Student Quick Tips

Use this Student Quick Tips guide for a quick and easy start with McGraw-Hill Connect. You'll get valuable tips on registering, doing assignments, and accessing resources, as well as information about the support center hours.

Getting Started

TIP: To get started in Connect, you will need the following:

- Your instructor's Connect Web Address

> Sample of Connect Web Address:
> http://www.mcgrawhillconnect.com/class/instructorname_section_name

- Connect Access Code

TIP: If you do not have an access code or have not yet secured your tuition funds, you can click "Free Trial" during registration. This trial will provide temporary Connect access (typically three weeks) and will remind you to purchase online access before the end of your trial.

Registration and Sign In

1. Go to the Connect Web Address provided by your instructor
2. Click on **Register Now**
3. Enter your email address

TIP: If you already have a McGraw-Hill account, you will be asked for your password and will not be required to create a new account.

4. Enter a registration code or choose **Buy Online** to purchase access online

(Continued: **Registration and Sign In**)

5. Follow the on-screen directions

TIP: Please choose your Security Question and Answer carefully. We will ask you for this information if you forget your password.

6. When registration is complete, click on **Go to Connect Now**
7. You are now ready to use **Connect**

Trouble Logging In?

- Ensure you are using the same email address you used during registration
- If you have forgotten your password, click on the "Forgot Password?" link at your Instructor's Connect Course Web Address
- When logged into Connect, you can update your account information (e.g. email address, password, and security question/answer) by clicking on the *"My Account"* link located at the top-right corner

Home (Assignments)

TIP: If you are unable to begin an assignment, verify the following:

- The assignment is available (start and due dates)
- That you have not exceeded the maximum number of attempts
- That you have not achieved a score of 100%
- If your assignment contains questions that require manual grading, you will not be able to begin your next attempt until your instructor has graded those questions

(Continued: **Home Assignments**)

TIP: Based on the assignment policy settings established by your Instructor, you may encounter the following limitations when working on your assignment(s):

- Ability to Print Assignment
- Timed assignments – once you begin a "*timed assignment*," the timer will not stop by design

TIP: "*Save & Exit*" vs. "*Submit*" button

- If you are unable to complete your assignment in one sitting, utilize the "*Save & Exit*" button to save your work and complete it at a later time
- Once you have completed your assignment, utilize the "*Submit*" button in order for your assignment to be graded

Library

TIP: The *Library* section of your Connect account provides shortcuts to various resources.

- If you purchased ConnectPlus, you will see an *eBook* link, which can also be accessed from the section information widget of the *Home* tab
- *Recorded Lectures* can be accessed if your instructor is using *Tegrity Campus* to capture lectures. You may also access recorded lectures when taking an assignment by clicking on the projector icon in the navigation bar
- Many McGraw-Hill textbooks offer additional resources such as narrated slides and additional problems, which are accessible through the *Student Resources* link

McGraw Hill connect™

Reports

TIP: Once you submit your assignment, you can view your available results in the *Reports* tab.

- If you see a dash (-) as your score, your instructor has either delayed or restricted your ability to see the assignment feedback
- Your instructor has the ability to limit the amount of information (e.g. questions, answers, scores) you can view for each submitted assignment

Need More Help?

CONTACT US ONLINE

Visit us at:

www.mcgrawhillconnect.com/support

Browse our support materials including tutorial videos and our searchable Connect knowledge base. If you cannot find an answer to your question, click on "Contact Us" button to send us an email.

GIVE US A CALL

Call us at:

1-800-331-5094

Our live support is available:

Mon-Thurs:	8 am – 11 pm CT
Friday:	8 am – 6 pm CT
Sunday:	6 pm – 11 pm CT

GENETICS 5E

ANALYSIS & PRINCIPLES

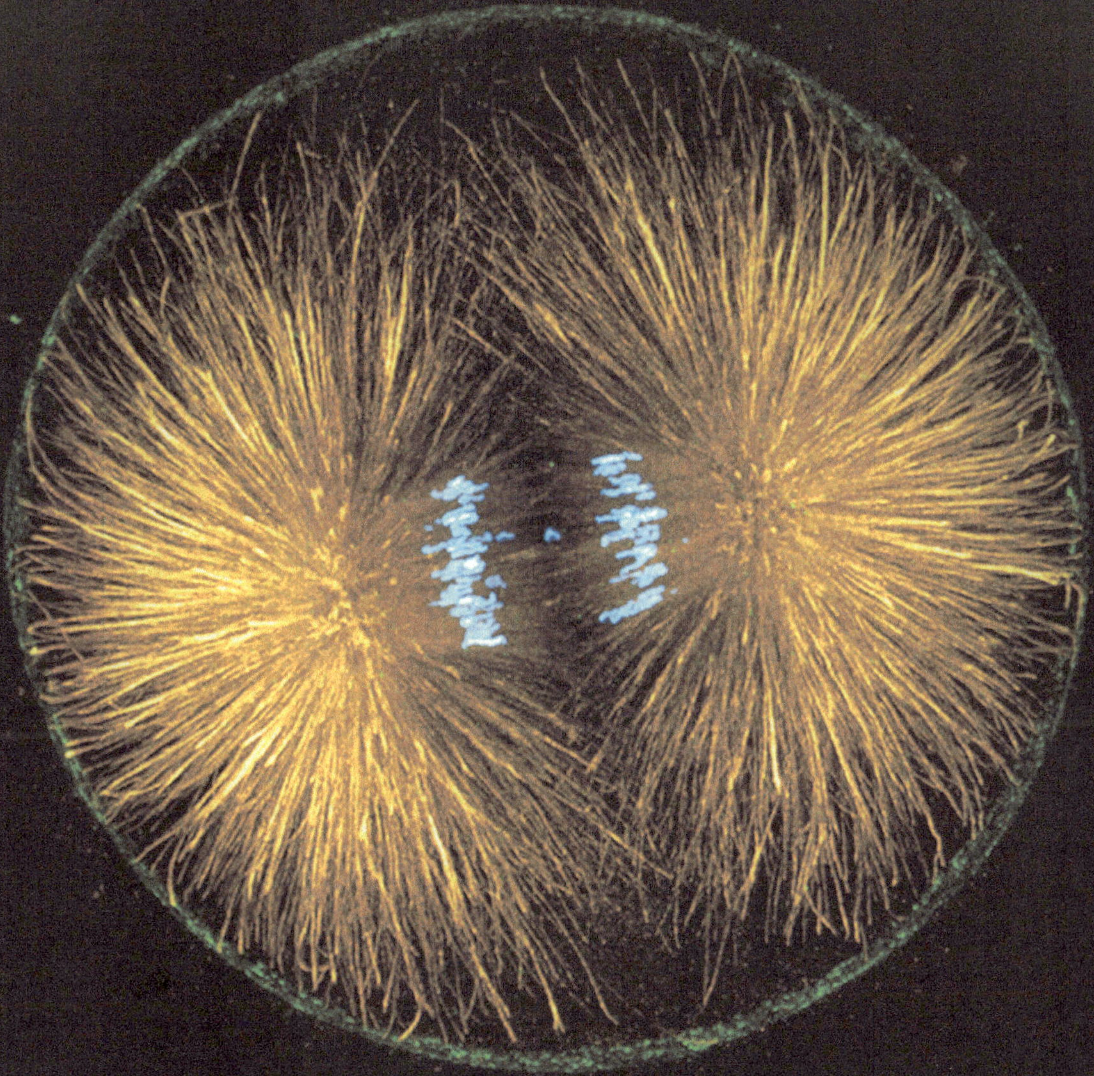

ROBERT J. BROOKER

University of Minnesota

BIO 184

CALIFORNIA STATE UNIVERSITY SACRAMENTO

Mc Graw Hill Education

1 2 3 4 5 6 7 8 9 0 KNG KNG 17 16 15 14

ISBN-13: 978-1-259-34304-9
ISBN-10: 1-259-34304-9

Learning Solutions Consultant: Julie Wolfe
Project Manager: James Doepke
Cover Photo Credits: Purple sea urchin zygote (Strongylocentrotus purpuratus) undergoing its first cell division © George von Dassow

BRIEF CONTENTS

TABLE OF CONTENTS

ABOUT THE AUTHOR

Robert J. Brooker is a professor in the Department of Genetics, Cell Biology, and Development at the University of Minnesota–Minneapolis. He received his B.A. in biology from Wittenberg University in 1978 and his Ph.D. in genetics from Yale University in 1983. At Harvard, he conducted postdoctoral studies on the lactose permease, which is the product of the *lacY* gene of the *lac* operon. He continues his work on transporters at the University of Minnesota. Dr. Brooker's laboratory primarily investigates the structure, function, and regulation of iron transporters found in bacteria and *C. elegans*. At the University of Minnesota he teaches undergraduate courses in biology, genetics, and cell biology.

DEDICATION

To my wife, Deborah, and our children, Daniel, Nathan, and Sarah

PREFACE

In the fifth edition of *Genetics: Analysis & Principles*, the content has been updated to reflect current trends in the field. In addition, the presentation of the content has been improved in a way that fosters active learning. As an author, researcher, and teacher, I want a textbook that gets students actively involved in learning genetics. To achieve this goal, I have worked with a talented team of editors, illustrators, and media specialists who have helped me to make the fifth edition of *Genetics: Analysis & Principles* a fun learning tool.

Overall, an effective textbook needs to accomplish four goals. First, it needs to provide comprehensive, accurate, and up-to-date content in its field. Second, it needs to expose students to the techniques and skills they will need to become successful in that field. Third, an effective textbook should have pedagogical features, such as formative assessment, that foster student learning. And finally, it should inspire students so they want to pursue that field as a career. The hard work that has gone into the fifth edition of *Genetics: Analysis & Principles* has been aimed at achieving all four of these goals!

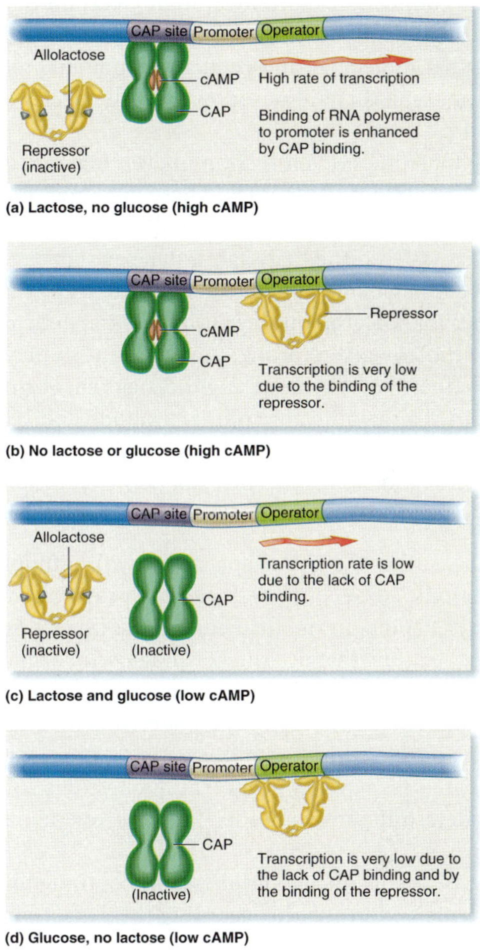

INTERACTIVE EXERCISE

FIGURE 14.8 The roles of the lac repressor and catabolite activator protein (CAP) in the regulation of the *lac* operon. This figure illustrates how the *lac* operon is regulated depending on its exposure to lactose or glucose.

Genes → Traits The mechanism of catabolite repression provides the bacterium with the trait of being able to choose between two sugars. When exposed to both glucose and lactose, the bacterium chooses glucose first. After the glucose is used up, it then expresses the genes necessary for lactose metabolism. This trait allows the bacterium to more efficiently use sugars from its environment.

CONCEPT CHECK: Why is it beneficial for the bacterium to regulate the *lac* operon with both a repressor protein and an activator protein?

FLIPPING THE CLASSROOM

A recent trend in science education is the phenomenon that is sometimes called "flipping the classroom." This phrase refers to the idea that some of the activities that used to be done in class are now done outside of class, and vice versa. For example, instead of spending the entire class time lecturing over textbook and other materials, some of the class time is spent engaging students in various activities, such as problem solving, working through case studies, and designing experiments. This approach is called active learning. For many instructors, the classroom has become more learner centered rather teacher centered. A learner-centered classroom provides a rich environment in which students can interact with each other and with their instructors. Instructors and fellow students often provide formative assessment—immediate feedback that helps each student understand if his or her learning is on the right track.

What are some advantages of active learning? Educational studies reveal that active learning usually promotes greater learning gains. In addition, active learning often focuses on skill development rather than on the memorization of facts that are easily forgotten. Students become trained to "think like scientists" and to develop a skill set that enables them to apply scientific reasoning. A common concern among instructors who are beginning to try out active learning is that they think they will have less time to teach and therefore will cover less material. However, this may not be the case. Although students may be provided with online lectures, "flipping the classroom" typically gives students more responsibility for understanding the textbook material on their own. Along these lines, *Genetics: Analysis & Principles*, fifth edition, is intended to provide students with a resource that can be effectively used outside of the classroom. Here are several of the key pedagogical features:

- Genes → Traits: Because genetics is such a broad discipline, ranging from the molecular level to populations, many instructors have told us that it is a challenge for students to see both "the forest and the trees." It is commonly mentioned that students often have trouble connecting the concepts they have learned in molecular genetics with the traits that occur at the level of a whole organism (i.e., What does transcription have to do with blue eyes?). To try to make this connection more meaningful, certain figure legends in each chapter, designated Genes → Traits, remind students that molecular and cellular phenomena ultimately lead to the traits that are observed in each species (see Figure 14.8).

- Learning Outcomes: Each section of every chapter begins with a set of learning outcomes. These outcomes help students understand what they should be able to do once they have mastered the material in that section.
- Formative Assessment: When students are expected to learn textbook material on their own, it is imperative that they are regularly given formative assessment so they can gauge whether they are mastering the material. Formative assessment is a major feature of this textbook and is bolstered by Connect—a state-of-the art digital assignment and assessment platform. In *Genetics: Analysis & Principles*, fifth edition, formative assessment is provided in multiple ways.

1. Each section of every chapter ends with multiple-choice questions. Also, compared with the previous edition, many chapters in the fifth edition are divided into more sections that are shorter in length. Formative assessment at the end of each section allows students to evaluate their mastery of the material before moving on to the next section.
2. Most figures have Concept Check questions so students can determine if they understand the key points in the figure.
3. Extensive end-of chapter questions continue to provide students with feedback regarding their mastery of the material.
4. Additional questions, including questions that pertain to every feature investigation, are found at the open-access companion website: www.mhhe.com/brookergenetics5e.
5. The textbook material is supported by digital learning tools found in Connect. Questions and activities are assignable in Connect, but students also have access to our valuable adaptive study tool, LearnSmart, offered for the first time with the fifth edition of *Genetics: Analysis & Principles.*

Overall, the pedagogy of *Genetics: Analysis & Principles*, fifth edition, has been designed to foster student learning. Instead of being a collection of "facts and figures," *Genetics: Analysis and Principles*, fifth edition, by Robert Brooker, is intended to be an engaging and motivating textbook in which formative assessment allows students to move ahead and learn the material in a productive way. We welcome your feedback so we can make future editions even better!

HOW WE ARE MEETING YOUR NEEDS

Text Organization

In surveying many genetics instructors, it became apparent that most people fall into two camps: "Mendel first" versus "Molecular first." I have taught genetics both ways. As a teaching tool, this textbook has been written with these different teaching strategies in mind. The organization and content lend themselves to various teaching formats.

Chapters 2 through 8 are largely inheritance chapters, whereas Chapters 26 through 28 examine population and quantitative genetics. The bulk of the molecular genetics is found in Chapters 9 through 25, although I have tried to weave a fair amount of molecular genetics into Chapters 2 through 8 as well. The information in Chapters 9 through 25 does not assume that a student has already covered Chapters 2 through 8. In fact, each chapter is written with the perspective that instructors may want to vary the order of their chapters to fit their students' needs.

For those who like to discuss inheritance patterns first, a common strategy would be to cover Chapters 1 through 8 first, and then possibly 26 through 28. (However, many instructors like to cover quantitative and population genetics at the end. Either way works fine.) The more molecular and technical aspects of genetics would then be covered in Chapters 9 through 25. Alternatively, if you like the "Molecular-first" approach, you would probably cover Chapter 1, then skip to Chapters 9 through 25, then return to Chapters 2 through 8, and then cover Chapters 26 through 28 at the end of the course. This textbook was written in such a way that either strategy works well.

Accuracy

Both the publisher and I acknowledge that inaccuracies can be a source of frustration for both the instructor and students. Therefore, throughout the writing and production of this textbook we have worked very hard to catch and correct errors during each phase of development and production.

In addition to input from reviewers, a development editor has gone through the material to check for accuracy in art and consistency between the text and art. With regard to the problem sets, the author personally checked every question and answer when the chapters were completed.

Feature Experiments

As in previous editions, each chapter (beginning with Chapter 2) incorporates one or two experiments that are presented according to the scientific method. These experiments are not "boxed off" from the rest of the chapter. Rather, they are integrated within the chapters and flow with the rest of the text. As you are reading the experiments, you will simultaneously explore the scientific method and the genetic principles that have been discovered using this approach. For students, I hope this textbook helps you to see the fundamental connection between scientific analysis and principles. For both students and instructors, I expect that this strategy makes genetics much more fun to explore.

Writing Style

Motivation in learning often stems from enjoyment. If you enjoy what you're reading, you are more likely to spend more time with it and focus your attention more crisply. The writing style of this book is meant to be interesting, down to earth, and easy to follow. Each section of every chapter begins with an overview of the contents of that section, usually with a table or figure that summarizes the broad points. The section then examines how those broad points were discovered experimentally, as well as explaining many of the finer scientific details. Important terms are introduced in a boldface font. These terms are also found at the end of the chapter and in the glossary.

A genetics book can be made interesting and inspiring in various ways. The subject matter itself is pretty amazing, so it's not difficult to build on that. In addition to describing the concepts and experiments in ways that motivate students, it is important to draw on examples that bring the concepts to life. In a genetics book, many of these examples come from the realm of medicine and medical research. This textbook contains lots of examples of human diseases that exemplify some of the underlying principles of genetics. Students often say they remember certain genetic concepts because they remember how defects in certain genes can cause disease. For example, defects in DNA repair genes cause a higher predisposition for developing cancer. In addition, I have provided interesting examples from the microbial and plant world. Finally, students are often fascinated by applications of genetics that affect their everyday lives. Because we frequently hear about genetics in the news, it's inspiring for students to learn the underlying basis for such technologies. Chapters 20 to 23 are devoted to genetic technologies, and applications of these and other technologies are found throughout this textbook. By the end of their genetics course, students should come away with a greater appreciation for the influence of genetics in their lives.

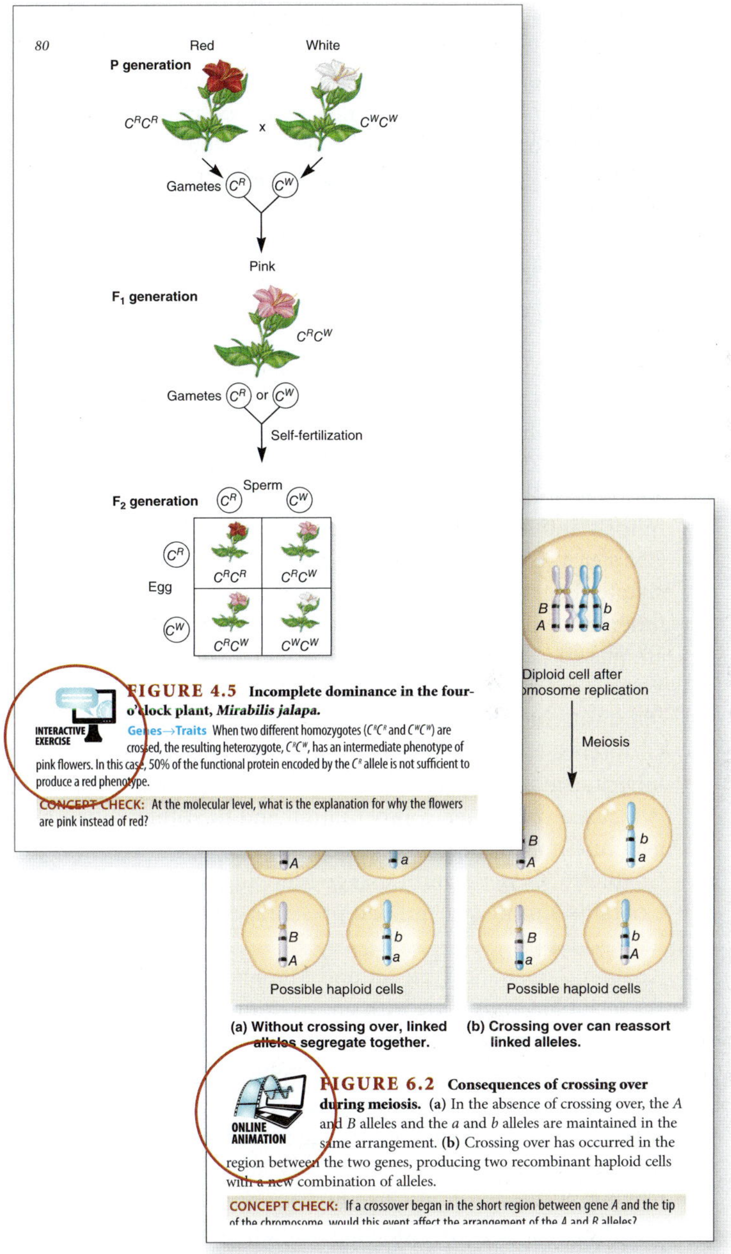

Interactive Exercises and Animations

Education specialists have crafted Interactive Exercises in which the students can make their own choices in problem-solving activities and predict what the outcomes will be. These Interactive Exercises are an excellent tool for helping students test their understanding of inheritance patterns and human genetic diseases, but additional exercises also explore molecular concepts.

Our media specialists have also created over 55 animations for a variety of genetic processes. These animations were made specifically for this textbook and use the art from the textbook. The animations make many of the figures in the textbook "come to life." Icons are found in figure legends where the concept is supported by an Interactive Exercise or animation. The Interactive Exercises and animations are available through the Presentation Tools table in Connect, allowing professors to incorporate both into their classroom discussions. The Interactive Exercises are a great jumping off point for Active Learning discussions.

SIGNIFICANT CONTENT CHANGES IN THE FIFTH EDITION

- Each section of every chapter begins with Learning Outcomes and ends with multiple-choice questions. The answers to the multiple-choice questions are in the back of the book.
- Concept Check questions have been added to the figure legends of hundreds of figures, enabling students to determine if they understand key points in those figures. The answers to these questions are also in the back of the book.
- Many chapters have been divided into more sections that are shorter in length. This helps students to see the big

picture of each topic and provides an opportunity to include more formative assessment.

- Although the overall length of the fifth edition is not longer than the fourth edition, two new chapters have been added to this edition: Chapter 16. Eukaryotic Gene Regulation II: Epigenetics and Regulation at the RNA Level; and Chapter 17. Genetics of Viruses.

Examples of Specific Content Changes to Individual Chapters

- Chapter 2. Mendelian Inheritance: Divided into shorter sections that end in questions to help students gauge their understanding of Mendel's laws.
- Chapter 3. Chromosome Transmission During Cell Division and Sexual Reproduction: Improvement in the figures of mitosis and meiosis (see Figures 3.8 and 3.11).
- Chapter 4. Extensions of Mendelian Inheritance: This revised chapter begins with an overview that compares different inheritance patterns, and then the inheritance patterns are placed in their own sections that end with formative assessment questions. This approach should help students see the similarities and differences among the various patterns.
- Chapter 5. Non-Mendelian Inheritance: Epigenetic inheritance is now divided into two sections. One section is focused on dosage compensation and the other concerns genomic imprinting.
- Chapter 6. Genetic and Linkage Mapping in Eukaryotes: Contains a more streamlined presentation of mapping in haploid fungi.
- Chapter 7. Genetic Transfer and Mapping in Bacteria and Bacteriophages: Begins with an overview that compares different types of genetic transfer between bacteria, and then each form of transfer is highlighted within its own section. Mapping in bacteriophages has been separated into sections that focus on intergenic complementation and intragenic mapping.
- Chapter 8. Variation in Chromosome Structure and Number: The fifth edition contains a more streamlined presentation of natural and experimental mechanisms that produce variation in chromosome number.
- Chapter 9. Molecular Structure of DNA and RNA: The four levels of DNA structure are introduced in an overview section and then the different levels of DNA structure are presented in their own sections, followed by formative assessment questions.
- Chapter 10. Chromosome Organization and Molecular Structure: The information on viruses has been moved to Chapter 17, which is a new chapter in the fifth edition. Improvements have been made to several figures that depict chromatin structure (see Figures 10.11 and 10.18).
- Chapter 11. DNA Replication: The figure illustrating a three-dimensional view of DNA replication has been improved (see Figure 11.12).
- Chapter 12. Gene Transcription and RNA Modification: An improved figure has been added, which shows how sigma factor binds into the major groove (see Figure 12.6). A summary table has been added at the end of the chapter that compares transcription and RNA modification between bacteria and eukaryotes.
- Chapter 13. Translation of mRNA: The relationship between the genetic code to protein synthesis and the experimental determination of the genetic code have been placed into two separate sections. A new table has been added that describes how certain antibiotics inhibit translation (see Table 13.9).
- Chapter 14. Gene Regulation in Bacteria: The material on bacteriophage gene regulation has been moved to Chapter 17 (Genetics of Viruses). The *lac* operon and *trp* operon are now discussed in their own separate sections.
- Chapter 15. Gene Regulation in Eukaryotes I: Transcriptional Regulation: The material on eukaryotic gene regulation is now divided into two chapters. The first one focuses on transcriptional regulation. A new section has been added on the ENCODE Project.
- Chapter 16. Gene Regulation in Eukaryotes II: Epigenetics and Regulation at the RNA Level: This chapter has three new sections that focus on epigenetics during development and environmental factors that cause epigenetic changes. It includes seven new figures and three new tables.
- Chapter 17. Genetics of Viruses: This chapter incorporates some material from the fourth edition, such as bacteriophage gene regulation, but largely includes new material on viral reproductive cycles and gene regulation in HIV. It has eight new figures.
- Chapter 18. Gene Mutation and DNA Repair: The figure concerning trinucleotide repeat expansion has been revised (see Figure 18.12).
- Chapter 19. Recombination and Transposition at the Molecular Level: The figure concerning the function of transposase has been revised into a two-part figure that shows how transposase causes the transposon to loop out (see Figure 19.12).
- Chapter 20. DNA Technologies: Based on reviewer feedback, the order of topics in this chapter has been revised. DNA sequencing and site-directed mutagenesis come directly after cloning methods.
- Chapter 21. Biotechnology: Various topics, such as the use of transgenic crops, have been updated.
- Chapter 22. Genomics I: Analysis of DNA: A new section has been added on metagenomics.
- Chapter 23. Genomics II: Functional Genomics, Proteomics, and Bioinformatics: Includes updates to the topic of bioinformatics.
- Chapter 24. Medical Genetics and Cancer: This chapter ends with a new section on personalized medicine.
- Chapter 25. Developmental Genetics: Improvements in the color scheme of several figures will help students better understand certain key points of development.
- Chapter 26. Population Genetics: The chapter now contains an overview of microevolution, and then natural selection, genetic drift, migration, nonrandom mating, and sources of new genetic variation are covered in their own separate sections.

- Chapter 27. Quantitative Genetics: This chapter in the fifth edition presents a more streamlined view of how quantitative loci are mapped. A section on the general features of heritability precedes a section on selective breeding.
- Chapter 28. Evolutionary Genetics: The cladistics method for constructing a phylogenetic tree is compared with the UPGMA method.

Suggestions Welcome!

It seems very appropriate to use the word *evolution* to describe the continued development of this textbook. I welcome any and all comments. The refinement of any science textbook requires input from instructors and their students. These include comments regarding writing, illustrations, supplements, factual content, and topics that may need greater or less emphasis. You are invited to contact me at:

Dr. Rob Brooker
Dept. of Genetics, Cell Biology, and Development
University of Minnesota
6-160 Jackson Hall
321 Church St.
Minneapolis, MN 55455
brook005@umn.edu
612-624-3053

ONLINE TEACHING AND LEARNING RESOURCES

Help Your Students Prepare for Class

Digital resources can help you achieve your instructional goals—making your students more responsible for learning outside of class by meeting your students where they live: on the go and online. Use the text and digital tools to empower students to come to class more prepared and ready to engage!

connect plus+ GENETICS

McGraw-Hill Connect Genetics provides online presentation, assignment, and assessment solutions. It connects your students with the tools and resources they'll need to achieve success. With **Connect Genetics** you can deliver assignments, quizzes, and tests online. A robust set of questions and activities is presented in the Question Bank and a separate set of questions to use for exams is presented in the Test Bank. As an instructor, you can edit existing questions and author entirely new problems. Track individual student performance—by question, by assignment, or in relation to the class overall—with detailed grade reports. Integrate grade reports easily with Learning Management Systems such as Blackboard and Canvas—and much more. **ConnectPlus Genetics** provides students with all the advantages of Connect Biology plus 24/7 online access to an eBook. This media-rich version of the book is available through the McGraw-Hill ConnectPlus platform and allows seamless integration of text, media, and assessments.

To learn more, visit **www.mcgrawhillconnect.com**.

Help Your Students by Making Assignments—Reading, Homework, and LearnSmart

Connect content can be assigned as homework for before class to help students with basic concepts so they can better understand classroom presentations and projects. Quizzes taken after class can also evaluate their comprehension. These assignments support the rich assessment presented in the text so that student and professors can gauge the level of understanding of concepts and the mastery of skills.

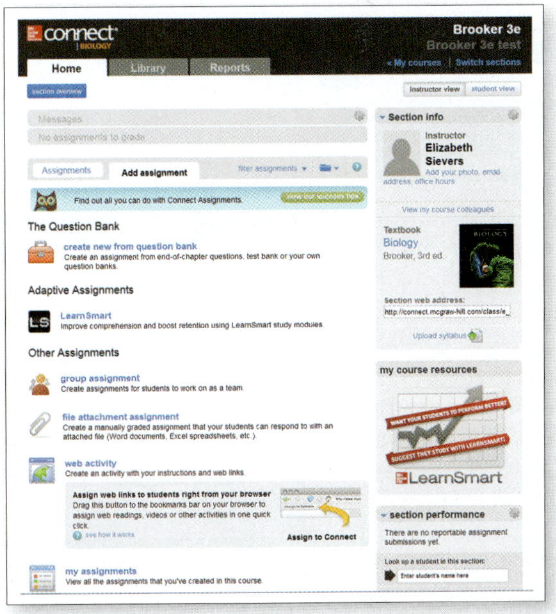

Assignments can include reading assignments from the ConnectPlus eBook or SmartBook, homework or quizzes, LearnSmart, your own web or short answer activities, and more.

The interactive eBook takes the reading experience to a new level with links to animations and interactive exercises that supplement the text.

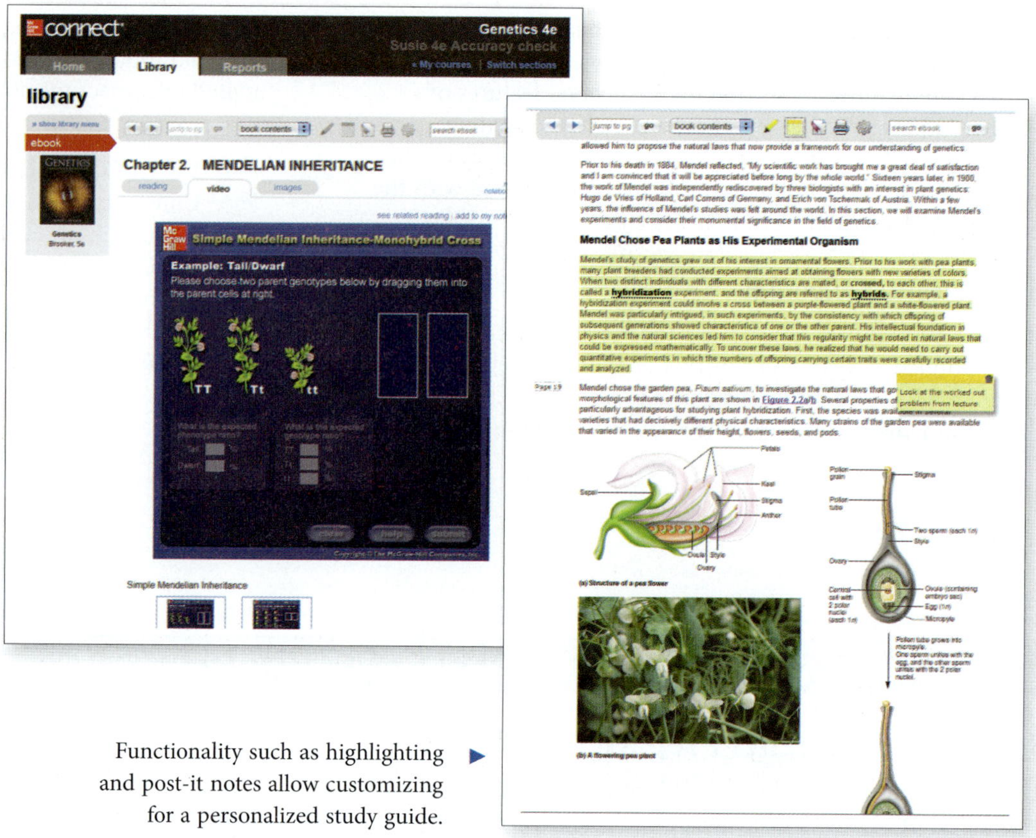

Functionality such as highlighting and post-it notes allow customizing for a personalized study guide.

SMARTBOOK™

Powered by an intelligent diagnostic and adaptive engine, **SmartBook** facilitates the reading process by identifying what content a student knows and doesn't know through adaptive assessments.

The Smartbook experience starts by previewing key concepts from the chapter and ensuring that you understand the big ideas.

SmartBook asks you questions that identify gaps in your knowledge. The reading experience then continuously adapts in response to the assessments, highlighting the material you need to review based on what you don't know.

The reports in SmartBook help identify topics where you need more work.

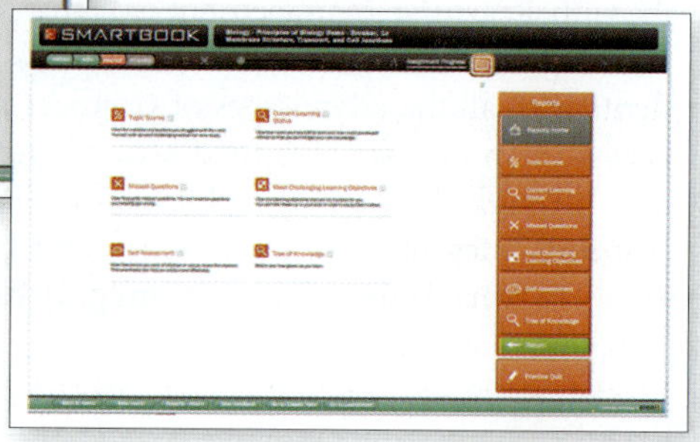

Help your students prepare for class and revisit content by assigning homework in Connect and LearnSmart.

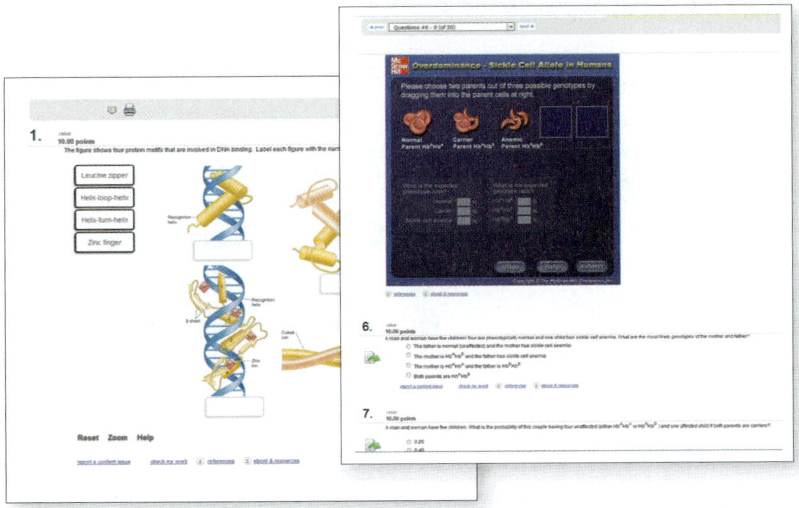

▲ Interactive and traditional questions help assess your knowledge of the material. Having achieved a base level of knowledge, you will get more out of your time in class.

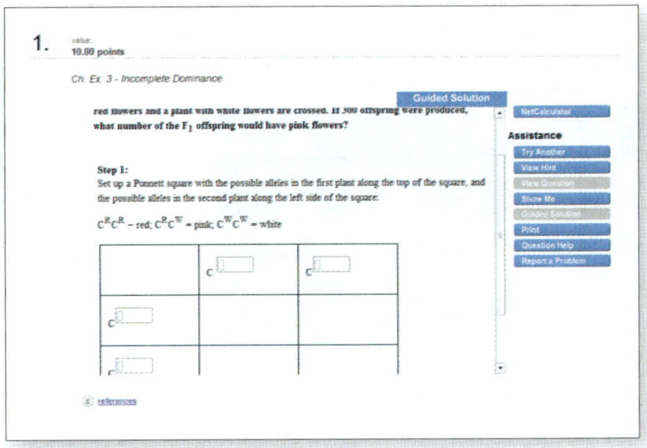

▲ Quantitative questions help you challenge your students to apply material they learn in class and in the book.

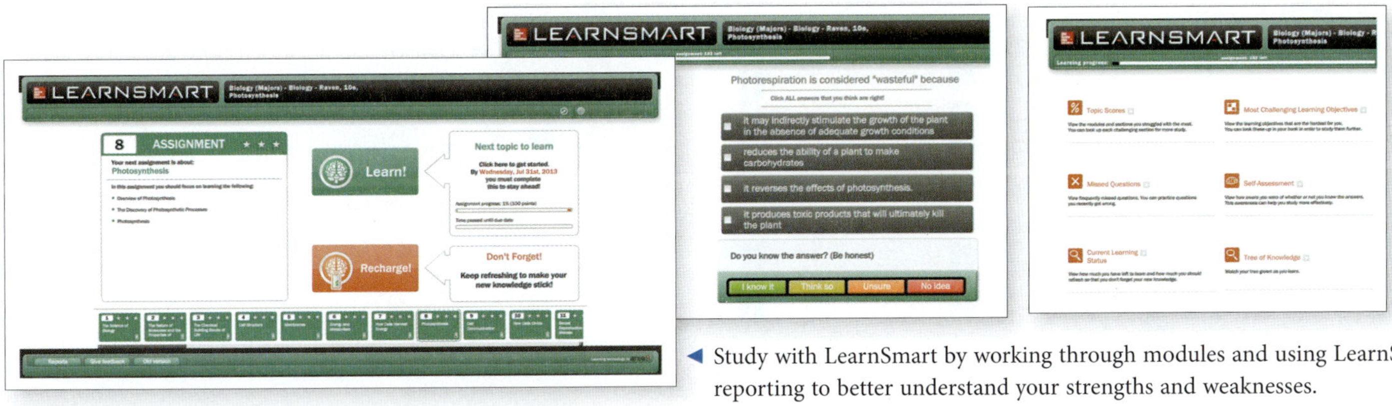

◄ Study with LearnSmart by working through modules and using LearnSmart's reporting to better understand your strengths and weaknesses.

Gauge your students' progress using reports in LearnSmart and Connect. Students can run these same reports in LearnSmart to track their own progress.

▲ The Tree of Knowledge tracks your progress, reporting on short-term successes and long-term retention.

▲ Download the LearnSmart app from iTunes or Google Play and work on LearnSmart from anywhere!

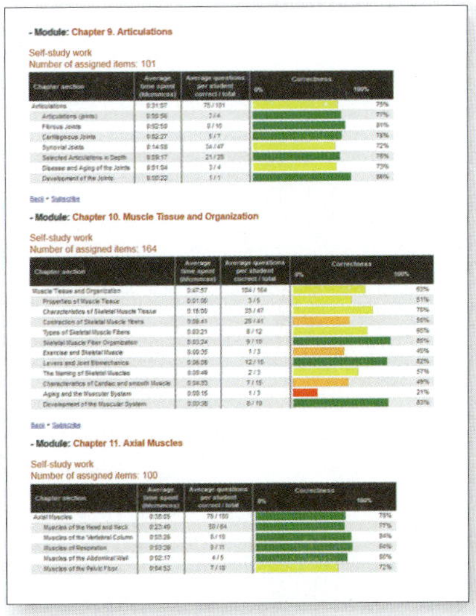

◄ Reports in Connect and LearnSmart help you monitor student assignments and performance, allowing for "just-in-time" teaching to clarify concepts that are more difficult for your students to understand.

Presentation Tools

Within Connect, you will find presentation materials to enhance your class all in one place.

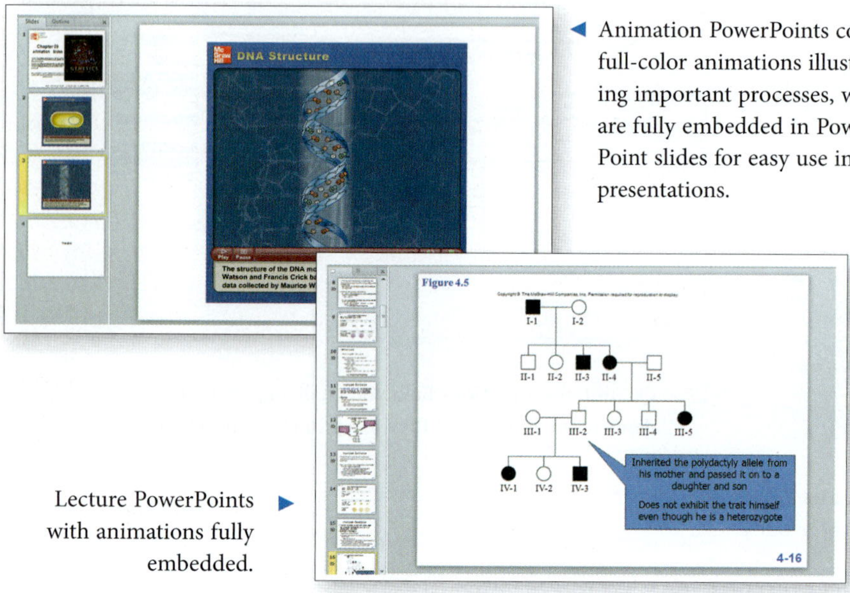

◄ Animation PowerPoints contain full-color animations illustrating important processes, which are fully embedded in PowerPoint slides for easy use in your presentations.

Lecture PowerPoints with animations fully embedded. ►

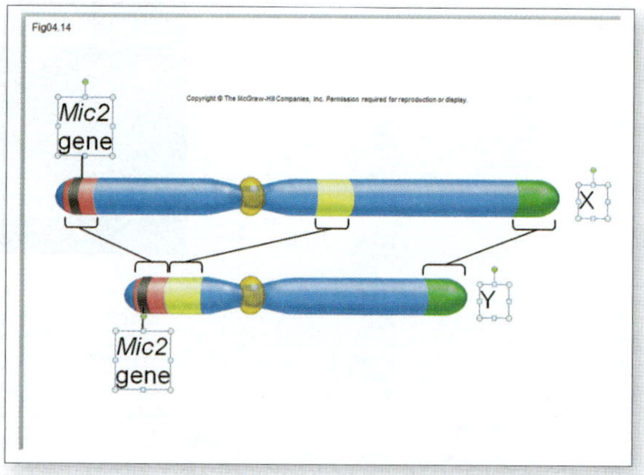

▲ FlexArt PowerPoints contain editable art from the text. For all figures, labels and leader lines are editable and some figures also have editable or stepped out art allowing you to customize your PowerPoint presentations.

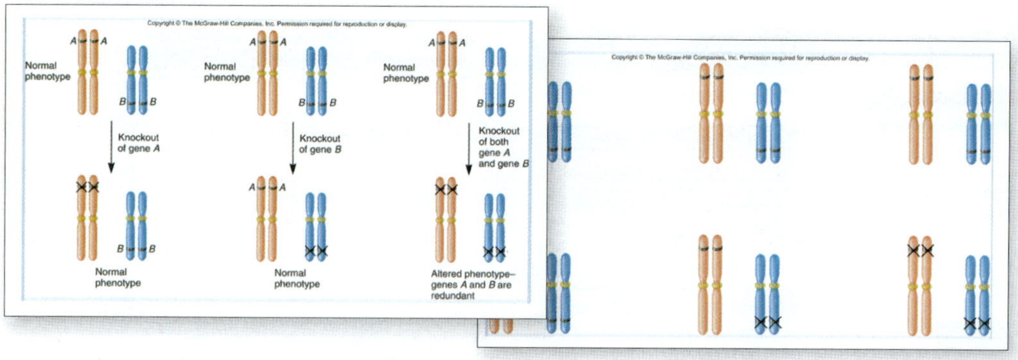

◄ Labeled and unlabeled JPEG files of all art and photos in the text to be readily incorporated into presentations, exams, or custom-made classroom materials.

Student Study Guide/Solutions Manual (ISBN: 0-07-767636-X)

The solutions to the end-of-chapter problems and questions aid students in developing their problem-solving skills by providing the steps for each solution. The Study Guide summarizes the main points in the chapter text, figures, and tables. It also contains concept-building exercises, self-help quizzes, and practice exams.

ACKNOWLEDGMENTS

The production of a textbook is truly a collaborative effort, and I am greatly indebted to a variety of people. All five editions of this textbook went through multiple rounds of rigorous revision that involved the input of faculty, students, editors, and educational and media specialists. Their collective contributions are reflected in the final outcome.

Deborah Brooker (Freelance Developmental Editor) meticulously read every chapter, analyzed every figure, and offered extensive feedback. Her attention to detail in this edition and previous editions has profoundly contributed to the accuracy and clarity of this textbook. I would also like to thank Linda Davoli (Freelance Copy Editor) for making grammatical improvements throughout the text and art, which has also improved the text's clarity.

I would like to acknowledge the many people at McGraw-Hill whose efforts are amazing. My highest praise goes to Rebecca Olson (Brand Manager) for her insights regarding the needs of genetics instructors and her skill at overseeing this project. I would also like to thank Elizabeth Sievers (Director of Development), who kept me on schedule and made sure that all of the pieces of the puzzle were in place. Other people at McGraw-Hill have played key roles in producing an actual book and the supplements that go along with it. In particular, Daryl Bruflodt (Content Project Manager) has done a superb job of managing the components that need to be assembled to produce a book. I would also like to thank John Leland (Photo Research Coordinator), who acted as an interface between me and the photo company. In addition, my gratitude goes to David Hash (Designer), who provided much input into the internal design of the book as well as created an awesome cover. Finally, I would like to thank Patrick Reidy (Executive Marketing Manager), whose major efforts begin when the fifth edition comes out!

I would also like to extend my thanks to Chris Black and everyone at Lachina Publishing Services, including the many artists who have played important roles in developing the art for the third, fourth, and fifth editions. Also, folks at Lachina Publishing Services worked with great care in the paging of the book, making sure that the figures and relevant text are as close to each other as possible. Likewise, the people at Photo Affairs, Inc. have done a great job of locating many of the photographs that have been used in the fifth edition.

Finally, I want to thank the many scientists who reviewed the chapters of this textbook. Their broad insights and constructive suggestions were an important factor that shaped its final content and organization. I am truly grateful for their time and effort.

REVIEWERS

Amy Abdulovic-Cui, *Augusta State University*
Steve Alas, *California State Polytechnic University, Pomona*
Harvey Babich, *Stern College for Women*
Laura Hill Bermingham, *University of Vermont*
Hector Biliran Jr., *Xavier University of Louisiana*
Mirjana Milosevic Brockett, *Georgia Institute of Technology*
Madhusudan Choudhary, *Sam Houston State University*
Reggie Cobb, *Nash Community College*
Dan Choffnes, *Carthage College*
Laurie Cotroneo, *Southern Utah University*
Robert S. Dotson, *Tulane University*
Robert Hinrichsen, *Indiana University of Pennsylvania*
Margaret Hollingsworth, *University at Buffalo*
Mitrick Johns, *Illinois University*
Ekaterina Kaverina, *East Tennessee State University*
Jesse Mager, *University of Massachusetts*
Norah R. McCabe, *Washington State University*
R. Deborah Overath, *Texas A&M University–Corpus Christi*
Thomas Peavy, *California State University–Sacramento*
Rongsun Pu, *Kean University*
Robert Rutledge, *DeSales University*
Julian Shull, *Appalachian State University*
Jeffry Shultz, *Louisiana Tech University*
Ronald Wagner, *Central Washington University*
Carey Waldburger, *William Paterson University*
Gary L. Walker, *Appalachian State University*
Jessica Wooten, *The University of Findlay*

A Visual Guide to
GENETICS: ANALYSIS & PRINCIPLES

Instructional Art

Each figure is carefully designed to follow closely with the text material.

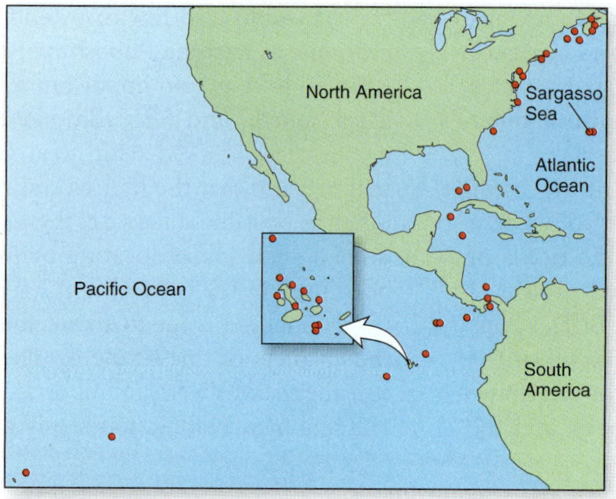

Every illustration was drawn with four goals in mind: completeness, clarity, consistency, and realism.

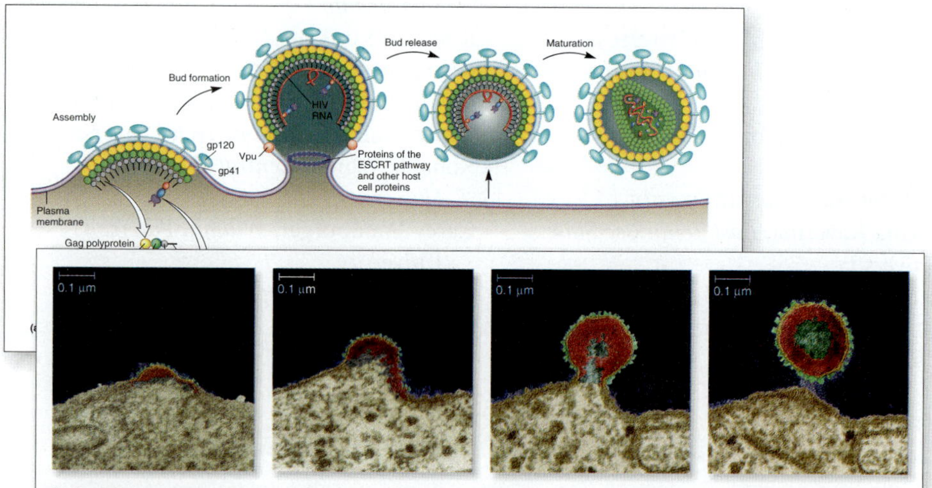

(b) Transmission electron micrograph of HIV assembly, budding, and maturation

The digitally rendered images have a vivid three-dimensional look that will stimulate a student's interest and enthusiasm.

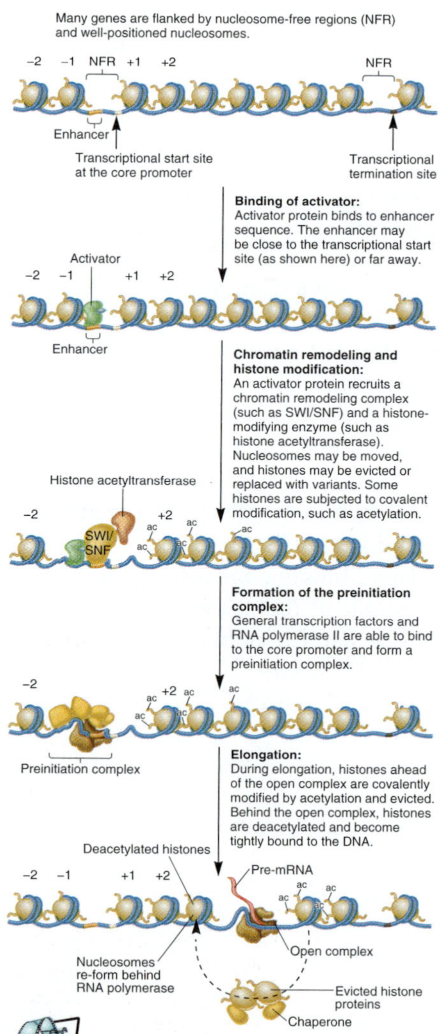

ONLINE ANIMATION

INTERACTIVE EXERCISE

FIGURE 15.13 A simplified model for the transcriptional activation of a eukaryotic gene. The activation involves changes in nucleosome arrangements and histone modifications.

CONCEPT CHECK: Explain why histone eviction is needed for the elongation phase of transcription.

Many figures are supported with online Interactive Exercises and Online Animations that allow students an opportunity to delve into problem-solving activities through the Interactive Exercises and explore dynamic processes through the Online Animations. Icons indicate which figures are supported with these online features.

Learning Through Experimentation

Each chapter (beginning with Chapter 2) incorporates one or two experiments that are presented according to the scientific method. These experiments are integrated within the chapters and flow with the rest of the textbook. As you read the experiments, you will simultaneously explore the scientific method and the genetic principles learned from this approach.

STEP 1: BACKGROUND OBSERVATIONS

Each experiment begins with a description of the information that led researchers to study an experimental problem. Detailed information about the researchers and the experimental challenges they faced help students to understand actual research.

STEP 2: HYPOTHESIS

The student is given a statement describing the possible explanation for the observed phenomenon that will be tested. The hypothesis section reinforces the scientific method and allows students to experience the process for themselves.

STEP 3: TESTING THE HYPOTHESIS

This section illustrates the experimental process, including the actual steps followed by scientists to test their hypothesis. Science comes alive for students with this detailed look at experimentation.

STEP 4: THE DATA

Actual data from the original research paper help students understand how real-life research results are reported. Each experiment's results are discussed in the context of the larger genetic principle to help students understand the implications and importance of the research.

STEP 5: INTERPRETING THE DATA

This discussion, which examines whether the experimental data supported or refuted the hypothesis, gives students an appreciation for scientific interpretation.

EXPERIMENT 17A

The Genome of Tobacco Mosaic Virus Is Composed of RNA

We now know that bacteria, archaea, protists, fungi, plants, and animals all use DNA as their genetic material. In 1956, Alfred Gierer and Gerhard Schramm isolated RNA from tobacco mosaic virus (TMV), which infects plant cells. When this purified RNA was applied to plant tissue, the plants developed the same types of lesions that occurred when they were exposed to intact TMVs. Gierer and Schramm correctly concluded that the viral genome of TMV is composed of RNA.

To further confirm that TMV uses RNA as its genetic material, Heinz Fraenkel-Conrat and Beatrice Singer conducted additional research that involved different strains of TMV. They focused their efforts on the wild-type strain and a mutant TMV strain called the Holmes ribgrass (HR) strain. The two strains differ in two ways. First, they cause significantly different symptoms when they infect plants. In particular, the wild-type strain produces a mottled area with yellow and green irregularly shaped lesions on infected leaves (see chapter-opening photo), whereas the HR strain often produces streaks along the veins and ringlike markings on other parts of the leaves. Second, the capsid protein in the HR strain has two amino acids (histidine and methionine), which are not found in the wild-type capsid protein.

Previous experiments had shown that purified TMV capsid proteins and purified TMV-RNA molecules can be mixed together and self-assemble into an intact virus. Such a procedure is referred to as a reconstitution experiment because intact viruses are made from their individual parts. In the experiment described in **Figure 17.2**, Fraenkel-Conrat and Singer mixed wild-type RNA with HR proteins or HR RNA with wild-type proteins and then placed the reconstituted viruses onto tobacco leaves. Following infection, they then observed the symptoms caused by the viruses and analyzed the amino acid composition of the proteins of newly made viruses that arose after the infection.

THE HYPOTHESIS

RNA is the genetic material of TMV.

TESTING THE HYPOTHESIS

FIGURE 17.2 **Evidence that RNA is the genetic material of TMV.**

Starting material: Purified preparations of RNA and proteins from wild-type TMV and from the Holmes ribgrass (HR) strain of TMV.

	Experimental level	Conceptual level
1. Mix together wild-type RNA and HR proteins or HR RNA and wild-type proteins. Allow time for the RNA and proteins to assemble into intact viruses. These are called reconstituted viruses.		

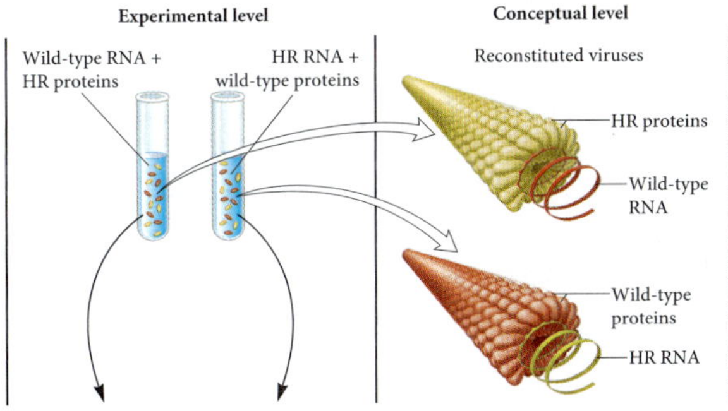

like wild-type

like the lesions of ribgrass strain.

-type proteins

proteins

a Table.

THE DATA

Composition of Reconstituted Virus Placed on Tobacco Leaves	Symptoms on Tobacco Leaves	Amino Acids Found in Newly Made Viral Proteins Following Infection: Methionine	Histidine
Wild-type RNA and HR protein	Like wild-type TMV	No	No
HR RNA and wild-type protein	Like HR TMV	Yes	Yes

Data adapted from H. Fraenkel-Conrat and B. Singer (1957) Virus reconstitution II. Combination of protein and nucleic acid from different strains. *Biochimica et Biophysica Acta 24*, 540–548.

INTERPRETING THE DATA

As seen in The Data, the outcome of infection depended on the RNA that was found in the reconstituted virus but not on the protein. If wild-type RNA was used, the leaves developed symptoms that were typical of wild-type TMV and the capsid proteins of newly made viruses lacked methionine or histidine. In contrast, if reconstituted virus had HR RNA, the symptoms were those of the HR TMV strain and the newly made capsid proteins contained both methionine and histidine. Taken together, these results are consistent with the hypothesis that the RNA component of TMV is its genetic material.

A self-help quiz involving this experiment can be found at www.mhhe.com/brookergenetics5e.

Formative Assessment Throughout Each Chapter

Formative assessment provides a means for students to gauge their learning. Genetics: Analysis & Principles, *5e, has incorporated three new pedagogical features that should help students navigate this textbook.*

LEARNING OUTCOMES

Each section begins with one or more Learning Outcomes. These allow a student to appreciate the skills and knowledge they will gain if they master the material.

17.4 HIV REPRODUCTIVE CYCLE

Learning Outcomes:

1. Outline the organization of the HIV genome.
2. Explain how HIV is reverse transcribed and integrated into the DNA of its host cell.
3. Describe the steps that lead to the formation of new HIV particles.

In the previous section, we focused on the reproductive cycles of phage λ, which infects bacteria. In this section, we will focus on HIV, which infects helper T cells found in humans. T cells are a type of lymphocyte that play a key role in cell-mediated immunity. T cells are distinguished from other lymphocytes by the presence of T-cell receptors in their plasma membrane. T-cell receptors are responsible for recognizing antigens—molecules that elicit an immune response. The activation of T cells to fight

cellular digestion by nucleases. The p6 protein facilitates the incorporation of certain proteins, such as Vpr (described later), into HIV particles.

- ***pol:* Enzymes Needed for Viral Replication and Viral Assembly.** The *pol* gene encodes a polyprotein that is cleaved into three enzymes: HIV protease, reverse transcriptase, and integrase. The HIV protease processes proteins made from the HIV genome so they can assemble into mature HIV particles. Reverse transcriptase is required for the production of DNA from the RNA genome of HIV. A portion of reverse transcriptase called RNase H digests RNA during reverse transcription. Integrase is used to incorporate this DNA into the host genome.
- ***vif, vpu:* Proteins That Promote Infectivity and Budding.** The Vif protein is incorporated into mature HIV particles and promotes infectivity by interacting with host-cell proteins. The Vpu protein promotes budding.
- ***vpr, rev, tat, nef:* Proteins with Regulatory Functions.** The Vpr protein has several functions including the transport of the HIV genome into the cell nucleus. The

protein is able to form a capsid structure that encloses two molecules of HIV RNA along with several different proteins (see Figure 17.11b). The nucleocapsid and p6 proteins are found inside the capsid. They protect the HIV RNA from nuclease digestion and have binding sites that promote the incorporation of other proteins into the capsid. For example, p6 has binding sites for Vpr.

As mentioned earlier, another polyprotein called the Gag-pol polyprotein plays a role in the maturation process, but is made in much lesser amounts than Gag polyprotein. As shown in Figure 17.15, a few copies of the Gag-pol polyprotein are incorporated into immature virus particles. When the Gag-pol polyprotein is cleaved by HIV protease, this releases more HIV protease, along with reverse transcriptase, and integrase. These proteins become captured within the capsid. As described in Figures 17.12 and 17.13, they are necessary for the reverse transcription and integration of the HIV genome in a newly infected host cell.

17.4 COMPREHENSION QUESTIONS

1. A viral protein that is needed to make HIV DNA is
   a. integrase. c. Vpr.
   b. reverse transcriptase. d. Gag polyprotein.
2. Which form of HIV RNA is packaged into HIV particles?
   a. Fully spliced c. Unspliced
   b. Incompletely spliced d. All of the above
3. After HIV components are made, what is the correct order of the stages that are needed to produce HIV particles?
   a. Maturation, budding, assembly
   b. Maturation, assembly, budding
   c. Assembly, budding, maturation
   d. Assembly, maturation, budding

COMPREHENSION QUESTIONS

Each section of every chapter ends with questions that help a student determine if they have learned the material and are ready to move on to the next section.

CONCEPT CHECK QUESTIONS

These are found within figure legends and help students determine their understanding of key points in the figure.

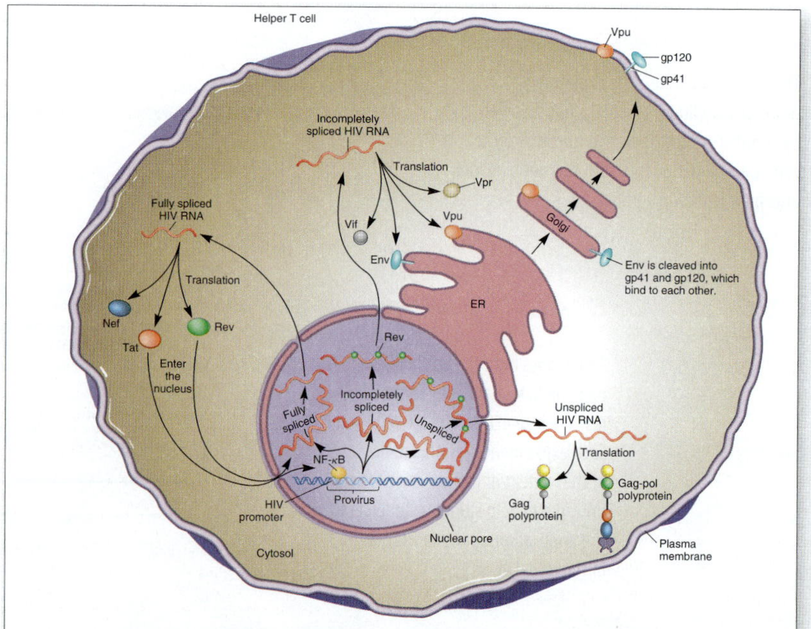

FIGURE 17.14 **Synthesis of new HIV components: HIV RNA and proteins.**

CONCEPT CHECK: Which form of HIV RNA, fully spliced, incompletely spiced, or unspliced, is needed during the early stage of synthesizing HIV components?

End-of-Chapter Support Materials

These study tools and problems are crafted to aid students in reviewing key information in the text and developing a wide range of problem-solving skills. They also develop a student's cognitive, writing, analytical, computational, and collaborative abilities.

KEY TERMS

Enhance student development of vital vocabulary necessary for the understanding and application of chapter content. Important terms are boldfaced throughout the chapter and page referenced at the end of each chapter for reflective study.

KEY TERMS

Page 393. epigenetics, epigenetic inheritance, transgenerational epigenetic inheritance, DNA methylation, chromatin remodeling, covalent histone modification, histone variants, feedback loops, noncoding RNAs
Page 394. *cis*-epigenetic mechanism, *trans*-epigenetic mechanism
Page 397. development, imprinting control region (ICR), differentially methylated region (DMR), de novo methylation
Page 398. maintenance methylation, X-chromosome inactivation (XCI), pluripotency factors, symmetry break
Page 400. trithorax group (TrxG), polycomb group (PcG), trimethylation, polycomb response element (PRE)
Page 404. oncogene
Page 405. alternative splicing
Page 406. constitutive exons, alternative exons, splicing factors, SR proteins, exon skipping
Page 407. polyA-binding protein, 3′-untranslated region (3′-UTR), AU-rich element (ARE)
Page 409. RNA interference (RNAi)
Page 410. microRNAs (miRNAs), short-interfering RNAs (siRNAs), RNA-induced silencing complex (RISC), processing body (P-body), iron regulatory protein (IRP), iron response element (IRE)

CHAPTER SUMMARY

Emphasizes the main concepts from each section of the chapter in a bulleted list form to provide students with a thorough review of the main topics covered.

CHAPTER SUMMARY

16.1 Overview of Epigenetics

- Epigenetics can be defined as the study of mechanisms that lead to changes in gene expression that are passed from cell to cell and are reversible but do not involve a change in the sequence of DNA. The transmission of epigenetic changes from one generation to the next is referred to as epigenetic inheritance.
- The most common types of molecular changes that underlie epigenetic control are DNA methylation, chromatin remodeling, covalent histone modification, the localization of histone variants, and feedback loops (see Table 16.1).
- Epigenetic changes can be established by transcription factors or noncoding RNAs (see Figure 16.1).
- Epigenetic changes may be maintained by *cis*- or *trans*-epigenetic mechanisms (see Figures 16.2, 16.3).
- Some epigenetic changes are programmed during development and others are caused by environmental agents (see Table 16.2).

PROBLEM SETS & INSIGHTS

Enhance student development

1. **Solved Problems:** Walk the student through the solutions of quantitative problems or provide explanations for the answers to more conceptual based questions.

PROBLEM SETS & INSIGHTS

Solved Problems

S1. Are the following examples best explained by genetic and/or epigenetic phenomena?

A. imprinting of the *Igf2* gene

B. variation in coat color in mice carrying the A^{vy} allele

C. formation of cancer cells

D. variation in flower color between different strains of pea plants, such as purple versus white

attached to DNA sites. After binding, the noncoding RNA can act as a bridge to attract other proteins to the site that cause epigenetic modifications, such as DNA methylation and covalent histone modifications.

S3. Prior to X-chromosome inactivation, what prevents the expression of the *Xist* gene?

Answer: Prior to X-chromosome inactivation, pluripotency factors promote the expression of the *Tsix* gene. The expression of the *Tsix* gene inhibits the expression of the *Xist* gene.

2. **Conceptual Questions:** Test the understanding of basic genetic principles. The student is given many questions with a wide range of difficulty. Some require critical thinking skills, and some require the student to write coherent essay questions.

Conceptual Questions

C1. Define epigenetics. Are all epigenetic changes passed from parent to offspring? Explain.

C2. List and briefly describe five types of molecular events that may underlie epigenetic gene regulation.

C3. Explain how epigenetic changes may be targeted to specific genes.

C4. What is the key difference between a *cis*- and *trans*-epigenetic mechanism that maintains an epigenetic modification? In Chapter 5, we considered genomic imprinting of the *Igf2* gene in which offspring express the copy of the gene they inherit from their father, but not the copy they inherit from their mother. Is this a *cis*- or *trans*-epigenetic mechanism?

C13. With regard to development, what would be the dire consequences if polycomb group complexes did not function properly?

C14. Using coat color in mice and the development of female honeybees as examples, how can dietary factors cause epigenetic modifications, leading to phenotypic effects?

C15. How can environmental agents that do not cause gene mutations contribute to cancer? Would these epigenetic changes be passed to offspring?

C16. Define alternative splicing. What are advantages and disadvantages of this process?

3. **Experimental Questions:** Test the ability to analyze data, design experiments, or appreciate the relevance of experimental techniques.

Experimental Questions

E1. A gene, which we will call gene *C*, can be epigenetically modified in such a way that its expression in some cells is permanently silenced. Describe how you could conduct cell fusion experiments to determine if a *cis*- or *trans*-epigenetic mechanism is responsible for maintaining the silencing of gene *C*.

E2. In the experiments described in Figure 16.7, explain the relationship between coat color and DNA methylation. How is coat color related to the diet of the mother?

E3. 5-Azocytidine is an inhibitor of DNA methyltransferase. If it were fed to female mice during pregnancy, explain how you think it would affect the coat color of offspring carrying the A^{vy} allele.

E4. A research study indicated that an agent in cigarette smoke caused the silencing of a tumor suppressor gene called *p53*. However, upon sequencing, no mutation was found in the DNA sequence for this gene. Give two possible explanations for these results.

E8. Chapter 20 describes a blotting method known as Northern blotting, in which a short segment of cloned DNA is used as a probe to detect RNA that is transcribed from a particular gene. The DNA probe, which is labeled, is complementary to the RNA that the researcher wishes to detect. After the probe DNA binds to the RNA within a blot of a gel, the RNA is visualized as a dark band. The method of Northern blotting can be used to determine the amount of a particular RNA transcribed in a given cell type. If one type of cell produces twice as much of a particular mRNA as another cell, the band appears twice as intense.

For this question, a researcher has a DNA probe complementary to the ferritin mRNA. This probe can be used to specifically detect the amount of ferritin mRNA on a gel. A researcher began with two flasks of human skin cells. One flask contained a very low concentration of iron, and the other flask had a high concentration of iron. The mRNA was isolated from these cells and then subjected

4. **Student Discussion/Collaboration Questions:** Encourage students to consider broad concepts and practical problems. Some questions require a substantial amount of computational activities, which can be worked on as a group.

Questions for Student Discussion/Collaboration

1. Go to the PubMed website and search words such as *epigenetic* and *cancer*. Scan through the journal articles you retrieve and make a list of environmental agents that may cause epigenetic changes that contribute to cancer.
2. Discuss the similarities and differences of phenotypic variation that are caused by epigenetic gene regulation versus variation in gene sequences (epigenetics versus genetics).
3. How are regulatory transcription factors (described in Chapter 15) and regulatory splicing factors (described in this chapter) similar in their mechanism of action? In your discussion, consider the domain structures of both types of proteins. How are they different?

Note: All answers appear at the website for this textbook; the answers to the even-numbered questions and all of the Concept Check and Comprehension Questions are in the back of the book.

CHAPTER OUTLINE

Carbon copy, the first cloned pet. *In 2002, the cat shown here, named Carbon copy (also called Copycat), was produced by cloning, a procedure described in Chapter 21.*

1 OVERVIEW OF GENETICS

Hardly a week goes by without a major news story involving a genetic breakthrough. The increasing pace of genetic discoveries has become staggering. The Human Genome Project is a case in point. This project began in the United States in 1990, when the National Institutes of Health and the Department of Energy joined forces with international partners to decipher the massive amount of information contained in our **genome**—the DNA found within all of our chromosomes (**Figure 1.1**). Working collectively, scientists from around the world produced a detailed series of maps that help geneticists navigate through human DNA. Remarkably, in only a decade, they determined the DNA sequence (read in the bases of A, T, G, and C) of over 90% of the human genome. The first draft of this sequence, published in 2001, showed that the human genome is approximately 3 billion nucleotide base pairs (bp) in length. The completed sequence, published in 2003, has an accuracy greater than 99.99%; less than one mistake was made in every 10,000 base pairs!

Studying the human genome allows us to explore fundamental details about ourselves at the molecular level. The results of the Human Genome Project are expected to shed considerable light on basic questions, like how many genes we have, how genes direct the activities of living cells, how species evolve, how single cells develop into complex tissues, and how defective genes cause disease. Furthermore, such understanding may lend itself to improvements in modern medicine by leading to better diagnoses of diseases and the development of new treatments for them.

The journey to unravel the mysteries within our genes has involved the invention of many new technologies. For example, researchers have developed genetic techniques to produce medicines, such as human insulin, that would otherwise be difficult or impossible to make. Human insulin is synthesized in strains of *Escherichia coli* bacteria that have been genetically altered by the addition of genes that encode the polypeptides that form this hormone. The bacteria are grown in a laboratory and make large amounts of human insulin. As discussed in Chapter 19, the insulin is purified and administered to many people with insulin-dependent diabetes.

New genetic technologies are often met with skepticism and sometimes even with disdain. An example is DNA fingerprinting, a molecular method to identify an individual based on a DNA sample (see Chapter 26). Though this technology is now relatively common in the area of forensic science, it was not always universally accepted. High-profile crime cases in the news cause us to realize that not everyone believes in DNA fingerprinting, in spite of its extraordinary ability to uniquely identify individuals.

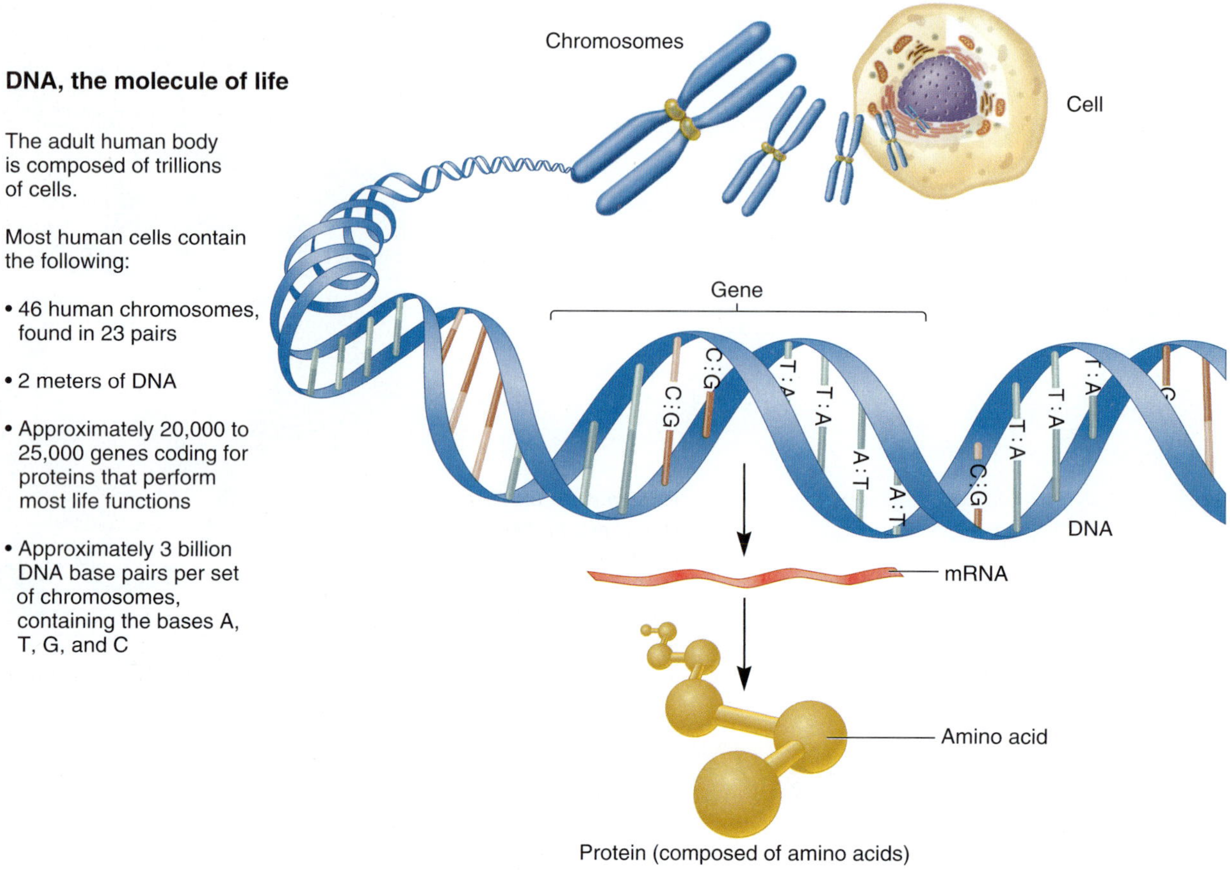

FIGURE 1.1 The human genome. The human genome is a complete set of human chromosomes. People have two sets of chromosomes—one from each parent—which are found in the cell nucleus. The Human Genome Project revealed that each set of chromosomes is composed of a DNA sequence that is approximately 3 billion nucleotide base pairs long. Estimates suggest that each set contains about 22,000 different genes. Most genes encode proteins. As discussed later, such genes are first transcribed into mRNA and then the mRNA is used to make proteins. This figure emphasizes the DNA found in the cell nucleus. Humans also have a small amount of DNA in their mitochondria, which has also been sequenced.

CONCEPT CHECK: How might a better understanding of our genes be used in the field of medicine?

A second controversial example is mammalian cloning. In 1997, Ian Wilmut and his colleagues created clones of sheep, using mammary cells from an adult animal (**Figure 1.2**). More recently, such cloning has been achieved in several mammalian species, including cows, mice, goats, pigs, and cats. In 2002, the first pet was cloned, a cat named Carbon copy, also known as Copycat (see photo at the beginning of the chapter). The cloning of mammals provides the potential for many practical applications. With regard to livestock, cloning would enable farmers to use cells from their best individuals to create genetically homogeneous herds. This could be advantageous in terms of agricultural yield, although such a genetically homogeneous herd may be more susceptible to certain diseases. However, people have become greatly concerned with the possibility of human cloning. This prospect has raised serious ethical questions. Within the past few years, legislation has been introduced that involves bans on human cloning.

Finally, genetic technologies provide the means to modify the traits of animals and plants in ways that would have been unimaginable just a few decades ago. **Figure 1.3a** illustrates a striking example in which scientists introduced a gene from jellyfish into mice. Certain species of jellyfish emit a "green glow" produced by a gene that encodes a bioluminescent protein called green fluorescent protein (GFP). When exposed to blue or ultraviolet (UV) light, the protein emits a striking green-colored light. Scientists were able to clone the *GFP* gene from a sample of jellyfish cells and then introduce this gene into laboratory mice. The green fluorescent protein is made throughout the cells of their bodies. As a result, their skin, eyes, and organs give off an eerie green glow when exposed to UV light. Only their fur does not glow.

The expression of green fluorescent protein allows researchers to identify particular proteins in cells or specific body parts. For example, Andrea Crisanti and colleagues have altered mosquitoes to express GFP only in the gonads of males (**Figure 1.3b**). This

FIGURE 1.2 The cloning of a mammal. The lamb on the left is Dolly, the first mammal to be cloned. She was cloned from the cells of a Finn Dorset (a white-faced sheep). The sheep on the right is Dolly's surrogate mother, a Blackface ewe. A description of how Dolly was produced is presented in Chapter 21.

Photo courtesy of The Roslin Institute, The University of Edinburgh

CONCEPT CHECK: What ethical issues may be associated with human cloning?

(a) GFP expressed in mice

(b) GFP expressed in the gonads of a male mosquito

FIGURE 1.3 The introduction of a jellyfish gene into laboratory mice and mosquitoes. (a) A gene that naturally occurs in jellyfish encodes a protein called green fluorescent protein (GFP). The *GFP* gene was cloned and introduced into mice. When these mice are exposed to UV light, GFP emits a bright green color. These mice glow green, just like the jellyfish! **(b)** The *GFP* gene was introduced next to a gene sequence that causes the expression of GFP only in the gonads of male mosquitoes. This allows researchers to identify and sort males from females.

CONCEPT CHECK: Why is it useful to sort male from female mosquitoes?

enables the researchers to identify and sort males from females. Why is this useful? Researchers can produce a population of mosquitoes and then sterilize the males. The ability to rapidly sort males and females makes it possible to release the sterile males without the risk of releasing additional females. The release of sterile males may be an effective means of controlling mosquito populations because females mate only once before they die. Mating with a sterile male prevents a female from producing offspring. In 2008, Osamu Shimomura, Martin Chalfie, and Roger Tsien received the Nobel Prize in chemistry for the discovery and the development of GFP, which has become a widely used tool in biology.

Overall, as we move forward in the twenty-first century, the excitement level in the field of genetics is high, perhaps higher than it has ever been. Nevertheless, the excitement generated by new genetic knowledge and technologies will also create many ethical and societal challenges. In this chapter, we begin with an overview of genetics and then explore the various fields of genetics and their experimental approaches.

1.1 THE MOLECULAR EXPRESSION OF GENES

Learning Outcomes:

1. Describe the biochemical composition of cells.
2. Explain how proteins are largely responsible for cell structure and function.
3. Outline how DNA stores the information to make proteins.

Genetics is the branch of biology that deals with heredity and variation. It stands as the unifying discipline in biology by allowing us to understand how life can exist at all levels of complexity, ranging from the molecular to the population level. Genetic variation is the root of the natural diversity that we observe among members of the same species as well as among different species.

Genetics is centered on the study of genes. A gene is classically defined as a unit of heredity. At the molecular level, a **gene** is a segment of DNA that produces a functional product. The functional product of most genes is a polypeptide, which is a linear sequence of amino acids that folds into units that constitute proteins. In addition, genes are commonly described according to the way they affect **traits,** which are the characteristics of an organism. In humans, for example, we speak of traits such as eye color, hair texture, and height. The ongoing theme of this textbook is the relationship between genes and traits. As an organism grows and develops, its collection of genes provides a blueprint that determines its traits.

In this section of Chapter 1, we examine the general features of life, beginning with the molecular level and ending with populations of organisms. As will become apparent, genetics is the common thread that explains the existence of life and its continuity from generation to generation. For most students, this chapter should serve as an overview of topics they have learned in other introductory courses such as General Biology. Even so, it is usually helpful to see the "big picture" of genetics before delving into the finer details that are covered in Chapters 2 through 28.

Living Cells Are Composed of Biochemicals

To fully understand the relationship between genes and traits, we need to begin with an examination of the composition of living organisms. Every cell is constructed from intricately organized chemical substances. Small organic molecules such as glucose and amino acids are produced from the linkage of atoms via chemical bonds. The chemical properties of organic molecules are essential for cell vitality in two key ways. First, the breaking of chemical bonds during the degradation of small molecules provides energy to drive cellular processes. A second important function of these small organic molecules is their role as the building blocks for the synthesis of larger molecules. Four important categories of larger cellular molecules are **nucleic acids** (i.e., DNA and RNA), **proteins, carbohydrates,** and **lipids.** Three of these—nucleic acids, proteins, and carbohydrates—form **macromolecules** that are composed of many repeating units of smaller building blocks. RNA, proteins, and carbohydrates can be made from hundreds or even thousands of repeating building blocks. DNA is the largest macromolecule found in living cells. A single DNA molecule can be composed of a linear sequence of hundreds of millions of nucleotides!

The formation of cellular structures relies on the interactions of molecules and macromolecules. For example, nucleotides are the building blocks of DNA, which is a constituent of chromosomes (**Figure 1.4**). In addition, DNA is associated with many proteins that provide organization to the structure of chromosomes. Within a eukaryotic cell, the chromosomes are contained in a compartment called the cell nucleus. The nucleus is bounded by a double membrane composed of lipids and proteins that shields the chromosomes from the rest of the cell. The

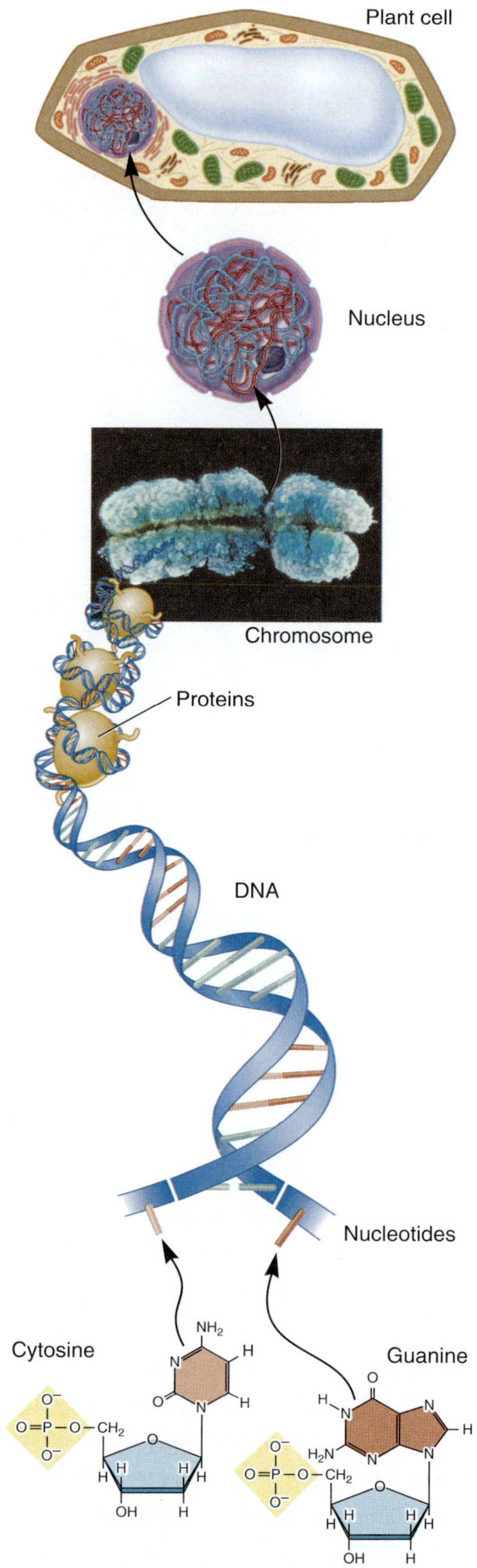

FIGURE 1.4 Molecular organization of a living cell. Cellular structures are constructed from smaller building blocks. In this example, DNA is formed from the linkage of nucleotides to produce a very long macromolecule. The DNA associates with proteins to form a chromosome. The chromosomes are located within a membrane-bound organelle called the nucleus, which, along with many different types of organelles, is found within a complete cell.

CONCEPT CHECK: Is DNA a small molecule, a macromolecule, or an organelle?

organization of chromosomes within a cell nucleus protects the chromosomes from mechanical damage and provides a single compartment for genetic activities such as gene transcription. As a general theme, the formation of large cellular structures arises from interactions among different molecules and macromolecules. These cellular structures, in turn, are organized to make a complete living cell.

Each Cell Contains Many Different Proteins That Determine Cell Structure and Function

To a great extent, the characteristics of a cell depend on the types of proteins that it makes. All of the proteins that a cell makes at a given time is called its **proteome.** The range of functions among different types of proteins is truly remarkable. Some proteins help determine the shape and structure of a given cell. For example, the protein known as tubulin assembles into large structures known as microtubules, which provide the cell with internal structure and organization. Other proteins are inserted into cell membranes and aid in the transport of ions and small molecules across the membrane. **Enzymes,** which accelerate chemical reactions, are a particularly important category of proteins. Some enzymes play a role in the breakdown of molecules or macromolecules into smaller units. These are known as catabolic enzymes and are important in the utilization of energy. Alternatively, anabolic enzymes and accessory proteins function in the synthesis of molecules and macromolecules throughout the cell. The construction of a cell greatly depends on its proteins involved in anabolism because these are required to synthesize all cellular macromolecules.

Molecular biologists have come to realize that the functions of proteins underlie the cellular characteristics of every organism. At the molecular level, proteins can be viewed as the active participants in the enterprise of life.

DNA Stores the Information for Protein Synthesis

The genetic material of living organisms is composed of a substance called **deoxyribonucleic acid,** abbreviated **DNA.** The DNA stores the information needed for the synthesis of all cellular proteins. In other words, the main function of the genetic blueprint is to code for the production of proteins in the correct cell, at the proper time, and in suitable amounts. This is an extremely complicated task because living cells make thousands of different proteins. Genetic analyses have shown that a typical bacterium can make a few thousand different proteins, and estimates among complex eukaryotic species range in the tens of thousands.

DNA's ability to store information is based on its structure. DNA is composed of a linear sequence of **nucleotides.** Each nucleotide contains one of four nitrogen-containing bases: adenine (A), thymine (T), guanine (G), or cytosine (C). The linear order of these bases along a DNA molecule contains information similar to the way that groups of letters of the alphabet represent words. For example, the "meaning" of the sequence of bases ATGGGCCTTAGC differs from that of TTTAAGCTTGCC. DNA sequences within most genes contain the information to direct the order of amino acids within **polypeptides** according to the **genetic code.** In the code, a three-base sequence specifies one particular **amino acid** among the 20 possible choices. One or more polypeptides form a functional protein. In this way, the DNA can store the information to specify the proteins made by an organism.

DNA Sequence	Amino Acid Sequence
ATG GGC CTT AGC	Methionine Glycine Leucine Serine
TTT AAG CTT GCC	Phenylalanine Lysine Leucine Alanine

In living cells, the DNA is found within large structures known as **chromosomes.** Figure 1.5 is a micrograph of the 46 chromosomes contained in a cell from a human male, otherwise known as a karyotype. The DNA of an average human

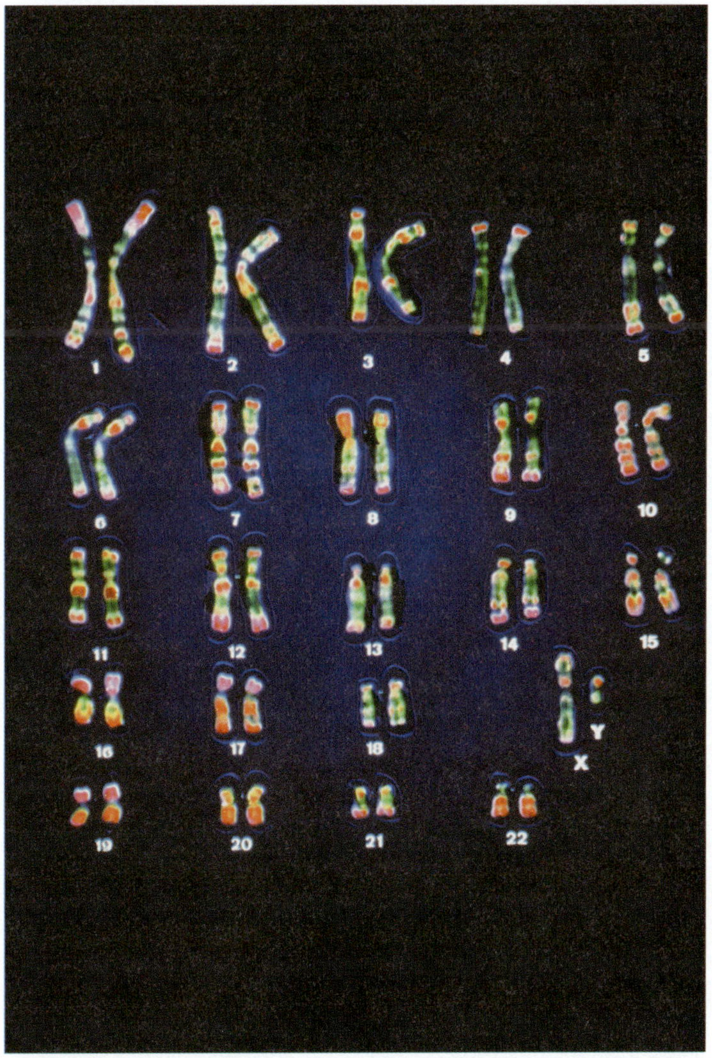

FIGURE 1.5 **A micrograph of the 46 chromosomes found in a cell from a human male.**

CONCEPT CHECK: Which types of macromolecules are found in chromosomes?

chromosome is an extraordinarily long, linear, double-stranded structure that contains well over a hundred million nucleotides. Along the immense length of a chromosome, the genetic information is parceled into functional units known as genes. An average-sized human chromosome is expected to contain about 1000 different genes.

The Information in DNA Is Accessed During the Process of Gene Expression

To synthesize its proteins, a cell must be able to access the information that is stored within its DNA. The process of using a gene sequence to affect the characteristics of cells and organisms is referred to as **gene expression.** At the molecular level, the information within genes is accessed in a stepwise process (Figure 1.6). In the first step, known as **transcription,** the DNA sequence within a gene is copied into a nucleotide sequence of **ribonucleic acid (RNA). Protein-encoding genes** (also called **structural genes**) carry the information for the amino acid sequence of a polypeptide. When a protein-encoding gene is transcribed, the first product is an RNA molecule known as **messenger RNA (mRNA).** During polypeptide synthesis—a process called **translation**—the sequence of nucleotides within the mRNA determines the sequence of amino acids in a polypeptide. One or more polypeptides then fold and assemble into a functional protein. The synthesis of functional proteins ultimately determines an organism's traits. As discussed further in Chapter 12 (look ahead to Figure 12.1), the pathway of gene expression from DNA to RNA to protein is called the **central dogma of genetics** (also called the central dogma of molecular biology). It forms a cornerstone of our understanding of genetics at the molecular level.

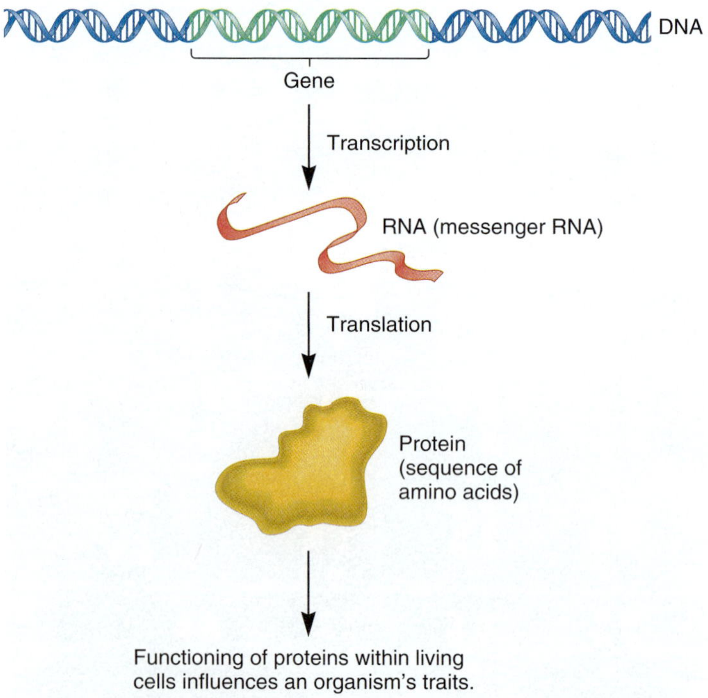

ONLINE ANIMATION

FIGURE 1.6 Gene expression at the molecular level. The expression of a gene is a multistep process. During transcription, one of the DNA strands is used as a template to make an RNA strand. During translation, the RNA strand is used to specify the sequence of amino acids within a polypeptide. One or more polypeptides produce a protein that functions within the cell, thereby influencing an organism's traits.

CONCEPT CHECK: Where is the information to make a polypeptide stored?

1.1 COMPREHENSION QUESTIONS

1. Which of the following is not a constituent of a cell's proteome?
   a. An enzyme
   b. A cytoskeletal protein
   c. A transport protein in the plasma membrane
   d. An mRNA
2. A gene is a segment of DNA that has the information to produce a functional product. The functional product of most genes is
   a. DNA.
   b. mRNA.
   c. a polypeptide.
   d. all of the above.
3. The function of the genetic code is to
   a. promote transcription.
   b. specify the amino acids within a polypeptide.
   c. alter the sequence of DNA.
   d. none of the above.
4. The process of transcription directly results in the synthesis of
   a. DNA.
   b. RNA.
   c. a polypeptide.
   d. all of the above.

1.2 THE RELATIONSHIP BETWEEN GENES AND TRAITS

Learning Outcomes:

1. Explain how the expression of genes leads to an organism's traits.
2. Define genetic variation.
3. Discuss the relationship between genes and traits.
4. Describe how genes are transmitted in sexually reproducing species.
5. Outline the process of evolution.

A trait is any characteristic that an organism displays. In genetics, we often focus our attention on **morphological traits** that affect the appearance, form, and structure of an organism. The color of a flower and the height of a pea plant are morphological traits. Geneticists frequently study these types of traits because they are easy to evaluate. For example, an experimenter can simply look at a plant and tell if it has red or white flowers. However, not all traits are morphological. **Physiological traits** affect the ability of an organism to function. For example, the rate at which a bacterium metabolizes a sugar such as lactose is a physiological trait. Like morphological traits, physiological traits are controlled, in part, by the expression of genes. **Behavioral traits** also affect the ways an organism responds to its environment. An example is the mating calls of bird species. In animals, the nervous system plays a key role in governing such traits. In this section, we will examine the relationship between the expression of genes and an organism's traits.

The Molecular Expression of Genes Leads to an Organism's Traits

A complicated, yet very exciting, aspect of genetics is that our observations and theories span four levels of biological organization: molecules, cells, organisms, and populations. This can make it difficult to appreciate the relationship between genes and traits. To understand this connection, we need to relate the following phenomena:

1. Genes are expressed at the **molecular level.** In other words, gene transcription and translation lead to the production of a particular protein, which is a molecular process.
2. Proteins often function at the **cellular level.** The function of a protein within a cell affects the structure and workings of that cell.
3. An organism's traits are determined by the characteristics of its cells. We do not have microscopic vision, yet when we view morphological traits, we are really observing the properties of an individual's cells. For example, a red flower has its color because the flower cells make a red pigment. The trait of red flower color is an observation at the **organism level.** Yet the trait is rooted in the molecular characteristics of the organism's cells.
4. A **species** is a group of organisms that maintains a distinctive set of attributes in nature. The occurrence of a trait within a species is an observation at the **population level.** Along with learning how a trait occurs, we also want to understand why a trait becomes prevalent in a particular species. In many cases, researchers discover that a trait predominates within a population because it promotes the reproductive success of the members of the population. This leads to the evolution of beneficial traits.

As a schematic example to illustrate the four levels of genetics, Figure 1.7 shows the trait of pigmentation in butterflies. One is dark-colored and the other is very light. Let's consider how we can explain this trait at the molecular, cellular, organism, and population levels.

At the molecular level, we need to understand the nature of the gene or genes that govern this trait. As shown in Figure 1.7a, a gene, which we will call the pigmentation gene, is responsible for the amount of pigment produced. The pigmentation gene exists in two different versions. Alternative versions of a specific gene are called **alleles.** In this example, one allele confers a dark pigmentation and the other causes a light pigmentation. Each of these alleles encodes a protein that functions as a pigment-synthesizing enzyme. However, the DNA sequences of

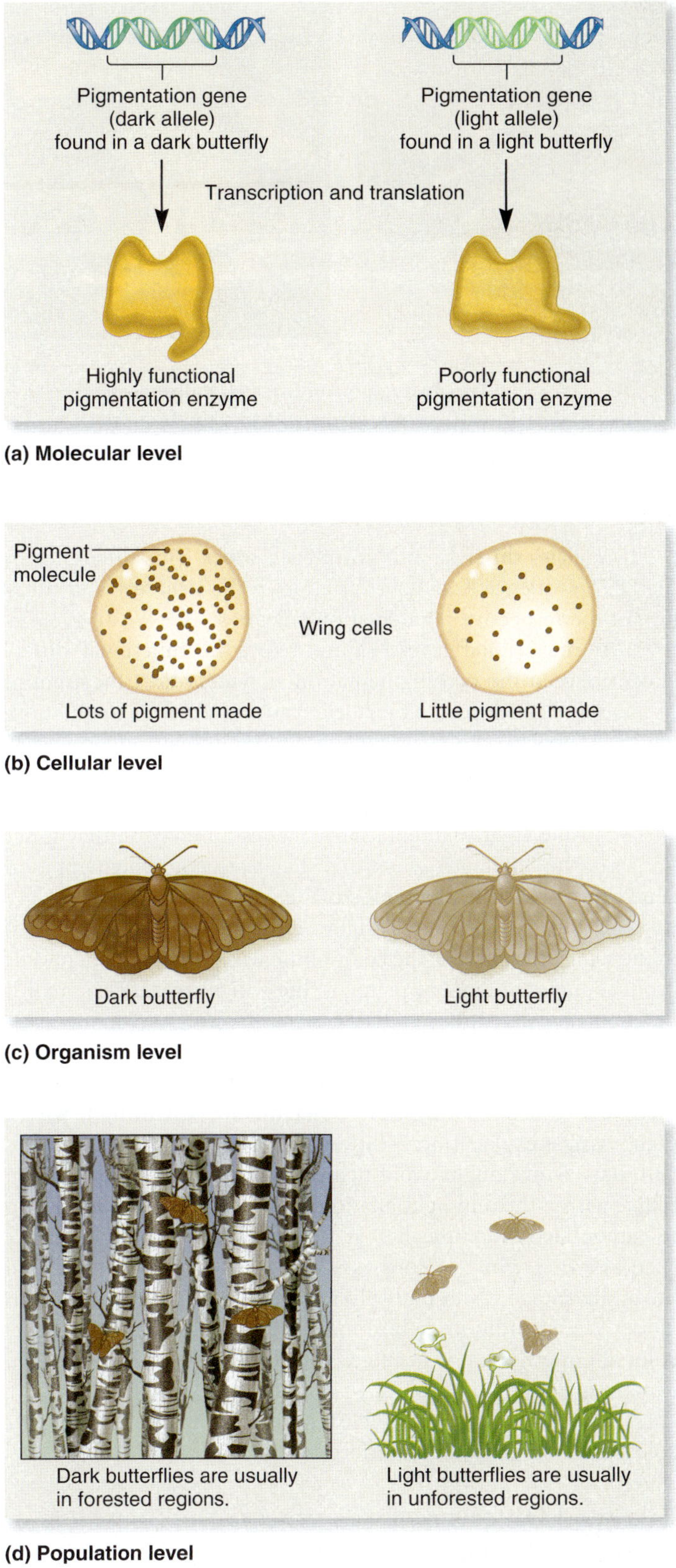

FIGURE 1.7 The relationship between genes and traits at the (a) molecular, (b) cellular, (c) organism, and (d) population levels.

CONCEPT CHECK: Which butterfly has a more active pigment-producing enzyme, the dark- or light-colored one?

FIGURE 1.8 **Two dyeing poison frogs (*Dendrobates tinctorius*) showing different morphs within a single species.**

CONCEPT CHECK: Why do these two frogs look so different?

the two alleles differ slightly from each other. This difference in the DNA sequence leads to a variation in the structure and function of the respective pigmentation enzymes.

At the cellular level (Figure 1.7b), the functional differences between the two pigmentation enzymes affect the amount of pigment produced. The allele causing dark pigmentation, which is shown on the left, encodes a protein that functions very well. Therefore, when this gene is expressed in the cells of the wings, a large amount of pigment is made. By comparison, the allele causing light pigmentation encodes an enzyme that functions poorly. Therefore, when this allele is the only pigmentation gene expressed, little pigment is made.

At the organism level (Figure 1.7c), the amount of pigment in the wing cells governs the color of the wings. If the pigment cells produce high amounts of pigment, the wings are dark-colored. If the pigment cells produce little pigment, the wings are light.

Finally, at the population level (Figure 1.7d), geneticists would like to know why a species of butterfly would contain some members with dark wings and other members with light wings. One possible explanation is differential predation. The butterflies with dark wings might avoid being eaten by birds if they happen to live within the dim light of a forest. The dark wings would help to camouflage the butterfly if it were perched on a dark surface such as a tree trunk. In contrast, the lightly colored wings would be an advantage if the butterfly inhabited a brightly lit meadow. Under these conditions, a bird may be less likely to notice a light-colored butterfly that is perched on a sunlit surface. A population geneticist might study this species of butterfly and find that the dark-colored members usually live in forested areas and the light-colored members reside in unforested regions.

Inherited Differences in Traits Are Due to Genetic Variation

In Figure 1.7, we considered how gene expression leads to variation in a trait of organisms, such as dark- versus light-colored butterflies. Variation in traits among members of the same species is very common. For example, some people have brown hair, and others have blond hair; some petunias have white flowers and some have purple flowers. These are examples of **genetic variation.** This term describes the differences in inherited traits among individuals within a population.

In large populations that occupy a wide geographic range, genetic variation can be quite striking. In fact, morphological differences have often led geneticists to misidentify two members of the same species as belonging to separate species. As an example, **Figure 1.8** shows two dyeing poison frogs that are members of the same species, *Dendrobates tinctorius.* They display dramatic differences in their markings. Such contrasting forms within a single species are termed **morphs.** You can easily imagine how someone might mistakenly conclude that these frogs are not members of the same species.

Changes in the nucleotide sequence of DNA underlie the genetic variation that we see among individuals. Throughout this textbook, we will routinely examine how variation in the genetic material results in changes in an organism's traits. At the molecular level, genetic variation can be attributed to different types of modifications.

1. Small or large differences can occur within gene sequences. When such changes initially occur, they are called **gene mutations.** Mutations result in genetic variation in which a gene is found in two or more alleles, as previously described in Figure 1.7. In many cases, gene mutations alter the expression or function of the protein that the gene encodes.
2. Major alterations can also occur in the structure of a chromosome. A large segment of a chromosome can be lost, rearranged, or reattached to another chromosome.
3. Variation may also occur in the total number of chromosomes. In some cases, an organism may inherit one too many or one too few chromosomes. In other cases, it may inherit an extra set of chromosomes.

Variations of sequences within genes are a common source of genetic variation among members of the same species. In humans, familiar examples of variation involve genes for eye color, hair texture, and skin pigmentation. Chromosome variation—a change in chromosome structure or number (or both)—is also found, but this type of change is often detrimental. Many human genetic disorders are the result of chromosomal alterations. The most common example is Down syndrome, which is due to the presence of an extra chromosome (**Figure 1.9a**). By comparison, chromosome variation in plants is common and often leads to plants with superior characteristics, such as increased resistance to disease. Plant breeders have frequently exploited this observation. Cultivated varieties of wheat, for example, have many more chromosomes than the wild species (**Figure 1.9b**).

Traits Are Governed by Genes and by the Environment

In our discussion thus far, we have considered the role that genes play in determining an organism's traits. Another critical factor is the **environment**—the surroundings in which an organism exists. A variety of factors in an organism's environment profoundly affect its morphological and physiological features. For example, a

(a)

(b)

FIGURE 1.9 Examples of chromosome variation. **(a)** A person with Down syndrome competing in the Special Olympics. This person has 47 chromosomes rather than the common number of 46, because she has an extra copy of chromosome 21. **(b)** A wheat plant. Cultivated wheat is derived from the contributions of three wild species with two sets of chromosomes each, producing an organism with six sets of chromosomes.

CONCEPT CHECK: Do these examples constitute variation in chromosome structure or variation in chromosome number?

person's diet greatly influences many traits such as height, weight, and even intelligence. Likewise, the amount of sunlight a plant receives affects its growth rate and the color of its flowers.

An interesting example of the interplay between genes and the environment involves the human genetic disease **phenylketonuria (PKU).** Humans have a gene that encodes an enzyme known as phenylalanine hydroxylase. Most people have two functional copies of this gene. People with one or two functional copies of the gene can eat foods containing the amino acid phenylalanine and metabolize it properly. A rare variation in the gene that encodes phenylalanine hydroxylase results in a nonfunctional version of this enzyme. Individuals with two copies of this rare, inactive allele cannot metabolize phenylalanine properly. When given a standard diet containing phenylalanine, individuals with this disorder are unable to break down this amino acid. Phenylalanine accumulates and is converted into phenylketones, which are detected in the urine. Individuals with PKU can manifest a variety of detrimental traits, including mental impairment, underdeveloped teeth, and foul-smelling urine. Fortunately, through routine newborn screening in the United States, PKU is now diagnosed early. Part of the treatment is a diet that restricts phenylalanine, which is present in high-protein foods such as eggs, meat, and dairy products. Restricting phenylalanine allows the affected child to develop normally. PKU provides a dramatic example of how the environment and an individual's genes can interact to influence the traits of the organism.

During Reproduction, Genes Are Passed from Parent to Offspring

Now that we have considered how genes and the environment govern the outcome of traits, we can turn to the issue of inheritance. How are traits passed from parents to offspring? The foundation for our understanding of inheritance came from Gregor Mendel's study of pea plants in the nineteenth century. His work revealed that the genetic determinants that govern traits, which we now call genes, are passed from parent to offspring as discrete units. We can predict the outcome of many genetic crosses based on Mendel's laws of inheritance.

The inheritance patterns identified by Mendel can be explained by the existence of chromosomes and their behavior during cell division. As in Mendel's pea plants, sexually reproducing species are commonly **diploid.** This means they contain two copies of each chromosome, one from each parent. The two copies are called **homologs** of each other. Because genes are located within chromosomes, diploid organisms have two copies of most genes. Humans, for example, have 46 chromosomes, which are found in homologous pairs (**Figure 1.10a**). With the exception of the sex chromosomes (X and Y), each homologous pair contains the same kinds of genes. For example, both copies of human chromosome 12 carry the gene that encodes phenylalanine hydroxylase, which was discussed previously. Therefore, an individual has two copies of this gene, which may or may not be identical alleles.

Most cells of the human body that are not directly involved in sexual reproduction contain 46 chromosomes. These cells are called **somatic cells.** In contrast, the **gametes**—sperm and egg cells—contain half that number (23) and are termed **haploid** (**Figure 1.10b**). The union of gametes during fertilization restores the diploid number of chromosomes. The primary advantage of sexual reproduction is that it enhances genetic variation. For example, a tall person with blue eyes and a short person with brown eyes may have short offspring with blue eyes or tall offspring with brown eyes. Therefore, sexual reproduction can result in new combinations of two or more traits that differ from those of either parent.

The Genetic Composition of a Species Evolves over the Course of Many Generations

As we have just seen, sexual reproduction has the potential to enhance genetic variation. This can be an advantage for a population of individuals as they struggle to survive and compete within their natural environment. The term **biological evolution,** or simply, **evolution,** refers to the phenomenon that the genetic makeup of a population changes from one generation to the next.

As suggested by Charles Darwin, the members of a species are in competition with one another for essential resources. Random genetic changes (i.e., mutations) occasionally occur within an individual's genes, and sometimes these changes lead to a modification of traits that promote reproductive success. For example,

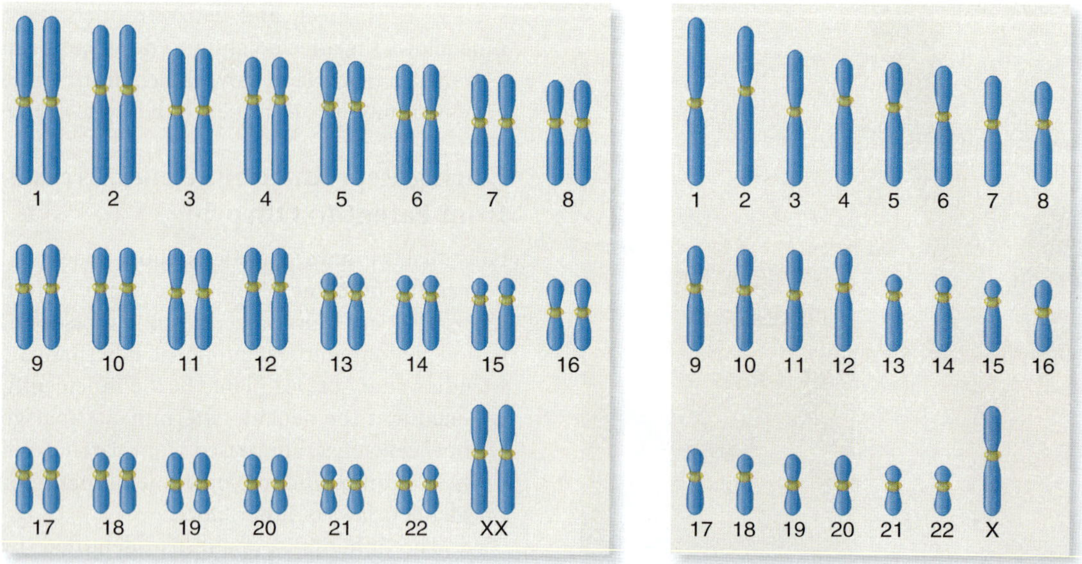

(a) Chromosomal composition found in human somatic cells of females (46 chromosomes)

(b) Chromosomal composition found in a human gamete (23 chromosomes)

FIGURE 1.10 **The complement of human chromosomes in somatic cells and gametes.** **(a)** A schematic drawing of the 46 chromosomes of a human. With the exception of the sex chromosomes, these are always found in homologous pairs. **(b)** The chromosomal composition of a gamete, which contains only 23 chromosomes, one from each pair. This gamete contains an X chromosome. Half of the gametes from human males contain a Y chromosome instead of the X chromosome.

CONCEPT CHECK: The leaf cells of a corn plant contain 20 chromosomes each. How many chromosomes are found in a gamete made by a corn plant?

over the course of many generations, random gene mutations have lengthened the snout and extended the tongue of the anteater, enabling it to probe into the ground and feed on ants. When a mutation creates a new allele that is beneficial, the allele may become prevalent in future generations because the individuals carrying the allele are more likely to reproduce and pass the beneficial allele to their offspring. This process is known as **natural selection.** In this way, a species becomes better adapted to its environment.

Over a long period of time, the accumulation of many genetic changes may lead to rather striking modifications in a species' characteristics. As an example, **Figure 1.11** depicts the evolution of the modern-day horse. Over time, a variety of morphological changes occurred, including an increase in size, fewer toes, and modified jaw structure. The changes can be attributed to natural selection. Over North America, where much of horse evolution occurred, large areas of dense forests were replaced with grasslands. The increase in size and changes in foot structure enabled horses to escape predators more easily and travel greater distances in search of food. Natural selection favored the changes seen in horses' teeth, because such changes allowed them to eat grasses and other vegetation that are more abrasive and require more chewing.

1.2 COMPREHENSION QUESTIONS

1. Gene expression can be viewed at which of the following levels?
   a. Molecular and cellular levels
   b. Organism level
   c. Population level
   d. All of the above
2. Variation in the traits of organisms may be attributable to
   a. gene mutations.
   b. alterations in chromosome structure.
   c. variation in chromosome number.
   d. all of the above.
3. A human skin cell has 46 chromosomes. A human sperm cell has
   a. 23. b. 46. c. 92. d. none of the above.
4. Evolutionary change caused by natural selection results in species with
   a. greater complexity.
   b. less complexity.
   c. greater reproductive success in their native environment.
   d. the ability to survive longer.

1.3 FIELDS OF GENETICS

Learning Outcomes:

1. Compare and contrast the three major fields of genetics: transmission, molecular, and population genetics.
2. Discuss how genetics is an experimental science.

Genetics is a broad discipline encompassing molecular, cellular, organism, and population biology. Many scientists who are interested in genetics have been trained in supporting disciplines such

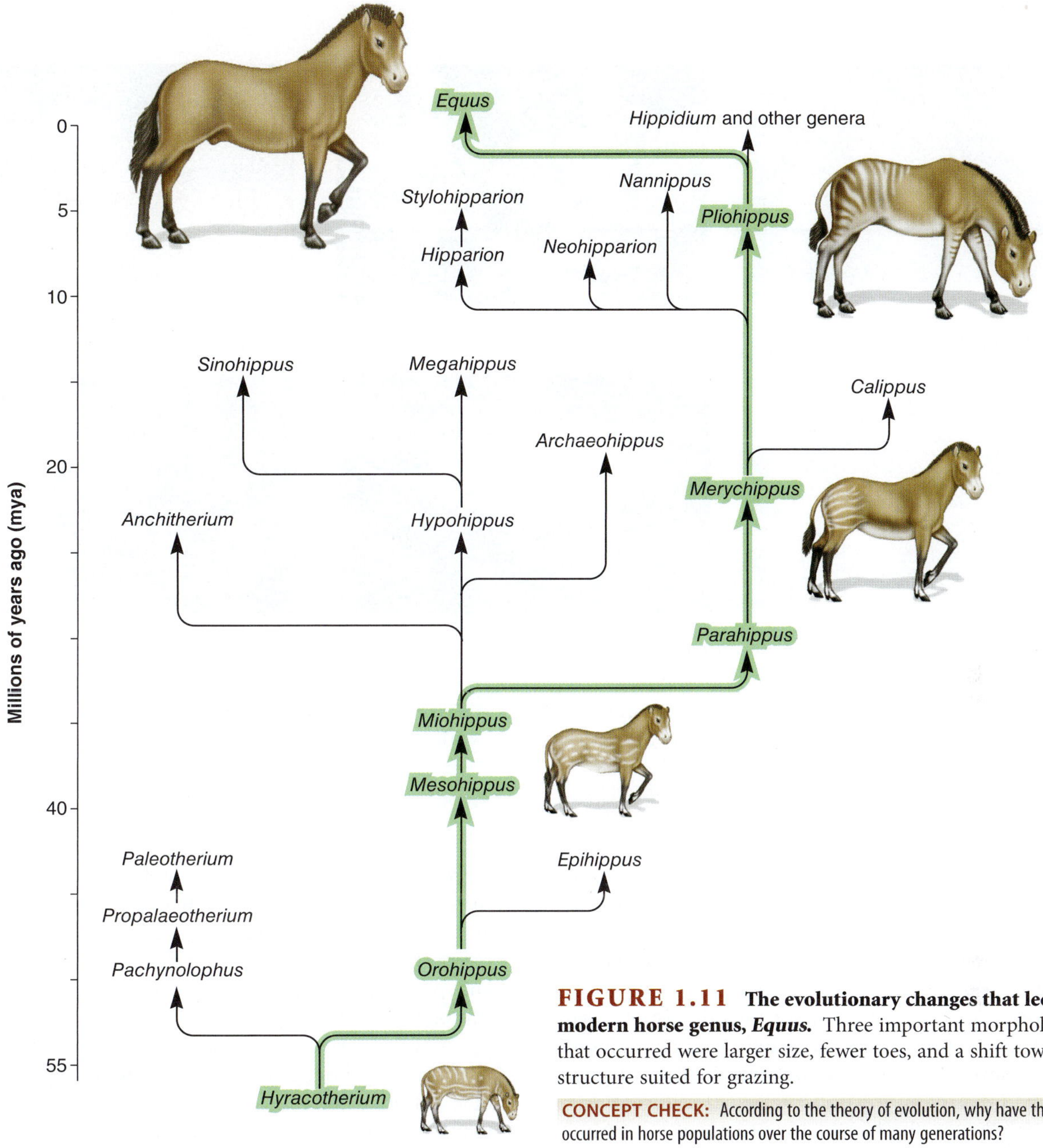

FIGURE 1.11 The evolutionary changes that led to the modern horse genus, *Equus*. Three important morphological changes that occurred were larger size, fewer toes, and a shift toward a jaw structure suited for grazing.

CONCEPT CHECK: According to the theory of evolution, why have these changes occurred in horse populations over the course of many generations?

as biochemistry, biophysics, cell biology, mathematics, microbiology, population biology, ecology, agriculture, and medicine. Experimentally, geneticists often focus their efforts on **model organisms**—organisms studied by many different researchers so they can compare their results and determine scientific principles that apply more broadly to other species. **Figure 1.12** shows some common examples, including *Escherichia coli* (a bacterium), *Saccharomyces cerevisiae* (a yeast), *Drosophila melanogaster* (fruit fly), *Caenorhabditis elegans* (a nematode worm), *Mus musculus* (mouse), and *Arabidopsis thaliana* (a flowering plant). Model organisms offer experimental advantages over other species. For example, *E. coli* is a very simple organism that can be easily grown in the laboratory. By limiting their work to a few such model organisms, researchers can more easily unravel the genetic mechanisms that govern the traits of a given species. Furthermore, the genes found in model organisms often function in a similar way to those found in humans.

The study of genetics has been traditionally divided into three areas—transmission, molecular, and population genetics—although overlap is found among them. In this section, we will examine the three fields of genetics.

Transmission Genetics Explores the Inheritance Patterns of Traits as They Are Passed from Parents to Offspring

A scientist working in the field of transmission genetics examines the relationship between the transmission of genes from parent to offspring and the outcome of the offspring's traits. For example, how can two brown-eyed parents produce a blue-eyed child? Or why do tall parents tend to produce tall children, but not always? Our modern understanding of transmission genetics began with the studies of Gregor Mendel. His work provided the

FIGURE 1.12 **Examples of model organisms studied by geneticists.** **(a)** *Escherichia coli* (a bacterium), **(b)** *Saccharomyces cerevisiae* (a yeast), **(c)** *Drosophila melanogaster* (fruit fly), **(d)** *Caenorhabditis elegans* (a nematode worm), **(e)** *Mus musculus* (mouse), and **(f)** *Arabidopsis thaliana* (a flowering plant).

CONCEPT CHECK: Can you think of another example of a model organism?

conceptual framework for transmission genetics. In particular, he originated the idea that factors, which we now call genes, are passed as discrete units from parents to offspring via sperm and egg cells. Since these pioneering studies in the 1860s, our knowledge of genetic transmission has greatly increased. Many patterns of genetic transmission are more complex than the simple Mendelian patterns that are described in Chapter 2. The additional complexities of transmission genetics are examined in Chapters 3 through 8.

Experimentally, the fundamental approach of a transmission geneticist is the **genetic cross**—the breeding of two selected individuals and the subsequent analysis of their offspring in an attempt to understand how traits are passed from parents to offspring. In the case of experimental organisms, the researcher chooses two parents with particular traits and then categorizes the offspring according to the traits they possess. In many cases, this analysis is quantitative in nature. For example, an experimenter may cross two tall pea plants and obtain 100 offspring that fall into two categories: 75 tall and 25 dwarf. As we will see in Chapter 2, the ratio of tall and dwarf offspring provides important information concerning the inheritance pattern of this trait.

Molecular Genetics Focuses on a Biochemical Understanding of the Hereditary Material

The goal of molecular genetics, as the name of the field implies, is to understand how the genetic material works at the molecular level. In other words, molecular geneticists want to understand the molecular features of DNA and how these features underlie the expression of genes. The experiments of molecular geneticists are usually conducted within the confines of a laboratory. Their efforts frequently progress to a detailed analysis of DNA, RNA, and proteins, using a variety of techniques that are described throughout Parts III, IV, and V of this textbook.

Molecular geneticists often study mutant genes that have abnormal function. This is called a **genetic approach** to the study of a research question. In many cases, researchers analyze the effects of gene mutations that eliminate the function of a gene. This type of mutation is called a **loss-of-function mutation,** and the resulting version of the gene is called a **loss-of-function allele.** By studying the effects of such mutations, the role of the functional, nonmutant gene is often revealed. For example, let's suppose that a particular plant species produces purple flowers. If a loss-of-function mutation within a given gene causes a plant of that species to produce white flowers, one would suspect the role of the functional gene involves the production of purple pigmentation.

Studies within molecular genetics interface with other disciplines such as biochemistry, biophysics, and cell biology. In addition, advances within molecular genetics have shed considerable light on the areas of transmission and population genetics. Our quest to understand molecular genetics has spawned a variety of modern molecular technologies and computer-based approaches. Furthermore, discoveries within molecular genetics have had widespread applications in agriculture, medicine, and biotechnology.

Population Genetics Is Concerned with Genetic Variation and Its Role in Evolution

The foundations of population genetics arose during the first few decades of the twentieth century. Although many scientists of this era did not accept the findings of Mendel or Darwin, the theories of population genetics provided a compelling way to connect the two viewpoints. Mendel's work and that of many succeeding geneticists gave insight into the nature of genes and how they are transmitted from parents to offspring. The work of Darwin provided a natural explanation for the variation in characteristics observed among the members of a species. To relate these two phenomena, population geneticists have developed mathematical theories to explain the prevalence of certain alleles within populations of individuals. This work helps us understand how processes such as natural selection have resulted in the prevalence of individuals that carry particular alleles.

Population geneticists are particularly interested in genetic variation and how that variation is related to an organism's environment. In this field, the frequencies of alleles within a population are of central importance.

Genetics Is an Experimental Science

Science is a way of knowing about our natural world. The science of genetics allows us to understand how the expression of our genes produces the traits that we possess. Researchers typically follow two general types of scientific approaches: hypothesis testing and discovery-based science. In **hypothesis testing,** also called the **scientific method,** scientists follow a series of steps to reach verifiable conclusions about the world. Although scientists arrive at their theories in different ways, the scientific method provides a way to validate (or invalidate) a particular hypothesis. Alternatively, research may also involve the collection of data without a preconceived hypothesis. For example, researchers might analyze the genes found in cancer cells to identify those that have become mutant. In this case, the scientists may not have a hypothesis about which particular genes may be involved. The collection and analysis of data without the need for a preconceived hypothesis is called **discovery-based science** or, simply, discovery science.

In traditional science textbooks, the emphasis often lies on the product of science. Namely, many textbooks are aimed primarily at teaching the student about the observations scientists have made and the hypotheses they have proposed to explain those observations. Along the way, the student is provided with many bits and pieces of experimental techniques and data. Likewise, this textbook also provides you with many observations and hypotheses. However, it attempts to go one step further. Each of the following chapters contains one or two experiments that have been "dissected" into five individual components to help you to understand the entire scientific process:

1. Background information is provided so you can appreciate what earlier observations were known prior to conducting the experiment.
2. Most experiments involve hypothesis testing. In those cases, the figure states the hypothesis the scientists were trying to test. In other words, what scientific question was the researcher trying to answer?
3. Next, the figure follows the experimental steps the scientist took to test the hypothesis. Each featured experiment contains two parallel illustrations labeled Experimental Level and Conceptual Level. The Experimental Level helps you to understand the techniques followed. The Conceptual Level helps you to understand what is actually happening at each step in the procedure.
4. The raw data for each experiment are then presented.
5. Last, an interpretation of the data is offered within the text.

The rationale behind this approach is that it will enable you to see the experimental process from beginning to end. As you read through the chapters, the experiments will help you to see the relationship between science and scientific theories.

As a student of genetics, you will be given the opportunity to involve your mind in the experimental process. As you are reading an experiment, you may find yourself thinking about different approaches and alternative hypotheses. Different people can view the same data and arrive at very different conclusions. As you progress through the experiments in this book, you will enjoy genetics far more if you try to develop your own skills at formulating hypotheses, designing experiments, and interpreting data. Also, some of the questions in the problem sets are aimed at refining these skills.

Finally, it is worthwhile to point out that science is a social discipline. As you develop your skills at scrutinizing experiments, it is fun to discuss your ideas with other people, including fellow students and faculty members. Keep in mind that you do not need to "know all the answers" before you enter into a scientific discussion. Instead, it is more rewarding to view science as an ongoing and never-ending dialogue.

1.3 COMPREHENSION QUESTIONS

1. Which of the following is *not* a model organism?
   a. *Mus musculus* (laboratory mouse)
   b. *Escherichia coli* (a bacterium)
   c. *Saccharomyces cerevisiae* (a yeast)
   d. *Sciurus carolinensis* (gray squirrel)

2. A person studying the rate of transcription of a particular gene is working in the field of
   a. molecular genetics.
   b. transmission genetics.
   c. population genetics.
   d. none of the above.

3. The scientific method involves which of the following?
   a. The collection of observations and the formulation of a hypothesis
   b. Experimentation
   c. Data analysis and interpretation
   d. All of the above

KEY TERMS

Page 1. genome
Page 3. genetics
Page 4. gene, traits, nucleic acids, proteins, carbohydrates, lipids, macromolecules
Page 5. proteome, enzymes, deoxyribonucleic acid (DNA), nucleotides, polypeptides, genetic code, amino acid, chromosomes
Page 6. gene expression, transcription, ribonucleic acid (RNA), protein-encoding genes, structural genes, messenger RNA (mRNA), translation, central dogma of genetics, morphological traits, physiological traits
Page 7. behavioral traits, molecular level, cellular level, organism level, species, population level, alleles
Page 8. genetic variation, morphs, gene mutations, environment
Page 9. phenylketonuria (PKU), diploid, homologs, somatic cells, gametes, haploid, biological evolution, evolution
Page 10. natural selection
Page 11. model organisms
Page 12. genetic cross, genetic approach, loss-of-function mutation, loss-of-function allele
Page 13. hypothesis testing, scientific method, discovery-based science

CHAPTER SUMMARY

- The complete genetic composition of a cell or organism is called a genome. The genome encodes all of the proteins a cell or organism can make. Many key discoveries in genetics are related to the study of genes and genomes (see Figures 1.1, 1.2, 1.3).

1.1 The Molecular Expression of Genes

- Living cells are composed of nucleic acids (DNA and RNA), proteins, carbohydrates, and lipids. The proteome largely determines the structure and function of cells (see Figure 1.4).
- DNA, which is found within chromosomes, stores the information to make proteins (see Figure 1.5).
- Most genes encode polypeptides that are units within functional proteins. Gene expression at the molecular level involves transcription to produce mRNA and translation to produce a polypeptide (see Figure 1.6).

1.2 The Relationship Between Genes and Traits

- Genetics, which governs an organism's traits, spans the molecular, cellular, organism, and population levels (see Figure 1.7).
- Genetic variation underlies variation in traits. In addition, the environment plays a key role (see Figures 1.8, 1.9).
- During reproduction, genetic material is passed from parents to offspring. In many species, somatic cells are diploid and have two sets of chromosomes, whereas gametes are haploid and have a single set (see Figure 1.10).
- Evolution refers to a change in the genetic composition of a population from one generation to the next (see Figure 1.11).

1.3 Fields of Genetics

- Genetics is traditionally divided into transmission genetics, molecular genetics, and population genetics, though overlap occurs among these fields. Many geneticists study model organisms (see Figure 1.12).
- Researchers in genetics carry out hypothesis testing or discovery-based science.

PROBLEM SETS & INSIGHTS

Solved Problems

S1. A human gene called the *CFTR* gene (for cystic fibrosis transmembrane regulator) encodes a protein that functions in the transport of chloride ions across the cell membrane. Most people have two copies of a functional *CFTR* gene and do not have cystic fibrosis. However, a mutant version of the *CFTR* gene is found in some people. If a person has two mutant copies of the gene, he or she develops the disease known as cystic fibrosis. Are the following examples a description of genetics at the molecular, cellular, organism, or population level?

A. People with cystic fibrosis have lung problems due to a buildup of thick mucus in their lungs.

B. The mutant *CFTR* gene encodes a defective chloride transporter.

C. A defect in the chloride transporter causes a salt imbalance in lung cells.

D. Scientists have wondered why the mutant *CFTR* gene is relatively common. In fact, it is the most common mutant gene that causes a severe disease in persons of Northern European descent. One possible explanation why CF is so common is that people who have one copy of the functional *CFTR* gene and one copy of the mutant gene may be more resistant to diarrheal diseases such as cholera. Therefore, even though individuals with two mutant copies are very sick, people with one mutant copy and one functional copy might have a survival advantage over people with two functional copies of the gene.

Answer:

A. Organism. This is a description of a trait at the level of an entire individual.

B. Molecular. This is a description of a gene and the protein it encodes.

C. Cellular. This is a description of how protein function affects the cell.

D. Population. This is a possible explanation why two alleles of the gene occur within a population.

S2. Explain the relationship between the following pairs of terms:

A. RNA and DNA

B. RNA and transcription

C. Gene expression and trait

D. Mutation and allele

Answer:

A. DNA is the genetic material. In a cell, DNA is used to make RNA. RNA is then used to specify a sequence of amino acids within a polypeptide.

B. Transcription is a process in which RNA is made using DNA as a template.

C. Genes are expressed at the molecular level to produce functional proteins. The functioning of proteins within living cells ultimately affects an organism's traits.

D. Alleles are alternative forms of the same gene. For example, a particular human gene affects eye color. The gene can exist as a blue allele or a brown allele. The difference between these two alleles is caused by a mutation. Research suggests that the brown allele was the first eye color allele in the human population. Within some ancestral person, however, a mutation occurred in the eye color gene that converted the brown allele to the blue allele. Now the human population has both the brown allele and the blue allele.

S3. In diploid species that carry out sexual reproduction, how are genes passed from generation to generation?

Answer: When a diploid individual makes haploid cells for sexual reproduction, the cells contain half the number of chromosomes. When two haploid cells (e.g., sperm and egg) combine with each other, a zygote is formed that begins the life of a new individual. This zygote has inherited half of its chromosomes and, therefore, half of its genes from each parent. This is how genes are passed from parents to offspring.

Conceptual Questions

C1. Pick any example of a genetic technology and describe how it has directly affected your life.

C2. At the molecular level, what is a gene? Where are genes located?

C3. Most genes encode proteins. Explain how the structure and function of proteins produce an organism's traits.

C4. Briefly explain how gene expression occurs at the molecular level.

C5. A human gene called the β-globin gene encodes a polypeptide that functions as a subunit of the protein known as hemoglobin. Hemoglobin is found within red blood cells; it carries oxygen. In human populations, the β-globin gene can be found as the common allele called the Hb^A allele, but it can also be found as the Hb^S allele. Individuals who have two copies of the Hb^S allele have the disease called sickle cell disease. Are the following examples a description of genetics at the molecular, cellular, organism, or population level?

A. The Hb^S allele encodes a polypeptide that functions slightly differently from the polypeptide encoded by the Hb^A allele.

B. If an individual has two copies of the Hb^S allele, that person's red blood cells take on a sickle shape.

C. Individuals who have two copies of the Hb^A allele do not have sickle cell disease, but they are not resistant to malaria. People who have one Hb^A allele and one Hb^S allele do not have sickle cell disease, and they are resistant to malaria. People who have two copies of the Hb^S allele have sickle cell disease, and this disease may significantly shorten their lives.

D. Individuals with sickle cell disease have anemia because their red blood cells are easily destroyed by the body.

C6. What is meant by the term "genetic variation"? Give two examples of genetic variation not discussed in Chapter 1. What causes genetic variation at the molecular level?

C7. What is the cause of Down syndrome?

C8. Your textbook describes how the trait of phenylketonuria (PKU) is greatly influenced by the environment. Pick a trait in your favorite plant and explain how genetics and the environment may play important roles.

C9. What is meant by the term "diploid"? Which cells of the human body are diploid, and which cells are not?

C10. What is a DNA sequence?

C11. What is the genetic code?

C12. Explain the relationships between the following pairs of genetic terms:

A. Gene and trait

B. Gene and chromosome

C. Allele and gene

D. DNA sequence and amino acid sequence

C13. With regard to biological evolution, which of the following statements is incorrect? Explain why.

A. During its lifetime, an animal evolves to become better adapted to its environment.

B. The process of biological evolution has produced species that are better adapted to their environments.

C. When an animal is better adapted to its environment, the process of natural selection makes it more likely for that animal to reproduce.

C14. What are the primary interests of researchers working in the following fields of genetics?

A. Transmission genetics

B. Molecular genetics

C. Population genetics

Experimental Questions

E1. What is a genetic cross?

E2. The technique known as DNA sequencing (described in Chapter 20) enables researchers to determine the DNA sequence of genes. Would this technique be used primarily by transmission geneticists, molecular geneticists, or population geneticists?

E3. Figure 1.5 shows a micrograph of chromosomes from a normal human cell. If you performed this type of experiment using cells from a person with Down syndrome, what would you expect to see?

E4. Many organisms are studied by geneticists. Of the following species, do you think it would be more likely for them to be studied by a transmission geneticist, a molecular geneticist, or a population geneticist? Explain your answer. Note: More than one answer may be possible.

A. Dogs

B. *E. coli*

C. Fruit flies

D. Leopards

E. Corn

E5. Pick any trait you like in any species of wild plant or animal. The trait must somehow vary among different members of the species (see Figure 1.7). Note: When picking a trait to answer this question, do not pick the trait of wing color in butterflies.

A. Discuss all of the background information that you already have (from personal observations) regarding this trait.

B. Propose a hypothesis that would explain the genetic variation within the species. For example, in the case of the butterflies, your hypothesis might be that the dark butterflies survive better in dark forests, and the light butterflies survive better in sunlit fields.

C. Describe the experimental steps you would follow to test your hypothesis.

D. Describe the possible data you might collect.

E. Interpret your data.

Note: All answers appear at the website for this textbook; the answers to the even-numbered questions and all of the Concept Check and Comprehension Questions are in the back of the book.

www.mhhe.com/brookergenetics5e

Visit the website for practice tests, answer keys, and other learning aids for this chapter. Enhance your understanding of genetics with our interactive exercises, quizzes, animations, and much more.

CHAPTER OUTLINE

The garden pea, studied by Mendel.

2 MENDELIAN INHERITANCE

An appreciation for the concept of heredity can be traced far back in human history. Hippocrates, a famous Greek physician, was the first person to provide an explanation for hereditary traits (ca. 400 B.C.E.). He suggested that "seeds" are produced by all parts of the body, which are then collected and transmitted to the offspring at the time of conception. Furthermore, he hypothesized that these seeds cause certain traits of the offspring to resemble those of the parents. This idea, known as pangenesis, was the first attempt to explain the transmission of hereditary traits from generation to generation.

For the next 2000 years, the ideas of Hippocrates were accepted by some and rejected by many. After the invention of the microscope in the late seventeenth century, some people observed sperm and thought they could see a tiny creature inside, which they termed a homunculus (little man). This homunculus was hypothesized to be a miniature human waiting to develop within the womb of its mother. Those who held that thought, known as spermists, suggested that only the father was responsible for creating future generations and that any resemblance between mother and offspring was due to influences within the womb. During the same time, an opposite school of thought also developed. According to the ovists, the egg was solely responsible for human characteristics. The only role of the sperm was to stimulate the egg onto its path of development. Of course, neither of these ideas was correct.

The first systematic studies of genetic crosses were carried out by Joseph Kölreuter from 1761 to 1766. In crosses between different strains of tobacco plants, he found that the offspring were usually intermediate in appearance between the two parents. This led Kölreuter to conclude that both parents make equal genetic contributions to their offspring. Furthermore, his observations were consistent with the blending hypothesis of inheritance. According to this idea, the factors that dictate hereditary traits can blend together from generation to generation. The blended traits would then be passed to the next generation. The popular view before the 1860s, which combined the notions of pangenesis and blending inheritance, was that hereditary traits were rather malleable and could change and blend over the course of one or two generations. However, the pioneering work of Gregor Mendel would prove instrumental in refuting this viewpoint.

In Chapter 2, we will first examine the outcome of Mendel's crosses in pea plants. We begin our inquiry into genetics here because the inheritance patterns observed in peas are fundamentally related to inheritance patterns found in other eukaryotic species, such as corn, fruit flies, mice, and humans. We will discover how Mendel's insights into the patterns of inheritance in pea plants revealed some simple rules that govern the process of inheritance. In Chapters 3 through 8, we will explore more complex patterns of inheritance and also consider the role that chromosomes play as the carriers of the genetic material.

In the last part of this chapter, we will examine some general concepts in probability and statistics. How are statistical methods useful? First, probability calculations allow us to predict the outcomes of simple genetic crosses, as well as the outcomes of more complicated crosses described in later chapters. In addition, we will learn how to use statistics to test the validity of genetic hypotheses that attempt to explain the inheritance patterns of traits.

2.1 MENDEL'S STUDY OF PEA PLANTS

Learning Outcomes:

1. Explain the characteristics of pea plants that make them a suitable organism to study genetically.
2. Outline the steps that Mendel followed to make crosses between different strains of pea plants.
3. List the seven characteristics of pea plants that Mendel chose to study.

Gregor Johann Mendel, born in 1822, is now remembered as the father of genetics (**Figure 2.1**). He grew up on a small farm in Hyncice (formerly Heinzendorf) in northern Moravia, which was then a part of Austria and is now a part of the Czech Republic. As a young boy, he worked with his father grafting trees to improve the family orchard. Undoubtedly, his success at grafting taught him that precision and attention to detail are important elements of success. These qualities would later be important in his experiments as an adult scientist. Instead of farming, however, Mendel was accepted into the Augustinian monastery of St. Thomas, completed his studies for the priesthood, and was ordained in 1847. Soon after becoming a priest, Mendel worked for a short time as a substitute teacher. To continue that role, he needed to obtain a teaching license from the government. Surprisingly, he failed the licensing exam due to poor answers in the areas of physics and natural history. Therefore, Mendel then enrolled at the University of Vienna to expand his knowledge in these two areas. Mendel's training in physics and mathematics taught him to perceive the world as an orderly place, governed by natural laws. In his studies, Mendel learned that these natural laws could be stated as simple mathematical relationships.

In 1856, Mendel began his historic studies on pea plants. For 8 years, he grew and crossed thousands of pea plants in a small 23- by 115-foot garden. He kept meticulously accurate records that included quantitative data concerning the outcome of his crosses. He published his work, entitled *Experiments on Plant Hybrids*, in 1866. This paper was largely ignored by scientists at that time, possibly because of its title, which did not reveal the key observations he made. Another reason his work went unrecognized could be tied

FIGURE 2.1 **Gregor Johann Mendel, the father of genetics.**

to a lack of understanding of chromosomes and their transmission, a topic we will discuss in Chapter 3. Nevertheless, Mendel's groundbreaking work allowed him to propose the natural laws that now provide a framework for our understanding of genetics.

Prior to his death in 1884, Mendel reflected, "My scientific work has brought me a great deal of satisfaction and I am convinced that it will be appreciated before long by the whole world." Sixteen years later, in 1900, the work of Mendel was independently rediscovered by three biologists with an interest in plant genetics: Hugo de Vries of Holland, Carl Correns of Germany, and Erich von Tschermak of Austria. Within a few years, the influence of Mendel's studies was felt around the world. In this section, we will examine Mendel's experimental approach.

Mendel Chose Pea Plants as His Experimental Organism

Mendel's study of genetics grew out of his interest in ornamental flowers. Prior to his work with pea plants, many plant breeders had conducted experiments aimed at obtaining flowers with new varieties of colors. When two distinct individuals with different characteristics are bred to each other, this is called a **cross,** or a **hybridization** experiment, and the offspring are referred to as **hybrids.** For example, a hybridization experiment could involve a cross between a purple-flowered plant and a white-flowered plant. Mendel was particularly intrigued by the consistency with which

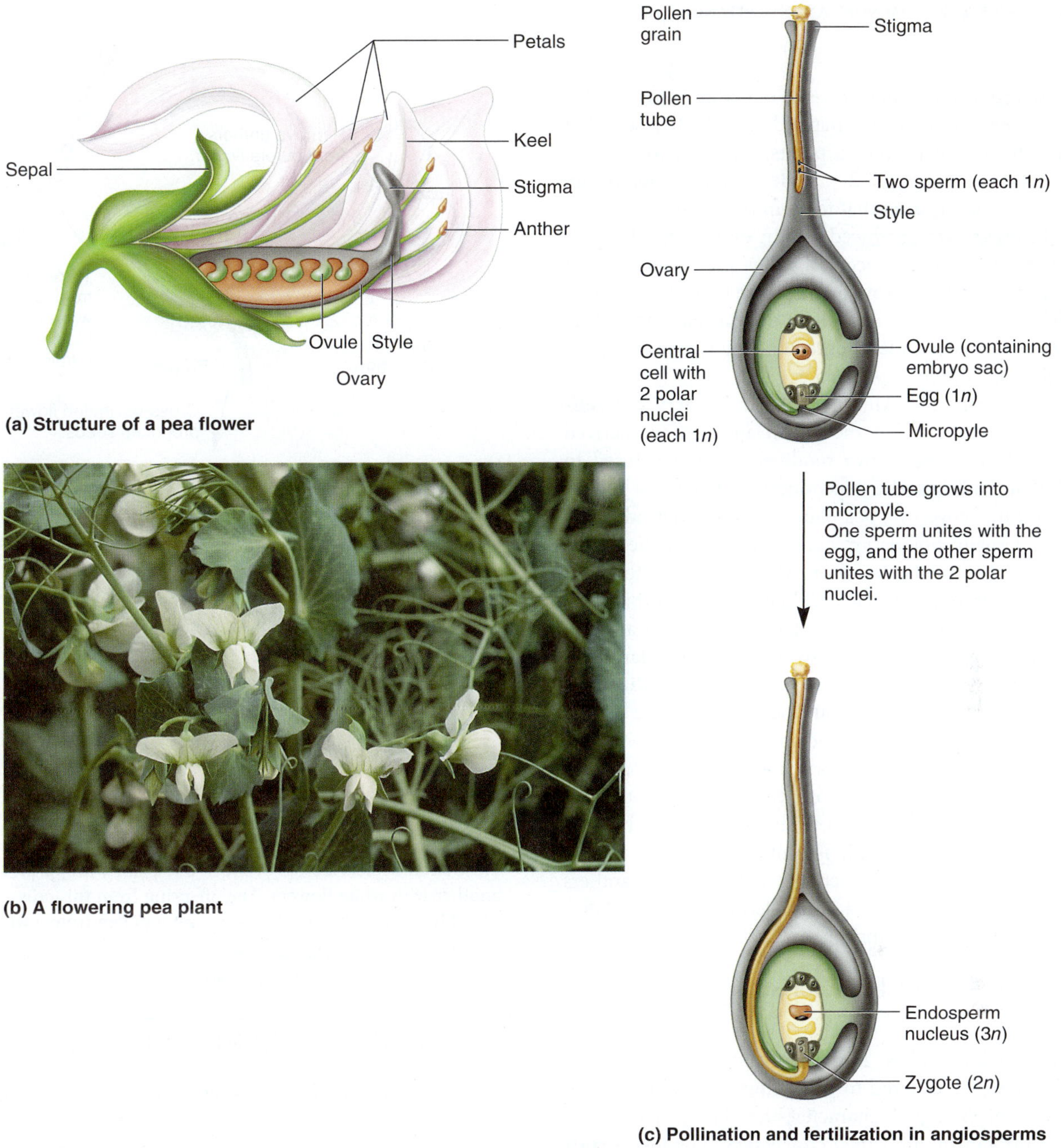

FIGURE 2.2 **Flower structure and pollination in pea plants.** **(a)** The pea flower can produce both pollen and egg cells. The pollen grains are produced within the anthers, and the egg cells are produced within the ovules that are contained within the ovary. The keel is two modified petals that are fused and enclose the anthers and ovaries. In this drawing, some of the keel is not shown so the internal reproductive structures of the flower can be seen. **(b)** Photograph of a flowering pea plant. **(c)** A pollen grain must first land on the stigma. After this occurs, the pollen sends out a long tube through which two sperm cells travel toward an ovule to reach an egg cell. The fusion between a sperm and an egg cell results in fertilization and creates a zygote. The second sperm fuses with a central cell containing two polar nuclei to create the endosperm. The endosperm provides a nutritive material for the developing embryo.

CONCEPT CHECK: Prior to fertilization, where is the male gamete located?

offspring of subsequent generations showed characteristics of one or the other parent. His educational training in physics and the natural sciences led him to consider that this regularity might be rooted in natural laws that could be expressed mathematically. To uncover these laws, he realized that he needed to carry out quantitative experiments in which the numbers of offspring carrying certain traits were carefully recorded and analyzed.

Mendel chose the garden pea, *Pisum sativum*, to investigate the natural laws that govern plant hybrids. The morphological features of this plant are shown in **Figure 2.2a** and **2.2b**. Several properties of this species were particularly advantageous for studying plant hybridization. First, the species was available in several varieties, which varied in height and in the appearance of their flowers, seeds, and pods.

A second important issue is the ease of making crosses. The term **gamete** is used to describe the haploid reproductive cells that fuse to form a zygote. In flowering plants, reproduction occurs by a pollination event (**Figure 2.2c**). Male gametes **(sperm)** are produced within **pollen grains** that form in the **anthers,** and the female gametes **(eggs)** are produced within **ovules** that form in the **ovaries.** For fertilization to occur, a pollen grain lands on the **stigma,** which stimulates the growth of a pollen tube. This enables sperm cells to enter the stigma and migrate toward an ovule. Fertilization takes place when a sperm enters the micropyle, an opening in the ovule wall, and fuses with an egg cell.

In some experiments, Mendel wanted to carry out **self-fertilization,** which means that the pollen and egg are derived from the same plant. In peas, two modified petals are fused to form a keel that encloses the reproductive structures of the plant. Because of this covering, pea plants naturally reproduce by self-fertilization. Usually, pollination occurs even before the flower opens. In other experiments, however, Mendel wanted to make crosses between different plants. How did he accomplish this goal? Fortunately, pea plants contain relatively large flowers that are easy to manipulate, making it possible to cross two particular plants and study the outcomes. This process, known as **cross-fertilization,** requires that the pollen from one plant be placed on the stigma of another plant. This procedure is shown in **Figure 2.3**. Mendel was able to pry open immature flowers and remove the anthers before they produced pollen. Therefore, these flowers could not self-fertilize. He then obtained pollen from another plant by gently touching its mature anthers with a paintbrush. Mendel applied this pollen to the stigma of the flower that already had its anthers removed. In this way, he was able to cross-fertilize his pea plants, thereby obtaining any type of hybrid he wanted.

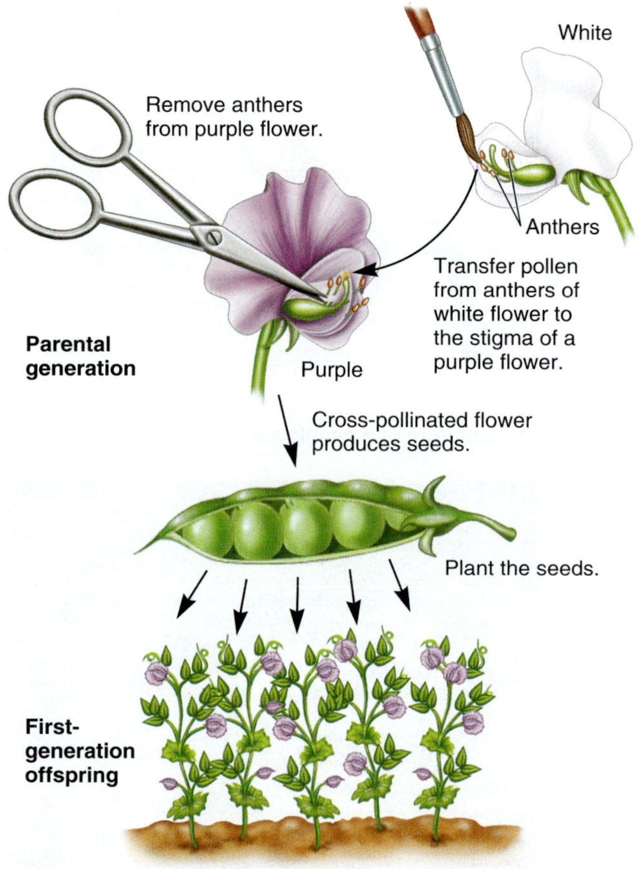

FIGURE 2.3 **How Mendel cross-fertilized pea plants.** This illustration depicts a cross between one plant with purple flowers and another with white flowers. The offspring from this cross are the result of pollination of the purple flower using pollen from a white flower.

CONCEPT CHECK: In this experiment, which plant, the white- or purple-flowered one, is providing the egg cells, and which is providing the sperm cells?

Mendel Studied Seven Characteristics That Bred True

When he initiated his studies, Mendel obtained several varieties of peas that were considered to be distinct. These plants had many different morphological characteristics. The general characteristics of an organism are called **characters.** The terms **trait** or **variant** are typically used to describe the specific properties of a character. For example, eye color is a character of humans and blue eyes is the trait (or variant) found in some people. Over the course of 2 years, Mendel tested his pea strains to determine if their characteristics bred true. This means that a trait did not vary in appearance from generation to generation. For example, if the seeds from a pea plant were yellow, the next generation would also produce yellow seeds. Likewise, if these offspring were allowed to self-fertilize, all of their offspring would also produce yellow seeds, and so on. A variety that continues to produce the same trait after several generations of self-fertilization is called a **true-breeding strain,** or **true-breeding line.**

Mendel next concentrated his efforts on the analysis of characters that were clearly distinguishable between different true-breeding lines. **Figure 2.4** illustrates the seven characters that Mendel eventually chose to follow in his breeding experiments. All seven were found in two variants. For example, one character he followed was height, which was found in two variants: tall and dwarf plants.

2.1 COMPREHENSION QUESTIONS

1. Experimental advantages of using pea plants include which of the following?
   a. They came in several different varieties.
   b. They were capable of self-fertilization.
   c. They were easy to cross.
   d. All of the above were advantages.
2. The term "cross" refers to an experiment in which
   a. the gametes come from different individuals.
   b. the gametes come from a single flower of the same individual.
   c. the gametes come from different flowers of the same individual.
   d. both a and c.
3. To avoid self-fertilization in his pea plants, Mendel had to
   a. spray the plants with a chemical that damaged the pollen.
   b. remove the anthers from immature flowers.
   c. grow the plants in a greenhouse that did not contain pollinators (e.g., bees).
   d. do all of the above.

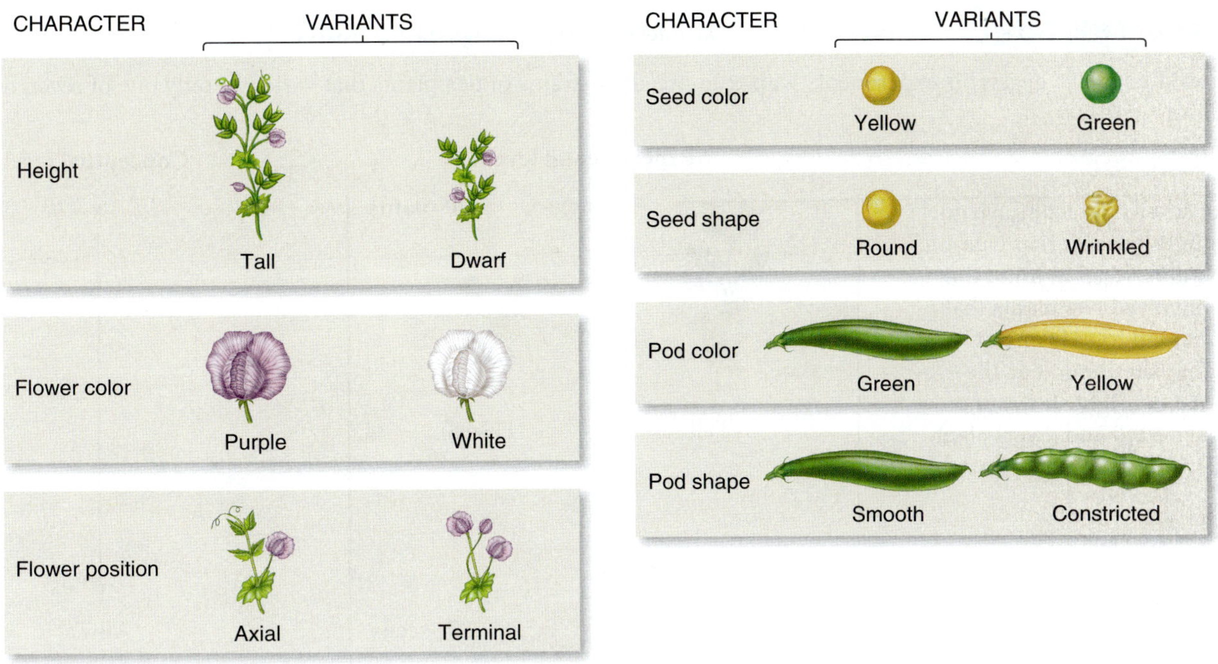

FIGURE 2.4 **An illustration of the seven characters that Mendel studied.** Each character was found as two variants that were decisively different from each other.

CONCEPT CHECK: What do we mean when we say a strain is true-breeding?

2.2 LAW OF SEGREGATION

Learning Outcomes:

1. Analyze Mendel's experiments involving single-factor crosses.
2. Define the law of segregation and explain how it is related to gamete formation and fertilization.
3. Predict the outcome of single-factor crosses using a Punnett square.

As discussed in the previous section, Mendel carried out self-fertilization or cross-fertilization experiments with his pea plants. In this section, we will examine how he studied the inheritance of characters by crossing the variants to each other. A cross in which an experimenter observes one character is called a **single-factor cross.** A cross between two parents with different variants for a given character produces single-character hybrids, also known as **monohybrids.** As you will learn, this type of experimental approach led Mendel to propose the law of segregation.

EXPERIMENT 2A

Mendel Followed the Outcome of a Single Character for Two Generations

Prior to conducting his studies, Mendel did not already have a hypothesis to explain the formation of hybrids. However, his educational background caused him to realize that a quantitative analysis of crosses may uncover mathematical relationships that would otherwise be mysterious. His experiments were designed to determine the relationships that govern hereditary traits. This rationale is called an **empirical approach.** Laws deduced from an empirical approach are known as empirical laws.

Mendel's experimental procedure is shown in Figure 2.5. He began with true-breeding plants that differed in a single character. These are termed the **parental generation,** or **P generation.** Crossing true-breeding parents to each other, called a P cross, produces the offspring that constitute the **F_1 generation,** for first filial generation. As seen in the data, all plants of the F_1 generation showed the trait of one parent but not the other. This prompted Mendel to follow the transmission of this character for one additional generation. To do so, the plants of the F_1 generation were allowed to self-fertilize to produce a second generation called the **F_2 generation,** for second filial generation.

THE GOAL (DISCOVERY-BASED SCIENCE)

Mendel speculated that the inheritance pattern for a single character may follow quantitative natural laws. The goal of this experiment was to uncover such laws.

ACHIEVING THE GOAL **FIGURE 2.5 Mendel's analysis of single-factor crosses.**

Starting material: Mendel began his experiments with true-breeding strains of pea plants that varied in only one of seven different characters (see Figure 2.4).

1. For each of seven characters, Mendel cross-fertilized two different true-breeding strains. Keep in mind that each cross involved two plants that differed in regard to only one of the seven characters studied. The illustration at the right shows one cross between a tall and dwarf plant. This is called a P (parental) cross.
2. Collect the F_1 generation seeds. The following spring, plant the seeds and allow the plants to grow. These are the plants of the F_1 generation.
3. Allow the F_1 generation plants to self-fertilize. This produces seeds that are part of the F_2 generation.
4. Collect the F_2 generation seeds and plant them the following spring to obtain the F_2 generation plants.
5. Analyze the traits found in each generation.

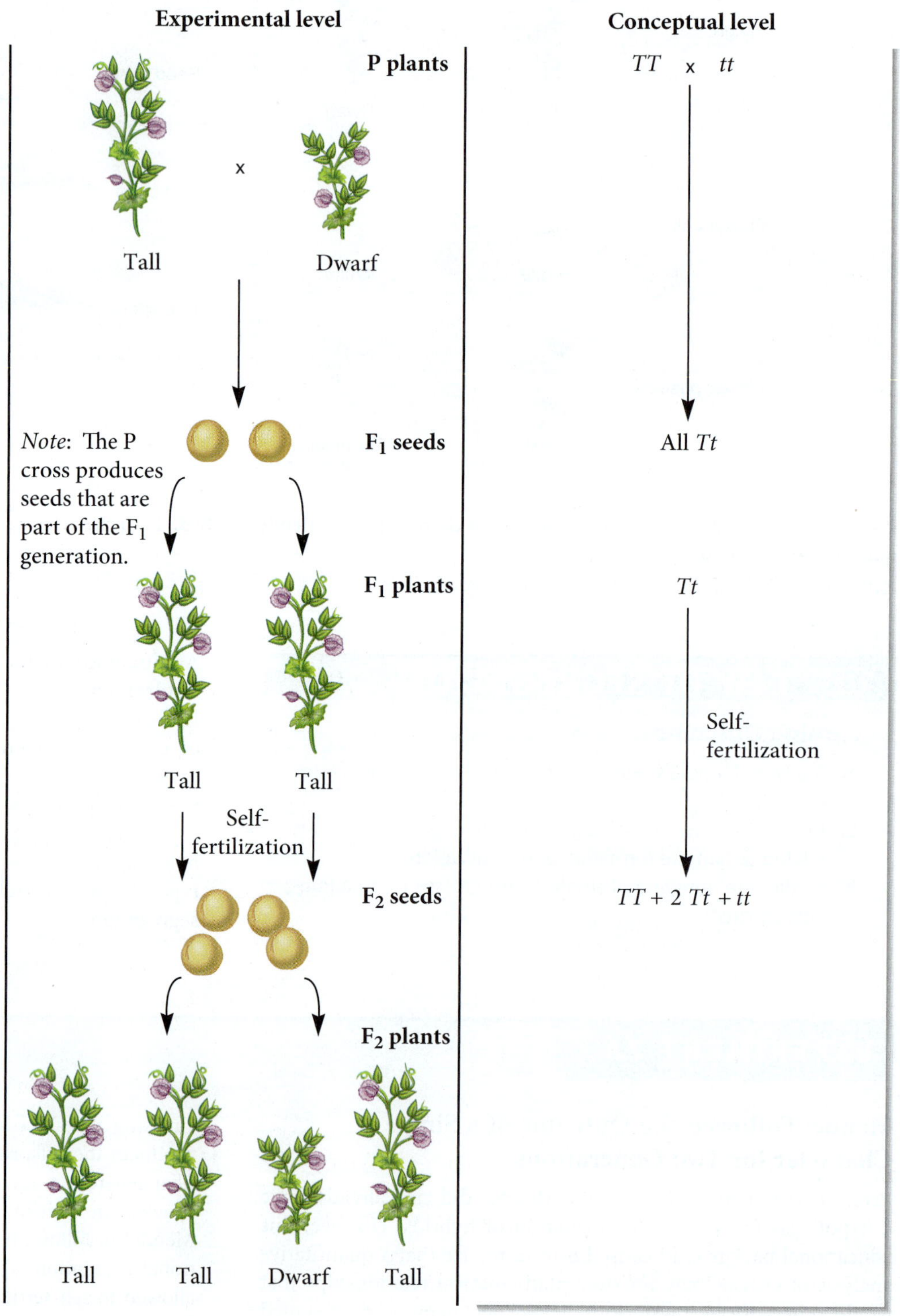

THE DATA

P cross	*F_1 generation*	*F_2 generation*	*Ratio*
Tall × dwarf stem	All tall	787 tall, 277 dwarf	2.84:1
Purple × white flowers	All purple	705 purple, 224 white	3.15:1
Axial × terminal flowers	All axial	651 axial, 207 terminal	3.14:1
Yellow × green seeds	All yellow	6,022 yellow, 2,001 green	3.01:1
Round × wrinkled seeds	All round	5,474 round, 1,850 wrinkled	2.96:1
Green × yellow pods	All green	428 green, 152 yellow	2.82:1
Smooth × constricted pods	All smooth	882 smooth, 299 constricted	2.95:1
Total	All dominant	14,949 dominant, 5,010 recessive	2.98:1

Data from Mendel, Gregor. 1866 Versuche über Plflanzenhybriden. *Verhandlungen des naturforschenden Vereines in Brünn, Bd IV für das Jahr 1865*, Abhandlungen, 3–47.

INTERPRETING THE DATA

The data shown in Figure 2.5 are the results of producing an F_1 generation via cross-fertilization and an F_2 generation via self-fertilization of the F_1 plants. A quantitative analysis of these data allowed Mendel to propose three important ideas:

1. Mendel's data argued strongly against a blending mechanism of heredity. In all seven cases, the F_1 generation displayed characteristics that were distinctly like one of the two parents rather than traits intermediate in character. His first proposal was that one variant for a particular character is **dominant** over another variant. For example, the variant of green pods is dominant to that of yellow pods. The term **recessive** is used to describe a variant that is masked by the presence of a dominant trait but reappears in subsequent generations. Yellow pods and dwarf stems are examples of recessive variants. They can also be referred to as recessive traits.
2. When a true-breeding plant with a dominant trait was crossed to a true-breeding plant with a recessive trait, the dominant trait was always observed in the F_1 generation. In the F_2 generation, most offspring displayed the dominant trait, but some showed the recessive trait. How did Mendel explain this observation? Because the recessive trait appeared in the F_2 generation, he made a second proposal—the genetic determinants of traits are passed along as "unit factors" from generation to generation. His data were consistent with a **particulate theory of inheritance,** in which the genetic determinants that govern traits are inherited as discrete units that remain unchanged as they are passed from parent to offspring. Mendel referred to the genetic determinants as unit factors, but we now call them genes.
3. When Mendel compared the numbers of dominant and recessive traits in the F_2 generation, he noticed a recurring pattern. Within experimental variation, he always observed approximately a 3:1 ratio between the dominant and the recessive trait. As described next, this quantitative approach allowed him to make a third proposal—genes **segregate** from each other during the process that gives rise to gametes.

A self-help quiz involving this experiment can be found at www.mhhe.com/brookergenetics5e.

Mendel's 3:1 Ratio Is Consistent with the Law of Segregation

Mendel's research was aimed at understanding the laws that govern the inheritance of traits. At that time, scientists did not understand the molecular composition of the genetic material or its mode of transmission during gamete formation and fertilization. We now know that the genetic material is composed of deoxyribonucleic acid (DNA), a component of chromosomes. Each chromosome contains hundreds to thousands of shorter segments that function as genes—a term that was originally coined by the Danish botanist Wilhelm Johannsen in 1909. A **gene** is defined as a "unit of heredity" that may influence the outcome of an organism's traits. Each of the seven characters that Mendel studied is influenced by a different gene.

Most eukaryotic species, such as pea plants and humans, have their genetic material organized into pairs of chromosomes. For this reason, eukaryotes have two copies of most genes. These copies may be the same or they may differ. The term **allele** (from the Latin *alius* meaning other) refers to an alternative version of a particular gene. For example, the height gene in pea plants is found as a tall allele and a dwarf allele. With this modern knowledge, the results shown in Figure 2.5 are consistent with the idea that each parent transmits only one copy of each gene (i.e., one allele) to each offspring. Using modern terminology, **Mendel's law of segregation** states the following:

> *The two copies of a gene segregate (or separate) from each other during transmission from parent to offspring.*

Therefore, only one copy of each gene is found in a gamete. At fertilization, two gametes combine randomly, potentially producing different allelic combinations.

Let's use Mendel's cross of tall and dwarf pea plants to illustrate how alleles are passed from parents to offspring (**Figure 2.6**). The letters *T* and *t* are used to represent the alleles

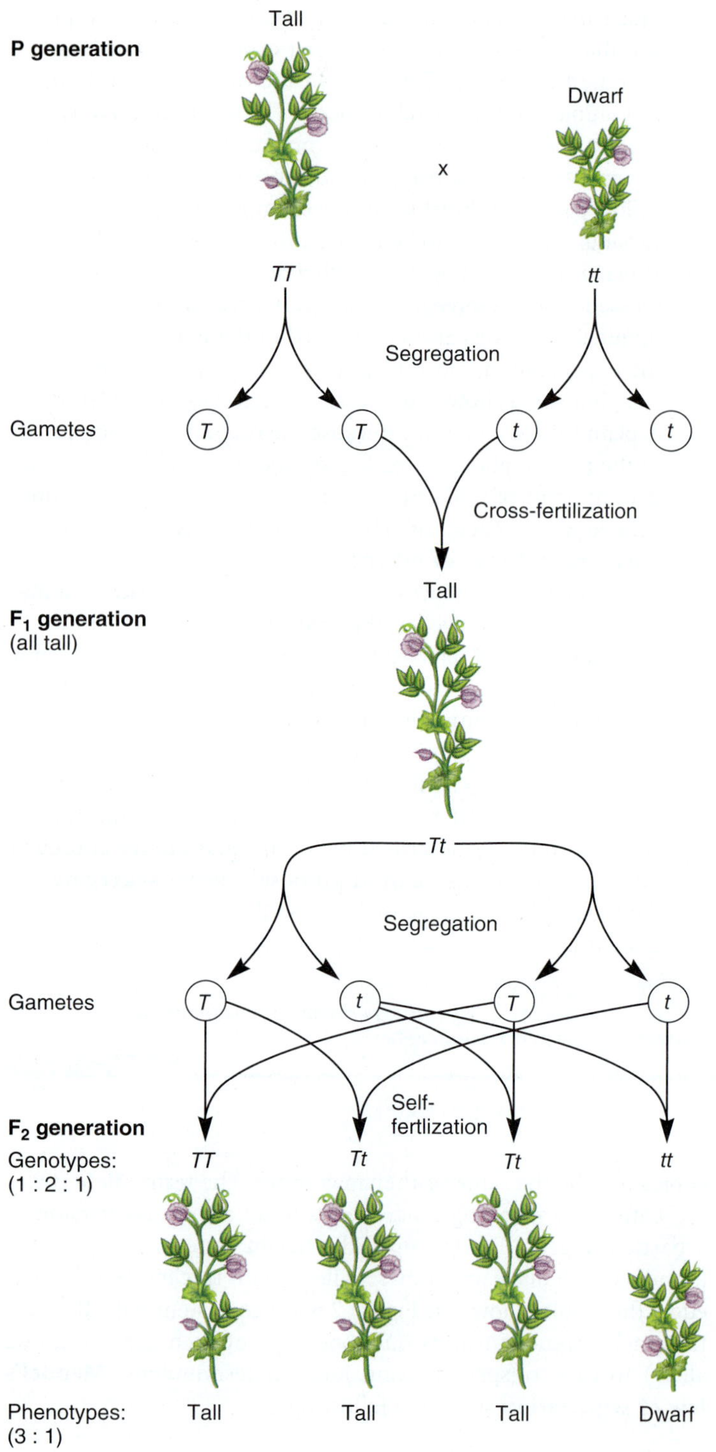

INTERACTIVE EXERCISE

FIGURE 2.6 Mendel's law of segregation. This illustration shows a cross between a true-breeding tall plant and a true-breeding dwarf plant and the subsequent segregation of the tall (*T*) and dwarf (*t*) alleles in the F_1 and F_2 generations.

CONCEPT CHECK: With regard to the *T* and *t* alleles, explain what the word "segregation" means.

of the gene that determines plant height. By convention, the uppercase letter represents the dominant allele (*T* for tall height, in this case), and the recessive allele is represented by the same letter in lowercase (*t*, for dwarf height). For the P cross, both parents are true-breeding plants. Therefore, each one has identical copies of the height gene. When an individual has two identical copies of a gene, the individual is said to be **homozygous** with respect to that gene. (The prefix *homo-* means like, and the suffix *-zygo* means pair.) In the P cross, the tall plant is homozygous for the tall allele *T*, and the dwarf plant is homozygous for the dwarf allele *t*. In contrast, the F_1 generation is **heterozygous,** with the genotype *Tt*, because every individual carries one copy of the tall allele and one copy of the dwarf allele. A heterozygous individual carries different alleles of a gene. (The prefix *hetero-* means different.)

The term **genotype** refers to the genetic composition of an individual. *TT* and *tt* are the genotypes of the P generation in this experiment. The term **phenotype** refers to an observable characteristic of an organism. In the P generation, the plants exhibit a phenotype that is either tall or dwarf. In some cases, plants have different genotypes yet the same phenotype. For example, both *TT* and *Tt* plants are tall.

The law of segregation predicts that the phenotypes of the F_2 generation will be tall and dwarf in a ratio of 3:1 (see Figure 2.6). The parents of the F_2 generation are heterozygous. Due to segregation, their gametes can carry either a *T* allele or a *t* allele, but not both. Following self-fertilization, *TT*, *Tt*, and *tt* are the possible genotypes of the F_2 generation (note that the genotype *Tt* is the same as *tT*). By randomly combining these alleles, the genotypes are produced in a 1:2:1 ratio. Because *TT* and *Tt* both produce tall phenotypes, a 3:1 phenotypic ratio is observed in the F_2 generation.

A Punnett Square Can Be Used to Predict the Outcome of Crosses

An easy way to predict the outcome of simple genetic crosses is to use a **Punnett square,** a method originally proposed by British geneticist Reginald Punnett. To construct a Punnett square, you must know the genotypes of the parents. With this information, the Punnett square enables you to predict the types of offspring the parents are expected to produce and in what proportions. We will follow a step-by-step description of the Punnett square approach using a cross of heterozygous tall plants as an example.

Step 1. *Write down the genotypes of both parents.* In this example, a heterozygous tall plant is crossed to another heterozygous tall plant. The plant providing the pollen is the male parent and the plant providing the eggs, the female parent.

Male parent: *Tt*

Female parent: *Tt*

Step 2. *Write down the possible gametes that each parent can make.* Remember that the law of segregation tells us that a gamete carries only one copy of each gene.

Male gametes: *T* or *t*

Female gametes: *T* or *t*

Step 3. *Create an empty Punnett square.* In the examples shown in this textbook, the number of columns equals the number of male gametes, and the number of rows equals the number of female gametes. Our example has two rows and two columns. Place the male gametes across the top of the Punnett square and the female gametes along the side.

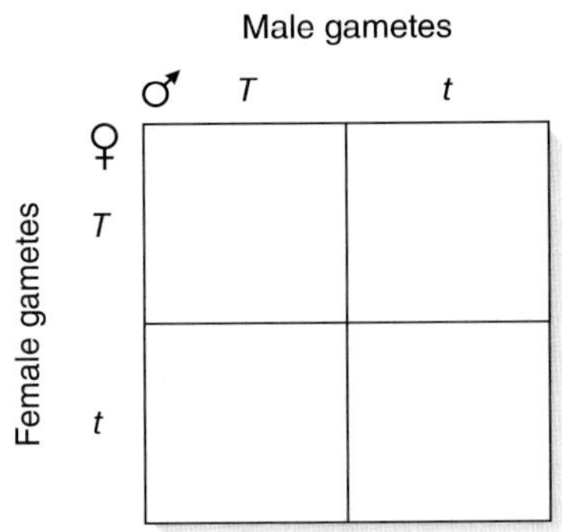

Step 4. *Fill in the possible genotypes of the offspring by combining the alleles of the gametes in the empty boxes.*

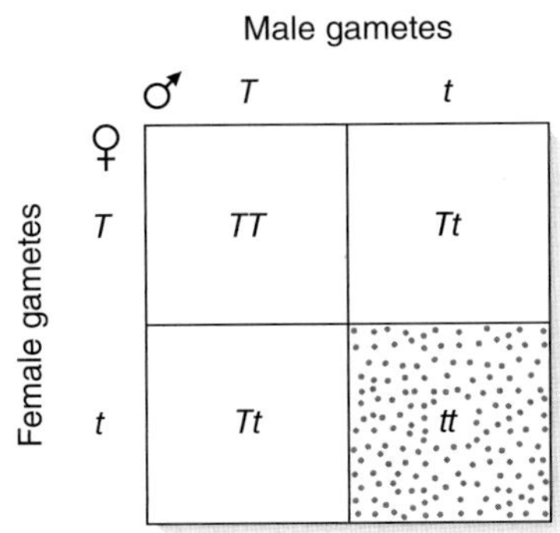

Step 5. *Determine the relative proportions of genotypes and phenotypes of the offspring.* The genotypes are obtained directly from the Punnett square. They are contained within the boxes that have been filled in. In this example, the genotypes are *TT*, *Tt*, and *tt* in a 1:2:1 ratio. To determine the phenotypes, you must know the dominant/recessive relationship between the alleles. For plant height, *T* (tall) is dominant to *t* (dwarf). The genotypes *TT* and *Tt* are tall, whereas the genotype *tt* is dwarf. Therefore, our Punnett square shows us that the ratio of phenotypes is 3:1, or 3 tall plants : 1 dwarf plant.

2.2 COMPREHENSION QUESTIONS

1. A pea plant is *Tt*. Which of the following statements is correct?
   a. Its genotype is *Tt*, and its phenotype is dwarf.
   b. Its phenotype is *Tt*, and its genotype is dwarf.
   c. Its genotype is *Tt*, and its phenotype is tall.
   d. Its phenotype is *Tt*, and its genotype is tall.
2. A *Tt* plant is crossed to a *tt* plant. What is the expected outcome for the ratio of offspring from this cross?
   a. 3 tall : 1 dwarf
   b. 1 tall : 1 dwarf
   c. 1 tall : 3 dwarf
   d. 2 tall : 1 dwarf

2.3 LAW OF INDEPENDENT ASSORTMENT

Learning Outcomes:

1. Analyze Mendel's experiments involving two-factor crosses.
2. Define the law of independent assortment.
3. Predict the outcome of two-factor crosses using a Punnett square.
4. Define a loss-of-function allele and explain why such alleles are useful to study.

Though his experiments described in Figure 2.5 revealed important ideas regarding a hereditary law, Mendel realized that additional insights might be uncovered if he conducted more complicated experiments. In this section, we will examine how he conducted crosses in which he simultaneously investigated the pattern of inheritance for two different characters. In other words, he carried out **two-factor crosses** in which he followed the inheritance of two different characters within the same groups of individuals. These experiments led to the formulation of a second law—the law of independent assortment.

EXPERIMENT 2B

Mendel Also Analyzed Crosses Involving Two Different Characters

To illustrate Mendel's work, we will consider an experiment in which one of the characters was seed shape, found in round or wrinkled variants; the second character was seed color, which existed as yellow and green variants. In this two-factor cross, Mendel followed the inheritance pattern for both characters simultaneously.

What results are possible from a two-factor cross? One possibility is that the genetic determinants for these two different characters are always linked to each other and inherited as a single unit (**Figure 2.7a**). If this were the case, the F_1 offspring could produce only two types of gametes, *RY* and *ry*. A second possibility is they are not linked and can assort themselves independently into gametes (**Figure 2.7b**). According to independent assortment, an F_1 offspring could produce four types of gametes, *RY*, *Ry*, *rY*, and *ry*. Keep in mind that the results of Figure 2.5 have already shown us that a gamete carries only one allele for each gene.

The experimental protocol for this two-factor cross is shown in **Figure 2.8**. Mendel began with two different strains of true-breeding pea plants that were different in seed shape and seed color. One plant was produced from seeds that were round and yellow; the other plant from seeds that were wrinkled and green. When these plants were crossed, the seeds, which contain the plant embryo, are considered part of the F_1 generation. As expected, the data revealed that the F_1 seeds displayed a phenotype of round and yellow. This was observed because round and yellow are dominant traits. It is the F_2 generation that supports the independent-assortment model and refutes the linkage model.

THE HYPOTHESES

The inheritance pattern for two different characters follows one or more quantitative natural laws. Two possible hypotheses are described in Figure 2.7.

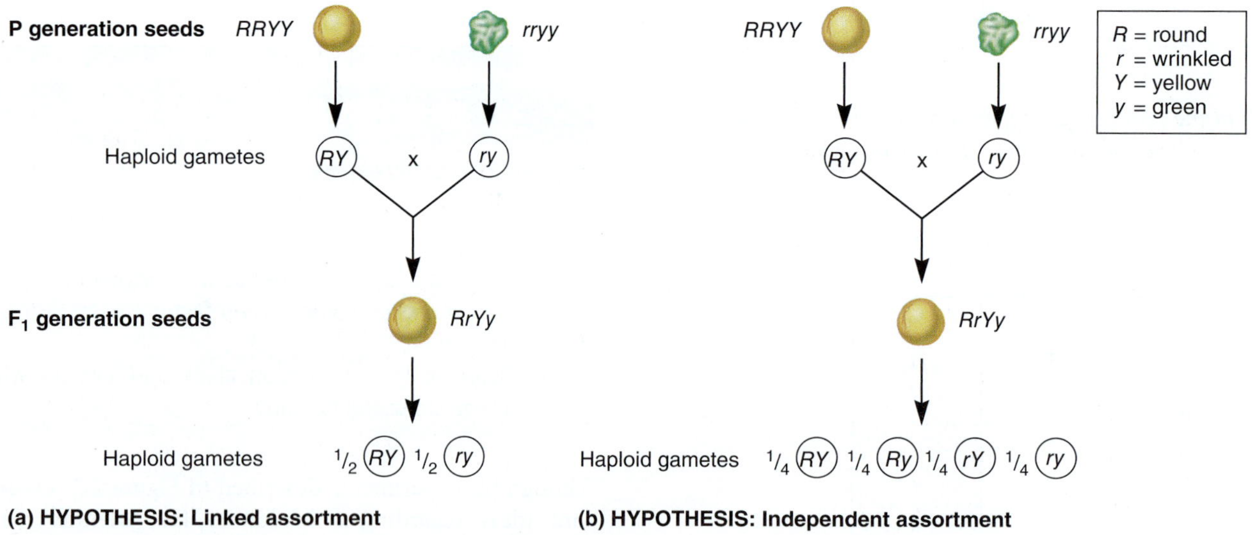

FIGURE 2.7 **Two hypotheses to explain how two different genes assort during gamete formation.** **(a)** According to the linked hypothesis, the two genes always stay associated with each other. **(b)** In contrast, the independent assortment hypothesis proposes that the two different genes randomly segregate into haploid cells.

CONCEPT CHECK: According to the linkage hypothesis shown here, what is linked? Are two different genes linked, or are two different alleles of the same gene linked, or both?

TESTING THE HYPOTHESES **FIGURE 2.8 Mendel's analysis of two-factor crosses.**

Starting material: In this experiment, Mendel began with two types of true-breeding strains of pea plants that were different with regard to two characters. One strain produced round, yellow seeds (*RRYY*); the other strain produced wrinkled, green seeds (*rryy*).

1. Cross the two true-breeding strains to each other. This produces F_1 generation seeds.
2. Collect many seeds and record their phenotype.
3. F_1 seeds are planted and grown, and the F_1 plants are allowed to self-fertilize. This produces seeds that are part of the F_2 generation.
4. Analyze the characteristics found in the F_2 generation seeds.

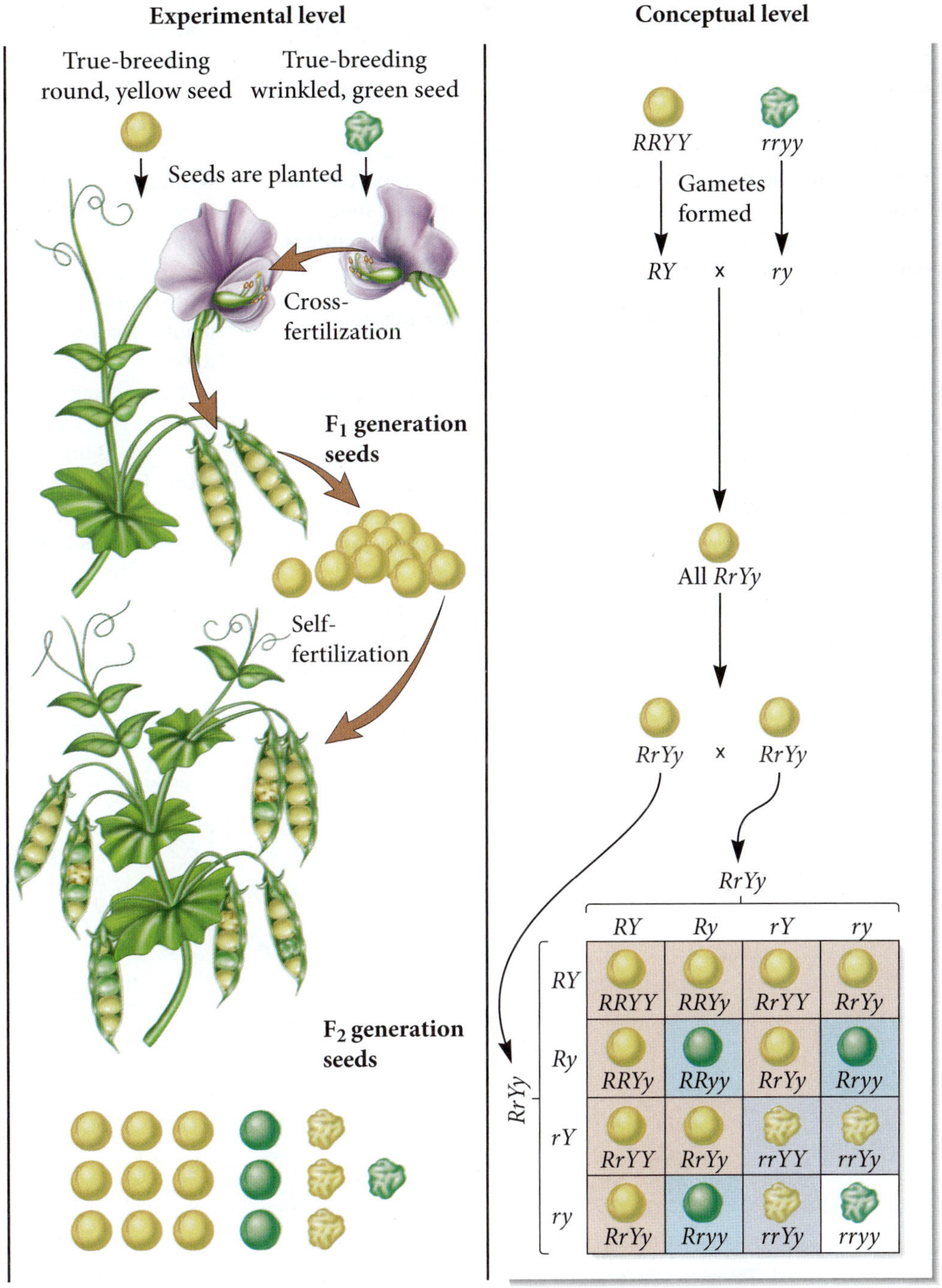

THE DATA

P cross	*F_1 generation*	*F_2 generation*
Round, yellow × wrinkled, green seeds	All round, yellow	315 round, yellow seeds 108 round, green seeds 101 wrinkled, yellow seeds 32 wrinkled, green seeds

INTERPRETING THE DATA

In addition to seeds that were like those of the parental generation, the F_2 generation also had seeds that were round and green and seeds that were wrinkled and yellow. These two categories of F_2 seeds are called **nonparentals** because these combinations of traits were not found in the true-breeding plants of the parental generation. The occurrence of nonparental variants contradicts the linkage model (see Figure 2.7a). According to the linkage model, the *R* and *Y* alleles should be linked together and so should the *r* and *y* alleles. If this were the case, the F_1 plants could produce gametes that are only *RY* or *ry*. These would combine to produce *RRYY* (round, yellow), *RrYy* (round, yellow), or *rryy* (wrinkled, green) in a 1:2:1 ratio. Nonparental seeds could, therefore, not be produced. However, Mendel did not obtain this result. Instead, he observed a phenotypic ratio of 9:3:3:1 in the F_2 generation.

Mendel's results from many two-factor crosses rejected the hypothesis of linked assortment and, instead, supported the hypothesis that different characters assort themselves independently. Using modern terminology, **Mendel's law of independent assortment** states the following:

> *Two different genes will randomly assort their alleles during the formation of haploid cells.*

In other words, the allele for one gene will be found within a resulting gamete independently of whether the allele for a different gene is found in the same gamete. In the example given in Figure 2.8, the round and wrinkled alleles are assorted into haploid gametes independently of the yellow and green alleles. Therefore, a heterozygous *RrYy* parent can produce four different gametes—*RY*, *Ry*, *rY*, and *ry*—in equal proportions.

In an F_1 self-fertilization experiment, any two gametes can combine randomly during fertilization. This allows for 4^2, or 16, possible offspring, although some offspring will be genetically identical to each other. As shown in **Figure 2.9**, these 16 possible combinations result in seeds with the following phenotypes: 9 round, yellow; 3 round, green; 3 wrinkled, yellow; and 1 wrinkled, green. This 9:3:3:1 ratio is the expected outcome when a plant that is heterozygous for both genes is allowed to self-fertilize. Mendel was clever enough to realize that the data for his two-factor experiments were close to a 9:3:3:1 ratio. In Figure 2.8, for example, his F_1 generation produced F_2 seeds with the following characteristics: 315 round, yellow seeds; 108 round, green seeds; 101 wrinkled, yellow seeds; and 32 wrinkled, green seeds. If we divide each of these numbers by 32 (the number of plants with wrinkled, green seeds), the phenotypic ratio of the F_2 generation is 9.8 : 3.4 : 3.2 : 1.0. Within experimental error, Mendel's data approximated the predicted 9:3:3:1 ratio for the F_2 generation.

The law of independent assortment held true for two-factor crosses involving all of the characters that Mendel studied in pea plants. However, in other cases, the inheritance pattern of two different genes is consistent with the linkage model described earlier in Figure 2.7a. In Chapter 6, we will examine the inheritance of genes that are linked because they are close to each other within the same chromosome. As we will see, linked genes do not assort independently.

An important consequence of the law of independent assortment is that a single individual can produce a vast array of genetically different gametes. As mentioned in Chapter 1, diploid species have pairs of homologous chromosomes, which may differ with respect to the alleles they carry. When an offspring receives a combination of alleles that differs from those in the parental generation, this phenomenon is termed **genetic recombination.**

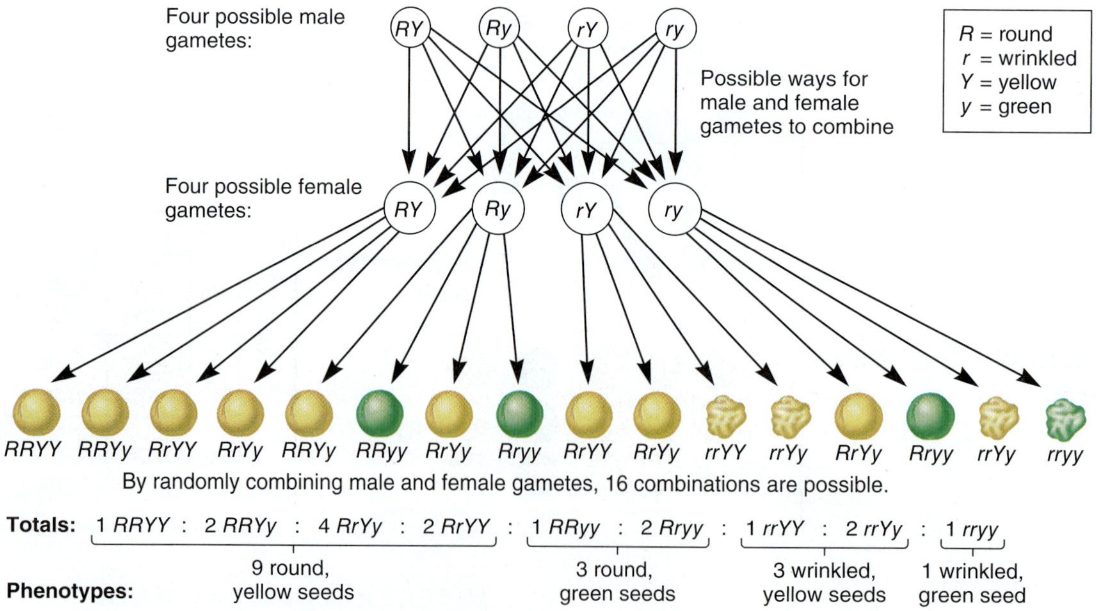

INTERACTIVE EXERCISE

FIGURE 2.9 **Mendel's law of independent assortment.**

Genes→Traits The cross is between two parents that are heterozygous for seed shape and seed color (*RrYy* × *RrYy*). Four types of male gametes are possible: *RY*, *Ry*, *rY*, and *ry*. Likewise, four types of female gametes are possible: *RY*, *Ry*, *rY*, and *ry*. These four types of gametes are the result of the independent assortment of the seed shape and seed color alleles relative to each other. During fertilization, any one of the four types of male gametes can combine with any one of the four types of female gametes. This results in 16 types of offspring, each one containing two copies of the seed shape gene and two copies of the seed color gene.

CONCEPT CHECK: Why does independent assortment promote genetic variation?

One mechanism that accounts for genetic recombination is independent assortment. A second mechanism, discussed in Chapter 6, is crossing over, which can reassort alleles that happen to be linked along the same chromosome.

The phenomenon of independent assortment is rooted in the random pattern by which the pairs of chromosomes assort themselves during the process of meiosis, a topic addressed in Chapter 3. When two different genes are found on different chromosomes, they randomly assort into haploid cells (look ahead to Figure 3.16). If a species contains a large number of chromosomes, this creates the potential for an enormous amount of genetic diversity. For example, human cells contain 23 pairs of chromosomes. These pairs separate and randomly assort into gametes during meiosis. The number of different gametes an individual can make equals 2^n, where n is the number of pairs of chromosomes. Therefore, humans can make 2^{23}, or over 8 million, possible gametes, due to independent assortment. The capacity to make so many genetically different gametes enables a species to produce a great diversity of individuals with different combinations of traits.

A self-help quiz involving this experiment can be found at www.mhhe.com/brookergenetics5e.

A Punnett Square Can Also Be Used to Solve Independent Assortment Problems

As already depicted in Figure 2.8, we can make a Punnett square to predict the outcome of crosses involving two or more genes that assort independently. Let's see how such a Punnett square is made by considering a cross between two plants that are heterozygous for height and seed color (**Figure 2.10**). This cross is *TtYy* × *TtYy*. When we construct a Punnett square for this cross, we must keep in mind that each gamete has a single allele for each of two genes. In this example, the four possible gametes from each parent are

TY, *Ty*, *tY*, and *ty*

In this two-factor cross, we need to make a Punnett square containing 16 boxes. The phenotypes of the resulting offspring are predicted to occur in a ratio of 9:3:3:1.

In crosses involving three or more genes, the construction of a single large Punnett square to predict the outcome of crosses becomes very unwieldy. For example, in a three-factor cross

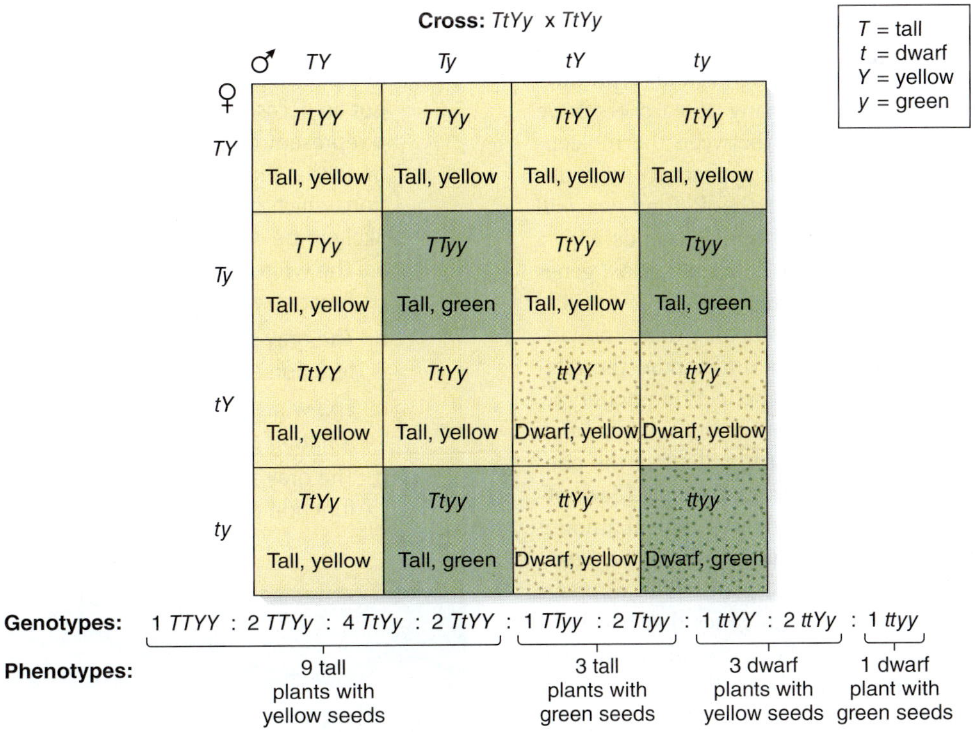

FIGURE 2.10 A Punnett square for a two-factor cross. The Punnett square shown here involves a cross between two pea plants that are heterozygous for height and seed color. The cross is *TtYy* × *TtYy*.

CONCEPT CHECK: If a parent is *Ttyy*, how many different types of gametes can it make?

between two pea plants that are *Tt Rr Yy*, each parent can make 2^3, or 8, possible gametes. Therefore, the Punnett square must contain $8 \times 8 = 64$ boxes. As a more reasonable alternative, we can consider each gene separately and then algebraically combine them by multiplying together the expected outcomes for each gene. Two such methods, termed the **multiplication method** and the **forked-line method,** are shown in solved problem S3 at the end of this chapter.

Independent assortment is also revealed by a **dihybrid testcross.** In this type of experiment, individuals that are heterozygous for two genes are crossed to individuals that are homozygous recessive for the same two genes. For example, individuals with a *TtYy* genotype could be crossed to *ttyy* plants. As shown below, independent assortment predicts a 1:1:1:1 ratio among the resulting offspring:

	TY	*Ty*	*tY*	*ty*
ty	*TtYy* Tall, yellow	*Ttyy* Tall, green	*ttYy* Dwarf, yellow	*ttyy* Dwarf, green

Modern Geneticists Are Often Interested in the Relationship Between the Molecular Expression of Genes and the Outcome of Traits

Mendel's work with pea plants was critically important because his laws of inheritance pertain to most eukaryotic organisms. These include corn, fruit flies, roundworms, mice, humans, and many others that transmit their genes through sexual reproduction. During the past several decades, many researchers have focused their attention on the relationship between the molecular expression of genes and the phenotypic appearance of traits. This theme will recur throughout the textbook (and we will draw attention to it by designating certain figure legends with a "Genes → Traits" label). As mentioned in Chapter 1, most genes encode proteins that function within living cells. The specific function of individual proteins affects the outcome of an individual's traits. A genetic approach can help us understand the relationship between a protein's function and its effect on phenotype. Most commonly, a geneticist will try to identify an individual that has a defective copy of a gene to see how that will affect the phenotype of the organism. These defective genes are called **loss-of-function alleles,** and they provide geneticists with a great amount of information. Unknowingly, Gregor Mendel had studied seven loss-of-function alleles among his strains of plants. The recessive characteristics in his pea plants were due to genes that had been rendered defective by a mutation. Such alleles are often inherited in a recessive manner, though this is not always the case.

How are loss-of-function alleles informative? In many cases, such alleles provide critical clues concerning the purpose of the protein's function within the organism. For example, we expect the gene affecting flower color (purple versus white) to encode a protein that is necessary for pigment production. This protein may function as an enzyme that is necessary for the synthesis of purple pigment. Furthermore, a reasonable guess is that the white allele is a loss-of-function allele that is unable to express this protein and therefore cannot make the purple pigment. To confirm this idea, a biochemist could analyze the petals from purple and white flowers and try to identify the protein that is defective or missing in the white petals but functionally active in the purple ones. The identification and characterization of this protein would provide a molecular explanation for this phenotypic characteristic.

2.3 COMPREHENSION QUESTIONS

1. A pea plant has the genotype *rrYy*. How many different types of gametes can it make and in what proportions?
   a. 1 *rr* : 1 *Yy*
   b. 1 *rY* : 1 *ry*
   c. 3 *rY* : 1 *ry*
   d. 1 *RY* : 1 *rY* : 1 *Ry* : 1 *ry*
2. A cross is made between a plant that is *RrYy* to a plant that is *rrYy*. What is the predicted outcome of the seed phenotypes?
   a. 9 round, yellow : 3 round, green : 3 wrinkled, yellow : 1 wrinkled, green
   b. 3 round, yellow : 3 round, green : 1 wrinkled, yellow : 1 wrinkled, green
   c. 3 round, yellow : 1 round, green : 3 wrinkled, yellow : 1 wrinkled, green
   d. 1 round, yellow : 1 round, green : 1 wrinkled, yellow : 1 wrinkled, green
3. In a population of wild squirrels, most of them have gray fur, but an occasional squirrel is completely white. If we let *P* and *p* represent dominant and recessive alleles, respectively, of a gene that encodes an enzyme necessary for pigment formation, which of the following statements do you think is most likely to be correct?
   a. The white squirrels are *pp*, and the *p* allele is a loss-of-function allele.
   b. The gray squirrels are *pp*, and the *p* allele is a loss-of-function allele.
   c. The white squirrels are *PP*, and the *P* allele is a loss-of-function allele.
   d. The gray squirrels are *PP*, and the *P* allele is a loss-of-function allele.

2.4 STUDYING INHERITANCE PATTERNS IN HUMANS

Learning Outcomes:

1. Describe the features of a pedigree.
2. Analyze a pedigree to determine if a trait or disease is dominant or recessive.

Before we end our discussion of simple Mendelian traits, let's address the question of how we can analyze inheritance patterns among humans. In his experiments, Mendel selectively made crosses and then analyzed a large number of offspring. When studying human traits, however, researchers cannot control parental crosses. Instead, they must rely on the information that is contained within family trees, or **pedigrees,** which are charts representing family relationships. This type of approach, known as **pedigree analysis,** is aimed at determining the type of inheritance pattern that a gene will follow. Although this method may be less definitive than the results described in Mendel's experiments, pedigree analyses often provide important clues concerning the pattern of inheritance of traits within human families. An expanded discussion of human pedigrees is found in Chapter 24, which concerns the inheritance patterns of many different human diseases.

In order to discuss the applications of pedigree analyses, we need to understand the organization and symbols of a pedigree (**Figure 2.11**). The oldest generation is at the top of the pedigree, and the most recent generation is at the bottom. Vertical lines connect each succeeding generation. A man (square) and woman (circle) who produce one or more offspring are directly connected by a horizontal line. A vertical line connects parents with their offspring. If parents produce two or more offspring, the group of siblings (brothers and sisters) is denoted by two or more individuals projecting from the same horizontal line.

When a pedigree involves the transmission of a human trait or disease, affected individuals are depicted by filled symbols (in this case, black) that distinguish them from unaffected individuals. Each generation is given a roman numeral designation, and individuals within the same generation are numbered from left to right. A few examples of the genetic relationships in Figure 2.11a are described here:

Individuals I-1 and I-2 are the grandparents of III-1, III-2, III-3, III-4, III-5, III-6, and III-7.

Individuals III-1, III-2, and III-3 are brother and sisters.

Individual III-4 is affected by a genetic disease.

The symbols shown in Figure 2.11 depict certain individuals, such as I-1, I-2, II-4, and II-5, as presumed heterozygotes because they are unaffected with a disease but produce homozygous offspring that are affected with a recessive genetic disease. However, in many pedigrees, such as those found in the problem sets at the end of the chapter, the inheritance pattern may not be known, so the symbols reflect only phenotypes. In most pedigrees, affected individuals are shown with filled symbols, and unaffected individuals, including those that might be heterozygous for a recessive disease, are depicted with unfilled symbols.

Pedigree analysis is commonly used to determine the inheritance pattern for human genetic diseases. Human geneticists are routinely interested in knowing whether a genetic disease is inherited as a recessive or dominant trait. One way to discern the dominant/recessive relationship between two alleles is by pedigree analysis. Genes that play a role in disease may exist as a normal allele or a mutant allele that causes disease symptoms. If the disease follows a simple Mendelian pattern of inheritance and is caused by a recessive allele, an individual must inherit two copies of the mutant allele to exhibit the disease. Therefore, a recessive pattern of inheritance makes two

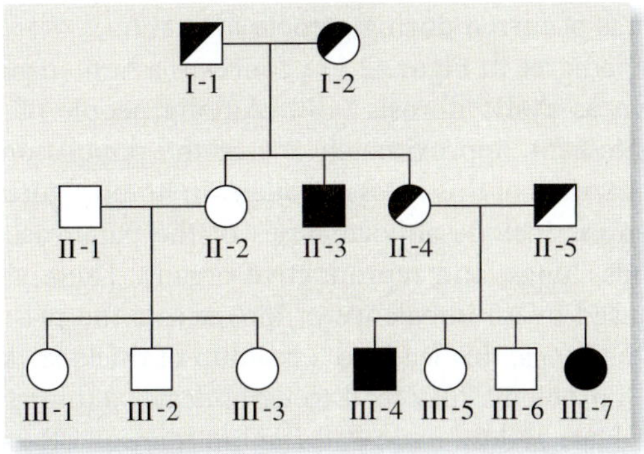

(a) Human pedigree showing cystic fibrosis

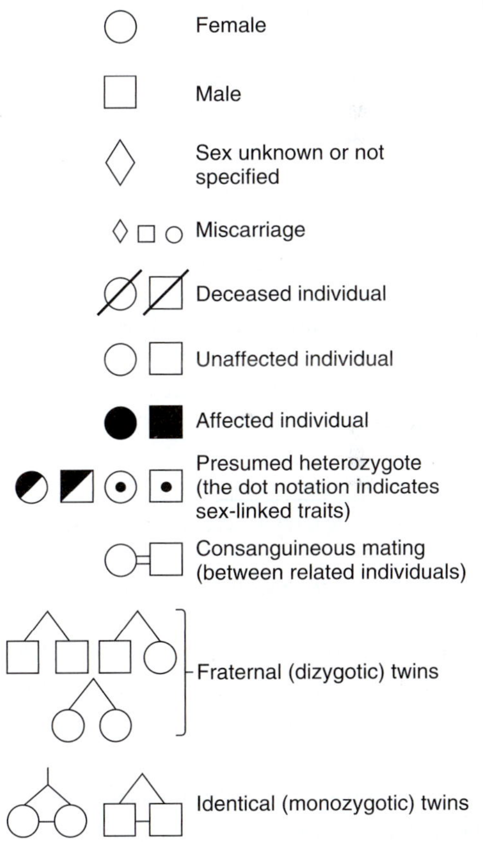

(b) Symbols used in a human pedigree

FIGURE 2.11 Pedigree analysis. **(a)** A family pedigree in which some of the members are affected with cystic fibrosis. Individuals I-1, I-2, II-4, and II-5 are depicted as presumed heterozygotes because they produce affected offspring. **(b)** The symbols used in pedigree analysis. Note: In most pedigrees shown in this textbook, such as those found in the problem sets, the heterozygotes are not shown as half-filled symbols. Most pedigrees throughout the book show individuals' phenotypes—unfilled symbols are unaffected individuals and filled symbols are affected individuals.

CONCEPT CHECK: What are the two different meanings of horizontal lines in a pedigree?

important predictions. First, two heterozygous normal individuals will, on average, have 1/4 of their offspring affected. Second, all offspring of two affected individuals will be affected. Alternatively, a dominant trait predicts that affected individuals will have inherited the gene from at least one affected parent (unless a new mutation has occurred during gamete formation).

The pedigree in Figure 2.11a concerns a human genetic disease known as cystic fibrosis (CF). Among people of Northern European descent, approximately 3% of the population are heterozygous carriers of this recessive allele. In homozygotes, the disease symptoms include abnormalities of the pancreas, intestine, sweat glands, lungs, and reproductive organs. These abnormalities are caused by an imbalance of ions across the plasma membrane. In the lungs, this leads to a buildup of thick, sticky mucus. Respiratory problems may lead to early death, although modern treatments have greatly increased the life span of CF patients. In the late 1980s, the gene for CF was identified. It encodes a protein called the cystic fibrosis transmembrane conductance regulator (CFTR). This protein regulates the ion balance across the cell membrane in tissues of the pancreas, intestine, sweat glands, and lungs. The mutant allele causing CF alters the encoded CFTR protein. The altered CFTR protein is not correctly inserted into the plasma membrane, resulting in a decreased function that causes the ionic imbalance. As seen in the pedigree, the pattern of affected and unaffected individuals is consistent with a recessive mode of inheritance. Two unaffected individuals can produce an affected offspring. Although not shown in this pedigree, a recessive mode of inheritance is also characterized by the observation that two affected individuals will produce 100% affected offspring. However, for human genetic diseases that limit survival or fertility (or both), there may never be cases where two affected individuals produce offspring.

2.4 COMPREHENSION QUESTIONS

1. Which of the following would *not* be observed in a pedigree if a genetic disorder was inherited in a recessive manner?
   a. Two unaffected parents have an affected offspring.
   b. Two affected parents have an unaffected offspring.
   c. One affected and one unaffected parent have an unaffected offspring.
   d. All of the above are possible for a recessive disorder.
2. In the pedigree shown here, which pattern(s) of inheritance is/are possible? Affected individuals are shown with a filled symbol.

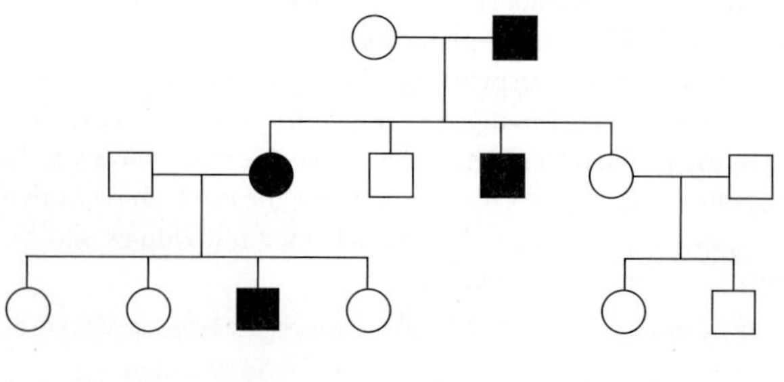

   a. Recessive
   b. Dominant
   c. Both recessive and dominant
   d. Neither recessive nor dominant

2.5 PROBABILITY AND STATISTICS

Learning Outcomes:

1. Define probability.
2. Predict the outcome of crosses using the product rule and binomial expansion equation.
3. Evaluate the validity of a hypothesis using a chi square test.

A powerful application of Mendel's work is that the laws of inheritance can be used to predict the outcome of genetic crosses. In agriculture, for example, plant and animal breeders are concerned with the types of offspring resulting from their crosses. This information is used to produce commercially important crops and livestock. In addition, people are often interested in predicting the characteristics of the children they may have. This may be particularly important to individuals who carry alleles that cause inherited diseases. Of course, we cannot see into the future and definitively predict what will happen. Nevertheless, genetic counselors can help couples to predict the likelihood of having an affected child. This probability is one factor that may influence a couple's decision whether to have children.

In this section, we will see how probability calculations are used in genetic problems to predict the outcome of crosses. We will consider two mathematical operations known as the product rule and the binomial expansion equation. These methods allow us to determine the probability that a cross between two individuals will produce a particular outcome. To apply these operations, we must have some knowledge regarding the genotypes of the parents and the pattern of inheritance of a given trait. These two operations are used to solve different types of problems:

- The product rule is used in problems in which the outcomes are independent of each other.
- The binomial expansion is used in problems having an unordered combination of outcomes.

Probability calculations can also be used in hypothesis testing. In many situations, a researcher would like to discern the genotypes and patterns of inheritance for traits that are not yet understood. A traditional approach to this problem is to conduct crosses and then analyze their outcomes. The proportions of offspring may provide important clues that allow the experimenter to propose a hypothesis, based on the quantitative laws of inheritance, that explains the transmission of the trait from parent to offspring. Statistical methods, such as the chi square test, can then be used to evaluate how well the observed data from crosses fit the expected data. We will end this chapter with an example that applies the chi square test to a genetic cross.

Probability Is the Likelihood That an Outcome Will Occur

The chance that an outcome will occur in the future is called the outcome's **probability.** For example, if you flip a coin, the probability is 0.50, or 50%, that the head side will be showing when the coin lands. Probability depends on the number of possible outcomes. In this case, two possible outcomes (heads or tails) are equally likely. This allows us to predict a 50% chance that a coin flip will produce heads. The general formula for probability (P) is

$$\text{Probability} = \frac{\text{Number of times a particular outcome occurs}}{\text{Total number of possible outcomes}}$$

$$P_{\text{heads}} = 1 \text{ heads} / (1 \text{ heads} + 1 \text{ tails}) = 1/2, \text{ or } 50\%$$

In genetic problems, we are often interested in the probability that a particular type of offspring will be produced. Recall that when two heterozygous tall pea plants (*Tt*) are crossed, the phenotypic ratio of the offspring is 3 tall to 1 dwarf. This information can be used to calculate the probability for either type of offspring.

$$\text{Probability} = \frac{\text{Number of individuals with a given phenotype}}{\text{Total number of individuals}}$$

$$P_{\text{tall}} = 3 \text{ tall} / (3 \text{ tall} + 1 \text{ dwarf}) = 3/4, \text{ or } 75\%$$

$$P_{\text{dwarf}} = 1 \text{ dwarf} / (3 \text{ tall} + 1 \text{ dwarf}) = 1/4, \text{ or } 25\%$$

The probability is 75% of obtaining a tall plant and 25% of obtaining a dwarf plant. When we add together the probabilities of all possible outcomes (tall and dwarf), we should get a sum of 100% (here, 75% + 25% = 100%).

A probability calculation allows us to predict the likelihood that an outcome will occur in the future. The accuracy of this prediction, however, depends to a great extent on the size of the sample. For example, if we toss a coin six times, our probability prediction suggests that 50% of the time we should get heads (i.e., three heads and three tails). In this small sample size, however, we would not be too surprised if we came up with four heads and two tails. Each time we toss a coin, there is a random chance that it will be heads or tails. The deviation between the observed and expected outcomes is called the **random sampling error.** In a small sample, the error between the predicted percentage of heads and the actual percentage observed may be quite large. By comparison, if we flipped a coin 1000 times, the percentage of heads would be fairly close to the predicted 50% value. In a larger sample, we expect the random sampling error to be a much smaller percentage. For example, the fairly large data sets produced by Mendel had a relatively small sampling error (refer back to Figure 2.5.)

The Product Rule Is Used to Predict the Probability of Independent Outcomes

We can use probability to make predictions regarding the likelihood of two or more independent outcomes from a genetic cross. When we say that outcomes are independent, we mean that the occurrence of one outcome does not affect the probability of another. As an example, let's consider a rare, recessive human trait known as congenital analgesia. Persons with this trait can distinguish between sharp and dull, and hot and cold, but do not perceive extremes of sensation as being painful. The first case of congenital analgesia, described in 1932, was a man who made his living entertaining the public as a "human pincushion."

For a phenotypically unaffected couple, each being heterozygous for the recessive allele causing congenital analgesia, we can ask the question, What is the probability that the couple's first three offspring will have congenital analgesia? To answer this question, the **product rule** is used. According to this rule,

> *The probability that two or more independent outcomes will occur is equal to the product of their individual probabilities.*

A strategy for solving this type of problem is shown here.

The Cross: *Pp* × *Pp* (where *P* is the common allele and *p* is the recessive congenital analgesia allele)

The Question: What is the probability that the couple's first three offspring will have congenital analgesia?

Step 1. *Calculate the individual probability of this phenotype.* As described previously, this is accomplished using a Punnett square.

The probability of an affected offspring is 1/4 (25%).

Step 2. *Multiply the individual probabilities.* In this case, we are asking about the first three offspring, and so we multiply 1/4 three times.

$$1/4 \times 1/4 \times 1/4 = 1/64 = 0.016, \text{ or } 1.6\%$$

Thus, the probability that the first three offspring will have this trait is 0.016. In other words, we predict that 1.6% of the time the first three offspring of a couple, each heterozygous for the recessive allele, will all have congenital analgesia. In this example, the phenotypes of the first, second, and third offspring are independent outcomes. In other words, the phenotype of the first offspring does not have an effect on the phenotype of the second or third offspring.

In the problem described here, we have used the product rule to determine the probability that the first three offspring will all have the same phenotype (congenital analgesia). We can also apply the rule to predict the probability of a sequence of outcomes that involves combinations of different offspring. For example, consider the question, What is the probability that the first offspring will be unaffected, the second offspring will have congenital analgesia, and the third offspring will be unaffected? Again, to solve this problem, begin by calculating the individual probability of each phenotype.

Unaffected = 3/4
Congenital analgesia = 1/4

The probability that these three phenotypes will occur in this specified order is

$$3/4 \times 1/4 \times 3/4 = 9/64 = 0.14, \text{ or } 14\%$$

This sequence of outcomes is expected to occur 14% of the time.

The product rule can also be used to predict the outcome of a cross involving two or more genes. Let's suppose an individual with the genotype *Aa Bb CC* was crossed to an individual with the genotype *Aa bb Cc*. We could ask the question, What is the probability that an offspring will have the genotype *AA bb Cc* ? If the three genes independently assort, the probability of inheriting alleles for each gene is independent of the other two genes. Therefore, we can separately calculate the probability of the desired outcome for each gene.

Cross: *Aa Bb CC* × *Aa bb Cc*

Probability that an offspring will be *AA* = 1/4

Probability that an offspring will be *bb* = 1/2

Probability that an offspring will be *Cc* = 1/2

We can use the product rule to determine the probability that an offspring will be *AA bb Cc* :

$P = (1/4)(1/2)(1/2) = 1/16 = 0.0625$, or 6.25%

The Binomial Expansion Equation Is Used to Predict the Probability of an Unordered Combination of Outcomes

A second type of prediction in genetics is to determine the probability that a certain proportion of offspring will be produced with particular characteristics; here they can be produced in an unspecified order. For example, we can consider a group of children produced by two heterozygous brown-eyed (*Bb*) individuals. We can ask the question, What is the probability that two out of five children will have blue eyes?

In this case, we are not concerned with the order in which the offspring are born. Instead, we are only concerned with the final numbers of blue-eyed and brown-eyed offspring. To solve this type of question, the **binomial expansion equation** can be used. This equation represents all of the possibilities for a given set of unordered outcomes.

$$P = \frac{n!}{x!(n-x)!} p^{x} q^{n-x}$$

where

P = the probability that the unordered outcome will occur

n = total number of outcomes

x = number of outcomes in one category (e.g., blue eyes)

p = individual probability of x

q = individual probability of the other category (e.g., brown eyes)

Note: In this case, $p + q = 1$.

The symbol ! denotes a factorial. $n!$ is the product of all integers from n down to 1. For example, $4! = 4 \times 3 \times 2 \times 1 = 24$. An exception is 0!, which equals 1.

The use of the binomial expansion equation is described next.

The Cross: *Bb* × *Bb*

The Question: What is the probability that two out of five offspring will have blue eyes?

Step 1. *Calculate the individual probabilities of the blue-eye and brown-eye phenotypes.* If we constructed a Punnett square, we would find the probability of blue eyes is 1/4 and the probability of brown eyes is 3/4:

$$p = 1/4$$

$$q = 3/4$$

Step 2. *Determine the number of outcomes in category x (in this case, blue eyes) versus the total number of outcomes.* In this example, the number of outcomes in category x is two blue-eyed children among a total number of five.

$$x = 2$$

$$n = 5$$

Step 3. *Substitute the values for p, q, x, and n in the binomial expansion equation.*

$$P = \frac{n!}{x!(n-x)!} p^{x} q^{n-x}$$

$$P = \frac{5!}{2!(5-2)!} (1/4)^2(3/4)^{5-2}$$

$$P = \frac{5 \times 4 \times 3 \times 2 \times 1}{(2 \times 1)(3 \times 2 \times 1)} (1/16)(27/64)$$

$$P = 0.26, \text{ or } 26\%$$

Thus, the probability is 0.26 that two out of five offspring will have blue eyes. In other words, 26% of the time we expect a *Bb* × *Bb* cross yielding five offspring to contain two blue-eyed children and three brown-eyed children.

When more than two outcomes are possible, we use a **multinomial expansion equation** to solve a problem involving an unordered number of outcomes. A general expression for this equation is

$$P = \frac{n!}{a!b!c! \ldots} p^{a} q^{b} r^{c} \ldots$$

where

P = the probability that the unordered number of outcomes will occur

n = total number of outcomes

$a + b + c + \ldots = n$

$p + q + r + \ldots = 1$

(p is the likelihood of a, q is the likelihood of b, r is the likelihood of c, and so on)

The multinomial expansion equation can be useful in many genetic problems where more than two combinations of offspring are possible. For example, this formula can be used to solve problems involving an unordered sequence of outcomes in a two-factor cross (see solved problem S6 at the end of this chapter).

The Chi Square Test Is Used to Test the Validity of a Genetic Hypothesis

We now look at a different issue in genetic problems, namely **hypothesis testing.** Our goal here is to determine if the data from genetic crosses are consistent with a particular pattern of inheritance. For example, a geneticist may study the inheritance of body color and wing shape in fruit flies over the course of two generations. The following question may be asked about the F_2 generation: Do the observed numbers of offspring agree with the predicted numbers based on Mendel's laws of segregation and independent assortment? As we will see in Chapters 3 through 8, not all traits follow a simple Mendelian pattern of inheritance. Some genes do not segregate and independently assort themselves the same way that Mendel's seven characters did in pea plants.

To distinguish inheritance patterns that obey Mendel's laws from those that do not, a conventional strategy is to make crosses and then quantitatively analyze the offspring. Based on the observed outcome, an experimenter may make a tentative hypothesis. For example, it may seem that the data are obeying Mendel's laws. Hypothesis testing provides an objective, statistical method to evaluate whether the observed data really agree with the hypothesis. In other words, we use statistical methods to determine whether the data that have been gathered from crosses are consistent with predictions based on quantitative laws of inheritance.

The rationale behind a statistical approach is to evaluate the **goodness of fit** between the observed data and the data that are predicted from a hypothesis. This is sometimes called a **null hypothesis** because it assumes there is no real difference between the observed and expected values. Any actual differences that occur are presumed to be due to random sampling error. If the observed and predicted data are very similar, we conclude that the hypothesis is consistent with the observed outcome. In this case, it is reasonable to accept the hypothesis. However, it should be emphasized that this does not prove that a hypothesis is correct. Statistical methods can never prove that a hypothesis is correct. They can provide insight as to whether or not the observed data seem reasonably consistent with the hypothesis. Alternative hypotheses, perhaps even ones that the experimenter has failed to realize, may also be consistent with the data. In some cases, statistical methods may reveal a poor fit between hypothesis and data. In other words, a high deviation would be found between the observed and expected values. If this occurs, the hypothesis is rejected. Hopefully, the experimenter can subsequently propose an alternative hypothesis that has a better fit with the data.

One commonly used statistical method to determine goodness of fit is the **chi square test** (often written χ^2). We can use the chi square test to analyze population data in which the members of the population fall into different categories. This is the kind of data we have when we evaluate the outcome of genetic crosses, because these usually produce a population of offspring that differ with regard to phenotypes. The general formula for the chi square test is

$$\chi^2 = \Sigma \frac{(O - E)^2}{E}$$

where

O = observed data in each category

E = expected data in each category based on the experimenter's hypothesis

Σ means to sum this calculation for each category. For example, if the population data fell into two categories, the chi square calculation would be

$$\chi^2 = \frac{(O_1 - E_1)^2}{E_1} + \frac{(O_2 - E_2)^2}{E_2}$$

We can use the chi square test to determine if a genetic hypothesis is consistent with the observed outcome of a genetic cross. The strategy described next provides a step-by-step outline for applying the chi square testing method. In this problem, the experimenter wants to determine if a two-factor cross is obeying Mendel's laws. The experimental organism is *Drosophila melanogaster* (the common fruit fly), and the two characters involve wing shape and body color. Straight wing shape and curved wing shape are designated by c^+ and c, respectively; gray body color and ebony body color are designated by e^+ and e, respectively. Note: In certain species, such as *Drosophila melanogaster*, the convention is to designate the common (wild-type) allele with a plus sign. Recessive mutant alleles are designated with lowercase letters and dominant mutant alleles with capital letters.

The Cross: A true-breeding fly with straight wings and a gray body ($c^+c^+e^+e^+$) is crossed to a true-breeding fly with curved wings and an ebony body (*ccee*). The flies of the F_1 generation are then allowed to mate with each other to produce an F_2 generation.

The Outcome:

F_1 generation:	All offspring have straight wings and gray bodies
F_2 generation:	193 straight wings, gray bodies
	69 straight wings, ebony bodies
	64 curved wings, gray bodies
	26 curved wings, ebony bodies
Total:	352

Step 1. *Propose a hypothesis that allows us to calculate the expected values based on Mendel's laws.* The F_1 generation suggests that the trait of straight wings is dominant to curved wings and gray body coloration is dominant to ebony. Looking at the F_2 generation, it appears that the data follow a 9:3:3:1 ratio. If so, this is consistent with an independent assortment of the two characters.

Based on these observations, the hypothesis is the following:

Straight (c^+) is dominant to curved (c), and gray (e^+) is dominant to ebony (e). The two characters segregate and assort independently from generation to generation.

Step 2. *Based on the hypothesis, calculate the expected values of the four phenotypes.* We first need to calculate the individual

TABLE 2.1
Chi Square Values and Probability

Degrees of Freedom	P · 0.99	0.95	0.80	0.50	0.20	Null Hypothesis Rejected 0.05	0.01
1.	0.000157	0.00393	0.0642	0.455	1.642	3.841	6.635
2.	0.020	0.103	0.446	1.386	3.219	5.991	9.210
3.	0.115	0.352	1.005	2.366	4.642	7.815	11.345
4.	0.297	0.711	1.649	3.357	5.989	9.488	13.277
5.	0.554	1.145	2.343	4.351	7.289	11.070	15.086
6.	0.872	1.635	3.070	5.348	8.558	12.592	16.812
7.	1.239	2.167	3.822	6.346	9.803	14.067	18.475
8.	1.646	2.733	4.594	7.344	11.030	15.507	20.090
9.	2.088	3.325	5.380	8.343	12.242	16.919	21.666
10.	2.558	3.940	6.179	9.342	13.442	18.307	23.209
15.	5.229	7.261	10.307	14.339	19.311	24.996	30.578
20.	8.260	10.851	14.578	19.337	25.038	31.410	37.566
25.	11.524	14.611	18.940	24.337	30.675	37.652	44.314
30.	14.953	18.493	23.364	29.336	36.250	43.773	50.892

From Fisher, R. A., and Yates, F. (1943) *Statistical Tables for Biological, Agricultural, and Medical Research*. Oliver and Boyd, London.

probabilities of the four phenotypes. According to our hypothesis, there should be a 9:3:3:1 ratio in the F_2 generation. Therefore, the expected probabilities are as follows:

9/16 = straight wings, gray bodies
3/16 = straight wings, ebony bodies
3/16 = curved wings, gray bodies
1/16 = curved wings, ebony bodies

The observed F_2 generation contained a total of 352 individuals. Our next step is to calculate the expected numbers of each type of offspring when the total equals 352. This is accomplished by multiplying each individual probability by 352.

9/16 × 352 = 198 (expected number with straight wings, gray bodies)

3/16 × 352 = 66 (expected number with straight wings, ebony bodies)

3/16 × 352 = 66 (expected number with curved wings, gray bodies)

1/16 × 352 = 22 (expected number with curved wings, ebony bodies)

Step 3. *Apply the chi square formula, using the data for the expected values that have been calculated in step 2.* In this case, the data include four categories, and thus the sum has four terms.

$$\chi^2 = \frac{(O_1 - E_1)^2}{E_1} + \frac{(O_2 - E_2)^2}{E_2} + \frac{(O_3 - E_3)^2}{E_3} + \frac{(O_4 - E_4)^2}{E_4}$$

$$\chi^2 = \frac{(193 - 198)^2}{198} + \frac{(69 - 66)^2}{66} + \frac{(64 - 66)^2}{66} + \frac{(26 - 22)^2}{22}$$

$$\chi^2 = 0.13 + 0.14 + 0.06 + 0.73 = 1.06$$

Step 4. *Interpret the calculated chi square value. This is done using a chi square table.*

Before interpreting the chi square value we obtained, we must understand how to use **Table 2.1**. The probabilities, called ***P* values,** listed in the chi square table allow us to determine the likelihood that the amount of variation indicated by a given chi square value is due to random chance alone, based on a particular hypothesis. For example, let's consider a value (0.00393) listed in row 1. (The meaning of the rows will be explained shortly.) Chi square values that are equal to or greater than 0.00393 are expected to occur 95% of the time when a hypothesis is correct. In other words, 95 out of 100 times we would expect that random chance alone would produce a deviation between the experimental data and hypothesized model that is equal to or greater than 0.00393. A low chi square value indicates a high probability that the observed deviations could be due to random chance alone. By comparison, chi square values that are equal to or greater than 3.841 are expected to occur less than 5% of the time due to random sampling error. If a high chi square value is obtained, an experimenter becomes suspicious that the high deviation has occurred because the hypothesis is incorrect.

A common convention is to reject the null hypothesis if the chi square value results in a probability that is less than 0.05 (less than 5%) or if the probability is less than 0.01 (less than 1%). These are sometimes called the 5% and 1% significance levels, respectively. Which level is better to choose? The choice is somewhat subjective. If you choose a 5% level rather than a 1% level, a disadvantage is that you are more likely to reject a null hypothesis that happens to be correct. Even so, choosing a 5% level rather than a 1% level has the advantage that you are less likely to accept an incorrect null hypothesis.

In our problem involving flies with straight or curved wings and gray or ebony bodies, we have calculated a chi square value of 1.06. Before we can determine the probability that this deviation would have occurred as a matter of random chance, we must first determine the degrees of freedom (*df*) in this experiment. The **degrees of freedom** is a measure of the number of categories that are independent of each other. When phenotype categories are derived from a Punnett square, it is typically $n - 1$, where n equals the total number of categories. In the preceding problem, $n = 4$ (the categories are the phenotypes: straight wings and gray body; straight wings and ebony body; curved wings and gray body; and curved wings and ebony body); thus, the degrees of freedom equals 3.* We now have sufficient information to interpret our chi square value of 1.06.

With $df = 3$, the chi square value of 1.06 we have obtained is slightly greater than 1.005, which gives a *P* value of 0.80, or 80%. What does this *P* value mean? If the hypothesis is correct, chi square values equal to or greater than 1.005 are expected to occur 80% of the time based on random chance alone. To reject the null hypothesis at the 5% significance level, the chi square would have to be greater than 7.815. Because it was actually far less than this value, we are inclined to accept that the null hypothesis is correct.

We must keep in mind that the chi square test does not prove a hypothesis is correct. It is a statistical method for evaluating whether the data and hypothesis have a good fit.

2.5 COMPREHENSION QUESTIONS

1. A cross is made between *AA Bb Cc Dd* and *Aa Bb cc dd* individuals. Rather than making a very large Punnett square, which statistical operation could you use to solve this problem, and what would be the probability of an offspring that is *AA bb Cc dd*?
   a. Product rule, 1/32
   b. Product rule, 1/4
   c. Binomial expansion, 1/32
   d. Binomial expansion, 1/4
2. In dogs, brown fur color (*B*) is dominant to white (*b*). A cross is made between two heterozygotes. If a litter contains six pups, what is the probability that half of them will be white?
   a. 0.066, or 6.6%
   b. 0.13, or 13%
   c. 0.25, or 25%
   d. 0.26, or 26%
3. Which of the following operations could be used for hypothesis testing?
   a. Product rule
   b. Binomial expansion
   c. Chi square test
   d. All of the above

KEY TERMS

Page 18. cross, hybridization, hybrids
Page 20. gamete, sperm, pollen grains, anthers, eggs, ovules, ovaries, stigma, self-fertilization, cross-fertilization, characters, trait, variant, true-breeding strain, true-breeding line
Page 21. single-factor cross, monohybrids, empirical approach, parental generation, P generation, F_1 generation, F_2 generation
Page 23. dominant, recessive, particulate theory of inheritance, segregate, gene, allele, Mendel's law of segregation
Page 24. homozygous, heterozygous, genotype, phenotype, Punnett square
Page 25. two-factor crosses
Page 27. nonparentals
Page 28. Mendel's law of independent assortment, genetic recombination
Page 30. multiplication method, forked-line method, dihybrid testcross, loss-of-function alleles
Page 31. pedigrees, pedigree analysis
Page 33. probability, random sampling error, product rule
Page 34. binomial expansion equation, multinomial expansion equation
Page 35. hypothesis testing, goodness of fit, null hypothesis, chi square test
Page 36. *P* values
Page 37. degrees of freedom

CHAPTER SUMMARY

- Early ideas regarding inheritance included pangenesis and blending inheritance. These ideas were later refuted by the work of Mendel.

2.1 Mendel's Study of Pea Plants

- Mendel chose pea plants as his experimental organism because it was easy to carry out self-fertilization or cross-fertilization experiments with these plants, and because pea plants were available in several varieties in which a character existed in two distinct variants (see Figures 2.1, 2.2, 2.3, 2.4).

*If our hypothesis already assumed that the law of segregation is obeyed, the degrees of freedom would be 1 (see Chapter 6).

2.2 Law of Segregation

- By conducting single-factor crosses, Mendel proposed three key ideas regarding inheritance. (1) Traits may be dominant or recessive. (2) Genes are passed unaltered from generation to generation. (3) The law of segregation states the following: The two copies of a gene segregate (or separate) from each other during transmission from parent to offspring (see Figures 2.5, 2.6).
- A Punnett square is used to deduce the outcome of single-factor crosses.

2.3 Law of Independent Assortment

- By conducting two-factor crosses, Mendel proposed the law of independent assortment: Two different genes will randomly assort their alleles during the formation of haploid cells (see Figures 2.8, 2.9).
- A Punnett square is used to predict the outcome of two-factor crosses (see Figure 2.10).

2.4 Studying Inheritance Patterns in Humans

- Human inheritance patterns are determined by analyzing family trees known as pedigrees (see Figure 2.11).

2.5 Probability and Statistics

- Probability is the number of times an outcome occurs divided by the total number of outcomes.
- According to the product rule, the probability of two or more independent outcomes is equal to the product of their individual probabilities. This rule can be used to predict the outcome of crosses involving two or more genes.
- The binomial expansion is used to predict the probability of an unordered combination of outcomes.
- The chi square test is used to test the validity of a hypothesis (see Table 2.1).

PROBLEM SETS & INSIGHTS

Solved Problems

S1. As described in this chapter, a human disease known as cystic fibrosis is inherited as a recessive trait. Two unaffected individuals have a first child with the disease. What is the probability that their next two children will not have the disease?

Answer: An unaffected couple has already produced an affected child. To be affected, the child must be homozygous for the disease allele and thus has inherited one copy from each parent. Therefore, because the parents are unaffected with the disease, we know that both of them must be heterozygous carriers for the recessive disease-causing allele. With this information, we can calculate the probability that they will produce an unaffected offspring. Using a Punnett square, this couple should produce a ratio of 3 unaffected : 1 affected offspring.

♀ \ ♂	*N*	*n*
N	*NN*	*Nn*
n	*Nn*	*nn*

N = common allele
n = cystic fibrosis allele

The probability of a single unaffected offspring is

$$P_{\text{unaffected}} = 3/(3 + 1) = 3/4, \text{ or } 75\%$$

To obtain the probability of getting two unaffected offspring in a row (i.e., in a specified order), we must apply the product rule.

$$3/4 \times 3/4 = 9/16 = 0.56, \text{ or } 56\%$$

The chance that their next two children will be unaffected is 56%.

S2. A cross was made between two heterozygous pea plants, *TtYy* × *TtYy*. The following Punnett square was constructed:

♀ \ ♂	*TT*	*Tt*	*Tt*	*tt*
YY	*TTYY*	*TtYY*	*TtYY*	*ttYY*
Yy	*TTYy*	*TtYy*	*TtYy*	*ttYy*
Yy	*TTYy*	*TtYy*	*TtYy*	*ttYy*
yy	*TTyy*	*Ttyy*	*Ttyy*	*ttyy*

Phenotypic ratio:

9 tall, yellow seeds : 3 tall, green seeds : 3 dwarf, yellow seeds : 1 dwarf, green seed

What is wrong with this Punnett square?

Answer: The outside of the Punnett square is supposed to contain the possible types of gametes. A gamete should contain one copy of each type of gene. Instead, the outside of this Punnett square contains two copies of one gene and zero copies of the other gene. The outcome happens to be correct (i.e., it yields a 9:3:3:1 ratio), but this is only a coincidence. The outside of the Punnett square must contain one copy of each type of gene. In this example, the correct possible types of gametes are *TY*, *Ty*, *tY*, and *ty* for each parent.

S3. A pea plant is heterozygous for three genes (*Tt Rr Yy*), where T = tall, t = dwarf, R = round seeds, r = wrinkled seeds, Y = yellow seeds, and y = green seeds. If this plant is self-fertilized, what are the predicted phenotypes of the offspring, and what fraction of the offspring will occur in each category?

Answer: You could solve this problem by constructing a large Punnett square and filling in the boxes. However, in this case, eight different male gametes and eight different female gametes are possible: *TRY*, *TRy*, *TrY*, *tRY*, *trY*, *Try*, *tRy*, and *try*. It would become rather tiresome to construct and fill in this Punnett square, which would contain 64 boxes. As an alternative, we can consider each gene separately and then algebraically combine them by multiplying together the expected phenotypic outcomes for each gene. In the cross *Tt Rr Yy* × *Tt Rr Yy*, the following Punnett squares can be made for each gene:

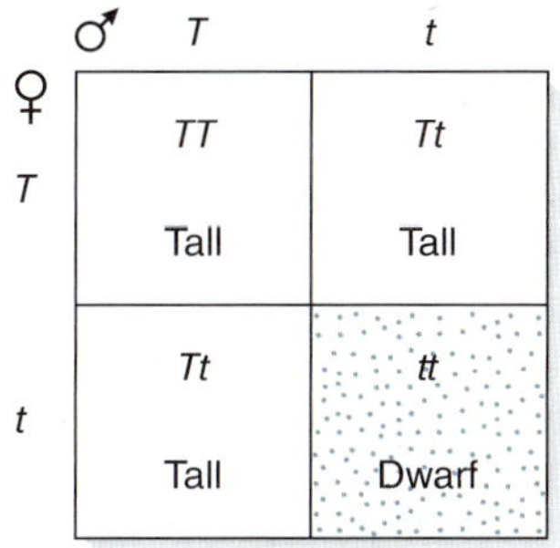

3 tall : 1 dwarf

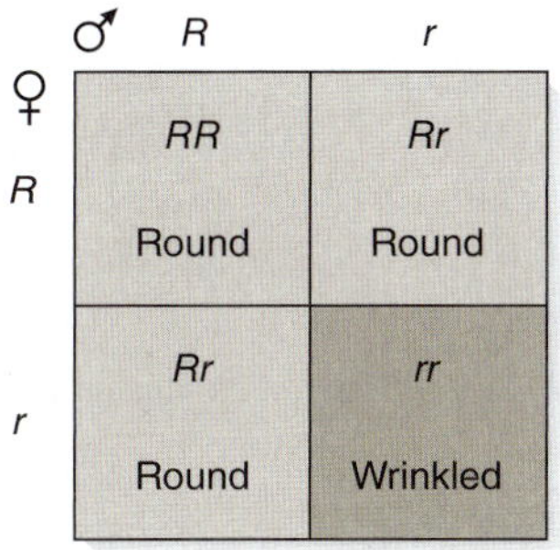

3 round : 1 wrinkled

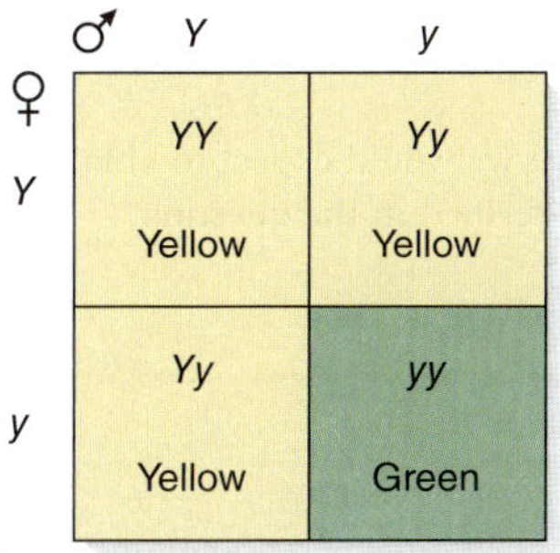

3 yellow : 1 green

Instead of constructing a large, 64-box Punnett square, we can use two similar ways to determine the phenotypic outcome of this three-factor cross. In the multiplication method, we simply multiply these three combinations together:

(3 tall +1 dwarf)(3 round + 1 wrinkled)(3 yellow + 1 green)

This multiplication operation can be done in a stepwise manner. First, multiply (3 tall + 1 dwarf) by (3 round + 1 wrinkled).

(3 tall + 1 dwarf)(3 round + 1 wrinkled) = 9 tall, round + 3 tall, wrinkled + 3 dwarf, round, + 1 dwarf, wrinkled

Next, multiply this product by (3 yellow + 1 green).

(9 tall, round + 3 tall, wrinkled + 3 dwarf, round + 1 dwarf, wrinkled) (3 yellow + 1 green) = 27 tall, round, yellow + 9 tall, round, green + 9 tall, wrinkled, yellow + 3 tall, wrinkled, green + 9 dwarf, round, yellow + 3 dwarf, round, green + 3 dwarf, wrinkled, yellow + 1 dwarf, wrinkled, green

Even though the multiplication steps are also somewhat tedious, this approach is much easier than making a Punnett square with 64 boxes, filling them in, deducing each phenotype, and then adding them up!

A second approach that is analogous to the multiplication method is the forked-line method**.** In this case, the genetic proportions are determined by multiplying together the probabilities of each phenotype.

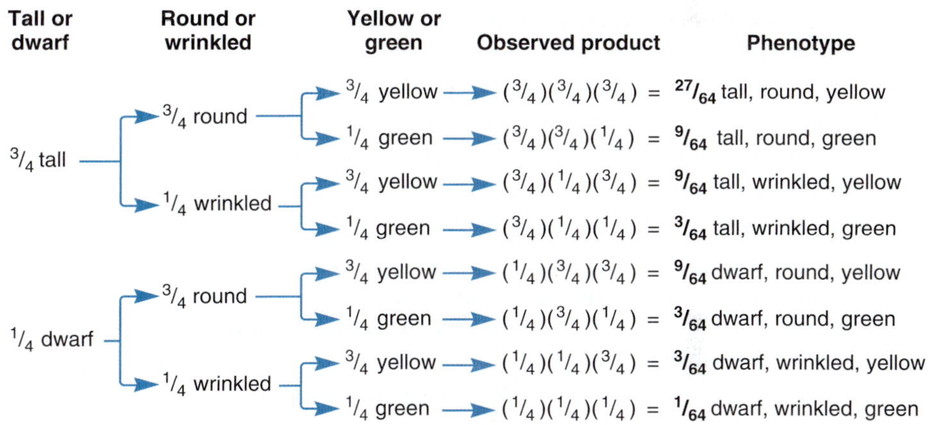

S4. For an individual expressing a dominant trait, how can you tell if it is a heterozygote or a homozygote?

Answer: One way is to conduct a testcross with an individual that expresses the recessive version of the same character. If the individual is heterozygous, half of the offspring will show the recessive trait, but if the individual is homozygous, none of the offspring will express the recessive trait.

Dd × *dd*			or	*DD* × *dd*
	↓			↓
1*Dd* (dominant trait)	:	1 *dd* (recessive trait)		All *Dd* (dominant trait)

Another way to determine heterozygosity involves a more careful examination of the individual at the cellular or molecular level. At the cellular level, the heterozygote may not look exactly like the homozygote. This phenomenon is described in Chapter 4. Also, gene cloning methods described in Chapter 20 can be used to distinguish between heterozygotes and homozygotes.

S5. In dogs, black fur color is dominant to white. Two heterozygous black dogs are mated. What would be the probability of the following combinations of offspring?

A. A litter of six pups, four with black fur and two with white fur.

B. A litter of six pups, the firstborn with white fur, and among the remaining five pups, two with white fur and three with black fur.

C. A first litter of six pups, four with black fur and two with white fur, and then a second litter of seven pups, five with black fur and two with white fur.

D. A first litter of five pups, four with black fur and one with white fur, and then a second litter of seven pups in which the firstborn is homozygous, the second born is black, and the remaining five pups are three black and two white.

Answer:

A. Because this is an unordered combination of outcomes, we use the binomial expansion equation, where $n = 6$, $x = 4$, $p = 0.75$ (probability of black), and $q = 0.25$ (probability of white).

The answer is 0.297, or 29.7%, of the time.

B. We use the product rule because the order is specified. The first pup is white and then the remaining five are born later. We also need to use the binomial expansion equation to determine the probability of the remaining five pups.

(probability of a white pup)(binomial expansion for the remaining five pups)

The probability of the white pup is 0.25. In the binomial expansion equation, $n = 5$, $x = 2$, $p = 0.25$, and $q = 0.75$.

The answer is 0.066, or 6.6%, of the time.

C. The order of the two litters is specified, so we need to use the product rule. We multiply the probability of the first litter times the probability of the second litter. We need to use the binomial expansion equation for each litter.

(binomial expansion of the first litter)(binomial expansion of the second litter)

For the first litter, $n = 6$, $x = 4$, $p = 0.75$, $q = 0.25$. For the second litter, $n = 7$, $x = 5$, $p = 0.75$, $q = 0.25$.

The answer is 0.092, or 9.2%, of the time.

D. The order of the litters is specified, so we need to use the product rule to multiply the probability of the first litter times the probability of the second litter. We use the binomial expansion equation to determine the probability of the first litter. The probability of the second litter is a little more complicated. The firstborn is homozygous. The probability of the first pup equals $0.25 + 0.25 = 0.5$. The probability of the second pup is 0.75, and we use the binomial expansion equation to determine the probability of the remaining pups.

(binomial expansion of first litter)([0.5][0.75][binomial expansion of second litter])

For the first litter, $n = 5$, $x = 4$, $p = 0.75$, $q = 0.25$. For the last five pups in the second litter, $n = 5$, $x = 3$, $p = 0.75$, $q = 0.25$.

The answer is 0.039, or 3.9%, of the time.

S6. A cross is made between two heterozygous tall plants with axial flowers (*TtAa*), where tall is dominant to dwarf and axial is dominant to terminal flowers. What is the probability that a group of five offspring will be composed of two tall plants with axial flowers, one tall plant with terminal flowers, one dwarf plant with axial flowers, and one dwarf plant with terminal flowers?

Answer:

Step 1. *Calculate the individual probabilities of each phenotype.* This can be accomplished using a Punnett square.

The phenotypic ratios are 9 tall with axial flowers, 3 tall with terminal flowers, 3 dwarf with axial flowers, and 1 dwarf with terminal flowers.

The probability of a tall plant with axial flowers is $9/(9 + 3 + 3 + 1) = 9/16$.

The probability of a tall plant with terminal flowers is $3/(9 + 3 + 3 + 1) = 3/16$.

The probability of a dwarf plant with axial flowers is $3/(9 + 3 + 3 + 1) = 3/16$.

The probability of a dwarf plant with terminal flowers is $1/(9 + 3 + 3 + 1) = 1/16$.

$p = 9/16$ $\quad r = 3/16$

$q = 3/16$ $\quad s = 1/16$

Step 2. *Determine the number of each type of outcome versus the total number of outcomes.*

$n = 5$

$a = 2$

$b = 1$

$c = 1$

$d = 1$

Step 3. *Substitute the values in the multinomial expansion equation.*

$$P = \frac{n!}{a!b!c!d!} p^a q^b r^c s^d$$

$$P = \frac{5!}{2!1!1!1!} (9/16)^2 (3/16)^1 (3/16)^1 (1/16)^1$$

$$P = 0.04, \text{ or } 4\%$$

This means that 4% of the time we would expect to obtain five offspring with the phenotypes described in the question.

Conceptual Questions

C1. Why did Mendel's work refute the idea of blending inheritance?

C2. What is the difference between cross-fertilization and self-fertilization?

C3. Describe the difference between genotype and phenotype. Give three examples. Is it possible for two individuals to have the same phenotype but different genotypes?

C4. With regard to genotypes, what is a true-breeding organism?

C5. How can you determine whether an organism is heterozygous or homozygous for a dominant trait?

C6. In your own words, describe Mendel's law of segregation. Do not use the word "segregation" in your answer.

C7. Based on genes in pea plants that we have considered in this chapter, which statement(s) is/are not correct?

A. The gene causing tall plants is an allele of the gene causing dwarf plants.

B. The gene causing tall plants is an allele of the gene causing purple flowers.

C. The alleles causing tall plants and purple flowers are dominant.

C8. In a cross between a heterozygous tall pea plant and a dwarf plant, predict the ratios of the offspring's genotypes and phenotypes.

C9. Do you know the genotype of an individual with a recessive trait and/or a dominant trait? Explain your answer.

C10. A cross is made between a pea plant that has constricted pods (a recessive trait; smooth is dominant) and is heterozygous for seed color (yellow is dominant to green) and a plant that is heterozygous for both pod texture and seed color. Construct a Punnett square that depicts this cross. What are the predicted outcomes of genotypes and phenotypes of the offspring?

C11. A pea plant that is heterozygous with regard to seed color (yellow is dominant to green) is allowed to self-fertilize. What are the predicted outcomes of genotypes and phenotypes of the offspring?

C12. Describe the significance of nonparentals with regard to the law of independent assortment. In other words, explain how the appearance of nonparentals refutes a linkage hypothesis.

C13. For the following pedigrees, describe what you think is the most likely inheritance pattern (dominant versus recessive). Explain your reasoning. Filled (black) symbols indicate affected individuals.

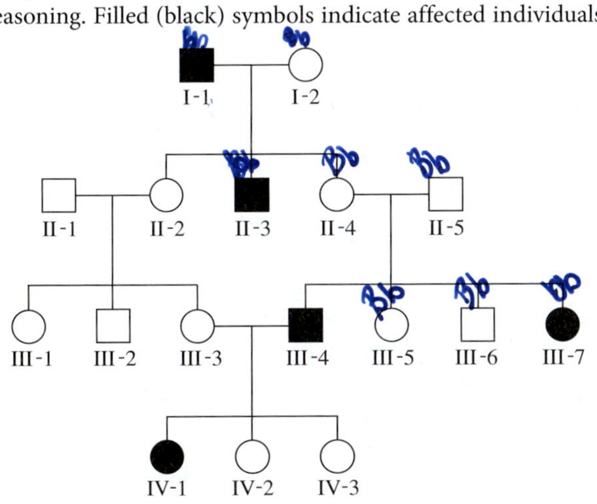

(a)

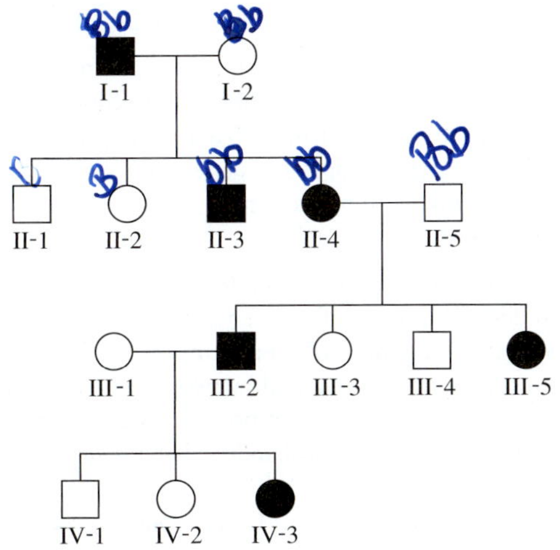

(b)

C14. Ectrodactyly, also known as "lobster claw syndrome," is a recessive disorder in humans. If a phenotypically unaffected couple produces an affected offspring, what are the following probabilities?

A. Both parents are heterozygotes.

B. An offspring is a heterozygote.

C. The next three offspring will be phenotypically unaffected.

D. Any two out of the next three offspring will be phenotypically unaffected.

C15. Identical twins are produced from the same sperm and egg (which splits after the first mitotic division), whereas fraternal twins are produced from separate sperm and separate egg cells. If two parents with brown eyes (a dominant trait) produce one twin boy with blue eyes, what are the following probabilities?

A. If the other twin is identical, he will have blue eyes.

B. If the other twin is fraternal, he or she will have blue eyes.

C. If the other twin is fraternal, he or she will transmit the blue eye allele to his or her offspring.

D. The parents are both heterozygotes.

C16. In cocker spaniels, solid coat color is dominant over spotted coat color. If two heterozygous dogs were crossed to each other, what would be the probability of the following combinations of offspring?

A. A litter of five pups, four with solid fur and one with spotted fur.

B. A first litter of six pups, four with solid fur and two with spotted fur, and then a second litter of five pups, all with solid fur.

C. A first litter of five pups, the firstborn with solid fur, and then among the next four, three with solid fur and one with spotted fur, and then a second litter of seven pups in which the firstborn is spotted, the second born is spotted, and the remaining five are composed of four solid and one spotted animal.

D. A litter of six pups, the firstborn with solid fur, the second born spotted, and among the remaining four pups, two with spotted fur and two with solid fur.

C17. A cross was made between a white male dog and two different black females. The first female gave birth to eight black pups, and the second female gave birth to four white and three black pups. What are the likely genotypes of the male parent and the two female parents? Explain whether you are uncertain about any of the genotypes.

C18. In humans, the allele for brown eye color (*B*) is dominant to blue eye color (*b*). If two heterozygous parents produce children, what are the following probabilities?

A. The first two children have blue eyes.

B. A total of four children, two with blue eyes and the other two with brown eyes.

C. The first child has blue eyes, and the next two have brown eyes.

C19. Albinism, a condition characterized by a partial or total lack of skin pigment, is a recessive human trait. If a phenotypically unaffected couple produce an albino child, what is the probability that their next child will be albino?

C20. A true-breeding tall plant was crossed to a dwarf plant. Tallness is a dominant trait. The F_1 individuals were allowed to self-fertilize. What are the following probabilities for the F_2 generation?

A. The first plant is dwarf.

B. The first plant is dwarf or tall.

C. The first three plants are tall.

D. For any seven plants, three are tall and four are dwarf.

E. The first plant is tall, and then among the next four, two are tall and the other two are dwarf.

C21. For pea plants with the following genotypes, list the possible gametes that the plant can make:

A. *TT Yy Rr*

B. *Tt YY rr*

C. *Tt Yy Rr*

D. *tt Yy rr*

C22. An individual has the genotype *Aa Bb Cc* and makes an abnormal gamete with the genotype *AaBc*. Does this gamete violate the law of independent assortment or the law of segregation (or both)? Explain your answer.

C23. In people with maple syrup urine disease, the body is unable to metabolize the amino acids leucine, isoleucine, and valine. One of the symptoms is that the urine smells like maple syrup. An unaffected couple produced six children in the following order: unaffected daughter, affected daughter, unaffected son, unaffected son, affected son, and unaffected son. The youngest unaffected son marries an unaffected woman and has three children in the following order: affected daughter, unaffected daughter, and unaffected son. Draw a pedigree that describes this family. What type of inheritance (dominant or recessive) would you propose to explain maple syrup urine disease?

C24. Marfan syndrome is a rare inherited human disorder characterized by unusually long limbs and digits plus defects in the heart (especially the aorta) and the eyes, among other symptoms. Following is a pedigree for this disorder. Affected individuals are shown with filled (black) symbols. What type of inheritance pattern do you think is the most likely?

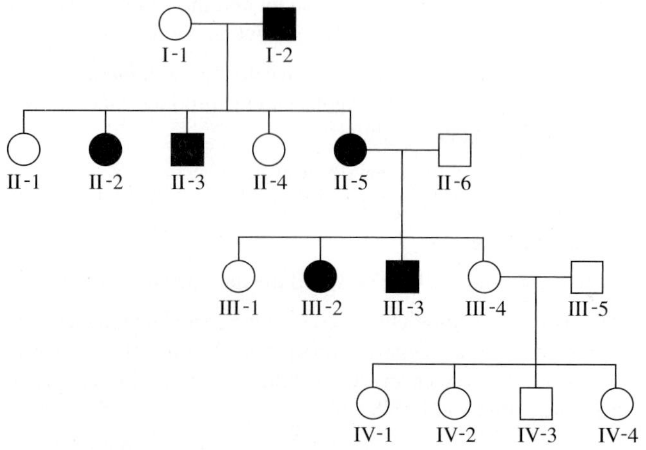

C25. A true-breeding pea plant with round and green seeds was crossed to a true-breeding plant with wrinkled and yellow seeds. Round and yellow seeds are the dominant traits. The F_1 plants were allowed to self-fertilize. What are the following probabilities for the F_2 generation?

A. An F_2 plant with wrinkled, yellow seeds.

B. Three out of three F_2 plants with round, yellow seeds.

C. Five F_2 plants in the following order: two have round, yellow seeds; one has round, green seeds; and two have wrinkled, green seeds.

D. An F_2 plant will not have round, yellow seeds.

C26. A true-breeding tall pea plant was crossed to a true-breeding dwarf plant. What is the probability that an F_1 individual will be true-breeding? What is the probability that an F_1 individual will be a true-breeding tall plant?

C27. What are the expected phenotypic ratios from the following cross: *Tt Rr yy Aa* × *Tt rr YY Aa*, where T = tall, t = dwarf, R = round, r = wrinkled, Y = yellow, y = green, A = axial, a = terminal; T, R, Y, and A are dominant alleles. Note: See solved problem S3 for help in answering this problem.

C28. When an abnormal organism has three copies of a chromosome and therefore contains three copies of the genes on that chromosome (instead of the normal number of two copies), the alleles for each gene usually segregate so that a gamete will contain one or two copies of the gene. Let's suppose that an abnormal pea plant has three copies of the chromosome that carries the height gene. Its genotype is *TTt*. The plant is also heterozygous for the seed color gene, *Yy*, which is found on a different chromosome. With regard to both genes, how many types of gametes can this plant make, and in what proportions? (Assume that it is equally likely that a gamete will contain one or two copies of the height gene.)

C29. Honeybees are unusual in that male bees (drones) have only one copy of each gene, but female bees have two copies of their genes. That is because drones develop from eggs that have not been fertilized by sperm cells. In bees, the trait of long wings is dominant over short wings, and the trait of black eyes is dominant over white eyes. If a drone with short wings and black eyes was mated to a queen bee that is heterozygous for both genes, what are the predicted genotypes and phenotypes of male and female offspring? What are the phenotypic ratios if we assume an equal number of male and female offspring?

C30. A pea plant that is dwarf with green, wrinkled seeds was crossed to a true-breeding plant that is tall with yellow, round seeds. The F_1 generation was allowed to self-fertilize. What types of gametes, and in what proportions, would the F_1 generation make? What would be the ratios of genotypes and phenotypes of the F_2 generation?

C31. A true-breeding plant with round and green seeds was crossed to a true-breeding plant with wrinkled and yellow seeds. The F_1 plants were allowed to self-fertilize. What is the probability of obtaining the following plants in the F_2 generation: two that have round, yellow seeds; one with round, green seeds; and two with wrinkled, green seeds?

C32. Wooly hair is a rare dominant trait found in people of Scandinavian descent in which the hair resembles the wool of a sheep. A male with wooly hair, who has a mother with straight hair, moves to an island that is inhabited by people who are not of Scandinavian descent. Assuming that no other Scandinavians immigrate to the island, what is the probability that a great-grandchild of this male will have wooly hair? (Hint: You may want to draw a pedigree to help you figure this out.) If this wooly-haired male has eight great-grandchildren, what is the probability that one out of eight will have wooly hair?

C33. Huntington disease is a rare dominant trait that causes neurodegeneration later in life. A man in his thirties, who already has three children, discovers that his mother has Huntington disease though his father is unaffected. What are the following probabilities?

A. That the man in his thirties will develop Huntington disease.

B. That his first child will develop Huntington disease.

C. That one out of three of his children will develop Huntington disease.

C34. A woman with achondroplasia (a dominant form of dwarfism) and a phenotypically unaffected man have seven children, all of whom have achondroplasia. What is the probability of producing such a family if this woman is a heterozygote? What is the probability that the woman is a heterozygote if her eighth child does not have this disorder?

Experimental Questions

E1. Describe three advantages of using pea plants as an experimental organism.

E2. Explain the technical differences between a cross-fertilization experiment versus a self-fertilization experiment.

E3. How long did it take Mendel to complete the experiment in Figure 2.5?

E4. For all seven characters described in the data of Figure 2.5, Mendel allowed the F_2 plants to self-fertilize. He found that when F_2 plants with recessive traits were crossed to each other, they always bred true. However, when F_2 plants with dominant traits were crossed, some bred true but others did not. A summary of Mendel's results is shown here.

The Ratio of True-Breeding and Non-True-Breeding Parents of the F_2 Generation

F_2 Parents	True-Breeding	Non-True-Breeding	Ratio
Round	193	372	1:1.93
Yellow	166	353	1:2.13
Gray	36	64	1:1.78
Smooth	29	71	1:2.45
Green	40	60	1:1.5
Axial	33	67	1:2.08
Tall	28	72	1:2.57
TOTAL:	525	1059	1:2.02

When considering the data in this table, keep in mind that it describes the characteristics of the F_2 generation parents that had displayed a dominant phenotype. These data were deduced by analyzing the outcome of the F_3 generation. Based on Mendel's laws, explain the 1:2 ratio obtained in these data.

E5. From the point of view of crosses and data collection, what are the experimental differences between a single-factor and a two-factor cross?

E6. As in many animals, albino coat color is a recessive trait in guinea pigs. Researchers removed the ovaries from an albino female guinea pig and then transplanted ovaries from a true-breeding black guinea pig. They then mated this albino female (with the transplanted ovaries) to an albino male. The albino female produced three offspring. What were their coat colors? Explain the results.

E7. The fungus *Melampsora lini* causes a disease known as flax rust. Different strains of *M. lini* cause varying degrees of the rust disease. Conversely, different strains of flax are resistant or sensitive to the various varieties of rust. The Bombay variety of flax is resistant to *M. lini*-strain 22 but sensitive to *M. lini*-strain 24. A strain of flax called 770B is just the opposite; it is resistant to strain 24 but sensitive to strain 22. When 770B was crossed to Bombay, all F_1 individuals were resistant to both strain 22 and strain 24. When F_1 individuals were self-fertilized, the following data were obtained:

43 resistant to strain 22 but sensitive to strain 24

9 sensitive to strain 22 and strain 24

32 sensitive to strain 22 but resistant to strain 24

110 resistant to strain 22 and strain 24

Explain the inheritance pattern for flax resistance and sensitivity to *M. lini* strains.

E8. For Mendel's data shown in Figure 2.8, conduct a chi square analysis to determine if the data agree with Mendel's law of independent assortment.

E9. Would it be possible to deduce the law of independent assortment from a single-factor cross? Explain your answer.

E10. In fruit flies, curved wings are recessive to straight wings, and ebony body is recessive to gray body. A cross was made between true-breeding flies with curved wings and gray bodies to flies with straight wings and ebony bodies. The F_1 offspring were then mated to flies with curved wings and ebony bodies to produce an F_2 generation.

A. Diagram the genotypes of this cross, starting with the parental generation and ending with the F_2 generation.

B. What are the predicted phenotypic ratios of the F_2 generation?

C. Let's suppose the following data were obtained for the F_2 generation:

114 curved wings, ebony body

105 curved wings, gray body

111 straight wings, gray body

114 straight wings, ebony body

Conduct a chi square analysis to determine if the experimental data are consistent with the expected outcome based on Mendel's laws.

E11. A recessive allele in mice results in an abnormally long neck. Sometimes, during early embryonic development, the abnormal neck causes the embryo to die. An experimenter began with a population of true-breeding normal mice and true-breeding mice with long necks. Crosses were made between these two populations to produce an F_1 generation of mice with normal necks. The F_1 mice were then mated to each other to obtain an F_2 generation. For the mice that were born alive, the following data were obtained:

522 mice with normal necks

62 mice with long necks

What percentage of homozygous mice (that would have had long necks if they had survived) died during embryonic development?

E12. The data in Figure 2.5 show the results of the F_2 generation for seven of Mendel's crosses. Conduct a chi square analysis to determine if these data are consistent with the law of segregation.

E13. Let's suppose you conducted an experiment involving genetic crosses and calculated a chi square value of 1.005. There were four categories of offspring (i.e., the degrees of freedom equaled 3). Explain what the 1.005 value means. Your answer should include the phrase "80% of the time."

E14. A tall pea plant with axial flowers was crossed to a dwarf plant with terminal flowers. Tall plants and axial flowers are dominant traits. The following offspring were obtained: 27 tall, axial flowers; 23 tall, terminal flowers; 28 dwarf, axial flowers; and 25 dwarf, terminal flowers. What are the genotypes of the parents?

E15. A cross was made between two strains of plants that are agriculturally important. One strain was disease-resistant but herbicide-sensitive; the other strain was disease-sensitive but herbicide-resistant. A plant breeder crossed the two plants and then allowed the F_1 generation to self-fertilize. The following data were obtained:

F_1 generation: All offspring are disease-sensitive and herbicide-resistant

F_2 generation:
157 disease-sensitive, herbicide-resistant
57 disease-sensitive, herbicide-sensitive
54 disease-resistant, herbicide-resistant
20 disease-resistant, herbicide-sensitive

Total: 288

Formulate a hypothesis that you think is consistent with the observed data. Test the goodness of fit between the data and your hypothesis using a chi square test. Explain what the chi square results mean.

E16. A cross was made between a plant that has blue flowers and purple seeds to a plant with white flowers and green seeds. The following data were obtained:

F_1 generation: All offspring have blue flowers with purple seeds

F_2 generation:
103 blue flowers, purple seeds
49 blue flowers, green seeds
44 white flowers, purple seeds
104 white flowers, green seeds

Total: 300

Start with the hypothesis that blue flowers and purple seeds are dominant traits and that the two genes assort independently. Calculate a chi square value. What does this value mean with regard to your hypothesis? If you decide to reject your hypothesis, which aspect of the hypothesis do you think is incorrect (i.e., the idea that blue flowers and purple seeds are dominant traits, or the idea that the two genes assort independently)?

Questions for Student Discussion/Collaboration

1. Consider this cross in pea plants: *Tt Rr yy Aa* × *Tt rr Yy Aa*, where *T* = tall, *t* = dwarf, *R* = round, *r* = wrinkled, *Y* = yellow, *y* = green, *A* = axial, *a* = terminal. What is the expected phenotypic outcome of this cross? Have one group of students solve this problem by making one big Punnett square, and have another group solve it by making four single-gene Punnett squares and using the product rule. Time each other to see who gets done first.

2. A cross was made between two pea plants, *TtAa* and *Ttaa*, where *T* = tall, *t* = dwarf, *A* = axial, and *a* = terminal. What is the probability that the first three offspring will be tall with axial flowers or dwarf with terminal flowers and the fourth offspring will be tall with axial flowers? Discuss what operation(s) (e.g., product rule or binomial expansion equation) you used and in what order they were used.

3. Consider this four-factor cross: *Tt Rr yy Aa* × *Tt RR Yy aa*, where *T* = tall, *t* = dwarf, *R* = round, *r* = wrinkled, *Y* = yellow, *y* = green, *A* = axial, *a* = terminal. What is the probability that the first three plants will have round seeds? What is the easiest way to solve this problem?

Note: All answers appear at the website for this textbook; the answers to the even-numbered questions and all of the Concept Check and Comprehension Questions are in the back of the book.

www.mhhe.com/brookergenetics5e

Visit the website for practice tests, answer keys, and other learning aids for this chapter. Enhance your understanding of genetics with our interactive exercises, quizzes, animations, and much more.

CHAPTER OUTLINE

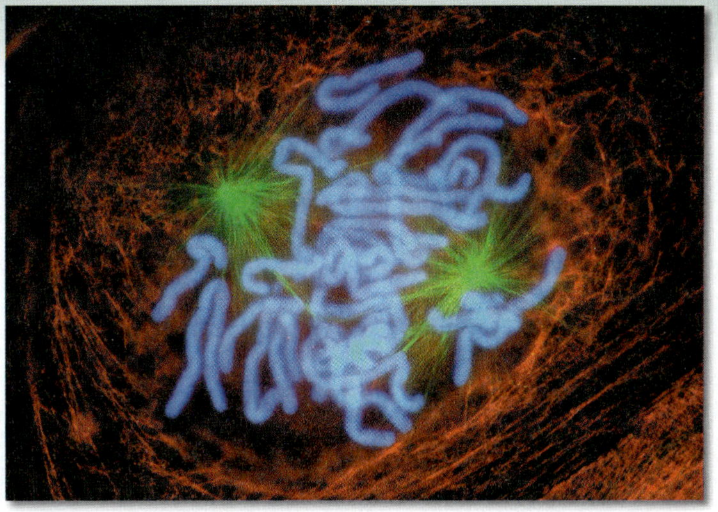

Chromosome sorting during cell division. *When eukaryotic cells divide, they replicate and sort their chromosomes (shown in blue), so that each cell receives the correct amount.*

3 CHROMOSOME TRANSMISSION DURING CELL DIVISION AND SEXUAL REPRODUCTION

In Chapter 2, we considered some patterns of inheritance that explain the passage of traits from parent to offspring. In this chapter, we begin by considering the general features of chromosomes and how they are observed under the microscope. We will then examine how bacterial and eukaryotic cells divide. During cell division, eukaryotic cells can sort their chromosomes in two different ways. One process called mitosis sorts chromosomes so that each daughter cell receives the same number and types of chromosomes as the original mother cell. A second process called meiosis results in daughter cells with half the number of chromosomes as the mother cell. Meiosis is needed for sexual reproduction in eukaryotic species, such as animals and plants. During sexual reproduction, gametes with half the number of chromosomes unite at fertilization.

Finally, we end this chapter by examining the chromosome theory of inheritance that explains the relationship between the transmission of chromosomes during sexual reproduction and the pattern of transmission of traits observed by Mendel. In addition, we will consider how chromosomes are involved with determining sex in certain species and how researchers were able to show that certain genes are located on sex chromosomes.

3.1 GENERAL FEATURES OF CHROMOSOMES

Learning Outcomes:

1. Define the term chromosome.
2. Outline key differences between prokaryotic and eukaryotic cells.
3. Describe the procedure for making a karyotype.
4. Compare and contrast the similarities and differences between homologous chromosomes.

The **chromosomes** are structures within living cells that contain the genetic material. The term chromosome—meaning colored body—refers to the microscopic observation of chromosomes after they have been stained with dyes. Genes are physically located within chromosomes. Biochemically, each chromosome contains a very long segment of DNA, which is the genetic material, and proteins, which are bound to the DNA and provide it with an organized structure. In eukaryotic cells, this complex between DNA and proteins is called **chromatin.** In this chapter, we will focus on the cellular mechanics of chromosome transmission during cell division to better understand the patterns of gene transmission that we considered in Chapter 2. In particular, we will examine how chromosomes are copied and sorted into newly made cells.

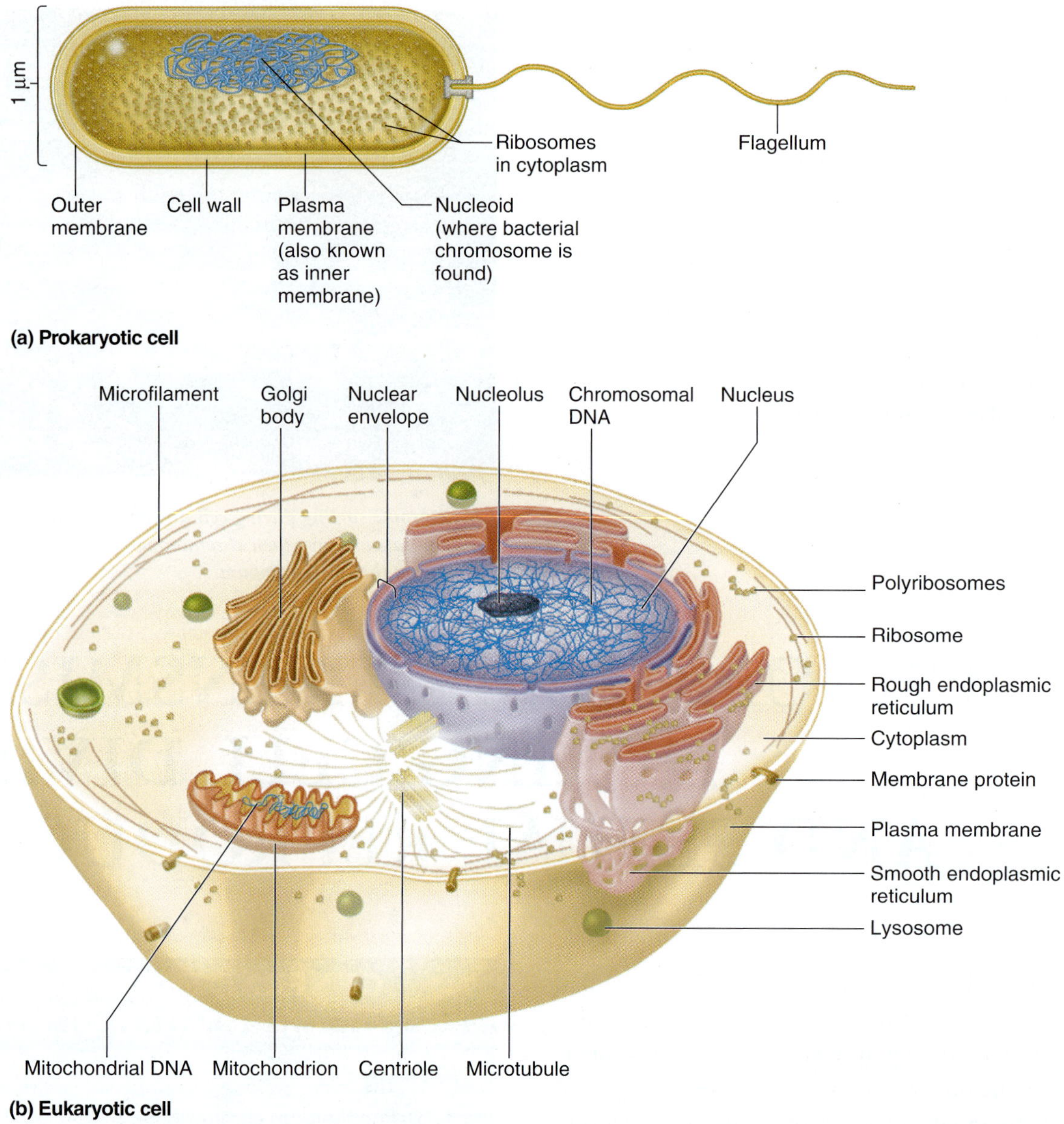

FIGURE 3.1 **The basic organization of cells.** **(a)** A prokaryotic cell. The example shown here is a typical bacterium, such as *Escherichia coli*, which has an outer membrane. **(b)** A eukaryotic cell. The example shown here is a typical animal cell.

CONCEPT CHECK: Eukaryotic cells exhibit compartmentalization. Define compartmentalization.

Before we begin a description of chromosome transmission, we need to consider the distinctive cellular differences between bacterial and eukaryotic species. Bacteria and archaea are referred to as **prokaryotes,** from the Greek meaning prenucleus, because their chromosomes are not contained within a membrane-bound nucleus of the cell. Prokaryotes usually have a single type of circular chromosome in a region of the cytoplasm called the **nucleoid** (**Figure 3.1a**). The cytoplasm is enclosed by a plasma membrane that regulates the uptake of nutrients and the excretion of waste products. Outside the plasma membrane is a rigid cell wall that protects the cell from breakage. Certain species of bacteria also have an outer membrane located beyond the cell wall.

Eukaryotes, from the Greek meaning true nucleus, include some simple species, such as single-celled protists and some fungi (such as yeast), and more complex multicellular species, such as plants, animals, and other fungi. The cells of eukaryotic species have internal membranes that enclose highly specialized compartments (**Figure 3.1b**). These compartments form membrane-bound **organelles** with specific functions. For example, the lysosomes play a role in the degradation of macromolecules. The endoplasmic reticulum and Golgi body play a role in protein modification and trafficking. A particularly conspicuous organelle is the **nucleus,** which is bounded by two membranes that constitute the nuclear envelope. Most of the genetic material is found within chromosomes, which are located in the nucleus. In addition to the nucleus, certain organelles in eukaryotic cells contain a small amount of their own DNA. These include the mitochondrion, which functions in ATP synthesis, and the chloroplast, in plant and algal cells, which functions in photosynthesis. The DNA found in these organelles is referred to as extranuclear or extrachromosomal DNA to distinguish it from the DNA that is found in the cell nucleus. We will examine the role of mitochondrial and chloroplast DNA in Chapter 5.

In this section, we will focus on the composition of chromosomes found in the nucleus of eukaryotic cells. As you will learn, eukaryotic species contain genetic material that comes in sets of linear chromosomes.

Eukaryotic Chromosomes Are Examined Cytologically to Yield a Karyotype

Insights into inheritance patterns have been gained by observing chromosomes under the microscope. **Cytogenetics** is the field of genetics that involves the microscopic examination of chromosomes. The most basic observation that a **cytogeneticist** can make is to examine the chromosomal composition of a particular cell. For eukaryotic species, this is usually accomplished by observing the chromosomes as they are found in actively dividing cells. When a cell is preparing to divide, the chromosomes become more tightly coiled, which shortens them and increases their diameter. The consequence of this shortening is that distinctive shapes and numbers of chromosomes become visible with a light microscope. Each species has a particular chromosome composition. For example, most human cells contain 23 pairs of chromosomes, for a total of 46. On rare occasions, some individuals may inherit an abnormal number of chromosomes or a chromosome with an abnormal structure. Such abnormalities can often be detected by a microscopic examination of the chromosomes within actively dividing cells. In addition, a cytogeneticist may examine chromosomes as a way to distinguish between two closely related species.

Figure 3.2a shows the general procedure for preparing human chromosomes to be viewed by microscopy. In this example, the cells were obtained from a sample of human blood; more specifically, the chromosomes within leukocytes (also called white blood cells) were examined. Blood cells are a type of **somatic cell.** This term refers to any cell of the body that is not a gamete or a precursor to a gamete. The **gametes** (sperm and egg cells or their precursors) are also called **germ cells.**

After the blood cells have been removed from the body, they are treated with chemicals that stimulate them to begin cell division and halt cell division during mitosis, which is described later in this chapter. As shown in Figure 3.2a, these actively dividing cells are subjected to centrifugation to concentrate them. The concentrated preparation is then mixed with a hypotonic solution that makes the cells swell. This swelling causes the chromosomes to spread out within the cell, thereby making it easier to see each individual chromosome. Next, the cells are treated with a fixative that chemically freezes them so that the chromosomes will no longer move around. The cells are then treated with a chemical dye that binds to the chromosomes and stains them. As discussed in greater detail in Chapter 8, this gives chromosomes a distinctive banding pattern that greatly enhances our ability to visualize and to uniquely identify them (also see Figure 8.1c, d). The cells are then placed on a slide and viewed with a light microscope.

In a cytogenetics laboratory, the microscopes are equipped with a camera that can photograph the chromosomes. In recent years, advances in technology have allowed cytogeneticists to view microscopic images on a computer screen (**Figure 3.2b**). On the computer screen, the chromosomes can be organized in a standard way, usually from largest to smallest. As seen in **Figure 3.2c**, the human chromosomes have been lined up, and a number is used to designate each type of chromosome. An exception would be the sex chromosomes, which are designated with the letters X and Y. An organized representation of the chromosomes within a cell is called a **karyotype.** A karyotype reveals how many chromosomes are found within an actively dividing somatic cell.

Eukaryotic Chromosomes Are Inherited in Sets

Most eukaryotic species are **diploid** or have a diploid phase to their life cycle, which means that each type of chromosome is a member of a pair. A diploid cell has two sets of chromosomes. In humans, most somatic cells have 46 chromosomes—two sets of 23 each. Other diploid species, however, have different numbers of chromosomes in their somatic cells. For example, the dog has 39 chromosomes per set (78 total), the fruit fly has 4 chromosomes per set (8 total), and the tomato has 12 per set (24 total).

When a species is diploid, the members of a pair of chromosomes are called **homologs;** each type of chromosome is found in a homologous pair. As shown in Figure 3.2c, for example, a human somatic cell has two copies of chromosome 1, two copies of chromosome 2, and so forth. Within each pair, the chromosome on the left is a homolog to the one on the right, and vice versa. In each pair, one chromosome was inherited from the mother and its homolog was inherited from the father. The two chromosomes in a homologous pair are nearly identical in size, have the same banding pattern, and contain a similar composition of genetic material. If a particular gene is found on one copy of a chromosome, it is also found on the other homolog. However, the two homologs may carry different versions of a given gene, which are called **alleles.** As discussed in Chapter 2, some alleles are dominant and mask the expression of recessive alleles. As an example, let's consider a gene in humans, called *OCA2*, which is one of a few different genes that affect eye color. The *OCA2* gene is located on chromosome 15 and comes in variants that result in brown, green, or blue eyes. In a person with brown eyes, one copy of chromosome 15 might carry a dominant brown allele, whereas its homolog could carry a recessive blue allele.

At the molecular level, how similar are homologous chromosomes? The answer is that the sequence of bases of one homolog would usually differ by less than 1% relative to the sequence of the other homolog. For example, the DNA sequence of chromosome 1 that you inherited from your mother would be greater than 99% identical to the sequence of chromosome 1 that you inherited from your father. Nevertheless, it should be emphasized that

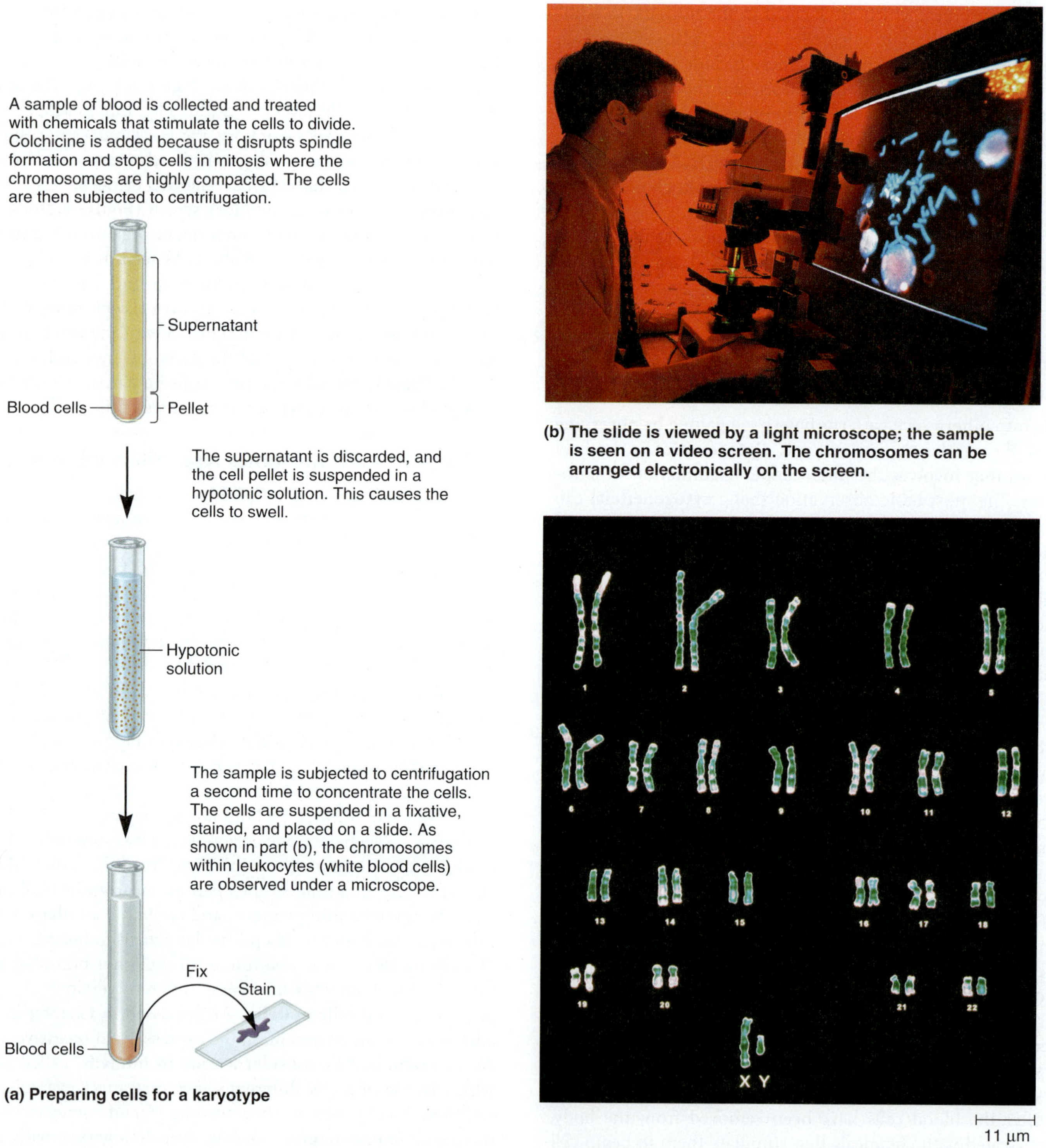

(a) Preparing cells for a karyotype

(b) The slide is viewed by a light microscope; the sample is seen on a video screen. The chromosomes can be arranged electronically on the screen.

(c) For a diploid human cell, two complete sets of chromosomes from a single cell constitute a karyotype of that cell.

FIGURE 3.2 **The procedure for making a human karyotype.**

CONCEPT CHECK: How do you think the end results would be affected if the cells were not treated with a hypotonic solution?

the sequences are not identical. The slight differences in DNA sequences provide the allelic differences in genes. Again, if we use the eye color gene as an example, a slight difference in DNA sequence distinguishes the brown, green, and blue alleles. It should also be noted that the striking similarities between homologous chromosomes do not apply to the pair of sex chromosomes—X and Y. These chromosomes differ in size and genetic composition. Certain genes that are found on the X chromosome are not found

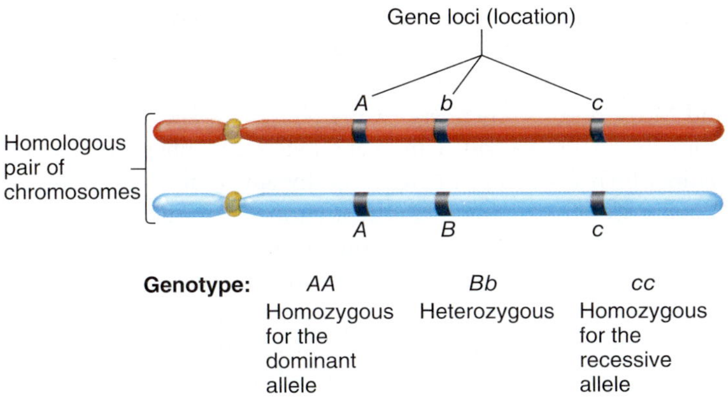

FIGURE 3.3 A comparison of homologous chromosomes. Each pair of homologous chromosomes carries the same types of genes, but, as shown here, the alleles may or may not be different.

CONCEPT CHECK: How are homologs similar to each other and how are they different?

on the Y chromosome, and vice versa. The X and Y chromosomes are not considered homologous chromosomes even though they do have short regions of homology.

Figure 3.3 considers two homologous chromosomes that are labeled with three different genes. An individual carrying these two chromosomes would be homozygous for the dominant allele of gene *A*, which means that both homologs carry the same allele. The individual would be heterozygous, *Bb*, for the second gene, which means the homologs carry different alleles. For the third gene, the individual is homozygous for a recessive allele, *c*. The physical location of a gene is called its **locus** (plural: **loci**). As seen in Figure 3.3, for example, the locus of gene *C* is toward one end of this chromosome, whereas the locus of gene *B* is more in the middle.

3.1 COMPREHENSION QUESTIONS

1. Which of the following is not found in a prokaryotic cell?
   a. Plasma membrane
   b. Ribosome
   c. Cell nucleus
   d. Cytoplasm
2. When preparing a karyotype, which of the following steps is conducted?
   a. Treat the cells with drugs that cause them to begin cell division.
   b. Treat the cells with a hypotonic solution that causes them to swell.
   c. Expose the cells to chemical dyes that bind to the chromosomes and stain them.
   d. All of the above.
3. How many sets of chromosomes are found in a human somatic cell, and how many chromosomes are within one set?
   a. 2 sets, with 23 in each set
   b. 23 sets, with 2 in each set
   c. 1 set, with 23 in each set
   d. 23 sets, with 1 in each set

3.2 CELL DIVISION

Learning Outcomes:

1. Describe the process of binary fission in bacteria.
2. List and outline the phases of the eukaryotic cell cycle.

Now that we have an appreciation for the chromosomal composition of living cells, we can consider how chromosomes are copied and transmitted when cells divide. One purpose of cell division is **asexual reproduction.** In this process, a preexisting cell divides to produce two new cells. By convention, the original cell is usually called the mother cell, and the new cells are the two daughter cells. When species are unicellular, the mother cell is judged to be one organism, and the two daughter cells are two new separate organisms. Asexual reproduction is how bacterial cells proliferate. In addition, certain unicellular eukaryotes, such as the amoeba and baker's yeast (*Saccharomyces cerevisiae*), can reproduce asexually.

A second important reason for cell division is **multicellularity.** Species such as plants, animals, most fungi, and some protists are derived from a single cell that has undergone repeated cellular divisions. Humans, for example, begin as a single fertilized egg; repeated cell divisions produce an adult with trillions of cells. The precise transmission of chromosomes during every cell division is critical so that all cells of the body receive the correct amount of genetic material.

In this section, we will consider how the process of cell division requires the duplication, organization, and distribution of the chromosomes. In bacteria, which have a single circular chromosome, the division process is relatively simple. Prior to cell division, bacteria duplicate their circular chromosome; they then distribute a copy into each of the two daughter cells. This process, known as binary fission, is described first. Eukaryotes have multiple numbers of chromosomes that occur as sets. Compared with bacteria, this added complexity requires a more complicated sorting process to ensure that each newly made cell receives the correct number and types of chromosomes. A mechanism known as mitosis entails the organization and distribution of eukaryotic chromosomes during cell division.

Bacteria Reproduce Asexually by Binary Fission

As discussed earlier (see Figure 3.1a), bacterial species are typically unicellular, although individual bacteria may associate with each other to form pairs, chains, or clumps. Unlike eukaryotes, which have their chromosomes in a separate nucleus, the circular chromosomes of bacteria are in direct contact with the cytoplasm.

The capacity of bacteria to divide is really quite astounding. Some species, such as *Escherichia coli*, a common bacterium of the intestine, can divide every 20 to 30 minutes. Prior to cell division, bacterial cells copy, or replicate, their chromosomal DNA. This produces two identical copies of the genetic material, as shown at the top of **Figure 3.4**. Following DNA replication, a bacterial cell divides into two daughter cells by a process known as **binary fission.** During this event, the two daughter cells become separated from each other by the formation of a septum. As seen in the figure, each cell receives a copy of the chromosomal genetic

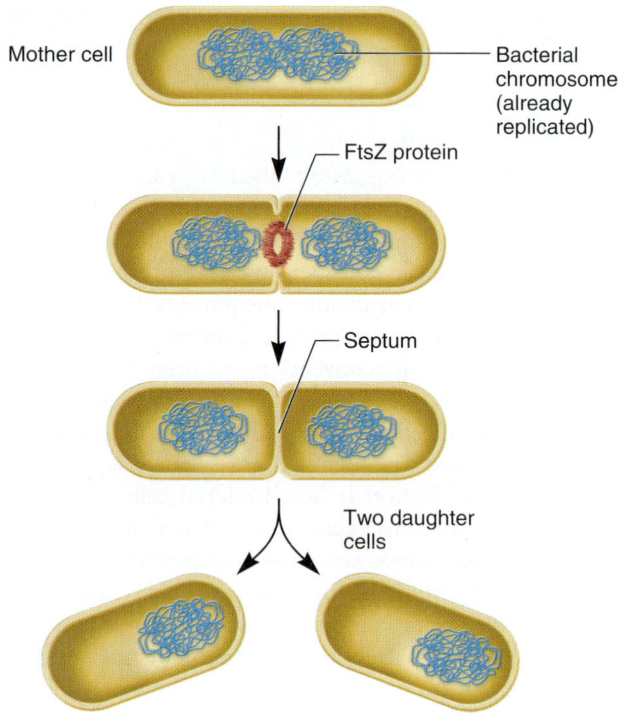

FIGURE 3.4 Binary fission: The process by which bacterial cells divide. Prior to division, the chromosome replicates to produce two identical copies. These two copies segregate from each other, with one copy going to each daughter cell.

CONCEPT CHECK: What is the function of the FtsZ protein during binary fission?

material. Except when rare mutations occur, the daughter cells are usually genetically identical because they contain exact copies of the genetic material from the mother cell.

Recent evidence has shown that bacterial species produce a protein called FtsZ, which is important in cell division. This protein assembles into a ring at the future site of the septum. FtsZ is thought to be the first protein to move to this division site, and it recruits other proteins that produce a new cell wall between the daughter cells. FtsZ is evolutionarily related to a eukaryotic protein called tubulin. As discussed later in this chapter, tubulin is the main component of microtubules, which play a key role in chromosome sorting in eukaryotes. Both FtsZ and tubulin form structures that provide cells with organization and play key roles in cell division.

Binary fission is an asexual form of reproduction because it does not involve genetic contributions from two different gametes. On occasion, bacteria can transfer small pieces of genetic material to each other, which is described in Chapter 7.

Eukaryotic Cells Progress Through a Cell Cycle to Produce Genetically Identical Daughter Cells

The common outcome of eukaryotic cell division is to produce two daughter cells that have the same number and types of chromosomes as the original mother cell. This requires a replication and division process that is more complicated than simple binary fission. Eukaryotic cells that are destined to divide progress through a series of phases known as the **cell cycle** (**Figure 3.5**). These phases are named G for gap, S for synthesis (of the genetic material), and M for mitosis. There are two G phases: G_1 and G_2. The term "gap" originally described the gaps between S phase and mitosis in which it was not microscopically apparent that significant changes were occurring in the cell. However, we now know that both gap phases are critical periods in the cell cycle that involve many molecular changes. In actively dividing cells, the G_1, S, and G_2 phases are collectively known as **interphase.** In addition, cells may remain permanently, or for long periods of time, in a phase of the cell cycle called G_0. A cell in the G_0 phase is either temporarily not progressing through the cell cycle or, in the case of terminally differentiated cells, such as most nerve cells in an adult mammal, will never divide again.

During the G_1 phase, a cell may prepare to divide. Depending on the cell type and the conditions that it encounters, a cell in the G_1 phase may accumulate molecular changes (e.g., produce new proteins) that cause it to progress through the rest of the cell cycle. When this occurs, cell biologists say that a cell has reached a **restriction point** and is committed on a pathway that leads to cell division. Once past the restriction point, the cell then advances to the S phase, during which the chromosomes are replicated. After replication, the two copies are called **chromatids.** They are joined to each other at a region of DNA called the **centromere** to form a unit known as a pair of **sister chromatids,** or a **dyad** (**Figure 3.6**). A single chromatid within a dyad is called a **monad.** An unreplicated chromosome can also be called a monad. The **kinetochore** is a group of proteins that are bound to the centromere. These proteins help to hold the sister chromatids together and also play a role in chromosome sorting, as discussed later.

When S phase is completed, a cell actually has twice as many chromatids as chromosomes in the G_1 phase. For example, a human cell in the G_1 phase has 46 distinct chromosomes, whereas in G_2, it would have 46 pairs of sister chromatids, for a total of 92 chromatids. The term chromosome can be a bit confusing because it originally meant a distinct structure that is observable with the microscope. Therefore, the term *chromosome* can refer to either a pair of sister chromatids (a dyad) during G_2 and early stages of M phase or to a structure that is observed at the end of M phase and during G_1, which is a monad and contains the equivalent of one chromatid (refer back to Figure 3.5).

During the G_2 phase, the cell accumulates the materials necessary for nuclear and cell division. It then progresses into the M phase of the cell cycle, when **mitosis** occurs. The primary purpose of mitosis is to distribute the replicated chromosomes, dividing one cell nucleus into two nuclei, so each daughter cell receives the same complement of chromosomes. For example, a human cell in the G_2 phase has 92 chromatids, which are found in 46 pairs. During mitosis, these pairs of chromatids are separated and sorted so each daughter cell receives 46 chromosomes.

Mitosis was first observed microscopically in the 1870s by the German biologist Walther Flemming, who coined the term mitosis (from the Greek *mitos*, meaning thread). He studied the dividing epithelial cells of salamander larvae and noticed that

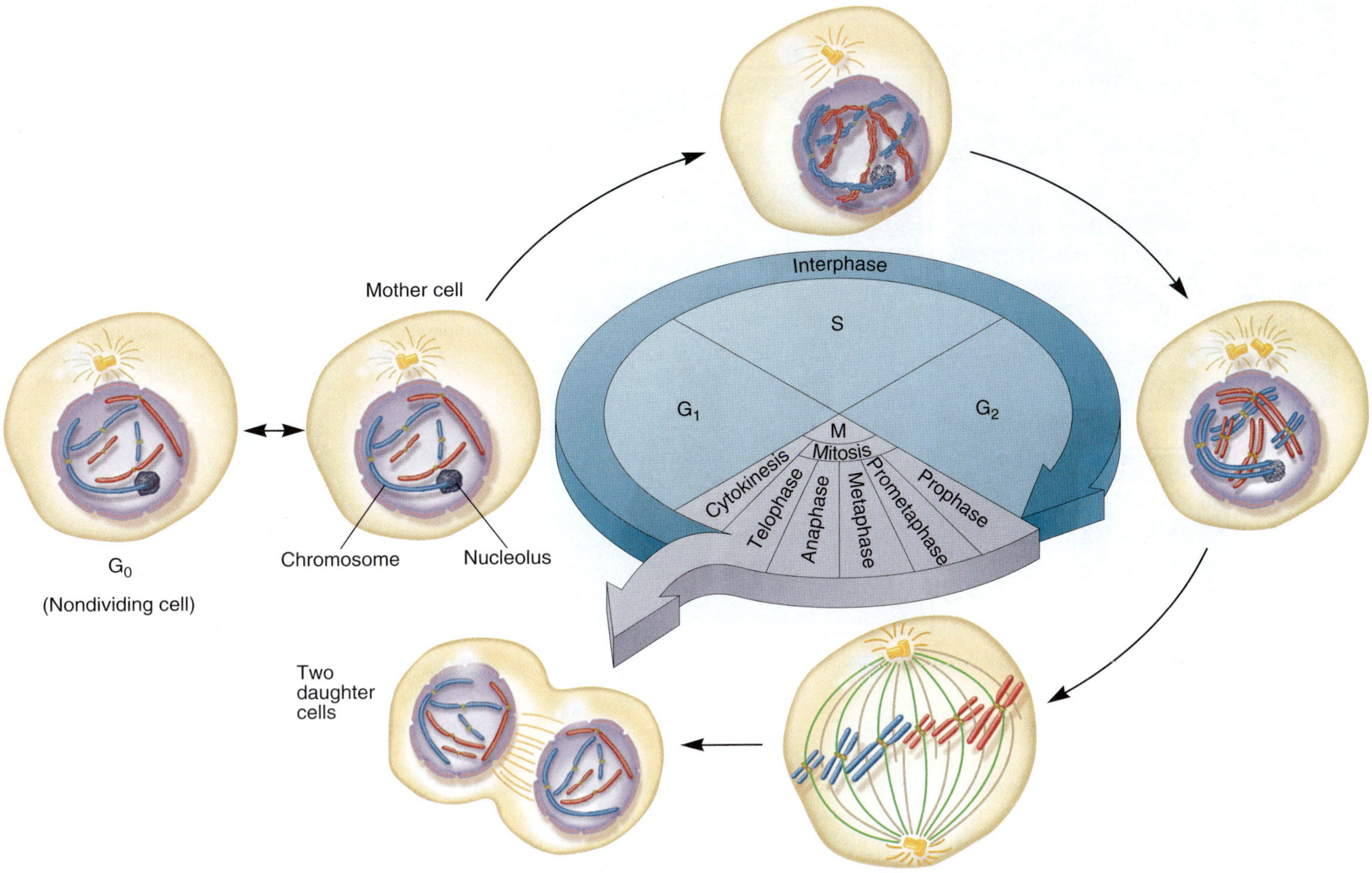

FIGURE 3.5 The eukaryotic cell cycle. Dividing cells progress through a series of phases, denoted G_1, S, G_2, and M phases. This diagram shows the progression of a cell through mitosis to produce two daughter cells. The original diploid cell had three pairs of chromosomes, for a total of six individual chromosomes. During S phase, these have replicated to yield 12 chromatids found in six pairs of sister chromatids. After mitosis and cytokinesis are completed, each of the two daughter cells contains six individual chromosomes, just like the mother cell. Note: The chromosomes in G_0, G_1, S, and G_2 phases are not actually condensed as shown here. In this drawing, they are shown partially condensed so they can be easily counted.

CONCEPT CHECK: What is the difference between the G_0 and G_1 phases?

chromosomes were constructed of two parallel "threads." These threads separated and moved apart, one going to each of the two daughter nuclei. By this mechanism, Flemming pointed out that the two daughter cells received an identical group of threads, of a quantity comparable to the number of threads in the parent cell.

3.2 COMPREHENSION QUESTIONS

1. Binary fission
   a. is a form of asexual reproduction.
   b. is a way for bacteria to reproduce.
   c. begins with a single mother cell and produces two genetically identical daughter cells.
   d. is all of the above.

2. Which of the following is the correct order of phases of the cell cycle?
   a. G_1, G_2, S, M
   b. G_1, S, G_2, M
   c. G_1, G_2, M, S
   d. G_1, S, M, G_2

3. What critical event occurs during the S phase of the eukaryotic cell cycle?
   a. Cells make a decision of whether or not to divide.
   b. DNA replication produces pairs of sister chromatids.
   c. The chromosomes condense.
   d. The single nucleus is divided into two nuclei.

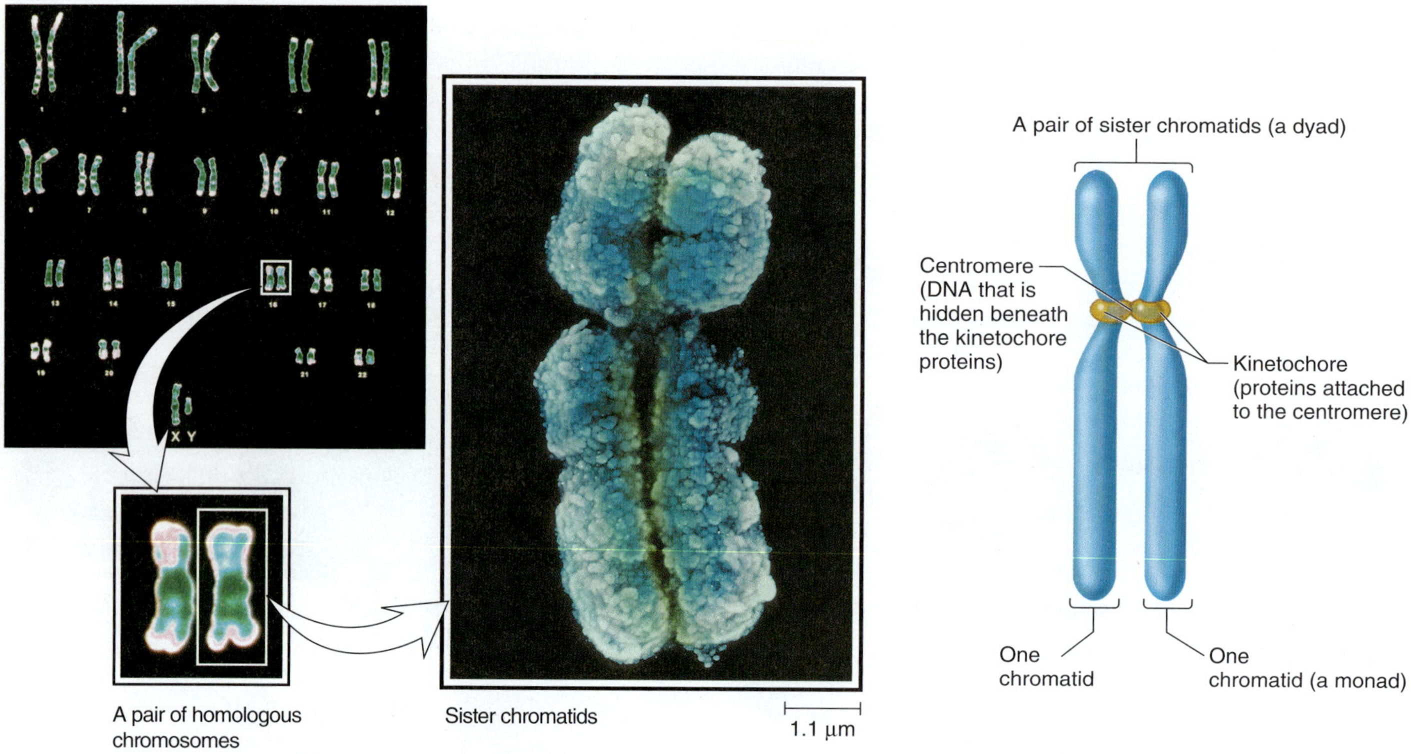

(a) Homologous chromosomes and sister chromatids

(b) Schematic drawing of sister chromatids

FIGURE 3.6 **Chromosomes following DNA replication.** **(a)** The photomicrograph on the right shows a chromosome in a form called a dyad, or pair of sister chromatids. This chromosome is in the metaphase stage of mitosis, which is described later in the chapter. Note: Each of the 46 chromosomes that are viewed in a human karyotype (upper left) is actually a pair of sister chromatids. Look closely at the white rectangular boxes in the two insets. **(b)** A schematic drawing of sister chromatids. This structure has two chromatids that lie side by side. As seen here, each chromatid is a distinct unit, called a monad. The two chromatids are held together by kinetochore proteins that bind to each other and to the centromeres of each chromatid.

CONCEPT CHECK: What is the difference between homologs versus the chromatids within a pair of sister chromatids?

3.3 MITOSIS AND CYTOKINESIS

Learning Outcomes:

1. Describe the structure and function of the mitotic spindle.
2. List and describe the phases of mitosis.
3. Outline the key differences between cytokinesis in animal and plant cells.

As we have seen, eukaryotic cell division involves a cell cycle in which the chromosomes are replicated and later sorted so each daughter cell receives the same amount of genetic material. This process ensures genetic consistency from one cell generation to the next. In this section, we will examine the stages of mitosis and cytokinesis in greater detail.

The Mitotic Spindle Apparatus Organizes and Sorts Eukaryotic Chromosomes

Before we discuss the events of mitosis, let's first consider the structure of the **mitotic spindle apparatus** (also known simply as the **mitotic spindle**), which is involved in the organization and sorting of chromosomes (**Figure 3.7**). The spindle apparatus is formed from **microtubule-organizing centers** (MTOCs), which are structures found in eukaryotic cells from which microtubules grow. Microtubules are produced from the rapid polymerization of tubulin proteins. In animal cells, the mitotic spindle is formed from two MTOCs called **centrosomes.** Each centrosome is located at a **spindle pole.** A pair of **centrioles** at right angles to each other is found within each centrosome of animal cells. However, centrosomes and centrioles are found in animal cells but not in all eukaryotic species. For example, plant cells do not have centrosomes. Instead, the nuclear envelope functions as an MTOC for spindle formation.

The mitotic spindle of a typical animal cell has three types of microtubules (see Figure 3.7). The **aster microtubules** emanate outward from the centrosome toward the plasma membrane. They are important for the positioning of the spindle apparatus within the cell and later in the process of cell division. The **polar microtubules** project toward the region where the chromosomes will be found during mitosis—the region between the two spindle poles. Polar microtubules that overlap with each other play a role in the separation of the two poles. They help to "push" the poles away from each other. Finally, the **kinetochore microtubules** have attachments to a kinetochore, which is a complex of proteins that is bound to the centromere of individual chromosomes.

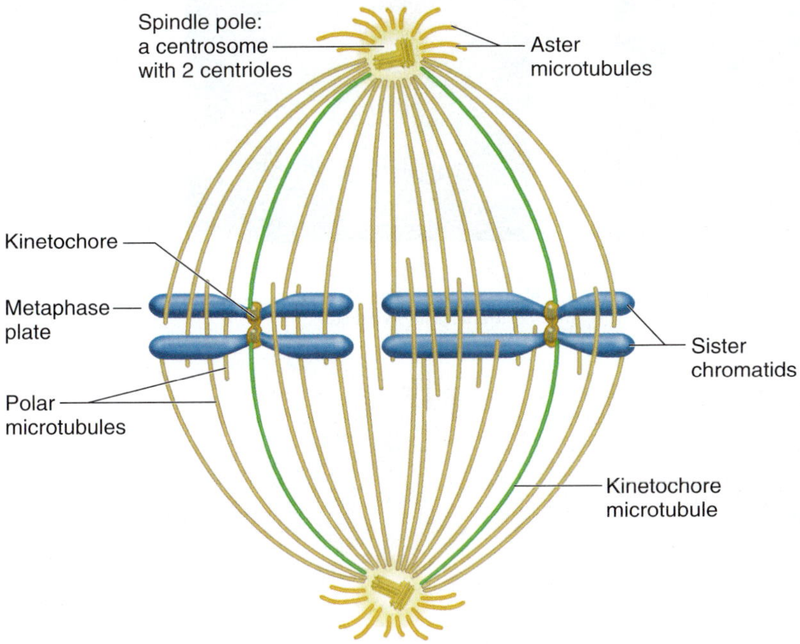

FIGURE 3.7 The structure of the mitotic spindle in a typical animal cell. During the cell cycle, a single centrosome duplicates in S phase and the two centrosomes separate at the beginning of M phase. The mitotic spindle is formed from microtubules that are rooted in the centrosomes. Each centrosome is located at a spindle pole. The aster microtubules emanate away from the region between the poles. They help to position the spindle within the cell and are used as reference points for cell division. However, the mitotic spindle formed in many species, such as plants, do not have astral microtubules. The polar microtubules project into the region between the two poles; they play a role in pole separation. The kinetochore microtubules are attached to the kinetochore of sister chromatids.

CONCEPT CHECK: Where are the two ends of a kinetochore microtubule?

The mitotic spindle allows cells to organize and separate chromosomes so each daughter cell receives the same complement of chromosomes. This sorting process, known as mitosis, is described next.

The Transmission of Chromosomes During the Division of Eukaryotic Cells Requires a Process Known as Mitosis

The process of mitosis is shown for a diploid animal cell in Figure 3.8. In the simplified diagrams shown along the bottom of this figure, the original mother cell contains six chromosomes; it is diploid ($2n$) and contains three chromosomes per set (n = 3). One set is shown in blue, and the homologous set is shown in red. As discussed next, mitosis is subdivided into phases known as prophase, prometaphase, metaphase, anaphase, and telophase.

Prophase Prior to mitosis, the cells are in interphase, during which the chromosomes are **decondensed**—less tightly compacted—and found in the nucleus (Figure 3.8a). At the start of mitosis, in **prophase,** the chromosomes have already replicated, resulting in 12 chromatids, joined as six pairs of sister chromatids (Figure 3.8b). As prophase proceeds, the nuclear membrane begins to dissociate into small vesicles and the nucleolus becomes less visible. At the same time, the chromatids become **condensed** into more compact structures that are readily visible by light microscopy. The two centrosomes move apart and the mitotic spindle begins to form.

Prometaphase As mitosis progresses from prophase to prometaphase, the centrosomes move to opposite ends of the cell and demarcate two spindle poles, one within each of the future daughter cells. During **prometaphase,** the nuclear membrane is completely disrupted into vesicles allowing the spindle fibers to interact with the sister chromatids (Figure 3.8c). How do sister chromatids become attached to the spindle? Initially, microtubules are rapidly formed and can be seen growing out from the two poles. As it grows, if the end of a microtubule happens to make contact with a kinetochore, its end is said to be "captured" and remains firmly attached to the kinetochore. This random process is how sister chromatids become attached to kinetochore microtubules. Alternatively, if the end of a microtubule does not collide with a kinetochore, the microtubule eventually depolymerizes and retracts to the centrosome. As the end of prometaphase nears, the kinetochore on a pair of sister chromatids is attached to kinetochore microtubules from opposite poles. As these events are occurring, the sister chromatids are seen to undergo jerky movements as they are tugged, back and forth, between the two poles. By the end of prometaphase, the mitotic spindle is completely formed.

Metaphase Eventually, the pairs of sister chromatids align themselves along a plane called the **metaphase plate.** As shown in Figure 3.8d, when this alignment is complete, the cell is in **metaphase** of mitosis. At this point, each pair of chromatids (each dyad) is attached to both poles by kinetochore microtubules. The pairs of sister chromatids have become organized into a single row along the metaphase plate. When this organizational process is finished, the chromatids can be equally distributed into two daughter cells.

Anaphase The next step in the division process occurs during **anaphase** (Figure 3.8e). At this stage, the connection that is responsible for holding the pairs of chromatids together is broken. (We will examine the process of sister chromatid cohesion and separation in more detail in Chapter 10.) Each chromatid or monad, now an individual chromosome, is linked to only one of the two poles. As anaphase proceeds, the chromosomes move toward the pole to which they are attached. This involves a shortening of the kinetochore microtubules. In addition, the two poles themselves move farther apart due to the elongation of the polar microtubules, which slide in opposite directions due to the actions of motor proteins.

Telophase During **telophase,** the chromosomes reach their respective poles and decondense. The nuclear membrane now re-forms to produce two separate nuclei. In Figure 3.8f, this has produced two nuclei that contain six chromosomes each. The nucleoli will also reappear.

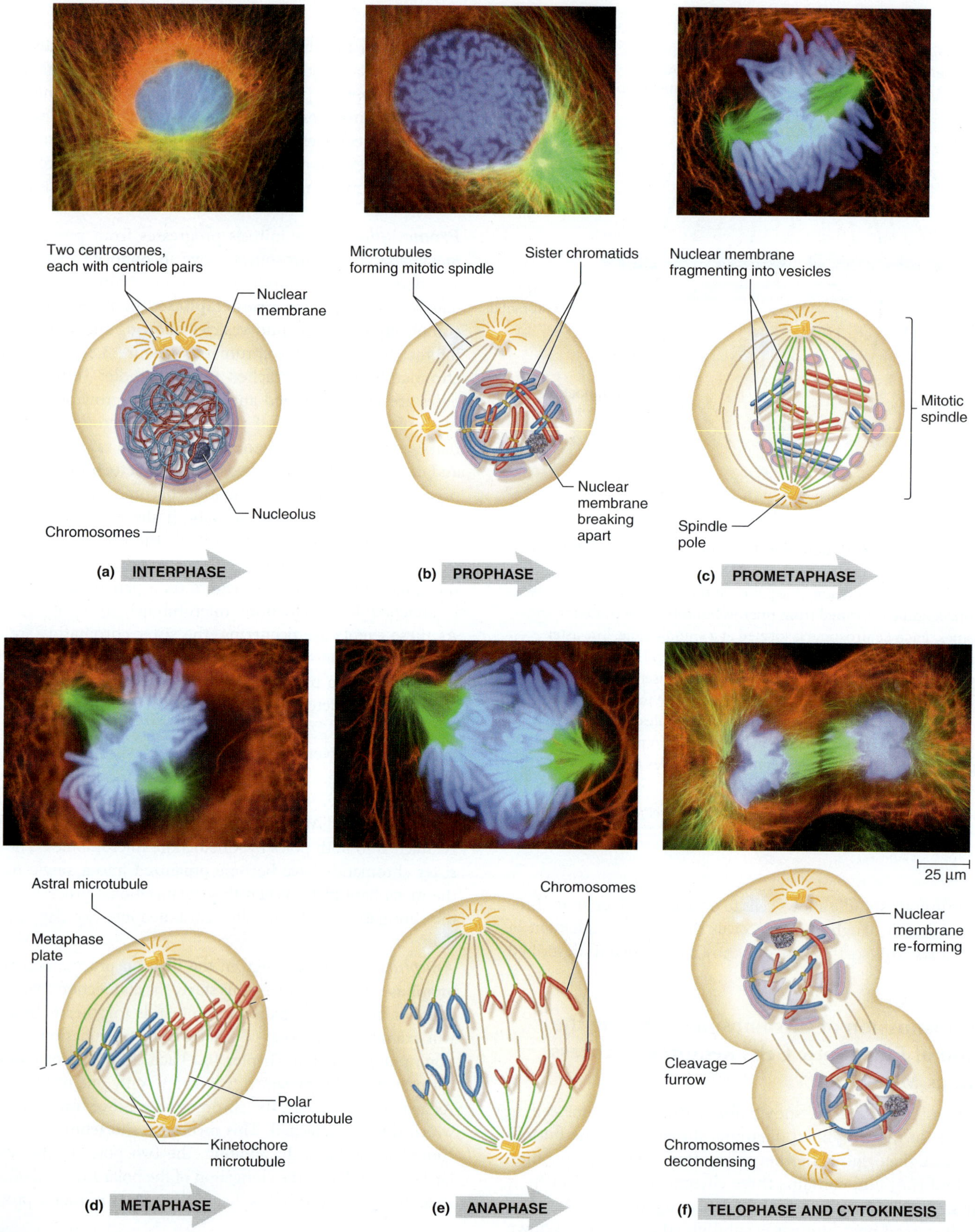

ONLINE ANIMATION

FIGURE 3.8 The process of mitosis in an animal cell. The top panel of photos illustrates cells of a fish embryo progressing through mitosis. The chromosomes are stained in blue and the spindle is green. The bottom panels are schematic drawings that emphasize the sorting and separation of the chromosomes. In this case, the original diploid cell had six chromosomes (three in each set). At the start of mitosis, these have already replicated into 12 chromatids. The final result is two daughter cells, each containing six chromosomes.

CONCEPT CHECK: During which phase are sister chromatids separated and sent to opposite poles?

Cytokinesis In most cases, mitosis is quickly followed by **cytokinesis,** in which the two nuclei are segregated into separate daughter cells. Likewise, cytokinesis also segregates cell organelles, such as mitochondria and chloroplasts, into daughter cells. In animal cells, cytokinesis begins shortly after anaphase. A contractile ring, composed of myosin motor proteins and actin filaments, assembles at the cytoplasmic surface of the plasma membrane. Myosin hydrolyzes ATP, which shortens the ring, thereby constricting the plasma membrane to form a **cleavage furrow** that ingresses, or moves inward (**Figure 3.9a**). Ingression continues until the cell is divided into two cells.

In plants, the two daughter cells are separated by the formation of a **cell plate** (**Figure 3.9b**). At the end of anaphase, Golgi-derived vesicles carrying cell wall materials are transported to the equator of a dividing cell. The fusion of these vesicles gives rise to the cell plate, which is a membrane-bound compartment. The cell plate begins in the middle of the cell and expands until it attaches to the mother cell wall. Once this attachment has taken place, the cell plate undergoes a process of maturation and eventually separates the mother cell into two daughter cells.

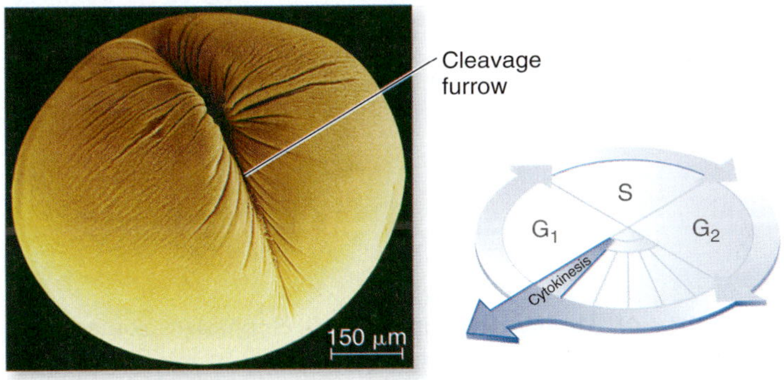

(a) Cleavage of an animal cell

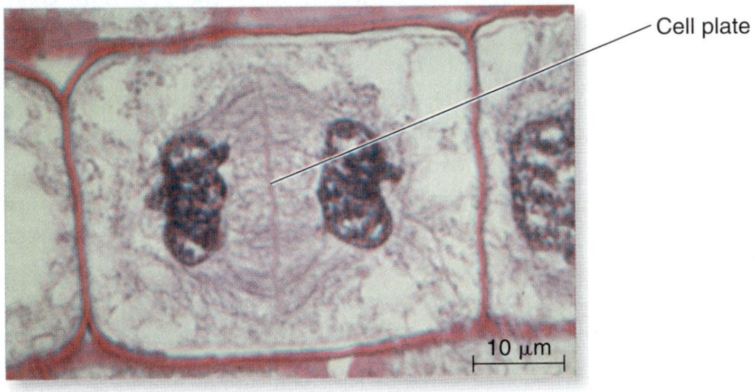

(b) Formation of a cell plate in a plant cell

FIGURE 3.9 Cytokinesis in an animal and plant cell. (a) In an animal cell, cytokinesis involves the formation of a cleavage furrow. (b) In a plant cell, cytokinesis occurs via the formation of a cell plate between the two daughter cells.

CONCEPT CHECK: What causes the cleavage furrow in an animal cell to ingress?

Outcome of Mitotic Cell Division Mitosis and cytokinesis ultimately produce two daughter cells having the same number of chromosomes as the mother cell. Barring rare mutations, the two daughter cells are genetically identical to each other and to the mother cell from which they were derived. The critical consequence of this sorting process is to ensure genetic consistency from one somatic cell to the next. The development of multicellularity relies on the repeated process of mitosis and cytokinesis. For diploid organisms that are multicellular, most of the somatic cells are diploid and genetically identical to each other.

3.3 COMPREHENSION QUESTIONS

1. What is the function of microtubules during mitosis?
   a. Aster microtubules are important for the positioning of the spindle apparatus within an animal cell and in the process of cell division.
   b. Polar microtubules play a role in the separation of the two poles.
   c. Kinetochore microtubules aid in the alignment of chromosomes during metaphase and the movement of individual chromosomes toward a pole during anaphase.
   d. All of the above
2. Which phase of mitosis is depicted in the drawing below?

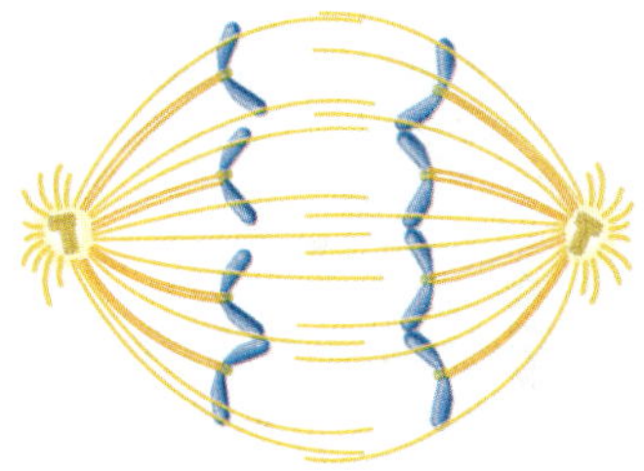

   a. Prophase
   b. Prometaphase
   c. Metaphase
   d. Anaphase
   e. Telophase

3.4 MEIOSIS

Learning Outcomes:

1. List and describe the phases of meiosis.
2. Compare and contrast the key differences between mitosis and meiosis.

In the previous section, we considered how a cell divides to produce two new cells with identical complements of genetic material. Diploid eukaryotic cells may also divide by an alternative process called **meiosis** (from the Greek meaning less). During meiosis, **haploid** cells, which contain a single set of chromosomes, are produced

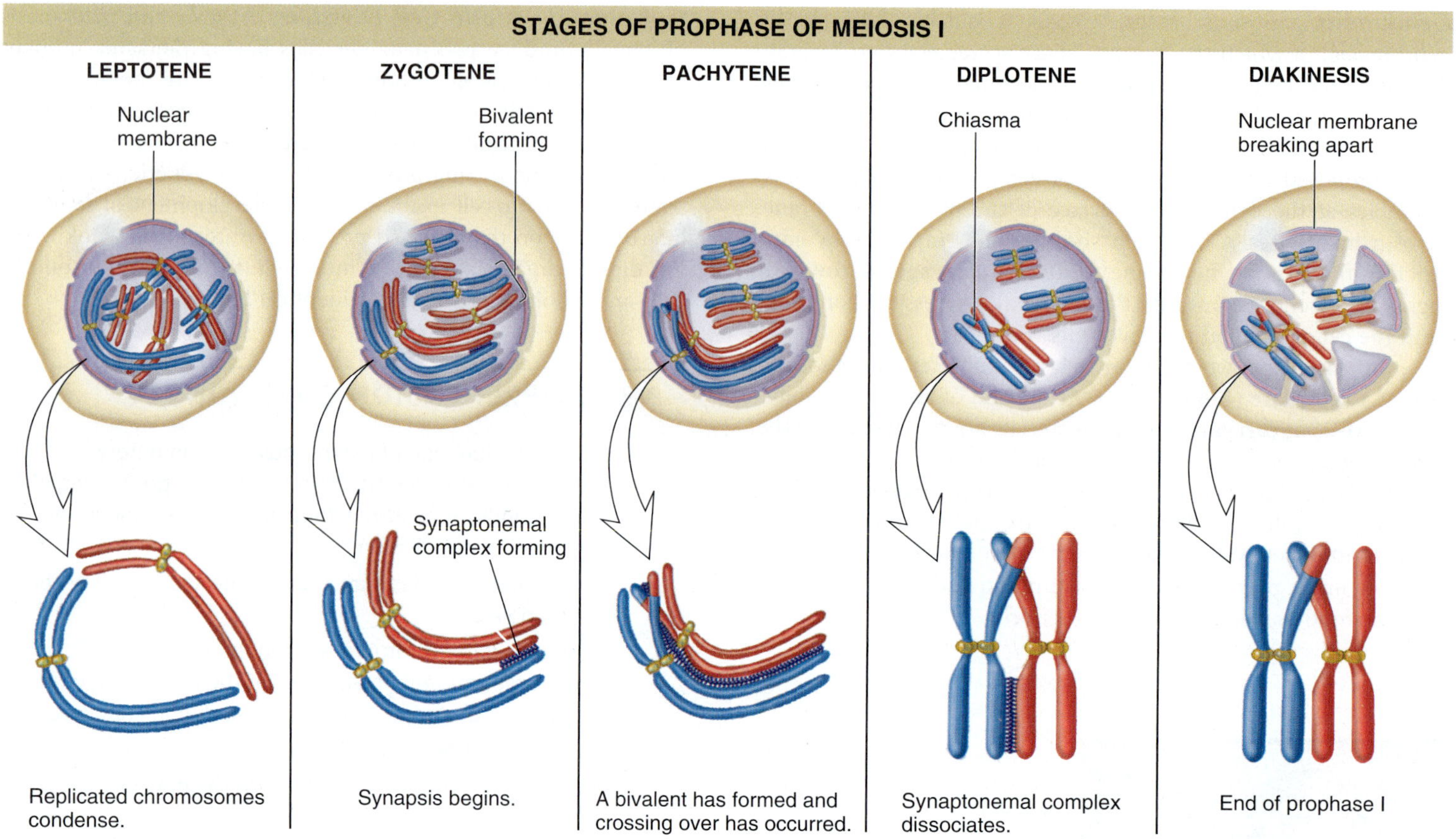

FIGURE 3.10 **The events that occur during prophase of meiosis I.**

CONCEPT CHECK: What is the end result of crossing over?

from a cell that was originally diploid. For this to occur, the chromosomes must be correctly sorted and distributed in a way that reduces the chromosome number to half its original value. For example, in humans, haploid gametes (sperm or egg cells) are produced by meiosis. Each gamete must receive half the total number of chromosomes, but not just any 23 chromosomes will do. A gamete must receive one chromosome from each of the 23 pairs. In this section, we will examine how the stages of meiosis lead to the formation of cells with a haploid complement of chromosomes.

Meiosis Produces Cells That Are Haploid

The process of meiosis bears striking similarities to mitosis. Like mitosis, meiosis begins after a cell has progressed through the G_1, S, and G_2 phases of the cell cycle. However, meiosis involves two successive divisions rather than one (as in mitosis). Prior to meiosis, the chromosomes are replicated in S phase to produce pairs of sister chromatids. This single replication event is then followed by two sequential cell divisions called meiosis I and II. As in mitosis, each of these is subdivided into prophase, prometaphase, metaphase, anaphase, and telophase.

Prophase of Meiosis I Figure 3.10 emphasizes some of the important events that occur during prophase of meiosis I, which is further subdivided into stages known as leptotene, zygotene, pachytene, diplotene, and diakinesis. During the **leptotene** stage, the replicated chromosomes begin to condense and become visible with a light microscope. Unlike mitosis, the **zygotene** stage of prophase of meiosis I involves a recognition process known as **synapsis,** in which the homologous chromosomes recognize each other and begin to align themselves. In most species, this involves the formation of a synaptonemal complex that forms between the homologous chromosomes. At **pachytene,** the homologs have become completely aligned. The associated chromatids are known as **bivalents.** Each bivalent contains two pairs of sister chromatids, or a total of four chromatids. A bivalent is also called a **tetrad** (from the prefix *tetra-*, meaning four) because it is composed of four chromatids—that is, four monads.

During the pachytene stage, when synapsis is complete, an event known as **crossing over** usually occurs. Crossing over involves a physical exchange of chromosome pieces. Depending on the size of the chromosome and the species, an average eukaryotic chromosome incurs a couple to a couple dozen crossovers.

During spermatogenesis in humans, for example, an average chromosome undergoes slightly more than two crossovers, whereas chromosomes in certain plant species may undergo 20 or more crossovers. Recent research has shown that crossing over is usually critical for the proper segregation of chromosomes. In fact, abnormalities in chromosome segregation may be related to a defect in crossing over. In a high percentage of people with Down syndrome, in which an individual has three copies of chromosome 21 instead of two, research has shown that the presence of the extra chromosome is associated with a lack of crossing over between homologous chromosomes.

In Figure 3.10, crossing over has occurred at a single site between two of the larger chromatids. The connection that results from crossing over is called a **chiasma** (plural: **chiasmata**), because it physically resembles the Greek letter chi, χ. We will consider the genetic consequences of crossing over in Chapter 6 and the molecular process of crossing over in Chapter 19. By the end of the **diplotene** stage, the synaptonemal complex has largely disappeared. The bivalent pulls apart slightly, and microscopically it becomes easier to see that it is actually composed of four chromatids. In the last stage of prophase of meiosis I, **diakinesis,** the synaptonemal complex completely disappears.

Prometaphase of Meiosis I Figure 3.10 has emphasized the pairing and crossing over that occurs during prophase of meiosis I. In **Figure 3.11**, we turn our attention to the general events in meiosis. Prophase of meiosis I is followed by prometaphase, in which the spindle apparatus is complete, and the chromatids are attached via kinetochore microtubules.

Metaphase of Meiosis I At metaphase of meiosis I, the bivalents (tetrads) are organized along the metaphase plate. Before we consider the rest of meiosis I, a particularly critical feature for you to appreciate is how the bivalents are aligned along the metaphase plate. In particular, the pairs of sister chromatids are aligned in a double row rather than a single row, as occurs in mitosis (refer back to Figure 3.8d). Furthermore, the arrangement of sister chromatids (dyads) within this double row is random with regard to the blue and red homologs. In Figure 3.11, one of the blue homologs is above the metaphase plate and the other two are below, whereas one of the red homologs is below the metaphase plate and the other two are above.

In an organism that produces many gametes, meiosis can produce many different arrangements of homologs—three blues above and none below, or none above and three below, and so on. As discussed later in this chapter, the random arrangement of homologs is consistent with Mendel's law of independent assortment. Because most eukaryotic species have several chromosomes per set, the sister chromatids can be randomly aligned along the metaphase plate in many possible ways. For example, consider humans, who have 23 chromosomes per set. The possible number of different, random alignments equals 2^n, where n equals the number of chromosomes per set. Thus, in humans, this would equal 2^{23}, or over 8 million, possibilities! Because the homologs are genetically similar but not identical, we see from this calculation that the random alignment of homologous chromosomes provides a mechanism to promote a vast amount of genetic diversity.

In addition to the random arrangement of homologs within a double row, a second distinctive feature of metaphase of meiosis I is the attachment of kinetochore microtubules to the sister chromatids (**Figure 3.12**). One pair of sister chromatids is linked to one of the poles, and the homologous pair is linked to the opposite pole. This arrangement is quite different from the kinetochore attachment sites during mitosis in which a pair of sister chromatids is linked to both poles (see Figure 3.8).

Anaphase of Meiosis I During anaphase of meiosis I, the two pairs of sister chromatids within a bivalent separate from each other (see Figure 3.11). However, the connection that holds sister chromatids together does not break. Instead, each joined pair of chromatids migrates to one pole, and the homologous pair of chromatids moves to the opposite pole. Another way of saying this is that the two dyads within a tetrad separate from each other and migrate to opposite poles.

Telophase of Meiosis I Finally, at telophase of meiosis I, the sister chromatids have reached their respective poles, and decondensation occurs in most, but not all, species. In many species, the nuclear membrane re-forms to produce two separate nuclei. The end result of meiosis I is two cells, each with three pairs of sister chromatids. It is thus a reduction division. The original diploid cell had its chromosomes in homologous pairs, but the two cells produced at the end of meiosis I are considered to be haploid; they do not have pairs of homologous chromosomes. The reduction division occurs because the connection holding the sister chromatids together does not break during anaphase.

Meiosis II The sorting events that occur during meiosis II are similar to those that occur during mitosis, but the starting point is different. For a diploid organism with six chromosomes, mitosis begins with 12 chromatids that are joined as six pairs of sister chromatids (refer back to Figure 3.8). In other words, mitosis begins with six dyads. By comparison, the two cells that begin meiosis II each have six chromatids that are joined as three pairs of sister chromatids; meiosis II begins with three dyads. Otherwise, the steps that occur during prophase, prometaphase, metaphase, anaphase, and telophase of meiosis II are analogous to a mitotic division.

Meiosis versus Mitosis If we compare the outcome of meiosis (see Figure 3.11) to that of mitosis (see Figure 3.8), the results are quite different. In these examples, mitosis produced two diploid daughter cells with six chromosomes each, whereas meiosis produced four haploid daughter cells with three chromosomes each. In other words, meiosis has halved the number of chromosomes per cell. **Table 3.1** describes key differences between mitosis and meiosis that account for the different outcomes of mitosis and meiosis.

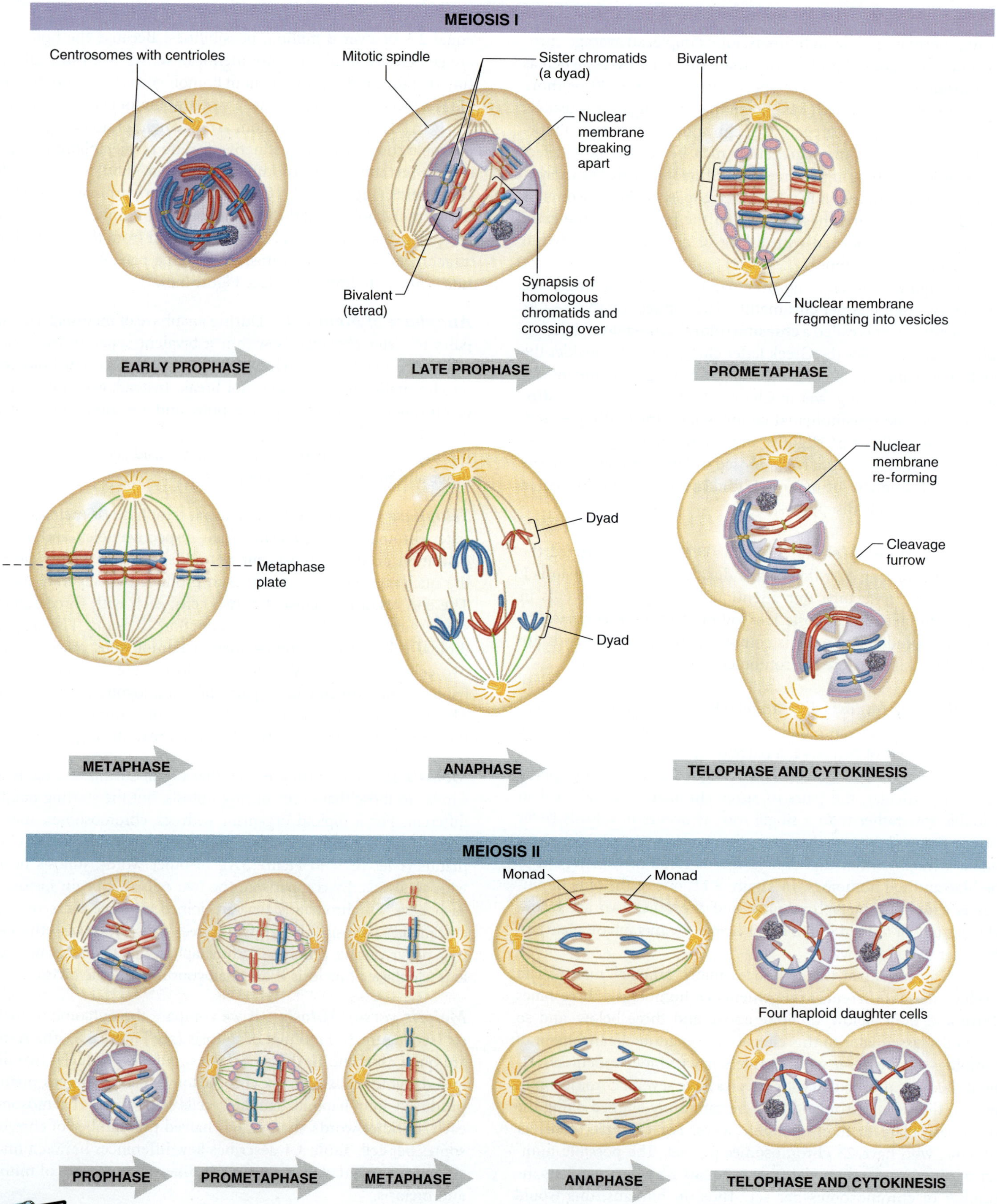

ONLINE ANIMATION

FIGURE 3.11 The stages of meiosis in an animal cell. See text for details.

CONCEPT CHECK: How do the four cells at the end of meiosis differ from the original mother cell?

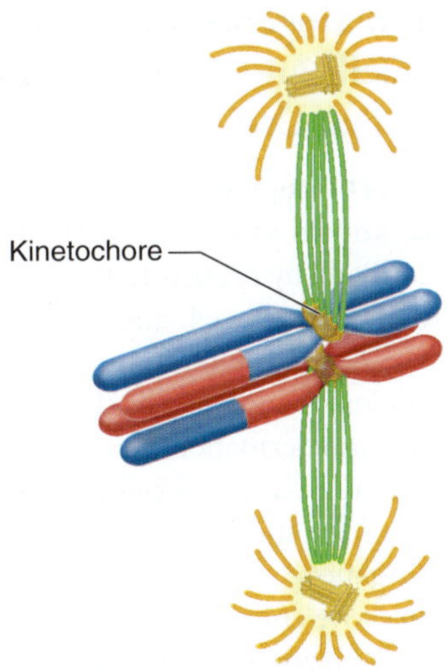

FIGURE 3.12 Attachment of the kinetochore microtubules to replicated chromosomes at metaphase of meiosis I. The kinetochore microtubules from a given pole are attached to one pair of chromatids in a bivalent, but not both. Therefore, each pair of sister chromatids is attached to only one pole.

CONCEPT CHECK: How is this attachment of chromosomes to kinetochore microtubules different from their attachment during metaphase of mitosis?

TABLE 3.1
A Comparison of Mitosis, Meiosis I, and Meiosis II

Phase	Event	Mitosis	Meiosis I	Meiosis II
Prophase	Synapsis	No	Yes	No
Prophase	Crossing over	Rarely	Commonly	Rarely
Prometaphase	Attachment to the poles	A pair of sister chromatids to both poles	A pair of sister chromatids to one pole	A pair of sister chromatids to both poles
Metaphase	Alignment along the metaphase plate	Sister chromatids	Bivalents	Sister chromatids
Anaphase	Separation of:	Sister chromatids	Bivalents	Sister chromatids
End result		Two diploid cells		Four haploid cells

With regard to alleles, the results of mitosis and meiosis are also different. The daughter cells produced by mitosis are genetically identical. However, the haploid cells produced by meiosis are not genetically identical to each other because they contain only one homologous chromosome from each pair. In Section 3.6, we will consider how the haploid cells may differ in the alleles that they carry on their homologous chromosomes.

3.4 COMPREHENSION QUESTIONS

1. When does crossing over usually occur, and what is the end result?
   a. It occurs during prophase of meiosis I, and the end result is the exchange of pieces between homologous chromosomes.
   b. It occurs during prometaphase of meiosis I, and the end result is the exchange of pieces between homologous chromosomes.
   c. It occurs during prophase of meiosis I, and the end result is the separation of sister chromatids.
   d. It occurs during prometaphase of meiosis I, and the end result is the separation of sister chromatids.
2. Which phase of meiosis is depicted in the drawing below?

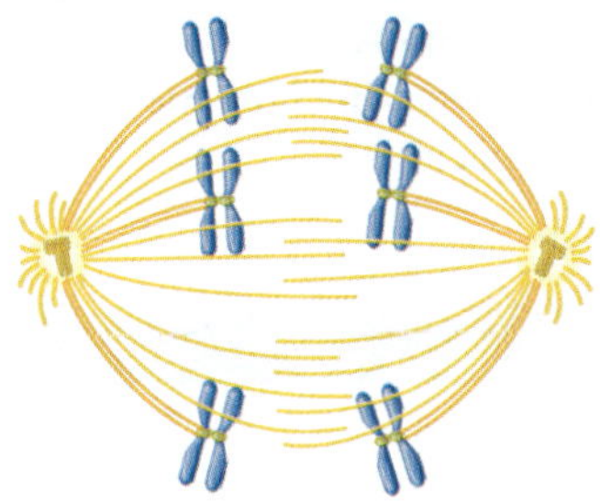

   a. Metaphase of meiosis I
   b. Metaphase of meiosis II
   c. Anaphase of meiosis I
   d. Anaphase of meiosis II

3.5 SEXUAL REPRODUCTION

Learning Outcomes:

1. Define sexual reproduction.
2. Describe how animals make sperm and egg cells.
3. Explain how plants alternate between haploid and diploid generations.

In the previous section, we considered how a diploid cell divides by meiosis to produce cells with half the genetic material of the original mother cell. This process is critical for sexual reproduction, which is a common way for eukaryotic organisms to produce offspring. During **sexual reproduction,** two gametes fuse with each other in the process of fertilization to begin the life of a new organism. Gametes are highly specialized cells that are produced by a process called **gametogenesis.** As discussed previously, gametes are typically haploid, which means they contain half the number of chromosomes of diploid cells. Haploid cells are represented by $1n$ and diploid cells by $2n$, where n refers to a set of chromosomes. A haploid gamete contains a single set of chromosomes, whereas a diploid cell has two sets. For example, a diploid human cell contains two sets of chromosomes, for a total of 46, but a gamete (sperm or egg cell) contains only a single set of 23 chromosomes.

Some simple eukaryotic species are **isogamous,** which means that their gametes are morphologically similar. Examples of isogamous organisms include many species of fungi and algae. Most eukaryotic species, however, are **heterogamous**—they produce two morphologically different types of gametes. Male gametes, or **sperm cells,** are relatively small and usually travel far distances to reach the female gamete—the **egg cell,** or **ovum.** The mobility of the sperm is an important characteristic, making it likely that it will come in close proximity to the egg cell. The sperm of most animal species contain a single flagellum that enables them to swim. The sperm of ferns and nonvascular plants, such as bryophytes, may have multiple flagella. In flowering plants, however, the sperm are contained within pollen grains. Pollen is a small mobile structure that can be carried by the wind or on the feet or hairs of insects. In flowering plants, sperm are delivered to egg cells via pollen tubes. Compared with sperm cells, the egg cell is usually very large and nonmotile. In animal species, the egg stores a large amount of nutrients to nourish the growing embryo. In this section, we will examine how sperm and egg cells are made in animal and plant species.

In Animals, Spermatogenesis Produces Four Haploid Sperm Cells and Oogenesis Produces a Single Haploid Egg Cell

In male animals, **spermatogenesis,** the production of sperm, occurs within glands known as the testes. The testes contain spermatogonial cells that divide by mitosis to produce two cells. One of these remains a spermatogonial cell, and the other cell becomes a primary spermatocyte. As shown in Figure 3.13a, the spermatocyte progresses through meiosis I and meiosis II to produce four haploid cells, which are known as spermatids. These cells then mature into sperm cells. The structure of a sperm cell includes a long flagellum and a head. The head of the sperm contains little more than a haploid nucleus and an organelle at its tip, known as an acrosome. The acrosome contains digestive enzymes that are released when a sperm meets an egg cell. These enzymes enable the sperm to penetrate the outer protective layers of the egg and gain entry into the egg cell's cytosol. In animal species without a mating season, sperm production is a continuous process in mature males. A mature human male, for example, produces several hundred million sperm each day.

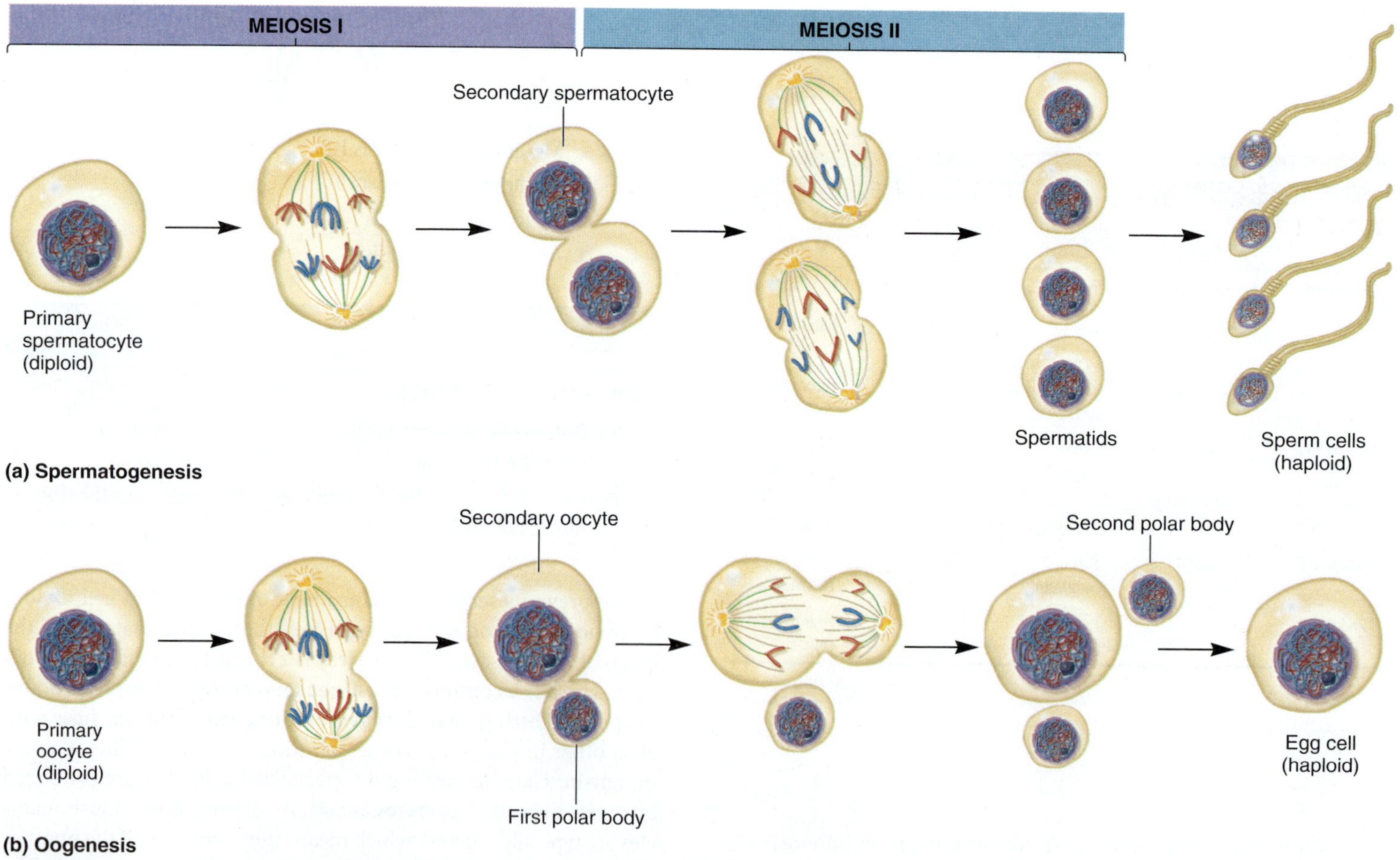

FIGURE 3.13 **Gametogenesis in animals.** (a) Spermatogenesis. A diploid spermatocyte undergoes meiosis to produce four haploid spermatids. These differentiate during spermatogenesis to become mature sperm. (b) Oogenesis. A diploid oocyte undergoes meiosis to produce one haploid egg cell and two or three polar bodies. For some species, the first polar body divides; in other species, it does not. Because of asymmetrical cytokinesis, the amount of cytoplasm the egg receives is maximized. The polar bodies degenerate.

CONCEPT CHECK: What are polar bodies?

In female animals, **oogenesis,** the production of egg cells, occurs within specialized diploid cells of the ovary known as oogonia. Quite early in the development of the ovary, the oogonia initiate meiosis to produce primary oocytes. For example, in humans, approximately 1 million primary oocytes per ovary are produced before birth. These primary oocytes are arrested—enter a dormant phase—at prophase of meiosis I, remaining at this stage until the female becomes sexually mature. Beginning at this stage, primary oocytes are periodically activated to progress through the remaining stages of oocyte development.

During oocyte maturation, meiosis produces only one cell that is destined to become an egg, as opposed to the four gametes produced from each primary spermatocyte during spermatogenesis. How does this occur? As shown in **Figure 3.13b**, the first meiotic division is asymmetrical and produces a secondary oocyte and a much smaller cell, known as a polar body. Most of the cytoplasm is retained by the secondary oocyte and very little by the polar body, allowing the oocyte to become a larger cell with more stored nutrients. The secondary oocyte then begins meiosis II. In mammals, the secondary oocyte is released from the ovary—an event called ovulation—and travels down the oviduct toward the uterus. During this journey, if a sperm cell penetrates the secondary oocyte, it is stimulated to complete meiosis II; the secondary oocyte produces a haploid egg and a second polar body. The haploid egg and sperm nuclei then unite to create the diploid nucleus of a new individual.

Plant Species Alternate Between Haploid (Gametophyte) and Diploid (Sporophyte) Generations

Most species of animals are diploid, and their haploid gametes are considered to be a specialized type of cell. By comparison, the life cycles of plant species alternate between haploid and diploid generations. The haploid generation is called the **gametophyte,** whereas the diploid generation is called the **sporophyte.** Certain cells in the sporophyte undergo meiosis and produce haploid cells called spores, which divide by mitosis to produce a gametophyte. In simpler plants, such as mosses, a haploid spore can produce a large multicellular gametophyte by repeated mitoses and cellular divisions. In flowering plants, however, spores develop into gametophytes that contain only a few cells. In this case, the organism that we think of as a "plant" is the sporophyte, whereas the gametophyte is very inconspicuous. In fact, the gametophytes of most plant species are small structures produced within the much larger sporophyte. Certain cells within the haploid gametophytes then become specialized as haploid gametes.

Figure 3.14 provides an overview of gametophyte development and gametogenesis in flowering plants. Meiosis occurs within cells found in two different structures of the sporophyte: the anthers and the ovaries, which produce male and female gametophytes, respectively. This diagram depicts a flower from an angiosperm, which is a plant that produces seeds within an ovary.

In the anther, diploid cells called microsporocytes undergo meiosis to produce four haploid microspores. These separate into individual microspores. In many angiosperms, each microspore

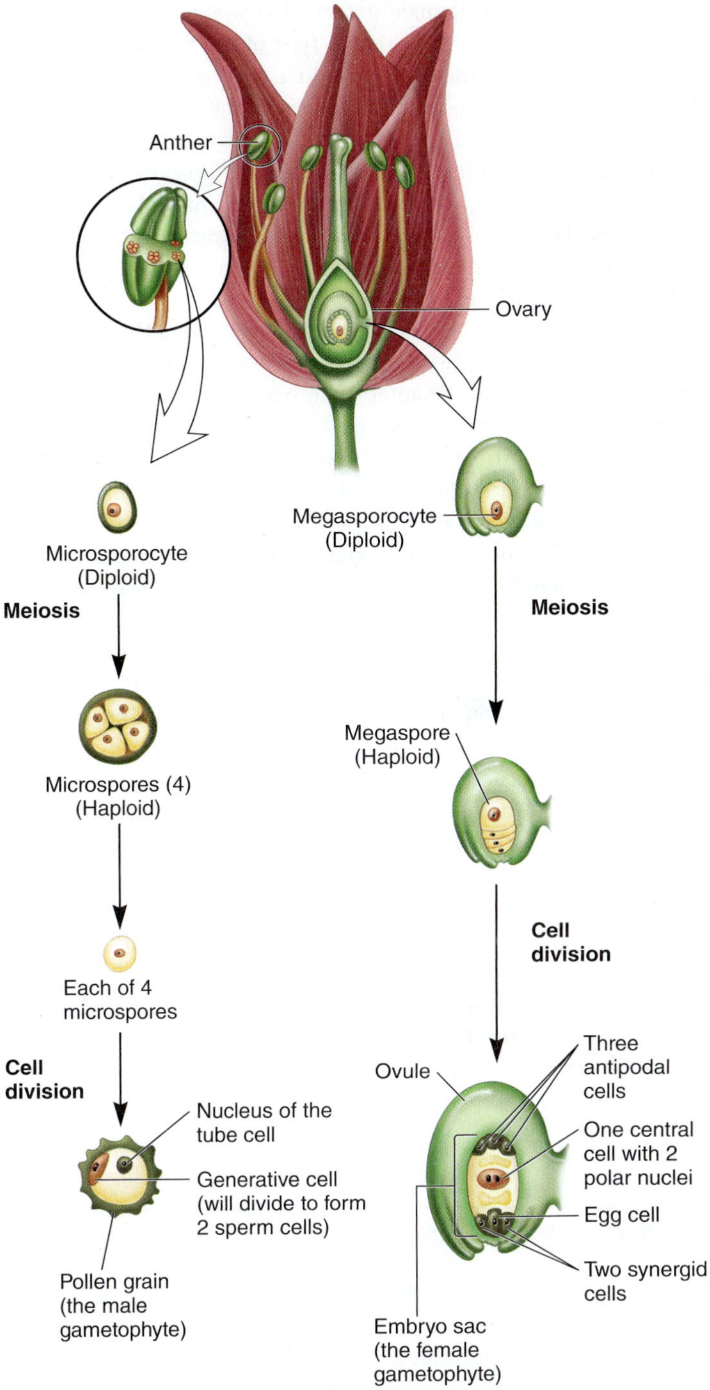

FIGURE 3.14 The formation of male and female gametes by the gametophytes of angiosperms (flowering plants).

CONCEPT CHECK: Are all of the cell nuclei in the embryo sac haploid or is just the egg haploid?

undergoes mitosis to produce a two-celled structure containing one tube cell and one generative cell, both of which are haploid. This structure differentiates into a **pollen grain,** which is the male gametophyte with a thick cell wall. Later, the generative cell undergoes mitosis to produce two haploid sperm cells. In most plant species, this mitosis occurs only if the pollen grain germinates—if it lands on a stigma and forms a pollen tube (refer back to Figure 2.2c).

By comparison, female gametophytes are produced within ovules found in the plant ovaries. A type of cell known as a megasporocyte undergoes meiosis to produce four haploid megaspores. Three of the four megaspores degenerate. The remaining haploid megaspore then undergoes three successive mitotic divisions accompanied by asymmetrical cytokinesis to produce seven individual cells—one egg, two synergids, three antipodals, and one central cell. This seven-celled structure, also known as the **embryo sac,** is the mature female gametophyte. Each embryo sac is contained within an ovule.

For fertilization to occur, specialized cells within the male and female gametophytes must meet. The steps of plant fertilization were described in Chapter 2. To begin this process, a pollen grain lands on a stigma (refer back to Figure 2.2c). This stimulates the tube cell to sprout a pollen tube that grows through the style and eventually makes contact with an ovule. As this is occurring, the generative cell undergoes mitosis to produce two haploid sperm cells. The sperm cells migrate through the pollen tube and eventually reach the ovule. One of the sperm enters the central cell, which contains the two polar nuclei. This results in a cell that is triploid ($3n$). This cell divides mitotically to produce **endosperm,** which acts as a food-storing tissue. The other sperm enters the egg cell. The egg and sperm nuclei fuse to create a diploid cell, the zygote, which becomes a plant embryo. Therefore, fertilization in flowering plants is actually a double fertilization. The result is that the endosperm, which uses a large amount of plant resources, will develop only when an egg cell has been fertilized. After fertilization is complete, the ovule develops into a seed, and the surrounding ovary develops into the fruit, which encloses one or more seeds.

When comparing animals and plants, it's interesting to consider how gametes are made. Animals produce gametes by meiosis. In contrast, plants produce gametes by mitosis. The gametophyte of plants is a haploid multicellular organism produced by mitotic cellular divisions of a haploid spore. Within the multicellular gametophyte, certain cells become specialized as gametes.

3.5 COMPREHENSION QUESTIONS

1. In animals, a key difference between spermatogenesis and oogenesis is that
   a. only oogenesis involves meiosis.
   b. only spermatogenesis involves meiosis.
   c. spermatogenesis produces four sperm, whereas oogenesis produces only one egg cell.
   d. None of the above.
2. Which of the following statements regarding plants is false?
   a. Meiosis within anthers produces spores that develop into pollen.
   b. Meiosis within ovules produces spores that develop into an embryo sac.
   c. The male gametophyte is a pollen grain, and the female gametophyte is an embyro sac.
   d. Meiosis directly produces sperm and egg cells in plants.

3.6 THE CHROMOSOME THEORY OF INHERITANCE AND SEX CHROMOSOMES

Learning Outcomes:

1. List the key tenets of the chromosome theory of inheritance.
2. Explain the relationship between meiosis and Mendel's laws of inheritance.
3. Outline different mechanisms of sex determination.
4. Analyze the results of Morgan's experiment, which showed that a gene affecting eye color in fruit flies is located on the X chromosome.

Thus far, we have considered how chromosomes are transmitted during cell division and gamete formation. In this section, we begin by examining how chromosomal transmission is related to the patterns of inheritance observed by Mendel. This relationship, known as the chromosome theory of inheritance, was a major breakthrough in our understanding of genetics because it established the framework for understanding how chromosomes carry and transmit the genetic determinants that govern the outcome of traits. This theory dramatically unfolded as a result of three lines of scientific inquiry (**Table 3.2**). One avenue concerned Mendel's breeding studies, in which he analyzed the transmission of traits from parent to offspring. A second line of inquiry involved the material basis for heredity. A Swiss botanist, Carl Nägeli, and a German biologist, August Weismann, championed the idea that a substance found in living cells is responsible for the transmission of traits from parents to offspring. Nägeli also suggested that both parents contribute equal amounts of this substance to their offspring. Several scientists, including Oscar Hertwig, Eduard Strasburger, and Walther Flemming, conducted studies suggesting that the chromosomes are the carriers of the genetic material. We now know the DNA within the chromosomes is the genetic material.

Finally, the third line of evidence involved the microscopic examination of the processes of fertilization, mitosis, and meiosis. Researchers became increasingly aware that the characteristics of organisms are rooted in the continuity of cells during the life of an organism and from one generation to the next. When the work of Mendel was rediscovered, several scientists noted striking parallels between the segregation and assortment of traits noted by Mendel and the behavior of chromosomes during meiosis. Among them were Theodore Boveri, a German biologist, and American geneticist Walter Sutton at Columbia University. They independently proposed the chromosome theory of inheritance, which was a milestone in our understanding of genetics. The principles of this theory are described at the beginning of this section.

The remainder of this section focuses on sex chromosomes. Even though an examination of meiosis provided compelling evidence that Mendel's laws could be explained by chromosome sorting, researchers still needed to correlate chromosome behavior with the inheritance of particular traits. Because sex chromosomes, such as the X and the Y chromosome, look very different under the microscope, and because many genes on the X chromosome are

TABLE 3.2

Chronology for the Development and Proof of the Chromosome Theory of Inheritance

1866	Gregor Mendel: Analyzed the transmission of traits from parents to offspring and showed that it follows a pattern of segregation and independent assortment.
1876–77	Oscar Hertwig and Hermann Fol: Observed that the nucleus of the sperm enters the egg during animal cell fertilization.
1877	Eduard Strasburger: Observed that the sperm nucleus of plants (and no detectable cytoplasm) enters the egg during plant fertilization.
1878	Walther Flemming: Described mitosis in careful detail.
1883	Carl Nägeli and August Weismann: Proposed the existence of a genetic material, which Nägeli called idioplasm and Weismann called germ plasm.
1883	Wilhelm Roux: Proposed that the most important event of mitosis is the equal partitioning of "nuclear qualities" to the daughter cells.
1883	Edouard van Beneden: Showed that gametes contain half the number of chromosomes and that fertilization restores the normal diploid number.
1884–85	Hertwig, Strasburger, and Weismann: Proposed that chromosomes are carriers of the genetic material.
1889	Theodore Boveri: Showed that enucleated sea urchin eggs that are fertilized by sperm from a different species develop into larva that have characteristics that coincide with the sperm's species.
1900	Hugo de Vries, Carl Correns, and Erich von Tschermak: Rediscovered Mendel's work, while analyzing inheritance patterns in other plants.
1901	Thomas Montgomery: Determined that maternal and paternal chromosomes pair with each other during meiosis.
1901	C. E. McClung: Discovered that sex determination in insects is related to differences in chromosome composition.
1902	Boveri: Showed that when sea urchin eggs were fertilized by two sperm, the abnormal development of the embryo was related to an abnormal number of chromosomes.
1903	Walter Sutton: Showed that even though the chromosomes seem to disappear during interphase, they do not actually disintegrate. Instead, he argued that chromosomes must retain their continuity and individuality from one cell division to the next.
1902–03	Boveri and Sutton: Independently proposed tenets of the chromosome theory of inheritance. Some historians primarily credit this theory to Sutton.
1910	Thomas Hunt Morgan: Showed that a genetic trait (i.e., white-eyed phenotype in *Drosophila*) was linked to a particular chromosome.
1913	E. Eleanor Carothers: Demonstrated that homologous pairs of chromosomes show independent assortment.
1916	Calvin Bridges: Studied chromosomal abnormalities as a way to confirm the chromosome theory of inheritance.

For a description of these experiments, the student is encouraged to read Voeller, B. R. (1968), *The chromosome theory of inheritance. Classic Papers in Development and Heredity.* New York: Appleton-Century-Crofts.

not on the Y chromosome, geneticists were able to correlate the inheritance of certain traits with the transmission of specific sex chromosomes. In particular, early studies identified genes on the X chromosome that govern eye color in fruit flies. This phenomenon, which is called **X-linked inheritance,** confirmed the idea that genes are found on chromosomes.

The Chromosome Theory of Inheritance Relates the Behavior of Chromosomes to the Mendelian Inheritance of Traits

According to the **chromosome theory of inheritance,** the inheritance patterns of traits can be explained by the transmission patterns of chromosomes during meiosis and fertilization. This theory is based on a few fundamental principles.

1. Chromosomes contain the genetic material that is transmitted from parent to offspring and from cell to cell.
2. Chromosomes are replicated and passed along, generation after generation, from parent to offspring. They are also passed from cell to cell during the development of a multicellular organism. Each type of chromosome retains its individuality during cell division and gamete formation.
3. The nuclei of most eukaryotic cells contain chromosomes that are found in homologous pairs—they are diploid. One member of each pair is inherited from the mother, the other from the father. At meiosis, one of the two members of each pair segregates into one daughter nucleus, and the homolog segregates into the other daughter nucleus. Gametes contain one set of chromosomes—they are haploid.
4. During the formation of haploid cells, different types of (nonhomologous) chromosomes segregate independently of each other.
5. Each parent contributes one set of chromosomes to its offspring. The maternal and paternal sets of homologous chromosomes are functionally equivalent; each set carries a full complement of genes.

The chromosome theory of inheritance allows us to see the relationship between Mendel's laws and chromosomal transmission. As shown in **Figure 3.15**, Mendel's law of segregation can be explained by the homologous pairing and segregation of chromosomes during meiosis. This figure depicts the behavior of a pair of homologous chromosomes in a pea plant that carry a gene for seed color. One of the chromosomes carries a dominant allele that confers yellow seed color, whereas the homologous chromosome carries a recessive allele that confers green color. A heterozygous individual passes only one of these alleles to each offspring. In other words, a gamete may contain the yellow allele or the green allele but not both. Because the homologs segregate during meiosis I and the sister chromatids separate during meiosis II, a gamete contains only one copy of each type of chromosome.

How is the law of independent assortment explained by the behavior of chromosomes? **Figure 3.16** considers the segregation of two types of chromosomes in a pea plant, each carrying a different gene. One pair of chromosomes carries the gene for seed color: the yellow (Y) allele is on one chromosome, and the green (y) allele

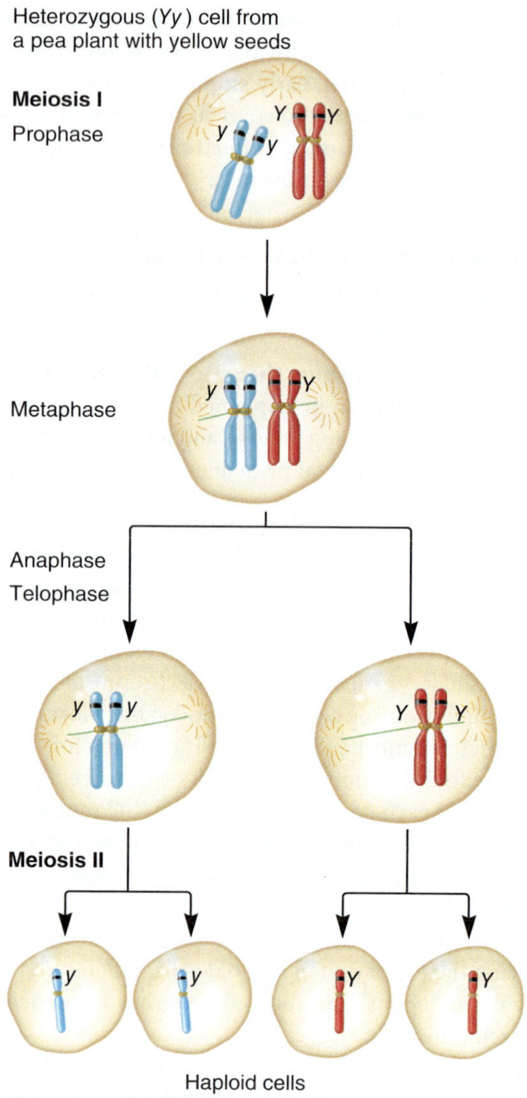

FIGURE 3.15 Mendel's law of segregation can be explained by the segregation of homologs during meiosis. The two copies of a gene are located on homologous chromosomes. In this example using pea seed color, the two alleles are *Y* (yellow) and *y* (green). During meiosis, the homologous chromosomes segregate from each other, leading to segregation of the two alleles into separate gametes.

Genes→Traits The gene for seed color exists in two alleles, *Y* (yellow) and *y* (green). During meiosis, the homologous chromosomes that carry these alleles segregate from each other. The resulting cells receive the *Y* or *y* allele, but not both. When two gametes unite during fertilization, the alleles they carry determine the traits of the resulting offspring.

CONCEPT CHECK: At which stage do homologous chromosomes separate from each other?

is on the homolog. The other pair of (smaller) chromosomes carries the gene for seed shape: one copy has the round (*R*) allele, and the homolog carries the wrinkled (*r*) allele. At metaphase of meiosis I, the different types of chromosomes have randomly aligned along the metaphase plate. As shown in Figure 3.16, this can occur in more than one way. On the left, the *y* allele has sorted with the *R* allele, whereas the *Y* allele has sorted with the *r* allele. On the right, the opposite situation has occurred. Therefore, the random alignment of chromatid pairs during meiosis I can lead to an independent assortment of genes that are found on nonhomologous chromosomes. As we will see in Chapter 6, this law is violated if two different genes are located close to one another on the same chromosome.

Sex Differences Often Correlate with the Presence of Sex Chromosomes

According to the chromosome theory of inheritance, chromosomes carry the genes that determine an organism's traits and are the basis of Mendel's law of segregation and independent assortment. Some early evidence supporting this theory involved the determination of sex. Many species are divided into male and female sexes. In 1901, Clarence McClung, who studied grasshoppers, was the first to suggest that male and female sexes are due to the inheritance of particular chromosomes. Since McClung's initial observations, we now know that a pair of chromosomes, called the **sex chromosomes,** determines sex in many different species. Some examples are described in Figure 3.17.

In the X-Y system of sex determination, which operates in mammals, the male has one X chromosome and one Y chromosome, whereas the female has two X chromosomes (Figure 3.17a). In this case, the male is called the **heterogametic sex.** Two types of sperm are produced: one that carries only the X chromosome, and another type that carries the Y. In contrast, the female is the **homogametic sex** because all eggs carry a single X chromosome. The 46 chromosomes carried by humans consist of 1 pair of sex chromosomes and 22 pairs of **autosomes**—chromosomes that are not sex chromosomes. In the human male, each of the four sperm produced during gametogenesis contains 23 chromosomes. Two sperm contain an X chromosome, and the other two have a Y chromosome. The sex of the offspring is determined by whether the sperm that fertilizes the egg carries an X or a Y chromosome.

What causes an offspring to develop into a male or female? One possibility is that two X chromosomes are required for female development. A second possibility is that the Y chromosome promotes male development. In the case of mammals, the second possibility is correct. This is known from the analysis of rare individuals who carry chromosomal abnormalities. For example, mistakes that occasionally occur during meiosis may produce an individual who carries two X chromosomes and one Y chromosome. Such an individual develops into a male. As discussed in Chapter 25, the *Sry* gene on the Y chromosome plays a key role in causing the development of male characteristics.

Other mechanisms of sex determination include the X-0, Z-W, and haplodiploid systems. The X-0 system of sex determination operates in many insects (Figure 3.17b). In such species, the male has one sex chromosome (the X) and is designated X0, whereas the female has a pair (two Xs). In other insect species, such as *Drosophila melanogaster*, the male is XY. For both types of insect species (i.e., X0 or XY males, and XX females), the ratio between X chromosomes and the number of autosomal sets determines sex. If a fly has one X chromosome and is diploid for the autosomes ($2n$), the ratio is 1/2, or 0.5. This fly will become a male even if it does not receive a Y chromosome. In contrast to mam-

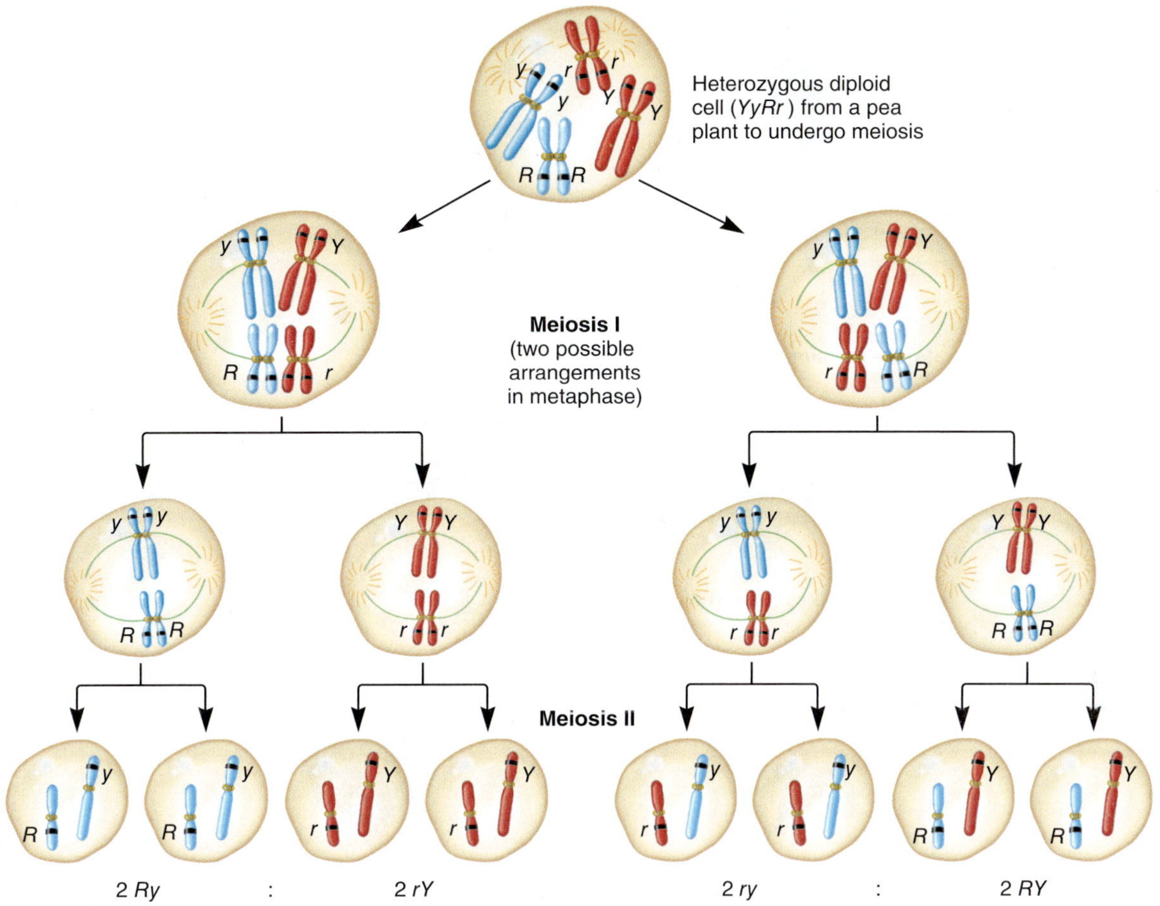

FIGURE 3.16 Mendel's law of independent assortment can be explained by the random alignment of bivalents during metaphase of meiosis I. This figure shows the assortment of two genes located on two different chromosomes in a pea plant, using seed color and shape as an example (*YyRr*). During metaphase of meiosis I, different possible arrangements of the homologs within bivalents can lead to different combinations of the alleles in the resulting gametes. For example, on the left, the recessive *y* allele has sorted with the dominant *R* allele; on the right, the dominant *Y* allele has sorted with the dominant *R* allele.

Genes→Traits Most species have several different chromosomes that carry many different genes. In this example, the gene for seed color exists in two alleles, *Y* (yellow) and *y* (green), and the gene for seed shape is found as *R* (round) and *r* (wrinkled) alleles. The two genes are found on different (nonhomologous) chromosomes. During meiosis, the homologous chromosomes that carry these alleles segregate from each other. In addition, the chromosomes carrying the *Y* or *y* alleles will independently assort from the chromosomes carrying the *R* or *r* alleles. As shown here, this provides a reassortment of alleles, potentially creating combinations of alleles that are different from the parental combinations. When two gametes unite during fertilization, the alleles they carry affect the traits of the resulting offspring.

CONCEPT CHECK: Let's suppose a pea plant is heterozygous for three genes, *Tt Yy Rr*, and each gene is on a different chromosome. How many different ways could the three homologous chromosomes line up during metaphase of meiosis I?

mals, the Y chromosome in the X-0 system does not determine maleness. If a fly receives two X chromosomes and is diploid, the ratio is 2/2, or 1.0, and the fly becomes a female.

For the Z-W system, which determines sex in birds and some fish, the male is ZZ and the female is ZW (Figure 3.17c). The letters Z and W are used to distinguish these types of sex chromosomes from those found in the X-Y pattern of sex determination of other species. In the Z-W system, the male is the homogametic sex, and the female is heterogametic.

Another interesting mechanism of sex determination, known as the haplodiploid system, is found in bees (Figure 3.17d). The male bee, called the drone, is produced from unfertilized haploid eggs. Female bees, both worker bees and queen bees, are produced from fertilized eggs and therefore are diploid.

Although sex in many species of animals is determined by chromosomes, other mechanisms are also known. In certain reptiles and fish, sex is controlled by environmental factors such as temperature. For example, in the American alligator (*Alligator mississippiensis*), temperature controls sex development. When fertilized eggs of this alligator are incubated at 33°C, nearly 100% of them produce male individuals. When the eggs are incubated at a temperature a few degrees below 33°C, they produce nearly all females, whereas at a temperature a few degrees above 33°C, they produce about 95% females.

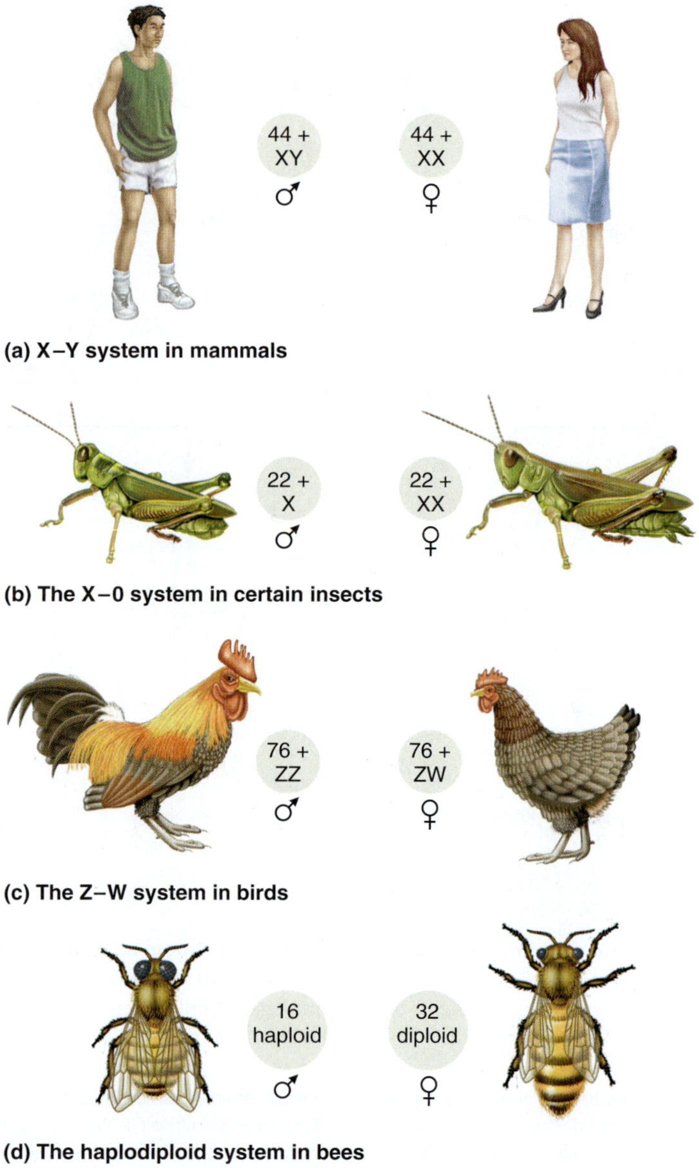

FIGURE 3.17 **Different mechanisms of sex determination in animals.** See text for a description.

CONCEPT CHECK: What is the difference between the X-Y and X-0 systems of sex determination?

EXPERIMENT 3A

Morgan's Experiments Showed a Connection Between a Genetic Trait and the Inheritance of a Sex Chromosome in *Drosophila*

In the early 1900s, American geneticist Thomas Hunt Morgan carried out a particularly influential study that confirmed the chromosome theory of inheritance. Morgan was trained as an embryologist, and much of his early research involved descriptive and experimental work in that field. He was particularly interested in ways that organisms change. He wrote, "The most distinctive problem of zoological work is the change in form that animals undergo, both in the course of their development from the egg (embryology) and in their development in time (evolution)." Throughout his life, he usually had dozens of different experiments going on simultaneously, many of them unrelated to each other. He jokingly said there are three kinds of experiments—those that are foolish, those that are damn foolish, and those that are worse than that!

In one of his most famous studies, Morgan engaged one of his graduate students to rear the fruit fly *Drosophila melanogaster* in the dark, hoping to produce flies whose eyes would atrophy from disuse and disappear in future generations. Even after many consecutive generations, however, the flies appeared to have no noticeable changes despite repeated attempts at inducing mutations by treatments with agents such as X-rays and radium. After two years, Morgan finally obtained an interesting result when a true-breeding line of *Drosophila* produced a male fruit fly with white eyes rather than the normal red eyes. Because this had been a true-breeding line of flies, this white-eyed male must have arisen from a new mutation that converted a red-eye allele (denoted w^+) into a white-eye allele (denoted w). Morgan is said to have carried this fly home with him in a jar, put it by his bedside at night while he slept, and then taken it back to the laboratory during the day.

Much like Mendel, Morgan studied the inheritance of this white-eye trait by making crosses and quantitatively analyzing their outcome. In the experiment described in **Figure 3.18**, he began with his white-eyed male and crossed it to a true-breeding red-eyed female. All of the F_1 offspring had red eyes, indicating that red is dominant to white. The F_1 offspring were then mated to each other to obtain an F_2 generation.

THE GOAL (DISCOVERY-BASED SCIENCE)

This is an example of discovery-based science rather than hypothesis testing. In this case, a quantitative analysis of genetic crosses may reveal the inheritance pattern for the white-eye allele.

ACHIEVING THE GOAL

FIGURE 3.18 Inheritance pattern of an X-linked trait in fruit flies.

Starting material: A true-breeding strain of red-eyed fruit flies plus one white-eyed male fly that was discovered in the culture.

1. Cross the white-eyed male to a true-breeding red-eyed female.
2. Record the results of the F_1 generation. This involves noting the eye color and sex of many offspring.
3. Cross F_1 offspring with each other to obtain F_2 offspring. Also record the eye color and sex of the F_2 offspring.
4. In a separate experiment, perform a testcross between a white-eyed male from the F_2 generation and a red-eyed female from the F_1 generation. Record the results.

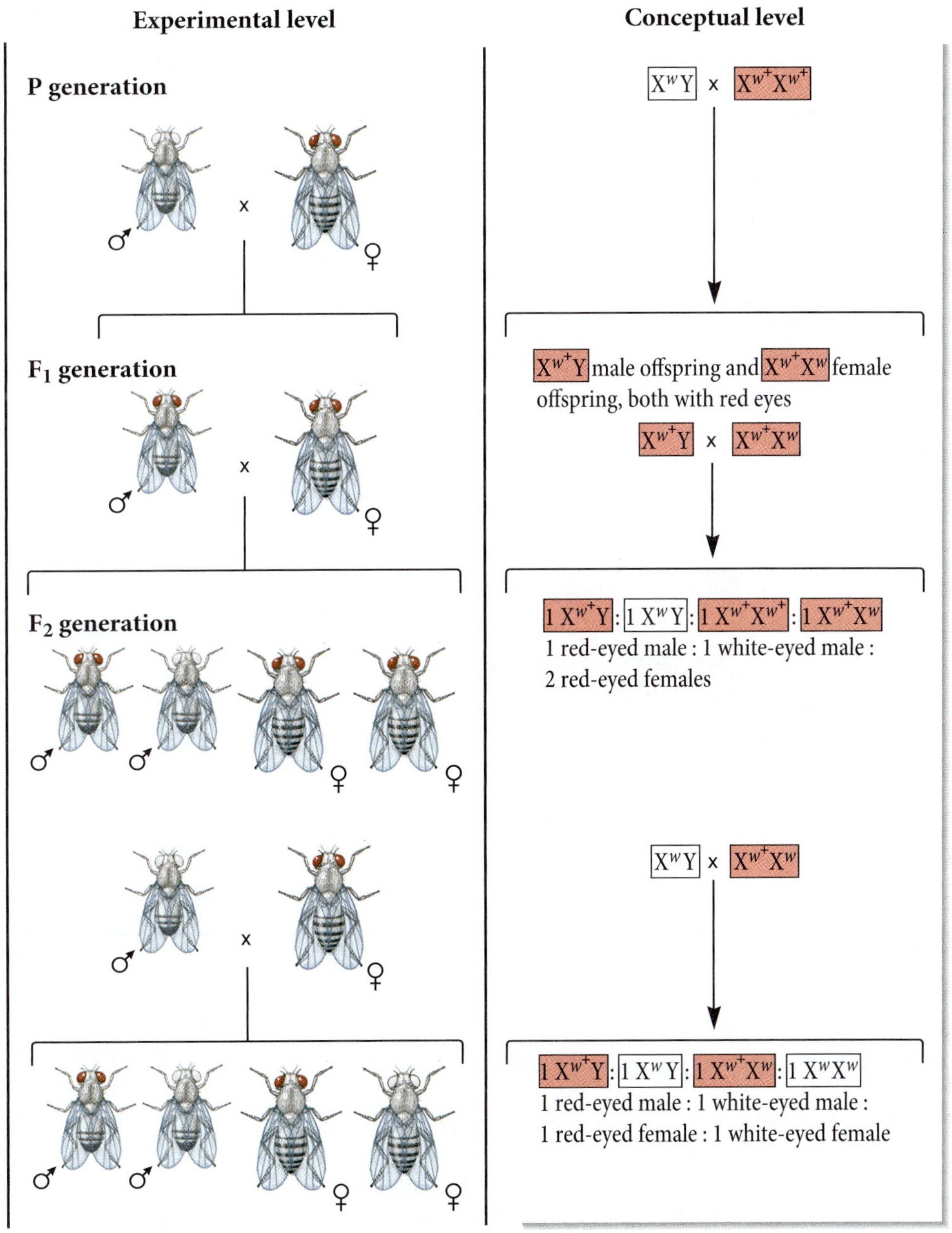

THE DATA

Cross	*Results*	
Original white-eyed male to a true-breeding red-eyed female	F_1 generation:	All red-eyed flies
F_1 male to F_1 female	F_2 generation:	2459 red-eyed females 1011 red-eyed males 0 white-eyed females 782 white-eyed males
F_2 white-eyed male to F_1 female	Testcross:	129 red-eyed females 132 red-eyed males 88 white-eyed females 86 white-eyed males

Data from Morgan, T.H. (1910) Sex limited inheritance in *Drosophila. Science* 32, 120–122.

INTERPRETING THE DATA

As seen in the data table, the F_2 generation consisted of 2459 red-eyed females, 1011 red-eyed males, and 782 white-eyed males. Most notably, no white-eyed female offspring were observed in the F_2 generation. These results suggested that the pattern of transmission from parent to offspring depends on the sex of the offspring and on the alleles that they carry. As shown in the Punnett square below, the data are consistent with the idea that the eye color alleles are located on the X chromosome.

F_1 male is $X^{w^+}Y$
F_1 female is $X^{w^+}X^w$

Male gametes

Female gametes	♂ X^{w^+}	Y
♀ X^{w^+}	$X^{w^+}X^{w^+}$ Red, female	$X^{w^+}Y$ Red, male
X^w	$X^{w^+}X^w$ Red, female	X^wY White, male

The Punnett square predicts that the F_2 generation will not have any white-eyed females. This prediction was confirmed experimentally. These results indicated that the eye color alleles are located on the X chromosome. Genes that are physically located within the X chromosome are called **X-linked genes,** or **X-linked alleles.** However, it should also be pointed out that the experimental ratio in the F_2 generation of red eyes to white eyes is (2459 + 1011):782, which equals 4.4:1. This ratio deviates significantly from the predicted ratio of 3:1. How can this discrepancy be explained? Later work revealed that the lower-than-expected number of white-eyed flies is due to their decreased survival rate.

Morgan also conducted a **testcross** (see step 4, Figure 3.18) in which an individual with a dominant phenotype and unknown genotype is crossed to an individual with a recessive phenotype. In this case, he mated an F_1 red-eyed female to an F_2 white-eyed male. This cross produced red-eyed males and females, and white-eyed males and females, in approximately equal numbers. The testcross data are also consistent with an X-linked pattern of inheritance. As shown in the following Punnett square, the testcross predicts a 1:1:1:1 ratio:

Testcross:
Male is X^wY
F_1 female is $X^{w^+}X^w$

Male gametes

Female gametes	♂ X^w	Y
♀ X^{w^+}	$X^{w^+}X^w$ Red, female	$X^{w^+}Y$ Red, male
X^w	X^wX^w White, female	X^wY White, male

The observed data are 129:132:88:86, which is a ratio of 1.5:1.5:1:1. Again, the lower-than-expected numbers of white-eyed males and females can be explained by a lower survival rate for white-eyed flies. In his own interpretation, Morgan concluded that R (red eye color) and X (a sex factor that is present in two copies in the female) are combined and have never existed apart. In other words, this gene for eye color is on the X chromosome. Morgan was the first geneticist to receive a Nobel Prize.

A self-help quiz involving this experiment can be found at www.mhhe.com/brookergenetics5e.

3.6 COMPREHENSION QUESTIONS

1. Which of the following is *not* one of the tenets of the chromosome theory of inheritance?
   a. Chromosomes contain the genetic material that is transmitted from parent to offspring and from cell to cell.
   b. Chromosomes are replicated and passed along, generation after generation, from parent to offspring.
   c. Chromosome replication occurs during the S phase of the cell cycle.
   d. Each parent contributes one set of chromosomes to its offspring.

2. A pea plant has the genotype *TtRr*. The independent assortment of these two genes occurs at ________ because chromosomes carrying the ________ alleles line up independently of the chromosomes carrying the __________ alleles.
   a. metaphase of meiosis I, *T* and *t*, *R* and *r*
   b. metaphase of meiosis I, *T* and *R*, *t* and *r*
   c. metaphase of meiosis II, *T* and *t*, *R* and *r*
   d. metaphase of meiosis II, *T* and *R*, *t* and *r*

3. In mammals, sex is determined by
   a. the *Sry* gene on the Y chromosome.
   b. having two copies of the X chromosome.
   c. having one copy of the X chromosome.
   d. both a and c.

4. An abnormal fruit fly has two sets of autosomes and is XXY. Such a fly would be
   a. a male.
   b. a female.
   c. a hermaphrodite.
   d. none of the above.

KEY TERMS

Page 45. chromosomes, chromatin
Page 46. prokaryotes, nucleoid, eukaryotes, organelles, nucleus
Page 47. cytogenetics, cytogeneticist, somatic cell, gametes, germ cells, karyotype, diploid, homologs, alleles
Page 49. locus, loci, asexual reproduction, multicellularity, binary fission
Page 50. cell cycle, interphase, restriction point, chromatids, centromere, sister chromatids, dyad, monad, kinetochore, mitosis
Page 52. mitotic spindle apparatus, mitotic spindle, microtubule-organizing centers, centrosomes, spindle pole, centrioles, aster microtubules, polar microtubules, kinetochore microtubules
Page 53. decondensed, prophase, condensed, prometaphase, metaphase plate, metaphase, anaphase, telophase
Page 55. cytokinesis, cleavage furrow, cell plate, meiosis, haploid
Page 56. leptotene, zygotene, synapsis, pachytene, bivalents, tetrad, crossing over
Page 57. chiasma, chiasmata, diplotene, diakinesis
Page 59. sexual reproduction, gametogenesis
Page 60. isogamous, heterogamous, sperm cells, egg cell, ovum, spermatogenesis
Page 61. oogenesis, gametophyte, sporophyte, pollen grain
Page 62. embryo sac, endosperm
Page 63. X-linked inheritance, chromosome theory of inheritance
Page 64. sex chromosomes, heterogametic sex, homogametic sex, autosomes
Page 68. X-linked genes, X-linked alleles, testcross

CHAPTER SUMMARY

3.1 General Features of Chromosomes

- Chromosomes are structures that contain the genetic material, which is DNA.
- Prokaryotic cells are simple and lack cell compartmentalization, whereas eukaryotic cells contain a cell nucleus and other compartments (see Figure 3.1).
- Chromosomes can be examined under the microscope. An organized representation of the chromosomes from a single cell is called a karyotype (see Figure 3.2).
- In eukaryotic species, the chromosomes are found in sets. Eukaryotic cells are often diploid, which means that each type of chromosome occurs in a homologous pair (see Figure 3.3).

3.2 Cell Division

- Bacteria divide by binary fission (see Figure 3.4).
- To divide, eukaryotic cells progress through a cell cycle (see Figure 3.5).

3.3 Mitosis and Cytokinesis

- Prior to cell division, eukaryotic chromosomes are replicated to form sister chromatids (see Figure 3.6).
- Chromosome sorting in eukaryotes is achieved via a spindle apparatus (see Figure 3.7).
- A common way for eukaryotic cells to divide is by mitosis and cytokinesis. Mitosis is divided into prophase, prometaphase, metaphase, anaphase, and telophase (see Figures 3.8, 3.9).

3.4 Meiosis

- Another way for eukaryotic cells to divide is via meiosis, which produces four haploid cells. During prophase of meiosis I, homologs synapse and crossing over may occur (see Figures 3.10–3.12).

3.5 Sexual Reproduction

- Animals produce gametes via spermatogenesis and oogenesis (see Figure 3.13).
- Plants exhibit alternation of generations between a diploid sporophyte and a haploid gametophyte. The gametophyte produces gametes (see Figure 3.14).

3.6 The Chromosome Theory of Inheritance and Sex Chromosomes

- The chromosome theory of inheritance describes how the transmission of chromosomes can explain Mendel's laws.
- Mendel's law of segregation is explained by the separation of homologs during meiosis (see Figure 3.15).
- Mendel's law of independent assortment is explained by the random alignment of different chromosomes during metaphase of meiosis I (see Figure 3.16).
- Mechanisms of sex determination in animals may involve differences in chromosome composition (see Figure 3.17).
- Morgan's work provided strong evidence for the chromosome theory of inheritance by showing that a gene affecting eye color in fruit flies is inherited on the X chromosome (see Figure 3.18).

PROBLEM SETS & INSIGHTS

Solved Problems

S1. A diploid cell has eight chromosomes, four per set. For the following diagram, in what phase of mitosis, meiosis I, or meiosis II is this cell?

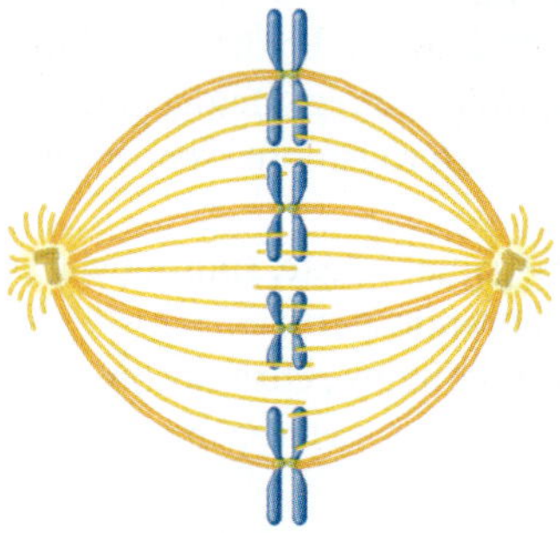

Answer: The cell is in metaphase of meiosis II. You can tell because the chromosomes are lined up in a single row along the metaphase plate, and the cell has only four pairs of sister chromatids. If it were mitosis, the cell would have eight pairs of sister chromatids.

S2. An unaffected woman (i.e., without disease symptoms) who is heterozygous for the X-linked allele causing Duchenne muscular dystrophy has children with a man with a normal allele. What are the probabilities of the following combinations of offspring?

A. An unaffected son

B. An unaffected son or daughter

C. A family of three children, all of whom are affected

Answer: The first thing we must do is construct a Punnett square to determine the outcome of the cross. *N* represents the normal allele, *n* the recessive allele causing Duchenne muscular dystrophy. The mother is heterozygous, and the father has the normal allele.

Male gametes

Female gametes ♀ \ ♂	X^N	Y
X^N	X^NX^N	X^NY
X^n	X^NX^n	X^nY

Phenotype ratio is

2 unaffected daughters :
1 unaffected son :
1 affected son

A. There are four possible children, one of whom is an unaffected son. Therefore, the probability of an unaffected son is 1/4.

B. 1/4 + 1/2 = 3/4.

C. You could use the product rule because the three offspring with the disorder are independent events: (1/4)(1/4)(1/4) = 1/64 = 0.016, or 1.6%.

S3. Among different plant species, both male and female gametophytes can be produced by single individuals or by separate sexes. In some species, such as the garden pea, a single individual can produce both male and female gametophytes. Fertilization takes place via self-fertilization or cross-fertilization. A plant species that has a single type of flower producing both pollen and eggs is termed a monoclinous plant. In other plant species, two different types of flowers produce either pollen or eggs. When both flower types are on a single individual, such a species is termed monoecious. It is most common for the "male flowers" to be produced near the top of the plant and the "female flowers" toward the bottom. Though less common, some species of plants are dioecious. For dioecious species, one individual makes either male flowers or female flowers, but not both.

Based on your personal observations of plants, try to give examples of monoclinous, monoecious, and dioecious plants. What would be the advantages and disadvantages of each?

Answer: Monoclinous plants—pea plant, tulip, and roses. The same flower produces pollen on the anthers and egg cells within the ovary.

Monoecious plants—corn and pine trees. In corn, the tassels are the male flowers and the ears result from fertilization within the female flowers. In pine trees, pollen is produced in cones near the top of the tree, and eggs cells are found in larger cones nearer the bottom.

Dioecious plants—holly and ginkgo trees. Certain individuals produce only pollen; others produce only eggs.

An advantage of being monoclinous or monoecious is that fertilization is relatively easy because the pollen and egg cells are produced on the same individual. This is particularly true for monoclinous plants. The proximity of the pollen to the egg cells makes it more likely for self-fertilization to occur. This is advantageous if the plant population is relatively sparse. On the other hand, a dioecious species can reproduce only via cross-fertilization. The advantage of cross-fertilization is that it enhances genetic variation. Over the long run, this can be an advantage because cross-fertilization is more likely to produce a varied population of individuals, some of which may possess combinations of traits that promote survival.

S4. To test the chromosome theory of inheritance, American geneticist Calvin Bridges made crosses involving the inheritance of X-linked traits. One of his experiments concerned two different X-linked genes affecting eye color and wing size. For the eye color gene, the red-eye allele (w^+) is dominant to the white-eye allele (w). A second X-linked trait is wing size; the allele called miniature is recessive to the normal allele. In this case, m represents the miniature allele and m^+ the normal allele. A male fly carrying a miniature allele on its single X chromosome has small (miniature) wings. A female must be homozygous, mm, in order to have miniature wings.

Bridges made a cross between $X^{w,m^+} X^{w,m^+}$ female flies (white eyes and normal wings) to $X^{w^+,m}$ Y male flies (red eyes and miniature wings). He then examined the eyes, wings, and sexes of thousands of offspring. Most of the offspring were females with red eyes and normal wings, and males with white eyes and normal wings. On rare occasions (approximately 1 out of 1700 flies), however, he also obtained female offspring with white eyes or males with red eyes. He also noted the wing shape in these flies and then cytologically examined their chromosome composition using a microscope. The following results were obtained:

Offspring	Eye Color	Wing Size	Sex Chromosomes
Expected females	Red	Normal	XX
Expected males	White	Normal	XY
Unexpected females (rare)	White	Normal	XXY
Unexpected males (rare)	Red	Miniature	X0

Data from: Bridges, C. B. (1916) "Non-disjunction as proof of the chromosome theory of heredity," *Genetics 1*, 1–52, 107–163.

Explain these data.

Answer: Remember that in fruit flies, the number of X chromosomes (not the presence of the Y chromosome) determines sex. As seen in the data, the flies with unexpected phenotypes were abnormal in their sex chromosome composition. The white-eyed female flies were due to the union between an abnormal XX female gamete and a normal Y male gamete. Likewise, the unexpected male offspring had only one X chromosome and no Y. These male offspring were due to the union between an abnormal egg without any X chromosome and a normal sperm containing one X chromosome. The wing size of the unexpected males was a particularly significant result. The red-eyed males showed a miniature wing size. As noted by Bridges, this means they inherited their X chromosome from their father rather than their mother. This observation provided compelling evidence that the inheritance of the X chromosome correlates with the inheritance of particular traits.

Conceptual Questions

C1. The process of binary fission begins with a single mother cell and ends with two daughter cells. Would you expect the mother and daughter cells to be genetically identical? Explain why or why not.

C2. What is a homolog? With regard to genes and alleles, how are homologs similar to and different from each other?

C3. What is a sister chromatid? Are sister chromatids genetically similar or identical? Explain.

C4. With regard to sister chromatids, which phase of mitosis is the organization phase, and which is the separation phase?

C5. A species is diploid and has three chromosomes per set. Draw what the chromosomes would look like in the G_1 and G_2 phases of the cell cycle.

C6. How does the attachment of kinetochore microtubules to the kinetochore differ in metaphase of meiosis I from metaphase of mitosis? Discuss what you think would happen if a sister chromatid was not attached to a kinetochore microtubule.

C7. For the following events, specify whether they occur during mitosis, meiosis I, or meiosis II:

A. Separation of conjoined chromatids within a pair of sister chromatids

B. Pairing of homologous chromosomes

C. Alignment of chromatids along the metaphase plate

D. Attachment of sister chromatids to both poles

C8. Describe the key events during meiosis that result in a 50% reduction in the amount of genetic material per cell.

C9. A cell is diploid and contains three chromosomes per set. Draw the arrangement of chromosomes during metaphase of mitosis and metaphase of meiosis I and II. In your drawing, make one set dark and the other lighter.

C10. The arrangement of homologs during metaphase of meiosis I is a random process. In your own words, explain what this means.

C11. A eukaryotic cell is diploid containing 10 chromosomes (5 in each set). For mitosis and meiosis, how many daughter cells would be produced, and how many chromosomes would each one contain?

C12. If a diploid cell contains six chromosomes (i.e., three per set), how many possible random arrangements of homologs could occur during metaphase of meiosis I?

C13. A cell has four pairs of chromosomes. Assuming that crossing over does not occur, what is the probability that a gamete will contain all of the paternal chromosomes? If n equals the number of chromosomes in a set, which of the following expressions can be used to calculate the probability that a gamete will receive all of the paternal chromosomes: $(1/2)^n$, $(1/2)^{n-1}$, or $n^{1/2}$?

C14. With regard to question C13, how would the phenomenon of crossing over affect the results? In other words, would the probability of a gamete inheriting only paternal chromosomes be higher or lower? Explain your answer.

C15. Eukaryotic cells must sort their chromosomes during mitosis so each daughter cell receives the correct number of chromosomes. Why don't bacteria need to sort their chromosomes?

C16. Why is it necessary for the chromosomes to condense during mitosis and meiosis? What do you think might happen if the chromosomes were not condensed?

C17. Nine-banded armadillos almost always give birth to four offspring that are genetically identical quadruplets. Explain how you think this happens.

C18. A diploid species has four chromosomes per set for a total of eight chromosomes in its somatic cells. Draw the cell as it would look in late prophase of meiosis II and prophase of mitosis. Discuss how prophase of meiosis II and prophase of mitosis differ from each other, and explain how the difference originates.

C19. Explain why the products of meiosis may not be genetically identical, whereas the products of mitosis are.

C20. The period between meiosis I and meiosis II is called interphase II. Does DNA replication take place during interphase II?

C21. List several ways in which telophase appears to be the reverse of prophase and prometaphase.

C22. Corn has 10 chromosomes per set, and the sporophyte of the species is diploid. If you performed a karyotype, what is the total number of chromosomes you would expect to see in the following types of cells?

A. A leaf cell

B. The sperm nucleus of a pollen grain

C. An endosperm cell after fertilization

D. A root cell

C23. The arctic fox has 50 chromosomes (25 per set), and the common red fox has 38 chromosomes (19 per set). These species can interbreed to produce viable but infertile offspring. How many chromosomes would the offspring have? What problems do you think may occur during meiosis that would explain the offspring's infertility?

C24. Let's suppose that a gene affecting pigmentation is found on the X chromosome (in mammals or insects) or the Z chromosome (in birds) but not on the Y or W chromosome. It is found on an autosome in bees. This gene exists in two alleles; *D* (dark), is dominant to *d* (light). What would be the phenotypic results of crosses between a true-breeding dark female and true-breeding light male, and the reciprocal crosses involving a true-breeding light female and true-breeding dark male, in the following species? Refer back to Figure 3.17 for the mechanism of sex determination in these species.

A. Birds

B. *Drosophila*

C. Bees

D. Humans

C25. Describe the cellular differences between male and female gametes.

C26. At puberty, the testes contain a finite number of cells and produce an enormous number of sperm cells during the life span of a male. Explain why testes do not run out of spermatogonial cells.

C27. Describe the timing of meiosis I and II during human oogenesis.

C28. Three genes (*A*, *B*, and *C*) are found on three different chromosomes. For the following diploid genotypes, describe all of the possible gamete combinations.

A. *Aa Bb Cc*

B. *AA Bb CC*

C. *Aa BB Cc*

D. *Aa bb cc*

C29. A phenotypically normal woman with an abnormally long chromosome 13 (and a normal homolog of chromosome 13) marries a phenotypically normal man with an abnormally short chromosome 11 (and a normal homolog of chromosome 11). What is the probability of producing an offspring that will have both a long chromosome 13 and a short chromosome 11? If such a child is produced, what is the probability that this child would eventually pass both abnormal chromosomes to one of his or her offspring?

C30. Assuming that such a fly would be viable, what would be the sex of a fruit fly with the following chromosomal composition?

A. One X chromosome and two sets of autosomes

B. Two X chromosomes, one Y chromosome, and two sets of autosomes

C. Two X chromosomes and four sets of autosomes

D. Four X chromosomes, two Y chromosomes, and four sets of autosomes

C31. What would be the sex of a human with the following numbers of sex chromosomes?

A. XXX

B. X (also described as X0)

C. XYY

D. XXY

Experimental Questions

E1. When studying living cells in a laboratory, researchers sometimes use drugs as a way to cause cells to remain in a particular phase of the cell cycle. For example, aphidicolin inhibits DNA synthesis in eukaryotic cells and causes them to remain in the G_1 phase because they cannot replicate their DNA. In what phase of the cell cycle—G_1, S, G_2, prophase, metaphase, anaphase, or telophase—would you expect somatic cells to stay if the following types of drug were added?

A. A drug that inhibits microtubule formation

B. A drug that allows microtubules to form but prevents them from shortening

C. A drug that inhibits cytokinesis

D. A drug that prevents chromosomal condensation

E2. In Morgan's experiments, which result do you think is the most convincing piece of evidence pointing to X-linkage of the eye color gene? Explain your answer.

E3. In his original studies of Figure 3.18, Morgan first suggested that the original white-eyed male had two copies of the white-eye allele. In this problem, let's assume that he meant the fly was X^wY^w instead of X^wY. Are his data in Figure 3.18 consistent with this hypothesis? What crosses would need to be made to rule out the possibility that the Y chromosome carries a copy of the eye color gene?

E4. How would you set up crosses to determine if a gene was Y linked versus X linked?

E5. Occasionally during meiosis, a mistake can happen whereby a gamete may receive zero or two sex chromosomes rather than one. Calvin Bridges made a cross between white-eyed female flies and red-eyed male flies. As you would expect, most of the offspring were red-eyed females and white-eyed males. On rare occasions, however, he found a white-eyed female or a red-eyed male. These rare flies were not due to new gene mutations, but instead were due to mistakes during meiosis in the parent flies. Consider the mechanism of sex determination in fruit flies and propose how this could happen. In your answer, describe the sex chromosome composition of these rare flies.

E6. Let's suppose that you have karyotyped a female fruit fly with red eyes and found that it has three X chromosomes instead of the normal two. Although you do not know its parents, you do know that this fly came from a mixed culture of flies in which some had red eyes, some had white eyes, and some had eosin eyes. Eosin is an allele of the same gene that has white and red alleles. Eosin is a pale orange color. The red allele is dominant and the white allele is recessive. The expression of the eosin allele, however, depends on the number of copies of the allele. When females have two copies of this allele, they have eosin eyes. When females are heterozygous for the eosin allele and white allele, they have light-eosin eyes. When females are heterozygous for the red allele and the eosin allele, they have red eyes. Males that have a single copy of eosin allele have light-eosin eyes.

You cross this female with a white-eyed male and count the number of offspring. You may assume that this unusual female makes half of its gametes with one X chromosome and half of its gametes with two X chromosomes. The following results were obtained:

	Females*	Males
Red eyes	50	11
White eyes	0	0
Eosin	20	0
Light-eosin	21	20

*A female offspring can be XXX, XX, or XXY.

Explain the 3:1 ratio between female and male offspring. What was the genotype of the original mother, which had red eyes and three X chromosomes? Construct a Punnett square that is consistent with these data.

E7. With regard to thickness and length, what do you think the chromosomes would look like if you microscopically examined them during interphase? How would that compare with their appearance during metaphase?

E8. White-eyed flies have a lower survival rate than red-eyed flies. Based on the data in Figure 3.18, what percentage of white-eyed flies survived compared with red-eyed flies, assuming 100% survival of red-eyed flies?

E9. A rare form of dwarfism that also included hearing loss was found to run in a particular family. It is inherited in a dominant manner. It was discovered that an affected individual had one normal copy of chromosome 15 and one abnormal copy of chromosome 15 that was unusually long. How would you determine if the unusually long chromosome 15 was causing this disorder?

E10. Discuss why crosses (i.e., the experiments of Mendel) and the microscopic observations of chromosomes during mitosis and meiosis were both needed to deduce the chromosome theory of inheritance.

E11. Female flies with white eyes and miniature wings (both X-linked recessive traits) were crossed to male flies with red eyes and normal wings. On rare occasions, female offspring were produced with white eyes. If we assume these females are due to errors in meiosis, what would be the most likely chromosomal composition of such flies? What would be their wing shape?

E12. Experimentally, how do you think researchers were able to determine that the Y chromosome causes maleness in mammals, whereas the ratio of X chromosomes to the sets of autosomes causes sex determination in fruit flies?

Questions for Student Discussion/Collaboration

1. In Figure 3.18, Morgan obtained a white-eyed male fly in a population containing many red-eyed flies that he thought were true-breeding. As mentioned in the experiment, he crossed this fly with several red-eyed sisters, and all the offspring had red eyes. But actually this is not quite true. Morgan observed 1237 red-eyed flies and 3 white-eyed males. Provide two or more explanations why he obtained 3 white-eyed males in the F_1 generation.
2. A diploid eukaryotic cell has 10 chromosomes (5 per set). As a group, take turns having one student draw the cell as it would look during a phase of mitosis, meiosis I, or meiosis II; then have the other students guess which phase it is.
3. Discuss the principles of the chromosome theory of inheritance. Which principles were deduced via light microscopy, and which were deduced from crosses? What modern techniques could be used to support the chromosome theory of inheritance?

Note: All answers appear at the website for this textbook; the answers to the even-numbered questions and all of the Concept Check and Comprehension Questions are in the back of the book.

www.mhhe.com/brookergenetics5e

Visit the website for practice tests, answer keys, and other learning aids for this chapter. Enhance your understanding of genetics with our interactive exercises, quizzes, animations, and much more.

CHAPTER OUTLINE

Inheritance patterns and alleles. *In the petunia, multiple alleles result in flowers with several different colors, such as the three shown here.*

4 EXTENSIONS OF MENDELIAN INHERITANCE

Many traits in eukaryotic species follow a **Mendelian inheritance** pattern. Such traits obey two laws: the law of segregation and the law of independent assortment. Furthermore, the genes that influence such traits are not altered (except by rare mutations) as they are passed from parent to offspring. The traits that are displayed by the offspring depend on the alleles they inherit and also on environmental factors. Until now, we have mainly considered traits that are affected by a single gene that is found in two different alleles. In these cases, one allele is dominant over the other. This type of inheritance is called **simple Mendelian inheritance** because the observed ratios in the offspring readily obey Mendel's laws. For example, when two different true-breeding pea plants are crossed (e.g., tall and dwarf) and the F_1 generation is allowed to self-fertilize, the F_2 generation shows a 3:1 phenotypic ratio of tall to dwarf offspring.

In Chapter 4, we will extend our understanding of Mendelian inheritance by first examining the transmission patterns for several traits that do not display a simple dominant/recessive relationship. Geneticists have discovered an amazing diversity of mechanisms by which alleles affect the outcome of traits. Many alleles don't produce the ratios of offspring that are expected from a simple Mendelian relationship. This does not mean that Mendel was wrong. Rather, the inheritance patterns of many traits are more complex and interesting than he had realized. In this chapter, we will examine how the outcome of a trait may be influenced by a variety of factors such as the level of protein expression, the sex of the individual, the presence of multiple alleles of a given gene, and environmental effects. We will also explore how two different genes can contribute to the outcome of a single trait.

In Chapters 5 and 6, we will examine eukaryotic inheritance patterns that are not considered Mendelian. For example, some genes are located in mitochondrial DNA and do not obey the law of segregation. Others are closely linked along the same chromosome and violate the law of independent assortment. In addition, some genes are altered (by DNA methylation) during gamete formation, which affects their expression in the offspring that inherit them. This results in an epigenetic pattern of inheritance that we will consider in Chapter 5.

4.1 OVERVIEW OF SIMPLE INHERITANCE PATTERNS

Learning Outcomes:

1. Compare and contrast the different types of inheritance patterns.
2. Describe the molecular mechanisms that account for different types of inheritance patterns involving single genes.

Before we delve more deeply into inheritance patterns, let's first compare a variety of inheritance patterns with each other (**Table 4.1**). We have already discussed two of these—simple dominant/recessive inheritance and X-linked inheritance in Chapter 2. These various patterns occur because the outcome of a trait may be governed by two or more alleles in several different ways. Geneticists want to understand Mendelian inheritance from two perspectives. One goal is to predict the outcome of crosses. Many of the inheritance patterns described in Table 4.1 do not produce a 3:1 phenotypic ratio when two heterozygotes produce offspring. A second goal is to understand how the molecular expression of genes can account for an individual's phenotype. In other words, what is the underlying relationship between molecular genetics—the expression of genes to produce functional proteins—and the traits of individuals that inherit the genes? The remaining sections of this chapter will explore several patterns of inheritance and consider their underlying molecular mechanisms.

4.1 COMPREHENSION QUESTION

1. Which of the following statements is *true*?
   a. Not all inheritance patterns follow a strict dominant/recessive relationship.
   b. Geneticists want to understand both inheritance patterns and the underlying molecular mechanisms that cause them to happen.
   c. Different inheritance patterns are explained by a variety of different molecular mechanisms.
   d. All of the above are true.

TABLE 4.1
Mendelian Inheritance Patterns Involving Single Genes

Type	Description
Simple Mendelian	**Inheritance:** This term is commonly applied to the inheritance of alleles that obey Mendel's laws and follow a strict dominant/recessive relationship. In Chapter 4, we will see that some genes are found in three or more alleles, making the relationship more complex. **Molecular:** 50% of the protein, produced by a single copy of the dominant (functional) allele in the heterozygote, is sufficient to produce the dominant trait.
Incomplete dominance	**Inheritance:** This pattern occurs when the heterozygote has a phenotype that is intermediate between either corresponding homozygote. For example, a cross between homozygous red-flowered and homozygous white-flowered parents will have heterozygous offspring with pink flowers. **Molecular:** 50% of the protein, produced by a single copy of the functional allele in the heterozygote, is not sufficient to produce the same trait as the homozygote making 100%.
Incomplete penetrance	**Inheritance:** In the case of dominant traits, this pattern occurs when a dominant phenotype is not expressed even though an individual carries a dominant allele. An example is an individual who carries the polydactyly allele but has a normal number of fingers and toes. **Molecular:** Even though a dominant allele is present, the protein encoded by the gene may not exert its effects. This can be due to environmental influences or due to other genes that may encode proteins that counteract the effects of the protein encoded by the dominant allele.
Overdominance	**Inheritance:** This pattern occurs when the heterozygote has a trait that is more beneficial than either homozygote. **Molecular:** Three common ways that heterozygotes gain benefits: (1) Their cells may have increased resistance to infection by microorganisms; (2) they may produce more forms of protein dimers, with enhanced function; or (3) they may produce proteins that function under a wider range of conditions.
Codominance	**Inheritance:** This pattern occurs when the heterozygote expresses both alleles simultaneously without forming an intermediate phenotype. For example, in blood typing, an individual carrying the *A* and *B* alleles will have an AB blood type. **Molecular:** The codominant alleles encode proteins that function slightly differently from each other, and the function of each protein in the heterozygote affects the phenotype uniquely.
X-linked	**Inheritance:** This pattern involves the inheritance of genes that are located on the X chromosome. In mammals and fruit flies, males are hemizygous for X-linked genes, whereas females have two copies. **Molecular:** If a pair of X-linked alleles shows a simple dominant/recessive relationship, 50% of the protein, produced by a single copy of the dominant allele in a heterozygous female, is sufficient to produce the dominant trait (in the female). Males have only one copy of X-linked genes and therefore express the copy they carry.
Sex-influenced inheritance	**Inheritance:** This pattern refers to the effect of sex on the phenotype of the individual. Some alleles are recessive in one sex and dominant in the opposite sex. **Molecular:** Sex hormones may regulate the molecular expression of genes. This can influence the phenotypic effects of alleles.
Sex-limited inheritance	**Inheritance:** This refers to traits that occur in only one of the two sexes. An example is breast development in mammals. **Molecular:** Sex hormones may regulate the molecular expression of genes. This can influence the phenotypic effects of alleles. In this case, sex hormones that are primarily produced in only one sex are essential to produce a particular phenotype.
Lethal alleles	**Inheritance:** An allele that has the potential of causing the death of an organism. **Molecular:** Lethal alleles are most commonly loss-of-function alleles that encode proteins that are necessary for survival. In some cases, the allele may be due to a mutation in a nonessential gene that changes a protein to function with abnormal and detrimental consequences.

4.2 DOMINANT AND RECESSIVE ALLELES

Learning Outcomes:

1. Define wild-type allele and genetic polymorphism.
2. Explain why loss-of-function alleles often follow a recessive pattern of inheritance.
3. Describe how alleles can exhibit incomplete penetrance and vary in their expressivity.

In Chapter 2, we examined patterns of inheritance that showed a simple dominant/recessive relationship. This means that a heterozygote exhibits the dominant trait. In this section, we will take a closer look at why an allele may be dominant or recessive and discuss how dominant alleles may not always exert their effects.

Recessive Mutant Alleles Often Cause a Reduction in the Amount or Function of the Encoded Proteins

For any given gene, geneticists refer to prevalent alleles in a natural population as **wild-type alleles.** In large populations, more than one wild-type allele may occur—a phenomenon known as **genetic polymorphism.** For example, Figure 4.1 illustrates a striking example of polymorphism in the elderflower orchid, *Dactylorhiza sambucina.* Throughout the range of this species in Europe, both yellow- and red-flowered individuals are prevalent. Both colors are considered wild type. At the molecular level, a wild-type allele typically encodes a protein that is made in the proper amount and functions normally. As discussed in Chapter 26, wild-type alleles tend to promote the reproductive success of organisms in their native environments.

In addition, random mutations occur in populations and alter preexisting alleles. Geneticists sometimes refer to these kinds of alleles as **mutant alleles** to distinguish them from the more common wild-type alleles. Because random mutations are more likely to disrupt gene function, mutant alleles are often defective in their ability to express a functional protein. Such mutant alleles tend to be rare in natural populations. They are typically, but not always, inherited in a recessive fashion.

FIGURE 4.1 An example of genetic polymorphism. Both yellow and red flowers are common in natural populations of the elderflower orchid, *Dactylorhiza sambucina*, and both are considered wild type.

CONCEPT CHECK: Why are both of these colors considered to be wild type?

Among Mendel's seven traits discussed in Chapter 2, the wild-type alleles are tall plants, purple flowers, axial flowers, yellow seeds, round seeds, green pods, and smooth pods (refer back to Figure 2.4). The mutant alleles are dwarf plants, white flowers, terminal flowers, green seeds, wrinkled seeds, yellow pods, and constricted pods. You may have already noticed that the seven wild-type alleles are dominant over the seven mutant alleles. Likewise, red eyes and normal wings are examples of wild-type alleles in *Drosophila*, and white eyes and miniature wings are recessive mutant alleles.

The idea that recessive mutant alleles usually cause a substantial decrease in the expression of a functional protein is supported by the analysis of many human genetic diseases. Keep in mind that a genetic disease is usually caused by a mutant allele. Table 4.2 lists several examples of human genetic diseases in

TABLE 4.2
Examples of Recessive Human Diseases

Disease	Protein That Is Produced by the Normal Gene*	Description
Phenylketonuria	Phenylalanine hydroxylase	Inability to metabolize phenylalanine. The disease can be prevented by following a phenylalanine-free diet. If the diet is not followed early in life, it can lead to severe mental impairment and physical degeneration.
Albinism	Tyrosinase	Lack of pigmentation in the skin, eyes, and hair.
Tay-Sachs disease	Hexosaminidase A	Defect in lipid metabolism. Leads to paralysis, blindness, and early death.
Sandhoff disease	Hexosaminidase B	Defect in lipid metabolism. Muscle weakness in infancy, early blindness, and progressive mental and motor deterioration.
Cystic fibrosis	Chloride transporter	Inability to regulate ion balance across epithelial cells. Leads to production of thick mucus and results in chronic lung infections, infertility, and organ malfunctions.
Lesch-Nyhan syndrome	Hypoxanthine-guanine phosphoribosyl transferase	Inability to metabolize purines, which are bases found in DNA and RNA. Leads to self-mutilation behavior, poor motor skills, and usually mental impairment and kidney failure.

*Individuals who exhibit the disease are either homozygous for a recessive allele or hemizygous (for X-linked genes in human males). The disease symptoms result from a defect in the amount or function of the normal protein.

which the recessive allele fails to produce a specific cellular protein in its active form. In many cases, molecular techniques have enabled researchers to clone these genes and determine the differences between the wild-type and mutant alleles. They have found that the recessive allele usually contains a mutation that causes a defect in the synthesis of a fully functional protein.

To understand why many defective mutant alleles are inherited recessively, we need to take a quantitative look at protein function. With the exception of sex-linked genes, diploid individuals have two copies of every gene. In a simple dominant/recessive relationship, the recessive allele does not affect the phenotype of the heterozygote. In other words, a single copy of the dominant allele is sufficient to mask the effects of the recessive allele. If the recessive allele cannot produce a functional protein, how do we explain the wild-type phenotype of the heterozygote? As described in **Figure 4.2**, a common explanation is that 50% of the functional protein is adequate to provide the wild-type phenotype. In this example, the *PP* homozygote and *Pp* heterozygote each make sufficient functional protein to yield purple flowers.

A second possible explanation for other genes is that the heterozygote actually produces more than 50% of the functional protein. Due to gene regulation, the expression of the normal gene may be increased or "up-regulated" in the heterozygote to compensate for the lack of function of the defective allele. The topic of gene regulation is discussed in Chapters 14 and 15.

Dominant Mutant Alleles Usually Exert Their Effects in One of Three Ways

Though dominant mutant alleles are much less common than recessive mutant alleles, they do occur in natural populations. How can a mutant allele be dominant over a wild-type allele? Three explanations account for most dominant mutant alleles:

Dominant (functional) allele: *P* (purple)
Recessive (defective) allele: *p* (white)

Genotype	*PP*	*Pp*	*pp*
Amount of functional protein P	100%	50%	0%
Phenotype	Purple	Purple	White
Simple dominant/recessive relationship			

FIGURE 4.2 A comparison of protein levels among homozygous (*PP* or *pp*) and heterozygous (*Pp*) genotypes.

Genes→Traits In a simple dominant/recessive relationship, 50% of the protein encoded by one copy of the dominant allele in the heterozygote is sufficient to produce the wild-type phenotype, in this case, purple flowers. A complete lack of the functional protein results in white flowers.

CONCEPT CHECK: Does a *PP* individual produce more of the protein encoded by the *P* gene than is necessary for the purple color?

a gain-of-function mutation, a dominant-negative mutation, or haploinsufficiency. Some dominant mutant alleles are due to **gain-of-function mutations.** For example, a mutant gene may be overexpressed or it may be expressed in the wrong cell type. A second category is **dominant-negative mutations** in which the protein encoded by the mutant gene acts antagonistically to the normal protein. In a heterozygote, the mutant protein counteracts the effects of the normal protein, thereby altering the phenotype. Finally, mutant alleles may affect phenotype via **haploinsufficiency,** in which the dominant mutant allele is a loss-of-function allele. Haploinsufficiency is used to describe patterns of inheritance in which a heterozygote (with one functional allele and one inactive allele) exhibits an abnormal or disease phenotype. An example in humans is polydactyly, which is discussed next.

Traits May Skip a Generation Due to Incomplete Penetrance and Vary in Their Expressivity

As we have seen, dominant alleles are expected to influence the outcome of a trait when they are present in heterozygotes. Occasionally, however, this may not occur. The phenomenon, called **incomplete penetrance,** is a situation in which an allele that is expected to cause a particular phenotype does not. **Figure 4.3a** illustrates a human pedigree for a dominant trait known as polydactyly. This trait causes the affected individual to have additional fingers or toes (or both) (**Figure 4.3b**). Polydactyly is due to an autosomal dominant allele—the allele is found in a gene located on an autosome (not a sex chromosome) and a single copy of this allele is sufficient to cause this condition. Sometimes, however, individuals carry the dominant allele but do not exhibit the trait. In Figure 4.3a, individual III-2 has inherited the polydactyly allele from his mother and passed the allele to a daughter and son. However, individual III-2 does not actually exhibit the trait himself, even though he is a heterozygote. In our polydactyly example, the dominant allele does not always "penetrate" into the phenotype of the individual. Alternatively, for recessive traits, incomplete penetrance would occur if a homozygote carrying the recessive allele did not exhibit the recessive trait. The measure of penetrance is described at the populational level. For example, if 60% of the heterozygotes carrying a dominant allele exhibit the trait, we would say that this trait is 60% penetrant. At the individual level, the trait is either present or not.

Another term used to describe the outcome of traits is the degree to which the trait is expressed, or its **expressivity.** In the case of polydactyly, the number of extra digits can vary. For example, one individual may have an extra toe on only one foot, whereas a second individual may have extra digits on both the hands and feet. Using genetic terminology, a person with several extra digits would have high expressivity of this trait, whereas a person with a single extra digit would have low expressivity.

How do we explain incomplete penetrance and variable expressivity? Although the answer may not always be understood, the range of phenotypes is often due to environmental influences and/or due to effects of modifier genes in which one or more genes alter the phenotypic effects of another gene. We will consider the issue of the environment next. The effects of modifier genes will be discussed later in the chapter.

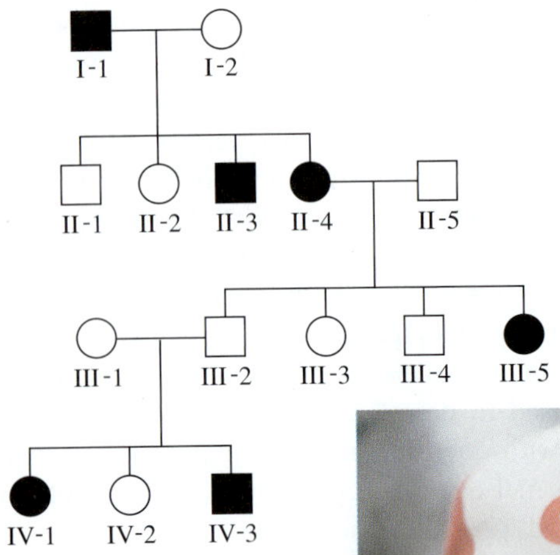

(a)

(b)

FIGURE 4.3 Polydactyly, a dominant trait that shows incomplete penetrance. **(a)** A family pedigree. Affected individuals are shown in black. Notice that offspring IV-1 and IV-3 have inherited the trait from a parent, III-2, who is heterozygous but does not exhibit polydactyly. **(b)** Antonio Alfonseca, a baseball player with polydactyly. His extra finger does not give him an advantage when pitching because it is small and does not touch the ball.

CONCEPT CHECK: Which individual(s) in this pedigree exhibit(s) incomplete penetrance?

4.2 COMPREHENSION QUESTIONS

1. Which of the following is *not* an example of a wild-type allele?
   a. Yellow-flowered elderflower orchid
   b. Red-flowered elderflower orchid
   c. A gray elephant
   d. An albino (white) elephant
2. Dominant alleles may result from mutations that cause
   a. the overexpression of a gene or its protein product.
   b. a protein to inhibit the function of a normal protein.
   c. a protein to be inactive and 50% of the normal protein is insufficient for a normal phenotype.
   d. all of the above.
3. Polydactyly is a condition in which a person has extra fingers and/or toes. It is caused by a dominant allele. If a person carries this allele but does not have any extra fingers or toes, this is an example of
   a. haploinsufficiency.
   b. a dominant negative mutation.
   c. incomplete penetrance.
   d. a gain-of-function mutation.

4.3 ENVIRONMENTAL EFFECTS ON GENE EXPRESSION

Learning Outcomes:

1. Discuss the role of the environment with regard to an individual's traits.
2. Define norm of reaction.

Throughout this book, our study of genetics tends to focus on the roles of genes in the outcome of traits. In addition to genetics, environmental conditions have a great effect on the phenotype of the individual. For example, the arctic fox (*Alopex lagopus*) goes through two color phases. During the cold winter, it is primarily white, but in the warmer summer, the fox is mostly brown (**Figure 4.4a**). This is an example of a **temperature-sensitive allele.** The phenotypic effects are dependent on the temperature.

A dramatic example of the relationship between environment and phenotype can be seen in the human genetic disease known as phenylketonuria (PKU). This autosomal recessive disease is caused by a defect in a gene that encodes the enzyme phenylalanine hydroxylase. Homozygous individuals with this defective allele are unable to metabolize the amino acid phenylalanine properly. When given a standard diet containing phenylalanine, which is found in most protein-rich foods, PKU individuals manifest a variety of detrimental traits including mental impairment, underdeveloped teeth, and foul-smelling urine. In contrast, when PKU is diagnosed early and patients follow a restricted diet free of phenylalanine, they develop normally (**Figure 4.4b**). Since the 1960s, testing methods have been developed that can determine if an individual is lacking the phenylalanine hydroxylase enzyme. These tests permit the identification of infants who have PKU, and their diets can then be modified before the harmful effects of phenylalanine ingestion have occurred. As a result of government legislation, more than 90% of infants born in the United States are now tested for PKU. This test prevents a great deal of human suffering and is also cost-effective. In the United States, the annual cost of PKU testing is estimated to be a few million dollars, whereas the cost of treating severely affected individuals with the disease would be hundreds of millions of dollars.

The examples of the arctic fox and PKU represent dramatic effects of very different environmental conditions. When considering the environment, geneticists often examine a range of conditions, rather than simply observing phenotypes under two different conditions. The term **norm of reaction** refers to the effects of environmental variation on a phenotype. Specifically,

(a) Arctic fox in winter and summer

(b) Healthy person with PKU

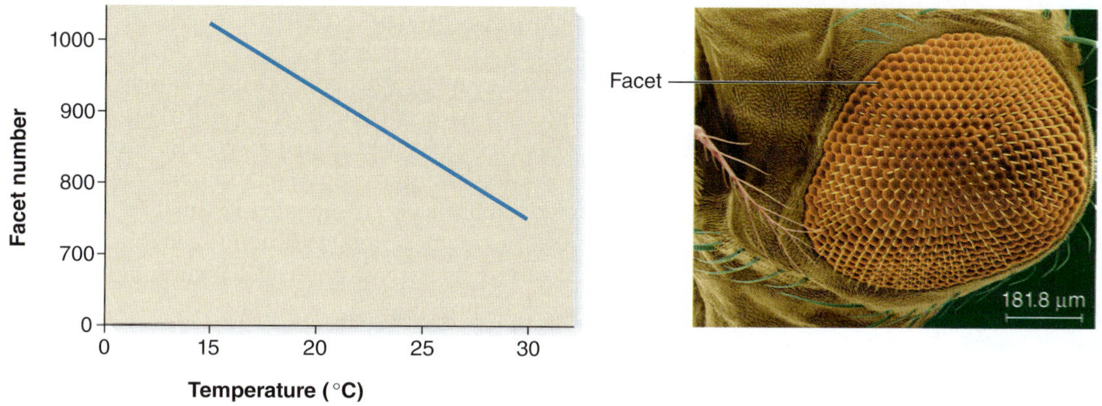

(c) Norm of reaction

FIGURE 4.4 **Variation in the expression of traits due to environmental effects.** (a) The arctic fox (*Alopex lagopus*) in the winter and summer. (b) A person with PKU who has followed a restricted diet and developed normally. (c) Norm of reaction. In this experiment, fertilized eggs from a population of genetically identical fruit flies (*Drosophila melanogaster*) were allowed to develop into adult flies at different environmental temperatures. This graph shows the relationship between temperature (an environmental factor) and facet number in the eyes of the resulting adult flies. The micrograph shows an eye of *D. melanogaster*.

CONCEPT CHECK: What are the two main factors that determine an organism's traits?

it is the phenotypic range seen in individuals with a particular genotype. To evaluate the norm of reaction, researchers begin with true-breeding strains that have the same genotypes and subject them to different environmental conditions. As an example, let's consider facet number in the eyes of fruit flies, *Drosophila melanogaster*. This species has compound eyes composed of many individual facets. **Figure 4.4c** shows the norm of reaction for facet number in genetically identical fruit flies that developed at different temperatures. As shown in the figure, the facet number varies with changes in temperature. At a lower temperature (15°C), the facet number is over 1000, whereas at a higher temperature (30°C), it is approximately 750.

4.3 COMPREHENSION QUESTION

1. The outcome of an individual's traits is controlled by
   a. genes.
   b. the environment.
   c. genes and the environment.
   d. neither genes nor the environment.

4.4 INCOMPLETE DOMINANCE, OVERDOMINANCE, AND CODOMINANCE

Learning Outcomes:

1. Predict the outcome of crosses involving incomplete dominance, overdominance, and codominance.
2. Explain the underlying molecular mechanisms of incomplete dominance, overdominance, and codominance.

Thus far, we have considered inheritance patterns that follow a simple dominant/recessive inheritance pattern. In these cases, the heterozygote exhibits a phenotype that is the same as a homozygote that carries two copies of the dominant allele but different from the homozygote carrying two copies of the recessive allele. In this section, we will examine three different inheritance patterns in which the heterozygote shows a phenotype that is different from both types of homozygotes.

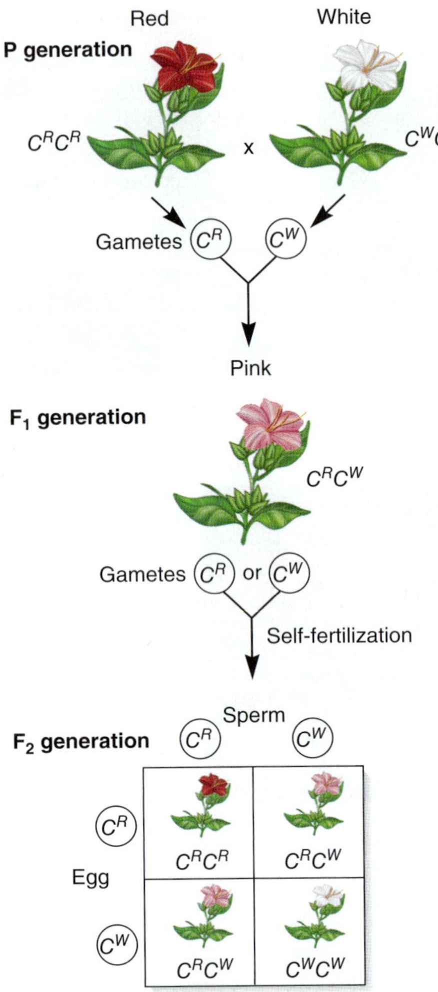

INTERACTIVE EXERCISE

FIGURE 4.5 Incomplete dominance in the four-o'clock plant, *Mirabilis jalapa*.

Genes→Traits When two different homozygotes (C^RC^R and C^WC^W) are crossed, the resulting heterozygote, C^RC^W, has an intermediate phenotype of pink flowers. In this case, 50% of the functional protein encoded by the C^R allele is not sufficient to produce a red phenotype.

CONCEPT CHECK: At the molecular level, what is the explanation for why the flowers are pink instead of red?

Incomplete Dominance Occurs When Two Alleles Produce an Intermediate Phenotype

Although many alleles display a simple dominant/recessive relationship, geneticists have also identified some cases in which a heterozygote exhibits **incomplete dominance**—a condition in which the phenotype is intermediate between the corresponding homozygous individuals. In 1905, the German botanist Carl Correns first observed this phenomenon in the color of the four-o'clock (*Mirabilis jalapa*). **Figure 4.5** describes Correns' experiment, in which a homozygous red-flowered four-o'clock plant was crossed to a homozygous white-flowered plant. The wild-type allele for red flower color is designated C^R and the white allele is C^W. As shown here, the offspring had pink flowers. When these F_1 offspring were allowed to self-fertilize, the F_2 generation consisted of 1/4 red-flowered plants, 1/2 pink-flowered plants, and 1/4 white-flowered plants. The pink plants in the F_2 generation were heterozygotes with an intermediate phenotype. As noted in the Punnett square in Figure 4.5, the F_2 generation displayed a 1:2:1 phenotypic ratio, which is different from the 3:1 ratio observed for simple Mendelian inheritance.

In Figure 4.5, incomplete dominance resulted in a heterozygote with an intermediate phenotype. At the molecular level, the allele that causes a white phenotype is expected to result in a lack of a functional protein required for pigmentation. Depending on the effects of gene regulation, the heterozygotes may produce only 50% of the normal protein, but this amount is not sufficient to produce the same phenotype as the C^RC^R homozygote, which may make twice as much of this protein. In this example, a reasonable explanation is that 50% of the functional protein cannot accomplish the same level of pigment synthesis that 100% of the protein can.

Finally, our opinion of whether a trait is dominant or incompletely dominant may depend on how closely we examine the trait in the individual. The more closely we look, the more likely we are to discover that the heterozygote is not quite the same as the wild-type homozygote. For example, Mendel studied the characteristic of pea seed shape and visually concluded that the *RR* and *Rr* genotypes produced round seeds and the *rr* genotype produced wrinkled seeds. The peculiar morphology of the wrinkled seed is caused by a large decrease in the amount of starch deposition in the seed due to a defective *r* allele. More recently, other scientists have dissected round and wrinkled seeds and examined their contents under the microscope. They have found that round seeds from heterozygotes actually contain an intermediate number of starch grains compared with seeds from the corresponding homozygotes (**Figure 4.6**). Within the seed, an intermediate amount of the functional protein is not enough to

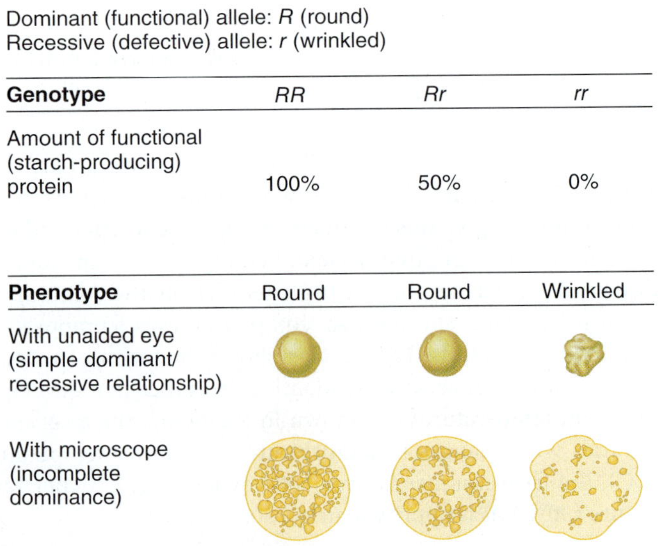

FIGURE 4.6 A comparison of phenotype at the macroscopic and microscopic levels.

Genes→Traits This illustration shows the effects of a heterozygote having only 50% of the functional protein needed for starch production. This seed appears to be as round as those of the homozygote carrying the *R* allele, but when examined microscopically, it has produced only half the amount of starch.

CONCEPT CHECK: At which level is incomplete dominance more likely to be observed—at the molecular/cellular level or at the organism level?

produce as many starch grains as in the homozygote carrying two copies of the *R* allele. Even so, at the level of our unaided eyes, heterozygotes produce seeds that appear to be round. With regard to phenotypes, the *R* allele is dominant to the *r* allele at the level of visual examination, but the *R* and *r* alleles show incomplete dominance at the level of starch biosynthesis.

Overdominance Occurs When Heterozygotes Have Superior Traits

As we have seen, the environment plays a key role in the outcome of traits. For certain genes, heterozygotes may display characteristics that are more beneficial for their survival in a particular environment. Such heterozygotes may be more likely to survive and reproduce. For example, a heterozygote may be larger, disease-resistant, or better able to withstand harsh environmental conditions. The phenomenon in which a heterozygote has greater reproductive success compared with either of the corresponding homozygotes is called **overdominance,** or **heterozygote advantage.**

A well-documented example involves a human allele that causes sickle cell disease in homozygous individuals. This disease is an autosomal recessive disorder in which the affected individual produces an altered form of the protein hemoglobin, which carries oxygen within red blood cells. Most people carry the Hb^A allele and make hemoglobin A. Individuals affected with sickle cell disease are homozygous for the Hb^S allele and produce only hemoglobin S. This causes their red blood cells to deform into a sickle shape under conditions of low oxygen concentration (**Figure 4.7a, b**). The sickling phenomenon causes the life span of these cells to be greatly shortened to only a few weeks compared with a normal span of four months, and therefore, anemia results. In addition, abnormal sickled cells can become clogged in the capillaries throughout the body, leading to localized areas of oxygen depletion. Such an event, called a crisis, causes pain and sometimes tissue and organ damage. For these reasons, the homozygous Hb^SHb^S individual usually has a shortened life span relative to an individual producing hemoglobin A.

In spite of the harmful consequences to homozygotes, the sickle cell allele has been found at a fairly high frequency among human populations that are exposed to malaria. The protist genus that causes malaria, *Plasmodium*, spends part of its life cycle within the *Anopheles* mosquito and another part within the red blood cells of humans who have been bitten by an infected mosquito. However, red blood cells of heterozygotes, Hb^AHb^S, are likely to rupture when infected by this parasite, thereby preventing the parasite from propagating. People who are heterozygous have better resistance to malaria than do Hb^AHb^A homozygotes, while not incurring the ill effects of sickle cell disease. Therefore, even though the homozygous Hb^SHb^S condition is detrimental, the greater survival of the heterozygote has selected for the presence of the Hb^S allele within populations where malaria is prevalent. When viewing survival in such a region, overdominance explains the prevalence of the sickle cell allele. In Chapter 26, we will consider the role that natural selection plays in maintaining alleles that are beneficial to the heterozygote but harmful to the homozygote.

Figure 4.7c illustrates the predicted outcome when two heterozygotes have children. In this example, 1/4 of the offspring are Hb^AHb^A (unaffected, not malaria-resistant), 1/2 are Hb^AHb^S (unaffected, malaria-resistant), and 1/4 are Hb^SHb^S (sickle cell disease). This 1:2:1 ratio deviates from a simple Mendelian 3:1 phenotypic ratio.

Overdominance is usually due to two alleles that produce proteins with slightly different amino acid sequences. How can we explain the observation that two protein variants in the heterozygote produce a more favorable phenotype? There are three common explanations. In the case of sickle cell disease, the phenotype is related to the infectivity of *Plasmodium* (**Figure 4.8a**). In the heterozygote, the infectious agent is less likely to propagate within red blood cells. Interestingly, researchers have speculated that other alleles in humans may confer disease resistance in the heterozygous condition but are detrimental in the homozygous state. These include PKU, in which the heterozygous fetus may be resistant to miscarriage caused by a fungal toxin, and Tay-Sachs disease, in which the heterozygote may be resistant to tuberculosis.

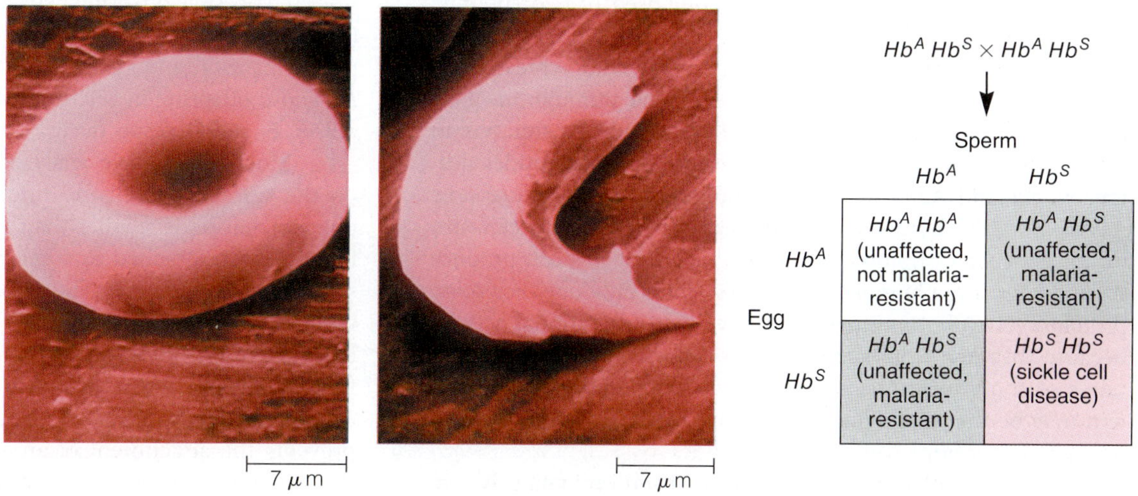

(a) Normal red blood cell

(b) Sickled red blood cell

(c) Example of sickle cell inheritance pattern

INTERACTIVE EXERCISE

FIGURE 4.7 Inheritance of sickle cell disease. A comparison of (a) a normal red blood cell and (b) one from a person with sickle cell disease. (c) The outcome of a cross between two heterozygous individuals.

CONCEPT CHECK: Why does the heterozygote have an advantage?

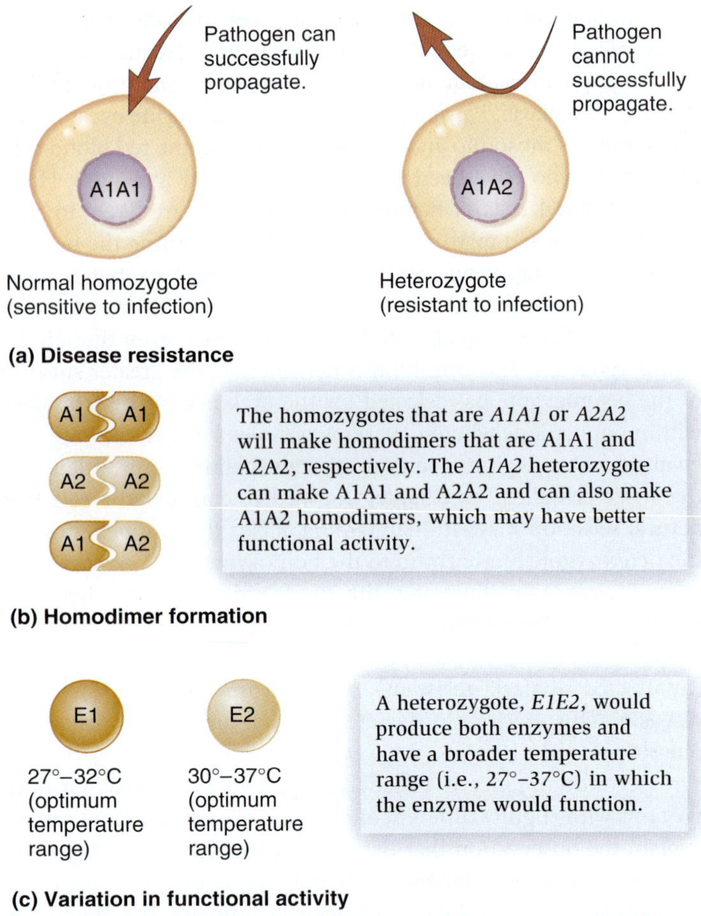

FIGURE 4.8 Three possible explanations for overdominance at the molecular level.

CONCEPT CHECK: Which of these three scenarios explains overdominance with regard to the sickle cell allele?

A second way to explain overdominance is related to the subunit composition of proteins. In some cases, a protein functions as a complex of multiple subunits; each subunit is composed of one polypeptide. A protein composed of two subunits is called a dimer. When both subunits are encoded by the same gene, the protein is a homodimer. The prefix homo- means that the subunits come from the same type of gene although the gene may exist in different alleles. **Figure 4.8b** considers a situation in which a gene exists in two alleles that encode polypeptides designated A1 and A2. Homozygous individuals can produce only A1A1 or A2A2 homodimers, whereas a heterozygote can also produce an A1A2 homodimer. Thus, heterozygotes can produce three forms of the homodimer, homozygotes only one. For some proteins, A1A2 homodimers may have better functional activity because they are more stable or able to function under a wider range of conditions. The greater activity of the homodimer protein may be the underlying reason why a heterozygote has characteristics superior to either homozygote.

A third molecular explanation of overdominance is that the proteins encoded by each allele exhibit differences in their functional activity. For example, suppose that a gene encodes a metabolic enzyme that can be found in two forms (corresponding to the two alleles), one that functions better at a lower temperature and the other that functions optimally at a higher temperature (**Figure 4.8c**). The heterozygote, which makes a mixture of both enzymes, may be at an advantage under a wider temperature range than either of the corresponding homozygotes.

Alleles of the ABO Blood Group Can Be Dominant, Recessive, or Codominant

Thus far, we have considered examples in which a gene exists in two different alleles. As researchers have probed genes at the molecular level within natural populations of organisms, they have discovered that most genes exist in **multiple alleles.** Within a population, genes are typically found in three or more alleles.

The ABO group of antigens, which determine blood type in humans, is an example of multiple alleles and illustrates yet another allelic relationship called codominance. To understand this concept, we first need to examine the molecular characteristics of human blood types. The plasma membranes of red blood cells have groups of interconnected sugars—oligosaccharides—that act as surface antigens (**Figure 4.9a**). Antigens are molecular structures that are recognized by antibodies produced by the immune system.

A person who is homozygous *ii* will have type O blood and produces a relatively short oligosaccharide, which is called H antigen. The *i* allele is recessive to both I^A and I^B. A homozygous I^AI^A or heterozygous I^Ai individual will have type A blood. The red blood cells of this individual will contain the surface antigen known as A. Similarly, a homozygous I^BI^B or heterozygous I^Bi individual will produce surface antigen B. As Figure 4.9a indicates, surface antigens A and B have different molecular structures. A person who is I^AI^B will have the blood type AB and express both surface antigens A and B. The phenomenon in which two alleles are both expressed in the heterozygous individual is called **codominance.** In this case, the I^A and I^B alleles are codominant to each other.

As an example of the inheritance of blood type, let's consider the possible offspring between two parents who are I^Ai and I^Bi (**Figure 4.9b**). The I^Ai parent makes I^A and *i* gametes, and the I^Bi parent makes I^B and *i* gametes. These combine to produce I^AI^B, I^Ai, I^Bi, and *ii* offspring in a 1:1:1:1 ratio. The resulting blood types are AB, A, B, and O, respectively.

Biochemists have analyzed the oligosaccharides produced on the surfaces of cells of differing blood types. In type O, the oligosaccharide is smaller than type A or type B because a sugar has not been attached to a specific site on the tree. This idea is schematically shown in Figure 4.9a. How do we explain this difference at the molecular level? The gene that determines ABO blood type encodes a type of enzyme called a glycosyl transferase that attaches a sugar to an oligosaccharide. The *i* allele carries a mutation that renders this enzyme inactive, which prevents the attachment of an additional sugar. By comparison, the two types of glycosyl transferase encoded by the I^A and I^B alleles have different structures in their active sites. The active site is the part of the protein that recognizes the sugar molecule

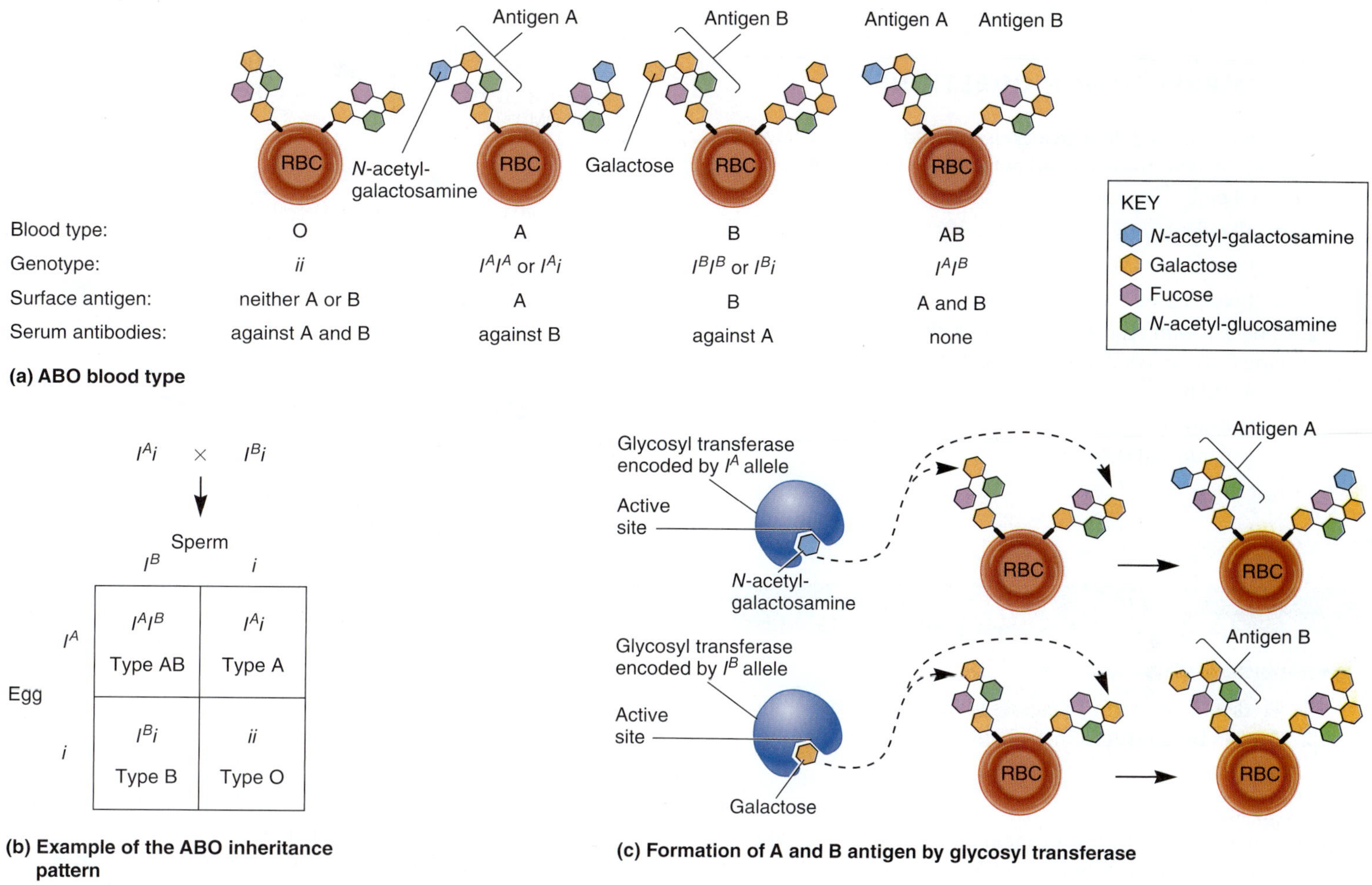

FIGURE 4.9 **ABO blood type.** **(a)** A schematic representation of blood type at the cellular level. Note: This is not drawn to scale. A red blood cell is much larger than the oligosaccharide on the surface of the cell. **(b)** The predicted offspring from parents who are I^Ai and I^Bi. **(c)** The glycosyl transferase encoded by the I^A and I^B alleles recognizes different sugars due to changes in its active site. The *i* allele results in a nonfunctional enzyme. The antigen produced by type O individuals is called H antigen. Humans of all blood types do not normally produce antibodies against H antigen.

INTERACTIVE EXERCISE

CONCEPT CHECK: Which allele is an example of a loss-of-function allele?

that will be attached to the oligosaccharide. The glycosyl transferase encoded by the I^A allele recognizes uridine diphosphate *N*-acetylgalactosamine (UDP-GalNAc) and attaches GalNAc to the oligosaccharide (**Figure 4.9c**). This produces the structure of surface antigen A. In contrast, the glycosyl transferase encoded by the I^B allele recognizes UDP-galactose and attaches galactose to the oligosaccharide. This produces the molecular structure of surface antigen B. A person with type AB blood makes both types of enzymes and thereby makes oligosaccharides with both types of sugar attached.

A small difference in the structure of the oligosaccharide, namely, a GalNAc in antigen A versus galactose in antigen B, explains why the two antigens are different from each other at the molecular level. These differences enable them to be recognized by different antibodies. A person who has blood type A makes antibodies to blood type B (refer back to Figure 4.9a). The antibodies against blood type B require a galactose in the oligosaccharide for their proper recognition. This person's antibodies will not recognize and destroy their own blood cells, but they will recognize and destroy the blood cells from a type B person.

With this in mind, let's consider why blood typing is essential for safe blood transfusions. The donor's blood must be an appropriate match with the recipient's blood. A person with type O blood has the potential to produce antibodies against both A and B antigens if they receive type A, type B, or type AB blood. After the antibodies are produced in the recipient, they will react with the donated blood cells and cause them to agglutinate (clump together). This is a life-threatening situation that causes the blood vessels to clog. Other incompatible combinations include a type A person receiving type B or type AB blood, and a type B person receiving type A or type AB blood. Because individuals with type AB blood do not produce antibodies to either A or B antigens, they can receive any type of blood and are known as universal recipients. By comparison, type O persons are universal donors because their blood can be given to type O, A, B, and AB people.

4.4 COMPREHENSION QUESTIONS

1. A pink-flowered four-o'clock is crossed to a red-flowered plant. What is the expected outcome for the offspring?
   a. All pink
   b. All red
   c. 1 red : 2 pink : 1 white
   d. 1 red : 1 pink
2. A person with type AB blood has a child with a person with type O blood. What are the possible blood types of the child?
   a. A and B
   b. A, B, and O
   c. A, B, AB, and O
   d. O only

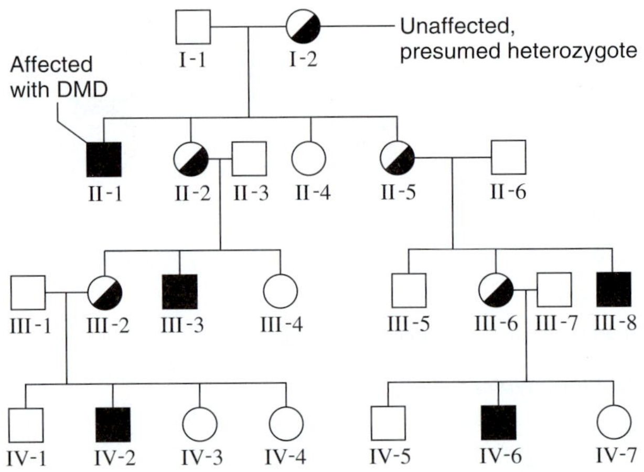

FIGURE 4.10 A human pedigree for Duchenne muscular dystrophy, an X-linked recessive trait. Affected individuals are shown with filled symbols. Females who are unaffected with the disease but have affected sons are presumed to be heterozygous carriers, as shown with half-filled symbols.

CONCEPT CHECK: What features of this pedigree indicate that the allele for Duchenne muscular dystrophy is X-linked?

4.5 X-LINKED INHERITANCE

Learning Outcomes:

1. Predict the outcome of crosses for X-linked inheritance.
2. Explain pseudoautosomal inheritance.

The term **sex chromosomes** refers to chromosomes that differ between males and females. In mammals and fruit flies, the sex chromosomes are designated X and Y. In Chapter 3, we considered experiments in fruit flies which showed that an eye-color gene is located on the X chromosome. When a gene is located on the X chromsome but not on the Y chromosome, it follows a pattern of transmission called **X-linked inheritance.** In this section, we will further explore the inheritance of traits in which genes are located on sex chromosomes.

The Inheritance Pattern of X-Linked Genes Can Be Revealed by Reciprocal Crosses

As discussed in Chapter 3, many species have males and females that differ in their sex chromosome composition. In mammals, for example, females are XX and males are XY. In such species, certain traits are governed by genes that are located on a sex chromosome. For these traits, the outcome of crosses depends on the genotypes and sexes of the parents and offspring.

As an example, let's consider a human disease known as Duchenne muscular dystrophy (DMD), which was first described by the French neurologist Guillaume Duchenne in the 1860s. Affected individuals show signs of muscle weakness as early as age 3. The disease gradually weakens the skeletal muscles and eventually affects the heart and breathing muscles. Survival is rare beyond the early 30s. The gene for DMD, found on the X chromosome, encodes a protein called dystrophin that is required inside muscle cells for structural support. Dystrophin is thought to strengthen muscle cells by anchoring elements of the internal cytoskeleton to the plasma membrane. Without it, the plasma membrane becomes permeable and may rupture.

DMD is inherited in an **X-linked recessive** pattern—the allele causing the disease is recessive and located on the X chromosome. In the pedigree shown in **Figure 4.10**, several males are affected by this disorder, as indicated by filled squares. The mothers of these males are presumed heterozygotes for the X-linked recessive allele. This recessive disorder is very rare among females because daughters would have to inherit a copy of the mutant allele from their mother and a copy from an affected father.

X-linked muscular dystrophy has also been found in certain breeds of dogs such as golden retrievers (**Figure 4.11a**). As in humans, the mutation occurs in the dystrophin gene, and the symptoms include severe weakness and muscle atrophy that begin at about 6 to 8 weeks of age. Many dogs that inherit this disorder die within the first year of life, though some can live 3 to 5 years and reproduce.

Figure 4.11b (left side) considers a cross between an unaffected female dog with two copies of the wild-type gene and a male dog with muscular dystrophy that carries the mutant allele and has survived to reproductive age. When setting up a Punnett square involving X-linked traits, we must consider the alleles on the X chromosome as well as the observation that males may transmit a Y chromosome instead of the X chromosome. The male makes two types of gametes, one that carries the X chromosome and one that carries the Y. The Punnett square must also include the Y chromosome even though this chromosome does not carry any X-linked genes. The X chromosomes from the female and male are designated with their corresponding alleles. When the Punnett square is filled in, it predicts the X-linked genotypes and sexes of the offspring. As seen on the left side of Figure 4.11b, none of the offspring from this cross are affected with the disorder, although all female offspring are carriers.

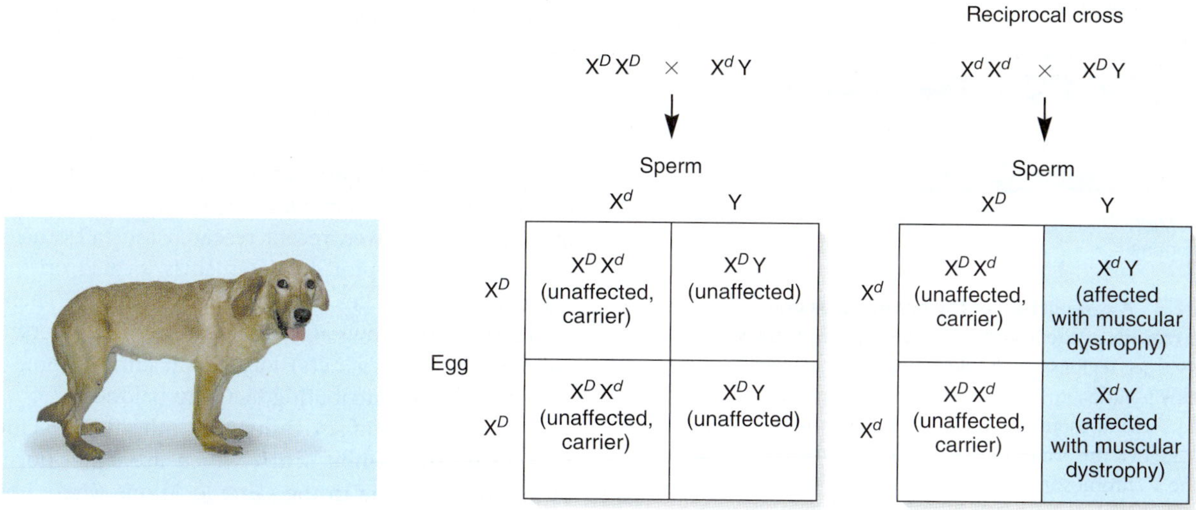

(a) Male golden retriever with X-linked muscular dystrophy

(b) Examples of X-linked muscular dystrophy inheritance patterns

INTERACTIVE EXERCISE

FIGURE 4.11 **X-linked muscular dystrophy in dogs.** **(a)** The male golden retriever shown here has the disease. **(b)** The left side shows a cross between an unaffected female and an affected male. The right shows a reciprocal cross between an affected female and an unaffected male. *D* represents the normal allele for the dystrophin gene, and *d* is the mutant allele that causes a defect in dystrophin function.

CONCEPT CHECK: Explain why the reciprocal cross yields a different result from the first cross.

The right side of Figure 4.11b shows a **reciprocal cross**—a second cross in which the sexes and phenotypes are reversed. In this case, an affected female animal is crossed to an unaffected male. This cross produces female offspring that are carriers and male offspring that are all affected with muscular dystrophy.

When comparing the two Punnett squares, the outcome of the reciprocal cross yielded different results. This is expected of X-linked genes, because the male transmits the gene only to female offspring, but the female transmits an X chromosome to both male and female offspring. Because the male parent does not transmit the X chromosome to his sons, he does not contribute to their X-linked phenotypes. This explains why X-linked traits do not behave equally in reciprocal crosses. Experimentally, the observation that reciprocal crosses do not yield the same results is an important clue that a trait may be X-linked.

Genes Located on Mammalian Sex Chromosomes Can Be Transmitted in an X-Linked, a Y-Linked, or a Pseudoautosomal Pattern

Our discussion of sex chromosomes has focused on genes that are located on the X chromosome but not on the Y chromosome. The term **sex-linked gene** refers to a gene that is found on one of the two types of sex chromosomes but not on both. Hundreds of X-linked genes have been identified in humans and other mammals.

The inheritance pattern of X-linked genes shows certain distinctive features. For example, males transmit X-linked genes only to their daughters, and sons receive their X-linked genes from their mothers. The term **hemizygous** is used to describe the single copy of an X-linked gene in the male. A male mammal is said to be hemizygous for X-linked genes. Because males of certain species, such as humans, have a single copy of the X chromosome, another distinctive feature of X-linked inheritance is that males are more likely to be affected by rare, recessive X-linked disorders.

By comparison, relatively few genes are located only on the Y chromosome. These few genes are called **holandric genes.** An example of a holandric gene is the *Sry* gene found in mammals. Its expression is necessary for proper male development. A Y-linked inheritance pattern is very distinctive—the gene is transmitted only from fathers to sons.

In addition to sex-linked genes, the X and Y chromosomes contain short regions of homology where the X and Y chromosomes carry the same genes. Along with several smaller regions, the human sex chromosomes have three large homologous regions (**Figure 4.12**). These regions, which are evolutionarily related, promote the necessary pairing of the X and Y chromosomes that occurs during meiosis I of spermatogenesis. Relatively few genes are located in these homologous regions. One example is a human gene called *Mic2*, which encodes a cell surface antigen. The *Mic2* gene is found on both the X and Y chromosomes. It follows a pattern of inheritance called **pseudoautosomal inheritance.** The term pseudoautosomal refers to the idea that the inheritance pattern of the *Mic2* gene is the same as the inheritance pattern of a gene located on an autosome even though the *Mic2* gene is actually located on the sex chromosomes. As in autosomal inheritance, males have two copies of pseudoautosomally inherited genes, and they can transmit the genes to both daughters and sons.

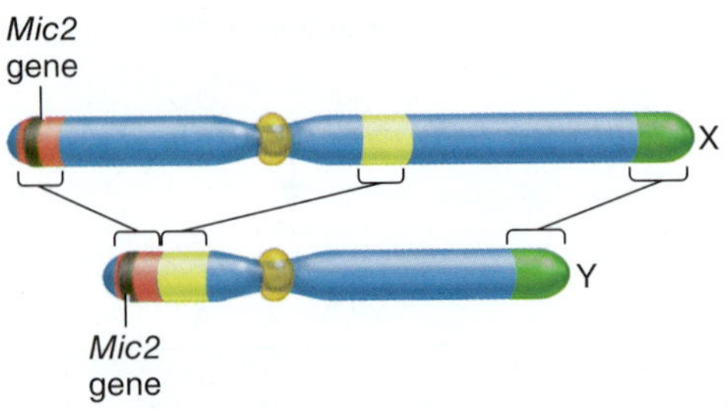

FIGURE 4.12 A comparison of the homologous and nonhomologous regions of the X and Y chromosome in humans. The brackets show three regions of homology between the X and Y chromosome. A few pseudoautosomal genes, such as *Mic2*, are found on both the X and Y chromosomes in these small regions of homology. Researchers estimate that the X chromosome contains between 900 and 1200 genes and the Y chromosome has between 70 and 300 genes.

CONCEPT CHECK: Why are the homologous regions of the X and Y chromosome important during meiosis?

4.5 COMPREHENSION QUESTIONS

1. A cross is made between a white-eyed female fruit fly and a red-eyed male. What would be the reciprocal cross?
   a. Female is X^wX^w and male is X^wY.
   b. Female is $X^{w+}X^{w+}$ and male is $X^{w+}Y$.
   c. Female is $X^{w+}X^{w+}$ and male is X^wY.
   d. Female is X^wX^w and male is $X^{w+}Y$.
2. Hemophilia is a blood clotting disorder in humans that follows an X-linked recessive pattern of inheritance. A man with hemophilia and woman without hemophilia have a daughter with hemophilia. If you let *H* represent the normal allele and *h* the hemophilia allele, what are the genotypes of the parents?
   a. Mother is X^HX^h and father is X^hY.
   b. Mother is X^hX^h and father is X^hY.
   c. Mother is X^hX^h and father is X^HY.
   d. Mother is X^HX^h and father is X^HY.

4.6 SEX-INFLUENCED AND SEX-LIMITED INHERITANCE

Learning Outcomes:

1. Compare and contrast sex-influenced inheritance and sex-limited inheritance.
2. Predict the outcome of crosses for sex-influenced inheritance.

As we have just seen, the transmission pattern of sex-linked genes depends on the sex of the parents and offspring. Sex can influence traits in other ways as well. The term **sex-influenced inheritance** refers to the phenomenon in which an allele is dominant in one sex but recessive in the opposite sex. Therefore, sex influence is a phenomenon of heterozygotes. Sex-influenced inheritance should not be confused with sex-linked inheritance. The genes that govern sex-influenced traits are autosomal, not on the X or Y chromosome. Researchers once thought that human pattern baldness, which is characterized by hair loss on the front and top of the head but not on the sides, is an example of sex-influenced inheritance. However, recent research indicates that mutations in the androgen receptor gene, which is located on the X chromosome, often play a key role in pattern baldness. Therefore, pattern baldness often follows an X-linked pattern of inheritance. Even so, variation in other gene(s) located on chromosome 20 (an autosome) can be a contributing factor to baldness.

An example of sex-influenced inheritance is found in cattle. Certain breeds exhibit scurs, which are small hornlike growths on the frontal bone in the same locations where horns (in other breeds of cattle) would grow (**Figure 4.13**). This trait appears to be controlled by a single gene that exists in *Sc* and *sc* alleles. The trait is dominant in males and recessive in females:

Genotype	Phenotype	
	Males	Females
ScSc	Scurs	Scurs
Scsc	Scurs	No scurs (hornless)
scsc	No scurs	No scurs

Another way in which sex affects an organism's phenotype is provided by **sex-limited inheritance,** in which a trait occurs in only one of the two sexes. Such genes are controlled by sex hormones or by the pathway that leads to male and female development, which is described in Chapter 25. The genes that affect

INTERACTIVE EXERCISE

FIGURE 4.13 Scurs in cattle, an example of a sex-influenced trait.

CONCEPT CHECK: What is the phenotype of a female cow that is heterozygous?

(a) Hen (b) Rooster

FIGURE 4.14 **Differences in the feathering pattern in female and male chickens, an example of sex-limited inheritance.**

CONCEPT CHECK: What is the molecular explanation for sex-limited inheritance?

sex-limited traits may be autosomal or X-linked. In humans, examples of sex-limited traits are the presence of ovaries in females and the presence of testes in males. Due to these two sex-limited traits, mature females only produce eggs, whereas mature males only produce sperm.

Sex-limited traits are responsible for **sexual dimorphism** in which members of the opposite sex have different morphological features. This phenomenon is common among many animal species and is often striking among various species of birds in which the male has more ornate plumage than the female. As shown in Figure 4.14, roosters have a larger comb and wattles and longer neck and tail feathers than do hens. These sex-limited features are never found in normal hens.

4.6 COMPREHENSION QUESTION

1. A female cow with scurs and a male bull with no scurs have an offspring. This offspring could be
   a. a female with scurs or a male with scurs.
   b. a female with no scurs or a male with scurs.
   c. a female with scurs or a male with no scurs.
   d. a female with no scurs or a male with no scurs.

4.7 LETHAL ALLELES

Learning Outcomes:

1. Describe the different types of lethal alleles.
2. Predict how lethal alleles may affect the outcome of a cross.

Let's now turn our attention to alleles that have the most detrimental effect on phenotype—those that result in death. An allele that has the potential to cause the death of an organism is called a **lethal allele.** These are usually inherited in a recessive manner. When the absence of a specific protein results in a lethal phenotype, the gene that encodes the protein is considered an **essential gene** for survival. Though it varies according to species, researchers estimate that approximately 1/3 of all genes are essential genes. By comparison, **nonessential genes** are not absolutely required for survival, although they are likely to be beneficial to the organism. A loss-of-function mutation in a nonessential gene will not usually cause death. On rare occasions, however, a nonessential gene may acquire a gain-of-function mutation that causes the gene product to be abnormally expressed in a way that may interfere with normal cell function and lead to a lethal phenotype. Therefore, not all lethal mutations occur in essential genes, although the great majority do.

Some lethal alleles may kill an organism only when certain environmental conditions prevail. Such **conditional lethal alleles** have been extensively studied in experimental organisms. For example, some conditional lethals cause an organism to die only in a particular temperature range. These alleles, called **temperature-sensitive (ts) lethal alleles,** have been observed in many organisms, including *Drosophila.* A ts lethal allele may be fatal for a developing larva at a high temperature (30°C), but the larva survives if grown at a lower temperature (22°C). Temperature-sensitive lethal alleles are typically caused by mutations that alter the structure of the encoded protein so it does not function correctly at the nonpermissive temperature or becomes unfolded and is rapidly degraded. Conditional lethal alleles may also be identified when an individual is exposed to a particular agent in the environment. For example, people with a defect in the gene that encodes the enzyme glucose-6-phosphate dehydrogenase (G6PD) have a negative reaction to the ingestion of fava beans. This can lead to an acute hemolytic syndrome with 10% mortality if not treated properly.

Finally, it is surprising that certain lethal alleles act only in some individuals. These are called **semilethal alleles.** Of course, any particular individual cannot be semidead. However, within a population, a semilethal allele will cause some individuals to die but not all of them. The reasons for semilethality are not always understood, but environmental conditions and the actions of other genes within the organism may help to prevent the detrimental effects of certain semilethal alleles. An example of a semilethal allele is the X-linked white-eyed allele in fruit flies, which is described in Chapter 3 (see Figure 3.18). Depending on the growth conditions, approximately 1/4 to 1/3 of the flies that carry this white-eyed trait die during early stages of development.

In some cases, a lethal allele may produce ratios that seemingly deviate from Mendelian ratios. An example is an allele in a breed of cats known as Manx, which originated on the Isle of Man (Figure 4.15a). The Manx cat is a heterozygote that carries a dominant mutant allele affecting the spine. This allele shortens the tail, resulting in a range of tail lengths from normal to tailless. When two heterozygous Manx cats are crossed to each other, the ratio of offspring is 1 normal to 2 Manx. How do we explain the 1:2 ratio? The answer is that 1/4 of the offspring are homozygous for the dominant mutant allele and they die during early embryonic development (Figure 4.15b). In this case, the Manx phenotype is dominant in heterozygotes, whereas the dominant mutant allele is lethal in the homozygous condition.

The time when a lethal allele exerts its effect can vary. Many lethal alleles prevent cell division and thereby cause an organism to die at a very early stage. Others may allow a short

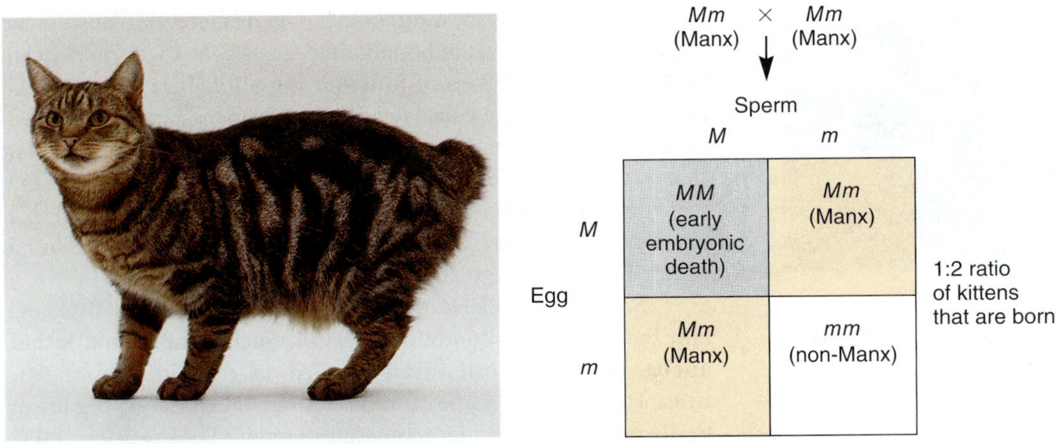

(a) A Manx cat

(b) Example of a Manx inheritance pattern

INTERACTIVE EXERCISE

FIGURE 4.15 **The Manx cat, which carries a lethal allele.** **(a)** Photo of a Manx cat, which typically has a shortened tail. **(b)** Outcome of a cross between two Manx cats. Animals that are homozygous for the dominant Manx allele (*M*) die during early embryonic development.

CONCEPT CHECK: Why do you think the heterozygote survives with developmental abnormalities, whereas the homozygote dies?

period of development before the organism dies. In the case of the Manx allele, the homozygote dies early in embryonic development. Certain inherited diseases in humans, such as Tay-Sachs (see Table 4.2), result in death during childhood. However, some lethal alleles may exert their effects later in life, or only under certain environmental conditions. For example, a human genetic disease known as Huntington disease is caused by a dominant allele. The disease is characterized by a progressive degeneration of the nervous system, dementia, and early death. The age when these symptoms appear, or the **age of onset,** is usually between 30 and 50.

4.7 COMPREHENSION QUESTION

1. The Manx phenotype in cats is caused by a dominant allele that is lethal in the homozygous state. A Manx cat is crossed to a normal (non-Manx) cat. What is the expected outcome for the surviving offspring?
   a. All Manx offspring
   b. All normal offspring
   c. 1 normal offspring : 1 Manx offspring
   d. 1 normal offspring : 2 Manx offspring

4.8 PLEIOTROPY

Learning Outcome:

1. Explain the phenomenon of pleiotropy.

Before ending our discussion of single-gene inheritance patterns, let's take a broader look at how a single gene may affect phenotype. Although we tend to discuss genes within the context of how they influence a single trait, most genes actually have multiple effects throughout a cell or throughout a multicellular organism. The multiple effects of a single gene on the phenotype of an organism is called **pleiotropy.** Pleiotropy occurs for several reasons, including the following:

1. The expression of a single gene can affect cell function in more than one way. For example, a defect in a microtubule protein may affect cell division and cell movement.
2. A gene may be expressed in different cell types in a multicellular organism.
3. A gene may be expressed at different stages of development.

In all or nearly all cases, the expression of a gene is pleiotropic with regard to the characteristics of an organism. The expression of any given gene influences the expression of many other genes in the genome, and vice versa. Pleiotropy is revealed when researchers study the effects of gene mutations. As an example of a pleiotropic mutation, let's consider cystic fibrosis, which is a recessive human disorder. In the late 1980s, the gene for cystic fibrosis was identified. It encodes a protein called the cystic fibrosis transmembrane conductance regulator (CFTR), which regulates ionic balance by allowing the transport of chloride ions (Cl^-) across epithelial cell membranes.

The mutation that causes cystic fibrosis diminishes the function of this Cl^- transporter, affecting several parts of the body in different ways. Because the movement of Cl^- affects water transport across membranes, the most severe symptom of cystic fibrosis is thick mucus in the lungs that occurs because of a water imbalance. In sweat glands, the normal Cl^- transporter has the function of recycling salt out of the glands and back into the skin before it can be lost to the outside world. Persons with cystic fibrosis have excessively salty sweat due to their inability to recycle salt back into their skin cells—a common test for cystic fibrosis is measurement of salt on the skin. Another effect is seen in the reproductive system of males who are homozygous for the cystic fibrosis allele. Males with cystic fibrosis may be infertile because the vas deferens, the tubules that transport sperm from the testes, may be absent or undeveloped. Presumably, a normally functioning Cl^- transporter is needed for the proper development of the vas deferens in the embryo. Taken together, we can see that a defect in CFTR has multiple effects throughout the body.

4.8 COMPREHENSION QUESTION

1. Which of the following is a possible explanation for pleiotropy?
   a. The expression of a single gene can affect cell function in more than one way.
   b. A gene may be expressed in different cell types in a multicellular organism.
   c. A gene may be expressed at different stages of development.
   d. All of the above are possible explanations.

4.9 GENE INTERACTIONS

Learning Outcomes:

1. Define gene interaction.
2. Predict the outcome of crosses involving epistasis, complementation, gene modifiers, and gene redundancy.
3. Describe examples that explain the molecular mechanisms of epistasis, complementation, gene modifier effects, and gene redundancy.

Thus far, we have considered the effects of single genes on the outcome of traits. This approach helps us to understand the various ways that alleles influence traits. Researchers often examine the effects of a single gene on the outcome of a single trait as a way to simplify the genetic analysis. For example, Mendel studied one gene that affected the height of pea plants—tall versus dwarf alleles. Actually, many other genes in pea plants also affect height, but Mendel did not happen to study variants in those other height genes. How then did Mendel study the effects of a single gene? The answer lies in the genotypes of his strains. Although many genes affect the height of pea plants, Mendel chose true-breeding strains that differed with regard to only one of those genes. As a hypothetical example, let's suppose that pea plants have 10 genes affecting height, which we will call *K*, *L*, *M*, *N*, *O*, *P*, *Q*, *R*, *S*, and *T*. The genotypes of two hypothetical strains of pea plants may be

Tall strain: *KK LL MM NN OO PP QQ RR SS* ***TT***
Dwarf strain: *KK LL MM NN OO PP QQ RR SS* ***tt***

In this example, the alleles affecting height differ at only a single gene. One strain is *TT* and the other is *tt*, and this accounts for the difference in their height. If we make crosses between these tall and dwarf strains, the genotypes of the F_2 offspring may differ with regard to only one gene; the other nine genes will be identical in all of them. This approach allows a researcher to study the effects of a single gene even though many genes may affect a single trait.

Researchers now appreciate that essentially all traits are affected by the contributions of many genes. Morphological features such as height, weight, growth rate, and pigmentation are all affected by the expression of many different genes in combination with environmental factors. In this section, we will further our understanding of genetics by considering how the allelic variants of two different genes affect a single trait. This phenomenon is known as **gene interaction.** Table 4.3 considers several examples in which two different genes interact to influence the outcome of particular traits. In this section, we will examine these examples in greater detail.

TABLE 4.3
Types of Mendelian Inheritance Patterns Involving Two Genes

Type	Description
Epistasis	An inheritance pattern in which the alleles of one gene mask the phenotypic effects of the alleles of a different gene.
Complementation	A phenomenon in which two parents that express the same or similar recessive phenotypes produce offspring with a wild-type phenotype.
Gene modifier effects	A phenomenon in which an allele of one gene modifies the phenotypic outcome of the alleles of a different gene.
Gene redundancy	A pattern in which the loss of function in a single gene has no phenotypic effect, but the loss of function of two genes has an effect. Functionality of only one of the two genes is necessary for a normal phenotype; the genes are functionally redundant.

A Cross Involving a Two-Gene Interaction Can Produce Four Distinct Phenotypes

The first case of two different genes interacting to affect a single trait was discovered by William Bateson and Reginald Punnett in 1906 while they were investigating the inheritance of comb morphology in chickens. Several common varieties of chickens have combs with different morphologies, as illustrated in **Figure 4.16a**. In their studies, Bateson and Punnett crossed a Wyandotte breed having a rose comb to a Brahma having a pea comb. All F_1 offspring had a walnut comb.

When these F_1 offspring were mated to each other, the F_2 generation consisted of chickens with four types of combs in the following phenotypic ratio: 9 walnut : 3 rose : 3 pea : 1 single comb. As we have seen in Chapter 2, a 9:3:3:1 ratio is obtained in the F_2 generation when the F_1 generation is heterozygous for two different genes and these genes assort independently. However, an important difference here is that we have four distinct categories of a single trait. Based on the 9:3:3:1 ratio, Bateson and Punnett reasoned that a single trait (comb morphology) was determined by two different genes.

R (rose comb) is dominant to *r*.
P (pea comb) is dominant to *p*.
R and *P* (walnut comb) are codominant.
rrpp produces a single comb.

As shown in the Punnett square of **Figure 4.16b**, each of the genes exists in two alleles, and the two genes show independent assortment.

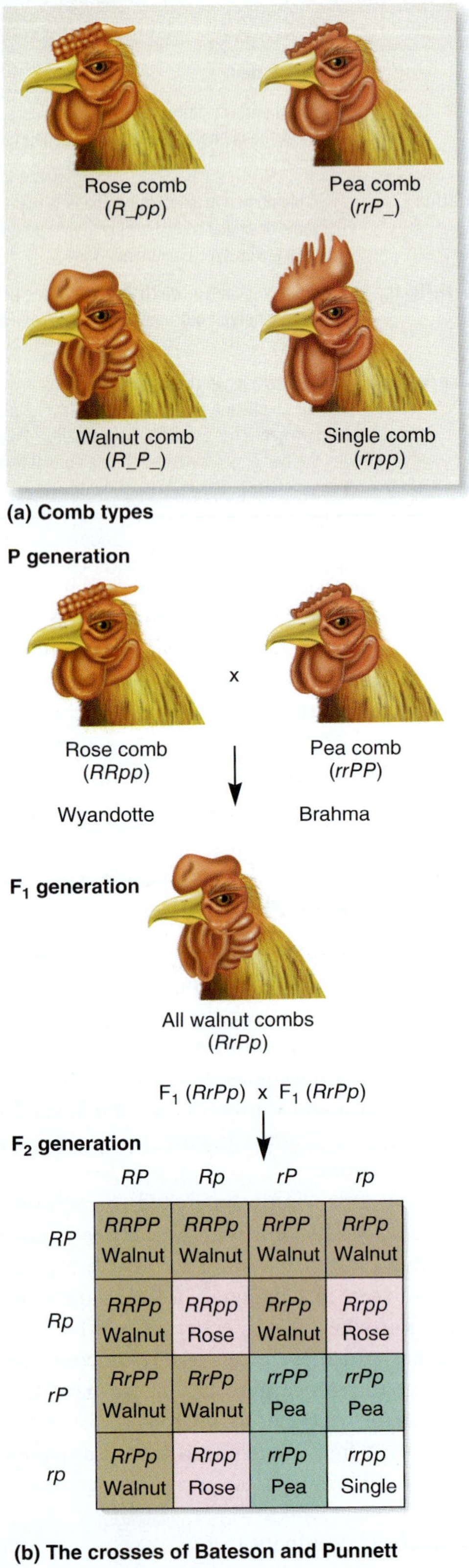

INTERACTIVE EXERCISE

FIGURE 4.16 Inheritance of comb morphology in chickens. This trait is influenced by two different genes, which each exist in two alleles. **(a)** Four phenotypic outcomes are possible. The underline symbol indicates that the allele could be either dominant or recessive. **(b)** The crosses of Bateson and Punnett examined the interaction of the two genes.

A Cross Involving a Two-Gene Interaction Can Produce Two Distinct Phenotypes Due to Epistasis

Bateson and Punnett also discovered an unexpected gene interaction when studying crosses involving the sweet pea, *Lathyrus odoratus.* The wild sweet pea has purple flowers. However, they obtained several true-breeding mutant varieties with white flowers. Not surprisingly, when they crossed a true-breeding purple-flowered plant to a true-breeding white-flowered plant, the F_1 generation contained all purple-flowered plants and the F_2 generation (produced by self-fertilization of the F_1 generation) consisted of purple- and white-flowered plants in a 3:1 ratio.

A surprising result came in an experiment where they crossed two different varieties of white-flowered plants (**Figure 4.17**). All of the F_1 generation plants had purple flowers! Bateson and Punnett then allowed the F_1 offspring to self-fertilize. The F_2 generation resulted in purple and white flowers in a ratio of 9 purple

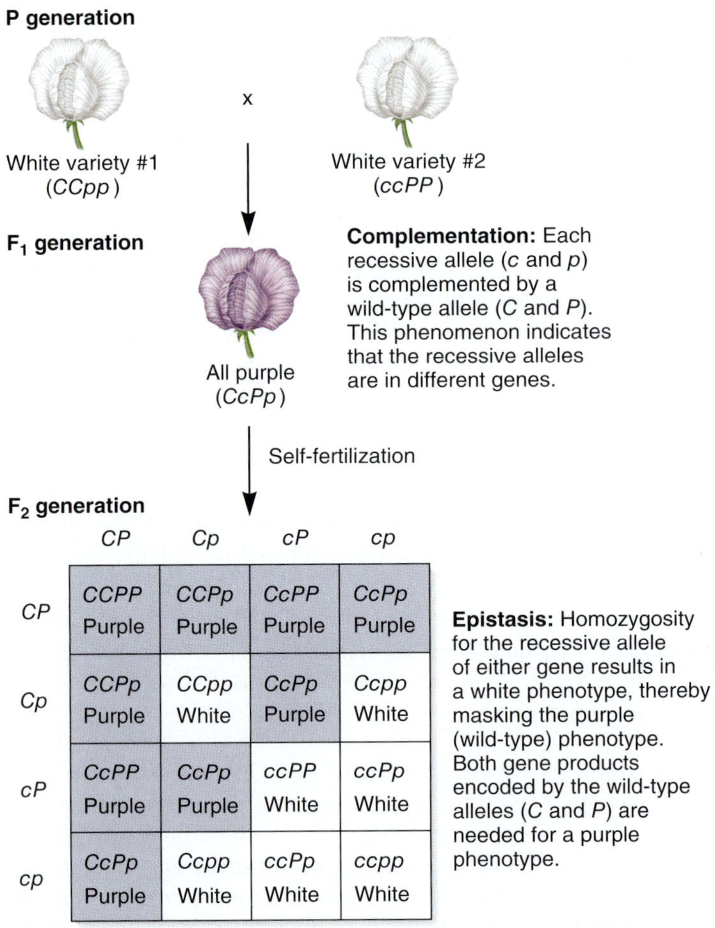

INTERACTIVE EXERCISE

FIGURE 4.17 A cross between two different white varieties of the sweet pea.

Genes→Traits The color of the sweet pea flower is controlled by two genes, which are epistatic to each other and show complementation. Each gene is necessary for the production of an enzyme required for pigment synthesis. The recessive allele of either gene encodes a defective enzyme. If an individual is homozygous recessive for either of the two genes, the purple pigment cannot be synthesized. This results in a white phenotype.

CONCEPT CHECK: What do the terms epistasis and complementation mean?

to 7 white. From this result, Bateson and Punnett deduced that two different genes were involved, with the following relationship:

> *C* (one purple-color-producing) allele is dominant to *c* (white).
>
> *P* (another purple-color-producing) allele is dominant to *p* (white).
>
> *cc* or *pp* masks the *P* or *C* alleles, producing white color.

When the alleles of one gene mask the phenotypic effects of the alleles of another gene, the phenomenon is called **epistasis.** Geneticists consider epistasis relative to a particular phenotype. If possible, geneticists use the wild-type phenotype as their reference phenotype when describing an epistatic interaction. In this case, purple flowers are wild type. Homozygosity for the white allele of one gene masks the expression of the purple-producing allele of another gene. In other words, the *cc* genotype is epistatic to a purple phenotype, and the *pp* genotype is also epistatic to a purple phenotype. At the level of genotypes, *cc* is epistatic to *PP* or *Pp*, and *pp* is epistatic to *CC* or *Cc*. This is an example of **recessive epistasis.** As seen in Figure 4.17, this epistatic interaction produces only two phenotypes—purple or white flowers—in a 9:7 ratio.

Epistatis often occurs because two (or more) different proteins participate in a common function. For example, two or more proteins may be part of an enzymatic pathway leading to the formation of a single product. To illustrate this idea, let's consider the formation of a purple pigment in the sweet pea.

$$\text{Colorless precursor} \xrightarrow{\text{Enzyme C}} \text{Colorless intermediate} \xrightarrow{\text{Enzyme P}} \textbf{Purple pigment}$$

In this example, a colorless precursor molecule must be acted on by two different enzymes to produce the purple pigment. Gene *C* encodes a functional protein called enzyme C, which converts the colorless precursor into a colorless intermediate. Two copies of the recessive allele (*cc*) result in a lack of production of this enzyme in the homozygote. Gene *P* encodes a functional enzyme P, which converts the colorless intermediate into the purple pigment. If an individual is homozygous for either recessive allele (*cc* or *pp*), it will not make any functional enzyme C or enzyme P, respectively. When one of these enzymes is missing, purple pigment cannot be made, and the flowers remain white.

The parental cross shown in Figure 4.17 illustrates another genetic phenomenon called **complementation.** This term refers to the production of offspring with a wild-type phenotype from parents that both display the same or similar recessive phenotype. In this case, purple-flowered F_1 offspring were obtained from two white-flowered parents. Complementation typically occurs because the recessive phenotype in the parents is due to homozgyosity at two different genes. In our sweet pea example, one parent is *CCpp* and the other is *ccPP*. In the F_1 offspring, the *C* and *P* alleles, which are wild-type and dominant, complement the *c* and *p* alleles, which are recessive. The offspring must have one wild-type allele of both genes to display the wild-type phenotype. Why is complementation an important experimental observation? When geneticists observe complementation in a genetic

F_1 generation *AaCc* × *AaCc* (Agouti)

F_2 generation

Egg \ Sperm	*AC*	*Ac*	*aC*	*ac*
AC	*AACC* Agouti	*AACc* Agouti	*AaCC* Agouti	*AaCc* Agouti
Ac	*AACc* Agouti	*AAcc* Albino	*AaCc* Agouti	*Aacc* Albino
aC	*AaCC* Agouti	*AaCc* Agouti	*aaCC* Black	*aaCc* Black
ac	*AaCc* Agouti	*Aacc* Albino	*aaCc* Black	*aacc* Albino

FIGURE 4.18 Inheritance pattern of coat color in rats involving a gene interaction between the agouti gene (*A* or *a*) and the colored gene (*C* or *c*).

cross, the results suggest that the recessive phenotype in the two parent strains is caused by mutant alleles in two different genes.

A Cross Involving a Two-Gene Interaction Can Produce Three Distinct Phenotypes

Thus far, we have observed two different gene interactions: one producing four phenotypes and the other producing two. Coat color in rodents provides an example that produces three phenotypes. If a true-breeding black rat is crossed to a true-breeding albino rat, the offspring have agouti coat color. The hairs exhibit black pigmentation at the tips that changes to brown pigmentation near the root. If two agouti animals of the F_1 generation are crossed to each other, they produce agouti, black, and albino offspring in a 9:3:4 ratio (**Figure 4.18**).

How do we explain this ratio? This cross involves two genes that are called *A* (for agouti) and *C* (for colored). The dominant *A* allele of the agouti gene encodes a protein that regulates hair color such that the pigmentation shifts from black (eumelanin) at the tips to yellow (pheomelanin) and black near the roots. The mixture of yellow and black produces brown. The recessive allele, *a*, inhibits the expression of yellow pigmentation and thereby results in black pigment production throughout the entire hair, when an animal is *aa*. The colored gene encodes tyrosinase, which is an enzyme needed for the first step in melanin synthesis. The *C* allele allows pigmentation to occur, whereas the *c* allele causes the loss of tyrosinase function. The *C* allele is dominant to the *c* allele; *cc* homozygotes are albino and have white coat color.

As shown at the top of Figure 4.18, the F_1 rats are heterozygous for the two genes. In this case, *C* is dominant to *c*, and *A* is dominant to *a*. If a rat has at least one copy of both dominant alleles, the result is agouti coat color. Let's consider agouti as our reference phenotype. In the F_2 generation, if a rat has a dominant

A allele but is *cc* homozygous, it will be albino and develop a white coat. The *c* allele is epistatic to *A* and masks pigment production.

By comparison, if an individual has a dominant *C* allele and is homozygous *aa*, the coat color is black. How can we view the effects of the *aa* genotype when an individual carries a *C* allele? Because the *aa* genotype actually masks yellow pigmentation, the black phenotype is viewed by most geneticists as an example of epistasis.

EXPERIMENT 4A

Bridges Observed an 8:4:3:1 Ratio Because the Cream-Eye Gene Modifies the X-Linked Eosin Allele But Not the Red or White Alleles

As we have seen, geneticists view epistasis as a situation in which the alleles of a given gene mask the phenotypic effects of the alleles of another gene. In some cases, however, two genes may interact to influence a particular phenotype, but the interaction of particular alleles seems to modify the phenotype, not mask it. This is called a **gene modifier effect**—the alleles of one gene modify the phenotypic effect of the alleles of a different gene.

Calvin Bridges discovered an early example in which one gene modifies the phenotypic effects of an X-linked eye color gene in *Drosophila*. As discussed in Chapter 3, the X-linked red allele (w^+) is dominant to the white allele (w). In addition to these two alleles, Bridges found another allele of this gene that he called eosin (*w-e*), which results in eyes that are a pale orange color. The red allele is dominant to the eosin allele. The expression of the eosin allele depends on the number of copies of the allele. When females have two copies of this allele, they have eosin eyes. When females are heterozygous for the eosin allele and white allele, they have light-eosin eyes. Similarly, males with a single copy of the eosin allele have light-eosin eyes.

Within a true-breeding culture of flies with eosin eyes, Bridges found a fly with cream-colored eyes. Bridges reasoned that this new eye color could be explained in two different ways. One possibility is that the cream-colored phenotype could be the result of a new mutation that changed the eosin allele into a cream allele. A second possibility is that a different gene may have incurred a mutation that modified the phenotypic expression of the eosin allele. This second possibility is an example of a gene interaction. To distinguish between these two possibilities, he carried out the crosses described in **Figure 4.19**. He crossed males with cream-colored eyes to wild-type females and then allowed the F_1 generation flies, which all had red eyes, to mate with each other. As shown in the data, all F_2 females had red eyes, but males had red eyes, light-eosin eyes, or cream eyes.

THE HYPOTHESES

Cream-colored eyes in fruit flies are due to the effect of an allele that is in the same gene as the eosin allele or in a second gene that modifies the expression of the eosin allele.

TESTING THE HYPOTHESES

FIGURE 4.19 **A gene interaction between the cream allele and eosin allele in fruit flies.**

INTERACTIVE EXERCISE

Starting material: From a culture of flies with eosin eyes, Bridges obtained a fly with cream-colored eyes and used it to produce a true-breeding culture of flies with cream-colored eyes. The allele was called *cream a* (c^a).

	Experimental level	Conceptual level
1. Cross males with cream-colored eyes to wild-type females.	P generation	$CCX^{w^+}X^{w^+}$ x $c^ac^aX^{w-e}Y$
2. Observe the F_1 offspring and then allow the offspring to mate with each other.	F_1 generation	Predicted outcome: $Cc^aX^{w^+}X^{w-e}$ and $Cc^aX^{w^+}Y$ Red-eyed females and males Allow F_1 offspring to mate.
3. Observe and record the eye color and sex of the F_2 generation.	F_2 generation (see The Data)	Predicted outcome: See Punnett square and The Data.

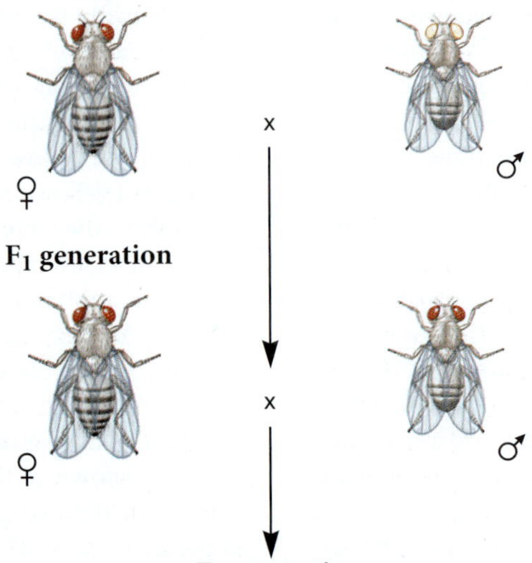

THE DATA

Cross	*Outcome*
P cross:	
Cream-eyed male × wild-type female	F_1: All red eyes
F_1 cross:	
F_1 brother × F_1 sister	F_2: 104 females with red eyes 47 males with red eyes 44 males with light-eosin eyes 14 males with cream eyes

Data from Calvin Bridges (1919) Specific modifiers of eosin eye color in *Drosophila melanogaster. J Exper Zool 28*, 337–384.

INTERPRETING THE DATA

To interpret these data, keep in mind that Bridges already knew that the eosin allele is X-linked. However, he did not know whether the cream allele was in the same gene as the eosin allele, in a different gene on the X chromosome, or on an autosome. The F_2 generation indicates that the cream allele is not in the same gene as the eosin allele. If the cream allele was in the same gene as the eosin allele, none of the F_2 males would have had light-eosin eyes; there would have been a 1:1 ratio of red-eyed males and cream-eyed males in the F_2 generation. This result was not obtained. Instead, the actual results are consistent with the idea that the male flies of the parental generation carried both the eosin and cream alleles. Therefore, Bridges concluded that the cream allele was an allele of a different gene.

One possibility is that the cream allele is an autosomal recessive allele. If so, we can let *C* represent the dominant allele (which does not modify the eosin phenotype) and c^a represent the cream allele that modifies the eosin color to cream. We already know that the eosin allele is X-linked and recessive to the red allele. The parental cross is expected to produce all red-eyed F_1 flies in which the males are $Cc^aX^{w+}Y$ and the females are $Cc^aX^{w+}X^{w\text{-}e}$. When these F_1 offspring are allowed to mate with each other, the Punnett square shown here predicts the following outcome:

$Cc^aX^{w^+}X^{w\text{-}e}$ × $Cc^aX^{w^+}Y$

Sperm

♀ Egg \ ♂ Sperm	CX^{w^+}	CY	$c^aX^{w^+}$	c^aY
CX^{w^+}	$CCX^{w^+}X^{w^+}$	$CCX^{w^+}Y$	$Cc^aX^{w^+}X^{w^+}$	$Cc^aX^{w^+}Y$
$CX^{w\text{-}e}$	$CCX^{w^+}X^{w\text{-}e}$	$CCX^{w\text{-}e}Y$	$Cc^aX^{w^+}X^{w\text{-}e}$	$Cc^aX^{w\text{-}e}Y$
$c^aX^{w^+}$	$Cc^aX^{w^+}X^{w^+}$	$Cc^aX^{w^+}Y$	$c^ac^aX^{w^+}X^{w^+}$	$c^ac^aX^{w^+}Y$
$c^aX^{w\text{-}e}$	$Cc^aX^{w^+}X^{w\text{-}e}$	$Cc^aX^{w\text{-}e}Y$	$c^ac^aX^{w^+}X^{w\text{-}e}$	$c^ac^aX^{w\text{-}e}Y$

Outcome:

1 $CCX^{w^+}X^{w^+}$: 1 $CCX^{w^+}X^{w\text{-}e}$: 2 $Cc^aX^{w^+}X^{w^+}$: 2 $Cc^aX^{w^+}X^{w\text{-}e}$: 1 $c^ac^aX^{w^+}X^{w^+}$: 1 $c^ac^aX^{w^+}X^{w\text{-}e}$ = **8 red-eyed females**

1 $CCX^{w^+}Y$: 2 $Cc^aX^{w^+}Y$: 1 $c^ac^aX^{w^+}Y$ = **4 red-eyed males**

1 $CCX^{w\text{-}e}Y$: 2 $Cc^aX^{w\text{-}e}Y$ = **3 light eosin-eyed males**

1 $c^ac^aX^{w\text{-}e}Y$ = **1 cream-eyed male**

This phenotypic outcome indicates that the specific modifier allele, c^a, can modify the phenotype of the eosin allele but not the red-eye allele. The eosin allele can be modified only when the c^a allele is homozygous. The predicted 8:4:3:1 ratio agrees reasonably well with Bridges' data.

A self-help quiz involving this experiment can be found at www.mhhe.com/brookergenetics5e.

Due to Gene Redundancy, Loss-of-Function Alleles May Have No Effect on Phenotype

During the past several decades, researchers have discovered new kinds of gene interactions by studying model organisms such as *Escherichia coli* (a bacterium), *Saccharomyces cerevisiae* (baker's yeast), *Arabidopsis thaliana* (a model plant), *Drosophila melanogaster* (fruit fly), *Caenorhabditis elegans* (a nematode worm), and *Mus musculus* (the laboratory mouse). The isolation of mutants that alter the phenotypes of these organisms has become a powerful tool for investigating gene function and has provided ways for researchers to identify new kinds of gene interactions. With the advent of modern molecular techniques (described in Chapters 20, 22, and 23), a common approach for investigating gene function is to intentionally produce loss-of-function alleles in a gene of interest. When a geneticist abolishes gene function by creating an organism that is homozygous for a loss-of-function allele, the resulting organism is said to have undergone a **gene knockout.**

Why are gene knockouts useful? The primary reason for making a gene knockout is to understand how that gene affects the structure and function of cells or the phenotypes of organisms. For example, if a researcher knocked out a particular gene in a mouse and the resulting animal was unable to hear, the researcher would suspect that the role of the functional gene is to promote the formation of ear structures that are vital for hearing.

Interestingly, by studying many gene knockouts in a variety of experimental organisms, geneticists have discovered that many

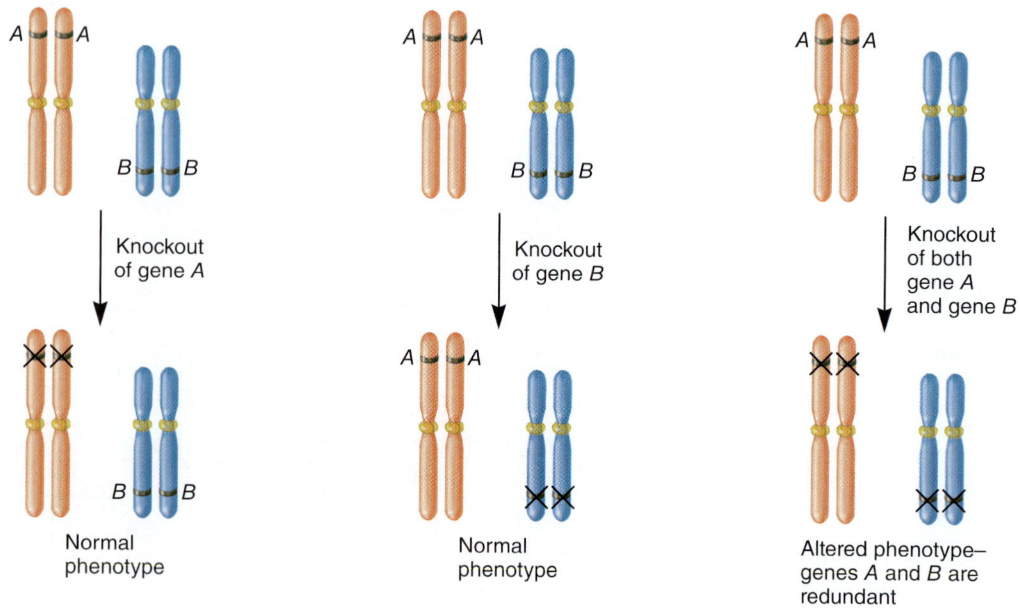

FIGURE 4.20 A molecular explanation for gene redundancy. To have a normal phenotype, an organism must have a functional copy of either gene *A* or gene *B*. If both gene *A* and gene *B* are knocked out, an altered phenotype occurs.

CONCEPT CHECK: Explain why a single gene knockout does not have an effect on the phenotype.

knockouts have no obvious effect on phenotype at the cellular level or the level of discernible traits. To explore gene function further, researchers may make two or more gene knockouts in the same organism. In some cases, gene knockouts in two different genes produce a phenotypic change even though the single knockouts have no effect (**Figure 4.20**). Geneticists may attribute this change to **gene redundancy**—the phenomenon that one gene can compensate for the loss of function of another gene.

Gene redundancy may be due to different underlying causes. One common reason is gene duplication. Certain genes have been duplicated during evolution, so a species may contain two or more copies of similar genes. These copies, which are not identical due to the accumulation of random changes during evolution, are called **paralogs** (look ahead to Figures 8.6 and 8.7). When one gene is missing, a paralog may be able to carry out the missing function. For example, genes *A* and *B* in Figure 4.20 could be paralogs of each other. Alternatively, gene redundancy may involve proteins that are involved in a common cellular function. When one of the proteins is missing due to a gene knockout, the function of another protein may be increased to compensate for the missing protein and thereby overcome the defect.

Let's explore the consequences of gene redundancy in a genetic cross. George Shull conducted one of the first studies that illustrated the phenomenon of gene redundancy. His work involved a weed known as shepherd's purse, a member of the mustard family. The trait he followed was the shape of the seed capsule, which is commonly triangular (**Figure 4.21**). Strains

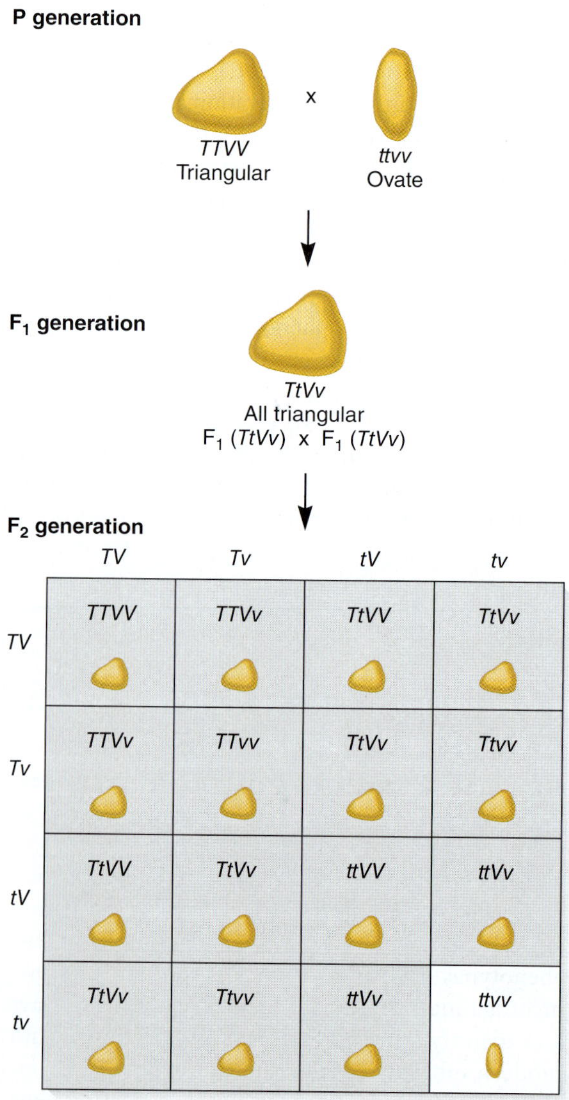

FIGURE 4.21 Inheritance of capsule shape in shepherd's purse, an example of gene redundancy. In this case, triangular shape requires a dominant allele in one of two genes, but not both. The *T* and *V* alleles are redundant.

CONCEPT CHECK: At the molecular level (with regard to loss-of-function alleles), explain why the *ttvv* homozygote has an ovate capsule phenotype.

producing smaller ovate capsules are due to loss-of-function alleles in two different genes (*ttvv*). The ovate strain is an example of a double gene knockout. When Shull crossed a true-breeding plant with triangular capsules to a plant having ovate capsules, the F_1 generation all had triangular capsules. When the F_1 plants were self-fertilized, a surprising result came in the F_2 generation. Shull observed a 15:1 ratio of plants having triangular capsules to ovate capsules. The result can be explained by gene redundancy. Having at least one functional copy of either gene (*T* or *V*) is sufficient to produce the triangular phenotype. *T* and *V* are functional alleles of redundant genes. Only one of them is necessary for a triangular shape. When the functions of both genes are knocked out, as in the *ttvv* homozygote, the capsule becomes smaller and ovate.

4.9 COMPREHENSION QUESTIONS

1. Two different strains of sweet peas are true-breeding and have white flowers. When these strains are crossed together, the F_1 offspring all have purple flowers. This phenomenon is called
   a. epistasis.
   b. complementation.
   c. incomplete dominance.
   d. incomplete penetrance.
2. If the F_1 offspring from question 1 are allowed to self-fertilize, what is the expected outcome for the F_2 offspring?
   a. All white
   b. All purple
   c. 3 purple to 1 white
   d. 9 purple to 7 white

KEY TERMS

Page 74. Mendelian inheritance, simple Mendelian inheritance
Page 76. wild-type alleles, genetic polymorphism, mutant alleles
Page 77. gain-of-function mutations, dominant-negative mutations, haploinsufficiency, incomplete penetrance, expressivity
Page 78. temperature-sensitive allele, norm of reaction
Page 80. incomplete dominance
Page 81. overdominance, heterozygote advantage
Page 82. multiple alleles, codominance
Page 84. sex chromosomes, X-linked inheritance, X-linked recessive
Page 85. reciprocal cross, sex-linked gene, hemizygous, holandric genes, pseudoautosomal inheritance
Page 86. sex-influenced inheritance, sex-limited inheritance
Page 87. sexual dimorphism, lethal allele, essential gene, nonessential genes, conditional lethal alleles, temperature-sensitive (ts) lethal alleles, semilethal alleles
Page 88. age of onset, pleiotropy
Page 89. gene interaction
Page 91. epistasis, recessive epistasis, complementation
Page 92. gene modifier effect
Page 93. gene knockout
Page 94. gene redundancy, paralogs

CHAPTER SUMMARY

- Mendelian inheritance patterns obey Mendel's laws.

4.1 Overview of Simple Inheritance Patterns

- Several inheritance patterns involving single genes differ from those observed by Mendel (see Table 4.1).

4.2 Dominant and Recessive Alleles

- Wild-type alleles are defined as those that are prevalent in a population. A gene that exists in two or more wild-type alleles is an example of genetic polymorphism (see Figure 4.1).
- Recessive alleles are often due to mutations that result in a reduction or loss of function of the encoded protein (see Figure 4.2 and Table 4.2).
- Dominant mutant alleles are most commonly caused by gain-of-function mutations, dominant negative mutations, or haploinsufficiency.
- Incomplete penetrance is a situation in which an allele that is expected to be expressed is not expressed (see Figure 4.3).
- Traits may vary in their expressivity.

4.3 Environmental Effects on Gene Expression

- The outcome of traits is influenced by the environment (see Figure 4.4).

4.4 Incomplete Dominance, Overdominance, and Codominance

- Incomplete dominance is an inheritance pattern in which the heterozygote has an intermediate phenotype (see Figure 4.5).
- Whether we judge an allele to be dominant or incompletely dominant may depend on how closely we examine the phenotype (see Figure 4.6).

- Overdominance is an inheritance pattern in which the heterozygote has greater reproductive success (see Figures 4.7, 4.8).
- Most genes exist in multiple alleles in a population. Some alleles, such as those that produce A and B blood antigens, are codominant (see Figure 4.9).

4.5 X-linked Inheritance

- X-linked inheritance patterns show differences between males and females, and are revealed in reciprocal crosses (see Figures 4.10, 4.11).
- The X and Y chromosomes carry different sets of genes, but they do have regions of short homology that can lead to pseudoautosomal inheritance (see Figure 4.12).

4.6 Sex-Influenced and Sex-Limited Inheritance

- For sex-influenced traits, heterozygous males and females have different phenotypes (see Figure 4.13).
- Sex-limited traits are expressed in only one sex, thereby resulting in sexual dimorphism (see Figure 4.14).

4.7 Lethal Alleles

- Lethal alleles most commonly occur in essential genes. Lethal alleles may result in inheritance patterns that yield unexpected ratios (see Figure 4.15).

4.8 Pleiotropy

- Single genes have pleiotropic effects.

4.9 Gene Interactions

- A gene interaction is the phenomenon in which two or more genes affect a single phenotype (see Table 4.3).
- Bateson and Punnett discovered the first case of a gene interaction affecting comb morphology in chickens (see Figure 4.16).
- Epistasis is a situation in which the allele of one gene masks the phenotypic expression of the alleles of a different gene. Complementation occurs when two individuals with similar phenotypes produce offspring with a wild-type phenotype (see Figures 4.17, 4.18).
- A gene modifier effect occurs when an allele of one gene modifies (but does not completely mask) the phenotypic effects of the alleles of a different gene. An example is the cream eye color in fruit flies observed by Bridges (see Figure 4.19).
- Two different genes may have redundant functions, which is revealed in a double gene knockout (see Figures 4.20, 4.21).

PROBLEM SETS & INSIGHTS

Solved Problems

S1. In humans, why are X-linked recessive traits more likely to occur in males than females?

Answer: Because a male is hemizygous for X-linked traits, the phenotypic expression of X-linked traits depends on only a single copy of the gene. When a male inherits a recessive X-linked allele, he will automatically exhibit the trait because he does not have another copy of the gene on the corresponding Y chromosome. This phenomenon is particularly relevant to the inheritance of recessive X-linked alleles that cause human disease. (Some examples are described in Chapter 24.)

S2. In Ayrshire cattle, the spotting pattern of the animals can be either red and white or mahogany and white. The mahogany and white pattern is caused by the allele *M*. The red and white phenotype is controlled by the allele *m*. When mahogany and white animals are mated to red and white animals, the following results are obtained:

Genotype	Phenotype	
	Females	**Males**
MM	Mahogany and white	Mahogany and white
Mm	Red and white	Mahogany and white
mm	Red and white	Red and white

Explain the pattern of inheritance.

Answer: The inheritance pattern for this trait is sex-influenced inheritance. The *M* allele is dominant in males but recessive in females, whereas the *m* allele is dominant in females but recessive in males.

S3. The following pedigree involves a single gene causing an inherited disease. If you assume that incomplete penetrance is *not* occurring, indicate which modes of inheritance are *not* possible. (Affected individuals are shown as filled symbols.)

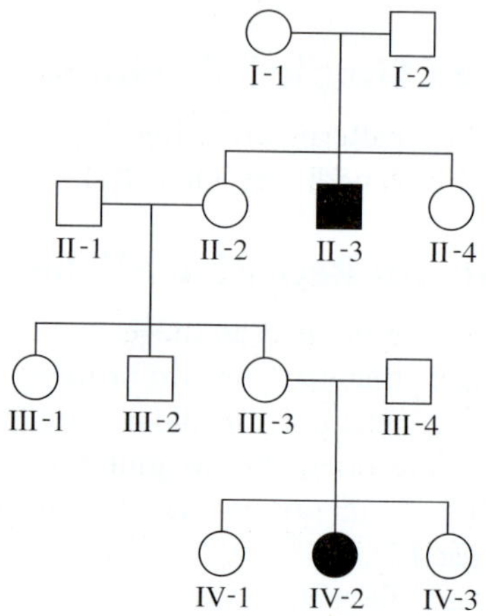

A. Recessive

B. Dominant

C. X-linked, recessive

D. Sex-influenced, dominant in females

E. Sex-limited, recessive in females

Answer:

A. It could be recessive.

B. It cannot be dominant (completely penetrant) because affected offspring have two unaffected parents.

C. It could not be X-linked recessive because individual IV-2 does not have an affected father.

D. It could not be sex-influenced, dominant in females because individual II-3 (who would have to be homozygous) has an unaffected mother (who would have to be heterozygous and affected).

E. It is not sex-limited because individual II-3 is an affected male and IV-2 is an affected female.

S4. Red-green color blindness is inherited as a recessive X-linked trait. What are the following probabilities?

A. A woman with phenotypically normal parents and a color-blind brother will have a color-blind son. Assume that she has no previous children.

B. A phenotypically normal woman, who has already had one color-blind son, will have a color-blind son.

C. A phenotypically normal woman, who has already had one color-blind son, and who is married to a color-blind man, will have a color-blind daughter.

Answer:

A. The woman's mother must have been a heterozygote. So there is a 50% chance that the woman is a carrier. If she has children, 1/4 (i.e., 25%) will be affected sons if she is a carrier. However, there is only a 50% chance that she is a carrier. We multiply 50% times 25%, which equals $0.5 \times 0.25 = 0.125$, or a 12.5% chance.

B. If she already had a color-blind son, then we know she must be a carrier, so the chance is 25%.

C. The woman is heterozygous and her husband is hemizygous for the color-blind allele. This couple will produce 1/4 offspring that are color-blind daughters. The rest are 1/4 carrier daughters, 1/4 normal sons, and 1/4 color-blind sons. The answer is 25%.

S5. Two pink-flowered four-o'clocks were crossed to each other. What are the following probabilities for the offspring?

A. A plant will be red-flowered.

B. The first three plants examined will be white.

C. A plant will be either white or pink.

D. A group of six plants composed of one pink, two whites, and three reds.

Answer: The first thing we need to do is construct a Punnett square to determine the individual probabilities for each type of offspring.

Because flower color is incompletely dominant, the cross is $Rr \times Rr$.

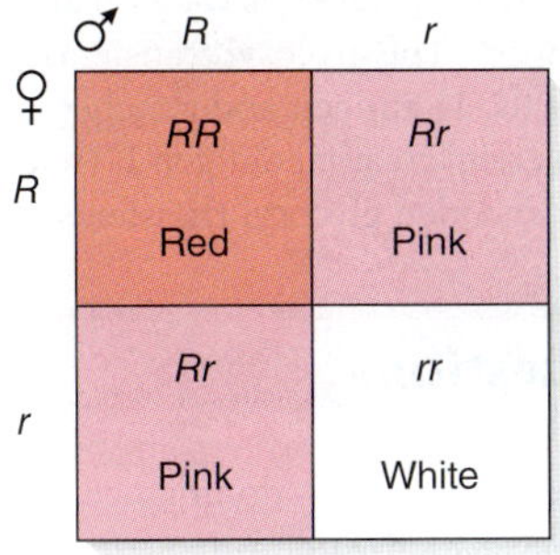

The phenotypic ratio is 1 red to 2 pink to 1 white. In other words, 1/4 are expected to be red, 1/2 pink, and 1/4 white.

A. The probability of a red-flowered plant is 1/4, which equals 25%.

B. Use the product rule.

$$1/4 \times 1/4 \times 1/4 = 1/64, \text{ or } 1.6\%$$

C. $1/4 + 1/2 = 3/4 = 75\%$

D. Use the multinomial expansion equation. See Chapter 2 for an explanation of the multinomial expansion equation. In this case, three phenotypes are possible.

$$P = \frac{n!}{a!b!c!}\, p^a q^b r^c$$

where

n = total number of offspring = 6
a = number of reds = 3
p = probability of reds = (1/4)
b = number of pinks = 1
q = probability of pinks = (1/2)
c = number of whites = 2
r = probability of whites = (1/4)

If we substitute these values into the equation,

$$P = \frac{6!}{3!1!2!}\,(1/4)^3(1/2)^1(1/4)^2$$
$$P = 0.029, \text{ or } 2.9\%$$

This means that 2.9% of the time we would expect to obtain six plants, three with red flowers, one with pink flowers, and two with white flowers.

S6. A woman with type O blood gives birth to a daughter with type B blood. The father of this daughter does not have type B or type AB blood. How is this possible?

Answer: This is an example of epistasis. In this case, the father could have been I^Bi, I^BI^B, or I^AI^B and passed the I^B allele to his daughter. In addition, the father also was homozgyous for a recessive, loss-of-function allele of a different gene (the *H* gene). This father is *hh*, whereas most people are *HH*. The functional *H* allele encodes a glycosyl transferase that is needed to attach the sugar fucose onto the oligosaccharide and thereby make H antigen, which is the antigen found in type O people.

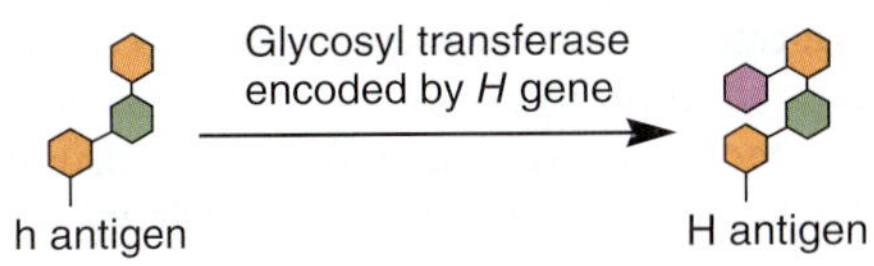

An individual who is *hh* is unable to attach this fucose and the resulting oligosaccharide, which is smaller, is called h antigen. Why is the *hh* genotype epistatic to I^B? The glycosyl transferase that is encoded by the I^B allele recognizes the H antigen and attaches a galactose sugar to the oligosaccharide, thereby creating antigen B (refer back to Figure 4.9c). However, the I^B-encoded glycosyl transferase does not recognize h antigen due to the missing fucose. Therefore, even though this father carries the I^B allele, the type B phenotype is masked by the *hh* genotype. Likewise, the *hh* genotype also masks the I^A allele. The *hh* genotype, which is very rare, is also called the Bombay genotype because the first case was reported in Mumbai (formerly Bombay).

Conceptual Questions

C1. Describe the differences among dominance, incomplete dominance, codominance, and overdominance.

C2. Discuss the differences among sex-influenced, sex-limited, and sex-linked inheritance. Give examples.

C3. What is meant by a gene interaction? How can a gene interaction be explained at the molecular level?

C4. Let's suppose a recessive allele encodes a completely defective protein. If the functional allele is dominant, what does that tell you about the amount of the functional protein that is sufficient to cause the phenotype? What if the allele shows incomplete dominance?

C5. A nectarine is a peach without the fuzz. The difference is controlled by a single gene that is found in two alleles, *D* and *d*. At the molecular level, would it make more sense to you that the nectarine is homozygous for a recessive allele or that the peach is homozygous for the recessive allele? Explain your reasoning.

C6. An allele in *Drosophila* produces a "star-eye" trait in the heterozygous individual. However, the star-eye allele is lethal in homozygotes. What would be the ratio and phenotypes of surviving flies if star-eyed flies were crossed to each other?

C7. A seed dealer wants to sell four-o'clock seeds that will produce only a single color of flowers (red, white, or pink). Explain how this should be done.

C8. The blood serum from one individual (let's call this person individual 1) is known to agglutinate the red blood cells from a second individual (individual 2). List the pairwise combinations of possible genotypes that individuals 1 and 2 could be. If individual 1 is the parent of individual 2, what are his or her possible genotypes?

C9. Which blood phenotypes (A, B, AB, and/or O) provide an unambiguous genotype? Is it possible for a couple to produce a family of children with all four blood types? If so, what would the genotypes of the parents have to be?

C10. A woman with type B blood has a child with type O blood. What are the possible genotypes and blood types of the father?

C11. A type A woman is the daughter of a type O father and type A mother. If she has children with a type AB man, what are the following probabilities?

A. A type AB child

B. A type O child

C. The first three children with type AB blood

D. A family composed of two children with type B blood and one child with type AB

C12. In Shorthorn cattle, coat color is controlled by a single gene that can exist as a red allele (*R*) or white allele (*r*). The heterozygotes (*Rr*) have a color called roan that looks less red than the *RR* homozygotes. However, when examined carefully, the roan phenotype in cattle is actually due to a mixture of completely red hairs and completely white hairs. Should this be called incomplete dominance, codominance, or something else? Explain your reasoning.

C13. In chickens, the Leghorn variety has white feathers due to an autosomal dominant allele. Silkies have white feathers due to a recessive allele in a second (different) gene. If a true-breeding white Leghorn is crossed to a true-breeding white Silkie, what is the expected phenotype of the F_1 generation? If members of the F_1 generation are mated to each other, what is the expected phenotypic outcome of the F_2 generation? Assume the chickens in the parental generation are homozygous for the white allele at one gene and homozygous for the brown allele at the other gene. In subsequent generations, nonwhite birds will be brown.

C14. Propose the most likely mode of inheritance (autosomal dominant, autosomal recessive, or X-linked recessive) for the following pedigrees. Affected individuals are shown with filled (black) symbols.

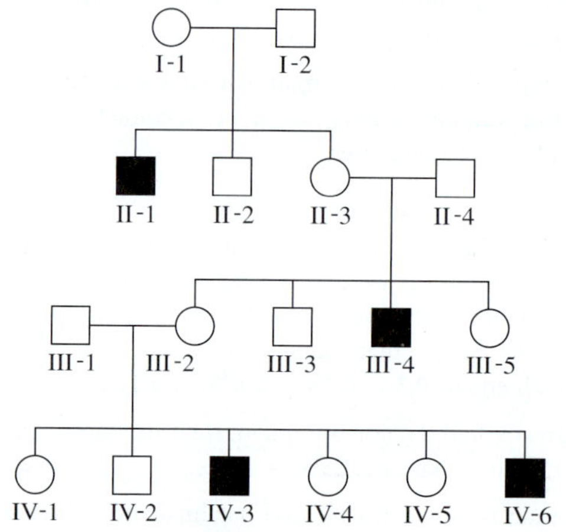

(a)

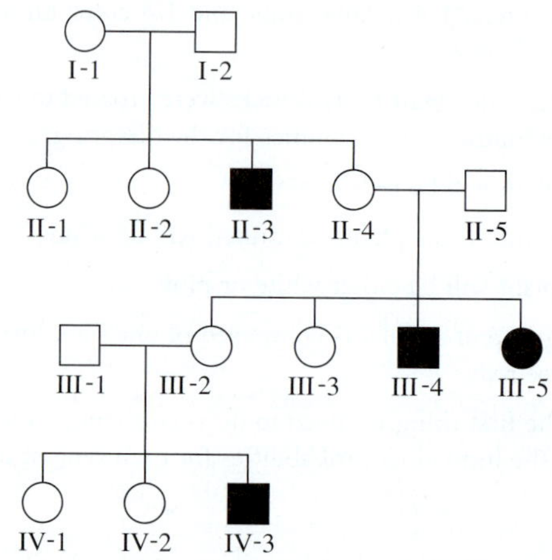

(b)

C15. A human disease known as vitamin D-resistant rickets is inherited as an X-linked dominant trait. If a male with the disease produces children with a female who does not have the disease, what is the expected ratio of affected and unaffected offspring?

C16. Hemophilia is an X-linked recessive trait in humans. If a heterozygous woman has children with an unaffected man, what is the probability of the following combinations of children?

A. An affected son

B. Four unaffected offspring in a row

C. An unaffected daughter or son

D. Two affected offspring out of five

C17. Incontinentia pigmenti is a rare, X-linked dominant disorder in humans characterized by swirls of pigment in the skin. If an affected female, who had an unaffected father, has children with an unaffected male, what would be the predicted ratios of affected and unaffected sons and daughters?

C18. With regard to scurs in cattle (a sex-influenced trait), a cow with no scurs whose mother had scurs had offspring with a bull with scurs whose father had no scurs. What are their probabilities of having the following combination of offspring?

A. Their first offspring will not have scurs.

B. Their first offspring will be a male with no scurs.

C. Their first three offspring will be females with no scurs.

C19. In rabbits, the color of body fat is controlled by a single gene with two alleles, designated *Y* and *y*. The outcome of this trait is affected by the diet of the rabbit. When raised on a standard vegetarian diet, the dominant *Y* allele confers white body fat, and the *y* allele confers yellow body fat. However, when raised on a xanthophyll-free diet, the homozygote *yy* animal has white body fat. If a heterozygous animal is crossed to a rabbit with yellow body fat, what are the proportions of offspring with white and yellow body fat when raised on a standard vegetarian diet? How do the proportions change if the offspring are raised on a xanthophyll-free diet?

C20. In cats, a temperature-sensitive allele produces the Siamese phenotype in which the cooler extremities are dark and the warmer trunk area is lighter. A Siamese cat that spends most of its time outside was accidentally injured in a trap and required several stitches in its right front paw. The veterinarian had to shave the fur from the paw and leg, which originally had rather dark fur. Later, when the fur grew back, it was much lighter than the fur on the other three legs. Do you think this injury occurred in the hot summer or cold winter? Explain your answer.

C21. A true-breeding male fly with light-eosin eyes is crossed to a white-eyed female that is heterozygous for the wild-type (*C*) and cream alleles (c^a). What are the expected proportions of their offspring?

C22. The trait of hen- versus cock-feathering is a sex-limited trait controlled by a single gene. Females always exhibit hen-feathering as do *HH* and *Hh* males. Only *hh* males show cock-feathering. Starting with two heterozygous fowl that are hen-feathered, explain how you would obtain a true-breeding line that always produced cock-feathered males.

C23. In the pedigree shown here for a trait determined by a single gene (affected individuals are shown in black), state whether it would be possible for the trait to be inherited in each of the following ways:

A. Recessive

B. X-linked recessive

C. Dominant, complete penetrance

D. Sex-influenced, dominant in males

E. Sex-limited

F. Dominant, incomplete penetrance

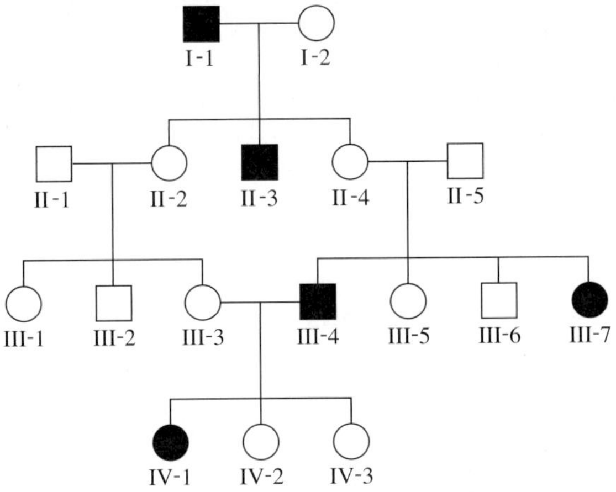

C24. The pedigree shown here also concerns a trait determined by a single gene (affected individuals are shown in black). Which of the following patterns of inheritance are possible?

A. Recessive

B. X-linked recessive

C. Dominant

D. Sex-influenced, recessive in males

E. Sex-limited

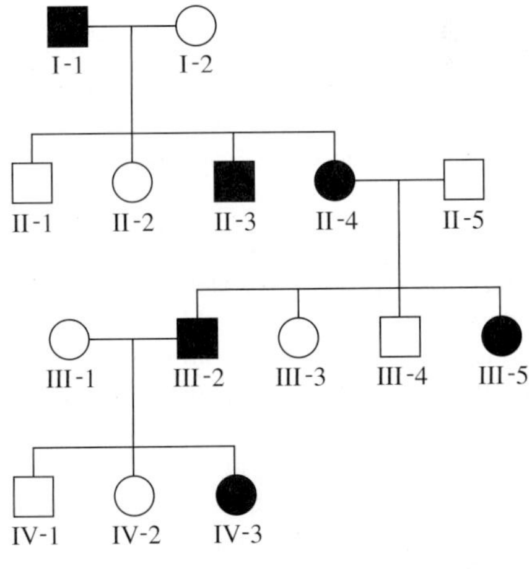

C25. Let's suppose you have pedigree data from thousands of different families involving a particular genetic disease. How would you decide whether the disease is inherited as a recessive trait as opposed to one that is dominant with incomplete penetrance?

C26. Compare phenotypes at the molecular, cellular, and organism levels for individuals who are homozygous for the hemoglobin allele, Hb^AHb^A, and the sickle cell allele, Hb^SHb^S.

C27. A very rare dominant allele that causes the little finger to be crooked has a penetrance value of 80%. In other words, 80% of heterozygotes carrying the allele will have a crooked little finger. If a homozygous unaffected person has children with a heterozygote carrying this mutant allele, what is the probability that an offspring will have little fingers that are crooked?

C28. A sex-influenced trait in humans is one that affects the length of the index finger. A "short" allele is dominant in males and recessive in females. Heterozygous males have an index finger that is significantly shorter than the ring finger. The gene affecting index finger length is located on an autosome. A woman with short index fingers has children with a man who has normal index fingers. They produce five children in the following order: female, male, male, female, male. The oldest female offspring marries a man with normal fingers and then has one daughter. The youngest male among the five children marries a woman with short index fingers, and then they have two sons. Draw the pedigree for this family. Indicate the phenotypes of every individual (filled symbols for individuals with short index fingers and open symbols for individuals with normal index fingers).

C29. Three coat-color patterns that occur in some breeds of horses are termed cremello (beige), chestnut (brown), and palomino (golden with light mane and tail). If two palomino horses are mated, they produce about 1/4 cremello, 1/4 chestnut, and 1/2 palomino offspring. In contrast, cremello horses and chestnut horses breed true. (In other words, two cremello horses will produce only cremello offspring, and two chestnut horses will produce only chestnut offspring.) Explain this pattern of inheritance.

Experimental Questions

E1. Mexican hairless dogs have little hair and few teeth. When a Mexican hairless is mated to another breed of dog, about half of the puppies are hairless. When two Mexican hairless dogs are mated to each other, about 1/3 of the surviving puppies have hair, and about 2/3 of the surviving puppies are hairless. However, about two out of eight puppies from this type of cross are born grossly deformed and do not survive. Explain this pattern of inheritance.

E2. In chickens, some varieties have feathered shanks (legs), but others do not. In a cross between a Black Langhans (feathered shanks) and Buff Rocks (unfeathered shanks), the shanks of the F_1 generation are all feathered. When members of the F_1 generation are crossed to each other, the F_2 generation contains chickens with feathered shanks to unfeathered shanks in a ratio of 15:1. Suggest an explanation for this result.

E3. In sheep, the formation of horns is a sex-influenced trait; the allele that results in horns is dominant in males and recessive in females. Females must be homozygous for the horned allele to have horns. A horned ram was crossed to a polled (unhorned) ewe, and the first offspring they produced was a horned ewe. What are the genotypes of the parents?

E4. A particular breed of dog can have long hair or short hair. When true-breeding long-haired animals were crossed to true-breeding short-haired animals, the offspring all had long hair. The F_2 generation produced a 3:1 ratio of long- to short-haired offspring. A second gene affects the texture of the hair. The two variants are wiry hair and straight hair. F_1 offspring from a cross of these two varieties all had wiry hair, and F_2 offspring showed a 3:1 ratio of wiry-haired to straight-haired puppies. Recently, a breeder of the short-, wiry-haired dogs found a female puppy that was albino. Similarly, another breeder of the long-, straight-haired dogs found a male puppy that was albino. Because the albino trait is always due to a recessive allele, the two breeders got together and mated the two dogs. Surprisingly, all of the puppies in the litter had black hair. How would you explain this result?

E5. In the clover butterfly, males are always yellow, but females can be yellow or white. In females, white is a dominant allele. Two yellow butterflies were crossed to yield an F_1 generation consisting of 50% yellow males, 25% yellow females, and 25% white females. Describe how this trait is inherited and specify the genotypes of the parents.

E6. The *Mic2* gene in humans is present on both the X and Y chromosome. Let's suppose the *Mic2* gene exists in a dominant *Mic2* allele, which results in normal surface antigen, and a recessive *mic2* allele, which results in defective surface antigen production. Using molecular techniques, it is possible to identify homozygous and heterozygous individuals. By following the transmission of the *Mic2* and *mic2* alleles in a large human pedigree, would it be possible to distinguish between pseudoautosomal inheritance and autosomal inheritance? Explain your answer.

E7. Duroc Jersey pigs are typically red, but a sandy variation is also seen. When two different varieties of true-breeding sandy pigs were crossed to each other, they produced F_1 offspring that were red. When these F_1 offspring were crossed to each other, they produced red, sandy, and white pigs in a 9:6:1 ratio. Explain this pattern of inheritance.

E8. As discussed in this chapter, comb morphology in chickens is governed by a gene interaction. Two walnut comb chickens were crossed to each other. They produced only walnut comb and rose comb offspring, in a ratio of 3:1. What are the genotypes of the parents?

E9. In certain species of summer squash, fruit color is determined by two interacting genes. A dominant allele, *W*, determines white color, and a recessive allele (*w*) allows the fruit to be colored. In a homozygous *ww* individual, a second gene determines fruit color: *G* (green) is dominant to *g* (yellow). A white squash and a yellow squash were crossed, and the F_1 generation yielded approximately 50% white fruit and 50% green fruit. What are the genotypes of the parents?

E10. Certain species of summer squash exist in long, spherical, or disk shapes. When a true-breeding long-shaped strain was crossed to a true-breeding disk-shaped strain, all of the F_1 offspring were disk-shaped. When the F_1 offspring were allowed to self-fertilize, the F_2 generation consisted of a ratio of 9 disk-shaped to 6 round-shaped to 1 long-shaped. Assuming the shape of summer squash is governed by two different genes, with each gene existing in two alleles, propose a mechanism to account for this 9:6:1 ratio.

E11. In a species of plant, two genes control flower color. The red allele (*R*) is dominant to the white allele (*r*); the color-producing allele (*C*) is dominant to the non-color-producing allele (*c*). You suspect that either an *rr* homozygote or a *cc* homozygote will produce white flowers. In other words, *rr* is epistatic to *C*, and *cc* is epistatic to *R*. To test your hypothesis, you allowed heterozygous plants (*RrCc*) to self-fertilize and counted the offspring. You obtained the following data: 201 plants with red flowers and 144 with white flowers. Conduct a chi square analysis to see if your observed data are consistent with your hypothesis.

E12. In *Drosophila*, red eyes is the wild-type phenotype. Several different genes (with each gene existing in two or more alleles) are known to affect eye color. One allele causes purple eyes, and a different allele causes sepia eyes. Both of these alleles are recessive compared with red eye color. When flies with purple eyes were crossed to flies with sepia eyes, all of the F_1 offspring had red eyes. When the F_1 offspring were allowed to mate with each other, the following data were obtained:

146 purple eyes

151 sepia eyes

50 purplish sepia eyes

444 red eyes

Explain this pattern of inheritance. Conduct a chi square analysis to see if the experimental data fit your hypothesis.

E13. As mentioned in Experimental Question E12, red eyes is the wild-type phenotype in *Drosophila*, and several different genes (with each gene existing in two or more alleles) affect eye color. One allele causes purple eyes, and a different allele causes vermilion eyes. The purple and vermilion alleles are recessive compared with red eye color. The following crosses were made, and the following data were obtained:

Cross 1: Males with vermilion eyes × females with purple eyes

354 offspring, all with red eyes

Cross 2: Males with purple eyes × females with vermilion eyes

212 male offspring with vermilion eyes

221 female offspring with red eyes

Explain the pattern of inheritance based on these results. What additional crosses might you make to confirm your hypothesis?

E14. Let's suppose you were looking through a vial of fruit flies in your laboratory and noticed a male fly that has pink eyes. What crosses would you make to determine if the pink allele is an X-linked gene? What crosses would you make to determine if the pink allele is an allele of the same X-linked gene that has white and eosin alleles? Note: The eosin allele is discussed in Figure 4.19.

E15. When examining a human pedigree, what features do you look for to distinguish between X-linked recessive inheritance and autosomal recessive inheritance? How would you distinguish X-linked dominant inheritance from autosomal dominant inheritance in a human pedigree?

E16. The cream allele is a modifier of eosin and the cream allele is autosomal. By comparison, the red and eosin alleles are X-linked. Based on these ideas, conduct a chi square analysis to determine if Bridges' data of Figure 4.19 agree with the predicted ratio of 8 red-eyed females, 4 red-eyed males, 3 light eosin-eyed males, and 1 cream-eyed male.

Questions for Student Discussion/Collaboration

1. Let's suppose a gene exists as a functional wild-type allele and a nonfunctional mutant allele. At the organism level (i.e., at the level of visible traits), the wild-type allele is dominant. In a heterozygote, discuss whether dominance occurs at the cellular or molecular level. Discuss examples in which the issue of dominance depends on the level of examination.
2. A true-breeding rooster with a rose comb, feathered shanks, and cock-feathering was crossed to a hen that is true-breeding for pea comb and unfeathered shanks but is heterozygous for hen-feathering. If you assume these genes assort independently, what is the expected outcome of the F_1 generation?
3. In oats, the color of the chaff is determined by a two-gene interaction. When a true-breeding black plant was crossed to a true-breeding white plant, the F_1 generation was composed of all black plants. When the F_1 offspring were crossed to each other, the ratio produced was 12 black to 3 gray to 1 white. First, construct a Punnett square that accounts for this pattern of inheritance. Which genotypes produce the gray phenotype? Second, at the level of protein function, how would you explain this type of inheritance?

Note: All answers appear at the website for this textbook; the answers to the even-numbered questions and all of the Concept Check and Comprehension Questions are in the back of the book.

www.mhhe.com/brookergenetics5e

Visit the website for practice tests, answer keys, and other learning aids for this chapter. Enhance your understanding of genetics with our interactive exercises, quizzes, animations, and much more.

CHAPTER OUTLINE

Shell coiling in the water snail, Lymnaea peregra. *In this species, some snails coil to the left, and others coil to the right. This is due to an inheritance pattern called the maternal effect.*

5 NON-MENDELIAN INHERITANCE

As we discussed in Chapter 4, Mendelian inheritance patterns involve genes that directly influence the outcome of an offspring's traits and obey Mendel's laws. To predict phenotype, we must consider several factors. These include the dominant/recessive relationship of alleles, gene interactions that may affect the expression of a single trait, and the roles that sex and the environment play in influencing the individual's phenotype. Once these factors are understood, we can predict the phenotypes of offspring from their genotypes. Genes that follow a Mendelian inheritance pattern conform to four rules:

1. The expression of the genes in the offspring directly influences their traits.
2. Except in the case of rare mutations, the genes are passed unaltered from generation to generation.
3. The genes obey Mendel's law of segregation.
4. For crosses involving two or more genes, the genes obey Mendel's law of independent assortment.

Most genes in eukaryotic species follow a Mendelian pattern of inheritance. However, many genes do not. In this and the following chapter, we will examine several additional and even bizarre types of inheritance patterns that deviate from a Mendelian pattern because one of these four rules is broken. We begin this chapter by analyzing a non-Mendelian pattern called maternal effect in which rule 1 is broken. The expression of maternal effect genes in the offspring does not influence their own traits. Instead, traits that follow a maternal effect inheritance pattern are influenced by the expression of genes in cells of the mother that contribute to the development of egg cells. We then turn our attention to epigenetic inheritance, which breaks rule number 2, because the genes are altered in the offspring. As we will see, in certain types of epigenetic inheritance, genes are methylated, which alters their expression. Finally, in the last section of this chapter, we will examine inheritance patterns that arise because some genetic material is not located in the cell nucleus. Certain organelles, such as mitochondria and chloroplasts, contain their own genetic material. We will consider examples in which traits are determined by genes within these organelles. These traits do not obey rule 3, the law of segregation. In Chapter 6, we will examine inheritance patterns that do not obey rule 4, the law of independent assortment.

5.1 MATERNAL EFFECT

Learning Outcomes:

1. Define maternal effect.
2. Predict the outcome of crosses for genes that exhibit a maternal effect pattern of inheritance.
3. Explain the molecular mechanism of maternal effect.

We begin by considering genes that have a **maternal effect.** For these genes the genotype of the mother directly determines the phenotype of her offspring. Surprisingly, for maternal effect genes, the genotypes of the father and offspring themselves do not affect the phenotype of the offspring. We will see that this phenomenon is explained by the accumulation of gene products that the mother provides to her developing oocytes (immature eggs).

The Genotype of the Mother Determines the Phenotype of the Offspring for Maternal Effect Genes

The first example of a maternal effect gene was studied in the 1920s by Arthur Boycott and involved morphological features of the water snail, *Lymnaea peregra*. In this species, the shell and internal organs can be arranged in either a right-handed (dextral) or left-handed (sinistral) direction. The dextral orientation is more common and is dominant to the sinistral orientation. Figure 5.1 describes the results of a genetic analysis carried out by Boycott. In this experiment, he began with two different true-breeding strains of snails with either a dextral or sinistral morphology. Many combinations of crosses produced results that could not be explained by a Mendelian pattern of inheritance. When a dextral female (*DD*) was crossed to a sinistral male (*dd*), all F_1 offspring were dextral. However, in the **reciprocal cross,** where a sinistral female (*dd)* was crossed to a dextral male (*DD*), all F_1 offspring were sinistral. Taken together, these results contradict a Mendelian pattern of inheritance.

How can we explain the unusual results obtained in Figure 5.1? Alfred Sturtevant proposed the idea that snail coiling is due to a maternal effect gene that exists as a dextral (*D*) or sinistral (*d*) allele. His conclusions were drawn from the inheritance patterns of the F_2 and F_3 generations. In this experiment, the genotype of the F_1 generation is expected to be heterozygous (*Dd*). When these F_1 individuals were crossed to each other, a genotypic ratio of 1 *DD* : 2 *Dd* : 1 *dd* is predicted for the F_2 generation. Because the *D* allele is dominant to the *d* allele, a 3:1 phenotypic ratio of dextral to sinistral snails should be produced according to a Mendelian pattern of inheritance. Instead of this predicted phenotypic ratio, however, the F_2 generation was composed of all dextral snails. This incongruity with Mendelian inheritance is due to the maternal effect. The phenotype of the offspring depended solely on the genotype of the mother. The F_1 mothers were *Dd*. The *D* allele in the mothers is dominant to the *d* allele and caused the offspring to be dextral, even if the offspring's genotype was *dd*. When the members of the

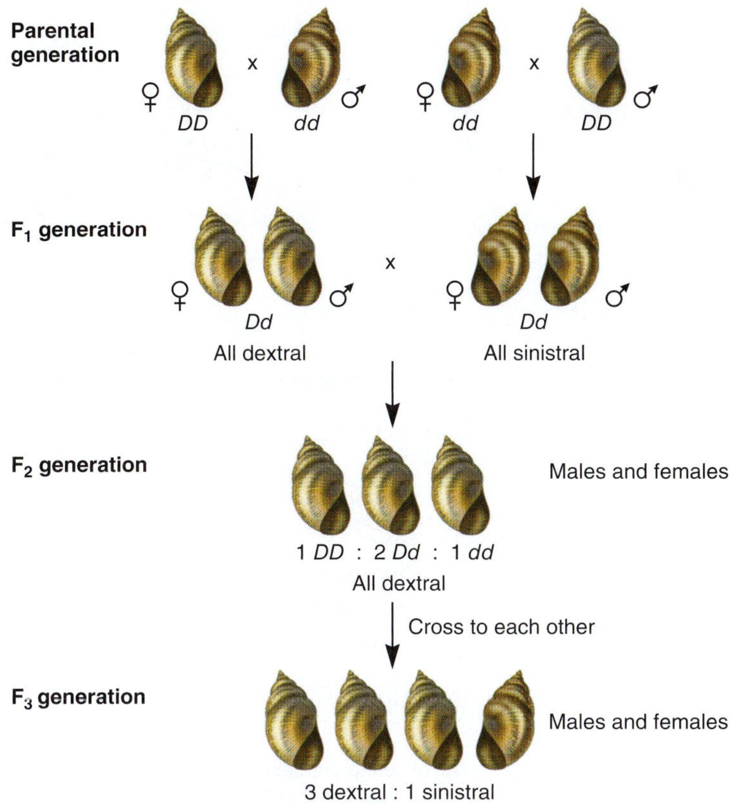

INTERACTIVE EXERCISE

FIGURE 5.1 Experiment showing the inheritance pattern of snail coiling. In this experiment, *D* (dextral) is dominant to *d* (sinistral). The genotype of the mother determines the phenotype of the offspring. This phenomenon is known as the maternal effect. In this case, a *DD* or *Dd* mother produces dextral offspring, and a *dd* mother produces sinistral offspring. The genotypes of the father and offspring do not affect the offspring's phenotype.

CONCEPT CHECK: Explain why all of the offspring in the F_2 generation are dextral even though some of them are *dd*.

F_2 generation were crossed, the F_3 generation exhibited a 3:1 ratio of dextral to sinistral snails. This ratio corresponds to the genotypes of the F_2 females, which were the mothers of the F_3 generation. The ratio of F_2 females was 1 *DD* : 2 *Dd* : 1 *dd*. The *DD* and *Dd* females produced dextral offspring, whereas the *dd* females produced sinistral offspring. This explains the 3:1 ratio of dextral and sinistral offspring in the F_3 generation.

Female Gametes Receive Gene Products from the Mother That Affect Early Developmental Stages of the Embryo

At the molecular and cellular level, the non-Mendelian inheritance pattern of maternal effect genes can be explained by the process of oogenesis in female animals (Figure 5.2a). As an animal oocyte matures, many surrounding maternal cells called nurse cells provide the oocyte with nutrients and other materials. In Figure 5.2a, a female is heterozygous for the snail-coiling maternal effect gene, with the alleles designated *D* and *d*. Depending on the outcome

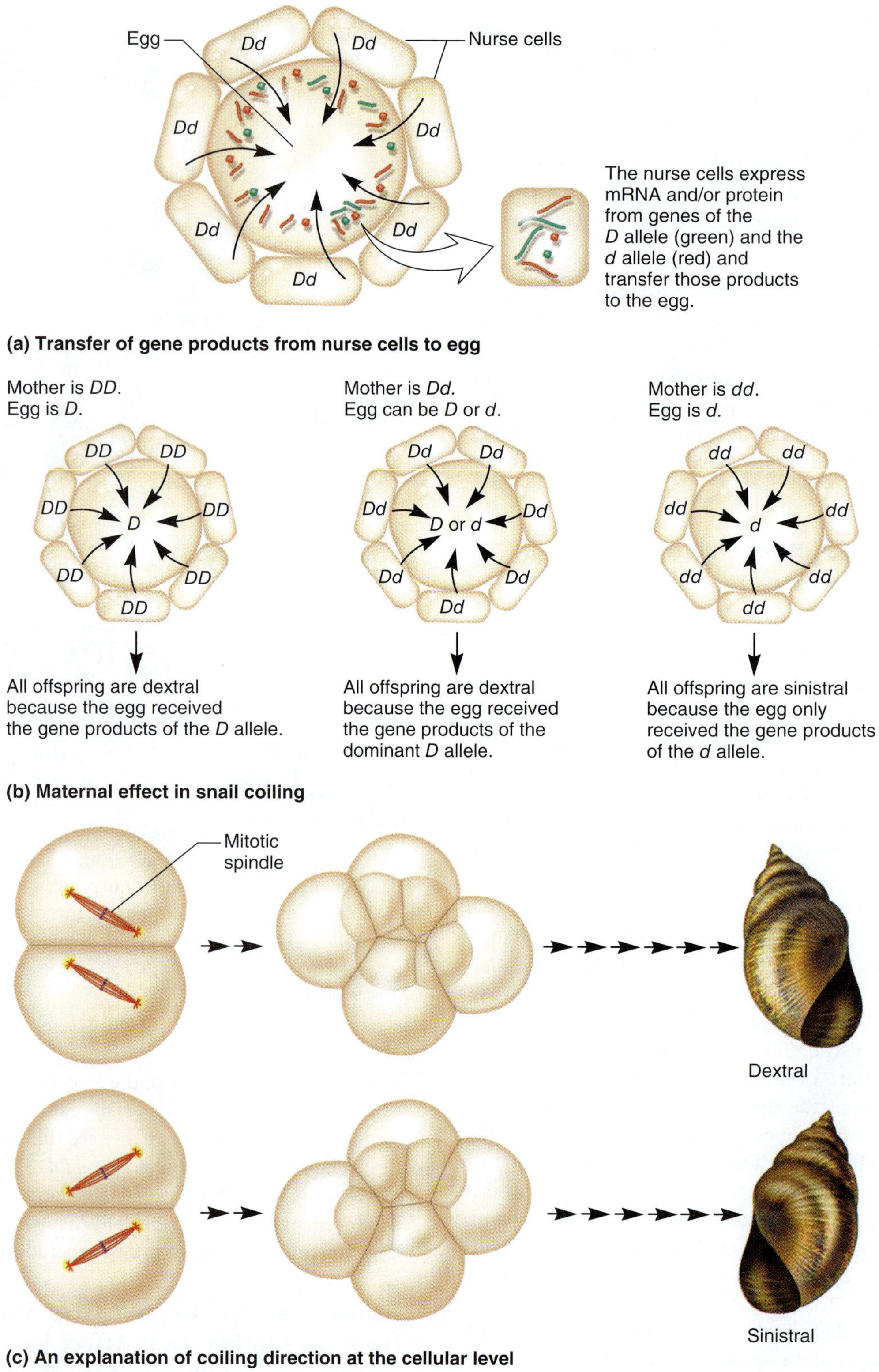

FIGURE 5.2 The mechanism of maternal effect in snail coiling. **(a)** Transfer of gene products from nurse cells to an oocyte. The nurse cells are heterozygous (*Dd*). Both the *D* and *d* alleles are activated in the nurse cells to produce *D* and *d* gene products (mRNA or proteins, or both). These products are transported into the cytoplasm of the oocyte, where they accumulate to significant amounts. **(b)** Explanation of the maternal effect in snail coiling. **(c)** The direction of snail coiling is determined by differences in the cleavage planes during early embryonic development.

Genes → Traits If the nurse cells are *DD* or *Dd*, they will transfer the *D* gene product to the oocyte and thereby cause the resulting offspring to be dextral. If the nurse cells are *dd*, only the *d* gene product will be transferred to the oocyte, so the resulting offspring will be sinistral.

CONCEPT CHECK: If a mother is heterozygous, *Dd*, which gene products will the oocyte receive?

of meiosis, the haploid oocyte may receive the *D* allele or the *d* allele, but not both. The surrounding nurse cells, however, produce both *D* and *d* gene products (mRNA and proteins). These gene products are then transported into the oocyte. As shown in Figure 5.2a, the oocyte has received both the *D* allele gene products and the *d* allele gene products. These gene products persist for a significant time after the egg has been fertilized and begins its embryonic development. In this way, the gene products of the nurse cells, which reflect the genotype of the mother, influence the early developmental stages of the embryo.

Now that we have an understanding of the relationship between oogenesis and maternal effect genes, let's reconsider the topic of snail coiling. As shown in **Figure 5.2b**, a female snail that is *DD* transmits only the *D* gene products to the oocyte. During the early stages of embryonic development, these gene products cause the embryo cleavage to occur in a way that promotes a right-handed body plan. A heterozygous female transmits both *D* and *d* gene products. Because the *D* allele is dominant, the maternal effect also causes a right-handed body plan. Finally, a *dd* mother contributes only *d* gene products that promote a left-handed body plan, even if the egg is fertilized by a sperm carrying a *D* allele. The sperm's genotype is irrelevant, because the expression of the sperm's gene would occur too late. The origin of dextral and sinistral coiling can be traced to the orientation of the mitotic spindle at the two- to four-cell stage of embryonic development. The dextral and sinistral snails develop as mirror images of each other (**Figure 5.2c**).

Since these initial studies, researchers have found that maternal effect genes encode proteins that are important in the early steps of embryogenesis. The accumulation of maternal gene products in the oocyte allows embryogenesis to proceed quickly after fertilization. Maternal effect genes often play a role in cell division, cleavage pattern, and body axis orientation. Therefore, defective alleles in maternal effect genes tend to have a dramatic effect on the phenotype of the individual, altering major features of morphology, often with dire consequences.

Our understanding of maternal effect genes has been greatly aided by their identification in experimental organisms such as *Drosophila melanogaster*. In such organisms with a short generation time, geneticists have successfully searched for mutant alleles that prevent the normal process of embryonic development. In *Drosophila*, geneticists have identified several maternal effect genes with profound effects on the early stages of development. The pattern of development of a *Drosophila* embryo occurs along axes, such as the anteroposterior axis and the dorsoventral axis. The proper development of each axis requires a distinct set of maternal gene products. For example, the maternal effect gene called *bicoid* produces a gene product that accumulates in a region of the oocyte that will become the anterior end of the embryo. After fertilization, the bicoid protein promotes the development of anterior structures in the developing embryo. If the mother is homozygous for a defective *bicoid* allele, the resulting embryos develop with two posterior ends. More recently, several maternal effect genes have been identified in mice and humans that are required for proper embryonic development. Chapter 25 examines the relationships among the actions of several maternal effect genes during embryonic development.

5.1 COMPREHENSION QUESTIONS

1. A female that coils to the left has offspring that coil to the right. What are the genotypes of the mother of these offspring and the maternal grandmother, respectively?
   a. *dd, DD*
   b. *Dd, Dd*
   c. *dd, Dd*
   d. *Dd, dd*
2. What is the molecular explanation for maternal effect?
   a. The father's gene is silenced at fertilization.
   b. During oogenesis, nurse cells transfer gene products to the oocyte.
   c. The gene products from nurse cells are needed during the very early stages of development.
   d. Both b and c are correct.

5.2 EPIGENETIC INHERITANCE: DOSAGE COMPENSATION

Learning Outcomes:

1. Define epigenetic inheritance.
2. Compare and contrast how dosage compensation is achieved in different animal species.
3. Describe the process of X-chromosome inactivation in mammals.
4. Explain how X-chromosome inactivation may affect the phenotype of female mammals.

Epigenetic inheritance is a pattern in which a modification occurs to a nuclear gene or chromosome that alters gene expression, but is not permanent over the course of many generations. As we will see, epigenetic inheritance patterns are the result of DNA and chromosomal modifications that occur during oogenesis, spermatogenesis, or early stages of embryogenesis. In this chapter, we will consider the general ways that epigenetic changes occur and how they affect traits. In Chapter 16, we will examine several molecular mechanisms of epigenetic inheritance in greater detail.

Once they are initiated, epigenetic changes alter the expression of particular genes in a way that may be fixed during an individual's lifetime. Therefore, epigenetic changes can permanently affect the phenotype of the individual. However, epigenetic modifications are not permanent over the course of many generations, and they do not change the actual DNA sequence. For example, a gene may undergo an epigenetic change that inactivates it for the lifetime of an individual. However, when this individual makes gametes, the gene may become activated and remain active during the lifetime of an offspring who inherits the gene.

In this section, we will examine a type of epigenetic inheritance called dosage compensation, which has the effect of offsetting differences in the number of sex chromosomes. One of the sex chromosomes is altered, with the result that males and females have similar levels of gene expression even though they do not possess the same complement of sex chromosomes. We will largely focus on mammals in which dosage compensation is initiated during the early stages of embryonic development.

Dosage Compensation Results in Similar Levels of Gene Expression Between the Sexes

Dosage compensation refers to the phenomenon in which the level of expression of many genes on the sex chromosomes (such as the X chromosome) is similar in both sexes even though males and females have a different complement of sex chromosomes. This term was coined in 1932 by Hermann Muller to explain the effects of eye color mutations in *Drosophila*. Muller observed that female flies homozygous for certain X-linked eye color alleles had a similar phenotype to hemizygous males. He noted that an X-linked gene conferring an apricot eye color produces a very similar phenotype in homozygous females and hemizygous males. In contrast, a female that has one copy of the apricot allele and a deletion of the apricot gene on the other X chromosome has eyes of paler color. Therefore, one copy of the allele in the female is not equivalent to one copy of the allele in the male. Instead, two copies of the allele in the female produce a phenotype that is similar to that produced by one copy in the male. In other words, the difference in gene dosage—two copies in females versus one copy in males—is being compensated for at the level of gene expression.

Since these initial studies, dosage compensation has been studied extensively in mammals, *Drosophila*, and *Caenorhabditis elegans* (a nematode). Depending on the species, dosage compensation occurs via different mechanisms (**Table 5.1**). Female mammals equalize the expression of X-linked genes by turning off one of their two X chromosomes. This process is known as **X-chromosome inactivation (XCI).** In *Drosophila*, the male accomplishes dosage compensation by doubling the expression of most X-linked genes. In *C. elegans*, the XX animal is a hermaphrodite that produces both sperm and egg cells, and an animal carrying a single X chromosome is a male that produces only sperm. The XX hermaphrodite diminishes the expression of X-linked genes on both X chromosomes to approximately 50% of that in the male.

In birds, the Z chromosome is a large chromosome, usually the fourth or fifth largest, which contains almost all of the known sex-linked genes. The W chromosome is generally a much smaller microchromosome containing a high proportion of repeat sequence DNA that does not encode genes. Males are ZZ and females are ZW. Several years ago, researchers studied the level of expression of a Z-linked gene that encodes an enzyme called aconitase. They discovered that males express twice as much aconitase as females do. These results suggested that dosage compensation does not occur in birds. More recently, the expression of hundreds of Z-linked genes has been examined in chickens. These newer results also suggest that birds lack a general mechanism of dosage compensation that controls the expression of most Z-linked genes. Even so, the pattern of gene expression between males and females was found to vary a great deal for certain Z-linked genes. Overall, the results suggest that some Z-linked genes may be dosage-compensated, but many of them are not.

TABLE 5.1
Mechanisms of Dosage Compensation Among Different Species

	Sex Chromosomes In:		
Species	**Females**	**Males**	**Mechanism of Compensation**
Placental mammals	XX	XY	One of the X chromosomes in the somatic cells of females is inactivated. In certain species, the paternal X chromosome is inactivated, and in other species, such as humans, either the maternal or paternal X chromosome is randomly inactivated throughout the somatic cells of females.
Marsupial mammals	XX	XY	The paternally derived X chromosome is inactivated in the somatic cells of females.
Drosophila melanogaster	XX	XY	The level of expression of genes on the X chromosome in males is increased twofold.
Caenorhabditis elegans	XX*	X0	The level of expression of genes on both X chromosomes in hermaphrodites is decreased to 50% levels compared with males.

*In *C. elegans*, an XX individual is a hermaphrodite, not a female.

Dosage Compensation Occurs in Female Mammals by the Inactivation of One X Chromosome

In 1961, Mary Lyon proposed that dosage compensation in mammals occurs by the inactivation of a single X chromosome in females. Liane Russell proposed the same idea around the same time. This proposal brought together two lines of study. The first type of evidence came from cytological studies. In 1949, Murray Barr and Ewart Bertram identified a highly condensed structure in the interphase nuclei of somatic cells in female cats that was not found in male cats. This structure became known as the **Barr body** (**Figure 5.3a**). In 1960, Susumu Ohno correctly proposed that the Barr body is a highly condensed X chromosome.

In addition to this cytological evidence, Lyon was also familiar with examples in which the coat color of mammals has a variegated pattern. **Figure 5.3b** is a photo of a calico cat, which is a female that is heterozygous for an X-linked gene that can occur as an orange or a black allele. (The white underside is due to a dominant allele in a different gene.) The orange and black patches are randomly distributed in different female individuals. The calico pattern does not occur in male cats, but similar kinds of mosaic patterns have been identified in the female mouse. Lyon suggested that both the Barr body and the calico pattern are the result of XCI in the cells of female mammals.

The mechanism of XCI, also known as the **Lyon hypothesis,** is schematically illustrated in **Figure 5.4**. This example involves a white and black variegated coat color found in certain

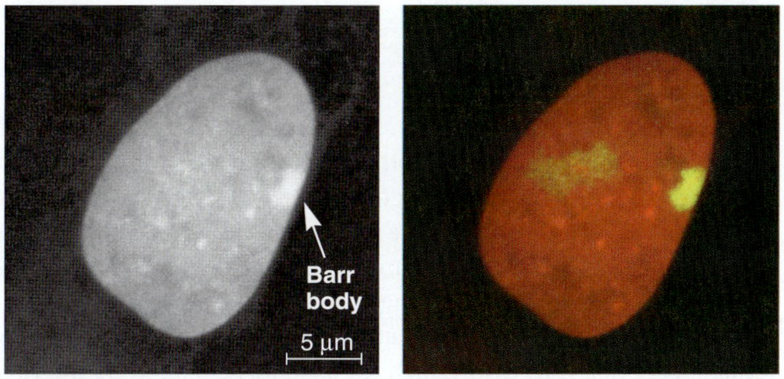

(a) Nucleus with a Barr body

(b) A calico cat

INTERACTIVE EXERCISE

FIGURE 5.3 X-chromosome inactivation in female mammals. **(a)** The left micrograph shows the Barr body on the periphery of a human nucleus after staining with a DNA-specific dye. Because it is compact, the Barr body is the most brightly staining. The right micrograph shows the same nucleus using a yellow fluorescent probe that recognizes the X chromosome. The Barr body is more compact than the active X chromosome, which is to the left of the Barr body. **(b)** The fur pattern of a calico cat.

Genes → Traits The pattern of black and orange fur on this cat is due to random X-chromosome inactivation during embryonic development. The orange patches of fur are due to the inactivation of the X chromosome that carries a black allele; the black patches are due to the inactivation of the X chromosome that carries the orange allele. In general, only heterozygous female cats can be calico. A rare exception would be a male cat (XXY) that has an abnormal composition of sex chromosomes.

CONCEPT CHECK: Why is the Barr body more brightly staining in a cell nucleus than the other chromosomes?

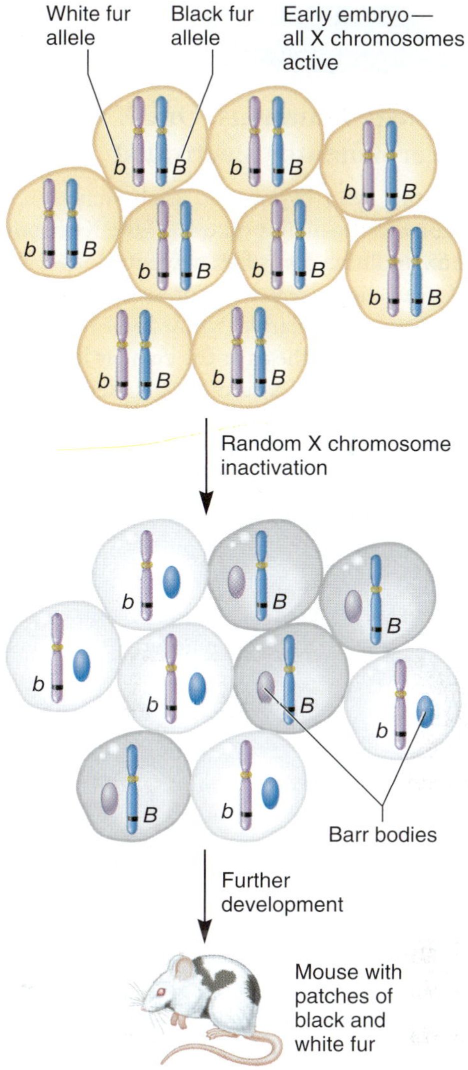

FIGURE 5.4 The mechanism of X-chromosome inactivation.

Genes → Traits The top of this figure represents a mass of several cells that compose the early embryo. Initially, both X chromosomes are active. At an early stage of embryonic development, random inactivation of one X chromosome occurs in each cell. This inactivation pattern is maintained as the embryo matures into an adult.

CONCEPT CHECK: At which stage of development does XCI initially occur?

strains of mice. As shown here, a female mouse has inherited an X chromosome from its mother that carries an allele conferring white coat color (X^b). The X chromosome from its father carries a black coat color allele (X^B). How can XCI explain a variegated coat pattern? Initially, both X chromosomes are active. However, at an early stage of embryonic development, one of the two X chromosomes is randomly inactivated in each somatic cell and becomes a Barr body. For example, one embryonic cell may have the X^B chromosome inactivated. As the embryo continues to grow and mature, this embryonic cell will divide and may eventually give rise to billions of cells in the adult animal. The epithelial (skin) cells that are derived from this embryonic cell will produce a patch of white fur because the X^B chromosome has been permanently inactivated. Alternatively, another embryonic cell may have the X^b chromosome inactivated. The epithelial cells derived from this embryonic cell will produce a patch of black fur. Because the primary event of XCI is a random process that occurs at an early stage of development, the result is an animal with some patches of white fur and other patches of black fur. This is the basis of the variegated phenotype.

During XCI, the chromosomal DNA becomes highly compacted into a Barr body, so most genes on the inactivated X chromosome cannot be expressed. When cell division occurs and the inactivated X chromosome is replicated, both copies remain highly compacted and inactive. Likewise, during subsequent cell divisions, XCI is passed along to all future somatic cells.

EXPERIMENT 5A

In Adult Female Mammals, One X Chromosome Has Been Permanently Inactivated

According to the Lyon hypothesis, each somatic cell of female mammals expresses the genes on one of the X chromosomes, but not both. If an adult female is heterozygous for an X-linked gene, only one of two alleles will be expressed in any given cell. In 1963, Ronald Davidson, Harold Nitowsky, and Barton Childs set out to test the Lyon hypothesis at the cellular level. To do so, they analyzed the expression of a human X-linked gene that encodes an enzyme involved with sugar metabolism known as glucose-6-phosphate dehydrogenase (G-6-PD).

Prior to the Lyon hypothesis, biochemists had found that individuals vary with regard to the G-6-PD enzyme. This variation can be detected when the enzyme is subjected to gel electrophoresis (see the Appendix for a description of gel electrophoresis). One *G-6-PD* allele encodes a G-6-PD enzyme that migrates very quickly during gel electrophoresis (the "fast" enzyme), whereas another *G-6-PD* allele produces an enzyme that migrates more slowly (the "slow" enzyme). As shown in **Figure 5.5**, when multiple skin samples are collected from a heterozygous adult female and mixed together, both types of enzymes are observed on a gel. In contrast, a hemizygous male produces either the fast or slow type, but not both. The difference in migration between the fast and slow G-6-PD enzymes is due to minor differences in the amino acid sequences of these enzymes. These minor differences do not significantly affect G-6-PD function, but they do enable geneticists to distinguish the proteins encoded by the two X-linked alleles.

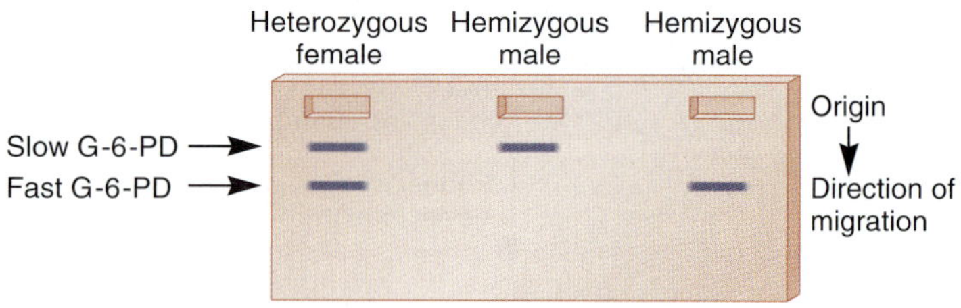

FIGURE 5.5 **Mobility of G-6-PD protein on gels.** *G-6-PD* exists as a fast allele that encodes a protein that migrates more quickly to the bottom of the gel and a slow allele that migrates more slowly. The protein encoded by the fast allele is closer to the bottom of the gel.

CONCEPT CHECK: Why do these two forms of G-6-PD migrate differently?

As shown in **Figure 5.6**, Davidson, Nitowsky, and Childs tested the Lyon hypothesis using cell culturing techniques. They removed several small samples of epithelial cells from a heterozygous female and grew them in the laboratory. When combined, these samples contained a mixture of both types of enzymes because the adult cells were derived from many different embryonic cells, some that had the slow allele inactivated and some that had the fast allele inactivated. In the experiment of Figure 5.6, these cells were sparsely plated onto solid growth media. After several days, each cell grew and divided to produce a **clone** of cells. All cells within a clone were derived from a single cell. The researchers reasoned that all cells within a single clone would express only one of the two *G-6-PD* alleles if the Lyon hypothesis was correct. Nine clones were grown in liquid cultures, and then the cells were lysed to release the G-6-PD proteins inside of them. The proteins were then subjected to sodium dodecyl sulfate (SDS) gel electrophoresis.

THE HYPOTHESIS

According to the Lyon hypothesis, an adult female who is heterozygous for the fast and slow *G-6-PD* alleles should express only one of the two alleles in any particular somatic cell and its descendants, but not both.

TESTING THE HYPOTHESIS

FIGURE 5.6 **Evidence that adult female mammals contain one X chromosome that has been permanently inactivated.**

Starting material: Several small skin samples taken from a woman who was heterozygous for the fast and slow alleles of *G-6-PD*.

	Experimental level	Conceptual level
1. Mince the tissue to separate the individual cells.	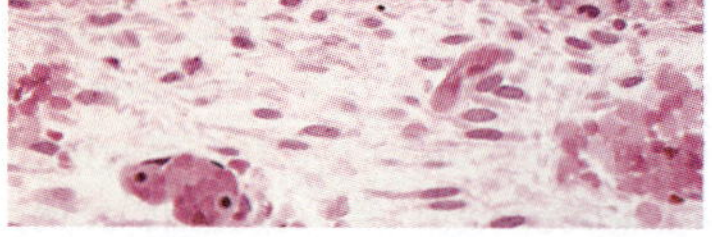 Mince.	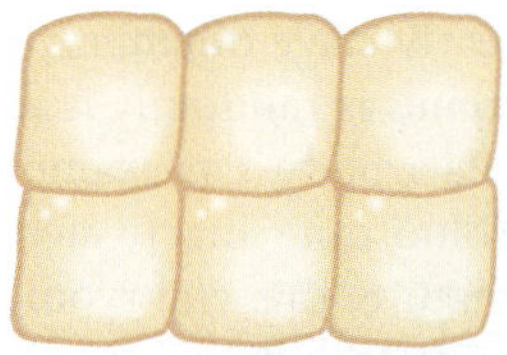 Mince and grow.

2. Grow the cells in a liquid growth medium and then plate (sparsely) onto solid growth medium. Each cell divides to form a clone of many cells.

3. Take nine isolated clones and grow in liquid cultures. (Only three are shown here.)

4. Take cells from the liquid cultures, lyse cells to obtain proteins, and subject to gel electrophoresis. (This technique is described in the Appendix.)

Note: As a control, lyse cells from step **1**, and subject the proteins to gel electrophoresis. The control is derived from several small skin samples from a heterozygous woman.

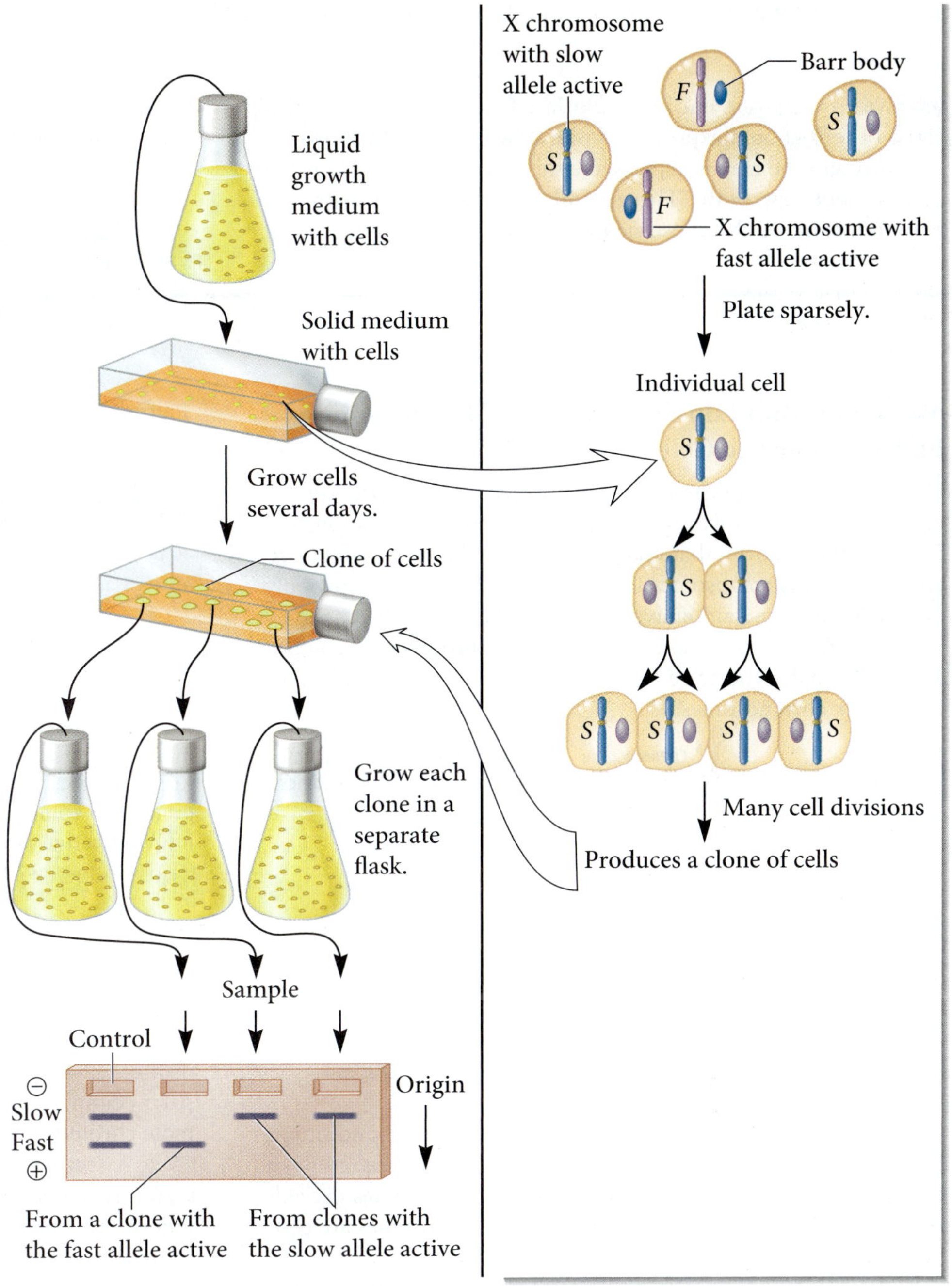

THE DATA

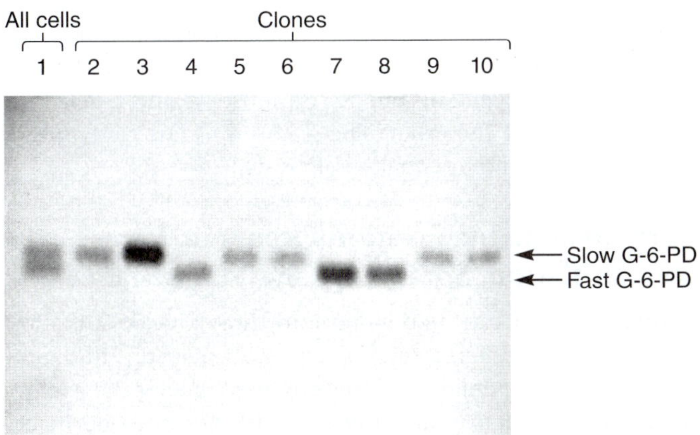

Data from Ronald G. Davidson, Harold M. Nitowsky, and Barton Childs (1963) Demonstration of two population of cells in the human female heterozygous for glucose-6-phosphate dehydrogenase variants. *Proc Natl Acad Sci USA 50*, 481–485.

INTERPRETING THE DATA

In the data shown in Figure 5.6, the control (lane 1) was a protein sample obtained from a mixture of epithelial cells from a heterozygous woman who produced both types of G-6-PD enzymes. Bands corresponding to the fast and slow enzymes were observed in this lane. As described in steps 2 to 4, this mixture of epithelial cells was also used to generate nine clones. The proteins obtained from these clones are shown in lanes 2 through 10. Each clone was a population of cells independently derived from a single epithelial cell. Because the epithelial cells were obtained from an adult female, the Lyon hypothesis predicts that each epithelial cell would already have one of its X chromosomes permanently inactivated and would pass this trait to its progeny cells. For example, suppose that an epithelial cell had inactivated the X chromosome that encodes the fast G-6-PD. If this cell was allowed to form a clone of cells on a plate, all cells in this clonal population would be expected to have the same X chromosome inactivated—the one encoding the fast G-6-PD. Therefore, this clone of cells should express only the slow G-6-PD. As shown in the data, all nine clones expressed either the fast or slow G-6-PD protein, but not both. These results are consistent with the hypothesis that X-chromosome inactivation has already occurred in any given epithelial cell and that this pattern of inactivation is passed to all of its progeny cells.

A self-help quiz involving this experiment can be found at www.mhhe.com/brookergenetics5e.

Mammals Maintain One Active X Chromosome in Their Somatic Cells

Since the Lyon hypothesis was confirmed, the genetic control of XCI has been investigated further by several laboratories. Research has shown that mammalian cells have the ability to count their X chromosomes in their somatic cells and allow only one of them to remain active. How was this determined? A key observation came from comparisons of the chromosome composition of people who were born with normal or abnormal numbers of sex chromosomes.

Phenotype	Chromosome Composition	Number of X Chromosomes	Number of Barr Bodies
Normal female	XX	2	1
Normal male	XY	1	0
Turner syndrome (female)	X0	1	0
Triple X syndrome (female)	XXX	3	2
Klinefelter syndrome (male)	XXY	2	1

In normal females, two X chromosomes are counted and one is inactivated, whereas in males, one X chromosome is counted and none inactivated. If the number of X chromosomes exceeds two, as in triple X syndrome, additional X chromosomes are converted to Barr bodies.

X-Chromosome Inactivation in Mammals Depends on the X-Inactivation Center and the *Xist* Gene

Although the genetic control of XCI is not entirely understood at the molecular level, a short region on the X chromosome called the **X-inactivation center (Xic)** is known to play a critical role (**Figure 5.7**). Eeva Therman and Klaus Patau discovered that if one of the two X chromosomes in a female is missing its Xic due to a chromosome mutation, a cell counts only one Xic and X-chromosome inactivation does not occur. Having two active X chromosomes is a lethal condition for a human female embryo.

Let's consider how the molecular expression of certain genes controls XCI. The expression of a specific gene within the Xic is required for the compaction of the X chromosome into a Barr body. This gene, discovered in 1991, is named *Xist* (for X-inactive specific transcript). The *Xist* gene on the inactivated X chromosome is active, which is unusual because most other genes on the inactivated X chromosome are silenced. The *Xist* gene product is an RNA molecule that does not encode a protein. Instead, the role of the *Xist* RNA is to coat the X chromosome and inactivate it. After coating, other proteins associate with the *Xist* RNA and promote chromosomal compaction into a Barr body.

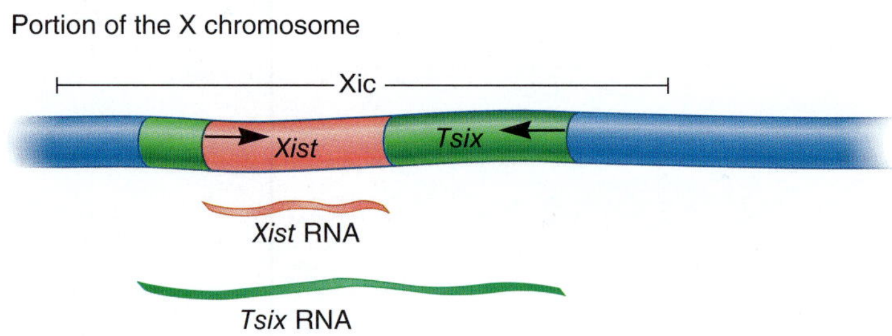

FIGURE 5.7 The X-inactivation center (Xic) of the X chromosome. The *Xist* gene is transcribed into RNA from the inactive X chromosome but not from the active X chromosome. This *Xist* RNA coats the inactive X chromosome and recruits proteins that promote its compaction. On the active X chromosome, the *Tsix* gene inhibits the transcription of the *Xist* gene.

CONCEPT CHECK: What is the function of the *Tsix* gene?

A second gene found within the Xic, designated *Tsix*, plays a role in preventing X-chromosome inactivation. As shown in Figure 5.7, the *Xist* and *Tsix* genes are overlapping and transcribed in opposite directions. (The name *Tsix* is *Xist* spelled backwards.) The expression of the *Tsix* gene inhibits the transcription of the *Xist* gene. On the active X chromosome, the *Tsix* gene is expressed, and the *Xist* gene is not. The opposite situation occurs on the inactive X chromosome—*Xist* is expressed, and *Tsix* is not. Researchers have studied heterozygous females that carry a normal *Tsix* gene on one X chromosome and a defective, mutant *Tsix* gene on the other. The X chromosome carrying the mutant *Tsix* gene is preferentially inactivated.

X-Chromosome Inactivation Occurs in Three Phases: Initiation, Spreading, and Maintenance

The process of XCI can be divided into three phases: initiation, spreading, and maintenance (**Figure 5.8**). During initiation, which occurs during embryonic development, one of the X chromosomes remains active, and the other is chosen to be inactivated. The molecular mechanism of X-chromosome inactivation is described in Chapter 16 (look ahead to Figure 16.5).

During the spreading phase, the chosen X chromosome is inactivated. This spreading requires the expression of the *Xist* gene. The *Xist* RNA coats the inactivated X chromosome and recruits proteins that promote compaction. This compaction involves DNA methylation and also the modification of histone proteins, which are described in Chapter 10. The spreading phase is so named because inactivation begins at the Xic and spreads in both directions along the X chromosome.

Once the initiation and spreading phases occur for a given X chromosome, the inactivated X chromosome is maintained as a Barr body during future cell divisions. When a cell divides, the Barr body is replicated, and both copies remain compacted. This maintenance phase continues from the embryonic stage through adulthood.

Some genes on the inactivated X chromosome are expressed in the somatic cells of adult female mammals. These genes are said to escape the effects of XCI. As mentioned, *Xist* is an example of a gene that is expressed from the highly condensed Barr body. In humans, up to a quarter of X-linked genes may escape inactivation to some degree. Many of these genes occur in clusters. Among these are the pseudoautosomal genes found on the X and Y chromosomes in the regions of homology described in Chapter 4. Dosage compensation is not necessary for X-linked pseudoautosomal genes because they are located on both the X and Y chromosomes. How are genes on the Barr body expressed? Although the mechanism is not understood, these genes may be found in localized regions where the chromatin is less tightly packed and able to be transcribed.

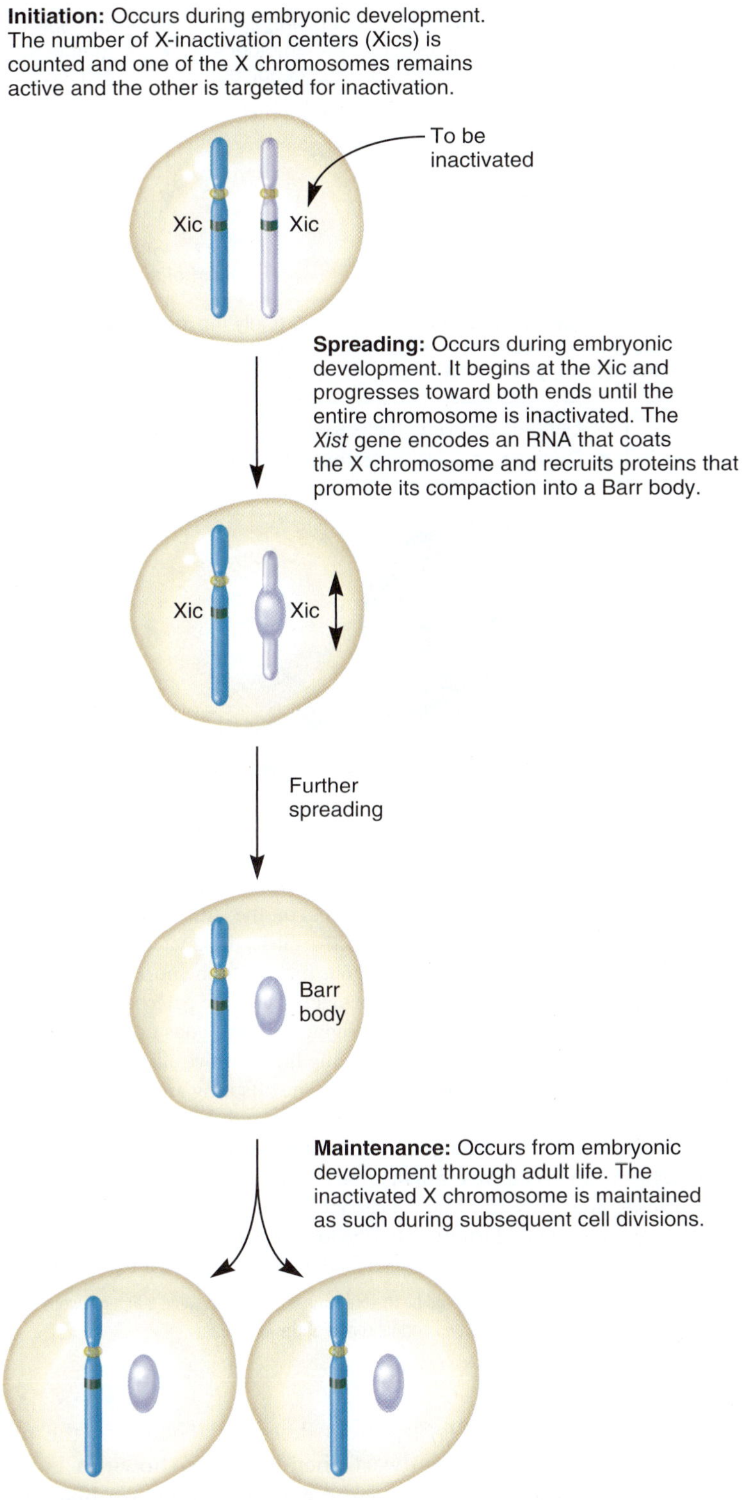

ONLINE ANIMATION

FIGURE 5.8 The function of the Xic during X-chromosome inactivation.

CONCEPT CHECK: Which of these phases occurs in an adult female?

5.2 COMPREHENSION QUESTIONS

1. In fruit flies, dosage compensation is achieved by
   a. X-chromosome inactivation.
   b. turning up the expression of genes on the single X chromosome twofold in the male.
   c. turning down the expression of genes on the two X chromosomes to one half in the female.
   d. all of the above.
2. According to the Lyon hypothesis,
   a. one of the X chromosomes is converted to a Barr body in somatic cells of female mammals.
   b. one of the X chromosomes is converted to a Barr body in all cells of female mammals.
   c. both of the X chromosomes are converted to Barr bodies in somatic cells of female mammals.
   d. both of the X chromosomes are converted to Barr bodies in all cells of female mammals.
3. Which of the following is *not* a phase of XCI?
   a. Initiation
   b. Spreading
   c. Maintenance
   d. Erasure

5.3 EPIGENETIC INHERITANCE: GENOMIC IMPRINTING

Learning Outcomes:

1. Define genomic imprinting.
2. Predict the outcome of crosses involving imprinted genes.
3. Explain the molecular mechanism of imprinting.

As we have just seen, dosage compensation changes the level of expression of many genes located on the X chromosome. We now turn to another epigenetic phenomenon known as genomic imprinting or simply imprinting. The term imprinting implies a type of marking process that has a memory. For example, newly hatched birds identify marks on their parents, which allows them to distinguish their parents from other individuals. The term **genomic imprinting** refers to an analogous situation in which a segment of DNA is marked, and that mark is retained and recognized throughout the life of the organism inheriting the marked DNA. Genomic imprinting happens prior to fertilization; it involves a change in a single gene or chromosome during gamete formation. Depending on whether the modification occurs during spermatogenesis or oogenesis, imprinting governs whether an offspring expresses a gene that has been inherited from its mother or father.

The Expression of an Imprinted Gene Depends on the Sex of the Parent from Which the Gene Was Inherited

The phenotypes that result from the expression of imprinted genes follow a non-Mendelian pattern of inheritance because the marking process causes the offspring to distinguish between maternally and paternally inherited alleles. Depending on how the genes are marked, the offspring expresses only one of the two alleles. This phenomenon is termed **monoallelic expression.**

To understand genomic imprinting, let's consider a specific example. In the mouse, a gene designated *Igf2* encodes a protein growth hormone called insulin-like growth factor 2. Imprinting occurs in a way that results in the expression of the paternal *Igf2* allele but not the maternal allele. The paternal allele is transcribed into RNA, but the maternal allele is transcriptionally silent. With regard to phenotype, a functional *Igf2* gene is necessary for normal size. A loss-of-function allele of this gene, designated *Igf2*$^-$, is defective in the synthesis of a functional Igf2 protein. This may cause a mouse to be a dwarf, but the dwarfism depends on whether the mutant allele is inherited from the male or female parent, as shown in **Figure 5.9**. On the left side, an offspring has inherited the *Igf2* allele from its father and the *Igf2*$^-$ allele from its mother. Due to imprinting, only the *Igf2* allele is expressed in the offspring. Therefore, this mouse grows to a normal size. Alternatively, in the reciprocal cross on the right side, an individual has inherited the *Igf2*$^-$ allele from its father and the *Igf2* allele from its mother. In this case, the *Igf2* allele is not expressed. In this mouse, the *Igf2*$^-$ allele would be transcribed into mRNA, but the

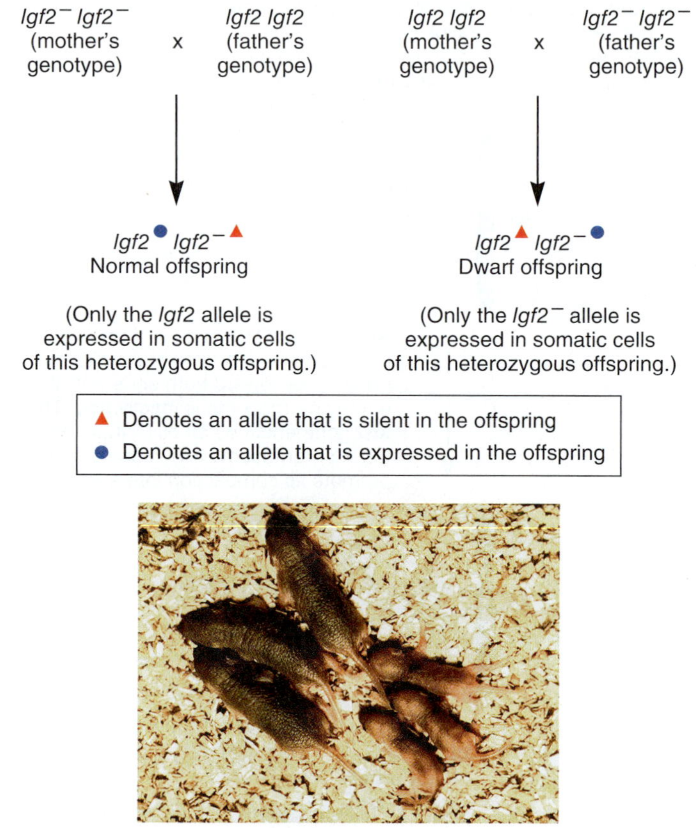

INTERACTIVE EXERCISE

FIGURE 5.9 An example of genomic imprinting in the mouse. In the cross on the left, a homozygous female carrying a defective allele, designated *Igf2*$^-$, is crossed to a homozygous male with the normal *Igf2* allele. An offspring is heterozygous and normal because the paternal allele is active. In the reciprocal cross on the right, a homozygous normal female is crossed to a homozygous male carrying the defective allele. In this case, the offspring is heterozygous and dwarf. The offspring are dwarf because the paternal allele is defective due to mutation and the maternal allele is not expressed. The photograph shows normal-size (left) and dwarf littermates (right) derived from a cross between a wild-type female and a heterozygous male (*Igf2 Igf2*$^-$) carrying one copy of a loss-of-function allele (courtesy of A. Efstratiadis). The loss-of-function allele was created using gene knockout methods described in Chapter 21 (see Figure 21.5).

CONCEPT CHECK: What would be the outcome of a cross between a heterozygous female and a male that carries two normal copies of the *Igf2* gene?

mutation renders the *Igf2* protein defective. Therefore, the offspring on the right has a dwarf phenotype. As shown in Figure 5.9, both offspring have the same genotype; they are heterozygous for the *Igf2* alleles (i.e., *Igf2 Igf2*$^-$). They are phenotypically different, however, because only the paternally inherited allele is expressed.

The Imprint Is Established During Gametogenesis

At the cellular level, imprinting is an epigenetic process that can be divided into three stages: (1) the establishment of the imprint during gametogenesis, (2) the maintenance of the imprint during embryogenesis and in adult somatic cells, and (3) the erasure and reestablishment of the imprint in the germ cells. These stages are described in **Figure 5.10**, which shows the imprinting of the *Igf2* gene. The two mice shown here have inherited the *Igf2* allele from their father and the *Igf2*$^-$ allele from their mother. Due to imprinting, both mice express the *Igf2* allele in their somatic cells, and the pattern of imprinting is maintained in the somatic cells throughout development. In the germline cells that give rise to gametes (i.e., sperm or eggs), the imprint is erased; it will be reestablished according to the sex of the animal. The female mouse on the left will transmit only transcriptionally inactive alleles to her offspring. The male mouse on the right will transmit transcriptionally active alleles. However, because this male is a heterozygote, it will transmit either a functionally active *Igf2* allele or a functionally defective mutant allele (*Igf2*$^-$). If this heterozygous male transmits the *Igf2* allele to an offspring, it will be normal sized. In contrast, if this male transmits an *Igf2*$^-$ allele, the offspring will be dwarf. Although an *Igf2*$^-$ allele inherited from a male mouse can be transcribed into mRNA, it will not produce a functional Igf2 protein, which is the reason for the dwarf phenotype.

As seen in Figure 5.10, genomic imprinting is permanent in the somatic cells of an animal, but the marking of alleles can be altered from generation to generation. For example, the female mouse on the left has an active copy of the *Igf2* allele, but any allele this female transmits to its offspring will be transcriptionally inactive.

Genomic imprinting occurs in several species, including numerous insects, mammals, and flowering plants. Imprinting may involve a single gene, a part of a chromosome, an entire chromosome, or even all of the chromosomes from one parent. Helen Crouse discovered the first example of imprinting, which involved an entire chromosome in the housefly, *Sciara coprophila*. In this species, the fly normally inherits three sex chromosomes, rather than two as in most other species. One X chromosome is inherited from the female, and two are inherited from the male. In male flies, both paternal X chromosomes are lost from somatic cells during embryogenesis. In female flies, only one of the paternal X chromosomes is lost. In both sexes, the maternally inherited X chromosome is never lost. These results indicate that either the maternal X chromosome is marked to promote its retention or paternal X chromosomes are marked to promote their loss.

Genomic imprinting can also be involved in the process of X-chromosome inactivation, described previously. In certain species, imprinting plays a role in the choice of the X chromosome that will be inactivated. For example, in marsupials, the paternal X chromosome is marked so it is the X chromosome that is always inactivated in the somatic cells of females. In marsupials, XCI is not random; the maternal X chromosome is always active.

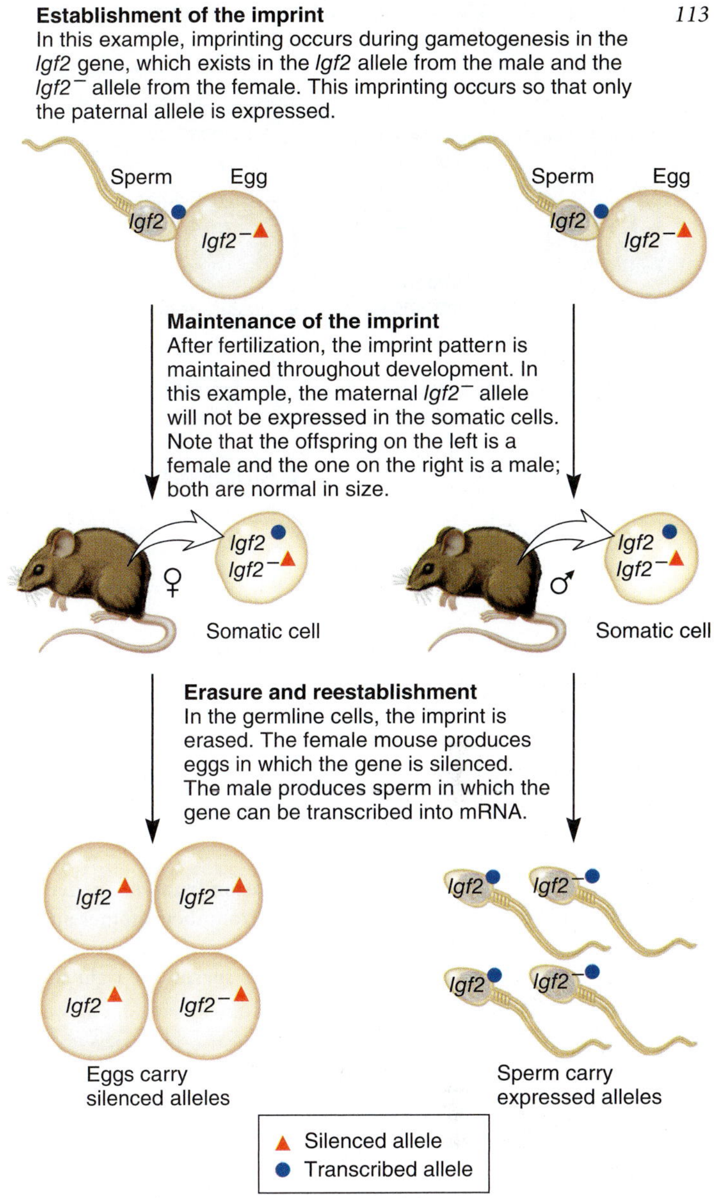

FIGURE 5.10 Genomic imprinting during gametogenesis. This example involves a mouse gene *Igf2*, which is found in two alleles designated *Igf2* and *Igf2*$^-$. The left side shows a female mouse that was produced from a sperm carrying the *Igf2* allele and an egg carrying the *Igf2*$^-$ allele. In the somatic cells of this female animal, the *Igf2* allele is active. However, when this female produces eggs, both alleles are transcriptionally inactive when they are transmitted to offspring. The right side of this figure shows a male mouse that was also produced from a sperm carrying the *Igf2* allele and an egg carrying the *Igf2*$^-$ allele. In the somatic cells of this male animal, the *Igf2* allele is active. The sperm from this male may carry either a functionally active *Igf2* allele or a functionally defective *Igf2*$^-$ allele. The allele inherited from the male is expressed, resulting in normal or dwarf offspring, respectively.

CONCEPT CHECK: Explain why the erasure phase of imprinting is necessary in eggs.

The Imprinting of Genes Is a Molecular Marking Process That Involves DNA Methylation

As we have seen, genomic imprinting must involve a marking process. A particular gene or chromosome must be marked differently during spermatogenesis versus oogenesis. After fertilization takes place, this differential marking affects the expression of particular genes. What is the molecular explanation for genomic imprinting? As discussed in Chapter 15, **DNA methylation**—the attachment of a methyl group onto a cytosine base—is a common way that eukaryotic genes may be regulated. Research indicates that genomic imprinting involves an **imprinting control region (ICR)** that is located near the imprinted gene (see Chapter 16,

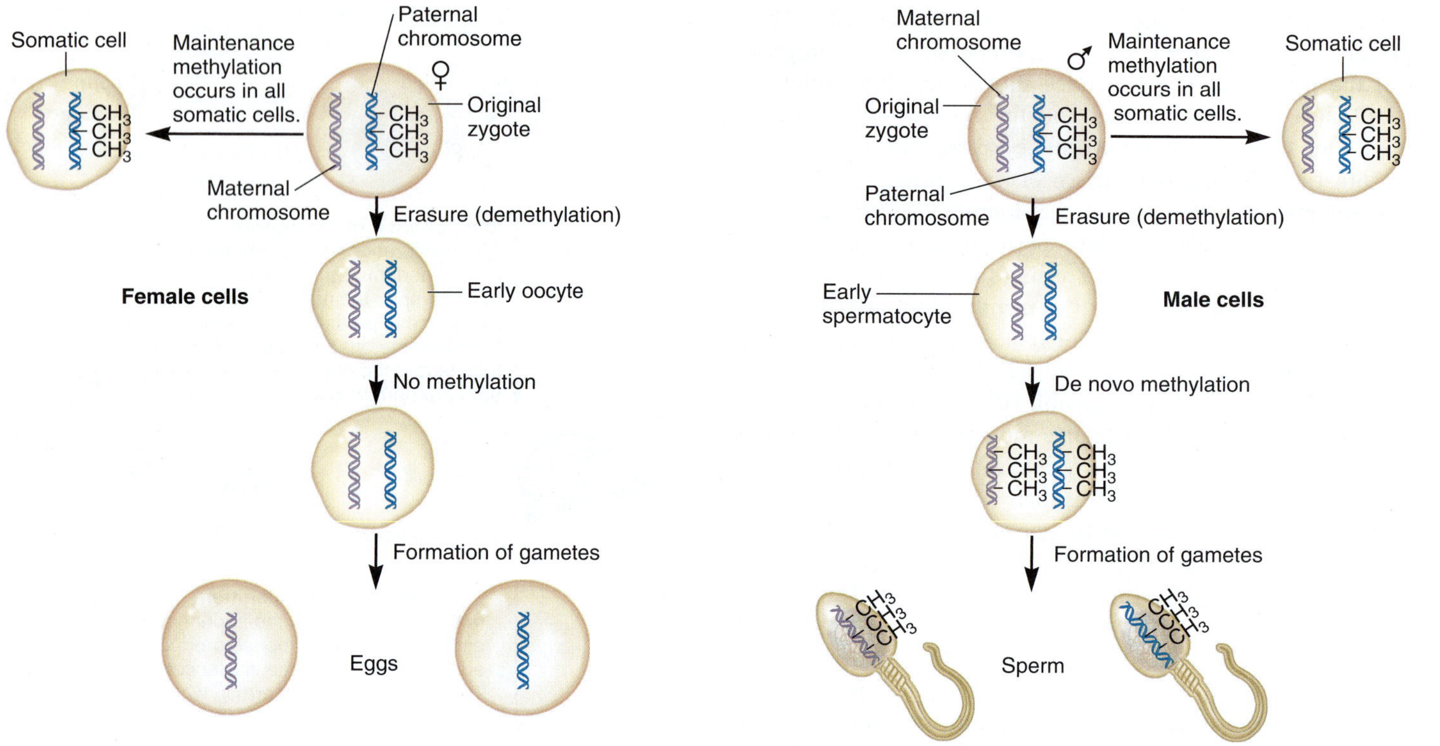

FIGURE 5.11 **The pattern of methylation from one generation to the next.** In this example, a female and a male offspring have inherited a nonmethylated gene and a methylated gene from their mother and father, respectively. Maintenance methylation retains the imprinting in somatic cells during embryogenesis and in adulthood. Demethylation occurs in cells that are destined to become gametes. In this example, de novo methylation occurs only in cells that are destined to become sperm (see right side). Haploid male gametes transmit a methylated gene, whereas haploid female gametes transmit an unmethylated gene.

CONCEPT CHECK: What is the difference between maintenance methylation and de novo methylation? In what cell types (somatic cells or germline cells) do they occur?

Figure 16.4). Depending on the particular gene, the ICR is methylated in the egg or the sperm, but not both. The ICR contains binding sites for one or more proteins that regulate the transcription of the imprinted gene. For most but not all imprinted genes, methylation causes an inhibition of transcription.

Let's now consider the methylation process from one generation to the next. In the example shown in **Figure 5.11**, the paternally inherited allele for a particular gene is methylated but the maternally inherited allele is not. A female (left side) and male (right side) have inherited an unmethylated gene from their mother and a methylated gene from their father. This pattern of imprinting is maintained in the somatic cells of both individuals. However, when the female makes gametes, the imprinting is erased during early oogenesis, so the female will pass an unmethylated gene to its offspring. In the male, the imprinting is also erased during early spermatogenesis, but then de novo (new) methylation occurs in both genes. Therefore, the male will transmit a methylated gene to its offspring.

Imprinting Plays a Role in the Inheritance of Certain Human Genetic Diseases

Several genes have been shown to be imprinted in humans (**Table 5.2**). Human diseases, such as Prader-Willi syndrome (PWS) and Angelman syndrome (AS), are influenced by imprinting. PWS is characterized by reduced motor function, obesity, and small hands and feet. Individuals with AS are thin and hyperactive, have unusual seizures and repetitive symmetrical muscle movements, and exhibit mental deficiencies. Most commonly, both PWS and AS involve a small deletion in human chromosome 15.

TABLE 5.2
Examples of Mammalian Genes That Are Imprinted*

Gene	Allele Expressed	Function
WT1	Maternal	Wilms tumor-suppressor gene; suppresses cell growth
INS	Paternal	Insulin; hormone involved in cell growth and metabolism
Igf2	Paternal	Insulin-like growth factor II; similar to insulin
Igf2R	Maternal	Receptor for insulin-like growth factor II
UBE3A	Maternal	Regulates protein degradation; involved in Angelman syndrome
SNRPN	Paternal	Splicing factor; involved in Prader-Willi syndrome
Gabrb	Maternal	Neurotransmitter receptor

*Researchers estimate that approximately 1–2% of human genes are subjected to genomic imprinting, but fewer than 100 have actually been demonstrated to be imprinted.

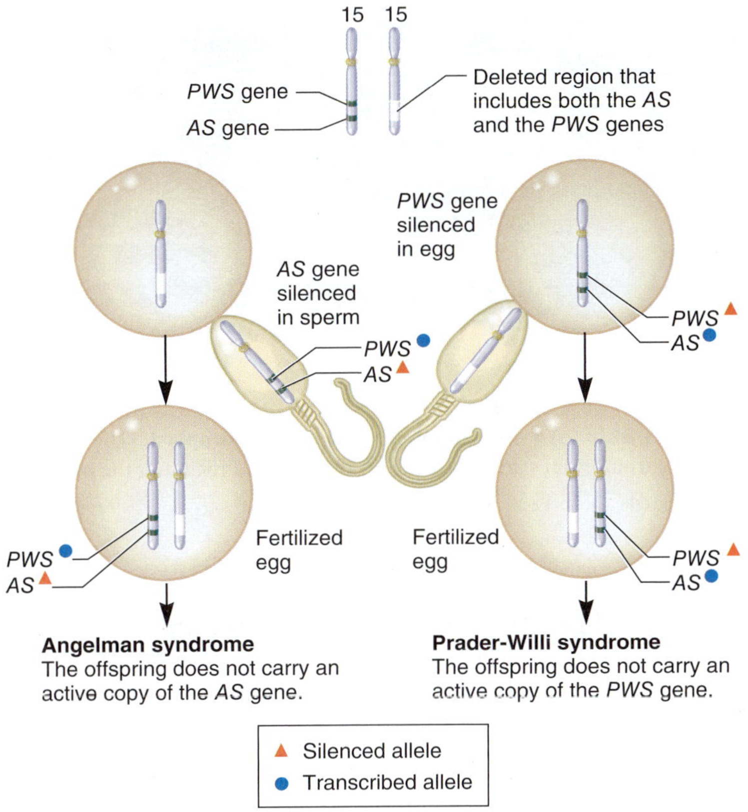

INTERACTIVE EXERCISE

FIGURE 5.12 The role of imprinting in the development of Angelman and Prader-Willi syndromes.

Genes → Traits A small region on chromosome 15 contains two different genes designated the *AS* gene and *PWS* gene in this figure. The *AS* gene is silenced during sperm formation and the *PWS* gene is silenced during egg formation. If a chromosome 15 deletion of this small region is inherited from the mother, Angelman syndrome occurs because the offspring does not inherit an active copy of the *AS* gene (left). Alternatively, if the chromosome 15 deletion is inherited from the father, Prader-Willi syndrome occurs because the offspring does not inherit an active copy of the *PWS* gene (right).

CONCEPT CHECK: On the left side of this figure, explain why the offspring does not have Prader-Willi syndrome but does have Angelman syndrome.

If this deletion is inherited from the mother, it leads to Angelman syndrome; if inherited from the father, it leads to Prader-Willi syndrome (**Figure 5.12**).

Researchers have discovered that this region of chromosome 15 contains closely linked but distinct genes that are maternally or paternally imprinted. AS results from the lack of expression of a single gene (*UBE3A*) that encodes a protein that regulates protein degradation. If the maternal allele is deleted, as in the left side of Figure 5.12, the individual will develop AS because she or he will not have an active copy of the *UBE3A* gene.

The gene(s) responsible for PWS has (have) not been definitively determined, although several imprinted genes are located in this region of chromosome 15. One gene that is thought to play a role in PWS is a gene designated *SNRPN*. This gene encodes a small nuclear ribonucleoprotein polypeptide N, which is part of a complex that controls RNA splicing. The maternal allele of *SNRPN* is silenced; only the paternal copy is active.

5.3 COMPREHENSION QUESTIONS

1. In mice, the copy of the *Igf2* gene that is inherited from the mother is never expressed in her offspring. This happens because the *Igf2* gene from the mother
   a. always undergoes a mutation that inactivates its function.
   b. is deleted during oogenesis.
   c. is deleted during embryonic development.
   d. is not transcribed in the somatic cells of the offspring.
2. A female mouse, *Igf2* $Igf2^-$, is crossed to a male that is also *Igf2* $Igf2^-$. The expected outcome of the offspring for this cross is
   a. all normal.
   b. all dwarf.
   c. 1 normal : 1 dwarf.
   d. 3 normal : 1 dwarf.
3. The marking process for genomic imprinting initially occurs during
   a. gametogenesis.
   b. fertilization.
   c. embryonic development.
   d. adulthood.
4. A female born with Angelman syndrome carries a deletion in the *AS* gene (i.e., the *UBE3A* gene). Which parent transmitted the deletion to her?
   a. Her father
   b. Her mother
   c. Either her mother or father

5.4 EXTRANUCLEAR INHERITANCE

Learning Outcomes:

1. Describe the general features of the mitochondrial and chloroplast genomes.
2. Predict the outcome of crosses involving extranuclear inheritance.
3. Explain how mutations in mitochondrial genes cause human diseases.
4. Evaluate the endosymbiosis theory.

Thus far, we have considered several types of non-Mendelian inheritance patterns. These include maternal effect genes, dosage compensation, and genomic imprinting. All of these inheritance patterns involve **nuclear genes**—genes located on chromosomes that are in the cell nucleus. Another cause of non-Mendelian inheritance patterns involves genes that are not located in the cell nucleus. In eukaryotic species, the most biologically important example of extranuclear inheritance is due to genetic material in cellular organelles. In addition to the cell nucleus, the mitochondria

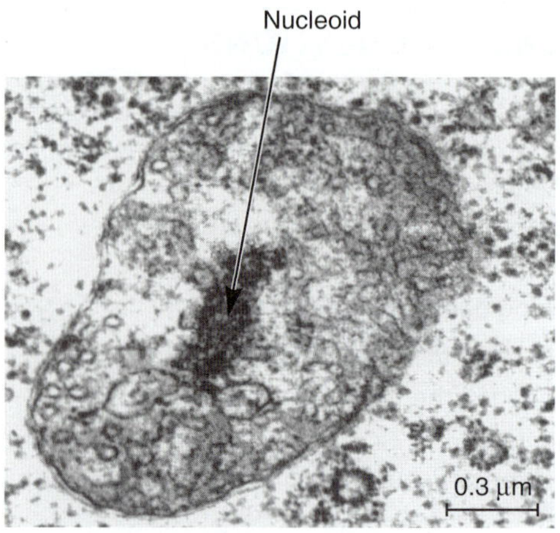

(a) Mitochondrion

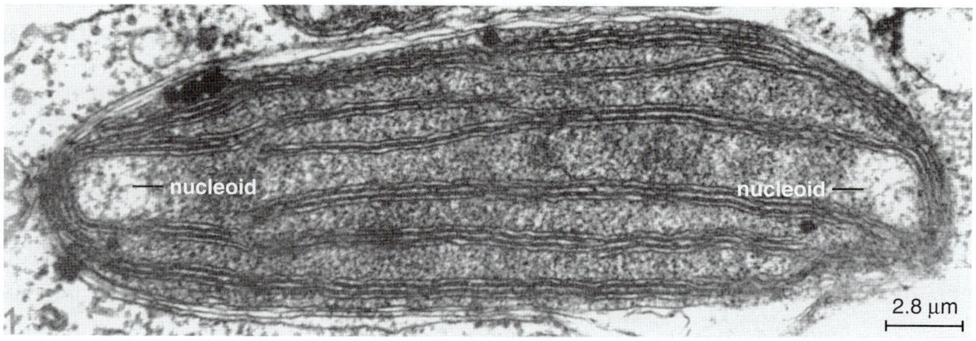

(b) Chloroplast

FIGURE 5.13 Nucleoids within (a) a mitochondrion and (b) a chloroplast. The mitochondrial and chloroplast chromosomes are found within the nucleoid region of the organelle.

CONCEPT CHECK: How is a nucleoid different from a cell nucleus?

and chloroplasts contain their own genetic material. Because these organelles are found within the cytoplasm of cells, the inheritance of organellar genetic material is called **extranuclear inheritance** (the prefix *extra-* means outside of) or **cytoplasmic inheritance.** In this section, we will examine the genetic composition of mitochondria and chloroplasts and explore the pattern of transmission of these organelles from parent to offspring.

Mitochondria and Chloroplasts Contain Circular Chromosomes with Many Genes

In 1951, Yukako Chiba was the first to suggest that chloroplasts contain their own DNA. He based his conclusion on the staining properties of a DNA-specific dye known as Feulgen. Researchers later developed techniques to purify organellar DNA. In addition, electron microscopy studies provided interesting insights into the organization and composition of mitochondrial and chloroplast chromosomes. More recently, the advent of molecular genetic techniques in the 1970s and 1980s has allowed researchers to determine the genome sequences of organellar DNAs. From these types of studies, the chromosomes of mitochondria and chloroplasts were found to resemble smaller versions of bacterial chromosomes.

The genetic material of mitochondria and chloroplasts is located inside the organelle in a region known as the **nucleoid** (Figure 5.13). The genome is a single circular chromosome (composed of double-stranded DNA), although a nucleoid contains several copies of this chromosome. In addition, a mitochondrion or chloroplast often has more than one nucleoid. In mice, for example, each mitochondrion has one to three nucleoids, with each nucleoid containing two to six copies of the circular mitochondrial genome. However, this number varies depending on the type of cell and the stage of development. In comparison, the chloroplasts of algae and higher plants tend to have more nucleoids per organelle. Table 5.3 describes the genetic composition of mitochondria and chloroplasts for a few selected species.

The sizes of mitochondrial and chloroplast genomes also vary greatly among different species. For example, a 400-fold variation is found in the sizes of mitochondrial chromosomes. In general, the mitochondrial genomes of animal species tend to be fairly small; those of fungi and protists are intermediate in size; and those of plant cells tend to be fairly large. Among algae and plants, substantial variation is also found in the sizes of chloroplast chromosomes.

TABLE 5.3
Genetic Composition of Mitochondria and Chloroplasts

Species	Organelle	Nucleoids per Organelle	Number of Chromosomes per Nucleoid
Tetrahymena	Mitochondrion	1	6–8
Mouse	Mitochondrion	1–3	2–6
Chlamydomonas	Chloroplast	5–6	~15
Euglena	Chloroplast	20–34	10–15
Flowering plants	Chloroplast	12–25	3–5

Data from: Gillham, N. W. (1994). *Organelle Genes and Genomes*. Oxford University Press, New York.

Figure 5.14 illustrates a map of human **mitochondrial DNA (mtDNA).** Each copy of the mitochondrial chromosome consists of a circular DNA molecule that is only 17,000 bp in length. This size is less than 1% of a typical bacterial chromosome. The human mtDNA carries relatively few genes. Thirteen genes encode proteins that function within the mitochondrion. In addition, mtDNA carries genes that encode ribosomal RNA and transfer RNA. These rRNAs and tRNAs are necessary for the synthesis of the 13 polypeptides that are encoded by the mtDNA. The primary role of mitochondria is to provide cells with the bulk of their adenosine triphosphate (ATP), which is used as an energy source to drive cellular reactions. These 13 polypeptides are subunits of proteins that function in a process known as oxidative phosphorylation, in which mitochondria use oxygen and synthesize ATP. However, mitochondria require many additional proteins to carry out oxidative phosphorylation and other mitochondrial functions. Most mitochondrial proteins are encoded

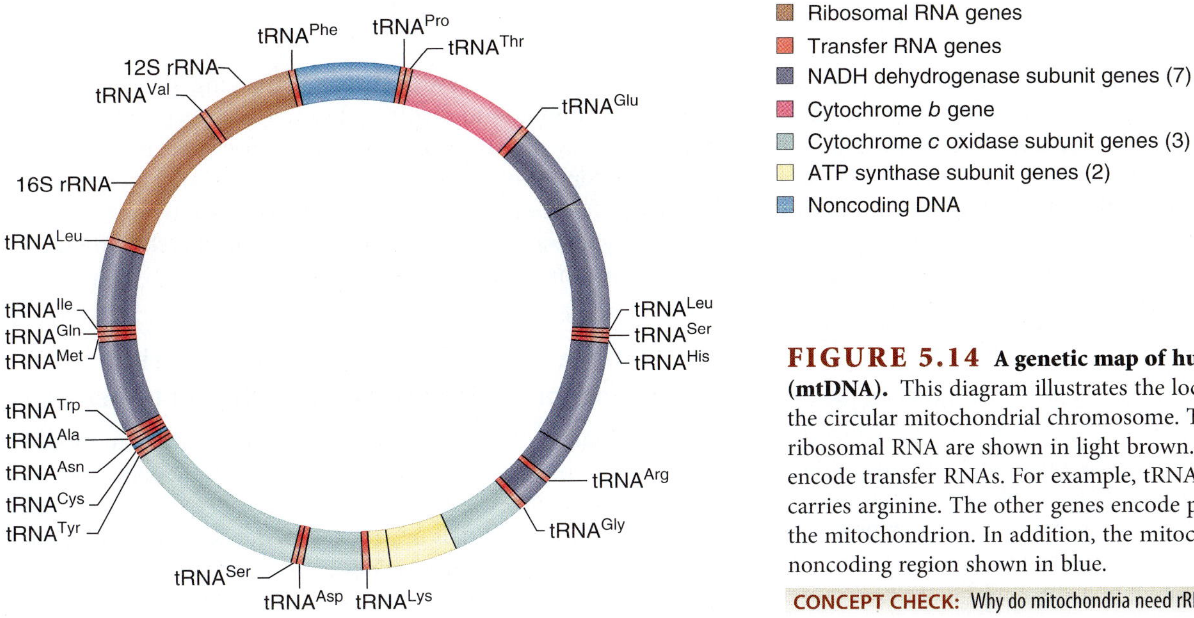

FIGURE 5.14 A genetic map of human mitochondrial DNA (mtDNA). This diagram illustrates the locations of many genes along the circular mitochondrial chromosome. The genes that encode ribosomal RNA are shown in light brown. The genes shown in red encode transfer RNAs. For example, tRNA^{Arg} encodes a tRNA that carries arginine. The other genes encode proteins that function within the mitochondrion. In addition, the mitochondrial genome has a noncoding region shown in blue.

CONCEPT CHECK: Why do mitochondria need rRNA and tRNA genes?

by genes within the cell nucleus. When these nuclear genes are expressed, the mitochondrial polypeptides are first synthesized outside the mitochondria in the cytosol of the cell. They are then transported into the mitochondria where they may associate with other polypeptides and become functional proteins.

Chloroplast genomes tend to be larger than mitochondrial genomes, and they have a correspondingly greater number of genes. A typical chloroplast genome is approximately 100,000 to 200,000 bp in length, which is about 10 times larger than the mitochondrial genome of animal cells. For example, the **chloroplast DNA (cpDNA)** of the tobacco plant is a circular DNA molecule that contains 156,000 bp of DNA and carries between 110 and 120 different genes. These genes encode ribosomal RNAs, transfer RNAs, and many proteins required for photosynthesis. As with mitochondria, many chloroplast proteins are encoded by genes found in the plant cell nucleus. These proteins contain chloroplast-targeting signals that direct them into the chloroplasts.

Extranuclear Inheritance Produces Non-Mendelian Results in Reciprocal Crosses

In diploid eukaryotic species, most genes within the nucleus obey a Mendelian pattern of inheritance because the homologous pairs of chromosomes segregate and independently assort during meiosis. Except for sex-linked traits, offspring inherit one copy of each gene from both the maternal and paternal parents. The sorting of chromosomes during meiosis explains the inheritance patterns of nuclear genes. By comparison, the inheritance of extranuclear genetic material does not display a Mendelian pattern. Mitochondria and chloroplasts are not sorted during meiosis and therefore do not segregate into gametes in the same way as nuclear chromosomes.

In 1909, Carl Correns discovered a trait that showed a non-Mendelian pattern of inheritance involving pigmentation in *Mirabilis jalapa* (the four-o'clock plant). Leaves can be green, white, or variegated with both green and white sectors. Correns demonstrated that the pigmentation of the offspring depended solely on the maternal parent (**Figure 5.15**). If the female parent had

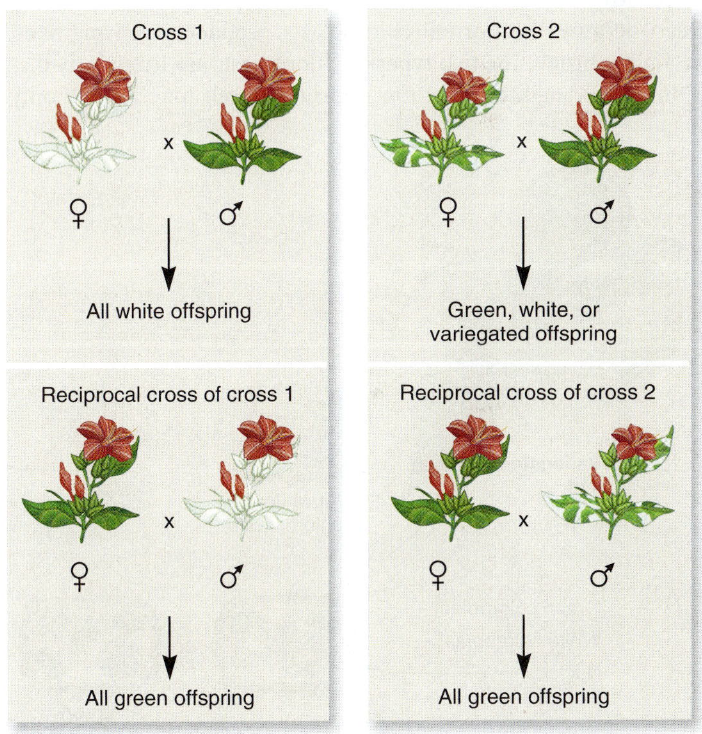

INTERACTIVE EXERCISE

FIGURE 5.15 Maternal inheritance in the four-o'clock plant, *Mirabilis jalapa*. The reciprocal crosses of four-o'clock plants by Carl Correns consisted of a pair of crosses between white-leaved and green-leaved plants, and a pair of crosses between variegated-leaved and green-leaved plants.

Genes → Traits In this example, the green phenotype is due to the synthesis of chlorophyll by normal chloroplasts, whereas the white phenotype is due to chloroplasts that carry a mutant allele that diminishes green pigmentation. The variegated phenotype is described in Figure 5.16. In the crosses shown here, the parent providing the eggs determines the phenotypes of the offspring. This is due to maternal inheritance. The egg contains the chloroplasts that are inherited by the offspring. (Note: The defective chloroplasts that give rise to white sectors are not completely defective in chlorophyll synthesis. Therefore, entirely white plants can survive, though they are smaller than green or variegated plants.)

CONCEPT CHECK: What is a reciprocal cross?

white pigmentation, all offspring had white leaves. Similarly, if the female was green, all offspring were green. When the female was variegated, the offspring could be green, white, or variegated.

The pattern of inheritance observed by Correns is a type of extranuclear inheritance called **maternal inheritance** (not to be confused with maternal effect). Chloroplasts are a type of plastid that makes chlorophyll, a green photosynthetic pigment. Maternal inheritance occurs because the chloroplasts are inherited only through the cytoplasm of the egg. The pollen grains of *M. jalapa* do not transmit chloroplasts to the offspring.

The phenotypes of leaves can be explained by the types of chloroplasts within the leaf cells. The green phenotype, which is the wild-type condition, is due to the presence of normal chloroplasts that make green pigment. By comparison, the white phenotype is due to a mutation in a gene within the chloroplast DNA that diminishes the synthesis of green pigment.

How does a variegated phenotype occur? **Figure 5.16** considers the leaf of a plant that began from a fertilized egg that contained both types of chloroplasts, a condition known as **heteroplasmy.** A leaf cell containing both types of chloroplasts is green because the normal chloroplasts produce green pigment. As a plant grows, the two types of chloroplasts are irregularly distributed to daughter cells. On occasion, a cell may receive only the chloroplasts that have a defect in making green pigment. Such a cell continues to divide and produces a sector of the plant that is entirely white. In this way, the variegated phenotype is produced. Similarly, if we consider the results of Figure 5.15, a variegated female parent may transmit green, white, or a mixture of these types of chloroplasts to the egg cell, thereby producing green, white, or variegated offspring, respectively.

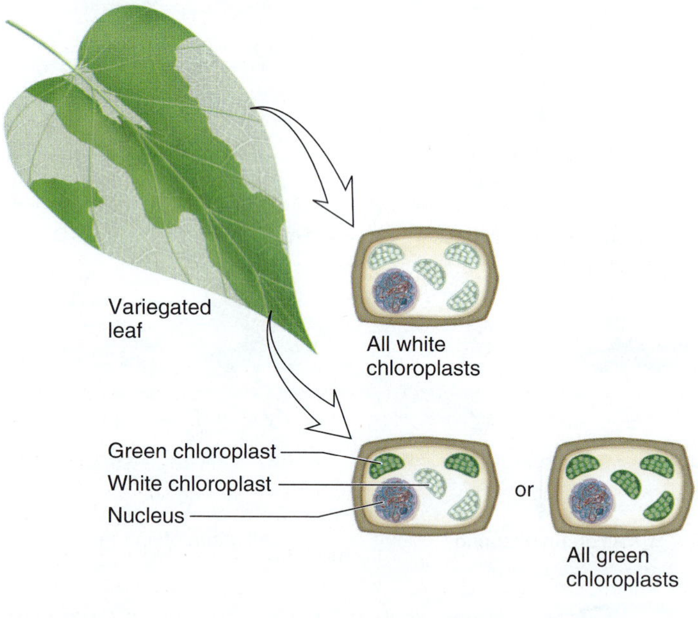

FIGURE 5.16 A cellular explanation of the variegated phenotype in *Mirabilis jalapa*. This plant inherited two types of chloroplasts—those that can produce green pigment and those that are defective. As the plant grows, the two types of chloroplasts are irregularly distributed to daughter cells. On occasion, a leaf cell may receive only the chloroplasts that are defective at making green pigment. Such a cell continues to divide and produces a sector of the leaf that is entirely white. Cells that contain both types of chloroplasts or cells that contain only green chloroplasts produce green tissue, which may be adjacent to a sector of white tissue. This is the basis for the variegated phenotype of the leaves.

CONCEPT CHECK: During growth, can a patch of tissue with a white phenotype give rise to a patch with a green phenotype? Explain.

The Pattern of Inheritance of Mitochondria and Chloroplasts Varies Among Different Species

The inheritance of traits via genetic material within mitochondria and chloroplasts is now a well-established phenomenon that geneticists have investigated in many different species. In **heterogamous** species, two kinds of gametes are made. The female gamete tends to be large and provides most of the cytoplasm to the zygote, whereas the male gamete is small and often provides little more than a nucleus. Therefore, mitochondria and chloroplasts are most often inherited from the maternal parent. However, this is not always the case. **Table 5.4** describes the inheritance patterns of mitochondria and chloroplasts in several selected species.

In species in which maternal inheritance is generally observed, the paternal parent may occasionally provide mitochondria via the sperm. This phenomenon is called **paternal leakage.** In the mouse, for example, approximately one to four paternal mitochondria are inherited for every 100,000 maternal mitochondria per generation of offspring. Most offspring do not inherit any paternal mitochondria, but a rare individual may inherit a mitochondrion from the sperm.

Many Human Diseases Are Caused by Mitochondrial Mutations

Researchers have identified many diseases that are due to mutations in the mitochondrial DNA. Such diseases can occur in two ways. In some cases, mitochondrial mutations are transmitted from mother to offspring. Human mtDNA is maternally inherited because it is transmitted from mother to offspring via the cytoplasm of the egg. Therefore, the transmission of inherited human mitochondrial diseases follows a maternal inheritance pattern. In addition, mitochondrial mutations may occur in somatic cells and accumulate as a person ages. Researchers have discovered that mitochondria are particularly susceptible to DNA damage. When more oxygen is consumed than is actually used to make ATP, mitochondria tend to produce free radicals that damage DNA. Unlike nuclear DNA, mitochondrial DNA has very limited repair abilities and almost no protective ability against free radical damage.

Table 5.5 describes several mitochondrial diseases that have been discovered in humans and are caused by mutations in mitochondrial genes. Over 200 diseases associated with defective mitochondria have been identified. These are usually chronic degenerative disorders that affect cells requiring a high level of ATP, such as nerve and muscle cells. For example, Leber hereditary optic neuropathy (LHON) affects the optic nerve and may lead to the progressive loss of vision in one or both eyes. LHON can be caused by a defective mutation in one of several different

TABLE 5.4

Transmission of Organelles Among Different Species

Species	Organelle	Transmission
S. cerevisiae (Yeast)	Mitochondria	Biparental inheritance
Molds	Mitochondria	Usually maternal inheritance; paternal inheritance has been found in the genus *Allomyces*
Chlamydomonas (Alga)	Mitochondria	*Chlamydomonas* exists in two mating types (mt^+ and mt^-). Inherited from the parent with the mt^- mating type
Chlamydomonas	Chloroplasts	Inherited from the parent with the mt^+ mating type
Angiosperms (Plants)	Mitochondria and chloroplasts	Often maternal inheritance, although biparental inheritance is found among some species
Gymnosperms (Plants)	Mitochondria and chloroplasts	Usually paternal inheritance
Mammals	Mitochondria	Maternal inheritance

TABLE 5.5

Examples of Human Mitochondrial Diseases

Disease	Mitochondrial Gene Mutated
Leber hereditary optic neuropathy	A mutation in one of several mitochondrial genes that encode respiratory chain proteins: ND1, ND2, CO1, ND4, ND5, ND6, and cytb.; tends to affect males more than females
Neurogenic muscle weakness	A mutation in the *ATPase6* gene that encodes a subunit of the mitochondrial ATP-synthetase, which is required for ATP synthesis
Mitochondrial encephalomyopathy, lactic acidosis, and strokelike episodes	A mutation in genes that encode tRNAs for leucine and lysine
Mitochondrial myopathy	A mutation in a gene that encodes a tRNA for leucine
Maternal myopathy and cardiomyopathy	A mutation in a gene that encodes a tRNA for leucine
Myoclonic epilepsy with ragged-red muscle fibers	A mutation in a gene that encodes a tRNA for lysine

mitochondrial genes. Researchers are still investigating how a defect in these mitochondrial genes produces the symptoms of this disease.

An important factor in mitochondrial disease is heteroplasmy, which means that a cell contains a mixed population of mitochondria. Within a single cell, some mitochondria may carry a disease-causing mutation but others may not. As cells divide, mutant and normal mitochondria randomly segregate into the resulting daughter cells. Some daughter cells may receive a high ratio of mutant to normal mitochondria, whereas others may receive a low ratio. The ratio of mutant to normal mitochondria must exceed a certain threshold value before disease symptoms are observed. Because of the random nature of heteroplasmy, the symptoms of mitochondrial diseases often vary widely within a given family. Some members that carry the mutant mitochondria may be asymptomatic, whereas others exhibit mild to severe symptoms.

Extranuclear Genomes of Chloroplasts and Mitochondria Evolved from an Endosymbiotic Relationship

The idea that the nucleus, chloroplasts, and mitochondria contain their own separate genetic material may at first seem puzzling. Wouldn't it be simpler to have all of the genetic material in one place in the cell? The underlying reason for distinct genomes of chloroplasts and mitochondria can be traced back to their evolutionary origin, which is thought to involve a symbiotic association.

A symbiotic relationship occurs when two different species live together in a close association. The symbiont is the smaller of the two species; the host is the larger. The term **endosymbiosis** describes a symbiotic relationship in which the symbiont actually lives inside (*endo-*, inside) the host. In 1883, Andreas Schimper proposed that chloroplasts were descended from an endosymbiotic relationship between cyanobacteria and eukaryotic cells. This idea, now known as the **endosymbiosis theory,** suggested that the ancient origin of chloroplasts was initiated when a cyanobacterium took up residence within a primordial eukaryotic cell (**Figure 5.17**). Over the course of evolution, the characteristics of the intracellular bacterial cell gradually changed to those of a chloroplast. In 1922, Ivan Wallin also proposed an endosymbiotic origin for mitochondria.

In spite of these hypotheses, the question of endosymbiosis was largely ignored until researchers in the 1950s discovered that chloroplasts and mitochondria contain their own genetic material. The issue of endosymbiosis was hotly debated after Lynn Margulis published a book entitled *Origin of Eukaryotic Cells* (1970). During the 1970s and 1980s, the advent of molecular genetic techniques allowed researchers to analyze genes from chloroplasts, mitochondria, bacteria, and eukaryotic nuclear genomes. They found that genes in chloroplasts and mitochondria are very similar to bacterial genes but not as similar to those found within the nucleus of eukaryotic cells. This observation provided strong support for the endosymbiotic origin of chloroplasts and mitochondria, which is now widely accepted.

The endosymbiosis theory proposes that the relationship provided eukaryotic cells with useful cellular characteristics. Chloroplasts were derived from cyanobacteria, a bacterial species that is capable of photosynthesis. The ability to carry out photosynthesis enabled algal and plant cells to use the energy from sunlight. By comparison, mitochondria are thought to have been derived from a different type of bacteria known as gram-negative nonsulfur purple bacteria. In this case, the endosymbiotic relationship enabled eukaryotic cells to synthesize greater amounts of ATP. It is less clear how the relationship would have been beneficial to cyanobacteria or purple bacteria, though the cytosol of a eukaryotic cell may have provided a stable environment with an adequate supply of nutrients.

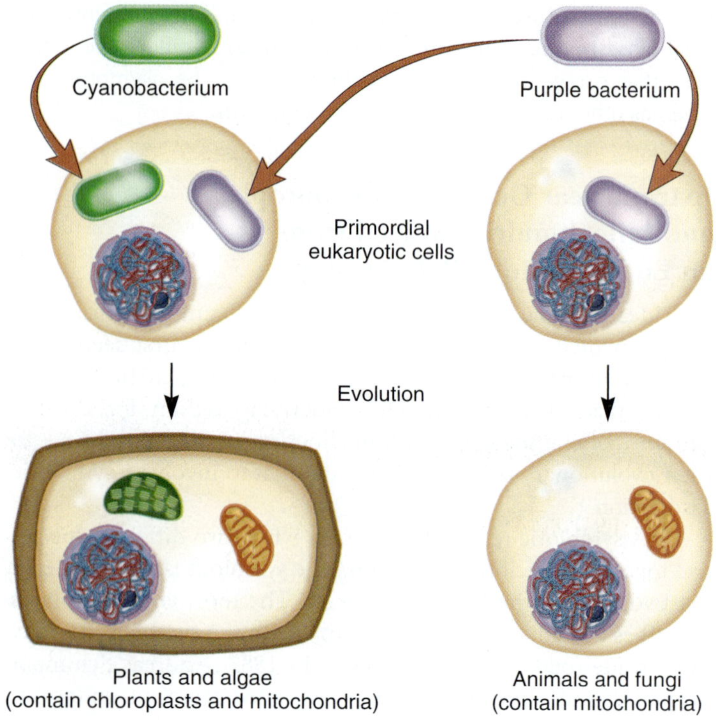

FIGURE 5.17 The endosymbiotic origin of chloroplasts and mitochondria. According to the endosymbiotic theory, chloroplasts descended from an endosymbiotic relationship between cyanobacteria and eukaryotic cells. This arose when a bacterium took up residence within a primordial eukaryotic cell. Over the course of evolution, the intracellular bacterial cell gradually changed its characteristics, eventually becoming a chloroplast. Similarly, mitochondria are derived from an endosymbiotic relationship between purple bacteria and eukaryotic cells.

CONCEPT CHECK: How have chloroplasts and mitochondria changed since the initial endosymbiosis events, which occurred hundreds of millions of years ago?

During the evolution of eukaryotic species, most genes that were originally found in the genome of the primordial cyanobacteria and purple bacteria have been lost or transferred from the organelles to the nucleus. The sequences of certain genes within the nucleus are consistent with their origin within an organelle. Such genes are more similar in their DNA sequence to known bacterial genes than to their eukaryotic counterparts. Therefore, researchers have concluded that these genes have been removed from the chloroplast and mitochondrial chromosomes and relocated to the nuclear chromosomes. This has occurred many times throughout evolution, so modern chloroplasts and mitochondria have lost most of the genes that are still found in present-day cyanobacteria and purple bacteria.

Most of this gene transfer occurred early in chloroplast and mitochondrial evolution. The functional transfer of mitochondrial genes seems to have ceased in animals, but gene transfer from chloroplasts and mitochondria to the nucleus continues to occur in plants at a low rate. The molecular mechanism of gene transfer is not entirely understood, but the direction of transfer is well established. During evolution, gene transfer has occurred primarily from the organelles to the nucleus. For example, about 1500 genes have been transferred from the mitochondrial genome to the nuclear genome of eukaryotes. Transfer of genes from the nucleus to the organelles has almost never occurred, although one example is known of a nuclear gene in plants that has been transferred to the mitochondrial genome. This unidirectional gene transfer from organelles to the nucleus partly explains why the organellar genomes now contain relatively few genes. In addition, gene transfer can occur between organelles. It can happen between two mitochondria, between two chloroplasts, and between a chloroplast and mitochondrion. Overall, the transfer of genetic material between the nucleus, chloroplasts, and mitochondria is an established phenomenon, although its biological benefits remain unclear.

5.4 COMPREHENSION QUESTIONS

1. Extranuclear inheritance occurs due to
   a. chromosomes that may become detached from the spindle during meiosis.
   b. genetic material that is found in chloroplasts and mitochondria.
   c. mutations that disrupt the integrity of the nuclear membrane.
   d. none of the above.
2. A cross is made between a green four-o'clock and a variegated four o'clock. If the variegated plant provides the pollen, the expected outcome of the offspring would be
   a. all plants with green leaves.
   b. 3 plants with green leaves to 1 plant with variegated leaves.
   c. 3 plants with green leaves to 1 plant with white leaves.
   d. some plants with green leaves, some with variegated leaves, and some with white leaves.
3. Some human diseases are caused by mutations in mitochondrial genes. Which of the following statements is *false*?
   a. Human mitochondrial diseases follow a maternal inheritance pattern.
   b. Mutations associated with mitochondrial diseases often affect cells with a high demand for ATP.
   c. The symptoms associated with mitochondrial diseases tend to improve with age.
   d. Heteroplasmy plays a key role in the severity of disease symptoms.
4. Chloroplasts and mitochondria evolved from an endosymbiotic relationship involving
   a. purple bacteria and cyanobacteria, respectively.
   b. cyanobacteria and purple bacteria, respectively.
   c. cyanobacteria.
   d. purple bacteria.

KEY TERMS

Page 103. maternal effect, reciprocal cross
Page 105. epigenetic inheritance
Page 106. dosage compensation, X-chromosome inactivation (XCI), Barr body, Lyon hypothesis
Page 108. clone
Page 110. X-inactivation center (Xic)
Page 112. genomic imprinting, monoallelic expression
Page 113. DNA methylation, imprinting control region (ICR)
Page 115. nuclear genes
Page 116. extranuclear inheritance, cytoplasmic inheritance, nucleoid, mitochondrial DNA (mtDNA)
Page 117. chloroplast DNA (cpDNA)
Page 118. maternal inheritance, heteroplasmy, heterogamous, paternal leakage
Page 119. endosymbiosis, endosymbiosis theory

CHAPTER SUMMARY

- Non-Mendelian inheritance refers to inheritance patterns that do not conform to (at least) one of the four rules of Mendelian inheritance.

5.1 Maternal Effect

- Maternal effect is an inheritance pattern in which the genotype of the mother determines the phenotype of the offspring. It occurs because gene products of maternal effect genes are transferred from nurse cells to the oocyte. These gene products affect early stages of development (see Figures 5.1, 5.2).

5.2 Epigenetic Inheritance: Dosage Compensation

- Epigenetic inheritance is a pattern in which a gene or chromosome is modified and gene expression is altered, but the modification is not permanent over the course of many generations.
- Dosage compensation often occurs in species that differ in their sex chromosomes (see Table 5.1).
- In mammals, the process of X-chromosome inactivation in females compensates for the single X chromosome found in males. The inactivated X chromosome is called a Barr body. The process can lead to a variegated phenotype, such as a calico cat (see Figures 5.3, 5.4).
- After it occurs during embryonic development, the pattern of X-chromosome inactivation is maintained when cells divide (see Figures 5.5, 5.6).
- X-chromosome inactivation is controlled by the X-inactivation center that contains the *Xist* and *Tsix* genes. XCI occurs as initiation, spreading, and maintenance phases (see Figures 5.7, 5.8).

5.3 Epigenetic Inheritance: Genomic Imprinting

- Genomic imprinting refers to a marking process in which an offspring expresses a gene that is inherited from one parent but not both (see Figures 5.9, 5.10).
- DNA methylation at imprinting control regions is the marking process that causes imprinting (see Figure 5.11).
- Human diseases such as Prader-Willi syndrome and Angelman syndrome are associated with genomic imprinting (see Figure 5.12, Table 5.2).

5.4 Extranuclear Inheritance

- Extranuclear inheritance involves the inheritance of genes that are found in mitochondria or chloroplasts.
- Mitochondria and chloroplasts carry circular chromosomes in a nucleoid region. These circular chromosomes contain relatively few genes compared with the number in the cell nucleus (see Figures 5.13–5.14, Table 5.3).
- Maternal inheritance occurs when organelles, such as mitochondria or chloroplasts, are transmitted via the egg (see Figure 5.15).
- Heteroplasmy for chloroplasts can result in a variegated phenotype (see Figure 5.16).
- The transmission patterns of mitochondria and chloroplasts vary among different species (see Table 5.4).
- Many diseases are caused by mutations in mitochondrial DNA (see Table 5.5).
- Chloroplasts and mitochondria were derived from an ancient endosymbiotic relationship (see Figure 5.17).

PROBLEM SETS & INSIGHTS

Solved Problems

S1. Our understanding of maternal effect genes has been greatly aided by their identification in experimental organisms such as *Drosophila melanogaster* and *Caenorhabditis elegans*. In experimental organisms with a short generation time, geneticists have successfully searched for mutant alleles that prevent the normal process of embryonic development. In many cases, the offspring die at early embryonic or larval stages. These are called maternal effect lethal alleles. How would a researcher identify a mutation that produced a recessive maternal effect lethal allele?

Answer: A maternal effect lethal allele can be identified when a phenotypically normal mother produces only offspring with gross developmental abnormalities. For example, let's call the normal allele *N* and the maternal effect lethal allele *n*. A cross between two flies that are heterozygous for a maternal effect lethal allele would produce 1/4 of the offspring with a homozygous genotype, *nn*. These flies are viable because of the maternal effect. Their mother would be *Nn* and provide the *n* oocyte with a sufficient amount of *N* gene product so the *nn* flies would develop properly. However, homozygous *nn* females cannot provide their oocytes with any normal gene product. Therefore, all of their offspring are abnormal and die during early stages.

S2. A maternal effect gene in *Drosophila*, called *torso*, is found as a recessive allele that prevents the correct development of anterior- and posterior-most structures. A wild-type male is crossed to a female of unknown genotype. This mating produces 100% larva that are missing their anterior- and posterior-most structures and therefore die during early development. What is the genotype and phenotype of the female fly in this cross? What are the genotypes and phenotypes of the female fly's parents?

Answer: Because this cross produces 100% abnormal offspring, the female fly must be homozygous for the abnormal *torso* allele. Even so, the female fly must be phenotypically normal in order to reproduce. This female fly had a mother that was heterozygous for a normal and abnormal *torso* allele and a father that was either heterozygous or homozygous for the abnormal *torso* allele.

$torso^+\ torso^-$ (grandmother) × $torso^+\ torso^-$ or $torso^-\ torso^-$ (grandfather)

↓

$torso^-\ torso^-$
(mother of 100% abnormal offspring)

This female fly is phenotypically normal because its mother was heterozygous and provided the gene products of the $torso^+$ allele from the nurse cells. However, this homozygous female will produce only abnormal offspring because it cannot provide them with the normal $torso^+$ gene products.

S3. A person is born with three X chromosomes and one Y chromosome (XXXY). Would you expect this person to be a male or female? How many Barr bodies would you expect in their somatic cells?

Answer: The person would be a male because the Y chromosome determines maleness in humans. He would have two Barr bodies in each cell because the counting of Xics allows just one X chromosome to remain active.

S4. An individual with Angelman syndrome produced an offspring with Prader-Willi syndrome. Why does this occur? What are the sexes of the parent with Angelman syndrome and the offspring with Prader-Willi syndrome?

Answer: These two different syndromes are most commonly caused by a small deletion in chromosome 15. In addition, genomic imprinting plays a role because genes in this deleted region are differentially imprinted, depending on sex. If this deletion is inherited from the paternal parent, the offspring develops Prader-Willi syndrome. Therefore, in this problem, the person with Angelman syndrome must have been a male because he produced a child with Prader-Willi syndrome. The child could be either a male or a female.

S5. Suppose that you are a horticulturist who has recently identified an interesting plant with variegated leaves. How would you determine if this trait is nuclearly or cytoplasmically inherited?

Answer: Make crosses and reciprocal crosses involving normal and variegated strains. In many species, chloroplast genomes are inherited maternally, although this is not always the case. In addition, a significant percentage of paternal leakage may occur. Nevertheless, when reciprocal crosses yield different outcomes, an organellar mode of inheritance is possibly at work.

Conceptual Questions

C1. Define the term epigenetic inheritance, and describe two examples.

C2. Describe the inheritance pattern of maternal effect genes. Explain how the maternal effect occurs at the cellular level. What are the expected functional roles of the proteins that are encoded by maternal effect genes?

C3. A maternal effect gene exists in a dominant *N* (normal) allele and a recessive *n* (abnormal) allele. What would be the ratios of genotypes and phenotypes for the offspring of the following crosses?

A. *nn* female × *NN* male

B. *NN* female × *nn* male

C. *Nn* female × *Nn* male

C4. A *Drosophila* embryo dies during early embryogenesis due to a recessive maternal effect allele called *bicoid*. The wild-type allele is designated $bicoid^+$. What are the genotypes and phenotypes of the embryo's mother and maternal grandparents?

C5. For Mendelian traits, the nuclear genotype (i.e., the alleles found on chromosomes in the cell nucleus) directly influences an offspring's traits. In contrast, for non-Mendelian inheritance patterns, the offspring's phenotype cannot be reliably predicted solely from its genotype. For the following traits, what do you need to know to predict the phenotypic outcome?

A. Dwarfism due to a mutant *Igf2* allele

B. Snail coiling direction

C. Leber hereditary optic neuropathy

C6. Suppose a maternal effect gene exists as a normal dominant allele and an abnormal recessive allele. A mother who is phenotypically abnormal produces all normal offspring. Explain the genotype of the mother.

C7. Suppose that a gene affects the anterior morphology in house flies and is inherited as a maternal effect gene. The gene exists in a normal allele, *H*, and a recessive allele, *h*, which causes a small head. A female fly with a normal head is mated to a true-breeding male with a small head. All of the offspring have small heads. What are the genotypes of the mother and offspring? Explain your answer.

C8. Explain why maternal effect genes exert their effects during the early stages of development.

C9. As described in Chapter 21, researchers have been able to clone mammals by fusing a cell having a diploid nucleus (i.e., a somatic cell) with an egg that has had its nucleus removed.

A. With regard to maternal effect genes, would the phenotype of such a cloned animal be determined by the animal that donated the egg or by the animal that donated the somatic cell? Explain.

B. Would the cloned animal inherit extranuclear traits from the animal that donated the egg or by the animal that donated the somatic cell? Explain.

C. In what ways would you expect this cloned animal to be similar to or different from the animal that donated the somatic cell? Is it fair to call such an animal a "clone" of the animal that donated the diploid nucleus?

C10. With regard to the numbers of sex chromosomes, explain why dosage compensation is necessary.

C11. What is a Barr body? How is its structure different from that of other chromosomes in the cell? How does the structure of a Barr body affect the level of X-linked gene expression?

C12. Among different species, describe three distinct strategies for accomplishing dosage compensation.

C13. Describe when X-chromosome inactivation occurs and how this leads to phenotypic results at the organism level. In your answer, you should explain why XCI causes results such as variegated coat patterns in mammals. Why do two different calico cats have their patches of orange and black fur in different places? Explain whether or not a variegated coat pattern due to XCI could occur in marsupials.

C14. Describe the molecular process of X-chromosome inactivation. This description should include the three phases of inactivation and the role of the Xic. Explain what happens to X chromosomes during embryogenesis, in adult somatic cells, and during oogenesis.

C15. On rare occasions, an abnormal human male is born who is somewhat feminized compared with normal males. Microscopic examination of the cells of one such individual revealed that he has a single Barr body in each cell. What is the chromosomal composition of this individual?

C16. How many Barr bodies would you expect to find in humans with the following abnormal compositions of sex chromosomes?

A. XXY

B. XYY

C. XXX

D. X0 (a person with just a single X chromosome)

C17. Certain forms of human color blindness are inherited as X-linked recessive traits. Hemizygous males are color-blind, but heterozygous females are not. However, heterozygous females sometimes have partial color blindness.

A. Discuss why heterozygous females may have partial color blindness.

B. Doctors identified an unusual case in which a heterozygous female was color-blind in her right eye but had normal color vision in her left eye. Explain how this might have occurred.

C18. A black female cat (X^BX^B) and an orange male cat (X^OY) were mated to each other and produced a male cat that was calico. Which sex chromosomes did this male offspring inherit from its mother and father? Remember that the presence of the Y chromosome determines maleness in mammals.

C19. What is the spreading phase of X-chromosome inactivation? Why do you think it is called a spreading phase? Discuss the role of the *Xist* gene in the spreading phase of XCI.

C20. When does the erasure and reestablishment phase of genomic imprinting occur? Explain why it is necessary to erase an imprint and then reestablish it in order to always maintain imprinting from the same sex of parent.

C21. In what types of cells would you expect de novo methylation to occur? In what cell types would it not occur?

C22. On rare occasions, people are born with a condition known as uniparental disomy. It happens when an individual inherits both copies of a chromosome from one parent and no copies from the other parent. This occurs when two abnormal gametes happen to complement each other to produce a diploid zygote. For example, an abnormal sperm that lacks chromosome 15 could fertilize an egg that contains two copies of chromosome 15. In this situation, the individual would be said to have maternal uniparental disomy 15 because both copies of chromosome 15 were inherited from the mother. Alternatively, an abnormal sperm with two copies of chromosome 15 could fertilize an egg with no copies. This is known as paternal uniparental disomy 15. If a female is born with paternal uniparental disomy 15, would you expect her to be phenotypically normal, have Angelman syndrome (AS), or have Prader-Willi syndrome (PWS)? Explain. Would you expect her to produce normal offspring or offspring affected with AS or PWS?

C23. Genes that cause Prader-Willi syndrome and Angelman syndrome are closely linked along chromosome 15. Although people with these syndromes do not usually reproduce, let's suppose that a couple produces two children with Angelman syndrome. The oldest child (named Pat) grows up and has two children with Prader-Willi syndrome. The second child (named Robin) grows up and has one child with Angelman syndrome.

A. Are Pat and Robin's parents both normal or does one of them have Angelman or Prader-Willi syndrome? If one of them has a disorder, explain why it is the mother or the father.

B. What are the sexes of Pat and Robin? Explain.

C24. How is the process of X-chromosome inactivation similar to genomic imprinting? How is it different?

C25. What is extranuclear inheritance? Describe three examples.

C26. What is a reciprocal cross? Suppose that a gene is found as a wild-type allele and a recessive mutant allele. What would be the expected outcomes of reciprocal crosses if a true-breeding normal individual was crossed to a true-breeding individual carrying the mutant allele? What would be the results if the gene is maternally inherited?

C27. Among different species, does extranuclear inheritance always follow a maternal inheritance pattern? Why or why not?

C28. Extranuclear inheritance often correlates with maternal inheritance. Even so, paternal leakage is not uncommon. What is paternal leakage? If a cross produced 200 offspring and the rate of mitochondrial paternal leakage was 3%, how many offspring would be expected to contain paternal mitochondria?

C29. Discuss the structure and organization of the mitochondrial and chloroplast genomes. How large are they, how many genes do they contain, and how many copies of the genome are found in each organelle?

C30. Explain the likely evolutionary origin of mitochondrial and chloroplast genomes. How have the sizes of the mitochondrial and chloroplast genomes changed since their origin? How has this occurred?

C31. Which of the following traits or diseases are determined by nuclear genes?

A. Snail coiling pattern

B. Prader-Willi syndrome

C. Leber hereditary optic neuropathy

C32. Acute murine leukemia virus (AMLV) causes leukemia in mice. This virus is easily passed from mother to offspring through the mother's milk. (Note: Even though newborn offspring acquire the virus, they may not develop leukemia until much later in life. Testing can determine if an animal carries the virus.) Describe how the formation of leukemia via AMLV resembles a maternal inheritance pattern. How could you determine that this form of leukemia is not caused by extranuclear inheritance?

C33. Describe how a biparental pattern of extranuclear inheritance would resemble a Mendelian pattern of inheritance for a particular gene. How would they differ?

Experimental Questions

E1. Figure 5.1 describes an example of a maternal effect gene. Explain how Sturtevant deduced a maternal effect gene based on the F_2 and F_3 generations.

E2. Discuss the types of experimental observations that Mary Lyon brought together in proposing her hypothesis concerning X-chromosome inactivation. In your own words, explain how these observations were consistent with her hypothesis.

E3. Chapter 20 describes three blotting methods (i.e., Southern blotting, Northern blotting, and Western blotting) used to detect specific genes and gene products. Southern blotting detects DNA, Northern blotting detects RNA, and Western blotting detects proteins. Suppose that a female fruit fly is heterozygous for a maternal effect gene, which we will call gene *B*. The female is *Bb*. The normal allele, *B*, encodes a functional mRNA that is 550 nucleotides long. A recessive allele, *b*, encodes a shorter mRNA that is 375 nucleotides long. (Allele *b* is due to a deletion within this gene.) How could you use one or more of these techniques to show that nurse cells transfer gene products from gene *B* to developing oocytes? You may assume that you can dissect the ovaries of fruit flies and isolate oocytes separately from nurse cells. In your answer, describe your expected results.

E4. As a hypothetical example, a trait in mice results in mice with long tails. You initially have a true-breeding strain with normal tails and a true-breeding strain with long tails. You then make the following types of crosses:

Cross 1: When true-breeding females with normal tails are crossed to true-breeding males with long tails, all F_1 offspring have long tails.

Cross 2: When true-breeding females with long tails are crossed to true-breeding males with normal tails, all F_1 offspring have normal tails.

Cross 3: When F_1 females from cross 1 are crossed to true-breeding males with normal tails, all offspring have normal tails.

Cross 4: When F_1 males from cross 1 are crossed to true-breeding females with long tails, half of the offspring have normal tails and half have long tails.

Explain the pattern of inheritance of this trait.

E5. You have a female snail that coils to the right, but you do not know its genotype. You may assume that right coiling (*D*) is dominant to left coiling (*d*). You also have male snails of known genotype. How would you determine the genotype of this female snail? In your answer, describe your expected results depending on whether the female is *DD*, *Dd*, or *dd*.

E6. On a camping trip, you find one male snail on a deserted island that coils to the right. However, in this same area, you find several shells (not containing living snails) that coil to the left. Therefore, you conclude that you are not certain of the genotype of this male snail. On a different island, you find a large colony of snails of the same species. All of these snails coil to the right, and every snail shell that you find on this second island coils to the right. With regard to the maternal effect gene that determines coiling pattern, how would you determine the genotype of the male snail that you found on the deserted island? In your answer, describe your expected results.

E7. Figure 5.6 describes the results of X-chromosome inactivation in mammals. If fast and slow alleles of glucose-6-phosphate dehydrogenase (G-6-PD) exist in other species, what would be the expected results of gel electrophoresis for a heterozygous female of the following species?

A. Marsupial

B. *Drosophila melanogaster*

C. *Caenorhabditis elegans* (Note: We are considering the hermaphrodite in *C. elegans* to be equivalent to a female.)

E8. Two male mice, which we will call male A and male B, are both phenotypically normal. Male A was from a litter that contained half phenotypically normal mice and half dwarf mice. The mother of male A was known to be homozygous for the normal *Igf2* allele. Male B was from a litter of eight mice that were all phenotypically normal. The parents of male B were a phenotypically normal male and a dwarf female. Male A and male B were put into a cage with two female mice that we will call female A and female B. Female A is dwarf, and female B is phenotypically normal. The parents of these two females were unknown, although it was known that they were from the same litter. The mice were allowed to mate with each other, and the following data were obtained:

Female A gave birth to three dwarf babies and four normal babies.

Female B gave birth to four normal babies and two dwarf babies.

Which male(s) mated with female A and female B? Explain.

E9. In the experiment of Figure 5.6, why does a clone of cells produce only one type of G-6-PD enzyme? What would you expect to happen if a clone was derived from an early embryonic cell? Why does the initial sample of tissue produce both forms of G-6-PD?

E10. Chapter 20 describes a blotting method known as Northern blotting that is used to determine the amount of mRNA produced by a particular gene. In this method, the amount of a specific mRNA produced by cells is detected as a band on a gel. If one type of cell produces twice as much of a particular mRNA as another cell, the band will appear twice as dark. Also, sometimes mutations affect the length of mRNA that is transcribed from a gene. For example, a small deletion within a gene may shorten an mRNA. Northern blotting also can discern the sizes of mRNAs.

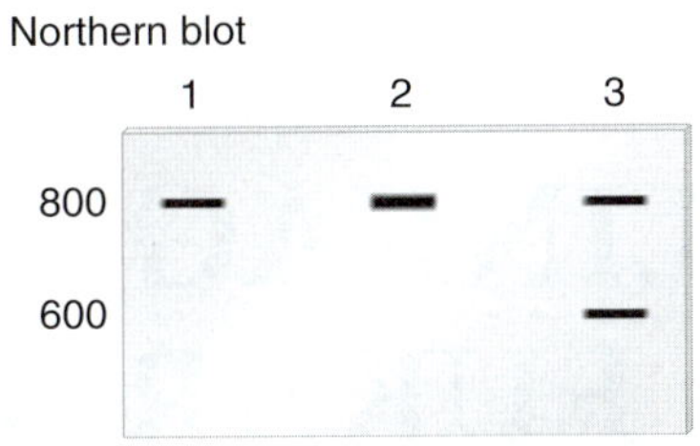

Lane 1 is a Northern blot of mRNA from cell type A that is 800 nucleotides long.

Lane 2 is a Northern blot of the same mRNA from cell type B. (Cell type B produces twice as much of this RNA as cell type A.)

Lane 3 shows a heterozygote in which one of the two genes has a deletion, which shortens the mRNA by 200 nucleotides.

Here is the question. Suppose an X-linked gene exists as two alleles: *B* and *b*. Allele *B* encodes an mRNA that is 750 nucleotides long, and allele *b* encodes a shorter mRNA that is 675 nucleotides long. Draw the expected results of a Northern blot using mRNA isolated from the same type of somatic cells taken from the following individuals:

A. First lane is mRNA from an X^bY male fruit fly.

Second lane is mRNA from an X^bX^b female fruit fly.

Third lane is mRNA from an X^BX^b female fruit fly.

B. First lane is mRNA from an X^BY male mouse.

Second lane is mRNA from an X^BX^b female mouse.*

Third lane is mRNA from an X^BX^B female mouse.*

C. First lane is mRNA from an X^B0 male *C. elegans*.

Second lane is mRNA from an X^BX^b hermaphrodite *C. elegans*.

Third lane is mRNA from an X^BX^B hermaphrodite *C. elegans*.

*The sample is taken from an adult female mouse. It is not a clone of cells. It is a tissue sample, like the one described in the experiment of Figure 5.6.

E11. A variegated trait in plants is analyzed using reciprocal crosses. The following results are obtained:

Variegated female × Normal male	Normal female × Variegated male
↓	↓
1024 variegated + 52 normal	1113 normal + 61 variegated

Explain this pattern of inheritance.

Questions for Student Discussion/Collaboration

1. Recessive maternal effect genes are identified in flies (for example) when a phenotypically normal mother cannot produce any normal offspring. Because all of the offspring are dead, this female fly cannot be used to produce a strain of heterozygous flies that could be used in future studies. How would you identify heterozygous individuals that are carrying a recessive maternal effect allele? How would you maintain this strain of flies in a laboratory over many generations?

2. According to the endosymbiosis theory, mitochondria and chloroplasts are derived from bacteria that took up residence within eukaryotic cells. At one time, prior to being taken up by eukaryotic cells, these bacteria were free-living organisms. However, we cannot take a mitochondrion or chloroplast out of a living eukaryotic cell and get it to survive and replicate on its own. Discuss why not.

Note: All answers appear at the website for this textbook; the answers to the even-numbered questions and all of the Concept Check and Comprehension Questions are in the back of the book.

www.mhhe.com/brookergenetics5e

Visit the website for practice tests, answer keys, and other learning aids for this chapter. Enhance your understanding of genetics with our interactive exercises, quizzes, animations, and much more.

CHAPTER OUTLINE

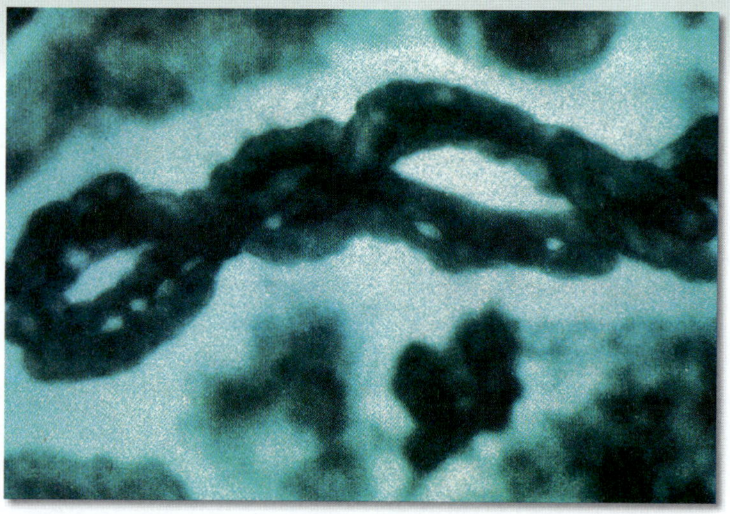

Crossing over during meiosis. *This event provides a way to reassort the alleles of genes that are located on the same chromosome.*

6 GENETIC LINKAGE AND MAPPING IN EUKARYOTES

In Chapter 2, we were introduced to Mendel's laws of inheritance. According to these principles, we expect that two different genes will segregate and independently assort themselves during the process that creates gametes. After Mendel's work was rediscovered at the turn of the twentieth century, chromosomes were identified as the cellular structures that carry genes. The chromosome theory of inheritance explained how the transmission of chromosomes is responsible for the passage of genes from parents to offspring.

When geneticists first realized that chromosomes contain the genetic material, they began to suspect that a conflict might sometimes occur between the law of independent assortment of genes and the actual behavior of chromosomes during meiosis. In particular, geneticists assumed that each species of organism must contain thousands of different genes, yet cytological studies revealed that most species have at most a few dozen chromosomes. Therefore, it seemed likely, and turned out to be true, that each chromosome would carry many hundreds or even thousands of different genes. The transmission of genes located close to each other on the same chromosome violates the law of independent assortment.

In this chapter, we will consider the pattern of inheritance that occurs when different genes are situated on the same chromosome. In addition, we will briefly explore how the data from genetic crosses are used to construct a **genetic map**—a diagram that describes the order of genes along a chromosome. Newer strategies for gene mapping are examined in Chapter 22. However, an understanding of traditional mapping studies, as described in this chapter, will strengthen our appreciation for these newer molecular approaches. More importantly, traditional mapping studies further illustrate how the location of two or more genes on the same chromosome can affect the transmission patterns from parents to offspring.

6.1 OVERVIEW OF LINKAGE

Learning Outcomes:

1. Define genetic linkage.
2. Explain how linkage affects the outcome of crosses.

In eukaryotic species, each linear chromosome contains a very long segment of DNA. A chromosome contains many individual functional units—called genes—that influence an organism's traits. A typical chromosome is expected to contain many hundreds or perhaps a few thousand different genes. The term

synteny means that two or more genes are located on the same chromosome. Genes that are syntenic are physically linked to each other, because each eukaryotic chromosome contains a single, continuous, linear molecule of DNA. **Genetic linkage** is the phenomenon in which genes that are close together on the same chromosome tend to be transmitted as a unit. Therefore, genetic linkage has an influence on inheritance patterns.

Chromosomes are sometimes called **linkage groups,** because a chromosome contains a group of genes that are physically linked together. In species that have been characterized genetically, the number of linkage groups equals the number of chromosome types. For example, human somatic cells have 46 chromosomes, which are composed of 22 types of autosomes that come in pairs plus one pair of sex chromosomes, the X and Y. Therefore, humans have 22 autosomal linkage groups, and an X chromosome linkage group, with males also having a Y chromosome linkage group. In addition, the human mitochondrial genome is another linkage group.

Geneticists are often interested in the transmission of two or more characters in a genetic cross. An experiment that follows the variants of two different characters in a cross is called a **two-factor cross;** one that follows three characters is a **three-factor cross;** and so on. The outcome of a two-factor or three-factor cross depends on whether or not the genes are linked to each other along the same chromosome. Next, we will examine how linkage affects the transmission patterns of two characters. Later sections will examine crosses involving three characters.

Bateson and Punnett Discovered Two Genes That Did Not Assort Independently

An early study indicating that some genes may not assort independently was carried out by William Bateson and Reginald Punnett in 1905. According to Mendel's law of independent assortment, a two-factor cross between two individuals, heterozygous for two genes, should yield a 9:3:3:1 phenotypic ratio among the offspring. However, a surprising result occurred when Bateson and Punnett conducted a cross in the sweet pea involving two different characters: flower color and pollen shape.

As seen in **Figure 6.1**, they began by crossing a true-breeding strain with purple flowers (*PP*) and long pollen (*LL*) to a strain with red flowers (*pp*) and round pollen (*ll*). This yielded an F_1 generation of plants that all had purple flowers and long pollen (*PpLl*). An unexpected result came from the F_2 generation. Even though the F_2 generation had four different phenotypic categories, the observed numbers of offspring did not conform to a 9:3:3:1 ratio. Bateson and Punnett found that the F_2 generation had a much greater proportion of the two phenotypes found in the parental generation—purple flowers with long pollen and red flowers with round pollen. Therefore, they suggested that the transmission of these two characters from the parental generation to the F_2 generation was somehow coupled and not easily assorted in an independent manner. However, Bateson and Punnett did not realize that this coupling was due to the linkage of the flower color gene and the pollen shape gene on the same chromosome.

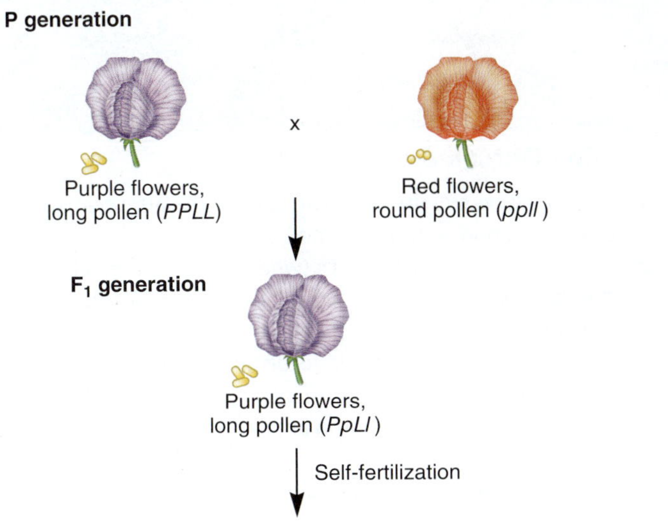

F_2 generation	Observed number	Ratio	Expected number	Ratio
Purple flowers, long pollen	296	15.6	240	9
Purple flowers, round pollen	19	1.0	80	3
Red flowers, long pollen	27	1.4	80	3
Red flowers, round pollen	85	4.5	27	1

FIGURE 6.1 An experiment of Bateson and Punnett with sweet peas, showing that independent assortment does not always occur. Note: The expected numbers are rounded to the nearest whole number.

Genes →Traits Two genes that govern flower color and pollen shape are found on the same chromosome. Therefore, the offspring tend to inherit the nonrecombinant pattern of alleles (*PL* or *pl*). Due to occasional crossing over, a lower percentage of offspring inherit a recombinant pattern of alleles (*Pl* or *pL*).

CONCEPT CHECK: Which types of offspring are found in excess based on Mendel's law of independent assortment?

6.1 COMPREHENSION QUESTIONS

1. Genetic linkage occurs because
   a. genes that are on the same chromosome may affect the same character.
   b. genes that are close together on the same chromosome tend to be transmitted together to offspring.
   c. genes that are on different chromosomes are independently assorted.
   d. none of the above.
2. In the experiment by Bateson and Punnett, which of the following observations suggested linkage in the sweet pea?
   a. A 9:3:3:1 ratio was observed in the F_2 offspring.
   b. A 9:3:3:1 ratio was not observed in the F_2 offspring.
   c. An unusually high number of F_2 offspring had phenotypes of the parental generation.
   d. Both b and c suggested linkage.

6.2 RELATIONSHIP BETWEEN LINKAGE AND CROSSING OVER

Learning Outcomes:

1. Describe how crossing over can change the arrangements of alleles along a chromosome.
2. Explain how the distance between linked genes affects the proportions of recombinant and nonrecombinant offspring.
3. Apply a chi square analysis to distinguish between linkage and independent assortment.
4. Analyze the data of Creighton and McClintock, and explain how it indicated that recombinant offspring carry chromosomes that are the result of crossing over.

Even though the alleles for different genes may be linked along the same chromosome, the linkage can be altered during meiosis. In diploid eukaryotic species, homologous chromosomes can exchange pieces with each other, a phenomenon called **crossing over.** This event occurs during prophase of meiosis I. As discussed in Chapter 3, the replicated chromosomes, known as sister chromatids, associate with the homologous sister chromatids to form a structure known as a **bivalent.** A bivalent is composed of two pairs of sister chromatids. In prophase of meiosis I, it is common for a sister chromatid of one pair to cross over with a sister chromatid from the homologous pair (refer back to Figure 3.10). In this section, we will consider how crossing over affects the pattern of inheritance for genes linked on the same chromosome.

Crossing Over May Produce Recombinant Genotypes

Figure 6.2 considers meiosis when two genes are linked on the same chromosome. Prior to meiosis, one of the chromosomes carries the *B* and *A* alleles, while the homolog carries the *b* and *a* alleles. In Figure 6.2a, no crossing over has occurred. Therefore, the resulting haploid cells contain the same combination of alleles as the original chromosomes. In this case, two haploid cells carry the dominant *B* and *A* alleles, and the other two carry the recessive *b* and *a* alleles. The arrangement of linked alleles has not been altered.

In contrast, Figure 6.2b illustrates what can happen when crossing over occurs. Two of the haploid cells contain combinations of alleles, namely *B* and *a* or *b* and *A*, which differ from those in the original chromosomes. In these two cells, the grouping of linked alleles has changed. An event such as this, leading to a new combination of alleles, is known as **genetic recombination.** The haploid cells carrying the *B* and *a*, or the *b* and *A*, alleles are called **recombinant** cells. Likewise, if such haploid cells were gametes that participated in fertilization, the resulting offspring are called recombinant offspring. These offspring can display combinations of traits that are different from those of either parent. In contrast, offspring that have inherited the same combination of alleles found in the chromosomes of their parents are known as **nonrecombinant** offspring.

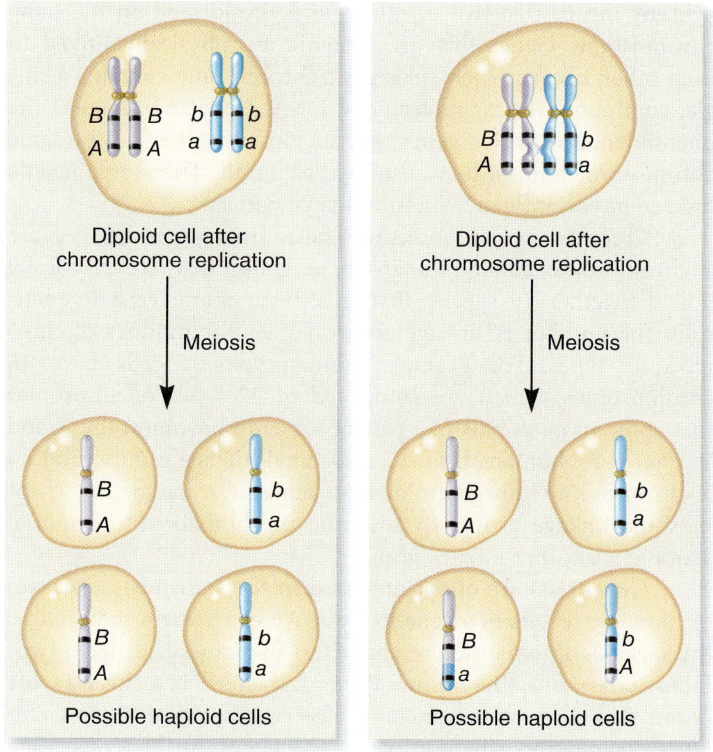

(a) Without crossing over, linked alleles segregate together. **(b) Crossing over can reassort linked alleles.**

ONLINE ANIMATION

FIGURE 6.2 Consequences of crossing over during meiosis. **(a)** In the absence of crossing over, the *A* and *B* alleles and the *a* and *b* alleles are maintained in the same arrangement. **(b)** Crossing over has occurred in the region between the two genes, producing two recombinant haploid cells with a new combination of alleles.

CONCEPT CHECK: If a crossover began in the short region between gene *A* and the tip of the chromosome, would this event affect the arrangement of the *A* and *B* alleles?

Morgan Provided Evidence for the Linkage of X-Linked Genes and Proposed That Crossing Over Between X Chromosomes Can Occur

The first direct evidence that different genes are physically located on the same chromosome came from the studies of Thomas Hunt Morgan in 1911, who investigated the inheritance pattern of different characters that had been shown to follow an X-linked pattern of inheritance. **Figure 6.3** illustrates an experiment involving three characters that Morgan studied. His parental crosses were female fruit flies that had yellow bodies (*yy*), white eyes (*ww*), and miniature wings (*mm*) mated to wild-type males. The wild-type alleles for these three genes are designated y^+ (gray body), w^+ (red eyes), and m^+ (long wings). As expected, the phenotypes of the F_1 generation were wild-type females, and males with yellow bodies, white eyes, and miniature wings. The linkage of these genes was revealed when the F_1 flies were mated to each other and the F_2 generation examined.

Instead of equal proportions of the eight possible phenotypes, Morgan observed a much higher proportion of the combinations of traits found in the parental generation. He observed

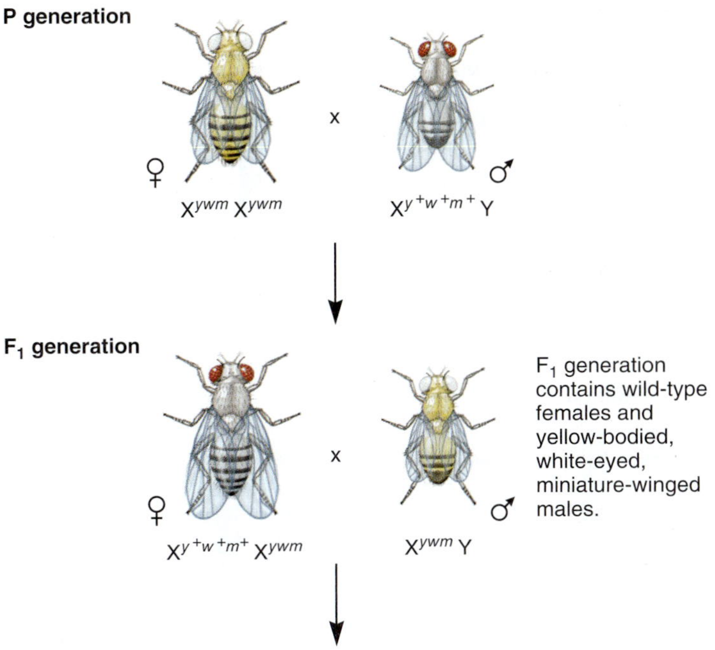

F_2 generation	Females	Males	Total
Gray body, red eyes, long wings	439	319	758
Gray body, red eyes, miniature wings	208	193	401
Gray body, white eyes, long wings	1	0	1
Gray body, white eyes, miniature wings	5	11	16
Yellow body, red eyes, long wings	7	5	12
Yellow body, red eyes, miniature wings	0	0	0
Yellow body, white eyes, long wings	178	139	317
Yellow body, white eyes, miniature wings	365	335	700

FIGURE 6.3 Morgan's three-factor crosses involving three X-linked traits in *Drosophila*.

Genes →Traits Three genes that govern body color, eye color, and wing length are all found on the X chromosome. Therefore, the offspring tend to inherit the nonrecombinant pattern of alleles (y^+ w^+ m^+ or y w m). Figure 6.5 explains how single and double crossovers can create a recombinant pattern of alleles.

CONCEPT CHECK: Of the eight possible phenotypic combinations in the F_2 generation, which ones are the product of a single crossover?

758 flies with gray bodies, red eyes, and long wings, and 700 flies with yellow bodies, white eyes, and miniature wings. Morgan's explanation for this higher proportion of a nonrecombinant pattern of traits was that all three genes are located on the X chromosome and, therefore, tend to be transmitted together as a unit.

However, to fully account for the data shown in Figure 6.3, Morgan needed to explain why a significant proportion of the F_2 generation had a recombinant pattern of traits. Along with the two nonrecombinant phenotypes, five other phenotypic combinations appeared that were not found in the parental generation. How did Morgan explain these data? He considered the studies conducted in 1909 of Belgian cytologist Frans Janssens, who observed chiasmata under the microscope and proposed that crossing over involves a physical exchange between homologous chromosomes. Morgan shrewdly realized that crossing over between homologous X chromosomes was consistent with his data. He assumed that crossing over did not occur between the X and Y chromosome and that these three genes are not found on the Y chromosome. With these ideas in mind, he hypothesized that the genes for body color, eye color, and wing length are all located on the same chromosome, namely, the X chromosome. Therefore, the alleles for all three characters are most likely to be inherited together. Due to crossing over, Morgan also proposed that the homologous X chromosomes (in the female) can exchange pieces of chromosomes and produce new (recombinant) patterns of alleles and recombinant patterns of traits in the F_2 generation.

To appreciate Morgan's proposals, let's simplify his data and consider only two of the three genes: those that affect body color and eye color. If we use the data from Figure 6.3, the following results were obtained:

Gray body, red eyes	1159	
Yellow body, white eyes	1017	
Gray body, white eyes	17	Recombinant offspring
Yellow body, red eyes	12	
Total	2205	

Figure 6.4 considers how Morgan's proposals could account for these data. The nonrecombinant offspring with gray bodies and red eyes or yellow body and white eyes were produced when no crossing over had occurred between the two genes (Figure 6.4a). This was the more common situation. By comparison, crossing over could alter the pattern of alleles along each chromosome and account for the recombinant offspring (Figure 6.4b). Why were there relatively few recombinant offspring? These two genes are very close together along the same chromosome, which makes it unlikely that a crossover would be initiated between them. As described next, the distance between two genes is an important factor that determines the relative proportions of recombinant offspring.

The Likelihood of Crossing Over Between Two Genes Depends on the Distance Between Them

For the experiment of Figure 6.3, Morgan also noticed a quantitative difference between recombinant outcomes involving body color and eye color versus eye color and wing length. This quantitative difference is revealed by reorganizing the data of Figure 6.3 by pairs of genes.

Gray body, red eyes	1159	
Yellow body, white eyes	1017	
Gray body, white eyes	17	Recombinant offspring
Yellow body, red eyes	12	
Total	2205	

Red eyes, long wings	770	
White eyes, miniature wings	716	
Red eyes, miniature wings	401	Recombinant offspring
White eyes, long wings	318	
Total	2205	

Morgan found a substantial difference between the numbers of recombinant offspring when pairs of genes were considered separately. Recombinant patterns involving only eye color and wing

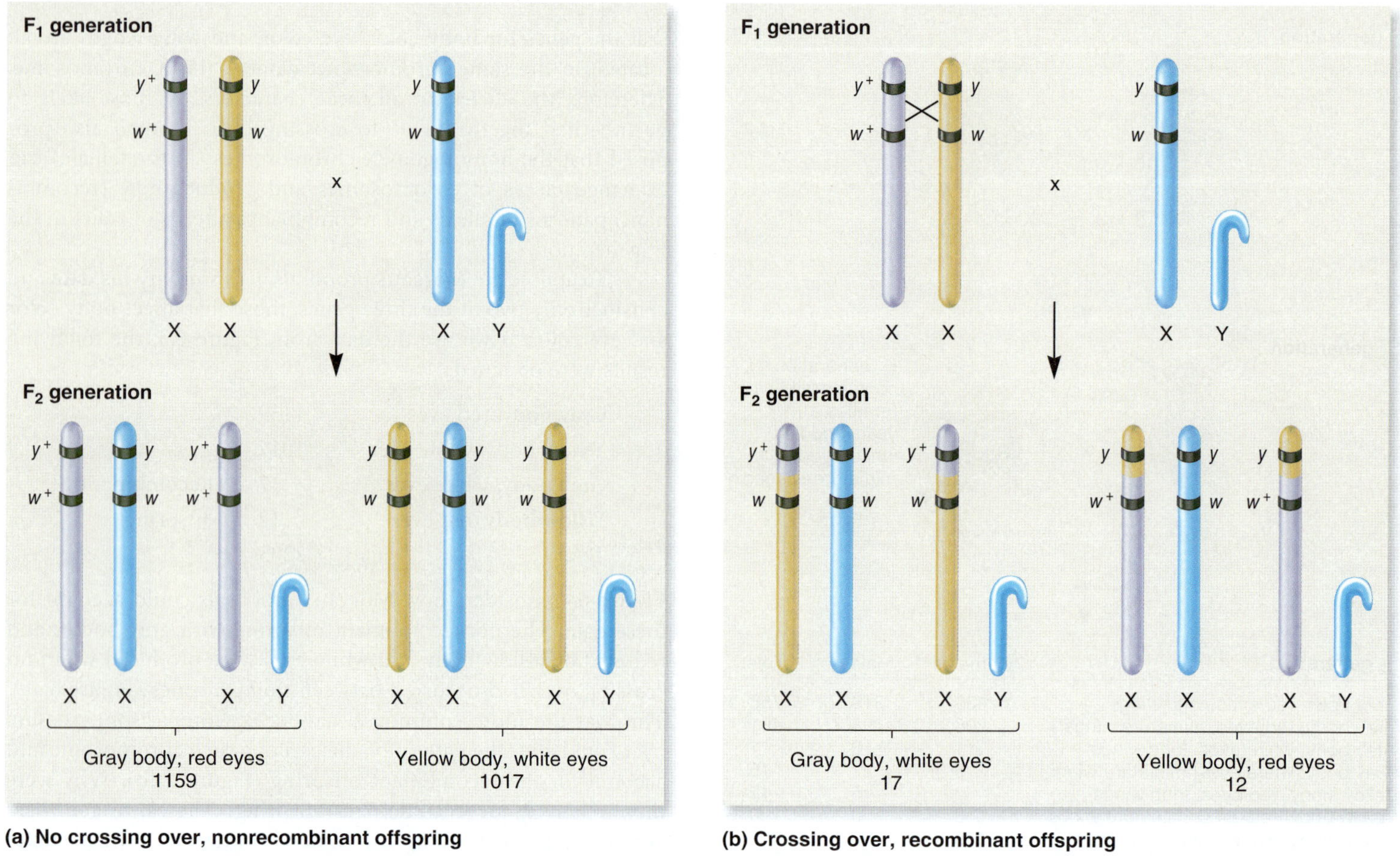

FIGURE 6.4 **Morgan's explanation for nonrecombinant and recombinant offspring.** As described in Chapter 3, crossing over actually occurs at the bivalent stage, but for simplicity, this figure shows only two X chromosomes (one of each homolog) rather than four chromatids, which would occur during the bivalent stage of meiosis. Also note that this figure shows only a portion of the X chromosome. A map of the entire X chromosome is shown in Figure 6.8.

CONCEPT CHECK: Why are the nonrecombinant offspring more common than the recombinant offspring?

length were fairly common—401 + 318 recombinant offspring. In sharp contrast, recombinant patterns for body color and eye color were quite rare—17 + 12 recombinant offspring.

How did Morgan explain these data? Another proposal that he made was that the likelihood of crossing over depends on the distance between two genes. If two genes are far apart from each other, crossing over is more likely to occur between them than it is between two genes that are close together.

Figure 6.5 illustrates the possible events that occurred in the F_1 female flies of Morgan's experiment. One of the X chromosomes carried all three dominant alleles; the other had all three recessive alleles. During oogenesis in the F_1 female flies, crossing over may or may not have occurred in this region of the X chromosome. If no crossing over occurred, the nonrecombinant phenotypes were produced in the F_2 offspring (Figure 6.5a). Alternatively, a crossover sometimes occurred between the eye color gene and the wing length gene to produce recombinant offspring with gray bodies, red eyes, and miniature wings or yellow bodies, white eyes, and long wings (Figure 6.5b). According to Morgan's proposal, such an event is fairly likely because these two genes are far apart from each other on the X chromosome. In contrast, he proposed that the body color and eye color genes are very close together, which makes crossing over between them an unlikely event. Nevertheless, it occasionally occurred, yielding offspring with gray bodies, white eyes, and miniature wings, or with yellow bodies, red eyes, and long wings (Figure 6.5c). Finally, it was also possible for two homologous chromosomes to cross over twice (Figure 6.5d). This double crossover is very unlikely. Among the 2205 offspring Morgan examined, he found only 1 fly with a gray body, white eyes, and long wings that could be explained by this phenomenon.

A Chi Square Analysis Can Be Used to Distinguish Between Linkage and Independent Assortment

Now that we have an appreciation for linkage and the production of recombinant offspring, let's consider how an experimenter can objectively decide whether two genes are linked or assort independently. In Chapter 2, we used a chi square analysis to evaluate the goodness of fit between a genetic hypothesis and the observed experimental data. This method can similarly be employed to determine if the outcome of a two-factor cross is consistent with linkage or independent assortment.

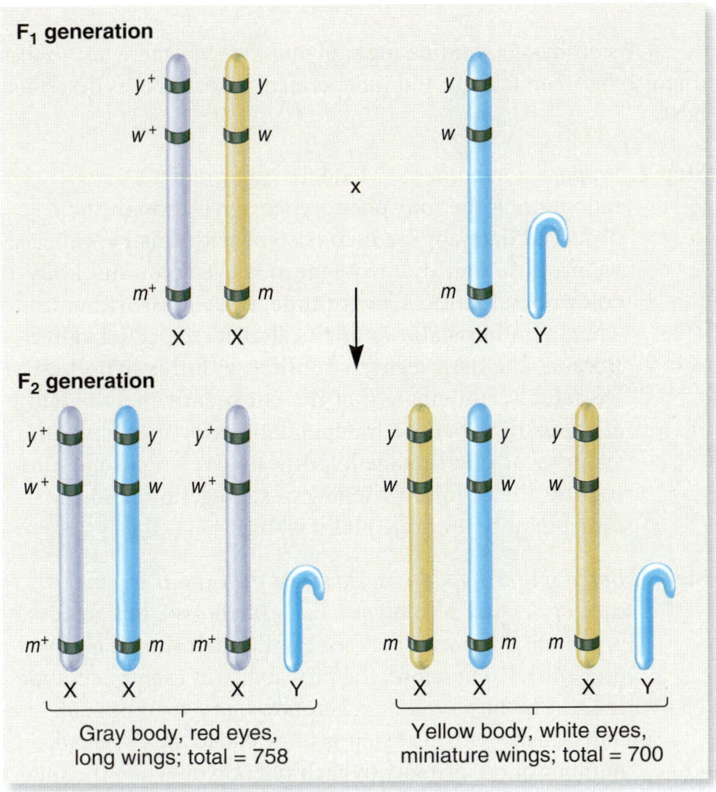

(a) No crossing over in this region, very common

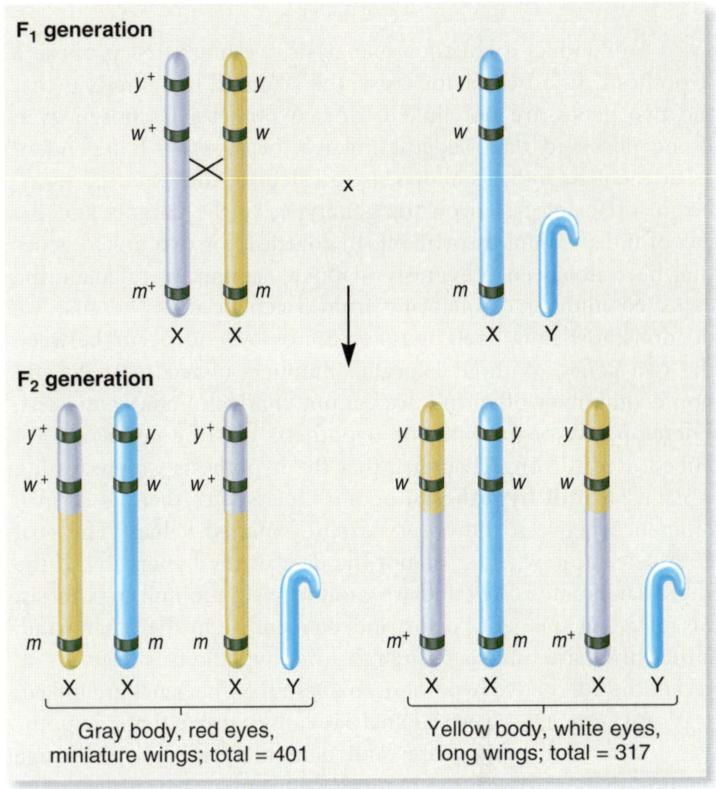

(b) Crossover between eye color and wing length genes, fairly common

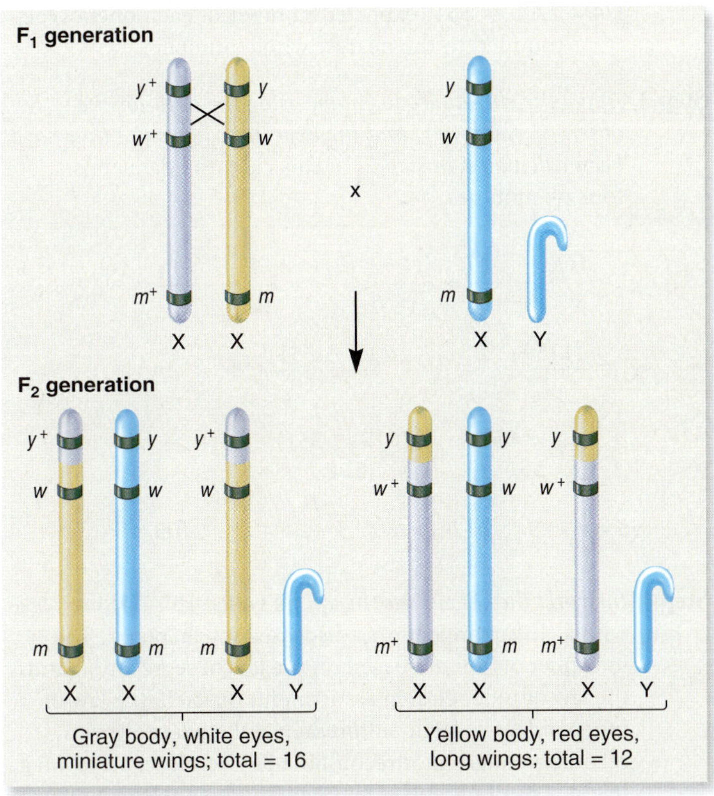

(c) Crosssover between body color and eye color genes, uncommon

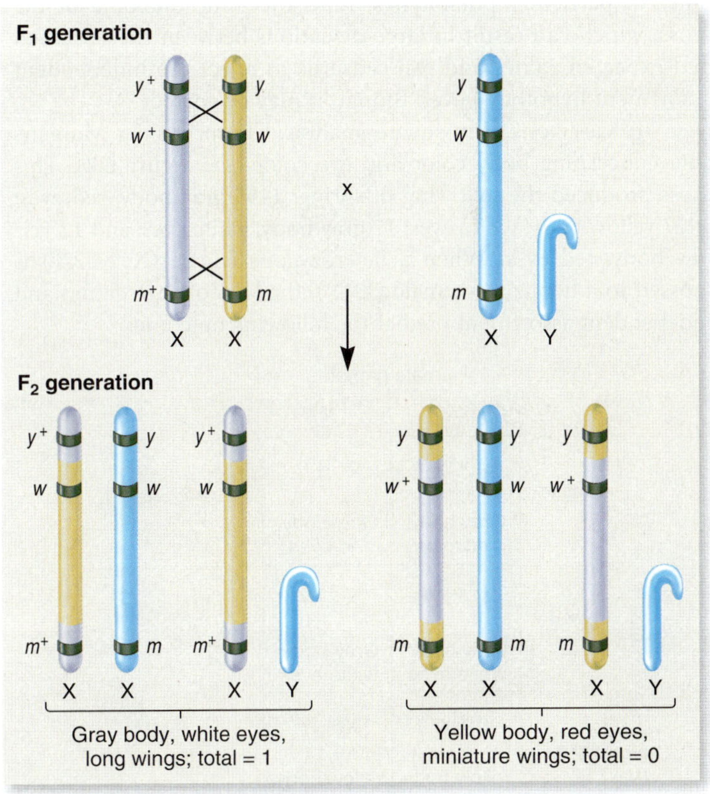

(d) Double crossover, very uncommon

FIGURE 6.5 **Morgan's explanation for different proportions of recombinant offspring.** Crossing over is more likely to occur between two genes that are relatively far apart than between two genes that are very close together. A double crossover is particularly uncommon.

CONCEPT CHECK: Why are the types of offspring described in part (b) more numerous than those described in part (c)?

To conduct a chi square analysis, we must first propose a hypothesis. In a two-factor cross, the standard hypothesis is that the two genes are not linked. This hypothesis is chosen even if the observed data suggest linkage, because an independent assortment hypothesis allows us to calculate the expected number of offspring based on the genotypes of the parents and the law of independent assortment. In contrast, for two linked genes that have not been previously mapped, we cannot calculate the expected number of offspring from a genetic cross because we do not know how likely it is for a crossover to occur between the two genes. Without expected numbers of recombinant and nonrecombinant offspring, we cannot conduct a chi square test. Therefore, we begin with the hypothesis that the genes are not linked. Recall from Chapter 2 that the hypothesis we are testing is called a **null hypothesis,** because it assumes there is no real difference between the observed and expected values. The goal is to determine whether or not the data fit the hypothesis. If the chi square value is low and we cannot reject the null hypothesis, we infer that the genes assort independently. On the other hand, if the chi square value is so high that our hypothesis is rejected, we accept the alternative hypothesis, namely, that the genes are linked.

Of course, a statistical analysis cannot prove that a hypothesis is true. If the chi square value is high, we accept the linkage hypothesis because we are assuming that only two explanations for a genetic outcome are possible: the genes are either linked or not linked. However, other factors, such as a decreased viability of particular phenotypes, may affect the outcome of the cross, which can result in large deviations between the observed and expected values and can cause us to reject the independent assortment hypothesis even though it may be correct.

To carry out a chi square analysis, let's reconsider Morgan's data concerning body color and eye color (see Figure 6.4). This cross produced the following offspring: 1159 gray body, red eyes; 1017 yellow body, white eyes; 17 gray body, white eyes; and 12 yellow body, red eyes. When a heterozygous female (X^{y+w+} X^{yw}) is crossed to a hemizygous male ($X^{yw}Y$), the laws of segregation and independent assortment predict the following outcome:

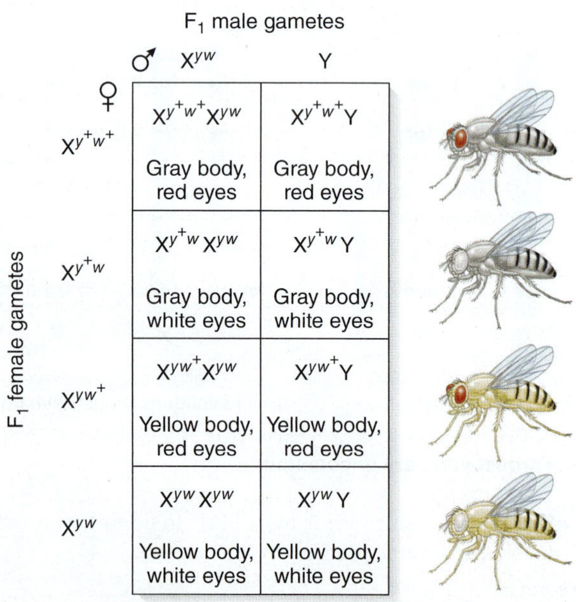

A step-by-step outline for applying the chi square test to distinguish between linkage and independent assortment is described next.

Step 1. *Propose a hypothesis.* Mendel's laws predict a 1:1:1:1 ratio among the four phenotypes. Even though the observed data appear inconsistent with this hypothesis, we propose that the two genes for eye color and body color obey Mendel's law of independent assortment. This hypothesis allows us to calculate expected values. Because the data seem to conflict with this hypothesis, we actually anticipate that the chi square analysis will allow us to reject the independent assortment hypothesis in favor of a linkage hypothesis. We are also assuming the alleles follow the law of segregation, and the four phenotypes are equally viable.

Step 2. *Based on the hypothesis, calculate the expected value of each of the four phenotypes.* Each phenotype has an equal probability of occurring (see the Punnett square given previously). Therefore, the probability of each phenotype is 1/4. The observed F_2 generation had a total of 2205 individuals. Our next step is to calculate the expected number of offspring with each phenotype when the total equals 2205; 1/4 of the offspring should be each of the four phenotypes:

$1/4 \times 2205 = 551$ (expected number of each phenotype, rounded to the nearest whole number)

Step 3. *Apply the chi square formula, using the data for the observed values (O) and the expected values (E) that have been calculated in step 2.* In this case, the data consist of four phenotypes.

$$\chi^2 = \frac{(O_1 - E_1)^2}{E_1} + \frac{(O_2 - E_2)^2}{E_2} + \frac{(O_3 - E_3)^2}{E_3} + \frac{(O_4 - E_4)^2}{E_4}$$

$$\chi^2 = \frac{(1159 - 551)^2}{551} + \frac{(17 - 551)^2}{551} + \frac{(12 - 551)^2}{551} + \frac{(1017 - 551)^2}{551}$$

$$\chi^2 = 670.9 + 517.5 + 527.3 + 394.1 = 2109.8$$

Step 4. *Interpret the calculated chi square value.* This is done with a chi square table, as discussed in Chapter 2. The four phenotypes are based on the law of segregation and the law of independent assortment. By itself, the law of independent assortment predicts only two categories, recombinant and nonrecombinant. Therefore, based on a hypothesis of independent assortment, the degree of freedom equals $n - 1$, which is $2 - 1$, or 1.

The calculated chi square value is enormous! This means that the deviation between observed and expected values is very large. With 1 degree of freedom, such a large deviation is expected to occur by chance alone less than 1% of the time (see Table 2.1). Therefore, we reject the hypothesis that the two genes assort independently. As an alternative, we can accept the hypothesis that the genes are linked. Even so, it should be emphasized that rejecting the null hypothesis does not necessarily mean that the linked hypothesis is correct. For example, some of the non-Mendelian inheritance patterns described in Chapter 5 can produce results that do not conform to independent assortment.

EXPERIMENT 6A

Creighton and McClintock Showed That Crossing Over Produced New Combinations of Alleles and Resulted in the Exchange of Segments Between Homologous Chromosomes

As we have seen, Morgan's studies were consistent with the hypothesis that crossing over occurs between homologous chromosomes to produce new combinations of alleles. To obtain direct evidence that crossing over can result in genetic recombination, Harriet Creighton and Barbara McClintock used an interesting strategy involving parallel observations. In studies conducted in 1931, they first made crosses involving two linked genes to produce nonrecombinant and recombinant offspring. Second, they used a microscope to view the structures of the chromosomes in the parents and in the offspring. Because the chromosomes had some unusual structural features, they could microscopically distinguish between the two homologous chromosomes within a pair. As we will see, this enabled them to correlate the occurrence of recombinant offspring with microscopically observable exchanges in segments of homologous chromosomes.

Creighton and McClintock focused much of their attention on the pattern of inheritance of traits in corn. This species has 10 different chromosomes per set, which are named chromosome 1, chromosome 2, chromosome 3, and so on. In previous cytological examinations of corn chromosomes, some strains were found to have an unusual chromosome 9 with a knob at one end. In addition, McClintock identified an abnormal version of chromosome 9 that also had an extra piece of chromosome 8 attached at the other end (**Figure 6.6a**). This chromosomal rearrangement is called a translocation.

Creighton and McClintock insightfully realized that this abnormal chromosome could be used to determine if two homologous chromosomes physically exchange segments as a result of crossing over. They knew that a gene was located near the knobbed end of chromosome 9 that provided color to corn kernels. This gene existed in two alleles, the dominant allele *C* (colored) and the recessive allele *c* (colorless). A second gene, located near the translocated piece from chromosome 8, affected the texture of the kernel endosperm. The dominant allele *Wx* caused starchy endosperm, and the recessive *wx* allele caused waxy endosperm. Creighton and McClintock reasoned that a crossover involving a normal chromosome 9 and a knobbed/translocated chromosome 9 would produce a recombinant chromosome that had either a knob or a translocation, but not both. These two types of chromosomes would be distinctly different from either of the original chromosomes (**Figure 6.6b**).

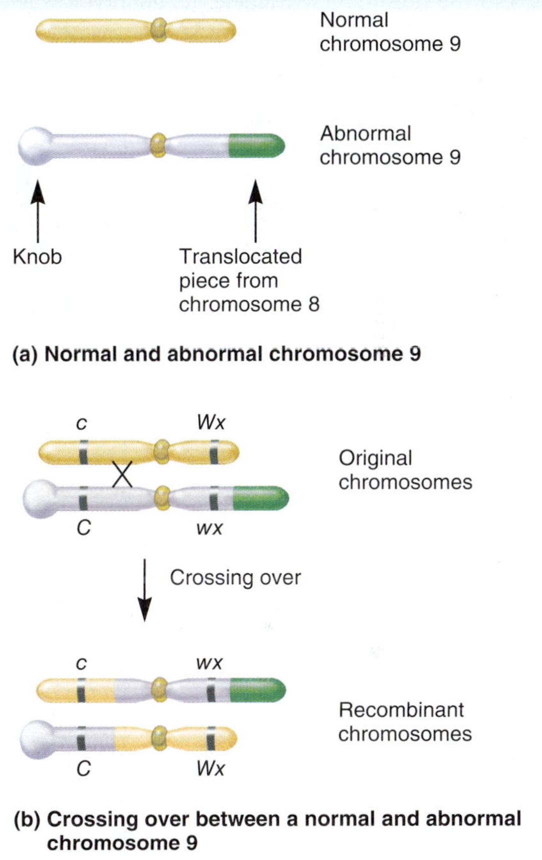

FIGURE 6.6 Crossing over between a normal and abnormal chromosome 9 in corn. **(a)** A normal chromosome 9 in corn is compared with an abnormal chromosome 9 that contains a knob at one end and a translocation at the opposite end. **(b)** A crossover produces a chromosome that contains only a knob at one end and another chromosome that contains only a translocation at the other end.

CONCEPT CHECK: In this experiment, what two types of characteristics can crossing over change? Hint: One type is seen only with a microscope, whereas the other type can be seen with the unaided eye.

As shown in the experiment of **Figure 6.7**, Creighton and McClintock began with a corn strain that carried an abnormal chromosome that had a knob at one end and a translocation at the other. Genotypically, this chromosome was *C wx*. The cytologically normal chromosome in this strain was *c Wx*. This corn plant, termed parent A, had the genotype *Cc Wx wx*. It was crossed to a strain called parent B that carried two cytologically normal chromosomes and had the genotype *cc Wx wx*.

The kernels of the corn resulting from this cross were analyzed in two ways. First, they examined the kernels to see if their phenotypes were colored versus colorless and starchy versus waxy. Second, the chromosomes in each kernel were observed under a microscope to determine their cytological appearance. Altogether, they analyzed a total of 25 kernels (see data of Figure 6.7).

THE HYPOTHESIS

Offspring with recombinant phenotypes are the product of a crossover. This crossover should produce recombinant chromosomes via an exchange of chromosomal segments between homologous chromosomes.

TESTING THE HYPOTHESIS

FIGURE 6.7 **Experimental correlation between genetic recombination and crossing over.**

Starting materials: Two different strains of corn. One strain (referred to as parent A) had an abnormal chromosome 9 (knobbed/translocation) with a dominant *C* allele and a recessive *wx* allele. It also had a cytologically normal copy of chromosome 9 that carried the recessive *c* allele and the dominant *Wx* allele. Its genotype was *Cc Wxwx*. The other strain (referred to as parent B) had two normal versions of chromosome 9. The genotype of this strain was *cc Wxwx*.

1. Cross the two strains described. The tassel is the pollen-bearing structure, and the silk (equivalent to the stigma and style) is the egg-bearing structure. Each fertilization event produces one kernel of corn.

2. Observe the kernels from this cross.

3. Microscopically examine chromosome 9 in the kernels.

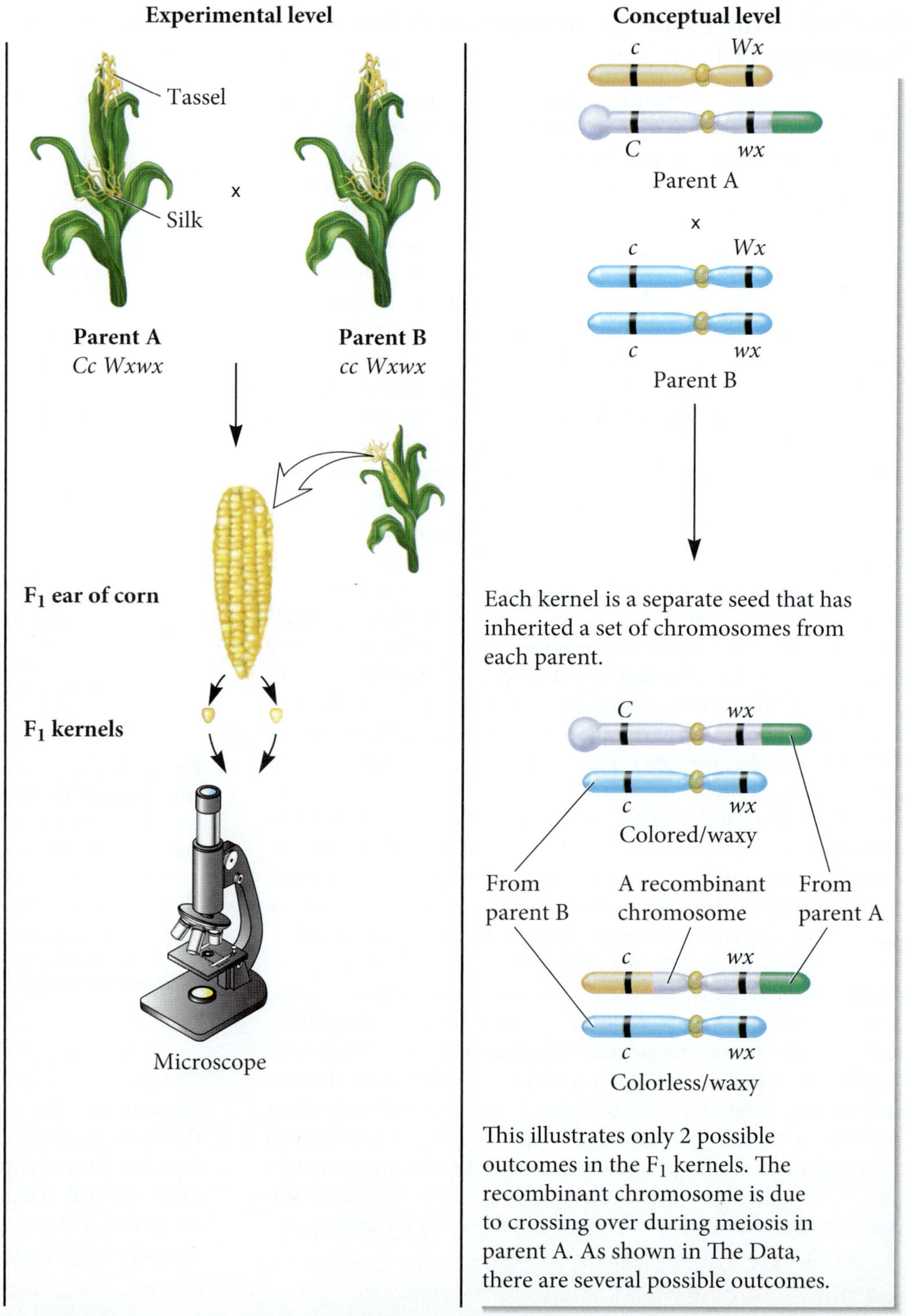

THE DATA

Phenotype of F_1 Kernel	*Number of Kernels Analyzed*	*Cytological Appearance of Chromosome 9 in F_1 Offspring**		*Did a Crossover Occur During Gamete Formation in Parent A?*
Colored/waxy	3	Knobbed/translocation: *C* *wx*	Normal: *c* *wx*	No
Colorless/starchy	11	Knobless/normal: *c* *Wx*	Normal: *c* *Wx* or *c* *wx*	No
Colorless/starchy	4	Knobless/translocation: *c* *wx*	Normal: *c* *Wx*	Yes
Colorless/waxy	2	Knobless/translocation: *c* *wx*	Normal: *c* *wx*	Yes
Colored/starchy	5	Knobbed/normal: *C* *Wx*	Normal: *c* *Wx* or *c* *wx*	Yes
Total	25			

*In this table, the chromosome on the left was inherited from parent A, and the blue chromosome on the right was inherited from parent B.
Data from Harriet B. Creighton and Barbara McClintock (1931) A Correlation of Cytological and Genetical Crossing-Over in *Zea Mays*. *Proc. Natl. Acad. Sci. USA 17*, 492–497.

INTERPRETING THE DATA

By combining the gametes in a Punnett square, the following types of offspring can be produced:

Parent B

Parent A: ♀ \ ♂	c Wx	c wx	
C wx	Cc Wxwx Colored, starchy*	Cc wxwx Colored, waxy	Nonrecombinant
c Wx	cc WxWx Colorless, starchy	cc Wxwx Colorless, starchy	Nonrecombinant
C Wx	Cc WxWx Colored, starchy	Cc Wxwx Colored, starchy	Recombinant
c wx	cc Wxwx Colorless, starchy	cc wxwx Colorless, waxy	Recombinant

*Note: Due to a small sampling size, this type of kernel was not observed.

In this experiment, the researchers were interested in whether or not crossing over had occurred in parent A, which was heterozygous for both genes. This parent could produce four types of gametes, but parent B could produce only two types.

As seen in the Punnett square, two of the phenotypic categories, colored, starchy (*Cc Wx wx* or *Cc Wx Wx*) and colorless, starchy (*cc Wx Wx* or *cc Wx wx*), were ambiguous because they could arise from a nonrecombinant and from a recombinant gamete. In other words, these phenotypes could be produced whether or not recombination occurred in parent A. Therefore, let's focus on the two unambiguous phenotypic categories: colored, waxy (*Cc wx wx*) and colorless, waxy (*cc wx wx*). The colored, waxy phenotype could happen only if recombination did not occur in parent A and if parent A passed the knobbed/translocated chromosome to its offspring. As shown in The Data, three kernels were obtained with this phenotype, and all of them had the knobbed/translocated chromosome. By comparison, the colorless, waxy phenotype could be obtained only if genetic recombination occurred in parent A and this parent passed on a chromosome 9 that was knobless and had a translocation. Two kernels were obtained with this phenotype, and both of them had the expected knobless chromosome with a translocation. Taken together, these results showed a perfect correlation between genetic recombination of alleles and the cytological presence of a chromosome displaying a genetic exchange of chromosomal pieces from parent A.

Overall, the observations described in this experiment were consistent with the idea that a crossover occurred in the region between the *C* and *wx* genes that involved an exchange of segments between two homologous chromosomes. These results supported the view that genetic recombination involves a physical exchange between homologous chromosomes. This microscopic evidence helped to convince geneticists that recombinant offspring arise from the physical exchange of segments of homologous chromosomes. As shown in the solved problem S4 at the end of this chapter, an experiment by Curt Stern was also consistent with the conclusion that crossing over between homologous chromosomes accounts for the formation of offspring with recombinant phenotypes.

A self-help quiz involving this experiment can be found at www.mhhe.com/brookergenetics5e.

6.2 COMPREHENSION QUESTIONS

1. With regard to linked genes along the same chromosome, which of the following statements is *false*?
   a. Crossing over is needed to produce nonrecombinant offspring.
   b. Crossing over is needed to produce recombinant offspring.
   c. Crossing over is more likely to separate alleles if they are far apart on the same chromosome.
   d. Crossing over that separates linked alleles occurs during prophase of meiosis I.

2. Morgan observed a higher number of recombinant offspring involving eye color and wing length (401 + 318) than recombinants for body color and eye color (17 + 12). These results occurred because
   a. the genes affecting eye color and wing length are farther apart on the X chromosome than are the genes affecting body color and eye color.
   b. the genes affecting eye color and wing length are closer together on the X chromosome than are the genes affecting body color and eye color.
   c. the gene affecting wing length is not on the X chromosome.
   d. the gene affecting body color is not on the X chromosome.

3. In a chi square analysis involving genes that may be linked, which of the following statements is correct?
   a. An independent assortment hypothesis is not proposed because the data usually suggest linkage.
   b. An independent assortment hypothesis is proposed because it allows you to calculate expected numbers of offspring.
   c. A large chi square value suggests that the observed and expected data are in good agreement.
   d. The null hypothesis is rejected when the chi square value is very low.

6.3 GENETIC MAPPING IN PLANTS AND ANIMALS

Learning Outcomes:

1. Describe why genetic mapping is useful.
2. Calculate the map distance between linked genes using data from a testcross.
3. Explain how interference affects the number of double crossovers.

The purpose of **genetic mapping,** also known as gene mapping or chromosome mapping, is to determine the linear order and distance of separation among genes that are linked to each other along the same chromosome. **Figure 6.8** illustrates a simplified genetic map of *Drosophila melanogaster*, depicting the locations of many different genes along the individual chromosomes. As shown here, each gene has its own unique **locus**—the site where the gene is found within a particular chromosome. For example, the gene designated *brown eyes* (*bw*), which affects eye color, is located near one end of chromosome 2. The gene designated *black body* (*b*), which affects body color, is found near the middle of the same chromosome.

Why is genetic mapping useful? First, it allows geneticists to understand the overall complexity and genetic organization of a particular species. The genetic map of a species portrays the underlying basis for the inherited traits an organism displays. In some cases, the known locus of a gene within a genetic map can help molecular geneticists to clone that gene and thereby obtain greater information about its molecular features. In addition, genetic maps are useful from an evolutionary point of view. A comparison of the genetic maps for different species can improve our understanding of the evolutionary relationships among them.

Along with these scientific uses, genetic maps have many practical benefits. For example, many human genes that play a role in human disease have been genetically mapped. This information can be used to diagnose and perhaps someday treat inherited human diseases. It can also help genetic counselors predict the likelihood that a couple will produce children with certain inherited diseases. In addition, genetic maps are gaining increasing importance in agriculture. A genetic map can provide plant and animal breeders with helpful information for improving agriculturally important strains through selective breeding programs.

In this section, we will examine traditional genetic mapping techniques that involve an analysis of crosses of individuals that are heterozygous for two or more genes. Determining the number of recombinant offspring due to crossing over provides a way to deduce the linear order of genes along a chromosome. As depicted in Figure 6.8, this linear arrangement of genes is known as a **genetic linkage map.** Such maps have been constructed for several plant species and certain species of animals, incuding *Drosophila*. Mapping has also been carried out on fungi and algae as described in Section 6.4. For many organisms, however, traditional mapping approaches are difficult due to long generation times or the inability to carry out experimental crosses (as in humans). Fortunately, many alternative methods of gene mapping have been developed to replace the need to carry out crosses. As described in Chapter 22, molecular approaches are increasingly used to map genes.

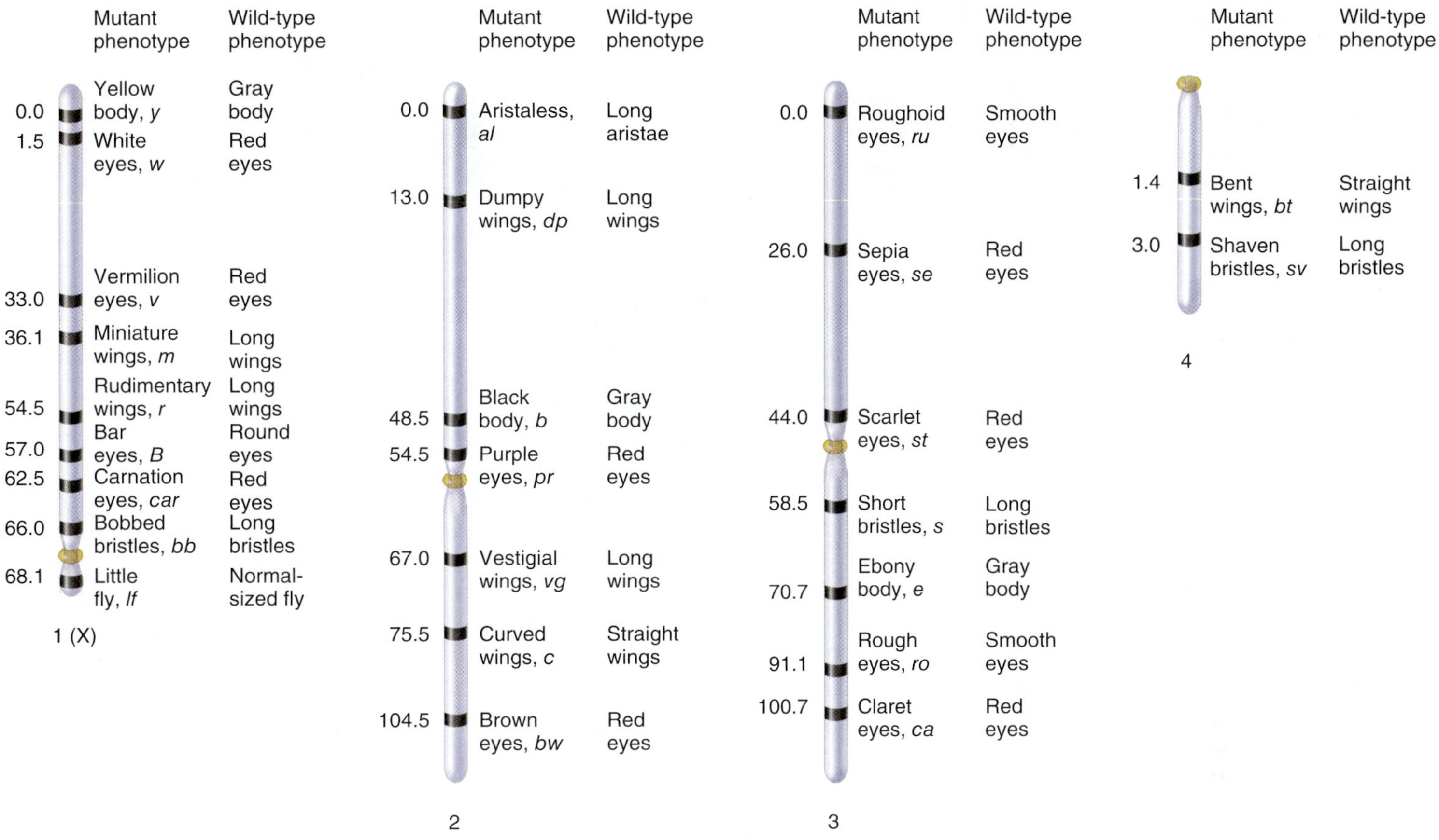

FIGURE 6.8 A simplified genetic linkage map of *Drosophila melanogaster*. This simplified map illustrates a few of the many thousands of genes that have been identified in this organism.

CONCEPT CHECK: List five reasons why genetic maps are useful.

A Testcross Is Conducted to Produce a Genetic Linkage Map

Genetic mapping allows us to estimate the relative distances between linked genes based on the likelihood that a crossover will occur between them. If two genes are very close together on the same chromosome, a crossover is unlikely to begin in the region between them. However, if two genes are very far apart, a crossover is more likely to be initiated in this region, thereby recombining the alleles of the two genes. Experimentally, the basis for genetic mapping is that the percentage of recombinant offspring is correlated with the distance between two genes. If two genes are far apart, many recombinant offspring will be produced. However, if two genes are close together, very few recombinant offspring will be observed.

To interpret a genetic mapping experiment, the experimenter must know if the characteristics of an offspring are due to crossing over during meiosis in a parent. This is accomplished by conducting a **testcross.** Most testcrosses are between an individual that is heterozygous for two or more genes and an individual that is recessive and homozygous for the same genes. The goal of the testcross is to determine if recombination has occurred during meiosis in the heterozygous parent. Therefore, genetic mapping is based on the level of recombination that occurs in just one parent—the heterozygote. In a testcross, new combinations of alleles cannot occur in the gametes of the other parent, which is homozygous for these genes.

Figure 6.9 illustrates how a testcross provides an experimental strategy to distinguish between recombinant and nonrecombinant offspring. This cross concerns two linked genes affecting bristle length and body color in fruit flies. The recessive alleles are *s* (short bristles) and *e* (ebony body), and the dominant (wild-type) alleles are s^+ (long bristles) and e^+ (gray body). One parent displays both recessive traits. Therefore, we know this parent is homozygous for the recessive alleles of the two genes (*ss ee*). The other parent is heterozygous for the linked genes affecting bristle length and body color. This parent was produced from a cross involving a true-breeding wild-type fly and a true-breeding fly with short bristles and an ebony body. Therefore, in this heterozygous parent, we know that the *s* and *e* alleles are located on one chromosome and the corresponding s^+ and e^+ alleles are located on the homologous chromosome.

Now let's take a look at the four possible types of offspring these parents can produce. The offspring's phenotypes are long bristles, gray body; short bristles, ebony body; long bristles, ebony body; and short bristles, gray body. All four types of offspring have inherited a chromosome carrying the *s* and *e* alleles from their homozygous parent (shown on the right in each pair of chromosomes). Focus your attention on the other chromosome. The offspring with long bristles and gray bodies have inherited a chromosome carrying the s^+ and e^+ alleles from the heterozygous parent. This chromosome is not the product of a crossover. The offspring with short bristles and ebony bodies

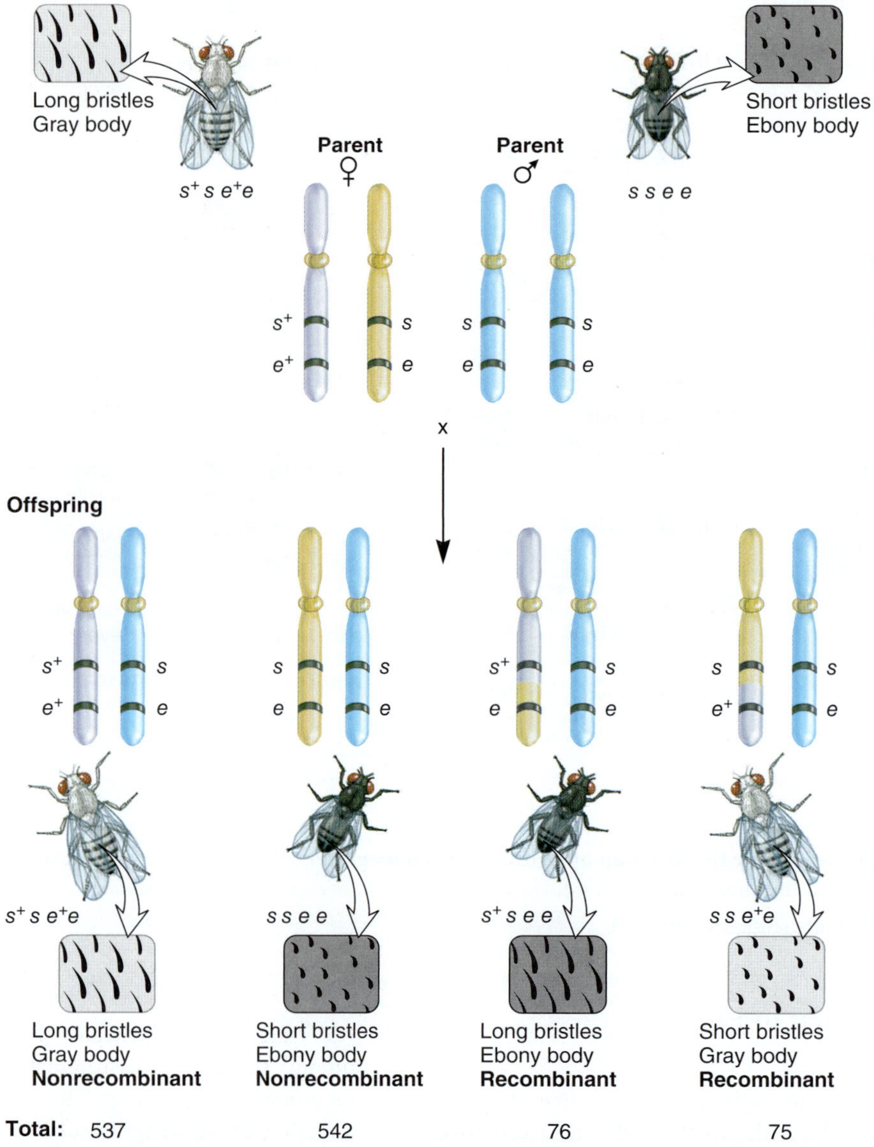

INTERACTIVE EXERCISE

FIGURE 6.9 Use of a testcross to distinguish between recombinant and nonrecombinant offspring. The cross involves one *Drosophila* parent that is heterozygous for both genes ($s^+s\ e^+e$) and the other parent that is homozygous recessive for short bristles (*ss*) and ebony body (*ee*). (Note: *Drosophila* geneticists normally designate the short allele as *ss* and a homozygous fly with short bristles as *ssss*. In this figure, the allele causing short bristles is designated with a single *s* to avoid confusion between the allele designation and the genotype of the fly. Also, crossing over does not occur during sperm formation in *Drosophila*, which is unusual among eukaryotes. Therefore, the heterozygote in a testcross involving *Drosophila* must be the female.)

CONCEPT CHECK: When and in which fly or flies did crossing over occur in order to produce the recombinant offspring?

have inherited a chromosome carrying the *s* and *e* alleles from the heterozygous parent. Again, this chromosome is not the product of a crossover.

The other two types of offspring, however, can be produced only if crossing over has occurred in the region between these two genes. Those with long bristles and ebony bodies or short bristles and gray bodies have inherited a chromosome that is the product of a crossover during meiosis in the heterozygous parent. A key point for you to observe is that the recombinant offspring of the F_2 generation must carry a chromosome that is the product of a crossover. As noted in Figure 6.9, the recombinant offspring are fewer in number than are the nonrecombinant offspring.

The amount of recombination can be used as an estimate of the physical distance between two genes on the same chromosome. The **map distance** is defined as the number of recombinant offspring divided by the total number of offspring, multiplied by 100. We can calculate the map distance between these two genes using this formula:

$$\text{Map distance} = \frac{\text{Number of recombinant offspring}}{\text{Total number of offspring}} \times 100$$

$$= \frac{76 + 75}{537 + 542 + 76 + 75} \times 100$$

$$= 12.3 \text{ map units}$$

The units of distance are called **map units (mu),** or sometimes **centiMorgans (cM)** in honor of Thomas Hunt Morgan. One map unit is equivalent to 1% recombinant offspring in a testcross. In this example, we would conclude that the *s* and *e* alleles are 12.3 mu apart from each other along the same chromosome.

Multiple crossovers set a quantitative limit on the relationship between map distance and the percentage of recombinant offspring. Even though two different genes can be on the same chromosome and more than 50 mu apart, a testcross is expected to yield a maximum of only 50% recombinant offspring (**Figure 6.10**). What accounts for this 50% limit? The answer lies in the pattern of multiple crossovers. As shown in the figure to solved problem S5 at the end of the chapter, a double crossover between two genes could involve four, three, or two chromatids, which would yield 100%, 50%, or 0% recombinants, respectively. Because all of these double crossovers are equally likely, we take the average of them to determine the maximum value. This average equals 50%. Therefore, when two different genes are more than 50 mu apart, they follow the law of independent assortment in a testcross and only 50% recombinants are observed.

Three-Factor Crosses Can Be Used to Determine the Order and Distance Between Linked Genes

Thus far, we have considered the construction of genetic maps using two-factor testcrosses to compute map distance. The data from three-factor crosses can yield additional information about map distance and gene order. In a three-factor cross, the experimenter crosses two individuals that differ in three characters. The following experiment outlines a common strategy for using three-factor crosses to map genes. In this experiment, the parental generation consists of fruit flies that differ in body color, eye color, and wing shape. We must begin with true-breeding lines so that we know which alleles are initially linked to each other on the same chromosome. In this example, all of the dominant alleles are linked on the same chromosome.

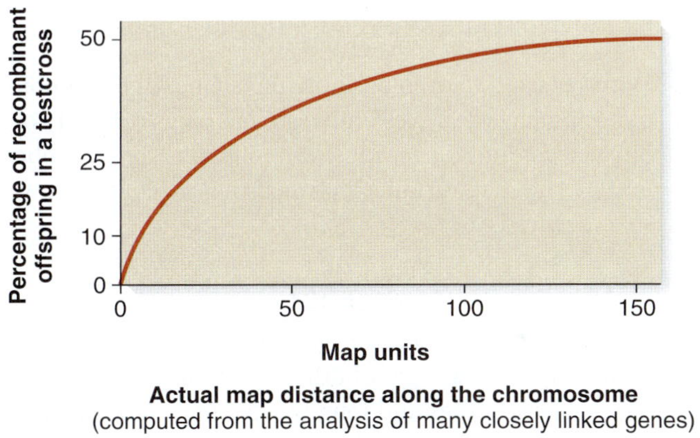

FIGURE 6.10 Relationship between the percentage of recombinant offspring in a testcross and the actual map distance between genes. The y-axis depicts the percentage of recombinant offspring that would be observed in a two-factor testcross. The actual map distance, shown on the x-axis, is calculated by analyzing the percentages of recombinant offspring from a series of many two-factor crosses involving closely linked genes. Even though two genes may be more than 50 mu apart, the percentage of recombinant offspring will not exceed 50%.

CONCEPT CHECK: What phenomenon explains why the maximum percentage of recombinant offspring does not exceed 50%?

Step 1. *Cross two true-breeding strains that differ with regard to three alleles.* In this example, we will cross a fly that has a black body (*bb*), purple eyes (*prpr*), and vestigial wings (*vgvg*) to a homozygous wild-type fly with a gray body (b^+b^+), red eyes (pr^+pr^+), and long wings (vg^+vg^+):

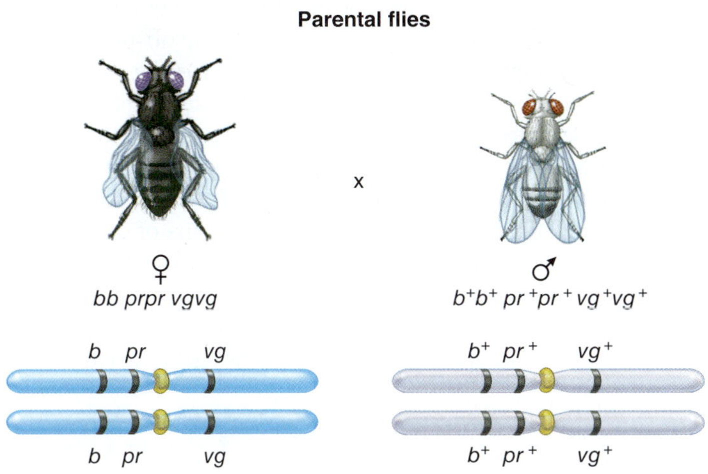

The goal in this step is to obtain F_1 individuals that are heterozygous for all three genes. In the F_1 heterozygotes, all dominant alleles are located on one chromosome, and all recessive alleles are on the other homologous chromosome.

Step 2. *Perform a testcross by mating F_1 female heterozygotes to male flies that are homozygous recessive for all three alleles (bb prpr vgvg).*

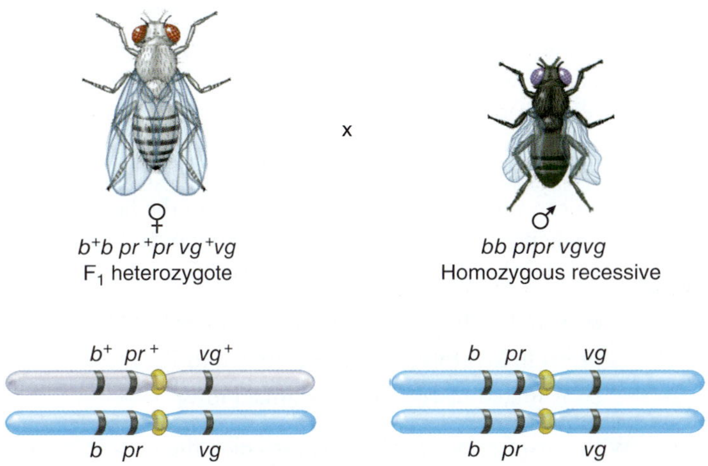

During gametogenesis in the heterozygous female F_1 flies, crossovers may produce new combinations of the three alleles.

Step 3. *Collect data for the F_2 generation.* As shown in **Table 6.1**, eight phenotypic combinations are possible. An analysis of the F_2 generation flies allows us to map these three genes. Because the three genes exist as two alleles each, we have 2^3, or 8, possible combinations of offspring. If these alleles assorted independently, all eight combinations would occur in equal proportions. However, we see that the proportions of the eight phenotypes are far from equal.

The genotypes of the parental generation correspond to the phenotypes gray body, red eyes, and long wings, and black body, purple eyes, and vestigial wings. In crosses involving linked genes, the nonrecombinant phenotypes occur most frequently in the offspring. The remaining six phenotypes are due to crossing over.

The double crossover is always expected to be the least frequent category of offspring. Two of the phenotypes—gray body, purple eyes, and long wings; and black body, red eyes, and vestigial wings—arose from a double crossover between two pairs of genes. Also, the combination of traits in the double crossover tells us which gene is in the middle. When a chromatid undergoes a double crossover, the gene in the middle becomes separated from the other two genes at either end.

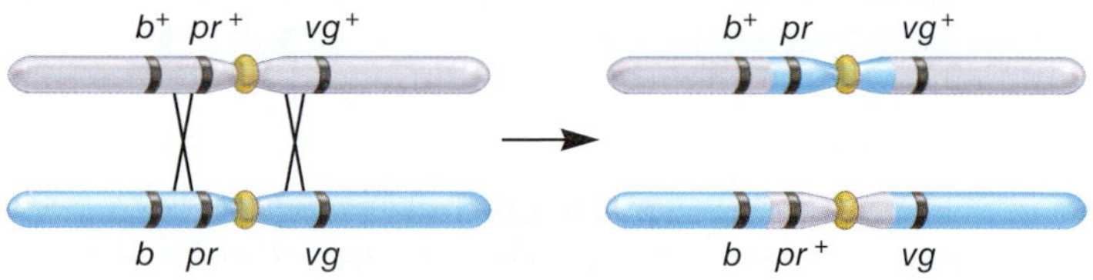

In the double-crossover categories, the recessive purple eye allele is separated from the other two recessive alleles. In the testcross, this yields flies with gray bodies, purple eyes, and long wings; or ones with black bodies, red eyes, and vestigial wings. This observation indicates that the gene for eye color lies between the genes for body color and wing shape.

Step 4. *Calculate the map distance between pairs of genes.* To do this, we need to understand which allele combinations are recombinant and which are nonrecombinant. The recombinant offspring are due to crossing over in the heterozygous female parent. If you look back at step 2, you can see the arrangement of alleles in the heterozygous female parent in the absence of crossing over. Let's consider this arrangement with regard to gene pairs:

b^+ is linked to pr^+, and b is linked to pr

pr^+ is linked to vg^+, and pr is linked to vg

b^+ is linked to vg^+, and b is linked to vg

With regard to body color and eye color, the recombinant offspring have gray bodies and purple eyes (2 + 30) or black bodies and red eyes (28 + 1). As shown along the right side of Table 6.1, these offspring were produced by crossovers in the female parents. The total number of these recombinant offspring is 61. The map distance between the body color and eye color genes is

$$\text{Map distance} = \frac{61}{944 + 61} \times 100 = 6.1 \text{ mu}$$

With regard to eye color and wing shape, the recombinant offspring have red eyes and vestigial wings (61 + 1) or purple eyes and long wings (2 + 60). The total number is 124. The map distance between the eye color and wing shape genes is

$$\text{Map distance} = \frac{124}{881 + 124} \times 100 = 12.3 \text{ mu}$$

With regard to body color and wing shape, the recombinant offspring have gray bodies and vestigial wings (61 + 30) or black bodies and long wings (28 + 60). The total number is 179. The map distance between the body color and wing shape genes is

$$\text{Map distance} = \frac{179}{826 + 179} \times 100 = 17.8 \text{ mu}$$

Step 5. *Construct the map.* Based on the map unit calculation, the body color (b) and wing shape (vg) genes are farthest apart. The eye color gene (pr) must lie in the middle. As mentioned earlier, this order of genes is also confirmed by the pattern of traits found in the double crossovers. To construct the map, we use the distances between the genes that are closest together.

TABLE 6.1

Data from a Three-Factor Cross (see step 2)

Phenotype	Number of Observed Offspring (males and females)	Chromosome Inherited from F_1 Female
Gray body, red eyes, long wings	411	b^+ pr^+ vg^+
Gray body, red eyes, vestigial wings	61	b^+ pr^+ vg
Gray body, purple eyes, long wings	2	b^+ pr vg^+
Gray body, purple eyes, vestigial wings	30	b^+ pr vg
Black body, red eyes, long wings	28	b pr^+ vg^+
Black body, red eyes, vestigial wings	1	b pr^+ vg
Black body, purple eyes, long wings	60	b pr vg^+
Black body, purple eyes, vestigial wings	412	b pr vg
Total	1005	

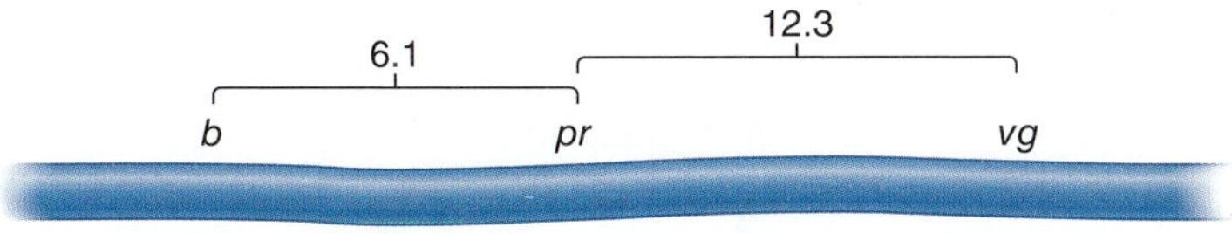

In our example, we have placed the body color gene first and the wing shape gene last. The data also are consistent with a map in which the wing shape gene comes first and the body color gene comes last. In detailed genetic maps, the locations of genes are mapped relative to the centromere.

You may have noticed that our calculations underestimate the distance between the body color and wing shape genes. We obtained a value of 17.8 mu even though the distance seems to be 18.4 mu when we add together the distance between body color and eye color genes (6.1 mu) and the distance between eye color and wing shape genes (12.3 mu). What accounts for this discrepancy? The answer is double crossovers. If you look at the data in Table 6.1, the offspring with gray bodies, purple eyes, and long wings or those with black bodies, red eyes, and vestigial wings are due to a double crossover. From a phenotypic perspective, these offspring are nonrecombinant with regard to the body color and wing shape alleles. Even so, we know they arose from a double crossover between these two genes. Therefore, we should consider these crossovers when calculating the distance between the body color and wing shape genes. In this case, three offspring (2 + 1) were due to double crossovers. The number of double crossovers (2 + 1) is multiplied by 2 and we add this number to our previous value of recombinant offspring:

$$\text{Map distance} = \frac{179 + 2(2 + 1)}{826 + 179} \times 100 = 18.4 \text{ mu}$$

Interference Can Influence the Number of Double Crossovers That Occur in a Short Region

In Chapter 2, we considered the product rule to determine the probability that two independent events will both occur. The product rule allows us to predict the expected likelihood of a double crossover, provided we know the individual probabilities of each single crossover. Let's reconsider the data of the three-factor testcross just described to see if the frequency of double crossovers is what we would expect based on the product rule. The map distance between between *b* and *pr* is 6.1 mu and the distance between *pr* and *vg* is 12.3 mu. Because map distances are calculated as the percentage of recombinant offspring, we divide the map distances by 100 to compute the crossover frequency between each pair of genes. In this case, the crossover frequency between *b* and *pr* is 0.061 and between *pr* and *vg* is 0.123. The product rule predicts

Expected likelihood of a double crossover =
$0.061 \times 0.123 = 0.0075$, or 0.75%

Expected number of offspring due to a double crossover, based on a total of 1005 offspring produced =
$1005 \times 0.0075 = 7.5$

In other words, we would expect about 7 or 8 offspring to be produced as a result of a double crossover. The observed number of offspring was only 3 (namely, 2 with gray bodies, purple eyes, and long wings, and 1 with a black body, red eyes, and vestigial wings). What accounts for the lower number? This lower-than-expected value is probably not due to random sampling error. Instead, the likely cause is a common genetic phenomenon known as **positive interference,** in which the occurrence of a crossover in one region of a chromosome decreases the probability that a second crossover will occur nearby. In other words, the first crossover interferes with the ability to form a second crossover in the immediate vicinity. To provide interference with a quantitative value, we first calculate the coefficient of coincidence (C), which is the ratio of the observed number of double crossovers to the expected number.

$$C = \frac{\text{Observed number of double crossovers}}{\text{Expected number of double crossovers}}$$

Interference (I) is expressed as

$$I = 1 - C$$

For the data of the three-factor testcross, the observed number of crossovers is 3 and the expected number is 7.5, so the coefficient of coincidence equals $3/7.5 = 0.40$. In other words, only 40% of the expected number of double crossovers were actually observed. The value for interference equals $1 - 0.4 = 0.60$, or 60%. This means that 60% of the expected number of crossovers did not occur. Because I has a positive value, this is called positive interference. Rarely, the outcome of a testcross yields a negative value for interference. A negative interference value suggests that a first crossover enhanced the rate of a second crossover in a nearby region. Although the molecular mechanisms that cause interference are not entirely understood, the number of crossovers in most organisms is regulated so that very few occur per chromosome. The reasons for positive and negative interference require further research.

6.3 COMPREHENSION QUESTIONS

Answer the multiple choice questions based on the following crosses:

Parental generation: True-breeding flies with red eyes and long wings were crossed to flies with white eyes and miniature wings. All F_1 offspring had red eyes and long wings.

The F_1 female flies were then crossed to males with white eyes and miniature wings. The following results were obtained for the F_2 generation:

129 red eyes, long wings
133 white eyes, miniature wings
71 red eyes, miniature wings
67 white eyes, long wings

1. What is/are the phenotype(s) of the recombinant offspring of the F_2 generation?
   a. Red eyes, long wings
   b. White eyes, miniature wings
   c. Red eyes, long wings and white eyes, miniature wings
   d. Red eyes, miniature wings and white eyes, long wings

2. The recombinant offspring of the F_2 generation were produced by crossing over that occurred
   a. during spermatogenesis in the parental generation males.
   b. during oogenesis in the parental generation females.
   c. during spermatogenesis in the F_1 males.
   d. during oogenesis in the F_1 females.
3. What is the map distance between these two genes?
   a. 32.3 mu
   b. 34.5 mu
   c. 16.2 mu
   d. 17.3 mu

6.4 GENETIC MAPPING IN HAPLOID EUKARYOTES

Learning Outcomes:

1. Explain the experimental advantage of genetic mapping in fungi.
2. Calculate the map distance between genes in fungi using tetrad analysis.

Before ending our discussion of genetic mapping, let's consider some pioneering studies that involved the genetic mapping of haploid organisms. You may find it surprising that certain species of simple eukaryotes, particularly unicellular algae and fungi, which spend part of their life cycle in the haploid state, have also been used in genetic mapping studies. The sac fungi, called ascomycetes, have been particularly useful to geneticists because of their unique style of sexual reproduction. In fact, much of our earliest understanding of genetic recombination came from the genetic analyses of fungi.

Fungi may be unicellular or multicellular organisms. Fungal cells are typically haploid (1*n*) and can reproduce asexually. In addition, fungi can also reproduce sexually by the fusion of two haploid cells to create a diploid zygote (2*n*) (**Figure 6.11**). The diploid zygote can then proceed through meiosis to produce four haploid cells, which are called **spores.** This group of four spores is known as a **tetrad** (not to be confused with a tetrad of four sister chromatids). In some species, meiosis is followed by a mitotic division to produce eight spores, known as an **octad.**

In ascomycete fungi and certain species of algae, the cells of a tetrad or octad are contained within a sac, which is called an **ascus** (plural: asci) in fungi. In other words, the products of a single meiotic division are contained within one sac. This mode of reproduction does not occur in other eukaryotic groups. Studies of fungi have been pivotal in our fundamental understanding of meiosis and crossing over. By comparison, the products of meiosis are produced differently in animals and plants. For example, in animals, oogenesis produces a single functional egg, and spermatogenesis occurs in the testes, where the resulting sperm become mixed with millions of other sperm.

Using a microscope, researchers can dissect asci and study the traits of each haploid spore. In this way, these organisms offer a unique opportunity for geneticists to identify and study all of the cells that are derived from a single meiotic division. In this section, we will consider how the analysis of asci can be used to map genes in fungi.

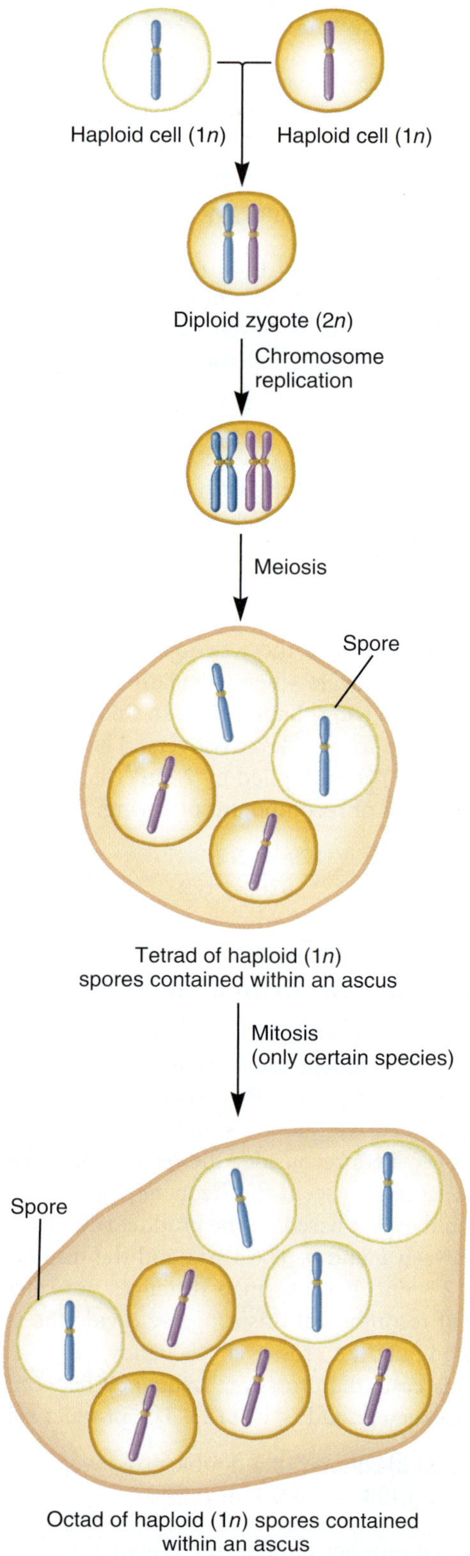

FIGURE 6.11 Sexual reproduction in ascomycetes. For simplicity, this diagram shows each haploid cell as having only one chromosome per haploid set. However, fungal species actually contain several chromosomes per haploid set.

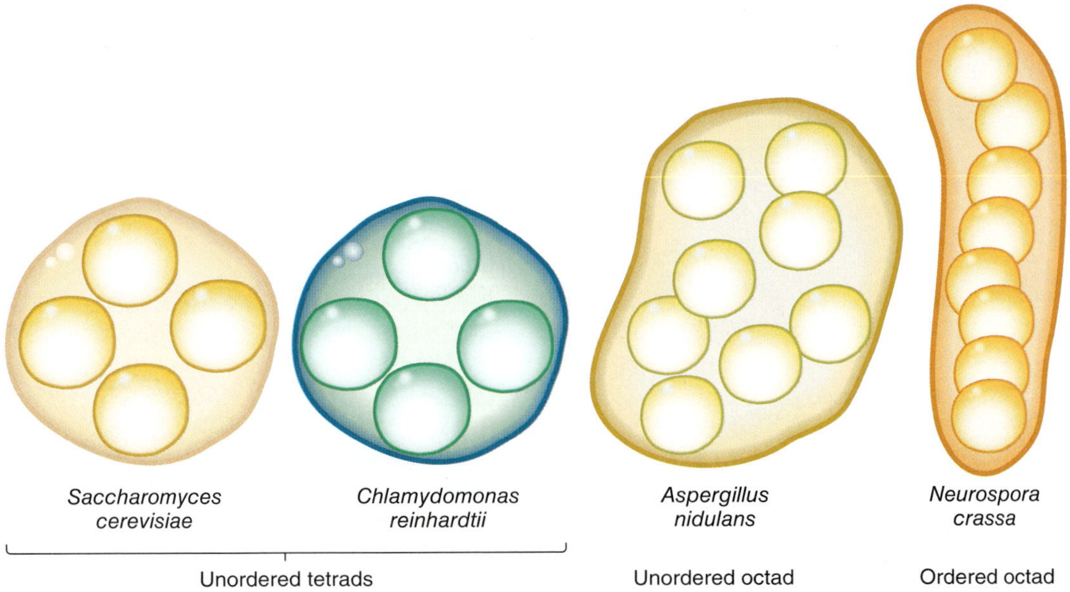

FIGURE 6.12 Arrangement of spores within asci of different species. *Saccharomyces cerevisiae* and *Chlamydomonas reinhardtii* (an alga) produce unordered tetrads, *Aspergillus nidulans* produces an unordered octad, and *Neurospora crassa* produces an ordered octad.

The Arrangement of Spores in an Ascus Can Be Ordered or Unordered

The arrangement of spores within an ascus varies from species to species (**Figure 6.12**). In some cases, the ascus provides enough space for the tetrads or octads of spores to randomly mix together. This creates an **unordered tetrad** or **octad.** These occur in fungal species such as *Saccharomyces cerevisiae* and *Aspergillus nidulans* and also in certain unicellular algae (*Chlamydomonas reinhardtii*). By comparison, other species of fungi produce a very tight ascus that prevents spores from randomly moving around, which results in an **ordered tetrad** or **octad.**

Unordered Tetrad Analysis Can Be Used to Map Genes in Two-Factor Crosses

Unordered tetrads contain a group of spores that are the product of meiosis and randomly arranged in an ascus. An experimenter can conduct a two-factor cross, remove the spores from each ascus, and determine the phenotypes of the spores. This analysis can determine if two genes are linked or assort independently. If two genes are linked, a tetrad analysis can also be used to compute map distance.

Figure 6.13 illustrates the possible outcomes starting with two haploid yeast strains. One strain carries the wild-type alleles ura^+ and arg^+, which are required for uracil and arginine biosynthesis, respectively. The other strain has defective alleles *ura-2* and *arg-3*; these result in yeast strains that require uracil and arginine in the growth medium. A diploid zygote with the genotype ura^+*ura-2* arg^+*arg-3* was produced from the fusion of haploid cells from these two strains. The diploid cell then proceeds through meiosis to produce four haploid cells. After the completion of meiosis, three distinct types of tetrads could be produced. One possibility is that the tetrad will contain four spores with the nonrecombinant arrangement of alleles. This ascus is said to have the **parental ditype (PD).** Alternatively, an ascus may have two nonrecombinant cells and two recombinant cells, which is called a **tetratype (T).** Finally, an ascus with a **nonparental ditype** (**NPD**) contains four cells with recombinant genotypes.

When two genes assort independently, the number of asci having a parental ditype is expected to equal the number having a nonparental ditype, thus yielding 50% recombinant spores. For linked genes, **Figure 6.14** illustrates the relationship between crossing over and the type of ascus that will result. If no crossing over occurs in the region between the two genes, the parental ditype will be produced (Figure 6.14a). A single crossover event produces a tetratype (Figure 6.14b). Double crossovers can yield a parental ditype, tetratype, or nonparental ditype, depending on the combination of chromatids that are involved (Figure 6.14c). A nonparental ditype is produced when a double crossover involves all four chromatids. A tetratype results from a three-chromatid crossover. Finally, a double crossover between the same two chromatids produces the parental ditype.

The data from a tetrad analysis can be used to calculate the map distance between two linked genes. As in conventional mapping, the map distance is calculated as the percentage of offspring that carry recombinant chromosomes. As mentioned, a tetratype contains 50% recombinant chromosomes; a nonparental ditype, 100%. Therefore, the map distance is computed as

$$\text{Map distance} = \frac{\text{NPD} + (1/2)\ (\text{T})}{\text{total number of asci}} \times 100$$

Over short map distances, this calculation provides a fairly reliable measure of distance. However, it does not adequately account for double crossovers. When two genes are far apart on the same chromosome, the calculated map distance using this equation underestimates the actual map distance due to double crossovers. Fortunately, a particular strength of tetrad analysis is that we can derive another equation that accounts for double crossovers, thereby providing a more accurate value for map distance. To begin this derivation, let's consider a more precise way to calculate map distance.

$$\text{Map distance} = \frac{\text{Single crossover tetrads} + (2)\ (\text{Double crossover tetrads})}{\text{Total number of asci}} \times 0.5 \times 100$$

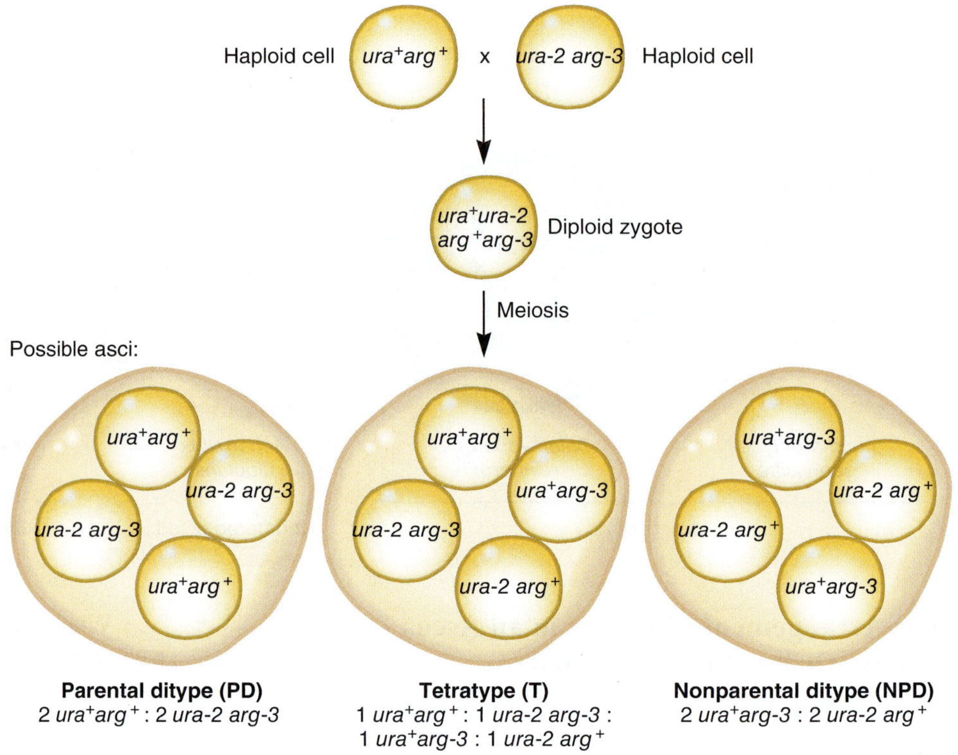

INTERACTIVE EXERCISE

FIGURE 6.13 **The assortment of two genes in an unordered tetrad.** If the tetrad contains 100% nonrecombinant cells, this ascus has the parental ditype (PD). If it contains 50% nonrecombinant and 50% recombinant cells, it is a tetratype (T). Finally, an ascus with 100% recombinant cells is called a nonparental ditype (NPD). This figure does not illustrate the chromosomal locations of the alleles. In this type of experiment, the goal is to determine whether the two genes are linked on the same chromosome.

This equation includes the number of single and double crossovers in the computation of map distance. The total number of crossovers equals the number of single crossovers plus 2 times the number of double crossovers. Overall, the tetrads that contain single and double crossovers also contain 50% nonrecombinant chromosomes. To calculate map distance, therefore, we divide the total number of crossovers by the total number of asci and multiply by 0.5 and 100.

To be useful, we need to relate this equation to the number of parental ditypes, nonparental ditypes, and tetratypes that are obtained by experimentation. To derive this relationship, we must consider the types of tetrads produced from no crossing over, a single crossover, and double crossovers. To do so, let's take another look at Figure 6.14. As shown there, the parental ditype and tetratype are ambiguous. The parental ditype can be derived from no crossovers or a double crossover; the tetratype can be derived from a single crossover or a double crossover. However, the nonparental ditype is unambiguous, because it can be produced only from a double crossover. We can use this observation as a way to determine the actual number of single and double crossovers. As seen in Figure 6.14, 1/4 of all the double crossovers are nonparental ditypes. Therefore, the total number of double crossovers equals four times the number of nonparental ditypes.

Next, we need to know the number of single crossovers. A single crossover yields a tetratype, but double crossovers can also yield a tetratype. Therefore, the total number of tetratypes overestimates the true number of single crossovers. Fortunately, we can compensate for this overestimation. Because two types of tetratypes are due to a double crossover, the actual number of tetratypes arising from a double crossover should equal 2NPD. Therefore, the true number of single crossovers is calculated as T – 2NPD.

Now we have accurate measures of both single and double crossovers. The number of single crossovers equals T – 2NPD, and the number of double crossovers equals 4NPD. We can substitute these values into our previous equation.

$$\text{Map distance} = \frac{(T - 2NPD) + (2)(4NPD)}{\text{Total number of asci}} \times 0.5 \times 100$$

$$= \frac{T + 6NPD}{\text{Total number of asci}} \times 0.5 \times 100$$

This equation provides a more accurate measure of map distance because it considers both single and double crossovers.

6.4 COMPREHENSION QUESTIONS

1. An ascus called an octad is the product of
   a. one meiotic division.
   b. two meiotic divisions.
   c. one meiotic division followed by one mitotic division.
   d. one mitotic division followed by one meiotic division.
2. One yeast strain is *lys*⁺ and *arg*⁺, whereas another strain is *lys-3* and *arg-2*. The two strains were crossed to each other and an ascus obtained from this cross has four spores with the following genotypes: *lys*⁺ *arg*⁺, *lys*⁺ *arg-2*, *lys-3 arg*⁺, and *lys-3 arg 2*. This ascus is a
   a. parental ditype.
   b. tetratype.
   c. nonparental ditype.
   d. could be b or c.

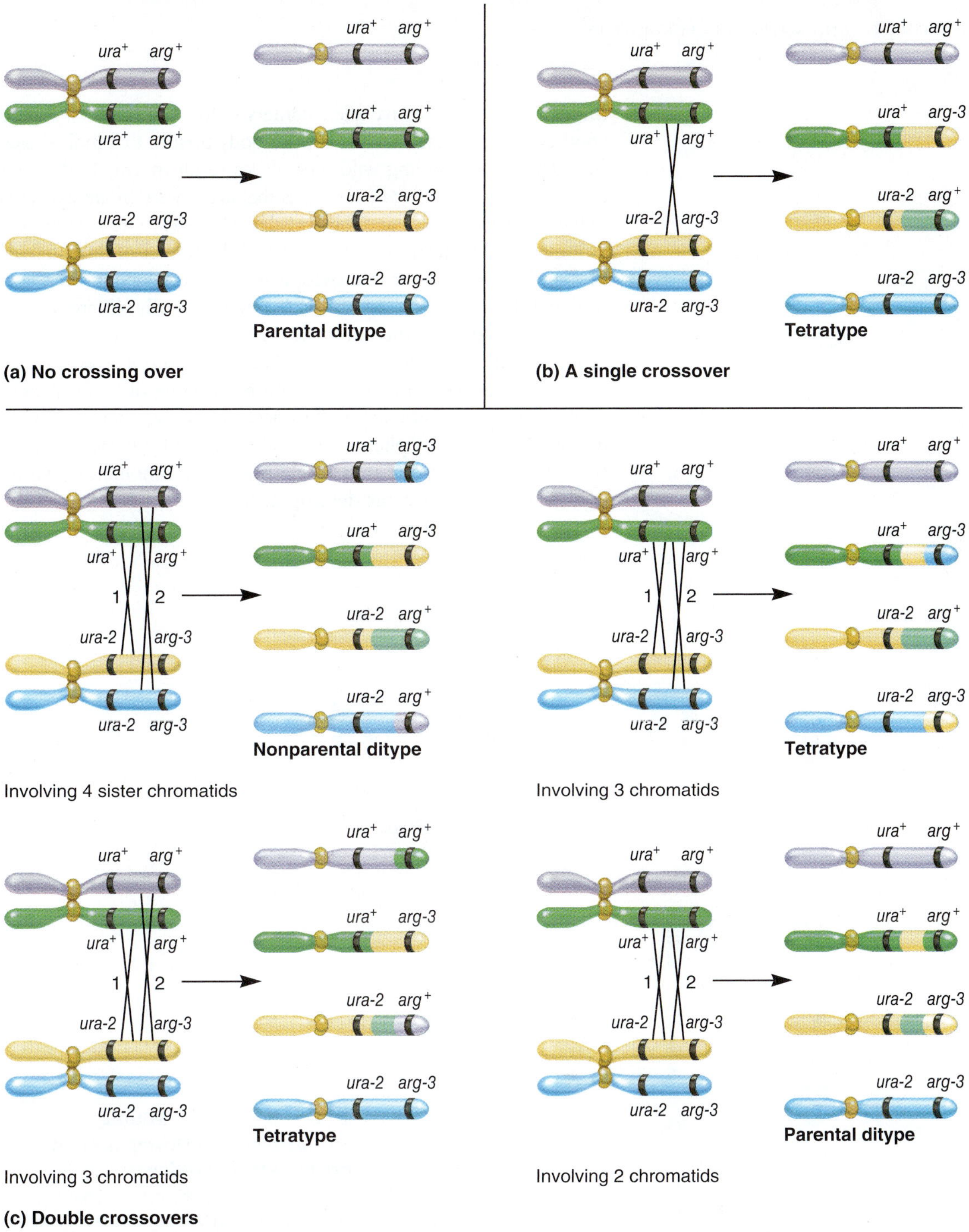

FIGURE 6.14 Relationship between crossing over and the production of the parental ditype, tetratype, and nonparental ditype for two linked genes. Note: For part **(c),** the outcome assumes that the crossover on the left (labeled 1) occurred before the crossover on the right (labeled 2).

6.5 MITOTIC RECOMBINATION

Learning Outcome:

1. Describe the process of mitotic recombination and explain how it can produce a twin spot.

Thus far, we have considered how the arrangement of linked alleles along a chromosome can be rearranged by crossing over. This event produces cells and offspring with a recombinant pattern of traits. In these cases, crossing over has occurred during meiosis, when the homologous chromosomes replicate and form bivalents.

In multicellular organisms, the union of egg and sperm is followed by many cellular divisions, which occur in conjunction with mitotic divisions of the cell nuclei. As discussed in Chapter 3, mitosis normally does not involve the homologous pairing of chromosomes to form a bivalent. Therefore, crossing over during mitosis is expected to occur much less frequently than during meiosis. Nevertheless, it does happen on rare occasions. Mitotic crossing over may produce a pair of recombinant chromosomes that have a new combination of alleles, an event known as **mitotic recombination.** If it occurs during an early stage of embryonic development, the daughter cells containing one recombinant chromosome and one nonrecombinant chromosome will continue to divide many times to produce a patch of tissue in the adult. This may result in a portion of tissue with characteristics different from those of the rest of the organism.

In 1936, Curt Stern identified unusual patches on the bodies of certain *Drosophila* strains. He was working with strains carrying X-linked alleles affecting body color and bristle morphology

(**Figure 6.15**). A recessive allele confers yellow body color (y), and another recessive allele causes shorter body bristles that look singed (sn). The corresponding wild-type alleles result in gray body color (y^+) and long bristles (sn^+). Females that are $y^+y\ sn^+sn$ are expected to have gray body color and long bristles. This was generally the case. However, when Stern carefully observed the bodies of these female flies under a low-power microscope, he occasionally noticed places in which two adjacent regions were phenotypically different from each other and also different from the rest of the body. These were called a twin spot. He concluded that twin spotting was too frequent to be explained by the random positioning of two independent single spots that happened to occur close together. How then did Stern explain the phenomenon of twin spotting? He proposed that twin spots are due to a single mitotic recombination within one cell during embryonic development.

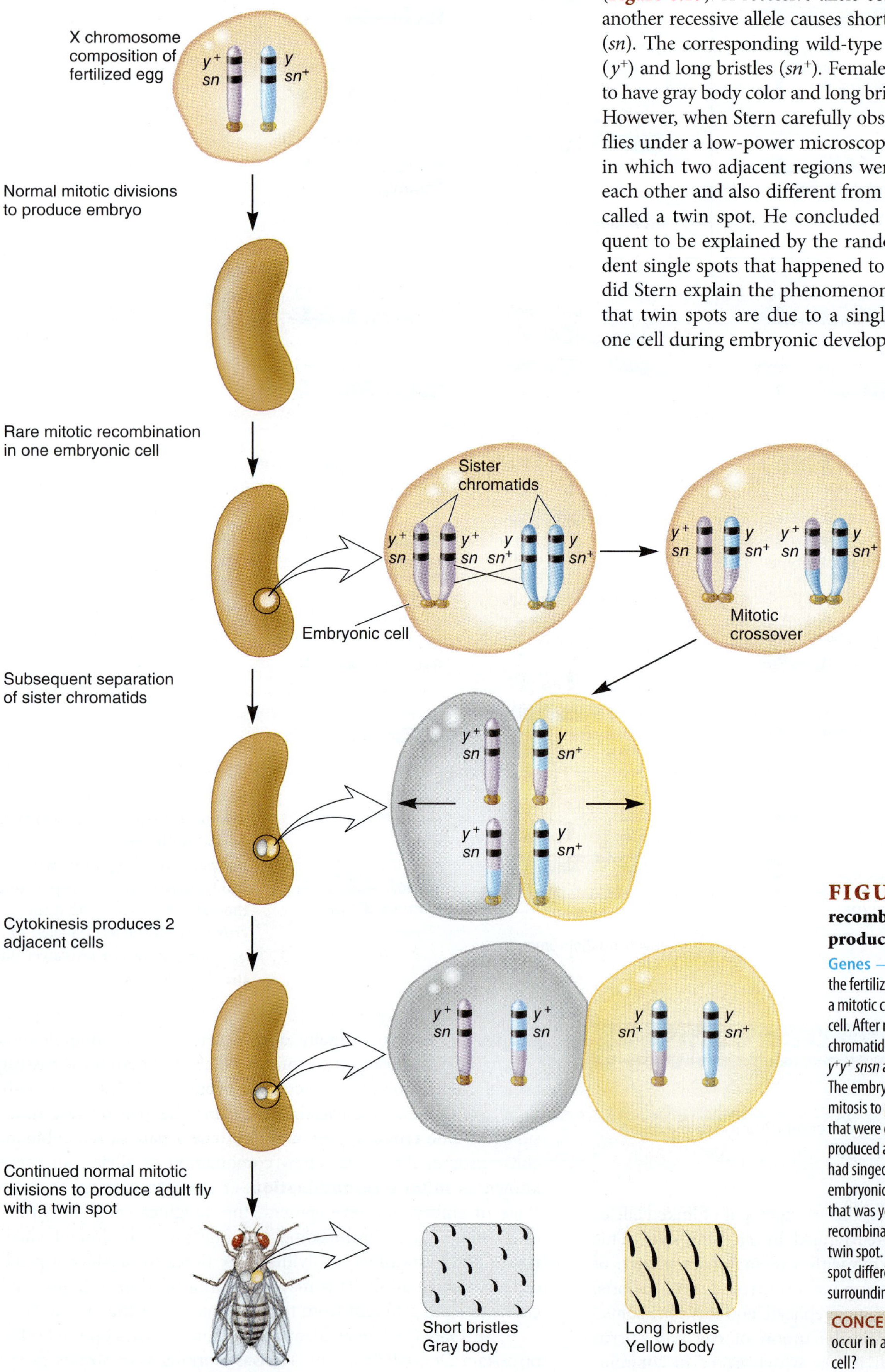

FIGURE 6.15 Mitotic recombination in *Drosophila* that produces twin spotting.

Genes → Traits In this illustration, the genotype of the fertilized egg was $y^+y\ sn^+sn$. During development, a mitotic crossover occurred in a single embryonic cell. After mitotic crossing over, the separation of the chromatids occurred, so one embryonic cell became $y^+y^+\ snsn$ and the adjacent sister cell became $yy\ sn^+sn^+$. The embryonic cells then continued to divide by normal mitosis to produce an adult fly. The cells in the adult that were derived from the $y^+y^+\ snsn$ embryonic cell produced a spot on the adult body that was gray and had singed bristles. The cells derived from the $yy\ sn^+sn^+$ embryonic cell produced an adjacent spot on the body that was yellow and had long bristles. In this case, mitotic recombination produced an unusual trait known as a twin spot. The two adjacent regions within the twin spot differed from each other and also differed from the surrounding tissue, which was gray with long bristles.

CONCEPT CHECK: Does mitotic recombination occur in a gamete (sperm or egg cell) or in a somatic cell?

As shown in Figure 6.15, the X chromosomes of the fertilized egg are y^+ *sn* and *y* sn^+. During development, a rare crossover can occur during mitosis to produce two adjacent daughter cells that are y^+y^+ *snsn* and *yy* sn^+sn^+. As embryonic development proceeds, the cell on the left continues to divide to produce many cells, eventually producing a patch on the body that has gray color with singed bristles. The daughter cell next to it produces a patch of yellow body color with long bristles. These two adjacent patches—a twin spot—are surrounded by cells that are y^+y sn^+sn and have gray color and long bristles. Twin spots provide evidence that mitotic recombination occasionally occurs.

6.5 COMPREHENSION QUESTION

1. The process of mitotic recombination involves the
   a. exchange of chromosomal regions between homologs during gamete formation.
   b. exchange of chromosomal regions between homologs during the division of somatic cells.
   c. reassortment of alleles that occurs at fertilization.
   d. reassortment of alleles that occurs during gamete formation.

KEY TERMS

Page 126. genetic map
Page 127. synteny, genetic linkage, linkage groups, two-factor cross, three-factor cross
Page 128. crossing over, bivalent, genetic recombination, recombinant, nonrecombinant
Page 132. null hypothesis
Page 136. genetic mapping, locus, genetic linkage map
Page 137. testcross
Page 138. map distance
Page 139. map units (mu), centiMorgans (cM)
Page 141. positive interference
Page 142. spores, tetrad, octad, ascus
Page 143. unordered tetrad, unordered octad, ordered tetrad, ordered octad, parental ditype (PD), tetratype (T), nonparental ditype (NPD)
Page 145. mitotic recombination

CHAPTER SUMMARY

6.1 Overview of Linkage

- Synteny refers to genes that are located on the same chromosome. Genetic linkage means that the alleles of two or more genes tend to be transmitted as a unit because they are relatively close together on the same chromosome.
- Bateson and Punnett discovered the first example of genetic linkage in sweet peas (see Figure 6.1).

6.2 Relationship Between Linkage and Crossing Over

- Crossing over during meiosis can alter the pattern of linked alleles along a chromosome (see Figure 6.2).
- Morgan discovered genetic linkage in *Drosophila* and proposed that recombinant offspring are produced by crossing over during meiosis (see Figures 6.3, 6.4).
- When genes are linked, the relative proportions of recombinant offspring depend on the distance between the genes (see Figure 6.5).
- A chi square analysis can be carried out to judge whether two genes assort independently.
- Creighton and McClintock were able to correlate the formation of recombinant offspring with the presence of chromosomes that had exchanged pieces due to crossing over (see Figures 6.6, 6.7).

6.3 Genetic Mapping in Plants and Animals

- A genetic linkage map is a diagram that portrays the order and relative spacing of genes along one or more chromosomes (see Figure 6.8).
- A testcross can be performed to map the distance between two or more genes (see Figure 6.9).
- Due to the effects of multiple crossovers, the map distance between two genes obtained from a testcross cannot exceed 50% (see Figure 6.10).
- The data from a three-factor cross can be used to map genes (see Table 6.1).
- Positive interference refers to the phenomenon that the number of double crossovers in a given region is less than expected based on the frequency of single crossovers.

6.4 Genetic Mapping in Haploid Eukaryotes

- Several haploid eukaryotes have been used in genetic mapping. Ascomycetes have the product of a single meiosis contained with an ascus (see Figure 6.11).
- Certain haploid species form unordered or ordered tetrads or octads (see Figure 6.12).
- The analysis of unordered tetrads in yeast is used to map the distance between two linked genes (see Figures 6.13, 6.14).

6.5 Mitotic Recombination

- Mitotic recombination occurs on rare occasions and may lead to twin spots (see Figure 6.15).

PROBLEM SETS & INSIGHTS

Solved Problems

S1. In the garden pea, orange pods (*orp*) are recessive to green pods (*Orp*), and sensitivity to pea mosaic virus (*mo*) is recessive to resistance to the virus (*Mo*). A plant with orange pods and sensitivity to the virus was crossed to a true-breeding plant with green pods and resistance to the virus. The F_1 plants were then testcrossed to plants with orange pods and sensitivity to the virus. The following results were obtained:

160 orange pods, virus sensitive
165 green pods, virus resistant
36 orange pods, virus resistant
39 green pods, virus sensitive
400 total

A. Conduct a chi square analysis to see if these genes are linked.

B. If they are linked, calculate the map distance between the two genes.

Answer:

A. Chi square analysis.

1. Our hypothesis is that the genes are not linked.
2. Calculate the predicted number of offspring based on the hypothesis. The testcross is

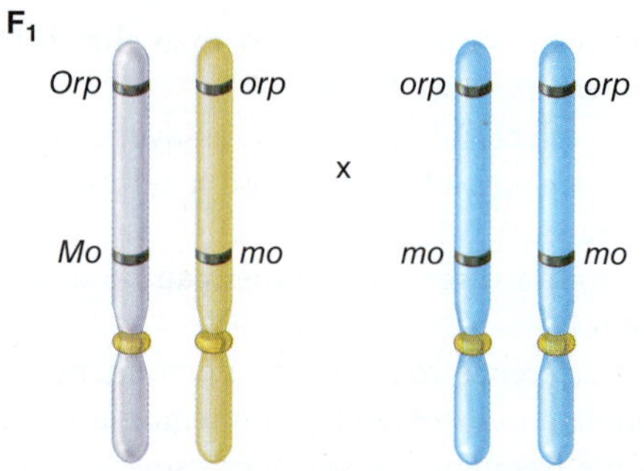

The predicted outcome of this cross under our hypothesis is a 1:1:1:1 ratio of plants with the four possible phenotypes. In other words, 1/4 should have the phenotype orange pods, virus sensitive; 1/4 should have green pods, virus resistant; 1/4 should have orange pods, virus resistant; and 1/4 should have green pods, virus sensitive. Because a total of 400 offspring were produced, our hypothesis predicts 100 offspring in each category.

3. Calculate the chi square.

$$\chi^2 = \frac{(O_1 - E_1)^2}{E_1} + \frac{(O_2 - E_2)^2}{E_2} + \frac{(O_3 - E_3)^2}{E_3} + \frac{(O_4 - E_4)^2}{E_4}$$

$$\chi^2 = \frac{(160 - 100)^2}{100} + \frac{(165 - 100)^2}{100} + \frac{(36 - 100)^2}{100} + \frac{(39 - 100)^2}{100}$$

$$\chi^2 = 36 + 42.3 + 41 + 37.2 = 156.5$$

4. Interpret the chi square value. The calculated chi square value is quite large. This indicates that the deviation between observed and expected values is very high. For 1 degree of freedom in Table 2.1, such a large deviation is expected to occur by chance alone less than 1% of the time. Therefore, we reject the hypothesis that the genes assort independently. As an alternative, we may infer that the two genes are linked.

B. Calculate the map distance.

$$\text{Map distance} = \frac{\text{(Number of recombinant offspring)}}{\text{Total number of offspring}} \times 100$$

$$= \frac{36 + 39}{36 + 39 + 160 + 165} \times 100$$

$$= 18.8 \text{ mu}$$

The genes are approximately 18.8 mu apart.

S2. Two recessive disorders in mice—droopy ears and flaky tail—are caused by genes that are located 6 mu apart on chromosome 3. A true-breeding mouse with normal ears (*De*) and a flaky tail (*ft*) was crossed to a true-breeding mouse with droopy ears (*de*) and a normal tail (*Ft*). The F_1 offspring were then crossed to mice with droopy ears and flaky tails. If this testcross produced 100 offspring, what is the expected outcome?

Answer: The testcross is

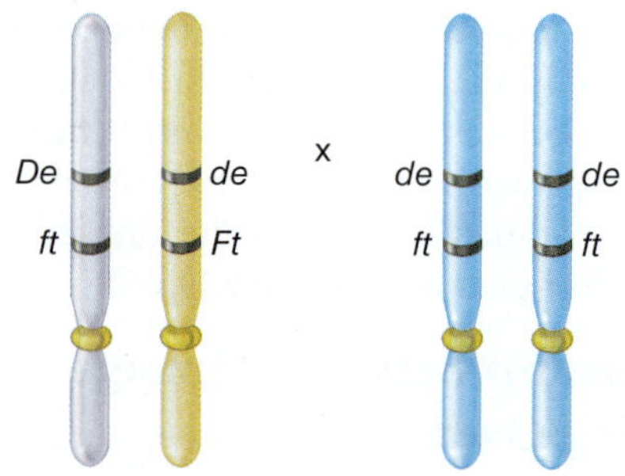

The nonrecombinant offspring are

Dede ftft Normal ears, flaky tail
dede Ftft Droopy ears, normal tail

The recombinant offspring are

dede ftft Droopy ears, flaky tail
Dede Ftft Normal ears, normal tail

Because the two genes are located 6 mu apart on the same chromosome, 6% of the offspring will be recombinants. Therefore, the expected outcome for 100 offspring is

3 droopy ears, flaky tail
3 normal ears, normal tail
47 normal ears, flaky tail
47 droopy ears, normal tail

S3. The following X-linked recessive traits are found in fruit flies: vermilion eyes are recessive to red eyes, miniature wings are recessive to long wings, and sable body is recessive to gray body. A cross was made between wild-type males with red eyes, long wings, and gray bodies to females with vermilion eyes, miniature wings, and sable bodies. The heterozygous females from this cross, which had red eyes, long wings, and gray bodies, were then crossed to males with vermilion eyes, miniature wings, and sable bodies. The following outcome was obtained:

Males and Females

1320 vermilion eyes, miniature wings, sable body

1346 red eyes, long wings, gray body

102 vermilion eyes, miniature wings, gray body

90 red eyes, long wings, sable body

42 vermilion eyes, long wings, gray body

48 red eyes, miniature wings, sable body

2 vermilion eyes, long wings, sable body

1 red eyes, miniature wings, gray body

A. Calculate the map distance between the three genes.

B. Is positive interference occurring?

Answer:

A. The first step is to determine the order of the three genes. We can do this by evaluating the pattern of inheritance in the double crossovers. The double crossover group occurs with the lowest frequency. Thus, the double crossovers are vermilion eyes, long wings, and sable body, and red eyes, miniature wings, and gray body. Compared with the nonrecombinant combinations of alleles (vermilion eyes, miniature wings, sable body and red eyes, long wings, gray body), the gene for wing length has been reassorted. Two flies have long wings associated with vermilion eyes and sable body, and one fly has miniature wings associated with red eyes and gray body. Taken together, these results indicate that the wing length gene is found in between the eye color and body color genes.

Eye color——wing length——body color

We now calculate the distance between eye color and wing length, and between wing length and body color. To do this, we consider the data according to gene pairs:

vermilion eyes, miniature wings = 1320 + 102 = 1422

red eyes, long wings = 1346 + 90 = 1436

vermilion eyes, long wings = 42 + 2 = 44

red eyes, miniature wings = 48 + 1 = 49

The recombinants are vermilion eyes, long wings and red eyes, miniature wings. The map distance between these two genes is

$(44 + 49)/(1422 + 1436 + 44 + 49) \times 100 = 3.2$ mu

Likewise, the other gene pair is wing length and body color.

miniature wings, sable body = 1320 + 48 = 1368

long wings, gray body = 1346 + 42 = 1388

miniature wings, gray body = 102 + 1 = 103

long wings, sable body = 90 + 2 = 92

The recombinants are miniature wings, gray body and long wings, sable body. The map distance between these two genes is

$(103 + 92)/(1368 + 1388 + 103 + 92) \times 100 = 6.6$ mu

With these data, we can produce the following genetic map:

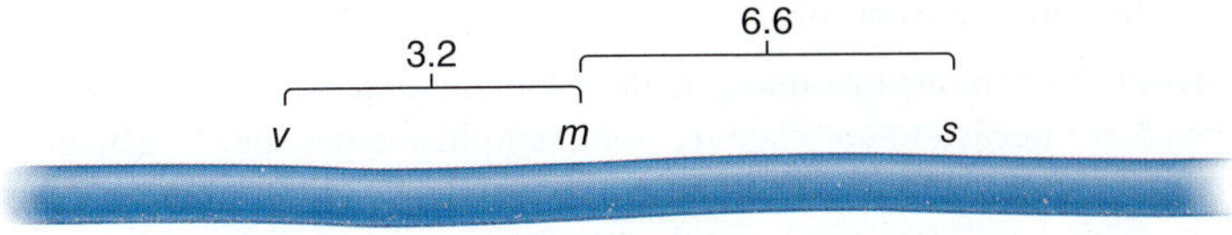

B. To calculate the interference value, we must first calculate the coefficient of coincidence.

$$C = \frac{\text{Observed number of double crossovers}}{\text{Expected number of double crossovers}}$$

Based on our calculation of map distances in part A, the percentage of single crossovers equals 3.2% (0.032) and 6.6% (0.066). The expected number of double crossovers equals 0.032×0.066, which is 0.002, or 0.2%. A total of 2951 offspring were produced. If we multiply 2951×0.002, we get 5.9, which is the expected number of double crossovers. The observed number was 3. Therefore,

$C = 3/5.9 = 0.51$

$I = 1 - C = 1 - 0.51 = 0.49$

In other words, approximately 49% of the expected double crossovers did not occur due to interference.

S4. Around the same time as the study of Creighton and McClintock, described in Figure 6.7, Curt Stern conducted similar experiments with *Drosophila.* He had strains of flies with microscopically detectable abnormalities in the X chromosome. In one case, the X chromosome was shorter than normal due to a deletion at one end. In another case, the X chromosome was longer than normal because an extra piece of the Y chromosome was attached at the other end of the X chromosome, where the centromere is located. He had female flies that had both abnormal chromosomes. On the short X chromosome, a recessive allele (*car*) was located that results in carnation-colored eyes, and a dominant allele (*B*) that causes bar-shaped eyes was also found on this chromosome. On the long X chromosome were located the wild-type alleles for these two genes (designated car^+ and B^+), which confer red eyes and round eyes, respectively. Stern realized that a crossover between the two X chromosomes in such female flies would result in recombinant chromosomes that would be cytologically distinguishable from the starting chromosomes. If a crossover occurred between the *B* and *car* genes on the X chromosome, this is expected to produce a normal-sized X chromosome and an abnormal chromosome with a deletion at one end and an extra piece of the Y chromosome at the other end.

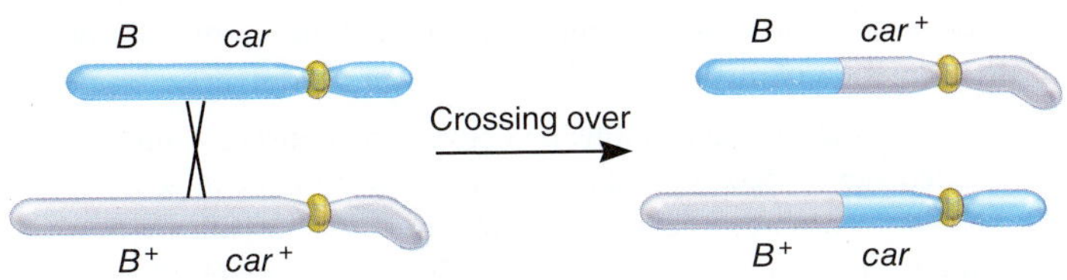

Stern crossed these female flies to male flies that had a normal-length X chromosome with the *car* allele and the allele for round eyes (*car* B^+). Using a microscope, he could discriminate between the morphologies of starting chromosomes—like those contained within the original parental flies—and recombinant chromosomes that may be found in the offspring. What would be the predicted phenotypes and chromosome characteristics in the offspring if crossing over did or did not occur between the X chromosomes in the female flies of this cross?

Answer: To demonstrate that genetic recombination is due to crossing over, Stern needed to correlate recombinant phenotypes (due to genetic recombination) with the inheritance of recombinant chromosomes (due to crossing over). Because he knew the arrangement of alleles in the female flies, he could predict the phenotypes of nonrecombinant and recombinant offspring. The male flies could contribute the *car* and B^+ alleles (on a cytologically normal X chromosome) or contribute a Y chromosome. In the absence of crossing over, the female flies could contribute a short X chromosome with the *car* and *B* alleles or a long X chromosome with the car^+ and B^+ alleles. If crossing over occurred in the region between these two genes, the female flies would contribute recombinant X chromosomes. One possible recombinant X chromosome would be normal-sized and carry the *car* and B^+ alleles, and the other recombinant X chromosome would be deleted at one end with a piece of the Y chromosome at the other end and carry the car^+ and *B* alleles. When combined with an X or Y chromosome from the males, the nonrecombinant offspring would have carnation, bar eyes or wild-type eyes; the recombinant offspring would have carnation, round eyes or red, bar eyes.

Female gametes ♀ \ Male gametes ♂	$car\,B^+$	Y	Phenotype	X chromosome from female
$car\,B$	$car\,B$ / $car\,B^+$	$car\,B$ / Y	Carnation, bar eyes	Short X chromosome
car^+B^+	car^+B^+ / $car\,B^+$	car^+B^+ / Y	Red, round eyes	Long X chromosome with a piece of Y
$car\,B^+$	$car\,B^+$ / $car\,B^+$	$car\,B^+$ / Y	Carnation, round eyes	Normal-sized X chromosome
car^+B	car^+B / $car\,B^+$	car^+B / Y	Red, bar eyes	Short X chromosome with a piece of Y

The results shown in the Punnett square are the actual results that Stern observed. His interpretation was that crossing over between homologous chromosomes—in this case, the X chromosome—accounts for the formation of offspring with recombinant phenotypes.

S5. A limit exists in the relationship between map distance and the percentage of recombinant offspring. Even though two genes on the same chromosome may be much more than 50 mu apart, we do not expect to obtain greater than 50% recombinant offspring in a testcross. You may be wondering why this is so. The answer lies in the pattern of multiple crossovers. At the pachytene stage of meiosis, a single crossover in the region between two genes produces only 50% recombinant chromosomes (see Figure 6.2b). Therefore, to exceed a 50% recombinant level, it would seem necessary to have multiple crossovers within the tetrad.

Let's suppose that two genes are far apart on the same chromosome. A testcross is made between a heterozygous individual, *AaBb*, and a homozygous individual, *aabb*. In the heterozygous individual, the dominant alleles (*A* and *B*) are linked on the same chromosome, and the recessive alleles (*a* and *b*) are linked on the same chromosome. Draw out all of the possible double crossovers (between two, three, or four chromatids) and determine the average number of recombinant offspring, assuming an equal probability of all of the double crossover possibilities.

Answer: A double crossover between the two genes could involve two chromatids, three chromatids, or four chromatids. The possibilities for all types of double crossovers are shown here:

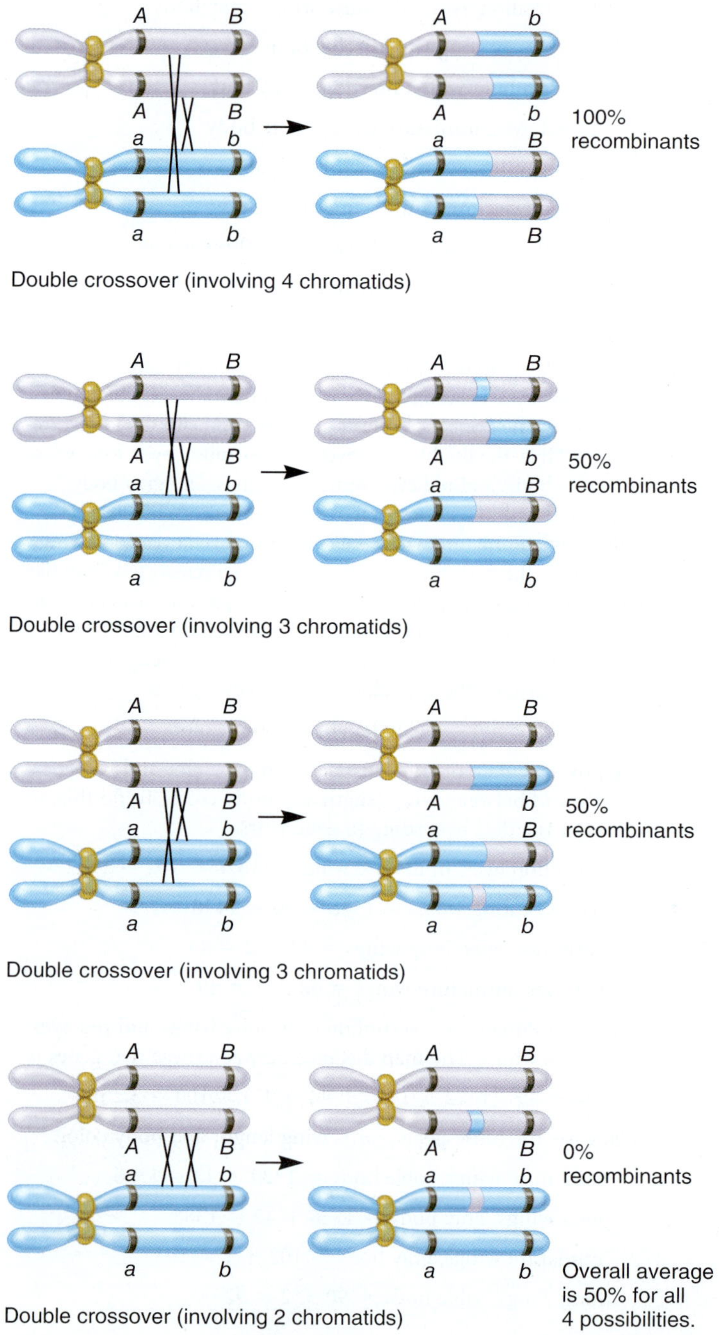

This drawing considers the situation where two crossovers are expected to occur in the region between the two genes. Because the tetrad is composed of two pairs of homologs, a double crossover between homologs could occur in several possible ways. In this illustration, the crossover on the right has occurred first. Because all of these double crossing over events are equally probable, we take the average of them to determine the maximum recombination. This average equals 50%.

Conceptual Questions

C1. What is the difference in meaning between the terms genetic recombination and crossing over?

C2. When applying a chi square approach in a linkage problem, explain why an independent assortment hypothesis is used.

C3. What is mitotic recombination? A heterozygous individual (*Bb*) with brown eyes has one eye with a small patch of blue. Provide two or more explanations for how the blue patch may have occurred.

C4. Mitotic recombination can occasionally produce a twin spot. Let's suppose an animal species is heterozygous for two genes that govern fur color and length: One gene affects pigmentation, with dark pigmentation (*A*) dominant to albino (*a*); the other gene affects hair length, with long hair (*L*) dominant to short hair (*l*). The two genes are linked on the same chromosome. Let's assume this animal is *AaLl*; *A* is linked to *l*, and *a* is linked to *L*. Draw the chromosomes labeled with these alleles, and explain how mitotic recombination could produce a twin spot with one spot having albino pigmentation and long fur, the other having dark pigmentation and short fur.

C5. A crossover has occurred in the bivalent shown here.

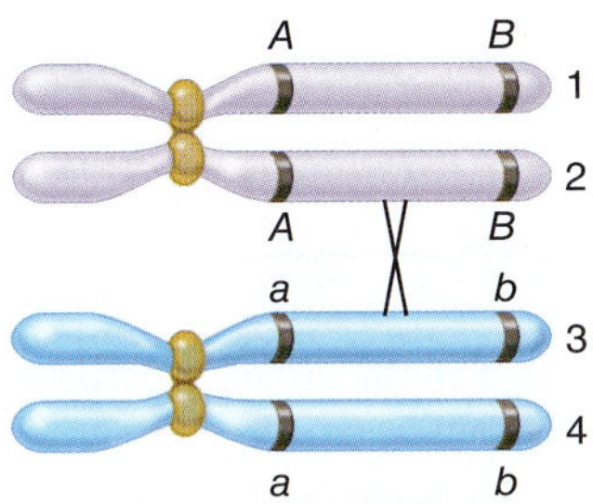

If a second crossover occurs in the same region between these two genes, which two chromatids would be involved to produce the following outcomes?

A. 100% recombinants

B. 0% recombinants

C. 50% recombinants

C6. A crossover has occurred in the bivalent shown here.

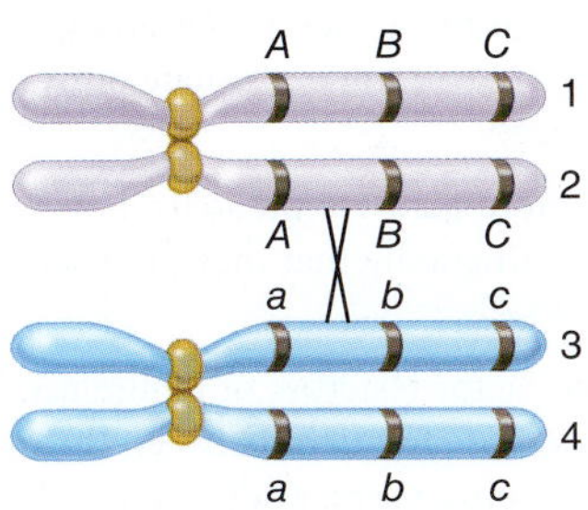

What is the outcome of this single crossover event? If a second crossover occurs somewhere between *A* and *C*, explain which two chromatids it would involve and where it would occur (i.e., between which two genes) to produce the types of chromosomes shown here:

A. *A B C*, *A b C*, *a B c*, and *a b c*

B. *A b c*, *A b c*, *a B C*, and *a B C*

C. *A B c*, *A b c*, *a B C*, and *a b C*

D. *A B C*, *A B C*, *a b c*, and *a b c*

C7. A diploid organism has a total of 14 chromosomes and about 20,000 genes per haploid genome. Approximately how many genes are in each linkage group?

C8. If you try to throw a basketball into a basket, the likelihood of succeeding depends on the size of the basket. It is more likely that you will get the ball into the basket if the basket is bigger. In your own words, explain how this analogy also applies to the idea that the likelihood of crossing over is greater when two genes are far apart than when they are close together.

C9. By conducting testcrosses, researchers have found that the sweet pea has seven linkage groups. How many chromosomes would you expect to find in leaf cells?

C10. In humans, a rare dominant disorder known as nail-patella syndrome causes abnormalities in the fingernails, toenails, and kneecaps. Researchers have examined family pedigrees with regard to this disorder and, within the same pedigree, also examined the individuals with regard to their blood types. (A description of blood genotypes is found in Chapter 4.) In the following pedigree, individuals affected with nail-patella disorder are shown with filled symbols. The genotype of each individual with regard to their ABO blood type is also shown. Does this pedigree suggest any linkage between the gene that causes nail-patella syndrome and the gene that causes blood type?

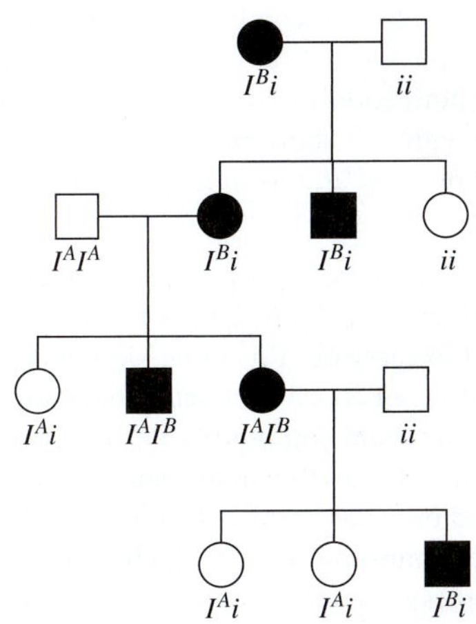

C11. When true-breeding mice with brown fur and short tails (*BBtt*) were crossed to true-breeding mice with white fur and long tails (*bbTT*), all F_1 offspring had brown fur and long tails. The F_1 offspring were crossed to mice with white fur and short tails. What are the possible phenotypes of the F_2 offspring? Which F_2 offspring are recombinant, and which are nonrecombinant? What are the ratios of the F_2 offspring if independent assortment is taking place? How are the ratios affected by linkage?

C12. Though we often think of genes in terms of the phenotypes they produce (e.g., curly leaves, flaky tail, brown eyes), the molecular function of most genes is to encode proteins. Many cellular proteins function as enzymes. The following table describes the map distances between six different genes that encode six different enzymes: *Ada*, adenosine deaminase; *Hao-1*, hydroxyacid oxidase-1; *Hdc*, histidine decarboxylase; *Odc-2*, ornithine decarboxylase-2; *Sdh-1*, sorbitol dehydrogenase-1; and *Ass-1*, arginosuccinate synthetase-1.

Map distances between two genes:

	Ada	***Hao-1***	***Hdc***	***Odc-2***	***Sdh-1***	***Ass-1***
Ada		14		8	28	
Hao-1	14		9		14	
Hdc		9		15	5	
Odc-2	8		15			63
Sdh-1	28	14	5			43
Ass-1				63	43	

Construct a genetic map that describes the locations of all six genes.

C13. If the likelihood of a single crossover in a particular chromosomal region is 10%, what is the theoretical likelihood of a double or triple crossover in that same region? How would positive interference affect these theoretical values?

C14. In most two-factor crosses involving linked genes, we cannot tell if a double crossover between the two genes has occurred because the offspring will inherit the nonrecombinant pattern of alleles. How does the inability to detect double crossovers affect the calculation of map distance? Is map distance underestimated or overestimated because of our inability to detect double crossovers? Explain your answer.

C15. Researchers have discovered that some regions of chromosomes are much more likely than others to cross over. We might call such a region a "hot spot" for crossing over. Let's suppose that two genes, gene *A* and gene *B*, are 5,000,000 bp apart on the same chromosome. Genes *A* and *B* are in a hot spot for crossing over. Two other genes, let's call them gene *C* and gene *D*, are also 5,000,000 bp apart but are not in a hot spot for recombination. If we conducted two-factor crosses to compute the map distance between genes *A* and *B*, and other two-factor crosses to compute the map distance between genes *C* and *D*, would the map distances be the same between *A* and *B* compared with *C* and *D*? Explain.

C16. Describe the unique features of ascomycetes that lend themselves to genetic analysis.

C17. In fungi, what is the difference between a tetrad and an octad? What cellular process occurs in an octad that does not occur in a tetrad?

C18. Explain the difference between an unordered versus an ordered octad.

Experimental Questions (Includes Most Mapping Questions)

E1. Figure 6.1 shows the first experimental results that indicated linkage between two different genes. Conduct a chi square analysis to confirm that the genes are really linked and the data could not be explained by independent assortment.

E2. In the experiment of Figure 6.7, the researchers followed the inheritance pattern of chromosomes that were abnormal at both ends to correlate genetic recombination with the physical exchange of chromosome pieces. Is it necessary to use a chromosome that is abnormal at both ends, or could the researchers have used a parental strain with two abnormal versions of chromosome 9, one with a knob at one end and its homolog with a translocation at the other end?

E3. The experiment of Figure 6.7 is not like a standard testcross, because neither parent is homozygous recessive for both genes. If you were going to carry out this same kind of experiment to verify that crossing over can explain the recombination of alleles of different genes, how would you modify this experiment to make it a standard testcross? For both parents, you should designate which alleles are found on an abnormal chromosome (i.e., knobbed, translocation chromosome 9) and which alleles are found on normal chromosomes.

E4. How would you determine that genes in mammals are located on the Y chromosome linkage group? Is it possible to conduct crosses (let's say in mice) to map the distances between genes along the Y chromosome? Explain.

E5. Explain the rationale behind a testcross. Is it necessary for one of the parents to be homozygous recessive for the genes of interest? In the heterozygous parent of a testcross, must all of the dominant alleles be linked on the same chromosome and all of the recessive alleles be linked on the homolog?

E6. In your own words, explain why a testcross cannot produce more than 50% recombinant offspring. When a testcross does produce 50% recombinant offspring, what do these results mean?

E7. Explain why the percentage of recombinant offspring in a testcross is a more accurate measure of map distance when two genes are close together. When two genes are far apart, is the percentage of recombinant offspring an underestimate or overestimate of the actual map distance?

E8. If two genes are more than 50 mu apart, how would you ever be able to show experimentally that they are located on the same chromosome?

E9. In Morgan's three-factor testcross of Figure 6.3, he realized that crossing over was more frequent between the eye color and wing length genes than between the body color and eye color genes. Explain how he determined this.

E10. Two genes are located on the same chromosome and are known to be 12 mu apart. An *AABB* individual was crossed to an *aabb* individual to produce *AaBb* offspring. The *AaBb* offspring were then crossed to *aabb* individuals.

A. If this cross produces 1000 offspring, what are the predicted numbers of offspring with each of the four genotypes: *AaBb*, *Aabb*, *aaBb*, and *aabb*?

B. What would be the predicted numbers of offspring with these four genotypes if the parental generation had been *AAbb* and *aaBB* instead of *AABB* and *aabb*?

E11. Two genes, designated *A* and *B*, are located 10 mu from each other. A third gene, designated *C*, is located 15 mu from *B* and 5 mu from *A*. The parental generation consisting of *AA bb CC* and *aa BB cc* individuals were crossed to each other. The F_1 heterozygotes were then testcrossed to *aa bb cc* individuals. If we assume no double crossovers occur in this region, what percentage of offspring would you expect with the following genotypes?

A. *Aa Bb Cc*

B. *aa Bb Cc*

C. *Aa bb cc*

E12. Two genes in tomatoes are 61 mu apart; normal fruit (*F*) is dominant to fasciated (flattened) fruit (*f*), and normal numbers of leaves (*Lf*) is dominant to leafy (*lf*). A true-breeding plant with normal leaves and fruit was crossed to a leafy plant with fasciated fruit. The F_1 offspring were then crossed to leafy plants with fasciated fruit. If this cross produced 600 offspring, what are the expected numbers of plants in each of the four possible categories: normal leaves, normal fruit; normal leaves, fasciated fruit; leafy, normal fruit; and leafy, fasciated fruit?

E13. In the tomato, three genes are linked on the same chromosome. Tall is dominant to dwarf, skin that is smooth is dominant to skin that is peachy, and fruit with a normal rounded tomato shape is dominant to flattened, or oblate shape. A plant that is true-breeding for the dominant traits was crossed to a dwarf plant with peachy skin and oblate fruit. The F_1 plants were then testcrossed to dwarf plants with peachy skin and oblate fruit. The following results were obtained:

151 tall, smooth, normal
33 tall, smooth, oblate
11 tall, peach, oblate
2 tall, peach, normal
155 dwarf, peach, oblate
29 dwarf, peach, normal
12 dwarf, smooth, normal
0 dwarf, smooth, oblate

Construct a genetic map that describes the order of these three genes and the distances between them.

E14. A trait in garden peas involves the curling of leaves. A two-factor cross was made by crossing a plant with yellow pods and curling leaves to a wild-type plant with green pods and normal leaves. All F_1 offspring had green pods and normal leaves. The F_1 plants were then crossed to plants with yellow pods and curling leaves. The following results were obtained:

117 green pods, normal leaves
115 yellow pods, curling leaves
78 green pods, curling leaves
80 yellow pods, normal leaves

A. Conduct a chi square analysis to determine if these two genes are linked.

B. If they are linked, calculate the map distance between the two genes. How accurate do you think this distance is?

E15. In mice, the gene that encodes the enzyme inosine triphosphatase is 12 mu from the gene that encodes the enzyme ornithine decarboxylase. Suppose you have identified a strain of mice homozygous for a defective inosine triphosphatase gene that does not produce any of this enzyme and is also homozygous for a defective ornithine decarboxylase gene. In other words, this strain of mice cannot make either enzyme. You crossed this homozygous recessive strain to a normal strain of mice to produce heterozygotes. The heterozygotes were then backcrossed to the strain that cannot produce either enzyme. What is the probability of obtaining a mouse that cannot make either enzyme?

E16. In the garden pea, several different genes affect pod characteristics. A gene affecting pod color (green is dominant to yellow) is approximately 7 mu away from a gene affecting pod width (wide is dominant to narrow). Both genes are located on chromosome 5. A third gene, located on chromosome 4, affects pod length (long is dominant to short). A true-breeding wild-type plant (green, wide, long pods) was crossed to a plant with yellow, narrow, short pods. The F_1 offspring were then testcrossed to plants with yellow, narrow, short pods. If the testcross produced 800 offspring, what are the expected numbers of the eight possible phenotypic combinations?

E17. A sex-influenced trait is dominant in males and causes bushy tails. The same trait is recessive in females. Fur color is not sex influenced. Yellow fur is dominant to white fur. A true-breeding female with a bushy tail and yellow fur was crossed to a white male without a bushy tail (i.e., a normal tail). The F_1 females were then crossed to white males without bushy tails. The following results were obtained:

Males	Females
28 normal tails, yellow	102 normal tails, yellow
72 normal tails, white	96 normal tails, white
68 bushy tails, yellow	0 bushy tails, yellow
29 bushy tails, white	0 bushy tails, white

A. Conduct a chi square analysis to determine if these two genes are linked.

B. If the genes are linked, calculate the map distance between them. Explain which data you used in your calculation.

E18. Three recessive traits in garden pea plants are as follows: yellow pods are recessive to green pods, bluish green seedlings are recessive to green seedlings, creeper (a plant that cannot stand up) is recessive to normal. A true-breeding normal plant with green pods and green seedlings was crossed to a creeper with yellow pods and bluish green seedlings. The F_1 plants were then crossed to creepers with yellow pods and bluish green seedlings. The following results were obtained:

2059 green pods, green seedlings, normal
151 green pods, green seedlings, creeper

281 green pods, bluish green seedlings, normal

15 green pods, bluish green seedlings, creeper

2041 yellow pods, bluish green seedlings, creeper

157 yellow pods, bluish green seedlings, normal

282 yellow pods, green seedlings, creeper

11 yellow pods, green seedlings, normal

Construct a genetic map that describes the map distance between these three genes.

E19. In mice, a trait called snubnose is recessive to a wild-type nose, a trait called pintail is dominant to a normal tail, and a trait called jerker (a defect in motor skills) is recessive to a normal gait. Jerker mice with a snubnose and pintail were crossed to normal mice, and then the F_1 mice were crossed to jerker mice that have a snubnose and normal tail. The outcome of this cross was as follows:

560 jerker, snubnose, pintail

548 normal gait, normal nose, normal tail

102 jerker, snubnose, normal tail

104 normal gait, normal nose, pintail

77 jerker, normal nose, normal tail

71 normal gait, snubnose, pintail

11 jerker, normal nose, pintail

9 normal gait, snubnose, normal tail

Construct a genetic map that describes the order and distance between these genes.

E20. In *Drosophila*, an allele causing vestigial wings is 12.5 mu away from another allele that causes purple eyes. A third gene that affects body color has an allele that causes black body color. This third gene is 18.5 mu away from the vestigial wings allele and 6 mu away from the allele causing purple eyes. The alleles causing vestigial wings, purple eyes, and black body are all recessive. The dominant (wild-type) traits are long wings, red eyes, and gray body. A researcher crossed wild-type flies to flies with vestigial wings, purple eyes, and black bodies. All F_1 flies were wild type. F_1 female flies were then crossed to male flies with vestigial wings, purple eyes, and black bodies. If 1000 offspring were observed, what are the expected numbers of the following types of flies?

Long wings, red eyes, gray body

Long wings, purple eyes, gray body

Long wings, red eyes, black body

Long wings, purple eyes, black body

Short wings, red eyes, gray body

Short wings, purple eyes, gray body

Short wings, red eyes, black body

Short wings, purple eyes, black body

Which kinds of flies can be produced only by a double crossover event?

E21. Three autosomal genes are linked along the same chromosome. The distance between gene *A* and *B* is 7 mu, the distance between *B* and *C* is 11 mu, and the distance between *A* and *C* is 4 mu. An individual who is *AA bb CC* was crossed to an individual who is *aa BB cc* to produce heterozygous F_1 offspring. The F_1 offspring were then crossed to homozygous *aa bb cc* individuals to produce F_2 offspring.

A. Draw the arrangement of alleles on the chromosomes in the parents and in the F_1 offspring.

B. Where would a crossover have to occur to produce an F_2 offspring that was heterozygous for all three genes?

C. If we assume that no double crossovers occur in this region, what percentage of F_2 offspring is likely to be homozygous for all three genes?

E22. Let's suppose that two different X-linked genes exist in mice, designated with the letters *N* and *L*. Gene *N* exists in a dominant, normal allele and in a recessive allele, *n*, that is lethal. Similarly, gene *L* exists in a dominant, normal allele and in a recessive allele, *l*, that is lethal. Heterozygous females are normal, but males that carry either recessive allele are born dead. Explain whether or not it would be possible to map the distance between these two genes by making crosses and analyzing the number of living and dead offspring. You may assume that you have strains of mice in which females are heterozygous for one or both genes.

E23. The alleles *his-5* and *lys-1*, found in baker's yeast, result in cells that require histidine and lysine for growth, respectively. A cross was made between two haploid yeast strains that are *his-5 lys-1* and his^+ lys^+. From the analysis of 818 individual tetrads, the following numbers of tetrads were obtained:

2 spores: *his-5* lys^+ + 2 spores: his^+ *lys-1* = 4

2 spores: *his-5 lys-1* + 2 spores: his^+ lys^+ = 502

1 spore: *his-5 lys-1* + 1 spore: *his-5* lys^+ + 1 spore: his^+ *lys-1* + 1 spore: his^+ lys^+ = 312

A. Compute the map distance between these two genes using the method of calculation that considers double crossovers and the one that does not. Which method gives a higher value? Explain why.

B. What is the frequency of single crossovers between these two genes?

C. Based on your answer to part B, how many NPDs are expected from this cross? Explain your answer. Is positive interference occurring?

Questions for Student Discussion/Collaboration

1. In mice, a dominant allele that causes a short tail is located on chromosome 2. On chromosome 3, a recessive allele causing droopy ears is 6 mu away from another recessive allele that causes a flaky tail. A recessive allele that causes a jerker (uncoordinated) phenotype is located on chromosome 4. A jerker mouse with droopy ears and a short, flaky tail was crossed to a normal mouse. All F_1 generation mice were phenotypically normal, except they had short tails. These F_1 mice were then testcrossed to jerker mice with droopy ears and long, flaky tails. If this cross produced 400 offspring, what would be the proportions of the 16 possible phenotypic categories?

2. In Chapter 3, we discussed the idea that the X and Y chromosomes have a few genes in common. These genes are inherited in a pseudoautosomal pattern. With this phenomenon in mind, discuss whether or not the X and Y chromosomes are really distinct linkage groups.
3. Mendel studied seven traits in pea plants, and the garden pea happens to have seven different chromosomes. It has been pointed out that Mendel was very lucky not to have conducted crosses involving two traits governed by genes that are closely linked on the same chromosome because the results would have confounded his theory of independent assortment. It has even been suggested that Mendel may not have published data involving traits that were linked! An article by Stig Blixt ("Why Didn't Gregor Mendel Find Linkage?" *Nature 256*:206, 1975) considers this issue. Look up this article and discuss why Mendel did not find linkage.

Note: All answers appear at the website for this textbook; the answers to the even-numbered questions and all of the Concept Check and Comprehension Questions are in the back of the book.

www.mhhe.com/brookergenetics5e

Visit the website for practice tests, answer keys, and other learning aids for this chapter. Enhance your understanding of genetics with our interactive exercises, quizzes, animations, and much more.

CHAPTER OUTLINE

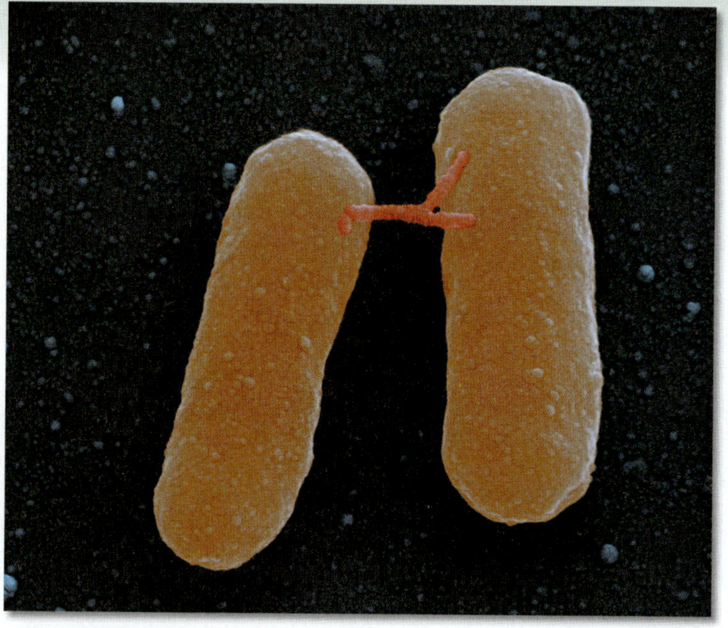

Conjugating bacteria. *The bacteria shown here are transferring genetic material by a process called conjugation.*

7 GENETIC TRANSFER AND MAPPING IN BACTERIA AND BACTERIOPHAGES

One reason researchers are so interested in bacteria and viruses is related to their impact on health. Infectious diseases caused by these agents are a leading cause of human death, accounting for a quarter to a third of deaths worldwide. The spread of infectious diseases is influenced by human behavior, and in recent times has been accelerated by increased trade and travel, and the inappropriate use of antibiotic drugs. Although the incidence of fatal infectious diseases in the United States is relatively low compared with the worldwide average, an alarming increase in more deadly strains of bacteria and viruses has occurred over the past few decades. Since 1980, the number of deaths in the United States due to infectious diseases has approximately doubled.

Thus far, our attention in Part II of this textbook has focused on genetic analyses of eukaryotic species such as plants and animals. As we have seen, these organisms are amenable to genetic studies for two reasons. First, differences, such as tall versus dwarf pea plants and red versus white eyes in *Drosophila*, provide readily discernible traits among different individuals. Second, because most eukaryotic species reproduce sexually, crosses can be made, and the pattern of transmission of traits from parent to offspring can be analyzed. The ability to follow allelic differences in a genetic cross is a basic tool in the genetic examination of eukaryotic species.

In Chapter 7, we turn our attention to the genetic analysis of bacteria. Like their eukaryotic counterparts, bacteria often possess allelic differences that affect their cellular traits. Common allelic variations among bacteria that are readily discernible involve traits such as sensitivity to antibiotics and differences in their nutrient requirements for growth. In these cases, the allelic differences are between different strains of bacteria, because any given bacterium is usually haploid for a particular gene. Throughout this chapter, we will consider interesting experiments that examine bacterial strains with allelic differences.

Compared with eukaryotes, another striking difference in bacterial species is their mode of reproduction. Because they reproduce asexually, researchers do not use crosses in the genetic analysis of bacterial species. Instead, they rely on a similar mechanism, called genetic transfer, in which a segment of bacterial DNA is transferred from one bacterium to another. In the first part of this chapter, we will explore the different routes of genetic transfer, and see how researchers have used genetic transfer to map the locations of genes along the chromosome of many bacterial species. In the later part of this chapter, we will examine **bacteriophages** (also known as **phages**), which are viruses that infect bacteria. Bacteriophages contain their own genetic material that governs the traits of the phage. As we will see, the genetic analysis of phages can yield a highly detailed genetic map of a short region of DNA. These types of analyses have provided researchers with insights regarding the structure and function of genes.

7.1 OVERVIEW OF GENETIC TRANSFER IN BACTERIA

Learning Outcome:

1. Compare and contrast the three mechanisms of genetic transfer in bacteria.

Genetic transfer is a process by which one bacterium transfers genetic material to another bacterium. Why is genetic transfer an advantage? Like sexual reproduction in eukaryotes, genetic transfer in bacteria is thought to enhance the genetic diversity of bacterial species. For example, a bacterial cell carrying a gene that provides antibiotic resistance may transfer this gene to another bacterial cell, allowing that bacterial cell to survive exposure to the antibiotic.

Bacteria can naturally transfer genetic material via three mechanisms (**Table 7.1**). The first route, known as **conjugation,** involves a direct physical interaction between two bacterial cells. One bacterium acts as a donor and transfers genetic material to a recipient cell. A second means of transfer is called **transduction.** This occurs when a virus infects a bacterium and then transfers bacterial genetic material from that bacterium to another. The last mode of genetic transfer is **transformation.** In this case, genetic material is released into the environment when a bacterial cell dies. This material then binds to a living bacterial cell, which can take it up. These three mechanisms of genetic transfer have been extensively investigated in research laboratories, and their molecular mechanisms continue to be studied with great interest. In later sections, we will examine these three systems of genetic transfer in greater detail.

We will also learn how genetic transfer between bacterial cells has provided unique ways to accurately map bacterial genes. The mapping methods described in this chapter have been largely replaced by molecular approaches described in Chapter 23. Even so, the mapping of bacterial genes serves to illuminate the mechanisms by which genes are transferred between bacterial cells and also helps us to appreciate the strategies of newer mapping approaches.

7.1 COMPREHENSION QUESTION

1. A form of genetic transfer that involves the uptake of a fragment of DNA from the environment is called
   a. conjugation.
   b. transduction.
   c. transformation.
   d. all of the above.

7.2 BACTERIAL CONJUGATION

Learning Outcomes:

1. Analyze the data of Lederberg and Tatum and the data of Davis, and explain how it indicated that some strains of bacteria can transfer genetic material via direct physical contact.
2. Outline the steps of conjugation via F factors.
3. Compare and contrast different types of plasmids.

As described briefly in Table 7.1, conjugation involves the direct transfer of genetic material from one bacterial cell to another. In this section, we will examine the steps in this process at the molecular and cellular levels.

Bacteria Can Transfer Genetic Material During Conjugation

The natural ability of some bacteria to transfer genetic material between each other was first recognized by Joshua Lederberg and Edward Tatum in 1946. They were studying strains of *Escherichia coli* that had different nutritional requirements for growth. A **minimal medium** is a growth medium that contains the essential nutrients for a wild-type (nonmutant) bacterial species to grow. Researchers often study bacterial strains that harbor mutations and cannot grow on a minimal medium. A strain that cannot synthesize a particular nutrient and needs that nutrient to be supplemented in its growth medium is called an **auxotroph.** For example, a strain that cannot make the amino acid methionine would not grow on a minimal medium because the minimal medium does not contain methionine. Such a strain would need to have methionine added to its growth medium and would be called a methionine auxotroph. By comparison, a strain that could make this amino acid would be termed a methionine prototroph. A **prototroph** does not need this nutrient in its growth medium.

The experiment in **Figure 7.1** considers one strain, designated met^- bio^- thr^+ leu^+ thi^+, which required one amino acid, methionine (met), and one vitamin, biotin (bio), to be added to its growth medium in order to grow. This strain did not require the amino acids threonine (thr) or leucine (leu), or the vitamin thiamine (thi) for growth. Another strain, designated met^+ bio^+ thr^- leu^- thi^-, had just the opposite requirements. It was an auxotroph for threonine, leucine, and thiamine, but a prototroph for methionine and biotin. These differences in nutritional requirements

TABLE 7.1

Three Mechanisms of Genetic Transfer Found in Bacteria

Mechanism		Description
Conjugation	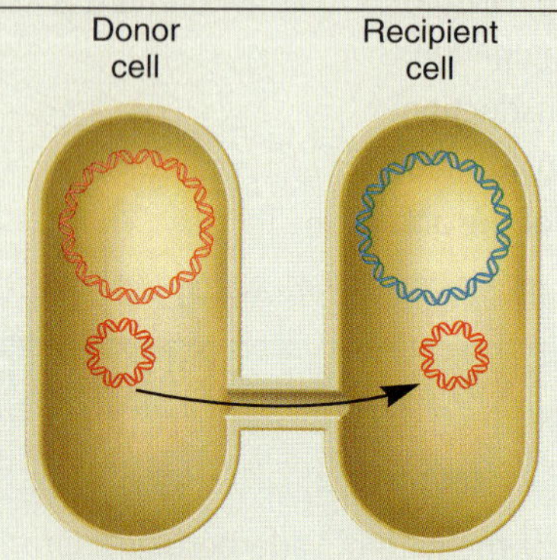	Requires direct contact between a donor and a recipient cell. The donor cell transfers a strand of DNA to the recipient. In the example shown here, DNA known as a plasmid is transferred to the recipient cell.
Transduction	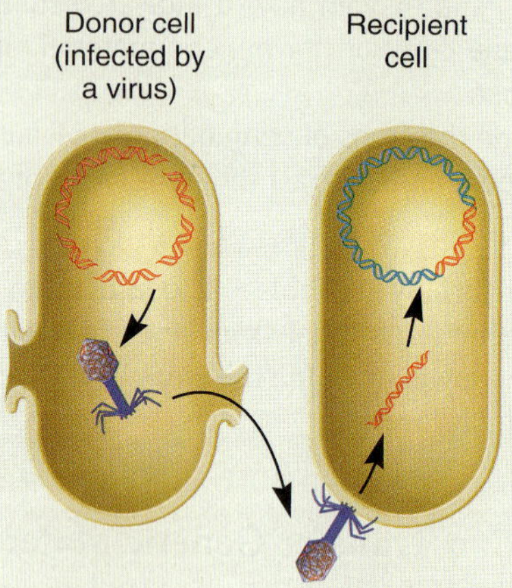	When a virus infects a donor cell, it incorporates a fragment of bacterial chromosomal DNA into a newly made virus particle. The virus then transfers this fragment of DNA to a recipient cell, which incorporates the DNA into its chromosome by recombination.
Transformation	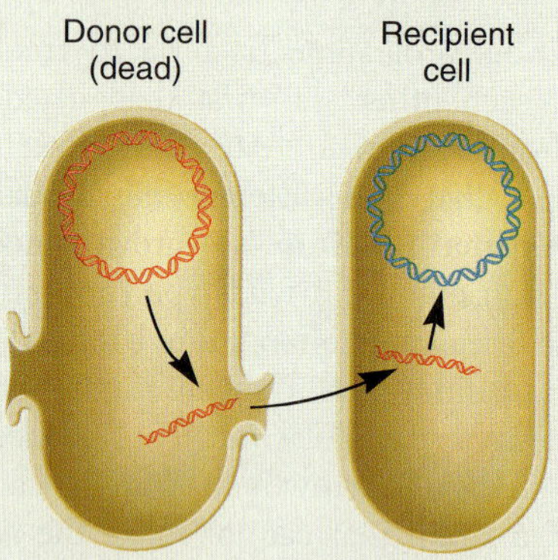	When a bacterial cell dies, it releases a fragment of its DNA into the environment. This DNA fragment is taken up by a recipient cell, which incorporates the DNA into its chromosome by recombination.

correspond to variations in the genetic material of the two strains. The first strain had two defective genes encoding enzymes necessary for methionine and biotin synthesis. The second strain contained three defective genes required to make threonine, leucine, and thiamine.

Figure 7.1 compares the results when the two strains were mixed together and when they were not mixed. Without mixing, about 100 million (10^8) $met^- \ bio^- \ thr^+ \ leu^+ \ thi^+$ cells were applied to plates on a growth medium lacking amino acids, biotin, and thiamine; no colonies were observed to grow. This result is expected because the medium did not contain methionine or biotin. Likewise, when 10^8 $met^+ \ bio^+ \ thr^- \ leu^- \ thi^-$ cells were plated, no colonies were observed because threonine, leucine, and thiamine were missing from this growth medium. However, when the two strains were mixed together and then 10^8 cells plated, approximately 10 bacterial colonies formed. Because growth occurred, the genotype of the cells within these colonies must have been $met^+ \ bio^+ \ thr^+ \ leu^+ \ thi^+$. How could this genotype occur? Because no colonies were observed on either plate in which the two strains were not mixed, Lederberg and Tatum concluded that it was not due to mutations that converted $met^- \ bio^-$ to $met^+ \ bio^+$ or to mutations that converted $thr^- \ leu^- \ thi^-$ to $thr^+ \ leu^+ \ thi^+$. Instead, they hypothesized that some genetic material was transferred between the two strains. One possibility is that the genetic

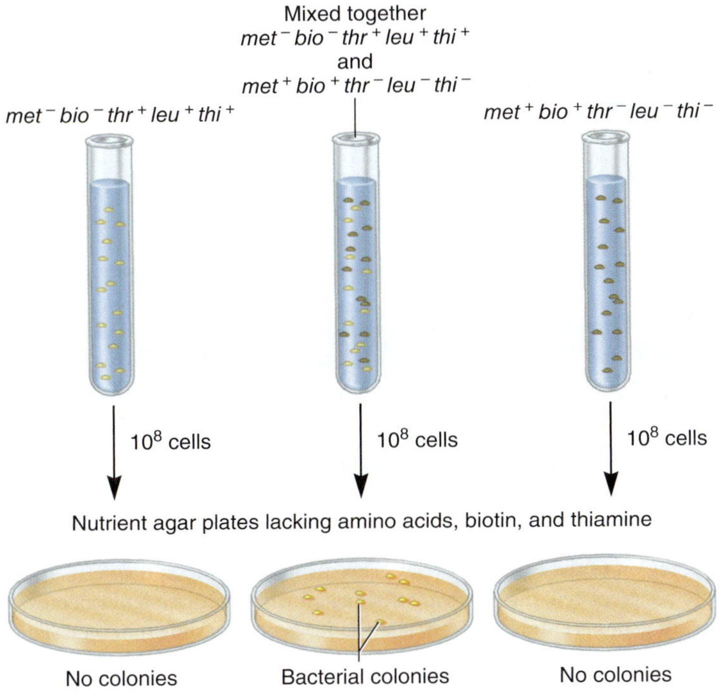

FIGURE 7.1 Experiment of Lederberg and Tatum demonstrating genetic transfer during conjugation in *E. coli*. When plated on a growth medium lacking amino acids, biotin, and thiamine, the met^- bio^- thr^+ leu^+ thi^+ or met^+ bio^+ thr^- leu^- thi^- strains were unable to grow. However, if they were mixed together and then plated, some colonies were observed. These colonies were due to the transfer of genetic material between these two strains by conjugation. Note: In bacteria, it is common to give genes a three-letter name (shown in italics) that is related to the function of the gene. A plus superscript ($^+$) indicates a functional gene, and a minus superscript ($^-$) indicates a mutation that has caused the gene or gene product to be inactive. In some cases, several genes have related functions. These may have the same three-letter name followed by different capital letters. For example, different genes involved with leucine biosynthesis may be called *leuA*, *leuB*, *leuC*, and so on. In the experiment described in Figure 7.1, the genes involved in leucine biosynthesis had not been distinguished, so the gene involved is simply referred to as leu^+ (for a functional gene) and leu^- (for a nonfunctional gene).

CONCEPT CHECK: Describe how gene transfer can explain the growth of colonies on the middle plate.

material providing the ability to synthesize methionine and biotin (met^+ bio^+) was transferred to the met^- bio^- thr^+ leu^+ thi^+ strain. Alternatively, the ability to synthesize threonine, leucine, and thiamine (thr^+ leu^+ thi^+) may have been transferred to the met^+ bio^+ thr^- leu^- thi^- cells. The results of this experiment did not distinguish between these two possibilities.

Conjugation Requires Direct Physical Contact

In 1950, Bernard Davis conducted experiments showing that two strains of bacteria must make physical contact with each other to transfer genetic material. The apparatus he used, known as a U-tube, is shown in **Figure 7.2**. At the bottom of the U-tube is a filter with pores small enough to allow the passage of genetic

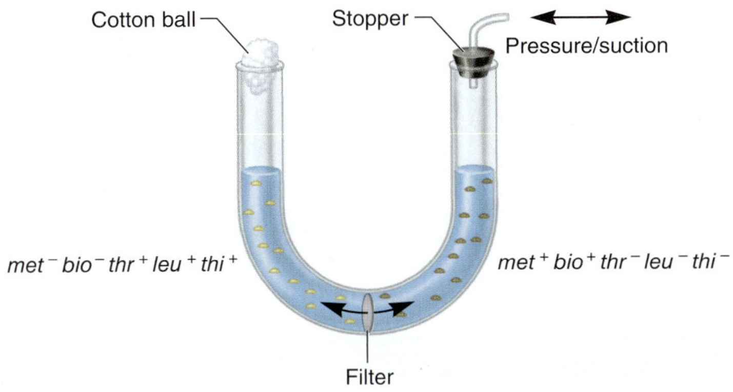

FIGURE 7.2 A U-tube apparatus like that used by Bernard Davis. The fluid in the tube is forced through the filter by alternating suction and pressure. However, the pores in the filter are too small for the passage of bacteria.

CONCEPT CHECK: With regard to studying the mechanism of conjugation, what is the purpose of using a U-tube?

material (i.e., DNA molecules) but too small to permit the passage of bacterial cells. On one side of the filter, Davis added a bacterial strain with a certain combination of nutritional requirements (the met^- bio^- thr^+ leu^+ thi^+ strain). On the other side, he added a different bacterial strain (the met^+ bio^+ thr^- leu^- thi^- strain). The application of alternating pressure and suction promoted the movement of liquid through the filter. Because the bacteria were too large to pass through the pores, the movement of liquid did not allow the two types of bacterial strains to mix with each other. However, any genetic material that was released from a bacterium could pass through the filter.

After incubation in a U-tube, bacteria from either side of the tube were placed on a medium that could select for the growth of cells that were met^+ bio^+ thr^+ leu^+ thi^+. The minimal medium lacked methionine, biotin, threonine, leucine, and thiamine, but contained all other nutrients essential for growth. In this case, no bacterial colonies grew on the plates. The experiment showed that, without physical contact, the two bacterial strains did not transfer genetic material to one another. The term conjugation is now used to describe the natural process of genetic transfer between bacterial cells that requires direct cell-to-cell contact. Many, but not all, species of bacteria can conjugate.

An F^+ Strain Transfers an F factor to an F^- Strain During Conjugation

We now know that certain donor strains of *E. coli* contain a small circular segment of genetic material known as an **F factor** (for fertility factor) in addition to their circular chromosome. Strains of *E. coli* that contain an F factor are designated F^+, and strains without F factors are termed F^-. F factors carry several genes that are required for conjugation to occur. For example, **Figure 7.3** shows the arrangement of genes on the F factor found in certain strains of *E. coli*. The functions of the proteins encoded by these genes are needed to transfer a strand of DNA from the donor cell to a recipient cell.

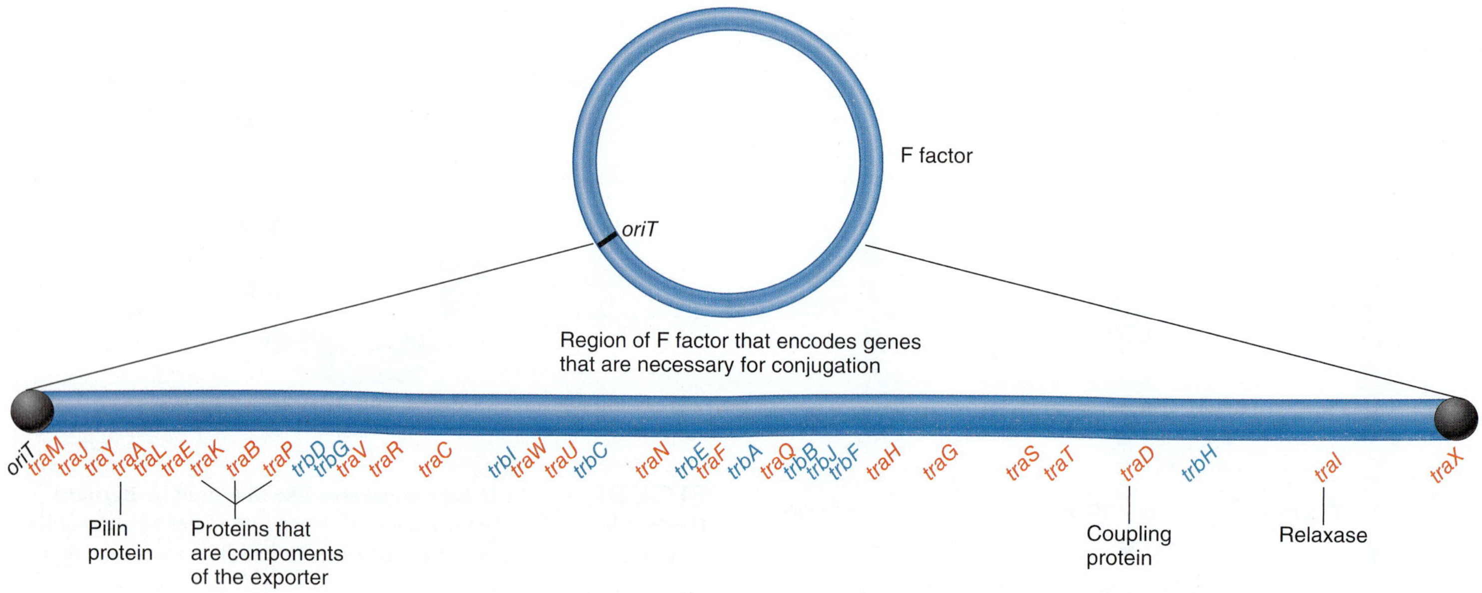

FIGURE 7.3 **Genes on the F factor that play a role during conjugation.** A region of the F factor carries genes that play a role in the conjugation process. Because they play a role in the transfer of DNA from donor to recipient cell, the genes are designated with the three-letter names of *tra* or *trb*, followed by a capital letter. (Note: the *tr* prefix refers to transfer.) The *tra* genes are shown in red, and the *trb* genes are shown in blue. The functions of a few examples are indicated. The origin of transfer is designated *oriT*.

CONCEPT CHECK: Would this circular DNA molecule be found in an F^+ or F^- cell?

Figure 7.4a describes the molecular events that occur during conjugation in *E. coli*. Contact between donor and recipient cells is a key step that initiates the conjugation process. **Sex pili** (singular: **pilus**) are made by F^+ strains (**Figure 7.4b**). The gene encoding the pilin protein (*traA*) is located on the F factor. The pili act as attachment sites that promote the binding of bacteria to each other. In this way, an F^+ strain makes physical contact with an F^- strain. In certain species, such as *E. coli*, long pili project from F^+ cells and attempt to make contact with nearby F^- cells. Once contact is made, the pili shorten, thereby drawing the donor and recipient cells closer together. A **conjugation bridge** is then formed between the two cells, which provides a passageway for DNA transfer.

The successful contact between a donor and recipient cell stimulates the donor cell to begin the transfer process. Genes within the F factor encode a protein complex called the **relaxosome.** This complex first recognizes a DNA sequence in the F factor known as the **origin of transfer** (see Figure 7.4a). Upon recognition, one DNA strand in the site is cut. The relaxosome also catalyzes the separation of the DNA strands, and only the cut DNA strand is transferred to the recipient cell. As the DNA strands separate, most of the proteins within the relaxosome are released, but one protein, called relaxase, remains bound to the end of the cut DNA strand. The complex between the single-stranded DNA and relaxase is called a **nucleoprotein** because it contains both nucleic acid (DNA) and protein (relaxase).

The next phase of conjugation involves the export of the nucleoprotein complex from the donor cell to the recipient cell. To begin this process, the DNA/relaxase complex is recognized by a coupling factor that promotes the entry of the nucleoprotein into the exporter, a complex of proteins that spans both inner and outer membranes of the donor cell. In bacterial species, this complex is formed from 10 to 15 different proteins that are encoded by genes within the F factor.

Once the DNA/relaxase complex is pumped out of the donor cell, it travels through the conjugation bridge and then into the recipient cell. As shown in Figure 7.4a, the other strand of the F factor DNA remains in the donor cell, where DNA replication restores the F factor DNA to its original double-stranded condition. After the recipient cell receives a single strand of the F factor DNA, relaxase catalyzes the joining of the ends of the linear DNA molecule to form a circular molecule. This single-stranded DNA is replicated in the recipient cell to become double-stranded. The result of conjugation is that the recipient cell has acquired an F factor, converting it from an F^- to an F^+ cell. The genetic composition of the donor strain has not changed.

Bacteria May Contain Different Types of Plasmids

Thus far, we have considered F factors, which are one type of DNA that can exist independently of the chromosomal DNA. The more general term for this type of DNA molecule is a **plasmid.** Most known plasmids are circular, although some are linear. Plasmids occur naturally in many strains of bacteria and in a few types of eukaryotic cells such as yeast. The smallest plasmids consist of just a few thousand base pairs (bp) and carry only a gene or two; the largest are in the range of 100,000 to 500,000 bp and carry several dozen or even hundreds of genes. Some plasmids, such as F factors, can integrate into the chromosome. These are also called **episomes.**

A plasmid has its own origin of replication that allows it to be replicated independently of the bacterial chromosome. The DNA sequence of the origin of replication influences how many copies of the plasmid are found within a cell. Some origins are said to be very strong because they result in many copies of the plasmid, perhaps as many as 100 per cell. Other origins of replication have sequences

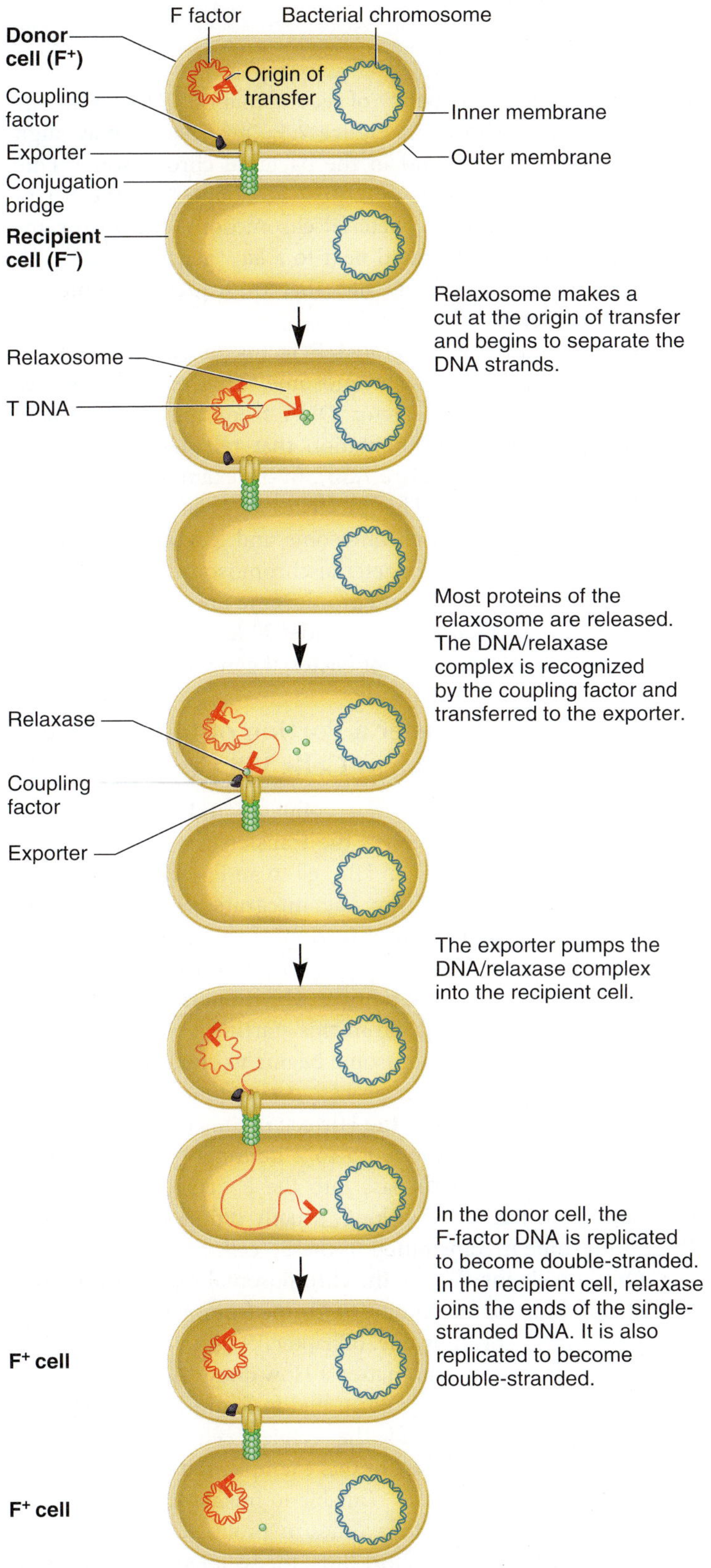

(a) Transfer of an F factor via conjugation

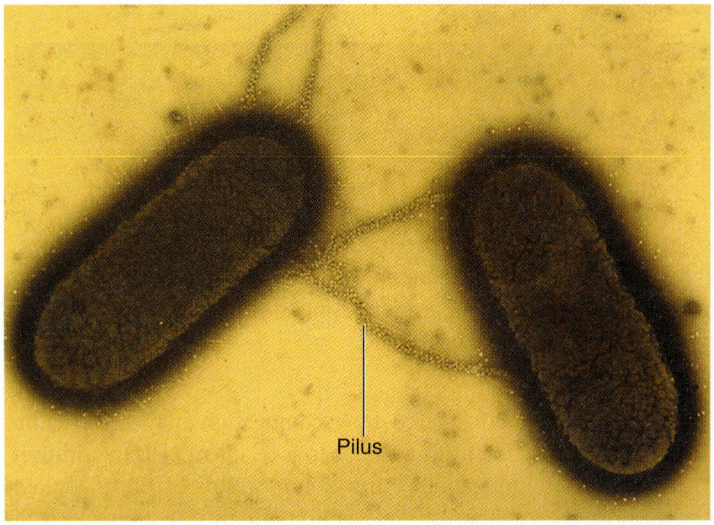

(b) Conjugating *E. coli*

ONLINE ANIMATION

FIGURE 7.4 The transfer of an F factor during bacterial conjugation. **(a)** The mechanism of transfer. The end result is that both cells have an F factor. **(b)** Two *E. coli* cells in the act of conjugation. The cell on the left is F^+, and the one on the right is F^-. The two cells make contact with each other via sex pili that are made by the F^+ cell.

CONCEPT CHECK: What are the functions of relaxase, coupling factor, and the exporter?

that are described as much weaker, in that the number of copies created is relatively low, such as one or two per cell.

Why do bacteria have plasmids? Plasmids are not usually necessary for bacterial survival. However, in many cases, certain genes within a plasmid provide some type of growth advantage to the cell. By studying plasmids in many different species, researchers have discovered that most plasmids fall into five different categories:

1. Fertility plasmids, also known as F factors, allow bacteria to conjugate with each other.
2. Resistance plasmids, also known as R factors, contain genes that confer resistance against antibiotics and other types of toxins.
3. Degradative plasmids carry genes that enable the bacterium to digest and utilize an unusual substance. For example, a degradative plasmid may carry genes that allow a bacterium to digest an organic solvent such as toluene.
4. Col-plasmids contain genes that encode colicins, which are proteins that kill other bacteria.
5. Virulence plasmids carry genes that turn a bacterium into a pathogenic strain.

7.2 COMPREHENSION QUESTIONS

1. A bacterial cell with an F factor conjugates with an F^- cell. Following conjugation, the two cells would be
   a. F^+.
   b. F^-.
   c. one F^+ and one F^-.
   d. none of the above.

2. Which of the following is a type of plasmid?
   a. F factor (fertility factor)
   b. R factor (resistance plasmid)
   c. Virulence plasmids
   d. All of the above

7.3 CONJUGATION AND MAPPING VIA *Hfr* STRAINS

Learning Outcomes:

1. Explain how an *Hfr* strain is produced.
2. Describe how an *Hfr* strain can transfer portions of the bacterial chromosome to recipient strains.
3. Construct a genetic map using data from conjugation experiments.

Thus far, we have considered how conjugation may involve the transfer of an F factor from a donor to a recipient cell. In addition to this form of conjugation, other donor strains of *E. coli*, called *Hfr* strains, are capable of conjugation. In this section, we will examine how *Hfr* strains are formed and how they transfer genes to recipient cells. We will also explore the use of *Hfr* strains to map genes along the *E. coli* chromosome.

Hfr Strains Contain an F Factor Integrated into the Bacterial Chromosome

Luca Cavalli-Sforza discovered a strain of *E. coli* that was very efficient at transferring many chromosomal genes to recipient F^- strains. Cavalli-Sforza designated this bacterial strain an **_Hfr_ strain** (for high frequency of recombination). How is an *Hfr* strain formed? As shown in **Figure 7.5a**, an F factor may align with a similar region found in the bacterial chromosome. Due to recombination, which is described in Chapter 19, the F factor may integrate into the bacterial chromosome. In this example, the F factor has integrated next to a lac^+ gene. F factors can integrate into several different sites that are scattered around the *E. coli* chromosome.

Occasionally, the integrated F factor in an *Hfr* strain is excised from the bacterial chromosome. This process involves the looping out of the F factor DNA from the chromosome, which is followed by recombination that releases the F factor from the chromosome (**Figure 7.5b**). In the example shown here, the excision is imprecise. This produces an F factor that carries a portion of the bacterial chromosome and leaves behind some of the F factor DNA in the bacterial chromosome. F factors that carry portions of the bacterial chromosome are called **F′ factors** (F prime factors). We will also consider F′ factors in Chapter 14 when we discuss mechanisms of bacterial gene regulation.

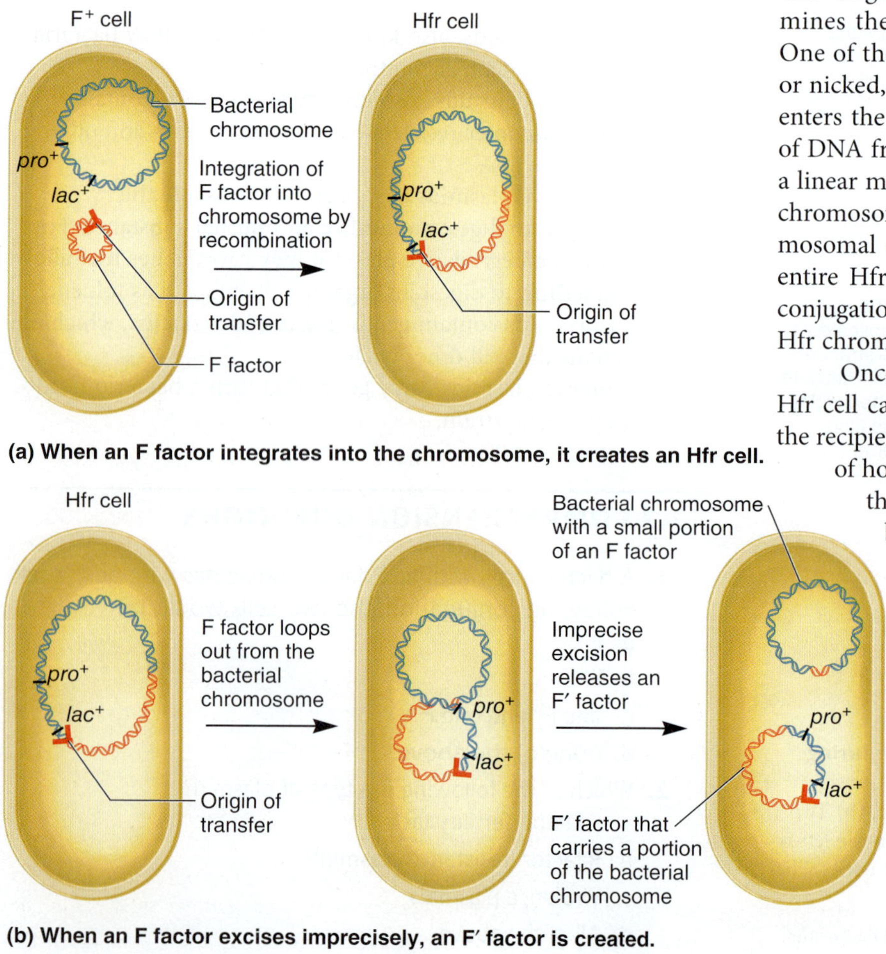

FIGURE 7.5 Integration of an F factor to form an Hfr cell and its subsequent excision to form an F′ factor. **(a)** An Hfr cell is created when an F factor integrates into the bacterial chromosome. **(b)** When an F factor is imprecisely excised, an F′ factor is created that carries a portion of the bacterial chromosome.

CONCEPT CHECK: How is an F′ factor different from an F factor?

Hfr Strains Can Transfer a Portion of the Bacterial Chromosome to Recipient Cells Via Conjugation

William Hayes determined that conjugation between an *Hfr* strain and an F^- strain involves the transfer of a portion of the bacterial chromosome from an Hfr cell to an F^- cell (**Figure 7.6**). The origin of transfer within the integrated F factor determines the starting point and direction of this transfer process. One of the DNA strands is cut at the origin of transfer. This cut, or nicked, site is the starting point at which the Hfr chromosome enters the F^- recipient cell. From this starting point, one strand of DNA from the Hfr chromosome begins to enter the F^- cell in a linear manner. The transfer process occurs in conjunction with chromosomal replication, so the Hfr cell retains its original chromosomal composition. About 1.5 to 2 hours is required for the entire Hfr chromosome to pass into the F^- cell. Because most conjugations do not last that long, usually only a portion of the Hfr chromosome is transmitted to the F^- cell.

Once inside the F^- cell, the chromosomal material from the Hfr cell can swap, or recombine, with the homologous region of the recipient cell's chromosome. (Chapter 19 describes the process of homologous recombination.) How does this process affect the recipient cell? As illustrated in Figure 7.6, this recombination may provide the recipient cell with a new combination of alleles. In this example, the recipient strain was originally lac^- (unable to metabolize lactose) and pro^- (unable to synthesize proline). If conjugation occurred for a short time, the recipient cell received a

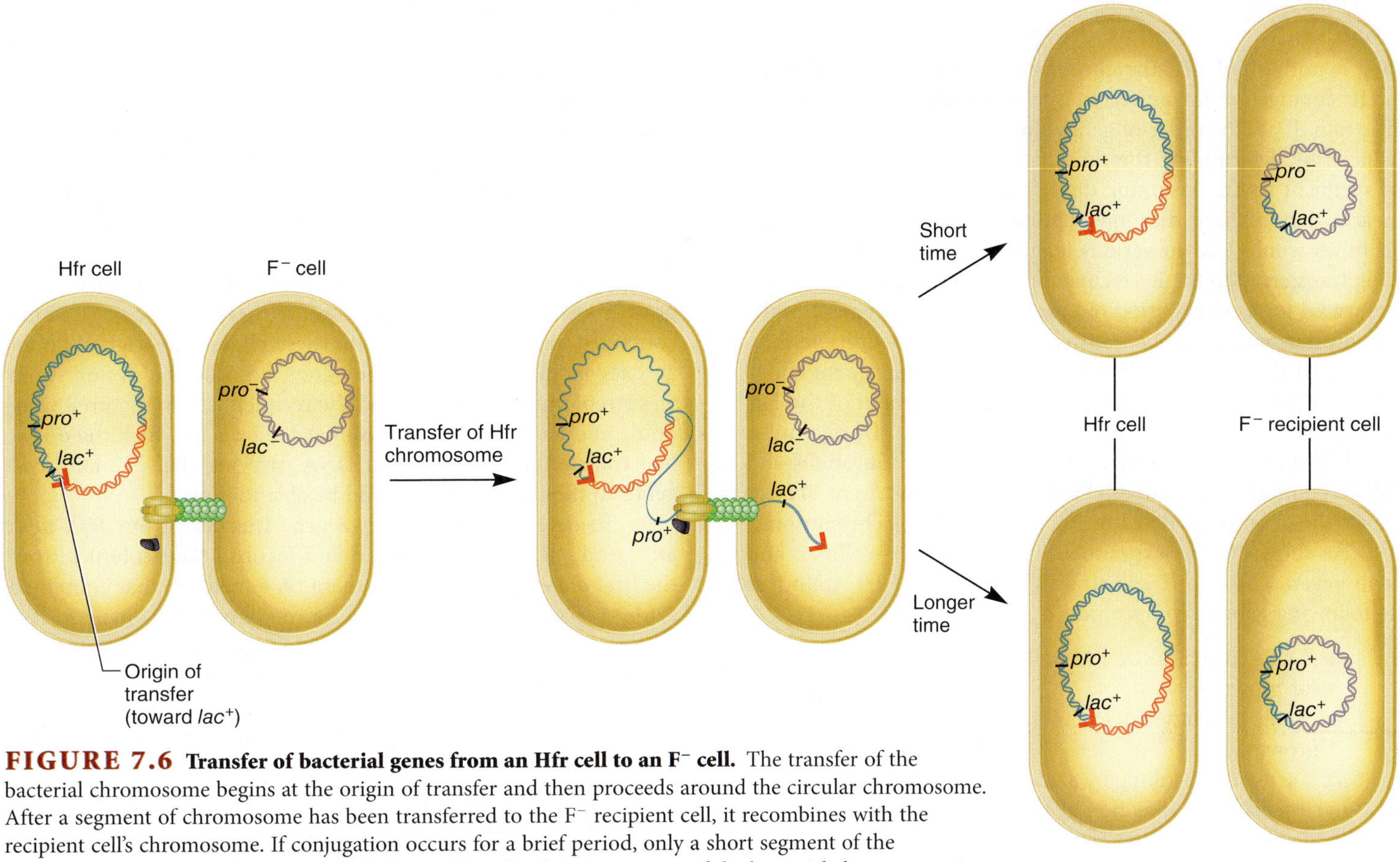

FIGURE 7.6 Transfer of bacterial genes from an Hfr cell to an F⁻ cell. The transfer of the bacterial chromosome begins at the origin of transfer and then proceeds around the circular chromosome. After a segment of chromosome has been transferred to the F⁻ recipient cell, it recombines with the recipient cell's chromosome. If conjugation occurs for a brief period, only a short segment of the chromosome is transferred. If conjugation is prolonged, a longer segment of the bacterial chromosome is transferred.

Genes→Traits The F⁻ recipient cell was originally *lac*⁻ (unable to metabolize lactose) and *pro*⁻ (unable to synthesize proline). If conjugation occurred for a short period of time, the recipient cell acquired *lac*⁺, allowing it to metabolize lactose. If conjugation occurred for a longer period of time, the recipient cell also acquired *pro*⁺, enabling it to synthesize proline.

CONCEPT CHECK: With regard to the timing of conjugation, explain why the recipient cell in the top right is *pro*⁻ whereas the recipient cell in the bottom right is *pro*⁺.

short segment of chromosomal DNA from the donor. In this case, the recipient cell has become *lac*⁺ but remains *pro*⁻. If the conjugation is prolonged, the recipient cell will receive a longer segment of chromosomal DNA from the donor. After a longer conjugation, the recipient becomes *lac*⁺ and *pro*⁺. As shown in Figure 7.6, an important feature of Hfr conjugation is that the bacterial chromosome is transferred linearly to the recipient strain. In this example, *lac*⁺ is always transferred first, and *pro*⁺ is transferred later.

In any particular *Hfr* strain, the origin of transfer has a specific orientation that promotes either a counterclockwise or clockwise transfer of genes. Among different *Hfr* strains, the origin of transfer may be located in different regions of the chromosome. Therefore, the order of gene transfer depends on the location and orientation of the origin of transfer. For example, another *Hfr* strain could have its origin of transfer next to *pro*⁺ and transfer *pro*⁺ first and *lac*⁺ later.

EXPERIMENT 7A

Conjugation Experiments Can Map Genes Along the *E. coli* Chromosome

The first genetic mapping experiments in bacteria were carried out by Elie Wollman and François Jacob in the 1950s. They were aware of previous microbiological studies concerning bacteriophages—viruses that bind to *E. coli* cells and subsequently infect them. These studies showed that bacteriophages can be sheared from the surface of *E. coli* cells if they are spun in a blender. In this treatment, the bacteriophages are detached from the surface of the bacterial cells, but the bacteria themselves remain healthy and viable. Wollman and Jacob reasoned that a blender treatment could be used to separate bacterial cells that were in the act of conjugation without killing them. This technique is known as an **interrupted mating.**

The rationale behind Wollman and Jacob's mapping strategy is that the time it takes for genes to enter a donor cell is directly related to their order along the bacterial chromosome.

They hypothesized that the chromosome of the donor strain in an Hfr conjugation is transferred in a linear manner to the recipient strain. If so, the order of genes along the chromosome can be deduced by determining the time it takes various genes to enter the recipient strain. Assuming the Hfr chromosome is transferred linearly, they realized that interruptions of conjugation at different times would lead to various lengths of the Hfr chromosome being transferred to the F^{-} recipient cell. If two bacterial cells had conjugated for a short period of time, only a small segment of the Hfr chromosome would be transferred to the recipient bacterium. However, if the bacterial cells were allowed to conjugate for a longer period before being interrupted, a longer segment of the Hfr chromosome could be transferred (see Figure 7.6). By determining which genes were transferred during short conjugations and which required longer times, Wollman and Jacob were able to deduce the order of particular genes along the *E. coli* chromosome.

As shown in the experiment of **Figure 7.7**, Wollman and Jacob began with two *E. coli* strains. The donor (*Hfr*) strain had the following genetic composition:

- *thr*$^{+}$: able to synthesize threonine, an essential amino acid for growth
- *leu*$^{+}$: able to synthesize leucine, an essential amino acid for growth
- *azi*^{s}: sensitive to killing by azide (a toxic chemical)
- *ton*^{s}: sensitive to infection by bacteriophage T1
- *lac*$^{+}$: able to metabolize lactose and use it for growth
- *gal*$^{+}$: able to metabolize galactose and use it for growth
- *str*^{s}: sensitive to killing by streptomycin (an antibiotic)

The recipient (F^{-}) strain had the opposite genotype: *thr*$^{-}$ *leu*$^{-}$ *azi*^{r} *ton*^{r} *lac*$^{-}$ *gal*$^{-}$ *str*^{r} (*r* = resistant). Before the experiment, Wollman and Jacob already knew the *thr*$^{+}$ gene was transferred first, followed by the *leu*$^{+}$ gene, and both were transferred relatively soon (5–10 minutes) after conjugation. Their main goal in this experiment was to determine the times at which the other genes (*azi*^{s} *ton*^{s} *lac*$^{+}$ *gal*$^{+}$) were transferred to the recipient strain. The transfer of the *str*^{s} gene was not examined because streptomycin was used to kill the donor strain following conjugation.

Before discussing the conclusions of this experiment, let's consider how Wollman and Jacob monitored gene transfer. To determine if particular genes had been transferred after conjugation, they took the conjugated cells and first plated them on a growth medium that lacked threonine (thr) and leucine (leu) but contained streptomycin (str). On these plates, the original donor and recipient strains could not grow because the donor strain was streptomycin-sensitive and the recipient strain required threonine and leucine. However, conjugated cells in which the donor had transferred chromosomal DNA carrying the *thr*$^{+}$ and *leu*$^{+}$ genes to the recipient cell would be able to grow.

To determine the order of gene transfer of the *azi*^{s}, *ton*^{s}, *lac*$^{+}$, and *gal*$^{+}$ genes, Wollman and Jacob picked colonies from the first plates and restreaked them on a medium that contained azide or bacteriophage T1 or on a medium that contained lactose or galactose as the sole source of energy for growth. The plates were incubated overnight to observe the formation of visible bacterial growth. Whether or not the bacteria could grow depended on their genotypes. For example, a cell that is *azi*^{s} cannot grow on a medium containing azide, and a cell that is *lac*$^{-}$ cannot grow on a medium containing lactose as the carbon source for growth. By comparison, a cell that is *azi*^{r} and *lac*$^{+}$ can grow on both types of media.

THE GOAL (DISCOVERY-BASED SCIENCE)

The chromosome of the donor strain in an Hfr conjugation is transferred in a linear manner to the recipient strain. The order of genes along the chromosome can be deduced by determining the time various genes take to enter the recipient strain.

ACHIEVING THE GOAL

FIGURE 7.7 The use of conjugation to map the order of genes along the *E. coli* chromosome.

Starting materials: The two *E. coli* strains already described, one *Hfr* strain (*thr*$^{+}$ *leu*$^{+}$ *azi*^{s} *ton*^{s} *lac*$^{+}$ *gal*$^{+}$ *str*^{s}) and one F^{-} (*thr*$^{-}$ *leu*$^{-}$ *azi*^{r} *ton*^{r} *lac*$^{-}$ *gal*$^{-}$ *str*^{r}).

Experimental level | **Conceptual level**

1. Mix together a large number of Hfr donor and F^{-} recipient cells.

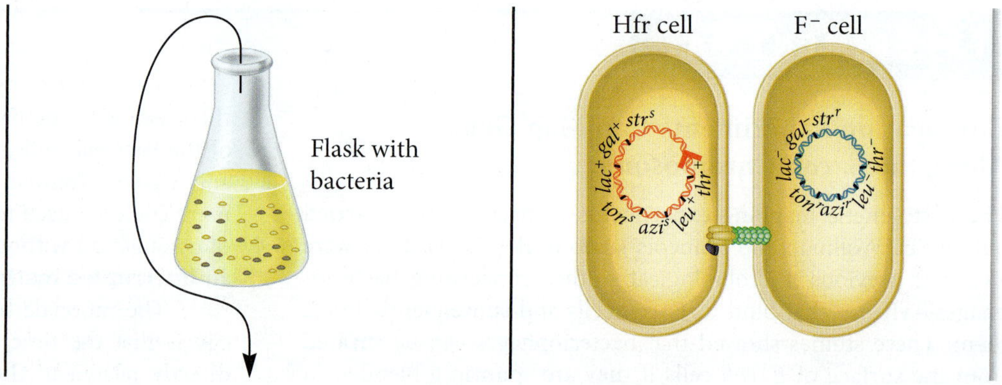

2. After different periods of time, take a sample of cells and interrupt conjugation in a blender.

3. Plate the cells on solid growth medium lacking threonine and leucine but containing streptomycin. Note: The general methods for growing bacteria in a laboratory are described in the Appendix.

4. Pick each surviving colony, which would have to be thr^+ leu^+ str^r, and test to see if it is sensitive to killing by azide, sensitive to infection by T1 bacteriophage, and able to metabolize lactose or galactose.

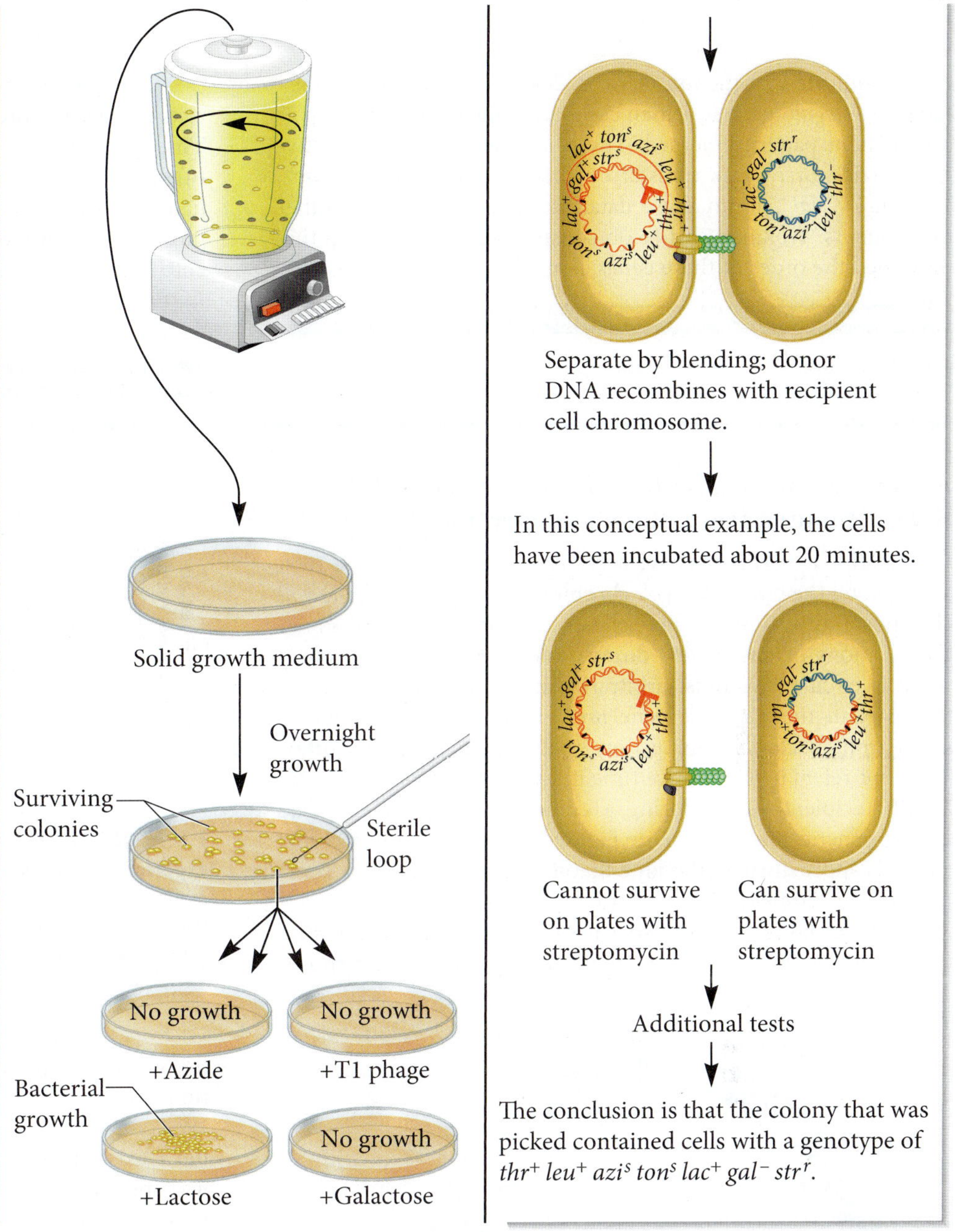

THE DATA

Minutes That Bacterial Cells Were Allowed to Conjugate Before Blender Treatment	Percent of Surviving Bacterial Colonies with the Following Genotypes: thr^+ leu^+	azi^s	ton^s	lac^+	gal^+
5	—*	—	—	—	—
10	100	12	3	0	0
15	100	70	31	0	0
20	100	88	71	12	0
25	100	92	80	28	0.6
30	100	90	75	36	5
40	100	90	75	38	20
50	100	91	78	42	27
60	100	91	78	42	27

*There were no surviving colonies within the first 5 minutes of conjugation.

Data from François Jacob and Elie Wollman (1961) *Sexuality and the Genetics of Bacteria.* Academic Press, New York.

INTERPRETING THE DATA

Now let's discuss the data shown in Figure 7.7. After the first plating, all colonies contained cells in which the thr^+ and leu^+ alleles had been transferred to the F^- recipient strain, which was already streptomycin-resistant. As seen in The Data, 5 minutes was not sufficient time to transfer the thr^+ and leu^+ alleles because no surviving colonies were observed. After 10 minutes or longer, however, surviving bacterial colonies with the thr^+ leu^+ genotype were obtained. To determine the order of the remaining genes (azi^s, ton^s, lac^+, and gal^+), each surviving colony was tested to see if it was sensitive to killing by azide, sensitive to infection by T1 bacteriophage, able to use lactose for growth, or able to use galactose for growth. The likelihood of colonies surviving depended on whether the azi^s, ton^s, lac^+, and gal^+ genes were close to the origin of transfer or farther away. For example, when cells were allowed to conjugate for 25 minutes, 80% carried the ton^s gene, whereas only 0.6% carried the gal^+ gene. These results indicate that the ton^s gene is closer to the origin of transfer than the gal^+ gene is. When comparing the data in Figure 7.7, a consistent pattern emerged. The gene that conferred sensitivity to azide (azi^s) was transferred first, followed by ton^s, lac^+, and finally, gal^+. From these data, as well as those from other experiments, Wollman and Jacob constructed a genetic map that described the order of these genes along the *E. coli* chromosome.

This work provided the first method for bacterial geneticists to map the order of genes along the bacterial chromosome. Throughout the course of their studies, Wollman and Jacob identified several different *Hfr* strains in which the origin of transfer had been integrated at different places along the bacterial chromosome. When they compared the order of genes among different *Hfr* strains, their results were consistent with the idea that the *E. coli* chromosome is circular (see solved problem S2).

A self-help quiz involving this experiment can be found at www.mhhe.com/brookergenetics5e.

A Genetic Map of the *E. coli* Chromosome Has Been Obtained from Many Conjugation Studies

Conjugation experiments have been used to map more than 1000 genes along the circular *E. coli* chromosome. A map of the *E. coli* chromosome is shown in **Figure 7.8**. This simplified map shows the locations of only a few dozen genes. Because the chromosome is circular, we must arbitrarily assign a starting point on the map, in this case the gene *thrA*. Researchers scale genetic maps from bacterial conjugation studies in units of **minutes.** This unit refers to the relative time it takes for genes to first enter an F^- recipient strain during a conjugation experiment. The *E. coli* genetic map shown in Figure 7.8 is 100 minutes long, which is approximately the time that it takes to transfer the complete chromosome during an Hfr conjugation.

The distance between two genes is determined by comparing their times of entry during a conjugation experiment. As shown in **Figure 7.9**, the time of entry is found by conducting conjugation experiments at different time intervals before interruption. We compute the time of entry by extrapolating the data back to the *x*-axis. In this experiment, the time of entry of the *lacZ* gene was approximately 16 minutes, and that of the *galE* gene was 25 minutes. Therefore, these two genes are approximately 9 minutes apart from each other along the *E. coli* chromosome.

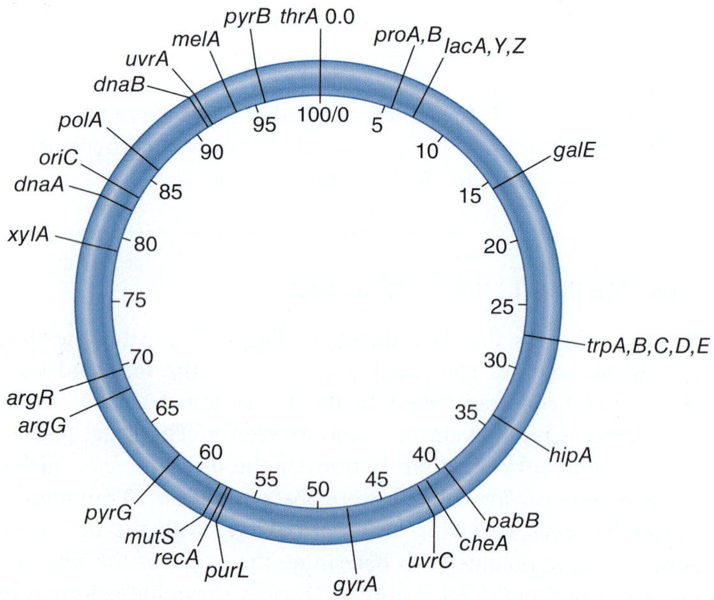

FIGURE 7.8 A simplified genetic map of the *E. coli* chromosome indicating the positions of several genes. *E. coli* has a circular chromosome with about 4300 different genes. This map shows the locations of several of them. The map is scaled in units of minutes, and it proceeds in a clockwise direction. The starting point on the map is the gene *thrA*.

CONCEPT CHECK: Why is the scale of this map in minutes?

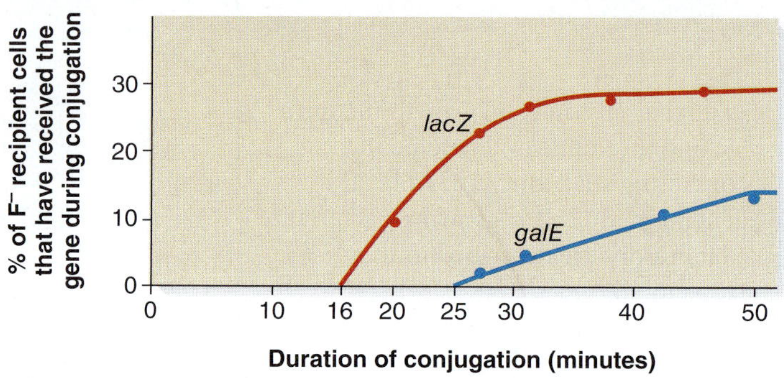

FIGURE 7.9 Time course of an interrupted *E. coli* conjugation experiment. By extrapolating the data back to the origin, the approximate time of entry of the *lacZ* gene is found to be 16 minutes; that of the *galE* gene, 25 minutes. Therefore, the distance between these two genes is 9 minutes.

CONCEPT CHECK: Which of these two genes is closer to the origin of transfer?

7.3 COMPREHENSION QUESTIONS

1. With regard to conjugation, a key difference between F^+ and Hfr cells is that an Hfr cell
   a. is unable to conjugate.
   b. transfers a plasmid to the recipient cell.
   c. transfers a portion of the bacterial chromosome to the recipient cell.
   d. becomes an F^- cell after conjugation.
2. In mapping experiments, _______ strains are conjugated to F^- strains. The distance between two genes is determined by comparing their __________________ during a conjugation experiment.
   a. F^+, times of entry
   b. *Hfr*, times of entry
   c. F^+, expression levels
   d. *Hfr*, expression levels

7.4 BACTERIAL TRANSDUCTION

Learning Outcomes:

1. Outline the steps of bacterial transduction.
2. Calculate the map distance between genes using data from a cotransduction experiment.

We now turn to a second method of genetic transfer, one that involves bacteriophages. Following infection of the cell, new viral particles are made, which are then released from the cell in an event called lysis. In this section, we will examine how mistakes can happen in the phage reproductive cycle that lead to the transfer of genetic material from one bacterial cell to another.

Bacteriophages Transfer Genetic Material from One Bacterial Cell to Another Via Transduction

Before we discuss the ability of bacteriophages to transfer genetic material between bacterial cells, let's consider some general features of a phage's reproductive cycle. Bacteriophages are composed of genetic material that is surrounded by a protein coat. As described in Chapter 17, certain bacteriophages bind to the surface of a bacterium and inject their genetic material into the bacterial cytoplasm. Depending on the specific type of virus and its growth conditions, a phage may follow a lytic cycle or a lysogenic cycle. During the **lytic cycle,** the bacteriophage directs the synthesis of many copies of the phage genetic material and coat proteins (look ahead to Figure 17.4, left side). These components then assemble to make new phages. When the synthesis and assembly of bacteriophages are completed, the bacterial host cell is lysed (broken apart), releasing the newly made phages into the environment.

When bacteriophages follow the lytic cycle, on rare occasions bacterial genes may be transferred from one bacterial cell to another. This process is called transduction. Examples of phages that can transfer bacterial chromosomal DNA from one bacterium to another are the P22 and P1 phages, which infect the bacterial species *Salmonella typhimurium* and *E. coli*, respectively.

How does a bacteriophage transfer bacterial chromosomal genes from one cell to another? As shown in **Figure 7.10**, when a phage infects a bacterial cell and follows the lytic cycle, the bacterial chromosome is digested into fragments of DNA. The phage DNA directs the synthesis of more phage DNA and proteins, which then assemble to make new phages. Occasionally, a mistake can happen in which a fragment of bacterial DNA assembles with phage proteins. This creates a phage that contains bacterial chromosomal DNA. When phage synthesis is completed, the bacterial cell is lysed and releases the newly made phage into the environment. Following release, this abnormal phage can bind to a living bacterial cell and inject its genetic material into the bacterium. The DNA fragment, which was derived from the chromosomal DNA of the first bacterium, can then recombine with the recipient cell's bacterial chromosome. In this case, the recipient bacterium has been changed from a cell that was *his*$^-$ (unable to synthesize histidine) to a cell that is *his*$^+$ (able to synthesize histidine).

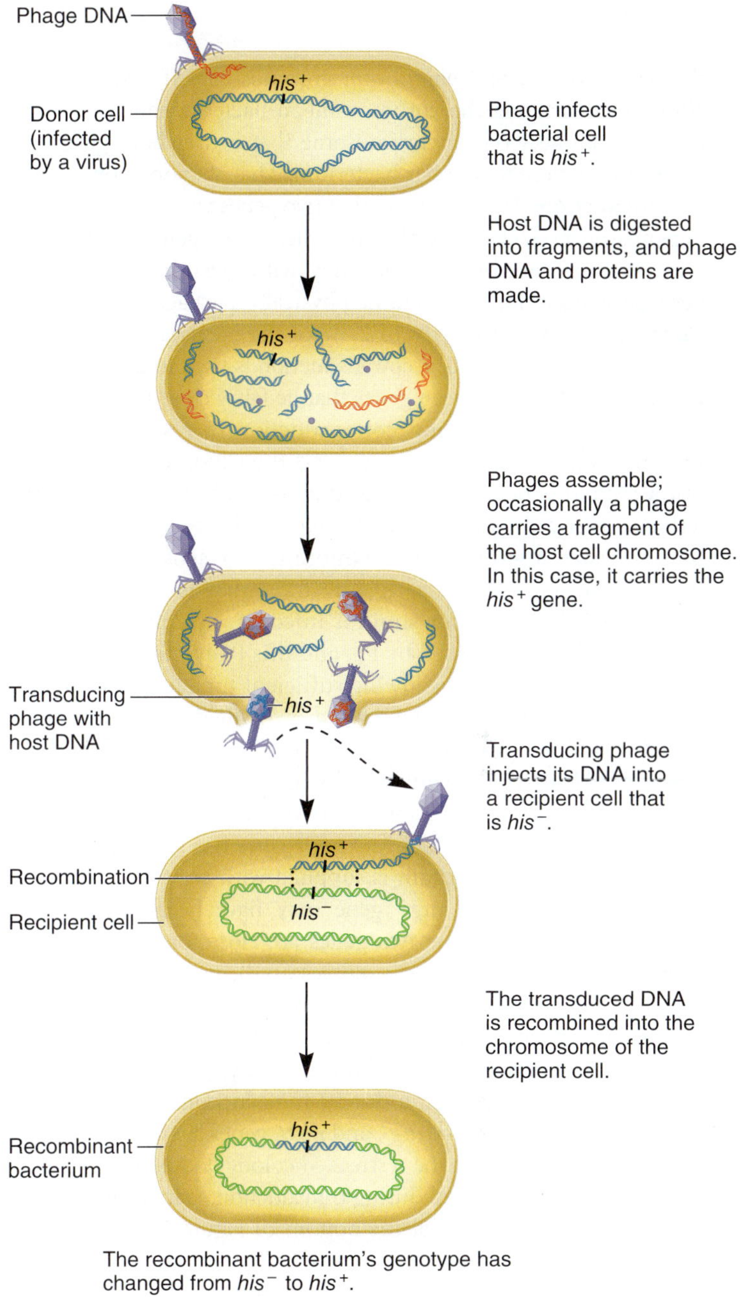

ONLINE ANIMATION

FIGURE 7.10 Transduction in bacteria.

Genes→Traits During transduction, a phage carries a segment of bacterial DNA from a donor to a recipient cell. In this case, the phage carried a segment of DNA with the *his*$^+$ gene and transferred this gene to a recipient cell that was originally *his*$^-$ (unable to synthesize histidine). Following transduction, the recipient cell became *his*$^+$, so it now can synthesize histidine.

CONCEPT CHECK: Transduction is sometimes described as a mistake in the bacteriophage reproductive cycle. Explain how it is a mistake.

Cotransduction Can Be Used to Map Genes That Are Within 2.5 Minutes of Each Other

Can transduction be used to map the distance between bacterial genes? The answer is yes, but only if the genes are relatively close together. During transduction, P1 phages cannot package pieces that are greater than 2% to 2.5% of the entire *E. coli* chromosome, and P22 phages cannot package pieces that are greater than 1% of the length of the *S. typhimurium* chromosome. If two genes are close together along the chromosome, a bacteriophage may package a single piece of the chromosome that carries both genes and transfer that piece to another bacterium. This phenomenon is called **cotransduction.** The likelihood that two genes will be cotransduced depends on how close together they lie. If two genes are far apart along a bacterial chromosome, they will never be cotransduced because a bacteriophage cannot physically package a DNA fragment that is larger than 1% to 2.5% of the bacterial chromosome. In genetic mapping studies, cotransduction is used to determine the order and distance between genes that lie fairly close to each other.

To map genes using cotransduction, a researcher selects for the transduction of one gene and then monitors whether or not a second gene is cotransduced along with it. As an example, let's consider a donor strain of *E. coli* that is arg^+ met^+ str^s (able to synthesize arginine and methionine but sensitive to killing by streptomycin) and a recipient strain that is arg^- met^- str^r (**Figure 7.11**). The donor strain is infected with phage P1. Some of the *E. coli* cells are lysed by P1, and this P1 lysate is mixed with the recipient cells. After allowing sufficient time for transduction, the recipient cells are plated on a growth medium that contains arginine and streptomycin but not methionine. Therefore, these plates select for the growth of cells in which the met^+ gene has been transferred to the recipient strain, but they do not select for the growth of cells in which the arg^+ gene has been transferred, because the growth medium is supplied with arginine.

Nevertheless, the arg^+ gene may have been cotransduced with the met^+ gene if the two genes are close together. To determine if any of the 50 bacterial colonies on the growth medium with arginine and streptomycin also carried the arg^+ gene, each colony can be picked up with a wire loop and restreaked on a medium that lacks arginine. If the cells from a colony grow, they must also have obtained the arg^+ gene during transduction. In other words, cotransduction of both the arg^+ and met^+ genes has occurred. Alternatively, if the cells from the restreaked colony do not grow, only the met^+ gene was transferred during transduction. Data from this type of experiment are shown at the bottom of Figure 7.11. These data indicate a cotransduction frequency of $21/50 = 0.42$, or 42%.

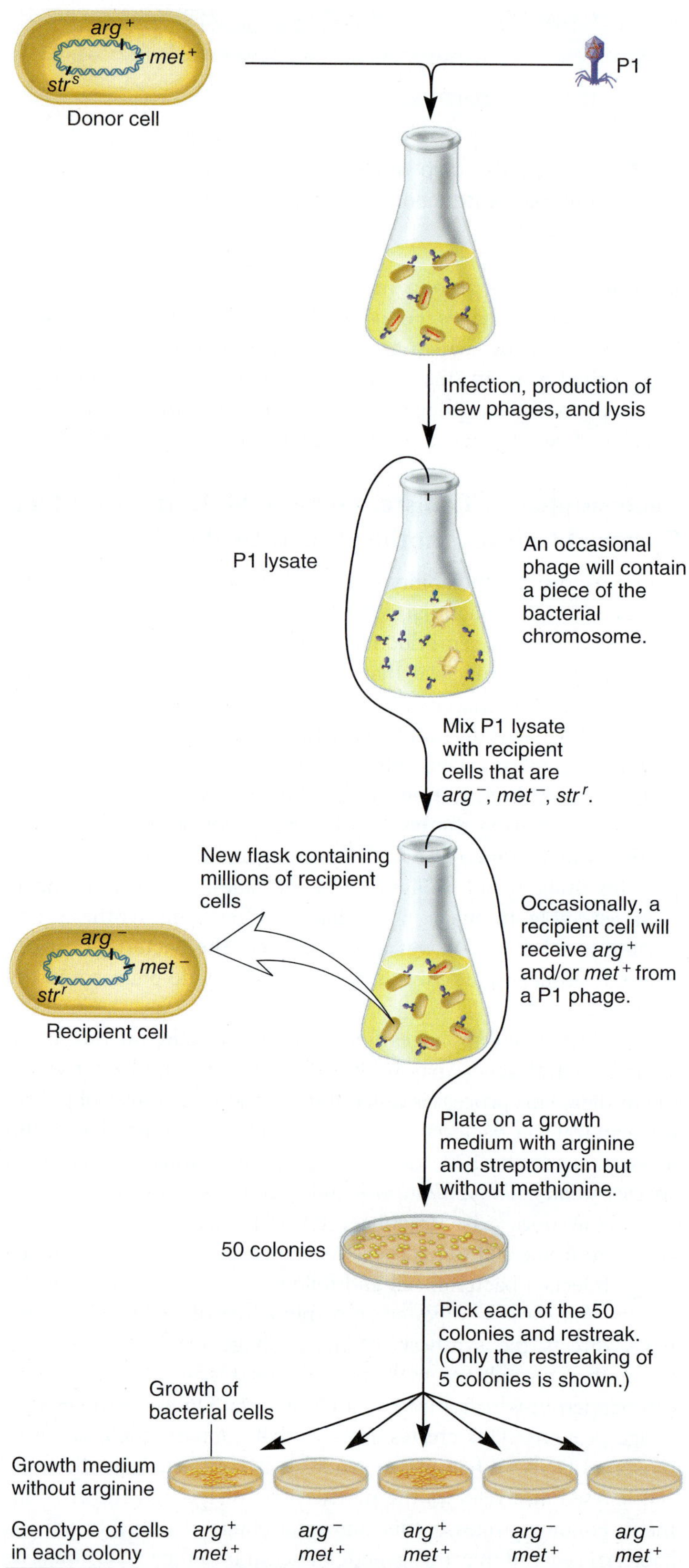

Results				
Selected gene	Nonselected gene	Number of colonies that grew on		Cotransduction frequency
		medium + arginine	medium − arginine	
met^+	arg^+	50	21	0.42

FIGURE 7.11 **The steps in a cotransduction experiment.** The donor strain, which is arg^+ met^+ str^s (able to synthesize arginine and methionine but sensitive to streptomycin), is infected with phage P1. Some of the cells are lysed by P1, and this P1 lysate is mixed with cells of the recipient strain, which are arg^- met^- str^r. P1 phages in this lysate may carry fragments of the donor cell's chromosome, and the P1 phage may inject that DNA into the recipient cells. To identify recipient cells that have received the met^+ gene from the donor strain, the recipient cells are plated on a growth medium that contains arginine and streptomycin but not methionine. To determine if the arg^+ gene has also been cotransduced, cells from each bacterial colony (on the plates containing arginine and streptomycin) are lifted with a sterile wire loop and streaked on a medium that lacks both amino acids. If the cells can grow, cotransduction of the arg^+ and met^+ genes has occurred. Alternatively, if the restreaked colony does not grow, it must have received only the met^+ gene during transduction.

CONCEPT CHECK: If these two genes were very far apart on the bacterial chromosome, how would the results have been different?

In 1966, Tai Te Wu derived a mathematical expression that relates cotransduction frequency with map distance obtained from conjugation experiments. This equation is

$$\text{Cotransduction frequency} = (1 - d/L)^3$$

where

d = distance between two genes in minutes
L = the size of the chromosomal pieces (in minutes) that the phage carries during transduction (For P1 transduction, this size is approximately 2% of the bacterial chromosome, which equals about 2 minutes.)

This equation assumes that the bacteriophage randomly packages pieces of the bacterial chromosome that are similar in size. Depending on the type of phage used in a transduction experiment, this assumption may not always be valid. Nevertheless, this equation has been fairly reliable in computing map distance for P1 transduction experiments in *E. coli*. We can use this equation to estimate the distance between the two genes described in Figure 7.11.

$$0.42 = (1 - d/2)^3$$

$$(1 - d/2) = \sqrt[3]{0.42}$$

$$1 - d/2 = 0.75$$

$$d/2 = 0.25$$

$$d = 0.5 \text{ minutes}$$

This equation tells us that the distance between the met^+ and arg^+ genes is approximately 0.5 minutes.

Historically, genetic mapping strategies in bacteria often involved data from both conjugation and transduction experiments. Conjugation has been used to determine the relative order and distance of genes, particularly those that are far apart along the chromosome. In comparison, transduction experiments can provide fairly accurate mapping data for genes that are relatively close together.

7.4 COMPREHENSION QUESTIONS

1. During transduction via P1,
   a. any small fragment of the bacterial chromosome may be transferred to another bacterium by a bacteriophage.
   b. only a specific fragment of DNA may be transferred to another bacterium by a bacteriophage.
   c. any small fragment of the bacterial chromosome may be transferred during conjugation.
   d. only a specific fragment of DNA may be transferred during conjugation.
2. Cotransduction may be used to map bacterial genes that are
   a. far apart on the bacterial chromosome.
   b. close together on the bacterial chromosome.
   c. both a and b.
   d. neither a or b.

7.5 BACTERIAL TRANSFORMATION

Learning Outcomes:

1. Outline the steps of bacterial transformation.
2. Explain how certain bacterial species preferentially take up DNA from their own species.

A third mechanism for the transfer of genetic material from one bacterium to another is known as transformation. This process was first discovered by Frederick Griffith in 1928 while he was working with strains of *Streptococcus pneumoniae* (see Chapter 9). During transformation, a living bacterial cell takes up DNA that was released from a dead bacterium. This DNA may then recombine into the living bacterium's chromosome, producing a bacterium with genetic material that it has received from the dead bacterium. In this section, we will examine the steps in the transformation process.

Transformation Involves the Uptake of DNA Molecules into Bacterial Cells

Transformation occurs as a natural process that has evolved in certain species of bacteria, in which case it is called **natural transformation.** This form of genetic transfer has been reported in a wide variety of bacterial species. Bacterial cells that are able to take up DNA are known as **competent cells.** Those that can take up DNA naturally carry genes that encode proteins called **competence factors.** These proteins facilitate the binding of DNA fragments to the cell surface, the uptake of DNA into the cytoplasm, and its subsequent incorporation into the bacterial chromosome. Temperature, ionic conditions, and nutrient availability can affect whether or not a bacterium is competent to take up genetic material from its environment. These conditions influence the expression of competence genes.

In recent years, geneticists have unraveled some of the steps that occur when competent bacterial cells are transformed by genetic material in their environment. **Figure 7.12** describes the steps of transformation. First, a large fragment of genetic material binds to the surface of the bacterial cell. Competent cells express DNA receptors that promote such binding. Before entering the cell, however, this large piece of chromosomal DNA must be cut into smaller fragments. This cutting is accomplished by an extracellular bacterial enzyme known as an endonuclease, which makes occasional random cuts in the long piece of chromosomal DNA. At this stage, the DNA fragments are composed of double-stranded DNA.

In the next step, the DNA fragment begins its entry into the bacterial cytoplasm. For this to occur, the double-stranded DNA interacts with proteins in the bacterial membrane. One of the DNA strands is degraded, and the other strand enters the bacterial cytoplasm via an uptake system, which is structurally similar to the one described for conjugation (as shown earlier in Figure 7.4a), but is involved with DNA uptake rather than export.

To be stably inherited, the DNA strand must be incorporated into the bacterial chromosome. If the DNA strand has a sequence that is similar to a region of DNA in the bacterial chromosome,

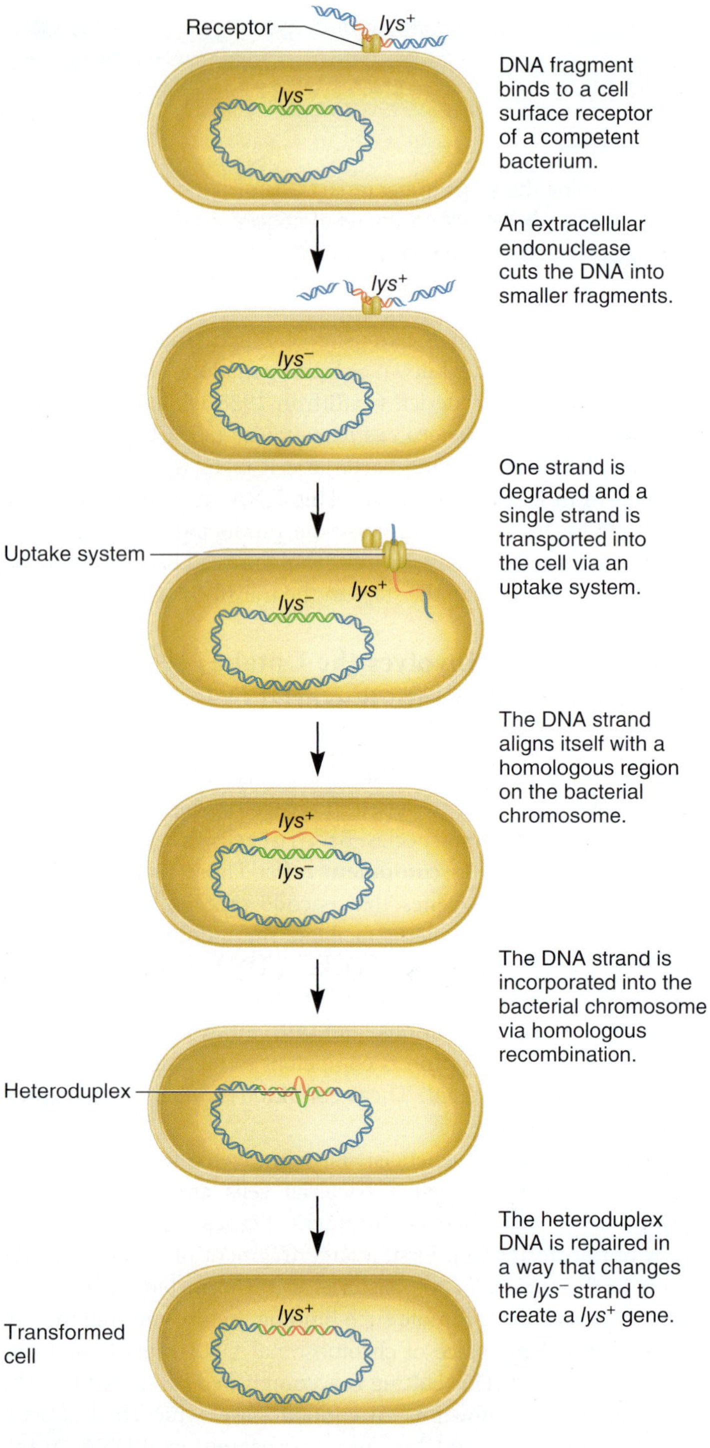

FIGURE 7.12 The steps of bacterial transformation. In this example, a fragment of DNA carrying a *lys*$^+$ gene enters the competent cell and recombines with the chromosome, transforming the bacterium from *lys*$^-$ to *lys*$^+$.

Genes → Traits Bacterial transformation can lead to new traits for the recipient cell. The recipient cell was *lys*$^-$ (unable to synthesize the amino acid lysine). Following transformation, it became *lys*$^+$. This result transforms the recipient bacterial cell into a cell that can synthesize lysine and grow on a medium that lacks this amino acid. Before transformation, the recipient *lys*$^-$ cell would not have been able to grow on a medium lacking lysine.

CONCEPT CHECK: If the recipient cell did not already have a *lys*$^-$ gene, could the *lys*$^+$ DNA become incorporated into the bacterial chromosome? Explain.

the DNA may be incorporated into the chromosome by a process known as **homologous recombination,** discussed in detail in Chapter 19. For this to occur, the single-stranded DNA aligns itself with the homologous location on the bacterial chromosome. In the example shown in Figure 7.12, the foreign DNA carries a functional *lys*$^+$ gene that aligns itself with a nonfunctional (mutant) *lys*$^-$ gene already present within the bacterial chromosome. The foreign DNA then recombines with one of the strands in the bacterial chromosome of the competent cell. In other words, the foreign DNA replaces one of the chromosomal strands of DNA, which is subsequently degraded. During homologous recombination, alignment of the *lys*$^-$ and the *lys*$^+$ alleles results in a region of DNA called a **heteroduplex** that contains one or more base sequence mismatches. However, the heteroduplex exists only temporarily. DNA repair enzymes in the recipient cell recognize the heteroduplex and repair it. In this example, the heteroduplex has been repaired by eliminating the mutation that caused the *lys*$^-$ genotype, thereby creating a *lys*$^+$ gene. Therefore, the recipient cell has been transformed from a *lys*$^-$ strain to a *lys*$^+$ strain. Alternatively, a DNA fragment that has entered a cell may not be homologous to any genes that are already found in the bacterial chromosome. In this case, the DNA strand may be incorporated at a random site in the chromosome. This process is known as **nonhomologous recombination** or **illegitimate recombination.**

Transformation has also been used to map many bacterial genes, using methods similar to the cotransduction experiments described earlier. If two genes are close together, the **cotransformation** frequency is expected to be high, whereas genes that are far apart have a cotransformation frequency that is very low or even zero. Like cotransduction, genetic mapping via cotransformation is used only to map genes that are relatively close together.

Transformation is also a commonly used laboratory method to get plasmid DNA into cells, such as *Escherichia coli*. This is called **artificial transformation** to distinguish it from the natural process. Different approaches can be used to achieve transformation. One method is to treat the cells with calcium chloride, which is followed by a brief period of high temperature (a heat shock). These conditions make the cells permeable to small DNA molecules. Another method, called electroporation, makes cells permeable to DNA by using an externally applied electric field. In Chapters 20 and 21, we will explore how plasmids are used in a variety of modern research methods and biotechnological applications.

Some Bacterial Species Have Evolved Ways to Take Up DNA from Their Own Species

Some bacteria preferentially take up DNA fragments from other bacteria of the same species or closely related species. How does this occur? Recent research has shown that the mechanism can vary among different species. In *Streptococcus pneumoniae*, the cells secrete a short peptide called the **competence-stimulating peptide (CSP).** When many *S. pneumoniae* cells are in the vicinity of one another, the concentration of CSP becomes high, which stimulates the cells to express the competence proteins needed for the uptake of DNA and its incorporation in the chromosome. Because competence requires a high external concentration of CSP, *S. pneuomoniae* cells are more likely to take up DNA from nearby *S. pneumoniae* cells that have died and released their DNA into the environment.

Other bacterial species promote the uptake of DNA among members of their own species via **DNA uptake signal sequences.** In the human pathogens *Neisseria meningitidis* (a causative agent of meningitis), *N. gonorrhoeae* (a causative agent of gonorrhea), and *Haemophilus influenzae* (a causative agent of ear, sinus, and respiratory infections), these sequences are found at many locations within their respective genomes. For example, *H. influenzae* contains approximately 1500 copies of the sequence 5′-AAGTGCGGT-3′ in its genome, and *N. meningitidis* contains about 1900 copies of the sequence 5′-GCCGTCTGAA-3′. DNA fragments that contain their own uptake signal sequence are preferentially taken up by these species instead of other DNA fragments. For example, *H. influenzae* is much more likely to take up a DNA fragment with the sequence 5′-AAGTGCGGT-3′. For this reason, transformation is more likely to involve DNA uptake between members of the same species.

7.5 COMPREHENSION QUESTIONS

1. From the following list, what is the correct order for the steps of transformation?
   1. Recombination with the bacterial chromosome
   2. Binding of a large DNA fragment to the surface of a bacterial cell
   3. Cutting a large DNA fragment into smaller pieces
   4. Uptake of DNA into the cytoplasm
   5. Degradation of one of the DNA strands

   a. 1, 2, 3, 4, 5
   b. 2, 3, 5, 4, 1
   c. 2, 3, 4, 5, 1
   d. 2, 5, 4, 3, 1
2. Some bacterial species preferentially take up DNA fragments from members of their own species. This can occur via
   a. competence-stimulating peptide (CSP).
   b. DNA uptake signal sequences.
   c. both a and b.
   d. none of the above.

7.6 MEDICAL RELEVANCE OF BACTERIAL GENETIC TRANSFER

Learning Outcomes:

1. Define horizontal gene transfer.
2. Explain the impact of bacterial horizontal gene transfer in medicine.

The term **horizontal gene transfer** refers to a process in which an organism incorporates genetic material from another organism without being the offspring of that organism. Conjugation, transformation, and transduction are examples of horizontal gene transfer. This process can occur between members of the same species or members of different species. A key reason why horizontal gene transfer is important is its medical relevance. One area of great concern is the topic of antibiotic resistance. Antibiotics are widely prescribed to treat bacterial infections in humans. They are also used in agriculture to control bacterial diseases in livestock. Unfortunately, the widespread use of antibiotics has increased the prevalence of antibiotic-resistant strains of bacteria, strains that have a selective advantage over those that are susceptible to antibiotics. Resistant strains carry genes that counteract the action of antibiotics in various ways. A resistance gene may encode a protein that breaks down the antibiotic, pumps it out of the cell, or prevents it from inhibiting cellular processes.

The term **acquired antibiotic resistance** refers to the common phenomenon of a previously susceptible strain becoming resistant to a specific antibiotic. This change may result from genetic alterations in the bacteria's own genome, but it is often due to the horizontal transfer of resistance genes from a resistant strain. As often mentioned in the news media, antibiotic resistance has increased dramatically worldwide over the past few decades, with resistant strains reported in almost all pathogenic strains of bacteria. As an example, some *Staphylococcus aureus* strains have developed resistance to methicillin and all penicillins. Evidence suggests that these so-called methicillin-resistant strains of *Staphlococcus aureus* (MRSA) acquired the methicillin-resistance gene by horizontal gene transfer, possibly from a strain of *Enterococcus faecalis*. MRSA strains cause skin infections that are more difficult to treat than "staph" infections caused by nonresistant strains of *S. aureus*.

7.6 COMPREHENSION QUESTION

1. Which of the following is an example of horizontal gene transfer?
   a. The transfer of a gene from one strain of *E. coli* to a different strain via conjugation
   b. The transfer of a gene from one strain of *E. coli* to a different strain via transduction
   c. The transfer of an antibiotic resistance gene from *E. coli* to *Salmonella typhimurium* via transformation
   d. All of the above

7.7 PLAQUE FORMATION AND INTERGENIC COMPLEMENTATION IN BACTERIOPHAGES

Learning Outcomes:

1. Describe how a plaque forms in the laboratory and how mutations may affect plaque morphology.
2. Explain how a complementation test can distinguish if two mutations are in the same bacteriophage gene or in different genes.

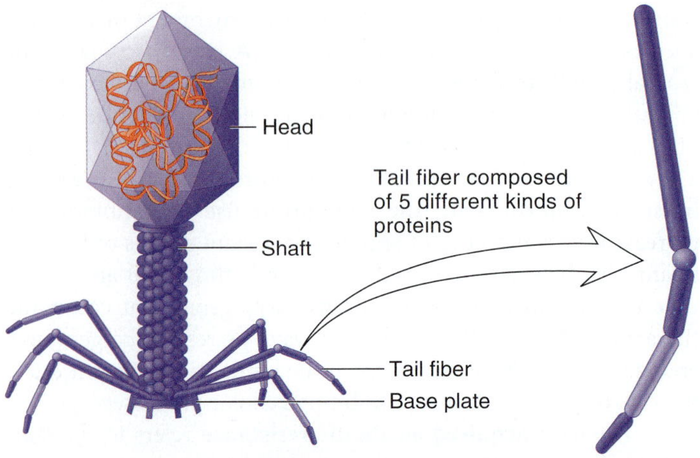

FIGURE 7.13 **Structure of T4 bacteriophage.**

Genes → Traits The inset to this figure shows the five different proteins, encoded by five different genes, that make up a tail fiber. The expression of these genes provides the bacteriophage with the trait of having tail fibers, which it uses to attach itself to the surface of a bacterium.

CONCEPT CHECK: Where is the genetic material located that determines the structure of the bacteriophage?

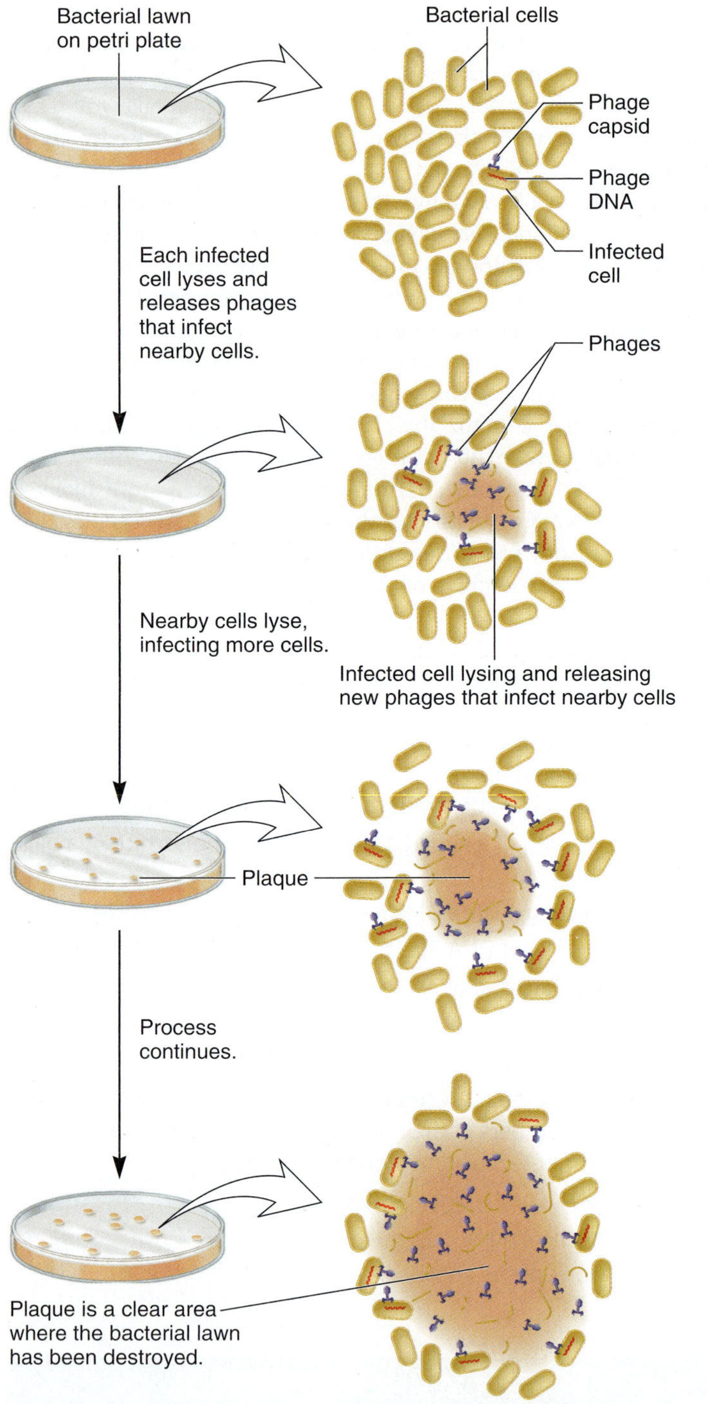

FIGURE 7.14 **The formation of phage plaques on a lawn of bacteria in the laboratory.** In this experiment, bacterial cells were mixed with a small number of lytic bacteriophages. In this particular example, 11 bacterial cells were initially infected by phages. The cells were poured onto petri plates containing nutrient agar for bacterial cell growth. Bacterial cells rapidly grow and divide to produce an opaque lawn of densely packed bacteria. The 11 infected cells lyse and release newly made bacteriophages. These bacteriophages then infect the nearby bacteria within the lawn. Likewise, these newly infected cells lyse and release new phages. By this repeated process, the area of cell lysis creates a clear zone known as a plaque. Note: In this textbook, *E. coli* cells are shown in a yellow/tan color. However, a lawn of *E. coli* is shown in white so the plaques are more readily visible.

CONCEPT CHECK: Over time, the plaques become larger in diameter. Explain why.

Biologists do not consider viruses to be living entities because they rely on a host cell for their existence and proliferation. Nevertheless, we can think of viruses as having traits because they have unique biological structures and functions. Each type of virus has its own genetic material, which may contain a few or many genes. In this section and the one following, we will focus our attention on a bacteriophage called T4, which is a virus that infects *E. coli*. Its genetic material contains several dozen different genes encoding proteins that carry out a variety of functions. For example, some of the genes encode proteins needed for the synthesis of new phages and the lysis of the host cell. Other genes encode the coat proteins that are found in the head, shaft, tail fibers, and base plate of the phage. **Figure 7.13** illustrates the structure of T4. As seen here, five different proteins bind to each other to form a tail fiber. The expression of T4 genes to make these proteins provides the bacteriophage with the trait of having tail fibers, enabling it to attach to the surface of a bacterium. In this section, we will examine the growth properties of bacteriophages in the laboratory and explore how mutations in bacteriophage genes are analyzed.

Lytic Bacteriophages Form Plaques Within a Lawn of Bacterial Cells

As they progress through the lytic cycle, bacteriophages ultimately produce new phages, which are released when the bacterial cell lyses (look ahead to Figure 17.4, left side). In the laboratory, researchers can visually observe the consequences of bacterial cell lysis in the following way (**Figure 7.14**). A sample of many bacterial cells and a few (in this case, 11) lytic bacteriophages are mixed together and then poured onto petri plates containing nutrient agar for bacterial cell growth. Bacterial cells that are not infected by a bacteriophage rapidly grow and divide to produce a "lawn" of bacteria. This lawn of bacteria is opaque—you cannot see through it to the underlying agar. In the experiment shown in Figure 7.14, eleven bacterial cells have been infected by bacteriophages and these infected cells are found at random locations in the lawn of uninfected bacteria. The infected cells lyse and release newly made bacteriophages. These bacteriophages then infect the nearby bacteria within the lawn. These cells eventually lyse and also release newly made phages. Over time, these repeated cycles of infection and lysis produce an observable clear area, or **plaque,** where the bacteria have been lysed around the original site where a phage infected a bacterial cell.

Phage Mutations Can Alter Plaque Morphology and Their Ability to Proliferate in Different Bacterial Strains

Because bacteriophages can be visualized only with an electron microscope, it would be rather difficult for geneticists to analyze mutations that affect phage morphology. However, some mutations in the bacteriophage's genetic material alter the ability of the phage to form plaques. Therefore, we can view the morphology of plaques as a trait of the bacteriophage. Because plaques are visible with the unaided eye, mutations affecting this trait lend themselves to a much easier genetic analysis. An example is a rapid-lysis mutant of bacteriophage T4, which tends to form unusually large plaques (**Figure 7.15**). The plaques are large because the mutant phages lyse the bacterial cells more rapidly than do the wild-type phages. Rapid-lysis mutants form large, clearly defined plaques, as opposed to wild-type bacteriophages, which produce smaller, fuzzy-edged plaques. Mutations in different bacteriophage genes can produce a rapid-lysis phenotype.

During the 1950s, Seymour Benzer studied one category of T4 phage mutants, designated *rII* (*r* stands for rapid lysis). Compared to wild-type T4 phage, these *rII* mutants showed the following properties:

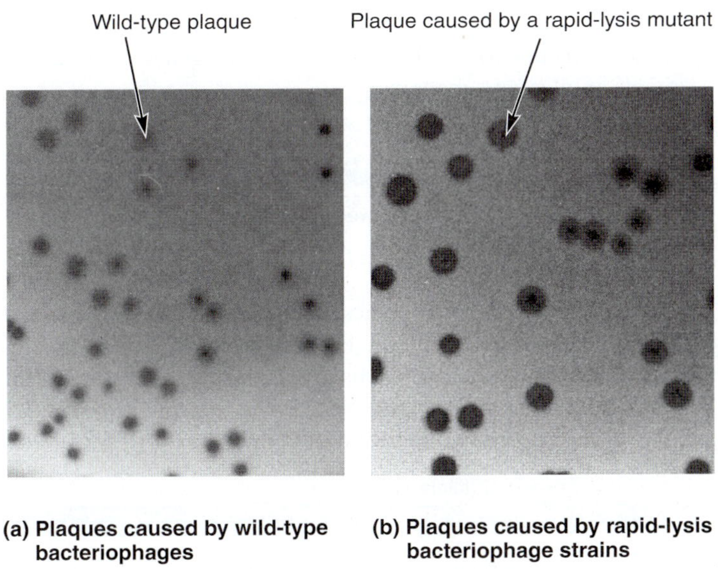

FIGURE 7.15 A comparison of plaques produced by the wild-type T4 bacteriophage and a rapid-lysis mutant.

Genes → Traits **(a)** These plaques were caused by the infection and lysis of *E. coli* cells by the wild-type T4 phage. **(b)** A mutation in a phage gene, called a rapid-lysis mutation, caused the phage to lyse *E. coli* cells more quickly. A phage carrying a rapid-lysis mutant allele yields much larger plaques with clearly defined edges.

Plaque formation in:

T4 strain	*E. coli* B	*E. coli* K12(λ)
Wild-type	Normal plaques	Normal plaques
rII mutants	Large plaques	No plaques

In the bacterial strain called *E. coli* B, *rII* phages produced abnormally large plaques. However, in *E. coli* K12(λ), the *rII* mutants were unable to produce plaques at all, for reasons that were not understood. As described next, these properties allowed Benzer to determine if different *rII* mutations were in the same phage gene or in different genes.

A Complementation Test Can Reveal If Mutations Are in the Same Gene or in Different Genes

In his experiments, Benzer was interested in a single trait, namely, the ability to form plaques. He had isolated many *rII* mutant strains that could form large plaques in *E. coli* B but could not produce plaques in *E. coli* K12(λ). To attempt gene mapping, which is described later, he needed to know if the various *rII* mutations were in the same gene or if they involved mutations in different genes. To accomplish this, he conducted a **complementation test.** In a complementation test, a researcher analyzes two mutant strains that exhibit a phenotype that is different from the wild-type strain. In Benzer's work, the mutant T4 strains were unable to form plaques in *E. coli* K12(λ), whereas the wild-type strain could.

The possible outcomes of a complementation test are shown in **Figure 7.16**. Bacterial cells were coinfected with two different *rII* mutant strains of T4 phage. In Figure 7.16a, this coinfection produced plaques in *E. coli* K12(λ). This result, which is called **complementation,** indicates that the mutations are in different genes. Complementation occurs because each mutant gene in one *rII* strain is complemented by the corresponding wild-type gene in the other strain. In this example, strain 1 carried a mutation in gene *A* and gene *B* was wild-type, whereas strain 2 carried a mutation in gene *B* and gene *A* was wild-type. Because the coinfected cells produce normal proteins that are encoded by the wild-type versions of both genes *A* and *B*, the coinfected cells are lysed in the same manner as if they were infected by the wild-type strain.

Another possible outcome is shown in Figure 7.16b, in which plaques do not form. This result, which is called **noncomplementation,** indicates that the two different *rII* phage strains carry mutations in the same gene. In this case, noncomplementation occurs because neither phage has a functional gene *A*.

By carefully considering the pattern of complementation and noncomplementation, Benzer found that the *rII* mutations occurred in two different genes, which were termed *rIIA* and *rIIB*. The identification of two distinct genes affecting plaque formation was a necessary step that preceded his intragenic mapping analysis, which is described in the following section. Benzer coined the term **cistron** to refer to the smallest genetic unit that gives a negative complementation test. In other words, if two mutations occur within the same cistron, they cannot complement each other. Since these studies, researchers have learned that a cistron is equivalent to a gene. In recent decades, the term gene has gained wide popularity but the term cistron

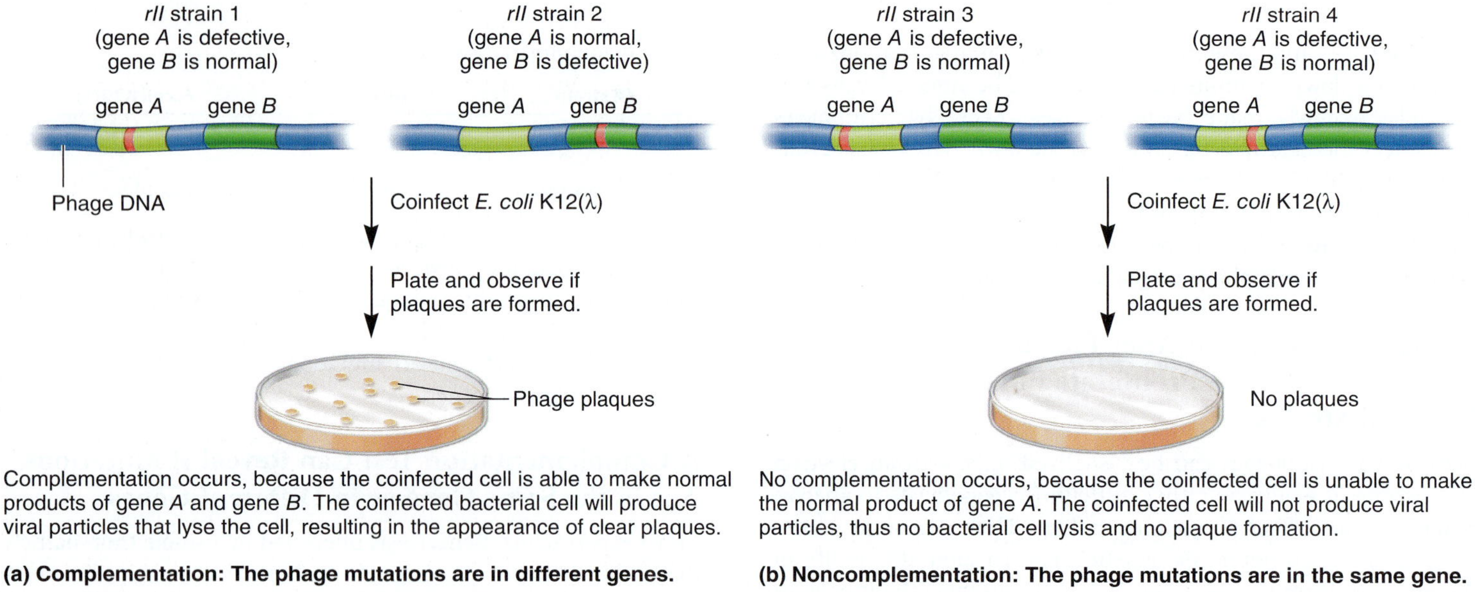

FIGURE 7.16 A comparison of noncomplementation and complementation. Four different T4 phage strains (designated 1 through 4) carry *rII* mutations in gene *A* or gene *B*. **(a)** If the *rII* mutations are in different genes, a coinfected cell can produce new phages and plaques will form. This result is called complementation because the defective genes in each *rII* strain are complemented by the corresponding wild-type genes. **(b)** If *E. coli* K12(λ) is coinfected with two *rII* phage strains with mutations in the same gene, noncomplementation will occur.

CONCEPT CHECK: With regard to the locations of mutations, explain the difference between noncomplementation and complementation.

is not commonly used. However, the term polycistronic is still used to describe bacterial mRNAs that carry two or more gene sequences, as described in Chapter 14.

7.7 COMPREHENSION QUESTIONS

1. A bacterial plaque is
   a. a clear region in a bacterial lawn where the bacteria have been lysed.
   b. a bacterial cell that has a phage attached to its surface.
   c. a dense region of bacterial growth.
   d. none of the above.
2. When two different alleles that affect the same trait exhibit complementation, an interpretation is that
   a. the alleles occur in the same gene.
   b. the alleles occur in two different genes.
   c. both alleles are dominant.
   d. one allele is dominant and the other is recessive.

7.8 INTRAGENIC MAPPING IN BACTERIOPHAGES

Learning Outcomes:

1. Distinguish between intragenic and intergenic mapping.
2. Explain the strategy of Benzer to conduct intragenic mapping.

In the previous section, we learned that Benzer identified T4 bacteriophage mutations that altered plaque morphology and he determined if such mutations were in the same or different genes via complementation. In this section, we will examine how he conducted a detailed type of genetic mapping known as **intragenic mapping.** How is this mapping different from mapping studies in eukaryotes and bacteria? The mapping studies in eukaryotes and bacteria are examples of **intergenic mapping.** The goal of intergenic mapping is to determine the distance between two different genes. For example, the determination of the distance between genes *A* and *B* is an example of intergenic mapping:

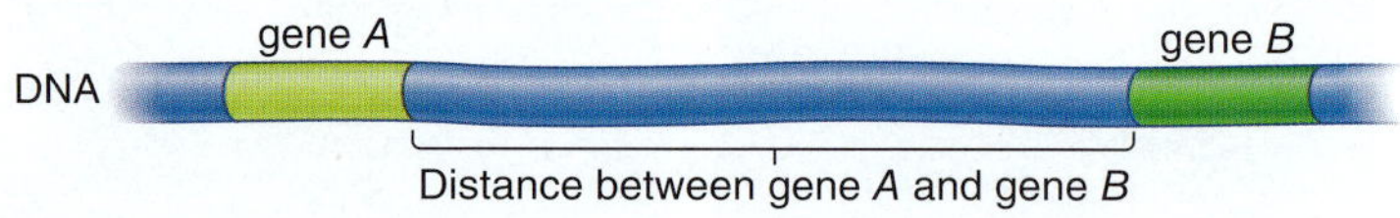

Intergenic mapping

In comparison, intragenic mapping seeks to establish distances between two or more mutations within the same gene. For example, in a population of viruses, gene *C* may exist as two different mutant alleles. One allele may be due to a mutation near the beginning of the gene; the second, to a mutation near the end:

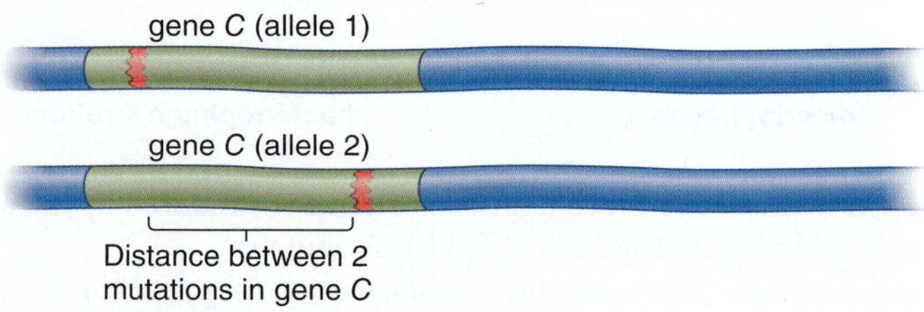

Intragenic mapping

In this section, we will explore the pioneering studies that led to the development of intragenic mapping and advanced our knowledge of gene function. Benzer's results showed that, rather than being an indivisible particle, a gene must be composed of a large structure that can be subdivided during intragenic crossing over.

Intragenic Maps Were Constructed By Analyzing Recombination Between Mutants Within the *rII* Region

As we saw in Figure 7.16, the ability of strains with mutations in two different genes to produce plaques after coinfection is due to complementation. Noncomplementation occurs when two different strains have mutations in the same gene. However, at an extremely low rate, two noncomplementing strains of viruses can produce an occasional plaque if intragenic recombination has taken place. For example, **Figure 7.17** shows a coinfection experiment between two phage strains that both contain *rII* mutations in gene *A*. These mutations are located at different places within the same gene. On rare occasions, a crossover may occur in the very short region between each mutation. This crossover produces a double-mutant gene *A* and a wild-type gene *A*. Because this event has produced a wild-type gene *A*, the function of the protein encoded by gene *A* is restored. Therefore, new phages can be made in *E. coli* K12(λ), resulting in the formation of a plaque.

Figure 7.18 describes the general strategy for intragenic mapping of *rII* phage mutations. Two different noncomplementing *rII* phage mutants (here, *r103* and *r104*) were mixed together in equal numbers and then coinfected into *E. coli* B. In this strain,

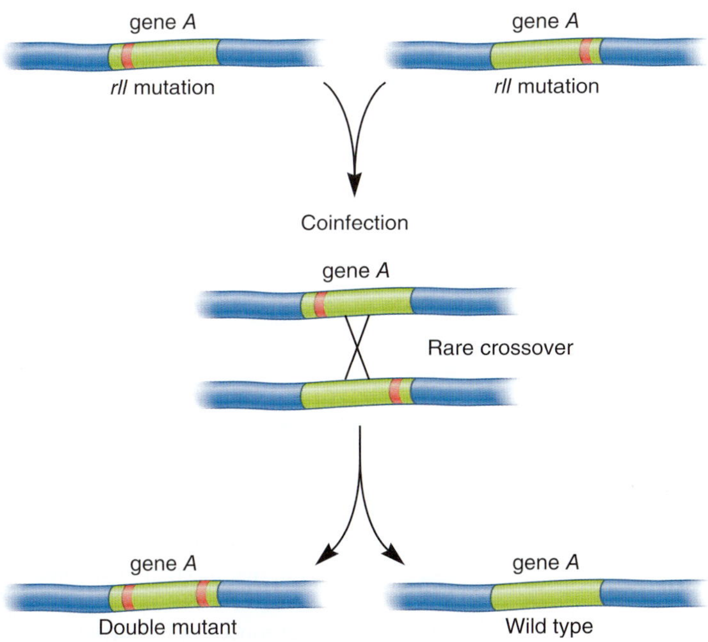

FIGURE 7.17 Intragenic recombination. Following coinfection, a rare crossover has occurred between the sites of the two mutations. This produces a wild-type phage with no mutations and a double-mutant phage with both mutations.

the *rII* mutants grew and propagated. Recall from Figure 7.17, when two different mutants coinfect the same cell, intragenic recombination can occur, producing wild-type phages and double-mutant phages. However, these intragenic recombinants were produced at a very low rate. Following coinfection and lysis of *E. coli* B, a new population of phages was isolated. This population was expected to contain predominantly nonrecombinant phages. However, due to intragenic recombination, it should also contain a very low percentage of wild-type phages and double-mutant phages (see Figure 7.17).

How could Benzer determine the number of rare phages that were produced by intragenic recombination? As you may recall from the previous section, a key observation is that *rII* mutant phages cannot grow in *E. coli* K12(λ). Following coinfection, he took this new population of phages and used some of them to infect *E. coli* B and some to infect *E. coli* K12(λ). After plating, the *E. coli* B infection was used to determine the total number of phages, because *rII* mutants as well as wild-type phages can produce plaques in this strain. The overwhelming majority of these phages were expected to be nonrecombinant phages. The *E. coli* K12(λ) infection was used to determine the number of rare intragenic recombinants that produce wild-type phages.

Figure 7.18 illustrates the great advantage of this experimental system in detecting a low percentage of recombinants. In the laboratory, phage preparations containing several billion phages per milliliter are readily made. Among billions of phages, a low percentage (e.g., 1 in every 1000) may be wild-type phages arising from intragenic recombination. The wild-type recombinants can produce plaques in *E. coli* K12(λ), whereas the *rII* mutant strains cannot. In other words, only the tiny fraction of wild-type recombinants would produce plaques in *E. coli* K12(λ).

The frequency of recombinant phages can be determined by comparing the number of wild-type phages, produced by intragenic recombination, and the total number of phages. As shown in Figure 7.18, the total number of phages can be deduced from the number of plaques obtained from the infection of *E. coli* B. In this experiment, the phage preparation was diluted by 10^{-8} (1:100,000,000), and 1 mL was used to infect *E. coli* B. Because this plate produced 66 plaques, the total number of phages in the original preparation was $66 \times 10^8 = 6.6 \times 10^9$, or 6.6 billion phages per milliliter. By comparison, the phage preparation used to infect *E. coli* K12(λ) was diluted by only 10^{-6} (1:1,000,000). This plate produced 11 plaques. Therefore, the number of wild-type phages was 11×10^6, which equals 11 million wild-type phages per milliliter.

As we have already seen in Chapter 6, genetic mapping distance is computed by dividing the number of recombinants by the total population (nonrecombinants and recombinants) times 100. In this experiment, intragenic recombination produced an equal number of two types of recombinants: wild-type phages and double-mutant phages. Only the wild-type phages are detected in the infection of *E. coli* K12(λ). Therefore, to obtain the total number of recombinants, the number of wild-type phages must be multiplied by 2. With all this information, we can use the following equation to compute the frequency of recombinants using the experimental approach described in Figure 7.18.

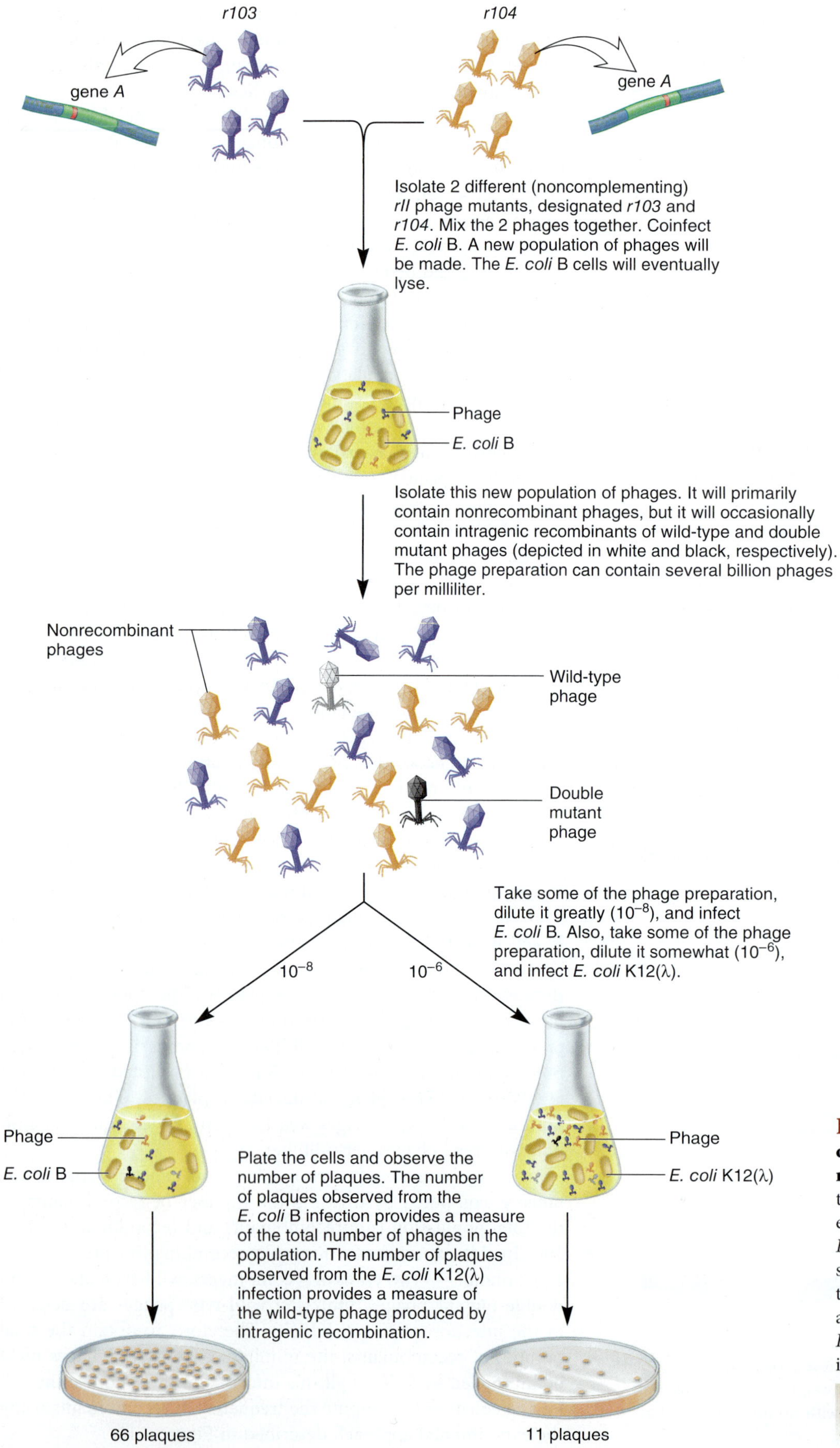

FIGURE 7.18 Benzer's method of intragenic mapping in the *rII* region. The coinfection carried out at the top of this figure was done using a great excess of phages, making it likely that the *E. coli* B cells will be coinfected with both strains. However, the infection experiment toward the bottom of the figure was done at a lower phage concentration such that an *E. coli* B cell or *E. coli* K12(λ) cell would be infected with a single phage.

CONCEPT CHECK: In the two flasks at the bottom of this figure, why was it necessary to infect *E. coli* K12(λ) with 100 times more phage compared to *E. coli* B?

$$\text{Frequency of recombinants} = \frac{2\ [\text{Wild-type plaques obtained in } E.\ coli\ \text{K12}(\lambda)]}{\text{Total number of plaques obtained in } E.\ coli\ \text{B}}$$

$$\text{Frequency of recombinants} = \frac{2(11 \times 10^{6})}{6.6 \times 10^{9}}$$

$$= 3.3 \times 10^{-3} = 0.0033$$

In this example approximately 3.3 recombinants were produced per 1000 phages.

The frequency of recombinants provides a measure of map distance. In eukaryotic mapping studies, we compute the map distance by multiplying the frequency of recombinants by 100 to give a value in map units (also known as centiMorgans). In the previous example, the map distance is 0.0033 × 100 = 0.33 map units. In this case, the map distance is between two mutations within the same gene. Like intergenic mapping, the frequency of intragenic recombinants is correlated with the distance between the two mutations; the farther apart they are, the higher the frequency of recombinants.

7.8 COMPREHENSION QUESTION

1. The goal of intragenic mapping is to determine the
   a. order and distance between different genes.
   b. order and distance between mutations in a single gene.
   c. locations of genes within a viral genome.
   d. insertion site of a viral genome into the host cell's chromosome.

KEY TERMS

Page 157. bacteriophages, phages, genetic transfer, conjugation, transduction, transformation, minimal medium, auxotroph, prototroph
Page 159. F factor
Page 160. sex pili (pilus), conjugation bridge, relaxosome, origin of transfer, nucleoprotein, plasmid, episomes
Page 162. *Hfr* strain, F′ factors
Page 163. interrupted mating
Page 166. minutes
Page 167. lytic cycle
Page 168. cotransduction
Page 169. natural transformation, competent cells, competence factors
Page 170. homologous recombination, heteroduplex, nonhomologous recombination, illegitimate recombination, cotransformation, artificial transformation, competence-stimulating peptide (CSP)
Page 171. DNA uptake signal sequences, horizontal gene transfer, acquired antibiotic resistance
Page 173. plaque, complementation test, complementation, noncomplementation, cistron
Page 174. intragenic mapping, intergenic mapping

CHAPTER SUMMARY

7.1 Overview of Genetic Transfer in Bacteria

- Three general mechanisms for genetic transfer in various species of bacteria are conjugation, transduction, and transformation (see Table 7.1).

7.2 Bacterial Conjugation

- Joshua Lederberg and Edward Tatum discovered conjugation in *E. coli* by analyzing auxotrophic strains (see Figure 7.1).
- Using a U-tube apparatus, Davis showed that conjugation requires cell-to-cell contact (see Figure 7.2).
- Certain strains of bacteria have F factors, which they can transfer via conjugation in a series of steps (see Figures 7.3, 7.4).

7.3 Conjugation and Mapping via *Hfr* Strains

- *Hfr* strains are formed when an F factor integrates into the bacterial chromosome. The imprecise excision can produce an F′ factor that carries a portion of the bacterial chromosome (see Figure 7.5).
- *Hfr* strains transfer a portion of the bacterial chromosome to a recipient cell during conjugation (see Figure 7.6).
- Wollman and Jacob showed that conjugation can be used to map the locations of genes along the bacterial chromosome, thereby creating a genetic map (see Figures 7.7–7.9).

7.4 Bacterial Transduction

- During transduction, a portion of a bacterial chromosome from a dead bacterium is transferred to a living recipient cell via a bacteriophage (see Figure 7.10).
- A cotransduction experiment can be used to map genes that are close together along a bacterial chromosome (see Figure 7.11).

7.5 Bacterial Transformation

- During transformation, a segment of DNA from a dead bacterium is taken up by a living bacterial cell and then incorporated into the bacterial chromosome (see Figure 7.12).

7.6 Medical Relevance of Bacterial Genetic Transfer

- Horizontal gene transfer is a process in which an organism incorporates genetic material from another organism without being the offspring of that organism. Conjugation, transformation, and transduction are examples of horizontal gene transfer.
- Horizontal gene transfer has involved the transfer of genes that confer antibiotic resistance. When pathogenic strains acquire such genes, they become more difficult to treat with commonly prescribed antibiotics.

7.7 Plaque Formation and Intergenic Complementation in Bacteriophages

- Bacteriophages are viruses that infect bacteria (see Figure 7.13).
- Some bacteriophages cause bacteria to lyse. This event leads to plaque formation when bacteria are plated as a lawn on a bacterial medium (see Figure 7.14).
- Benzer studied a type of bacteriophage called T4. He identified mutant strains that caused rapid lysis, thereby leading to larger plaques (see Figure 7.15).
- Benzer identified strains of T4 that could not form plaques in *E. coli* K12(λ). He conducted complementation tests to determine if the mutations in these strains were in the same phage gene or in different genes (see Figure 7.16).

7.8 Intragenic Mapping in Bacteriophages

- Benzer devised a method to study intragenic recombination, which allowed him to map the distance between two mutations within the same gene (see Figures 7.17, 7.18).

PROBLEM SETS & INSIGHTS

Solved Problems

S1. In *E. coli*, the gene $bioD^+$ encodes an enzyme involved in biotin synthesis, and $galK^+$ encodes an enzyme involved in galactose utilization. An *E. coli* strain that contained wild-type versions of both genes was infected with P1, and then a P1 lysate was obtained. This lysate was used to transduce (infect) a strain that was $bioD^-$ and $galK^-$. The cells were plated on a medium containing galactose as the sole carbon source for growth to select for transduction of the $galK^+$ gene. This medium also was supplemented with biotin. The colonies were then restreaked on a medium that lacked biotin to see if the $bioD^+$ gene had been cotransduced. The following results were obtained:

Selected Gene	Nonselected Gene	Number of Colonies That Grew On: Galactose + Biotin	Galactose − Biotin	Cotransduction Frequency
$galK^+$	$bioD^+$	80	10	0.125

How far apart are these two genes?

Answer: We can use the cotransduction frequency to calculate the distance between the two genes (in minutes) using the equation

$$\text{Cotransduction frequency} = (1 - d/2)^3$$

$$0.125 = (1 - d/2)^3$$

$$1 - d/2 = \sqrt[3]{0.125}$$

$$1 - d/2 = 0.5$$

$$d/2 = 1 - 0.5$$

$$d = 1.0 \text{ minute}$$

The two genes are approximately 1 minute apart on the *E. coli* chromosome.

S2. By conducting conjugation experiments between *Hfr* and recipient strains, Wollman and Jacob mapped the order of many bacterial genes. Throughout the course of their studies, they identified several different *Hfr* strains in which the F factor DNA had been integrated at different places along the bacterial chromosome. A sample of their experimental results is shown in the following table:

Hfr Strain	Origin	Order of Transfer of Several Different Bacterial Genes: First								Last
H	O	*thr*	*leu*	*azi*	*ton*	*pro*	*lac*	*gal*	*str*	*met*
1	O	*leu*	*thr*	*met*	*str*	*gal*	*lac*	*pro*	*ton*	*azi*
2	O	*pro*	*ton*	*azi*	*leu*	*thr*	*met*	*str*	*gal*	*lac*
3	O	*lac*	*pro*	*ton*	*azi*	*leu*	*thr*	*met*	*str*	*gal*
4	O	*met*	*str*	*gal*	*lac*	*pro*	*ton*	*azi*	*leu*	*thr*
5	O	*met*	*thr*	*leu*	*azi*	*ton*	*pro*	*lac*	*gal*	*str*
6	O	*met*	*thr*	*leu*	*azi*	*ton*	*pro*	*lac*	*gal*	*str*
7	O	*ton*	*azi*	*leu*	*thr*	*met*	*str*	*gal*	*lac*	*pro*

A. Explain how these results are consistent with the idea that the bacterial chromosome is circular.

B. Draw a map that shows the order of genes and the locations of the origins of transfer among these different *Hfr* strains.

Answer:

A. In comparing the data among different *Hfr* strains, the order of the nine genes was always the same or the reverse of the same order. For example, *HfrH* and *Hfr4* transfer the same genes but their orders are reversed relative to each other. In addition, the *Hfr* strains showed an overlapping pattern of transfer with regard to the origin. For example, *Hfr1* and *Hfr2* had the same order of genes, but *Hfr1* began with *leu* and ended with *azi*, whereas *Hfr2* began with *pro* and ended with *lac*. From these findings, Wollman and Jacob concluded that the origin of transfer had been inserted at different points within a circular *E. coli* chromosome in different *Hfr* strains. They also concluded that the origin can be inserted in either orientation, so the direction of gene transfer can be clockwise or counterclockwise around the circular bacterial chromosome.

B. A genetic map consistent with these results is shown here.

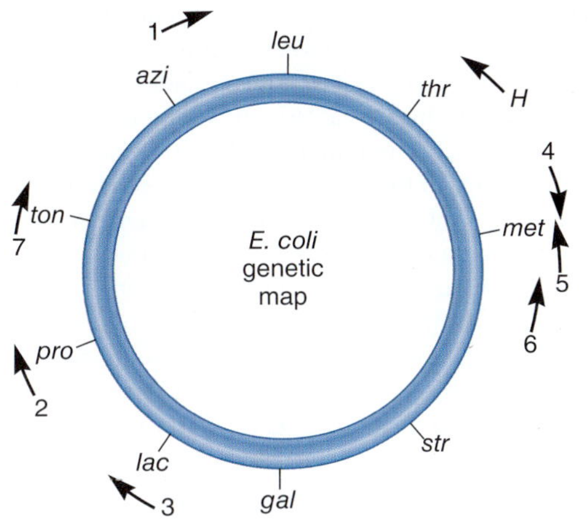

S3. An *Hfr* strain that is $leuA^+$ and $thiL^+$ was mixed with a strain that is $leuA^-$ and $thiL^-$. In the data points shown here, the conjugation was interrupted, and the percentage of recombinants for each gene was determined by streaking on a medium that lacked either leucine or thiamine. The results are shown in the following graph.

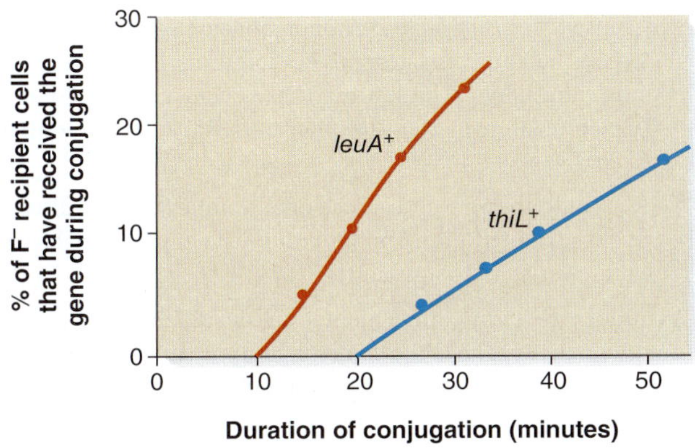

What is the map distance (in minutes) between these two genes?

Answer: This problem is solved by extrapolating the data points to the *x*-axis to determine the time of entry. For $leuA^+$, they extrapolate back to 10 minutes. For $thiL^+$, they extrapolate back to 20 minutes. Therefore, the distance between the two genes is approximately 10 minutes.

S4. Genetic transfer via transformation can also be used to map genes along the bacterial chromosome. In this approach, fragments of chromosomal DNA are isolated from one bacterial strain and used to transform another strain. The experimenter examines the transformed bacteria to see if they have incorporated two or more different genes. For example, the DNA may be isolated from a donor *E. coli* bacterium that has functional copies of the *araB* and *leuD* genes. Let's call these genes $araB^+$ and $leuD^+$ to indicate the genes are functional. These two genes are required for arabinose metabolism and leucine synthesis, respectively. To map the distance between these two genes via transformation, a recipient bacterium is used that is $araB^-$ and $leuD^-$. Following transformation, the recipient bacterium may become $araB^+$ and $leuD^+$. This phenomenon is called cotransformation because two genes from the donor bacterium have been transferred to the recipient via transformation. In this type of experiment, the recipient cell is exposed to a fairly low concentration of donor DNA, making it unlikely that the recipient bacterium will take up more than one fragment of DNA. Therefore, under these conditions, cotransformation is likely only when two genes are fairly close together and are found on one fragment of DNA.

In a cotransformation experiment, a researcher has isolated DNA from an $araB^+$ and $leuD^+$ donor strain. This DNA was transformed into a recipient strain that was $araB^-$ and $leuD^-$. Following transformation, the cells were plated on a medium containing arabinose and leucine. On this medium, only bacteria that are $araB^+$ can grow. The bacteria can be either $leuD^+$ or $leuD^-$ because leucine is provided in the medium. Colonies that grew on this medium were then restreaked on a medium that contained arabinose but lacked leucine. Only $araB^+$ and $leuD^+$ cells could grow on these secondary plates. Following this protocol, a researcher obtained the following results:

Number of colonies growing on a medium containing arabinose plus leucine: 57

Number of colonies that grew when restreaked on a medium containing arabinose medium without leucine: 42

What is the map distance between these two genes? Note: This problem can be solved using the strategy of a cotransduction experiment except that the researcher must determine the average size of DNA fragments that are taken up by the bacterial cells. This would correspond to the value of *L* in a cotransduction experiment.

Answer: As mentioned, the basic principle of gene mapping via cotransformation is identical to the method of gene mapping via cotransduction described in this chapter. One way to calculate the map distance is to use the same equation that we used for cotransduction data, except that we substitute cotransformation frequency for cotransduction frequency.

$$\text{Cotransformation frequency} = (1 - d/L)^3$$

(Note: Cotransformation is not quite as accurate as cotransduction because the sizes of chromosomal pieces tend to vary significantly from experiment to experiment, so the value of *L* is not quite as reliable. Nevertheless, cotransformation has been used extensively to map the order and distance between closely linked genes along the bacterial chromosome.)

The researcher needs to experimentally determine the value of *L* by running the DNA on a gel and estimating the average size of the DNA fragments. Let's assume they are about 2% of the bacterial chromosome, which, for *E. coli*, would be about 80,000 bp in length. So *L* equals 2 minutes, which is the same as 2%.

$$\text{Cotransformation frequency} = (1 - d/L)^3$$

$$42/57 = (1 - d/2)^3$$

$$d = 0.2 \text{ minutes}$$

The distance between *araB* and *leuD* is approximately 0.2 minutes.

Conceptual Questions

C1. The terms conjugation, transduction, and transformation are used to describe three different natural forms of genetic transfer between bacterial cells. Briefly discuss the similarities and differences among these processes.

C2. Conjugation is sometimes called "bacterial mating." Is it a form of sexual reproduction? Explain.

C3. If you mix together an equal number of F^+ and F^- cells, how would you expect the proportions to change over time? In other words, do you expect an increase in the relative proportions of F^+ or of F^- cells? Explain your answer.

C4. What is the difference between an F^+ and an *Hfr* strain? Which type of strain do you expect to transfer many bacterial genes to recipient cells?

C5. What is the role of the origin of transfer during F^+- and Hfr-mediated conjugation? What is the significance of the direction of transfer in Hfr-mediated conjugation?

C6. What is the role of sex pili during conjugation?

C7. Think about the structure and transmission of F factors and discuss how you think F factors may have originated.

C8. Each species of bacteria has its own distinctive cell surface. The characteristics of the cell surface play an important role in processes such as conjugation and transduction. For example, certain strains of *E. coli* have pili on their cell surface. These pili enable *E. coli* to conjugate with other *E. coli*, and the pili also enable certain bacteriophages (such as M13) to bind to the surface of *E. coli* and gain entry into the cytoplasm. With these ideas in mind, explain which forms of genetic transfer (i.e., conjugation, transduction, and transformation) are more likely to occur between different species of bacteria. Discuss some of the potential consequences of interspecies genetic transfer.

C9. Briefly describe the lytic and lysogenic cycles of bacteriophages. In your answer, explain what a prophage is.

C10. What is cotransduction? What determines the likelihood that two genes will be cotransduced?

C11. When bacteriophage P1 causes *E. coli* to lyse, the resulting material is called a P1 lysate. What type of genetic material would be found in most of the P1 phages in the lysate? What kind of genetic material is occasionally found within a P1 phage?

C12. As described in Figure 7.10, host DNA is hydrolyzed into small pieces, which are occasionally assembled with phage proteins, creating a phage with bacterial chromosomal DNA. If the breakage of the chromosomal DNA is not random (i.e., it is more likely to break at certain spots as opposed to other spots), how might nonrandom breakage affect cotransduction frequency?

C13. Describe the steps that occur during bacterial transformation. What is a competent cell? What factors may determine whether a cell will be competent?

C14. Which bacterial genetic transfer process does not require recombination with the bacterial chromosome?

C15. Researchers who study the molecular mechanism of transformation have identified many proteins in bacteria that function in the uptake of DNA from the environment and its recombination into the host cell's chromosome. This means that bacteria have evolved molecular mechanisms for the purpose of transformation by extracellular DNA. What advantage(s) does a bacterium derive from importing DNA from the environment and/or incorporating it into its chromosome?

C16. Antibiotics such as tetracycline, streptomycin, and bacitracin are small organic molecules that are synthesized by particular species of bacteria. Microbiologists have hypothesized that the reason why certain bacteria make antibiotics is to kill other species that occupy the same environment. Bacteria that produce an antibiotic may be able to kill competing species. This provides more resources for the antibiotic-producing bacteria. In addition, bacteria that have the genes necessary for antibiotic biosynthesis contain genes that confer resistance to the same antibiotic. For example, tetracycline is made by the soil bacterium *Streptomyces aureofaciens*. Besides the genes that are needed to make tetracycline, *S. aureofaciens* also contains genes that confer tetracycline resistance; otherwise, it would kill itself when it makes tetracycline. In recent years, however, many other species of bacteria that do not synthesize tetracycline have acquired the genes that confer tetracycline resistance. For example, certain strains of *E. coli* carry tetracycline-resistance genes, even though *E. coli* does not synthesize tetracycline. When these genes were analyzed at the molecular level, it was found that they are evolutionarily related to the genes in *S. aureofaciens*. This observation indicates that the genes from *S. aureofaciens* have been transferred to *E. coli*.

A. What form of genetic transfer (i.e., conjugation, transduction, or transformation) would be the most likely mechanism of interspecies gene transfer?

B. Because *S. aureofaciens* is a nonpathogenic soil bacterium and *E. coli* is an enteric bacterium, do you think it was direct gene transfer, or do you think it may have occurred in multiple steps (i.e., from *S. aureofaciens* to other bacterial species and then to *E. coli*)?

C. How could the widespread use of antibiotics to treat diseases have contributed to the proliferation of many bacterial species that are resistant to antibiotics?

C17. What does the term complementation mean? If two different mutations that produce the same phenotype can complement each other, what can you conclude about the locations of each mutation?

C18. Intragenic mapping is sometimes called interallelic mapping. Explain why the two terms mean the same thing. In your own words, explain what an intragenic map is.

C19. As discussed in Chapter 12, genes are composed of a sequence of nucleotides. A typical gene in a bacteriophage is a few hundred or a few thousand nucleotides in length. If two different strains of bacteriophage T4 have a mutation in the *rIIA* gene that gives a rapid-lysis phenotype, yet they never produce wild-type phages by intragenic recombination when they are coinfected into *E. coli* B, what would you conclude about the locations of the mutations in the two different T4 strains?

Experimental Questions

E1. In the experiment of Figure 7.1, a met^- bio^- thr^+ leu^+ thi^+ cell could become met^+ bio^+ thr^+ leu^+ thi^+ by a (rare) double mutation that converts the met^- bio^- genes into met^+ bio^+. Likewise, a met^+ bio^+ thr^- leu^- thi^- cell could become met^+ bio^+ thr^+ leu^+ thi^+ by three mutations that convert the thr^- leu^- thi^- genes into thr^+ leu^+ thi^+. From the results of Figure 7.1, how do you know that the occurrence of 10 met^+ bio^+ thr^+ leu^+ thi^+ colonies is not due to these types of rare double or triple mutations?

E2. In the experiment of Figure 7.1, Joshua Lederberg and Edward Tatum could not discern whether met^+ bio^+ genetic material was transferred to the met^- bio^- thr^+ leu^+ thi^+ strain or if thr^+ leu^+ thi^+ genetic material was transferred to the met^+ bio^+ thr^- leu^- thi^- strain. Let's suppose that one strain is streptomycin-resistant (say, met^+ bio^+ thr^- leu^- thi^-) and the other strain is sensitive to streptomycin. Describe an experiment that could determine whether the met^+ bio^+ genetic material was transferred to the met^- bio^- thr^+ leu^+ thi^+ strain or if the thr^+ leu^+ thi^+ genetic material was transferred to the met^+ bio^+ thr^- leu^- thi^- strain.

E3. Explain how a U-tube apparatus can distinguish between genetic transfer involving conjugation and genetic transfer involving transduction. Do you think a U tube could be used to distinguish between transduction and transformation?

E4. What is an interrupted mating experiment? What type of experimental information can be obtained from this type of study? Why is it necessary to interrupt mating?

E5. In a conjugation experiment, what is meant by the time of entry? How is the time of entry determined experimentally?

E6. In your laboratory, you have an F^- strain of *E. coli* that is resistant to streptomycin and is unable to metabolize lactose, but it can metabolize glucose. Therefore, this strain can grow on a medium that contains glucose and streptomycin, but it cannot grow on a medium containing lactose. A researcher has sent you two *E. coli* strains in two separate tubes. One strain, let's call it strain *A*, has an F′ factor (an F prime factor) that carries the genes that are required for lactose metabolism. On its chromosome, it also has the genes that are required for glucose metabolism. However, it is sensitive to streptomycin. This strain can grow on a medium containing lactose or glucose, but it cannot grow if streptomycin is added to the medium. The second strain, let's call it strain *B*, is an F^- strain. On its chromosome, it has the genes that are required for lactose and glucose metabolism. Strain *B* is also sensitive to streptomycin. Unfortunately, when strains *A* and *B* were sent to you, the labels had fallen off the tubes. Describe how you could determine which tubes contain strain *A* and strain *B*.

E7. As mentioned in solved problem S2, origins of transfer can be located in many different locations, and their direction of transfer can be clockwise or counterclockwise. Let's suppose a researcher conjugated six different *Hfr* strains that were thr^+ leu^+ ton^s str^r azi^s lac^+ gal^+ pro^+ met^+ to an F^- strain that was thr^- leu^- ton^r str^s azi^r lac^- gal^- pro^- met^-, and obtained the following results:

Strain	Order of Gene Transfer
1	ton^s azi^s leu^+ thr^+ met^+ str^r gal^+ lac^+ pro^+
2	leu^+ azi^s ton^s pro^+ lac^+ gal^+ str^r met^+ thr^+
3	lac^+ gal^+ str^r met^+ thr^+ leu^+ azi^s ton^s pro^+
4	leu^+ thr^+ met^+ str^r gal^+ lac^+ pro^+ ton^s azi^s
5	ton^s pro^+ lac^+ gal^+ str^r met^+ thr^+ leu^+ azi^s
6	met^+ str^r gal^+ lac^+ pro^+ ton^s azi^s leu^+ thr^+

Draw a circular map of the *E. coli* chromosome and describe the locations and orientations of the origins of transfer in these six *Hfr* strains.

E8. An *Hfr* strain that is $hisE^+$ and $pheA^+$ was mixed with a strain that is $hisE^-$ and $pheA^-$. The conjugation was interrupted and the percentage of recombinants for each gene was determined by streaking on a medium that lacked either histidine or phenylalanine. The following results were obtained:

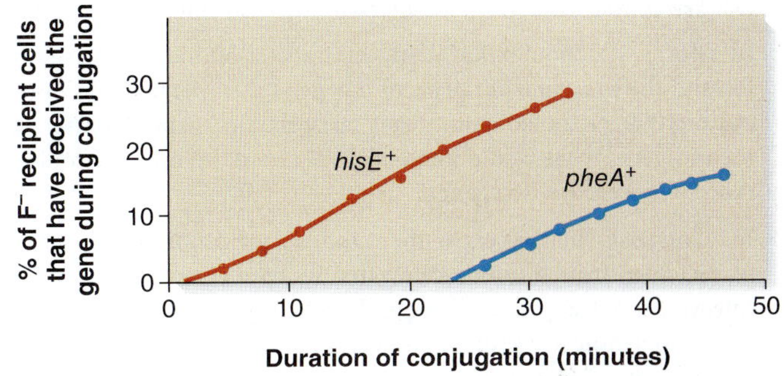

A. Determine the map distance (in minutes) between these two genes.

B. In a previous experiment, it was found that *hisE* is 4 minutes away from the gene *pabB*. *PheA* was shown to be 17 minutes from this gene. Draw a genetic map describing the locations of all three genes.

E9. Acridine orange is a chemical that inhibits the replication of F factor DNA but does not affect the replication of chromosomal DNA, even if the chromosomal DNA contains an Hfr. Let's suppose that you have an *E. coli* strain that is unable to metabolize lactose and has an F factor that carries a streptomycin-resistant gene. You also have an F^- strain of *E. coli* that is sensitive to streptomycin and has the genes that allow the bacterium to metabolize lactose. This second strain can grow on a lactose-containing medium. How would you generate an *Hfr* strain that is resistant to streptomycin and can metabolize lactose? (Hint: F factors occasionally integrate into the chromosome to become *Hfr* strains, and occasionally *Hfr* strains excise their DNA from the chromosome to become F^+ strains that carry an F′ factor.)

E10. In a P1 transduction experiment, the P1 lysate contains phages that carry pieces of the host chromosomal DNA, but the lysate also contains broken pieces of chromosomal DNA (see Figure 7.10). If a P1 lysate is used to transfer chromosomal DNA to another bacterium, how could you show experimentally that the recombinant bacterium has been transduced (i.e., taken up a P1 phage with a piece of chromosomal DNA inside) versus transformed (i.e., taken up a piece of chromosomal DNA that is not within a P1 phage coat)?

E11. Can you devise an experimental strategy to get P1 phage to transduce the entire λ genome from one strain of bacterium to another strain? (Note: The general features of phage lambda's reproductive cycle are described in Chapter 17.) Phage λ has a genome size of 48,502 nucleotides (about 1% of the size of the *E. coli* chromosome) and can follow the lytic or lysogenic reproductive cycle. Growth of *E. coli* on minimal growth medium favors the lysogenic reproductive cycle, whereas growth on a rich medium and/or under UV light promotes the lytic cycle.

E12. Let's suppose a new strain of P1 has been identified that packages larger pieces of the *E. coli* chromosome. This new P1 strain packages pieces of the *E. coli* chromosome that are 5 minutes long. If two genes are 0.7 minutes apart along the *E. coli* chromosome, what would be the cotransduction frequency using a normal strain of P1 and using this new strain of P1 that packages large pieces? What would be the experimental advantage of using this new P1 strain?

E13. If two bacterial genes are 0.6 minutes apart on the bacterial chromosome, what frequency of cotransductants would you expect to observe in a P1 transduction experiment?

E14. In an experiment involving P1 transduction, the cotransduction frequency was 0.53. How far apart are the two genes?

E15. In a cotransduction experiment, the transfer of one gene is selected for and the presence of the second gene is then determined. If 0 out of 1000 P1 transductants that carry the first gene also carry the second gene, what would you conclude about the minimum distance between the two genes?

E16. In a cotransformation experiment (see solved problem S4), DNA was isolated from a donor strain that was $proA^+$ and $strC^+$ and sensitive to tetracycline. (The *proA* and *strC* genes confer the ability to synthesize proline and confer streptomycin resistance, respectively.) A recipient strain is $proA^-$ and $strC^-$ and is resistant to tetracycline. After transformation, the bacteria were first streaked on a medium containing proline, streptomycin, and tetracycline. Colonies were then restreaked on a medium containing streptomycin and tetracycline. (Note: Each type of medium had carbon and nitrogen sources for growth.) The following results were obtained:

> 70 colonies grew on the medium containing proline, streptomycin, and tetracycline, but only 2 of these 70 colonies grew when restreaked on the medium containing streptomycin and tetracycline but lacking proline.

A. If we assume the average size of the DNA fragments is 2 minutes, how far apart are these two genes?

B. What would you expect the cotransformation frequency to be if the average size of the DNA fragments was 4 minutes and the two genes are 1.4 minutes apart?

E17. If you took a pipette tip and removed a phage plaque from a petri plate, what would it contain?

E18. As shown in Figure 7.16, phages with *rII* mutations cannot produce plaques in *E. coli* K12(λ), but wild-type phages can. From an experimental point of view, explain why this observation is so significant.

E19. In the experimental strategy described in Figure 7.18, explain why it was necessary to dilute the phage preparation used to infect *E. coli* B so much more than the phage preparation used to infect *E. coli* K12(λ).

E20. Here are data from several complementation experiments, involving rapid-lysis mutations in genes *rIIA* and *rIIB*. The strain designated *L51* is known to have a mutation in *rIIB*.

Phage Mixture	Complementation
L91 and L65	No
L65 and L62	No
L33 and L47	Yes
L40 and L51	No
L47 and L92	No
L51 and L47	Yes
L51 and L92	Yes
L33 and L40	No
L91 and L92	Yes
L91 and L33	No

List which groups of mutations are in the *rIIA* gene and which groups are in the *rIIB* gene.

E21. A researcher has several different strains of T4 phage with single mutations in the same gene. In these strains, the mutations render the phage temperature-sensitive. This means that temperature-sensitive phages can propagate when the bacterium (*E. coli*) is grown at 32°C but cannot propagate themselves when *E. coli* is grown at 37°C. Think about Benzer's strategy for intragenic mapping and propose an experimental strategy to map the temperature-sensitive mutations.

Questions for Student Discussion/Collaboration

1. Discuss the advantages of the genetic analysis of bacteria and bacteriophages. Make a list of the types of allelic differences among bacteria and phages that are suitable for genetic analyses.
2. Complementation occurs when two defective alleles in two different genes are found within the same organism and produce a normal phenotype. What other examples of complementation have we encountered in previous chapters of this textbook?

Note: All answers appear at the website for this textbook; the answers to the even-numbered questions and all of the Concept Check and Comprehension Questions are in the back of the book.

www.mhhe.com/brookergenetics5e

Visit the website for practice tests, answer keys, and other learning aids for this chapter. Enhance your understanding of genetics with our interactive exercises, quizzes, animations, and much more.

CHAPTER OUTLINE

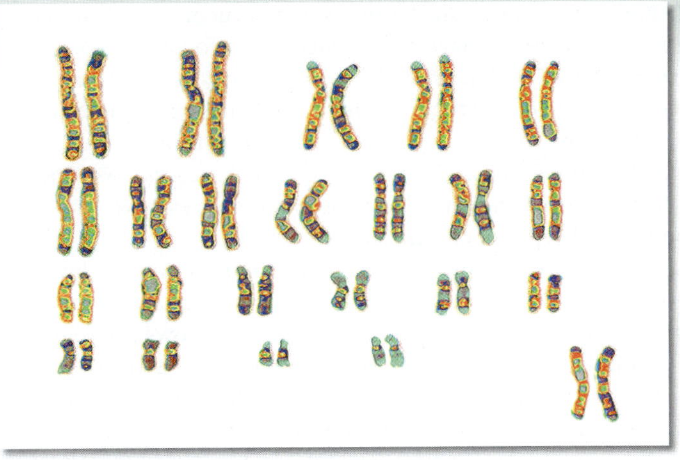

The chromosome composition of humans. *Somatic cells in humans contain 46 chromosomes, which come in 23 pairs.*

8 VARIATION IN CHROMOSOME STRUCTURE AND NUMBER

The term **genetic variation** refers to genetic differences among members of the same species or between different species. Throughout Chapters 2–7, we have focused primarily on variation in specific genes, which is called **allelic variation.** In Chapter 8, our emphasis will shift to larger types of genetic changes that affect the structure or number of eukaryotic chromosomes. These larger alterations may affect the expression of many genes and thereby influence phenotypes. Variation in chromosome structure and number are of great importance in the field of genetics because they are critical in the evolution of new species and have widespread medical relevance. In addition, agricultural geneticists have discovered that such variation can lead to the development of new crops, which may be hardier and more productive.

We begin Chapter 8 by exploring how the structure of a eukaryotic chromosome can be modified, either by altering the total amount of genetic material or by rearranging the order of genes along a chromosome. Such changes may often be detected microscopically. The rest of the chapter is concerned with changes in the total number of chromosomes. We will explore how variation in chromosome number occurs and consider examples in which it has significant phenotypic consequences. We will also examine how changes in chromosome number can be induced through experimental treatments and how these approaches have applications in research and agriculture.

8.1 MICROSCOPIC EXAMINATION OF EUKARYOTIC CHROMOSOMES

Learning Outcome:

1. Describe the characteristics that are used to classify and identify chromosomes.

To identify changes in chromosome structure and number, researchers need to have a reference point for a normal set of chromosomes. To determine what the normal chromosomes of a species look like, a **cytogeneticist**—a scientist who studies chromosomes microscopically—examines the chromosomes from several members of a given species. In most cases, two phenotypically normal individuals of the same species have the same number and types of chromosomes.

Because chromosomes are more compact and readily visible during cell division, a cytogeneticist usually analyzes chromosomes in actively dividing cells. **Figure 8.1a** shows micrographs of chromosomes from three species: a human, a fruit fly, and a corn plant. As seen here, a human has 46 chromosomes (23 pairs), a fruit fly has 8 chromosomes (4 pairs), and corn has 20 chromosomes (10 pairs). Except for the sex chromosomes, which differ between males and females, most members of the same species have very similar chromosomes. For example, the overwhelming majority of people have 46 chromosomes in their somatic cells. By comparison, the chromosomal compositions of distantly related species, such as humans and fruit flies, may be very different. A total of 46 chromosomes is normal for humans, whereas 8 chromosomes is the norm for fruit flies.

Cytogeneticists have various ways to classify and identify chromosomes. The three most commonly used features are location of the centromere, size, and banding patterns that are revealed when the chromosomes are treated with stains. As shown in **Figure 8.1b**, chromosomes are classified as **metacentric** (in which the centromere is near the middle), **submetacentric** (in which the centromere is slightly off center), **acrocentric** (in which the centromere is significantly off center but not at the end), and **telocentric** (in which the centromere is at one end). Because the centromere is never exactly in the center of a chromosome, each chromosome has a short arm and a long arm. For human chromosomes, the short arm is designated with the letter p (for the French, petit), and the long arm is designated with the next letter in the alphabet, q. In the case of telocentric chromosomes, the short arm may be nearly nonexistent.

Figure 8.1c shows a human karyotype. The procedure for making a karyotype is described in Chapter 3 (see Figure 3.2). A **karyotype** is a micrograph in which all of the chromosomes within a single cell have been arranged in a standard fashion. When preparing a karyotype, the chromosomes are aligned with the short arms on top and the long arms on the bottom. By convention, the chromosomes are numbered roughly according to their size, with the largest chromosomes having the smallest numbers. For example, human chromosomes 1, 2, and 3 are relatively large, whereas 21 and 22 are the two smallest. An exception to the numbering system involves the sex chromosomes, which are designated with letters (for humans, X and Y).

Because different chromosomes often have similar sizes and centromeric locations (e.g., compare human chromosomes 8, 9, and 10), geneticists must use additional methods to accurately identify each type of chromosome within a karyotype. For detailed identification, chromosomes are treated with stains to produce characteristic banding patterns. Several different staining procedures are used by cytogeneticists to identify specific chromosomes. An example is **G banding,** which is shown in Figure 8.1c. In this procedure, chromosomes are treated with mild heat or with proteolytic enzymes that partially digest chromosomal proteins. When exposed to the dye called Giemsa (named after its inventor Gustav Giemsa), some chromosomal regions bind the dye heavily and produce a dark band. In other regions, the stain hardly binds at all and a light band results. Though the mechanism of staining is not completely understood, the dark bands are thought to represent regions that are more tightly compacted. As shown in Figure 8.1c and d, the alternating pattern of G bands is a unique feature for each chromosome.

In the case of human chromosomes, approximately 300 G bands can usually be distinguished during metaphase. A larger number of G bands (in the range of 800) can be observed in prometaphase chromosomes because they are less compacted than metaphase chromosomes. **Figure 8.1d** shows the conventional numbering system that is used to designate G bands along a set of human chromosomes. The left chromatid in each pair of sister chromatids shows the expected banding pattern during metaphase, and the right chromatid shows the banding pattern as it would appear during prometaphase.

Why is the banding pattern of eukaryotic chromosomes useful? First, when stained, individual chromosomes can be distinguished from each other, even if they have similar sizes and centromeric locations. For example, compare the differences in banding patterns between human chromosomes 8 and 9 (see Figure 8.1d). These differences permit us to distinguish these two chromosomes even though their sizes and centromeric locations are very similar. Banding patterns are also used to detect changes in chromosome structure. As discussed next, chromosomal rearrangements or changes in the total amount of genetic material are more easily detected in banded chromosomes. Also, chromosome banding can be used to assess evolutionary relationships between species. Research studies have shown that the similarity of chromosome banding patterns is a good measure of genetic relatedness.

8.1 COMPREHENSION QUESTIONS

1. A chromosome that is metacentric has its centromere
   a. at the very tip.
   b. near one end, but not at the very tip.
   c. near the middle.
   d. at two distinct locations.
2. Staining eukaryotic chromosomes is useful because it makes it possible to
   a. distinguish chromosomes that are similar in size and centromeric locations.
   b. identify changes in chromosome structure.
   c. explore evolutionary relationships among different species.
   d. do all of the above.

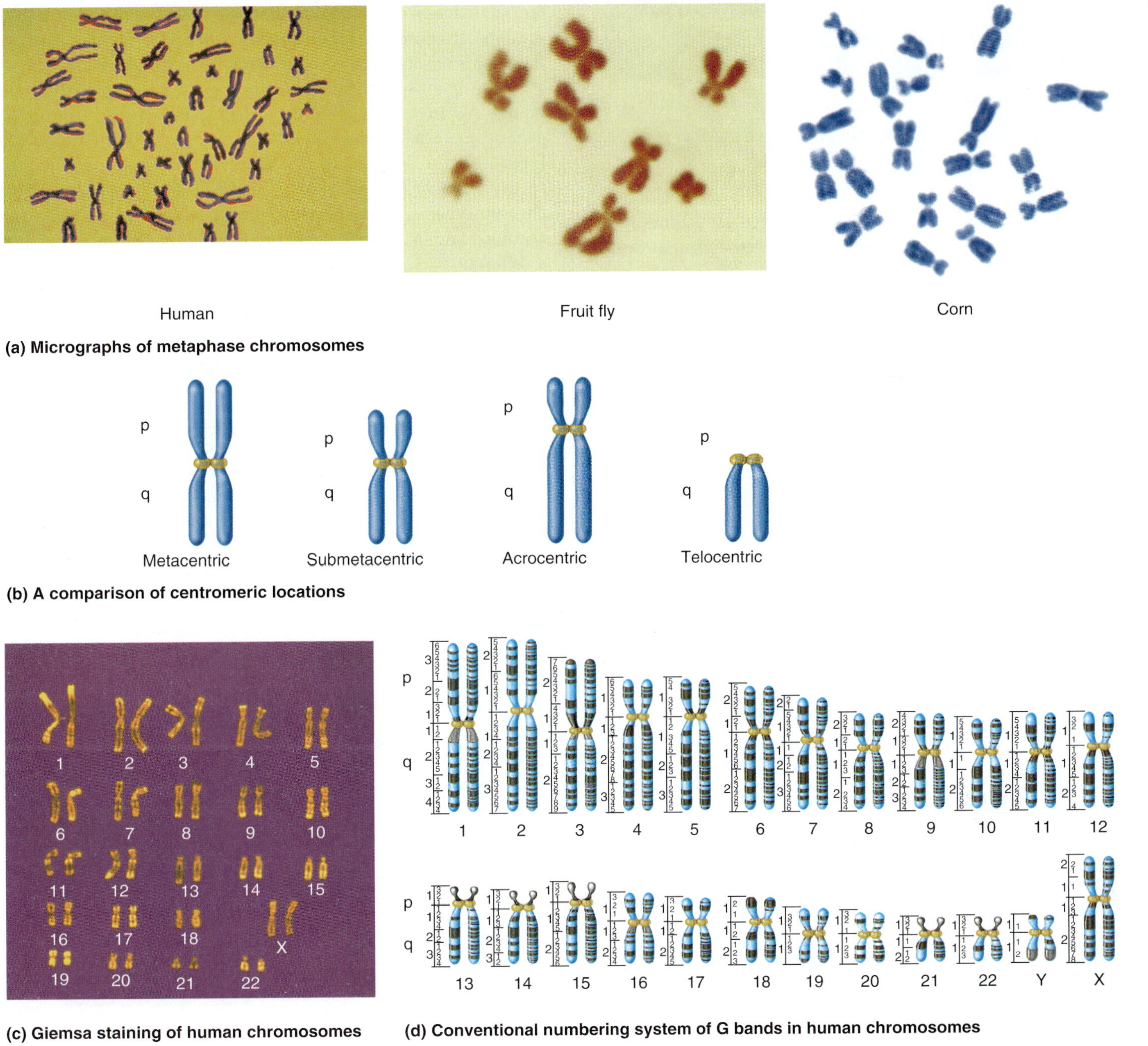

FIGURE 8.1 Features of normal chromosomes. **(a)** Micrographs of chromosomes from a human, a fruit fly, and corn. **(b)** A comparison of centromeric locations. Centromeres can be metacentric, submetacentric, acrocentric (near one end), or telocentric (at the end). **(c)** Human chromosomes that have been stained with Giemsa. This micrograph has been colorized to more easily discern the banding patterns. **(d)** The conventional numbering of bands in Giemsa-stained human chromosomes. The numbering is divided into broad regions, which then are subdivided into smaller regions. The numbers increase as the region gets farther away from the centromere. For example, if you take a look at the left chromatid of chromosome 1, the uppermost dark band is at a location designated p35. The banding patterns of chromatids change as the chromatids condense. The left chromatid of each pair of sister chromatids shows the banding pattern of a chromatid in metaphase, and the right side shows the banding pattern as it would appear in prometaphase. Note: In prometaphase, the chromatids are less compacted than in metaphase.

CONCEPT CHECK: Why is it useful to stain chromosomes?

8.2 CHANGES IN CHROMOSOME STRUCTURE: AN OVERVIEW

Learning Outcome:

1. Compare and contrast the four types of changes in chromosome structure.

Now that we understand that chromosomes typically come in a variety of shapes and sizes, let's consider how the structures of normal chromosomes can be modified by mutation. In some cases, the total amount of genetic material within a single chromosome can be decreased or increased significantly. Alternatively, the genetic material in one or more chromosomes may be rearranged without affecting the total amount of material. As shown in **Figure 8.2**, these mutations are categorized as deletions, duplications, inversions, and translocations.

Deletions and duplications are changes in the total amount of genetic material within a single chromosome. In Figure 8.2, human chromosomes are labeled according to their normal G-banding patterns. When a **deletion** occurs, a segment of chromosomal material is missing. In other words, the affected chromosome is deficient in a significant amount of genetic material. The term **deficiency** is also used to describe a missing region of a chromosome. In contrast, a **duplication** occurs when a section of a chromosome is repeated compared with a normal chromosome.

Inversions and translocations are chromosomal rearrangements. An **inversion** involves a change in the direction of the genetic material along a single chromosome. For example, in Figure 8.2c, a segment of one chromosome has been inverted, so the order of four G bands is opposite to that of the original chromosome. A **translocation** occurs when one segment of a chromosome becomes attached to a different chromosome or to a different part of the same chromosome. A **simple translocation** occurs when a single piece of chromosome is attached to another chromosome. In a **reciprocal translocation,** two different types of chromosomes exchange pieces, thereby producing two abnormal chromosomes carrying translocations.

8.2 COMPREHENSION QUESTION

1. A change in chromosome structure that does *not* involve a change in the total amount of genetic material is
   a. a deletion.
   b. a duplication.
   c. an inversion.
   d. none of the above.

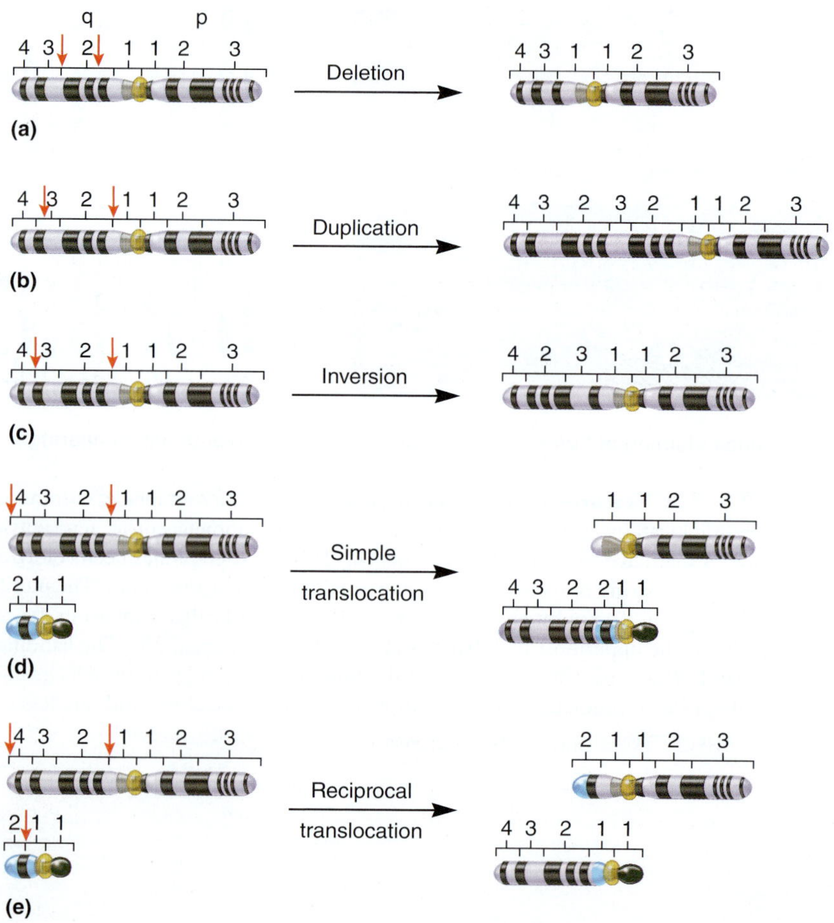

ONLINE ANIMATION

FIGURE 8.2 Types of changes in chromosome structure. The large chromosome shown throughout is human chromosome 1. The smaller chromosome seen in (d) and (e) is human chromosome 21. The red arrows refer to the ends of the affected portion. **(a)** A deletion occurs that removes a large portion of the q2 region, indicated by the red arrows. **(b)** A duplication occurs that doubles the q2–q3 region. **(c)** An inversion occurs that inverts the q2–q3 region. **(d)** The q2–q4 region of chromosome 1 is translocated to chromosome 21. A region of a chromosome cannot be inserted directly to the tip of another chromosome because telomeres at the tips of chromosomes prevent such an event. In this example, a small piece at the end of chromosome 21 must be removed for the q2–q4 region of chromosome 1 to be attached to chromosome 21. **(e)** The q2–q4 region of chromosome 1 is exchanged with the q1–q2 region of chromosome 21.

CONCEPT CHECK: Which of these changes alter the total amount of genetic material?

8.3 DELETIONS AND DUPLICATIONS

Learning Outcomes:

1. Explain how deletions and duplications occur.
2. Describe how deletions and duplications may affect the phenotype of an organism.
3. Define copy number variation.
4. Interpret the results of an experiment that uses the technique of comparative genomic hybridization.

As we have seen, deletions and duplications alter the total amount of genetic material within a chromosome. In this section, we will examine how these changes occur, how the changes are detected experimentally, and how they affect the phenotypes of the individuals that inherit them.

The Loss of Genetic Material in a Deletion Tends to Be Detrimental to an Organism

A chromosomal deletion occurs when a chromosome breaks in one or more places and a fragment of the chromosome is lost. In **Figure 8.3a**, a normal chromosome has broken into two separate pieces. The piece without the centromere is lost and degraded. This event produces a chromosome with a **terminal deletion.** In **Figure 8.3b**, a chromosome has broken in two places to produce three chromosomal fragments. The central fragment is lost, and the two outer pieces reattach to each other. This process has created a chromosome with an **interstitial deletion.** Delections can also happen when recombination takes place at incorrect locations between two homologous chromosomes. The products of this type of aberrant recombination event are one chromosome with a deletion and another chromosome with a duplication. This process is examined later in this chapter.

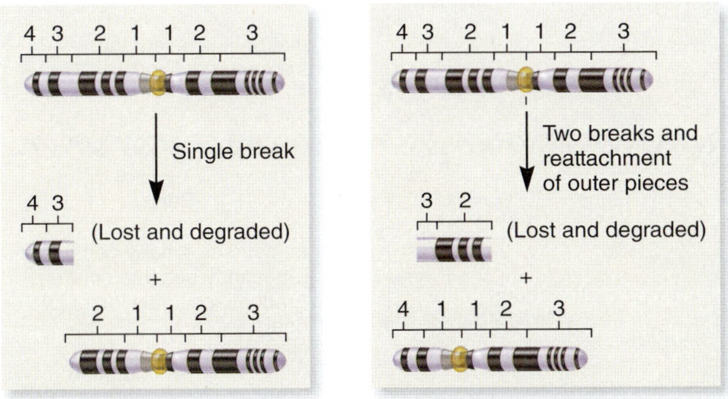

(a) Terminal deletion (b) Interstitial deletion

FIGURE 8.3 Production of terminal and interstitial deletions. This illustration shows the production of deletions in human chromosome 1.

CONCEPT CHECK: Why is a chromosomal fragment without a centromere lost and then subsequently degraded?

(a) Chromosome 5 (b) A child with cri-du-chat syndrome

FIGURE 8.4 Cri-du-chat syndrome. (a) Chromosome 5 from the karyotype of an individual with this disorder. A section of the short arm of one copy of chromosome 5 is missing. **(b)** An affected individual.

Genes → Traits Compared with an individual who has two copies of each gene on chromosome 5, an individual with cri-du-chat syndrome has only one copy of the genes that are located within the missing segment. This genetic imbalance (one versus two copies of many genes on chromosome 5) causes the phenotypic characteristics of this disorder, which include a catlike cry in infancy, short stature, characteristic facial anomalies (e.g., a triangular face, almond-shaped eyes, broad nasal bridge, and low-set ears), and microencephaly (a smaller than normal brain).

The phenotypic consequences of a chromosomal deletion depend on the size of the deletion and whether it includes genes or portions of genes that are vital to the development of the organism. When deletions have a phenotypic effect, they are usually detrimental. Larger deletions tend to be more harmful because more genes are missing. Many examples are known in which deletions have significant phenotypic influences. For example, a human genetic disease known as cri-du-chat, or Lejeune, syndrome is caused by a deletion in a segment of the short arm of human chromosome 5 (**Figure 8.4a**). Individuals who carry a single copy of this abnormal chromosome along with a normal chromosome 5 display an array of abnormalities, including mental deficiencies, unique facial anomalies, and an unusual catlike cry in infancy, which is the meaning of the French name for the syndrome (**Figure 8.4b**). Two other human genetic diseases, Angelman syndrome and Prader-Willi syndrome, which are described in Chapter 5, are due to a deletion in chromosome 15.

Duplications Tend to Be Less Harmful than Deletions

Duplications result in extra genetic material. They are usually caused by abnormal events during recombination. Under normal circumstances, crossing over occurs at analogous sites between homologous chromosomes. On rare occasions, a crossover may occur

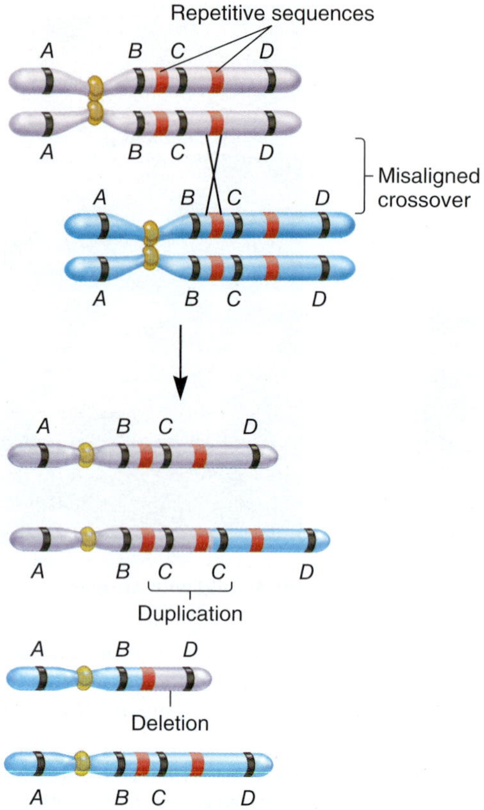

FIGURE 8.5 Nonallelic homologous recombination, leading to a duplication and a deletion. Repetitive sequences, shown in red, have promoted the misalignment of homologous chromosomes. A crossover has occurred at sites between genes *C* and *D* in one chromatid and between genes *B* and *C* in another chromatid. After crossing over is completed, one chromatid contains a duplication, and the other contains a deletion.

CONCEPT CHECK: What is the underlying cause of nonallelic homologous recombination?

at misaligned sites on the homologs (**Figure 8.5**). What causes the misalignment? In some cases, a chromosome may carry two or more homologous segments of DNA that have identical or similar sequences. These are called **repetitive sequences** because they occur multiple times. An example of repetitive sequences are transposable elements, which are described in Chapter 19. In Figure 8.5, the repetitive sequence on the right (in the upper homolog) has lined up with the repetitive sequence on the left (in the lower homolog). A misaligned crossover then occurs. This is called **nonallelic homologous recombination** because it has occurred at homologous sites (i.e., repetitive sequences), but the alleles of neighboring genes are not properly aligned. The result is that one chromatid has a duplication and another chromatid has a deletion. In Figure 8.5, the chromosome with the extra genetic material carries a **gene duplication,** because the number of copies of gene *C* has been increased from one to two. In most cases, gene duplications happen as rare, sporadic events during the evolution of species.

Like deletions, the phenotypic consequences of duplications tend to be correlated with size. Duplications are more likely to have phenotypic effects if they involve a large piece of the chromosome. In general, small duplications are less likely to have harmful effects than are deletions of comparable size. This observation suggests that having only one copy of a gene is more harmful than having three copies. In humans, relatively few well-defined syndromes are caused by small chromosomal duplications. An example is Charcot-Marie-Tooth disease (type 1A), a peripheral neuropathy characterized by numbness in the hands and feet that is caused by a small duplication on the short arm of chromosome 17.

Duplications Provide Additional Material for Gene Evolution, Sometimes Leading to the Formation of Gene Families

In contrast to the gene duplication that causes Charcot-Marie-Tooth disease, the majority of small chromosomal duplications have no phenotypic effect. Nevertheless, they are vitally important because they provide raw material for the addition of more genes into a species' chromosomes. Over the course of many generations, this can lead to the formation of a **gene family** consisting of two or more genes in a particular species that are similar to each other. As shown in **Figure 8.6**, the members of a gene family are derived from the same ancestral gene. Over time, two copies of an ancestral gene can accumulate different mutations. Therefore, after many generations, the two genes will be similar but not identical. During evolution, this type of event can occur several times, creating a family of many similar genes.

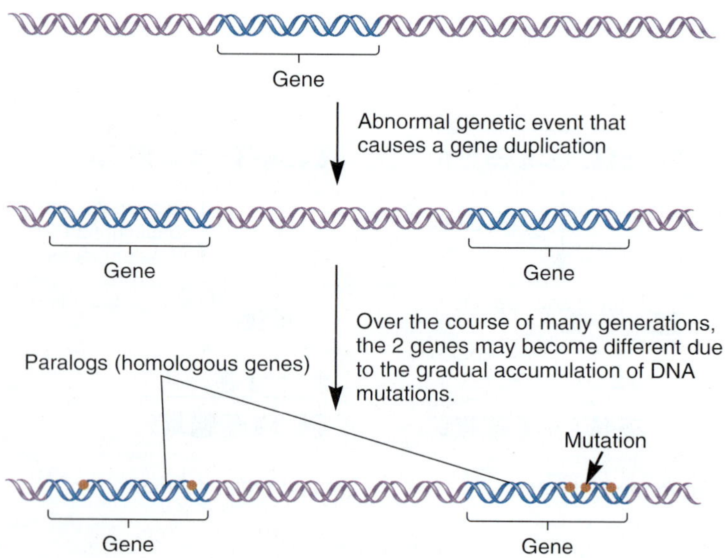

ONLINE ANIMATION

FIGURE 8.6 Gene duplication and the evolution of paralogs. An abnormal crossover event like the one described in Figure 8.5 leads to a gene duplication. Over time, each gene accumulates different mutations.

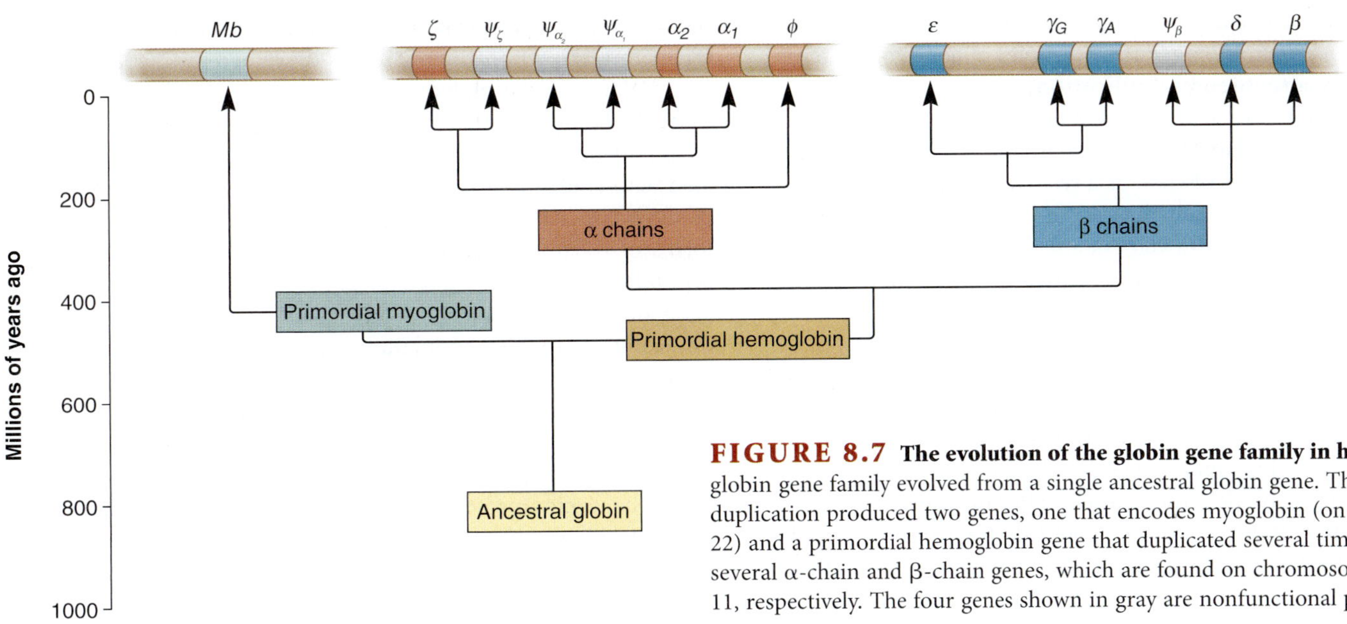

FIGURE 8.7 The evolution of the globin gene family in humans. The globin gene family evolved from a single ancestral globin gene. The first gene duplication produced two genes, one that encodes myoglobin (on chromosome 22) and a primordial hemoglobin gene that duplicated several times to produce several α-chain and β-chain genes, which are found on chromosomes 16 and 11, respectively. The four genes shown in gray are nonfunctional pseudogenes.

When two or more genes are derived from a single ancestral gene, the genes are said to be **homologous.** Homologous genes within a single species are called **paralogs** and constitute a gene family. A well-studied example of a gene family is shown in Figure 8.7, which illustrates the evolution of the globin gene family found in humans. The globin genes encode polypeptides that are subunits of proteins that function in oxygen binding. One such protein is hemoglobin found in red blood cells. The globin gene family is composed of 14 paralogs that were originally derived from a single ancestral globin gene. According to an evolutionary analysis, the ancestral globin gene first duplicated about 500 mya and became separate genes encoding myoglobin and the hemoglobin group of genes. The primordial hemoglobin gene duplicated into an α-chain gene and a β-chain gene, which subsequently duplicated to produce several genes located on chromosomes 16 and 11, respectively. Currently, 14 globin genes are found on three different human chromosomes.

Why is it advantageous to have a family of globin genes? Although all globin polypeptides are subunits of proteins that play a role in oxygen binding, the accumulation of different mutations in the various family members has produced globins that are more specialized in their function. For example, myoglobin is better at binding and storing oxygen in muscle cells, and the hemoglobins are better at binding and transporting oxygen via the red blood cells. Also, different globin genes are expressed during different stages of human development. The ζ- and ε-globin genes are expressed very early in embryonic life, whereas the α-globin and γ-globin genes are expressed during the second and third trimesters of gestation. Following birth, the α-globin gene remains turned on, but the γ-globin genes are turned off and the β-globin gene is turned on. These differences in the expression of the globin genes reflect the differences in the oxygen transport needs of humans during the embryonic, fetal, and postpartum stages of life.

Copy Number Variation Is Relatively Common Among Members of the Same Species

As we have seen, deletions and duplication can alter the number of copies of a given gene. In recent years, geneticists have analyzed the occurrence of deletions and duplications within modern populations. The term **copy number variation (CNV)** refers to a type of structural variation in which a segment of DNA, which is 1000 bp or more in length, commonly exhibits copy number differences among members of the same species. In other words, copy number variation is a phenomenon that occurs at the populational level. One possibility is that some members of a species may carry a chromosome that is missing a particular gene or part of a gene. Alternatively, a CNV may involve a duplication. For example, some members of a diploid species may have one copy of gene *A* on both homologs of a chromosome, and thereby have two copies of the gene (Figure 8.8). By comparison, other members of the same species might have one copy of gene *A* on a

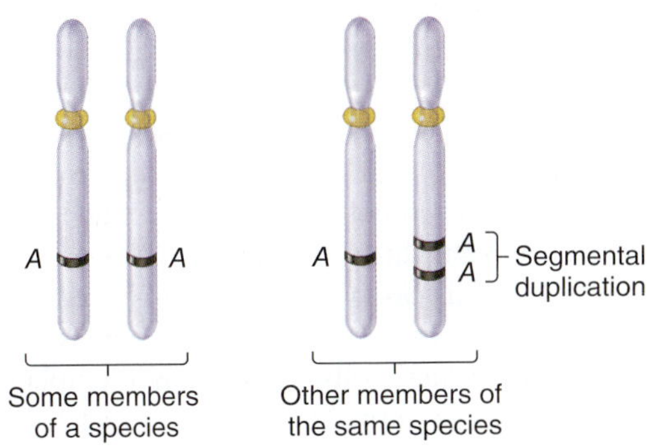

FIGURE 8.8 An example of copy number variation. Among members of the same species, some individuals carry two copies of gene *A* (left side), whereas others carry three copies (right side).

particular chromosome and two copies on its homolog for a total of three copies. The homolog with two copies of gene *A* is said to have undergone a **segmental duplication.**

In the past 10 years, researchers have discovered that copy number variation is relatively common in animal and plant species. Though the analysis of CNV is a relatively new area of investigation, researchers estimate that 0.1%–10% of a genome may show CNV within a typical species of animal or plant. In humans, approximately 0.4% of the genomes of two unrelated individuals typically differ with respect to copy number.

Most CNV is inherited and has happened in the past, but it may also be caused by new mutations. A variety of mechanisms may bring about copy number variation. One common cause is nonallelic homologous recombination, which was described earlier in Figure 8.5. This type of event can produce a chromosome with a duplication or deletion, thereby altering the copy number of genes. Researchers also speculate that the proliferation of transposable elements may increase the copy number of DNA segments. A third mechanism that underlies CNV may involve errors in DNA replication, which is described in Chapter 11.

What are the phenotypic consequences of CNV? In many cases, CNV has no obvious phenotypic consequences. However, recent medical research is revealing that some CNV is associated with specific human diseases. For example, particular types of CNV are associated with schizophrenia, autism, and certain forms of learning disabilities. In addition, CNV may affect susceptibility to infectious diseases. An example is the human *CCL3* gene that encodes a chemokine protein, which is involved in immunity. In human populations, the copy number of this gene varies from one to six. In people infected with HIV (human immunodeficiency virus), copy number variation of *CCL3* may affect the progression of AIDS (acquired immune deficiency syndrome). Individuals with a higher copy number of *CCL3* produce more chemokine protein and often show a slower advancement of AIDS. Finally, another reason why researchers are interested in CNV is its relationship to cancer, which is described next.

EXPERIMENT 8A

Comparative Genomic Hybridization Is Used to Detect Chromosome Deletions and Duplications

As we have seen, chromosome deletions and duplications may influence the phenotypes of individuals who inherit them. One very important reason why researchers have become interested in these types of chromosomal changes is related to cancer. As discussed in Chapter 24, chromosomal deletions and duplications have been associated with many types of human cancers. Though such changes may be detectable by traditional chromosomal staining and karyotyping methods, small deletions and duplications may be difficult to detect in this manner. Fortunately, researchers have been able to develop more sensitive methods for identifying changes in chromosome structure.

In 1992, Anne Kallioniemi, Daniel Pinkel, and colleagues devised a method called **comparative genomic hybridization (CGH).** This technique is largely used to determine if cancer cells have changes in chromosome structure, such as deletions or duplications. To begin this procedure, DNA is isolated from a test sample, which in this case was a sample of breast cancer cells, and also from cells of normal breast tissue (**Figure 8.9**). The DNA from the breast cancer cells was used as a template to make green fluorescent DNA, and the DNA from normal cells was used to make red fluorescent DNA. These green or red DNA molecules averaged 800 bp in length and were made from sites that were scattered all along each chromosome. The green and red DNA molecules were then denatured by heat treatment. Equal amounts of the two fluorescently labeled DNA samples were mixed together and applied to normal metaphase chromosomes from white blood cells in which the DNA had also been denatured. Because the fluorescently labeled DNA fragments and the metaphase chromosomes had both been denatured, the fluorescently labeled DNA strands can bind to complementary regions on the metaphase chromosomes. This process is called **hybridization** because the DNA from one sample (a green or red DNA strand) forms a double-stranded region with a DNA strand from another sample (an unlabeled metaphase chromosome). Following hybridization, the metaphase chromosomes were visualized using a fluorescence microscope, and the images were analyzed by a computer that can determine the relative intensities of green and red fluorescence. What are the expected results? If a chromosomal region in the breast cancer cells and the normal cells are present in the same amount, the ratio between green and red fluorescence should be 1. If a chromosomal region is deleted in the breast cancer cell line, the ratio will be less than 1, or if a region is duplicated, it will be greater than 1.

THE GOAL (DISCOVERY-BASED SCIENCE)

Deletions or duplications in cancer cells can be detected by comparing the ability of fluorescently labeled DNA from cancer cells and normal cells to bind (hybridize) to normal metaphase chromosomes.

ACHIEVING THE GOAL **FIGURE 8.9** **The use of comparative genomic hybridization to detect deletions and duplications in cancer cells.**

Starting materials: Breast cancer cells and normal cells.

1. Isolate DNA from human breast cancer cells and normal breast cells. This involves breaking open the cells and isolating the DNA by chromatography. (See Appendix for a description of chromatography.)

2. Label the breast cancer DNA with a green fluorescent molecule and the normal DNA with a red fluorescent molecule. This is done by using the DNA from step 1 as a template, and incorporating fluorescently labeled nucleotides into newly made DNA strands.

3. The DNA strands are then denatured by heat treatment. Mix together equal amounts of fluorescently labeled DNA and add it to a preparation of metaphase chromosomes from white blood cells. The procedure for preparing metaphase chromosomes is described in Figure 3.2. The metaphase chromosomes were also denatured.

4. Allow the fluorescently labeled DNA to hybridize to the metaphase chromosomes.

5. Visualize the chromosomes with a fluorescence microscope. Analyze the amount of green and red fluorescence along each chromosome with a computer.

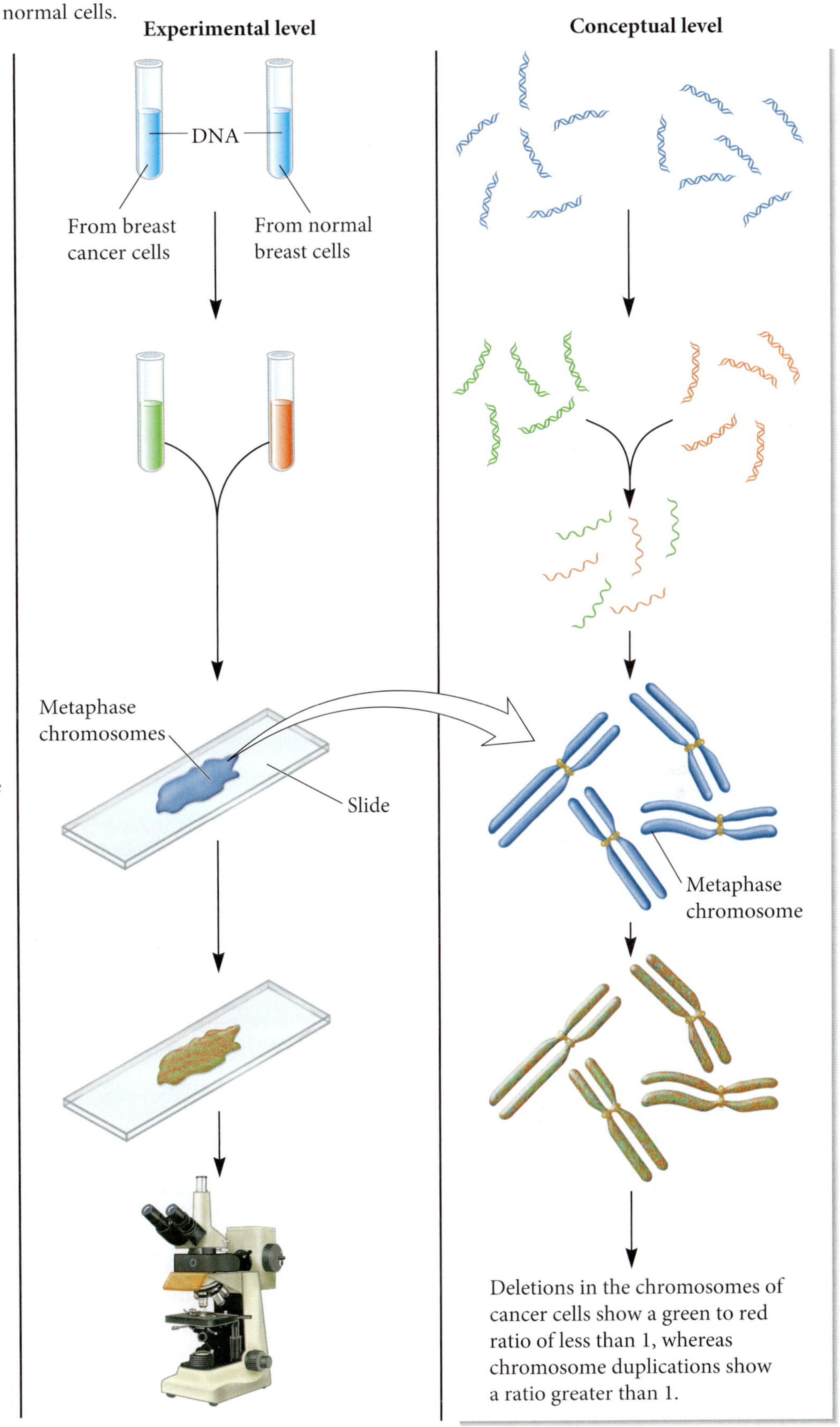

THE DATA

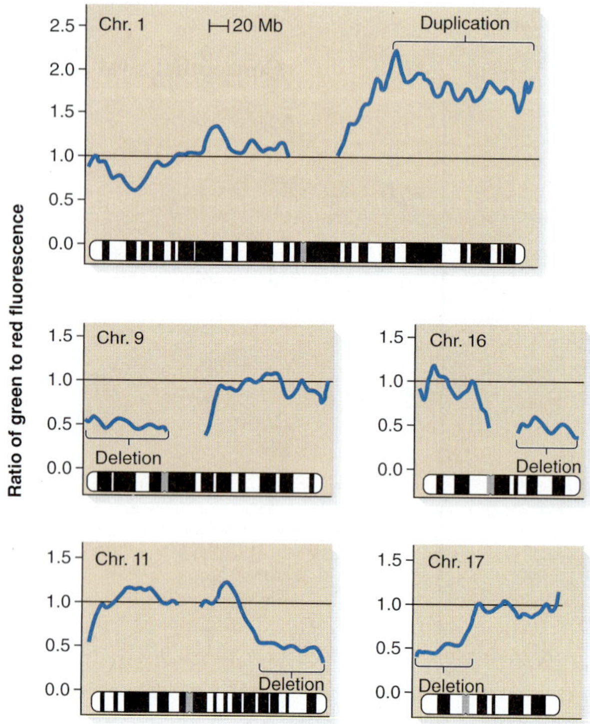

Note: Unlabeled repetitive DNA was also included in this experiment to decrease the level of nonspecific, background labeling. This repetitive DNA also prevents labeling near the centromere. As seen in The Data, regions in the chromosomes where the curves are missing are due to the presence of highly repetitive sequences near the centromere. Data from A. Kallioniemi, O. P. Kallioniemi, D. Sudar, et al. (1992) Comparative genomic hybridization for molecular cytogenetic analysis of solid tumors. *Science 258*, 818–821.

INTERPRETING THE DATA

The Data of Figure 8.9 show the ratio of green (cancer DNA) to red (normal DNA) fluorescence along five different metaphase chromosomes. Chromosome 1 shows a large duplication, as indicated by the ratio of 2. One interpretation of this observation is that both copies of chromosome 1 carry a duplication. In comparison, chromosomes 9, 11, 16, and 17 have regions with a value of 0.5. This value indicates that one of the two chromosomes of these four types in the cancer cells carries a deletion, but the other chromosome does not. (A value of 0 would indicate both copies of a chromosome had deleted the same region.) Overall, these results illustrate how this technique can be used to map chromosomal duplications and deletions in cancer cells.

This method is named comparative genomic hybridization because a comparison is made between the ability of two DNA samples (cancer versus normal cells) to hybridize to an entire genome. In this case, the entire genome is in the form of metaphase chromosomes. As discussed in Chapter 22, the fluorescently labeled DNAs can be hybridized to a DNA microarray instead of metaphase chromosomes. This newer method, called array comparative genomic hybridization (aCGH), is gaining widespread use in the analysis of cancer cells.

A self-help quiz involving this experiment can be found at www.mhhe.com/brookergenetics5e.

8.3 COMPREHENSION QUESTIONS

1. Which of the following statements is correct?
   a. If a deletion and a duplication are the same size, the deletion is more likely to be harmful.
   b. If a deletion and a duplication are the same size, the duplication is more likely to be harmful.
   c. If a deletion and a duplication are the same size, the likelihood of causing harm is about the same.
   d. A deletion is always harmful, whereas a duplication is always beneficial.
2. With regard to gene duplications, which of the following statement(s) is/are correct?
   a. Gene duplications may be caused by nonallelic homologous recombination.
   b. Large gene duplications are more likely to be harmful than smaller ones.
   c. Gene duplications are responsible for creating gene families that encode proteins with similar and specialized functions.
   d. All of the above statements are correct.

8.4 INVERSIONS AND TRANSLOCATIONS

Learning Outcomes:

1. Define pericentric and paracentric inversions.
2. Diagram how inversion heterozygotes produce abnormal chromosomes due to crossing over.
3. Explain two mechanisms that result in reciprocal translocations.
4. Describe how reciprocal translocations align during meiosis and how they segregate.

As discussed earlier in this chapter, inversions and translocations are types of chromosomal rearrangements. In this section, we will explore how they occur and how they may affect an individual's phenotype or fertility.

Inversions Often Occur Without Phenotypic Consequences

A chromosome with an inversion contains a segment that has been flipped to the opposite direction. Geneticists classify inversions according to the location of the centromere. If the centromere

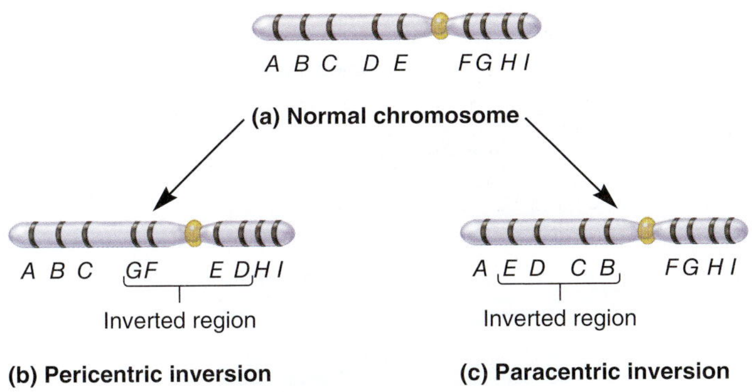

FIGURE 8.10 **Types of inversions.** (a) Depicts a normal chromosome with the genes ordered from *A* through *I*. A pericentric inversion (b) includes the centromere, whereas a paracentric inversion (c) does not.

lies within the inverted region of the chromosome, the inverted region is known as a **pericentric inversion** (**Figure 8.10b**). Alternatively, if the centromere is found outside the inverted region, the inverted region is called a **paracentric inversion** (**Figure 8.10c**).

When a chromosome contains an inversion, the total amount of genetic material remains the same as in a normal chromosome. Therefore, the great majority of inversions do not have any phenotypic consequences. In rare cases, however, an inversion can alter the phenotype of an individual. Whether or not this occurs is related to the boundaries of the inverted segment. When an inversion occurs, the chromosome is broken in two places, and the center piece flips around to produce the inversion. If either breakpoint occurs within a vital gene, the function of the gene is expected to be disrupted, possibly producing a phenotypic effect. For example, some people with hemophilia (type A) have inherited an X-linked inversion in which the breakpoint has inactivated the gene for factor VIII—a blood-clotting protein. In other cases, an inversion (or translocation) may reposition a gene on a chromosome in a way that alters its normal level of expression. This is a type of **position effect**—a change in phenotype that occurs when the position of a gene changes from one chromosomal site to a different location. This topic is also discussed in Chapter 18 (see Figures 18.2 and 18.3).

Because inversions seem like an unusual genetic phenomenon, it is perhaps surprising that they are found in human populations in significant numbers. About 2% of the human population carries inversions that are detectable with a light microscope. In most cases, such individuals are phenotypically normal and live their lives without knowing they carry an inversion. In a few cases, however, an individual with an inversion chromosome may produce offspring with phenotypic abnormalities. This event may prompt a physician to request a microscopic examination of the individual's chromosomes. In this way, phenotypically normal individuals may discover they have a chromosome with an inversion. Next, we will examine how an individual carrying an inversion may produce offspring with phenotypic abnormalities.

Inversion Heterozygotes May Produce Abnormal Chromosomes Due to Crossing Over

An individual carrying one copy of a normal chromosome and one copy of an inverted chromosome is known as an **inversion heterozygote.** Such an individual, though possibly phenotypically normal, may have a high probability of producing haploid cells that are abnormal in their total genetic content.

The underlying cause of gamete abnormality is the phenomenon of crossing over within the inverted region. During meiosis I, pairs of homologous sister chromatids synapse with each other. **Figure 8.11** illustrates how this occurs in an inversion heterozygote. For the normal chromosome and inversion chromosome to synapse properly, an **inversion loop** must form to permit the homologous genes on both chromosomes to align next to each other despite the inverted sequence. If a crossover occurs within the inversion loop, highly abnormal chromosomes are produced. A crossover is more likely to occur in this region if the inversion is large. Therefore, individuals carrying large inversions are more likely to produce abnormal gametes.

The consequences of this type of crossover depend on whether the inversion is pericentric or paracentric. Figure 8.11a describes a crossover in the inversion loop when one of the homologs has a pericentric inversion in which the centromere lies within the inverted region of the chromosome. This event consists of a single crossover that involves only two of the four sister chromatids. Following the completion of meiosis, this single crossover yields two abnormal chromosomes. Both of these abnormal chromosomes have a segment that is deleted and a different segment that is duplicated. In this example, one of the abnormal chromosomes is missing genes *H* and *I* and has an extra copy of genes *A*, *B*, and *C*. The other abnormal chromosome has the opposite situation; it is missing genes *A*, *B*, and *C* and has an extra copy of genes *H* and *I*. These abnormal chromosomes may result in gametes that are inviable. Alternatively, if these abnormal chromosomes are passed to offspring, they are likely to produce phenotypic abnormalities, depending on the amount and nature of the duplicated and deleted genetic material. A large deletion is likely to be lethal.

Figure 8.11b shows the outcome of a crossover involving a paracentric inversion in which the centromere lies outside the inverted region. This single crossover event produces a very strange outcome. One product is a piece of chromosome without any centromere—an **acentric fragment,** which is lost and degraded in subsequent cell divisions. The other product is a **dicentric** chromosome that contains two centromeres. The region of the chromosome connecting the two centromeres is a **dicentric bridge.** The dicentric chromosome is a temporary condition. If the two centromeres try to move toward opposite poles during anaphase, the dicentric bridge will be forced to break at some random location. Therefore, the net result of this crossover is to produce one normal chromosome, one chromosome with an inversion, and two chromosomes that contain deletions. These two chromosomes with deletions result from the breakage of the dicentric chromosome.

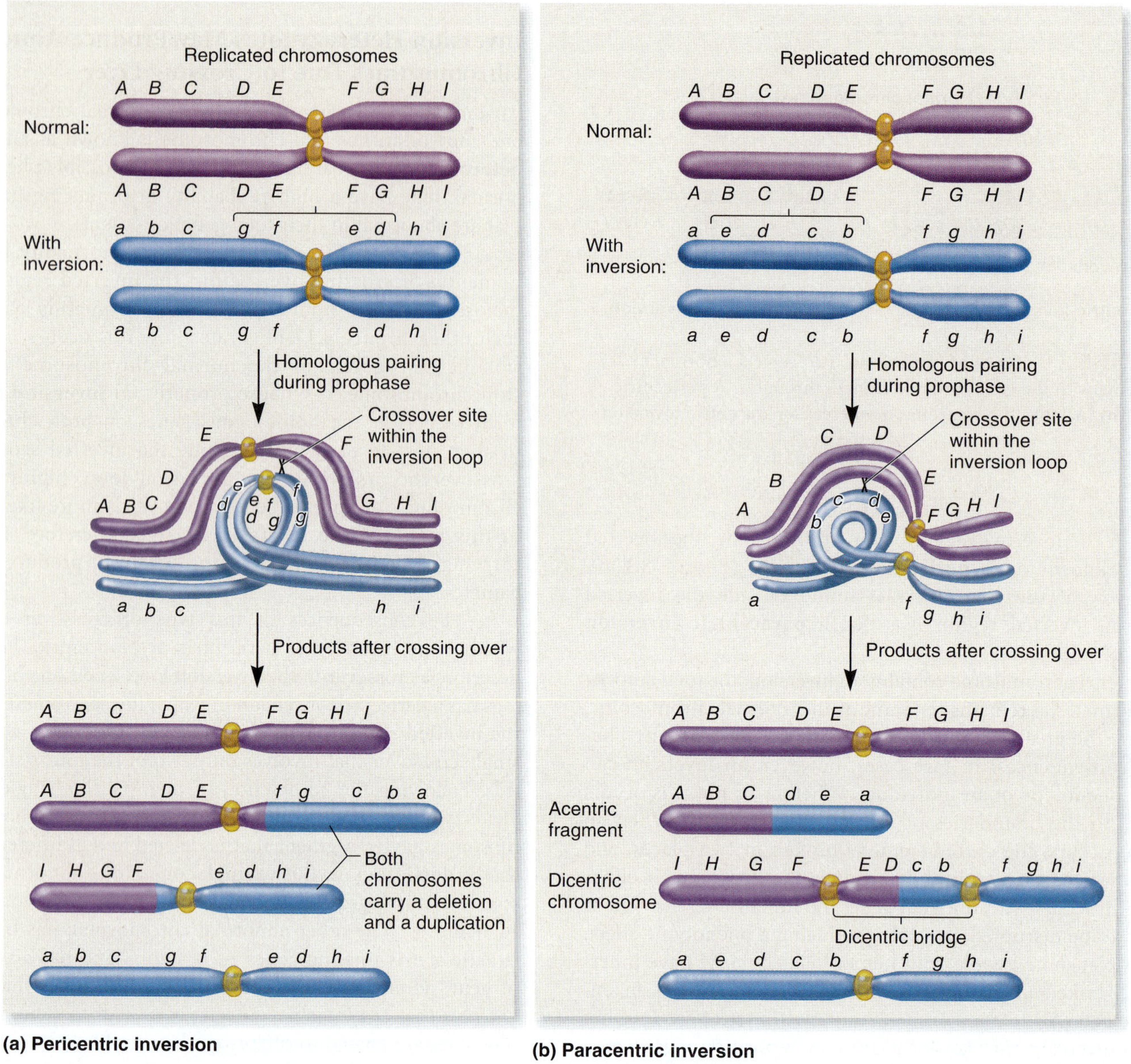

FIGURE 8.11 The consequences of crossing over in the inversion loop. (a) Crossover within a pericentric inversion. (b) Crossover within a paracentric inversion.

CONCEPT CHECK: Explain why these chromosomes can synapse only if an inversion loop forms.

Translocations Involve Exchanges Between Different Chromosomes

Another type of chromosomal rearrangement is a translocation in which a piece from one chromosome is attached to another chromosome. Eukaryotic chromosomes have telomeres, which tend to prevent translocations from occurring. As described in Chapters 10 and 11, **telomeres**—specialized repeated sequences of DNA—are found at the ends of normal chromosomes. Telomeres allow cells to identify where a chromosome ends and prevent the attachment of chromosomal DNA to the natural ends of a chromosome.

If cells are exposed to agents that cause chromosomes to break, the broken ends lack telomeres and are said to be reactive—a reactive end readily binds to another reactive end. If a single chromosome break occurs, DNA repair enzymes will usually recognize the two reactive ends and join them back together; the chromosome is repaired properly. However, if multiple chromosomes are broken, the reactive ends may be joined incorrectly to produce abnormal chromosomes (Figure 8.12a). This is one mechanism that causes reciprocal translocations to occur. A second mechanism occurs when two nonhomologous chromosomes cross over (Figure 8.12b).

The reciprocal translocations we have considered thus far are also called **balanced translocations** because the total amount of genetic material is not altered. Like inversions, balanced translocations usually occur without any phenotypic consequences because the individual has a normal amount of genetic material. In a few cases, balanced translocations can result in position

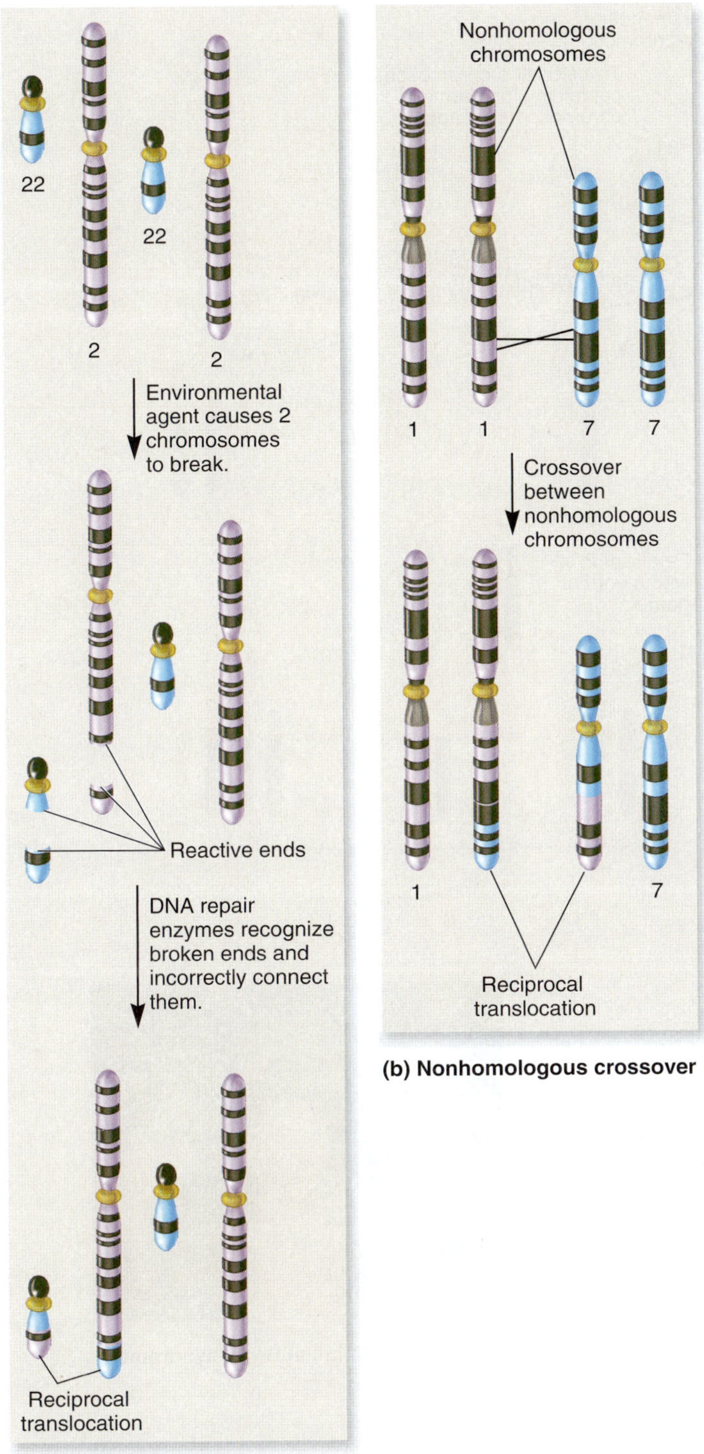

(a) Chromosomal breakage and DNA repair

(b) Nonhomologous crossover

FIGURE 8.12 Two mechanisms that cause a reciprocal translocation. **(a)** When two different chromosomes break, the reactive ends are recognized by DNA repair enzymes, which attempt to reattach them. If two different chromosomes are broken at the same time, the broken pieces may reattach incorrectly. **(b)** A nonhomologous crossover has occurred between chromosome 1 and chromosome 7. This crossover yields two chromosomes that carry translocations.

CONCEPT CHECK: Which of these two mechanisms might be promoted by the presence of the same transposable element in many places in a species genome?

effects similar to those that can occur in inversions. In addition, carriers of a balanced translocation are at risk of having offspring with an **unbalanced translocation,** in which significant portions of genetic material are duplicated and/or deleted. Unbalanced translocations are generally associated with phenotypic abnormalities or even lethality.

Let's consider how a person with a balanced translocation may produce gametes and offspring with an unbalanced translocation. An inherited human syndrome known as familial Down syndrome provides an example. Due to a translocation, a person may have one copy of chromosome 14, one copy of chromosome 21, and one copy of a chromosome that is a fusion between chromosome 14 and 21 (**Figure 8.13a**). The individual has a normal phenotype because the total amount of genetic material is present (with the exception of the short arms of these chromosomes that do not carry vital genetic material). During meiosis, these three types of chromosomes replicate and segregate from each other. However, because the three chromosomes cannot segregate evenly, six possible types of gametes may be produced. One gamete is normal, and one is a balanced carrier of a translocated chromosome. The four gametes to the right, however, are unbalanced, either containing too much or too little material from chromosome 14 or 21. The unbalanced gametes may be nonviable, or they could combine with a normal gamete. The three offspring on the right will not survive. In comparison, the unbalanced gamete that carries chromosome 21 and the fused chromosome results in an offspring with familial Down syndrome (also see karyotype in **Figure 8.13b**). Such an offspring has three copies of the genes that are found on the long arm of chromosome 21. **Figure 8.13c** shows a person with this disorder. She has characteristics similar to those of an individual who has the more prevalent form of Down syndrome, which is due to three entire copies of chromosome 21. We will examine this common form of Down syndrome later in this chapter.

The abnormal chromosome that occurs in familial Down syndrome is an example of a **Robertsonian translocation,** named after William Robertson, who first described this type of fusion in grasshoppers. This type of translocation arises from breaks near the centromeres of two nonhomologous acrocentric chromosomes. In the example shown in Figure 8.13, the long arms of chromosomes 14 and 21 have fused, creating one large single chromosome; the two short arms are lost. This type of translocation between two nonhomologous acrocentric chromosomes is the most common type of chromosome rearrangement in humans, occurring at a frequency of approximately 1 in 900 live births. In humans, Robertsonian translocations involve only the acrocentric chromosomes 13, 14, 15, 21, and 22.

Individuals with Reciprocal Translocations May Produce Abnormal Gametes Due to the Segregation of Chromosomes

As we have seen, individuals who carry balanced translocations have a greater risk of producing gametes with unbalanced combinations of chromosomes. Whether or not this occurs depends on the segregation pattern during meiosis I (**Figure 8.14**). In this example, the parent carries a reciprocal translocation and is

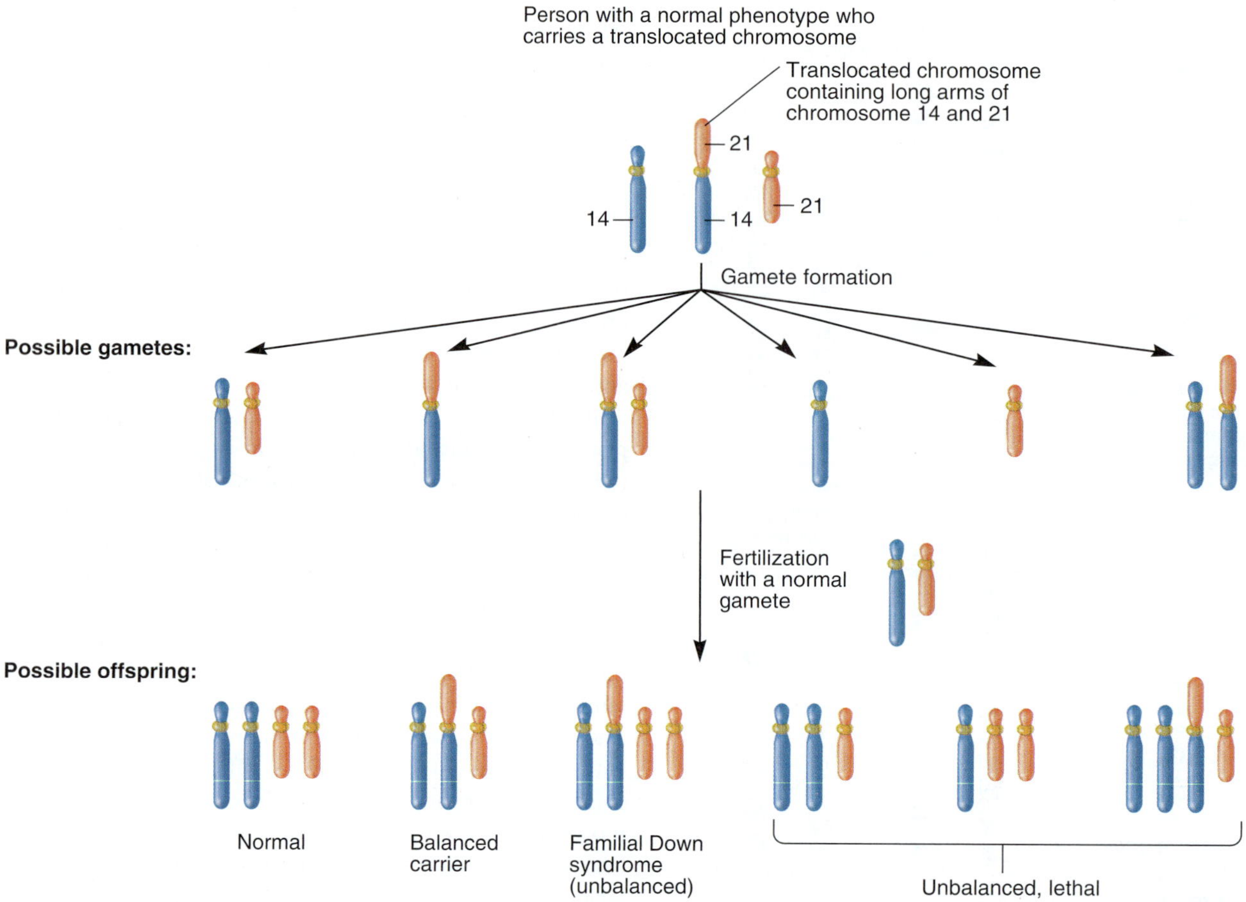

(a) Possible transmission patterns

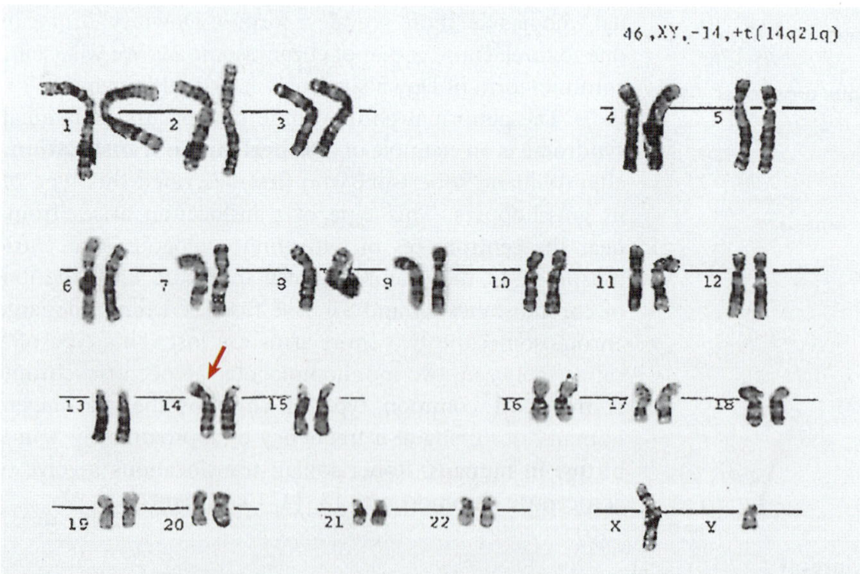

(b) Karyotype of a male with familial Down syndrome

(c) Child with familial Down syndrome

FIGURE 8.13 **Transmission of familial Down syndrome.** **(a)** Potential transmission of familial Down syndrome. The individual with the chromosome composition shown at the top of this figure may produce a gamete carrying chromosome 21 and a fused chromosome containing the long arms of chromosomes 14 and 21. Such a gamete can give rise to an offspring with familial Down syndrome. **(b)** The karyotype of an individual with familial Down syndrome. This karyotype shows that the long arm of chromosome 21 has been translocated to chromosome 14 (see arrow). In addition, the individual also carries two normal copies of chromosome 21. **(c)** An individual (in the center) with this disorder.

CONCEPT CHECK: If these segregation patterns are equally likely, what is the probability that a gamete made from this individual will result in a viable offspring with a normal phenotype?

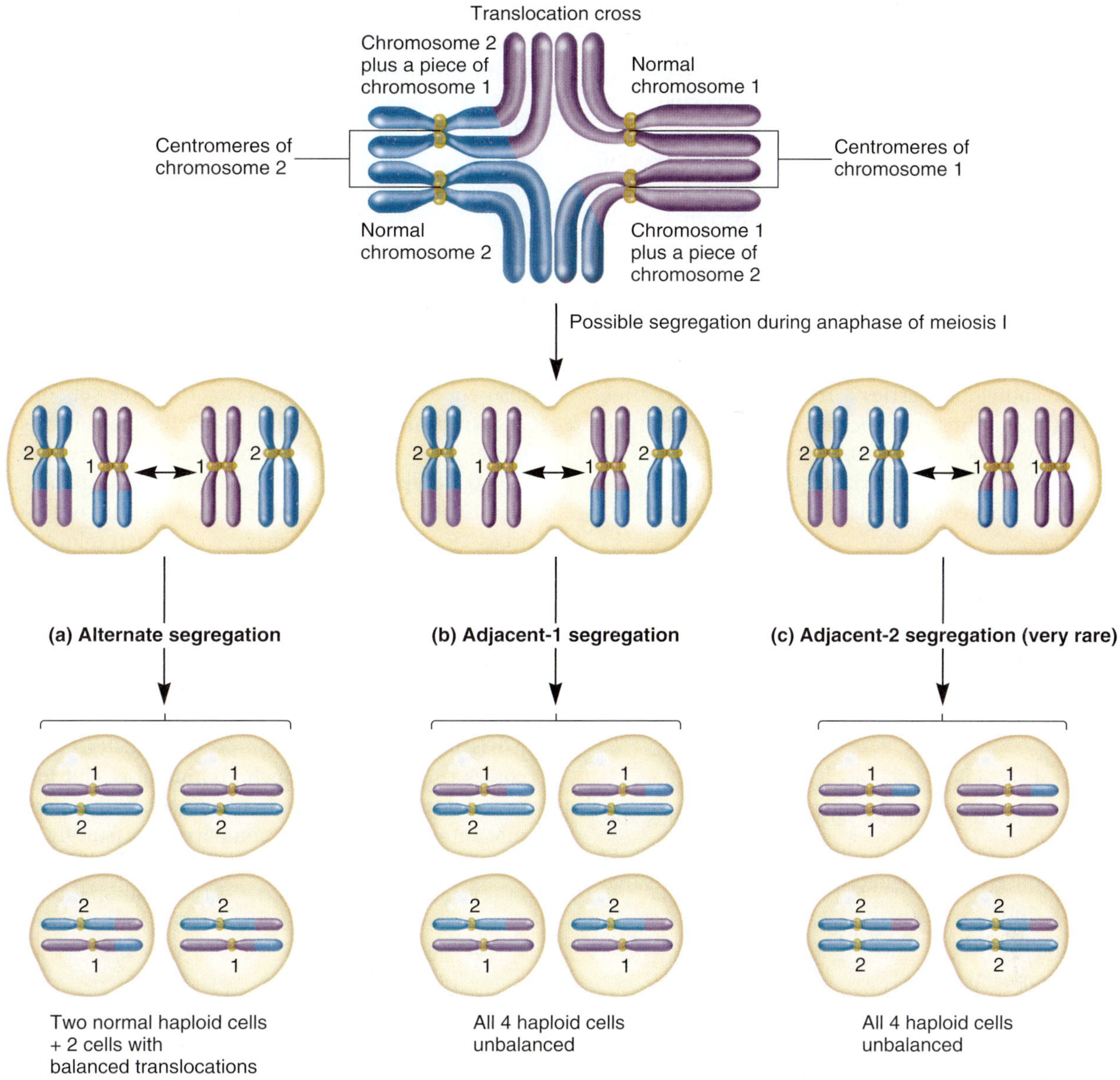

FIGURE 8.14 Meiotic segregation of a reciprocal translocation. Follow the numbered centromeres through each process. **(a)** Alternate segregation gives rise to balanced haploid cells, whereas **(b)** adjacent-1 segregation and **(c)** adjacent-2 segregation produce haploid cells with an unbalanced amount of genetic material.

CONCEPT CHECK: Explain why these chromosomes form a translocation cross during prophase of meiosis I.

likely to be phenotypically normal. During meiosis, the homologous chromosomes attempt to synapse with each other. Because of the translocations, the pairing of homologous regions leads to the formation of an unusual structure that contains four pairs of sister chromatids (i.e., eight chromatids), termed a **translocation cross.**

To understand the segregation of translocated chromosomes, pay close attention to the centromeres, which are numbered in Figure 8.14. For these translocated chromosomes, the expected segregation pattern is governed by the centromeres. Each haploid cell should receive one centromere located on chromosome 1 and one centromere located on chromosome 2. This occurs in two ways. One possibility is alternate segregation. As shown in Figure 8.14a, the chromosomes diagonal to each other within the translocation cross sort into the same cell. One daughter cell receives two normal chromosomes, and the other cell gets two translocated chromosomes. Following meiosis II, four haploid cells are produced: two have normal chromosomes, and two have reciprocal (balanced) translocations.

Another possible segregation pattern is called adjacent-1 segregation (Figure 8.14b). In this case, adjacent chromosomes (one of each type of centromere) sort into the same cell. Each daughter cell receives one normal chromosome and one translocated chromosome. After meiosis II is completed, four haploid cells are produced, all of which are genetically unbalanced because part of one chromosome is missing and part of another is duplicated. If these haploid cells give rise to gametes that unite with a normal gamete, the zygote is expected to be abnormal genetically and possibly phenotypically.

On very rare occasions, adjacent-2 segregation can occur (Figure 8.14c). In this case, the centromeres do not segregate as they should. One daughter cell has received both copies of the centromere on chromosome 1; the other, both copies of the centromere on chromosome 2. This rare segregation pattern also yields four abnormal haploid cells that contain an unbalanced combination of chromosomes.

Alternate and adjacent-1 segregation patterns are the likely outcomes when an individual carries a reciprocal translocation. Depending on the sizes of the translocated segments, both types may be equally likely to occur. In many cases, the haploid cells from adjacent-1 segregation are not viable, thereby lowering the fertility of the parent. This condition is called **semisterility.**

8.4 COMPREHENSION QUESTIONS

1. A paracentric inversion
   a. includes the centromere within the inverted region.
   b. does not include the centromere within the inverted region.
   c. has two adjacent inverted regions.
   d. has an inverted region at the very end of a chromosome.
2. Due to crossing over within an inversion loop, a heterozygote with a pericentric inversion may produce gametes that
   a. carry a deletion.
   b. carry a duplication.
   c. carry a translocation.
   d. include both a and b.
3. A mechanism that may cause a translocation is
   a. the joining of reactive ends when two different chromosomes break.
   b. crossing over between nonhomologous chromosomes.
   c. crossing over between homologous chromosomes.
   d. both a and b.

8.5 CHANGES IN CHROMOSOME NUMBER: AN OVERVIEW

Learning Outcomes:

1. Define euploid and aneuploid.
2. Compare and contrast polyploidy and aneuploidy.

As we saw in Sections 8.1–8.4, chromosome structure can be altered in a variety of ways. Likewise, the total number of chromosomes can vary. Eukaryotic species typically contain several chromosomes that are inherited as one or more sets. Variations in chromosome number can be categorized in two ways: variation in the number of sets of chromosomes and variation in the number of particular chromosomes within a set.

Organisms that are **euploid** have a chromosome number that is an exact multiple of a chromosome set. In *Drosophila melanogaster*, for example, a single set is composed of 4 different chromosomes. The species is diploid, having two sets of 4 chromosomes each (**Figure 8.15a**). A normal fruit fly is euploid because 8 chromosomes divided by 4 chromosomes per set equals two exact sets. On rare occasions, an abnormal fruit fly can be produced with 12 chromosomes, containing three sets of 4 chromosomes each. This alteration in euploidy produces a **triploid** fruit fly with 12 chromosomes. Such a fly is also euploid because it has exactly three sets of chromosomes. A **tetraploid** fly with 4 sets of chromosomes is also euploid. Organisms with three or more sets of chromosomes are also called **polyploid** (**Figure 8.15b**). Geneticists use the letter n to represent a set of chromosomes. A diploid organism is referred to as $2n$, a triploid organism as $3n$, a tetraploid organism as $4n$, and so on.

A second way in which chromosome number can vary is by **aneuploidy.** Such variation involves an alteration in the number of particular chromosomes, so the total number of chromosomes is not an exact multiple of a set. For example, an abnormal fruit fly could contain nine chromosomes instead of eight because it has three copies of chromosome 2 instead of the normal two copies (**Figure 8.15c**). Such an animal is said to have trisomy 2 or to be **trisomic.** Instead of being perfectly diploid ($2n$), a trisomic animal is $2n + 1$. By comparison, a fruit fly could be lacking a single chromosome, such as chromosome 1, and contain a total of seven chromosomes ($2n - 1$). This animal is **monosomic** and is described as having monosomy 1.

8.5 COMPREHENSION QUESTION

1. Humans have 23 chromosomes per set. A person with 45 chromosomes could be described as being
   a. euploid.
   b. aneuploid.
   c. monoploid.
   d. trisomic.

8.6 VARIATION IN THE NUMBER OF CHROMOSOMES WITHIN A SET: ANEUPLOIDY

Learning Outcomes:

1. Explain why aneuploidy usually has a detrimental effect on phenotype.
2. Describe examples of aneuploidy in humans.

In this section, we will consider several examples of aneuploidy. As you will learn, this is generally regarded as an abnormal condition that usually has a negative effect on phenotype.

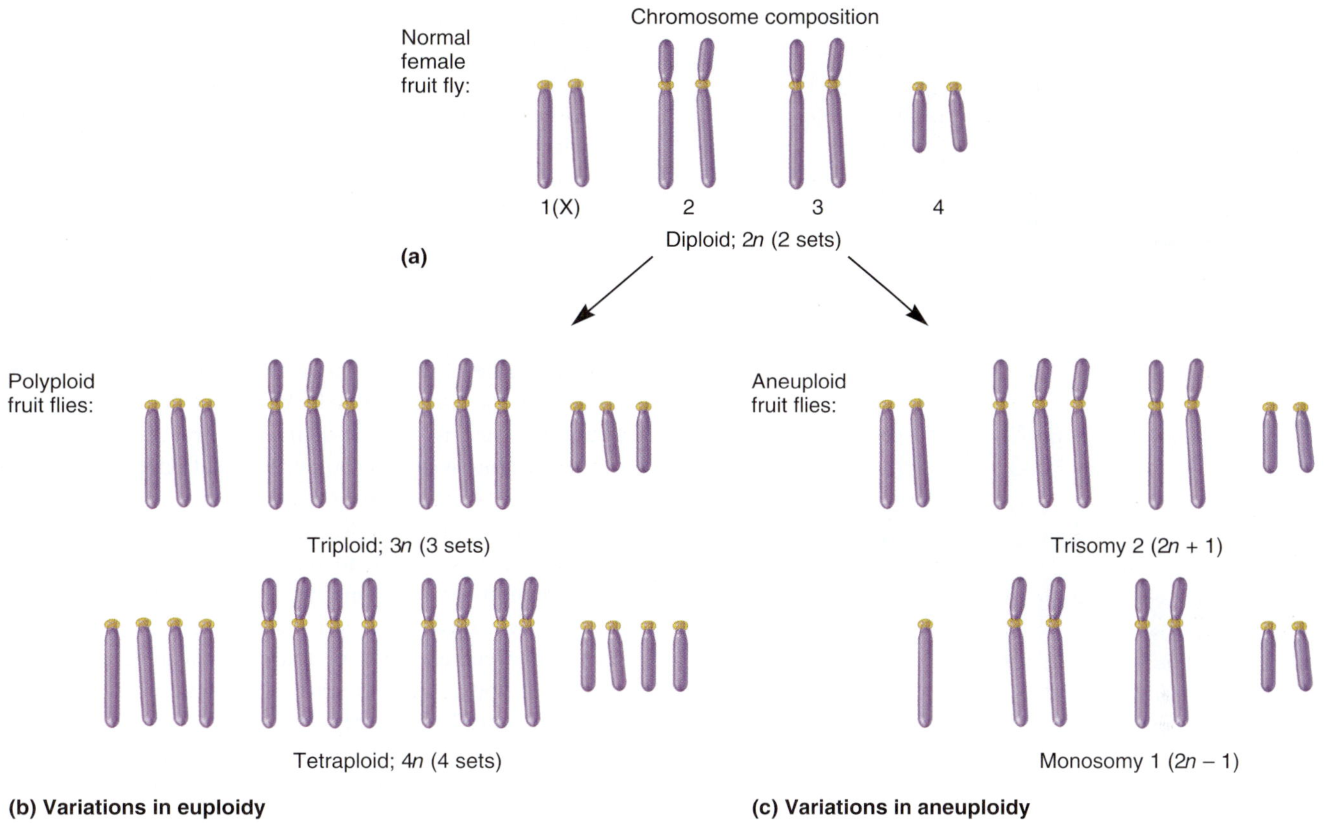

FIGURE 8.15 **Types of variation in chromosome number.** **(a)** Depicts the normal diploid number of chromosomes in *Drosophila*. **(b)** Examples of polyploidy. **(c)** Examples of aneuploidy.

CONCEPT CHECK: What adjectives can be used to describe a fruit fly that has a total of seven chromosomes because it is missing one copy of chromosome 3?

Aneuploidy Causes an Imbalance in Gene Expression That Is Often Detrimental to the Phenotype of the Individual

The phenotype of every eukaryotic species is influenced by thousands of different genes. In humans, for example, a single set of chromosomes contains approximately 22,000 different genes. To produce a phenotypically normal individual, the expression of thousands of genes requires intricate coordination. In the case of humans and other diploid species, evolution has resulted in a developmental process that works correctly when somatic cells have two copies of each chromosome. In other words, when a human is diploid, the balance of gene expression among many different genes usually produces a person with a normal phenotype.

Aneuploidy commonly causes an abnormal phenotype. To understand why, let's consider the relationship between gene expression and chromosome number in a species that has three pairs of chromosomes (**Figure 8.16**). The level of gene expression is influenced by the number of genes per cell. Compared with a diploid cell, if a gene is carried on a chromosome that is present in three copies instead of two, more of the gene product is typically made. For example, a gene present in three copies instead of two may produce 150% of the gene product, though that number may vary due to effects of gene regulation. Alternatively, if only one copy of that gene is present due to a missing chromosome, less of the gene product is usually made, perhaps only 50%. Therefore, in trisomic and monosomic individuals, an imbalance occurs between the level of gene expression on the chromosomes found in pairs versus the ones with extra or missing copies.

At first glance, the difference in gene expression between euploid and aneuploid individuals may not seem terribly dramatic. Keep in mind, however, that a eukaryotic chromosome carries hundreds or even thousands of different genes. Therefore, when an organism is trisomic or monosomic, many gene products occur in excessive or deficient amounts. This imbalance among many genes appears to underlie the abnormal phenotypic effects that aneuploidy frequently causes. In most cases, these effects are detrimental and produce an individual that is less likely to survive than a euploid individual.

Aneuploidy in Humans Causes Abnormal Phenotypes

A key reason why geneticists are so interested in aneuploidy is its relationship to certain inherited disorders in humans. Even though most people are born with a normal number of chromosomes (i.e., 46), alterations in chromosome number occur fairly

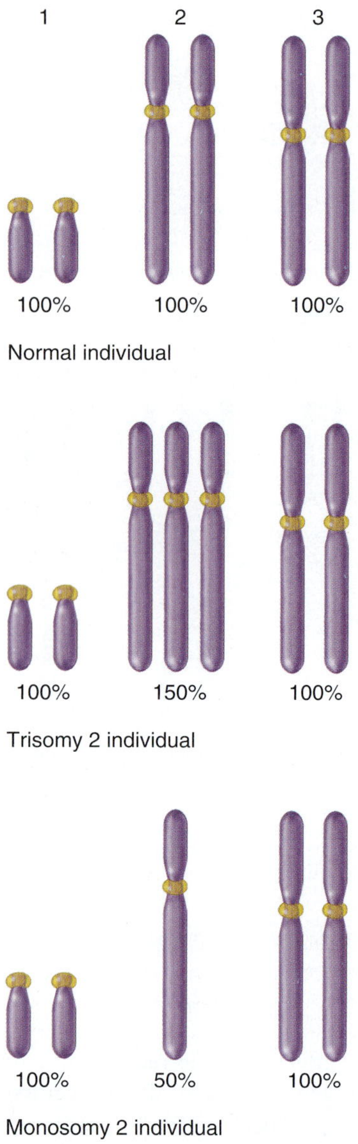

FIGURE 8.16 Imbalance of gene products in trisomic and monosomic individuals. Aneuploidy of chromosome 2 (i.e., trisomy and monosomy) leads to an imbalance in the amount of gene products from chromosome 2 compared with the amounts from chromosomes 1 and 3.

CONCEPT CHECK: Describe the imbalance in gene products that would occur in monosomy 2.

frequently during gamete formation. About 5%–10% of all fertilized human eggs result in an embryo with an abnormality in chromosome number! In most cases, these abnormal embryos do not develop properly and result in a spontaneous abortion very early in pregnancy. Approximately 50% of all spontaneous abortions are due to alterations in chromosome number.

In some cases, an abnormality in chromosome number produces an offspring that survives to birth or longer. Several human disorders involve abnormalities in chromosome number. The most common are trisomies of chromosomes 13, 18, or 21, and abnormalities in the number of the sex chromosomes (**Table 8.1**). Most of the known trisomies involve chromosomes that are relatively small—chromosome 13, 18, or 21—and carry fewer genes than the larger chromosomes. Trisomies of the other human autosomes and monosomies of all autosomes are presumed to produce a lethal phenotype, and many have been found in spontaneously aborted embryos and fetuses. For example, all possible human trisomies have been found in spontaneously aborted embryos except trisomy 1. It is believed that trisomy 1 is lethal at such an early stage that it prevents the successful implantation of the embryo.

Variation in the number of X chromosomes, unlike that of other large chromosomes, is often nonlethal. The survival of trisomy X individuals may be explained by X-chromosome inactivation, which is described in Chapter 5. In an individual with more than one X chromosome, all additional X chromosomes are converted to Barr bodies in the somatic cells of adult tissues. In an individual with trisomy X, for example, two out of three X chromosomes are converted to inactive Barr bodies. Unlike the level of expression for autosomal genes, the normal level of expression for most X-linked genes is from a single X chromosome. In other words, the correct level of mammalian gene expression results from two copies of each autosomal gene and one copy of each X-linked gene. This explains how the expression of X-linked genes in males (XY) can be maintained at the same levels as in females (XX). It may also explain why trisomy X is not a lethal condition.

The phenotypic effects noted in Table 8.1 involving X chromosome abnormalities may be due to the expression of X-linked genes prior to embryonic X-chromosome inactivation or to

TABLE 8.1
Aneuploid Conditions in Humans

Condition	Frequency	Syndrome	Characteristics
Autosomal			
Trisomy 13	1/15,000	Patau	Mental and physical deficiencies, wide variety of defects in organs, large triangular nose, early death
Trisomy 18	1/6000	Edward	Mental and physical deficiencies, facial abnormalities, extreme muscle tone, early death
Trisomy 21	1/800	Down	Mental deficiencies, abnormal pattern of palm creases, slanted eyes, flattened face, short stature
Sex Chromosomal			
XXY	1/1000	Klinefelter	Sexual immaturity (no sperm [males]), breast swelling
XYY	1/1000	Jacobs	Tall and thin (males)
XXX	1/1500	Triple X	Tall and thin, menstrual irregularity (females)
XO	1/5000	Turner	Short stature, webbed neck, sexually undeveloped (females)

the expression of genes on the inactivated X chromosome. As described in Chapter 5, pseudoautosomal genes and some other genes on the inactivated X chromosome are expressed in humans. Having three copies or one copy of the X chromosome results in an over- or underexpression of these X-linked genes, respectively.

Human abnormalities in chromosome number are influenced by the age of the parents. Older parents are more likely to produce children with abnormalities in chromosome number. Down syndrome provides an example. The common form of this disorder is caused by the inheritance of three copies of chromosome 21. The incidence of Down syndrome rises with the age of either parent. In males, however, the rise occurs relatively late in life, usually past the age when most men have children. By comparison, the likelihood of having a child with Down syndrome rises dramatically with age during a woman's reproductive years (**Figure 8.17**). This syndrome was first described by the English physician John Langdon Down in 1866. The association between maternal age and Down syndrome was later discovered by L. S. Penrose in 1933, even before the chromosomal basis for the disorder was identified by the French scientist Jérôme Lejeune in 1959. Down syndrome is most commonly caused by **nondisjunction,** which means that the chromosomes do not segregate properly. (Nondisjunction is discussed later in this chapter.) In this case, nondisjunction of chromosome 21 most commonly occurs during meiosis I in the oocyte.

Different hypotheses have been proposed to explain the relationship between maternal age and Down syndrome. One popular idea suggests that it may be due to the age of the oocytes. Human primary oocytes are produced within the ovary of the female fetus prior to birth and are arrested at prophase of meiosis I and remain in this stage until the time of ovulation. Therefore, as a woman ages, her primary oocytes have been in prophase I for a progressively longer period of time. This added length of time may contribute to an increased frequency of nondisjunction. About 5% of the time, Down syndrome is due to an extra paternal chromosome.

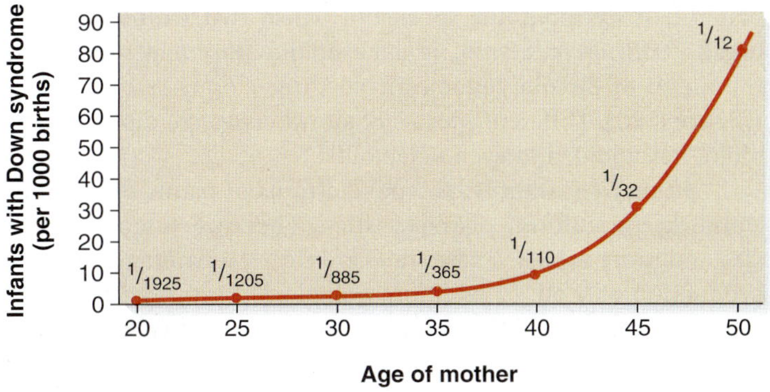

FIGURE 8.17 The incidence of Down syndrome births according to the age of the mother. The *y*-axis shows the number of infants born with Down syndrome per 1000 live births, and the *x*-axis plots the age of the mother at the time of birth. The data points indicate the fraction of live offspring born with Down syndrome.

8.6 COMPREHENSION QUESTIONS

1. In a trisomic individual, such as a person with trisomy 21 (Down syndrome), a genetic imbalance occurs because
   a. genes on chromosome 21 are overexpressed.
   b. genes on chromosome 21 are underexpressed.
   c. genes on the other chromosomes are overexpressed.
   d. genes on the other chromosomes are underexpressed.
2. Humans that survive with aneuploidy usually have incorrect numbers of chromosome 13, 18, or 21 or the sex chromosomes. A possible explanation why these abnormalities permit survival is because
   a. the chromosomes have clusters of genes that aid in embryonic growth.
   b. the chromosomes are small and carry relatively few genes.
   c. X-chromosome inactivation.
   d. both b and c.

8.7 VARIATION IN THE NUMBER OF SETS OF CHROMOSOMES

Learning Outcomes:

1. Describe examples in animals that involve variation in euploidy.
2. Define endopolyploidy.
3. Outline the process of polytene chromosome formation.
4. Discuss the effects of polyploidy among plant species and its impact in agriculture.

We now turn our attention to changes in the number of sets of chromosomes, referred to as variations in euploidy. In some cases, such changes are detrimental. However, in many species, particularly plants, additional sets of chromosomes are very common and are often beneficial with regard to an individual's phenotype.

Variations in Euploidy Occur Naturally in a Few Animal Species

Most species of animals are diploid. In some cases, changes in euploidy are not well tolerated. For example, polyploidy in mammals is generally a lethal condition. However, many examples of naturally occurring variations in euploidy occur. In **haplodiploid** species, which include many species of bees, wasps, and ants, one of the sexes is haploid, usually the male, and the other is diploid. For example, male bees, which are called drones, contain a single set of chromosomes. They are produced from unfertilized eggs. By comparison, female bees are produced from fertilized eggs and are diploid.

Examples of vertebrate polyploid animals have been discovered. Interestingly, on several occasions, animals that are morphologically very similar can be found as a diploid species as

(a) ***Hyla chrysoscelis***

(b) ***Hyla versicolor***

FIGURE 8.18 Differences in euploidy in two closely related frog species. The frog in **(a)** is diploid, whereas the frog in **(b)** is tetraploid.

Genes → Traits Though similar in appearance, these two species differ in their number of chromosome sets. At the level of gene expression, this observation suggests that the number of copies of each gene (two versus four) does not critically affect the phenotype of these two species.

well as a separate polyploid species. This situation occurs among certain amphibians and reptiles. **Figure 8.18** shows photographs of a diploid and a tetraploid ($4n$) frog. As you can see, they look indistinguishable from each other. Their difference can be revealed only by an examination of the chromosome number in the somatic cells of the animals and by mating calls—*H. chrysoscelis* has a faster trill rate than *H. versicolor*.

Variations in Euploidy Can Occur in Certain Tissues Within an Animal

Thus far, we have considered variations in chromosome number that occur at fertilization, so all somatic cells of an individual contain this variation. Euploidy may also change after fertilization. In many animals, certain tissues of the body display normal variations in the number of sets of chromosomes. For example, the cells of the human liver are typically polyploid. Liver cells contain nuclei that can be triploid, tetraploid, and even octaploid ($8n$). The occurrence of polyploid tissues or cells in organisms that are otherwise diploid is known as **endopolyploidy.** What is the biological significance of endopolyploidy? One possibility is that the increase in chromosome number in certain cells may enhance their ability to produce specific gene products that are needed in great abundance.

An unusual example of natural variation in the ploidy of somatic cells occurs in *Drosophila* and some other insects. *Drosophila* cells contain eight chromosomes (two sets of four chromosomes each; see Figure 8.15a). Within certain tissues, such as the salivary glands, the homologous chromosomes synapse with each other and undergo repeated rounds of chromosome replication without separating from each other. For example, in the salivary gland cells of *Drosophila*, the homologous chromosomes double approximately nine times ($2^9 = 512$). As shown in **Figure 8.19a**, repeated rounds of chromosomal replication produce a bundle of chromosomes that lie together in a parallel fashion. During this process, the four types of chromosomes aggregate to form an enormous **polytene chromosome** with several arms (**Figure 8.19b**). The central point where the chromosomes aggregate is known as the **chromocenter** (**Figure 8.19c**). Each of the four types of chromosome is attached to the chromocenter near its centromere. The X and Y and chromosome 4 are telocentric, and chromosomes 2 and 3 are metacentric. Therefore, the X and Y and chromosome 4 have a single arm projecting from the chromocenter, whereas chromosomes 2 and 3 have two arms. Polytene chromosomes were first observed by E. G. Balbiani in 1881. Later, in the 1930s, Theophilus Painter and colleagues recognized that the large size of polytene chromosomes provided geneticists with unique opportunities to study chromosome structure and gene organization.

Variations in Euploidy Are Common in Plants

We now turn our attention to variations of euploidy that occur in plants. Compared with animals, plants more commonly exhibit polyploidy. Among ferns and seed plants, at least 30%–35% of species are polyploid. Polyploidy is also important in agriculture. Many of the fruits and grains we eat are produced from polyploid plants. For example, the species of wheat that we use to make bread, *Triticum aestivum*, is a hexaploid ($6n$) that arose from the union of diploid genomes from three closely related species (**Figure 8.20**). Different species of strawberries are diploid, tetraploid, hexaploid, and even octaploid!

In many instances, polyploid strains of plants display outstanding agricultural characteristics. They are often larger in size and more robust. These traits are clearly advantageous in the production of food. In addition, polyploid plants tend to exhibit a greater adaptability, which allows them to withstand harsher environmental conditions. Also, polyploid ornamental plants often produce larger flowers than their diploid counterparts.

Polyploid plants having an odd number of chromosome sets, such as triploids ($3n$) or pentaploids ($5n$), usually cannot reproduce. Why are they sterile? The sterility arises because they produce highly aneuploid gametes. During prophase of meiosis I, homologous pairs of sister chromatids form bivalents. However, organisms with an odd number of chromosome sets, such as three,

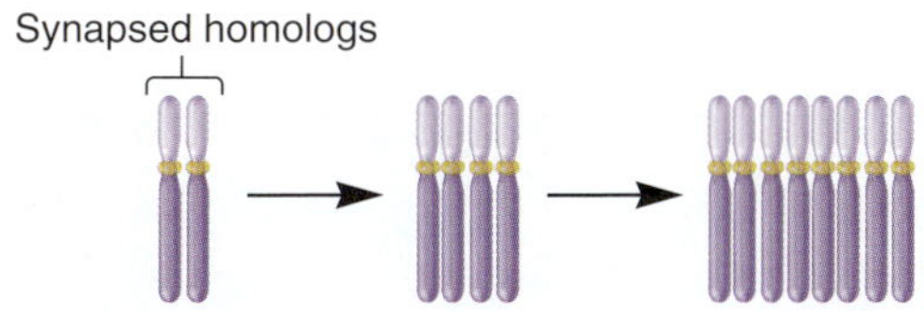

(a) Repeated chromosome replication produces polytene chromosome.

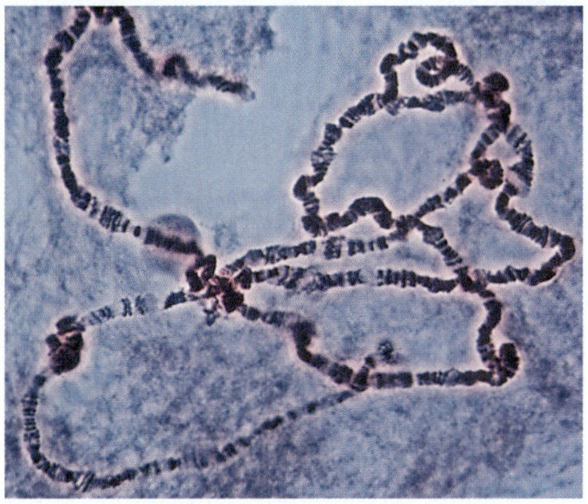

(b) A polytene chromosome

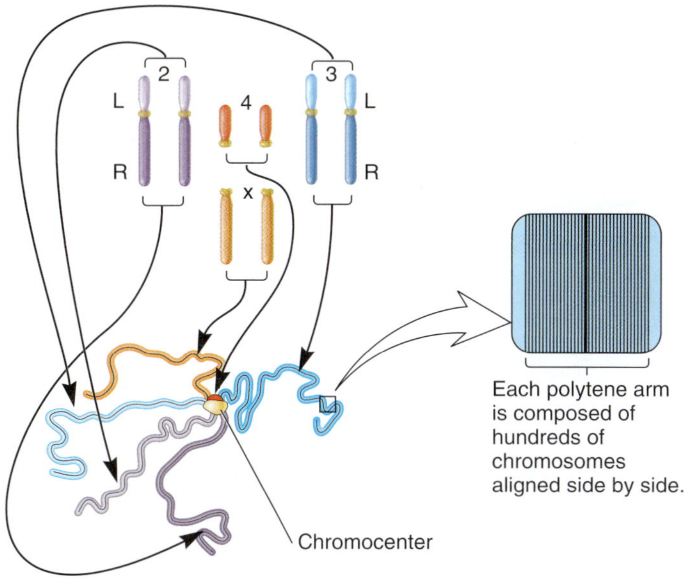

(c) Relationship between a polytene chromosome and regular *Drosophila* chromosomes

FIGURE 8.19 Polytene chromosomes in *Drosophila*. **(a)** A schematic illustration of the formation of a polytene chromosome. Homologous chromosomes synapse and undergo several rounds of replication without separating from each other. This results in a bundle of chromosomes that are parallel to each other. Note: This replication does not occur in highly condensed, heterochromatic DNA near the centromere. **(b)** A photograph of a polytene chromosome. **(c)** This drawing shows the relationship between the four pairs of chromosomes and the formation of a polytene chromosome in the salivary gland. The heterochromatic regions of the chromosomes aggregate at the chromocenter, and the arms of the chromosomes project outward. In chromosomes with two arms, the short arm is labeled L and the long arm is labeled R.

CONCEPT CHECK: Approximately how many copies of chromosome 2 are found in a polytene chromosome?

display an unequal separation of homologous chromosomes during anaphase of meiosis I (**Figure 8.21**). An odd number cannot be divided equally between two daughter cells. For each type of chromosome, a daughter cell randomly gets either one or two copies. For example, one daughter cell might receive one copy of chromosome 1, two copies of chromosome 2, two copies of chromosome 3, one copy of chromosome 4, and so forth. For a triploid species containing many different chromosomes in a set, meiosis is very unlikely to produce a daughter cell that is euploid. If we assume that a daughter cell receives either one copy or two copies of each kind of chromosome, the probability that meiosis will produce a cell that is perfectly haploid or diploid is $(1/2)^{n-1}$, where n is the number of chromosomes in a set. As an example, in a triploid organism containing 20 chromosomes per set, the

Cultivated wheat, a hexaploid species

FIGURE 8.20 Example of a polyploid plant. Cultivated wheat, *Triticum aestivum*, is a hexaploid. It was derived from three different diploid species of grasses that originally were found in the Middle East and were cultivated by ancient farmers in that region.

Genes → Traits An increase in chromosome number from diploid to tetraploid or hexaploid affects the phenotype of the individual. In the case of many plant species, a polyploid individual is larger and more robust than its diploid counterpart. This suggests that having additional copies of each gene is somewhat better than having two copies of each gene. This phenomenon in plants is rather different from the situation in animals. Tetraploidy in animals may have little effect (as in Figure 8.18b), but it is also common for polyploidy in animals to be detrimental.

CONCEPT CHECK: What are some common advantages of polyploidy in plants?

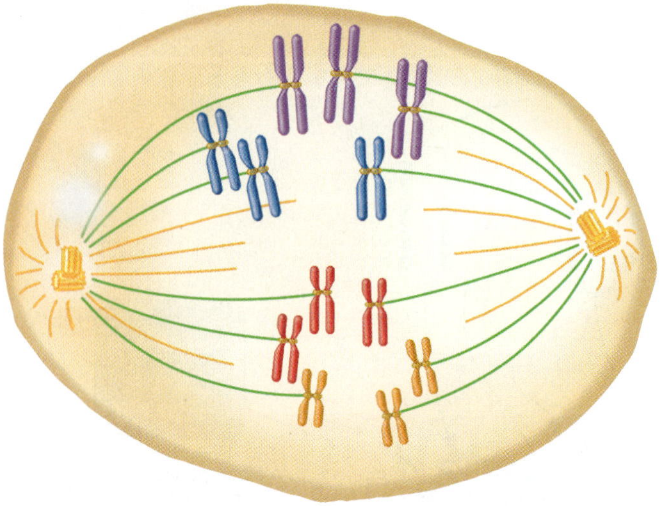

FIGURE 8.21 Schematic representation of anaphase of meiosis I in a triploid organism containing three sets of four chromosomes. In this example, the homologous chromosomes (three each) do not evenly separate during anaphase. Each cell receives one copy of some chromosomes and two copies of other chromosomes. This produces aneuploid gametes.

CONCEPT CHECK: Explain why a triploid individual is usually infertile.

probability of producing a haploid or diploid cell is 0.000001907, or 1 in 524,288. Thus, meiosis is almost certain to produce cells that contain one copy of some chromosomes and two copies of the others. This high probability of aneuploidy underlies the reason for triploid sterility.

Though sterility is generally a detrimental trait, it can be desirable agriculturally because it may result in a seedless fruit. For example, domestic bananas and seedless watermelons are triploid varieties. The domestic banana was originally derived from a seed-producing diploid species and has been asexually propagated by humans via cuttings. The small black spots in the center of a domestic banana are degenerate seeds. In the case of flowers, the seedless phenotype can also be beneficial. Seed producers such as Burpee have developed triploid varieties of flowering plants such as marigolds. Because the triploid marigolds are sterile and unable to set seed, more of their energy goes into flower production. According to Burpee, "They bloom and bloom, unweakened by seed bearing."

8.7 COMPREHENSION QUESTIONS

1. The term *endopolyploidy* refers to the phenomenon of having
   a. too many chromosomes.
   b. extra chromosomes inside the cell nucleus.
   c. certain cells of the body with extra sets of chromosomes.
   d. extra sets of chromosomes in gametes.
2. In agriculture, an advantage of triploid plants is that they are
   a. more fertile.
   b. often seedless.
   c. always disease-resistant.
   d. all of the above.

8.8 NATURAL AND EXPERIMENTAL MECHANISMS THAT PRODUCE VARIATION IN CHROMOSOME NUMBER

Learning Outcomes:

1. Describe how meiotic and mitotic nondisjunction occur and their possible phenotypic consequences.
2. Compare and contrast autopolyploidy, alloploidy, and allopolyploidy.
3. Explain why allotetraploids are more likely to be fertile than are allodiploids.
4. Describe how colchicine is used to produce polyploid species.

As we have seen, variations in chromosome number are fairly widespread and usually have a significant effect on the phenotypes of plants and animals. For these reasons, researchers have wanted to understand the cellular mechanisms that cause variations in chromosome number.

In some cases, a change in chromosome number is the result of nondisjunction, which refers to an event in which the chromosomes do not segregate properly. **Meiotic nondisjunction,** which is an improper separation of chromosomes during meiosis, produces haploid cells that have too many or too few chromosomes. If such a cell gives rise to a gamete that fuses with a normal gamete during fertilization, the resulting offspring will have an abnormal chromosome number in all of its cells. An abnormal nondisjunction event also may occur after fertilization in one of the somatic cells of the body. This mechanism is known as **mitotic nondisjunction.** When this occurs during embryonic stages of development, it may lead to a patch of tissue in the organism that has an altered chromosome number.

Another common way in which the chromosome number of an organism can vary is by interspecies crosses. An **alloploid** organism contains sets of chromosomes from two or more different species. This term refers to the occurrence of chromosome sets (ploidy) from the genomes of different (allo) species. In this section, we will examine these three natural mechanisms in greater detail, and also explore how chromosome number can be changed experimentally.

Meiotic Nondisjunction Can Produce Aneuploidy or Polyploidy

The process of meiosis is described in Chapter 3. Nondisjunction can occur during anaphase of meiosis I or meiosis II. If it happens during meiosis I, an entire bivalent migrates to one pole (**Figure 8.22a**). Following the completion of meiosis, the four resulting haploid cells produced from this event are abnormal. If nondisjunction occurs during anaphase of meiosis II (**Figure 8.22b**), the net result is two abnormal and two normal haploid cells. If a gamete carrying an extra chromosome unites with a normal gamete, the offspring will be trisomic. Alternatively,

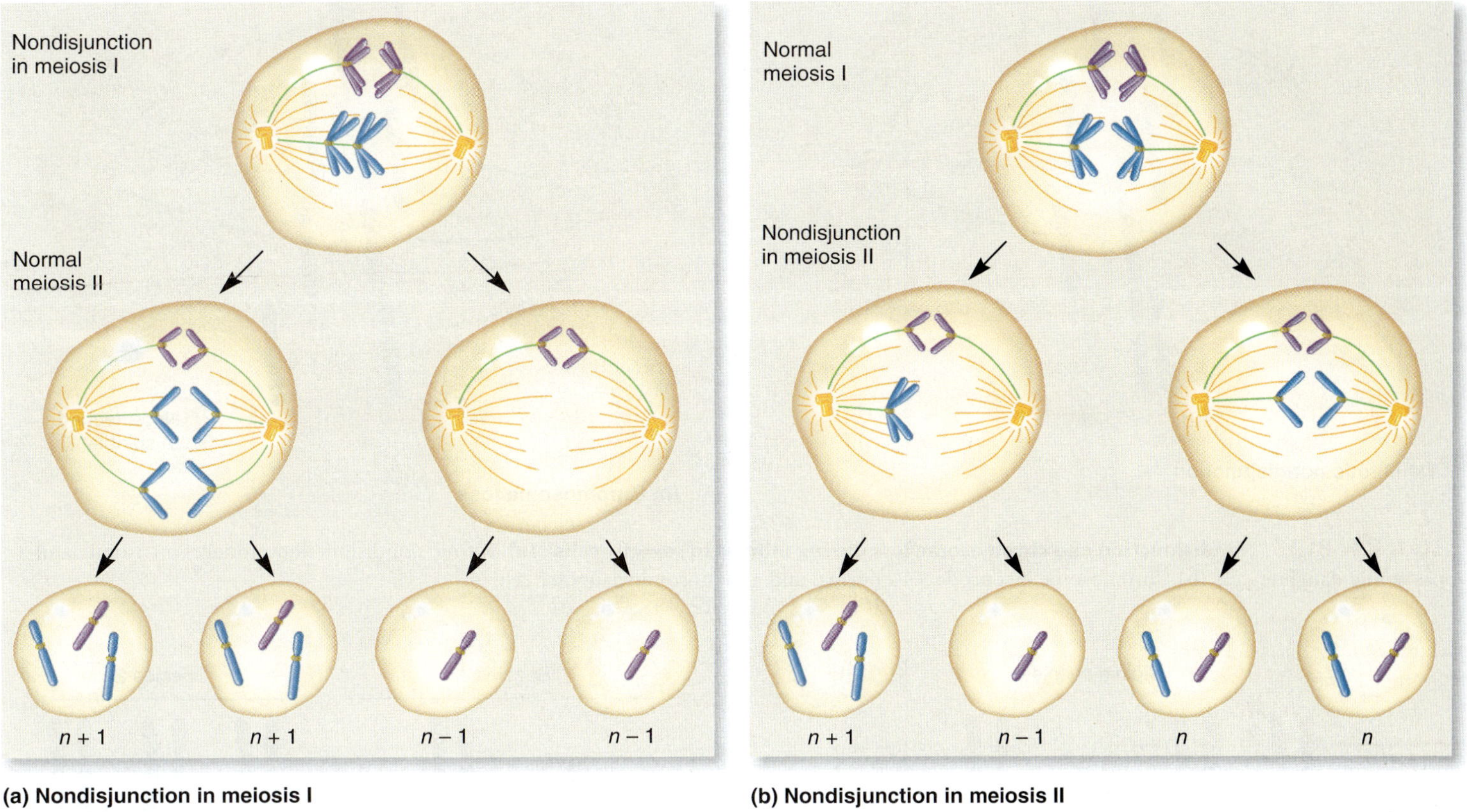

FIGURE 8.22 **Nondisjunction during meiosis I and II.** The chromosomes shown in purple are behaving properly during meiosis I and II, so each cell receives one copy of this chromosome. The chromosomes shown in blue are not disjoining correctly. In **(a)**, nondisjunction occurred in meiosis I, so the resulting four cells receive either two copies of the blue chromosome or zero copies. In **(b)**, nondisjunction occurred during meiosis II, so one cell has two blue chromosomes and another cell has zero. The remaining two cells are normal.

CONCEPT CHECK: Explain what the word nondisjunction means.

if a gamete that is missing a chromosome is viable and participates in fertilization, the resulting offspring is monosomic for the missing chromosomes.

In rare cases, all of the chromosomes undergo nondisjunction and migrate to one of the daughter cells, an event called **complete nondisjunction.** The daughter cell that does not receive any chromosomes is nonviable. In contrast, the daughter cell receiving all of the chromosomes may complete meiosis and form two diploid cells, which give rise to diploid gametes. If a diploid gamete participates in fertilization with a normal haploid gamete, a triploid individual is produced. Therefore, complete nondisjunction can result in individuals that are polyploid.

Mitotic Nondisjunction or Chromosome Loss Can Produce a Patch of Tissue with an Altered Chromosome Number

Abnormalities in chromosome number occasionally occur after fertilization takes place. In this case, the abnormal event happens during mitosis rather than meiosis. One possibility is that the sister chromatids separate improperly, so one daughter cell receives three copies of a chromosome, whereas the other daughter cell gets only one (**Figure 8.23a**). Alternatively, the sister chromatids could separate during anaphase of mitosis, but one of the chromosomes could be improperly attached to the spindle, so it does not migrate to a pole (**Figure 8.23b**). A chromosome will be degraded if it is left outside the nucleus when the nuclear membrane re-forms. In this case, one of the daughter cells has two copies of that chromosome, whereas the other has only one.

When genetic abnormalities occur after fertilization, the organism contains a subset of cells that are genetically different from those of the rest of the organism. This condition is referred to as **mosaicism.** The size and location of the mosaic region depend on the timing and location of the original abnormal event. If a genetic alteration happens very early in the embryonic development of an organism, the abnormal cell will be the precursor for a large section of the organism.

Changes in Euploidy Occur by Autopolyploidy, Alloploidy, and Allopolyploidy

Different mechanisms account for changes in the number of chromosome sets among natural populations of plants and animals (**Figure 8.24**). As previously mentioned, complete nondisjunction, due to a general defect in the spindle apparatus, can produce an individual with one or more extra sets of chromosomes. This individual is known as an **autopolyploid** (Figure 8.24a).

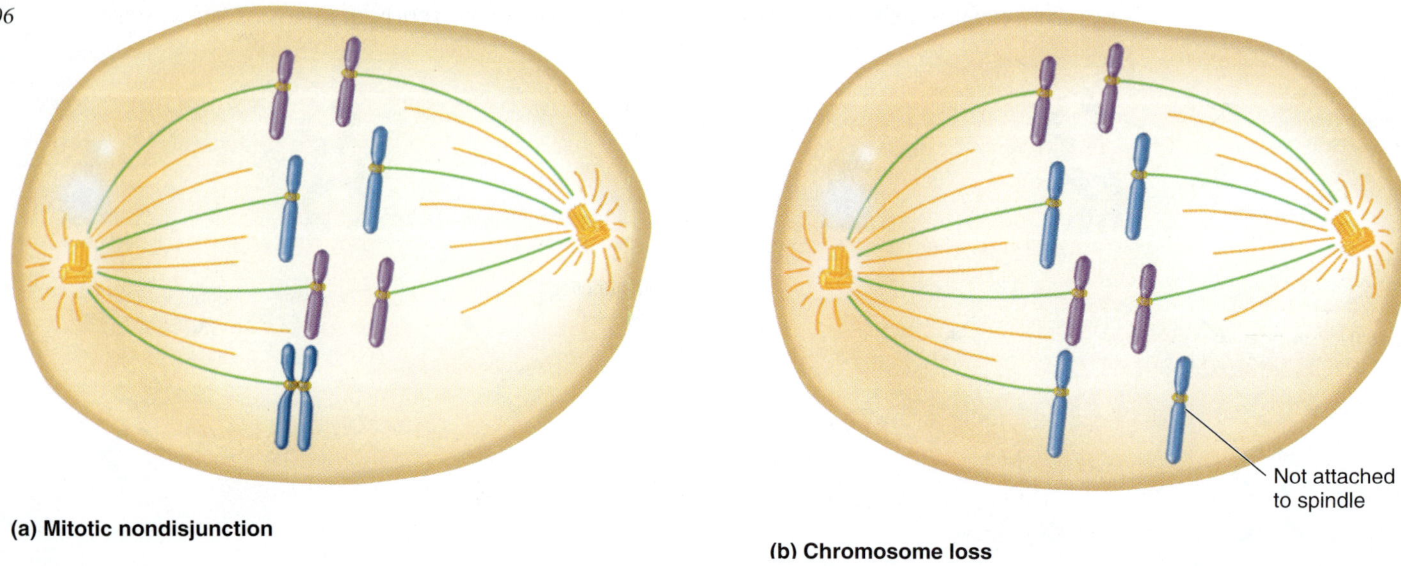

FIGURE 8.23 Nondisjunction and chromosome loss during mitosis in somatic cells. **(a)** Mitotic nondisjunction produces a trisomic and a monosomic daughter cell. **(b)** Chromosome loss produces a normal and a monosomic daughter cell.

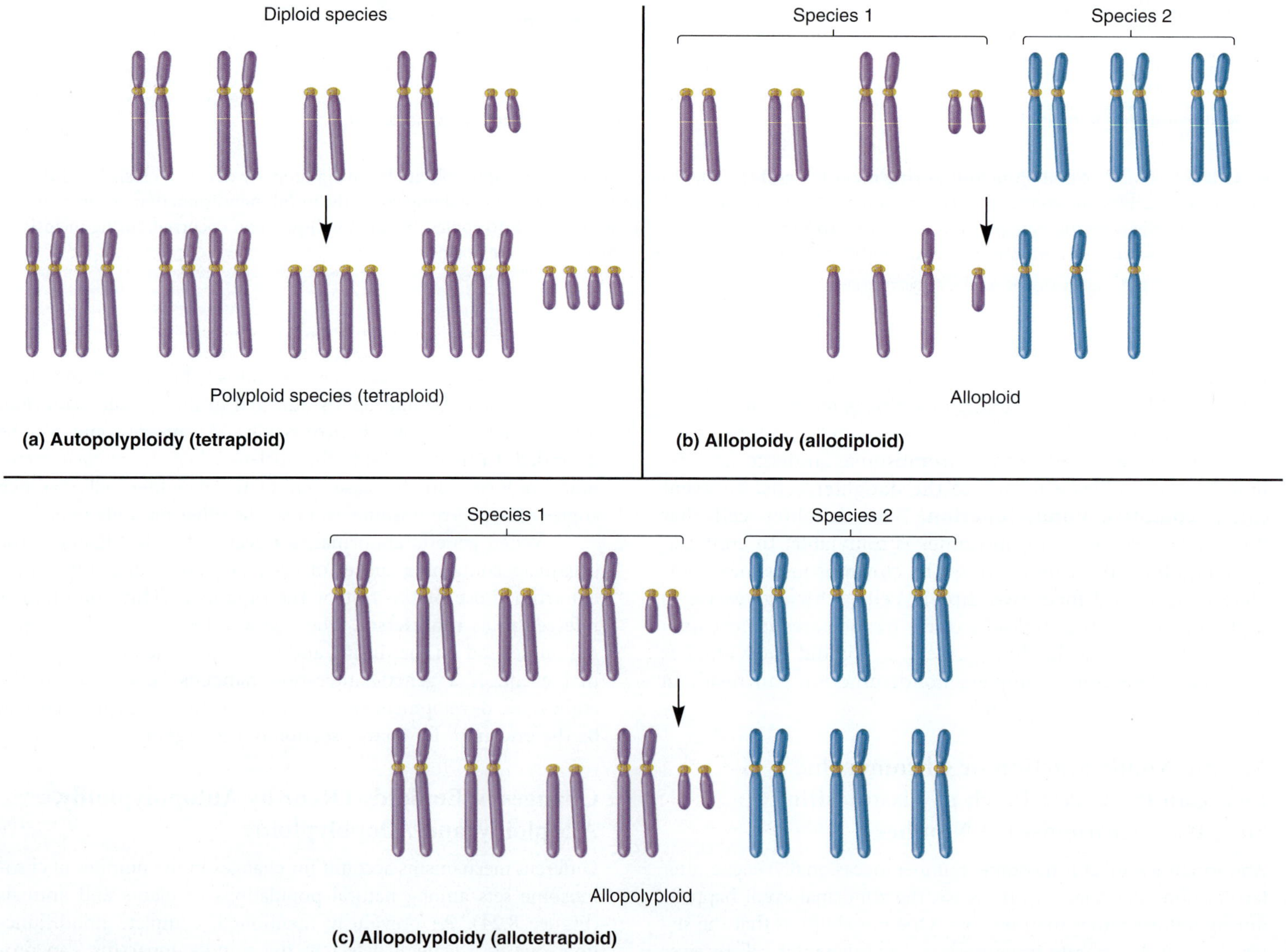

FIGURE 8.24 A comparison of autopolyploidy, alloploidy, and allopolyploidy.

CONCEPT CHECK: What is the key difference between autopolyploidy and allopolyploidy?

The prefix auto- (meaning self) and term polyploid (meaning many sets of chromosomes) refer to an increase in the number of chromosome sets within a single species.

A more common mechanism that changes chromosome number, called **alloploidy,** is a result of interspecies crosses (Figure 8.24b). An alloploid that has one set of chromosomes from two different species is called an **allodiploid.** This event is most likely to occur between species that are close evolutionary relatives. For example, closely related species of grasses may interbreed to produce allodiploids. As shown in Figure 8.24c, an **allopolyploid** contains two (or more) sets of chromosomes from two (or more) species. In this case, the **allotetraploid** contains two complete sets of chromosomes from two different species, for a total of four sets. In nature, allotetraploids usually arise from allodiploids. This can occur when a somatic cell in an allodiploid undergoes complete nondisjunction to create an allotetraploid cell. In plants, such a cell can continue to grow and produce a section of the plant that is allotetraploid. If this part of the plant produces seeds by self-pollination, the seeds give rise to allotetraploid offspring. Cultivated wheat (refer back to Figure 8.20) is a plant in which two species must have interbred to create an allotetraploid, and then a third species interbred with the allotetraploid to create an allohexaploid.

Allodiploids Formed from Closely Related Species Carry Homeologous Chromosomes

When comparing two closely related species, the number and types of chromosomes are typically very similar. **Figure 8.25** shows a karyotype of an allodiploid that was produced by an interspecies mating between the roan antelope (*Hippotragus equinus*) and the sable antelope (*Hippotragus niger*). As seen here, these two closely related species have the same number of chromosomes. The sizes and banding patterns of the chromosomes show that they correspond to one another. For example, chromosome 1 from both species is fairly large, has very similar banding patterns, and carries many of the same genes. Evolutionarily related chromosomes from two different species are called **homeologous** chromosomes (not to be confused with homologous chromosomes). This allodiploid is fertile because the homeologous chromosomes can properly synapse during meiosis to produce haploid gametes.

In some cases, an allodiploid may not be fertile because the chromosomes from each species do not pair up properly during meiosis. By comparison, allotetraploids are often times fertile because there are two sets of chromosomes from both species, which allows all of the chromosomes to pair up during meiosis.

Experimental Treatments Can Promote Polyploidy

Because autopolyploid and allopolyploid plants often exhibit desirable traits, the development of polyploids is of considerable interest among plant breeders. Experimental studies on the ability of environmental agents to promote polyploidy began in the early 1900s. Since that time, various agents have been shown to promote nondisjunction, which thereby lead to polyploidy. These include abrupt temperature changes during the initial stages of seedling growth and the treatment of plants with chemical agents that interfere with the formation of the spindle apparatus.

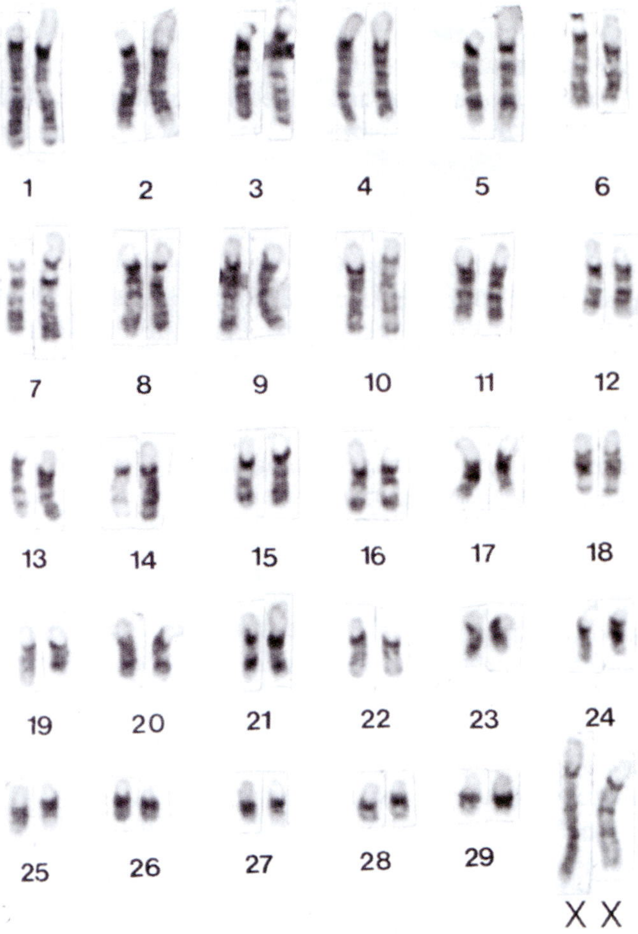

FIGURE 8.25 The karyotype of a hybrid animal produced from two closely related antelope species. In each chromosome pair in this karyotype, one chromosome was inherited from the roan antelope (*Hippotragus equinus*) and the other from the sable antelope (*Hippotragus niger*). When it is possible to distinguish the chromosomes from the two species, the roan chromosomes are shown on the left side of each pair.

The drug colchicine is one such commonly used agent for promoting polyploidy. Once inside the cell, colchicine binds to tubulin (a protein found in the spindle apparatus) and thereby interferes with normal chromosome segregation during mitosis or meiosis. In 1937, Alfred Blakeslee and Amos Avery applied colchicine to plant tissue and, at high doses, were able to cause complete mitotic nondisjunction and produce polyploidy in plant cells. Colchicine can be applied to seeds, young embryos, or rapidly growing regions of a plant (**Figure 8.26**). This application may produce aneuploidy, which is usually an undesirable outcome, but it often produces polyploid cells, which may grow faster than the surrounding diploid tissue. In a diploid plant, colchicine may cause complete mitotic nondisjunction, yielding tetraploid ($4n$) cells. As the tetraploid cells continue to divide, they generate a sector that is often morphologically distinguishable from the rest of the plant. For example, a tetraploid stem may have a larger diameter and produce larger leaves and flowers.

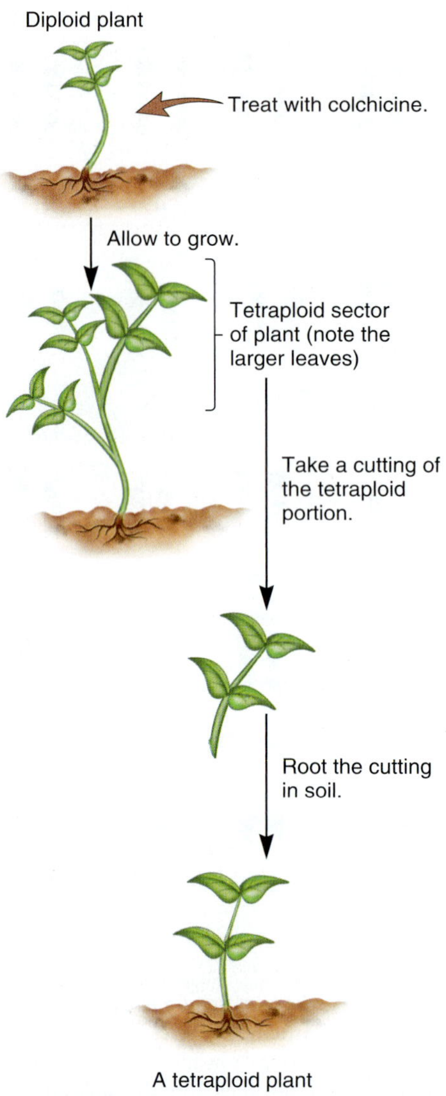

FIGURE 8.26 Use of colchicine to promote polyploidy in plants. Colchicine interferes with the mitotic spindle apparatus and promotes mitotic nondisjunction. If complete nondisjunction occurs in a diploid cell, a tetraploid daughter cell will be formed. Such a tetraploid cell may continue to divide and produce a segment of the plant with more robust characteristics. This segment may be cut from the rest of the plant and rooted. In this way, a tetraploid plant can be propagated.

Because individual plants can be propagated asexually from pieces of plant tissue (i.e., cuttings), the polyploid portion of the plant can be removed, treated with the proper growth hormones, and grown as a separate plant. Alternatively, the tetraploid region of a plant may have flowers that produce seeds by self-pollination. For many plant species, a tetraploid flower produces diploid pollen and eggs, which can combine to produce tetraploid offspring. In this way, colchicine can be used to provide a straightforward method of producing polyploid strains of plants.

8.8 COMPREHENSION QUESTIONS

1. In a diploid species, complete nondisjunction during meiosis I may produce a viable cell that is
   a. trisomic.
   b. haploid.
   c. diploid.
   d. triploid.
2. The somatic cells of an allotetraploid usually contain
   a. one set of chromosomes from four different species.
   b. two sets of chromosomes from two different species.
   c. four sets of chromosomes from one species.
   d. one set of chromosomes from two different species.

KEY TERMS

Page 183. genetic variation, allelic variation
Page 184. cytogeneticist, metacentric, submetacentric, acrocentric, telocentric, karyotype, G banding
Page 186. deletion, deficiency, duplication, inversion, translocation, simple translocation, reciprocal translocation
Page 187. terminal deletion, interstitial deletion
Page 188. repetitive sequences, nonallelic homologous recombination, gene duplication, gene family
Page 189. homologous, paralogs, copy number variation (CNV)
Page 190. segmental duplication, comparative genomic hybridization (CGH), hybridization
Page 193. pericentric inversion, paracentric inversion, position effect, inversion heterozygote, inversion loop, acentric fragment, dicentric, dicentric bridge
Page 194. telomeres, balanced translocations
Page 195. unbalanced translocation, Robertsonian translocation
Page 197. translocation cross
Page 198. semisterility, euploid, triploid, tetraploid, polyploid, aneuploidy, trisomic, monosomic
Page 201. nondisjunction, haplodiploid
Page 202. endopolyploidy, polytene chromosome, chromocenter
Page 204. meiotic nondisjunction, mitotic nondisjunction, alloploid
Page 205. complete nondisjunction, mosaicism, autopolyploid
Page 207. alloploidy, allodiploid, allopolyploid, allotetraploid, homeologous

CHAPTER SUMMARY

8.1 Microscopic Examination of Eukaryotic Chromosomes

- Among different species, natural variation exists with regard to chromosome structure and number. Three features of chromosomes that aid in their identification are centromere location, size, and banding patterns (see Figure 8.1).

8.2 Changes in Chromosome Structure: An Overview

- Within a species, variations in chromosome structure include deletions, duplications, inversions, and translocations (see Figure 8.2).

8.3 Deletions and Duplications

- Chromosome breaks can create terminal and interstitial deletions. Some deletions are associated with human genetic disorders such as cri-du-chat syndrome (see Figures 8.3, 8.4).
- Nonallelic homologous recombination creates gene duplications and deletions. Over time, gene duplications can lead to the formation of gene families, such as the globin gene family (see Figures 8.5–8.7).
- Copy number variation (CNV) is fairly common within a species. In humans, CNV is associated with certain diseases (see Figure 8.8).
- Comparative genomic hybridization is one technique for detecting chromosome deletions and duplications. It is used in the analysis of cancer cells (see Figure 8.9).

8.4 Inversions and Translocations

- Inversions can be pericentric or paracentric. In an inversion heterozygote, crossing over within the inversion loop creates deletions and duplications in the resulting chromosomes (see Figures 8.10, 8.11).
- Two mechanisms that may produce translocations are (1) chromosome breakage with subsequent repair and (2) nonhomologous crossing over (see Figure 8.12).
- Familial Down syndrome is due to a translocation between chromosome 14 and 21 (see Figure 8.13).
- Due to the formation of a translocation cross, individuals that carry a balanced translocation may have a high probability of producing unbalanced gametes (see Figure 8.14).

8.5 Changes in Chromosome Number: An Overview

- Chromosome number variation may involve changes in the number of sets (euploidy) or changes in the number of particular chromosomes within a set (aneuploidy) (see Figure 8.15).

8.6 Variation in the Number of Chromosomes Within a Set: Aneuploidy

- Aneuploidy is often detrimental due to an imbalance in gene expression. Down syndrome, an example of an aneuploidy in humans, increases in frequency with maternal age (see Figures 8.16, 8.17, Table 8.1).

8.7 Variation in the Number of Sets of Chromosomes

- Among animals, variation in euploidy is relatively rare, though it does occur. Some tissues within an animal may exhibit polyploidy. An example is the polytene chromosomes found in salivary cells of *Drosophila* (see Figures 8.18, 8.19).
- Polyploidy in plants is relatively common and has found many advantages in agriculture. Triploid plants are usually seedless because they cannot segregate their chromosomes evenly during meiosis (see Figures 8.20, 8.21).

8.8 Natural and Experimental Mechanisms That Produce Variations in Chromosome Number

- Meiotic nondisjunction, mitotic nondisjunction, and chromosome loss result in changes in chromosome number (see Figures 8.22, 8.23).
- Autopolyploidy is an increase in the number of sets within a single species. Interspecies matings result in alloploids. Alloploids with multiple sets of chromosomes from each species are allopolyploids (see Figures 8.24, 8.25).
- The drug colchicine promotes nondisjunction and is used to make polyploid plants (see Figure 8.26).

PROBLEM SETS & INSIGHTS

Solved Problems

S1. Describe how a gene family is produced. Discuss the common and unique features of the members of the globin gene family.

Answer: A gene family is produced when a single gene is copied one or more times by a gene duplication event. This duplication may occur by an abnormal (misaligned) crossover, which produces a chromosome with a deletion and another chromosome with a gene duplication.

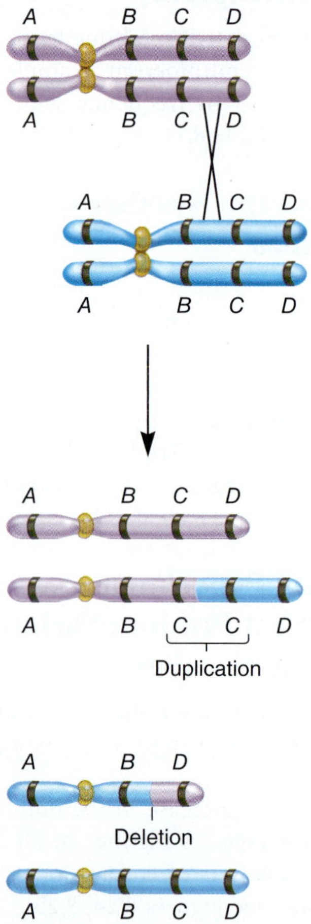

Over time, this type of duplication may occur several times to produce many copies of a particular gene. In addition, translocations may move the duplicated genes to other chromosomes, so the members of the gene family may be dispersed among several different chromosomes. Eventually, each member of a gene family will accumulate mutations, which may subtly alter their function.

All members of the globin gene family bind oxygen. Myoglobin tends to bind it more tightly; therefore, it is good at storing oxygen. Hemoglobin binds it more loosely, so it can transport oxygen throughout the body (via red blood cells) and release it to the tissues that need oxygen. The polypeptides that form hemoglobins are expressed in red blood cells, whereas the myoglobin gene is expressed in many different cell types. The expression pattern of the globin genes changes during different stages of development. The ε- and ζ-globin genes are expressed in the early embryo. They are turned off near the end of the first trimester, and then the α- and γ-globin genes are expressed. Following birth, the γ-globin genes are silenced, and the β-globin gene is expressed for the rest of a person's life. These differences in the expression of the globin genes reflect the differences in the oxygen transport needs of humans during the different stages of life. Overall, the evolution of gene families has resulted in gene products that are better suited to a particular tissue or stage of development. This has allowed a better "fine-tuning" of human traits.

S2. An inversion heterozygote has the following inverted chromosome:

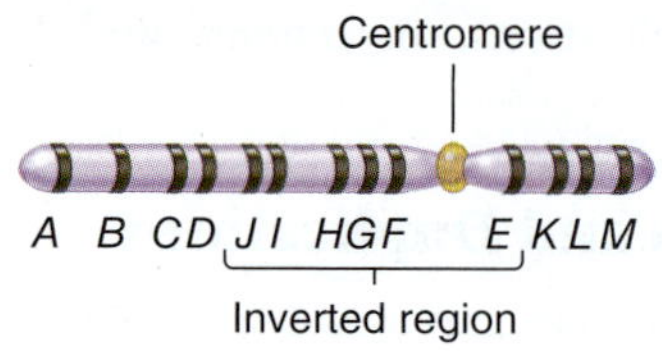

What is the result if a crossover occurs between genes *F* and *G* on one inversion and one normal chromosome?

Answer: The resulting product is four chromosomes. One chromosome is normal, one is an inversion chromosome, and two chromosomes have duplications and deletions. The two duplicated and deficient chromosomes are shown here:

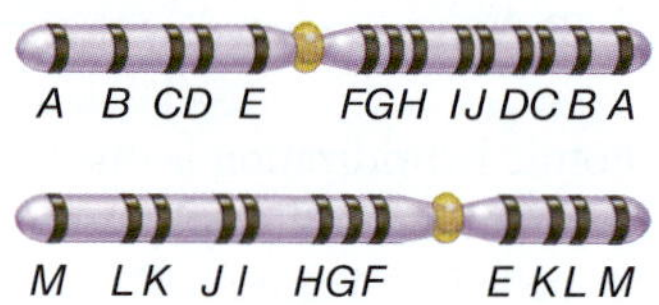

S3. In humans, the number of chromosomes per set equals 23. Even though the following conditions are lethal, what would be the total number of chromosomes for the following individuals?

A. Trisomy 22

B. Monosomy 11

C. Triploid individual

Answer:

A. 47 (the diploid number, 46, plus 1)

B. 45 (the diploid number, 46, minus 1)

C. 69 (3 times 23)

S4. A diploid species with 44 chromosomes (i.e., 22/set) is crossed to another diploid species with 38 chromosomes (i.e., 19/set). How many chromosomes are produced in an allodiploid or allotetraploid from this cross? Would you expect the offspring to be sterile or fertile?

Answer: An allodiploid would have $22 + 19 = 41$ chromosomes. This individual would likely be sterile, because not all of the chromosomes have homeologous partners to pair with during meiosis. This yields aneuploidy, which usually causes sterility. An allotetraploid would have $44 + 38 = 82$ chromosomes. Because each chromosome has a homologous partner, the allotetraploid would likely be fertile.

S5. Pseudodominance occurs when a single copy of a recessive allele is phenotypically expressed because the second copy of the gene has been deleted from the homologous chromosome; the individual is hemizygous for the recessive allele. As an example, we can consider the notch phenotype in *Drosophila*, which is an X-linked trait. Fruit flies with this condition have wings with a notched appearance at their edges. Female flies that are heterozygous for this mutation have notched wings; homozygous females and hemizygous males are unable to survive. The notched phenotype is due to a defect in a single gene called *Notch* (*N*). Geneticists studying fruit flies with this phenotype have discovered that one X chromosome of some of the mutant flies has a small deletion that includes the *Notch* gene as well as a few genes on either side of it. The mutation causing this phenotype in other flies is confined within the *Notch* gene itself. A genetic analysis can distinguish between notched fruit flies carrying a deletion versus those that carry a single-gene mutation. This is possible because the *Notch* gene happens to be located next to the red/white eye color gene on the X chromosome. How would you distinguish between a notched phenotype due to a deletion that included the *Notch* gene and the adjacent eye color gene versus a notch phenotype due to a small mutation only within the *Notch* gene itself?

Answer: To determine if the *Notch* mutation is due to a deletion, red-eyed females with the notched phenotype can be crossed to white-eyed males. Only the daughters with notched wings need to be analyzed. If they have red eyes, this means the *Notch* mutation has not deleted the red-eye allele from the X chromosome. Alternatively, if they have white eyes, this indicates the red-eye allele has been deleted from the X chromosome that carries the *Notch* mutation. In this case, the white-eye allele is expressed because it is present in a single copy in a female fly with two X chromosomes. This phenomenon is pseudodominance.

S6. Albert Blakeslee began using the Jimson weed (*Datura stramonium*) as an experimental organism to teach his students the laws of Mendelian inheritance. Although this plant has not gained widespread use in genetic studies, Blakeslee's work provided a convincing demonstration that changes in chromosome number affect the phenotype of organisms. Blakeslee's assistant, B. T. Avery, identified a Jimson weed mutant that he called "globe" because the capsule is more rounded than normal. In genetic crosses, he found that the globe mutant had a peculiar pattern of inheritance. The globe trait was passed to about 25% of the offspring when the globe plants were allowed to self-fertilize. Unexpectedly, about 25% of the offspring also had the globe phenotype when globe plants were pollinated by a normal plant. In contrast, when pollen from a globe plant was used to pollinate a normal plant, less than 2% of the offspring had the globe phenotype. This non-Mendelian pattern of inheritance caused Blakeslee and his colleagues to investigate the nature of this trait further. We now know the globe phenotype is due to trisomy 11. Can you explain this unusual pattern of inheritance knowing it is due to trisomy 11?

Answer: This unusual pattern of inheritance of these aneuploid strains can be explained by the viability of euploid versus aneuploid gametes or gametophytes. An individual that is trisomic has a 50% chance of producing an egg or sperm that will inherit an extra chromosome and a 50% chance that a gamete will be normal. Blakeslee's results indicate that when a pollen grain inherited an extra copy of chromosome 11, it was almost always nonviable and unable to produce an aneuploid offspring. However, a significant percentage of aneuploid eggs were viable, so some (25%) of the offspring were aneuploid. Because an aneuploid plant should produce a 1:1 ratio between euploid and aneuploid eggs, the observation that only 25% of the offspring were aneuploid also indicates that about half of the female gametophytes or aneuploid eggs from such gametophytes were also nonviable. Overall, these results provide compelling evidence that imbalances in chromosome number can alter reproductive viability and also cause significant phenotypic consequences.

Conceptual Questions

C1. Which changes in chromosome structure cause a change in the total amount of genetic material, and which do not?

C2. Explain why small deletions and duplications are less likely to have a detrimental effect on an individual's phenotype than large ones. If a small deletion within a single chromosome happens to have a phenotypic effect, what would you conclude about the genes in this region?

C3. How does a chromosomal duplication occur?

C4. What is a gene family? How are gene families produced over time? With regard to gene function, what is the biological significance of a gene family?

C5. Following a gene duplication, two genes will accumulate different mutations, causing them to have slightly different sequences. In Figure 8.7, which pair of genes would you expect to have more similar sequences, α_1 and α_2 or $\psi_{\alpha 1}$ and α_2? Explain your answer.

C6. Two chromosomes have the following order of genes:

Normal: *A B C* centromere *D E F G H I*

Abnormal: *A B G F E D* centromere *C H I*

Does the abnormal chromosome have a pericentric or paracentric inversion? Draw a sketch showing how these two chromosomes would pair during prophase of meiosis I.

C7. An inversion heterozygote has the following inverted chromosome:

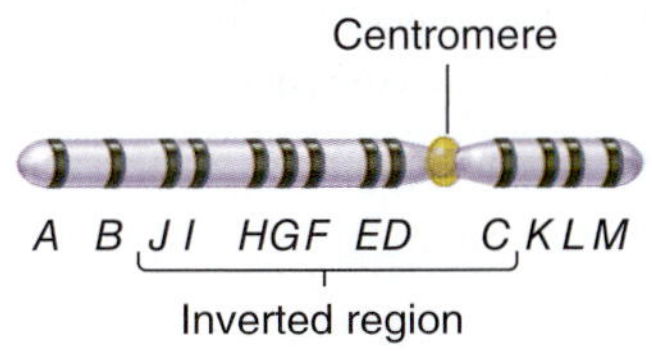

What would be the products if a crossover occurred between genes *H* and *I* on one inverted and one normal chromosome?

C8. An inversion heterozygote has the following inverted chromosome:

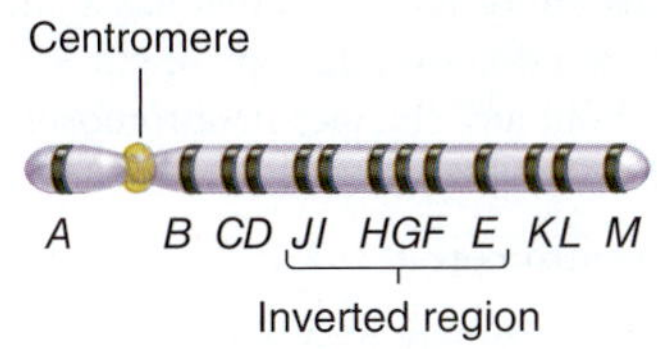

What would be the products if a crossover occurred between genes *H* and *I* on one inverted and one normal chromosome?

C9. Explain why inversions and reciprocal translocations do not usually cause a phenotypic effect. In a few cases, however, they do. Explain how.

C10. An individual has the following reciprocal translocation:

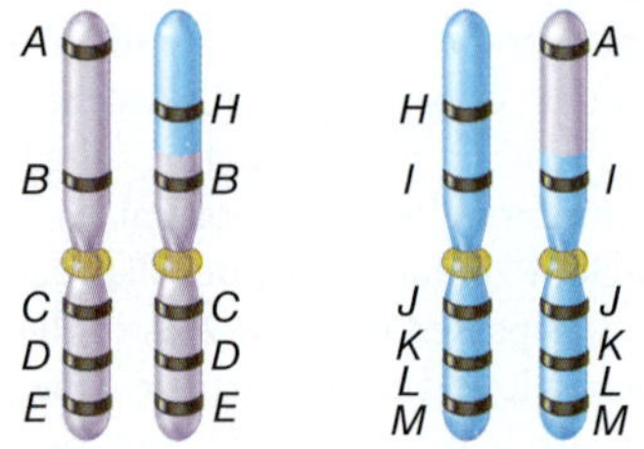

What would be the outcome of alternate and adjacent-1 segregation?

C11. A phenotypically normal individual has the following combinations of abnormal chromosomes:

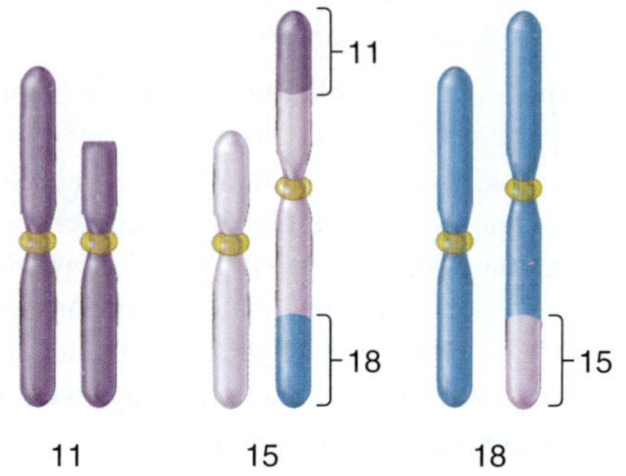

The normal chromosomes are shown on the left of each pair. Suggest a series of events (breaks, translocations, crossovers, etc.) that may have produced this combination of chromosomes.

C12. Two phenotypically normal parents produce a phenotypically abnormal child in which chromosome 5 is missing part of its long arm but has a piece of chromosome 7 attached to it. The child also has one normal copy of chromosome 5 and two normal copies of chromosome 7. With regard to chromosomes 5 and 7, what do you think are the chromosomal compositions of the parents?

C13. With regard to the segregation of centromeres, why is adjacent-2 segregation less frequent than alternate or adjacent-1 segregation?

C14. Which of the following types of chromosomal changes would you expect to have phenotypic consequences? Explain your choices.

A. Pericentric inversion

B. Reciprocal translocation

C. Deletion

D. Unbalanced translocation

C15. Explain why a translocation cross occurs during metaphase of meiosis I when a cell contains a reciprocal translocation.

C16. A phenotypically abnormal individual has a phenotypically normal father with an inversion on one copy of chromosome 7 and a normal mother without any changes in chromosome structure. The order of genes along chromosome 7 in the father is as follows:

R T D M centromere *P U X Z C* (normal chromosome 7)

R T D U P centromere *M X Z C* (inverted chromosome 7)

The phenotypically abnormal offspring has a chromosome 7 with the following order of genes:

R T D M centromere *P U D T R*

With a sketch, explain how this chromosome was formed. In your answer, explain where the crossover occurred (i.e., between which two genes).

C17. A diploid fruit fly has eight chromosomes. How many total chromosomes would be found in the following flies?

A. Tetraploid

B. Trisomy 2

C. Monosomy 3

D. $3n$

E. $4n + 1$

C18. A person is born with one X chromosome, zero Y chromosomes, trisomy 21, and two copies of the other chromosomes. How many chromosomes does this person have altogether? Explain whether this person is euploid or aneuploid.

C19. Two phenotypically unaffected parents produce two children with familial Down syndrome. With regard to chromosomes 14 and 21, what are the chromosomal compositions of the parents?

C20. Aneuploidy is typically detrimental, whereas polyploidy is sometimes beneficial, particularly in plants. Discuss why you think this is the case.

C21. Explain how aneuploidy, deletions, and duplications cause genetic imbalances. Why do you think that deletions and monosomies are more detrimental than duplications and trisomies?

C22. Female fruit flies homozygous for the X-linked white-eye allele are crossed to males with red eyes. On very rare occasions, an offspring is a male with red eyes. Assuming these rare offspring are not due to a new mutation in one of the mother's X chromosomes that converted the white-eye allele into a red-eye allele, explain how this red-eyed male arose.

C23. A cytogeneticist has collected tissue samples from members of the same butterfly species. Some of the butterflies were located in Canada, and others were found in Mexico. Upon karyotyping, the cytogeneticist discovered that chromosome 5 of the Canadian butterflies had a large inversion compared with the Mexican butterflies. The Canadian butterflies were inversion homozygotes, whereas the Mexican butterflies had two normal copies of chromosome 5.

A. Explain whether a mating between the Canadian and Mexican butterflies would produce phenotypically normal offspring.

B. Explain whether the offspring of a cross between Canadian and Mexican butterflies would be fertile.

C24. Why do you think that human trisomies 13, 18, and 21 can survive but the other trisomies are lethal? Even though X chromosomes are large, aneuploidies of this chromosome are also tolerated. Explain why.

C25. A zookeeper has collected a male and a female lizard that look like they belong to the same species. They mate with each other and produce phenotypically normal offspring. However, the offspring are sterile. Suggest one or more explanations for their sterility.

C26. What is endopolyploidy? What is its biological significance?

C27. What is mosaicism? How is it produced?

C28. Explain how polytene chromosomes of *Drosophila melanogaster* are produced and how they form a six-armed structure.

C29. Describe some of the advantages of polyploid plants. What are the consequences of having an odd number of chromosome sets?

C30. While conducting field studies on a chain of islands, you decide to karyotype two phenotypically identical groups of turtles, which are found on different islands. The turtles on one island have 24 chromosomes, but those on another island have 48 chromosomes. How would you explain this observation? How do you think the turtles with 48 chromosomes came into being? If you mated the two types of turtles together, would you expect their offspring to be phenotypically normal? Would you expect them to be fertile? Explain.

C31. A diploid fruit fly has eight chromosomes. Which of the following terms should not be used to describe a fruit fly with four sets of chromosomes?

A. Polyploid

B. Aneuploid

C. Euploid

D. Tetraploid

E. $4n$

C32. Which of the following terms should not be used to describe a human with three copies of chromosome 12?

A. Polyploid

B. Triploid

C. Aneuploid

D. Euploid

E. $2n + 1$

F. Trisomy 12

C33. The kidney bean, *Phaseolus vulgaris*, is a diploid species containing a total of 22 chromosomes in somatic cells. How many possible types of trisomic individuals could be produced in this species?

C34. The karyotype of a young girl who is affected with familial Down syndrome revealed a total of 46 chromosomes. Her older brother, however, who is phenotypically unaffected, actually had 45 chromosomes. Explain how this could happen. What would you expect to be the chromosomal number in the parents of these two children?

C35. A triploid plant has 18 chromosomes (i.e., 6 chromosomes per set). If we assume a gamete has an equal probability of receiving one or two copies of each of the six types of chromosome, what are the odds of this plant producing a monoploid or a diploid gamete? What are the odds of producing an aneuploid gamete? If the plant is allowed to self-fertilize, what are the odds of producing a euploid offspring?

C36. Describe three naturally occurring ways that chromosome number can change.

C37. Meiotic nondisjunction is much more likely than mitotic nondisjunction. Based on this observation, would you conclude that meiotic nondisjunction is usually due to nondisjunction during meiosis I or meiosis II? Explain your reasoning.

C38. A woman who is heterozygous, *Bb*, has brown eyes. *B* (brown) is a dominant allele, and *b* (blue) is recessive. In one of her eyes, however, there is a patch of blue color. Give three different explanations for how this might have occurred.

C39. What is an allodiploid? What factor determines the fertility of an allodiploid? Why are allotetraploids more likely to be fertile?

C40. What are homeologous chromosomes?

C41. Meiotic nondisjunction usually occurs during meiosis I. What is not separating properly: bivalents or sister chromatids? What is not separating properly during mitotic nondisjunction?

C42. Table 8.1 shows that Turner syndrome occurs when an individual inherits one X chromosome but lacks a second sex chromosome. Can Turner syndrome be due to nondisjunction during oogenesis, spermatogenesis, or both? If a phenotypically normal couple has a color-blind child (due to a recessive X-linked allele) with Turner syndrome, did nondisjunction occur during oogenesis or spermatogenesis in this child's parents? Explain your answer.

C43. Male honeybees, which are monoploid, produce sperm by meiosis. Explain what unusual event (compared with other animals) must occur during spermatogenesis in honeybees to produce sperm. Does this unusual event occur during meiosis I or meiosis II?

Experimental Questions

E1. What is the main goal of comparative genome hybridization? Explain how the ratio of green to red fluorescence provides information about chromosome structure.

E2. Let's suppose a researcher conducted comparative genomic hybridization (see Figure 8.9) and accidentally added twice as much (red) DNA from normal cells. What green-to-red ratio would you expect in a region from a chromosome from a cancer cell that carried a duplication on both chromosomal copies? What ratio would be observed for a region that was deleted on just one of the chromosomes from cancer cells?

E3. With regard to the analysis of chromosome structure, explain the experimental advantage that polytene chromosomes offer. Discuss why changes in chromosome structure are more easily detected in polytene chromosomes than in ordinary (nonpolytene) chromosomes.

E4. Describe how colchicine can be used to alter chromosome number.

E5. Describe the steps you would take to produce a tetraploid plant from a diploid plant.

E6. It is an exciting time to be a plant breeder because so many options are available for the development of new types of agriculturally useful plants. Let's suppose you wish to develop a seedless tomato that could grow in a very hot climate and is resistant to a viral pathogen that commonly infects tomato plants. At your disposal, you have a seed-bearing tomato strain that is heat-resistant and produces great-tasting tomatoes. You also have a wild strain of tomato plants (which have lousy-tasting tomatoes) that is resistant to the viral pathogen. Suggest a series of steps you might follow to produce a great-tasting, seedless tomato that is resistant to heat and the viral pathogen.

E7. What is a G band? Discuss how G bands are useful in the analysis of chromosome structure.

E8. A female fruit fly has one normal X chromosome and one X chromosome with a deletion. The deletion is in the middle of the X chromosome and is about 10% of the entire length of the X chromosome. If you stained and observed the chromosomes of this female fly in salivary gland cells, draw a picture of the polytene arm of the X chromosome. Explain your drawing.

Questions for Student Discussion/Collaboration

1. A chromosome involved in a reciprocal translocation also has an inversion. In addition, the cell contains two normal chromosomes.

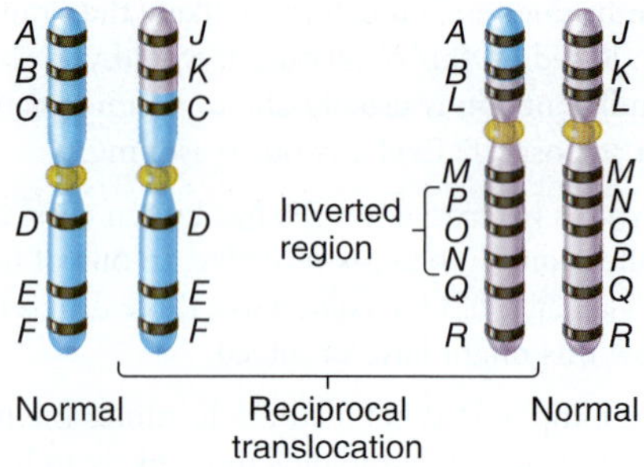

Make a drawing that shows how these chromosomes will pair during metaphase of meiosis I.

2. Besides the ones mentioned in this textbook, look for other examples of variations in euploidy. Perhaps you might look in more advanced textbooks concerning population genetics, ecology, or the like. Discuss the phenotypic consequences of these changes.

3. Cell biology textbooks often discuss cellular proteins encoded by genes that are members of a gene family. Examples of such proteins include myosins and glucose transporters. Look through a cell biology textbook and identify some proteins encoded by members of gene families. Discuss the importance of gene families at the cellular level.

4. Discuss how variation in chromosome number has been useful in agriculture.

Note: All answers appear at the website for this textbook; the answers to the even-numbered questions and all of the Concept Check and Comprehension Questions are in the back of the book.

www.mhhe.com/brookergenetics5e

Visit the website for practice tests, answer keys, and other learning aids for this chapter. Enhance your understanding of genetics with our interactive exercises, quizzes, animations, and much more.

CHAPTER OUTLINE

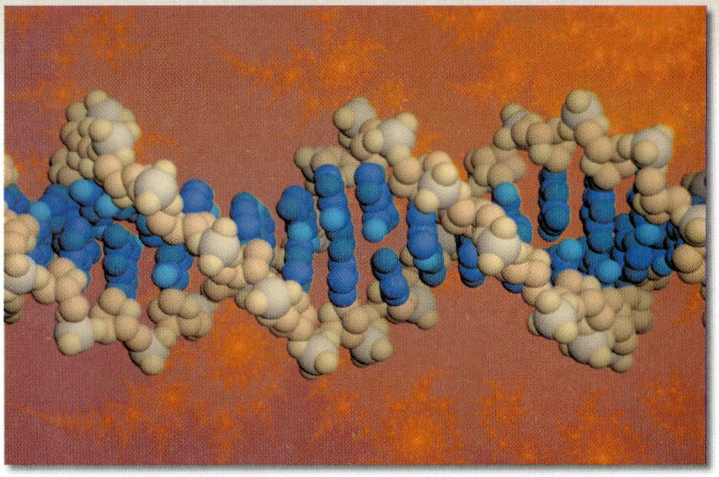

A molecular model showing the structure of the DNA double helix.

9 MOLECULAR STRUCTURE OF DNA AND RNA

In Chapters 2 through 8, we focused on the relationship between the inheritance of genes and chromosomes and the outcome of an organism's traits. In Chapter 9, we will shift our attention to **molecular genetics**—the study of DNA structure and function at the molecular level. An exciting goal of molecular genetics is to use our knowledge of DNA structure to understand how DNA functions as the genetic material. Using molecular techniques, researchers have determined the organization of many genes. This information, in turn, has helped us understand how the expression of genes governs the outcome of an individual's inherited traits.

The past several decades have seen dramatic advances in techniques and approaches to investigating and even to altering the genetic material. These advances have greatly expanded our understanding of molecular genetics and also have provided key insights into the mechanisms underlying transmission and population genetics. Molecular genetic technology is also widely used in supporting disciplines such as biochemistry, cell biology, and microbiology.

To a large extent, our understanding of genetics comes from our knowledge of the molecular structure of **DNA (deoxyribonucleic acid)** and **RNA (ribonucleic acid).** In this chapter, we will begin by considering classic experiments that showed that DNA is the genetic material. We will then survey the molecular features of DNA and RNA that underlie their function.

9.1 IDENTIFICATION OF DNA AS THE GENETIC MATERIAL

Learning Outcomes:

1. Describe the four criteria that the genetic material must fulfill.
2. Analyze the results of (1) Griffith, (2) Avery, MacLeod, and McCarty, and (3) Hershey and Chase, and explain how they indicate that DNA is the genetic material.

To fulfill its role, the genetic material must meet four criteria.

1. **Information:** The genetic material must contain the information necessary to construct an entire organism. In other words, it must provide the blueprint for determining the inherited traits of an organism.
2. **Transmission:** During reproduction, the genetic material must be passed from parents to offspring.
3. **Replication:** Because the genetic material is passed from parents to offspring, and from mother cell to daughter cells during cell division, it must be copied.
4. **Variation:** Within any species, a significant amount of phenotypic variability occurs. For example, Mendel studied several characteristics in pea plants that varied among different strains. These included height (tall versus

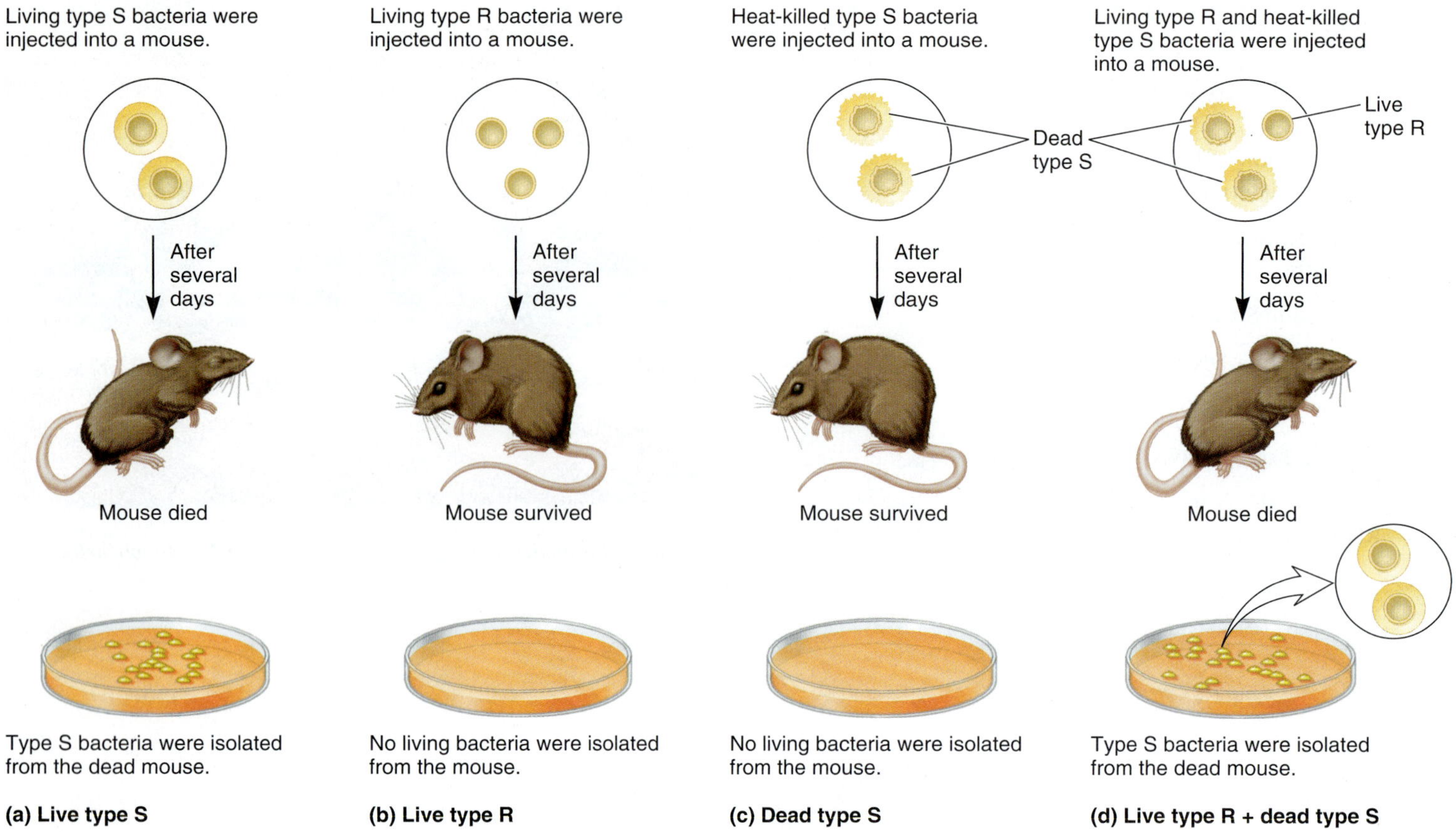

FIGURE 9.1 **Griffith's experiments on genetic transformation in *S. pneumoniae*.**

CONCEPT CHECK: Explain why the mouse in part (d) died.

dwarf) and seed color (yellow versus green). Therefore, the genetic material must also vary in ways that can account for the known phenotypic differences within each species.

In the 1900s, the data of many geneticists including those of Mendel were consistent with these four properties of genetic material. However, the experimental study of genetic crosses cannot, by itself, identify the chemical nature of the genetic material.

In the 1880s, August Weismann and Carl Nägeli championed the idea that a chemical substance within living cells is responsible for the transmission of traits from parents to offspring. The chromosome theory of inheritance was developed, and experimentation demonstrated that the chromosomes are the carriers of the genetic material (see Chapter 3). Nevertheless, the story was not complete because chromosomes contain both DNA and proteins. Also, RNA is found in the vicinity of chromosomes. Therefore, further research was needed to precisely identify the genetic material. In this section, we will examine the first experimental approaches to achieve this goal.

Experiments with *Streptococcus* Suggested That DNA Is the Genetic Material

Some early work in microbiology was important in developing an experimental strategy for identifying the genetic material. Frederick Griffith studied a type of bacterium known then as pneumococci and now classified as *Streptococcus pneumoniae*. Certain strains of *S. pneumoniae* secrete a polysaccharide capsule, whereas other strains do not. When streaked onto petri plates containing a solid growth medium, capsule-secreting strains have a smooth colony morphology, whereas those strains unable to secrete a capsule have a rough appearance.

The different forms of *S. pneumoniae* also affect their virulence, or ability to cause disease. When smooth strains of *S. pneumoniae* infect a mouse, the capsule allows the bacteria to escape attack by the mouse's immune system. As a result, the bacteria can grow and eventually kill the mouse. In contrast, the nonencapsulated (rough) bacteria are destroyed by the animal's immune system.

In 1928, Griffith conducted experiments that involved the injection of live and/or heat-killed bacteria into mice. He then observed whether or not the bacteria caused a lethal infection. Griffith was working with two strains of *S. pneumoniae*, a type S (S for smooth) and a type R (R for rough). When injected into a live mouse, the type S bacteria proliferated within the mouse's bloodstream and ultimately killed the mouse (**Figure 9.1a**). Following the death of the mouse, Griffith found many type S bacteria within the mouse's blood. In contrast, when type R bacteria were injected into a mouse, the mouse lived (**Figure 9.1b**). To verify that the proliferation of the smooth bacteria was causing the death of the mouse, Griffith killed the smooth bacteria with heat treatment before injecting them into the mouse. In this case, the mouse also survived (**Figure 9.1c**).

The critical and unexpected result was obtained in the experiment outlined in **Figure 9.1d**. In this experiment, live type R bacteria were mixed with heat-killed type S bacteria. As shown here, the mouse died. Furthermore, extracts from tissues of the

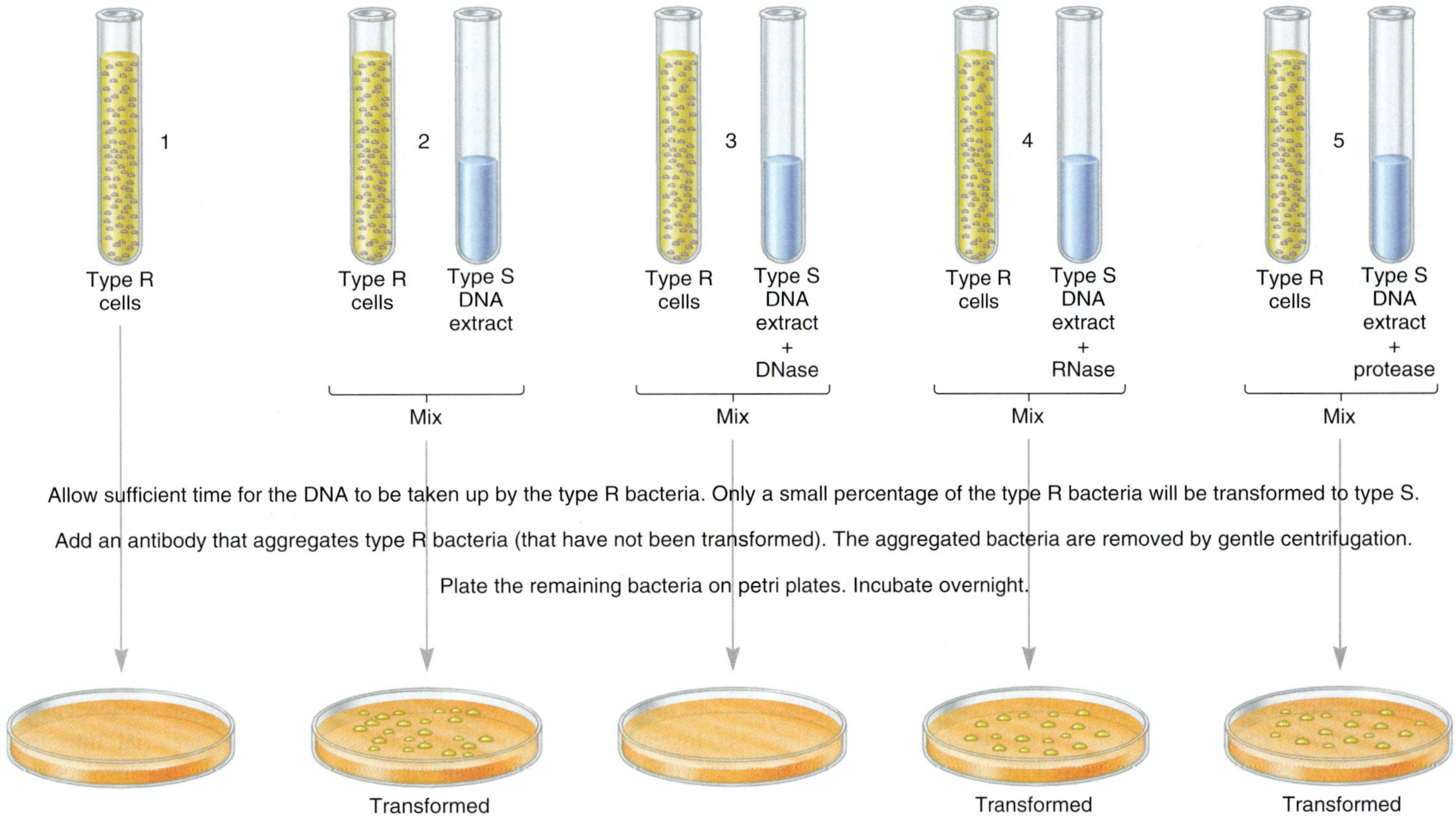

FIGURE 9.2 Experiments of Avery, MacLeod, and McCarty to identify the transforming principle. Samples of *Streptococcus pneumoniae* cells were either not exposed to a type S DNA extract (experiment 1, left side) or exposed to a type S DNA extract (experiments 2–5). Experiments 3, 4, and 5 also contained DNase, RNase, or protease, respectively. After incubation, the cells were exposed to antibodies, which are molecules that can specifically recognize the molecular structure of macromolecules. In this experiment, the antibodies recognized the cell surface of type R bacteria and caused them to clump together. The clumped bacteria were removed by a gentle centrifugation step. Only the bacteria that were not recognized by the antibody (namely, the type S bacteria) remained in the supernatant. The cells in the supernatant were plated on solid growth media. After overnight incubation, visible colonies may be observed.

CONCEPT CHECK: What was the purpose of adding RNase and protease to a DNA extract?

dead mouse were found to contain living type S bacteria! What can account for these results? Because living type R bacteria alone could not proliferate and kill the mouse (Figure 9.1b), the interpretation of the data in Figure 9.1d is that something from the dead type S bacteria was transforming the type R bacteria into type S bacteria. Griffith called this process **transformation,** and the unidentified substance causing this to occur was termed the transforming principle. The steps of bacterial transformation are described in Chapter 7 (see Figure 7.12).

At this point, let's look at what Griffith's observations mean in genetic terms. The transformed bacteria acquired the *information* to make a capsule. Among different strains, *variation* exists in the ability to create a capsule and to cause mortality in mice. The genetic material that is necessary to create a capsule must be *replicated* so it can be *transmitted* from mother to daughter cells during cell division. Taken together, these observations are consistent with the idea that the formation of a capsule is governed by the bacteria's genetic material, meeting the four criteria described previously. Griffith's experiments showed that some genetic material from the dead bacteria had been transferred to the living bacteria and provided them with a new trait. However, Griffith did not know what the transforming substance was.

Important scientific discoveries often take place when researchers recognize that someone else's experimental observations can be used to address a particular scientific question. Oswald Avery, Colin MacLeod, and Maclyn McCarty realized that Griffith's observations could be used as part of an experimental strategy to identify the genetic material. They asked the question, What substance is being transferred from the dead type S bacteria to the live type R?

At the time of these experiments in the 1940s, researchers already knew that DNA, RNA, proteins, and carbohydrates are major constituents of living cells. To separate these components and to determine if any of them was the genetic material, Avery, MacLeod, and McCarty used established biochemical purification procedures and prepared bacterial extracts from type S strains containing each type of these molecules. After many repeated attempts with different types of extracts, they discovered that only one of the extracts, namely, the one that contained purified DNA from type S bacteria, was able to convert the type R bacteria into type S. As shown in **Figure 9.2**, when this extract was mixed with type R bacteria, some of the bacteria were converted to type S. However, if no DNA extract was added, no type S bacterial colonies were observed on the petri plates.

A biochemist might point out that a DNA extract may not be 100% pure. In fact, any purified extract might contain small traces of some other substances. Therefore, one can argue that a small amount of contaminating material in the DNA extract might actually be the genetic material. The most likely contaminating substances in this case would be RNA or protein. To further verify that the DNA in the extract was indeed responsible for the transformation, Avery, MacLeod, and McCarty treated samples of the DNA extract with enzymes that digest DNA (called **DNase**), RNA (**RNase**), or protein (**protease**) (see Figure 9.2). When the DNA extracts were treated with RNase or protease, they still converted type R bacteria into type S. These results indicated that any RNA or protein in the extract was not acting as the genetic material. However, when the extract was treated with DNase, it lost its ability to convert type R into type S bacteria. These results indicated that the degradation of the DNA in the extract by DNase prevented conversion of type R to type S. This interpretation is consistent with the hypothesis that DNA is the genetic material. A more elegant way of saying this is that "DNA is the transforming principle."

EXPERIMENT 9A

Hershey and Chase Provided Evidence That DNA Is the Genetic Material of T2 Phage

A second experimental approach indicating that DNA is the genetic material came from the studies of Alfred Hershey and Martha Chase in 1952. Their research centered on the study of a virus known as T2. This virus infects *Escherichia coli* bacterial cells and is therefore known as a **bacteriophage,** or simply a **phage.** As shown in **Figure 9.3**, the external structure of the T2 phage, called the capsid or phage coat, consists of a head, sheath, tail fibers, and base plate. Biochemically, the phage coat is composed entirely of protein, which includes several different polypeptides. DNA is found inside the head of the T2 capsid. From a molecular point of view, this virus is rather simple, because it is composed of only two types of macromolecules: DNA and proteins.

Although the viral genetic material contains the blueprint to make new viruses, a virus itself cannot synthesize new viruses. Instead, a virus must introduce its genetic material into the cytoplasm of a living cell. In the case of T2, this first involves the attachment of its tail fibers to the bacterial cell wall and the subsequent injection of its genetic material into the cytoplasm of the cell (**Figure 9.4**). The phage coat remains attached on the outside of the bacterium and does not enter the cell. After the entry of the viral genetic material, the bacterial cytoplasm provides all of the machinery necessary to make viral proteins and DNA. The viral proteins and DNA assemble to make new viruses that are subsequently released from the cell by lysis (i.e., cell breakage).

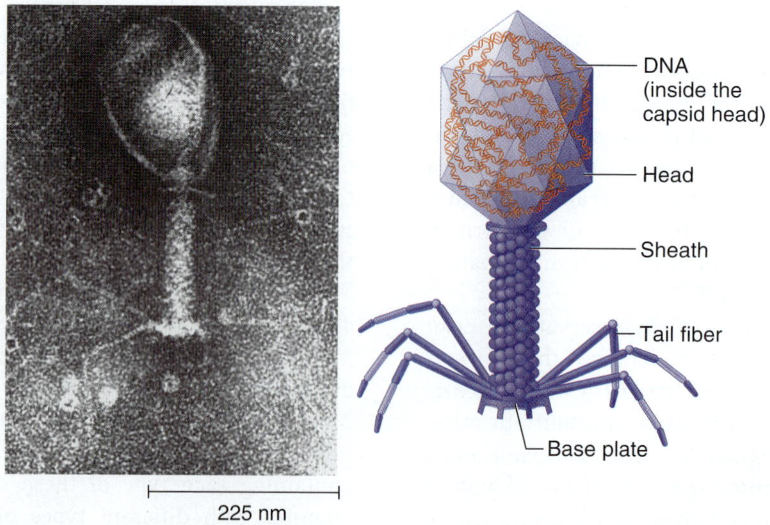

FIGURE 9.3 Structure of the T2 bacteriophage. The T2 bacteriophage is composed of a phage coat, or capsid, with genetic material inside the head of the capsid. The capsid is divided into regions called the head, sheath, tail fibers, and base plate. The components of the capsid are composed of proteins. The genetic material is composed of DNA.

Genes → Traits The genetic material of a bacteriophage contains many genes, which provide the blueprint for making new viruses. When the bacteriophage injects its genetic material into a bacterium, these genes are activated and direct the host cell to make new bacteriophages, as described in Figure 9.4.

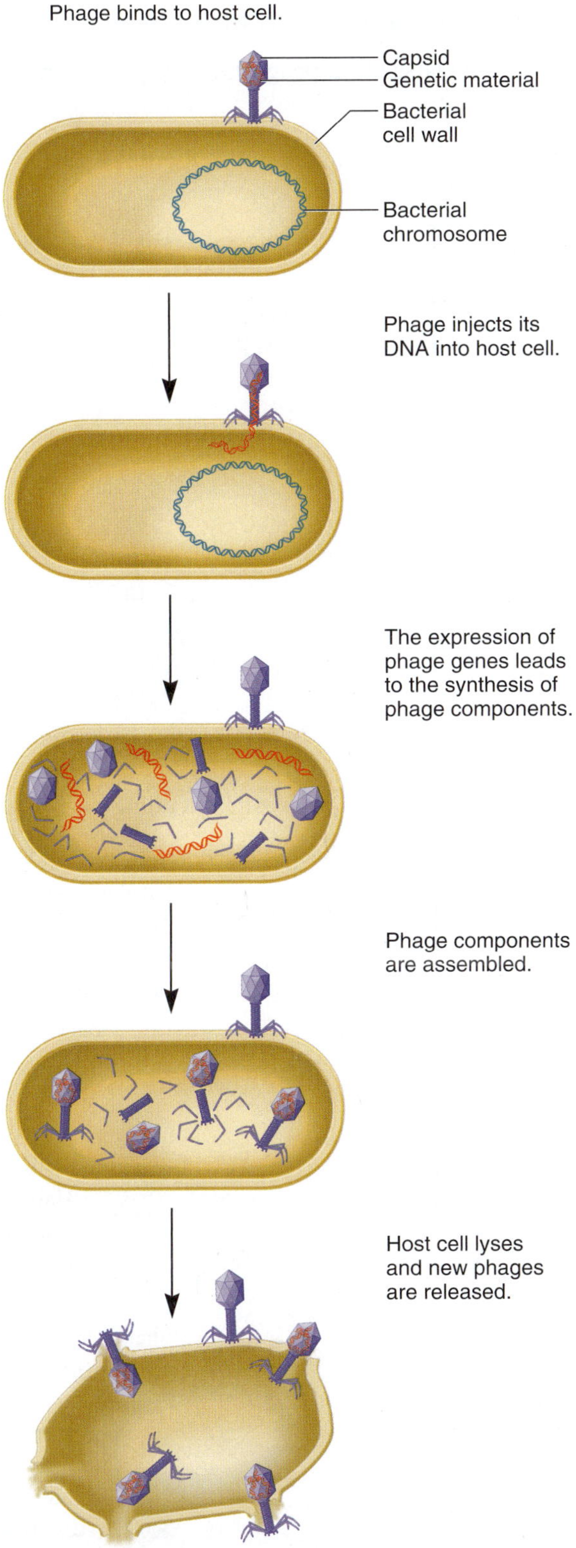

FIGURE 9.4 Reproductive cycle of the T2 bacteriophage.

ONLINE ANIMATION

CONCEPT CHECK: For infection to occur, what has to happen to the T2 genetic material?

To verify that DNA is the genetic material of T2, Hershey and Chase devised a method to separate the phage coat, which is attached to the outside of the bacterium, from the genetic material, which is injected into the cytoplasm. They were aware of microscopy experiments by Thomas Anderson showing that the T2 phage attaches itself to the outside of a bacterium by its tail fibers. Hershey and Chase reasoned that this is a fairly precarious attachment that could be disrupted by subjecting the bacteria to high shear forces, such as those produced in a kitchen blender. Their method was to expose bacteria to T2 phage, allowing sufficient time for the viruses to attach to bacteria and inject their genetic material. They then sheared the phage coats from the surface of the bacteria by a blender treatment. In this way, the phages' genetic material, which had been injected into the cytoplasm of the bacterial cells, could be separated from the phage coats that were sheared away.

Hershey and Chase used radioisotopes to distinguish proteins from DNA. Sulfur atoms are found in proteins but not in DNA, whereas phosphorus atoms are found in DNA but not in phage proteins. Therefore, ^{35}S (a radioisotope of sulfur) and ^{32}P (a radioisotope of phosphorus) were used to specifically label phage proteins and DNA, respectively. To make radiolabeled phages, they grew *E. coli* cells in media that contained ^{35}S or ^{32}P and then infected the *E. coli* cells with T2 phages. In this way, the newly made phages were radiolabeled with ^{35}S or ^{32}P.

In the experiment described in **Figure 9.5**, Hershey and Chase began with *E. coli* cells, ^{35}S-labeled phage, and ^{32}P-labeled phage. In one flask, *E. coli* cells were mixed with ^{35}S-labeled phage and, in the other flask, the cells were mixed with ^{32}P-labeled phage. The phages were given sufficient time to inject their genetic material into the cells, and then samples were taken from each flask and subjected to a shearing force using a blender. This treatment removed the phage coats from the surface of the bacterial cells. The samples were then subjected to centrifugation at a speed that caused the heavier bacterial cells to form a pellet at the bottom of the tube, whereas the lighter phage coats remained in the supernatant, the liquid found above the pellet. The amount of radioactivity in the supernatant (emitted from either ^{35}S or ^{32}P) was determined using a Geiger counter.

THE HYPOTHESIS

Only the genetic material of the phage is injected into the bacterium. Isotope labeling will reveal if it is DNA or protein.

TESTING THE HYPOTHESIS **FIGURE 9.5 Evidence that DNA is the genetic material of T2 bacteriophage.**

Starting materials: The starting materials were *E. coli* cells and two preparations of radiolabeled T2 phage. One phage preparation had phage proteins labeled with ^{35}S, and the other preparation had phage DNA labeled with ^{32}P.

1. Grow bacterial cells. Divide into two flasks.
2. Into one flask, add ^{35}S-labeled phage; in the second flask, add ^{32}P-labeled phage.
3. Allow infection to occur.
4. Agitate solutions in blenders for different lengths of time to shear the empty phages off the bacterial cells.
5. Centrifuge at 10,000 rpm.
6. The heavy bacterial cells sediment to the pellet, while the lighter phages remain in the supernatant. (See Appendix for explanation of centrifugation.)
7. Count the amount of both types of radioisotope in the supernatant with a Geiger counter. Compare it with the starting amount.

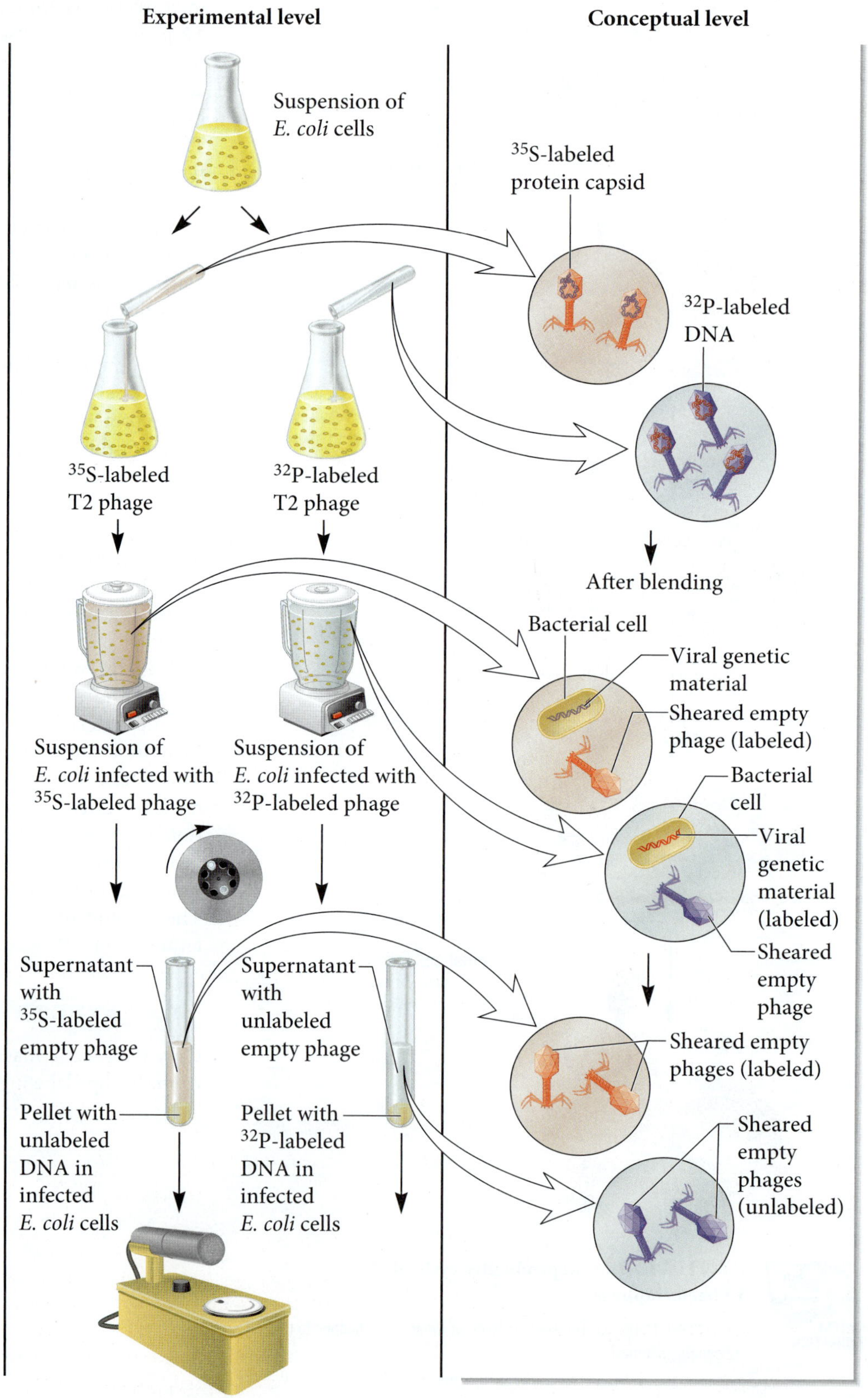

THE DATA

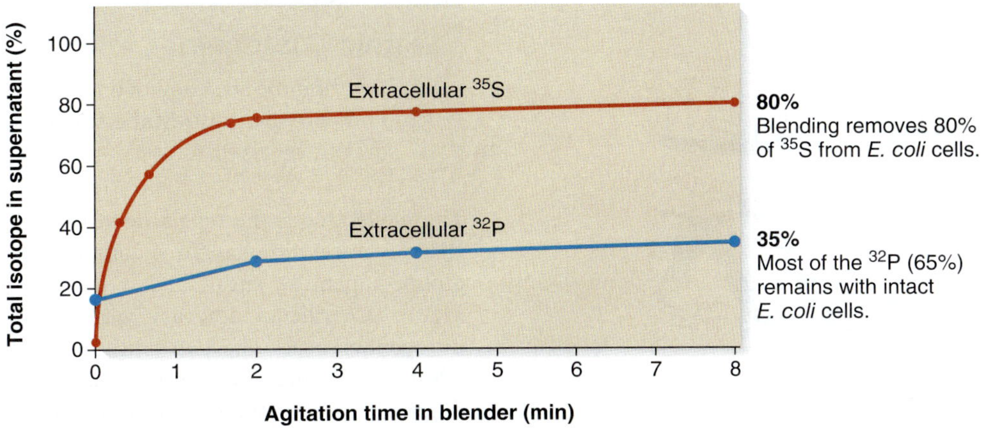

Data from A. D. Hershey and Martha Chase (1952) Independent Functions of Viral Protein and Nucleic Acid in Growth of Bacteriophage. *Journal of General Physiology* 36, 39–56.

INTERPRETING THE DATA

As seen in The Data, most of the ^{35}S isotope was found in the supernatant. Because the shearing force was expected to remove the phage coat, this result indicates that the empty phages contain primarily protein. By comparison, only about 35% of the ^{32}P was found in the supernatant following shearing. Therefore, most of the DNA was located within the bacterial cells in the pellet. These results are consistent with the idea that DNA is injected into the bacterial cytoplasm during infection, which would be the expected result if DNA is the genetic material.

By themselves, the results described in Figure 9.5 were not conclusive evidence that DNA is the genetic material. For example, you may have noticed that less than 100% of the phage protein was found in the supernatant. Therefore, some of the phage protein could have been introduced into the bacterial cells (and could function as the genetic material). Nevertheless, the results of Hershey and Chase were consistent with the conclusion that the genetic material is DNA rather than protein. Overall, their studies of the T2 phage were quite influential in convincing the scientific community that DNA is the genetic material.

We now know that bacteria, archaea, protists, fungi, plants, and animals all use DNA as their genetic material. Many viruses, such as T2, also use DNA as their genetic material. However, as discussed in Chapter 17, some viruses use RNA, rather than DNA, as their genetic material.

A self-help quiz involving this experiment can be found at www.mhhe.com/brookergenetics5e.

9.1 COMPREHENSION QUESTIONS

1. In the experiment of Avery, McLeod, and McCarty, the addition of RNase and protease to the DNA extracts
   a. prevented the conversion of type S bacteria into type R bacteria.
   b. allowed the conversion of type S bacteria into type R bacteria.
   c. prevented the conversion of type R bacteria into type S bacteria.
   d. allowed the conversion of type R bacteria into type S bacteria.
2. In the Hershey and Chase experiment involving T2 phage, after blender treatment and centrifugation
   a. most of the ^{32}P was in the supernatant and most of the ^{35}S was in the pellet.
   b. most of the ^{35}S was in the supernatant and most of the ^{32}P was in the pellet.
   c. equal amounts of ^{32}P and ^{35}S were in the supernatant and pellet.
   d. none of the above was observed.

9.2 OVERVIEW OF DNA AND RNA STRUCTURE

Learning Outcomes:

1. Define nucleic acid.
2. Describe the four levels of complexity of DNA.

DNA and its molecular cousin, RNA, are known as **nucleic acids.** This term is derived from the discovery of DNA by Friedrich Miescher in 1869. He identified a novel phosphorus-containing substance from the nuclei of white blood cells found in waste surgical bandages. He named this substance nuclein. As the structure of DNA and RNA became better understood, it was

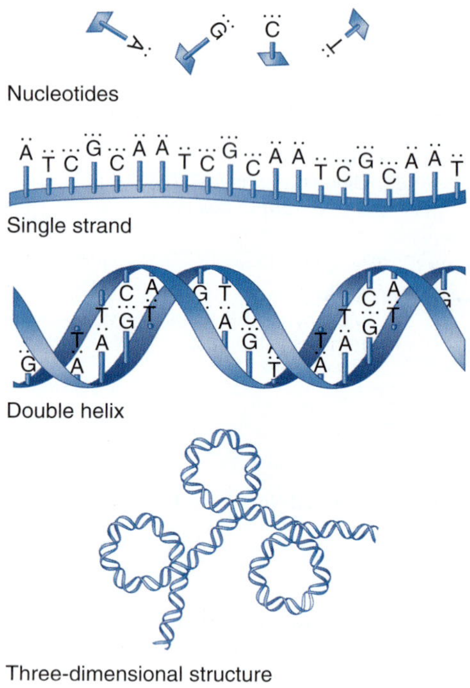

FIGURE 9.6 **Levels of nucleic acid structure.**

determined that they are acidic molecules, which means they release hydrogen ions (H^+) in solution and have a net negative charge at neutral pH. Thus, the name nucleic acid was coined.

Geneticists, biochemists, and biophysicists have been interested in the molecular structure of nucleic acids for several decades. Both DNA and RNA are macromolecules composed of smaller building blocks. To fully appreciate their structures, we need to consider four levels of complexity (**Figure 9.6**):

1. **Nucleotides** form the repeating structural unit of nucleic acids.
2. Nucleotides are linked together in a linear manner to form a **strand** of DNA or RNA.
3. Two strands of DNA (and sometimes strands of RNA) interact with each other to form a **double helix.**
4. The three-dimensional structure of DNA results from the folding and bending of the double helix. Within living cells, DNA is associated with a wide variety of proteins that influence its structure. Chapter 10 examines the roles of these proteins in creating the three-dimensional structure of DNA found within chromosomes.

9.2 COMPREHENSION QUESTION

1. When going from simple to complex, which of the following is the proper order for DNA?
   a. Nucleotide, double helix, DNA strand, chromosome
   b. Nucleotide, chromosome, double helix, DNA strand
   c. Nucleotide, DNA strand, double helix, chromosome
   d. Chromosome, nucleotide, DNA strand, double helix

9.3 NUCLEOTIDE STRUCTURE

Learning Outcomes:

1. Describe the structure of a nucleotide.
2. Compare and contrast the structure of nucleotides found in DNA versus RNA.

The nucleotide is the repeating structural unit of both DNA and RNA. A nucleotide has three components: at least one phosphate group, a pentose sugar, and a nitrogenous base. As shown in **Figure 9.7**, nucleotides vary with regard to the sugar and the nitrogenous base. The two types of sugars are **deoxyribose** and **ribose,** which are found in DNA and RNA, respectively. The five different bases are subdivided into two categories: the **purines** and the **pyrimidines.** The purine bases, **adenine (A)** and **guanine (G),** contain a double-ring structure; the pyrimidine bases, **thymine (T), cytosine (C),** and **uracil (U),** contain a single-ring structure. The sugar in DNA is always deoxyribose. In RNA, the sugar is always ribose. Also, the base thymine is not found in RNA. Rather, uracil is found in RNA instead of thymine. Adenine, guanine, and cytosine occur in both DNA and RNA. As noted in Figure 9.7, the bases and sugars have a standard numbering system. The nitrogen and carbon atoms found in the ring structure of the bases are given numbers 1 through 9 for the purines and 1 through 6 for the pyrimidines. In comparison, the five carbons found in the sugars are designated with primes, such as 1′, to distinguish them from the numbers found in the bases.

Figure 9.8 shows the repeating unit of nucleotides found in DNA and RNA. The locations of the attachment sites of the base and phosphate to the sugar molecule are important to the nucleotide's function. In the sugar ring, carbon atoms are numbered in a clockwise direction, beginning with a carbon atom adjacent to the ring oxygen atom. The fifth carbon is outside the ring structure. In a single nucleotide, the base is always attached to the 1′ carbon atom, and one or more phosphate groups are attached at the 5′ position. As discussed later, the –OH group attached to the 3′ carbon is important in allowing nucleotides to form covalent linkages with each other.

The terminology used to describe nucleic acid units is based on three structural features: the type of sugar, the type of base, and the number of phosphate groups. When a sugar is attached to only a base, this pair is a **nucleoside.** If ribose is attached to adenine, this nucleoside is called adenosine (**Figure 9.9**). Nucleosides composed of ribose and guanine, cytosine, or uracil are named guanosine, cytidine, and uridine, respectively. Nucleosides made of deoxyribose and adenine, guanine, thymine or cytosine are called deoxyadenosine, deoxyguanosine, deoxythymidine, and deoxycytidine, respectively.

The covalent attachment of one or more phosphate molecules to a nucleoside creates a nucleotide. One or more phosphate groups are attached to a sugar via an ester bond. If a nucleotide contains ribose, adenine, and one phosphate, it is named adenosine monophosphate, abbreviated AMP. A nucleotide composed of ribose, adenine, and three phosphate groups is called adenosine triphosphate, or ATP. By comparison, a nucleotide made of deoxyribose, adenine, and three phosphate groups is referred to as deoxyadenosine triphosphate (dATP).

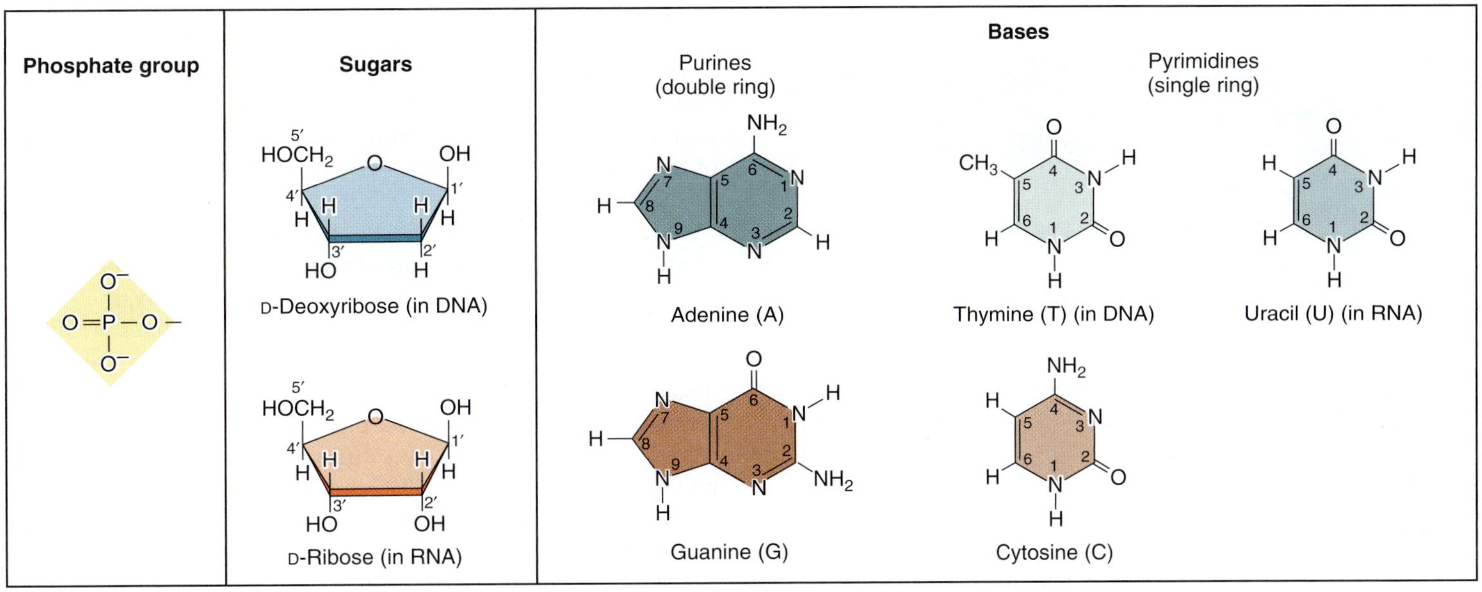

FIGURE 9.7 **The components of nucleotides.** The three building blocks of a nucleotide are one or more phosphate groups, a sugar, and a base. The bases are categorized as purines (adenine and guanine) and pyrimidines (thymine, cytosine, and uracil). Note: the location of a double bond in phosphate is not fixed. Because the sharing of electrons between phosphorus and the oxygen atoms is delocalized, phosphate exists in multiple resonance structures.

INTERACTIVE EXERCISE

CONCEPT CHECK: Which of these components are not found in DNA?

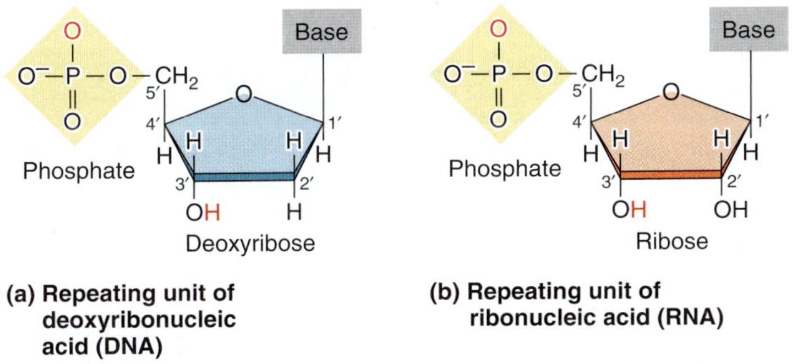

FIGURE 9.8 **The structure of nucleotides found in (a) DNA and (b) RNA.** DNA contains deoxyribose as its sugar and the bases A, T, G, and C. RNA contains ribose as its sugar and the bases A, U, G, and C. In a DNA or RNA strand, the oxygen on the 3′ carbon is linked to the phosphorus atom of phosphate in the adjacent nucleotide. The two atoms (O and H) shown in red are found within individual nucleotides but not when nucleotides are joined together to make strands of DNA and RNA.

INTERACTIVE EXERCISE

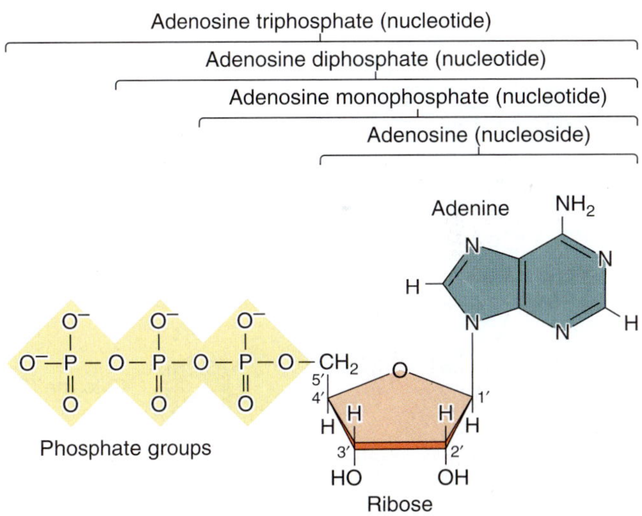

FIGURE 9.9 **A comparison between the structures of an adenine-containing nucleoside and nucleotides.**

INTERACTIVE EXERCISE

9.3 COMPREHENSION QUESTIONS

1. Which of the following could be the components of a single nucleotide found in DNA?
   a. Deoxyribose, adenine, and thymine
   b. Ribose, phosphate, and cytosine
   c. Deoxyribose, phosphate, and thymine
   d. Ribose, phosphate, and uracil

2. A key difference between the nucleotides found in DNA versus RNA is that
   a. DNA has phosphate, but RNA does not.
   b. DNA has deoxyribose, but RNA has ribose.
   c. DNA has thymine, but RNA has uracil.
   d. both b and c are correct.

9.4 STRUCTURE OF A DNA STRAND

Learning Outcome:

1. Describe the structural features of a DNA strand.

A strand of DNA or RNA has nucleotides that are linked to each other in a linear fashion. **Figure 9.10** depicts a short strand of DNA with four nucleotides. A few structural features are worth noting. A phosphate group connects two sugar molecules via ester bonds. For this reason, the linkage in DNA or RNA strands is called a **phosphodiester linkage.** The phosphates and sugar molecules form the **backbone** of a DNA or RNA strand. The bases project from the backbone. The backbone is negatively charged due to a negative charge on each phosphate.

A second important structural feature is the orientation of the nucleotides. As mentioned, the carbon atoms in a sugar molecule are numbered in a particular way. In Figure 9.10, the 5′ carbons in every sugar molecule are above the 3′ carbons. Therefore, a strand has a **directionality** because all sugar molecules have the same orientation. In this example, the direction of the strand is 5′ to 3′ when going from top to bottom.

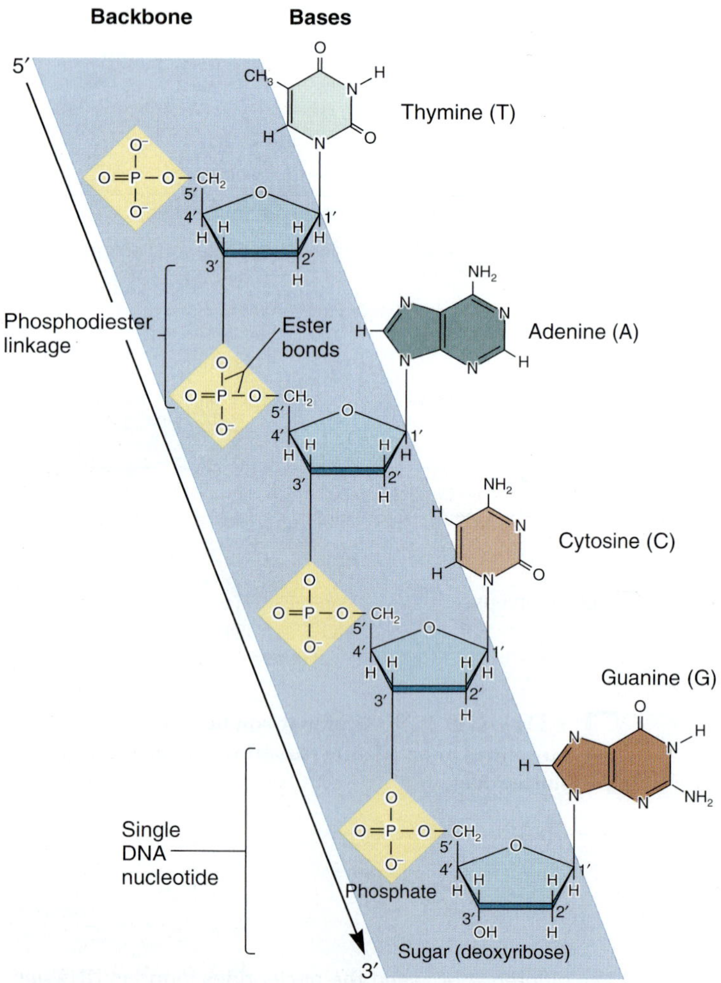

FIGURE 9.10 A short strand of DNA containing four nucleotides. Nucleotides are covalently linked together to form a strand of DNA.

CONCEPT CHECK: Which components of a nucleotide form the backbone of a DNA strand?

A critical aspect regarding DNA and RNA structure is that a strand contains a specific sequence of bases. In Figure 9.10, the sequence of bases is thymine–adenine–cytosine–guanine, abbreviated TACG. Furthermore, to show the directionality, the strand should be abbreviated 5′–TACG–3′. The nucleotides within a strand are covalently attached to each other via phosphodiester linkages, so the sequence of bases cannot shuffle around and become rearranged. Therefore, the sequence of bases in a DNA strand remains the same over time, except in rare cases when mutations occur. As we will see throughout this textbook, the sequence of bases within DNA and RNA is the defining feature that allows them to carry information.

9.4 COMPREHENSION QUESTION

1. In a DNA strand, a phosphate connects a 3′ carbon atom in one deoxyribose to
   a. a 5′ carbon in an adjacent deoxyribose.
   b. a 3′ carbon in an adjacent deoxyribose.
   c. a base in an adjacent nucleotide.
   d. none of the above.

9.5 DISCOVERY OF THE DOUBLE HELIX

Learning Outcome:

1. Outline the key experiments that led to the discovery of the DNA double helix.

A major discovery in molecular genetics was made in 1953 by James Watson and Francis Crick. At that time, DNA was already known to be composed of nucleotides. However, it was not understood how the nucleotides are bonded together to form the structure of DNA. Watson and Crick committed themselves to determining the structure of DNA because they felt this knowledge was needed to understand the functioning of genes. Other researchers, such as Rosalind Franklin and Maurice Wilkins, shared this view. Before we examine the characteristics of the double helix, let's consider the events that provided the scientific framework for Watson and Crick's breakthrough.

A Few Key Events Led to the Discovery of the Double-Helix Structure

One method that proved important in the discovery of the structure of the DNA double helix was model building. In the early 1950s, Linus Pauling proposed that regions of proteins can fold into a secondary structure known as an α helix (**Figure 9.11a**). To elucidate this structure, Pauling built large models by linking together simple ball-and-stick units (**Figure 9.11b**). By carefully scaling the objects in his models, he could visualize if

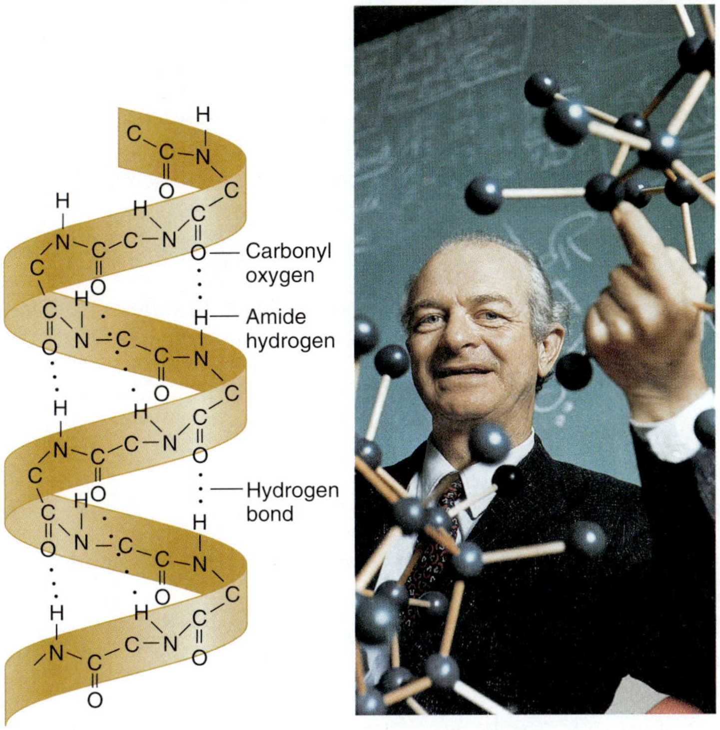

(a) An α helix in a protein **(b) Linus Pauling**

FIGURE 9.11 Linus Pauling and the α-helix protein structure. **(a)** An α helix is a secondary structure found in proteins. This structure emphasizes the polypeptide backbone (shown as a tan ribbon), which is composed of amino acids linked together in a linear fashion. Hydrogen bonding between hydrogen and oxygen atoms stabilizes the helical conformation. **(b)** Linus Pauling with a ball-and-stick model.

CONCEPT CHECK: Why is modeling useful?

(a) Rosalind Franklin

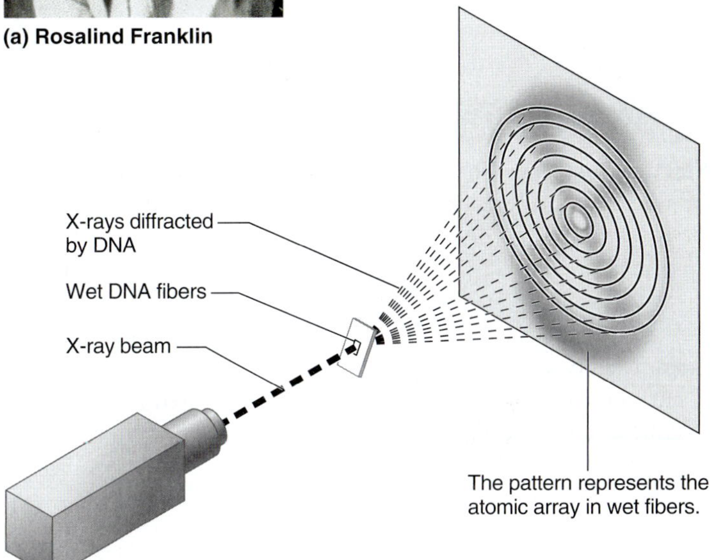

(b) X-ray diffraction of wet DNA fibers

FIGURE 9.12 X-ray diffraction of DNA.

atoms fit together properly in a complicated three-dimensional structure. As we will see, Watson and Crick also used a ball-and-stick approach to solve the structure of the DNA double helix. Interestingly, they were well aware that Pauling might figure out the structure of DNA before they did. This stimulated a rivalry between the researchers. It is worth noting that the use of models is still an important tool in understanding structural features of biomolecules. However, modern-day molecular geneticists usually construct their three-dimensional models on computers.

A second important development that led to the elucidation of the double helix was X-ray diffraction data. When a purified substance, such as DNA, is subjected to X-rays, it produces a well-defined diffraction pattern if the molecule is organized into a regular structural pattern. An interpretation of the diffraction pattern (using mathematical theory) can ultimately provide information concerning the structure of the molecule. Rosalind Franklin (**Figure 9.12a**), working in the same laboratory as Maurice Wilkins, used X-ray diffraction to study wet DNA fibers. Franklin made marked advances in X-ray diffraction techniques while working with DNA. She adjusted her equipment to produce an extremely fine beam of X-rays. She extracted finer DNA fibers than ever before and arranged them in parallel bundles. Franklin also studied the fibers' reactions to humid conditions. The diffraction pattern of Franklin's DNA fibers is shown in **Figure 9.12b**. This pattern suggested several structural features of DNA. First, it was consistent with a helical structure. Second, the diameter of the helical structure was too wide to be only a single-stranded helix. Finally, the diffraction pattern indicated that the helix contains about 10 base pairs (bp) per complete turn. These observations were instrumental in solving the structure of DNA.

EXPERIMENT 9B

Chargaff Found That DNA Has a Biochemical Composition in Which the Amount of A Equals T and the Amount of G Equals C

Another piece of information that led to the discovery of the double-helix structure came from the studies of Erwin Chargaff. In the 1940s and 1950s, he pioneered many of the biochemical techniques for the isolation, purification, and measurement of nucleic acids from living cells. This was no trivial undertaking, because the biochemical composition of living cells is complex. At the time of Chargaff's work, researchers already knew that the building blocks of DNA are nucleotides containing the bases adenine, thymine, guanine, or cytosine. Chargaff analyzed the base composition of DNA, which was isolated from many different species. He expected that the results might provide important clues concerning the structure of DNA.

Chargaff's experimental protocol is described in **Figure 9.13**. He began with various types of cells as starting material. The chromosomes were extracted from cells and then treated with protease to separate the DNA from chromosomal proteins. The DNA was then treated with a strong acid, which cleaved the bonds between the sugars and bases, thereby releasing the individual bases from the DNA strands. This mixture of bases was then subjected to paper chromatography to separate the four types. The amounts of the four bases were determined spectroscopically.

THE GOAL (DISCOVERY-BASED SCIENCE)

An analysis of the base composition of DNA in different organisms may reveal important features about the structure of DNA.

ACHIEVING THE GOAL

FIGURE 9.13 **An analysis of base composition among different DNA samples.**

Starting material: The following types of cells were obtained: *Escherichia coli*, *Streptococcus pneumoniae*, yeast, turtle red blood cells, salmon sperm cells, chicken red blood cells, and human liver cells.

1. For each type of cell, extract the chromosomal material. This can be done in a variety of ways, including the use of high salt, detergent, or mild alkali treatment. Note: The chromosomes contain both DNA and protein.

2. Remove the protein. This can be done in several ways, including treament with protease.

3. Hydrolyze the DNA to release the bases from the DNA strands. A common way to do this is by strong acid treatment.

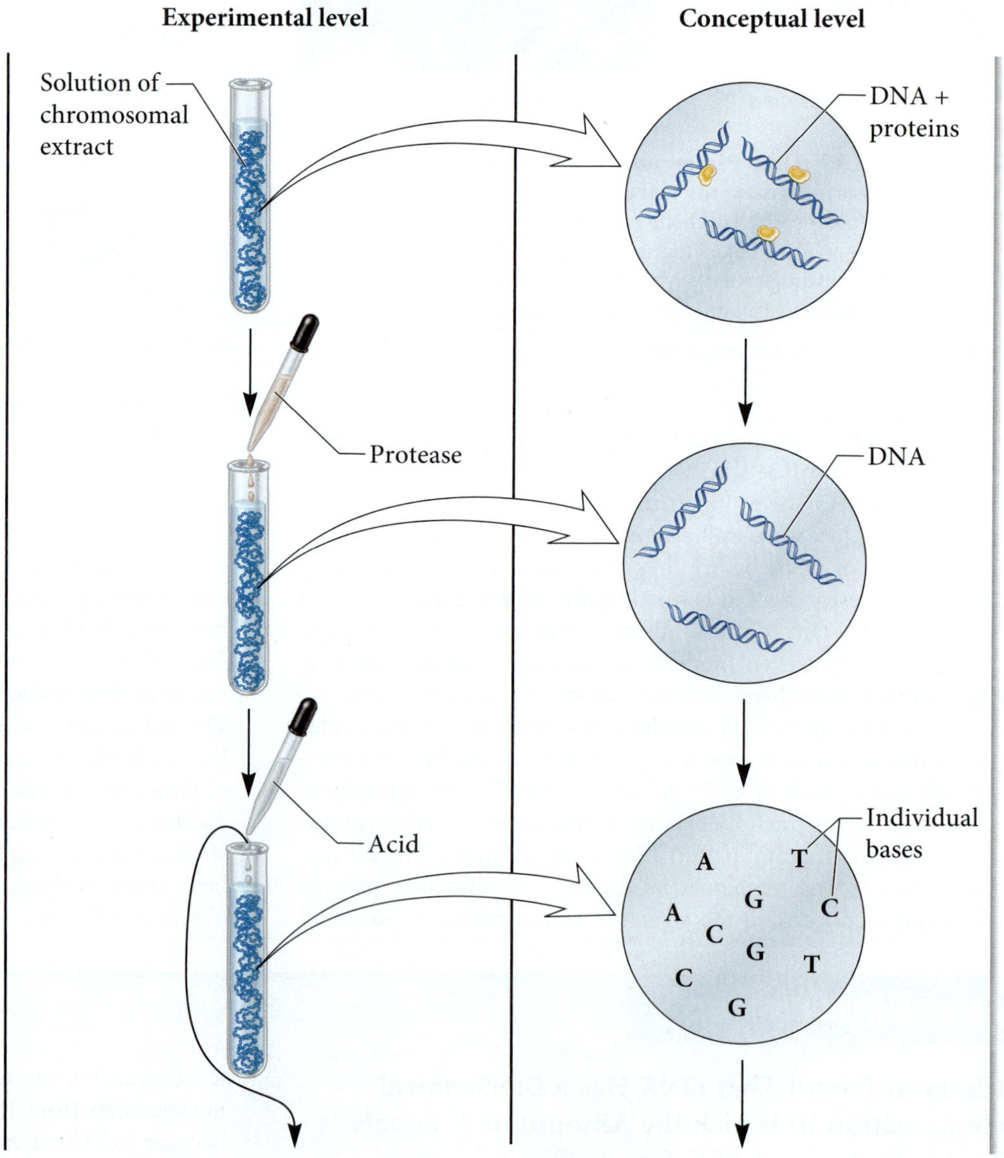

4. Separate the bases by chromatography. Paper chromatography provides an easy way to separate the four types of bases. (The technique of chromatography is described in the Appendix.)

5. Extract bands from paper into solutions and determine the amounts of each base by spectroscopy. Each base will absorb light at a particular wavelength. By examining the absorption profile of a sample of base, it is then possible to calculate the amount of the base. (Spectroscopy is described in the Appendix.)

6. Compare the base content in the DNA from different organisms.

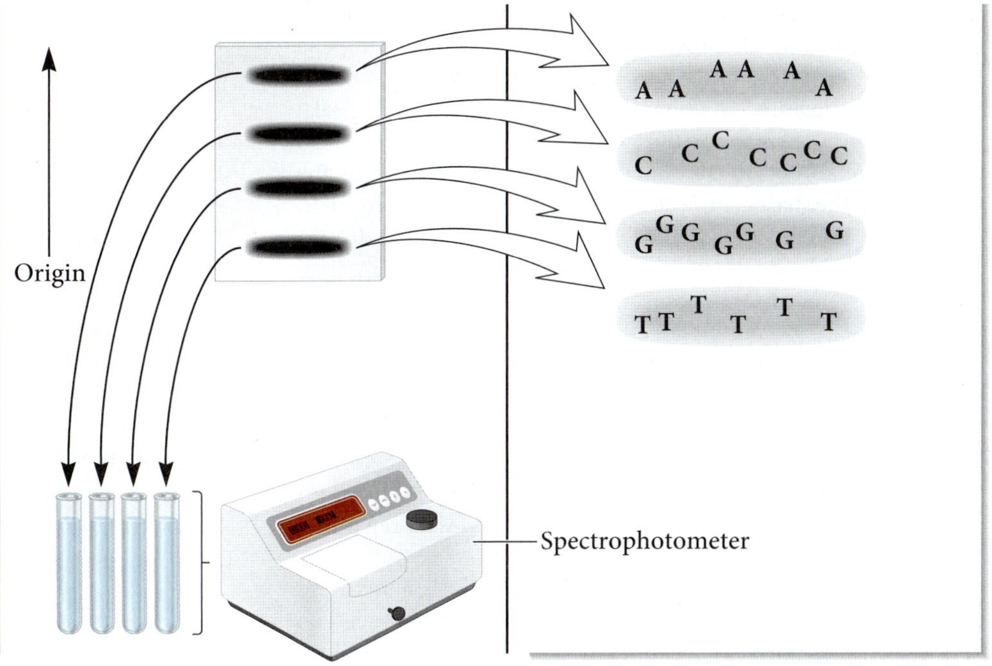

THE DATA

Base Content in the DNA from a Variety of Organisms*

	Percentage of Bases (based on molarity)			
Organism	*Adenine*	*Thymine*	*Guanine*	*Cytosine*
Escherichia coli	26.0	23.9	24.9	25.2
Streptococcus pneumoniae	29.8	31.6	20.5	18.0
Yeast	31.7	32.6	18.3	17.4
Turtle red blood cells	28.7	27.9	22.0	21.3
Salmon sperm	29.7	29.1	20.8	20.4
Chicken red blood cells	28.0	28.4	22.0	21.6
Human liver cells	30.3	30.3	19.5	19.9

*When the base compositions from different tissues within the same species were measured, similar results were obtained. These data were compiled from several sources. See E. Chargaff and J. Davidson, Eds. (1995) *The Nucleic Acids*. Academic Press, New York.

INTERPRETING THE DATA

The Data shown in Figure 9.13 are only a small sampling of Chargaff's results. During the late 1940s and early 1950s, Chargaff published many papers concerned with the chemical composition of DNA from biological sources. Hundreds of measurements were made. The compelling observation was that the amount of adenine was similar to that of thymine, and the amount of guanine was similar to cytosine. The idea that the amount of A in DNA equals the amount of T, and the amount of G equals C, is known as **Chargaff's rule.**

These results were not sufficient to propose a model for the structure of DNA. However, they provided the important clue that DNA is structured so that each molecule of adenine interacts with a thymine, and each molecule of guanine interacts with a cytosine. A DNA structure in which A binds to T, and G to C, would explain the equal amounts of A and T, and G and C observed in Chargaff's experiments. As we will see, this observation became crucial evidence that Watson and Crick used to elucidate the structure of the double helix.

A self-help quiz involving this experiment can be found at www.mhhe.com/brookergenetics5e.

Watson and Crick Deduced the Double-Helical Structure of DNA

Thus far, we have examined key pieces of information used to determine the structure of DNA. In particular, the X-ray diffraction work of Franklin suggested a helical structure composed of two or more strands with 10 bases per turn. In addition, Chargaff's work indicated that the amount of A equals T, and the amount of G equals C. Furthermore, Watson and Crick were familiar with Pauling's success in using ball-and-stick models to deduce the secondary structure of proteins. With these key observations, they set out to solve the structure of DNA.

Watson and Crick assumed DNA is composed of nucleotides that are linked together in a linear fashion. They also assumed the chemical linkage between two nucleotides is always the same. With these ideas in mind, they tried to build ball-and-stick models that incorporated the known experimental observations. During this time, Rosalind Franklin had produced even clearer X-ray diffraction patterns, which provided greater detail concerning the relative locations of the bases and backbone of DNA. This major breakthrough suggested a two-strand interaction that was helical.

In their model building, Watson and Crick's emphasis shifted to models containing the two backbones on the outside of the model, with the bases projecting toward each other. At first, a structure was considered in which the bases form hydrogen bonds with the identical base in the opposite strand (A to A, T to T, G to G, and C to C). However, the model building revealed that the bases could not fit together this way. The final hurdle was overcome when it was realized that the hydrogen bonding of adenine to thymine was structurally similar to that of guanine to cytosine. With an interaction between A and T and between G and C, the ball-and-stick models showed that the two strands would fit together properly. This ball-and-stick model, shown in **Figure 9.14**, was consistent with all of the known data regarding DNA structure.

For their work, Watson, Crick, and Maurice Wilkins were awarded the 1962 Nobel Prize in physiology or medicine. The contribution of Rosalind Franklin to the discovery of the double helix was also critical and has been acknowledged in several books and articles. Franklin was independently trying to solve the structure of DNA. However, Wilkins, who worked in the same laboratory, shared Franklin's X-ray data with Watson and Crick, presumably without her knowledge. This provided important information that helped them solve the structure of DNA, which was published in the journal *Nature* in April 1953. Though she was not given credit in the original publication of the double-helix structure, Franklin's key contribution became known in later years. Unfortunately, however, Rosalind Franklin died in 1958, and so could not share in the Nobel Prize because it is not awarded posthumously.

FIGURE 9.14 Watson and Crick and their model of the DNA double helix. James Watson is shown here on the left and Francis Crick on the right, next to the molecular model they originally proposed for the double helix. Each strand contains a sugar-phosphate backbone. In opposite strands, A hydrogen bonds to T, and G hydrogen bonds with C.

9.5 COMPREHENSION QUESTIONS

1. Evidence or approaches that led to the discovery of the DNA double helix include
   a. the determination of structures using ball-and-stick models.
   b. the X-ray diffraction data of Franklin.
   c. the base composition data of Chargaff.
   d. all of the above.
2. Chargaff's analysis of the base composition of DNA is consistent with base pairing between
   a. A and G, and T and C.
   b. A and A, G and G, T and T, and C and C.
   c. A and T, and G and C.
   d. A and C, and T and G.

9.6 STRUCTURE OF THE DNA DOUBLE HELIX

Learning Outcomes:

1. Outline the key structural features of the DNA double helix.
2. Compare and contrast B DNA and Z DNA.
3. Describe how triplex DNA is formed.

As we have seen, the discovery of the DNA double helix required different kinds of evidence. In this section, we will examine its structure in greater detail.

The Molecular Structure of the DNA Double Helix Has Several Key Features

The general structural features of the double helix are shown in **Figure 9.15**. In a DNA double helix, two DNA strands are twisted together around a common axis to form a structure that resembles a spiral staircase. This double-stranded structure is stabilized by **base pairs (bp)**—pairs of bases in opposite strands that are hydrogen bonded to each other. As you move past 10 bp, you have gone 360° around the backbone. The linear distance of a complete turn is 3.4 nm; each base pair traverses 0.34 nm.

A distinguishing feature of the hydrogen bonding between base pairs is its specificity. An adenine base in one strand hydrogen bonds with a thymine base in the opposite strand, or a guanine

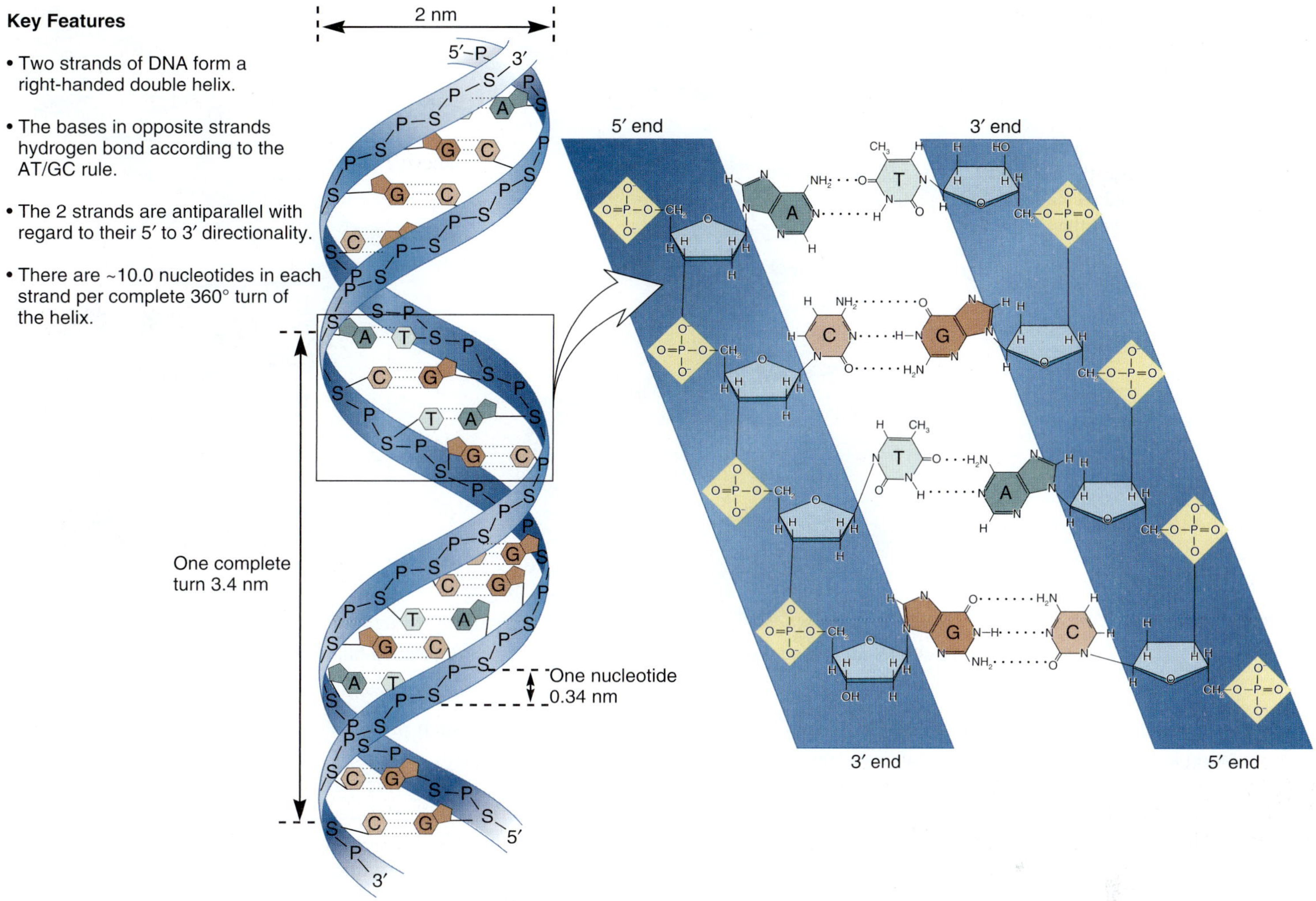

ONLINE ANIMATION

FIGURE 9.15 Key features of the structure of the double helix. Note: In the drawing on the left and in the inset, the planes of the bases and sugars are shown parallel to each other in order to depict the hydrogen bonding between the bases. In an actual DNA molecule, the bases would be rotated about 90° so the planes of the bases would be facing each other, as shown in Figure 9.16a.

CONCEPT CHECK: What holds the DNA strands together?

base hydrogen bonds with a cytosine. This **AT/GC rule** explained the earlier data of Chargaff, showing that the DNA from many organisms contains equal amounts of A and T, and equal amounts of G and C (see Figure 9.13). The AT/GC rule indicates that purines (A and G) always bond with pyrimidines (T and C). This keeps the width of the double helix relatively constant. As noted in Figure 9.15, three hydrogen bonds occur between G and C but only two between A and T. For this reason, DNA sequences with a high proportion of G and C tend to form more stable double-stranded structures.

The AT/GC rule implies that we can predict the sequence in one DNA strand if the sequence in the opposite strand is known. For example, let's consider a DNA strand with the sequence of 5′–ATGGCGGATTT–3′. The opposite strand would have to be 3′–TACCGCCTAAA–5′. Using genetic terms, we would say that these two sequences are **complementary** to each other or that the two sequences exhibit complementarity. In addition, you may have noticed that the sequences are labeled with 5′ and 3′ ends. These numbers designate the direction of the DNA backbones. The direction of DNA strands is depicted in the inset to Figure 9.15. When going from the top of this figure to the bottom, one strand is running in the 5′ to 3′ direction, and the other strand is 3′ to 5′. This opposite orientation of the two DNA strands is referred to as an **antiparallel** arrangement. An antiparallel structure was initially proposed in the models of Watson and Crick.

Figure 9.16a is a schematic model that emphasizes certain molecular features of DNA structure. The bases in this model are depicted as flat rectangular structures that hydrogen bond in pairs. (The hydrogen bonds are the dotted lines.) Although the bases are not actually rectangular, they do form flattened planar structures. Within DNA, the bases are oriented so the flattened regions are facing each other, an arrangement referred to as base stacking. In other words, if you think of the bases as flat plates, these plates are stacked on top of each other in the

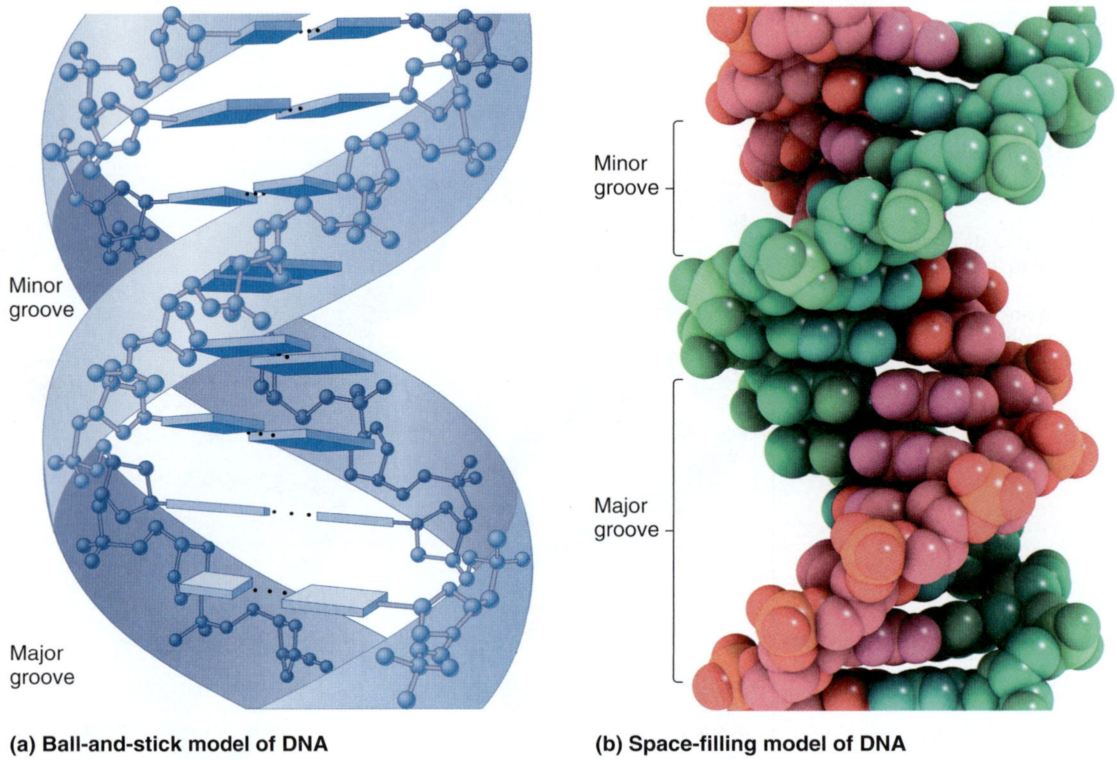

ONLINE ANIMATION

FIGURE 9.16 **Two models of the double helix.** **(a)** Ball-and-stick model of the double helix. The deoxyribose-phosphate backbone is shown in detail, whereas the bases are depicted as flattened rectangles. **(b)** Space-filling model of the double helix.

CONCEPT CHECK: Describe the major and minor grooves.

double-stranded DNA structure. Along with hydrogen bonding, base stacking is a structural feature that stabilizes the double helix by excluding water molecules. The helical structure of the DNA backbone depends on the hydrogen bonding between base pairs and also on base stacking.

By convention, the direction of the DNA double helix shown in Figure 9.16a spirals in a direction that is called "right-handed." To understand this terminology, imagine that a double helix is laid on your desk; one end of the helix is close to you, and the other end is at the opposite side of the desk. As it spirals away from you, a right-handed helix turns in a clockwise direction. By comparison, a left-handed helix spirals in a counterclockwise manner. Both strands in Figure 9.16a spiral in a right-handed direction.

Figure 9.16b is a space-filling model for DNA in which the atoms are represented by spheres. This model emphasizes the surface features of DNA. Note that the backbone—composed of sugar and phosphate groups—is on the outermost surface. In a living cell, the backbone has the most direct contact with water. In contrast, the bases are more internally located within the double-stranded structure. Biochemists use the term **grooves** to describe the indentations where the atoms of the bases are in contact with the surrounding water. As you travel around the DNA helix, the structure of DNA has two grooves: the **minor groove** and the **major groove.**

As discussed in later chapters, proteins can bind to DNA and affect its conformation and function. For example, some proteins hydrogen bond to the bases within the major groove. This hydrogen bonding can be very precise so that a protein interacts with a particular sequence of bases. In this way, a protein recognizes a specific gene and affect its ability to be transcribed. We will consider such proteins in Chapters 12, 14, and 15. Alternatively, other proteins bind to the DNA backbone. For example, histone proteins, which are discussed in Chapter 10, form ionic interactions with the negatively charged phosphates in the DNA backbone. The histones are important for the proper compaction of DNA in eukaryotic cells and also play a role in gene transcription.

DNA Forms Alternative Types of Double Helices

The DNA double helix can form different types of structures. **Figure 9.17** compares the structures of **B DNA** and **Z DNA.** The highly detailed structures shown here were deduced by X-ray crystallography on short segments of DNA. B DNA is the predominant form of DNA in living cells though some is found in a Z DNA conformation. B DNA is a right-handed helix, whereas Z DNA is left-handed. In addition, the helical backbone in Z DNA appears to zigzag slightly as it winds itself around the double-helical structure. The numbers of base pairs per 360° turn are 10.0 in B and 12.0 in Z DNA. In B DNA, the bases tend to be centrally located, and the hydrogen bonds between base pairs occur relatively perpendicular to the central axis. In contrast, the bases in Z DNA are substantially tilted relative to the central axis.

The ability of the predominant B DNA to adopt a Z-DNA conformation depends on various factors. At high ionic strength (i.e., high salt concentration), formation of a Z-DNA conformation is favored by a sequence of bases that alternates between purines and pyrimidines. One such sequence is

5′-GCGCGCGCG-3′
3′-CGCGCGCGC-5′

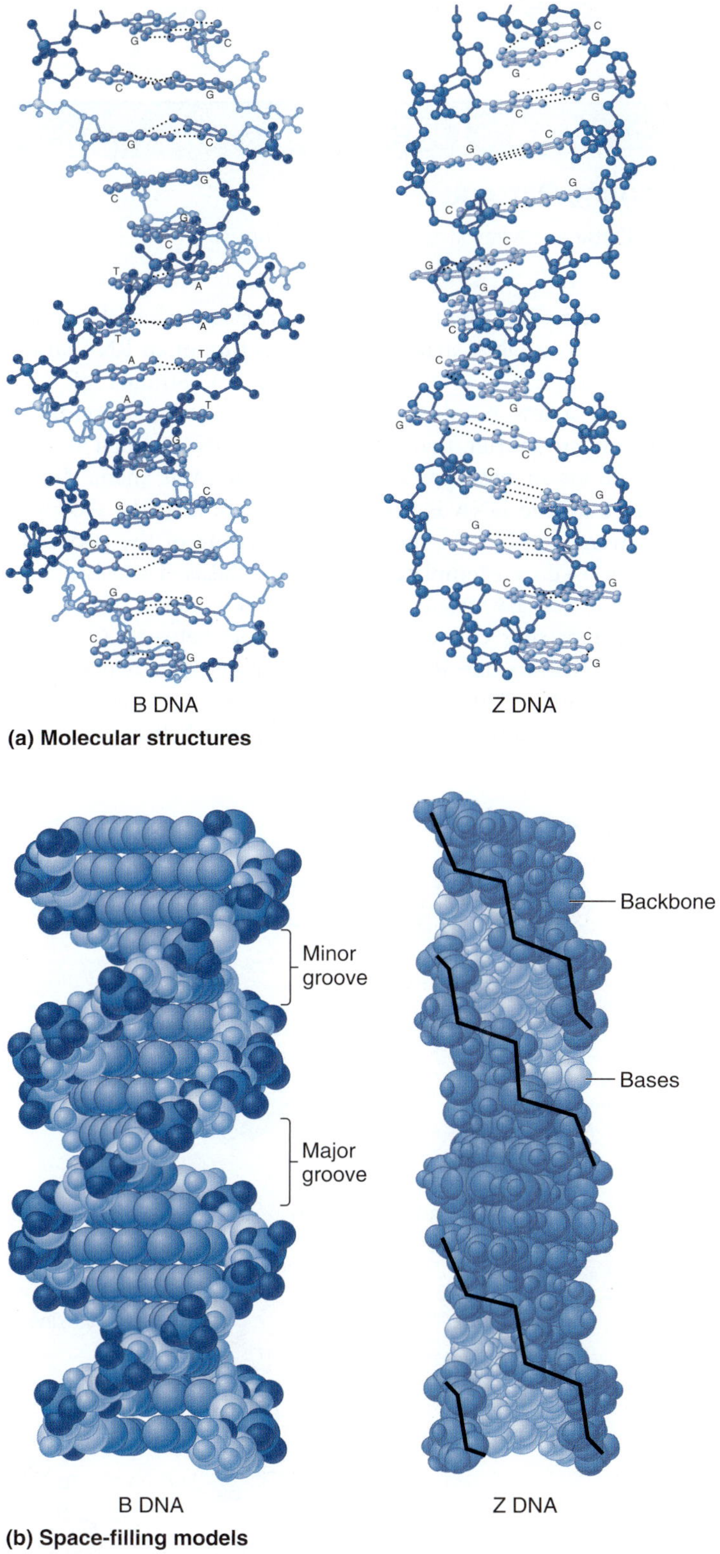

FIGURE 9.17 Comparison of the structures of B DNA and Z DNA. **(a)** The highly detailed structures shown here were deduced by X-ray crystallography performed on short segments of DNA. In contrast to the less detailed structures obtained from DNA wet fibers, the diffraction pattern obtained from the crystallization of short segments of DNA provides much greater detail concerning the exact placement of atoms within a double-helical structure. Alexander Rich, Richard Dickerson, and their colleagues were the first researchers to crystallize a short piece of DNA. **(b)** Space-filling models of the B-DNA and Z-DNA structures. In the case of Z DNA, the black lines connect the phosphate groups in the DNA backbone. As seen here, they travel along the backbone in a zigzag pattern.

CONCEPT CHECK: What are the structural differences between B DNA and Z DNA?

At lower ionic strength, the methylation of cytosine bases favors Z-DNA formation. Cytosine **methylation** occurs when a cellular enzyme attaches a methyl group ($-CH_3$) to the cytosine base. In addition, negative supercoiling (a topic discussed in Chapter 10) favors the Z-DNA conformation.

What is the biological significance of Z DNA? Accumulating evidence suggests a possible biological role for Z DNA in the process of transcription. Recent research has identified cellular proteins that specifically recognize Z DNA. In 2005, Alexander Rich and colleagues reported that the Z-DNA-binding region of one such protein played a role in regulating the transcription of particular genes. In addition, other research has suggested that Z DNA may play a role in chromosome structure by affecting the level of compaction.

DNA Can Form a Triple Helix, Called Triplex DNA

A surprising discovery made in 1957 by Alexander Rich, David Davies, and Gary Felsenfeld was that DNA can form a triple-helical structure called **triplex DNA.** This triplex was formed in vitro using pieces of DNA that were made synthetically. Although this result was interesting, it seemed to have little, if any, biological relevance.

About 30 years later, however, interest in triplex DNA was renewed by the observation that triplex DNA can form in vitro by mixing natural double-stranded DNA and a third short strand that is synthetically made. The synthetic strand binds into the major groove of the naturally occurring double-stranded DNA (**Figure 9.18**). As shown here, an interesting feature of triplex DNA formation is that it is sequence-specific. In other words, the synthetic third strand incorporates itself into a triple helix due to specific interactions between the synthetic DNA and the biological DNA. The pairing rules are that a thymine in the synthetic DNA hydrogen bonds at an AT pair in the biological DNA and that a cytosine in the synthetic DNA hydrogen bonds at a GC pair.

The formation of triplex DNA has been implicated in several cellular processes, including recombination, which is described in Chapter 19. In addition, researchers are interested in triplex DNA due to its potential as a tool for specifically inhibiting particular genes. As shown in Figure 9.18, the synthetic DNA strand binds into the major groove according to specific base-pairing rules. Therefore, researchers can design a synthetic DNA to recognize the base sequence found in a particular gene. When the synthetic DNA binds to a gene, it inhibits transcription. In addition, the synthetic DNA can contain reactive groups that cause mutations in a gene, thereby inactivating it. Researchers are excited about the possibility of using such synthetic DNA to silence the expression of particular genes. For example, this approach could be used to silence genes that become overactive in cancer cells. However, further research is needed to develop effective ways to promote the uptake of synthetic DNAs into the appropriate target cells.

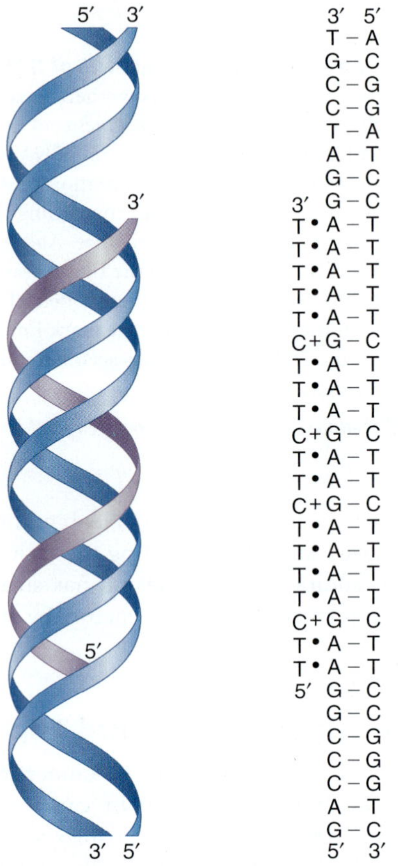

(a) Ribbon model **(b) Example sequence of bases**

FIGURE 9.18 The structure of triplex DNA. (a) As seen in the ribbon model, the third, synthetic strand binds within the major groove of the double-stranded structure. **(b)** Within triplex DNA, the third strand hydrogen bonds according to the rule T to AT, and C to GC. The cytosine bases in the third strand are protonated (i.e., positively charged).

9.6 COMPREHENSION QUESTIONS

1. Which of the following is not a feature of the DNA double helix?
   a. It obeys the AT/GC rule.
   b. The DNA strands are antiparallel.
   c. The structure is stabilized by base stacking.
   d. All of the above are correct features of a DNA double helix.
2. A groove in the DNA refers to
   a. the indentations where the atoms of the bases are in contact with the surrounding water.
   b. the interactions between bases in the DNA.
   c. the spiral structure of the DNA.
   d. all of the above.
3. A key difference between B DNA and Z DNA is that
   a. B DNA is right-handed, whereas Z DNA is left-handed.
   b. B DNA obeys the AT/GC rule, whereas Z DNA does not.
   c. Z DNA allows ribose in its structure, whereas B DNA uses deoxyribose.
   d. Z DNA allows uracil in its structure, whereas B DNA uses thymine.

9.7 RNA STRUCTURE

Learning Outcome:

1. Outline the key structural features of RNA.

Let's now turn our attention to RNA structure, which has many similarities with DNA structure. The structure of an RNA strand is much like a DNA strand (**Figure 9.19**). Strands of RNA are usually a few hundred to several thousand nucleotides in length—much shorter than chromosomal DNA, which is typically millions of base pairs long. When RNA is made during transcription, the DNA is used as a template to make a copy of single-stranded RNA. In most cases, only one of the two DNA strands is used as a template for RNA synthesis. Therefore, only one complementary strand of RNA is usually made.

Base pairing between A and U and between G and C may occur within one RNA molecule or between two separate RNA molecules. This base pairing causes short segments of RNA to form a double-stranded region that is helical. As shown

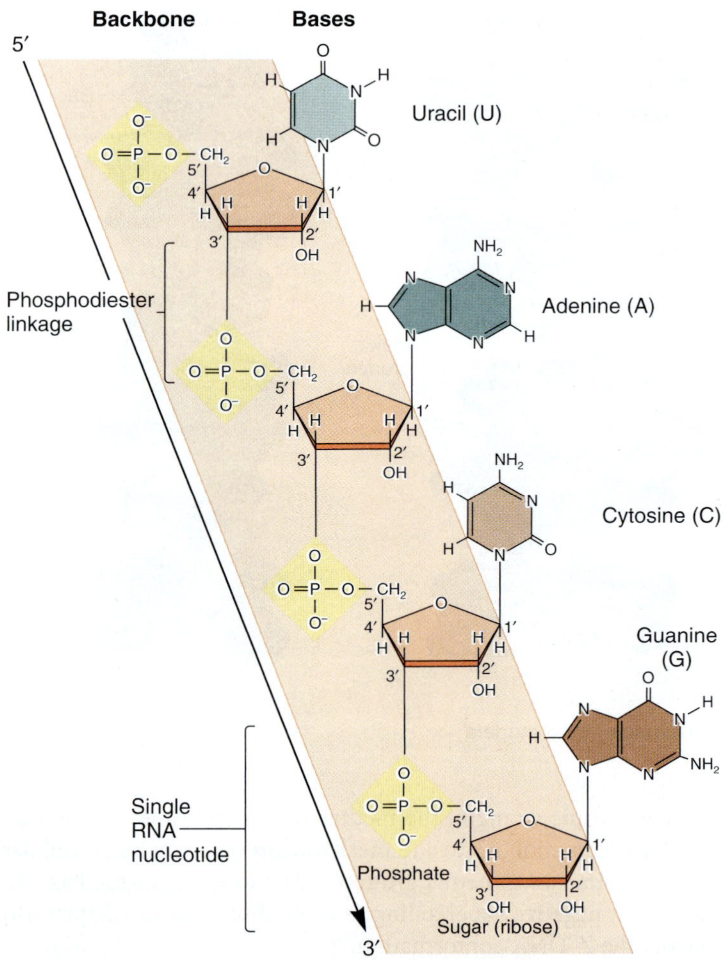

FIGURE 9.19 A strand of RNA. This structure is very similar to a DNA strand (see Figure 9.10), except that the sugar is ribose instead of deoxyribose, and uracil is substituted for thymine.

CONCEPT CHECK: What types of bonds hold nucleotides together in an RNA strand?

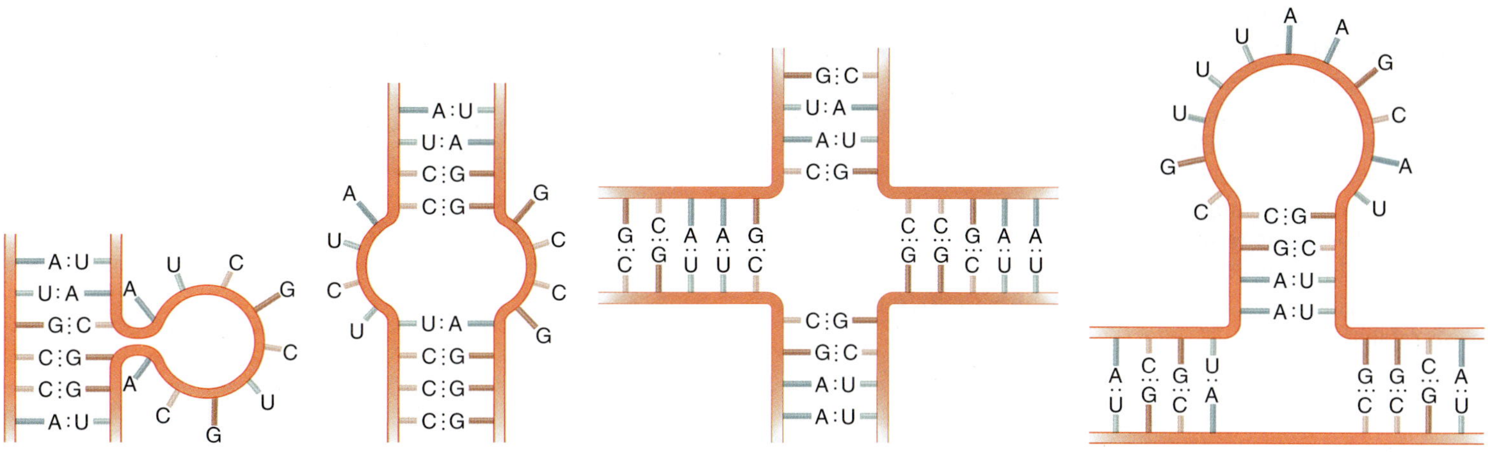

(a) Bulge loop (b) Internal loop (c) Multibranched junction (d) Stem-loop

FIGURE 9.20 Possible structures of RNA molecules. The double-stranded regions are depicted by connecting hydrogen bonds between complementary bases. Double-stranded regions can form within a single RNA molecule or between two separate RNA molecules. Note: Though not shown here, double-stranded regions are in a helical conformation.

CONCEPT CHECK: What are the base-pairing rules in RNA?

in **Figure 9.20**, different arrangements of base pairing are possible, which result in structures called bulge loops, internal loops, multibranched junctions, and stem-loops (also called hairpins). These structures contain regions of complementarity punctuated by regions of noncomplementarity. As shown in Figure 9.20, the complementary regions are held together by hydrogen bonds between base pairs, whereas the noncomplementary regions have their bases projecting away from the double-stranded region.

Many factors contribute to the structure of RNA molecules. These include hydrogen bonding between base pairs, stacking between bases, and hydrogen bonding between bases and backbone regions. In addition, interactions with ions, small molecules, and large proteins may influence RNA structure. **Figure 9.21** depicts the structure of a transfer RNA molecule known as $tRNA^{Phe}$, which is a tRNA molecule that carries the amino acid phenylalanine. It was the first naturally occurring RNA to have its structure elucidated. This RNA molecule has several double-stranded and single-stranded regions. RNA double helices are antiparallel and right-handed, with 11–12 bp per turn. In a living cell, the various regions of an RNA molecule fold and interact with each other to produce the three-dimensional structure.

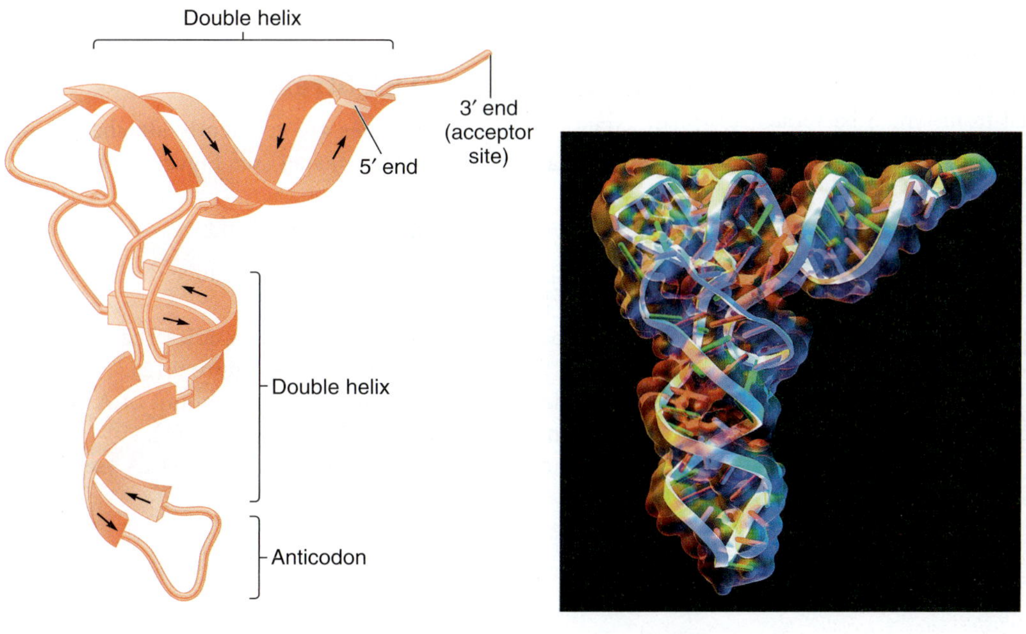

(a) Ribbon model (b) Space-filling model

FIGURE 9.21 The structure of $tRNA^{Phe}$, the transfer RNA molecule that carries phenylalanine. (a) The double-stranded regions of the molecule are shown as antiparallel ribbons. (b) A space-filling model of $tRNA^{Phe}$.

The folding of RNA into a three-dimensional structure is important for its function. For example, as discussed in Chapter 13, a tRNA molecule has two key functional sites—an anticodon and a 3′ acceptor site—that play important roles in translation. In a folded tRNA molecule, these sites are exposed on the surface of the molecule and can perform their roles (see Figure 9.21a). Many other examples are known in which RNA folding is key to its structure and function. These include the folding of ribosomal RNAs (rRNAs), which are important components of ribosomes, and ribozymes, which are RNA molecules with catalytic function.

9.7 COMPREHENSION QUESTION

1. A double-stranded region of RNA
   a. forms a helical structure.
   b. obeys the AU/GC rule.
   c. may promote the formation of different structures, such as bulge loops and stem-loops.
   d. does all of the above.

KEY TERMS

Page 215. molecular genetics, DNA (deoxyribonucleic acid), RNA (ribonucleic acid)
Page 217. transformation
Page 218. DNase, RNase, protease, bacteriophage, phage
Page 221. nucleic acids
Page 222. nucleotides, strand, double helix, deoxyribose, ribose, purines, pyrimidines, adenine (A), guanine (G), thymine (T), cytosine (C), uracil (U), nucleoside
Page 224. phosphodiester linkage, backbone, directionality
Page 227. Chargaff's rule
Page 228. base pairs (bp)
Page 229. AT/GC rule, complementary, antiparallel
Page 230. grooves, major groove, minor groove, B DNA, Z DNA
Page 231. methylation, triplex DNA

CHAPTER SUMMARY

- Molecular genetics is the study of DNA structure and function at the molecular level.

9.1 Identification of DNA as the Genetic Material

- To fulfill its role, genetic material must meet four criteria: information, transmission, replication, and variation.
- Griffith showed that the genetic material from type S bacteria could transform type R bacteria into type S (see Figure 9.1).
- Avery, MacLeod, and McCarty discovered that the transforming substance is DNA (see Figure 9.2).
- Hershey and Chase determined that the genetic material of T2 phage is DNA (see Figures 9.3–9.5).

9.2 Overview of DNA and RNA Structure

- DNA and RNA are types of nucleic acids.
- In DNA, nucleotides are linked together to form strands, which then form a double helix that is found within chromosomes (see Figure 9.6).

9.3 Nucleotide Structure

- A nucleotide is composed of one or more phosphates, a sugar, and a base. The purine bases are adenine and guanine, whereas the pyrimidine bases are thymine (DNA only), cytosine, and uracil (RNA only) (see Figures 9.7–9.9).

9.4 Structure of a DNA Strand

- In a DNA strand, nucleotides are covalently attached to one another via phosphodiester linkages (see Figure 9.10).

9.5 Discovery of the Double Helix

- Pauling used ball-and-stick models to deduce the structure of an α helix in a protein (see Figure 9.11).
- Franklin performed X-ray diffraction studies that helped to determine the structure of DNA (see Figure 9.12).
- Chargaff determined that, in DNA, the amount of A equals T and the amount of G equals C (see Figure 9.13).
- Watson and Crick deduced the structure of DNA (see Figure 9.14).

9.6 Structure of the DNA Double Helix

- DNA is a right-handed double helix in which adenine (A) hydrogen bonds to thymine (T) and guanine (G) hydrogen bonds to cytosine (C). The two strands are antiparallel and contain about 10 bp per turn (see Figure 9.15).
- The spiral structure of DNA has a major groove and a minor groove (see Figure 9.16).
- B DNA is the major form of DNA found in living cells. Z DNA is an alternative conformation for DNA (see Figure 9.17).
- Under certain conditions, DNA can form a triple-helix structure that obeys specific base-pairing rules (see Figure 9.18).

9.7 RNA Structure

- RNA is also a strand of nucleotides (see Figure 9.19).
- RNA can form double-stranded helical regions, and it folds into a three-dimensional structure (see Figures 9.20, 9.21).

PROBLEM SETS & INSIGHTS

Solved Problems

S1. A hypothetical sequence at the beginning of an mRNA molecule is

5′–AUUUGCCCUAGCAAACGUAGCAAACG . . . rest of the coding sequence

Using two out of the three underlined sequences, draw two possible models for potential stem-loop structures at the 5′ end of this mRNA.

Answer:

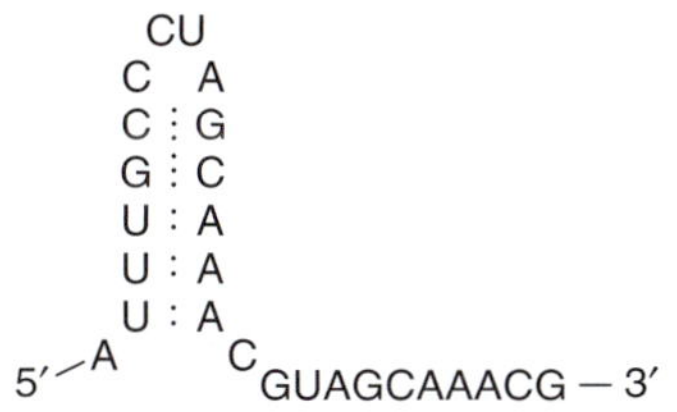

or

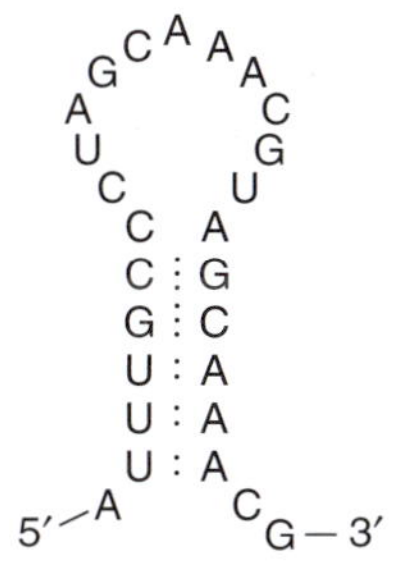

S2. Describe the experimental evidence that led Watson and Crick to the discovery of the DNA double helix.

Answer:

1. The chemical structure of single nucleotides was understood by the 1950s.
2. Watson and Crick assumed DNA is composed of nucleotides linked together in a linear fashion to form a strand. They also assumed the chemical linkage between two nucleotides is always the same.
3. Franklin's diffraction patterns suggested several structural features. First, it was consistent with a helical structure. Second, the diameter of the helical structure was too wide to be only a single-stranded helix. Finally, the line spacing on the diffraction pattern indicated the helix contains about 10 bp per complete turn.
4. In the chemical analysis of the DNA from different species, Chargaff's work indicated that the amount of adenine equals the amount of thymine, and the amount of cytosine equals that of guanine.
5. In the early 1950s, Linus Pauling proposed that regions of proteins fold into a secondary structure known as an α helix. To discover this, Pauling built large models by linking together simple ball-and-stick units. In this way, he could determine if atoms fit together properly in a complicated three-dimensional structure. A similar approach was used by Watson and Crick to solve the structure of the DNA double helix.

S3. Within living cells, many different proteins play important functional roles by binding to DNA and RNA. As described throughout your textbook, the dynamic interactions between nucleic acids and proteins lie at the heart of molecular genetics. Some proteins bind to DNA (or RNA) but not in a sequence-specific manner. For example, histones are proteins important in the formation of chromosome structure. In this case, the positively charged histone proteins bind to the negatively charged phosphate groups in DNA. In addition, several other proteins interact with DNA but do not require a specific nucleotide sequence to carry out their function. For example, DNA polymerase, which catalyzes the synthesis of new DNA strands, does not bind to DNA in a sequence-dependent manner. By comparison, many other proteins do interact with nucleic acids in a sequence-dependent fashion. This means that a specific sequence of bases can provide a structure that is recognized by a particular protein. Throughout the textbook, the functions of many of these proteins will be described. Some examples include transcription factors that affect the rate of transcription, proteins that bind to centromeres, and proteins that bind to origins of replication. With regard to the three-dimensional structure of DNA, where would you expect DNA-binding proteins to bind if they recognize a specific base sequence? What about DNA-binding proteins that do not recognize a base sequence?

Answer: DNA-binding proteins that recognize a base sequence must bind into the major or minor groove of the DNA, which is where the bases are accessible to a DNA-binding protein. Most DNA-binding proteins, which recognize a base sequence, fit into the major groove. By comparison, other DNA-binding proteins, such as histones, which do not recognize a base sequence, bind to the DNA backbone.

S4. The formation of a double-stranded structure must obey the rule that adenine hydrogen bonds to thymine (or uracil) and cytosine hydrogen bonds to guanine. Based on your previous understanding of genetics (from this course or a general biology course), discuss reasons why complementarity is an important feature of DNA and RNA structure and function.

Answer: Note: Many of the topics described next are discussed in Chapters 10 through 13. One way that complementarity underlies function is that it provides the basis for the synthesis of new strands of DNA and RNA. During replication, the synthesis of new DNA strands occurs in such a way that adenine hydrogen bonds to thymine, and cytosine hydrogen bonds to guanine. In other words, the molecular feature of a complementary double-stranded structure makes it possible to produce exact copies of DNA. Likewise, the ability to transcribe DNA into RNA is based on complementarity. During transcription, one strand of DNA is used as a template to make a complementary strand of RNA.

In addition to the synthesis of new strands of DNA and RNA, complementarity is important in other ways. As mentioned in this chapter, the folding of RNA into a particular structure is driven by the hydrogen bonding of complementary regions. This event is necessary to produce functionally active tRNA molecules. Likewise, stem-loop structures also occur in other types of RNA. For example, the rapid formation of stem-loop structures is known to occur as RNA is being transcribed and to affect the termination of transcription.

A third way that complementarity can be functionally important is that it can promote the interaction of two separate RNA molecules. During translation, codons in mRNA bind to the anticodons in tRNA (see Chapter 13). This binding is due to complementarity. For example, if a codon is 5′–AGG–3′, the anticodon is 3′–UCC–5′. This type of specific interaction between codons and anticodons is an important step that enables the nucleotide sequence in mRNA to code for an amino acid sequence within a protein. In addition, many other examples of RNA-RNA interactions are known and will be described throughout this textbook.

S5. An important feature of triplex DNA formation is that it is sequence-specific. The synthetic third strand incorporates itself into a triple helix, so a thymine in the synthetic DNA binds near an AT pair in the biological DNA, and a cytosine in the synthetic DNA binds near a GC pair. From a practical point of view, this opens the possibility of synthesizing a short strand of DNA that forms a triple helix at a particular target site. For example, if the sequence of a particular gene is known, researchers can make a synthetic piece of DNA that forms a triple helix somewhere within that gene according to the T to AT, and C to GC rule. Triplex DNA formation is known to inhibit gene transcription. In other words, when the synthetic DNA binds within the DNA of a gene, the formation of triplex DNA prevents that gene from being transcribed into RNA. Discuss how this observation might be used to combat diseases.

Answer: Triplex DNA formation opens the exciting possibility of designing synthetic pieces of DNA to inhibit the expression of particular genes. Theoretically, such a tool could be used to combat viral diseases or to inhibit the growth of cancer cells. To combat a viral disease, a synthetic DNA could be made that specifically binds to an essential viral gene, thereby preventing viral proliferation. To inhibit cancer, a synthetic DNA could be made to bind to an oncogene. (Note: As described in Chapter 24, an oncogene is a gene that promotes cancerous growth.) Inhibition of an oncogene could prevent cancer. At this point, a primary obstacle in applying this approach is devising a method of getting the synthetic DNA into living cells.

Conceptual Questions

C1. What is the meaning of the term *genetic material*?

C2. After the DNA from type S bacteria is exposed to type R bacteria, list all of the steps that you think must occur for the bacteria to start making a capsule.

C3. Look up the meaning of the word transformation in a dictionary and explain whether it is an appropriate word to describe the transfer of genetic material from one organism to another.

C4. What are the building blocks of a nucleotide? With regard to the 5′ and 3′ positions on a sugar molecule, how are nucleotides linked together to form a strand of DNA?

C5. Draw the structure of guanine, guanosine, and deoxyguanosine triphosphate.

C6. Draw the structure of a phosphodiester linkage.

C7. Describe how bases interact with each other in the double helix. This discussion should address the issues of complementarity, hydrogen bonding, and base stacking.

C8. If one DNA strand is 5′–GGCATTACACTAGGCCT–3′, what is the sequence of the complementary strand?

C9. What is meant by the term *DNA sequence*?

C10. Make a side-by-side drawing of two DNA helices: one with 10 bp per 360° turn and the other with 15 bp per 360° turn.

C11. Discuss the differences in the structural features of B DNA and Z DNA.

C12. What parts of a nucleotide (namely, phosphate, sugar, and/or bases) occupy the major and minor grooves of double-stranded DNA, and what parts are found in the DNA backbone? If a DNA-binding protein does not recognize a specific nucleotide sequence, do you expect that it recognizes the major groove, the minor groove, or the DNA backbone? Explain.

C13. List the structural differences between DNA and RNA.

C14. Draw the structure of deoxyribose and number the carbon atoms. Describe the numbering of the carbon atoms in deoxyribose with regard to the directionality of a DNA strand. In a DNA double helix, what does the term antiparallel mean?

C15. Write out a sequence of an RNA molecule that could form a stem-loop with 24 nucleotides in the stem and 16 nucleotides in the loop.

C16. Compare the structural features of a double-stranded RNA structure with those of a DNA double helix.

C17. Which of the following DNA double helices would be more difficult to separate into single-stranded molecules by treatment with heat, which breaks hydrogen bonds?

A. GGCGTACCAGCGCAT
   CCGCATGGTCGCGTA

B. ATACGATTTACGAGA
   TATGCTAAATGCTCT

Explain your choice.

C18. What structural feature allows DNA to store information?

C19. Discuss the structural significance of complementarity in DNA and in RNA.

C20. An organism has a G + C content of 64% in its DNA. What are the percentages of A, T, G, and C?

C21. Let's suppose you have recently identified an organism that was scraped from an asteroid that hit the earth. (Fortunately, no one was injured.) When you analyze this organism, you discover that its DNA is a triple helix, composed of six different nucleotides: A, T, G, C, X, and Y. You measure the chemical composition of the bases and find the following amounts of these six bases: A = 24%, T = 23%, G = 11%, C = 12%, X = 21%, Y = 9%. What rules would you propose that govern triplex-DNA formation in this organism? Note: There is more than one possibility.

C22. On further analysis of the DNA described in conceptual question C21, you discover that the triplex DNA in this alien organism is composed of a double helix, with the third helix wound within the major groove (just like the DNA in Figure 9.18). How would you propose that this DNA is able to replicate itself? In your answer, be specific about the base pairing rules within the double helix and which part of the triplex DNA would be replicated first.

C23. A DNA-binding protein recognizes the following double-stranded sequence:

5′–GCCCGGGC–3′
3′–CGGGCCCG–5′

This type of double-stranded structure could also occur within the stem region of an RNA stem-loop molecule. Discuss the structural differences between RNA and DNA that might prevent this DNA-binding protein from recognizing a double-stranded RNA molecule.

C24. Within a protein, certain amino acids are positively charged (e.g., lysine and arginine), some are negatively charged (e.g., glutamate and aspartate), some are polar but uncharged, and some are nonpolar. If you knew that a DNA-binding protein was recognizing the DNA backbone rather than base sequences, which amino acids in the protein would be good candidates for interacting with the DNA?

C25. In what ways are the structures of an α helix in proteins and the DNA double helix similar, and in what ways are they different?

C26. A double-stranded DNA molecule contains 560 nucleotides. How many complete turns would be found in this double helix?

C27. As the minor and major grooves of the DNA wind around a DNA double helix, do they ever intersect each other, or do they always run parallel to each other?

C28. What chemical group (phosphate group, hydroxyl group, or a nitrogenous base) is found at the 3′ end of a DNA strand? What group is found at the 5′ end?

C29. The base composition of an RNA virus was analyzed and found to be 14.1% A, 14.0% U, 36.2% G, and 35.7% C. Would you conclude that the viral genetic material is single-stranded RNA or double-stranded RNA?

C30. The genetic material found within some viruses is single-stranded DNA. Would this genetic material contain equal amounts of A and T and equal amounts of G and C?

C31. A medium-sized human chromosome contains about 100 million bp. If the DNA were stretched out in a linear manner, how long would it be?

C32. A double-stranded DNA molecule is 1 cm long, and the percentage of adenine is 15%. How many cytosines would be found in this DNA molecule?

C33. Could single-stranded DNA form a stem-loop structure? Why or why not?

C34. As described in Chapter 15, the methylation of cytosine bases can have an important effect on gene expression. For example, the methylation of cytosines may inhibit the transcription of genes. A methylated cytosine base has the following structure:

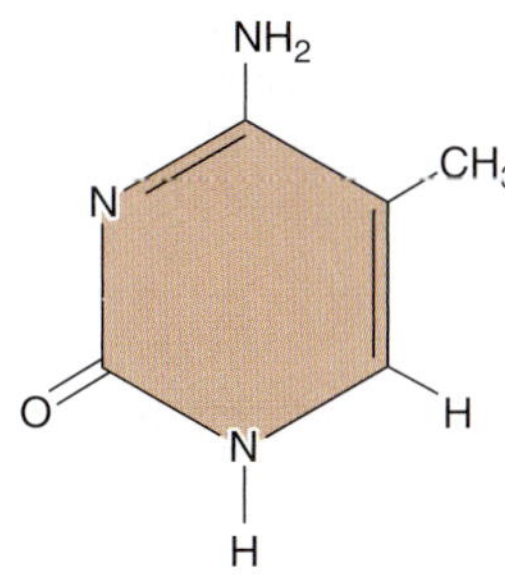

Would you expect the methylation of cytosine to affect the hydrogen bonding between cytosine and guanine in a DNA double helix? Why or why not? (Hint: See Figure 9.15 for help.) Take a look at solved problem S3 and speculate as to how methylation could affect gene expression.

C35. An RNA molecule has the following sequence:

Region 1 Region 2 Region 3
5′–CAUCCAUCCAUUCCCCAUCCGAUAAGGGGAAUGGAUCCGAAUGGAUAAC–3′

Parts of region 1 can form a stem-loop with region 2 and with region 3. Can region 1 form a stem-loop with region 2 and region 3 at the same time? Why or why not? Which stem-loop would you predict to be more stable: a region 1/region 2 interaction or a region 1/region 3 interaction? Explain your choice.

Experimental Questions

E1. Genetic material acts as a blueprint for an organism's traits. Explain how Griffith's experiments indicated that genetic material was being transferred to the type R bacteria.

E2. With regard to the experiment described in Figure 9.2, answer the following:

A. List several possible reasons why only a small percentage of the type R bacteria was converted to type S.

B. Explain why an antibody was used to remove the bacteria that are not transformed. What would the results look like, in all five cases, if the antibody/centrifugation step had not been included in the experimental procedure?

C. The DNA extract was treated with DNase, RNase, or protease. Why was this done? (In other words, what were the researchers trying to demonstrate?)

E3. An interesting trait that some bacteria exhibit is resistance to killing by antibiotics. For example, certain strains of bacteria are resistant to tetracycline, whereas other strains are sensitive to tetracycline. Describe an experiment you would carry out to demonstrate that tetracycline resistance is an inherited trait encoded by the DNA of the resistant strain.

E4. With regard to the experiment of Figure 9.5, answer the following:

A. Provide possible explanations why some of the DNA is in the supernatant.

B. Plot the results if the radioactivity in the pellet, rather than in the supernatant, had been measured.

C. Why were ^{32}P and ^{35}S chosen as radioisotopes to label the phages?

D. List possible reasons why less than 100% of the phage protein was removed from the bacterial cells during the shearing process.

E5. Does the experiment of Figure 9.5 rule out the possibility that RNA is the genetic material of T2 phage? Explain your answer. If it does not, could you modify Hershey and Chase's approach to show that DNA and not RNA is the genetic material of T2 bacteriophage? Note: It is possible to specifically label DNA or RNA by providing bacteria with radiolabeled thymine or uracil, respectively.

E6. In this chapter, we considered two experiments—one by Avery, MacLeod, and McCarty and the second by Hershey and Chase—that indicated DNA is the genetic material. Evaluate the strengths and weaknesses of the two approaches. Which experimental approach did you find the most convincing? Why?

E7. The type of model building used by Pauling and Watson and Crick involved the use of ball-and-stick units. Now we can do model building on a computer screen. Even though you may not be familiar with this approach, discuss potential advantages of using computers in molecular model building.

E8. With regard to Chargaff's experiment described in Figure 9.13, answer the following:

A. What is the purpose of paper chromatography?

B. Explain why it is necessary to remove the bases in order to determine the base composition of DNA.

C. Would Chargaff's experiments have been convincing if they had been done on only one species? Discuss.

E9. Gierer and Schramm exposed plant tissue to purified RNA from tobacco mosaic virus, and the plants developed the same types of lesions as if they had been exposed to the virus itself. What would be the results if the RNA was treated with DNase, RNase, or protease prior to its exposure to the plant tissue?

Questions for Student Discussion/Collaboration

1. Try to propose structures for a genetic material that are substantially different from the double helix. Remember that the genetic material must have a way to store information and a way to be faithfully replicated.

2. How might you provide evidence that DNA is the genetic material in mice?

Note: All answers appear at the website for this textbook; the answers to the even-numbered questions and all of the Concept Check and Comprehension Questions are in the back of the book.

www.mhhe.com/brookergenetics5e

Visit the website for practice tests, answer keys, and other learning aids for this chapter. Enhance your understanding of genetics with our interactive exercises, quizzes, animations, and much more.

CHAPTER OUTLINE

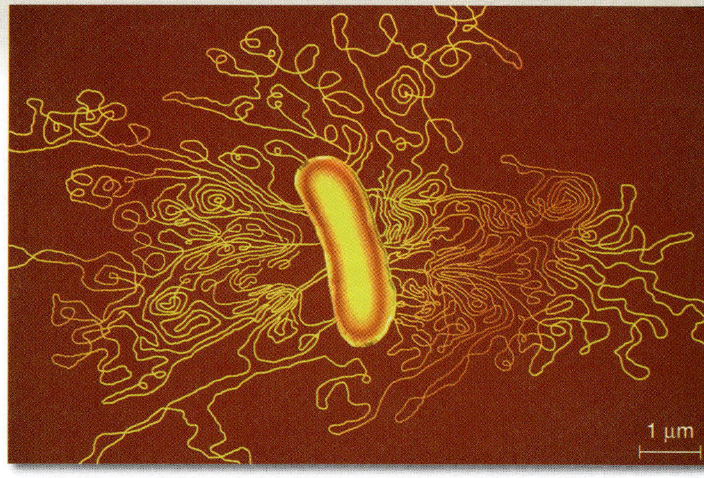

Structure of a bacterial chromosome. *Electron micrograph of a bacterial chromosome, which has been released from a bacterial cell.*

10 CHROMOSOME ORGANIZATION AND MOLECULAR STRUCTURE

Chromosomes are the structures within cells that contain the genetic material. The term **genome** refers to the entire complement of genetic material in an organism or species. For bacteria, the genome is typically a single circular chromosome. For eukaryotes, genetic material is found in different cellular compartments. The nuclear genome refers to one haploid set of chromosomes that resides in the cell nucleus. In the case of humans, this includes 22 autosomes, the X chromosome, and (in males) the Y chromosome. Eukaryotes also have a mitochondrial genome, and plants have a chloroplast genome.

The primary function of the genetic material is to store the information needed to produce the characteristics of an organism. As we saw in Chapter 9, the sequence of bases in a DNA molecule can store information. To fulfill their role at the molecular level, chromosomal sequences facilitate four important processes: (1) the synthesis of RNA and cellular proteins, (2) the replication of chromosomes, (3) the proper segregation of chromosomes, and (4) the compaction of chromosomes so they can fit within living cells. In this chapter, we will examine the general organization of the genetic material within bacterial and eukaryotic chromosomes. In addition, the molecular mechanisms that account for the packaging of the genetic material in bacteria and eukaryotic cells will be described.

10.1 ORGANIZATION OF SITES ALONG BACTERIAL CHROMOSOMES

Learning Outcome:

1. Describe the organization of sites along bacterial chromosomes.

Bacterial chromosomal DNA is usually a circular molecule, though some bacteria have linear chromosomes (**Figure 10.1**). Although bacteria usually contain a single type of chromosome, more than one copy of that chromosome may be found within one bacterial cell. A typical chromosome is a few million base pairs (bp) in length. For example, the chromosome of *Escherichia coli* has approximately 4.6 million bp, and the *Haemophilus influenzae* chromosome has roughly 1.8 million bp. A bacterial chromosome

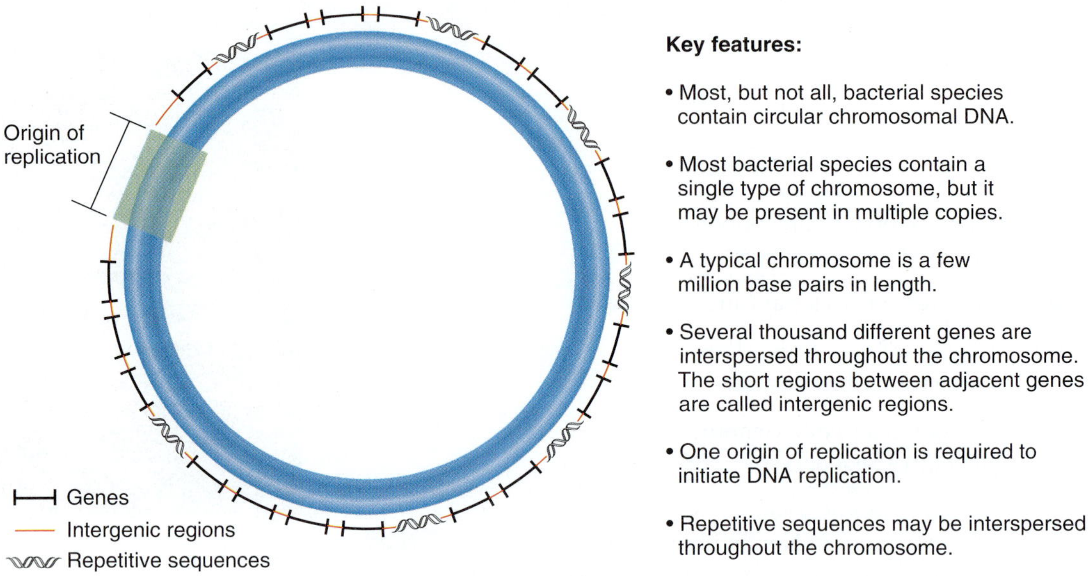

FIGURE 10.1 Organization of sequences in bacterial chromosomal DNA.

CONCEPT CHECK: What types of sequences constitute most of a bacterial genome?

commonly has a few thousand different genes, which are interspersed throughout the entire chromosome. **Protein-encoding genes** (also called structural genes) account for the majority of bacterial DNA. The nontranscribed regions of DNA located between adjacent genes are termed **intergenic regions.**

Other sequences in chromosomal DNA influence DNA replication, gene transcription, and chromosome structure. For example, bacterial chromosomes have one **origin of replication** (see Figure 10.1)**,** a sequence that is a few hundred nucleotides in length. This nucleotide sequence functions as an initiation site for the assembly of several proteins required for DNA replication. Also, a variety of **repetitive sequences** have been identified in many bacterial species. These sequences are found in multiple copies and are usually interspersed within the intergenic regions throughout the bacterial chromosome. Repetitive sequences may play a role in a variety of genetic processes, including DNA folding, DNA replication, gene regulation, and genetic recombination. As discussed in Chapter 19, some repetitive sequences are transposable elements that can move throughout the genome.

10.1 COMPREHENSION QUESTION

1. A bacterial chromosome typically contains
   a. a few thousand genes.
   b. one origin of replication.
   c. some repetitive sequences.
   d. all of the above.

10.2 STRUCTURE OF BACTERIAL CHROMOSOMES

Learning Outcomes:

1. Outline the two processes that make a bacterial chromosome more compact.
2. Describe how DNA gyrase alters DNA supercoiling.

Inside a bacterial cell, a chromosome is highly compacted and found within a region of the cell known as a **nucleoid.** Depending on the growth conditions and phase of the cell cycle, bacteria

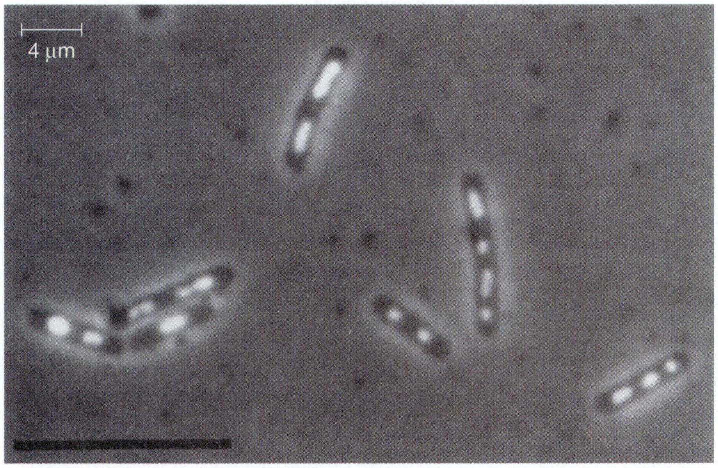

FIGURE 10.2 The localization of nucleoids within *Bacillus subtilis* bacteria. The nucleoids are fluorescently labeled and seen as bright, oval-shaped regions within the bacterial cytoplasm. Note that two or more nucleoids are found within each cell. Some of the cells seen here are in the process of dividing.

CONCEPT CHECK: How many nucleoids are in the bacterial cell on the far right?

may have one to four identical chromosomes per cell. In addition, the number of copies varies depending on the bacterial species. As shown in **Figure 10.2**, each chromosome occupies its own distinct nucleoid region within the cell. Unlike the eukaryotic nucleus, the bacterial nucleoid is not a separate cellular compartment surrounded by a membrane. Rather, the DNA in a nucleoid is in direct contact with the cytoplasm of the cell. In this section, we will explore the structure of bacterial chromosomes and how they are compacted to fit within a nucleoid.

The Formation of Chromosomal Loops Helps Make the Bacterial Chromosome More Compact

To fit within the bacterial cell, the chromosomal DNA must be compacted about 1000-fold. Part of this compaction process involves the formation of **loop domains** within the bacterial

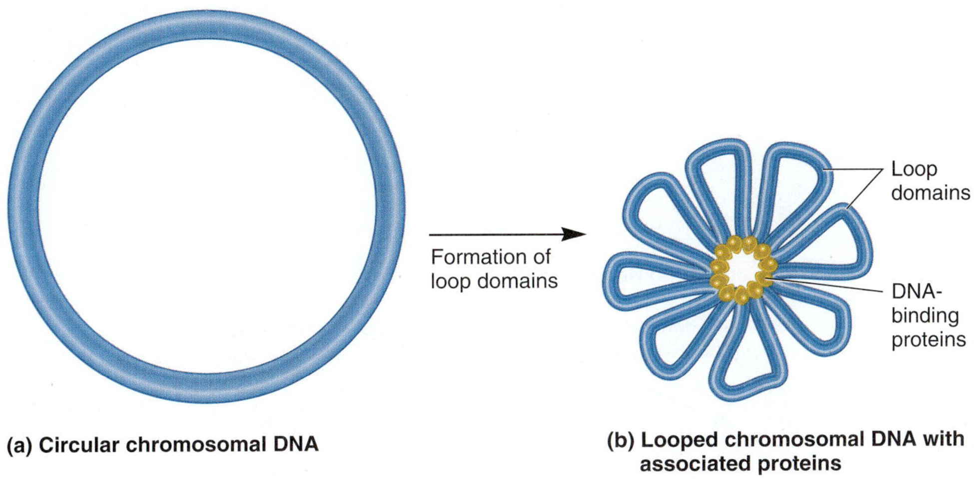

ONLINE ANIMATION

FIGURE 10.3 The formation of loop domains within the bacterial chromosome. To promote compaction, **(a)** the large, circular chromosomal DNA is organized into **(b)** smaller, loop domains with associated proteins.

CONCEPT CHECK: What holds the loop domains in place?

chromosome (**Figure 10.3**). As its name suggests, a loop domain is a segment of chromosomal DNA folded into a structure that resembles a loop. DNA-binding proteins anchor the base of the loops in place. The number of loops varies according to the size of the bacterial chromosome and the species. In *E. coli*, a chromosome has 50–100 loop domains with about 40,000–80,000 bp of DNA in each loop. This looped structure compacts the circular chromosome about 10-fold.

DNA Supercoiling Further Compacts the Bacterial Chromosome

Because DNA is a long thin molecule, twisting forces can dramatically change its conformation. This effect is similar to twisting a rubber band. If twisted in one direction, a rubber band eventually coils itself into a compact structure as it absorbs the energy applied by the twisting motion. Because the two strands within DNA already coil around each other, the formation of additional coils due to twisting forces is referred to as **DNA supercoiling** (**Figure 10.4**).

How do twisting forces affect DNA structure? **Figure 10.5** illustrates four possibilities. In Figure 10.5a, a double-stranded DNA molecule with five complete turns is anchored between two plates. In this hypothetical example, the ends of the DNA molecule cannot rotate freely. Both underwinding and overwinding of the DNA double helix can cause supercoiling of the helix. Because B DNA is a right-handed helix, underwinding is a left-handed twisting motion, and overwinding is a right-handed twist. Along the left side of Figure 10.5, one of the plates has been given a left-handed turn that tends to unwind the helix. As the helix absorbs this force, two things can happen. The underwinding motion can cause fewer turns (Figure 10.5b) or cause a negative supercoil to form (Figure 10.5c). On the right side of Figure 10.5, one of the plates has been given a right-handed turn, which overwinds the double helix. This can lead to either more turns (Figure 10.5d) or the formation of a positive supercoil (Figure 10.5e). The DNA conformations shown in Figure 10.5a, c, and e differ only with regard to supercoiling. These three DNA conformations are referred to as **topoisomers** of each other. The DNA conformations shown in Figure 10.5b and d are not structurally stable and do not occur in living cells.

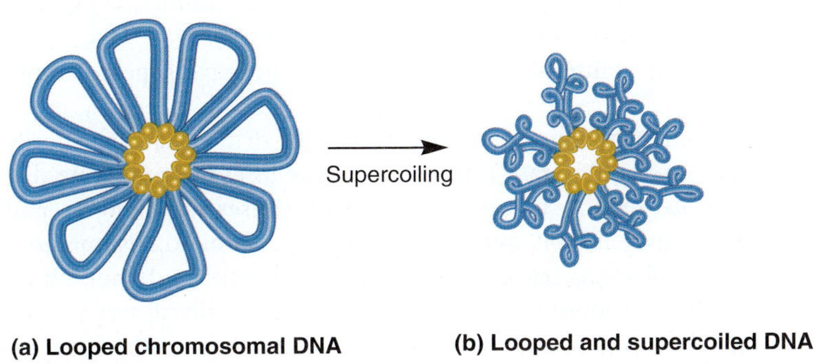

ONLINE ANIMATION

FIGURE 10.4 DNA supercoiling leads to further compaction of the looped chromosomal DNA. (a) The looped chromosomal DNA becomes much more compacted due to **(b)** supercoiling within the loops.

Chromosome Function Is Influenced by DNA Supercoiling

The chromosomal DNA in living bacteria is negatively supercoiled. In the chromosome of *E. coli*, about one negative supercoil occurs per 40 turns of the double helix. Negative supercoiling has several important consequences. As already mentioned, the supercoiling of chromosomal DNA makes it much more compact (see Figure 10.4). Therefore, supercoiling helps to greatly decrease the size of the bacterial chromosome. In addition, negative supercoiling also affects DNA function. To understand how, remember that negative supercoiling is due to an underwinding force on the DNA. Therefore, negative supercoiling creates tension on the DNA strands that may be released by DNA strand separation (**Figure 10.6**). Although most of the chromosomal DNA is negatively supercoiled and compact, the force of negative supercoiling may promote DNA strand separation in small regions. This enhances genetic activities such as replication and transcription that require the DNA strands to be separated.

How does bacterial DNA become supercoiled? In 1976, Martin Gellert and colleagues discovered the enzyme **DNA gyrase,** also known as topoisomerase II. This enzyme, which contains four subunits (two A and two B subunits), introduces negative

FIGURE 10.5 Schematic representation of DNA supercoiling. In this example, the DNA in **(a)** is anchored between two plates and given a twist as noted by the arrows. A left-handed twist (underwinding) could produce either **(b)** fewer turns or **(c)** a negative supercoil. A right-handed twist (overwinding) produces **(d)** more turns or **(e)** a positive supercoil. The structures shown in (b) and (d) are unstable.

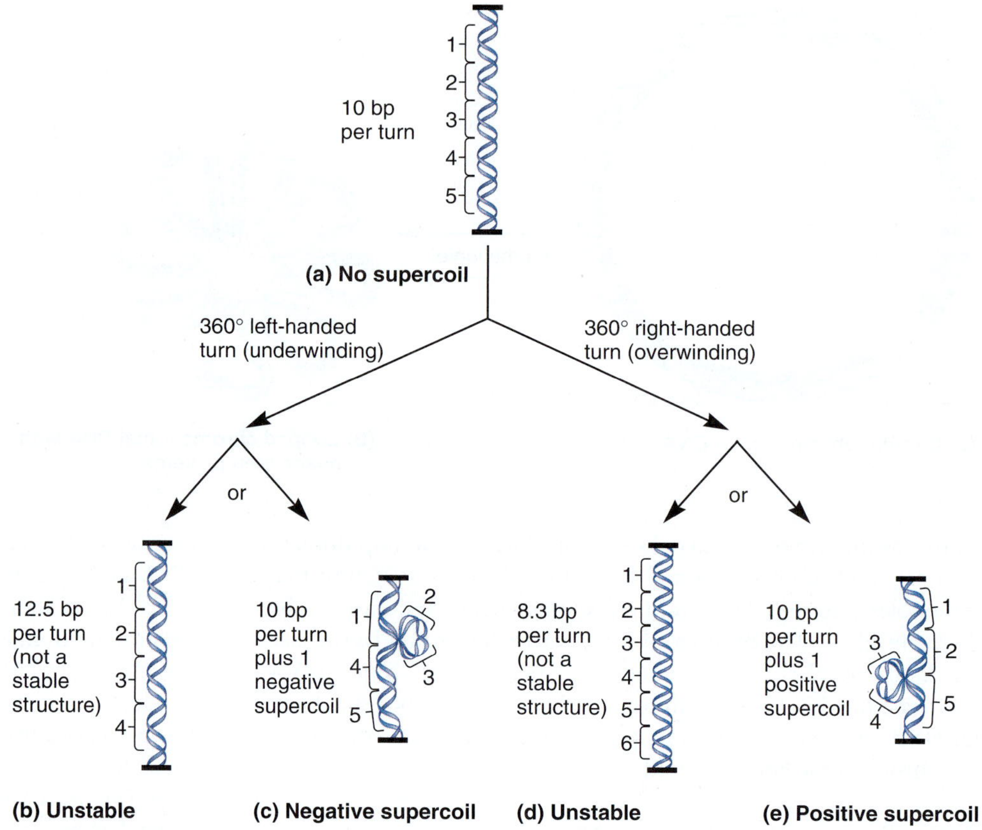

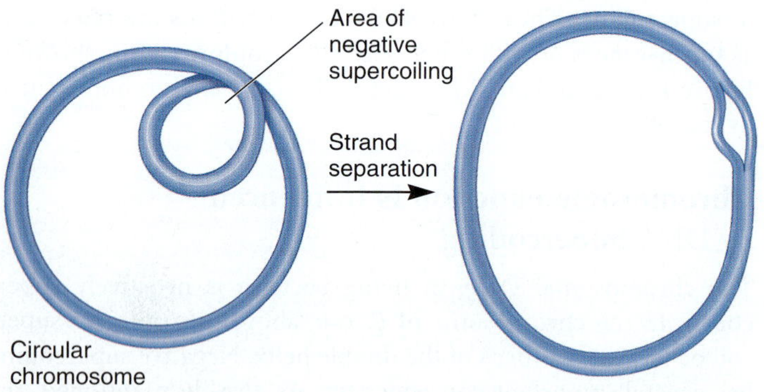

FIGURE 10.6 Negative supercoiling promotes strand separation.

CONCEPT CHECK: Why is strand separation beneficial?

supercoils (or relaxes positive supercoils) using energy from ATP (**Figure 10.7a**). To alter supercoiling, DNA gyrase has two sets of jaws that allow it to grab onto two regions of DNA. One of the DNA regions is grabbed by the lower jaws and then is wrapped in a right-handed direction around the two A subunits. The upper jaws then clamp onto another region of DNA. The DNA in the lower jaws is cut in both strands, and the other region of DNA is then released from the upper jaws and passed through this double-stranded break. To complete the process, the double-stranded break is ligated back together. The net result is that two negative supercoils have been introduced into the DNA molecule (**Figure 10.7b**). In addition, DNA gyrase can untangle DNA molecules. For example, as discussed in Chapter 11, circular DNA molecules are sometimes intertwined following DNA replication (see Figure 11.14). Such interlocked molecules can be separated via DNA gyrase.

A second type of enzyme, **topoisomerase I,** relaxes negative supercoils. This enzyme can bind to a negatively supercoiled region and introduce a break in one of the DNA strands. After one DNA strand has been broken, the DNA molecule rotates to relieve the tension that is caused by negative supercoiling. This rotation relaxes negative supercoiling. The broken strand is then repaired. The competing actions of DNA gyrase and topoisomerase I govern the overall supercoiling of the bacterial DNA.

The ability of gyrase to introduce negative supercoils into DNA is critical for bacterial survival. For this reason, much research has been aimed at identifying drugs that specifically block bacterial gyrase function as a way to cure or alleviate diseases caused by bacteria. Two main classes—quinolones and coumarins—inhibit gyrase and other bacterial topoisomerases, thereby blocking bacterial cell growth. These drugs do not inhibit eukaryotic topoisomerases, which are structurally different from their bacterial counterparts. This finding has been the basis for the production of many drugs with important antibacterial applications. An example is ciprofloxacin (known also by the brand name Cipro), which is used to treat a wide spectrum of bacterial diseases, including anthrax.

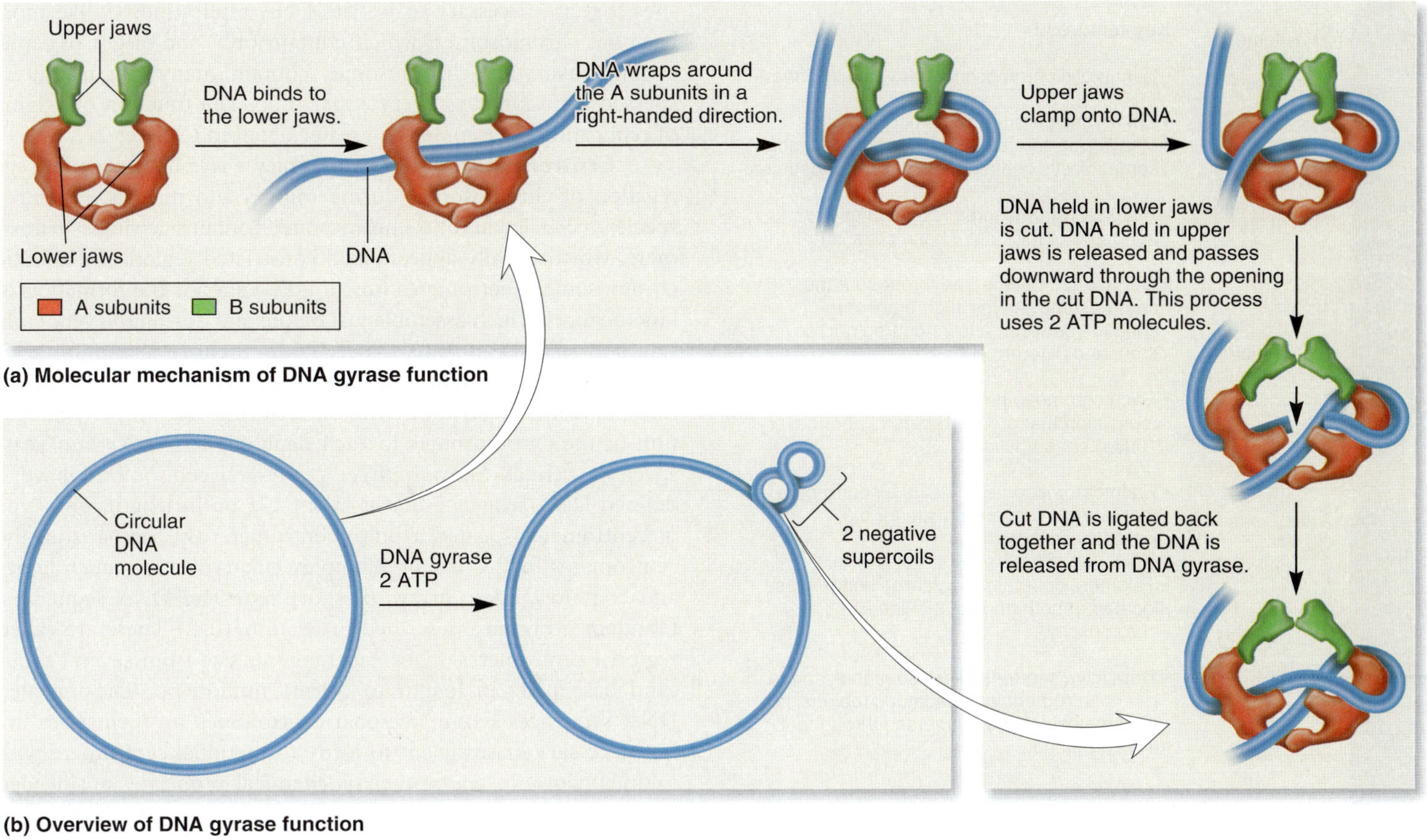

ONLINE ANIMATION

FIGURE 10.7 The action of DNA gyrase. **(a)** DNA gyrase, also known as topoisomerase II, is composed of two A and two B subunits. The lower jaws and then the upper jaws bind to two regions of DNA. The lower region is wrapped around the A subunits, which then cleave this DNA. The unbroken segment of DNA is released from the upper jaws and passes through the break. The break is repaired. The B subunits capture the energy from the hydrolysis of 2 ATP molecules to catalyze this process. **(b)** The result is that two negative turns have been introduced into the DNA molecule.

CONCEPT CHECK: In your own words, describe the step that requires the use of ATP.

10.2 COMPREHENSION QUESTIONS

1. Mechanisms that make the bacterial chromosome more compact include
   a. the formation of loop domains.
   b. DNA supercoiling.
   c. crossing over.
   d. both a and b.
2. Negative supercoiling may enhance activities like transcription and DNA replication because it
   a. allows the binding of proteins to the major groove.
   b. promotes DNA strand separation.
   c. makes the DNA more compact.
   d. causes all of the above.
3. DNA gyrase
   a. promotes negative supercoiling.
   b. relaxes positive supercoils.
   c. cuts DNA strands as part of its function.
   d. does all of the above.

10.3 ORGANIZATION OF SITES ALONG EUKARYOTIC CHROMOSOMES

Learning Outcome:

1. Describe the organization of sites along a eukaryotic chromosome.

Eukaryotic species have one or more sets of chromosomes in the cell nucleus; each set is composed of several different linear chromosomes (refer back to Figure 8.1). Humans, for example, have two sets of 23 chromosomes each, for a total of 46. Each eukaryotic chromosome contains a long, linear DNA molecule that is typically tens of millions to hundreds of millions of base pairs in length (**Figure 10.8**).

A single chromosome usually has a few hundred to several thousand different genes. A typical eukaryotic gene is several thousand to tens of thousands of base pairs in length. In less complex eukaryotes such as yeast, genes are relatively small,

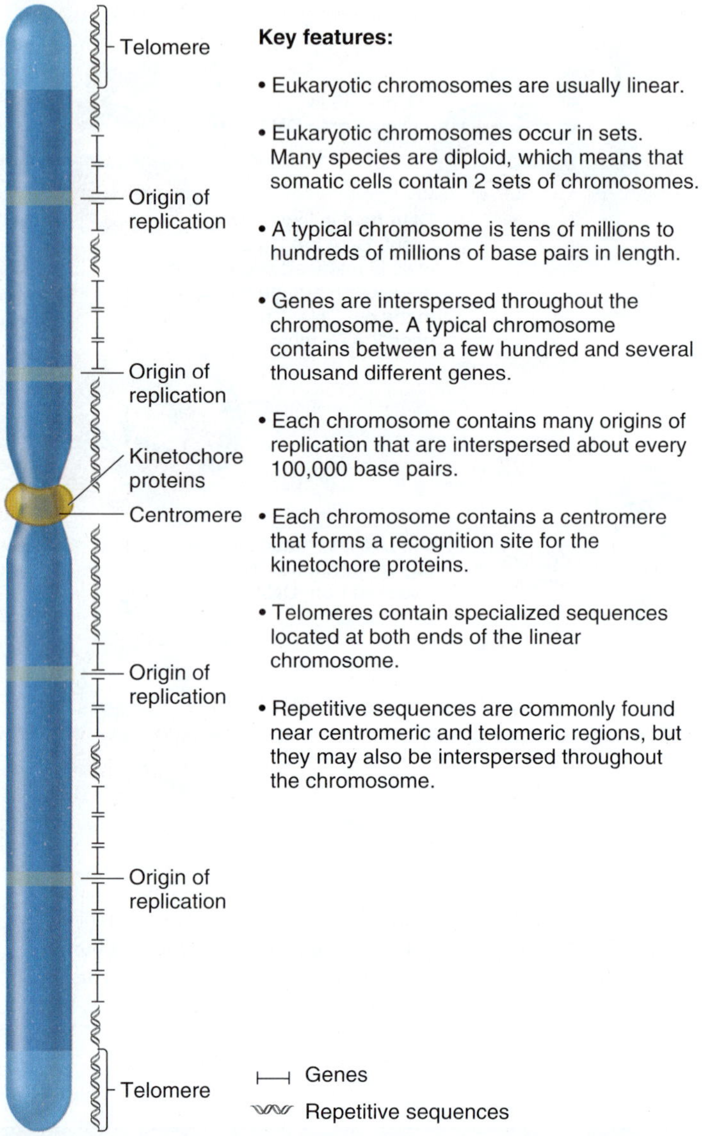

FIGURE 10.8 Organization of eukaryotic chromosomes.

CONCEPT CHECK: What are some differences between the types of sequences found in eukaryotic chromosomes versus bacterial chromosomes?

often several hundred to a few thousand base pairs long. In more complex eukaryotes such as mammals and flowering plants, protein-encoding genes tend to be much longer due to the presence of **introns**—noncoding intervening sequences. As described in Chapter 12, **exons** are regions of an RNA molecule that remain after splicing has removed the introns. The size of introns ranges from less than 100 bp to more than 10,000 bp. Therefore, the presence of large introns can greatly increase the lengths of eukaryotic genes.

In addition to genes, chromosomes contain three types of regions that are required for chromosomal replication and segregation: origins of replication, centromeres, and telomeres. As mentioned previously, origins of replication are chromosomal sites that are necessary to initiate DNA replication. Unlike most bacterial chromosomes, which contain only one origin of replication, eukaryotic chromosomes contain many origins, interspersed approximately every 100,000 bp. The function of origins of replication is discussed in greater detail in Chapter 11.

Centromeres are regions that play a role in the proper segregation of chromosomes during mitosis and meiosis. For most species, each eukaryotic chromosome contains a single centromere, which usually appears as a constricted region of a mitotic chromosome. Centromeres function as a site for the formation of kinetochores, which assemble just before and during the very early stages of mitosis and meiosis. The **kinetochore** is composed of a group of proteins that link the centromere to the spindle apparatus during mitosis and meiosis, ensuring the proper segregation of the chromosomes to each daughter cell. In certain yeast species, such as *Saccharomyces cerevisiae*, centromeres have a defined DNA sequence that is about 125 bp in length. This type of centromere is called a point centromere. By comparison, the centromeres found in more complex eukaryotes are much larger and contain tandem arrays of short repetitive DNA sequences. (Tandem arrays are described in Section 10.4.) These are called regional centromeres. They can range in size from several thousand base pairs in length to over 1 million bp. The repeated DNA sequences within regional centromeres by themselves are not necessary or sufficient to form a functional centromere with a kinetochore. Instead, other biochemical properties are needed to make a functional centromere. For example, a distinctive feature of all eukaryotic centromeres is that histone protein H3 is replaced with a histone variant called CENP-A. (Histone variants are described in Chapter 15.) However, researchers are still trying to identify all of the biochemical properties that distinguish regional centromeres, and understand how these properties are transmitted during cell division.

At the ends of linear chromosomes are found specialized regions known as **telomeres.** Telomeres serve several important functions in the replication and stability of the chromosome. As discussed in Chapter 8, telomeres prevent chromosomal rearrangements such as translocations. In addition, they prevent chromosome shortening in two ways. First, the telomeres protect chromosomes from digestion via enzymes called exonucleases that recognize the ends of DNA. Second, an unusual form of DNA replication occurs at the telomere to ensure that eukaryotic chromosomes do not become shortened with each round of DNA replication (see Chapter 11).

10.3 COMPREHENSION QUESTION

1. The chromosomes of eukaryotes typically contain
   a. a few hundred to several thousand different genes.
   b. multiple origins of replication.
   c. a centromere.
   d. telomeres at their ends.
   e. all of the above.

10.4 SIZES OF EUKARYOTIC GENOMES AND REPETITIVE SEQUENCES

Learning Outcomes:

1. Describe the variation in size of eukaryotic genomes.
2. Define *repetitive sequence* and explain how this type of sequence affects genome sizes.

The total amount of DNA in cells of eukaryotic species is usually much greater than in bacterial cells. This enables eukaryotic genomes to contain many more genes than their bacterial counterparts. In this section, we will examine the sizes of eukaryotic genomes and consider how repetitive sequences may greatly contribute to their overall size.

The Sizes of Eukaryotic Genomes Vary Substantially

Different eukaryotic species vary dramatically in the size of their genomes (**Figure 10.9a**; note that this is a log scale). In many cases, this variation is not related to the complexity of the species. For example, two closely related species of salamander, *Plethodon richmondi* and *Plethodon larselli*, differ considerably in genome size (**Figure 10.9b**, **c**). The genome of *P. larselli* is more than twice as large as the genome of *P. richmondi*. However, the genome of *P. larselli* probably doesn't contain more genes. How do we explain the difference in genome size? The additional DNA in *P. larselli* is due to the accumulation of repetitive DNA sequences present in many copies. In some species, these repetitive sequences have accumulated to enormous levels. Such highly repetitive sequences do not encode proteins, and their function remains a matter of controversy and great interest. The structure and significance of repetitive DNA will be discussed next.

The Genomes of Eukaryotes Contain Sequences That Are Unique, Moderately Repetitive, or Highly Repetitive

The term **sequence complexity** refers to the number of times a particular base sequence appears throughout the genome. Unique or nonrepetitive sequences are those found once or a few times within the genome. Protein-encoding genes are typically unique sequences of DNA. The vast majority of proteins in eukaryotic cells are encoded by genes present in one or a few copies. In the case of humans, unique sequences make up roughly 41% of the entire genome (**Figure 10.10**). These unique sequences include the protein-encoding regions of genes (2%), introns (24%), and unique regions that are not found within genes (15%).

Moderately repetitive sequences are found a few hundred to several thousand times in the genome. In a few cases, moderately repetitive sequences are multiple copies of the same gene. For example, the genes that encode ribosomal RNA (rRNA) are found in many copies. Ribosomal RNA is necessary for the functioning of ribosomes. Cells need a large amount of rRNA for making ribosomes, and this is accomplished by having multiple copies of the genes that encode rRNA. Likewise, the genes encoding histone proteins are also found in multiple copies because a large number of histone proteins are needed for the structure of chromosomes. In addition, other types of functionally important sequences are moderately repetitive. For example, moderately repetitive sequences may play a role in the regulation of gene transcription and translation. By comparison, some moderately repetitive sequences do not play a functional role and are derived from **transposable elements (TEs)**—segments of DNA that have the ability to move within a genome. This category of repetitive sequences is discussed in greater detail in Chapter 19.

Highly repetitive sequences are found tens of thousands or even millions of times throughout a genome. Each copy of a highly repetitive sequence is relatively short, ranging from a few

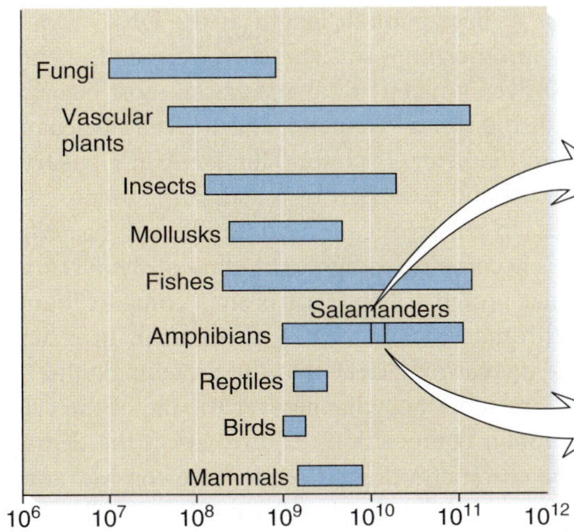

(a) Genome sizes (nucleotide base pairs per haploid genome)

(b) *Plethodon richmondi*

(c) *Plethodon larselli*

FIGURE 10.9 Haploid genome sizes among groups of eukaryotic species. **(a)** Ranges of genome sizes among different groups of eukaryotes. Data from T. Ryan Gregory, et al. (2007) Eukaryotic genome size databases. *Nucleic Acids Res.* 35:D332–D338. **(b)** A species of salamander, *Plethodon richmondi*, and **(c)** a close relative, *Plethodon larselli*. The genome of *P. larselli* is over twice as large as that of *P. richmondi*.

Genes→Traits The two species of salamander shown here have very similar traits, even though the genome of *P. larselli* is over twice as large as that of *P. richmondi*. However, the genome of *P. larselli* is not likely to contain more genes. Rather, the additional DNA is due to the accumulation of short repetitive DNA sequences that do not code for genes and are present in many copies.

CONCEPT CHECK: What are two reasons for the wide variation in genome sizes among eukaryotic species?

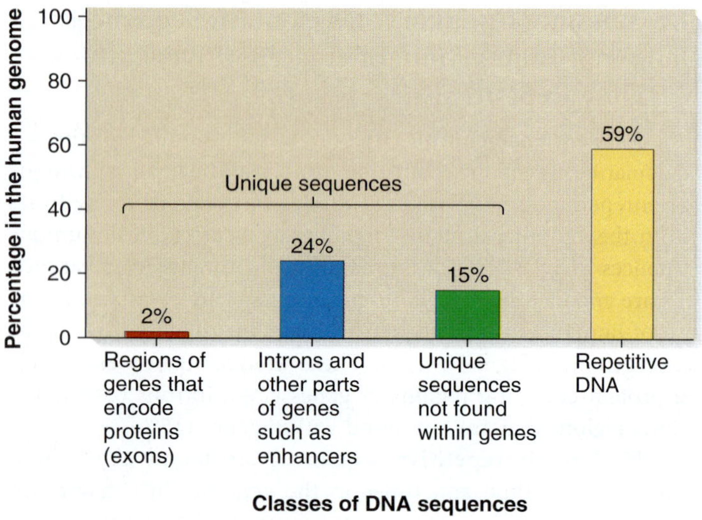

FIGURE 10.10 **Relative amounts of unique and repetitive DNA sequences in the human genome.**

nucleotides to several hundred in length. A widely studied example is the *Alu* family of sequences found in humans and other primates. The *Alu* sequence is approximately 300 bp long. This sequence derives its name from the observation that it contains a site for cleavage by a restriction enzyme known as *Alu*I. (The function of restriction enzymes is described in Chapter 20.) The *Alu* sequence is present in about 1,000,000 copies in the human genome. It represents about 10% of the total human DNA and occurs approximately every 5000–6000 bp! Evolutionary studies suggest that the *Alu* sequence arose 65 mya from a section of a single ancestral gene known as the *7SL RNA* gene. Since that time, this gene has become a type of TE called a **retroelement,** which is transcribed into RNA, copied into DNA, and inserted into the genome. Over the past 65 million years, the *Alu* sequence has been copied and inserted into the human genome many times and is now present in about 1,000,000 copies.

Repetitive sequences, like the *Alu* family, are interspersed throughout the genome. However, some moderately and highly repetitive sequences are clustered together in a **tandem array,** also known as tandem repeats. In a tandem array, a very short nucleotide sequence is repeated many times in a row. In *Drosophila*, for example, 19% of the chromosomal DNA is highly repetitive DNA found in tandem arrays. An example is shown here.

AATATAATATAATATAATATAATATATAATAT
TTATATTATATTATATTATATTATATATTATA

In this particular tandem array, two related sequences, AATAT and AATATAT, are repeated. As mentioned earlier, tandem arrays of short sequences are commonly found in centromeric regions of chromosomes and can be quite long, sometimes more than 1,000,000 bp in length!

What is the functional significance of highly repetitive sequences? Whether they have any significant function is controversial. Some experiments in *Drosophila* indicate that highly repetitive sequences may be important in the proper segregation of chromosomes during meiosis. It is not yet clear if highly repetitive DNA plays the same role in other species. The sequences within highly repetitive DNA vary greatly from species to species. Likewise, the amount of highly repetitive DNA can vary a great deal even among closely related species (as noted earlier in Figure 10.9).

10.4 COMPREHENSION QUESTION

1. Which of the following is an example of a moderately repetitive sequence?
   a. rRNA genes
   b. Most protein-encoding genes
   c. Both a and b
   d. None of the above

10.5 STRUCTURE OF EUKARYOTIC CHROMOSOMES IN NONDIVIDING CELLS

Learning Outcomes:

1. Define chromatin.
2. Outline the structures of nucleosomes, the 30-nm fiber, and radial loop domains.
3. Analyze Noll's results and explain how they support the beads-on-a-string model.
4. Describe the structure of a chromosome territory.

A distinguishing feature of eukaryotic cells is that their chromosomes are located within a separate cellular compartment known as the **nucleus.** The DNA within a typical eukaryotic chromosome is a single, linear double-stranded molecule that may be hundreds of millions of base pairs in length. If the DNA from a single set of human chromosomes was stretched from end to end, the length would be over 1 meter! By comparison, most eukaryotic cells are only 10–100 μm in diameter, and the cell nucleus is only about 2–4 μm in diameter. Therefore, the DNA in a eukaryotic cell must be folded and packaged by a staggering amount to fit inside the nucleus. In eukaryotic chromosomes, as in bacterial chromosomes, this is accomplished by the binding of the DNA to many different cellular proteins. The DNA-protein complex found within eukaryotic chromosomes is termed **chromatin.** In recent years, it has become increasingly clear that the proteins bound to chromosomal DNA are subject to change over the life of the cell. These changes in protein composition, in turn, affect the degree of compaction of the chromatin. In this section, we consider how chromosomes are compacted during interphase—the period of the cell cycle that includes the G_1, S, and G_2 phases. In the following section, we examine the additional compaction that is necessary to produce the highly condensed chromosomes found in M phase.

Linear DNA Wraps Around Histone Proteins to Form Nucleosomes, the Repeating Structural Unit of Chromatin

The repeating structural unit within eukaryotic chromatin is the **nucleosome**—a double-stranded segment of DNA wrapped around an octamer of histone proteins (**Figure 10.11a**). Each octamer contains eight histone subunits: two copies each of four different histone proteins. The DNA is negatively supercoiled over the surface of this octamer; it makes 1.65 negative superhelical turns around a histone octamer. The amount of DNA required to wrap around the histone octamer is 146 or 147 bp. At its widest point, a single nucleosome is about 11 nm in diameter.

The chromatin of eukaryotic cells contains a repeating pattern in which the nucleosomes are connected by linker regions of DNA that vary in length from 20–100 bp, depending on the species and cell type. It has been suggested that the overall structure of connected nucleosomes resembles beads on a string. This structure shortens the length of the DNA molecule about sevenfold.

Each of the **histone proteins** consists of a globular domain and a flexible, charged amino terminus called an amino terminal tail. Histone proteins are very basic proteins because they contain a large number of positively charged lysine and arginine amino acids. The arginines, in particular, play a major role in binding to the DNA. Arginines within the histone proteins form electrostatic and hydrogen-bonding interactions with the phosphate groups along the DNA backbone. The octamer of histones contains two molecules each of four different histone proteins: H2A, H2B, H3, and H4. These are called the core histones. In 1997, Timothy Richmond and colleagues determined the structure of a nucleosome by X-ray crystallography (**Figure 10.11b**).

Another histone, H1, is found in most eukaryotic cells and is called the linker histone. It binds to the DNA in the linker region between nucleosomes and may help to compact adjacent nucleosomes (**Figure 10.11c**). The linker histones are less tightly bound to the DNA than are the core histones. In addition, nonhistone proteins bound to the linker region play a role in the organization and compaction of chromosomes, and their presence may affect the expression of nearby genes.

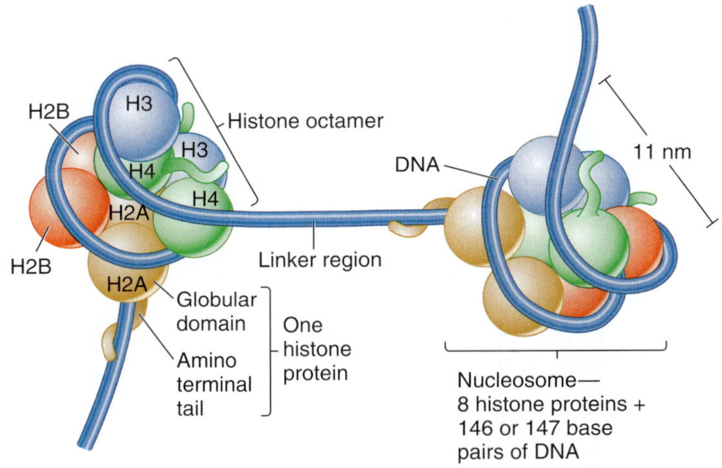

(a) Nucleosomes showing core histone proteins

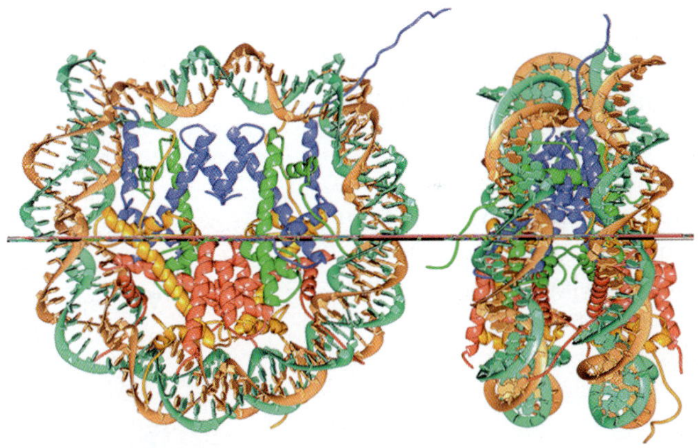

(b) Molecular model for nucleosome structure

(Image courtesy of Timothy J. Richmond. Reprinted by permission from Macmillan Publishers Ltd. Luger K., Mader, AW, Richmond, RK, Sargent, DF, Richmond, TJ [1997] Crystal structure of the nucleosome core particle at 2.8 Å resolution. *Nature 389*:6648, 251–260.) This drawing shows two views of a nucleosome that are at right angles to each other. The horizontal bar passes through the center of the nucleosome in each view.

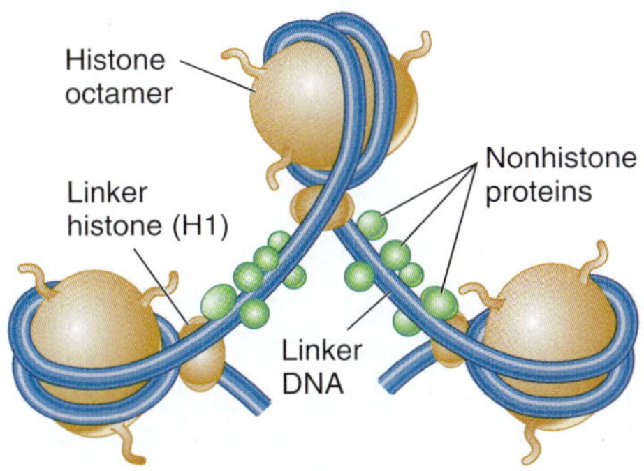

(c) Nucleosomes showing linker histones and nonhistone proteins

FIGURE 10.11 Nucleosome structure. (a) A nucleosome consists of 146 or 147 bp of DNA wrapped around an octamer of core histone proteins. **(b)** A model for the structure of a nucleosome as determined by X-ray crystallography. **(c)** The linker region of DNA connects adjacent nucleosomes. The linker histone H1 and nonhistone proteins also bind to this linker region.

CONCEPT CHECK: What is the diameter of a nucleosome?

EXPERIMENT 10A

The Repeating Nucleosome Structure Is Revealed by Digestion of the Linker Region

The model of nucleosome structure was originally proposed by Roger Kornberg in 1974. He based his proposal on several observations. Biochemical experiments had shown that chromatin contains a ratio of one molecule of each of the four core histones (namely, H2A, H2B, H3, and H4) per 100 bp of DNA. Approximately one H1 protein was found per 200 bp of DNA. In addition, purified core histone proteins were observed to bind to each other via specific pairwise interactions. Subsequent X-ray diffraction studies showed that chromatin is composed of a repeating pattern of smaller units. Finally, electron microscopy of chromatin fibers revealed a diameter of approximately 11 nm. Taken together, these observations led Kornberg to propose a model in which the DNA double helix is wrapped around an octamer of core histone proteins. Including the linker region, this involves about 200 bp of DNA.

Markus Noll decided to test Kornberg's model by digesting chromatin with DNase I, an enzyme that cuts the DNA backbone. He then accurately measured the molecular mass of the DNA fragments by gel electrophoresis. Noll assumed that the linker region of DNA is more accessible to DNase I and, therefore, DNase I is more likely to make cuts in the linker region than in the 146-bp region that is tightly bound to the core histones. If this is correct, incubation with DNase I is expected to make cuts in the linker region, thereby producing DNA pieces approximately 200 bp in length. The size of the DNA fragments may vary somewhat because the linker region is not of constant length and because the cut within the linker region may occur at different sites.

Figure 10.12 describes Noll's experimental protocol. He began with nuclei from rat liver cells and incubated them with low, medium, or high concentrations of DNase I. The DNA was extracted into an aqueous phase and then loaded onto an agarose gel that separated the fragments according to their molecular mass. The DNA fragments within the gel were stained with a UV-sensitive dye, ethidium bromide, which made it possible to view the DNA fragments under UV illumination.

THE HYPOTHESIS

This experiment seeks to test the beads-on-a-string model for chromatin structure. According to this model, DNase I should preferentially cut the DNA in the linker region, thereby producing DNA pieces that are about 200 bp in length.

TESTING THE HYPOTHESIS

FIGURE 10.12 **DNase I cuts chromatin into repeating units containing 200 bp of DNA.**

Starting material: Nuclei from rat liver cells.

1. Incubate the nuclei with low, medium, and high concentrations of DNase I. The conceptual level illustrates a low DNase I concentration.

2. Isolate the DNA. This involves dissolving the nuclear membrane with detergent and treating the sample with the organic solvent phenol.

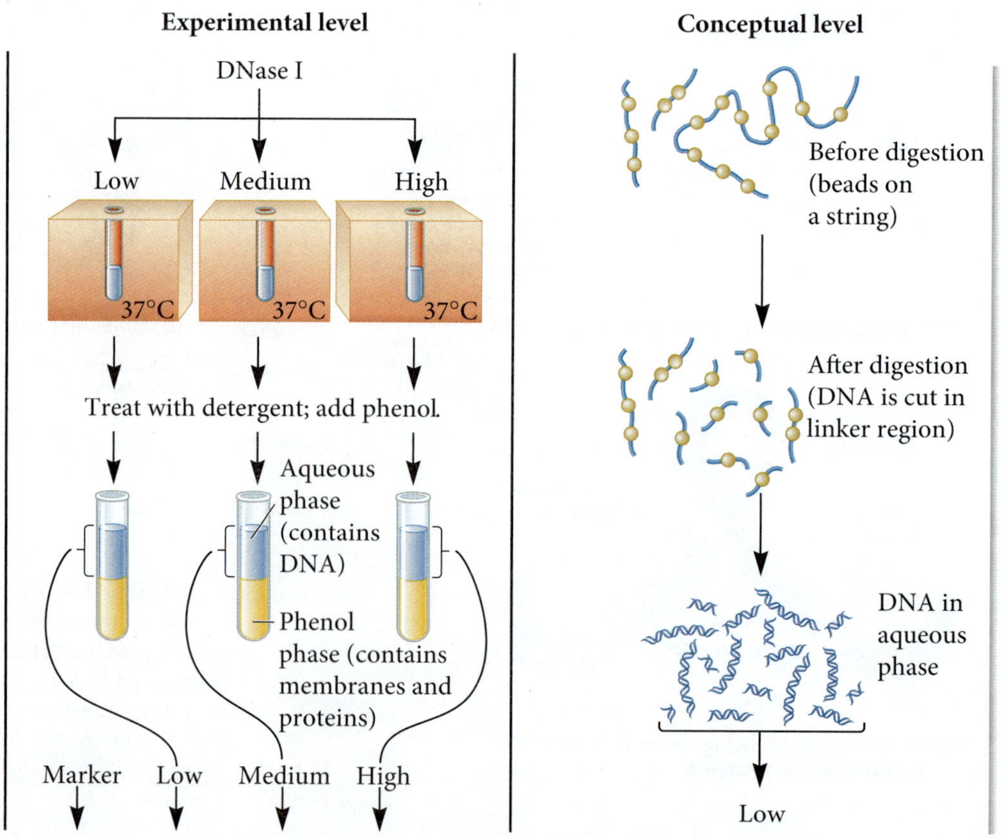

3. Load the DNA into a well of an agarose gel and run the gel to separate the DNA pieces according to size. On this gel, also load DNA fragments of known molecular mass (marker lane).

4. Visualize the DNA fragments by staining the DNA with ethidium bromide, a dye that binds to DNA and is fluorescent when excited by UV light.

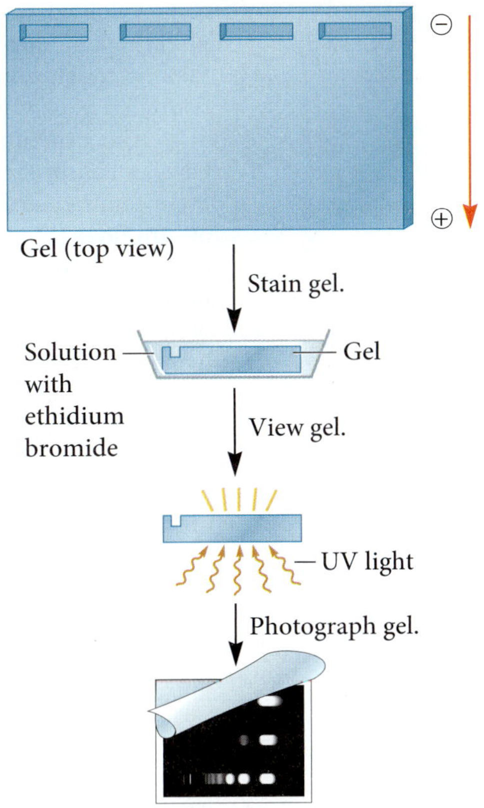

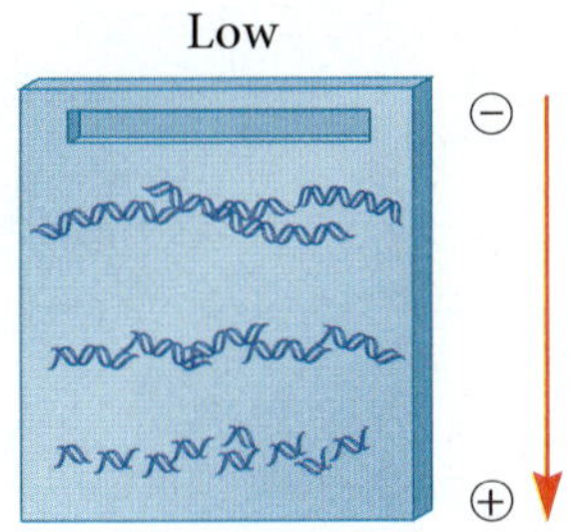

THE DATA

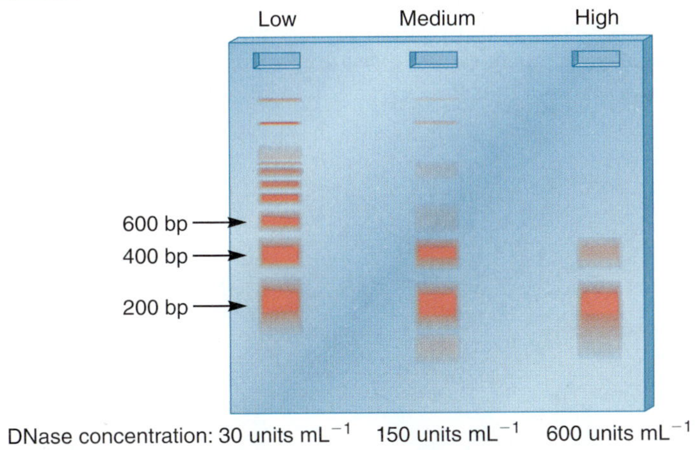

Note: The marker lane is omitted from this drawing.

INTERPRETING THE DATA

As shown in The Data of Figure 10.12, at a high DNase I concentration, the entire sample of chromosomal DNA was digested into fragments of approximately 200 bp in length. This result is predicted by the beads-on-a-string model. Furthermore, at a low or medium DNase I concentration, longer pieces were observed, and these were in multiples of 200 bp (400, 600, etc.). How do we explain these longer pieces? They occurred because occasional linker regions remained uncut at a low or medium DNase I concentration. For example, if one linker region was not cut, a DNA piece would contain two nucleosomes and be 400 bp in length. If two consecutive linker regions were not cut, this would produce a piece with three nucleosomes containing about 600 bp of DNA. Taken together, these results strongly supported the nucleosome model for chromatin structure.

A self-help quiz involving this experiment can be found at www.mhhe.com/brookergenetics5e.

Nucleosomes Become Closely Associated to Form a 30-nm Fiber

In eukaryotic chromatin, nucleosomes associate with each other to form a more compact structure that is 30 nm in diameter. Evidence for the packaging of nucleosomes was obtained in the microscopy studies of Fritz Thoma in 1977. Chromatin samples were exposed to a solid resin that could bind to histone H1 and remove it from the DNA. However, the removal of H1 depended on the NaCl concentration. A moderate salt solution (100 mM NaCl) removed H1, but a solution with no added NaCl did not remove H1. Both types of samples were then observed with an electron microscope. At moderate salt concentrations

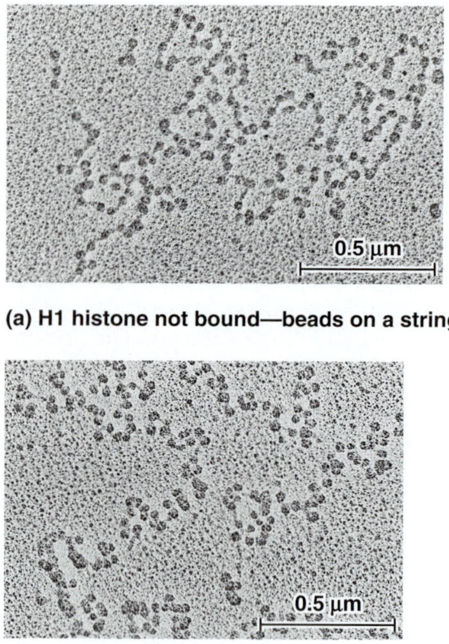

(a) H1 histone not bound—beads on a string

(b) H1 histone bound to linker region—nucleosomes more compact

FIGURE 10.13 The nucleosome structure of eukaryotic chromatin as viewed by electron microscopy. The chromatin in **(a)** has been treated with moderate salt concentrations to remove the linker histone, H1. It exhibits the classic beads-on-a-string morphology. The chromatin in **(b)** has been incubated without added NaCl and shows a more compact morphology.

(**Figure 10.13a**), the chromatin exhibited the classic beads-on-a-string morphology. Without added NaCl (when H1 is expected to remain bound to the DNA), these "beads" associated with each other into a more compact conformation (**Figure 10.13b**). These results suggest that the nucleosomes are packaged into a more compact unit and that H1 has a role in the packaging and compaction of nucleosomes. However, the precise role of H1 in chromatin compaction remains unclear. Recent data suggest that the core histones also play a key role in the compaction and relaxation of chromatin.

The experiment of Figure 10.13 and other experiments have established that nucleosome units are organized into a more compact structure that is 30 nm in diameter, known as the **30-nm fiber** (**Figure 10.14a**). The 30-nm fiber shortens the total length of DNA another sevenfold. The structure of the 30-nm fiber has proven difficult to determine because the conformation of the DNA may be substantially altered when it is extracted from living cells. Most models for the 30-nm fiber fall into two main classes. The solenoid model suggests a helical structure in which contact between nucleosomes produces a symmetrically compact structure within the 30-nm fiber (**Figure 10.14b**). This type of model is still favored by some researchers in the field. However, experimental data also suggest that the 30-nm fiber may not form such a regular structure. Instead, an alternative zigzag model, advocated by Rachel Horowitz, Christopher Woodcock, and others, is based on techniques such as cryoelectron microscopy (electron microscopy at low temperature). According to the zigzag model, linker regions within the 30-nm structure are variably bent and twisted, and little face-to-face contact occurs between nucleosomes (**Figure 10.14c**). At this level of compaction, the overall picture of chromatin that emerges is an irregular, fluctuating, three-dimensional zigzag structure with stable nucleosome units connected by deformable linker regions. In 2005, Timothy Richmond and colleagues were the first to solve the crystal structure of a segment of DNA containing multiple nucleosomes, in this case four. The structure with four nucleosomes revealed that the linker DNA zigzags back and forth between each nucleosome, a feature consistent with the zigzag model.

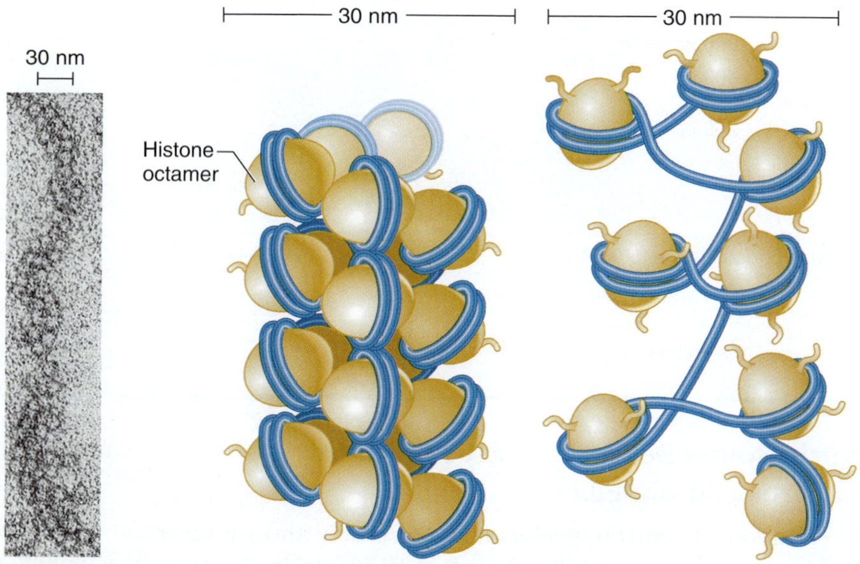

(a) Micrograph of a 30-nm fiber (b) Solenoid model (c) Zigzag model

FIGURE 10.14 The 30-nm fiber. (a) A photomicrograph of the 30-nm fiber. **(b)** In the solenoid model, the nucleosomes are packed in a helical configuration. **(c)** In the zigzag model, the linker DNA forms a more irregular structure, and less contact occurs between adjacent nucleosomes. The zigzag model is consistent with more recent data regarding chromatin conformation.

CONCEPT CHECK: Describe the distinguishing features of the solenoid model and zigzag model.

Chromosomes Are Further Compacted by Anchoring the 30-nm Fiber into Radial Loop Domains Along the Nuclear Matrix

Thus far, we have examined two mechanisms that compact eukaryotic DNA. These involve the wrapping of DNA within nucleosomes and the arrangement of nucleosomes to form a 30-nm fiber. Taken together, these two events shorten the DNA nearly 50-fold. A third level of compaction involves interactions between the 30-nm fibers and a filamentous network of proteins in the nucleus called the **nuclear matrix.** As shown in **Figure 10.15a**, the nuclear matrix consists of two parts. The **nuclear lamina** is a collection of filaments that line the inner nuclear membrane. These filaments are composed of intermediate filament proteins. The second part is an **internal nuclear matrix,** which is connected to the nuclear lamina and fills the interior of the nucleus. The internal nuclear matrix, whose structure and functional role remain controversial, is hypothesized to be an intricate fine network of irregular protein filaments plus many other proteins that bind to these filaments. Even when the chromatin is extracted from the nucleus, the internal nuclear matrix may remain intact (**Figure 10.15b**, **c**). However, the matrix should not be considered a static structure. Research indicates that the protein composition of the internal nuclear matrix is very dynamic and complex, consisting of dozens or perhaps hundreds of different proteins. The protein composition varies depending on species, cell type, and environmental conditions. This complexity has made it difficult to propose models regarding its overall organization. Further research is necessary to understand the structure and dynamic nature of the internal nuclear matrix.

The proteins of the nuclear matrix are involved in compacting the DNA into **radial loop domains,** similar to those described for the bacterial chromosome. During interphase, chromatin is organized into loops, often 25,000–200,000 bp in size, which are anchored to the nuclear matrix. The chromosomal DNA of eukaryotic species contains sequences called

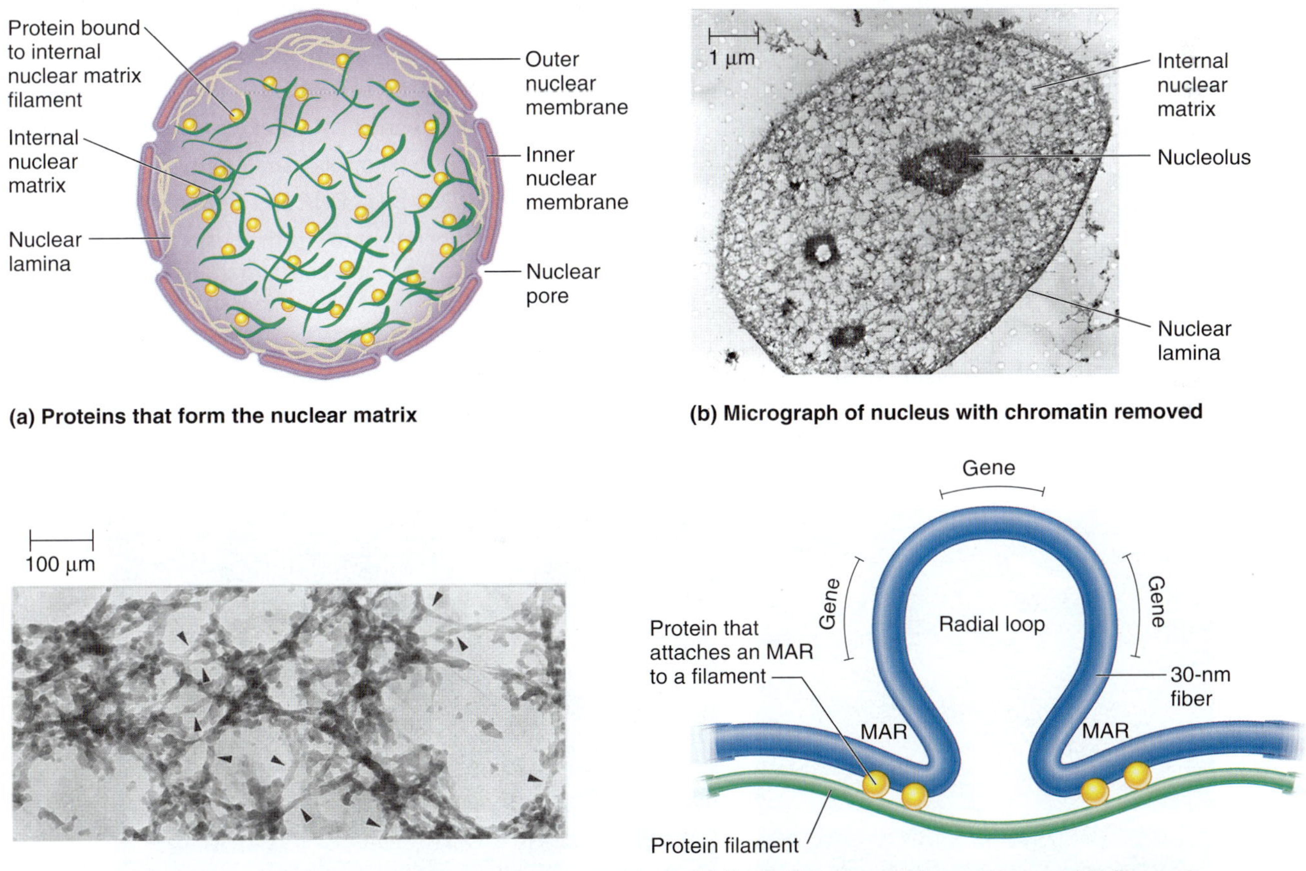

(a) Proteins that form the nuclear matrix

(b) Micrograph of nucleus with chromatin removed

(c) Micrograph showing a close-up of nuclear matrix

(d) Radial loop bound to a nuclear matrix filament

FIGURE 10.15 **Structure of the nuclear matrix.** **(a)** This schematic drawing shows the arrangement of the matrix within a cell nucleus. The nuclear lamina (depicted as yellow filaments) is a collection of filamentous proteins that line the inner nuclear membrane. The internal nuclear matrix is composed of protein filaments (depicted in green) that are interconnected. These filaments also have many other proteins associated with them (depicted in orange). **(b)** An electron micrograph of the nuclear matrix during interphase after the chromatin has been removed. The internal nuclear matrix is seen as a meshwork of protein filaments. **(c)** At higher magnification, the protein filaments are more easily seen (arrowheads point at filaments). **(d)** The matrix-attachment regions (MARs), which contain a high percentage of A and T bases, bind to the nuclear matrix and create radial loops. This causes a greater compaction of eukaryotic chromosomal DNA.

CONCEPT CHECK: What is the function of the nuclear matrix?

matrix-attachment regions (MARs) or **scaffold-attachment regions (SARs),** which are interspersed at regular intervals throughout the genome. The MARs bind to specific proteins in the nuclear matrix, thus forming chromosomal loops (**Figure 10.15d**).

Why is the attachment of radial loops to the nuclear matrix important? In addition to compaction, the nuclear matrix serves to organize the chromosomes within the nucleus. Each chromosome in the cell nucleus is located in a discrete **chromosome territory.** As shown in studies by Thomas Cremer, Christoph Cremer, and others, these territories can be viewed when interphase cells are exposed to multiple fluorescent molecules that recognize specific sequences on particular chromosomes. **Figure 10.16** illustrates an experiment in which chicken cells were exposed to a mixture of probes that recognize multiple sites along several of the larger chromosomes found in this species (*Gallus gallus*). Figure 10.16a shows the chromosomes in metaphase. The probes label each type of metaphase chromosome with a different color. Figure 10.16b shows the use of the same probes during interphase, when the chromosomes are less condensed and found in the cell nucleus. As seen here, each chromosome occupies its own distinct territory. The binding of each chromosome to the nuclear matrix is thought to play a key role in forming these chromosome territories.

Before ending the topic of interphase chromosome compaction, let's consider how the compaction level of interphase chromosomes may vary. This variability can be seen with a light microscope and was first observed by the German cytologist Emil Heitz in 1928. He coined the term **heterochromatin** to describe the tightly compacted regions of chromosomes. These regions of the chromosome are transcriptionally inactive. By comparison, the less condensed regions, known as **euchromatin,** usually are capable of gene transcription. In euchromatin, the 30-nm fiber forms radial loop domains. In heterochromatin, these radial loop domains become compacted even further.

Figure 10.17 illustrates the distribution of heterochromatin and euchromatin in a typical eukaryotic chromosome during interphase. The chromosome contains regions of both heterochromatin and euchromatin. Heterochromatin is most abundant in the centromeric regions of the chromosomes and, to a lesser extent, in the telomeric regions. The term **constitutive heterochromatin** refers to chromosomal regions that are always heterochromatic and permanently inactive with regard to transcription. Constitutive heterochromatin usually contains highly repetitive DNA sequences, such as tandem repeats, rather than gene sequences. **Facultative heterochromatin** refers to chromatin that can occasionally interconvert between heterochromatin and euchromatin. An example of facultative heterochromatin occurs in female mammals when one of the two X chromosomes is converted to a heterochromatic Barr body. As discussed in Chapter 5, most of the genes on the Barr body are transcriptionally inactive. The conversion of one X chromosome to heterochromatin occurs during embryonic development in the somatic cells of the body.

10.5 COMPREHENSION QUESTIONS

1. What are the components of a single nucleosome?
   a. About 146 bp of DNA and four histone proteins
   b. About 146 bp of DNA and eight histone proteins
   c. About 200 bp of DNA and four histone proteins
   d. About 200 bp of DNA and eight histone proteins
2. In Noll's experiment to test the beads-on-a-string model, exposure of nuclei to a low concentration of DNase I resulted in
   a. a single band of DNA with a size of approximately 200 bp.
   b. several bands of DNA in multiples of 200 bp.
   c. a single band of DNA with a size of 100 bp.
   d. several bands of DNA in multiples of 100 bp.

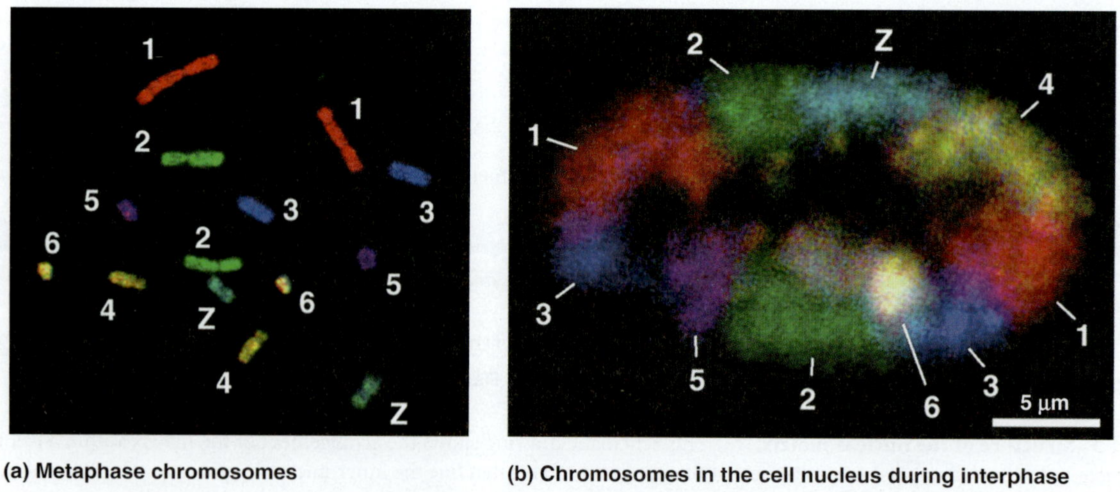

FIGURE 10.16 Chromosome territories in the cell nucleus. **(a)** Several metaphase chromosomes from the chicken were labeled with chromosome-specific probes. Each of seven types of chicken chromosomes (i.e., 1, 2, 3, 4, 5, 6, and Z) is labeled a different color. **(b)** The same probes were used to label interphase chromosomes in the cell nucleus. Each of these chromosomes occupies its own distinct, nonoverlapping territory within the cell nucleus. (Note: Chicken cells are diploid, with two copies of each chromosome.)

CONCEPT CHECK: What is a chromosome territory?

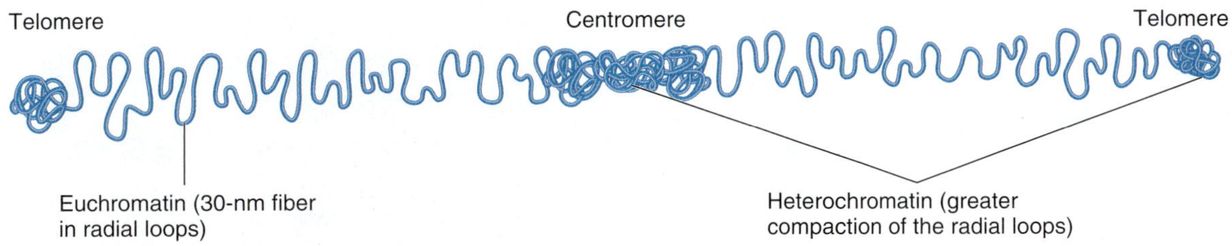

FIGURE 10.17 Chromatin structure during interphase. Heterochromatic regions are more highly condensed and tend to be localized in centromeric and telomeric regions.

CONCEPT CHECK: Would you expect to find active genes in heterochromatic or euchromatic regions?

3. With regard to the 30-nm fiber, a key difference between the solenoid and zigzag models is
   a. the solenoid model suggests a helical structure.
   b. the zigzag model suggests a more irregular pattern of nucleosomes.
   c. the zigzag model does not contain nucleosomes.
   d. both a and b are correct.
4. A chromosome territory is a region
   a. along a chromosome where many genes are clustered.
   b. along a chromosome where the nucleosomes are close together.
   c. in a cell nucleus where a single chromosome is located.
   d. in a cell nucleus where multiple chromosomes are located.

10.6 STRUCTURE OF EUKARYOTIC CHROMOSOMES DURING CELL DIVISION

Learning Outcomes:

1. Describe the levels of compaction that lead to a metaphase chromosome.
2. Explain the functions of condensin and cohesin.

As described in Chapter 2, when eukaryotic cells prepare to divide, the chromosomes become very condensed or compacted. This aids in their proper sorting during metaphase. The highly compacted chromosomes that are viewed during metaphase are called metaphase chromosomes. In this section, we will examine the compaction level of metaphase chromosomes and discuss certain proteins that are involved in the process.

Metaphase Chromosomes Are Highly Compacted

Figure 10.18 illustrates the compaction levels of eukaryotic chromosomes. The first level involves the wrapping of DNA around histone octamers to form nucleosomes. The next level is the formation of a 30-nm fiber in which nucleosomes form a zigzag or spiral structure via histone H1 and nonhistone proteins that bind to the DNA (Figure 10.18b). The 30-nm fibers then form radial loops by anchoring to protein filaments of the nuclear matrix. The average distance that the loops radiate from the protein filaments is approximately 300 nm. The level of compaction shown in Figure 10.18c is found in euchromatin, which is the predominant form of chromatin in nondividing cells.

When cells prepare to divide, the protein filaments come closer together and form a more compact scaffold for anchoring the radial loops (Figure 10.18d). This additional level of compaction greatly shortens the overall length of a chromosome and produces a diameter of approximately 700 nm, which is the compaction level found in heterochromatin. By the end of prophase, sister chromatids are entirely heterochromatic. Two parallel chromatids have a diameter of approximately 1400 nm and a much shorter length than interphase chromosomes. The overall size of a metaphase chromosome is much smaller than a chromosome territory found in the cell nucleus during interphase (refer back to Figure 10.16).

Figure 10.19a illustrates the roles of histone and nonhistone proteins in the compaction of a metaphase chromosome. Figure 10.19a shows a normal metaphase chromosome in which the radial loops are in a very compact configuration. In **Figure 10.19b**, a metaphase chromosome was treated with a high concentration of salt to remove both the core and linker histones. As seen here, the highly compact configuration is lost, but the bottoms of the elongated loops remain attached to the scaffold, which is composed of nonhistone proteins. A black arrow points to an elongated DNA strand emanating from the darkly staining scaffold. Remarkably, the scaffold retains the shape of the original metaphase chromosome even though the DNA strands have become greatly elongated. These results illustrate that the structure of metaphase chromosomes is determined by nuclear matrix proteins (which form a scaffold) and by histones (which are needed to compact the radial loops).

Condensin and Cohesin Promote the Formation of Metaphase Chromosomes

Researchers are trying to understand the steps that lead to the formation and organization of metaphase chromosomes. During the past several years, studies in yeast and in frog oocytes have been aimed at the identification of proteins that promote the conversion of interphase chromosomes into metaphase chromosomes.

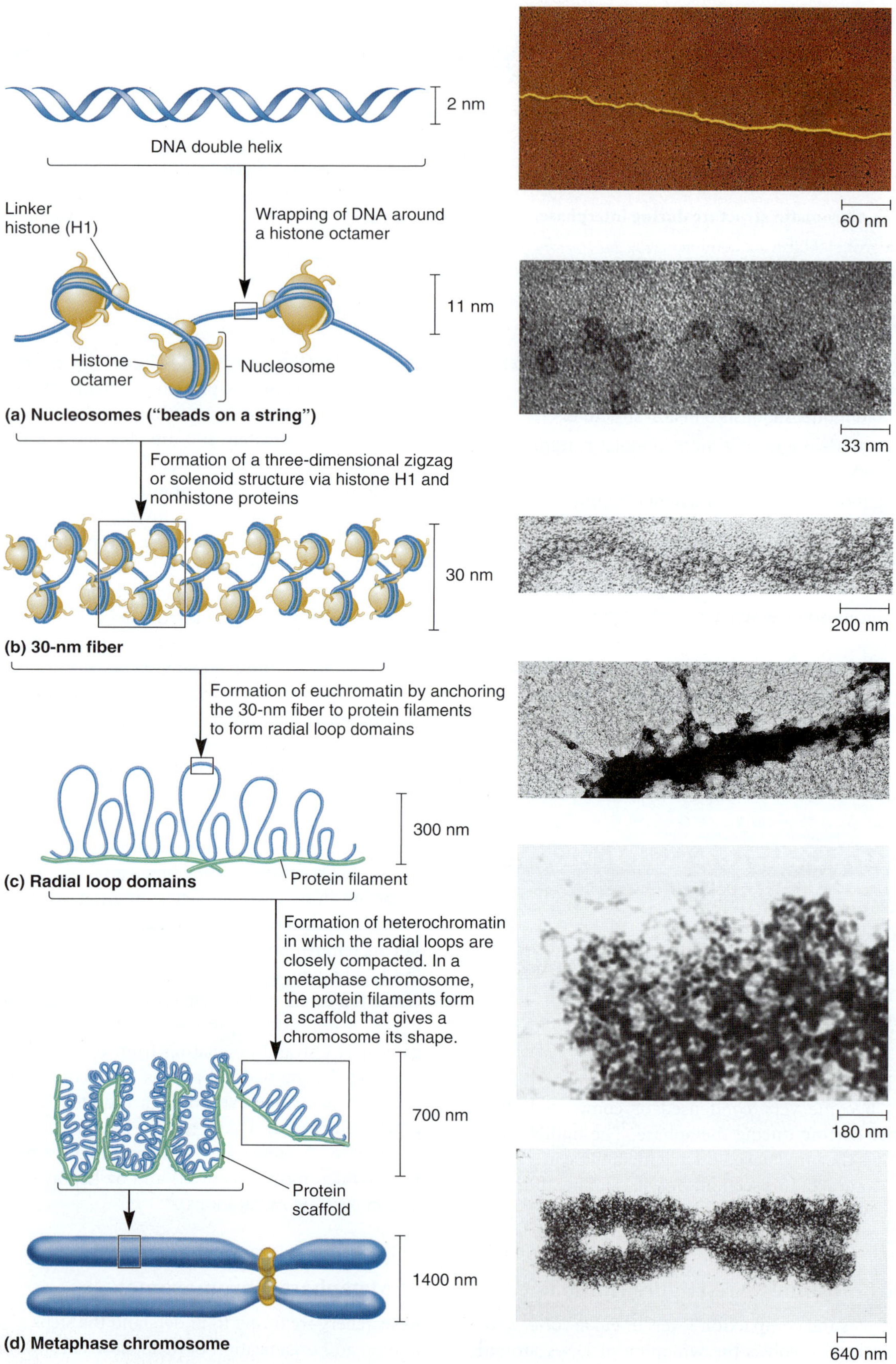

FIGURE 10.18 **The steps in eukaryotic chromosomal compaction leading to the metaphase chromosome.**

CONCEPT CHECK: Describe what structural changes convert a region that is 300 nm in diameter to one that is 700 nm in diameter.

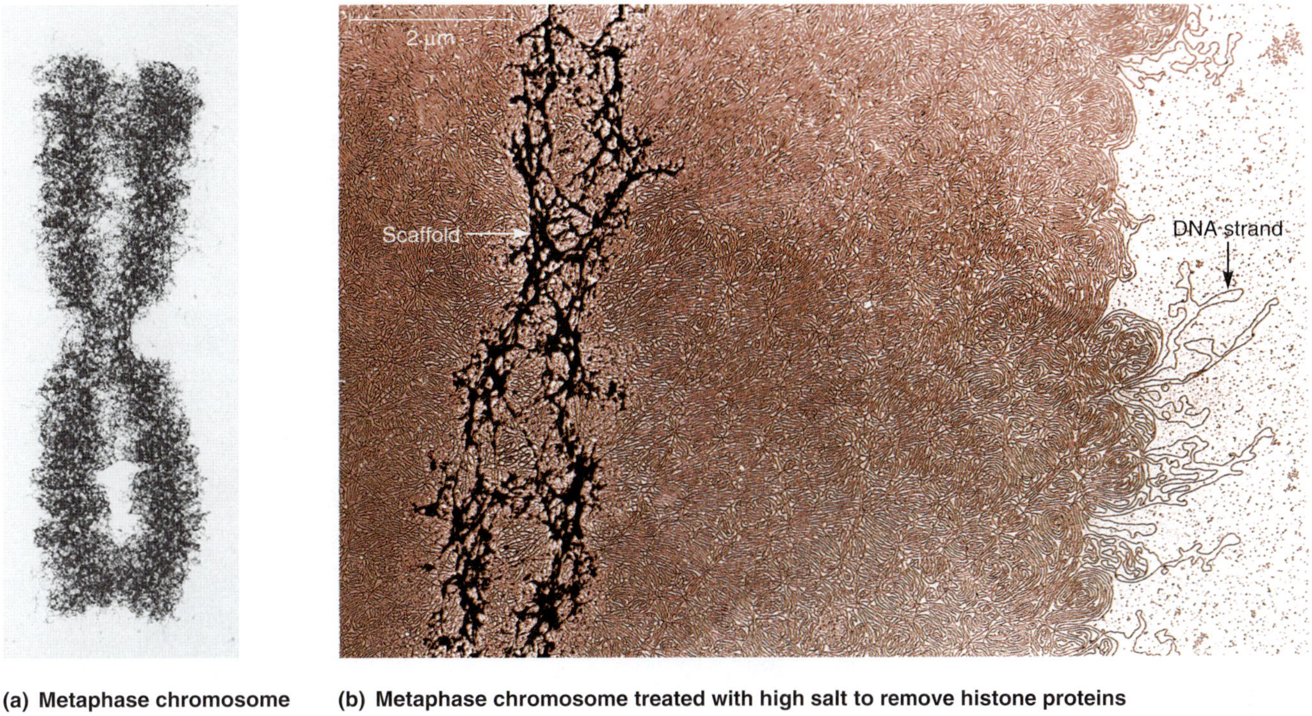

FIGURE 10.19 The importance of histone proteins and scaffolding proteins in the compaction of eukaryotic chromosomes. (**a**) A metaphase chromosome. (**b**) A metaphase chromosome following treatment with high salt concentration to remove the histone proteins. The black arrow on the right points to an elongated strand of DNA. The white arrow on the left points to the scaffold (composed of nonhistone proteins), which anchors the bases of the radial loops. As shown earlier in Figure 10.18d, this scaffold contains protein filaments.

In yeast, mutants have been characterized that have alterations in the condensation or the segregation of chromosomes. Similarly, biochemical studies using frog oocytes resulted in the purification of protein complexes that promote chromosomal condensation or sister chromatid alignment. These two lines of independent research produced the same results. Researchers discovered two multiprotein complexes called **condensin** and **cohesin,** which play a critical role in chromosomal condensation and sister chromatid alignment, respectively.

Condensin and cohesin are two completely distinct complexes, but both contain a category of proteins called **SMC proteins.** SMC stands for structural maintenance of chromosomes. These proteins use energy from ATP to catalyze changes in chromosome structure. Together with topoisomerases, SMC proteins have been shown to promote major changes in DNA structure. An emerging theme is that SMC proteins actively fold, tether, and manipulate DNA strands. The monomers, which are connected at a hinge region, have two long coiled arms with a head region that binds ATP (**Figure 10.20**). The length of each monomer is about 50 nm, which is equivalent to approximately 150 bp of DNA.

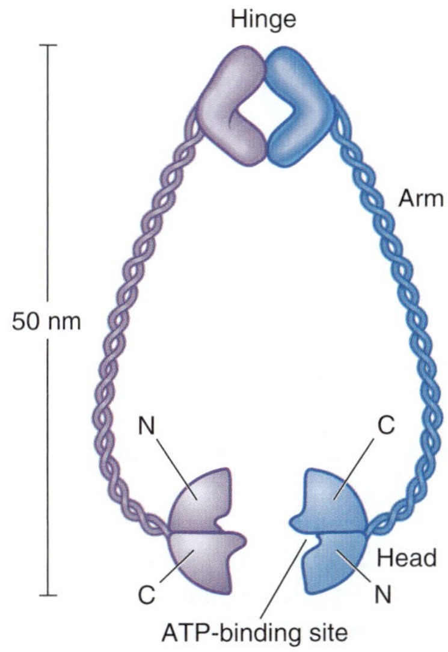

FIGURE 10.20 The structure of SMC proteins. This figure shows the generalized structure of SMC proteins, which are dimers consisting of hinge, arm, and head regions. The head regions bind and hydrolyze ATP. Condensin and cohesin have additional protein subunits not shown here. N indicates the amino terminus. C indicates the carboxyl terminus.

As their names suggest, condensin and cohesin play different roles in metaphase chromosome structure. Prior to M phase, condensin is found outside the nucleus (**Figure 10.21**). However, as M phase begins, the nuclear envelope breaks apart and condensin is observed to coat the individual chromatids as euchromatin is condensed into heterochromatin. Although the mechanism of

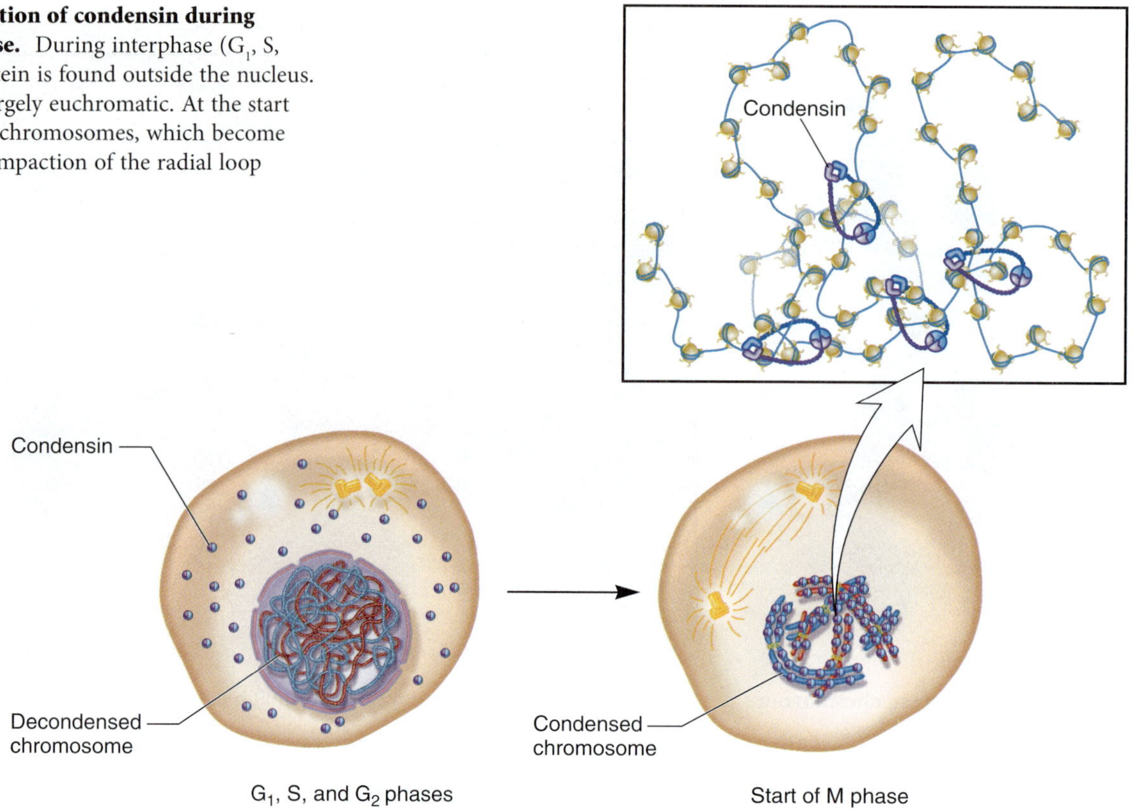

FIGURE 10.21 The localization of condensin during interphase and the start of M phase. During interphase (G_1, S, and G_2), most of the condensin protein is found outside the nucleus. The interphase chromosomes are largely euchromatic. At the start of M phase, condensin binds to the chromosomes, which become heterochromatic due to a greater compaction of the radial loop domains.

compaction is not entirely understood, recent evidence suggests that condensin proteins form a ring around the DNA. Compaction occurs because multiple condensin proteins bring chromatin loops closer together and hold them in place (see inset to Figure 10.21).

In comparison, the function of cohesin is to promote the binding (i.e., cohesion) between sister chromatids. After S phase and until the middle of prophase, sister chromatids remain attached to each other along their length. As shown in **Figure 10.22**, this attachment is promoted by cohesin, which is found along the entire length of each chromatid. In certain species, such as mammals, cohesins located along the chromosome arms are released during prophase, which allows the arms to separate. However, some cohesins remain attached, primarily to the centromeric regions, leaving the centromeric region as the main linkage before anaphase. At anaphase, the cohesins bound to the centromere are rapidly degraded by a protease aptly named separase, thereby allowing sister chromatid separation.

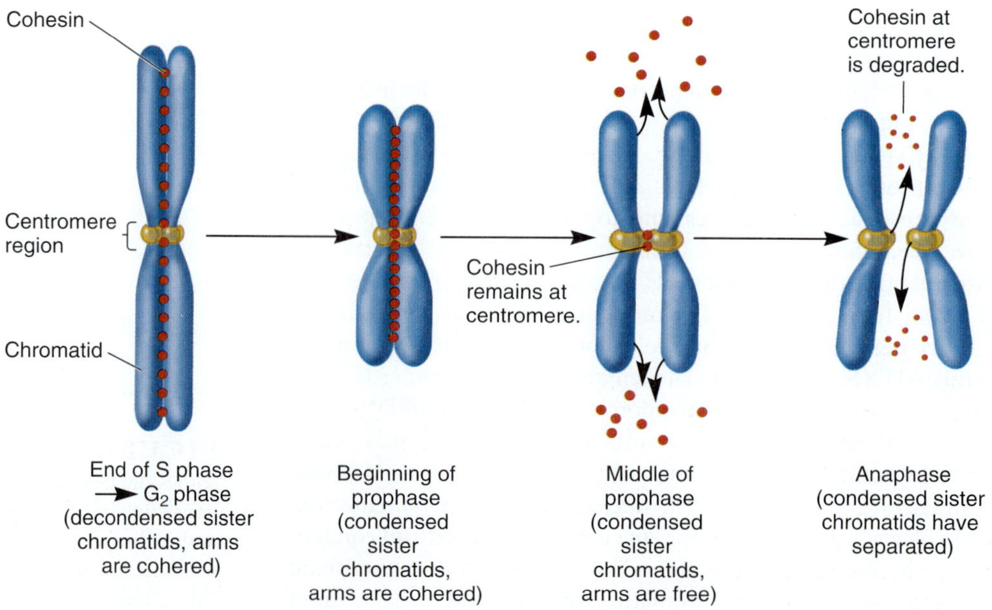

ONLINE ANIMATION

FIGURE 10.22 The alignment of sister chromatids via cohesin. After S phase is completed, many cohesin complexes bind along each chromatid, thereby facilitating their attachment to each other. During the middle of prophase, cohesin is released from the chromosome arms, but some cohesin remains in the centromeric regions. At anaphase, the remaining cohesin complexes are rapidly degraded by separase, which promotes sister chromatid separation.

CONCEPT CHECK: Describe what happens to cohesin from the beginning of prophase to anaphase.

10.6 COMPREHENSION QUESTIONS

1. The compaction leading to a metaphase chromosome involves which of the following?
   a. The formation of nucleosomes
   b. The formation of the 30-nm fiber
   c. Anchoring and further compaction of the radial loops
   d. All of the above
2. The role of cohesin is to
   a. make chromosomes more compact.
   b. allow for the replication of chromosomes.
   c. hold sister chromatids together.
   d. promote the separation of sister chromatids.

KEY TERMS

Page 239. chromosomes, genome
Page 240. protein-encoding genes, intergenic regions, origin of replication, repetitive sequences, nucleoid, loop domains
Page 241. DNA supercoiling, topoisomers, DNA gyrase
Page 242. topoisomerase I
Page 244. introns, exons, centromeres, kinetochore, telomeres
Page 245. sequence complexity, moderately repetitive sequences, transposable elements (TEs), highly repetitive sequences
Page 246. retroelement, tandem array, nucleus, chromatin
Page 247. nucleosome, histone proteins
Page 250. 30-nm fiber
Page 251. nuclear matrix, nuclear lamina, internal nuclear matrix, radial loop domains
Page 252. matrix-attachment regions (MARs), scaffold-attachment regions (SARs), chromosome territory, heterochromatin, euchromatin, constitutive heterochromatin, facultative heterochromatin
Page 255. condensin, cohesin, SMC proteins

CHAPTER SUMMARY

- Chromosomes contain the genetic material, which is DNA. A genome refers to a complete set of genetic material.

10.1 Organization of Sites Along Bacterial Chromosomes

- Bacterial chromosomes are typically circular and carry an origin of replication and a few thousand genes (see Figure 10.1).

10.2 Structure of Bacterial Chromosomes

- Bacterial chromosomes are found in the nucleoid region of a bacterial cell (see Figure 10.2).
- Bacterial chromosomes are made more compact by the formation of loop domains and DNA supercoiling (see Figures 10.3–10.5).
- Negative DNA supercoiling promotes DNA strand separation (see Figure 10.6).
- DNA gyrase (topoisomerase II) is a bacterial enzyme that introduces negative supercoils. Topoisomerase I relaxes negative supercoils (see Figure 10.7).

10.3 Organization of Sites Along Eukaryotic Chromosomes

- Eukaryotic chromosomes are usually linear and contain a centromere, telomeres, multiple origins of replication, and many genes (see Figure 10.8).

10.4 Sizes of Eukaryotic Genomes and Repetitive Sequences

- The genome sizes of eukaryotes vary greatly. Some of this variation is due to the accumulation of repetitive sequences (see Figure 10.9).
- The human genome contains about 41% unique sequences and 59% repetitive sequences (see Figure 10.10).

10.5 Structure of Eukaryotic Chromosomes in Nondividing Cells

- The term chromatin refers to the DNA-protein complex found within eukaryotic chromosomes.
- Eukaryotic DNA wraps around an octamer of histone proteins to form nucleosomes (see Figure 10.11).
- Noll tested Kornberg's nucleosome model by digesting eukaryotic chromatin with varying concentrations of DNase I (see Figure 10.12).
- The linker histone, H1, plays a role in nucleosome compaction (see Figure 10.13).
- Nucleosomes are further compacted to form a 30-nm fiber. Solenoid and zigzag models have been proposed (see Figure 10.14).
- Chromatin is further compacted by the attachment of 30-nm fibers to protein filaments to form radial loop domains (see Figure 10.15).

- Within the cell nucleus, each eukaryotic chromosome occupies its own unique chromosome territory (see Figure 10.16).
- In nondividing cells, each chromosome has highly compacted regions called heterochromatin and less compacted regions called euchromatin (see Figure 10.17).

10.6 Structure of Eukaryotic Chromosomes During Cell Division

- Chromatin compaction occurs due to nucleosome formation, the formation of a 30-nm fiber, and radial loop formation. A metaphase chromosome is completely heterochromatic due to the further compaction of radial loops (see Figure 10.18).
- Both histone and nonhistone proteins are important for the compaction of metaphase chromosomes (see Figure 10.19).
- Condensin and cohesion contain SMC proteins that promote chromosome compaction and sister chromatid cohesion, respectively (see Figures 10.20–10.22).

PROBLEM SETS & INSIGHTS

Solved Problems

S1. What are the general features of the organization of sites along a bacterial chromosome?

Answer:

1. A bacterial chromosome is usually circular and is usually of one type.
2. It typically contains a few million base pairs of DNA.
3. Several thousand genes are found interspersed along the chromosome; short intergenic regions are found between genes.
4. One origin of replication is present.
5. Repetitive sequences are often present, but they are not as abundant as those found in eukaryotic species.

S2. Let's suppose a bacterial DNA molecule is given a left-handed twist. How does this affect the structure and function of the DNA?

Answer: A left-handed twist is negative supercoiling. Negative supercoiling makes the bacterial chromosome more compact. It also promotes DNA functions that involve strand separation, including gene transcription and DNA replication.

S3. To hold bacterial DNA in a more compact configuration, specific proteins must bind to the DNA and stabilize its conformation (as shown in Figure 10.3). Several different proteins are involved in this process. These proteins have been collectively referred to as "histone-like" due to their functional similarity to the histone proteins found in eukaryotes. Based on your knowledge of eukaryotic histone proteins, what biochemical properties would you expect from bacterial histone-like proteins?

Answer: The histone-like proteins have the properties expected for proteins involved in DNA folding. They are all small proteins found in relative abundance within the bacterial cell. In some cases, the histone-like proteins are biochemically similar to eukaryotic histones. For example, they tend to be basic (positively charged) and bind to DNA in a non-sequence-dependent fashion. However, other proteins appear to bind to bacterial DNA at specific sites in order to promote DNA bending.

S4. Describe two key reasons why eukaryotic genomes vary greatly in their sizes.

Answer: One reason is that more complex eukaryotes, such as mammals, have more genes than simpler eukaryotes, such as yeast. Another reason is that the amount of repetitive DNA can vary greatly among eukaryotic species.

S5. Outline the ways that a eukaryotic chromosome may be compacted.

Answer:

1. The DNA is wrapped around histones to form nucleosomes.
2. The nucleosomes are brought closer together to form a 30-nm fiber.
3. The 30-nm fiber is anchored to protein filaments to form radial loop domains.
4. In a metaphase chromosome, the protein filaments form a more compact scaffold, which results in a chromosome that is much shorter.

Conceptual Questions

C1. What is a bacterial nucleoid? With regard to cellular membranes, what is the difference between a bacterial nucleoid and a eukaryotic nucleus?

C2. In Part II of this textbook, we considered inheritance patterns for diploid eukaryotic species. Bacteria frequently contain two or more nucleoids. With regard to genes and alleles, how is a bacterium that contains two nucleoids similar to a diploid eukaryotic cell, and how is it different?

C3. Describe the two main mechanisms by which the bacterial DNA becomes compacted.

C4. As described in Chapter 9, one bp of DNA is approximately 0.34 nm in length. A bacterial chromosome is about 4 million bp in length and is organized into about 100 loops that are about 40,000 bp in length.

A. If it was stretched out linearly, how long (in micrometers) would one loop be?

B. If a bacterial chromosomal loop is circular, what would be its diameter? (Note: Circumference = πD, where D is the diameter of the circle.)

C. Is the diameter of the circular loop calculated in part B small enough to fit inside a bacterium? The dimensions of the bacterial cytoplasm, such as *E. coli*, are roughly 0.5 μm wide and 1.0 μm long.

C5. Why is DNA supercoiling called supercoiling rather than just coiling? Why is positive supercoiling called overwinding and negative supercoiling called underwinding? How would you define the terms positive and negative supercoiling for Z DNA (described in Chapter 9)?

C6. Coumarins and quinolones are two classes of drugs that inhibit bacterial growth by directly inhibiting DNA gyrase. Discuss two reasons why inhibiting DNA gyrase also inhibits bacterial growth.

C7. Take two pieces of string that are approximately 10 inches each, and create a double helix by wrapping them around each other to make 10 complete turns. Tape one end of the strings to a table, and now twist the strings three times (360° each time) in a right-handed direction. Note: As you are looking down at the strings from above, a right-handed twist is in the clockwise direction.

A. Did the three turns create more or fewer turns in your double helix? How many turns are now in your double helix?

B. Is your double helix right-handed or left-handed? Explain your answer.

C. Did the three turns create any supercoils?

D. If you had coated your double helix with rubber cement and allowed the cement to dry before making the three additional right-handed turns, would the rubber cement make it more or less likely for the three turns to create supercoiling? Would a pair of cemented strings be more or less like a real DNA double helix than an uncemented pair of strings? Explain your answer.

C8. Try to explain the function of DNA gyrase with a drawing.

C9. How are two topoisomers different from each other? How are they the same?

C10. On rare occasions, a chromosome can suffer a small deletion that removes the centromere. When this occurs, the chromosome usually is not found within subsequent daughter cells. Explain why a chromosome without a centromere is not transmitted very efficiently from mother to daughter cells. (Note: If a chromosome is located outside the nucleus after telophase, it is degraded.)

C11. What is the function of a centromere? At what stage of the cell cycle would you expect the centromere to be the most important?

C12. Describe the characteristics of highly repetitive DNA.

C13. Describe the structures of a nucleosome and a 30-nm fiber.

C14. Beginning with the G_1 phase of the cell cycle, describe the level of compaction of the eukaryotic chromosome. How does the level of compaction change as the cell progresses through the cell cycle? Why is it necessary to further compact the chromatin during mitosis?

C15. If you assume the average length of linker DNA is 50 bp, approximately how many nucleosomes are found in the haploid human genome, which contains 3 billion bp?

C16. Draw a picture depicting the binding between the nuclear matrix and MARs.

C17. Compare heterochromatin and euchromatin. What are the differences between them?

C18. Compare the structure and cell localization of chromosomes during interphase and M phase.

C19. What types of genetic activities occur during interphase? Explain why these activities cannot occur during M phase.

C20. Let's assume the linker DNA averages 54 bp in length. How many molecules of H2A would you expect to find in a DNA sample that is 46,000 bp in length?

C21. In Figure 10.13, what are we looking at in part b? Is this an 11-nm fiber, a 30-nm fiber, or a 300-nm fiber? Does this DNA come from a cell during M phase or interphase?

C22. What are the roles of the core histone proteins compared with the role of histone H1 in the compaction of eukaryotic DNA?

C23. A typical eukaryotic chromosome found in humans contains about 100 million bp of DNA. As described in Chapter 9, one bp of DNA has a linear length of 0.34 nm.

A. What is the linear length of the DNA for a typical human chromosome in micrometers?

B. What is the linear length of a 30-nm fiber of a typical human chromosome?

C. Based on your calculation of part B, would a typical human chromosome fit inside the nucleus (with a diameter of 5 μm) if the 30-nm fiber was stretched out in a linear manner? If not, explain how a typical human chromosome fits inside the nucleus during interphase.

C24. Which of the following terms should *not* be used to describe a Barr body?

A. Chromatin

B. Euchromatin

C. Heterochromatin

D. Chromosome

E. Genome

C25. Discuss the differences in the compaction levels of metaphase chromosomes compared with interphase chromosomes. When would you expect gene transcription and DNA replication to take place, during M phase or interphase? Explain why.

C26. What is an SMC protein? Describe two examples.

Experimental Questions

E1. Two circular DNA molecules, which we will call molecule A and molecule B, are topoisomers of each other. When viewed under the electron microscope, molecule A appears more compact than molecule B. The level of gene transcription is much lower for molecule A. Which of the following three possibilities could account for these observations?

First possibility: Molecule A has three positive supercoils, and molecule B has three negative supercoils.

Second possibility: Molecule A has four positive supercoils, and molecule B has one negative supercoil.

Third possibility: Molecule A has zero supercoils, and molecule B has three negative supercoils.

E2. Let's suppose you have isolated DNA from a cell and viewed it under a microscope. It looks supercoiled. What experiment would you perform to determine if it is positively or negatively supercoiled? In your answer, describe your expected results. You may assume that you have purified topoisomerases at your disposal.

E3. We seem to know more about the structure of eukaryotic chromosomal DNA than bacterial DNA. Discuss why you think this is so, and list several experimental procedures that have yielded important information concerning the compaction of eukaryotic chromatin.

E4. In Noll's experiment of Figure 10.12, explain where the DNase I cuts the DNA. Why were the bands on the gel in multiples of 200 bp at lower DNase I concentrations?

E5. When chromatin is treated with a moderate salt concentration, the linker histone H1 is removed (see Figure 10.13a). Higher salt concentration removes the rest of the histone proteins (see Figure 10.19b). If the experiment of Figure 10.12 were carried out after the DNA was treated with moderate or high salt, what would be the expected results?

E6. Let's suppose you have isolated chromatin from some bizarre eukaryote with a linker region that is usually 300–350 bp in length. The nucleosome structure is the same as in other eukaryotes. If you digested this eukaryotic organism's chromatin with a high concentration of DNase I, what would be your expected results?

E7. If you were given a sample of chromosomal DNA and asked to determine if it is bacterial or eukaryotic, what experiment would you perform, and what would be your expected results?

E8. Consider how histone proteins bind to DNA and then explain why a high salt concentration can remove histones from DNA (as shown in Figure 10.19b).

E9. In Chapter 22, the technique of fluorescence in situ hybridization (FISH) is described. This is another method for examining sequence complexity within a genome. In this method, a DNA sequence, such as a particular gene sequence, can be detected within an intact chromosome by using a DNA probe that is complementary to the sequence. For example, let's consider the β-globin gene, which is found on human chromosome 11. A probe complementary to the β-globin gene binds to the β-globin gene and shows up as a brightly colored spot on human chromosome 11. In this way, researchers can detect where the β-globin gene is located within a set of chromosomes. Because the β-globin gene is unique and because human cells are diploid (i.e., have two copies of each chromosome), a FISH experiment shows two bright spots per cell; the probe binds to each copy of chromosome 11. What would you expect to see if you used the following types of probes?

A. A probe complementary to the *Alu*I sequence

B. A probe complementary to a tandemly repeated sequence near the centromere of the X chromosome

Questions for Student Discussion/Collaboration

1. Bacterial and eukaryotic chromosomes are very compact. Discuss the advantages and disadvantages of having a compact structure.
2. The prevalence of highly repetitive sequences seems rather strange to many geneticists. Do they seem strange to you? Why or why not? Discuss whether or not you think they have an important function.
3. Discuss and make a list of the similarities and differences between bacterial and eukaryotic chromosomes.

Note: All answers appear at the website for this textbook; the answers to the even-numbered questions and all of the Concept Check and Comprehension Questions are in the back of the book.

www.mhhe.com/brookergenetics5e

Visit the website for practice tests, answer keys, and other learning aids for this chapter. Enhance your understanding of genetics with our interactive exercises, quizzes, animations, and much more.

CHAPTER OUTLINE

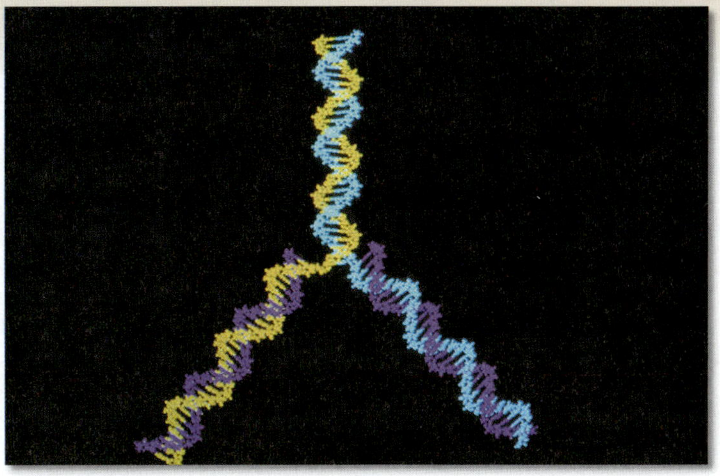

A model for DNA undergoing replication. *This molecular model shows a DNA replication fork, the site where new DNA strands are made. In this model, the original DNA is yellow and blue. The newly made strands are purple.*

11

DNA REPLICATION

As discussed throughout Chapters 2–8, genetic material is transmitted from parent to offspring and from cell to cell. For transmission to occur, the genetic material must be copied. During this process, known as **DNA replication,** the original DNA strands are used as templates for the synthesis of new DNA strands. We will begin Chapter 11 with a consideration of the structural features of the double helix that underlie the replication process. Then we examine how chromosomes are replicated within living cells, addressing the following questions: where does DNA replication begin, how does it proceed, and where does it end? We first consider bacterial DNA replication and examine how DNA replication occurs within living cells, and then we turn our attention to the unique features of the replication of eukaryotic DNA. At the molecular level, it is rather remarkable that the replication of chromosomal DNA occurs very quickly, very accurately, and at the appropriate time in the life of the cell. For this to happen, many cellular proteins play vital roles. In this chapter, we will examine the mechanism of DNA replication and consider the functions of several proteins involved in the process.

11.1 STRUCTURAL OVERVIEW OF DNA REPLICATION

Learning Outcomes:

1. Describe the structural features of DNA that enable it to be replicated.
2. Analyze the experiment of Meselson and Stahl and explain how the results were consistent with the semiconservative model of DNA replication.

Because they bear directly on the replication process, let's begin by recalling a few important structural features of the double helix from Chapter 9. The double helix is composed of two DNA strands, and the individual building blocks of each strand are nucleotides. The nucleotides contain one of four bases: adenine, thymine, guanine, or cytosine. The double-stranded structure is held together by base stacking and by hydrogen bonding between the bases in opposite strands. A critical feature of the double-helix structure is that adenine forms two hydrogen bonds with thymine, and guanine forms three hydrogen bonds with cytosine. This rule, known as the AT/GC rule, is the basis for the complementarity of the base sequences in double-stranded DNA.

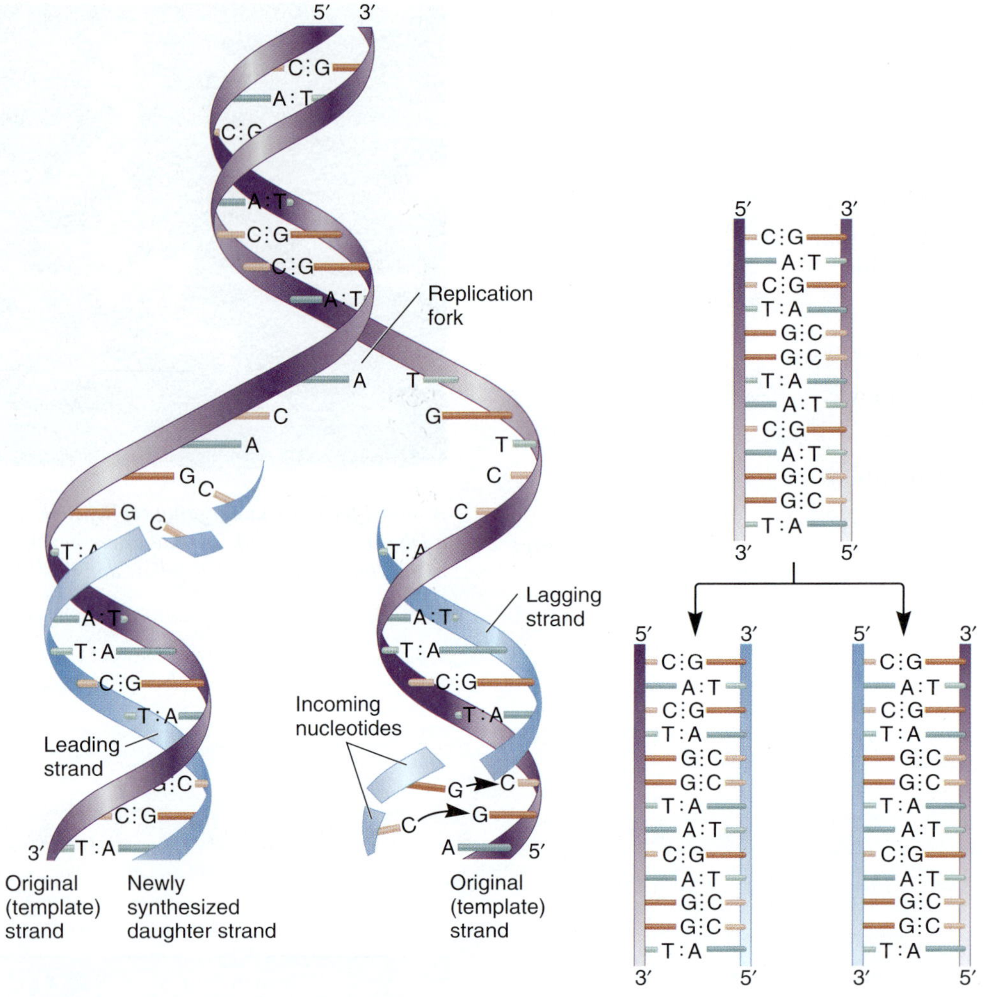

(a) The mechanism of DNA replication

(b) The products of replication

ONLINE ANIMATION

FIGURE 11.1 The structural basis for DNA replication. **(a)** The mechanism of DNA replication as originally proposed by Watson and Crick. As discussed later, the synthesis of one newly made strand (the leading strand) occurs in the direction toward the replication fork, whereas the synthesis of the other newly made strand (the lagging strand) occurs in small segments away from the replication fork. **(b)** DNA replication produces two copies of DNA with the same sequence as the original DNA molecule.

CONCEPT CHECK: What features of DNA structure enable it to be replicated?

Another feature worth noting is that the strands within a double helix have an antiparallel alignment. This directionality is determined by the orientation of sugar molecules within the sugar-phosphate backbone. If one strand is running in the 5′ to 3′ direction, the complementary strand is running in the 3′ to 5′ direction. The issue of directionality will be important when we consider the function of the enzymes that synthesize new DNA strands. In this section, we will consider how the structure of the DNA double helix provides the basis for DNA replication.

Existing DNA Strands Act as Templates for the Synthesis of New Strands

As shown in **Figure 11.1a**, DNA replication relies on the complementarity of DNA strands according to the AT/GC rule. During the replication process, the two complementary strands of DNA come apart and serve as **template strands,** or **parental strands,** for the synthesis of two new strands of DNA. After the double helix has separated, individual nucleotides have access to the template strands. Hydrogen bonding between individual nucleotides and the template strands must obey the AT/GC rule. To complete the replication process, a covalent bond is formed between the phosphate of one nucleotide and the sugar of the previous nucleotide. The two newly made strands are referred to as the **daughter strands.** Note that the base sequences are identical in both double-stranded molecules after replication (**Figure 11.1b**). Therefore, DNA is replicated so both copies retain the same information—the same base sequence—as the original molecule.

EXPERIMENT 11A

Three Different Models Were Proposed That Described the Net Result of DNA Replication

Scientists in the late 1950s had considered three different mechanisms to explain the net result of DNA replication. These mechanisms are shown in **Figure 11.2**. The first is referred to as a **conservative model.** According to this hypothesis, both strands of parental DNA remain together following DNA replication. In this model, the original arrangement of parental strands is completely conserved, while the two newly made daughter strands also remain together following replication. The second is called a **semiconservative model.** In this mechanism, the double-stranded DNA is half conserved following the replication process. In other words, the newly made double-stranded DNA contains one parental strand and one daughter strand. The third, called the **dispersive model,** proposes that segments of parental DNA and newly made DNA are interspersed in both strands following the replication process. Only the semiconservative model shown in Figure 11.2b is actually correct.

In 1958, Matthew Meselson and Franklin Stahl devised a method to experimentally distinguish newly made daughter strands from the original parental strands. Their technique involved labeling DNA with a heavy isotope of nitrogen. Nitrogen, which is found in the bases of DNA, occurs in both a heavy (^{15}N) and light (^{14}N) form. Prior to their experiment, they grew *Escherichia coli* cells in the presence of ^{15}N for many generations. This produced a population of cells in which all of the DNA was heavy-labeled. At the start of their experiment, shown in **Figure 11.3** (generation 0), they switched the bacteria to a medium that contained only ^{14}N and then collected samples of cells after various time points. Under the growth conditions they employed, the cells replicated their DNA and divided into daughter cells every 30 minutes. After each cell doubling, the new daughter cells were viewed as part of a new generation. Because the bacteria were doubling in a medium that contained only ^{14}N, all of the newly made DNA strands were labeled with light nitrogen, but the original strands remained in the heavy form.

Meselson and Stahl then analyzed the density of the DNA by centrifugation, using a cesium chloride (CsCl) gradient. (The procedure of gradient centrifugation is described in the Appendix.) If both DNA strands contained ^{14}N, the DNA had a light density and remained near the top of the tube. If one strand contained ^{14}N and the other strand contained ^{15}N, the DNA was half-heavy and had an intermediate density. Finally, if both strands contained ^{15}N, the DNA was heavy and moved closer to the bottom of the centrifuge tube.

THE HYPOTHESIS

Based on Watson's and Crick's ideas, the hypothesis was that DNA replication is semiconservative. Figure 11.2 also shows two alternative models.

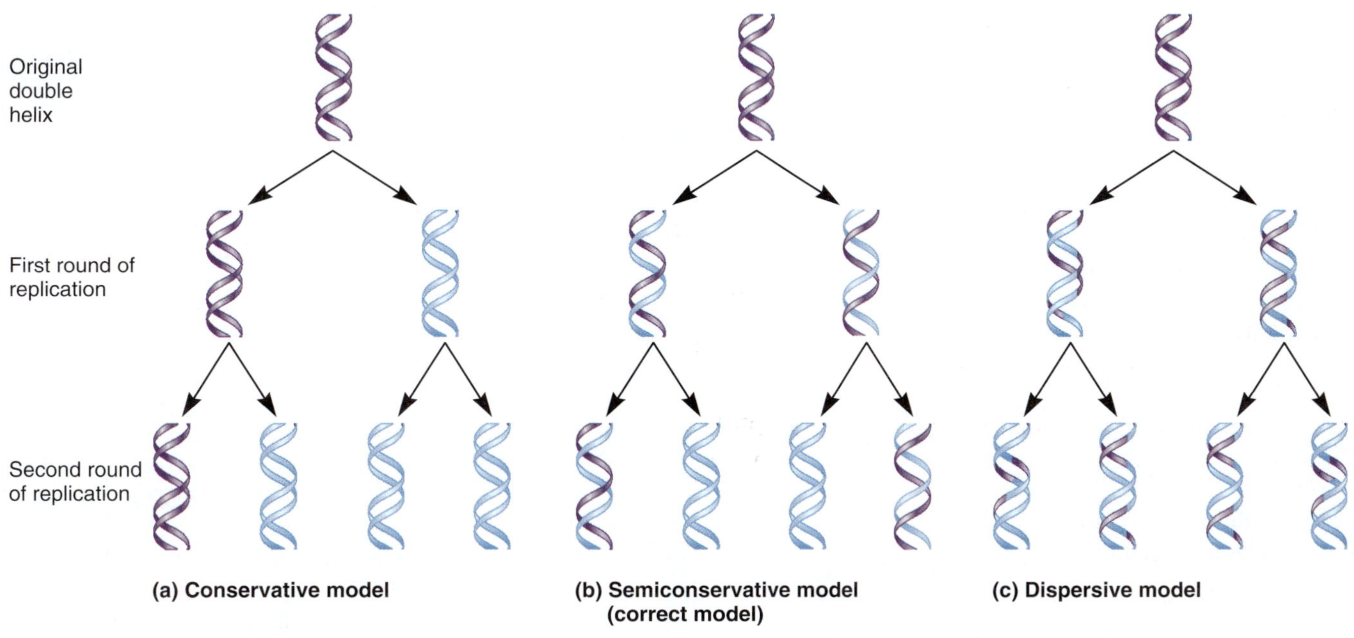

FIGURE 11.2 Three possible models for DNA replication. The two original parental DNA strands are shown in purple, and the newly made strands after one and two generations are shown in light blue.

TESTING THE HYPOTHESIS **FIGURE 11.3** **Evidence that DNA replication is semiconservative.**

Starting material: A strain of *E. coli* that has been grown for many generations in the presence of ^{15}N. All of the nitrogen in the DNA is labeled with ^{15}N.

1. Add an excess of ^{14}N-containing compounds to the growth medium so all of the newly made DNA will contain ^{14}N.
2. Incubate the cells for various lengths of time. Note: The ^{15}N-labeled DNA is shown in purple and the ^{14}N-labeled DNA is shown in blue.
3. Lyse the cells by the addition of lysozyme and detergent, which disrupt the bacterial cell wall and cell membrane, respectively.
4. Load a sample of the lysate onto a CsCl gradient. (Note: The average density of DNA is around 1.7 g/cm^3, which is well isolated from other cellular macromolecules.)
5. Centrifuge the gradients until the DNA molecules reach their equilibrium densities.
6. DNA within the gradient can be observed under a UV light.

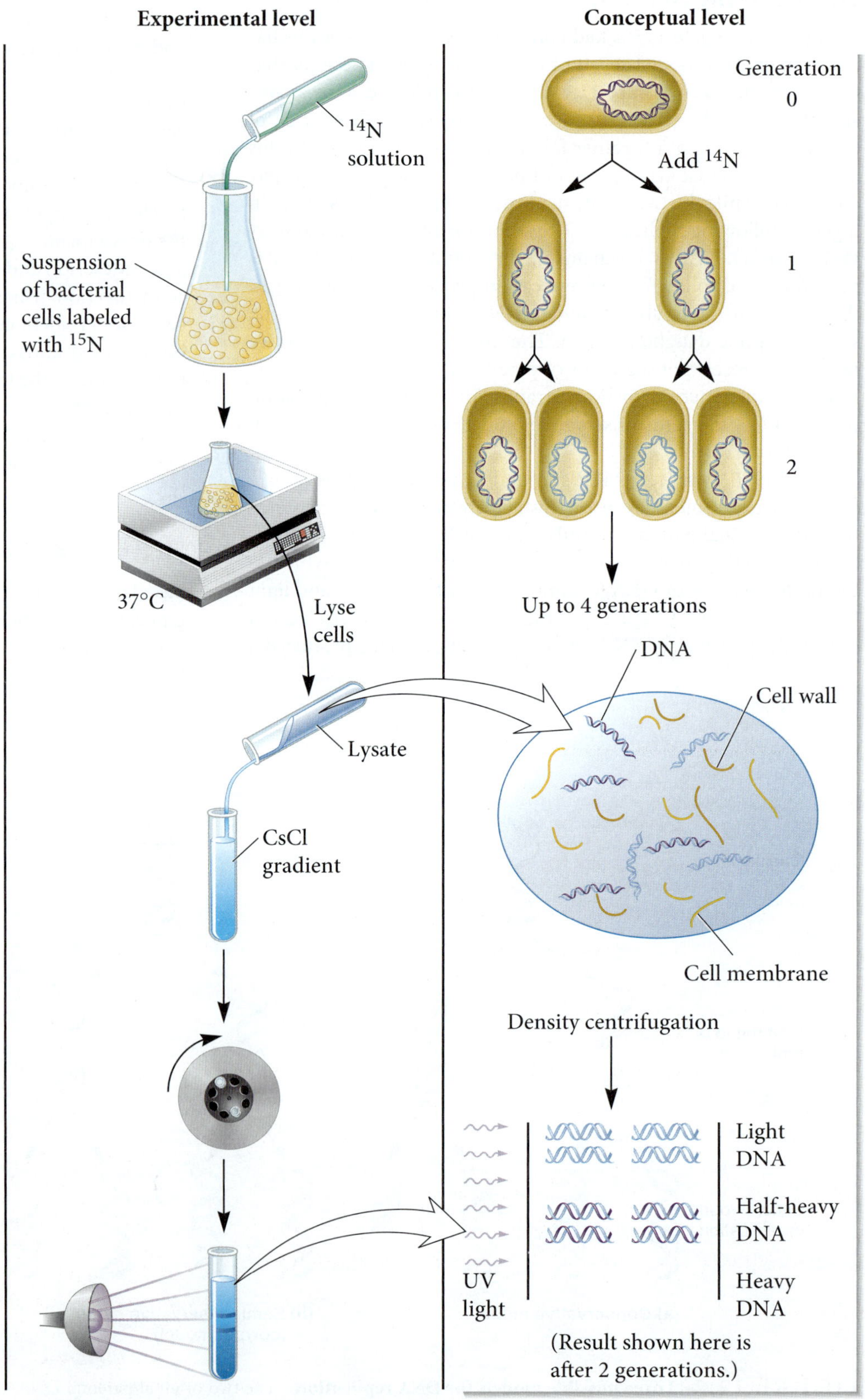

THE DATA

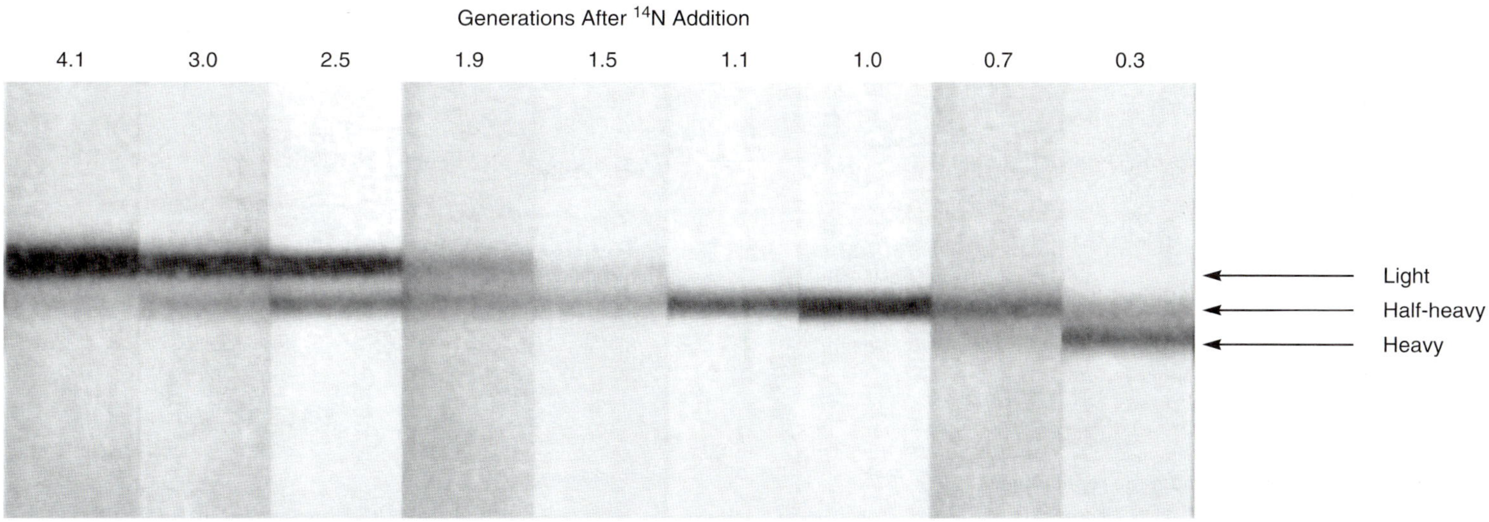

Data from: Meselson, M. and Stahl, F.W. (1958) The Replication of DNA in *Escherichia coli*. *Proc Natl Acad Sci USA* 44: 671–682.

INTERPRETING THE DATA

As seen in The Data, after one round of DNA replication (i.e., one generation), all of the DNA sedimented at a density that was half-heavy. Which of the three models is consistent with this result? Both the semiconservative and dispersive models are consistent. In contrast, the conservative model predicts two separate DNA types: a light type and a heavy type. Because all of the DNA had sedimented as a single band, this model was disproved. According to the semiconservative model, the replicated DNA would contain one original strand (a heavy strand) and a newly made daughter strand (a light strand). Likewise, in a dispersive model, all of the DNA should have been half-heavy after one generation as well. To determine which of these two remaining models is correct, therefore, Meselson and Stahl had to investigate future generations.

After approximately two rounds of DNA replication (i.e., 1.9 generations), a mixture of light DNA and half-heavy DNA was observed. This result was consistent with the semiconservative model of DNA replication, because some DNA molecules should contain all light DNA, and other molecules should be half-heavy (see Figure 11.2b). The dispersive model predicts that after two generations, the heavy nitrogen would be evenly dispersed among four strands, each strand containing 1/4 heavy nitrogen and 3/4 light nitrogen (see Figure 11.2c). However, this result was not obtained. Instead, the results of the Meselson and Stahl experiment provided compelling evidence in favor of only the semiconservative model for DNA replication.

A self-help quiz involving this experiment can be found at www.mhhe.com/brookergenetics5e.

11.1 COMPREHENSION QUESTIONS

1. The complementarity of DNA strands is based on
   a. the chemical properties of a phosphodiester linkage.
   b. the binding of proteins to the DNA.
   c. the AT/GC rule.
   d. none of the above.
2. To make a new DNA strand, which of the following is necessary?
   a. A template strand
   b. Nucleotides
   c. Heavy nitrogen
   d. Both a and b
3. The model that correctly describes the process of DNA replication is
   a. the conservative model.
   b. the semiconservative model.
   c. the dispersive model.
   d. all of the above.

11.2 BACTERIAL DNA REPLICATION: THE FORMATION OF TWO REPLICATION FORKS AT THE ORIGIN OF REPLICATION

Learning Outcomes:

1. Describe the key features of a bacterial origin of replication.
2. Explain how DnaA protein initiates DNA replication.

Thus far, we have considered how a complementary, double-stranded structure underlies the ability of DNA to be copied. In addition, the experiment of Meselson and Stahl showed that DNA replication results in two double helices, each one containing an original parental strand and a newly made daughter strand. We now turn our attention to how DNA replication occurs within

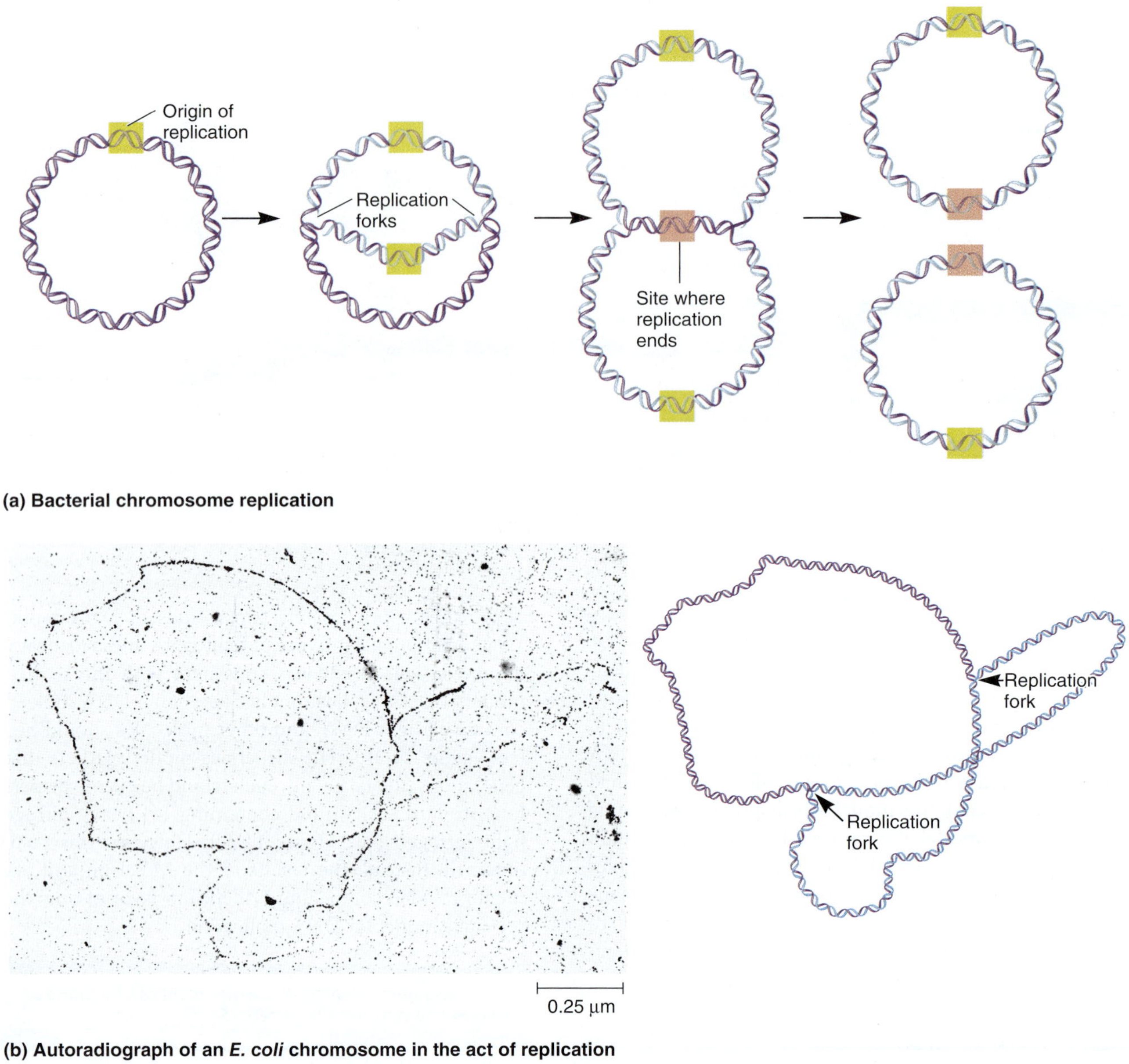

FIGURE 11.4 The process of bacterial chromosome replication. (a) An overview of the process of bacterial chromosome replication. **(b)** A replicating *E. coli* chromosome visualized by autoradiography and transmission electron microscopy (TEM). This chromosome was radiolabeled by growing bacterial cells in media containing radiolabeled deoxythymidine. The diagram at the right shows the locations of the two replication forks. The chromosome is about one-third replicated. New strands are shown in blue.

living cells. Much research has focused on understanding the process in the bacterium *Escherichia coli.* In this section, we will examine how bacterial DNA replication begins.

Bacterial Chromosomes Contain a Single Origin of Replication

Figure 11.4 presents an overview of the process of bacterial chromosomal replication. The site on the bacterial chromosome where DNA synthesis begins is known as the **origin of replication.** Bacterial chromosomes have a single origin of replication. The synthesis of new daughter strands is initiated within the origin and proceeds in both directions, or **bidirectionally,** around the bacterial chromosome. This means that two **replication forks** move in opposite directions outward from the origin. A replication fork is the site where the parental DNA strands have separated and new daughter strands are being made. Eventually, these replication forks meet each other on the opposite side of the bacterial chromosome to complete the replication process.

Replication Is Initiated by the Binding of DnaA Protein to the Origin of Replication

Considerable research has focused on the origin of replication in *E. coli.* This origin is named *oriC* for origin of Chromosomal replication (**Figure 11.5**). Three types of DNA sequences are found within *oriC:* an AT-rich region, DnaA box sequences, and GATC methylation sites.

FIGURE 11.5 The sequence of *oriC* in *E. coli*. The AT-rich region is composed of three similar sequences that are 13 bp long and highlighted in blue. The five DnaA boxes are highlighted in orange. The GATC methylation sites are underlined.

CONCEPT CHECK: What are the functions of the AT-rich region and DnaA boxes?

DNA replication is initiated by the binding of **DnaA proteins** to sequences within the origin known as **DnaA box sequences.** The DnaA box sequences serve as recognition sites for the binding of the DnaA proteins. When DnaA proteins are in their ATP-bound form, they bind to the five DnaA boxes in *oriC* to initiate DNA replication. DnaA proteins also bind to each other to form a complex (**Figure 11.6**). With the aid of other DNA-binding proteins, such as HU and IHF, this causes the DNA to bend around the complex of DnaA proteins and results in the separation of the AT-rich region. Because only two hydrogen bonds form between AT base pairs, whereas three hydrogen bonds occur between G and C, the DNA strands are more easily separated at an AT-rich region.

Following separation of the AT-rich region, the DnaA proteins, with the help of DnaC proteins, recruit **DNA helicase** proteins to this site. DNA helicase is also known as DnaB protein. When a DNA helicase encounters a double-stranded region, it breaks the hydrogen bonds between the two strands, thereby generating two single strands. Two DNA helicases begin strand separation within the *oriC* region and continue to separate the DNA

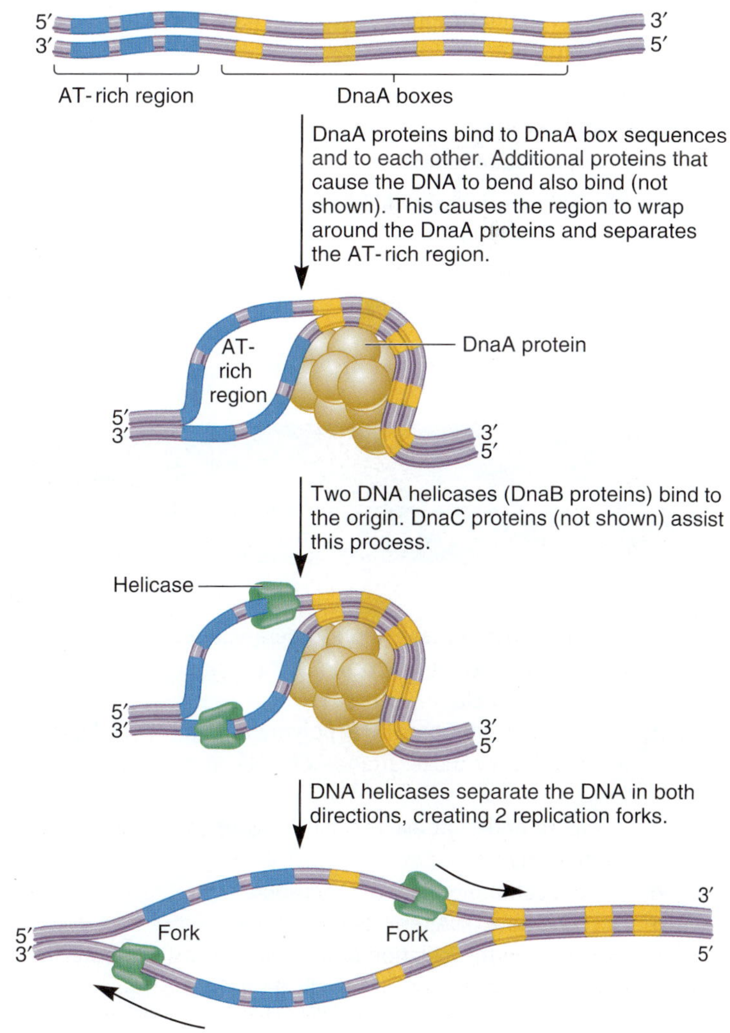

INTERACTIVE EXERCISE

ONLINE ANIMATION

FIGURE 11.6 The events that occur at *oriC* to initiate the DNA replication process. To initiate DNA replication, DnaA proteins bind to sequences in the five DnaA boxes, which causes the DNA strands to separate at the AT-rich region. DnaA and DnaC proteins then recruit DNA helicase (DnaB) into this region. Each DNA helicase is composed of six subunits, which form a ring around one DNA strand and travel in the 5′ to 3′ direction. As shown here, the movement of two DNA helicases separates the DNA strands beyond the *oriC* region.

CONCEPT CHECK: How many replication forks are formed at the origin?

strands beyond the origin. These proteins use the energy from ATP hydrolysis to catalyze the separation of the double-stranded parental DNA. In *E. coli*, DNA helicases bind to single-stranded DNA and travel along the DNA in a 5′ to 3′ direction to keep the replication fork moving. As shown in Figure 11.6, the action of DNA helicases promotes the movement of two replication forks outward from *oriC* in opposite directions. This initiates the replication of the bacterial chromosome in both directions, an event termed **bidirectional replication.**

The GATC methylation sites within *oriC* are involved with regulating DNA replication. These sites are methylated by an enzyme known as DNA adenine methyltransferase (Dam). Prior to DNA replication, the GATC sites are methylated in both strands. This full methylation facilitates the initiation of DNA replication at the origin. Following DNA replication, the newly made strands are not methylated, because adenine rather than methyladenine is incorporated into the daughter strands. The initiation of DNA replication at the origin does not readily occur until after it has become fully methylated. Because it takes several minutes for Dam to methylate the GATC sites within this region, DNA replication does not occur again too quickly.

11.2 COMPREHENSION QUESTIONS

1. A site in a chromosome where DNA replication begins is
   a. a promoter.
   b. an origin of replication.
   c. an operator.
   d. a replication fork.
2. The origin of replication in *E. coli* contains
   a. an AT-rich region.
   b. DnaA box sequences.
   c. GATC methylation sites.
   d. all of the above.

11.3 BACTERIAL DNA REPLICATION: SYNTHESIS OF NEW DNA STRANDS

Learning Outcomes:

1. Describe how helicase, topoisomerase, and single-strand binding protein are important for the unwinding of the DNA double helix.
2. Outline how primase, DNA polymerase, and DNA ligase are needed to make strands of DNA at the replication fork.
3. Compare and contrast the synthesis of the leading and lagging strands.
4. List the components of the replisome.
5. Describe how DNA replication is terminated.
6. Explain how the isolation of mutants was instrumental in our understanding of DNA replication.

As we have seen, bacterial DNA replication begins at the origin of replication. The synthesis of new DNA strands is a stepwise process in which many cellular proteins participate. In this section, we will examine how DNA strands are made at a replication fork.

Several Proteins Are Required for DNA Replication at the Replication Fork

Figure 11.7 provides an overview of the molecular events that occur as one of the two replication forks moves around the bacterial chromosome, and **Table 11.1** summarizes the functions of the major proteins involved in *E. coli* DNA replication.

Unwinding of the Helix Let's begin with strand separation. To act as a template for DNA replication, the strands of a double helix must separate. As mentioned previously, the function of DNA helicase is to break the hydrogen bonds between base pairs and thereby unwind the strands; this action generates positive supercoiling ahead of each replication fork. As shown in Figure 11.7, an enzyme known as **topoisomerase II,** also called **DNA gyrase,** travels in front of DNA helicase and alleviates positive supercoiling.

After the two parental DNA strands have been separated and the supercoiling relaxed, they must be kept that way until the complementary daughter strands have been made. What prevents the DNA strands from coming back together? DNA replication requires **single-strand binding proteins** that bind to the strands of parental DNA and prevent them from re-forming a double helix. In this way, the bases within the parental strands are kept in an exposed condition that enables them to hydrogen bond with individual nucleotides.

Synthesis of RNA Primers via Primase The next event in DNA replication involves the synthesis of short strands of RNA (rather than DNA) called **RNA primers.** These strands of RNA are synthesized by the linkage of ribonucleotides via an enzyme known as **primase.** This enzyme synthesizes short strands of RNA, typically 10–12 nucleotides in length. These short RNA strands start, or prime, the process of DNA replication. In the **leading strand,** a single primer is made at the origin of replication. In the **lagging strand,** multiple primers are made. As discussed later, the RNA primers are eventually removed.

Synthesis of DNA via DNA Polymerase An enzyme known as **DNA polymerase** is responsible for synthesizing the DNA of the leading and lagging strands. This enzyme catalyzes the formation of covalent bonds between adjacent nucleotides and thereby makes the new daughter strands. In *E. coli*, five distinct proteins function as DNA polymerases and are designated polymerase I, II, III, IV, and V. DNA polymerases I and III are involved in normal DNA replication, whereas DNA polymerases II, IV, and V play a role in DNA repair and the replication of damaged DNA.

DNA polymerase III is responsible for most of the DNA replication. It is a large enzyme consisting of 10 different subunits that play various roles in the DNA replication process (**Table 11.2**). The α subunit catalyzes the bond formation between adjacent nucleotides, and the remaining nine subunits fulfill other functions. The

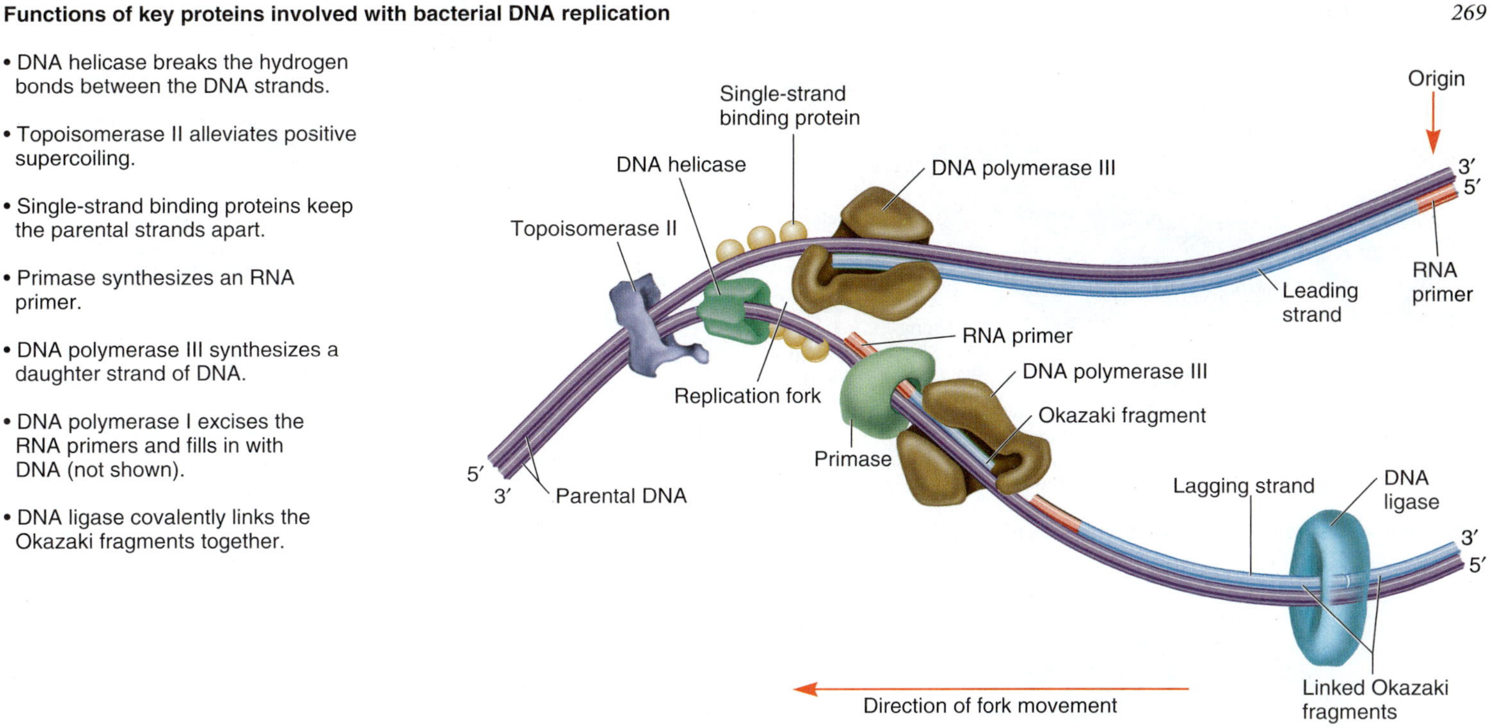

INTERACTIVE EXERCISE

ONLINE ANIMATION

FIGURE 11.7 **The proteins involved with bacterial DNA replication.**
Note: The drawing of DNA polymerase III depicts the catalytic subunit that synthesizes DNA.

CONCEPT CHECK: Why is primase needed for DNA replication?

complex of all 10 subunits together is called DNA polymerase III holoenzyme. By comparison, DNA polymerase I is composed of a single subunit. Its role during DNA replication is to remove the RNA primers and fill in the vacant regions with DNA.

Though the various DNA polymerases in *E. coli* and other bacterial species vary in their subunit composition, several common structural features have emerged. The catalytic subunit of all DNA polymerases has a structure that resembles a human hand. As shown in **Figure 11.8**, the template DNA is threaded through the palm of the hand; the thumb and fingers are wrapped around the DNA. The incoming deoxyribonucleoside triphosphates (dNTPs) enter the catalytic site, bind to the template strand according to the AT/GC rule, and then are covalently attached to the 3′ end of the growing strand. DNA polymerase also contains a 3′ exonuclease site that removes mismatched bases, as described later.

As researchers began to unravel the function of DNA polymerase, two features seemed unusual (**Figure 11.9**). DNA polymerase cannot begin DNA synthesis by linking together the first two individual nucleotides. Rather, this type of enzyme can elongate only a preexisting strand starting with an RNA primer or

TABLE 11.1
Proteins Involved in *E. coli* DNA Replication

Common Name	Function
DnaA proteins	Bind to DnaA box sequences within the origin to initiate DNA replication
DnaC proteins	Aid DnaA in the recruitment of DNA helicase to the origin
DNA helicase (DnaB)	Separates double-stranded DNA
Topoisomerase II (DNA gyrase)	Removes positive supercoiling ahead of the replication fork
Single-strand binding proteins	Bind to single-stranded DNA and prevent it from re-forming a double-stranded structure
Primase	Synthesizes short RNA primers
DNA polymerase III	Synthesizes DNA in the leading and lagging strands
DNA polymerase I	Removes RNA primers, fills in gaps with DNA
DNA ligase	Covalently attaches adjacent Okazaki fragments
Tus	Binds to ter sequences and prevents the advancement of the replication fork

TABLE 11.2
Subunit Composition of DNA Polymerase III Holoenzyme from *E. coli*

Subunit(s)	Function
α	Synthesizes DNA
ε	3′ to 5′ proofreading (removes mismatched nucleotides)
θ	Accessory protein that stimulates the proofreading function
β	Clamp protein, which allows DNA polymerase to slide along the DNA without falling off
τ, γ, δ, δ', ψ, and χ	Clamp loader complex, involved with helping the clamp protein bind to the DNA

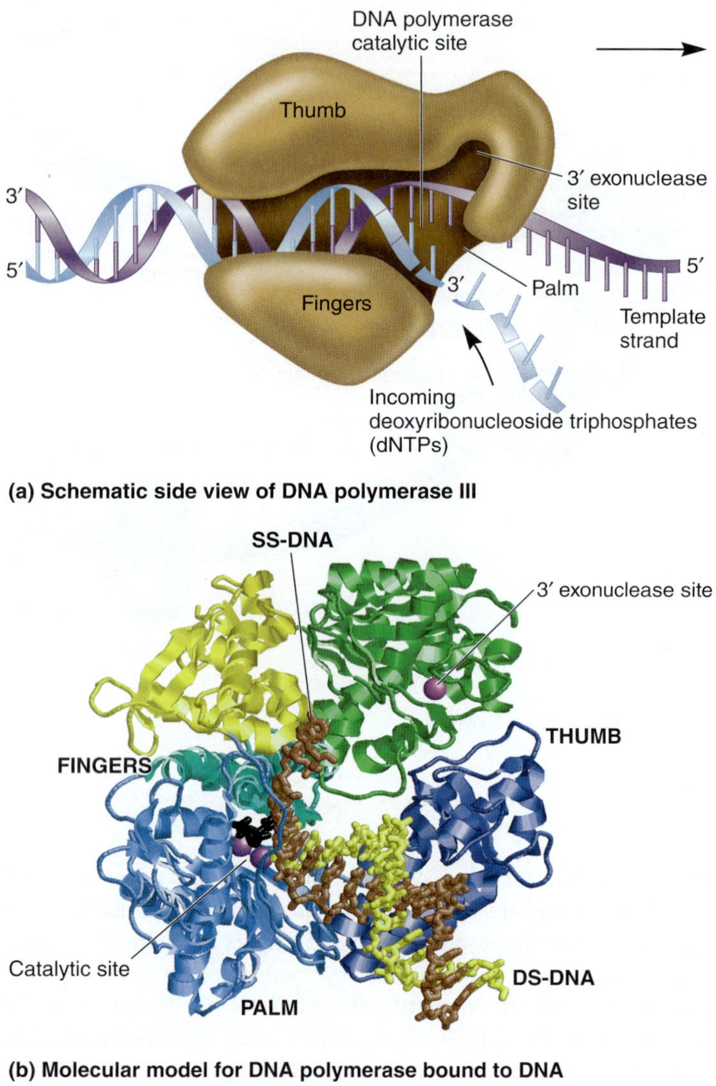

FIGURE 11.8 **The action of DNA polymerase.** **(a)** DNA polymerase slides along the template strand as it synthesizes a new strand by connecting deoxyribonucleoside triphosphates (dNTPs) in a 5′ to 3′ direction. The catalytic subunit of DNA polymerase resembles a hand that is wrapped around the template strand. In this regard, the movement of DNA polymerase along the template strand is similar to a hand that is sliding along a rope. **(b)** The molecular structure of DNA polymerase I. This model shows a portion of DNA polymerase I that is bound to DNA. This molecular structure depicts a different view of DNA polymerase compared to part (a), which is a schematic side view.

CONCEPT CHECK: Is the template strand read in the 5′ to 3′ or the 3′ to 5′ direction?

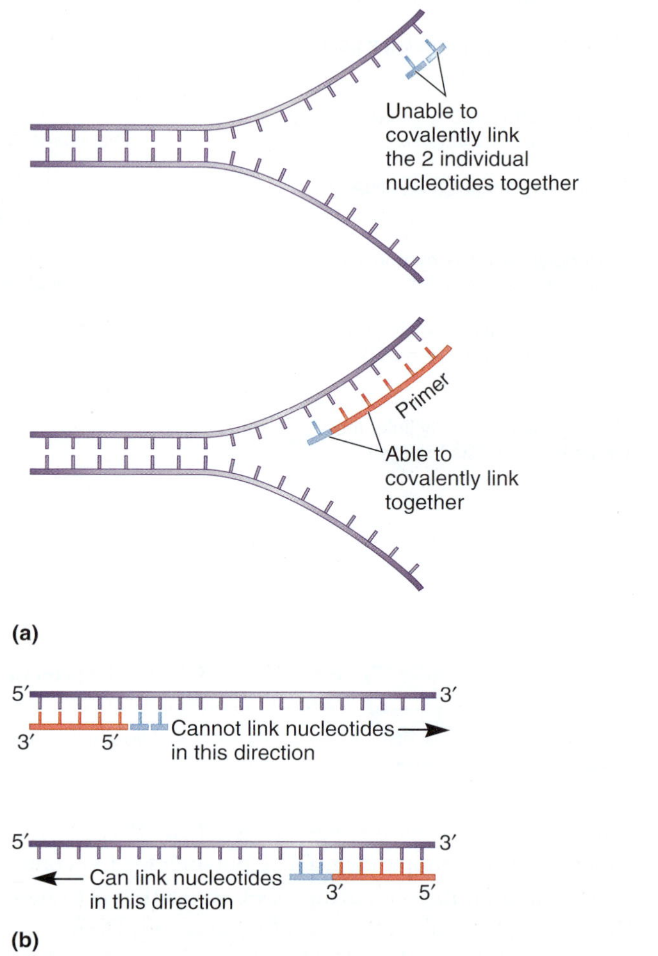

FIGURE 11.9 **Unusual features of DNA polymerase function.** **(a)** DNA polymerase can elongate a strand only from an RNA primer or an existing DNA strand. **(b)** DNA polymerase can attach nucleotides only in a 5′ to 3′ direction. Note the template strand is read in the opposite, 3′ to 5′, direction.

an existing DNA strand (Figure 11.9a). A second unusual feature is the directionality of strand synthesis. DNA polymerase can attach nucleotides only in the 5′ to 3′ direction, not in the 3′ to 5′ direction (Figure 11.9b).

The Synthesis of the Leading and Lagging Strands Is Distinctly Different

Due to the two unusual features described in Figure 11.9, the synthesis of the leading and lagging strands shows distinctive differences (**Figure 11.10**). The synthesis of RNA primers by primase allows DNA polymerase III to begin the synthesis of complementary daughter strands of DNA. DNA polymerase III catalyzes the attachment of nucleotides to the 3′ end of each primer, in a 5′ to 3′ direction. In the leading strand, one RNA primer is made at the origin, and then DNA polymerase III attaches nucleotides in a 5′ to 3′ direction as it slides toward the opening of the replication fork. The synthesis of the leading strand is continuous.

In the lagging strand, the synthesis of DNA also elongates in a 5′ to 3′ manner, but it does so in the direction away from the replication fork. In the lagging strand, RNA primers repeatedly initiate the synthesis of short segments of DNA; the synthesis is discontinuous. The length of these fragments in bacteria is typically 1000–2000 nucleotides. In eukaryotes, the fragments are shorter: 100–200 nucleotides. Each fragment contains a short RNA primer at the 5′ end, which is made by primase. The remainder of the fragment is a strand of DNA made by DNA polymerase. The DNA fragments made in this manner are known as **Okazaki fragments,** after Reiji and Tuneko Okazaki, who initially discovered them in the late 1960s.

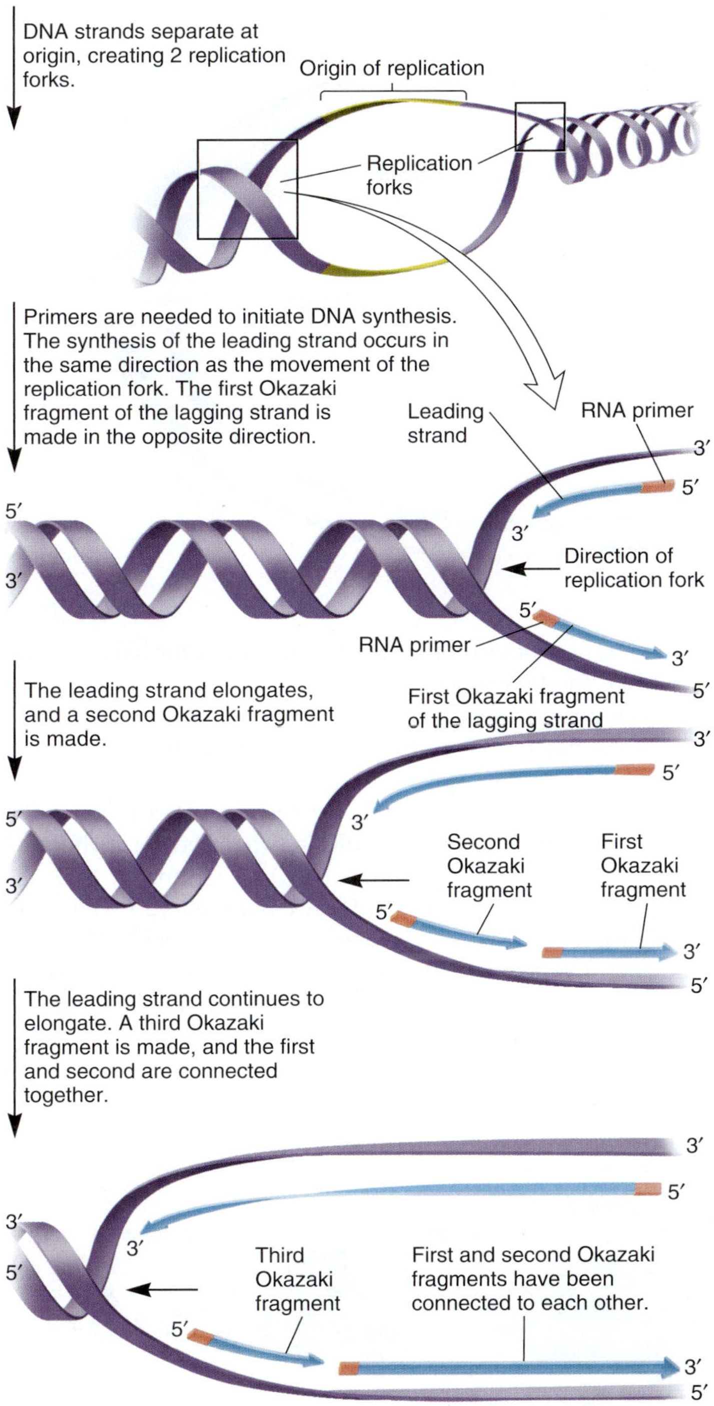

FIGURE 11.10 **The synthesis of DNA at the replication fork.**

ONLINE ANIMATION

CONCEPT CHECK: Describe the differences between the synthesis of the leading and lagging strands.

To complete the synthesis of Okazaki fragments within the lagging strand, three additional events must occur: removal of the RNA primers, synthesis of DNA in the area where the primers have been removed, and the covalent attachment of adjacent fragments of DNA (see Figure 11.10 and refer back to Figure 11.7). In *E. coli*, the RNA primers are removed by the action of DNA polymerase I. This enzyme has a 5′ to 3′ exonuclease activity, which means that DNA polymerase I digests away the RNA primers in a 5′ to 3′ direction, leaving a vacant area. DNA polymerase I then synthesizes DNA to fill in this region. It uses the 3′ end of an adjacent Okazaki fragment as a primer. For example, in Figure 11.10, DNA polymerase I would remove the RNA primer from the first Okazaki fragment and then synthesize DNA in the vacant region by attaching nucleotides to the 3′ end of the second Okazaki fragment.

After the vacant region has been completely filled in, a covalent bond is still missing between the last nucleotide added by DNA polymerase I (on the second Okazaki fragment) and the adjacent DNA strand of the first Okazaki fragment (see Figure 11.10). An enzyme known as **DNA ligase** catalyzes a covalent bond between adjacent Okazaki fragments to complete the replication process in the lagging strand (also refer back to Figure 11.7). In *E. coli*, DNA ligase requires NAD^+ to carry out this reaction, whereas the DNA ligases found in archaea and eukaryotes require ATP.

Now that we understand how the leading and lagging strands are made, **Figure 11.11** shows how new strands are constructed from a single origin of replication. To the left of the origin, the top strand is made continuously, whereas to the right of the origin it is made in Okazaki fragments. By comparison, the synthesis of the bottom strand is just the opposite. To the left of the origin it is made in Okazaki fragments and to the right of the origin the synthesis is continuous.

Certain Enzymes of DNA Replication Bind to Each Other to Form a Complex

Figure 11.12 provides a more three-dimensional view of the DNA replication process. DNA helicase and primase are physically bound to each other to form a complex known as a **primosome.** This complex leads the way at the replication fork. The primosome tracks along the DNA, separating the parental strands and synthesizing RNA primers at regular intervals along the lagging strand. By acting within a complex, the actions of DNA helicase and primase can be better coordinated.

The primosome is physically associated with two DNA polymerase holoenzymes to form a **replisome.** As shown in Figure 11.12, two DNA polymerase III proteins act in concert to replicate the leading and lagging strands. The term **dimeric DNA polymerase** is used to describe two DNA polymerase holoenzymes that move as a unit toward a replication fork. For this to occur, the lagging strand is looped out with respect to the DNA polymerase that synthesizes the lagging strand. This loop allows the lagging-strand polymerase to make DNA in a 5′ to 3′ direction yet move toward the opening of the replication fork. Interestingly, when this DNA polymerase reaches the end of an Okazaki fragment, it must be released from the template DNA and "hop" to the RNA primer that is closest to the fork. The clamp loader complex (see Table 11.2), which is part of DNA polymerase holoenzyme, then reloads the holoenzyme at the site where the next RNA primer has been made. Similarly, after primase synthesizes an RNA primer in the 5′ to 3′ direction, it must hop over the primer and synthesize the next primer closer to the replication fork.

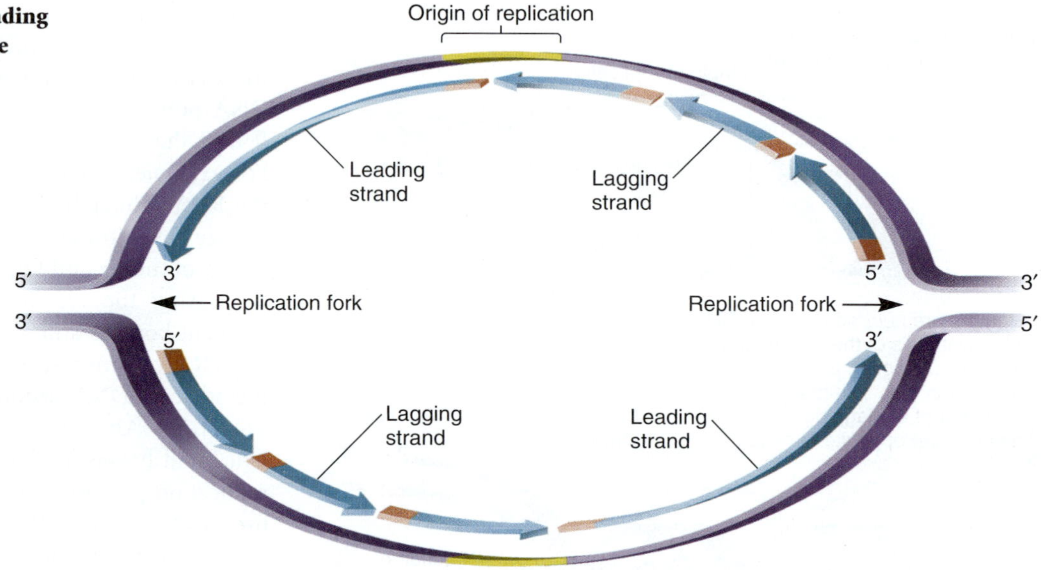

FIGURE 11.11 **The synthesis of leading and lagging strands outward from a single origin of replication.**

Replication Is Terminated When the Replication Forks Meet at the Termination Sequences

On the opposite side of the *E. coli* chromosome from *oriC* is a pair of **termination sequences** called ter sequences. A protein known as the termination utilization substance (Tus) binds to the ter sequences and stops the movement of the replication forks. As shown in **Figure 11.13**, one of the ter sequences, designated T1, prevents the advancement of the fork moving left to right, but allows the movement of the other fork (see the inset to Figure 11.13). Alternatively, T2 prevents the advancement of the fork moving right to left, but allows the advancement of the other fork. In any given cell, only one ter sequence is required to stop the advancement of one replication fork, and then the other fork ends its synthesis of DNA when it reaches the halted replication fork. In other words, DNA replication ends when oppositely advancing forks meet, usually at T1 or T2. Finally, DNA ligase covalently links the two daughter strands, creating two circular, double-stranded molecules.

After DNA replication is completed, one last problem may exist. DNA replication often results in two intertwined DNA molecules known as **catenanes** (**Figure 11.14**). Fortunately, catenanes are only transient structures in DNA replication. In *E. coli*, topoisomerase II introduces a temporary break into the DNA strands and then rejoins them after the strands have become unlocked. This allows the catenanes to be separated into individual circular molecules.

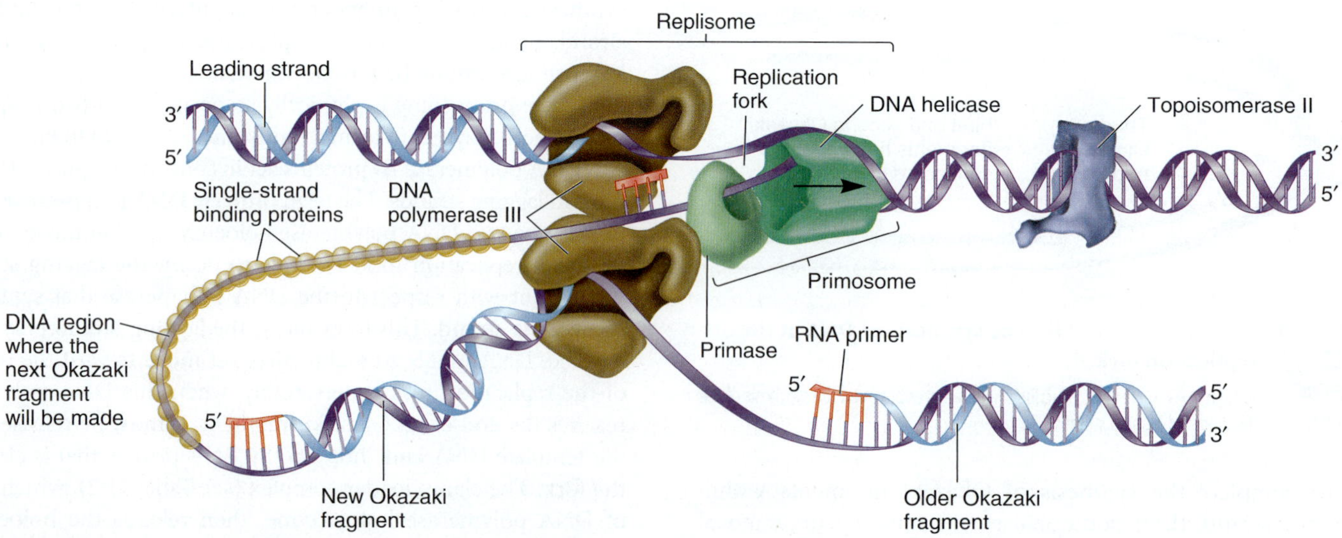

FIGURE 11.12 **A three-dimensional view of DNA replication.** DNA helicase and primase associate together to form a primosome. The primosome associates with two DNA polymerase holoenzymes to form a replisome.

CONCEPT CHECK: What is the advantage of having the replication machinery in a complex?

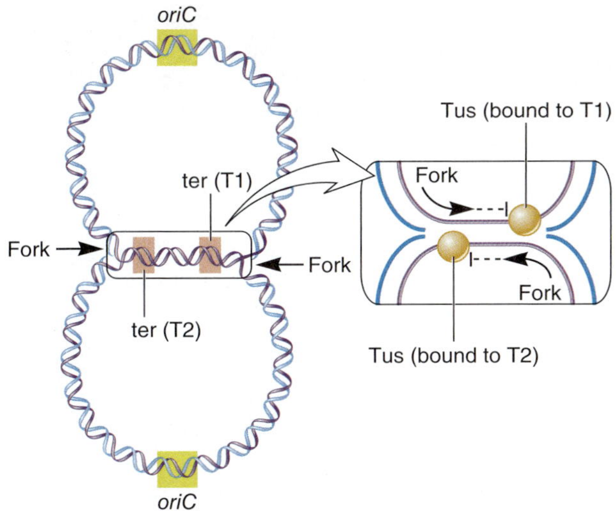

FIGURE 11.13 The termination of DNA replication. Two sites in the bacterial chromosome, shown with rose-colored rectangles, are ter sequences designated T1 and T2. The T1 site prevents the further advancement of the fork moving left to right, and T2 prevents the advancement of the fork moving right to left. As shown in the inset, the binding of Tus prevents the replication forks from proceeding past the ter sequences in a particular direction.

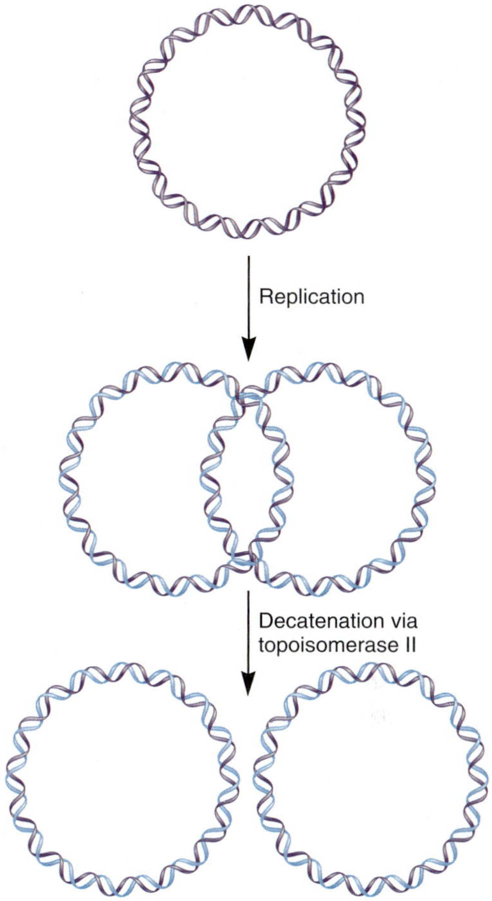

FIGURE 11.14 Separation of catenanes. DNA replication can result in two intertwined chromosomes called catenanes. These catenanes are separated by the action of topoisomerase II.

CONCEPT CHECK: Is DNA strand breakage needed for catenane separation?

The Isolation of Mutants Has Been Instrumental to Our Understanding of DNA Replication

Thus far, we have considered how a variety of proteins play a role in the replication of bacterial DNA. An important experimental approach that led to the identification of most of these proteins involved the isolation of mutants in *E. coli* that have abnormalities in DNA replication. The first such mutant, which was identified by Paula DeLucia and John Cairns in 1969, was in the gene that encodes DNA polymerase I.

Because DNA replication is vital for cell division, mutations that block DNA replication prevent cell growth. For this reason, if researchers want to identify loss-of-function mutations in vital genes, they must screen for **conditional mutants.** One type of conditional mutant is a **temperature-sensitive (ts) mutant.** For example, a ts mutant might grow at 30°C (the permissive temperature) but fail to grow at a higher temperature, such as 42°C, which is chosen by the experimenter. The higher temperature at which a mutant strain fails to grow is called the nonpermissive temperature. The mutant strain fails to grow because the higher temperature inactivates the function of the protein encoded by the mutant gene.

Figure 11.15 shows a general strategy for the isolation of ts mutants. Researchers expose bacterial cells to a mutagen that increases the likelihood of mutations. The mutagenized cells are plated on growth media and incubated at the permissive temperature. The colonies are then replica plated onto two plates: one incubated at the permissive temperature and one at the nonpermissive temperature. (The technique of replica plating is described in Chapter 18, Figure 18.6.) As seen in Figure 11.15, this method enables researchers to identify ts mutations that are unable to grow at the nonpermissive temperature.

With regard to the study of DNA replication, researchers analyzed a large number of ts mutants to discover if any of them had a defect in DNA replication. For example, one could expose a ts mutant to radiolabeled thymine (a base that is incorporated into DNA), shift to the nonpermissive temperature, and determine if a mutant strain could make radiolabeled DNA. Because *E. coli* has many vital genes not involved with DNA replication, only a small subset of ts mutants would be expected to have mutations in genes that encode proteins that are critical to the replication process. Therefore, researchers had to screen many thousands of ts mutants to identify the few involved in DNA replication. This approach is sometimes called a "brute force" genetic screen.

Table 11.3 summarizes some of the genes that were identified using this type of strategy. The genes were originally designated with the name dna, followed by a capital letter that generally refers to the order in which they were discovered. When shifted to a nonpermissive temperature, certain mutants showed a rapid arrest of DNA synthesis. These so-called rapid-stop mutations were found in genes that encode proteins needed for DNA synthesis. By comparison, other mutants were able to complete their

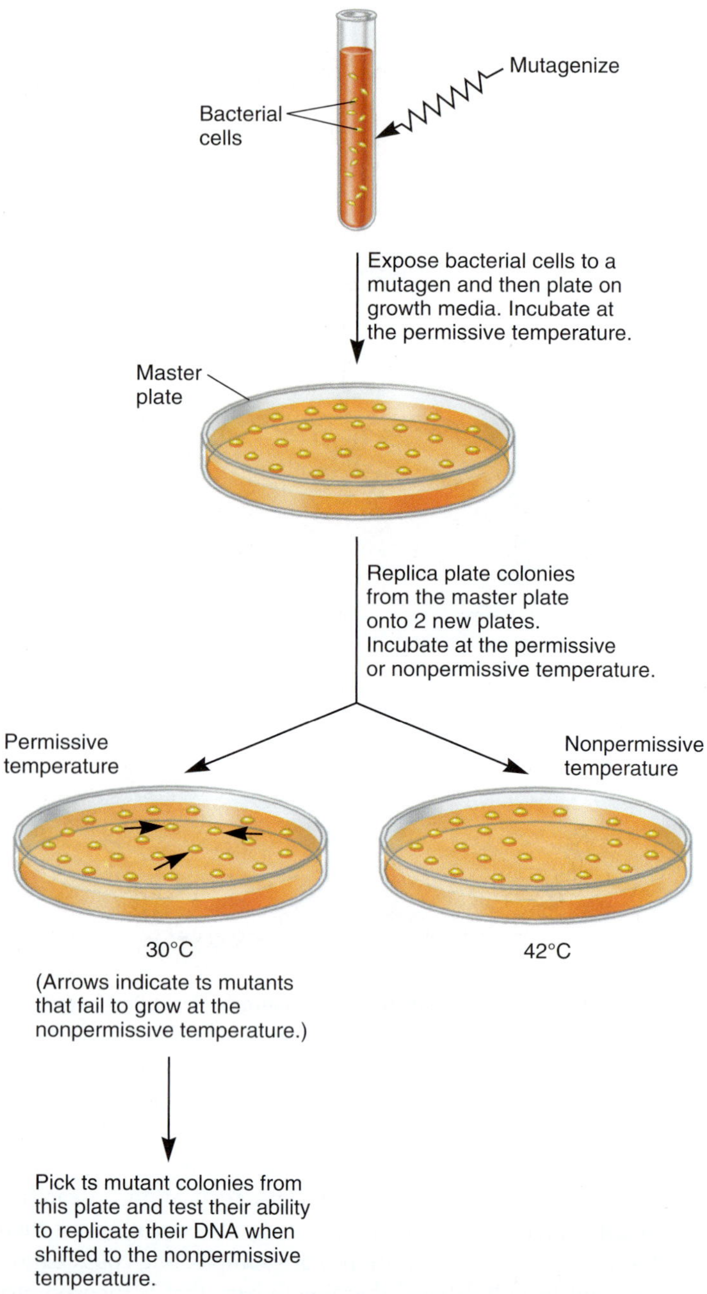

FIGURE 11.15 A strategy to identify ts mutations in vital genes. In this approach, bacteria are mutagenized, which increases the likelihood of mutation, and then grown at the permissive temperature. Colonies are then replica plated and grown at both the permissive and nonpermissive temperatures. (Note: The procedure of replica plating is shown in Chapter 18, Figure 18.6.) Ts mutants fail to grow at the nonpermissive temperature. The appropriate colonies are picked from the master plate, grown at the permissive temperature, and analyzed to see if DNA replication is altered at the nonpermissive temperature.

current round of replication but could not start another round. These slow-stop mutants involved genes that encode proteins needed for the initiation of replication at the origin. In later studies, the proteins encoded by these genes were purified, and their functions were studied in vitro. This work contributed greatly to our modern understanding of DNA replication at the molecular level.

TABLE 11.3
Examples of ts Mutants Involved in DNA Replication in *E. coli*

Gene Name	Protein Function
Rapid-Stop Mutants	
dnaE	α subunit of DNA polymerase III; synthesizes DNA
dnaX	τ subunit of DNA polymerase III; part of the clamp loader complex and also promotes the dimerization of two DNA polymerase III proteins together at the replication fork
dnaN	β subunit of DNA polymerase III; functions as a clamp protein that makes DNA polymerase a processive enzyme
dnaZ	γ subunit of DNA polymerase III; helps the β subunit bind to the DNA
dnaG	Primase; needed to make RNA primers
dnaB	Helicase; needed to unwind the DNA strands during replication
Slow-Stop Mutants	
dnaA	DnaA protein that recognizes the DnaA boxes at the origin
dnaC	DnaC protein that recruits DNA helicase to the origin

11.3 COMPREHENSION QUESTIONS

1. The enzyme known as ________ uses ________ and separates the DNA strands at the replication fork.
   a. helicase, ATP
   b. helicase, GTP
   c. gyrase, ATP
   d. gyrase, GTP
2. In the lagging strand, DNA is made in the direction ________ the replication fork and is made as ________.
   a. toward, one continuous strand
   b. away from, one continuous strand
   c. toward, Okazaki fragments
   d. away from, Okazaki fragments

11.4 BACTERIAL DNA REPLICATION: CHEMISTRY AND ACCURACY

Learning Outcomes:

1. Describe how nucleotides are connected to a growing DNA strand.
2. Define processivity.
3. Explain the proofreading function of DNA polymerase.

In Sections 11.2 and 11.3, we examined the origin of replication and considered how DNA strands are synthesized. In this section, we will take a closer look at the process in which nucleotides are attached to a growing DNA strand and discuss how DNA replication is amazingly accurate.

DNA Polymerase III Is a Processive Enzyme That Uses Deoxyribonucleoside Triphosphates

Let's now turn our attention to other enzymatic features of DNA polymerase. As shown in **Figure 11.16**, DNA polymerases catalyze the covalent attachment between the phosphate in one nucleotide and the sugar in the previous nucleotide. The formation of this covalent bond requires an input of energy. Prior to bond formation, the nucleotide about to be attached to the growing strand is a deoxyribonucleoside triphosphate (dNTP). It contains three phosphate groups attached at the 5′–C atom of deoxyribose. The dNTP first enters the catalytic site of DNA polymerase and binds to the template strand according to the AT/GC rule. Next, the 3′–OH group on the previous nucleotide reacts with the phosphate group (PO_4^{2-}) adjacent to the sugar on the incoming nucleotide. This reaction is highly exergonic and results in a covalent bond between the sugar at the 3′ end of the DNA strand and the PO_4^{2-} of the incoming nucleotide. The formation of this covalent bond causes the newly made strand to grow in the 5′ to 3′ direction. As shown in Figure 11.16, pyrophosphate (PP_i) is released and broken down into 2 phosphates.

DNA polymerase catalyzes the covalent attachment of nucleotides with great speed. In *E. coli*, DNA polymerase III attaches approximately 750 nucleotides per second! DNA polymerase III can catalyze the synthesis of the daughter strands so quickly because it is a **processive enzyme.** This means it does not dissociate from the growing strand after it has catalyzed the covalent joining of two nucleotides. Rather, as depicted in Figure 11.8a, it remains clamped to the DNA template strand and slides along the template as it catalyzes the synthesis of the daughter strand. The β subunit of the holoenzyme, also known as the clamp protein, promotes the association of the holoenzyme with the DNA as it glides along the

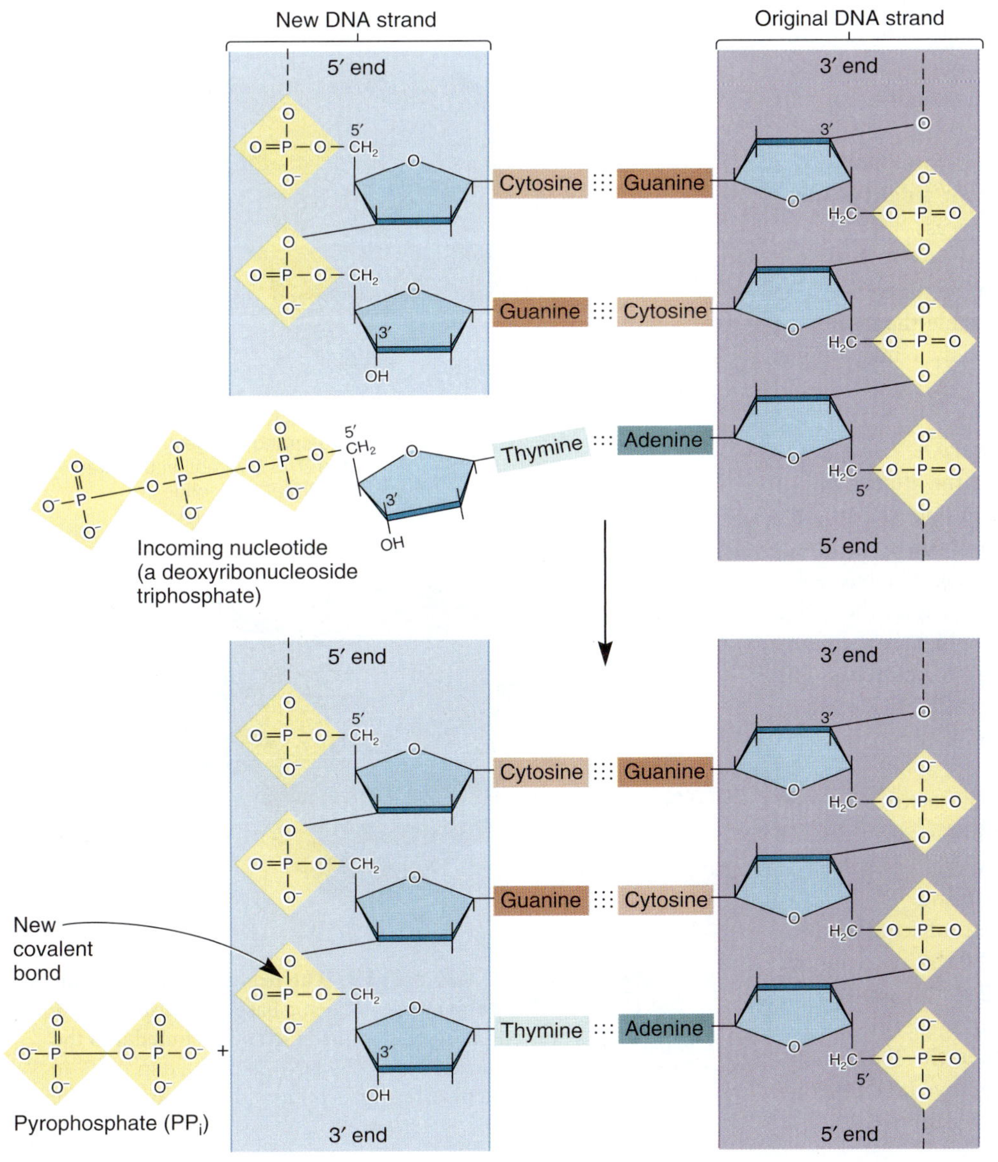

FIGURE 11.16 The enzymatic action of DNA polymerase. An incoming deoxyribonucleoside triphosphate (dNTP) is cleaved to form a nucleoside monophosphate and pyrophosphate (PP_i). The energy released from this exergonic reaction allows the nucleoside monophosphate to form a covalent bond at the 3′ end of the growing strand. This reaction is catalyzed by DNA polymerase. PP_i is released.

CONCEPT CHECK: Does the oxygen in the newly made ester bond come from the phosphate or from the sugar?

template strand (refer back to Table 11.2). The β subunit forms a dimer in the shape of a ring; the hole of the ring is large enough to accommodate a double-stranded DNA molecule, and its width is about one turn of DNA. A complex of several subunits functions as a clamp loader that allows the DNA polymerase holoenzyme to initially clamp onto the DNA.

The effects of processivity are really quite remarkable. In the absence of the β subunit, DNA polymerase can synthesize DNA at a rate of only about 20 nucleotides per second and typically falls off the DNA template after approximately 10 nucleotides have been linked together. By comparison, when the β subunit is present, as in the holoenzyme, the synthesis rate is approximately 750 nucleotides per second. In the leading strand, DNA polymerase III has been estimated to synthesize a segment of DNA that is over 500,000 nucleotides in length before it inadvertently falls off.

The Fidelity of DNA Replication Is Ensured by Three Mechanisms

With replication occurring so rapidly, one might imagine that mistakes could happen in which the wrong nucleotide is incorporated into the growing daughter strand. Although mistakes do happen during DNA replication, they are extraordinarily rare. In the case of DNA synthesis via DNA polymerase III, only 1 mistake per 100 million nucleotides is made. Therefore, DNA synthesis occurs with a high degree of accuracy, or **fidelity.** Three factors account for this remarkably accurate process.

Stability of Base Pairing The hydrogen bonding between G and C or between A and T is much more stable than between mismatched pairs. Due to this stability alone, only 1 mistake per 1000 nucleotides would be made.

Structure of the Active Site of DNA Polymerase The active site of DNA polymerase preferentially catalyzes the attachment of nucleotides when the correct bases are located in opposite strands. Helix distortions caused by mispairing usually prevent an incorrect nucleotide from properly occupying the active site of DNA polymerase. By comparison, the correct nucleotide occupies the active site with precision and promotes induced fit, which is a conformational change in the enzyme that is necessary for catalysis. The inability of incorrect nucleotides to promote induced fit decreases the error rate to a range of 1 in 100,000–1 million.

Proofreading DNA polymerase decreases the error rate even further by the enzymatic removal of mismatched nucleotides. As shown in Figure 11.17, DNA polymerase can identify a mismatched nucleotide and remove it from the daughter strand. This occurs by exonuclease cleavage of the bonds between adjacent nucleotides at the 3′ end of the newly made strand. The ability to remove mismatched bases by this mechanism is called the **proofreading function** of DNA polymerase. Proofreading occurs by the removal of nucleotides in the 3′ to 5′ direction at the 3′ exonuclease site. After the mismatched nucleotide is removed, DNA polymerase resumes DNA synthesis in the 5′ to 3′ direction.

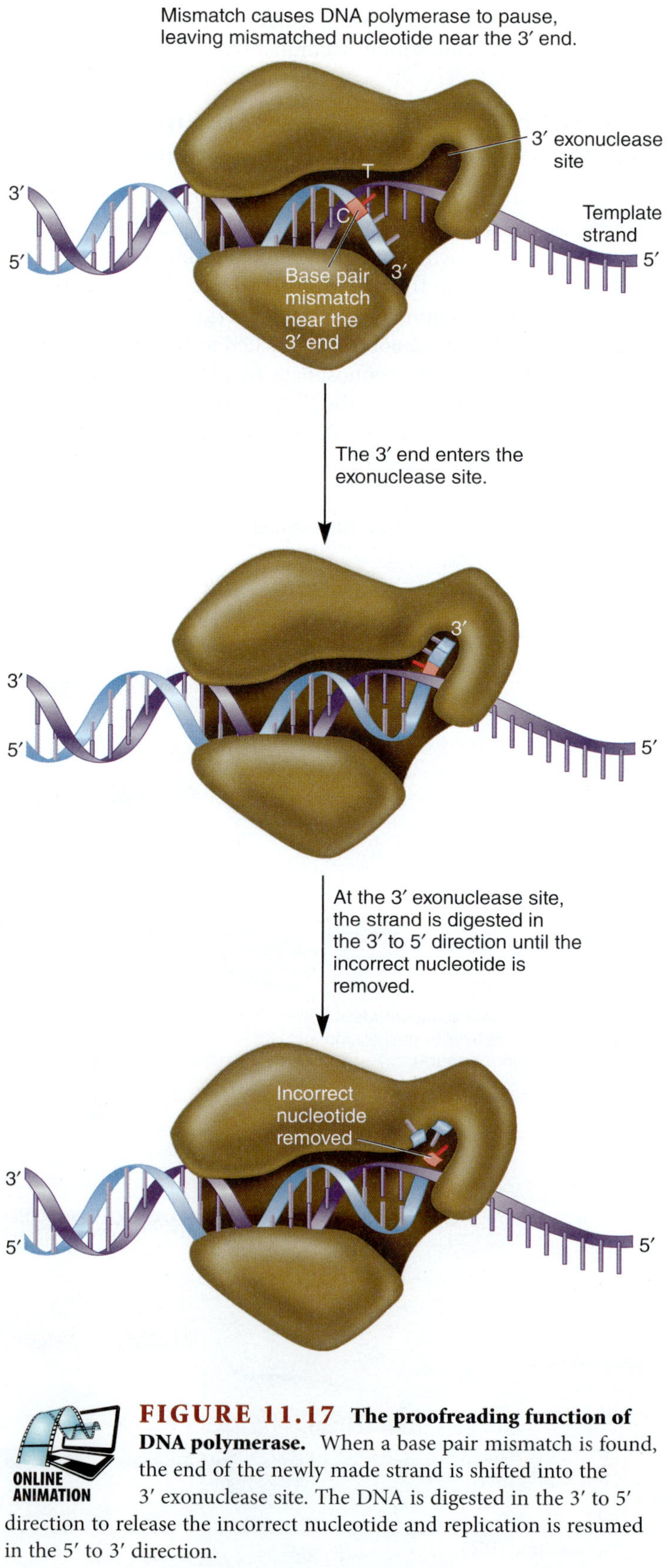

FIGURE 11.17 The proofreading function of DNA polymerase. When a base pair mismatch is found, the end of the newly made strand is shifted into the 3′ exonuclease site. The DNA is digested in the 3′ to 5′ direction to release the incorrect nucleotide and replication is resumed in the 5′ to 3′ direction.

11.4 COMPREHENSION QUESTIONS

1. DNA polymerase III is a processive enzyme, which means that
   a. it does not dissociate from the growing strand after it has attached a nucleotide to the 3′ end.
   b. it makes a new strand very quickly.
   c. it proceeds toward the opening of the replication fork.
   d. it copies DNA with relatively few errors.
2. The proofreading function of DNA polymerase involves the recognition of a _________ and the removal of a short segment of DNA in the ___________ direction.
   a. missing base, 5′ to 3′
   b. base mismatch, 5′ to 3′
   c. missing base, 3′ to 5′
   d. base mismatch, 3′ to 5′

11.5 EUKARYOTIC DNA REPLICATION

Learning Outcomes:

1. Compare and contrast origins of replication in bacteria and eukaryotes.
2. Outline the functions of different DNA polymerases in eukaryotes.
3. Describe how RNA primers are removed in eukaryotes.
4. Explain how DNA replication occurs at the ends of eukaryotic chromosomes.

Eukaryotic DNA replication is not as well understood as bacterial replication. Much research has been carried out on a variety of experimental organisms, particularly yeast and mammalian cells. Many of these studies have found extensive similarities between the general features of DNA replication in prokaryotes and eukaryotes. For example, DNA helicases, topoisomerases, single-strand binding proteins, primases, DNA polymerases, and DNA ligases—the types of bacterial enzymes described in Table 11.1—have also been identified in eukaryotes. Nevertheless, at the molecular level, eukaryotic DNA replication appears to be substantially more complex. These additional intricacies of eukaryotic DNA replication are related to several features of eukaryotic cells. In particular, eukaryotic cells have larger, linear chromosomes, the chromatin is tightly packed within nucleosomes, and cell-cycle regulation is more complicated. This section emphasizes some of the unique features of eukaryotic DNA replication.

Initiation Occurs at Multiple Origins of Replication on Linear Eukaryotic Chromosomes

Because eukaryotes have long, linear chromosomes, multiple origins of replication are needed so the DNA can be replicated within a reasonable length of time. In 1968, Joel Huberman and Arthur Riggs provided evidence for multiple origins of replication by adding a radiolabeled nucleoside (^{3}H-deoxythymidine) to a culture of actively dividing cells, followed by a chase with nonlabeled deoxythymidine. The radiolabeled deoxythymidine was taken up by the cells and incorporated into their newly made DNA strands for a brief period. The chromosomes were then isolated from the cells and subjected to autoradiography. As seen in Figure 11.18, radiolabeled segments were interspersed among nonlabeled segments. This result is consistent with the hypothesis that eukaryotic chromosomes contain multiple origins of replication.

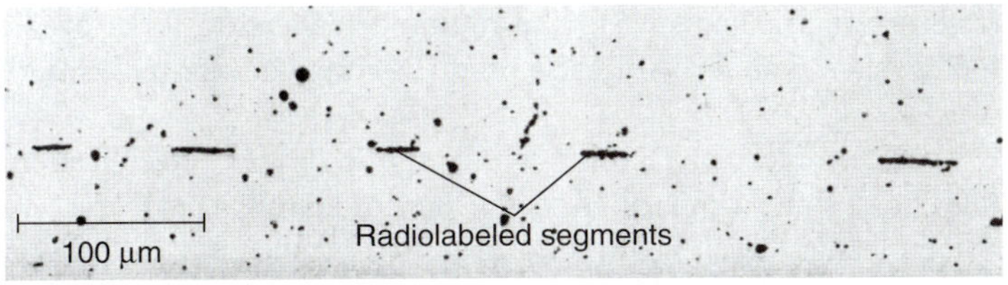

FIGURE 11.18 Evidence for multiple origins of replication in eukaryotic chromosomes. In this experiment, cells were given a pulse-chase of ^{3}H-deoxythymidine and unlabeled deoxythymidine. The chromosomes were isolated and subjected to autoradiography. In this micrograph, radiolabeled segments were interspersed among nonlabeled segments, indicating that eukaryotic chromosomes contain multiple origins of replication.

As shown schematically in Figure 11.19, DNA replication proceeds bidirectionally from many origins of replication. As discussed in Chapter 3, this occurs during S phase of the cell cycle

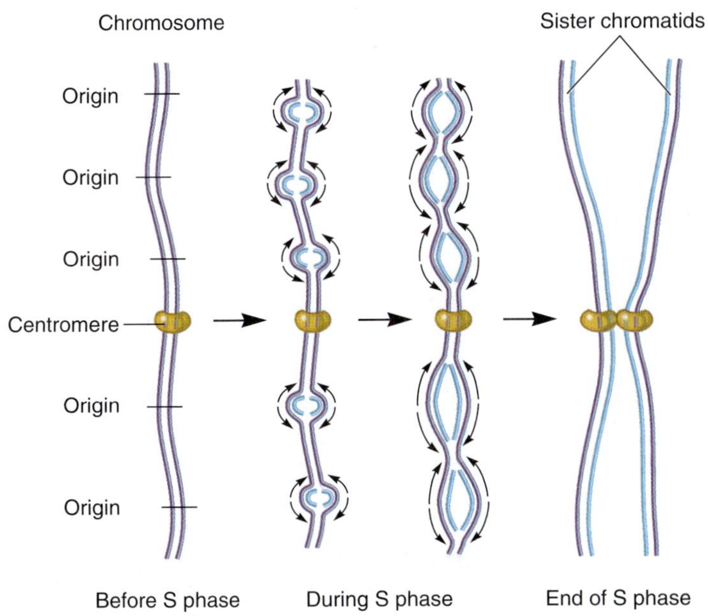

FIGURE 11.19 The replication of eukaryotic chromosomes. Replication begins from multiple origins of replication, and the replication forks move bidirectionally to replicate the DNA. Eventually, all of the replication forks will merge. The net result is two sister chromatids attached to each other at the centromere.

CONCEPT CHECK: Why do eukaryotes need multiple origins of replication?

(refer back to Figure 3.5). The multiple replication forks eventually make contact with each other to complete the replication process.

The molecular features of eukaryotic origins of replication may have some similarities to the origins found in bacteria. At the molecular level, eukaryotic origins of replication have been extensively studied in the yeast *Saccharomyces cerevisiae*. In this organism, several replication origins have been identified and sequenced. They have been named **ARS elements** (for autonomously replicating sequence). ARS elements, which are about 50 bp in length, are necessary to initiate chromosome replication. ARS elements contain a higher percentage of A and T bases than the rest of the chromosomal DNA. In addition, they have a copy of the ARS consensus sequence, ATTTAT(A or G)TTTA, along with additional elements that enhance origin function. This arrangement is similar to bacterial origins, which also have an AT-rich region and specific elements, such as DnaA boxes.

In *S. cerevisiae*, origins of replication are determined primarily by their DNA sequences. In more complex eukaryotes such as animals, the critical features that define origins of replication are not completely understood. In many species, origins are not determined by particular DNA sequences but instead occur at specific sites along a chromosome due to features of chromatin structure such as histone modifications.

DNA replication in eukaryotes begins with the assembly of a **prereplication complex (preRC),** which is composed of many different proteins. Part of the preRC is a group of six proteins called the **origin recognition complex (ORC)** that acts as the first initiator of eukaryotic DNA replication. ORC first binds to the origin of replication and other proteins of the preRC then bind, including a group of proteins called **MCM helicase.**[1] The binding of MCM helicase at the origin completes a process called **DNA replication licensing.** Those origins with MCM helicase are able to begin the process of DNA synthesis.

Eukaryotes Contain Several Different DNA Polymerases

Eukaryotes have many types of DNA polymerases. For example, mammalian cells have well over a dozen different DNA polymerases (**Table 11.4**). Four of these, designated α (alpha), ε (epsilon), δ (delta), and γ (gamma), have the primary function of replicating DNA. DNA polymerase γ functions in the mitochondria to replicate mitochondrial DNA, whereas α, ε, and δ are involved with DNA replication in the cell nucleus.

DNA polymerase α is the only eukaryotic polymerase that associates with primase. The functional role of the DNA polymerase α/primase complex is to synthesize a short RNA-DNA primer of approximately 10 RNA nucleotides followed by 20–30 DNA nucleotides. This short RNA-DNA strand is then used by DNA polymerase ε or δ for the processive elongation of the DNA strands. For this to happen, the DNA polymerase α/primase complex dissociates

[1] MCM is an abbreviation for minichromosome maintenance. The genes encoding MCM proteins were originally identified in mutant yeast strains that are defective in the maintenance of minichromosomes in the cell. MCM proteins have since been shown to play a role in DNA replication.

TABLE 11.4
Eukaryotic DNA Polymerases

Polymerase Types*	Function
α	Initiates DNA replication in conjunction with primase
ε	Replication of the leading strand
δ	Replication of the lagging strand
γ	Replication of mitochondrial DNA
η, κ, ι, ξ (lesion-replicating polymerases)	Replication of damaged DNA
α, β, δ, ε, σ, λ, μ, φ, θ, η	DNA repair or other functions†

*The designations are those of mammalian enzymes.
†Many DNA polymerases have dual functions. For example, DNA polymerases α, δ, and ε are involved in the replication of normal DNA and also play a role in DNA repair. In cells of the immune system, certain genes that encode antibodies (i.e., immunoglobulin genes) undergo a phenomenon known as hypermutation. This increases the variation in the kinds of antibodies the cells can make. Certain polymerases in this list, such as η, may play a role in hypermutation of immunoglobulin genes.

from the replication fork and is exchanged for DNA polymerase ε or δ. This exchange is called a **polymerase switch.** DNA polymerase ε is involved with leading-strand synthesis, whereas DNA polymerase δ is responsible for lagging-strand synthesis.

What are the functions of the other DNA polymerases? Several of them play an important role in DNA repair, a topic that will be examined in Chapter 18. DNA polymerase β, which has been studied for several decades, is not involved in the replication of normal DNA, but plays an important role in removing incorrect bases from damaged DNA. More recently, several additional DNA polymerases have been identified. Although their precise roles have not been elucidated, many are in a category called **lesion-replicating polymerases.** When DNA polymerase α, δ, and ε encounter abnormalities in DNA structure, such as abnormal bases or crosslinks, they may be unable to replicate over the aberration. When this occurs, lesion-replicating polymerases are attracted to the damaged DNA and have special properties that enable them to synthesize a complementary strand over the abnormal region. The various types of lesion-replicating polymerases may be able to replicate over different kinds of DNA damage.

Flap Endonuclease Removes RNA Primers During Eukaryotic DNA Replication

Another key difference between bacterial and eukaryotic DNA replication is the way that RNA primers are removed. As discussed earlier in this chapter, bacterial RNA primers are removed by DNA polymerase I. By comparison, a DNA polymerase enzyme does not play this role in eukaryotes. Instead, an enzyme called flap endonuclease is primarily responsible for RNA primer removal.

Flap endonuclease gets its name because it removes small pieces of RNA flaps that are generated by the action of DNA polymerase δ. In the diagram shown in **Figure 11.20**, DNA polymerase δ elongates the left Okazaki fragment until it runs into the RNA primer of the adjacent Okazaki fragment on the right. This causes a portion of the RNA primer to form a short flap,

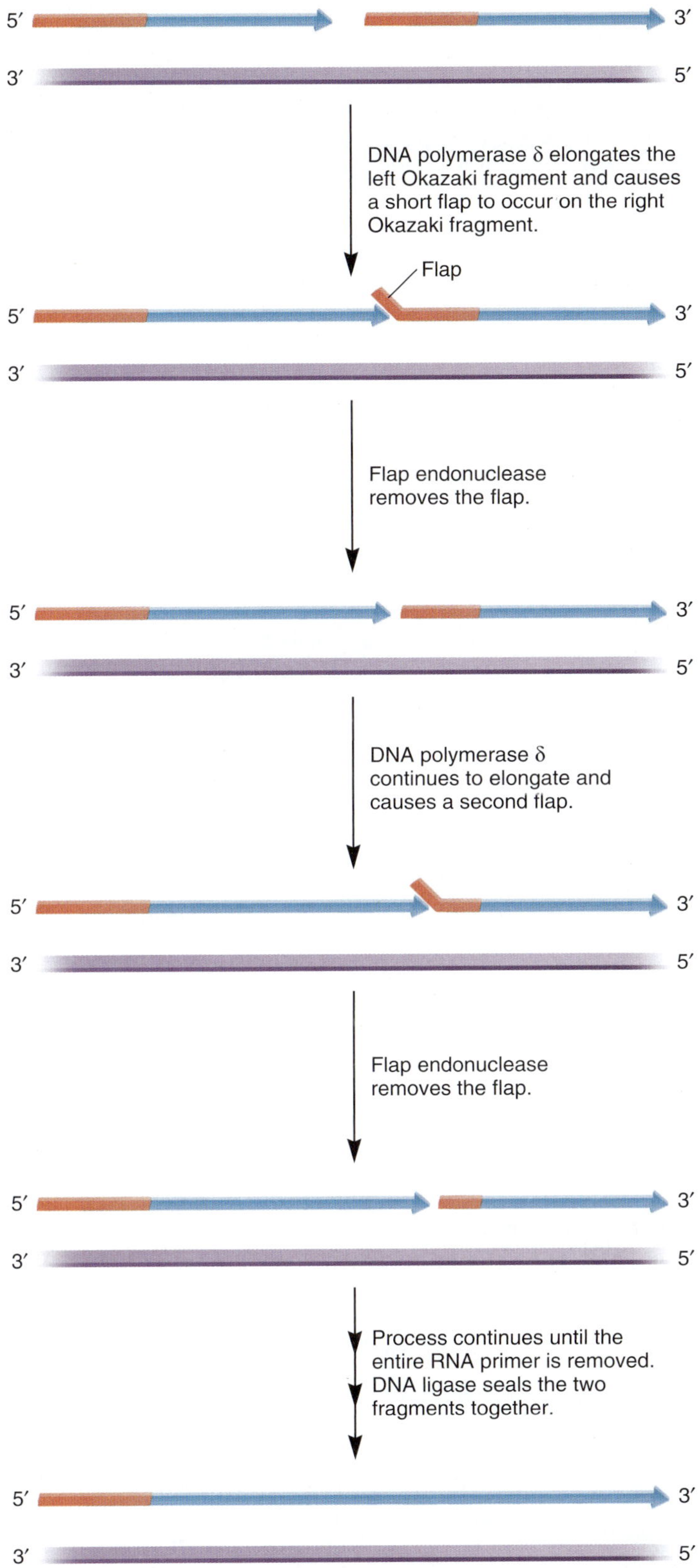

FIGURE 11.20 **Removal of an RNA primer by flap endonuclease.**

which is removed by flap endonuclease. As DNA polymerase δ continues to elongate the DNA, short flaps continue to be generated, which are sequentially removed by flap endonuclease. Eventually, all of the RNA primer is removed, and DNA ligase seals the DNA fragments together.

Though flap endonuclease is thought to be the primary pathway for RNA primer removal in eukaryotes, it cannot remove a flap that is too long. In such cases, a long flap is cleaved by an enzyme called Dna2 nuclease/helicase. This enzyme can cut a long flap, thereby generating a short flap. The short flap is then removed via flap endonuclease.

The Ends of Eukaryotic Chromosomes Are Replicated by Telomerase

Linear eukaryotic chromosomes contain telomeres at both ends. The term **telomere** refers to the telomeric sequences within the DNA and the specific proteins that are bound to those sequences. Telomeres consist of a tandemly repeated sequence and a 3′ overhang region that is 12–16 nucleotides in length (**Figure 11.21**).

The tandem array that occurs within the telomere has been studied in a wide variety of eukaryotic organisms. A common feature is that the telomeric sequence contains several guanine nucleotides and often many thymine nucleotides (**Table 11.5**). Depending on the species and the cell type, this sequence can be tandemly repeated up to several hundred times in the telomere region.

One reason why telomeric repeat sequences are needed is because DNA polymerase is unable to replicate the 3′ ends of DNA strands. Why is DNA polymerase unable to replicate this region? The answer lies in the two unusual enzymatic features of this enzyme. As discussed previously, DNA polymerase synthesizes DNA only in a 5′ to 3′ direction, and it cannot link together the first two individual nucleotides; it can elongate only preexisting strands. These two features of DNA polymerase function pose a problem at the 3′ ends of linear chromosomes. As shown in **Figure 11.22**, the 3′ end of a DNA strand cannot be replicated by DNA polymerase because a primer cannot be made upstream from this point. Therefore, if this problem were not solved, the chromosome would become progressively shorter with each round of DNA replication.

To prevent the loss of genetic information due to chromosome shortening, additional DNA sequences are attached to the ends of telomeres. In 1984, Carol Greider and Elizabeth Blackburn discovered an enzyme called **telomerase** that prevents chromosome shortening. It recognizes the sequences at the ends of eukaryotic chromosomes and synthesizes additional repeats of telomeric sequences. They received the 2009 Nobel Prize in physiology or medicine for their discovery.

Figure 11.23 shows the interesting mechanism by which telomerase works. Telomere lengthening occurs in three phases.

Binding of Telomerase Telomerase contains both protein subunits and RNA. The RNA part of telomerase contains a sequence complementary to the DNA sequence found in the telomeric repeat. This allows telomerase to bind to the 3′ overhang region of the telomere.

Polymerization Following binding, the RNA sequence beyond the binding site functions as a template for the synthesis of a 6-nucleotide sequence at the end of the DNA strand. This is

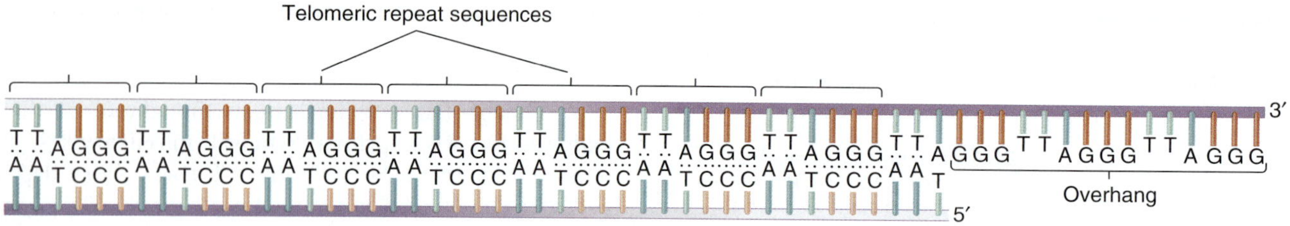

FIGURE 11.21 **General structure of telomeric sequences.** The telomere DNA consists of a tandemly repeated sequence and a 12- to 16-nucleotide overhang.

ONLINE ANIMATION

TABLE 11.5
Telomeric Repeat Sequences Within Selected Organisms

Group	Example	Telomeric Repeat Sequence
Mammals	Humans	TTAGGG
Slime molds	*Physarum, Didymium*	TTAGGG
	Dictyostelium	$AG_{(1-8)}$
Filamentous fungi	*Neurospora*	TTAGGG
Budding yeast	*Saccharomyces cerevisiae*	$TG_{(1-3)}$
Ciliates	*Tetrahymena*	TTGGGG
	Paramecium	TTGGG(T/G)
	Euplotes	TTTTGGGG
Flowering plants	*Arabidopsis*	TTTAGGG

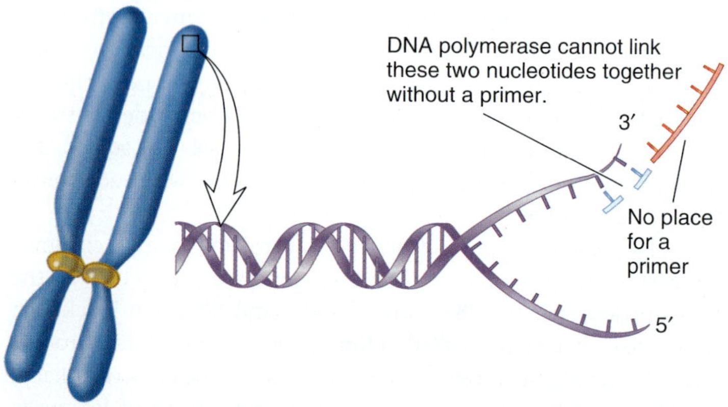

FIGURE 11.22 **The replication problem at the ends of linear chromosomes.** DNA polymerase cannot synthesize a DNA strand that is complementary to the 3′ end because a primer cannot be made upstream from this site.

called polymerization, because it is analogous to the function of DNA polymerase. Telomere lengthening is catalyzed by two identical protein subunits of telomerase called **telomerase reverse transcriptase (TERT).** TERT's name indicates that it catalyzes the reverse of transcription; it uses an RNA template to synthesize DNA.

Translocation Following polymerization, telomerase then moves—a process called translocation—to the new end of this DNA strand and attaches another 6 nucleotides to the end.

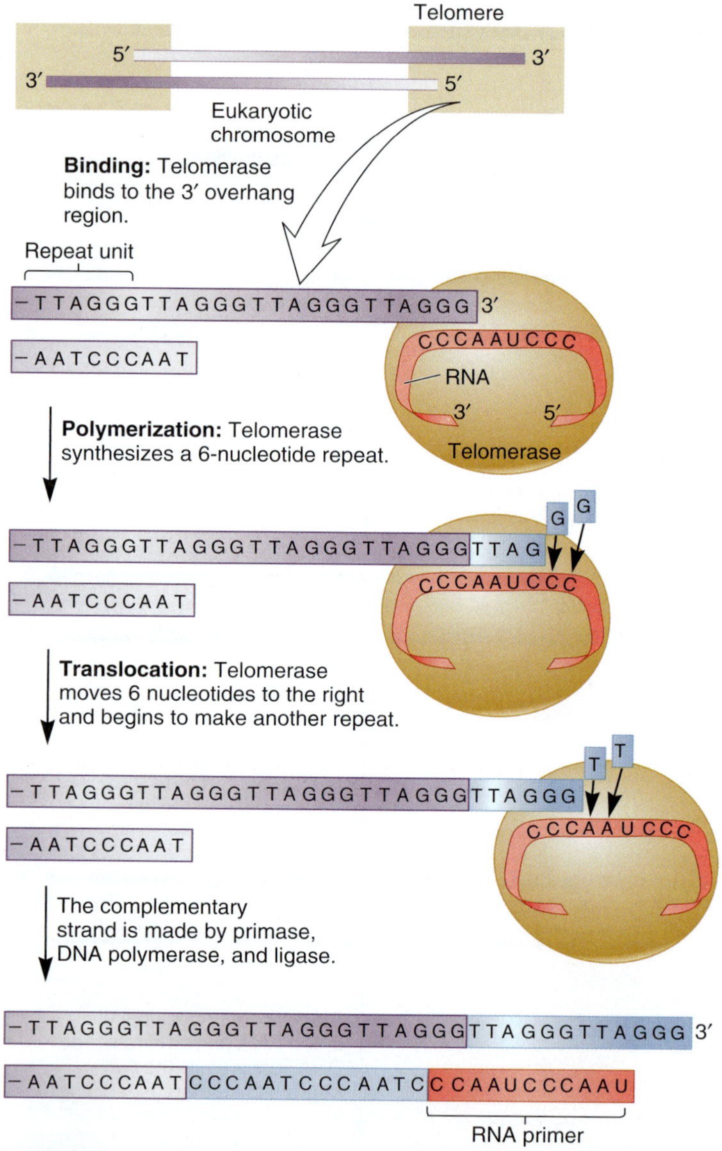

ONLINE ANIMATION

FIGURE 11.23 **The enzymatic action of telomerase.** A short, 3-nucleotide segment of RNA within telomerase causes it to bind to the 3′ overhang. The adjacent part of the RNA is used as a template to make a short, 6-nucleotide repeat of DNA. After the repeat is made, telomerase moves 6 nucleotides to the right and then synthesizes another repeat. This process is repeated many times to lengthen the top strand shown in this figure. The bottom strand is made by DNA polymerase, using an RNA primer at the end of the chromosome that is complementary to the telomeric repeat sequence in the top strand. DNA polymerase fills in the region, which is sealed by ligase. Note: Though not shown in the figure, the RNA primer is eventually removed, which leaves a 3′ overhang.

CONCEPT CHECK: How many times would telomerase have to bind to a different site in the telomere to make a segment of DNA that is 36 nucleotides in length?

This binding-polymerization-translocation cycle occurs many times in a row, thereby greatly lengthening the 3′ end of the DNA strand in the telomeric region. The complementary strand is synthesized by primase, DNA polymerase, and DNA ligase, as described earlier in this chapter. The RNA primer is later removed, which leaves a 3′ overhang.

Telomere Length May Play a Role in Aging and Cancer

In actively dividing somatic cells, telomeres tend to shorten with age. For example, the average length of telomere DNA in human blood cells is about 8,000 bp at birth, but can be as short as 1500 bp in an elderly person. This decrease occurs because the activity of telomerase decreases as a person ages. When telomeres are too short, the cells become **senescent,** which means they lose their ability to divide. Researchers are investigating whether the shortening of telomeres is simply a sign of aging, like gray hair, or if the shortening is a programmed process that contributes to aging. In 1998, Andrea Bodnar and her colleagues inserted a gene that encodes a highly active telomerase into human cells grown in the laboratory. The expression of telomerase prevented telomere shortening and senescence!

If telomeres tend to shorten with each cell division, how do cancer cells keep dividing? Though some types of cancer cells have very short telomeres, it is common for cancer cells to carry mutations that increase the activity of telomerase, thereby preventing telomere shortening. This increase in telomerase function is thought to prevent senescence. In the laboratory, when cancer cells are treated with drugs that inhibit telomerase, they often stop dividing. For this reason, researchers are interested in whether such drugs could be used to combat cancer. However, additional research is needed before such drugs can be used. A major concern is how they could affect normal cells of the body, possibly accelerating senescence.

11.5 COMPREHENSION QUESTIONS

1. In eukaryotes, DNA replication is initiated at an origin of replication by
   a. DnaA proteins.
   b. the origin recognition complex.
   c. DNA polymerase δ.
   d. MCM helicase.
2. Which of the following statements regarding DNA polymerases in eukaryotes is *not* correct?
   a. DNA polymerase α synthesizes a short RNA-DNA primer.
   b. DNA polymerases ε and δ synthesize most of the leading and lagging, respectively.
   c. Lesion-replicating DNA polymerases can replicate over damaged DNA.
   d. All of the above statements are correct.
3. In eukaryotes, RNA primers are primarily removed by
   a. DNA polymerase I.
   b. DNA polymerase α.
   c. flap endonuclease.
   d. helicase.
4. To synthesize DNA, what does telomerase use as a template?
   a. It uses the DNA in the 3′ overhang.
   b. It uses RNA that is a component of telomerase.
   c. No template is used.
   d. Both a and b are correct.

KEY TERMS

Page 261. DNA replication
Page 262. template strands, parental strands, daughter strands
Page 263. conservative model, semiconservative model, dispersive model
Page 266. origin of replication, bidirectionally, replication forks
Page 267. DnaA proteins, DnaA box sequences, DNA helicase
Page 268. bidirectional replication, topoisomerase II, DNA gyrase, single-strand binding proteins, RNA primers, primase, leading strand, lagging strand, DNA polymerase
Page 270. Okazaki fragments
Page 271. DNA ligase, primosome, replisome, dimeric DNA polymerase
Page 272. termination sequences, catenanes
Page 273. conditional mutants, temperature-sensitive (ts) mutant
Page 275. processive enzyme
Page 276. fidelity, proofreading function
Page 278. ARS elements, prereplication complex (preRC), origin recognition complex (ORC), MCM helicase, DNA replication licensing, polymerase switch, lesion-replicating polymerases, flap endonuclease
Page 279. telomere, telomerase
Page 280. telomerase reverse transcriptase (TERT)
Page 281. senescent

CHAPTER SUMMARY

- DNA replication is the process in which existing DNA strands are used to make new DNA strands.

11.1 Structural Overview of DNA Replication

- DNA replication occurs when the strands of DNA unwind and each strand is used as a template to make a new strand according to the AT/GC rule. The resulting DNA molecules have the same base sequence as the original DNA (see Figure 11.1).

- By labeling DNA with heavy and light isotopes of nitrogen and using centrifugation, Meselson and Stahl showed that DNA replication is semiconservative (see Figures 11.2, 11.3).

11.2 Bacterial DNA Replication: The Formation of Two Replication Forks at the Origin of Replication

- Bacterial DNA replication begins at a single origin of replication and proceeds bidirectionally around the circular chromosome (see Figure 11.4).
- In *E. coli*, DNA replication is initiated when DnaA proteins bind to sequences in five DnaA boxes at the origin of replication. This causes the AT-rich region to unwind. DNA helicases then promote the movement of two forks (see Figures 11.5, 11.6).

11.3 Bacterial DNA Replication: Synthesis of New DNA Strands

- At each replication fork, DNA helicase unwinds the DNA and topoisomerase alleviates positive supercoiling. Single-strand binding proteins coat the DNA to prevent the strands from coming back together. Primase synthesizes RNA primers and DNA polymerase synthesizes complementary strands of DNA. DNA ligase seals the gaps between Okazaki fragments (see Figure 11.7, Table 11.1).
- DNA polymerase III in *E. coli* is an enzyme with several subunits. The catalytic subunit wraps around the DNA like a hand (see Figure 11.8, Table 11.2).
- DNA polymerase enzymes need a primer to synthesize DNA and make new DNA strands in a 5′ to 3′ direction (see Figure 11.9).
- During DNA synthesis, the leading strand is made continuously in the direction toward the replication fork, whereas the lagging strand is made as Okazaki fragments in the direction away from the fork (see Figures 11.10, 11.11).
- The primosome is a complex between helicase and primase. The replisome is a complex between the primosome and dimeric DNA polymerase (see Figure 11.12).
- In *E. coli*, DNA replication is terminated at ter sequences (see Figure 11.13).
- Following DNA replication, interlocked catenanes sometimes need to be unlocked via topoisomerase II (see Figure 11.14).
- The isolation and characterization of temperature-sensitive mutants was a useful strategy for identifying proteins involved with DNA replication (see Figure 11.15, Table 11.3).

11.4 Bacterial DNA Replication: Chemistry and Accuracy

- DNA polymerase III is a processive enzyme that uses deoxynucleoside triphosphates to make new DNA strands (see Figure 11.16).
- The high fidelity of DNA replication is a result of (1) the stability of hydrogen bonding between the correct bases, (2) the phenomenon of induced fit, and (3) the proofreading ability of DNA polymerase (see Figure 11.17).

11.5 Eukaryotic DNA Replication

- Eukaryotic chromosomes contain multiple origins of replication. Part of the prereplication complex (preRC) is formed from a group of six proteins called the origin recognition complex (ORC). The binding of MCM helicase completes a process called DNA replication licensing (see Figures 11.18, 11.19).
- Eukaryotes have several different DNA polymerases with specialized roles. Different types of DNA polymerases switch with each other during the process of DNA replication (see Table 11.4).
- Flap endonuclease is an enzyme that removes RNA primers from Okazaki fragments (see Figure 11.20).
- The ends of eukaryotic chromosomes contain telomeres, which are composed of tandemly repeated sequences and proteins (see Figure 11.21, Table 11.5).
- DNA polymerase is unable to replicate the very end of a eukaryotic chromosome (see Figure 11.22).
- Telomerase uses a short RNA molecule as a template to add repeat sequences onto telomeres (see Figure 11.23).

PROBLEM SETS & INSIGHTS

Solved Problems

S1. Describe three ways to account for the high fidelity of DNA replication. Discuss the quantitative contributions of each of the three ways.

Answer:

First: AT and GC pairs are preferred in the double-helix structure. This provides fidelity to around 1 mistake per 1000.

Second: Induced fit by DNA polymerase prevents covalent bond formation unless the proper nucleotides are in place. This increases fidelity another 100- to 1000-fold, to about 1 error in 100,000–1 million.

Third: Exonuclease proofreading increases fidelity another 100- to 1000-fold, to about 1 error per 100 million nucleotides added.

S2. What do you think would happen if the ter sequences were deleted from the bacterial DNA?

Answer: Instead of meeting at the ter sequences, the two replication forks would meet somewhere else. This would depend on how fast they were moving. For example, if one fork was advancing faster than the other, they would meet closer to where the slower-moving fork started. In fact, researchers have actually conducted this experiment. Interestingly, *E. coli* without the ter sequences seemed to survive just fine.

S3. Summarize the steps that occur in the process of chromosomal DNA replication in *E. coli.*

Answer:

Step 1. DnaA proteins bind to the origin of replication, resulting in the separation of the AT-rich region.

Step 2. DNA helicase breaks the hydrogen bonds between the DNA strands, topoisomerases alleviate positive supercoiling, and single-strand binding proteins keep the parental strands from rejoining.

Step 3. Primase synthesizes one RNA primer in the leading strand and many RNA primers in the lagging strand. DNA polymerase III synthesizes the daughter strands of DNA. In the lagging strand, many short segments of DNA (Okazaki fragments) are made. DNA polymerase I removes the RNA primers and fills in with DNA, and DNA ligase covalently links the Okazaki fragments together.

Step 4. The processes described in steps 2 and 3 continue until the two replication forks reach the ter sequences on the other side of the circular bacterial chromosome.

Step 5. Topoisomerases unravel the intertwined chromosomes, if necessary.

S4. If a strain of *E. coli* overproduced the Dam enzyme, how would that affect the DNA replication process? Would you expect such a strain to have more or fewer chromosomes per cell than a normal strain of *E. coli*? Explain why.

Answer: If a strain overproduced the Dam enzyme, DNA would replicate more rapidly. The GATC methylation sites in the origin of replication have to be fully methylated for DNA replication to occur. Immediately after DNA replication, a delay occurs before the next round of DNA replication because the two copies of newly replicated DNA are hemimethylated. A strain that overproduces Dam would rapidly convert the hemimethylated DNA into fully methylated DNA and more quickly allow the next round of DNA replication to occur. For this reason, the overproducing strain might have more copies of the *E. coli* chromosome per cell because it would not have a long delay in DNA replication.

Conceptual Questions

C1. What key structural features of the DNA molecule underlie its ability to be faithfully replicated?

C2. With regard to DNA replication, define the term *bidirectional replication.*

C3. Which of the following statements is *not* true? Explain why.

A. A DNA strand can serve as a template strand on many occasions.

B. Following semiconservative DNA replication, one strand is a newly made daughter strand and the other strand is a parental strand.

C. A DNA double helix may contain two strands of DNA that were made at the same time.

D. A DNA double helix obeys the AT/GC rule.

E. A DNA double helix could contain one strand that is 10 generations older than its complementary strand.

C4. The compound known as nitrous acid is a reactive chemical that replaces amino groups ($-NH_2$) with keto groups ($=O$). When nitrous acid reacts with the bases in DNA, it can change cytosine to uracil and change adenine to hypoxanthine. A DNA double helix has the following sequence:

TTGGATGCTGG
AACCTACGACC

A. What would be the sequence of this double helix immediately after reaction with nitrous acid? Let the letter H represent hypoxanthine and U represent uracil.

B. Let's suppose this DNA was reacted with nitrous acid. The nitrous acid was then removed, and the DNA was replicated for two generations. What would be the sequences of the DNA products after the DNA had replicated twice? Your answer should contain the sequences of four double helices. Note: During DNA replication, uracil hydrogen bonds with adenine, and hypoxanthine hydrogen bonds with cytosine.

C5. One way that bacterial cells regulate DNA replication is by GATC methylation sites within the origin of replication. Would this mechanism work if the DNA was conservatively (rather than semiconservatively) replicated?

C6. The chromosome of *E. coli* contains 4.6 million bp. How long will it take to replicate its DNA? Assuming DNA polymerase III is the primary enzyme involved and this enzyme can actively proofread during DNA synthesis, how many base pair mistakes will be made in one round of DNA replication in a bacterial population containing 1000 bacteria?

C7. Here are two strands of DNA.

—————————————DNA polymerase→

———————————————————————————

The one on the bottom is a template strand, and the one on the top is being synthesized by DNA polymerase in the direction shown by the arrow. Label the 5′ and 3′ ends of the top and bottom strands.

C8. A DNA strand has the following sequence:

5′–GATCCCGATCCGCATACATTTACCAGATCACCACC–3′

In which direction would DNA polymerase slide along this strand (from left to right or from right to left)? If this strand was used as a template by DNA polymerase, what would be the sequence of the newly made strand? Indicate the 5′ and 3′ ends of the newly made strand.

C9. List and briefly describe the three types of sequences within bacterial origins of replication that are functionally important.

C10. As shown in Figure 11.5, five DnaA boxes are found within the origin of replication in *E. coli*. Take a look at these five sequences carefully.

A. Are the sequences of the five DnaA boxes very similar to each other? (Hint: Remember that DNA is double-stranded; think about these sequences in the forward and reverse direction.)

B. What is the most common sequence for the DnaA box? In other words, what is the most common base in the first position, second position, and so on until the ninth position? The most common sequence is called the consensus sequence.

C. The *E. coli* chromosome is about 4.6 million bp long. Based on random chance, is it likely that the consensus sequence for a DnaA box occurs elsewhere in the *E. coli* chromosome? If so, why aren't there multiple origins of replication in *E. coli*?

C11. Obtain two strings of different colors (e.g., black and white) that are the same length. A length of 20 inches is sufficient. Tie a knot at one end of the black string, and tie a knot at one end of the white string. Each knot designates the 5′ end of your strings. Make a double helix with your two strings. Now tape one end of the double helix to a table so that the tape is covering the knot on the black string.

A. Pretend your hand is DNA helicase and use your hand to unravel the double helix, beginning at the end that is not taped to the table. Should your hand be sliding along the white string or the black string?

B. As in Figure 11.12, imagine that your two hands together form a dimeric replicative DNA polymerase. Unravel your two strings halfway to create a replication fork. Grasp the black string with your left hand and the white string with your right hand. Your thumbs should point toward the 5′ end of each string. You need to loop one of the strings so that one of the DNA polymerases can synthesize the lagging strand. With such a loop, dimeric replicative DNA polymerase can move toward the replication fork and synthesize both DNA strands in the 5′ to 3′ direction. In other words, with such a loop, your two hands can touch each other with both of your thumbs pointing toward the fork. Should the black string be looped, or should the white string be looped?

C12. Sometimes DNA polymerase makes a mistake, and the wrong nucleotide is added to the growing DNA strand. With regard to pyrimidines and purines, two general types of mistakes are possible. The addition of an incorrect pyrimidine instead of the correct pyrimidine (e.g., adding cytosine where thymine should be added) is called a transition. If a pyrimidine is incorrectly added to the growing strand instead of purine (e.g., adding cytosine where an adenine should be added), this type of mistake is called a transversion. If a transition or transversion is not detected by DNA polymerase, a mutation is created that permanently changes the DNA sequence. Though both types of mutations are rare, transition mutations are more frequent than transversion mutations. Based on your understanding of DNA replication and DNA polymerase, offer three explanations why transition mutations are more common.

C13. A short genetic sequence, which may be recognized by primase, is repeated many times throughout the *E. coli* chromosome. Researchers have hypothesized that primase may recognize this sequence as a site to begin the synthesis of an RNA primer for DNA replication. The *E. coli* chromosome is roughly 4.6 million bp in length. How many copies of the primase recognition sequence would be necessary to replicate the entire *E. coli* chromosome?

C14. Single-strand binding proteins keep the two parental strands of DNA separated from each other until DNA polymerase has an opportunity to replicate the strands. Suggest how single-strand binding proteins keep the strands separated and yet do not impede the ability of DNA polymerase to replicate the strands.

C15. The ability of DNA polymerase to digest a DNA strand from one end is called its exonuclease activity. Exonuclease activity is used to digest RNA primers and also to proofread a newly made DNA strand. Note: DNA polymerase I does not change direction while it is removing an RNA primer and synthesizing new DNA. It does change direction during proofreading.

A. In which direction, 5′ to 3′ or 3′ to 5′, is the exonuclease activity occurring during the removal of RNA primers and during the proofreading and removal of mistakes following DNA replication?

B. Figure 11.17 shows a drawing of the 3′ exonuclease site. Do you think this site would be used by DNA polymerase I to remove RNA primers? Why or why not?

C16. In the following drawing, the top strand is the template DNA, and the bottom strand shows the lagging strand prior to the action of DNA polymerase I. The lagging strand contains three Okazaki fragments. The RNA primers have not yet been removed.

The top strand is the template DNA.

```
3'————————————————————————————————————————5'
5'*************____***************____*************______3'
   RNA primer     ↑ RNA primer ↑        RNA primer
|——————————————||——————————————||——————————————|
   Left Okazaki     Middle Okazaki     Right Okazaki
   fragment         fragment           fragment
```

A. Which Okazaki fragment was made first, the one on the left or the one on the right?

B. Which RNA primer would be the first one to be removed by DNA polymerase I, the primer on the left or the primer on the right? For this primer to be removed by DNA polymerase I and for the gap to be filled in, is it necessary for the Okazaki fragment in the middle to have already been synthesized? Explain.

C. Let's consider how DNA ligase connects the left Okazaki fragment with the middle Okazaki fragment. After DNA polymerase I removes the middle RNA primer and fills in the gap with DNA, where does DNA ligase function? See the arrows on either side of the middle RNA primer. Is ligase needed at the left arrow, at the right arrow, or both?

D. When connecting two Okazaki fragments, DNA ligase uses NAD^+ or ATP as a source of energy to catalyze this reaction. Explain why DNA ligase needs another source of energy to connect two nucleotides, but DNA polymerase needs nothing more than the incoming nucleotide and the existing DNA strand. Note: You may want to refer to Figure 11.16 to answer this question.

C17. Describe the three important functions of DnaA protein.

C18. Draw a picture that illustrates how DNA helicase works.

C19. What is an Okazaki fragment? In which strand of DNA are Okazaki fragments found? Based on the properties of DNA polymerase, why is it necessary to make these fragments?

C20. Discuss the similarities and differences in the synthesis of DNA in the lagging and leading strands. What is the advantage of a primosome and a replisome as opposed to having all replication enzymes functioning independently of each other?

C21. Explain the proofreading function of DNA polymerase.

C22. What is a processive enzyme? Explain why this is an important feature of DNA polymerase.

C23. What enzymatic features of DNA polymerase prevent it from replicating one of the DNA strands at the ends of linear chromosomes? Compared with DNA polymerase, how is telomerase different in its ability to synthesize a DNA strand? What does telomerase use as its template for the synthesis of a DNA strand? How does the use of this template result in a telomere sequence that is tandemly repetitive?

C24. As shown in Figure 11.23, telomerase attaches additional DNA, 6 nucleotides at a time, to the ends of eukaryotic chromosomes. However, it makes only one DNA strand. Describe how the opposite strand is replicated.

C25. If a eukaryotic chromosome has 25 origins of replication, how many replication forks does it have at the beginning of DNA replication?

C26. A diagram of a linear chromosome is shown here. The end of each strand is labeled with an A, B, C, or D. Which ends could not be replicated by DNA polymerase? Why not?

5′–A————————————————B–3′
3′–C————————————————D–5′

C27. As discussed in Chapter 17, some viruses contain RNA as their genetic material. Certain RNA viruses can exist as a provirus in which the viral genetic material has been inserted into the chromosomal DNA of the host cell. For this to happen, the viral RNA must be copied into a strand of DNA. An enzyme called reverse transcriptase, encoded by the viral genome, copies the viral RNA into a complementary strand of DNA. The strand of DNA is then used as a template to make a double-stranded DNA molecule. This double-stranded DNA molecule is then inserted into the chromosomal DNA, where it may exist as a provirus for a long period of time.

A. How is the function of reverse transcriptase similar to the function of telomerase?

B. Unlike DNA polymerase, reverse transcriptase does not have a proofreading function. How might this affect the proliferation of the virus?

C28. Telomeres contain a 3′ overhang region, as shown in Figure 11.21. Does telomerase require a 3′ overhang to replicate the telomere region? Explain.

Experimental Questions

E1. Answer the following questions that pertain to the experiment of Figure 11.3.

A. What would be the expected results if the Meselson and Stahl experiment were carried out for four or five generations?

B. What would be the expected results of the Meselson and Stahl experiment after three generations if the mechanism of DNA replication was dispersive?

C. As shown in The Data, explain why three different bands (i.e., light, half-heavy, and heavy) can be observed in the CsCl gradient.

E2. An absentminded researcher follows the steps of Figure 11.3, and when the gradient is viewed under UV light, the researcher does not see any bands at all. Which of the following mistakes could account for this observation? Explain how.

A. The researcher forgot to add ^{14}N-containing compounds.

B. The researcher forgot to add lysozyme.

C. The researcher forgot to turn on the UV lamp.

E3. Figure 11.4b shows an autoradiograph of a replicating bacterial chromosome. If you analyzed many replicating chromosomes, what types of information could you learn about the mechanism of DNA replication?

E4. As described in Table 11.3, what is the difference between a rapid-stop and a slow-stop mutant? What are different roles of the proteins that are defective in rapid-stop and slow-stop mutants?

E5. The technique of dideoxy sequencing of DNA is described in Chapter 20. The technique relies on the use of dideoxyribonucleotides (shown in Figures 20.11 and 20.12). A dideoxyribonucleotide has a hydrogen atom attached to the 3′-carbon atom instead of an —OH group. When a dideoxyribonucleotide is incorporated into a newly made strand, the strand cannot grow any longer. Explain why.

E6. Another technique described in Chapter 20 is polymerase chain reaction (PCR) (see Figures 20.5 and 20.6). This method is based on our understanding of DNA replication. In this method, a small amount of double-stranded template DNA is mixed with a high concentration of primers. Nucleotides and DNA polymerase are also added. The template DNA strands are separated by heat treatment, and when the temperature is lowered, the primers bind to the single-stranded DNA, and then DNA polymerase replicates the DNA. This increases the amount of DNA made from the primers. This cycle of steps (i.e., heat treatment, lower temperature, allowing DNA replication to occur) is repeated again and again. Because the cycles are repeated many times, this method is called a chain reaction. It is called polymerase chain reaction because DNA polymerase is the enzyme needed to increase the amount of DNA with each cycle. In a PCR experiment, the template DNA is placed in a tube, and the primers, nucleotides, and DNA polymerase are added to the tube. The tube is then placed in a machine called a thermocycler, which raises and lowers the temperature. During one cycle, the temperature is raised (e.g., to 95°C) for a brief period and then lowered (e.g., to 60°C) to allow the primers to bind. The sample is then incubated at a slightly higher temperature for a few minutes to allow DNA replication to proceed. In a typical PCR experiment, the tube may be left in the thermocycler for 25–30 cycles. The total time for a PCR experiment is a few hours.

A. Why is DNA helicase not needed in a PCR experiment?

B. How is the sequence of each primer important in a PCR experiment? Do the two primers recognize the same strand or opposite strands?

C. The DNA polymerase used in PCR experiments is a DNA polymerase isolated from thermophilic bacteria. Why is this kind of polymerase used?

D. If a tube initially contained 10 copies of double-stranded DNA, how many copies of double-stranded DNA (in the region flanked by the two primers) would be obtained after 27 cycles?

Questions for Student Discussion/Collaboration

1. The complementarity of double-stranded DNA is the underlying reason that DNA can be faithfully copied. Propose alternative chemical structures that could be faithfully copied.
2. Compare and contrast DNA replication in bacteria and eukaryotes.
3. DNA replication is fast, virtually error-free, and coordinated with cell division. Discuss which of these three features you think is the most important.

Note: All answers appear at the website for this textbook; the answers to the even-numbered questions and all of the Concept Check and Comprehension Questions are in the back of the book.

www.mhhe.com/brookergenetics5e

Visit the website for practice tests, answer keys, and other learning aids for this chapter. Enhance your understanding of genetics with our interactive exercises, quizzes, animations, and much more.

CHAPTER OUTLINE

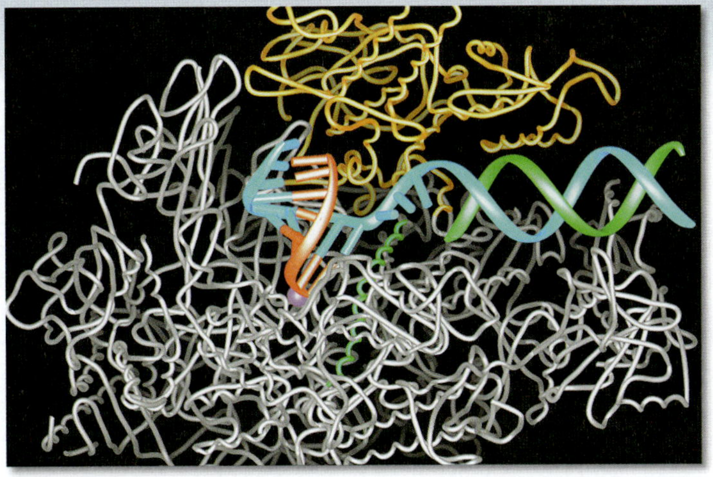

A molecular model showing the enzyme RNA polymerase in the act of sliding along the DNA and synthesizing a copy of RNA.

12 GENE TRANSCRIPTION AND RNA MODIFICATION

The primary function of the genetic material, which is DNA, is to store the information necessary to create a living organism. The information is contained within units called genes. At the molecular level, a **gene** is defined as a segment of DNA that is used to make a functional product, either an RNA molecule or a polypeptide. How is the information within a gene accessed? The first step in this process is called **transcription,** which literally means the act or process of making a copy. In genetics, this term refers to the process of synthesizing RNA from a DNA template (**Figure 12.1**). The structure of DNA is not altered as a result of transcription. Rather, the DNA base sequence has only been accessed to make a copy in the form of RNA. Therefore, the same DNA can continue to store information. DNA replication, which was discussed in Chapter 11, provides a mechanism for copying that information so it can be transmitted to new daughter cells and from parent to offspring.

Protein-encoding genes (also called **structural genes**) carry the information for the amino acid sequence of a polypeptide. When a protein-encoding gene is transcribed, the first product is an RNA molecule known as **messenger RNA (mRNA).** During polypeptide synthesis—a process called **translation**—the sequence of nucleotides within the mRNA determines the sequence of amino acids in a polypeptide. One or more polypeptides then assemble into a functional protein. The structures and functions of proteins ultimately determines an organism's traits. The model depicted in Figure 12.1, which is called the **central dogma of genetics** (also called the central dogma of molecular biology), was first enunciated by Francis Crick in 1958. It forms a cornerstone of our understanding of genetics at the molecular level. The flow of genetic information occurs from DNA to mRNA to polypeptide.

In this chapter, we begin to study the molecular steps in gene expression, with an emphasis on transcription and the modifications that may occur to an RNA transcript after it has been made. Chapter 13 will examine the process of translation, and Chapters 14, 15, and 16 will focus on how the level of gene expression is regulated at the molecular level.

12.1 OVERVIEW OF TRANSCRIPTION

Learning Outcomes:

1. Describe the organization of a protein-encoding gene and its mRNA transcript.
2. Outline the three stages of transcription.
3. Describe the various functions of RNA transcripts.

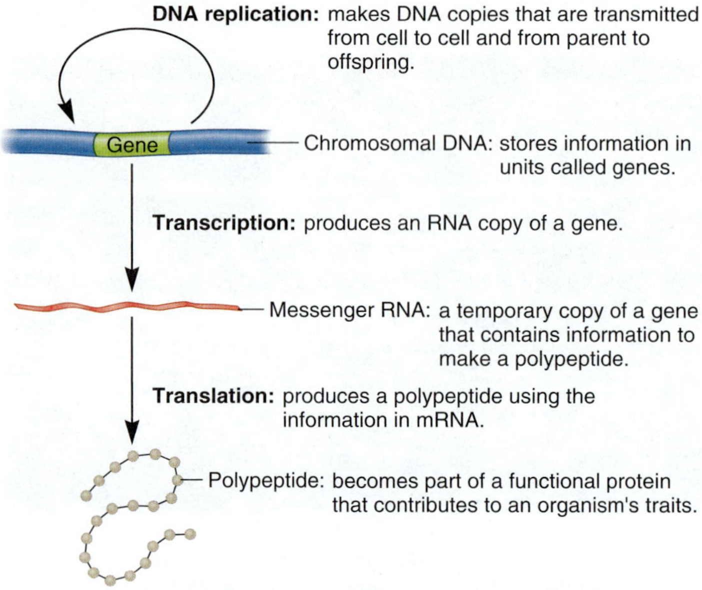

FIGURE 12.1 The central dogma of genetics. The usual flow of genetic information is from DNA to mRNA to polypeptide. Note: The direction of informational flow shown in this figure is the most common direction found in living organisms, but rare exceptions do occur. For example, RNA viruses use an enzyme called reverse transcriptase to make a copy of DNA from RNA.

One key concept important in the process of transcription is that specific base sequences define the beginning and ending of a gene and also play a role in regulating the level of RNA synthesis. In this section, we begin by examining the sequences that determine where transcription starts and ends, and then we also briefly consider DNA sequences, called regulatory sites, that influence whether a gene is turned on or off. The functions of regulatory sites will be examined in greater detail in Chapters 14, 15, and 16. A second important concept is the role of proteins in transcription. DNA sequences, in and of themselves, just exist. For genes to be actively transcribed, proteins must recognize particular DNA sequences and act on them in a way that affects the transcription process. In the later part of this section, we will consider how proteins participate in the general steps of transcription and the types of RNA transcripts that can be made.

Gene Expression Requires Base Sequences That Perform Different Functional Roles

At the molecular level, **gene expression** is the overall process by which the information within a gene is used to produce a functional product, such as a polypeptide. Along with environmental factors, the molecular expression of genes determines an organism's traits. Figure 12.2 shows a common organization of base sequences needed to create a protein-encoding gene that functions in a bacterium such as *E. coli*. The promoter and terminator are base sequences used during gene transcription. Specifically, the **promoter** provides a site for beginning transcription, and the **terminator** specifies the end of transcription. These two sequences cause RNA synthesis to occur within a defined location. As shown in Figure 12.2, the DNA is transcribed into RNA from the end of the promoter to the end of the terminator. As described later, the base sequence in the RNA transcript is complementary to the **template strand** of DNA. The opposite strand of DNA is the **nontemplate strand.** For protein-encoding genes, the nontemplate

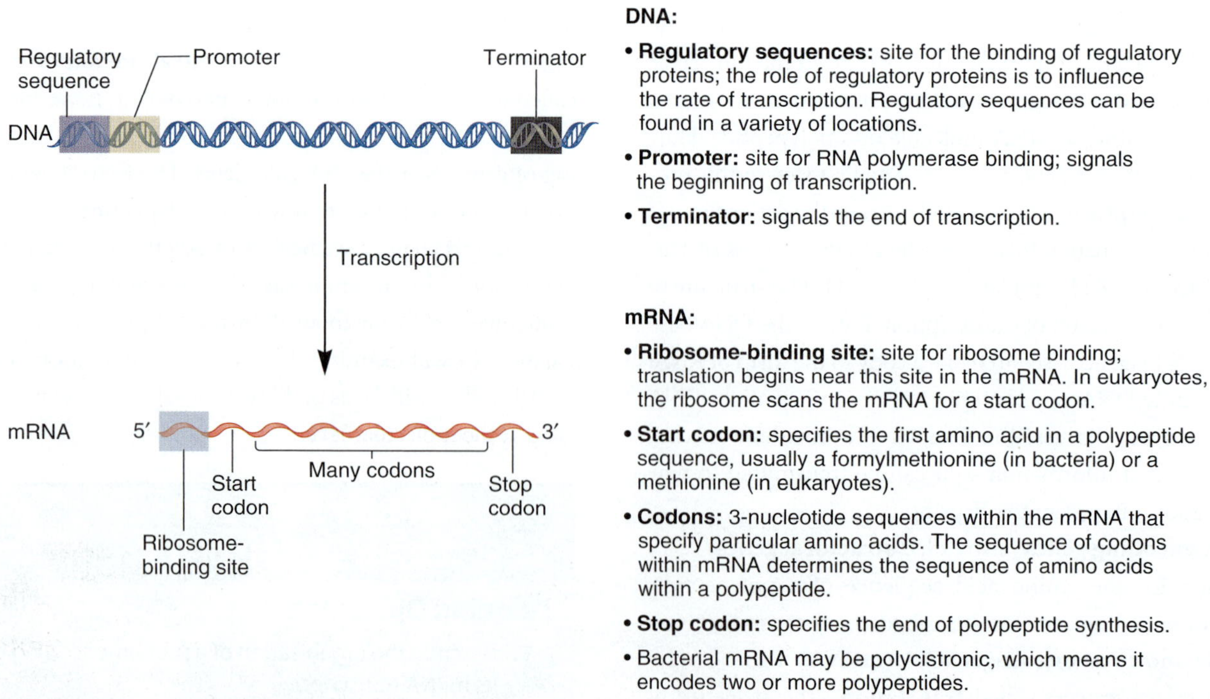

FIGURE 12.2 Organization of sequences of a bacterial gene and its mRNA transcript. This figure depicts the general organization of sequences that are needed to create a functional gene that encodes an mRNA.

CONCEPT CHECK: If a mutation changed the start codon into a stop codon, would this mutation affect the length of the RNA? Explain.

strand is also called the **coding strand,** or sense strand, because its sequence is the same as the transcribed mRNA that encodes a polypeptide, except that the DNA has T's in places where the mRNA contains U's. By comparison, the template strand is also called the noncoding strand, or antisense strand.

A category of proteins called **transcription factors** recognizes base sequences in the DNA and controls transcription. Some transcription factors bind directly to the promoter and facilitate transcription. Other transcription factors recognize **regulatory sequences** or **regulatory elements**—short stretches of DNA involved in the regulation of transcription. Certain transcription factors bind to such regulatory sequences and increase the rate of transcription, whereas others inhibit transcription.

As shown at the bottom of Figure 12.2, base sequences within an mRNA are used during the translation process. In bacteria, a short sequence within the mRNA, the **ribosome-binding site** (also known as the **Shine-Dalgarno sequence**), provides a location for a ribosome to bind and begin translation. The bacterial ribosome recognizes this site because it is complementary to a sequence in ribosomal RNA. In addition, mRNA contains a series of **codons,** read as groups of three nucleotides, which contain the information for a polypeptide's sequence. The first codon, which is very close to the ribosome-binding site, is the **start codon.** This is followed by many more codons that dictate the sequence of amino acids within the synthesized polypeptide. Finally, a **stop codon** signals the end of translation. Chapter 13 will examine the process of translation in greater detail.

The Three Stages of Transcription Are Initiation, Elongation, and Termination

Transcription occurs in three stages: **initiation; elongation,** or synthesis of the RNA transcript; and **termination** (Figure 12.3). These steps involve protein-DNA interactions in which proteins such as **RNA polymerase,** the enzyme that synthesizes RNA, interact with DNA sequences. What causes transcription to begin? The initiation stage in the transcription process is a recognition step. The sequence of bases within the promoter region is recognized by one or more transcription factors. The specific binding of transcription factors to the promoter identifies the starting site for transcription.

Transcription factors and RNA polymerase first bind to the promoter region when the DNA is in the form of a double helix. For transcription to occur, the DNA strands must be separated. This allows one of the two strands to be used as a template for the synthesis of a complementary strand of RNA. This synthesis occurs as RNA polymerase slides along the DNA, forming a small bubble-like structure known as the open promoter complex, or simply as the **open complex.** Eventually, RNA polymerase reaches a terminator, which causes both RNA polymerase and the newly made RNA transcript to dissociate from the DNA.

RNA Transcripts Have Different Functions

Once they are made, RNA transcripts play different functional roles (Table 12.1). Well over 90% of all genes are protein-encoding genes, which are transcribed into mRNA. For protein-encoding

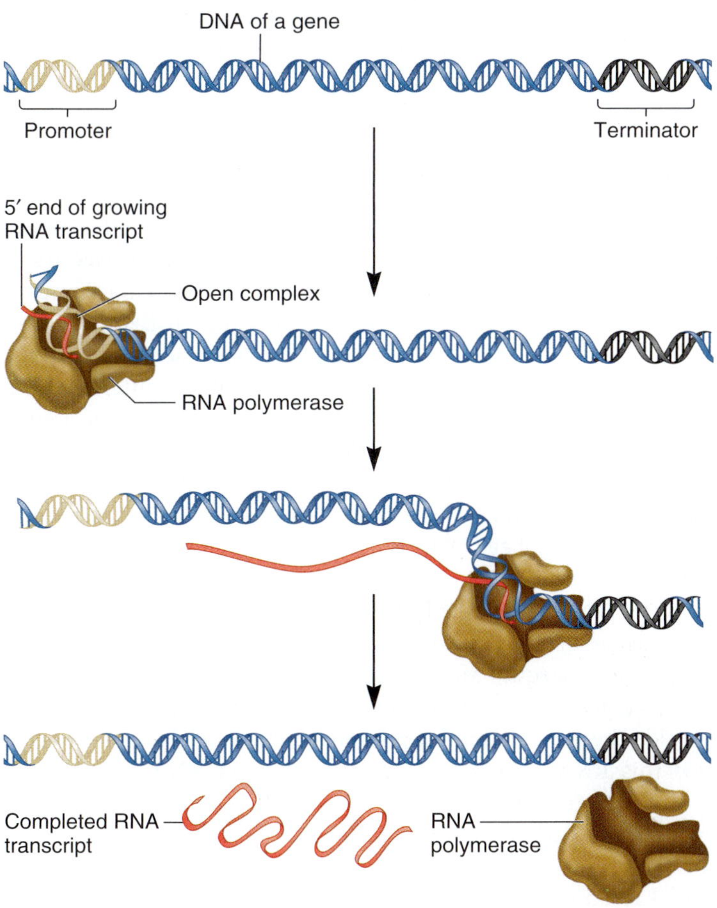

Initiation: The promoter functions as a recognition site for transcription factors (not shown). The transcription factor(s) enables RNA polymerase to bind to the promoter. Following binding, the DNA is denatured into a bubble known as the open complex.

Elongation/synthesis of the RNA transcript: RNA polymerase slides along the DNA in an open complex to synthesize RNA.

Termination: A terminator is reached that causes RNA polymerase and the RNA transcript to dissociate from the DNA.

ONLINE ANIMATION

FIGURE 12.3 Stages of transcription.

Genes→Traits The ability of genes to produce an organism's traits relies on the molecular process of gene expression. Transcription is the first step in gene expression. During transcription, the gene's sequence within the DNA is used as a template to make a complementary copy of RNA. In Chapter 13, we will examine how the sequence in mRNA is translated into a polypeptide. After polypeptides are made within a living cell, they fold into functional proteins that govern an organism's traits.

TABLE 12.1
Functions of RNA Molecules

Type of RNA	Description
mRNA	Messenger RNA (mRNA) encodes the sequence of amino acids within a polypeptide. In bacteria, some mRNAs encode a single polypeptide, and others are polycistronic—a single mRNA encodes two or more polypeptides. In most species of eukaryotes, each mRNA usually encodes a single polypeptide. However, in some species, such as *Caenorhabditis elegans* (a nematode worm), polycistronic mRNAs are relatively common.
tRNA	Transfer RNA (tRNA) is necessary for the translation of mRNA. The structure and function of transfer RNA are outlined in Chapter 13.
rRNA	Ribosomal RNA (rRNA) is necessary for the translation of mRNA. Ribosomes are composed of both rRNAs and protein subunits. The structure and function of ribosomes are examined in Chapter 13.
miRNA	MicroRNAs (miRNAs) are short RNA molecules that are involved in gene regulation in eukaryotes (see Chapter 16).
scRNA	Small cytoplasmic RNA (scRNA) is found in the cytoplasm of bacteria and eukaryotes. In bacteria, scRNA is needed for protein secretion. An example in eukaryotes is 7S RNA, which is necessary in the targeting of proteins to the endoplasmic reticulum. It is a component of a complex known as signal recognition particle (SRP), which is composed of 7S RNA and six different protein subunits.
RNA of RNaseP	RNaseP is a catalyst necessary in the processing of tRNA molecules. The RNA is the catalytic component. RNaseP is composed of a 350- to 410-nucleotide RNA and one protein subunit.
snRNA	Small nuclear RNA (snRNA) is necessary in the splicing of eukaryotic pre-mRNA. snRNAs are components of a spliceosome, which is composed of both snRNAs and protein subunits. The structure and function of spliceosomes are examined later in this chapter.
Telomerase RNA	The enzyme telomerase, which is involved in the replication of eukaryotic telomeres, is composed of an RNA molecule and protein subunits.
snoRNA	Small nucleolar RNA (snoRNA) is necessary in the processing of eukaryotic rRNA transcripts. snoRNAs are also associated with protein subunits. In eukaryotes, snoRNAs are found in the nucleolus, where rRNA processing and ribosome assembly occur.
Viral RNAs	Some types of viruses use RNA as their genome, which is packaged within the viral capsid.

genes, mRNAs are made first, but the final, functional products are polypeptides that are components of proteins. The remaining types of RNAs described in Table 12.1 do not contain the information to make proteins. The RNA transcripts from such genes have various important cellular functions. For these genes, the functional product is the RNA. In some cases, the RNA transcript becomes part of a complex that contains both protein subunits and one or more RNA molecules. Examples of protein-RNA complexes include ribosomes, signal recognition particles, RNaseP, spliceosomes, and telomerase.

12.1 COMPREHENSION QUESTIONS

1. Which of the following base sequences is used during transcription?
   a. Promoter and terminator
   b. Start and stop codons
   c. Ribosome-binding site
   d. Both a and b
2. The three stages of transcription are
   a. initiation, ribosome-binding, and termination.
   b. elongation, ribosome-binding, and termination.
   c. initiation, elongation, and termination.
   d. initiation, regulation, and termination.
3. Which of the following statements is false?
   a. All genes encode polypeptides that function as units within proteins.
   b. Some genes encode RNAs that are never translated.
   c. Most genes encode mRNAs.
   d. Some RNAs function within protein-RNA complexes.

12.2 TRANSCRIPTION IN BACTERIA

Learning Outcomes:

1. Describe the characteristics of a bacterial promoter.
2. Explain how RNA polymerase transcribes a bacterial gene.
3. Compare and contrast two mechanisms for transcriptional termination in bacteria.

Our molecular understanding of gene transcription initially came from studies involving bacteria and bacteriophages. Several early investigations focused on the production of viral RNA after bacteriophage infection. The first suggestion that RNA is derived from the transcription of DNA was made by Elliot Volkin and Lazarus Astrachan in 1956. When the researchers exposed *E. coli* cells to T2 bacteriophage, they observed that the RNA made immediately after infection had a base composition substantially different from the base composition of RNA prior to infection. Furthermore, the base composition after infection was very similar to the base composition in the T2 DNA, except that the RNA contained uracil instead of thymine. These results were consistent with the idea that the bacteriophage DNA is used as a template for the synthesis of bacteriophage RNA.

In 1960, Matthew Meselson and François Jacob found that proteins are synthesized on ribosomes. One year later, Jacob and his colleague Jacques Monod proposed that a certain type of RNA acts as a genetic messenger (from the DNA to the ribosome) to provide the information for protein synthesis. They hypothesized that this RNA, which they called messenger RNA (mRNA), is transcribed from the sequence within DNA and then directs the synthesis of particular polypeptides. In the early 1960s, this proposal was remarkable, considering that it was made before the actual isolation and characterization of the mRNA molecules in vitro. In 1961, the hypothesis was confirmed by Sydney Brenner in collaboration with Jacob and Meselson. They found that when a virus infects a bacterial cell, a virus-specific RNA is made that rapidly associates with preexisting ribosomes in the cell.

Since these pioneering studies, a great deal has been learned about the molecular features of bacterial gene transcription. Much of our knowledge comes from studies of *E. coli*. In this section, we will examine the three steps in the gene transcription process as they occur in bacteria.

A Promoter Is a Short Sequence of DNA That Is Necessary to Initiate Transcription

The type of DNA sequence known as the promoter gets its name from the idea that it "promotes" gene expression. More precisely, this sequence of bases directs the exact location for the initiation of transcription. Most of the promoter region is located just ahead of or upstream from the site where transcription of a gene actually begins. By convention, the bases in a promoter sequence are numbered in relation to the **transcriptional start site** (Figure 12.4). This site is the first base used as a template for transcription and is denoted +1. The bases preceding this site are numbered in a negative direction. No base is numbered zero. Therefore, most of the promoter region is labeled with negative numbers that describe the number of bases preceding the beginning of transcription.

Although the promoter may encompass a region several dozen nucleotides in length, short sequences are particularly critical for promoter recognition. By comparing the sequence of DNA bases within many promoters, researchers have learned that certain sequences of bases are necessary to create a functional promoter. In many promoters found in *E. coli* and similar species, two sequences, which are located at approximately the −35 and −10 site in the promoter region, are particularly important (see Figure 12.4).

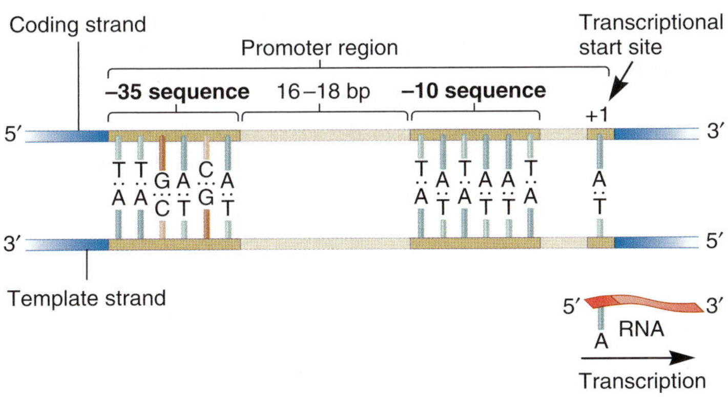

FIGURE 12.4 The conventional numbering system of promoters. The first nucleotide that acts as a template for transcription is designated +1. The numbering of nucleotides to the left of this spot is in a negative direction, whereas the numbering to the right is in a positive direction. For example, the nucleotide that is immediately to the left of the +1 nucleotide is numbered −1, and the nucleotide to the right of the +1 nucleotide is numbered +2. There is no zero nucleotide in this numbering system. In many bacterial promoters, sequence elements at the −35 and −10 regions play a key role in promoting transcription.

The sequence in the top DNA strand at the −35 site is 5′–TTGACA–3′, and the one at the −10 site is 5′–TATAAT–3′. The TATAAT sequence is called the **Pribnow box** after David Pribnow, who initially discovered it in 1975.

Sequences within DNA, such as those found in promoters or regulatory elements, vary among different genes. The most commonly occurring bases within a specific type of sequence is called the **consensus sequence.** As an example, let's consider how sequences may vary at the −35 and −10 sites among different genes. Figure 12.5 illustrates the sequences found in several different *E. coli* promoters. The consensus sequence for this group of promoters is shown at the bottom. This sequence is efficiently recognized by proteins that initiate transcription. For many bacterial genes, a strong correlation is found between the maximal rate of transcription and the degree to which the −35 and −10 regions agree with their consensus sequences.

Bacterial Transcription Is Initiated When RNA Polymerase Holoenzyme Binds at a Promoter Sequence

Thus far, we have considered the DNA sequences that constitute a functional promoter. Let's now turn our attention to the proteins that recognize those sequences and carry out the transcription process. The enzyme that catalyzes the synthesis of RNA is

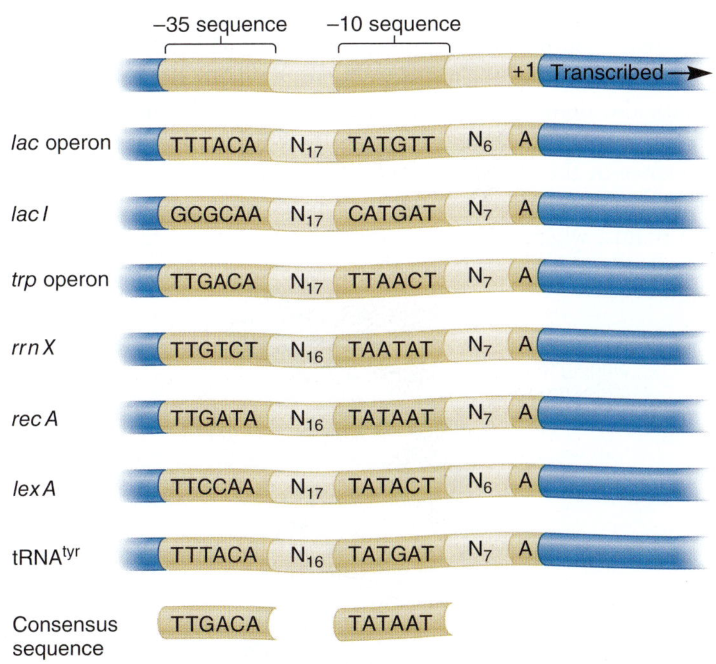

FIGURE 12.5 Examples of −35 and −10 sequences within several *E. coli* promoters. This figure shows the −35 and −10 sequences for one DNA strand found in seven different *E. coli* promoters. The consensus sequence is shown at the bottom. The spacer regions contain the designated number of nucleotides between the −35 and −10 region or between the −10 region and the transcriptional start site. For example, N_{17} means there are 17 nucleotides between the end of the −35 region and the beginning of the −10 region.

CONCEPT CHECK: What does the term *consensus sequence* mean?

RNA polymerase. In *E. coli*, the **core enzyme** is composed of five subunits, $\alpha_2\beta\beta'\omega$. The association of a sixth subunit, **sigma (σ) factor,** with the core enzyme is referred to as RNA polymerase **holoenzyme.** The different subunits within the holoenzyme play distinct functional roles. The two α subunits are important in the proper assembly of the holoenzyme and in the process of binding to DNA. The β and β′ subunits are also needed for binding to the DNA and carry out the catalytic synthesis of RNA. The ω (omega) subunit is important for the proper assembly of the core enzyme. The holoenzyme is required to initiate transcription; the primary role of σ factor is to recognize the promoter. Proteins, such as σ factor, that influence the function of RNA polymerase are types of transcription factors.

After RNA polymerase holoenzyme is assembled into its six subunits, it binds loosely to the DNA and then slides along the DNA, much as a train rolls down the tracks. How is a promoter identified? When the holoenzyme encounters a promoter sequence, σ factor recognizes the bases at both the −35 and −10 regions. The σ-factor protein contains a structure called a **helix-turn-helix motif** that can bind tightly to these regions. Alpha (α) helices within the protein fit into the major groove of the DNA double helix and form hydrogen bonds with the bases. This phenomenon of molecular recognition is shown in **Figure 12.6**. Hydrogen bonding occurs between nucleotides in the −35 and −10 regions of the promoter and amino acid side chains in the helix-turn-helix structure of σ factor.

As shown in **Figure 12.7**, the process of transcription is initiated when σ factor within the holoenzyme has bound to the promoter region to form a **closed complex.** For transcription to begin, the double-stranded DNA must then be unwound into an open complex. This unwinding first occurs at the TATAAT sequence in the −10 region, which contains only AT base pairs, as shown in Figure 12.4. AT base pairs form only two hydrogen bonds, whereas GC pairs form three. Therefore, DNA in an AT-rich region is more easily separated because fewer hydrogen bonds must be broken. A short strand of RNA is made within the open complex, and then σ factor is released from the core enzyme. The release of σ factor marks the transition to the elongation phase of transcription. The core enzyme may now slide down the DNA to synthesize a strand of RNA.

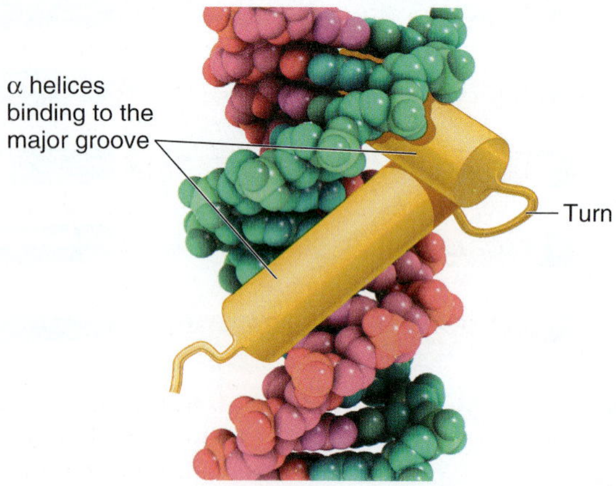

FIGURE 12.6 The binding of σ-factor protein to the promoter region. The σ-factor protein contains two α helices connected by a turn, termed a helix-turn-helix motif. Two α helices of the protein fit within the major groove of the DNA. Amino acids within the α helices form hydrogen bonds with the bases in the DNA. The DNA strands are shown in red and green.

CONCEPT CHECK: Why is it necessary for portions of sigma (σ)-factor protein to fit into the major groove?

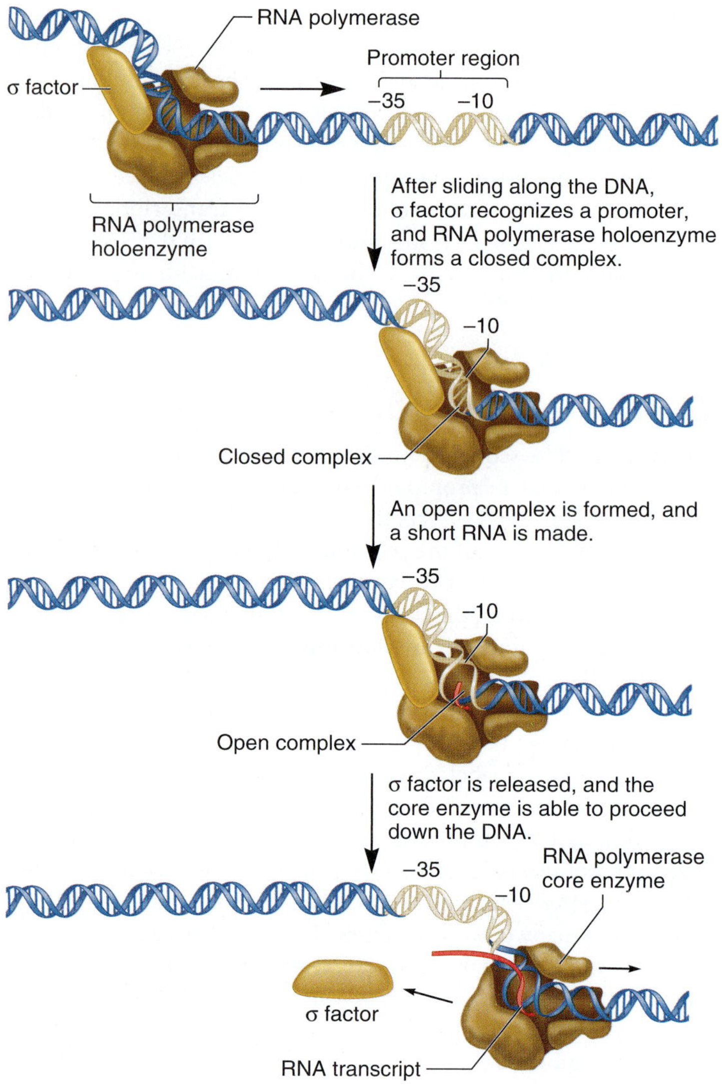

ONLINE ANIMATION

FIGURE 12.7 The initiation stage of transcription in bacteria. The σ-factor subunit of RNA polymerase holoenzyme recognizes the −35 and −10 regions of the promoter. The DNA unwinds in the −10 region to form an open complex, and a short RNA is made. σ factor then dissociates from the holoenzyme, and RNA polymerase core enzyme proceeds down the DNA to synthesize RNA, forming an open complex as it goes.

CONCEPT CHECK: What feature of the −10 sequence makes it easy to unwind?

The RNA Transcript Is Synthesized During the Elongation Stage

After the initiation stage of transcription is completed, the RNA transcript is made during the elongation stage. During the synthesis of the RNA transcript, RNA polymerase moves along the DNA,

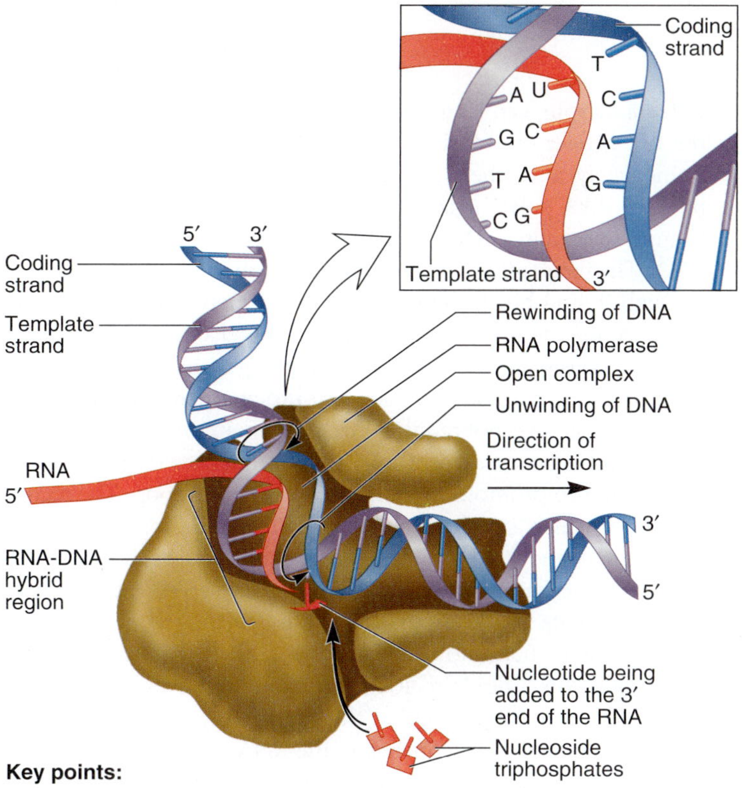

Key points:

- RNA polymerase slides along the DNA, creating an open complex as it moves.
- The DNA strand known as the template strand is used to make a complementary copy of RNA, resulting in an RNA-DNA hybrid.
- RNA polymerase moves along the template strand in a 3′ to 5′ direction, and RNA is synthesized in a 5′ to 3′ direction using nucleoside triphosphates as precursors. Pyrophosphate is released (not shown).
- The complementarity rule is the same as the AT/GC rule except that U is substituted for T in the RNA.

ONLINE ANIMATION

FIGURE 12.8 Synthesis of the RNA transcript.

causing it to unwind (**Figure 12.8**). As previously mentioned, the DNA strand used as a template for RNA synthesis is called the template strand. As it moves along the DNA, the open complex formed by the action of RNA polymerase is approximately 17 bp long. On average, the rate of RNA synthesis is about 43 nucleotides per second! Behind the open complex, the DNA rewinds back into a double helix.

As described in Figure 12.8, the chemistry of transcription by RNA polymerase is similar to the replication of DNA via DNA polymerase, which is discussed in Chapter 11. RNA polymerase always connects nucleotides in the 5′ to 3′ direction. During this process, RNA polymerase catalyzes the formation of a bond between the 5′ PO_4^{2-} group on one nucleotide and the 3′—OH group on the previous nucleotide. The complementarity rule is similar to the AT/GC rule, except that uracil substitutes for thymine in the RNA. In other words, RNA synthesis obeys an $A_{DNA}U_{RNA}$/ $T_{DNA}A_{RNA}$/ $G_{DNA}C_{RNA}$/$C_{DNA}G_{RNA}$ rule.

When considering the transcription of multiple genes within a chromosome, the direction of transcription and the DNA strand used as a template varies among different genes. **Figure 12.9** shows three genes adjacent to each other within a chromosome. Genes *A* and *B* are transcribed from left to right, using the bottom DNA strand as a template. By comparison, gene *C* is transcribed from right to left and uses the top DNA strand as a template. Note that in all three cases, the template strand is read in the 3′ to 5′ direction, and the synthesis of the RNA transcript occurs in a 5′ to 3′ direction. Within a given gene, only the template strand is used for RNA synthesis.

Transcription Is Terminated by Either an RNA-Binding Protein or an Intrinsic Terminator

The end of RNA synthesis is referred to as termination. Prior to termination, the hydrogen bonding between the DNA and RNA within the open complex is of central importance in preventing dissociation of RNA polymerase from the template strand. Termination occurs when this short RNA-DNA hybrid region is forced to separate, thereby releasing RNA polymerase as well as the newly made RNA transcript. In *E. coli*, two different mechanisms for termination have been identified. For certain genes, an RNA-binding protein known as **ρ (rho)** is responsible for terminating transcription, in a mechanism called ρ-dependent termination. For other genes, termination does not require the involvement of the ρ protein—a process referred to as ρ-independent termination.

In **ρ-dependent termination,** the termination process requires two components. First, a site in the DNA, called the *rut* site (or rho utilization site), encodes a sequence in the RNA that acts as a recognition site for the binding of the ρ protein (**Figure 12.10**).

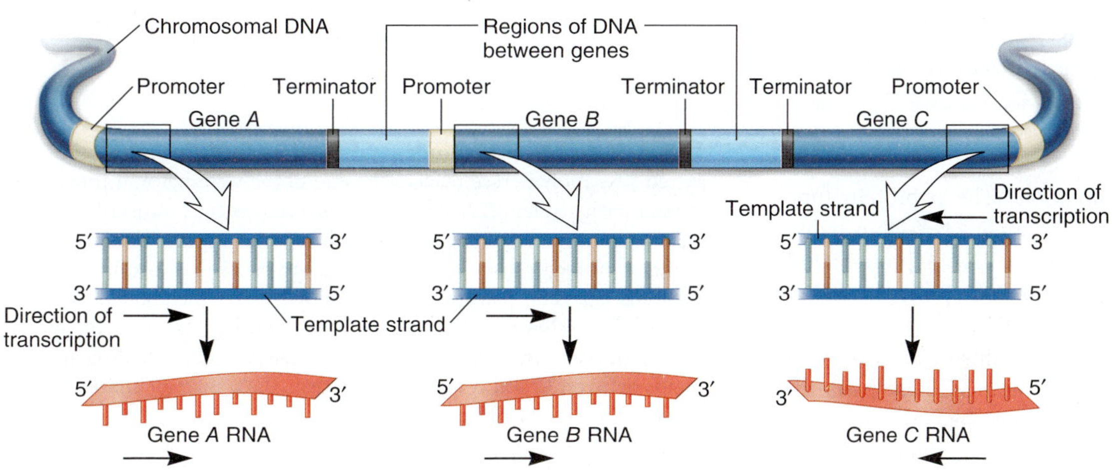

FIGURE 12.9
The transcription of three different genes found in the same chromosome. RNA polymerase synthesizes each RNA transcript in a 5′ to 3′ direction, sliding along a DNA template strand in a 3′ to 5′ direction. However, the use of the template strand varies from gene to gene. For example, genes *A* and *B* use the bottom strand, but gene *C* uses the top strand.

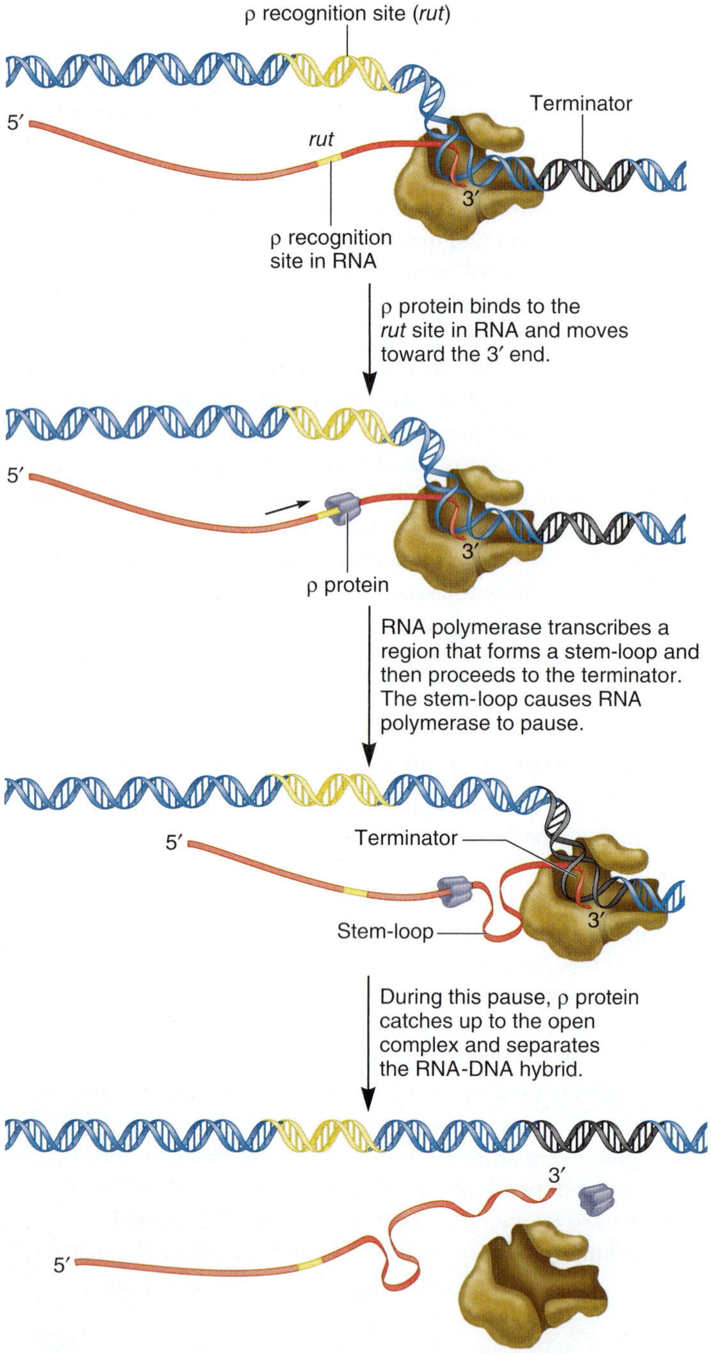

ONLINE ANIMATION

FIGURE 12.10 **ρ-dependent termination.**

CONCEPT CHECK: What would be the consequences if a mutation removed the *rut* site from this RNA molecule?

How does ρ protein facilitate termination? The ρ protein functions as a helicase, an enzyme that can separate RNA-DNA hybrid regions. After the *rut* site is synthesized in the RNA, ρ protein binds to the RNA and moves in the direction of RNA polymerase. The second component of ρ-dependent termination is the site where termination actually takes place. At this terminator site, the DNA encodes an RNA sequence containing several GC base pairs that form a stem-loop. RNA synthesis terminates several nucleotides beyond this stem-loop. As discussed in Chapter 9, a stem-loop, also called a hairpin, can form due to complementary sequences within the RNA (refer back to Figure 9.20). This stem-loop forms almost immediately after the RNA sequence is synthesized and quickly binds to RNA polymerase. This binding results in a conformational change that causes RNA polymerase to pause in its synthesis of RNA. The pause allows ρ protein to catch up to the stem-loop, pass through it, and break the hydrogen bonds between the DNA and RNA within the open complex. When this occurs, the completed RNA strand is separated from the DNA along with RNA polymerase.

Let's now turn our attention to **ρ-independent termination**—a process that does not require the ρ protein. In this case, the terminator involves two adjacent nucleotide sequences (**Figure 12.11**). One sequence promotes the formation of a stem-loop. The second sequence, which is downstream from the stem-loop, is a uracil-rich sequence located at the 3′ end of the RNA. As shown in Figure 12.11, the formation of the stem-loop causes RNA polymerase to pause in its synthesis of RNA. This pausing is stabilized by other proteins that bind to RNA polymerase. For example, a protein called NusA binds to RNA polymerase and promotes pausing at stem-loop sequences. At the precise time RNA polymerase pauses, the uracil-rich sequence in the RNA transcript is bound to the DNA template strand. As previously mentioned, the hydrogen

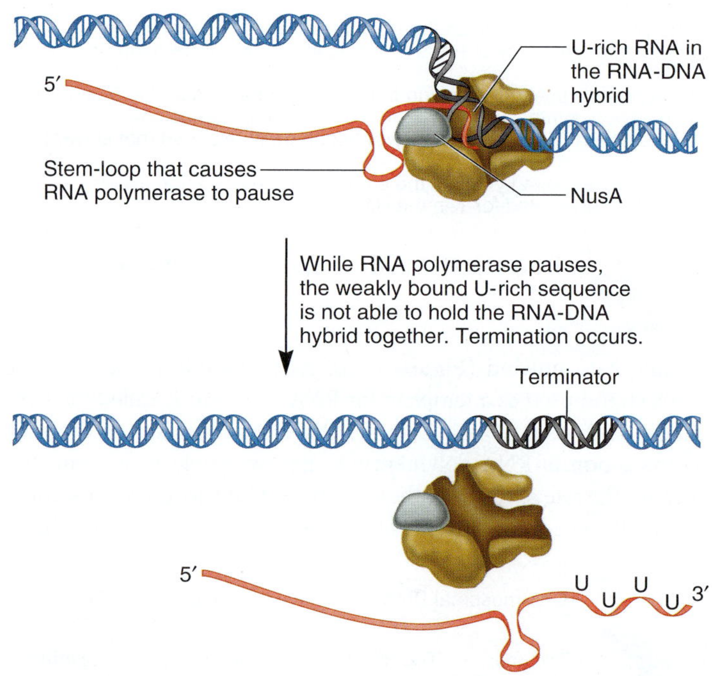

FIGURE 12.11 **ρ-independent or intrinsic termination.** When RNA polymerase reaches the end of the terminator, it transcribes a uracil-rich sequence. As this uracil-rich sequence is transcribed, a stem-loop forms just upstream from the open complex. The formation of this stem-loop causes RNA polymerase to pause in its synthesis of the transcript. This pausing is stabilized by NusA, which binds near the region where RNA exits the open complex. During this pause, the uracil-rich sequence in the RNA is bound to the DNA. Because hydrogen bonds between U and A are relatively weak, the transcript and RNA polymerase dissociate from the DNA.

CONCEPT CHECK: Why is NusA important for this process?

bonding of RNA to DNA keeps RNA polymerase clamped onto the DNA. However, the binding of this uracil-rich sequence to the DNA template strand is relatively weak, causing the RNA transcript to spontaneously dissociate from the DNA and cease further transcription. Because this process does not require the ρ protein to physically remove the RNA transcript from the DNA, it is also referred to as **intrinsic termination.** In *E. coli*, about half of the genes show intrinsic termination, and the other half are terminated by ρ protein.

12.2 COMPREHENSION QUESTIONS

1. With regard to a promoter, a transcriptional start site is
   a. located at the −35 position and recognized by σ factor.
   b. located at the −35 position and is where the first base is used as a template for transcription.
   c. located at the +1 position and recognized by σ factor.
   d. located at the +1 position and is where the first base is used as a template for transcription.
2. For the following five sequences, what is the consensus sequence?

   5′–GGGAGCG–3′
   5′–GAGAGCG–3′
   5′–GAGTGCG–3′
   5′–GAGAACG–3′
   5′–GAGAGCA–3′

   a. 5′–GGGAGCG–3′
   b. 5′–GAGAGCG–3′
   c. 5′–GAGTGCG–3′
   d. 5′–GAGAACG–3′
3. Sigma factor (σ) is needed during which stage(s) of transcription?
   a. Initiation c. Termination
   b. Elongation d. All of the above
4. A uracil-rich sequence occurs at the end of the RNA in
   a. ρ-dependent termination. c. both a and b.
   b. ρ-independent termination. d. none of the above.

12.3 TRANSCRIPTION IN EUKARYOTES

Learning Outcomes:

1. List the functions of the three types of RNA polymerases in eukaryotes.
2. Describe the characteristics of a eukaryotic promoter for a protein-encoding gene.
3. Explain how general transcription factors and RNA polymerase assemble at the promoter and form an open complex.
4. Compare and contrast two possible mechanisms for transcriptional termination in eukaryotes.

Many of the basic features of gene transcription are very similar in bacterial and eukaryotic species. Much of our understanding of transcription has come from studies in *Saccharomyces cerevisiae* (baker's yeast) and other eukaryotic species, including mammals. In general, gene transcription in eukaryotes is more complex than that of their bacterial counterparts. Eukaryotic cells are larger and contain a variety of compartments known as organelles. This added level of cellular complexity dictates that eukaryotes require many more proteins and consequently have many more protein-coding genes. In addition, most eukaryotic species are multicellular, being composed of many different cell types. Multicellularity adds the requirement that genes be transcribed in the correct type of cell and during the proper stage of development. Therefore, in any given species, the transcription of the thousands of different genes that an organism possesses requires appropriate timing and coordination. In this section, we will examine the basic features of gene transcription in eukaryotes. We will focus on the proteins that are needed to make an RNA transcript. In addition, an important factor that affects eukaryotic gene transcription is chromatin structure. Eukaryotic gene transcription requires changes in the positions and structures of nucleosomes. However, because these changes are important for regulating transcription, they are described in Chapter 15 rather than this chapter.

Eukaryotes Have Multiple RNA Polymerases That Are Structurally Similar to the Bacterial Enzyme

The genetic material within the nucleus of a eukaryotic cell is transcribed by three different RNA polymerase enzymes, designated RNA polymerase I, II, and III. What are the roles of these enzymes? Each of the three RNA polymerases transcribes different categories of genes. RNA polymerase I transcribes all of the genes that encode ribosomal RNA (rRNA) except for the 5S rRNA. RNA polymerase II plays a major role in transcription because it transcribes all of the protein-encoding genes. It is responsible for the synthesis of all mRNA and also transcribes certain snRNA genes, which are needed for RNA splicing. RNA polymerase III transcribes all tRNA genes, the 5S rRNA gene, and microRNA genes.

All three RNA polymerases are structurally very similar to each other and are composed of many subunits. They contain two large catalytic subunits similar to the β and β′ subunits of bacterial RNA polymerase. The structures of RNA polymerase from a few different species have been determined by X-ray crystallography. A remarkable similarity also exists between the bacterial enzyme and its eukaryotic counterparts. **Figure 12.12a** compares the structures of a bacterial RNA polymerase with RNA polymerase II from yeast. As seen here, both enzymes have a very similar structure. Also, it is very exciting that this structure provides a way to envision how the transcription process works. As seen in **Figure 12.12b**, DNA enters the enzyme through the jaw and lies on a surface within RNA polymerase termed the bridge. The part of the enzyme called the clamp is thought to control the movement of the DNA through RNA polymerase. A wall in the enzyme forces the RNA-DNA hybrid to make a right-angle turn. This bend facilitates the ability of nucleotides to bind to the template strand. Mg^{2+} is located at the catalytic site, which is precisely at the 3′ end of the growing RNA strand. Nucleoside triphosphates (NTPs) enter the

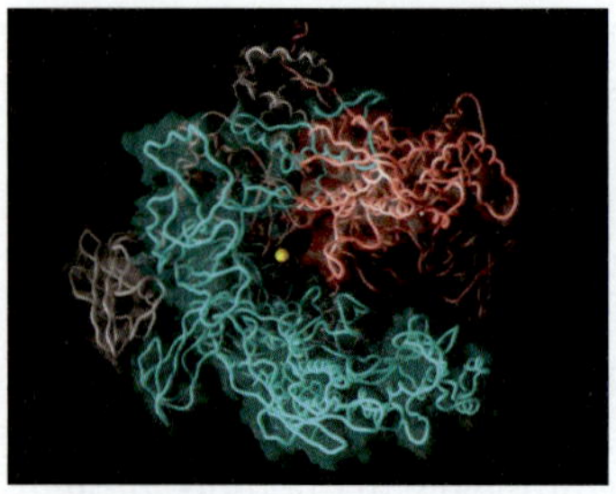

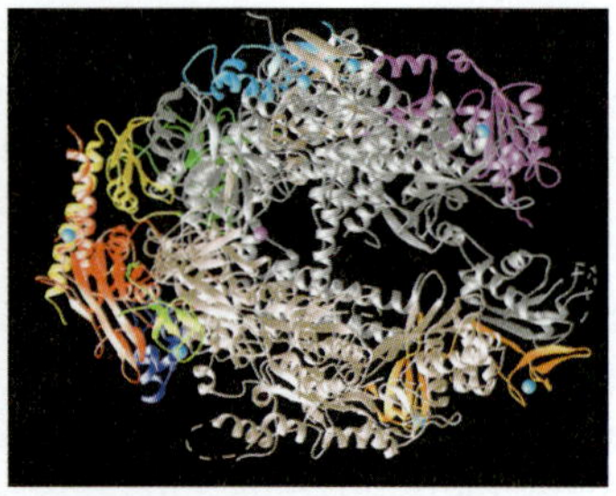

(a) Structure of a bacterial RNA polymerase **Structure of a eukaryotic RNA polymerase II (yeast)**

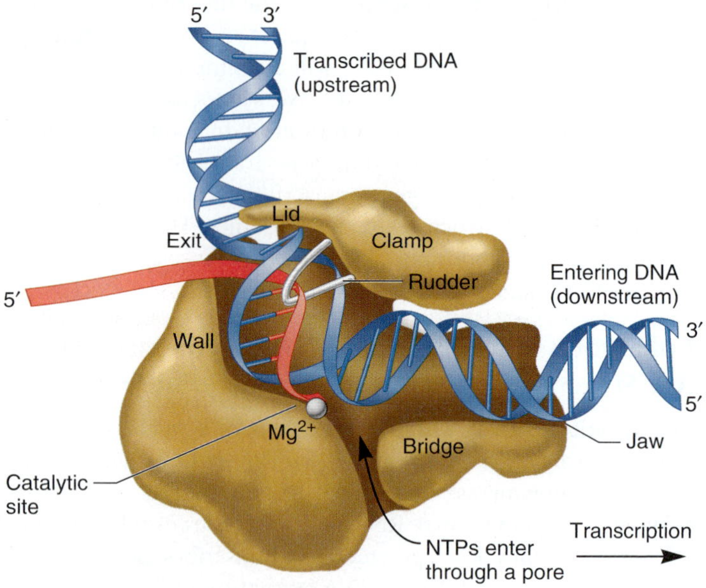

(b) Schematic structure of RNA polymerase

FIGURE 12.12 Structure and molecular function of RNA polymerase. **(a)** A comparison of the crystal structures of a bacterial RNA polymerase (left) to a eukaryotic RNA polymerase II (right). The bacterial enzyme is from *Thermus aquaticus*. The eukaryotic enzyme is from *Saccharomyces cerevisiae.* **(b)** A mechanism for transcription based on the crystal structure. In this diagram, the direction of transcription is from left to right. The double-stranded DNA enters the polymerase along a bridge surface that is between the jaw and clamp. At a region termed the wall, the RNA-DNA hybrid is forced to make a right-angle turn, which enables nucleotides to bind to the template strand. Mg^{2+} is located at the catalytic site. Nucleoside triphosphates (NTPs) enter the catalytic site via a pore region and bind to the template DNA. At the catalytic site, the nucleotides are covalently attached to the 3′ end of the RNA. As RNA polymerase slides down the template, a small region of the protein termed the rudder separates the RNA-DNA hybrid. The DNA and single-stranded RNA then exit under a small lid.

catalytic site via a pore region. The correct nucleotide binds to the template DNA and is covalently attached to the 3′ end. As RNA polymerase slides down the template, a rudder, which is about 9 bp away from the 3′ end of the RNA, forces the RNA-DNA hybrid apart. The DNA and the single-stranded RNA then exit under a small lid.

Eukaryotic Genes Have a Core Promoter and Regulatory Elements

For transcription to occur at a reasonable rate, most eukaryotic genes have two features: a core promoter and regulatory elements. **Figure 12.13** shows a common pattern of sequences found in protein-encoding genes. The **core promoter** is a relatively short DNA sequence that is necessary for transcription to take place. It typically consists of a TATAAA sequence called the **TATA box** and the transcriptional start site, where transcription begins. The TATA box, which is usually about 25 bp upstream from a transcriptional start site, is important in determining the precise starting point for transcription. If it is missing from the core promoter, the transcription start site becomes undefined, and transcription may start at a variety of different locations. The core promoter, by itself, produces a low level of transcription. This is termed **basal transcription.**

Regulatory elements are short DNA sequences that affect the ability of RNA polymerase to recognize the core promoter and begin the process of transcription. These elements are recognized by transcription factors—proteins that bind to regulatory elements and influence the rate of transcription. There are two categories of regulatory elements. Activating sequences, known as **enhancers,** are needed to stimulate transcription. In the absence of enhancer sequences, most eukaryotic genes have very low levels of basal transcription. Under certain conditions, it may also be necessary to prevent transcription of a given gene. This occurs via **silencers**—DNA sequences that are recognized by transcription factors that inhibit transcription. As seen in Figure 12.13, a common location for regulatory elements is the −50 to −100 region. However, the locations of regulatory elements vary considerably among different eukaryotic genes. These elements can be far away from the core promoter yet strongly influence the ability of RNA polymerase to initiate transcription.

DNA sequences such as the TATA box, enhancers, and silencers exert their effects only over a particular gene. They are called ***cis*-acting elements.** The term *cis* comes from chemistry nomenclature meaning "next to." *Cis*-acting elements, though possibly far away from the core promoter, are always found within the same chromosome as the genes they regulate. By comparison, the regulatory transcription factors that bind to such elements are called ***trans*-acting factors** (the term *trans* means "across from"). The transcription factors that control the expression of a gene are themselves encoded by genes; regulatory genes that encode transcription factors may be far away from the genes they control, even on a different chromosome. When a gene encoding a *trans*-acting factor is expressed, the transcription factor protein can diffuse into the cell nucleus and bind to its appropriate *cis*-acting element. Let's now turn our attention to the function of such proteins.

Transcription of Eukaryotic Protein-Encoding Genes Is Initiated When RNA Polymerase II and General Transcription Factors Bind to a Promoter Sequence

Thus far, we have considered the DNA sequences that play a role in the promoter region of eukaryotic protein-encoding genes. By studying transcription in a variety of eukaryotic species, researchers

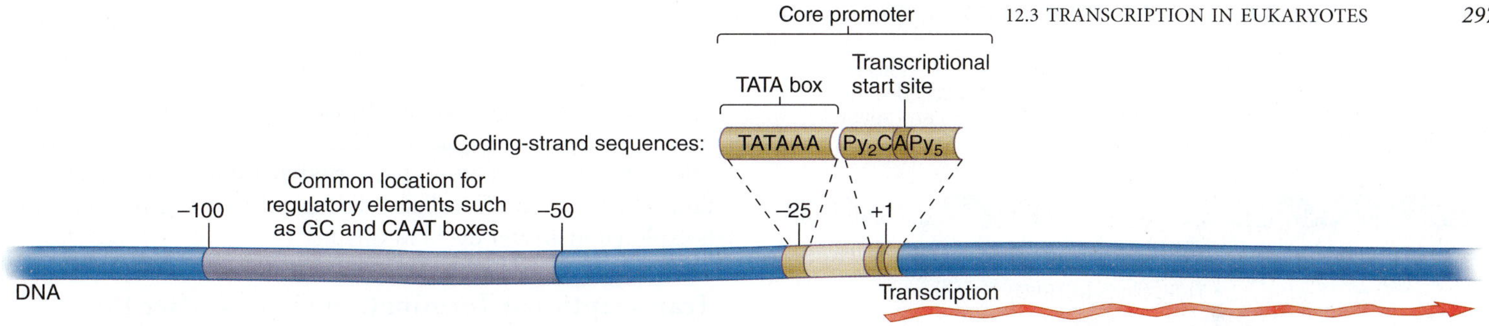

FIGURE 12.13 **A common pattern for the promoter of protein-encoding genes recognized by RNA polymerase II.** The start site usually occurs at adenine (A); two pyrimidines (Py: cytosine or thymine) and a cytosine (C) are to the left of this adenine, and five pyrimidines (Py) are to the right. A TATA box is approximately 25 bp upstream. However, the sequences that constitute eukaryotic promoters are quite diverse, and not all protein-encoding genes have a TATA box. Regulatory elements, such as GC or CAAT boxes, vary in their locations but are often found in the −50 to −100 region. The core promoters for RNA polymerase I and III are quite different. A single upstream regulatory element is involved in the binding of RNA polymerase I to its promoter, whereas two regulatory elements, called A and B boxes, facilitate the binding of RNA polymerase III.

CONCEPT CHECK: What is the functional role of the TATA box?

have discovered that three categories of proteins are needed for basal transcription at the core promoter: RNA polymerase II, general transcription factors, and a complex called mediator (Table 12.2).

Five different proteins called **general transcription factors (GTFs)** are always needed for RNA polymerase II to initiate transcription of protein-encoding genes. Figure 12.14 describes the assembly of GTFs and RNA polymerase II at the TATA box. As shown here, a series of interactions leads to the formation of the open complex. Transcription factor IID (TFIID) first binds to the TATA box and thereby plays a critical role in the recognition of the core promoter. TFIID is composed of several subunits, including TATA-binding protein (TBP), which directly binds to the TATA box, and several other proteins called TBP-associated factors (TAFs). After TFIID binds to the TATA box, it associates with TFIIB. TFIIB promotes the binding of RNA polymerase II and TFIIF. Lastly, TFIIE and TFIIH bind to the complex. This completes the assembly of proteins to form a closed complex, also known as a **preinitiation complex.**

TFIIH plays a major role in the formation of the open complex. TFIIH has several subunits that perform different functions. Certain subunits act as helicases, which break the hydrogen bonding between the double-stranded DNA and are needed to initially form an open complex. Another subunit hydrolyzes ATP and phosphorylates a domain in RNA polymerase II known as the carboxyl terminal domain (CTD). Phosphorylation of the CTD releases the contact between RNA polymerase II and TFIIB. Next, TFIIB, TFIIE, and TFIIH dissociate, and RNA polymerase II is free to proceed to the elongation stage of transcription.

In vitro, when researchers mix together TFIID, TFIIB, TFIIF, TFIIE, TFIIH, RNA polymerase II, and a DNA sequence containing a TATA box and transcriptional start site, the DNA is transcribed into RNA. Therefore, these components are referred to as the **basal transcription apparatus.** In a living cell, however, additional components regulate transcription and allow it to proceed at a reasonable rate.

In addition to GTFs and RNA polymerase II, another component required for transcription is a large protein complex termed mediator. This complex was discovered by Roger Kornberg and colleagues in 1990. In 2006, Kornberg was awarded the Nobel Prize in chemistry for his studies regarding the molecular basis of eukaryotic transcription. **Mediator** derives its name from the observation that it mediates interactions between RNA polymerase II and regulatory transcription factors that bind to enhancers or silencers. It serves as an interface between RNA

TABLE 12.2

Proteins Needed for Transcription via the Core Promoter of Eukaryotic Protein-Encoding Genes

RNA polymerase II: The enzyme that catalyzes the linkage of nucleotides in the 5′ to 3′ direction, using DNA as a template. Essentially all eukaryotic RNA polymerase II proteins are composed of 12 subunits. The two largest subunits are structurally similar to the β and β′ subunits found in *E. coli* RNA polymerase.

General transcription factors:

- **TFIID:** Composed of TATA-binding protein (TBP) and other TBP-associated factors (TAFs). Recognizes the TATA box of eukaryotic protein-encoding gene promoters.
- **TFIIB:** Binds to TFIID and then enables RNA polymerase II to bind to the core promoter. Also promotes TFIIF binding.
- **TFIIF:** Binds to RNA polymerase II and plays a role in its ability to bind to TFIIB and the core promoter. Also plays a role in the ability of TFIIE and TFIIH to bind to RNA polymerase II.
- **TFIIE:** Plays a role in the formation or the maintenance (or both) of the open complex. It may exert its effects by facilitating the binding of TFIIH to RNA polymerase II and regulating the activity of TFIIH.
- **TFIIH:** A multisubunit protein that has multiple roles. First, certain subunits act as helicases and promote the formation of the open complex. Other subunits phosphorylate the carboxyl terminal domain (CTD) of RNA polymerase II, which releases its interaction with TFIIB, thereby allowing RNA polymerase II to proceed to the elongation phase.

Mediator: A multisubunit complex that mediates the effects of regulatory transcription factors on the function of RNA polymerase II. Though mediator typically has certain core subunits, many of its subunits vary, depending on the cell type and environmental conditions. The ability of mediator to affect the function of RNA polymerase II is thought to occur via the CTD of RNA polymerase II. Mediator can influence the ability of TFIIH to phosphorylate CTD, and subunits within mediator itself have the ability to phosphorylate CTD. Because CTD phosphorylation is needed to release RNA polymerase II from TFIIB, mediator plays a key role in the ability of RNA polymerase II to switch from the initiation to the elongation stage of transcription.

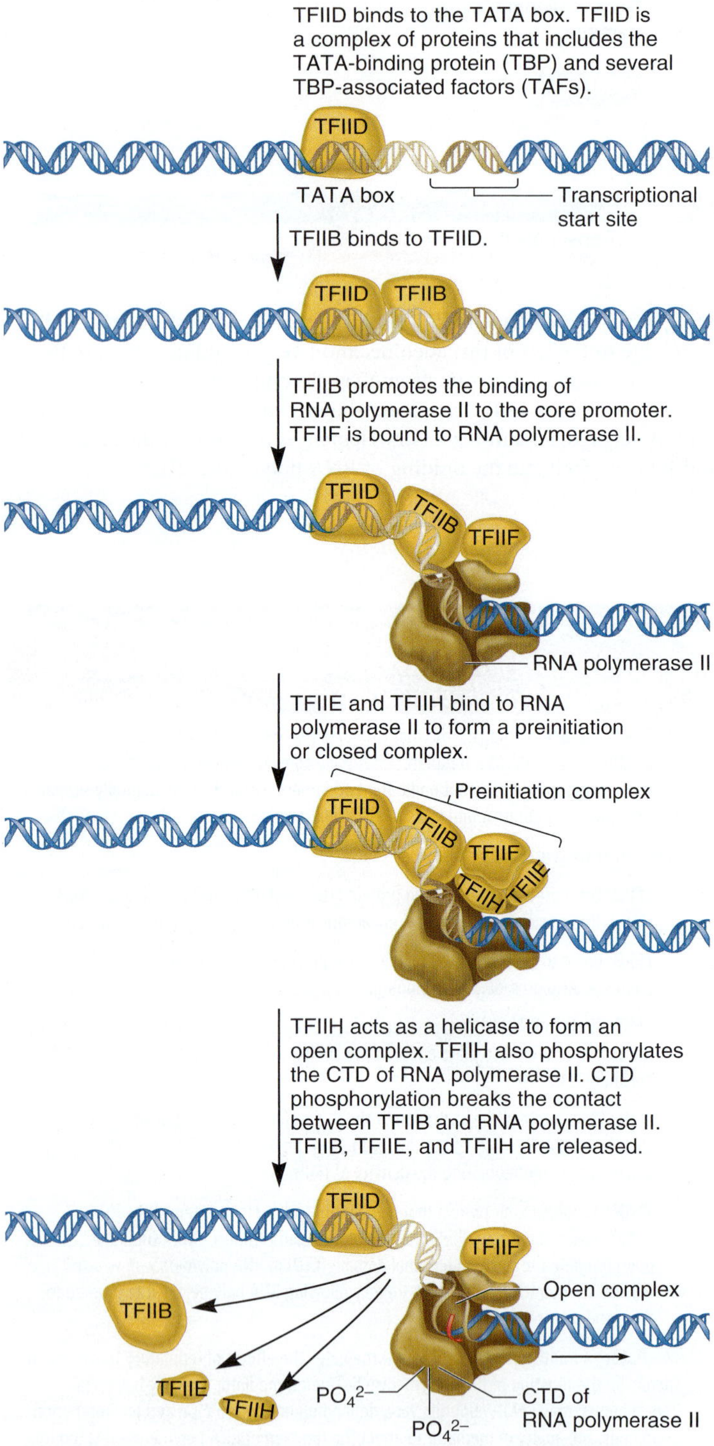

FIGURE 12.14 Steps leading to the formation of the open complex. The DNA is first denatured at the TATA box to form an open complex. In this diagram, the open complex has moved to the transcriptional start site, which is usually about 25 bp away from the TATA box.

CONCEPT CHECK: Why is carboxyl terminal domain (CTD) phosphorylation functionally important?

polymerase II and many diverse regulatory signals. The subunit composition of mediator is quite complex and variable. The core subunits form an elliptically shaped complex that partially wraps around RNA polymerase II. Mediator itself may phosphorylate the CTD of RNA polymerase II, and it may regulate the ability of TFIIH to phosphorylate the CTD. Therefore, it plays a pivotal role in the switch between transcriptional initiation and elongation. The function of mediator during eukaryotic gene regulation is explored in greater detail in Chapter 15.

Transcriptional Termination Occurs After the 3′ End of the Transcript Is Cleaved near the Polyadenylation Signal Sequence

As discussed later in this chapter, eukaryotic mRNAs are modified by cleavage near their 3′ end and the subsequent attachment of a string of adenine nucleotides (look ahead at Figure 12.24). This process, which is called polyadenylation, requires the transcription of a polyadenylation signal sequence that directs the cleavage of the mRNA. Transcription via RNA polymerase II typically terminates about 500–2000 nucleotides downstream from the polyadenylation signal.

Figure 12.15 shows a simplified scheme for the transcriptional termination of a protein-encoding gene. After RNA polymerase II has transcribed the polyadenylation signal sequence, the RNA is cleaved just downstream from this sequence. This cleavage occurs before transcriptional termination. Two models have been proposed for transcriptional termination. According to the allosteric model, RNA polymerase II becomes destabilized after it has transcribed the polyA signal sequence, and it eventually dissociates from the DNA. This destabilization may be caused by the release of proteins that function as elongation factors or by the binding of proteins that function as termination factors. A second model, called the torpedo model, suggests that RNA polymerase II is physically removed from the DNA. According to this model, the region of RNA that is still being transcribed and downstream from the polyadenylation signal sequence is cleaved by an exonuclease that degrades the transcript in the 5′ to 3′ direction. When the exonuclease catches up to RNA polymerase II, this causes RNA polymerase II to dissociate from the DNA.

Which of these two models is correct? Additional research is needed, but the results of studies over the past few years have provided evidence that the two models are not mutually exclusive. Therefore, both mechanisms may play a role in transcriptional termination.

12.3 COMPREHENSION QUESTIONS

1. Which RNA polymerase in eukaryotes is responsible for the transcription of genes that encode proteins?
   a. RNA polymerase I
   b. RNA polymerase II
   c. RNA polymerase III
   d. All of the above
2. An enhancer is a ______________ that ____________ the rate of transcription.
   a. *trans*-acting factor, increases
   b. *trans*-acting factor, decreases
   c. *cis*-acting element, increases
   d. *cis*-acting element, decreases

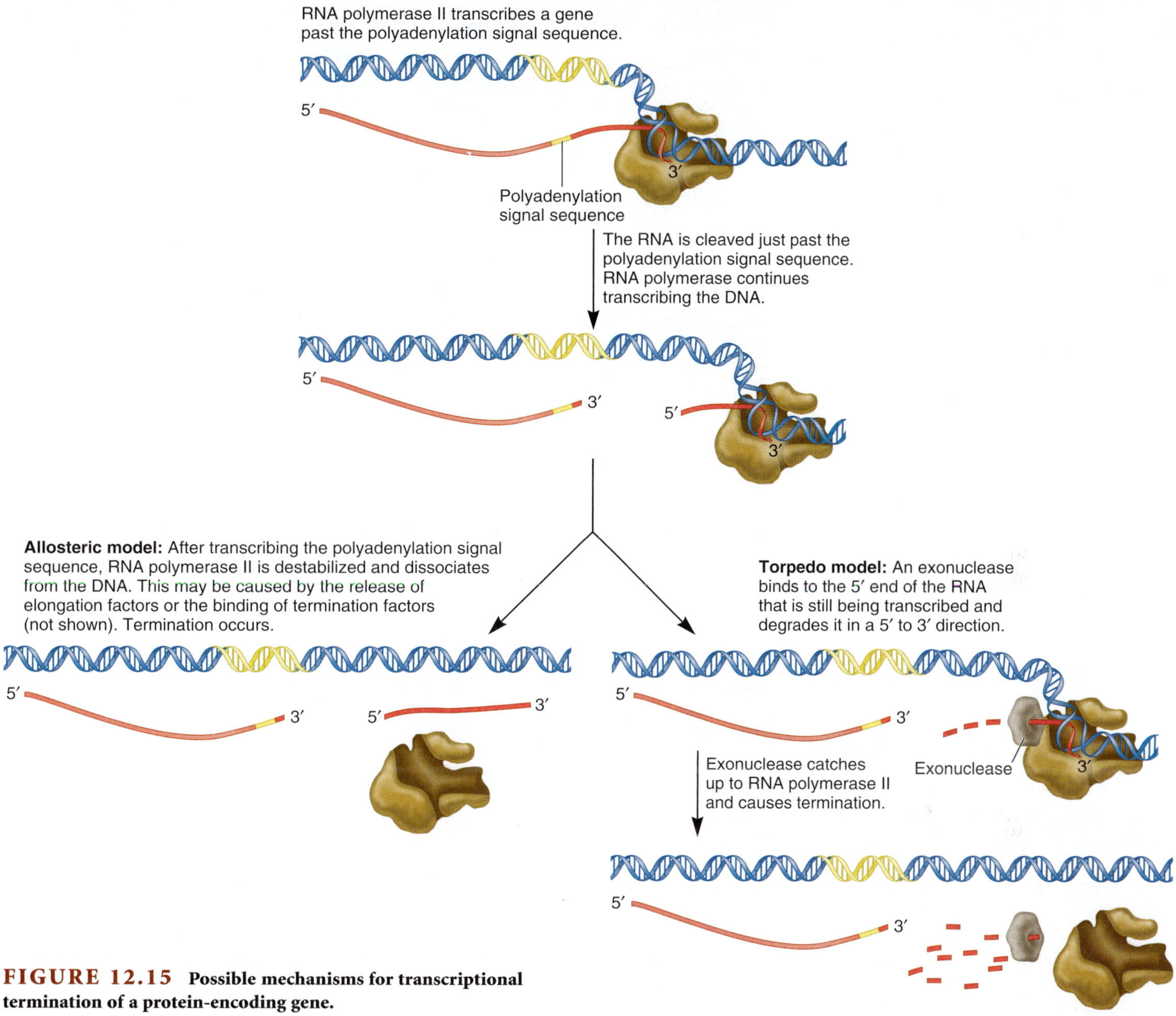

FIGURE 12.15 **Possible mechanisms for transcriptional termination of a protein-encoding gene.**

3. The basal transcription apparatus is composed of
   a. five general transcription factors.
   b. RNA polymerase II.
   c. a DNA sequence containing a TATA box and transcriptional start site.
   d. all of the above.
4. With regard to transcriptional termination in eukaryotes, which model suggests that RNA polymerase is physically removed from the DNA?
   a. Allosteric model
   b. Torpedo model
   c. Both models
   d. Neither model

12.4 RNA MODIFICATION

Learning Outcomes:

1. List the different types of RNA modifications.
2. Describe the processing of ribosomal RNAs and tRNAs.
3. Analyze the microscopy experiment of Leder and explain how it verified the existence of introns.
4. Compare and contrast different mechanisms of RNA splicing.
5. Explain how eukaryotic mRNAs are modified to have a cap and a tail.
6. Describe the process of RNA editing.

TABLE 12.3

Modifications That May Occur to RNAs

Modification	Description	Occurrence
Processing 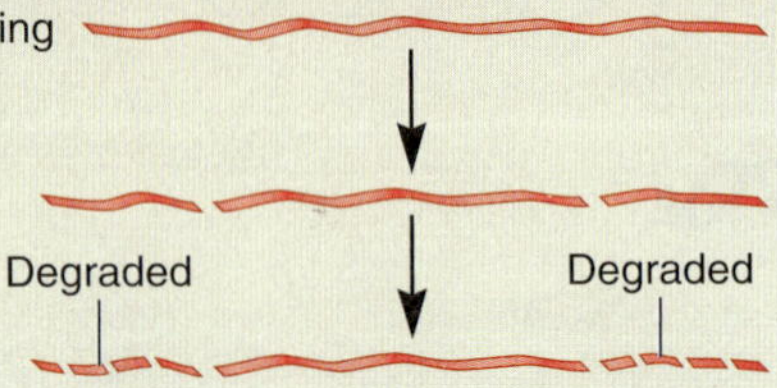	The cleavage of a large RNA transcript into smaller pieces. One or more of the smaller pieces becomes a functional RNA molecule.	Processing occurs in both prokaryotic and eukaryotic rRNAs and tRNAs.
Splicing 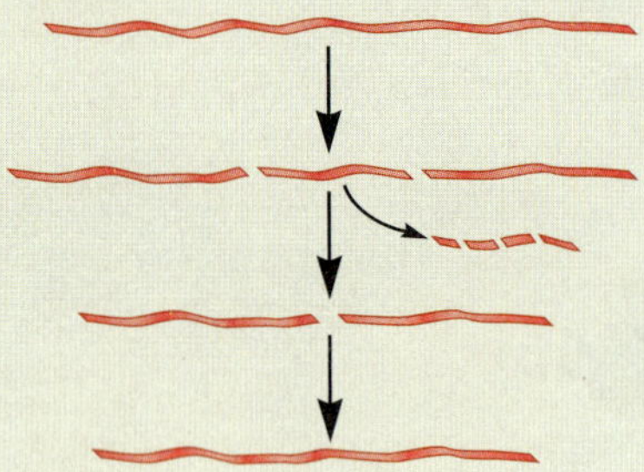	Splicing involves both cleavage and joining of RNA molecules. The RNA is cleaved at two sites, which allows an internal segment of RNA, known as an intron, to be removed. After the intron is removed, the two ends of the RNA molecules are joined together.	Splicing is common among eukaryotic pre-mRNAs, and it also occurs occasionally in rRNAs, tRNAs, and a few bacterial RNAs.
5′ capping 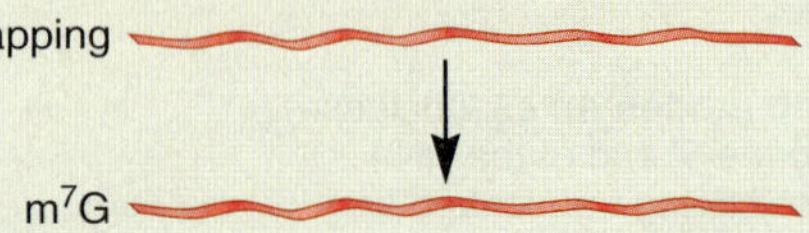	The attachment of a 7-methylguanosine cap (m^7G) to the 5′ end of mRNA. The cap plays a role in the splicing of introns, the exit of mRNA from the nucleus, and the binding of mRNA to the ribosome.	Capping occurs on eukaryotic mRNAs.
3′ polyA tailing 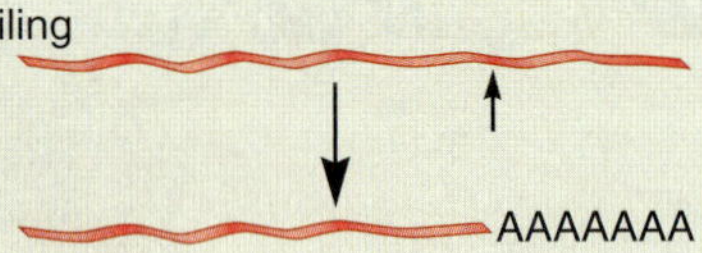	The attachment of a string of adenine-containing nucleotides to the 3′ end of mRNA at a site where the mRNA is cleaved (see upward arrow). It is important for RNA stability and translation in eukaryotes.	Occurs on eukaryotic mRNAs and occasionally occurs on bacterial RNAs.
RNA editing 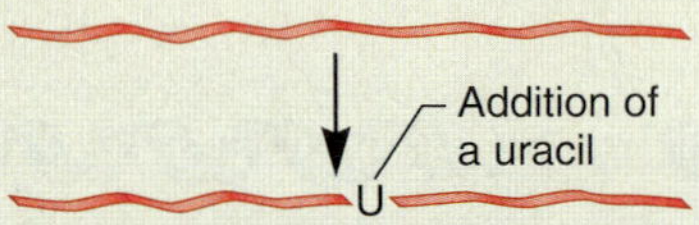	The change of the base sequence of an RNA after it has been transcribed.	Occurs occasionally among eukaryotic RNAs.
Base modification 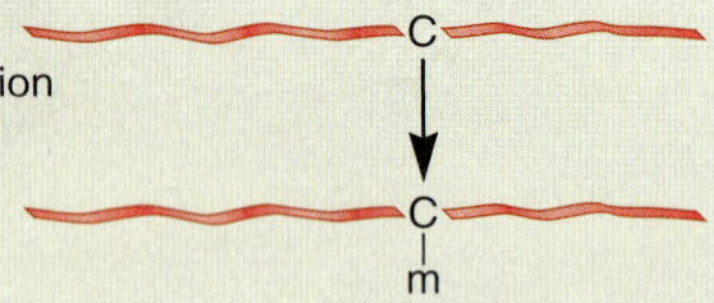	The covalent modification of a base within an RNA molecule.	As discussed in Chapter 13, base modification commonly occurs in tRNA molecules found in both prokaryotes and eukaryotes. C—m indicates methylation of cytosine.

During the 1960s and 1970s, studies in bacteria established the physical structure of the gene. The analysis of bacterial genes showed that the sequence of DNA within the coding strand corresponds to the sequence of nucleotides in the mRNA, except that T is replaced with U. During translation, the sequence of codons in mRNA is then read, providing the instructions for the correct amino acid sequence in a polypeptide. The correspondence between the sequence of codons in the DNA coding strand and the amino acid sequence of the polypeptide has been termed the **colinearity** of gene expression.

The situation dramatically changed in the late 1970s, when the tools became available to study eukaryotic genes at the molecular level. The scientific community was astonished by the discovery that eukaryotic protein-encoding genes are not always colinear with their functional mRNAs. Instead, the coding sequences within many eukaryotic genes are separated by DNA sequences that are not translated into protein. The coding sequences are found within **exons,** which are regions that are contained within functional mRNA. By comparison, the sequences that are found between the exons are called **intervening sequences,** or **introns.** During transcription, a **pre-mRNA** is made corresponding to the entire gene sequence that is transcribed. To become a functional, or mature, mRNA, the sequences in the pre-mRNA that correspond to the introns are removed and the exons are connected, or spliced, together. This process is called **RNA splicing** (Table 12.3). Since the 1970s, research has revealed that splicing is a common genetic phenomenon in eukaryotic species. Splicing occurs occasionally in bacteria as well.

Aside from splicing, research has also shown that RNA transcripts are modified in several other ways. Table 12.3 describes several types of RNA modifications. For example, rRNAs and tRNAs are synthesized as long transcripts that are processed into smaller functional pieces. In addition, most mature mRNAs in eukaryotes have a cap attached to their 5′ end and a tail attached at their 3′ end. In this section, we will examine the molecular mechanisms that account for these types of RNA modifications and consider why they are functionally important.

Some Large RNA Transcripts Are Cleaved into Smaller Functional Transcripts

For some genes, the RNA transcript initially made during gene transcription is processed or cleaved into smaller pieces. As an example, **Figure 12.16** shows the processing of mammalian ribosomal RNA. The ribosomal RNA gene is transcribed by RNA polymerase I to make a long primary transcript, known as 45S rRNA. The term 45S refers to the sedimentation characteristics of this transcript in Svedberg units. Following the synthesis of the 45S rRNA, cleavage occurs at several points to produce three fragments, termed 18S, 5.8S, and 28S rRNA. These functional rRNA molecules play a key role in forming the structure of the ribosome. In eukaryotes, the cleavage of 45S rRNA into smaller rRNAs and the assembly of ribosomal subunits occur in a site within the cell nucleus known as the **nucleolus.**

The production of tRNA molecules requires processing via exonucleases and endonucleases. An **exonuclease** cleaves a covalent bond between two nucleotides at one end of a strand. Starting at one end, an exonuclease digests a strand, one nucleotide at a time. Some exonucleases begin this digestion only from the 3′ end, traveling in the 3′ to 5′ direction, whereas others begin only at the 5′ end and digest in the 5′ to 3′ direction. By comparison, an **endonuclease** cleaves the bond between two adjacent nucleotides within a strand.

Like ribosomal RNA, tRNAs are synthesized as large precursor tRNAs that must be cleaved to produce mature, functional tRNAs that carry amino acids. This processing has been studied extensively in *E. coli*. The details of tRNA processing vary among different tRNAs. **Figure 12.17** shows an example, which involves the action of two endonucleases and one exonuclease. The precursor tRNA is recognized by RNaseP, which is an endonuclease that cuts the precursor tRNA. The action of RNaseP produces the correct 5′ end of the mature tRNA. A different endonuclease cleaves the precursor tRNA to remove a 170-nucleotide segment from the 3′ end. Next, an exonuclease, called RNaseD, binds to the 3′ end and digests the RNA in the 3′ to 5′ direction. All tRNAs have a CCA sequence at their 3′ end. In some tRNAs, the CCA sequence is encoded in the tRNA gene. However, for most tRNAs in prokaryotes and eukaryotes, the CCA sequence is added to the tRNA by an enzyme called tRNA nucleotidyltransferase. Finally, certain bases in tRNA molecules may be covalently modified to alter their structure. The functional importance of modified bases in tRNAs is discussed in Chapter 13.

As researchers studied tRNA processing, they discovered certain features that were very unusual and exciting, changing the way biologists view the actions of catalysts. RNaseP has been found to be a catalyst that contains both RNA and protein subunits. In 1983, Sidney Altman and colleagues made the surprising discovery that the RNA portion of RNaseP, not the protein subunit, contains the catalytic ability to cleave the precursor tRNA. RNaseP is an example of a **ribozyme,** an RNA molecule with catalytic activity. Prior to the study of RNaseP and the identification of self-splicing RNAs (discussed later), biochemists had staunchly believed that only protein molecules could function as biological catalysts.

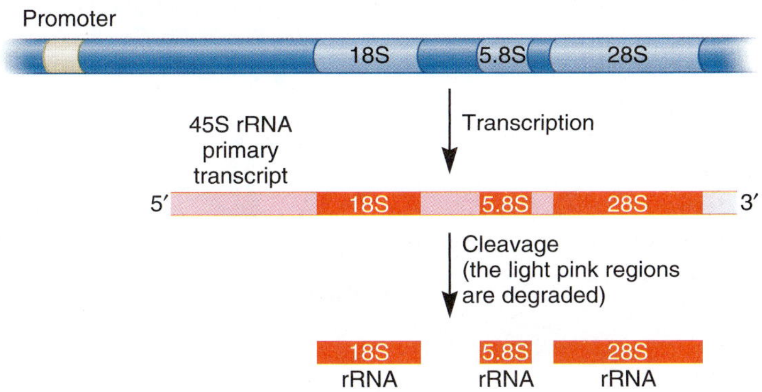

FIGURE 12.16 The processing of ribosomal RNA in eukaryotes. The large ribosomal RNA gene is transcribed into a long 45S rRNA primary transcript. This transcript is cleaved to produce 18S, 5.8S, and 28S rRNA molecules, which become associated with protein subunits in the ribosome. This processing occurs within the nucleolus of the cell.

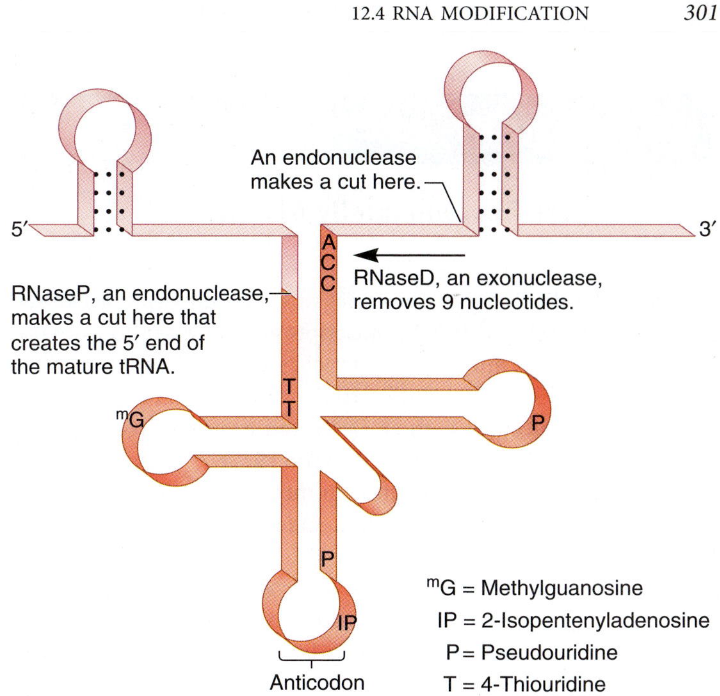

FIGURE 12.17 The processing of a precursor tRNA molecule in *E. coli*. RNaseP is an endonuclease that makes a cut that creates the 5′ end of the mature tRNA. At the 3′ end, another endonuclease makes a cut, which removes a 170-nucleotide segment. Next, the exonuclease RNaseD removes 9 nucleotides at the 3′ end. For most tRNAs, the CCA sequence at the 3′ end is added by tRNA nucleotidyltransferase. In addition to these steps, several bases are modified to other bases as schematically indicated. A similar type of precursor tRNA processing occurs in eukaryotes.

CONCEPT CHECK: What is the difference between an endonuclease and an exonuclease?

EXPERIMENT 12A

Introns Were Experimentally Identified via Microscopy

Although the discovery of ribozymes was very surprising, the observation that rRNA and tRNA transcripts are processed to a smaller form did not seem unusual to geneticists and biochemists, because the cleavage of RNA was similar to the cleavage that can occur for other macromolecules such as DNA and proteins. In sharp contrast, when splicing was detected in the 1970s, it was a novel concept. RNA is cleaved at two sites, an intron is removed, and—in a unique step—the remaining fragments are hooked back together again.

Eukaryotic introns were first detected by comparing the base sequence of viral genes and their mRNA transcripts during viral infection of mammalian cells by adenovirus. This research was carried out in 1977 by two groups headed by Philip Sharp and Richard Roberts. This pioneering observation led to the next question: Are introns a peculiar phenomenon that occurs only in viral genes, or are they found in eukaryotic genes as well?

In the late 1970s, several research groups, including those of Pierre Chambon, Bert O'Malley, and Philip Leder, investigated the presence of introns in eukaryotic protein-encoding genes. The experiments of Leder used electron microscopy to identify introns in the β-globin gene. β globin is a polypeptide that is a subunit of hemoglobin, the protein that carries oxygen in red blood cells. To detect introns within the gene, Leder considered the possible effects of mRNA binding to a gene. **Figure 12.18a** considers the situation in which a gene does not contain an intron. In this experiment, a segment of double-stranded chromosomal DNA containing a gene was first denatured into single-stranded molecules. The single-stranded

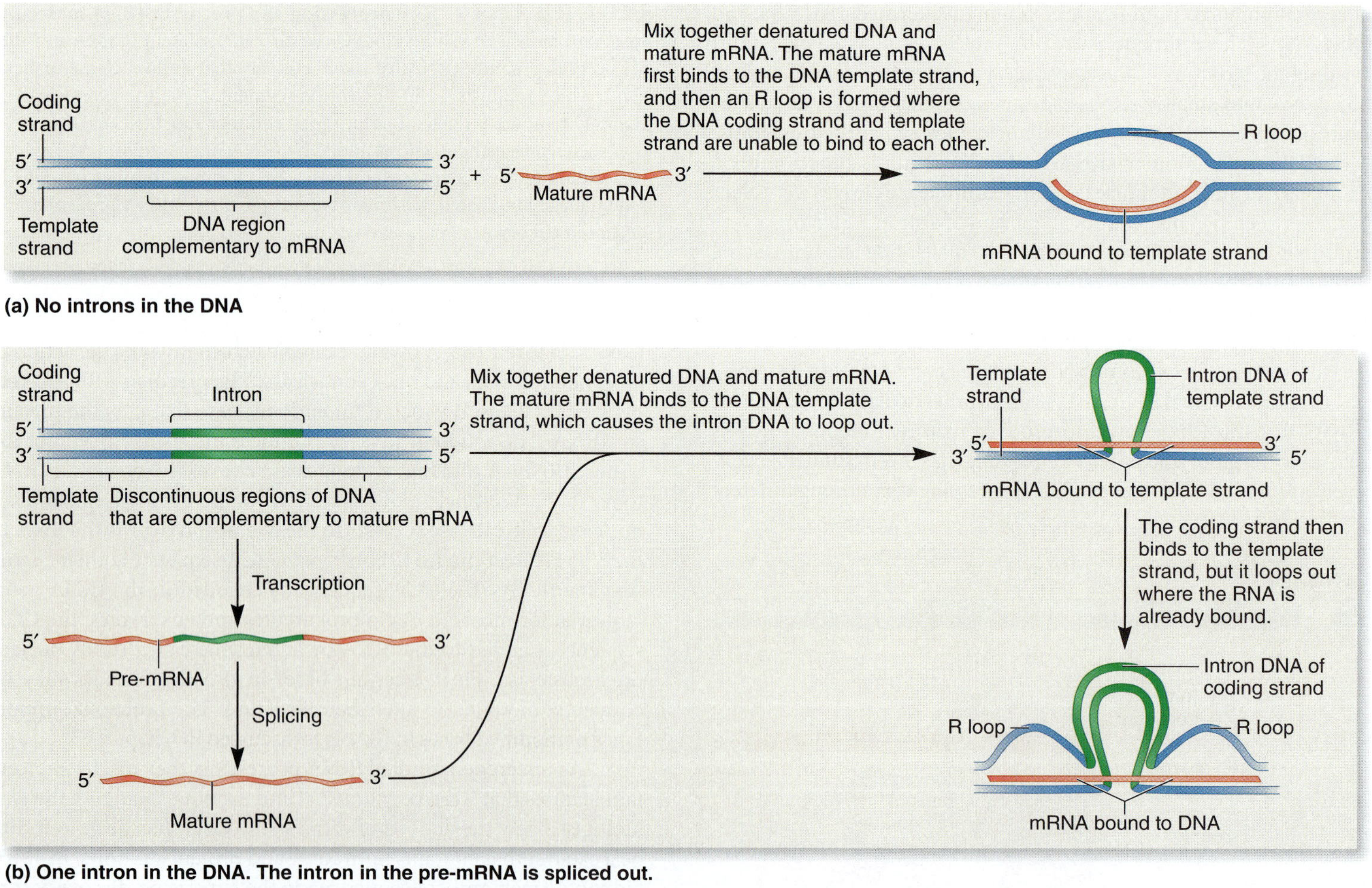

FIGURE 12.18 Hybridization of mRNA to double-stranded DNA. In this experiment, the DNA is denatured and mixed with mRNA under conditions that favor an RNA-DNA hybrid. Later, the conditions are changed, which allows the DNA to form double-stranded regions. **(a)** If the DNA does not contain an intron, the binding of the mRNA to the template strand prevents the two strands of DNA from forming a double helix and results in the formation of one R loop. **(b)** When mRNA hybridizes to a gene containing one intron and then the DNA strands are allowed to bind to each other, two R loops will form that are separated by a double-stranded DNA region. The intervening double-stranded region occurs because an intron has been spliced out of the mRNA and the mRNA cannot hybridize to this segment of the gene.

DNA was then mixed with mature mRNA encoded by that gene. This is done under conditions that favor the formation of an RNA-DNA hybrid rather than the formation of double-stranded DNA. Because the mRNA is complementary to the template strand of the DNA, the template strand and the mRNA can bind to each other to form a hybrid molecule, an event called **hybridization.** Later, the conditions are changed, which allows the DNA strands to bind to each other (see right side of Figure 12.18a). However, if the mRNA is already bound to the template strand of DNA, it prevents the two DNA strands from forming a double helix in that region. In the absence of any introns, the single-stranded DNA forms a loop. Because the RNA has displaced one of the DNA strands, this structure is known as an RNA displacement loop, or **R loop,** as shown in Figure 12.18a.

In contrast, Leder and colleagues realized that a different type of R loop structure would form if a gene contained an intron (**Figure 12.18b**). When mRNA is hybridized to a region of a gene containing one intron and then the other DNA strand is allowed to bind, two single-stranded R loops form that are separated by a double-stranded DNA region. The intervening double-stranded region occurs because an intron has been spliced out of the mature mRNA, so the mRNA cannot hybridize to this segment of the gene.

As shown in steps 1 through 4 of **Figure 12.19**, this hybridization approach was used to identify introns within the β-globin gene. Following hybridization, the samples were placed on a microscopy grid, shadowed with heavy metal, and then observed by electron microscopy.

THE HYPOTHESIS

The mouse β-globin gene contains one or more introns.

TESTING THE HYPOTHESIS

FIGURE 12.19 **RNA hybridization to the β-globin gene reveals an intron.**

Starting material: A cloned fragment of chromosomal DNA that contains the mouse β-globin gene.

1. Isolate mature mRNA for the mouse β-globin gene. Note: β-globin mRNA is abundant in reticulocytes, which are immature red blood cells.

2. Mix together the β-globin mRNA and cloned DNA of the β-globin gene.

3. Separate the double-stranded DNA and allow the mRNA to hybridize. This is done using 70% formamide, at 52°C, for 16 hours.

4. Dilute the sample to decrease the formamide concentration. This allows the DNA to re-form a double-stranded structure. Note: The DNA cannot form a double-stranded structure in regions where the mRNA has already hybridized.

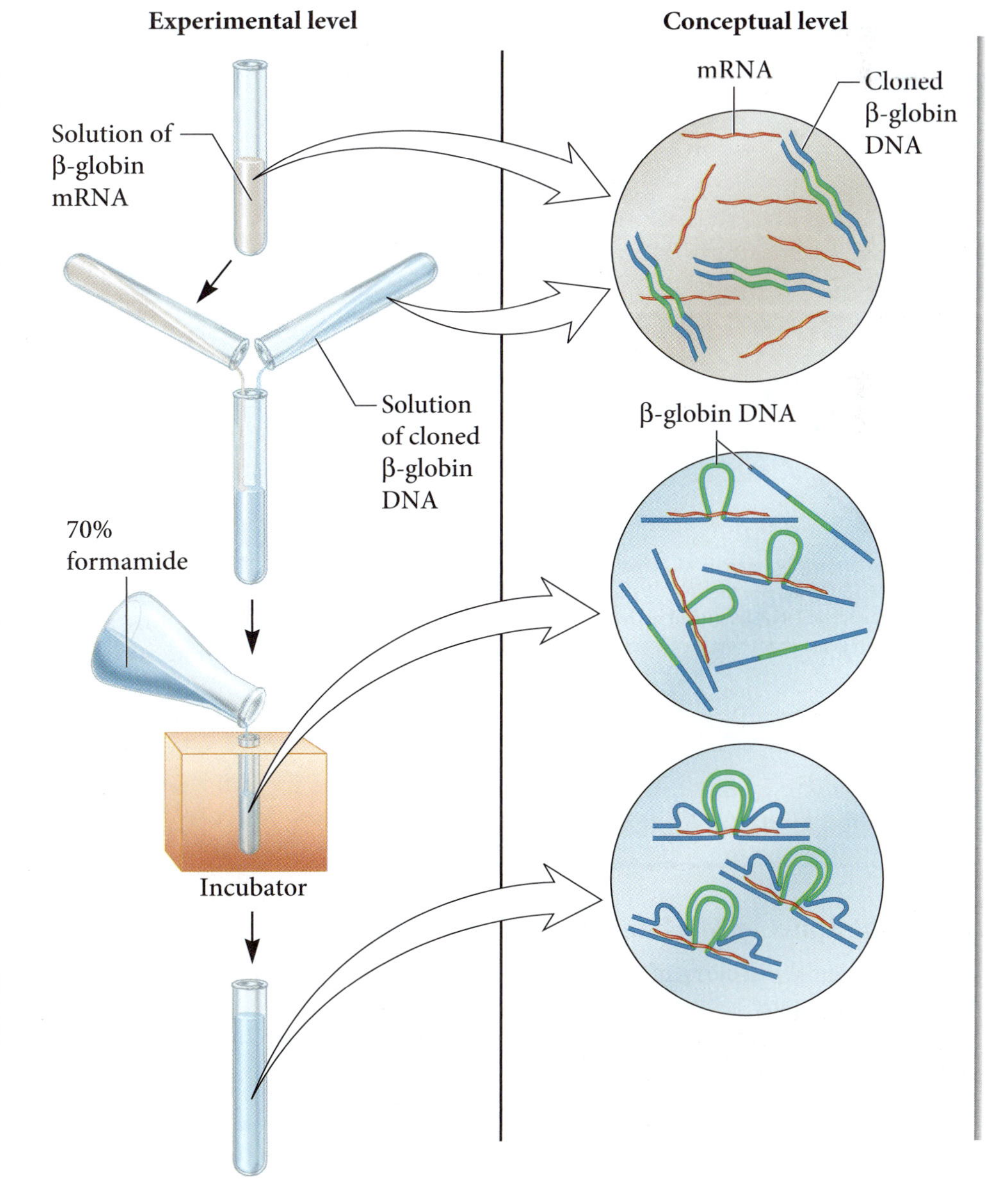

5. Spread the sample onto a microscopy grid.

6. Stain with uranyl acetate and shadow with heavy metal. Note: The technique of electron microscopy is described in the Appendix.

7. View the sample under the electron microscope.

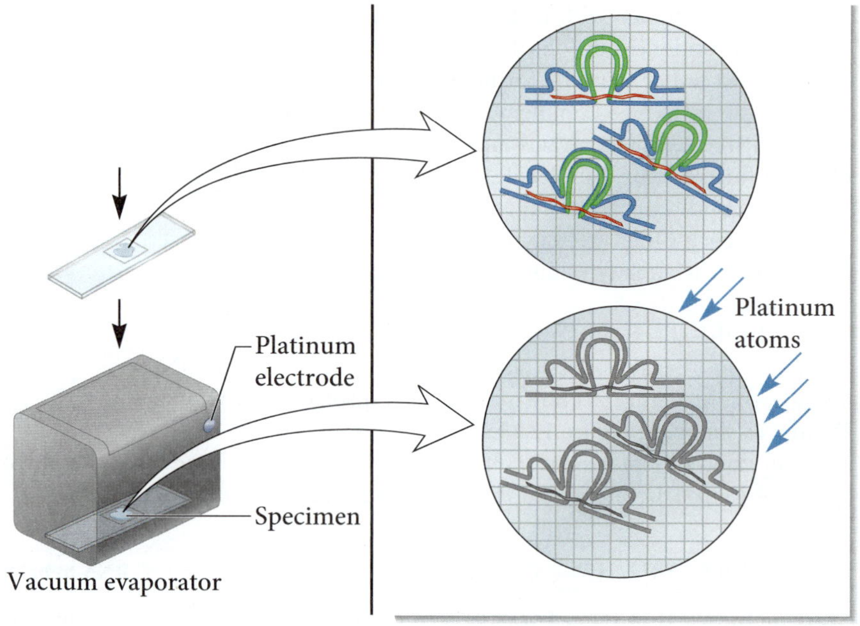

THE DATA

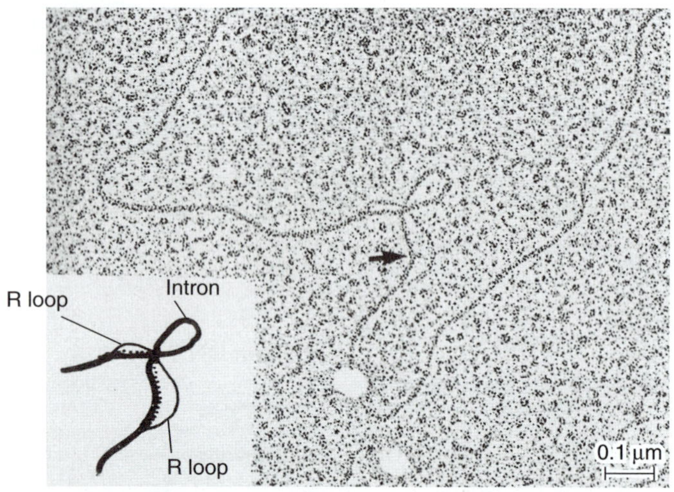

*Data from: Tilghman, S. M., Tiemeier, D. C., Seidman, J. G., Peterlin, B. M., Sullivan, M., Maizel, J. V., and Leder, P. (1978) Intervening sequence of DNA identified in the structural portion of a mouse beta-globin gene. *Proc. Natl. Acad. Sci. USA* 75:725–729.

INTERPRETING THE DATA

As seen in the electron micrograph, the β-globin mRNA hybridized to the DNA of the β-globin gene, which resulted in the formation of two R loops separated by a double-stranded DNA region. These data were consistent with the idea that the DNA of the β-globin gene contains an intron. Similar results were obtained by Chambon and O'Malley for other protein-encoding genes. Since these initial discoveries, introns have been found in many eukaryotic genes. The prevalence and biological significance of introns are discussed later in this chapter.

Since the late 1970s, DNA sequencing methods have permitted an easier and more precise way of detecting introns. Researchers can clone a fragment of chromosomal DNA that contains a particular gene. This is called a **genomic clone.** In addition, mRNA can be used as a starting material to make a copy of DNA known as **complementary DNA (cDNA).** The cDNA does not contain introns, because it is made from mRNA that had its introns removed via splicing. In contrast, if a gene contains introns, a genomic clone of that gene also contains introns. Therefore, a comparison of the DNA sequences of a genomic clone and a cDNA clone can provide direct evidence that a particular gene contains introns. Compared with genomic DNA, the cDNA is missing base sequences that were removed during splicing.

A self-help quiz involving this experiment can be found at www.mhhe.com/brookergenetics5e.

Different Splicing Mechanisms Remove Introns

Since the original discovery of introns, the investigations of many research groups have shown that most protein-encoding genes in complex eukaryotes contain one or more introns. Less commonly, introns can occur within tRNA and rRNA genes. At the molecular level, different RNA splicing mechanisms have been identified. In the three examples shown in **Figure 12.20**, splicing leads to the removal of the intron RNA and the covalent connection of the exon RNA.

The splicing among **group I** and **group II introns** occurs via **self-splicing**—splicing that does not require the aid of other catalysts. Instead, the RNA functions as its own ribozyme. Group I and II differ in the ways that the introns are removed and the exons are connected. Group I introns that occur within the rRNA of *Tetrahymena* (a protist) have been studied extensively by Thomas Cech and colleagues. In this organism, the splicing process involves the binding of a single guanosine to a guanosine-binding

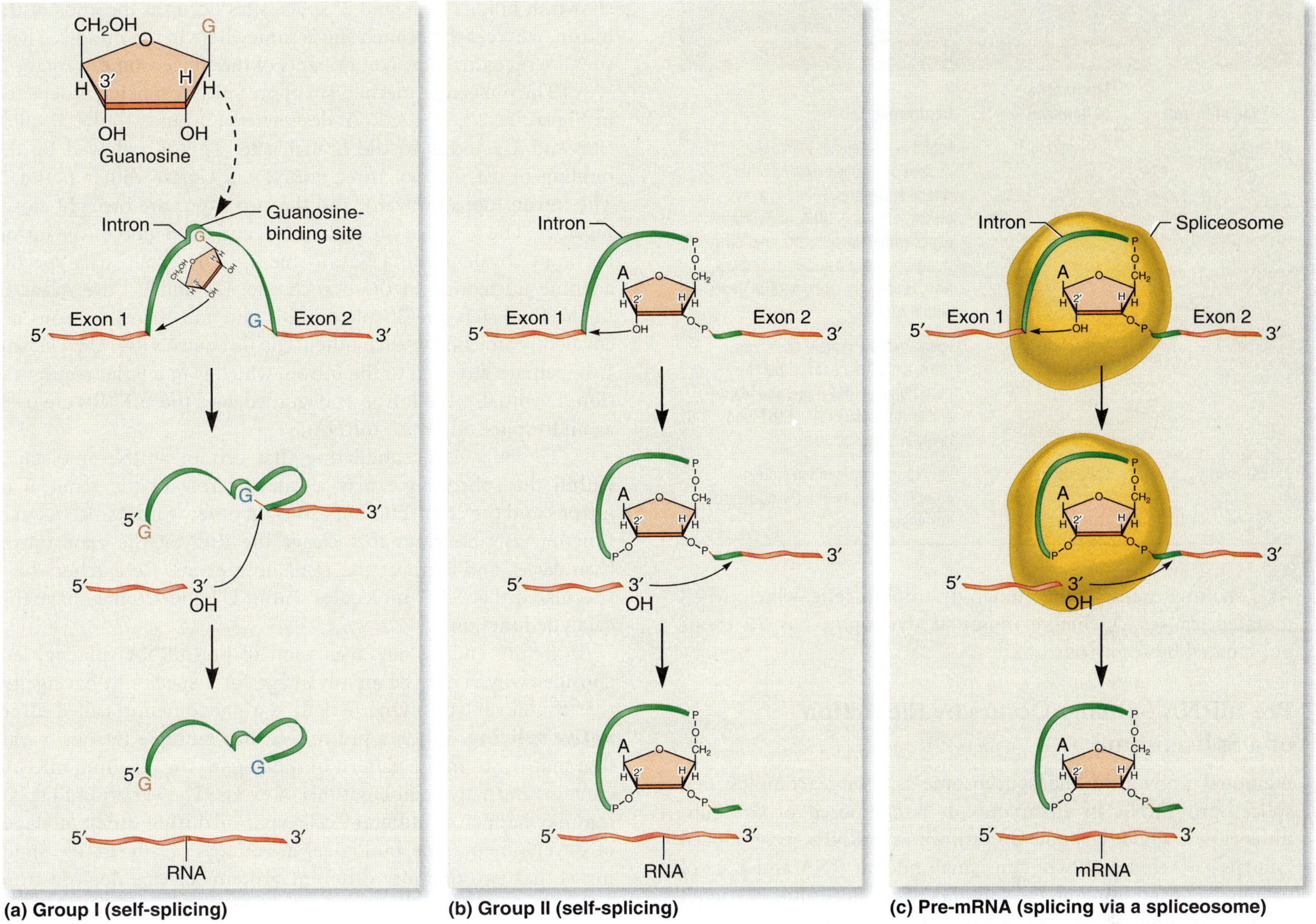

FIGURE 12.20 Mechanisms of RNA splicing. Group I and II introns are self-splicing. **(a)** The splicing of group I introns involves the binding of a free guanosine to a site within the intron, leading to the cleavage of RNA at the 3′ end of exon 1. The bond between a different nucleotide in the intron strand (in this case, guanine) and the 5′ end of exon 2 is cleaved. The 3′ end of exon 1 then forms a covalent bond with the 5′ end of exon 2. **(b)** In group II introns, a similar self-splicing mechanism occurs, except that the 2′—OH group on an adenine nucleotide (already within the intron) begins the catalytic process. **(c)** Pre-mRNA splicing requires the aid of a multicomponent structure known as the spliceosome.

CONCEPT CHECK: Which of these three mechanisms is very common in eukaryotes?

site within the intron (Figure 12.20a). This guanosine breaks the bond between the first exon and the intron. The guanosine becomes attached to the 5′ end of the intron. The 3′—OH group of exon 1 then breaks the bond next to a different nucleotide (in this example, a G) that lies at the boundary between the end of the intron and exon 2; exon 1 forms a covalent bond with the 5′ end of exon 2. The intron RNA is subsequently degraded. In this example, the RNA molecule functions as its own ribozyme, because it splices itself without the aid of a catalytic protein.

In group II introns, a similar splicing mechanism occurs, except the 2′—OH group on ribose found in an adenine nucleotide already within the intron strand begins the catalytic process (Figure 12.20b). Experimentally, group I and II self-splicing can occur in vitro without the addition of any proteins. However, in a living cell, proteins known as **maturases** often enhance the rate of splicing of group I and II introns.

In eukaryotes, the transcription of protein-encoding genes produces a long transcript known as **pre-mRNA,** which is made in the nucleus. This pre-mRNA is usually altered by splicing and other modifications before it exits the nucleus. Unlike group I and II introns, which may undergo self-splicing, pre-mRNA splicing requires the aid of a multicomponent structure known as the **spliceosome.** As discussed shortly, this is needed to recognize the intron boundaries and to properly remove it.

Table 12.4 describes the occurrence of introns among the genes of different species. The splicing of group I and II introns is relatively uncommon. By comparison, pre-mRNA splicing is a widespread phenomenon among complex eukaryotes. In mammals and flowering plants, most protein-encoding genes have at least one intron that can be located anywhere within the gene. An average human gene has about eight introns. In some cases, a single gene can have a great number of introns.

TABLE 12.4
Occurrence of Introns

Type of Intron	Mechanism of Removal	Occurrence
Group I	Self-splicing	Found in rRNA genes within the nucleus of *Tetrahymena* and other simple eukaryotes. Found in a few protein-encoding, tRNA, and rRNA genes within the mitochondrial DNA (fungi and plants) and in chloroplast DNA. Found very rarely in tRNA genes within bacteria.
Group II	Self-splicing	Found in a few protein-encoding, tRNA, and rRNA genes within the mitochondrial DNA (fungi and plants) and in chloroplast DNA. Also found rarely in bacterial genes.
Pre-mRNA	Spliceosome	Very commonly found in protein-encoding genes within the nucleus of eukaryotes.

As a striking example, the human dystrophin gene, which, when mutated, causes Duchenne muscular dystrophy, has 79 exons punctuated by 78 introns.

Pre-mRNA Splicing Occurs by the Action of a Spliceosome

As noted previously, the spliceosome is a large complex that splices pre-mRNA in eukaryotes. It is composed of five subunits (U1, U2, U4, U5, and U6) known as **snRNPs** (pronounced "snurps"). Each snRNP contains small nuclear RNA and a set of proteins. During splicing, the subunits of a spliceosome carry out several functions. First, spliceosome subunits bind to an intron sequence and precisely recognize the intron-exon boundaries. In addition, the spliceosome must hold the pre-mRNA in the correct configuration to ensure the splicing together of the exons. And finally, the spliceosome catalyzes the chemical reactions that cause the introns to be removed and the exons to be covalently linked.

Intron RNA is defined by particular sequences within the intron and at the intron-exon boundaries. The consensus sequences for the splicing of mammalian pre-mRNA are shown in **Figure 12.21**. These sequences serve as recognition sites for the binding of the spliceosome. The bases most commonly found at these sites—those that are highly conserved evolutionarily—are shown in bold. The 5′ and 3′ splice sites occur at the ends of the intron, whereas the branch site is somewhere in the middle. These sites are recognized by components of the spliceosome.

The molecular mechanism of pre-mRNA splicing is depicted in **Figure 12.22**. The snRNP designated U1 binds to the 5′ splice site, and U2 binds to the branch site. This is followed by the binding of a trimer of three snRNPs: a U4/U6 dimer plus U5. The intron loops outward, and the two exons are brought closer together. The 5′ splice site is then cut, and the 5′ end of the intron becomes covalently attached to the 2′—OH group of a specific adenine nucleotide in the branch site. U1 and U4 are released. In the final step, the 3′ splice site is cut, and then the exons are covalently attached to each other. The three snRNPs—U2, U5, and U6—remain attached to the intron, which is in a lariat configuration. Eventually, the intron is degraded, and the snRNPs are used again to splice other pre-mRNAs.

Evidence is accumulating that certain snRNA molecules within the spliceosome play a catalytic role in the removal of introns and the connection of exons. In other words, snRNAs may function as ribozymes that cleave the RNA at the exon-intron boundaries and connect the remaining exons. Researchers have speculated that RNA molecules within U2 and U6 may have this catalytic function.

Before ending our discussion of pre-mRNA splicing, let's consider why it may be an advantage for a species to have genes that contain introns. One benefit is a phenomenon called **alternative splicing.** When a pre-mRNA has multiple introns, variation may occur in the pattern of splicing, so the resulting mRNAs contain alternative combinations of exons. The variation in splicing may happen in different cell types or during different stages of development. The biological advantage of alternative splicing is that two or more different proteins can be derived from a single gene. This allows an organism to carry fewer genes in its genome. The molecular mechanism of alternative splicing is examined in Chapter 16. It involves the actions of proteins (not shown in Figure 12.22) that influence whether or not U1 and U2 can begin the splicing process.

The Ends of Eukaryotic Pre-mRNAs Have a 5′ Cap and a 3′ Tail

In addition to splicing, pre-mRNAs in eukaryotes are also subjected to modifications at their 5′ and 3′ ends. At their 5′ end, most mature mRNAs have a 7-methylguanosine covalently attached—an event known as **capping.** Capping occurs while the pre-mRNA is being

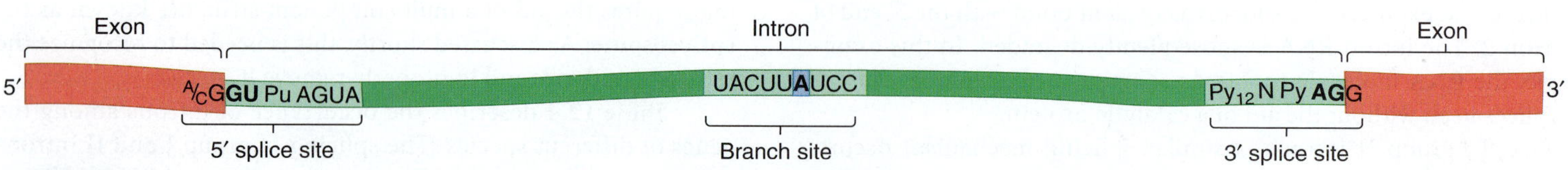

ONLINE ANIMATION

FIGURE 12.21 Consensus sequences for pre-mRNA splicing in complex eukaryotes. Consensus sequences exist at the intron-exon boundaries and at a branch site found within the intron itself. The adenine nucleotide shown in blue in this figure corresponds to the adenine nucleotide at the branch site in Figure 12.22. The nucleotides shown in bold are highly conserved. Designations: A/C = A or C, Pu = purine, Py = pyrimidine, N = any of the four bases.

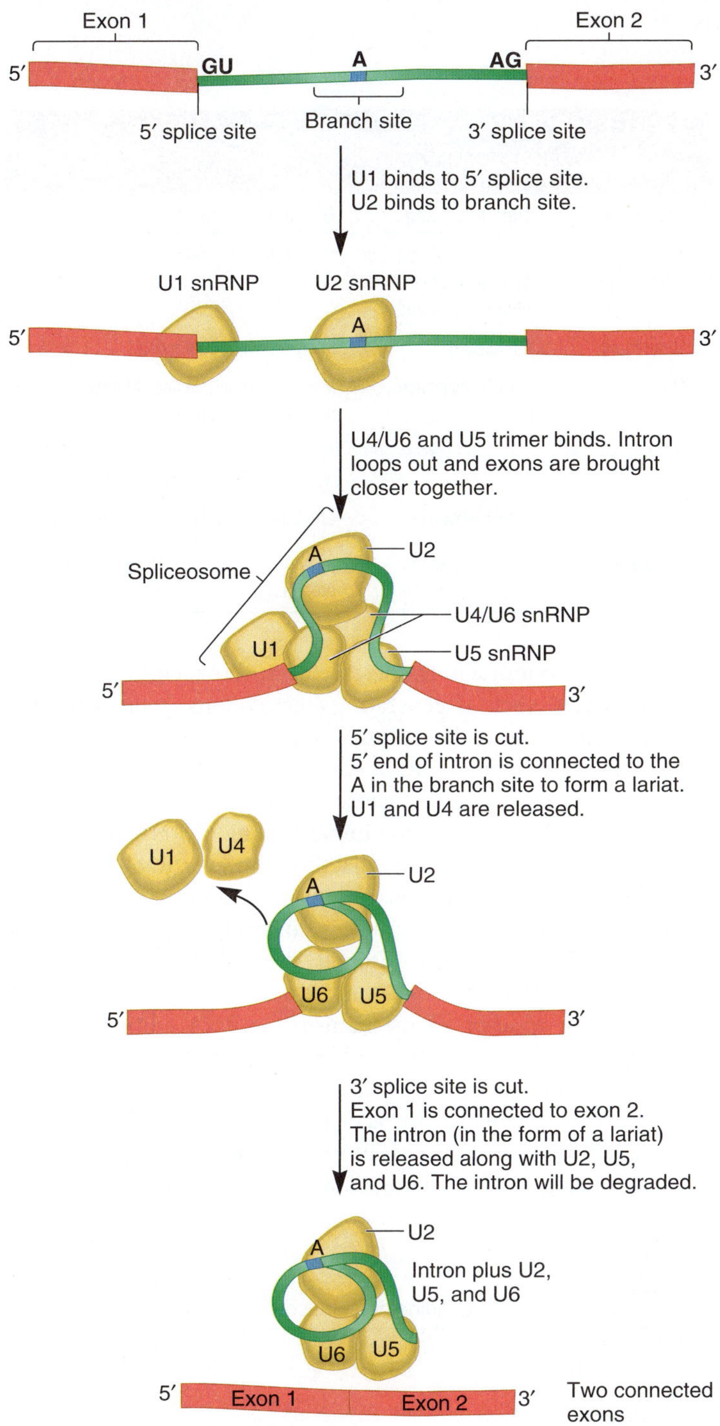

FIGURE 12.22 **Splicing of pre-mRNA via a spliceosome.**

ONLINE ANIMATION

CONCEPT CHECK: Describe the roles of snRNPs in the splicing process.

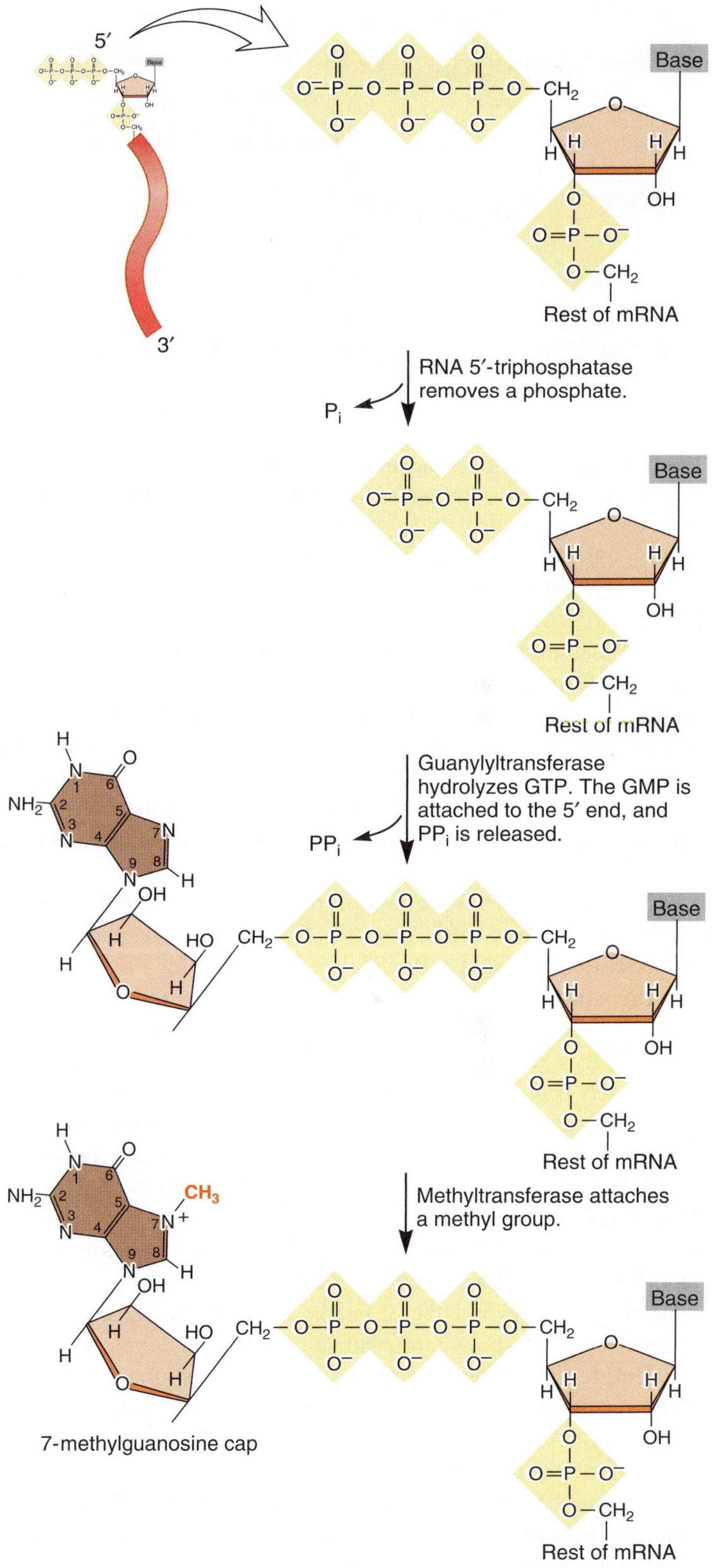

FIGURE 12.23 **Attachment of a 7-methylguanosine cap to the 5′ end of mRNA.** When the transcript is about 20–25 nucleotides in length, RNA 5′-triphosphatase removes one of the three phosphates, and then a second enzyme, guanylyltransferase, attaches GMP to the 5′ end. Finally, a methyltransferase attaches a methyl group to the guanine base.

CONCEPT CHECK: What are three functional roles of the 7-methlyguanosine cap?

made by RNA polymerase II, usually when the transcript is only 20–25 nucleotides in length. As shown in **Figure 12.23**, it is a three-step process. The nucleotide at the 5′ end of the transcript has three phosphate groups. First, an enzyme called RNA 5′-triphosphatase removes one of the phosphates, and then a second enzyme, guanylyltransferase, hydrolyzes guanosine triphosphate (GTP) to attach a guanosine monophosphate (GMP) to the 5′ end. Finally, a methyltransferase attaches a methyl group to a nitrogen at position 7 in the guanine base.

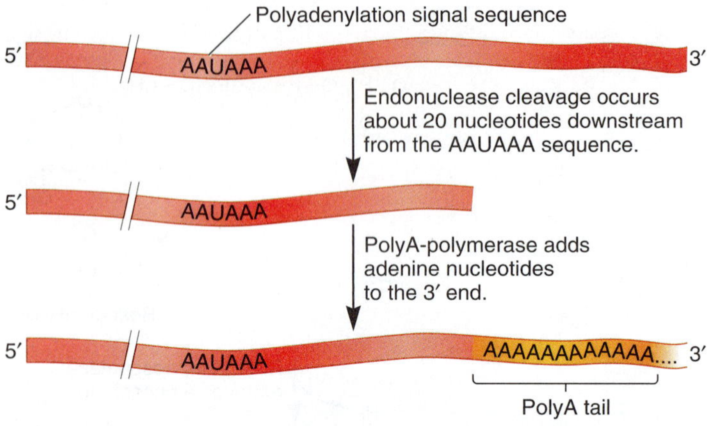

FIGURE 12.24 Attachment of a polyA tail. First, an endonuclease cuts the RNA at a location that is 11–30 nucleotides after the AAUAAA polyadenylation sequence, making the RNA shorter at its 3′ end. Adenine-containing nucleotides are then attached, one at a time, to the 3′ end by the enzyme polyA-polymerase.

What are the functions of the 7-methylguanosine cap? The cap structure is recognized by cap-binding proteins, which perform various roles. For example, cap-binding proteins are required for the proper exit of most mRNAs from the nucleus. Also, the cap structure is recognized by initiation factors that are needed during the early stages of translation. Finally, the cap structure may be important in the efficient splicing of introns, particularly the first intron located nearest the 5′ end.

Let's now turn our attention to the 3′ end of the RNA molecule. Most mature mRNAs have a string of adenine nucleotides, referred to as a **polyA tail,** which is important for mRNA stability, the exit of mRNA from the nucleus, and in the synthesis of polypeptides. The polyA tail is not encoded in the gene sequence. Instead, it is added enzymatically after the pre-mRNA has been completely transcribed—a process termed **polyadenylation.**

The steps required to synthesize a polyA tail are shown in Figure 12.24. To acquire a polyA tail, the pre-mRNA contains a polyadenylation signal sequence near its 3′ end. In mammals, the consensus sequence is AAUAAA. An endonuclease recognizes the signal sequence and cleaves the pre-mRNA at a location that is about 20 nucleotides beyond the 3′ end of the AAUAAA sequence. The fragment beyond the 3′ cut is degraded. Next, an enzyme known as polyA-polymerase attaches many adenine-containing nucleotides. The length of the polyA tail varies among different mRNAs; the maximum length is typically around 250 nucleotides. A long polyA tail facilitates mRNA export from the nucleus, stability of mRNA in the cytosol, and translation.

Some bacterial RNAs are also polyadenylated. However, in bacteria, polyadenylation targets the RNA to a structure called the degradosome where it is degraded.

The Nucleotide Sequence of RNA Can Be Modified by RNA Editing

The term **RNA editing** refers to a change in the nucleotide sequence of an RNA molecule that involves additions or deletions of particular bases, or a conversion of one type of base to another, such as a cytosine to a uracil. In the case of mRNAs, editing can have various effects, such as generating start codons, generating stop codons, and changing the coding sequence for a polypeptide.

TABLE 12.5
Examples of RNA Editing

Organism	Type of Editing	Found In
Trypanosomes (protozoa)	Primarily additions but occasionally deletions of uracil nucleotides	Many mitochondrial mRNAs
Slime mold	C additions	Many mitochondrial mRNAs
Plants	C-to-U conversion	Many mitochondrial and chloroplast mRNAs, tRNAs, and rRNAs
Mammals	C-to-U conversion	Apoliproprotein B mRNA, and NFI mRNA, which encodes a tumor-suppressor protein
	A-to-I conversion	Glutamate receptor mRNA, many tRNAs
Drosophila	A-to-I conversion	mRNA for calcium and sodium channel proteins

The phenomenon of RNA editing was first discovered in trypanosomes, the protists that cause sleeping sickness. As with the discovery of RNA splicing, the initial finding of RNA editing was met with great skepticism. Since that time, however, RNA editing has been shown to occur in various organisms and in a variety of ways, although its functional significance is slowly emerging (Table 12.5). In the specific case of trypanosomes, the editing process involves the addition or deletion of one or more uracil nucleotides.

A more widespread mechanism for RNA editing involves changes of one type of base to another. In this form of editing, a base in the RNA is deaminated, meaning an amino group is removed from the base. When cytosine is deaminated, uracil is formed, and when adenine is deaminated, inosine is formed (Figure 12.25). Inosine is recognized as guanine during translation.

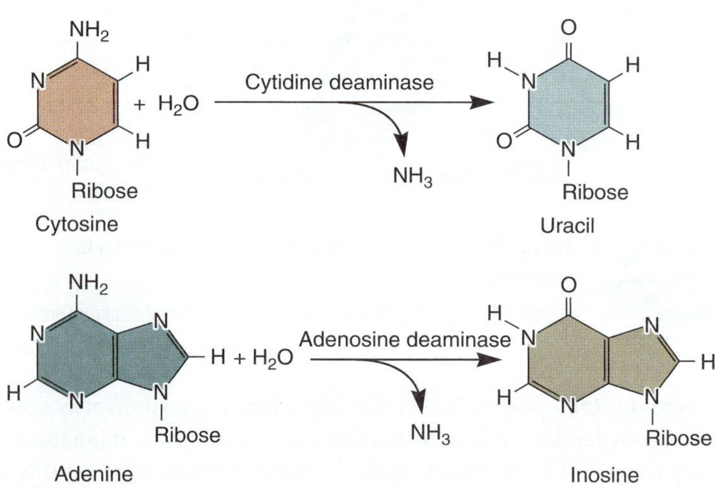

FIGURE 12.25 RNA editing by deamination. Cytidine deaminase removes an amino group from cytosine, thereby creating uracil. Adenosine deaminase removes an amino group from adenine to make inosine.

CONCEPT CHECK: What is a functional consequence of RNA editing?

An example of RNA editing occurs in mammals involving an mRNA that encodes a protein called apolipoprotein B. In the liver, translation of an unedited mRNA produces apolipoprotein B-100, a protein that is essential for the transport of cholesterol in the blood. In intestinal cells, the mRNA may be edited so that a single C is changed to a U. What is the significance of this base substitution? This change converts a glutamine codon (CAA) to a stop codon (UAA), resulting in a shorter apolipoprotein. In this case, RNA editing produces an apolipoprotein B with an altered structure. Therefore, RNA editing can produce two proteins from the same gene, much like the phenomenon of alternative splicing.

How widespread is RNA editing that involves C to U and A to I substitutions? In invertebrates such as *Drosophila*, researchers estimate that 50–100 RNAs are edited for the purpose of changing the RNA coding sequence. In mammals, the RNAs from fewer than 25 genes are currently known to be edited.

12.4 COMPREHENSION QUESTIONS

1. Which of the following are examples of RNA modification?
   a. Splicing
   b. 7-Methylguanosine cap
   c. PolyA tail
   d. All of the above
2. A ribozyme is
   a. a complex between RNA and a protein.
   b. an RNA than encodes a protein that functions as an enzyme.
   c. an RNA molecule with catalytic function.
   d. a protein that degrades RNA molecules.
3. Which of the following statements about the spliceosome is false?
   a. A spliceosome splices pre-mRNA molecules.
   b. A spliceosome removes exons from RNA molecules.
   c. A spliceosome is composed of snRNPs.
   d. A spliceosome recognizes the exon-intron boundaries and the branch site.
4. Which of the following is a function of the 7-methylguanosine cap?
   a. Exit of mRNA from the nucleus
   b. Efficient splicing of pre-mRNA
   c. Initiation of translation
   d. All of the above

12.5 A COMPARISON OF TRANSCRIPTION AND RNA MODIFICATION IN BACTERIA AND EUKARYOTES

Learning Outcome:

1. Compare and contrast the processes of transcription and RNA modification in bacteria and eukaryotes.

Throughout this chapter, we have considered the process of transcription and RNA modification. Many similarities have been noted between bacteria and eukaryotes. However, these processes are more complex in eukaryotes than their bacterial counterparts. Table 12.6 summarizes many of the key differences.

12.5 COMPREHENSION QUESTION

1. Which of the following is *not* a key difference between bacteria and eukaryotes?
   a. Initiation of transcrption requires more proteins in eukaryotes.
   b. A 7-methylguanosine cap is added only to eukaryotic mRNAs.
   c. Splicing is common in complex eukaryotes but not in bacteria.
   d. All of the above are key differences.

TABLE 12.6

Key Differences Between Transcription and RNA Modification in Bacteria and Eukaryotes*

Component	Bacteria	Eukaryotes
Promoter	Consists of a −35 and −10 sequence	For protein-encoding genes, the core promoter often consists of a TATA box and a transcriptional start site.
RNA polymerase	A single RNA polymerase	Three types of RNA polymerases; RNA polymerase II transcribes protein-encoding genes.
Initiation	σ factor is needed for promoter recognition.	Five general transcription factors assemble at the core promoter.
Elongation	Requires the release of σ factor	Mediator controls the switch to the elongation phase.
Termination	ρ-dependent or ρ-independent	According to the allosteric or torpedo model
Splicing	Very rare; self-splicing	Commonly occurs in protein-encoding pre-mRNAs in complex eukaryotes via a spliceosome; self-splicing occurs rarely.
Capping	Does not occur	Addition of 7-methylguanosine cap
Tailing	Added to 3′ end; promotes degradation	Added to the 3′ end; promotes stability
RNA editing	Does not occur	Occurs occasionally

*Note: This table does not include the process of gene regulation, which is described in Chapters 14, 15, and 16.

KEY TERMS

Page 287. gene, transcription, protein-encoding genes, structural genes, messenger RNA (mRNA), translation, central dogma of genetics
Page 288. gene expression, promoter, terminator, template strand, nontemplate strand
Page 289. coding strand, transcription factors, regulatory sequences, regulatory elements, ribosome-binding site, Shine-Dalgarno sequence, codons, start codon, stop codon, initiation, elongation, termination, RNA polymerase, open complex
Page 291. transcriptional start site, Pribnow box, consensus sequence
Page 292. core enzyme, sigma (σ) factor, holoenzyme, helix-turn-helix motif, closed complex
Page 293. ρ (rho), ρ-dependent termination
Page 294. ρ-independent termination
Page 295. intrinsic termination
Page 296. core promoter, TATA box, basal transcription, enhancers, silencers, *cis*-acting elements, *trans*-acting factors
Page 297. general transcription factors (GTFs), preinitiation complex, basal transcription apparatus, mediator
Page 300. colinearity, exons, intervening sequences, introns, pre-mRNA, RNA splicing
Page 301. nucleolus, exonuclease, endonuclease, ribozyme
Page 303. hybridization, R loop
Page 304. genomic clone, complementary DNA (cDNA), group I introns, group II introns, self-splicing
Page 305. maturases, pre-mRNA, spliceosome
Page 306. snRNPs, alternative splicing, capping
Page 308. polyA tail, polyadenylation, RNA editing

CHAPTER SUMMARY

- According to the central dogma of genetics, DNA is transcribed into mRNA, and mRNA is translated into a polypeptide. DNA replication allows the DNA to be passed from cell to cell and from parent to offspring (see Figure 12.1).

12.1 Overview of Transcription

- A gene is an organization of DNA sequences. A promoter signals the start of transcription, and a terminator signals the end. Regulatory sequences control the rate of transcription. For genes that encode polypeptides, the gene sequence also specifies a start codon, a stop codon, and many codons in between. Bacterial genes also specify a ribosomal binding site (see Figure 12.2).
- Transcription occurs in three phases called initiation, elongation, and termination (see Figure 12.3).
- RNA transcripts have several different functions (see Table 12.1).

12.2 Transcription in Bacteria

- Many bacterial promoters have sequence elements at the −35 and −10 regions. The transcriptional start site is at +1 (see Figures 12.4, 12.5).
- During the initiation phase of transcription in *E. coli*, sigma (σ) factor, which is bound to RNA polymerase, binds into the major groove of DNA and recognizes sequence elements at the promoter. This process forms a closed complex. Following the formation of an open complex, σ factor is released (see Figures 12.6, 12.7).
- During the elongation phase of transcription, RNA polymerase slides along the DNA and maintains an open complex as it goes. RNA is made in the 5′ to 3′ direction according to complementary base pairing (see Figure 12.8).
- In a given chromosome, the use of the template strand varies from gene to gene (see Figure 12.9).
- Transcriptional termination in *E. coli* occurs by a rho (ρ)-dependent or ρ-independent mechanism (see Figures 12.10, 12.11).

12.3 Transcription in Eukaryotes

- Eukaryotes use RNA polymerase I, II, and III to transcribe different categories of genes. Prokaryotic and eukaryotic RNA polymerases have similar structures (see Figure 12.12).
- Eukaryotic promoters have a core promoter and regulatory elements such as enhancers and silencers (see Figure 12.13).
- Transcription of protein-encoding genes in eukaryotes requires RNA polymerase II, five general transcription factors, and mediator. The five general transcription factors and RNA polymerase assemble together to form an open complex (see Table 12.2, Figure 12.14).
- Transcriptional termination of RNA polymerase II may occur via an allosteric model or a torpedo model (see Figure 12.15).

12.4 RNA Modification

- RNA transcripts can be modified in a variety of ways, which include processing, splicing, 5′ capping, 3′ polyA tailing, RNA editing, and base modification (see Table 12.3).
- Certain RNA molecules such as ribosomal RNAs and precursor tRNAs are processed to smaller, functional molecules via cleavage steps (see Figures 12.16, 12.17).
- Leder and colleagues identified introns in a globin gene using microscopy (see Figures 12.18, 12.19).
- Group I and II introns are removed by self-splicing. Pre-mRNA introns are removed via a spliceosome (see Table 12.4, Figure 12.20).

- The spliceosome is a multicomponent structure that recognizes intron sequences and removes them from pre-mRNA (see Figures 12.21, 12.22).
- In eukaryotes, mRNA is given a 7-methylguanosine cap at the 5′ end and a polyA tail at the 3′ end (see Figures 12.23, 12.24).
- RNA editing changes the base sequence of an RNA after it has been synthesized (see Table 12.5, Figure 12.25).

12.5 A Comparison of Transcription and RNA Modification in Bacteria and Eukaryotes

- Several key differences have been found when comparing transcription and RNA modification in bacteria and eukaryotes (see Table 12.6).

PROBLEM SETS & INSIGHTS

Solved Problems

S1. Describe the important events that occur during the three stages of gene transcription in bacteria. What proteins play critical roles in the three stages?

Answer: The three stages are initiation, elongation, and termination.

Initiation: RNA polymerase holoenzyme slides along the DNA until σ factor recognizes a promoter. σ factor binds tightly to this sequence, forming a closed complex. The DNA is then denatured to form a bubble-like structure known as the open complex.

Elongation: RNA polymerase core enzyme slides along the DNA, synthesizing RNA as it goes. The α subunits of RNA polymerase keep the enzyme bound to the DNA, while the β subunits are responsible for binding and for the catalytic synthesis of RNA. The ω (omega) subunit is also important for the proper assembly of the core enzyme. During elongation, RNA is made according to the AU/GC rule, with nucleotides being added in the 5′ to 3′ direction.

Termination: RNA polymerase eventually reaches a sequence at the end of the gene that signals the end of transcription. In ρ-independent termination, the properties of the termination sequences in the DNA are sufficient to cause termination. In ρ-dependent termination, the ρ (rho) protein recognizes a sequence within the RNA, binds there, and travels toward RNA polymerase. When the formation of a stem-loop causes RNA polymerase to pause, ρ catches up and separates the RNA-DNA hybrid region, releasing RNA polymerase.

S2. What is the difference between a protein-encoding gene and one that is not?

Answer: Protein-encoding genes encode mRNAs that are translated into polypeptides. Other genes encode RNAs that are never translated. Products of these other genes include tRNA and rRNA, which function during translation; microRNA, which is involved in gene regulation; 7S RNA, which is part of a complex known as SRP; scRNA, small cytoplasmic RNA found in bacteria; the RNA of RNaseP; telomerase RNA, which is involved in telomere replication; snoRNA, which is involved in rRNA processing; and snRNA, which is a component of spliceosome. In many cases, the RNA from such genes becomes part of a complex composed of RNA molecules and protein subunits.

S3. When RNA polymerase transcribes DNA, only one of the two DNA strands is used as a template. Take a look at Figure 12.4 and explain how RNA polymerase determines which DNA strand is the template strand.

Answer: The binding of σ factor and RNA polymerase depends on the sequence of the promoter. RNA polymerase binds to the promoter in such a way that the −35 sequence TTGACA and the −10 sequence TATAAT are within the coding strand, whereas the −35 sequence AACTGT and the −10 sequence ATATTA are within the template strand.

S4. A lariat contains a closed loop and a linear end. An intron has the following sequence: 5′–GUPuAGUA–60 nucleotides–UACUUAUCC–100 nucleotides–Py_{12}NPyAG–3′. Which sequence would be found within the closed loop of the lariat, the 60-nucleotide sequence or the 100-nucleotide sequence?

Answer: The 60-nucleotide sequence would be found in the closed loop. The closed loop is the region between the 5′ splice site and the branch site.

Conceptual Questions

C1. Genes may be protein-encoding genes that encode polypeptides, or they may not.

A. Describe three examples of genes that are not protein-encoding genes.

B. For protein-encoding genes, one DNA strand is called the template strand, and the complementary strand is called the coding strand. Are these two terms appropriate for genes that do not encode proteins? Explain.

C. Do genes that do not encode proteins have a promoter and terminator?

C2. In bacteria, what event marks the end of the initiation stage of transcription?

C3. What is the meaning of the term consensus sequence? Give an example. Describe the locations of consensus sequences within bacterial promoters. What are their functions?

C4. What is the consensus sequence of the following six DNA sequences?

GGCATTGACT
GCCATTGTCA
CGCATAGTCA
GGAAATGGGA
GGCTTTGTCA
GGCATAGTCA

C5. Mutations in bacterial promoters may increase or decrease the rate of gene transcription. Promoter mutations that increase transcription are termed up-promoter mutations, and those that decrease transcription are termed down-promoter mutations. As shown in Figure 12.5, the sequence of the −10 region of the promoter for the *lac* operon is TATGTT. Would you expect the following mutations to be up-promoter or down-promoter mutations?

A. TATGTT to TATATT

B. TATGTT to TTTGTT

C. TATGTT to TATGAT

C6. According to the examples shown in Figure 12.5, which positions of the −35 sequence (i.e., first, second, third, fourth, fifth, or sixth) are more tolerant of changes? Do you think these positions play a more or less important role in the binding of σ factor? Explain why.

C7. In Chapter 9, we considered the dimensions of the double helix (see Figure 9.15). In an α helix of a protein, there are 3.6 amino acids per complete turn. Each amino acid advances the α helix by 0.15 nm; a complete turn of an α helix is 0.54 nm in length. As shown in Figure 12.6, two α helices of a transcription factor occupy the major groove of the DNA. According to Figure 12.6, estimate the number of amino acids that bind to this region. How many complete turns of the α helices occupy the major groove of DNA?

C8. A mutation within a gene sequence changes the start codon to a stop codon. How will this mutation affect the transcription of this gene?

C9. What is the subunit composition of bacterial RNA polymerase holoenzyme? What are the functional roles of the different subunits?

C10. At the molecular level, describe how σ factor recognizes a bacterial promoter. Be specific about the structure of σ factor and the type of chemical bonding.

C11. Let's suppose a DNA mutation changes the consensus sequence at the −35 location in a way that inhibits σ factor binding. Explain how a mutation could inhibit the binding of σ factor to the DNA. Look at Figure 12.5 and describe two specific base substitutions you think would inhibit the binding of σ factor. Explain why your base substitutions would have this effect.

C12. What is the complementarity rule that governs the synthesis of an RNA molecule during transcription? An RNA transcript has the following sequence:

5′–GGCAUGCAUUACGGCAUCACACUAGGGAUC–3′

What is the sequence of the template and coding strands of the DNA that encodes this RNA? On which side (5′ or 3′) of the template strand is the promoter located?

C13. Describe the movement of the open complex along the DNA.

C14. Describe what happens to the chemical bonding interactions when transcriptional termination occurs. Be specific about the type of chemical bonding.

C15. Discuss the differences between ρ-dependent and ρ-independent termination.

C16. In Chapter 11, we discussed the function of DNA helicase, which is involved in DNA replication. Discuss how the functions of ρ-protein and DNA helicase are similar and how they are different.

C17. Discuss the similarities and differences between RNA polymerase (described in this chapter) and DNA polymerase (described in Chapter 11).

C18. Mutations that occur at the end of a gene may alter the sequence of the gene and prevent transcriptional termination.

A. What types of mutations would prevent ρ-independent termination?

B. What types of mutations would prevent ρ-dependent termination?

C. If a mutation prevented transcriptional termination at the end of a gene, where would gene transcription end? Or would it end?

C19. If the following RNA polymerases were missing from a eukaryotic cell, what types of genes would not be transcribed?

A. RNA polymerase I

B. RNA polymerase II

C. RNA polymerase III

C20. What sequence elements are found within the core promoter of protein-encoding genes in eukaryotes? Describe their locations and specific functions.

C21. For each of the following transcription factors, how would eukaryotic transcriptional initiation be affected if it were missing?

A. TFIIB

B. TFIID

C. TFIIH

C22. Describe the allosteric and torpedo models for transcriptional termination of RNA polymerase II. Which model is more similar to ρ-dependent termination in bacteria and which model is more similar to ρ-independent termination?

C23. Which eukaryotic transcription factor(s) shown in Figure 12.14 play(s) an equivalent role to σ factor found in bacterial cells?

C24. The initiation phase of eukaryotic transcription via RNA polymerase II is considered an assembly and disassembly process. Which types of biochemical interactions—hydrogen bonding, ionic bonding, covalent bonding, and/or hydrophobic interactions—would you expect to drive the assembly and disassembly process? How would temperature and salt concentration affect assembly and disassembly?

C25. A eukaryotic protein-encoding gene contains two introns and three exons: exon 1–intron 1–exon 2–intron 2–exon 3. The 5′ splice site at the boundary between exon 2 and intron 2 has been eliminated by a small deletion in the gene. Describe how the pre-mRNA encoded by this mutant gene would be spliced. Indicate which introns and exons would be found in the mRNA after splicing occurs.

C26. Describe the processing events that occur during the production of tRNA in *E. coli*.

C27. Describe the structure and function of a spliceosome. Speculate why the spliceosome subunits contain snRNA. In other words, what do you think is/are the functional role(s) of snRNA during splicing?

C28. What is the unique feature of ribozyme function? Give two examples described in this chapter.

C29. What does it mean to say that gene expression is colinear?

C30. What is meant by the term self-splicing? What types of introns are self-splicing?

C31. In eukaryotes, what types of modifications occur to pre-mRNAs?

C32. What is alternative splicing? What is its biological significance?

C33. The processing of ribosomal RNA in eukaryotes is shown in Figure 12.16. Why is this called cleavage or processing but not splicing?

C34. In the splicing of group I introns shown in Figure 12.20, does the 5′ end of the intron have a phosphate group? Explain.

C35. According to the mechanism shown in Figure 12.22, several snRNPs play different roles in the splicing of pre-mRNA. Identify the snRNP that recognizes the following sites:

A. 5′ splice site

B. 3′ splice site

C. Branch site

C36. After the intron (which is in a lariat configuration) is released during pre-mRNA splicing, a brief moment occurs before the two exons are connected to each other. Which snRNP(s) holds the exons in place so they can be covalently connected to each other?

Experimental Questions

E1. A research group has sequenced the cDNA and genomic DNA from a particular gene. The cDNA is derived from mRNA, so it does not contain introns. Here are the DNA sequences.

cDNA:

5′–ATTGCATCCAGCGTATACTATCTCGGGCCCAATTAATGCCA-

GCGGCCAGACTATCACCCAACTCGGTTACCTACTAGTATATC-

CCATATACTAGCATATATTTTACCCATAATTTGTGTGTGGGTATA-

CAGTATAATCATATA–3′

Genomic DNA (contains one intron):

5′-ATTGCATCCAGCGTATACTATCTCGGGCCCAATTAATGCCAG

CGGCCAGACTATCACCCAACTCGGCCCACCCCCCAGGTTTA-

CACAGTCATACCATACATACAAAAATCGCAGTTACTTATCCCA-

AAAAAACCTAGATACCCCACATACTATTAACTCTTTCTTTCTAG-

GTTACCTACTAGTATATCCCATATACTAGCATATATTTTAC-

CCATAATTTGTGTGTGGGTATACAGTATAATCATATA–3′

Indicate where the intron is located. Does the intron contain the normal consensus splice site sequences based on those described in Figure 12.21? Underline the splice site sequences, and indicate whether or not they fit the consensus sequence.

E2. What is an R loop? In an R loop experiment, to which strand of DNA does the mRNA bind, the coding strand or the template strand?

E3. If a gene contains three introns, draw what it would look like in an R loop experiment.

E4. Chapter 20 describes a technique known as Northern blotting that can be used to detect RNA transcribed from a particular gene. In this method, a specific RNA is detected using a short segment of cloned DNA as a probe. The DNA probe, which is labeled, is complementary to the RNA that the researcher wishes to detect. After the probe DNA binds to the RNA, the RNA is visualized as a labeled (dark) band after the RNA is run on a gel. As shown here, the method of Northern blotting can be used to determine the amount of a particular RNA transcribed in a given cell type. If one type of cell produces twice as much of a particular mRNA as occurs in another cell, the band will appear twice as intense. Also, the method can distinguish if alternative RNA splicing has occurred to produce an RNA that has a different molecular mass.

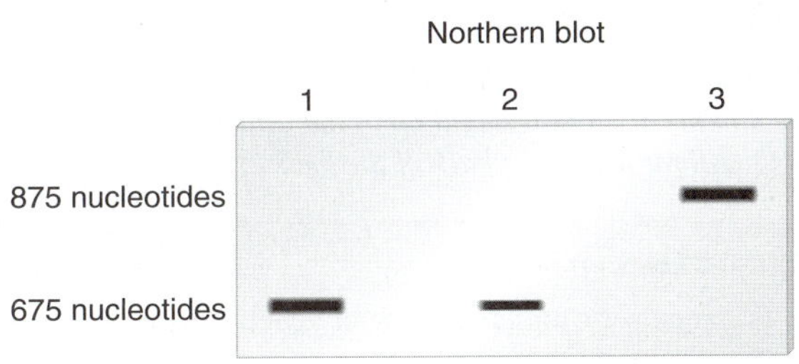

Lane 1 is a sample of RNA isolated from nerve cells.

Lane 2 is a sample of RNA isolated from kidney cells. Nerve cells produce twice as much of this RNA as do kidney cells.

Lane 3 is a sample of RNA isolated from spleen cells. Spleen cells produce an alternatively spliced version of this RNA that is about 200 nucleotides longer than the RNA produced in nerve and kidney cells.

Let's suppose a researcher was interested in the effects of mutations on the expression of a particular protein-encoding gene in eukaryotes. The gene has one intron that is 450 nucleotides long. After this intron is removed from the pre-mRNA, the mRNA transcript is 1100 nucleotides in length. Diploid somatic cells have two copies of this gene. Make a drawing that shows the expected results of a Northern blot using mRNA from the cytosol of somatic cells, which were obtained from the following individuals:

Lane 1: A normal individual

Lane 2: A homozygote for a deletion that removes the −50 to −100 region of the gene that encodes this mRNA

Lane 3: A heterozygote in which one gene is normal and the other gene has a deletion that removes the −50 to −100 region

Lane 4: A homozygote for a mutation that introduces an early stop codon into the middle of the coding sequence of the gene

Lane 5: A homozygote for a three-nucleotide deletion that removes the AG sequence at the 3′ splice site

E5. A gel retardation assay can be used to study the binding of proteins to a segment of DNA. This method is described in Chapter 20. When a protein binds to a segment of DNA, it retards the movement of the DNA through a gel, so the DNA appears at a higher point in the gel (see the following).

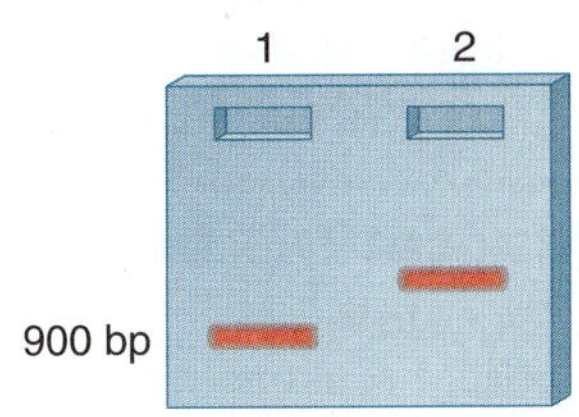

Lane 1: 900-bp fragment alone

Lane 2: 900-bp fragment plus a protein that binds to the 900-bp fragment

In this example, the segment of DNA is 900 bp in length, and the binding of a protein causes the DNA to appear at a higher point in the gel. If this 900-bp fragment of DNA contains a eukaryotic promoter for a protein-encoding gene, draw a gel that shows the relative locations of the 900-bp fragment under the following conditions:

Lane 1: 900 bp plus TFIID

Lane 2: 900 bp plus TFIIB

Lane 3: 900 bp plus TFIID and TFIIB

Lane 4: 900 bp plus TFIIB and RNA polymerase II

Lane 5: 900 bp plus TFIID, TFIIB, and RNA polymerase II/TFIIF

E6. As described in Chapter 20 and in experimental question E5, a gel retardation assay can be used to determine if a protein binds to DNA. This method can also determine if a protein binds to RNA. In the combinations described here, would you expect the migration of the RNA to be retarded due to the binding of a protein?

A. mRNA from a gene that is terminated in a ρ-independent manner plus ρ protein

B. mRNA from a gene that is terminated in a ρ-dependent manner plus ρ protein

C. Pre-mRNA from a protein-encoding gene that contains two introns plus the snRNP called U1

D. Mature mRNA from a protein-encoding gene that contains two introns plus the snRNP called U1

E7. The technique of DNA footprinting is described in Chapter 20. If a protein binds over a region of DNA, it will protect the DNA in that region from digestion by DNase I. To carry out a DNA footprinting experiment, a researcher has a sample of a cloned DNA fragment. The fragments are exposed to DNase I in the presence and absence of a DNA-binding protein. Regions of the DNA fragment not covered by the DNA-binding protein will be digested by DNase I, producing a series of bands on a gel. Regions of the DNA fragment not digested by DNase I (because a DNA-binding protein is preventing DNase I from gaining access to the DNA) will be revealed, because a region of the gel will not contain any bands.

In the DNA footprinting experiment shown here, a researcher began with a sample of cloned DNA 300 bp in length. This DNA contained a eukaryotic promoter for RNA polymerase II. For the sample loaded in lane 1, no proteins were added. For the sample loaded in lane 2, the 300-bp fragment was mixed with RNA polymerase II plus TFIID and TFIIB.

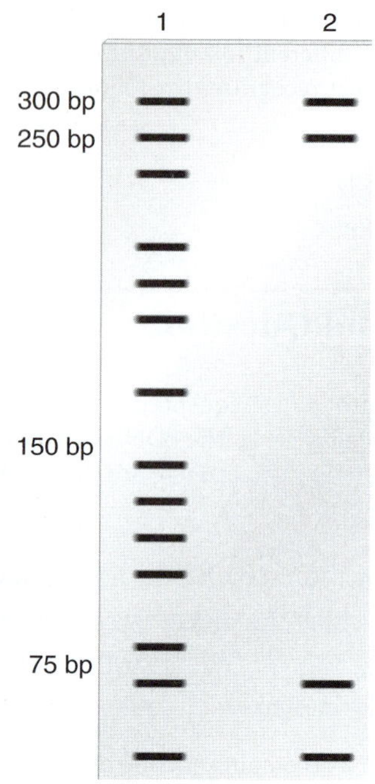

A. How long of a region of DNA is "covered up" by the binding of RNA polymerase II and the transcription factors?

B. Describe how this binding would occur if the DNA was within a nucleosome structure. (Note: The structure of nucleosomes is described in Chapter 10.) Do you think that the DNA is in a nucleosome structure when RNA polymerase and transcription factors are bound to the promoter? Explain why or why not.

E8. As described in Table 12.1, several different types of RNA are made, especially in eukaryotic cells. Researchers are sometimes interested in focusing their attention on the transcription of protein-encoding genes in eukaryotes. Such researchers want to study mRNA. One method that is used to isolate mRNA is column chromatography. (Note: See the Appendix for a general description of chromatography.) Researchers can covalently attach short pieces of DNA that contain stretches of thymine (i.e., TTTTTTTTTTTT) to the column matrix. This is called a poly-dT column. When a cell extract is poured over the column, mRNA binds to the column, but other types of RNA do not.

A. Explain how you would use a poly-dT column to obtain a purified preparation of mRNA from eukaryotic cells. In your description, explain why mRNA binds to this column and what you would do to release the mRNA from the column.

B. Can you think of ways to purify other types of RNA, such as tRNA or rRNA?

Questions for Student Discussion/Collaboration

1. Based on your knowledge of introns and pre-mRNA splicing, discuss whether or not you think alternative splicing fully explains the existence of introns. Can you think of other possible reasons to explain the existence of introns?

2. Discuss the types of RNA transcripts and the functional roles they play. Why do you think some RNAs form complexes with protein subunits?

Note: All answers appear at the website for this textbook; the answers to the even-numbered questions and all of the Concept Check and Comprehension Questions are in the back of the book.

www.mhhe.com/brookergenetics5e

Visit the website for practice tests, answer keys, and other learning aids for this chapter. Enhance your understanding of genetics with our interactive exercises, quizzes, animations, and much more.

CHAPTER OUTLINE

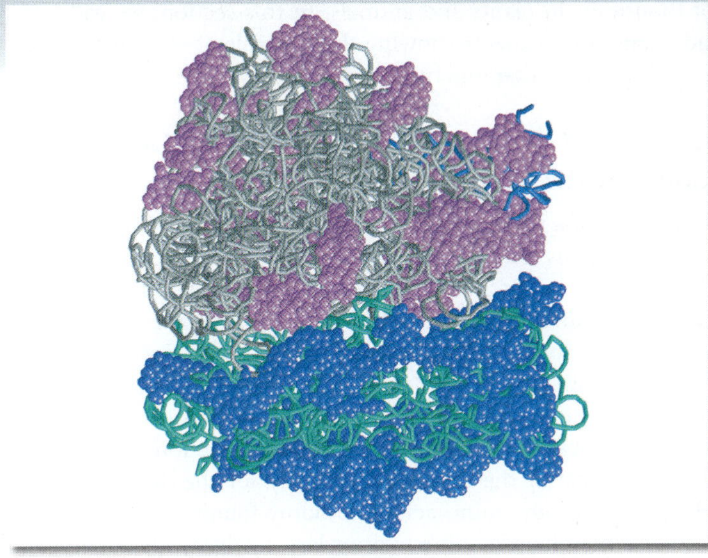

A molecular model for the structure of a ribosome. *This is a model of ribosome structure based on X-ray crystallography. Ribosomes are needed to synthesize polypeptides, using mRNA as a template. A detailed description of this model is discussed later in Figure 13.13.*

13 TRANSLATION OF mRNA

The synthesis of proteins occurs via the translation of the sequence of codons within mRNA into a sequence of amino acids that constitute a polypeptide. In this chapter, we will explore the current state of knowledge regarding translation, with an eye toward the specific molecular interactions responsible for this process. During the past few decades, the combined efforts of geneticists, cell biologists, and biochemists have advanced our understanding of translation in profound ways. Even so, many questions remain unanswered, and this topic continues to be an exciting area of investigation.

We will begin this chapter by considering classic experiments which revealed that some genes encode proteins that function as enzymes. Next, we examine how the genetic code is used to decipher the information within mRNA to produce a polypeptide with a specific amino acid sequence. The rest of this chapter is devoted to an understanding of translation at the molecular level as it occurs in living cells. This will involve an examination of the cellular components—including many different proteins, RNAs, and small molecules—needed for the translation process. We will consider the structure and function of tRNA molecules, which act as the translators of the genetic information within mRNA, and then examine the composition of ribosomes. Finally, we will explore the stages of translation and examine differences between translation in bacterial cells and eukaryotic cells.

13.1 THE GENETIC BASIS FOR PROTEIN SYNTHESIS

Learning Outcomes:

1. Explain how the work of Garrod indicated that some genes encode enzymes.
2. Analyze the experiments of Beadle and Tatum, and explain how their results were consistent with the idea that certain genes encode a single enzyme.

Proteins, which are composed of one or more polypeptides, are critically important as active participants in cell structure and function. The primary role of DNA is to store the information needed for the synthesis of all the proteins that an organism makes. As we discussed in Chapter 12, genes that encode the amino acid sequence of a polypeptide are known as **protein-encoding genes** (also called **structural genes**). The RNA transcribed from a protein-encoding gene is called **messenger RNA (mRNA).** The main function of the genetic material is to encode the production of cellular proteins in the correct cell, at the proper time, and in suitable amounts. This is an extremely complicated task because living cells make thousands of different proteins. Genetic analyses have shown that a typical bacterium can make a few thousand different proteins, and estimates for eukaryotes range from several thousand in simple eukaryotic organisms, such as yeast, to tens of thousands in plants and animals. In this section, we will consider early experiments showing that the role of some genes is to encode proteins that function as enzymes.

Archibald Garrod Proposed That Some Genes Code for the Production of Enzymes

The idea that a relationship exists between genes and the production of proteins was first suggested at the beginning of the twentieth century by Archibald Garrod, a British physician. Prior to Garrod's studies, biochemists had studied many metabolic pathways within living cells. These pathways consist of a series of metabolic conversions of one molecule to another, each step catalyzed by a specific enzyme. Each enzyme is a distinctly different protein that catalyzes a particular chemical reaction. **Figure 13.1** illustrates part of the metabolic pathway for the degradation of phenylalanine, an amino acid commonly found in human diets. The enzyme phenylalanine hydroxylase catalyzes the conversion of phenylalanine to tyrosine, and a different enzyme (tyrosine aminotransferase) converts tyrosine into *p*-hydroxyphenylpyruvic acid, and so on. In all of the steps shown in Figure 13.1, a specific enzyme catalyzes a single type of chemical reaction.

Garrod studied patients who had defects in their ability to metabolize certain compounds. He was particularly interested in the inherited disease known as **alkaptonuria.** In this disorder, the patient's body accumulates abnormal levels of homogentisic acid (also called alkapton), which is excreted in the urine, causing it to appear black on exposure to air. In addition, the disease is characterized by bluish black discoloration of cartilage and skin (ochronosis). Garrod proposed that the accumulation of homogentisic acid in these patients is due to a missing enzyme, namely, homogentisic acid oxidase (see Figure 13.1).

How did Garrod realize that certain genes encode enzymes? He already knew that alkaptonuria is an inherited trait that follows an autosomal recessive pattern of inheritance. Therefore, an individual with alkaptonuria must have inherited the mutant (defective) gene that causes this disorder from both parents. From these observations, Garrod proposed that a relationship exists between the inheritance of the trait and the inheritance of a defective enzyme. Namely, if an individual inherited the mutant gene (which causes a loss of enzyme function), she or he would not produce any normal enzyme and would be unable to metabolize homogentisic acid. Garrod described alkaptonuria as an **inborn error of metabolism.** This hypothesis was the first suggestion that a connection exists between the function of genes and the production of enzymes. At the turn of the century, this idea was particularly insightful, because the structure and function of the genetic material were completely unknown.

Beadle and Tatum's Experiments with *Neurospora* Led Them to Propose the One-Gene/One-Enzyme Hypothesis

In the early 1940s, American geneticists George Beadle and Edward Tatum were also interested in the relationship among genes, enzymes, and traits. They developed an experimental system for investigating the connection between genes and the production

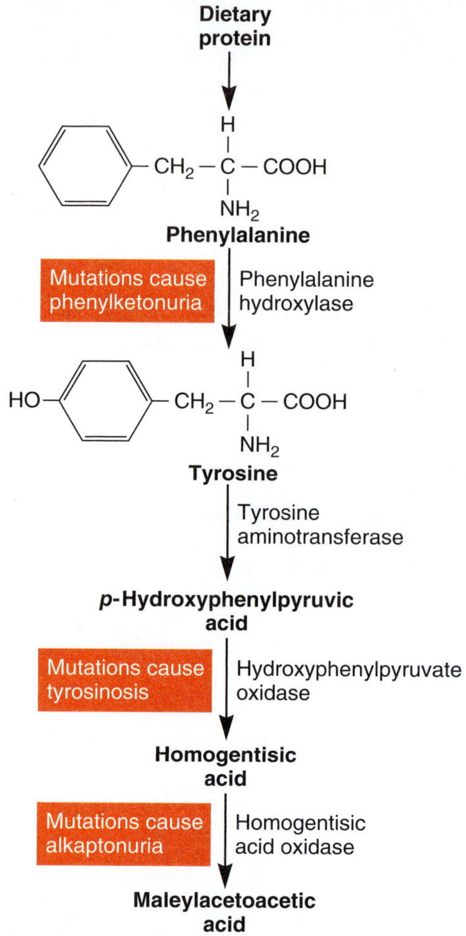

FIGURE 13.1 The metabolic pathway of phenylalanine breakdown. This diagram shows part of the pathway of phenylalanine metabolism, which consists of specific enzymes (shown to the right of the arrows) that successively convert one molecule to another. Certain human genetic diseases (shown in red boxes) are caused when enzymes in this pathway are missing or defective.

Genes → Traits A person who inherits two defective copies of the gene that encodes homogentisic acid oxidase cannot convert homogentisic acid into maleylacetoacetic acid. He or she accumulates large amounts of homogentisic acid in the urine and has other symptoms of the disease known as alkaptonuria. Similarly, a person with two defective copies of the gene encoding phenylalanine hydroxylase is unable to synthesize the enzyme phenylalanine hydroxylase and has the disease called phenylketonuria (PKU).

CONCEPT CHECK: Which disease occurs when homogentisic acid oxidase is defective?

of particular enzymes. Consistent with the ideas of Garrod, the underlying assumption behind their approach was that a relationship exists between genes and the production of enzymes. However, the quantitative nature of this relationship was unclear. In particular, they asked the question, "Does one gene control the production of one enzyme, or does one gene control the synthesis of many enzymes involved in a complex biochemical pathway?"

At the time of their studies, many geneticists were trying to understand the nature of the gene by studying morphological traits. However, Beadle and Tatum realized that morphological traits are likely to be based on systems of biochemical reactions so complex as to make analysis exceedingly difficult. Therefore, they turned their genetic studies to the analysis of simple nutritional requirements in *Neurospora crassa*, a common bread mold. *Neurospora* can be easily grown in the laboratory and has few nutritional requirements: a

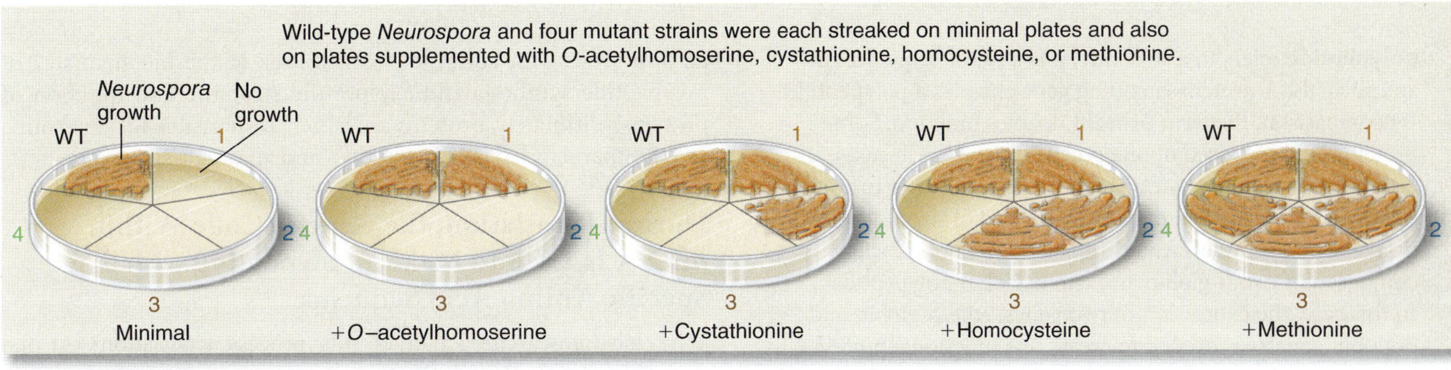

(a) Growth of strains on minimal and supplemented growth media

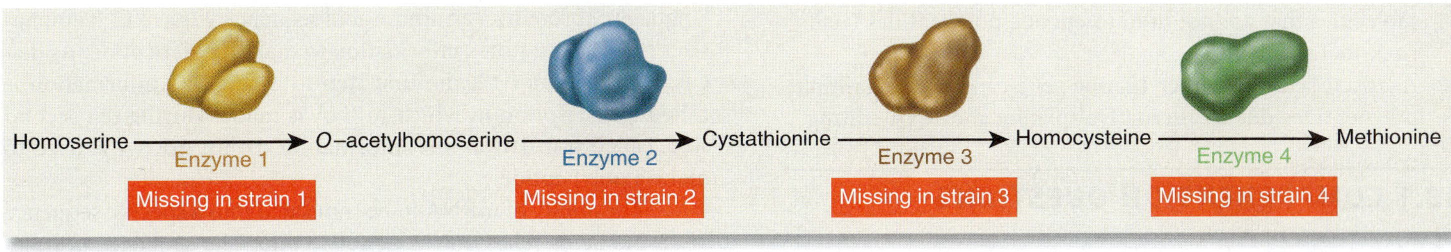

(b) Simplified pathway for methionine biosynthesis

FIGURE 13.2 An example of an experiment that supported Beadle and Tatum's one-gene/one-enzyme hypothesis. **(a)** Growth of wild-type (WT) and mutant strains of *Neurospora crassa* on minimal media or in the presence of *O*-acetylhomoserine, cystathionine, homocysteine, or methionine. **(b)** A simplified pathway for methionine biosynthesis. Note: The first compound in this pathway, homoserine, is made by *Neurospora* via enzymes and precursor molecules not discussed in this experiment.

CONCEPT CHECK: What enzymatic function is missing in the strain 2 mutants?

carbon source (sugar), inorganic salts, and the vitamin biotin. Normal *Neurospora* cells produce many different enzymes that can synthesize the organic molecules, such as amino acids and other vitamins, which are essential for growth.

Beadle and Tatum wanted to understand how enzymes are controlled by genes. They reasoned that a mutation in a gene, causing a defect in an enzyme needed for the cellular synthesis of an essential molecule, would prevent that mutant strain from growing on minimal media, which contain only a carbon source, inorganic salts, and biotin. In their original study of 1941, Beadle and Tatum exposed *Neurospora* cells to X-rays, which caused mutations to occur, and studied the growth of the resulting cells. By plating the cells on media with or without vitamins, they were able to identify mutant strains that required vitamins for growth. In each case, a single mutation resulted in the requirement for a single type of vitamin in the growth media.

This early study by Beadle and Tatum led to additional research to study enzymes involved with the synthesis of other substances, including the amino acid methionine. They first isolated several different mutant strains that required methionine for growth. They hypothesized that each mutant strain might be blocked at only a single step in the consecutive series of reactions that lead to methionine synthesis. To test this hypothesis, the mutant strains were examined for their ability to grow in the presence of *O*-acetylhomoserine, cystathionine, homocysteine, or methionine. *O*-Acetylhomoserine, cystathionine, and homocysteine are intermediates in the synthesis of methionine from homoserine. A simplified depiction of the results is shown in **Figure 13.2a**. The wild-type strain could grow on minimal growth media that contained the minimum set of nutrients that is required for growth. The minimal media did not contain *O*-acetylhomoserine, cystathionine, homocysteine, or methionine. Based on their growth properties, the mutant strains that had been originally identified as requiring methionine for growth could be placed into four groups designated strains 1, 2, 3, and 4 in this figure. A strain 1 mutant was missing enzyme 1, needed for the conversion of homoserine into *O*-acetylhomoserine. The cells could grow only if *O*-acetylhomoserine, cystathionine, homocysteine, or methionine was added to the growth medium. A strain 2 mutant was missing the second enzyme in this pathway that is needed for the conversion of *O*-acetylhomoserine into cystathionine, and a strain 3 mutant was unable to convert cystathionine into homocysteine. Finally, a strain 4 mutant could not make methionine from homocysteine. Based on these results, the researchers could order the enzymes into a biochemical pathway as depicted in **Figure 13.2b**. Taken together, the analysis of these mutants allowed Beadle and Tatum to conclude that a single gene controlled the synthesis of a single enzyme. This was referred to as the **one-gene/one-enzyme hypothesis.**

In later decades, this hypothesis had to be modified in four ways:

1. Enzymes are only one category of cellular proteins. All proteins are encoded by genes, and many of them do not function as enzymes.
2. Some proteins are composed of two or more different polypeptides. Therefore, it is more accurate to say that a protein-encoding gene encodes a polypeptide. The term

polypeptide refers to a structure; it is a linear sequence of amino acids. A protein-encoding gene encodes a polypeptide. By comparison, the term **protein** denotes function. Some proteins are composed of one polypeptide. In such cases, a single gene does encode a single protein. In other cases, however, a functional protein is composed of two or more different polypeptides. An example is hemoglobin, which is composed of two α-globin and two β-globin polypeptides. In this case, the expression of two genes—the α-globin and β-globin genes—is needed to create one functional protein.

3. Many genes do not encode polypeptides. As discussed in Chapter 12, several types of genes specify functional RNA molecules that are not used to encode polypeptides (refer back to Table 12.1).
4. As discussed in Chapter 12, one gene can encode multiple polypeptides due to alternative splicing and RNA editing.

13.1 COMPREHENSION QUESTIONS

1. An inborn error of metabolism is caused by
   a. a mutation in a gene that causes an enzyme to be inactive.
   b. a mutation in a gene that occurs in somatic cells.
   c. the consumption of foods that disrupt metabolic processes.
   d. all of the above.
2. The reason why Beadle and Tatum observed four different categories of mutants that could not grow on media without methionine is because
   a. the enzyme involved in methionine biosynthesis is composed of four different subunits.
   b. the enzyme involved in methionine biosynthesis is present in four copies in the *Neurospora* genome.
   c. four different enzymes are involved in a pathway for methionine biosynthesis.
   d. a lack of methionine biosynthesis can inhibit *Neurospora* growth in four different ways.

13.2 THE RELATIONSHIP BETWEEN THE GENETIC CODE AND PROTEIN SYNTHESIS

Learning Outcomes:

1. Outline how the information within DNA is used to make mRNA and a polypeptide.
2. Explain the function of the genetic code.
3. List a few exceptions to the genetic code.
4. Describe the four levels of protein structure.
5. Compare and contrast the functions of proteins.

Thus far, we have considered experiments that led to the conclusion that some genes encode enzymes. The sequence of a protein-encoding gene provides a template for the synthesis of mRNA. In turn, the mRNA contains the information to synthesize a polypeptide. In this section, we will examine the general features of the genetic code—the sequence of bases in a codon that specifies an amino acid. In addition, we will look at the biochemistry of polypeptide synthesis and explore the structure and function of proteins. Ultimately, proteins are largely responsible for determining the characteristics of living cells and an organism's traits.

During Translation, the Genetic Code Within mRNA Is Used to Make a Polypeptide with a Specific Amino Acid Sequence

Why have researchers named this process translation? At the molecular level, **translation** involves an interpretation of one language—the language of mRNA, a nucleotide sequence—into the language of proteins—an amino acid sequence. **Figure 13.3** emphasizes how a gene stores information to make a polypeptide. As discussed in Chapter 12, the first step to access this information is called transcription in which mRNA is made. During the second step, called translation, the information within mRNA is used to make a polypeptide.

The ability of mRNA to be translated into a specific sequence of amino acids relies on the **genetic code.** The sequence of bases within an mRNA molecule provides coded information that is read in groups of three nucleotides known as codons (see Figure 13.3). The sequence of three bases in most codons specifies a particular amino acid. These codons are termed **sense codons.** For example, the codon AGC specifies the amino acid serine. The codon AUG, which specifies methionine, is used as a **start codon;** it is usually the first codon that begins a polypeptide sequence. The AUG codon can also be used to specify additional methionines within the coding sequence. Finally, three codons, UAA, UAG, and UGA, which are known as **stop codons,** are used to end the process of translation. Stop codons are also known as **termination codons** or **nonsense codons.**

The codons in mRNA are recognized by the anticodons in transfer RNA (tRNA) molecules (see Figure 13.3). **Anticodons** are 3-nucleotide sequences that are complementary to codons in mRNA. The tRNA molecules carry the amino acids that are specified by the codons in the mRNA. In this way, the order of codons in mRNA dictates the order of amino acids within a polypeptide.

The details of the genetic code are shown in **Table 13.1.** Because polypeptides are composed of 20 different kinds of amino acids, a minimum of 20 codons is needed to specify each type. With four types of bases in mRNA (A, U, G, and C), a genetic code containing two bases in a codon would not be sufficient because it would have only 4^2, or 16, possible types. By comparison, a three-base codon system can specify 4^3, or 64, different codons. Because the number of possible codons exceeds 20—which is the number of different types of amino acids—the genetic code is termed **degenerate.** This means that more than one codon can specify the same amino acid. For example, the codons GGU, GGC, GGA, and GGG all specify the amino acid glycine. Such codons are termed **synonymous codons.**

The start codon (AUG) defines the **reading frame** of an mRNA—a sequence of codons determined by reading the bases in groups of three, beginning with the start codon. This concept is best understood with a few examples. The mRNA sequence shown below encodes a short polypeptide with 7 amino acids:

5′–AUGCCCGGAGGCACCGUCCAAU–3′
Met–Pro–Gly–Gly–Thr–Val–Gln

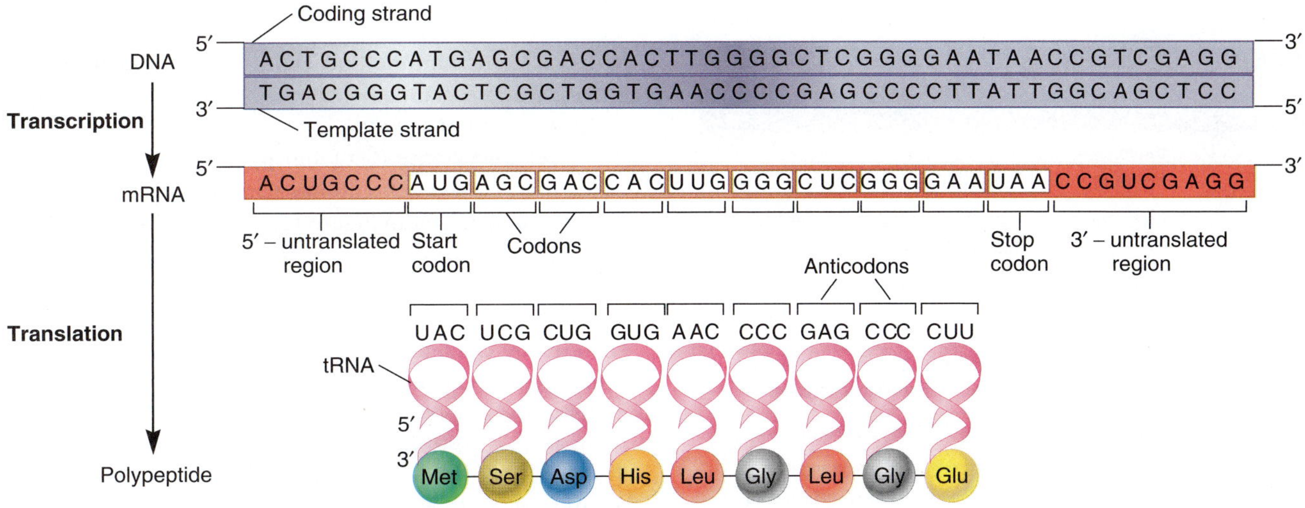

FIGURE 13.3 The relationships among the DNA coding sequence, mRNA codons, tRNA anticodons, and amino acids in a polypeptide. The sequence of nucleotides within DNA is transcribed to make a complementary sequence of nucleotides within mRNA. This sequence of nucleotides in mRNA is translated into a sequence of amino acids of a polypeptide. tRNA molecules act as intermediates in this translation process.

CONCEPT CHECK: Describe the role of DNA in the synthesis of a polypeptide.

TABLE 13.1

The Genetic Code

First base	Second base: U	Second base: C	Second base: A	Second base: G	Third base
U	UUU Phenylalanine (Phe) UUC Phenylalanine (Phe) UUA Leucine (Leu) UUG Leucine (Leu)	UCU Serine (Ser) UCC Serine (Ser) UCA Serine (Ser) UCG Serine (Ser)	UAU Tyrosine (Tyr) UAC Tyrosine (Tyr) UAA Stop codon UAG Stop codon	UGU Cysteine (Cys) UGC Cysteine (Cys) UGA Stop codon UGG Tryptophan (Trp)	U C A G
C	CUU Leucine (Leu) CUC Leucine (Leu) CUA Leucine (Leu) CUG Leucine (Leu)	CCU Proline (Pro) CCC Proline (Pro) CCA Proline (Pro) CCG Proline (Pro)	CAU Histidine (His) CAC Histidine (His) CAA Glutamine (Gln) CAG Glutamine (Gln)	CGU Arginine (Arg) CGC Arginine (Arg) CGA Arginine (Arg) CGG Arginine (Arg)	U C A G
A	AUU Isoleucine (Ile) AUC Isoleucine (Ile) AUA Isoleucine (Ile) AUG Methionine (Met); start codon	ACU Threonine (Thr) ACC Threonine (Thr) ACA Threonine (Thr) ACG Threonine (Thr)	AAU Asparagine (Asn) AAC Asparagine (Asn) AAA Lysine (Lys) AAG Lysine (Lys)	AGU Serine (Ser) AGC Serine (Ser) AGA Arginine (Arg) AGG Arginine (Arg)	U C A G
G	GUU Valine (Val) GUC Valine (Val) GUA Valine (Val) GUG Valine (Val)	GCU Alanine (Ala) GCC Alanine (Ala) GCA Alanine (Ala) GCG Alanine (Ala)	GAU Aspartic acid (Asp) GAC Aspartic acid (Asp) GAA Glutamic acid (Glu) GAG Glutamic acid (Glu)	GGU Glycine (Gly) GGC Glycine (Gly) GGA Glycine (Gly) GGG Glycine (Gly)	U C A G

If we remove one base (C) adjacent to the start codon, this changes the reading frame to produce a different polypeptide sequence:

5′–AUGCCGGAGGCACCGUCCAAU–3′

Met–Pro–Glu–Ala–Pro–Ser–Asn

Alternatively, if we remove three bases (CCC) next to the start codon, the resulting polypeptide has the same reading frame as the first polypeptide, though one amino acid (Pro, proline) has been deleted:

5′–AUGGGAGGCACCGUCCAAU–3′

Met–Gly–Gly–Thr–Val–Gln

Exceptions to the Genetic Code Are Known to Occur, Which Includes the Incorporation of Selenocysteine and Pyrrolysine into Polypeptides

From the analysis of many different species, including bacteria, archaea, protists, fungi, plants, and animals, researchers have found that the genetic code is nearly universal. However, a few exceptions to the genetic code have been noted (Table 13.2). The eukaryotic organelles known as mitochondria have their own DNA, which encodes a few protein-encoding genes. In mammals, the mitochondrial genetic code contains differences such

TABLE 13.2

Examples of Exceptions to the Genetic Code

Codon	Universal Meaning	Exception*
AUA	Isoleucine	Methionine in yeast and mammalian mitochondria
UGA	Stop	Tryptophan in mammalian mitochondria
CUU, CUC, CUA, CUG	Leucine	Threonine in yeast mitochondria
AGA, AGG	Arginine	Stop codon in ciliated protozoa and in yeast and mammalian mitochondria
UAA, UAG	Stop	Glutamine in ciliated protozoa
UGA	Stop	Encodes selenocysteine in certain genes found in bacteria, archaea, and eukaryotes
UAG	Stop	Encodes pyrrolysine in certain genes found in methane-producing archaea

*Several other exceptions, sporadically found among various species, are also known.

as AUA = methionine and UGA = tryptophan. Also, in mitochondria and certain ciliated protists, AGA and AGG specify stop codons instead of arginine.

Selenocysteine (Sec) and **pyrrolysine** (Pyl) are sometimes called the 21st and 22nd amino acids in polypeptides. Their structures are shown later in Figure 13.5f. Selenocysteine is found in several enzymes involved in oxidation-reduction reactions in bacteria, archaea, and eukaryotes. Pyrrolysine is found in a few enzymes of methane-producing archaea. Selenocysteine and pyrrolysine are encoded by UGA and UAG codons, respectively, which normally function as stop codons. Like the standard 20 amino acids, selenocysteine and pyrrolysine are bound to tRNAs that specifically carry them to the ribosome for their incorporation into polypeptides. The anticodon of the tRNA that carries selenocysteine is complementary to a UGA codon, and the tRNA that carries pyrrolysine has an anticodon that is complementary to UAG.

How are UGA and UAG codons "recoded" to specify the incorporation of selenocysteine or pyrrolysine, respectively? In the case of selenocysteine, a UGA codon is followed by a sequence called the selenocysteine insertion sequence (SECIS), which forms a stem-loop. In bacteria, a SECIS may be located immediately following the UGA codon, whereas in archaea and eukaryotes the SECIS may be further downstream in the 3′-untranslated region of the mRNA. The SECIS is recognized by proteins that favor the binding of a UGA codon to a tRNA carrying selenocysteine instead of the binding of release factors that are needed for polypeptide termination. Similarly, pyrrolysine incorporation may involve sequences downstream from a UAG codon that form a stem-loop.

A Polypeptide Has Directionality from Its Amino-Terminus to Its Carboxyl-Terminus

Let's now turn our attention to polypeptide biochemistry. Polypeptide synthesis has a directionality that parallels the order of codons in the mRNA. As a polypeptide is made, a **peptide bond** is formed between the carboxyl group in the last amino acid of the polypeptide and the amino group in the amino acid being added. As shown in **Figure 13.4a**, this occurs via a condensation reaction that releases a water molecule. The newest amino acid added to a growing polypeptide always has a free carboxyl group. **Figure 13.4b** compares the sequence of a very short polypeptide

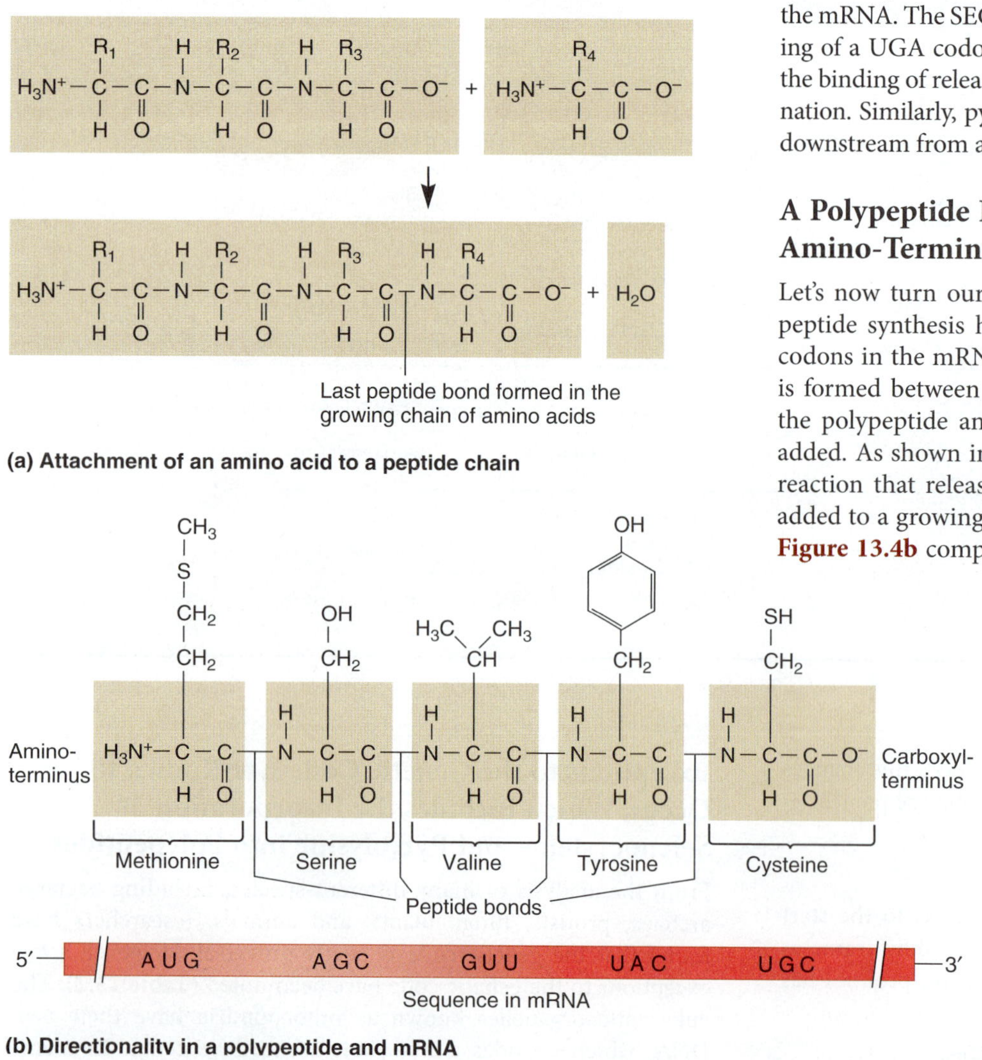

FIGURE 13.4 **The directionality of polypeptide synthesis.** **(a)** An amino acid is connected to a polypeptide via a condensation reaction that releases a water molecule. The letter R is a general designation for an amino acid side chain. **(b)** The first amino acid in a polypeptide (usually methionine) is located at the amino-terminus, and the last amino acid is at the carboxyl-terminus. Thus, the directionality of amino acids in a polypeptide is from the amino-terminus to the carboxyl-terminus, which corresponds to the 5′ to 3′ orientation of codons in mRNA.

with the mRNA that encodes it. The first amino acid is said to be at the **N-terminus,** or **amino-terminus,** of the polypeptide. An amino group (NH_3^+) is found at this site. The term N-terminus refers to the presence of a nitrogen atom (N) at this end. The first amino acid is specified by a codon that is near the 5′ end of the mRNA. By comparison, the last amino acid in a completed polypeptide is located at the **C-terminus,** or **carboxyl-terminus.** A carboxyl group (COO^-) is always found at this site in the polypeptide. This last amino acid is specified by a codon that is closer to the 3′ end of the mRNA.

The Amino Acid Sequences of Polypeptides Determine the Structure and Function of Proteins

Now that we understand how mRNAs encode polypeptides, let's consider the structure and function of the gene product, namely, polypeptides. **Figure 13.5** shows the 20 different amino acids that are most commonly found within polypeptides. Each amino acid contains a unique **side chain,** or **R group,** that has its own particular chemical properties. For example, aliphatic and aromatic amino acids are relatively nonpolar, which means they are

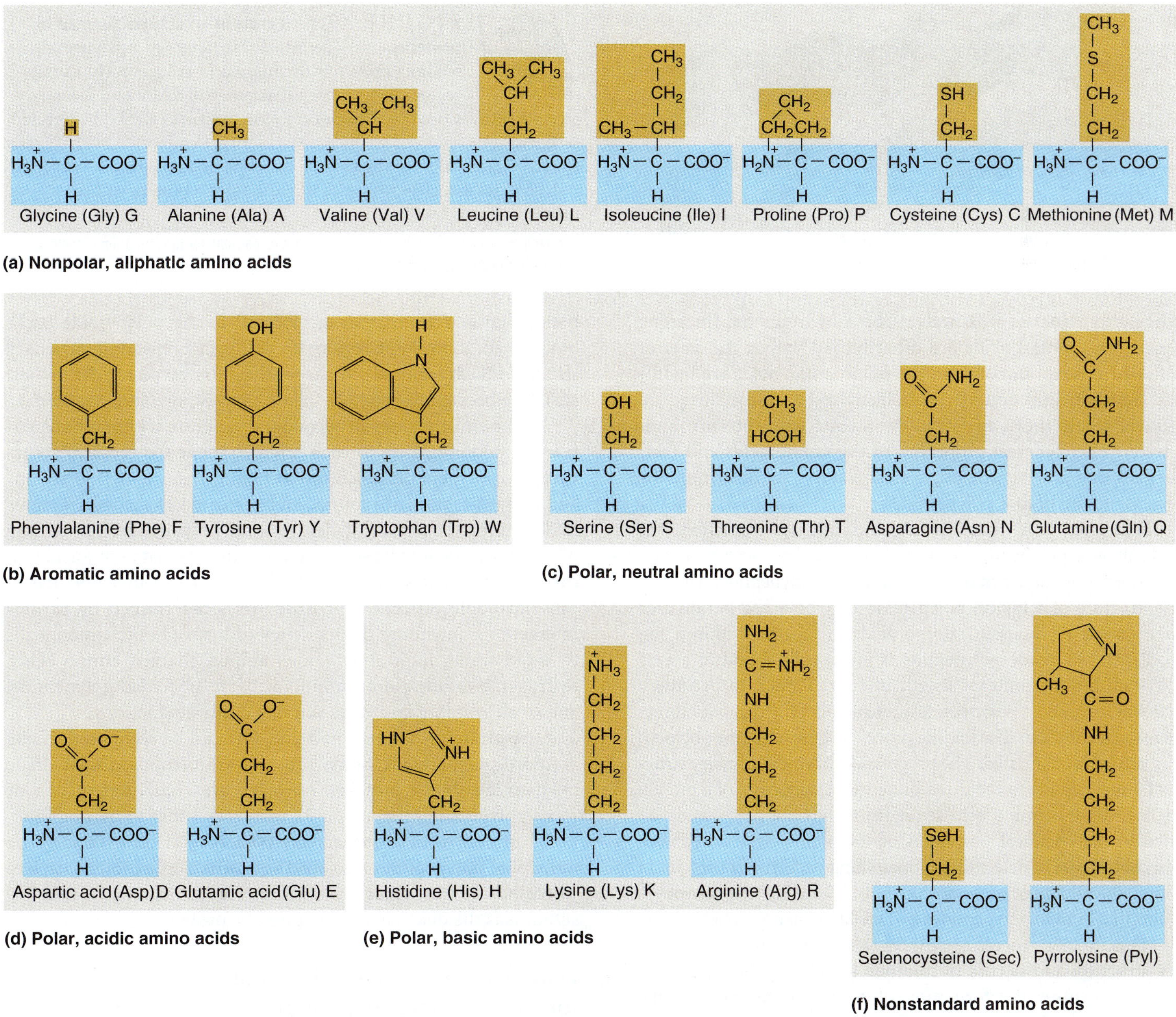

FIGURE 13.5 The amino acids that are incorporated into polypeptides during translation. Parts (a) through (e) show the 20 standard amino acids, and part (f) shows two amino acids that are occasionally incorporated into polypeptides by the use of stop codons (see Table 13.2). The structures of amino acid side chains can also be covalently modified after a polypeptide is made, a phenomenon called posttranslational modification.

CONCEPT CHECK: Which two amino acids do you think are the least soluble in water?

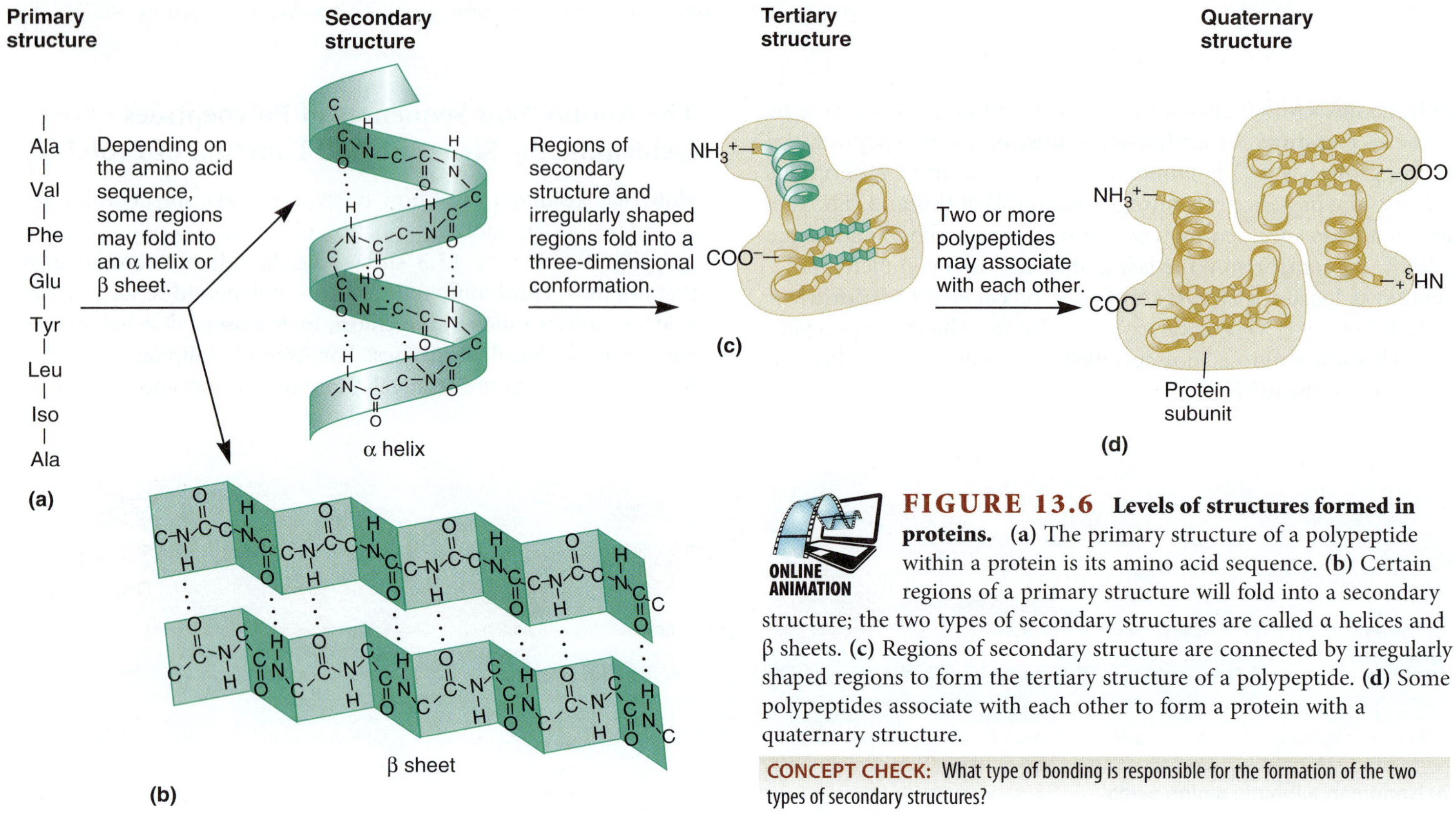

FIGURE 13.6 Levels of structures formed in proteins. (a) The primary structure of a polypeptide within a protein is its amino acid sequence. (b) Certain regions of a primary structure will fold into a secondary structure; the two types of secondary structures are called α helices and β sheets. (c) Regions of secondary structure are connected by irregularly shaped regions to form the tertiary structure of a polypeptide. (d) Some polypeptides associate with each other to form a protein with a quaternary structure.

CONCEPT CHECK: What type of bonding is responsible for the formation of the two types of secondary structures?

less likely to associate with water. These hydrophobic (meaning water-fearing) amino acids are often buried within the interior of a folded protein. In contrast, the polar amino acids are hydrophilic (water-loving) and are more likely to be on the surface of a protein, where they can favorably interact with the surrounding water. The chemical properties of the amino acids and their sequences in a polypeptide are critical factors that determine the unique structure of that polypeptide.

Following gene transcription and mRNA translation, the end result is a polypeptide with a defined amino acid sequence. This sequence is the **primary structure** of a polypeptide. The primary structure of a typical polypeptide may be a few hundred or even a couple of thousand amino acids in length. Within a living cell, a newly made polypeptide is not usually found in a long linear state for a significant length of time. Rather, to become a functional unit, most polypeptides quickly adopt a compact three-dimensional structure. The folding process begins while the polypeptide is still being translated. The progression from the primary structure of a polypeptide to the three-dimensional structure of a protein is dictated by the amino acid sequence within the polypeptide. In particular, the chemical properties of the amino acid side chains play a central role in determining the folding pattern of a protein. In addition, the folding of some polypeptides is aided by **chaperones**—proteins that bind to polypeptides and facilitate their proper folding.

This folding process of polypeptides is governed by the primary structure and occurs in multiple stages (**Figure 13.6**). The first stage involves the formation of a regular, repeating shape known as a **secondary structure.** The two main types of secondary structures are the **α helix** and the **β sheet** (Figure 13.6b). A single polypeptide may have some regions that fold into an α helix and other regions that fold into a β sheet. Secondary structures within polypeptides are primarily stabilized by the formation of hydrogen bonds between atoms that are located in the polypeptide backbone. In addition, some regions do not form a repeating secondary structure. Such regions have shapes that look very irregular in their structure because they do not follow a repeating folding pattern.

The short regions of secondary structure within a polypeptide are folded relative to each other to make the **tertiary structure** of a polypeptide. As shown in Figure 13.6c, α-helical regions and β-sheet regions are connected by irregularly shaped segments to determine the tertiary structure of the polypeptide. The folding of a polypeptide into its secondary and then tertiary conformation can usually occur spontaneously because it is a thermodynamically favorable process. The structure is determined by various interactions, including the tendency of hydrophobic amino acids to avoid water, ionic interactions among charged amino acids, hydrogen bonding among amino acids in the folded polypeptide, and weak bonding known as van der Waals interactions.

A protein is a functional unit that can be composed of one or more polypeptides. Some proteins are composed of a single polypeptide. Many proteins, however, are composed of two or more polypeptides that associate with each other to make a functional protein with a **quaternary structure** (Figure 13.6d). The individual polypeptides are called **subunits** of the protein, each of which has its own tertiary structure. The association of multiple subunits is the quaternary structure of a protein.

Proteins Are Primarily Responsible for the Characteristics of Living Cells and an Organism's Traits

Why is the genetic material largely devoted to storing the information to make proteins? To a great extent, the characteristics of a cell depend on the types of proteins that it makes. In turn, the

TABLE 13.3
Functions of Selected Cellular Proteins

Function	Examples
Cell shape and organization	Tubulin: Forms cytoskeletal structures known as microtubules
Transport	Sodium channels: Transport sodium ions across the nerve cell membranes
	Hemoglobin: Transports oxygen in red blood cells
Movement	Myosin: Involved in muscle cell contraction
Cell signaling	Insulin: A hormone that influences cell metabolism and growth
	Insulin receptor: Recognizes insulin and initiates a cell response
Cell surface recognition	Integrins: Bind to large extracellular proteins
Enzymes	Hexokinase: Phosphorylates glucose during the first step in glycolysis
	β-Galactosidase: Cleaves lactose into glucose and galactose
	Glycogen synthetase: Uses glucose molecules as building blocks to synthesize a large carbohydrate known as glycogen
	RNA polymerase: Uses ribonucleotides as building blocks to synthesize RNA
	DNA polymerase: Uses deoxyribonucleotides as building blocks to synthesize DNA

traits of multicellular organisms are determined by the properties of their cells. Proteins perform a variety of roles that are critical to the life of cells and to the morphology and function of organisms. Table 13.3 describes several examples of how proteins function.

13.2 COMPREHENSION QUESTIONS

1. What is the genetic code?
   a. The relationship between a three-base codon sequence and an amino acid
   b. The entire base sequence of an mRNA molecule
   c. The entire sequence from the promoter to the terminator of a gene
   d. The binding of tRNA to mRNA
2. The reading frame begins with a ______________ and is read ______________.
   a. promoter, one base at a time
   b. promoter, in groups of three bases
   c. start codon, one base at a time
   d. start codon, in groups of three bases
3. The fourth codon in an mRNA-coding sequence is GGG, which specifies glycine. If we assume that no amino acids are removed from the polypeptide, which of the following statements is correct?
   a. The third amino acid from the N-terminus is glycine.
   b. The fourth amino acid from the N-terminus is glycine.
   c. The third amino acid from the C-terminus is glycine.
   d. The fourth amino acid from the C-terminus is glycine.
4. A type of secondary structure found in proteins is
   a. an α helix.
   b. a β sheet.
   c. both a and b.
   d. none of the above.

13.3 EXPERIMENTAL DETERMINATION OF THE GENETIC CODE

Learning Outcome:

1. Compare and contrast the experiments of (1) Nirenberg and Matthaei, (2) Khorana, and (3) Nirenberg and Leder that were instrumental in deciphering the genetic code.

In the previous section, we examined how the genetic code determines the amino acid sequence of a polypeptide. In this section, we will consider the experimental approaches that deduced the genetic code.

EXPERIMENT 13A

Synthetic RNA Helped to Decipher the Genetic Code

How did scientists determine the functions of the 64 codons of the genetic code? During the early 1960s, three research groups headed by Marshall Nirenberg, Severo Ochoa, and H. Gobind Khorana set out to decipher the genetic code. Though they used different methods, all of these groups used synthetic mRNA in their experimental approaches to "crack the code." We first consider the work of Nirenberg and his colleagues. Prior to their studies, several laboratories had already determined that extracts from bacterial cells, containing a mixture of components including ribosomes, tRNAs, and other factors required for translation, are able to synthesize polypeptides if mRNA and amino acids are added. This mixture is termed an in vitro, or **cell-free, translation system.** If radiolabeled amino acids are added to a cell-free translation system, the synthesized polypeptides are radiolabeled and easy to detect.

To decipher the genetic code, Nirenberg and colleagues needed to gather information regarding the relationship between mRNA composition and polypeptide composition. To accomplish this goal, they made mRNA molecules of a known base composition, added them to a cell-free translation system, and then analyzed the amino acid composition of the resultant polypeptides. For example, if an mRNA molecule consisted of a string of adenine-containing nucleotides (e.g., 5′–AAAAAAAAAAAAAAAA–3′), researchers could add this polyA mRNA to a cell-free translation system and ask the question, "Which amino acid is specified by a codon that contains only adenine nucleotides?" (As Table 13.1 shows, it is lysine.)

Before discussing the details of this type of experiment, let's consider how the synthetic mRNA molecules were made. To synthesize mRNA, an enzyme known as polynucleotide phosphorylase was used. In the presence of excess ribonucleoside diphosphates, also called nucleoside diphosphates (NDPs), this enzyme catalyzes the covalent linkage of nucleotides to make a polymer of RNA. Because it does not use a DNA template, the order of the nucleotides is random. For example, if nucleotides containing two different bases, such as uracil and guanine, are added, then polynucleotide phosphorylase makes a random polymer containing both nucleotides (5′–GGGUGUGUGGUGGGUG–3′). An experimenter can control the amounts of the nucleotides that are added. For example, if 70% G and 30% U are mixed together with polynucleotide phosphorylase, the predicted amounts of the codons within the random polymer are as follows:

Codon Possibilities	*Percentage in the Random Polymer*
GGG	$0.7 \times 0.7 \times 0.7 = 0.34 =$ 34%
GGU	$0.7 \times 0.7 \times 0.3 = 0.15 =$ 15%
GUU	$0.7 \times 0.3 \times 0.3 = 0.06 =$ 6%
UUU	$0.3 \times 0.3 \times 0.3 = 0.03 =$ 3%
UUG	$0.3 \times 0.3 \times 0.7 = 0.06 =$ 6%
UGG	$0.3 \times 0.7 \times 0.7 = 0.15 =$ 15%
UGU	$0.3 \times 0.7 \times 0.3 = 0.06 =$ 6%
GUG	$0.7 \times 0.3 \times 0.7 = 0.15 =$ 15%
	100%

By controlling the amounts of the NDPs in the phosphorylase reaction, the relative amounts of the possible codons can be predicted.

The first experiment that demonstrated the ability to synthesize polypeptides from synthetic mRNA was performed by Marshall Nirenberg and J. Heinrich Matthaei in 1961. As shown in **Figure 13.7**, a cell-free translation system was placed into 20 different tubes. An mRNA template made via polynucleotide phosphorylase was then added into each tube. In this example, the mRNA was made using 70% G and 30% U. Next, the 20 amino acids were added to each tube, but each tube differed with regard to the type of radiolabeled amino acid. For example, radiolabeled glycine would be found in only 1 of the 20 tubes. The tubes were incubated a sufficient length of time to allow translation to occur. The newly made polypeptides were then precipitated by treatment with trichloroacetic acid. This step precipitates polypeptides but not amino acids. The contents of each tube were then subjected to filtration. The precipitated polypeptides were captured on a filter, whereas amino acids that had not been incorporated into polypeptides remained in solution and passed through the filter. Finally, the amount of radioactivity captured on the filter was determined by liquid scintillation counting.

THE GOAL (DISCOVERY-BASED SCIENCE)

The researchers assumed that the sequence of bases in mRNA determines the incorporation of specific amino acids into a polypeptide. The purpose of this experiment was to provide information that would help to decipher the relationship between base composition and particular amino acids.

ACHIEVING THE GOAL

FIGURE 13.7 Elucidation of the genetic code.

Starting material: A cell-free translation system that can synthesize polypeptides if mRNA and amino acids are added. The translation system contained ribosomes, tRNAs, and other factors required for translation. These other factors included enzymes that attach amino acids to tRNA molecules.

1. Place the cell-free translation system into 20 tubes.

2. To each tube, add random mRNA polymers of G and U made via polynucleotide phosphorylase using 70% G and 30% U.

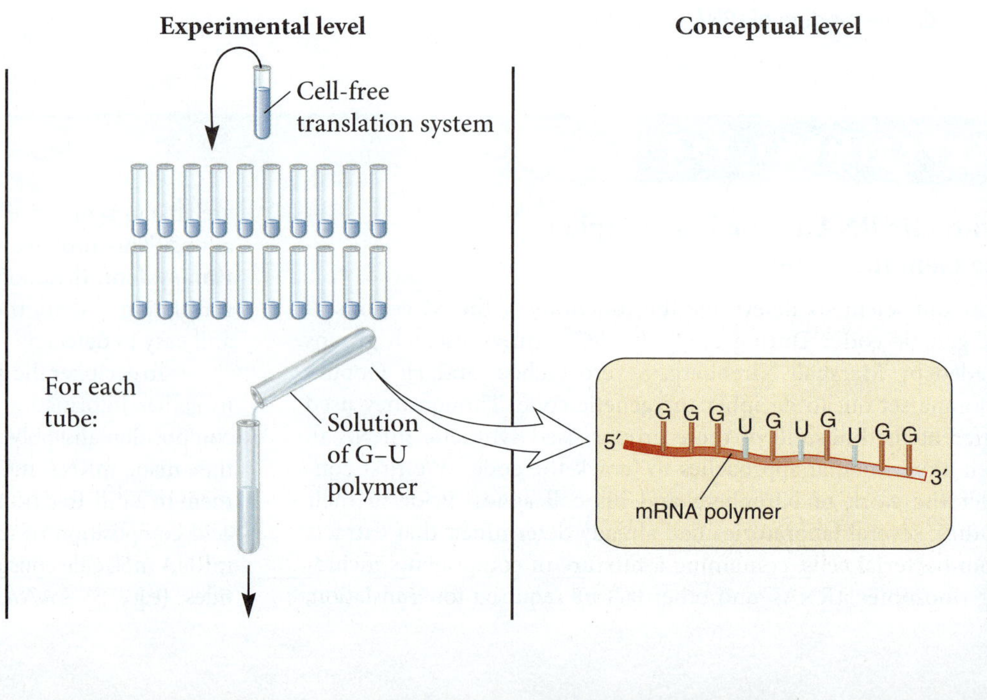

3. Add a different radiolabeled amino acid to each tube, and add the other 19 nonradiolabeled amino acids. In this example, glycine was radiolabeled.

4. Incubate for 60 minutes to allow translation to occur.

5. Add 15% trichloroacetic acid (TCA), which precipitates polypeptides but not amino acids.

6. Capture the precipitated polypeptides on a filter. Note: Amino acids that were not incorporated into polypeptides pass through the filter.

7. Count the radioactivity on the filter in a scintillation counter (see the Appendix for a description).

8. Calculate the amount of radiolabeled amino acids in the precipitated polypeptides.

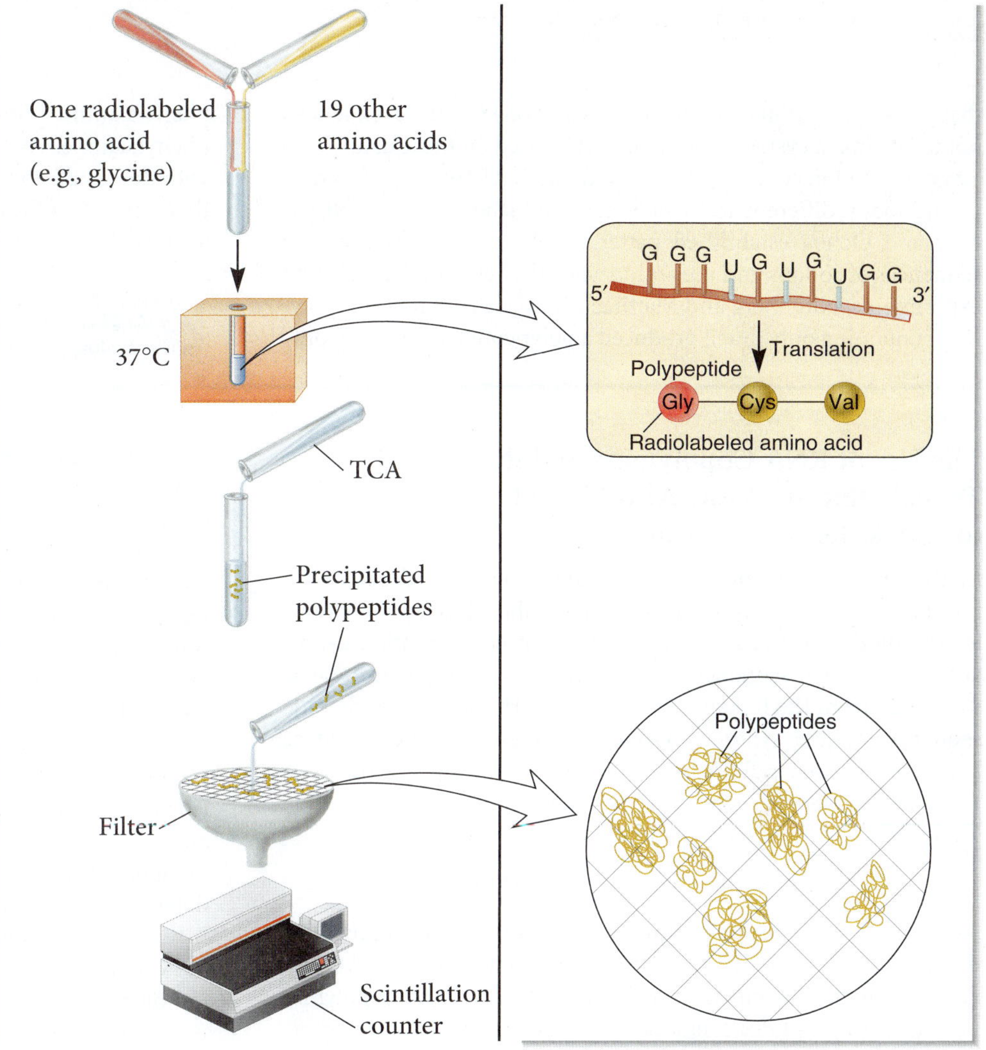

THE DATA

Radiolabeled Amino Acid Added	*Relative Amount of Radiolabeled Amino Acid Incorporated into Translated Polypeptides (% of total)*	*Radiolabeled Amino Acid Added*	*Relative Amount of Radiolabeled Amino Acid Incorporated into Translated Polypeptides (% of total)*
Alanine	0	Leucine	6
Arginine	0	Lysine	0
Asparagine	0	Methionine	0
Aspartic acid	0	Phenylalanine	3
Cysteine	6	Proline	0
Glutamic acid	0	Serine	0
Glutamine	0	Threonine	0
Glycine	49	Tryptophan	15
Histidine	0	Tyrosine	0
Isoleucine	0	Valine	21

Adapted from Nirenberg, Marshall W., and Matthaei, J.H. (1961) The dependence of cell-free protein synthesis in *E. coli* upon naturally occurring or synthetic polyribonucleotides. *Proc Natl Acad Sci USA 47*, 1588–1602.

INTERPRETING THE DATA

According to the calculation previously described, codons should occur in the following percentages: 34% GGG, 15% GGU, 6% GUU, 3% UUU, 6% UUG, 15% UGG, 6% UGU, and 15% GUG. We now know that the value of 49% for glycine is due to two codons: GGG (34%) and GGU (15%). The 6% cysteine is due to UGU, and so on. It is important to realize that the genetic code was not deciphered in a single experiment such as the one described here. Furthermore, this kind of experiment yields information regarding only the nucleotide content of codons, not the specific order of bases within a single codon. For example, this experiment indicates that a cysteine codon contains two U's and one G. However, it does not tell us that a cysteine codon is UGU. Based on these data alone, a cysteine codon could be UUG, GUU, or UGU. However, by comparing many different RNA polymers, the laboratories of Nirenberg and Ochoa established patterns between the specific base sequences of codons and the amino acids they encode. In their first experiments, Nirenberg and Matthaei showed that a random polymer containing only uracil produced a polypeptide containing only phenylalanine. From this result, they inferred that UUU specifies phenylalanine. As shown in The Data, this idea is also consistent with the results involving a random 70% G and 30% U polymer. In this case, 3% of the codons will be UUU. Likewise, 3% of the amino acids within the polypeptides were found to be phenylalanine.

A self-help quiz involving this experiment can be found at www.mhhe.com/brookergenetics5e.

The Use of RNA Copolymers and the Triplet-Binding Assay Also Helped to Crack the Genetic Code

In the 1960s, H. Gobind Khorana and colleagues developed a novel method for synthesizing RNA. First, they developed a chemical method to make DNA molecules that had repeating 2-, 3-, and 4-nucleotide sequences. This is called a copolymer, because it is made from the linkage of smaller di-, tri-, or tetra-nucleotide sequences. These synthetic DNA molecules were then used as templates to make RNA molecules with repeating sequences. For example, RNA molecules with the repeating trinucleotide sequence 5′–AUC–3′ were made

5′—AUCAUCAUCAUCAUCAUCAUCAUCAUCAUC—3′

Depending on the reading frame, such a copolymer contains three different codons: AUC, UCA, and CAU. Using a cell-free translation system like the one described in Figure 13.7, such a copolymer produced polypeptides containing isoleucine, serine, and histidine. **Table 13.4** summarizes some of the copolymers that were made using this approach and the amino acids that were incorporated into polypeptides.

Finally, another method that helped to decipher the genetic code also involved the chemical synthesis of short RNA molecules. In 1964, Marshall Nirenberg and Philip Leder discovered that an RNA molecule composed of 3 nucleotides—a triplet—could cause a ribosome to bind a tRNA. In other words, the RNA triplet acted like a codon.

As an example, in one experiment the researchers began with a sample of ribosomes that were mixed with 5′–CCC–3′ triplets. Portions of this sample were then added to 20 different tubes that had tRNAs with different radiolabeled amino acids. For example, one tube contained radiolabeled histidine, a second tube had radiolabeled proline, a third tube contained radiolabeled glycine, and so on. Only one radiolabeled amino acid was added to each tube. After allowing sufficient time for tRNAs to bind to the ribosomes, the samples were filtered; only the large ribosomes and anything bound to them were trapped on the filter (**Figure 13.8**). Unbound tRNAs passed through the filter. Next, the researchers determined the amount of radioactivity trapped on each filter. If the filter contained a large amount of radioactivity, the results indicated that the added triplet encoded the amino acid that was radiolabeled.

Using the triplet-binding assay, Nirenberg and Leder were able to establish relationships between particular RNA triplet sequences and the binding of tRNAs carrying specific (radiolabeled) amino acids. In the case of the 5′–CCC–3′ triplet, they determined that tRNAs carrying radiolabeled proline were bound to the ribosomes. Unfortunately, in some cases, a triplet could not promote sufficient tRNA binding to yield unambiguous results. Nevertheless, the triplet-binding assay was an important tool in the identification of the majority of codons.

TABLE 13.4

Examples of Copolymers That Were Analyzed by Khorana and Colleagues

Synthetic RNA*	Codon Possibilities	Amino Acids Incorporated into Polypeptides
UC	UCU, CUC	Serine, leucine
AG	AGA, GAG	Arginine, glutamic acid
UG	UGU, GUG	Cysteine, valine
AC	ACA, CAC	Threonine, histidine
UUC	UUC, UCU, CUU	Phenylalanine, serine, leucine
AAG	AAG, AGA, GAA	Lysine, arginine, glutamic acid
UUG	UUG, UGU, GUU	Leucine, cysteine, valine
CAA	CAA, AAC, ACA	Glutamine, asparagine, threonine
UAUC	UAU, AUC, UCU, CUA	Tyrosine, isoleucine, serine, leucine
UUAC	UUA, UAC, ACU, CUU	Leucine, tyrosine, threonine

*The synthetic RNAs were made using DNA templates that were composed of copolymers.

13.3 COMPREHENSION QUESTIONS

1. Let's suppose a researcher mixed together nucleotides with the following percentage of bases: 30% G, 30% C, and 40% A. If RNA was made via polynucleotide phosphorylase, what percentage of the codons would be 5′-GGC-3′?
   a. 30%
   b. 9%
   c. 2.7%
   d. 0%
2. In the triplet-binding assay of Nirenberg and Leder, an RNA triplet composed of three bases was able to cause the
   a. translation of a polypeptide.
   b. binding of a tRNA carrying the appropriate amino acid.
   c. termination of translation.
   d. release of the amino acid from the tRNA.

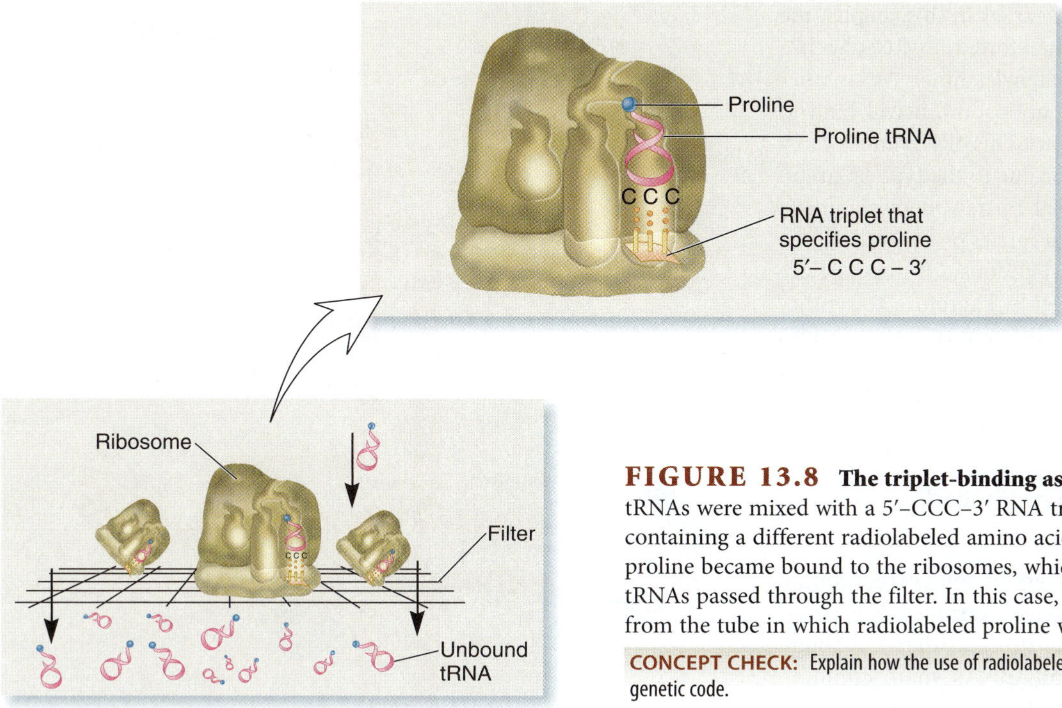

FIGURE 13.8 The triplet-binding assay. In this experiment, ribosomes and tRNAs were mixed with a 5′–CCC–3′ RNA triplet in 20 separate tubes, with each tube containing a different radiolabeled amino acid (not shown). Only tRNAs carrying proline became bound to the ribosomes, which became trapped on the filter. Unbound tRNAs passed through the filter. In this case, radioactivity was trapped on the filter only from the tube in which radiolabeled proline was added.

CONCEPT CHECK: Explain how the use of radiolabeled amino acids in this procedure helped to reveal the genetic code.

13.4 STRUCTURE AND FUNCTION OF tRNA

Learning Outcomes:

1. Describe the specificity between a tRNA and a codon in mRNA.
2. Describe the key structural features of a tRNA molecule.
3. Explain how an amino acid is attached to a tRNA via aminoacyl-tRNA synthetase.
4. Outline the wobble rules.

Thus far, we have considered the general features of translation and surveyed the structure and functional significance of proteins. The rest of this chapter is devoted to a molecular understanding of translation as it occurs in living cells. Biochemical studies of protein synthesis began in the 1950s. As work progressed toward an understanding of translation, research revealed that certain types of RNA molecules are involved in the incorporation of amino acids into growing polypeptides. Francis Crick proposed the **adaptor hypothesis.** According to this idea, the position of an amino acid within a polypeptide is determined by the binding between the mRNA and an adaptor molecule carrying a specific amino acid. Later, work by Paul Zamecnik and Mahlon Hoagland suggested that the adaptor molecule is tRNA. During translation, a tRNA has two functions: (1) It recognizes a three-base codon sequence in mRNA, and (2) it carries an amino acid specific for that codon. In this section, we will examine the general function of tRNA molecules. We begin by exploring the important structural features that underlie tRNA function.

The Function of a tRNA Depends on the Specificity Between the Amino Acid It Carries and Its Anticodon

The adaptor hypothesis proposes that tRNA molecules recognize the codons within mRNA and carry the correct amino acids to the site of polypeptide synthesis. During mRNA-tRNA recognition, the anticodon in a tRNA molecule binds to a codon in mRNA in an antiparallel manner and according to the AU/GC rule (**Figure 13.9**). For example, if the anticodon in the tRNA is 3′–AAG–5′, it will bind to a 5′–UUC–3′ codon. Importantly, the anticodon in the tRNA corresponds to the amino acid that it carries.

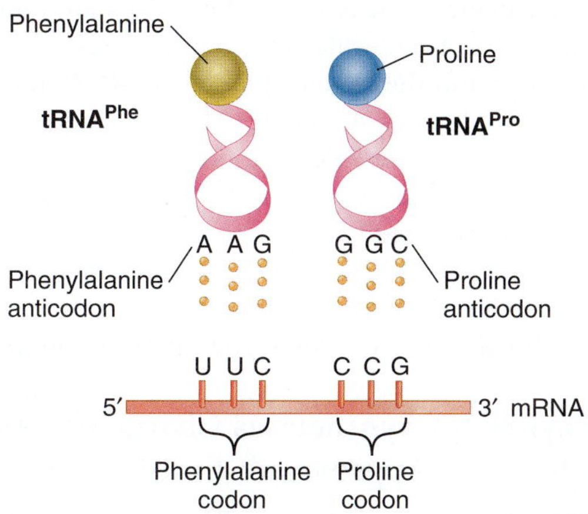

FIGURE 13.9 Recognition between tRNAs and mRNA. The anticodon in the tRNA binds to a complementary sequence in the mRNA. At its 3′ end, the tRNA carries the amino acid that corresponds to the codon in the mRNA via the genetic code.

CONCEPT CHECK: What are the two key functional sites of a tRNA molecule?

According to the genetic code, described earlier in this chapter, the UUC codon specifies phenylalanine. Therefore, the tRNA with a 3′–AAG–5′ anticodon must carry a phenylalanine. As another example, if the tRNA has a 3′–GGC–5′ anticodon, it will bind to a 5′–CCG–3′ codon that specifies proline. This tRNA must carry proline. tRNA molecules are named according to the type of amino acid they carry. For example, a tRNA that carries phenylalanine is described as $tRNA^{Phe}$, whereas a tRNA that carries proline is $tRNA^{Pro}$.

Recall that the genetic code has 64 codons. Of these, 61 are sense codons that specify the 20 amino acids. Therefore, to synthesize proteins, a cell must produce many different tRNA molecules having specific anticodon sequences. To do so, the chromosomal DNA contains many distinct tRNA genes that encode tRNA molecules with different sequences.

Common Structural Features Are Shared by All tRNAs

To understand how tRNAs act as carriers of the correct amino acids during translation, researchers have examined the structural characteristics of these molecules in great detail. Though a cell makes many different tRNAs, all tRNAs share common structural features. As originally proposed by Robert W. Holley in 1965, the secondary structure of tRNAs exhibits a cloverleaf pattern. A tRNA has three stem-loops, a few variable sites, and an acceptor stem with a 3′ single-stranded region (**Figure 13.10**). The acceptor stem is where an amino acid becomes attached to a tRNA (see inset). All tRNA molecules have a sequence, CCA, at their 3′ ends. These three nucleotides are usually added enzymatically by the enzyme tRNA nucleotidyltransferase after the tRNA is made. A conventional numbering system for the nucleotides within a tRNA molecule begins at the 5′ end and proceeds toward the 3′ end. Among different types of tRNA molecules, the variable sites (shown in blue) can differ in the number of nucleotides they contain. The anticodon is located in the second loop region.

The three-dimensional, or tertiary, structure of tRNA molecules involves additional folding of the secondary structure. In the tertiary structure of tRNA, the stem-loops are folded into a much more compact molecule. The ability of RNA molecules to form stem-loops and the tertiary folding of tRNA molecules are described in Chapter 9 (see Figure 9.21). In addition to the normal A, U, G, and C bases, tRNA molecules commonly contain modified bases. For example, Figure 13.10 illustrates a tRNA that contains several modified bases. Among many different species, researchers have found that more than 80 different base modifications can occur in tRNA molecules. We will explore the significance of modified bases in codon recognition later in this chapter.

Aminoacyl-tRNA Synthetases Charge tRNAs by Attaching the Appropriate Amino Acid

To function correctly, each type of tRNA must have the appropriate amino acid attached to its 3′ end. How does an amino acid get attached to the correct tRNA? Enzymes in the cell known as **aminoacyl-tRNA synthetases** catalyze the attachment of amino acids to tRNA molecules. Cells produce 20 different aminoacyl-tRNA synthetase enzymes, 1 for each of the 20 distinct amino

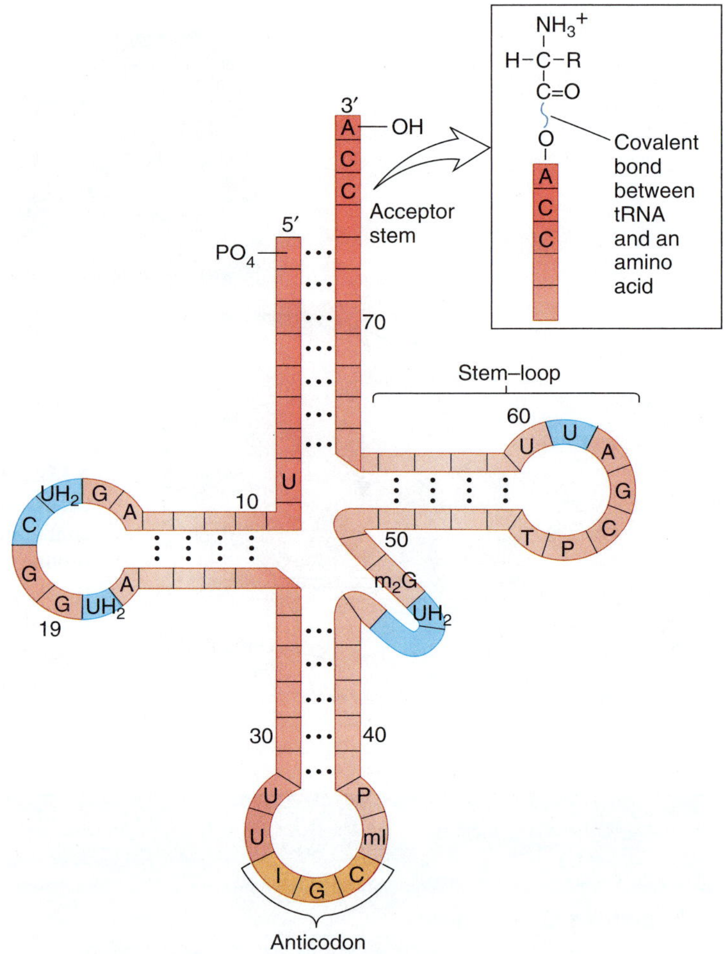

FIGURE 13.10 **Secondary structure of tRNA.** The conventional numbering of nucleotides begins at the 5′ end and proceeds toward the 3′ end. In all tRNAs, the nucleotides at the 3′ end contain the sequence CCA. Certain locations have additional nucleotides not found in all tRNA molecules. These variable sites are shown in blue. The figure also shows the locations of a few modified bases specifically found in a yeast tRNA that carries alanine. The modified bases are as follows: UH_2 = dihydrouridine, I = inosine, mI = methylinosine, P = pseudouridine, m_2G = dimethylguanosine, and T = ribothymidine. The inset shows an amino acid covalently attached to the 3′ end of a tRNA. Note: The 5′ to 3′ orientation of the anticodon in this drawing (left to right) is opposite to that of tRNAs shown in other drawings, such as Figure 13.9.

acids. Each aminoacyl-tRNA synthetase is named for the specific amino acid it attaches to tRNA. For example, alanyl-tRNA synthetase recognizes a tRNA with an alanine anticodon—$tRNA^{Ala}$—and attaches an alanine to it.

Aminoacyl-tRNA synthetases catalyze a chemical reaction involving three different molecules: an amino acid, a tRNA molecule, and ATP. In the first step of the reaction, a synthetase recognizes a specific amino acid and also ATP (**Figure 13.11**). The ATP is hydrolyzed, and AMP becomes attached to the amino acid; pyrophosphate is released. During the second step, the correct tRNA binds to the synthetase. The amino acid becomes covalently attached to the 3′ end of the tRNA molecule at the acceptor stem,

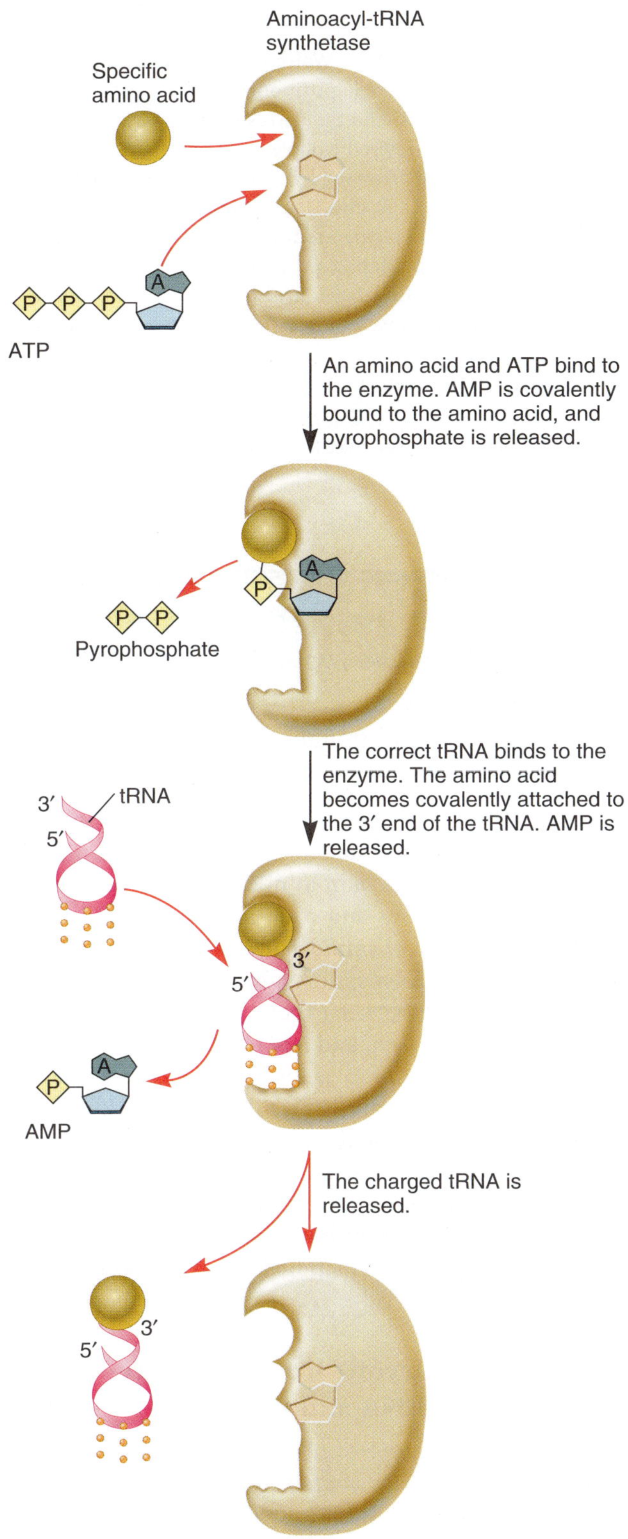

ONLINE ANIMATION

FIGURE 13.11 Catalytic function of aminoacyl-tRNA synthetase. Aminoacyl-tRNA synthetase has binding sites for a specific amino acid, ATP, and a particular tRNA. In the first step, the enzyme catalyzes the covalent attachment of AMP to an amino acid, yielding an activated amino acid. In the second step, the activated amino acid is attached to the appropriate tRNA. The charged tRNA is then released.

CONCEPT CHECK: What is the difference between a charged tRNA versus an uncharged tRNA?

and AMP is released. Finally, the tRNA with its attached amino acid is released from the enzyme. At this stage, the tRNA is called a **charged tRNA** or an **aminoacyl-tRNA.** In a charged tRNA molecule, the amino acid is attached to the 3′ end of the tRNA by a covalent bond (see Figure 13.10 inset).

The ability of the aminoacyl-tRNA synthetases to recognize tRNAs has sometimes been called the "second genetic code." This recognition process is necessary to maintain the fidelity of genetic information. The frequency of error for aminoacyl-tRNA synthetases is less than 10^{-4}. In other words, the wrong amino acid is attached to a tRNA less than once in 10,000 times! As you might expect, the anticodon region of the tRNA is usually important for precise recognition by the correct aminoacyl-tRNA synthetase. In studies of *Escherichia coli* synthetases, 17 of the 20 types of aminoacyl-tRNA synthetases recognize the anticodon region of the tRNA. However, other regions of the tRNA are also important recognition sites. These include the acceptor stem and bases in the stem-loops. Also, most aminoacyl-tRNA synthetases have a proofreading ability. They can enzymatically remove the wrong amino acid from the 3′ end.

As mentioned previously, tRNA molecules frequently contain bases within their structure that have been chemically modified. These modified bases have important effects on tRNA function. For example, modified bases within tRNA molecules affect the rate of translation and the recognition of tRNAs by aminoacyl-tRNA synthetases. Positions 34 and 37 contain the largest variety of modified nucleotides; position 34 is the first base in the anticodon that matches the third base in the codon of mRNA. As discussed next, a modified base at position 34 can have important effects on codon-anticodon recognition.

Mismatches That Follow the Wobble Rules Occur at the Third Position in Codon-Anticodon Pairing

After considering the structure and function of tRNA molecules, let's reexamine some subtle features of the genetic code. As discussed earlier, the genetic code is degenerate, which means that more than one codon can specify the same amino acid. Degeneracy usually occurs at the third position in the codon. For example, valine is specified by GUU, GUC, GUA, and GUG. In all four cases, the first two bases are always G and U. The third base, however, can be U, C, A, or G. To explain this pattern of degeneracy, Francis Crick proposed in 1966 that it is due to "wobble" at the third position in the codon-anticodon recognition process. According to the **wobble rules,** the first two positions pair strictly according to the AU/GC rule. However, the third position can tolerate certain types of mismatches (**Figure 13.12**). This proposal suggested that the base at the third position in the codon does not have to hydrogen bond as precisely with the corresponding base in the anticodon.

Because of the wobble rules, some flexibility is observed in the recognition between a codon and anticodon during the process of translation. When two or more tRNAs that differ at the wobble base are able to recognize the same codon, these are termed **isoacceptor tRNAs.** As an example, tRNAs with an anticodon of 3′–CCA–5′ or 3′–CCG–5′, which both carry glycine, can recognize a codon with the sequence of 5′–GGU–3′. In addition,

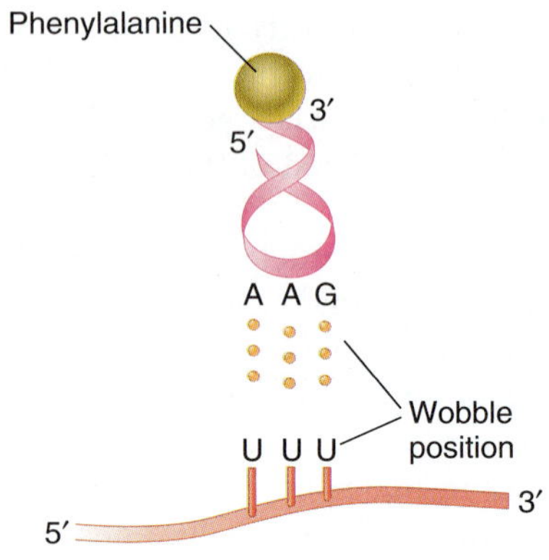

(a) Location of wobble position

Third base of mRNA codon:	Base in anticodon can be:
A	U, I, xm^5s^2U, xm^5Um, Um, xm^5U, xo^5U, k^2C
U	A, G, U, I, xo^5U
G	C, A, U, xo^5U
C	G, A, I

(b) Revised wobble rules

FIGURE 13.12 Wobble position and base-pairing rules. **(a)** The wobble position occurs at the third base in the mRNA codon and the corresponding base in the anticodon. **(b)** The revised wobble rules are slightly different from those originally proposed by Crick. The standard bases found in RNA are G, C, A, and U. In addition, the structures of bases in tRNAs may be modified. Some modified bases that may occur in the wobble position in tRNA are I = inosine; xm^5s^2U = 5-methyl-2-thiouridine; xm^5Um = 5-methyl-2′-*O*-methyluridine; Um = 2′-*O*-methyluridine; xm^5U = 5-methyluridine; xo^5U = 5-hydroxyuridine; k^2C = lysidine (a cytosine derivative).

CONCEPT CHECK: How do the wobble rules affect the total number of different tRNAs that are needed to carry out translation?

the wobble rules enable a single type of tRNA to recognize more than one codon. For example, a tRNA with an anticodon sequence of 3′–AAG–5′, which carries phenylalanine, can recognize a 5′–UUC–3′ and a 5′–UUU–3′ codon. The 5′–UUC–3′ codon is a perfect match with this tRNA. The 5′–UUU–3′ codon is mismatched according to the standard RNA-RNA hybridization rules (namely, G in the anticodon is mismatched to U in the codon), but the two can fit according to the wobble rules described in Figure 13.12. The ability of a single tRNA to recognize more than one codon makes it unnecessary for a cell to make 61 different tRNA molecules with anticodons that are complementary to the 61 possible sense codons. *E. coli* cells, for example, make a population of tRNA molecules that have just 40 different anticodon sequences.

13.4 COMPREHENSION QUESTIONS

1. If a tRNA has an anticodon with the sequence 5′-CAG-3′, which amino acid does it carry?
   a. Aspartic acid
   b. Leucine
   c. Valine
   d. Glutamine
2. The anticodon of a tRNA is located in the
   a. 3′ single-stranded region of the acceptor stem.
   b. loop of the first stem-loop.
   c. loop of the second stem-loop.
   d. loop of the third stem-loop.
3. An enzyme known as __________ attaches an amino acid to the __________ of a tRNA, thereby producing __________.
   a. aminoacyl-tRNA synthetase, anticodon, a charged tRNA
   b. aminoacyl-tRNA synthetase, 3′ single-stranded region of the acceptor stem, a charged tRNA
   c. polynucleotide phosphorylase, anticodon, a charged tRNA
   d. polynucleotide phosphorylase, anticodon, an aminoacyl-tRNA

13.5 RIBOSOME STRUCTURE AND ASSEMBLY

Learning Outcome:

1. Outline the structural features of ribosomes.

In Section 13.4, we examined how the structure and function of tRNA molecules are important in translation. According to the adaptor hypothesis, tRNAs bind to mRNA due to complementarity between the anticodons and codons. Concurrently, the tRNA molecules have the correct amino acid attached to their 3′ ends.

To synthesize a polypeptide, additional events must occur. In particular, the bond between the 3′ end of the tRNA and the amino acid must be broken, and a peptide bond must be formed between the adjacent amino acids. To facilitate these events, translation occurs on a macromolecular complex known as the **ribosome.** The ribosome can be thought of as the macromolecular arena where translation takes place.

In this section, we will begin by outlining the biochemical compositions of ribosomes in bacterial and eukaryotic cells. We will then examine the key functional sites on ribosomes for the translation process.

Bacterial and Eukaryotic Ribosomes Are Assembled from rRNA and Proteins

Bacterial cells have one type of ribosome that is found within the cytoplasm. Eukaryotic cells contain biochemically distinct ribosomes in different cellular locations. The most abundant type of eukaryotic ribosome functions in the cytosol, which is the region of the cell that is inside the plasma membrane but outside the membrane-bound organelles. In addition to the cytosolic ribosomes, all eukaryotic cells have ribosomes within the mitochondria. Plant and algal cells also have ribosomes in their chloroplasts. The compositions of mitochondrial and chloroplast ribosomes are quite different from that of the cytosolic ribosomes. Unless otherwise noted, the term eukaryotic ribosome refers to ribosomes in the cytosol, not to those found within organelles. Likewise, the description of eukaryotic translation refers to translation via cytosolic ribosomes.

Each ribosome is composed of structures called the large and small subunits. This term is perhaps misleading because each ribosomal subunit itself is formed from the assembly of many different proteins and RNA molecules called ribosomal RNA or rRNA. In bacterial ribosomes, the 30S subunit is formed from the assembly of 21 different ribosomal proteins and a 16S rRNA molecule; the 50S subunit contains 34 different proteins and 5S and 23S rRNA molecules (**Table 13.5**). With regard to the ribosomal subunits, the designations of 30S and 50S refer to the rate that these subunits sediment when subjected to a centripetal force. This rate is described as a sedimentation coefficient in Svedberg units (S), in honor of Theodor Svedberg, who invented the ultracentrifuge. Together, the 30S and 50S subunits form a 70S ribosome. (Note: Svedberg units do not add up linearly.) In bacteria, the ribosomal proteins and rRNA molecules are synthesized in the cytoplasm, and the ribosomal subunits are assembled there.

The synthesis of eukaryotic rRNA occurs within the nucleus, and the ribosomal proteins are made in the cytosol, where translation takes place. The 40S subunit is composed of 33 proteins and an 18S rRNA; the 60S subunit is made of 49 proteins and 5S, 5.8S, and 28S rRNAs (see Table 13.5). The assembly of the rRNAs and ribosomal proteins to make the 40S and 60S subunits occurs within the **nucleolus,** a region of the nucleus specialized for this purpose. The 40S and 60S subunits are then exported into the cytosol, where they associate to form an 80S ribosome during translation.

Components of Ribosomal Subunits Form Functional Sites for Translation

In recent years, many advances have been made toward understanding ribosomes on a molecular level. Microscopic and biophysical methods have been used to study ribosome structure. An electron micrograph of bacterial ribosomes in the act of translation is shown in **Figure 13.13a**. In this example, many ribosomes are translating a single mRNA. The term **polyribosome,** or **polysome,** is used to describe an mRNA transcript that has many bound ribosomes in the act of translation.

More recently, a few research groups have succeeded in crystallizing ribosomal subunits, and even intact ribosomes. This is an amazing technical feat, because it is difficult to find the right conditions under which large macromolecules will form highly ordered crystals. **Figure 13.13b** is a schematic drawing of the crystal structure of bacterial ribosomal subunits. The overall shape of each subunit is largely determined by the structure of the rRNAs, which constitute most of the mass of the ribosome. The interface between the 30S and 50S subunits is primarily composed of rRNA. Ribosomal proteins cluster on the outer surface of the ribosome and on the periphery of the interface.

During bacterial translation, the mRNA lies on the surface of the 30S subunit within a space between the 30S and 50S subunits. As the polypeptide is being synthesized, it exits through a channel within the 50S subunit (**Figure 13.13c**). Ribosomes contain discrete sites where tRNAs bind and the polypeptide is synthesized. In 1964, James Watson was the first to propose a two-site model for tRNA binding to the ribosome. These sites are known

TABLE 13.5
Composition of Bacterial and Eukaryotic Ribosomes

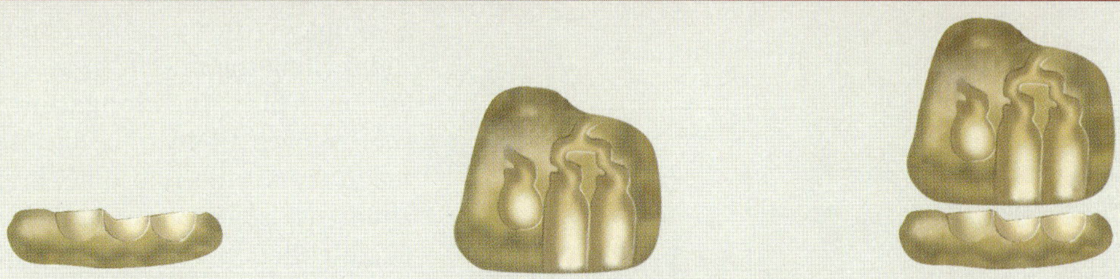

	Small subunit	Large subunit	Assembled ribosome
Bacterial			
Sedimentation coefficient	30S	50S	70S
Number of proteins	21	34	55
rRNA	16S rRNA	5S rRNA, 23S rRNA	16S rRNA, 5S rRNA, 23S rRNA
Eukaryotic			
Sedimentation coefficient	40S	60S	80S
Number of proteins	33	49	82
rRNA	18S rRNA	5S rRNA, 5.8S rRNA, 28S rRNA	18S rRNA, 5S rRNA, 5.8S rRNA, 28S rRNA

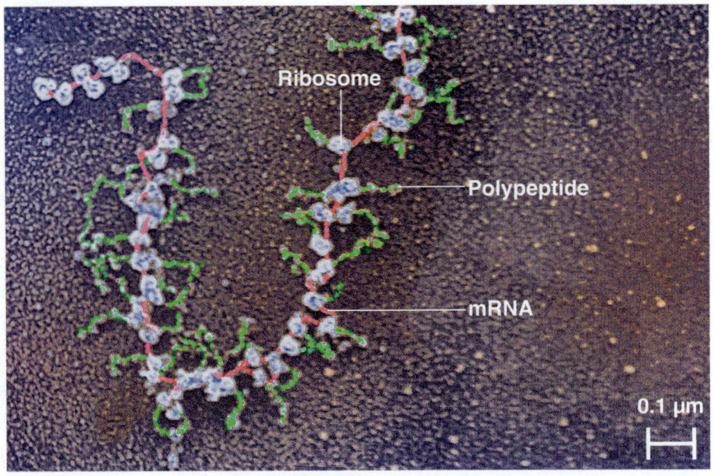

(a) Ribosomes in the act of translation

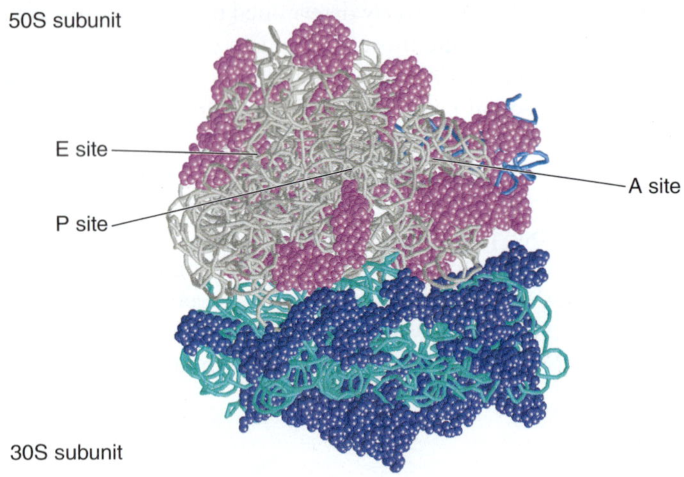

(b) Bacterial ribosome model based on X-ray diffraction studies

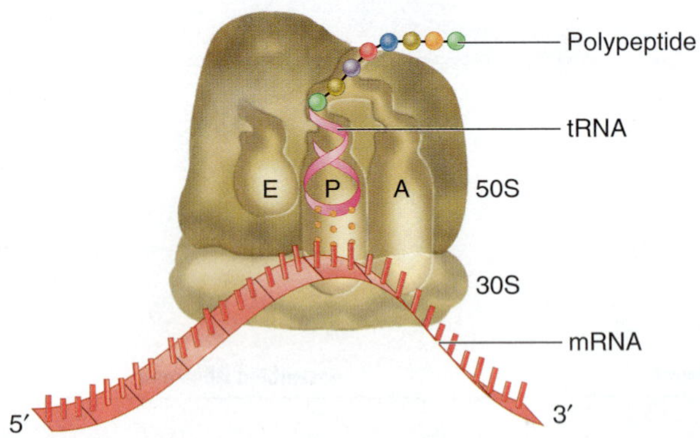

(c) Model for ribosome structure

FIGURE 13.13 **Ribosomal structure.** **(a)** Electron micrograph of ribosomes in the act of translation. Ribosomes are blue, mRNA is red, and polypeptides are green. **(b)** Crystal structure of the 50S and 30S subunits in bacteria. The rRNA is shown in gray strands (50S subunit) and turquoise strands (30S subunit), and proteins are shown in magenta (50S subunit) and navy blue (30S subunit). **(c)** A model depicting the sites where tRNA and mRNA bind to an intact ribosome. The mRNA lies on the surface of the 30S subunit. The E, P, and A sites are formed at the interface between the large and small subunits. The growing polypeptide exits through a hole in the 50S subunit.

as the **peptidyl site (P site)** and **aminoacyl site (A site).** In 1981, Knud Nierhaus, Hans Sternbach, and Hans-Jörg Rheinberger proposed a three-site model. This model incorporated the observation that uncharged tRNA molecules bind to a site on the ribosome that is distinct from the P and A sites. This third site is now known as the **exit site (E site).** The locations of the E, P, and A sites are shown in Figure 13.13c. Next, we will examine the roles of these sites during the three stages of translation.

13.5 COMPREHENSION QUESTIONS

1. Each ribosomal subunit is composed of
   a. multiple proteins.
   b. rRNA.
   c. tRNA.
   d. both a and b.
2. The site(s) on a ribosome where tRNA molecules may be located include the
   a. A site.
   b. P site.
   c. E site.
   d. all of the above.

13.6 STAGES OF TRANSLATION

Learning Outcomes:

1. Outline the three stages of translation.
2. Describe the steps that occur during the initiation, elongation, and termination stages of translation.
3. Compare and contrast bacterial and eukaryotic translation.

Like transcription, the process of translation can be viewed as occurring in three stages: initiation, elongation, and termination. **Figure 13.14** presents an overview of these stages, which are similar between bacteria and eukaryotes. During **initiation,** the ribosomal subunits, mRNA, and the first tRNA assemble to form a complex. After the initiation complex is formed, the ribosome slides along the mRNA in the 5′ to 3′ direction, moving over the codons. This is the **elongation** stage of translation. As the ribosome moves, tRNA molecules sequentially bind to the mRNA at the A site in the ribosome, bringing with them the appropriate amino acids. Therefore, amino acids are linked in the order dictated by the codon sequence in the mRNA. Finally, a stop codon is reached, signaling the **termination** of translation. At this point, disassembly occurs, and the newly made polypeptide is released. In this section, we will examine the components required for the translation process and consider their functional roles during the three stages of translation.

The Initiation Stage Involves the Binding of mRNA and the Initiator tRNA to the Ribosomal Subunits

During initiation, an mRNA and the first tRNA bind to the ribosomal subunits. A specific tRNA functions as the **initiator tRNA,** which recognizes the start codon in the mRNA. In bacteria, the initiator tRNA, which is designated $tRNA^{fMet}$, carries a methionine that has been covalently modified to *N*-formylmethionine. In this

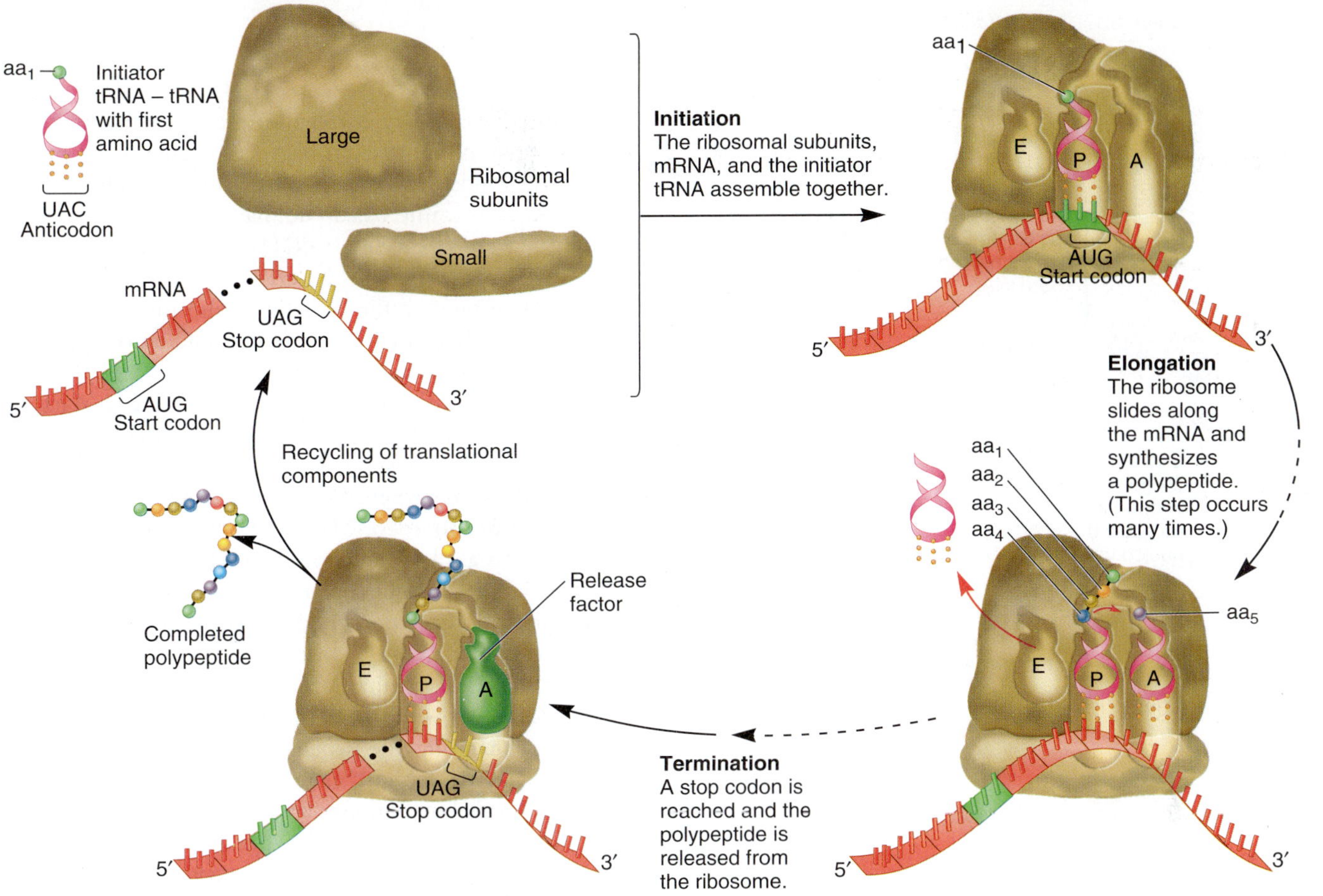

FIGURE 13.14 **Overview of the stages of translation.** Note: In this and succeeding figures in this chapter, the ribosomes are drawn schematically to emphasize different aspects of the translation process. The structures of ribosomes are described in Figure 13.13.

Genes → Traits The ability of genes to produce an organism's traits relies on the molecular process of gene expression. During translation, the codon sequence within mRNA (which is derived from a gene sequence during transcription) is translated into a polypeptide sequence. After polypeptides are made within a living cell, they function as components of proteins to govern an organism's traits. For example, once the β-globin polypeptide is made, it functions within the hemoglobin protein and provides red blood cells with the ability to carry oxygen, a vital trait for survival. Translation allows functional proteins to be made within living cells.

CONCEPT CHECK: Explain how mRNA plays a role in all three stages.

modification, a formyl group (—CHO) is attached to the nitrogen atom in methionine after the methionine has been attached to the tRNA.

Figure 13.15 describes the initiation stage of translation in bacteria during which the mRNA, $tRNA^{fMet}$, and ribosomal subunits associate with each other to form an initiation complex. The formation of this complex requires the participation of three initiation factors: IF1, IF2, and IF3. First, IF1 and IF3 bind to the 30S subunit, which prevents the association of the 50S subunit. Next, the mRNA binds to the 30S subunit. This binding is facilitated by a 9-nucleotide sequence within the bacterial mRNA called the **Shine-Dalgarno sequence.** The location of this sequence is shown in Figure 13.15 and in more detail in **Figure 13.16**. How does the Shine-Dalgarno sequence facilitate the binding of mRNA to the ribosome? The Shine-Dalgarno sequence is complementary to a short sequence within the 16S rRNA, which promotes the hydrogen bonding of the mRNA to the 30S subunit.

The next step requires IF2, which has GTP bound to it. IF2 promotes the binding of the initiator tRNA ($tRNA^{fMet}$) to the mRNA already bound to the 30S subunit (see Figure 13.15). The $tRNA^{fMet}$ binds to the start codon, which is typically a few nucleotides downstream from the Shine-Dalgarno sequence. The start codon is usually AUG, but in some cases it can be GUG or UUG. Even when the start codon is GUG (which normally encodes valine) or UUG (which normally encodes leucine), the first amino acid in the polypeptide is still a formylmethionine because only a $tRNA^{fMet}$ can initiate translation. During or after translation of the entire polypeptide, the formyl group or the entire *N*-formylmethionine may be removed. Therefore, some polypeptides may not have *N*-formylmethionine as their first amino acid. As noted in Figure 13.15, the $tRNA^{fMet}$ binds to the P site on the ribosome. IF1 is thought to occupy a portion of the A site, thereby preventing the binding of $tRNA^{fMet}$ to the A site during initiation. By comparison, during the elongation stage that is discussed later, all of the other tRNAs initially bind to the A site.

After the mRNA and $tRNA^{fMet}$ have become bound to the 30S subunit, IF1 and IF3 are released, and then IF2 hydrolyzes its GTP and is also released. This allows the 50S ribosomal subunit to associate with the 30S subunit. After translation is completed, IF1 binding is necessary to dissociate the 50S and 30S ribosomal subunits so that the 30S subunit can reinitiate with another mRNA molecule.

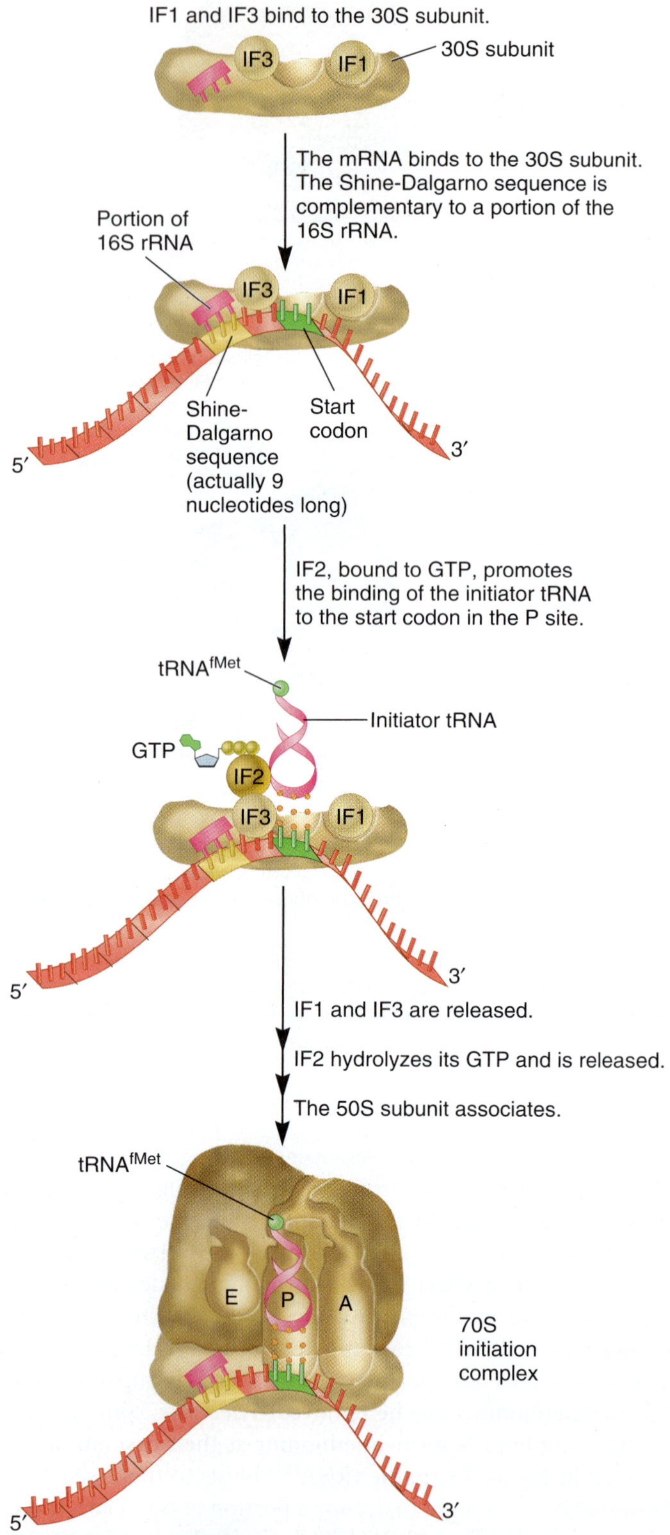

INTERACTIVE EXERCISE

ONLINE ANIMATION

FIGURE 13.15 The initiation stage of translation in bacteria.

In eukaryotes, the assembly of the initiation complex bears similarities to that in bacteria. However, as described in **Table 13.6**, additional factors are required for the initiation process. Note that the initiation factors are designated eIF (for eukaryotic Initiation Factor) to distinguish them from bacterial initiation factors.

The initiator tRNA in eukaryotes carries methionine rather than *N*-formylmethionine, as in bacteria. A eukaryotic initiation factor, eIF2, binds directly to $tRNA^{Met}$ to recruit it to the 40S subunit. Eukaryotic mRNAs do not have a Shine-Dalgarno sequence. How then are eukaryotic mRNAs recognized by the ribosome? The mRNA is recognized by eIF4, which is a multiprotein complex that recognizes the 7-methylguanosine cap at the 5′ end of the mRNA. eIF4 then facilitates the binding of the 5′ end of the mRNA to the 40S ribosomal subunit.

The identification of the correct AUG start codon in eukaryotes differs from that in bacteria. After the initial binding of mRNA to the ribosome, the next step is to locate an AUG start codon that is somewhere downstream from the 7-methylguanosine cap structure. In 1986, Marilyn Kozak proposed that the ribosome begins at the 5′ end and then scans along the mRNA in the 3′ direction in search of an AUG start codon. In many, but not all, cases, the ribosome uses the first AUG codon that it encounters as a start codon. When a start codon is identified, eIF5 causes the release of the other initiation factors, which enables the 60S subunit to associate with the 40S subunit.

By analyzing the sequences of many eukaryotic mRNAs, researchers have found that not all AUG codons near the 5′ end of mRNA can function as start codons. In some cases, the scanning ribosome passes over the first AUG codon and chooses an AUG farther down the mRNA. The sequence of bases around the AUG codon plays an important role in determining whether or not it is selected as the start codon by a scanning ribosome. The consensus sequence for optimal start codon recognition in more complex eukaryotes, such as vertebrates and vascular plants, is shown here.

						Start	*Codon*		
G	C	C	(A/G)	C	C	A	U	G	G
−6	−5	−4	−3	−2	−1	+1	+2	+3	+4

Aside from an AUG codon itself, a guanine at the +4 position and a purine, preferably an adenine, at the −3 position are the most important sites for start codon selection. These rules for optimal translation initiation are called **Kozak's rules.**

Polypeptide Synthesis Occurs During the Elongation Stage

During the elongation stage of translation, amino acids are added, one at a time, to a growing polypeptide (**Figure 13.17**). Even though this process is rather complex, it occurs with remarkable speed. Under normal cellular conditions, a polypeptide can elongate at a rate of 15–20 amino acids per second in bacteria and 2–6 amino acids per second in eukaryotes!

To begin elongation, a charged tRNA brings a new amino acid to the ribosome so it can be attached to the end of the growing polypeptide. At the top of Figure 13.17, which describes bacterial translation, a short polypeptide is attached to the tRNA located at the P site of the ribosome. A charged tRNA carrying a single amino acid binds to the A site. This binding occurs because the anticodon in the tRNA is complementary to the codon in the mRNA. The hydrolysis of GTP by the elongation factor, EF-Tu,

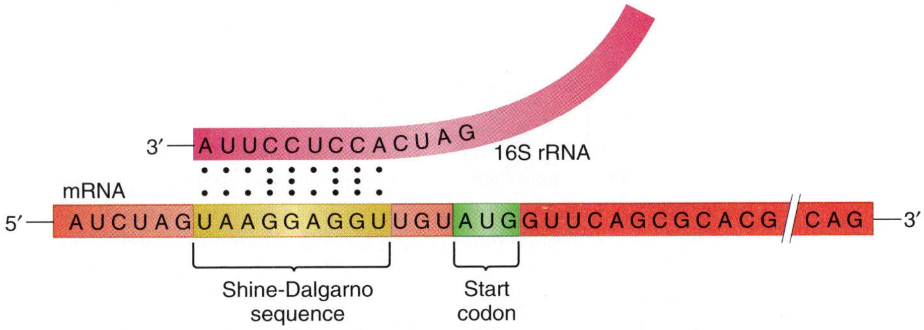

FIGURE 13.16 The locations of the Shine-Dalgarno sequence and the start codon in bacterial mRNA. The Shine-Dalgarno sequence is complementary to a sequence in the 16S rRNA. It hydrogen bonds with the 16S rRNA to promote initiation. The start codon is typically a few nucleotides downstream from the Shine-Dalgarno sequence.

CONCEPT CHECK: Why does a bacterial mRNA bind specifically to the small ribosomal subunit?

TABLE 13.6
A Simplified Comparison of Translational Protein Factors in Bacteria and Eukaryotes

Bacterial Factors	Eukaryotic Factors*	Function
Initiation Factors		
	eIF4	Recognizes the 7-methylguanosine cap at the 5′-end of mRNA and facilitates the binding of the mRNA to the small ribosomal subunit
IF1, IF3	eIF1, eIF3, eIF6	Prevent the association between the small and large ribosomal subunits and favor their dissociation
IF2	eIF2	Promotes the binding of the initiator tRNA to the small ribosomal subunit
	eIF5	Helps to dissociate the initiation factors, which allows the two ribosomal subunits to assemble
Elongation Factors		
EF-Tu	eEF1α	Involved in the binding of tRNAs to the A site
EF-Ts	eEF1βγ	Nucleotide exchange factors required for the functioning of EF-Tu and eEF1α, respectively
EF-G	eEF2	Required for translocation
Release Factors		
RF1, RF2	eRF1	Recognize a stop codon and trigger the cleavage of the polypeptide from the tRNA
RF3	eRF3	GTPases that are also involved in termination

*Eukaryotic translation factors are typically composed of multiple proteins.

provides energy for the binding of a tRNA to the A site. In addition, the 16S rRNA, which is a component of the small 30S ribosomal subunit, plays a key role by ensuring the proper recognition between the mRNA and correct tRNA. The 16S rRNA can detect when an incorrect tRNA is bound at the A site and will prevent elongation until the mispaired tRNA is released from the A site. This phenomenon, termed the **decoding function** of the ribosome, is important in maintaining high fidelity of mRNA translation. An incorrect amino acid is incorporated into a growing polypeptide at a rate of approximately one mistake per 10,000 amino acids, or 10^{-4}.

The next step of elongation is the **peptidyl transfer** reaction—the polypeptide is removed from the tRNA in the P site and transferred to the amino acid at the A site. This transfer is accompanied by the formation of a peptide bond between the amino acid at the A site and the polypeptide, lengthening the polypeptide by one amino acid. The peptidyl transfer reaction is catalyzed by a component of the 50S subunit known as **peptidyl transferase,** which is composed of several proteins and rRNA. Interestingly, based on the crystal structure of the 50S subunit, Thomas Steitz, Peter Moore, and their colleagues concluded that the 23S rRNA (a component of peptidyl transferase)—not a ribosomal protein—catalyzes bond formation between adjacent amino acids that occurs during peptidyl transfer. In other words, the ribosome is a ribozyme!

After the peptidyl transfer reaction is complete, the ribosome moves, or translocates, to the next codon in the mRNA. This moves the tRNAs at the P and A sites to the E and P sites, respectively. Finally, the uncharged tRNA exits the E site. You should notice that the next codon in the mRNA is now exposed in the unoccupied A site. At this point, a charged tRNA can enter the empty A site, and the same series of steps adds the next amino acid to the growing polypeptide. As you may have realized, the A, P, and E sites are named for the role of the tRNA that is usually found there. The A site binds an aminoacyl-tRNA (also called a charged tRNA), the P site usually contains the peptidyl-tRNA (a tRNA with an attached peptide), and the E site is where the uncharged tRNA exits.

Termination Occurs When a Stop Codon Is Reached in the mRNA

The final stage of translation, known as termination, occurs when a stop codon is reached in the mRNA. In most species, the three stop codons are UAA, UAG, and UGA. The stop codons are not recognized by a tRNA with a complementary sequence. Instead, they are recognized by proteins known as **release factors** (see Table 13.6). Interestingly, the three-dimensional structures of release factor proteins are "molecular mimics" that resemble the structure of tRNAs. Such proteins can specifically bind to a stop

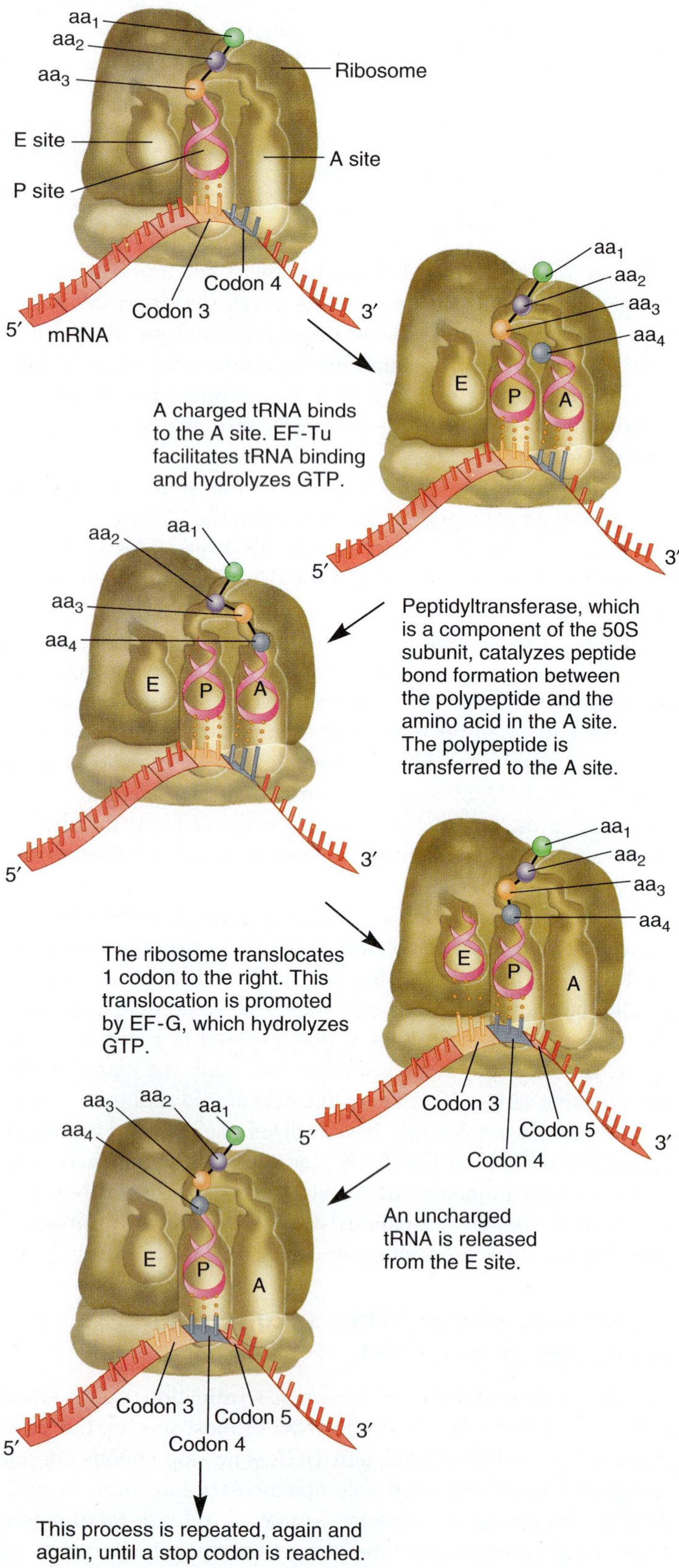

INTERACTIVE EXERCISE ONLINE ANIMATION

FIGURE 13.17 The elongation stage of translation in bacteria. This process begins with the binding of an incoming charged tRNA. The hydrolysis of GTP by EF-Tu provides the energy for the binding of the tRNA to the A site. A peptide bond is then formed between the amino acid at the A site and the last amino acid in the growing polypeptide. This moves the polypeptide to the A site. The ribosome then translocates in the 3′ direction so the two tRNAs are moved to the E and P sites. The tRNA carrying the polypeptide is now back in the P site. This translocation requires the hydrolysis of GTP via EF-G. The uncharged tRNA in the E site is released from the ribosome. Now the process is ready to begin again. Each cycle of elongation causes the polypeptide to grow by one amino acid.

CONCEPT CHECK: What is the role of peptidyl transferase during the elongation stage?

codon sequence. In bacteria, RF1 recognizes UAA and UAG, and RF2 recognizes UAA and UGA. A third release factor, RF3, is also required. In eukaryotes, a single release factor, eRF1, recognizes all three stop codons and eRF3 is also required for termination.

Figure 13.18 illustrates the termination stage of translation in bacteria. At the top of this figure, the completed polypeptide is attached to a tRNA in the P site. A stop codon is located at the A site. In the first step, RF1 or RF2 binds to the stop codon at the A site and RF3 (not shown) binds at a different location on the ribosome. After RF1 (or RF2) and RF3 have bound, the bond between the polypeptide and the tRNA is hydrolyzed. The polypeptide and tRNA are then released from the ribosome. The final step in translational termination is the disassembly of ribosomal subunits, mRNA, and the release factors.

Bacterial Translation Can Begin Before Transcription Is Completed

Microscopic, biochemical, and genetic studies have shown that the translation of a bacterial protein-encoding gene begins before the mRNA transcript is completed. In other words, as soon as an mRNA strand is long enough, a ribosome attaches to the 5′ end and begins translation, even before RNA polymerase has reached the transcriptional termination site within the gene. This phenomenon is termed the coupling between transcription and translation in bacterial cells. Note that coupling of these processes does not usually occur in eukaryotes, because transcription takes place in the nucleus, whereas translation occurs in the cytosol.

Bacterial and Eukaryotic Translation Show Similarities and Differences

Throughout this chapter, we have compared translation in bacteria and eukaryotic organisms. The general steps of translation are similar in all forms of life, but we have also seen some striking differences between bacteria and eukaryotes. **Table 13.7** compares translation between these groups.

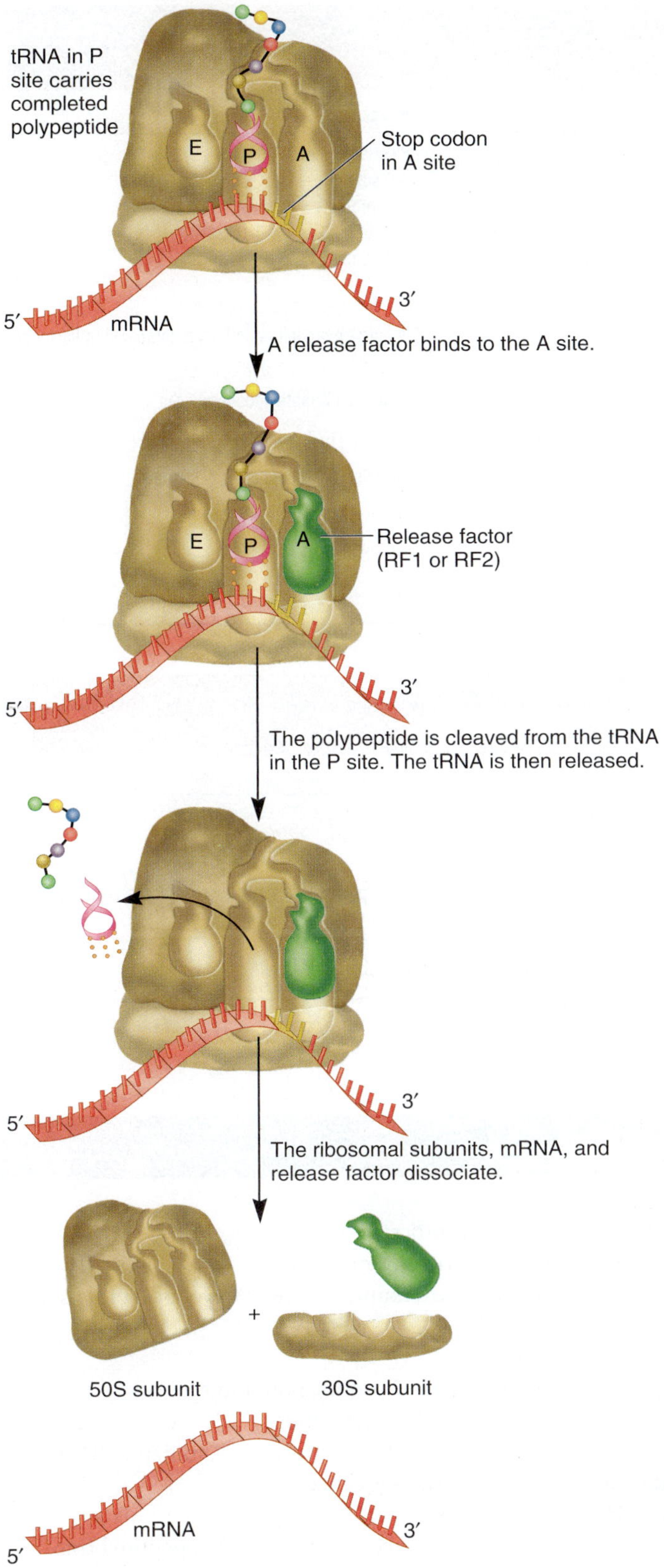

INTERACTIVE EXERCISE ONLINE ANIMATION

FIGURE 13.18 The termination stage of translation in bacteria. When a stop codon is reached, a release factor (either RF1 or RF2) binds to the A site. (RF3 binds elsewhere and uses GTP to facilitate the termination process.) The polypeptide is cleaved from the tRNA in the P site and released. The tRNA is released, and the rest of the components disassemble.

CONCEPT CHECK: Explain why release factors are called "molecular mimics."

Antibiotics That Inhibit Bacterial Translation Are Used to Treat Bacterial Diseases

Many different diseases that affect people and domesticated animals are caused by pathogenic bacteria. An **antibiotic** is any substance produced by a microorganism that inhibits the growth of other microorganics, such as pathogenic bacteria. Most antibiotics are small organic molecules, less than 2000 daltons (Da). In some cases, antibiotics exert their effect because they inhibit or interfere with bacterial translation. Because the components of translation differ somewhat between bacteria and eukaryotes, some antibiotics inhibit bacterial translation without affecting eukaryotic translation. Therefore, they can be used to treat bacterial infections in humans, pets, and livestock. **Table 13.8** describes a few examples.

13.6 COMPREHENSION QUESTIONS

1. During the initiation stage of translation in bacteria, which of the following event(s) occur(s)?
   a. IF1 and IF3 bind to the 30S subunit.
   b. The mRNA binds to the 30S subunit and $tRNA^{fMet}$ binds to the start codon in the mRNA.
   c. IF2 hydrolyzes its GTP and is released; the 50S subunit binds to the 30S subunit.
   d. All of the above events occur.
2. The Kozak rules determine
   a. the choice of the start codon in complex eukaryotes.
   b. the choice of the start codon in bacteria.
   c. the site in the mRNA where translation ends.
   d. how fast the mRNA is translated.
3. During the peptidyl transferase reaction, the polypeptide, which is attached to a tRNA in the ___________, becomes bound via _________ to an amino acid attached to a tRNA in the ___________.
   a. A site, several hydrogen bonds, P site
   b. A site, a peptide bond, P site
   c. P site, a peptide bond, A site
   d. P site, several hydrogen bonds, A site
4. A release factor is referred to as a "molecular mimic" because its structure is similar to
   a. a ribosome.
   b. an mRNA.
   c. a tRNA.
   d. an elongation factor.

TABLE 13.7
A Comparison of Bacterial and Eukaryotic Translation

	Bacterial	Eukaryotic
Ribosome composition:	70S ribosomes: 30S subunit— 21 proteins + 1 rRNA 50S subunit— 34 proteins + 2 rRNAs	80S ribosomes: 40S subunit— 33 proteins + 1 rRNA 60S subunit— 49 proteins + 3 rRNAs
Initiator tRNA:	$tRNA^{fmet}$	$tRNA^{Met}$
Formation of the initiation complex:	Requires IF1, IF2, and IF3	Requires more initiation factors compared to bacterial initiation
Initial binding of mRNA to the ribosome:	Requires a Shine-Dalgarno sequence	Requires a 7-methylguanosine cap
Selection of a start codon:	AUG, GUG, or UUG located just downstream from the Shine-Dalgarno sequence	According to Kozak's rules
Elongation rate:	Typically 15–20 amino acids per second	Typically 2–6 amino acids per second
Termination:	Requires RF1, RF2, and RF3	Requires eRF1 and eRF3
Location of translation:	Cytoplasm	Cytosol
Coupled to transcription:	Yes	No

TABLE 13.8
Mechanisms of Inhibition of Bacterial Translation via Selected Antibiotics

Antibiotic	Description
Chloraphenical	Blocks elongation by acting as competitive inhibitor of the peptidyltransferase complex.
Erythromycin	Binds to the 23S RNA and blocks elongation by interfering with the translocation step.
Puromycin	Binds to the A site and causes premature release of the polypeptide. This early termination of translation results in polypeptides that are shorter than normal.
Tetracycline	Blocks elongation by inibiting the binding of aminoacyl-tRNAs to the ribosome.
Streptomycin	Interferes with normal pairing between aminoacyl-tRNAs and codons. This causes misreading, thereby producing abnormal proteins.

KEY TERMS

Page 315. protein-encoding genes, structural genes, messenger RNA (mRNA)
Page 316. alkaptonuria, inborn error of metabolism
Page 317. one-gene/one-enzyme hypothesis
Page 318. polypeptide, protein, translation, genetic code, sense codons, start codon, stop codons, termination codons, nonsense codons, anticodons, degenerate, synonymous codons, reading frame
Page 320. selenocysteine, pyrrolysine, peptide bond
Page 321. N-terminus, amino-terminus, C-terminus, carboxyl-terminus, side chain, R group
Page 322. primary structure, chaperones, secondary structure, α helix, β sheet, tertiary structure, quaternary structure, subunits
Page 323. cell-free translation system
Page 327. adaptor hypothesis
Page 328. aminoacyl-tRNA synthetases
Page 329. charged tRNA, aminoacyl-tRNA, wobble rules, isoacceptor tRNAs
Page 330. ribosome
Page 331. nucleolus, polyribosome, polysome
Page 332. peptidyl site (P site), aminoacyl site (A site), exit site (E site), initiation, elongation, termination, initiator tRNA
Page 333. Shine-Dalgarno sequence
Page 334. Kozak's rules
Page 335. decoding function, peptidyl transfer, peptidyl transferase, release factors
Page 337. antibiotic

CHAPTER SUMMARY

- Cellular proteins are made via the translation of mRNA.

13.1 The Genetic Basis for Protein Synthesis

- Garrod studied the disease called alkaptonuria and suggested that some genes encode enzymes (see Figure 13.1).
- Beadle and Tatum studied *Neurospora* mutants that were altered in their nutritional requirements and hypothesized that one gene encodes one enzyme. This hypothesis was later modified because (1) some proteins are not enzymes; (2) some proteins are composed of two or more different polypeptides; (3) some genes encode RNAs that are not translated into polypeptides; and (4) some mRNAs are alternatively spliced (see Figure 13.2).

13.2 The Relationship Between the Genetic Code and Protein Synthesis

- During translation, the sequence of codons in mRNA is used via tRNA molecules to make a polypeptide with a specific amino acid sequence (see Figure 13.3).
- The genetic code refers to the relationship between the three-base codons in the mRNA and the amino acids that are incorporated into a polypeptide. One codon (AUG) is a start codon, which determines the reading frame of the mRNA. Three codons (UAA, UAG, and UGA) function as stop codons (see Table 13.1).
- The genetic code is largely universal, but some exceptions are known to occur (see Table 13.2).
- A polypeptide is made by the formation of peptide bonds between adjacent amino acids. Each polypeptide has a directionality from its amino-terminus to its carboxyl-terminus that parallels the arrangement of codons in mRNA in the 5′ to 3′ direction (see Figure 13.4).
- Amino acids differ in their side chain structure (see Figure 13.5).
- Protein structure can be viewed at different levels, which include primary structure (sequence of amino acids), secondary structure (repeating folding patterns such as the α helix and the β sheet), tertiary structure (additional folding), and quaternary structure (the binding of multiple subunits to each other) (see Figure 13.6).
- Proteins carry out a variety of functions. The structures and functions of proteins are largely responsible for an organism's traits (see Table 13.3).

13.3 Experimental Determination of the Genetic Code

- Nirenberg and colleagues used synthetic RNA and a cell-free translation system to decipher the genetic code (see Figure 13.7).
- Other methods to decipher the genetic code included the synthesis of copolymers by Khorana and the triplet-binding assay of Nirenberg and Leder (see Table 13.4, Figure 13.8).

13.4 Structure and Function of tRNA

- The anticodon in a tRNA binds to a codon in mRNA in an antiparallel manner and obeys the AU/GC rule. The tRNA carries a specific amino acid that corresponds to the codon in the mRNA according to the genetic code (see Figure 13.9).
- The secondary structure of tRNA resembles a cloverleaf. The anticodon is in the second loop and the amino acid is attached to the 3′ end (see Figure 13.10).
- Aminoacyl-tRNA synthetases are a group of enzymes that attach the correct amino acid to a tRNA. The resulting tRNA is called a charged tRNA or an aminoacyl-tRNA (see Figure 13.11).
- Mismatches are allowed between the pairing of tRNAs and mRNA according to the wobble rules (see Figure 13.12).

13.5 Ribosome Structure and Assembly

- Ribosomes are the site of polypeptide synthesis. The small and large subunit of ribosomes are composed of rRNAs and multiple proteins (see Table 13.5).
- A ribosome contains an A (aminoacyl), P (peptidyl), and E (exit) site, which are occupied by tRNA molecules (see Figure 13.13).

13.6 Stages of Translation

- The three stages of translation are initiation, elongation, and termination (see Figure 13.14).
- During the initiation stage of translation, the mRNA, initiator tRNA, and ribosomal subunits assemble. Initiation factors are involved in the process. In bacteria, the Shine-Dalgarno sequence promotes the binding of the mRNA to the small ribosomal subunit (see Figures 13.15, 13.16, Table 13.6).
- Start codon selection in complex eukaryotes follows Kozak's rules.
- During elongation, tRNAs bring amino acids to the A site and a series of peptidyl transfer reactions creates a polypeptide. At each step, the polypeptide is transferred from the P site to the A site. The uncharged tRNAs are released from the E site. Elongation factors are involved in this process (see Figure 13.17).
- During termination, a release factor binds to a stop codon in the A site. This promotes the cleavage of the polypeptide from the tRNA and the subsequent disassembly of the tRNA, ribosomal subunits, mRNA, and release factor (see Figure 13.18).
- Bacterial translation can begin before transcription is completed.
- Bacterial and eukaryotic translation show many similarities and differences (see Table 13.7).
- Antibiotics that inhibit translation are used to treat bacterial diseases (see Table 13.8).

PROBLEM SETS & INSIGHTS

Solved Problems

S1. The first amino acid in a purified bacterial protein is methionine. The start codon in the mRNA is GUG, which codes for valine. Why isn't the first amino acid formylmethionine or valine?

Answer: During polypeptide synthesis, the first amino acid carried by the initiator tRNA is formylmethionine. This occurs even when the start codon is GUG (valine) or UUG (leucine). The formyl group can be later removed to yield methionine as the first amino acid.

S2. A tRNA has the anticodon sequence 3′–CAG–5′. What amino acid does it carry?

Answer: Because the anticodon is 3′–CAG–5′, it would bind to a codon with the sequence 5′–GUC–3′. According to the genetic code, this codon specifies the amino acid valine. Therefore, this tRNA must carry valine at its acceptor stem.

S3. In eukaryotic cells, the assembly of ribosomal subunits occurs in the nucleolus. As discussed in Chapter 12 (see Figure 12.16), a single 45S rRNA transcript is cleaved to produce the three rRNA fragments—18S, 5.8S, and 28S rRNA—that play a key role in creating the structure of the ribosome. The genes that encode the 45S precursor are found in multiple copies (i.e., they are moderately repetitive). The segments of chromosomes that contain the 45S rRNA genes align themselves at the center of the nucleolus. This site is called the nucleolar-organizing center. In this region, active transcription of the 45S gene takes place. Briefly explain how the assembly of the ribosomal subunits occurs.

Answer: In the nucleolar-organizing center, the 45S RNA is cleaved to the 18S, 5.8S, and 28S rRNAs. The other components of the ribosomal subunits, 5S rRNA and ribosomal proteins, must also be imported into the nucleolar region. Because proteins are made in the cytosol, they must enter the nucleus through the nuclear pores. When all components are present, they assemble into 40S and 60S ribosomal subunits. Following assembly, the ribosomal subunits exit the nucleus through the nuclear pores and enter the cytosol.

S4. The use of antibiotics has been of great importance in the battle against many infectious diseases caused by microorganisms. For many antibiotics, their mode of action is to inhibit the translation process within bacterial cells. Certain antibiotics selectively bind to bacterial ribosomes but do not inhibit eukaryotic ribosomes. Their ability to inhibit translation can occur at different steps in the translation process (see Table 13.8). How does inhibition of translation by antibiotics such as tetracycline prevent bacterial growth?

Answer: Because bacterial ribosomes have a different protein and rRNA composition than eukaryotic ribosomes, certain drugs can recognize these different components, bind specifically to bacterial ribosomes, and interfere with the process of translation. In other words, the surface of a bacterial ribosome must be somewhat different from the surface of a eukaryotic ribosome so that the drugs bind to the surface of only bacterial ribosomes. If a bacterial cell is exposed to tetracycline or other antibiotics, it cannot synthesize new polypeptides. Because polypeptides form functional proteins needed for processes such as cell division, the bacterium is unable to grow and proliferate.

Conceptual Questions

C1. An mRNA has the following sequence:

5′–GGCGAUGGGCAAUAAACCGGGCAGUAAGC–3′

Identify the start codon and determine the complete amino acid sequence that would be translated from this mRNA.

C2. What does it mean when we say that the genetic code is degenerate? Discuss the universality of the genetic code.

C3. According to the adaptor hypothesis, are the following statements true or false?

A. The sequence of anticodons in tRNA directly recognizes codon sequences in mRNA, with some room for wobble.

B. The amino acid attached to the tRNA directly recognizes codon sequences in mRNA.

C. The amino acid attached to the tRNA affects the binding of the tRNA to a codon sequence in mRNA.

C4. In bacteria, researchers have isolated strains that carry mutations within tRNA genes. These mutations can change the sequence of the anticodon. For example, a normal $tRNA^{Trp}$ gene would encode a tRNA with the anticodon 3′–ACC–5′. A mutation could change this sequence to 3′–CCC–5′. When this mutation occurs, the tRNA still carries a tryptophan at its 3′ acceptor stem, even though the anticodon sequence has been altered.

A. How would this mutation affect the synthesis of polypeptides within the bacterium?

B. What does this mutation tell you about the recognition between tryptophanyl-tRNA synthetase and $tRNA^{Trp}$? Does the enzyme primarily recognize the anticodon or not?

C5. The covalent attachment of an amino acid to a tRNA is an endergonic reaction. In other words, it requires an input of energy for the reaction to proceed. Where does the energy come from to attach amino acids to tRNA molecules?

C6. The wobble rules for tRNA-mRNA pairing are shown in Figure 13.12. If we assume that the tRNAs do not contain modified bases, what is the minimum number of tRNAs needed to efficiently recognize the codons for the following types of amino acids?

A. Leucine

B. Methionine

C. Serine

C7. How many different sequences of mRNA could encode a peptide with the sequence proline-glycine-methionine-serine?

C8. If a tRNA molecule carries a glutamic acid, what are the two possible anticodon sequences that it could contain? Be specific about the 5′ and 3′ ends.

C9. A tRNA has an anticodon sequence 3′–GGU–5′. What amino acid does it carry?

C10. If a tRNA has an anticodon sequence 3′–CCI–5′, what codon(s) can it recognize?

C11. Describe the anticodon of a single tRNA that could recognize the codons 5′–AAC–3′ and 5′–AAU–3′. How would this tRNA need to be modified for it to also recognize 5′–AAA–3′?

C12. Describe the structural features that all tRNA molecules have in common.

C13. In the tertiary structure of tRNA, where is the anticodon region relative to the attachment site for the amino acid? Are they adjacent to each other?

C14. What is the role of aminoacyl-tRNA synthetase? The ability of aminoacyl-tRNA synthetases to recognize tRNAs has sometimes been called the "second genetic code." Why has the function of this type of enzyme been described this way?

C15. What is an activated amino acid?

C16. Discuss the significance of modified bases within tRNA molecules.

C17. How and when does formylmethionine become attached to the initiator tRNA in bacteria?

C18. Is it necessary for a cell to make 61 different tRNA molecules, corresponding to the 61 codons for amino acids? Explain your answer.

C19. List the components required for translation. Describe the relative sizes of these different components. In other words, which components are small molecules, macromolecules, or assemblies of macromolecules?

C20. Describe the components of eukaryotic ribosomal subunits and where the assembly of the subunits occurs within living cells.

C21. The term subunit can be used in a variety of ways. Compare the use of the term subunit in proteins versus ribosomal subunit.

C22. Do the following events during bacterial translation occur primarily within the 30S subunit, within the 50S subunit, or at the interface between these two ribosomal subunits?

A. mRNA-tRNA recognition

B. Peptidyl transfer reaction

C. Exit of the polypeptide from the ribosome

D. Binding of initiation factors IF1, IF2, and IF3

C23. What are the three stages of translation? Discuss the main events that occur during these three stages.

C24. Describe the sequence in bacterial mRNA that promotes recognition by the 30S subunit.

C25. For each of the following initiation factors, how would eukaryotic initiation of translation be affected if it were missing?

A. eIF2 B. eIF4 C. eIF5

C26. How does a eukaryotic ribosome select its start codon? Describe the sequences in eukaryotic mRNA that provide an optimal context for a start codon.

C27. For each of the following sequences, rank them in order (from best to worst) as sequences that could be used to initiate translation according to Kozak's rules.

GACGCCAUGG

GCCUCCAUGC

GCCAUCAAGG

GCCACCAUGG

C28. Explain the functional roles of the A, P, and E sites during translation.

C29. An mRNA has the following sequence:

5′—AUG UAC UAU GGG GCG UAA—3′

Describe the amino acid sequence of the polypeptide that would be encoded by this mRNA. Be specific about the amino-terminus and carboxyl-terminus.

C30. Which steps during the translation of bacterial mRNA involve an interaction between complementary strands of RNA?

C31. What is the function of the nucleolus?

C32. In which of the ribosomal sites, the A site, P site, and/or E site, could the following be found?

A. A tRNA without an amino acid attached

B. A tRNA with a polypeptide attached

C. A tRNA with a single amino acid attached

C33. What is a polysome?

C34. According to Figure 13.17, explain why the ribosome translocates along the mRNA in a 5′ to 3′ direction rather than a 3′ to 5′ direction.

C35. Lactose permease of *E. coli* is a protein composed of a single polypeptide that is 417 amino acids in length. By convention, the amino acids within a polypeptide are numbered from the amino-terminus to the carboxyl-terminus. Are the following questions about lactose permease true or false?

A. Because the 64th amino acid is glycine and the 68th amino acid is aspartic acid, the codon for glycine, 64, is closer to the 3′ end of the mRNA than the codon for aspartic acid, 68.

B. The mRNA that encodes lactose permease must be greater than 1241 nucleotides in length.

C36. An mRNA encodes a polypeptide that is 312 amino acids in length. The 53rd codon in this polypeptide is a tryptophan codon. A mutation in the gene that encodes this polypeptide changes this tryptophan codon into a stop codon. How many amino acids would be in the resulting polypeptide: 52, 53, 259, or 260?

C37. Explain what is meant by the coupling of transcription and translation in bacteria. Does coupling occur in bacterial and/or eukaryotic cells? Explain.

Experimental Questions

E1. In the experiment of Figure 13.7, what would be the predicted amounts of amino acids incorporated into polypeptides if the RNA was a random polymer containing 50% C and 50% G?

E2. Polypeptides can be translated in vitro. Would a bacterial mRNA be translated in vitro by eukaryotic ribosomes? Would a eukaryotic mRNA be translated in vitro by bacterial ribosomes? Why or why not?

E3. Discuss how the elucidation of the structure of the ribosome can help us to understand its function.

E4. Describe the structure of a polysome, which is depicted in Figure 13.13a.

E5. Chapter 20 describes a blotting method known as Western blotting that can be used to detect the production of a polypeptide that is translated from a particular mRNA. In this method, a protein is detected with an antibody that specifically recognizes and binds to its amino acid sequence. The antibody acts as a probe to detect the presence of the protein. In a Western blotting experiment, a mixture of cellular proteins is separated using gel electrophoresis according to their molecular masses. After the antibody has bound to the protein of interest within a blot of a gel, the protein is visualized as a dark band. For example, an antibody that recognizes the β-globin polypeptide could be used to specifically detect the β-globin polypeptide in a blot. As shown here, the method of Western blotting can be used to determine the amount and relative size of a particular protein that is produced in a given cell type.

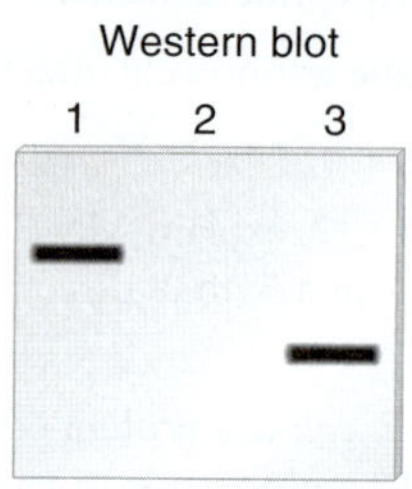

Lane 1 is a sample of proteins isolated from normal red blood cells.

Lane 2 is a sample of proteins isolated from kidney cells. Kidney cells do not produce β globin.

Lane 3 is a sample of proteins isolated from red blood cells from a patient with β-thalassemia. This patient is homozygous for a mutation that results in the shortening of the β-globin polypeptide.

Now here is the question. A protein called troponin contains 334 amino acids. Because each amino acid weighs 120 daltons (Da) (on average), the molecular mass of this protein is about 40,000 Da, or 40 kDa. Troponin functions in muscle cells, and it is not expressed in nerve cells. Draw the expected results of a Western blot for the following samples:

Lane 1: Proteins isolated from muscle cells

Lane 2: Proteins isolated from nerve cells

Lane 3: Proteins isolated from the muscle cells of an individual who is homozygous for a mutation that introduces a stop codon at codon 177

E6. The technique of Western blotting can be used to detect proteins that are translated from a particular mRNA. This method is described in Chapter 20 and also in experimental question E5. Let's suppose a researcher was interested in the effects of mutations on the expression of a protein-encoding gene that encodes a protein we will call protein X. This protein is expressed in skin cells and contains 572 amino acids. Its molecular mass is approximately 68,600 Da, or 68.6 kDa. Make a drawing that shows the expected results of a Western blot using proteins isolated from the skin cells obtained from the following individuals:

Lane 1: A normal individual

Lane 2: An individual who is homozygous for a deletion, which removes the promoter for this gene

Lane 3: An individual who is heterozygous in which one gene is normal and the other gene has a mutation that introduces an early stop codon at codon 421

Lane 4: An individual who is homozygous for a mutation that introduces an early stop codon at codon 421

Lane 5: An individual who is homozygous for a mutation that changes codon 198 from a valine codon into a leucine codon

E7. The protein known as tyrosinase is needed to make certain types of pigments. Tyrosinase is composed of a single polypeptide with 511 amino acids. Because each amino acid weighs 120 Da (on average), the molecular mass of this protein is approximately 61,300 Da, or 61.3 kDa. People who carry two defective copies of the tyrosinase gene have the condition known as albinism. They are unable to make pigment in the skin, eyes, and hair. Western blotting can be used to detect proteins that are translated from a particular mRNA. This method is described in Chapter 20 and also in experimental question E5. Skin samples were collected from a pigmented individual (lane 1) and from three unrelated albino individuals (lanes 2, 3, and 4) and subjected to a Western blot analysis using an antibody that recognizes tyrosinase. Explain the possible cause of albinism in the three albino individuals.

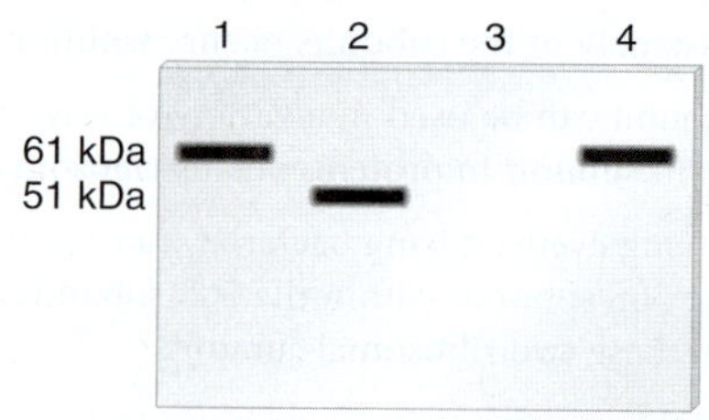

E8. Although 61 codons specify the 20 amino acids, most species display a codon bias. This means that certain codons are used much more frequently than other codons. For example, UUA, UUG, CUU, CUC, CUA, and CUG all specify leucine. In yeast, however, the UUG codon is used to specify leucine approximately 80% of the time.

A. The experiment of Figure 13.7 shows the use of an in vitro or cell-free translation system. In this experiment, the RNA, which was used for translation, was chemically synthesized. Instead of using a chemically synthesized RNA, researchers can isolate mRNA from living cells and then add the mRNA to the cell-free translation system. If a researcher isolated mRNA from kangaroo cells and then added it to a cell-free translation system that came from yeast cells, how might the phenomenon of codon bias affect the production of proteins?

B. Discuss potential advantages and disadvantages of codon bias for translation.

Questions for Student Discussion/Collaboration

1. Discuss why you think the ribosome needs to contain so many proteins and rRNA molecules. Does it seem like a waste of cellular energy to make such a large structure so that translation can occur?
2. Discuss and make a list of the similarities and differences in the events that occur during the initiation, elongation, and termination stages of transcription (see Chapter 12) and translation discussed in this chapter.
3. Which events during translation involve molecular recognition between base sequences within different RNAs? Which events involve recognition between different protein molecules?

Note: All answers appear at the website for this textbook; the answers to the even-numbered questions and all of the Concept Check and Comprehension Questions are in the back of the book.

www.mhhe.com/brookergenetics5e

Visit the website for practice tests, answer keys, and other learning aids for this chapter. Enhance your understanding of genetics with our interactive exercises, quizzes, animations, and much more.

CHAPTER OUTLINE

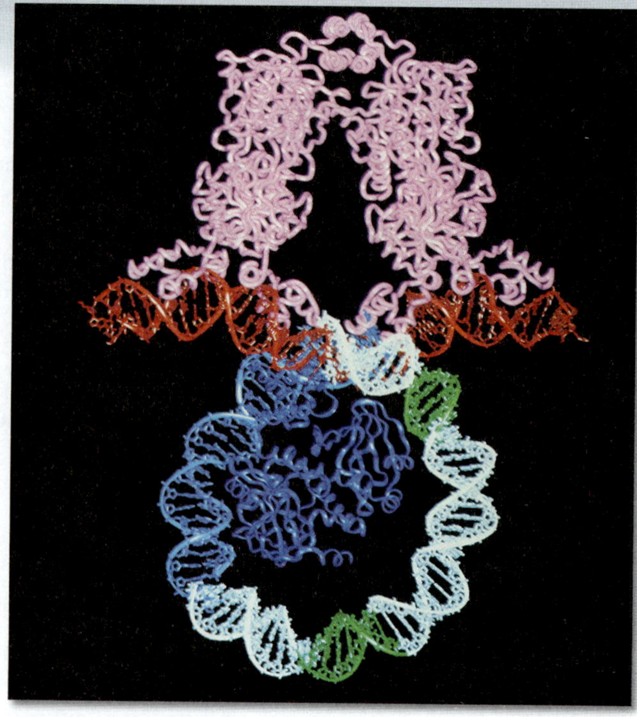

A model showing the binding of a genetic regulatory protein to DNA, which results in a DNA loop. *The model shown here involves the lac repressor protein found in E. coli binding to the operator site in the lac operon.*

14 GENE REGULATION IN BACTERIA

Chromosomes of bacteria, such as *Escherichia coli*, contain a few thousand different genes. **Gene regulation** is the phenomenon in which the level of gene expression can vary under different conditions. In comparison, unregulated genes have essentially constant levels of expression in all conditions over time. Unregulated genes are also called **constitutive genes.** Frequently, constitutive genes encode proteins that are continuously needed for the survival of the bacterium. In contrast, the majority of genes are regulated so that the proteins they encode can be produced at the proper times and in the proper amounts.

A key benefit of gene regulation is that the encoded proteins are produced only when they are required. Therefore, the cell avoids wasting valuable energy making proteins it does not need. From the viewpoint of natural selection, this enables an organism such as a bacterium to compete as efficiently as possible for limited resources. Gene regulation is particularly important because bacteria exist in an environment that is frequently changing with regard to temperature, nutrients, and many other factors. The following are a few common processes regulated at the genetic level:

1. ***Metabolism:*** Some proteins function in the metabolism of small molecules. For example, certain enzymes are needed for a bacterium to metabolize particular sugars. These enzymes are required only when the sugars are present in the bacterium's environment.
2. ***Response to environmental stress:*** Certain proteins help a bacterium to survive environmental stress such as osmotic shock or heat shock. These proteins are required only when the bacterium is confronted with the stress.
3. ***Cell division:*** Some proteins are needed for cell division. These are necessary only when the bacterial cell is getting ready to divide.

The expression of protein-encoding genes, which encode polypeptides, ultimately leads to the production of functional proteins. As we saw in Chapters 12 and 13, gene expression is a multistep process that proceeds from transcription to translation, and it may involve posttranslational effects on protein structure and function. As shown in **Figure 14.1**, gene regulation can occur at any of these steps in the pathway of gene expression. In this chapter, we will examine the molecular mechanisms of these types of gene regulation.

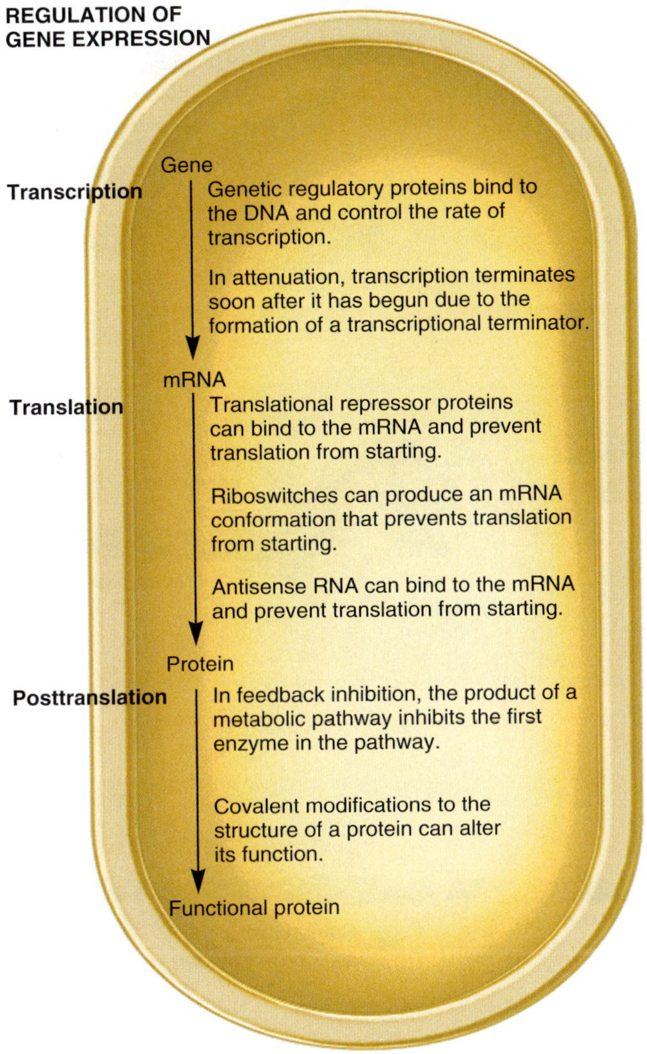

FIGURE 14.1 **Common points where regulation of gene expression occurs in bacteria.**

CONCEPT CHECK: What is an advantage of gene regulation?

14.1 OVERVIEW OF TRANSCRIPTIONAL REGULATION

Learning Outcomes:

1. Describe the function of activators and repressors.
2. Explain how small effector molecules affect the function of activators and repressors.

In bacteria, the most common way to regulate gene expression is by influencing the rate at which transcription is initiated. Although we frequently refer to genes as being "turned on or off," it is more accurate to say that the level of gene expression is increased or decreased. At the level of transcription, this means that the rate of RNA synthesis can be increased or decreased.

In most cases, transcriptional regulation involves the actions of regulatory proteins that can bind to the DNA and affect the rate of transcription of one or more nearby genes. Two types of regulatory proteins are common. A **repressor** is a regulatory protein that binds to the DNA and inhibits transcription, whereas an **activator** is a regulatory protein that increases the rate of transcription. Transcriptional regulation by a repressor protein is termed **negative control,** and regulation by an activator protein is considered to be **positive control.**

In conjunction with regulatory proteins, small effector molecules often play a critical role in transcriptional regulation. However, small effector molecules do not bind directly to the DNA to alter transcription. Rather, an effector molecule exerts its effects by binding to a repressor or activator. The binding of the effector molecule causes a conformational change in the regulatory protein and thereby influences whether or not the protein can bind to the DNA. Genetic regulatory proteins that respond to small effector molecules typically have two binding sites. One site is where the protein binds to the DNA; the other is the binding site for the effector molecule.

Regulatory proteins are given names describing how they affect transcription when they are bound to the DNA (repressor or activator). In contrast, small effector molecules are given names that describe how they affect transcription when they are present in the cell at a sufficient concentration to exert their effect (Figure 14.2). An **inducer** is a small effector molecule that causes transcription to increase. An inducer may accomplish this in two ways: It could bind to a repressor protein and prevent it from binding to the DNA, or it could bind to an activator protein and cause it to bind to the DNA. In either case, the transcription rate is increased. Genes that are regulated in this manner are called **inducible genes.**

Alternatively, the presence of a small effector molecule may inhibit transcription. This can also occur in two ways. A **corepressor** is a small molecule that binds to a repressor protein, thereby causing the protein to bind to the DNA. An **inhibitor** binds to an activator protein and prevents it from binding to the DNA. Both corepressors and inhibitors act to reduce the rate of transcription. Therefore, the genes they regulate are termed **repressible genes.** Unfortunately, this terminology can be confusing because a repressible gene could be controlled by an activator protein, or an inducible gene could be controlled by a repressor protein.

14.1 COMPREHENSION QUESTIONS

1. A repressor is a ___________ that ___________ transcription.
   a. small effector molecule, inhibits
   b. small effector molecule, enhances
   c. regulatory protein, inhibits
   d. regulatory protein, enhances
2. Which of the following conditions would cause transcription to be activated?
   a. A repressor plus an inducer
   b. A repressor plus a corepressor
   c. An activator plus an inhibitor
   d. None of the above

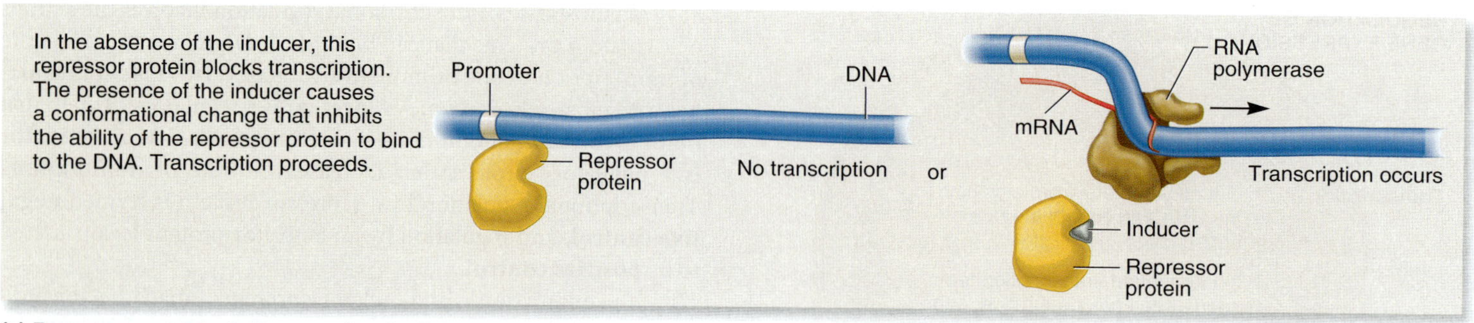

(a) Repressor protein, inducer molecule, inducible gene

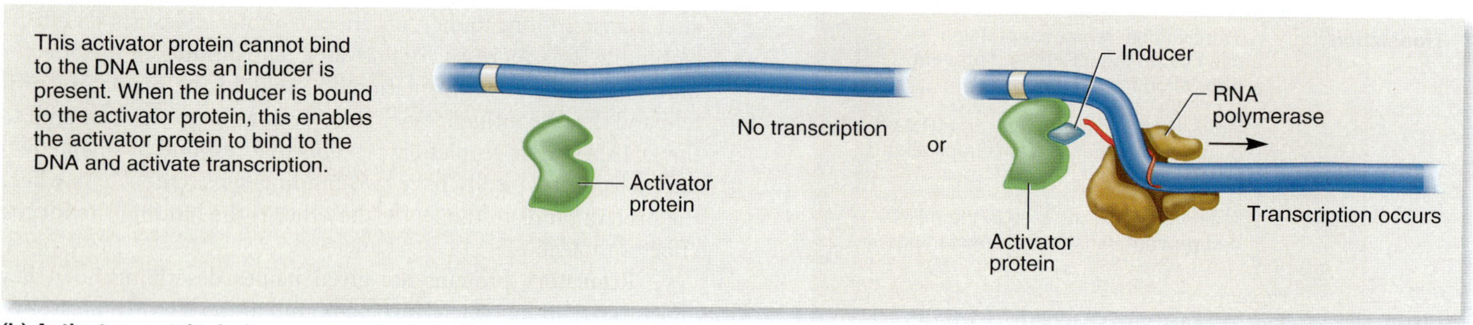

(b) Activator protein, inducer molecule, inducible gene

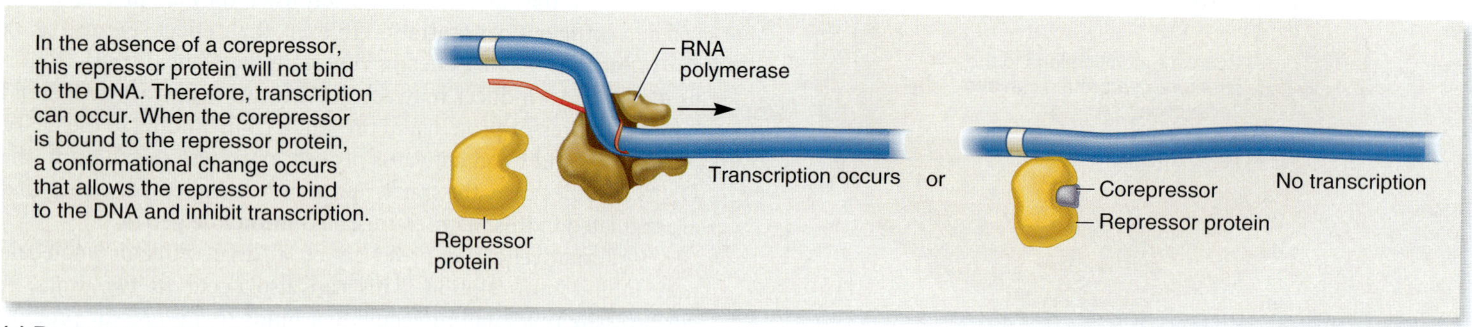

(c) Repressor protein, corepressor molecule, repressible gene

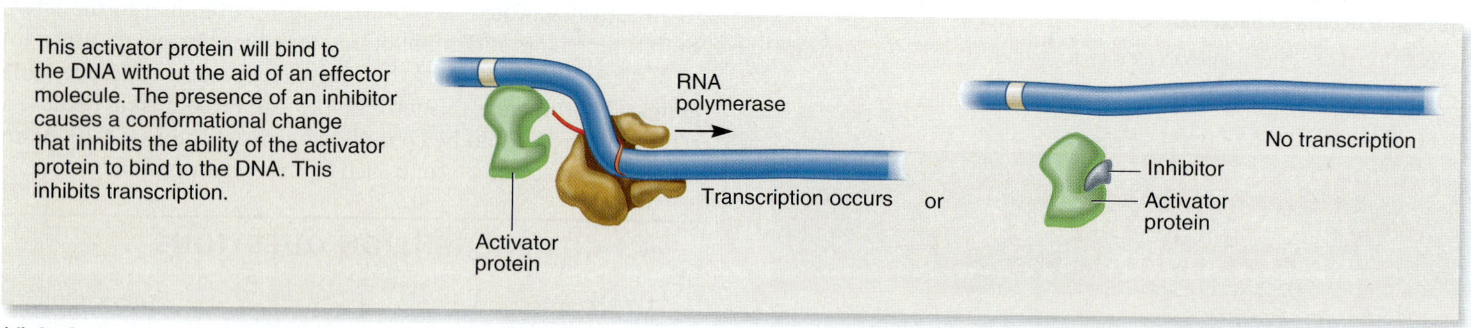

(d) Activator protein, inhibitor molecule, repressible gene

FIGURE 14.2 Binding sites on a genetic regulatory protein. In these examples, a regulatory protein has two binding sites: one for DNA and one for a small effector molecule. The binding of the small effector molecule changes the conformation of the regulatory protein, which alters the DNA-binding site structure, thereby influencing whether the protein can bind to the DNA.

CONCEPT CHECK: Which of these are genetic regulatory proteins and which are small effector molecules?

14.2 REGULATION OF THE *lac* OPERON

Learning Outcomes:

1. Describe the organization of the *lac* operon.
2. Explain how the *lac* operon is regulated by lac repressor and by catabolite activator protein.
3. Analyze the results of Jacob, Monod, and Pardee and explain how they indicated that the *lacI* gene encodes a diffusible repressor protein.

We will now turn our attention to a specific example of gene regulation that is found in *E. coli*. This example involves genes that play a role in the utilization of lactose, which is a sugar found in milk. As you will learn, the regulation of these genes involves a repressor protein and also an activator protein.

The Phenomenon of Enzyme Adaptation Is Due to the Synthesis of Cellular Proteins

Our initial understanding of gene regulation can be traced back to the creative minds of François Jacob and Jacques Monod at the Pasteur Institute in Paris, France. Their research into genes and gene regulation stemmed from an interest in the phenomenon known as **enzyme adaptation,** which had been identified at the turn of the twentieth century. Enzymes are composed of proteins. Enzyme adaptation refers to the observation that a particular enzyme appears within a living cell only after the cell has been exposed to the substrate for that enzyme. When a bacterium is not exposed to a particular substance, it does not make the enzymes needed to metabolize that substance.

To investigate this phenomenon, Jacob and Monod focused their attention on lactose metabolism in *E. coli*. Key experimental observations that led to an understanding of this genetic system are listed here.

1. The exposure of bacterial cells to lactose increased the levels of lactose-utilizing enzymes by 1000- to 10,000-fold.
2. Antibody and labeling techniques revealed that the increase in the activity of these enzymes was due to the increased synthesis of the proteins that form the enzymes.
3. The removal of lactose from the environment caused an abrupt termination in the synthesis of the enzymes.
4. The analysis of mutations in the *lac* operon revealed that each protein involved with lactose utilization is encoded by a separate gene.

These critical observations indicated to Jacob and Monod that enzyme adaptation is due to the synthesis of specific proteins in response to lactose in the environment. Next, we will examine how Jacob and Monod discovered that this phenomenon is due to the interactions between genetic regulatory proteins and small effector molecules. In other words, we will see that enzyme adaptation is due to the transcriptional regulation of genes.

The *lac* Operon Encodes Proteins Involved in Lactose Metabolism

In bacteria, it is common for a few genes to be arranged together in an **operon**—a group of two or more genes under the transcriptional control of a single promoter. An operon encodes a **polycistronic mRNA,** an RNA that contains the sequences of two or more genes. Why do operons occur in bacteria? One biological advantage of an operon organization is that it allows a bacterium to coordinately regulate a group of two or more genes that are involved with a common functional goal; the expression of the genes occurs as a single unit. For transcription to take place, an operon is flanked by a **promoter** that signals the beginning of transcription and a **terminator** that specifies the end of transcription. Two or more genes are found between these two sequences.

Figure 14.3a shows the organization of the genes in *E. coli* involved in lactose utilization and their transcriptional regulation. Two distinct transcriptional units are present. The first unit, known as the *lac* operon, contains a CAP site; *lac* promoter (*lacP*); operator site (*lacO*); three protein-encoding genes, *lacZ*, *lacY*, and *lacA*; and a terminator. *LacZ* encodes the enzyme β-galactosidase, an enzyme that cleaves lactose into galactose and glucose. As a side reaction, β-galactosidase also converts a small percentage of lactose into allolactose, a structurally similar sugar (**Figure 14.3b**). As we will see later, allolactose acts as a small effector molecule for regulating the *lac* operon. The *lacY* gene encodes lactose permease, a membrane protein required for the active transport of lactose into the cytoplasm of the bacterium. The *lacA* gene encodes galactoside transacetylase, an enzyme that covalently modifies lactose and lactose analogs. Although the functional necessity of the transacetylase remains unclear, the acetylation of nonmetabolizable lactose analogs may prevent their toxic buildup within the bacterial cytoplasm.

The CAP site and the operator site are short DNA segments that function in gene regulation. The **CAP site** is a DNA sequence recognized by an activator protein called the **catabolite activator protein (CAP).** The **operator site** (also known simply as the **operator**) is a sequence of bases that provides a binding site for a repressor protein called the lac repressor.

A second transcriptional unit is the *lacI* gene (see Figure 14.3a), which is not part of the *lac* operon. The *lacI* gene has its own promoter called the *i* promoter and is constitutively expressed at fairly low levels. The *lacI* gene encodes the **lac repressor,** a protein that regulates the *lac* operon by binding to the operator site and repressing transcription. The lac repressor functions as a homotetramer, a protein composed of four identical subunits. Only a small amount of the lac repressor protein is needed to repress the *lac* operon. The mechanism of action of the lac repressor is described next.

The *lac* Operon Is Regulated by a Repressor Protein

The *lac* operon can be transcriptionally regulated in more than one way. The first mechanism that we will examine is inducible and under negative control. As shown in **Figure 14.4**, this form of

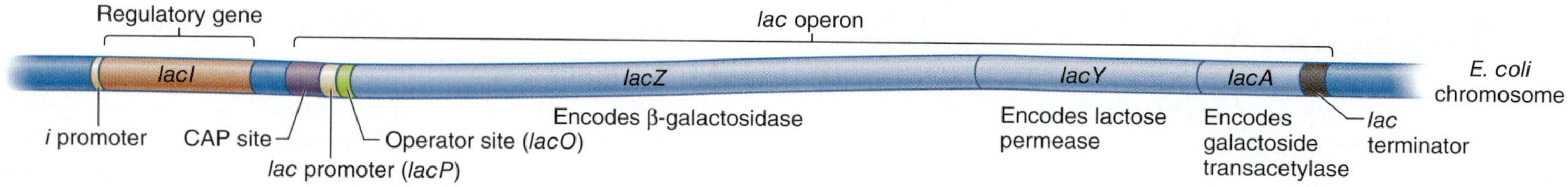

(a) Organization of DNA sequences in the *lac* region of the *E. coli* chromosome

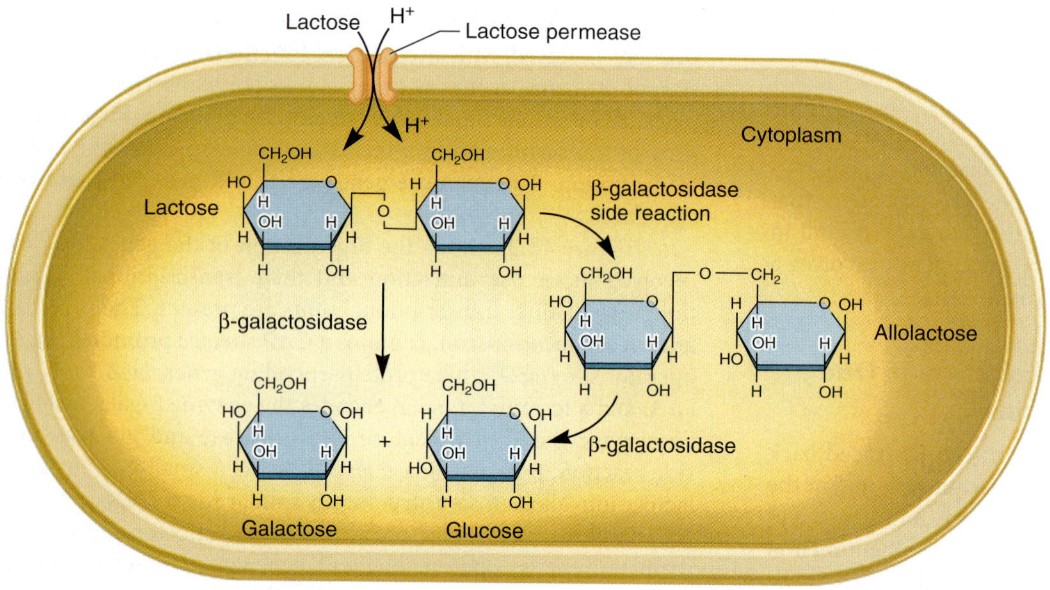

(b) Functions of lactose permease and β-galactosidase

ONLINE ANIMATION

FIGURE 14.3 Organization of the lac operon and other genes involved with lactose metabolism in *E. coli.* **(a)** The CAP site is the binding site for the catabolite activator protein (CAP). The operator site is a binding site for the lac repressor. The promoter (*lacP*) is responsible for the transcription of the *lacZ, lacY,* and *lacA* genes as a single unit, which ends at the *lac* terminator. The *i* promoter is responsible for the transcription of the *lacI* gene. **(b)** Lactose permease allows the uptake of lactose into the bacterial cytoplasm. It cotransports lactose with H^+. Because bacteria maintain an H^+ gradient across their cytoplasmic membrane, this cotransport permits the active accumulation of lactose against a gradient. β-Galactosidase is a cytoplasmic enzyme that cleaves lactose and related compounds into galactose and glucose. As a minor side reaction, β-galactosidase also converts lactose into allolactose. Allolactose can also be broken down into galactose and glucose.

CONCEPT CHECK: Which genes are under the control of the *lac* promoter?

regulation involves the lac repressor protein, which binds to the sequence of nucleotides found within the *lac* operator site. Once bound, the lac repressor prevents RNA polymerase from transcribing the *lacZ*, *lacY*, and *lacA* genes (Figure 14.4a). The binding of the repressor to the operator site is a reversible process. In the absence of allolactose, the lac repressor is bound to the operator site most of the time.

The ability of the lac repressor to bind to the operator site depends on whether or not allolactose is bound to it. Each of the repressor protein's four subunits has a single binding site for allolactose, the inducer. How does a small molecule like allolactose exert its effects? When four molecules of allolactose bind to the repressor, a conformational change occurs that prevents the lac repressor from binding to the operator site. Under these conditions, RNA polymerase is now free to transcribe the operon (Figure 14.4b). In genetic terms, we would say that the operon has been **induced.** The action of a small effector molecule, such as allolactose, is called **allosteric regulation.** The functioning of allosteric proteins, such as the lac repressor, is controlled by effector molecules that bind to the proteins' **allosteric sites.** In the case of the lac repressor, the binding of allolactose alters the function of lac repressor because it prevents it from binding to the DNA.

Rare mutations in the *lacI* gene that alter the regulation of the *lac* operon reveal that each subunit of the lac repressor has a region that binds to the DNA and another region that contains the allolactose-binding site. As discussed later in Figure 14.7, researchers have identified $lacI^-$ mutations that result in the constitutive expression of the *lac* operon, which means that it is expressed in both the presence and the absence of lactose. Such mutations may result in an inability to synthesize any repressor protein, or they may produce a repressor protein that is unable to bind to the DNA at the *lac* operator site. If the lac repressor is unable to bind to the DNA, the *lac* operon cannot be repressed. By comparison, $lacI^S$ mutations have the opposite effect: The *lac* operon cannot be induced even in the presence of lactose. These mutations, which are called super-repressor mutations, alter the region of the lac repressor that binds allolactose. The mutation usually results in a lac repressor protein that cannot bind allolactose. If the lac repressor is unable to bind allolactose, it will remain bound to the *lac* operator site and therefore induction cannot occur.

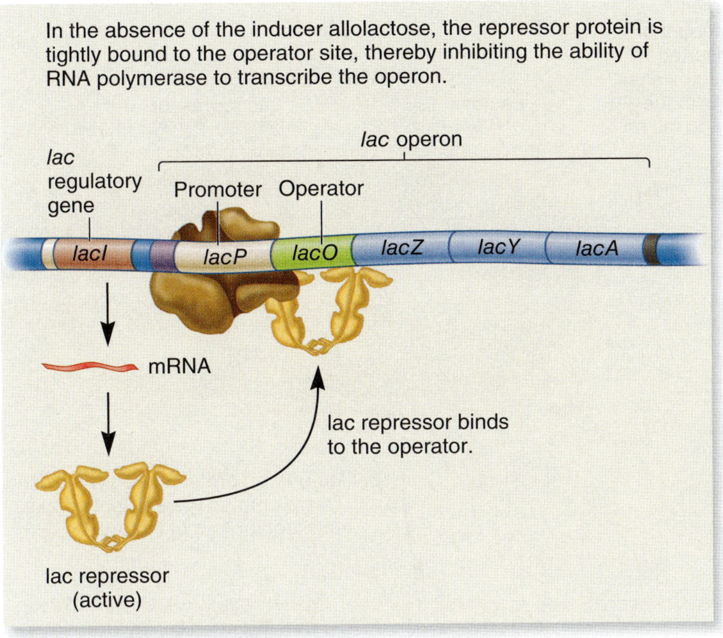

(a) No lactose in the environment

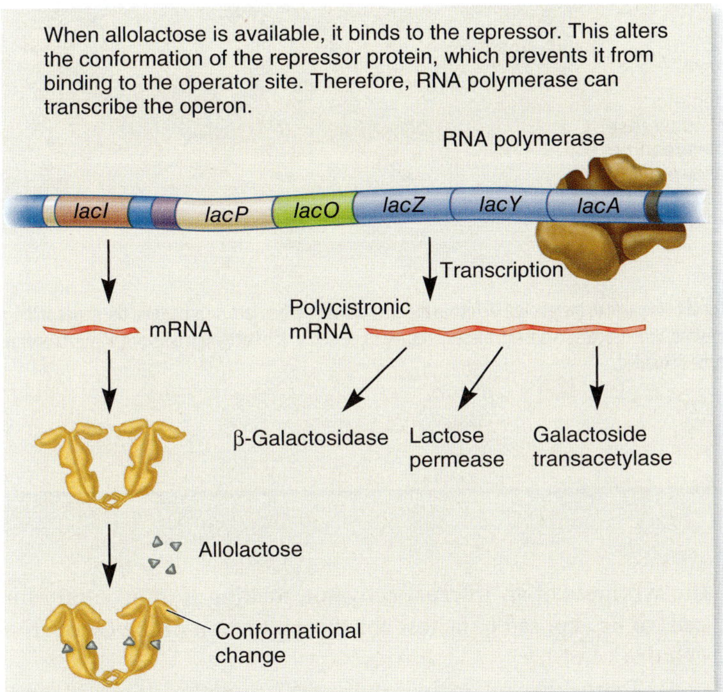

(b) Lactose present

ONLINE ANIMATION

FIGURE 14.4 Mechanism of induction of the *lac* operon. Note: The CAP site is not labeled in these drawings.

The Regulation of the *lac* Operon Allows a Bacterium to Respond to Environmental Change

To better appreciate *lac* operon regulation at the cellular level, let's consider the process as it occurs over time. **Figure 14.5** illustrates the effects of external lactose on the regulation of the *lac* operon. In the absence of lactose, no inducer (allolactose) is available to bind to the lac repressor. Therefore, the lac repressor binds to the operator site and inhibits transcription. In reality, the repressor does not completely inhibit transcription, so very small amounts of β-galactosidase, lactose permease, and galactoside transacetylase are made. However, the levels are far too low for the bacterium to readily use lactose. When the bacterium is exposed to lactose, a small amount can be transported into the cytoplasm via lactose permease, and β-galactosidase converts some of it to allolactose. As this occurs, the cytoplasmic level of allolactose gradually rises; eventually, allolactose binds to the lac repressor. The binding of allolactose promotes a conformational change that prevents the repressor from binding to the *lac* operator site, thereby allowing transcription of the *lacZ*, *lacY*, and *lacA* genes to occur. Translation of the encoded polypeptides produces the proteins needed for lactose uptake and metabolism.

To understand how the induction process is shut off when lactose in the environment has been depleted, let's consider the interaction between allolactose and the lac repressor. The lac repressor has a measurable affinity for allolactose. The binding of allolactose to the lac repressor is reversible. The likelihood that allolactose will bind to the repressor depends on the allolactose concentration. During induction of the operon, the concentration of allolactose (the inducer) rises and approaches the affinity for the repressor protein. This makes it likely that allolactose will bind to the lac repressor, thereby causing it to be released from the operator site.

The intracellular concentration of allolactose remains high as long as lactose is available in the environment. However, when lactose is depleted from the environment, the concentration of allolactose also becomes lower due to the action of metabolic enzymes. Eventually, the concentration of allolactose drops below its affinity for the repressor. At this point, the lac repressor is unlikely to be bound to allolactose. When allolactose is released, the lac repressor returns to the conformation that binds to the operator site. In this way, the binding of the repressor shuts down the *lac* operon when lactose is depleted from the environment. After repression occurs, the mRNA and proteins encoded by the *lac* operon are eventually degraded (see Figure 14.5).

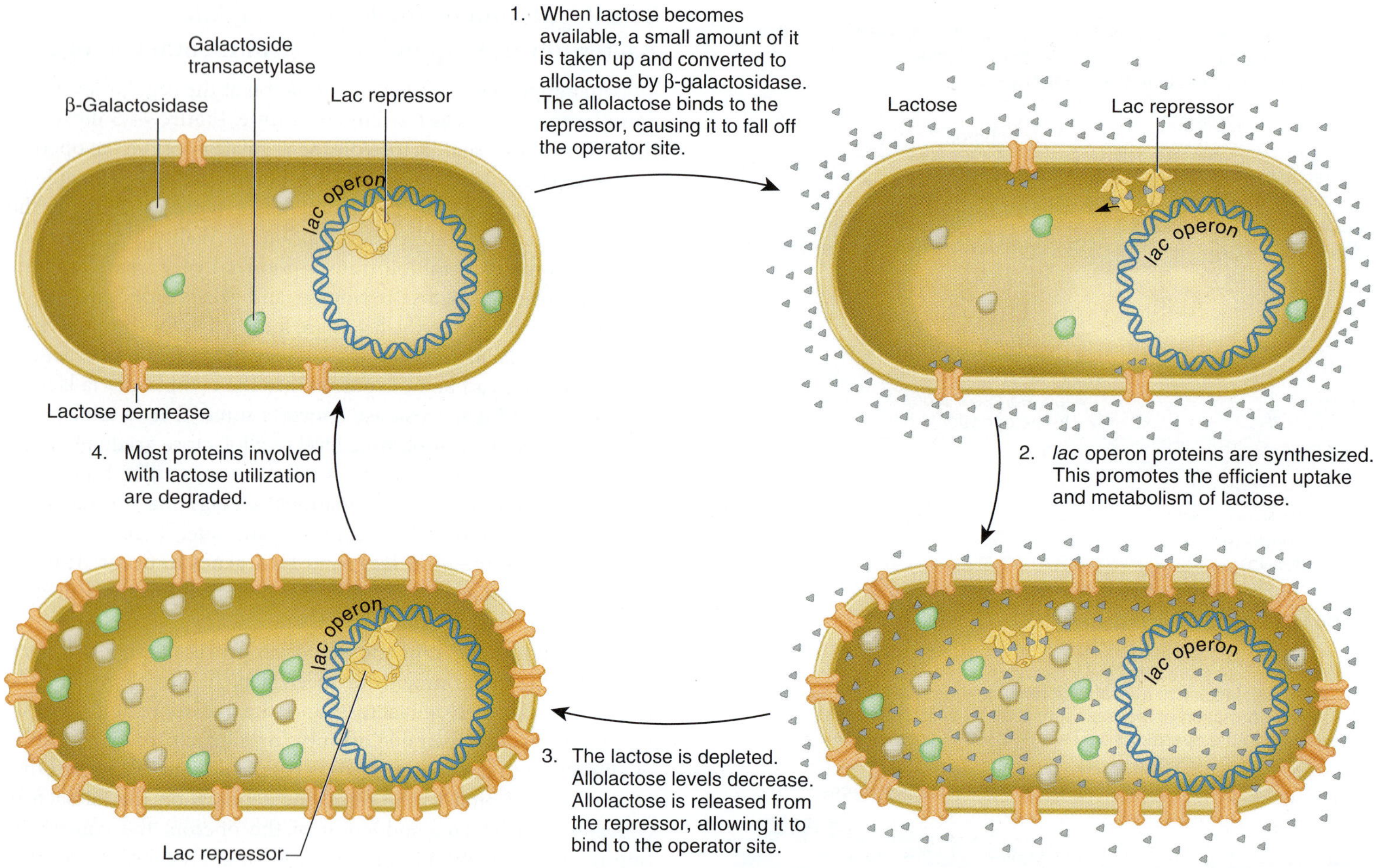

FIGURE 14.5 **The cycle of *lac* operon induction and repression.**

Genes → Traits The genes and the genetic regulation of the *lac* operon provide the bacterium with the trait of being able to metabolize lactose in the environment. When lactose is present, the genes of the *lac* operon are induced, and the proteins needed for the efficient uptake and metabolism of lactose are synthesized. When lactose is absent, these genes are repressed so the bacterium does not waste its energy expressing them. Note: The proteins involved with lactose utilization are fairly stable, but they will eventually be degraded.

CONCEPT CHECK: Under what conditions is the lac repressor bound to the *lac* operon?

EXPERIMENT 14A

The *lacI* Gene Encodes a Diffusible Repressor Protein

Now that we have an understanding of the *lac* operon, let's consider one of the experimental approaches that was used to elucidate its regulation. In the 1950s, Jacob, Monod, and their colleague Arthur Pardee had identified a few rare mutant strains of bacteria that had abnormal lactose adaptation. As mentioned earlier, one type of mutant, designated *lacI*$^-$, resulted in the constitutive expression of the *lac* operon even in the absence of lactose. As shown in **Figure 14.6a**, the correct explanation is that a loss-of-function mutation in the *lacI* gene prevented the lac repressor protein from binding to the *lac* operator site and inhibiting transcription. At the time of their work, however, the function of the lac repressor was not yet known. Instead, the researchers incorrectly hypothesized that the *lacI*$^-$ mutation resulted in the synthesis of an internal activator, making it unnecessary for cells to be exposed to lactose for the expression of the *lac* operon (**Figure 14.6b**).

To further explore the nature of this mutation, Jacob, Monod, and Pardee applied a genetic approach. In order to understand their approach, let's briefly consider the process of bacterial conjugation (described in Chapter 7). The earliest studies of Jacob, Monod, and Pardee in 1959 involved matings between recipient cells, termed F$^-$, and donor cells, which were *Hfr* strains that transferred a portion of the bacterial chromosome. Later experiments in 1961 involved the transfer of circular segments of DNA known as F factors. We consider the latter type of experiment here. Sometimes an F factor also carries genes that were originally found within the bacterial chromosome. These types of F factors are called F′ factors (F prime factors). In their studies, Jacob, Monod, and Pardee identified F′ factors that carried the *lacI* gene

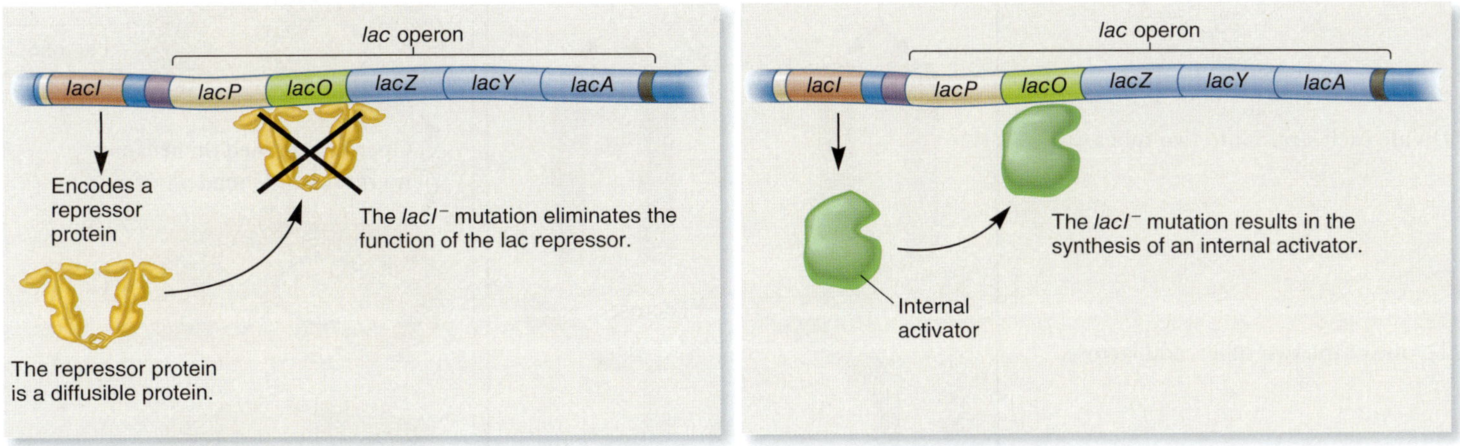

(a) Correct explanation **(b) Internal activator hypothesis**

FIGURE 14.6 Alternative hypotheses to explain how a *lacI*$^-$ mutation could cause the constitutive expression of the *lac* operon. **(a)** The correct explanation in which the *lacI*$^-$ mutation eliminates the function of the lac repressor protein, which prevents it from repressing the *lac* operon. **(b)** The hypothesis of Jacob, Monod, and Pardee. In this case, the *lacI*$^-$ mutation would result in the synthesis of an internal activator that turns on the *lac* operon.

and the *lac* operon. These F′ factors can be transferred from one cell to another by bacterial conjugation. A strain of bacteria containing F′ factor genes is called a **merozygote,** or partial diploid.

The production of merozygotes was instrumental in allowing Jacob, Monod, and Pardee to elucidate the function of the *lacI* gene. This experimental approach has two key points. First, the two *lacI* genes in a merozygote may be different alleles. For example, the *lacI* gene on the chromosome may be a *lacI*$^-$ allele that causes constitutive expression, whereas the *lacI* gene on the F′ factor may be normal. Second, the genes on the bacterial chromosome and the genes on the F′ factor are not physically adjacent to each other. As we now know, the expression of a normal *lacI* gene on an F′ factor produces repressor proteins that can diffuse within the cell and eventually bind to the operator site of the *lac* operon located on the chromosome and also to the operator site on an F′ factor.

Figure 14.7 shows one experiment of Jacob, Monod, and Pardee in which they analyzed a *lacI*$^-$ mutant strain that was already known to constitutively express the *lac* operon and compared it with the corresponding merozygote. The merozygote had a *lacI*$^-$ mutant gene on the chromosome and a normal *lacI* gene on an F′ factor. Each strain was grown and then divided into two tubes. In one tube, lactose was omitted; the other tubes contained lactose in order to determine if lactose was needed to induce the expression of the operon. The cells were lysed by sonication to release β-galactosidase from the cells, and then a lactose analog, β-*o*-nitrophenylgalactoside (β-ONPG), was added. This molecule is colorless, but β-galactosidase cleaves it into a product that has a yellow color. Therefore, the amount of yellow color produced in a given amount of time is a measure of the amount of β-galactosidase being expressed from the *lac* operon.

THE HYPOTHESIS

The *lacI*$^-$ mutation results in the synthesis of an internal activator.

TESTING THE HYPOTHESIS

FIGURE 14.7 Evidence that the *lacI* gene encodes a diffusible repressor protein.

Starting material: The genotype of the mutant strain was *lacI*$^-$ *lacZ*$^+$ *lacY*$^+$ *lacA*$^+$. The merozygote strain had an F′ factor that was *lacI*$^+$ *lacZ*$^+$ *lacY*$^+$ *lacA*$^+$. The F′ factor had been introduced into the mutant strain via conjugation.

1. Grow mutant strain and merozygote strain separately.

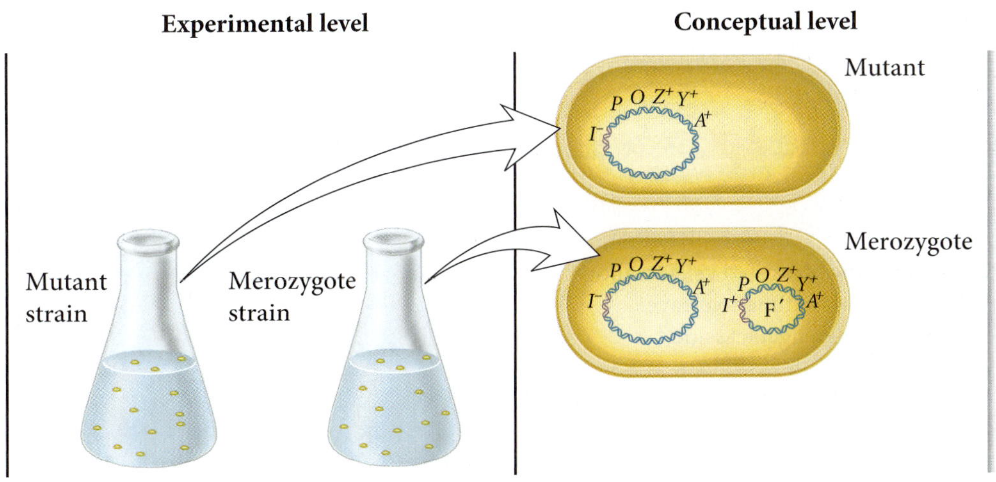

2. Divide each strain into two tubes.
3. To one of the two tubes, add lactose.
4. Incubate the cells long enough to allow *lac* operon induction.
5. Lyse the cells with a sonicator. This allows β-galactosidase to escape from the cells.
6. Add β-*o*-nitrophenylgalactoside (β-ONPG). This is a colorless compound. If β-galactosidase is present, it will cleave the compound to produce galactose and *o*-nitrophenol (O-NP). *o*-Nitrophenol has a yellow color. The deeper the yellow color, the more β-galactosidase was produced.
7. Incubate the sonicated cells to allow β-galactosidase time to cleave β-*o*-nitrophenylgalactoside.

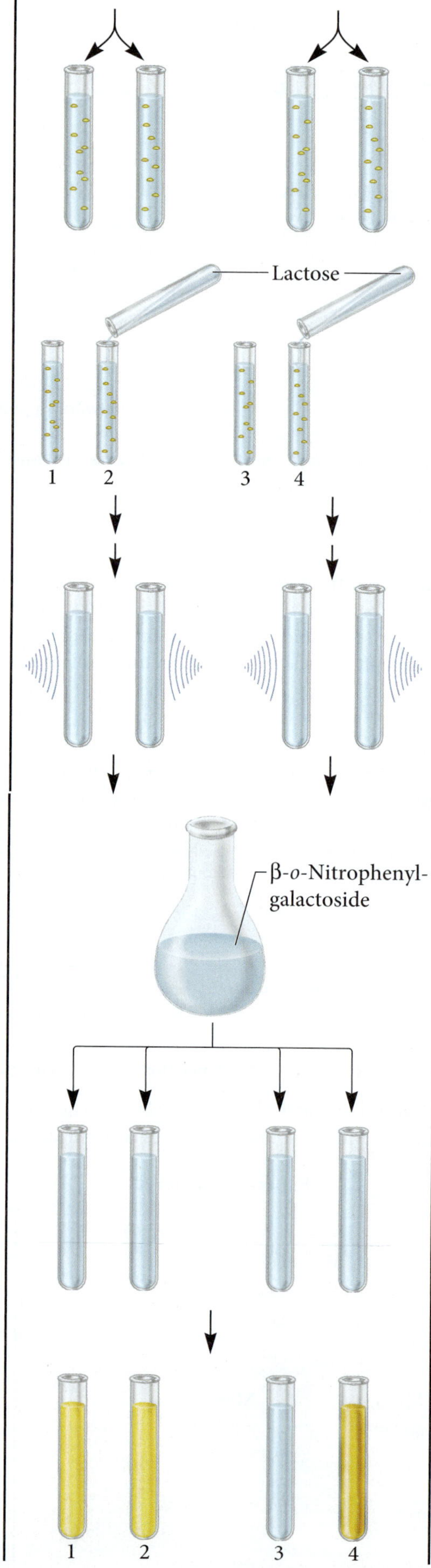

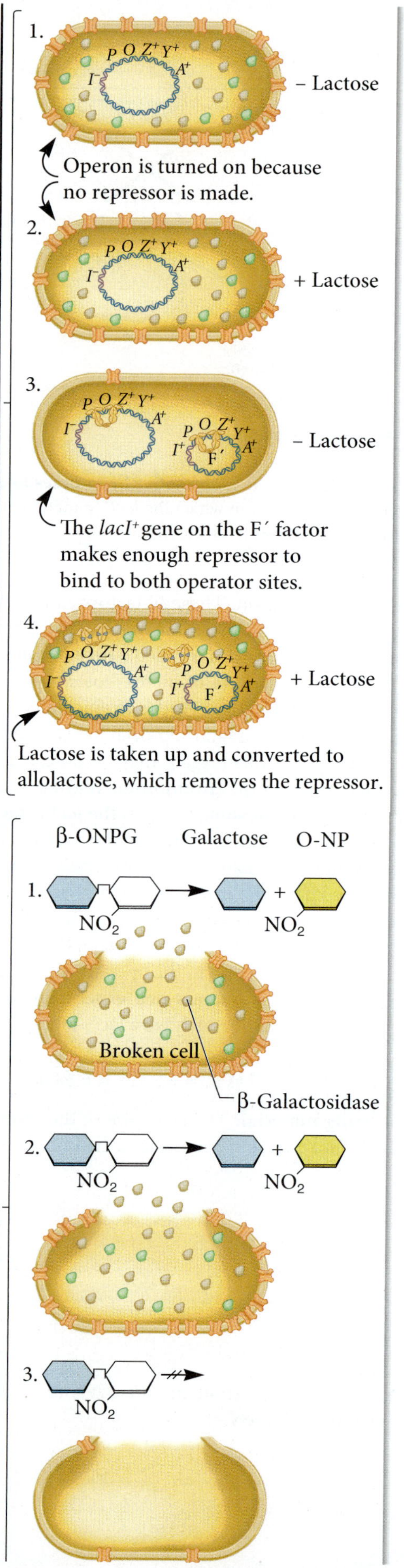

8. Measure the yellow color produced with a spectrophotometer. (See the Appendix for a description of spectrophotometry.)

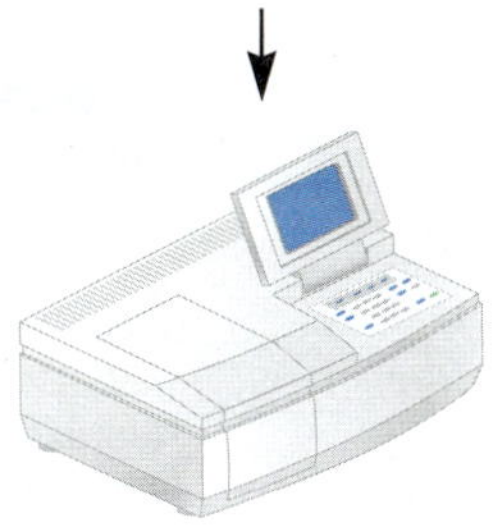

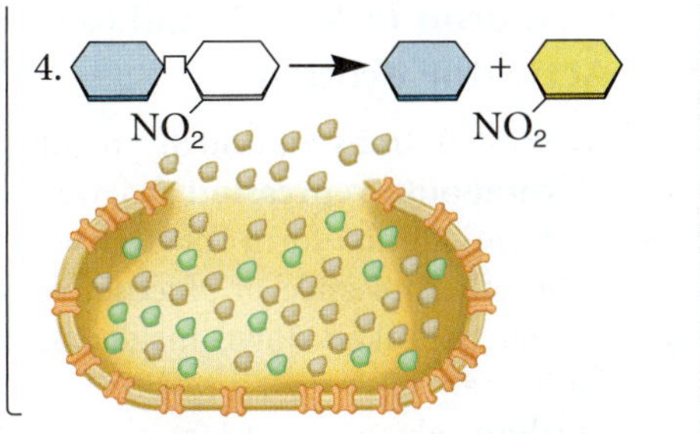

THE DATA

Strain	*Addition of Lactose*	*Amount of β-Galactosidase (percentage)*
Mutant	No	100
Mutant	Yes	100
Merozygote	No	<1
Merozygote	Yes	220

Data from F. Jacob, and J. Monod (1961) Genetic regulatory mechanisms in the synthesis of proteins. *J Mol Biol. 3*, 318–356.

INTERPRETING THE DATA

As seen in The Data, the yellow production in the original mutant strain was the same in either the absence or the presence of lactose. This result is expected because the expression of β-galactosidase in the *lacI*⁻ mutant strain was already known to be constitutive. In other words, the presence of lactose was not needed to induce the operon due to a defective *lacI* gene. In the merozygote strain, however, a different result was obtained. In the absence of lactose, the *lac* operons were repressed—even the operon on the bacterial chromosome. How do we explain these results? Because the normal *lacI* gene on the F′ factor was not physically located next to the chromosomal *lac* operon, this result is consistent with the idea that the *lacI* gene codes for a repressor protein that can diffuse throughout the cell and bind to any *lac* operon. The hypothesis that the *lacI*⁻ mutation resulted in the synthesis of an internal activator was rejected. If that hypothesis had been correct, the merozygote strain would have still made an internal inducer, and the *lac* operons in the merozygote would have been expressed in the absence of lactose. This result was not obtained.

The interactions between regulatory proteins and DNA sequences illustrated in this experiment have led to the definition of two genetic terms. A ***trans*-effect** is a form of genetic regulation that can occur even though two DNA segments are not physically adjacent. The action of the lac repressor on the *lac* operon is a *trans*-effect. A regulatory protein, such as the lac repressor, is called a ***trans*-acting factor.** In contrast, a ***cis*-acting element** is a DNA segment that must be adjacent to the gene(s) that it regulates, and it is said to have a ***cis*-effect** on gene expression. The *lac* operator site is an example of a *cis*-acting element. A *trans*-effect is mediated by genes that encode regulatory proteins, whereas a *cis*-effect is mediated by DNA sequences that are binding sites for regulatory proteins.

TABLE 14.1

A Comparison of Loss-of-Function Mutations in the *lacI* Gene Versus the Operator Site

Chromosome	F′ factor	Expression of the *lac* Operon (%) With Lactose	Without Lactose
Wild type	None	100	<1
lacI⁻	None	100	100
lacO⁻	None	100	100
lacI⁻	*lacI*⁺ and a normal *lac* operon	200	<1
lacO⁻	*lacI*⁺ and a normal *lac* operon	200	100

Jacob and Monod also isolated constitutive mutants that affected the operator site, *lacO*. **Table 14.1** summarizes the effects of mutations based on their locations in the *lacI* regulatory gene versus *lacO* and their analysis in merozygotes. As seen here, a loss-of-function mutation in a gene encoding a repressor protein has the same effect as a mutation in an operator site that cannot bind a repressor protein. In both cases, the genes of the *lac* operon are constitutively expressed.

In a merozygote, however, the results are quite different. When a *lacI*⁺ gene and a normal *lac* operon carried on a F′ factor are introduced into a cell harboring a defective (mutant) *lacI* gene on the bacterial chromosome, the *lacI*⁺ gene on the F′ factor can regulate both operons. In contrast, when a *lacI*⁺ gene and a normal *lac* operon carried on an F′ factor are introduced into a cell with a defective (mutant) operator site on the bacterial chromosome, the *lac* operon on the chromosome continues to be expressed without lactose present. This occurs because the repressor cannot bind to the defective operator site on the chromosome. Overall, a mutation in a *trans*-acting factor can be complemented by the introduction of a second gene with normal function. However, a mutation in a *cis*-acting element is not affected by the introduction of a normal *cis*-acting element into the cell.

A self-help quiz involving this experiment can be found at www.mhhe.com/brookergenetics5e.

The *lac* Operon Is Also Regulated by an Activator Protein

The *lac* operon is transcriptionally regulated in a second way, known as **catabolite repression** (as we shall see, a somewhat imprecise term). This form of transcriptional regulation is influenced by the presence of glucose, which is a catabolite—a substance that is broken down inside the cell. The presence of glucose ultimately leads to repression of the *lac* operon. When exposed to both glucose and lactose, *E. coli* cells first use glucose, and catabolite repression prevents the use of lactose. Why is this an advantage? The explanation is efficiency. The bacterium does not have to express all of the genes necessary for both glucose and lactose metabolism. If the glucose is used up, catabolite repression is alleviated, and the bacterium then expresses the *lac* operon. The sequential use of two sugars by a bacterium, known as **diauxic growth,** is a common phenomenon among many bacterial species. Glucose, a more commonly encountered sugar, is metabolized preferentially, and then a second sugar is metabolized after glucose is depleted from the environment.

Glucose, however, is not itself the small effector molecule that binds directly to a genetic regulatory protein. Instead, this form of regulation involves a small effector molecule, **cyclic-AMP (cAMP),** which is produced from ATP via an enzyme known as adenylyl cyclase. When a bacterium is exposed to glucose, the transport of glucose into the cell stimulates a signaling pathway that causes the intracellular concentration of cAMP to decrease because the pathway inhibits adenylyl cyclase, the enzyme needed for cAMP synthesis. The effect of cAMP on the *lac* operon is mediated by an activator protein called the catabolite activator protein (CAP). CAP is composed of two subunits, each of which binds one molecule of cAMP.

Figure 14.8 considers how the interplay between the lac repressor and CAP determines whether the *lac* operon is expressed in the presence or absence of lactose and/or glucose. When only lactose is present, allolactose and cAMP levels are high (Figure 14.8a). Allolactose binds to the lac repressor and prevents it from binding to the DNA. The cAMP binds to CAP, and then CAP binds to the CAP site and stimulates the ability of RNA polymerase to begin transcription. Under these conditions, transcription proceeds at a high rate. In the absence of both lactose and glucose, cAMP levels are also high (Figure 14.8b). However, the binding of the lac repressor inhibits transcription even though CAP is bound to the DNA. Therefore, the transcription rate is very low.

Figure 14.8c shows the situation in which both sugars are present. The presence of lactose causes the lac repressor to be inactive, which prevents it from binding to the operator site. Even so, the presence of glucose decreases cAMP levels so that cAMP is released from CAP, which prevents CAP from binding to the CAP site. Because CAP is not bound to the CAP site, the transcription of the *lac* operon is low in the presence of both sugars. Finally, Figure 14.8d illustrates what happens when only glucose is present. The transcription of the *lac* operon is very low because the lac repressor is bound to the operator site and CAP is not bound to the CAP site due to low cAMP levels.

The effect of glucose, called catabolite repression, may seem like a puzzling way to describe this process because this regulation involves the action of an inducer (cAMP) and an activator

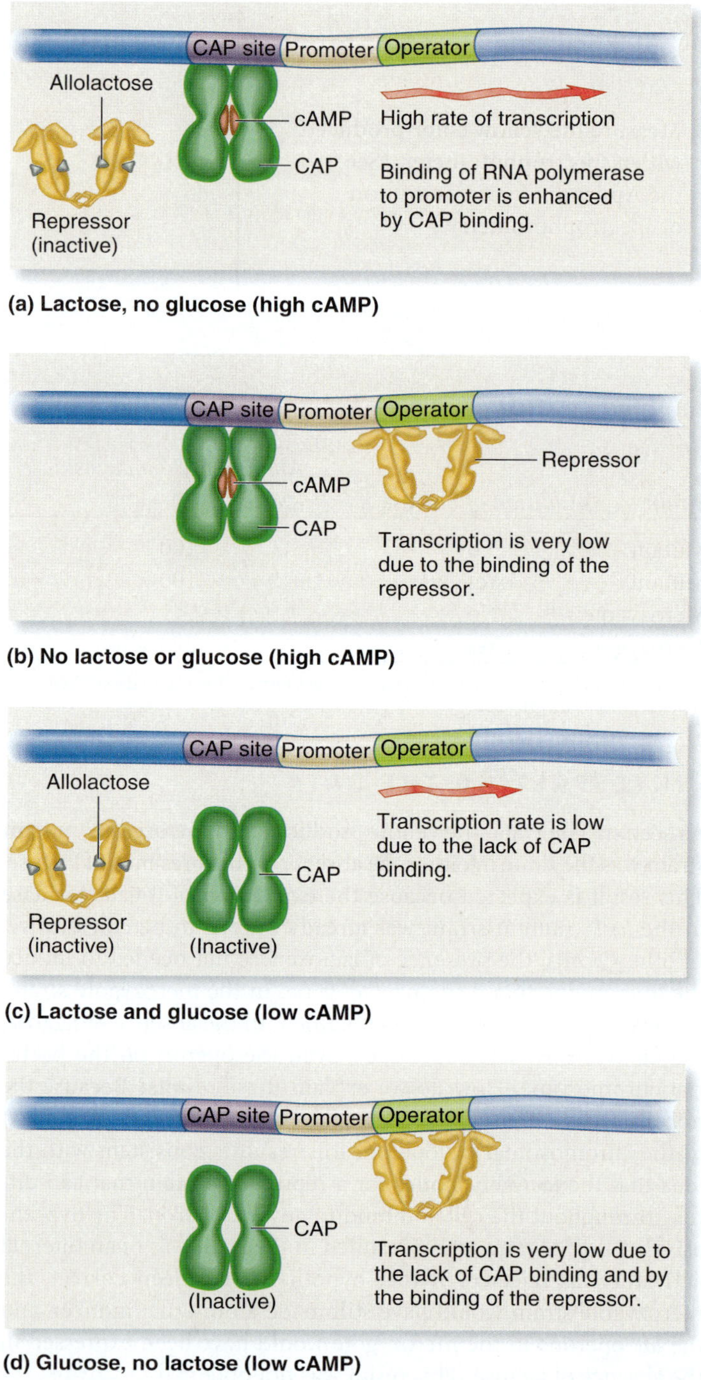

INTERACTIVE EXERCISE

FIGURE 14.8 The roles of the lac repressor and catabolite activator protein (CAP) in the regulation of the *lac* operon. This figure illustrates how the *lac* operon is regulated depending on its exposure to lactose or glucose.

Genes → Traits The mechanism of catabolite repression provides the bacterium with the trait of being able to choose between two sugars. When exposed to both glucose and lactose, the bacterium chooses glucose first. After the glucose is used up, it then expresses the genes necessary for lactose metabolism. This trait allows the bacterium to more efficiently use sugars from its environment.

CONCEPT CHECK: Why is it beneficial for the bacterium to regulate the *lac* operon with both a repressor protein and an activator protein?

protein (CAP), not a repressor. The term was coined before the action of the cAMP-CAP complex was understood. At that time, the primary observation was that glucose (a catabolite) inhibited (repressed) lactose metabolism.

Many bacterial promoters that transcribe genes involved in the breakdown of other sugars, such as maltose, arabinose, and melibiose, also have binding sites for CAP. Therefore, when glucose levels are high, these operons are inhibited. This promotes diauxic growth.

Further Studies Have Revealed That the *lac* Operon Has Three Operator Sites for the lac Repressor

Our traditional view of the regulation of the *lac* operon has been modified as we gained a greater understanding of the molecular process. In particular, detailed genetic and crystallographic studies have shown that the binding of the lac repressor is more complex than originally realized. The site in the *lac* operon commonly called the operator site was first identified by mutations that prevented lac repressor binding. These mutations, called *lacO*$^-$ or *lacO*^{C} mutants, resulted in the constitutive expression of the *lac* operon even in strains that make a normal lac repressor protein. The *lacO*^{C} mutations were localized in the *lac* operator site, which is now known as O_1. This led to the view that a single operator site was bound by the lac repressor to inhibit transcription, as in Figure 14.4.

In the late 1970s and 1980s, two additional operator sites were identified. As shown at the top of **Figure 14.9**, these sites are called O_2 and O_3. O_1 is the operator site slightly downstream from the promoter. O_2 is located farther downstream in the *lacZ* coding sequence, and O_3 is located slightly upstream from the promoter. The O_2 and O_3 operator sites were initially called pseudo-operators, because substantial repression occurred in the absence of either one of them. However, studies by Benno Müller-Hill and his colleagues revealed a surprising result. As shown in Figure 14.9 (fourth example down), if both O_2 and O_3 are missing, repression is dramatically reduced even when O_1 is present.

How were these results interpreted? The data of Figure 14.9 supported a hypothesis that the lac repressor must bind to two operator sites to repress the *lac* operon. The data indicate that the lac repressor can bind to O_1 and O_2, or to O_1 and O_3, but not to O_2 and O_3. If either O_2 or O_3 is missing, maximal repression is not achieved because it is less likely for the repressor to bind when only two operator sites are present. When O_1 is missing, even in the presence of the other operator sites, repression is nearly abolished because the lac repressor cannot bind to O_2 and O_3. Look at Figure 14.9 and you will notice that the operator sites are a fair distance from each other. For this reason, the researchers proposed that the binding of the lac repressor to two operator sites requires the DNA to form a loop. A loop in the DNA brings the operator sites closer together, thereby facilitating the binding of the repressor protein (**Figure 14.10a**).

In 1996, the proposal that the lac repressor binds to two operator sites was confirmed by studies in which the lac repressor was crystallized by Mitchell Lewis and his colleagues. The crystal structure of the lac repressor has provided exciting insights into its mechanism of action. As mentioned earlier in this chapter, the lac repressor is a tetramer of four identical subunits. The crystal structure revealed that each dimer within the tetramer recognizes one operator site. **Figure 14.10b** is a molecular model illustrating the binding of the lac repressor to the O_1 and O_3 sites. The amino acid side chains in the protein interact directly with bases in the major groove of the DNA helix. This is how genetic regulatory proteins recognize specific DNA sequences. Because each dimer within the tetramer recognizes a single operator site, the association of two dimers to form a tetramer requires that the two operator sites be close to each other. For this to occur, a loop must form in the DNA. The formation of this loop dramatically inhibits the ability of RNA polymerase to slide past the O_1 site and transcribe the operon.

Figure 14.10b also shows the binding of the cAMP-CAP complex to the CAP site (see the dark blue protein within the loop). A particularly striking observation is that the binding of the cAMP-CAP complex to the DNA causes a bend in the DNA structure. When the repressor is active—not bound to allolactose—the cAMP-CAP complex facilitates the formation of a loop in which the lac repressor binds to the O_1 and O_3 sites. When the repressor is inactive, this bending also appears to be important in the ability of RNA polymerase to initiate transcription slightly downstream from the bend.

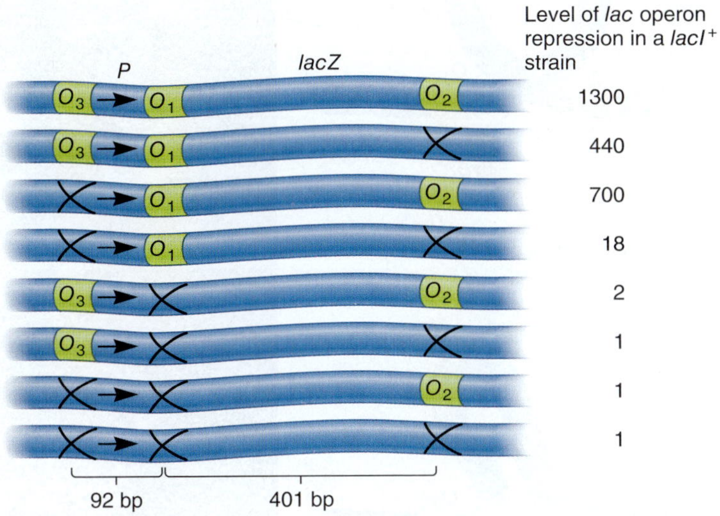

FIGURE 14.9 The identification of three *lac* operator sites. The top of this figure shows the locations of three *lac* operator sites, designated O_1, O_2, and O_3. O_1 is the *lac* operator site shown in previous figures. The arrows depict the starting site for transcription. Defective operator sites are indicated with an X. When all three operator sites are present, the repression of the *lac* operon is 1300-fold; this means there is 1/1300 the level of expression compared with an induced *lac* operon. This figure also shows the amount of repression when one or more operator sites are removed. A repression value of 1.0 indicates that no repression is occurring. In other words, a value of 1.0 indicates constitutive expression.

CONCEPT CHECK: Which data provide the strongest evidence that O_1 is not the only operator site?

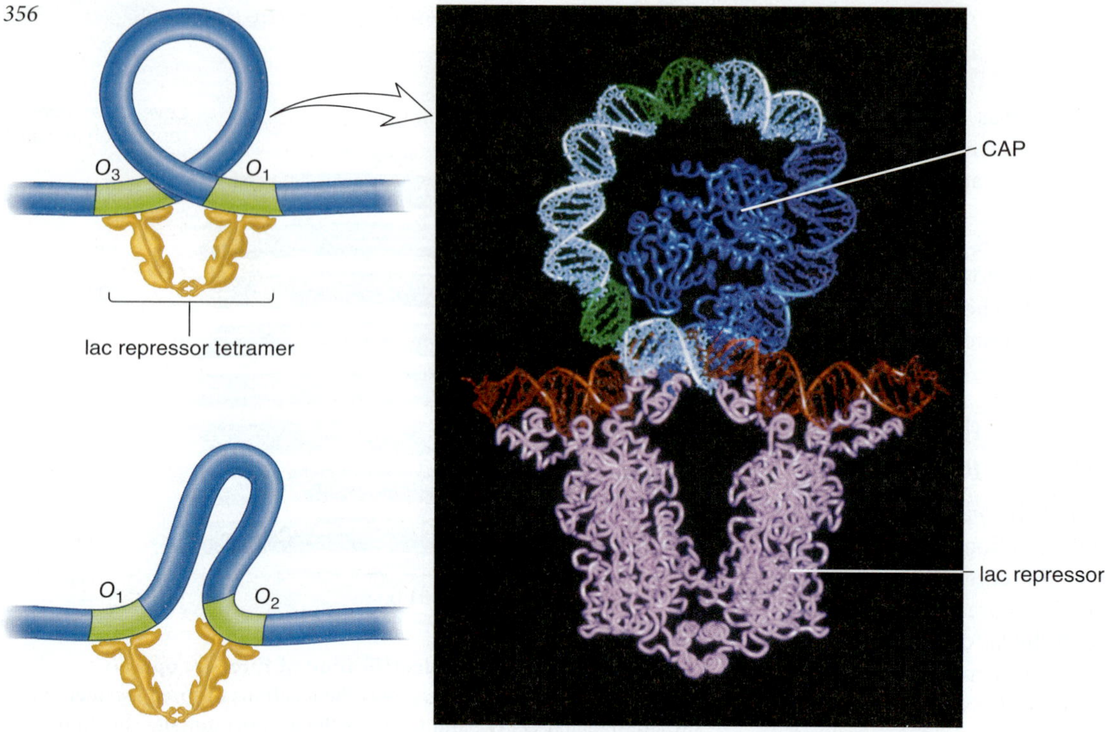

(a) DNA loops caused by the binding of the lac repressor

(b) Proposed model of the lac repressor binding to O_1 and O_3 based on crystallography

FIGURE 14.10 The binding of the lac repressor to two operator sites. **(a)** The binding of the lac repressor protein to the O_1 and O_3 or to the O_1 and O_2 operator sites. Because the two sites are far apart, a loop must form in the DNA. **(b)** A molecular model for the binding of the lac repressor to O_1 and O_3. Each repressor dimer binds to one operator site, so the repressor tetramer brings the two operator sites together. This causes the formation of a DNA loop in the intervening region. Note that the DNA loop contains the −35 and −10 regions (shown in green), which are recognized by σ factor of RNA polymerase. This loop also contains the binding site for the cAMP-CAP complex, which is the blue protein within the loop.

14.2 COMPREHENSION QUESTIONS

1. What is an operon?
   a. A site in the DNA where a regulatory protein binds
   b. A group of genes under the control of a single promoter
   c. An mRNA that encodes several genes
   d. All of the above
2. The binding of ________ to the lac repressor causes the lac repressor to ________ to the operator site, thereby ________ transcription.
   a. glucose, bind, inhibiting
   b. allolactose, bind, inhibiting
   c. glucose, not bind, increasing
   d. allolactose, not bind, increasing
3. On its chromosome, an *E. coli* cell is *lacI*$^-$ *lacZ*$^+$ *lacY*$^+$ *lacA*$^+$. It has an F′ factor that is *lacI*$^+$ *lacZ*$^+$ *lacY*$^+$ *lacA*$^+$. What is the expected level of expression of the *lac* operon genes (*lacZ*$^+$ *lacY*$^+$ *lacA*$^+$) in the absence of lactose?
   a. Both *lac* operons would be expressed.
   b. Neither *lac* operon would be expressed.
   c. Only the chromosomal *lac* operon would be expressed.
   d. Only the *lac* operon on the F′ factor would be expressed.
4. How does exposing an *E. coli* cell to glucose affect the regulation of the *lac* operon via CAP?
   a. cAMP binds to CAP and transcription is increased.
   b. cAMP binds to CAP and transcription is decreased.
   c. cAMP does not bind to CAP and transcription is increased.
   d. cAMP does not bind to CAP and transcription is decreased.

14.3 REGULATION OF THE *trp* OPERON

Learning Outcomes:

1. Describe the organization of the *trp* operon.
2. Explain how the *trp* operon is regulated by the trp repressor and by attenuation.

We now turn our attention to a second operon in *E. coli*, called the *trp* operon, which encodes enzymes involved with the synthesis of the amino acid tryptophan. Like the *lac* operon, the *trp* operon is regulated by a repressor protein. In addition, you will learn that the *trp* operon is regulated by another mechanism called **attenuation** in which transcription begins, but is stopped prematurely, or "attenuated," before most of the *trp* operon is transcribed.

The *trp* Operon Is Regulated by a Repressor Protein and Also by Attenuation

The *trp* operon (pronounced "trip") contains five genes, designated *trpE*, *trpD*, *trpC*, *trpB*, and *trpA*, which encode enzymes involved in tryptophan biosynthesis (**Figure 14.11a**). In addition, two genes, *trpR* or *trpL*, play a role in the regulation of the *trp* operon. The *trpL* gene is part of the *trp* operon, whereas the *trpR* gene has its own promoter and is not part of the *trp* operon.

Let's first consider how the *trp* operon is regulated by the *trpR* gene, which encodes the **trp repressor** protein. When tryptophan levels within the cell are very low, the trp repressor cannot bind to the operator site. Under these conditions, RNA polymerase transcribes the *trp* operon (see Figure 14.11a). In this way,

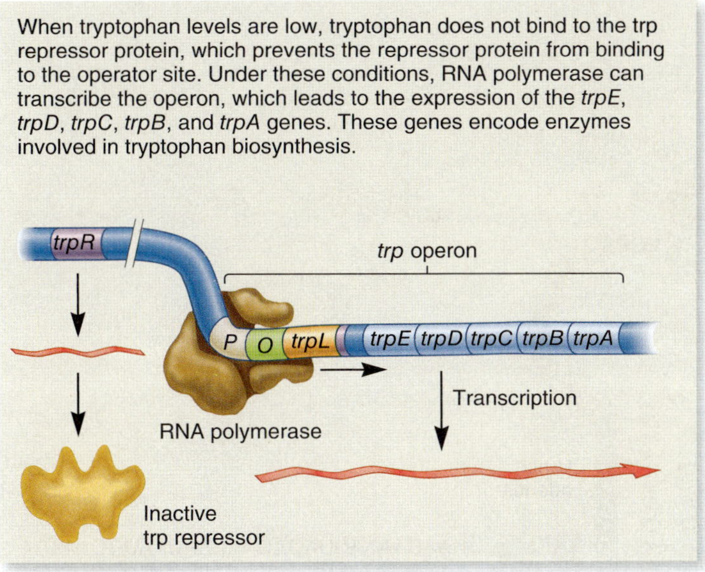

(a) Low tryptophan levels, transcription of the entire *trp* operon occurs

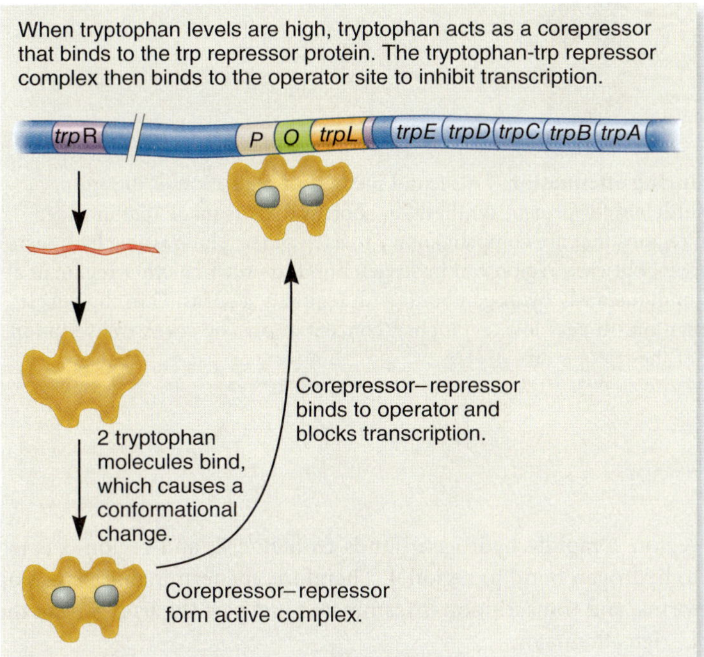

(b) High tryptophan levels, repression occurs

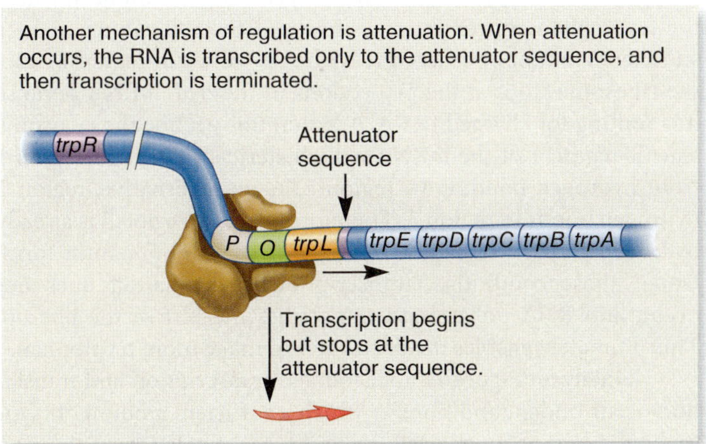

(c) High tryptophan levels, attenuation occurs

FIGURE 14.11 Organization of the *trp* operon and regulation via the trp repressor protein and attenuation.

CONCEPT CHECK: How does tryptophan affect the function of the trp repressor?

the cell expresses the genes required for the synthesis of tryptophan. When the tryptophan levels within the cell become high, tryptophan acts as a corepressor that binds to the trp repressor protein. This causes a conformational change in the trp repressor that allows it to bind to the *trp* operator site (**Figure 14.11b**). This inhibits the ability of RNA polymerase to transcribe the operon. Therefore, when a high level of tryptophan is present within the cell—when the cell does not need to make more tryptophan—the *trp* operon is turned off.

The *trp* Operon Is Also Regulated by Attenuation

In the 1970s, after the action of the trp repressor was elucidated, Charles Yanofsky and coworkers made a few unexpected observations. Mutant strains were found that lacked the trp repressor protein. Surprisingly, these mutant strains could still inhibit the expression of the *trp* operon when the tryptophan levels were high. In addition, *trp* operon mutations were identified in which a region including the *trpL* gene was missing from the operon. These mutations resulted in higher levels of expression of the other genes in the *trp* operon. As is often the case, unusual observations can lead scientists into productive avenues of study. By pursuing this research further, Yanofsky discovered a second regulatory mechanism in the *trp* operon, called attenuation, that is mediated by the region that includes the *trpL* gene (**Figure 14.11c**).

During attenuation, transcription actually begins, but it is terminated before the entire mRNA is made. A segment of DNA, termed the **attenuator sequence,** is important in facilitating this termination. When attenuation occurs, the mRNA from the *trp* operon is made as a short piece that terminates shortly past the *trpL* gene (see Figure 14.11c). Because this short mRNA has been terminated before RNA polymerase has transcribed the *trpE*, *trpD*, *trpC*, *trpB*, and *trpA* genes, it will not encode the proteins required for tryptophan biosynthesis. In this way, attenuation inhibits the further production of tryptophan in the cell.

The segment of the *trp* operon immediately downstream from the operator site plays a critical role during attenuation. The first gene in the *trp* operon is the *trpL* gene. The mRNA made from the *trpL* gene contains codons for 14 amino acids that form the trp leader peptide (**Figure 14.12**). One key feature of attenuation is that two tryptophan (Trp) codons are found within the mRNA. As we will see later, these two codons provide a way for the bacterium to sense whether or not it has sufficient tryptophan to synthesize its proteins.

A second key feature that underlies attenuation is that the mRNA made from the *trpL* gene has four regions that are complementary to each other, which causes the mRNA to form stem-loops. Different combinations of stem-loops are possible due to interactions among these four regions (see the color key in Figure 14.12).

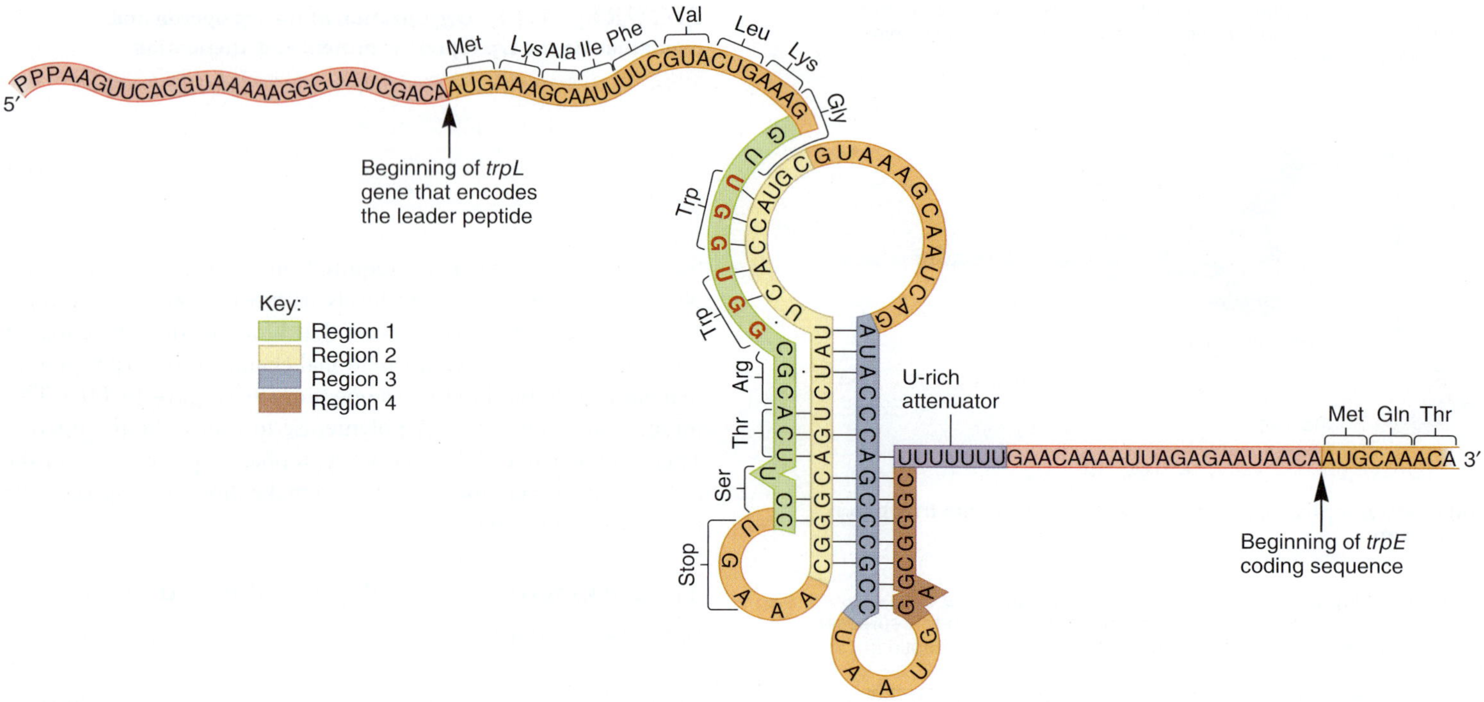

FIGURE 14.12 Sequence of the *trpL* mRNA produced during attenuation. A second method of regulation of the *trp* operon is attenuation, which occurs when tryptophan levels are sufficient for protein synthesis. A short mRNA is made that includes the *trpL* coding sequence. As shown here, this mRNA has several regions that are complementary to each other. The possible hydrogen bonding between regions 1 and 2, 2 and 3, and 3 and 4 is also shown, but each region can hydrogen bond to only one other region at a given time. For example, if region 2 is hydrogen bonded to region 1, it cannot simultanteously hydrogen bond with region 3. The last U in the purple attenuator sequence is the last nucleotide that would be transcribed during attenuation. At very low tryptophan concentrations, however, transcription occurs beyond the end of *trpL* and proceeds through the *trpE* gene and the rest of the *trp* operon.

ONLINE ANIMATION

CONCEPT CHECK: What type of bonding interaction causes the stem-loops to form?

Region 2 is complementary to region 1 and also to region 3. Region 3 is complementary to region 2 as well as to region 4. Therefore, three stem-loops are possible: 1–2, 2–3, and 3–4. Even so, keep in mind that a particular segment of RNA can participate in the formation of only one stem-loop. For example, if region 2 forms a stem-loop with region 1, it cannot (at the same time) form a stem-loop with region 3. Alternatively, if region 2 forms a stem-loop with region 3, then region 3 cannot form a stem-loop with region 4. Though three stem-loops are possible, the 3–4 stem-loop is functionally unique. The 3–4 stem-loop together with the U-rich attenuator sequence acts as an intrinsic terminator—a ρ-independent terminator, as described in Chapter 12. Therefore, the formation of the 3–4 stem-loop causes RNA polymerase to pause, and the U-rich sequence dissociates from the DNA. This terminates transcription at the U-rich attenuator. In comparison, if region 3 forms a stem-loop with region 2, transcription will not be terminated because a 3–4 stem-loop cannot form.

Conditions that favor the formation of the 3–4 stem-loop rely on the translation of the *trpL* gene and on the amount of tryptophan in the cell. As shown in **Figure 14.13**, three scenarios are possible. On some occasions, translation is not coupled with transcription (Figure 14.13a). As the *trpL* gene is being transcribed, region 1 rapidly hydrogen bonds to region 2, and region 3 is left to hydrogen bond to region 4. Therefore, the terminator stem-loop forms, and transcription is terminated just past the *trpL* gene at the U-rich attenuator.

In bacteria, transcription and translation are often coupled, which means that translation begins as transcription is occurring. In Figure 14.13b, this coupling happens under conditions in which the tryptophan concentration is low and the cell cannot make a sufficient amount of charged $tRNA^{Trp}$. As we see in Figure 14.13b, the ribosome stops at the Trp codons in the *trpL* mRNA because it is waiting for charged $tRNA^{Trp}$. When this occurs, the ribosome shields region 1 of the mRNA, which sterically prevents region 1 from hydrogen bonding to region 2. As an alternative, region 2 hydrogen bonds to region 3. Therefore, because region 3 is already hydrogen bonded to region 2, the 3–4 stem-loop cannot form. Under these conditions, transcriptional termination does not occur, and RNA polymerase transcribes the rest of the operon. This ultimately enables the bacterium to make more tryptophan.

Finally, in Figure 14.13c, coupled transcription and translation occur under conditions in which a sufficient amount of tryptophan is present in the cell. In this case, translation of the *trpL* mRNA progresses to its stop codon, where the ribosome pauses.

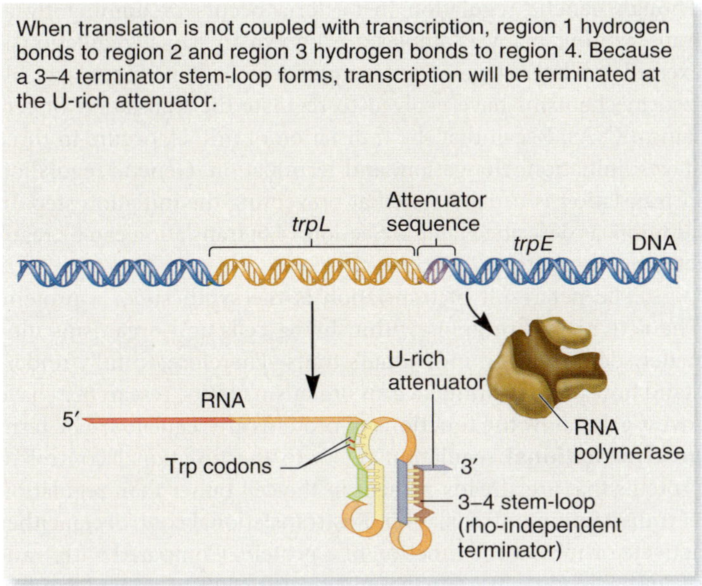

(a) No translation, 1–2 and 3–4 stem-loops form

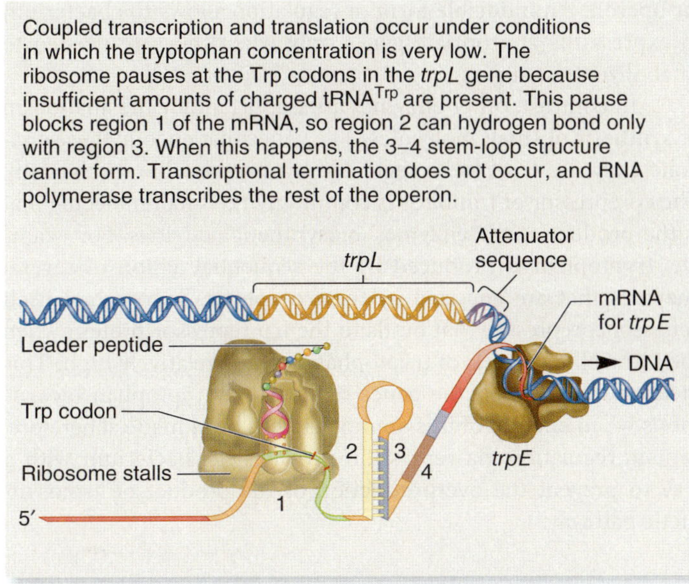

(b) Low tryptophan levels, 2–3 stem-loop forms

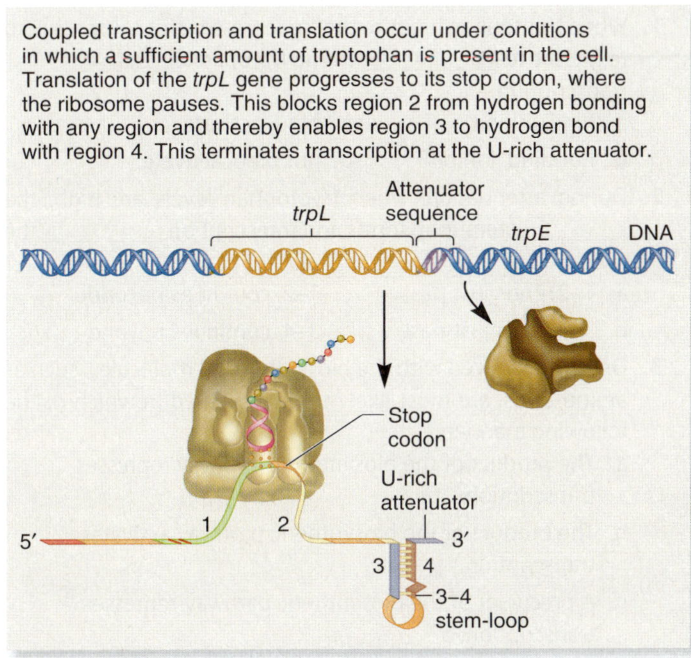

(c) High tryptophan levels, 3–4 stem-loop forms

ONLINE ANIMATION

FIGURE 14.13 Possible stem-loops formed from *trpL* mRNA under different conditions of translation. Attenuation occurs in parts **(a)** and **(c)** due to the formation of a 3–4 stem-loop.

CONCEPT CHECK: Explain how the presence of tryptophan favors the formation of the 3–4 stem-loop.

The pausing at the stop codon prevents region 2 from hydrogen bonding with any region, thereby enabling region 3 to hydrogen bond with region 4. As in Figure 14.13a, this terminates transcription. Of course, keep in mind that the *trpL* mRNA contains two tryptophan codons. For the ribosome to smoothly progress to the *trpL* stop codon, enough charged $tRNA^{Trp}$ must be available to translate this mRNA. It follows that the bacterium must have a sufficient amount of tryptophan. Under these conditions, the rest of the transcription of the operon is terminated.

Attenuation is a mechanism for regulating transcription, which is found in several other operons involved with amino acid biosynthesis. In all cases, the mRNAs that encode leader peptides have codons for the amino acid that is synthesized by the enzymes encoded by the operon. For example, the mRNA for the leader peptide of the histidine operon has seven histidine codons, and the mRNA for the leader peptide of the leucine operon has four leucine codons. Like the *trp* operon, these other operons have alternative stem-loops, one of which is a transcriptional terminator.

Inducible Operons Usually Encode Catabolic Enzymes, and Repressible Operons Encode Anabolic Enzymes

Thus far, we have seen that bacterial genes can be transcriptionally regulated in a positive or negative way—and sometimes both. The *lac* operon is an inducible operon that is regulated by sugar molecules. By comparison, the *trp* operon is a repressible operon regulated by tryptophan, a corepressor that binds to the repressor and turns the operon off. In addition, an abundance of charged $tRNA^{Trp}$ in the cytoplasm can turn the *trp* operon off via attenuation.

By studying the genetic regulation of many operons, geneticists have discovered a general trend concerning inducible versus repressible regulation. When the genes in an operon encode proteins that function in the catabolism or breakdown of a substance, they are usually regulated in an inducible manner. The substance to be broken down or a related compound often acts as the inducer. For example, allolactose acts as an inducer of the *lac* operon. An inducible form of regulation allows the bacterium to express the appropriate genes only when they are needed to catabolize these sugars.

In contrast, other enzymes are important for the anabolism or synthesis of small molecules. The genes that encode these anabolic enzymes tend to be regulated by a repressible mechanism. The corepressor or inhibitor is commonly the small molecule that is the product of the enzymes' biosynthetic activities. For example, tryptophan is produced by the sequential action of several enzymes that are encoded by the *trp* operon. Tryptophan itself acts as a corepressor that binds to the trp repressor protein when the intracellular levels of tryptophan become relatively high. This mechanism turns off the genes required for tryptophan biosynthesis when enough of this amino acid has been made. Therefore, genetic regulation via repression provides the bacterium with a way to prevent the overproduction of the product of a biosynthetic pathway.

14.3 COMPREHENSION QUESTIONS

1. When tryptophan binds to the trp repressor, this causes the trp repressor to __________ to the *trp* operator and ________ transcription.
   a. bind, inhibit
   b. not bind, inhibit
   c. bind, activate
   d. not bind, activate
2. During attenuation, when tryptophan levels are high, the ________ stem-loop forms and transcription _________ the *trpL* gene.
   a. 1–2, ends just past
   b. 3–4, ends just past
   c. 1–2, continues beyond
   d. 3–4, continues beyond
3. Operons involved with the biosynthesis of molecules, such as amino acids, are most likely to be regulated in which of the following manners?
   a. The product of the biosynthetic pathway represses transcription.
   b. The product of the biosynthetic pathway activates transcription.
   c. A precursor of the biosynthetic pathway represses transcription.
   d. A precursor of the biosynthetic pathway activates transcription.

14.4 TRANSLATIONAL AND POSTTRANSLATIONAL REGULATION

Learning Outcomes:

1. Explain how translational regulatory proteins and antisense RNAs regulate translation.
2. Summarize how feedback inhibition and posttranslational modifications regulate protein function.

Though genetic regulation in bacteria occurs predominantly at transcription, many examples are known in which regulation is exerted at a later stage in gene expression. In some cases, specialized mechanisms have evolved to regulate the translation of certain mRNAs. Recall that the translation of mRNA occurs in three stages: initiation, elongation, and termination. Genetic regulation of translation is usually aimed at preventing the initiation step. In addition, as described later in Section 14.5, translation can be regulated by riboswitches.

The net result of translation is the synthesis of a protein. The activities of proteins within living cells and organisms ultimately determine an individual's traits. Therefore, to fully understand how proteins influence an organism's traits, researchers have investigated how the functions of proteins are regulated. The term **posttranslational** regulation refers to the functional control of proteins that are already present in the cell rather than regulation of transcription or translation. Posttranslational control can either activate or inhibit the function of a protein. Compared with transcriptional or translational regulation, posttranslational control can be relatively fast, occurring in a matter of seconds, which is an important advantage. In contrast, transcriptional and translational regulation typically require several minutes or even hours to take effect because these two mechanisms involve the synthesis and turnover of mRNA and polypeptides. In this section, we will examine some of the ways that bacteria can regulate the initiation of translation, as well as ways that protein function can be regulated posttranslationally.

Repressor Proteins and Antisense RNA Can Inhibit Translation

For some bacterial genes, the translation of mRNA is regulated by the binding of proteins or other RNA molecules that influence the ability of ribosomes to translate the mRNA into a polypeptide. A **translational regulatory protein** recognizes sequences within the mRNA, similar to the way that transcription factors recognize DNA sequences. In most cases, translational regulatory proteins act to inhibit translation. These are known as **translational repressors.**

When a translational repressor protein binds to the mRNA, it can inhibit translational initiation in one of two ways. One possibility is that it can bind in the vicinity of the Shine-Dalgarno sequence and/or the start codon. As discussed in Chapter 13, the Shine-Dalgarno sequence is needed for the mRNA to bind to a ribosome and the start codon is the site where translation begins. A translational repressor can bind to either of these two sites, thereby sterically blocking the ribosome's ability to initiate translation in this region.

Alternatively, a repressor protein may bind to the mRNA outside the Shine-Dalgarno/start codon region but stabilize an mRNA secondary structure that prevents initiation. For example, a translational repressor could stabilize an mRNA secondary structure in which the Shine-Dalgarno sequence is inaccessible. This would prevent the mRNA from binding to the ribosome, thereby blocking translation. Translational repression is also a form of genetic regulation found in eukaryotic species, and we will consider specific examples in Chapter 16.

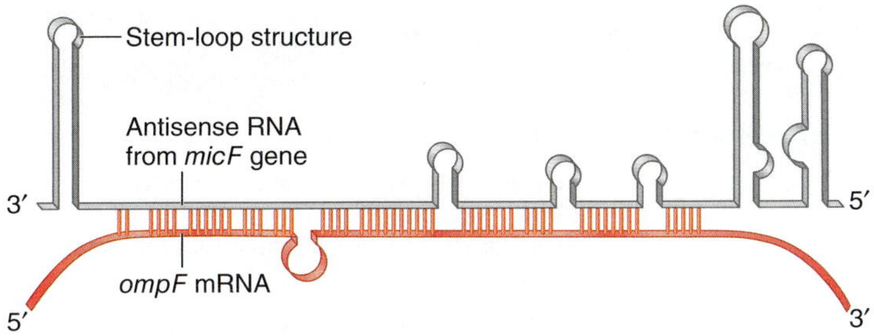

FIGURE 14.14 The double-stranded RNA structure formed between the *micF* antisense RNA and the *ompF* mRNA. Because they have regions that are complementary to each other, the *micF* antisense RNA binds to the *ompF* mRNA to form a double-stranded structure in which the *ompF* mRNA cannot be translated.

CONCEPT CHECK: How does *micF* antisense RNA affect the translation of *ompF* mRNA?

A second way to regulate translation is via the synthesis of **antisense RNA,** an RNA strand that is complementary to a strand of mRNA. To understand this form of genetic regulation, let's consider a trait known as osmoregulation, which is essential for the survival of most bacteria. Osmoregulation refers to the ability to control the amount of water inside the cell. Because the solute concentrations in the external environment may rapidly change between hypotonic and hypertonic conditions, bacteria must have an osmoregulation mechanism to maintain their internal cell volume. Otherwise, they would be susceptible to the harmful effects of swelling or shrinking.

In *E. coli*, an outer membrane protein encoded by the *ompF* gene is important in osmoregulation. At low osmolarity, the ompF protein is preferentially produced, whereas at high osmolarity, its synthesis is decreased. The expression of another gene, known as *micF*, is responsible for inhibiting the expression of the *ompF* gene at high osmolarity. As shown in **Figure 14.14**, the inhibition occurs because the RNA transcribed from the *micF* gene is complementary to the *ompF* mRNA; it is an antisense strand of RNA. When the *micF* gene is transcribed, its RNA product binds to the *ompF* mRNA via hydrogen bonding between their complementary regions. The binding of the *micF* RNA to the *ompF* mRNA prevents the *ompF* mRNA from being translated. The RNA transcribed from the *micF* gene is called antisense RNA because it is complementary to the *ompF* mRNA, which is a sense strand of mRNA that encodes a protein. The *micF* RNA does not encode a protein.

Posttranslational Regulation Can Occur via Feedback Inhibition and Covalent Modifications

Let's now turn our attention to ways that protein function is regulated posttranslationally. A common mechanism to regulate the activity of metabolic enzymes is **feedback inhibition.** The synthesis of many cellular molecules such as amino acids, vitamins, and nucleotides occurs via the action of a series of enzymes that convert precursor molecules to a particular product. The final product in a metabolic pathway then inhibits an enzyme that acts early in the pathway.

Figure 14.15 depicts feedback inhibition in a metabolic pathway. Enzyme 1 is an example of an **allosteric enzyme,** an enzyme that contains two different binding sites. (The lac repressor is an allosteric protein, but not an enzyme.) The catalytic site is responsible for the binding of the substrate and its conversion to intermediate 1. The second site is a regulatory or allosteric site. This site binds the final product of the metabolic pathway. When bound to the regulatory site, the final product inhibits the catalytic ability of enzyme 1.

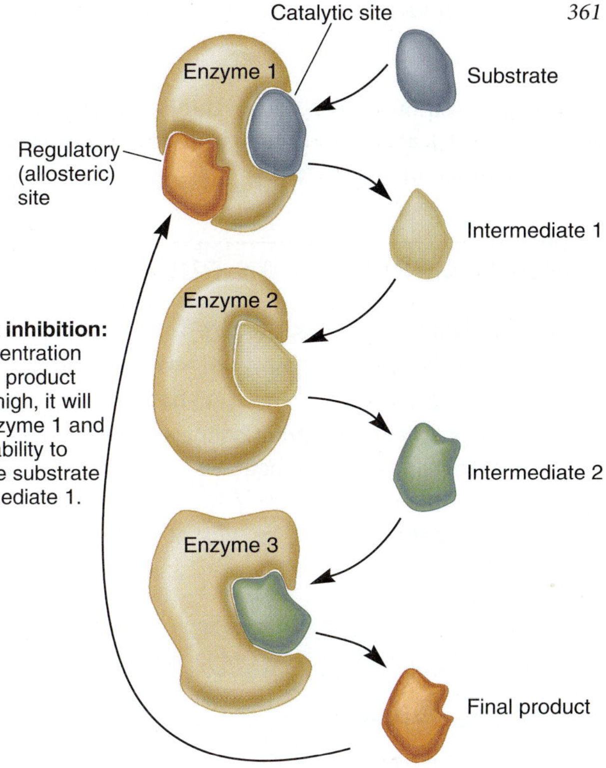

FIGURE 14.15 Feedback inhibition in a metabolic pathway. The substrate is converted to a product by the sequential action of three different enzymes. Enzyme 1 has a catalytic site that recognizes the substrate, and it also has a regulatory site (also called an allosteric site) that recognizes the final product. When the final product binds to the regulatory site, it inhibits enzyme 1.

CONCEPT CHECK: Why is feedback inhibition an advantage to the bacterium?

To appreciate feedback inhibition at the cellular level, we can consider the relationship between the product concentration and the regulatory site on enzyme 1. As the final product is made within the cell, its concentration gradually increases. Once the final product concentration has reached a level that is similar to its affinity for enzyme 1, the final product is likely to bind to the regulatory site on enzyme 1 and inhibit its function. In this way, the net result is that the final product of a metabolic pathway inhibits the further synthesis of more product. Under these conditions, the concentration of the final product has reached a level sufficient for the purpose of the cell.

A second strategy to control the function of proteins is by the covalent modification of their structure, a process called **posttranslational covalent modification** (look ahead to Figure 23.4b). Certain types of modifications are involved primarily in the assembly and construction of a functional protein. These alterations include proteolytic processing; disulfide bond formation; and the attachment of prosthetic groups, sugars, or lipids. These are typically irreversible changes required to produce a functional protein. In contrast, other types of modifications, such as phosphorylation (—PO_4), acetylation (—$COCH_3$), and methylation (—CH_3), are often reversible modifications that transiently affect the function of a protein.

14.4 COMPREHENSION QUESTIONS

1. Translation can be regulated by
   a. translational repressors. c. attenuation.
   b. antisense RNA. d. both a and b.
2. An example of a posttranslational covalent modification that may regulate protein function is
   a. phosphorylation. c. methylation.
   b. acetylation. d. all of the above.

14.5 RIBOSWITCHES

Learning Outcome:

1. Explain how riboswitches can regulate transcription and translation.

In 2001 and 2002, researchers in a few different laboratories discovered a mechanism of gene regulation called a **riboswitch.** In this form of regulation, an RNA molecule can exist in two different secondary conformations. The conversion from one conformation to the other is due to the binding of a small molecule. As described in **Table 14.2**, a riboswitch can regulate transcription, translation, RNA stability, and splicing.

Riboswitches are widespread in bacteria. Researchers estimate that 3–5% of all bacterial genes may be regulated by riboswitches. In bacteria, they are involved in regulating genes associated with the biosynthesis of purines, amino acids, vitamins, and other essential molecules. Riboswitches are also found in archaea, algae, fungi, and plants. In this section, we will examine two examples of riboswitches in bacteria. The first example shows how riboswitches can regulate transcription and the second example involves translational regulation.

TABLE 14.2
Types of Riboswitches

Type of Regulation	Description
Transcription	The 5′ region of an mRNA may exist in one conformation that forms a ρ-independent terminator, which causes attenuation of transcription. The other conformation does not form a terminator and is completely transcribed.
Translation	The 5′ region of an mRNA may exist in one conformation in which the Shine-Dalgarno sequence cannot be recognized by the ribosome, whereas the other conformation has an accessible Shine-Dalgarno sequence that allows the mRNA to be translated.
RNA stability	One mRNA conformation may be stable, whereas the other conformation acts as ribozyme that causes self-degradation.
Splicing	In eukaryotes, one pre-mRNA conformation may be spliced in one way, whereas another conformation is spliced in a different way.

A Riboswitch Can Regulate Transcription

Thiamin, also called vitamin B_1, is an important organic molecule for bacteria, archaea, and eukaryotes. The active form of this vitamin is thiamin pyrophosphate (TPP). TPP is an essential coenzyme for the functioning of a variety of enzymes, such as certain enzymes in the citric acid cycle. In bacteria, TPP is made via biosynthetic enzymes that are encoded in the bacterial genome. For example, in *Bacillus subtilis*, the majority of genes involved in TPP synthesis are found within the *thi* operon that contains seven genes.

Certain genes that encode TPP biosynthetic enzymes, such as those of the *thi* operon in *B. subtilis*, are regulated by a riboswitch that controls transcription (**Figure 14.16**). As the polycistronic mRNA for the *thi* operon is being made, the 5′ end quickly folds into a secondary structure. When TPP levels are low, the secondary structure has a stem-loop called an **antiterminator,** which prevents the formation of the terminator stem-loop. Therefore, under these conditions, transcription of the entire *thi* operon occurs. In this way, the bacterium is able to make more TPP, which is in short supply. By comparison, when TPP levels are high, TPP binds to the RNA as it is being made and causes a change in its secondary structure. As shown on the right side of Figure 14.16, a terminator stem-loop forms instead of the antiterminator stem-loop.

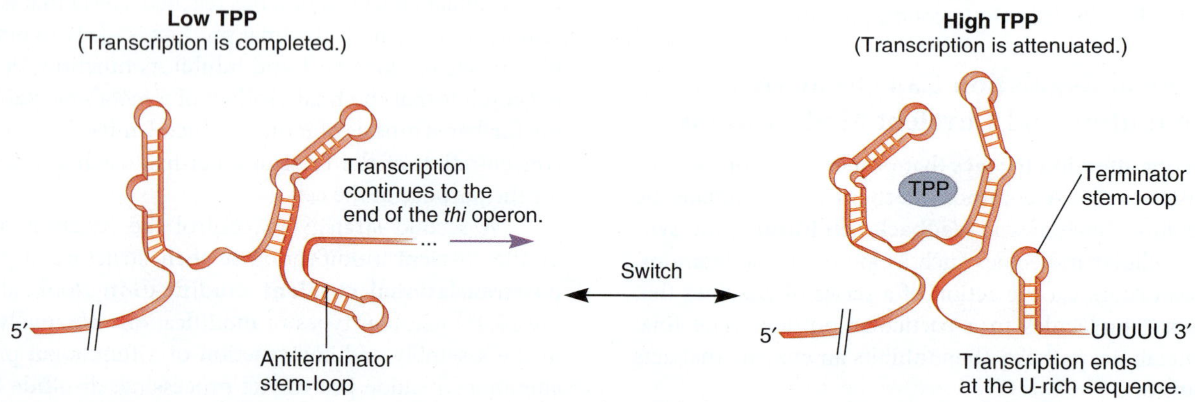

FIGURE 14.16 **A TPP riboswitch in *B. subtilis* that regulates transcription of the *thi* operon.**

CONCEPT CHECK: Which RNA conformation favors transcription—the form with the antiterminator stem-loop or the form with the terminator stem-loop?

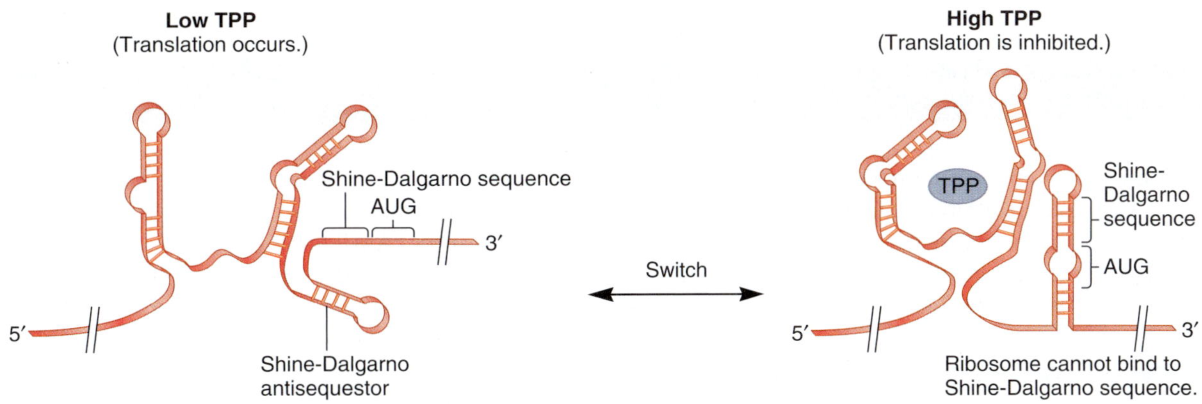

FIGURE 14.17 A TPP riboswitch in *E. coli* that regulates translation of the *thiMD* operon. Note: In both cases, the mRNA for the *thiMD* operon is completely transcribed. The double slashes near the 5′ and 3′ ends of the RNA indicate that this figure is not showing a large portion of the mRNA from this operon.

CONCEPT CHECK: Which RNA conformation favors translation—the form with the Shine-Dalgarno antisequestor or the form in which the Shine-Dalgarno sequence is within a stem-loop?

As described in Chapter 13, this is an example of a ρ-independent terminator. The formation of this terminator abruptly stops transcription, thereby inhibiting the production of the enzymes that are needed to make more TPP. Similar to the *trp* operon discussed earlier in this chapter, this is an example of attenuation. (Compare Figures 14.13a and the right side of Figure 14.16.)

A Riboswitch Can Regulate Translation

As we have just seen, the regulation of TPP biosynthetic enzymes in *B. subtilis* occurs via a riboswitch that controls transcription. However, in gram-negative bacteria, such as *Escherichia coli*, the regulation of TPP biosynthetic enzymes occurs via a riboswitch that controls translation.

Figure 14.17 shows how a riboswitch can regulate translation. In *E. coli*, the *thiMD* operon encodes two enzymes involved with TPP biosynthesis. When TPP levels are low, the 5′ end of the mRNA folds into a structure that contains a stem-loop called the Shine-Dalgarno antisequestor. When this stem-loop forms, the Shine-Dalgarno sequence is accessible to ribosome binding. Therefore, the mRNA is translated when TPP is in short supply. By comparison, when TPP levels are high, TPP binds to the RNA and causes a change in its secondary structure. As shown on the right side of Figure 14.17, a stem-loop forms that contains the Shine-Dalgarno sequence and the start codon. The formation of this stem-loop sequesters the Shine-Dalgarno sequence, thereby preventing ribosomal binding. This blocks the translation of the enzymes that are needed to make more TPP.

When comparing Figures 14.16 and 14.17, it is worth noting that the 5′ end of the *thiMD* mRNA of *E. coli* can exist in two different secondary structures that greatly resemble those that occur at the 5′ end of the *thi* operon of *B. subtilis*. However, the effects on regulation are quite different. The TPP riboswitch in *E. coli* controls translation, whereas the TPP riboswitch in *B. subtilis* controls transcription.

14.5 COMPREHENSION QUESTION

1. For a riboswitch that controls transcription, the binding of a small molecule such as TPP controls whether the RNA
   a. has an antiterminator or terminator stem-loop.
   b. has a Shine-Dalgarno antisequestor or the Shine-Dalgarno sequence in a stem-loop.
   c. is degraded from its 5′ end.
   d. has both a and b.

KEY TERMS

Page 344. gene regulation, constitutive genes
Page 345. repressor, activator, negative control, positive control, inducer, inducible genes, corepressor, inhibitor, repressible genes
Page 347. enzyme adaptation, operon, polycistronic mRNA, promoter, terminator, CAP site, catabolite activator protein (CAP), operator site, operator, lac repressor
Page 348. induced, allosteric regulation, allosteric sites
Page 351. merozygote
Page 353. *trans*-effect, *trans*-acting factor, *cis*-acting element, *cis*-effect
Page 354. catabolite repression, diauxic growth, cyclic-AMP (cAMP)
Page 356. attenuation, trp repressor
Page 357. attenuator sequence
Page 360. posttranslational, translational regulatory protein, translational repressors
Page 361. antisense RNA, feedback inhibition, allosteric enzyme, posttranslational covalent modification
Page 362. riboswitch, antiterminator

CHAPTER SUMMARY

- Gene regulation is the phenomenon in which the level of gene expression can vary under different conditions. By comparison, constitutive genes are expressed at relatively constant levels. Gene regulation can occur at transcription, at translation, or posttranslationally (see Figure 14.1).

14.1 Overview of Transcriptional Regulation

- Repressors and activators are proteins that exert negative control or positive control, respectively. Small effector molecules control the function of repressors and activators. These include inducers, corepressors, and inhibitors (see Figure 14.2).

14.2 Regulation of the *lac* Operon

- Enzyme adaptation is the phenomenon in which the enzymes that metabolize a molecule are made only when the molecule is present in the environment.
- The *lac* operon encodes a polycistronic mRNA for proteins that are involved with the uptake and metabolism of lactose (see Figure 14.3).
- The lac repressor protein binds to the *lac* operator and inhibits transcription. Allolactose binds to the repressor and causes a conformational change that prevents the repressor from binding to the operator. This event induces transcription (see Figure 14.4).
- The regulation of the *lac* operon enables the bacterium *E. coli* to respond to changes in the level of lactose in its environment (see Figure 14.5).
- Jacob and Monod constructed merozygotes to show that the *lacI* gene encodes a diffusible repressor protein (see Figures 14.6, 14.7).
- Mutations in the *lacI* gene and the *lac* operator site may have different effects in a merozygote (see Table 14.1).
- CAP is an activator protein of the *lac* operon. It binds to the CAP site when cAMP levels are high. The presence of glucose causes cAMP levels to decrease, thereby inhibiting the *lac* operon (see Figure 14.8).
- The lac repressor, which is a tetramer, binds to two operator sites, either to O_1 and O_2, or to O_1 and O_3 (see Figures 14.9, 14.10).

14.3 Regulation of the *trp* Operon

- The *trp* operon is regulated by the trp repressor that binds to the *trp* operator when tryptophan, a corepressor, is present (see Figure 14.11).
- A second way that the *trp* operon is regulated is via attenuation in which the formation of a terminator stem-loop causes early termination of transcription (see Figures 14.12, 14.13).
- Inducible operons typically encode catabolic enzymes, whereas repressible operons often encode anabolic enzymes.

14.4 Translational and Posttranslational Regulation

- Translation can be regulated by translational regulatory proteins and by antisense RNA (see Figure 14.14).
- Posttranslational control of protein function may involve feedback inhibition and posttranslational covalent modifications (see Figure 14.15).

14.5 Riboswitches

- A riboswitch is a form of regulation in which an RNA exists in two different secondary structures. The conversion from one secondary structure to the other is due to the binding of a small molecule.
- A riboswitch can regulate transcription, translation, RNA stability, and splicing (see Table 14.2, Figures 14.16, 14.17).

PROBLEM SETS & INSIGHTS

Solved Problems

S1. Researchers have identified mutations in the promoter region of the *lacI* gene that make it more difficult for the *lac* operon to be induced. These are called *lacI*^{Q} mutants, because a greater Quantity of lac repressor protein is made. Explain why an increased transcription of the *lacI* gene makes it more difficult to induce the *lac* operon.

Answer: An increase in the amount of lac repressor protein makes it easier to repress the *lac* operon. When the cells are exposed to lactose, allolactose levels slowly rise. Some of the allolactose binds to the lac repressor protein and causes it to be released from the operator site. If many more lac repressor proteins are found within the cell, more allolactose is needed to ensure that no unoccupied repressor proteins can repress the operon.

S2. Explain how the pausing of the ribosome in the presence or absence of tryptophan affects the formation of a terminator stem-loop.

Answer: The key issue is the location where the ribosome stalls. In the absence of tryptophan, it stalls over the Trp codons in the *trpL* mRNA. Stalling at this site shields region 1 in the attenuator region. Because region 1 is unavailable to hydrogen bond with region 2, region 2 hydrogen bonds with region 3. Therefore, region 3 cannot form a terminator stem-loop with region 4. Alternatively, if tryptophan levels in the cell are sufficient, the ribosome pauses over the stop codon in the *trpL* RNA. In this case, the ribosome shields region 2. Therefore, regions 3 and 4 hydrogen bond with each other to form a terminator stem-loop, which abruptly halts the continued transcription of the *trp* operon.

S3. In bacteria, it is common for two or more genes to be arranged together in an operon. Discuss the arrangement of genetic sequences within an operon. What is the biological advantage of an operon organization?

Answer: An operon contains several different DNA sequences that play specific roles. For transcription to take place, an operon is flanked by a promoter to signal the beginning of transcription and a terminator to signal the end of transcription. Two or more genes are found between these two sequences. A key feature of an operon is that the expression of the genes occurs as a unit. When transcription takes place, a polycistronic mRNA is made that encodes all of the genes.

In order to control the ability of RNA polymerase to transcribe an operon, an additional DNA sequence, known as the operator site, is usually present. The base sequence within the operator site can serve as a binding site for genetic regulatory proteins called activators or repressors. The advantage of an operon organization is that it allows a bacterium to coordinately regulate a group of genes with a common function. For example, an operon may contain a group of genes involved in lactose breakdown or a group of genes involved in tryptophan synthesis. The genes within an operon usually encode proteins within a common metabolic pathway or cellular function.

S4. The sequential use of two sugars by a bacterium is known as diauxic growth. It is a common phenomenon among many bacterial species. When glucose is one of the two sugars available, it is typical that the bacterium metabolizes glucose first, and then a second sugar after the glucose has been used up. Among *E. coli* and related species, diauxic growth is regulated by intracellular cAMP levels and the catabolite activator protein (CAP). Summarize the effects of glucose and lactose on the ability of the lac repressor and the cAMP-CAP complex to regulate the *lac* operon.

Answer: In the absence of lactose, the lac repressor has the dominating effect of shutting off the *lac* operon. Even when glucose is also absent and the cAMP-CAP complex is formed, the presence of the bound lac repressor prevents the expression of the *lac* operon. The effects of the cAMP-CAP complex are exerted only in the presence of lactose. When lactose is present and glucose is absent, the cAMP-CAP complex acts to enhance the rate of transcription. However, when both lactose and glucose are present, the inability of CAP to bind to the *lac* operon decreases the rate of transcription. The table shown here summarizes these effects.

***lac* Operon Regulation**

Sugar Present	Transcription of the *lac* Operon
None	The operon is turned off due to the dominating effect of the lac repressor protein.
Lactose	The operon is maximally turned on. The repressor protein is removed from the operator site, and the cAMP-CAP complex is bound to the CAP site.
Glucose	The operon is turned off due to the dominating effect of the lac repressor protein.
Lactose and glucose	The expression of the *lac* operon is greatly decreased. The lac repressor is removed from the operator site, and the catabolite activator protein is not bound to the CAP site. The absence of CAP at the CAP site makes it difficult for RNA polymerase to begin transcription. However, a little more transcription occurs under these conditions than in the absence of lactose, when the repressor is bound.

Conceptual Questions

C1. What is the difference between a constitutive gene and a regulated gene?

C2. In general, why is it important to regulate genes? Discuss examples of situations in which it would be advantageous for a bacterial cell to regulate genes.

C3. If a gene is repressible and under positive control, describe what kind of effector molecule and regulatory protein are involved. Explain how the binding of the effector molecule affects the regulatory protein.

C4. Transcriptional regulation often involves a regulatory protein that binds to a segment of DNA and a small effector molecule that binds to the regulatory protein. Do the following terms apply to a regulatory protein, a segment of DNA, or a small effector molecule?

A. Repressor
B. Inducer
C. Operator site
D. Corepressor
E. Activator
F. Attenuator
G. Inhibitor

C5. An operon is repressible—a small effector molecule turns off transcription. Which combinations of small effector molecules and regulatory proteins could be involved?

A. An inducer plus a repressor

B. A corepressor plus a repressor

C. An inhibitor plus an activator

D. An inducer plus an activator

C6. Some mutations have a *cis*-effect, whereas others have a *trans*-effect. Explain the molecular differences between *cis*- and *trans*-mutations. Which type of mutation (*cis* or *trans*) can be complemented in a merozygote experiment?

C7. What is enzyme adaptation? From a genetic point of view, how does it occur?

C8. In the *lac* operon, how would gene expression be affected if one of the following segments was missing?

A. *lac* operon promoter

B. Operator site

C. *lacA* gene

C9. If an abnormal repressor protein could still bind allolactose but the binding of allolactose did not alter the conformation of the repressor protein, how would the expression of the *lac* operon be affected?

C10. What is diauxic growth? Explain the roles of cAMP and the catabolite activator protein in this process.

C11. Mutations may have an effect on the expression of the *lac* operon and the *trp* operon. Would the following mutations have a *cis*- or *trans*-effect on the expression of the protein-encoding genes in the operon?

A. A mutation in the operator site that prevents the lac repressor from binding to it

B. A mutation in the *lacI* gene that prevents the lac repressor from binding to DNA

C. A mutation in *trpL* that prevents attenuation

C12. Would a mutation that inactivated the lac repressor and prevented it from binding to the *lac* operator site result in the constitutive expression of the *lac* operon under all conditions? Explain. What is the disadvantage to the bacterium of having a constitutive *lac* operon?

C13. What is meant by the term attenuation? Is it an example of gene regulation at the level of transcription or translation? Explain your answer.

C14. As described in Figure 14.12, four regions within the *trpL* mRNA can form stem-loops. Let's suppose that mutations have been previously identified that prevent the ability of a particular region to form a stem-loop with a complementary region. For example, a region 1 mutant cannot form a 1–2 stem-loop, but it can still form a 2–3 or 3–4 stem-loop. Likewise, a region 4 mutant can form a 1–2 or 2–3 stem-loop but not a 3–4 stem-loop. Under the following conditions, would attenuation occur?

A. Region 1 is mutant, tryptophan is high, and translation is not occurring.

B. Region 2 is mutant, tryptophan is low, and translation is occurring.

C. Region 3 is mutant, tryptophan is high, and translation is not occurring.

D. Region 4 is mutant, tryptophan is low, and translation is not occurring.

C15. As described in Chapter 13, enzymes known as aminoacyl-tRNA synthetases are responsible for attaching amino acids to tRNAs. Let's suppose that tryptophanyl-tRNA synthetase was partially defective at attaching tryptophan to tRNA; its activity was only 10% of that found in a normal bacterium. How would attenuation of the *trp* operon be affected? Would it be more or less likely to be attenuated? Explain your answer.

C16. The 3–4 stem-loop and U-rich attenuator found in the *trp* operon (see Figure 14.12) is an example of ρ-independent termination. The function of ρ-independent terminators is described in Chapter 12. Would you expect attenuation to occur if the tryptophan levels were high and mutations changed the attenuator sequence from UUUUUUUU to UGGUUGUC? Explain why or why not.

C17. Mutations in tRNA genes can create tRNAs that recognize stop codons. Because stop codons are sometimes called nonsense codons, the types of mutations that affect tRNAs are called nonsense suppressors. For example, a normal $tRNA^{Gly}$ has an anticodon sequence CCU that recognizes a glycine codon in mRNA (GGA) and puts in a glycine during translation. However, a mutation in the gene that encodes $tRNA^{Gly}$ could change the anticodon to ACU. This mutant $tRNA^{Gly}$ would still carry glycine, but it would recognize the stop codon UGA. Would this mutation affect attenuation of the *trp* operon? Explain why or why not. Note: To answer this question, you need to look carefully at Figure 14.12 and see if you can identify any stop codons that may exist beyond the UGA stop codon that is found after region 1.

C18. Translational control is usually aimed at preventing the initiation of translation. With regard to cellular efficiency, why do you think this is the case?

C19. What is antisense RNA? How does it affect the translation of a complementary mRNA?

C20. A species of bacteria can synthesize the amino acid histidine so it does not require histidine in its growth medium. A key enzyme, which we will call histidine synthetase, is necessary for histidine biosynthesis. When these bacteria are given histidine in their growth media, they stop synthesizing histidine intracellularly. Based on this observation alone, propose three different regulatory mechanisms to explain why histidine biosynthesis ceases when histidine is in the growth medium. To explore this phenomenon further, you measure the amount of intracellular histidine synthetase protein when cells are grown in the presence and absence of histidine. In both conditions, the amount of this protein is identical. Which mechanism of regulation would be consistent with this observation?

C21. Using three examples, describe how allosteric sites are important in the function of genetic regulatory proteins.

C22 In what ways are the actions of the lac repressor and trp repressor similar and how are they different with regard to their binding to operator sites, their effects on transcription, and the influences of small effector molecules?

C23. Transcriptional repressor proteins (e.g., lac repressor), antisense RNA, and feedback inhibition are three different mechanisms that turn off the expression of genes and gene products. Which of these three mechanisms would be most effective in each of the following situations?

A. Shutting down the synthesis of a polypeptide

B. Shutting down the synthesis of mRNA

C. Shutting off the function of a protein

For your answers in parts A–C that have more than one mechanism, which mechanism would be the fastest or the most efficient?

Experimental Questions

E1. Answer the following questions that pertain to the experiment of Figure 14.7.

A. Why was β-ONPG used? Why was no yellow color observed in one of the four tubes? Can you propose alternative methods to measure the level of expression of the *lac* operon?

B. The optical density values were twice as high for the mated strain as for the parent strain. Why was this result obtained?

E2. Chapter 20 describes a blotting method known as Northern blotting, which can be used to detect RNA transcribed from a particular gene or a particular operon. In this method, a specific RNA is detected by using a short segment of cloned DNA as a probe. The DNA probe is complementary to the RNA that the researcher wishes to detect. After the probe DNA binds to the RNA within a blot of a gel, the RNA is visualized as a dark band. For example, a DNA probe complementary to the mRNA of the *lac* operon could be used to specifically detect the *lac* operon mRNA on a gel blot. As shown here, the method of Northern blotting can be used to determine the amount of a particular RNA transcribed under different types of growth conditions. In this Northern blot, bacteria containing a normal *lac* operon were grown under different types of conditions, and then the mRNA was isolated from the cells and subjected to Northern blotting, using a probe that is complementary to the mRNA of the *lac* operon.

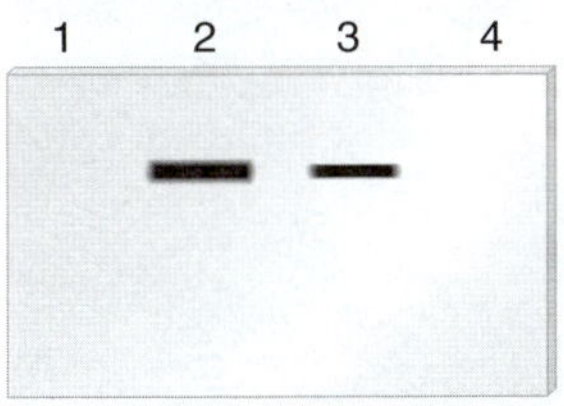

Lane 1. Growth in media containing glucose
Lane 2. Growth in media containing lactose
Lane 3. Growth in media containing glucose and lactose
Lane 4. Growth in media that doesn't contain glucose or lactose

Based on your understanding of the regulation of the *lac* operon, explain these results. Which is more effective at shutting down the *lac* operon, the binding of the lac repressor or the removal of CAP? Explain your answer based on the results shown in the Northern blot.

E3. As described in experimental question E2 and also in Chapter 20, the technique of Northern blotting can be used to detect the transcription of RNA. Draw the results you would expect from a Northern blot if bacteria were grown in media that contained lactose (and no glucose) but had the following mutations:

Lane 1. Normal strain

Lane 2. Strain with a mutation that inactivates the lac repressor

Lane 3. Strain with a mutation that prevents allolactose from binding to the lac repressor

Lane 4. Strain with a mutation that inactivates CAP

How would your results differ if these bacterial strains were grown in media that did not contain lactose or glucose?

E4. An absentminded researcher follows the protocol described in Figure 14.7 and (at the end of the experiment) does not observe any yellow color in any of the tubes. Yikes! Which of the following mistakes could account for this observation?

A. Forgot to sonicate the cells

B. Forgot to add lactose to two of the tubes

C. Forgot to add β-ONPG to the four tubes

E5. Explain how the data shown in Figure 14.9 indicate that two operator sites are necessary for repression of the *lac* operon. What would the results have been if all three operator sites were required for the binding of the lac repressor?

E6. A mutant strain has a defective *lac* operator site that results in the constitutive expression of the *lac* operon. Outline an experiment you would carry out to demonstrate that the operator site must be physically adjacent to the genes that it influences. Based on your knowledge of the *lac* operon, describe the results you would expect.

E7. Let's suppose you have isolated a mutant strain of *E. coli* in which the *lac* operon is constitutively expressed. To understand the nature of this defect, you create a merozygote in which the mutant strain contains an F′ factor with a normal *lac* operon and a normal *lacI* gene. You then compare the mutant strain and the merozygote with regard to their β-galactosidase activities in the presence and absence of lactose. You obtain the following results:

	Addition of Lactose	Amount of β-Galactosidase (percentage of mutant strain in the presence of lactose)
Mutant	No	100
Mutant	Yes	100
Merozygote	No	100
Merozygote	Yes	200

Explain the nature of the defect in the mutant strain.

Questions for Student Discussion/Collaboration

1. Discuss the advantages and disadvantages of genetic regulation at the different levels described in Figure 14.1.
2. As you look at Figure 14.10, discuss possible "molecular ways" that the cAMP-CAP complex and lac repressor may influence RNA polymerase function. In other words, try to explain how the bending and looping in DNA may affect the ability of RNA polymerase to initiate transcription.

Note: All answers appear at the website for this textbook; the answers to the even-numbered questions and all of the Concept Check and Comprehension Questions are in the back of the book.

www.mhhe.com/brookergenetics5e

Visit the website for practice tests, answer keys, and other learning aids for this chapter. Enhance your understanding of genetics with our interactive exercises, quizzes, animations, and much more.

CHAPTER OUTLINE

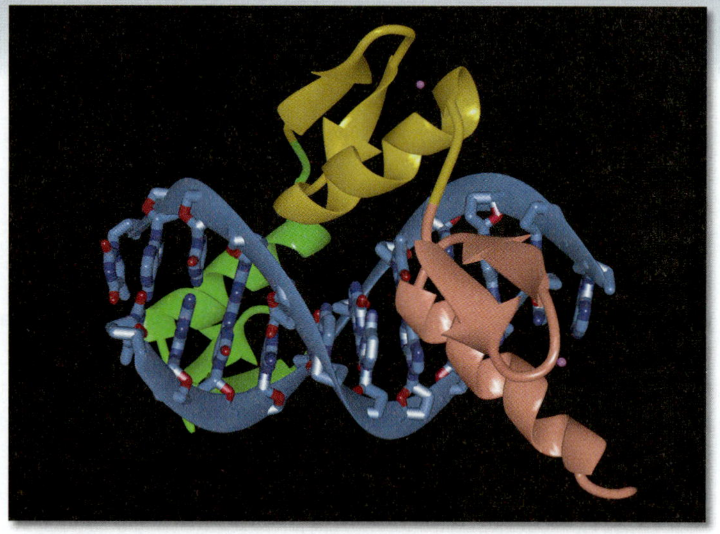

Certain proteins, known as regulatory transcription factors, have the ability to bind into the major groove of DNA and regulate gene transcription.

15 GENE REGULATION IN EUKARYOTES I: TRANSCRIPTIONAL REGULATION

Gene regulation refers to the phenomenon that the level of gene expression can be controlled so that genes can be expressed at high or low levels. The ability to regulate genes provides many benefits to eukaryotic organisms—a category that includes protists, fungi, plants, and animals. Like their prokaryotic counterparts, eukaryotic cells need to adapt to changes in their environment. For example, eukaryotic cells respond to changes in nutrient availability by enzyme adaptation, much as prokaryotic cells do. Eukaryotic cells also respond to environmental stresses such as ultraviolet (UV) radiation by inducing genes that provide protection against harmful environmental agents. An example is the ability of humans to develop a tan. The tanning response helps to protect a person's skin cells against the damaging effects of UV rays.

Among plants and animals, multicellularity and a more complex cell structure also demand a much greater level of gene regulation. The life cycle of complex eukaryotic organisms involves the progression through developmental stages to achieve a mature organism. Some genes are expressed only during early stages of development, such as the embryonic stage, whereas others are expressed only in the adult. In addition, complex eukaryotic species are composed of many different tissues that contain a variety of cell types. Gene regulation is necessary to ensure the differences in structure and function among distinct cell types. It is amazing that the various cells within a multicellular organism usually contain the same genetic material, yet phenotypically may look quite different. For example, the appearance of a human nerve cell seems about as similar to a muscle cell as an amoeba is to a paramecium. In spite of these phenotypic differences, a human nerve cell and muscle cell actually contain the same complement of human chromosomes. Nerve and muscle cells look strikingly different because of gene regulation rather than differences in DNA content. Many genes are expressed in the nerve cell and not the muscle cell, and vice versa.

The molecular mechanisms that underlie gene regulation in eukaryotes bear many similarities to the ways that bacteria regulate their genes. As in bacteria, regulation in eukaryotes can occur at any step in the pathway of gene expression (**Figure 15.1**). In this chapter, we will focus on gene regulation at the level of transcription, an important form of control. In addition, research in the

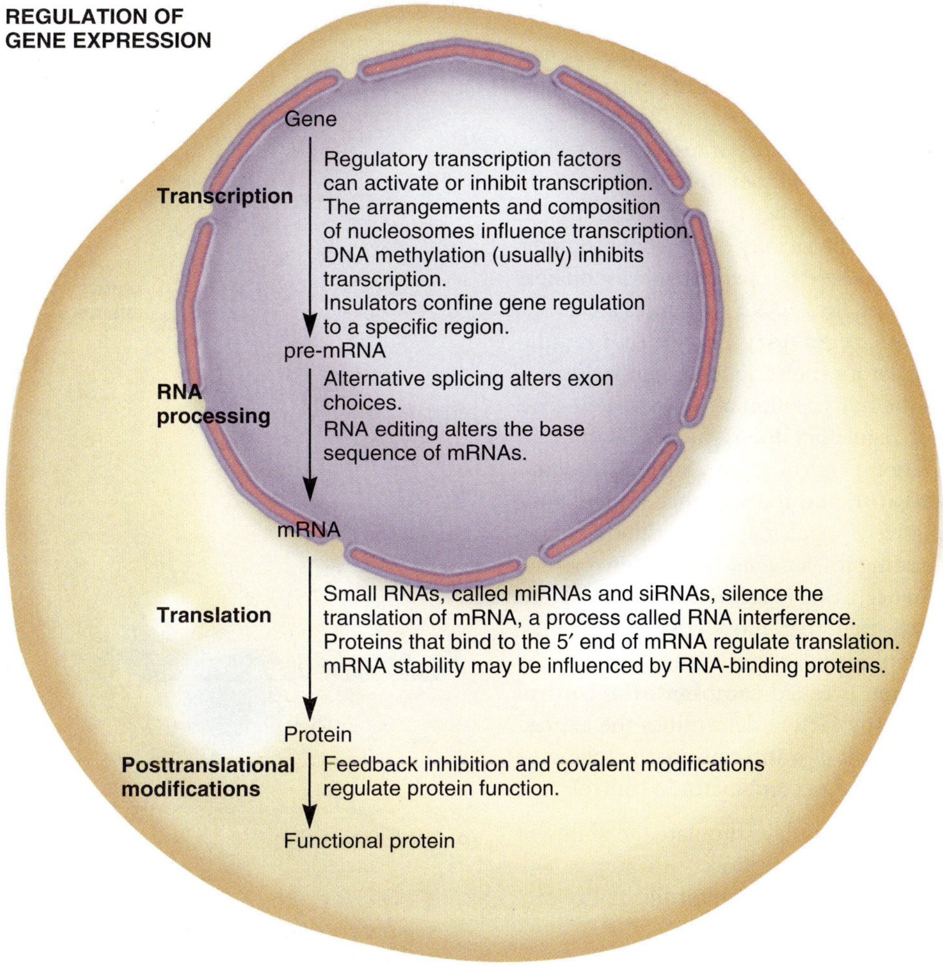

FIGURE 15.1 **Levels of gene expression commonly subject to regulation.**

CONCEPT CHECK: Which of these levels is the most energy-efficient way to regulate gene expression?

past few decades has revealed that eukaryotic organisms frequently regulate gene expression at points other than transcription. We will discuss some well-studied examples in which genes are regulated at these other control points in Chapter 16.

15.1 REGULATORY TRANSCRIPTION FACTORS

Learning Outcomes:

1. Distinguish between general and regulatory transcription factors.
2. List the factors that contribute to combinatorial control.
3. Describe how a regulatory transcription factor binds to a regulatory element.
4. Explain how a regulatory transcription factor exerts its effects via TFIID or mediator.
5. Describe three ways that the function of a regulatory transcription factor can be modulated.
6. Outline the steps whereby glucocorticoid receptors and CREB proteins regulate genes.

The term **transcription factor** is broadly used to describe proteins that influence the ability of RNA polymerase to transcribe a given gene. We will focus our attention on transcription factors that affect the ability of RNA polymerase to begin the transcription process. Such transcription factors regulate the binding of the transcriptional complex to the core promoter and/or control the switch from the initiation to the elongation stage of transcription. Two categories of transcription factors play a key role in these processes. In Chapter 12, we considered **general transcription factors,** which are required for the binding of RNA polymerase to the core promoter and its progression to the elongation stage. General transcription factors are necessary for any transcription to occur. In addition, eukaryotic cells possess a diverse array of **regulatory transcription factors** that serve to regulate the rate of transcription of target genes. The importance of transcription factors is underscored by the number of genes that encode this category of proteins. A sizeable portion of the genomes of complex organisms, such as plants and animals, is devoted to the process of gene regulation. For example, in *Arabidopsis thaliana*, a plant that is used as a model organism by plant geneticists, approximately 5% of its genome encodes transcription factors. This species has more than 1500 different genes that encode proteins that regulate the transcription of other genes. Similarly, 2–3% of the genes in the human genome encode transcription factors.

As discussed in this section, regulatory transcription factors exert their effects by influencing the ability of RNA polymerase to begin transcription of a particular gene. They typically recognize *cis*-acting elements that are located in the vicinity of the core promoter. These DNA sequences are analogous to the operator sites found near bacterial promoters. In eukaryotes, these DNA sequences are generally known as **control elements, regulatory elements,** or **regulatory sequences.** When a regulatory transcription factor binds to a regulatory element, it affects the transcription of an associated gene. For example, the binding of regulatory transcription factors may enhance the rate of transcription (**Figure 15.2a**). Such a transcription factor is termed an **activator,** and the sequence it binds to is called an **enhancer.** Alternatively, regulatory transcription factors may act as **repressors** by binding to elements called **silencers** and preventing transcription from occurring (**Figure 15.2b**).

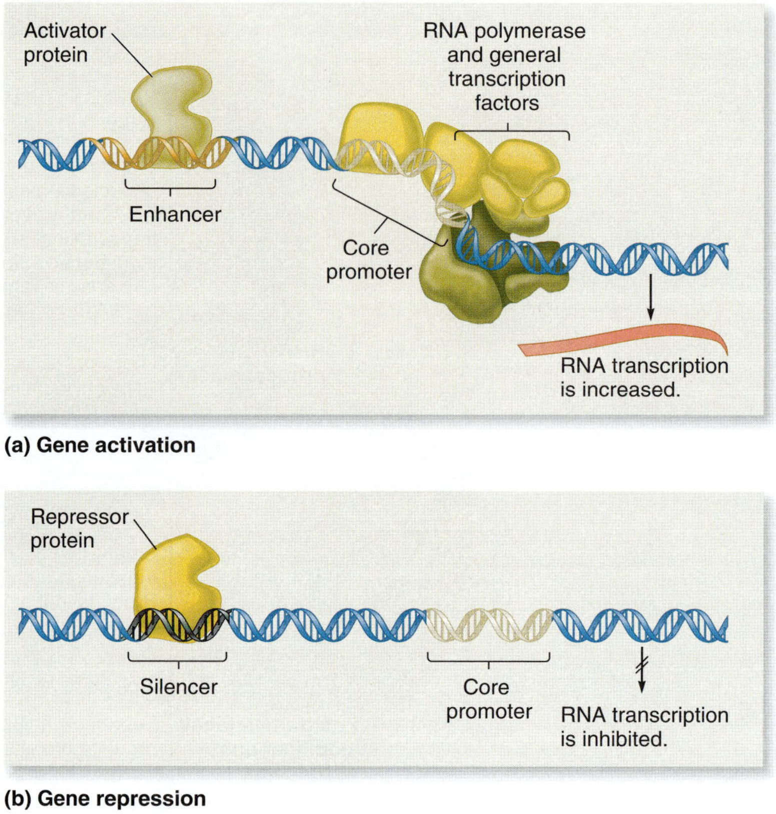

FIGURE 15.2 Overview of transcriptional regulation by regulatory transcription factors. These proteins can act as either **(a)** an activator to increase the rate of transcription or **(b)** a repressor to decrease the rate of transcription.

Researchers have discovered that most eukaryotic genes, particularly those found in multicellular species, are regulated by many factors. This phenomenon is called **combinatorial control** because the combination of many factors determines the expression of any given gene. At the level of transcription, the following are common factors that contribute to combinatorial control:

1. One or more activator proteins may stimulate the ability of RNA polymerase to initiate transcription.
2. One or more repressor proteins may inhibit the ability of RNA polymerase to initiate transcription.
3. The function of activators and repressors may be modulated in a variety of ways, including the binding of small effector molecules, protein-protein interactions, and covalent modifications.
4. Regulatory proteins may alter the composition or arrangements of nucleosomes in the vicinity of a promoter, thereby affecting transcription.
5. DNA methylation may inhibit transcription, either by preventing the binding of an activator protein or by recruiting proteins that change the structure of chromatin in a way that inhibits transcription.

All five of these factors can contribute to the regulation of a single gene, or possibly only three or four will play a role. In most cases, transcriptional regulation is aimed at controlling the initiation of transcription at the promoter. In this section, we will survey the first three factors that can provide combinatorial control of transcription. In the sections to follow, we will consider the fourth and fifth factors.

Structural Features of Regulatory Transcription Factors Allow Them to Bind to DNA

Genes that encode general and regulatory transcription factor proteins have been identified and sequenced from a wide variety of eukaryotic species, including yeast, plants, and animals. Several different families of evolutionarily related transcription factors have been discovered. In recent years, the molecular structures of transcription factor proteins have become an area of intense research. Transcription factor proteins contain regions, called **domains,** that have specific functions. For example, one domain of a transcription factor may have a DNA-binding function, and another may provide a binding site for a small effector molecule. When a domain or portion of a domain has a very similar structure in many different proteins, such a structure is called a **motif.**

Figure 15.3 depicts several different domain structures found in transcription factor proteins. The protein secondary structure known as an α helix is frequently found in transcription factors. Why is the α helix common in such proteins? The explanation is that the α helix is the proper width to bind into the major groove of the DNA double helix. In helix-turn-helix and helix-loop-helix motifs, an α helix called the recognition helix makes contact with and recognizes a base sequence along the major groove of the DNA (Figure 15.3a and b). Recall that the major groove is a region of the DNA double helix where the bases contact the surrounding water in the cell. Hydrogen bonding between the amino acid side chains in an α helix and nucleotide bases in the DNA is one way that a transcription factor binds to a specific DNA sequence. In addition, the recognition helix often contains many positively charged amino acids (e.g., arginine and lysine) that favorably interact with the DNA backbone, which is negatively charged.

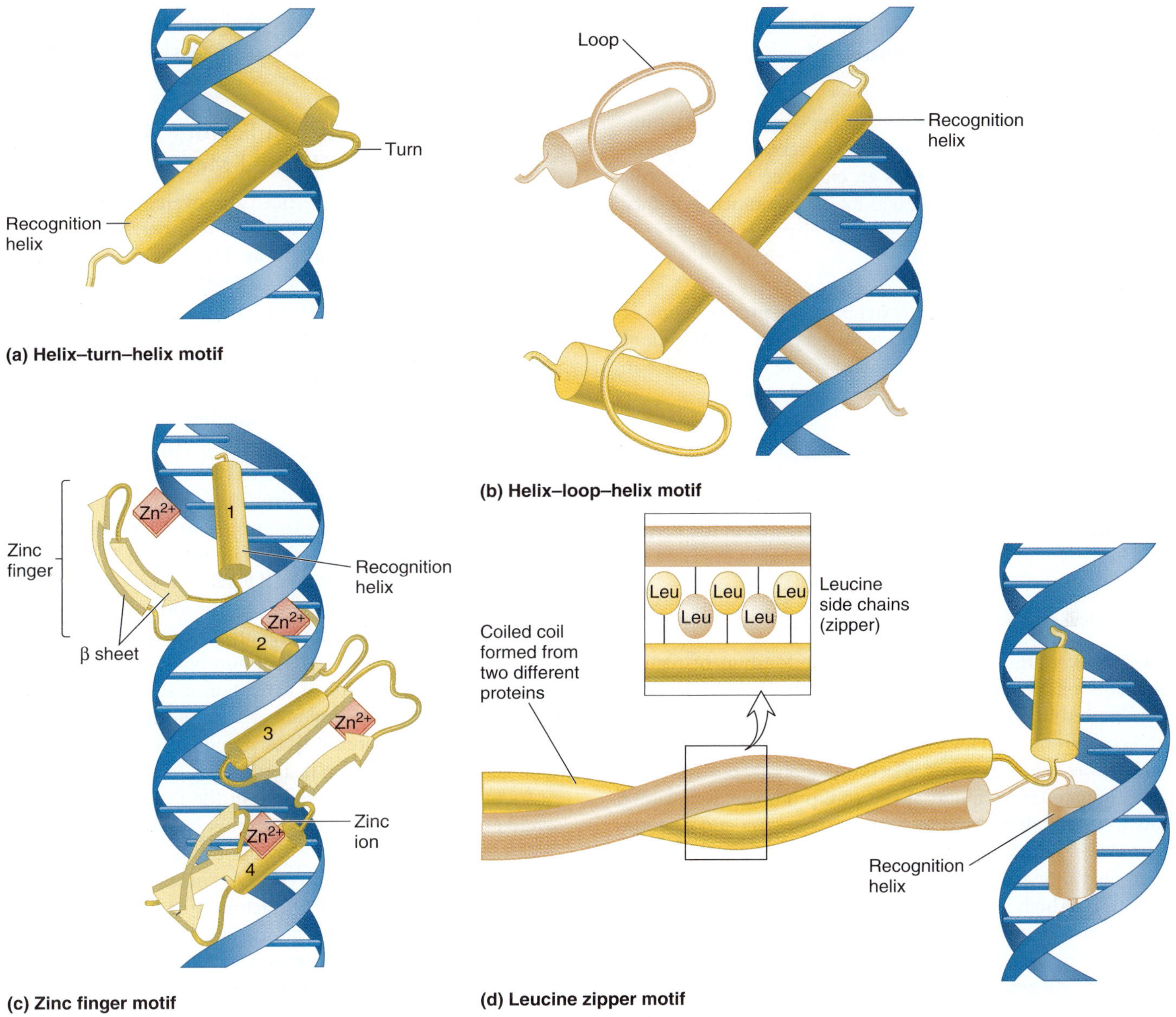

FIGURE 15.3 Structural motifs found in transcription factor proteins. Certain types of protein secondary structure are found in many different transcription factors. In this figure, α helices are shown as cylinders and β sheets as flattened arrows. **(a)** Helix-turn-helix motif: Two α helices are connected by a turn. The α helices bind to the DNA within the major groove. **(b)** Helix-loop-helix motif: A short α helix is connected to a longer α helix by a loop. In this illustration, a dimer is formed from the interactions of two helix-loop-helix motifs, and the longer helices are binding to the DNA. **(c)** Zinc finger motif: Each zinc finger is composed of one α helix and two antiparallel β sheets. A zinc ion (Zn^{2+}), shown in red, holds the zinc finger together. This illustration shows four zinc fingers in a row. **(d)** Leucine zipper motif: The leucine zipper promotes the dimerization of two transcription factor proteins. Two α helices (termed a coiled coil) are intertwined via the leucines (see inset).

CONCEPT CHECK: Explain how an α helix is able to function as a recognition helix in a transcription factor protein.

A zinc finger motif is composed of one α helix and two β-sheet structures that are held together by a zinc (Zn^{2+}) metal ion (Figure 15.3c). The zinc finger also can recognize DNA sequences within the major groove.

A second interesting feature of certain motifs is that they promote protein dimerization. The leucine zipper (Figure 15.3d) and helix-loop-helix motif (see Figure 15.3b) mediate protein dimerization. For example, Figure 15.3d depicts the dimerization and DNA binding of two proteins that have several leucine amino acids (a zipper). Alternating leucines in both proteins interact ("zip up"), resulting in protein dimerization. Two identical transcription factor proteins may come together to form a **homodimer,** or two different transcription factors can form a **heterodimer.** As discussed later, the dimerization of transcription factors can be an important way to modulate their function.

Regulatory Transcription Factors Recognize Regulatory Elements That Function as Enhancers or Silencers

As mentioned previously, when the binding of a regulatory transcription factor to a regulatory element increases transcription, the regulatory element is known as an enhancer. Such elements can stimulate transcription 10- to 1000-fold, a phenomenon known as **up regulation.** Alternatively, regulatory elements that serve to inhibit transcription are called silencers, and their action is called **down regulation.**

Many regulatory elements are **orientation-independent,** or **bidirectional.** This means that the regulatory element functions in the forward or reverse direction. For example, let's consider an enhancer with a forward orientation as shown below:

5′—GATA—3′
3′—CTAT—5′

This enhancer is also bound by a regulatory transcription factor and enhances transcription even when it is oriented in the reverse direction:

5′—TATC—3′
3′—ATAG—5′

Striking variation is also found in the location of regulatory elements relative to a gene's promoter. Regulatory elements are often located in a region within 200 base pairs (bp) upstream from the core promoter. However, they can be quite distant, even 100,000 bp away, yet exert strong effects on the ability of RNA polymerase to initiate transcription at the core promoter! Regulatory elements were first discovered by Susumu Tonegawa and coworkers in the 1980s. While studying genes that play a role in immunity, they identified a region that is far away from the core promoter, but is needed for high levels of transcription to take place. In some cases, regulatory elements are located downstream from the promoter site and may even be found within introns, the noncoding parts of genes. As you may imagine, the variation in regulatory element orientation and location profoundly complicates the efforts of geneticists to identify the regulatory elements that affect the expression of any given gene.

Regulatory Transcription Factors May Exert Their Effects Through TFIID and Mediator

Different mechanisms have been discovered that explain how a regulatory transcription factor binds to a regulatory element and thereby affects gene transcription. The net effect of a regulatory transcription factor is to influence the ability of RNA polymerase II to transcribe a given gene. However, most regulatory transcription factors do not bind directly to RNA polymerase. How then do most regulatory transcription factors exert their effects?

Some regulatory transcription factors bind to a regulatory element and then influence the function of TFIID. As discussed in Chapter 12, **TFIID** is a general transcription factor that binds to the TATA box and is needed to recruit RNA polymerase II to the core promoter. Activator proteins are expected to enhance the ability of TFIID to initiate transcription. One possibility is that activator proteins could help TFIID bind to the TATA box or they could enhance the function of TFIID in a way that facilitates its ability to recruit RNA polymerase II. In some cases, activator proteins exert their effects by interacting with **coactivators**—proteins that increase the rate of transcription but do not directly bind to the DNA itself. This type of activation is shown in Figure 15.4a. Coactivators typically contain a **transactivation domain** that promotes the activation of RNA polymerase, often by interacting with general transcription factors.

In contrast, repressors inhibit the function of TFIID. They could exert their effects by preventing the binding of TFIID to the TATA box (Figure 15.4b) or by inhibiting the ability of TFIID to recruit RNA polymerase II to the core promoter.

A second way that regulatory transcription factors control RNA polymerase II is via mediator—a protein complex discovered by Roger Kornberg and colleagues in 1990. The term **mediator** refers to the observation that it mediates the interaction

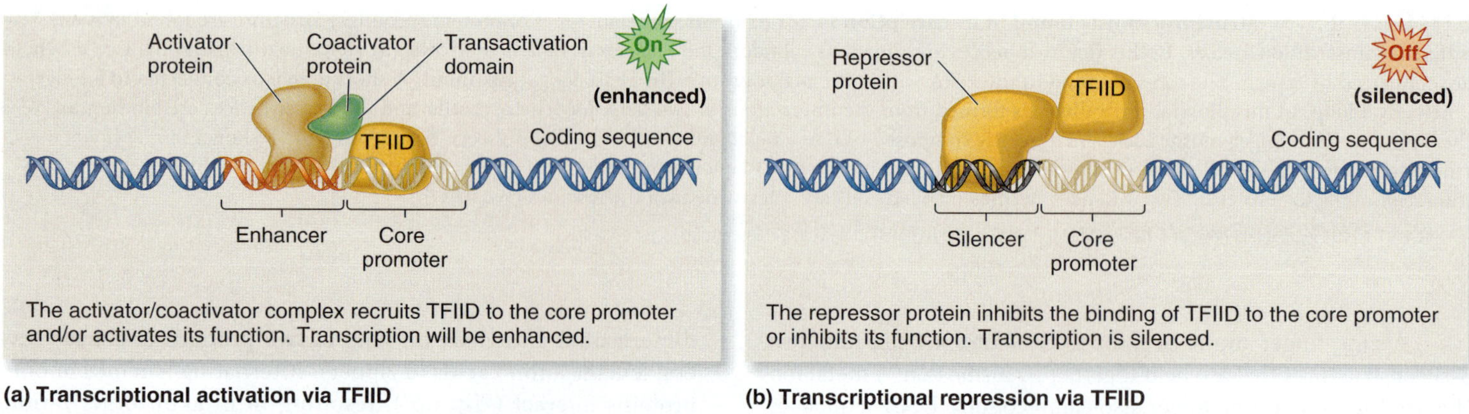

FIGURE 15.4 Effects of regulatory transcription factors on TFIID. (a) Some activators stimulate the function of TFIID, thereby enhancing transcription. In this example, the activator interacts with a coactivator that directly binds to TFIID and stimulates its function. (b) Regulatory transcription factors may also function as repressors. In the example shown here, the repressor binds to a silencer and inhibits the ability of TFIID to bind to the TATA box at the core promoter.

CONCEPT CHECK: If a repressor prevents TFIID from binding to the TATA box, why does this inhibit transcription?

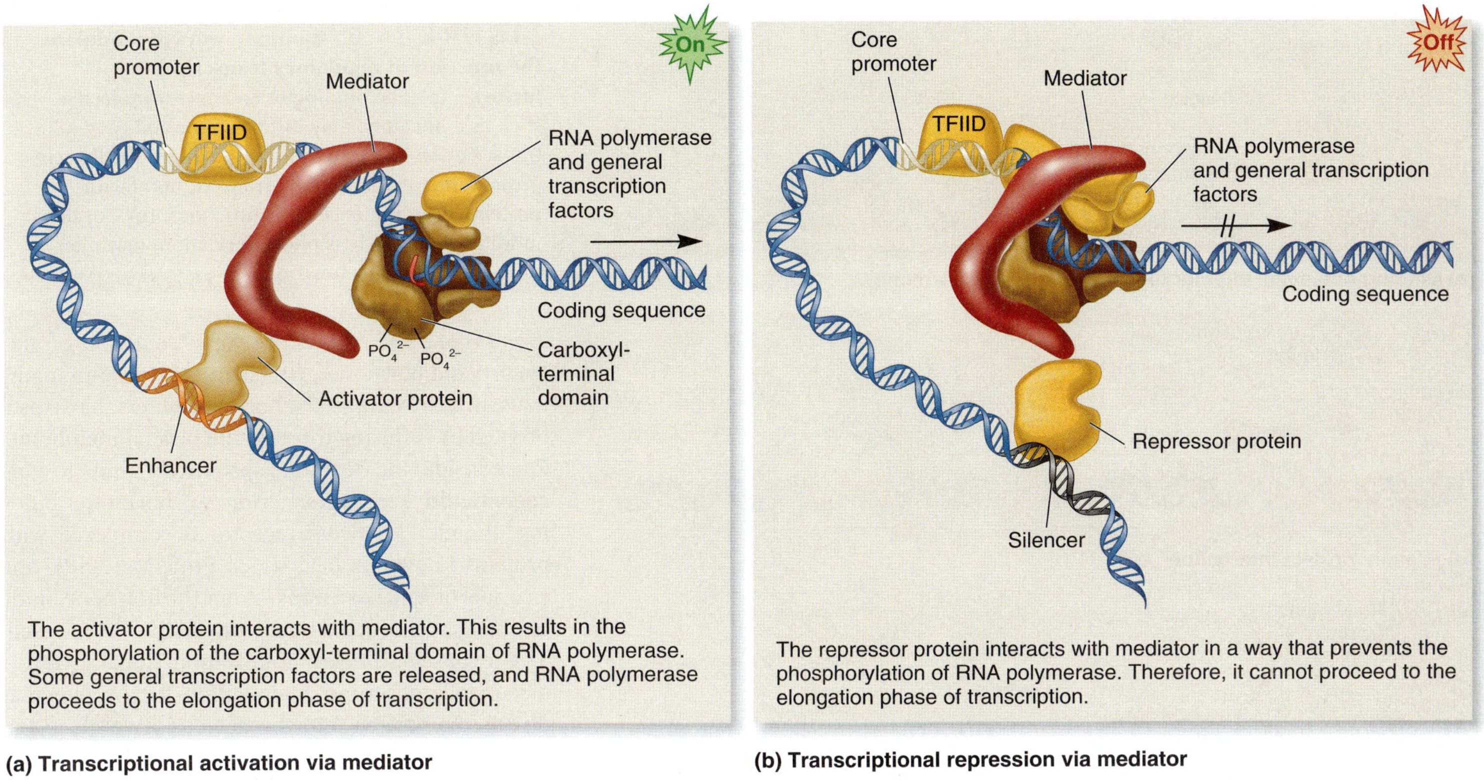

FIGURE 15.5 **Effects of an activator protein and a repressor protein on mediator.**

CONCEPT CHECK: When an activator protein interacts with mediator, how does this affect the function of RNA polymerase?

between RNA polymerase II and regulatory transcription factors. As discussed in Chapter 12, mediator controls the ability of RNA polymerase II to progress to the elongation stage of transcription via phosphorylation of the carboxyl-terminal domain. Transcriptional activators stimulate the ability of mediator to cause the phosphorylation of the carboxyl-terminal domain, thereby facilitating the switch between the initiation and elongation stages. In contrast, repressors have the opposite effect. In the example shown in **Figure 15.5a**, an activator binds to a distant enhancer element. The activator protein and mediator are brought together by the formation of a loop within the intervening DNA. Alternatively, a repressor protein may prevent mediator from allowing RNA polymerase to proceed to the elongation phase of transcription (**Figure 15.5b**).

A third way that regulatory transcription factors influence transcription is by recruiting proteins to the promoter region that affect nucleosome positions and compositions. For example, certain transcriptional activators recruit proteins to the promoter region that alter chromatin structure. We will return to this topic in Section 15.2.

The Function of Regulatory Transcription Factor Proteins Can Be Modulated in Three Ways

Thus far, we have considered the structures of regulatory transcription factors and the molecular mechanisms that account for their abilities to control transcription. The functions of the regulatory transcription factors themselves must also be modulated. Why is this necessary? The answer is that the genes they control must be turned on at the proper time, in the correct cell type, and under the appropriate environmental conditions. Therefore, eukaryotes have evolved different ways to modulate the functions of these proteins.

The functions of regulatory transcription factor proteins are controlled in three common ways: through (1) the binding of a small effector molecule, (2) protein-protein interactions, and (3) covalent modifications. **Figure 15.6** depicts these three mechanisms of modulating regulatory transcription factor function. Usually, one or more of these modulating effects are important in determining whether a transcription factor binds to the DNA or influences transcription by RNA polymerase. For example, a small effector molecule may bind to a regulatory transcription factor and promote its binding to DNA (Figure 15.6a). We will see that steroid hormones function in this manner. Another important way is via protein-protein interactions (Figure 15.6b). The formation of homodimers and heterodimers is a fairly common means of controlling transcription. Finally, the function of a regulatory transcription factor can be affected by covalent modifications, such as the attachment of a phosphate group (Figure 15.6c). As discussed later, the phosphorylation of activators can control their ability to stimulate transcription.

Steroid Hormones Exert Their Effects by Binding to a Regulatory Transcription Factor

Now that we have a general understanding of the structure and function of regulatory transcription factors, let's turn our attention to specific examples that illustrate how they carry out their roles

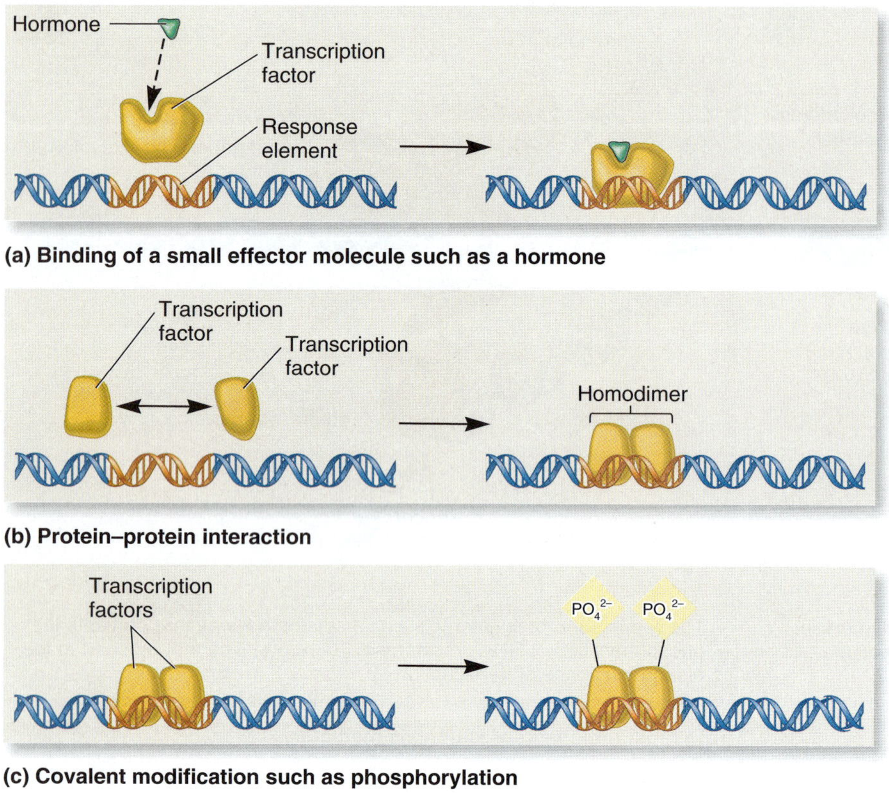

FIGURE 15.6 Common ways to modulate the function of regulatory transcription factors. **(a)** The binding of an effector molecule such as a hormone may influence the ability of a transcription factor to bind to the DNA. **(b)** Protein-protein interactions among transcription factor proteins may influence their functions. **(c)** Covalent modifications, such as phosphorylation, may alter transcription factor function.

within living cells. Our first example is a category that responds to steroid hormones. This type of regulatory transcription factor is known as a **steroid receptor,** because the steroid hormone binds directly to the protein.

The ultimate action of a steroid hormone is to affect gene transcription. In animals, steroid hormones act as signaling molecules that are synthesized by endocrine glands and secreted into the bloodstream. The hormones are then taken up by cells that respond to the hormones in different ways. For example, glucocorticoid hormones influence nutrient metabolism in most body cells. Other steroid hormones, such as estrogen and testosterone, are called gonadocorticoids because they influence the growth and function of the gonads.

Figure 15.7 shows the stepwise action of glucocorticoid hormones, which are produced in mammals. In this example, the hormone enters the cytosol of a cell by diffusing through the plasma membrane. Once inside, the hormone specifically binds to **glucocorticoid receptors.** Prior to hormone binding, the glucocorticoid receptor is complexed with proteins known as heat shock proteins (HSP), one example of which is HSP90. After the hormone binds to the glucocorticoid receptor, HSP90 is released. This exposes a nuclear localization signal (NLS)—a sequence of amino acids within the protein that directs the protein into the nucleus. Two glucocorticoid receptors form a homodimer and then travel through a nuclear pore into the nucleus.

How does the glucocorticoid receptor regulate the expression of particular genes? In the nucleus, the glucocorticoid receptor homodimer binds to DNA sites with the following consensus sequence:

5′—AGRACA—3′
3′—TCYTGT—5′

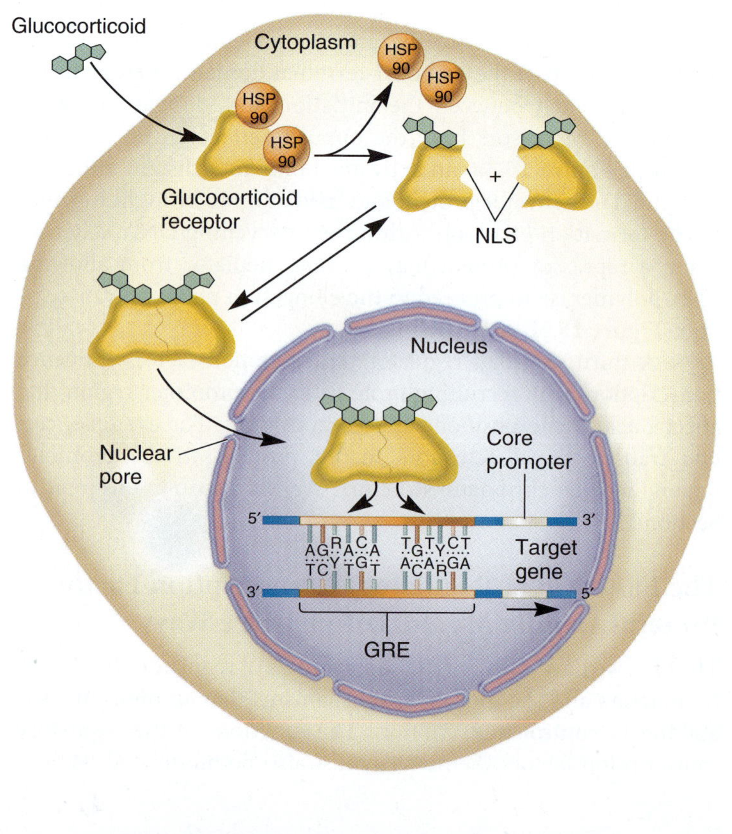

FIGURE 15.7 The action of glucocorticoid hormones. Once inside the cell, the glucocorticoid hormone binds to the glucocorticoid receptor, releasing it from a heat shock protein known as HSP90. This exposes a nuclear localization signal (NLS). Two glucocorticoid receptors then form a dimer and travel into the nucleus, where the dimer binds to a glucocorticoid response element (GRE) that is next to a particular gene. The binding of the glucocorticoid receptors to the GRE activates the transcription of the adjacent target gene.

ONLINE ANIMATION

Genes→Traits Glucocorticoid hormones are produced by the endocrine glands in response to fasting and activity. They enable the body to regulate its metabolism properly. When glucocorticoids are produced, they are taken into cells and bind to the glucocorticoid receptors. This eventually leads to the activation of genes that encode proteins involved in the synthesis of glucose, the breakdown of proteins, and the mobilization of fats.

CONCEPT CHECK: Explain why the glucocorticoid receptor binds next to the core promoter of some genes, but not next to the core promoter of most genes.

where R is a purine and Y is a pyrimidine. A glucocorticoid response element (GRE) contains two of these sequences in opposite directions (Figure 15.7). A GRE is found next to many genes and functions as an enhancer. The binding of the glucocorticoid receptor homodimer to a GRE activates the transcription of the nearby gene, eventually leading to the synthesis of the encoded protein.

Mammalian cells usually have a large number of glucocorticoid receptors within the cytoplasm. Because GREs are located near dozens of different genes, the uptake of many hormone molecules can activate many glucocorticoid receptors, thereby stimulating the transcription of many different genes. For this reason, a cell can respond to the presence of the hormone in a very complex way. Glucocorticoid hormones stimulate many genes that encode proteins involved in several different cellular processes, including the synthesis of glucose, the breakdown of proteins, and the mobilization of fats. Although the genes are not physically adjacent to each other, the regulation of multiple genes via glucocorticoid hormones is much like the ability of bacterial operons to simultaneously control the expression of several genes.

The CREB Protein Is an Example of a Regulatory Transcription Factor Modulated by Covalent Modification

As we have just seen, steroid hormones function as signaling molecules that bind directly to regulatory transcription factors to alter their function. This enables a cell to respond to a hormone by up regulating a particular set of genes. Most extracellular signaling molecules, however, do not enter the cell or bind directly to transcription factors. Instead, most signaling molecules must bind to receptors in the plasma membrane. This binding activates the receptor and may lead to the synthesis of an intracellular signal that causes a cellular response. One type of cellular response is to affect the transcription of particular genes within the cell.

As our second example of regulatory transcription factor function within living cells, we will examine the **cAMP response element-binding protein (CREB protein).** The CREB protein is a regulatory transcription factor that becomes activated in response to cell-signaling molecules that cause an increase in the cytoplasmic concentration of the molecule cyclic adenosine monophosphate (cAMP). The CREB protein, which is composed of two identical subunits, recognizes a DNA site that has two adjacent copies of the following consensus sequence:

5′—TGACGTCA—3′
3′—ACTGCAGT—5′

This response element, which is found near many different genes, is called the **cAMP response element (CRE).**

Figure 15.8 shows the steps leading to the activation of the CREB protein. A wide variety of hormones, growth factors, neurotransmitters, and other signaling molecules bind to plasma membrane receptors to initiate an intracellular response. In this case, the response involves the production of a second messenger, cAMP. The extracellular signaling molecule itself is considered the primary messenger. When the signaling molecule binds to the receptor, it activates a G protein that subsequently activates the enzyme adenylyl cyclase. The activated adenylyl cyclase catalyzes the synthesis of cAMP. The cAMP molecule then binds to a second enzyme, protein kinase A, and activates it. Protein kinase A travels into the nucleus and phosphorylates several different cellular proteins, including CREB protein. This phosphorylation activates the CREB protein by allowing it to bind to CREB-binding protein (CBP), which functions as a coactivator. The binding of CBP leads to the transcription activation of RNA polymerase. In contrast, unphosphorylated CREB protein can still bind to CREs but does not lead to the activation of RNA polymerase.

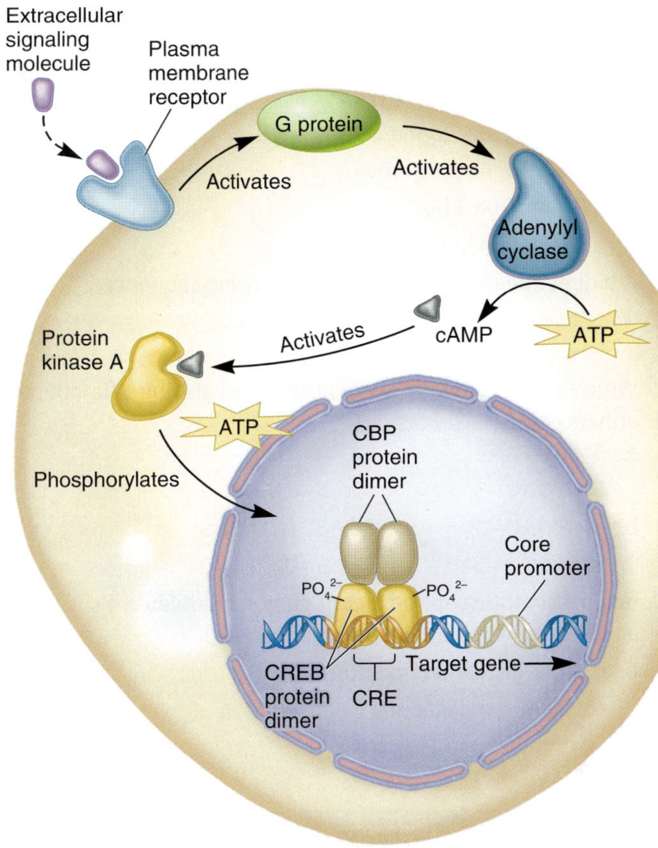

FIGURE 15.8 The activity of the CREB protein. An extracellular signaling molecule binds to a receptor in the plasma membrane, thereby activating a G protein, which then activates adenylyl cyclase, leading to the synthesis of cAMP. Next, cAMP binds to protein kinase A, which activates it. Protein kinase A then travels into the nucleus and phosphorylates the CREB protein. Once phosphorylated, the CREB protein acts as a transcriptional activator by promoting the binding of CBP, which is a coactivator.

15.1 COMPREHENSION QUESTIONS

1. Combinatorial control refers to the phenomenon that
   a. transcription factors always combine with each other when regulating genes.
   b. the combination of many factors determines the expression of any given gene.
   c. small effector molecules and regulatory transcription factors are found in many different combinations.
   d. genes and regulatory transcription factors must combine with each other during gene regulation.

2. A regulatory transcription factor protein typically contains __________ that binds to the _________ of the DNA.
   a. an α helix, backbone
   b. an α helix, major groove
   c. a β sheet, backbone
   d. a β sheet, major groove
3. A bidirectional enhancer has the following sequence:

   5′—GTCA—3′
   3′—CAGT—5′

   Which of the following sequences would also be a functional enhancer?
   a. 5′—ACTG—3′
      3′—TGAC—5′
   b. 5′—TGAC—3′
      3′—ACTG—5′
   c. 3′—GTCA—5′
      5′—CAGT—3′
   d. 3′—TGAC—5′
      5′—ACTG—3′
4. Regulatory transcription factors can be modulated by
   a. the binding of small effector molecules.
   b. protein-protein interactions.
   c. covalent modifications.
   d. all of the above.

15.2 CHROMATIN REMODELING, HISTONE VARIATION, AND HISTONE MODIFICATION

Learning Outcomes:

1. Describe how chromatin-remodeling complexes alter nucleosomes.
2. Define histone variant and explain why they are functionally important.
3. Explain how histone modifications affect transcription.
4. Outline the procedure of chromatin immunoprecipitation sequencing and explain how it was used to identify nucleosome-free regions that flank eukaryotic genes.
5. Summarize the steps that occur for transcriptional activation of a eukaryotic gene.

The term ATP-dependent chromatin remodeling, or simply **chromatin remodeling,** refers to dynamic changes in the structure of chromatin that occur during the life of a cell. These changes range from local alterations in the positioning of one or a few nucleosomes to larger changes that affect chromatin structure over a longer distance. Chromatin remodeling is carried out by ATP-dependent chromatin-remodeling complexes, which are a set of diverse multiprotein machines that reposition and restructure nucleosomes.

In eukaryotes, changes in nucleosome position and histone composition are key features of gene regulation. Such changes in chromatin structure affect the ability of transcription factors to gain access to and bind their target sequences in the promoter region. If the chromatin is in a **closed conformation,** transcription may be difficult or impossible. By comparison, chromatin that is in an **open conformation** is more easily accessible to transcription factors and RNA polymerase so that transcription can occur. Although the closed and open conformations may be affected by the relative compaction of a chromosomal region, researchers have recently determined that the precise positioning of nucleosomes at or near promoters and the composition of the histones within nucleosomes often play a key role in eukaryotic gene regulation. In this section, we will examine molecular mechanisms that explain how changes in chromatin structure control the regulation of eukaryotic genes.

Chromatin Remodeling Complexes Alter the Positions and Compositions of Nucleosomes

In recent years, geneticists have been trying to identify the steps that promote the interconversion between the closed and open conformations of chromatin. Nucleosomes have been shown to change position in cells that normally express a particular gene compared with cells in which the gene is inactive. For example, in reticulocytes that express the β-globin gene, an alteration in nucleosome positioning occurs in the promoter region from nucleotide −500 to +200. This alteration is thought to be an important step in gene activation. Based on the analysis of many genes, researchers have discovered that a key role of some transcriptional activators is to orchestrate changes in chromatin structure from the closed to the open conformation by altering nucleosomes.

One way to change chromatin structure is through **ATP-dependent chromatin remodeling.** In this process, the energy of ATP hydrolysis is used to drive a change in the positions and/or compositions of nucleosomes, thereby making the DNA more or less amenable to transcription. Therefore, chromatin remodeling is important for both the activation and repression of transcription.

The remodeling process is carried out by a protein complex that recognizes nucleosomes and uses ATP to alter their configuration. All remodeling complexes have a catalytic ATPase subunit that is similar to other motor proteins, called **DNA translocases,** that move along the DNA. Eukaryotes have multiple families of chromatin remodelers. Though their names may differ depending on the species, common chromatin-remodeling complex families are referred to as the SWI/SNF-family, the ISWI-family, the INO80-family, and the Mi-2-family. The names of these remodelers sometimes refer to the effects of mutations in genes that encode remodeling proteins. For example, the abbreviations SWI and SNF refer to the effects that occur in yeast when these remodeling enzymes are defective. SWI mutants are defective in mating-type switching, and SNF mutations create a sucrose nonfermenting phenotype.

How do chromatin remodelers change chromatin structure? Three effects are possible. One result of ATP-dependent chromatin remodeling is a change in the position of nucleosomes (**Figure 15.9a**). This may involve a shift in nucleosomes to a new location or a change in the relative spacing of nucleosomes over a long stretch of DNA. A second effect is that remodelers may evict histone octamers from the DNA, thereby creating gaps where nucleosomes are not found (**Figure 15.9b**). A third possibility is that remodelers may change the composition of nucleosomes by removing standard histones and replacing them with histone variants (**Figure 15.9c**). The functions of histone variants are described next.

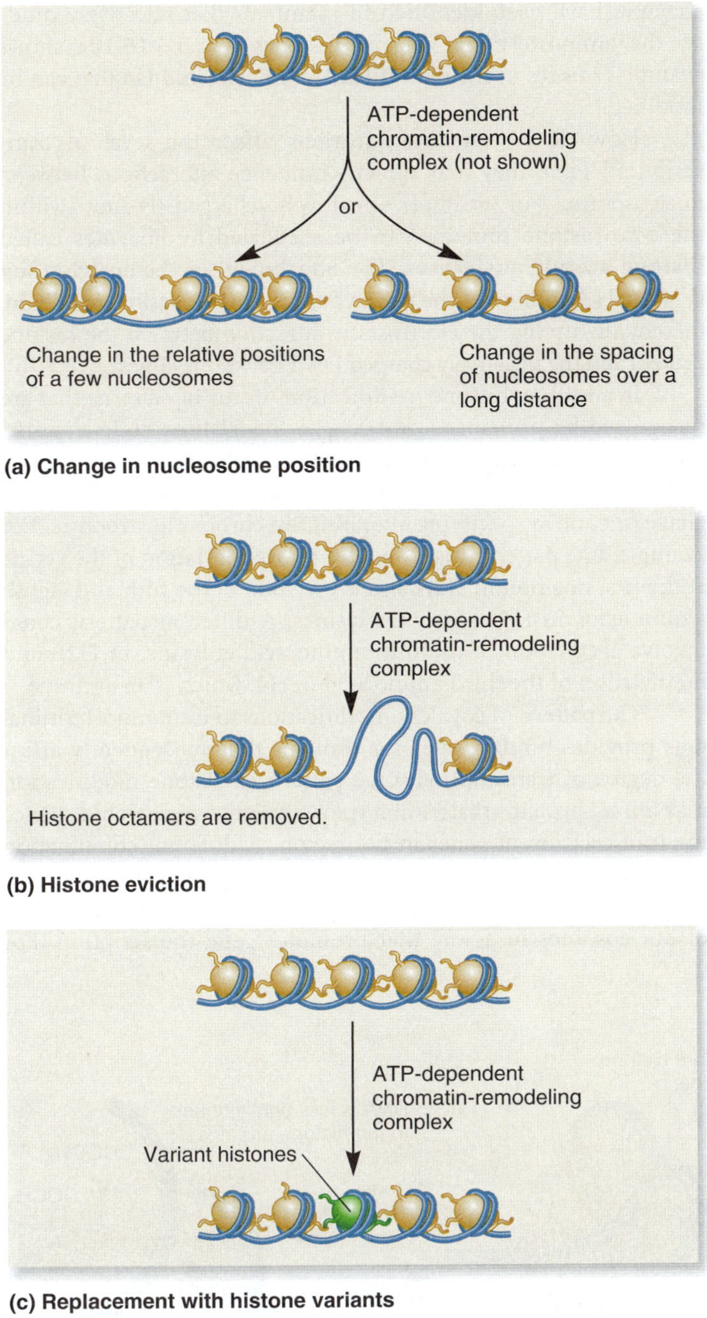

FIGURE 15.9 **ATP-dependent chromatin remodeling.** The top part of each illustration shows five nucleosomes. Chromatin-remodeling complexes may **(a)** change the locations of nucleosomes, **(b)** remove histone octamers from the DNA, or **(c)** replace core histones with histone variants.

CONCEPT CHECK: How might nucleosome eviction affect transcription?

Histone Variants Play Specialized Roles in Chromatin Structure and Function

The genes, which encode histones H1, H2A, H2B, H3, and H4, are moderately repetitive. The total number of histone genes varies from species to species. As an example, the human genome contains over 70 histone genes that have been produced by gene duplication events during evolution. Most of the histone genes encode standard histone proteins. However, a few have accumulated mutations that change the amino acid sequence of the histone proteins. These are called **histone variants.** Among eukaryotic species, histone variants have been identified for histone H1, H2A, H2B, and H3 genes, but not for the H4 gene.

What are the consequences of histone variation? Certain histone variants play specialized roles in chromatin structure and function. In all eukaryotes, histone variants are incorporated into a subset of nucleosomes to create functionally specialized regions of chromatin. In most cases, the standard histones are incorporated into the nucleosomes while new DNA is synthesized during S phase of the cell cycle. Later, some of the standard histones are replaced by histone variants via chromatin-remodeling complexes.

Table 15.1 describes the standard histones and a few histone variants that are found in humans. A key role of many histone variants is to regulate the structure of chromatin, thereby influencing

TABLE 15.1
Standard Human Histones and Examples of Histone Variants

Histone	Type	Number of Genes in Humans	Function
H1	Standard	11	Standard linker histone*
H1^0	Variant	1	Linker histone associated with chromatin compaction and gene repression
H2A	Standard	15	Standard core histone
MacroH2A	Variant	1	Core histone that is abundant on the inactivated X chromosome in female mammals. Plays a role in chromatin compaction.
H2A.Z	Variant	1	Core histone that is usually found in nucleosomes that flank the transcriptional start site of promoters. Plays a role in gene transcription.
H2A.Bbd	Variant	1	Core histone that promotes open chromatin. Plays a role in gene activation.
H2A.X	Variant	1	Plays a role in DNA repair
H2B	Standard	17	Standard core histone
spH2B	Variant	1	Core histone found in the telomeres of sperm cells
H3	Standard	10	Standard core histone
cenH3	Variant	1	Core histone found at centromeres. Involved with the binding of kinetochore proteins.
H3.3	Variant	2	Core histone that promotes open chromatin. Plays a role in gene activation.
H4	Standard	14	Standard core histone

*H1 in mammals is found in five subtypes.

gene transcription. Such variants can have opposite effects. The incorporation of histone H2A.Bbd into a chromosomal region where a particular gene is found favors gene activation. In contrast, the incorporation of histone H1^0 represses gene expression.

Although our focus in this chapter is on gene regulation, it is worth noting that histone variants play other important roles. For example, histone cenH3 (also called CENP-A), which is a variant of histone H3, is found at the centromeres of each chromosome and functions in the binding of kinetochore proteins. Histone cenH3 is required for the proper segregation of eukaryotic chromosomes. Other histone variants are primarily found at specialized sites in certain cells. Histone macroH2A is found along the inactivated X chromosome in female mammals, whereas spH2B is found in the telomeres in sperm cells. Finally, certain histone variants appear to play a role in DNA repair. For example, histone H2A.X becomes phosphorylated where a double-stranded DNA break occurs. This phosphorylation is thought to be important for the proper repair of that break.

The Histone Code Also Controls Gene Transcription

As described in Chapter 10, each of the core histone proteins consists of a globular domain and a flexible, charged amino-terminus called an amino-terminal tail (refer back to Figure 10.11a). The DNA wraps around the globular domains, and the amino-terminal tails protrude from the chromatin. Particular amino acids in the amino-terminal tails of standard histones and histone variants are subject to several types of covalent modifications, including acetylation, methylation, and phosphorylation. Over 50 different enzymes have been identified in mammals that selectively modify the amino-terminal tails of histones. **Figure 15.10a** shows examples of sites in the tails of H2A, H2B, H3, and H4 that can be modified.

How do histone modifications affect the level of transcription? First, they may directly influence interactions between nucleosomes. For example, positively charged lysines within the core histone proteins can be acetylated by enzymes called **histone acetyltransferases.** The attachment of the acetyl group ($—COCH_3$) eliminates the positive charge on the lysine side chain, thereby disrupting the electrostatic attraction between the histone protein and the negatively charged DNA backbone (**Figure 15.10b**).

In addition, histone modifications occur in patterns that are recognized by proteins. According to the **histone code hypothesis,** proposed by Brian Strahl, C. David Allis, and Bryan Turner in 2000, the pattern of histone modification acts much like a language or code in specifying alterations in chromatin structure. For example, one pattern might involve phosphorylation of the serine at the first position in H2A and acetylation of the fifth and eighth amino acids in H4, which are lysines. A different pattern could involve acetylation of the fifth amino acid, a lysine, in H2B and methylation of the third amino acid in H4, which is an arginine.

The pattern of covalent modifications to the amino-terminal tails provides binding sites for proteins that subsequently affect the degree of transcription. One pattern of histone modification may attract proteins that inhibit transcription, which would silence the transcription of genes in the region. A different combination of histone modifications may attract proteins, such as chromatin-remodeling enzymes, which would serve to alter the positions of nucleosomes in a way that promotes gene transcription. For

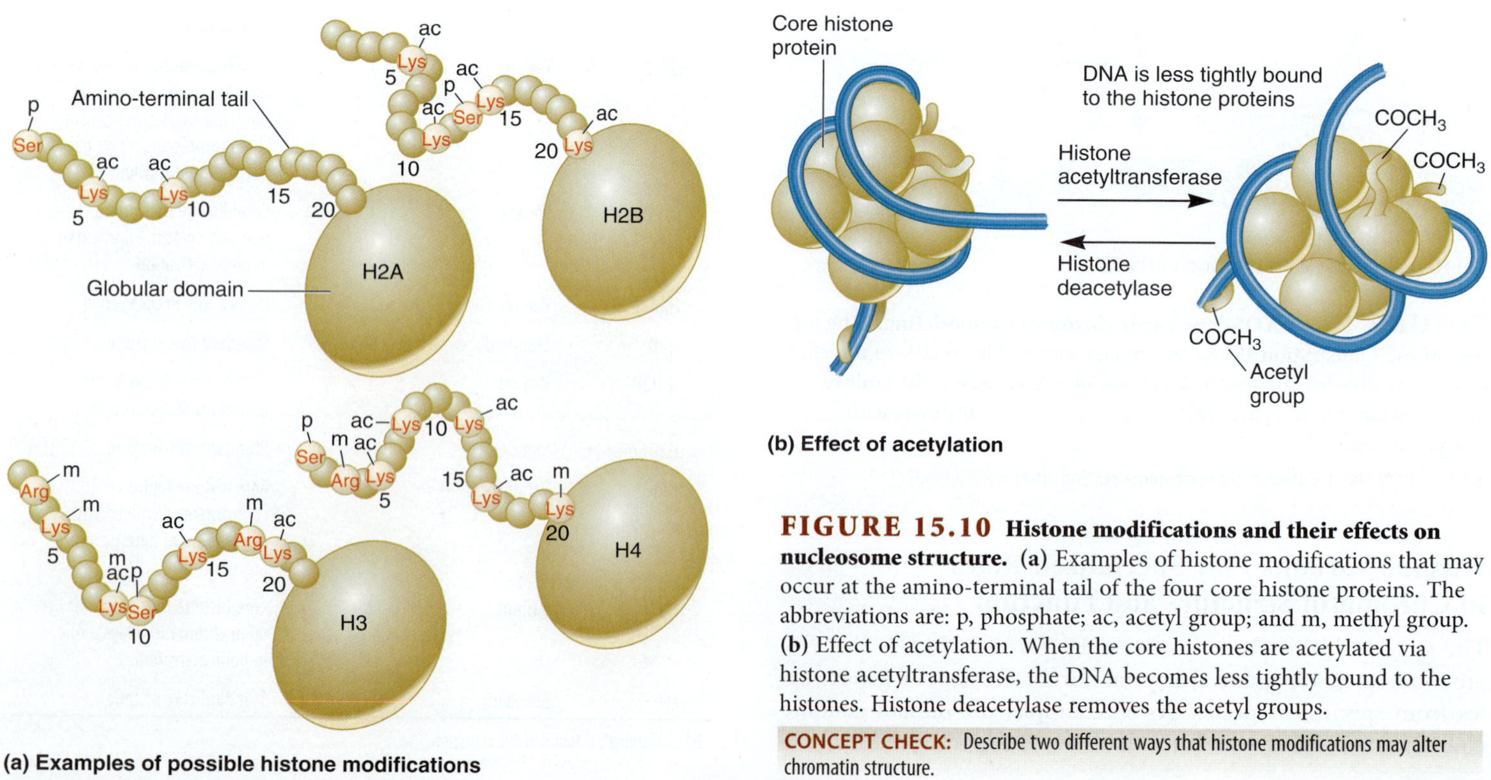

FIGURE 15.10 Histone modifications and their effects on nucleosome structure. **(a)** Examples of histone modifications that may occur at the amino-terminal tail of the four core histone proteins. The abbreviations are: p, phosphate; ac, acetyl group; and m, methyl group. **(b)** Effect of acetylation. When the core histones are acetylated via histone acetyltransferase, the DNA becomes less tightly bound to the histones. Histone deacetylase removes the acetyl groups.

CONCEPT CHECK: Describe two different ways that histone modifications may alter chromatin structure.

example, the acetylation of histones attracts certain chromatin remodelers that shift nucleosomes or evict histone octamers, thereby aiding in the transcription of genes. Overall, the histone code is thought to play an important role in determining whether the information within the genomes of eukaryotic species is accessed. Researchers are trying to decipher the effects of the covalent modifications that make up the histone code.

Chromatin Immunoprecipitation Sequencing Allows Researchers to Determine the Precise Locations of Nucleosomes Throughout Entire Genomes

Thus far, we have learned that nucleosomes can change in three ways. First, their locations can be altered along a DNA molecule either by a change in their spacing or the removal of histone octamers. Second, histones exist in different variants that play specialized roles. And third, histones are subject to covalent modifications at their amino-terminal tails.

Within the last 15 years researchers have been able to map the locations of specific nucleosomes within a genome. This allows for the determination of (1) where nucleosomes are located, (2) where histone variants are found, and (3) where covalent modifications of histones occur. This amazing achievement has been possible in large part by using an approach called **chromatin immunoprecipitation sequencing (ChIP-Seq).**

As shown in **Figure 15.11**, ChIP-Seq begins with living cells. The cells are treated with formaldehyde, which covalently links proteins to the DNA. Such a linkage is called a crosslink. Next, the cells are broken open, and the chromatin is exposed to a high concentration of micrococcal nuclease (MNase), which is an enzyme that digests DNA. MNase digests the linker regions between nucleosomes, but it is unable to cleave DNA that is attached to the core histone proteins. Following this digestion, what remains is a collection of millions of nucleosomes that have DNA attached to them.

The nucleosomes can be removed from this mixture using antibodies that specifically recognize histone proteins. All nucleosomes are removed if the antibody recognizes a histone such as H4, which is found in all nucleosomes and does not exist as a histone variant. Alternatively, specific types of nucleosomes can be removed from this mixture using antibodies that recognize particular histones. For example, an antibody could be added that recognizes a histone variant. Also, antibodies can be used that recognize a histone that has been covalently modified in a certain way. For example, a researcher could use an antibody that recognizes an acetylated histone.

The antibodies, which are attached to heavy beads, are added and they recognize specific nucleosomes. The material is then centrifuged and the DNA and proteins attached to the beads form a pellet at the bottom of the centrifuge tube. This step is called **immunoprecipitation.** Next, the crosslinks between the histones and DNA are broken, and the addition of proteases partially digests the core histones, thereby removing the histone octamers from DNA. The DNA is then subjected to gel electrophoresis. Only those DNA fragments that are about 150 bp in length are saved. This is the length of DNA that wraps around one nucleosome. At this stage, the researcher has thousands or millions of DNA fragments that are about 150 bp in length. Short oligonucleotides, also called linkers, are added to the ends of the DNA fragments, which facilitates their ability to be amplified by the process of polymerase chain reaction (PCR) and subjected to DNA sequencing. (PCR and DNA sequencing are described in Chapter 20.)

How are the DNA sequences analyzed? The ChIP-Seq method is used on species in which the entire genome has already been sequenced, such as yeast, *Drosophila*, humans, and *Arabidopsis*. A genome map describes the locations of DNA sequences along each chromosome in a genome. Using computer software, the sequences obtained via ChIP-Seq can be matched to identical sequences on a genome map. Because the genome map also shows the locations of genes, this method allows researchers to determine the relative positions of nucleosomes from the beginning of a gene to the end, as described next.

Eukaryotic Genes Are Flanked by Nucleosome-Free Regions and Well-Positioned Nucleosomes

Studies using ChIP-Seq, including those in yeast, *Drosophila*, humans, and other eukaryotic species have revealed that many eukaryotic genes show a common pattern of nucleosome organization (**Figure 15.12**). For active genes or those genes that can be activated, the transcriptional start site at the core promoter is found in a **nucleosome-free region (NFR),** which is a segment of DNA that is missing histones. The NFR is typically 150 bp in length. Although the NFR may be required for transcription, it is not, by itself, sufficient for gene activation. At any given time in the life of a cell, many genes that contain an NFR are not being actively transcribed.

The NFR at the transcriptional start site is flanked by two well-positioned nucleosomes that are termed the −1 and +1 nucleosomes. In yeast, the transcriptional start site (TSS) is usually at the boundary between the NFR and the +1 nucleosome. However, in animals, the TSS is about 60 bp farther upstream into the NFR. The +1 nucleosome typically contains histone variants H2A.Z and H3.3. Depending on the species and the gene, these variants may also be found in the −1 nucleosome and in some of the nucleosomes that immediately follow the +1 nucleosome in the transcribed region. For example, the +2 nucleosome is likely to contain H2A.Z but not as likely as the +1 nucleosome. Similarly, the +3 nucleosome is likely to contain H2A.Z, but not as likely as the +2 nucleosome.

The nucleosomes downstream from the +1 nucleosome tend to be evenly spaced near the beginning of a eukaryotic gene, but their spacing becomes less regular farther downstream. The ends of many eukaryotic genes appear to have a well-positioned nucleosome that is followed by an NFR. This arrangement at the ends of genes may be important for transcriptional termination.

Transcriptional Activation Involves Changes in Nucleosome Positions, Composition, and Histone Modifications

A key role of certain transcriptional activators is to recruit chromatin-remodeling enzymes and histone-modifying enzymes to the promoter region. Though the order of recruitment may differ among specific transcriptional activators, this appears to be critical

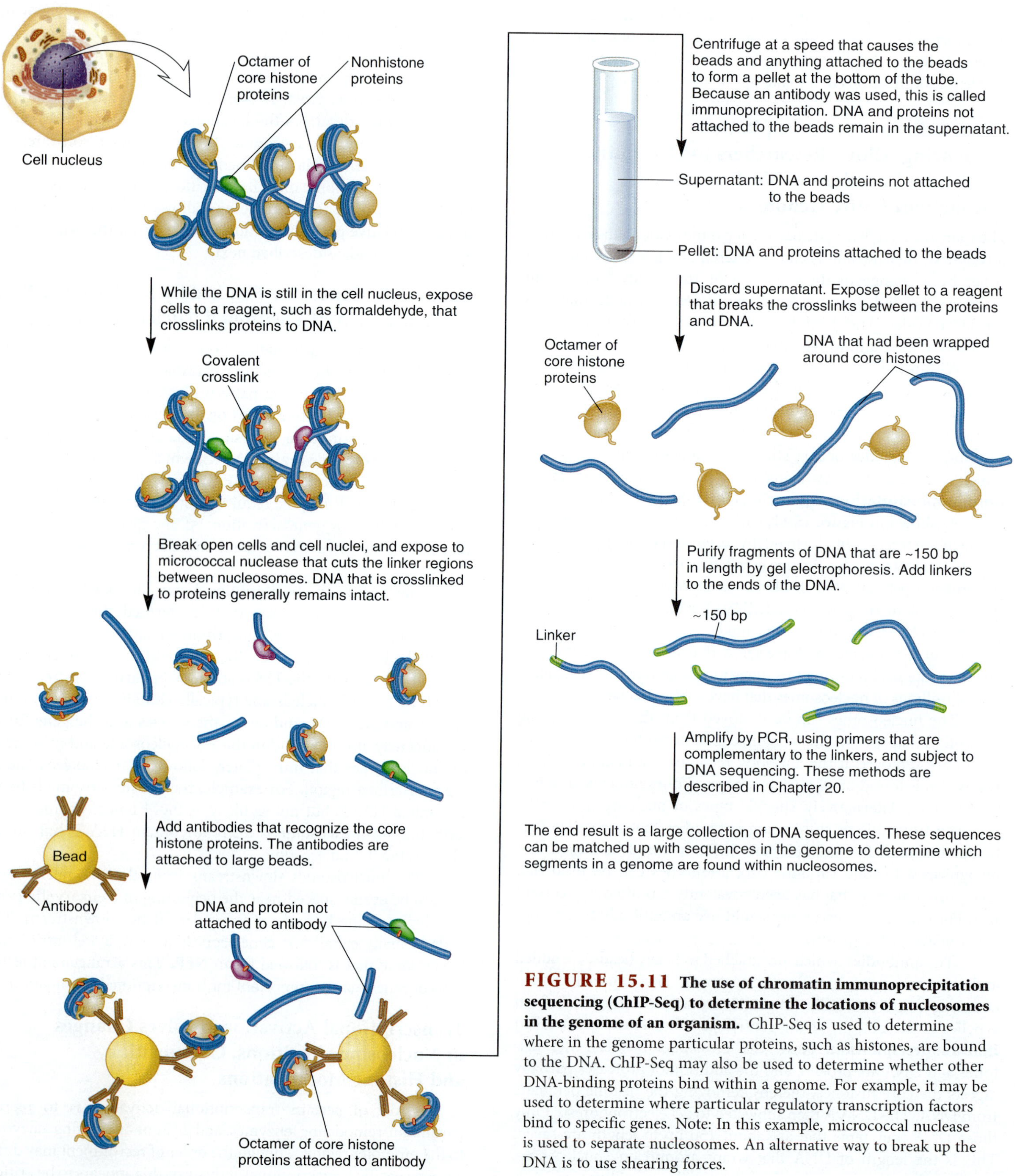

FIGURE 15.11 The use of chromatin immunoprecipitation sequencing (ChIP-Seq) to determine the locations of nucleosomes in the genome of an organism. ChIP-Seq is used to determine where in the genome particular proteins, such as histones, are bound to the DNA. ChIP-Seq may also be used to determine whether other DNA-binding proteins bind within a genome. For example, it may be used to determine where particular regulatory transcription factors bind to specific genes. Note: In this example, micrococcal nuclease is used to separate nucleosomes. An alternative way to break up the DNA is to use shearing forces.

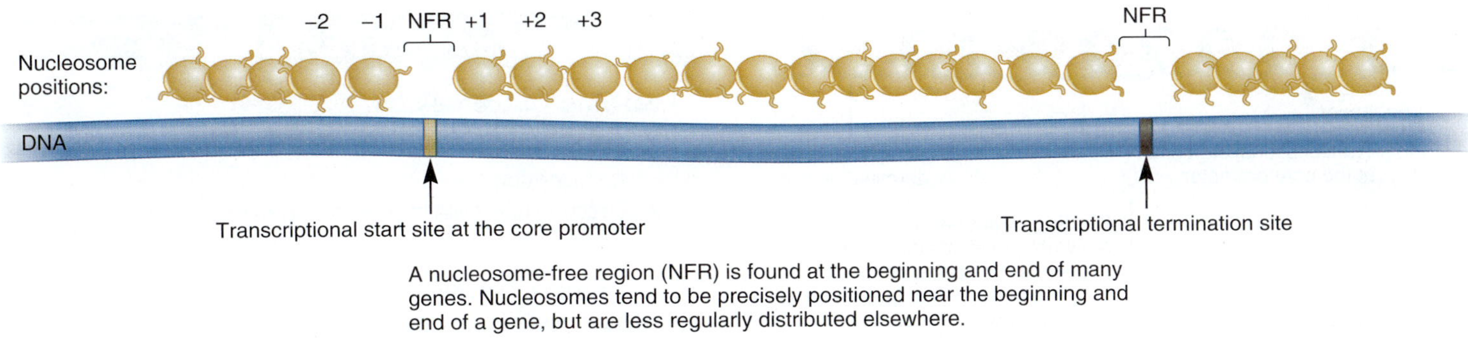

FIGURE 15.12 **Nucleosome arrangements and composition in the vicinity of a protein-encoding gene.**

CONCEPT CHECK: Why is an NFR needed at the core promoter for transcription to occur?

for transcriptional initiation and elongation. **Figure 15.13** presents a general scheme for how transcriptional activators may facilitate transcription in a eukaryotic gene, such as a gene found in yeast.

As discussed earlier, a gene that is able to be activated has an NFR at the transcriptional start site. In the scenario shown in Figure 15.13, a transcriptional activator binds to an enhancer in the NFR. The activator then recruits a chromatin-remodeling complex and a histone-modifying enzyme to this region. The chromatin remodelers may shift nucleosomes or temporarily evict histone octamers from the promoter region. Nucleosomes containing the histone variant H2A.Z, which are typically found at the +1 nucleosome, are thought to be more easily removed from the DNA than those containing the standard histone H2A. Histone-modifying enzymes, such as histone acetyltransferase, covalently modify histone proteins and may affect nucleosome contact with the DNA. The actions of chromatin remodelers and histone-modifying enzymes facilitate the binding of general transcription factors and RNA polymerase II to the core promoter, thereby allowing the formation of a preinitiation complex (see Figure 15.13).

Further changes in chromatin structure are necessary for elongation to occur. RNA polymerase II cannot transcribe DNA that is tightly wrapped in nucleosomes. For transcription to occur, histone octamers are evicted or destabilized so that RNA polymerase II can pass. Evicted histone proteins are transferred to histone chaperones, which are proteins that bind histones and aid in the assembly of histone octamers. Assembled histone octamers are then placed back on the DNA behind the moving RNA polymerase II (see Figure 15.13).

Histone-modifying enzymes also play a key role in histone removal and replacement during the elongation phase of transcription. Histone-modifying enzymes have been found to travel with RNA polymerase II during the elongation phase of transcription. These include enzymes that carry out histone acetylation, H3 methylation, and H2B ubiquitination. These modifications facilitate histone removal ahead of the traveling RNA polymerase II. Behind RNA polymerase II, histone deacetylase removes the acetyl groups, thereby favoring the binding of histones to the DNA to form nucleosomes. The re-formation of nucleosomes behind RNA polymerase II is thought to be critical to maintaining the fidelity of transcription. If nucleosome re-formation did not occur properly, transcriptional initiation might occur at multiple points in a gene, thereby producing faulty transcripts.

Changes in the relative amounts of histone variants also occur within actively transcribed genes. For example, histone variant H3.3 is often found in the transcribed regions of genes, but is less common in silent genes. H3.3 may facilitate the eviction and replacement of nucleosomes ahead and behind RNA polymerase II, respectively. Also, genes with very high levels of transcription may be largely devoid of nucleosomes, because multiple RNA polymerases are transcribing them at the same time.

15.2 COMPREHENSION QUESTIONS

1. A chromatin-remodeling complex may
   a. change the locations of nucleosomes.
   b. evict nucleosomes from DNA.
   c. replace standard histones with histone variants.
   d. do all of the above.
2. According to the histone code hypothesis, the pattern of histone modifications acts like a language that
   a. influences chromatin structure.
   b. promotes transcriptional termination.
   c. inhibits the elongation of RNA polymerase.
   d. does all of the above.
3. Which of the following characteristics is typical of a eukaryotic gene that can be transcribed?
   a. The core promoter is wrapped around a nucleosome.
   b. The core promoter is found in a nucleosome-free region.
   c. The terminator is wrapped around a nucleosome.
   d. None of the above are correct.
4. Transcriptional activation of eukaryotic genes involves which of the following events?
   a. Changes in nucleosome locations
   b. Changes in histone composition within nucleosomes
   c. Changes in histone modifications
   d. All of the above

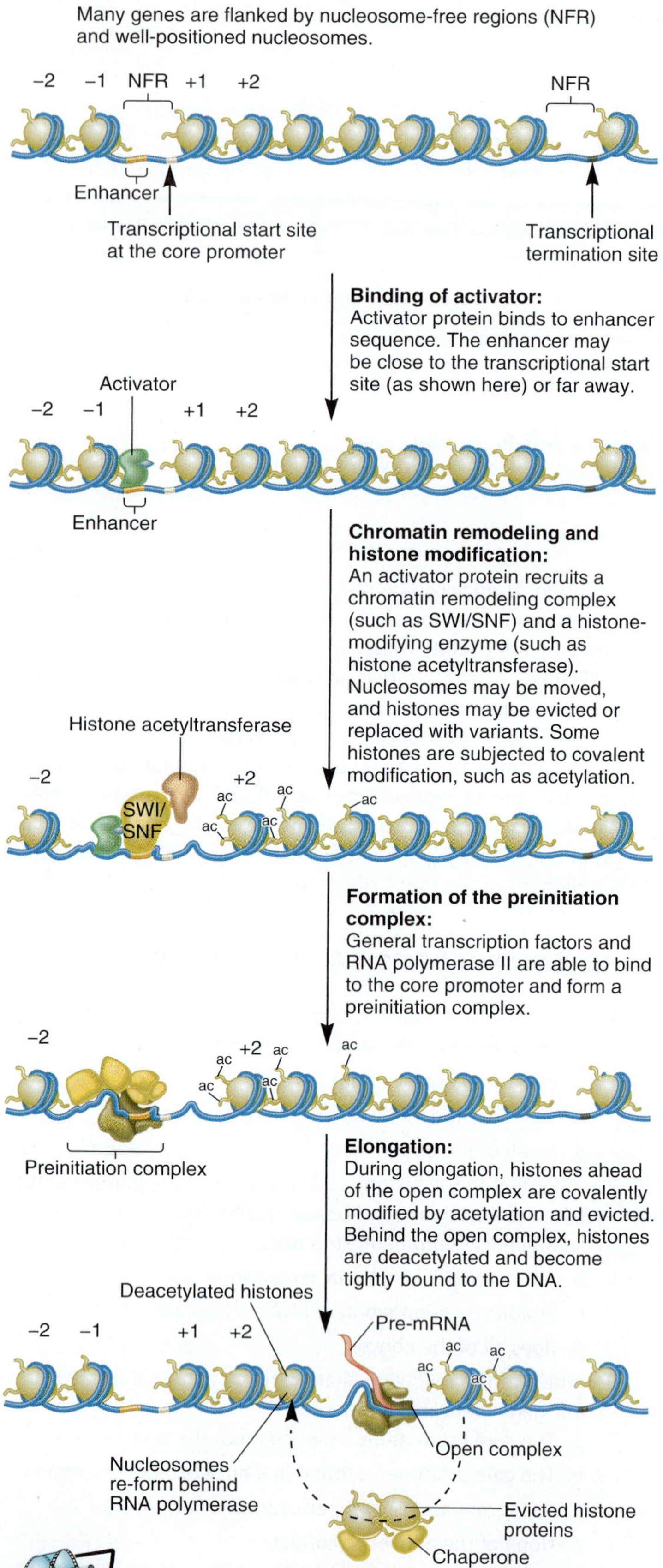

ONLINE ANIMATION

FIGURE 15.13 A simplified model for the transcriptional activation of a eukaryotic gene. The activation involves changes in nucleosome arrangements and histone modifications.

INTERACTIVE EXERCISE

CONCEPT CHECK: Explain why histone eviction is needed for the elongation phase of transcription.

15.3 DNA METHYLATION

Learning Outcomes:

1. Define DNA methylation and explain how it affects transcription.
2. Describe how DNA methylation is heritable.

We now turn our attention to a regulatory mechanism that usually silences gene expression. DNA structure can be modified by the covalent attachment of methyl groups, a mechanism called **DNA methylation.** This process is common in some, but not all, eukaryotic species. For example, yeast and *Drosophila* have little or no detectable methylation of their DNA, whereas DNA methylation in vertebrates and plants is relatively abundant. In mammals, approximately 2 to 7% of the DNA is methylated. In this section, we will examine how DNA methylation occurs and how it controls gene expression.

DNA Methylation Occurs on the Cytosine Base and Usually Inhibits Gene Transcription

As shown in Figure 15.14, eukaryotic DNA methylation occurs via an enzyme called **DNA methyltransferase,** which attaches a methyl group to the number 5 position of the cytosine base, forming 5-methylcytosine. The sequence that is methylated is shown here.

```
  CH3
  |
5′–CG–3′
3′–GC–5′
     |
    CH3
```

Note that this sequence contains cytosines in both strands. Methylation of the cytosine in both strands is termed full methylation, whereas methylation of only one strand is called hemimethylation.

DNA methylation usually inhibits the transcription of eukaryotic genes, particularly when it occurs in the vicinity of the promoter. In vertebrates and plants, **CpG islands** occur near many promoters of genes. (Note: CpG refers to a dinucleotide of C and G in DNA that is connected by a phosphodiester linkage.) These CpG islands are commonly 1000–2000 bp in length and contain a high number of CpG sites. In the case of **housekeeping genes**—genes that encode proteins required in most cells of a multicellular organism—the cytosine bases in the CpG islands are unmethylated. Therefore, housekeeping genes tend to be expressed in most cell types. By comparison, other genes are highly regulated and may be expressed only in a particular cell type. These are **tissue-specific genes.** In some cases, it has been found that the expression of such genes may be silenced by the methylation of CpG islands. In general, unmethylated CpG islands are correlated with active genes, whereas suppressed genes contain methylated CpG islands. In this way, DNA methylation is thought to play an important role in the silencing of tissue-specific genes to prevent them from being expressed in the wrong tissue.

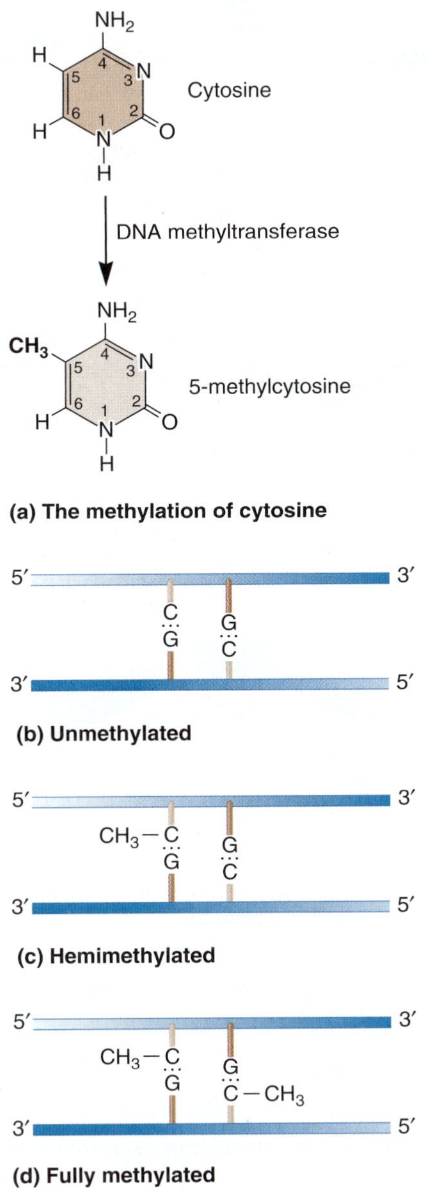

FIGURE 15.14 DNA methylation on cytosine bases. **(a)** Methylation occurs via an enzyme known as DNA methyltransferase, which attaches a methyl group to the number 5 carbon on cytosine. The CG sequence can be **(b)** unmethylated, **(c)** hemimethylated, or **(d)** fully methylated.

Methylation can affect transcription in two ways. First, methylation of CpG islands may prevent or enhance the binding of regulatory transcription factors to the promoter region. For example, methylated CG sequences could prevent the binding of an activator protein to an enhancer element, presumably by the methyl group protruding into the major groove of the DNA (**Figure 15.15a**). The inability of an activator protein to bind to the DNA would inhibit the initiation of transcription. However, CG methylation does not slow down the movement of RNA polymerase along a gene. In vertebrates and plants, coding regions downstream from the core promoter usually contain methylated CG sequences, but these do not hinder the elongation phase of transcription. This suggests that methylation must occur in the vicinity of the promoter to have an effect on transcription.

A second way that methylation inhibits transcription is via proteins known as **methyl-CpG-binding proteins,** which bind methylated sequences (**Figure 15.15b**). These proteins contain a domain called the methyl-binding domain that specifically recognizes a methylated CpG island. Once bound to the DNA, the methyl-CpG-binding protein recruits other proteins to the region that inhibit transcription. For example, methyl-CpG-binding proteins may recruit histone deacetylase to a methylated CpG island near a promoter. Histone deacetylation removes acetyl groups from the histone proteins, which makes it more difficult for nucleosomes to be removed from the DNA. In this way, deacetylation tends to inhibit transcription.

DNA Methylation Is Heritable

Methylated DNA sequences are inherited during cell division. Experimentally, if fully methylated DNA is introduced into a plant or vertebrate cell, the DNA will remain fully methylated even in subsequently produced daughter cells. However, if the same sequence of nonmethylated DNA is introduced into a cell, it will remain nonmethylated in the daughter cells. These observations indicate that the pattern of methylation is retained following DNA replication and, therefore, is inherited in future daughter cells.

How can methylation be inherited from cell to cell? **Figure 15.16** illustrates a molecular model that explains this process, which was originally proposed by Arthur Riggs, Robin Holliday, and J. E. Pugh. The DNA in a particular cell may become methylated by **de novo methylation**—the methylation of DNA that was previously unmethylated. When a fully methylated segment of DNA replicates in preparation for cell division, the newly made daughter strands contain unmethylated cytosines. Such DNA is said to be hemimethylated. This hemimethylated DNA is efficiently recognized by DNA methyltransferase, which makes it fully methylated. This process is called **maintenance methylation,** because it preserves the methylated condition in future cells. However, maintenance methylation does not act on unmethylated DNA. Overall, maintenance methylation appears to be an efficient process that routinely occurs within vertebrate and plant cells. By comparison, de novo methylation and demethylation are infrequent and highly regulated events. According to this view, the initial methylation or demethylation of a given gene can be regulated so that it occurs in a specific cell type or stage of development. Once methylation has occurred, it can then be transmitted from mother to daughter cells via maintenance methylation.

The methylation mechanism shown in Figure 15.16 can explain the phenomenon of genomic imprinting, which is described in Chapters 5 and 16. In this case, specific genes are methylated during oogenesis or spermatogenesis, but not both. Following fertilization, the pattern of methylation is maintained in the offspring. For example, if a gene is methylated only during spermatogenesis, the allele that is inherited from the father will be methylated in the somatic cells of the offspring, but the maternal allele will remain unmethylated. Along these lines, geneticists are also eager to determine how

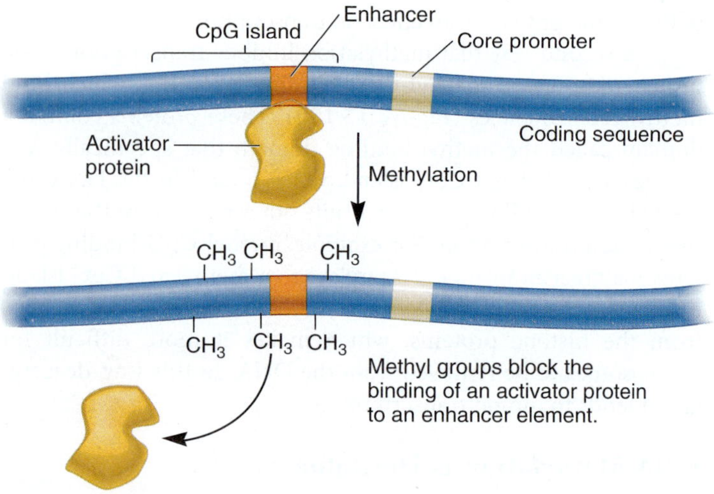

(a) Methylation inhibits the binding of an activator protein.

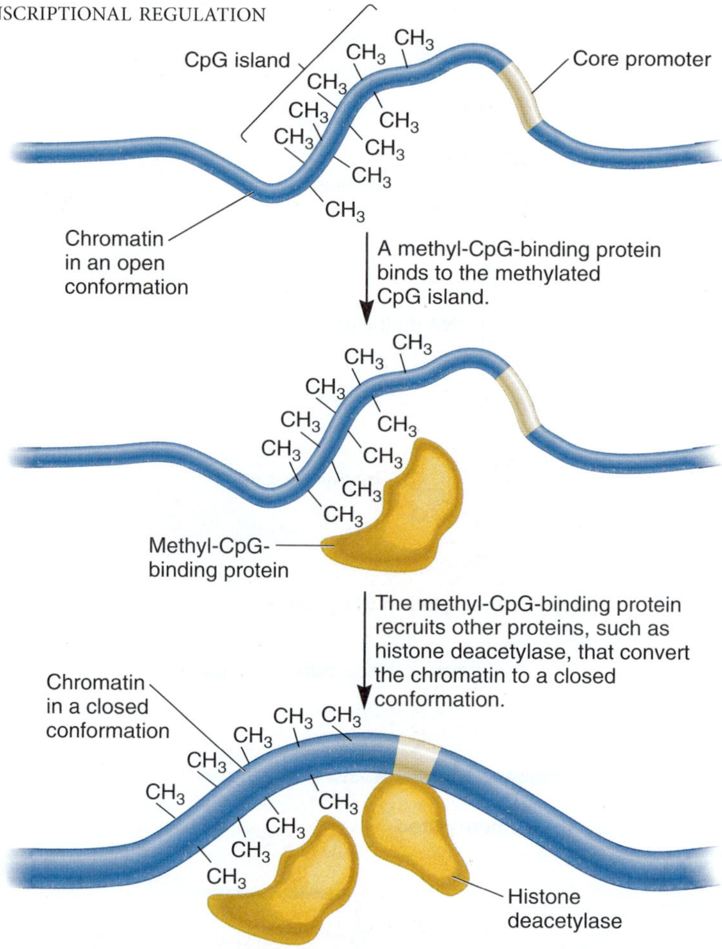

(b) Methyl-CpG-binding protein recruits other proteins that change the chromatin to a closed conformation.

FIGURE 15.15 Transcriptional silencing via methylation. (a) The methylation of a CpG island may inhibit the binding of a transcriptional activator protein to the promoter region. (b) The binding of a methyl-CpG-binding protein to a CpG island may lead to the recruitment of other proteins, such as histone deacetylase, that convert chromatin to a closed conformation and thus suppress transcription.

CONCEPT CHECK: Explain why the events shown in part (a) inhibit transcription.

variations in DNA methylation patterns may be important for cell differentiation. It may be a key way to silence genes in different cell types. However, additional research is necessary to understand how specific genes may be targeted for de novo methylation or demethylation in specific cell types or during different stages of development.

15.3 COMPREHENSION QUESTIONS

1. How can methylation affect transcription?
   a. It may prevent the binding of regulatory transcription factors.
   b. It may enhance the binding of regulatory transcription factors.
   c. It may attract the binding of methyl-CpG-binding proteins, which inhibit transcription.
   d. All of the above are possible.
2. The process in which completely unmethylated DNA becomes methylated is called
   a. maintenance methylation.
   b. de novo methylation.
   c. primary methylation.
   d. demethylation.

15.4 INSULATORS

Learning Outcomes:

1. Define insulator.
2. Explain two ways that insulators exert their effects.

Thus far, we have considered how regulatory transcription factors, chromatin remodeling, and DNA methylation can regulate gene transcription. As we have seen, eukaryotic gene regulation involves changes in chromatin structure that may occur over a relatively long distance. In addition, gene regulation often involves regulatory elements such as enhancers that are far away from the promoters they control.

In eukaryotes, adjacent genes usually exhibit different patterns of gene regulation. Because eukaryotic gene regulation can occur over long distances, a bewildering aspect of such regulation is its ability to control a particular gene but not affect neighboring genes. How are the effects of chromatin remodeling and histone modifications constrained so they only affect a particular gene and not its neighbors? How is an enhancer prevented from controlling multiple genes?

An **insulator** is a segment of DNA that functions as a boundary between two genes. An insulator is so named because it protects, or "insulates," a gene from the regulatory effects of a neighboring gene.

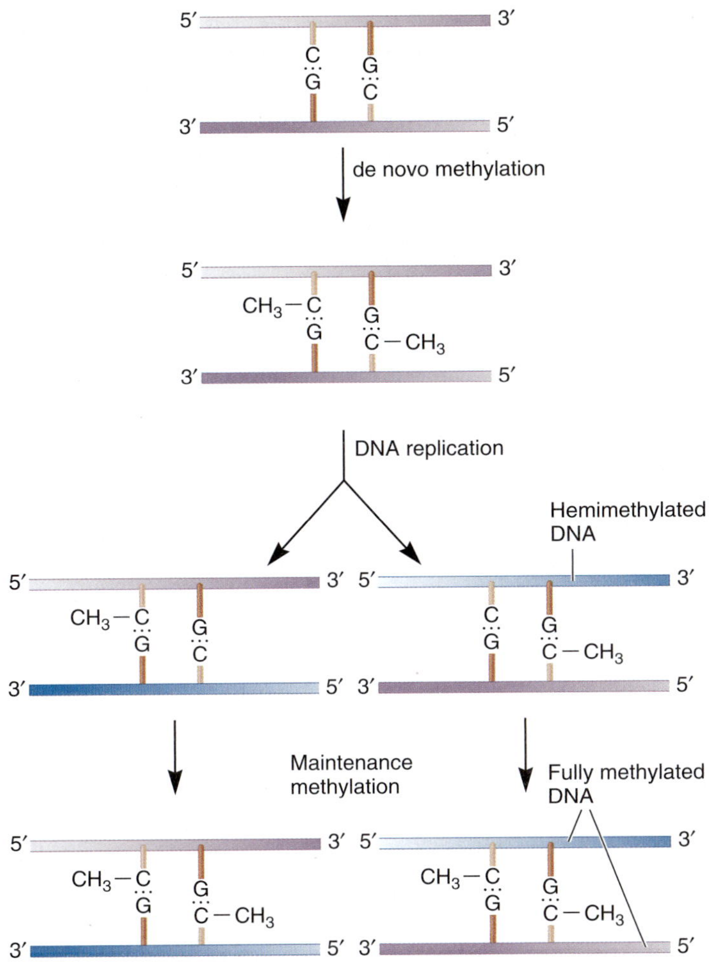

FIGURE 15.16 A molecular model for the inheritance of DNA methylation. The DNA initially undergoes de novo methylation, which is a rare, highly regulated event. Once this occurs, DNA replication produces hemimethylated DNA molecules, which are then fully methylated by DNA methyltransferase. This process, called maintenance methylation, is a routine event that is expected to occur for all hemimethylated DNA.

CONCEPT CHECK: What is the difference between de novo methylation and maintenance methylation?

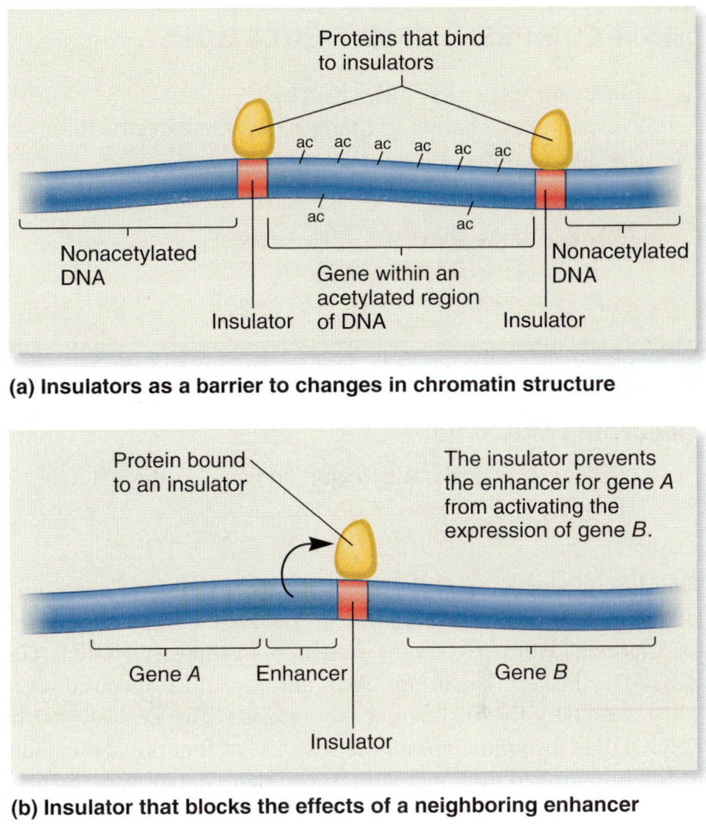

FIGURE 15.17 Potential effects of insulators. **(a)** Some insulators act as a barrier to enzymes, such as histone deacetylase, that move along the DNA and change chromatin structure or composition. The label ac refers to acetyl groups attached to histones. **(b)** Other insulators prevent enhancers of an adjacent gene from exerting their effects.

CONCEPT CHECK: Why are insulators important for gene regulation in eukaryotes?

Insulators typically perform two roles. One role is to act as a barrier to chromatin-remodeling or histone-modifying enzymes. As an example, let's consider the effects of histone deacetylase, which removes acetyl groups from core histone proteins, thereby favoring a closed chromatin conformation that is transcriptionally silent. Histone deacetylase may act over a long region of chromatin. How is the action of histone deacetylase controlled so that not all genes in a chromosomal region are silenced? In the example of Figure 15.17a, a gene is found in a chromosomal region in which most of the histones are not acetylated due to the action of histone deacetylase. However, this gene is flanked by two insulators that allow the region where the gene is located to be highly acetylated and transcriptionally activated. In this example, the insulators act as barriers to the action of histone deacetylase. The mechanisms that enable insulators to act as barriers are not well understood. However, such insulators often bind proteins that recruit histone-modifying enzymes or chromatin-remodeling enzymes to the region. For example, an insulator could bind a protein that recruits histone acetyltransferase, which would favor the acetylation of core histones.

A second role of insulators is to block the effects of enhancers that exert their effects on neighboring genes. As discussed earlier in this chapter, enhancers may stimulate the expression of genes that are relatively far away. To prevent an enhancer of one gene from activating the expression of an adjacent gene, an insulator may be located in between them. In the example shown in Figure 15.17b, the enhancer can activate the expression of gene *A* but the protein bound to the insulator prevents the enhancer from exerting its effects on gene *B*. In other words, gene *B* is insulated from the effects of this enhancer. How does this occur? Although the mechanism of enhancer blocking may vary among different genes, one mechanism is chromosome looping, which is described in Chapter 16 (see Figure 16.4).

15.4 COMPREHENSION QUESTION

1. Insulators may exert their effect by
   a. acting as a barrier to changes in chromatin structure.
   b. blocking the effects of neighboring enhancers.
   c. both a and b.
   d. none of the above.

15.5 THE ENCODE PROJECT

Learning Outcome:

1. Outline the goals and major findings of the ENCODE Project.

The Encyclopedia of DNA Elements (ENCODE) Consortium is an international collaboration of research groups funded by the National Human Genome Research Institute (NHGRI). The ENCODE Project began in 2003 and by 2012 involved over 400 researchers in 30 different laboratories. The central goal of ENCODE is to build a comprehensive list of functional elements in the human genome, including those that control gene expression. Therefore, a major goal of the ENCODE Project is to identify elements that are involved in gene regulation. All data generated from this project are released into public databases.

Investigators employ a variety of strategies and methods to identify functional elements, which include the following:

- Isolate and sequence RNA molecules transcribed from the human genome.
- Identify DNA binding sites for many transcription factors.
- Map DNA methylation sites.
- Identify sites of histone modification.
- Map sites where DNaseI can cleave the DNA. Note: DNaseI usually cleaves DNA at sites where regulatory transcription factors bind, but it doesn't readily cleave DNA that is wrapped around histones.

In 2012, some initial results of the ENCODE Project were published in a set of 30 papers found in the scientific journals *Nature*, *Genome Biology*, and *Genome*. More than 80% of the human genome sequence was linked to a biological function. This result was surprising because scientists had previously thought that most of the human genome was "junk DNA" with no function. In addition, they mapped more than 4 million regulatory regions where proteins specifically interact with the DNA. While it will take time to unravel the roles of these regulatory regions, these results underscore the complexity and precision of gene regulation. The ENCODE Project has revealed that a gene's regulation is far more complex than previously thought, being influenced by multiple regulatory regions both near and far from the gene itself.

Another important finding from the ENCODE Project is related to human disease. Genetic variation contributes to the development of many human diseases, including diabetes, asthma, heart disease, and others. For many such diseases, most DNA sequence changes that are associated with disease do not lie within the core promoter or coding sequences of genes. Such genetic variation is thought to lie within regulatory elements that have yet to be identified and whose function is unknown. As we begin to understand the roles of the 4 million regulatory regions identified by the ENCODE Project, researchers expect that genetic variation at many of these regions will be shown to affect gene expression, thereby contributing to the development of certain types of diseases. Therefore, the results of the ENCODE Project may lead to a better understanding of human diseases and aid in the development of new drugs and therapies to treat those diseases.

15.5 COMPREHENSION QUESTION

1. The overall goal of the ENCODE Project is
   a. to sequence the entire genome from many different people.
   b. to identify all of the functional elements in the human genome.
   c. to study the expression of selected human genes.
   d. none of the above.

KEY TERMS

Page 368. gene regulation
Page 369. transcription factor, general transcription factors, regulatory transcription factors
Page 370. control elements, regulatory elements, regulatory sequences, activator, enhancer, repressors, silencers, combinatorial control, domains, motif
Page 371. homodimer, heterodimer
Page 372. up regulation, down regulation, orientation-independent, bidirectional, TFIID, coactivators, transactivation domain, mediator
Page 374. steroid receptor, glucocorticoid receptors
Page 375. cAMP response element-binding protein (CREB protein), cAMP response element (CRE)
Page 376. chromatin remodeling, closed conformation, open conformation, ATP-dependent chromatin remodeling, DNA translocases
Page 377. histone variants
Page 378. histone acetyltransferases, histone code hypothesis
Page 379. chromatin immunoprecipitation sequencing (ChIP-Seq), immunoprecipitation, nucleosome-free region (NFR)
Page 382. DNA methylation, DNA methyltransferase, CpG islands, housekeeping genes, tissue-specific genes
Page 383. methyl-CpG-binding proteins, de novo methylation, maintenance methylation
Page 384. insulator
Page 388. promoter bashing

CHAPTER SUMMARY

- Gene regulation refers to the phenomenon that the level of gene expression can be controlled so that genes can be expressed at high or low levels. Gene regulation can occur at any step in the pathway of gene expression (see Figure 15.1).

15.1 Regulatory Transcription Factors

- Regulatory transcription factors can be activators that bind to enhancers and increase transcription, or they can be repressors that bind to silencers and inhibit transcription (see Figure 15.2).
- Combinatorial control means that the transcription of a gene is controlled by a variety of factors.
- Transcription factor proteins contain specific domains that may be involved in a variety of processes such as DNA binding and protein dimerization (see Figure 15.3).
- Control or response elements are usually orientation-independent.
- Regulatory transcription factors may exert their effects by interacting with TFIID (a general transcription factor) or mediator (see Figures 15.4, 15.5).
- The functions of regulatory transcription factors can be modulated by small effector molecules, protein-protein interactions, and covalent modifications (see Figure 15.6).
- Steroid hormones, such as glucocorticoids, bind to receptors that function as transcriptional activators (see Figure 15.7).
- The CREB protein activates transcription when it has become phosphorylated by protein kinase A (see Figure 15.8).

15.2 Chromatin Remodeling, Histone Variation, and Histone Modification

- Chromatin remodeling occurs via ATP-dependent chromatin-remodeling complexes that alter the positions and compositions of nucleosomes (see Figure 15.9).
- Histone variants, which have amino acid sequences that differ slightly from the standard histones, play specialized roles in chromatin structure and function (see Table 15.1).
- The amino-terminal tails of histones are subject to covalent modifications that act as a histone code for the binding of proteins that affect chromatin structure and gene expression (see Figure 15.10).
- Chromatin immunoprecipitation sequencing (ChIP-Seq) is a method that can be used to determine the precise locations of nucleosomes throughout an entire genome (see Figure 15.11).
- Many eukaryotic genes are flanked by nucleosome-free regions (NFRs) with well-positioned nucleosomes (see Figure 15.12).
- Gene transcription involves changes in nucleosome positions, composition, and histone modifications (see Figure 15.13).

15.3 DNA Methylation

- DNA methylation is the attachment of a methyl group to a cytosine base via DNA methyltransferase (see Figure 15.14).
- The methylation of CpG islands near promoters usually silences transcription. Methylation may affect the binding of regulatory transcription factors or it may inhibit transcription via methyl-CpG-binding proteins (see Figure 15.15).
- Maintenance methylation can preserve a methylation pattern following cell division (see Figure 15.16).

15.4 Insulators

- An insulator is a segment of DNA that functions as a boundary that protects one gene from the regulatory effects of a neighboring gene. Insulators may act as a barrier to changes in chromatin structure or block the effects of neighboring enhancers (see Figure 15.17).

15.5 The ENCODE Project

- The ENCODE Project is an international collaboration of research groups with a common goal of identifying the functional elements in the human genome.

PROBLEM SETS & INSIGHTS

Solved Problems

S1. Describe how the arrangements and compositions of nucleosomes may result in a closed conformation that prevents gene transcription.

Answer: Nucleosome composition and arrangements are important for transcriptional initiation and elongation. The locations of nucleosomes may physically prevent general transcription factors or RNA polymerase from binding to the DNA. Also, nucleosome composition and certain types of covalent modifications may prevent RNA polymerase from forming an open complex, which is necessary to begin transcription. These types of changes are also important for transcription to proceed to the elongation phase.

S2. Eukaryotic regulatory elements are often orientation-independent and can function in a variety of locations. Explain the meaning of this statement.

Answer: Orientation independence means that the regulatory element can function in the forward or reverse direction. In addition, regulatory elements can function at a variety of locations that may be upstream or downstream from the core promoter. A loop in the DNA must form to bring the regulatory transcription factors bound at the regulatory elements and the core promoter in close proximity with one another.

S3. To gain a molecular understanding of how the glucocorticoid receptor works, geneticists have attempted to "dissect" the protein to identify smaller domains that play specific functional roles.

Using recombinant DNA techniques described in Chapter 20, particular segments in the coding region of the glucocorticoid receptor gene can be removed. The altered gene can then be expressed in a living cell to see if functional aspects of the receptor have been changed or lost. For example, the removal of the portion of the gene encoding the carboxyl-terminal half of the protein causes a loss of glucocorticoid binding. These results indicate that the carboxyl-terminal portion contains a domain that functions as a glucocorticoid-binding site. The following figure illustrates the locations of several functional domains within the glucocorticoid receptor relative to the entire amino acid sequence.

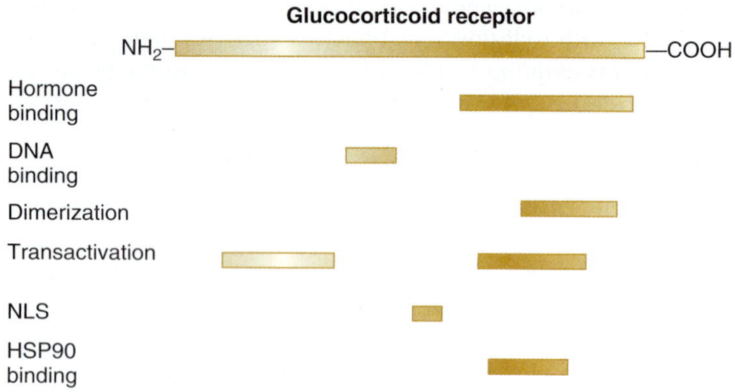

Based on your understanding of the mechanism of glucocorticoid receptor function described in Figure 15.7, explain the functional roles of the different domains in the glucocorticoid receptor.

Answer: The hormone-binding domain is located in the carboxyl-terminal half of the protein. This part of the protein also contains the region necessary for HSP90 binding and receptor dimerization. A nuclear localization sequence (NLS) is located near the center of the protein. After hormone binding, the NLS is exposed on the surface of the protein and allows it to be targeted to the nucleus. The DNA-binding domain, which contains zinc fingers, is also centrally located in the primary amino acid sequence. Zinc fingers promote DNA binding to the major groove. Finally, two separate regions of the protein, one in the amino-terminal half and one in the carboxyl-terminal half, are necessary for the transactivation of RNA polymerase. If these domains are removed from the receptor, it can still bind to the DNA, but it cannot activate transcription.

S4. A common approach to identify genetic sequences that play a role in the transcriptional regulation of a gene is the strategy sometimes called **promoter bashing.** This approach requires gene cloning methods, which are described in Chapter 20. A clone is obtained that has the coding region for a protein-encoding gene as well as the region that is upstream from the core promoter. This upstream region is likely to contain genetic regulatory elements such as enhancers and silencers. The diagram shown here depicts a cloned DNA region that contains the upstream region, the core promoter, and the coding sequence for a protein that is expressed in human liver cells. The upstream region may be several thousand base pairs in length.

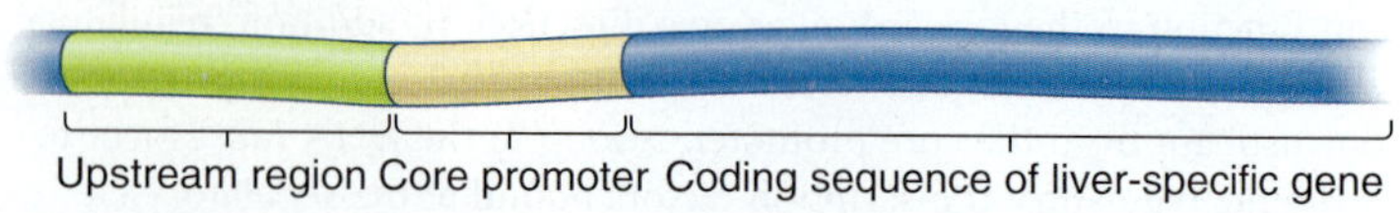

To determine if promoter bashing has an effect on transcription, it is helpful to have an easy way to measure the level of gene expression. One way to accomplish this is to swap the coding sequence of the gene of interest with the coding sequence of another gene. For example, the coding sequence of the *lacZ* gene, which encodes β-galactosidase, is frequently swapped because it is easy to measure the activity of β-galactosidase using an assay for its enzymatic activity. The *lacZ* gene is called a "reporter gene" because its activity is easy to measure. As shown here, the coding sequence of the *lacZ* gene has been swapped with the coding sequence of the liver-specific gene. In this new genetic construct, the expression and transcriptional regulation of the *lacZ* gene are under the control of the core promoter and upstream region of the liver-specific gene.

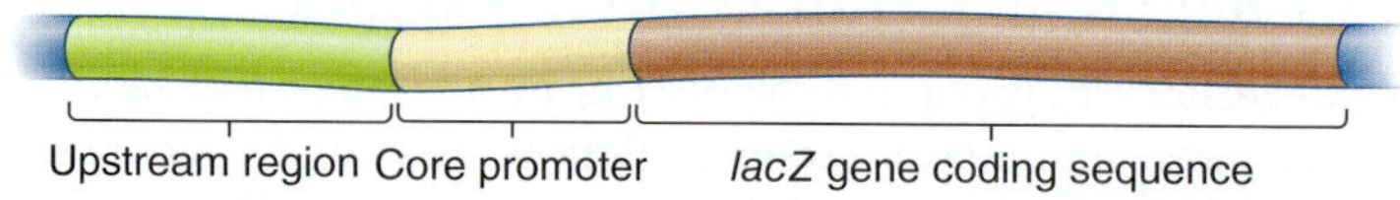

Now comes the "bashing" part of the experiment. Different segments of the upstream region are deleted, and then the DNA is transformed into living cells. In this case, the researcher would probably transform the DNA into liver cells, because those are the cells where the gene is normally expressed. The last step is to measure the β-galactosidase activity in the transformed liver cells.

In the diagram shown here, the upstream region and the core promoter have been divided into five regions, labeled A–E.

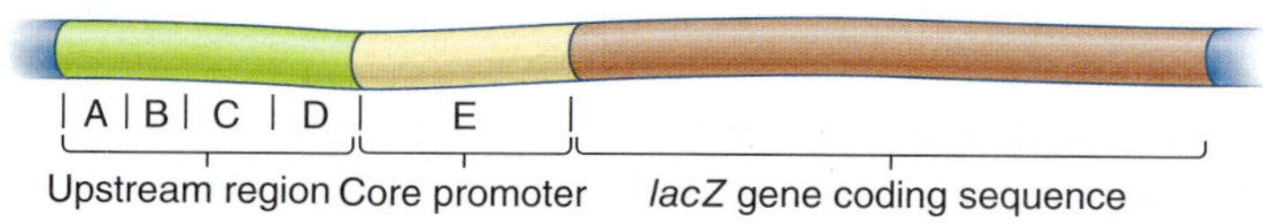

One of these regions was deleted (i.e., bashed out), and the rest of the DNA segment was transformed into liver cells. The data shown here are from this experiment.

Region Deleted	Percentage of β-Galactosidase Activity*
None	100
A	100
B	330
C	100
D	5
E	<1

*The amount of β-galactosidase activity in the cells carrying an undeleted upstream and promoter region was assigned a value of 100%. The amounts of activity in the cells carrying a deletion were expressed relative to this 100% value.

Explain what these results mean.

Answer: The amount of β-galactosidase activity found in liver cells that do not carry a deletion reflects the amount of expression under normal circumstances. If the core promoter (region E) is deleted, very little expression is observed. This is expected because a core promoter is needed for transcription. If enhancers are deleted, the activity should be less than 100%. It appears that one or more enhancers are found in region D. If a silencer is deleted, the activity should be above 100%. From the data shown, it appears that one or more silencers are found in region B. Finally, if a deletion has no effect, the region may not contain any regulatory elements. This was observed for regions A and C.

Note: The deletion of an enhancer has an effect on β-galactosidase activity only if the cell is expressing the regulatory transcription factor that binds to the enhancer and activates transcription. Likewise, the deletion of a silencer has an effect only if the cell is expressing the repressor protein that binds to the silencer and inhibits transcription. In the problem described, the liver cells must be expressing the activator(s) and repressor(s) that recognize the regulatory elements found in regions D and B, respectively.

Conceptual Questions

C1. Discuss the common points of control in eukaryotic gene regulation.

C2. Discuss the structure and function of regulatory elements. Where are they located relative to the core promoter?

C3. What is meant by the term transcription factor modulation? List three general ways that this can occur.

C4. What are the functions of transcriptional activator proteins and repressor proteins? Explain how they work at the molecular level.

C5. Are the following statements true or false?

A. An enhancer is a type of regulatory element.

B. A core promoter is a type of regulatory element.

C. Regulatory transcription factors bind to regulatory elements.

D. An enhancer may cause the down regulation of transcription.

C6. Transcription factors usually contain one or more motifs that play key roles in their function. What is the function of the following motifs?

A. Helix-turn-helix

B. Zinc finger

C. Leucine zipper

C7. The binding of an effector molecule, protein-protein interactions, and covalent modifications are three common ways to modulate the activities of transcription factors. Which of these three mechanisms are used by steroid receptors and the CREB protein?

C8. Describe the steps that occur for the glucocorticoid receptor to bind to a GRE.

C9. Let's suppose a mutation in the glucocorticoid receptor does not prevent the binding of the glucocorticoid hormone to the protein but prevents the ability of the receptor to activate transcription. Make a list of all the possible defects that may explain why transcription cannot be activated.

C10. Explain how phosphorylation affects the function of the CREB protein.

C11. A particular drug inhibits the protein kinase that is responsible for phosphorylating the CREB protein. How would this drug affect the following events?

A. The ability of the CREB protein to bind to CREs

B. The ability of extracellular hormones to enhance cAMP levels

C. The ability of the CREB protein to stimulate transcription

D. The ability of the CREB protein to dimerize

C12. The glucocorticoid receptor and the CREB protein are two examples of transcriptional activators. These proteins bind to response elements and activate transcription. (Note: The answers to this question are not directly described in this chapter. You have to rely on your understanding of the functioning of other proteins that are modulated by the binding of effector molecules, such as the lac repressor.)

A. How would the function of the glucocorticoid receptor be shut off?

B. What type of enzyme would be needed to shut off the activation of transcription by the CREB protein?

C13. Transcription factors such as the glucocorticoid receptor and the CREB protein form homodimers and activate transcription. Other transcription factors form heterodimers. For example, a transcription factor known as myogenic bHLH forms a heterodimer with a protein called the E protein. This heterodimer activates the transcription of genes that promote muscle cell differentiation. However, when myogenic bHLH forms a heterodimer with a protein called the Id protein, transcriptional activation does not occur. (Note: Id stands for Inhibitor of differentiation.) Which of the following possibilities best explains this observation? Only one possibility is correct.

	Myogenic bHLH	E Protein	Id Protein
Possibility 1			
DNA-binding domain:	Yes	No	No
Leucine zipper:	Yes	No	Yes
Possibility 2			
DNA-binding domain:	Yes	Yes	No
Leucine zipper:	Yes	Yes	Yes
Possibility 3			
DNA-binding domain:	Yes	No	Yes
Leucine zipper:	Yes	No	No

C14. An enhancer, located upstream from a gene, has the following sequence:

5′–GTAG–3′
3′–CATC–5′

This enhancer is orientation-independent. Which of the following sequences also works as an enhancer?

A. 5′–CTAC–3′
3′–GATG–5′

B. 5′–GATG–3′
3′–CTAC–5′

C. 5′–CATC–3′
3′–GTAG–5′

C15. The DNA-binding domain of each CREB protein subunit recognizes the sequence 5′–TGACGTCA–3′. As a matter of random chance, how often would you expect this sequence to occur in the human genome, which contains approximately 3 billion base pairs? Actually, only a few dozen genes are activated by the CREB protein. Does the value of a few dozen agree with the number of random sites found in the human genome? If the number of random sites in the human genome is much higher than a few dozen, provide at least one explanation why the CREB protein is not activating more than a few dozen genes.

C16. Solved problem S3 shows the locations of domains in the glucocorticoid receptor relative to the amino- and carboxyl-terminal ends of the protein. Make a drawing that illustrates the binding of a glucocorticoid receptor dimer to the DNA. In your drawing, label the amino and carboxyl ends, the hormone-binding domains, the DNA-binding domains, the dimerization domains, and the transactivation domains.

C17. The gene that encodes the enzyme called tyrosine hydroxylase is known to be up regulated by the CREB protein. Tyrosine hydroxylase is expressed in nerve cells and is involved in the synthesis of catecholamine, a neurotransmitter. The exposure of cells to adrenaline normally up regulates the transcription of the tyrosine hydroxylase gene. A mutant cell line was identified in which the tyrosine hydroxylase gene was not up regulated when exposed to adrenaline. List all the possible mutations that could explain this defect. How would you explain the defect if only the tyrosine hydroxylase gene was not up regulated by the CREB protein, whereas other genes having CREs were properly up regulated in response to adrenaline in this cell line?

C18. Briefly describe three ways that ATP-dependent chromatin-remodeling complexes may change chromatin structure.

C19. What is a histone variant?

C20. Explain how the acetylation of core histones may loosen chromatin packing.

C21. What is meant by the term histone code? With regard to gene regulation, what is the proposed role of the histone code?

C22. What is a nucleosome-free region? Where are they typically found in a genome? How are nucleosome-free regions thought to be functionally important?

C23. Histones are thought to be displaced as RNA polymerase is transcribing a gene. What would be the potentially harmful consequences if histones were not put back into a gene after RNA polymerase had passed?

C24. What is an insulator? Describe two different ways that insulators may exert their effects.

C25. What is DNA methylation? When we say that DNA methylation is heritable, what do we mean? How is it passed from a mother to a daughter cell?

C26. Let's suppose that a vertebrate organism carries a mutation that causes some cells that normally differentiate into nerve cells to differentiate into muscle cells. A molecular analysis of this mutation revealed that it was in a gene that encodes a methyltransferase. Explain how an alteration in a methyltransferase could produce this phenotype.

C27. What is a CpG island? Where would you expect one to be located? How does the methylation of CpG islands affect gene expression?

Experimental Questions

E1. Briefly describe the method of chromatin immunoprecipitation sequencing (ChIP-Seq). How is it used to determine nucleosome positions within a genome?

E2. Researchers can isolate a sample of cells, such as skin fibroblasts, and grow them in the laboratory. This procedure is called a cell culture. A cell culture can be exposed to a sample of DNA. If the cells are treated with agents that make their membranes permeable to DNA, the cells may take up the DNA and incorporate the DNA into their chromosomes. This process is called transformation or transfection. Scientists have transformed human skin fibroblasts with methylated DNA and then allowed the fibroblasts to divide for several cellular generations. The DNA in the daughter cells was then isolated, and the segment that corresponded to the transformed DNA was examined. This DNA segment in the daughter cells was also found to be methylated. However, if the original skin fibroblasts were transformed with unmethylated DNA, the DNA found in the daughter cells was also unmethylated. With regard to the transformed DNA, do fibroblasts perform de novo methylation, maintenance methylation, or both? Explain your answer.

E3. Restriction enzymes, described in Chapter 20, are enzymes that recognize a particular DNA sequence and cleave the DNA (along the DNA backbone) at that site. The restriction enzyme known as *Not*I recognizes the sequence

5′–GCGGCCGC–3′
3′–CGCCGGCG–5′

However, if the cytosines in this sequence have been methylated, *Not*I will not cleave the DNA at this site. For this reason, *Not*I is commonly used to investigate the methylation state of CpG islands.

A researcher has studied a gene, which we will call gene *T*, that is found in corn. This gene encodes a transporter involved in the uptake of phosphate from the soil. A CpG island is located near the core promoter of gene *T*. The CpG island has a single *Not*I site. The arrangement of gene *T* is shown here.

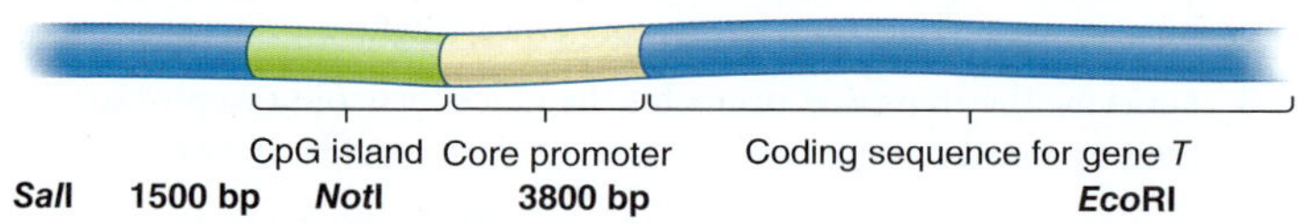

A *Sal*I restriction site is located upstream from the CpG island, and an *Eco*RI restriction site is located near the end of the coding sequence for gene *T*. The distance between the *Sal*I and *Not*I sites is 1500 bp, and the distance between the *Not*I and *Eco*RI sites is 3800 bp. No other sites for *Sal*I, *Not*I, or *Eco*RI are found in this region.

Here is the question. Let's suppose a researcher has isolated DNA samples from four different tissues in a corn plant. These include the leaf, the tassel, a section of stem, and a section of root. The DNA was then digested with all three restriction enzymes, separated by gel electrophoresis, and then probed with a DNA fragment complementary to the gene *T* coding sequence. The results are shown here.

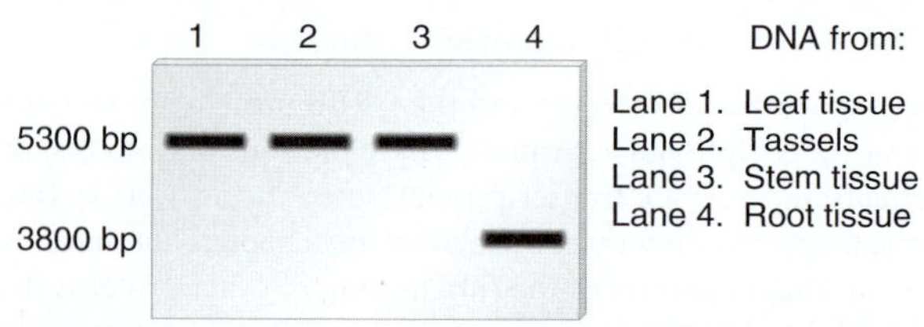

In which type of tissue is the CpG island methylated? Does this make sense based on the function of the protein encoded by gene *T*?

E4. You will need to understand solved problem S4 before answering this question. A muscle-specific gene was cloned and then subjected to promoter bashing. As shown here, six regions, labeled A–F, were deleted, and then the DNA was transformed into muscle cells.

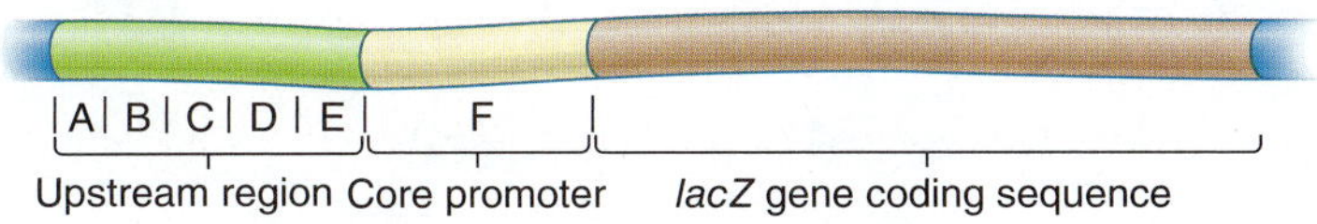

The data shown here are from this experiment.

Region Deleted	Percentage of β-Galactosidase Activity
None	100
A	20
B	330
C	100
D	5
E	15
F	<1

Explain these results.

E5. You will need to understand solved problem S4 before answering this question. A gene that is normally expressed in pancreatic cells was cloned and then subjected to promoter bashing. As shown here, four regions, labeled A–D, were individually deleted, and then the DNA was transformed into pancreatic cells or into kidney cells.

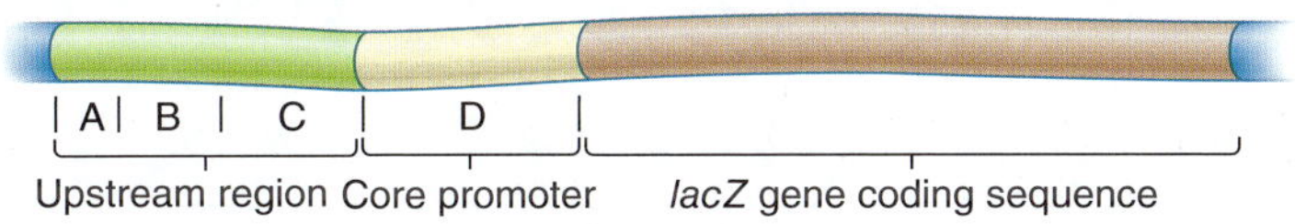

The data shown in the following table are from this experiment (see upper right column).

Region Deleted	Cell Type Transformed	Percentage of β-Galactosidase Activity
None	Pancreatic	100
A	Pancreatic	5
B	Pancreatic	100
C	Pancreatic	100
D	Pancreatic	<1
None	Kidney	<1
A	Kidney	<1
B	Kidney	100
C	Kidney	<1
D	Kidney	<1

If we assume that the upstream region has one silencer and one enhancer, answer the following questions:

A. Where are the silencer and enhancer located?

B. Why don't we detect the presence of the silencer in the pancreatic cells?

C. Why isn't this gene normally expressed in kidney cells?

E6. A gel retardation assay can be used to determine if a protein binds to a segment of DNA. When a segment of DNA is bound by a protein, its mobility will be retarded, and the DNA band will appear higher in the gel. In the gel retardation assay shown here, a cloned gene fragment that is 750 bp in length contains a regulatory element that is recognized by a transcription factor called protein X. Previous experiments have shown that the presence of hormone X results in transcriptional activation by protein X. The results of a gel retardation assay are shown here.

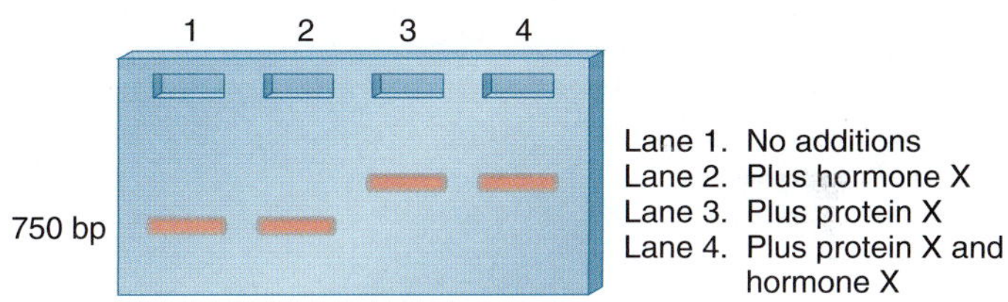

Explain the action of hormone X.

Questions for Student Discussion/Collaboration

1. Explain how DNA methylation could be used to regulate gene expression in a tissue-specific way. When and where would de novo methylation occur, and when would demethylation occur? What would occur in the germ line?

2. Enhancers can be almost anywhere and affect the transcription of a gene. Let's suppose you have a gene cloned on a piece of DNA, and the DNA fragment is 50,000 bp in length. Using cloning methods described in Chapter 20, you can cut out short segments from this 50,000-bp fragment and then reintroduce the smaller fragments into a cell that can express the gene. You would like to know if any enhancers are within the 50,000-bp region that may affect the expression of the gene. Discuss the most efficient strategy you can think of to trim your 50,000-bp fragment, thereby locating enhancers. You can assume that the coding sequence of the gene is in the center of the 50,000-bp fragment and that you can trim the 50,000-bp fragment into any size piece you want using molecular techniques described in Chapter 20.

Note: All answers appear at the website for this textbook; the answers to the even-numbered questions and all of the Concept Check and Comprehension Questions are in the back of the book.

www.mhhe.com/brookergenetics5e

Visit the website for practice tests, answer keys, and other learning aids for this chapter. Enhance your understanding of genetics with our interactive exercises, quizzes, animations, and much more.

CHAPTER OUTLINE

Epigenetics and the environment. *Female honeybees that are fed royal jelly throughout the entire larval stage and into adulthood develop into queen bees. The larger queen bee is shown with a blue disk labeled 68. By comparison, those larvae that do not receive this diet become smaller worker bees. These differences in development are caused by epigenetic modifications.*

16 GENE REGULATION IN EUKARYOTES II: EPIGENETICS AND REGULATION AT THE RNA LEVEL

In Chapter 15, we began our discussion of gene regulation in eukaryotes with an examination of regulatory transcription factors. We also explored mechanisms that alter chromatin or DNA structure and thereby affect gene expression. These mechanisms included chromatin remodeling, histone variants, the covalent modification of histones, and DNA methylation. In this chapter, we will continue our discussion of eukaryotic gene regulation by examining epigenetics at the molecular level, currently one of the hottest topics in molecular genetics. Epigenetics is a field of genetics that explores changes in gene expression that may be permanent over the course of an individual's life but are not permanent over the course of multiple generations. Epigenetic changes are responsible for the establishment and maintenance of gene activation or repression. Such changes enable cells to "remember" past events, such as developmental alterations in embryonic cells or prior exposure to environmental agents. In Chapter 5, we considered two patterns of inheritance—X-chromosome inactivation and genomic imprinting—which are explained by epigenetic gene regulation.

In this chapter, we will begin with an overview of the types of epigenetic modifications that affect gene expression. We will then examine how some epigenetic modifications are programmed to occur during development, and others are the result of environmental factors. The remaining part of this chapter is concerned with regulation that occurs after transcription. We will examine some well-studied examples in which the expression of mRNAs are regulated after they are made.

16.1 OVERVIEW OF EPIGENETICS

Learning Outcomes:

1. Define epigenetics and epigenetics inheritance.
2. Outline the types of molecular changes that underlie epigenetic gene regulation.

3. Distinguish *cis*- and *trans*-mechanisms that maintain epigenetic changes.
4. Compare and contrast epigenetic changes that are programmed during development versus those that are caused by environmental agents.

The term epigenetics was first coined by Conrad Waddington in 1941. The prefix epi- means "over." In the past few decades, researchers have used this term to describe certain types of variation in gene expression that are not based on variation in DNA sequences. How do geneticists distinguish epigenetic events from other types of gene regulation, such as those described in Chapters 14 and 15? In epigenetic gene regulation, an initial event causes a change in gene expression. For example, DNA methylation may inhibit transcription. For this to be an epigenetic phenomenon, this change must be passed from cell to cell and does not involve a change in the sequence of DNA. Therefore, a key feature of epigenetic gene regulation is the long-term maintenance of a change in gene expression. However, epigenetic changes are also reversible from one generation to the next. For example, a gene that is silenced in one individual may be active in the offspring from that individual. Although researchers are still debating the proper definition, one way to define **epigenetics** is the study of mechanisms that lead to changes in gene expression that can be passed from cell to cell and are reversible, but do not involve a change in the sequence of DNA.

With regard to transmission, epigenetic changes may be inherited from cell to cell, and some are passed from parent to offspring. In multicellular species that reproduce via gametes (i.e., sperm and egg cells), an epigenetic change that is passed from parent to offspring is called **epigenetic inheritance,** or **transgenerational epigenetic inheritance.** For example, as we learned in Chapter 5, genomic imprinting is an epigenetic change that is passed from parent to offspring. However, not all epigenetic changes fall into this category. For example, an organism may be exposed to an environmental agent that causes an epigenetic change in a lung cell that is subsequently transmitted from cell to cell and promotes lung cancer. Such a change would not be transmitted to offspring.

In this section, we will begin with an examination of the molecular changes that cause epigenetic gene regulation. We will then consider how such changes may be programmed into an organism's development or caused by environmental agents.

Different Types of Molecular Changes Underlie Epigenetic Gene Regulation

The molecular mechanisms that promote epigenetic gene regulation are the subject of a large amount of recent research. The most common types of molecular changes that underlie epigenetic control are **DNA methylation, chromatin remodeling, covalent histone modification,** the localization of **histone variants,** and **feedback loops** (Table 16.1). These types of changes can also be involved in transient (nonepigenetic) gene regulation. The details of the first four mechanisms were examined in Chapter 15. In Sections 16.2 and 16.3, we will explore specific examples in which epigenetic gene regulation occurs by these five mechanisms. In some cases, epigenetic changes stimulate the transcription of a given gene and in other cases, they repress gene transcription.

Epigenetic Changes May Be Targeted to Specific Genes by Transcription Factors or Noncoding RNAs

How are specific genes or chromosomes targeted for the types of epigenetic changes described in Table 16.1? The answer to this question is not well understood, but researchers are beginning to uncover a few types of mechanisms. In some cases, transcription factors may bind to a specific gene and initiate a series of events that leads to an epigenetic modification. For example, particular transcription factors in stem cells initiate epigenetic modifications that cause the cells to follow a specific pathway of development. For this to occur, the transcription factors recognize specific sites in the genome and recruit proteins to those sites, such as histone-modifying enzymes and DNA methyltransferase. This recruitment leads to epigenetic changes, such as changes in chromatin structure and DNA methylation. A simplified illustration of this process is shown in **Figure 16.1a**.

In other cases, **noncoding RNAs**—RNAs that do not encode proteins—are involved in establishing an epigenetic modification. In plants, for example, noncoding RNAs can promote DNA methylation at specific sites. This RNA-dependent DNA methylation has been shown to regulate the expression of specific genes. The noncoding RNAs are thought to act as bridges between specific sites in the DNA and proteins that alter chromatin or DNA structure, such as histone-modifying enzymes and DNA methyltransferase (**Figure 16.1b**).

TABLE 16.1
Molecular Mechanisms That Underlie Epigenetic Gene Regulation

Type of Modification	Description
DNA methylation	Methyl groups may be attached to cytosine bases in DNA. When this occurs near promoters, transcription is often inhibited.
Chromatin remodeling	Nucleosomes may be moved to new locations or evicted. When such changes occur in the vicinity of promoters, the level of transcription may be altered. Also, larger scale changes in chromatin structure may occur such as those that happen during X-chromosome inactivation in female mammals.
Covalent histone modification	Specific amino acid side chains found in the amino terminal tails of histones can be covalently modified. For example, they can be acetylated or phosphorylated. Such modifications may enhance or inhibit transcription.
Localization of histone variants	Histone variants may become localized to specific locations, such as near the promoters of genes, and affect transcription.
Feedback loop	The activation of a gene that encodes a transcription factor may result in a feedback loop in which that transcription factor continues to stimulate its own expression.

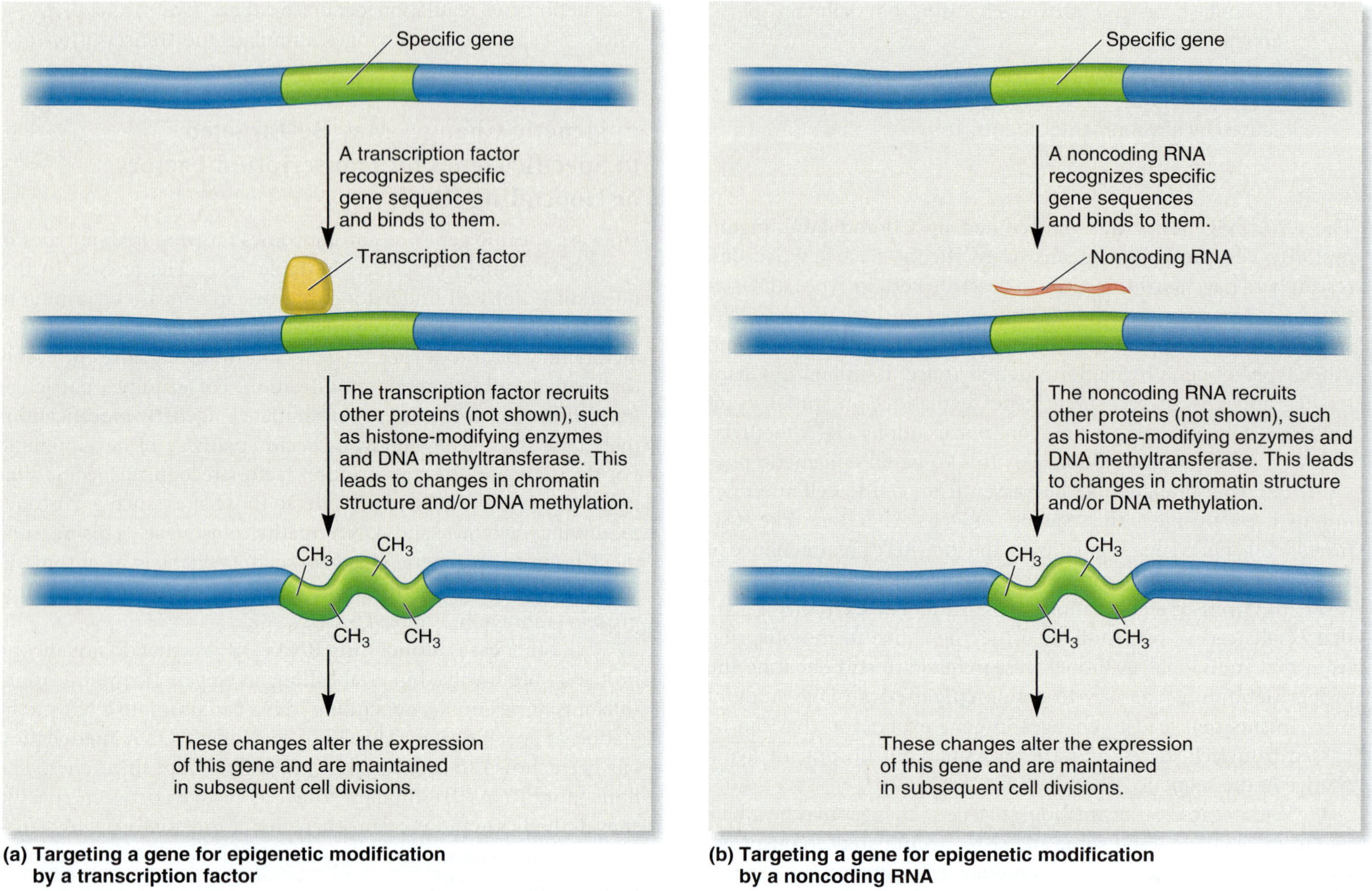

FIGURE 16.1 Establishing epigenetic modifications. Two common ways are **(a)** via transcription factors and **(b)** via noncoding RNA.

Epigenetic Changes May Be Maintained by *Cis*- or *Trans*-Epigenetic Mechanisms

By studying epigenetic inheritance from parents to offspring and by conducting cell fusion experiments, researchers have discovered that the types of epigenetic changes described in Table 16.1 can be maintained in two general ways, called ***cis*- and *trans*-epigenetic mechanisms.** Researchers can distinguish between these two mechanisms by studying the epigenetic modification of a specific gene that occurs in multiple copies within a given cell (**Figure 16.2**). In a *cis*-epigenetic mechanism, the epigenetic change at a given site is maintained only at that site; it does not affect the expression of the same gene located elsewhere in the cell nucleus. For example, if one copy of gene *B* is modified by DNA methylation and the other copy is not, a *cis*-epigenetic mechanism will maintain this pattern from one cell division to the next (see Figure 16.2). Genomic imprinting and X-chromosome inactivation, which are discussed in Chapter 5, are examples of *cis*-epigenetic mechanisms.

By comparison, some epigenetic phenomena are explained by *trans*-epigenetic mechanisms that are maintained by diffusible proteins, an example of which is a feedback loop. In this mechanism, an epigenetic change is established by activating a gene that encodes a transcription factor. After the transcription factor is initially made, it stimulates its own expression. Furthermore, if the gene that encodes this transcription factor is present in two copies in the cell, both copies will be activated because the transcription factor is a diffusible protein and many of these proteins are made when the gene is expressed. This pattern in which both copies are activated will be maintained during cell division. The transcription factor may also turn on other genes in the cell that encode proteins that affect cell structure and function. In this way, a *trans*-epigenetic mechanism may have a phenotypic effect. *Trans*-epigenetic mechanisms are more commonly found in prokaryotes and single-celled eukaryotes.

Experimentally, researchers can distinguish between *cis*- and *trans*-epigenetic mechanisms by conducting cell fusion experiments. In the example shown in **Figure 16.3**, one cell has gene *B* epigenetically modified so it is transcriptionally activated, whereas this same gene in another cell is not. When the cells are fused, two different outcomes are possible. According to a *cis*-epigenetic mechanism, the epigenetic modification will be maintained only for the copy of gene *B* that was originally modified. The other copy will remain inactive. This pattern will be maintained following cell division. By comparison, if a *trans*-epigenetic mechanism occurs, both copies of gene *B* will be expressed because the fused

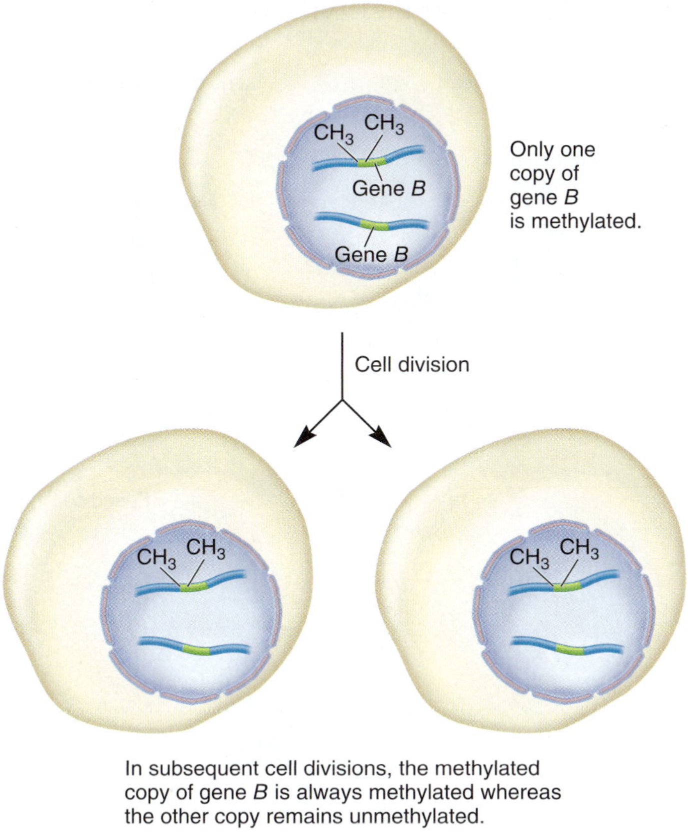

FIGURE 16.2 Pattern of transmission of a *cis*-epigenetic mechanism that maintains an epigenetic modification.

CONCEPT CHECK: Explain how DNA methylation could follow a *cis*-epigenetic mechanism of transmission.

cell will contain enough of the transcription factor protein to stimulate both copies. This pattern will also be maintained following cell division.

Epigenetic Gene Regulation May Occur as a Programmed Developmental Change or Be Caused by Environmental Agents

Many epigenetic modifications that regulate gene expression are programmed changes that occur at specific stages of development (**Table 16.2**). For example, in Chapter 5, we examined genomic imprinting and X-chromosome inactivation. Genomic imprinting of the *Igf2* gene occurs during gametogenesis—the maternal allele is silenced, whereas the paternal allele is active. By comparison, X-chromosome inactivation occurs during embryogenesis in female mammals. In early embryonic cells, one of the X chromosomes of a female is inactivated and forms a Barr body, whereas the other remains active. This pattern is maintained as the cells divide and eventually form an adult organism. Similarly, the differentiation of specific cell types, such as muscle cells and neurons, involves epigenetic modifications. During embryonic development, certain genes undergo epigenetic changes that affect their expression throughout the rest of development. For example, in an embryonic cell that is destined to give rise to muscle cells, a large number of genes that should not be expressed in muscle cells undergo epigenetic modifications that prevent their expression; such changes persist through adulthood.

As described in Table 16.2, a particularly exciting discovery in the field of epigenetics is that a wide range of environmental agents have epigenetic effects, although researchers are often uncertain whether such environmental agents are responsible for altering phenotype. Even so, many recent studies have suggested that environmentally induced changes in an organism's characteristics are rooted in epigenetic changes that alter gene expression. For example, several studies have indicated that temperature changes have epigenetic effects. In certain species of flowering plants, a process known as vernalization occurs in which flowering or germination in the spring is caused by colder temperatures during the previous winter. Researchers studying this process in *Arabidopsis* have discovered that vernalization involves covalent histone modifications of specific genes, which persist from winter to spring. A second environmental factor that can have an epigenetic effect is diet. A striking example is found in honey bees (*Apis mellifora*). Female bees have two alternative body types—queen bees and worker bees (see chapter-opening photo). Their distinct body types are caused by dietary differences. Only larvae that are persistently fed royal jelly develop into queens. Researchers have determined that patterns of DNA methylation are quite different between queen and worker bees and that the methylation affects the expression of many genes. A third factor that is of great interest to many geneticists is how environmental toxins can cause epigenetic changes. In humans, exposure to tobacco smoke has been shown to alter DNA methylation and covalent histone modifications of specific genes in lung cells. As discussed in Section 16.3, such changes may play a role in the development of cancer.

16.1 COMPREHENSION QUESTIONS

1. Which of the following are examples of epigenetic changes that alter gene expression?
   a. Chromatin remodeling
   b. Covalent histone modification
   c. Localization of histone variants
   d. DNA methylation
   e. Feedback loops
   f. All of the above
2. An epigenetic modification to a specific gene may be initially established by
   a. a transcription factor.
   b. a noncoding RNA.
   c. both a and b.
   d. none of the above.
3. In one cell, gene *C* is expressed, whereas in another cell, gene *C* is inactive. After the cells are fused experimentally, both copies of gene *C* are expressed. This observation could be explained by
   a. a *cis*-epigenetic mechanism.
   b. a *trans*-epigenetic mechanism.
   c. DNA methylation.
   d. both a and b.
4. Epigenetic changes may
   a. be programmed during development.
   b. be caused by environmental changes.
   c. involve changes in the DNA sequence of a gene.
   d. be both a and b.

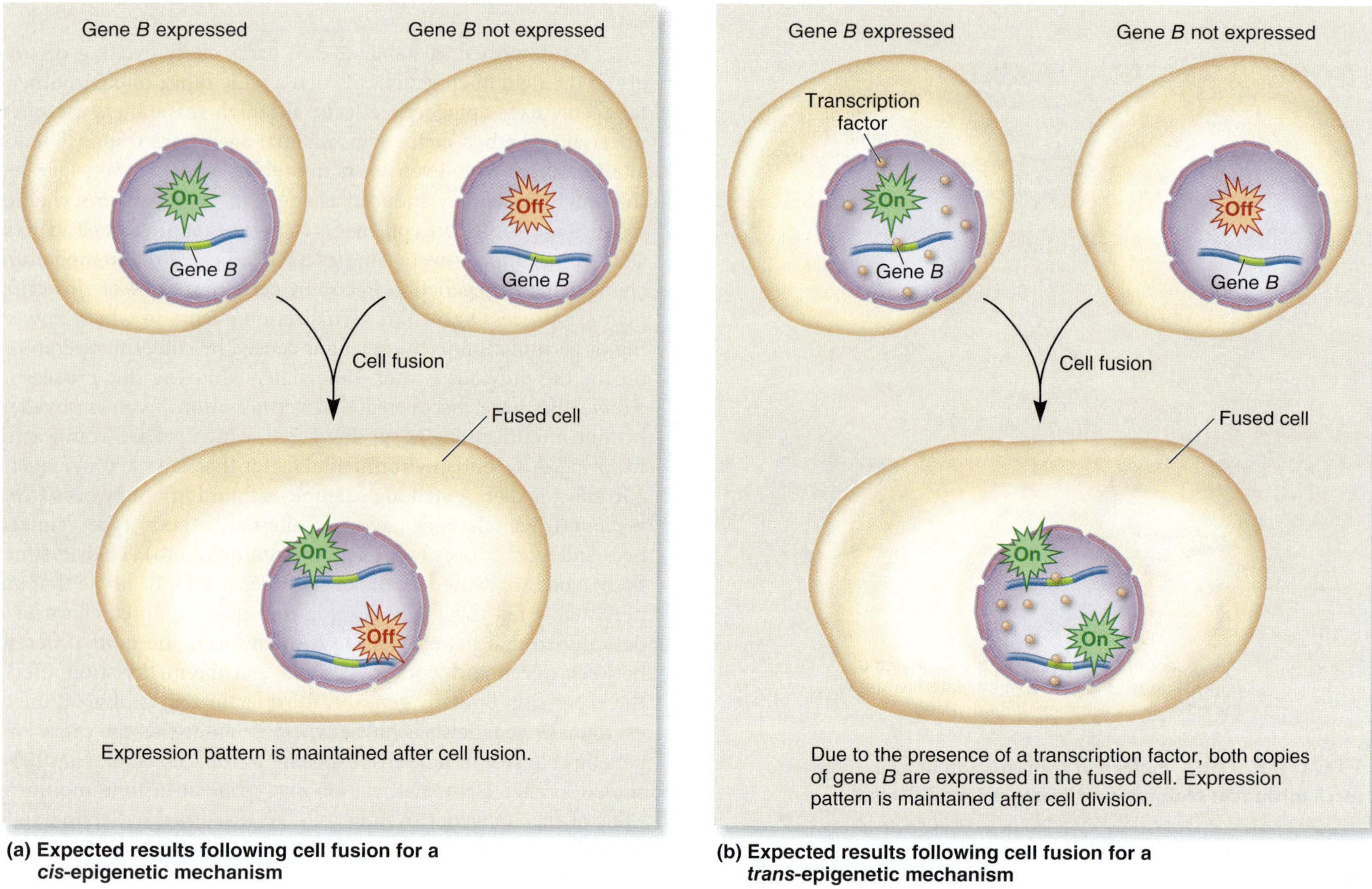

FIGURE 16.3 The use of cell-fusion experiments to distinguish *cis*- and *trans*-epigenetic mechanisms. This simplified example shows only one gene from each cell. Most eukaryotic cells are diploid, so two copies of a gene would be found in each cell prior to fusion. **(a)** In a *cis*-epigenetic mechanism, the genes in the fused cell retain their original pattern of expression. **(b)** In a *trans*-epigenetic mechanism, diffusible transcription factors, shown as yellow balls, in this figure stimulate the expression of both genes.

CONCEPT CHECK: Which of these patterns applies to the imprinting of the *Igf2* gene, described in Chapter 5?

TABLE 16.2

Factors That Promote Epigenetic Changes

Factor	Examples
Programmed Changes During Development	
Genomic imprinting	Certain genes, such as *Igf2* described in Chapter 5, undergo different patterns of DNA methylation during oogenesis and spermatogenesis. Such patterns affect whether the maternal or paternal allele is expressed in offspring.
X-chromosome inactivation	As described in Chapter 5, X-chromosome inactivation occurs during embryogenesis in female mammals.
Cell differentiation	The differentiation of cells into particular cell types involves epigenetic changes such as DNA methylation and covalent histone modification.
Environmental Agents	
Temperature	In flowering plants, cold winter temperatures cause specific types of covalent histone modifications that are thought to affect the expression of specific genes the following spring. This process may be necessary for germination and flowering in the spring.
Diet	The different diets of queen and worker bees alter DNA methylation patterns, which affect the expression of many genes. Such effects may underlie the different body types of queen and worker bees.
Toxins	Cigarette smoke contains a variety of toxins that affect DNA methylation and covalent histone modifications in lung cells. These epigenetic changes may play a role in the development of lung cancer. In addition, metals, such as chromium and cadmium, and certain chemicals found in pesticides and herbicides, cause epigenetic changes that can affect gene expression.

16.2 EPIGENETICS AND DEVELOPMENT

Learning Outcomes:

1. Describe the mechanism of genomic imprinting of the *Igf2* gene in mammals.
2. Outline the process of X-chromosome inactivation.
3. Explain how epigenetic modifications are involved in developmental changes that lead to the formation of specific cell types.

Beginning with gametes—sperm and egg cells—**development** in multicellular species involves a series of genetically programmed stages in which a fertilized egg becomes an embryo and eventually develops into an adult. These stages are discussed in Chapter 25. Over the past few decades, researchers have determined that epigenetic changes play key roles in the process of development in animals and plants. At the molecular level, cells in the adult are able to "remember" events that happened much earlier in development. For example, in Chapter 5, we examined genomic imprinting. For the *Igf2* gene in mammals, an offspring expresses the *Igf2* gene that was inherited from the father, but not the copy that was inherited from the mother. From an epigenetic perspective, the cells in the adult are able to "remember" an event that occurred during gamete formation. Likewise, during embryonic development, cells become destined to embark on a pathway that leads to particular cell types. For example, an embryonic cell may give rise to a lineage of daughter cells that become a group of muscle cells. The muscle cells in the adult are remembering an event that occurred during embryonic development. In this section, we will explore the epigenetic mechanisms that explain how cells can remember events that occurred during specific stages of development.

Genomic Imprinting Occurs During Gamete Formation

As mentioned, we considered the imprinting of the *Igf2* gene in mice as an example of epigenetic inheritance (refer back to Figure 5.9). This gene encodes a protein that is required for proper growth.

The molecular mechanism of *Igf2* imprinting is due to different patterns of methylation during oogenesis and spermatogenesis (**Figure 16.4**). The *Igf2* gene is located next to another gene called *H19*. The function of the *H19* gene is not well understood, but it appears to play a role in some forms of cancer. Methylation may occur at a site called the **imprinting control region (ICR)** that is located between the *H19* and *Igf2* genes. A second site called a **differentially methylated region (DMR)** may also be methylated. During oogenesis (see upper right side of Figure 16.4), methylation does not occur at either site. A protein called the CTC-binding factor (CTCF) binds to a DNA sequence containing CTC (cytosine-thymine-cytosine) that is found in both the ICR and DMR. The CTCFs bound to these sites then bind to each other to form a loop in the DNA. How does this loop affect the expression of *Igf2*? To understand how, we need to consider the effects of an enhancer that is located next to the *H19* gene. Even though it is fairly far away, this enhancer can stimulate transcription of the *Igf2* gene. However, when a loop forms due to the interactions between two CTCFs, the enhancer is prevented from stimulating *Igf2*. Under these conditions, the maternal allele of *Igf2* is turned off.

Alternatively, if the ICR and DMR are methylated, which occurs during sperm formation, CTCFs are unable to bind to these sites (see bottom right of Figure 16.4). This prevents loop formation, which allows the enhancer to stimulate the *Igf2* gene. Therefore, the paternally inherited *Igf2* allele is transcriptionally activated. The methylation that occurs during sperm formation is **de novo methylation,** which is the methylation of a completely

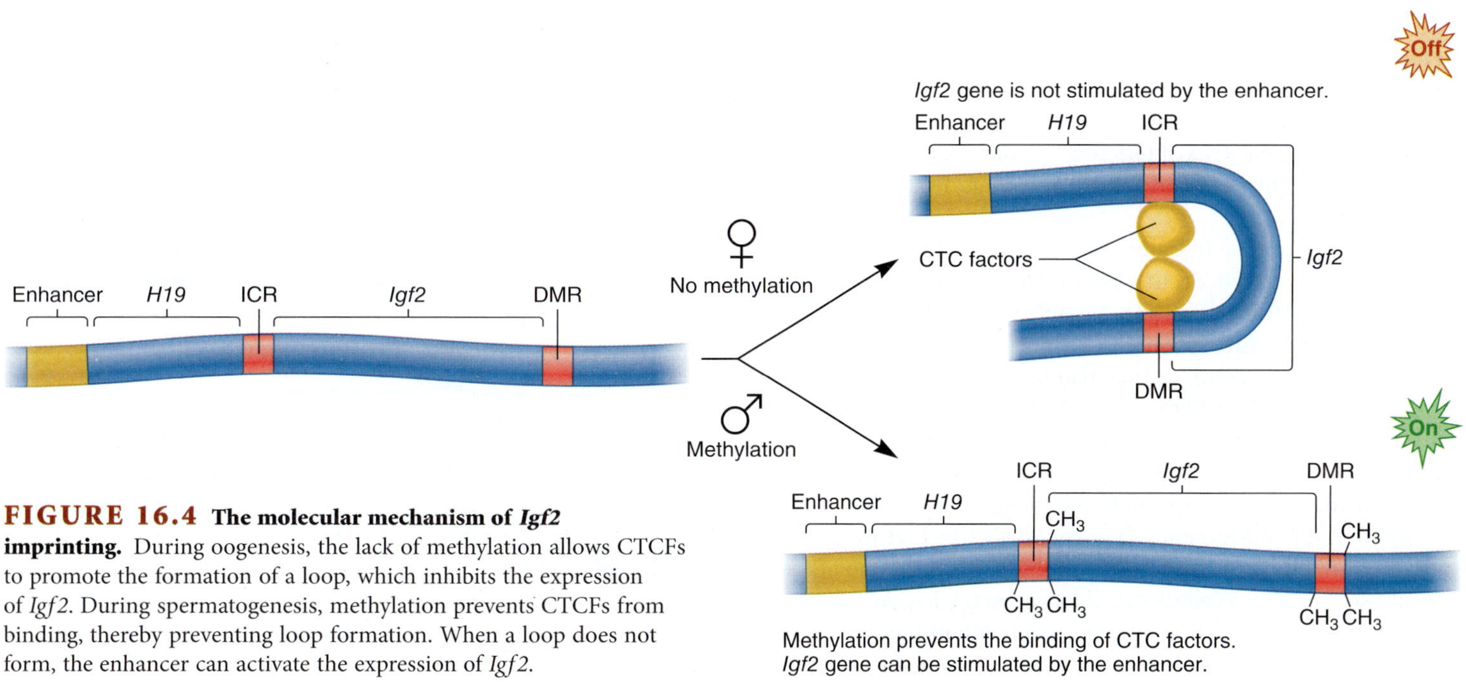

FIGURE 16.4 The molecular mechanism of *Igf2* imprinting. During oogenesis, the lack of methylation allows CTCFs to promote the formation of a loop, which inhibits the expression of *Igf2*. During spermatogenesis, methylation prevents CTCFs from binding, thereby preventing loop formation. When a loop does not form, the enhancer can activate the expression of *Igf2*.

unmethylated site. Following this de novo methylation, this methylation pattern is maintained in somatic cells of offspring due to **maintenance methylation,** which is the methylation of hemimethylated sites (refer back to Figure 15.16).

As we have just seen, DNA methylation causes the *Igf2* gene to be transcriptionally active. However, it is worth noting that DNA methylation more commonly has the opposite effect. As discussed in Chapter 15, DNA methylation at CpG islands in the vicinity of promoters for other genes usually inhibits their transcription.

X-Chromosome Inactivation in Mammals Occurs During Embryogenesis

As described in Chapter 5, **X-chromosome inactivation (XCI)** occurs in female mammals. One of the two X chromosomes in somatic cells is inactivated and becomes a condensed Barr body. Because females are XX and males are XY, the process of XCI achieves dosage compensation—both females and males express a single copy of most X-linked genes. In addition, XCI is responsible for certain traits such as the calico coat pattern observed in certain breeds of female cats (refer back to Figure 5.3).

XCI is an epigenetic phenomenon. During early embryonic development in female mammals, one of the X chromosomes in each somatic cell is randomly chosen for XCI. After this occurs, the same X chromosome is maintained in an inactivated state during subsequent cell divisions (refer back to Figure 5.4). A region of the X chromosome called the X inactivation center (Xic) plays a key role in this process (**Figure 16.5a**). Within the Xic are two genes called *Xist*, for X inactive-specific transcript, and *Tsix*. *Xist* is expressed from the inactivated X chromosome, whereas *Tsix* is expressed from the active X chromosome. The two genes are transcribed in opposite directions. (Note: *Tsix* is *Xist* spelled backwards; this naming refers to the opposite direction of transcription of the two genes.)

Figure 16.5b shows a simplified diagram of how XCI occurs at the molecular level. Prior to XCI, transcription factors called **pluripotency factors** stimulate the expression of *Tsix*. The expression of the *Tsix* gene from both X chromosomes inhibits the expression of the *Xist* gene in multiple ways. For example, the *Tsix* RNA inhibits *Xist* RNA via RNA interference, which is described later in this chapter. At this very early embryonic stage, both X chromosomes are active (Xa).

A key event that initiates XCI is the pairing of the two X chromosomes, which occurs briefly (for less than an hour). This pairing, which happens only in embryonic cells, is required for choosing one X chromosome to be inactivated. Pairing begins at the *Tsix* gene via a complex that also includes the pluripotency factors and CTCFs (see Figure 16.5b). As you may recall from Figure 16.4, CTCFs bind to unmethylated sequences; in this case, they bind to an unmethylated sequence in the Xic. Due to the thermodynamics of protein-protein interactions, researchers have proposed that the pluripotency factors and CTCFs that were previously bound to both X chromosomes shift entirely to one of the X chromosomes. This event is called a **symmetry break.** Because the pluripotency factors stimulate the expression of the *Tsix* gene, the X chromosome to which they shift is chosen as the active X chromosome (Xa). By comparison, on the other X chromosome, the *Tsix* gene is not expressed, which permits the expression of the *Xist* gene. This chromosome is the one that becomes the inactivated X chromosome (Xi).

The next phase of X-chromosome inactivation, called the spreading phase, involves the coating of Xi with *Xist* RNA, which is a 17-kb noncoding RNA (in humans) that is transcribed exclusively from Xi. This RNA consists of six repetitive sequences (called repeats A–F). Repeat A is located at the 5′ end and is necessary for later events that lead to the formation of a Barr body. Repeat C is necessary for the binding of *Xist* RNA to Xi. The process of spreading begins at the Xic. A transcription factor bound to the Xic initially binds to repeat C in one *Xist* RNA molecule, tethering it to the Xic. After this occurs, subsequent *Xist* RNAs bind to each other and to a protein, called hnRNP-U, which binds to numerous AT-rich DNA sequences along Xi.

After the spreading phase, the first repeat sequence (repeat A) within the *Xist* RNA recruits proteins to Xi, thereby leading to epigenetic changes. For example, *Xist* RNA recruits protein complexes to Xi that cause covalent modifications of specific sites in histone tails. These histone marks are recognized by other proteins that promote changes in chromatin structure. In addition, a histone variant, called macroH2A, is incorporated into nucleosomes at many sites along Xi. Finally, another key event that occurs after *Xist* RNA coating is the recruitment of DNA methyltransferases to Xi, which leads to DNA methylation. Collectively, covalent histone modifications, the incorporation of macroH2A into nucleosomes, and the methylation of many CpG islands are thought to play key roles in the silencing of genes on Xi and its compaction into a Barr body. These epigenetic changes in Xi are then maintained in subsequent cell divisions.

The Development of Body Parts and the Formation of Specific Cell Types in Multicellular Organisms Involves Epigenetic Gene Regulation

As described in Chapter 25, development involves a series of genetically programmed stages in which a fertilized egg becomes a multicellular adult with well-defined body parts composed of

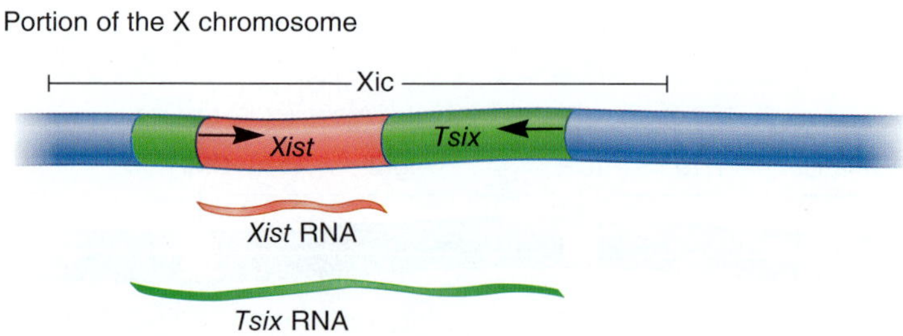

(a) The X-inactivation center (Xic)

ONLINE ANIMATION

FIGURE 16.5 The process of X-chromosome inactivation. Note: This is a simplified description of XCI. Several more proteins that are not described in this figure are involved in the process.

CONCEPT CHECK: What is the role of the symmetry break? Does it occur in embryonic cells, in adult somatic cells, or both?

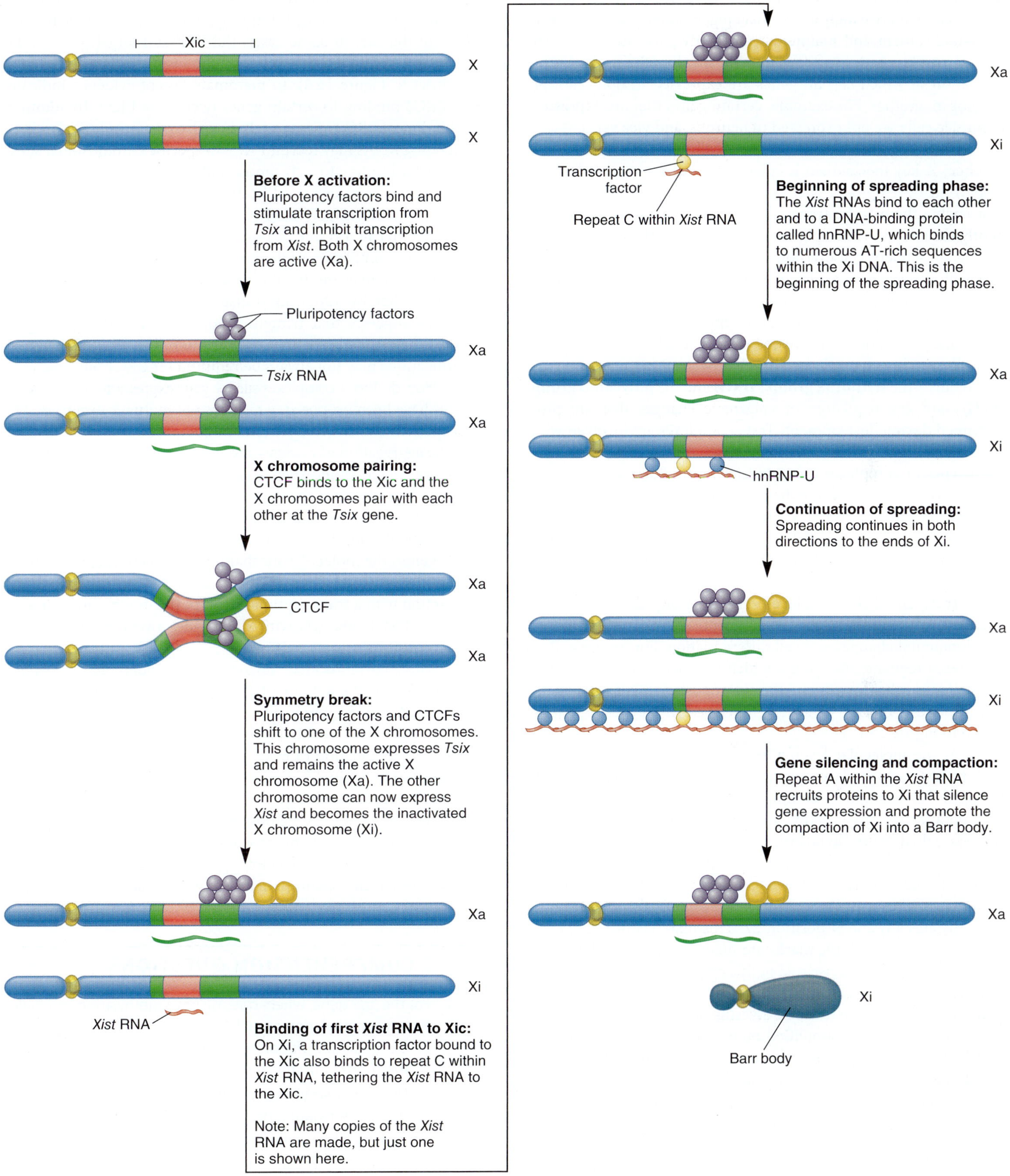

(b) Mechanism of X inactivation during embryonic development in mammals

specific types of cells. Over the past three decades, researchers have become increasingly aware that epigenetics plays a key role in the establishment and maintenance of body parts and particular cell types. The process of development involves differential gene regulation in which certain genes are expressed in one cell type but not in another. For example, certain genes that are expressed in muscle cells are not expressed in neurons, and vice versa.

How are genes activated in one cell type and repressed in another? A key mechanism is epigenetic gene regulation. During embryonic development, many genes undergo epigenetic changes that enable them to be transcribed or cause them to be permanently repressed. Such epigenetic changes are then transmitted during subsequent cell divisions. For example, an embryonic cell that will eventually give rise to muscle tissue is programmed to undergo epigenetic modifications that will enable the transcription of muscle-specific genes and repress the transcription of genes that should not be expressed in muscle cells.

Researchers have discovered that two competing complexes of proteins—the **trithorax group (TrxG)** and the **polycomb group (PcG)**—are key regulators of epigenetic changes that are programmed during development. TrxG complexes are involved with gene activation, whereas PcG complexes cause gene repression. Both types of complexes are found in multicellular species, such as animals and plants, where they are required for proper development. TrxG and PcG complexes were discovered in genetic studies of *Drosophila*. The names trithorax and polycomb refer to altered body parts associated with mutations in genes that encode TrxG and PcG proteins, respectively.

TrxG and PcG complexes regulate many different genes, particularly those that encode transcription factors that control developmental changes and cell differentiation. For example, PcG complexes regulate *Hox* genes, which are involved in specifying the structures that form along the anteroposterior axis in animals. The functions of *Hox* genes are described in Chapter 25 (look ahead to Figure 25.17).

At the molecular level, a key function of specific proteins within TrxG and PcG complexes is the covalent modifcation of histones. For example, a component of TrxG recognizes histone H3 and attaches three methyl groups to a lysine at position 4—a process called **trimethylation.** (Note: This is methylation of a histone protein, not methylation of DNA.) This mark is abbreviated H3K4me3 and is called an activating mark. By comparison, a component of certain PcG complexes recognizes histone H3 and trimethylates a lysine at position 27. This mark (H3K27me3) is a repressive mark. Multiple marks are made by TrxG and PcG proteins. However, the precise role of this epigenetic marking process in gene activation or gene repression is not completely understood and is an area of active research investigation.

Figure 16.6 is a simplified molecular model of how a gene may be targeted for epigenetic silencing by PcG complexes. Keep in mind that some of these steps are not entirely understood and that the details of silencing steps may not be the same among different genes. This model considers the actions of two different types of PcG complexes, which are named Polycomb Repressive Complex 1 and 2 (PRC1 and PRC2). The first step involves the binding of PRC2 to a chromosomal site that is near a gene controlled by PcG complexes. In *Drosophila*, PRC2 binds to a DNA element called a **polycomb response element (PRE).** The response element is initially recognized by a PRE-binding protein, which then recruits PRC2 to the site. In some cases, the PRE and target gene are far apart and brought close together by the formation of a DNA loop (see inset to Figure 16.6). In mammals, recent evidence indicates that PRC2 binding to certain genes occurs via PREs. In addition, the binding of PRC2 in mammals may also occur at CpG islands, or noncoding RNAs may recruit PRC2 to specific chromosomal sites.

After binding to a chromosomal site, a protein within PRC2 catalyzes the covalent modification of histone H3 (see Figure 16.6). As mentioned, certain PcG complexes recognize histone H3 and trimethylate a lysine at position 27. The effect of trimethylation is not completely understood, but different effects are possible. First, it may inhibit the binding of RNA polymerase to a promoter, thereby inhibiting transcription. Second, experimental evidence suggests that trimethylation promotes the binding of PRC1, which can also inhibit transcription. However, H3K27me3 modifications may not always be required for PRC1 binding.

How do PRC1 complexes silence gene expression? Three mechanisms have been proposed, which are not mutually exclusive. The first mechanism involves chromatin compaction. PRC1 can catalyze the aggregation of nucleosomes into a more compact knot-like structure, which would silence gene expression (see Figure 16.6). A second mechanism involves another covalent modification—the attachment of a ubiquitin molecule onto histone H2A. Though this covalent modification is associated with the silencing of many genes, the molecular mechanism by which it represses transcription is not understood. Finally, a third possibility is a direct interaction with a transcription factor. In the model shown in Figure 16.6, PRC1 interacts with TFIID (described in Chapter 15), thereby inhibiting transcription.

After chromosomal regions have undergone epigenetic changes due to the actions of PcG complexes, these changes are maintained during subsequent cell divisions. In this way, epigenetic changes that occur during embryonic development can be transmitted to a population of cells that gives rise to a particular type of tissue, such as muscle tissue. The molecular mechanism(s) of how such epigenetic changes are maintained during cell divisions is not well understood. One possibility is that PcG complexes may remain bound to their chromosomal sites during the process of DNA replication, thereby facilitating covalent histone modification and changes in chromatin structure after replication has occurred.

16.2 COMPREHENSION QUESTIONS

1. For the *Igf2* gene, where does de novo methylation and maintenance methylation occur?
   a. De novo methylation occurs in sperm, and maintenance methylation occurs in egg cells.
   b. De novo methylation occurs in egg cells, and maintenance methylation occurs in sperm cells.
   c. De novo methylation occurs in sperm, and maintenance methylation occurs in somatic cells of offspring.
   d. De novo methylation occurs in egg cells, and maintenance methylation occurs in somatic cells of offspring.

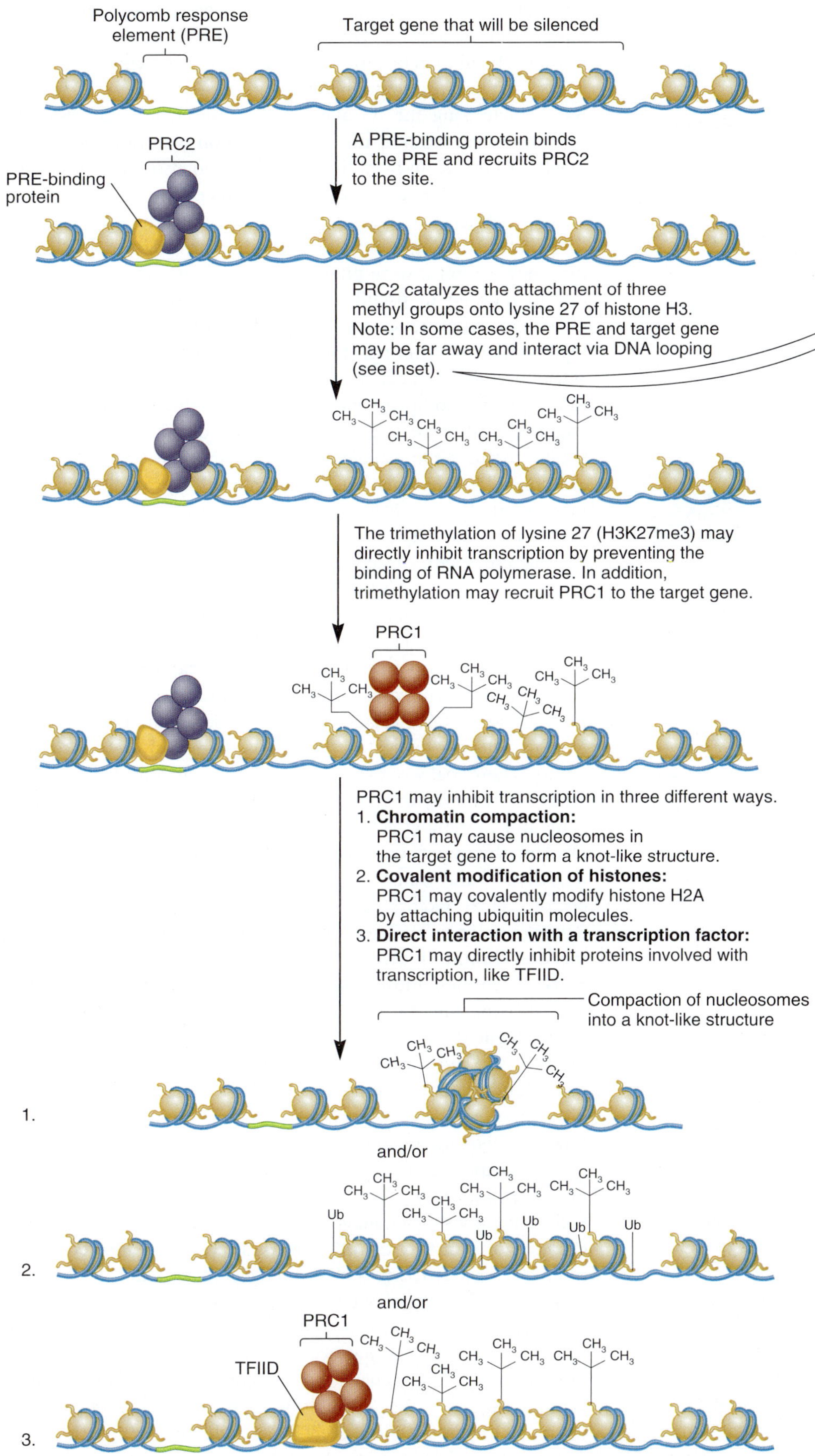

ONLINE ANIMATION

FIGURE 16.6 A simplified model of epigenetic silencing of a gene by polycomb group complexes.

CONCEPT CHECK: Describe how the compaction of nucleosomes into a knot-like structure could silence gene expression.

2. For XCI to occur, where are the *Xist* and *Tsix* genes expressed?
   a. *Xist* is expressed only on Xa, and *Tsix* is expressed only on Xi.
   b. *Xist* is expressed only on Xi, and *Tsix* is expressed only on Xa.
   c. *Xist* is expressed only on Xa, and *Tsix* is expressed only on Xa.
   d. *Xist* is expressed only on Xi, and *Tsix* is expressed only on Xi.
3. Which of the following possibilities could explain how PcG complexes are able to silence genes?
   a. The compaction of nucleosomes
   b. The attachment of ubiquitin to histone proteins
   c. The direct inhibition of transcriptional components, such as TFIID
   d. All of the above

16.3 EPIGENETICS AND ENVIRONMENTAL AGENTS

Learning Outcomes:

1. Explain how coat color in mice is epigenetically modified by dietary factors.
2. Describe how some environmental agents may contribute to the development of cancer without causing mutations.

One of the most active fields in genetics is studying how certain environmental agents cause epigenetic changes that affect gene expression. Two areas that have received a great deal of attention are the effects of diet on epigenetic modifications and the potential effects of toxic agents, such as carcinogens—cancer-causing agents. In this section, we will consider examples of both types.

Environmental Agents at Early Stages of Development May Cause Epigenetic Changes That Affect Phenotype

A striking example of how the environment can promote epigenetic changes is illustrated by studies of the *Agouti* gene (also designated *A*) found in mice. This gene encodes the Agouti signaling peptide that controls the deposition of yellow pigment in developing hairs. In wild-type mice (*AA*), the expression of this gene promotes the synthesis of pheomelanin, a yellow pigment. During the growth of a hair, melanocytes (pigment-producing cells) within a hair follicle initially make eumelanin, which is black. The transient expression of the *Agouti* gene causes the cells to express pheomelanin. The melanoncytes then revert back to making black pigment. The result is a band of yellow pigment sandwiched between layers of black pigment, which gives a brown color. The yellow pigment is not synthesized near the tip of the hair, so the hair of wild-type mice is brown with black tips.

Researchers have identified many mutations that affect the expression of the *Agouti* gene. For example, mice that are homozygous for a loss-of-function mutation (*aa*) have black fur because pheomelanin is not made. Alternatively, a gain-of-function mutation that causes the *Agouti* gene to be overexpressed results in a mouse with yellow fur. One such mutation is designated A^{vy} (*A* refers to *Agouti*, *v* refers to viable, and *y* refers to yellow. The letter *v* was used because some mutations in the *Agouti* gene are lethal). By characterizing the A^{vy} allele at the molecular level, researchers determined that it is due to the insertion of a transposable element (TE) upstream from the normal promoter of the *Agouti* gene (**Figure 16.7a**). (TEs are described in Chapter 19.) The TE carries an active promoter that causes the overexpression of the *Agouti* gene.

An intriguing observation of mice carrying the A^{vy} allele is that they exhibit a wide phenotypic variation, ranging from yellow to mottled to pseudo-agouti (**Figure 16.7b**). Why should mice with the same genotype show such a wide range of phenotypic variation? Although the answer is not entirely understood, researchers have speculated that TEs are particularly sensitive to epigenetic modifications. In the case of the A^{vy} allele, epigenetic modifications may affect the function of the promoter within the TE that is responsible for overexpressing the *Agouti* gene. For example, DNA methylation could inhibit this promoter. Furthermore, a variety of environmental factors may cause such epigenetic changes to occur. The sensitivity of TEs to epigenetic modifications together with variation in environmental factors may explain the phenotypic variation seen in these mice.

One environmental factor that may affect epigenetic modification is diet. With regard to the A^{vy} allele, the exposure of pregnant female mice to different types of diets can have a significant effect on the phenotypes of the resulting offspring. For example, in 2003, Robert Waterland and Randy Jirtle conducted a study in which they investigated the effects of certain dietary supplements. Their goal was to determine if nutrients that are known to affect DNA methylation would alter the expression of the *Agouti* gene and thereby affect coat color. For DNA to be methylated, DNA methyltransferase removes a methyl group from *S*-adenosyl methionine and transfers it to a cytosine base in DNA. A variety of dietary factors can increase the synthesis of *S*-adenosyl methionine in cells. These include folic acid, vitamin B_{12}, betaine, and choline chloride. Waterland and Jirtle took female mice and divided them into a control group that was fed a normal diet and an experimental group that was fed a diet supplemented with folic acid, vitamin B_{12}, betaine, and choline chloride. Both groups were fed their respective diets before and during pregnancy and up to the stage of weaning. Offspring carrying the A^{vy} allele were then analyzed with regard to their coat color and levels of DNA methylation.

As expected, a range of coat colors was observed among the offspring (**Figure 16.7c**). However, the offspring of females that had been fed a supplemented diet tended to have darker coats. For example, over 25% of the offspring with heavily mottled coats had mothers that were fed a supplemented diet (black bars), whereas less than 10% had mothers that were given a normal diet (white bars).

The coat colors of the offspring largely correlated with the degree of methylation that occurred at CpG islands in the TE—offspring with darker coat color had greater levels of DNA methylation (**Figure 16.7d**). How do we explain these results? In the mice that are more yellow, the TE has undergone very little methylation. Therefore, the promoter would remain active, thereby

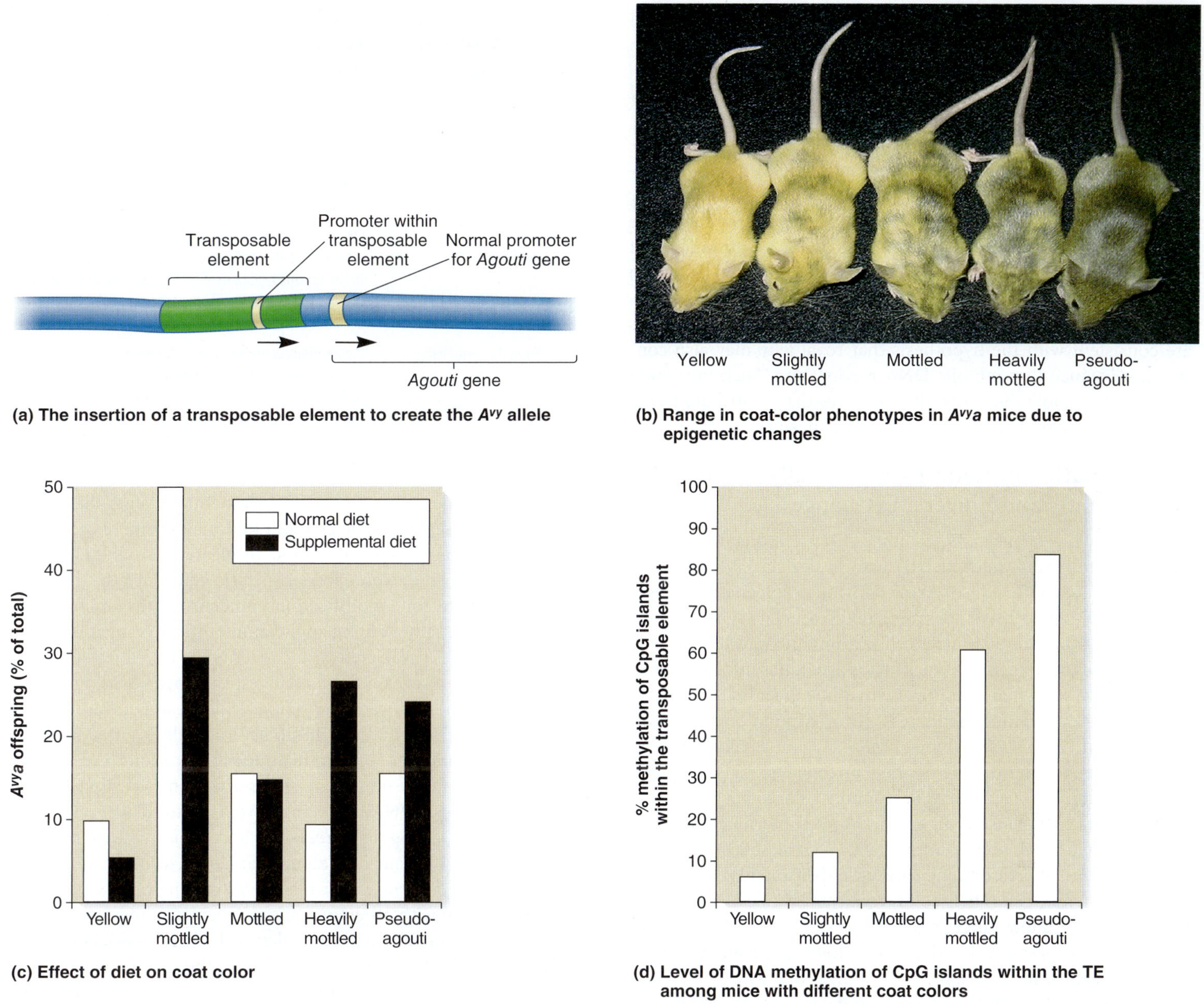

FIGURE 16.7 Dietary effects on coat color in mice. **(a)** A mutation in the *Agouti* gene, designated A^{vy}, is caused by the insertion of a TE upstream from the normal *Agouti* promoter. The TE promoter is very active and causes the overexpression of the *Agouti* gene. **(b)** Mice carrying the A^{vy} allele exhibit a range of phenotypes. The mice shown here are heterozygotes, $A^{vy}a$; they carry the mutant A^{vy} allele and a loss-of-function allele (a). **(c)** Effects of diet supplementation on coat color. White bars represent offspring from females given a normal diet and black bars represent offspring from females given a supplemented diet. **(d)** DNA methylation patterns among mice with different coat colors. The samples to determine DNA methylation were obtained from cells in the tail.

Data from R. A. Waterland and R. L. Jirtle (2003) Transposable Elements: Targets for Early Nutritional Effects on Epigenetic Gene Regulation. *Mol. Cell. Biol.* *23*, 5293–5300.

Genes → Traits The mice shown in part (b) are genetically identical. Their differences in coat color are due to epigenetic modifications that occur during early stages of development.

leading to the transcription of the *Agouti* gene and the overproduction of yellow pigment. By contrast, the TE in the darker mice has undergone extensive methylation. Such methylation would inhibit the overexpression of the *Agouti* gene, thereby preventing the overproduction of yellow pigment and resulting in darker fur.

Evidence that diet may affect DNA methylation also comes from the study of honey bees (*Apis mellifera*). Female honey bees are of two types: queen bees and worker bees (see chapter-opening photo). Queens are larger, live for years, and produce up to 2000 eggs each day. By comparison, the smaller worker bees are sterile, typically live only for weeks, and engage in specialized types of work, which includes the cleaning and constructing of comb cells, nurturing larvae, guarding the hive entrance, and foraging for pollen and nectar. The striking differences between queen and worker bees are largely caused by differences in their diets. Certain worker bees, called nurse bees, produce royal jelly from glands in their mouths. All female larvae are initially fed royal jelly, but those that are bathed in royal jelly throughout their entire larval development and feed on it into adulthood become queens. In contrast, female larvae that are weaned at an early stage of development and switched to a diet of pollen and nectar become worker bees.

In 2008, a study conducted by Ryszard Maleszka and colleagues indicated that DNA methylation may play a role in controlling developmental pathways that result in queen or worker bee morphologies. Bee larvae were fed a diet that should produce worker bees. These larvae were injected with a substance that inhibits DNA methyltransferase. The result was that most of them became queen bees with fully developed ovaries! While other factors may contribute to the development of queens, these results are consistent with the hypothesis that royal jelly may also contain a substance that inhibits DNA methylation. Such inhibition is thought to allow the expression of genes that contribute to the development of traits that are observed in queen bees.

The Formation of Cancer Cells Usually Involves Both Genetic and Epigenetic Changes

Thus far, we have considered examples in which dietary environmental agents exert their effects during early stages of development. Researchers are also discovering that epigenetic changes can occur in adult cells, contributing to diseases such as cancer. As described in Chapter 24, cancer cells exhibit uncontrolled growth, which is due to two general types of abnormalities in gene expression. First, certain genes are overactive in cancer cells. In many cases, this is due to increased rates of transcription or the expression of a gene in the wrong type of cell. The abnormally high level of expression of these genes, called **oncogenes,** causes cellular changes that promote cancer. For example, a higher expression of an oncogene may cause a higher rate of cell division. A second general type of change in cancer cells is a decrease in the expression of tumor suppressor genes. Because the proteins encoded by tumor suppressor genes help to prevent cancer, a decrease in their expression may allow cancer to occur.

Over the past few decades, researchers have come to understand that cancer involves a series of changes that usually increases the expression of multiple oncogenes and decreases the expression of multiple tumor suppressor genes. Such changes in the level of gene expression may be caused by genetic and epigenetic mechanisms. As described in Chapter 24, several genetic mechanisms are involved. For example, a mutation may occur in the promoter of a gene and increase its level of transcription, thereby converting a normal gene into an oncogene. Alternatively, a mutation in a tumor suppressor gene could inactivate the function of the encoded protein, thereby promoting cancer.

In addition to genetic changes, multiple epigenetic changes can foster the development of cancer cells (Table 16.3). A particularly common change is hypermethylation—an abnormally high level of methylation, typically at CpG islands. This occurrence may promote cancer by inhibiting the expression of tumor suppressor genes. In addition, changes in the covalent modification of histones and chromatin remodeling are common in cancer cells. Such changes could increase the expression of oncogenes or decrease the expression of tumor suppressor genes.

TABLE 16.3
Epigenetic Changes Associated with Cancer

Type of Modification	Description
DNA methylation	The hypermethylation of CpG islands is a common occurrence in cancer cells. This may promote cancer by inhibiting tumor suppressor genes. Also, other chromosomal regions may exhibit the opposite phenomenon, hypomethylation. A decrease in DNA methylation could activate oncogenes or affect genomic imprinting.
Covalent histone modification	The covalent modification of histones has been shown to be altered at specific genes in cancer cells. Depending on the specific type of modification, such changes could increase the expression of oncogenes or inhibit the expression of tumor suppressor genes.
Chromatin remodeling	The locations of nucleosomes have been shown to be altered in cancer cells. Depending on how the nucleosomes are rearranged, such changes could increase the expression of oncogenes or inhibit the expression of tumor suppressor genes.

Experimentally, many environmental agents have been shown to cause epigenetic changes that may promote cancer. For example, certain chemicals found in cigarette smoke have been shown to alter the expression of specific genes by affecting DNA methylation, histone modifications, and chromatin remodeling.

A potentially exciting application of the investigation of epigenetic changes associated with cancer is the development of new drugs that may reverse these changes, thereby providing another treatment option for cancer patients. Researchers are actively investigating drugs that may inhibit cancer cells by affecting either DNA methylation or covalent histone modifications. For example, inhibitors of DNA methyltransferase, the enzyme that attaches methyl groups to DNA, are being developed to treat certain forms of cancer including leukemia—a cancer of white blood cells. Drugs such as 5-azacytidine and decitabine, which inhibit DNA methyltransferase, have shown some promising results when used in conjunction with other anticancer drugs. In some cases, clinical improvement in patients with leukemia has been associated with a decrease in DNA methylation. Although the specific reason(s) for patient improvement are not understood at the molecular level, one possibility is that the lower level of DNA methylation has reversed the inhibition of tumor suppressor genes.

16.3 COMPREHENSION QUESTIONS

1. When mice carrying the A^{vy} allele exhibit a darker coat color, this phenotype is thought to be caused by dietary factors that result in
   a. a greater level of DNA methylation and a decrease in the expression of the *Agouti* gene.
   b. a lower level of DNA methylation and a decrease in the expression of the *Agouti* gene.
   c. a greater level of DNA methylation and the overexpression of the *Agouti* gene.
   d. a lower level of DNA methylation and the overexpression of the *Agouti* gene.

2. Which of the following types of epigenetic changes may promote cancer?
   a. DNA methylation
   b. Covalent modification of histones
   c. Chromatin remodeling
   d. All of the above

16.4 REGULATION OF RNA SPLICING, RNA STABILITY, AND TRANSLATION

Learning Outcomes:

1. Define alternative splicing and explain the role of splicing factors in this process.
2. Describe how sequences within mRNAs affect their stability.
3. Analyze experimental evidence that double-stranded RNA is more potent at inhibiting mRNA than is antisense RNA.
4. Outline the steps of RNA interference.
5. Explain how the translation of mRNA may be controlled by RNA-binding proteins.

In Chapter 15 and this chapter, we have surveyed a variety of mechanisms that regulate the level of gene transcription in eukaryotes. These mechanisms control the rate of RNA transcription from a given gene. In addition, eukaryotic gene expression is commonly regulated after the RNA is made (**Table 16.4**). One such mechanism of RNA regulation is via splicing. In this section, we will consider how the splicing of a pre-mRNA can be regulated so it occurs in more than one way, leading to the production of polypeptides with different amino acid sequences.

A second strategy for regulating mRNA is by controlling its stability. The concentration of any given mRNA is greatly affected by its stability or half-life. Factors that increase mRNA stability are expected to maintain a higher concentration of that RNA molecule, whereas factors that decrease its stability will result in a lower concentration. We will explore how sequences within mRNA molecules can greatly affect their stability. In addition, we will consider a mechanism called RNA interference that can target mRNA molecules for degradation.

A third way to regulate specific mRNAs is by controlling their ability to be translated into a polypeptide. We will end this section by examining how RNA-binding proteins can recognize specific sequences within mRNAs and influence the ability of ribosomes to translate them.

Alternative Splicing Regulates Which Exons Occur in a Mature mRNA, Allowing Different Polypeptides to Be Made from the Same Gene

When it was first discovered, the phenomenon of splicing seemed like a wasteful process. During transcription, energy is used to synthesize intron sequences, which are subsequently removed by spliceosomes (refer back to Figure 12.22). Making and then removing introns uses a significant amount of energy. This observation intrigued many geneticists, because natural selection tends to eliminate wasteful processes. Therefore, the researchers expected to find that pre-mRNA splicing has one or more important biological roles. In recent years, one very important biological advantage has become apparent. This is **alternative splicing,** which refers to the phenomenon that a pre-mRNA can be spliced in more than one way.

What is the advantage of alternative splicing? To understand the biological effects of alternative splicing, remember that the sequence of amino acids within a polypeptide determines the structure and function of a protein. Alternative splicing produces two or more polypeptides with differences in their amino acid sequences, leading to possible changes in their functions. In most cases, the alternative versions of the protein have similar functions, because most of their amino acid sequences are identical to each other. Nevertheless, alternative splicing produces differences in amino acid sequences that provide each polypeptide with its own unique characteristics. Because alternative splicing allows two or more different polypeptide sequences to be derived from a single gene, some geneticists have speculated that an important advantage of this process is that it allows an organism to carry fewer genes in its genome.

The degree of splicing and alternative splicing varies greatly among different species. Baker's yeast (*Saccharomyces cerevisiae*), for example, contains about 6300 genes and approximately 300

TABLE 16.4
Regulation of mRNA Molecules

Effect	Description
Pre-mRNA splicing	Certain pre-mRNAs are spliced in more than one way, leading to polypeptides that have different amino acid sequences. This phenomenon, called alternative splicing, is often cell-specific so that a protein can be fine-tuned to function in a particular cell type. It is an important form of gene regulation in multicellular eukaryotic species.
RNA stability	The amount of mRNA is greatly influenced by its half-life. A long polyA tail on mRNAs promotes their stability due to the binding of polyA-binding protein. Some mRNAs with a relatively short half-life contain sequences that act as destabilizing elements that target them for rapid destruction. Such mRNAs are stabilized by specific RNA-binding proteins that usually bind near the 3′ end.
RNA interference	Double-stranded RNA can mediate the degradation of homologous mRNAs in the cell. This is a mechanism of gene regulation. Also, it probably provides eukaryotic cells with protection from invasion by certain types of viruses and may prevent the movement of transposable elements.
Regulation of translation of specific RNAs	Some mRNAs are regulated by RNA-binding proteins that inhibit the ability of the ribosomes to initiate translation. These proteins usually bind at the 5′ end of the mRNA, thereby preventing the ribosome from binding.

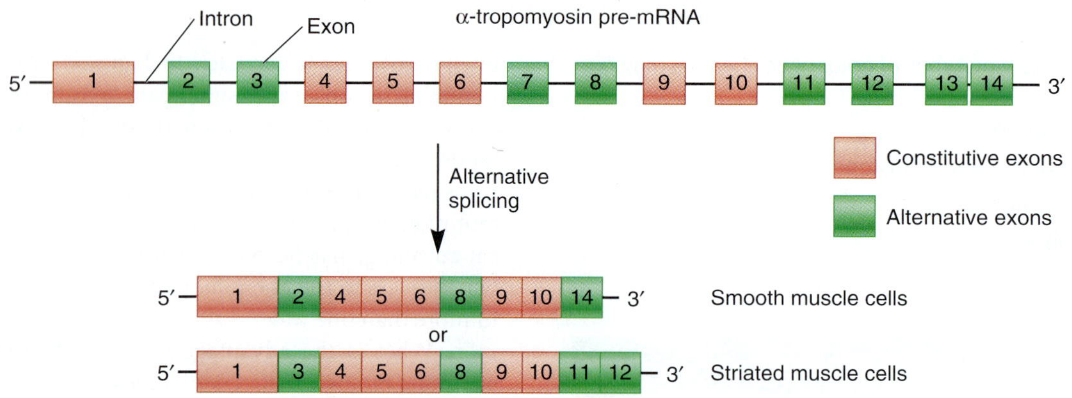

FIGURE 16.8 Alternative ways that the rat α-tropomyosin pre-mRNA can be spliced. The top part of this figure depicts the structure of the rat α-tropomyosin pre-mRNA. Exons are shown as colored boxes, and introns are illustrated as connecting black lines. The lower part of the figure depicts the mature mRNA of smooth and striated muscle cells. The constitutive exons (red) are found in the mature α-tropomyosin mRNAs in all cell types. Alternative exons are also found in mRNAs but they vary from one cell type to another. Though exon 8 in this figure is found in both mature mRNAs, it is considered an alternative exon because it is not found in the mature α-tropomyosin mRNAs of some other cell types.

Genes → Traits α-tropomyosin functions in the regulation of cell contraction in muscle and nonmuscle cells. Alternative splicing of the pre-mRNA provides a way to vary contractibility in different types of cells by modifying the function of α-tropomyosin. As shown here, the alternatively spliced versions of the pre-mRNA produce α-tropomyosin proteins that differ from each other in their structure (i.e., amino acid sequence). These alternatively spliced versions vary in function to meet the needs of the cell type in which they are found. For example, the sequence of exons 1–2–4–5–6–8–9–10–14 produces an α-tropomyosin protein that functions suitably in smooth muscle cells. Overall, alternative splicing affects the traits of an organism by allowing a single gene to encode several versions of a protein, each optimally suited to the cell type in which it is made.

(i.e., approximately 5%) encode pre-mRNAs that are spliced. Of these, only a few have been shown to be alternatively spliced. Therefore, in this unicellular eukaryote, alternative splicing is not a major mechanism for generating protein diversity. In comparison, complex multicellular organisms seem to rely on alternative splicing to a great extent. Humans have approximately 22,000 different genes, and most of these contain one or more introns. Recent estimates suggest that about 70% of all human pre-mRNAs are alternatively spliced. Furthermore, certain pre-mRNAs are alternatively spliced to an extraordinary extent. Some pre-mRNAs can be alternatively spliced to produce dozens of different mRNAs. This provides a much greater potential for human cells to create protein diversity.

Figure 16.8 considers an example of alternative splicing for a gene that encodes a protein known as α-tropomyosin, which functions in the regulation of cell contraction. It is located along the thin filaments found in smooth muscle cells, such as those in the uterus and small intestine, and in striated muscle cells that are found in cardiac and skeletal muscle. α-tropomyosin is also synthesized in many types of nonmuscle cells but in lower amounts. Within a multicellular organism, different types of cells must regulate their contractibility in subtly different ways. One way this may be accomplished is by the production of different forms of α-tropomyosin.

The intron-exon structure of the rat α-tropomyosin pre-mRNA and two alternative ways that the pre-mRNA can be spliced are described in Figure 16.8. The pre-mRNA contains 14 exons, 6 of which are **constitutive exons** (shown in red), which are always found in the mature mRNA from all cell types. Presumably, constitutive exons encode polypeptide segments of the α-tropomyosin protein that are necessary for its general structure and function. The mature mRNA also contains other exons, called **alternative exons** (shown in green), which vary from one cell type to another. The polypeptide sequences encoded by alternative exons may subtly change the function of α-tropomyosin to meet the needs of the cell type in which it is found. For example, Figure 16.8 shows the predominant splicing products found in smooth muscle cells and striated muscle cells. Exon 2 encodes a segment of the α-tropomyosin protein that alters its function to make it suitable for smooth muscle cells. By comparison, the α-tropomyosin mRNA found in striated muscle cells does not include exon 2. Instead, this mRNA contains exon 3, which is more suitable for that cell type.

Alternative splicing is not a random event. Rather, the specific pattern of splicing is regulated in any given cell. The molecular mechanism for the regulation of alternative splicing involves proteins known as **splicing factors,** which play a key role in the choice of particular splice sites. **SR proteins** are a type of splicing factor. They contain a domain at their carboxyl-terminal end that is rich in serines (S) and arginines (R) and is involved in protein-protein recognition. They also contain an RNA-binding domain at their amino-terminal end.

As discussed in Chapter 12, components of a spliceosome recognize the 5′- and 3′-splice sites and then remove the intervening intron. The function of splicing factors is to modulate the ability of a spliceosome to choose 5′- and 3′-splice sites. This can occur in two ways. Some splicing factors act as repressors that inhibit the ability of the spliceosome to recognize a splice site. In **Figure 16.9a**, a splicing repressor binds to a 3′-splice site and prevents the spliceosome from recognizing the site. Instead, the spliceosome binds to the next available 3′-splice site. The splicing repressor causes exon 2 to be spliced out of the mature mRNA, an event called **exon skipping.** Alternatively, other splicing factors enhance the ability of the spliceosome to recognize particular splice sites. In **Figure 16.9b**, splicing enhancers bind to the 3′- and 5′-splice sites that flank exon 3, which results in the inclusion of exon 3 in the mature mRNA.

Alternative splicing occurs because each cell type has its own characteristic concentration of many kinds of splicing factors. Furthermore, splicing factors may be regulated by the binding of small effector molecules, protein-protein interactions, and covalent modifications. Overall, the differences in the composition of splicing factors and the regulation of their activities form the basis for alternative splicing decisions.

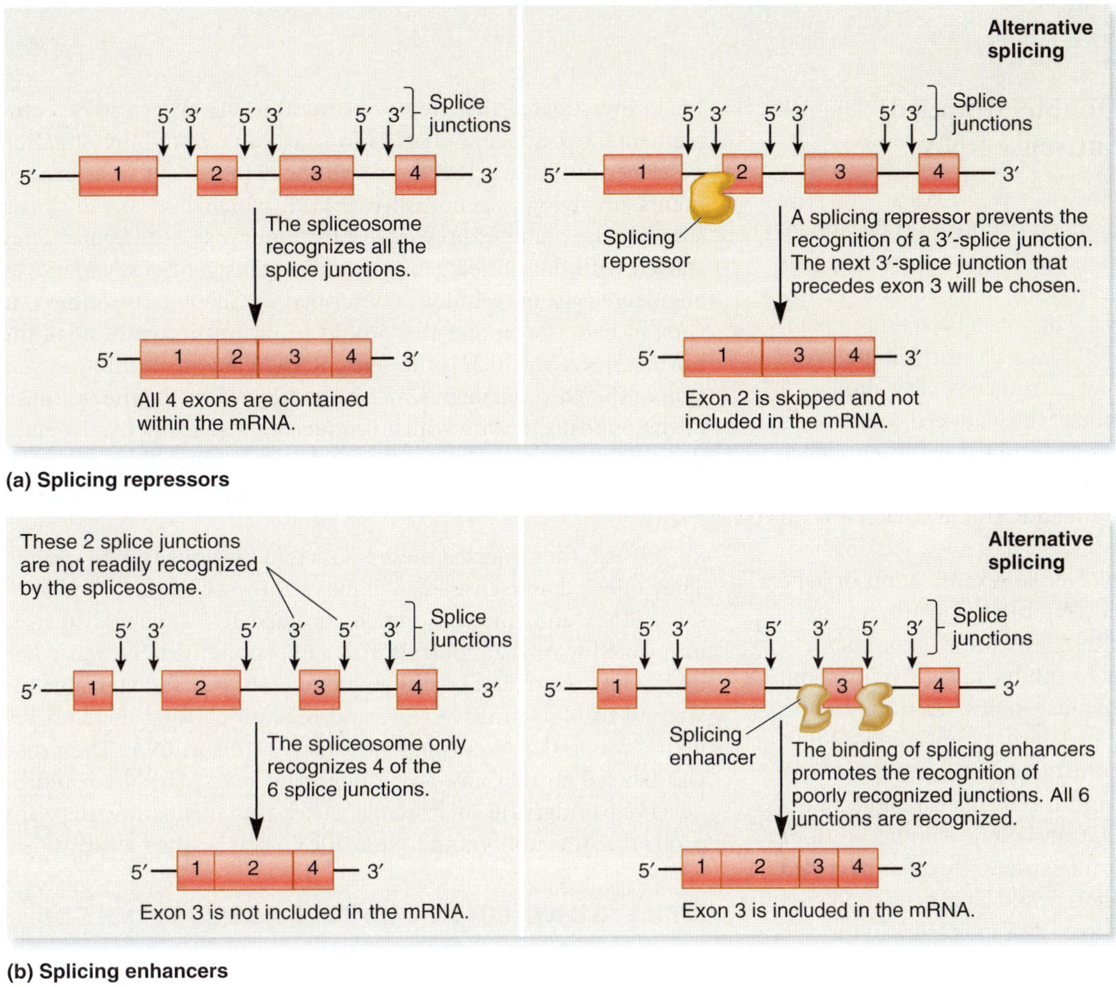

FIGURE 16.9 The roles of splicing factors during alternative splicing. **(a)** Splicing factors can act as repressors to prevent the recognition of splice sites. In this example, the presence of a splicing repressor causes exon 2 to be skipped and thus not included in the mRNA. **(b)** Other splicing factors can enhance the recognition of splice sites. In this example, splicing enhancers promote the recognition of sites that flank exon 3, thereby causing its inclusion in the mRNA.

CONCEPT CHECK: A pre-mRNA is recognized by just one splicing repressor that binds to the 3′ end of the third intron. The third intron is located between exon 3 and exon 4. After splicing is complete, would you expect the mRNA to contain exon 3 and/or exon 4 in the presence of the splicing repressor?

The Stability of mRNA Influences mRNA Concentration

In eukaryotes, the stability of mRNAs varies considerably. Certain mRNAs have very short half-lives (e.g., several minutes), whereas others persist for many days or even months. In some cases, the stability of an mRNA is regulated so that its half-life is shortened or lengthened. A change in the stability of mRNA can greatly influence the concentration of that mRNA molecule in the cell. In this way, mechanisms that control mRNA stability dramatically affect gene expression.

Various factors play a role in mRNA stability. One important structural feature is the length of the polyA tail. As you may recall from Chapter 12, most newly made mRNAs contain a polyA tail that averages 200 nucleotides in length. The polyA tail is recognized by **polyA-binding protein.** The binding of this protein enhances RNA stability. However, as an mRNA ages, its polyA tail tends to be shortened by the action of cellular exonucleases. Once it becomes less than 10 to 30 adenosines in length, polyA-binding protein can no longer bind, and the mRNA is rapidly degraded by exo- and endonucleases.

Certain mRNAs, particularly those with short half-lives, contain sequences that act as destabilizing elements. Although these destabilizing elements can be located anywhere within the mRNA, they are most commonly found near the 3′ end beyond the stop codon in a region known as the **3′-untranslated region (3′-UTR).** An example of a destabilizing element is the **AU-rich element (ARE)** that is found in many short-lived mRNAs in mammals (**Figure 16.10**). This element, which contains the consensus sequence AUUUA, is recognized by different types of proteins. Some proteins bind to the ARE and protect the mRNA from degradation. Other proteins bind to the ARE and destabilize the mRNA, which leads to its degradation by endonucleases.

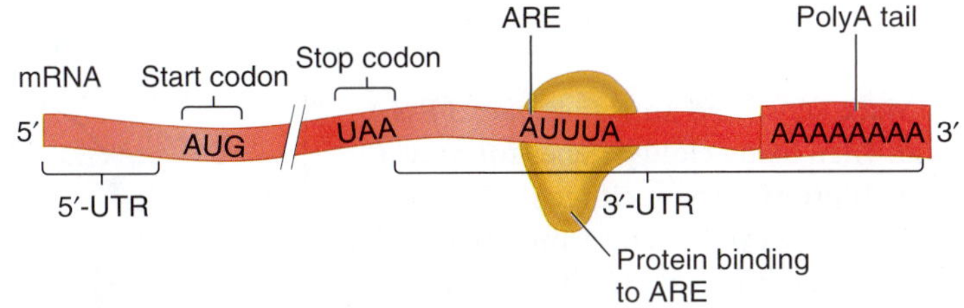

FIGURE 16.10 The location of AU-rich elements (AREs) within mRNAs of mammals. One or more AREs are commonly found within the 3′-UTRs (untranslated regions) of mRNAs with short half-lives. The 5′-UTR is the untranslated region of the mRNA that precedes the start codon. Proteins bind to AREs and usually cause the mRNA to be rapidly degraded.

EXPERIMENT 16A

Fire and Mello Show That Double-Stranded RNA Is More Potent than Antisense RNA at Silencing mRNA

Specific RNAs can be targeted for degradation or translational inhibition by a mechanism involving double-stranded RNA. This mechanism was discovered by research in plants and the nematode worm *Caenorhabditis elegans*. For example, certain plant viruses produce double-stranded RNA as part of their life cycle. In addition, these plant viruses may carry genes very similar to genes that already exist in the genome of the plant cell. When such viruses infect plant cells, researchers determined that they silence the expression of the plant gene that is similar to the viral gene. Later work revealed that double-stranded RNA plays a key role in this silencing process.

The study described here involved an examination of gene expression in *Caenorhabditis elegans*. Prior to this work, researchers had often introduced antisense RNA (RNA that is complementary to mRNA) into cells as a way to inhibit mRNA translation. Because antisense RNA is complementary to mRNA, the antisense RNA binds to the mRNA, thereby preventing translation. Oddly, in other experiments, researchers introduced sense RNA (RNA with the same sequence as mRNA) into cells instead, and, unexpectedly, this also inhibited mRNA translation. Another curious observation was that the effects of antisense RNA often persisted for a very long time, much longer than would have been predicted by the relatively short half-lives of most RNA molecules in the cell. These two unusual observations caused Andrew Fire, Craig Mello, and colleagues to investigate how the introduction of RNA into cells inhibits mRNA.

They used *C. elegans* as their experimental organism because it was relatively easy to inject with RNA and the expression of many of its genes had already been established. In 1998, Fire and Mello investigated the effects of several RNAs known to be complementary to specific mRNAs in *C. elegans*. In the investigation described in **Figure 16.11**, we will focus on one of their experiments involving a gene called *mex-3*, which had already been shown to be highly expressed in early embryos of *C. elegans*. They started with the cloned gene for *mex-3*. This gene was genetically engineered using techniques described in Chapter 20 so that one version had a promoter that would result in the synthesis of the normal mRNA, which is the sense RNA. They made another version of the *mex-3* gene in which a promoter directed the synthesis of the opposite strand, which produced antisense RNA. To make these RNAs in vitro, they added RNA polymerase and nucleotides to these two types of cloned genes to make sense or antisense RNA.

Next, they injected these RNAs into the gonads of *C. elegans*. They injected antisense RNA alone or they mixed sense and antisense RNA and injected double-stranded RNA. They also used uninjected worms as controls. After injection, the RNA was taken up by eggs, which later developed into embryos. To determine the amount of *mex-3* mRNA, Fire and Mello incubated the embryos with a probe that was complementary to this mRNA. The probe was labeled so that any probe bound to *mex-3* mRNA could be observed under the microscope. After this incubation step, any probe that was not bound to this mRNA was washed away.

THE GOAL (DISCOVERY-BASED SCIENCE)

The goal was to further understand how the experimental injection of RNA was responsible for the silencing of particular mRNAs.

ACHIEVING THE GOAL

FIGURE 16.11 **Injection of antisense and double-stranded RNA into *C. elegans* to compare their effects on mRNA silencing.**

Starting material: The researchers used *C. elegans* as their model organism. They also had the cloned *mex-3* gene, which had been previously shown to be highly expressed in the embryo.

1. Make sense and antisense *mex-3* RNA in vitro using cloned genes for *mex-3* with promoters on either side of the gene. RNA polymerase and nucleotides are added to synthesize the RNAs.

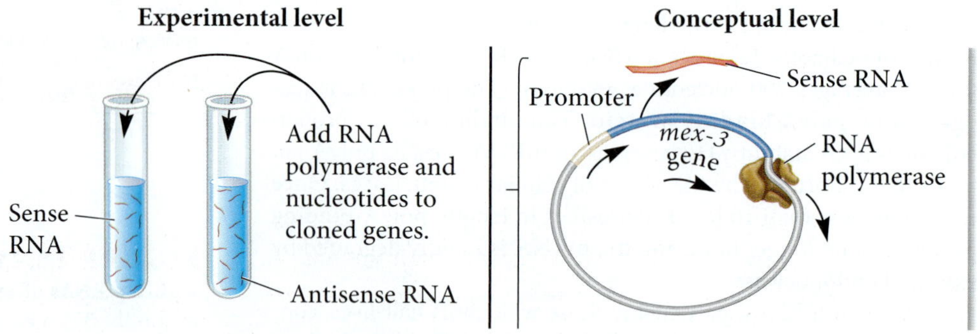

2. Inject either *mex-3* antisense RNA or a mixture of *mex-3* sense and antisense RNA into the gonads of *C. elegans*. This RNA is taken up by the eggs and early embryos. As a control, do not inject any RNA.

3. Incubate and then subject early embryos to *in situ* hybridization. In this method, a labeled probe is added that is complementary to *mex-3* mRNA. If cells express *mex-3*, the mRNA in the cells will bind to the probe and become labeled. After incubation with a labeled probe, the cells are washed to remove unbound probe.

4. Observe embryos under the microscope.

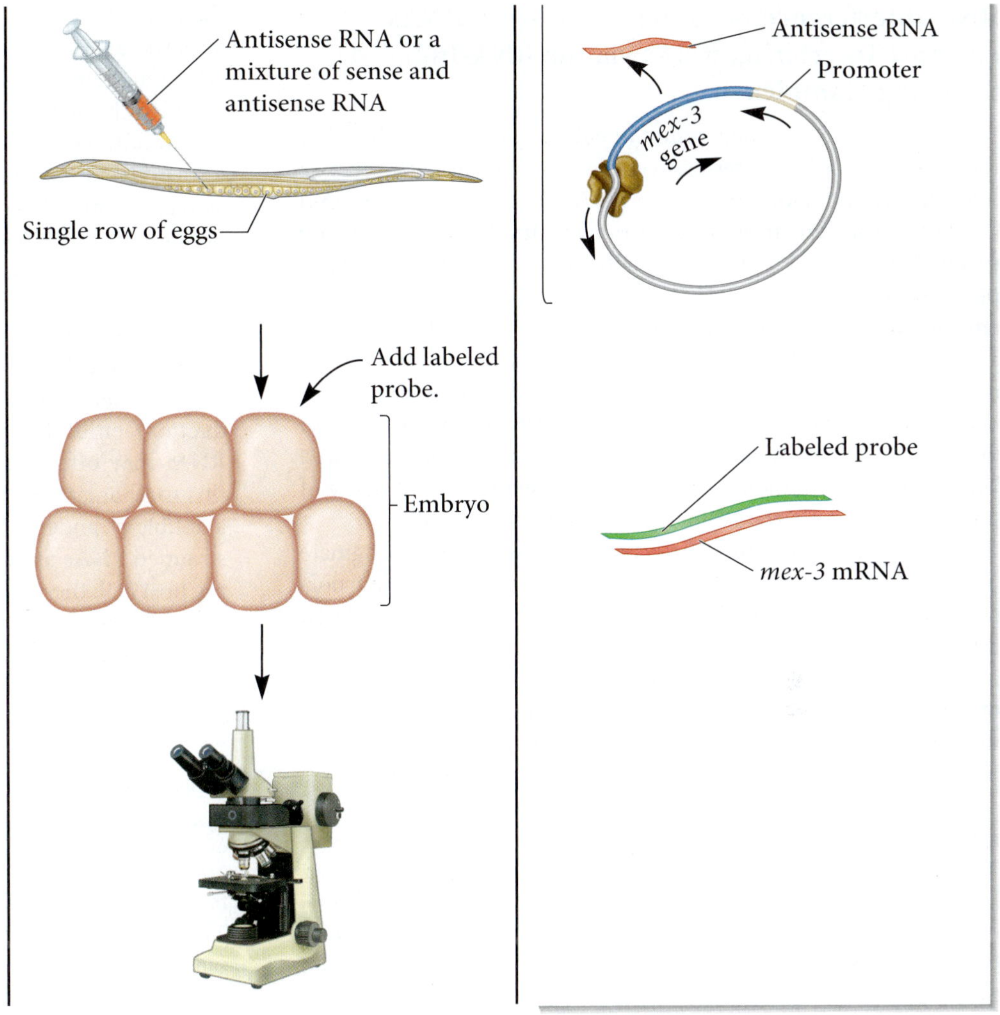

THE DATA

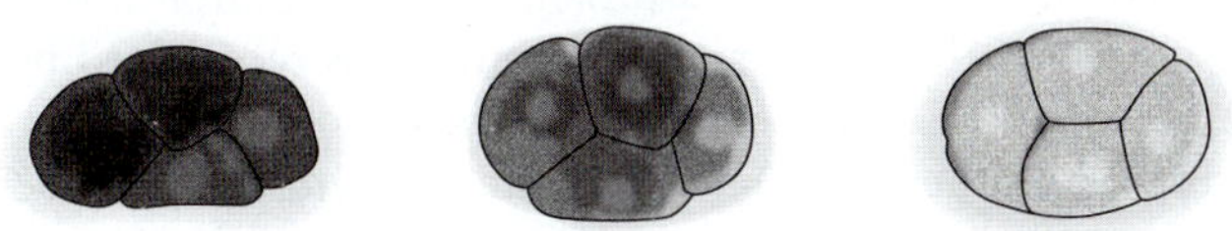

Control

Injected with *mex-3* antisense RNA

Injected with double-stranded RNA (both *mex-3* sense and antisense RNA)

Data adapted from A. Fire, S. Xu, M. K. Montgomery, et al. (1998) Potent and specific genetic interference by double-stranded RNA in *Caenorhabditis elegans*. *Nature 391*, 806–811.

INTERPRETING THE DATA

As seen in the data of Figure 16.11, the control embryos were very darkly labeled. These results indicated that the control embryos contained a high amount of *mex-3* mRNA, which was known from previous research. In the embryos that had received antisense RNA, *mex-3* mRNA levels were decreased, but detectable, as shown by light labeling. Remarkably, in embryos that received double-stranded RNA, no *mex-3* mRNA was detected! These results indicated that double-stranded RNA is more potent at silencing mRNA than is antisense RNA. In this case, the double-stranded RNA caused the *mex-3* mRNA to be degraded. They used the term **RNA interference (RNAi)** to describe the phenomenon in which double-stranded RNA causes the silencing of mRNA. This surprising observation led researchers to investigate the underlying molecular mechanism that accounts for this phenomenon, as described next.

A self-help quiz involving this experiment can be found at www.mhhe.com/brookergenetics5e.

RNA Interference Is Mediated by MicroRNAs or Short-Interfering RNAs via the RNA-Induced Silencing Complex

RNA interference is widely found in eukaryotic species. The RNA that promotes RNA interference can come from two sources: microRNAs and short-interfering RNAs. **MicroRNAs (miRNAs)** are RNAs that are transcribed from eukaryotic genes that are endogenous—genes that are normally found in the genome. They play key roles in regulating gene expression, particularly during embryonic development in animals and plants. Most commonly, a single type of miRNA inhibits the translation of several different mRNAs. An miRNA and an mRNA hydrogen bond to each other because they have sequences that are partially complementary. In humans, estimates suggest that over 1000 genes encode miRNAs. These miRNAs are thought to regulate the expression of many of the mRNAs that are transcribed by the human genome.

By comparison, **short-interfering RNAs (siRNAs)** originate from sources that are exogenous, which means they are not normally made by cells. The siRNAs can come from viruses that infect a cell, or researchers can make siRNAs to study gene function experimentally. In most cases, siRNAs are a perfect match to a single type of mRNA. The functioning of siRNAs is thought to play a key role in preventing certain types of viral infections. In addition, siRNAs have become important experimental tools in molecular biology.

How do miRNAs and siRNAs cause the silencing of specific mRNAs? **Figure 16.12** shows how an miRNA or siRNA leads to RNA interference. In this example, the miRNA is first synthesized as a pri-miRNA (for primary-miRNA) in the nucleus. Due to complementary base pairing, the pri-miRNA forms a stem-loop structure with long single-stranded 5′ and 3′ ends. It is processed in the nucleus by an RNase called Drosha and a dsRNA-binding protein called DGCR8. The result is a 70-nucleotide RNA molecule that is called a pre-miRNA (not to be confused with pri-miRNA). The pre-miRNA is then exported from the nucleus with the aid of a dsRNA-binding protein called exportin 5.

As shown in Figure 16.12, siRNAs do not go through the processing events that occur in the nucleus. Instead, pre-siRNAs are derived from viral RNAs or they may be made by researchers and taken up by cells. For example, in the work of Fire and Mello described in Experiment 16A, the double-stranded *mex-3* RNA is an example of a pre-siRNA. In Figure 16.12, the pre-siRNA is formed from two complementary RNA molecules that base-pair with each other.

In the cytosol, both pre-miRNAs and pre-siRNAs are cut by an endonuclease called dicer (see Figure 16.12). This releases a double-stranded RNA molecule that is typically 20–25 bp long. This double-stranded RNA associates with proteins to form a complex called the **RNA-induced silencing complex (RISC).** One of the RNA strands is degraded. The remaining single-stranded miRNA or siRNA is complementary to specific mRNAs that will be silenced. The miRNA or siRNA allows the RISC to recognize and hydrogen bond to such mRNA molecules.

After binding to an mRNA, two different things may happen. In some cases, the RISC may inhibit translation. This is more common for miRNAs, which often are only partially complementary to their target mRNAs. The RISC-mRNA complex is sent to a cellular structure called a **processing body (P-body),** where it may be stored and later reused. The mRNA may also be degraded in the P-body. Alternatively, the RISC may immediately direct the degradation of the mRNA. This usually occurs for siRNAs that typically are a perfect match to their target mRNA. In either case, the expression of the mRNA is silenced. The effect is termed RNA interference because the miRNA or siRNA interferes with the proper expression of an mRNA. In 2006, Fire and Mello received the Nobel Prize in physiology or medicine for their discovery of this mechanism.

RNA interference is believed to have at least three benefits. First, this mechanism represents an important form of gene regulation. When genes encoding pri-miRNAs are turned on, the production of miRNAs silences the expression of other genes. Second, RNAi may offer a defense against certain viruses. In particular, this mechanism may inhibit the proliferation of viruses that have a double-stranded RNA genome and viruses that produce double-stranded RNAs as part of their reproductive cycles. After entering the host cell, the viral RNA would be degraded, as in Figure 16.12, and the cell would survive the infection. Third, researchers have speculated that RNAi may also play a role in the silencing of genes within certain transposable elements (TEs), which are discussed in Chapter 19.

RNA-Binding Proteins Bind to Specific mRNAs and Regulate Translation and RNA Stability

Specific mRNAs are sometimes regulated by RNA-binding proteins that directly affect translational initiation or RNA stability. The regulation of iron assimilation provides a well-studied example in which both of these phenomena occur. Before discussing this form of translational control, let's consider the biology of iron metabolism.

Iron is an essential element for the survival of living organisms because it is required for the function of many different enzymes. The pathway by which mammalian cells take up iron is depicted in **Figure 16.13**. Iron (Fe^{3+}) ingested by an animal becomes bound to transferrin, a protein that carries iron through the bloodstream. The transferrin-Fe^{3+} complex is recognized by a transferrin receptor on the surface of cells; the complex binds to the receptor and then is transported into the cytosol by endocytosis to form an endocytic vesicle. The iron within the vesicle is then released from transferrin and transported out of the vesicle. At this stage, Fe^{3+} may bind to enzymes that require iron for their activity. Alternatively, if too much iron is present, the excess iron is stored within a hollow, spherical protein known as ferritin. The storage of excess iron within ferritin helps to prevent the toxic buildup of too much iron within the cell.

Because iron is a vital yet potentially toxic substance, animals have evolved an interesting way to regulate its assimilation. The two mRNAs that encode ferritin and the transferrin receptor are both influenced by an RNA-binding protein known as the **iron regulatory protein (IRP).** How does IRP exert its effects? This protein binds to a regulatory element within these two mRNAs known as the **iron response element (IRE).** The ferritin mRNA has an IRE in its 5′-untranslated region (5′-UTR).

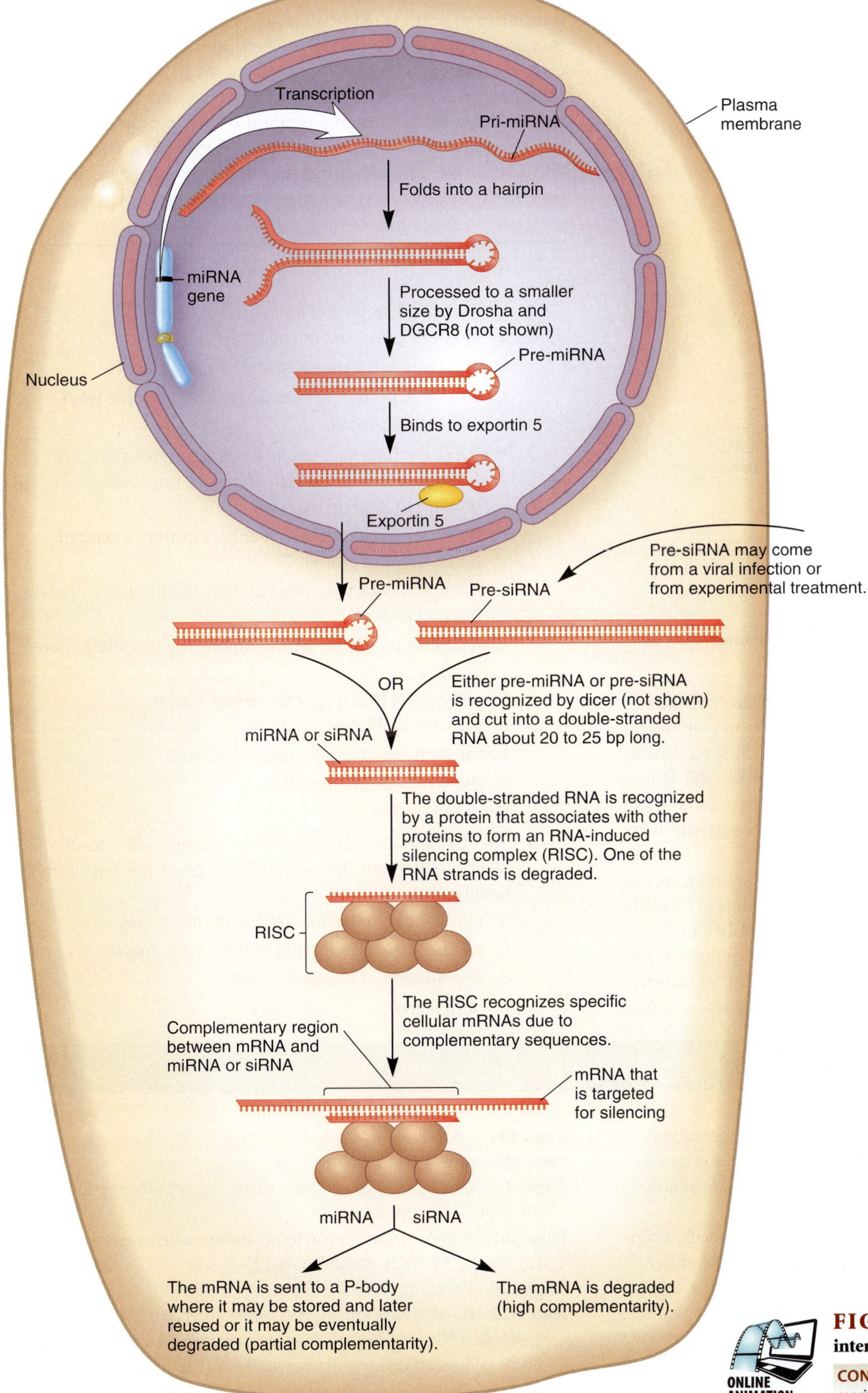

ONLINE ANIMATION

FIGURE 16.12 **Mechanism of RNA interference.**

CONCEPT CHECK: Explain why RISC binds to a specific mRNA. What type of bonding occurs?

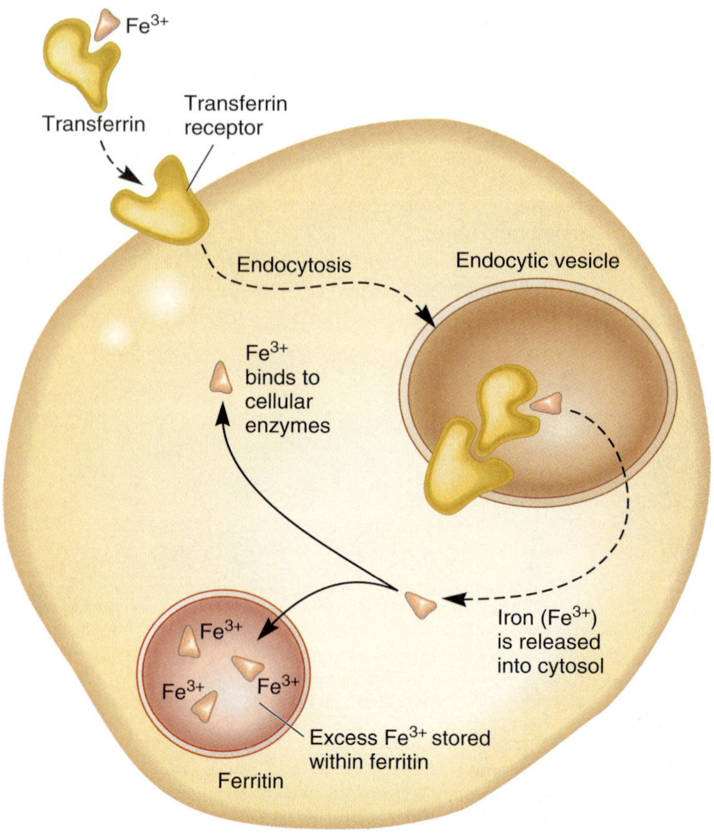

FIGURE 16.13 **The uptake of iron (Fe^{3+}) into mammalian cells.**

When IRP binds to this IRE, it inhibits the translation of the ferritin mRNA (**Figure 16.14a**, left). However, when iron is abundant in the cytosol, the iron binds directly to IRP and causes a conformational change that prevents it from binding to the IRE. Under these conditions, the ferritin mRNA is translated to make more ferritin protein (Figure 16.14a, right), which prevents the toxic buildup of iron within the cytosol.

The transferrin receptor mRNA also contains an iron response element, but it is located in the 3′-UTR. When IRP binds to this IRE, it does not inhibit translation. Instead, the binding of IRP increases the stability of the mRNA by blocking the action of endonucleases that degrade the RNA. This leads to increased amounts of transferrin receptor mRNA within the cell when the cytosolic levels of iron are very low (**Figure 16.14b**, left). Under these conditions, more transferrin receptor is made. This promotes the uptake of iron when it is in short supply. In contrast, when iron is abundant within the cytosol, the iron binds to IRP, which causes IRP to dissociate from transferrin receptor mRNA. After this occurs, the mRNA becomes rapidly degraded because the removal of IRP exposes sites that are recognized by endonucleases (Figure 16.14b, right). This leads to a decrease in the amount of transferrin receptor, thereby helping to prevent the uptake of too much iron into the cell.

16.4 COMPREHENSION QUESTIONS

1. What is an advantage of alternative splicing?
   a. Multiple genes can encode the same protein.
   b. The same gene can encode multiple proteins.
   c. The same protein can be digested in multiple ways.
   d. All of the above are true.
2. Which of the following would decrease mRNA stability?
   a. A longer polyA tail, which enhances the binding of polyA-binding protein
   b. A shorter polyA tail, which enhances the binding of polyA-binding protein
   c. A longer polyA tail, which inhibits the binding of polyA-binding protein
   d. A shorter polyA tail, which inhibits the binding of polyA-binding protein
3. The process of RNA interference may lead to
   a. the degradation of an mRNA.
   b. the inhibition of translation of an mRNA.
   c. the synthesis of an mRNA.
   d. both a and b.
4. The binding of iron regulatory protein (IRP) to the iron response element (IRE) in the 5′ region of the ferritin mRNA results in the
   a. inhibition of translation of the ferritin mRNA.
   b. stimulation of translation of the ferritin mRNA.
   c. degradation of the ferritin mRNA.
   d. both a and c.

KEY TERMS

Page 393. epigenetics, epigenetic inheritance, transgenerational epigenetic inheritance, DNA methylation, chromatin remodeling, covalent histone modification, histone variants, feedback loops, noncoding RNAs

Page 394. *cis*-epigenetic mechanism, *trans*-epigenetic mechanism

Page 397. development, imprinting control region (ICR), differentially methylated region (DMR), de novo methylation

Page 398. maintenance methylation, X-chromosome inactivation (XCI), pluripotency factors, symmetry break

Page 400. trithorax group (TrxG), polycomb group (PcG), trimethylation, polycomb response element (PRE)

Page 404. oncogene

Page 405. alternative splicing

Page 406. constitutive exons, alternative exons, splicing factors, SR proteins, exon skipping

Page 407. polyA-binding protein, 3′-untranslated region (3′-UTR), AU-rich element (ARE)

Page 409. RNA interference (RNAi)

Page 410. microRNAs (miRNAs), short-interfering RNAs (siRNAs), RNA-induced silencing complex (RISC), processing body (P-body), iron regulatory protein (IRP), iron response element (IRE)

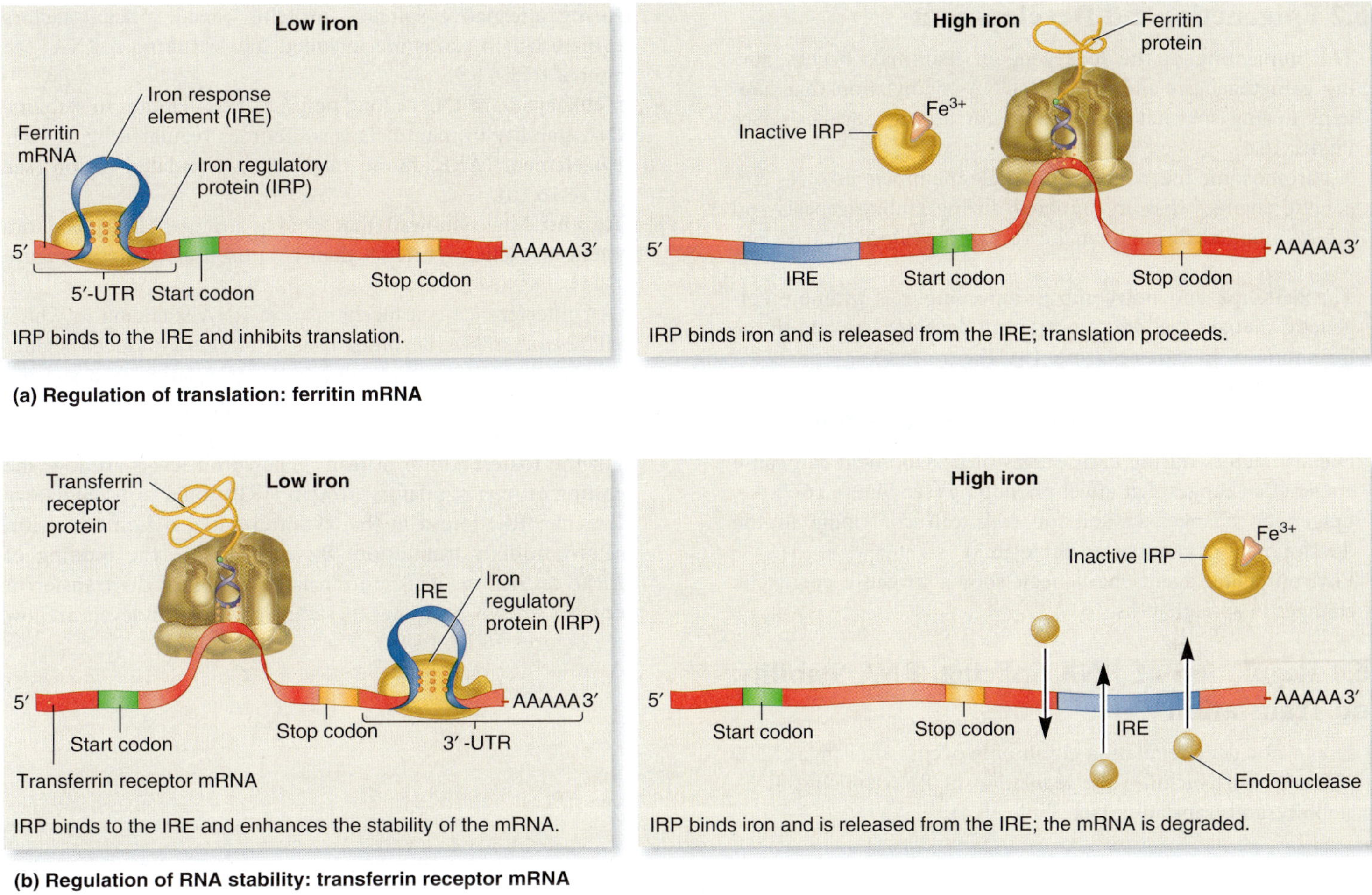

FIGURE 16.14 The regulation of iron assimilation genes by IRP and IRE. **(a)** When the Fe^{3+} concentration is low, the binding of the iron regulatory protein (IRP) to the iron response element (IRE) in the 5′-UTR of ferritin mRNA inhibits translation (left). When Fe^{3+} concentration is high and Fe^{3+} binds to IRP, the IRP is removed from the ferritin mRNA, so translation can proceed (right). **(b)** The binding of IRP to IRE in the 3′-UTR of the transferrin receptor mRNA enhances the stability of the mRNA and leads to a higher concentration of this mRNA. Therefore, more transferrin receptor proteins are made when the Fe^{3+} concentration is low (left). When the Fe^{3+} concentration is high and this metal binds to IRP, the IRP dissociates from the IRE, and the transferrin receptor mRNA is rapidly degraded (right).

Genes → Traits This form of translational control allows cells to use iron appropriately. When the cellular concentration of iron is low, the stability of the transferrin receptor mRNA is increased, thereby enhancing the ability of cells to synthesize more receptors and take up more iron. Also, the translation of ferritin mRNA is inhibited, because ferritin is not needed to store excess iron. By comparison, when the cellular concentration of iron is high, the translation of ferritin mRNA is enhanced. This leads to the synthesis of ferritin and prevents the toxic buildup of iron. When iron is high, the transferrin receptor mRNA is degraded, which decreases further uptake of iron.

CONCEPT CHECK: If a mutation prevented IRP from binding to the IRE in the ferritin mRNA, how would the mutation affect the regulation of ferritin synthesis? Do you think there would be too much or too little ferritin?

CHAPTER SUMMARY

16.1 Overview of Epigenetics

- Epigenetics can be defined as the study of mechanisms that lead to changes in gene expression that are passed from cell to cell and are reversible but do not involve a change in the sequence of DNA. The transmission of epigenetic changes from one generation to the next is referred to as epigenetic inheritance.
- The most common types of molecular changes that underlie epigenetic control are DNA methylation, chromatin remodeling, covalent histone modification, the localization of histone variants, and feedback loops (see Table 16.1).
- Epigenetic changes can be established by transcription factors or noncoding RNAs (see Figure 16.1).
- Epigenetic changes may be maintained by *cis*- or *trans*-epigenetic mechanisms (see Figures 16.2, 16.3).
- Some epigenetic changes are programmed during development and others are caused by environmental agents (see Table 16.2).

16.2 Epigenetics and Development

- The imprinting of the *Igf2* gene in mammals occurs during gametogenesis and involves DNA methylation that happens during spermatogenesis but not during oogenesis (see Figure 16.4).
- X-chromosome inactivation in female mammals involves epigenetic changes that are initiated during embryogenesis and are maintained throughout the rest of development (see Figure 16.5).
- The trithorax and polycomb group complexes promote epigenetic changes that are important in development and in the formation of specific cell types (see Figure 16.6).

16.3 Epigenetics and Environmental Agents

- Dietary factors during early stages of development can cause epigenetic changes that affect phenotype (see Figure 16.7).
- Epigenetic changes in somatic cells can contribute to the development of cancer (see Table 16.3).
- Environmental agents have been shown to cause epigenetic changes in somatic cells.

16.4 Regulation of RNA Splicing, RNA Stability, and Translation

- Eukaryotic gene regulation commonly occurs after the mRNA is made. This includes the regulation of RNA splicing, RNA stability, and translation (see Table 16.4).
- During alternative splicing, proteins called splicing factors regulate which exons are included in a resulting mRNA (see Figures 16.8, 16.9).
- In eukaryotic mRNA, a long polyA tail contributes to stability. RNA stability in mammals is sometimes regulated by an AU-rich element (ARE) found in the 3′-untranslated region (see Figure 16.10).
- Fire and Mello showed that double-stranded RNA is more potent at silencing mRNA than is antisense RNA (see Figure 16.11).
- RNA interference is a mechanism of RNA silencing in which miRNA or siRNA becomes part of an RNA-induced silencing complex (RISC) that inhibits the translation of a specific mRNA or causes its degradation (see Figure 16.12).
- The regulation of iron uptake and storage is needed to prevent the toxic buildup of iron. When iron levels are low, the binding of iron regulatory protein (IRP) to an iron regulatory element (IRE) found in the 5′-untranslated region of ferritin mRNA inhibits translation. By comparison, the binding of IRP to an IRE in the 3′-untranslated region of the transferrin receptor mRNA promotes its stability when iron levels are low (see Figures 16.13, 16.14).

PROBLEM SETS & INSIGHTS

Solved Problems

S1. Are the following examples best explained by genetic and/or epigenetic phenomena?

A. imprinting of the *Igf2* gene

B. variation in coat color in mice carrying the A^{vy} allele

C. formation of cancer cells

D. variation in flower color between different strains of pea plants, such as purple versus white

E. X-chromosome inactivation

Answer:

A. epigenetic

B. epigenetic

C. usually both genetic and epigenetic

D. genetic; it is due to variation in DNA sequences

E. epigenetic

S2. Explain how a noncoding RNA could play a role in establishing an epigenetic modification.

Answer: A noncoding RNA can bind to a specific site in a chromosome due to complementary base pairing or it may bind to proteins that are attached to DNA sites. After binding, the noncoding RNA can act as a bridge to attract other proteins to the site that cause epigenetic modifications, such as DNA methylation and covalent histone modifications.

S3. Prior to X-chromosome inactivation, what prevents the expression of the *Xist* gene?

Answer: Prior to X-chromosome inactivation, pluripotency factors promote the expression of the *Tsix* gene. The expression of the *Tsix* gene inhibits the expression of the *Xist* gene.

S4. As shown in the following diagram, a pre-mRNA contains 7 exons. If a splicing repressor binds at the site of the arrow (just before exon 5), what exons would be contained in the mature mRNA?

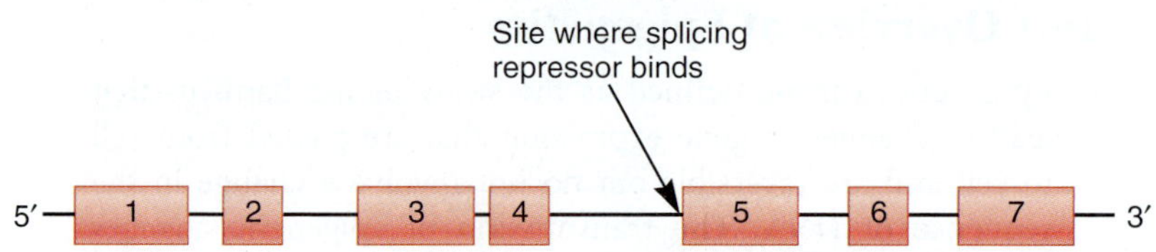

Answer: The splicing repressor would cause exon skipping. In this case, it would skip exon 5, so the mature mRNA would contain exons 1, 2, 3, 4, 6, and 7.

S5. What are the two alternative ways that IRP can affect gene expression at the RNA level?

Answer: The ferritin mRNA has an IRE in its 5′-UTR. When IRP binds to this IRE, it inhibits the translation of the ferritin mRNA. This decreases the amount of ferritin protein, which is not needed when iron levels are low. However, when iron is abundant in the cytosol, the iron binds directly to IRP and prevents it from binding to the IRE. This allows the ferritin mRNA to be translated, producing more ferritin protein. An IRE is also located in the transferrin receptor mRNA in the 3′-UTR. When IRP binds to this IRE, it increases the stability of the mRNA, which leads to an increase in the amount of transferrin receptor mRNA within the cell when the cytosolic levels of iron are very low. Under these conditions, more transferrin receptor is made to promote the uptake of iron, which is in short supply. When the iron is found in abundance within the cytosol, IRP is removed from the mRNA, and the mRNA becomes rapidly degraded. This leads to a decrease in the amount of the transferrin receptor.

Conceptual Questions

C1. Define epigenetics. Are all epigenetic changes passed from parent to offspring? Explain.

C2. List and briefly describe five types of molecular events that may underlie epigenetic gene regulation.

C3. Explain how epigenetic changes may be targeted to specific genes.

C4. What is the key difference between a *cis*- and *trans*-epigenetic mechanism that maintains an epigenetic modification? In Chapter 5, we considered genomic imprinting of the *Igf2* gene in which offspring express the copy of the gene they inherit from their father, but not the copy they inherit from their mother. Is this a *cis*- or *trans*-epigenetic mechanism?

C5. Epigenetic modifications may be programmed during development or they be caused by environmental agents. Of the following examples, which are programmed during development and which are caused by environmental agents?

A. imprinting of the *Igf2* gene in mammals

B. development of queen and worker honeybees

C. X-chromosome inactivation in mammals

D. development of cancer

C6. Explain how DNA methylation and the formation of a DNA loop controls the expression of the *Igf2* gene in mammals. How is this gene imprinted so that only the paternal copy is expressed in offspring?

C7. Let's suppose a mutation removes the ICR next to the *Igf2* gene. If this mutation was inherited from the mother, would the *Igf2* gene (from the mother) be silenced or expressed? Explain.

C8. Outline the molecular steps in the process of X-chromosome inactivation (XCI). Which step plays a key role in choosing the X chromosome that will remain active versus the one that will become inactivated?

C9. Following X-chromosome inactivation, most of the genes on the inactivated X chromosome are silenced. Explain how. Name one gene that is not silenced.

C10. In general, explain how epigenetic modifications are an important mechanism for developmental changes that lead to specialized body parts and cell types. How do the trithorax and polycomb group complexes participate in this process?

C11. What are the contrasting roles of trithorax and polycomb group complexes during development in animals and plants?

C12. Describe the molecular steps by which polycomb group complexes cause epigenetic gene silencing.

C13. With regard to development, what would be the dire consequences if polycomb group complexes did not function properly?

C14. Using coat color in mice and the development of female honeybees as examples, how can dietary factors cause epigenetic modifications, leading to phenotypic effects?

C15. How can environmental agents that do not cause gene mutations contribute to cancer? Would these epigenetic changes be passed to offspring?

C16. Define alternative splicing. What are advantages and disadvantages of this process?

C17. What is the function of a splicing factor? Explain how splicing factors can regulate the cell-specific splicing of mRNAs.

C18. Figure 16.8 shows the products of alternative splicing for the α-tropomyosin pre-mRNA. Let's suppose that smooth muscle cells produce splicing factors that are not produced in other cell types. Explain where you think such splicing factors bind and how they influence the splicing of the α-tropomyosin pre-mRNA.

C19. Let's suppose a person is homozygous for a mutation in the *IRP* gene that changed the structure of the iron regulatory protein in such a way that it could not bind iron, but it could still bind to IREs. How would this mutation affect the regulation of ferritin and transferrin receptor mRNAs? Do you think such a person would need more iron in his or her diet than normal individuals? Do you think that excess iron in the diet would be more toxic than it would be for normal individuals? Explain your answer.

C20. In response to potentially toxic substances (e.g., high levels of iron), eukaryotic cells often use translational regulatory mechanisms to prevent cell death, rather than using transcriptional regulatory mechanisms. Explain why.

C21. What are the advantages and disadvantages of mRNAs with a short half-life compared with mRNAs with a long half-life?

C22. What is the phenomenon of RNA interference (RNAi)? During RNAi, explain how the double-stranded RNA is processed and how it leads to the silencing of a complementary mRNA.

C23. With regard to RNAi, what are three possible sources for double-stranded RNA?

C24. What is the relationship between mRNA stability and mRNA concentration? What factors affect mRNA stability?

C25. Describe how the binding of the iron regulatory protein affects the mRNAs for ferritin and the transferrin receptor. How does iron influence this process?

Experimental Questions

E1. A gene, which we will call gene C, can be epigenetically modified in such a way that its expression in some cells is permanently silenced. Describe how you could conduct cell fusion experiments to determine if a *cis-* or *trans-*epigenetic mechanism is responsible for maintaining the silencing of gene C.

E2. In the experiments described in Figure 16.7, explain the relationship between coat color and DNA methylation. How is coat color related to the diet of the mother?

E3. 5-Azocytidine is an inhibitor of DNA methyltransferase. If it were fed to female mice during pregnancy, explain how you think it would affect the coat color of offspring carrying the A^{vy} allele.

E4. A research study indicated that an agent in cigarette smoke caused the silencing of a tumor suppressor gene called *p53*. However, upon sequencing, no mutation was found in the DNA sequence for this gene. Give two possible explanations for these results.

E5. Let's suppose you were interested in developing drugs to prevent epigenetic changes that may contribute to cancer. What cellular proteins would be the target of your drugs? What possible side effects might your drugs cause?

E6. In Experiment 16A by Fire and Mello, were they injecting pre-miRNA or pre-siRNA? Explain.

E7. Explain how the data of Fire and Mello suggested that double-stranded RNA is responsible for the silencing of the *mex-3* mRNA.

E8. Chapter 20 describes a blotting method known as Northern blotting, in which a short segment of cloned DNA is used as a probe to detect RNA that is transcribed from a particular gene. The DNA probe, which is labeled, is complementary to the RNA that the researcher wishes to detect. After the probe DNA binds to the RNA within a blot of a gel, the RNA is visualized as a dark band. The method of Northern blotting can be used to determine the amount of a particular RNA transcribed in a given cell type. If one type of cell produces twice as much of a particular mRNA as another cell, the band appears twice as intense.

For this question, a researcher has a DNA probe complementary to the ferritin mRNA. This probe can be used to specifically detect the amount of ferritin mRNA on a gel. A researcher began with two flasks of human skin cells. One flask contained a very low concentration of iron, and the other flask had a high concentration of iron. The mRNA was isolated from these cells and then subjected to Northern blotting, using a probe complementary to the ferritin mRNA. The sample loaded in lane 1 was from the cells grown in a low concentration of iron, and the sample in lane 2 was from the cells grown in a high concentration of iron. Three Northern blots are shown here, but only one of them is correct. Based on your understanding of ferritin mRNA regulation, which blot (a, b, or c) would be your expected result? Explain. Which blot (a, b, or c) would be your expected result if the gel had been probed with a DNA segment complementary to the transferrin receptor mRNA?

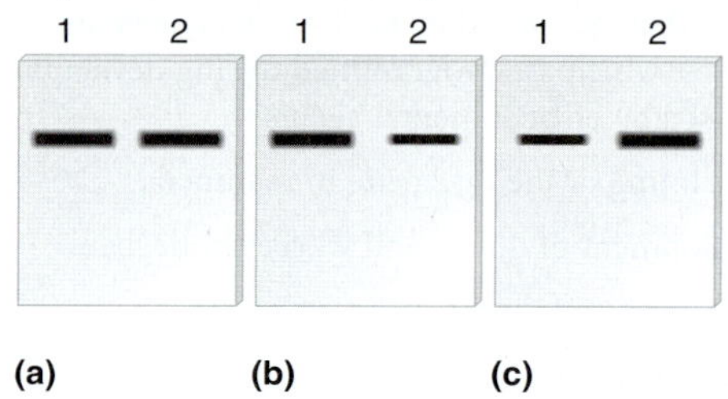

Questions for Student Discussion/Collaboration

1. Go to the PubMed website and search words such as *epigenetic* and *cancer*. Scan through the journal articles you retrieve and make a list of environmental agents that may cause epigenetic changes that contribute to cancer.
2. Discuss the similarities and differences of phenotypic variation that are caused by epigenetic gene regulation versus variation in gene sequences (epigenetics versus genetics).
3. How are regulatory transcription factors (described in Chapter 15) and regulatory splicing factors (described in this chapter) similar in their mechanism of action? In your discussion, consider the domain structures of both types of proteins. How are they different?

Note: All answers appear at the website for this textbook; the answers to the even-numbered questions and all of the Concept Check and Comprehension Questions are in the back of the book.

www.mhhe.com/brookergenetics5e

Visit the website for practice tests, answer keys, and other learning aids for this chapter. Enhance your understanding of genetics with our interactive exercises, quizzes, animations, and much more.

CHAPTER OUTLINE

A plant infected with tobacco mosaic virus.

17 GENETICS OF VIRUSES

Viruses are nonliving, infectious particles with nucleic acid genomes. Viruses are considered nonliving because they do not exhibit all of the properties associated with living organisms. For example, viruses are not composed of cells, and by themselves, they do not carry out metabolism, use energy, maintain homeostasis, or even reproduce. A virus or its genetic material must be taken up by a living cell to replicate.

The first virus to be discovered was tobacco mosaic virus (TMV), a virus that infects many species of plants. On the leaves, TMV causes mosaic-like patterns in which normal-colored patches are interspersed with light green or yellowish patches (see chapter-opening photo). TMV damages leaves, flowers, and fruit, but almost never kills the plant. In 1883, the German scientist Adolf Mayer determined that this disease could be spread by spraying the sap from one plant onto another. By subjecting this sap to filtration, the Russian scientist Dmitri Ivanovski demonstrated that the disease-causing agent was not a bacterium. Sap that had been passed through filters with pores small enough to trap bacterial cells was still able to spread the disease. At first, some researchers suggested the agent was a chemical toxin. However, the Dutch botanist Martinus Beijerinck ruled out this possibility by showing that sap could continue to transmit the disease after many plant generations. A toxin would have been diluted after many generations, whereas Beijerinck's results indicated the disease agent was multiplying in the plant. Around the same time, animal viruses were discovered in connection with a livestock infection called foot-and-mouth disease. In 1900, the first human virus—the virus that causes yellow fever—was identified.

Since these early studies, microbiologists, geneticists, and molecular biologists have taken a great interest in the structure, genetic composition, and replication of viruses. All organisms are susceptible to infection by viruses. Once a cell is infected, the genetic material of a virus orchestrates a series of events that ultimately leads to the production of new virus particles. In this chapter, we will first examine the structure and genetic composition of viruses and explore the general features of viral reproductive cycles. We will then take a closer look at the reproductive cycle of a virus called phage λ, which infects *E. coli* cells, and then examine the reproductive cycle of human immunodeficiency virus (HIV).

17.1 VIRUS STRUCTURE AND GENETIC COMPOSITION

Learning Outcomes:

1. Compare and contrast viruses with regard to host range, structure, and genomes.
2. Analyze the results of an experiment indicating that the genome of tobacco mosaic virus is RNA.

Researchers have identified and studied over 4000 different types of viruses. In this section, we will survey the basic features of viruses and consider an early experiment that showed that some viruses use RNA as their genetic material.

Viruses Differ in Their Host Range, Structure, and Genome Composition

Although all viruses share some similarities, such as small size and the reliance on a living cell for replication, they vary greatly in their characteristics, including their host range, structure, and genome composition. Some of the major differences are described next, and characteristics of selected viruses are shown in **Table 17.1**.

Host Range A cell that is infected by a virus is called a **host cell,** and a species that can be infected by a specific virus is called a host species for that virus. Viruses differ greatly in their **host range**—the number of species they can infect. Table 17.1 lists a few examples of viruses with widely different ranges of host species. Tobacco mosaic virus has a broad host range, being known to infect over 150 different species of plants. By comparison, other viruses have a narrow host range, with some infecting only a single species. Furthermore, a virus may infect only certain cell types in a host species. For example, HIV specifically infects a type of lymphocyte called a helper T cell.

Structure Although the existence of viruses was postulated in the 1890s, viruses were not observed until the 1930s, when the electron microscope was invented. Viruses cannot be resolved by even the best light microscope. Most of them are smaller than the wavelength of visible light. Viruses range in size from about 20–400 nm in diameter (1 nanometer = 10^{-9} meters). For comparison, a typical bacterium is 1000 nm in diameter, and the diameter of most eukaryotic cells is 10 to 1000 times that of a bacterium. Adenoviruses, which cause infections of the respiratory and gastrointestinal tracts, have an average diameter of 75 nm. Over 50 million adenoviruses could fit into an average-sized human cell.

What are the common structural features of all viruses? As shown in **Figure 17.1**, all viruses have a protein coat called a **capsid,** which encloses its genome and consists of one or more molecules of nucleic acid. Capsids are composed of one or several different protein subunits called capsomers. Capsids have a variety of shapes, including helical and polyhedral. Figure 17.1a shows the structure of TMV, which has a helical capsid made of identical capsomers. Figure 17.1b shows an adenovirus, which has a polyhedral capsid. Protein fibers with a terminal knob are located at the corners of the polyhedral capsid. Many viruses that infect animal cells, such as the influenza virus shown in Figure 17.1c, have a **viral envelope** enclosing the capsid. The envelope consists of a lipid bilayer that is derived from the plasma membrane of the host cell and is embedded with virally encoded spike glycoproteins, also called spikes or peplomers. In addition to encasing and protecting the genetic material, the capsid and envelope enable viruses to infect their hosts. In many viruses, the capsids or envelopes have specialized proteins, including protein fibers with a knob (Figure 17.1b) and spike glycoproteins (Figure 17.1c), that help them bind to the surface of a host cell. Viruses that infect bacteria, called **bacteriophages,** or **phages,** may have more complex capsids, with accessory structures used for anchoring the virus to a host cell and injecting the viral nucleic acid (Figure 17.1d). As discussed later, the tail fibers of such bacteriophages attach the virus to the bacterial cell wall.

Genome Composition The genetic material in a virus is called a **viral genome.** The composition of viral genomes varies markedly among different types of viruses, as suggested by the examples in Table 17.1. The nucleic acid of some viruses is DNA, whereas in others it is RNA, referred to as DNA viruses and RNA

TABLE 17.1
Hosts and Characteristics of Selected Viruses

Virus or Group of Viruses	Host	Effect on Host	Nucleic Acid*	Genome Size (kb)†	Number of Genes‡
Phage fd	*E. coli*	Slows growth	ssDNA	6.4	10
Phage λ	*E. coli*	Can exist harmlessly in the host cell or cause lysis	dsDNA	48.5	36
Phage T4	*E. coli*	Causes lysis	dsDNA	169	288
Phage Qß	*E. coli*	Slows growth	ssRNA	4.2	4
Tobacco mosaic virus (TMV)	Many plants	Causes mottling and necrosis of leaves and other plant parts	ssRNA	6.4	6
Baculoviruses	Insects	Usually kill the insect	dsDNA	133.9	154
Parvovirus	Mammals	Causes respiratory, flulike symptoms	ssDNA	5.0	5
Influenza virus	Mammals	Causes classic flu symptoms—fever, cough, sore throat, and headache	ssRNA	13.5	11
Epstein-Barr virus	Humans	Causes mononucleosis, with fever, sore throat, and fatigue	dsDNA	172	80
Adenovirus	Humans	Causes respiratory symptoms and diarrhea	dsDNA	34	35
Herpes simplex type II	Humans	Causes blistering sores around the genital region	dsDNA	158.4	77
HIV	Humans	Causes AIDS, an immunodeficiency syndrome that can lead to death	ssRNA	9.7	9

*ss refers to single stranded and ds refers to double stranded.
†Number of thousands of nucleotides or nucleotide base pairs.
‡This number refers to the number of protein-encoding units. In some cases, two or more proteins are made from a single gene due to events such as protein processing.

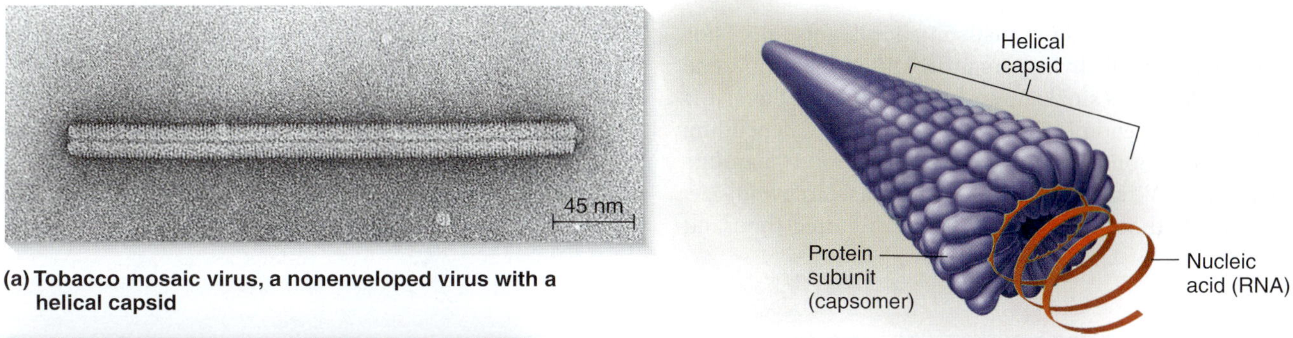

(a) Tobacco mosaic virus, a nonenveloped virus with a helical capsid

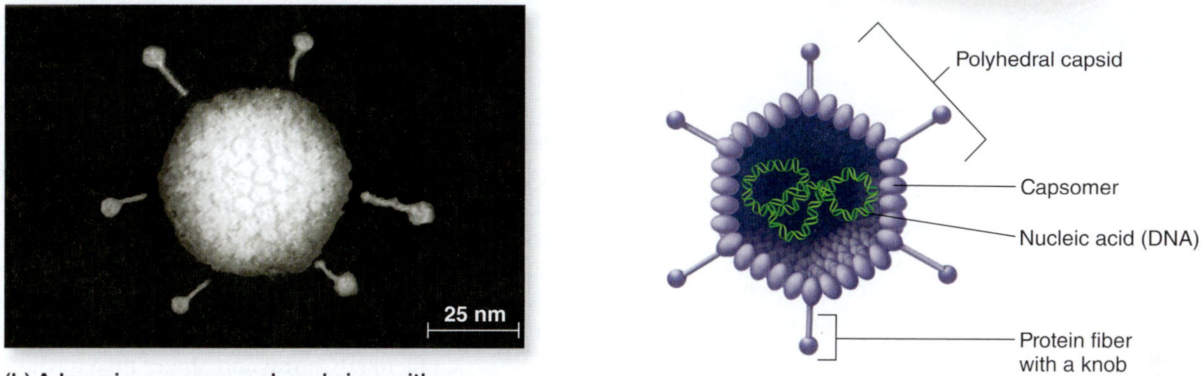

(b) Adenovirus, a nonenveloped virus with a polyhedral capsid and protein fibers with a knob

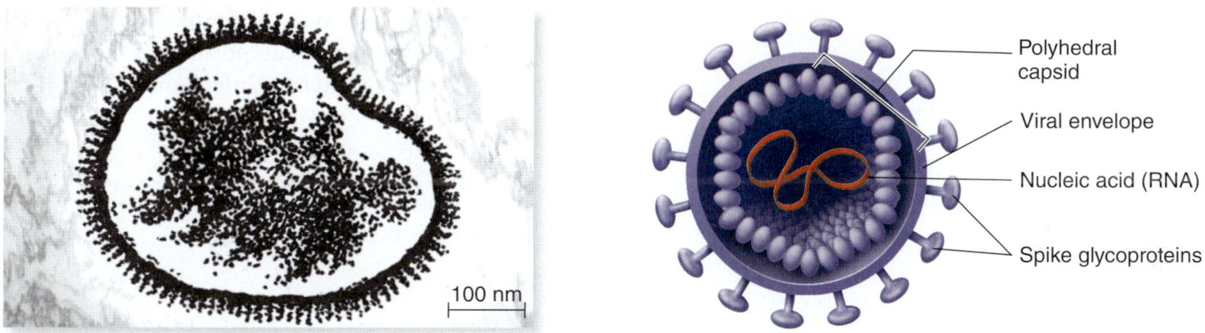

(c) Influenza virus, an enveloped virus with spikes

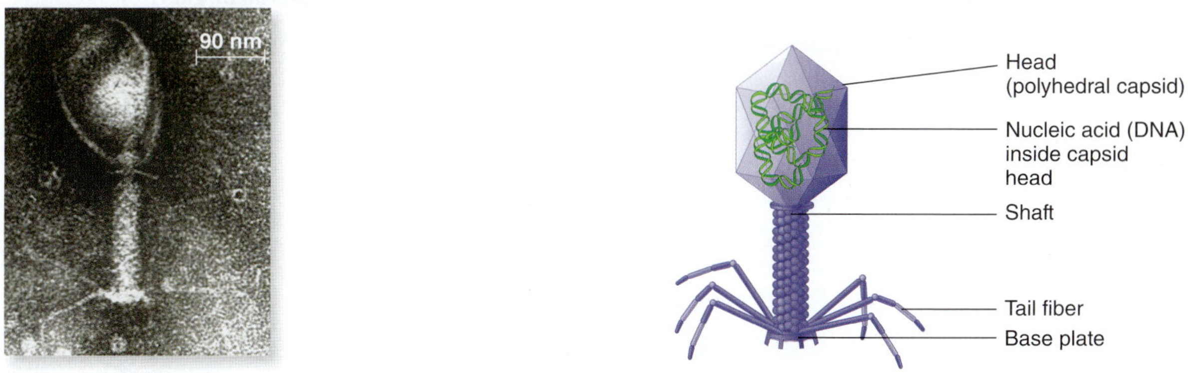

(d) T4, a bacteriophage

FIGURE 17.1 Variations in the structure of viruses, as shown by transmission electron microscopy. All viruses contain nucleic acid (DNA or RNA) surrounded by a protein capsid. They may or may not have an outer envelope surrounding the capsid. **(a)** Tobacco mosaic virus (TMV) has a helical capsid made of 2130 identical protein subunits (capsomeres), arranged around a strand of RNA. **(b)** Adenovirus has a polyhedral capsid containing protein fibers with a knob. **(c)** Many animal viruses, including the influenza virus, have an envelope composed of a lipid bilayer and spike glycoproteins. The lipid bilayer is obtained from the host cell when the virus buds from the plasma membrane. **(d)** Some bacteriophages, such as T4, have capsids with accessory structures, such as tail fibers and base plates, that facilitate invasion of a bacterial cell.

CONCEPT CHECK: What features vary among different types of viruses?

viruses, respectively. It is striking that some viruses use RNA for their genome, whereas all living organisms use DNA. In some viruses, the nucleic acid is single-stranded, whereas in others, it is double-stranded. The genome can be linear or circular, depending on the type of virus. Some kinds of viruses have more than one copy of the genome. Viral genomes also vary considerably in size, ranging from a few thousand to more than a hundred thousand nucleotides in length (see Table 17.1). For example, the genomes of some simple viruses, such as phage Qβ, are only a few thousand nucleotides in length and contain only a few genes. Other viruses, particularly those with a complex structure, such as phage T4, contain many more genes. These extra genes encode many different proteins that are involved in the formation of the elaborate structure shown in Figure 17.1d.

EXPERIMENT 17A

The Genome of Tobacco Mosaic Virus Is Composed of RNA

We now know that bacteria, archaea, protists, fungi, plants, and animals all use DNA as their genetic material. In 1956, Alfred Gierer and Gerhard Schramm isolated RNA from tobacco mosaic virus (TMV), which infects plant cells. When this purified RNA was applied to plant tissue, the plants developed the same types of lesions that occurred when they were exposed to intact TMVs. Gierer and Schramm correctly concluded that the viral genome of TMV is composed of RNA.

To further confirm that TMV uses RNA as its genetic material, Heinz Fraenkel-Conrat and Beatrice Singer conducted additional research that involved different strains of TMV. They focused their efforts on the wild-type strain and a mutant TMV strain called the Holmes ribgrass (HR) strain. The two strains differ in two ways. First, they cause significantly different symptoms when they infect plants. In particular, the wild-type strain produces a mottled area with yellow and green irregularly shaped lesions on infected leaves (see chapter-opening photo), whereas the HR strain often produces streaks along the veins and ringlike markings on other parts of the leaves. Second, the capsid protein in the HR strain has two amino acids (histidine and methionine), which are not found in the wild-type capsid protein.

Previous experiments had shown that purified TMV capsid proteins and purified TMV-RNA molecules can be mixed together and self-assemble into an intact virus. Such a procedure is referred to as a reconstitution experiment because intact viruses are made from their individual parts. In the experiment described in **Figure 17.2**, Fraenkel-Conrat and Singer mixed wild-type RNA with HR proteins or HR RNA with wild-type proteins and then placed the reconstituted viruses onto tobacco leaves. Following infection, they then observed the symptoms caused by the viruses and analyzed the amino acid composition of the proteins of newly made viruses that arose after the infection.

THE HYPOTHESIS

RNA is the genetic material of TMV.

TESTING THE HYPOTHESIS

FIGURE 17.2 Evidence that RNA is the genetic material of TMV.

Starting material: Purified preparations of RNA and proteins from wild-type TMV and from the Holmes ribgrass (HR) strain of TMV.

1. Mix together wild-type RNA and HR proteins or HR RNA and wild-type proteins. Allow time for the RNA and proteins to assemble into intact viruses. These are called reconstituted viruses.

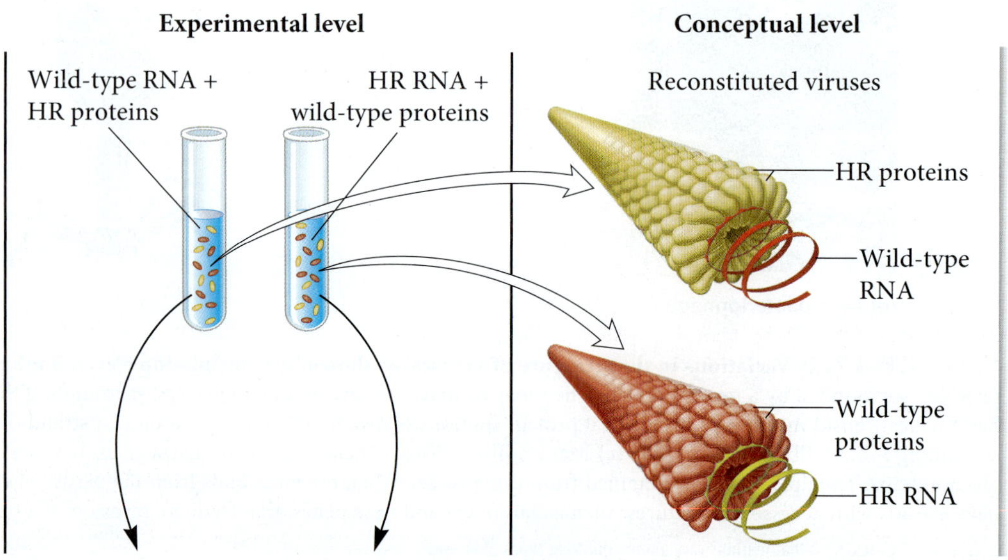

2. Inoculate a small amount of reconstituted viruses onto healthy tobacco leaves. Allow time for infection to occur.

3. Observe the types of lesions that form on the leaves.

4. Take plant tissue containing viral lesions and isolate newly made viral proteins. This is done by extracting the protein with mild alkali.

5. Determine the amino acid composition of the newly made viral proteins. This involves hydrolyzing the proteins into individual amino acids and then separating the amino acids by chromatography.

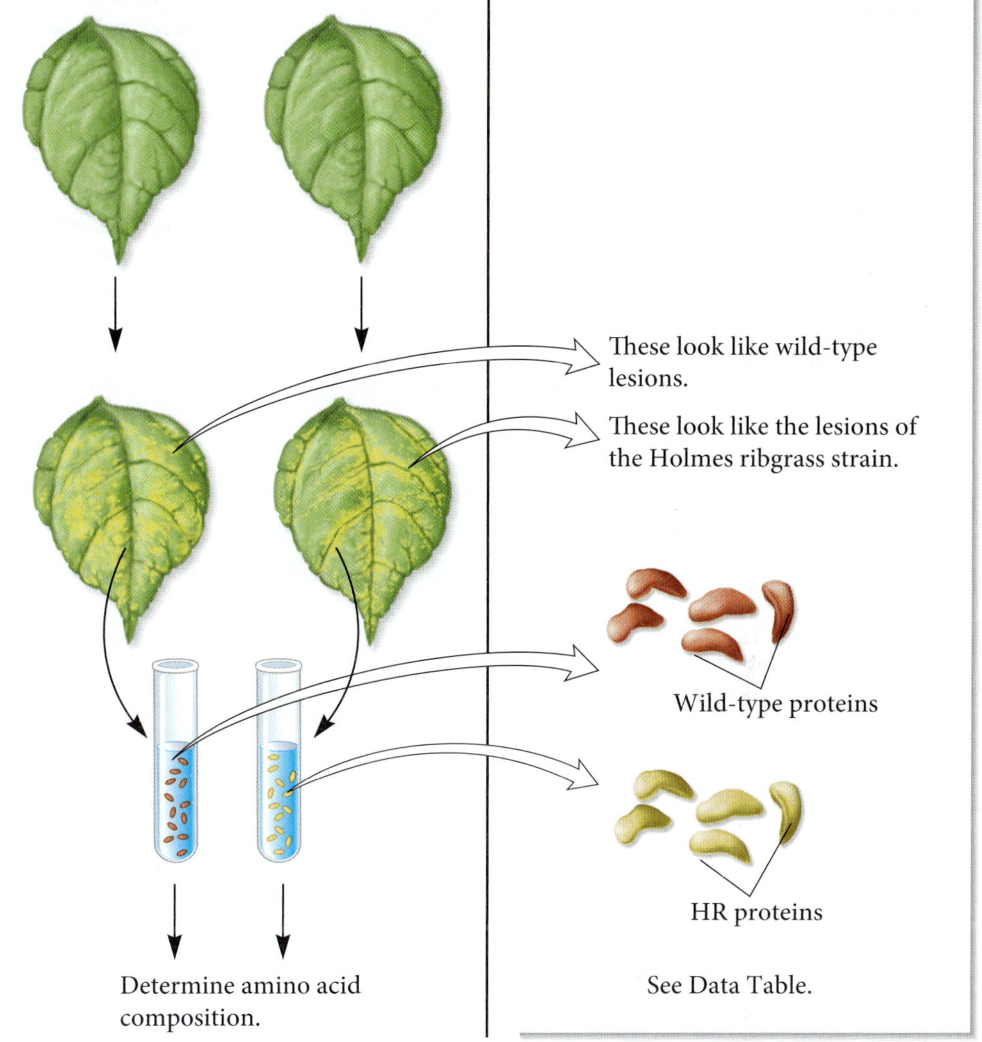

THE DATA

Composition of Reconstituted Virus Placed on Tobacco Leaves	*Symptoms on Tobacco Leaves*	*Amino Acids Found in Newly Made Viral Proteins Following Infection:*	
		Methionine	*Histidine*
Wild-type RNA and HR protein	Like wild-type TMV	No	No
HR RNA and wild-type protein	Like HR TMV	Yes	Yes

Data adapted from H. Fraenkel-Conrat and B. Singer (1957) Virus reconstitution II. Combination of protein and nucleic acid from different strains. *Biochimica et Biophysica Acta 24*, 540–548.

INTERPRETING THE DATA

As seen in The Data, the outcome of infection depended on the RNA that was found in the reconstituted virus but not on the protein. If wild-type RNA was used, the leaves developed symptoms that were typical of wild-type TMV and the capsid proteins of newly made viruses lacked methionine or histidine. In contrast, if reconstituted virus had HR RNA, the symptoms were those of the HR TMV strain and the newly made capsid proteins contained both methionine and histidine. Taken together, these results are consistent with the hypothesis that the RNA component of TMV is its genetic material.

A self-help quiz involving this experiment can be found at www.mhhe.com/brookergenetics5e.

17.1 COMPREHENSION QUESTIONS

1. What are common features found in all viruses?
   a. An envelope
   b. DNA
   c. Nucleic acid surrounded by a protein capsid
   d. All of the above
2. Viral genomes can be
   a. DNA or RNA.
   b. single-stranded or double-stranded.
   c. linear or circular.
   d. all of the above.

17.2 OVERVIEW OF VIRAL REPRODUCTIVE CYCLES

Learning Outcomes:

1. Compare and contrast the reproductive cycles of phage λ and HIV.
2. Define latency and explain how it occurs for phage λ and HIV.
3. Describe the properties of emerging viruses.

When a virus infects a host cell, the expression of viral genes leads to a series of steps, called a **viral reproductive cycle,** that results in the production of new viruses. The details of the steps may differ greatly among various types of viruses, and even the same virus may have the capacity to follow alternative cycles. Even so, by studying the reproductive cycles of hundreds of different viruses, researchers have determined that the reproductive cycle consists of five or six basic steps. In this section, we will examine the basic steps of viral reproductive cycles and consider how new viruses can quickly spread through a population.

Viruses Follow a Reproductive Cycle That Leads to the Synthesis of New Viruses

To illustrate the general features of viral reproductive cycles, Figure 17.3 considers the basic steps for two viruses. Figure 17.3a shows the cycle of **phage λ** (lambda), a bacteriophage with double-stranded DNA as its genome, and Figure 17.3b depicts the cycle of **human immunodeficiency virus (HIV),** an enveloped animal virus containing two copies of single-stranded RNA. The descriptions that follow compare the reproductive cycles of these two very different viruses.

Step 1: Attachment In the first step of a viral reproductive cycle, the virus must attach to the surface of a host cell. This attachment is usually specific for one or just a few types of cells because proteins in the virus recognize and bind to specific molecules on the cell surface. In the case of phage λ, the phage tail fibers bind to proteins in the outer bacterial cell membrane of *E. coli* cells. In the case of HIV, spike glycoproteins in the viral envelope bind to receptors in the plasma membrane of human blood cells called helper T cells.

Step 2: Entry After attachment, the viral genome enters the host cell. Attachment of phage λ stimulates a conformational change in the phage coat proteins, so the shaft contracts, and the phage injects its DNA into the bacterial cytoplasm. In contrast, the envelope of HIV fuses with the plasma membrane of the host cell, so both the capsid and its contents are released into the cytosol. Some of the HIV capsid proteins are then removed by host-cell enzymes, a process called uncoating. This releases the two copies of viral RNA and molecules of enzymes called reverse transcriptase and integrase into the cytosol. The functions of these enzymes are described later. Once a viral genome has entered the cell, specific viral genes may be expressed immediately due to the action of host-cell enzymes and ribosomes. Expression of these key genes leads quickly to either step 3 or step 4 of the reproductive cycle, depending on the type of virus. The genome of some viruses, including both phage λ and HIV, can integrate into a chromosome of the host cell. For such viruses, the cycle may proceed from step 2 to step 3 as described next, delaying the production of new viruses. Alternatively, the cycle may proceed directly from step 2 to step 4 and quickly lead to the production of new viruses.

Step 3: Integration Viruses capable of integration carry a gene that encodes an enzyme called **integrase.** For some viruses, such as phage λ, the integrase gene within the phage genome is expressed soon after entry, which leads to the synthesis of the integrase protein. For other viruses, such as HIV, integrase proteins are packaged into newly made viruses, which are released during the uncoating process. The function of integrase is to cut the host's chromosomal DNA and integrate the viral genome into the chromosome. In the case of phage λ, the double-stranded DNA that entered the cell can be directly integrated into the double-stranded DNA of the chromosome. Once integrated, the phage DNA in a bacterium is called a **prophage.** While it exists as a prophage, this type of viral reproductive cycle is called the **lysogenic cycle.** As discussed later, new phages are not made during the lysogenic cycle, and the host cell is not destroyed. On occasion, a prophage can be excised from the bacterial chromosome and proceed to step 4.

How can an RNA virus, such as HIV, integrate its genome into the host-cell DNA? For this to occur, the viral genome must be copied into DNA. HIV accomplishes this by means of a viral enzyme called **reverse transcriptase,** which is carried within the capsid and released into the host cell along with the viral RNA. Reverse transcriptase uses the viral RNA strand to make a complementary copy of DNA, and it then uses the DNA strand as a template to make double-stranded viral DNA. This process is called reverse transcription because it is the reverse of the usual transcription process, in which a DNA strand is used to make a complementary strand of RNA. The viral double-stranded DNA enters the host-cell nucleus and is integrated into a host chromosome via integrase. Once integrated, the viral DNA in a eukaryotic cell is called a **provirus.** Viruses that follow this mechanism are called retroviruses.

Step 4: Synthesis of Viral Components The production of new viruses by a host cell involves the replication of the viral genome and the synthesis of viral proteins that make up the capsid. In the case of bacteriophage λ that has been integrated into the host chromosome, the prophage must be excised before synthesis of new viral components can occur. A viral enzyme called excisionase plays a role in the excision process. Following excision, host-cell enzymes make many copies of the phage DNA and transcribe the genes within these copies into mRNA. Host-cell ribosomes translate this viral mRNA into viral proteins. The expression of phage genes also leads to the degradation of the host chromosomal DNA.

In the case of HIV, the DNA provirus is not excised from the host chromosome. Instead, it is transcribed in the nucleus to produce many copies of viral RNA. These viral RNA molecules enter the cytosol, where they are used to make viral proteins and serve as the genome for new viral particles.

Step 5: Viral Assembly After all of the necessary components have been synthesized, they must be assembled into new viruses. Some viruses with a simple structure self-assemble, meaning that viral components spontaneously bind to each other to form a complete virus particle. An example of a self-assembling virus is TMV, which we examined earlier (see Figure 17.1a). The capsid proteins assemble around the RNA genome, which becomes trapped inside the hollow capsid (see solved problem S1 at the end of the chapter). This assembly process can occur in vitro if purified capsid proteins and RNA are mixed together. Other viruses, including the two shown in Figure 17.3, do not self-assemble. The correct assembly of phage λ requires the help of noncapsid proteins not found in the completed phage particle. Some of these noncapsid proteins function as enzymes that modify capsid proteins, and others serve as scaffolding for the assembly of the capsid. The assembly of HIV components occurs at the plasma membrane. Capsid proteins assemble around two molecules of viral RNA and molecules of reverse transcriptase and integrase. As this is occurring, the virus acquires its outer envelope in a budding process.

Step 6: Release The last step of a viral reproductive cycle is the release of new viruses from the host cell. The release of bacteriophages is a dramatic event. Because bacteria are surrounded by a rigid cell wall, the phages must burst, or lyse, their host cell in order to escape. After the phages have been assembled, a phage-encoded enzyme called lysozyme digests the bacterial cell wall, causing the cell to burst. Lysis releases many new phages into the environment, where they can infect other bacteria and begin the cycle again. Collectively, steps 1, 2, 4, 5, and 6 are called the **lytic cycle** because they lead to cell lysis.

The release of enveloped viruses from an animal cell is far less dramatic. This type of virus escapes by a mechanism called budding that does not lyse the cell. A portion of the host-cell plasma membrane enfolds the viral capsid and eventually buds from the cell surface.

Latency in Bacteriophages As we saw in step 3, viruses can integrate their genomes into a host chromosome. In some cases, the prophage or provirus may remain inactive, or **latent,** for a long time. Most of the viral genes are silent during latency, and the viral reproductive cycle does not progress to step 4. Latency in bacteriophages is also called **lysogeny.** When this occurs, both the prophage and its host cell are said to be lysogenic. When a lysogenic bacterium prepares to divide, it copies the prophage DNA along with its own DNA, so each daughter cell inherits a copy of the prophage. The prophage DNA can be replicated repeatedly in this way without killing the host cell or producing new phage particles. As mentioned earlier, this is called the lysogenic cycle.

Many bacteriophages can alternate between lysogenic and lytic cycles (**Figure 17.4**). A bacteriophage that may spend some of its time in the lysogenic cycle is called a **temperate phage.** Phage λ is an example of a temperate phage. Upon infection, it can either enter the lysogenic cycle or proceed directly to the lytic cycle. Other phages, called **virulent phages,** have only lytic cycles. The genome of a virulent phage is not capable of integration into a host chromosome.

Latency in Human Viruses Latency among human viruses can occur in two different ways. For HIV, latency occurs because the virus has integrated into the host genome and may remain dormant for long periods of time. Alternatively, the genomes of other viruses can exist as an **episome**—a genetic element that can replicate independently of the chromosomal DNA but also can occasionally integrate into chromosomal DNA. Examples of viral genomes that exist as episomes include different types of herpesviruses that cause cold sores (herpes simplex type I), genital herpes (herpes simplex type II), and chickenpox (varicella-zoster virus). A person infected with a given type of herpesvirus may have periodic outbreaks of disease symptoms when the virus switches from the latent, episomal form to the active form that produces new virus particles.

As an example, let's consider the herpesvirus called varicella-zoster virus. The initial infection by this virus causes chickenpox, after which the virus may remain latent for many years as an episome. The disease called shingles usually occurs decades later, when varicella-zoster virus switches from the latent state and starts making new virus particles. It usually occurs in adults over the age of 60 or in people with weakened immune systems. Shingles begins as a painful rash that eventually erupts into blisters. The blisters follow the path of the nerve cells that carry the latent varicella-zoster virus. The blisters often form a ring around the back of the patient's body. The term shingles is derived from a Latin word meaning girdle, referring to the observation that the blisters girdle a part of the body.

Emerging Viruses, Such as HIV, Have Arisen Recently and May Rapidly Spread Through a Population

A primary reason researchers have been interested in viral reproductive cycles is the ability of many viruses to cause diseases in humans and other hosts. Some examples of human disease-causing viruses were presented earlier in Table 17.1. **Emerging viruses** are viruses that have arisen recently and are more likely to cause infection than are previous strains. Such viruses may lead to a significant loss of human life and often cause public alarm.

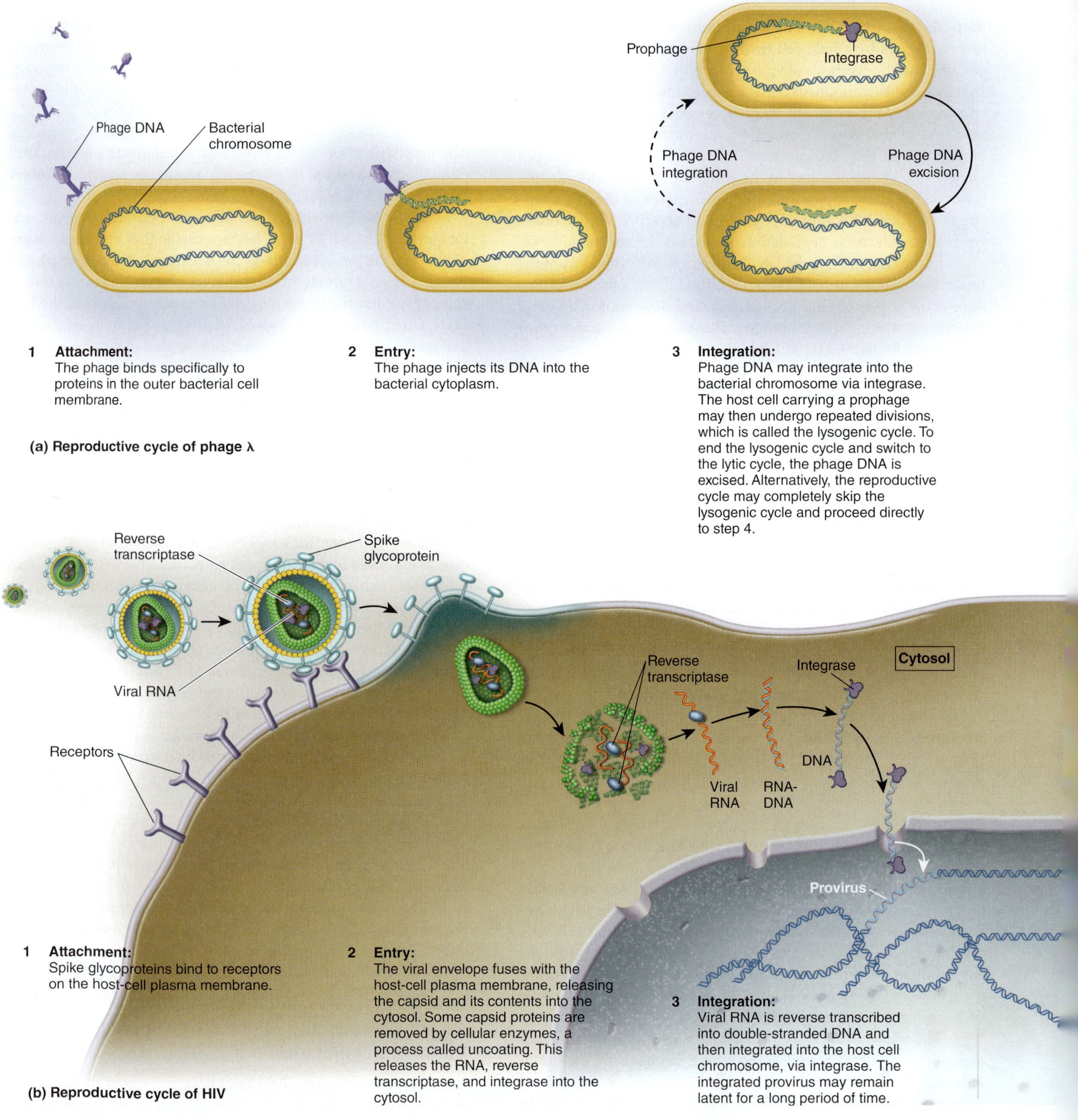

FIGURE 17.3 Comparison of the steps of two viral reproductive cycles. (**a**) The reproductive cycles of phage λ, a bacteriophage with a double-stranded DNA genome. Note: As described in Chapter 19, Figure 19.6, the phage DNA is in a circular form when it integrates into the bacterial chromosome. (**b**) The reproductive cycle of HIV, an enveloped animal virus with a single-stranded RNA genome.

CONCEPT CHECK: During which step may a virus remain latent?

4 **Synthesis of viral components:** In the lytic cycle, phage DNA directs the synthesis of viral components. During this process, the phage DNA circularizes, and the host chromosomal DNA is digested into fragments.

5 **Viral assembly:** Phage components are assembled with the help of noncapsid proteins to make many new phages.

6 **Release:** The viral enzyme called lysozyme causes cell lysis, and new phages are released from the broken cell.

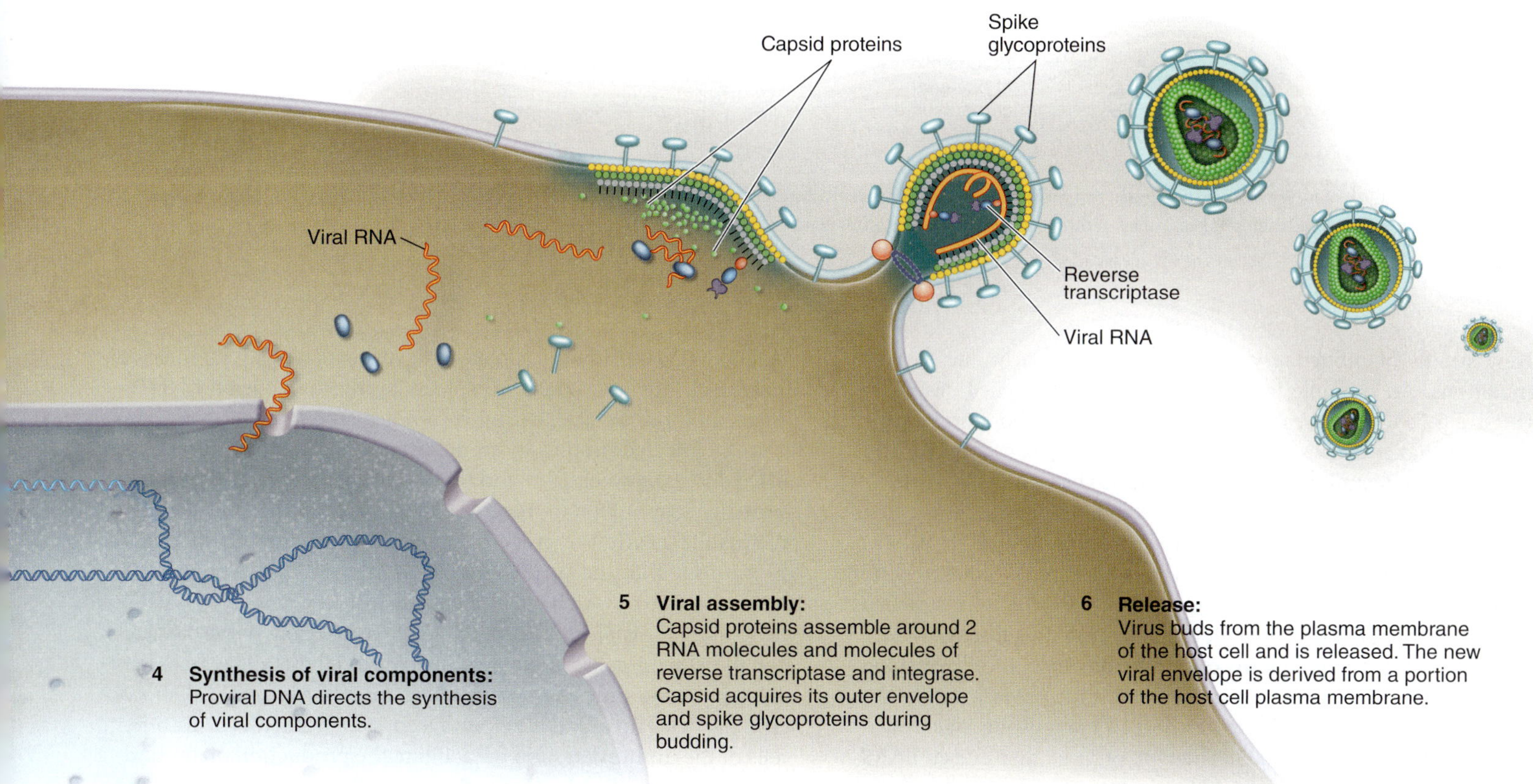

4 **Synthesis of viral components:** Proviral DNA directs the synthesis of viral components.

5 **Viral assembly:** Capsid proteins assemble around 2 RNA molecules and molecules of reverse transcriptase and integrase. Capsid acquires its outer envelope and spike glycoproteins during budding.

6 **Release:** Virus buds from the plasma membrane of the host cell and is released. The new viral envelope is derived from a portion of the host cell plasma membrane.

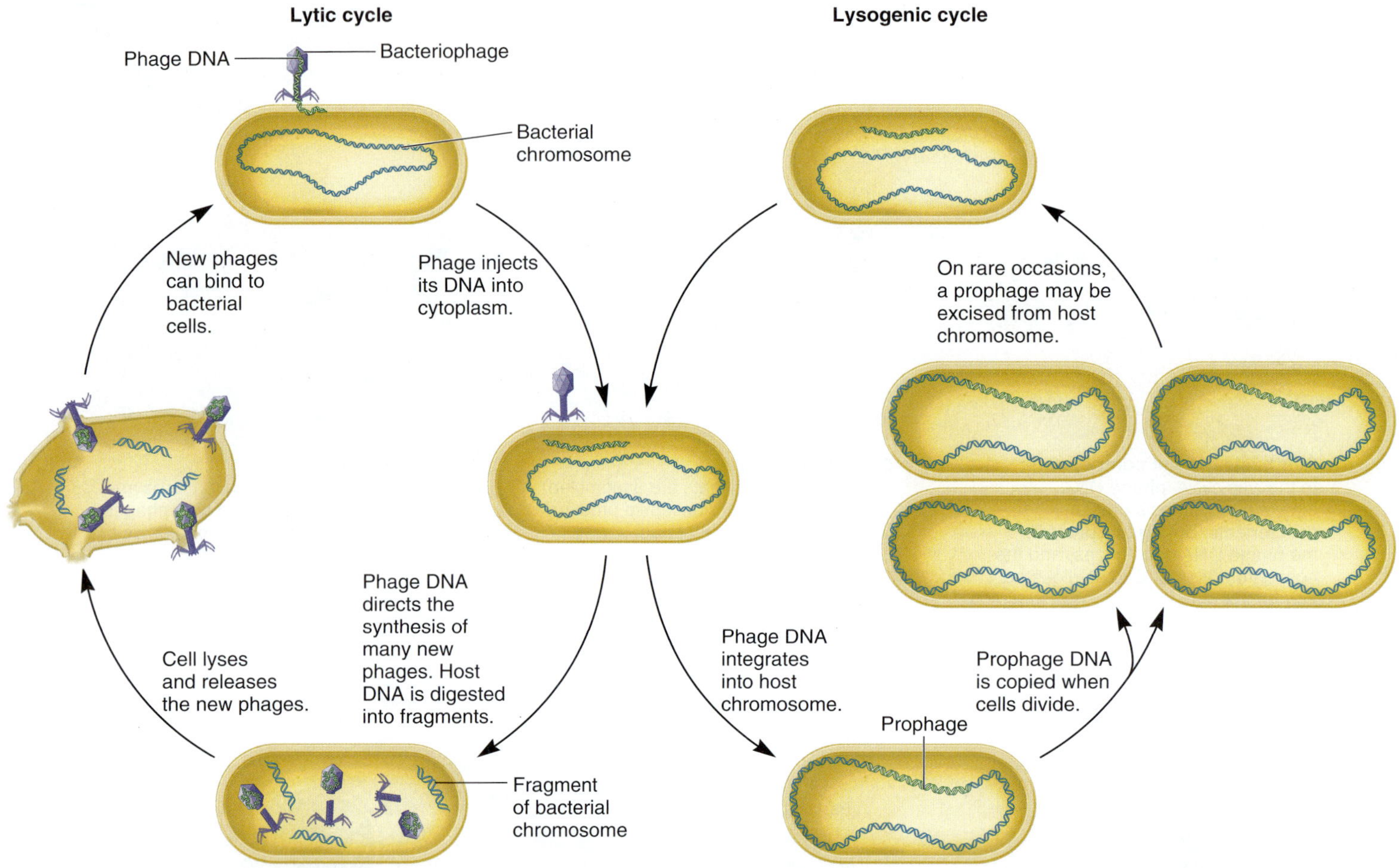

FIGURE 17.4 The lytic and lysogenic reproductive cycles of certain bacteriophages. Some bacteriophages, known as virulent phages, can follow only a lytic cycle. Other phages, called temperate phages, can follow both cycles.

CONCEPT CHECK: Which cycle produces new phage particles?

New strains of influenza virus arise fairly regularly due to new mutations. An example is the strain H1N1, also called swine flu (**Figure 17.5**). It was called swine flu because laboratory testing revealed that it carries two genes that are normally found in flu viruses that infect pigs in Europe and Asia. The Centers for Disease Control and Prevention (CDC) recommends vaccination as the first and most important step in preventing infection by H1N1 as well as other strains of influenza. However, in the United States, despite vaccination campaigns that aim to minimize deaths from seasonal influenza, over 30,000 people die annually from this disease.

Another emerging virus caused a potentially life-threatening illness called severe acute respiratory syndrome (SARS). This RNA virus was first identified by researchers in Hong Kong, the United States, and Germany in 2003. The type of virus that caused SARS is called a coronavirus. The SARS coronavirus is believed to have originated in bats and acquired the ability to infect humans and livestock. Symptoms in people included a high fever, headache, a dry cough, and pneumonia. The SARS outbreak ended in 2004.

During the past few decades, the most devastating example of an emerging virus has been human immunodeficiency virus (HIV), the causative agent of acquired immune deficiency syndrome (AIDS). HIV is primarily spread by sexual contact with an infected individual, but it can also be spread by the transfusion of HIV-infected blood, by the sharing of needles among drug users, and from an infected mother to an unborn child. The total number of AIDS deaths between 1981 and the end of 2013 was over 30 million; more than 0.6 million of these deaths occurred in the United States. Worldwide, nearly 1 in every 100 adults between the ages of 15 and 49 is infected. In the U.S., about 55,000 new HIV infections occur each year, 70% of which are in men and 30% in women.

The devastating effects of AIDS result from viral destruction of helper T cells, a type of white blood cell that plays an essential role in the immune system of mammals. **Figure 17.6** shows HIV particles invading a helper T cell, which interact with other cells of the immune system to facilitate the production of antibodies and other molecules that target and kill foreign invaders of the body. When large numbers of helper T cells are destroyed by HIV, the function of the immune system is seriously compromised, and the individual becomes susceptible to opportunistic infections—diseases that would not normally occur in a healthy person. For example, *Pneumocystis jiroveci*, a fungus that causes pneumonia, is easily destroyed by a healthy immune system. However, in people

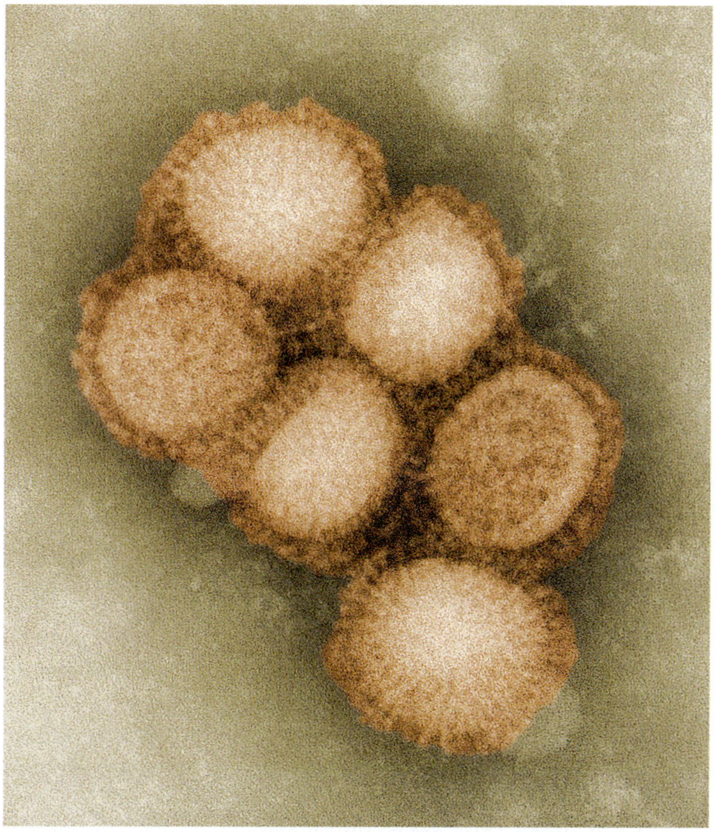

FIGURE 17.5 A micrograph of influenza virus, strain H1N1. This virus, which infects people, is a relatively new strain of influenza virus that causes classic flu symptoms. This is a colorized transmission electron micrograph.

with AIDS, infection by this fungus can be fatal. An insidious feature of HIV replication, described in Figure 17.3b, is that reverse transcriptase, the enzyme that copies the RNA genome into DNA, lacks a proofreading function. In Chapter 11, we discussed how DNA polymerase can identify and remove mismatched nucleotides in newly synthesized DNA. Because reverse transcriptase lacks this function, it makes more errors and thereby tends to create many mutant strains of HIV. This undermines the ability of the body to combat HIV because mutant strains may not be destroyed by the body's defenses. In addition, mutant strains of HIV may be resistant to antiviral drugs.

17.2 COMPREHENSION QUESTIONS

1. Which of the following is a common order of steps for a viral reproductive cycle? Note: Integration is an optional step.
   a. Entry, integration, attachment, synthesis of viral components, viral assembly, viral release
   b. Entry, integration, synthesis of viral components, viral assembly, attachment, viral release
   c. Attachment, entry, integration, synthesis of viral components, viral assembly, viral release
   d. Attachment, entry, integration, viral assembly, synthesis of viral components, viral release

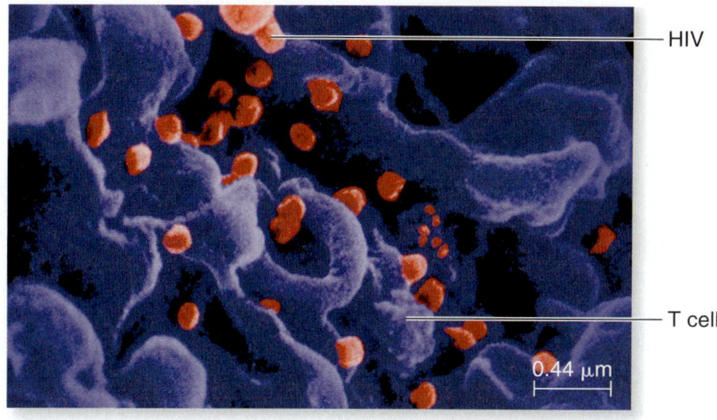

FIGURE 17.6 Micrograph of HIV invading a human helper T cell. This is a colorized scanning electron micrograph. The surface of the helper T cell is purple, and HIV particles are red.

CONCEPT CHECK: What is meant by the term "emerging virus"?

2. An example of an emerging virus is
   a. a coronavirus that causes SARS.
   b. HIV that causes AIDS.
   c. a strain of influenza virus called H1N1 that causes the flu.
   d. all of the above.

17.3 BACTERIOPHAGE λ REPRODUCTIVE CYCLE

Learning Outcomes:

1. Compare and contrast the lytic and lysogenic cycles of phage λ.
2. Explain how gene regulation determines the choice between the lytic and lysogenic cycles.

During the past several decades, the reproduction of viruses has presented an interesting and challenging problem for geneticists to investigate. The study of bacteriophages has greatly advanced our basic knowledge of how regulatory proteins work. In this section, we will largely focus on the function of bacteriophage genes that encode regulatory proteins that control the choice between the lytic and lysogenic cycles of phage λ, which was discovered by Esther Lederberg in 1951. The general features of its reproductive cycles were described earlier in Figure 17.3a. Since its discovery, phage λ has been investigated extensively and has provided geneticists with a model on which to base our understanding of viral proliferation.

Phage λ Can Follow a Lytic or Lysogenic Cycle

Phage λ binds to the surface of a bacterium and injects its genetic material into the bacterial cytoplasm. After this occurs, the phage proceeds along one of two alternative cycles, known as the lytic cycle and the lysogenic cycle, which were described earlier in this chapter (see Figure 17.4). During the lytic cycle, the information encoded by the bacteriophage DNA directs the synthesis of many copies of the phage genetic material and coat proteins that are then assembled to make new phages. When synthesis and assembly are completed, the bacterial host cell is lysed, and the newly made phages are released into the environment.

Alternatively, in the lysogenic cycle, the phage does not direct the synthesis of new phages and does not kill the bacterial cell that acts as its host. During the lysogenic cycle, phage λ integrates its genetic material into the chromosome of the bacterium. As mentioned, this integrated phage DNA is known as a prophage. A prophage can exist in a dormant state for a long time, during which no new bacteriophages are made. When a bacterium containing a lysogenic prophage divides to produce two daughter cells, it copies the prophage DNA along with its own DNA (see Figure 17.4). Therefore, both daughter cells inherit the prophage. At some later time, a prophage may become activated to excise itself from the bacterial chromosome and enter the lytic cycle. The phage DNA then directs the synthesis of new phages and eventually lyses the host cell.

Inside the virus particle, phage λ DNA is linear. After injection into the bacterium, the two ends of the DNA become covalently attached to each other to form a circular piece of DNA. **Figure 17.7** shows the genome of phage λ. The organization of the genes reflects the two alternative cycles of this virus. The genes shaded in orange are transcribed very soon after infection. As we will see, the expression pattern of these early genes determines the choice between the lytic or lysogenic cycle.

If the lysogenic cycle prevails, the integrase (*int*) gene is turned on. The integrase gene encodes an enzyme that integrates the λ DNA into the bacterial chromosome. (The mechanism of phage λ integration is described in Chapter 19.) If the lysogenic cycle is not chosen, the genes on the right side and bottom of the figure are transcribed. These genes are necessary for the synthesis of new phages. They encode replication proteins, coat proteins (that form the phage head, shaft, and tail fibers), proteins involved in coat assembly, proteins involved in packaging the DNA into the phage head, and enzymes that cause the bacterium to lyse.

The Choice Between the Lysogenic or Lytic Cycle Depends on the Relative Levels of the cII and Cro Proteins

Now that we understand phage λ reproductive cycles and genome organization, let's examine how the decision is made between the lysogenic and lytic cycles. This choice depends on the actions of regulatory proteins and how they affect transcription. **Table 17.2** summarizes the key regulatory elements and proteins involved in this process.

Soon after λ DNA enters the bacterial cell, two promoters—designated P_L and P_R—are used for transcription. This initiates a competition between the lysogenic and lytic cycles (**Figure 17.8**).

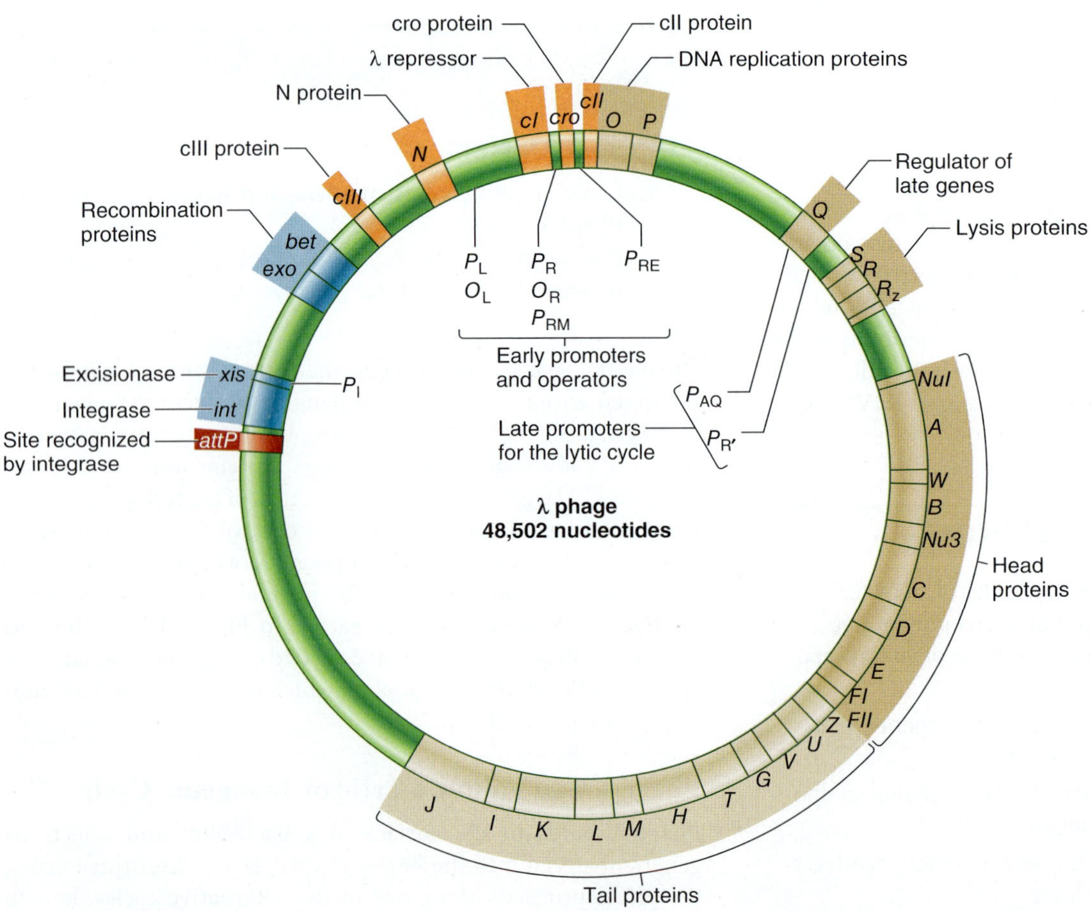

FIGURE 17.7 The genome of phage λ. The genes shaded in orange encode regulatory proteins that determine whether the lysogenic or lytic cycle prevails. Genes shaded in blue encode proteins necessary for the lysogenic cycle. The attP site, shaded in red, is recognized by integrase. The genes shaded in dark and light tan encode proteins required for the lytic cycle.

Initially, transcription from P_L and P_R results in the synthesis of two short RNA transcripts that encode two proteins called the N protein and the cro protein, both of which are regulatory proteins. We will examine the function of cro protein later, which is involved in the lytic cycle.

The N protein is a regulatory protein with an interesting function that we have not yet considered. Its function, known as **antitermination,** is to prevent transcriptional termination. The N protein inhibits termination at three sites, designated t_L, t_{R1}, and t_{R2}. The N protein binds to RNA polymerase and prevents transcriptional termination when these sites are being transcribed. When the N protein prevents termination at t_{R1} and t_{R2}, the transcript from P_R is extended to include the *cII*, *O*, *P*, and *Q* genes. The *cII* gene encodes an activator protein, the *O* and *P* genes encode enzymes needed for the initiation of λ DNA synthesis, and the *Q* gene encodes another antiterminator that is required for the lytic cycle. When the N protein prevents termination at t_L, the transcript from P_L is extended to include the *int*, *xis*, and *cIII* genes. The *int* gene encodes integrase, which is involved with integrating λ DNA into the *E. coli* chromosome, and the *xis* gene encodes excisionase, which plays a role in excising the λ DNA if a switch is made from the lysogenic to the lytic cycle. The *cIII* gene encodes the cIII protein, which inhibits a cellular protease and thereby makes cII less vulnerable to protease digestion.

As shown at the bottom of Figure 17.8, if the cII protein accumulates to sufficient levels, the lysogenic cycle is favored. We will explore its function in greater detail later. Alternatively, if the level of the cro protein becomes high, the lytic cycle occurs. What environmental factors determine whether the lysogenic or lytic cycle prevails? A critical issue is that the cII protein is easily degraded by a cellular protease that is produced by *E. coli.* Whether or not this protease is made at high levels depends on the environmental conditions. If the growth conditions are very favorable, such as a rich growth medium, the intracellular protease levels are relatively high, and the cII protein tends to be degraded. The cro protein is not sensitive to such protease degradation and slowly accumulates to high levels. Therefore, environmental conditions that are favorable for growth promote the lytic cycle. This makes sense, because a sufficient supply of nutrients is necessary to synthesize new bacteriophages.

Alternatively, starvation conditions favor the lysogenic cycle. When nutrients are limited, the protease level in the cell is relatively low. Under these conditions, the cII protein builds up much more quickly than the cro protein. This event favors the lysogenic cycle. From the perspective of the bacteriophage, lysogeny may be a better choice under starvation conditions because nutrients may be insufficient for the production of new λ phages.

The λ Repressor and Integrase Control the Lysogenic Cycle

Let's now consider how the lysogenic cycle is followed (**Figure 17.9a**). If the level of cII becomes high, the cII protein activates two different promoters in the λ genome, P_{RE} and P_I. When

TABLE 17.2
Genetic Regulatory Elements and Proteins of Phage λ

Type	Description
Promoters	
P_L	Promoter for the *N* and *cIII* genes, which encode N protein and cIII proteins, respectively
P_R	Promoter for the *cro*, *cII*, *O*, and *P* genes; the *cro* and *cII* genes encode cro protein and cII protein, respectively, whereas the *O* and *P* genes encode proteins involved with the replication of λ DNA
P_{RE}	Promoter for the *cI* gene, which encodes λ repressor; this promoter is used to establish the lysogenic cycle
P_{RM}	Promoter for the *cI* gene, which encodes λ repressor; this promoter is used to maintain the lysogenic cycle
P_I	Promoter for *int* and *xis* genes, which encode integrase and excisionase, respectively
$P_{R'}$	Promoter for a large operon that encodes many of the proteins necessary for the lytic cycle
Operators	
O_L	Operator that controls P_L
O_R	Operator that controls P_R and P_{RM}
Regulatory Proteins	
N protein	Promotes antitermination by binding to RNA polymerase and allowing transcription past t_L, t_{R1}, and t_{R2}; leads to the transcription of the *cIII*, *cII*, *O*, and *P* genes
cII protein	Favors the lysogenic cycle; binds to P_{RE} and P_I and activates their transcription
λ repressor	Establishes and maintains the lysogenic cycle; binds to O_L and inhibits transcription from P_L; binds to O_R and inhibits transcription from P_R and activates transcription from P_{RM}
cro protein	Favors the lytic cycle; binds to O_L and inhibits transcription from P_L; binds to O_R and inhibits transcription from P_{RM} and later from P_R
Q protein	Promotes antitermination by binding to RNA polymerase and allowing the transcription from $P_{R'}$ that is needed for the lytic cycle

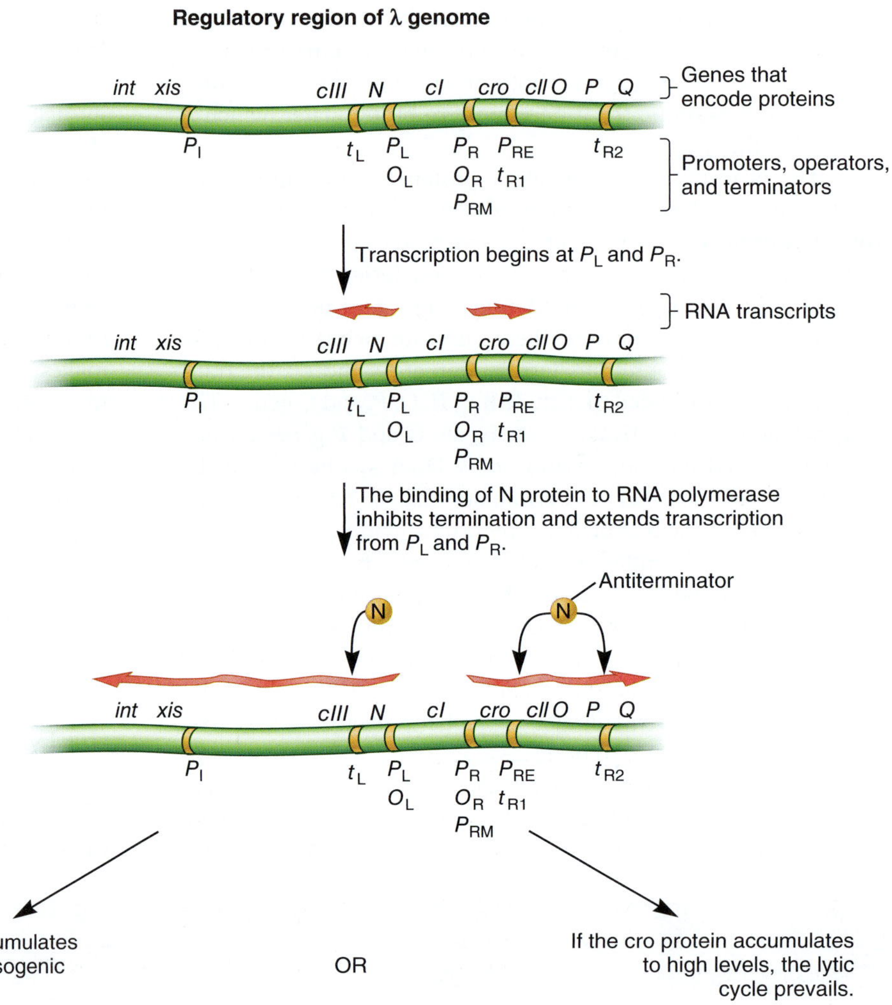

FIGURE 17.8 The events that lead to the beginning of the lysogenic and lytic cycles of phage λ. This figure shows the region of the phage λ genome that regulates the choice between the lysogenic and lytic cycles. DNA is shown in green. The names of genes that encode proteins are shown above the DNA. Promoters, operators, and terminators are shown below the DNA. In this figure and Figure 17.9, RNA transcripts are shown in red. The key regulatory proteins are indicated as spheres.

Immediately after infection, P_L and P_R are used to make two short mRNAs. These mRNAs encode two early proteins, designated N and cro. The N protein is an antiterminator that prevents transcriptional termination at three sites in the RNA (t_L, t_{R1}, and t_{R2}). This allows the transcription of several genes, which include *int*, *xis*, *cIII*, *cII*, *O*, *P*, and *Q*. The expression of the *O*, *P*, and *Q* genes is necessary only for the lytic cycle. If the lysogenic cycle is chosen, transcription of these genes is abruptly inhibited.

CONCEPT CHECK: Let's suppose a drug inhibits the function of the N protein. Would such a drug favor the lysogenic cycle, favor the lytic cycle, or prevent both cycles from occurring?

the cII protein binds to P_{RE}, it turns on the transcription of *cI*, a gene that encodes the λ repressor. The cII protein also activates the *int* gene by binding to P_I. The λ repressor and integrase proteins promote the lysogenic cycle. When the λ repressor is made in sufficient quantities, it binds to operator sites (O_L and O_R) that are adjacent to P_L and P_R. When the λ repressor is bound to O_R, it inhibits the expression of the genes required for the lytic cycle. As described in Chapter 19 (see Figure 19.6), the integrase protein recognizes the attP site and catalyzes the insertion of the λ DNA into a specific site in the *E. coli* chromosome.

Notice in Figure 17.9a that the binding of the λ repressor to O_R inhibits the expression of *cII*. This may seem counterintuitive, because the cII protein was initially required to bind to P_{RE} and activate the *cI* gene that encodes the λ repressor. You may be thinking that the inhibition of the *cII* gene would eventually prevent the expression of the *cI* gene and ultimately stop the synthesis of the λ repressor protein. What prevents the inhibition of *cI* gene expression? The explanation is that the *cI* gene has two promoters: P_{RE}, which is activated by the cII protein, and a second promoter called P_{RM}.

Transcription from P_{RE} occurs at the beginning of the lysogenic cycle. P_{RE} gets its name because the use of this promoter results in the expression of the λ Repressor during the Establishment of the lysogenic cycle. The transcript made from P_{RE} is very stable and quickly leads to a buildup of the λ repressor protein. This causes an abrupt inhibition of the lytic cycle because the binding of the λ repressor protein to O_R blocks the P_R promoter. Later in the lysogenic cycle, it is no longer necessary to make a large amount of the λ repressor. At this point, the use of the P_{RM} promoter is sufficient to make enough Repressor protein to Maintain the lysogenic cycle. Interestingly, the P_{RM} promoter is activated by the λ repressor protein. The λ repressor was named when it was understood that it repressed the lytic cycle. Later studies revealed that it also activates its own transcription from P_{RM}.

After the lysogenic cycle is established, certain environmental conditions favor induction to the lytic cycle. For example, exposure to UV light promotes induction. In this case, a protein known as recA detects the DNA damage from UV light and is activated to become a mediator of protein cleavage. RecA protein mediates the cleavage of the λ repressor, thereby inactivating it. This allows transcription from P_R and eventually leads to the accumulation of the cro protein, which favors the lytic cycle. Under these conditions, it may be advantageous for λ to make new phages and lyse the cell, because the exposure to UV light may have already damaged the bacterium to the point where further bacterial growth and division are prevented.

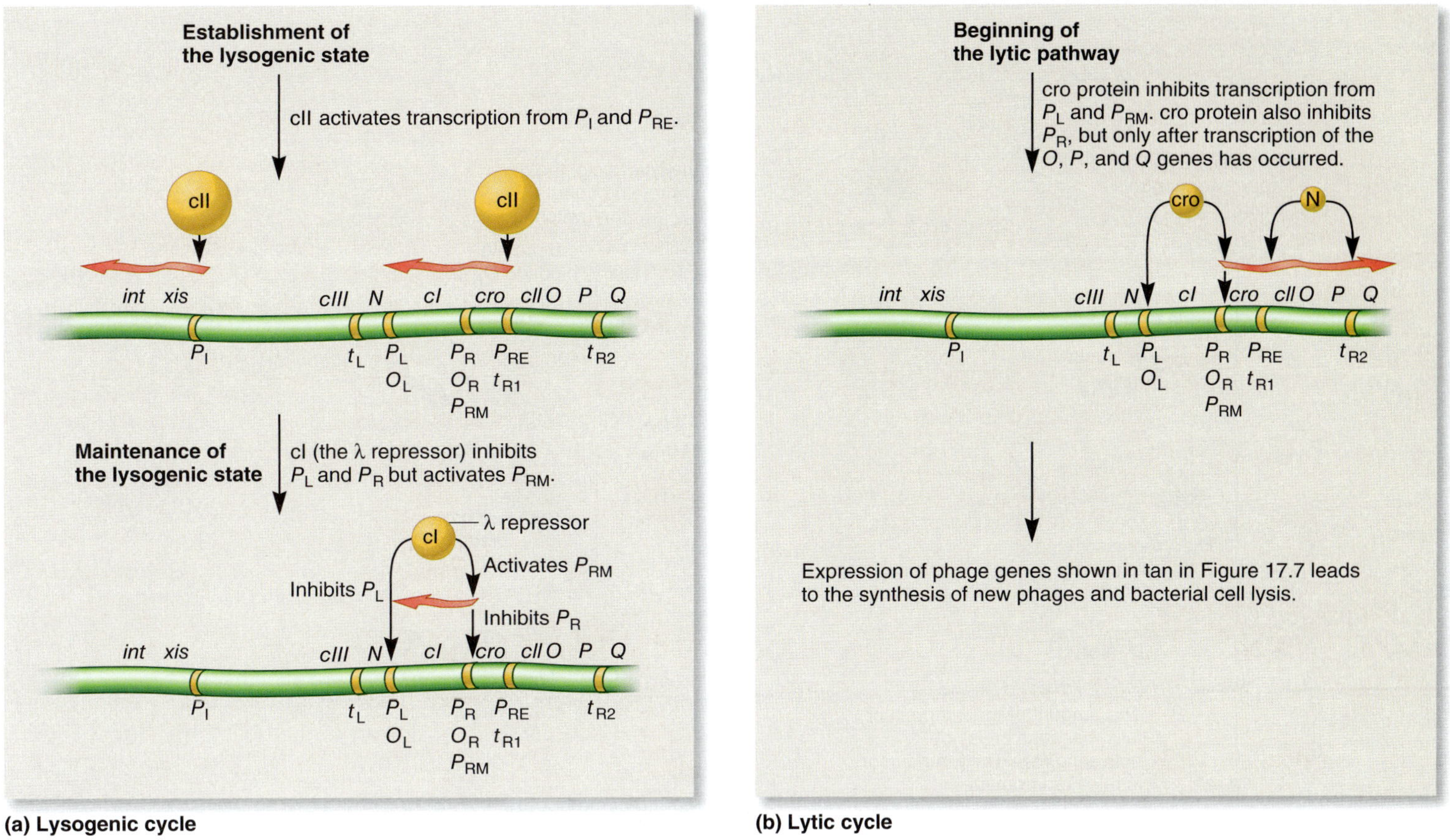

FIGURE 17.9 Phage λ gene regulation during the lysogenic and lytic cycles. (**a**) Gene regulation pattern that leads to the lysogenic cycle. (**b**) Steps that lead to the lytic cycle.

Genes→Traits The ability to choose between two alternative reproductive cycles can be viewed as a trait of this bacteriophage. As described here, the choice between the two cycles depends on the pattern of gene regulation.

CONCEPT CHECK: What is the function of P_{RM}?

The Lytic Cycle Depends on the Action of the Cro Protein

If the activity of the cro protein exceeds the cII protein, the lytic cycle prevails (**Figure 17.9b**). As mentioned, an early step in the expression of λ genes is the transcription from P_R to produce the cro protein. If the concentration of the cro protein builds to sufficient levels, it will bind to two operators: O_L and O_R. The binding of cro to O_L inhibits transcription from P_L; the binding of cro to O_R has two effects. First, the binding of cro protein to O_R inhibits transcription from P_{RM} in the leftward direction. This inhibition prevents the expression of the *cI* gene, which encodes the λ repressor; the λ repressor is needed to maintain the lysogenic state. Therefore, the λ repressor cannot successfully shut down transcription from P_R.

Second, the binding of the cro protein to O_R inhibits transcription from P_R in the rightward direction. However, this inhibition occurs after the transcription of the *O*, *P*, and *Q* genes. The O and P proteins are necessary for the replication of the phage DNA. The Q protein is an antiterminator protein that permits transcription through another promoter, designated $P_{R'}$ (i.e., $P_{R\ prime}$, not to be confused with P_R). The $P_{R'}$ promoter controls a very large operon that encodes the proteins necessary for the phage coat, the assembly of the coat proteins, the packaging of the λ DNA, and the lysis of the bacterial cell (refer back to Figure 17.7). These proteins are made toward the end of the lytic cycle. The expression of these late genes leads to the synthesis and assembly of many new λ phages that are released from the bacterial cell when it lyses.

O_R Acts as a Genetic Switch Between the Lysogenic and Lytic Cycles

Before we end this section on the λ reproductive cycles, let's consider how O_R acts as a genetic switch between the two cycles. Depending on the binding of genetic regulatory proteins to this region, the switch can be turned to favor the lysogenic or lytic cycle. How does this switch work? To understand the mechanism, we need to take a closer look at O_R (**Figure 17.10**).

O_R has three operators, designated O_{R3}, O_{R2}, and O_{R1}. These operators control two promoters called P_{RM} and P_R that transcribe in opposite directions. The λ repressor protein or the cro protein can bind to any or all of the three operators. The binding of these two proteins at these sites governs the switch between the lysogenic and lytic cycles. Two critical issues influence this binding

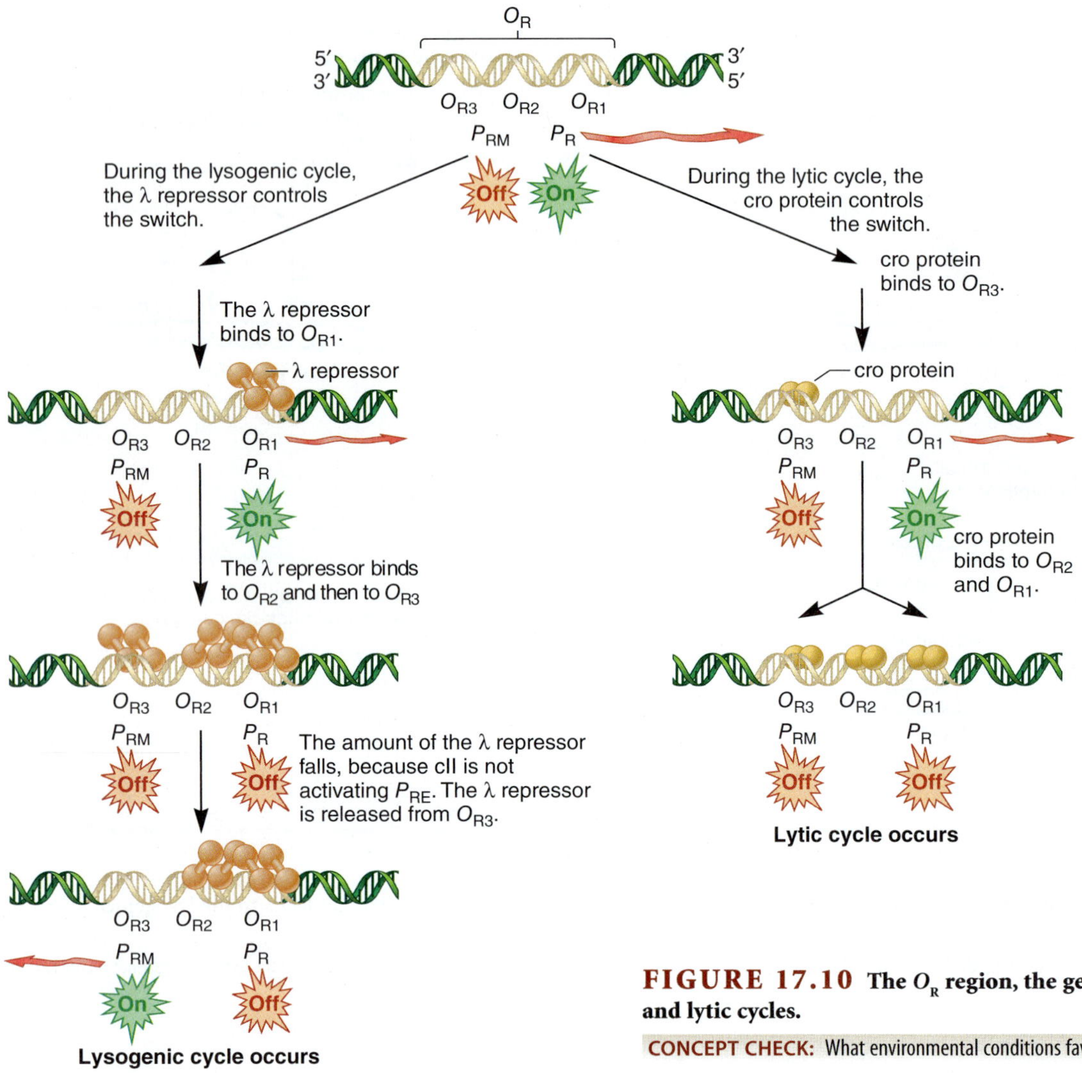

FIGURE 17.10 **The O_R region, the genetic switch between the lysogenic and lytic cycles.**

CONCEPT CHECK: What environmental conditions favor a switch to the lytic cycle?

event. The first is the relative affinities that the regulatory proteins have for these operators. The second is the concentrations of the λ repressor protein and the cro protein within the cell.

Let's consider how an increasing concentration of the λ repressor protein can switch on the lysogenic cycle and switch off the lytic cycle (Figure 17.10, left side). λ repressor binds with highest affinity to O_{R1}, followed by O_{R2} and then O_{R3}. As the concentration of the λ repressor builds within the cell, a dimer of the λ repressor protein first binds to O_{R1} because it has the highest affinity for this operator. Next, a second λ repressor dimer binds to O_{R2}. This occurs very rapidly, because the binding of the first dimer to O_{R1} favors the binding of a second dimer to O_{R2}. This is called a cooperative interaction. The binding of the λ repressor to O_{R1} and O_{R2} inhibits transcription from P_R and switches off the lytic cycle. Early in the lysogenic cycle, the λ repressor protein concentration may become so high that it occupies O_{R3}.

Eventually, the λ repressor concentration begins to drop, because the binding of the λ repressor to O_{R1} and O_{R2} inhibits P_R and thereby decreases the synthesis of cII. Recall that cII initially activates the λ repressor gene from P_{RE}. As the λ repressor concentration gradually falls, it is first removed from O_{R3}. This allows transcription from P_{RM}. As mentioned earlier, the binding of λ repressor at only O_{R1} and O_{R2} acts as an activator of P_{RM}. The ability of the λ repressor to activate its own transcription allows the switch to the lysogenic cycle to be maintained.

In the lytic cycle (Figure 17.10, right side), the binding of the cro protein controls the switch. The cro protein has its highest affinity for O_{R3} and has a similar affinity for O_{R2} and O_{R1}. Under conditions that favor the lytic cycle, the cro protein accumulates, and a cro dimer first binds to O_{R3}. This blocks transcription from P_{RM}, thereby switching off the lysogenic cycle. Later in the lytic cycle, the cro protein concentration continues to rise, so eventually it binds to O_{R2} and O_{R1}. This inhibits transcription from P_R that is not needed in the later stages of the lytic cycle.

Genetic switches, like the one just described for phage λ, represent an important form of genetic regulation. As we have seen, a genetic switch can be used to control two alternative reproductive cycles of a bacteriophage. In addition, genetic switches are also important in the developmental pathways of bacteria and eukaryotes. As we will examine in Chapter 25, they are key events in the initiation of cell differentiation during development. Studies of the phage λ life cycle have provided fundamental information with which to understand how these other switches can operate at the molecular level.

17.3 COMPREHENSION QUESTIONS

1. A mutation in phage λ results in 10-fold greater transcription of the *cII* gene. How do you think this mutation would affect the reproductive cycles?
   a. The lytic cycle would be favored.
   b. The lysogenic cycle would be favored.
   c. Neither cycle could occur.
   d. None of the above would happen.
2. The *cI* gene that encodes the λ repressor has two promoters designated P_{RE} and P_{RM}. Which of the following statements is false?
   a. P_{RE} is activated by the cII protein.
   b. P_{RE} is activated by the λ repressor.
   c. P_{RM} is turned on after P_{RE}.
   d. P_{RM} is needed to maintain the lysogenic cycle.

17.4 HIV REPRODUCTIVE CYCLE

Learning Outcomes:

1. Outline the organization of the HIV genome.
2. Explain how HIV is reverse transcribed and integrated into the DNA of its host cell.
3. Describe the steps that lead to the formation of new HIV particles.

In the previous section, we focused on the reproductive cycles of phage λ, which infects bacteria. In this section, we will focus on HIV, which infects helper T cells found in humans. T cells are a type of lymphocyte that play a key role in cell-mediated immunity. T cells are distinguished from other lymphocytes by the presence of T-cell receptors in their plasma membrane. T-cell receptors are responsible for recognizing antigens—molecules that elicit an immune response. The activation of T cells to fight infection is initiated by the activation of T-cell receptors. Over time, the destruction of helper T cells by HIV compromises immune system function, thereby leading to an inability to fight infections or kill cancer cells.

An understanding of the reproductive cycle of HIV has been critical in the development of drugs to combat the virus. For example, azidothymidine (AZT), which is used to prevent the proliferation of HIV, was developed using knowledge of HIV reproduction. In this section, we will explore how HIV infects helper T cells and how new HIV particles are made.

The HIV-1 Genome Has Nine Genes

Researchers have identified two different strains of HIV called HIV-1 and HIV-2. We will focus on HIV-1, which is much more common. The organization of the HIV genome is described in **Figure 17.11a**. The HIV-1 genome found within virus particles is composed of single-stranded RNA that is 9749 nucleotides long; each end has long-terminal repeat sequences (LTRs); the repeat sequences are the same at both ends. The HIV genome contains nine genes, but some of these genes encode more than one protein. Most of the viral proteins are found within mature HIV particles (**Figure 17.11b**).

A great deal of research has been aimed at determining the functions of HIV proteins. Many of the HIV proteins have multiple functions during the viral reproductive cycle. The following is an overview of HIV genes and some of the best characterized functions of the proteins they encode:

- ***gag:* Proteins Used for Viral Assembly and Capsid Formation.** The *gag* gene encodes the Gag polyprotein, which is cleaved into four different proteins, called matrix protein (p17), capsid protein (p24), nucleocapsid protein (p7), and p6. The p17 protein acts as a matrix to anchor the capsid to the viral envelope. The p24 protein is the major capsid protein. The p7 protein is a nucleocapsid protein that binds to the viral RNA and protects it from cellular digestion by nucleases. The p6 protein facilitates the incorporation of certain proteins, such as Vpr (described later), into HIV particles.
- ***pol:* Enzymes Needed for Viral Replication and Viral Assembly.** The *pol* gene encodes a polyprotein that is cleaved into three enzymes: HIV protease, reverse transcriptase, and integrase. The HIV protease processes proteins made from the HIV genome so they can assemble into mature HIV particles. Reverse transcriptase is required for the production of DNA from the RNA genome of HIV. A portion of reverse transcriptase called RNase H digests RNA during reverse transcription. Integrase is used to incorporate this DNA into the host genome.
- ***vif, vpu:* Proteins That Promote Infectivity and Budding.** The Vif protein is incorporated into mature HIV particles and promotes infectivity by interacting with host-cell proteins. The Vpu protein promotes budding.
- ***vpr, rev, tat, nef:* Proteins with Regulatory Functions.** The Vpr protein has several functions including the transport of the HIV genome into the cell nucleus. The Rev and Tat proteins are involved in the expression of HIV genes. The Nef protein affects the expression of several host-cell genes.
- ***env:* Proteins That Are Part of the Viral Envelope.** The *env* gene encodes a protein that is cleaved into two membrane glycoproteins, gp41 and gp120, which are components of the viral envelope.

HIV RNA Is Reverse Transcribed into Double-Stranded DNA

As briefly described in Figure 17.3, the reproductive cycle of HIV involves reverse transcription and integration of viral DNA into the host-cell's genome. The synthesis of double-stranded DNA from viral RNA occurs via reverse transcriptase. This viral enzyme is found within the HIV particle and is released during the uncoating process. We will now take a closer look at how viral RNA is used to make double-stranded DNA.

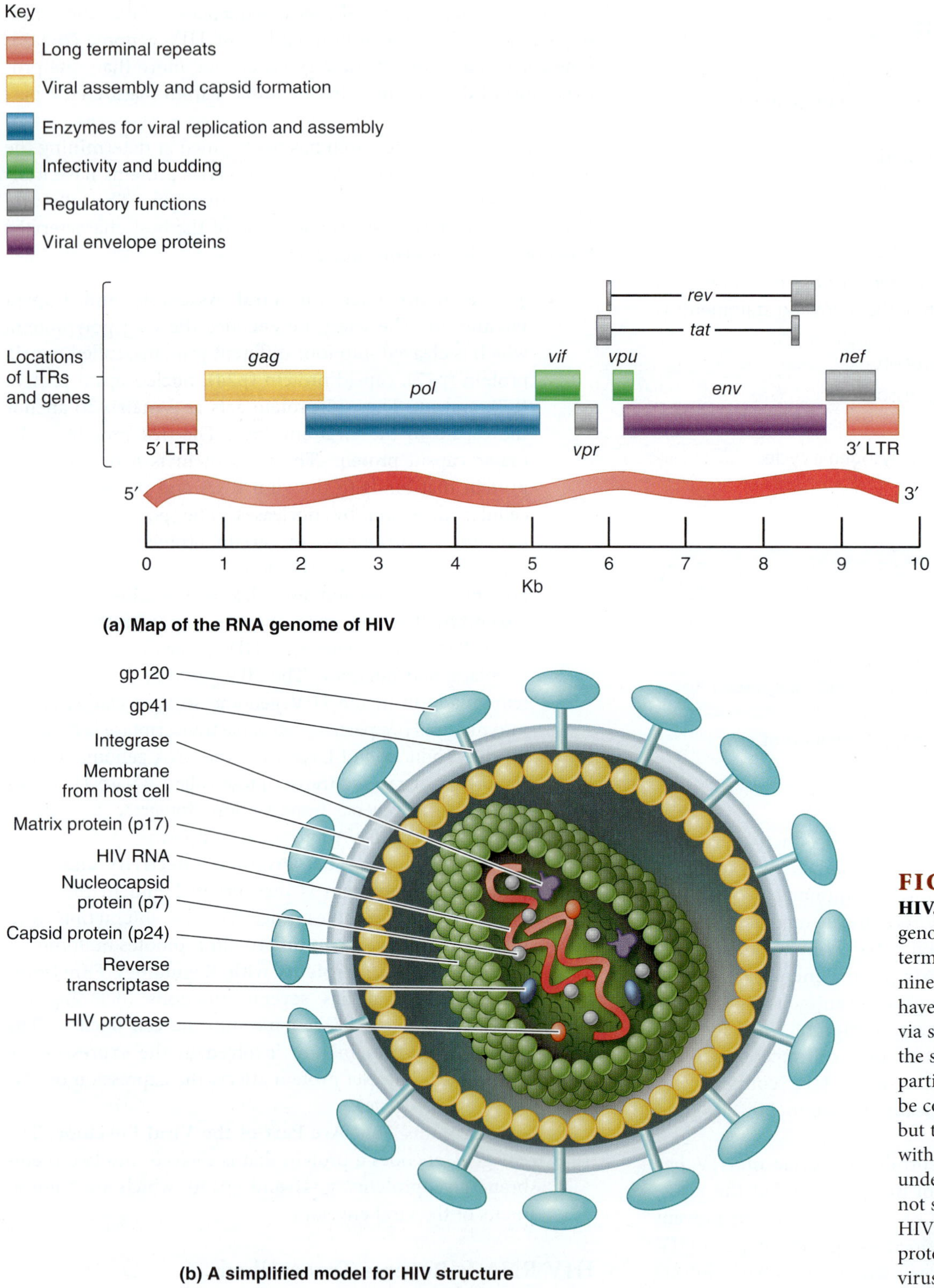

FIGURE 17.11 Structure of HIV. (a) Organization of the RNA genome of HIV. The HIV RNA has a long terminal repeat at each end and carries nine genes. Note: The *rev* and *tat* genes have two exons that become connected via splicing. **(b)** A simplified model for the structure of HIV, also called an HIV particle. Most HIV proteins are thought to be contained within mature virus particles, but their relative amounts and locations within the particle are not completely understood. For simplicity, this figure does not show all of the proteins found in an HIV particle. In addition, several host-cell proteins are incorporated into a mature virus.

The top of **Figure 17.12** shows a drawing of HIV RNA that emphasizes regions that are involved in the synthesis of double-stranded HIV DNA. Let's first focus on the ends of the RNA. Identical repeat sequences (LTRs) are found at both ends of the RNA. Throughout Figure 17.12, these are labeled simply as Repeats. Next to the 5′ repeats is a unique sequence (U5) and then a site called the primer binding site (PBS), which is a site for the binding of a primer that will initiate the synthesis of HIV DNA. Next to the 3′ repeats is a unique sequence (U3) and a site called PPT (polypurine tract), which also plays a role in viral DNA synthesis. The drawing at the top of Figure 17.12 also shows the locations of the *gag*, *pol*, and *env* genes, but leaves out the other six HIV genes for simplicity.

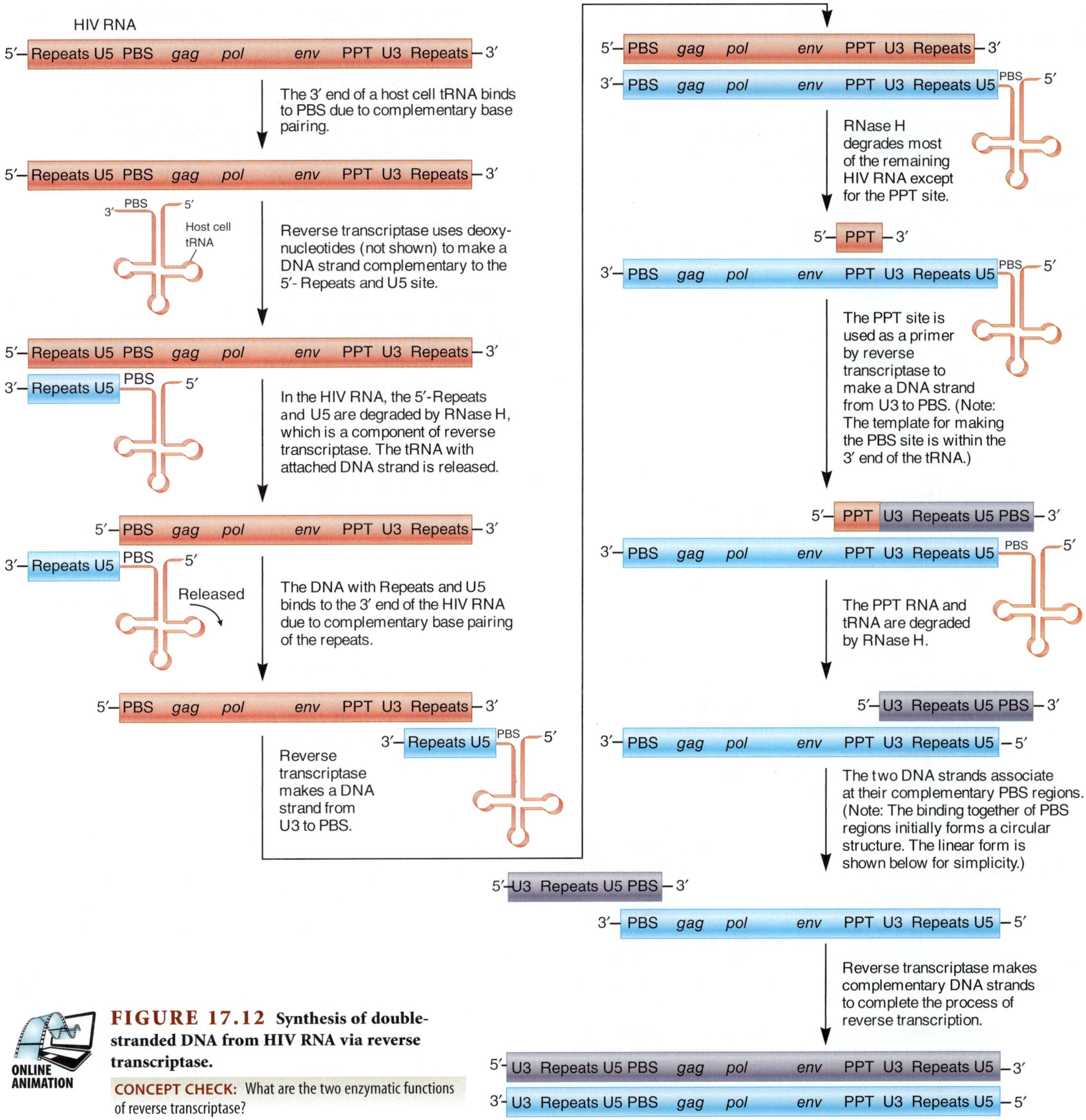

FIGURE 17.12 Synthesis of double-stranded DNA from HIV RNA via reverse transcriptase.

CONCEPT CHECK: What are the two enzymatic functions of reverse transcriptase?

Let's now consider the steps in the reverse transcription of HIV. Once HIV enters a host cell and uncoating occurs, the viral RNA is recognized by a host tRNA that binds to it at PBS. The 3′ end of the tRNA acts as a primer for DNA synthesis. Reverse transcriptase first synthesizes the 5′ repeats and U5 into DNA. After this occurs, the region of the RNA strand encompassing the repeats and U5 is degraded by RNase H, which is a component of reverse transcriptase. The tRNA with an attached DNA strand containing the repeats and U5 is released. Because the 5′ and 3′ repeats are the same, the newly made DNA with the repeats and U5 (shown in blue) can bind to the 3′ repeats in the RNA (shown in red) due to complementary base pairing. Reverse transcriptase then synthesizes the rest of the viral genome, except for the 5′ repeats and U5, which were previously degraded.

Next, the remaining viral RNA except for the PPT site is degraded by RNase H. The RNA at the PPT site acts as a primer to synthesize the region encompassing U3, the repeats, U5, and PBS. This occurs via reverse transcriptase, which also can use a DNA template to make a complementary DNA strand. At this stage, all of the RNA is degraded, which includes the RNA at the PPT site and the tRNA that initiated the synthesis of HIV DNA.

The short purple DNA strand binds to the longer blue DNA strand at their complementary PBS regions. Though not shown in Figure 17.12, this results in a circular structure. Reverse transcriptase then uses the 3′ ends of both DNA strands to complete the synthesis of viral DNA.

The bottom of Figure 17.12 shows the double-stranded DNA of HIV after DNA synthesis is completed. Note that the ends of the double-stranded DNA are not identical to the ends of the single-stranded RNA. In particular, U3 and U5 are found at both ends in the DNA but not in the RNA.

Double-Stranded HIV DNA Is Integrated into the Host Genome

The enzyme called integrase is involved with inserting the double-stranded HIV DNA into the genome of the host cell. Like reverse transcriptase, integrase is found within HIV particles and is released during the uncoating process.

A simplified description of HIV integration is shown in **Figure 17.13**. For integration to occur, a dimer of integrase first binds to specific sequences in the repeats at each end of the HIV DNA. This causes the HIV DNA to form a circular structure (see first inset). Next, integrase makes cuts that remove two nucleotides at both 3′ ends. These steps occur in the cytosol. Other proteins associate with integrase and the HIV DNA to form a **preintegration complex.** One such protein is Vpr, which facilitates the transport of the preintegration complex into the nucleus of the host cell through the nuclear pores. The preintegration complex then binds to a site in a chromosome of the host cell. At this site of integration, integrase makes staggered cuts in the host-cell DNA and then joins the 3′ ends of the HIV DNA strands to the 5′ ends of the host-cell DNA strands. The two nucleotides at the 5′ ends of the HIV DNA, which do not base pair with host-cell DNA, are removed. Due to the staggered cuts and the removal of the two nucleotides at the 5′ ends of the HIV DNA, small gaps occur between the 5′ ends of the viral DNA strands and the 3′ ends of host DNA. These gaps are filled in by DNA gap repair synthesis, which is described in Chapter 19. At this stage, the double-stranded HIV DNA is fully integrated into a site in the host-cell genome. In this condition, it is referred to as a provirus.

The Production of New HIV Particles Begins with the Synthesis of HIV RNA and Viral Proteins

After the double-stranded HIV DNA has been integrated into the host-cell genome, it may remain there in a dormant state for a very long time—several years or more. As mentioned previously, this condition is called latency. To switch from the latent state and start making new virus particles, certain transcription factors must transcribe the HIV provirus. In particular, a transcription factor called NF-κB plays a key role in activating the transcription of the HIV provirus. NF-κB is usually present in the cytosol of T cells in an inactive state. How does NF-κB become active? If an antigen binds to a T-cell receptor, this initiates a signal cascade that causes the activation of NF-κB and its movement into the nucleus. Once in the nucleus, NF-κB binds to the HIV-proviral DNA and stimulates transcription. In this way, those T cells that are currently fighting infection are most likely to produce new HIV particles and to be killed.

Figure 17.14 shows a simplified version of how new HIV components are synthesized. The 5′-LTR of HIV contains a single promoter and the binding sites for several transcription factors, including NF-κB. These transcription factors stimulate the transcription of HIV RNA from the promoter. The HIV RNA contains several 5′ and 3′ splicing sites so that HIV RNA can be spliced in more than 30 different ways or it may not be spliced at all. Though the splicing process is complex, we can view splicing as occurring in three general ways, as described next.

Fully Spliced HIV RNA Early in the process of synthesizing new HIV components, the HIV RNA is fully spliced, which removes two introns. This form of HIV RNA is able to exit the nucleus and is used for the translation of the Nef, Tat, and Rev proteins—the Nef, Tat, and Rev proteins are early gene products. The Nef protein maintains T-cell activation and inhibits the production of host-cell proteins that are involved with cellular defense against viruses. The Tat and Rev proteins enter the nucleus, where the Tat protein greatly increases the level of transcription of HIV RNA. The function of the Rev protein is described next.

Incompletely Spliced HIV RNA Incompletely spliced RNA is also produced during the transcription of HIV RNA, but eukaryotic cells possess mechanisms that prevent RNAs containing introns from exiting the cell nucleus. After the Rev protein is translated from the fully spliced HIV RNA, it enters the nucleus. Multiple copies of the Rev protein bind to incompletely spliced versions of the HIV RNA. This allows these RNAs to exit the nucleus. Incompletely spliced versions of HIV RNA are needed for the translation of Vif, Env, Vpu, and Vpr proteins. The Env and Vpu proteins are integral membrane proteins that are initially inserted into the endoplasmic reticulum (ER) membrane and move to the Golgi and then to the plasma membrane via membrane vesicles. During its transport, the Env protein is cleaved into two components called gp41 and gp120. The gp41 protein is a transmembrane protein that anchors gp120 to the membrane (see Figure 17.11b). Vpu is needed for the release of HIV particles from the host cell.

Unspliced HIV RNA Rev proteins also bind to unspliced HIV RNA to allow its export from the nucleus. Unspliced HIV RNA has two key functions. First, unspliced HIV RNA is packaged into HIV particles. Second, it is used for the translation of Gag and Gag-pol polyproteins. The Gag-pol polyprotein is a single polyprotein in which the Gag polyprotein is connected to the Pol polyprotein. How is the Gag-pol polyprotein made? On rare occasions the stop codon at the end of *gag* is not recognized. This readthrough past

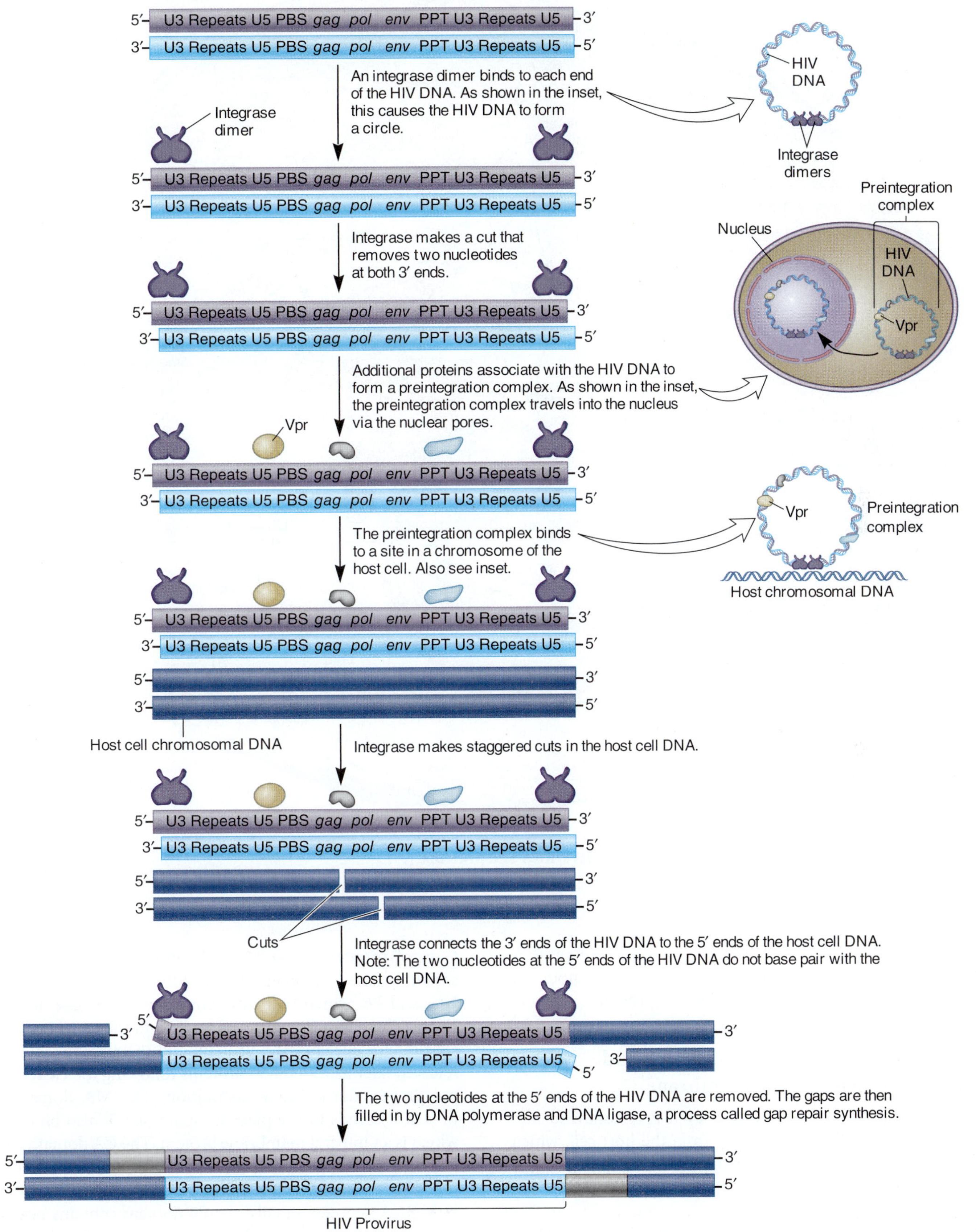

ONLINE ANIMATION

FIGURE 17.13 Integration of the double-stranded HIV DNA into the host-cell genome. Note: This is a simplified version of HIV integration that focuses on the role of integrase. Other proteins are necessary for integration.

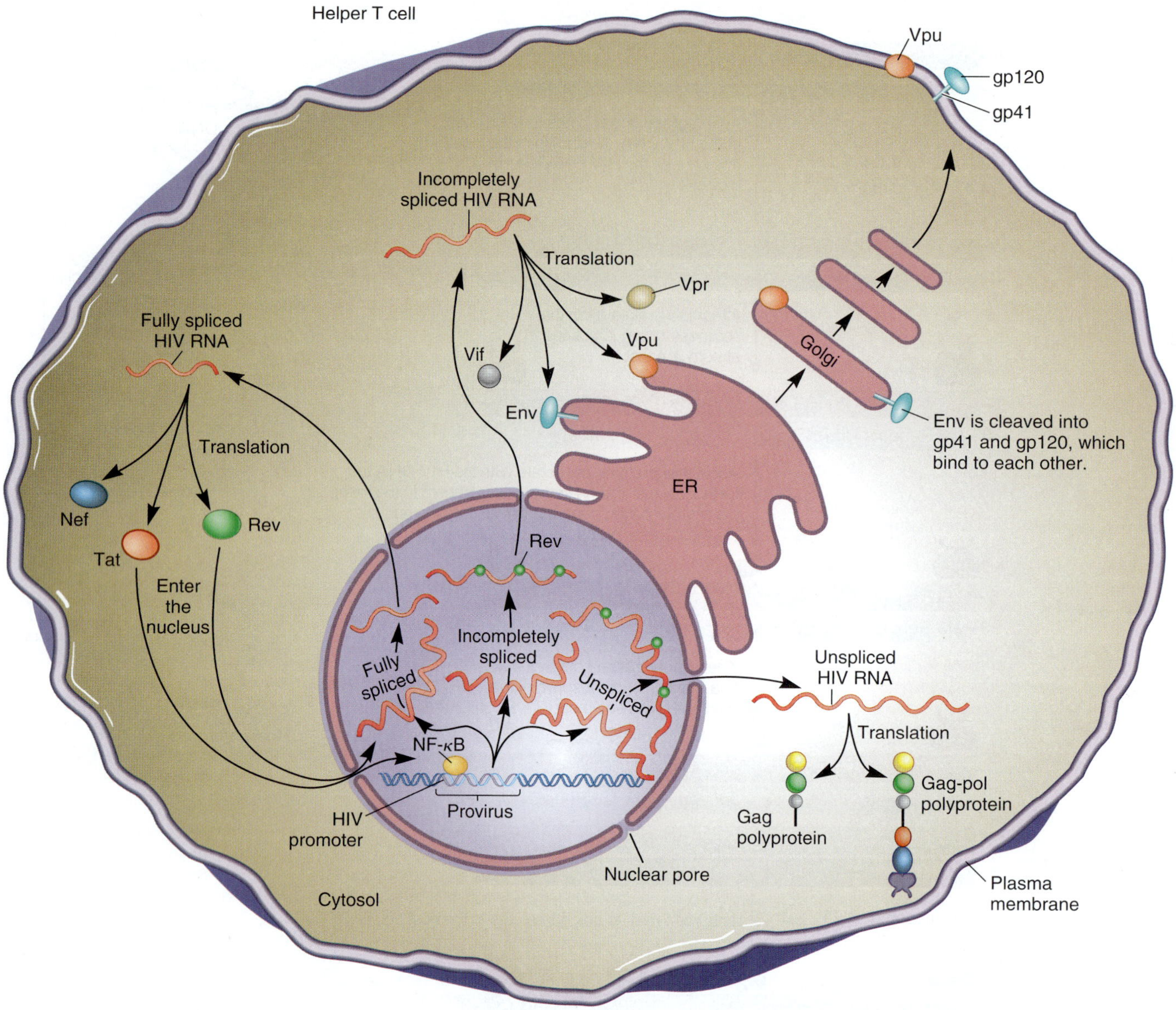

FIGURE 17.14 **Synthesis of new HIV components: HIV RNA and proteins.**

CONCEPT CHECK: Which form of HIV RNA, fully spliced, incompletely spiced, or unspliced, is needed during the early stage of synthesizing HIV components?

the stop codon results in the synthesis of Gag-pol polyprotein. However, because the readthrough is a rare event, the amount of Gag-pol polyprotein is much less than Gag polyprotein.

HIV Components Are Assembled and Bud from the Host-Cell Plasma Membrane

The final event in the HIV reproductive cycle involves the formation of HIV particles that are released from the host cell, which is a helper T cell. This process occurs in three overlapping stages: assembly, budding, and maturation.

Assembly As we have seen, the expression of HIV-proviral DNA leads to the synthesis of viral components, including unspliced HIV RNA and viral proteins. The assembly of HIV particles occurs at the host-cell plasma membrane and is mediated by the Gag polyprotein. **Figure 17.15** shows a simplified model for HIV assembly. The Gag polyprotein has four domains called MA, CA, NC, and P6, which play different roles in the assembly process. (Note: These four domains correspond to the matrix protein, capsid protein, nucleocapsid protein, and p6, respectively.) They are released later as individual proteins following the cleavage of Gag polyprotein during viral maturation. The MA domain (shown in yellow) binds to the plasma membrane. It also binds to gp41, which is an integral membrane protein. The CA domain (shown in green) facilitates protein–protein interactions required for immature viral assembly. The NC domain (shown in gray) captures the HIV RNA genome. Finally, the P6 domain contains binding sites for several other proteins that will be contained within virus particles, such as Vpr. The P6 domain also recognizes Vpu and cellular proteins that are required for budding, but are not contained within mature HIV particles.

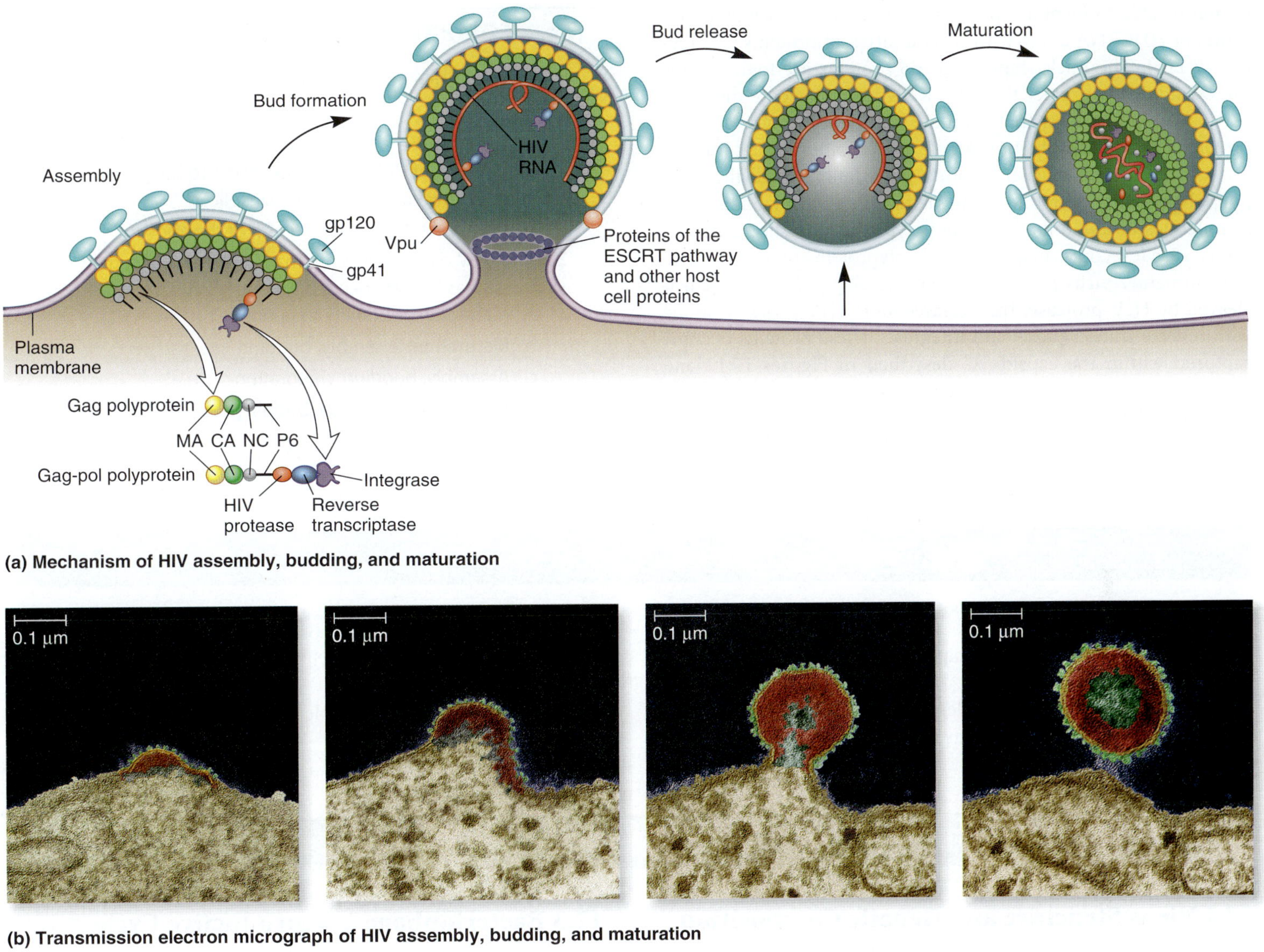

FIGURE 17.15 Assembly, budding, and maturation of new HIV particles. Part **(a)** is a schematic drawing and part **(b)** shows a colorized transmission electron micrograph of the process. Note: The colors in parts **(a)** and **(b)** don't match each other.

Budding As seen in Figure 17.15, as assembly occurs, the interactions of many Gag polyproteins results in the formation of a spherical structure that protrudes from the plasma membrane. The release of this bud requires Vpu (an HIV protein) and host-cell proteins, including ones that are components of the ESCRT (endosomal sorting complex required for transport) pathway. The P6 domain of Gag polyprotein recognizes certain proteins of the ESCRT pathway. This initiates a series of events that leads to membrane fusion at the neck of the bud, thereby releasing the bud from the surface of the cell. The released bud is an immature HIV particle.

The HIV protein called Vpu is also important for budding by enhancing the release of immature HIV particles from the cell surface. Vpu exerts its effects by inhibiting a protein known as tetherin (not shown in Figure 17.15). In the absence of Vpu, tetherin binds to the membrane of the viral envelope and tethers it to the cell surface. When Vpu inhibits tetherin, this allows immature HIV particles to be released.

Maturation The final stage of the HIV reproductive cycle is the maturation of immature HIV particles into mature virus particles. As a polyprotein, Gag promotes the assembly of the components that constitute an HIV particle. However, for a capsid to be formed and surround the HIV RNA, the Gag polyprotein must be cleaved into its individual components. HIV protease plays a key role in this maturation process. HIV protease cleaves Gag polyprotein at multiple sites, which releases MA, CA, NC, and P6 as individual proteins. The matrix protein lines the inside of the HIV envelope, helping to anchor the gp41/gp120 spikes to the envelope. Once released from the Gag polyprotein, the capsid protein is able to form a capsid structure that encloses two molecules of HIV RNA along with several different proteins (see Figure 17.11b). The nucleocapsid and p6 proteins are found inside the capsid. They protect the HIV RNA from nuclease digestion and have binding sites that promote the incorporation of other proteins into the capsid. For example, p6 has binding sites for Vpr.

As mentioned earlier, another polyprotein called the Gag-pol polyprotein plays a role in the maturation process, but is made in much lesser amounts than Gag polyprotein. As shown in Figure 17.15, a few copies of the Gag-pol polyprotein are incorporated into immature virus particles. When the Gag-pol polyprotein is cleaved by HIV protease, this releases more HIV protease, along with reverse transcriptase, and integrase. These proteins become captured within the capsid. As described in Figures 17.12 and 17.13, they are necessary for the reverse transcription and integration of the HIV genome in a newly infected host cell.

17.4 COMPREHENSION QUESTIONS

1. A viral protein that is needed to make HIV DNA is
   a. integrase.
   b. reverse transcriptase.
   c. Vpr.
   d. Gag polyprotein.
2. Which form of HIV RNA is packaged into HIV particles?
   a. Fully spliced
   b. Incompletely spliced
   c. Unspliced
   d. All of the above
3. After HIV components are made, what is the correct order of the stages that are needed to produce HIV particles?
   a. Maturation, budding, assembly
   b. Maturation, assembly, budding
   c. Assembly, budding, maturation
   d. Assembly, maturation, budding

KEY TERMS

Page 417. virus

Page 418. host cell, host range, capsid, viral envelope, bacteriophages, phages, viral genome

Page 422. viral reproductive cycle, phage λ, human immunodeficiency virus (HIV), integrase, prophage, lysogenic cycle, reverse transcriptase, provirus

Page 423. lytic cycle, latent, lysogeny, temperate phage, virulent phages, episome, emerging viruses

Page 429. antitermination

Page 436. preintegration complex

CHAPTER SUMMARY

17.1 Virus Structure and Genetic Composition

- Tobacco mosaic virus (TMV) was the first virus to be discovered. It infects many species of plants.
- Viruses vary with regard to their host range, structure, and genome composition (see Table 17.1, Figure 17.1).
- The production of reconstituted viruses confirmed that the genome of TMV is composed of RNA (see Figure 17.2).

17.2 Overview of Viral Reproductive Cycles

- The viral reproductive cycle consists of five or six basic steps, including attachment, entry, integration, synthesis, assembly, and release (see Figure 17.3).
- Some bacteriophages can alternate between two reproductive cycles: the lysogenic cycle and lytic cycle (see Figure 17.4).
- Emerging viruses have arisen recently and are more likely to cause infection (see Figure 17.5).
- The disease AIDS is caused by a virus called human immunodeficiency virus (HIV). The virus is a retrovirus whose reproductive cycle involves the integration of the viral genome into a chromosome in a helper T cell (see Figure 17.6).

17.3 Bacteriophage λ Reproductive Cycle

- Phage λ has different sets of genes that allow it to follow a lysogenic or lytic cycle (see Figure 17.7).
- The expression of the cII protein favors the lysogenic cycle, whereas the expression of cro protein favors the lytic cycle (see Figures 17.8, 17.9).
- O_R has three operators that are binding sites for the λ repressor and cro protein. O_R acts as a genetic switch between the lysogenic and lytic cycles (see Figure 17.10).

17.4 HIV Reproductive Cycle

- The HIV genome is an RNA molecule that contains nine genes, which are needed to make HIV particles (Figure 17.11).
- Following infection, reverse transcriptase catalyzes the synthesis of HIV DNA and integrase is used to insert the HIV DNA into the host-cell genome (Figures 17.12, 17.13).
- The transcription of HIV DNA leads to the synthesis of HIV RNA and viral proteins (Figure 17.14).
- HIV components are assembled at the host-cell plasma membrane. Buds are released that mature into HIV particles (Figure 17.15).

PROBLEM SETS & INSIGHTS

Solved Problems

S1. Discuss how the components of viruses assemble to make new virus particles.

Answer: Some viruses with a simple structure self-assemble, meaning that viral components spontaneously bind to each other to form a complete virus particle. An example of a self-assembling virus is tobacco mosaic virus (TMV), which we examined earlier. The capsid proteins assemble around the RNA genome, which becomes trapped inside the hollow capsid.

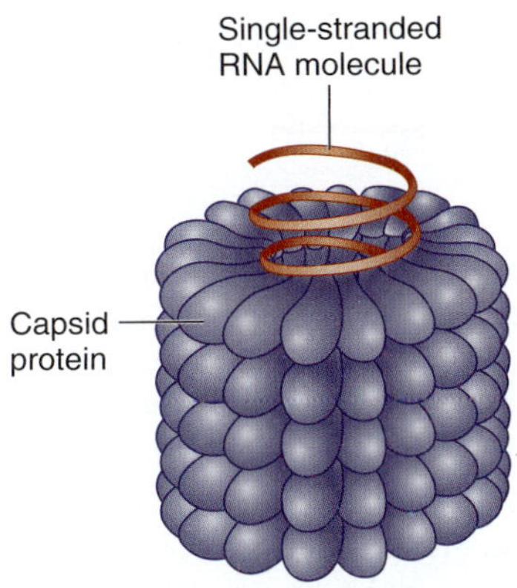

Other viruses do not self-assemble. For example, the assembly of phage λ requires the help of noncapsid proteins not found in the completed phage particle. Some of these noncapsid proteins function as enzymes that modify capsid proteins, whereas others serve as scaffolding for the assembly of the capsid. The formation of HIV particles occurs in three stages. First, HIV RNA and viral proteins assemble at the plasma membrane. Next, the forming virus acquires its outer envelope in a budding process. After the bud is released from the cell, it matures into a virus particle.

S2. From the perspective of the bacteriophage, what is the advantage of having both a lytic and lysogenic cycle?

Answer: For a phage such as λ that can follow either cycle, environmental conditions influence whether it follows the lytic or lysogenic cycle. If nutrients are readily available, phage λ usually proceeds directly to the lytic cycle after its DNA enters the cell. This allows the phage to rapidly proliferate. Alternatively, if nutrients are in short supply, the lysogenic cycle is usually favored because sufficient nutrients may not be available to make new viruses. If more nutrients become available later, the prophage may become activated, and the viral reproductive cycle will switch to the lytic cycle. Therefore, under more favorable conditions, new viruses are made and released.

S3. What are three different mechanisms of viral latency?

Answer: Some bacteriophages enter the lysogenic cycle in which the phage DNA becomes incorporated into the bacterial chromosome as a prophage. Eukaryotic viruses can become latent in other ways. Retroviruses, which have an RNA genome, are reverse transcribed and the viral DNA is incorporated into a chromosome as a provirus. Alternatively, herpesviruses can exist as episomes that remain latent for long periods of time.

S4. What is an emerging virus? Give three examples. How do you think that emerging viruses arise?

Answer: Emerging viruses are viruses that have arisen recently or are more likely to cause infection. Examples include HIV, the corona virus that causes SARS, and the influenza strain called H1N1. Emerging viruses arise by new mutations. Such mutations may alter the ability of a virus to infect its host or alter its ability to cause disease symptoms.

Conceptual Questions

C1. Discuss why viruses are considered nonliving.

C2. What structural features are common to all viruses? Which features are found only in certain types of viruses?

C3. What are the similarities and differences among viral genomes?

C4. What is a viral envelope? Describe how it is made.

C5. What do the terms *host cell* and *host range* mean?

C6. Describe why the attachment step in a viral reproductive cycle is usually specific for one or just a few cell types.

C7. Compare and contrast the entry step of the viral reproductive cycle of phage λ and HIV.

C8. What is the role of reverse transcriptase in the reproductive cycle of HIV and other retroviruses?

C9. Describe how lytic bacteriophages are released from their host cells.

C10. What is the difference between a temperate phage versus a virulent phage?

C11. What is a prophage, a provirus, and an episome? What is their common role in a viral reproductive cycle?

C12. What key features distinguish the lytic from the lysogenic cycles?

C13. With regard to promoting the lytic or lysogenic cycle, what would happen if the following genes were missing from the λ genome?

A. *cro*

B. *cI*

C. *cII*

D. *int*

E. *cII* and *cro*

C14. How do the λ repressor and the cro protein affect the transcription from P_R and P_{RM}? Explain where these proteins are binding to cause their effects.

C15. In your own words, explain why it is necessary for the *cI* gene to have two promoters. What would happen if it had only P_{RE}?

C16. Figure 17.10 shows a genetic switch that controls the choice between the lytic and lysogenic cycles of phage λ. What is a genetic switch? Compare the roles of a genetic switch and a simple operator site (like the one found in the *lac* operon) in gene regulation.

C17. Describe the process of reverse transcription of the HIV RNA.

C18. Why is a host-cell tRNA needed for reverse transcription?

C19. Explain the role of RNase H (a component of reverse transcriptase) during the synthesis of HIV DNA.

C20. Describe how HIV DNA is integrated into a chromosome of the host cell.

C21. What is the role of Vpr protein during the process of HIV DNA integration?

C22. Why is gap repair synthesis needed during HIV DNA integration?

C23. Compare and contrast the roles of fully spliced, incompletely spliced, and unspliced HIV RNA. Which type is needed in the early stages of HIV proliferation, and which is needed in later stages?

C24. Describe the role of Gag polyprotein during the assembly of HIV components at the plasma membrane.

C25. How does an HIV particle acquire its envelope?

C26. Explain the role of HIV protease during the process of HIV maturation.

Experimental Questions

E1. Discuss how researchers determined that TMV is a virus that causes damage to plants.

E2. What technique must be used to visualize a virus?

E3. What is a reconstituted virus?

E4. Following the infection of reconstituted viruses on tobacco leaves, what two characteristics did Fraenkel-Conrat and Singer analyze? Explain how their results were consistent with the idea that the RNA of TMV is responsible for the traits of the virus.

E5. Certain drugs to combat human viral diseases affect spike glycoproteins in the viral envelope. Discuss how you think that such drugs may prevent viral infection.

E6. Other drugs that inhibit HIV proliferation are inhibitors of HIV protease. Explain how these drugs would help to stop the spread of HIV.

E7. A researcher identified a mutation in P_R of phage λ that causes its transcription rate to be increased 10-fold. Do you think this mutation would favor the lytic or lysogenic cycle? Explain your answer.

E8. Experimentally, when an *E. coli* bacterium already has a λ prophage integrated into its chromosome, another λ phage cannot usually infect the cell and establish the lysogenic or lytic cycle. Based on your understanding of the genetic regulation of the λ life cycles, why do you think the other phage would be unsuccessful?

E9. A bacterium is exposed to a drug that inhibits the N protein. What would you expect to happen if the bacterium was later infected by phage λ? Would phage λ follow the lytic cycle, the lysogenic cycle, or neither? Explain your answer.

E10. This question combines your knowledge of conjugation (described in Chapter 7) and the genetic regulation that directs the phage λ reproductive cycles. When researchers mix donor *Hfr* strains with recipient F^- bacteria that are lysogenic for phage λ, the conjugated cells survive normally. However, if donor *Hfr* strains that are lysogenic for phage λ conjugate with recipient F^- bacteria that do not contain any phage λ, the recipient cells often lyse, due to the induction of λ into the lytic cycle. Based on your knowledge of the regulation of the two λ cycles, explain this experimental observation.

Questions for Student Discussion/Collaboration

1. Discuss the properties of emerging viruses. What are the challenges associated with combating them?
2. Certain environmental conditions such as UV light are known to activate lysogenic λ prophages and cause them to progress into the lytic cycle. UV light initially causes the repressor protein to be proteolytically degraded. Make a flow diagram that describes the subsequent events that would lead to the lytic cycle. Note: The *xis* gene codes for an enzyme that is necessary to excise the λ prophage from the *E. coli* chromosome. The integrase enzyme is also necessary to excise the λ prophage.
3. Browse the Internet and determine which drugs are used to treat people with AIDS. Which proteins do these drugs affect? Discuss how an understanding of the HIV reproductive cycle has been helpful in developing a treatment to combat AIDS.

Note: All answers appear at the website for this textbook; the answers to the even-numbered questions and all of the Concept Check and Comprehension Questions are in the back of the book.

www.mhhe.com/brookergenetics5e

Visit the website for practice tests, answer keys, and other learning aids for this chapter. Enhance your understanding of genetics with our interactive exercises, quizzes, animations, and much more.

CHAPTER OUTLINE

The effects of a mutation. *A mutation during embryonic development has caused this sheep to have a black spot on its side.*

18 GENE MUTATION AND DNA REPAIR

The primary function of DNA is to store information for the synthesis of proteins. A key aspect of the gene expression process is that the DNA itself does not normally change. This allows DNA to function as a permanent storage unit. However, on relatively rare occasions, a mutation can occur. The term **mutation** refers to a heritable change in the genetic material. This means that the structure of DNA has been changed permanently, and this alteration can be passed from mother to daughter cells during cell division. If a mutation occurs in reproductive cells, it may also be passed from parent to offspring.

The topic of mutation is centrally important in all fields of genetics, including molecular genetics, Mendelian inheritance, and population genetics. Mutations provide the allelic variation that we have discussed throughout this textbook. For example, phenotypic differences, such as tall versus dwarf pea plants, are due to mutations that alter the expression of particular genes. With regard to their phenotypic effects, mutations can be beneficial, neutral, or detrimental. On the positive side, mutations are essential to the continuity of life. They provide the variation that enables species to adapt to their environment via natural selection. Mutations are the foundation for evolutionary change. On the negative side, however, new mutations are much more likely to be harmful rather than beneficial to the individual. The genes within each species have evolved to work properly. They have functional promoters, coding sequences, terminators, and so on, that allow the genes to be expressed. Random mutations are more likely to disrupt these sequences rather than improve their function. For example, many inherited human diseases result from mutated genes. In addition, diseases such as skin and lung cancer can be caused by environmental agents that are known to cause DNA mutations. For these and many other reasons, understanding the molecular nature of mutations is a deeply compelling area of research. In this chapter, we will consider the nature of mutations and their consequences on gene expression at the molecular level.

Because mutations can be quite harmful, species have evolved several ways to repair damaged DNA. DNA repair systems reverse DNA damage before it results in a mutation that could potentially have negative consequences. DNA repair systems have been studied extensively in many organisms, particularly *Escherichia coli*, yeast, mammals, and plants. A variety of systems repair different types of DNA lesions. We will conclude the chapter by examining the ways that several of these DNA repair systems operate.

18.1 EFFECTS OF MUTATIONS ON GENE STRUCTURE AND FUNCTION

Learning Outcomes:

1. Define point mutation.
2. Describe how mutations within the coding sequence of a gene may alter a polypeptide's structure.
3. Explain how mutations within noncoding sequences may alter gene function.
4. Compare and contrast intragenic and intergenic suppressors.
5. Explain how changes in chromosome structure may affect gene expression.
6. Distinguish between germ-line and somatic mutations.

How do mutations affect phenotype? To answer this question, we must appreciate how changes in DNA structure can ultimately affect gene function. Much of our understanding of mutation has come from the study of experimental organisms, such as bacteria, yeast, and *Drosophila*. Researchers can expose these organisms to environmental agents that cause mutation and then study the consequences of the induced mutations. In addition, because these organisms have a short generation time, researchers can investigate the effects of mutation when they are passed from cell to cell and from parent to offspring.

As discussed in Chapter 8, changes in chromosome structure and number are important occurrences within natural populations of eukaryotic organisms. These types of changes are considered to be mutations because the genetic material has been altered in a way that can be inherited. Changes in chromosome structure and number usually affect the expression of many genes. In comparison, a mutation in a single gene is a relatively small change in DNA structure. In this section, we will be primarily concerned with the ways that mutations affect the molecular and phenotypic expression of single genes. We will also consider how the timing of mutations during an organism's development has important consequences.

Gene Mutations Are Molecular Changes in the DNA Sequence of a Gene

A gene mutation occurs when the sequence of the DNA within a gene is altered in a permanent way. It can involve a base substitution or a removal or addition of one or more base pairs.

A **point mutation** is a change in a single base pair within the DNA. For example, the DNA sequence shown here has been altered by a **base substitution,** in which one base is substituted for another.

```
                      ↓
5′–AACGCTAGATC–3′ → 5′–AACGCGAGATC–3′
3′–TTGCGATCTAG–5′   3′–TTGCGCTCTAG–5′
```

A change of a pyrimidine to another pyrimidine, such as C to T, or a purine to another purine, such as A to G, is called a **transition.** This type of mutation is more common than a **transversion,** in which a purine and a pyrimidine are interchanged. The example just shown is a transversion (a change from T to G in the top strand).

Besides base substitutions, a short sequence of DNA may be deleted from or added to the chromosomal DNA:

```
                          ↓
5′–AACGCTAGATC–3′ → 5′–AACGCTC–3′
3′–TTGCGATCTAG–5′   3′–TTGCGAG–5′
                    (Deletion of 4 bp)

                       _↓_
5′–AACGCTAGATC–3′ → 5′–AACAGTCGCTAGATC–3′
3′–TTGCGATCTAG–5′   3′–TTGTCAGCGATCTAG–5′
                    (Addition of 4 bp)
```

As we will see next, small deletions or additions to the sequence of a gene can significantly affect the function of the encoded protein.

Gene Mutations Can Alter the Coding Sequence Within a Gene

How might a mutation within the coding sequence of a protein-encoding gene affect the amino acid sequence of the polypeptide that is encoded by the gene? **Table 18.1** describes the possible effects of point mutations. **Silent mutations** are those that do not alter the amino acid sequence of the polypeptide even though the nucleotide sequence has changed. Because the genetic code is degenerate, silent mutations can occur in certain bases within a codon, such as the third base, so the specific amino acid is not changed. In contrast, **missense mutations** are base substitutions in which an amino acid change does occur. An example of a missense mutation is the one that causes the human disease known as sickle cell disease. This disease involves a mutation in the β-globin gene, which alters the polypeptide sequence so the sixth amino acid is changed from a glutamic acid to valine. This single amino acid substitution alters the structure and function of the hemoglobin protein. One consequence of this alteration is that under conditions of low oxygen, the red blood cells assume a sickle shape (**Figure 18.1**). In this case, a single amino acid substitution has a profound effect on the phenotype of cells and can even cause a serious disease.

Nonsense mutations involve a change from a normal codon to a stop codon. This terminates the translation of the polypeptide earlier than expected, producing a truncated polypeptide (see Table 18.1). When a nonsense mutation occurs in a bacterial operon, it may also inhibit the expression of downstream genes. This phenomenon, termed **polarity,** is described in solved problem S4 at the end of this chapter. Finally, **frameshift mutations** involve the addition or deletion of a number of nucleotides that is not divisible by 3. Because the codons are read in multiples of 3, this shifts the reading frame. The translation of the mRNA then results in a completely different amino acid sequence downstream from the mutation.

Except for silent mutations, new mutations are more likely to produce polypeptides that have reduced rather than enhanced function. For example, when nonsense mutations produce polypeptides that are substantially shorter, the shorter polypeptides are unlikely to function properly. Likewise, frameshift mutations dramatically alter the amino acid sequence of polypeptides and

TABLE 18.1

Consequences of Point Mutations Within the Coding Sequence

Type of Change	Mutation in the DNA	Example*	Amino Acids Altered	Likely Effect on Protein Function
None	None	5′–A–T–G–A–C–C–G–A–C–C–C–G–A–A–A–G–G–G–A–C–C–3′ Met – Thr – Asp – Pro – Lys – Gly – Thr –	None	None
Silent	Base substitution	↓ 5′–A–T–G–A–C–C–G–A–C–C–C–C–A–A–A–G–G–G–A–C–C–3′ Met – Thr – Asp – Pro – Lys – Gly – Thr –	None	None
Missense	Base substitution	↓ 5′–A–T–G–C–C–C–G–A–C–C–C–G–A–A–A–G–G–G–A–C–C–3′ Met – Pro – Asp – Pro – Lys – Gly – Thr –	One	Neutral or inhibitory
Nonsense	Base substitution	↓ 5′–A–T–G–A–C–C–G–A–C–C–C–G–T–A–A–G–G–G–A–C–C–3′ Met – Thr – Asp – Pro – STOP!	Many	Inhibitory
Frameshift	Addition/deletion	↓ 5′–A–T–G–A–C–C–G–A–C–G–C–C–G–A–A–A–G–G–G–A–C–C–3′ Met – Thr – Asp – Ala – Glu – Arg – Asp –	Many	Inhibitory

*DNA sequence in the coding strand. Note that this sequence is the same as the mRNA sequence except that the RNA contains uracil (U) instead of thymine (T). For the nonmutant sequence, the complementary sequence of the other strand would be: 3′–T–A–C–T–G–G–C–T–G–G–G–C–T–T–T–C–C–C–T–G–G–3′. The 3-base codons are shown in alternating black and red colors. Mutations are shown in green. Changes in the amino acid sequence are shown in blue.

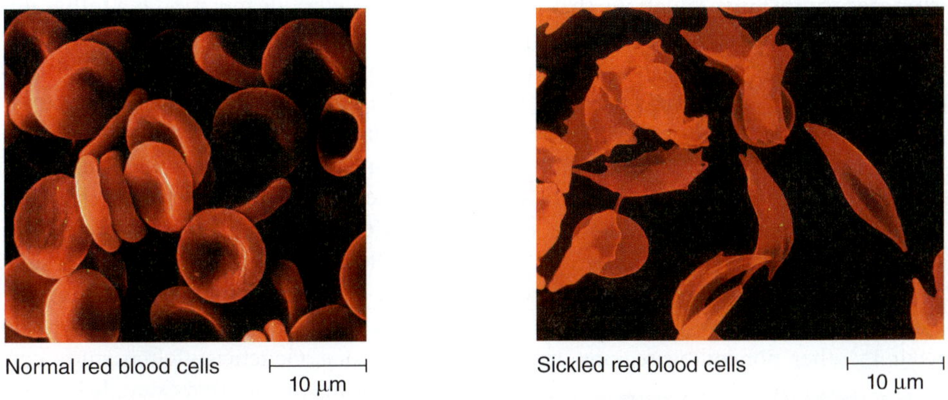

(a) Micrographs of red blood cells

NORMAL: NH_2 – VALINE – HISTIDINE – LEUCINE – THREONINE – PROLINE – GLUTAMIC ACID – GLUTAMIC ACID...

SICKLE CELL: NH_2 – VALINE – HISTIDINE – LEUCINE – THREONINE – PROLINE – VALINE – GLUTAMIC ACID...

(b) A comparison of the amino acid sequence between normal β globin and sickle cell β globin

FIGURE 18.1 Missense mutation in sickle cell disease. **(a)** Normal red blood cells (left) and sickled red blood cells (right). **(b)** A comparison of the amino acid sequence of the normal β-globin polypeptide and the polypeptide encoded by the sickle cell allele. This figure shows only a portion of the polypeptide sequence, which is 146 amino acids long. In β globin, the first methionine at the amino terminus is removed. As seen here, a missense mutation changes the sixth amino acid from a glutamic acid to a valine.

Genes → Traits A missense mutation alters the structure of β globin, which is a subunit of hemoglobin, the oxygen-carrying protein in the red blood cells. When an individual is homozygous for this allele, this missense mutation causes the red blood cells to sickle under conditions of low oxygen concentration. The sickling phenomenon is a description of the trait at the cellular level. At the organism level, the sickled cells can clog the capillaries, causing painful episodes, called crises, that can result in organ damage. The shortened life span of the red blood cells leads to symptoms of anemia.

are thereby likely to disrupt function. Missense mutations are less likely to alter function because they involve a change of a single amino acid within polypeptides that typically contain hundreds of amino acids. When a missense mutation has no detectable effect on protein function, it is referred to as a **neutral mutation.** A missense mutation that substitutes an amino acid with a similar chemistry as the original amino acid is likely to be neutral or nearly neutral. For example, a missense mutation that substitutes a glutamic acid for an aspartic acid may be neutral because both amino acids are negatively charged and have similar side chain structures. Silent mutations are also considered a type of neutral mutation.

Mutations can occasionally produce a polypeptide with an enhanced ability to function. Although these favorable mutations are relatively rare, they may result in an organism with a greater likelihood of surviving and reproducing. If this is the case, natural selection may cause such a favorable mutation to increase in frequency within a population. This topic will be discussed later in this chapter and also in Chapter 26.

Gene Mutations Can Occur Outside of the Coding Sequence and Influence Gene Expression

Thus far, we have focused our attention on mutations in the coding regions of genes and their effects on polypeptide structure and protein function. In previous chapters, we learned how various sequences outside the coding sequence play important roles during the process of gene expression. A mutation can occur within noncoding sequences, thereby affecting gene expression (**Table 18.2**). For example, a mutation may alter the sequence within the core promoter of a gene. Promoter mutations that increase transcription are termed **up promoter mutations.** Mutations that make a sequence more like the consensus sequence are likely to be up promoter mutations. In contrast, a **down promoter mutation** causes the promoter to become less like the consensus sequence, decreasing its affinity for transcription factors and decreasing the transcription rate.

In Chapter 14, we considered how mutations can affect regulatory elements. For example, mutations in the *lac* operator site, called *lacO*^{C} mutations, prevent the binding of the lac repressor protein. This causes the *lac* operon to be constitutively expressed even in the absence of lactose. As noted in Table 18.2, mutations can also occur in other noncoding regions of a gene and alter gene expression in a way that may affect phenotype. For example, in Chapter 16, we discussed how an iron regulatory element (IRE) plays a role in regulating translation and RNA stability (refer back to Figure 16.14). Mutations in the untranslated regions of mRNA—the 5′-UTR and 3′-UTR—may alter the sequence of the IRE, thereby affecting the mRNA's ability to be translated or affecting its stability. In addition, mutations in eukaryotic genes can alter splice recognition sequences, thereby affecting the order or number of exons contained within mRNA.

TABLE 18.2

Possible Consequences of Gene Mutations Outside of the Coding Sequence

Sequence	Effect of Mutation
Promoter	May increase or decrease the rate of transcription
Regulatory element/operator site	May disrupt the ability of the gene to be properly regulated
5′-UTR/3′-UTR	May alter the ability of mRNA to be translated; may alter mRNA stability
Splice recognition sequence	May alter the ability of pre-mRNA to be properly spliced

Gene Mutations Are Also Given Names That Describe How They Affect the Wild-Type Genotype and Phenotype

Thus far, several genetic terms have been introduced that describe the molecular effects of mutations. Genetic terms are also used to describe the effects of mutations relative to a wild-type genotype or phenotype. In a natural population, a **wild type** is a relatively prevalent genotype. Many or most genes exist as multiple alleles so that a population may have two or more wild-type alleles. A mutation may change a wild-type genotype by altering the DNA sequence of a gene. When such a mutation is rare in a population, it is generally referred to as a **mutant allele.** A reverse mutation, more commonly called a **reversion,** changes a mutant allele back to a wild-type allele.

Another way to describe a mutation is based on its influence on the wild-type phenotype. Mutants are often characterized by their differential ability to survive. As mentioned, a neutral mutation does not alter protein function, so it does not affect survival or reproductive success. A **deleterious mutation,** however, decreases the chances of survival and reproduction. The extreme example of a deleterious mutation is a **lethal mutation,** which results in death to a cell or organism. On the other hand, a **beneficial mutation** enhances the survival or reproductive success of an organism. In some cases, an allele may be either deleterious or beneficial, depending on the genotype and/or the environmental conditions. An example is the sickle cell allele. In the homozygous state, the sickle cell allele lessens the chances of survival. However, an individual who is heterozygous for the sickle cell allele and wild-type allele has an increased chance of survival due to malarial resistance.

Finally, some mutations are called **conditional mutants** because they affect the phenotype only under a defined set of conditions. Geneticists often study conditional mutants in microorganisms; a common example is a temperature-sensitive (*ts*) mutant. A bacterium that has a *ts* mutation grows normally in one temperature range—the permissive temperature range—but exhibits defective growth at a different temperature range—the nonpermissive range. For example, an *E. coli* strain carrying a *ts* mutation may be able to grow from 33°–38°C but not between 40°–42°C, whereas a wild-type strain can grow at either temperature range.

Suppressor Mutations Reverse the Phenotypic Effects of Another Mutation

A second mutation sometimes affects the phenotypic expression of a first mutation. As an example, let's consider a mutation that causes an organism to grow very slowly. A second mutation at another site in the organism's DNA may restore the normal growth rate, converting the mutant back to the wild-type phenotype. Geneticists call these second-site mutations **suppressors,** or **suppressor mutations.** This name is meant to indicate that a suppressor mutation acts to suppress the phenotypic effects of another mutation. A suppressor mutation differs from a reversion, because it occurs at a different site in the DNA from the first mutation.

Suppressor mutations are classified according to their relative locations with regard to the mutation they suppress. **Table 18.3** provides a few examples. When the second mutant site is within the same gene as the first, the mutation is termed an **intragenic suppressor.** This type of suppressor often involves a change in protein structure that compensates for an abnormality in protein structure caused by the first mutation.

Researchers often isolate suppressor mutations by setting up an experiment in which cells or organisms carrying the suppressor mutation are easy to identify. For example, Robert Brooker and colleagues have isolated many intragenic suppressors in the *lacY* gene of *E. coli*, which encodes the lactose permease described in Chapter 14. They began with single mutations that altered amino acids and prevented lactose transport. When cells harboring these single mutations were placed on media in which lactose was the sole carbon source for growth, such mutants failed to grow. However, rare suppressor mutations were easy to identify because they allowed the cells to grow and form visible colonies. By isolating many such suppressors and analyzing their locations, the researchers determined that certain regions in the protein are critical for conformational changes required for lactose transport.

Alternatively, a suppressor mutation can be in a different gene from the first mutation—an **intergenic suppressor.** How do intergenic suppressors work? These suppressor mutations usually involve a change in the expression of one gene that compensates for a loss-of-function mutation affecting another gene (see Table 18.3). For example, a first mutation may cause one protein to be partially or completely defective. An intergenic suppressor mutation in a different protein-encoding gene might overcome this defect by altering the structure of a second protein so it could take over the functional role of the defective protein. Alternatively, intergenic suppressors may involve proteins that participate in a common cellular pathway. When a first mutation affects the activity of a protein, a suppressor mutation could alter the function of a second protein involved in this pathway, thereby overcoming the defect in the first.

In some cases, intergenic suppressors involve multimeric proteins, with each subunit encoded by a different gene. A mutation in one subunit that inhibits function may be compensated for by a mutation in another subunit. Another type of intergenic suppressor is one that involves mutations in genetic regulatory proteins such as transcription factors. When a first mutation causes a protein to be defective, a suppressor mutation may occur in a gene that encodes a transcription factor. The mutant transcription factor transcriptionally activates another gene that can compensate for the loss-of-function mutation in the first gene.

Less commonly, intergenic suppression involves mutations in nonprotein-encoding genes that alter the translation of particular codons. Examples are suppressor tRNA mutants, which have been identified in bacteria (see solved problem S1). Suppressor tRNA mutants have a change in the anticodon region that causes the tRNA to behave contrary to the genetic code. Nonsense suppressors are mutant tRNAs that recognize a stop codon and, instead of stopping translation, put an amino acid into the growing polypeptide chain. This type of mutant tRNA can suppress a nonsense mutation in the mRNA from a protein-encoding gene. However, such bacterial strains grow poorly because they may also suppress stop codons in the mRNAs from normal genes.

Changes in Chromosome Structure Can Affect the Expression of a Gene

Thus far, we have considered small changes in the DNA sequence of particular genes. A change in chromosome structure can also be associated with an alteration in the expression of single genes. Quite commonly, an inversion or translocation has no obvious phenotypic consequence. However, in 1925, Alfred Sturtevant was the first to recognize that chromosomal rearrangements in *Drosophila* can influence phenotypic expression (namely, eye morphology). In some cases, a chromosomal rearrangement may affect a gene because a chromosomal **breakpoint**—the region where two chromosome pieces break and rejoin with other chromosome pieces—occurs within a gene. A breakpoint in the middle of a gene is very likely to inhibit gene function because it separates the gene into two pieces.

In other cases, a gene may be left intact, but its expression may be altered when it is moved to a new location. When this occurs, the change in gene location is said to have a **position effect.** How do position effects alter gene expression? Researchers have discovered two common explanations. **Figure 18.2** depicts a schematic example in which a piece of one chromosome has been inverted or translocated to a different chromosome. One possibility is that a gene may be moved next to regulatory sequences for a different gene, such as silencers or enhancers, that influence the expression of the relocated gene (Figure 18.2a). Alternatively, a chromosomal rearrangement may reposition a gene from a less condensed, or euchromatic chromosome, where it is active, to a very highly condensed, or heterochromatic chromosome. (See Chapter 10 for a description of euchromatin and heterochromatin.) When the gene is moved to a heterochromatic region, its expression may be turned off (Figure 18.2b). This second type of position effect may produce a variegated phenotype in which the expression of the gene is variable. For genes that affect pigmentation, this produces a variegated appearance rather than an even color.

Figure 18.3 shows a position effect that alters eye color in *Drosophila*. Figure 18.3a depicts a normal red-eyed fruit fly, and Figure 18.3b shows a mutant fly that has inherited a chromosomal rearrangement in which a gene affecting eye color has been relocated to a heterochromatic region of a chromosome. The variegated appearance of the eye occurs because the degree of heterochromatin formation varies across different regions of the eye. In cells where heterochromatin formation has turned off the eye color gene, a white phenotype occurs, but in other cells, this region remains euchromatic and produces a red phenotype.

TABLE 18.3

Examples of Suppressor Mutations

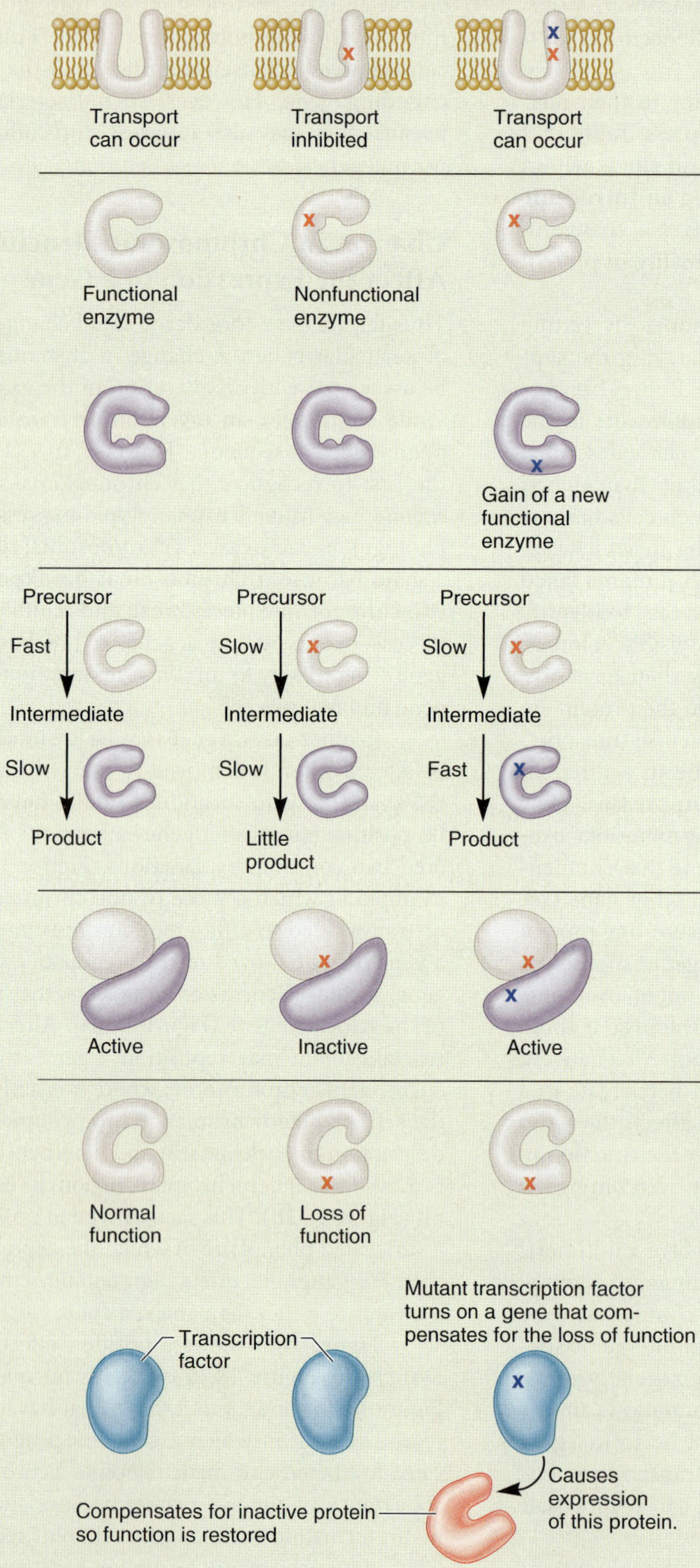

Type	No Mutation	First Mutation (x)	Second Mutation (x)	Description
Intragenic	Transport can occur	Transport inhibited	Transport can occur	A first mutation disrupts normal protein function, and a suppressor mutation affecting the same protein restores function. In this example, the first mutation inhibits transport function, and the second mutation restores it.
Intergenic Redundant function	Functional enzyme	Nonfunctional enzyme	Gain of a new functional enzyme	A first mutation inhibits the function of a protein, and a second mutation alters a different protein to carry out that function. In this example, the proteins function as enzymes.
Common pathway	Precursor → Fast → Intermediate → Slow → Product	Precursor → Slow → Intermediate → Slow → Little product	Precursor → Slow → Intermediate → Fast → Product	Two or more different proteins may function as enzymes in a common pathway. A mutation that causes a defect in one enzyme may be compensated for by a mutation that increases the function of a different enzyme in the same pathway.
Multimeric protein	Active	Inactive	Active	A mutation in a gene encoding one protein subunit that inhibits function may be suppressed by a mutation in a gene that encodes a different subunit. The double mutant has restored function.
Transcription factor	Normal function	Loss of function	Mutant transcription factor turns on a gene that compensates for the loss of function. Causes expression of this protein. Compensates for inactive protein so function is restored	A first mutation causes loss of function of a particular protein. A second mutation may alter a transcription factor and cause it to activate the expression of another gene. This other gene encodes a protein that can compensate for the loss of function caused by the first mutation.

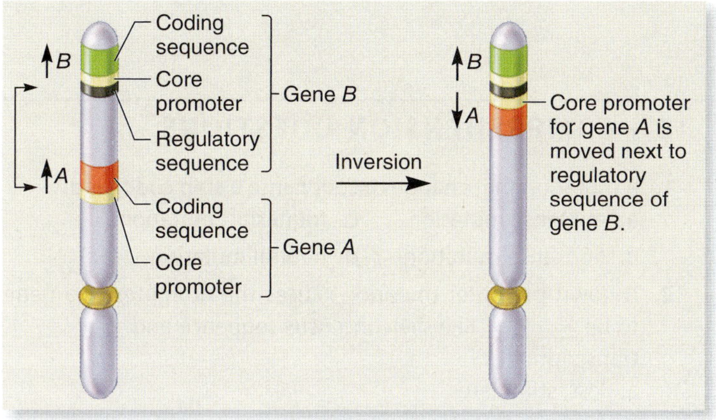

(a) Position effect due to regulatory sequences

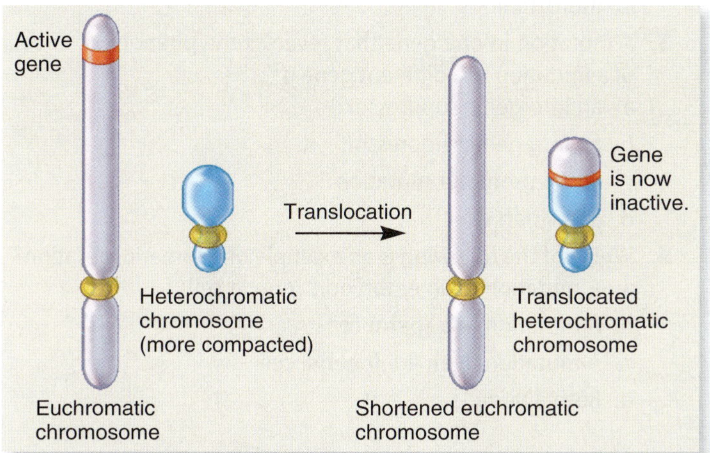

(b) Position effect due to translocation to a heterochromatic chromosome

FIGURE 18.2 Causes of position effects. **(a)** A chromosomal inversion has repositioned the core promoter of gene *A* next to the regulatory sequences for gene *B*. Because regulatory sequences are often bidirectional, the regulatory sequences for gene *B* may regulate the transcription of gene *A*. **(b)** A translocation has moved a gene from a euchromatic to a heterochromatic chromosome. This type of position effect inhibits the expression of the relocated gene.

CONCEPT CHECK: Explain what the term *position effect* means.

Mutations Can Occur in Germ-Line or Somatic Cells

In this section, we have considered many different ways that mutations affect gene expression. For multicellular organisms, the timing of mutations also plays an important role. A mutation can occur very early in life, such as in a gamete or a fertilized egg, or it may occur later in life, such as in the embryonic or adult stages. The exact time when mutations occur can be important with regard to the severity of the genetic effect and whether they are passed from parent to offspring.

Geneticists classify the cells of animals into two types: germ-line and somatic cells. The term **germ line** refers to cells that give rise to the gametes such as eggs and sperm. A **germ-line mutation** can occur directly in a sperm or egg cell, or it can occur in a precursor cell that produces the gametes. If a mutant gamete participates in fertilization, all cells of the resulting offspring will contain

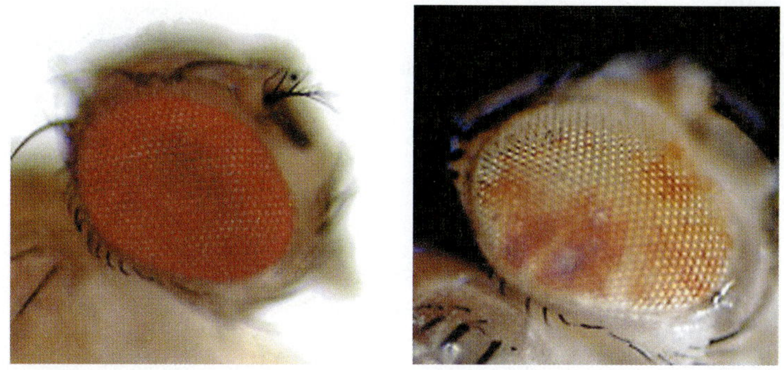

(a) Normal eye **(b) Variegated eye**

FIGURE 18.3 A position effect that alters eye color in *Drosophila*. **(a)** A normal red eye. **(b)** An eye in which an eye color gene has been relocated to a heterochromatic chromosome. This can inactivate the gene in some cells and produces a variegated phenotype.

Genes → Traits Variegated eye color occurs because the degree of heterochromatin formation varies throughout different regions of the eye. In some cells, heterochromatin formation occurs and turns off the eye color gene, thereby leading to the white phenotype. In other patches of cells, the region containing the eye color allele remains euchromatic, yielding a red phenotype.

CONCEPT CHECK: Has the DNA sequence of the eye color gene been changed in part (b) compared with part (a)? How do we explain the phenotypic difference?

the mutation (Figure 18.4a). Likewise, when an individual with a germ-line mutation produces gametes, the mutation may be passed along to future generations of offspring.

The **somatic cells** are all cells of the body excluding the germ-line cells. Examples include muscle cells, nerve cells, and skin cells. Mutations can also happen within somatic cells at early or late stages of development. Figure 18.4b illustrates the consequences of a mutation that took place during the embryonic stage. In this example, a **somatic mutation** has occurred within a single embryonic cell. As the embryo grows, this single cell is the precursor for many cells of the adult organism. Therefore, in the adult, a portion of the body contains the mutation. The size of the affected region depends on the timing of the mutation. In general, the earlier the mutation occurs during development, the larger the region. An individual that has somatic regions that are genotypically different from each other is called a **genetic mosaic.**

Figure 18.5 is a photo of an individual with a somatic mutation that occurred during an early stage of development. In this case, the child has a patch of white hair, but the rest of his hair is pigmented. Presumably, a single mutation initially occurred in this individual in an embryonic cell that ultimately gave rise to a patch of scalp that produced the white hair. Although a patch of white hair is not a harmful phenotypic effect, mutations during early stages of life can be quite detrimental, especially if they disrupt essential developmental processes. Therefore, even though it is smart to avoid environmental agents that cause mutations during all stages of life, the possibility of somatic mutations is a rather compelling reason to avoid them during the very early stages of life such as embryonic development, infancy, and early childhood. For example, the possibility of somatic mutations in an embryo is a reason why women are advised to avoid getting X-rays during pregnancy.

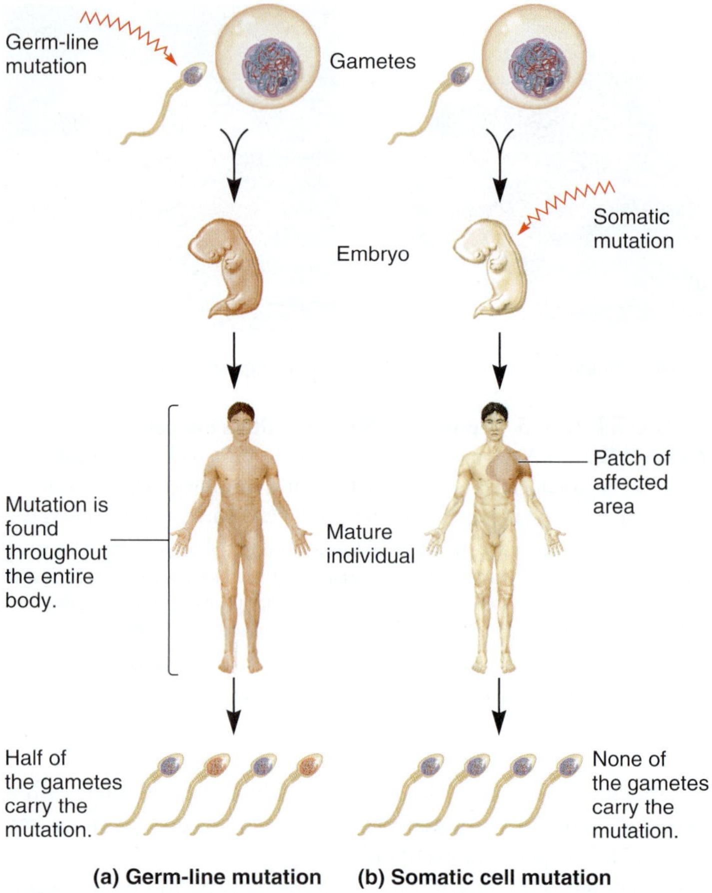

FIGURE 18.4 **The effects of germ-line versus somatic mutations.**

FIGURE 18.5 **Example of a somatic mutation.**

Genes → Traits This child has a patch of white hair because a somatic mutation occurred in a single cell during embryonic development that prevented pigmentation of the hair. This cell continued to divide to produce a patch of white hair.

CONCEPT CHECK: Can this trait be passed to offspring?

18.1 COMPREHENSION QUESTIONS

1. A mutation that changes a codon into a stop codon is a
   a. missense mutation.
   b. nonsense mutation.
   c. frameshift mutation.
   d. neutral mutation.
2. A down promoter mutation causes the promoter of a gene to be _______ like the consensus sequence and _________ transcription.
   a. less, stimulates
   b. more, stimulates
   c. less, inhibits
   d. more, inhibits
3. A mutation in one gene that reverses the phenotypic effects of a mutation in a different gene is
   a. an intergenic suppressor.
   b. an intragenic suppressor.
   c. an up promoter mutation.
   d. a position effect.
4. Which of the following is an example of a somatic mutation?
   a. A mutation in an embryonic muscle cell
   b. A mutation in a sperm cell
   c. A mutation in an adult nerve cell
   d. Both a and c

18.2 RANDOM NATURE OF MUTATIONS

Learning Outcome:

1. Analyze the results of the Lederbergs and explain how they are consistent with the random mutation theory.

For a couple of centuries, biologists had questioned whether heritable changes occur purposefully as a result of behavior or exposure to particular environmental conditions or whether they are events that happen randomly. In the nineteenth century, the naturalist Jean-Baptiste Lamarck proposed that physiological events—such as the use or disuse of muscles—determine whether traits are passed along to offspring. For example, his hypothesis suggested that an individual who practiced and became adept at a physical activity and developed muscular legs would pass that characteristic on to the next generation. The alternative point of view is that genetic variation exists in a population as a matter of random chance, and natural selection results in the differential reproductive success of organisms that are better adapted to their environments. Those individuals who, by chance, happen to have mutations that turn out to be beneficial will be more likely to survive and pass these genes to their offspring. These opposing ideas of the nineteenth century—one termed *physiological adaptation* and the other termed *random mutation*—were tested in bacterial studies in the 1940s and 1950s, one of which is described here.

EXPERIMENT 18A

Replica-Plating Experiments of the Lederbergs Confirmed the Random Mutation Theory

To distinguish between the physiological adaptation and random mutation hypotheses, Joshua and Esther Lederberg developed a technique known as **replica plating** in the 1950s. As shown in **Figure 18.6**, they plated a large number of bacteria onto a master plate that did not contain any selective agent (namely, the T1 phage). A sterile piece of velvet cloth was lightly touched to this plate in order to pick up a few bacterial cells from each colony. This replica was then transferred to two secondary plates that contained an agent that selected for the growth of bacterial cells with a particular mutation. In the example shown in Figure 18.6, the secondary plates contained T1 bacteriophages, which lyse the wild-type cells.

THE HYPOTHESIS

Mutations are random events.

TESTING THE HYPOTHESIS

FIGURE 18.6 Evidence that mutations are random events.

Starting material: They began with a wild-type strain of *E. coli*. They also had a preparation of T1 bacteriophage.

1. Place individual bacterial cells onto growth media.
2. Incubate overnight to allow the formation of bacterial colonies. This is called the master plate.
3. Press a velvet cloth (wrapped over a cylinder) onto the master plate, and then lift gently to obtain a replica of each bacterial colony. Press the replica onto 2 secondary plates that contain T1 phage. Incubate overnight to allow growth of mutant cells.

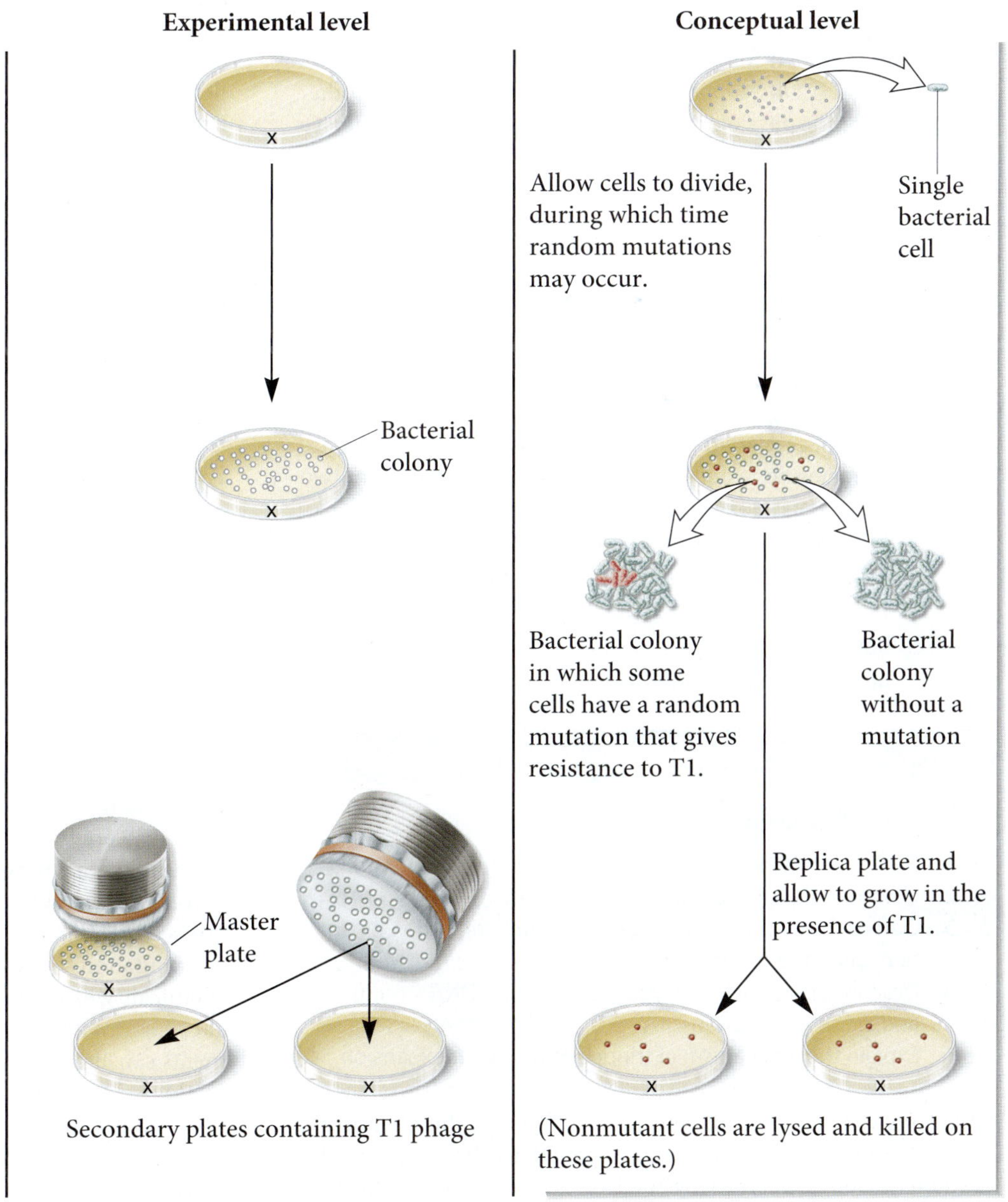

THE DATA

Note: The black × indicates the alignment of the velvet and the plates. Data adapted from J. Lederberg and E. M. Lederberg (1952) Replica plating and indirect selection of bacterial mutants. *J Bacteriol 63*, 399–406.

INTERPRETING THE DATA

On the secondary plates, only those mutant cells that are resistant to lysis by T1 phage (*ton*^{r}) could grow. A few colonies were observed. Strikingly, they occupied the same location on each plate. These results indicated that the mutations conferring *ton*^{r} occurred randomly while the cells were growing on the nonselective master plate. The presence of the T1 phage in the secondary plates simply selected for the growth of previously occurring *ton*^{r} mutants. These results supported the random mutation hypothesis. In contrast, the physiological adaptation hypothesis would have predicted that *ton*^{r} bacterial mutants would occur after exposure to the selective agent on the secondary plates. If that had been the case, the colonies would not be expected to arise in identical locations on different secondary plates but rather in different locations.

The results of the Lederbergs supported the random mutation hypothesis, now known as the **random mutation theory.** According to this theory, mutation is a random process—it can occur in any gene and does not require the exposure of an organism to an environmental condition that causes specific types of mutations to happen. For example, exposure to T1 phage does not cause *ton*^{r} mutations. In some cases, a random mutation may provide a mutant organism with an advantage, such as resistance to T1 phage. Although such mutations occur as a matter of random chance, growth conditions may select for organisms that happen to carry them.

As researchers have learned more about mutation at the molecular level, the view that mutations are a totally random process has required some modification. Within the same individual, some genes mutate at a much higher rate than other genes. Why does this happen? Some genes are larger than others, which provides a greater chance for mutation. Also, the relative locations of genes within a chromosome may cause some genes to be more susceptible to mutation than others. Even within a single gene, **hot spots** are usually found—certain regions of a gene that are more likely to mutate than other regions.

A self-help quiz involving this experiment can be found at www.mhhe.com/brookergenetics5e.

18.2 COMPREHENSION QUESTION

1. In the replica-plating experiments of the Lederbergs, bacterial colonies appeared at the same locations in each of two secondary plates because
   a. T1 phage caused the mutations to happen.
   b. the mutations occurred on the master plate prior to T1 exposure and prior to replica plating.
   c. both a and b are true.
   d. none of the above are true.

18.3 SPONTANEOUS MUTATIONS

Learning Outcomes:

1. Distinguish between spontaneous and induced mutations.
2. List examples of spontaneous mutations.
3. Outline how mutations arise by depurination, deamination, and tautomeric shifts.
4. Describe how reactive oxygen species alter DNA structure and cause mutation.
5. Explain the mechanism of trinucleotide repeat expansion.

Mutations can have a wide variety of effects on the phenotypic expression of genes. For this reason, geneticists have expended a great deal of effort identifying the causes of mutations. This task has been truly challenging, because a staggering number of agents can alter the structure of DNA, thereby causing mutation. Geneticists categorize the cause of mutation in one of two ways. **Spontaneous mutations** are changes in DNA structure that result from natural biological or chemical processes, whereas **induced mutations** are caused by environmental agents (**Table 18.4**). In this section, we will focus on spontaneous mutations and the following section will concern induced mutations.

Many causes of spontaneous mutations are examined in other chapters throughout this textbook. As discussed in Chapter 8, abnormalities in crossing over can produce mutations such as deletions, duplications, translocations, and inversions. In addition, abnormal chromosomal segregation during meiosis can cause changes in chromosome number. In Chapter 11, we learned that DNA polymerase can make a mistake during DNA replication by putting the wrong base in a newly synthesized daughter strand. Errors in DNA replication are usually infrequent except in certain viruses, such as HIV, that have relatively high rates of spontaneous mutations. Also, normal metabolic processes may produce chemicals within the cell that can react directly with the DNA and alter its structure. As discussed in Chapter 19, transposable elements can alter gene sequences by inserting themselves into genes. In this section, we will explore mechanisms, such as depurination, deamination, and tautomeric shifts, which explain how some spontaneous mutations can occur. We will also consider how oxidative stress and trinucleotide repeat expansions cause mutations.

TABLE 18.4
Causes of Mutations

Common Causes of Mutations	Description
Spontaneous	
Aberrant recombination	Abnormal crossing over may cause deletions, duplications, translocations, and inversions (see Chapter 8).
Aberrant segregation	Abnormal chromosomal segregation may cause aneuploidy or polyploidy (see Chapter 8).
Errors in DNA replication	A mistake by DNA polymerase may cause a base pair mismatch (see Chapter 11).
Toxic metabolic products	The products of normal metabolic processes, such as reactive oxygen species, may be chemically reactive agents that can alter the structure of DNA.
Transposable elements	Transposable elements can insert themselves into the sequence of a gene (see Chapter 19).
Depurination	On rare occasions, the linkage between purines (i.e., adenine and guanine) and deoxyribose can spontaneously break. If not repaired, it can lead to mutation.
Deamination	Cytosine and 5-methylcytosine can spontaneously deaminate to create uracil or thymine.
Tautomeric shifts	Spontaneous changes in base structure can cause mutations if they occur immediately prior to DNA replication.
Induced	
Chemical agents	Chemical substances may cause changes in the structure of DNA.
Physical agents	Physical phenomena such as UV light and X-rays can damage the DNA.

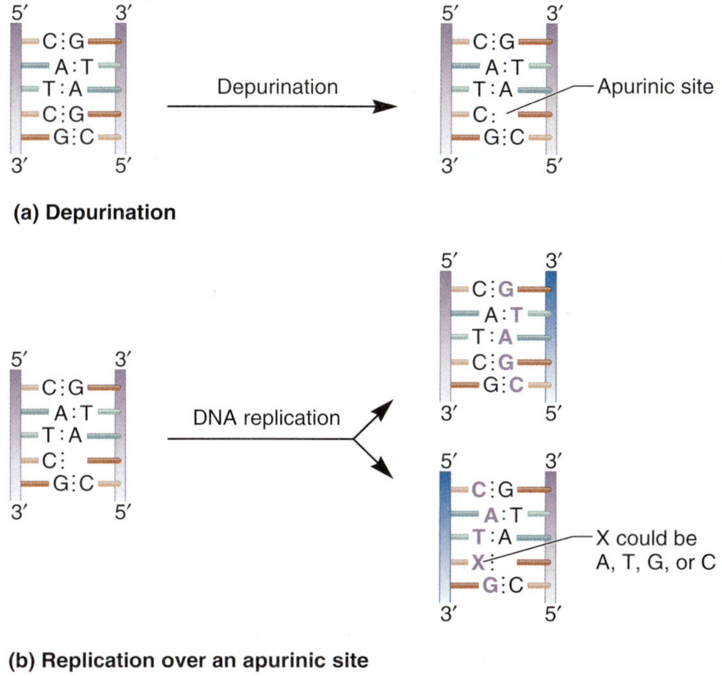

FIGURE 18.7 Spontaneous depurination. **(a)** The bond between guanine and deoxyribose is broken, thereby releasing the base. This leaves an apurinic site in the DNA. **(b)** If an apurinic site remains in the DNA as it is being replicated, any of the four nucleotides can be added to the newly made strand.

CONCEPT CHECK: When DNA replication occurs over an apurinic site, what is the probability that a mutation will occur?

Spontaneous Mutations Can Arise by Depurination, Deamination, and Tautomeric Shifts

How can molecular changes in DNA structure cause mutation? Our first examples concern changes that can occur spontaneously, albeit at a low rate. The most common type of chemical change that occurs naturally is **depurination,** which involves the removal of a purine (adenine or guanine) from the DNA. The covalent bond between deoxyribose and a purine base is somewhat unstable and occasionally undergoes a spontaneous reaction with water that releases the base from the sugar, thereby creating an **apurinic site** (**Figure 18.7a**). In a typical mammalian cell, approximately 10,000 purines are lost from the DNA in a 24-hour period at 37°C. The rate of loss is higher if the DNA is exposed to agents that cause certain types of base modifications such as the attachment of alkyl groups (methyl or ethyl groups). Fortunately, as discussed later in this chapter, apurinic sites are recognized by DNA base excision repair enzymes, which can repair the site. If the repair system fails, however, a mutation may result during subsequent rounds of DNA replication. What happens at an apurinic site during DNA replication? Because a complementary base is not present to specify the incoming base for the new strand, any of the four bases are added to the new strand in the region that is opposite the apurinic site (**Figure 18.7b**). This may lead to a new mutation.

A second spontaneous lesion that may occur in DNA is the **deamination** of cytosines. The other bases are not readily deaminated. As shown in **Figure 18.8a**, deamination involves the removal of an amino group from the cytosine base. This produces uracil. As discussed later, DNA repair enzymes can recognize uracil as an inappropriate base within DNA and subsequently remove it. However, if such repair does not take place, a mutation may result because uracil hydrogen bonds with adenine during DNA replication. Therefore, if a DNA template strand has uracil instead of cytosine, a newly made strand will incorporate adenine into the daughter strand instead of guanine.

Figure 18.8b shows the deamination of 5-methylcytosine. As discussed in Chapter 15, the methylation of cytosine occurs in many eukaryotic species, forming 5-methylcytosine. It also occurs in prokaryotes. If 5-methylcytosine is deaminated, the resulting base is thymine, which is a normal constituent of DNA. Therefore, this poses a problem for DNA repair because DNA repair proteins cannot distinguish which is the incorrect base—the thymine that was produced by deamination or the guanine in the opposite strand that originally base-paired with the methylated cytosine. For this reason, methylated cytosine bases tend to produce hotspots for

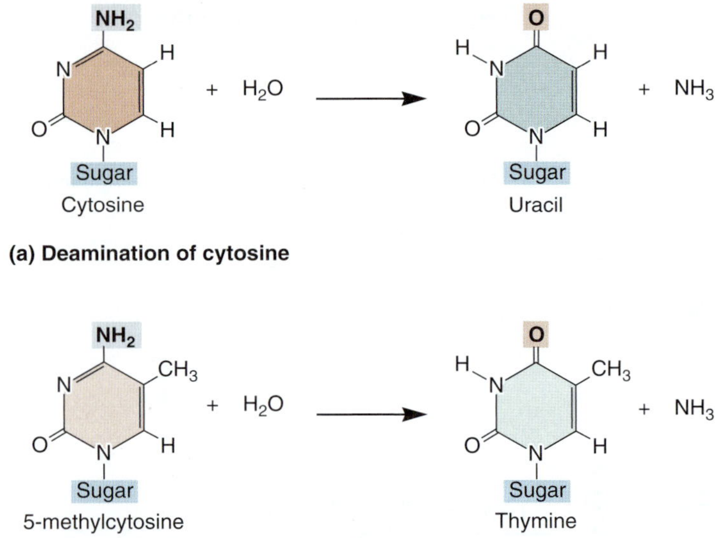

(a) Deamination of cytosine

(b) Deamination of 5-methylcytosine

FIGURE 18.8 Spontaneous deamination of cytosine and 5-methylcytosine. **(a)** The deamination of cytosine produces uracil. **(b)** The deamination of 5-methylcytosine produces thymine.

CONCEPT CHECK: Which of these two changes is more difficult for DNA repair enzymes to fix correctly? Explain why.

mutation. As an example, researchers analyzed 55 spontaneous mutations that occurred within the *lacI* gene of *E. coli* and determined that 44 of them involved changes at sites that were originally occupied by a methylated cytosine base.

A third way that mutations may arise spontaneously involves a temporary change in base structure called a **tautomeric shift.** In this case, the **tautomers** are bases, which exist in keto and enol or amino and imino forms. These forms can interconvert by a chemical reaction that involves the migration of a hydrogen atom and a switch of a single bond and an adjacent double bond. The common, stable form of guanine and thymine is the keto form; the common form of adenine and cytosine is the amino form (**Figure 18.9a**). At a low rate, G and T can interconvert to an enol form, and A and C can change to an imino form. Though the amounts of the enol and imino forms of these bases are relatively small, they can cause a mutation because these rare forms of the bases do not conform to the AT/GC rule of base pairing. Instead, if one of the bases is in the enol or imino form, hydrogen bonding will promote TG and CA base pairs, as shown in **Figure 18.9b**.

How does a tautomeric shift cause a mutation? The answer is that it must occur immediately prior to DNA replication. When DNA is double-stranded, the base pairing usually holds the bases in their more stable forms. After the strands unwind, however, a tautomeric shift may occur. In the example shown in **Figure 18.9c**, a thymine base in the template strand has undergone a tautomeric shift just prior to the replication of the complementary daughter strand. During replication, the daughter strand incorporates a guanine opposite this thymine, creating a base mismatch. This mismatch could be repaired via the proofreading function of DNA polymerase or via a mismatch repair system (discussed later in this chapter). However, if these repair mechanisms do not happen, the next round of DNA replication will produce a double helix with a CG base pair, whereas the correct base pair should be TA. As shown in the right side of Figure 18.9c, one of four daughter cells inherits this CG mutation.

Oxidative Stress May Also Lead to DNA Damage and Mutation

Aerobic organisms use oxygen as a terminal acceptor of their electron transport chains. **Reactive oxygen species (ROS),** such as hydrogen peroxide, superoxide, and hydroxyl radical, are products of oxygen metabolism in all aerobic organisms. In eukaryotes, ROS are naturally produced as unwanted by-products of energy production in mitochondria. They may also be produced during certain types of immune responses and from a variety of detoxification reactions in the cell. If ROS accumulate, they can damage cellular molecules, including DNA, proteins, and lipids. To prevent this from happening, cells use a variety of enzymes, such as superoxide dismutase and catalase, to prevent the buildup of ROS. In addition, small molecules, such as vitamin C, may act as antioxidants. Likewise, certain foods contain chemicals that act as antioxidants. Colorful fruits and vegetables, including grapes, blueberries, cranberries, citrus fruits, spinach, broccoli, beets, beans, red peppers, carrots, and strawberries, are usually high in antioxidants. In humans, the overaccumulation of ROS has been implicated in a wide variety of medical conditions, including cardiovascular disease, Alzheimer disease, chronic fatigue syndrome, and aging. However, the production of ROS is not always harmful. ROS are produced by the immune system as a means of killing pathogens. In addition, some ROS are used in cell signaling.

Oxidative stress refers to an imbalance between the production of ROS and an organism's ability to break them down. If ROS overaccumulate, one particularly harmful consequence is **oxidative DNA damage,** which refers to changes in DNA structure that are caused by ROS. DNA bases are very susceptible to oxidation. Guanine bases are particularly vulnerable to oxidation, which can lead to several different oxidized guanine products. The most thoroughly studied guanine oxidation product is 7,8-dihydro-8-oxoguanine, which is commonly known as 8-oxoguanine (8-oxoG) (**Figure 18.10**). Researchers often measure the amount of 8-oxoG in a sample of DNA to determine the extent of oxidative stress. Why are oxidized bases harmful? In the case of 8-oxoG, it base pairs with adenine during DNA replication, causing mutations in which a GC base pair becomes a TA base pair. This is a transversion mutation.

Although oxidative DNA damage can occur spontaneously, it also results from environmental agents, such as ultraviolet light, X-rays, and many chemicals, including those found in cigarette smoke. Later in this chapter, we will discuss such environment agents and their abilities to cause mutation.

DNA Sequences Known as Trinucleotide Repeats Are Hotspots for Mutation

Researchers have discovered that several human genetic diseases are caused by an unusual form of mutation known as **trinucleotide repeat expansion (TNRE).** The term refers to the phenomenon

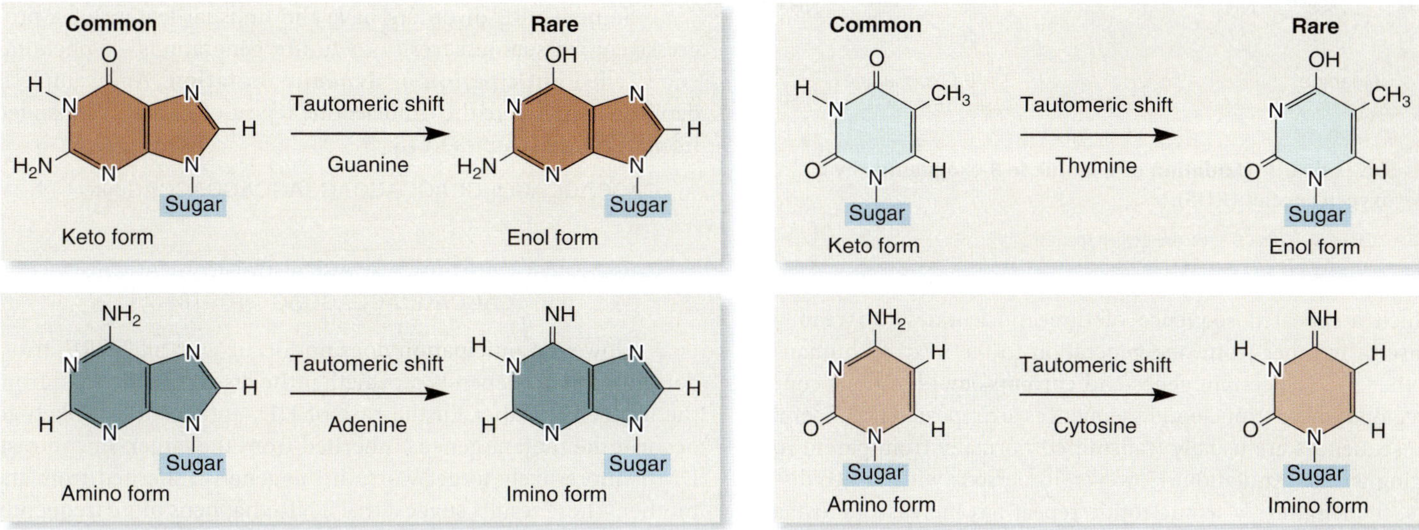

(a) Tautomeric shifts that occur in the 4 bases found in DNA

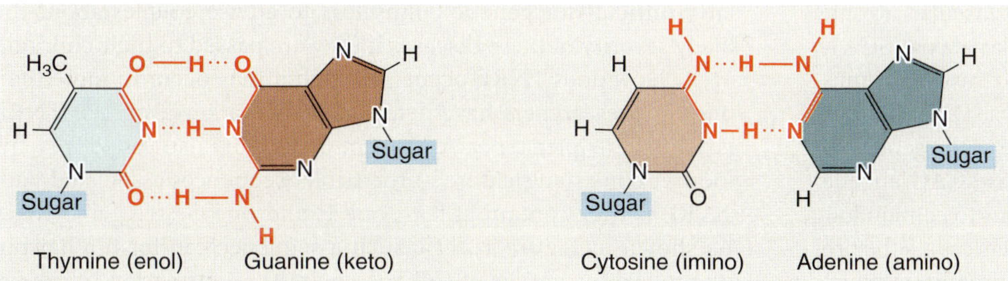

(b) Mis-base pairing due to tautomeric shifts

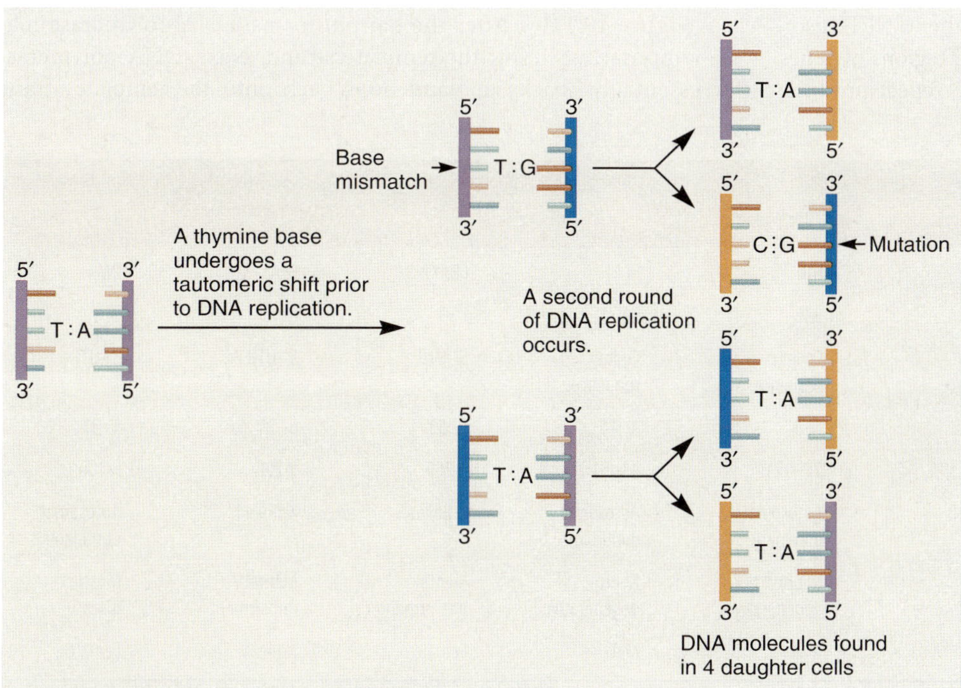

(c) Tautomeric shifts and DNA replication can cause mutation

FIGURE 18.9 Tautomeric shifts and their ability to cause mutation. (a) The common forms of the bases are shown on the left, and the rare forms produced by a tautomeric shift are shown on the right. **(b)** On the left, the rare enol form of thymine pairs with the common keto form of guanine (instead of adenine); on the right, the rare imino form of cytosine pairs with the common amino form of adenine (instead of guanine). **(c)** A tautomeric shift occurred in a thymine base just prior to replication, causing the formation of a TG base pair. If not repaired, a second round of replication will lead to the formation of a permanent CG mutation. Note: A tautomeric shift is a very temporary situation. During the second round of replication, the thymine base that shifted prior to the first round of DNA replication is likely to have shifted back to its normal form. Therefore, during the second round of replication, an adenine base is found opposite this thymine.

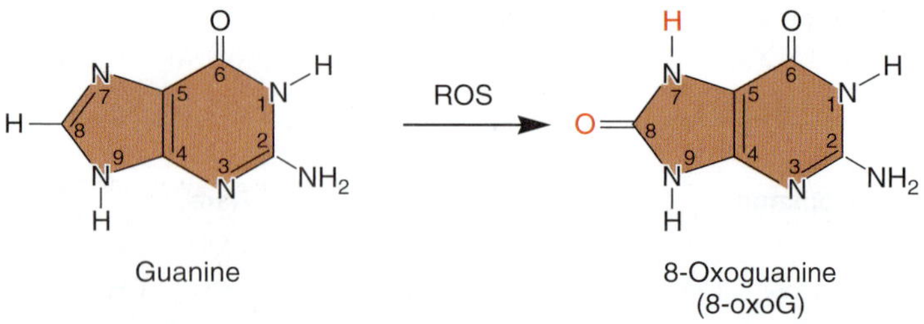

FIGURE 18.10 Oxidation of guanine to 8-oxoguanine by a reactive oxygen species (ROS).

CONCEPT CHECK: What is a reactive oxygen species?

in which a repeated sequence of three nucleotides can readily increase in number from one generation to the next. In humans and other species, certain genes and chromosomal locations contain regions where trinucleotide sequences are repeated in tandem. These sequences are usually transmitted normally from parent to offspring without mutation. However, in persons with TNRE disorders, the length of a trinucleotide repeat has increased above a certain critical size, thereby causing disease symptoms.

Table 18.5 describes several human diseases that involve these types of expansions: spinal and bulbar muscular atrophy (SBMA), Huntington disease (HD), spinocerebellar ataxia (SCA1), fragile X syndromes (FRAXA and FRAXE), and myotonic muscular dystrophy (also called dystrophia myotonica, DM). In some cases, the expansion is within the coding sequence of the gene. Typically, such an expansion is a CAG repeat. Because CAG encodes glutamine, these repeats cause the encoded proteins to contain long tracts of glutamine. The presence of glutamine tracts causes the proteins to aggregate. This aggregation of proteins or protein fragments carrying glutamine repeats is correlated with the progression of the disease. However, recent evidence suggests that the aggregated proteins may not cause the disease symptoms. In other TNRE disorders, the expansions are located in the noncoding regions of genes. In the case of the two fragile X syndromes, the repeat produces CpG islands that become methylated. As discussed in Chapter 15, DNA methylation can silence gene transcription. For DM, it has been hypothesized that these expansions cause abnormal changes in RNA structure, which produce disease symptoms.

Some TNRE disorders have the unusual feature of a progressively worsening severity in future generations—a phenomenon called **anticipation** or **dynamic mutation.** An example is depicted here, where the trinucleotide repeat of CAG has expanded from 11 tandem copies to 18.

CAGCAGCAGCAGCAGCAGCAGCAGCAGCAGCAG $n = 11$

to

CAGCAGCAGCAGCAGCAGCAGCAGCAGCAGCAGCAGCAG
CAGCAGCAGCAGCAG $n = 18$

However, anticipation does not occur with all TNRE disorders and usually depends on whether the disease is inherited from the mother or father. In the case of HD, anticipation is likely to occur if the mutant gene is inherited from the father. In contrast, DM is more likely to get worse if the gene is inherited from the mother. These results suggest that TNRE happens more frequently during oogenesis or spermatogenesis, depending on the particular gene involved. The phenomenon of anticipation makes it particularly difficult for genetic counselors to advise couples about the likely severity of these diseases if they are passed to their children.

How does TNRE occur? Though it may occur in more than one way, researchers have determined that a key aspect of TNRE is that the triplet repeat can form a hairpin, also called a stem-loop. A consistent feature of the triplet sequences associated with TNRE is they contain at least one C and one G (see Table 18.5). As shown in **Figure 18.11a**, such a sequence can form a hairpin due to the formation of CG base pairs. The formation of a hairpin during DNA replication can lead to an increase in the length of a DNA region if it occurs in the newly made daughter strand (**Figure 18.11b**). After the hairpin forms, DNA polymerase may temporarily slip off the template strand. Next, DNA polymerase essentially "backs up" and hops back onto the template strand

TABLE 18.5
TNRE Disorders

Disease	SBMA	HD	SCA1	FRAXA	FRAXE	DM
Repeat Sequence	CAG	CAG	CAG	CGG	GCC	CTG
Location of Repeat	Coding sequence	Coding sequence	Coding sequence	5′-UTR	5′-UTR	3′-UTR
Number of Repeats in Unaffected Individuals	11–33	6–37	6–44	6–53	6–35	5–37
Number of Repeats in Affected Individuals	36–62	27–121	43–81	>200	>200	>200
Pattern of Inheritance	X-linked	Autosomal dominant	Autosomal dominant	X-linked	X-linked	Autosomal dominant
Disease Symptoms	Neuro-degenerative	Neuro-degenerative	Neuro-degenerative	Mental impairment	Mental impairment	Muscle disease
Anticipation*	None	Male	Male	Female	None	Female

*Indicates the parent in which anticipation occurs most prevalently.
SBMA, spinal and bulbar muscular atrophy; HD, Huntington disease; SCA1, spinocerebellar ataxia; FRAXA and FRAXE, fragile X syndromes; DM, *dystrophia myotonica* (myotonic muscular dystrophy).

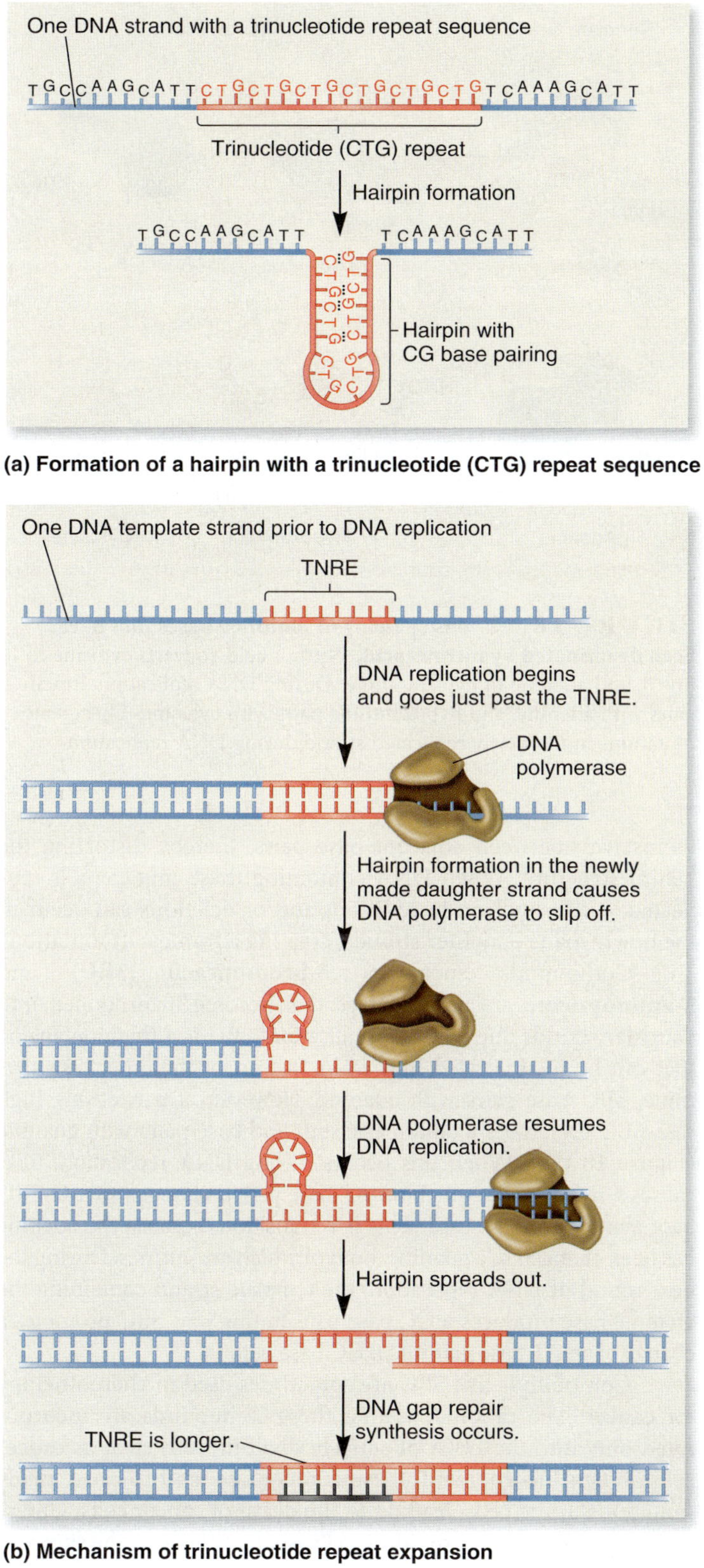

FIGURE 18.11 Proposed mechanism of trinucleotide repeat expansion (TNRE). **(a)** Trinucleotide repeats can form hairpin structures due to CG base pairing. **(b)** Formation of a trinucleotide repeat expansion.

and resumes DNA replication from the end of the hairpin. When this occurs, DNA polymerase is synthesizing most of the hairpin region twice.

Hairpins within DNA are short-lived. The hairpin can spread out, which leaves a gap in the opposite strand. This gap is later filled in by DNA polymerase and ligase as shown by the black DNA in Figure 18.11b. The end result is that the TNRE has become longer. When the trinucleotide repeat sequence is abnormally long, such expansions may frequently occur during gamete formation, and therefore future generations may have trinucleotide repeat sequences that are even longer than their parents.

18.3 COMPREHENSION QUESTIONS

1. Which of the following is *not* an example of a spontaneous mutation?
   a. A mutation caused by an error in DNA replication
   b. A mutation caused by a tautomeric shift
   c. A mutation caused by UV light
   d. None of the above
2. A point mutation could be caused by
   a. depurination.
   b. deamination.
   c. tautomeric shift.
   d. all of the above.
3. One way that TNRE may occur involves the formation of __________ that disrupts ______________.
   a. a double-strand break, chromosome segregation
   b. an apurinic site, DNA replication
   c. a hairpin, DNA replication
   d. a free radical, DNA structure

18.4 INDUCED MUTATIONS

Learning Outcomes:

1. Define mutagen.
2. Distinguish between chemical and physical mutagens and provide examples.
3. Define mutation rate.
4. Analyze the results of an Ames test.

As we have seen, spontaneous mutations occur in a wide variety of ways. They result from natural biological and chemical processes. In this section, we turn our attention to induced mutations that are caused by environmental agents. These agents can be either chemical or physical. They enter the cell and lead to changes in DNA structure. Agents known to alter the structure of DNA and thereby cause mutations are called **mutagens.** In this section, we will explore several mechanisms by which mutagens alter the structure of DNA. We will also consider laboratory tests that can identify potential mutagens.

TABLE 18.6
Examples of Mutagens

Mutagen	Effect(s) on DNA Structure
Chemical	
Nitrous acid	Deaminates bases
Nitrogen mustard	Alkylates bases
Ethyl methanesulfonate	Alkylates bases
Proflavin	Intercalates within DNA helix
5-Bromouracil	Functions as a base analog
2-Aminopurine	Functions as a base analog
Physical	
X-rays	Cause base deletions, single-stranded nicks in DNA, crosslinking, chromosomal breaks, and oxidized bases
UV light	Promotes pyrimidine dimer formation, such as thymine dimers

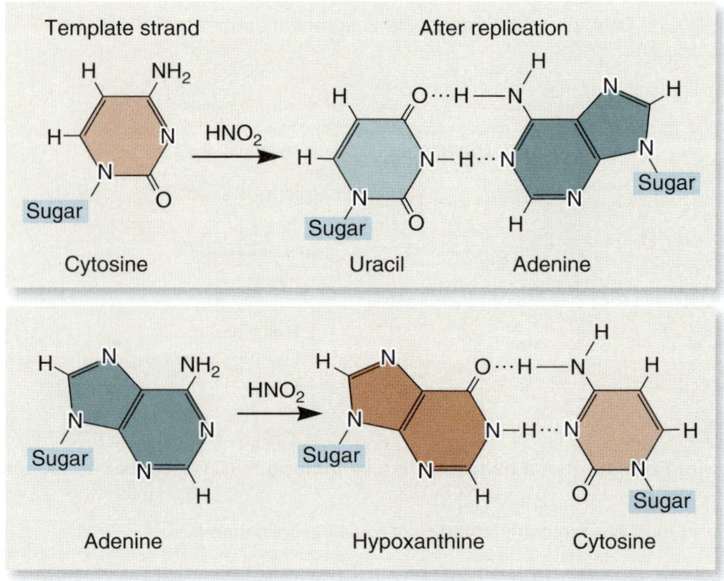

FIGURE 18.12 Base pairing of modified bases that have been deaminated by nitrous acid. Nitrous acid converts cytosine to uracil, and adenine to hypoxanthine. During DNA replication, uracil pairs with adenine, and hypoxanthine pairs with cytosine. This creates mutations in the newly replicated strand during DNA replication.

Mutagens Alter DNA Structure in Different Ways

Over the past few decades, researchers have found that an enormous array of agents act as mutagens that permanently alter the structure of DNA. We often hear in the news media that we should avoid these agents in our foods and living environment. For example, we use products such as sunscreens to help us avoid the mutagenic effects of ultraviolet (UV) rays. The public is concerned about mutagens for two important reasons. First, mutagenic agents are often involved in the development of human cancers. In addition, because new mutations may be deleterious, people want to avoid mutagens to prevent gene mutations that may have harmful effects on their future offspring.

Mutagenic agents are usually classified as chemical or physical mutagens. Examples of both types of agents are listed in Table 18.6. In some cases, chemicals that are not mutagenic can be altered to a mutagenically active form after being ingested. Cellular enzymes such as oxidases have been shown to activate some mutagens.

Mutagens alter the structure of DNA in various ways. Some mutagens act by covalently modifying the structure of bases. For example, **nitrous acid** (HNO_2) replaces amino groups with keto groups ($=NH_2$ to $=O$), a process called deamination. This changes cytosine to uracil and adenine to hypoxanthine. When this altered DNA replicates, mutations can occur because these modified bases pair differently than the original bases. For example, uracil pairs with adenine, and hypoxanthine pairs with cytosine (Figure 18.12).

Other chemical mutagens also disrupt the appropriate pairing between nucleotides by alkylating bases within the DNA. During alkylation, methyl or ethyl groups are covalently attached to the bases. Examples of alkylating agents include **nitrogen mustard** (a type of mustard gas) and **ethyl methanesulfonate (EMS).** Mustard gas was used as a chemical weapon during World War I. Such agents severely damage the skin, eyes, mucous membranes, lungs, and blood-forming organs.

Some mutagens exert their effects by directly interfering with the DNA replication process. For example, **acridine dyes** such as **proflavin** contain flat structures that intercalate, or insert, themselves, between adjacent base pairs, thereby distorting the helical structure. When DNA containing these mutagens is replicated, single-nucleotide additions and/or deletions can occur in the newly made daughter strands, creating a frameshift mutation.

Compounds such as **5-bromouracil (5BU)** and **2-aminopurine** are base analogs that become incorporated into daughter strands during DNA replication. 5BU is a thymine analog that can be incorporated into DNA instead of thymine. Like thymine, 5BU base-pairs with adenine. However, at a relatively high rate, 5BU undergoes a tautomeric shift and base-pairs with guanine (Figure 18.13a). When this occurs during DNA replication, 5BU causes a mutation in which an AT base pair is changed to a G-5BU base pair (Figure 18.13b). This is a transition, because the adenine has been changed to a guanine, both of which are purines. During the next round of DNA replication, the template strand containing the guanine base produces a GC base pair. In this way, 5BU promotes a change of an AT base pair into a GC base pair.

Compounds like 5BU are sometimes used in chemotherapy for cancer. The rationale is that these compounds are incorporated only into the DNA of actively dividing cells such as cancer cells. When incorporated, these compounds tend to cause many mutations in the cells, leading to the death of cancer cells. Unfortunately, other actively dividing cells, such as those in the skin and the lining of the digestive tract, also incorporate 5BU, which leads to unwanted side effects of chemotherapy, such as hair loss and a diminished appetite.

DNA molecules are also sensitive to physical agents such as radiation. In particular, radiation of short wavelength and high energy, known as ionizing radiation, can alter DNA structure. This type of radiation includes X-rays and gamma rays. Ionizing radiation

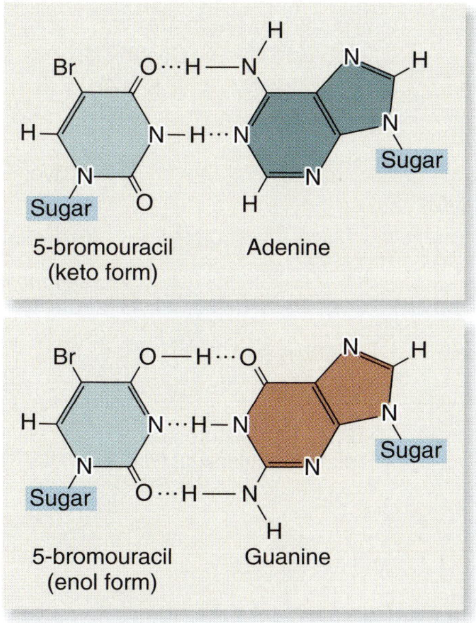

(a) Base pairing of 5BU (a thymine analog) with adenine or guanine

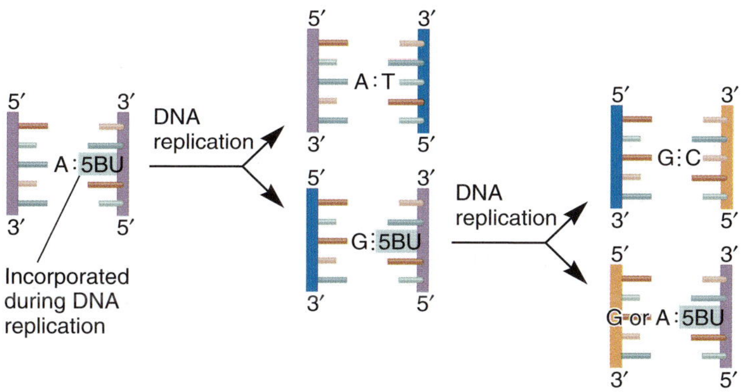

(b) How 5BU causes a mutation in a base pair during DNA replication

FIGURE 18.13 Base pairing of 5-bromouracil and its ability to cause mutation. **(a)** 5BU is a thymine analog. In its keto form, 5BU bonds with adenine; in its enol form, it bonds with guanine. **(b)** During DNA replication, guanine may be incorporated into a newly made strand by pairing with 5BU. After a second round of replication, the DNA contains a GC base pair instead of the original AT base pair.

CONCEPT CHECK: Does 5-bromouracil cause a transition or a transversion?

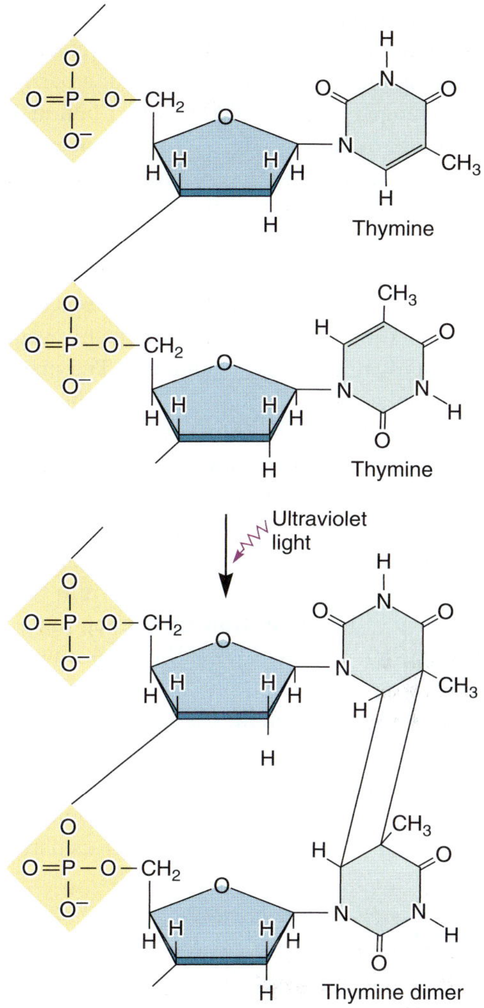

FIGURE 18.14 Formation and structure of a thymine dimer.

CONCEPT CHECK: In people, what is a common cause of thymine dimer formation and in what cell type(s) would it be most likely to occur?

can penetrate deeply into biological tissues, where it produces chemically reactive molecules known as free radicals. These molecules alter the structure of DNA in a variety of ways. Exposure to high doses of ionizing radiation results in base deletions, oxidized bases, single-stranded nicks in DNA, crosslinking, and even chromosomal breaks.

Nonionizing radiation, such as UV light, contains less energy, and so it penetrates only the surface of an organism, such as the skin. Nevertheless, UV light is known to cause DNA mutations. As shown in Figure 18.14, the energy in UV light causes the formation of crosslinked **thymine dimers.** Thymine dimers do not base pair properly during DNA replication. This can result in a mutation when that DNA strand is replicated. Plants, in particular, must have effective ways of preventing UV damage because they are exposed to sunlight throughout the day. Sun tanning greatly increases a person's exposure to UV light, raising the potential for thymine dimers and mutation. This explains the higher incidence of skin cancer among people who have been exposed to large amounts of sun during their lifetime. Because of the known link between skin cancer and sun exposure, people now apply sunscreen to their skin to prevent the harmful effects of UV light. Most sunscreens contain organic compounds, such as oxybenzone, which absorb UV light, and/or opaque ingredients, such as zinc oxide, that reflect UV light.

The Mutation Rate Is the Likelihood of a New Mutation

Because mutations occur spontaneously and may be induced by environmental agents, geneticists are greatly interested in learning how prevalent they are. The term **mutation rate** is the likelihood that a gene will be altered by a new mutation. This rate is commonly expressed as the number of new mutations in a given gene per cell generation. The spontaneous mutation rate for a particular gene is typically in the range of 1 in 100,000 to 1 in 1 billion, or 10^{-5}–10^{-9} per cell generation. In addition, mutations can occur at other sites in a genome, not only in genes. For example, people usually carry about 100–200 new mutations in their entire genome compared with their parents. Most of these are single-nucleotide changes that do not occur within the coding sequence of genes. With a human genome size of approximately 3,200,000,000 bp, these numbers tell us that a mutation is a relatively infrequent event. However, the mutation rate is not a constant number. The presence of certain environmental agents, such as X-rays, can increase the rate of induced mutations to a much higher value than the spontaneous mutation rate. In addition, mutation rates vary substantially from species to species and even within different strains of the same species. One explanation for this variation is that there are many different causes of mutations (refer back to Table 18.4).

The rate of new mutation is different from the concept of mutation frequency. The **mutation frequency** for a gene is the number of mutant genes divided by the total number of copies of that gene within a population. If 1 million bacteria were plated and 10 were found to be mutant, the mutation frequency would be 1 in 100,000, or 10^{-5}. The mutation frequency is an important genetic concept, particularly in the field of population genetics. As we will see in Chapter 26, mutation frequencies may rise above the mutation rate due to evolutionary factors such as natural selection and genetic drift.

Testing Methods Can Determine If an Agent Is a Mutagen

To determine if an agent is mutagenic, researchers use testing methods that monitor whether or not an agent increases the mutation rate. Many different kinds of tests have been used to evaluate mutagenicity. One common method is the **Ames test,** which was developed by Bruce Ames. This test uses strains of a bacterium, *Salmonella typhimurium*, that cannot synthesize the amino acid histidine. These strains contain a point mutation within a gene that encodes an enzyme required for histidine biosynthesis. The mutation renders the enzyme inactive. Therefore, the bacteria cannot grow on petri plates unless histidine has been added to the growth medium. However, a second mutation—a reversion—may occur that restores the ability to synthesize histidine. In other words, a second mutation can cause a reversion back to the wild-type condition. The Ames test monitors the rate at which this second mutation occurs, thereby indicating whether an agent increases the mutation rate above the spontaneous rate.

Figure 18.15 outlines the steps in the Ames test. The suspected mutagen is mixed with a rat liver extract and a strain of *Salmonella* that cannot synthesize histidine. A mutagen may require activation by cellular enzymes, which are provided by the rat liver extract. This step improves the ability of the test to identify agents that may cause mutation in mammals. After the incubation period, a large number of bacteria are then plated on a growth medium that does not contain histidine. The *Salmonella* strain is not expected to grow on these plates. However, if a mutation has occurred that allows the strain to

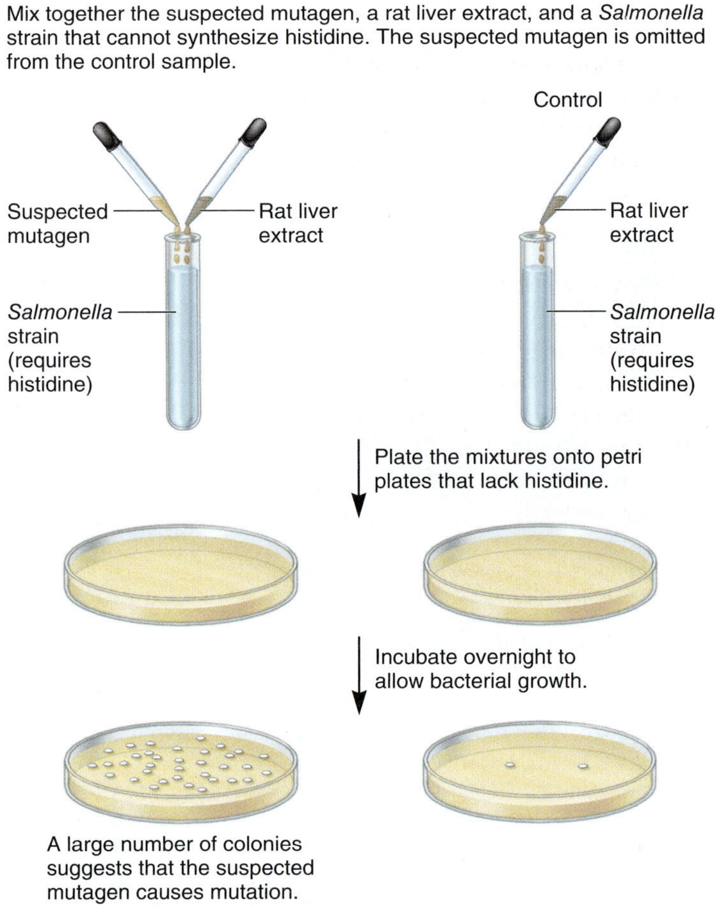

FIGURE 18.15 **The Ames test for mutagenicity.**

CONCEPT CHECK: What is the purpose of the rat liver extract in this procedure?

synthesize histidine, it can grow on these plates to form a visible bacterial colony. To estimate the mutation rate, the colonies that grow on the media are counted and compared with the total number of bacterial cells that were originally streaked on the plate. For example, if 10,000,000 bacteria were plated and 10 growing colonies were observed, the rate of mutation would be 10 out of 10,000,000; this equals 1 in 10^6, or simply 10^{-6}. As a control, bacteria that have not been exposed to the mutagen are also tested, because a low level of spontaneous mutations is expected to occur.

How do we judge if an agent is a mutagen? Researchers compare the mutation rate in the presence and absence of the suspected mutagen. The experimental approach shown in Figure 18.15 is conducted several times. If statistics reveal that the mutation rate in the experimental and control samples are significantly different, researchers may tentatively conclude that the agent may be a mutagen. Many studies have been conducted in which researchers used the Ames test to compare the urine from cigarette smokers to that from nonsmokers. This research has shown that the urine from smokers contains much higher levels of mutagens.

As discussed in solved problem S3 at the end of this chapter, several commercially available *his*$^-$ strains of *Salmonella* carry different kinds of mutations within the coding sequence. This makes it possible to determine if a mutagen causes transitions, transversions, or frameshift mutations.

18.4 COMPREHENSION QUESTIONS

1. Nitrous acid replaces amino groups with keto groups, a process called
   a. alkylation.
   b. deamination.
   c. depurination.
   d. crosslinking.
2. A mutagen that is a base analog is
   a. ethyl methanesulfonate (EMS).
   b. 5-bromouracil.
   c. UV light.
   d. proflavin.
3. In the Ames test, a ____________ number of colonies is observed if a substance ________ a mutagen, compared with the control that is not exposed to the suspected mutagen.
   a. significantly higher, is
   b. significantly higher, is not
   c. significantly lower, is
   d. significantly lower, is not

18.5 DNA REPAIR

Learning Outcomes:

1. Compare and contrast the different types of DNA repair mechanisms.
2. Describe how specialized DNA polymerases are able to synthesize DNA over a damaged region.

Because most mutations are deleterious, DNA repair systems are vital to the survival of all organisms. If DNA repair systems did not exist, spontaneous and environmentally induced mutations would be so prevalent that few species, if any, would survive. The necessity of DNA repair systems becomes evident when they are missing. Bacteria contain several different DNA repair systems. Yet, when even a single system is absent, the bacteria have a much higher rate of mutation. In fact, the rate of mutation is so high that these bacterial strains are sometimes called mutator strains. Likewise, in humans, an individual who is defective in only a single DNA repair system may manifest various disease symptoms, including a higher risk of skin cancer. This increased risk is due to the inability to repair UV-induced mutations.

Living cells produce several DNA repair systems that can fix different types of DNA alterations (**Table 18.7**). Each repair system is composed of one or more proteins that play specific roles in the repair mechanism. In most cases, DNA repair is a multistep process. First, one or more proteins in the DNA repair system detect an irregularity in DNA structure. Next, the abnormality is removed by the action of DNA repair enzymes. Finally, normal DNA is synthesized via DNA replication enzymes. In this section, we will examine several different repair systems that have been characterized in bacteria, yeast, mammals, and plants. Their diverse ways of repairing DNA underscore the extreme necessity for the structure of DNA to be maintained properly.

TABLE 18.7
Common Types of DNA Repair Systems

System	Description
Direct repair	An enzyme recognizes an incorrect alteration in DNA structure and directly converts it back to a correct structure.
Base excision repair and nucleotide excision repair	An abnormal base or nucleotide is first recognized and removed from the DNA, and a segment of DNA in this region is excised. Then the complementary DNA strand is used as a template to synthesize a normal DNA strand.
Mismatch repair	Similar to excision repair except that the DNA defect is a base pair mismatch in the DNA, not an abnormal nucleotide. The mismatch is recognized, and a segment of DNA in this region is removed. The parental strand is used to synthesize a normal daughter strand of DNA.
Homologous recombination	Occurs at double-strand breaks or when DNA repair damage causes a gap in synthesis during DNA replication. The strands of a normal sister chromatid are used to repair a damaged sister chromatid.
Nonhomologous end joining	Occurs at double-strand breaks. The broken ends are recognized by proteins that keep the ends together; the broken ends are eventually rejoined.

Damaged Bases Can Be Directly Repaired

In a few cases, the covalent modification of nucleotides by mutagens can be reversed by specific DNA repair enzymes. As discussed earlier in this chapter, UV light causes the formation of thymine dimers. Bacteria, fungi, most plants, and some animals produce an enzyme called **photolyase** that recognizes thymine dimers and splits them, which returns the DNA to its original condition (**Figure 18.16a**). Photolyase contains two light-sensitive cofactors. The repair mechanism itself requires light and is known as **photoreactivation.** This process directly restores the structure of DNA. Because plants are exposed to sunlight throughout the day, photolyase is a critical DNA repair enzyme for many plant species.

A protein known as **alkyltransferase** can remove methyl or ethyl groups from guanine bases that have been mutagenized by alkylating agents such as nitrogen mustard and EMS. This protein is called alkyltransferase because it transfers the methyl or ethyl group from the base to a cysteine side chain within the alkyltransferase protein (**Figure 18.16b**). Surprisingly, this permanently inactivates alkyltransferase, which means it can be used only once!

Base Excision Repair Removes a Damaged Base

A second type of repair system, called **base excision repair (BER),** is primarily responsible for removing non-helix-distorting changes that affect the structure of individual bases. BER involves the function of a category of enzymes known as **DNA *N*-glycosylases.** This type of enzyme can precisely recognize a site in the DNA where an abnormal base is located. Living organisms produce

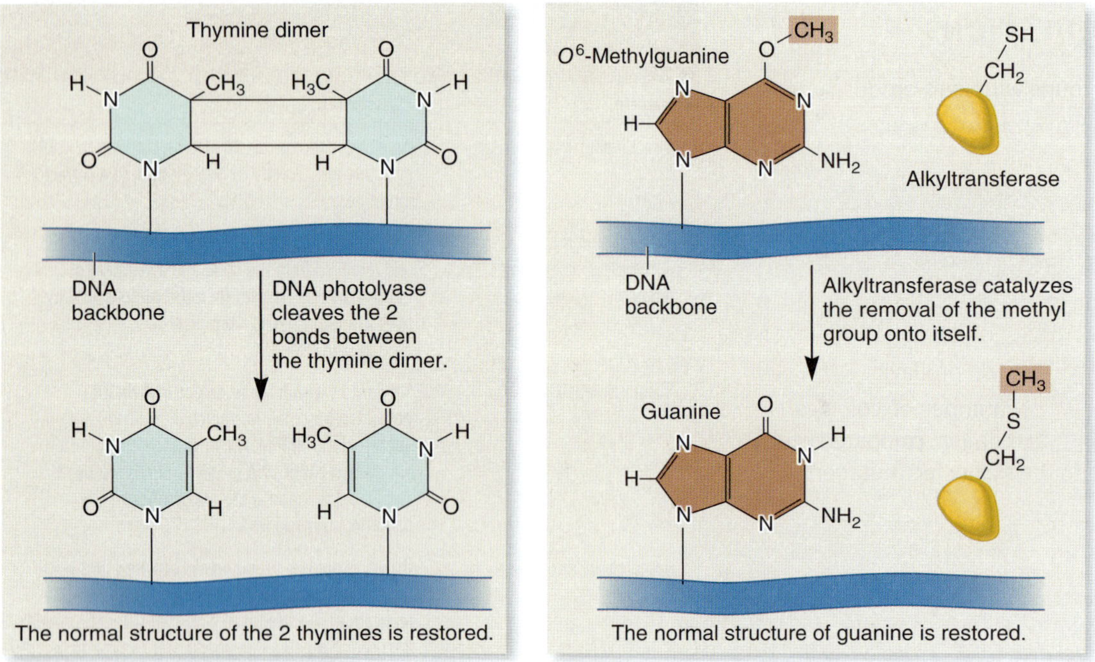

(a) Direct repair of a thymine dimer

(b) Direct repair of a methylated base

ONLINE ANIMATION

FIGURE 18.16 **Direct repair of damaged bases in DNA.** **(a)** The repair of thymine dimers by photolyase. **(b)** The repair of methylguanine by the transfer of the methyl group to alkyltransferase.

CONCEPT CHECK: Which of these repair systems is particularly valuable to plants?

multiple types of DNA *N*-glycosylases, each recognizing particular types of abnormal base structures. Depending on the DNA *N*-glycosylase, this repair system can eliminate abnormal bases such as uracil, 3-methyladenine, 7-methylguanine, and thymine dimers. BER is particularly important for the repair of oxidative DNA damage.

Figure 18.17 illustrates the general steps involved in DNA repair via *N*-glycosylase. In this example, the DNA contains a uracil in its sequence. This could have happened spontaneously or by the action of a chemical mutagen. After *N*-glycosylase recognizes a uracil within the DNA, the glycosylase flips the altered base out of the double helix and then cleaves the bond between the base and the sugar in the DNA backbone. This cleavage releases the uracil base and leaves behind an apyrimidinic site. This abnormality is recognized by a second enzyme, **AP endonuclease,** which makes a cut (a nick) on the 5′ side.

Following this cut by AP endonuclease, one of three things can happen. (1) In some species such as *E. coli*, DNA polymerase I, which has a 5′ to 3′ exonuclease activity, removes a DNA segment containing the abnormal region and, at the same time, replaces it with normal nucleotides. This process is called **nick translation** (although DNA replication, not mRNA translation, actually occurs). Alternatively, in eukaryotic species such as humans, the DNA is repaired in two possible ways. (2) DNA polymerase β has the enzymatic ability to remove a site that is missing a base, and then insert a nucleotide with the correct base in its place. (3) Another possibility is that DNA polymerase δ or ε can synthesize a short segment of DNA, which displaces a short segment of the original DNA, thereby generating a flap. The flap is then removed by flap endonuclease, which is described in Chapter 11. In all three cases, the final step is carried out by DNA ligase, which makes a covalent connection in the DNA backbone.

Nucleotide Excision Repair Systems Remove Segments of Damaged DNA

Another important mechanism for DNA repair is the **nucleotide excision repair (NER)** system, which is often used to repair bulky, helix-distorting lesions. This type of system can repair many different types of DNA damage, including thymine dimers, chemically modified bases, missing bases, and certain types of crosslinks. In NER, several nucleotides in the damaged strand are removed from the DNA, and the intact strand is used as a template for resynthesis of a normal complementary strand. NER is found in all eukaryotes and prokaryotes, although its molecular mechanism is better understood in prokaryotic species.

In *E. coli*, the NER system requires four key proteins, designated UvrA, UvrB, UvrC, and UvrD, plus the help of DNA polymerase and DNA ligase. UvrA, B, C, and D recognize and remove a short segment of a damaged DNA strand. They are named Uvr because they are involved in Ultraviolet light repair of thymine dimers, although the UvrA–D proteins are also important in repairing chemically damaged DNA.

Figure 18.18 outlines the steps involved in the *E. coli* NER system. A protein complex consisting of two UvrA molecules and one UvrB molecule tracks along the DNA in search of damaged DNA. Such DNA has a distorted double helix, which is sensed by the UvrA/UvrB complex. When a damaged segment is identified, the two UvrA proteins are released, and UvrC binds to the site. The UvrC protein makes cuts in the damaged strand on both sides of the damaged site. Typically, the damaged strand is cut 8 nucleotides from the 5′ end of the damage and 4–5 nucleotides away from the 3′ end. After this process, UvrD, which is a helicase, recognizes the region and separates the two strands of DNA. This releases a short DNA segment that contains the damaged region, and UvrB and UvrC are also released. Following the excision of the damaged DNA, DNA polymerase fills in the gap, using the undamaged strand as a template. Finally, DNA ligase makes the covalent connection between the newly made DNA and the original DNA strand.

In eukaryotes, NER systems are thought to operate similarly to bacterial NER systems, though more proteins are involved. Several human diseases are due to inherited defects in genes involved in NER. These include xeroderma pigmentosum (XP), Cockayne syndrome (CS), and PIBIDS. (PIBIDS is an acronym for a syndrome with symptoms that include photosensitivity; ichthyosis, a skin abnormality; brittle hair; impaired intelligence; decreased fertility; and short stature.) A common characteristic in all three syndromes is an increased sensitivity to sunlight because of an inability to repair UV-induced lesions. **Figure 18.19** is a photograph of a child with XP. Such individuals have pigmentation abnormalities, many premalignant lesions, and a high predisposition to skin cancer. They may also develop early degeneration of the nervous system.

Genetic analyses of patients with XP, CS, and PIBIDS have revealed that these syndromes result from defects in a variety of different genes that encode NER proteins. For example, XP can be caused by defects in any of seven different NER genes. In all cases, individuals have a defective NER pathway. In recent years, several human NER genes have been successfully cloned and sequenced.

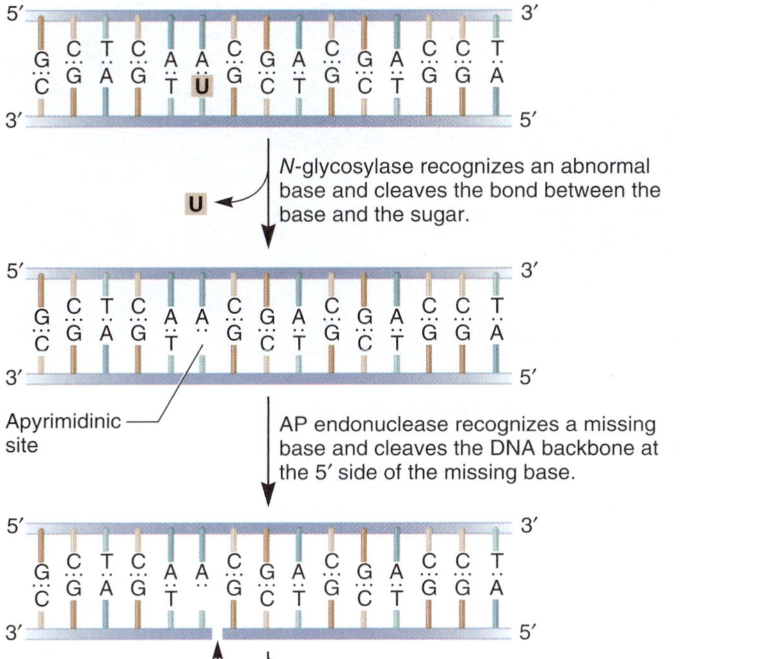

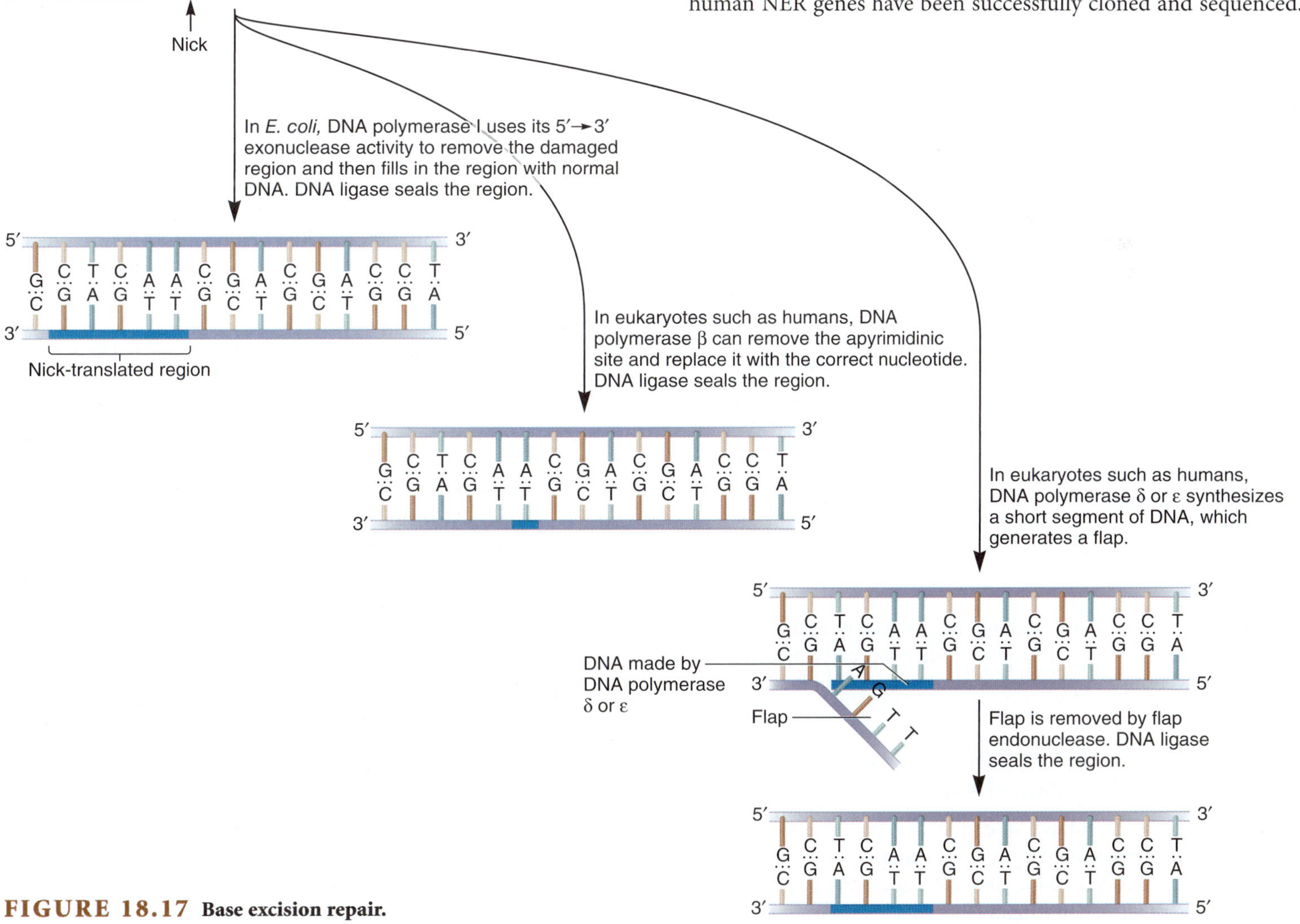

FIGURE 18.17 **Base excision repair.**

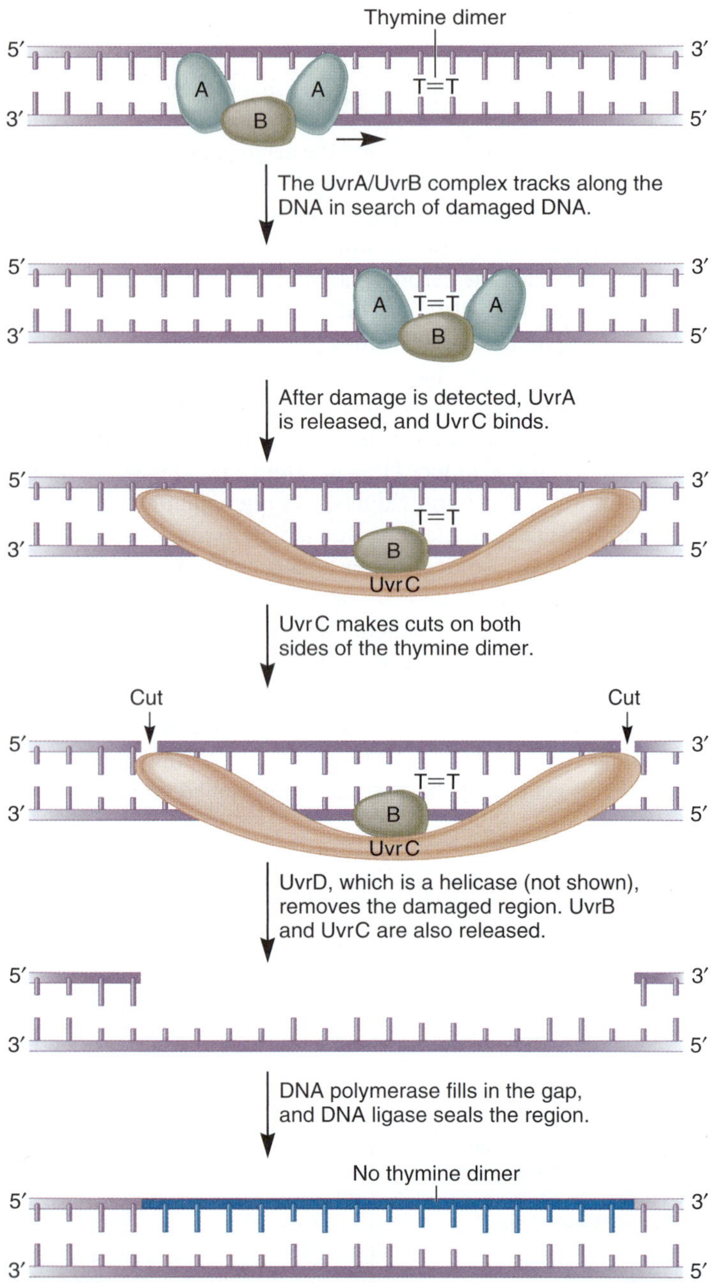

ONLINE ANIMATION

FIGURE 18.18 **Nucleotide excision repair in *E. coli*.**

CONCEPT CHECK: Explain why cuts are made on both sides of the DNA damage.

Although more research is needed to completely understand the mechanism of DNA repair, the identification of NER genes has helped unravel the complexities of NER pathways in human cells.

Mismatch Repair Systems Recognize and Correct a Base Pair Mismatch

Thus far, we have considered several DNA repair systems that recognize abnormal nucleotide structures within DNA, including thymine dimers, alkylated bases, and the presence of uracil in the DNA. Another type of abnormality that should not occur in DNA is

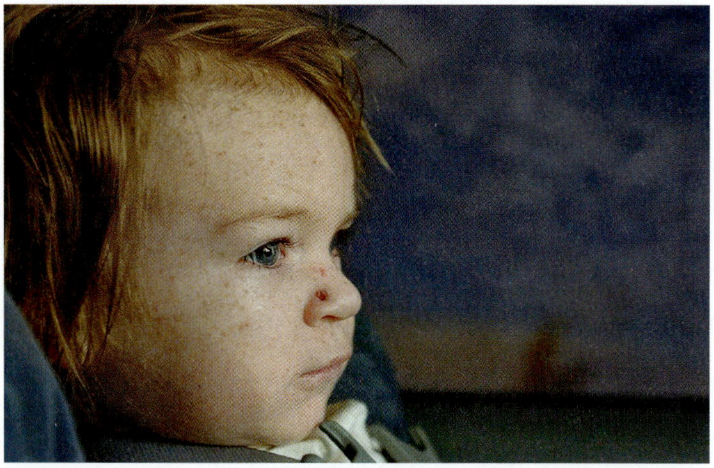

FIGURE 18.19 **An individual affected with xeroderma pigmentosum.**

Genes → Traits This disease involves a defect in genes that are involved with nucleotide excision repair (NER). Xeroderma pigmentosum can be caused by defects in seven different NER genes. Affected individuals have an increased sensitivity to sunlight because of an inability to repair UV-induced DNA lesions. In addition, they may also have pigmentation abnormalities, many premalignant lesions, and a high predisposition to skin cancer.

a **base pair mismatch.** The structure of the DNA double helix obeys the AT/GC rule of base pairing. During the normal course of DNA replication, however, an incorrect nucleotide may be added to the growing strand by mistake. This produces a mismatch between a nucleotide in the parental and the newly made strand. Various DNA repair mechanisms can recognize and remove this mismatch. For example, as described in Chapter 11, DNA polymerase has a 3′ to 5′ proofreading ability that detects mismatches and removes them. However, if this proofreading ability fails, cells have additional DNA repair systems that detect base mismatches and fix them. An interesting DNA repair system that exists in all species is the **mismatch repair system.**

In the case of a base mismatch, how does a DNA repair system determine which base to remove? If the mismatch is due to an error in DNA replication, the newly made daughter strand contains the incorrect base, whereas the parental strand is normal. Therefore, an important aspect of mismatch repair is that it specifically repairs the newly made strand rather than the parental template strand. Prior to DNA replication, the parental DNA has already been methylated. Immediately after DNA replication, some time must pass before a newly made strand is methylated. Therefore, newly replicated DNA is hemimethylated—only the parental DNA strand is methylated. Hemimethylation provides a way for a DNA repair system to distinguish between the parental DNA strand and the daughter strand.

The molecular mechanism of mismatch repair has been studied extensively in *E. coli*. As shown in **Figure 18.20**, three proteins, designated MutS, MutL, and MutH, detect the mismatch and direct the removal of the mismatched base from the newly made strand. These proteins are named Mut because their absence leads to a much higher mutation rate than occurs in normal strains of *E. coli*. The role of MutS is to locate mismatches. Once a mismatch is detected, MutS forms a complex with MutL. MutL

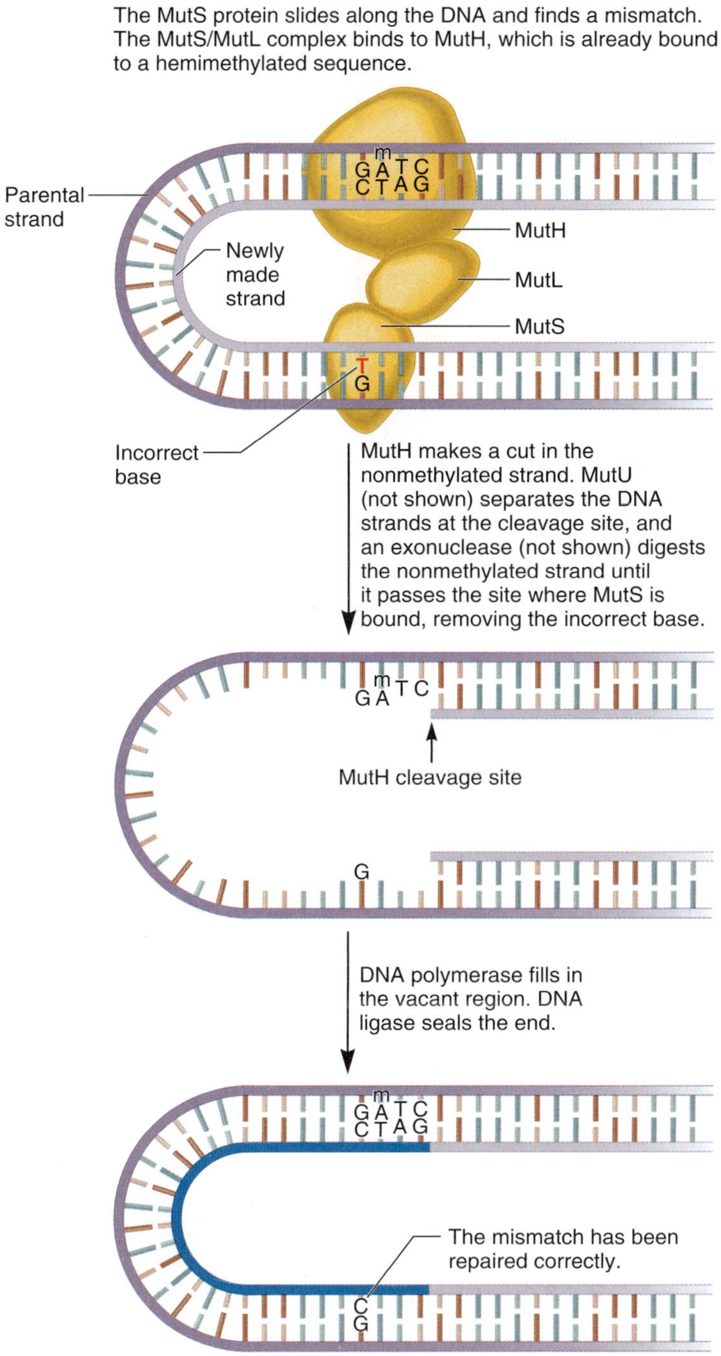

FIGURE 18.20 **Mismatch repair in *E. coli*.** In this example, the methylated adenine is designated with an m.

ONLINE ANIMATION

CONCEPT CHECK: Which of these three proteins is responsible for ensuring that the mismatched base in the newly made daughter strand is the one that is removed?

acts as a linker that binds to MutH, forming a loop in the DNA. The role of MutH is to identify the methylated strand of DNA, which is the nonmutated parental strand. The formation of the MutS/MutL/MutH complex stimulates MutH, which is already bound to a hemimethylated site, to make a cut in the newly made, nonmethylated DNA strand. After the strand is cut, MutU, which functions as a helicase, separates the strands, and an exonuclease then digests the nonmethylated DNA strand in the direction of the mismatch and proceeds beyond the site where MutS is located. This leaves a gap in the daughter strand that is repaired by DNA polymerase and DNA ligase. The net result is that the mismatch has been corrected by removing the incorrect region in the daughter strand and then resynthesizing the correct sequence using the parental DNA as a template.

Eukaryotic species also have homologs to MutS and MutL, along with many other proteins that are needed for mismatch repair. However, a MutH homolog has not been identified in eukaryotes, and the mechanism by which the eukaryotic mismatch repair system distinguishes between the parental and daughter strands is not well understood. As with defects in nucleotide excision repair systems, mutations in the human mismatch repair system are associated with particular types of cancer. For example, mutations in two human mismatch repair genes, *hMSH2* and *hMLH1*, play a role in the development of a type of colon cancer known as hereditary nonpolyposis colorectal cancer.

Double-Strand Breaks Can Be Repaired by Homologous Recombination Repair and by Nonhomologous End Joining

Of the many types of DNA damage that can occur within living cells, the breakage of chromosomes—called a DNA double-strand break (DSB)—is perhaps the most dangerous. DSBs can be caused by ionizing radiation (X-rays or gamma rays), chemical mutagens, and certain drugs used for chemotherapy. In addition, reactive oxygen species that are the by-products of aerobic metabolism cause double-strand breaks. Surprisingly, researchers estimate that naturally occurring double-strand breaks in a typical human cell occur at a rate of between 10–100 breaks per cell per day! Such breaks are harmful in a variety of ways. First, DSBs can result in chromosomal rearrangements such as inversions and translocations (refer back to Figure 8.2). In addition, DSBs can lead to terminal or interstitial deletions (refer back to Figure 8.3). Such genetic changes have the potential to result in detrimental phenotypic effects.

How are DSBs repaired? The two main mechanisms are **homologous recombination repair (HRR)** and **nonhomologous end joining (NHEJ).** HRR occurs when homologous DNA strands, usually from a sister chromatid, are used to repair a DSB in the other sister chromatid (**Figure 18.21**). First, the DSB is processed by the short digestion of DNA strands at the break site. This processing event is followed by the exchange of DNA strands between the broken and unbroken sister chromatids. The unbroken strands are then used as templates to synthesize DNA in the region where the break occurred. Finally, the crisscrossed strands are resolved, which means they are broken and then rejoined in a way that produces separate chromatids. Because sister chromatids are genetically identical, an advantage is that homologous recombination repair can be an error-free mechanism for repairing a DSB. A disadvantage is that sister chromatids are available only during the S and G_2 phases of the cell cycle in eukaryotes or following DNA replication in bacteria. Although sister chromatids are strongly preferred, HRR may also

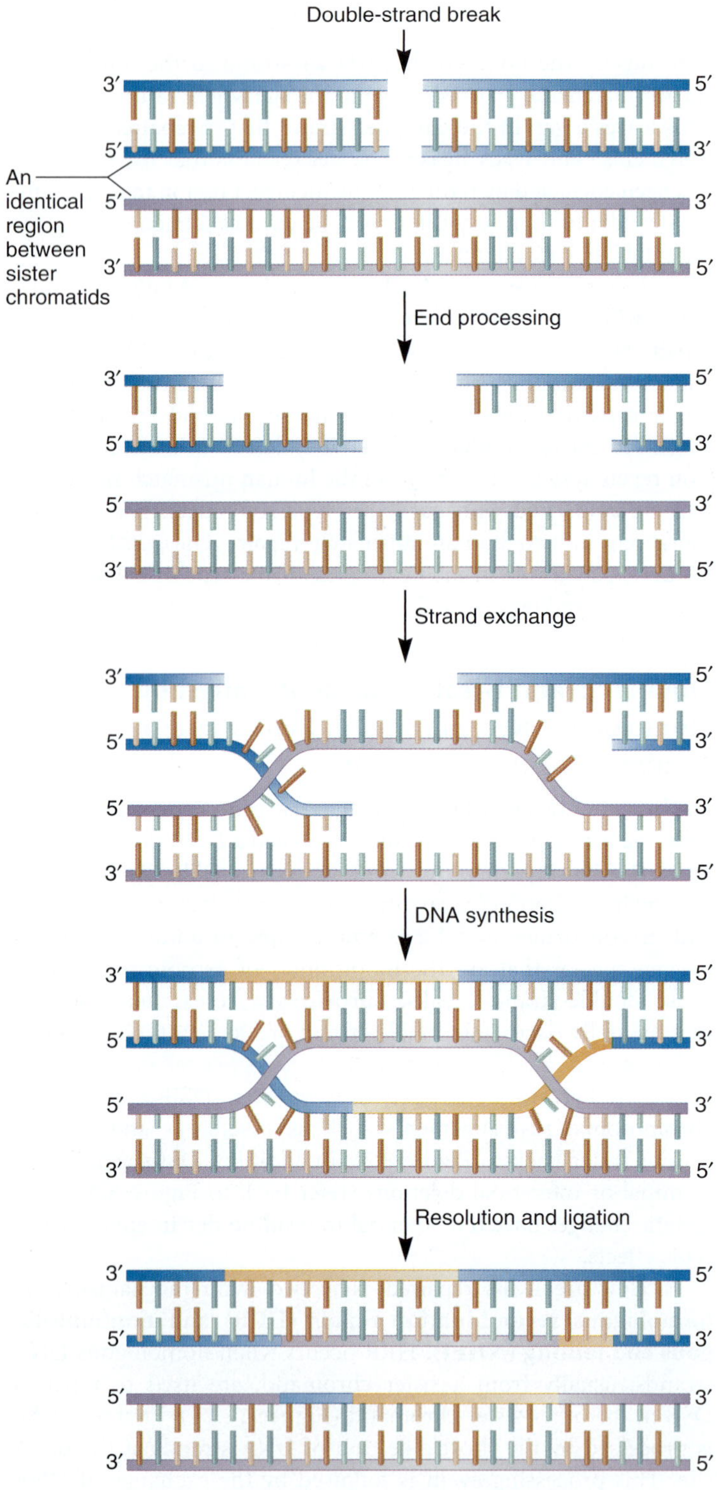

FIGURE 18.21 DNA repair of a double-strand break via homologous recombination repair.

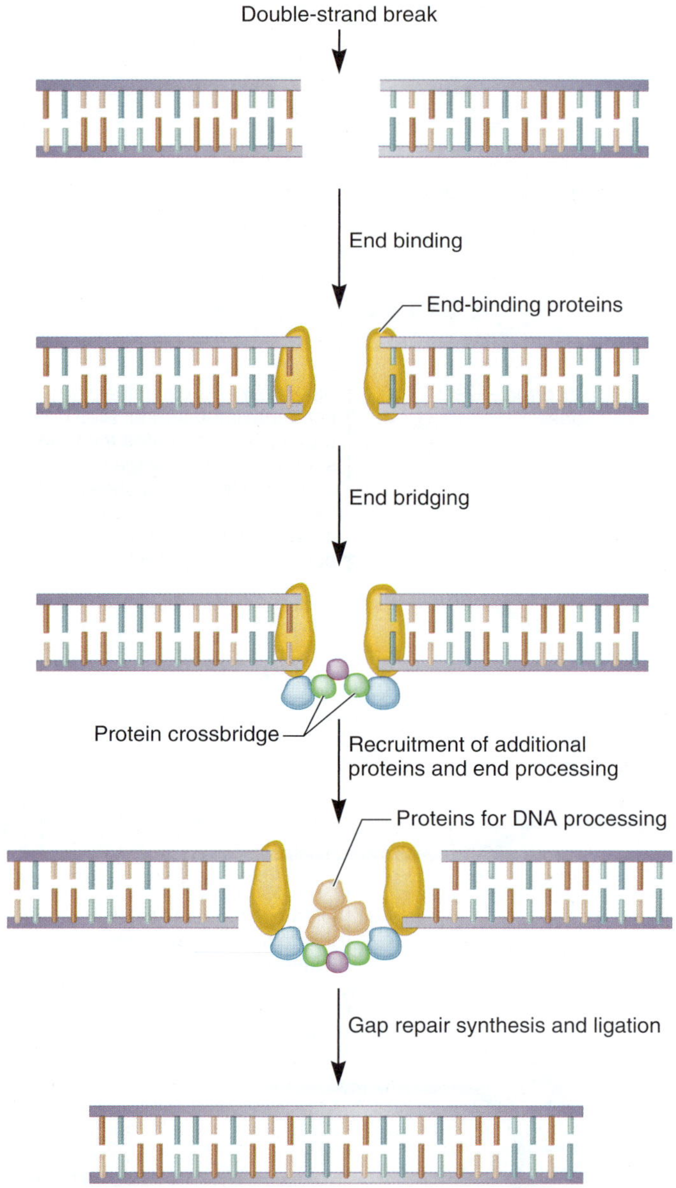

FIGURE 18.22 DNA repair of a double-strand break via nonhomologous end joining.

CONCEPT CHECK: What is an advantage and a disadvantage of this repair system?

occur between homologous regions that are not identical. Therefore, HRR may occasionally happen when sister chromatids are unavailable. The proteins involved in homologous recombination repair are described in Chapter 19 (see Table 19.1).

During nonhomologous end joining, the two broken ends of DNA are simply pieced back together (**Figure 18.22**). This mechanism requires the participation of several proteins that play key roles in the process. First, the DSB is recognized by end-binding proteins. These proteins then recognize additional proteins that form a crossbridge that prevents the two ends from drifting apart. Next, additional proteins are recruited to the region that may process the ends of the broken chromosome by digesting particular DNA strands. This processing may result in the deletion of a small amount of genetic material from the region. Finally, any gaps are filled in via DNA polymerase, and the DNA ends are ligated together. One advantage of NHEJ is that it doesn't involve the participation of a sister chromatid, so it can occur at any stage of the cell cycle. However, a disadvantage is that NHEJ can result in a small deletion in the region that has been repaired.

Damaged DNA May Be Replicated by Translesion DNA Polymerases

Despite the efficient action of numerous repair systems that remove lesions in DNA in an error-free manner, it is inevitable that some lesions may escape these repair mechanisms. Such lesions may be present when DNA is being replicated. If so, replicative DNA polymerases, such as polIII in *E. coli*, are highly sensitive to geometric distortions in DNA and are unable to replicate through DNA lesions. During the past decade, researchers have discovered that cells are equipped with specialized DNA polymerases that assist the replicative DNA polymerase during the process of **translesion synthesis (TLS)**—the synthesis of DNA over a template strand that harbors some type of DNA damage. These lesion-replicating polymerases, which are also described in Chapter 11 (see Table 11.4), contain an active site with a loose, flexible pocket that can accommodate aberrant structures in the template strand. When a replicative DNA polymerase encounters a damaged region, it is swapped with a lesion-replicating polymerase.

A negative consequence of translesion synthesis is low fidelity. Due to their flexible active site, lesion-replicating polymerases are much more likely to incorporate the wrong nucleotide into a newly made daughter strand. The mutation rate is typically in the range of 10^{-2}–10^{-3}. This phenomenon is called **error-prone replication.** By comparison, replicative DNA polymerases are highly intolerant of the geometric distortions imposed on DNA by the incorporation of incorrect nucleotides, and consequently, they incorporate wrong nucleotides with a very low frequency of approximately 10^{-8}. In other words, they copy DNA with a high degree of fidelity.

18.5 COMPREHENSION QUESTIONS

1. The function of photolyase is to repair
   a. double-strand breaks. c. thymine dimers.
   b. apurinic sites. d. all of the above.
2. Which of the following DNA repair systems may involve the removal of a segment of a DNA strand?
   a. Base excision repair
   b. Nucleotide excision repair
   c. Mismatch repair
   d. All of the above
3. In nucleotide excision repair in *E. coli*, the function of the UvrA/UvrB complex is to
   a. detect DNA damage.
   b. make cuts on both sides of the damage.
   c. remove the damaged piece of DNA.
   d. replace the damaged DNA with undamaged DNA.
4. Double-strand breaks can be repaired by
   a. homologous recombination repair (HRR).
   b. nonhomologous end joining (NHEJ).
   c. nucleotide excision repair (NER).
   d. both a and b.
5. An advantage of translesion DNA polymerases is they can replicate ________________, but a disadvantage is they ________________.
   a. very quickly, have low fidelity
   b. over damaged DNA, have low fidelity
   c. when resources are limited, are very slow
   d. over damaged DNA, are very slow

KEY TERMS

Page 443. mutation
Page 444. point mutation, base substitution, transition, transversion, silent mutations, missense mutations, nonsense mutations, polarity, frameshift mutations
Page 446. neutral mutation, up promoter mutation, down promoter mutation, wild type, mutant allele, reversion, deleterious mutation, lethal mutation, beneficial mutation, conditional mutants
Page 447. suppressors, suppressor mutations, intragenic suppressor, intergenic suppressor, breakpoint, position effect
Page 449. germ line, germ-line mutation, somatic cells, somatic mutation, genetic mosaic
Page 451. replica plating
Page 452. random mutation theory, hot spots, spontaneous mutations, induced mutations
Page 453. depurination, apurinic site, deamination
Page 454. tautomeric shift, tautomers, reactive oxygen species (ROS), oxidative stress, oxidative DNA damage, trinucletoide repeat expansion (TNRE)
Page 456. anticipation, dynamic mutation
Page 457. mutagens
Page 458. nitrous acid, nitrogen mustard, ethyl methanesulfonate (EMS), acridine dyes, proflavin, 5-bromouracil (5BU), 2-aminopurine
Page 459. thymine dimers, mutation rate
Page 460. mutation frequency, Ames test
Page 461. photolyase, photoreactivation, alkyltransferase, base excision repair (BER), DNA *N*-glycosylases
Page 462. AP endonuclease, nick translation, nucleotide excision repair (NER)
Page 464. base pair mismatch, mismatch repair system
Page 465. homologous recombination repair (HRR), nonhomologous end joining (NHEJ)
Page 467. translesion synthesis (TLS), error-prone replication
Page 469. polar mutation

CHAPTER SUMMARY

- A mutation is a heritable change in the genetic material.

18.1 Effects of Mutations on Gene Structure and Function

- A point mutation is a change in a single base pair. These can be transitions or transversions.
- Silent, missense, nonsense, and frameshift mutations may occur within the coding region of a gene (see Table 18.1, Figure 18.1).
- Mutations may also occur within noncoding regions of a gene and affect gene expression (see Table 18.2).
- Suppressor mutations reverse the phenotypic effects of another mutation. They can be intragenic or intergenic (see Table 18.3).
- Changes in chromosome structure can have a position effect that alters gene expression (see Figures 18.2, 18.3).
- With regard to timing, mutations can occur in the germ line or in somatic cells (see Figures 18.4, 18.5).

18.2 Random Nature of Mutations

- The experiments performed by Lederberg and Lederberg were consistent with the random mutation theory (see Figure 18.6).

18.3 Spontaneous Mutations

- Spontaneous mutations result from natural biological or chemical processes, whereas induced mutations are caused by environmental agents (see Table 18.4).
- Three common ways that mutations can arise spontaneously is by depurination, deamination, and tautomeric shifts (see Figures 18.7–18.9).
- Reactive oxygen species (ROS) can cause spontaneous mutations by oxidizing bases in DNA (see Figure 18.10).
- In individuals with a trinucleotide repeat expansion (TNRE), a trinucleotide repeat increases above a certain critical size and becomes prone to frequent expansion. This type of mutation, called a dynamic mutation, is responsible for certain types of human diseases. The repeats can expand due to hairpin formation during DNA replication (see Table 18.5, Figure 18.11).

18.4 Induced Mutations

- A mutagen is an agent that can cause a mutation. Researchers have identified many different mutagens that change DNA structure in a variety of ways (see Table 18.6, Figures 18.12–18.14).
- Mutation rate is the likelihood that a new mutation will occur. Mutation frequency is the number of mutant genes divided by the total number of genes in a population.
- The Ames test is used to determine if an agent is a mutagen (see Figure 18.15).

18.5 DNA Repair

- All species have a variety of DNA repair systems to avoid the harmful effects of mutations (see Table 18.7).
- Photolyase and alkyltransferase can directly repair certain types of damaged bases (see Figure 18.16).
- Base excision repair recognizes and removes a damaged base (see Figure 18.17).
- Nucleotide excision repair removes damaged bases and damaged segments of DNA. Several human inherited diseases involve defects in nucleotide excision repair (see Figures 18.18, 18.19).
- Base mismatch repair recognizes a base mismatch and removes the segment of the DNA strand with the incorrect base (see Figure 18.20).
- Double-strand breaks can be repaired by homologous recombination repair and by nonhomologous end joining (see Figures 18.21, 18.22).
- Damaged DNA may be replicated by translesion DNA polymerases that are error prone.

PROBLEM SETS & INSIGHTS

Solved Problems

S1. Mutant tRNAs may act as nonsense and missense suppressors. At the molecular level, explain how you think these suppressors work.

Answer: A suppressor is a second-site mutation that suppresses the phenotypic effects of a first mutation. Intergenic suppressor mutations in tRNA genes can act as nonsense or missense suppressors. For example, let's suppose a first mutation puts a stop codon into a protein-encoding gene. A second mutation in a tRNA gene can alter the anticodon region of a tRNA so that the anticodon recognizes a stop codon but inserts an amino acid at this site. A missense suppressor is a mutation in a tRNA gene that changes the anticodon so that it puts in the wrong amino acid at a codon that is not a stop codon. These mutant tRNAs are termed missense tRNAs. For example, a tRNA that normally recognizes glutamic acid may incur a mutation that changes its anticodon sequence so it recognizes a glycine codon instead. Like nonsense suppressors, missense suppressors can be produced by mutations in the anticodon region of tRNAs so the tRNA recognizes an incorrect codon. Alternatively, missense suppressors can also be produced by mutations in aminoacyl-tRNA synthetases that cause them to attach the incorrect amino acid to a tRNA.

S2. If the rate of mutation for a particular gene is 10^{-5} per cell generation, how many new mutations per gene would you expect in a population of 1 million bacteria?

Answer: If we multiply the mutation rate times the number of bacteria ($10^{-5} \times 10^{6}$), we obtain a value of 10 new mutations in this gene per million bacteria. This answer is correct, but it is an oversimplification of mutation rate. For any given gene, the mutation rate is based on a probability that an event will occur. Therefore, when we consider a particular population of bacteria, we should be aware that the actual rate of new mutation would vary. Even though the rate may be 10^{-5}, we would not be surprised if a population of 1 million bacteria had 9 or 11 new mutations in this gene instead of the expected number of 10. We would be surprised if it had 5000 new mutations in this gene, because this value would deviate much too far from our expected number.

S3. In the Ames test, several *Salmonella* strains are used that contain different types of mutations within the gene that encodes an enzyme necessary for histidine biosynthesis. These mutations include transversions, transitions, and frameshift mutations. Why do you think it would be informative to test a mutagen with these different strains?

Answer: Different types of mutagens have different effects on DNA structure. For example, if a mutagen caused transversions, an experimenter would want to use a *Salmonella* strain in which a transversion would convert a *his*$^{-}$ strain into a *his*$^{+}$ strain. This type of strain would make it possible to detect the effects of the mutagen.

S4. In Chapter 14, we discussed how bacterial genes can be arranged in an operon structure in which a polycistronic RNA contains the coding sequences for two or more genes. For genes in an operon, a relatively short distance occurs between the stop codon in the first gene and the start codon in the next gene. In addition to the ribosome-binding sequence at the beginning of the first gene, a ribosome-binding sequence is also found at the beginning of the second gene and at the beginning of all subsequent genes. After the ribosome has moved past the stop codon in the first gene, it quickly encounters a ribosome-binding site in the next gene. This prevents the complete disassembly of the ribosome as would normally occur after a stop codon, and leads to the efficient translation of the second gene.

Jacob and Monod, who studied mutations in the *lac* operon, discovered a mutation in the *lacZ* gene that had an unusual effect. This mutation resulted in a shorter version of the β-galactosidase protein, and it also prevented the expression of the *lacY* gene, even though the *lacY* gene was perfectly normal. Explain this mutation.

Answer: This mutation introduced a stop codon much earlier in the *lacZ* gene sequence. This type of mutation, which affects the translation of a gene downstream in an operon, is referred to as a **polar mutation,** and the phenomenon is called polarity. The explanation is that the early nonsense codon causes the ribosome to disassemble before it has a chance to reach the next gene sequence in the operon. The ribosome-binding sequence that precedes the second gene is efficiently recognized only by a ribosome that has traveled to the normal stop codon in the first gene.

S5. A reverse mutation or a reversion is a mutation that returns a mutant codon back to a codon that gives a wild-type phenotype. At the DNA level, this can be an exact reversion or an equivalent reversion.

GAG (glutamic acid)	First → mutation	GTG (valine)	Exact → reversion	GAG (glutamic acid)
GAG (glutamic acid)	First → mutation	GTG (valine)	Equivalent → reversion	GAA (glutamic acid)
GAG (glutamic acid)	First → mutation	GTG (valine)	Equivalent → reversion	GAT (aspartic acid)

An equivalent reversion produces a protein equivalent to the wild type in structure and function. This can occur in two ways. In some cases, the reversion produces the wild-type amino acid (in this case, glutamic acid), but it uses a different codon than the wild-type gene. Alternatively, an equivalent reversion may substitute an amino acid structurally similar to the wild-type amino acid. In our example, an equivalent reversion has changed valine to an aspartic acid. Because aspartic and glutamic acids are structurally similar—they are acidic amino acids—this type of reversion can restore the wild-type structure and function.

Now here is the question. The template strand within the coding sequence of a gene has the following sequence:

3′—TACCCCTTCGACCCCGGA—5′

This template produces an mRNA,

5′—AUGGGGAAGCUGGGGCCA—3′

that encodes a polypeptide with the following sequence: methionine–glycine–lysine–leucine–glycine–proline.

A mutation changes the template strand to:

3′—TACCCCTACGACCCCGGA—5′

After the first mutation, another mutation occurs to change this sequence again. Would the following second mutations be an exact reversion, an equivalent reversion, or neither?

A. 3′—TACCCCTCCGACCCCGGA—5′

B. 3′—TACCCCTTCGACCCCGGA—5′

C. 3′—TACCCCGACGACCCCGGA—5′

Answer:

A. This is probably an equivalent reversion. The third codon, which encodes a lysine in the normal gene, is now an arginine codon. Arginine and lysine are both basic amino acids, so the polypeptide would probably function normally.

B. This is an exact reversion.

C. The third codon, which is a lysine in the normal gene, has been changed to a leucine codon. It is difficult to say if this would be an equivalent reversion or not. Lysine is a basic amino acid and leucine is a nonpolar, aliphatic amino acid. The protein may still function normally with a leucine at the third codon, or it may function abnormally. You would need to test the function of the protein to determine if this was an equivalent reversion or not.

Conceptual Questions

C1. For each of the following mutations, is it a transition, transversion, addition, or deletion? The original DNA strand is

5′—GGACTAGATAC—3′

(Note: Only the coding DNA strand is shown.)

A. 5′—GAACTAGATAC—3′

B. 5′—GGACTAGAGAC—3′

C. 5′—GGACTAGTAC—3′

D. 5′—GGAGTAGATAC—3′

C2. A gene mutation changes an AT base pair to a GC pair. This causes a gene to encode a truncated protein that is nonfunctional. An organism that carries this mutation cannot survive at high temperatures. Make a list of all the genetic terms that could be used to describe this type of mutation.

C3. What does a suppressor mutation suppress? What is the difference between an intragenic and an intergenic suppressor?

C4. How would each of the following types of mutations affect the amount of functional protein that is expressed from a gene?

A. Nonsense

B. Missense

C. Up promoter mutation

D. Mutation that affects splicing

C5. X-rays strike a chromosome in a living cell and ultimately cause the cell to die. Did the X-rays produce a mutation? Explain why or why not.

C6. Lactose permease is encoded by the *lacY* gene of the *lac* operon. A mutation occurred at codon 64 that changed the normal glycine codon into a valine codon. The mutant lactose permease is unable to function. However, a second mutation, which changes codon 50 from an alanine codon to a threonine codon, is able to restore function. Are the following terms appropriate or inappropriate to describe this second mutation?

A. Reversion

B. Intragenic suppressor

C. Intergenic suppressor

D. Missense mutation

C7. Nonsense suppressors, which are described in solved problem S1, tend to be very inefficient at their job of allowing readthrough of a stop codon. How would it affect the cell if they were efficient at their job?

C8. Are each of the following mutations silent, missense, nonsense, or frameshift mutations? The original DNA strand is 5′–ATGGGACTAGATACC–3′. (Note: Only the coding strand is shown; the first codon is methionine.)

A. 5′—ATGGGTCTAGATACC—3′

B. 5′—ATGCGACTAGATACC—3′

C. 5′—ATGGGACTAGTTACC—3′

D. 5′—ATGGGACTAAGATACC—3′

C9. In Chapters 12 through 16, we discussed many sequences that are outside the coding sequence and are important for gene expression. Look up two of these sequences and write them out. Explain how a mutation could change these sequences, thereby altering gene expression.

C10. Explain two ways that a chromosomal rearrangement can cause a position effect.

C11. Is a random mutation more likely to be beneficial or harmful? Explain your answer.

C12. Which of the following mutations could be appropriately described as a position effect?

A. A point mutation at the −10 position in the promoter region prevents transcription.

B. A translocation places the coding sequence for a muscle-specific gene next to an enhancer that is turned on in nerve cells.

C. An inversion flips a gene from the long arm of chromosome 17 (which is euchromatic) to the short arm (which is heterochromatic).

C13. As discussed in Chapter 24, most forms of cancer are caused by environmental agents that produce mutations in somatic cells. Is an individual with cancer considered a genetic mosaic? Explain why or why not.

C14. Discuss the consequences of a germ-line versus a somatic mutation.

C15. Draw and explain how alkylating agents alter the structure of DNA.

C16. Explain how a mutagen can interfere with DNA replication to cause a mutation. Give two examples.

C17. What type of mutation (transition, transversion, or frameshift) would you expect each of the following mutagens to cause?

A. Nitrous acid

B. 5-Bromouracil

C. Proflavin

C18. Explain what happens to the sequence of DNA during trinucleotide repeat expansion (TNRE). If someone was mildly affected with a TNRE disorder, what issues would be important when considering possible effects in future offspring?

C19. Distinguish between spontaneous and induced mutations. Which are more harmful? Which are avoidable?

C20. Are mutations random events? Explain your answer.

C21. Give an example of a mutagen that can change cytosine to uracil. Which DNA repair system(s) would be able to repair this defect?

C22. If a mutagen causes bases to be removed from nucleotides within DNA, what repair system would fix this damage?

C23. Trinucleotide repeat expansions (TNREs) are associated with several different human inherited diseases. Certain types of TNREs produce a long stretch of glutamines (an amino acid) within the encoded protein. This long stretch of glutamines somehow inhibits the function of the protein, thereby causing a disorder. When a TNRE exerts its detrimental effect by producing a glutamine stretch, are the following statements true or false?

A. The TNRE is within the coding sequence of the gene.

B. The TNRE prevents RNA polymerase from transcribing the gene properly.

C. The trinucleotide sequence is CAG.

D. The trinucleotide sequence is CCG.

C24. With regard to TNRE, what is meant by the term anticipation?

C25. What is the difference between the mutation rate and the mutation frequency?

C26. Achondroplasia is a rare form of dwarfism. It is caused by an autosomal dominant mutation within a single gene. Among 1,422,000 live births, the number of babies born with achondroplasia was 31. Among those 31 babies, 18 of them had one parent with achondroplasia. The remaining babies had two unaffected parents. What is the mutation frequency for this disorder among these 1,422,000 babies? What is the mutation rate for achondroplasia?

C27. A segment of DNA has the following sequence:

```
TTGGATGCTGG
AACCTACGACC
```

A. What would be the sequence immediately after reaction with nitrous acid? Let the letters H represent hypoxanthine and U represent uracil.

B. Let's suppose this DNA was reacted with nitrous acid. The nitrous acid was then removed, and the DNA was replicated for two generations. What would be the sequences of the DNA products after the DNA had replicated two times? Your answer should contain the sequences of four double helices.

C28. In the treatment of cancer, the basis for many types of chemotherapy and radiation therapy is that mutagens are more effective at killing dividing cells than nondividing cells. Explain why. What are possible harmful side effects of chemotherapy and radiation therapy?

C29. An individual carries a somatic mutation that changes a lysine codon into a glutamic acid codon. Prior to acquiring this mutation, the individual had been exposed to UV light, proflavin, and 5-bromouracil. Which of these three agents would be the most likely to have caused this somatic mutation? Explain your answer.

C30. Which of the following examples is likely to be caused by a somatic mutation?

A. A purple flower has a small patch of white tissue.

B. One child, in a family of seven, is an albino.

C. One apple tree, in a very large orchard, produces its apples 2 weeks earlier than any of the other trees.

D. A 60-year-old smoker develops lung cancer.

C31. How would nucleotide excision repair be affected if one of the following proteins was missing? Describe the condition of the DNA if repair had been attempted in the absence of the protein.

A. UvrA

B. UvrC

C. UvrD

D. DNA polymerase

C32. During mismatch repair, why is it necessary to distinguish between the template strand and the newly made daughter strand? How is this accomplished?

C33. What are the two main mechanisms by which cells repair double-strand breaks? Briefly describe each one.

C34. With regard to the repair of double-strand breaks, what are the advantages and disadvantages of homologous recombination repair versus nonhomologous end joining?

C35. When DNA *N*-glycosylase recognizes thymine dimers, it detects only the thymine located on the 5′ side of the thymine dimer as being abnormal. Draw and explain the steps whereby a thymine dimer is repaired by the consecutive actions of DNA *N*-glycosylase, AP endonuclease, and DNA polymerase.

C36. What is the underlying genetic defect that causes xeroderma pigmentosum? How can the symptoms of this disease be explained by the genetic defect?

C37. Three common ways to repair changes in DNA structure are nucleotide excision repair, mismatch repair, and homologous recombination repair. Which of these three mechanisms would be used to fix the following types of DNA changes?

A. A change in the structure of a base caused by a mutagen in a nondividing eukaryotic cell

B. A change in DNA sequence caused by a mistake made by DNA polymerase

C. A thymine dimer in the DNA of an actively dividing bacterial cell

C38. Discuss the similarities and differences between the nucleotide excision repair and mismatch repair systems.

C39. In *E. coli*, a methyltransferase enzyme encoded by the *dam* gene recognizes the sequence 5′–GATC–3′ and attaches a methyl group to the N^6 position of adenine. *E. coli* strains that have the *dam* gene deleted are known to have a higher spontaneous mutation rate than normal strains. Explain why.

Experimental Questions

E1. Explain how the technique of replica plating supports the random mutation theory but conflicts with the physiological adaptation hypothesis.

E2. Outline how you would use the technique of replica plating to show that antibiotic resistance is due to random mutations.

E3. From an experimental point of view, is it better to use haploid or diploid organisms for mutagen testing? Consider the Ames test when preparing your answer.

E4. How would you modify the Ames test to discover physical mutagens? Would it be necessary to add the rat liver extract? Explain why or why not.

E5. During an Ames test, bacteria were exposed to a potential mutagen. Also, as a control, another sample of bacteria was not exposed to the mutagen. In both cases, 10 million bacteria were plated and the following results were obtained:

No mutagen: 17 colonies

With mutagen: 2017 colonies

Calculate the mutation rate in the presence and absence of the mutagen. How much does the mutagen increase the rate of mutation?

E6. Richard Boyce and Paul Howard-Flanders conducted an experiment that provided biochemical evidence that thymine dimers are removed from DNA by a DNA repair system. In their studies, bacterial DNA was radiolabeled so the amount of radioactivity reflected the amount of thymine dimers. The DNA was then subjected to UV light, causing the formation of thymine dimers. When radioactivity was found in the soluble fraction, thymine dimers had been excised from the DNA by a DNA repair system. But when the radioactivity was in the insoluble fraction, the thymine dimers had been retained within the DNA. The following table illustrates some of their results involving a normal strain of *E. coli* and a mutant strain that was very sensitive to killing by UV light:

Strain	Treatment	Radioactivity in the Insoluble Fraction (cpm)	Radioactivity in the Soluble Fraction (cpm)
Normal	No UV	<100	<40
Normal	UV-treated, incubated 2 hours at 37°C	357	940
Mutant	No UV	<100	<40
Mutant	UV-treated, incubated 2 hours at 37°C	890	<40

Adapted from R. P. Boyce and P. Howard-Flanders (1964) Release of ultraviolet light-induced thymine dimers from DNA in *E. coli* K-12. *Proc Natl Acad Sci USA 51*, 293–300.

Explain the results found in this table. Why is the mutant strain sensitive to UV light?

Questions for Student Discussion/Collaboration

1. In *E. coli*, a variety of mutator strains have been identified in which the spontaneous rate of mutation is much higher than in normal strains. Make a list of the types of abnormalities that could cause a strain of bacteria to become a mutator strain. Which abnormalities do you think would give the highest rate of spontaneous mutation?
2. Discuss the times in a person's life when it would be most important to avoid mutagens. Which parts of a person's body should be most protected from mutagens?
3. A large amount of research is aimed at studying mutation. However, there is not an infinite amount of research money. Where would you put your money for mutation research?

A. Testing of potential mutagens

B. Investigating molecular effects of mutagens

C. Investigating DNA repair mechanisms

D. Some other place

Note: All answers appear at the website for this textbook; the answers to the even-numbered questions and all of the Concept Check and Comprehension Questions are in the back of the book.

www.mhhe.com/brookergenetics5e

Visit the website for practice tests, answer keys, and other learning aids for this chapter. Enhance your understanding of genetics with our interactive exercises, quizzes, animations, and much more.

Speckled corn kernels. *This speckling phenotype is due to the movement of DNA segments called transposable elements.*

CHAPTER OUTLINE

19 RECOMBINATION AND TRANSPOSITION AT THE MOLECULAR LEVEL

In this chapter, we will examine a variety of molecular processes in which segments of chromosomal DNA become rearranged. **Homologous recombination** is the process whereby DNA segments that are similar or identical to each other break and rejoin to form a new combination. As described in Chapter 3, homologous recombination occurs when chromosomes cross over during meiosis. Not only does homologous recombination enhance genetic diversity, it also helps to repair DNA and ensures the proper segregation of chromosomes.

We will also examine ways that nonhomologous segments of DNA may recombine with each other. During **site-specific recombination,** nonhomologous DNA segments are recombined at specific sites. This type of recombination occurs when certain viruses integrate their genomes into host cell DNA and also happens within genes that encode antibody polypeptides. In addition, we will consider a widespread form of recombination known as **transposition.** As you will learn, small segments of DNA, called transposable elements, can move themselves to multiple locations within the chromosomal DNA.

From a molecular viewpoint, homologous recombination, site-specific recombination, and transposition are important mechanisms for DNA rearrangement. These processes involve a series of steps that direct the breakage and rejoining of DNA fragments and require various cellular proteins. The past few decades have seen many exciting advances in our understanding of recombination at the molecular level. In this chapter, we will consider the general concepts of recombination and examine molecular models that explain how recombination occurs.

19.1 HOMOLOGOUS RECOMBINATION

Learning Outcomes:

1. Describe the Holliday model and double-strand break model for homologous recombination.
2. Explain how gene conversion can occur via mismatch repair and DNA gap repair synthesis.

Homologous recombination involves an exchange between DNA segments that are similar or identical in their DNA sequences. Eukaryotic chromosomes that have similar or identical sequences frequently participate in crossing over during meiosis I and

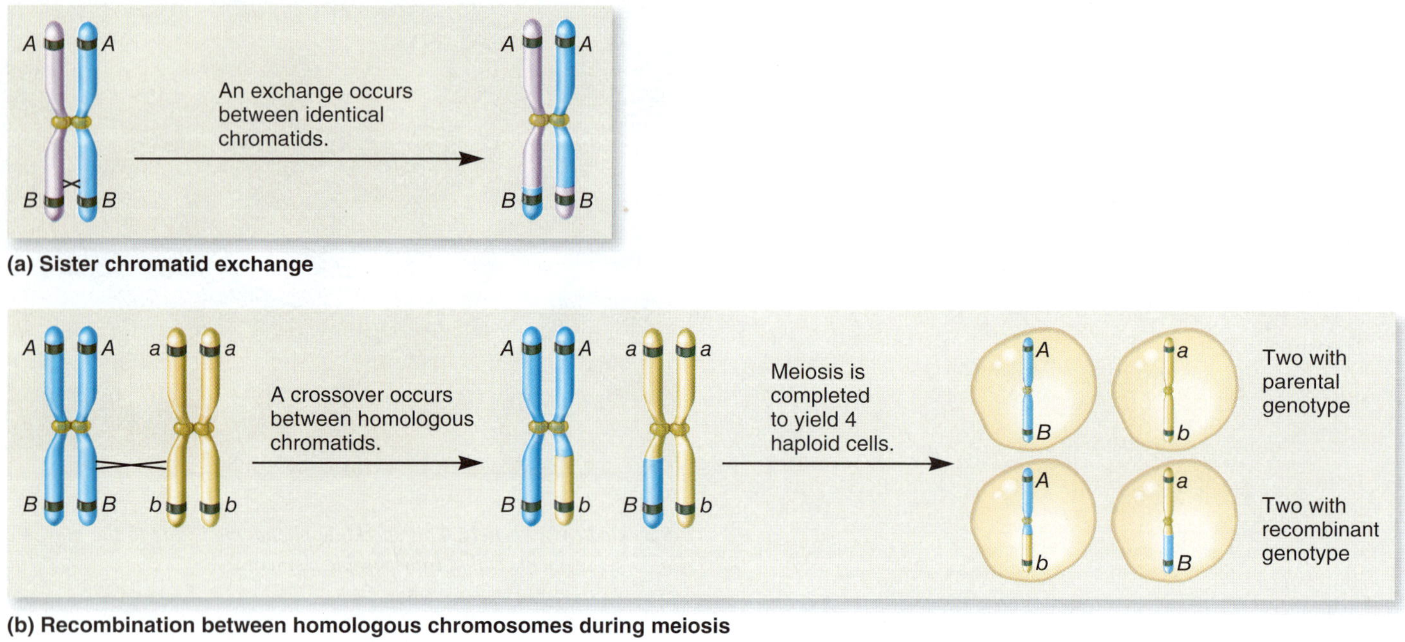

FIGURE 19.1 **Two types of homologous recombination in eukaryotes.** **(a)** Homologous recombination between sister chromatids. **(b)** Homologous recombination between homologous chromatids. This second form of homologous recombination may lead to a new combination of alleles, which is called a recombinant (or nonparental) genotype.

Genes→Traits Homologous recombination is particularly important when we consider the relationships between multiple genes and multiple traits. For example, if the X chromosome in a female fruit fly carried alleles for red eyes and gray body and its homolog carried alleles for white eyes and yellow body, homologous recombination could produce recombinant chromosomes that carry alleles for red eyes and yellow body, or alleles for white eyes and gray body. Therefore, new combinations of two or more alleles can arise when homologous recombination takes place.

CONCEPT CHECK: What is the advantage of genetic recombination, which is depicted in part b?

occasionally during mitosis. Crossing over involves the alignment of a pair of homologous chromosomes, followed by the breakage of two chromatids at analogous locations, and the subsequent exchange of the corresponding segments (refer to Figure 3.10).

Figure 19.1 shows two types of homologous recombination that may occur between replicated chromosomes in a diploid species. When it takes place between sister chromatids, the process is called **sister chromatid exchange (SCE).** Because sister chromatids are genetically identical, SCE does not produce a new combination of alleles (Figure 19.1a). By comparison, crossing over may occur between homologous chromosomes during meiosis. As shown in Figure 19.1b, this form of homologous recombination may produce a new combination of alleles in the resulting chromosomes. In this second case, homologous recombination has resulted in **genetic recombination,** which refers to the shuffling of genetic material to create a new genetic combination that differs from the original. Homologous recombination is an important mechanism to foster genetic recombination.

Bacteria are usually haploid. They do not have pairs of homologous chromosomes. Even so, bacteria also can undergo homologous recombination. How can the exchange of DNA segments occur in a haploid organism? First, bacteria may have more than one copy of a chromosome per cell, though the copies are usually identical. These copies can exchange genetic material via homologous recombination. Second, during DNA replication, the replicated regions may also undergo homologous recombination. In bacteria, homologous recombination is particularly important in the repair of DNA segments that have been damaged. In this section, we will focus our attention on the molecular mechanisms that underlie homologous recombination.

The Holliday Model Describes a Molecular Mechanism for the Recombination Process

We now turn our attention to genetic exchange that occurs between homologous chromosomes. Perhaps it is surprising that the first molecular model of homologous recombination did not come from a biochemical analysis of DNA or from electron microscopy studies. Instead, it was deduced from the outcome of genetic crosses in fungi.

As discussed in Chapter 6, geneticists have learned a great deal from the analysis of fungal asci, because an ascus is a sac that contains the products of a single meiosis. When two haploid fungi that differ at a single gene are crossed to each other, the ascus is expected to contain an equal proportion of each genotype. For example, if a pigmented strain of *Neurospora* producing orange spores is crossed to an albino strain producing white spores, the resulting group of eight cells, or octad, should contain four orange spores and four white spores. In 1934, H. Zickler noticed that unequal proportions of the spores sometimes occurred within asci. He occasionally observed octads with six orange spores and two white spores, or six white spores and two orange spores.

Zickler used the term **gene conversion** to describe the phenomenon in which one allele is converted to the allele on the homologous chromosome. Subsequent studies by several researchers confirmed this phenomenon in yeast and *Neurospora*. Gene conversion occurred at too high a rate to be explained by new mutations. In addition, research showed that gene conversion often occurs in a chromosomal region where a crossover has taken place.

Based on studies involving gene conversion, Robin Holliday proposed a model in 1964 to explain the molecular steps that occur during homologous recombination. We will first consider the steps in the Holliday model and then consider more recent models. Later, we will examine how the Holliday model can explain the phenomenon of gene conversion.

The **Holliday model** is shown in **Figure 19.2a**. At the beginning of the process, two homologous chromatids are aligned with each other. According to the model, a break or nick occurs at identical sites in one strand of each of the two homologous chromatids. The strands then invade the opposite helices and base-pair with the complementary strands. This event is followed by a covalent linkage to create a **Holliday junction.** The cross in the Holliday junction can migrate in a lateral direction. As it does so, a DNA strand in one helix is swapped for a DNA strand in the other helix. This process is called **branch migration** because the branch connecting the two double helices migrates laterally. Because the DNA sequences in the homologous chromosomes are similar but may not be identical, the swapping of the DNA strands during branch migration may produce a **heteroduplex,** a region in the double-stranded DNA that contains base mismatches.

The final two steps in the recombination process are collectively called **resolution** because they involve the breakage and rejoining of two DNA strands to create two separate chromosomes. In other words, the entangled DNA strands become resolved into two separate structures. The bottom left side of Figure 19.2a shows the Holliday junction viewed from two different planes. If breakage occurs in the same two DNA strands that were originally nicked at the beginning of this process, the subsequent joining of strands produces nonrecombinant chromosomes with a heteroduplex region. Alternatively, if breakage occurs in the strands that were not originally nicked, the rejoining process results in recombinant chromosomes, also with a heteroduplex region.

The Holliday model can account for the general properties of recombinant chromosomes formed during eukaryotic meiosis. As mentioned, the original model was based on the results of crosses in fungi in which the products of meiosis are contained within a single ascus. Nevertheless, molecular research in many other organisms has supported the central tenets of the model. Particularly convincing evidence came from electron microscopy studies in which recombination structures could be visualized. **Figure 19.2b** shows an electron micrograph of two DNA fragments in the process of recombination. This structure has been called a chi form because its shape is similar to the Greek letter χ (chi).

More Recent Models Have Refined the Molecular Steps of Homologous Recombination

As more detailed studies of homologous recombination have become available, certain steps in the Holliday model have been reconsidered. In particular, more recent models have modified the initiation phase of recombination. Researchers now suggest that homologous recombination is not likely to involve nicks at identical sites in one strand in each of two homologous chromatids. Instead, it is more likely for a DNA helix to incur a break in both strands of one chromatid or a single nick. Both of these types of changes have been shown to initiate homologous recombination. Therefore, newer models have tried to incorporate these experimental observations. In 1975, a model proposed by Matthew Meselson and Charles Radding hypothesized that a single nick in one DNA strand initiates recombination. A second model, proposed by Jack Szostak, Terry Orr-Weaver, Rodney Rothstein, and Franklin Stahl, suggests that a double-strand break initiates the recombination process. This is called the **double-strand break model.** Though recombination may occur via more than one mechanism, research suggests that double-strand breaks commonly promote homologous recombination during meiosis and during DNA repair.

Figure 19.3 shows the general steps in the double-strand break model. As seen here, the top chromatid has experienced a double-strand break. A small region near the break is degraded, which generates a single-stranded DNA segment that can invade the intact bottom chromatid. The strand displaced by the invading segment forms a structure called a displacement loop (D-loop). After the D-loop is formed, two regions have a gap in the DNA. How is the problem fixed? DNA synthesis occurs in the relatively short gaps where a DNA strand is missing. This DNA synthesis is called **DNA gap repair synthesis.** Once this is completed, two Holliday junctions are produced. Depending on the way these are resolved, the end result is nonrecombinant or recombinant chromosomes containing a short heteroduplex. In eukaryotes such as yeast, evidence suggests that certain proteins bound to Holliday junctions may regulate the resolution step in a way that favors the formation of recombinant chromosomes rather than nonrecombinant chromosomes.

Various Proteins Are Necessary to Facilitate Homologous Recombination

The homologous recombination process requires the participation of many proteins that catalyze different steps in the recombination pathway. Homologous recombination is found in all species, and the types of proteins that participate in the steps outlined in Figure 19.3 are very similar. The cells of any given species may have more than one molecular mechanism to carry out homologous recombination. This process is best understood in *Escherichia coli*. **Table 19.1** summarizes some of the *E. coli* proteins that play critical roles in one recombination pathway found in this species. Though it is beyond the scope of this textbook to describe them, *E. coli* has other pathways to carry out homologous recombination.

Before ending this discussion of the molecular mechanism of homologous recombination, let's consider recombinational events during meiosis in eukaryotic cells. As described in Chapter 3, crossing over between homologous chromosomes is an important event during prophase of meiosis I. An intriguing question is, How are crossover sites chosen between two homologous chromosomes? Although the answer is not entirely understood, molecular studies in two different yeast species, *Saccharomyces cerevisiae* and *Schizosaccharomyces pombe*, suggest

FIGURE 19.2 The Holliday model for homologous recombination. The Holliday model is adapted from Holliday R. (1964) A mechanism for gene conversion in fungi. *Genet Res* 5, 282–304.

INTERACTIVE EXERCISE

ONLINE ANIMATION

CONCEPT CHECK: Explain why a heteroduplex region may be produced after branch migration occurs.

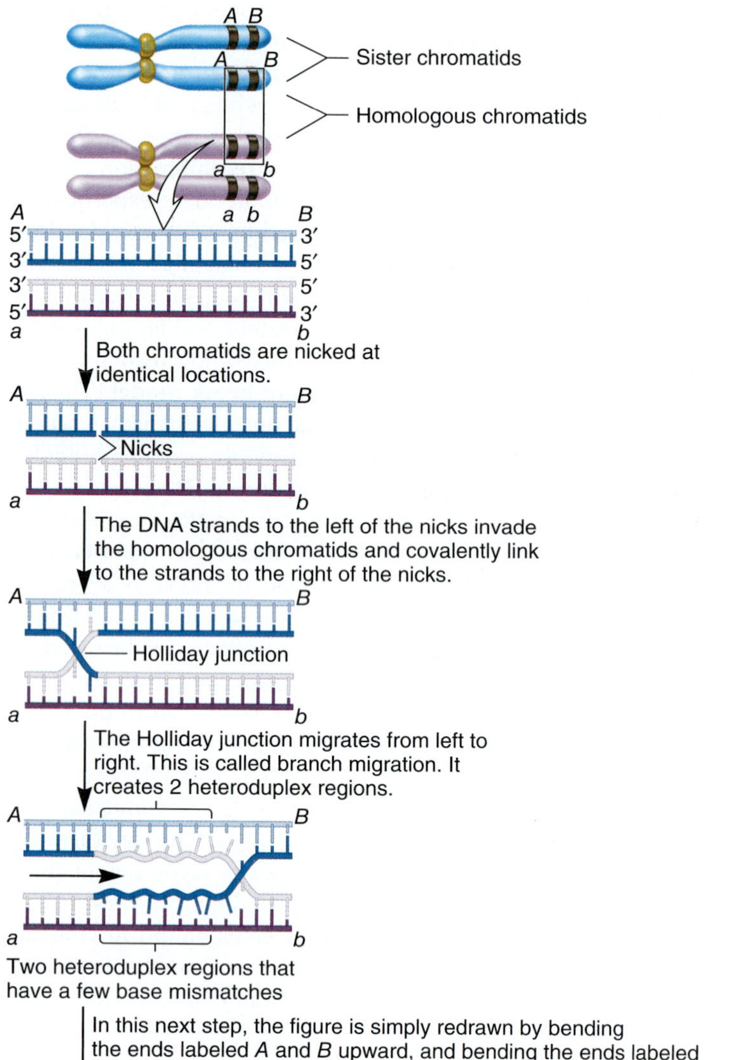
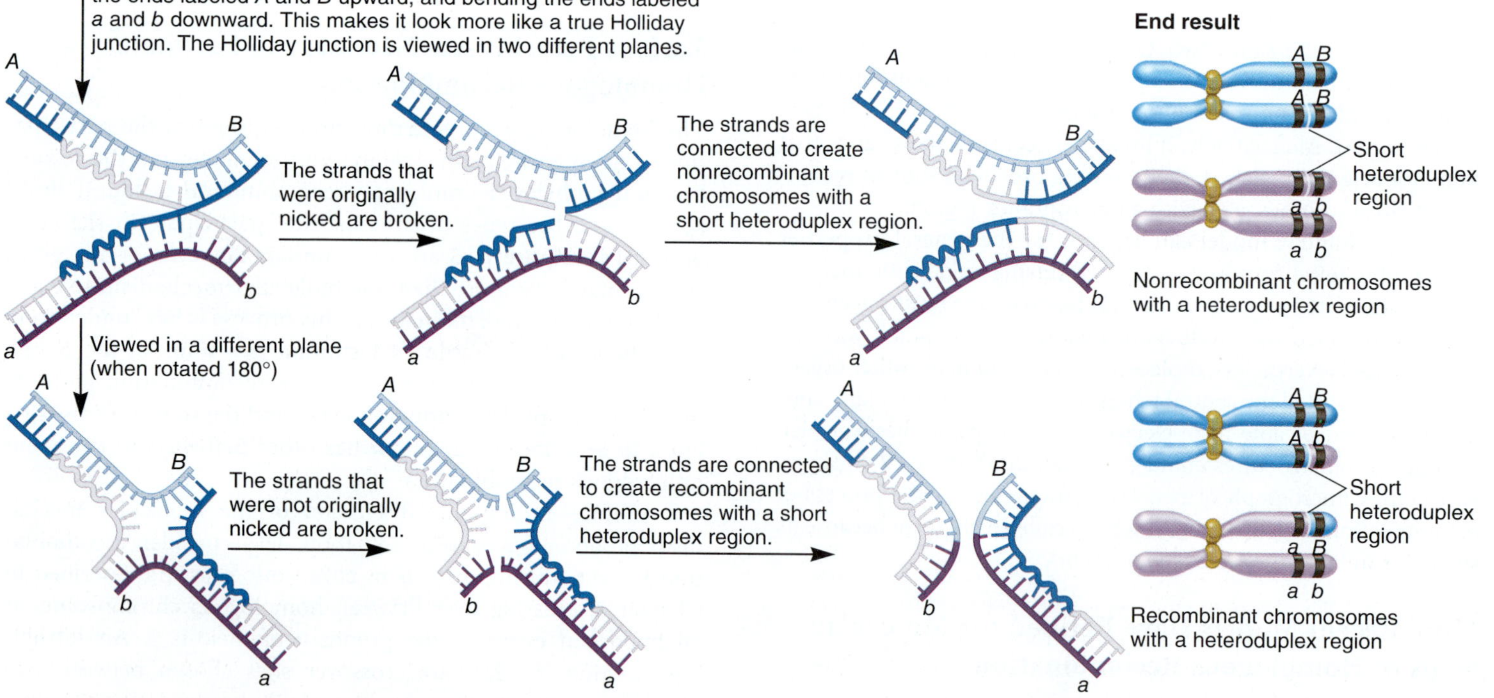

(a) The Holliday model for homologous recombination

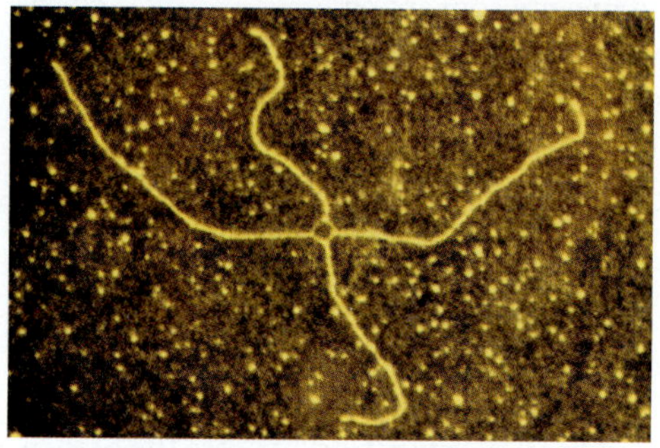

(b) Micrograph of a Holliday junction

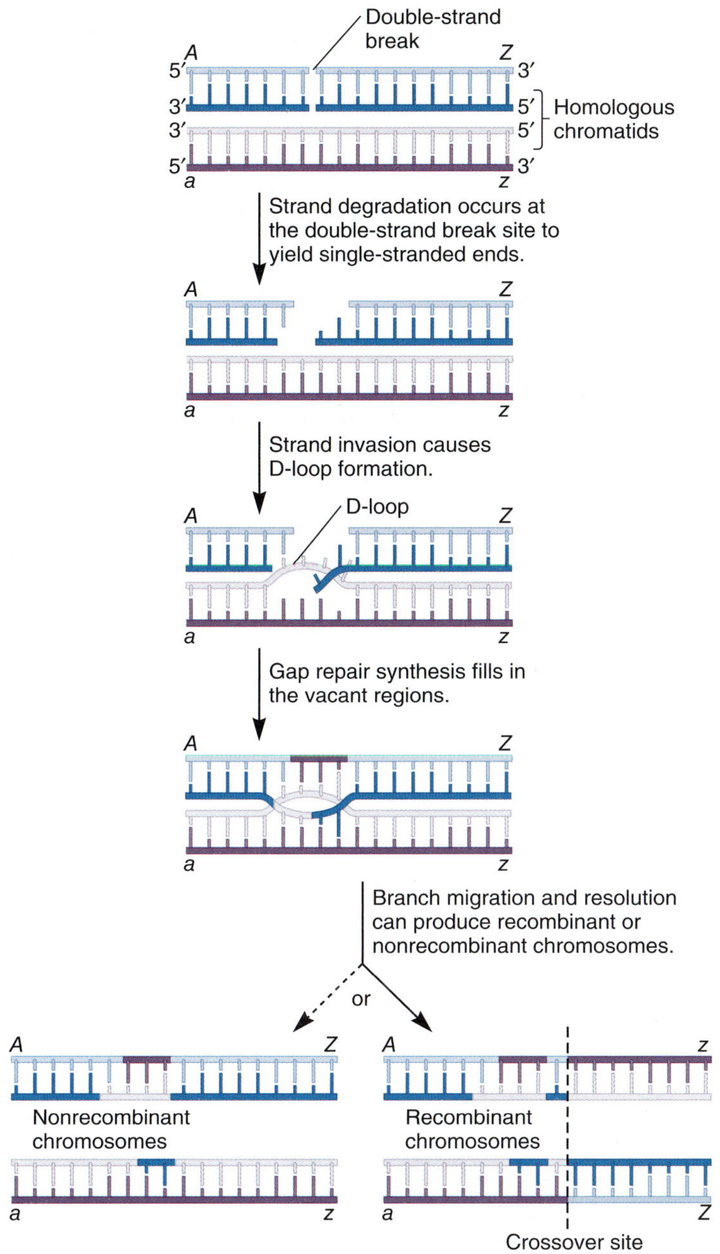

FIGURE 19.3 A simplified version of the double-strand break model. For simplicity, this illustration does not include the formation of heteroduplexes. The dashed arrow indicates that the pathway to the left may be less favored.

CONCEPT CHECK: Describe the structure and location of a D-loop.

that double-strand breaks initiate the homologous recombination that occurs during meiosis. In other words, double-strand breaks create sites where a crossover will occur. In *S. cerevisiae*, the formation of DNA double-strand breaks that initiate meiotic recombination requires at least 10 different proteins. One particular protein, termed Spo11 protein, is thought to be instrumental in cleaving the DNA, thereby creating a double-strand break. However, the roles of the other proteins, and the interactions among them, are not completely understood. Once a double-strand break is made, homologous recombination can then occur according to the model described in Figure 19.3.

TABLE 19.1

***E. coli* Proteins That Play a Role in Homologous Recombination**

Protein	Description
RecBCD	A complex of three proteins that tracks along the DNA and recognizes double-strand breaks. The complex partially degrades the double-stranded regions to generate single-stranded regions that can participate in strand invasion. RecBCD is also involved in loading RecA onto single-stranded DNA. In addition, RecBCD can create single-strand breaks that are used to initiate homologous recombination.
Single-strand binding protein	Coats broken ends of chromosomes and prevents excessive strand degradation.
RecA	Binds to single-stranded DNA and promotes strand invasion, which enables homologous strands to find each other. It also promotes the displacement of the complementary strand to generate a D-loop.
RuvABC	This protein complex binds to Holliday junctions. RuvAB promotes branch migration. RuvC is an endonuclease that cuts the crossed or uncrossed strands to resolve Holliday junctions into separate chromosomes.
RecG	RecG protein can also promote branch migration of Holliday junctions.

Gene Conversion May Result from DNA Mismatch Repair or DNA Gap Repair

As mentioned earlier, homologous recombination can lead to gene conversion, in which one allele is converted to the allele on the homologous chromosome. The original Holliday model was based on this phenomenon.

How can homologous recombination account for gene conversion? Researchers have identified two possible ways this can occur. One mechanism involves DNA mismatch repair, a topic that was described in Chapter 18. To understand how this works, let's take a closer look at the heteroduplexes formed during branch migration of a Holliday junction (see Figure 19.2a). A heteroduplex contains a DNA strand from each of the two original parental chromosomes. The two parental chromosomes may contain an allelic difference within this region. In other words, this short region may contain DNA sequence differences. If this is the case, the heteroduplex formed after branch migration will contain an area of base mismatch. Gene conversion occurs when recombinant chromosomes are repaired and result in two copies of the same allele.

As shown in **Figure 19.4**, mismatch repair of a heteroduplex may result in gene conversion. In this example, the two chromosomes had different alleles due to a single base pair difference in their DNA sequences, as shown at the top of the figure. During recombination, branch migration has occurred across this region, thereby creating two heteroduplexes with base mismatches. As described in Chapter 18, DNA mismatches will be recognized by DNA repair systems and repaired to a double helix that obeys the AT/GC rule. These two mismatches can be repaired in four possible ways. As shown here, two possibilities produce no gene conversion, whereas the other two lead to gene conversion.

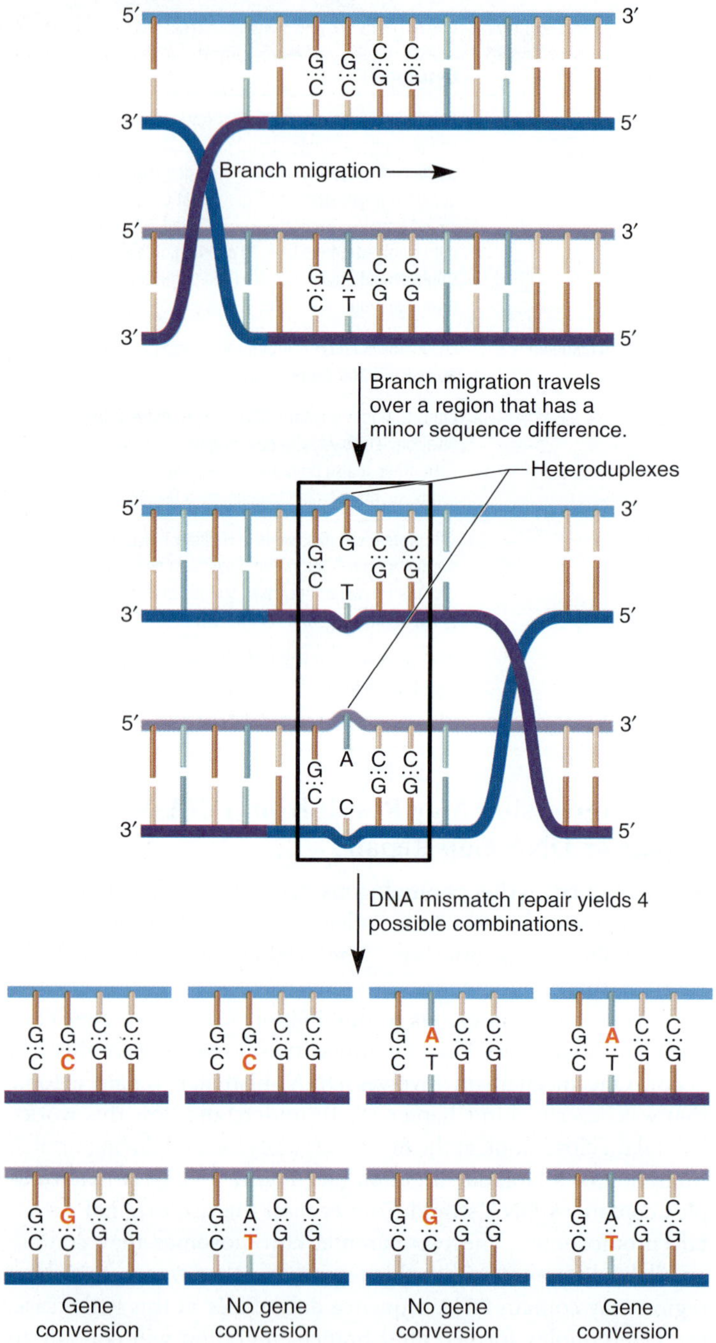

FIGURE 19.4 Gene conversion by DNA mismatch repair. A branch migrates past a homologous region that contains slightly different DNA sequences. This produces two heteroduplexes: DNA double helices with mismatches. The mismatches can be repaired in four possible ways by the mismatch repair system described in Chapter 18. Two of these ways result in gene conversion. The repaired base is shown in red.

A second mechanism for gene conversion occurs via DNA gap repair synthesis. **Figure 19.5** illustrates how gap repair synthesis can lead to gene conversion according to the double-strand break model. The top chromatid, which carries the recessive *b* allele, has suffered a double-strand break in this gene. A gap is created by the digestion of the DNA in the double helix. This digestion eliminates the *b* allele. The two template strands used in gap

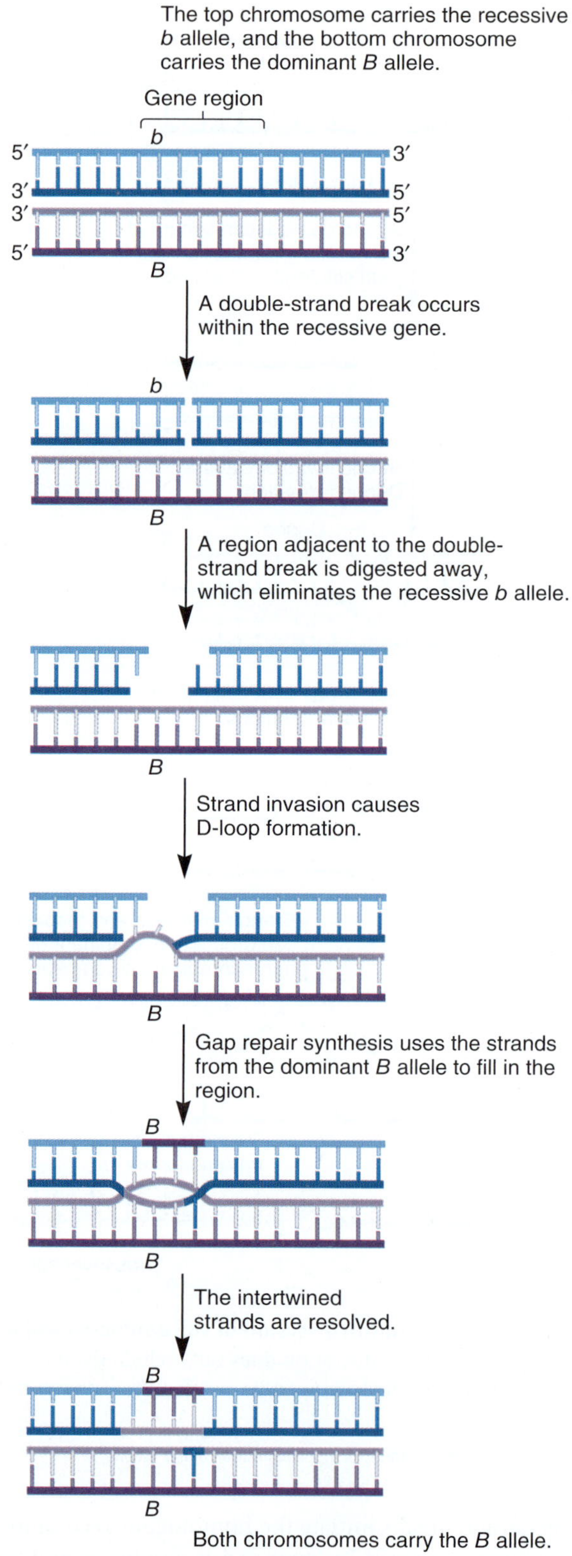

FIGURE 19.5 Gene conversion by gap repair synthesis in the double-strand break model. A gene is found in two alleles, designated *b* and *B*. A double-strand break occurs in the DNA encoding the *b* allele. Both of these DNA strands are digested away, thereby eliminating the *b* allele. A complementary DNA strand encoding the *B* allele migrates to this region and provides the template to synthesize a double-stranded region. Following resolution, both DNA double helices carry the *B* allele.

CONCEPT CHECK: Explain what happened to the *b* allele that allowed gene conversion to occur.

repair synthesis are from the chromatid carrying the dominant *B* allele. This helix carries the dominant *B* allele. Therefore, after gap repair synthesis takes place, the top chromatid carries the *B* allele, as does the bottom one. Gene conversion has changed the recessive *b* allele to a dominant *B* allele.

19.1 COMPREHENSION QUESTIONS

1. Homologous recombination refers to the exchange between DNA segments that are
   a. similar or identical in their DNA sequences.
   b. in close proximity to one another.
   c. broken due to ionizing radiation.
   d. misaligned along a chromosome.
2. During the molecular process of homologous recombination between homologous chromosomes,
   a. a Holliday junction forms.
   b. branch migration occurs.
   c. a heteroduplex region forms.
   d. all of the above occur.
3. A key difference between the original Holliday model and the double-strand break model is the way that
   a. the DNA strands are initially broken.
   b. branch migration occurs.
   c. a heteroduplex is formed.
   d. resolution occurs.
4. Which of the following mechanisms can cause gene conversion?
   a. DNA mismatch repair
   b. DNA gap repair
   c. The resolution of a Holliday junction
   d. Both a and b

19.2 SITE-SPECIFIC RECOMBINATION

Learning Outcomes:

1. Describe how phage λ is inserted into the *E. coli* chromosome.
2. Explain how antibody diversity is generated in vertebrates via site-specific recombination.

Thus far, we have examined recombination between segments of DNA that are similar or identical to each other. Site-specific recombination is another mechanism by which DNA fragments can recombine to make new genetic combinations. The recombination sites are relatively short DNA sequences (a dozen or so base pairs in length) that provide a specific location where recombination will occur. Chromosome breakage and reunion occur at these defined locations to create a recombinant chromosome. For this to occur, these sites are recognized by specialized enzymes that catalyze the breakage and rejoining of DNA fragments within the sites.

Certain viruses use site-specific recombination to insert their viral DNA into their host cell's chromosome. This process has been examined extensively in bacteriophage λ. In addition, mammalian genes that encode antibody polypeptides are rearranged by site-specific recombination that enables the generation of a diverse array of antibodies. In this section, we will consider both mechanisms.

The Integration of Viral Genomes Can Occur by Site-Specific Recombination

The reproductive cycle of some viruses involves the integration of viral DNA into host cell DNA. Certain bacteriophages, for example, can integrate their viral DNA into the bacterial chromosome, creating a prophage. The prophage can exist in a latent, or **lysogenic,** state for many generations. The two cycles of phage λ—lytic and lysogenic—are described in Chapter 17. Phage λ infects *E. coli*. The integration of this phage into the host chromosome is well understood and occurs by a mechanism involving site-specific recombination (**Figure 19.6**).

Integration of the λ DNA into the *E. coli* chromosome requires sequences known as **attachment sites.** As shown at the top of Figure 19.6, a common sequence within the attachment sites is identical in the λ DNA and the *E. coli* chromosome. The attP sequence is in the λ DNA, and the attB sequence is in the *E. coli* chromosome. An enzyme known as integrase is encoded by a gene in the λ DNA. Several integrase proteins recognize the attP and attB sequences and bring them close together. Integrase then makes staggered cuts in both the λ and *E. coli* attachment sites. The strands are then exchanged, and the ends are ligated together. In this way, the phage DNA is integrated into the host cell chromosome. As a prophage, the λ DNA may remain latent for many generations. Certain environmental conditions, such as exposure to UV light, may alter gene expression in a way that causes the excision of the prophage from the host DNA. Excision also requires integrase, which catalyzes the reverse reaction, as well as a second protein known as excisionase.

Antibody Diversity in the Immune System Is Produced by Site-Specific Recombination

Antibodies, or **immunoglobulins (Igs),** are proteins produced by the immune system of vertebrates. Their function is to recognize foreign agents, such as viruses and bacteria, and target them for destruction. A foreign substance that elicits an immune response is called an **antigen.** Antibodies recognize sites within antigens. The recognition between an antibody and antigen is very specific.

B cells, a type of lymphocyte, are responsible for producing antibodies within the immune system of mammals. Antibodies with different polypeptide sequences are generated by an unusual mechanism in which the genes that encode antibody polypeptides are cut and reconnected by site-specific recombination. The random nature of this "cut-and-paste" mechanism provides great

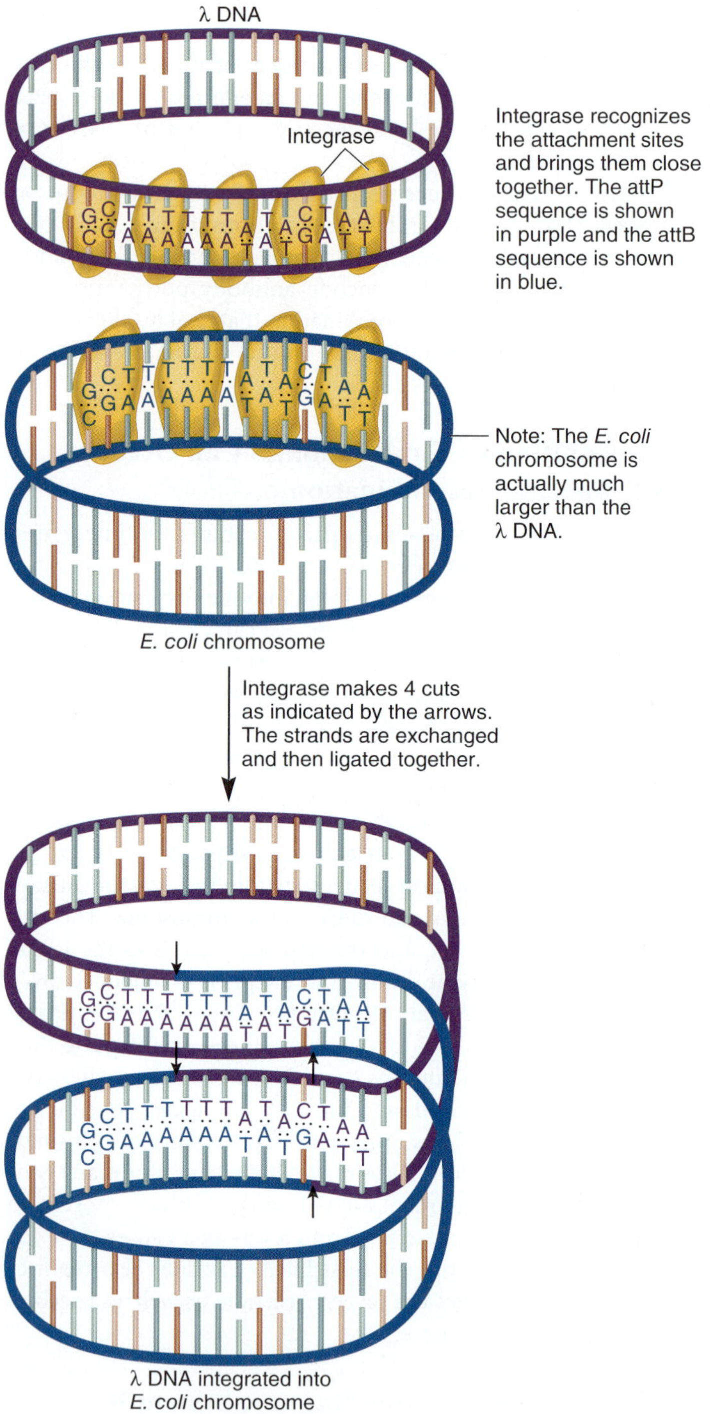

FIGURE 19.6 The integration of λ DNA into the *E. coli* chromosome. The attP sequence in the λ DNA attaches to the attB sequence in the *E. coli* chromosome. As noted here, the attP and attB sequences are identical and thereby provide recognition sites for site-specific recombination.

variation in the amino acid sequences of the antibodies that are produced by different B cells. Because the differences in the amino acid sequences of antibodies enable them to recognize different antigens, site-specific recombination plays an important role in the ability of the immune system to identify an impressive variety of antigens found in invaders, such as viruses and bacteria, and thereby target them for destruction. With this mechanism, only a few antibody genes are needed to produce millions of different antibodies within a large population of B cells.

Antibodies are tetrameric proteins composed of two heavy polypeptide chains and two light chains. One type of light chain is the κ (kappa) light chain, which is found in different classes of antibodies such as immunoglobulin G (IgG). Mammals have multiple copies of the gene that encodes the light and heavy chains of antibodies. As an example, the top of **Figure 19.7** shows the organization of one precursor gene for the κ light chain that is found in the mouse. This type of gene is sometimes called a precursor gene because its structure is subsequently altered by site-specific recombination. As shown on the upper left side of Figure 19.7, the precursor gene has approximately 300 regions known as variable (V) sequences or domains. In addition, the gene contains four different joining (J) sequences and a single constant (C) sequence. Each variable domain or joining domain encodes a different amino acid sequence.

Figure 19.7 shows the site-specific recombination of a light-chain precursor gene in the mouse. During the maturation of B cells, the κ light-chain precursor gene is cut and rejoined so that one variable domain becomes adjacent to a joining domain. At the end of every V domain and the beginning of every J domain is located a **recombination signal sequence** that functions as a recognition site for site-specific recombination between the V and J regions. The recombination event is initiated by two proteins called **RAG1** and **RAG2.** RAG is an abbreviation for recombination-activating gene. These proteins recognize recombination signal sequences and generate two double-strand breaks: one at the end of a V domain and one at the beginning of a J domain. For example, in the recombination event shown in Figure 19.7, RAG1 and RAG2 have made cuts at the end of variable domain number 78 and the beginning of joining domain number 2. The intervening region is lost, and the two ends are then joined to each other. The connection phase of this process is catalyzed by **nonhomologous end-joining (NHEJ) proteins,** which are described in Chapter 18. The fusion process may not be entirely precise, so a few nucleotides can be added or lost at the junction between the variable and joining domains. This imprecision further accentuates the diversity in antibody genes.

Following transcription, the fused VJ region is contained within a pre-mRNA transcript that is then spliced to connect the J and C domains. After this has occurred in a given B cell, it produces only the particular κ light chain encoded by the specific VJ fusion domain and the constant domain.

The heavy-chain polypeptides are produced by a similar recombination mechanism. Though mammals have multiple copies of heavy-chain genes, a typical gene in mice may have about 500 variable domains and four joining domains. In addition, a heavy-chain gene encodes several diversity (D) domains, which are found between the variable and joining domains. The recombination first involves the connection of a D and J domain, followed by the connection of a V and DJ domain. The same proteins that catalyze VJ fusion are involved in the recombination with the heavy-chain gene. Collectively, this process is called **V(D)J recombination.** The D is in parentheses because this type of domain is found only in the heavy-chain genes, not in the light-chain genes.

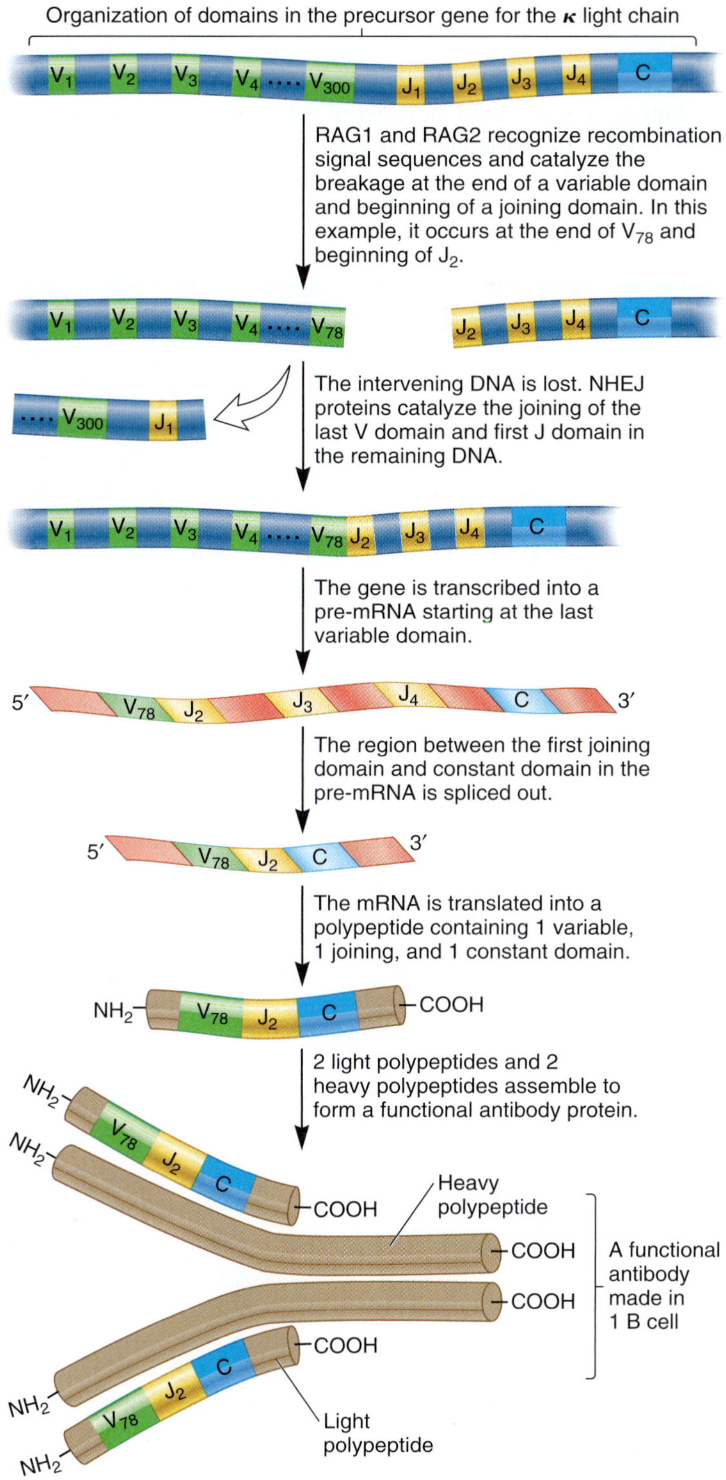

FIGURE 19.7 **Site-specific recombination within the precursor gene that encodes the κ light chain for immunoglobulin G (IgG) proteins.**

The recombination process within immunoglobulin genes produces an enormous diversity in polypeptides. Even though it occurs at specific junctions within the antibody gene, the recombination is fairly random with regard to the particular V, D, or J domains that are joined.

Overall, the possible number of functional antibodies that can be produced by V(D)J recombination is rather staggering. For example, if we assume that any of the 300 different variable sequences can be spliced next to any of the four joining sequences, this results in 1200 possible light-chain combinations. If we also assume that a heavy-chain gene has 500 variable regions, 12 diversity domains, and 4 joining regions, the number of heavy-chain possibilities is $500 \times 12 \times 4 = 24{,}000$. Because any light-chain/heavy-chain combination is possible, this yields $1200 \times 24{,}000 = 28{,}800{,}000$ possible antibody molecules from the random recombination within two precursor genes! The diversity is actually higher because mammals have multiple copies of nonidentical precursor genes for the light and heavy chains of antibodies.

19.2 COMPREHENSION QUESTIONS

1. With regard to phage λ, the role of attachment sites is to
   a. provide a site where the phage can align with the *E. coli* chromosome and become inserted there.
   b. attach the phage to the surface of the bacterium.
   c. promote the movement of the phage from one bacterium to another.
   d. facilitate the attachment of phage DNA to phage coat proteins.
2. During site-specific recombination that occurs for an antibody gene, the protein(s) that catalyze the joining of V and J domains to each other is/are
   a. RAG1 and RAG2.
   b. nonhomologous end-joining (NHEJ) proteins.
   c. RecA.
   d. RecBCD.

19.3 TRANSPOSITION

Learning Outcomes:

1. Analyze the results of McClintock and explain how they revealed the existence of transposable elements.
2. Describe the organization of sequences within different types of transposable elements.
3. Explain how transposons and retroelements move to new locations in a genome.
4. Discuss the effects of transposable elements on gene function.
5. Describe how transposable elements are used as experimental tools.

The last form of recombination we will consider is transposition. In some ways, transposition resembles the site-specific recombination we examined for phage λ. In that case, a segment of λ DNA was able to integrate itself into the *E. coli* chromosome. Transposition also involves the integration of small segments of DNA into the chromosome. Transposition, though, can occur at many different locations within the genome. The DNA segments that transpose themselves are known as **transposable elements (TEs).** TEs have sometimes been referred to as "jumping genes" because they are inherently mobile.

Transposable elements were first identified by Barbara McClintock in the early 1950s from her classic studies with corn plants. Since that time, geneticists have discovered many different types of TEs in organisms as diverse as bacteria, fungi, plants, and animals. The advent of molecular technology has allowed scientists to understand more about the characteristics of TEs that enable them to be mobile. In this section, we examine the characteristics of TEs and explore the mechanisms that explain how they move. We will also discuss the biological significance of TEs and their uses as experimental tools.

EXPERIMENT 19A

McClintock Found That Chromosomes of Corn Plants Contain Loci That Can Move

Barbara McClintock began her scientific career as a student at Cornell University. Her interests quickly became focused on the structure and function of the chromosomes of corn plants, an interest that continued for the rest of her life. She spent countless hours examining corn chromosomes under the microscope. She was technically gifted and had a theoretical mind that could propose ideas that conflicted with conventional wisdom.

During her long career as a scientist, McClintock identified many unusual features of corn chromosomes. She noticed that chromosome 9 in one strain of corn had the strange characteristic of tending to break at a fairly high rate at the same site. McClintock termed this a **mutable site,** or locus. This observation initiated a 6-year study concerned with highly unstable chromosomal locations. In 1951, at the end of her study, McClintock proposed that these sites are actually locations where TEs have been inserted into the chromosomes. At the time of McClintock's studies, such an idea was entirely unorthodox.

McClintock focused her efforts on the relationship between a mutable locus and its phenotypic effects on corn kernels. Chromosome 9 with a mutable locus also carried several genes that affected the phenotype of corn kernels. Each gene existed in (at least) one dominant and one recessive allele. The mutable locus was termed *Ds* (for dissociation), because the locus was known to frequently cause chromosomal breaks. In the chromosome shown here, the *Ds* locus is located next to several genes affecting kernel traits.

Three genes are located in this region that exist as two or more alleles:

1. *C* is an allele for normal kernel color (dark red), *c* is a recessive allele of the same gene that causes a colorless kernel, and C^I is a third allele of this gene that is dominant to the *C* allele and causes a colorless kernel.
2. *Sh* is an allele that produces normal endosperm, whereas *sh* is a recessive allele that causes shrunken endosperm. (Note: The endosperm is the storage material in the kernel that is used by the plant embryo to provide energy for growth.)
3. *Wx* is the allele that produces normal starch in the endosperm, and *wx* is a recessive allele that produces endosperm with a stiff, waxy texture.

During her intensive work, in which she studied corn chromosomes under the microscope, McClintock identified strains of corn in which the *Ds* locus was found in different places within the corn genome. She could determine the location of *Ds* because the movement of *Ds* out of its original site to a new site occasionally caused chromosome breakage at the original site (**Figure 19.8**). Keep in mind that the endosperm of a kernel is triploid because it is derived from the fusion of two maternal haploid nuclei and one paternal haploid nucleus (refer back to Chapter 2, Figure 2.2c). The kernel shown in Figure 19.8 was produced by a cross in which the pollen carried the top chromosome—C^I *Sh Wx Ds*—whereas the two maternal chromosomes are *C sh wx*. This kernel is expected to be colorless, because the C^I allele is dominant and causes a colorless phenotype. However, as the kernel grows by cell division, the movement of the *Ds* locus out of its original site may occasionally cause a chromosome to break, and the distal part of this chromosome is lost. This chromosome breakage may happen in several cells, which continue to divide and grow as the kernel becomes larger. This process produces a sectoring phenotype—patches of cells occur in the kernel that are red, shrunken, and waxy.

By analyzing many kernels, McClintock was sometimes able to identify the new site where *Ds* had moved. In one case, she determined that *Ds* had moved out of its original location and inserted into the middle of the *C* allele.

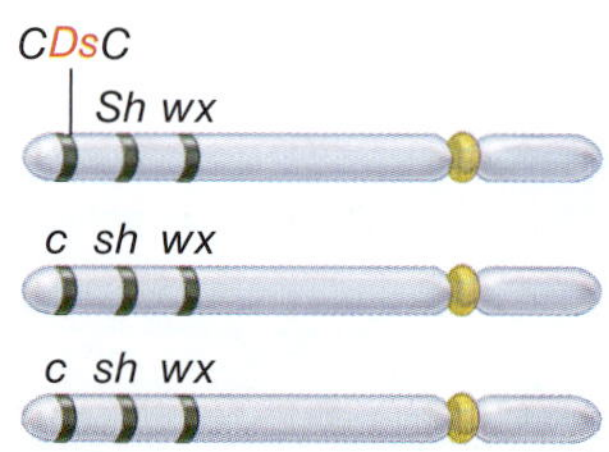

This strain had the opposite phenotype—a colorless kernel with red patches. How did McClintock explain this phenotype? She hypothesized the following:

1. The colorless background was due to a transposition of *Ds* into the *C* allele, which would inactivate it.
2. The red sectoring occurred because *Ds* occasionally transposed out of the *C* allele during kernel growth. This event would reconnect the *C* allele, resulting in a red patch. In this case, the formation of patches, or sectoring, was due to the movement of *Ds* out of its location within the *C* gene and the rejoining of the two ends, not simply due to chromosome breakage.

To test her hypothesis, McClintock carried out the experiment shown in **Figure 19.9**. As seen in steps 1 and 2, her cross produced kernels with a colorless background with red sectoring. The red sectors occurred because the *Ds* element left the *C* gene

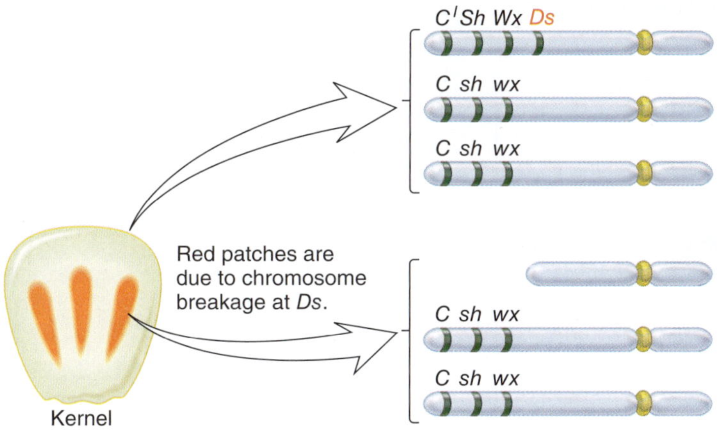

FIGURE 19.8 The sectoring trait in corn kernels.

Genes→Traits This kernel is expected to be colorless, because the *C'* allele is dominant. On occasion, though, the movement of *Ds* may cause a chromosome to break, thereby losing the *C'*, *Sh*, and *Wx* alleles. Such a cell continues to divide and produces a patch of daughter cells that are red, shrunken, and waxy. Therefore, this sectoring trait arises from the loss of genes that occurs when the movement of *Ds* causes chromosome breakage. Note: The movement of *Ds* does not usually cause chromosome breakage. In most cases, the movement of *Ds* out of a site is followed by nonhomologous end joining (described in Chapter 18) in which the chromosome pieces are connected together.

in a few cells during the growth of the kernel. However, on rare occasions, a kernel was completely red. This suggested that the *Ds* element had transposed out of the *C* gene prior to kernel growth, probably during pollen formation. Therefore, all of the cells in this kernel would have a red phenotype. (Note: Every kernel in an ear of corn is equivalent to a distinct offspring, because each is produced from the union of haploid cells from the male and female gametophytes.)

THE HYPOTHESIS

The transposition of the *Ds* element into the normal *C* gene prevents kernel pigmentation. When the *Ds* element transposes back out of the *C* gene, the normal *C* allele is restored, which gives a red phenotype.

TESTING THE HYPOTHESIS

FIGURE 19.9 Evidence for transposable elements in corn.

Starting material: The plant contributing the female gamete was homozygous for a chromosome carrying *c sh Wx*. The male pollen of the corn plant was homozygous for a chromosome in which the *Ds* element had moved into the *C* gene: *CDsC Sh wx*.

1. Cross a plant that is homozygous for *c*, *sh*, and *Wx* to a plant that is homozygous for *CDsC*, *Sh*, *wx*.

2. In this cross, most kernels will have a colorless background with red sectoring. On rare occasions, *Ds* may have transposed during male gametophyte formation, producing pollen that gives rise to a completely red kernel.

3. Identify rare kernels that are completely red.

4. Germinate the solid red kernels.

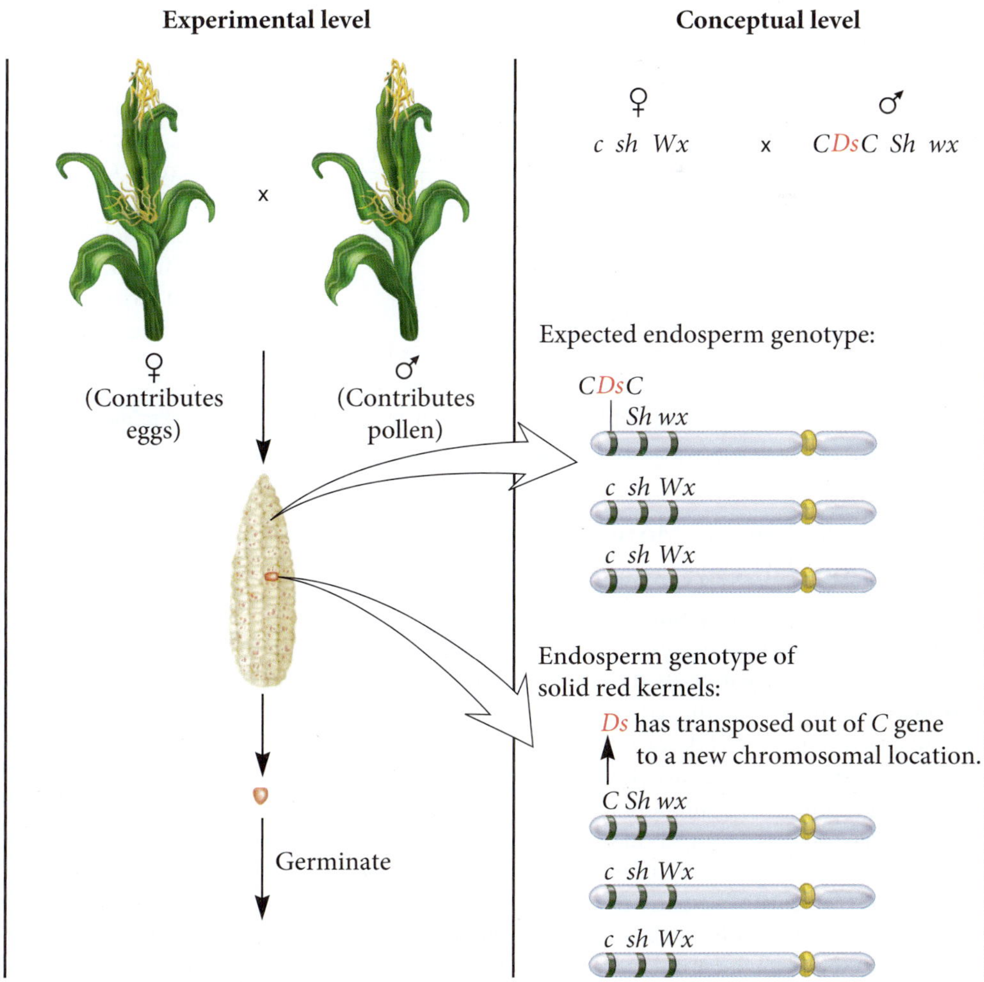

5. Conduct crosses to determine the new chromosomal location of *Ds* in the plants derived from the solid red kernels. Note: This can be done using strains of plants with many chromosomal markers and following the strategy that was described at the beginning of this experiment. When observing the results of these crosses, determine whether the *C* gene is still mutable. (Is red sectoring occurring?)

Make crosses and observe kernels.

THE DATA

Strain	*Kernels Produced*	*Location of Ds*	*Mutability?*
Offspring from the parental cross (see step 2)	Usually a colorless background with red sectoring	*Ds* was within the *C* gene.	Yes, red sectoring occurred.
Offspring that grew from red kernels (see step 3)	All red kernels	*Ds* had moved out the *C* gene to a new chromosomal location.	No, the *C* gene was stable; no sectoring was observed.

Data summarized from: McClintock, B. (1951) Chromosome organization and genic expression. *Cold Spring Harb Symp Quant Biol* 16:13–47.

INTERPRETING THE DATA

McClintock's hypothesis made two predictions. In a red kernel, *Ds* should have moved to a new location. Second, the restored *C* allele should no longer be mutable because the *Ds* had been removed. By conducting the appropriate crosses, McClintock found that in the offspring that grew from a solid red kernel, the *Ds* locus had moved out of the *C* gene to a new location (see The Data of Figure 19.9). In addition, the restored *C* gene behaved normally. In other words, the *C* gene was no longer highly mutable; the kernels did not show a sectoring phenotype. Taken together, the results are consistent with the hypothesis that the *Ds* locus can move around the corn genome by transposition.

When McClintock published these results in 1951, they were met with great skepticism. Some geneticists of that time were unable to accept the idea that the genetic material was susceptible to frequent rearrangement. Instead, they believed that the genetic material was always very stable and permanent in its structure. Over the next several decades, however, the scientific community came to realize that TEs are a widespread phenomenon. Much like Gregor Mendel and Charles Darwin, Barbara McClintock was clearly ahead of her time. She was awarded the Nobel Prize in physiology or medicine in 1983, more than 30 years after her original discovery.

A self-help quiz involving this experiment can be found at www.mhhe.com/brookergenetics5e.

Transposable Elements Move by Different Transposition Pathways

Since Barbara McClintock's pioneering studies, many different TEs have been found in bacteria, fungi, plants, and animals. Different types of transposition pathways have been identified. In **simple transposition,** the TE is removed from its original site and transferred to a new target site (**Figure 19.10a**). This mechanism is also called a cut-and-paste mechanism because the element is cut out of its original site and pasted into a new one. Transposable elements that move via simple transposition are widely found in bacterial and eukaryotic species. Such TEs are also called **transposons.**

Another category of elements moves via an RNA intermediate. This form of transposition, termed **retrotransposition,** is found in eukaryotic species, where it is very common (**Figure 19.10b**). Transposable elements that move via retrotransposition are known as **retrotransposons** or **retroelements.** In retrotransposition, the element is transcribed into RNA. An enzyme called reverse transcriptase uses the RNA as a template to synthesize a DNA molecule that is integrated into a new region of the genome. Retrotransposons increase in number during retrotransposition.

Each Type of Transposable Element Has a Characteristic Pattern of DNA Sequences

Research on TEs from many species has established that the DNA sequences within them are organized in several different ways. **Figure 19.11** describes a few of those ways, although many variations are possible. All TEs are flanked by **direct repeats (DRs),** also called **target-site duplications,** which are identical

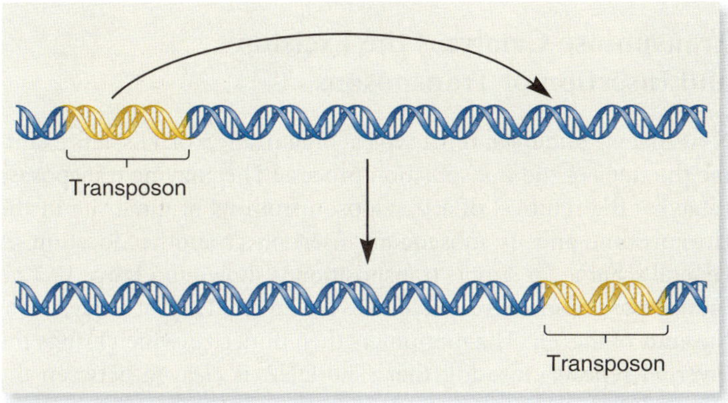

(a) Simple transposition

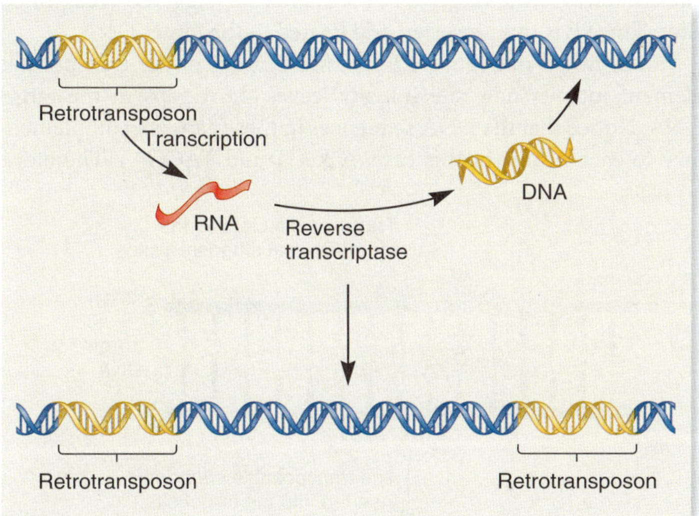

(b) Retrotransposition

FIGURE 19.10 Different mechanisms of transposition.

CONCEPT CHECK: Which of these mechanisms causes the TE to increase in number?

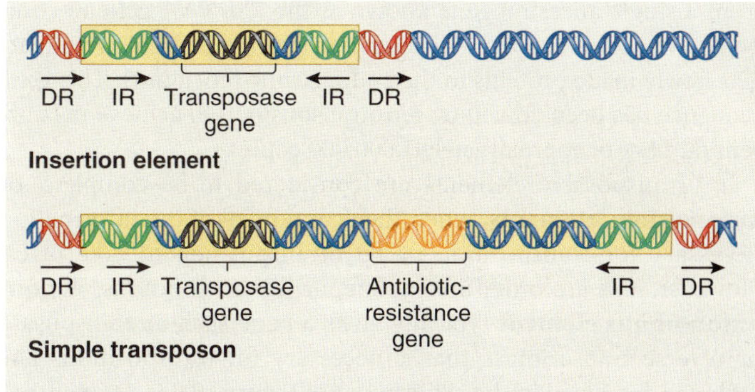

(a) Elements that move by simple transposition

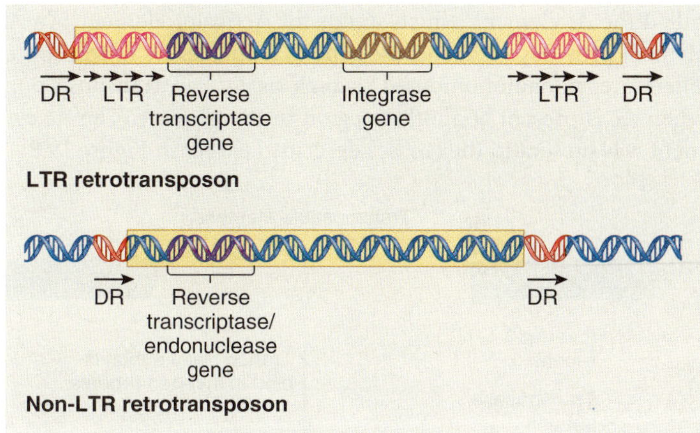

(b) Elements that move by retrotransposition (via an RNA intermediate)

FIGURE 19.11 Common organizations of transposable elements. Direct repeats (DRs) are found within the host DNA. Inverted repeats (IRs) are at the ends of some transposable elements. Long terminal repeats (LTRs) are regions containing a large number of tandem repeats.

nucleotide sequences that are oriented in the same direction and repeated. Direct repeats are adjacent to both ends of the element. The simplest TE is known as an **insertion element (IS element).** As shown in Figure 19.11a, an IS element has two important characteristics. First, both ends of the element contain **inverted repeats (IRs).** Inverted repeats are DNA sequences that are identical (or very similar) but run in opposite directions, such as the following:

```
5′-CTGACTCTT-3′     and     5′-AAGAGTCAG-3′
3′-GACTGAGAA-5′             3′-TTCTCAGTC-5′
```

Depending on the particular element, the lengths of inverted repeats range from 9–40 bp in length. In addition, IS elements may contain a central region that encodes the enzyme **transposase,** which catalyzes the transposition event. By comparison, a **simple transposon** carries one or more genes that are not required for a transposon to move. For example, the simple transposon shown in Figure 19.11a carries an antibiotic resistance gene.

The organization of retrotransposons varies greatly. They are categorized based on their evolutionary relationship to retroviruses. As described in Chapter 17, retroviruses are RNA viruses that make a DNA copy that integrates into the host's genome. The **LTR retrotransposons** are evolutionarily related to known retroviruses. These TEs have retained the ability to move around the genome, though, in most cases, they do not produce mature viral particles. LTR retrotransposons are so named because they contain **long terminal repeats (LTRs)** at both ends of the element (Figure 19.11b). The LTRs are typically a few hundred base pairs in length. Like their viral counterparts, LTR retrotransposons encode virally related proteins, such as reverse transcriptase and integrase, that are needed for the retrotransposition process.

By comparison, **non-LTR retrotransposons** appear less like retroviruses in their sequence. They may contain a gene that encodes a protein that has both reverse transcriptase and endonuclease function (see Figure 19.11b). As discussed later, these functions are needed for retrotransposition. Some non-LTR retrotransposons are evolutionarily derived from normal eukaryotic genes. For example, the *Alu* family of repetitive sequences found in humans is derived from a single ancestral gene known as the *7SL RNA* gene (a component of the complex called signal recognition particle, which targets newly made proteins to the endoplasmic reticulum). This gene sequence has been copied by retrotransposition to achieve the current number of approximately 1,000,000 copies.

Transposable elements are considered to be complete, or **autonomous elements,** when they contain all of the information necessary for transposition or retrotransposition to take place. However, TEs are often incomplete, or nonautonomous. A **nonautonomous element** typically lacks a gene such as transposase or reverse transcriptase that is necessary for transposition. The *Ds* locus described in the experiment of Figure 19.9 is a nonautonomous element, because it lacks a transposase gene. An element that is similar to *Ds* but contains a functional transposase gene is called the *Ac* element, which stands for Activator element. An *Ac* element provides a transposase gene that enables *Ds* to transpose. Therefore, nonautonomous TEs such as *Ds* can transpose only when *Ac* is present at another region in the genome. The *Ac* element was present in the crosses described earlier in Figure 19.9.

Transposase Catalyzes the Excision and Insertion of Transposons

Now that we understand the typical organization of TEs, let's examine the steps of the transposition process. The enzyme transposase catalyzes the removal of a transposon from its original site in the chromosome and its subsequent insertion at another location. A general scheme for simple transposition is shown in **Figure 19.12a**. Transposase monomers first bind to the inverted repeat sequences at the ends of the TE. The monomers then dimerize, which brings the inverted repeats close together. The DNA is cleaved between the inverted and direct repeats, excising the TE from its original site within the chromosome. Transposase carries the TE to a new site and cleaves the target DNA sequence at staggered recognition sites. The TE is then inserted and ligated to the target DNA.

As noted in **Figure 19.12b**, the ligation of the transposable element into its new site initially leaves short gaps in the target DNA. Notice that the DNA sequences in these gaps are complementary to each other (in this case, ATGCT and TACGA). Therefore,

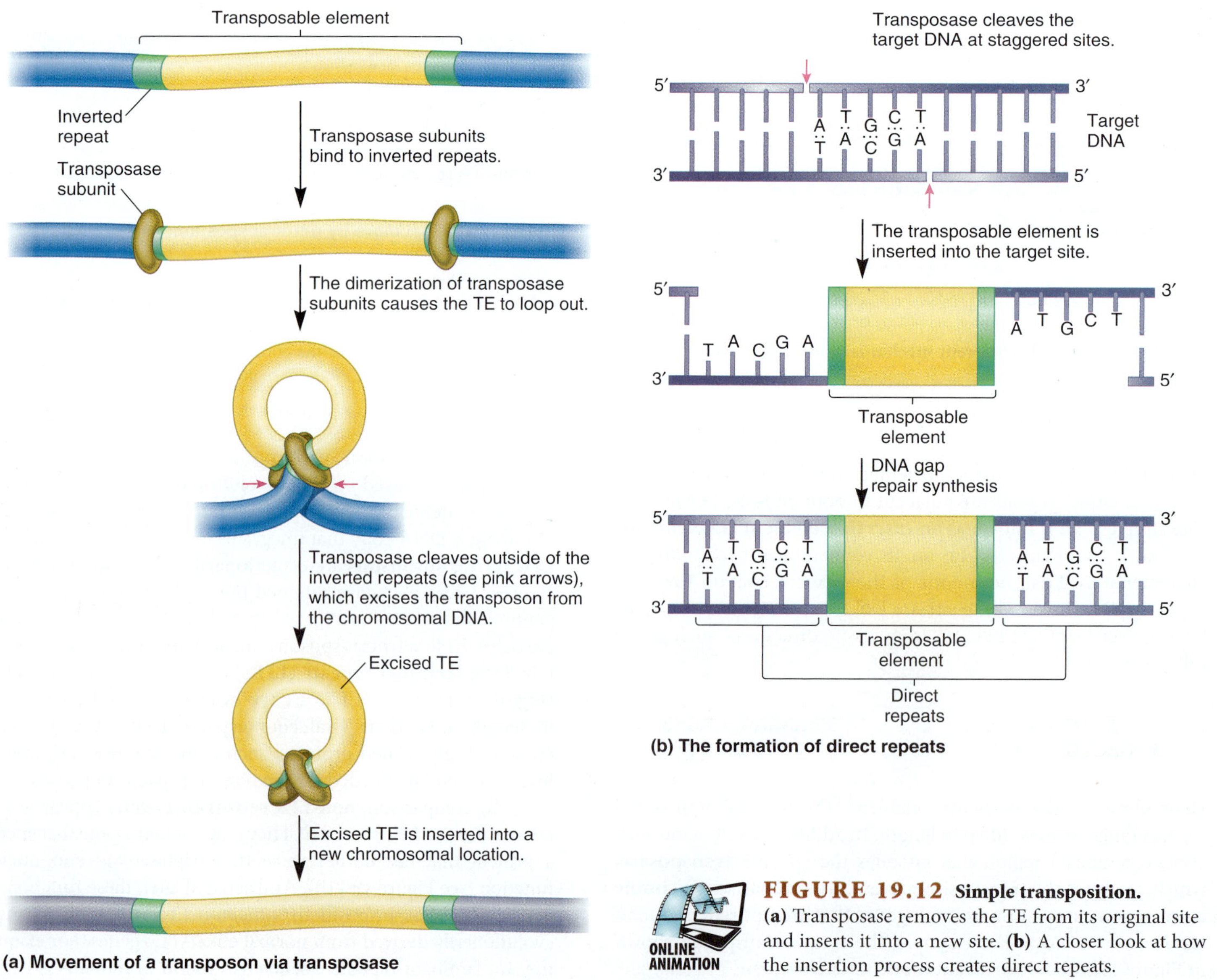

ONLINE ANIMATION

FIGURE 19.12 Simple transposition.
(a) Transposase removes the TE from its original site and inserts it into a new site. **(b)** A closer look at how the insertion process creates direct repeats.

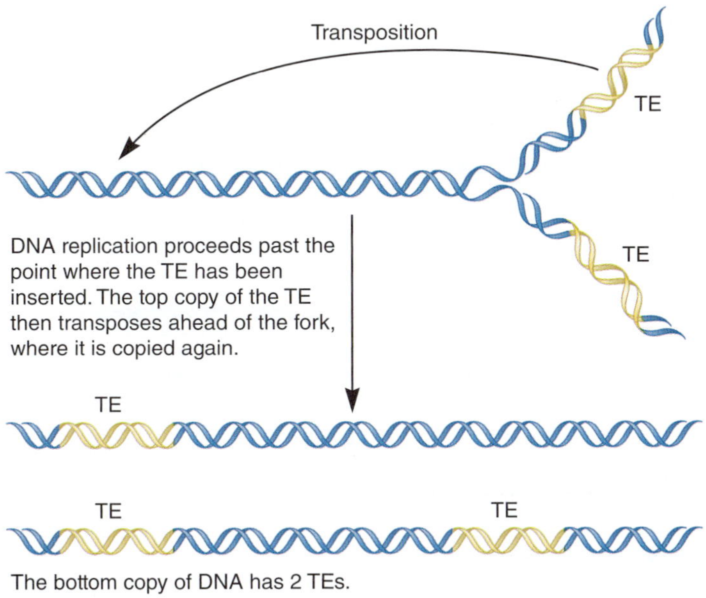

FIGURE 19.13 Increase in transposable element (TE) copy number via simple transposition. In this example, a TE that has already been replicated transposes to a new site that has not yet replicated. Following the completion of DNA replication, the TE has increased in number.

when they are filled in by DNA gap repair synthesis, this produces direct repeats that flank both ends of the TE. These direct repeats are common features found adjacent to all TEs (see Figure 19.11).

Although the transposition process depicted in Figure 19.12 does not directly alter the number of TEs, simple transposition is known to increase their number in genomes, in some cases to fairly high levels. How can this happen? The answer is that transposition often occurs around the time of DNA replication (**Figure 19.13**). After a replication fork has passed a region containing a TE, two TEs will be found behind the fork—one in each of the replicated regions. One of these TEs could then transpose from its original location into a region ahead of the replication fork. After the replication fork has passed this second region and DNA replication is completed, two TEs will be found in one of the chromosomes and one TE in the other chromosome. In this way, simple transposition can lead to an increase in TEs. We will discuss the biological significance of transposon proliferation later in this chapter.

Retrotransposons Use Reverse Transcriptase for Retrotransposition

Thus far, we have considered how DNA elements—transposons—can move throughout the genome. By comparison, retrotransposons use an RNA intermediate in their transposition mechanism. Let's begin with LTR retrotransposons. As shown in **Figure 19.14**, the movement of LTR retrotransposons requires two key enzymes: reverse transcriptase and integrase. In this example, the cell already contains a retrotransposon known as *Ty* within its genome. This retrotransposon is transcribed into RNA. In a series of steps, **reverse transcriptase** uses this RNA as a template to synthesize a double-stranded DNA molecule. The long terminal repeats (LTRs) at the ends of the double-stranded DNA are then recognized by **integrase,** which catalyzes the insertion of the DNA into the target chromosomal DNA. The integration of retrotransposons can

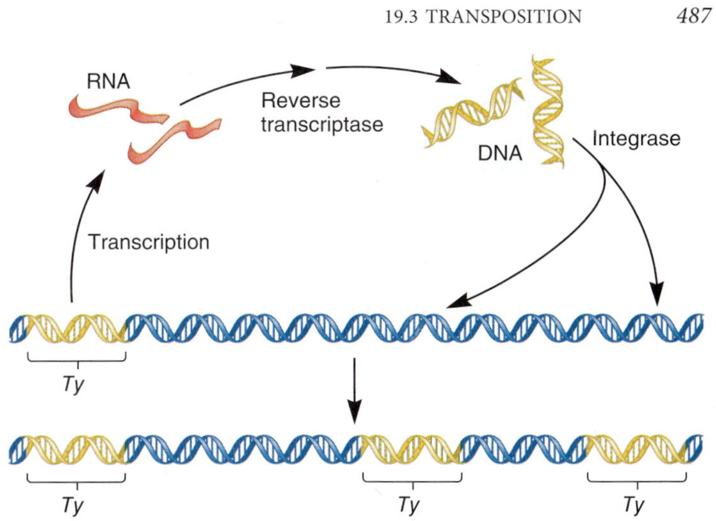

FIGURE 19.14 Retrotransposition of an LTR retrotransposon.

CONCEPT CHECK: What is the function of reverse transcriptase?

occur at many locations within the genome. Furthermore, because a single retrotransposon can be copied into many RNA transcripts, retrotransposons may accumulate rapidly within a genome.

The currently accepted model for the replication and integration of non-LTR retrotransposons is called **target-site primed reverse transcription (TPRT).** As shown in **Figure 19.15**, the retrotransposon is first copied into RNA with a polyA tail. The target DNA site is recognized by an endonuclease, which may be encoded by the retrotransposon. This endonuclease recognizes a consensus sequence of 5′-TTTTA-3′, and initially cuts just one of the DNA strands. The 3′-polyA tail of the retrotransposon RNA binds to this nicked site due to AT base pairing. Reverse transcriptase then uses the target DNA as a primer and makes a DNA copy of the RNA. This is the process of target-site-primed reverse transcription. To be fully integrated into the target DNA, the endonuclease makes a second cut in the other DNA strand usually about 7–20 nucleotides away from the first cut. The retrotransposon DNA is then ligated into the target site within a chromosome, perhaps by nonhomologous end joining. The mechanism for synthesis of the other DNA strand of the retrotransposon is not completely understood. It could occur via DNA gap repair described in Chapter 18.

Transposable Elements May Have Important Influences on Mutation and Evolution

Over the past few decades, researchers have found that TEs probably occur in the genomes of all species. **Table 19.2** describes a few TEs that have been studied in great detail. As discussed in Chapter 10, the genomes of eukaryotic species typically contain moderately and highly repetitive sequences. In some cases, these repetitive sequences are due to the proliferation of TEs. In mammals, for example, **LINEs** are long interspersed elements that are usually 1000–10,000 bp in length and found in a few thousand to several hundred thousand copies. In humans, a particular family of related LINEs called LINE-1, or L1, is found in about 500,000 copies and represents about 17% of the total human DNA! By comparison, **SINEs** are short interspersed elements that are less than

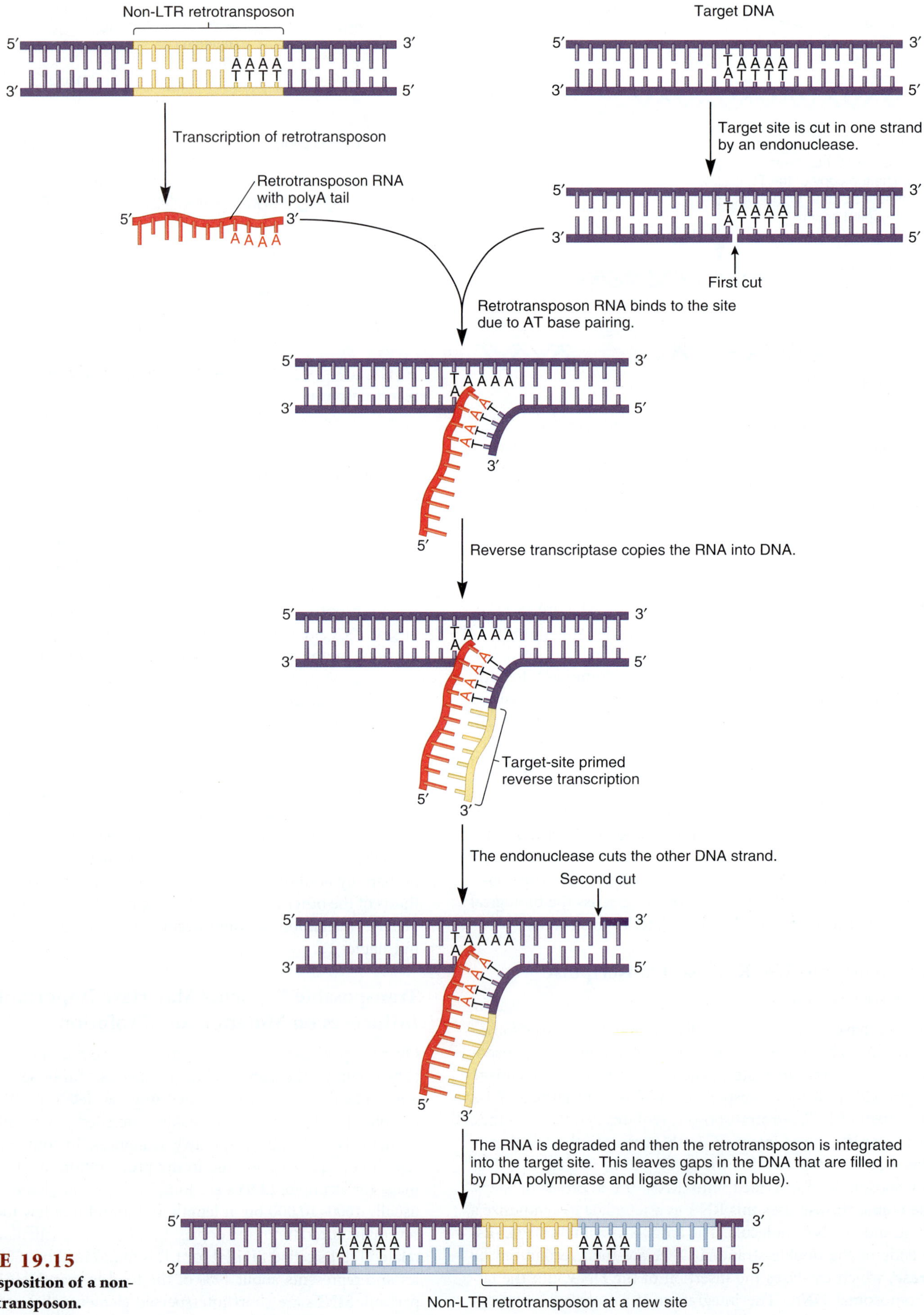

FIGURE 19.15
Retrotransposition of a non-LTR retrotransposon.

TABLE 19.2
Examples of Transposable Elements

Element	Type	Approximate Length (bp)	Description
Bacterial			
IS1	Transposon	768	An insertion element that is commonly found in five to eight copies in *E. coli.*
Tn10	Transposon	9300	One of many different bacterial transposons that carries antibiotic resistance.
Tn951	Transposon	16,600	A transposon that provides bacteria with genes that allow them to metabolize lactose.
Yeast			
Ty element	Retrotransposon	6300	Found in *S. cerevisiae* at about 35 copies per genome.
Fruit Fly			
P element	Transposon	500–3000	A transposon that may be found in 30–50 copies in P strains of *Drosophila.* It is absent from M strains.
Copia-like element	Retrotransposon	5000–8000	A family of *copia*-like elements found in *Drosophila,* which vary slightly in their lengths and sequences. Typically, each family member is found at about 5–100 copies per genome.
Humans			
Alu sequence	Retrotransposon	300	A SINE that is abundantly interspersed throughout the human genome.
L1	Retrotransposon	6500	A LINE found in about 500,000 copies in the human genome.
Plants			
Ac/Ds	Transposon	4500	*Ac* is an autonomous transposon found in corn and other plant species. It carries a transposase gene. *Ds* is a nonautonomous version that lacks a functional transposase gene.
Opie	Retrotransposon	9000	A retrotransposon found in plants that is related to the *copia*-like elements found in animals.

500 bp in length. A specific example of a SINE is the *Alu* sequence, present in about 1 million copies in the human genome. About 10% of the human genome is composed of this particular TE.

LINEs and SINEs continue to proliferate in the human genome but at a fairly low rate. In about 1 live birth in 100, an *Alu* or an L1 (or both) sequence has been inserted into a new site in the human genome. On rare occasions, a new insertion can disrupt a gene and cause phenotypic abnormalities. For example, new insertions of L1 or *Alu* sequences into particular genes have been shown, on occasion, to be associated with diseases such as hemophilia, muscular dystrophy, and breast and colon cancer.

The relative abundance of TEs varies widely among different species. As shown in **Table 19.3**, TEs can be quite prevalent in amphibians, mammals, and flowering plants, but tend to be less abundant in simpler organisms such as bacteria and yeast. The biological significance of TEs in the evolution of prokaryotic and eukaryotic species remains a matter of debate. According to the **selfish DNA hypothesis,** TEs exist because they have characteristics that allow them to multiply within the chromosomal DNA of living cells. In other words, they resemble parasites in the sense that they inhabit a cell without offering any selective advantage to the organism. They can proliferate as long as they do not harm the organism to the extent that they significantly disrupt survival.

TABLE 19.3
Abundance of TEs in the Genomes of Selected Species

Species	Percentage of the Total Genome Composed of Transposable Elements*
Frog (*Xenopus laevis*)	77
Corn (*Zea mays*)	60
Human (*Homo sapiens*)	45
Mouse (*Mus musculus*)	40
Fruit fly (*Drosophila melanogaster*)	20
Nematode (*Caenorhabditis elegans*)	12
Yeast (*Saccharomyces cerevisiae*)	4
Bacterium (*Escherichia coli*)	0.3

*In some cases, the abundance of TEs may vary somewhat among different strains of the same species. The values reported here are typical values.

Alternatively, other geneticists have argued that transpositional events are often deleterious. Therefore, TEs would be eliminated from the genome by natural selection if they did not also offer a compensating advantage. Several potential advantages

TABLE 19.4

Possible Consequences of Transposition

Consequence	Cause
Chromosome Structure	
Chromosome breakage	Excision of a TE.
Chromosomal rearrangements	Homologous recombination between TEs located at different positions in the genome.
Gene Expression	
Mutation	Incorrect excision of TEs.
Gene inactivation	Insertion of a TE into a gene.
Alteration in gene regulation	Transposition of a gene next to regulatory sequences or the transposition of regulatory sequences next to a gene.
Alteration in the exon content of a gene	Insertion of exons into the coding sequence of a gene via TEs. This phenomenon is called exon shuffling.

have been suggested. For example, TEs may cause greater genetic variability by promoting recombination. In addition, bacterial TEs often carry an antibiotic resistance gene that provides the organism with a survival advantage. Researchers have also suggested that transposition may cause the insertion of exons from one gene into another gene, thereby producing a new gene with novel function(s). This phenomenon, called **exon shuffling,** is described in Chapter 26. Also, as discussed in Chapter 15, the ENCODE Project has revealed that much of the noncoding DNA in humans may play a role in gene regulation, which could include segments of DNA that are derived from transposable elements.

This controversy remains unresolved, but it is clear that TEs can rapidly enter the genome of an organism and proliferate quickly. In *Drosophila melanogaster*, for example, a TE known as a P element was probably introduced into this species in the 1950s. Laboratory stocks of *D. melanogaster* collected prior to this time do not contain P elements. Remarkably, in the last 60 years, the P element has expanded throughout *D. melanogaster* populations worldwide. The only strains without the P element are laboratory strains collected prior to the 1950s. This observation underscores the surprising ability of TEs to infiltrate a population of organisms.

Transposable elements have a variety of effects on chromosome structure and gene expression (**Table 19.4**). Many of these outcomes are likely to be harmful. Usually, transposition is a relatively rare event that occurs only in a few individuals under certain conditions, such as exposure to radiation. Strains that harbor transposons often produce inhibitors that greatly decrease their movement. When transposition occurs at a high rate, it is likely to be detrimental. For example, in *D. melanogaster*, if M strain females that lack P elements and lack an inibitor of transposase are crossed with males that contain numerous P elements (P strain males), the egg cells allow the P elements to transpose at a high rate. The resulting hybrid offspring exhibit a variety of abnormalities, which include a high rate of sterility, mutation, and chromosome breakage. This deleterious outcome, which is called **hybrid dysgenesis,** occurs because the P elements were able to insert into a variety of locations in the genome.

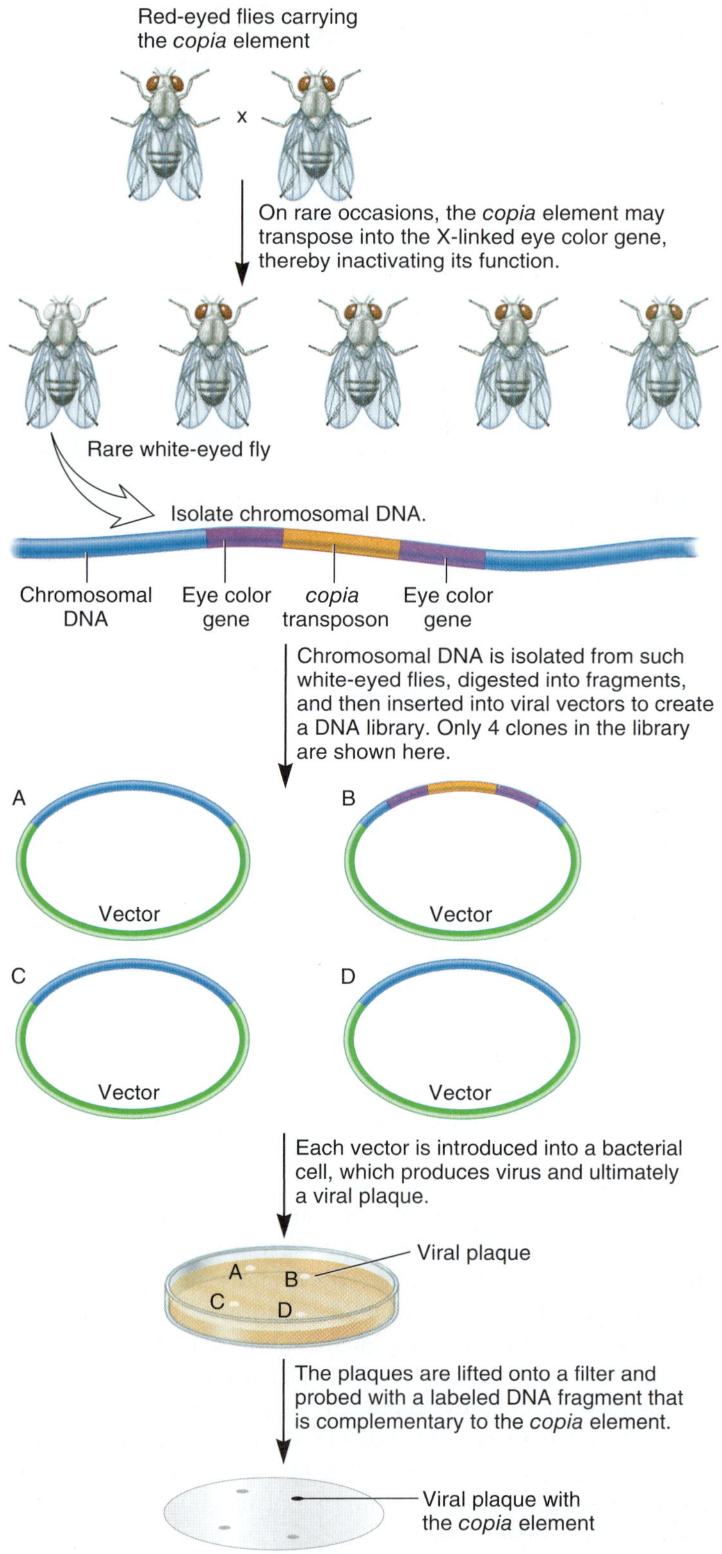

FIGURE 19.16 The procedure of transposon tagging.

Genes→Traits A white-eyed fly may occur due to the insertion of a transposable element into a gene that confers red eye color. The wild-type eye color gene encodes a protein that is necessary for red pigment production. When a TE inserts into this gene, it disrupts the coding sequence, thereby causing the gene to produce a nonfunctional protein. Therefore, no red pigment can be made, and a white-eye phenotype results. In many cases, transposons affect the phenotypes of organisms by inactivating individual genes.

Transposons Have Become Important Tools in Molecular Biology

The unique and unusual features of transposons have made them an important experimental tool in molecular biology. For researchers, the introduction of a transposon into a cell is a convenient way to abolish the expression of a particular gene. If a transposon "hops" into a gene, it is likely to inactivate the gene's function. This phenomenon is called **insertional mutagenesis.**

Aside from inactivating gene function, insertional mutagenesis is used to clone a particular gene in an approach known as **transposon tagging.** An early example of transposon tagging involved an X-linked gene in *Drosophila* that affects eye color. This X-linked gene can exist in the wild-type (red) allele and a loss-of-function allele that causes a white-eye phenotype. In 1981, Paul Bingham, in collaboration with Robert Levis and Gerald Rubin, used transposon tagging to clone this gene (**Figure 19.16**). Prior to their cloning work, a wild-type strain of *Drosophila* had been characterized that carried a transposable element called *copia*. From this red-eyed strain, a white-eyed strain was obtained in which the *copia* element had transposed into a region on the X chromosome that corresponded to where the eye color gene mapped. The researchers reasoned that the white-eye phenotype could be due to the insertion of the *copia* element into the wild-type gene, thereby inactivating it.

To clone the eye color gene, chromosomal DNA from this white-eyed strain was isolated, digested with restriction enzymes, and inserted into vectors that were derived from a naturally occurring virus (viral vectors). This procedure creates a DNA library, a collection of clones consisting of vectors that contain different pieces of chromosomal DNA (described in Chapter 20). If a transposon had "jumped" into the eye color gene, vectors that contain this gene will also contain the transposon sequence. In other words, the presence of the transposon tags the eye color gene. Each vector is introduced into a bacterial cell, which produces a virus and ultimately a viral plaque, a clear area where the bacteria have been lysed. A labeled fragment of DNA that is complementary to the transposon sequence can be used as a probe to identify plaques that also contain the eye color gene. The method of using a probe to screen a DNA library is also described in Chapter 20. In the example of Figure 19.16, the method of transposon tagging was successful at cloning an eye color gene in *Drosophila*.

19.3 COMPREHENSION QUESTIONS

1. Which of the following types of transposable elements involve an RNA intermediate?
   a. Insertion elements
   b. Simple transposons
   c. Retroelements
   d. All of the above
2. The function of transposase is
   a. to recognize inverted repeats.
   b. to remove a TE from its original site.
   c. to insert a TE into a new site.
   d. all of the above.
3. According to the selfish DNA hypothesis, TEs exist because
   a. they offer the host a selective advantage.
   b. they have characteristics that allow them to multiply within the chromosomal DNA of living cells.
   c. they promote the expression of certain beneficial genes.
   d. all of the above.

KEY TERMS

Page 473. homologous recombination, site-specific recombination, transposition
Page 474. sister chromatid exchange (SCE), genetic recombination
Page 475. gene conversion, Holliday model, Holliday junction, branch migration, heteroduplex, resolution, double-strand break model, DNA gap repair synthesis
Page 479. lysogenic, attachment sites, antibodies, immunoglobulins (Igs), antigen
Page 480. recombination signal sequence, RAG1, RAG2, nonhomologous end-joining (NHEJ) proteins, V(D)J recombination
Page 481. transposable elements (TEs)
Page 482. mutable site
Page 484. simple transposition, transposons, retrotransposition, retrotransposons, retroelements, direct repeats (DRs), target-site duplications
Page 485. insertion element (IS element), inverted repeats (IRs), transposase, simple transposon, LTR retrotransposons, long terminal repeats (LTRs), non-LTR retrotransposons
Page 486. autonomous elements, nonautonomous element
Page 487. reverse transcriptase, integrase, target-site primed reverse transcription (TPRT), LINEs, SINEs
Page 489. selfish DNA hypothesis
Page 490. exon shuffling, hybrid dysgenesis
Page 491. insertional mutagenesis, transposon tagging

CHAPTER SUMMARY

19.1 Homologous Recombination

- Homologous recombination involves an exchange between DNA segments that are similar or identical in their DNA sequences. It can occur between sister chromatids or between homologous chromosomes (see Figure 19.1).
- The Holliday model describes the molecular steps that occur during homologous recombination between homologous chromosomes (see Figure 19.2).
- The initiation of homologous recombination is thought to usually occur via a double-strand break (see Figure 19.3).
- Several different proteins are involved with homologous recombination (see Table 19.1).
- Two different mechanisms, DNA mismatch repair and gap repair synthesis, can explain how gene conversion may occur during homologous recombination (see Figures 19.4, 19.5).

19.2 Site-Specific Recombination

- Site-specific recombination is the insertion of DNA segments at specific sites in a chromosome or the rearrangements of particular sites within genes.
- Some viral genomes, such as phage λ, are inserted into their host chromosome via site-specific recombination (see Figure 19.6).
- Antibody precursor genes are rearranged via site-specific recombination to produce a vast array of different antibodies (see Figure 19.7).

19.3 Transposition

- Barbara McClintock discovered the phenomenon of transposition in which a segment of DNA called a transposable element can move to multiple sites in a genome (see Figures 19.8, 19.9).
- Transposable elements can move via simple transposition (cut and paste) and retrotransposition (see Figure 19.10).
- Each type of transposable element has its own pattern of DNA sequences, which always includes direct repeats (see Figure 19.11).
- Transposase catalyzes the excision of transposons and their insertion at a new site (see Figure 19.12).
- Simple transposition can increase the copy number of a transposon if it occurs just after a transposon has been replicated (see Figure 19.13).
- Retrotransposition of LTR retrotransposons occurs via reverse transcriptase and integrase, whereas retrotransposition of non-LTR retrotransposons occurs via target-site primed reverse transcription (see Figures 19.14, 19.15).
- Many different transposons are found among living organisms. Their abundance varies among different species (see Tables 19.2, 19.3).
- Transposition can have a variety of effects on chromosome structure and gene expression (see Table 19.4).
- Transposon tagging is a method to clone and identify a particular gene in a genome (see Figure 19.16).

PROBLEM SETS & INSIGHTS

Solved Problems

S1. Zickler was the first person to demonstrate gene conversion by observing unusual ratios in *Neurospora* octads. At first, it was difficult for geneticists to believe these results because they seemed to contradict the Mendelian concept that alleles do not physically interact with each other. However, work by Mary Mitchell provided convincing evidence that gene conversion actually takes place. She investigated three different genes in *Neurospora*. One *Neurospora* strain had three mutant alleles: *pdx-1* (pyridoxine-requiring), *pyr-1* (pyrimidine-requiring), and *col-4* (a mutation that affected growth morphology). The *pdx-l* gene had been previously shown to map between the *pyr-1* and *col-4* genes. As shown here, Mitchell crossed this strain to a wild-type *Neurospora* strain:

$$pyr\text{-}1\ pdx\text{-}1\ col\text{-}4 \times pyr\text{-}1^+\ pdx\text{-}1^+\ col\text{-}4^+$$

She first analyzed many octads with regard to their requirement for pyridoxine. Out of 246 octads, two had an aberrant ratio in which two spores were *pdx-1,* and six were *pdx-1*$^+$. These same spores were then analyzed with regard to the other two genes. In both cases, the aberrant asci gave a normal 4:4 ratio of *pyr-1* : *pyr-1*$^+$ and *col-4* : *col-4*$^+$. Explain these results.

Answer: These results can be explained by gene conversion. The gene conversion took place in a limited region of the chromosome (within the *pdx-1* gene), but it did not affect the flanking genes (*pyr-1* and *col-4*) located on either side of the *pdx-1* gene. In the asci containing two *pdx-1* alleles and six *pdx-1*$^+$ alleles, a crossover occurred during meiosis I in the region of the *pdx-1* gene. Gene conversion changed the *pdx-1* allele into the *pdx-1*$^+$ allele. This gene conversion could have occurred by two mechanisms. If branch migration occurred across the *pdx-1* gene, a heteroduplex may have formed, which could be repaired by DNA mismatch repair, as described in Figure 19.4. In the aberrant asci with two *pdx-1* and six *pdx-1*$^+$ alleles, the *pdx-1* allele was converted to *pdx-1*$^+$. Alternatively, gene conversion of *pdx-1* into *pdx-1*$^+$ could have taken place via gap repair synthesis, as described in Figure 19.5. In this case, the *pdx-1* allele would have been digested away, and the DNA encoding the *pdx-1*$^+$ allele would have migrated into the digested region and provided a template to make a copy of the *pdx-1*$^+$ allele. (Note: Since this pioneering work, additional studies have shed considerable light concerning the phenomenon of gene conversion. It occurs at a fairly high rate in fungi, approximately 0.1–1% of the time. It is not due to new mutations occurring during meiosis.)

S2. Recombination involves the pairing of identical or similar sequences, followed by crossing over and the resolution of the intertwined helices. On rare occasions, the direct repeats or the inverted repeats within a single TE can align and undergo homologous recombination. What are the consequences when the direct repeats recombine? What are the consequences when the inverted repeats recombine?

Answer:

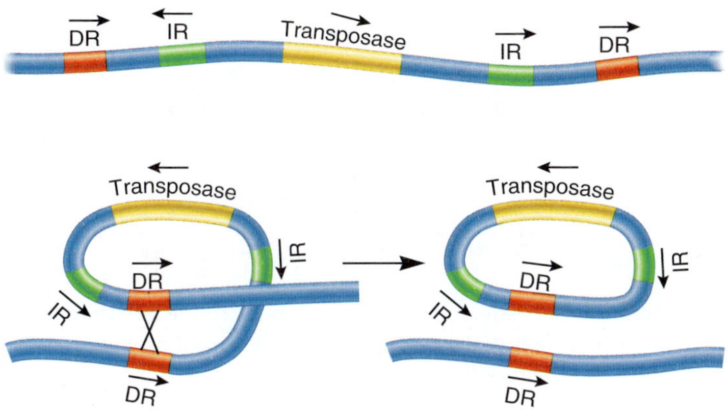

Most of the transposable element has been excised.

(a) Recombination between direct repeats

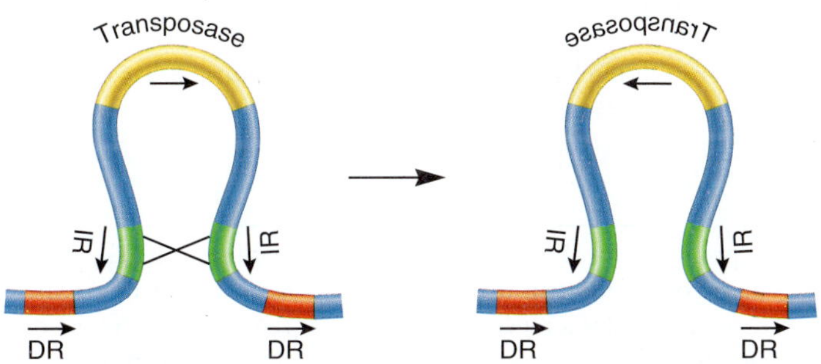

The sequence within the transposable element has been inverted.
Note that the transposase gene has changed to the opposite direction.

(b) Recombination between inverted repeats

S3. A schematic drawing of an uncrossed Holliday junction is shown here. One chromatid is shown in red, and the homologous chromatid is shown in blue. The red chromatid carries a dominant allele labeled *A* and a recessive allele labeled *b*, whereas the blue chromatid carries a recessive allele labeled *a* and a dominant allele labeled *B*.

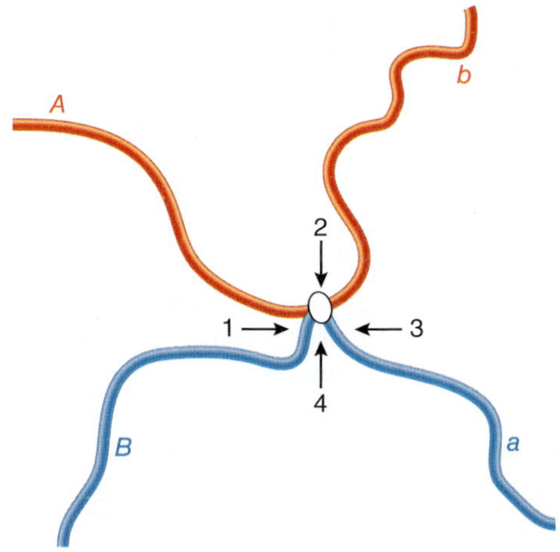

Where would the breakage of the crossed strands have to occur to get recombinant chromosomes? Would it have to occur at sites 1 and 3, or at sites 2 and 4? What would be the genotypes of the two recombinant chromosomes?

Answer: Breakage would have to occur at the arrows labeled 2 and 4. This would connect the *A* allele with the *B* allele. The *a* allele in the other homolog would become connected with the *b* allele. In other words, one chromosome would be *AB*, and the homolog would be *ab*.

Conceptual Questions

C1. Describe the similarities and differences between homologous recombination involving sister chromatid exchange (SCE) and that involving homologs. Would you expect the same types of proteins to be involved in both processes? Explain.

C2. The molecular mechanism of SCE is similar to homologous recombination between homologs except that the two segments of DNA are sister chromatids instead of homologous chromatids. If branch migration occurs during SCE, will a heteroduplex be formed? Explain why or why not. Can gene conversion occur during sister chromatid exchange?

C3. Which steps in the double-strand break model for recombination would be inhibited if the following proteins were missing? Explain the function of each protein required for the step that is inhibited.

A. RecBCD

B. RecA

C. RecG

D. RuvABC

C4. What two molecular mechanisms can explain the phenomenon of gene conversion? Would both occur in the double-strand break model?

C5. Is homologous recombination an example of mutation? Explain.

C6. What are recombinant chromosomes? How do they differ from the original parental chromosomes from which they are derived?

C7. In the Holliday model for homologous recombination (see Figure 19.2), the resolution steps can produce recombinant or nonrecombinant chromosomes. Explain how this can occur.

C8. What is gene conversion?

C9. Make a list of the differences between the Holliday model and the double-strand break model.

C10. In recombinant chromosomes, where is gene conversion likely to take place: near the breakpoint or far away from the breakpoint? Explain.

C11. What events does the RecA protein facilitate?

C12. According to the double-strand break model, does gene conversion necessarily involve DNA mismatch repair? Explain.

C13. What type of DNA structure is recognized by RecG and RuvABC? Do you think these proteins recognize DNA sequences? Be specific about what type(s) of molecular recognition these proteins can perform.

C14. Briefly describe three ways that antibody diversity is produced.

C15. Describe the function of RAG1 and RAG2 proteins and NHEJ proteins.

C16. According to the scenario shown in Figure 19.7, how many segments of DNA (one, two, or three) are removed during site-specific recombination within the gene that encodes the κ (kappa) light chain for IgG proteins? How many segments are spliced out of the pre-mRNA?

C17. Describe the role that integrase plays during the insertion of λ DNA into the host chromosome.

C18. If you were examining a sequence of chromosomal DNA, what characteristics would cause you to believe that the DNA contained a transposable element?

C19. For insertion elements and simple transposons, what is the function of the inverted repeat sequences during transposition?

C20. Why does transposition always produce direct repeats in the chromosomal DNA?

C21. Which types of TEs have the greatest potential for proliferation: insertion elements, simple transposons, or retrotransposons? Explain your choice.

C22. Do you consider TEs to be mutagens? Explain.

C23. Let's suppose that a species of mosquito has two different types of simple transposons that we will call X elements and Z elements. The X elements appear quite stable. When analyzing a population of 100 mosquitoes, every mosquito has six X elements, and they are always located in the same chromosomal locations among different individuals. In contrast, the Z elements seem to "move around" quite a bit. Within the same 100 mosquitoes, the number of Z elements ranges from 2–14, and the locations of the Z elements tend to vary considerably among different individuals. Explain how one simple transposon can be stable and another simple transposon can be mobile, within the same group of individuals.

C24. This chapter describes different types of TEs, including insertion elements, simple transposons, LTR retrotransposons, and non-LTR retrotransposons. Which of these four types of TEs would have the following features?

A. Require reverse transcriptase to transpose

B. Require transposase to transpose

C. Are flanked by direct repeats

D. Have inverted repeats

C25. What features distinguish a transposon from a retrotransposon? How are their sequences different, and how are their mechanisms of transposition different?

C26. Solved problem S2 illustrates the consequences of crossing over between the direct and inverted repeats within a single TE. The drawing here shows the locations of two copies of the same TE within a single chromosome. The chromosome is depicted according to its G banding pattern. (Note: G bands are illustrated in Figure 8.1.) The direct and inverted repeats are labeled 1, 2, 3, and 4, from left to right.

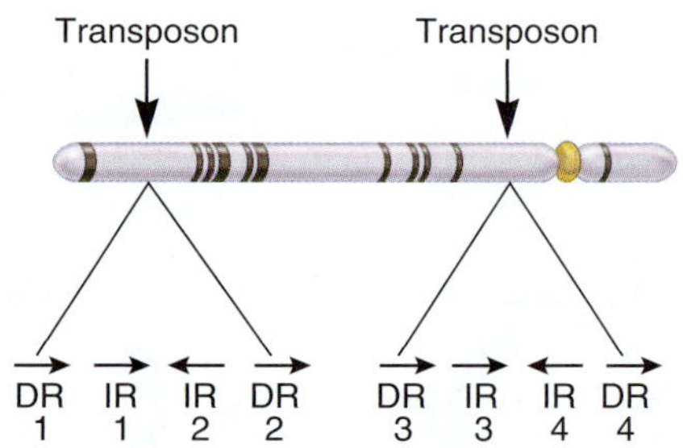

Draw the end result of recombination between the following sequences. Your drawing should include the banding pattern of the resulting chromosome.

A. DR-1 and DR-4

B. IR-1 and IR-4

C27. What is the difference between an autonomous and a nonautonomous TE? Is it possible for nonautonomous TEs to move? If yes, explain how.

C28. An operon in the bacterium *Salmonella typhimurium* has the following arrangement:

DR IR   IR DR

↓ ↓ promoter→ ↓ ↓ *H2 rH1*

The promoter for this operon is contained within a TE. The *H2* gene encodes a protein that is part of the bacterial flagellum. The *rH1* gene encodes a repressor protein that represses the *H1* gene, which is found at another location in the bacterial chromosome. The *H1* gene also encodes a flagellar protein. When the promoter is found in the arrangement shown here, the *H2/rH1* operon is turned on. This results in flagella that contain the H2 protein. The H1 protein is not made because the *rH1* repressor prevents the transcription of the *H1* gene. At a frequency of approximately 1 in 10,000 (which is much higher than the spontaneous mutation rate), this strain of bacterium can "switch" its expression so that *H2* is turned off and *H1* is turned on. Bacteria that have *H1* turned on and *H2* turned off can also switch back to having *H2* turned on and *H1* turned off. This switch also occurs at a frequency of about 1 in 10,000. Based on your understanding of transposons and recombination, explain how switching occurs in *Salmonella typhimurium*. Hint: Take a look at solved problem S2.

C29. The occurrence of multiple transposons within the genome of organisms has been suggested as a possible cause of chromosomal rearrangements such as deletions, translocations, and inversions. How could the occurrence of transposons promote these kinds of structural rearrangements?

Experimental Questions

E1. Briefly explain how McClintock determined that *Ds* was occasionally moving from one chromosomal location to another. Discuss the type of data she examined to arrive at this conclusion.

E2. In the data of Figure 19.9, is the solid red phenotype due to chromosome breakage or due to the excision of a TE and chromosomal rejoining? Explain how you have arrived at your conclusion.

E3. As discussed in the experiment of Figure 19.9, the presence of a transposon can create a mutable site or locus that is subject to frequent chromosome breakage. Why do you think a transposon creates a mutable site? If chromosome breakage occurs, do you think the transposon has moved somewhere else? How would you experimentally determine if it has?

E4. In your own words, explain the term transposon tagging.

E5. Tumor-suppressor genes are normal human genes that prevent uncontrollable cell growth. Starting with a normal laboratory human cell line, describe how you could use transposon tagging to identify tumor-suppressor genes. Note: When a TE hops into a tumor-suppressor gene, it may cause uncontrolled cell growth. This is detected as a large clump of cells among a normal monolayer of cells.

E6. Gerald Rubin and Allan Spradling devised a method of introducing a transposon into *Drosophila*. This approach has been important for the transposon tagging of many *Drosophila* genes. They began with a P element that had been cloned on a plasmid. (Note: Methods of cloning are described in Chapter 20.) Using cloning methods, they inserted the wild-type allele for the *rosy* gene into the P element in this plasmid. The recessive allele, *rosy*, results in a rosy eye color, while the wild-type allele, *rosy*$^+$, produces red eyes. This plasmid also has an intact transposase gene. The cloned DNA is shown here.

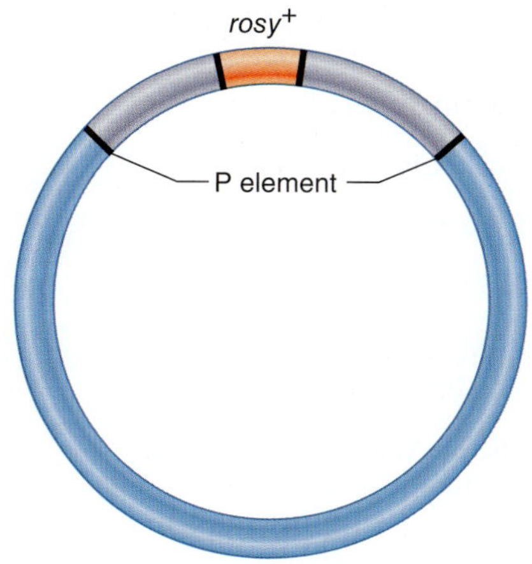

They used a micropipette to inject this DNA into regions of embryos that would later become reproductive cells. These embryos were originally homozygous for the recessive *rosy* allele. However, the P element carrying the *rosy*$^+$ allele could "hop" out of the plasmid and into a chromosome of the cells that were destined to become germ cells (i.e., sperm or egg cells). After they had matured to adults, these flies were then mated to flies that were homozygous for the recessive *rosy* allele. If offspring inherited a chromosome carrying the P element with the *rosy*$^+$ gene, such offspring would have red eyes. Therefore, the phenotype of red eyes provided a way to identify offspring that had a P element insertion.

Now here is the question. Let's suppose you were interested in identifying genes that play a role in wing development. Outline the experimental steps you would follow, using the plasmid with the P element containing the *rosy*$^+$ gene, as a way to transposon tag genes that play a role in wing development. Note: You should assume that the inactivation of a gene involved in wing development would cause an abnormality in wing shape. Also keep in mind that most P element insertions inactivate genes and may be inherited in a recessive manner.

Questions for Student Discussion/Collaboration

1. Make a list of the similarities and differences among homologous recombination, site-specific recombination, and transposition.

2. If homologous recombination could not occur, what would be the harmful and beneficial consequences?

3. Based on your current knowledge of genetics, discuss whether or not you think the selfish DNA hypothesis is correct.

Note: All answers appear at the website for this textbook; the answers to the even-numbered questions and all of the Concept Check and Comprehension Questions are in the back of the book.

www.mhhe.com/brookergenetics5e

Visit the website for practice tests, answer keys, and other learning aids for this chapter. Enhance your understanding of genetics with our interactive exercises, quizzes, animations, and much more.

CHAPTER OUTLINE

- 20.1 Gene Cloning Using Vectors
- 20.2 Polymerase Chain Reaction
- 20.3 DNA Sequencing
- 20.4 Site-Directed Mutagenesis
- 20.5 Blotting Methods to Detect Genes and Gene Products
- 20.6 Methods for Analyzing DNA- and RNA-Binding Proteins

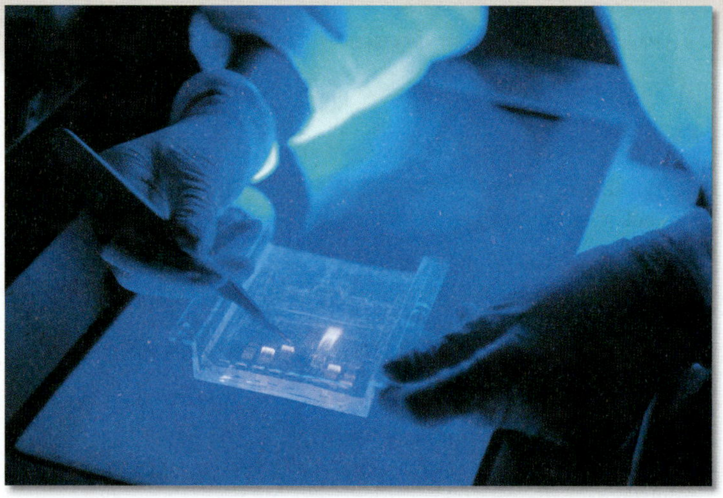

Detection of DNA bands on a gel. *DNA can be cut into fragments that can be separated via gel electrophoresis and then observed by staining with a dye called ethidium bromide. The cutting and pasting of DNA fragments allows researchers to clone genes.*

20 DNA TECHNOLOGIES

In this chapter, we will focus on methods aimed at manipulating and analyzing DNA at the molecular level. In some cases, DNA technologies involve the cutting and pasting of DNA segments to produce new arrangements. **Recombinant DNA technology** is the use of in vitro molecular techniques that manipulate fragments of DNA to produce new arrangements. In the early 1970s, the first successes in making recombinant DNA molecules were accomplished independently by two groups at Stanford University: David Jackson, Robert Symons, and Paul Berg; and Peter Lobban and A. Dale Kaiser. Both groups were able to isolate and purify pieces of DNA in a test tube and then covalently link DNA fragments from two different sources. In other words, they constructed molecules called **recombinant DNA molecules.** Shortly thereafter, researchers were able to introduce such recombinant DNA molecules into living cells. Once inside a host cell, the recombinant molecules are replicated to produce many identical copies of a gene—a process called gene cloning.

DNA technologies, such as gene cloning, have enabled geneticists to probe relationships between gene sequences and phenotypic consequences, and have thereby been fundamental to our understanding of gene structure and function. Researchers in molecular genetics use a variety of different methods to study gene structure and gene expression. Also, many practical applications of DNA technologies have been developed, including exciting advances such as gene therapy, screening for human diseases, recombinant vaccines, and the production of transgenic plants and animals in agriculture, in which a cloned gene from one species is transferred to some other species. Transgenic organisms have also been important in basic research.

In this chapter, we will focus primarily on the use of DNA technologies as a way to further our understanding of gene structure and function. We will look at the materials and molecular techniques used in gene cloning and explore polymerase chain reaction (PCR), which can make many copies of DNA within a defined region. We will then discuss how scientists analyze and alter DNA sequences through the techniques of DNA sequencing (a method that enables researchers to determine the base sequence of a DNA strand) and site-directed mutagenesis (a procedure that allows researchers to make mutations within a cloned segment of DNA). Finally, we will examine techniques for identifying specific genes or gene products, as well as methods for detecting the binding of proteins to DNA or RNA sequences.

In Chapter 21, we will consider many of the practical applications that have arisen as a result of these technologies. Chapters 22 and 23 are devoted to genomics, the molecular analysis of many genes and even the entire genome of a species.

20.1 GENE CLONING USING VECTORS

Learning Outcomes:

1. Outline the procedure for cloning a gene into a vector.
2. Describe how cDNA is made.
3. Compare and contrast a genomic library versus a cDNA library.

Molecular biologists want to understand how the molecules within living cells contribute to cell structure and function. Because proteins are the workhorses of cells and because they are the products of genes, many molecular biologists focus their attention on the structure and function of proteins or the genes that encode them. Researchers may focus their efforts on the study of just one or perhaps a few different genes or proteins. At the molecular level, this poses a daunting task. In all species, any given cell can express hundreds or thousands of different proteins, making the study of any single gene or protein akin to a "needle-in-a-haystack" exploration. To overcome this truly formidable obstacle, researchers frequently take the approach of cloning the genes that encode their proteins of interest. The term **gene cloning** refers to the process of making many copies of a gene. The laboratory methods necessary to clone a gene were devised during the early 1970s. Since then, many technical advances have enabled gene cloning to become a widely used procedure among scientists, including geneticists, cell biologists, biochemists, plant biologists, microbiologists, evolutionary biologists, clinicians, and biotechnologists.

Table 20.1 summarizes some of the common uses of gene cloning. In modern molecular biology, the diversity of uses for gene cloning is remarkable. For this reason, gene cloning has provided the foundation for critical technical advances in a variety of disciplines, including molecular biology, genetics, cell biology, biochemistry, and medicine. In this and the following section, we will examine the two general strategies used to make copies of a gene: the insertion of a gene into a vector that is then propagated in living cells, and cloning via polymerase chain reaction. Later sections in this chapter, as well as Chapters 21–23, will consider many of the uses of gene cloning that are described in Table 20.1.

Cloning Experiments May Involve Two Kinds of DNA Molecules: Chromosomal DNA and Vector DNA

If a scientist wants to clone a particular gene, a common source of the gene is the chromosomal DNA of the species that carries the gene. For example, if the goal is to clone the rat β-globin gene, this gene is found within the chromosomal DNA of rat cells. In this case, therefore, the rat's chromosomal DNA is one type of DNA needed in a cloning experiment. To prepare chromosomal DNA, an experimenter first obtains cellular tissue from the organism of interest. The preparation of chromosomal DNA then involves the breaking open of cells and the extraction and purification of the DNA using biochemical techniques such as chromatography and centrifugation (see the Appendix for a description of these techniques).

Let's begin our discussion of gene cloning by considering a recombinant DNA technology in which a gene is removed from its native site within a chromosome and inserted into a smaller

TABLE 20.1
Some Uses of Gene Cloning

Technique	Description
DNA sequencing	Cloned genes provide enough DNA to subject the gene to DNA sequencing (described later in this chapter). The sequence of the gene can reveal the gene's promoter, regulatory sequences, and coding sequence. DNA sequencing is also important in the identification of alleles that cause cancer and inherited human diseases.
Site-directed mutagenesis	A cloned gene can be manipulated to change its DNA sequence. Mutations within genes help to identify gene sequences, such as promoters and regulatory elements. The study of a mutant gene can also help to elucidate its normal function and how its expression may affect the roles of other genes. Mutations in the coding sequence may reveal which amino acids are important for a protein's structure and function.
Gene probes	Labeled DNA strands from a cloned gene can be used as probes for identifying similar or identical genes or RNA. These methods, known as Southern and Northern blot analysis, are described later in this chapter. Probes can also be used to localize genes within intact chromosomes (see Chapter 22).
Expression of cloned genes	Cloned genes can be introduced into a different cell type or different species. The expression of cloned genes has many uses: *Research* 1. The expression of a cloned gene can help to elucidate its cellular function. 2. The coding sequence of a gene can be placed next to an active promoter and then introduced into a culture of cells that will express a large amount of the protein. This greatly aids in the purification of large amounts of protein that may be needed for biochemical or biophysical studies. *Biotechnology* 1. Cloned genes can be introduced into bacteria to make pharmaceutical products such as insulin (see Chapter 21). 2. Cloned genes can be introduced into plants and animals to make transgenic species with desirable traits (see Chapter 21). *Clinical trials* 1. Cloned genes have been used in clinical trials involving gene therapy (see Chapter 21).

segment of DNA known as a **vector**—a small DNA molecule that can replicate independently of host cell chromosomal DNA and produce many identical copies of an inserted gene. The purpose of vector DNA is to act as a carrier of the DNA segment that is to be cloned. (The term vector comes from a Latin term meaning carrier.) In cloning experiments, a vector may carry a small segment of chromosomal DNA, perhaps only a single gene. By comparison, a chromosome carries many more genes, perhaps a few hundred or thousand. Like a chromosome, a vector is replicated when it resides within a living cell; a cell that harbors a vector is called a **host cell.** When a vector is replicated within a host cell, the DNA that it carries is also replicated.

The vectors commonly used in gene-cloning experiments were derived originally from two natural sources: plasmids or viruses. Most vectors are **plasmids,** which are small circular pieces of DNA. As discussed in Chapter 7, plasmids are found naturally in many strains of bacteria and occasionally in eukaryotic cells. Many naturally occurring plasmids carry genes that confer resistance to antibiotics or other toxic substances. These plasmids are called **R factors.** Some of the plasmids used in modern cloning experiments were derived from R factors.

Plasmids also contain a DNA sequence, known as an **origin of replication,** that is recognized by the replication enzymes of the host cell and allow it to be replicated. The sequence of the origin of replication determines whether or not the vector can replicate in a particular type of host cell. Some plasmids have origins of replication with a broad host range. Such a plasmid can replicate in the cells of many different species. Alternatively, many vectors used in cloning experiments have a limited host cell range. In cloning experiments, researchers must choose a vector that replicates in the appropriate cell type(s) for their experiments. For example, if researchers want a cloned gene to be propagated in *Escherichia coli*, the vector they employ must have an origin of replication that is recognized by this species of bacterium. The origin of replication also determines the copy number of a plasmid. Some plasmids are said to have strong origins because they can achieve a high copy number—perhaps 100–200 copies of the plasmid per cell. Others have weaker origins, whereby only one or two plasmids are found per cell.

Commercially available plasmids have been genetically engineered for effective use in cloning experiments. They contain unique sites where geneticists can easily insert pieces of DNA. Another useful feature is that cloning vectors often contain resistance genes, which provide host cells with the ability to grow in the presence of a toxic substance. Such a gene is called a **selectable marker** because the expression of the gene selects for the growth of the host cells. Many plasmid vectors are derived from R factors and carry selectable markers, such as genes that confer antibiotic resistance to the host cell. For example, the gene amp^R encodes an enzyme known as β-lactamase. This enzyme degrades ampicillin, an antibiotic that normally kills bacteria. Bacteria containing the amp^R gene can grow on media containing ampicillin because they can degrade it. In a cloning experiment in which the amp^R gene is found within the plasmid, the growth of cells in the presence of ampicillin identifies bacteria that contain the plasmid. These bacteria can grow and form visible colonies on solid growth media. In contrast, those cells that do not contain the plasmid are ampicillin-sensitive and do not grow.

An alternative type of vector used in cloning experiments is a viral vector. As discussed in Chapter 17, viruses infect living cells and propagate themselves by taking control of the host cell's metabolic machinery. When a chromosomal gene is inserted into a viral genome, the gene is replicated whenever the viral DNA is replicated. Therefore, viruses can be used as vectors to carry other pieces of DNA. When a virus is used as a vector, the researcher may analyze viral plaques rather than bacterial colonies. The characteristics of viral plaques are described in Chapter 7 (see Figure 7.14).

Molecular biologists may choose from hundreds of different vectors to use in their cloning experiments. **Table 20.2** provides a general description of several different types of vectors that are commonly used to clone small segments of DNA. In addition, other types of vectors, such as cosmids, bacterial artificial chromosomes (BACs), and yeast artificial chromosomes (YACs), are used to clone large pieces of DNA. These vectors are described in detail in Chapter 22. Vectors designed to introduce genes into plants and animals are discussed in Chapter 21.

TABLE 20.2

Some Vectors Used in Cloning Experiments

Example	Type	Description
pBluescript	Plasmid	A type of vector like the one shown in Figure 20.2. It is used to clone small segments of DNA and propagate them in *E. coli.*
YEp24	Plasmid	This plasmid is an example of a **shuttle vector,** which can replicate in two different host species, *E. coli* and *Saccharomyces cerevisiae.* It carries origins of replication for both species.
λgt11	Viral	This vector is derived from the bacteriophage λ, which is described in Chapter 17. λgt11 also contains a promoter from the *lac* operon. When fragments of DNA are inserted next to this promoter, the DNA is expressed in *E. coli.* This is an example of an **expression vector.** An expression vector is designed to clone the coding sequence of genes so they are transcribed and translated correctly.
SV40	Viral	This virus naturally infects mammalian cells. Genetically altered derivatives of the SV40 viral DNA are used as vectors for the cloning and expression of genes in mammalian cells that are grown in the laboratory.
Baculovirus	Viral	This virus naturally infects insect cells. In a laboratory, insect cells can be grown in liquid media. Unlike many other types of eukaryotic cells, insect cells often express large amounts of proteins that are encoded by cloned genes. When researchers want to make a large amount of a protein, they can clone the gene that encodes the protein into baculovirus and then purify the protein from insect cells.

Enzymes Are Used to Cut DNA into Pieces and Join the Pieces Together

A key step in a cloning experiment is the insertion of chromosomal DNA into a plasmid or viral vector. This requires the cutting and pasting of DNA fragments. To cut DNA, researchers use enzymes known as **restriction endonucleases,** or **restriction enzymes.** The restriction enzymes used in cloning experiments bind to a specific base sequence and then cleave the DNA backbone at two defined locations, one in each strand. Proposed by Werner Arber in the 1960s and discovered by Hamilton Smith and Daniel Nathans in the 1970s, restriction enzymes are made naturally by many different species of bacteria and protect bacterial cells from invasion by foreign DNA, particularly that of bacteriophages. The bacterial DNA is typically methylated to protect it from digestion by the restriction enzyme.

Figure 20.1 shows the role of a restriction enzyme, called *Eco*RI, in producing a recombinant DNA molecule. Certain types of restriction enzymes are useful in cloning because they digest DNA into fragments with "sticky ends." As shown in Figure 20.1,

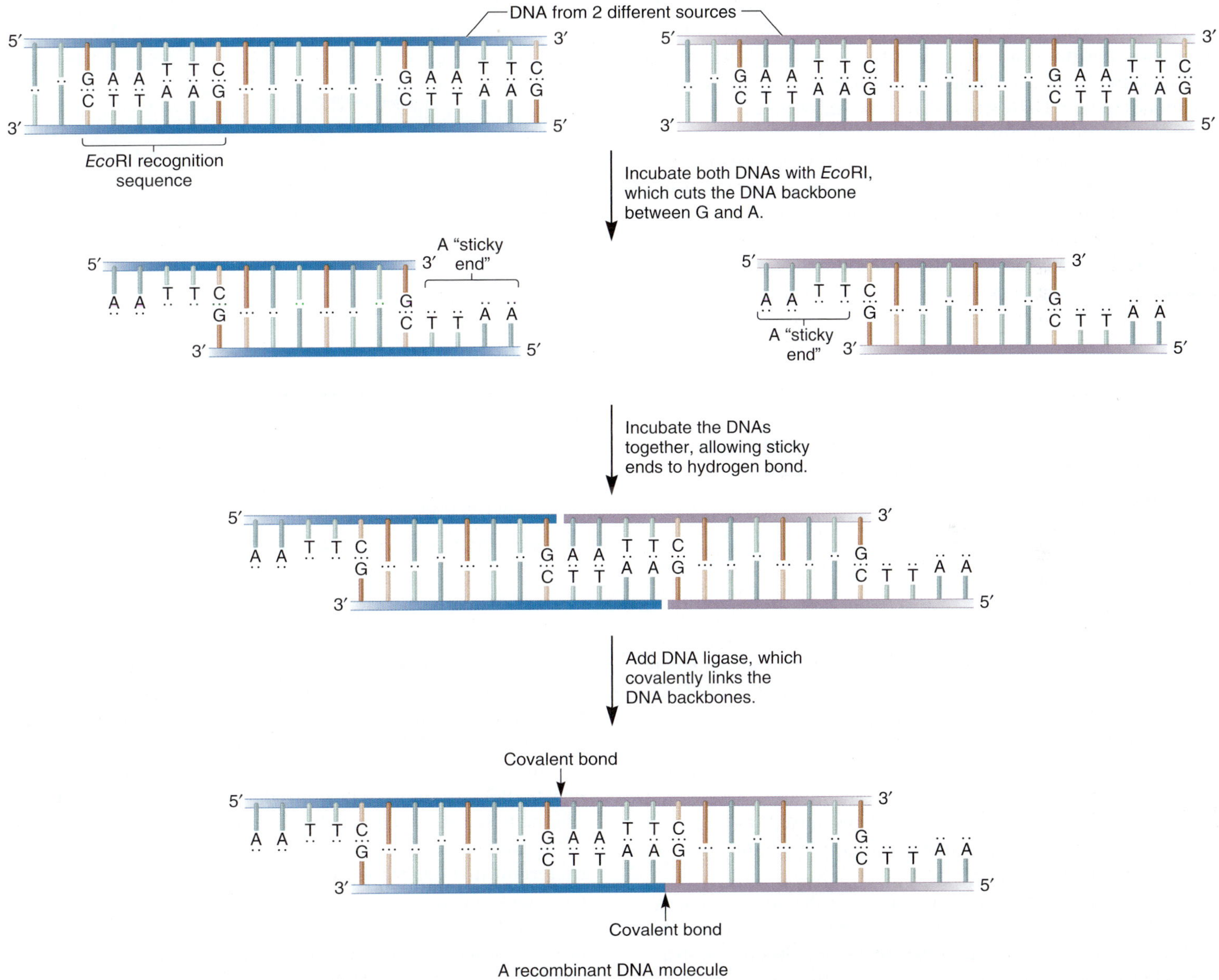

FIGURE 20.1 The action of a restriction enzyme and the production of recombinant DNA. The restriction enzyme *Eco*RI binds to a specific sequence, in this case 5′–GAATTC–3′. It then cleaves the DNA backbone between G and A, producing DNA fragments. The single-stranded ends of different DNA fragments can hydrogen bond with each other, because they have complementary sequences. The enzyme DNA ligase then catalyzes the formation of covalent bonds in the DNA backbones of the fragments.

CONCEPT CHECK: Prior to the action of DNA ligase, how many hydrogen bonds are holding these two DNA fragments together?

the sticky ends are single-stranded regions of DNA that can hydrogen bond to a complementary sequence of DNA from a different source. The ends of two different DNA pieces hydrogen bond to each other because of their complementary sticky ends.

The hydrogen bonding between the sticky ends of DNA fragments promotes a temporary interaction between the two fragments. However, this interaction is not stable because it involves only a few hydrogen bonds between complementary bases. How can this interaction be made more permanent? The answer is that the sugar-phosphate backbones within the DNA strands must be covalently linked together. Experimentally, this linkage is catalyzed by the addition of **DNA ligase.** (Note: As described in Chapter 11, DNA ligase plays a natural role in DNA replication.) Figure 20.1 illustrates the action of DNA ligase, which catalyzes covalent bond formation in the sugar-phosphate backbones of both DNA strands after the sticky ends have hydrogen bonded with each other.

Currently, several hundred different restriction enzymes from various bacterial species have been identified and are available commercially to molecular biologists. **Table 20.3** gives a few examples. Restriction enzymes usually recognize sequences that are **palindromic;** that is, the sequence in one strand is identical when read in the opposite direction in the complementary strand. For example, the sequence recognized by *Eco*RI is 5′–GAATTC–3′ in the top strand. Read in the opposite direction in the bottom strand, this sequence is also 5′–GAATTC–3′.

TABLE 20.3
Some Restriction Enzymes Used in Gene Cloning

Restriction Enzyme*	Bacterial Source	Sequence Recognized†
*Bam*HI	*Bacillus amyloliquefaciens* H	↓ 5′–GGATCC–3′ 3′–CCTAGG–5′ ↑
*Sau3A*I	*Staphylococcus aureus* 3A	↓ 5′–GATC–3′ 3′–CTAG–5′ ↑
*Eco*RI	*Escherichia coli* RY13	↓ 5′–GAATTC–3′ 3′–CTTAAG–5′ ↑
*Nae*I	*Nocardia aerocolonigenes*	↓ 5′–GCCGGC–3′ 3′–CGGCCG–5′ ↑
*Pst*I	*Providencia stuartii*	↓ 5′–CTGCAG–3′ 3′–GACGTC–5′ ↑

*Restriction enzymes are named according to the species in which they are found. The first three letters are italicized because they indicate the genus and species names. Because a species may produce more than one restriction enzyme, the enzymes are designated I, II, III, and so on, to indicate the order in which they were discovered in a given species. Some restriction enzymes, like *Eco*RI, produce a sticky end with a 5′ overhang (see Figure 20.1), whereas others, such as *Pst*I, produce a 3′ overhang. However, not all restriction enzymes cut DNA to produce sticky ends. For example, the enzyme *Nae*I cuts DNA to produce blunt ends.

†The arrows show the locations in the upper and lower DNA strands where the restriction enzymes cleave the DNA backbone.

Gene Cloning Involves the Insertion of DNA Fragments into Vectors, Which Are Then Propagated Within Host Cells

Now that we are familiar with the materials, let's outline the general strategy for a typical cloning experiment. In the procedure shown in **Figure 20.2**, the goal is to clone a chromosomal gene of interest into a plasmid vector that already carries the *amp*^{R} gene. To begin this experiment, the chromosomal DNA is isolated and digested with a restriction enzyme. This enzyme cuts the chromosomes into many small fragments. The plasmid DNA is also cut with the same restriction enzyme. However, the plasmid has only one unique site for the restriction enzyme. After cutting, the plasmid has two ends that are complementary to the sticky ends of the chromosomal DNA fragments. The digested chromosomal DNA and plasmid DNA are mixed together and incubated under conditions that promote the binding of these complementary sticky ends.

DNA ligase is then added to catalyze the covalent linkage between DNA fragments. In some cases, the two ends of the vector simply ligate back together, restoring the vector to its original structure. This is called a recircularized vector. In other cases, a fragment of chromosomal DNA may become ligated to both ends of the vector. In this way, a segment of chromosomal DNA has been inserted into the vector. The vector containing a piece of chromosomal DNA is referred to as a **recombinant vector.**

Following ligation, the DNA is introduced into living cells treated with agents that render them permeable to DNA molecules. Cells that can take up DNA from the extracellular medium are called **competent cells.** This step in the procedure is commonly called **transformation.** Only a very small percentage of bacterial cells actually take up a plasmid. How can an experimenter distinguish between bacterial cells that have taken up a plasmid versus those that have not? In the experiment shown in Figure 20.2, a plasmid is introduced into bacterial cells that were originally sensitive to ampicillin. The bacteria are then streaked onto plates containing bacterial growth media and ampicillin. A bacterium that has taken up a plasmid carrying the *amp*^{R} gene continues to divide and forms a bacterial colony containing tens of millions of cells. Because each cell within a single colony is derived from the same original cell, all cells within a colony contain the same type of plasmid DNA.

In the experiment shown in Figure 20.2, how can the experimenter distinguish between bacterial colonies that contain a recircularized vector versus those with a recombinant vector carrying a piece of chromosomal DNA? As shown here, the chromosomal DNA has been inserted into a region of the vector that contains the *lacZ* gene, which encodes the enzyme β-galactosidase (see Chapter 14). The insertion of chromosomal DNA into the

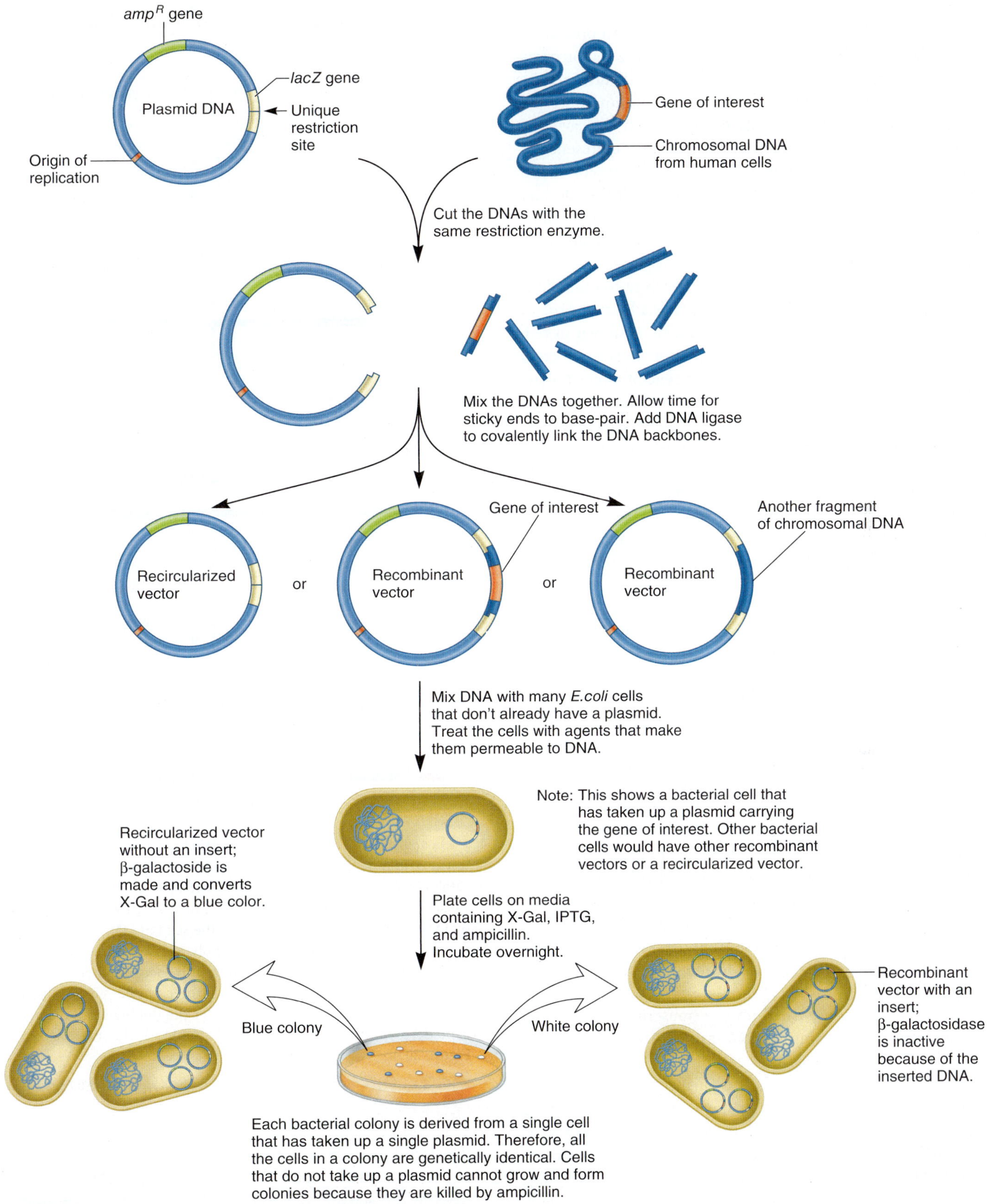

INTERACTIVE EXERCISE

ONLINE ANIMATION

FIGURE 20.2 The steps in gene cloning. Note: X-Gal refers to the colorless compound 5-bromo-4-chloro-3-indolyl-β-D-galactoside. It is converted by β-galactosidase into a blue dye. IPTG is an abbreviation for isopropyl-β-D-thiogalactopyranoside, which is a nonmetabolizable lactose analog that can induce the *lac* promoter.

CONCEPT CHECK: Explain the role of the selectable marker gene in this experiment.

vector disrupts the *lacZ* gene so it is no longer able to produce a functional enzyme. By comparison, a recircularized vector has a functional *lacZ* gene. The functionality of *lacZ* is determined by providing the growth medium with a colorless compound, X-Gal (5-bromo-4-chloro-3-indolyl-β-D-galactoside), which is cleaved by β-galactosidase into a blue dye. Bacteria grown in the presence of X-Gal and IPTG (an inducer of the *lacZ* gene) form blue colonies if they have a functional *lacZ* gene and white colonies if they do not. In this experiment, therefore, bacterial colonies containing recircularized vectors form blue colonies, whereas colonies containing recombinant vectors are white.

In the example of Figure 20.2, one of the white colonies contains cells with a recombinant vector that carries a human gene of interest; the segment containing the human gene is shown in red. The goal of gene cloning is to produce an enormous number of copies of a recombinant vector that carry the gene of interest. During transformation, a single bacterial cell usually takes up a single copy of a recombinant vector. However, two subsequent events lead to the amplification of the cloned gene. First, because the vector has an origin of replication, the bacterial host cell replicates the recombinant vector to produce many identical copies per cell. Second, the bacterial cells divide approximately every 20 minutes. Following overnight growth, a population of many millions of bacteria is obtained. Each of these bacterial cells contains many copies of the cloned gene. For example, a bacterial colony may be composed of 10 million cells, with each cell containing 50 copies of the recombinant vector. Therefore, this bacterial colony would contain 500 million copies of the cloned gene!

cDNA Can Be Made from mRNA via Reverse Transcriptase

In the discussion of gene cloning described in Figure 20.2, chromosomal DNA and plasmid DNA were used as the material to clone genes. Alternatively, a sample of RNA can provide a starting point for cloning DNA. As described in Chapter 17, the enzyme **reverse transcriptase** uses RNA as a template to make a complementary strand of DNA. This enzyme is encoded in the genome of retroviruses and provides a way for retroviruses to copy their RNA genome into DNA molecules that integrate into the host cell's chromosomes. Likewise, reverse transcriptase is encoded in some retrotransposons and is needed in the retrotransposition of such elements.

Researchers can use purified reverse transcriptase in a strategy for cloning genes, using mRNA as the starting material (**Figure 20.3**). To begin this experiment, mRNAs, which naturally contain a polyA tail at their 3′ ends, are purified from a sample of cells. The mRNAs are mixed with primers composed of a string of thymine-containing nucleotides. This short strand of DNA, or **oligonucleotide,** is called a poly-dT primer. The poly-dT primers are complementary to the 3′ end of mRNAs. Reverse transcriptase and deoxyribonucleotides (dNTPs) are then added to make a DNA strand that is complementary to the mRNA. One way to make the other DNA strand is to use RNaseH, which partially digests the RNA, generating short RNAs that are used as primers by DNA polymerase to make a second DNA strand that is complementary to

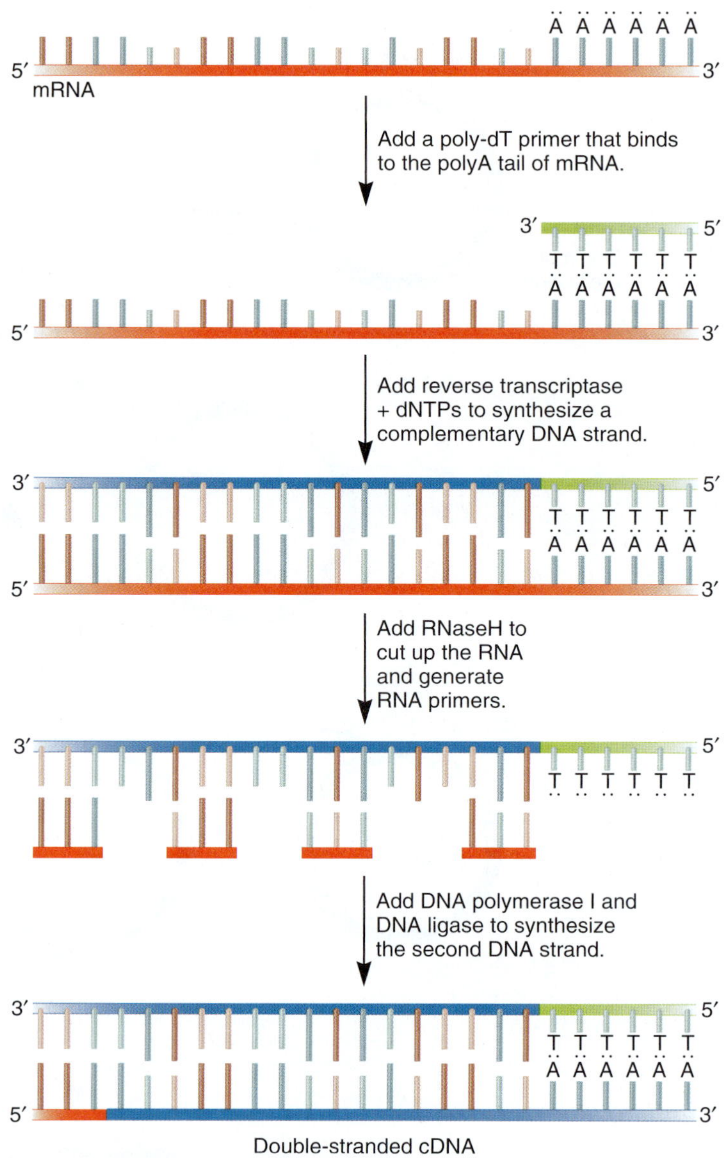

FIGURE 20.3 **Synthesis of cDNA.** A poly-dT primer binds to the 3′ end of eukaryotic mRNAs. Reverse transcriptase then catalyzes the synthesis of a complementary DNA strand (cDNA). RNaseH digests the mRNA into short pieces that are used as primers by DNA polymerase I to synthesize the second DNA strand. The 5′ to 3′ exonuclease function of DNA polymerase I removes all of the RNA primers except the one at the 5′ end (because there is no primer upstream from this site). This RNA primer can be removed by the subsequent addition of an RNase. After the double-stranded cDNA is made, it can then be inserted into a vector as described later in Figure 20.4b.

CONCEPT CHECK: Explain the meaning of the term reverse transcriptase.

the strand made by reverse transcriptase. Finally, DNA ligase seals any nicks in this second DNA strand. When DNA is made from RNA as the starting material, the DNA is called **complementary DNA (cDNA).** The term originally referred to the single strand of DNA that is complementary to the RNA template. However, cDNA now refers to any DNA, whether it is single- or double-stranded, that is made using RNA as the starting material.

Why is cDNA cloning useful? From a research perspective, an important advantage of cDNA is that it lacks introns, which are often found in eukaryotic genes. Because introns can be quite large, it is much simpler to insert cDNAs into vectors if researchers want to focus their attention on the coding sequence of a gene. For example, if the primary goal is to determine the coding sequence of a protein-encoding gene, a researcher inserts cDNA into a vector and then determines the DNA sequence of the insert, as described later in this chapter. Similarly, if a scientist wants to express an encoded protein of interest in a cell that does not splice out the introns properly (e.g., in a bacterial cell), it is necessary to make cDNA clones of the respective gene.

A DNA Library May Be Constructed Using Genomic DNA or cDNA

In a typical cloning experiment that involves the use of vectors (see Figure 20.2), the treatment of the chromosomal DNA with restriction enzymes yields tens of thousands of different DNA fragments. Therefore, after the DNA fragments are ligated individually to vectors, the researcher has a collection of recombinant vectors, with each vector containing a particular fragment of chromosomal DNA. A collection of recombinant vectors is known as a **DNA library.** When the starting material is chromosomal DNA, the library is called a **genomic library** (**Figure 20.4a**). The library shown here uses a plasmid vector. Alternatively, a viral vector could be used, which would result in viral plaques rather than bacterial colonies.

It is also common for researchers to make a **cDNA library** that contains recombinant vectors with cDNA inserts. **Figure 20.4b** illustrates how cDNAs are made and inserted into vectors. As described earlier in Figure 20.3, cDNA is first made via reverse transcriptase. To insert the cDNAs into vectors, short oligonucleotides called linkers are attached to the cDNAs via DNA ligase. The linkers contain DNA sequences with a unique site for a restriction enzyme. After the linkers are attached to the cDNAs, the cDNAs and the vectors are cut with restriction enzymes and then ligated to each other. This produces a cDNA library.

20.1 COMPREHENSION QUESTIONS

1. Which of the following may be used as a vector in a gene-cloning experiment?
   a. mRNA
   b. Plasmid
   c. Virus
   d. Both b and c
2. The restriction enzymes used in gene-cloning experiments ___________, which generates sticky ends that can ___________.
   a. cut the DNA, enter bacterial cells
   b. cut the DNA, hydrogen bond with complementary sticky ends
   c. methylate DNA, enter bacterial cells
   d. methylate DNA, hydrogen bond with complementary sticky ends
3. Which of the following is the proper order of steps in a gene-cloning experiment involving vectors?
   1. Add DNA ligase.
   2. Incubate the chromosomal DNA and the vector DNA with a restriction enzyme.
   3. Introduce the DNA into living cells.
   4. Mix the chromosomal DNA and vector DNA together.

   a. 1, 2, 3, 4
   b. 2, 3, 1, 4
   c. 2, 4, 1, 3
   d. 1, 2, 4, 3
4. The function of reverse transcriptase is to
   a. copy RNA into DNA.
   b. copy DNA into RNA.
   c. translate RNA into protein.
   d. translate DNA into protein.
5. A collection of recombinant vectors that contain fragments of chromosomal DNA is called
   a. a genomic library.
   b. a cDNA library.
   c. a Southern blot.
   d. both a and b.

20.2 POLYMERASE CHAIN REACTION

Learning Outcomes:

1. Describe the three steps of a PCR cycle.
2. Explain how reverse-transcriptase PCR is carried out.
3. Outline the method of real-time PCR and discuss why it is used.

In our previous discussion of gene cloning, the DNA of interest was inserted into a vector, which then was introduced into a host cell. The replication of the vector within the host cell, and the proliferation of the host cells, led to the production of many copies of the DNA. Another way to copy DNA, without the aid of vectors and host cells, is a technique called **polymerase chain reaction (PCR),** which was developed by Kary Mullis in 1985. In this section, we begin with a general description of PCR and then examine how it can be used to quantitate the amount of DNA or RNA in a biological sample.

Each Cycle of PCR Involves Three Steps: Denaturation, Primer Annealing, and Primer Extension

The PCR method is used to make large of amounts of DNA in a defined region that is flanked by two **primers.** The primers are oligonucleotides, which are short segments of DNA, usually about 15–20 nucleotides in length. As shown in **Figure 20.5a**, the starting material of a PCR experiment can be a complex mixture of DNA. The two primers bind to specific sites in the DNA because their bases are complementary at these sites. The end result of PCR is that the region that is flanked by the primers, which contains the gene of interest, is amplified. The term amplification means that many copies of the region have been made. In other words, the region between the two primers has been cloned.

FIGURE 20.4 The construction of a DNA library. **(a)** To make a genomic library, chromosomal DNA is first digested to produce many fragments. The fragment containing the gene of interest is highlighted in red. Following ligation, each vector contains a different piece of chromosomal DNA. **(b)** To make a cDNA library, oligonucleotide linkers that contain a restriction site are attached to the cDNAs, so they can be inserted into vectors.

CONCEPT CHECK: What is an advantage of making a cDNA library rather than a genomic library?

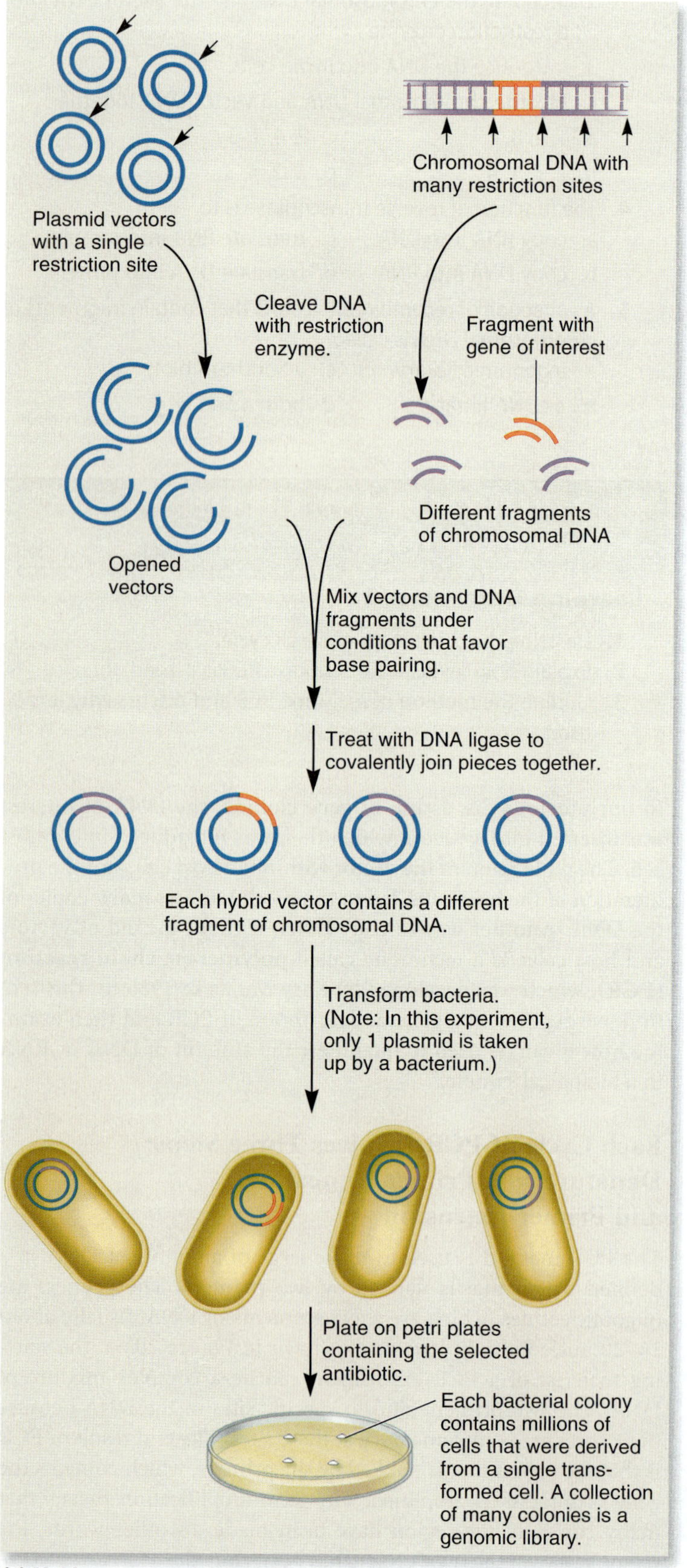

(a) Making a genomic library

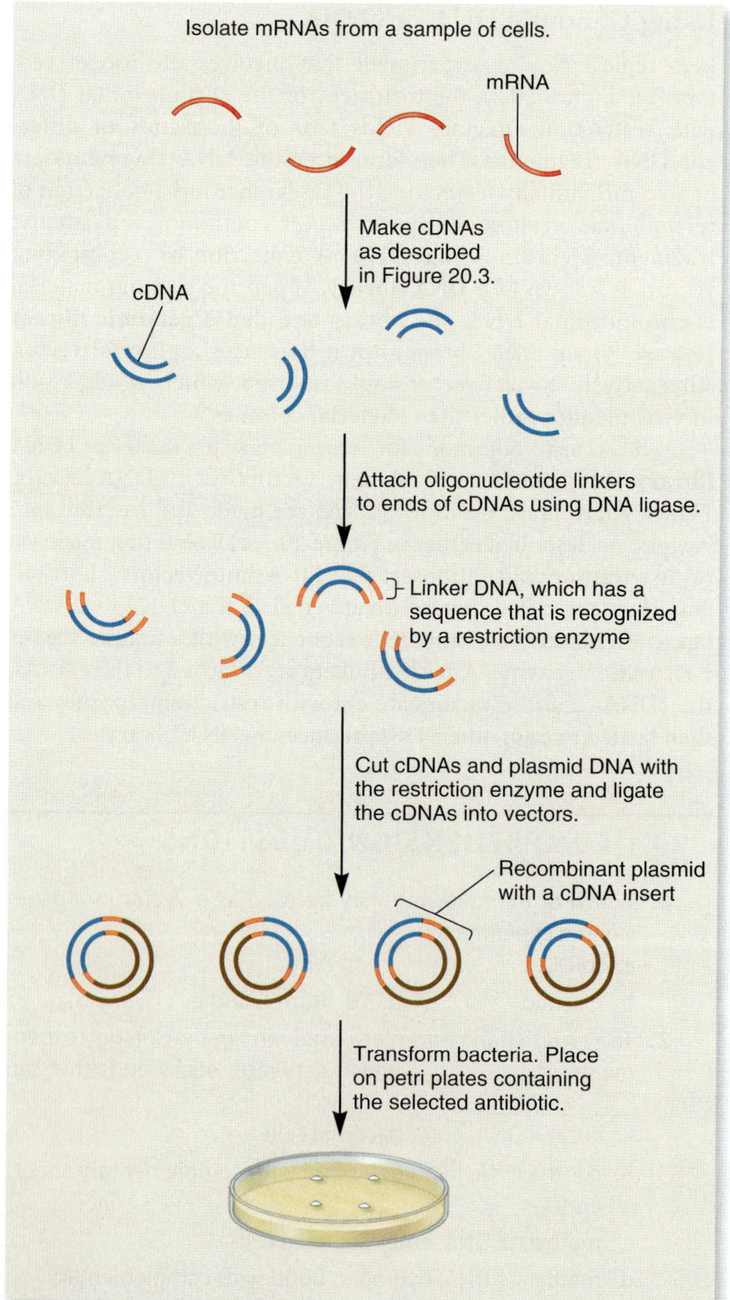

(b) Making a cDNA library

The primers are key reagents in a PCR experiment. They are made chemically. They are not typically made in research or clinical laboratories, but instead are made at university or industrial facilities. Researchers or clinicians simply order the primers they need with a specified sequence. How does someone choose the primer sequences? In most cases, PCR is conducted on DNA samples in which a scientist already knows the DNA sequences that flank the region of interest. Because the DNA sequence of the entire genome has already been determined for many species, sequence information can come from the genome database of the species of interest. For example, if you wanted to clone a human gene or a portion of human gene, such as the β-globin gene, you would look up that gene sequence in the human genome database and decide where you want your two primers to bind. You would then order primers with sequences that are complementary to those sites.

Let's now consider the reagents that are needed for a PCR experiment. The two reagents that we have already considered in Figure 20.5a are a sample of DNA that contains a gene of interest, which is called **template DNA,** and primers. In addition, deoxyribonucleoside triphosphates (dNTPs) are needed, as is a thermostable form of DNA polymerase such as ***Taq*** **polymerase,** isolated from the bacterium *Thermus aquaticus*. A thermostable form of DNA polymerase is necessary because PCR involves heating steps that inactivate most other natural forms of DNA polymerase (which are thermolabile, or readily denatured by heat).

As outlined in **Figure 20.5b**, PCR involves three steps: denaturation, primer annealing, and primer extension. To make copies of the DNA, the template DNA is first denatured by heat treatment, causing the strands to separate (also see Online Animation). As the temperature is lowered, oligonucleotide primers bind

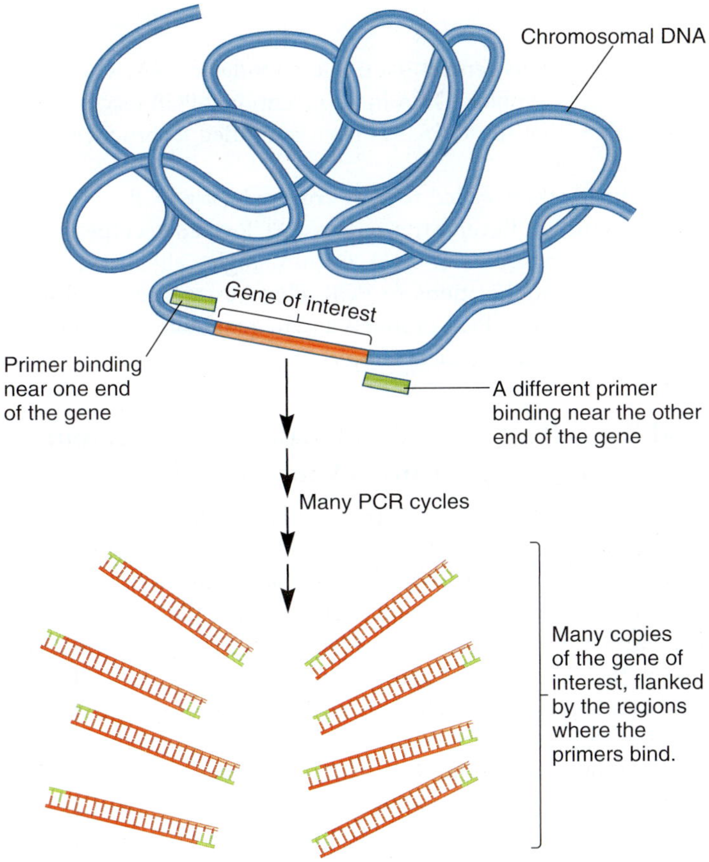

(a) The outcome of a PCR experiment

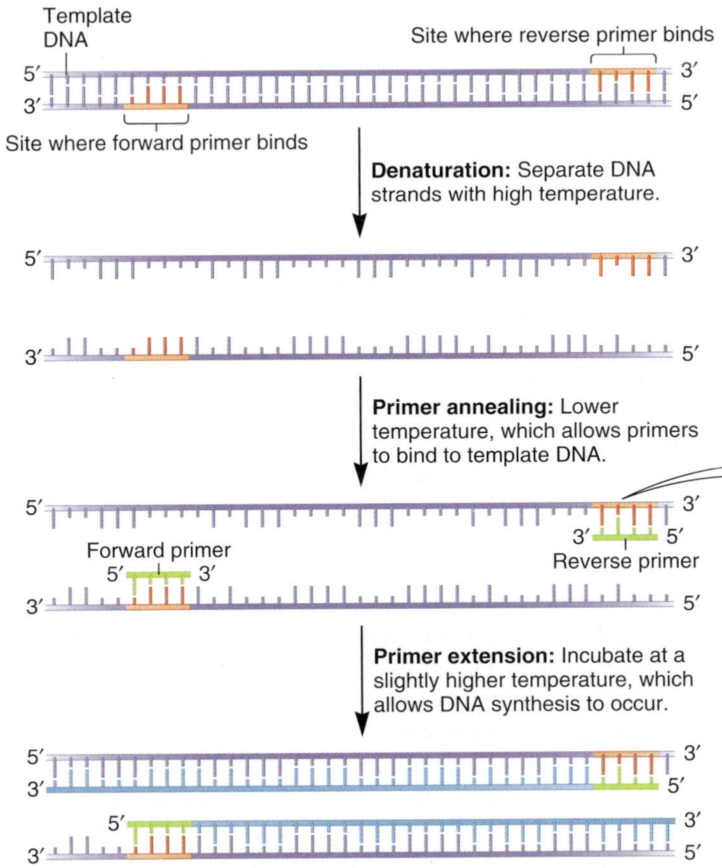

(b) The 3 steps of a PCR cycle

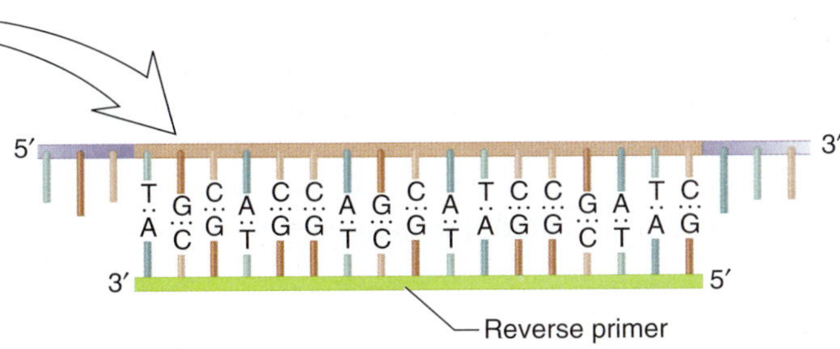

ONLINE ANIMATION

FIGURE 20.5 The technique of polymerase chain reaction (PCR). **(a)** Because the two primers anneal to specific sites, PCR amplifies a defined region of DNA that is located between the two primers. **(b)** During each cycle of PCR, three steps occur: denaturation, primer annealing, and primer extension. The primers used in actual PCR experiments are usually about 15–20 nucleotides in length. The region between the two primers is typically hundreds of nucleotides in length, not just several nucleotides as shown here.

to the DNA in a process called **annealing.** Once the primers have annealed, the temperature is raised slightly, and *Taq* polymerase catalyzes the synthesis of complementary DNA strands in the 5′ to 3′ direction, starting at the primers. This process, which is called **primer extension,** doubles the amount of the template DNA. The three steps shown in Figure 20.5b constitute one cycle of a PCR reaction. The sequential process of denaturation—primer annealing—primer extension is then repeated for many cycles to double the amount of template DNA many times in a row. This method is called a chain reaction because the products of each previous reaction (the newly made DNA strands) are used as reactants (the template strands) in subsequent reactions.

Figure 20.6 follows a PCR experiment through four cycles. Because the region of interest is doubled with each cycle, the end result of four cycles is 2^4, or 16 copies of the region of interest. Because the starting material contains long strands of chromosomal DNA, some of the products of a PCR experiment contain the region of interest plus some additional DNA at either end. However, after many cycles, the products that contain only the region of interest greatly predominate in the mixture.

PCR is carried out in a machine, known as a **thermocycler,** that automates the timing of each cycle. The experimenter mixes the DNA sample, dNTPs, *Taq* polymerase, and an excess amount of primers together in a single tube. The tube is placed in a thermocycler, and the experimenter sets the machine to operate within a defined temperature range and number of cycles. During each cycle, the thermocycler increases the temperature to denature the DNA strands and then lowers the temperature to allow annealing and extension to take place. Typically, each cycle lasts 2–3 minutes and is then repeated. A typical PCR run is likely to involve 20–30 cycles of replication and takes a couple of hours to complete. The PCR technique can amplify the amount of DNA by a staggering amount. Assuming 100% efficiency, the intervening region between the two primers increases 2^{20}-fold after 20 cycles. This is approximately a million-fold!

An important advantage of PCR is that it can amplify a particular region of DNA from a very complex mixture of template DNA. For example, if a researcher uses two primers that anneal to the human β-globin gene, PCR can amplify just the β-globin gene from a DNA sample that contains all of the human chromosomes!

Alternatively, PCR can be used to amplify a sample of chromosomal DNA nonspecifically. A nonspecific approach uses a mixture of short PCR primers with many different random sequences. These primers anneal randomly throughout the genome and amplify most of the chromosomal DNA. Nonspecific DNA amplification is used to increase the total amount of DNA in very small samples, such as blood stains found at crime scenes.

Reverse Transcriptase PCR Is Used to Amplify RNA

PCR is also used to detect and quantitate the amount of specific RNAs in living cells. To accomplish this goal, RNA is isolated from a sample and mixed with reverse transcriptase, a primer that binds near the 3′ end of the RNA of interest, and deoxyribonucleotides (**Figure 20.7**). This generates a single-stranded cDNA, which then can be used as template DNA in a conventional PCR reaction. The end result is that the RNA has been amplified to produce many copies of DNA.

This method, called **reverse transcriptase PCR,** is extraordinarily sensitive. Reverse transcriptase PCR can detect the expression of small amounts of RNA from a single cell! As discussed next, certain modifications to PCR allow researchers to observe the accumulation of PCR products and to quantitate the amount of DNA or RNA in a biological sample.

Real-Time PCR Is Used to Quantitate the Amount of a Specific Gene or mRNA in a Sample

In some applications of PCR, the goal is to obtain large amounts of a DNA region. To determine if PCR is successful, a researcher typically runs a sample of DNA on a gel, stains the DNA with ethidium bromide (EtBr), and then observes the gel under UV light, which causes EtBr to fluoresce. If a band of the correct size is seen, the experiment is likely to have been successful. This PCR approach is sometimes called endpoint analysis, because the success of the experiment is judged after PCR is completed.

By comparison, **real-time PCR** technology allows a researcher to follow the amount of a specific PCR product in "real time" as PCR is taking place in a thermocycler. Because the PCR products are ultimately derived from the template DNA that was initially added to the reaction, this approach allows researchers to determine how much DNA, such as the DNA that encodes a specific gene, was originally in the sample before PCR was conducted. Similarly, if the starting material is mRNA that is reverse-transcribed into DNA, real-time PCR can be used to determine how much mRNA from a specific gene was in a sample. This provides a way to quantitatively measure gene expression.

How do researchers determine the amount of a PCR product during real-time PCR? The procedure is carried out in a thermocycler that has the capacity to measure changes in the level of fluorescence that is emitted from probes that are added to the PCR mixture. The fluorescence given off by the probe depends on the amount of the PCR product. Several probes have been developed. We will consider one type called TaqMan.

The TaqMan probe is an oligonucleotide that has a reporter molecule at one end and a quencher molecule at the other (**Figure 20.8a**). The oligonucleotide is complementary to a site within the PCR product of interest. The reporter molecule emits fluorescence at a certain wavelength, but that fluorescence is largely absorbed by the nearby quencher. Therefore, the close proximity of the reporter molecule to the quencher molecule prevents the detection of fluorescence from the reporter molecule.

As shown in **Figure 20.8b**, a primer and TaqMan probe both anneal to the template DNA during the primer annealing step. During the primer extension step, the 5′ to 3′ exonuclease activity of *Taq* polymerase cleaves the oligonucleotide in the TaqMan probe into individual nucleotides, thereby separating the reporter from the quencher. This allows the reporter to emit (unquenched) fluorescence that can be measured within the

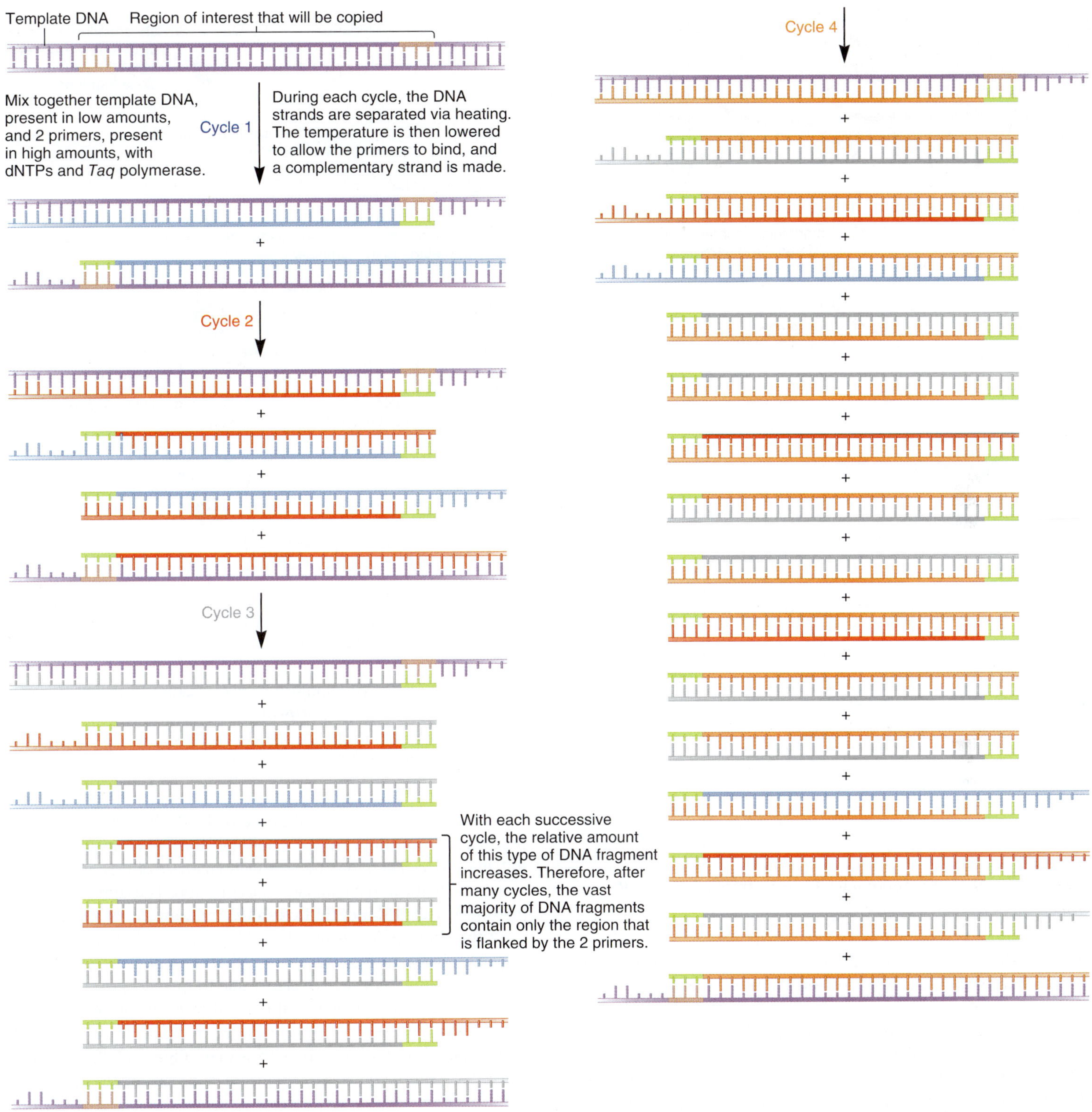

ONLINE ANIMATION

FIGURE 20.6 The technique of PCR carried out for four cycles. During each cycle, oligonucleotides that are complementary to the ends of the targeted DNA sequence (green) bind to the DNA and act as primers for the synthesis of this DNA region. Note: The original DNA strands are purple, and those strands made during the first, second, third, and fourth cycles are blue, red, gray, and orange, respectively.

CONCEPT CHECK: After four cycles of PCR, which type of PCR product predominates? Explain why.

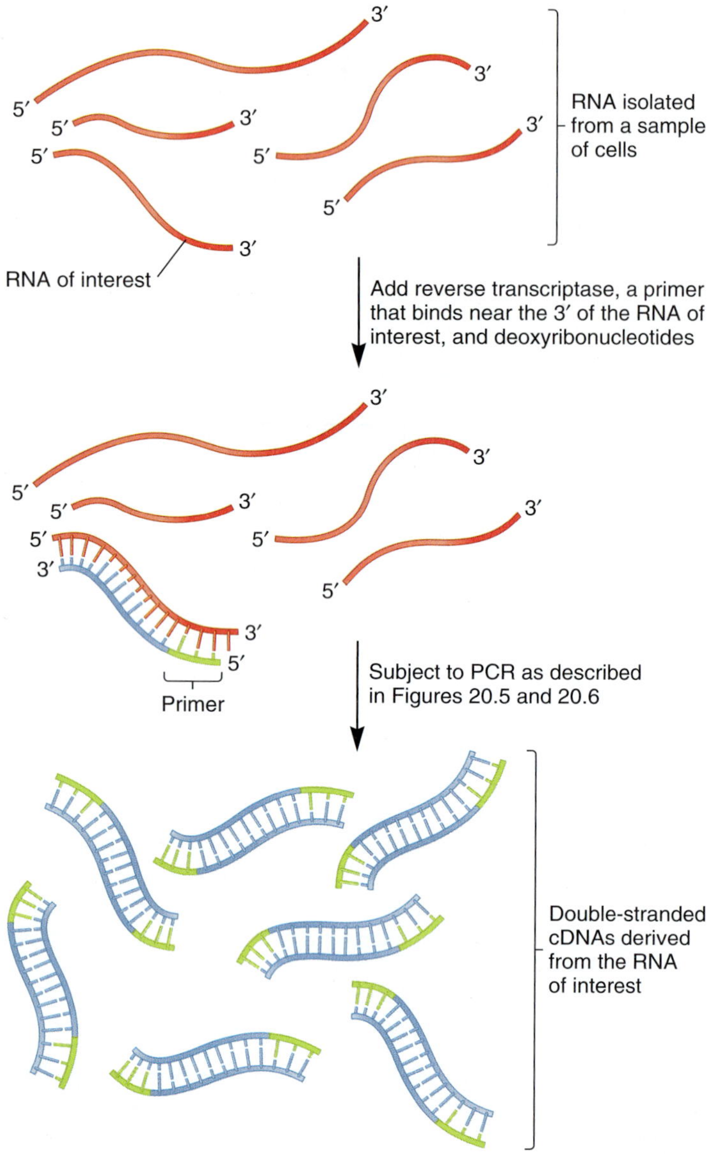

FIGURE 20.7 The technique of reverse transcriptase PCR.

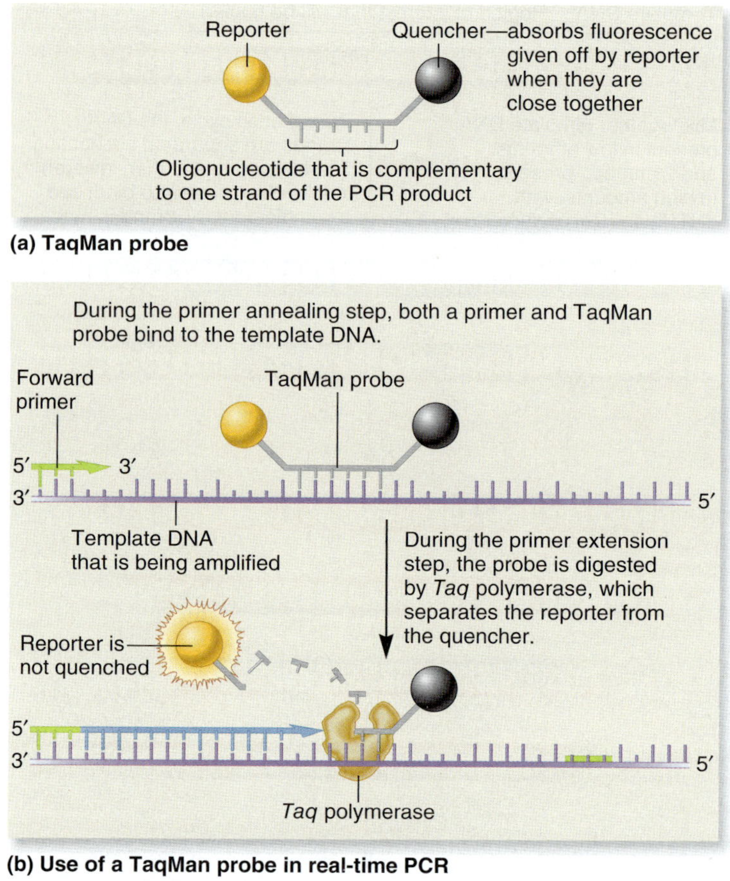

FIGURE 20.8 An example of a probe, called TaqMan, which is used in a real-time PCR experiment.

CONCEPT CHECK: What needs to happen so the reporter molecule can emit fluorescence that is not quenched?

thermocycler. As PCR products accumulate, more and more of the TaqMan probes are digested, and therefore the level of fluorescence increases.

Figure 20.9a considers a real-time PCR experiment as it occurs over the course of many cycles, such as 20–40. Real-time PCR goes through three main phases. Initially, when the amount of PCR product is small and reagents are not limiting, the synthesis of the PCR product is close to 100% efficiency—the amount of product nearly doubles with each cycle. This exponential accumulation is difficult to detect in the earliest cycles because the amount of PCR product is small. Therefore, the amount of TaqMan probe that is cleaved is also small. The second phase is approximately linear as PCR products continue to accumulate, but the reaction efficiency falls as reagents become limiting. Finally, in the third phase, the accumulation of PCR products reaches a plateau as one or more reagents are used up.

During the exponential phase of PCR, the amount of PCR product is proportional to the amount of starting template that was initially added. Therefore, the exponential phase of PCR is analyzed to quantitate the initial template concentration. The commonly used method is called the **cycle threshold method (C_t method).** The cycle threshold (C_t) is reached when the accumulation of fluorescence is significantly greater than the background level. During the early cycles, the fluorescence signal due to the background level of fluorescence is greater than that derived from the amplification of the PCR product. Once the C_t value is exceeded, the exponential accumulation of product can be measured. **Figure 20.9b** considers three PCR runs in which the template DNA was initially added at a high, medium, or low concentration. When the initial concentration of the template DNA is higher, the C_t is reached at an earlier amplification cycle.

To determine the amount of starting template DNA, the sample of interest that has an unknown amount of starting template DNA is compared with some type of standard, which will be described shortly. Real-time PCR involves the coamplification of two templates: the sample of interest and the standard. The two types of PCR products can be monitored simultaneously with different-colored fluorescent molecules.

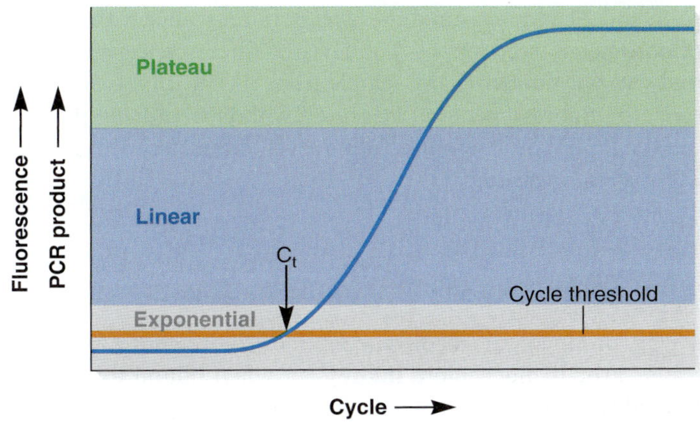

(a) Phases of PCR

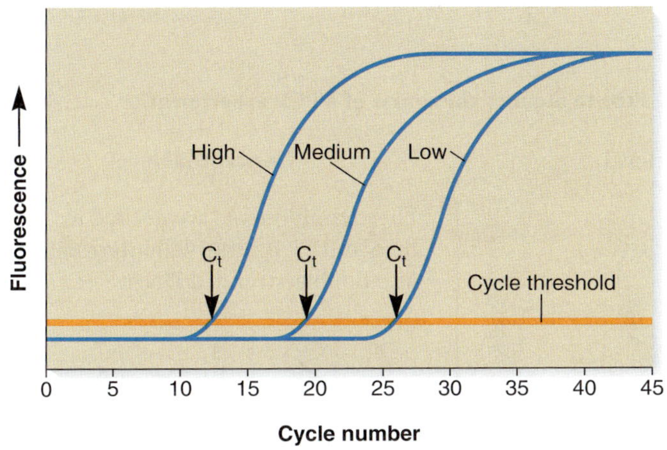

(b) Real-time PCR at high, medium, and low concentrations of the starting template DNA

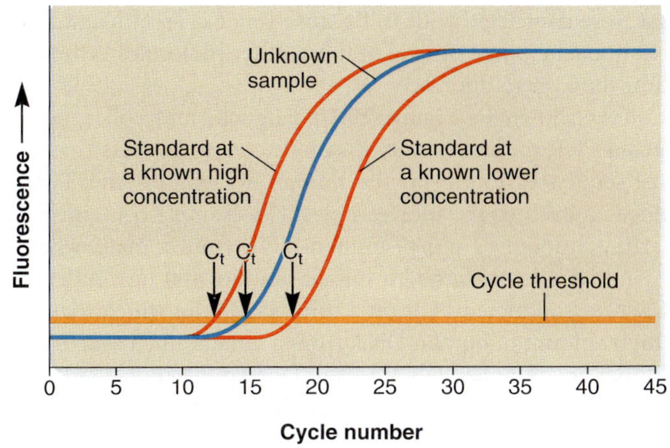

(c) A comparison between an unknown sample and standards of known concentrations

FIGURE 20.9 Examples of data that are obtained from a real-time PCR experiment. **(a)** The three phases that occur in a typical PCR experiment. **(b)** PCR carried out at three different starting template concentrations. **(c)** A comparison between a sample with an unknown DNA concentration and a standard at a high and low concentration.

What types of standards are commonly used? One possibility is that a standard of known concentration is added to the PCR mixture. For example, plasmid DNA carrying a specific gene can be added to the PCR mixture in known amounts, and the amplification of this plasmid-encoded gene provides a standard. By comparing the C_t values of the standard and the unknown sample, researchers can determine the concentration of the unknown sample. This is schematically shown in **Figure 20.9c**, but is actually done via computer software. Alternatively, researchers may use an internal standard in which another gene that is already present in the sample is also amplified. For example, a gene that encodes the cytoskeletal protein called actin may be used. This relative quantitation method is somewhat simpler. The amount of unknown template DNA of interest is expressed relative to the internal standard.

EXPERIMENT 20A

Early Attempts at Monitoring the Course of PCR Used Ethidium Bromide as a Probe

The first studies that showed the feasibility of real-time PCR were carried out by Russel Higuchi and colleagues in the early 1990s. To monitor the synthesis of PCR products, the researchers chose EtBr, which is a fluorescent dye that is used to stain DNA (see chapter-opening photograph). EtBr intercalates between the stacked bases in double-stranded DNA and becomes highly fluorescent when exposed to UV light. Why is EtBr more fluorescent when bound to double-stranded DNA? Although the answer is not entirely understood, researchers speculate that it is related to the hydrophobic environment between the bases. By moving into this hydrophobic environment, EtBr sheds any water molecules that were associated with it. Because water is an efficient fluorescent quencher, the removal of these water molecules is thought to allow the EtBr to fluoresce.

As shown in **Figure 20.10**, Higuchi and colleagues conducted PCR on four different samples using primers that recognize sequences found on the human Y chromosome. EtBr was added to each of the four samples. One sample contained a low starting amount (2 ng) of human DNA from a male, whereas a second sample contained a higher amount (60 ng). As controls, they also conducted PCR on a third sample in which they added a high amount of human DNA from a female and a fourth sample without any template DNA added. During the course of PCR, the researchers removed the tubes from the thermocycler at cycles 17, 21, 25, and 29 and measured the level of fluorescence with a spectrofluorometer (see step 4). The level of fluorescence was measured without removing any sample from the PCR tube. In other words, the fluorescence was measured without removing the cap from the PCR tube. The background level of fluorescence prior to PCR was also measured.

THE HYPOTHESIS

As PCR products are made, the level of fluorescence increases because EtBr is much more fluorescent when bound to double-stranded DNA.

TESTING THE HYPOTHESIS

FIGURE 20.10 **Using EtBr to monitor the course of a PCR experiment.**

1. Into each of four tubes, add primers that recognize DNA sequences on the human Y chromosome, dNTPs, and *Taq* polymerase. Also add ethidium bromide (EtBr) to the tubes.

2. Into tubes 1 and 2, add a low and high amount of human male DNA, respectively. Into tube 3, add a high amount of human female DNA. Tube 4 is a control that does not contain any template DNA.

3. Allow PCR to proceed as described in Figures 20.5 and 20.6. The template DNA should be amplified only in the tubes containing human male DNA because the primers are complementary to DNA sequences on the Y chromosome.

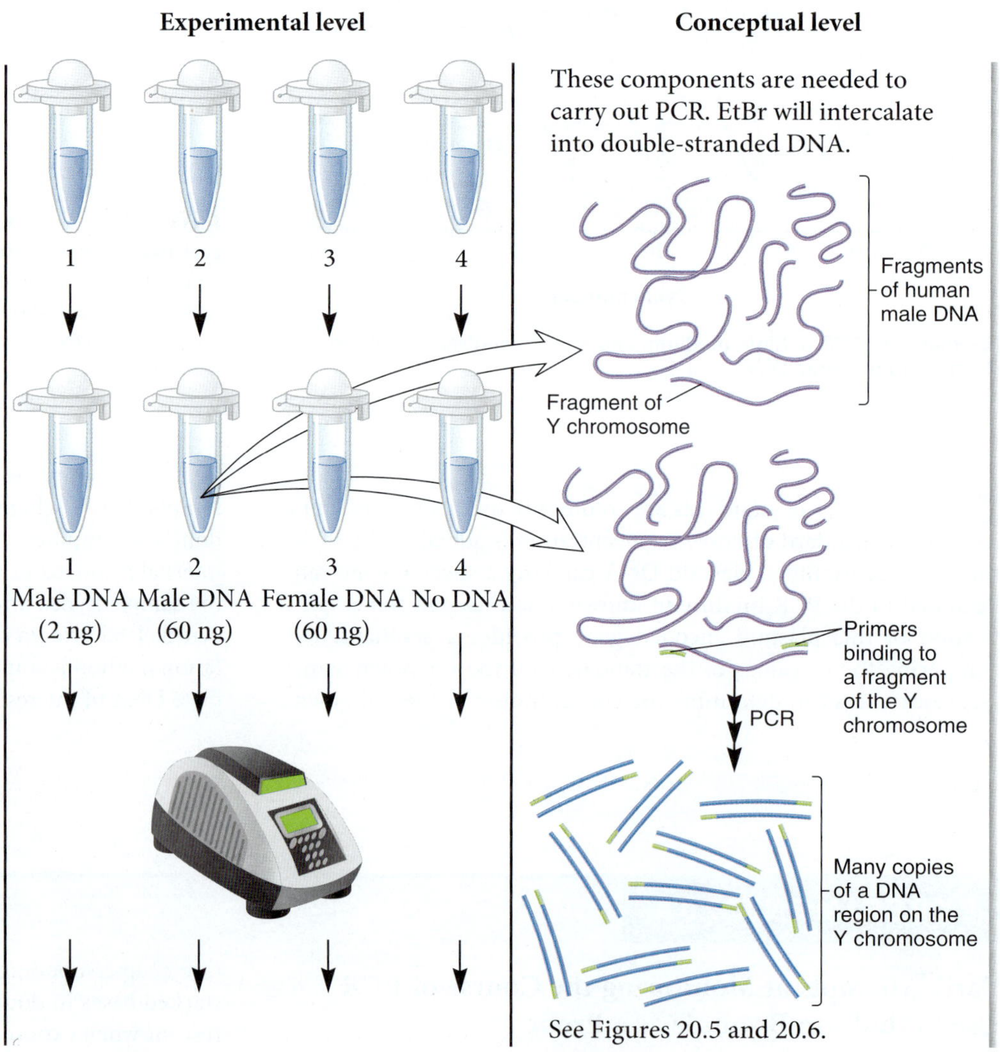

4. At cycles 17, 21, 25, and 29, remove the PCR tubes from the thermocycler and measure the level of fluorescence with a spectrofluorometer. Note: The tubes remained capped while the fluorescence was being measured.

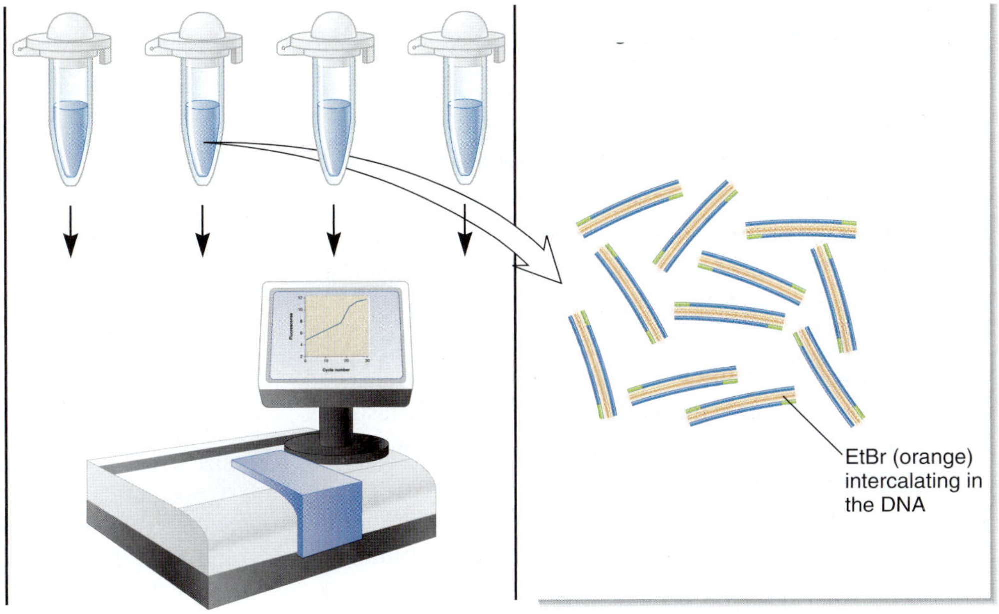

THE DATA

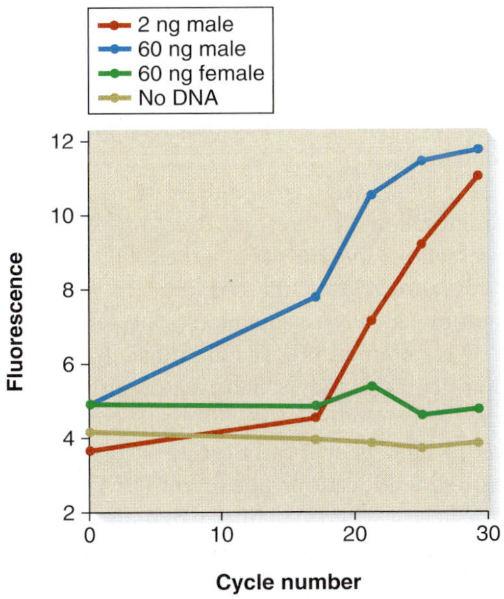

Data from R. Higuchi, G. Dollinger, P.S. Walsh, and R. Griffith (1992) Simultaneous amplification and detection of specific DNA sequences. *Bio/Technol 10*, 413–417.

INTERPRETING THE DATA

As shown in The Data, the level of fluorescence increased during the course of their PCR experiments when human DNA from a male was included in the reaction. The level of fluorescence increased more rapidly when a high concentration was added compared to a low concentration. In the two controls with human DNA from a female or when no template DNA was added, the fluorescence level did not increase. Because the fluorescence was measured without opening the tubes, these results indicated that EtBr fluorescence could be used to monitor the progress of a PCR experiment.

In other experiments not described in Figure 20.10, the researchers connected their spectrofluorometer directly to their thermocycler using a fiber-optic device. In this way, they were able to continuously monitor a PCR experiment. This was the first real-time monitoring of a PCR experiment. These pivotal experiments led to the development of thermocyclers with the built-in ability to monitor fluorescence.

Although EtBr is not typically used to monitor PCR experiments, other intercalators that are more highly fluorescent have found widespread use. For example, a cyanine dye called SYBR Green I is frequently used in real-time PCR experiments instead of probes such as TaqMan, which was described earlier in Figure 20.8.

A self-help quiz involving this experiment can be found at www.mhhe.com/brookergenetics5e.

20.2 COMPREHENSION QUESTIONS

1. In one PCR cycle, the correct order of steps is
   a. primer annealing, primer extension, denaturation.
   b. primer annealing, denaturation, primer extension.
   c. denaturation, primer annealing, primer extension.
   d. denaturation, primer extension, primer annealing.
2. In reverse transcriptase PCR, the starting biological material is
   a. chromosomal DNA.
   b. mRNA.
   c. proteins.
   d. all of the above.
3. During real-time PCR, the synthesis of PCR products is analyzed
   a. at the very end of the reaction by gel electrophoresis.
   b. at the very end of the reaction by fluorescence that is emitted within the thermocycler.
   c. during the PCR cycles by gel electrophoresis.
   d. during the PCR cycles by fluorescence that is emitted within the thermocycler.

20.3 DNA SEQUENCING

Learning Outcome:

1. Outline the steps in automated DNA sequencing via the dideoxy-sequencing method.

As we have seen throughout this textbook, our knowledge of genetics can be largely attributed to an understanding of DNA structure and function. The feature that underlies all aspects of inherited traits is the DNA sequence. For this reason, analyzing DNA sequences is a powerful approach for understanding genetics. A technique called **DNA sequencing** enables researchers to determine the base sequence of DNA found in genes and other chromosomal regions. It is one of the most important tools for exploring genetics at the molecular level. Molecular geneticists often want to determine DNA base sequences as a first step toward understanding the function and expression of genes. For example, the investigation of genetic sequences has been vital to our understanding of promoters, regulatory elements, and the genetic code itself. Likewise, an examination of sequences has facilitated our understanding of origins of replication, centromeres, telomeres, and transposable elements.

During the 1970s, two methods for DNA sequencing were devised. One method, developed by Allan Maxam and Walter Gilbert, involved the base-specific chemical cleavage of DNA. Another method, developed by Frederick Sanger and colleagues, is known as **dideoxy sequencing.** Because it became the more popular method of DNA sequencing, we will consider the dideoxy method here. In addition, Chapter 22 considers some newer methods of sequencing DNA that are not based on the Sanger dideoxy method. These newer methods are commonly used in projects aimed at determining the DNA sequence of entire genomes.

The dideoxy procedure of DNA sequencing is based on our knowledge of DNA replication but uses a clever twist. As described in Chapter 11, DNA polymerase connects adjacent deoxyribonucleotides by catalyzing a covalent bond between the 5′ phosphate

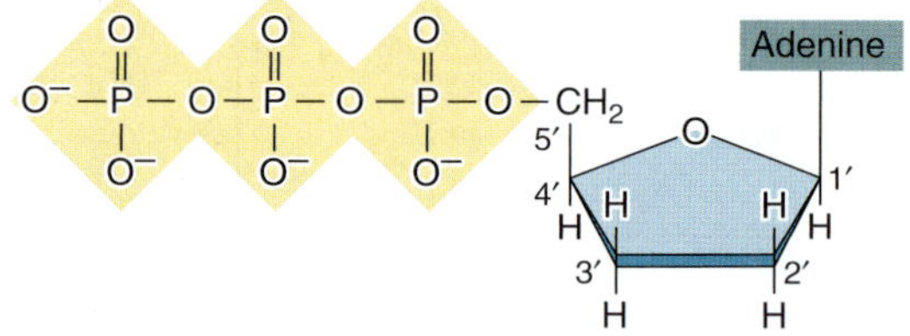

2′, 3′-Dideoxyadenosine triphosphate (ddA)

FIGURE 20.11 **The structure of a dideoxyribonucleotide.** Note that the 3′ group is a hydrogen rather than an –OH group. For this reason, another nucleotide cannot be attached at the 3′ position.

on one nucleotide and the 3′ –OH group on the previous nucleotide (refer back to Figure 11.16). Chemists, though, can synthesize deoxyribonucleotides that are missing the –OH group at the 3′ position (**Figure 20.11**). These synthetic nucleotides are called **dideoxyribonucleotides (ddNTPs).** (Note: The prefix dideoxy- indicates that two (di) oxygens (oxy) are removed (de) from this sugar compared with ribose; ribose has –OH groups at both the 2′ and 3′ positions.) Sanger reasoned that if a dideoxyribonucleotide is added to a growing DNA strand, the strand can no longer grow because the dideoxyribonucleotide is missing the 3′ –OH group. The incorporation of a dideoxyribonucleotide into a growing strand is therefore referred to as **chain termination.**

To detect the incorporation of dideoxyribonucleotides during DNA replication, the newly made DNA strands must be labeled in some way. When dideoxy sequencing was first invented, researchers labeled the newly made DNA with radioisotopes, which allowed them to be detected via autoradiography. This traditional DNA sequencing method has been replaced by **automated DNA sequencing** in which each type of dideoxyribonucleotide is labeled with a different-colored fluorescent dye. For example, a common way to fluorescently label ddNTPs is the following: ddA is green, ddT is red, ddG is yellow, and ddC is blue.

For automated DNA sequencing, the segment of DNA to be sequenced is obtained in large amounts. This can be accomplished using gene cloning, which was described earlier in this chapter. In the example in **Figure 20.12**, the segment of DNA to be sequenced, which we will call the target DNA, was cloned into a vector at a defined location. The target DNA was inserted next to a site in the vector where a primer will bind, which is called the primer-annealing site. The aim of the experiment is to determine the base sequence of the target DNA. In the experiment shown in Figure 20.12, the recombinant vector DNA has been previously denatured into single strands, usually via heat treatment. Only the strand needed for DNA sequencing is shown here.

Let's now consider the steps that are involved in DNA sequencing (Figure 20.12). First, a sample containing many copies of the single-stranded recombinant vector is mixed with many primers that will bind to the primer-annealing site. The primer binds to the DNA because the primer and primer-annealing site are complementary to each other. All four types of deoxyribonucleotides are added at a high concentration and all four dideoxyribonucleotides (ddA, ddT, ddG, or ddC), which are fluorescently labeled, are added at a low concentration. DNA polymerase is then added, which causes the synthesis of strands that are complementary to the target DNA sequence.

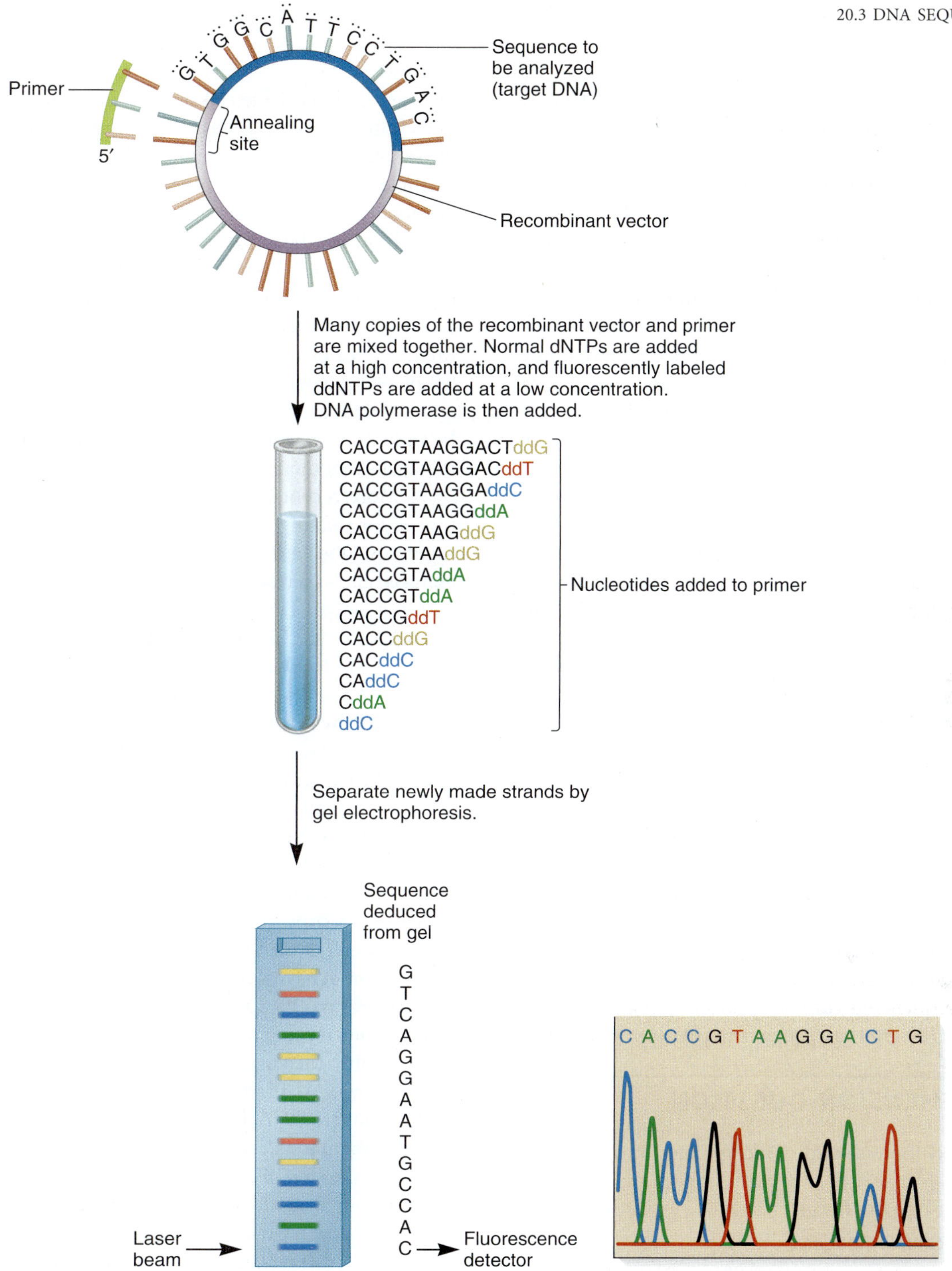

(a) Automated DNA sequencing **(b) Output from automated sequencing**

INTERACTIVE EXERCISE

FIGURE 20.12 The protocol for DNA sequencing by the dideoxy method. **(a)** The method begins with DNA in which the target DNA has been inserted into a vector. For the primer to bind, the recombinant vector must be denatured into single-stranded DNA at the beginning of the experiment. Only the strand needed for DNA sequencing is shown here. This diagram schematically depicts a series of bands on a gel; the four colors of the bands occur because each type of dideoxyribonucleotide is labeled with a different-colored fluorescent molecule. As each band passes a laser, the fluorescent dye is excited by the laser beam, and the fluorescence emission is recorded by a fluorescence detector. The detector reads the level of fluorescence at four wavelengths corresponding to the four dyes. **(b)** As shown in the printout, the peaks of fluorescence correspond to the DNA sequence that is complementary to the target DNA.

DNA synthesis continues until a dideoxyribonucleotide is incorporated into a growing strand. For example, chain termination can occasionally occur at the sixth or thirteenth position of the newly synthesized DNA strand if a ddT becomes incorporated at either of these sites. Note that the complementary A base is found at the sixth and thirteenth position in the target DNA. Therefore, we expect to make DNA strands that terminate at the sixth or thirteenth positions and have a ddT at their ends. Because these DNA strands contain a ddT, they are fluorescently labeled in red. Alternatively, ddA causes chain termination at the second, seventh, eighth, or eleventh positions because a complementary T base is found at the corresponding positions in the target strand. Strands that are terminated with ddA are fluorescently labeled in green.

After the samples have been incubated for several minutes, mixtures of DNA strands of different lengths are made, depending on the number of nucleotides attached to the primer. These DNA strands are separated according to their lengths by running them on a slab gel or more commonly by running them through a gel-filled capillary tube. The shorter strands move to the bottom of the gel more quickly than the longer strands. Because we know the color of each dideoxyribonucleotide, we also know which base is at the very end of each DNA strand separated on the gel. Therefore, we can deduce the DNA sequence that is complementary to the target DNA by "reading" which base is at the end of every DNA strand and matching this sequence with the length of the strand. Reading the base sequence, from bottom to top, is much like climbing a ladder of bands. For this reason, the sequence obtained by this method is referred to as a **sequencing ladder.**

Theoretically, it is possible to read this sequence directly from the gel. From a practical perspective, however, it is faster and more efficient to automate the procedure using a laser and fluorescence detector. As the gel is running, each band passes the laser and the laser beam excites the fluorescent dye. The fluorescence detector records the amount of fluorescence emission from the excited dye. The detector reads the level of fluorescence at four wavelengths, corresponding to the four different-colored dyes. An example of the printout from the fluorescence detector is shown in Figure 20.12b. As seen here, the peaks of fluorescence correspond to the DNA sequence that is complementary to the target DNA. Note that ddG is usually labeled with a yellow dye, but it is converted to black ink on the printout shown in Figure 20.12b for ease of reading. Though improvements in automated sequencing continue to be made, a typical sequencing run can provide a DNA sequence that is approximately 700–900 bases long, and perhaps even longer.

20.3 COMPREHENSION QUESTION

1. When a dideoxyribonucleotide is incorporated into a growing DNA strand,
   a. the strand elongates faster.
   b. the strand cannot elongate.
   c. the strand becomes more susceptible to DNase I cleavage.
   d. none of the above occurs.

20.4 SITE-DIRECTED MUTAGENESIS

Learning Outcomes:

1. Describe the method of site-directed mutagenesis.
2. List a few reasons why this method is useful.

To understand how the genetic material functions, researchers often analyze mutations that alter the normal DNA sequence, thereby affecting the expression of genes and the outcome of traits. For example, geneticists have discovered that many inherited human diseases, such as sickle cell disease and hemophilia, involve mutations within specific genes. These mutations provide insight into the function of the genes in unaffected individuals. Hemophilia, for example, is caused by mutations in genes that encode blood clotting factors.

Because the analysis of mutations can provide important information about normal genetic processes, researchers often wish to obtain mutant organisms. As discussed in Chapter 18, mutations can arise spontaneously or can be induced by environmental agents. Mendel's pea plants are a classic example of allelic strains with different phenotypes that arose from spontaneous mutations. X-rays and UV light are physical agents that can cause induced mutations. In addition, experimental organisms can be treated with chemical mutagens that increase the rate of mutations.

More recently, researchers have developed molecular techniques for making mutations within cloned genes or other DNA segments. One widely used method, known as **site-directed mutagenesis,** allows a researcher to produce a mutation at a specific site within a cloned DNA segment. For example, if a DNA sequence is 5′–AAATTTCTTTAAA–3′, a researcher can use site-directed mutagenesis to change it to 5′–AAATTTGTTTAAA–3′. In this case, the researcher deliberately changed the seventh base from a C to a G. Why is this method useful? The site-directed mutant can then be introduced into a living organism to see how the mutation affects the expression of a gene, the function of a protein, and the phenotype of an organism.

The first successful attempts at site-directed mutagenesis involved changes in the sequences of viral genomes. These studies were conducted in the 1970s. Mark Zoller and Michael Smith also developed a protocol for the site-directed mutagenesis of DNA that has been cloned into a viral vector. Since these early studies, many approaches have been used to achieve site-directed mutagenesis. **Figure 20.13** describes the general steps in the procedure. Prior to this experiment, the DNA was denatured into single strands; only the single strand needed for site-directed mutagenesis is shown. As in PCR, this single-stranded DNA is referred to as the template DNA, because it is used as a template to synthesize a complementary strand.

As shown in Figure 20.13, an oligonucleotide primer is allowed to hybridize or anneal to the template DNA. The primer, typically 20 or so nucleotides in length, is synthesized chemically. (A shorter version of the primer is shown in Figure 20.13 for simplicity.) The scientist designs the base sequence of the primer. The primer has two important characteristics. First, most of the sequence of the primer is complementary to the site in the DNA where the mutation is to be made. However, a second feature is that the primer contains a region of mismatch where the primer and template DNA are not complementary. The mutation occurs in this mismatched region. For this reason, site-directed mutagenesis is sometimes referred to as oligonucleotide-directed mutagenesis.

After the primer and template have annealed, the complementary strand is synthesized by adding deoxyribonucleoside triphosphates (dNTPs), DNA polymerase, and DNA ligase. This yields a double-stranded molecule that contains a mismatch only at the desired location. This double-stranded DNA can then be

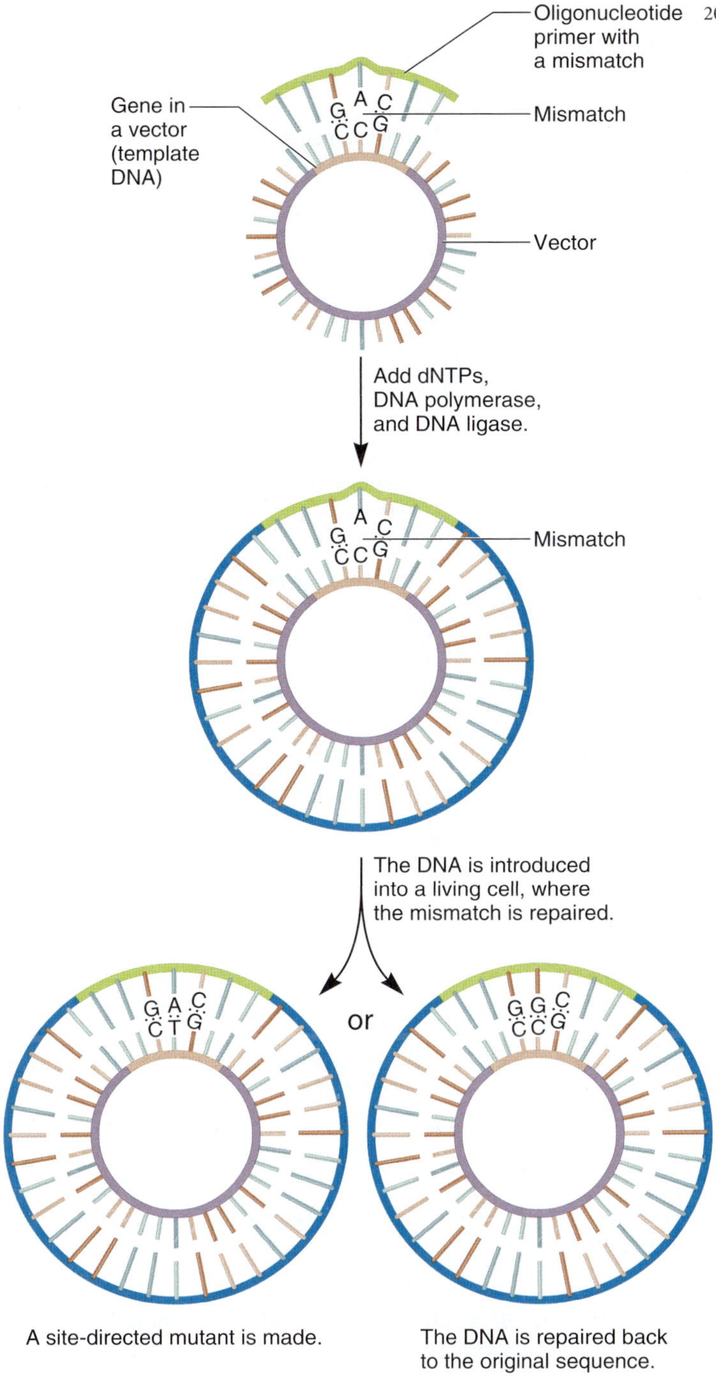

FIGURE 20.13 The method of site-directed mutagenesis.

Genes→Traits To examine the relationship between genes and traits, researchers can alter gene sequences via site-directed mutagenesis. The altered gene can then be introduced into a living organism to examine how the mutation affects the organism's traits. For example, a researcher could introduce a nonsense mutation into the middle of the *lacY* gene in the *lac* operon. If this site-directed mutant was introduced into an *E. coli* bacterium that did not a have a normal copy of the *lacY* gene, the bacterium would be unable to use lactose. These results indicate that a functional *lacY* gene is necessary for bacteria to have the trait of lactose utilization.

CONCEPT CHECK: Describe three possible uses of site-directed mutagenesis.

introduced into a bacterial cell. Within the cell, the DNA mismatch is likely to be repaired (see Chapter 18). Depending on which base is replaced, this may produce the mutant sequence or the original sequence. Clones containing the desired mutation are identified by DNA sequencing and used for further studies.

After a site-directed mutation has been made within a cloned gene, its consequences are analyzed by introducing the mutant gene into a living cell or organism. As described earlier, recombinant vectors containing cloned genes can be introduced into bacterial cells. Following transformation into a bacterium, a researcher can study the differences in function between the mutant and wild-type genes and the proteins they encode. Similarly, mutant genes made via site-directed mutagenesis can be introduced into plants and animals, as discussed in Chapter 21.

20.4 COMPREHENSION QUESTION

1. The purpose of site-directed mutagenesis is to
   a. determine if a protein binds to a DNA segment.
   b. determine the sequence of a segment of DNA.
   c. alter the sequence of a segment of DNA.
   d. determine if a gene is expressed.

20.5 BLOTTING METHODS TO DETECT GENES AND GENE PRODUCTS

Learning Outcomes:

1. Explain how a specific gene can be detected using colony hybridization.
2. Compare and contrast the techniques of Southern, Northern, and Western blotting and explain why each is used.

Researchers often want to focus their studies on a single gene. This presents a problem because chromosomal DNA contains thousands of different genes. For this reason, researchers need methods for specifically identifying a gene within a mixture of many other genes or DNA fragments. Similarly, when studying gene expression at the molecular level, scientists also need techniques for the identification of gene products, such as the RNA that is transcribed from a particular gene or the protein that is encoded by an mRNA. In this section, we will consider the methodology and uses of several common techniques used to detect specific genes, mRNAs, and proteins.

A DNA Library May Be Screened by Colony Hybridization to Identify a Cloned Gene

In many cases, the goal of a gene-cloning experiment is to obtain many copies of a specific gene or region of a gene. In cloning via PCR, this is achieved by selecting primers that flank the region of interest. This allows a researcher to specifically clone a gene of interest without cloning other segments of chromosomal DNA. However, when cloning is done using vectors, many different segments of chromosomal DNA are cloned during the construction of a DNA library. For example, let's suppose that a geneticist wishes to clone the rat β-globin gene. To begin a cloning experiment, chromosomal DNA is isolated from rat cells. This chromosomal DNA is then digested with a restriction enzyme, yielding thousands of DNA fragments. The chromosomal fragments are then ligated to vector DNA and transformed into bacterial cells to make a DNA library. Unfortunately, only a small percentage of the recombinant vectors, perhaps one in a few thousand, actually contains the rat β-globin gene. For this reason, researchers need some way of identifying those rare bacterial colonies within a DNA library that happen to contain the cloned gene of interest, in this case, a colony that contains the rat β-globin gene.

Figure 20.14 describes one method for identifying a bacterial colony that contains the gene of interest. This procedure is referred to as **colony hybridization.** The master plate shown at the top of Figure 20.14 has many bacterial colonies. Each bacterial colony is composed of bacterial cells containing a recombinant vector with a different piece of rat chromosomal DNA. The goal is to identify a colony that contains the gene of interest, in this case, the rat β-globin gene. To do so, a nylon membrane is laid gently onto the master plate. After the membrane is lifted, some cells from each colony are attached to it. In this way, the membrane contains a replica of the colonies on the master plate.

The nylon membrane is then treated with detergent, which dissolves the bacterial cell membranes and makes the DNA accessible. The DNA within the cells is then fixed (i.e., adhered) to the nylon membrane and denatured with NaOH. The membrane is then submerged in a solution containing a probe, which can be an oligonucleotide, a fragment of DNA from a cloned gene, or a specific RNA. In this case, the probe is single-stranded DNA with a base sequence that is complementary to one of the DNA strands of the rat β-globin gene. The probe is labeled in some way for ease of detection. For example, it could be labeled with a dye. The probe is given time to hybridize to the DNA on the membrane, which has already been denatured into single strands. If a bacterial colony contains the rat β-globin gene, the probe will hybridize to the DNA in this colony. Most bacterial colonies are not expected to contain this gene. The unbound probe is then washed away. If the DNA within a bacterial colony did hybridize to the probe, a labeled spot (shown in purple) will appear at the corresponding location. The pattern of spots on the membrane is compared with the master plate to identify bacterial colonies carrying the rat β-globin gene. Following the identification of labeled colonies, the experimenter can pick them from the plate, and grow bacteria containing the cloned gene.

A scientist might follow several different strategies to obtain a probe for a colony hybridization experiment. In some cases, the gene of interest may have already been cloned. A piece of the cloned gene then can be used as a labeled probe. However, in many circumstances, a scientist may attempt to clone a gene that has never been cloned before. In such a case, one strategy is to use a probe that likely has a sequence similar to that of the gene of interest. For example, if the goal is to clone the β-globin gene from a South American rodent, it is expected that this gene is similar to the rat β-globin gene, which has already been cloned. Therefore, the cloned rat β-globin gene can be used as a DNA probe to "fish out" the β-globin gene from a similar species.

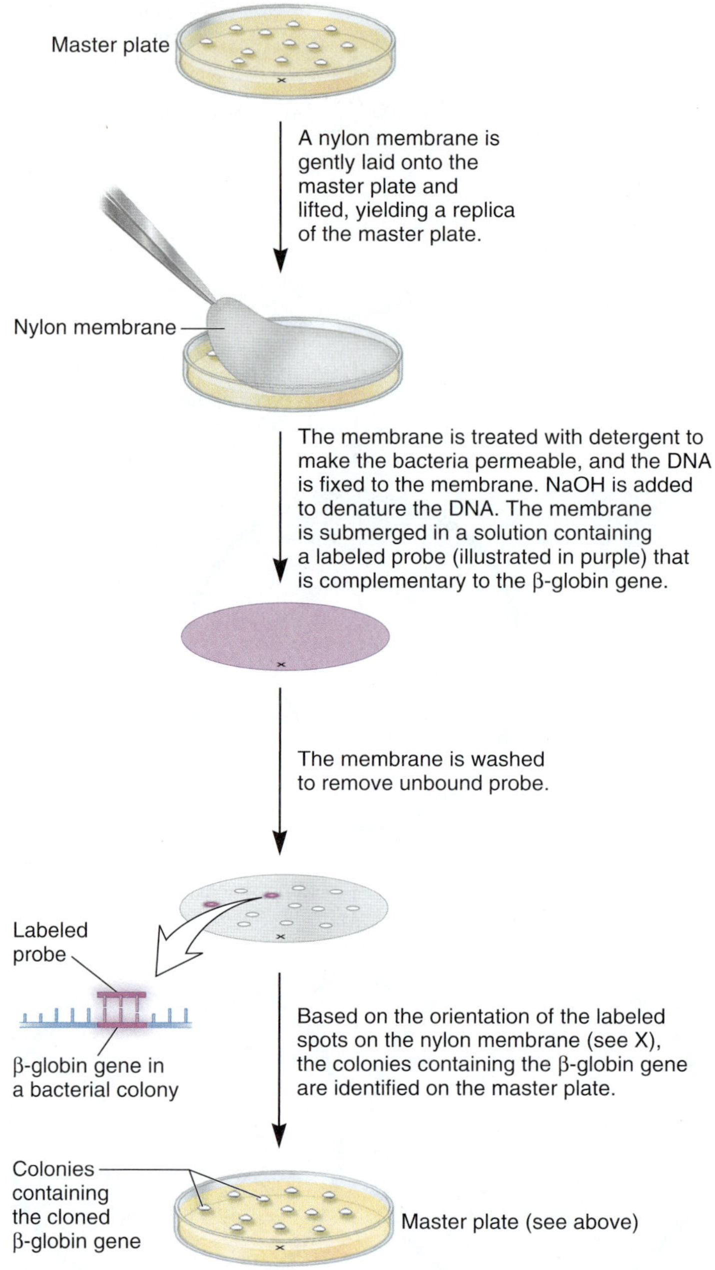

FIGURE 20.14 A colony hybridization experiment. Note: If the vector used was a viral vector, the master plate would have viral plaques. The steps shown here could also be followed, in which case the experiment would be called plaque hybridization.

Southern Blotting Is Used to Detect DNA Sequences

The technique of **Southern blotting** is used to detect the presence of a particular gene sequence within a mixture of many chromosomal DNA fragments. This method, developed by Edwin Southern in 1975, has several uses. It can determine the copy number of a gene within the genome of an organism. For example, Southern blotting has revealed that rRNA genes are found in multiple copies within a genome, whereas many protein-encoding genes are unique. In the study of human genetic diseases, Southern blotting can also detect small gene deletions that cannot be distinguished under the light microscope (see solved problem S4 at the end of this chapter). A common use of Southern blotting is to identify gene families. As discussed in Chapter 8, a gene family is a group of two or more genes derived from the same ancestral gene. The members of a gene family have similar but not identical DNA sequences; they are homologous genes. As we will see, Southern blotting can distinguish the homologous members of a gene family. Similarly, Southern blotting can identify homologous genes in different species. This is called a zoo blot. Finally, as discussed in Chapter 21, researchers may produce transgenic organisms that carry new genes or modified genes. Southern blotting may be used to verify that transgenic organisms carry such genes.

Prior to a Southern blotting experiment, a gene or fragment of a gene of interest has been cloned. This cloned DNA is labeled (e.g., with a dye) in vitro, and then the labeled DNA is used as a probe to detect the presence of the gene or a homologous gene within a mixture of many DNA fragments. The basis for a Southern blotting experiment is that two DNA fragments bind to each other only if they have complementary sequences.

Figure 20.15a shows the Southern blotting procedure. The goal of this experiment is to determine if a sample of chromosomal DNA contains a segment of DNA that is complementary to a specific probe. To begin the experiment, the chromosomal DNA is isolated and digested with a restriction enzyme. Because the restriction enzyme cuts the chromosomal DNA at many different sites within the chromosomes, this step produces thousands of DNA pieces of different sizes. Depending on the location of the restriction sites, each gene is contained on one or more DNA pieces with specific masses. The chromosomal pieces are separated according to their masses by gel electrophoresis, typically using a gel made of agarose. The DNA pieces within the gel are denatured by soaking the gel in an NaOH solution, and then the DNA is transferred (i.e., blotted) to a nylon membrane.

As shown in **Figure 20.15b**, the traditional way to transfer the DNA to the membrane is to lay the gel on blotting paper in a transfer solution. A nylon membrane is placed on top of the gel, and then additional blotting paper and dry paper towels are layered above the nylon membrane. A glass plate and heavy weight are placed on the very top. The liquid from the transfer solution is drawn upward, carrying the DNA from the gel to the nylon membrane. When the DNA enters the nylon membrane, it remains bound to it. Alternatively, another way to transfer DNA is via a transfer apparatus in which the gel is placed on top of a nylon membrane and then an electric field causes the DNA fragments to electrophoretically move from the gel into the membrane (**Figure 20.15c**).

The procedures shown in Figure 20.15b and c result in a nylon membrane that contains many unlabeled DNA fragments that have been separated according to mass and then denatured. The nylon membrane is incubated in an oven or exposed to UV light to permanently attach the DNA fragments to the membrane. The next step is to determine if any of these unlabeled fragments from the chromosomal DNA contain sequences that are complementary to a labeled probe. The nylon membrane, which has the unlabeled chromosomal DNA attached to it, is submerged in a solution containing the labeled probe. If the labeled probe and a fragment of chromosomal DNA are complementary, they will hydrogen bond to each other. Any unbound labeled DNA is then washed away. Locations where the labeled probe has bound appear as labeled bands (shown in purple).

Important variables in the Southern blotting procedure are the temperature and ionic strength of the hybridization and wash steps. If the procedure is done at very high temperatures and/or at low salt concentrations, the probe DNA and chromosomal fragment must be very complementary—nearly a perfect match—in order to hybridize. This condition is called **high stringency.** Conditions of high stringency are used to detect close or exact complementarity between the probe and a chromosomal DNA fragment. However, if the temperature is lower and/or the salt concentration is higher, DNA sequences that are not perfectly complementary may hybridize to the probe. This is called **low stringency.** Conditions of low stringency are used to detect homologous genes with DNA sequences that are similar but not identical to the cloned gene being used as a probe. In the results shown in Figure 20.15a, conditions of high stringency reveal that the gene of interest is unique in the genome. At low stringency, however, two other bands are detected. These results suggest that this gene is a member of a gene family composed of three distinct members.

Northern Blotting Is Used to Detect RNA

As discussed earlier in this chapter, reverse transcriptase PCR and real-time PCR are used to detect and quantitate the amount of RNA in a biological sample. Another approach, known as **Northern blotting,** is also used to identify a specific RNA within a mixture of many RNA molecules. (Note: Even though Southern blotting was named after Edwin Southern, Northern blotting was not named after anyone called Northern! It was originally termed reverse-Southern blotting and later became known as Northern blotting.) Northern blotting is used to investigate the transcription of genes at the molecular level. This method can determine if a specific gene is transcribed in a particular cell type, such as nerve or muscle cells, or at a particular stage of development, such as fetal or adult cells. Also, Northern blotting can reveal if a pre-mRNA transcript is alternatively spliced into two or more mRNAs of different sizes.

From a technical viewpoint, Northern blotting is rather similar to Southern blotting with a few important differences. RNA is the starting material and is extracted and purified from living cells. This RNA can be isolated from a particular cell type under a given set of conditions or during a particular stage of development. Any particular cell produces thousands of different types of RNA molecules, because cells express many genes at any given time. After the RNA is extracted from cells and purified, it is loaded onto an agarose gel that separates the RNA transcripts according to their size. The RNAs within the gel are then blotted onto a nylon membrane and probed with a labeled fragment of DNA from a cloned gene. RNAs that are complementary to the probe DNA are detected as labeled bands (shown in purple).

In **Figure 20.16**, the probe was complementary to an mRNA that encodes a protein called tropomyosin. In lane 1, the RNA was isolated from smooth muscle cells, in lane 2 from striated muscle

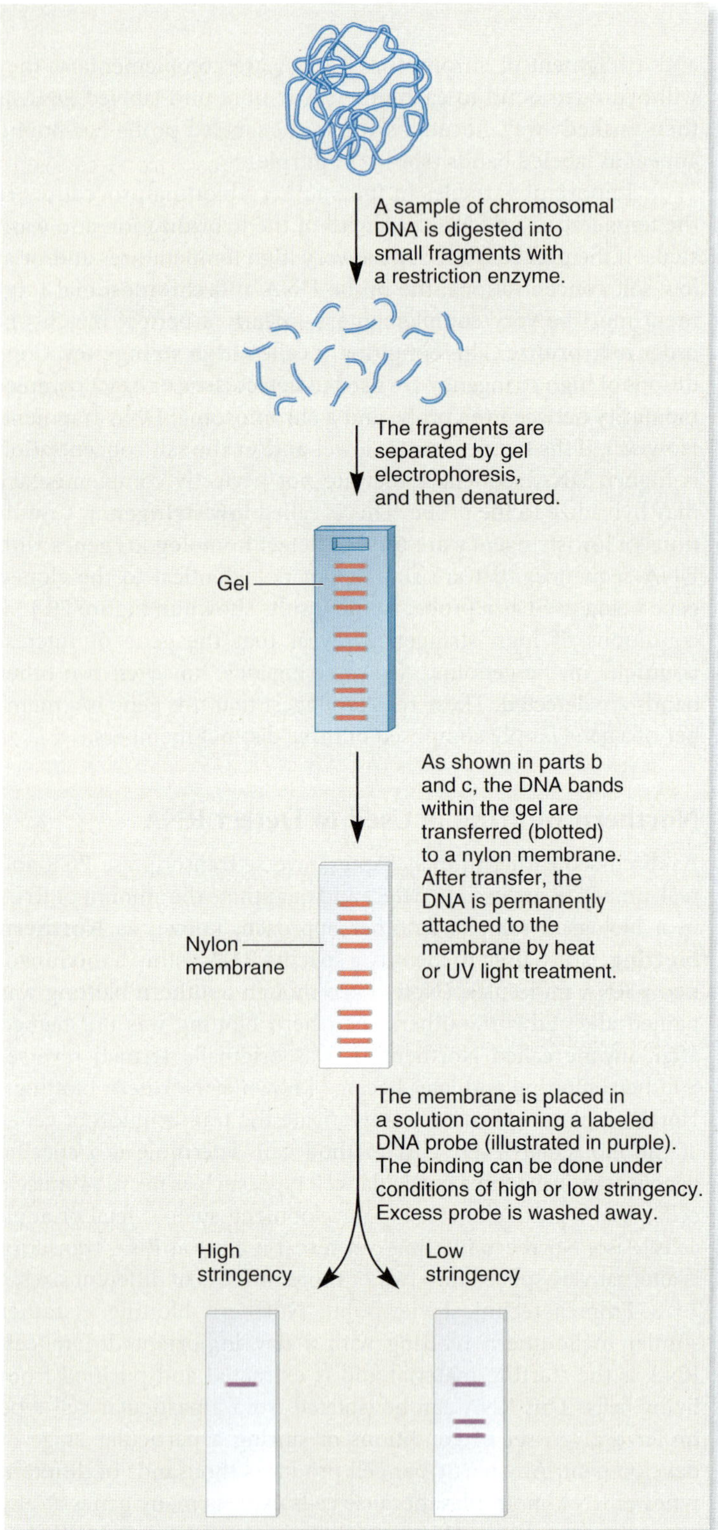

(a) The steps in Southern blotting

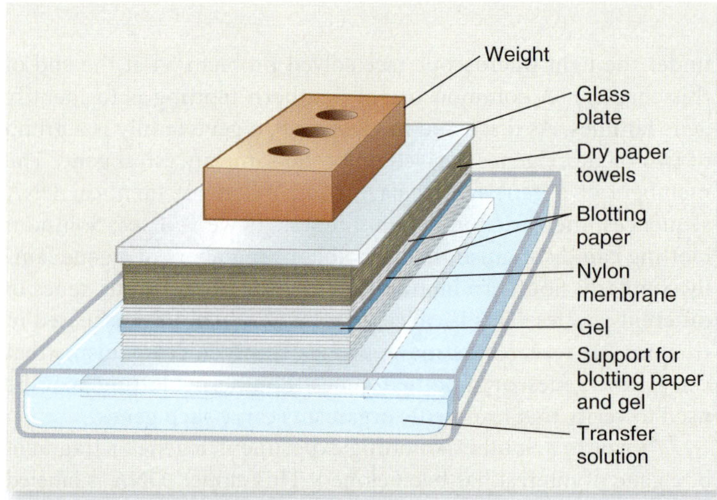

(b) The transfer step (traditional method)

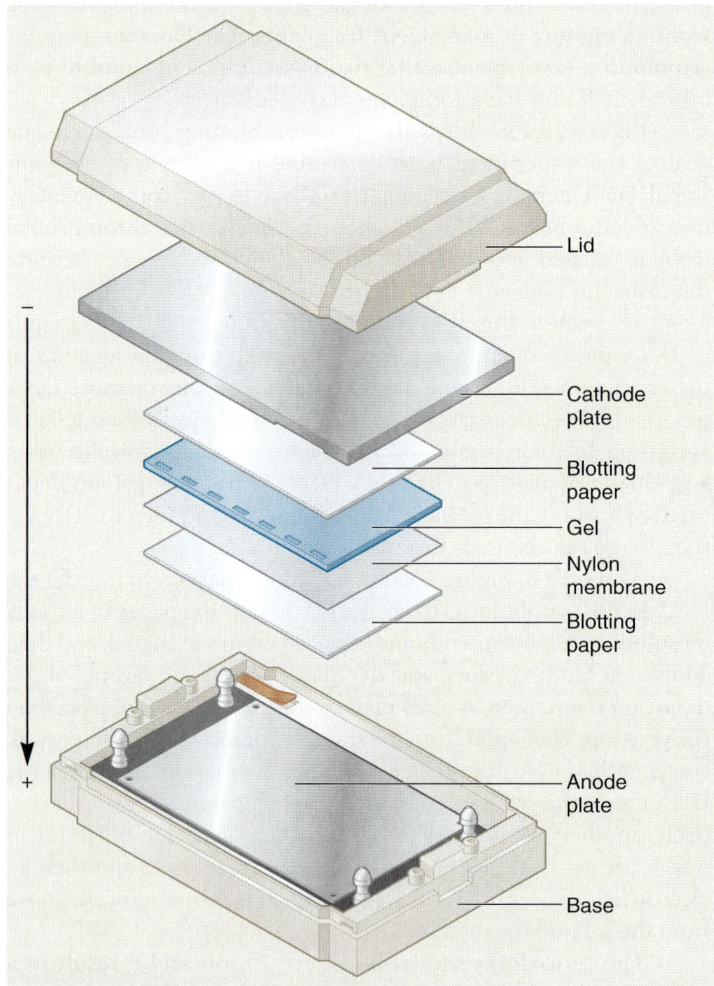

(c) The transfer step via electrophoresis

FIGURE 20.15 The technique of Southern blotting. **(a)** Chromosomal DNA is isolated from a sample of cells and then digested with a restriction enzyme to yield many DNA fragments. The fragments are then separated by gel electrophoresis according to molecular mass and denatured with NaOH. The bands within the gel are transferred to a nylon membrane, a step known as blotting. The blot is then placed in a solution containing a labeled DNA probe (shown in purple). When the DNA probe is hybridized to the mixture of chromosomal DNA fragments at high temperature and/or low ionic strength (i.e., high stringency), the probe recognizes only a single band. When hybridization is performed at lower temperature and/or higher ionic strength (i.e., low stringency), this band and two additional bands are produced. **(b)** The transfer step of DNA blotting (traditional method). **(c)** The transfer step of DNA blotting via electrophoresis.

CONCEPT CHECK: What type of information is obtained under conditions of low stringency?

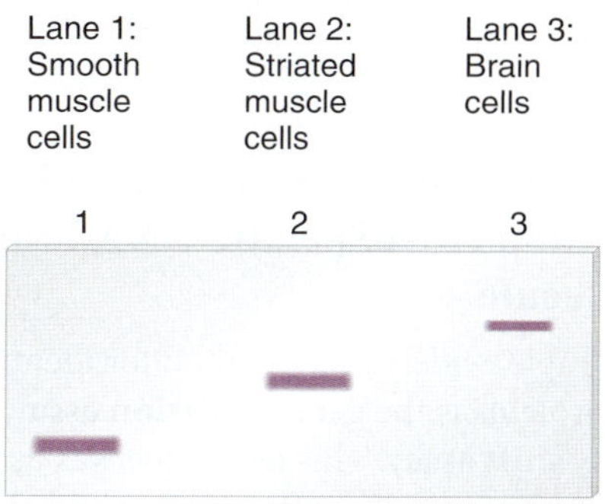

FIGURE 20.16 The results of Northern blotting. RNA is isolated from a sample of cells and then separated by gel electrophoresis. The separated RNA bands are blotted to a nylon membrane and then placed in a solution containing a labeled DNA probe that is complementary to tropomyosin mRNA. RNA molecules that are complementary to the probe appear as labeled bands (shown in purple).

cells, and in lane 3 from brain cells. As seen here, smooth and striated muscle cells contain a large amount of this mRNA. This result is expected because tropomyosin plays a role in the regulation of cell contraction. By comparison, brain cells have much less of this mRNA. In addition, we see that the molecular masses of the three mRNAs are different among the three cell types. This observation indicates that the pre-mRNA is alternatively spliced to contain different combinations of exons.

Western Blotting Is Used to Detect Proteins

For protein-encoding genes, the end result of gene expression is the synthesis of proteins. A particular protein within a mixture of many different protein molecules can be identified by **Western blotting,** a third detection procedure. This method can determine if a specific protein is made in a particular cell type or at a particular stage of development. Technically, this procedure is also similar to Southern and Northern blotting.

In a Western blotting experiment, proteins are extracted from living cells. As with RNA, any given cell produces many different proteins at any time, because it is expressing many different genes. After the proteins have been extracted from the cells, they are loaded onto a gel that separates them by molecular mass. To perform the separation step, the proteins are first dissolved in sodium dodecyl sulfate (SDS), a detergent that denatures proteins and coats them with negative charges. The negatively charged proteins are then separated in a gel made of polyacrylamide. This method of separating proteins is called SDS-PAGE (polyacrylamide gel electrophoresis).

Following SDS-PAGE, the proteins within the gel are blotted to a nylon membrane. The next step is to use a probe that recognizes a specific protein of interest. An important difference between Western blotting and either Southern or Northern blotting is the use of an **antibody** as a probe, rather than a labeled DNA strand. Antibodies bind to sites known as **epitopes.** An epitope has a three-dimensional structure that is recognized by an antibody. The term **antigen** refers to any molecule that is recognized by an antibody. An antigen contains one or more epitopes. In the case of proteins, an epitope is a short sequence of amino acids. Because the amino acid sequence is a unique feature of each protein, any given antibody specifically recognizes a particular protein. In a Western blot, this is called the primary antibody for that protein (**Figure 20.17a**).

After the primary antibody has been given sufficient time to recognize the protein of interest, any unbound primary antibody is washed away, and a secondary antibody is added. A secondary antibody is an antibody that specifically recognizes and binds to a region in the primary antibody. Secondary antibodies, which may be labeled or conjugated to an enzyme, are used for convenience, because secondary antibodies are available commercially. In general, it is easier for researchers to obtain these antibodies from commercial sources rather than labeling their own primary antibodies. In a Western blotting experiment, the secondary antibody provides a way to detect the protein of interest in a gel blot. For example, it is common for the secondary antibody to be linked to the enzyme alkaline phosphatase. When the colorless compound, XP (5-bromo-4-chloro-3-indolyl phosphate), is added to the

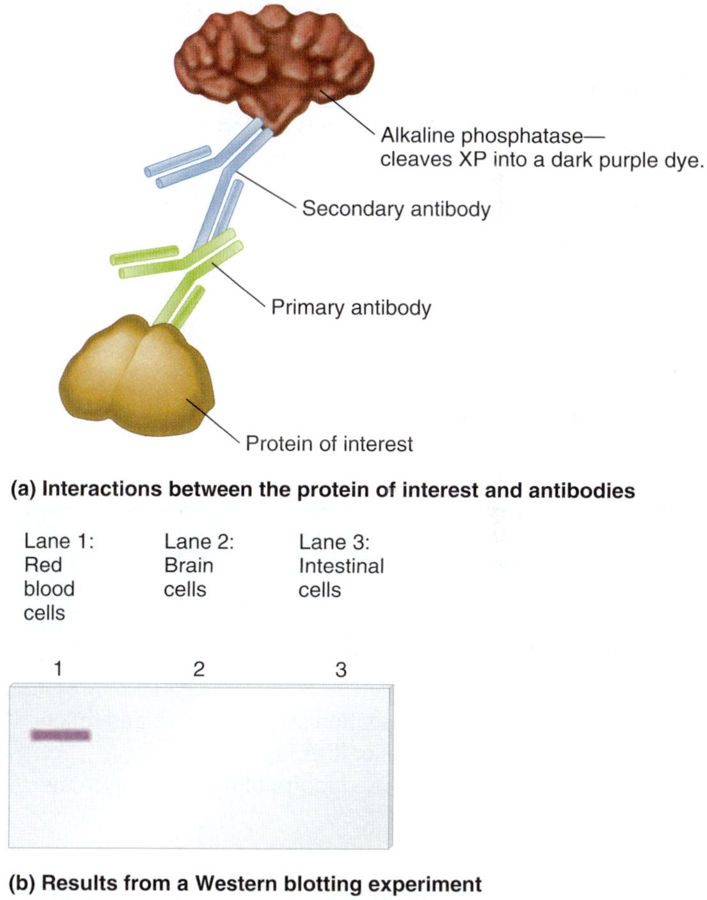

FIGURE 20.17 Western blotting. **(a)** After blotting, a primary antibody is added that binds to the protein of interest. Then a secondary antibody is added that binds to the primary antibody. In this example, the secondary antibody is also attached to an enzyme called alkaline phosphatase. When XP is added, alkaline phosphatase cleaves XP to a compound with a dark color (shown in purple). **(b)** The dark purple band indicates where the primary antibody has recognized the protein of interest. In lane 1, proteins were isolated from mouse red blood cells. As seen here, the β-globin polypeptide is made in these cells. By comparison, lanes 2 and 3 were samples of protein from brain cells and intestinal cells, respectively, which do not synthesize β globin.

CONCEPT CHECK: What is the purpose of using a secondary antibody?

blotting solution, alkaline phosphatase converts the compound to a dark purple dye (Figure 20.17a). Because the secondary antibody binds to the primary antibody, a protein band that is recognized by the primary antibody becomes dark purple (nearly black).

In the Western blot shown in **Figure 20.17b**, the primary antibody recognized β globin. In lane 1, proteins were isolated from mouse red blood cells. The labeled band indicates that the β-globin polypeptide is made in these cells. By comparison, the proteins loaded into lanes 2 and 3 were from brain cells and intestinal cells, respectively. The absence of any bands indicates that these cell types do not synthesize β globin.

20.5 COMPREHENSION QUESTIONS

1. Which of the following is a hybridization method that uses a labeled nucleic acid probe, such as a labeled fragment of DNA?
   a. Southern blotting
   b. Northern blotting
   c. Western blotting
   d. Both a and b
2. Which of the following methods is used to detect a specific RNA within a mixture of many different RNAs?
   a. Southern blotting
   b. Northern blotting
   c. Western blotting
   d. None of the above
3. During Western blotting, the primary antibody recognizes
   a. the secondary antibody.
   b. the protein of interest.
   c. an mRNA of interest.
   d. a specific fragment of chromosomal DNA.

20.6 METHODS FOR ANALYZING DNA- AND RNA-BINDING PROTEINS

Learning Outcomes:

1. Describe how a gel retardation assay is used to determine if a protein binds to DNA or RNA.
2. Outline the steps in DNase I footprinting and analyze the results.

In addition to detecting the presence of genes and gene products using blotting techniques, researchers often want to study the binding of proteins to specific sites on a DNA or RNA molecule. For example, the molecular investigation of transcription factors requires methods for identifying interactions between transcription factor proteins and specific DNA sequences. In Chapter 15, we considered the technique of chromatin immunoprecipitation sequencing (ChIP-seq) that can be used to identify where particular proteins, such as histones, bind to the DNA (refer back to Figure 15.11). In this section, we will consider two additional methods, called the gel retardation assay and DNase I footprinting, which are used to study protein-DNA interactions. Gel retardation is also used to study protein-RNA interactions.

The Gel Retardation Assay Is Used to Determine if a Protein Binds to a Specific DNA Fragment or RNA Molecule

A technically simple, widely used method for identifying DNA- or RNA-binding proteins is the **gel retardation assay,** also known as the **gel mobility shift assay.** This technique was used originally to study interactions between specific proteins and rRNA molecules and quickly became popular after its success in studying protein-DNA interactions in the *lac* operon. Now it is commonly used as a technique for detecting interactions between RNA-binding proteins and mRNAs and between transcription factors and DNA regulatory elements.

In the case of DNA-binding proteins, the technical basis for a gel retardation assay is that the binding of a protein to a DNA fragment retards the fragment's ability to move through a polyacrylamide or agarose gel. During electrophoresis, DNA fragments are pulled through the gel matrix toward the bottom of the gel by a voltage gradient. Smaller molecules migrate more quickly through a gel matrix than do larger ones. A key aspect of this procedure is that the binding of a protein to a DNA fragment retards the DNA's rate of movement through the gel matrix, because a protein-DNA complex has a higher mass. When comparing a DNA fragment alone versus that same DNA fragment that is found within a protein-DNA complex, the complex is shifted to a higher band than the DNA alone, because the complex migrates more slowly to the bottom of the gel (**Figure 20.18**). The bands are visualized by staining the DNA with a dye such as EtBr. To increase the sensitivity of the gel retardation assay, the DNA can also be labeled with a fluorescent molecule.

A gel retardation assay must be carried out under nondenaturing conditions. This means that the buffers and gel cannot cause the unfolding of proteins or the separation of the DNA double helix. This is necessary so the proteins and DNA retain their proper structure and are thereby able to bind to each other. The nondenaturing conditions of a gel retardation assay differ from the more common SDS gel electrophoresis, in which the proteins are denatured by the detergent SDS.

DNase I Footprinting Shows the Detailed Interactions Between a Protein and DNA

Another method for studying protein-DNA interactions is **DNase I footprinting,** a technique described by David Galas and Albert Schmitz in 1978. A DNase I footprinting experiment attempts to identify one or more regions of DNA that interact with a DNA-binding protein. Compared with a gel retardation assay, DNase I footprinting provides more detailed information about the interactions between a protein and DNA. One drawback is that DNase I footprinting is a more complicated technique than a gel retardation assay.

To understand the basis of a DNase I footprinting experiment, we need to consider the interactions among three types of molecules: a fragment of DNA, DNA-binding proteins, and reagents that alter DNA structure. As an example, let's examine the binding of RNA polymerase to a bacterial promoter, a topic discussed in Chapter 12. When RNA polymerase holoenzyme binds to the promoter

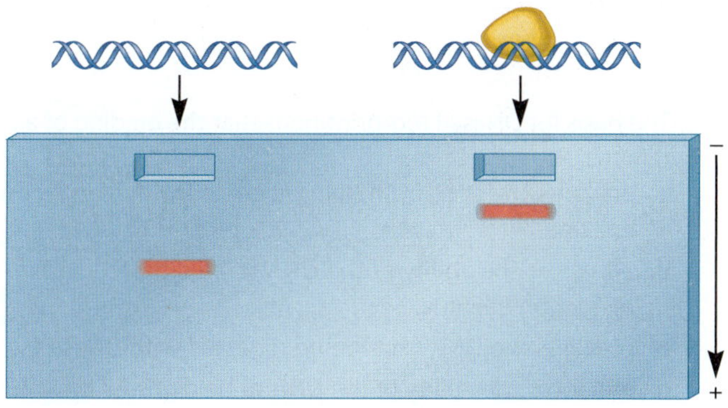

FIGURE 20.18 The results of a gel retardation assay. The binding of a protein to a labeled fragment of DNA retards its rate of movement through a gel. For the results shown in the lane on the right, if the concentration of the DNA fragment was higher than the concentration of the protein, there would be two bands: one band with protein bound (at a higher molecular mass) and one band without protein bound (corresponding to the band found in the left lane).

to form a closed complex, it binds tightly to the −35 and −10 promoter regions, but the protein covers up an even larger region of the DNA. Therefore, holoenzyme bound at the promoter prevents other molecules from gaining access to this region of the DNA. The enzyme DNase I, which cleaves covalent bonds in the DNA backbone, is used as a reagent for determining if a DNA region has a protein bound to it. Galas and Schmitz reasoned that DNase I cannot cleave the DNA at locations where a protein is bound. In this example, it is expected that RNA polymerase holoenzyme will bind to a promoter and protect this DNA region from DNase I cleavage.

Figure 20.19 shows the results of a DNase I footprinting experiment. In this experiment, a sample of many identical DNA fragments, all of which are 150 bp in length, were labeled at only one end. The sample of fragments was then divided into two tubes: tube A, which did not contain any holoenzyme, and tube B, which contained RNA polymerase holoenzyme. DNase I was then added to both tubes. The tubes were incubated long enough for DNase I to cleave the DNA at a single site in each DNA fragment. Each tube contained many 150-bp DNA fragments, and the cutting in any DNA strand by DNase I occurred randomly. Therefore, the DNase I treatment should produce a mixture of many smaller DNA fragments. A key point, however, is that DNase I cannot cleave the DNA in a region where RNA polymerase holoenzyme is bound. After DNase I treatment, the DNA fragments within the two tubes were separated by gel electrophoresis, which reveals a series of labeled bands.

In the absence of RNA polymerase holoenzyme (tube A), DNase I cleaves the 150-bp fragments randomly at any single location. Therefore, a continuous range of sizes occurs (see Figure 20.19). However, if we look at the gel lane from tube B, no bands are observed in the size range from 25–105 nucleotides. Why are these bands missing? The answer is that DNase I cannot cleave the DNA within the region where the holoenzyme is bound. The middle portion of the 150-bp fragment contains a promoter sequence that binds the RNA polymerase holoenzyme. Along the right side of the gel, the bases are numbered according to their position within the gene. (The site labeled +1 is where transcription begins.) As seen here, RNA polymerase covers up a fairly large region (its "footprint") of about 80 bp, from the −50 region to the +30 region.

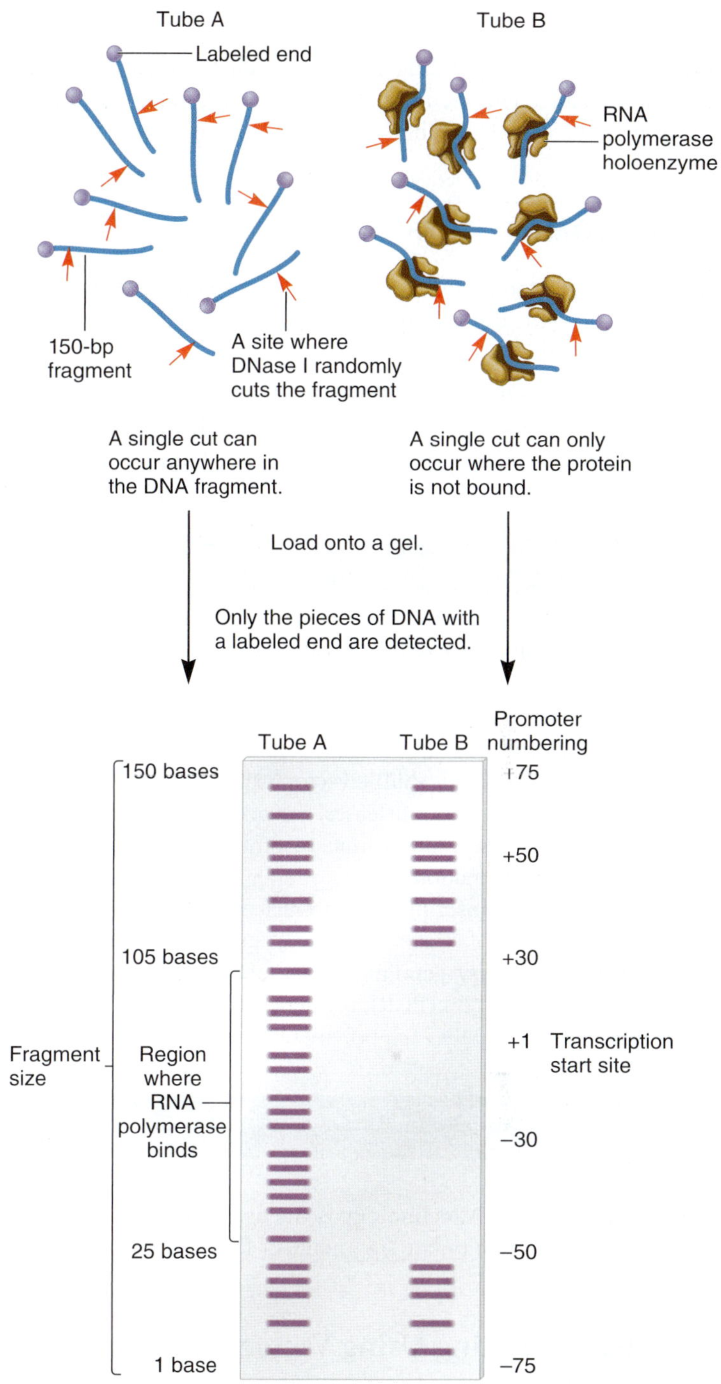

FIGURE 20.19 A DNase I footprinting experiment. Both tubes contained 150-bp fragments of DNA that were incubated with DNase I. Tube B also contained RNA polymerase holoenzyme. The binding of RNA polymerase holoenzyme protected a region of about 80 bp (namely, the −50 region to the +30 region) from DNase I digestion. Note: The promoter numbering convention shown here is the same as that described previously in Figure 12.4.

CONCEPT CHECK: How does the binding of a protein to DNA influence the ability of DNase I to cleave the DNA?

As illustrated in this experiment, DNase I footprinting can identify the DNA region that interacts with a DNA-binding protein. In addition to RNA polymerase-promoter binding, DNase I footprinting has been used to identify the binding sites for many other types of DNA-binding proteins, such as eukaryotic transcription factors. This technique has greatly facilitated our understanding of protein-DNA interactions.

20.6 COMPREHENSION QUESTIONS

1. In a gel retardation assay, the binding of a protein to DNA
   a. prevents the DNA from being digested with a restriction enzyme.
   b. causes the DNA to migrate more slowly through a gel.
   c. causes the DNA to migrate more quickly through a gel.
   d. inhibits the expression of any genes within the DNA.
2. The basis for DNase I footprinting is that the binding of a protein to DNA
   a. prevents the DNA from being digested with a restriction enzyme.
   b. enhances the ability of the DNA to be digested with a restriction enzyme.
   c. prevents the DNA from being digested with DNase I.
   d. enhances the ability of the DNA to be digested by DNase I.

KEY TERMS

Page 496. recombinant DNA technology, recombinant DNA molecules
Page 497. gene cloning
Page 498. vector, host cell, plasmids, R factors, origin of replication, selectable marker, shuttle vector, expression vector
Page 499. restriction endonucleases, restriction enzymes
Page 500. DNA ligase, palindromic, recombinant vector, competent cells, transformation
Page 502. reverse transcriptase, oligonucleotide, complementary DNA (cDNA)
Page 503. DNA library, genomic library, cDNA library, polymerase chain reaction (PCR), primers
Page 505. template DNA, *Taq* polymerase
Page 506. annealing, primer extension, thermocycler, reverse transcriptase PCR, real-time PCR
Page 508. cycle threshold method (C_t method)
Page 512. DNA sequencing, dideoxy sequencing, dideoxyribonucleotides (ddNTPs), chain termination, automated DNA sequencing
Page 514. sequencing ladder, site-directed mutagenesis
Page 516. colony hybridization, Southern blotting
Page 517. high stringency, low stringency, Northern blotting
Page 519. Western blotting, antibody, epitopes, antigen
Page 520. gel retardation assay, gel mobility shift assay, DNase I footprinting

CHAPTER SUMMARY

- Recombinant DNA technology is the use of in vitro molecular techniques to manipulate fragments of DNA and produce new arrangements.

20.1 Gene Cloning Using Vectors

- Gene cloning has many uses including DNA sequencing, site-directed mutagenesis, the use of genes as probes, and the expression of cloned genes (see Table 20.1).
- Cloning vectors are derived from plasmids or viruses (see Table 20.2).
- Restriction enzymes, also called restriction endonucleases, cut chromosomal DNA and vector DNA to produce sticky ends that will hydrogen bond with each other. DNA ligase is needed to make a covalent link in the DNA backbone (see Figure 20.1, Table 20.3).
- Gene cloning using vectors involves the insertion of the gene into a vector and then its propagation in a living cell such as *E. coli* (see Figure 20.2).
- Complementary DNA (cDNA) is made via reverse transcriptase by starting with mRNA (see Figure 20.3).
- A DNA library is a collection of recombinant vectors. The inserts can be chromosomal DNA or cDNA (see Figure 20.4).

20.2 Polymerase Chain Reaction

- Polymerase chain reaction (PCR) uses oligonucleotide primers to copy a specific region of DNA. Each cycle of PCR involves three steps: denaturation, primer annealing, and primer extension (see Figures 20.5, 20.6).
- Reverse transcriptase PCR begins with mRNA to study gene expression (see Figure 20.7).
- Real-time PCR monitors PCR as it occurs in a thermocycler. It is used to quantitatively measure the amount of starting DNA or mRNA in a sample. Real-time PCR uses fluorescent probes to follow the PCR reaction. A sample that has an unknown amount of DNA is compared with a standard (see Figures 20.8, 20.9).
- Higuchi and colleagues carried out the first real-time PCR experiment, using ethidium bromide as a fluorescent probe (see Figure 20.10).

20.3 DNA Sequencing

- A commonly used method of DNA sequencing, called dideoxy sequencing, uses fluorescently labeled dideoxyribonucleotides that cause chain termination (see Figures 20.11, 20.12).

20.4 Site-Directed Mutagenesis

- In the method of site-directed mutagenesis, an oligonucleotide with one or more mismatches directs a mutation into a specific region of DNA (see Figure 20.13).

20.5 Blotting Methods to Detect Genes and Gene Products

- Colony hybridization is used to identify colonies that carry a gene or cDNA of interest (see Figure 20.14).
- Southern blotting uses a labeled DNA probe to detect the presence of a particular gene sequence within a mixture of many gene sequences. At low stringency, gene families may be detected (see Figure 20.15).
- Northern blotting uses a labeled DNA probe to detect a specific RNA within a mixture of many different RNAs (see Figure 20.16).
- Western blotting uses antibodies to detect a specific protein within a mixture of many different proteins (see Figure 20.17).

20.6 Methods for Analyzing DNA- and RNA-Binding Proteins

- A gel retardation assay can determine if a protein binds to a specific DNA fragment or RNA molecule because the binding of the protein slows down the movement of the DNA or RNA through a gel (see Figure 20.18).
- DNase I footprinting can determine the regions of a DNA molecule that are bound by a protein (see Figure 20.19).

PROBLEM SETS & INSIGHTS

Solved Problems

S1. RNA was isolated from four different cell types and probed with labeled DNA strands from a cloned gene that is called gene *X*. The results are shown here.

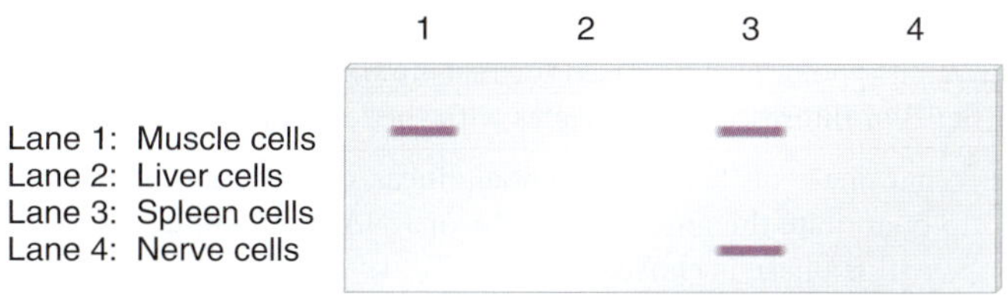

Explain the results of this experiment.

Answer: In this Northern blot, a labeled band appears in those lanes where RNA was isolated from muscle and spleen cells but not from liver and nerve cells. These results indicate that the muscle and spleen cells express a significant amount of RNA from gene *X*, but the liver and nerve cells do not. The muscle cells show a single band, whereas the spleen cells show this band plus a second band of lower molecular mass. An interpretation of these results is that the spleen cells can alternatively splice the RNA to produce a second RNA containing fewer exons.

S2. In the Western blotting experiment shown here, proteins were extracted from red blood cells obtained from tissue samples at different stages of human development. An equal amount of total cellular proteins was added to each lane. The primary antibody recognizes the β-globin polypeptide that is found in the hemoglobin protein. Explain these results.

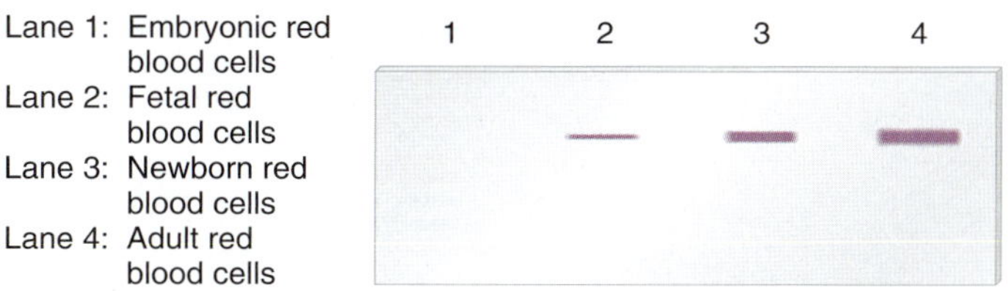

Answer: As shown here, the amount of β globin increases during development. Little detectable β globin is produced during embryonic development. The amount increases significantly during fetal development and becomes maximal in the adult. These results indicate that the β-globin gene is "turned on" in later stages of development, leading to the synthesis of the β-globin polypeptide. This experiment illustrates how a Western blot can provide information concerning the relative amount of a specific protein within living cells.

S3. A DNA strand is 3′–ATACGACTAGTCGGGACCATATC–5′. If the primer in a dideoxy-sequencing reaction anneals just to the left of this sequence, draw what the sequencing ladder would look like.

Answer:

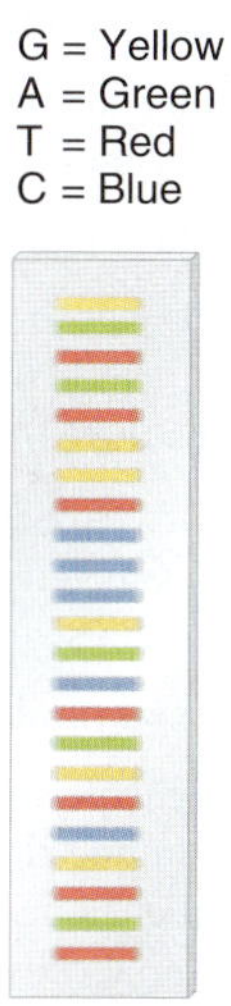

S4. The human genetic disease phenylketonuria (PKU) involves a defect in a gene that encodes the enzyme phenylalanine hydroxylase. It is inherited as a recessive autosomal disorder. Using a strand of DNA from the nonmutant phenylalanine hydroxylase gene as a probe, a Southern blot was carried out on a PKU patient, one of her parents, and an unaffected, unrelated person. In this example, the DNA fragments were subjected to acrylamide gel electrophoresis, rather than agarose gel electrophoresis, because acrylamide gel electrophoresis is better able to detect small deletions within genes. The following results were obtained:

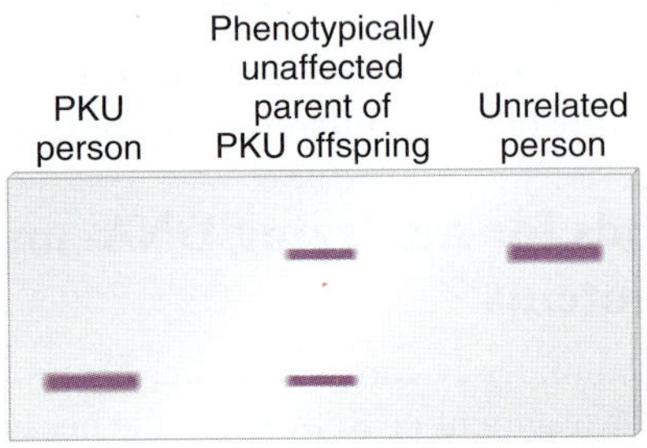

Suggest an explanation for these results.

Answer: In the affected person, the PKU defect is caused by a small deletion within the PKU gene. The unaffected parent is heterozygous for the nonmutant gene and the deletion. The PKU-affected person carries only the deletion, which runs at a lower molecular mass than the nonmutant gene. The unrelated person carries two copies of the nonmutant (nondeleted) gene.

Conceptual Questions

C1. Discuss three important advances that have resulted from gene cloning.

C2. What is a restriction enzyme? What structure does it recognize? What type of chemical bond does it cleave? Be as specific as possible.

C3. Write a double-stranded sequence that is 20 bases long and is palindromic.

C4. What is cDNA? In eukaryotes, how does cDNA differ from genomic DNA?

C5. Explain and draw the structural feature of a dideoxyribonucleotide that causes chain termination.

Experimental Questions

E1. What is the functional significance of sticky ends in a cloning experiment? What type of bonding makes the ends sticky?

E2. Table 20.3 describes the cleavage sites of five different restriction enzymes. After these restriction enzymes have cleaved the DNA, four of them produce sticky ends that can hydrogen bond with complementary sticky ends, as shown in Figure 20.1. The efficiency of sticky ends binding together depends on the number of hydrogen bonds; more hydrogen bonds makes the ends "stickier" and more likely to stay attached. Rank these four restriction enzymes in Table 20.3 (from best to worst) with regard to the efficiency of their sticky ends binding to each other.

E3. Describe the important features of cloning vectors. Explain the purpose of selectable marker genes in cloning experiments.

E4. How does gene cloning produce many copies of a gene?

E5. In your own words, describe the series of steps necessary to clone a gene. Your answer should include the use of a probe to identify a bacterial colony that contains the cloned gene of interest.

E6. What is a recombinant vector? How is a recombinant vector constructed? Explain how X-Gal is used in a method of identifying recombinant vectors that contain segments of chromosomal DNA.

E7. What is a DNA library? Do you think this is an appropriate name?

E8. Some vectors used in cloning experiments contain bacterial promoters that are adjacent to unique cloning sites. This makes it possible to insert a gene sequence next to the bacterial promoter and express the gene in bacterial cells. These are called expression vectors. If you wanted to express a eukaryotic protein in bacterial cells, would you clone genomic DNA or cDNA into the expression vector? Explain your choice.

E9. Why is a thermostable form of DNA polymerase (e.g., *Taq* polymerase) used in PCR? Is it necessary to use a thermostable form of DNA polymerase in the techniques of dideoxy DNA sequencing or site-directed mutagenesis?

E10. Starting with a sample of RNA that contains the mRNA for the β-globin gene, explain how you could create many copies of the β-globin cDNA using reverse transcriptase PCR.

E11. What type of probe is used for real-time PCR? Explain how the level of fluorescence correlates with the level of PCR product.

E12. What phase of PCR (exponential, linear, or stationary) is analyzed to quantitate the amount of DNA or RNA in a sample? Explain why this phase is chosen.

E13. DNA sequencing can help us to identify mutations within genes. The following data are derived from an experiment in which a normal gene and a mutant gene have been sequenced:

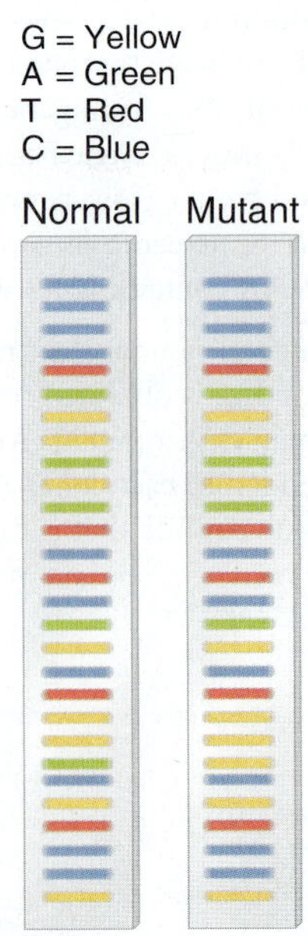

Locate and describe the mutation.

E14. A sample of DNA was subjected to automated DNA sequencing as shown here.

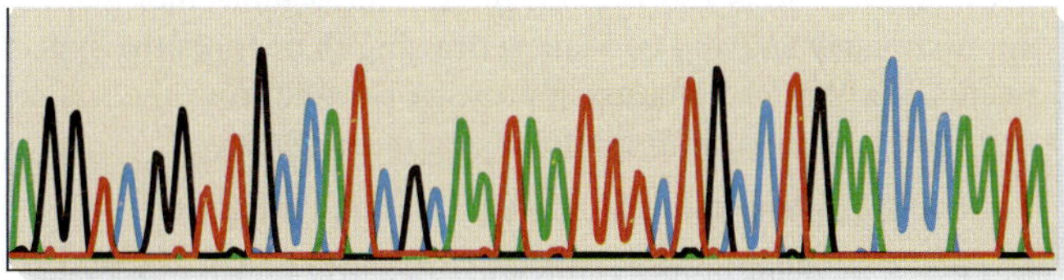

G = Black T = Red
A = Green C = Blue

What is the sequence of this DNA segment?

E15. A portion of the coding sequence of a cloned gene is shown here:

5′–GCCCCCGATCTACATCATTACGGCGAT–3′
3′–CGGGGGCTAGATGTAGTAATGCCGCTA–5′

This portion of the gene encodes a polypeptide with the amino acid sequence alanine–proline–aspartic acid–leucine–histidine–histidine–tyrosine–glycine–aspartic acid. Using the method of site-directed mutagenesis, a researcher wants to change the leucine codon into an arginine codon, with an oligonucleotide that is 19 nucleotides long. What is the sequence of the oligonucleotide that should be made? You should designate the 5′ and 3′ ends of the oligonucleotide in your answer. Note: The mismatch should be in the middle of the oligonucleotide, and a 1-base mismatch is preferable over a 2- or 3-base mismatch. Use the bottom strand as the template strand for this site-directed mutagenesis experiment.

E16. Let's suppose you want to use site-directed mutagenesis to investigate a DNA sequence that functions as a response element for hormone binding. From previous work, you have narrowed down the response element to a sequence of DNA that is 20 bp in length with the following sequence:

5′–GGACTGACTTATCCATCGGT–3′
3′–CCTGACTGAATAGGTAGCCA–5′

As a strategy to pinpoint the actual response element sequence, you decide to make 10 different site-directed mutants and then analyze their effects by a gel retardation assay. What mutations would you make? What results would you expect to obtain?

E17. Site-directed mutagenesis is also used to explore the structure and function of proteins. For example, changes can be made to the coding sequence of a gene to determine how alterations in the amino acid sequence affect the function of a protein. Let's suppose that you are interested in the functional importance of one glutamic acid residue within a protein you are studying. By site-directed mutagenesis, you make mutant proteins in which this glutamic acid codon has been changed to other codons. You then test the encoded mutant proteins for functionality. The results are as follows:

	Functionality (%)
Normal protein	100
Mutant proteins containing	
Tyrosine	5
Phenylalanine	3
Aspartic acid	94
Glycine	4

From these results, what would you conclude about the functional significance of the glutamic acid residue within the protein?

E18. Let's suppose you have recently cloned a gene, which we will call gene *X*, from corn. You use a labeled DNA strand from this cloned gene to probe genomic DNA from corn in a Southern blot experiment under conditions of low and high stringency. The following results were obtained:

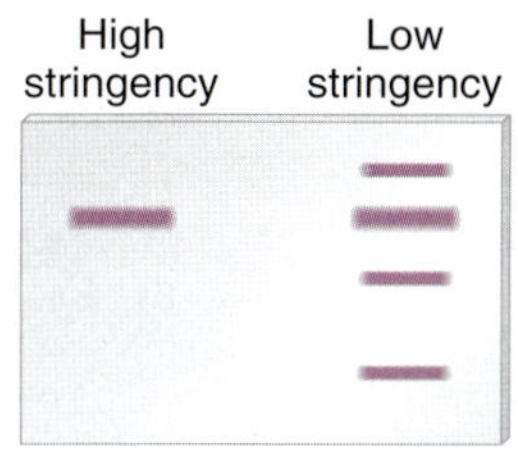

What do these results mean?

E19. Southern and Northern blotting depend on the phenomenon of hybridization. In these two techniques, explain why hybridization occurs. Which member of the hybrid is labeled?

E20. In Southern, Northern, and Western blotting, what is the purpose of gel electrophoresis?

E21. What is the purpose of a Northern blotting experiment? What types of information can it tell you about the transcription of a gene?

E22. Let's suppose an X-linked gene in mice exists as two alleles, which we will call *B* and *b*. X-chromosome inactivation, a process in which one X chromosome is turned off, occurs in the somatic cells of female mammals (see Chapter 5). Allele *B* encodes an mRNA that is 900 nucleotides long, whereas allele *b* contains a small deletion that shortens the mRNA to a length of 825 nucleotides. Draw the expected results of a Northern blot using mRNA isolated from somatic tissue of the following mice:

Lane 1. mRNA from an X^bY male mouse

Lane 2. mRNA from an X^bX^b female mouse

Lane 3. mRNA from an X^BX^b female mouse. Note: The sample taken from the female mouse is not from a clone of cells. It is from a tissue sample, like the one shown at the beginning of the experiment of Figure 5.6.

E23. The method of Northern blotting is used to determine the amount and size of a particular RNA transcribed in a given cell type. Alternative splicing (discussed in Chapter 16) produces mRNAs from the same gene that have different lengths. A Northern blot is shown here using a DNA probe that is complementary to the mRNA encoded by a particular gene. The mRNA in lanes 1 through 4 was isolated from different cell types, and equal amounts of total cellular mRNA were added to each lane.

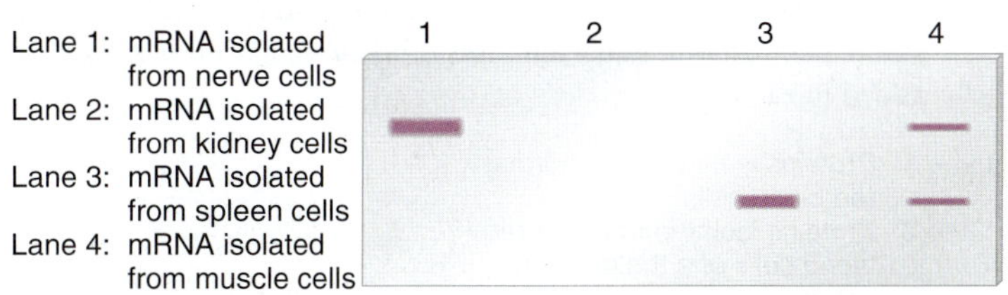

Explain these results.

E24. Southern blotting can be used to detect the presence of repetitive sequences, such as transposable elements, that are present in multiple copies within the chromosomal DNA of an organism. (Note: Transposable elements are described in Chapter 19.) In the Southern blot shown here, chromosomal DNA was isolated from three different strains of baker's yeast, digested with a restriction enzyme, run on a gel, blotted, and then labeled with a probe that is complementary to a transposable element called the *Ty* element.

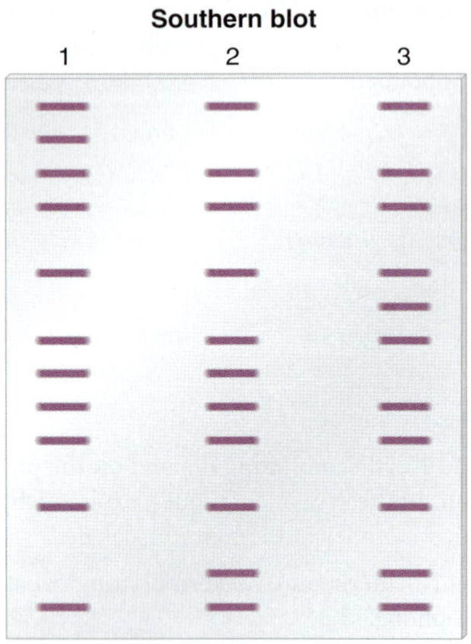

Explain, in a general way, why the banding patterns are not the same in lanes 1, 2, and 3.

E25. In Chapter 8, Figure 8.7 describes the evolution of the globin gene family. All genes in this family are homologous to each other, though the degree of sequence similarity varies depending on the time of divergence. Genes that have diverged more recently have sequences that are more similar. For example, the α_1 and α_2 genes have DNA sequences that are more similar to each other than to the α_1 and ξ genes. In a Southern blotting experiment, the degree of sequence similarity can be discerned by varying the stringency of hybridization. At high temperature (i.e., high stringency), the probe recognizes genes that are only a perfect or very close match. At a lower temperature, however, homologous genes with lower degrees of similarities can be detected because slight mismatches are tolerated. If a Southern blot was conducted on a sample of human chromosomal DNA and a probe was used that was a perfect match to the β-globin gene, rank the following genes (from those that are detected at high stringency down to those that are only detected at low stringency) as they would appear in a Southern blot experiment: *Mb*, α_1, β, γ_A, δ, and ε.

E26. In the Western blot shown here, proteins were isolated from red blood cells and muscle cells from two different individuals. One individual was unaffected, and the other individual suffered from a disease known as thalassemia, which involves a defect in hemoglobin. In the Western blot, the gel blot was exposed to an antibody that recognizes β globin, which is one of the polypeptides that constitute hemoglobin. Equal amounts of total cellular proteins were added to each lane.

Lane 1: Proteins isolated from normal red blood cells
Lane 2: Proteins isolated from the red blood cells of a thalassemia patient
Lane 3: Proteins isolated from normal muscle cells
Lane 4: Proteins isolated from the muscle cells of a thalassemia patient

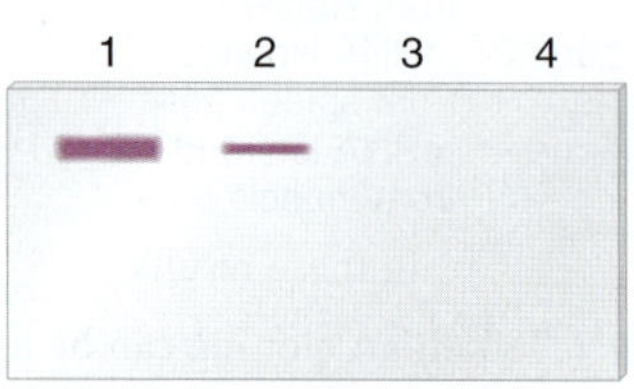

Explain these results.

E27. Let's suppose a researcher was interested in the effects of mutations on the expression of a protein-encoding gene for a protein that is 472 amino acids in length. This protein is expressed in leaf cells of *Arabidopsis thaliana.* Because the average molecular mass of an amino acid is 120 Da, this protein has a molecular mass of approximately 56,640 Da. Make a drawing that shows the expected results of a Western blot using proteins isolated from the leaf cells that were obtained from the following individuals:

Lane 1. A plant homozygous for a nonmutant gene

Lane 2. A plant homozygous for a deletion that removes the promoter for this gene

Lane 3. A heterozygous plant in which one gene is nonmutant and the other gene has a mutation that introduces an early stop codon at codon 112

Lane 4. A plant homozygous for a mutation that introduces an early stop codon at codon 112

Lane 5. A plant homozygous for a mutation that changes codon 108 from a phenylalanine codon into a leucine codon

E28. If you wanted to know if a protein was made during a particular stage of development, what technique would you choose?

E29. Explain the basis for using an antibody as a probe in a Western blotting experiment.

E30. Starting with pig cells and a probe that is a labeled DNA strand from the human β-globin gene, describe how you would clone the β-globin gene from pigs. You may assume that you have available all of the materials needed in a cloning experiment. How would you confirm that a putative clone really contained the β-globin gene?

E31. A cloned gene fragment contains a regulatory element that is recognized by a regulatory transcription factor. Previous experiments have shown that the presence of a hormone results in transcriptional activation by this transcription factor. To study this effect, you conduct a gel retardation assay and obtain the following results:

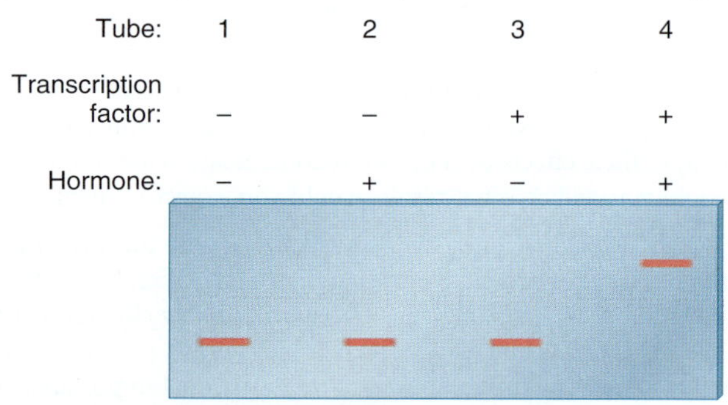

Explain the action of the hormone.

E32. Describe the rationale behind a gel retardation assay.

E33. Certain hormones, such as epinephrine, can increase the levels of cAMP within cells. Let's suppose you pretreat cells with or without epinephrine and then prepare a cell extract that contains the CREB protein (see Chapter 15 for a description of the CREB protein). You then use a gel retardation assay to analyze the ability of the CREB protein to bind to a DNA fragment containing a cAMP response element (CRE). Describe what the expected results would be.

E34. A gel retardation assay can be used to study the binding of proteins to a segment of DNA. In the experiment shown here, a gel retardation assay was used to examine the requirements for the binding of RNA polymerase II (from eukaryotic cells) to the promoter of a protein-encoding gene. The assembly of general transcription factors and RNA polymerase II at the core promoter is described in Chapter 12 (Figure 12.14). In this experiment, the segment of DNA containing a promoter sequence was 1100 bp in length. The fragment was mixed with various combinations of proteins and then subjected to a gel retardation assay.

Lane 1: No proteins added
Lane 2: TFIID
Lane 3: TFIIB
Lane 4: RNA polymerase II
Lane 5: TFIID + TFIIB
Lane 6: TFIID + RNA polymerase II
Lane 7: TFIID + TFIIB + RNA polymerase II

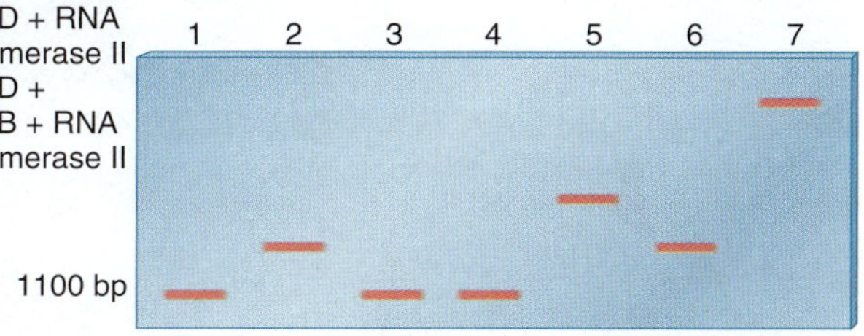

Explain which proteins (TFIID, TFIIB, or RNA polymerase II) are able to bind to this DNA fragment by themselves. Which transcription factors (i.e., TFIID or TFIIB) are needed for the binding of RNA polymerase II?

E35. As described in Chapter 15 (Figures 15.7 and 15.8), certain regulatory transcription factors bind to DNA and activate RNA polymerase II. When glucocorticoid binds to the glucocorticoid receptor (a regulatory transcription factor), this changes the conformation of the receptor and allows it ultimately to bind to the DNA. The glucocorticoid receptor binds to a DNA sequence called a glucocorticoid response element (GRE). In contrast, other regulatory transcription factors, such as the CREB protein, do not require hormone binding in order to bind to DNA. The CREB protein can bind to the DNA in the absence of any hormone, but it does not activate RNA polymerase II unless the CREB protein is phosphorylated. (Phosphorylation is stimulated by certain hormones.) The CREB protein binds to a DNA sequence called a cAMP response element (CRE). With these ideas in mind, draw the expected results of a gel retardation assay conducted on the following samples:

Lane 1. A 600-bp fragment containing a GRE, plus the glucocorticoid receptor

Lane 2. A 600-bp fragment containing a GRE, plus the glucocorticoid receptor, plus glucocorticoid hormone

Lane 3. A 600-bp fragment containing a GRE, plus the CREB protein

Lane 4. A 700-bp fragment containing a CRE, plus the CREB protein

Lane 5. A 700-bp fragment containing a CRE, plus the CREB protein, plus a hormone (such as epinephrine) that causes the phosphorylation of the CREB protein

Lane 6. A 700-bp fragment containing a CRE, plus the glucocorticoid receptor, plus glucocorticoid hormone

E36. In the technique of DNase I footprinting, the binding of a protein to a region of DNA protects that region from digestion by DNase I by blocking the ability of DNase I to gain access to the DNA. In the DNase I footprinting experiment shown here, a researcher began with a sample of cloned DNA 400 bp in length. This DNA contained a eukaryotic promoter for RNA polymerase II. The assembly of general transcription factors and RNA polymerase II at the core promoter is described in Chapter 12 (see Figure 12.14). For the sample loaded in lane 1, no proteins were added. For the sample loaded in lane 2, the 400-bp fragment was mixed with RNA polymerase II plus TFIID and TFIIB.

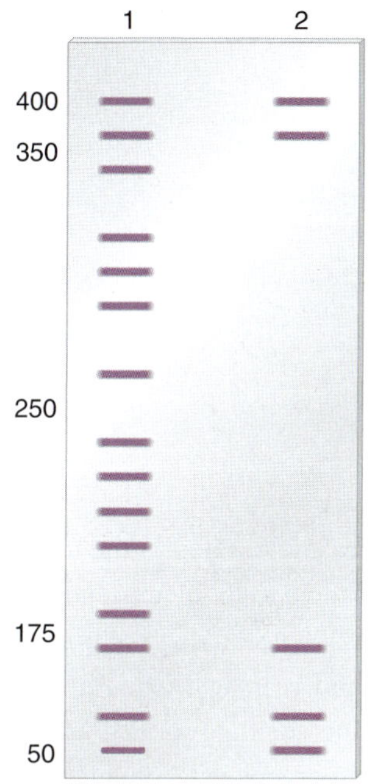

Which region of this 400-bp fragment of DNA is bound by RNA polymerase II and TFIID and TFIIB?

E37. Explain the rationale behind a DNase I footprinting experiment.

Questions for Student Discussion/Collaboration

1. Discuss and make a list of some of the reasons why it would be informative for a geneticist to determine the amount of a gene product. Use specific examples of known genes (e.g., β globin and other genes) when making your list.

2. Make a list of possible research questions that could be answered using site-directed mutagenesis.

Note: All answers appear at the website for this textbook; the answers to the even-numbered questions and all of the Concept Check and Comprehension Questions are in the back of the book.

www.mhhe.com/brookergenetics5e

Visit the website for practice tests, answer keys, and other learning aids for this chapter. Enhance your understanding of genetics with our interactive exercises, quizzes, animations, and much more.

CHAPTER OUTLINE

The sheep named Dolly, which was cloned using genetic material from a somatic cell.

21 BIOTECHNOLOGY

Biotechnology is broadly defined as the use of living organisms or their by-products in the development of goods or processes that are beneficial to humans. Biotechnology is not a new phenomenon. It began several thousand years ago when humans began to domesticate animals and plants for the production of food. Since that time, many species of microorganisms, animals, and plants have become routinely utilized by people. More recently, the term biotechnology has become associated with molecular genetics. Since the 1970s, molecular genetic tools have provided novel ways to make use of living organisms for products and services. As discussed in Chapter 20, recombinant DNA techniques can be used to genetically engineer microorganisms. Recombinant methods also enable the introduction of genetic material into animals and plants, resulting in **genetically modified organisms (GMOs).** An organism that has received genetic material from a different species is called a **transgenic organism.** A gene from one species that is introduced into another species is called a **transgene.**

In the 1980s, court rulings made it possible to patent recombinant organisms such as transgenic animals and plants. This was one factor that contributed to the growth of many biotechnology industries. In this chapter, we will examine how molecular techniques have expanded our knowledge of the genetic characteristics of commercially important species. We will also discuss examples in which recombinant microorganisms and transgenic animals and plants have been given characteristics that are useful in the treatment of disease or in agricultural production. These include recombinant bacteria that make human insulin, transgenic livestock that produce human proteins in their milk, and transgenic corn that is resistant to insects. In addition, the topics of mammalian cloning and stem cell research are examined from a technical point of view. Likewise, the current and potential use of human gene therapy—the introduction of cloned genes into living cells in the treatment of a disease—will be addressed. In the process, we will also touch upon some of the ethical issues associated with these technologies.

21.1 USES OF MICROORGANISMS IN BIOTECHNOLOGY

Learning Outcomes:

1. Explain how bacteria are genetically engineered to produce human insulin.
2. Define biological control and bioremediation, and describe how microorganisms may play a role in these two processes.

TABLE 21.1

Common Uses of Microorganisms

Application	Examples
Production of medicines	Antibiotics
	Synthesis of human insulin in recombinant *E. coli*
Food fermentation	Cheese, yogurt, vinegar, wine, and beer
Biological control	Control of plant diseases, insect pests, and weeds
	Symbiotic nitrogen fixation
	Prevention of frost formation
Bioremediation	Cleanup of environmental pollutants such as petroleum hydrocarbons and synthetics that are difficult to degrade

TABLE 21.2

Examples of Medical Agents Produced by Recombinant Microorganisms

Drug	Action	Treatment
Insulin	A hormone that promotes glucose uptake	Diabetes
Tissue plasminogen activator (TPA)	Dissolves blood clots	Heart attacks and other vascular occlusions
Superoxide dismutase	Antioxidant	Heart attacks and to minimize tissue damage
Factor VIII	Blood-clotting factor	Certain types of hemophilias
Renin inhibitor	Lowers blood pressure	Hypertension
Erythropoietin	Stimulates the production of red blood cells	Anemia

Microorganisms are used to benefit humans in various ways (**Table 21.1**). In this section, we will examine how molecular genetic tools have become increasingly important for improving our use of microorganisms. Such tools can produce recombinant microorganisms with genes that have been manipulated in vitro. Why are recombinant organisms useful? Recombinant techniques can improve strains of microorganisms and have even yielded strains that make products not normally produced by microorganisms. For example, human genes have been introduced into bacteria to produce medically important products such as insulin and human growth hormone. As discussed in this section, several recombinant strains are in widespread use. However, in some areas of biotechnology and in some parts of the world, the commercialization of recombinant strains has proceeded very slowly. This is particularly true for applications in which recombinant microorganisms are used to produce food products or where they are released into the environment. In such cases, safety and environmental concerns along with negative public perceptions have slowed or even halted the commercial use of recombinant microorganisms. Nevertheless, molecular genetic research continues, and many biotechnologists expect an expanding use of recombinant microbes in the future.

Many Important Medicines Are Produced by Recombinant Microorganisms

During the 1970s, geneticists became aware of the great potential of recombinant DNA technology to produce therapeutic agents for treating certain human diseases. Healthy individuals have many different genes that encode short-peptide and longer-polypeptide hormones. Diseases can result when an individual is unable to produce these hormones.

In 1976, Robert Swanson and Herbert Boyer formed Genentech Inc. The aspiration of this company was to engineer bacteria to synthesize useful products, particularly peptide and polypeptide hormones. Their first contract was with researchers Keiichi Itakura and Arthur Riggs. They were able to engineer a bacterial strain that produced somatostatin, a human hormone that inhibits the secretion of a number of other hormones, including growth hormone, insulin, and glucagon. Somatostatin was not chosen for its commercial potential. Instead, it was chosen because the researchers thought it would be technically less difficult to produce than other hormones. Somatostatin is very small (only 14 amino acids long), so it requires a short coding sequence; in addition it can be detected easily. Since this pioneering work, recombinant DNA technology has been used to develop bacterial strains that synthesize several other medical agents, a few of which are described in **Table 21.2**.

In 1982, the U.S. Food and Drug Administration approved the sale of the first genetically engineered drug, human insulin, which was produced by Genentech and marketed by Eli Lilly. In nondiabetic individuals, insulin is produced by the β cells of the pancreas. Insulin functions to regulate several physiological processes, particularly the uptake of glucose into fat and muscle cells. Persons with insulin-dependent diabetes cannot synthesize an adequate amount of insulin due to a loss of their β cells. Prior to 1982, insulin was isolated from the pancreases removed from cattle and pigs. Unfortunately, in some cases, diabetic individuals became allergic to such insulin and had to use expensive combinations of insulin from other animals and human cadavers. Today, people with diabetes can use genetically engineered human insulin to treat their disease.

Insulin is a hormone composed of two different polypeptide chains, called the A and B chains. To make this hormone using bacteria, the coding sequence of either the A or B chain is placed next to the coding sequence of a native *E. coli* protein, β-galactosidase (**Figure 21.1**). This creates a fusion protein comprising β-galactosidase and either the A or B chain. This step is necessary because the A and B chains are rapidly degraded when expressed in bacterial cells by themselves. The fusion proteins, however, are not. How are the two fusion proteins used to make human insulin? After the fusion proteins are expressed in bacteria, they can be purified and then treated with cyanogen bromide (CNBr), which cleaves after a methionine that is found at the junction between β-galactosidase and the A or B chain. This cleavage step separates β-galactosidase from the A or B chain. The A and B chains are then purified and mixed together under conditions in which they refold

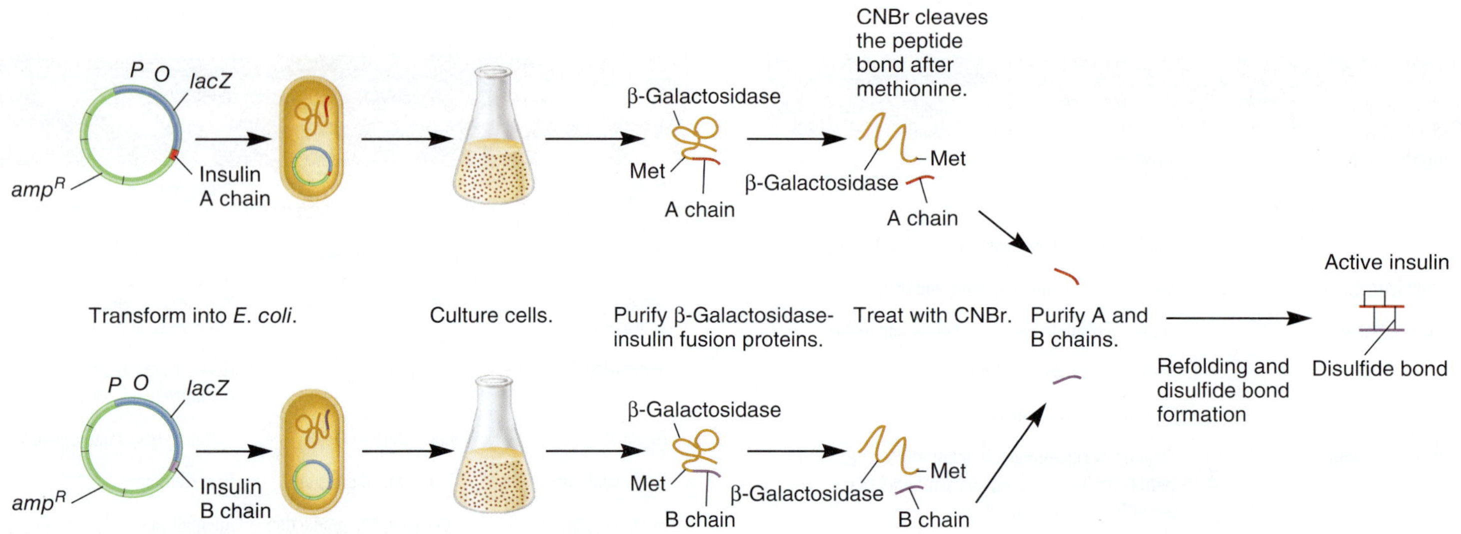

FIGURE 21.1 The use of bacteria to make human insulin.

Genes → Traits The synthesis of human insulin is not a trait that bacteria normally possess. However, genetic engineers can introduce the genetic sequences that encode the A and B chains of human insulin via recombinant DNA technology, yielding bacteria that make these polypeptides as fusion proteins with β-galactosidase. Cyanogen bromide (CNBr) treatment releases the A and B polypeptides, which are then purified and mixed together under conditions in which they refold and form functional human insulin.

CONCEPT CHECK: What is the purpose of using CNBr in this experiment?

and form disulfide bonds with each other to make an active insulin hormone, which has the same structure as the human-made hormone.

Bacterial Species Can Be Used as Biological Control Agents

The term **biological control** refers to the use of living organisms or their products to alleviate plant diseases. During the past 20 years, interest in the biological control of plant diseases and insect pests as an alternative to chemical pesticides has increased. Biological control agents can prevent disease in several ways. In some cases, nonpathogenic microorganisms are used to compete effectively against pathogenic strains for nutrients or space. In other cases, microorganisms may produce a toxin that inhibits other pathogenic microorganisms or insects without harming the plant.

Biological control can also involve the use of microorganisms living in the field. A successful example is the use of *Agrobacterium radiobacter* to prevent crown gall disease caused by *Agrobacterium tumefaciens*. The disease gets its name from the large swellings (galls) produced by the plant in response to the bacteria. *A. radiobacter* produces agrocin 84, an antibiotic that kills *A. tumefaciens*. Molecular geneticists have determined that *A. radiobacter* contains a plasmid with genes that encode enzymes responsible for agrocin 84 synthesis, which is a small molecule. The plasmid also carries genes that confer resistance to this antibiotic. Unfortunately, this plasmid is occasionally transferred from *A. radiobacter* to *A. tumefaciens* during interspecies conjugation. When this occurs, *A. tumefaciens* gains resistance to agrocin 84. Researchers have identified *A. radiobacter* strains in which this plasmid has been altered genetically to prevent its transfer during conjugation. This conjugation-deficient strain is now used commercially worldwide to prevent crown gall disease.

Another biological control agent is *Bacillus thuringiensis*, usually referred to as Bt (pronounced "bee-tee"). This naturally occurring bacterium produces toxins that are lethal to many caterpillars and beetles that feed on a wide variety of food crops and ornamental plants. Bt is generally harmless to plants and other animals, such as humans, and does not usually harm beneficial insects that act as pollinators. Therefore, it is viewed as an environmentally friendly pesticide. Commercially, Bt is sold in a powder form that is used as a dust or mixed with water as a foliage spray. Bt is then dusted or sprayed on plants that are under attack by caterpillars or beetles so the pests will ingest the bacteria as they eat the leaves, flowers, or fruits. The toxins produced by Bt bring about paralysis of the insect's digestive tract, causing it to stop feeding within hours and die within a few days. Geneticists have cloned the genes that encode Bt toxins, which are proteins. As discussed later, such genes have been introduced into crops, such as corn, to produce transgenic plants resistant to insect attack.

Microorganisms Can Reduce Environmental Pollutants

The term **bioremediation** refers to the use of living organisms or their products to decrease pollutants in the environment. As its name suggests, this is a biological remedy for pollution. During bioremediation via microorganisms, enzymes produced by a microorganism modify a toxic pollutant by altering or transforming its structure. This event is called **biotransformation.** In many cases, biotransformation results in **biodegradation,** in which the toxic pollutant is degraded, yielding less complex, nontoxic metabolites. Alternatively, biotransformation without biodegradation can also occur. For example, toxic heavy metals can often be rendered less toxic by oxidation or reduction reactions carried out by microorganisms. Another way to alter the toxicity of organic pollutants is by promoting polymerization. In many cases, polymerized toxic compounds are less likely to leach from the soil and, therefore, are less environmentally toxic than their parent compounds.

Since the early 1900s, microorganisms have been used in the treatment and degradation of sewage. More recently, the field of bioremediation has expanded into the treatment of hazardous and refractory chemical wastes—chemicals that are difficult to degrade and usually associated with industrial activity. These pollutants include petroleum hydrocarbons, halogenated organic compounds, pesticides, herbicides, and organic solvents. Many new applications that use microorganisms to degrade these pollutants are being tested. The field of bioremediation has been fostered, to a large extent, by better knowledge of how pollutants are degraded by microorganisms, the identification of new and useful strains of microbes, and the ability to enhance bioremediation through genetic engineering.

Molecular genetic technology is key in identifying genes that encode enzymes involved in bioremediation. The characterization of the relevant genes greatly enhances our understanding of how microbes can modify toxic pollutants. In addition, recombinant strains created in the laboratory can be more efficient at degrading certain types of pollutants.

In 1980, in a landmark case (*Diamond v. Chakrabarty*), the U.S. Supreme Court ruled that a live, recombinant microorganism is patentable as a "manufacture or composition of matter." The first recombinant microorganism to be patented was an "oil-eating" bacterium that contained a laboratory-constructed plasmid. This strain can oxidize the hydrocarbons commonly found in petroleum. It grew faster on crude oil than did any of the natural isolates tested. However, it has not been a commercial success because this recombinant strain metabolizes only a limited number of toxic compounds, a fraction of the more than 3000 actually present in crude oil. Unfortunately, the recombinant strain did not degrade many higher-molecular-weight compounds, which tend to persist in the environment.

Thus far, most bioremediation has involved the use of natural microorganisms rather than recombinant ones. Currently, bioremediation should be considered a developing industry. Many studies are currently underway aimed at elucidating the mechanisms whereby microorganisms degrade toxic pollutants. In the future, recombinant microorganisms may provide an effective way to decrease the levels of toxic chemicals within our environment. However, this approach will require careful studies to demonstrate that recombinant organisms are effective at reducing pollutants and are safe and able to survive when released into the environment.

21.1 COMPREHENSION QUESTIONS

1. Which of the following is a use of a microorganism in biotechnology?
   a. Production of medicines
   b. Food fermentation
   c. Biological control
   d. All of the above
2. What is the key reason why the A and B chains of insulin are made as fusion proteins with β-galactosidase?
   a. To make purification easier
   b. To prevent their degradation
   c. To be secreted from the cell
   d. All of the above
3. Which of the following was the first living organism to be patented?
   a. A strain of *E. coli* that makes somatostatin
   b. A strain of *E. coli* that makes insulin
   c. An oil-eating bacterium
   d. A strain of *B. thuringiensis* that makes an insecticide

21.2 GENETICALLY MODIFIED ANIMALS

Learning Outcomes:

1. Distinguish between gene replacement and gene addition.
2. Explain how gene replacements and gene knockins are made in mice, and list some of their important uses.
3. Outline how transgenic livestock are used to make human medicines in their milk.

As mentioned at the beginning of this chapter, transgenic organisms contain recombinant DNA from another species that has been integrated into their genome. A dramatic example of this is shown in Figure 21.2. In this case, the gene that encodes human growth hormone was introduced into the genome of a mouse. The larger mouse on the right is a transgenic mouse that expresses the human growth hormone gene at a very high level.

The production of transgenic animals is a relatively new area of biotechnology. In recent years, a few transgenic species have reached the stage of commercialization. Many researchers believe that this technology holds great promise for innovations in biotechnology. However, the degree to which this potential may be realized depends, in part, on the public's concern about the production and consumption of transgenic species.

In this section, we begin by examining the mechanisms through which cloned DNA becomes integrated into the chromosomal DNA of animal cells. We will then explore the current techniques used to produce transgenic animals and their potential uses.

The Integration of a Cloned Gene into a Chromosome Can Result in Gene Replacement or Gene Addition

In Chapter 20, we considered methods for cloning genes. A common approach is to insert a chromosomal gene into a vector and then propagate the vector in living microorganisms such as bacteria or yeast cells. Cloned genes can also be introduced into animal and plant cells. However, to be inherited stably from generation

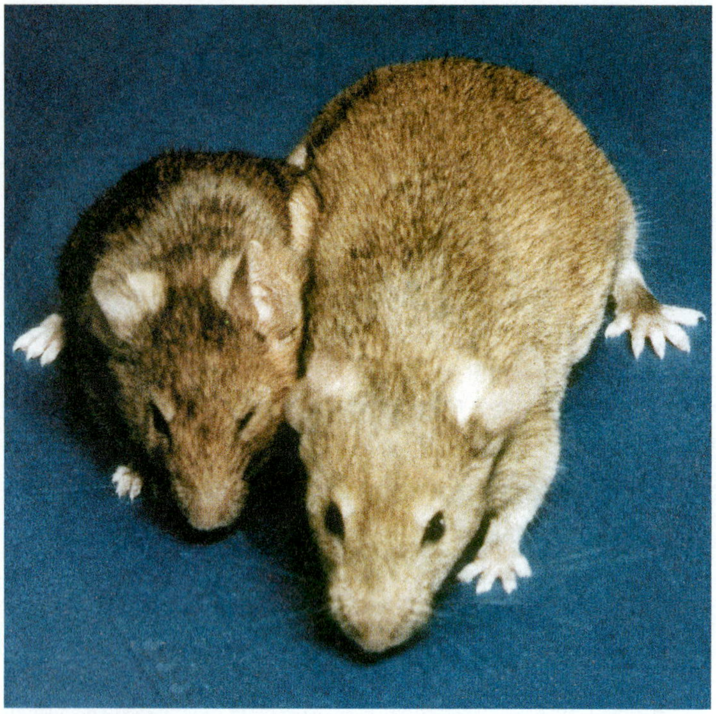

FIGURE 21.2 A comparison between a normal mouse and a transgenic mouse that carries a growth hormone gene.

Genes → Traits The transgenic mouse (on the right) carries a gene encoding human growth hormone. The introduction of the human gene into the mouse's genome has caused it to grow larger.

to generation, the cloned gene must become integrated into one (or more) of the chromosomes that reside in the cell nucleus. The integration of cloned DNA into a chromosome occurs by recombination, which is described in Chapter 19.

Figure 21.3 illustrates how a cloned gene can integrate into a chromosome by recombination. If the genome of the host cell carries the same type of gene and if the cloned gene is swapped with the normal chromosomal gene by homologous recombination, then the cloned gene will replace the normal gene within the chromosome (Figure 21.3a). This process is termed **gene replacement.** If the cloned gene has been rendered inactive and replaces the normal gene, researchers can study how the loss of normal gene function affects the organism. This is called a **gene knockout.** More commonly, a cloned gene may be introduced into a cell and become integrated into the genome by nonhomologous recombination in a process known as **gene addition** (Figure 21.3b). As shown here, both the cloned and normal genes are present following gene addition.

Researchers may also introduce a cloned gene that is not already present in a particular species' genome. An interesting example of this type of gene addition involves the production of aquarium fish that "glow," the aptly named GloFish. The **GloFish** is a brand of transgenic zebrafish (*Danio rerio*) that glows with bright green, red, or yellow fluorescent color (Figure 21.4). How were these fish produced? In 1999, Zhiyuan Gong and his colleagues started with a gene from jellyfish that encodes a green fluorescent protein (GFP) and inserted it into the zebrafish genome by gene addition, causing the zebrafish to glow green. By placing the gene next to a gene promoter that would turn the gene on in the presence of certain environmental toxins, their long-term goal was to develop a fish that could be used to detect water pollution.

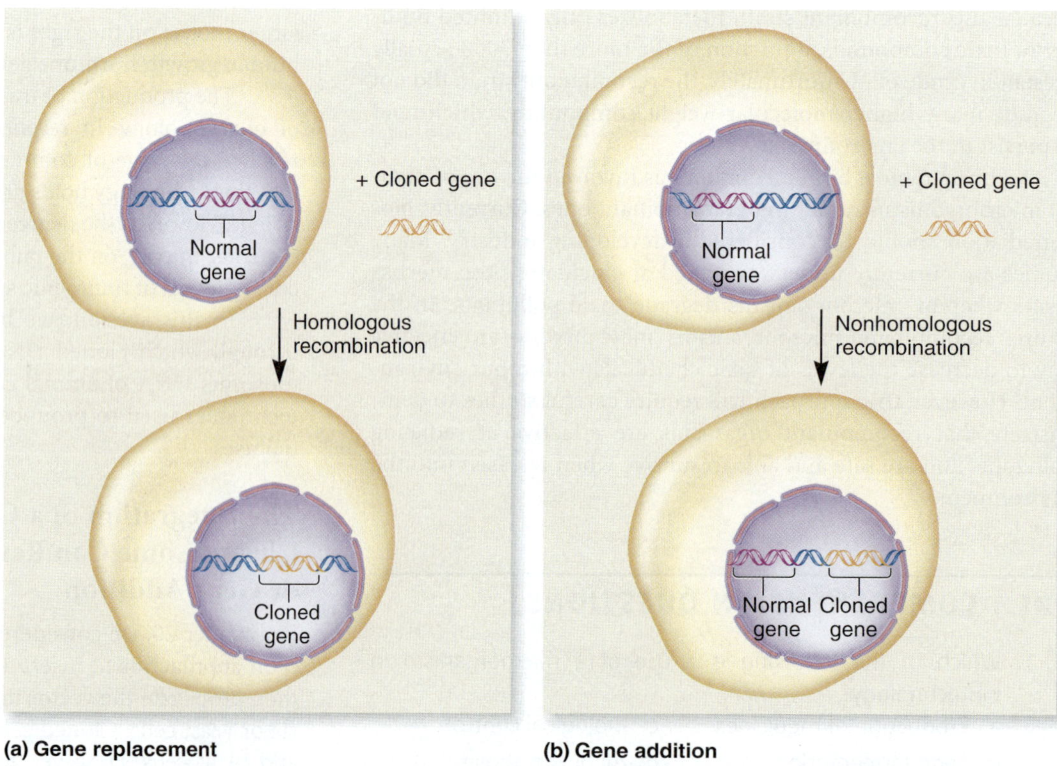

FIGURE 21.3
The introduction of a cloned gene into a cell can lead to gene replacement or gene addition. In these two examples, the cloned gene is similar to a normal gene already present in the genome of a cell. **(a)** When the cloned gene undergoes homologous recombination and replaces the normal gene, this is called gene replacement. **(b)** Alternatively, the cloned gene may recombine nonhomologously at some other chromosomal location, leading to gene addition.

CONCEPT CHECK: What is the difference between gene addition and gene replacement?

FIGURE 21.4 The use of gene addition to produce fish that glow. The aquarium fish shown here, which are named GloFish, are transgenic organisms that have received a gene from jellyfish or sea corals that encodes a fluorescent protein, causing them to glow green, red, or yellow.

However, as a first step in this process, they initially placed the *GFP* gene next to a promoter that was continually expressed. The researchers subsequently collaborated with a company to market the fish for aquarium use. They also developed a red fluorescent zebrafish by adding a gene from a sea coral, and a yellow fluorescent zebrafish by adding a variant of the jellyfish gene. In 2003, the GloFish became the first genetically modified organism to be sold as a pet. GloFish have been successfully marketed in several countries, including the United States, although the sale of GloFish is banned in California.

Molecular Biologists Can Produce Mice That Contain Gene Replacements

In bacteria and yeast, which have relatively small genomes, homologous recombination between cloned genes and the host cell chromosome occurs at a relatively high rate, so gene replacement is commonly achieved. Gene replacement is useful when a researcher or biotechnologist wants to make a gene mutation in the laboratory and then introduce the mutant gene into a living organism to compare the effects of the normal and mutant genes on the phenotype. However, in more complex eukaryotes with very large genomes, the introduction of cloned genes into cells is much more likely to result in gene addition rather than gene replacement. For example, when a cloned gene is introduced into a mouse cell, it undergoes homologous recombination less than 0.1% of the time. Most of the time (over 99.9%), gene addition occurs.

To produce mice with gene replacements, molecular biologists have devised laboratory procedures to preferentially select cells in which homologous recombination has occurred. In the approach shown in **Figure 21.5**, the goal is to produce a gene knockout. The gene of interest was cloned so that it can be manipulated in vitro. The cloned gene is shown at the top of Figure 21.5. The cloning procedure involves two selectable marker genes that influence whether or not mouse embryonic cells can grow in the presence of certain drugs. First, the gene of interest is inactivated by inserting a neomycin resistance gene (called Neo^R) into the center of its coding sequence. Neo^R provides cells with resistance to neomycin. Next, a thymidine kinase gene, designated *TK*, is inserted adjacent to the gene of interest but not within the gene itself. The *TK* gene renders cells sensitive to killing by a drug called gancyclovir.

After the cloned gene has been modified in vitro, it is introduced into mouse embryonic stem cells. How do researchers identify the cells in which homologous recombination has occurred? When the cells are grown in the presence of neomycin and gancyclovir, most nonhomologous recombinants are killed because they also carry the *TK* gene. In contrast, homologous recombinants in which the normal gene has recombined (via a double crossover) with the cloned gene contain only the Neo^R gene, so they are resistant to both drugs. The surviving embryonic cells can then be injected into blastocysts, early embryos that are obtained from a pregnant mouse. In the example shown in Figure 21.5, the embryonic cells are from a mouse with dark fur, and the blastocysts are from a mouse with white fur. The embryonic cells can mix with the blastocyst cells to create a **chimera,** an organism that contains cells from two different individuals. To identify chimeras, the injected blastocysts are reimplanted into the uterus of a female mouse and allowed to develop. When this mouse gives birth, chimeras are easily identified because they contain patches of white and dark fur (see Figure 21.5).

Some chimeric animals will produce sperm or eggs that carry the inactivated gene. Such chimeric animals are mated to other mice to produce offspring that carry the mutant gene in all of the cells of their bodies. Because mice are diploid, researchers must make two or more subsequent crosses to create a strain of mice that contains both copies of the mutant target gene. However, homozygous strains for a gene knockout cannot be produced if the mutant gene is lethal in the homozygous state.

An alternative method for producing genetically modified mice is to inject the desired gene into a fertilized egg. To conduct this type of experiment, researchers obtain mouse eggs and fertilize them in vitro. Immediately following fertilization, a period of time occurs in which the haploid sperm pronucleus has yet to fuse with the haploid egg pronucleus. The cloned DNA is usually injected into the sperm pronucleus, because it is larger. The two pronuclei then naturally fuse to form the diploid nucleus of a zygote. The zygote begins to divide and is introduced into the uterus of a female

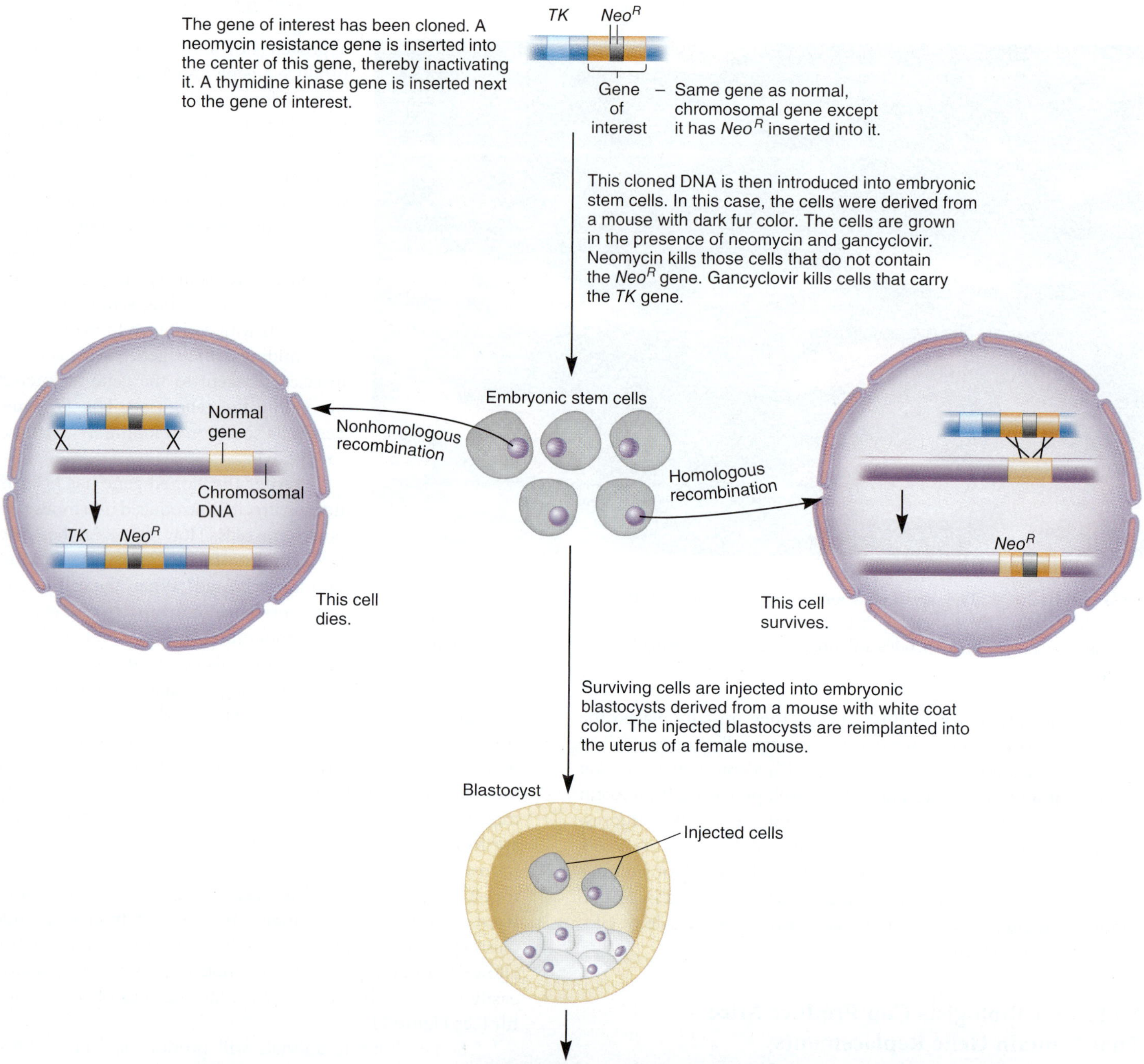

Following birth, chimeric mice are identified as those that contain a coat with both dark and white fur. The appropriate crosses are made in order to produce mice that have two copies of the inactivated gene.

Chimeric offspring

FIGURE 21.5 Producing a gene replacement in mice. The bottom of this figure shows a photograph of a chimeric mouse. Note the patches of black and white fur.

CONCEPT CHECK: What is the purpose of the *Neo*^{R} gene?

mouse, where it becomes implanted and grows. As discussed next, genetically modified mice are used in basic research and to study human diseases.

Gene Knockouts and Knockins Are Produced in Mice to Understand Gene Function and Human Disease

As we have seen, researchers can replace a normal mouse gene with one that has been inactivated by the insertion of an antibiotic resistance gene. As mentioned, when a mouse is homozygous for an inactivated gene, this condition is called a gene knockout. The inactive mutant gene has replaced both copies of the normal gene. In other words, the function of the normal gene has been "knocked out." By creating gene knockouts, researchers can study how the loss of normal gene function affects the organism. Gene knockouts frequently have specific effects on the phenotype of a mouse, which helps researchers to determine that the function of a gene is critical within a particular organ or during a specific stage of development.

In many cases, however, a gene knockout produces no obvious phenotypic effect. One explanation is that a single gene may make such a small contribution to an organism's phenotype that its loss may be difficult to detect. Alternatively, another possible explanation for a lack of observable phenotypic change in a knockout mouse may involve **gene redundancy.** This means that when one type of gene is inactivated, another gene with a similar function may be able to compensate for the inactive gene. A third explanation for why a gene knockout has no obvious effect is that the effects of the knockout may be observed only under certain types of environmental conditions.

A particularly exciting avenue of gene knockout research is its application in the study and treatment of human disease. How is this useful? Knocking out the function of a gene may provide clues about what that gene normally does. Because humans share many genes with mice, observing the characteristics of knockout mice gives researchers information that can be used to better understand how a similar gene mutation may cause or contribute to a disease in humans. Examples of research areas in which knockout mice have been useful include the study of cancer, obesity, heart disease, diabetes, and many inherited disorders.

In contrast to knockouts, researchers may introduce genes into the mouse genome to study the effects of gene overexpression or to examine the effects of particular alleles, such as those that may cause disease in humans. To accomplish this, researchers can produce a **gene knockin.** A gene knockin is a gene addition in which a gene of interest has been added to a particular site in the mouse genome (Figure 21.6). In this example, the cloned gene is inserted into the middle of a segment of DNA from a noncritical site in the mouse genome. The noncritical segment is very long, which allows the gene of interest to be targeted to that specific, noncritical integration site by homologous recombination after it is introduced into mouse cells. Such gene knockins tend to result in a more consistent level of expression of the gene of interest than gene additions that may occur randomly in another place in the genome. Also, because the insertion does not interfere with a critical mouse locus, the researcher can be more certain that any resulting phenotypic effect is due to the expression of the gene of interest.

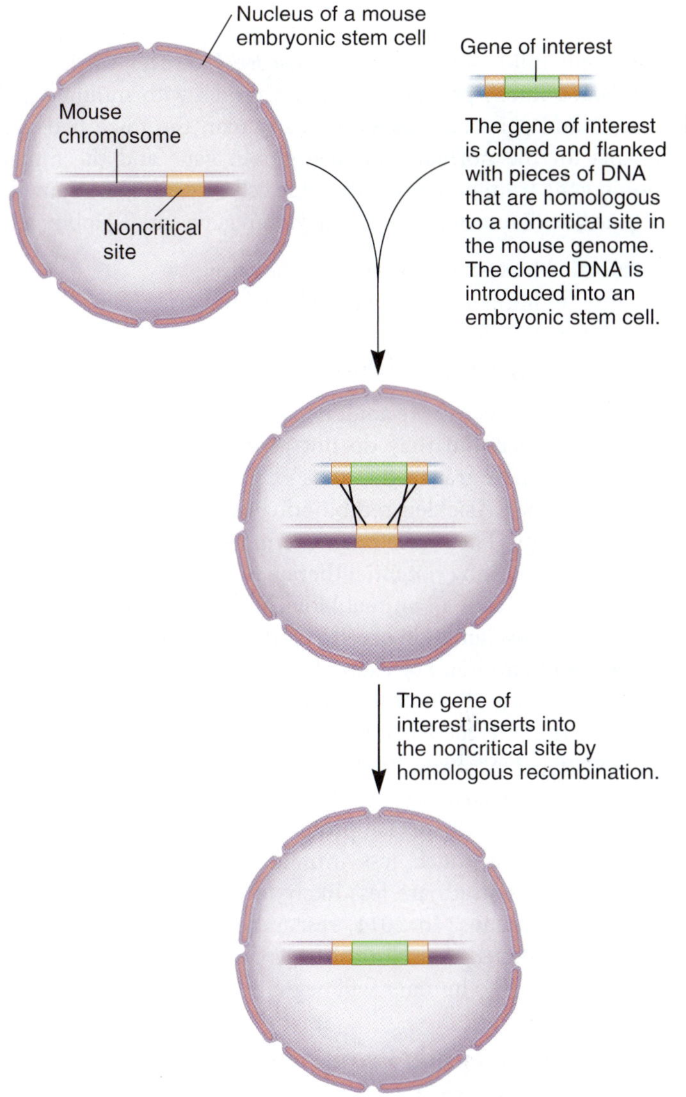

FIGURE 21.6 Producing a gene knockin in mice. This simplified diagram illustrates the strategy of producing a gene knockin. The gene of interest is cloned, and the cloned gene is flanked by DNA sequences that are homologous to a noncritical site in the mouse genome. When introduced into a mouse embryonic stem cell, the gene inserts into the genome by homologous recombination, which is an example of a targeted gene addition.

CONCEPT CHECK: What is the difference between a gene knockout and a gene knockin?

To study human diseases, researchers have produced strains of transgenic mice that harbor both gene knockouts and gene knockins. A strain of mice engineered to carry a mutation that is analogous to a disease-causing mutation in a human gene is termed a **mouse model.** As an example, let's consider sickle cell disease, which is due to a mutation in the human β-globin gene (refer back to Figure 4.7). This gene encodes a polypeptide called β globin; adult hemoglobin is composed of both α-globin and β-globin polypeptides. When researchers produced a gene knockin by introducing the mutant human β-globin gene into mice, the resulting mice showed only mild symptoms of the disease. However, Chris Pászty and Edward Rubin produced a mouse model with multiple gene knockins and gene knockouts. In particular, the mice had gene knockins for the normal human α-globin gene and the mutant β-globin gene from patients with sickle cell disease. The strain also had gene knockouts of the mouse α-globin gene and β-globin gene:

normal human α-globin gene knockin;
mutant human β-globin gene knockin;
mouse α-globin gene knockout;
mouse β-globin gene knockout.

Therefore, these mice make adult hemoglobin just like people with sickle cell disease, but they do not produce any normal mouse hemoglobin. These transgenic mice exhibit the major features of sickle cell disease—sickled red blood cells, anemia, and multiple organ pathology. They have been useful as a model for studying the disease and testing potential therapies.

Another avenue of animal biotechnology is **xenotransplantation**—the transplantation of cells, tissues, or organs from one animal species to another. For example, work is currently under way to produce genetically modified pigs whose organs are resistant to the rejection mechanisms that occur following transplantation. Strains of pigs have been made in which the gene that encodes 1,3 α-galactosyl transferase has been knocked out. This enzyme attaches sugars to cell surface proteins. When it is knocked out, the cell surface is much less immunogenic and therefore pig organs from this strain are less likely to be rejected when transplanted into humans. In 2011, researchers at the University of Pittsburgh predicted that clinical trials for the transplantation of such pig organs into humans will begin in a few years.

Biotechnology Holds Promise in Producing Transgenic Livestock

The technology for creating transgenic mice has been extended to other animals, and much research is under way to develop transgenic species of livestock, including fish, sheep, pigs, goats, and cattle. For some farmers, the ability to modify the characteristics of livestock via the introduction of cloned genes is an exciting prospect.

A novel avenue of research involves the production of medically important proteins in the mammary glands of livestock. This approach is sometimes called **molecular pharming.** (The term is also used to describe the manufacture of medical products by agricultural plants.) As shown in **Table 21.3**, several human proteins have been successfully produced in the milk of domestic livestock. Compared with the production of proteins in bacteria, one advantage of molecular pharming is that certain proteins are more likely to function properly when expressed in mammals. This may be due to covalent modifications, such as the attachment of carbohydrate groups, which occur in eukaryotes but not in bacteria. In addition, certain proteins may be degraded rapidly or folded improperly when expressed in bacteria. Furthermore, the yield of recombinant proteins in milk can be quite large. Dairy cows, for example, produce about 10,000 L of milk per year per cow. In some cases, a transgenic cow can produce approximately 1 g of the transgenic protein per liter of milk.

TABLE 21.3
Proteins That Are Produced in the Milk of Domestic Animals

Protein	Host	Use
Lactoferrin	Cattle	Used as an iron supplement in infant formula
Tissue plasminogen activator (TPA)	Goat	Dissolves blood clots
Antibodies	Cattle	Used to combat specific infectious diseases
α_1-Antitrypsin	Sheep	Treatment of emphysema
Factor IX	Sheep	Treatment of certain inherited forms of hemophilia
Insulin-like growth factor	Cattle	Treatment of diabetes

To introduce a human gene into an animal so that the encoded protein will be secreted into its milk, the gene is inserted next to a milk-specific promoter. Eukaryotic genes often are expressed in a tissue-specific fashion. In mammals, certain genes are expressed specifically within the mammary gland so that their protein product is secreted into the milk. Examples of milk-specific genes include genes that encode milk proteins such as β-lactoglobulin, casein, and whey acidic protein. To express a human gene that encodes a specific protein, such as a hormone, into a domestic animal's milk, the promoter for a milk-specific gene is linked to the coding sequence for the human gene (**Figure 21.7**). The DNA is then injected into an oocyte, where it is integrated into the genome. The fertilized oocyte is then implanted into the uterus of a female animal, which later gives birth to a transgenic offspring. If the offspring is a female, the protein hormone encoded by the human gene is expressed within the mammary gland and secreted into the milk. The milk can then be obtained from the animal, and the human hormone isolated.

21.2 COMPREHENSION QUESTIONS

1. When a cloned gene integrates into a genome by nonhomologous recombination, the end result is
   a. gene addition.
   b. gene replacement.
   c. a gene knockout.
   d. both a and b.
2. What is a chimera?
   a. A transgenic organism
   b. An organism that contains cells from two different individuals
   c. An organism that is produced by a gene knockin
   d. None of the above
3. One strategy for producing a protein in the milk of a cow is to place the coding sequence of the gene of interest next to a ___________ and then inject the gene into a ___________.
   a. *lac* operon promoter, cow oocyte
   b. β-lactoglobulin promoter, cow oocyte
   c. *lac* operon promoter, cow mammary cell
   d. β-lactoglobulin promoter, cow mammary cell

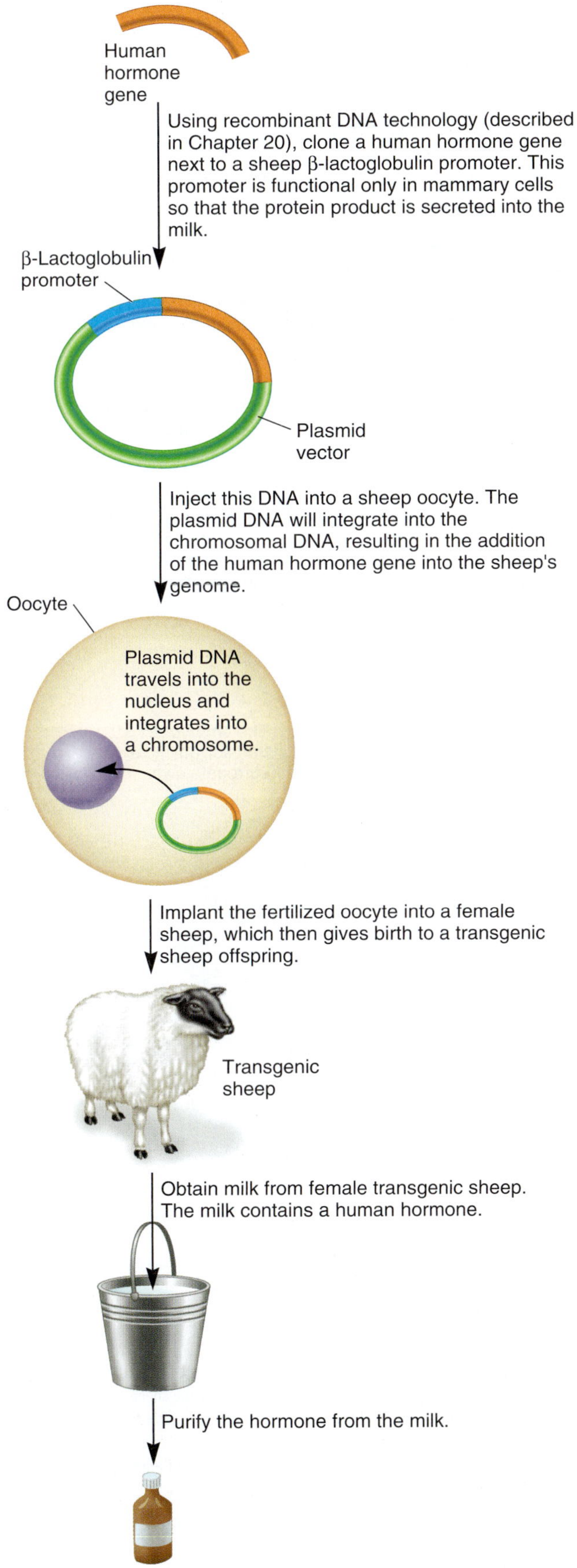

FIGURE 21.7 Strategy for expressing human genes in a domestic animal's milk. The β-lactoglobulin gene is normally expressed in mammary cells, whereas the human hormone gene is not. To express the human hormone gene in milk, the promoter from the milk-specific gene in sheep is linked to the coding sequence of the human hormone gene. In addition to the promoter, a short signal sequence may also be necessary so the protein is secreted from the mammary cells and into the milk.

Genes → Traits By using genetic engineering, researchers can give sheep the trait of producing a human hormone in their milk. This hormone can be purified from the milk and used to treat humans.

CONCEPT CHECK: Why is a β-lactoglobulin promoter used?

21.3 REPRODUCTIVE CLONING AND STEM CELLS

Learning Outcomes:

1. Outline the steps of reproductive cloning in mammals.
2. Define stem cells and describe their two key properties.
3. Distinguish between totipotent, pluripotent, multipotent, and unipotent stem cells.
4. List examples of how stem cells have the potential to treat human diseases.

The previous section focused on the area of biotechnology in which cloned genes are introduced into animals. Another aspect of biotechnology involves the cloning of whole organisms or the manipulation of stem cells. In this section, we will consider mammalian cloning and stem cell research. These topics have received enormous public attention due to the complex ethical issues they raise.

Researchers Have Succeeded in Cloning Mammals from Somatic Cells

The term "cloning" has several different meanings. In Chapter 20, we discussed gene cloning, which involves methods that produce many copies of a gene. The cloning of an entire organism is a different matter. **Reproductive cloning** refers to methods that produce two or more genetically identical individuals. This happens occasionally in nature; identical twins are genetic clones that began from the same fertilized egg. Similarly, researchers can take mammalian embryos at an early stage of development (e.g., the two-cell to eight-cell stage), separate the cells, implant them into the uterus, and obtain multiple births of genetically identical individuals.

In the case of plants, cloning is an easier undertaking, as we will explore later in the chapter. Plants can be cloned from somatic cells. In most cases, it is relatively easy to take a cutting from a plant, expose it to growth hormones, and obtain a separate plant that is genetically identical to the original. However, this approach has not been possible with mammals. For several decades, scientists believed that chromosomes within the somatic cells of mammals had incurred irreversible genetic changes that render them unsuitable for cloning. However, this hypothesis has proven to be incorrect. In 1997, Ian Wilmut and his colleagues announced that a sheep, named Dolly, had been cloned using the genetic material from somatic cells.

How was Dolly produced? As shown in **Figure 21.8**, the researchers removed mammary cells from an adult female sheep and grew them in the laboratory. The researchers then extracted the nucleus from an egg cell of a different sheep and used electrical pulses to fuse the diploid mammary cell with the enucleated egg cell. After fusion, the zygote began embryonic development, and the resulting embryo was implanted into the uterus of a surrogate mother sheep. One hundred and forty-eight days later, Dolly was born.

Although Dolly was clearly a clone of the donor female sheep, tests conducted when she was 3 years old suggested that she was "genetically older" than her actual age indicated. As mammals age, chromosomes in somatic cells tend to shorten from the telomeres—the ends of eukaryotic chromosomes. Therefore, older individuals have shorter chromosomes in their somatic cells than younger ones do. This shortening does not seem to occur in the cells of the germ line, however. When researchers analyzed the chromosomes in Dolly's somatic cells when she was about 3 years old, the lengths of her chromosomes were consistent with a sheep that was significantly older, say, 9–10 years old. The sheep that donated the somatic cell that produced Dolly was 6 years old, and her mammary cells had been grown in culture for several cell doublings before a mammary cell was fused with an oocyte. This led researchers to postulate that Dolly's shorter telomeres were a result of chromosome shortening in the somatic cells of the sheep that donated the nucleus. In 2003, the Roslin Institute announced the decision to euthanize 6-year-old Dolly after an examination showed progressive lung disease. Her death has raised concerns among experts that the techniques used to produce Dolly could have caused premature aging.

With regard to telomere length, research in mice and cattle has shown different results; the telomeres of cloned mice appear to be the correct length for their age. For example, cloning was conducted on mice via the method described in Figure 21.8 for six consecutive generations. The cloned mice of the sixth generation had appropriate telomere lengths. Further research is necessary to determine if cloning via somatic cells has an effect on the length of telomeres in subsequent generations. However, other studies in mice point to various types of genetic flaws in cloned animals. For example, Rudolf Jaenisch and his colleagues used DNA microarray technology (described in Chapter 22) to analyze the transcription patterns of over 10,000 genes in cloned mice. As much as 4% of those genes were not expressed normally. Furthermore, research has shown that cloned mice die at a younger age than their naturally bred counterparts.

Mammalian cloning is still at an early stage of development. Nevertheless, the breakthrough of creating Dolly has shown that it is technically possible. In recent years, cloning from somatic cells has been achieved in several mammalian species, including sheep, cattle, mice, goats, and pigs. In 2002, the first pet was cloned. She was named Carbon Copy, also called Copy Cat (**Figure 21.9**).

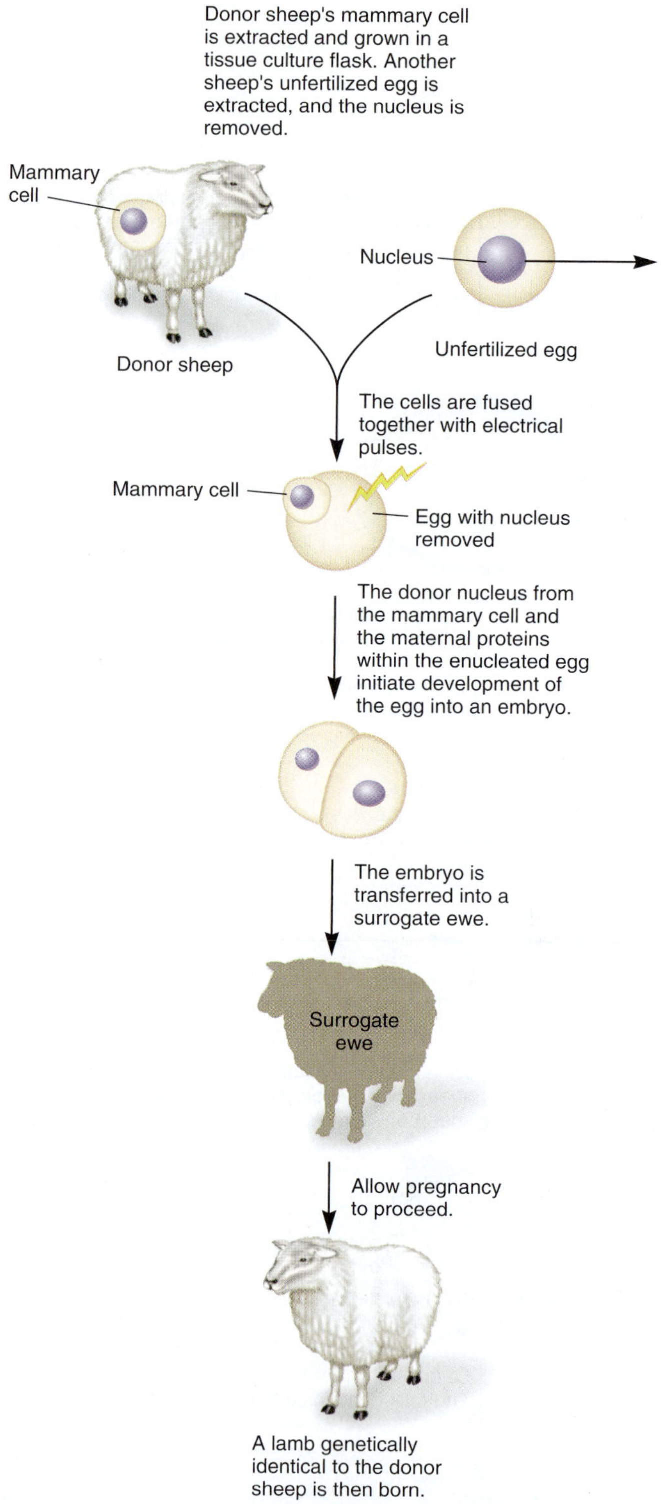

FIGURE 21.8 Protocol for the successful cloning of a sheep.
Genes → Traits Dolly was (almost) genetically identical to the sheep that donated a mammary cell to create her. Dolly and the donor sheep were (almost) genetically identical in the same way that identical twins are; they carried the same set of genes and looked remarkably similar. However, they may have had minor genetic differences due to possible variation in their mitochondrial DNA and may have exhibited some phenotypic differences due to maternal effect or imprinted genes.

CONCEPT CHECK: In the protocol, why is the nucleus of the oocyte removed?

FIGURE 21.9 Carbon Copy, the first cloned pet. The animal shown here was produced using a procedure similar to the one shown in Figure 21.8.

CONCEPT CHECK: Is Carbon Copy a transgenic animal?

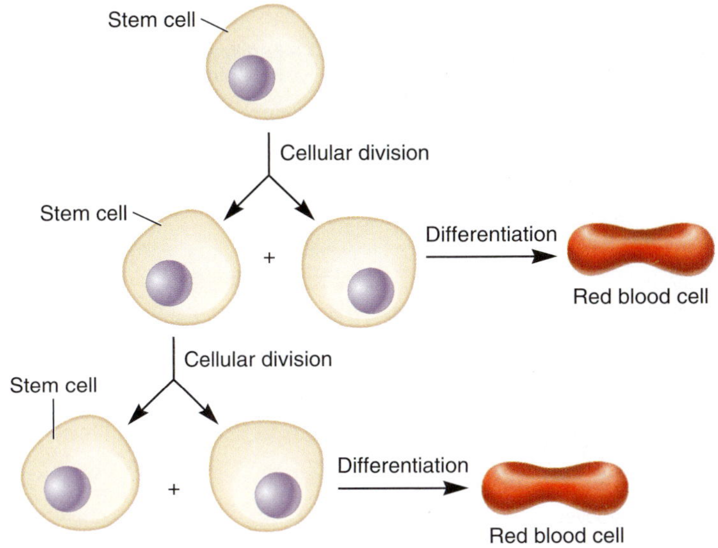

FIGURE 21.10 Growth pattern of stem cells. The two main traits that stem cells exhibit are an ability to divide and an ability to differentiate. When a stem cell divides, one of the two cells remains a stem cell, and the other daughter cell differentiates into a specialized cell type.

CONCEPT CHECK: Explain why stem cells are not depleted during the life of an organism.

Mammalian cloning may potentially have many practical applications. Cloning livestock would enable farmers to use the somatic cells from their best individuals to create genetically homogeneous herds, which could increase agricultural yield. A possible disadvantage, however, could be that such a genetically homogeneous herd may be more susceptible to rare diseases.

Though some people are concerned about the practical uses of cloning agricultural species, a majority have become very concerned with the possibility of human cloning. This prospect has raised a host of serious ethical questions. For example, some people feel that it is morally wrong and threatens the basic fabric of parenthood and family. Others feel that it is a technology that could offer a new avenue for reproduction, one that could be offered to infertile couples, for example. In the public sector, the sentiment toward human cloning has been generally negative. Many countries have issued an all-out ban on human cloning, but others permit limited research in this area. Because the technology for cloning exists, our society will continue to wrestle with the legal and ethical aspects of cloning as it applies not only to animals but also to people.

Stem Cells Have the Ability to Divide and Differentiate into Different Cell Types

Stem cells supply the various kinds of cells that construct our bodies from a fertilized egg. In adults, stem cells also replenish worn-out or damaged cells. To accomplish this task, stem cells have two common characteristics. First, they have the capacity to divide, and second, they can differentiate into one or more specialized cell types. As shown in **Figure 21.10**, the two daughter cells produced from the division of a stem cell can have different fates. One of the cells may remain an undifferentiated stem cell, while the other daughter cell can differentiate into a specialized cell type. With this type of asymmetrical division/differentiation pattern, the population of stem cells remains relatively constant, yet the stem cells provide a population of specialized cells. In the adult, this type of mechanism is needed to replenish cells that have a finite life span, such as skin epithelial cells and red blood cells.

In mammals, stem cells are commonly categorized according to their developmental stage and their ability to differentiate (**Figure 21.11**). The ultimate stem cell is the fertilized egg, which, via multiple cellular divisions, can give rise to an entire organism. A fertilized egg is considered **totipotent,** because it can give rise to all cell types in the adult organism.

The early mammalian embryo contains **embryonic stem cells (ES cells),** which are found in the inner cell mass of the blastocyst. The blastocyst is the stage of embryonic development prior to uterine implantation—the preimplantation embryo. ES cells are **pluripotent,** which means they can differentiate into almost every cell type of the body. However, a single ES cell has lost the ability to produce an entire, intact individual. During the early fetal stage of development, the germ-line cells found in the gonads also are pluripotent. These cells are called **embryonic germ cells (EG cells).**

As mentioned, adults also contain stem cells, but these are thought to be multipotent or unipotent. A **multipotent** stem cell can differentiate into several cell types but far fewer than an ES cell. For example, hematopoietic stem cells (HSCs) found in the bone marrow supply cells that populate two different tissues,

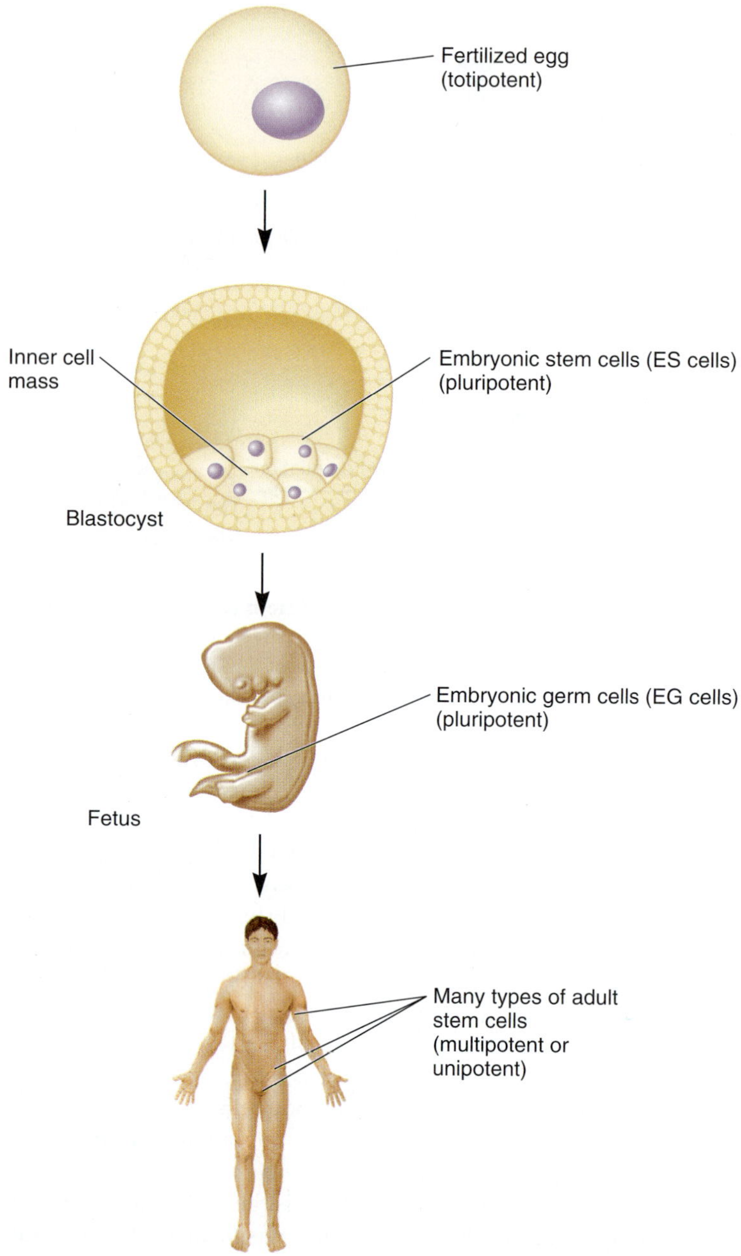

FIGURE 21.11 Occurrence of stem cells at different stages of human development.

namely, the blood and lymphoid tissues (**Figure 21.12**). Furthermore, each of these tissues contains several cell types. Multipotent HSCs follow a pathway in which cell division produces a myeloid progenitor cell, which then differentiates into a red blood cell, megakaryocyte, basophil, monocyte, eosinophil, neutrophil, or dendritic cell. Alternatively, an HSC follows a path in which it becomes a lymphoid progenitor cell, which then differentiates into a T cell, B cell, natural killer cell, or dendritic cell. Other stem cells found in the adult seem to be **unipotent.** For example, primordial germ cells in the testis differentiate only into a single cell type, the sperm.

Stem Cells Have the Potential to Treat a Variety of Diseases

Interest in stem cells centers on two main areas. Because stem cells have the capacity to differentiate into multiple cell types, the study of stem cells may help us to understand basic genetic mechanisms that underlie the process of development, the details of which are described in Chapter 25. A second compelling reason why people have become interested in stem cells is their potential to treat human diseases or injuries that cause cell and tissue damage. This application has already become a reality in certain cases. For example, bone marrow transplantation is used to treat patients with certain forms of cancers. Such patients may be given radiation treatments that destroy their immune systems. When these patients are injected with bone marrow from a healthy person, the stem cells within the transplanted marrow have the ability to proliferate and differentiate within their bodies and provide them with a functioning immune system.

Renewed interest in the use of stem cells in the potential treatment of many other diseases was fostered in 1998 by studies of two separate teams, headed by James Thomson and John Gearhart, showing that embryonic cells, either ES or EG cells, can be successfully propagated in the laboratory. As mentioned, ES and EG cells are pluripotent and therefore have the capacity to produce many different kinds of tissue. As shown in **Table 21.4**, embryonic cells could potentially be used to treat a wide variety of diseases associated with cell and tissue damage. By comparison, it would be difficult, based on our current knowledge, to treat these diseases with adult stem cells because of the inability to locate most types of adult stem cells within the body and successfully grow them in the laboratory. Even HSCs are elusive. In the bone marrow, about 1 cell in 10,000 is a stem cell, yet that is enough to populate all of the blood and lymphoid cells of the body. The stem cells of most other adult tissues are equally difficult to locate, if not more so. In addition, with the exception of stem cells in the blood, other types of stem cells in the adult body are difficult to remove in sufficient numbers for transplantation. By comparison, ES and EG cells are easy to identify and have the great advantage of rapid growth in the laboratory. For these reasons, ES and EG cells offer a greater potential for transplantation, based on our current knowledge of stem cell biology.

For ES or EG cells to be used in transplantation, researchers need to derive methods that cause them to differentiate into the appropriate type of tissue. For example, if the goal was to repair a spinal cord injury, ES or EG cells would need the appropriate cues that cause them to differentiate into neural tissue. At present, much research is needed to understand and potentially control the fate of ES or EG cells. Currently, research indicates that a complex variety of factors determines the developmental fates of stem cells. These include internal factors within the stem cells themselves, as well as external factors such as the properties of neighboring cells and the presence of hormones and growth factors in the environment.

From an ethical perspective, the primary issue that raises debate is the source of the stem cells for research and potential treatments. Most ES cells have been derived from human

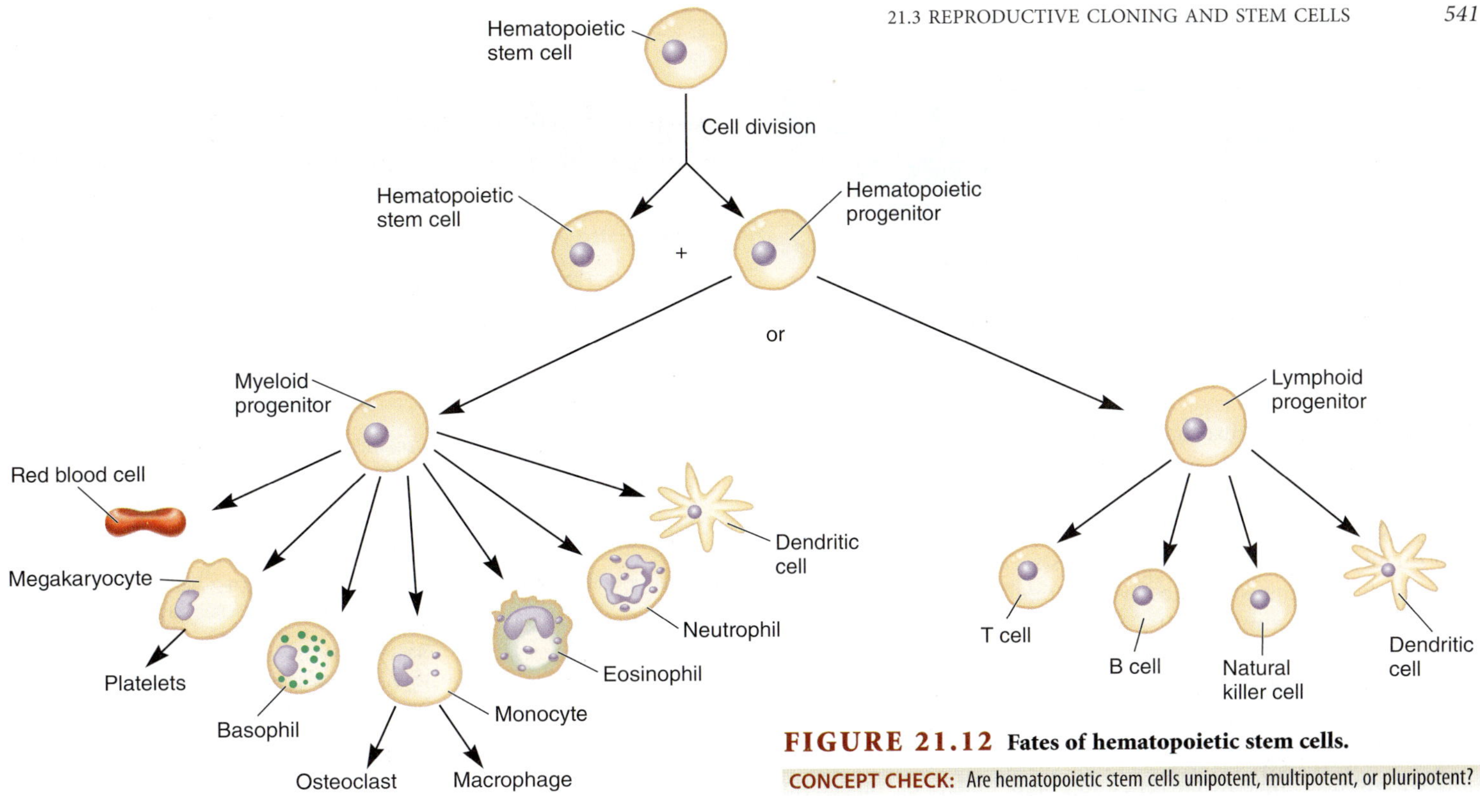

FIGURE 21.12 Fates of hematopoietic stem cells.

CONCEPT CHECK: Are hematopoietic stem cells unipotent, multipotent, or pluripotent?

TABLE 21.4
Potential Uses of Stem Cells to Treat Diseases

Cell/Tissue Type	Disease Treatment
Neural	Implantation of cells into the brain to treat Parkinson's disease
	Treatment of injuries such as those to the spinal cord
Skin	Treatment of burns and other types of skin disorders
Cardiac	Repair of heart damage associated with heart attacks
Cartilage	Repair of joints damaged by injury or arthritis
Bone	Repair of damaged bone or replacement with new bone
Liver	Repair or replacement of liver tissue that has been damaged by injury or disease
Skeletal muscle	Repair or replacement of damaged muscle

embryos that were produced from in vitro fertilization and were subsequently not used. Most EG cells are obtained from aborted fetuses. Some feel that it is morally wrong to use such tissue in research and/or the treatment of disease or they fear that this technology could lead to intentional abortions for the sole purpose of obtaining fetal tissues for transplantation. Alternatively, others feel that the embryos and fetuses that provide the ES and EG cells are not going to become living individuals, and therefore, it is beneficial to study these cells and use them in a positive way to treat human diseases and injury. It is not clear whether these two opposing viewpoints can reach a common ground.

If stem cells could be obtained from adult cells and propagated in the laboratory, an ethical dilemma may be avoided because most people do not have serious moral objections to current procedures such as bone marrow transplantation. In 2006, work by Shinya Yamanaka and colleagues showed that adult mouse fibroblasts (a type of connective tissue cell) could become pluripotent via the injection of four different genes that encode transcription factors. In 2007, Yamanaka's laboratory and two other research groups showed that such induced pluripotent stem cells (iPS) can differentiate into all cell types when injected into mouse blastocysts and grown into baby mice. Though further research is still needed, these recent results indicate that adult cells can be reprogrammed to become embryonic stem cells.

21.3 COMPREHENSION QUESTIONS

1. During mammalian reproductive cloning, __________ is fused with __________.
   a. a somatic cell, a stem cell
   b. a somatic cell, an egg cell
   c. a somatic cell, an enucleated egg cell
   d. an enucleated somatic cell, an egg cell
2. Which of the following is a key feature of stem cells?
   a. They have the ability to divide.
   b. They have the ability to differentiate.
   c. They are always pluripotent.
   d. Both a and b.

21.4 GENETICALLY MODIFIED PLANTS

Learning Outcomes:

1. List examples of transgenic plants that are useful to people.
2. Outline the steps to make transgenic plants using *Agrobacterium tumefaciens*.

For centuries, agriculture has relied on selective breeding programs to produce plants and animals with desirable characteristics. For agriculturally important species, this often means the production of strains that are larger, disease-resistant, and yield high-quality food. Agricultural scientists can now complement traditional breeding strategies with modern molecular genetic approaches. In the mid-1990s, genetically modified crops first became commercialized. Since that time, their use has progressively increased. In 2012, over 30% of all agricultural crops were transgenic. Worldwide, more than 100 million hectares (247 million acres) of transgenic crops were planted. In this section, we will discuss some current and potential uses of transgenic plants in agriculture, and examine the methods that scientists follow to make them.

Transgenic Plants Can Be Given Characteristics That Are Agriculturally Useful

Various traits can be modified in transgenic plants (**Table 21.5**). Most commonly, transgenic research has sought to produce plant strains resistant to herbicides, disease, and insects. For example, transgenic plants highly tolerant of particular herbicides have been made. The Monsanto Company has produced transgenic plant strains tolerant of glyphosate, the active agent in the herbicide Roundup. The herbicide remains effective against weeds, but the herbicide-resistant crop is spared (**Figure 21.13**).

Another important approach is to make plant strains that are disease-resistant. In many cases, virus-resistant plants have been developed by introducing a gene that encodes a viral coat protein. When the plant cells express the viral coat protein, they become resistant to infection by that pathogenic virus.

A very successful example of the use of transgenic plants has involved the introduction of genes from *Bacillus thuringiensis* (Bt). As discussed earlier in this chapter, this bacterium produces toxins that kill certain types of caterpillars and beetles and has been widely used as a biological control agent for several decades. These toxins are proteins encoded in the genome of *B. thuringiensis*. Researchers have succeeded in cloning toxin genes from *B. thuringiensis* and transferring those genes into plants. Such Bt

TABLE 21.5
Traits That Have Been Modified in Transgenic Plants

Trait	Examples
Plant Protection	
Resistance to herbicides	Transgenic plants can express proteins that render them resistant to particular herbicides (see Figure 21.13).
Resistance to viral, bacterial, and fungal pathogens	Transgenic plants that express the pokeweed antiviral protein are resistant to a variety of viral pathogens.
Resistance to insects	Transgenic plants that express the CryIA protein from *Bacillus thuringiensis* are resistant to a variety of insects (see Figure 21.14).
Plant Quality	
Improvement in storage	Transgenic plants can express antisense RNA that silences a gene involved in fruit softening.
Change in plant composition	Transgenic strains of canola have been altered with regard to oil composition; the seeds of the Brazil nut have been rendered methionine-rich via transgenic technology.
New Products	
Biodegradable plastics	Transgenic plants have been made that can synthesize polyhydroxyalkanoates, which are used as biodegradable plastics.
Vaccines	Transgenic plants have been modified to produce vaccines in their leaves against many human and animal diseases, including hepatitis B, cholera, and malaria.
Pharmaceuticals	Transgenic plants have been made that produce a variety of medicines, including human interferon-α (to fight viral diseases and cancer), human epidermal growth factor (for wound repair), and human aprotinin (for reducing blood loss during transplantation surgery).
Antibodies	Human antibodies have been made in transgenic plants to battle various diseases such as non-Hodgkin lymphoma.

FIGURE 21.13 Transgenic plants that are resistant to glyphosate.

Genes → Traits This field of soybean plants has been treated with glyphosate. The plants on the left have been genetically engineered to contain a herbicide-resistant gene, which makes them resistant to killing by glyphosate. By comparison, the dead plants in the row with the orange stick do not have this gene.

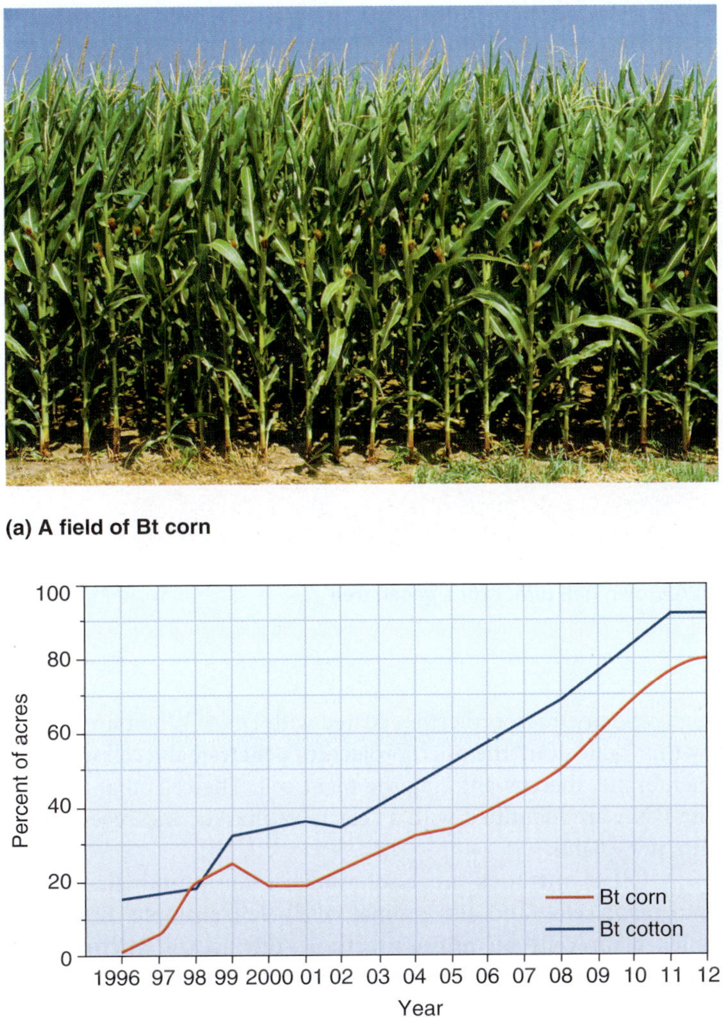

(a) A field of Bt corn

(b) Bt corn and Bt cotton usage since 1996

FIGURE 21.14 The production of Bt crops. **(a)** A field of Bt corn. These corn plants carry an endotoxin gene from *Bacillus thuringiensis* that provides them with resistance to insects such as corn borers, which are a major pest of corn plants. **(b)** A graph showing the increase in usage of Bt corn and Bt cotton in the United States since their commercial introduction in 1996.

varieties of plants produce the toxins themselves and therefore are resistant to many types of caterpillars and beetles. Examples of commercialized crops include Bt corn (**Figure 21.14a**) and Bt cotton. Since their introduction in 1996, the commercial use of these two Bt crops has steadily increased (**Figure 21.14b**).

The introduction of transgenic plants into agriculture has been strongly opposed by some people. What are the perceived risks? One potential risk is that transgenes in commercial crops could endanger native species. For example, Bt crops may kill pollinators of native species. Another worry is that the planting of transgenic crops could potentially lead to the proliferation of resistant insects. To prevent this from happening, researchers are producing transgenic strains that carry more than one toxin gene, which makes it more difficult for insect resistance to arise. Despite these and other concerns, many farmers are embracing transgenic crops, and their use continues to rise.

Agrobacterium tumefaciens and Other Methods Are Used to Make Transgenic Plants

As we have seen, the introduction of cloned genes into embryonic cells can produce transgenic animals. The production of transgenic plants is somewhat easier, because some somatic cells are totipotent, which means they are capable of developing into an entire organism. Therefore, a transgenic plant can be made by the introduction of cloned genes into somatic tissue, such as the tissue of a leaf. After the cells of a leaf have become transgenic, an entire plant can be regenerated by the treatment of the leaf with plant growth hormones, which cause it to form roots and shoots.

Molecular biologists can use the bacterium *Agrobacterium tumefaciens*, which naturally infects plant cells, to produce transgenic plants. A plasmid from the bacterium, known as the Ti plasmid (Tumor-inducing plasmid), carries genes that cause tumor formation after a plant has been infected (**Figure 21.15a**). In particular, a segment of the plasmid DNA, known as **T DNA** (for transferred DNA), is naturally transferred from the bacterium to the infected plant cells. The T DNA from the Ti plasmid is integrated into the chromosomal DNA of the plant cell by recombination. After this occurs, genes within the T DNA that encode plant growth hormones cause uncontrolled plant cell growth. This produces a cancerous plant growth known as a crown gall tumor (**Figure 21.15b**).

Because *A. tumefaciens* inserts its T DNA into the chromosomal DNA of plant cells, it can be used as a vector to introduce cloned genes into plants. Molecular geneticists have been able to modify the Ti plasmid to make this an efficient process. Such vectors are called **T-DNA vectors.** The T-DNA genes that cause the development of a gall have been identified. Fortunately for genetic engineers, when these genes are removed, the T DNA is still taken up into plant cells and integrated within the plant chromosomal DNA. However, a gall tumor does not form. In addition, geneticists have inserted selectable marker genes into the T DNA to allow selection of plant cells that have taken up the T DNA. A gene that provides resistance to the antibiotic kanamycin (*kan*^{R}) is a commonly used selectable marker. The T-DNA vectors used in cloning experiments are also modified to contain unique restriction sites for the convenient insertion of any gene.

Figure 21.16 shows the general strategy for producing transgenic plants via T-DNA-mediated gene transfer. A gene of interest is inserted into a genetically engineered T-DNA vector and then transformed into *A. tumefaciens*. Plant cells are exposed to the transformed *A. tumefaciens*. After allowing time for infection, the plant cells are exposed to the antibiotics carbenicillin and kanamycin. Carbenicillin kills *A. tumefaciens*, and kanamycin kills any plant cells that have not taken up the T DNA with the antibiotic resistance gene. Therefore, the only surviving cells are those plant cells that have integrated the T DNA into their genome. Because the T DNA also contains the cloned gene of interest, the selected plant cells are expected to have received this cloned gene as well. The cells are then transferred to a medium that contains the plant growth hormones necessary for the regeneration of entire plants. These plants can then be analyzed to verify that they are transgenic plants containing the cloned gene.

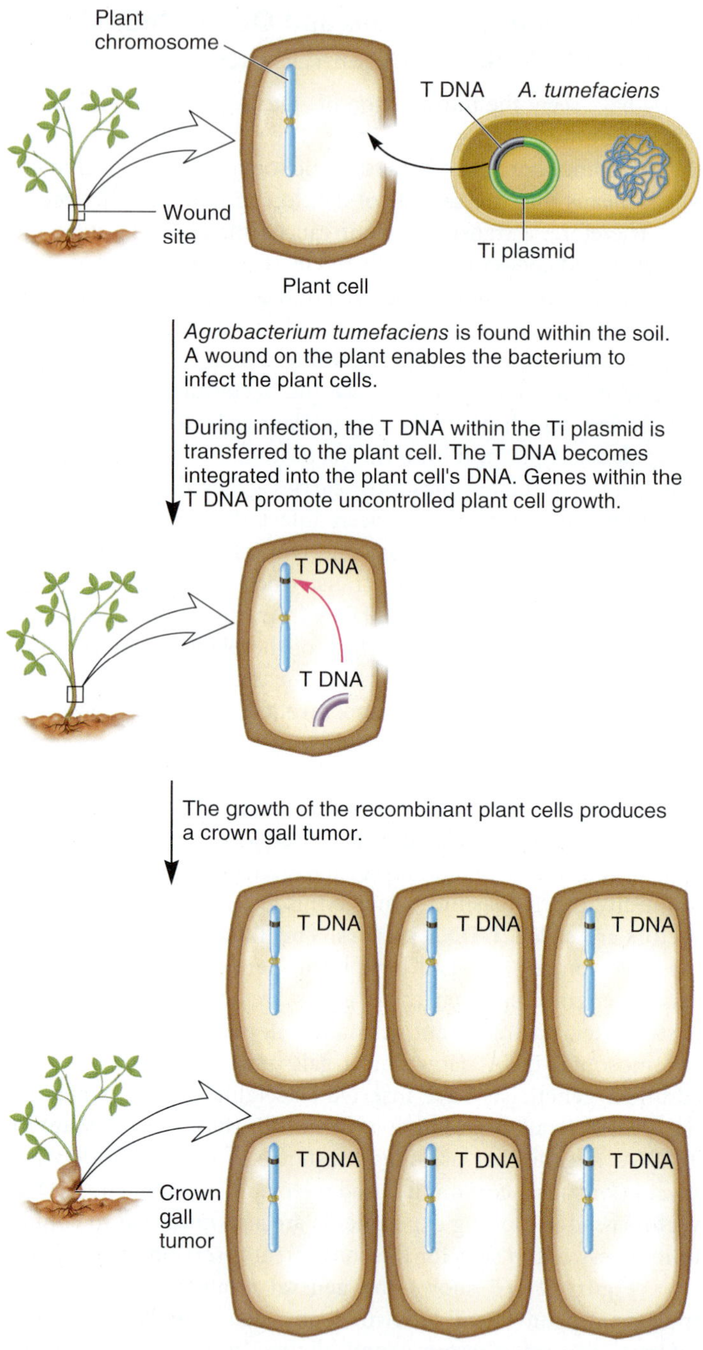

(a) The production of a crown gall tumor by *A. tumefaciens* infection

(b) A crown gall tumor on a pecan tree

FIGURE 21.15 ***Agrobacterium tumefaciens* infecting a plant and causing a crown gall tumor.**

A. tumefaciens infects a wide range of plant species, including most dicotyledonous plants, most gymnosperms, and some monocotyledonous plants. However, not all plant species are infected by this bacterium. Fortunately, other methods are available for introducing genes into plant cells. Another common way to produce transgenic plants is an approach known as **biolistic gene transfer.** In this method, plant cells are bombarded with high-velocity microprojectiles coated with DNA. When fired upon by this "gene gun," the microprojectiles penetrate the cell wall and membrane, thereby entering the plant cell. The cells that take up the DNA are identified with a selectable marker and regenerated into new plants.

Other methods are also available for introducing DNA into plant cells (and also animal cells). For example, DNA can enter plant cells by **microinjection**—the use of microscopic-sized needles—or by **electroporation**—the use of electrical current to create temporary pores in the plasma membrane. Because the rigid plant cell wall is a difficult barrier for DNA entry, other approaches involve the use of protoplasts, which are plant cells that have had their cell walls removed. DNA can be introduced into protoplasts using a variety of methods, including treatment with polyethylene glycol and calcium phosphate.

21.4 COMPREHENSION QUESTIONS

1. When *A. tumefaciens* is used to make a transgenic plant, a gene of interest is inserted into a __________, which is first transformed into __________. The __________ is then transferred to a plant.
   a. viral vector, *E. coli*, gene of interest
   b. T-DNA vector, *A. tumefaciens*, T DNA carrying the gene of interest
   c. T-DNA vector, *A. tumefaciens*, gene of interest alone
   d. T-DNA vector, *E. coli*, gene of interest alone

2. In addition to T-DNA vectors, other technologies to produce transgenic plants include
   a. biolistic gene transfer.
   b. microinjection.
   c. electroporation.
   d. all of the above.

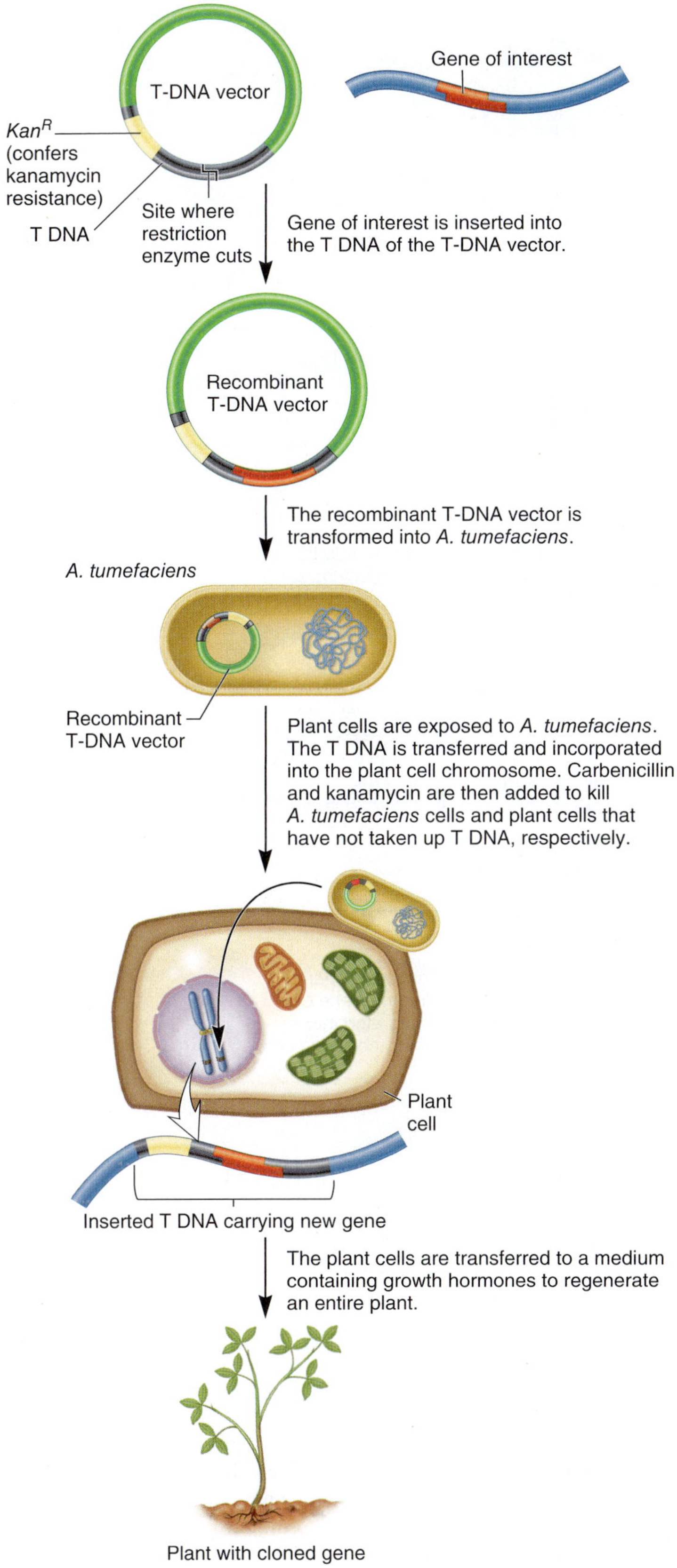

ONLINE ANIMATION

FIGURE 21.16 **The transfer of genes into plants using a T-DNA vector from *A. tumefaciens.***

CONCEPT CHECK: Which portion of a T-DNA vector is transferred to a plant?

21.5 HUMAN GENE THERAPY

Learning Outcomes:

1. Define gene therapy.
2. Analyze the results of the first gene therapy study involving adenosine deaminase deficiency.
3. Discuss potential side effects of gene therapy.

Because some mutations can alter gene function and thereby cause disease, geneticists are actively pursuing the goal of using normal, cloned genes to compensate for defects in mutant genes. **Gene therapy** is the introduction of cloned genes into somatic cells to treat disease. It is a potential method for treating a wide variety of illnesses.

Many current research efforts in gene therapy are aimed at alleviating inherited human diseases. Over 7000 human genetic diseases are known to involve a single gene abnormality. Familiar examples include cystic fibrosis, sickle cell disease, and hemophilia. In addition, gene therapies have also been aimed at treating diseases such as cancer and cardiovascular disease, which may occur later in life. Some scientists are even pursuing research that will use gene therapy to combat infectious diseases such as AIDS. Even though a large amount of research has already been conducted, success has been limited, and relatively few patients have been treated with gene therapy. Nevertheless, a few results have been somewhat promising. Table 21.6 describes several types of diseases that are being investigated as potential targets for gene therapy. In this section, we will examine the approaches to gene therapy and how it may be used to treat human disease.

Gene Therapy Involves the Introduction of Cloned Genes into Human Cells

A key step in gene therapy is the introduction of a cloned gene into the somatic cells of people. This is a challenging aspect of gene therapy, because it requires that many cells of a person's body take

TABLE 21.6
Future Prospects in Gene Therapy

Type of Disease	Treatment of
Blood	Sickle cell disease, hemophilia, severe combined immunodeficiency disease (SCID)
Metabolic	Glycogen storage diseases, lysosomal storage diseases, phenylketonuria
Muscular	Duchenne muscular dystrophy, myotonic muscular dystrophy
Lung	Cystic fibrosis
Cancer	Brain tumors, breast cancer, colorectal cancer, malignant melanoma, ovarian cancer, several other types of malignancies
Cardiovascular	Atherosclerosis, essential hypertension
Infectious	AIDS, possibly other viral diseases that involve latent infections

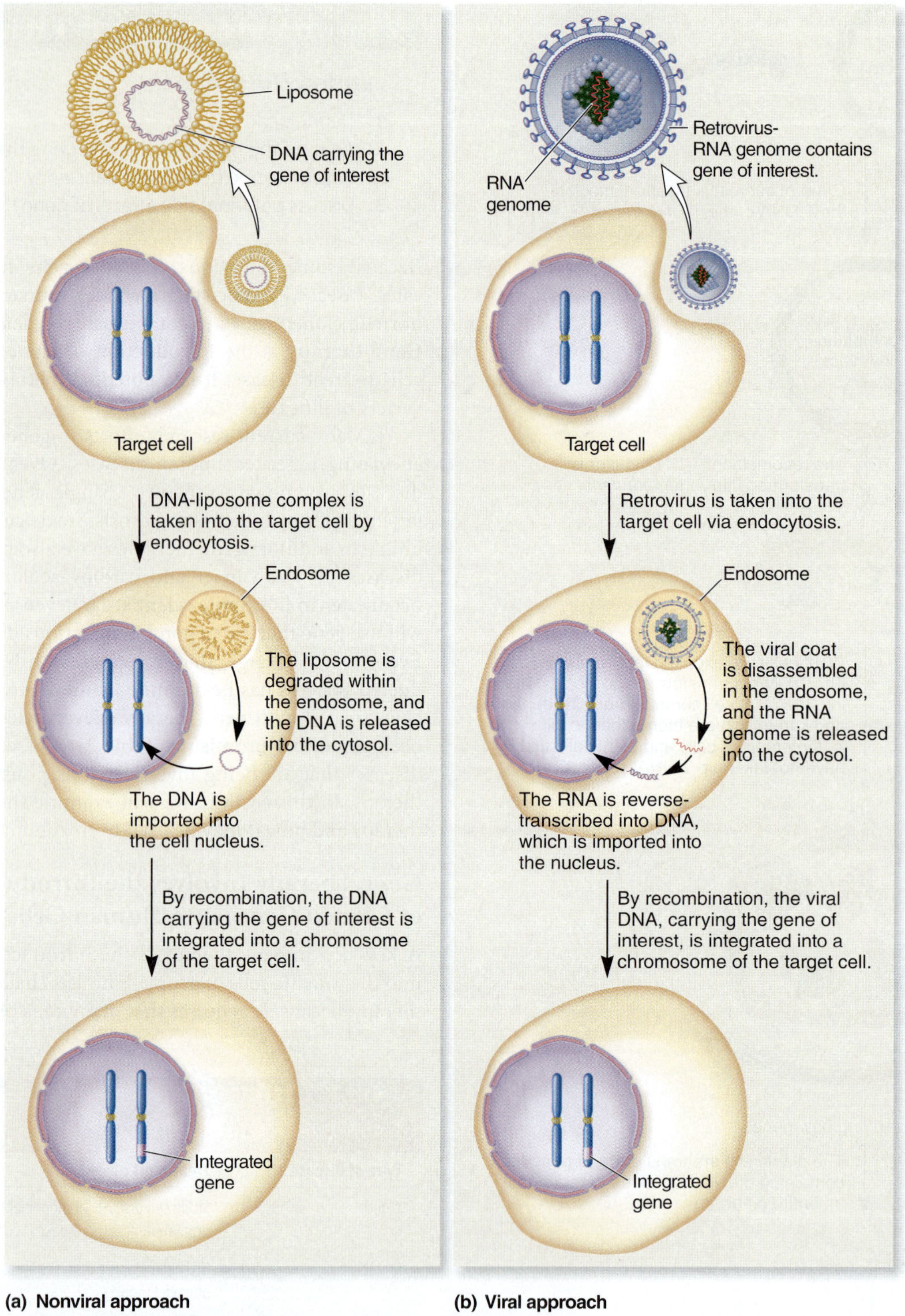

FIGURE 21.17 **Methods of gene transfer used in gene therapy.** **(a)** In this example, the DNA containing the gene of interest is complexed with cationic liposomes. These complexes are taken into cells by endocytosis, in which a portion of the plasma membrane invaginates and creates an intracellular vesicle known as an endosome. After it is released from the endosome, the DNA may then integrate into the chromosomal DNA via recombination. **(b)** In this example, the gene of interest is inserted into a retrovirus. When the retrovirus infects a cell via endocytosis, the RNA genome is reverse-transcribed into double-stranded DNA, which then integrates into the chromosome via recombination. Viruses used in gene therapy have been genetically altered so they cannot proliferate after entry into the target cell.

CONCEPT CHECK: What are advantages and disadvantages of these two methods?

up the cloned gene and express it. Otherwise, the disease symptoms will not be corrected. The techniques to transfer a cloned gene into human cells can be categorized as nonviral and viral gene transfer methods. The most common nonviral technique involves the use of **liposomes,** which are lipid vesicles (**Figure 21.17a**). The DNA containing the gene of interest is complexed with liposomes that carry a positive charge (i.e., cationic liposomes). The DNA-liposome complexes are taken into cells via endocytosis, in which a portion of the plasma membrane invaginates and creates an intracellular vesicle known as an endosome; the liposome is degraded within the endosome. The DNA is then released into the cytosol, imported into the nucleus, and then integrated into a chromosome of the target cell by recombination. An advantage of gene transfer via liposomes is that the liposomes do not elicit an immune response. A disadvantage is that the efficiency of gene transfer may be very low.

A second way to transfer genes into human cells is via viruses. Commonly used viruses for gene therapy include retroviruses, adenoviruses, and parvoviruses. The genetic modification of these viral genomes has led to the development of gene therapy vectors with a capacity to infect cells or tissues, much like the ability of wild-type viruses to infect cells. However, in contrast to wild-type viruses, gene therapy viral vectors have been genetically engineered so they can no longer replicate within target cells. Nevertheless, the genetically engineered viruses are naturally taken up by cells via endocytosis (**Figure 21.17b**). The viral coat disassembles, and the viral genome is released into the cytosol. In the case of retroviruses, the genome is RNA, which is reverse-transcribed into DNA. The viral DNA, which carries a gene of interest, is imported into the nucleus and is then integrated into a chromosome of the target cell by recombination.

A key advantage of viral vectors is their ability to efficiently transfer cloned genes to a variety of human cell types. However, a major disadvantage of viral-mediated gene therapy is the potential to evoke an undesirable immune response when injected into a patient. The inflammatory responses induced by adenovirus particles, for example, can be very strong and even fatal. Therefore, much effort has been aimed at preventing the inflammatory responses mediated by virus particles. These include the application of immunosuppressive drugs and the generation of less immunogenic viral vectors by further genetic modification within the viral genome.

EXPERIMENT 21A

Adenosine Deaminase Deficiency Was the First Inherited Disease Treated with Gene Therapy

Adenosine deaminase (ADA) is an enzyme involved in purine metabolism. If both copies of the *ADA* gene are defective, deoxyadenosine accumulates within the cells of the individual. At high concentrations, deoxyadenosine is particularly toxic to lymphocytes in the immune system, namely, T cells and B cells. In affected individuals, the destruction of T and B cells leads to a form of severe combined immunodeficiency (SCID). If left untreated, SCID is typically fatal at an early age (generally, 1–2 years old), because the immune system of these individuals is severely compromised and cannot fight infections.

Three approaches can be used to treat ADA deficiency. In some cases, a patient may receive a bone marrow transplant from a compatible donor. A second method is to treat SCID patients with purified ADA that is coupled to polyethylene glycol (PEG). This PEG-ADA is taken up by lymphocytes and can correct the ADA deficiency. Unfortunately, these two approaches are not always available and/or successful. A third, more recent approach is to treat ADA patients with gene therapy.

On September 14, 1990, the first human gene therapy was approved for a young girl suffering from ADA deficiency. This work was carried out by a large team of researchers composed of R. Michael Blaese, Kenneth Culver, W. French Anderson, and colleagues. Prior to this clinical trial, the normal gene for ADA had been cloned into a retroviral vector that can infect lymphocytes. The general aim of this therapy was to remove lymphocytes from the blood of the young girl with SCID, introduce the normal *ADA* gene into her cells, and then return them to her bloodstream.

Figure 21.18 outlines the protocol for the experimental treatment. Lymphocytes (i.e., T cells) were removed from the patient and cultured in a laboratory. The lymphocytes were then transfected with a nonpathogenic retrovirus that had been genetically engineered to contain the normal *ADA* gene. During the life cycle of a retrovirus, the retroviral genetic material is inserted into the host cell's DNA. Therefore, because this retrovirus contained the normal *ADA* gene, this gene also was inserted into the chromosomal DNA of the girl's lymphocytes. After this had occurred in the laboratory, the cells were reintroduced back into the patient. This approach is called an **ex vivo approach** because the genetic manipulations occur outside the body, and the products are reintroduced into the body.

THE HYPOTHESIS

Infecting lymphocytes with a retrovirus containing the normal *ADA* gene will correct the inherited deficiency of the mutant *ADA* gene in patients with ADA deficiency.

TESTING THE HYPOTHESIS

FIGURE 21.18 The first human gene therapy for adenosine deaminase deficiency by Blaese and colleagues.

Starting material: A retrovirus carrying the normal *ADA* gene.

1. Remove *ADA*-deficient lymphocytes from the patient with severe combined immunodeficiency disease (SCID).
2. Culture the cells in a laboratory.
3. Transfect the cells with a retrovirus that contains the normal *ADA* gene. Retroviruses insert their DNA into the host cell chromosome as part of their reproductive cycle.
4. Infuse the *ADA*-gene-corrected lymphocytes back into the SCID patient.

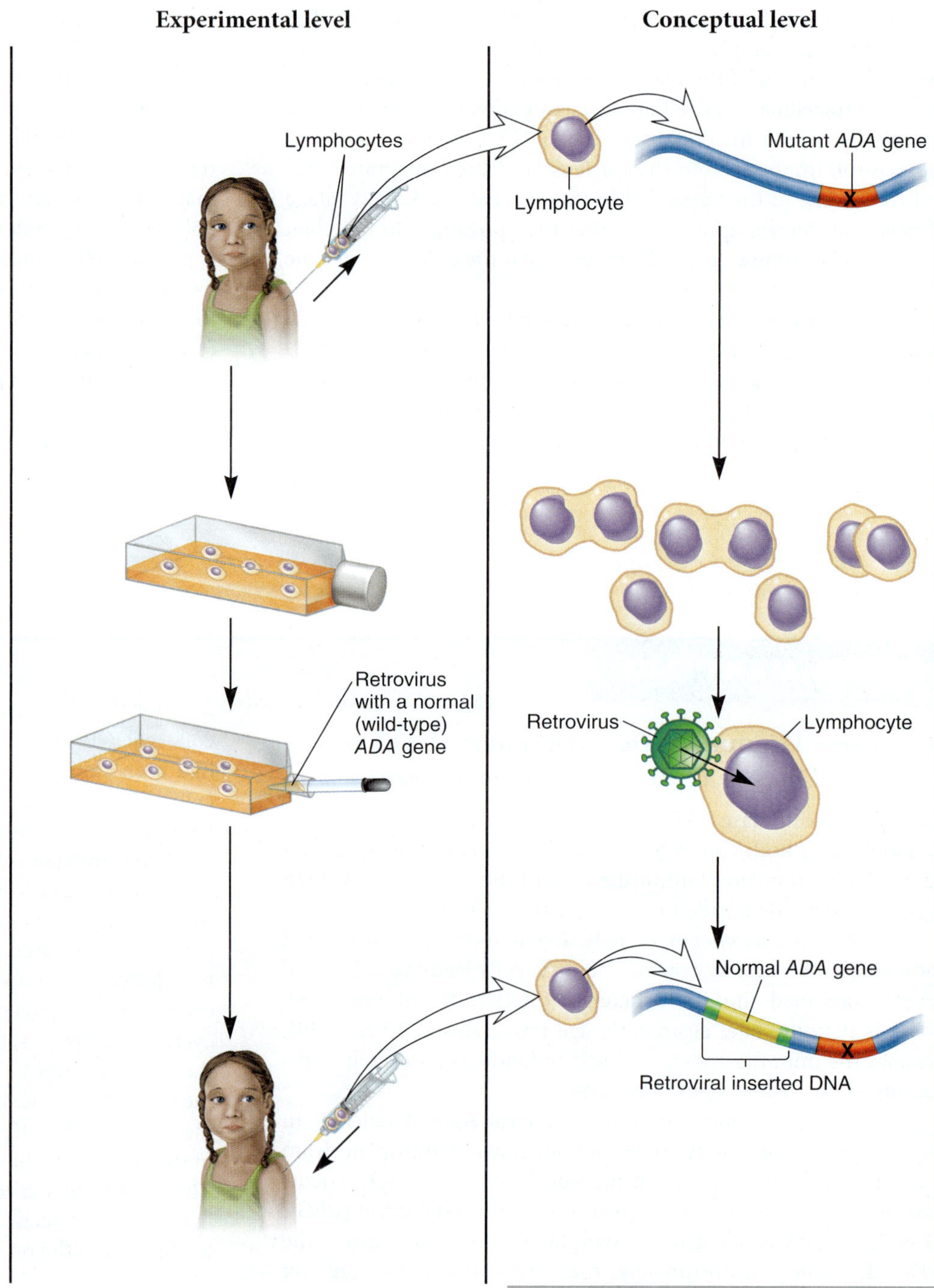

THE DATA

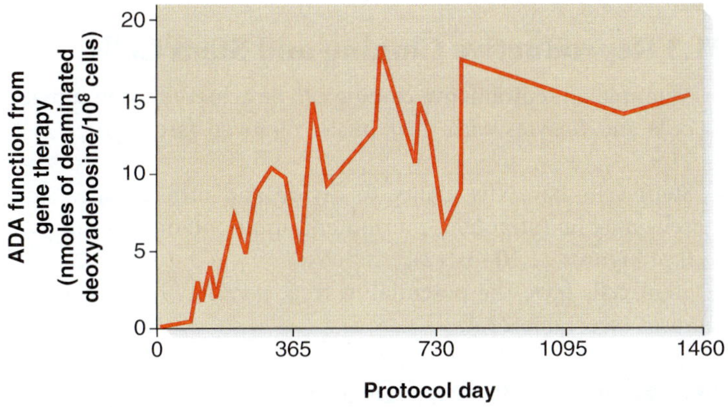

Data from: R.M. Blaese, K.W. Culver, A.D. Miller, et al. (1995) T lymphocyte-directed gene therapy for ADA-SCID: Initial trial results after 4 years. *Science 270*, 475–480.

INTERPRETING THE DATA

In this clinical trial, two patients were enrolled, and a third patient was later treated in Japan. Was the treatment a success? The Data show the level of ADA function for one patient involved in this study. The patient began with negligible ADA function. Even after several years, significantly higher levels of ADA were observed. However, because the patients also received a low dose of PEG-ADA treatment, researchers could not determine whether or not gene transfer into T cells by itself was of significant clinical benefit.

Another form of SCID, termed SCID-X1, is inherited as an X-linked trait. SCID-X1 is characterized by a block in T-cell growth and differentiation. This block is caused by mutations in the gene encoding the γ_c cytokine receptor, which plays a key role in the recognition of signals that are needed to promote the growth, survival, and differentiation of T cells. A gene therapy trial for SCID-X1 similar to the trial shown in Figure 21.18 was initiated in 2000 in which a normal γ_c cytokine receptor gene was cloned into a retroviral vector and then introduced into SCID-X1 patients' lymphocytes. The lymphocytes were then reintroduced back into their bodies. At a 10-month follow-up, T cells expressing the normal γ_c cytokine receptor were detected in two patients. Most importantly, the T-cell counts in these two patients had risen to levels that were comparable to those in normal individuals. This clinical trial was the first clear demonstration that gene therapy can offer clinical benefit, providing in these cases what seemed to be a complete correction of the disease phenotype. However, in a French study involving 10 SCID-X1 patients, an unexpected and serious side effect occurred. Within 3 years of gene therapy treatment, 3 out of the 10 treated children developed leukemia—a form of cancer involving the proliferation of white blood cells. In these cases, the disease was caused by the integration of the retroviral vector next to a particular gene in the patients' genomes. The development of leukemia in these patients has halted many clinical trials involving gene therapy.

A self-help quiz involving this experiment can be found at www.mhhe.com/brookergenetics5e.

21.5 COMPREHENSION QUESTIONS

1. A method of introducing a cloned gene into cells for gene therapy is via

a. liposomes.
b. retroviral vectors.
c. T-DNA vectors.
d. both a and b.

2. Which of the following best describes the approach that was used in the first gene therapy trial for SCID?

a. The normal ADA gene was introduced by injecting liposomes directly into the patients' bodies.
b. Lymphocytes were removed from a SCID patient, the normal *ADA* gene was transferred into the lymphocytes via liposomes, and then the lymphocytes were returned to the patient's body.
c. Lymphocytes were removed from a SCID patient, the normal *ADA* gene was transferred into the lymphocytes via a retrovirus, and then the lymphocytes were returned to the patient's body.
d. None of the above.

KEY TERMS

Page 528. biotechnology, genetically modified organisms (GMOs), transgenic organism, transgene
Page 530. biological control, bioremediation, biotransformation, biodegradation
Page 532. gene replacement, gene knockout, gene addition, GloFish
Page 533. chimera
Page 535. gene redundancy, gene knockin, mouse model
Page 536. xenotransplantation, molecular pharming
Page 537. reproductive cloning
Page 539. stem cells, totipotent, embryonic stem cells (ES cells), pluripotent, embryonic germ cells (EG cells), multipotent
Page 540. unipotent
Page 543. T DNA, T-DNA vectors
Page 544. biolistic gene transfer, microinjection, electroporation
Page 545. gene therapy
Page 547. liposomes, ex vivo approach

CHAPTER SUMMARY

- Biotechnology is broadly defined as the use of living organisms or their by-products in the development of goods or processes that are beneficial to humans. Genetically modified organisms have received genetic material via recombinant DNA technology.

21.1 Uses of Microorganisms in Biotechnology

- Microorganisms have several uses in biotechnology (see Table 21.1).
- Recombinant microorganisms are used to make human hormones, such as insulin (see Table 21.2, Figure 21.1).
- Microorganisms may be used as biological control agents.
- Microorganisms may be used in bioremediation, which is the method of using living organisms to decrease pollutants in the environment.

21.2 Genetically Modified Animals

- A transgenic organism carries a gene from another species (see Figure 21.2).
- The integration of a cloned gene into a chromosome may result in gene replacement or gene addition (see Figures 21.3, 21.4).
- Gene knockouts and knockins in mice are made to study gene function and to produce mouse models for human diseases (see Figures 21.5, 21.6).
- Transgenic livestock may produce human proteins, such as hormones, in their milk (see Table 21.3, Figure 21.7).

21.3 Reproductive Cloning and Stem Cells

- Mammalian reproductive cloning can be achieved using somatic cells and oocytes with their nuclei removed (see Figures 21.8, 21.9).
- Stem cells have the ability to divide and differentiate. Stem cells may be totipotent, pluripotent, multipotent, or unipotent (see Figures 21.10–21.12).
- Stem cells have the potential to treat a variety of human diseases (see Table 21.4).

21.4 Genetically Modified Plants

- Researchers have made many transgenic plants that have traits that are useful to humans, including herbicide and pesticide resistance (see Table 21.5, Figures 21.13, 21.14).
- *Agrobacterium tumefaciens* transfers T DNA to plants. Researchers have used T-DNA vectors to make genetically modified plants (see Figures 21.15, 21.16).

21.5 Human Gene Therapy

- Human gene therapy is the introduction of cloned genes into cells in an attempt to treat a disease. Genes may be introduced via liposomes or viruses (see Table 21.6, Figure 21.17).
- The first human gene therapy trial was aimed at treating adenosine deaminase deficiency (see Figure 21.18).

PROBLEM SETS & INSIGHTS

Solved Problems

S1. Which of the following would appropriately be described as a transgenic organism?

A. The sheep "Dolly," which was produced by cloning

B. A sheep that produces human α_1-antitrypsin in its milk

C. Bt corn

D. A hybrid strain of corn produced from crossing two inbred strains of corn (The inbred strains were not transgenic.)

Answer:

A. No, Dolly was not produced using recombinant techniques. Pieces of DNA were not cut and combined in a new way.

B. Yes

C. Yes

D. No, the hybrids simply contain chromosomal genes from two different parental strains.

S2. Describe the strategy for producing human proteins in the milk of livestock.

Answer: Milk proteins are encoded by genes with promoters and regulatory sequences that direct the expression of these genes within the cells of the mammary gland. To get other proteins expressed in the mammary gland, the strategy is to link the promoter and regulatory sequences from a milk-specific gene to the coding sequence of the gene that encodes the human protein of interest. In some cases, it is also necessary to add a signal sequence to the amino-terminal end of the target protein. A signal sequence is a short polypeptide that directs the secretion of a protein from a cell. If the target protein does not already have a signal sequence, it is possible to use a signal sequence from a milk-specific gene to promote the secretion of the target protein from the mammary cells and into the milk. During this process, the signal sequence is cleaved from the secreted protein.

S3. With regard to genetically modified organisms, describe one example that has not been successful and two that have.

Answer: One unsuccessful example is the oil-eating bacteria. Two very successful examples include bacteria that make human insulin and Bt varieties of agricultural crops.

Conceptual Questions

C1. What is a recombinant microorganism? Discuss examples.

C2. A conjugation-deficient strain of *A. radiobacter* is used to combat crown gall disease. Explain how this bacterium prevents the disease and the advantage of a conjugation-deficient strain.

C3. What is bioremediation? What is the difference between biotransformation and biodegradation?

C4. What is a biological control agent? Briefly describe two examples.

C5. As described in Table 21.2, several medical agents are now commercially produced by genetically engineered microorganisms. Discuss the advantages and disadvantages of making these agents this way.

C6. What is a mouse model for human disease?

C7. What is a transgenic organism? Describe three examples.

C8. What part of the *A. tumefaciens* DNA gets transferred to the genome of a plant cell during infection?

C9. Explain the difference between gene addition and gene replacement. Are the following descriptions examples of gene addition or gene replacement?

A. A mouse model to study cystic fibrosis

B. Introduction of a pesticide-resistance gene into corn using the T-DNA vector of *A. tumefaciens*

C10. As described in Chapter 5, not all inherited traits are determined by nuclear genes (i.e., genes located in the cell nucleus) that are expressed during the life of an individual. In particular, maternal effect genes and mitochondrial genes are notable exceptions. With these ideas in mind, let's consider the cloning of sheep (e.g., Dolly).

A. With regard to maternal effect genes, is the phenotype of such a cloned animal determined by the animal that donated the enucleated egg or by the animal that donated the somatic cell nucleus? Explain.

B. Does the cloned animal inherit extranuclear traits from the animal that donated the egg or from the animal that donated the somatic cell? Explain.

C. In what ways would you expect this cloned animal to be similar to or different from the animal that donated the somatic cell? Is it accurate to call such an animal a "clone" of the animal that donated the nucleus?

C11. Discuss some of the worthwhile traits that can be modified in transgenic plants.

C12. Discuss the concerns that some people have with regard to the uses of genetically engineered organisms.

Experimental Questions

E1. Recombinant bacteria can produce hormones that are normally produced in humans. Briefly describe how this is accomplished.

E2. *Bacillus thuringiensis* makes toxins that kill insects. This toxin must be applied several times during the growth season to prevent insect damage. As an alternative to repeated applications, one strategy is to apply bacteria directly to leaves. However, *B. thuringiensis* does not survive very long in the field. Other bacteria, such as *Pseudomonas syringae*, do survive. Propose a way to alter *P. syringae* so that it could be used as an insecticide. Discuss advantages and disadvantages of this approach compared with the repeated applications of the insecticide from *B. thuringiensis*.

E3. In the experiment of Figure 21.1, why was it necessary to link the coding sequence for the A or B chains to the sequence for β-galactosidase? How were the A or B chains separated from β-galactosidase after the fusion protein was synthesized in *E. coli*?

E4. Explain how it is possible to select for homologous recombination in mice. What phenotypic marker is used to readily identify chimeric mice?

E5. To produce transgenic plants, plant tissue is exposed to *Agrobacterium tumefaciens* and then grown in media containing kanamycin, carbenicillin, and plant growth hormones. Explain the purpose behind each of these three agents. What would happen if you left out the kanamycin?

E6. List and briefly describe five methods for the introduction of cloned genes into plants.

E7. What is a gene knockout? Is an animal or plant with a gene knockout a heterozygote or homozygote? What might you conclude if a gene knockout does not have a phenotypic effect?

E8. In the study of plants and animals, it is relatively common for researchers to identify genes using molecular techniques without knowing the function of the gene. In the case of mice, the function of the gene can be investigated by making a gene knockout. A knockout that causes a phenotypic change in the mouse may provide an important clue regarding the function of a gene. For example, a gene knockout that produced an albino mouse would indicate that the knocked-out gene probably plays a role in pigment formation. The experimental strategy of first identifying a gene based on its molecular properties and then investigating its function by making a knockout is called reverse genetics. Explain how this approach is opposite (or "in reverse") to the conventional way that geneticists study the function of genes.

E9. According to the methods described in Figure 21.5, can homologous recombination that results in gene replacement cause the integration of both the *TK* and Neo^R genes? Explain why or why not. Describe how the *TK* gene and Neo^R gene are used in a selection scheme that favors gene replacement.

E10. What is a chimera? How are chimeras made?

E11. Evidence (see P.G. Shiels, A.J. Kind, K.H. Campbell, et al. (1999) Analysis of telomere lengths in cloned sheep. *Nature 399*, 316–317) suggests that Dolly may have been "genetically older" than her actual age would have suggested. As mammals age, the chromosomes in somatic cells tend to shorten from the telomeres.

Therefore, older individuals have shorter chromosomes in their somatic cells than do younger ones. When researchers analyzed the chromosomes in the somatic cells of Dolly when she was about 3 years old, the lengths of her chromosomes were consistent with a sheep that was significantly older, say, 9–10 years old. (Note: As described in the chapter, the sheep that donated the somatic cell that produced Dolly was 6 years old, and her mammary cells had been grown in culture for several cell doublings before a mammary cell was fused with an oocyte.)

A. Suggest an explanation why Dolly's chromosomes seemed so old.

B. Let's suppose that a female sheep (like Dolly), which was produced via reproductive cloning, was mated at age 11 to a normal male sheep and then gave birth to a lamb named Molly. When Molly was 8 years old, a sample of somatic cells was analyzed. How old would you expect Molly's chromosomes to appear, based on the phenomenon of telomere shortening? Explain your answer.

C. Discuss how the observation of chromosome shortening, which was observed in Dolly, might affect the popularity of reproductive cloning.

E12. When transgenic organisms are made, the transgene may integrate into multiple sites within the genome. Furthermore, the integration site may influence the expression of the gene. For example, if a transgene integrates into a heterochromatic (i.e., highly condensed) region of a chromosome, the transgene may not be expressed. For these reasons, it is important for geneticists to analyze transgenic organisms with regard to the number of transgene insertions and the expression levels of the transgenes. Chapter 20 describes three methods (Southern blotting, Northern blotting, and Western blotting) that can be used to detect genes and gene products. Which of these techniques would you use to determine the number of transgenes in a transgenic plant or animal? Why is it important to know the number of transgenes? (Hint: You may want to use a transgenic animal or plant as breeding stock to produce many more transgenic animals or plants.) Which technique would you use to determine the expression levels of transgenes?

E13. What is molecular pharming? Compared with the production of proteins by bacteria, why might it be advantageous?

E14. What is reproductive cloning? Are identical twins in humans considered to be clones? With regard to agricultural species, what are some potential advantages to reproductive cloning?

E15. Researchers have identified a gene in humans that (when mutant) causes severe dwarfism and mental impairment. This disorder is inherited in an autosomal recessive manner, and the mutant allele is known to be a loss-of-function mutation. The same gene has been found in mice, although a mutant version of the gene has not been discovered in mice. To develop drugs and an effective therapy to treat this disorder in humans, it would be experimentally useful to have a mouse model. In other words, it would be desirable to develop a strain of mice that carry the mutant allele in the homozygous condition. Experimentally, how would you develop such a strain?

E16. Treatment of adenosine deaminase (ADA) deficiency is an example of ex vivo gene therapy. Why is this therapy called ex vivo? Can ex vivo gene therapy be used to treat all inherited diseases? Explain.

E17. Several research studies are under way that involve the use of gene therapies to inhibit the growth of cancer cells. As discussed in Chapter 24, oncogenes are mutant genes that are overexpressed and cause cancer. New gene therapies are aimed at silencing an oncogene by producing antisense RNA that recognizes the mRNA transcribed from an oncogene. Based on your understanding of antisense RNA (described in Chapters 14 and 16), explain how this strategy would prevent the growth of cancer cells.

Questions for Student Discussion/Collaboration

1. Discuss the advantages and disadvantages of gene therapy. Because a limited amount of funding is available for gene therapy research, make a priority list of the three top diseases for which you would fund research. Discuss your choices.

2. A commercially available strain of *P. syringae* marketed as Frostban B is used to combat frost damage. This is a naturally occurring strain that carries a loss-of-function mutation in a gene that encodes a protein that promotes frost formation. In addition, researchers have used recombinant DNA techniques to eliminate the function of this gene. Discuss the advantages and disadvantages of using the nonrecombinant strain compared with a recombinant version.

3. Make a list of the types of traits you would like to see altered in transgenic plants and animals. Suggest ways (i.e., what genes would you use?) to accomplish these alterations.

Note: All answers appear at the website for this textbook; the answers to the even-numbered questions and all of the Concept Check and Comprehension Questions are in the back of the book.

www.mhhe.com/brookergenetics5e

Visit the website for practice tests, answer keys, and other learning aids for this chapter. Enhance your understanding of genetics with our interactive exercises, quizzes, animations, and much more.

CHAPTER OUTLINE

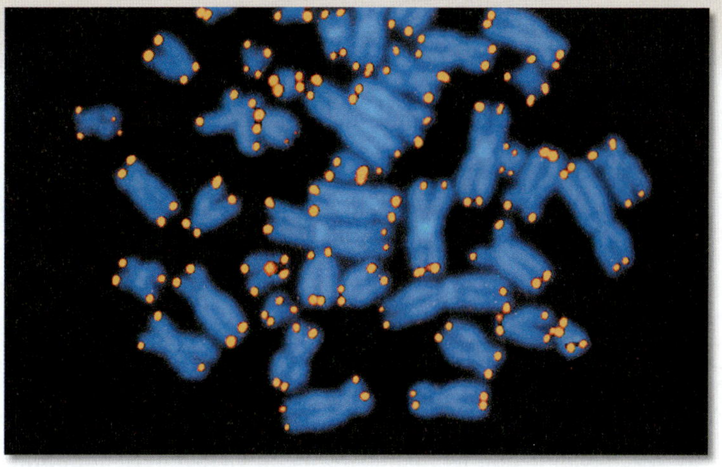

Labeling the ends of chromosomes. *In this micrograph, the telomeric sequences at the ends of chromosomes are labeled with an orange fluorescent probe, and the rest of the chromosomes are labeled in blue. This method, called fluorescent in situ hybridization, allows geneticists to identify particular sequences within intact chromosomes.*

22 GENOMICS I: ANALYSIS OF DNA

The term **genome** refers to the total genetic composition of an organism or species. For example, the nuclear genome of humans is composed of 22 different autosomes and an X chromosome and (in males) a Y chromosome. In addition, humans have a mitochondrial genome composed of a single circular chromosome.

As genetic technology has progressed over the past few decades, researchers have gained an increasing ability to analyze the composition of genomes as a whole unit. The term **genomics** refers to the molecular analysis of the entire genome of a species. Genome analysis is a molecular dissection process applied to a complete set of chromosomes. Segments of chromosomes are analyzed in progressively smaller pieces, the locations of which are known on the intact chromosomes. This is the mapping phase of genome analysis. The mapping of the genome ultimately progresses to the determination of the complete DNA sequence for all of a species' chromosomes.

In 1995, a team of researchers headed by J. Craig Venter and Hamilton Smith completed the first entire DNA sequence of the bacterial genome for *Haemophilus influenzae*. As discussed later in this chapter, its genome is composed of a single circular chromosome 1.83 million base pairs (bp) in length and contains approximately 1743 genes. In 1996, the first entire DNA sequence of a eukaryote, *Saccharomyces cerevisiae* (baker's yeast), was completed. The yeast genome contains 16 linear chromosomes, which have a combined length of about 12.1 million bp and contain approximately 6300 genes. Since that time, complete genome sequences from many prokaryotes and eukaryotes have been determined.

In this chapter, we will focus on methods aimed at determining the sequences within a species' genome. This process may begin with the mapping of regions along a species' chromosomes. It is finished when the complete DNA sequence is known. We will consider three mapping strategies—cytogenetic mapping, linkage mapping, and physical mapping—and the approaches used to carry them out. Then we will explore genome-sequencing projects—research endeavors that have the ultimate goal of determining the sequence of DNA bases of the entire genome of a given species. We will examine the methods, goals, and results of these large undertakings, which include the Human Genome Project.

Once a genome sequence is known, researchers can examine, at the level of many genes, how the components of a genome interact to produce the traits of an organism. This approach is called **functional genomics.** Ultimately, a long-term goal of researchers is to determine the roles of all cellular proteins to produce the characteristics of particular cell types and the traits of complete organisms. This is the research area known as **proteomics.** In this chapter, we explore genomics at the level of DNA segments and sequences. In Chapter 23, we will consider the exciting prospects that functional genomics and proteomics have to offer regarding our current and future understanding of genetics.

22.1 OVERVIEW OF CHROMOSOME MAPPING

Learning Outcomes:

1. Define mapping.
2. Distinguish among cytogenetic, linkage, and physical mapping.

In genetics, the term **mapping** refers to the experimental process of determining the relative locations of genes or other segments of DNA along individual chromosomes. Researchers may follow three general approaches to mapping a chromosome: cytogenetic, linkage, and physical mapping strategies. Before we discuss these approaches in detail, let's compare them. **Cytogenetic mapping** (also called cytological mapping) is aimed at determining the locations of specific sequences, such as gene sequences, within chromosomes that are viewed microscopically. When stained, each chromosome of a given species has a characteristic banding pattern, and genes are mapped cytogenetically relative to a band location. By comparison, in Chapter 6, we considered how genetic crosses are conducted to map the relative locations of genes within a chromosome. Such studies, known as **linkage mapping,** use the frequency of genetic recombination between different genes to determine their relative spacing and order along a chromosome. In eukaryotes, linkage mapping involves crosses among organisms that are heterozygous for two or more genes. The number of recombinant offspring provides a relative measure of the distance between genes, which is computed in map units (mu). Finally, a third approach is **physical mapping** in which DNA-cloning techniques are used to determine the location of and distance between genes and other DNA regions. In a physical map, the distances are computed as the number of base pairs between genes.

A **genetic map,** or **chromosome map,** is a chart that describes the relative locations of genes or other DNA segments along a chromosome. The term **locus** (plural, **loci**) refers to the site within a genetic map where a specific gene or other DNA segment is found. Figure 22.1 compares genetic maps that show the loci for two X-linked genes, *sc* (scute, a gene affecting bristle morphology) and *w* (a gene affecting eye color), in *Drosophila melanogaster*. In the cytogenetic map at the top, the *sc* gene is located at band 1A8, and the *w* gene is located at band 3B6. In the linkage map in the middle, genetic crosses indicate that the two genes are approximately 1.5 map units (mu) apart. The physical map at the bottom shows that the two genes are approximately 2.4×10^6 bp apart along the X chromosome. Correlations among cytogenetic, linkage, and physical maps often vary from species to species and from one region of the chromosome to another. For example, a distance of 1 mu may correspond to 1–2 million bp in one region of the chromosome, but other regions may recombine at a much lower rate, so a distance of 1 mu may be a much longer physical segment of DNA.

22.1 COMPREHENSION QUESTION

1. What type of mapping relies on microscopy?
   a. Cytogenetic mapping
   b. Linkage mapping
   c. Physical mapping
   d. All of the above

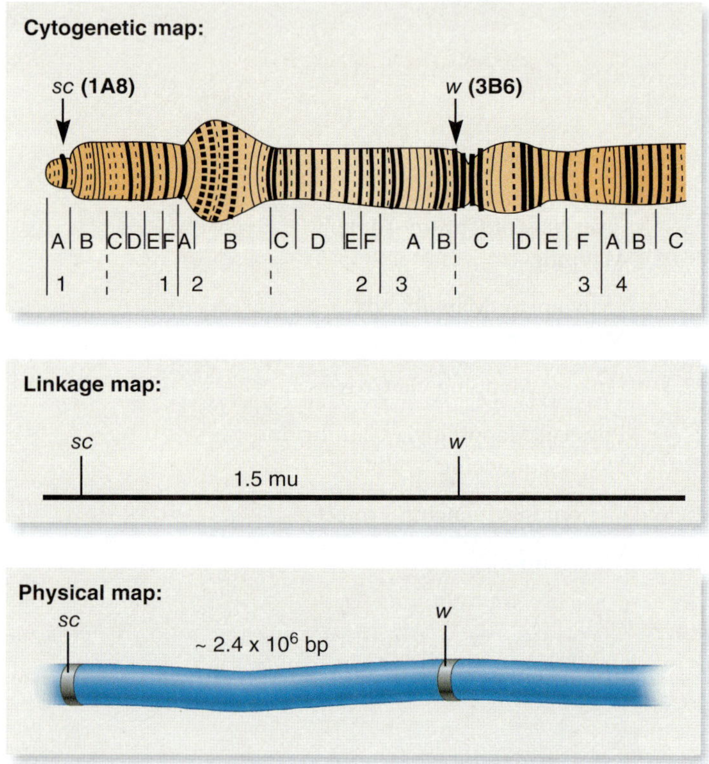

FIGURE 22.1 A comparison of cytogenetic, linkage, and physical maps. Each of these maps shows the distance between the *sc* and *w* genes along the X chromosome in *Drosophila melanogaster*. The cytogenetic map is that of the polytene chromosome.

CONCEPT CHECK: What is a genetic map?

22.2 CYTOGENETIC MAPPING VIA MICROSCOPY

Learning Outcomes:

1. Outline the method of in situ hybridization.
2. Describe the technique of chromosome painting and explain why it is used.

Cytogenetic mapping determines the location of specific sequences within chromosomes that are viewed microscopically. It is commonly used in eukaryotes, which have very large chromosomes relative to those of bacteria. Microscopically, eukaryotic chromosomes can be distinguished from one another by their size, centromeric location, and banding patterns (refer to Chapter 8, Figure 8.1). By treating chromosomes with particular dyes, a discrete banding pattern is obtained. Cytogeneticists use this banding pattern as a way to describe specific regions along a chromosome. In this section, we will explore techniques that are aimed at producing cytogenetic maps.

A Goal of Cytogenetic Mapping Is to Determine the Location of a Gene Along an Intact Chromosome

Cytogenetic mapping can localize a particular gene to a site within a chromosomal banding pattern. For example, let's consider the cystic fibrosis transmembrane regulator (CFTR) protein that is defective in people with cystic fibrosis. The human *CFTR* gene that encodes this protein is located on chromosome 7, at a specific site in the q3 region.

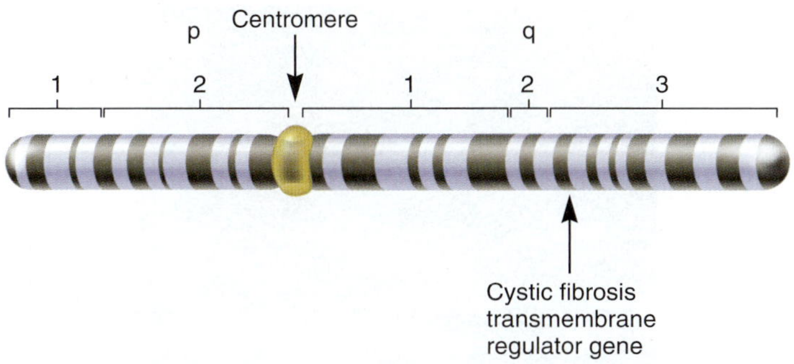

Cytogenetic mapping may be used as a first step in the localization of genes in plants and animals. However, because it relies on light microscopy, cytogenetic analysis has a fairly crude limit of resolution. In most species, cytogenetic mapping is accurate only within limits of approximately 5 million bp along a chromosome. In species that have large polytene chromosomes, such as *Drosophila*, the resolution is much better. A common strategy used by geneticists is to roughly locate a gene by cytogenetic analysis and then determine its location more precisely by the physical mapping methods described later.

In Situ Hybridization Localizes Genes Along Particular Chromosomes

The technique of **in situ hybridization** is widely used to cytogenetically map the locations of genes or other DNA sequences within large eukaryotic chromosomes. The term in situ (from the Latin for "in place") indicates that the procedure is conducted on chromosomes that are being held in place—adhered to a surface.

To map a gene via in situ hybridization, researchers use a probe to detect the location of the gene within a set of chromosomes. If the gene of interest has been cloned previously, as described in Chapter 20, the DNA of the cloned gene can be used as a probe. Because a DNA strand from a cloned gene, which is a very small piece of DNA relative to a chromosome, hybridizes only to its complementary sequence on a particular chromosome, this technique provides a way of localizing the gene of interest. For example, let's consider the gene that causes the white-eye phenotype in *Drosophila* when it carries a loss-of-function mutation. This gene has already been cloned. If a single-stranded piece of this cloned DNA is mixed with *Drosophila* chromosomes in which the DNA has been denatured, it will bind only to the X chromosome at the location corresponding to the site of the eye color gene.

The most common method of in situ hybridization uses fluorescently labeled DNA probes and is referred to as **fluorescence in situ hybridization (FISH).** Figure 22.2 describes the steps of the

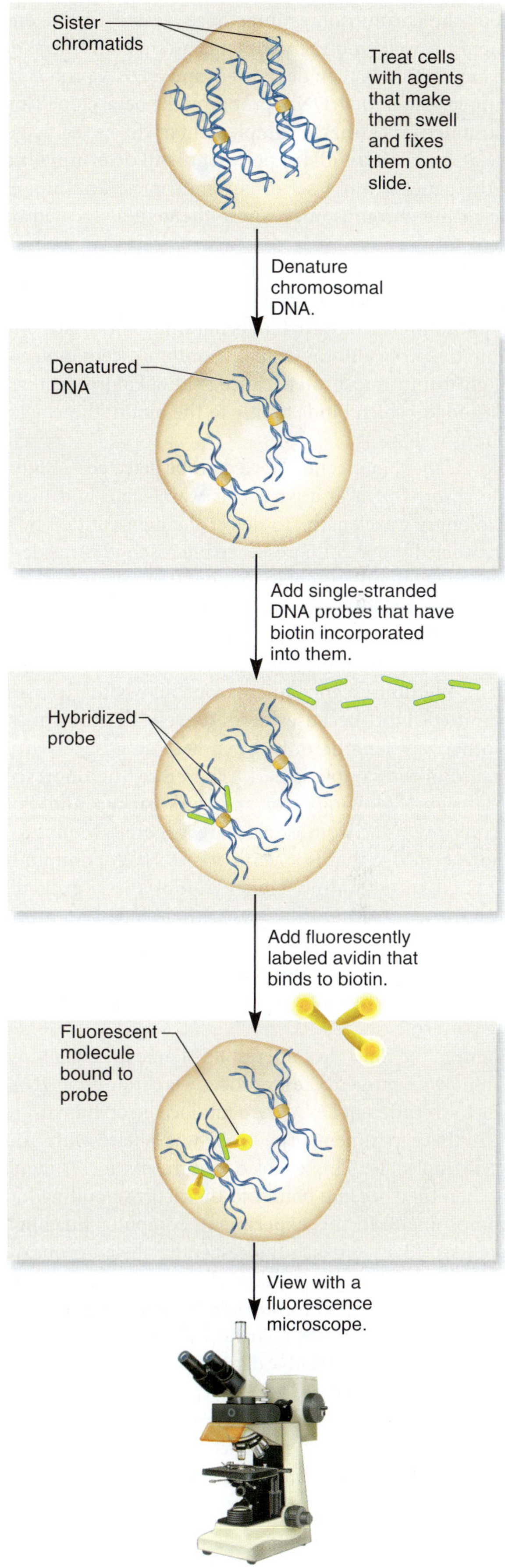

ONLINE ANIMATION

FIGURE 22.2 The technique of fluorescence in situ hybridization (FISH). The probe hybridizes to the denatured chromosomal DNA only at specific, complementary sites in the genome. Note that the chromosomes are highly condensed metaphase chromosomes that have already replicated. These are sister chromatids. Therefore, each X-shaped chromosome actually contains two copies of a particular gene. Because the sister chromatids are identical, a probe that recognizes one sister chromatid will also bind to the other.

CONCEPT CHECK: Why does the probe bind to a specific site on a chromosome?

FISH procedure. The cells, which are prepared using a technique that keeps the chromosomes intact, are treated with agents that cause them to swell, and their contents are fixed to the slide. The chromosomal DNA is then denatured, and a DNA probe is added. For example, the added DNA probe might be a cloned piece of single-stranded DNA that is complementary to a specific gene. In this case, the goal of a FISH experiment is to determine the location of the gene within a set of chromosomes. The probe binds to a site in the chromosomes where the gene is located because the probe and chromosomal gene line up and hydrogen bond with each other. To detect where the probe has bound to a chromosome, the probe is subsequently tagged with a fluorescent molecule. This is usually accomplished by first incorporating biotin-labeled nucleotides into the probe. Biotin, a small, nonfluorescent molecule, has a very high affinity for a protein called avidin. Fluorescently labeled avidin is added, which binds tightly to the biotin, thereby labeling the probe as well.

How is the fluorescently labeled probe detected? A fluorescent molecule absorbs light at a particular wavelength and then emits light at a longer wavelength. To detect the light emitted by a fluorescently labeled probe, a fluorescence microscope is needed. Such a microscope contains filters that allow the passage of light only within a defined wavelength range. The sample is illuminated at the wavelength of light that is absorbed by the fluorescent molecule. The fluorescent molecule then emits light at a longer wavelength and the fluorescence microscope allows the transmission of the emitted light. Because filters prevent the transmission of light of other wavelengths, only the emitted light is viewed and the background of the sample is dark. Therefore, the fluorescence is seen as a brightly glowing color on a dark background. For most FISH experiments, chromosomes are generally counterstained by a fluorescent dye that is specific for DNA. A commonly used dye is DAPI (4′,6-diamidino-2-phenylindol) that is excited by UV light. This provides all of the DNA with a blue background. The results of a FISH experiment are then compared with a sample of chromosomes that have been stained with Giemsa to produce banding, so the location of a probe can be mapped relative to the banding pattern.

Figure 22.3 illustrates the results of an experiment involving six different DNA probes. The six probes were strands of DNA corresponding to six different DNA segments located on human chromosome 5. In this experiment, each probe was labeled with a different fluorescent molecule. This enabled researchers to distinguish the probes when they became bound to their corresponding locations on chromosome 5. In this experiment, computer-imaging methods were used to assign each fluorescently labeled probe a different color. Using this method, called **chromosome painting,** FISH discerns the sites along chromosome 5 corresponding to the six different probes. In a visual, colorful way, FISH was used here to determine the order and relative distances between six specific sites along a single chromosome. FISH is commonly used in genetics and cell biology research, and its use has become more widespread in clinical applications. For example, clinicians may use FISH to detect changes in chromosome structure, such as deletions, duplications, and translocations, which may occur in patients with genetic disorders.

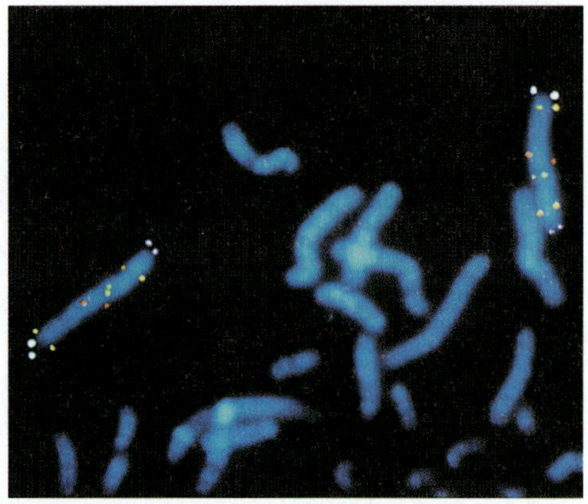

FIGURE 22.3 Chromosome painting via fluorescence in situ hybridization. In this experiment, six different probes were used to locate six different sites along human chromosome 5. The colors are due to computer imaging of the fluorescence emission; they are not the actual colors of the fluorescent labels. Two spots are usually seen at each site because the probe binds to both sister chromatids.

22.2 COMPREHENSION QUESTION

1. The technique of fluorescence in situ hybridization involves the use of a ___________ that hybridizes to a ___________.
   a. radiolabeled probe, band on a gel
   b. radiolabeled probe, specific site on an intact chromosome
   c. fluorescent probe, band on a gel
   d. fluorescent probe, specific site on an intact chromosome

22.3 LINKAGE MAPPING VIA CROSSES

Learning Outcomes:

1. Define molecular marker.
2. Explain the use of molecular markers in mapping studies.

Let's now turn to linkage mapping, which relies on the frequency of recombinant offspring to determine the distance between sites located along the same chromosome. We already considered linkage mapping methods in Chapter 6, where allelic differences between genes were used to map the relative locations of those genes along a chromosome by conducting testcrosses. In this section, we will focus on the use of molecular markers to map genes.

Linkage Mapping Can Use Molecular Markers

As an alternative to relying on allelic differences between genes, geneticists have realized that regions of DNA that do not encode genes can be used as markers along a chromosome. A **molecular marker** is a segment of DNA found at a specific site along a chromosome with properties that enable it to be uniquely recognized using molecular tools such as polymerase chain reaction (PCR) and gel electrophoresis. As with alleles, molecular markers may be **polymorphic;** that is, within a population, they may vary from individual to individual. Therefore, the distances between linked molecular markers can be determined from the outcomes of crosses. Using molecular techniques, researchers have found it easier to identify many molecular markers within a given species' genome rather than identifying many allelic differences among individuals. For this reason, geneticists have increasingly turned to molecular markers as points of reference along genetic maps. As described in **Table 22.1**, many different kinds of molecular markers are used by geneticists.

Researchers have constructed detailed genetic maps in which a series of many molecular markers have been identified along each chromosome of certain species. These species include humans, model organisms, agricultural species, and many others. Why are molecular markers useful? One key reason is that molecular markers can be used to determine the approximate location of an unknown gene that causes a human disease. Clinical geneticists sometimes follow the transmission patterns of polymorphic molecular markers in family pedigrees to identify genes that, when they are mutated, cause human disease. The discovery of a particular marker in those who have the disease can indicate that the marker is close to the disease-causing allele (see Chapter 24, Figures 24.6, 24.7). This may help researchers identify the gene by cloning methods, such as chromosome walking, which we will examine later in this chapter.

In addition, molecular markers may help researchers identify the locations of genes involved in quantitative traits, such as fruit yield and meat weight, that are valuable in agriculture. The use of molecular markers to identify such genes is described in Chapter 27 (see Figure 27.5). Genetic maps with a large number of markers are used by evolutionary biologists to determine patterns of genetic variation within a species and the evolutionary relatedness of different species.

As an example of a map using molecular markers, **Figure 22.4** shows a simplified map of two chromosomes found in the plant *Arabidopsis thaliana*, which is one of the favorite model organisms of plant molecular geneticists. Many RFLPs, which are described in Table 22.1, have been mapped to different locations along the *Arabidopsis* chromosomes.

TABLE 22.1

Common Types of Molecular Markers

Marker	Description
Restriction fragment length polymorphism (RFLP)	A site in a genome where the distance between two restriction sites varies among different individuals. These sites are identified by restriction enzyme digestion of chromosomal DNA and the use of Southern blotting.
Amplified restriction fragment length polymorphism (AFLP)	The same as an RFLP except that the site is amplified via PCR instead of isolating the chromosomal DNA.
Microsatellite	A site in the genome that contains many short sequences that are repeated many times in a row. The total length is usually in the size range of 50–200 bp, and their lengths may be polymorphic within a population. They are isolated via PCR. Microsatellites are also called short tandem repeats (STRs) and simple sequence repeats (SSRs).
Single-nucleotide polymorphism (SNP)	A site in the genome where a single nucleotide is polymorphic among different individuals. These sites occur commonly in all genomes, and they are gaining greater use in the mapping of disease-causing alleles and in the mapping of genes that contribute to quantitative traits that are valuable in agriculture (see Chapter 27).
Sequence-tagged site (STS)	This is a general term to describe any molecular marker that is found at a unique site in the genome and is amplified by PCR. AFLPs, microsatellites, and SNPs can provide sequence-tagged sites within a genome.

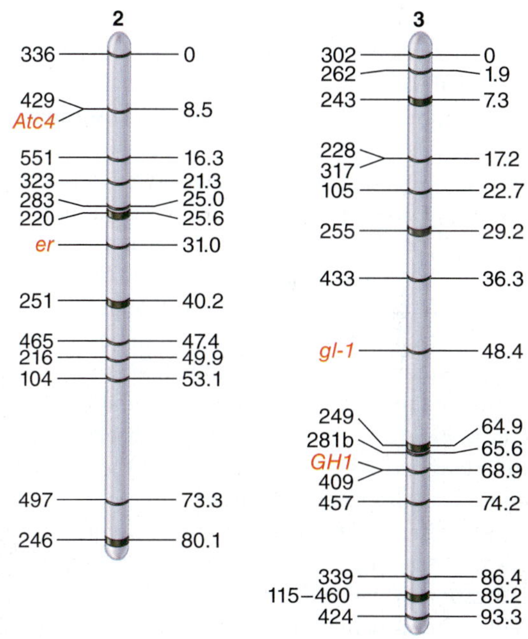

FIGURE 22.4 An RFLP linkage map of two chromosomes in *Arabidopsis thaliana*. This plant has five different chromosomes, but only the maps of chromosomes 2 and 3 are shown here. This map shows the locations of many RFLP markers. The numbers along the left side of each chromosome designate the locations of RFLP markers. The numbers along the right side of each chromosome are the map distances in map units (mu). For example, the RFLPs at map positions 16.3 and 40.2 on chromosome 2 are designated 551 and 251, respectively, and are 23.9 mu apart. The top marker at the end of each chromosome was arbitrarily assigned as the starting point (zero) for each chromosome. In addition, the map shows the locations of a few known genes (shown in red): *Atc4* = actin, *er* = erecta, *gl-1* = glabra-1, and *GH1* = acetolactate synthase.

Linkage Mapping Commonly Uses Molecular Markers Called Microsatellites

To make a highly refined map of a genome, many different polymorphic sites must be identified and their transmission followed from parent to offspring over many generations. RFLPs were among the first molecular markers studied by geneticists. More recently, other molecular markers have been used because they are easier to generate via PCR. As an example, let's consider **microsatellites,** which are short repetitive sequences that are abundantly interspersed throughout a species' genome and tend to vary in length among different individuals. Microsatellites usually contain di-, tri-, tetra-, or pentanucleotide sequences that are repeated many times in a row. For example, the most common microsatellite encountered in humans is a dinucleotide sequence $(CA)_n$, where n ranges from 5 to more than 50. In other words, this dinucleotide sequence can be tandemly repeated 5–50 or more times. The $(CA)_n$ microsatellite is found, on average, about every 10,000 bases in the human genome. Researchers have identified thousands of different DNA segments that contain $(CA)_n$ microsatellites, located at many distinct sites within the human genome. Using primers complementary to the unique DNA sequences that flank a specific $(CA)_n$ region, a particular microsatellite can be amplified by PCR. In other words, the PCR primers copy only a particular microsatellite, but not the thousands of others that are interspersed throughout the genome (**Figure 22.5**).

When a pair of PCR primers copies a single site within a set of chromosomes, the amplified region is called a **sequence-tagged site (STS).** When DNA is collected from a haploid cell, an STS produces only a single band on a gel. In a diploid species, an individual has two copies of a given STS. When an STS contains a microsatellite, the two PCR products may be identical and result in a single band on a gel if the region is the same length in both copies (i.e., if the individual is homozygous for the microsatellite). However, if an individual has two copies that differ in the number of repeats in the microsatellite sequence (i.e., if the individual is heterozygous for the microsatellite), the two PCR products obtained will be different in length (as in Figure 22.5). The two bands observed in this figure differ in their lengths because the same site on the two homologous chromosomes contains different numbers of repeat sequences.

Microsatellites that have length polymorphisms allow researchers to follow their transmission from parent to offspring. PCR amplification of particular microsatellites provides a strategy in the genetic analysis of human pedigrees, as shown in **Figure 22.6**. Prior to this analysis, a unique segment of DNA containing a microsatellite had been identified. Using PCR primers complementary to this microsatellite's unique flanking segments, two parents and their three offspring were tested for the inheritance of this microsatellite. A small sample of cells was obtained from each individual and subjected to PCR amplification, as described in Figure 22.5. The amplified PCR products were then analyzed by high-resolution gel electrophoresis, which detects small differences in the lengths of DNA fragments. The mother's PCR products were 154 and 150 bp in length; the father's were 146 and 140 bp. Their first offspring inherited the 154-bp microsatellite from the mother and the 146-bp one from the father, the second inherited the 150-bp microsatellite from the mother and the 146-bp one from the father, and the third inherited the 150-bp one from the mother and the 140-bp

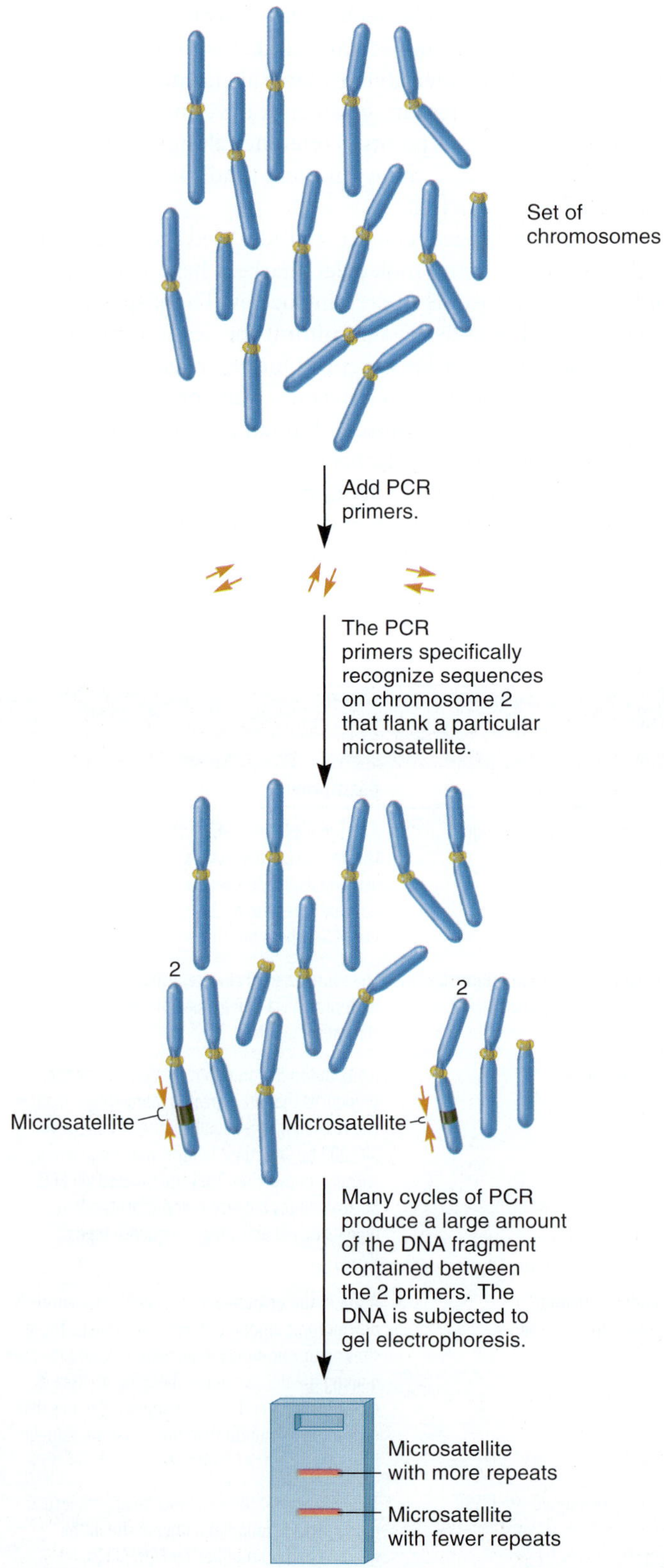

FIGURE 22.5 **Identifying a microsatellite using PCR primers.**

CONCEPT CHECK: What causes microsatellites to be polymorphic?

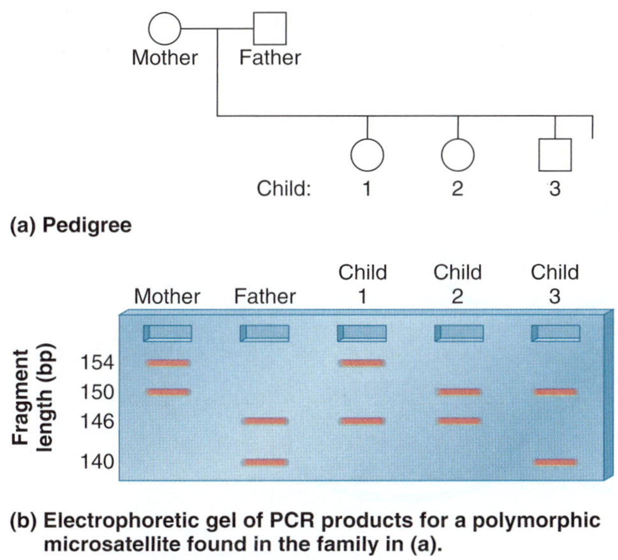

FIGURE 22.6 Inheritance pattern of a polymorphic microsatellite in a human pedigree. This microsatellite is autosomal.

microsatellite from the father. As shown in the figure, the transmission of polymorphic microsatellites is relatively easy to follow from generation to generation.

The simple pedigree analysis shown in Figure 22.6 illustrates the general method used to follow the transmission of a single microsatellite that is polymorphic in length. In linkage studies, the goal is to follow the transmission of many different microsatellites to determine those that are linked along the same chromosome versus those that are not. Those that are not linked will independently assort from generation to generation. Those that are linked tend to be transmitted together to the same offspring. In a large pedigree, it is possible to identify cases in which linked microsatellites have segregated due to crossing over. The frequency of crossing over provides a measure of the map distance, in this case between different microsatellites. This approach can help researchers obtain a finely detailed linkage map of the human chromosomes without having to depend on alleles of closely linked genes that affect phenotype.

Pedigree analysis involving STSs, such as polymorphic microsatellites, enables researchers to identify the location of disease-causing alleles. The assumption behind this approach is that a disease-causing allele had its origin in a single individual known as a **founder,** who lived many generations ago. Since that time, the allele has spread throughout portions of the human population. A second assumption is that the founder is likely to have had a polymorphic molecular marker that lies somewhere near the mutant allele. This is a reasonable assumption, because all people carry many polymorphic markers throughout their genomes. If a polymorphic marker lies very close to the disease-causing allele, it is unlikely that a crossover will occur in the intervening region. Therefore, such a polymorphic marker may be linked to the disease-causing allele for many generations. By following the transmission of many polymorphic markers within large family pedigrees, it may be possible to determine that particular markers are found in people who carry specific disease-causing alleles. (An example is shown in Chapter 24, Figure 24.7.) After the identification of a closely linked marker, a disease-causing allele can be identified using a technique called chromosome walking, which is described later in this chapter.

22.3 COMPREHENSION QUESTIONS

1. A molecular marker is a _______________ found at a specific site on a chromosome that has properties that allow it to be ____________.
   a. colored dye, visualized via microscopy
   b. colored dye, visualized on a gel
   c. segment of DNA, uniquely identified using molecular tools
   d. segment of DNA, visualized via microscopy
2. Which of the following is an example of a molecular marker?
   a. RFLP
   b. Microsatellite
   c. Single-nucleotide polymorphism
   d. All of the above
3. To map the distance between molecular markers via crosses, the markers must be
   a. polymorphic.
   b. monomorphic.
   c. fluorescently labeled.
   d. on different chromosomes.

22.4 PHYSICAL MAPPING VIA CLONING AND DNA SEQUENCING

Learning Outcomes:

1. Define contig.
2. Describe the uses of vectors known as YACs, BACs, and PACs.
3. Explain the technique of chromosome walking and why it is used.

We now turn our attention to methods aimed at establishing a physical map of a species' genome. Physical mapping typically involves the cloning of many pieces of chromosomal DNA. The cloned DNA fragments are then characterized by size (i.e., their length in base pairs), as well as the genes they contain and their relative locations along a chromosome.

As described later in this chapter, eukaryotic genomes are very large; the *Drosophila* genome is roughly 175 million bp long, and the human genome is approximately 3 billion bp in length.

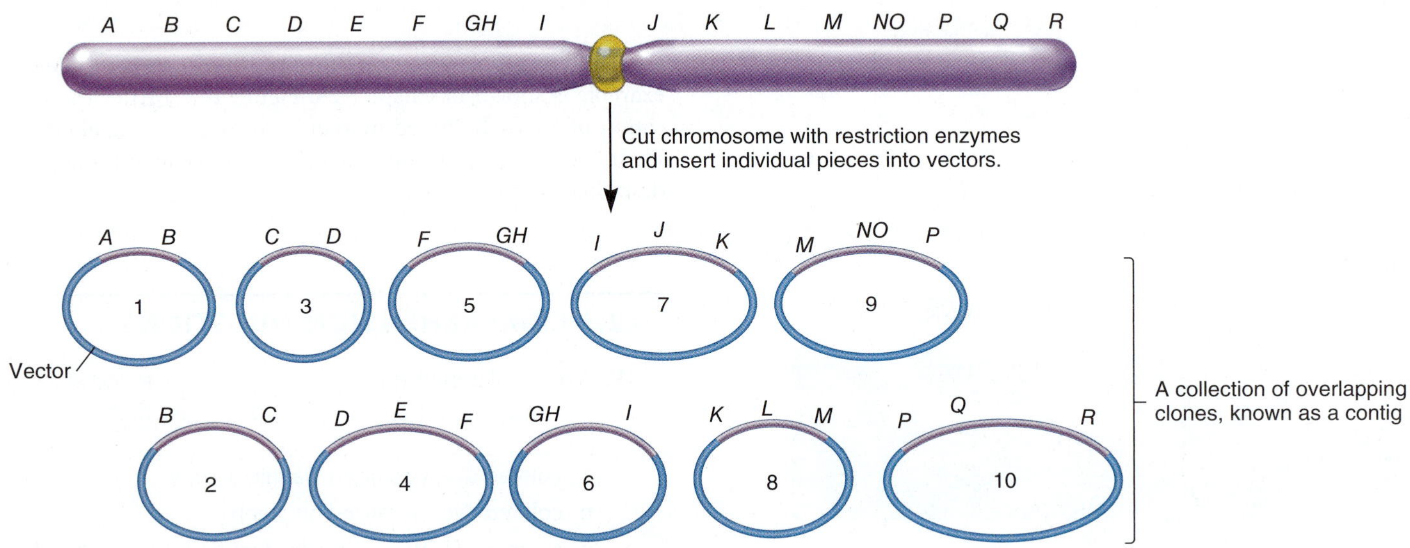

FIGURE 22.7 The construction of a contig. Large pieces of chromosomal DNA are cloned into vectors, and their order is determined by the identification of overlapping regions. For example, clone 3 ends with gene *D*, and clone 4 begins with gene *D*; clone 4 ends with gene *F*, and clone 5 begins with gene *F*.

CONCEPT CHECK: What is a contig?

When making a physical map of a genome via cloning, researchers must characterize many DNA clones that contain much smaller pieces of the genome. In this section, we examine the general strategies used in creating a physical map of a species' genome. We will also consider how physical mapping information can be used to clone genes.

A Physical Map of a Chromosome Is Constructed by Creating a Contiguous Series of Clones That Span a Chromosome

As discussed in Chapter 20, a DNA library contains a collection of recombinant vectors in which each vector contains a particular fragment of chromosomal DNA. Such a library is obtained by digesting chromosomal DNA into small fragments. Initially, researchers do not know how the fragments of DNA in a DNA library relate to DNA segments in intact chromosomes. Therefore, one goal in physical mapping is to determine the relative locations of the cloned chromosomal fragments from a DNA library as they would occur in an intact chromosome. In other words, the members of the library must be organized according to their actual locations along a chromosome. This can be achieved by identifying regions within clones that overlap with each other. To obtain a complete physical map of a chromosome via cloning, researchers need a series of clones that contain contiguous, overlapping pieces of chromosomal DNA. Such a collection of clones is known as a **contig** (Figure 22.7). A contig represents a physical map of a chromosome. As discussed later, cloning vectors known as BACs and cosmids are commonly used in the construction of a contig.

Different experimental strategies can be used to align the members of a contig. The general approach is to identify clones that contain overlapping regions. Historically, Southern blotting was used to determine if two different clones contain an overlapping region. In the example shown in Figure 22.7, the DNA from clone 1 could be labeled in vitro and then used as a probe in a Southern blot of the other clones shown in this figure. Clone 1 would hybridize to clone 2, because they share identical DNA sequences in the overlapping region. Similarly, clone 2 could be used as a probe to show that it hybridizes to clone 1 and clone 3. By conducting Southern blots between many combinations of clones, researchers can determine which clones have common overlapping regions and thereby order them as they would occur along the chromosome.

Alternatively, other methods for ordering the members of a contig involve the use of molecular markers. For example, if many STSs have already been identified along a chromosome via linkage or cytogenetic analysis, the STSs can be used as markers to order the members of a contig. Another approach is to subject the members of a library to a restriction enzyme, which produces a restriction digestion pattern for each member. The patterns of fragments are then analyzed by computer programs, which can identify regions that are potentially overlapping.

An ultimate goal of physical mapping procedures is to obtain a complete contig for each type of chromosome within a full set. For example, in the case of humans, a complete physical map requires a contig for each of the 22 autosomes and for the X and Y chromosomes. Geneticists can also correlate cloned DNA fragments in a contig with locations along a chromosome obtained from linkage or cytogenetic mapping. For example, a member of a contig may contain a gene, RFLP, or STS that was previously mapped by linkage analysis. Figure 22.8 considers a situation in which two members of a contig carry genes previously mapped by linkage analysis to be approximately 1.5 mu apart on chromosome 11. In this example, clone 2 has an insert that carries gene *A*, and clone 7 has an insert that carries gene *B*. Because a contig is composed of overlapping members, a researcher can align the contig along chromosome 11, starting with gene *A* and gene *B* as reference points. In this example, genes *A* and *B* serve as markers that identify the location of specific clones within the contig and align clones 2 and 7 within the cytogenetic map.

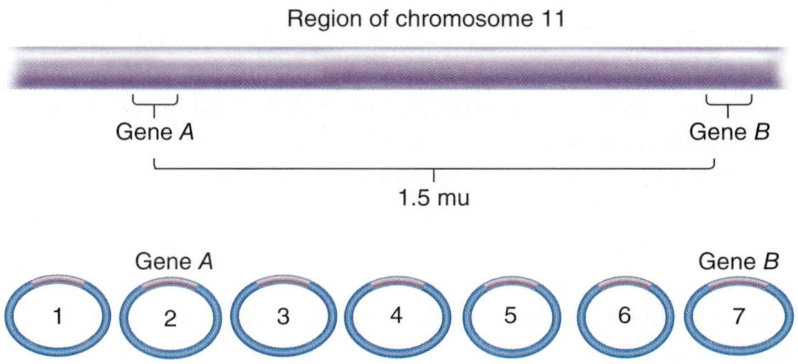

FIGURE 22.8 The use of genetic markers to align a contig. In this example, gene *A* and gene *B* had been mapped previously to specific regions of chromosome 11. Gene *A* was found within the insert of clone 2, gene *B* within the insert of clone 7. This made it possible to align the contig using gene *A* and gene *B* as genetic markers (i.e., reference points) along chromosome 11.

YAC, BAC, and PAC Vectors Are Used to Clone Large Segments of DNA

For large eukaryotic genomes, researchers often begin with vectors that can accept chromosomal DNA inserts of very large size. By having large insert sizes, a contig is more easily constructed and aligned because fewer recombinant vectors are needed. In general, most plasmid and viral vectors can accommodate inserts only a few thousand to perhaps tens of thousands of nucleotides in length. If a plasmid or viral vector has a DNA insert that is too large, it will have difficulty with DNA replication and is likely to suffer deletions in the insert.

By comparison, other cloning vectors, known as **artificial chromosomes,** can accommodate much larger sizes of DNA inserts. As their name suggests, they behave like chromosomes when inside living cells. The first type to be made was the **yeast artificial chromosome (YAC),** which was developed by David Burke, Georges Carle, and Maynard Olson in 1987. An insert within a YAC can be several hundred thousand to perhaps 2 million bp in length. For an average human chromosome, a few hundred YACs are sufficient to create a contig with fragments that span the entire length of the chromosome. By comparison, it would take thousands or even tens of thousands of recombinant plasmid vectors to create such a contig.

At the molecular level, YACs have structural similarities to normal eukaryotic chromosomes, yet have characteristics that make them suitable for cloning. The general structure of a YAC vector is shown at the top left of Figure 22.9. The YAC vector contains two telomeres (*TEL*), a centromere (*CEN*), a bacterial origin of replication (*ORI*), a yeast origin of replication (known as an *ARS*, for autonomous replication sequence), selectable markers, and a site where a large piece of chromosomal DNA can be inserted. Without an insert, the circular form of this vector can replicate in *E. coli*. After a large fragment has been inserted, the linear form of the vector can replicate in yeast.

FIGURE 22.9 The use of a YAC vector in DNA cloning. Note: *Eco*RI and *Bam*HI both cut DNA to yield sticky ends. In this diagram, only the *Eco*RI ends are shown with a notch to emphasize that these are the sticky ends that bind to the chromosomal DNA.

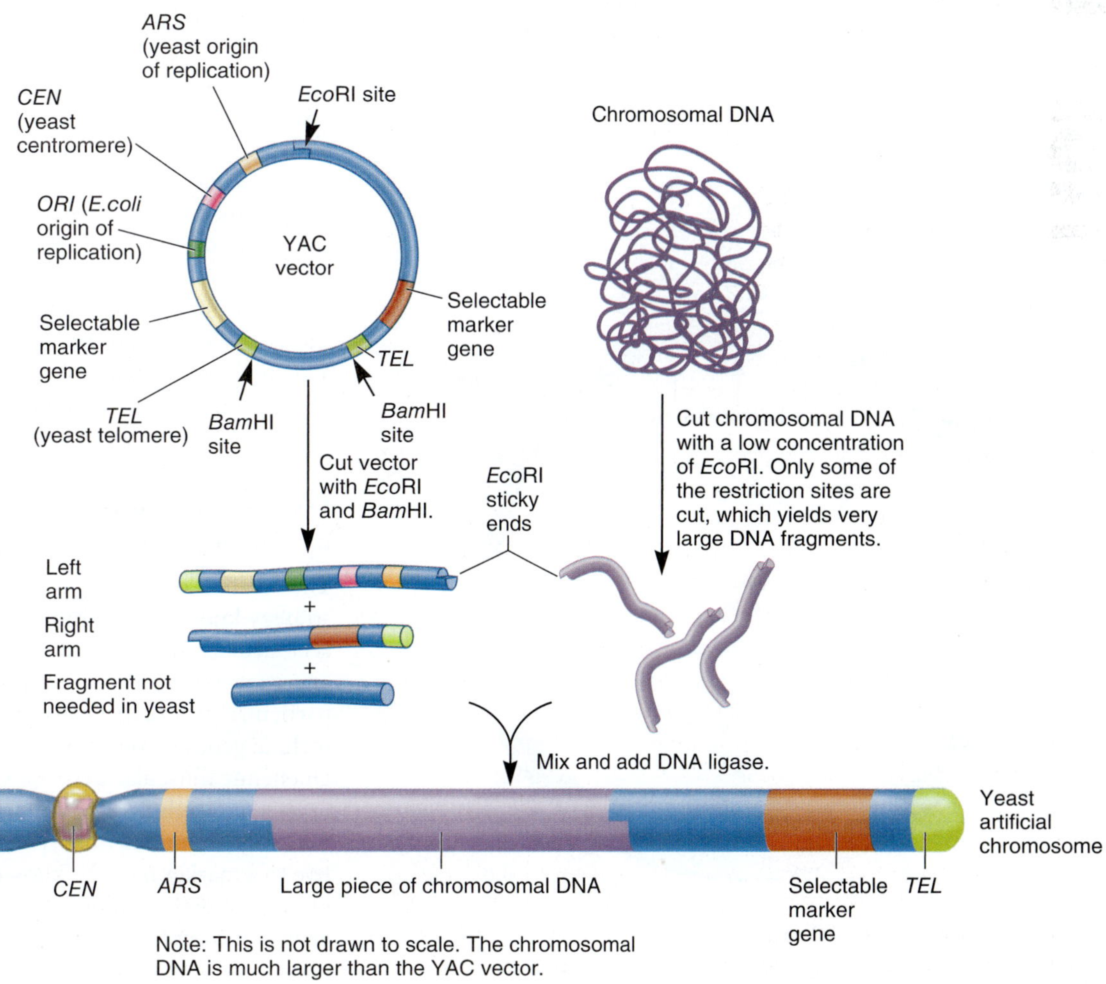

In the procedure shown in Figure 22.9, chromosomal DNA is digested with the restriction enzyme *Eco*RI at a low concentration so that only some of the restriction sites are cut. This partial digestion results in only occasional cleavage of the chromosomal DNA to yield very large DNA fragments.

The circular YAC vector is also digested with *Eco*RI and a second restriction enzyme, *Bam*HI. This digestion yields two arms of the YAC and a third fragment that is not needed in yeast. The YAC arms are then ligated to the large fragments of chromosomal DNA and transformed into yeast cells. When ligation occurs in the desired way, a large piece of chromosomal DNA is inserted between both arms of the YAC. Because each arm contains a different selectable marker, it is possible to select for the growth of yeast cells that carry a YAC with both arms.

Newer types of cloning vectors called **bacterial artificial chromosomes (BACs)** and **P1 artificial chromosomes (PACs)** have been constructed. BACs were developed from bacterial F factors, which are described in Chapter 7, and PACs were developed from P1 bacteriophage DNA. BACs and PACs typically can contain inserts up to 300,000 bp and sometimes larger. These vectors are somewhat easier to use than YACs because the DNA is inserted into a circular molecule and transformed into *E. coli*. BACs and PACs are more commonly used than YACs for the cloning of large DNA fragments.

Figure 22.10 shows a simplified drawing of a BAC cloning vector. The vector contains several genes that function in vector replication and segregation. The origin of replication is designated *oriS*, and the *repE* gene encodes a protein essential for replication at oriS. The *parA*, *parB*, and *parC* genes encode proteins required for the proper segregation of the vector into daughter cells. A chloramphenicol resistance gene, cm^R, provides a way to select for cells that have taken up the vector based on their ability to grow in the presence of the antibiotic chloramphenicol. The vector also contains unique restriction enzyme sites, such as *Hind*III, *Bam*HI, and *Sph*I, for the insertion of large fragments of DNA. These sites are located within the *lacZ* gene, which encodes the enzyme β-galactosidase. Vectors with DNA inserts can be determined by plating cells on media containing the compound X-Gal (as described for plasmid vectors in Chapter 20, Figure 20.2).

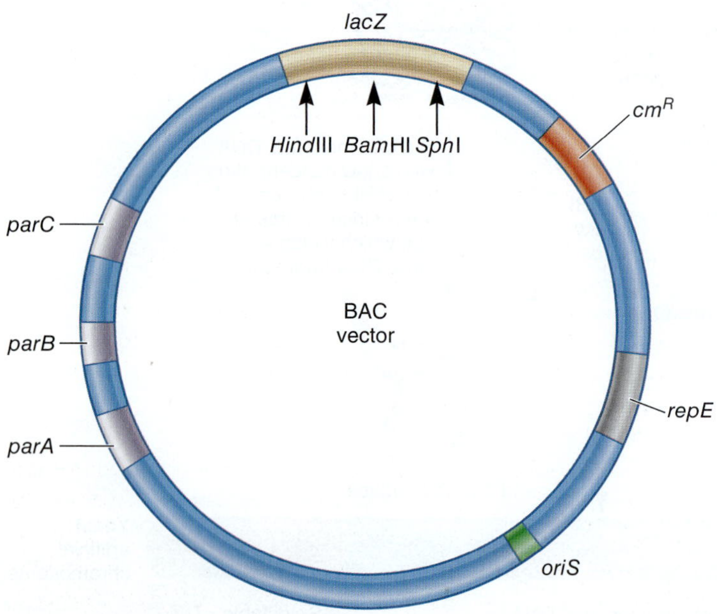

FIGURE 22.10 **A BAC vector.**

CONCEPT CHECK: What is the main advantage of YACs, BACs, and PACs?

YAC, BAC, and PAC cloning vectors have been very useful in the construction of contigs that span long segments of chromosomes. They have been used as the first step in creating a rough physical map of a genome. Although this is an important step in physical mapping, the large insert sizes make them difficult to use in gene-cloning and sequencing experiments. Therefore, libraries containing hybrid vectors with smaller insert sizes are needed. In many cases, a type of cloning vector called a cosmid is used. A **cosmid** is a hybrid between a plasmid vector and phage λ; its DNA can replicate in a cell like a plasmid or be packaged into a protein coat like a phage. Cosmid vectors typically can accept DNA fragments that are tens of thousands of base pairs in length.

Researchers Can Make Genetic Maps of Large Eukaryotic Chromosomes

Figure 22.11 illustrates a comparison among cytogenetic, linkage, and physical maps of human chromosome 16. Actually, this is a very simplified map of chromosome 16. A much more detailed map is available, although it would take well over 10 pages of this textbook to print it! The top of this figure shows the banding pattern of this chromosome. Underneath the banded chromosome are molecular markers that have been mapped by linkage analysis. These same markers have been localized to particular regions of the chromosome by cytogenetic mapping. Though not shown in this figure, a complete contig of this chromosome has also been produced by generating a series of overlapping YACs. Figure 22.11 shows the location of only one YAC and a cosmid contig for that YAC is shown beneath it. In addition, an STS is found within the region and provides a molecular marker for cosmids N16Y1-19 and N16Y1-10, as well as YAC N16Y1.

Positional Cloning Can Be Achieved by Chromosome Walking

The creation of a contig bears many similarities to a gene-cloning strategy known as **positional cloning,** a strategy in which a gene is cloned based on its mapped position along a chromosome. This approach has been successful in the cloning of many human genes, particularly those that cause genetic diseases when mutated. These include genes involved in cystic fibrosis, Huntington disease, and Duchenne muscular dystrophy.

One method of positional cloning is known as **chromosome walking.** To initiate this type of experiment, a gene's position relative to a marker must be known from mapping studies. For example, a gene may be known to be fairly close to a previously mapped gene or molecular marker. This provides a starting point to molecularly "walk" toward the gene of interest.

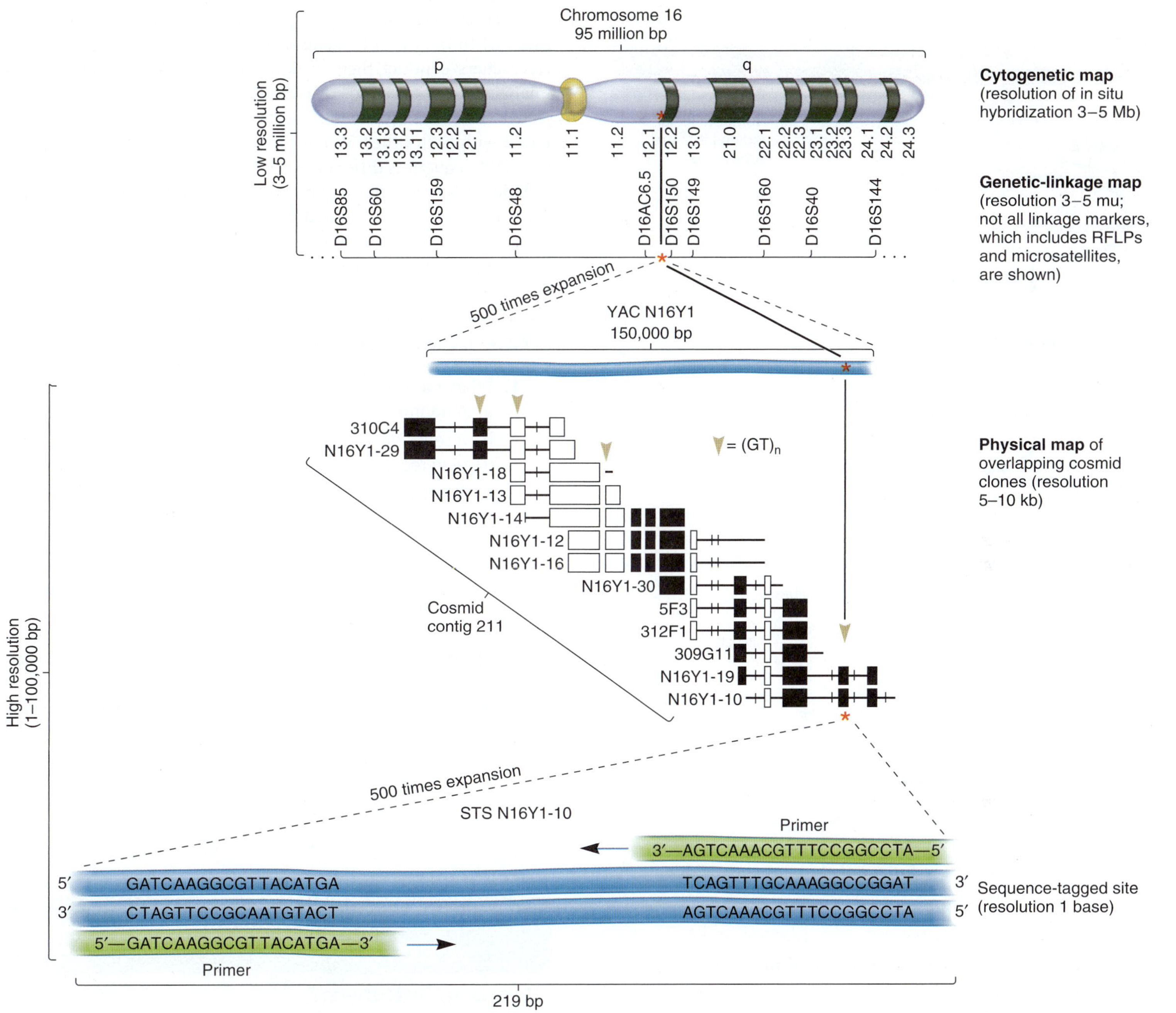

FIGURE 22.11 A correlation of cytogenetic, linkage, and physical maps of human chromosome 16. The top part shows the cytogenetic map of human chromosome 16 according to its G banding pattern. A very simple linkage map of molecular markers (D16S85, D16S60, etc.) is aligned below the cytogenetic map. A correlation between the linkage map and a segment of the physical map is shown below the linkage map. A YAC clone designated YAC N16Y1 is located between markers D16AC6.5 and D16S150 on the linkage map. Pieces of DNA from this YAC were subcloned into cosmid vectors, and the cosmids (310C4, N16Y1-29, N16Y1-18, etc.) were aligned relative to each other. One of the cosmids (N16Y1-10) was sequenced, and this sequence was used to generate an STS shown at the bottom of the figure.

Figure 22.12 considers a chromosome walk in which the goal is to locate a gene that we will call gene *A*. In this example, linkage mapping studies have revealed that gene *A* is relatively close to another gene, called gene *B*, that was previously cloned. Gene *A* and gene *B* have been deduced from genetic crosses to be approximately 1 mu apart. To begin this chromosome walk, a cloned DNA fragment that contains gene *B* and flanking sequences can be used as a starting point.

To walk from gene *B* to gene *A*, a series of library screening methods are followed. In this example, the starting materials are a cosmid library and a clone containing gene *B*. A small piece of DNA from the first cosmid vector containing gene *B* is inserted into another vector. This is called **subcloning.** The subcloned DNA is labeled and used as a probe to screen a cosmid library. This enables the researchers to identify a second clone that extends into the region that is closer to gene *A*. A subclone from this second clone

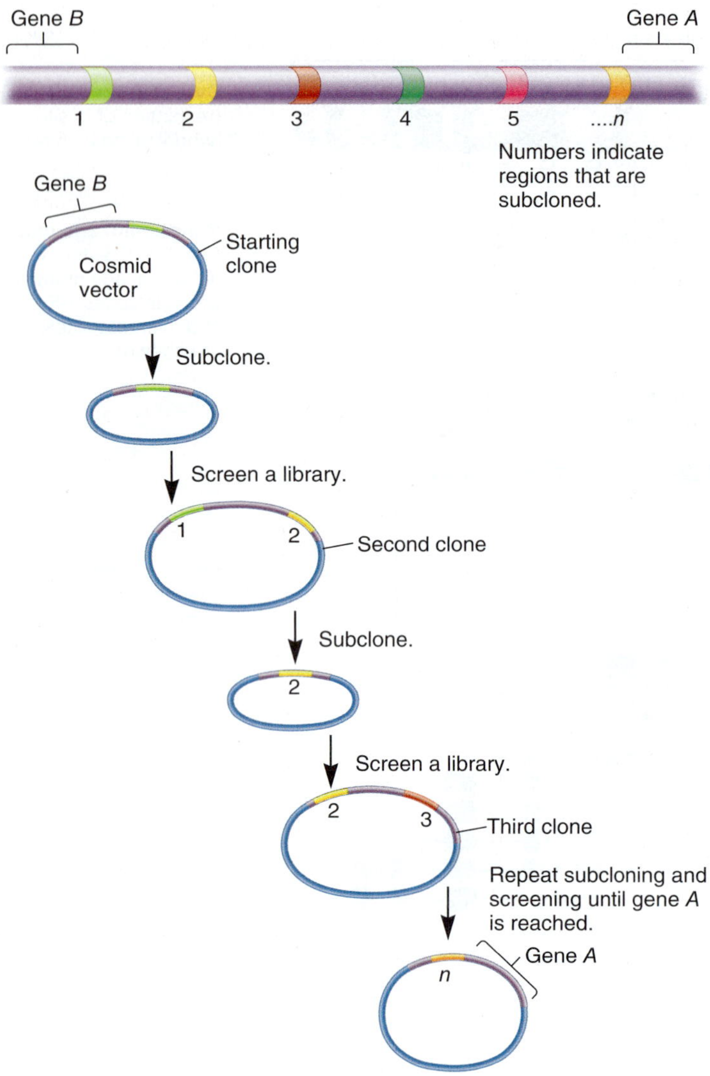

FIGURE 22.12 **The technique of chromosome walking.**

CONCEPT CHECK: Why is this technique used?

is then used to screen the library a second time. This allows the researchers to identify a clone that is even closer to gene *A*. This repeated pattern of subcloning and library screening is used to reach gene *A*. The term chromosome walking is an appropriate description of this technique, because each clone takes you a step closer to the gene of interest. When starting at gene *B* in Figure 22.12, researchers also want to have markers to the left of gene *B* to ensure they are not walking in the wrong direction.

The number of steps required to reach the gene of interest depends on the distance between the starting and ending points and on the sizes of the DNA inserts in the library. If the two points are 1 mu apart, they are expected to be approximately 1 million bp apart, although the correlation between map units and physical distances can vary greatly. In a typical walking experiment, each clone might have an average insert size of 50,000 bp. Therefore, it takes about 20 walking steps to reach the gene of interest. Researchers want to locate starting points in a chromosome walking experiment that are as close as possible to the gene they wish to identify.

How do researchers know when they have reached a gene of interest? In the case of a gene that causes a disease when mutant, researchers conduct their walking steps on DNA from both an unaffected and an affected individual. Each set of clones is subjected to DNA sequencing, and those DNA sequences are compared with each other. When the researchers reach a spot where the DNA sequences differ between the unaffected and the affected individual, such a site may be within the gene of interest. However, this has to be confirmed by sequencing the region from several unaffected and affected individuals to be certain the change in DNA sequence is correlated with the disease.

22.4 COMPREHENSION QUESTIONS

1. What is a contig?
   a. A fragment of DNA that has been inserted into a vector
   b. A series of vectors that contain inserts that have overlapping regions of chromosomal DNA
   c. A method of identifying a disease-causing allele
   d. A method of sequencing DNA

2. A vector that can carry a large fragment of chromosomal DNA is a
   a. YAC.
   b. BAC.
   c. PAC.
   d. all of the above.

3. Chromosomal walking is a type of _______________ in which a researcher begins at a specific site on a chromosome and analyzes _______________ until the gene of interest is reached.
   a. DNA sequencing, a series of subclones
   b. positional cloning, a series of subclones
   c. DNA sequencing, bands on a gel
   d. positional cloning, bands on a gel

22.5 GENOME-SEQUENCING PROJECTS

Learning Outcomes:

1. Describe the shotgun method for sequencing an entire genome.
2. List the goals of the Human Genome Project.
3. Outline different methods of DNA sequencing.
4. Compare and contrast the sizes of different species' genomes.

Genome-sequencing projects are research endeavors with the ultimate goal of determining the sequence of DNA bases of the entire genome of a given species. Such projects involve many participants, including scientists who isolate DNA and perform DNA-sequencing reactions, as well as theoreticians who gather the DNA sequence information and assemble it into a long DNA sequence for each chromosome. For bacteria and archaea, which usually have just one chromosome, the genome sequence is that of a single chromosome.

For eukaryotes, each chromosome must be sequenced. Thus, for humans, the genome sequence includes sequences of 22 autosomes, 2 sex chromosomes, and the mitochondrial genome.

In just a couple of decades, our ability to map and sequence genomes has improved dramatically. As of 2014, the complete genome sequences have been obtained for hundreds of different species, including over 300 prokaryotes and 100 eukaryotes. Considering that the first genome sequence was generated in 1995, the progress of genome-sequencing projects since then has been truly remarkable! In this section, we examine the approaches that researchers follow when tackling such a large project. We will also survey some of the general goals of the Human Genome Project, the largest of its kind, and compare the results from the genome sequencing of various species.

EXPERIMENT 22A

Venter, Smith, and Colleagues Sequenced the First Genome in 1995

The first genome to be entirely sequenced was that of the bacterium *Haemophilus influenzae*. This bacterium causes a variety of diseases in humans, including respiratory illnesses and bacterial meningitis. *H. influenzae* has a relatively small genome consisting of approximately 1.8 Mb of DNA in a single circular chromosome.

When sequencing an entire genome, researchers must consider factors such as genome size, the efficiency of the methods used to sequence DNA, and the costs of the project. Since genome-sequencing projects began in the 1990s, researchers have learned that the most efficient and inexpensive way to sequence genomes is via an approach called **shotgun sequencing,** in which DNA fragments to be sequenced are randomly generated from larger DNA fragments. In this method, genomic DNA is isolated and broken into smaller DNA fragments, typically 1500 bp or longer in length. The researchers then randomly sequence such fragments from the genome. As a matter of chance, some of the fragments overlap, as shown schematically in **Figure 22.13**. The DNA sequences in two different fragments are identical in the overlapping region. This allows researchers to order them as they are found in the intact chromosome. An advantage of shotgun DNA sequencing is that it does not require extensive mapping, which can be very time-consuming and expensive. A disadvantage is that researchers waste some time sequencing the same region of DNA more times than needed.

To obtain a complete sequence of a genome with the shotgun approach, how do researchers decide how many fragments to sequence? We can calculate the probability that a base will not be sequenced using this approach with the following equation:

$$P = e^{-m}$$

where

P is the probability that a base will be left unsequenced

e is the base of the natural logarithm; $e = 2.72$

m is the number of bases sequenced divided by the total genome size

For example, in the case of *H. influenzae* with a genome size of 1.8 Mb (i.e., 1.8 million bp), if researchers sequenced 9 Mb, $m = 5$ (i.e., 9.0 Mb divided by 1.8 Mb):

$$P = e^{-m} = e^{-5} = 0.0067, \text{ or } 0.67\%$$

This means that if we randomly sequence 9.0 Mb, which is five times the length of a single genome, we are likely to miss only 0.67% of the genome. With a genome size of 1.8 Mb, we would miss about 12,000 nucleotides out of approximately 1,800,000. Such missed sequences are typically on small DNA fragments that—as a matter of random chance—did not happen to be sequenced. The missing links in the genome can be sequenced later using mapping methods such as chromosome walking.

The general protocol conducted by J. Craig Venter, Hamilton Smith, and colleagues in this discovery-based investigation is described in **Figure 22.14**. This is a shotgun DNA sequencing approach. The researchers isolated chromosomal DNA from *H. influenzae* and used sound waves to break the DNA into small fragments of approximately 2000 bp in length. These fragments were randomly cloned into vectors, allowing the DNA to be propagated in *E. coli*. Each *E. coli* clone carried a vector with a different piece of DNA from *H. influenzae*. The researchers then subjected many of these clones to the procedure of DNA sequencing. They sequenced a total of approximately 10.8 Mb of DNA.

THE GOAL (DISCOVERY-BASED SCIENCE)

The goal was to obtain the entire genome sequence of *Haemophilus influenzae*. This information reveals its genome size as well as which genes the organism has.

```
                                            Overlapping region
TTACGGTACCAGTTACAAATTCCAGACCTAGTACC
AATGCCATGGTCAATGTTTAAGGTCTGGATCATGG
                        GACCTAGTACCGGACTTATTCGATCCCCAATTTTGCAT
                        CTGGATCATGGCCTGAATAAGCTAGGGGTTAAAACGTA
```

FIGURE 22.13 **A comparison of two DNA fragments that contain an overlapping region.**

ACHIEVING THE GOAL

FIGURE 22.14 The use of shotgun DNA sequencing to determine the first genome sequence of a bacterial species.

Starting materials: A strain of *H. influenzae.*

1. Purify DNA from a strain of *H. influenzae*. This involves breaking the cells open by adding phenol and chloroform. Most protein and lipid components go into the phenol-chloroform phase. DNA remains in the aqueous (water) phase, which is removed and used in step 2.

2. Sonicate the DNA to break it into small fragments of about 2000 bp in length.

3. Clone the DNA fragments into vectors. The procedures for cloning are described in Chapter 20. This produces a DNA library.

4. Subject many clones to the procedure of dideoxy DNA sequencing, also described in Chapter 20. A total of 10.8 Mb was sequenced.

5. Use tools of bioinformatics, described in Chapter 23, to identify various types of genes in the genome.

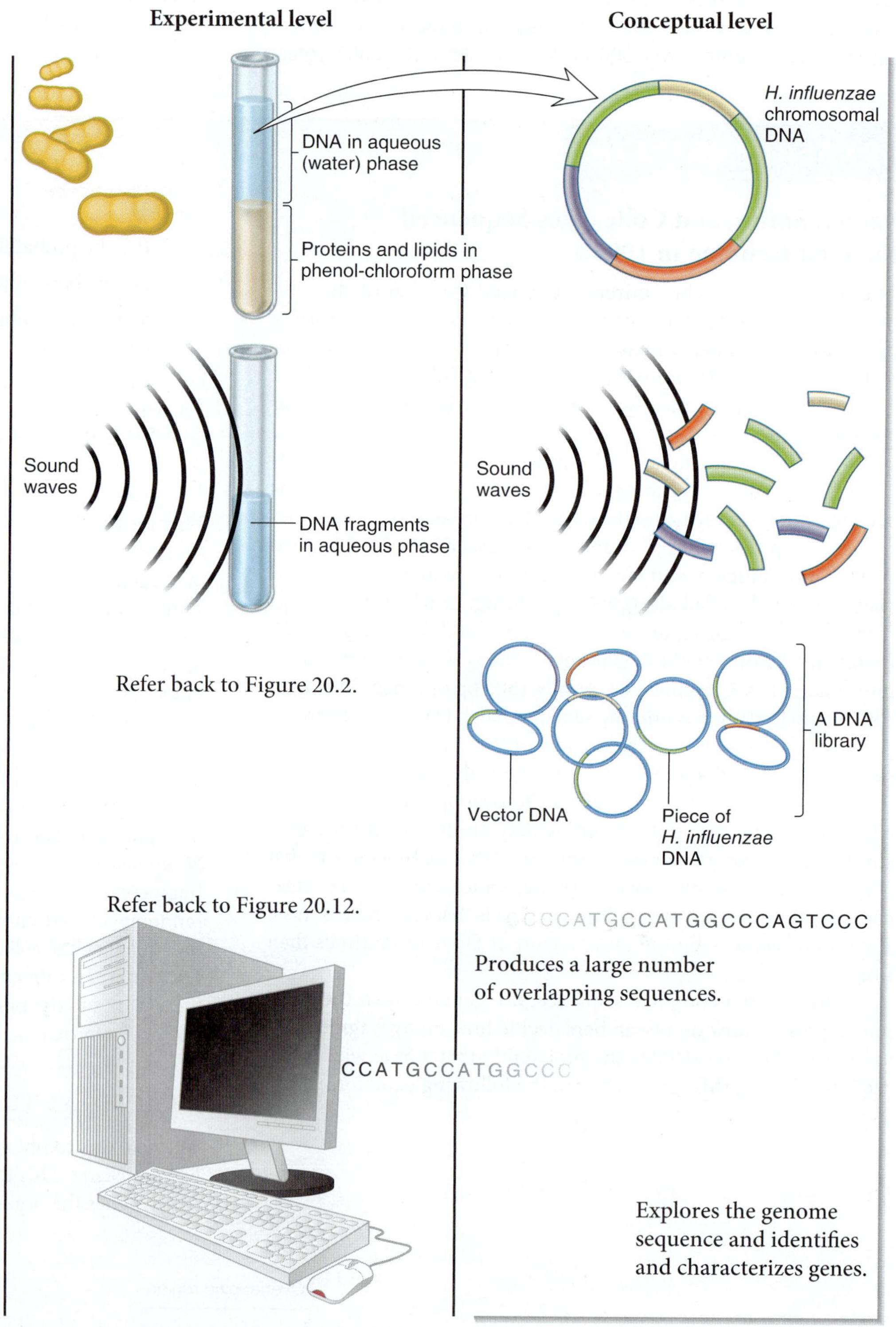

THE DATA

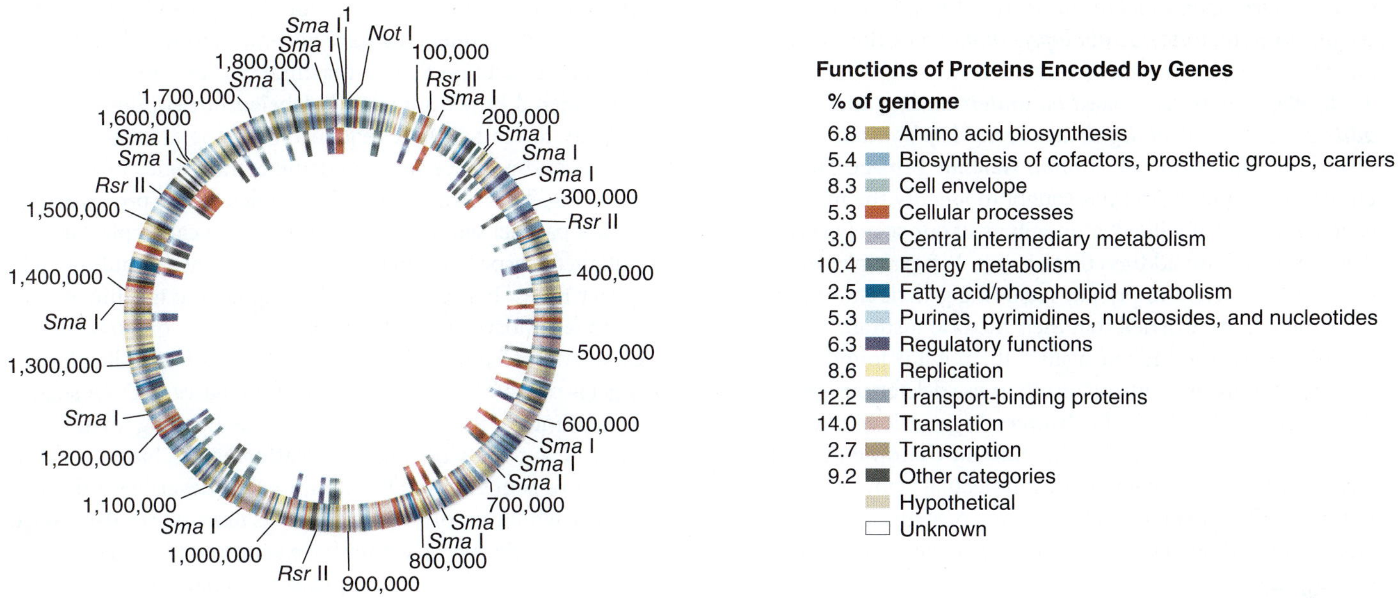

INTERPRETING THE DATA

The outcome of this genome-sequencing project was a very long DNA sequence. In 1995, Venter, Smith, and colleagues published the entire DNA sequence of *H. influenzae.* The researchers then analyzed the genome sequence using a computer to obtain information about the properties of the genome. Questions they asked included, How many genes does the genome contain and what are the likely functions of those genes? In Chapter 23, we will learn how scientists can answer such questions with the use of computers. The Data of Figure 22.14 summarize the results that the researchers obtained. The *H. influenzae* genome is composed of 1,830,137 bp of DNA. The computer analysis predicted 1743 genes. Based on their similarities to known genes in other species, the researchers also predicted the functions of nearly two-thirds of these genes. The diagram shown in The Data from Figure 22.14 places genes in various categories based on the predicted function of their encoded proteins. These results gave the first comprehensive "genome picture" of a living organism!

A self-help quiz involving this experiment can be found at www.mhhe.com/brookergenetics5e.

The Human Genome Project Was the Largest Genome-Sequencing Project in History

Due to its large size, the sequencing of the human genome was an enormous undertaking. Scientists had been discussing how to undertake this project since the mid-1980s. In 1988, the National Institutes of Health established an Office of Human Genome Research, with James Watson as its first director. The **Human Genome Project,** which officially began on October 1, 1990, was a 13-year effort coordinated and funded by the U.S. Department of Energy and the National Institutes of Health. The human DNA that was used in the Human Genome Project was obtained from several volunteers; their identity was not revealed to protect their privacy. From its outset, the Human Genome Project had the following goals:

1. ***To obtain a genetic linkage map of the human genome.*** This has involved the identification of millions of genetic markers and their localization along the autosomes and sex chromosomes.
2. ***To obtain a physical map of the human genome.*** This has required the cloning of many segments of chromosomal DNA into BACs, YACs, and cosmids.
3. ***To obtain the DNA sequence of the entire human genome.*** The first (nearly complete) sequence was published in February 2001. This was considered a first draft. A second draft was published in 2003, and the completed maps and sequences for all of the human chromosomes were published by 2006. The entire genome is approximately 3 billion bp in length. If the entire human genome were typed in a textbook like this, it would be nearly 1 million pages long!
4. ***To develop technology for the management of human genome information.*** The amount of information obtained from this project is staggering, to say the least. The Human Genome Project developed user-friendly tools to provide scientists easy access to up-to-date information obtained from the project. The Human Genome Project also developed analytical tools for interpreting genome information.

5. ***To analyze the genomes of model organisms.*** These include bacterial species (e.g., *Escherichia coli* and *Bacillus subtilis*), *Drosophila melanogaster* (fruit fly), *Caenorhabditis elegans* (a nematode), *Arabidopsis thaliana* (a flowering plant), and *Mus musculus* (mouse).
6. ***To develop programs focused on understanding and addressing the ethical, legal, and social implications of the results obtained from the Human Genome Project.*** The Human Genome Project has sought to identify the major genetic issues that will affect members of our society and to develop policies to address these issues. For example, what is an individual's right to privacy regarding genetic information? Some people are worried that their medical insurance company may discriminate against them if it is found that they carry a disease-causing or otherwise deleterious gene.
7. ***To develop technological advances in genetic methodologies.*** Some of the efforts of the Human Genome Project have involved improvements in molecular genetic technology such as gene cloning, contig construction, and DNA sequencing. The project has also been aimed at developing computer technology for data processing, storage, and the analysis of sequence information (see Chapter 23).

A great benefit expected from the characterization of the human genome is the ability to identify and study the sequences of our genes. Mutations in many different genes are known to be correlated with human diseases, which include cancer, heart disease, and many other abnormalities. The identification of mutant genes that cause inherited diseases was a strong motivation for the Human Genome Project. A detailed genetic and physical map has made it considerably easier for researchers to locate such genes. Furthermore, a complete DNA sequence of the human genome provides researchers with insight into the types of proteins encoded by these genes. The cloning and sequencing of disease-causing alleles is expected to play an increasingly important role in the diagnosis and treatment of disease.

Innovations in DNA Sequencing Have Made It Faster and Less Expensive

Since it was invented in the 1970s, technological advances in DNA sequencing have been aimed at making it faster and less expensive. The Human Genome Project, which began in the early 1990s, was originally estimated to cost about \$3 billion to sequence a single genome. However, cost reductions due to innovations in DNA-sequencing technology drove the actual cost down to about \$300 million. By the end of the project, researchers estimated that if they were starting again, they could have sequenced the genome for less than \$50 million. The project took about 13 years to complete. In 2007, researchers undertook the sequencing of James Watson's genome, which cost less than \$1 million. By 2011, the cost of sequencing a human genome had been reduced to \$5000 or less. By 2015, the sequencing of a single human genome will cost less than \$1000. Such innovation will make it feasible to sequence an individual's genome as a routine diagnostic procedure.

The ability to rapidly sequence large amounts of DNA is referred to as **high-throughput sequencing.** Different types of technological advances have made this possible. First, different aspects of DNA sequencing have become automated so that samples can be processed rapidly in a machine. For example, in Chapter 20, we considered how fluorescently labeled nucleotides can automate the ability to read a DNA sequence by using a fluorescence detector. A second advance in sequencing technologies involves parallel sequencing. This means multiple samples can be processed at once. The first parallel sequencing machines, also called platforms, could simultaneously perform many sequencing runs via multiple gel-filled capillary tubes. For example, DNA-sequencing machines produced by Applied Biosystems, which rely on Sanger dideoxy sequencing (described in Chapter 20), run 96 capillaries in parallel. Each capillary gel is capable of producing 500–1000 bases of DNA sequence.

Although the Sanger dideoxy sequencing method is still in use, newer high-throughput platforms are based on different methods of sequencing DNA. Table 22.2 describes a few of these methods, which are often referred to as **next-generation sequencing technologies** because they have superseded the Sanger dideoxy method for large sequencing projects. What sets next-generation sequencers apart from conventional Sanger dideoxy sequencing? One key technological advance is the ability to process thousands or even millions of sequence reads in parallel rather than only 96 at a time. This massively parallel throughput may require only one or two instrument runs to complete the sequencing of an entire prokaryotic genome. Also, next-generation sequencers are able to use samples that contain mixtures of DNA fragments that have not been subjected to the conventional vector-based cloning. By comparison, shotgun methods of DNA sequencing using capillary sequencing involve DNA-cloning steps (see Figure 22.14). The elimination of such cloning steps saves a great deal of time and money.

The newer sequencing platforms employ a complex interplay of enzymology, chemistry, high-resolution optics, and new approaches to processing the data. These instruments allow for easy sample preparation steps prior to DNA sequencing. Most of them involve strategies in which fragmented DNA is immobilized in a fixed position and repeatedly exposed to reagents. Some sequencing platforms use PCR to amplify the DNA, whereas others actually read single DNA molecules.

As an example of next-generation sequencing, Figure 22.15 considers the technology called **pyrosequencing,** which relies on the release of pyrophosphate. This method was developed by Pal Nyren and Mostafa Ronaghi in 1996 and is the basis for the Roche 454/FLX Pyrosequencer. Samples, such as genomic DNA, are broken into small 300- to 800-bp fragments. Short oligonucleotides called adaptors are then linked to the 5′ and 3′ ends of the DNA fragments. (Note: The red adaptor contains a sequence that is complementary to an oligonucleotide attached to the beads and the green adaptor is complementary to primers used in the DNA sequencing reaction.) The DNA is then denatured into single strands. The mixture of single-stranded DNA fragments with adaptors is called the sample library.

The single-stranded DNA fragments are then attached to beads via the adaptors. Initially, just one DNA strand is attached per bead. The beads are emulsified in an oil-water mixture so that

TABLE 22.2

Examples of Next-Generation DNA-Sequencing Technologies

Technology*	DNA Preparation	Enzyme(s) Used	Detection
Roche 454/FLX Pyrosequencer	DNA fragments are bound to small beads, which are dropped into tiny wells in a picotiter plate.	DNA polymerase, ATP sulfurylase, luciferase, apyrase	Pyrophosphate release activates luciferase, which breaks down luciferin and gives off light.
Illumina/Solexa Genome Analyzer	DNA fragments are bound to a flow cell surface..	DNA polymerase	Four different fluorescently labeled nucleotides are detected.
DNA Nanoball Sequencing	Linkers are attached to small fragments of DNA that are amplified by rolling circle DNA replication into DNA nanoballs. The nanoballs are then bound to a flow cell surface.	DNA polymerase, restriction enzymes, DNA ligase	Four different fluorescently labeled nucleotides are detected.
Single-Molecule DNA Sequencers			
Pacific Biosciences SMRT	DNA fragments and DNA polymerase are trapped within tiny holes on a thin metal film.	DNA polymerase	The growth of individual DNA molecules is monitored by fluorescence imaging.
Helicos Biosciences tSMS	DNA fragments are bound to a flow cell surface..	DNA polymerase	The growth of individual DNA molecules is monitored by fluorescence imaging.
ZS Genetics TEM	DNA is labeled with heavier elements, such as iodine or bromine.	None	The DNA sequence of single DNA molecules is read via transmission electron microscopy (TEM).
Ion Semiconductor Sequencing	DNA fragments are bound to a semiconductor chip.	DNA polymerase	When a nucleotide is incorporated into a growing strand, pyrophosphate and a hydrogen ion are released. The release of the hydrogen ion is detected.

*Also includes the company name associated with the technology.

each bead becomes localized into a single droplet. The mixture also contains PCR reagents, including primers that are complementary to the adaptors. During this step, the single-stranded DNA on the bead becomes amplified into many identical products, which also become attached to the beads. At this stage, each bead contains about one million copies of a particular DNA segment.

Next, each bead is placed into a well of a picotiter plate; the diameter of each well is large enough to accommodate only one bead. Sequencing reagents, which include primers that are complementary to the green adaptors, are added to the wells. The picotiter plate acts like a flow cell in which the beads are stuck in the wells and pure solutions of nucleotides can be flowed over the top of them in a stepwise fashion. The synthesis of DNA is monitored in real time. Therefore, this type of method is referred to as **sequencing by synthesis (SBS)** because it involves the identification of each nucleotide immediately after its incorporation into a DNA strand by DNA polymerase.

When a solution has been added that contains one particular nucleotide, how does the machine determine if that nucleotide has been added to a growing DNA strand? The pyrosequencing method relies on the functions of other enzymes, ATP sulfurylase, luciferase, and apyrase, along with additional molecules called adenosine 5′-phosphosulfate and luciferin, which are also included in the sequencing reaction. If a nucleotide has been incorporated into a DNA strand, pyrophosphate is released. ATP sulfurylase uses that pyrophosphate along with adenosine 5′-phosphosulfate to make ATP. The ATP is then used by luciferase to break down luciferin. This reaction gives off light, which is detected by a camera in the sequencing machine. Therefore, light is given off only when a nucleotide is incorporated into a DNA strand. Unincorporated nucleotides and ATP are degraded by apyrase, and the reaction can start again by flowing a new solution containing a particular nucleotide over the picotiter plate. By sequentially adding solutions with only one of the four possible nucleotides (A, T, G, or C), the sequence of the DNA strand can be monitored in real time by determining when light is given off.

Many Genome Sequences Have Been Determined

The amazing advances in DNA-sequencing technology have enabled researchers to determine the complete genome sequences of hundreds of species. By 2014, over 300 prokaryotic and 100 eukaryotic genomes had been sequenced, including more than two dozen mammalian genomes.

How do researchers decide which genomes to sequence? Motivation behind genome-sequencing projects comes from a variety of sources. For example, basic research scientists can greatly benefit from a genome sequence. It allows them to know which genes a given species has, and it aids in the cloning and characterization of such genes. This has been the impetus for genome projects involving model organisms such as *Escherichia coli, Saccharomyces cerevisiae, Arabidopsis thaliana, Caenorhabditis elegans, Drosophila melanogaster,* and the mouse. A second impetus for genome sequencing has involved human disease. As noted previously, researchers expect that the sequencing of the human genome will aid in the identification of genes that, when mutant, play a role

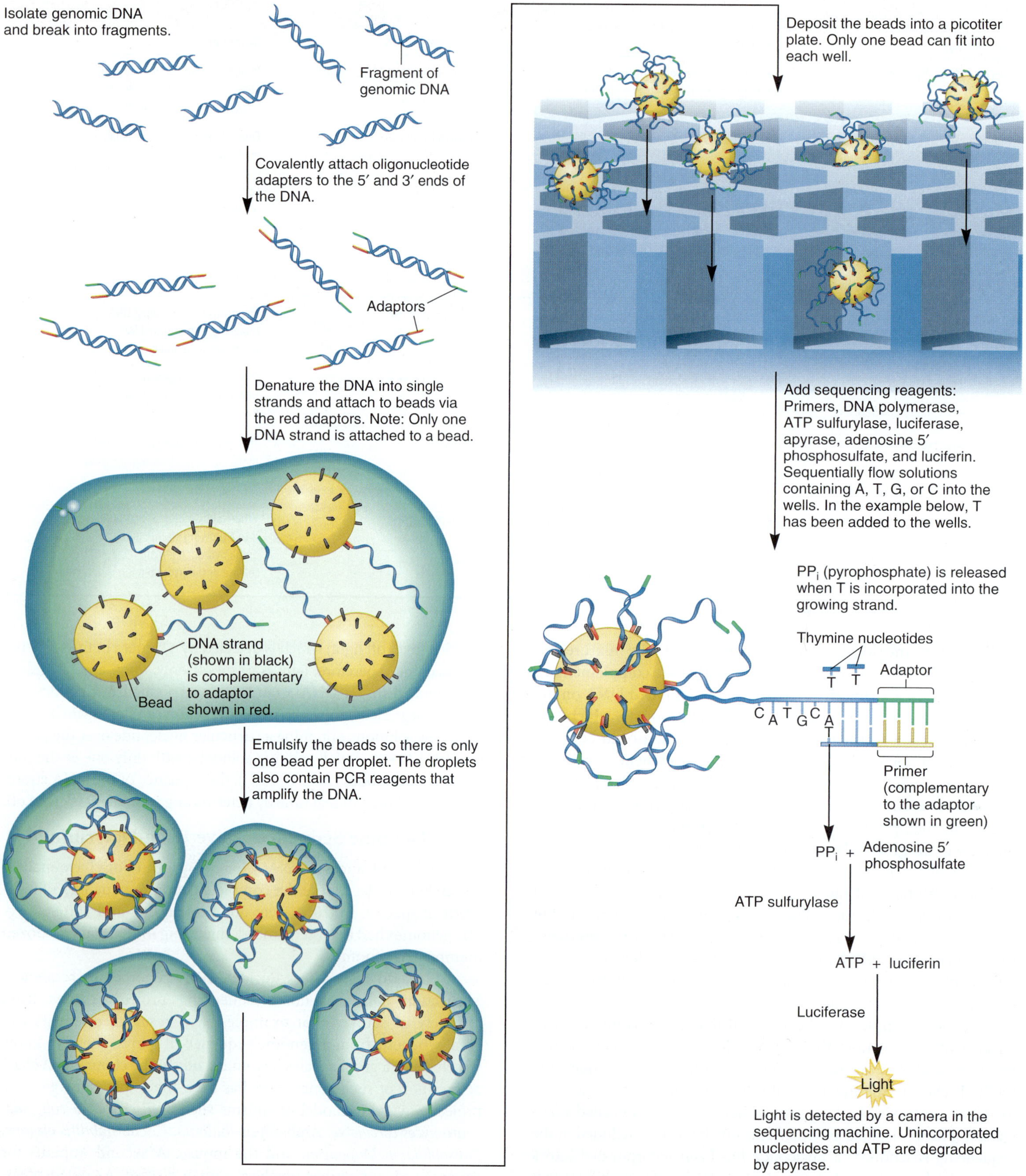

FIGURE 22.15 **Pyrosequencing, an example of a newer high-throughput sequencing technology.**

CONCEPT CHECK: Is this a "sequencing by synthesis" method? Explain.

TABLE 22.3
Examples of Genomes That Have Been Sequenced

Species	Genome Size (bp)*	Number of Genes†	Description
Prokaryotic genomes			
Bacteria			
Mycoplasma genitalium	580,000	521	Bacterial inhabitant of the human genital tract with a very small genome
Helicobacter pylori	1,668,000	1590	Bacterial inhabitant of the stomach that may cause gastritis, peptic ulcer, and gastric cancer
Mycobacterium tuberculosis	4,412,000	4294	Bacterial species that causes tuberculosis
Escherichia coli	4,639,000	4377	Widespread bacterial inhabitant of the gut of animals; also a model research organism
Archaea			
Thermoplasma volcanium	1,580,000	1494	Archaea with an optimal growth temperature of 60°C and pH optimum of <2.0
Pyrococcus abyssi	1,760,000	1765	Archaea that was originally isolated from samples taken close to a hot spring situated 3500 m deep in the southeast Pacific. Optimal growth conditions of 103°C and 200 atm.
Sulfolobus solfataricus	2,990,000	2977	Archaea found in terrestrial volcanic hot springs with optimum growth occurring at a temperature of 75–80°C and pH 2–3
Eukaryotic genomes			
Protists			
Plasmodium falciparum	22,900,000	5268	Parasitic protist that causes malaria in humans
Entamoeba histolytica	23,800,000	9938	Amoeba that causes dysentery in humans
Fungi			
Saccharomyces cerevisiae	12,100,000	6294	Baker's yeast—a structurally simple eukaryotic species that has been extensively studied by researchers to understand eukaryotic genetics and cell biology
Neurospora crassa	40,000,000	10,082	Common bread mold and also a structurally simple eukaryotic species that has been extensively studied by researchers
Plants			
Arabidopsis thaliana	142,000,000	26,000	A small flowering plant of the mustard family used as a model organism by plant biologists
Oryza sativa	440,000,000	40,000	Rice—cereal grain with a relatively small genome. It is very important worldwide as a food crop.
Populus trichocarpa	550,000,000	45,555	Balsam poplar—a tree with a relatively small genome
Animals			
Caenorhabditis elegans	97,000,000	19,000	Nematode worm that has been used as a model organism to study animal development
Drosophila melanogaster	175,000,000	14,000	Fruit fly—a model organism used to study many genetic phenomena, including development
Anopheles gambiae	278,000,000	13,683	Mosquito that carries the malaria parasite, *Plasmodium falciparum*
Canis familiaris	2,400,000,000	22,000	Dog—a common house pet
Mus musculus	2,500,000,000	22,000	House mouse—a rodent and a model organism studied by researchers
Pan troglodytes	3,100,000,000	22,000	Chimpanzee—a primate and the closest living relative to humans
Homo sapiens	3,200,000,000	22,000	Human

*In some cases, the values indicate the estimated amount of DNA. For eukaryotic genomes, DNA sequencing is considered completed in the euchromatic regions. The DNA in certain heterochromatic regions cannot be sequenced, and the total amount is difficult to estimate.
†The numbers of genes were predicted using computer methods described in Chapter 23. These numbers should be considered as estimates of the total gene number.

in disease. Likewise, the decision to sequence many bacterial, protist, and fungal genomes has been related to the role of these species in infectious diseases. Hundreds of microbial genomes have been sequenced. Many of them are from species that are pathogenic in humans. The sequencing of such genomes may help us understand which genes play a role in the infection process.

In addition, the genomes of agriculturally important species have been the subject of genome sequencing. An understanding of a species' genome may aid in the development of new strains of livestock and plant species that have improved traits from an agricultural perspective. Finally, genome-sequencing projects help us to better understand the evolutionary relationships among living species. This approach, called **comparative genomics,** uses information from genome projects to understand the genetic variation among different populations. We will explore this topic in Chapter 28.

Table 22.3 describes the results of several genome-sequencing projects that have been completed. Newly completed genome sequences are emerging rapidly, particularly those of microbial species. As we obtain more genome sequences, it becomes progressively more interesting to compare them to each other as a way to understand the process of evolution. As we will discover in

Chapter 23, the field of functional genomics enables researchers to probe the roles of many genes as they interact to generate the phenotypic traits of the species that have them.

22.5 COMPREHENSION QUESTIONS

1. Shotgun sequencing is a method of DNA sequencing in which
   a. the DNA fragments to be sequenced are randomly generated from larger DNA fragments.
   b. the sequencing reactions are carried out in rapid succession.
   c. the samples to be sequenced are rapidly generated by PCR.
   d. all of the above occur.
2. Which of the following was *not* a goal of the Human Genome Project?
   a. To obtain the DNA sequence of the entire human genome
   b. To successfully clone a mammal
   c. To develop technology for the management of human genome information
   d. To analyze the genomes of model organisms
3. A prokaryotic genome is about 4 million bp in length. About how many genes would you expect it to contain?
   a. 400 b. 4000 c. 40,000 d. 400,000

22.6 METAGENOMICS

Learning Outcomes:

1. Define metagenomics.
2. Describe the general strategy of metagenomics.
3. Outline the uses of metagenomics.

Most microorganisms that exist in soil, water, and the intestinal tract of humans have not been successfully grown in the laboratory because researchers may not understand their growth requirements and because some microbes require the presence of a complex microbial community to survive. In the past, such "unculturable microbes" were difficult to study. In the 1980s and early 1990s, Norman Pace and colleagues showed that 16S rRNA genes could be analyzed from samples of different microbes using PCR. Many of the 16S rRNA genes that were sequenced from environmental samples were found to be different from 16S rRNA genes of bacteria that had been grown in the laboratory. Therefore, this work revealed that environmental samples contain an abundance of unculturable microbes. Though such studies were exciting, a limitation of PCR is that it amplifies specific genes, leaving the rest of the genome of unculturable microbes unknown.

This limitation was overcome by the development of **metagenomics,** which is the study of a complex mixture of genetic material obtained from an environmental sample. The term **metagenome** refers to a collection of genes from a particular environmental sample. Such a sample can be analyzed in a way analogous to the study of a single genome. **Figure 22.16** outlines a general strategy for metagenomics. First, researchers obtain a sample from the environment. This could be a soil sample, a water sample from the ocean, or a fecal sample from a person. After the sample is filtered, the cells within the sample are lysed and the DNA is extracted and purified. During this procedure, the purified DNA is sheared and contains fragments of different sizes. The DNA fragments are then randomly inserted into a cloning vector and transformed into a host cell. The result is a DNA library of thousands or tens of thousands of cells, each carrying a DNA fragment from the metagenome. All cells together constitute the metagenomic library. The members of the library are then subjected to shotgun DNA sequencing to identify the genes they may contain. Note: The cloning step may be skipped if certain types of DNA sequencing methods, such as pyrosequencing, are used.

What are the uses of metagenomics? Some of the common applications of metagenomics are described next.

1. ***Human medicine:*** Various places in the body, such as the mouth and intestines, support a complex array of microorganisms. Metagenomics is being used to characterize these populations of organisms and to study changes in the relative compositions of microorganisms that may be associated with different diseases.
2. ***Agriculture:*** The metagenomic analysis of soil samples has revealed an astonishing complexity of microorganisms in the soil. As researchers learn more about which microbes facilitate plant growth, such knowledge may be used to improve agricultural yields.
3. ***Bioremediation:*** The types of microorganisms found in soil and water have a great effect on the decomposition of pollutants in the environment. Metagenomics may play a key role in the identification of microorganisms that can break down specific types of pollutants in the environment.
4. ***Biotechnology:*** Microorganisms are capable of synthesizing a vast array of chemicals, some of which are useful to humans. An example is antibiotics that are used to treat bacterial infections. Such chemicals are made by enzymes encoded by microbial genes. Metagenomics is being used to identify such genes in soil and aquatic microorganisms as a way to discover possible products that microorganisms can make.
5. ***Global change:*** Microorganisms carry out about half of the photosynthesis that takes place on Earth and are key participants in the cycling of various elements such as carbon, phosphorus, and nitrogen. Metagenomics is helping us understand the complexity of microbial communities in these processes.
6. ***Identification of viruses:*** Environmental samples are analyzed to identify viruses that infect humans and other organisms.
7. ***Aquatic biology:*** The metagenomic analysis of water samples from rivers, freshwater lakes, and oceans has revealed a much greater complexity of microbial communities than previously thought. An example of one such study is described next.

The first extensive large-scale environmental sequencing project, called the Global Ocean Sampling Expedition, was carried out by J. Craig Venter and colleagues. It began in 2003 as a pilot project to

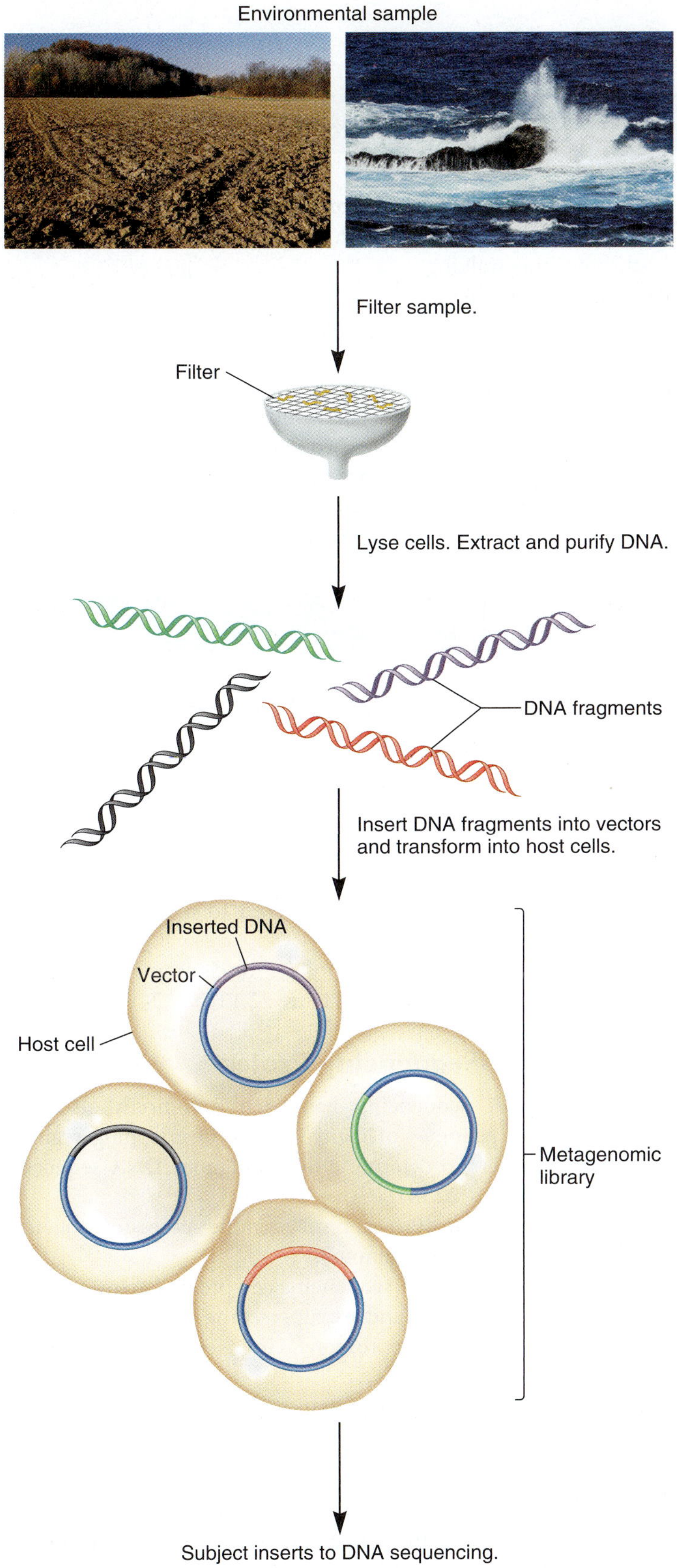

FIGURE 22.16 The general strategy of metagenomics.

sample the microbial population of the nutrient-limited Sargasso Sea, a region of the Atlantic Ocean close to Bermuda. This site was chosen because it was thought to contain a relatively simple population of microbes compared with other aquatic locations. The samples from the Sargasso Sea revealed several hundred new proteorhodopsin

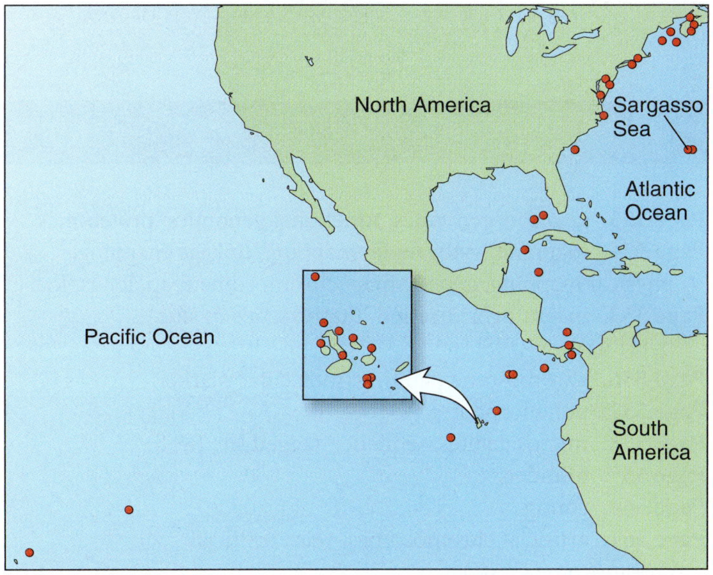

FIGURE 22.17 Sampling sites of the Global Ocean Sampling Expedition. The sampling sites are shown as red dots. The first two sites were in the Sargasso Sea. The expedition then headed to Nova Scotia and proceeded southwest, ending in the Pacific Ocean. (From Rusch, D.B., Halpern, A.L., Sutton, G., Heidelberg, K.B., Williamson, S., et al. (2007) The Sorcerer II Global Ocean Sampling Expedition: Northwest Atlantic through easterntropical Pacific. *PLoS Biol* 5(3): e77. doi:10.1371/journal.pbio.0050077)

light-harvesting genes. This discovery may help us understand their role in energy metabolism under low nutrient conditions. In addition, over 1800 new species of microorganisms were identified.

Venter announced the full expedition in 2004. In the spirit of Darwin's travels on the *Beagle*, Venter's 95-foot sailboat (the *Sorcerer II*) was outfitted as a research vessel. Using a sample size of 200 liters, the team traveled over 32,000 miles, sampling every 200 miles; they collected about 40 different samples (**Figure 22.17**). After being frozen and shipped to a land-based research center, the samples were sequenced using shotgun techniques. 7.7 million sequencing runs were performed yielding over 6 billion bp of DNA sequence information. Most of the DNA sequences were unique, reflecting the incredible diversity in naturally occurring microbial populations. Though the expedition yielded many exciting results, one of the key findings that emerged involved the identification of species variants, called subtypes, that exist alongside each other. As communities adapt and change, previous research had suggested that certain subtypes tend to dominate and outcompete others. In contrast, this expedition found many closely related species and subtypes that were living in the same environment. Further research will be needed to explain this paradox.

22.6 COMPREHENSION QUESTION

1. Metagenomics is aimed at
   a. determining the complete genome sequence of newly identified microorganisms.
   b. mapping the genes along chromosomes of newly identified microorganisms.
   c. determining the sequence of DNA fragments in environmental samples.
   d. determining the functions of all of the genes in a given species' genome.

KEY TERMS

Page 553. genome, genomics, functional genomics, proteomics
Page 554. mapping, cytogenetic mapping, linkage mapping, physical mapping, genetic map, chromosome map, locus, loci
Page 555. in situ hybridization, fluorescence in situ hybridization (FISH)
Page 556. chromosome painting, molecular marker
Page 557. polymorphic
Page 558. microsatellites, sequence-tagged site (STS)
Page 559. founder
Page 560. contig
Page 561. artificial chromosomes, yeast artificial chromosome (YAC)
Page 562. bacterial artificial chromosomes (BACs), P1 artificial chromosomes (PACs), cosmid, positional cloning, chromosome walking
Page 563. subcloning
Page 564. genome-sequencing projects
Page 565. shotgun sequencing
Page 567. Human Genome Project
Page 568. high-throughput sequencing, next-generation sequencing technologies, pyrosequencing
Page 569. sequencing by synthesis (SBS)
Page 571. comparative genomics
Page 572. metagenomics, metagenome

CHAPTER SUMMARY

- The genome is the total genetic composition of an organism or species. Genomics refers to the molecular analysis of a genome as a whole unit. Functional genomics is aimed at studying the expression of genes, whereas proteomics is focused on the structures and functions of proteins.

22.1 Overview of Chromosome Mapping

- The term mapping refers to the process of determining the relative locations of genes or other DNA segments along a chromosome, thereby producing a genetic or chromosome map. The process may involve cytogenetic, linkage, or physical mapping techniques (see Figure 22.1).

22.2 Cytogenetic Mapping via Microscopy

- Cytogenetic mapping attempts to determine the locations of particular genes relative to a banding pattern of a chromosome.
- Fluorescence in situ hybridization is a commonly used method for mapping genes and other segments along a chromosome (see Figures 22.2, 22.3).

22.3 Linkage Mapping via Crosses

- Linkage mapping often relies on molecular markers to map the locations of genes along chromosomes (see Table 22.1).
- Restriction fragment length polymorphisms and microsatellites are used as molecular markers (see Figures 22.4–22.6).

22.4 Physical Mapping via Cloning and DNA Sequencing

- A physical map of a chromosome is made by producing a contiguous collection of clones called a contig, which spans an entire chromosome (see Figures 22.7, 22.8).
- YACs, BACs, PACs, and cosmids are vectors that can carry a long segment of DNA and are used to make contigs (see Figures 22.9–22.11).
- Chromosome walking is a technique for cloning a gene that is close to a known marker (see Figure 22.12).

22.5 Genome-Sequencing Projects

- Shotgun DNA sequencing has been commonly used to sequence the genomes of many species. Overlapping regions allow researchers to determine the contiguous DNA sequence (see Figures 22.13, 22.14).
- The Human Genome Project resulted in the successful sequencing of the entire human genome.
- Newer methods of DNA sequencing can process many samples of DNA simultaneously and are superseding the Sanger dideoxy method. An example is pyrosequencing (see Table 22.2, Figure 22.15).
- Since 1995, the genomes of hundreds of species have been sequenced (see Table 22.3).

22.6 Metagenomics

- Metagenomics is the study of a complex mixture of genetic material obtained from an environmental sample (see Figures 22.16, 22.17).

PROBLEM SETS & INSIGHTS

Solved Problems

S1. The distance between two molecular markers that are linked along the same chromosome can be determined by analyzing the outcome of crosses. This can be done in humans by analyzing the members of a pedigree. However, the accuracy of linkage mapping in human pedigrees is fairly limited because the number of people in most families is relatively small. As an alternative, researchers can analyze a population of sperm, produced from a single male, and compute linkage distance in this manner. As an example, let's suppose a male is heterozygous for two polymorphic STSs. STS-1 exists in two sizes: 234 bp and 198 bp. STS-2 also exists in two sizes: 423 bp and 322 bp. A sample of sperm was collected from this man, and individual sperm were placed into 40 separate tubes. In other words, there was one sperm in each tube. Believe it or not, PCR is sensitive enough to allow analysis of DNA in a single sperm! Into each of the 40 tubes were added the primers that amplify STS-1 and STS-2, and then the samples were subjected to PCR. The following results were obtained.

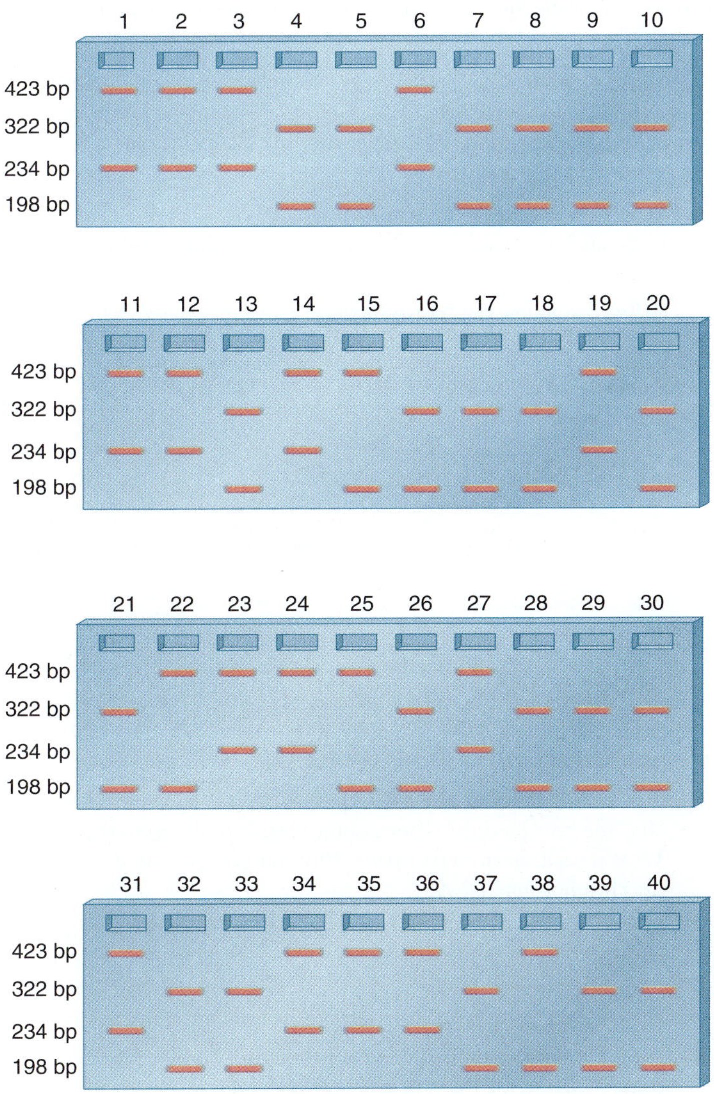

A. What is the arrangement of these two STSs in this individual?

B. What is the map distance between STS-1 and STS-2?

Answer: Keep in mind that mature sperm are haploid, so they have only one copy of STS-1 and one copy of STS-2.

A. If we look at the 40 lanes, most of the lanes (i.e., 36 of them) have either the 234-bp and 423-bp STSs or the 198-bp and 322-bp ones. This is the arrangement of STSs in this male. One chromosome has STS-1 that is 234 bp and STS-2 that is 423 bp, and the homologous chromosome has STS-1 that is 198 bp and STS-2 that is 322 bp.

B. There are four recombinant sperm, shown in lanes 15, 22, 25, and 38.

$$\text{Map distance} = \frac{4}{40} \times 100 = 10.0 \text{ mu}$$

Note: This is a relatively easy experiment compared with a pedigree analysis, which would involve contacting lots of relatives and collecting samples from each of them.

S2. An RFLP marker is located 1 million bp away from a gene of interest. Your goal is to start at this RFLP marker and walk to this gene. The average insert size in the library is 55,000 bp, and the average overlap at each end is 5000 bp. Approximately how many steps will it take to get there?

Answer: Each step is only 50,000 bp (i.e., 55,000 minus 5000), because you have to subtract the overlap between adjacent fragments, which is 5000 bp, from the average inset size. Therefore, it will take about 20 steps to go 1 million bp.

S3. Does a molecular marker have to be polymorphic to be useful in mapping studies? Does a molecular marker have to be polymorphic to be useful in linkage mapping (i.e., involving family pedigree studies or genetic crosses)? Explain why or why not.

Answer: A molecular marker does not have to be polymorphic to be useful in physical or cytological mapping studies. Many sequence-tagged sites (STSs) that are used in physical or cytogenetic studies are monomorphic. Monomorphic markers can provide landmarks in such mapping studies.

In linkage mapping studies, however, a marker must be polymorphic to be useful. Polymorphic molecular markers can be RFLPs, microsatellites, or SNPs. To compute map distances in linkage analysis, individuals must be heterozygous for two or more markers (or genes). For experimental organisms, heterozygotes are testcrossed to homozygotes, and then the number of recombinant offspring and nonrecombinant offspring are determined. For markers that do not assort independently (i.e., linked markers), the map distance is computed as the number of recombinant offspring divided by the total number of offspring times 100.

S4. The technique of chromosome walking involves the stepwise cloning of adjacent DNA fragments until the gene of interest is reached. This method was used to clone the *CFTR* gene that encodes the cystic fibrosis transmembrane regulator protein. When mutant, this causes cystic fibrosis (CF), an autosomal recessive disorder in humans. Linkage mapping studies first indicated that the *CFTR* gene was located between two markers designated MET and D7S8. Later mapping studies put the *CFTR* gene between IRP and D7S8. Walking was initiated from the D7S122 marker.

MET	D7S122	IRP	*CFTR* gene	D7S8

In this study, the researchers restriction mapped this region using two enzymes, *Not*I and *Xho*I. The locations of sites are shown here:

MET	D7S122	IRP	*CFTR* gene	D7S8

```
↑A↑     B ↑C↑  D ↑E ↑F ↑
N X       X  X    X  X   N
```

N = *Not*I, X = *Xho*I

How would the researchers know they were walking toward the *CFTR* gene and not away from it?

From J.M. Rommens, M.C. Iannuzzi, B. Kerem, et al. (1989) Identification of the cystic fibrosis gene: Chromosome walking and jumping. *Science 245,* 1059–1065.

Answer: The general answer is that they had to determine if they were walking toward the IRP marker or toward the MET marker.

As depicted in the previous diagram, digestion of the DNA with *Not*I would produce a large DNA fragment (including regions A–F) that contained both the D7S122 and IRP markers. Because the D7S122 and IRP markers were available as probes, they could confirm that this region contained both markers by Southern blotting of this *Not*I fragment to both markers. *Xho*I cuts this region into smaller pieces. The piece labeled B carried the D7S122 marker, and the piece labeled D carried the IRP marker. Again, this could be confirmed by Southern blotting. This map enabled the researchers to determine in which direction to walk. They began at fragment B (which contained the D7S122 marker) and walked toward the adjacent fragment labeled C. They then walked to the fragment labeled D, which contains the IRP marker. If they had walked toward the fragment labeled A, they would then have walked to the MET marker. This is the incorrect direction, because they knew from their linkage mapping that IRP is closer to the *CFTR* gene than to MET.

Conceptual Questions

C1. A person with a rare genetic disease has a sample of her chromosomes subjected to in situ hybridization using a probe that is known to recognize band p11 on chromosome 7. Even though her chromosomes look cytologically normal, the probe does not bind to this person's chromosomes. How would you explain these results? How would you use this information to positionally clone the gene that is related to this disease?

C2. For each of the following, decide if it could be appropriately described as a genome:

A. The *E. coli* chromosome

B. Human chromosome 11

C. A complete set of 10 chromosomes in corn

D. A copy of the single-stranded RNA packaged into human immunodeficiency virus (HIV)

C3. Which of the following statements are true about molecular markers?

A. All molecular markers are segments of DNA that carry specific genes.

B. A molecular marker is a segment of DNA that is found at a specific location in a genome.

C. We can follow the transmission of a molecular marker by analyzing the phenotype (i.e., the individual's bodily characteristics) of offspring.

D. We can follow the transmission of molecular markers using molecular techniques such as gel electrophoresis.

E. An STS is a molecular marker.

Experimental Questions

E1. Would the following methods be described as linkage, cytogenetic, or physical mapping?

A. Fluorescence in situ hybridization (FISH)

B. Conducting dihybrid crosses to compute map distances

C. Chromosome walking

D. Examination of polytene chromosomes in *Drosophila*

E. Use of RFLPs in crosses

F. Using BACs and cosmids to construct a contig

E2. In an in situ hybridization experiment, what is the relationship between the base sequence of the probe DNA and the site on the chromosomal DNA where the probe binds?

E3. Describe the technique of in situ hybridization. Explain how it can be used to map genes.

E4. The cells from a malignant tumor were subjected to in situ hybridization using a probe that recognizes a unique sequence on chromosome 14. The probe was detected only once in each of these cells. Explain these results and speculate on their significance with regard to the malignant characteristics of these cells.

E5. Figure 22.2 describes the technique of FISH. Why is it necessary to "fix" the cells (and the chromosomes inside of them) to the slides? What does it mean to fix them? Why is it necessary to denature the chromosomal DNA?

E6. Explain how the use of DNA probes with different fluorescence emission wavelengths can be used in a single FISH experiment to map the locations of two or more genes. This method is called chromosome painting. Explain why this is an appropriate term.

E7. A researcher is interested in a gene found on human chromosome 21. Describe the expected results of a FISH experiment using a probe that is complementary to this gene. How many spots would you see if the probe was used on a sample from an individual with 46 chromosomes versus an individual with Down syndrome?

E8. What is a contig? Explain how you would determine that two clones in a contig are overlapping.

E9. Contigs are often made using BAC or cosmid vectors. What are the advantages and disadvantages of these two types of vectors? Which type of contig would you make first, a BAC or cosmid contig? Explain.

E10. Describe the molecular features of a BAC cloning vector. What is the primary advantage of a BAC over a plasmid or viral vector?

E11. A woman has been married to two different men and produced five children. This group is analyzed with regard to three different STSs: STS-1 is 146 and 122 bp; STS-2 is 102 and 88 bp; and STS-3 is 188 and 204 bp. The mother is homozygous for all three STSs: STS-1 = 122, STS-2 = 88, and STS-3 = 188. Father 1 is homozygous for STS-1 = 122 and STS-2 = 102, and heterozygous for STS-3 = 188/204. Father 2 is heterozygous for STS-1 = 122/146, STS-2 = 88/102, and homozygous for STS-3 = 204. The five children have the following results:

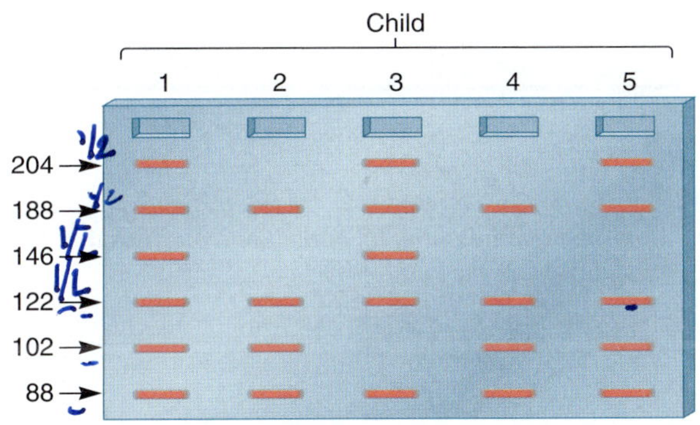

Which children can you definitely assign to father 1 or father 2?

E12. An experimenter used primers to nine different STSs to test their presence along five different BAC clones. The results are shown here.

Alignment of STSs and BACs

	STSs								
	1	2	3	4	5	6	7	8	9
BACs									
1	−	−	−	−	−	+	+	−	+
2	+	−	−	−	+	−	−	+	−
3	−	−	−	+	−	+	−	−	−
4	−	+	+	−	−	−	−	+	−
5	−	−	+	−	−	−	−	−	+

Make a contig map that describes the alignment of the five BACs.

E13. In the Human Genome Project, researchers have collected linkage data from many crosses in which the male was heterozygous for markers and many crosses where the female was heterozygous for markers. The distance between the same two markers, computed in map units or centiMorgans, is different between males and females. In other words, the linkage maps for human males and females are not the same. Propose an explanation for this discrepancy. Do you think the sizes of chromosomes (excluding the Y chromosome) in human males and females are different? How could physical mapping resolve this discrepancy?

E14. Take a look at solved problem S1. Let's suppose a male is heterozygous for two polymorphic sequence-tagged sites. STS-1 exists in two sizes: 211 bp and 289 bp. STS-2 also exists in two sizes: 115 bp and 422 bp. A sample of sperm was collected from this man, and individual sperm were placed into 30 separate tubes. Into each of the 30 tubes were added the primers that amplify STS-1 and STS-2, and then the samples were subjected to PCR. The following results were obtained:

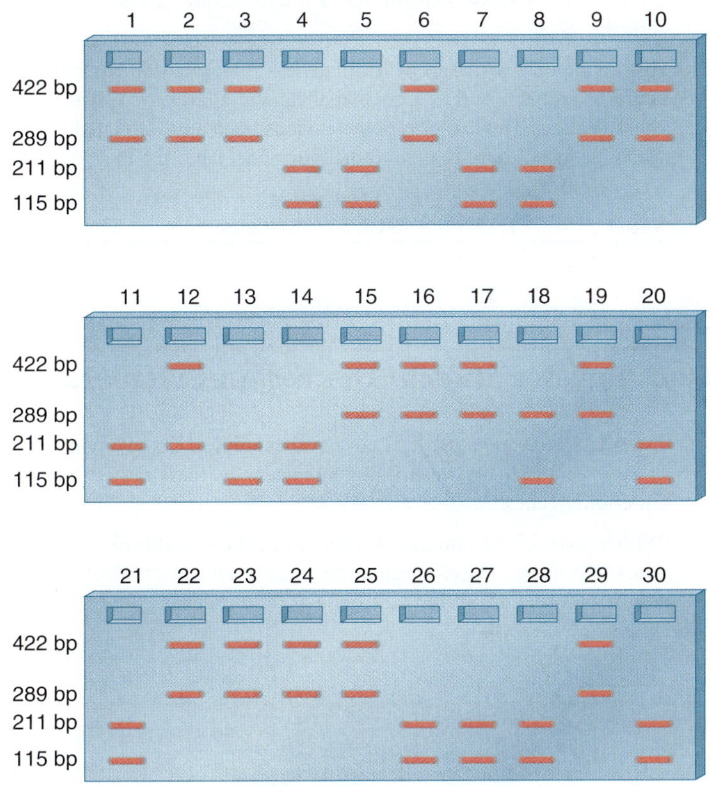

A. What is the arrangement of these STSs in this individual?

B. What is the linkage distance between STS-1 and STS-2?

C. Could this approach of analyzing a population of sperm be applied to RFLPs?

E15. Compared with a conventional plasmid, what additional sequences are required in a YAC vector so it can behave like an artificial chromosome? Describe the importance of each required sequence.

E16. When conducting physical mapping studies, place the following methods in their most logical order:

A. Clone large fragments of DNA to make a BAC library.

B. Determine the DNA sequence of subclones from a cosmid library.

C. Subclone BAC fragments to make a cosmid library.

D. Subclone cosmid fragments for DNA sequencing.

E17. Four cosmid clones, which we will call cosmid A, B, C, and D, were subjected to a Southern blot in pairwise combinations. The insert size of each cosmid was also analyzed. The following results were obtained:

Cosmid	Insert Size (bp)	Hybridized to?
A	6000	C
B	2200	C, D
C	11,500	A, B, D
D	7000	B, C

Draw a map that shows the order of the inserts within these four cosmids.

E18. What is an STS? How are STSs generated experimentally? What are the uses of STSs? Explain how a microsatellite can be a polymorphic STS.

E19. A human gene, which we will call gene *X*, is located on chromosome 11 and is found as a normal allele and a recessive disease-causing allele. The location of gene *X* has been approximated on the map shown here that contains four STSs, labeled STS-1, STS-2, STS-3, and STS-4.

STS-1 STS-2 STS-3 Gene *X* STS-4

A. Explain the general strategy of positional cloning.

B. If you applied the approach of positional cloning to clone gene *X*, where would you begin? As you progressed in your cloning efforts, how would you know if you were walking toward or away from gene *X*?

C. How would you know you had reached gene *X*? (Keep in mind that gene *X* exists as a normal allele and a disease-causing allele.)

E20. Describe how you would clone a gene by positional cloning. Explain how a (previously made) contig would make this task much easier.

E21. A bacterium has a genome size of 4.4 Mb. If a researcher carries out shotgun DNA sequencing and sequences a total of 19 Mb, what is the probability that a base will be left unsequenced? What percentage of the total genome will be left unsequenced?

E22. Discuss the advantages of next-generation sequencing technologies.

E23. What is meant by "sequence by synthesis"?

E24. Outline the general strategy of metagenomics.

Questions for Student Discussion/Collaboration

1. What is a molecular marker? Give two examples. Discuss why it is easier to locate and map many molecular markers rather than functional genes.
2. Which goals of the Human Genome Project do you think are the most important? Why? Discuss the types of ethical problems that might arise as a result of identifying all of our genes.

Note: All answers appear at the website for this textbook; the answers to the even-numbered questions and all of the Concept Check and Comprehension Questions are in the back of the book.

www.mhhe.com/brookergenetics5e

Visit the website for practice tests, answer keys, and other learning aids for this chapter. Enhance your understanding of genetics with our interactive exercises, quizzes, animations, and much more.

CHAPTER OUTLINE

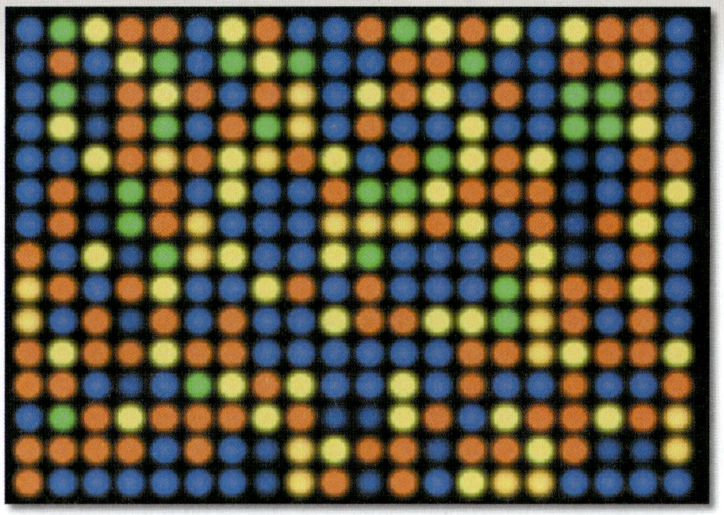

A DNA microarray for measuring gene expression at the genome level. *Each spot on the array corresponds to a specific gene. The color of the spots, which occurs via computer imaging techniques, indicates the amount of RNA transcribed from that gene. A DNA microarray allows researchers to simultaneously analyze the expression of many genes.*

23 GENOMICS II: FUNCTIONAL GENOMICS, PROTEOMICS, AND BIOINFORMATICS

In Chapter 22, we learned that genomics involves the mapping of an entire genome and, eventually, the determination of a species' complete DNA sequence. The amount of information found within a species' genome is enormous. The goal of **functional genomics** is to understand the roles of genetic sequences—DNA, RNA, and amino acid sequences—in a given species. In most cases, functional genomics is aimed at understanding gene function. At the genomic level, researchers can study genes as groups. For example, the information gained from a genome-sequencing project can help researchers study entire metabolic pathways. This provides a description of the ways in which gene products interact to carry out cellular processes. In addition, a study of genetic sequences can help to identify regions that play particular functional roles. For example, an analysis of genome sequences from certain species of bacteria helped to identify particular DNA sequences that promote the uptake of DNA during bacterial transformation.

Because most genes encode proteins, a goal of many molecular biologists is to understand the functional roles of all the proteins a species produces. The entire collection of proteins a given cell or organism makes is called its **proteome,** and the study of the functions and interactions of these proteins is termed **proteomics.** An objective of researchers in the field of proteomics is to understand the interplay among many proteins as they function to create cells and, ultimately, the traits of a given species.

From a research perspective, functional genomics and proteomics can be broadly categorized in two ways: experimental and computational. The experimental approach involves the study of groups of genes or proteins using molecular techniques in the laboratory. In the first two sections in this chapter, we will focus on these techniques. In the last section, we will consider bioinformatics. As a very general definition, **bioinformatics** describes the use of computers, mathematical tools, and statistical techniques to record, store, and analyze biological information. We often think of bioinformatics in the context of examining genetic data, such as DNA sequences. Even so, bioinformatics can also be applied to information from various sources, such as patient statistics and scientific literature. This rapidly developing branch of biology is highly interdisciplinary, incorporating principles from mathematics, statistics, information science, chemistry, and physics. We will consider the field of bioinformatics and discover how this area has provided great insights into the subjects of functional genomics and proteomics.

23.1 FUNCTIONAL GENOMICS

Learning Outcomes:

1. Describe the composition of a DNA microarray and how it is used.
2. Analyze microarray data regarding coordinate gene regulation.
3. Explain how microarrays are used in conjunction with chromatin immunoprecipitation.
4. Define gene knockouts and explain why they are useful.

Though the rapid sequencing of genomes, particularly the human genome, has generated great excitement in the field of genetics, many would argue that an understanding of genome function is fundamentally more interesting. In the past, our ability to study genes involved many of the techniques described in Chapter 20, such as gene cloning, Northern blotting, gel retardation assays, and site-directed mutagenesis. These approaches continue to provide a solid foundation for our understanding of gene function. More recently, genome-sequencing projects have enabled researchers to consider gene function at a more complex level. We now can analyze groups of many genes simultaneously to determine how they work as integrated units to produce the characteristics of cells and the traits of complete organisms. In this section, we examine the use of DNA microarrays, which enable researchers to monitor the expression of thousands of genes simultaneously. We will also discuss why researchers are producing gene knockout collections in which each gene within a given species is inactivated in order to understand gene function at the genomic level.

A DNA Microarray Can Identify Genes That Are Transcribed

Researchers have developed an exciting new technology, called a **DNA microarray** (also called a **gene chip**), that makes it possible to monitor the expression of thousands of genes simultaneously. A DNA microarray is a small silica, glass, or plastic slide that is dotted with many different sequences of DNA, each corresponding to a short sequence within a known gene. For example, one spot in a microarray may correspond to a sequence within the β-globin gene, whereas another could correspond to a gene that encodes actin, which is a cytoskeletal protein. A single slide may contain tens of thousands of different spots in an area the size of a postage stamp. The relative location of each spotted gene is known.

How are microarrays made? Some are produced by spotting different samples of DNA onto a slide, much like the way an ink-jet printer works. For many species, researchers know the entire genome sequence. With this information, they can make primers that flank any given gene and use PCR to synthesize the DNA from a specific gene. The DNA segments from many different gene sequences are then individually spotted onto the slide. Such DNA segments are typically 500–5000 nucleotides in length, and a few thousand to tens of thousands are spotted to make a single array. Alternatively, other microarrays contain shorter DNA segments—oligonucleotides—that are directly synthesized on the surface of the slide. In this case, the DNA sequence at a given spot is produced by selectively controlling the growth of an oligonucleotide using narrow beams of light. Such oligonucleotides are typically 25–30 nucleotides in length. Hundreds of thousands of different spots can be found on a single array. Overall, the technology of making DNA microarrays is quite amazing.

Once a DNA microarray has been made, it is used as a hybridization tool, as shown in **Figure 23.1**. In this experiment, RNA was isolated from a sample of cells and then used to make fluorescently labeled cDNA. The labeled cDNAs were then layered onto a DNA microarray. Those cDNAs that are complementary to the DNAs in the microarray hybridize, thereby remaining bound to the microarray. The array is then washed with a buffer to remove any unbound cDNAs and placed in a microscopic device called a laser scanner, which produces higher-resolution images than a conventional optical microscope. The device scans each pixel—the smallest element in a visual image—and after correction for local background, the final fluorescence intensity for each spot is obtained by averaging across the pixels in each spot. This results in a group of fluorescent spots at defined locations in the microarray. High fluorescence intensity in a particular spot means that a large amount of cDNA in the sample hybridized to the DNA at that location. Because the DNA sequence of each spot is already known, a fluorescent spot identifies cDNAs that are complementary to those DNA sequences. Furthermore, because the cDNA was generated from mRNA, this technique identifies RNAs that have been made in a particular cell type under a given set of conditions.

The technology of DNA microarrays has found many important uses (**Table 23.1**). Thus far, its most common use is for studying gene expression patterns. This helps us understand how genes are regulated in a cell-specific manner and how environmental conditions can induce or repress the transcription of genes. In some cases, microarrays can even help identify which genes encode the proteins that participate in a complicated metabolic pathway. Microarrays can also be used as identification tools. For example, gene expression patterns can aid in the categorization of tumor types. Such identification is used to determine the best course of clinical treatment for a patient. Instead of using labeled cDNA, researchers can also hybridize labeled genomic DNA to a microarray. This can be used to identify mutant alleles in a population and to detect deletions and duplications. In addition, this technology is proving useful in correctly identifying closely related bacterial species and subspecies. Finally, microarrays can be used to study DNA-protein interactions as described later in this section.

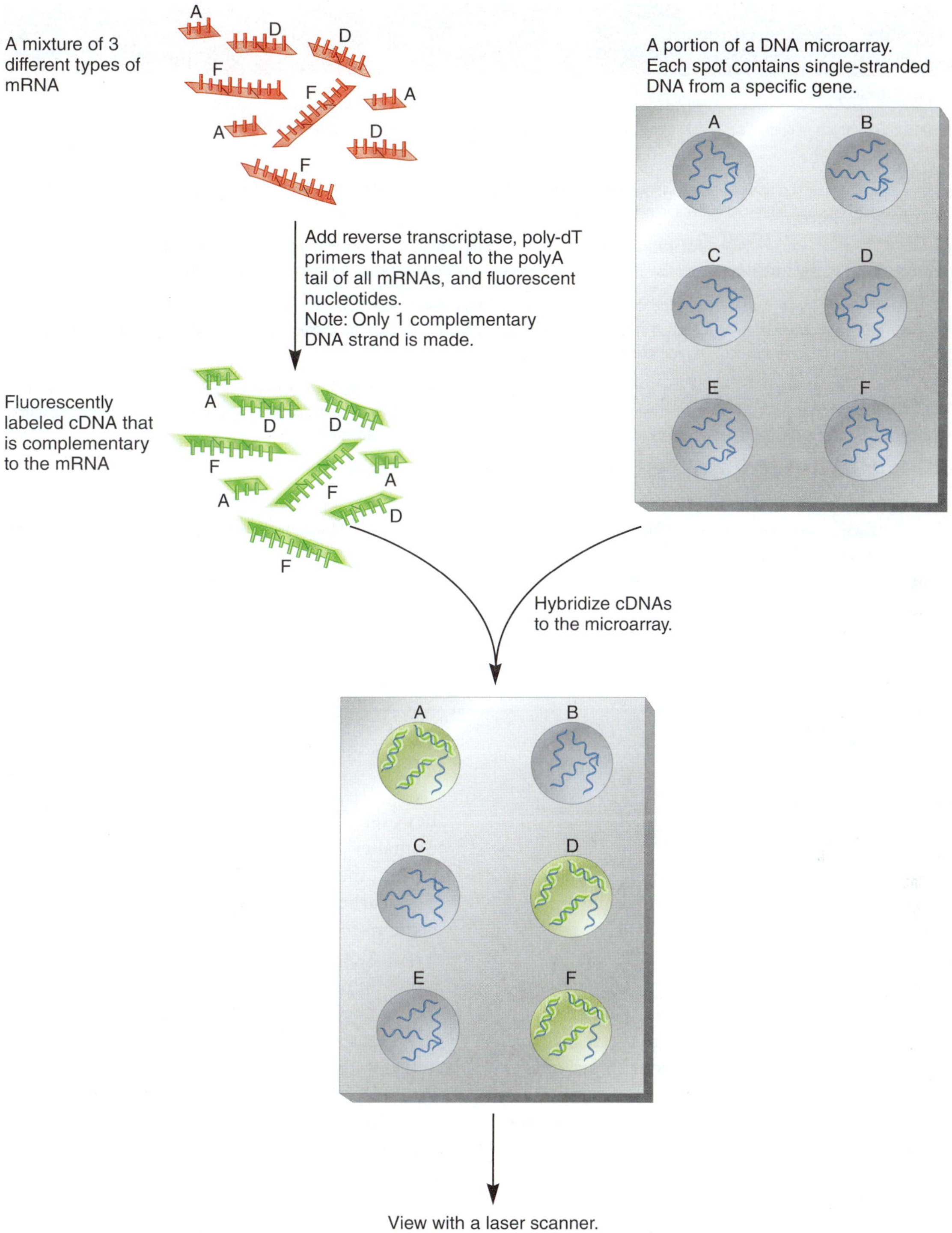

ONLINE ANIMATION

FIGURE 23.1 Using a DNA microarray to study gene expression. A mixture of mRNAs isolated from a sample of cells are used to create cDNAs that are fluorescently labeled. The cDNAs are applied to the microarray, and any unbound cDNAs are washed away. In this simplified example, three cDNAs specifically hybridize to spots on the microarray. In an actual experiment, there are typically hundreds or thousands of different cDNAs and tens of thousands of different spots on the array. After hybridization, some spots become fluorescent and can be visualized using a laser scanner.

CONCEPT CHECK: Explain how this experiment provides information regarding the expression of genes.

TABLE 23.1

Applications of DNA Microarrays

Application	Description
Cell-specific gene expression	A comparison of microarray data using cDNAs derived from RNA of different cell types can identify genes that are expressed in a cell-specific manner.
Gene regulation	Environmental conditions play an important role in gene regulation. A comparison of microarray data may reveal genes that are induced under one set of conditions and repressed under another.
Elucidation of metabolic pathways	Genes that encode proteins that participate in a common metabolic pathway are often expressed in a parallel manner. This can be revealed from a microarray analysis (as described in Figure 23.2). This application overlaps with the study of gene regulation via microarrays.
Tumor profiling	Different types of cancer cells exhibit striking differences in their profiles of gene expression. This can be revealed by a DNA microarray analysis. This approach is gaining widespread use as a method of subclassifying tumors that are sometimes morphologically indistinguishable. Tumor profiling may provide information that can improve a patient's clinical treatment.
Genetic variation	A mutant allele may not hybridize to a spot on a microarray as well as a wild-type allele. Therefore, microarrays are gaining widespread use as a tool for detecting genetic variation. They have been used to identify disease-causing alleles in humans and mutations that contribute to quantitative traits in plants and other species. In addition, microarrays are used to detect chromosomal deletions and duplications.
Microbial strain identification	Microarrays can distinguish between closely related bacterial species and subspecies.
DNA-protein binding	Chromatin immunoprecipitation, which is described in Figure 23.3, can be used with DNA microarrays to determine where in the genome a particular protein binds to the DNA.

EXPERIMENT 23A

The Coordinate Regulation of Many Genes Is Revealed by a DNA Microarray Analysis

One way that cells respond to environmental changes is via the coordinate regulation of genes. Under one set of environmental conditions, a particular set of genes may be induced, whereas under other conditions, those same genes may be repressed. In the past, researchers have been able to study this type of gene regulation using tools that analyze the expression of a few genes at a time. The advent of microarrays, however, has made it possible to study the expression of the whole genome under different sets of environmental conditions. In other words, microarrays can be comparative—researchers can compare gene expression patterns under different sets of conditions. In addition, as described in Chapter 24, microarrays can be used to compare gene expressions in different cell types, such as normal and cancer cells.

One of the first studies using a comparative approach involved the analysis of the yeast genome. The genome of baker's yeast, *Saccharomyces cerevisiae*, was the first eukaryotic genome to be sequenced. It encodes approximately 6300 different genes. An important process in the growth of yeast cells, as well as cells of other species, is the ability to metabolize carbon sources using different metabolic pathways. When yeast cells have glucose available, they preferentially carry out glycolysis, which breaks down glucose to smaller products. When the glucose is used up and if oxygen is present, these smaller products are subsequently broken down via the tricarboxylic acid cycle (TCA, or citric acid, cycle) that occurs in the mitochondria. The process of switching from glycolysis to the TCA cycle, called a diauxic shift, involves major changes in the expression of genes involved with carbohydrate metabolism. The goal of the experiment described in **Figure 23.2** was to identify genes that are induced and repressed as yeast cells shift from glycolysis to the TCA cycle. It was carried out by Joseph DeRisi, Vishwanath Iyer, and Patrick Brown in 1997.

As shown in Figure 23.2, yeast cells were initially given glucose as their carbon source for growth and then allowed to grow for several hours. Over time, the glucose was used up, and the cells shifted from glycolysis to the TCA cycle. At various time points, samples of cells were removed, and the RNA was isolated. The RNA was then exposed to reverse transcriptase, poly-dT primers, and deoxyribonucleotides, one of which was fluorescently labeled. This created fluorescently labeled cDNAs.

To determine the relative changes in RNA synthesis, two different fluorescent dyes were used. The RNA collected at the first time point—when glucose was at a high level—was used to make cDNA that contained a green fluorescent dye. The RNA collected at later time points was used to make cDNAs that contained a red fluorescent dye. A sample of green cDNA (from the first time point) was then mixed with a sample of red cDNA (from later time points), and the mixture was hybridized to a microarray containing about 6200 yeast genes. A laser scanner was used to measure the amount of red fluorescence and green fluorescence at each spot in the microarray. The fluorescence ratio (red fluorescent units divided by green fluorescent units) provided a way to quantitatively determine how the expression of each gene was changing. For example, if the ratio was high, this means that a gene was being induced as glucose levels fell, because the amount of red cDNA would be higher than the amount of green cDNA. Alternatively, if the red-to-green ratio was low, this means that a gene was being repressed as glucose levels fell.

THE GOAL (DISCOVERY-BASED SCIENCE)

A diauxic switch from glycolysis to the TCA cycle involves the induction of certain genes and the repression of other genes. The goal was to identify such groups of genes.

ACHIEVING THE GOAL

FIGURE 23.2 The use of a DNA microarray to study carbohydrate metabolism in yeast.

Starting material: A commonly used strain of *Saccharomyces cerevisiae* (baker's yeast) was used in this study. The researchers had made a DNA microarray containing (nearly) all of the known yeast genes.

Experimental level | **Conceptual level**

1. Inoculate yeast into media containing 2% glucose. Grow for up to 21 hours.

2. Beginning at 9 hours after inoculation, take out samples of cells and isolate mRNA. This involves breaking open cells and running the cell contents over a poly-dT column under high-salt conditions. Since mRNA has a polyA tail, it will bind to the beads in this column while other cell components flow through. The purified mRNA can then be released from the beads by adding a buffer to the top of the column that contains a low concentration of salt. This breaks the interaction between the poly-dT and polyA tails. The mRNA is then collected in a tube. In this experiment, mRNA samples were isolated at approximately 2-hour intervals, beginning at 9 hours and ending at 21 hours.

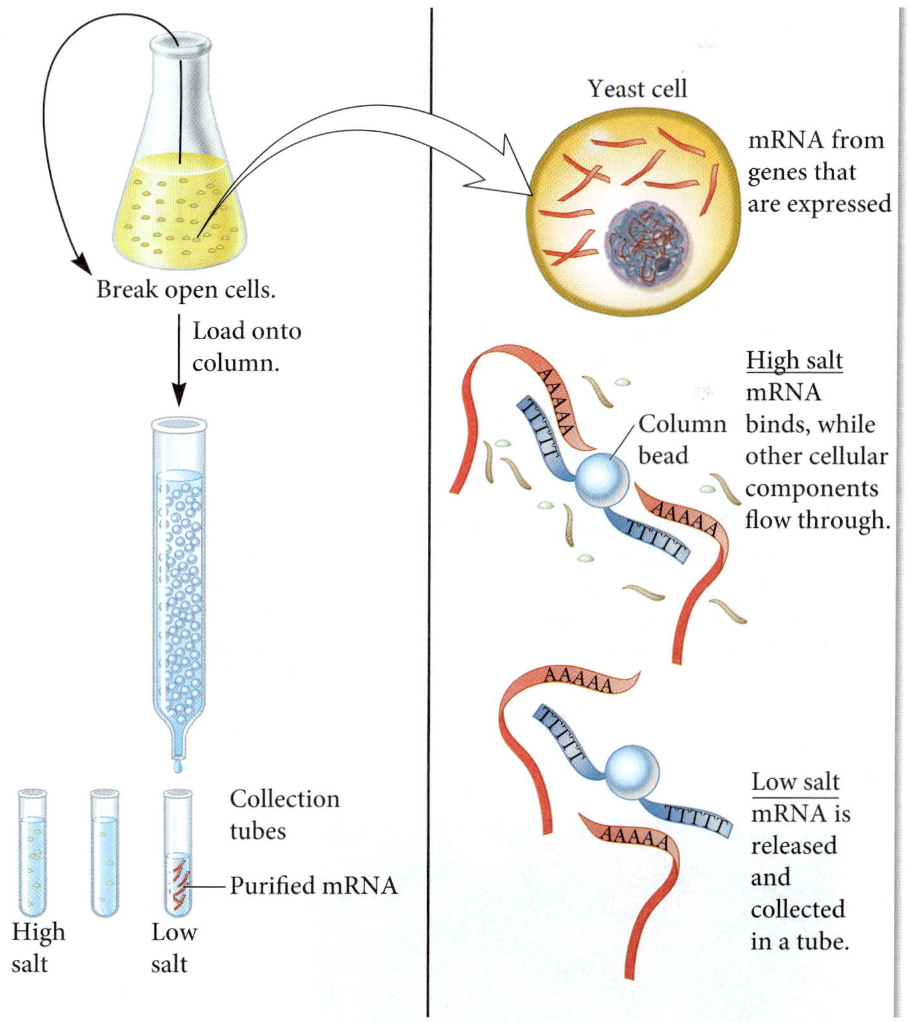

3. Add reverse transcriptase, poly-dT primers, and fluorescently labeled nucleotides to make complementary strands of fluorescently labeled cDNA. Note: The sample at 9 hours was used to make green cDNA, while mRNA samples collected at later time points were used to make red cDNA.

4. At each time point, mix together the cDNA derived from that time (i.e., red cDNA) with cDNA derived from the 9-hour time point (i.e., green cDNA).

5. Hybridize the mixture to the yeast DNA microarray as described in Figure 23.1.

6. Examine the DNA microarray with a laser scanner. The data are then analyzed by a computer, which can correlate expression levels among different genes. (This is a cluster analysis.)

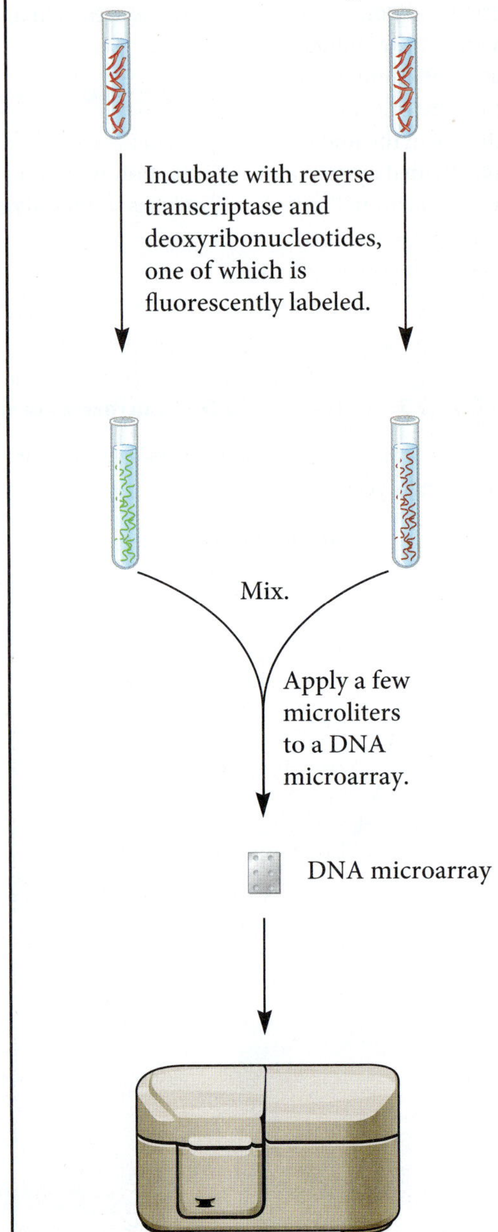

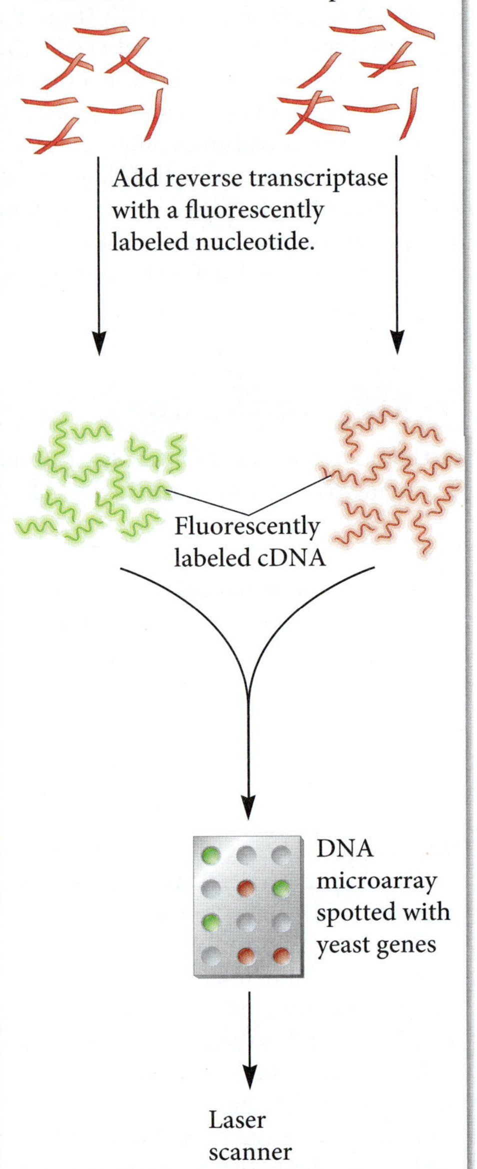

THE DATA

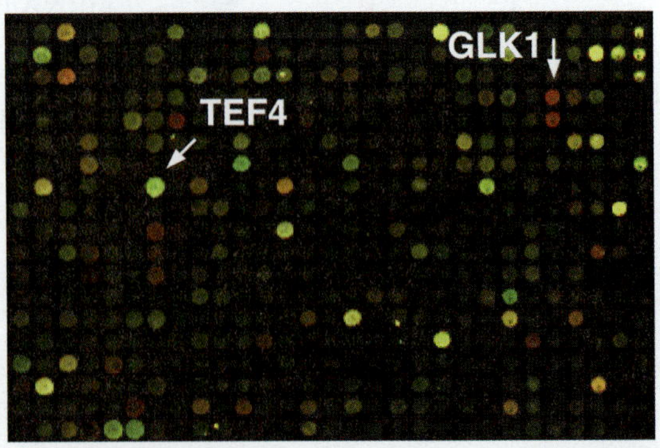

Green spots are genes expressed early in growth, whereas red spots are expressed later. Yellow spots are expressed more evenly. Spots that are barely visible indicate genes that are not substantially expressed under these growth conditions. [Data from J.L. DeRisi, V.R. Iyer, and P.O. Brown (1997) Exploring the metabolic and genetic control of gene expression on a genomic scale. *Science 278,* 680–686.]

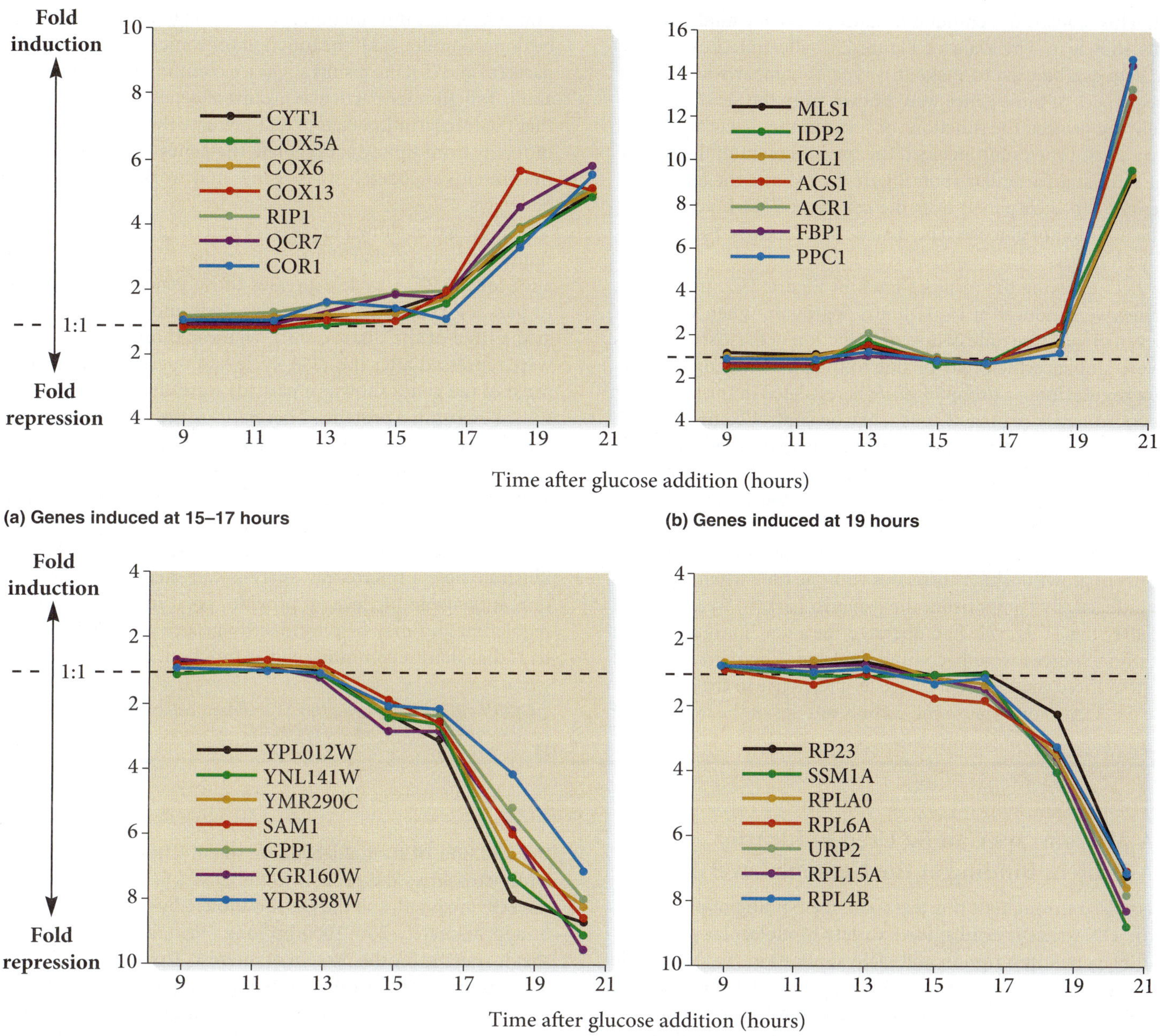

(a) Genes induced at 15–17 hours **(b) Genes induced at 19 hours**

(c) Genes repressed at 15–17 hours **(d) Genes repressed at 19 hours**

INTERPRETING THE DATA

A portion of one microarray is shown at the beginning of The Data of Figure 23.2. As seen here, the array shows many spots, some of which are relatively green and some of which are relatively red. The green spots indicate genes that are expressed at higher levels during early stages of growth when glucose levels are high. An example is a gene designated *TEF4*, which is involved with protein synthesis. Red spots identify genes that are expressed when the glucose is depleted. The gene *GLK1* encodes an enzyme that phosphorylates glucose but is expressed only when glucose levels are low. In addition, many spots are not fluorescent, indicating there was not much cDNA in the sample to hybridize to the DNA strands in those locations. These spots correspond to genes that are not greatly expressed under either condition. Judging from the numbers of red and green spots, the shift from glycolysis to the TCA cycle involved a great amount of gene induction and repression. By determining the red-to-green ratio at each spot, it was found that 710 genes were induced by at least a factor of 2, whereas 1030 genes were repressed by a factor of 2 or more. Therefore, it was found that a diauxic shift involves a staggering amount of gene regulation. Of the 6200 yeast genes analyzed, 1740, or roughly 28%, appeared to be regulated as a result of a diauxic shift.

Because the gene sequence in each spot was known, the next step was to relate the levels of gene expression based on the microarray data to specific genes. A common goal is to identify genes whose pattern of expression seems to strongly correlate with each other. This statistical technique is called a **cluster analysis.** The graphs shown in The Data of Figure 23.2 illustrate how microarray data can be used to make this type of comparison. Though the sequences of these genes were known from the yeast genome-sequencing project, the functions of some of these genes were not known. Each time point involved the measurement of the red-to-green ratio at a particular spot. Figure 23.2 shows the analysis of 28 genes (7 in each panel) over the course of 9–21 hours following the addition of glucose. The diauxic shift occurred at approximately 15 hours.

Let's examine the results shown in Figure 23.2. In part (a), the genes were induced at 15–17 hours of growth. When looking at the seven graphs, the time course of induction of these seven genes was very similar, which may be due to the phenomenon of **coordinate regulation**—a group of genes is regulated in the same way. Such coordinately regulated genes may be controlled by the same transcription factor(s) and may participate in a common metabolic response to the induction of the TCA cycle. In fact, several of the genes in part (a) have already been studied, and they are known to play a role in the ability of the mitochondria to run the TCA cycle. This makes sense because the diauxic shift occurred around 15 hours of growth, which correlates with the time when the genes were induced. By comparison, the genes in part (b) were induced later, after the TCA cycle had operated for a few hours and when the carbon sources were becoming depleted. Therefore, it would seem that these genes were induced as a response to the operation of the TCA cycle or because the carbon sources in the media were low. Why is this information useful? Though further work needs to be done to understand the functions of some of the genes shown in parts (a) and (b), the results suggest a common regulation and metabolic function for particular groups of genes. It seems more likely that the transcription factors that regulate the genes in part (a) are different from those that regulate the genes in part (b). Likewise, the proteins encoded by these genes may work in different cellular pathways.

The genes shown in parts (c) and (d) illustrate a similar phenomenon, except that the switch to the TCA cycle or the operation of the TCA cycle represses those genes. The genes shown in part (c) were active during glycolysis and then were repressed at the time of the diauxic shift, whereas those shown in part (d) were repressed after the TCA cycle had operated for a few hours. Most of the genes shown in part (d) were already known to function as ribosomal proteins. Therefore, as the carbon sources in the media were depleted, these results suggest that one of the cellular responses is to diminish the synthesis of ribosomes, which, in turn, slows down the rate of protein synthesis. It seems that the yeast cells were trying to conserve energy at this late stage of growth.

Overall, the results shown in the experiment of Figure 23.2 illustrate how a microarray analysis can shed light on gene function at the genomic level. It provides great insight regarding gene regulation and may help to identify groups of proteins (i.e., clusters) that share a common cellular function.

A self-help quiz involving this experiment can be found at www.mhhe.com/brookergenetics5e.

DNA Microarrays Can Be Used to Identify DNA-Protein Binding at the Genome Level

As discussed throughout this textbook, the binding of proteins to specific DNA sites is critical for a variety of molecular processes, including gene transcription and DNA replication. To study these processes, researchers have devised a variety of techniques for identifying whether or not specific proteins bind to particular sites in the DNA. For example, the techniques of gel retardation and DNaseI footprinting, which are described in Chapter 20, are used for this purpose (see Figures 20.18 and 20.19).

More recently, a newer approach called **chromatin immunoprecipitation (ChIP)** has gained widespread use in the analysis of DNA-protein interactions. This method can determine whether proteins can bind to a particular region of DNA. A distinguishing feature of this method is that it analyzes DNA-protein interactions as they occur in the chromatin of living cells. In contrast, a gel retardation assay and DNaseI footprinting are in vitro techniques, which typically use cloned DNA and purified proteins.

In Chapter 15, we considered how chromatin immunoprecipitation can be used in conjunction with DNA sequencing to determine the sites in the genome where nucleosomes are found (refer back to Figure 15.11). **Figure 23.3** describes the steps of chromatin immunoprecipitation when it is used in conjunction with or without microarrays. Proteins that are noncovalently bound to DNA can be more tightly attached to the chromatin by the addition of formaldehyde or some other method that covalently crosslinks the protein to the DNA. Following crosslinking, the cells are lysed, and the DNA is broken by sonication into pieces of approximately 200–1000 bp long. Next, an antibody is added that is specific for the protein of interest. To conduct a ChIP assay, a researcher must already suspect that a particular protein binds to the DNA and has previously had an antibody made against that protein. As discussed in Chapter 19, antibodies are molecules that specifically bind to antigens. In this case, the antigen is the protein of interest that is thought to be a DNA-binding protein. In Figure 23.3, the antibodies are attached to heavy beads. The antibodies bind to DNA-protein complexes and cause the complexes to form a pellet following centrifugation. (See the Appendix for a description of centrifugation.) Because an antibody is made by the immune system of an animal, this step is called immunoprecipitation.

The next step is to identify the DNA to which the protein is covalently crosslinked. To do so, the protein is removed by treatment with chemicals that break the covalent crosslinks. Because the DNA is usually present in very low amounts, researchers must amplify the DNA to analyze it. This is done using PCR (described in Chapter 20). If researchers already suspect that a protein binds to a known DNA region, they can use PCR primers that specifically flank the DNA region (see bottom left side of Figure 23.3). If a PCR product is obtained, the protein of interest must have been bound (either directly or through other proteins) to this DNA site in living cells.

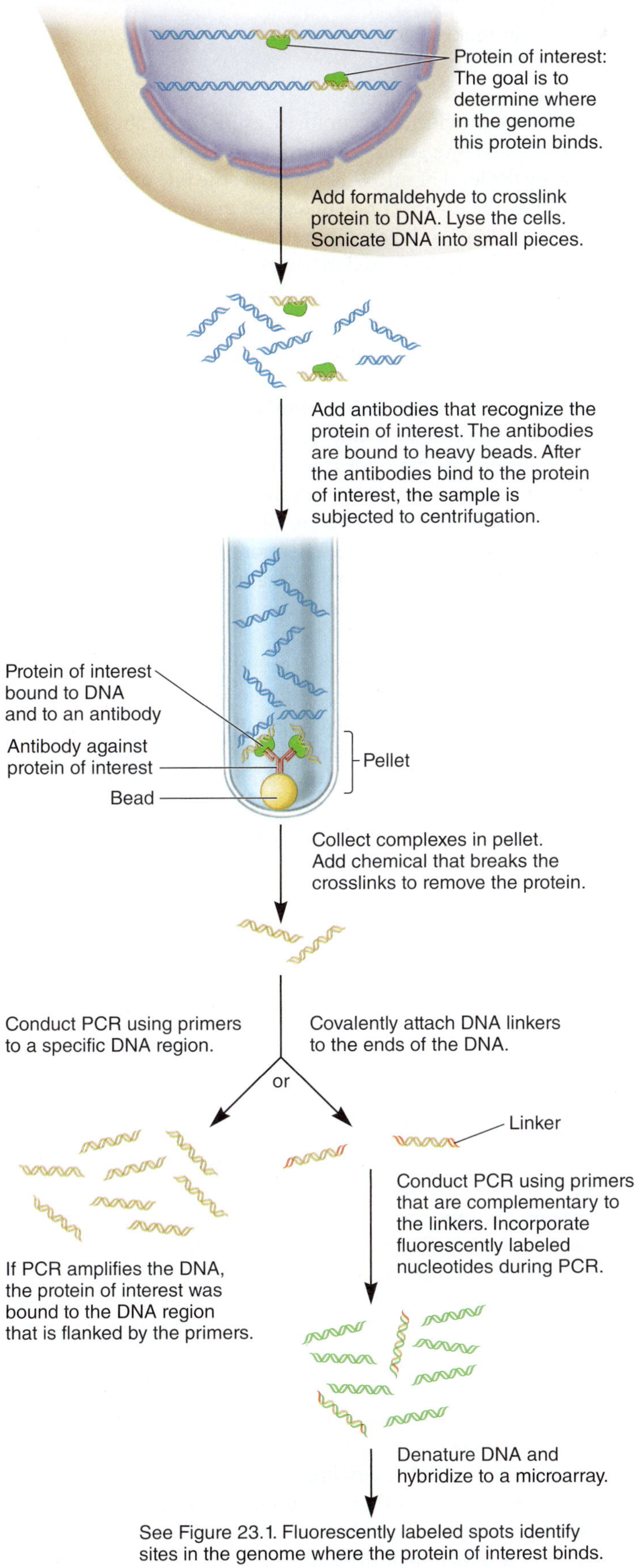

FIGURE 23.3 Chromatin Immunoprecipitation (ChIP) used with and without a DNA microarray. The bottom left shows ChIP without a microarray. The bottom right shows ChIP with a microarray, also called a ChiP-chip assay.

CONCEPT CHECK: Why is an antibody used in this experiment?

Alternatively, a researcher may want to determine where the protein of interest binds across the whole genome. In this case, a DNA microarray can be used (see bottom right side of Figure 23.3). Because a DNA microarray is found on a chip, this is called a **ChIP-chip assay.** The ends of the precipitated DNA are first ligated to short DNA pieces called linkers. PCR primers are then added that are complementary to the linkers and, therefore, amplify the DNA regions between the linkers. During PCR, the DNA is fluorescently labeled. The labeled DNA is then hybridized to a DNA microarray. Because the DNA was isolated using an antibody against the protein of interest, the fluorescent spots on the microarray identify sites in the genome where the protein binds. In this way, researchers may be able to determine where a protein binds to locations in the genome, even if those site(s) had not been previously determined by other methods.

Gene Knockout Collections Allow Researchers to Study Gene Function at the Genomic Level

One broad goal of functional genomics is to determine the functions of all of the genes in a species' genome. Because each species has thousands of different genes, this is a very complicated task. One approach to achieving this goal is to produce collections of organisms of the same species in which each strain has one of its genes knocked out. For example, in *E. coli*, which has 4377 different genes, a complete knockout collection would be composed of 4377 different strains, with a different gene knocked out in each strain. A **gene knockout** means that the gene has been altered in a way that inactivates its function.

Why are knockout collections useful? Consider, for example, the phenotype produced by a particular gene knockout, which causes deafness in mice. Such a result suggests that the function of the normal gene is to promote hearing. Geneticists may also produce knockouts involving two or more genes to understand how the protein products of genes participate in a particular cellular pathway or contribute to a complex trait. In addition, gene knockouts in mice are used in the study of inherited human diseases. For example, as discussed in Chapter 21, gene knockouts have been used to study sickle cell disease.

Knockout collections are made in different ways. One way is via transposable elements. When a transposable element hops into a gene, it often inactivates its function. Another way to produce knockouts is via homologous recombination. In Chapter 21, we examined how mice gene replacement can produce gene knockouts (refer back to Figure 21.5). In 2006, the National Institutes of Health (NIH) launched the Knockout Mouse Project. The goal of this program is to build a comprehensive and publicly available resource comprising a collection of mouse embryonic stem cells (ES cells) containing a loss-of-function mutation in every gene in the mouse genome. The NIH Knockout Mouse Project collaborates with other large-scale efforts to produce mouse knockouts in Canada, called the North American Conditional Mouse Mutagenesis Project (NorCOMM), and in Europe, called the European Conditional Mouse Mutagenesis Program (EUCOMM). The collective goal of these programs is to create at least one loss-of-function mutation in each of the approximately 22,000 genes in the mouse genome. In addition, knockout collections are currently available for other model organisms, including *E. coli*, *Saccharomyces cerevisiae*, and *Caenorhabditis elegans*.

23.1 COMPREHENSION QUESTIONS

1. A DNA microarray is a slide that is dotted with
   a. mRNA from a sample of cells.
   b. fluorescently labeled cDNA.
   c. known sequences of DNA.
   d. known cellular proteins.
2. To study gene expression from a particular sample of cells, what material is usually hybridized to a DNA microarray?
   a. Fluorescently labeled mRNA
   b. Fluorescently labeled cDNA
   c. Radiolabeled mRNA
   d. Radiolabeled cDNA
3. The purpose of a ChIP-chip assay is to determine
   a. the expression levels of particular genes in the genome.
   b. the amount of a specific protein that is made in a given cell type.
   c. the sites in a genome where a particular protein binds.
   d. all of the above.
4. A gene knockout is a gene
   a. whose function has been inactivated.
   b. that has been transferred to a different species.
   c. that has been moved to a new location in the genome.
   d. that has been eliminated from a species during evolution.

23.2 PROTEOMICS

Learning Outcomes:

1. List reasons why the proteome is larger than the genome.
2. Outline the techniques of 2D-gel electrophoresis and tandem mass spectroscopy and explain why they are used.
3. Describe two different types of protein microarrays and discuss their uses.

Thus far, we have considered ways to characterize the genome of a given species and study its function. Because most genes encode proteins, a logical next step is to examine the functional roles of the proteins that a species can make. As mentioned, this field is called proteomics, and the entire collection of a cell's or organism's proteins is its proteome.

Genomics represents only the first step in our comprehensive understanding of protein structure and function. Researchers often use genomic information to initiate proteomic studies, but such information must be followed up with research that involves the direct analysis of proteins. For example, as discussed in Section 23.1, a DNA microarray may provide insights into the transcription of particular genes under a given set of conditions. However, mRNA levels may not provide an accurate measure of the abundance of a protein encoded by a given gene. Protein levels are greatly affected, not only by the level of mRNA, but also by the rate of mRNA translation and by the turnover rate of a given protein. Therefore, DNA microarray data must be corroborated using other methods, such as Western blotting (discussed in Chapter 20), which directly determine the abundance of a protein in a given cell type.

A second way genomic data can explain the workings of the proteome is via homology. As discussed later in this chapter, homology between the genes of different species can be used to predict protein structure and/or function when the structure or function of the protein in one of the species is already known. However, homology may not provide direct information regarding the regulation of protein structure and function. Also, it may not reveal potential types of protein-protein interactions in which a given protein may participate. Therefore, even though homology is profoundly useful as a guiding tool, a full understanding of protein function must involve the direct analysis of proteins as they are found in living cells.

As we move forward in the twenty-first century, the study of proteomes represents a key challenge facing molecular biologists. Much like genomic research, proteomics will require the collective contributions of many research scientists, as well as improvements in technologies that are aimed at unraveling the complexities of the proteome. In this section, we will discuss the phenomena that increase protein diversity beyond genetic diversity. In addition, we will see how the techniques of two-dimensional gel electrophoresis and mass spectrometry can isolate and identify cellular proteins, and explore the use of protein microarrays to study protein expression and function.

The Proteome Is Much Larger than the Genome

From the sequencing and analysis of an entire genome, researchers can identify all or nearly all of the genes from a given species. The entire proteome of a species, however, is larger than the genome, and its actual size is somewhat more difficult to determine. What phenomena account for the larger size of the proteome? First, changes in pre-mRNAs may ultimately affect the resulting amino acid sequence of a protein. The most important alteration that occurs commonly in eukaryotic species is **alternative splicing** (**Figure 23.4a**). For many genes, a single pre-mRNA can be spliced into more than one version. The splicing is often cell-specific, or it may be related to environmental conditions. As discussed in Chapter 16, alternative splicing is widespread, particularly among complex multicellular organisms. It can lead to the production of several or perhaps dozens of different polypeptide sequences from the same pre-mRNA, which greatly increases the number of potential proteins in the proteome. Similarly, the phenomenon of **RNA editing,** a change in the base sequence of RNA after it has been transcribed (see Chapter 12), can lead to changes in the coding sequence of an mRNA. However, RNA editing is much less common than alternative splicing.

Another process that greatly diversifies the composition of a proteome is the phenomenon of **posttranslational covalent modification** (**Figure 23.4b**). Certain types of modifications are involved with the assembly and construction of a functional protein. These alterations include proteolytic processing, disulfide bond formation, and the attachment of prosthetic groups, sugars, or lipids. These are

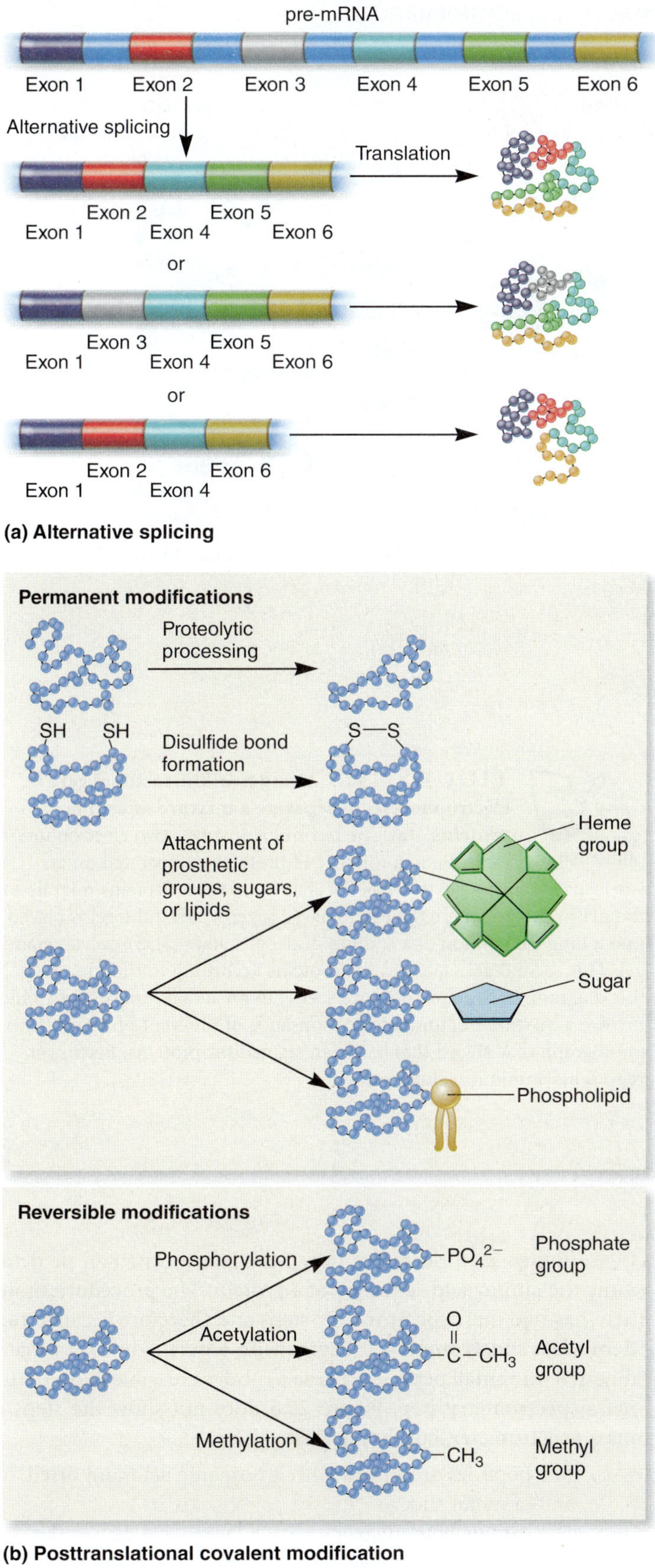

FIGURE 23.4 Cellular mechanisms that increase protein diversity. **(a)** During alternative splicing, the pattern of exons that remains in a mature mRNA can be different, creating multiple types of transcripts from the same gene. **(b)** In posttranslational covalent modification, after a protein is made, it can be modified in a variety of ways, some of which are permanent and some reversible.

CONCEPT CHECK: Explain how these mechanisms affect protein diversity.

typically irreversible changes that are necessary to produce a functional protein. Other types of modifications, such as phosphorylation, acetylation, and methylation, are often reversible modifications that transiently affect the function of a protein. A protein may be subject to several different types of covalent modification, which can greatly increase the number of different forms of a particular protein found in a cell at any given time.

Two-Dimensional Gel Electrophoresis Is Used to Separate a Mixture of Cellular Proteins

As we have just learned, the proteome is usually much larger than a species' genome. Even so, any given cell within a complex multicellular organism produces only a subset of the proteins found in the proteome of a species. For example, the human genome has approximately 22,000 different genes, yet a muscle cell expresses only a subset of those genes at significant levels, perhaps 12,000 or so. The proteins a cell makes depend primarily on the cell type, the stage of development, and the environmental conditions. An objective of researchers in the field of proteomics is the identification and functional characterization of all the proteins a cell type can make. Because cells produce thousands of different proteins, this is a daunting task. Nevertheless, along with genomic research, the past decade has seen important advances in our ability to isolate and identify cellular proteins.

A common technique in the field of proteomics is **two-dimensional (2D) gel electrophoresis.** This technique can separate hundreds or even thousands of different proteins within a cell extract. The steps in this procedure are shown in Figure 23.5. As its name suggests, the technique involves two different gel electrophoresis procedures. A sample of cells is lysed, and the proteins are loaded onto the top of a tube gel that separates proteins according to their net charge at a given pH. A protein migrates to the point in the gel where its net charge is zero. This is termed **isoelectric focusing.** After the tube gel has run, it is laid horizontally on top of a polyacrylamide slab gel that contains sodium dodecyl sulfate (SDS). The SDS coats each protein with negative charges and denatures them. Proteins in the slab gel are separated according to their molecular mass. Smaller proteins move toward the bottom of the gel more quickly than larger ones. After the slab gel has run, the proteins within the gel can be stained with a dye. As seen in Figure 23.5, the end result is a collection of spots, each of which corresponds to a unique cellular protein, with proteins of a larger mass remaining higher in the gel. The resolving power of 2D-gel electrophoresis is extraordinary. Proteins that differ by a single charged amino acid can be resolved as two distinct spots using this method.

Various approaches can be followed to identify spots on a gel that may be of interest to researchers. One possibility is that a given cell type may show a few very large spots that are not found when proteins are analyzed from other cell types. The relative abundance of such spots may indicate that a particular protein is important to that cell's structure or function. Secondly, certain spots on a 2D gel may be seen only under a given set of conditions. For example, a researcher may be interested in the effects of a hormone on the function of a particular cell type. Two-dimensional gel electrophoresis could be conducted on a sample in which the cells had not been exposed to the hormone versus a sample in which they

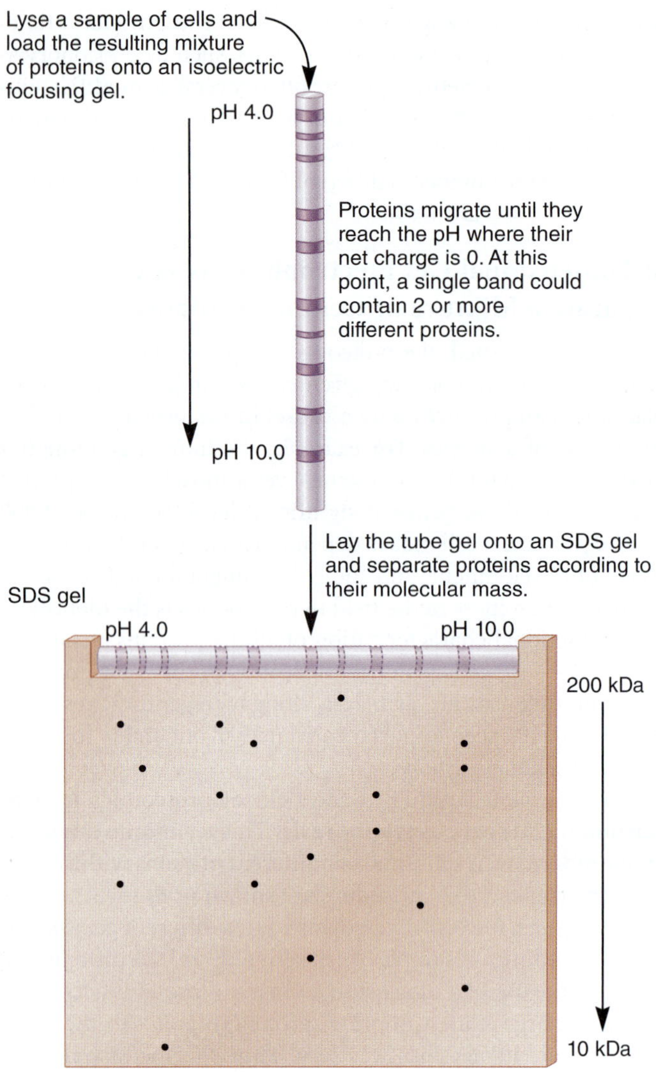

(a) The technique of two-dimensional gel electrophoresis

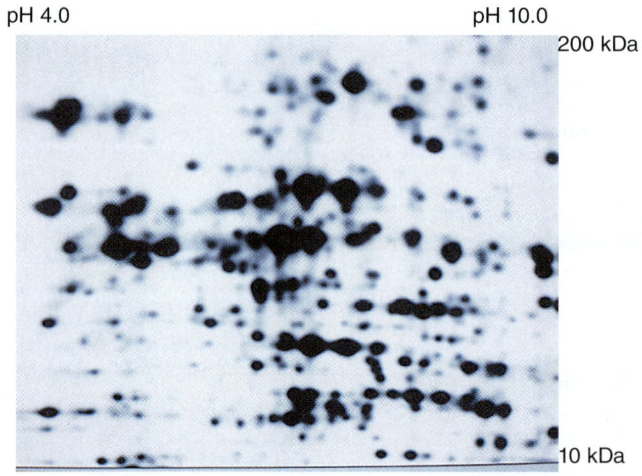

(b) A two-dimensional gel that has been stained to visualize proteins

ONLINE ANIMATION

FIGURE 23.5 Using two-dimensional gel electrophoresis to separate a mixture of cellular proteins. **(a)** The technique involves two electrophoresis steps. First, a mixture of proteins is separated on an isoelectric focusing gel that has the shape of a tube. Proteins migrate to the pH in the gel where their net charge is zero. This tube gel is placed into a long well on top of a sodium dodecyl sulfate (SDS)-polyacrylamide gel. This second gel separates the proteins according to their mass. In this diagram, only a few spots are seen, but an actual experiment would involve a mixture of hundreds or thousands of different proteins. **(b)** A photograph of a 2D gel that has been stained for proteins. Each spot represents a unique cellular protein.

had. This may reveal particular spots that are present only when the cells are exposed to a given hormone. Finally, abnormal cells often express proteins that are not found in normal cells. This is particularly true for cancer cells. A researcher may compare normal and cancer cells via 2D-gel electrophoresis to identify proteins expressed only in cancer cells.

Mass Spectrometry Is Used to Identify Proteins

Two-dimensional gel electrophoresis may be used as the first step in the separation of cellular proteins. The next goal is to correlate a given spot on a 2D gel with a particular protein. To accomplish this goal, a spot on a 2D gel can be cut out of the gel to obtain a tiny amount of the protein within the spot. In essence, the 2D-gel electrophoresis procedure purifies a small amount of the cellular protein of interest. The next step is to identify that protein. This can be accomplished via **mass spectrometry,** a technique that measures the mass of a molecule, such as a peptide fragment.

Figure 23.6 describes how mass spectrometry can determine the amino acid sequence of a protein. The procedure shown here has two mass spectrometry steps and therefore is called **tandem mass spectrometry.** It begins with a purified protein that is digested into small peptides. These peptides are subjected to a first mass spectrometry step. Figure 23.6 does not show the steps in mass spectrometry, but they are listed here:

1. The peptides are mixed with an organic acid and dried onto a metal slide.
2. The sample is then struck with a laser. This causes the peptides to become ejected from the slide in the form of an ionized gas in which the peptide contains one or more positive charges.
3. The charged peptides are then accelerated via an electric field and fly toward a detector. The time of flight is determined by their mass and net charge, which provide an extremely accurate way of determining the mass of a peptide.

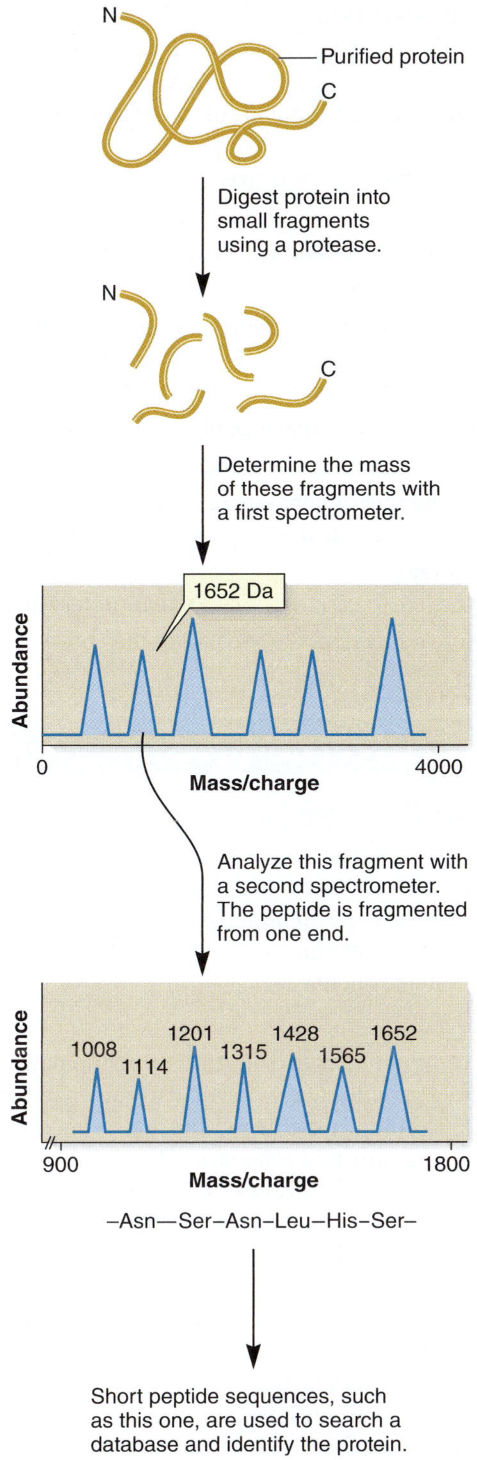

FIGURE 23.6 The use of tandem mass spectrometry to determine the amino acid sequence of a peptide and identify a protein.

CONCEPT CHECK: What is the purpose of tandem mass spectrometry?

In Figure 23.6, the first mass spectrometry step has determined the masses of six different peptides; we will focus on the second peptide that has a mass of 1652 daltons (Da). This 1652-Da peptide is then broken down into many smaller fragments. The mixture of smaller fragments is analzyed by a second mass spectrometry step. The second step reveals the amino acid sequence of the peptide, because the masses of all 20 amino acids are known. For example, as shown at the bottom of Figure 23.6, the starting peptide had a mass of 1652 Da. When one amino acid at the end was removed, the smaller peptide had a mass that was 87 Da less (i.e., 1565 Da); this indicates that a serine is at one end of the peptide, because the mass of serine within a polypeptide chain is 87 Da. When two amino acids were removed at one end, the mass was 224 Da less; this corresponds to the removal of one serine (87 Da) and one histidine (137 Da). When three amino acids were removed and the mass was decreased by 337 Da, this corresponds to the removal of one serine (87 Da), one histidine (137 Da), and one leucine (113 Da). Therefore, from these measurements, we conclude that the amino acid sequence from one end of the 1652-Da peptide is serine-histidine-leucine.

How does this information lead to the identification of a specific protein? As discussed in Chapter 22, the genome sequences of many species have already been determined. This information has allowed researchers to predict the amino acid sequences of most proteins that such species make and enter those sequences into a database. Using computer software described later in this chapter, the amino acid sequences obtained by tandem mass spectrometry can be used as query sequences to search a large database that contains a collection of protein sequences. This program may locate a match between the experimental sequences and a specific protein within a particular species. In this way, tandem mass spectrometry makes it possible to identify the purified protein.

Tandem mass spectrometry can also identify protein covalent modifications. For example, if an amino acid within a peptide was phosphorylated, the mass of the peptide is increased by the mass of a phosphate group. This increase in mass can be detected via tandem mass spectrometry.

Protein Microarrays Are Used to Study Protein Expression and Function

Earlier in this chapter, we learned about DNA microarrays, which have gained widespread use to study gene expression at the RNA level. The technology to make DNA microarrays is also being applied to make **protein microarrays.** In this type of technology, proteins, rather than DNA molecules, are spotted onto a glass or silica slide. The development of protein microarrays is more challenging because proteins are much more easily damaged by the manipulations that occur during microarray formation. For example, the three-dimensional structure of a protein may be severely damaged by drying, which usually occurs during the formation of a microarray. This has created additional challenges for researchers who are developing the technology of protein microarrays. In addition, the synthesis and purification of proteins tend to be more time-consuming than the production of DNA, which can be amplified by PCR or directly synthesized on the microarray itself. In spite of these technical difficulties, the last few years have seen progress in the production and uses of protein microarrays (**Table 23.2**).

The two common types of protein microarray analyses are antibody microarrays and functional protein microarrays. The purpose of an **antibody microarray** is to study protein expression. As discussed in Chapter 19, antibodies are proteins that recognize antigens. One type of antigen that an antibody can recognize is a short peptide sequence found within another protein. Therefore, an antibody can specifically recognize a cellular protein. Researchers

TABLE 23.2
Some Applications of Protein Microarrays

Application	Description
Protein expression	An antibody microarray can measure protein expression because each antibody in a given spot recognizes a specific amino acid sequence. This can be used to study the expression of proteins in a cell-specific manner. It can also be used to determine how environmental conditions affect the levels of particular proteins.
Protein function	The substrate specificity and enzymatic activities of groups of proteins can be analyzed by exposing a functional protein microarray to a variety of substrates.
Protein-protein interactions	The ability of two proteins to interact with each other can be determined by exposing a functional protein microarray to fluorescently labeled proteins.
Pharmacology	The ability of drugs to bind to cellular proteins can be determined by exposing a functional protein microarray to different kinds of labeled drugs. This can help to identify the proteins within a cell to which a given drug may bind.

and commercial laboratories can produce different antibodies; each antibody recognizes a specific peptide sequence. These can be spotted onto a microarray. Cellular proteins can then be isolated, fluorescently labeled, and exposed to the antibody microarray. When a given protein is recognized by an antibody on the microarray, it is captured by the antibody and remains bound to that spot. The level of fluorescence at a given spot indicates the amount of a protein that is recognized by a particular antibody.

The other type of array is a **functional protein microarray.** To make this type of array, researchers must purify cellular proteins and then spot them onto a microarray. The microarray can then be analyzed with regard to specific kinds of protein function. In 2000, for example, Heng Zhu, Michael Snyder, and colleagues purified 119 proteins from yeast that were known to function as protein kinases. These kinds of proteins attach phosphate groups onto other cellular proteins. A microarray was made consisting of different possible proteins that may or may not be phosphorylated by these 119 kinases, and then the array was exposed to each of the kinases in the presence of radiolabeled ATP. By following the incorporation of phosphate into the array, they determined the protein specificity of each kinase. On a much larger scale, the same group of researchers purified 5800 different yeast proteins and spotted them onto a microarray. The array was then exposed to fluorescently labeled calmodulin, which is a regulatory protein that binds calcium ions. Several proteins in the microarray were found to bind calmodulin. Although some of these were already known to be regulated by calmodulin, other proteins in the array that had not been previously known to bind calmodulin were identified.

23.2 COMPREHENSION QUESTIONS

1. Which of the following is a reason why the proteome of a eukaryotic cell is usually much larger than its genome?
   a. Alternative splicing
   b. RNA editing
   c. Posttranslational covalent modifications
   d. All of the above
2. During 2D-gel electrophoresis, proteins are separated based on
   a. their net charge at a given pH.
   b. their mass.
   c. their ability to bind to a specific resin.
   d. both a and b.
3. The technique of tandem mass spectrometry is used to determine
   a. the amino acid sequence of a peptide fragment.
   b. the nucleotide sequence of a segment of RNA.
   c. the nucleotide sequence of a segment of DNA.
   d. the number of genes in a species' genome.
4. Which of the following can be analyzed using a protein microarray?
   a. Protein expression
   b. Protein function
   c. Protein-protein interactions
   d. All of the above

23.3 BIOINFORMATICS

Learning Outcomes:

1. Define database.
2. Outline different strategies to identify gene sequences.
3. Describe the BLAST program and explain why it is used.
4. Explain how genetic sequences are used to predict RNA and protein structures.

Geneticists use computers to collect, store, manipulate, and analyze data. Molecular genetic data are particularly amenable to computer analysis because they come in the form of sequences, such as a DNA, RNA, or amino acid sequence. The ability of computers to analyze data at a rate of millions or even billions of operations per second has made it possible to solve problems concerning genetic information that were thought intractable a few decades ago.

In recent years, the marriage between genetics and computational tools and approaches has yielded an important branch of science known as bioinformatics. In this section, we first consider the fundamental concepts that underlie the analysis of genetic sequences. We then explore how these methods are used to provide insights into functional genomics and proteomics. Chapter 28 will describe applications of bioinformatics in the area of evolutionary biology. In addition, you may wish to actually run computer programs, which are widely available at university and government websites (e.g., see www.ncbi.nlm.nih.gov/Tools). This type of hands-on learning will help you to see how the computer has become a valuable tool to analyze genetic data.

Sequence Files Are Analyzed by Computer Programs

Most people are familiar with **computer programs,** which consist of a defined series of operations that can manipulate and analyze data in a desired way. For example, a computer program might be designed to take a DNA sequence and translate it into an amino acid sequence. A first step in the computer analysis of genetic data is the creation of a **computer data file** to store the data. This file is simply a collection of information in a form suitable for storage and manipulation on a computer. In genetic studies, a computer data file might contain an experimentally obtained DNA, RNA, or amino acid sequence. For example, a file could contain the DNA sequence of one strand of the *lacY* gene from *Escherichia coli*, as shown here. The numbers to the left represent the base number in the sequence file.

```
    1 ATGTACTATT TAAAAAACAC AAACTTTTGG ATGTTCGGTT TATTCTTTTT
   51 CTTTTACTTT TTTATCATGG GAGCCTACTT CCCGTTTTTC CCGATTTGGC
  101 TACATGACAT CAACCATATC AGCAAAGTG ATACGGGTAT TATTTTGCC
  151 GCTATTTCTC TGTTCTCGCT ATTATTCCAA CCGCTGTTTG GTCTGCTTC
  201 TGACAAACTC GGGCTGCGCA AATACCTGCT GTGGATTATT ACGGCATGT
  251 TAGTCATGTT TGCGCCGTTC TTTATTTTTA TCTTCGGGCC ACTGTTACAA
  301 TACAACATTT TAGTAGGATC GATTGTTGGT GGTATTTATC TAGGCTTTTG
  351 TTTTAACGCC GGTGCGCCAG CAGTAGAGGC ATTTATTGAG AAAGTCAGCC
  401 GTCGCAGTAA TTTCGAATTT GGTCGCGCGC GGATGTTTGG CTGTGTTGGC
  451 TGGGCGCTGT GTGCCTGAT TGTCGGCATC ATGTTCACCA TCAATAATCA
  501 GTTTGTTTTC TGGCTGGGCT CTGGCTGTGC ACTCATCCTC GCCGTTTTAC
  551 TCTTTTTCGC CAAAACGGAT GCGCCCTCTT CTGCCACGGT TGCCAATGCG
  601 GTAGGTGCCA ACCATTCGGC ATTTAGCCTT AAGTGGCAC TGGAACTGTT
  651 CAGACAGCCA AAACTGTGGT TTTGTCTACT GTATGTTATT GGCGTTTCCT
  701 GCACCTACGA TGTTTTTGAC CAACAGTTTG CTAATTTCTT TACTTCGTTC
  751 TTTGCTACCG GTGAACAGGG TACGCGGGTA TTTGGCTACG TAACGACAAT
  801 GGGCGCAATTA CTTAACGCCT CGATTATCTT CTTGCGCCA CTGATCATTA
  851 ATCGCATCGG TGGGAAAAAC GCCCTGCTGC TGGCTGGCAC TATTATCTCT
  901 CTACGTATTA TTGGCTCATC GTTCGCCACC TCAGCGCTGG AAGTGGTTAT
  951 TCTGAAAACG CTGCATATGT TTGAAGTACC GTTCCTGCTG GTGGGCTGCT
1,001 TTAAATATAT TACCAGCCAG TTTGAAGTGC GTTTTTCAGC GACGATTTAT
1,051 CTGGTCTGTT TCTGCTTCTT TAAGCAACTG GCATGATTT TTATGTCTGT
1,101 ACTGGCGGGC AATATGTATG AAAGCATCGG TTTCCAGGGC GCTTATCTGG
1,151 TGCTGGGTCT GGTGGCGCTG GGCTTCACCT TAATTTCCGT GTTCACGCTT
1,201 AGCGGCCCCG GCCCGCTTTC CCTGCTGCGT CGTCAGGTGA ATGAAGTCGC
1,251 TTAA
```

To store data in a computer data file, a scientist creates the file and enters the data, either by hand or, as is now more common, by laboratory instruments such as densitometers and fluorometers. These instruments have the capability to read data, such as a sequencing ladder, and enter the DNA sequence information directly into a computer file.

The purpose of making a computer file that contains a genetic sequence is to take advantage of the swift speed with which computers can analyze this information. Genetic sequence data in a computer file can be investigated in many different ways, corresponding to the many questions a researcher might ask about the sequence and its functional significance. These include the following:

1. Does a sequence contain a gene?
2. Where are functional sequences such as promoters, regulatory sites, and splice sites located within a particular gene?
3. Does a sequence encode a polypeptide? If so, what is the amino acid sequence of the polypeptide?
4. Does a sequence predict certain structural features for RNA or proteins? For example, what is the secondary structure of an RNA sequence or polypeptide sequence?
5. Is a sequence homologous to any other known sequences?
6. What is the evolutionary relationship between two or more genetic sequences?

To answer these and many other questions, different computer programs have been written to analyze genetic sequences in particular ways. As an example, let's consider a computer program aimed at translating a DNA sequence into an amino acid sequence and consider how it might work in practice. The geneticist—the user—has a DNA sequence file that she or he wants to have translated into an amino acid sequence. The user is sitting at a computer that is connected to a program that can translate a DNA sequence into an amino acid sequence. The program is opened and the user provides the program with a DNA sequence file to be translated (see the sequence of the *lacY* gene to the left). The user also specifies which portion of the sequence should be translated. In this case, the user decides to begin the translation at the first nucleotide in the sequence file and end the translation at nucleotide number 1254. The user wants the program to translate the sequence in all three forward reading frames and to show the longest reading frame—the longest amino acid sequence that is uninterrupted by a stop codon. This translated sequence is saved in a file. The contents of the file are shown here.

```
  1 MYYLKNTNFW MFGLFFFFYF FIMGAYFPFF PIWLHDINHI SKSDTGIIFA
 51 AISLFSLLFQ PLFGLLSDKL GLRKYLLWII TGMLVMFAPF FIFIFGPLLQ
101 YNILVGSIVG GIYLGFCFNA GAPAVEAFIE KVSRRSNFEF GRARMFGCVG
151 WALCASIVGI MFTINNQFVF WLGSGCALIL AVLLFFAKTD APSSATVANA
201 VGANHSAFSL KLALELFRQP KLWFLSLYVI GVSCTYDVFD QQFANFFTSF
251 FATGEQGTRV FGYVTTMGEL LNASIMFFAP LIINRIGGKN ALLLAGTIMS
301 VRIIGSSFAT SALEVVIKLT LHMFEVPFLL VGCFKYITSQ FEVRFSATIY
351 LVCFCFFKQL AMIFMSVLAG NMYESIGFQG AYLVLGLVAL GFTLISVFTL
401 SGPGPLSLLR RQVNEVA
```

In this file, which was created by a computer program, each of the 20 amino acids is given a single-letter abbreviation (see Figure 13.5).

Why is such a program useful? The advantages of running this program are speed and accuracy. It can translate a relatively long genetic sequence within milliseconds. By comparison, it would probably take you a few hours to look up each codon in the genetic code table and write out the sequence in the correct order. If you visit a website and actually run a program like this, you will discover that such a program can translate a genetic sequence into six reading frames—three forward and three reverse. This is useful if a researcher does not know where the start codon is located and/or does not know the direction of the coding sequence.

In genetic research, large software packages typically contain many computer programs that can analyze genetic sequences in different ways. For example, one program can translate a DNA sequence into an amino acid sequence, and another program can locate introns within genes. These software packages are found at universities, government facilities, hospitals, and industries. At such locations, a central computer with substantial memory and high-speed computational abilities runs the software, and individuals can connect to this central computer via their personal computer (e.g., laptop). Many such programs are freely available on the Internet. As mentioned, www.ncbi.nlm.nih.gov/Tools has a variety of useful programs.

The Scientific Community Has Collected Sequence Files and Stored Them in Large Computer Databases

In Chapter 20, we considered how researchers can clone and sequence genes. Likewise, in Chapter 22, we learned how scientists are investigating the genetic sequences of entire genomes from several species, including humans. An enormous amount of genetic information has been generated by researchers. For example, at its completion, the Human Genome Project had produced more data than any other undertaking in the history of biology. With these advances, scientists realize that another critical use of computers is simply to store the staggering amount of data produced from genetic research.

A large number of computer data files collected and stored in a single location is called a **database.** In addition to genetic sequences, the files within databases are **annotated,** which means they contain additional information such as a concise description of a sequence, the name of the organism from which this sequence was obtained, and the function of the encoded protein, if it is known. The file may also describe other features of significance and cite a published scientific journal reference that describes the sequence.

The scientific community has collected the genetic information from thousands of research labs and created several large databases. **Table 23.3** describes some of the major genetic databases in use worldwide. These databases enable researchers to access and compare genetic sequences obtained by many laboratories. Later, we will learn how researchers use databases to analyze genetic sequences.

The databases described in Table 23.3 collect genetic information from many different species. Scientists have also created more specialized databases, called **genome databases,** that focus on the genetic sequences and characteristics of a single species. Genome databases have been created for species of bacteria (e.g., *E. coli*), yeast (e.g., *Saccharomyces cerevisiae*), worms (e.g., *Caenorhabditis elegans*), fruit flies (e.g., *Drosophila melanogaster*), plants (e.g., *Arabidopsis thaliana*), and mammals (e.g., mice and humans). The primary aim of genome databases is to organize the information from sequencing and mapping projects for a single species. Genome databases identify an organism's known genes and describe their map location within the genome. In addition, a genome database may provide information concerning gene alleles, bibliographic information, a directory of researchers who study the species, and other pertinent information.

TABLE 23.3

Examples of Major Computer Databases

Type	Description
Nucleotide sequence	DNA sequence data are collected into three internationally collaborating databases: GenBank (a U.S. database), EMBL (European Molecular Biology Laboratory Nucleotide Sequence Database), and DDBJ (DNA Databank of Japan). These databases receive sequence and sequence annotation data from genome projects, sequencing centers, individual scientists, and patent offices. These databases are accessed via the Internet.
Amino acid sequence	Amino acid sequence data are collected into a few international databases including Swissprot (Swiss protein database), PIR (Protein Information Resource), Genpept (translated peptide sequences from the GenBank database), and TrEMBL (Translated sequences from the EMBL database).
Three-dimensional structure	PDB (Protein Data Bank) collects the three-dimensional structures of biological macromolecules with an emphasis on protein structure. These are primarily structures that have been determined by X-ray crystallography and nuclear magnetic resonance (NMR). These structures are stored in files that can be viewed on a computer with the appropriate software.
Protein motifs	Prosite is a database containing a collection of amino acid sequence motifs that are characteristic of a protein family, domain structure, or certain posttranslational modifications. Pfam is a database of protein families with multiple amino acid sequence alignments.
Gene expression data	Gene Expression Omnibus (GEO) contains data regarding the expression patterns of genes within a data set, such as the data obtained from microarrays or ChIP-chip assays.

Different Computational Strategies Can Identify Functional Genetic Sequences

At the molecular level, the function of the genetic material is based largely on specific genetic sequences that play distinct roles. For example, codons are three-base sequences that specify particular amino acids, and promoters are sequences that provide a binding site for RNA polymerase to initiate transcription. Computer programs can be designed to scan very long sequences, such as those obtained from genome-sequencing projects, and locate meaningful features within them. To illustrate this concept, let's first consider the following sequence file, which contains an alphabetic sequence of 54 letters:

Sequence file:

GJTRLLAMAQLHEOGYLTOBWENTMNMTORXXXTGOODNTHEQ
ALLRTLSTORE

Now let's compare how three different computer programs might analyze this sequence to identify meaningful features. The goal of the first program is to locate all of the English words within this sequence. If we ran this program, we would obtain the following result:

GJTRLLAMAQLHEOGYLTOBWENTMNMTORXXXTGOODNTHEQ
ALLRTLSTORE

In this case, a computer program has identified locations where the sequence of letters forms a word. Several words (which are underlined) have been located within this sequence.

A second computer program could be aimed at locating a series of words that are organized in the correct order to form a grammatically logical English sentence. If we used our sequence file and ran this program, we would obtain the following result:

GJTRLLAMAQLHEOGYLTOBWENTMNMTORXXXTGOODNTHEQ
ALLRTLSTORE

The second program has identified five words that form a logical sentence.

Finally, a computer program might be used to identify patterns of letters, rather than words. For example, a computer program could locate a pattern of five letters that occurs in both the forward and reverse directions. If we applied this program to our sequence file, we would obtain the following:

GJTRLLAMAQLHEOGYLTOBWENTMNMTORXXXTGOODNTHEQ
ALLRTLSTORE

In this case, the program has identified a pattern where five letters are found in both the forward and reverse directions.

In the three previous examples, we can distinguish between **sequence recognition** (as in our first example) and **pattern recognition** (as in our third example). In sequence recognition, the program has the information that a specific sequence of symbols has a specialized meaning. This information must be supplied to the computer program. For example, the first program has access to the information from a dictionary with all known English words. With this information, the first program can identify sequences of letters that make words. By comparison, the third program does not rely on specialized sequence information. Rather, it is looking for a pattern of symbols that can occur within any group of symbol arrangements.

Overall, the simple programs we have considered illustrate three general types of identification strategies:

1. ***Locate specialized sequences within a very long sequence.*** A specialized sequence with a particular meaning or function is called a **sequence element** or **motif.** As in the first example, the computer program has a list of predefined sequence elements and can identify such elements within a sequence of interest.
2. ***Locate an organization of sequences.*** As shown in the second program, this could be an organization of sequence elements. Alternatively, it could be an organization of a pattern of symbols.
3. ***Locate a pattern of symbols.*** The third program is an example of locating a pattern of symbols.

The great power of computer analysis is that these types of operations can be performed with great speed and accuracy on sequences that may be enormously long.

Now that we understand the general ways that computer programs identify sequences, let's consider specific examples. As we have discussed throughout this textbook, many short nucleotide sequences play specialized roles in the structure or function of genetic material. A geneticist may want to locate a short sequence element within a longer nucleotide sequence in a data file. For example, a sequence of chromosomal DNA might be tens of thousands of nucleotides in length, and a geneticist may want to know whether a sequence element, such as a TATA box, is found at one or more sites within the chromosomal DNA. To do so, a researcher could visually examine the long chromosomal DNA sequence in search of a TATA box. Of course, this process would be tedious and prone to error. By comparison, the appropriate computer program can locate a sequence element in seconds. Therefore, computers are very useful for this type of application. Table 23.4 lists some examples of sequence elements that can be identified by computer analysis.

By comparing the amino acid sequences and known functions of many different proteins, researchers have also found amino acid sequence motifs that carry out specialized functions within proteins. For example, researchers have determined that the amino acid sequence motif asparagine–X–serine (where X is any amino acid except proline) within eukaryotic proteins is a glycosylation site (i.e., it may have a carbohydrate attached to it). The Prosite database (refer back to Table 23.3) contains a collection of all amino acid sequence motifs known to be functionally important. Researchers can use computer programs to determine whether an amino acid sequence contains any of the motifs found in the Prosite database. This may help them to understand the function of a newly discovered protein.

TABLE 23.4

Short Sequence Elements That Can Be Identified by Computer Analysis

Type of Sequence	Examples*
Promoter	Many *E. coli* promoters contain TTGACA (−35 site) and TATAAT (−10 site). Eukaryotic core promoters may contain CAAT boxes, GC boxes, TATA boxes, etc.
Response elements	Glucocorticoid response element (AGRACA), cAMP response element (GTGACGTRA)
Start codon	ATG
Stop codons	TAA, TAG, TGA
Splice site	GTRAGT————YNYTRAC(Y)$_n$AG
Polyadenylation signal	AATAAA
Highly repetitive sequences	Relatively short sequences that are repeated many times throughout a genome
Transposable elements	Often characterized by a pattern in which direct repeats flank inverted repeats

*The sequences shown in this table would be found in the DNA. For gene sequences, only the coding strand is shown. R = purine (A or G); Y = pyrimidine (T or C); N = A, T, G, or C; U in RNA = T in DNA.

Several Computer-Based Approaches Can Identify Genes Within a Nucleotide Sequence

A typical gene that encodes a protein contains a promoter, followed by a start codon, a coding sequence, a stop codon, and a transcriptional termination site. In addition, most genes contain regulatory sequences (e.g., eukaryotic response elements or prokaryotic operator sites), and eukaryotic genes are likely to contain introns. After researchers have sequenced a long segment of chromosomal DNA, they frequently want to know if the sequence contains any genes. In an attempt to answer this question, geneticists can use computer programs that are aimed at identifying genes in long genomic DNA sequences.

How do computer programs identify a gene within a long DNA sequence? Such programs may employ several different strategies. A **search-by-signal** approach relies on known sequences such as promoters, start and stop codons, and splice sites to help predict whether or not a DNA sequence contains a protein-encoding gene. The program tries to locate a region that contains a promoter sequence, followed by a start codon, a coding sequence, a stop codon, and a transcriptional terminator.

A second strategy is a **search-by-content** approach. The goal here is to identify sequences with a nucleotide content that differs significantly from a random distribution. Within protein-coding genes, this occurs primarily due to preferential codon usage.

Although there are 64 codons, most organisms display a **codon bias** within the coding regions of genes. This means that certain codons are used much more frequently than others. For example, UUA, UUG, CUU, CUC, CUA, and CUG all specify leucine. In yeast, however, 80% of the leucine codons are UUG. Codon bias allows organisms to more efficiently rely on a smaller population of tRNA molecules. A search-by-content strategy, therefore, attempts to locate coding sequences by identifying regions where the nucleotide content displays a known codon bias.

A third way to locate genes within a DNA sequence is to examine translational reading frames. Recall that the reading frame is a sequence of codons determined by reading bases in groups of three. In a new DNA sequence, researchers must consider that the reading of codons (in groups of three nucleotides) could begin with the first nucleotide (reading frame 1), the second nucleotide (reading frame 2), or the third nucleotide (reading frame 3). An **open reading frame (ORF)** is a region of a nucleotide sequence that does not contain any stop codons. Because most proteins are several hundred amino acids in length, a relatively long reading frame is required to encode them. In bacteria, such long ORFs are contained within genes. In **Figure 23.7**, a bacterial DNA sequence has been translated in all three reading frames. Only one of the three reading frames (3) contains a very long ORF without any stop codons, suggesting that this DNA sequence encodes a protein. In eukaryotic genes, the approach of identifying a long ORF may not always be successful, because the ORFs may be interrupted by multiple introns, which makes the ORFs relatively short.

Even though computer programs are a valuable tool, they do not always accurately predict gene sequences. In particular, it is often difficult for programs to predict the correct start codon and the precise intron-exon boundaries. In some cases, computer programs may even suggest that a region encodes a gene when it does not. Therefore, although a bioinformatics approach is a relatively easy tool for identifying potential genes, it should not be viewed as a definitive method. The confirmation that a DNA region encodes an actual gene requires laboratory experimentation to show that it is truly transcribed into RNA.

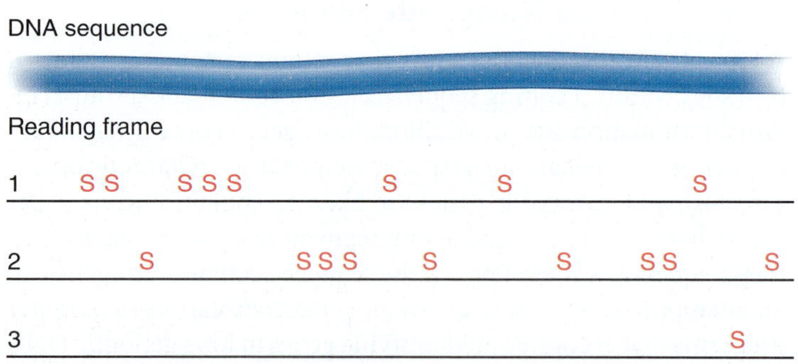

FIGURE 23.7 Translation of a bacterial DNA sequence in all three reading frames. The three lines represent the translation of a gene sequence in each of three forward reading frames; the reading frames proceed from left to right. The letter S indicates the location of a stop codon. Reading frame 3 has a very long open reading frame, suggesting that it may be the reading frame for a protein-encoding gene. Reading frames 1 and 2 are not likely to be the reading frames for a protein-encoding gene, because they contain many stop codons. During the cloning of DNA, the orientation of a gene may become flipped so that the coding sequence is inverted. Therefore, when analyzing many cloned DNA fragments, six reading frames (i.e., three forward and three reverse) are evaluated. Only the three forward frames are shown here.

Homologous Genes Are Derived from the Same Ancestral Gene

Let's now turn our attention to the uses of computer technology to identify genes that are evolutionarily related. The ability to sequence DNA allows geneticists to examine evolutionary relationships at the molecular level. This has become an extremely powerful tool in the field of genomics. When comparing genetic sequences, researchers frequently find two or more similar sequences. For example, the sequence of the *lacY* gene that encodes the lactose permease in *E. coli* is similar to that of the *lacY* gene that encodes the lactose permease in another bacterium, *Klebsiella pneumoniae*. When segments of the two *lacY* genes are lined up, approximately 78% of their bases are a perfect match (**Figure 23.8a**).

In this case, the two sequences are similar because the genes are **homologous,** meaning they have been derived from the same ancestral gene. This idea is shown schematically in **Figure 23.8b**. An ancestral *lacY* gene was located in a bacterium that preceded the evolutionary divergence between *E. coli* and *K. pneumoniae*. After these two species diverged from each other, their *lacY* genes accumulated distinct mutations that produced somewhat different base sequences for this gene. Therefore, in these two species of bacteria, the *lacY* genes are similar but not identical. When two homologous genes are found in different species and they continue to serve the same function, they are termed **orthologs.**

Two or more homologous genes can also be found within a single species. As discussed in Chapter 8, abnormal gene duplications may happen several times during evolution, which results in multiple copies of a gene. These multiple copies of a gene within a single species are called **paralogs.** A **gene family** consists of two or more paralogs within the genome of a single species. During evolution, the functions of paralogs may change. For example, the globin genes have become more specialized to function at different stages of mammalian development (see Chapter 8, Figure 8.7).

When a gene family occurs, the concept of orthologs becomes more complex. Again, let's consider the globin gene family found in mammals. Researchers would say that the β-globin gene in humans is an ortholog to the β-globin gene found in mice. Likewise, α-globin genes found in both species are considered orthologs. However, the α-globin gene in humans is not called an ortholog of the β-globin gene in mice, though they could be called homologous. The most closely related genes in two different species are considered orthologs.

A Simple Dot Matrix Can Compare the Degree of Similarity Between Two Sequences

A computer program can evaluate the similarity between two sequences by constructing a matrix. In a general way, **Figure 23.9** illustrates the use of a simple dot matrix. As shown in Figure 23.9a, the sequence GENETICSISCOOL is compared with itself. Each

```
              151                                                  200
E. coli       TCTTTTTCTT TTACTTTTTT ATCATGGGAG CCTACTTCCC GTTTTTCCCG
K. pneumoniae TCTTTTTCTT TTACTATTTC ATTATGTCAG CCTACTTTCC TTTTTTTCCG

              201                                                  250
              ATTTGGCTAC ATGACATCAA CCATATCAGC AAAAGTGATA CGGGTATTAT
              GTGTGGCTGG CGGAAGTTAA CCATTTAACC AAAACCGAGA CGGGTATTAT
```

(a) A comparison of DNA sequences in a portion of the *lacY* gene from *E. coli* and *K. pneumoniae*

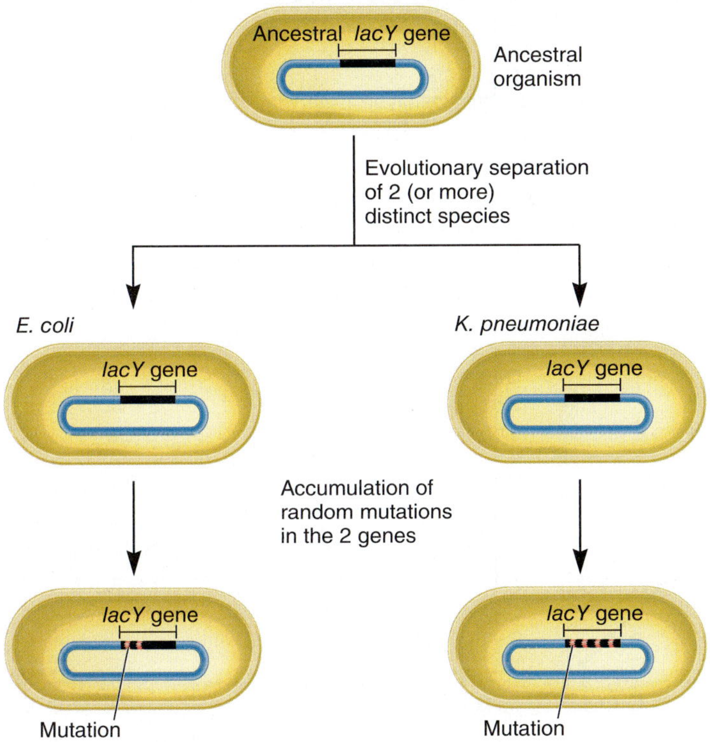

(b) The formation of homologous genes in two species, known as orthologs

FIGURE 23.8 The origin of homologous *lacY* genes in *Escherichia coli* and *Klebsiella pneumoniae*. **(a)** A comparison of the DNA sequences of a short region of the *lacY* gene from *E. coli* and *K. pneumoniae*. Bases that are different between the two genes are shown in red. **(b)** This figure emphasizes a single gene within an ancestral organism. During evolution, the ancestral organism diverged into many different species, including *E. coli* and *K. pneumoniae*. After this divergence, the *lacY* gene in the two species accumulated mutations, yielding *lacY* genes with somewhat different sequences.

Genes→Traits After two species diverge evolutionarily, their genes accumulate different random mutations. This example concerns the *lacY* gene, which encodes lactose permease. In both species, the function of lactose permease is to transport lactose into the cell. The *lacY* gene in these two species has accumulated different mutations that alter the amino acid sequence of the protein. Researchers have determined that these two species transport lactose at significantly different rates. Therefore, the changes in gene sequences have affected the ability of these two species to transport lactose.

CONCEPT CHECK: Explain why the sequences of these two genes are similar to each other but not identical.

point in the grid corresponds to one position of each sequence. The matrix allows all such pairs to be compared simultaneously. Dots are placed where the same letter occurs in both sequences. The key observation is that regions of similarity are distinguished by the occurrence of many dots along a diagonal line within the dot matrix. In contrast, Figure 23.9b compares two unrelated sequences: GENETICSISCOOL and THECOURSEISFUN. In this comparison, no diagonal line is seen. In some cases, two sequences may be related to each other but differ in length. Figure 23.9c compares the sequences GENETICSISCOOL and GENETICSISVERYCOOL. In this example, the second sequence is four letters longer than the first. In the dot matrix, two diagonal lines occur. To align these two lines, a gap must be created in the shorter sequence. Figure 23.9d shows the insertion of a gap that aligns the two sequences so that the two diagonal lines fall along the same line.

Overall, Figure 23.9 illustrates two important features of dot matrix methods. First, regions of homology are recognized by a series of dots that lie along a diagonal line. Second, gaps can be inserted to align sequences of unequal length but that still have some similarity. These same concepts hold true when genetic sequences are compared with each other. Unfortunately, for anything but the shortest genetic sequences, a simple dot matrix approach is inadequate. Instead, dynamic programming methods are used to identify similarities between genetic sequences. These methods, originally proposed by Saul Needleman and Christian Wunsch, are theoretically similar to a dot matrix, but they involve complex mathematical operations beyond the scope of this textbook.

In their original work of 1970, Needleman and Wunsch demonstrated that whale myoglobin and human β hemoglobin have similar sequences. Since then, this approach has been extended to compare more than two genetic sequences. Newer computer programs can align several genetic sequences and sensibly put in gaps. This produces a **multiple-sequence alignment.**

To illustrate the usefulness of a multiple-sequence alignment, let's use the general methods of Needleman and Wunsch and apply them to the globin gene family. Hemoglobin, a protein found in red blood cells, is responsible for carrying oxygen through the bloodstream. In humans, nine paralogous globin genes are functionally expressed. (There are also four pseudogenes that are not expressed and one myoglobin gene.) The nine globin genes fall into two categories: the α chains and the β chains. The α-chain genes are α_1, α_2, θ, and ξ; the β-chain genes are β, δ, γ_A, γ_G, and ε. Each hemoglobin protein is composed of two α chains and two β chains.

Because the globin genes are expressed at different stages of human development, the composition of hemoglobin changes during the course of growth. For example, the ξ and ε genes are expressed during early embryonic development, whereas the α_1, α_2, and β genes are expressed in the adult.

Insights into the structure and function of the hemoglobin polypeptides can be gained by comparing their sequences. In Figure 23.10, the amino acid sequences of the human globin polypeptides are compared in a multiple-sequence alignment using dynamic programming methods. An inspection of a multiple-sequence alignment may reveal important features concerning the similarities and differences within a gene family. In this alignment, dots are shown where it is necessary to create gaps to keep the amino acid sequences aligned. As we can see, the sequence similarity is very high between α_1, α_2, θ, and ζ. In fact, the amino acid sequences encoded by the α_1 and α_2 genes are identical. This suggests that the four types of α chains likely carry out very similar functions. Likewise, the β chains encoded by the β, δ, γ_A, γ_G, and ε genes are very similar to each other. In the globin gene family, the α chains are much more similar to each other than they are to the β chains, and vice versa.

In general, amino acids that are highly conserved within a gene family are more likely to be important functionally. The arrows in the multiple-sequence alignment point to histidine amino acids that are conserved in all nine members of the globin gene family. These histidines, which are highlighted in red, are involved in the necessary function of binding the heme molecule to the globin polypeptides.

Overall, the alignment shown in Figure 23.10 illustrates the type of information that can be derived from a multiple-sequence alignment. In this case, multiple-sequence alignment has shown that a group of nine genes falls into two closely related subgroups. The alignment has also identified particular amino acids within the proteins' sequences that are highly conserved. This conservation is consistent with these amino acids having an important role in this protein's function.

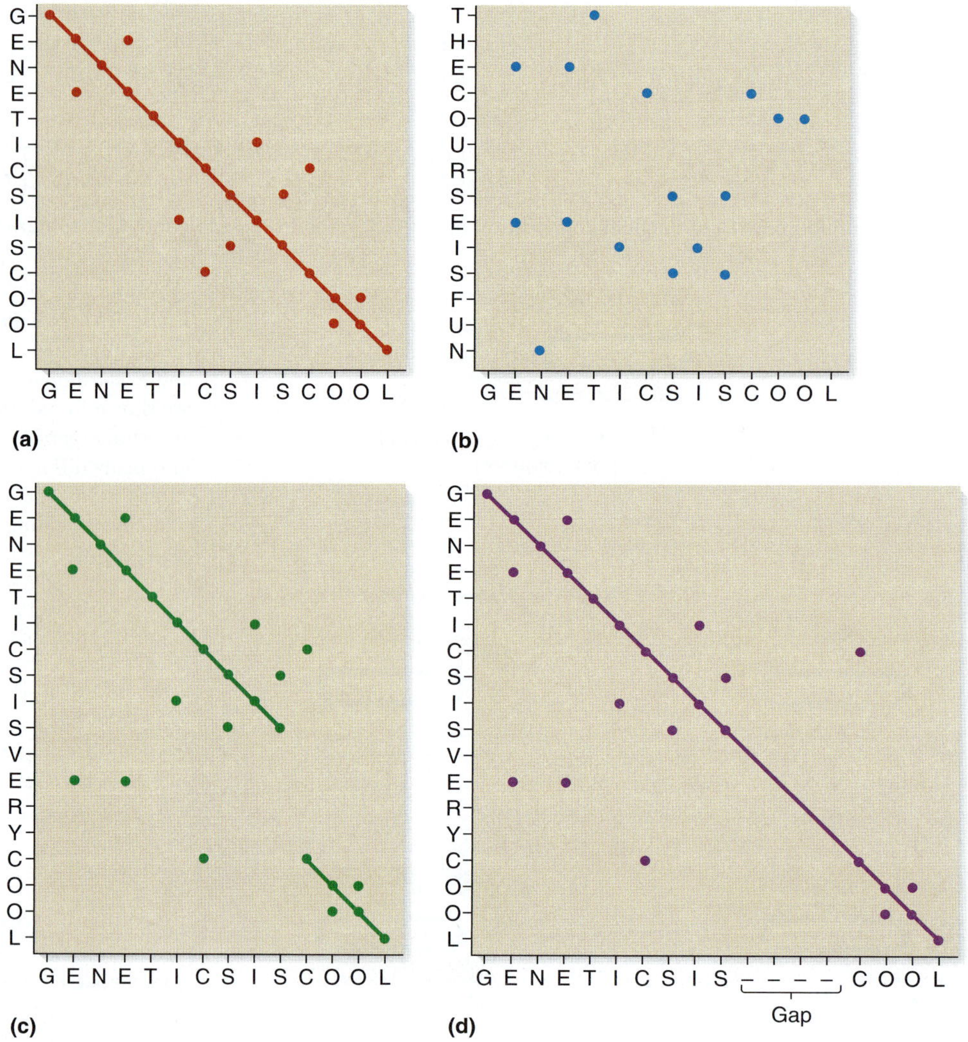

FIGURE 23.9 The use of dot matrices to evaluate similarity between two sequences. On a dot matrix, many points that fall on a diagonal line indicate regions of similarity.

A Database Can Be Searched to Identify Homologous Sequences

Homologous genes usually encode proteins that carry out similar or identical functions. As we have just considered, the members of the globin gene family are all involved with carrying and transporting oxygen. Likewise, the *lacY* genes in *E. coli* and *K. pneumoniae* both encode lactose permease, a protein that transports lactose across the bacterial cell membrane.

A strong correlation is typically found between homology and function. How is this relationship useful with regard to bioinformatics? In many cases, the first indication of the function of a newly determined sequence is through homology to known sequences in a database. An example is the *CFTR* gene that is altered in cystic fibrosis patients. After the *CFTR* gene was identified in humans, a database search revealed that it is homologous to several genes found in other species. Moreover, a few of the homologous genes were already known to encode proteins that function in the transport of ions and small molecules across the plasma membrane. This observation provided an important clue that cystic fibrosis involves a defect in ion transport.

The ability of computer programs to identify homology between genetic sequences provides a powerful tool for predicting the function of genetic sequences. In 1990, Stephen Altschul, David Lipman, and colleagues developed a program called **BLAST** (for **basic local alignment search tool**). The BLAST program has been described by many geneticists as the single most important bioinformatic tool. This computer program starts with a particular genetic sequence and then locates homologous sequences within a large database.

To see how the BLAST program works, let's consider the human enzyme phenylalanine hydroxylase, which functions in the metabolism of phenylalanine, an amino acid. Recessive mutations in the gene that encodes this enzyme are responsible for the disease called phenylketonuria (PKU). The computational experiment shown in Table 23.5 started with the amino acid sequence of this protein and used the BLAST program to search the Swissprot database, which contains millions of different protein sequences.

```
                  Gap
             1                                                          50
α1 alpha-1   VLSP.ADKTN VKAAWGKVGA HAGEGAEAL  ERMFLSFPTT KTYFPHF.DL
α2 alpha-2   VLSP.ADKTN VKAAWGKVGA HAGEGAEAL  ERMFLSFPTT KTYFPHF.DL
θ theta      ALSA.EDRAL VRALWKKLGS NVGVYTTEAL ERTFLAFPAT KTYFSHL.DL
ζ zeta       SLTK.TERTI IVSMWAKIST QADTIGTETL ERLFLSHPQT KTYFPHF.DL
β beta       VHLTPEEKSA VTALWGKV.. NVDEVGGEAL GRLLVVYPWT QRFFESFGDL
δ delta      VHLTPEEKSA VNALWGKV.. NVDAVGGEAL GRLLVVYPWT QRFFESFGDL
γA gamma-A   GHFTEEDKAT ITSLWGKV.. NVEDAGGEAL GRLLVVYPWT QRFFESFGDL
γG gamma-G   GHFTEEDKAT ITSLWGKV.. NVEDAGGEAL GRLLVVYPWT QRFFESFGDL
ε epsilon    VHFTAEEKAA VTSLWSKM.. NVEEAGGEAL GRLLVVYPWT QRFFESFGDL
                                 Gap

                    Gap
             51                                                         100
α1 alpha-1   SHGSA..... QVKGHGKKVA DALTNAVAHV DDMPNALSAL SDLHAHKLRV
α2 alpha-2   SHGSA..... QVKGHGKKVA DALTNAVAHV DDMPNALSAL SDLHAHKLRV
θ theta      SPGSS..... QVKAHGQKVA DALSLAVERL DDLPHALSAL SHLHACQLRV
ζ zeta       HPGSA..... QLRAHGSKVV AAVGDAVKSI DDIGGALSKL SELHAYQLRV
β beta       STPDAVMGNP KVKAHGKKVL GAFSDGLAHL DNLKGTFATL SELHCDKLHV
δ delta      SSPDAVMGNP KVKAHGKKVL GAFSDGLAHL DNLKGTFSQL SELHCDKLHV
γA gamma-A   SSASAIMGNP KVKAHGKKVL TSLGDAIKHL DDLKGTFAQL SELHCDKLHV
γG gamma-G   SSASAIMGNP KVKAHGKKVL TSLGDAIKHL DDLKGTFAQL SELHCDKLHV
ε epsilon    SSPSAILGNP KVKAHGKKVL TSFGDAIKNM DNLKPAFAKL SELHCDKLHV

             101                                                     148
α1 alpha-1   DPVNFKLLSH CLLVTLAAHL PAEFTPAVHA SLDKFLASVS TVLTSKYR
α2 alpha-2   DPVNFKLLSH CLLVTLAAHL PAEFTPAVHA SLDKFLASVS TVLTSKYR
θ theta      DPASFQLLGH CLLVTLARHL PGDFSPALQA SLDKFLSHSVI SALVSEYR
ζ zeta       DPVNFKLLSH CLLVTLAARF PADFTAEAHA AWDKFLSVVS SVLTEKYR
β beta       DPENFRLLGN VLVCVLAHHF GKEFTPPVQA AYQKVVAGVA NALAHKYH
δ delta      DPENFRLLGN VLVCVLARNF GKEFTPQMQA AYQKVVAGVA NALAHKYH
γA gamma-A   DPENFRLLGN VLVCVLAIHF GKEFTPEVQA SWQKMVTAVA SALSSRYH
γG gamma-G   DPENFRLLGN VLVCVLAIHF GKEFTPEVQA SWQKMVTAVA SALSSRYH
ε epsilon    DPENFRLLGN VMVIILATHF GKEFTPEVQA AWQKLVSAVA IALAHKYH
```

FIGURE 23.10 A multiple sequence alignment among selected members of the globin gene family in humans.

TABLE 23.5

Results from a BLAST Program Comparing Human Phenylalanine Hydroxylase with Database Sequences

Match*	% of Identical Amino Acids†	Species	Function of Sequence‡	E-value
1	100	Human (*Homo sapiens*)	Phenylalanine hydroxylase	0
2	99	Orangutan (*Pongo pygmaeus*)	Phenylalanine hydroxylase	0
3	92	Mouse (*Mus musculus*)	Phenylalanine hydroxylase	0
4	92	Rat (*Rattus norvegicus*)	Phenylalanine hydroxylase	0
5	83	Chicken (*Gallus gallus*)	Phenylalanine hydroxylase	0
6	78	Western clawed frog (*Xenopus tropicalis*)	Phenylalanine hydroxylase	0
7	75	Zebrafish (*Danio rerio*)	Phenylalanine hydroxylase	0
8	72	Japanese pufferfish (*Takifugu rubripes*)	Phenylalanine hydroxylase	0
9	62	Fruit fly (*Drosophila melanogaster*)	Phenylalanine hydroxylase	10^{-154}
10	57	Nematode (*Caenorhabditis elegans*)	Phenylalanine hydroxylase	10^{-141}

*The 10 examples shown here were randomly chosen from the results of a BLAST program using human phenylalanine hydroxylase as the starting sequence.
†The number indicates the percentage of amino acids that are identical to the amino acid sequence of human phenylalanine hydroxylase.
‡In some cases, the function of the sequence was determined by biochemical assay. In other cases, the function was inferred due to the high degree of sequence similarity with other species.

The BLAST program determines which sequences in this database are the closest matches to the amino acid sequence of human phenylalanine hydroxylase. Table 23.5 shows a portion of the results—10 selected matches to human phenylalanine hydroxylase that were identified by the program. Because this enzyme is found in nearly all eukaryotic species, the program identified phenylalanine hydroxylase from many different species. The column to the right of the match number shows the percentage of amino acids that are identical between the species indicated and the human sequence. Because the human phenylalanine hydroxylase sequence is already in the Swissprot database, the closest match of human phenylalanine hydroxylase is to itself (100%). The next nine sequences are in order of similarity. The most similar sequence is from the orangutan (99%), a close relative of humans. This is followed by two mammals, the mouse and rat, and then four vertebrates that are not mammals. The ninth and tenth best matches are from *Drosophila* and *C. elegans*, which are invertebrates.

As shown in the right column of Table 23.5, the relationship between the query sequence and each matching sequence is given an E-value (Expect value). The **E-value** represents the number of times that the match or a better one would be expected to occur purely by random chance in a search of the entire database. An E-value that is very small indicates that the similarity between the query sequence and the matching sequence is unlikely to have occurred by random chance. Instead, researchers would accept the hypothesis that the two sequences are homologous, which means they are derived from the same ancestral sequence.

E-values depend greatly on several parameters, such as the length of the query sequence, the number of gaps between the query sequence and the matching sequence, and the database size. As a general rule, if the E-value is less than 1×10^{-50}, the match is very similar to the query sequence and is likely to be homologous. (Values that are much less than 10^{-100} are reported as zero by the BLAST program.) If the value lies between 1×10^{-50} and 1×10^{-10}, the match, or part of it, is likely to be homologous. If the value is between 1×10^{-10} and 1×10^{-2}, the match has a significant chance of being related to the query sequence, whereas values between 1 and 1×10^{-2} have a relatively low probability of being homologous. Values above 1 are usually not evolutionarily related. As seen in Table 23.5, all of the E-values are below 1×10^{-140}, which suggests that all of these matches are homologous to the query sequence.

The order of the matches follows the evolutionary relatedness of the various species to humans. The similarity between any two sequences is related to the time that has passed since they diverged from a common ancestor. Among the species listed in Table 23.5, the human sequence is most similar to the orangutan, a closely related primate. The next most similar sequences are found in other mammals, followed by other vertebrates, and finally invertebrates.

Genetic Sequences Can Be Used to Predict the Structure of RNA and Proteins

Another area in which bioinformatics has influenced functional genomics and proteomics is the topic of structure prediction. The function of macromolecules such as RNA and proteins relies on their three-dimensional structure, which, in turn, depends on the linear sequences of their building blocks. In the case of RNA, this means a linear sequence of nucleotides; proteins are composed of a linear sequence of amino acids. Currently, the three-dimensional structure of macromolecules is determined primarily through the use of biophysical techniques, such as X-ray crystallography and nuclear magnetic resonance (NMR). These methods are technically difficult and very time-consuming. By comparison, DNA sequencing, which can be used to deduce RNA and amino acid sequences, requires much less effort. Therefore, because the three-dimensional structure of macromolecules depends ultimately on the linear sequence of their building blocks, it would be far easier if we could predict the structure (and function) of RNAs and proteins from their sequence of building blocks.

RNA molecules typically are folded into a secondary structure, which commonly contains double-stranded regions. This secondary structure is further folded and twisted to adopt a tertiary conformation. Such structural features of RNA molecules are functionally important. For example, the folding of RNA into secondary structures, such as stem-loops, affects transcriptional termination and other regulatory events. Therefore, geneticists are interested in the secondary and tertiary structures that RNA molecules can adopt.

Many approaches are available for investigating RNA structure. In addition to biophysical and biochemical techniques, computer modeling of RNA structure has become an important tool. Modeling programs consider different types of information. For example, the known characteristics of RNA secondary structure, such as the ability to form double-stranded regions, provide parameters for use in a modeling program.

A comparative approach can also be used in RNA structure prediction. This method assumes that RNAs of similar function and sequence have a similar structure. For example, the genes that encode certain types of RNAs, such as the 16S rRNA that makes up most of the small ribosomal subunit, have been sequenced from many different species. Among different species, the 16S rRNAs have similar but not identical sequences. Computer programs can compare many different 16S rRNA sequences to aid in the prediction of secondary structure. **Figure 23.11** illustrates a secondary structural model for 16S rRNA from *E. coli* based on a comparative sequence analysis of many bacterial 16S rRNA sequences. This large RNA contains 45 stem-loops. As you can imagine, it would be rather difficult to deduce such a model without the aid of a computer! In addition, RNA secondary structure prediction may be aimed at predicting the lowest energy state of a folded molecule. This approach, called free energy minimization, is also related to the base sequence of an RNA molecule.

Structure prediction is also used in the area of proteomics. As described in Chapter 13, proteins contain repeating secondary structural patterns known as α helices and β sheets. Several computer-based approaches attempt to predict secondary structure from the primary amino acid sequence. These programs base their predictions on different types of parameters. Some programs rely on the physical and energetic properties of the amino acids and the polypeptide backbone. More commonly, however, secondary structure predictions are based on the statistical frequency of amino acids within secondary structures that have already been crystallized.

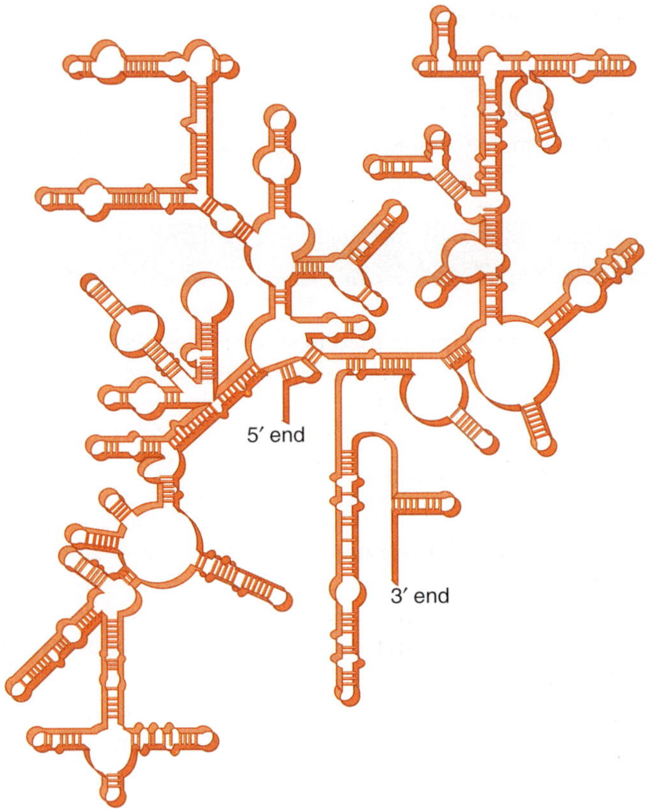

FIGURE 23.11 **A secondary structural model for *E. coli* 16S rRNA.**

CONCEPT CHECK: Would you expect the secondary structure of 16S rRNA from two closely related bacterial species to be similar? Explain.

For example, Peter Chou and Gerald Fasman compiled X-ray crystallographic data to calculate the likelihood that an amino acid will be found in an α helix or a β sheet. Certain amino acids, such as glutamic acid and alanine, are likely to be found in an α helix; others, such as valine and isoleucine, are more likely to be found in a β sheet. Such information can be used to predict whether a sequence of amino acids within a protein is likely to be folded in an α-helix or β-sheet conformation. Secondary structure prediction is correct for approximately 60–70% of all sequences. Although this degree of accuracy is promising, it is generally not sufficient to predict protein secondary structure reliably. Therefore, one must be cautious in interpreting the results of a secondary structure prediction program.

Although the three-dimensional structure of a protein is extremely difficult to predict solely from its amino acid sequence, researchers have had some success in predicting tertiary structure using a comparative approach. This strategy requires that the protein of interest be homologous to another protein, the tertiary structure of which already has been solved by X-ray crystallography. In this situation, the crystal structure of the known protein can be used as a starting point to model the three-dimensional structure of the protein of interest. This approach is known as **homology-based modeling.** As an example, two research groups independently predicted the three-dimensional structure of a protein, now known as HIV protease, that is encoded by a gene in HIV (see Chapter 17). The protein is homologous to other proteases with structures that had been solved by X-ray crystallography. Two similar models of the HIV protease were predicted before its actual crystal structure was determined. Both models turned out to be fairly accurate representations of the actual structure, which was later solved by X-ray crystallography. Knowledge of the structure of HIV protease helped scientists synthesize compounds known as HIV protease inhibitors, which are an important part of drug therapy for treating HIV/AIDS.

23.3 COMPREHENSION QUESTIONS

1. The identification of a stop codon for a particular gene is an example of
   a. sequence recognition.
   b. pattern recognition.
   c. both a and b.
   d. none of the above.
2. Homologous genes
   a. are derived from the same ancestral gene.
   b. are likely to carry out the same or similar functions.
   c. have similar DNA sequences.
   d. exhibit all of the above.
3. The BLAST program begins with a particular genetic sequence and
   a. translates it into an amino acid sequence.
   b. determines if it contains one or more genes.
   c. identifies homologs within a database.
   d. does all of the above.

KEY TERMS

Page 579. functional genomics, proteome, proteomics, bioinformatics
Page 580. DNA microarray, gene chip
Page 586. cluster analysis, coordinate regulation, chromatin immunoprecipitation (ChIP)
Page 587. ChIP-chip assay, gene knockout
Page 588. alternative splicing, RNA editing, posttranslational covalent modification
Page 589. two-dimensional (2D) gel electrophoresis, isoelectric focusing
Page 590. mass spectrometry, tandem mass spectrometry
Page 591. protein microarrays, antibody microarray
Page 592. functional protein microarray, computer programs
Page 593. computer data file
Page 594. database, annotated, genome databases
Page 595. sequence recognition, pattern recognition, sequence element, motif, search-by-signal, search-by-content
Page 596. codon bias, open reading frame (ORF), homologous, orthologs, paralogs, gene family

CHAPTER SUMMARY

23.1 Functional Genomics

- The goal of functional genomics is to understand the role of genetic sequences in a given species.
- A microarray is a slide dotted with many DNA sequences. It is used to study the expression of many genes simultaneously and also has other uses (see Figure 23.1, Table 23.1).
- A microarray can be used to study the coordinate regulation of groups of genes—a process called cluster analysis (see Figure 23.2).
- In a ChIP-chip assay, chromatin immunoprecipitation is used in conjunction with a microarray to study DNA-protein interactions at the genomic level (see Figure 23.3).
- Researchers are producing gene knockout collections for certain species, such as mice, to determine the functions of genes at the genomic level.

23.2 Proteomics

- The proteome is the entire collection of proteins that a cell or organism makes. The study of the function and interactions of many proteins is called proteomics.
- The proteome is much larger than the genome due to alternative splicing, RNA editing, and posttranslational covalent modifications (see Figure 23.4).
- Two-dimensional gel electrophoresis separates a complex mixture of proteins (see Figure 23.5).
- Tandem mass spectrometry identifies short amino acid sequences within purified proteins. These short sequences can be used to identify a protein (see Figure 23.6).
- Protein microarrays are used to study protein expression, protein function, protein-protein interactions, and protein-drug interactions (see Table 23.2).

23.3 Bioinformatics

- Bioinformatics involves the use of computers, mathematical tools, and statistical techniques to record, store, and analyze biological information, such as DNA sequences.
- Sequence files are analyzed by computer programs.
- Researchers have collected genetic sequences and compiled them in large databases (see Table 23.3).
- Different computational strategies, such as sequence recognition and pattern recognition, identify functional genetic sequences. Sequence recognition may identify sequence elements or motifs (see Table 23.4).
- Genes may be identified by computational strategies such as search-by-signal or search-by-content approaches. Searching for a long open reading frame may also identify a gene (see Figure 23.7).
- Homologous genes are derived from the same ancestral gene. They can be orthologs (genes in different species) or paralogs (genes in the same species) (see Figure 23.8).
- A dot matrix can be constructed to evaluate the degree of similarity between short sequences. Researchers may produce a multiple-sequence alignment to compare the sequences of several homologous genes (see Figures 23.9, 23.10).
- The BLAST program identifies homologous sequences that are found within a database (see Table 23.5).
- Computer programs may also use genetic sequences and try to predict the structure of RNA or proteins (see Figure 23.11).

PROBLEM SETS & INSIGHTS

Solved Problems

S1. To answer this question, you will need to look back at the evolution of the globin gene family, which is shown in Chapter 8, Figure 8.7. Throughout the evolution of this gene family, mutations have occurred that have resulted in globin polypeptides with similar but significantly different amino acid sequences. If we look at the sequence alignment in Figure 23.10, we can make logical guesses regarding the timing of mutations, based on a comparison of the amino acid sequences of family members. What is/are the most probable time(s) that mutations occurred to produce the following amino acid differences? Note: You need to examine the alignment in Figure 23.10 and the evolutionary time scale in Figure 8.7 to answer this question.

A. Val-111 and Cys-111

B. Met-112 and Leu-112

C. Ser-141, Asn-141, Ile-141, and Thr-141

Answer:

A. We do not know if the original globin gene encoded a cysteine or valine at codon 111. The mutation could have changed cysteine to valine or valine to cysteine. The mutation probably occurred after the duplication that produced the α-globin family and β-globin family (about 300 mya) but before the gene duplications that occurred in the last 200 million years to produce the multiple copies of the globin genes on chromosome 11 and chromosome 16. Therefore, all of the globin genes on chromosome 11 have a valine at codon 111, and all of the globin genes on chromosome 16 have a cysteine.

B. Met-112 occurs only in the ε-globin polypeptide; all of the other globin polypeptides contain a leucine at position 112. Therefore, the primordial globin gene probably contained a leucine codon at position 112. After the gene duplication that produced the ε-globin gene, a mutation occurred that changed this leucine codon into a methionine codon. This would have occurred since the evolution of primates (i.e., within the last 10–20 million years).

C. When we look at the possible codons at position 141 (i.e., Ser-141, Asn-141, Ile-141, and Thr-141), we notice that a serine codon is found in θ globin, ξ globin, and γ globin. Because the θ- and ξ-globin genes are found on chromosome 16 and the γ-globin genes are found on chromosome 11, it is probable that serine is the primordial codon and that the other codons (asparagine, isoleucine, and threonine) arose later by mutation of the serine codon. If this is correct, the Thr-141 codon arose before the gene duplication that produced the α-globin genes. The Asn-141 and Ile-141 mutations arose after the gene duplications that produced the γ-globin genes. Therefore, the Thr-141, Asn-141, and Ile-141 arose since the evolution of primates (i.e., within the last 10–20 million years).

S2. Using a comparative sequence analysis, the secondary structures of rRNAs have been predicted. Among many homologous rRNAs, one stem-loop usually has the following structure:

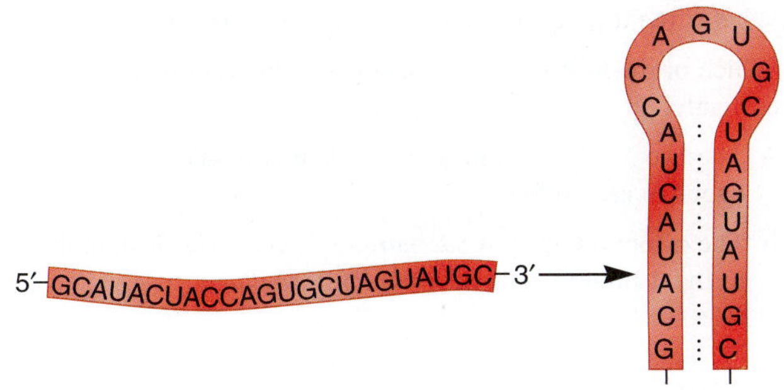

You have sequenced the DNA that encodes a homologous rRNA from a new species and have obtained most of its sequence, but you cannot read the last five bases on your sequencing gel.

5′–GCATTCTACCAGTGCTAG?????–3′

Of course, you will eventually repeat this experiment to determine the last five bases. However, before you get around to doing this, what do you expect will be the sequence of the last five bases?

Answer: AATGC–3′

This will also form a similar stem-loop structure.

S3. How can codon bias be used to search for protein-encoding genes within uncharacterized genetic sequences?

Answer: Most species exhibit a bias in the codons they use within the coding sequence of protein-encoding genes. This causes the base content within coding sequences to differ significantly from that of noncoding DNA regions. By knowing the codon bias for a particular species, researchers can use a computer to locate regions that display this bias and thereby identify what are likely to be the coding regions of protein-encoding genes.

Conceptual Questions

C1. Discuss the meaning of the following terms: genomics, functional genomics, and proteomics.

C2. Discuss the reasons why the proteome is larger than the genome of a given species.

C3. What is a database? What types of information are stored within a database? Where does the information come from? Discuss the objectives of a genome database.

C4. Besides the examples listed in Table 23.4, list five types of short sequences that a geneticist might want to locate within a DNA sequence.

C5. Discuss the distinction between sequence recognition and pattern recognition.

C6. A multiple sequence alignment of five homologous proteins is shown here:

```
  1                                                  50
1 MLAFLNQVRK PTLDLPLEVR RKMWFKPFM. QSYLVVFIGY LTMYLIRKNF
2 MLAFLNQVRK PTLDLALDVR RKMWFKPFM. QSYLVVFIGY LTMYLIRKNF
3 MLPFLKAPAD APL.MTDKYE IDARYRYWRR HILLTIWLGY ALFYFTRKSF
4 MLSFLKAPAN APL.ITDKHE VDARYRYWRR HILITIWLGY ALFYFTRKSF
5 MLSIFKPAPH KAR.LPAA.E IDPTYRRLRW QIFLGIFFGY AAYYLVRKNF

  51                                                100
1 NIAQNDMIST YGLSMTQLGM IGLGFSITYG VGKTLVSYYA DGKNTKQFLP
2 NIAQNDMIST YGLSMTELGM IGLGFSITYG VGKTLVSYYA DGKNTKQFLP
3 NAAVPEILAN GVLSRSDIGL LATLFYITYG VSKFVSGIVS DRSNARYFMG
4 NAAAPEILAS GILTRSDIGL LATLFYITYG VSKFVSGIVS DRSNARYFMG
5 ALAMPYLVEQ .GFSRGDLGF ALSGISIAYG FSKFIMGSVS DRSNPRVFLP
```

Discuss some of the interesting features that this alignment reveals.

C7. What is the difference between similarity and homology?

C8. Which of the following statements uses the term homologous correctly?

A. The two X chromosomes in female mammalian cells are homologous to each other.

B. The α-tubulin gene in *Saccharomyces cerevisiae* is homologous to the α-tubulin gene in *Arabidopsis thaliana*.

C. The promoter of the *lac* operon is homologous to the promoter of the *trp* operon.

D. The *lacY* gene of *E. coli* and *Klebsiella pneumoniae* are approximately 60% homologous to each other.

C9. When comparing (i.e., aligning) two or more genetic sequences, it is sometimes necessary to put in gaps. Explain why. Discuss two changes (i.e., two types of mutations) that could happen during the evolution of homologous genes that would explain the occurrence of gaps in multiple sequence alignments.

Experimental Questions

E1. With regard to DNA microarrays, answer the following questions:

A. What is attached to the slide? Be specific about the number of spots, the lengths of DNA fragments, and the origin of the DNA fragments.

B. What is hybridized to the microarray?

C. How is hybridization detected?

E2. In the experiment of Figure 23.2, explain how the ratio of red-to-green fluorescence provides information regarding gene regulation.

E3. What is meant by the term cluster analysis? How is this approach useful?

E4. For 2D-gel electrophoresis, what physical properties of proteins promote their separation in the first dimension and the second dimension?

E5. Can 2D-gel electrophoresis be used as a purification technique? Explain.

E6. Explain how tandem mass spectroscopy is used to determine the sequence of a peptide. Once a peptide sequence is known, how is this information used to determine the sequence of the entire protein?

E7. Describe the two general types of protein microarrays. What are their possible applications?

E8. Discuss the strategies that can be used to identify a protein-encoding gene using bioinformatics.

E9. What is a motif? Why is it useful for computer programs to identify functional motifs within amino acid sequences?

E10. Discuss why it is useful to search a database to identify sequences that are homologous to a newly determined sequence.

E11. The secondary structure of 16S rRNA has been predicted using a computer-based sequence analysis. In general terms, discuss what type of information is used in a comparative sequence analysis, and explain what assumptions are made concerning the structure of homologous RNAs.

E12. Discuss the basis for secondary structure prediction in proteins. How reliable is it?

E13. To reliably predict the tertiary structure of a protein based on its amino acid sequence, what type of information must be available?

E14. In this chapter, we considered a computer program that can translate a DNA sequence into a polypeptide sequence. A researcher has a sequence file that contains the amino acid sequence of a polypeptide and runs a program that is opposite to this program. This other program is called BACKTRANSLATE. It takes an amino acid sequence file and determines the sequence of DNA that would encode such a polypeptide. How does this program work? In other words, what genetic principles underlie this program? What type of sequence file would this program generate: a nucleotide sequence or an amino acid sequence? Would the BACKTRANSLATE program produce only a single sequence file? Explain why or why not.

E15. In this chapter, we considered a computer program that translates a DNA sequence into a polypeptide sequence. Instead of running this program, a researcher could simply look the codons up in a genetic code table and determine the sequence by hand. What are the advantages of running this program rather than doing it the old-fashioned way by hand?

E16. To identify the following types of genetic occurrences, would a program use sequence recognition, pattern recognition, or both?

A. Whether a segment of *Drosophila* DNA contains a P element (which is a specific type of transposable element)

B. Whether a segment of DNA contains a stop codon

C. In a comparison of two DNA segments, whether there is an inversion in one segment compared with the other segment

D. Whether a long segment of bacterial DNA contains one or more genes

E17. The goal of many computer programs is to identify sequence elements within a long segment of DNA. What is a sequence element? Give two examples. How is the specific sequence of a sequence element determined? In other words, is it determined by the computer program or by genetic studies? Explain.

E18. Take a look at the multiple sequence alignment in Figure 23.10 of the globin polypeptides from amino acids 101 to 148.

A. Which of these amino acids are likely to be most important for globin structure and function? Explain why.

B. Which are likely to be least important?

E19. See solved problem S1 before answering this question. Based on the sequence alignment in Figure 23.10, what is/are the most probable time(s) that mutations occurred in the human globin gene family to produce the following amino acid differences?

A. His-119 and Arg-119

B. Gly-121 and Pro-121

C. Glu-103, Val-103, and Ala-103

E20. Here is a short nucleotide sequence within a gene. Via the Internet (e.g., see www.ncbi.nlm.nih.gov/Tools), determine what gene this sequence is found within. Also, determine the species in which this gene sequence is found.

5′–GGGCGCAATTACTTAACGCCTCGATTATCTTCTTGC
GCCACTGATCATTA–3′

E21. Take a look at solved problem S1 and the codon table found in Chapter 13 (Table 13.1). Assuming that a mutation involving a single-base change is more likely than a double-base change, propose how the Asn-141, Ile-141, and Thr-141 codons arose. In your answer, describe which of the six possible serine codons is/are likely to be the primordial serine codon of the globin gene family and how that codon changed to produce the Asn-141, Ile-141, and Thr-141 codons.

E22. Membrane proteins often have transmembrane regions that span the membrane in an α-helical conformation. These transmembrane segments are about 20 amino acids long and usually contain amino acids with nonpolar (i.e., hydrophobic) amino acid side chains. Researchers can predict whether a polypeptide sequence has transmembrane segments based on the occurrence of segments that contain 20 nonpolar amino acids. To do so, each amino acid is assigned a hydropathy value, based on the chemistry of its amino acid side chain. Amino acids with very nonpolar side chains are given a high (positive) value, whereas amino acids that are charged and/or polar are given low (negative) values. The hydropathy values usually range from about +4 to −4.

Computer programs have been devised that scan the amino acid sequence of a polypeptide and calculate values based on the hydropathy values of the amino acid side chains. The program usually scans a window of seven amino acids and assigns an average hydropathy value. For example, the program would scan amino acids 1–7 and give an average value, then it would scan 2–8 and give a value, then it would scan 3–9 and give a value, and so on, until it reached the end of the polypeptide sequence (i.e., until it reached the carboxyl-terminus).

The program then produces a figure, known as a hydropathy plot, which describes the average hydropathy values throughout the entire polypeptide sequence. An example of a hydropathy plot is shown here.

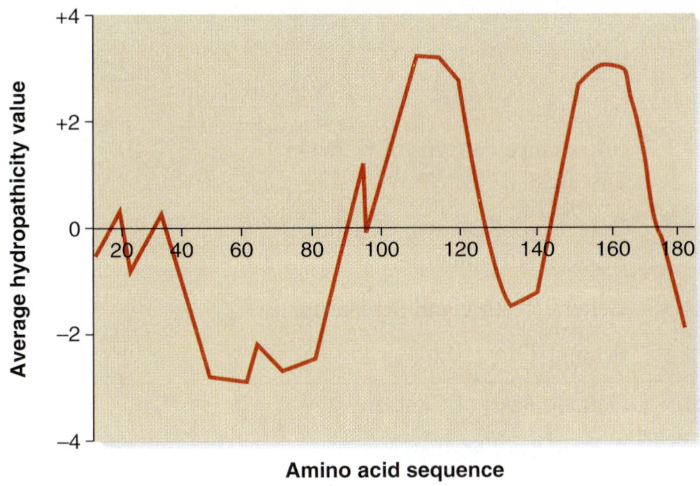

A. How many transmembrane segments are likely in this polypeptide?

B. Draw the structure of this polypeptide if it were embedded in the plasma membrane. Assume that the amino terminus is found in the cytoplasm of the cell.

E23. Explain how a computer program can predict RNA secondary structure. What underlying genetic concept is used by the program to predict secondary structure?

E24. Are the following statements about protein structure prediction true or false?

A. The prediction of secondary structure relies on information regarding the known occurrence of amino acid residues in α helices or β sheets from X-ray crystallographic data.

B. The prediction of secondary structure is highly accurate, nearly 100% correct.

C. To predict the tertiary structure of a protein based on its amino acid sequence, it is necessary that the protein of interest is homologous to another protein whose tertiary structure is already known.

Questions for Student Discussion/Collaboration

1. Let's suppose you are in charge of organizing and publicizing a genomic database for the mouse genome. Make a list of innovative strategies you would initiate to make the mouse genome database useful and effective.

2. Let's suppose a 5-year-old told you that she was interested in pursuing a career studying the three-dimensional structure of proteins. (Okay, so she's a bit precocious.) Would you advise her to become a geneticist, a mathematical theoretician, or a biophysicist?

3. If you have access to the necessary computer software, make a sequence file and analyze it in the following ways: What is the translated sequence in all three reading frames? What is the longest open reading frame? Is the sequence homologous to any known sequences? If so, does this provide any clues about the function of the sequence?

Note: All answers appear at the website for this textbook; the answers to the even-numbered questions and all of the Concept Check and Comprehension Questions are in the back of the book.

www.mhhe.com/brookergenetics5e

Visit the website for practice tests, answer keys, and other learning aids for this chapter. Enhance your understanding of genetics with our interactive exercises, quizzes, animations, and much more.

CHAPTER OUTLINE

Cigarette smoking and lung cancer. *Cigarette smoke contains chemicals that are known to mutate genes in the cells of a person's lungs, thereby leading to lung cancer. Lung cancer remains the top cause of cancer death in the United States, with 87% of deaths linked to smoking.*

24 MEDICAL GENETICS AND CANCER

Genetic insight is expected to bring about revolutionary changes in medical practices. In fact, changes are already under way. Currently, several hundred genetic tests are in clinical use, with many more under development. Most of these tests detect mutations associated with rare genetic disorders that follow Mendelian inheritance patterns. These include Duchenne muscular dystrophy, cystic fibrosis, sickle cell disease, and Huntington disease. In addition, genetic tests are available to detect the predisposition to develop certain forms of cancer. As discussed in Chapter 22, DNA sequencing technologies are progressing to the point where the sequencing of a person's entire genome will be inexpensive enough to practice as a routine diagnostic procedure. Many people in the medical field expect that this advance will usher in the era of personalized medicine—the use of information about a patient's genotype and other clinical data in order to select a medication, therapy or preventative measure that is specifically suited to that patient. We will explore this topic at the end of the chapter.

Approximately 12,000 genetic diseases are known to afflict people, but this is almost certainly an underestimate. The website Online Mendelian Inheritance in Man (OMIM) is a compendium of human genes and phenotypes (see www.ncbi.nlm.nih.gov/omim/). OMIM contains information on all known disorders that are inherited in a Mendelian manner. Most of the genetic disorders discussed in the first part of this chapter are the direct result of a mutation in one gene. However, many diseases, including common medical disorders such as diabetes, asthma, and mental illness, have a complex pattern of inheritance involving several genes. In these cases, a single mutant gene does not determine whether a person has a disease. Instead, a number of genes may each make a subtle contribution to a person's susceptibility to a disease. Unraveling these complexities will be a challenge for some time to come. The availability of the human genome sequence, discussed in Chapter 22, will be of great help.

In this chapter, we will focus our attention on ways that mutant genes contribute to human disease. In the first part of the chapter, we explore the molecular basis of several genetic disorders and their patterns of inheritance. We also examine how genetic testing can determine if an individual carries a defective allele. We will then consider cancer, a disease that involves the uncontrolled growth of somatic cells. We will examine the underlying genetic basis for cancer and discuss the roles that many different genes may play in the development of this disease.

24.1 INHERITANCE PATTERNS OF GENETIC DISEASES

Learning Outcomes:

1. List seven factors that suggest a disease may have a genetic component.
2. Analyze human pedigrees and be able to distinguish autosomal recessive, autosomal dominant, X-linked recessive, and X-linked dominant patterns.
3. Define locus heterogeneity and explain how it can confound pedigree analysis.

Human genetics is a topic that is hard to resist. Almost everyone who looks at a newborn is tempted to speculate whether the baby resembles the mother, the father, or perhaps a distant relative. In this section, we will focus primarily on the inheritance of human genetic diseases rather than common traits found in the general population. Even so, the study of human genetic diseases often provides insights about such common traits. The disease hemophilia illustrates this point. Hemophilia is a condition in which the blood does not clot properly. By analyzing people with this disorder, researchers have identified genes that participate in the process of blood clotting. The study of hemophilia has helped to identify a clotting pathway involving several different proteins. Therefore, as with the study of mutants in model organisms such as *Drosophila*, mice, and yeast, when we study the inheritance of genetic diseases, we often learn a great deal about the genetic basis for normal physiological processes as well.

Because thousands of human diseases have an underlying genetic basis, human genetic analysis is of great medical importance. In this section, we will examine the causes and inheritance patterns of human genetic diseases that result from defects in single genes. As you will learn, the mutant genes that cause these diseases often follow simple Mendelian inheritance patterns.

A Genetic Basis for a Human Disease May Be Suggested from a Variety of Observations

When we view the characteristics of people, we usually think that some traits are inherited, whereas others are caused by environmental factors. For example, when the facial features of two related individuals look strikingly similar, we think that this similarity has a genetic basis. The profound resemblance between identical twins is an obvious example. By comparison, other traits are governed by the environment. If we see a person with purple hair, we likely suspect that he or she has used hair dye as opposed to showing an unusual genetic trait.

For human diseases, geneticists would like to know the relative contributions from genetics and the environment. Is a disease caused by a pathogenic microorganism, a toxic agent in the environment, or a faulty gene? Unlike the case with experimental organisms, we cannot conduct human crosses to determine the genetic basis for diseases. Instead, we must rely on analyzing the occurrence of a disease in families that already exist. As described in the following list, several observations are consistent with the idea that a disease is caused, at least in part, by the inheritance of mutant genes. When the occurrence of a disease correlates with several of these observations, a geneticist becomes increasingly confident that the disease has a genetic basis.

1. ***An individual who exhibits a disease is more likely to have genetic relatives with the disorder than people in the general population.*** For example, someone with cystic fibrosis is more likely to have relatives with this disease than would a randomly chosen member of the general population.
2. ***Identical twins share the disease more often than nonidentical twins.*** Identical twins, also called **monozygotic (MZ) twins,** are genetically identical to each other, because they were formed from the same sperm and egg. By comparison, nonidentical twins, also called fraternal, or **dizygotic (DZ) twins,** are formed from separate pairs of sperm and egg cells. Fraternal twins share, on average, 50% of their genetic material, the same as any two siblings. When a disorder has a genetic component, both identical twins are more likely to exhibit the disorder than are fraternal twins.

   Geneticists evaluate a disorder's **concordance,** the degree to which it is inherited, by calculating the percentage of twin pairs in which both twins exhibit the disorder relative to pairs where only one twin shows the disorder. Theoretically, for diseases caused by a single gene, concordance among identical twins should be 100%. For fraternal twins, concordance for dominant disorders is expected to be 50%, assuming only one parent is heterozygous for the disease. For recessive diseases, concordance among fraternal twins would be 25% if we assume both parents are heterozygous carriers. However, the actual concordance values observed for most single-gene disorders are usually less than such theoretical values for a variety of reasons. Some disorders are not completely penetrant, meaning that the symptoms associated with the disorder are not always produced. Also, one twin may have a disorder due to a new mutation that occurred after fertilization; it would be very unlikely for the other twin to have the same mutation.
3. ***The disease does not spread to individuals sharing similar environmental situations.*** Inherited disorders cannot spread from person to person. The only way genetic diseases can be transmitted is from parent to offspring.
4. ***Different populations tend to have different frequencies of the disease.*** Due to evolutionary factors, the frequencies of traits usually vary among different populations of humans. For example, the frequency of sickle cell disease is highest among certain African and Asian populations and relatively low in other parts of the world (see Chapter 26, Figure 26.9).

5. ***The disease tends to develop at a characteristic age.*** Many genetic disorders exhibit a characteristic **age of onset** at which the disease appears. Some mutant genes exert their effects during embryonic and fetal development, so their effects are apparent at birth. Other genetic disorders tend to develop much later in life.
6. ***The human disorder may resemble a disorder that is already known to have a genetic basis in an animal.*** In animals, where we can conduct experiments, various traits are known to be governed by genes. For example, the albino phenotype is found in many animals as well as in humans (**Figure 24.1**).
7. ***A correlation is observed between a disease and a mutant human gene or a chromosomal alteration.*** A particularly convincing piece of evidence of a disease having a genetic basis is the identification of altered genes or chromosomes that occur only in people exhibiting the disorder. When comparing two individuals, one with a disease and one without, we expect to see differences in their genetic material if the disorder has a genetic component. Alterations in gene sequences are determined by gene cloning and DNA-sequencing techniques (see Chapter 20). Also, changes in chromosome structure and number can be detected by the microscopic examination of chromosomes (see Chapter 8).

Inheritance Patterns of Human Diseases May Be Determined via Pedigree Analysis

When a human disorder is caused by a mutation in a single gene, the pattern of inheritance can be deduced by analyzing human pedigrees. How is this accomplished? A geneticist must obtain data from many large pedigrees containing several individuals who exhibit the disorder and then follow its pattern of inheritance from generation to generation. To appreciate the basic features of pedigree analysis, we will examine a few large pedigrees that involve diseases inherited in different ways. You may wish to review Chapter 2 (see Figure 2.11) for the organization and symbols of pedigrees.

Autosomal Recessive Inheritance The pedigree shown in **Figure 24.2** concerns a genetic disorder called Tay-Sachs disease (TSD), first described by Warren Tay, a British ophthalmologist, and Bernard Sachs, an American neurologist, in the 1880s.

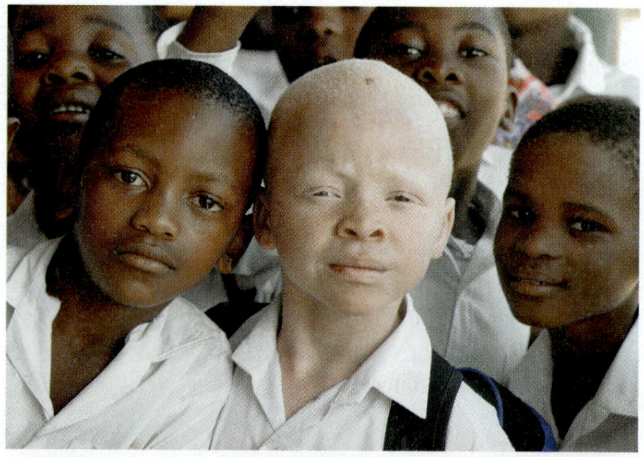

FIGURE 24.1 The albino phenotype in a human and a wildebeest.

Genes→Traits Certain enzymes (encoded by genes) are necessary for the production of pigment. A homozygote with two defective alleles in these pigmentation genes exhibits an albino phenotype. This phenotype can occur in humans and other animals.

Affected individuals appear healthy at birth but then develop neurodegenerative symptoms at 4–6 months of age. The primary characteristics are cerebral degeneration, blindness, and loss of motor function. Individuals with TSD typically die in the third or fourth year of life. This disease is particularly prevalent in Ashkenazi (eastern European) Jewish populations, in which it has a frequency of about 1 in 3600 births, which is over 100 times more frequent than in most other human populations.

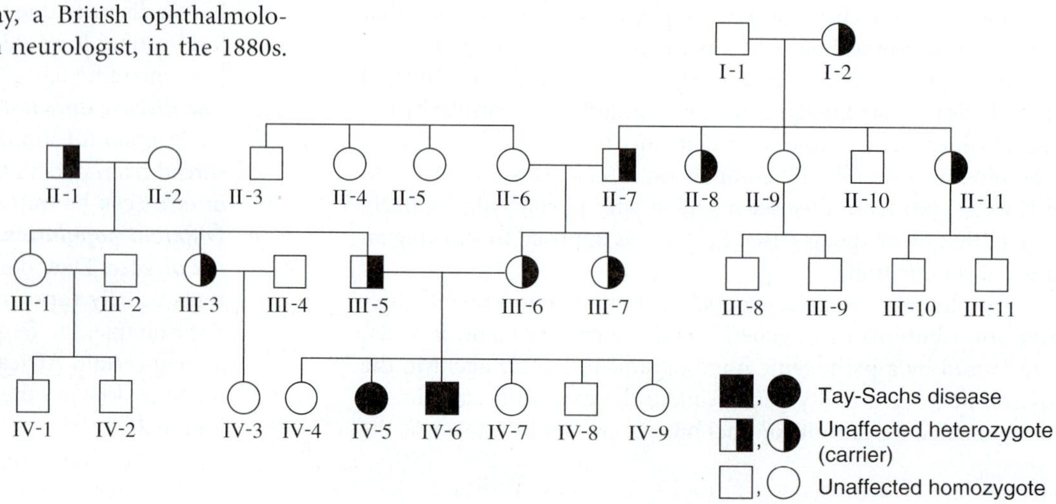

INTERACTIVE EXERCISE

FIGURE 24.2 A family pedigree of Tay-Sachs disease, indicating recessive inheritance. Heterozygous carriers were determined by genetic testing.

CONCEPT CHECK: What feature(s) of this pedigree indicate recessive inheritance?

At the molecular level, the mutation that causes TSD is in a gene that encodes the enzyme hexosaminidase A (HexA). HexA is responsible for the breakdown of a category of lipids called G_{M2}-gangliosides, which are prevalent in the cells of the central nervous system. A defect in the ability to break down this lipid leads to its excessive accumulation in neurons and eventually causes the neurodegenerative symptoms characteristic of TSD. This defect in lipid breakdown was recognized long before the *hexA* gene was identified as being defective in these patients.

As illustrated in Figure 24.2, Tay-Sachs disease is inherited in an autosomal recessive manner. The four common features of autosomal recessive inheritance are as follows:

1. ***Frequently, an affected offspring has two unaffected parents.*** For rare recessive traits, the parents are usually unaffected, meaning they do not exhibit the disease. For deleterious alleles that cause early death or infertility, the two parents must be unaffected. This is always the case in TSD.
2. ***When two unaffected heterozygotes have children, the percentage of affected children is (on average) 25%.***
3. ***Two affected individuals have 100% affected children.*** This observation can be made only when a recessive trait produces fertile, viable individuals. In the case of TSD, the affected individual dies in early childhood, and so it is not possible to observe crosses between two affected people.
4. ***The trait occurs with the same frequency in both sexes.***

Autosomal recessive inheritance is a common mode of transmission for genetic disorders, particularly those that involve defective enzymes. Human recessive alleles are often caused by mutations that result in a loss of function in the encoded enzyme. In the case of Tay-Sachs disease, a heterozygous carrier has approximately 50% of the functional enzyme, which is sufficient for a normal phenotype. However, for other genetic diseases, the level of functional protein may vary due to the effects of gene regulation. Thousands of human genetic diseases are inherited in a recessive manner, and in many cases, the mutant genes have been identified. A few of these diseases are described in **Table 24.1**.

Autosomal Dominant Inheritance Now let's examine a human pedigree involving an autosomal dominant disease (**Figure 24.3**). In this example, the affected individuals have a disorder called Huntington disease. The major symptoms of this disease, which usually occurs during middle age, are due to the degeneration of certain types of neurons in the brain, leading to personality changes, dementia, and early death. In 1993, the gene involved in Huntington disease was identified and sequenced. It encodes a protein called huntingtin that is expressed in neurons but is also found in some cells not affected in Huntington disease. In persons with this disorder, a mutation called a trinucleotide repeat expansion produces a polyglutamine tract—many glutamines in a row—within the huntingtin protein (refer back to Table 18.5). This causes an aggregation of the protein in neurons. However, additional research is needed to understand the molecular relationship between the abnormality in the huntingtin protein and the disease symptoms. Five common features of autosomal dominant inheritance are as follows:

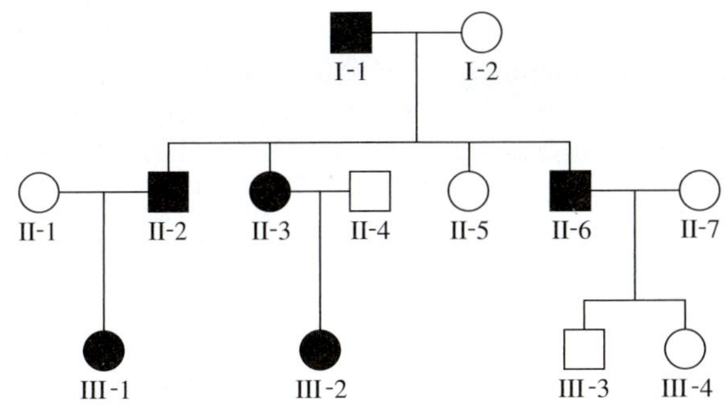

INTERACTIVE EXERCISE

FIGURE 24.3 A family pedigree of Huntington disease, indicating dominant inheritance. Because the dominant allele is rare, most affected individuals are heterozygotes. Rare cases of people who are homozygous for the disease-causing allele have been reported. Such individuals tend to have more severe symptoms.

CONCEPT CHECK: What feature(s) of this pedigree indicate dominant inheritance?

1. ***An affected offspring usually has one or both affected parents.*** However, this is not always the case. Some dominant traits show incomplete penetrance (see Chapter 4), so a heterozygote may not exhibit the trait even though it may be passed to offspring who do exhibit the trait. Also, a dominant mutation may occur during gametogenesis, so two unaffected parents may produce an affected offspring.

TABLE 24.1
Examples of Human Disorders Inherited in an Autosomal Recessive Manner

Disorder	Chromosomal Location of Gene	Gene Product	Effect of Disease-Causing Allele
Albinism (type I)	11q	Tyrosinase	Inability to synthesize melanin, resulting in white skin, hair, etc.
Cystic fibrosis (CF)	7q	CF transmembrane conductance regulator	Water imbalance in tissues of the pancreas, intestine, sweat glands, and lungs due to impaired ion transport; leads to lung damage
Phenylketonuria (PKU)	12q	Phenylalanine hydroxylase	Foul-smelling urine, neurological abnormalities, mental impairment; may be remedied by diet modification at birth
Sickle cell disease	11p	β Globin	Anemia, blockages in blood circulation
Tay-Sachs disease (TSD)	15q	Hexosaminidase A	Progressive neurodegeneration

TABLE 24.2
Examples of Human Disorders Inherited in an Autosomal Dominant Manner

Disorder	Chromosomal Location of Gene	Gene Product	Effects of Disease-Causing Allele
Aniridia	11p	Pax6 transcription factor	An absence of the iris of the eye, leading to visual impairment and sometimes blindness
Achondroplasia	4p	Fibroblast growth factor receptor-3	A common form of dwarfism associated with a defect in the growth of long bones
Marfan syndrome	15q	Fibrillin-1	Tall and thin individuals with abnormalities in the skeletal, ocular, and cardiovascular systems due to a weakening in the elasticity of certain body tissues
Familial hypercholesterolemia	19p	LDL receptor	Very high serum levels of low-density lipoprotein (LDL), a predisposing factor in heart disease
Huntington disease	4p	Huntingtin	Neurodegeneration that occurs relatively late in life, usually in middle age

2. ***An affected individual with only one affected parent is expected to produce 50% affected offspring (on average).***
3. ***Two affected, heterozygous individuals have (on average) 25% unaffected offspring.***
4. ***The trait occurs with the same frequency in both sexes.***
5. ***For most dominant, disease-causing alleles, the homozygote is more severely affected with the disorder. In some cases, a dominant allele may be lethal in the homozygous condition.***

Numerous autosomal dominant diseases have been identified in humans (Table 24.2). The three common explanations for dominant disorders are haploinsufficiency, a gain-of-function mutation, or a dominant-negative mutation. Let's consider examples of all three types.

Haploinsufficiency is the phenomenon in which a person has only a single functional copy of a gene, and that single functional copy does not produce a normal phenotype. In these disorders, 50% of the normal protein is not sufficient to produce a normal phenotype. Haploinsufficiency shows a dominant pattern of inheritance because a heterozygote (with one normal allele and one inactive allele) has the disease. An example is aniridia, which is a rare disorder that results in an absence of the iris of the eye. Aniridia leads to visual impairment and blindness in severe cases.

A second category of dominant disorders involves **gain-of-function mutations.** Such mutations change the gene product so that it "gains" a new or abnormal function. An example of such a disorder is achondroplasia, which is characterized by abnormal bone growth that results in short stature with relatively short arms and legs. This disorder is caused by a mutation that occurs in the fibroblast growth factor receptor-3 gene. In achondroplasia, the mutant form of the receptor is overactive. This overactivity disrupts the normal signaling pathway and leads to severely shortened bones.

A third category of dominant disorders is characterized by **dominant-negative mutations** in which the altered gene product acts antagonistically to the normal gene product. In humans, Marfan syndrome, which is due to a mutation in the *fibrillin-1* gene, is an example. The *fibrillin-1* gene encodes a glycoprotein that is a structural component of the extracellular matrix that provides structure and elasticity to tissues. The mutant gene encodes a fibrillin-1 protein that antagonizes the effects of the normal protein, thereby weakening the elasticity of certain body parts. For example, the walls of the major arteries such as the aorta, the large artery that leaves the heart, are often affected.

X-Linked Recessive Inheritance Let's now turn to another inheritance pattern common in humans that is called X-linked recessive inheritance (Table 24.3). X-linked recessive inheritance of diseases poses a special problem for males. Why are males more likely to be affected? Most X-linked genes lack a counterpart on the Y chromosome. Males are hemizygous—have a single copy—for these genes. Therefore, a female heterozygous for an X-linked recessive gene passes this trait on to 50% of her sons, as shown in the following Punnett square for hemophilia. In this example, X^H carries the wild-type allele, whereas $X^{h\text{-}A}$ is the chromosome that carries a mutant allele causing hemophilia.

♀ \ ♂	X^H	Y
X^H	X^HX^H Unaffected female	X^HY Unaffected male
$X^{h\text{-}A}$	$X^HX^{h\text{-}A}$ Carrier female	$X^{h\text{-}A}Y$ Male with hemophilia

As mentioned previously, hemophilia is a disorder in which the blood cannot clot properly after an injury. For individuals with this trait, a minor cut may bleed for a very long time, and small injuries can lead to serious internal bleeding because internal broken capillaries may leak blood profusely before they are repaired. Hemophilia A, also called classic hemophilia, is caused by a defect in an X-linked gene that encodes the protein clotting factor VIII. This

TABLE 24.3

Examples of Human Disorders Inherited in an X-linked Recessive Manner

Disorder	Gene Product	Effects of Disease-Causing Allele
Duchenne muscular dystrophy	Dystrophin	Progressive degeneration of muscles that begins in early childhood
Hemophilia A	Clotting factor VIII	Defect in blood clotting
Hemophilia B	Clotting factor IX	Defect in blood clotting
Androgen insensitivity syndrome	Androgen receptor	Missing male steroid hormone receptor; XY individuals have external features that are feminine but internally have undescended testes and no uterus

disease has also been called the "royal disease," because it affected many members of European royal families. The pedigree shown in Figure 24.4 illustrates the prevalence of hemophilia A among the descendants of Queen Victoria of England. The pattern of X-linked recessive inheritance is revealed by the following observations:

1. ***Males are much more likely to exhibit the trait.***
2. ***Mothers of affected males often have brothers or fathers who are affected with the same trait.***
3. ***Daughters of affected males produce, on average, 50% affected sons.***

X-Linked Dominant Inheritance Relatively few genetic disorders in humans follow an X-linked dominant inheritance pattern. In most cases, males are more severely affected than females, probably because females carry an X chromosome with a normal

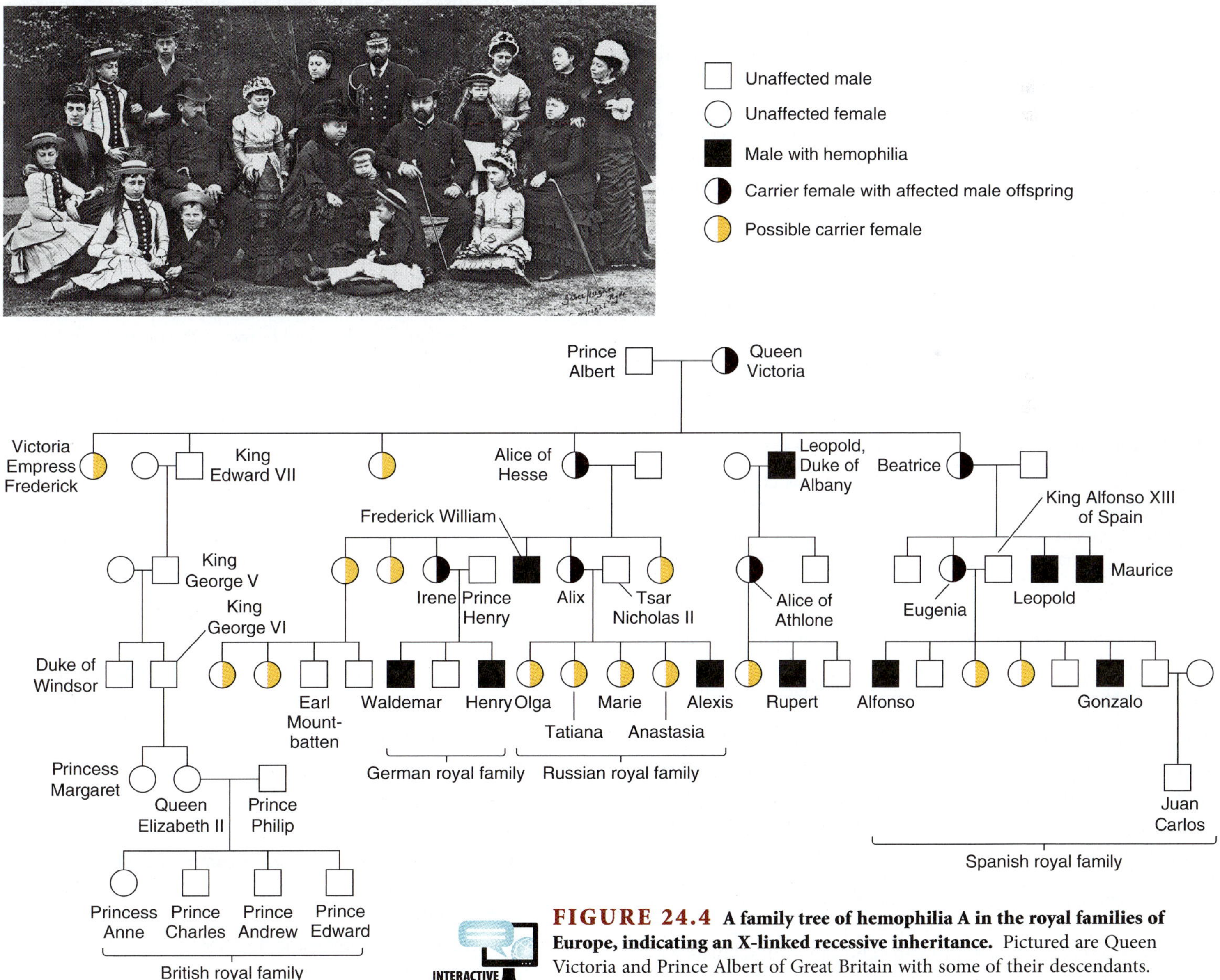

INTERACTIVE EXERCISE

FIGURE 24.4 A family tree of hemophilia A in the royal families of Europe, indicating an X-linked recessive inheritance. Pictured are Queen Victoria and Prince Albert of Great Britain with some of their descendants.

CONCEPT CHECK: What feature(s) of this pedigree indicate X-linked recessive inheritance?

TABLE 24.4
Examples of Human Disorders Inherited in an X-linked Dominant Manner

Disorder	Gene Product	Effects of Disease-Causing Allele
Vitamin D-resistant rickets	Metallopeptidase	Defects in bone mineralization at the sites of bone growth or remodeling, leading to bone deformity and stunted growth in children
Rett syndrome	Methyl-CpG-binding protein-2	A neurodevelopmental disorder that includes a deceleration of head growth and small hands and feet; fatal in males
Aicardi syndrome	Unknown	Characterized by the partial or complete absence of a key structure in the brain called the corpus callosum, and the presence of retinal abnormalities; fatal in males
Incontinentia pigmenti	NFκβ essential modulator	Characterized by morphological and pigmentation abnormalities in the skin, hair, teeth and nails; fatal in males

copy of the gene in question. In most of the X-linked dominant disorders listed in **Table 24.4**, male embryos die at an early stage of development so that most individuals exhibiting the disorder are females. Also, due to their dominant nature and severity, persons with some of the disorders listed in Table 24.4 do not reproduce. Therefore, these dominant disorders, which include Rett syndrome and Aicardi syndrome, are not passed from parent to offspring. Instead, they are caused by new mutations that occur during gamete formation or early embryogenesis. For those X-linked dominant disorders in which the offspring can reproduce, the following pattern is often observed:

1. ***Females are much more likely to exhibit the trait when it is lethal to males.***
2. ***Affected mothers have a 50% chance of passing the trait to daughters.*** Note: Affected mothers also have a 50% chance of passing the trait to sons, but for many of these disorders, affected sons are not observed because of lethality.

Many Genetic Disorders Exhibit Locus Heterogeneity

Hemophilia, which we considered earlier in this chapter, can be used to illustrate another concept in genetics called **locus heterogeneity.** This term refers to the phenomenon in which a particular type of disease may be caused by mutations in two or more different genes. For example, blood clotting involves the participation of several different proteins that take part in a cellular cascade that leads to the formation of a clot. Hemophilia is usually caused by a defect in one of three different clotting factors. In hemophilia A, also called classic hemophilia, a protein called factor VIII is missing. Hemophilia B is a deficiency in a different clotting factor, called factor IX. Both factor VIII and factor IX are encoded by different genes on the X chromosome. These two types of hemophilia show an X-linked recessive pattern of inheritance. By comparison, hemophilia C is due to a factor XI deficiency. The gene encoding factor XI is found on chromosome 4, and this form of hemophilia follows an autosomal recessive pattern of inheritance.

Unfortunately, locus heterogeneity may greatly confound pedigree analysis. For example, a human pedigree might contain individuals with X-linked hemophilia and other individuals with hemophilia C. A geneticist who assumed all affected individuals had defects in the same gene would be unable to explain the resulting pattern of inheritance. For disorders such as hemophilia, pedigree analysis is not a major problem because the biochemical basis for this disease is well understood. However, for rare diseases that are poorly understood at the molecular level, locus heterogeneity may profoundly obscure the pattern of inheritance.

24.1 COMPREHENSION QUESTIONS

1. Which of the following would *not* be consistent with the idea that a disorder has a genetic component?
   a. The disorder is more likely to occur in a person's relatives.
   b. The disorder can spread to individuals sharing similar environments.
   c. The disorder tends to develop at a characteristic age.
   d. A correlation is observed between the disorder and a mutant gene.
2. Assuming complete penetrance, which type of inheritance pattern is consistent with the pedigree shown here?

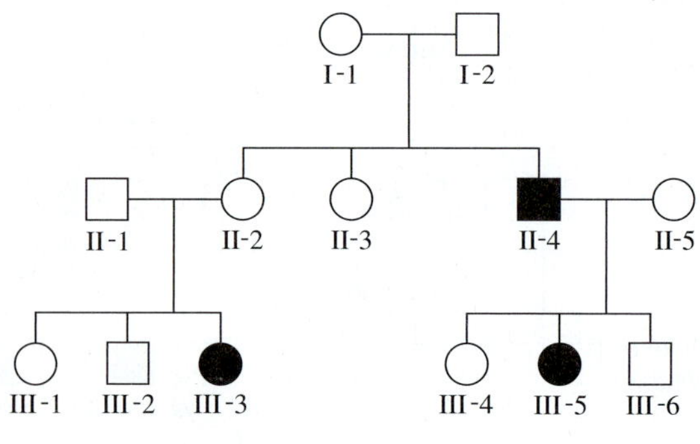

   a. Autosomal recessive
   b. Autosomal dominant
   c. X-linked recessive
   d. X-linked dominant
3. Which of the following is *not* a common explanation for a dominant disorder?
   a. Haploinsufficiency
   b. A change in chromosome number
   c. A gain-of-function mutation
   d. A dominant-negative mutation

4. Locus heterogeneity refers to a genetic disorder that
   a. has a heterogeneous phenotype.
   b. is caused by mutations in two or more different genes.
   c. involves a structural change in multiple chromosomes.
   d. is inherited from both parents.

24.2 DETECTION OF DISEASE-CAUSING ALLELES VIA HAPLOTYPES

Learning Outcomes:

1. Define haplotype.
2. Explain how haplotypes are analyzed to identify disease-causing alleles in humans.

Because mutant genes are known to play a role in thousands of diseases, researchers have devoted great effort to identifying alleles associated with genetic diseases. In this section, we will explore various approaches used to identify mutant alleles that cause disease.

Haplotypes Exhibit Genetic Variation

To identify disease-causing alleles, researchers often rely on the known locations of genes and molecular markers along chromosomes that have been characterized in human populations. A disease-causing allele may be identified due to its proximity to another known gene or its proximity to molecular markers.

As discussed in Chapter 22, researchers can characterize chromosomes at the molecular level and determine the precise locations of genes and molecular markers along each chromosome. During the course of evolution, new mutations arise that alter the DNA sequences of genes and molecular markers. For this reason, homologous chromosomes exhibit gene differences (i.e., allelic variation) and show variation in their molecular markers.

As an example, **Figure 24.5** considers a pair of homologous chromosomes from two different individuals and focuses on four sites (called 1, 2, 3, and 4) that occur at particular locations along those chromosomes. These sites could be within particular genes or they could be molecular markers used in mapping studies. In this drawing, each site is also given a letter designation (A, B, or C) depending on the variation in the DNA sequence at the site. In individual 1, sites 1, 2, and 4 differ at one bp between the homologs. In individual 2, all four sites differ at one or two bp. Also note that individuals 1 and 2 differ with regard to some of these sites.

The term **haplotype,** which is a contraction for haploid genotype, refers to the linkage of alleles or molecular markers along a single chromosome. In Figure 24.5, the haplotype for these four sites are shown at the bottom of each chromosome. For example, the haplotype of the left homolog in individual 2 is 1A 2B 3B 4C.

Because mutations are rare events, haplotypes do not dramatically change from one generation to the next due to new mutations. By comparison, haplotypes are more likely to change over the course of a few generations due to crossing over. However, the likelihood of changing a haplotype depends on the distance between the alleles or molecular markers. If two sites are far apart, a crossover is more likely to alter their pattern than if they are close together. If sites 1, 2, 3, and 4 were very close together along this chromosome, the haplotypes shown in this figure would be likely to stay the same after a few generations. For example, a great-great-great grandchild of individual 2 may inherit the haplotype 1A 2B 3B 4C or 1C 2C 3C 4A. In contrast, the inheritance of either haplotype would be much less likely if the sites are far apart and could frequently recombine by crossing over.

Haplotype Association Studies Are Conducted to Identify Disease-Causing Alleles

How do geneticists identify genes that cause disease when they are mutant? Although a variety of approaches may be followed, the hunt often begins with family pedigrees. The goal is to localize a disease-causing allele to a small region on a chromosome that is distinguished by its haplotype. This approach is based on two assumptions:

1. The disease-causing allele had its origin in a single individual known as a **founder,** who lived many generations ago. Since that time, the allele has spread throughout portions of the human population.

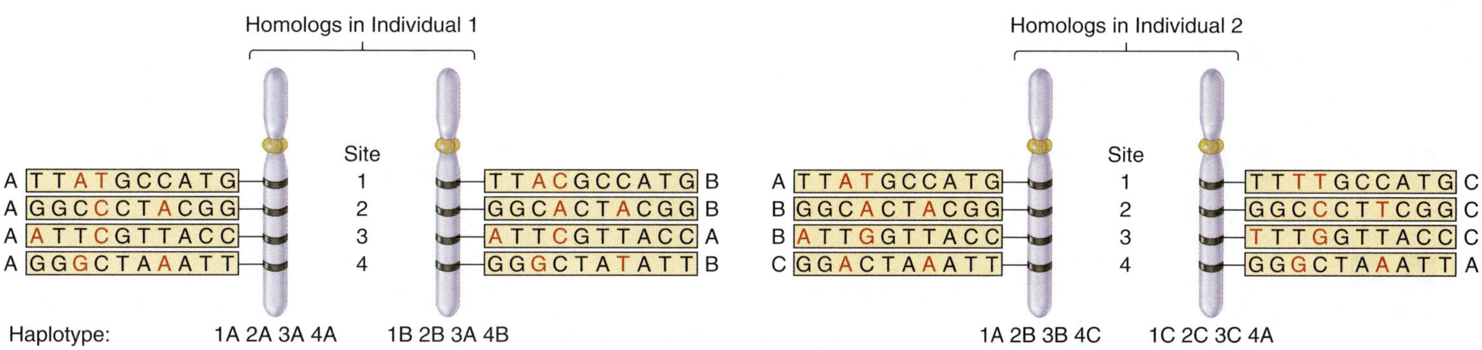

FIGURE 24.5 A schematic representation of haplotypes along a human chromosome. This example considers four sites along a chromosome that may exist in different versions designated A, B, or C. Bases that differ are shown in red.

CONCEPT CHECK: What is a haplotype?

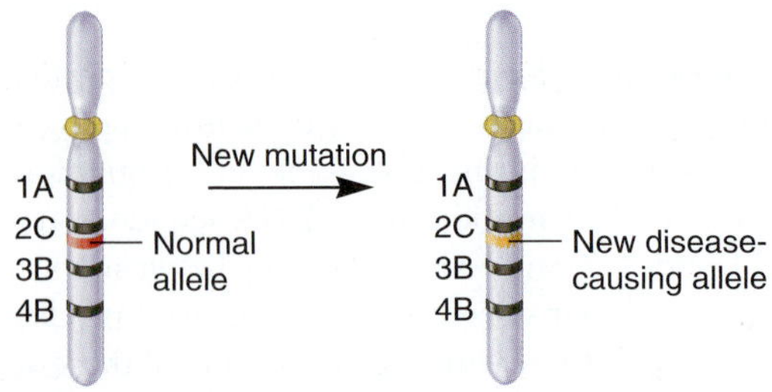

FIGURE 24.6 The occurrence of a new mutation in a founder. In this example, the new mutation occurred in a region with a specific haplotype: 1A 2C 3B 4B. Due to their close linkage, the same haplotype is likely to be found in individuals of succeeding generations who have inherited the mutant allele.

2. When the disease-causing allele originated in the founder, it occurred in a region with a particular haplotype. The haplotype is not likely to have changed over the course of several generations if the disease-causing allele and markers in this region are very close together.

By comparing the transmission patterns of many molecular markers with the occurrence of an inherited disease, researchers can pinpoint particular markers that are closely linked to the disease-causing mutant allele.

To understand this concept, Figure 24.6 shows a situation in which an individual—a founder—has incurred a new mutation that results in a disease-causing allele. The molecular markers designated 1A, 2C, 3B, and 4B are close to the location of this mutant allele. In succeeding generations, the disease-causing allele would more likely be present in individuals with haplotype 1A 2C 3B 4B than in those without this haplotype. Furthermore, people who have inherited the disease-causing allele would be particularly likely to inherit the 2C marker because it is the closest to the disease-causing allele. Due to their close proximity, it would be unlikely that a crossover would separate the 2C marker and the disease-causing allele from each other and create a different haplotype. When alleles and molecular markers are associated with each other at a frequency that is significantly higher than expected by random chance, they are said to exhibit **linkage disequilibrium.** The phenomenon of linkage disequilibrium is common when a disease-causing allele arose in a founder and the allele is closely linked to other markers along a chromosome.

Haplotype Association Studies Identified the Mutant Gene That Causes Huntington Disease

As an example of haplotype association studies, let's consider Huntington disease. Nancy Wexler, whose own mother died of this disorder, has studied Huntington disease among a population of related individuals in Venezuela. To date, over 15,000 individuals have been analyzed. Figure 24.7 shows a simplified version of a large family pedigree from this Venezuelan population in which many individuals were affected with Huntington disease. A specific molecular marker, called G8, is found in four different versions, named A, B, C, or D. In this Venezuelan population, pedigree

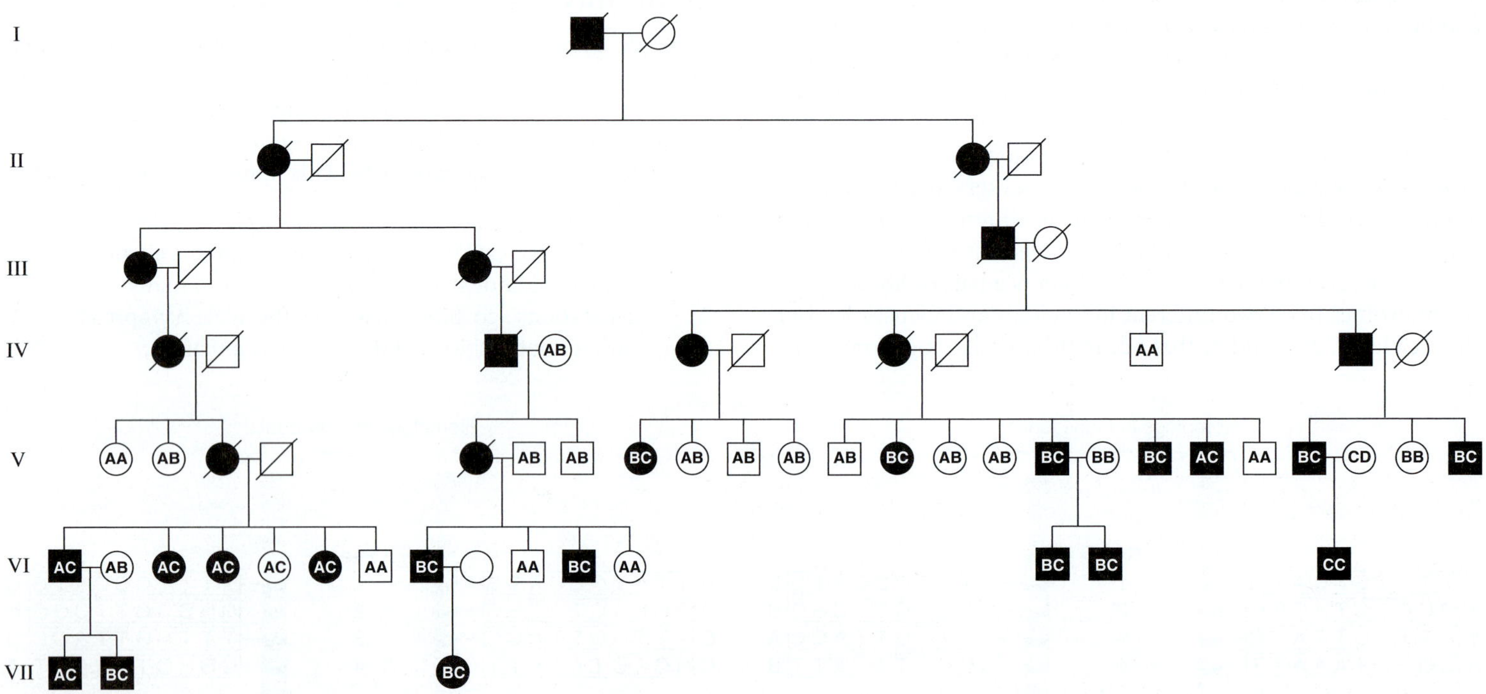

FIGURE 24.7 The transmission pattern of a molecular marker for Huntington disease. Affected members are shown with black symbols. The letters within each symbol indicate the forms of the G8 marker (A, B, C, or D) the individual carries. Affected individuals always carry the C version of this marker. In rare cases, an unaffected individual may also carry the G8-C marker. Symbols with slashes indicate deceased individuals.

(Data from Gusella J.F., Wexler N.S., Conneally P.M., et al. (1983) A polymorphic DNA marker genetically linked to Huntington's disease. *Nature 306*, 234–238.)

CONCEPT CHECK: Explain the connection between the founder and the G8-C marker.

analysis revealed that the G8-C marker, which is located near the tip of the short arm of chromosome 4, is almost always associated with the mutant gene causing Huntington disease. In other words, the G8-C marker is closely linked to the *Huntington* allele.

Once a disease-causing mutant gene has been localized to a short chromosomal region, the next step is to determine which gene in the region is responsible for the disease. Modern haplotype association studies typically localize a gene to a chromosome region that is about 1 Mb (1,000,000 bp) in length. One way to identify a disease-causing allele is chromosome walking (refer back to Figure 22.12). This is how the *Huntingtin* gene was identified. However, the technique of chromosome walking is no longer widely used because the entire human genome has been sequenced and most genes have been identified. Therefore, researchers can analyze the 1-Mb region where a gene has been mapped to determine if a mutant gene in this region is responsible for a disease. In the human genome, a 1-Mb region usually contains about 5–10 different genes, though the number can vary greatly. Therefore, mapping does not definitively tell researchers which gene may play a role in human disease, but it usually narrows down the list to a few candidate genes.

How does a researcher determine which of the candidate genes is the correct one? To further narrow down the list, researchers may also consider biological function. As the scientific community explores the functions of genes experimentally, the data are published in the research literature and placed into databases. In some cases, this information may help to narrow down the list of candidate genes. For example, if the disease of interest is neurological, researchers may discover that only certain genes in the mapped region are expressed in neurons. Also, researchers may compare data from other organisms. If a mutant gene in a mouse causes neurological problems and a human homolog of the mouse gene is found in the mapped region, this human gene would be a good candidate for being responsible for the human disease symptoms.

After researchers have narrowed down the number of candidate genes to as short a list as possible, the next phase is to sequence the candidate gene(s) from many affected and unaffected individuals, using the DNA-sequencing methods described in Chapters 20 or 22. The goal is to identify a gene in which affected individuals always carry a mutation. Identifying a gene that has a genetic change found in affected individuals is strong evidence that the candidate gene causes the disease symptoms.

Why is it useful to identify disease-causing alleles? In many cases, the identification of these alleles helps us to understand how genes contribute to pathogenesis and may even aid in developing strategies aimed at the treatment of the disease. As described in Section 24.3, the identification of such genes may also result in the development of genetic tests that can enable people to determine if they are carrying disease-causing alleles.

The International HapMap Project Is an International Effort to Identify Haplotypes in Human Populations

The genetic sequences of different people are usually very similar. When the chromosomes of two different humans are compared, their DNA sequences differ at about one in every 1200 bases. As described earlier in Figure 24.5, differences in individual bases are by far the most common type of genetic variation. Such differences, which are known as single nucleotide polymorphisms (SNPs), were also described in Chapter 18. The **International HapMap Project** is a worldwide effort to identify such SNPs and other types of human variation (see http://hapmap.ncbi.nlm.nih.gov/). Researchers estimate that approximately 10 million SNPs are commonly found in the human genome.

This project is producing an extensive **HapMap**—a catalog of common genetic variants that occur in human beings. It describes what these variants are, where they are located in the human genome, and how they are distributed among human populations throughout the world. The International HapMap Project is not using the information in the HapMap to understand connections between particular genetic variants and diseases. Instead, the goal of the project is to provide the HapMap so that other researchers can find links between genetic variants and the risk of developing specific diseases. Such links are expected to lead to new methods of diagnosing and treating illnesses.

24.2 COMPREHENSION QUESTIONS

1. What is a haplotype?
   a. A species with one set of chromosomes
   b. A cell with one set of chromosomes
   c. The linkage of alleles or molecular markers along a chromosome
   d. All of the above
2. Haplotype association studies are aimed at the identification of a particular _________ based on _____________.
   a. chromosome, an abnormality in its structure
   b. chromosome, the arrangement of molecular markers
   c. gene, its linkage to other genes or molecular markers
   d. gene, chromosomal rearrangements

24.3 GENETIC TESTING AND SCREENING

Learning Outcomes:

1. Compare and contrast genetic testing versus genetic screening.
2. List different testing methods for genetic abnormalities.
3. Describe how genetic testing can be conducted before birth.

Because genetic abnormalities occur in the human population at a significant level, people have sought ways to determine whether individuals carry disease-causing alleles or other types of genetic abnormalities. The term **genetic testing** refers to the use of testing methods to determine if an individual carries a genetic abnormality. By comparison, the term **genetic screening** refers to population-wide genetic testing. In this section, we will examine both approaches.

Genetic Testing Is Used to Identify Many Inherited Human Diseases

Table 24.5 describes several different testing strategies. In many cases, single-gene mutations that affect the function of proteins can be examined at the protein level. If a gene encodes an enzyme, biochemical assays to measure that enzyme's activity may be available. As mentioned earlier, Tay-Sachs disease involves a defect in the enzyme hexosaminidase A (HexA). Enzymatic assays for this enzyme involve the use of an artificial substrate in which 4-methylumbelliferone (MU) is covalently linked to *N*-acetylglucosamine (GlcNAc). HexA cleaves this covalent bond and releases MU, which is fluorescent.

$$\underset{\text{(nonfluorescent)}}{\text{MU–GlcNAc}} \xrightarrow{\text{HexA}} \underset{\text{(fluorescent)}}{\text{MU}} + \text{GlcNAc}$$

To perform this assay, a small sample of cells is collected and incubated with MU–GlcNAc, and the fluorescence is measured. Individuals affected with Tay-Sachs, who do not produce the HexA enzyme, produce little or no fluorescence, whereas individuals who are homozygous for the normal *hexA* allele produce a high level of fluorescence. Heterozygotes, who have 50% HexA activity, produce intermediate levels of fluorescence.

TABLE 24.5
Testing Methods for Genetic Abnormalities

Method	Description
Protein Level	
Biochemical	As described in this chapter for Tay-Sachs disease, the enzymatic activity of a protein can be assayed in vitro.
Immunological	The presence of a protein can be detected using antibodies that specifically recognize that protein. Western blotting is an example of this type of technique (see Chapter 20).
DNA or Chromosomal Level	
DNA sequencing	If the normal gene has already been identified and sequenced, it is possible to design PCR primers that can amplify the gene from a sample of cells. The amplified DNA segment can then be subjected to DNA sequencing, as described in Chapter 20.
In situ hybridization	A DNA probe that hybridizes to a particular gene or gene segment can be used to determine if the gene is present, absent, or altered in an individual. The technique of fluorescence *in situ* hybridization (FISH) is described in Chapter 22 (see Figure 22.2).
Karyotyping	The chromosomes from a sample of cells can be stained and then analyzed microscopically for abnormalities in chromosome structure and number (see Figure 3.2).
DNA microarrays	Microarrays are used to detect polymorphisms found in the human population that are associated with diseases.

An alternative and more common approach is to detect single-gene mutations at the DNA level. To apply this testing strategy, researchers must have previously identified the mutant gene using molecular techniques. The identification of many human genes, such as those involved in Duchenne muscular dystrophy, cystic fibrosis, and Huntington disease, has made it possible to test for affected individuals or those who may be carriers of these diseases. Table 24.5 describes several ways to test for gene mutations. These laboratory techniques are described in Chapters 20 and 22.

Many human genetic abnormalities involve changes in chromosome number and/or structure. In fact, changes in chromosome number are a common class of human genetic abnormality. Most of these result in spontaneous abortions. However, approximately 1 in 200 live births are aneuploid—have an abnormal number of chromosomes (see Chapter 8, Table 8.1). About 5% of infant and childhood deaths are related to such genetic abnormalities. Changes in chromosome number and many changes in chromosome structure can be detected by karyotyping the chromosomes with a light microscope.

Genetic Screening Identifies Genetic Abnormalities at the Population Level

In the United States, genetic screening for certain disorders has become common medical practice. For example, pregnant women older than 35 often have tests conducted to see if their fetuses are carrying chromosomal abnormalities. As discussed in Chapter 8, these tests are indicated because the rate of such defects increases with the age of the mother. Another example is the widespread screening for phenylketonuria (PKU). An inexpensive test can determine if newborns have this disease. Those who test positive can then be given a low-phenylalanine diet to avoid PKU's devastating effects.

Genetic screening also has been conducted on specific populations in which a genetic disease is prevalent. For example, in 1971, community-based screening for heterozygous carriers of Tay-Sachs disease was begun among specific Ashkenazi Jewish populations. With the use of this screening, over the course of one generation, the incidence of TSD births was reduced by 90%. For most rare genetic abnormalities, however, genetic screening is not routine practice. Rather, genetic testing is performed only when a family history reveals a strong likelihood that a couple may produce an affected child. This typically involves a couple that already has an affected child or has other relatives with a genetic disease.

Genetic testing and screening are medical practices with many social and ethical dimensions. For example, people must decide whether or not they want to make use of available tests, particularly when the disease in question has no cure. For example, Huntington disease typically does not affect people until their 50s and can last 20 years. People who learn they are carriers of genetic diseases such as Huntington disease can be devastated by the news. Some argue that people have a right to know about their genetic makeup; others assert that it does more harm than good. Another issue is privacy. Who should have access to personal genetic information, and how could it be used? Could routine genetic testing lead to discrimination by employers or medical insurance companies? In the coming years, we will gain an ever-increasing awareness of

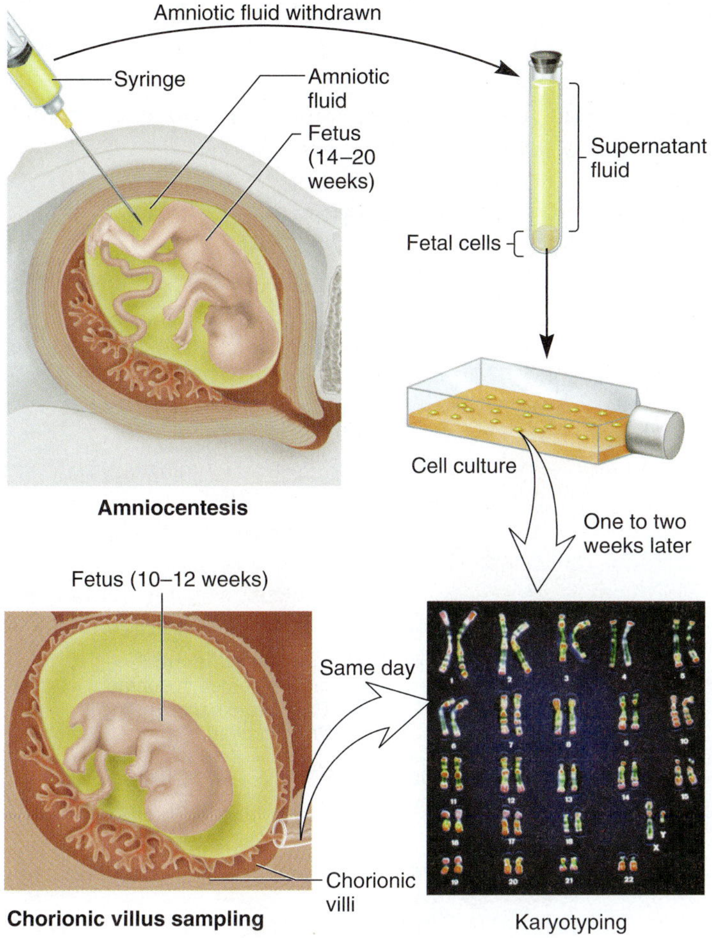

FIGURE 24.8 Techniques for determining genetic abnormalities during pregnancy. In amniocentesis, amniotic fluid is withdrawn, and fetal cells are collected by centrifugation. The cells are then allowed to grow in a laboratory culture medium for one to two weeks prior to karyotyping. In chorionic villus sampling, a small piece of the chorion is removed. These cells can be prepared directly for karyotyping.

our genetic makeup and the underlying causes of genetic diseases. As a society, establishing guidelines for the uses of genetic testing will be a necessary, yet very difficult, task.

Genetic Testing Can Be Performed Prior to Birth

Genetic testing can be performed during pregnancy, which may affect a woman's decision to terminate that pregnancy. The two common ways of obtaining cellular material from a fetus for the purpose of genetic testing are **amniocentesis** and **chorionic villus sampling.** In amniocentesis, a doctor removes amniotic fluid containing fetal cells using a needle that is passed through the abdominal wall (Figure 24.8). Because relatively few fetal cells are found in the amniotic fluid, the cells are grown in the laboratory for 1–2 weeks and then karyotyped to determine the number of chromosomes per cell and whether changes in chromosome structure have occurred. In chorionic villus sampling, a small piece of the chorion (the fetal part of the placenta) is removed, and a karyotype is prepared directly from the collected cells. Chorionic villus sampling can be performed earlier during pregnancy than amniocentesis, usually around the tenth week to twelfth week, compared with the fourteenth to twentieth week for amniocentesis, and results are available sooner. Weighed against these advantages, however, is that chorionic villus sampling may pose a slightly greater risk of causing a miscarriage.

Another method of genetic screening prior to birth is called **preimplantation genetic diagnosis (PGD).** This approach, which is conducted before pregnancy even occurs, involves the genetic testing of embryos that have been produced by **in vitro fertilization (IVF)**—a process in which sperm and egg are combined outside of the mother's body. The testing is typically done to check for a specific genetic abnormality, such as the allele that causes Huntington disease. PGD can also determine if an embryo contains the correct number of chromosomes (also called aneuploidy screening).

PGD is done by removing one or two cells usually at about the eight-cell stage, which occurs 3 days after fertilization. This process is called embryo biopsy or blastomere biopsy. Molecular techniques described in Table 24.5 are then conducted on the removed cell(s) to either check for a particular genetic disease or determine the chromosome composition. The testing can usually be completed in a day or so. Depending on the outcome of the results, a decision can be made whether or not to transfer the embryo into the uterus of the prospective mother in hopes of implantation and the eventual birth of a baby. In most cases, only embryos that do not harbor genetic abnormalities are used. As with the genetic screening of adults, the screening of embryos and fetuses raises many ethical questions.

24.3 COMPREHENSION QUESTIONS

1. Which of the following is *not* a method used in genetic testing?
   a. Chromosome walking c. *In situ* hybridization
   b. DNA sequencing d. DNA microarrays
2. Which of the following prenatal genetic testing methods is done in conjunction with in vitro fertilization?
   a. Amniocentesis
   b. Chorionic villus sampling
   c. Preimplantation genetic diagnosis
   d. All of the above

24.4 PRIONS

Learning Outcomes:

1. Define prion.
2. Explain how prions cause disease.

We now turn to an unusual mechanism in which agents known as prions cause disease. As shown in Table 24.6, prions cause several types of neurodegenerative diseases affecting humans and livestock, including mad cow disease. Recent evidence has shown that prions also exist in yeast. In the 1960s, British researchers Tikvah Alper and John Stanley Griffith discovered that preparations from animals with certain neurodegenerative diseases remained infectious even after exposure to radiation that would destroy any DNA or RNA. They suggested that the infectious agent was a protein.

TABLE 24.6
Neurodegenerative Diseases Caused by Prions*

Disease	Description
Infectious Diseases	
Kuru	A human disease that was once common in New Guinea. It begins with a loss of coordination, usually followed by dementia.
Scrapie	A disease of sheep and pigs characterized by intense itching in which the animals tend to scrape themselves against trees, followed by neurodegeneration
Mad cow disease	Begins with changes in posture and temperament, followed by loss of coordination and neurodegeneration
Human Inherited Diseases	
Creutzfeldt-Jakob disease	Characterized by loss of coordination and dementia
Gerstmann-Straüssler-Scheinker disease	Characterized by loss of coordination and dementia
Familial fatal insomnia	Begins with sleeping and autonomic nervous system disturbances followed by insomnia and dementia

*All of these diseases are eventually fatal.

Furthermore, Alper and Griffith speculated that the protein usually preferred one folding pattern but could sometimes misfold and then catalyze other proteins to do the same. In the early 1970s, Stanley Prusiner, moved by the death of a patient from a neurodegenerative disease, began to search for the causative agent. In 1982, he isolated a disease-causing agent composed entirely of protein, which he called a **prion.** The term emphasizes the prion's unusual character as a proteinaceous infectious agent. Before the discovery of prions, all known infectious agents such as viruses and bacteria contained their own genetic material (either DNA or RNA).

Prion-related diseases arise from the ability of a prion protein to exist in two conformational states: a normal form PrP^{C}, which does not cause disease, and an abnormal form, PrP^{Sc}, which does. (Note: The superscript C refers to the normal conformation, and the superscript Sc refers to the abnormal conformation, such as the one found in the disease called scrapie.) The gene encoding a prion protein (*PrP*) is found in humans and other mammals, and the protein is expressed at low levels in certain types of cells such as neurons. The abnormal conformation of the prion protein can come from two sources. An individual can be infected with the abnormal protein by ingesting the abnormal protein. For example, someone may eat products from an animal that had the disease. Alternatively, some people carry alleles of the *PrP* gene that cause their prion protein to convert spontaneously to the abnormal conformation at a very low rate. These individuals have an inherited predisposition to develop a prion-related disease. An example of an inherited prion disease is familial fatal insomnia (see Table 24.6).

What is the molecular mechanism through which prions cause disease? As shown in **Figure 24.9**, the abnormal conformation, PrP^{Sc}, acts as a catalyst to convert normal prion proteins within the cell to the misfolded conformation. As a prion disease progresses, the PrP^{Sc} proteins form dense aggregates in the cells of the brain and peripheral nervous tissues. This deposition is correlated

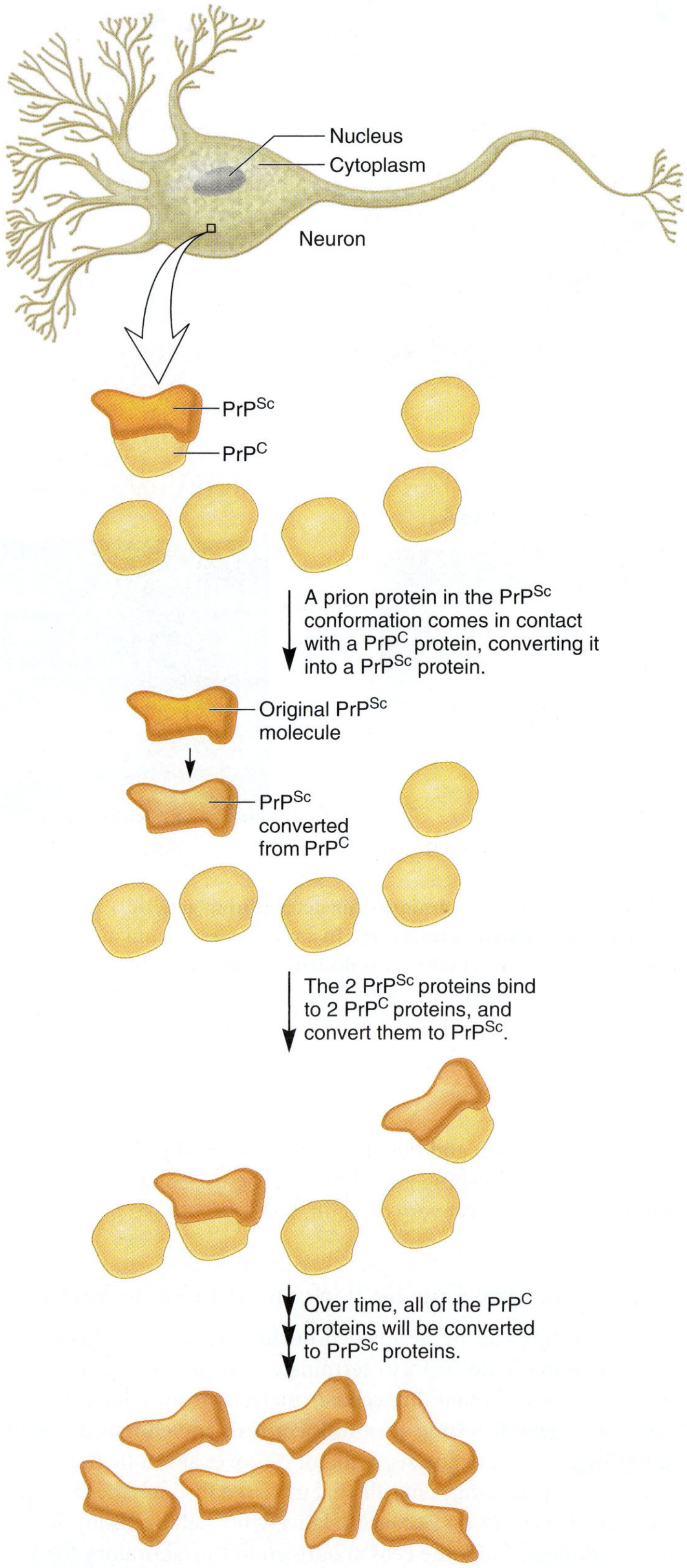

ONLINE ANIMATION

FIGURE 24.9 A proposed molecular mechanism of prion diseases. The PrP^{Sc} protein catalyzes the conversion of PrP^{C} to PrP^{Sc}. Over time, the PrP^{Sc} conformation accumulates to high levels, leading to symptoms of the prion disease.

CONCEPT CHECK: Where does the PrP^{Sc} protein come from?

with the disease symptoms affecting the nervous system. Some of the abnormal prion proteins are also excreted from infected cells and travel through the nervous system to infect other cells.

24.4 COMPREHENSION QUESTION

1. A prion is a disease-causing agent composed of
   a. cells.
   b. nucleic acid and protein coat.
   c. protein alone.
   d. nucleic acid alone.

24.5 GENETIC BASIS OF CANCER

Learning Outcomes:

1. Describe the key characteristics of cancer.
2. Compare and contrast oncogenes versus tumor-suppressor genes.
3. Describe the genetic changes that convert proto-oncogenes into oncogenes or eliminate the function of tumor-suppressor genes.
4. Summarize the different ways that oncogenes or the loss of tumor-suppressor genes can lead to cancer.

Cancer is a disease characterized by uncontrolled cell division. It is a genetic disease at the cellular level. More than 100 kinds of human cancers have been identified, and they are classified according to the type of cell that has become cancerous. Though cancer is a diverse collection of many diseases, some characteristics are common to all cancers.

1. Most cancers originate in a single cell. This single cell, and its line of daughter cells, usually undergoes a series of genetic changes that accumulate during cell division. In this regard, a cancerous growth can be considered to be **clonal** in origin. A hallmark of a cancer cell is that it divides to produce two daughter cancer cells.
2. At the cellular and genetic levels, cancer usually is a multistep process that begins with a precancerous genetic change—a **benign** growth—and is followed by additional genetic changes that lead to cancerous cell growth (Figure 24.10).
3. When cells have become cancerous, their growth is described as **malignant.** Cancer cells are **invasive**—they can invade healthy tissues—and **metastatic**—they can migrate to other parts of the body and cause secondary tumors.

In the United States, approximately 1 million people are diagnosed with cancer each year, and about half that number will die from the disease. In 5–10% of all cases, a predisposition to develop the cancer is an inherited trait. We will examine some inherited forms of cancer later in this chapter. Most cancers, though, perhaps 90–95%, are not passed from parent to offspring. Rather, cancer is usually an acquired condition that typically occurs later in life. Although some cancers are caused by spontaneous mutations and viruses, at least 80% of all human cancers are related to exposure to agents that promote genetic changes in somatic cells. An environmental agent that causes cancer is called a **carcinogen.** Most carcinogens, such as UV light and certain chemicals, are mutagens that alter the DNA in a way that affects the function of normal genes. If the DNA is permanently modified in somatic cells, such changes will be transmitted during cell division. These DNA alterations can lead to effects on gene expression that ultimately affect cell division and thereby lead to cancer.

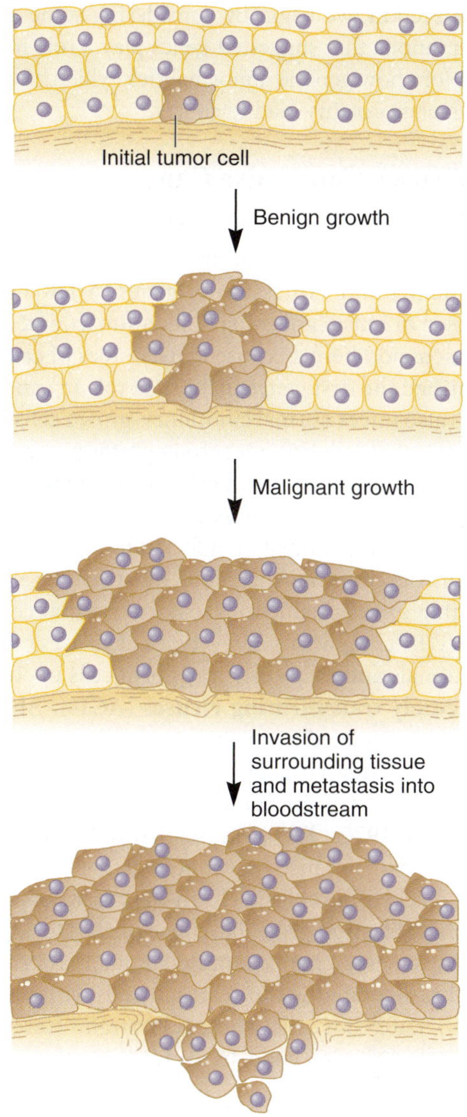

FIGURE 24.10 Progression of cellular growth leading to cancer.

Genes→Traits In a healthy individual, an initial gene mutation converts a normal cell into a tumor cell. This tumor cell divides to produce a benign tumor. Additional genetic changes in the tumor cells may occur, leading to malignant growth. At a later stage in malignancy, the tumor cells invade surrounding tissues, and some malignant cells may metastasize by traveling through the bloodstream to other parts of the body where they can grow and cause secondary tumors. As a trait, cancer can be viewed as a series of genetic changes that eventually lead to uncontrolled cell growth.

In this section, we begin by considering some early experimental observations that suggested genes play a role in cancer. We will then explore how genetic abnormalities, which affect the functions of particular proteins, lead to cancer.

Certain Viruses Can Cause Cancer by Carrying Oncogenes into the Cell

As mentioned, most types of cancers are caused by mutagens that alter the structure and expression of genes. A few viruses, however, are known to cause cancer in plants, animals, and humans. We begin our discussion here, because early studies of cancer-causing viruses identified the first genes that play a role in cancer. Many of these viruses also infect normal laboratory-grown cells and convert them into malignant cells. In cancer biology, the process of converting a normal cell into a malignant cell is called **transformation.**

Most cancer-causing viruses are not very potent at inducing cancer. An organism must be infected for a long time before a tumor actually develops. Furthermore, most viruses are inefficient at transforming or are unable to transform normal cells grown in the laboratory. By comparison, a few types of viruses rapidly induce tumors in animals and efficiently transform cells in culture. These are called **acutely transforming viruses (ACTs).** About 40 such viruses have been isolated from chickens, turkeys, mice, rats, cats, and monkeys. The first virus of this type, the Rous sarcoma virus (RSV), was isolated from chicken sarcomas by Peyton Rous in 1911.

During the 1970s, RSV research led to the identification of the first gene known to cause cancer. Researchers investigated RSV by infecting chicken fibroblast cells in the laboratory. This causes the chicken fibroblasts to grow like cancer cells. During the course of their studies, researchers identified mutant RSV strains that infected and proliferated within chicken cells without transforming them into malignant cells. These RSV strains were determined to contain a defective viral gene. In contrast, in other strains where this gene is functional, cancer occurs. This viral gene was designated *src* for sarcoma, the type of cancer it causes. The *src* gene was the first example of an **oncogene,** a gene that promotes cancer.

Since these early studies of RSV, many other retroviruses carrying oncogenes have been investigated. The characterization of such oncogenes has led to the identification of several genes with cancer-causing potential. In addition to retroviruses, several viruses with DNA genomes cause tumors, and some of these are known to cause cancer in humans (Table 24.7). Researchers estimate that up to 15% of all human cancers are associated with viruses.

TABLE 24.7
Examples of Viruses That May Cause Cancer

Virus	Description
Retroviruses	
Rous sarcoma virus	Causes sarcomas in chickens
Hardy-Zuckerman-4 feline sarcoma virus	Causes sarcomas in cats
DNA Viruses	
Hepatitis B, SV40, polyomavirus	Causes liver cancer in several species, including humans
Papillomavirus	Causes benign tumors and malignant carcinomas in several species, including humans. Causes cervical cancer in humans.
Herpesvirus	Causes carcinoma in frogs and T-cell lymphoma in chickens. A human herpesvirus, Epstein-Barr virus, is a causative agent in Burkitt lymphoma, which occurs primarily in immunosuppressed individuals such as AIDS patients.

EXPERIMENT 24A

DNA Isolated from Malignant Mouse Cells Can Transform Normal Mouse Cells into Malignant Cells

The study of retroviruses and other tumor-producing viruses led to the identification of a few dozen viral oncogenes, which were the first genes implicated in causing cancer. However, most cancers are not caused by viruses but instead are due to environmental mutagens, which cause mutations that alter the expression of normal cellular genes. Therefore, researchers also wanted to identify human genes that have been altered in a way that leads to malignancy. Methods developed in the study of viral oncogenes proved valuable in this search. Particularly useful were studies conducted in 1971 by Miroslav Hill and Jana Hillova. These experiments demonstrated that the purified DNA of RSV-infected cells could be taken up by chicken fibroblasts, transforming them into malignant cells.

In 1979, Robert Weinberg and colleagues wanted to determine if chromosomal DNA purified from cells that had become malignant due to exposure to mutagens could transform normal cells into malignant cells. If so, this would be the first step in the identification of chromosomally located genes that had been converted to oncogenes. Before we discuss this experiment, let's consider how a researcher can identify malignant cells in a laboratory. A widely used assay relies on the ability to recognize a clump of transformed cells as a distinct **focus,** or pile of raised cells, on a culture dish (Figure 24.11). Unlike normal cells, which grow as a single layer, or monolayer, on culture dishes, a malignant focus piles up over the monolayer to form a mass of cells. The focus of cells is a clone derived from a single cell that had become malignant and then proliferated. At the microscopic level, the malignant cells also have altered morphologies.

Now that we understand how malignant cells are identified in a laboratory, let's examine the steps of Weinberg and coworkers' experiment. They began with several malignant mouse cell lines that had been previously characterized, along with normal mouse cell lines. As shown in Figure 24.12, they isolated the DNA from the

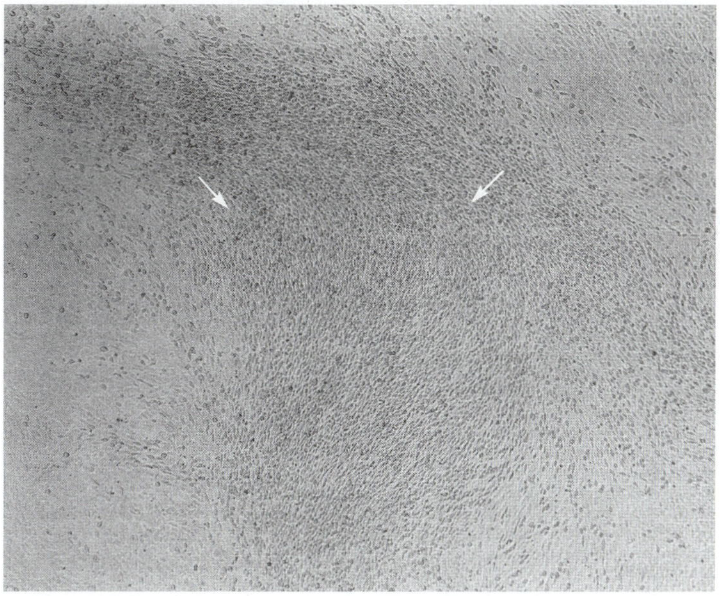

FIGURE 24.11 The malignant growth of cells. The growth of normal cells leads to the formation of a monolayer, while certain transformed cells grow as a focus (see arrows) or a raised pile of cells.

malignant and normal cells and then separately mixed this DNA with normal mouse fibroblast cells. Calcium phosphate was added to promote the uptake of the DNA by the mouse fibroblast cells. Once inside the cells, the DNA was incorporated into the chromosomes of the cells by recombination. The cells were incubated for 2–3 weeks, and then the researchers examined the plates for the growth of malignant foci of cells. In other words, they looked to see if any of the normal mouse cells had been transformed into malignant cells.

THE HYPOTHESIS

DNA isolated from malignant cells will be taken up by normal mouse fibroblast cells and will transform them into malignant cells.

TESTING THE HYPOTHESIS

FIGURE 24.12 Identification of chromosomal oncogenes.

Starting material: Several mouse cell lines. Some of the cell lines were malignant due to exposure to chemical or physical mutagens, whereas others were normal. It was known that none of the cell lines in this experiment were infected with oncogenic viruses.

1. Extract the chromosomal DNA from malignant or normal cell lines.

   Homogenize the cells to break them open.

   Isolate the nuclei by differential centrifugation. See the Appendix.

   Dissolve the nuclear membrane with detergent, and isolate the chromosomal DNA.

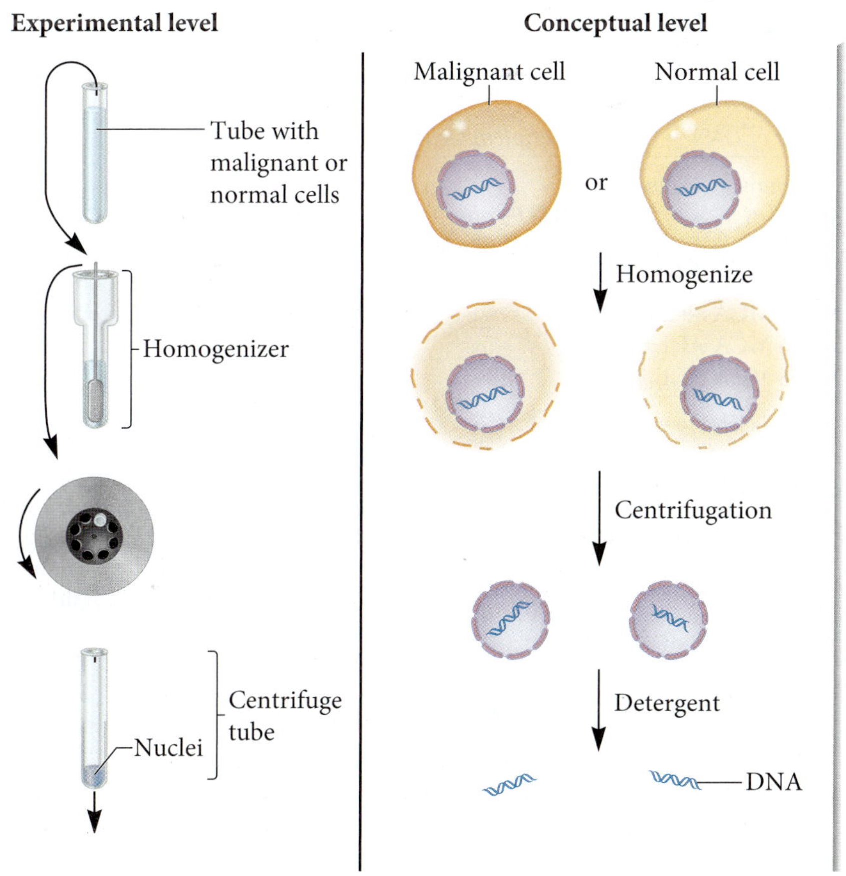

2. Mix the DNA from the malignant cells or from the normal cells with normal mouse fibroblast cells that are growing on a tissue culture plate.

3. Add a buffer containing calcium phosphate. This buffer makes the cells permeable to DNA.

4. Incubate for 2–3 weeks.

5. Examine the plates under a light microscope for cells growing as transformed foci.

   Note: Normal cells in culture grow as a monolayer and very rarely form foci.

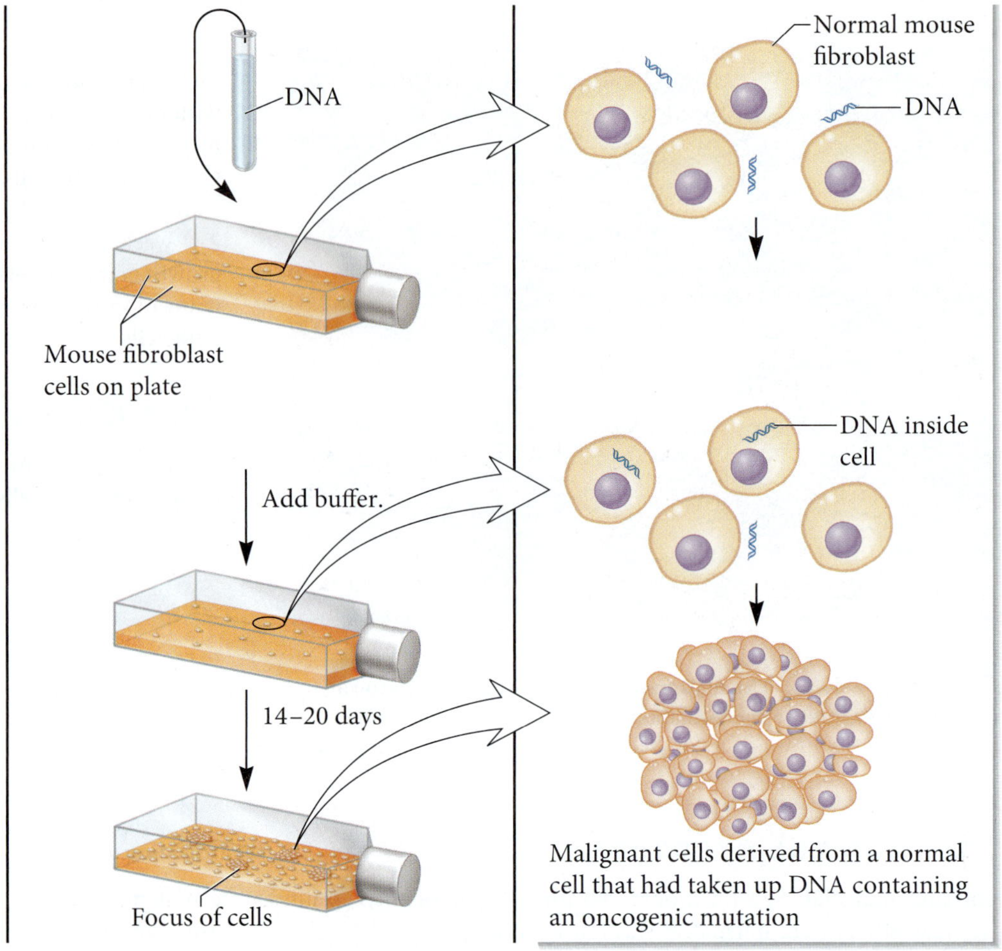

THE DATA

Source of DNA	*Recipient Cells*	*Number of Malignant Foci Found on 12 Plates*
Malignant Cell Lines		
MC5-5-0	NIH3T3 (normal fibroblasts)	48*
MCA16	"	5
MB66 MCA ad 36	"	8
MB66 MCA ACL 6	"	0
MB66 MCA ACL 13	"	0
Normal Cell Lines		
NIH3T3	"	1
C3H10T1/2	"	0

*In this experiment, two of the plates were contaminated, so this is 48 foci on 10 plates.

Data from C. Shih, B.Z. Shilo, M.P. Goldfarb, et al. (1979) Passage of phenotypes of chemically transformed cells via transfection of DNA and chromatin. *Proc. Natl. Acad. Sci. USA. 76*, 5714–5718.

INTERPRETING THE DATA

As shown in The Data, the DNA isolated from some (but not all) malignant cell lines could transform normal mouse cells, which then proliferated and produced malignant foci. These results are consistent with the hypothesis that oncogenes had been taken up and expressed in the normal mouse cells, converting them into malignant cells. By comparison, the DNA isolated from normal cells did not cause a significant amount of transformation. From these experiments, it is unclear why the DNA from two of the malignant cell lines, MB66 MCA ACL 6 and MB66 MCA ACL 13, could not transform the normal mouse cells. One possibility is that some malignancies are caused by dominant oncogenes, and others involve genes that act recessively. Recessive genes would be unable to transform normal mouse cells that already contain the normal (nonmalignant) dominant allele. Later, we will learn how another category of genes involved in cancer, called tumor-suppressor genes, acts recessively.

Just 2 years later, in 1981, the laboratories of Weinberg and Geoffrey Cooper identified the first oncogene in humans. They isolated chromosomal DNA from human bladder carcinoma cells and used it to transform mouse cells in vitro. This chromosomal DNA contained a human oncogene that was the result of a mutation. The identity of the gene was eventually determined by transposon tagging, as described in solved problem S2 at the end of this chapter. This observation paved the way for the isolation of many human oncogenes.

A self-help quiz involving this experiment can be found at www.mhhe.com/brookergenetics5e.

Oncogenes Have Gain-of-Function Mutations That May Affect Proteins Involved in Cell Division Pathways

As described in Chapter 3, eukaryotic cells destined to divide progress through a series of stages known as the cell cycle. You may want to refer back to Figure 3.5. The phases consist of G_1 (first gap), S (synthesis of DNA, the genetic material), G_2 (second gap), and M phase (mitosis and cytokinesis). The G_1 phase is a period in a cell's life when it may become committed to divide. Depending on the conditions, a cell in the G_1 phase may accumulate molecular changes that cause it to progress through the rest of the cell cycle. When this occurs, cell biologists say that a cell has reached a special control point called the restriction point. The commitment to divide is based on a variety of factors. For example, environmental conditions, such as the presence of sufficient nutrients, are important for cell division. In addition, multicellular organisms rely on signaling molecules to coordinate cell division throughout the body. These signaling molecules are often called **growth factors** because they promote cell division.

As researchers began to identify oncogenes, they wanted to understand how these mutant genes promote abnormal cell division. In parallel with cancer research, cell biologists have studied the roles that normal cellular proteins play in cell division. As mentioned, the cell cycle is regulated in part by growth factors that bind to cell surface receptors and initiate a cascade of cellular events that lead eventually to cell division. **Figure 24.13** considers a hormone called epidermal growth factor (EGF), which is secreted from endocrine cells and stimulates epidermal cells, such as skin cells, to divide. As seen here, EGF binds to its receptor, leading to the activation of an intracellular signaling pathway. This pathway, also known as a signal cascade, leads to a change in gene transcription. In other words, the transcription of specific genes is activated in response to the growth factor. Once these genes are transcribed and translated, the gene products function to promote the progression through the cell cycle. Figure 24.13 is just one example of a pathway between a growth factor and gene activation. Eukaryotic species produce many different growth factors, and the signaling pathways are often more complex than the one shown here.

What is the relationship between normal genes and oncogenes? A normal, nonmutated gene that has the potential to become an oncogene is termed a **proto-oncogene.** To become an oncogene, a proto-oncogene must incur a mutation that causes its expression to be abnormally active. This is another example of a gain-of-function mutation, which we discussed earlier in this chapter in the context of dominant mutations. A gain-of-function mutation that produces an oncogene typically has one of three possible effects:

1. The amount of the encoded protein is greatly increased.
2. A change occurs in the structure of the encoded protein that causes it to be overly active.
3. The encoded protein is expressed in a cell type where it is not normally expressed.

The mutations that convert proto-oncogenes into oncogenes have been analyzed in many types of cancers. Oncogenes commonly encode proteins that function in cell growth signaling pathways (**Table 24.8**). These include growth factors, growth factor receptors, intracellular signaling proteins, and transcription factors. As an example, mutations that alter the amino acid sequence

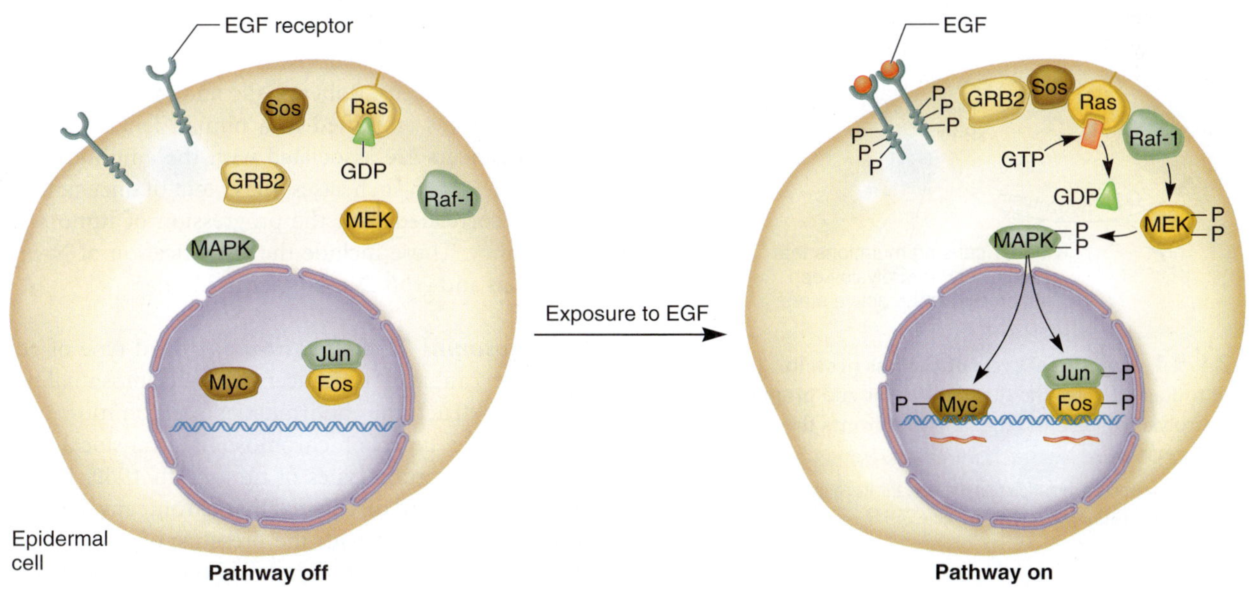

FIGURE 24.13 The activation of a cell-signaling pathway by a growth factor. In this example, epidermal growth factor (EGF) binds to two EGF receptors, causing them to dimerize and phosphorylate each other. An intracellular protein called GRB2 is attracted to the phosphorylated EGF receptor, and it is subsequently bound by another protein called Sos. The binding of Sos to GRB2 enables Sos to activate a protein called Ras. This activation involves the release of GDP and the binding of GTP. The activated Ras/GTP complex then activates Raf-1, which is a protein kinase. Raf-1 phosphorylates MEK, and then MEK phosphorylates MAPK. More than one MAPK may be involved. Finally, the phosphorylated form of MAPK activates transcription factors, such as Myc, Jun, and Fos. This leads to the transcription of genes, which encode proteins that promote cell division.

TABLE 24.8
Examples of Proto-Oncogenes That Can Mutate into Oncogenes

Gene	Cellular Function
Growth Factors*	
sis	Platelet-derived growth factor
int-2	Fibroblast growth factor
Growth Factor Receptors	
erbB	Growth factor receptor for EGF (epidermal growth factor)
fms	Growth factor receptor for NGF (nerve growth factor)
Intracellular Signaling Proteins	
ras	GTP/GDP-binding protein
raf	Serine/threonine kinase
src	Tyrosine kinase
abl	Tyrosine kinase
Transcription Factors	
myc	Transcription factor
jun	Transcription factor
fos	Transcription factor

*The genes described in this table are found in humans as well as other vertebrate species. Many of these genes were initially identified in retroviruses. Most of the genes have been given three-letter names that are abbreviations for the type of cancer the oncogene causes or the type of virus in which the gene was first identified.

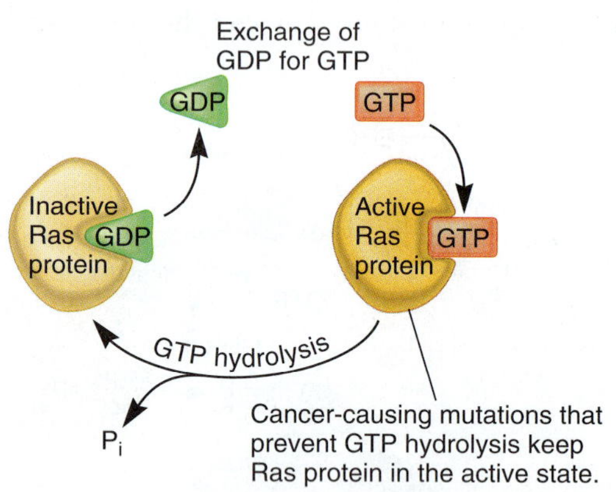

FIGURE 24.14 Functional cycle of the Ras protein. The binding of GTP to Ras activates the function of Ras and promotes cell division. The hydrolysis of GTP to GDP and P_i converts the active form of Ras to an inactive form.

CONCEPT CHECK: How would a mutation that prevents the Ras protein from hydrolyzing GTP affect the EFG pathway described in Figure 24.13?

of the Ras protein have been shown to cause functional abnormalities. The Ras protein is a GTPase, which hydrolyzes GTP to GDP + P_i (**Figure 24.14**). Therefore, after it has been activated, the Ras protein returns to its inactive state by hydrolyzing GTP. Certain mutations that convert the normal *ras* gene into an oncogene decrease the ability of the Ras protein to hydrolyze GTP. This results in a greater amount of the active GTP-bound form of the Ras protein. In this way, such mutations keep the signaling pathway turned on, thereby stimulating the cell to divide.

Genetic Changes in Proto-Oncogenes Convert Them to Oncogenes

How do specific genetic alterations convert proto-oncogenes into oncogenes? By isolating and studying oncogenes at the molecular level, researchers have discovered four main ways this occurs. These changes can be categorized as missense mutations, gene amplifications (i.e., an increase in copy number), chromosomal translocations, and viral integrations.

Missense Mutations As mentioned previously, changes in the structure of the Ras protein can cause it to become permanently activated. These changes are caused by a missense mutation in the *ras* gene. The human genome contains four different but evolutionarily related *ras* genes: *ras*H, *ras*N, *ras*K-4a, and *ras*K-4b. All four homologous genes encode proteins with very similar amino acid sequences containing a total of 188 or 189 amino acids. Missense mutants in these normal *ras* genes are associated with particular forms of cancer. For example, a missense mutation in *ras*H that changes the twelfth amino acid in the protein from a glycine to a valine is responsible for the conversion of *ras*H into an oncogene. Experimentally, chemical carcinogens have been shown to cause these missense mutations and thereby lead to cancer.

Gene Amplification Another genetic event that may occur in cancer cells is gene amplification, or an abnormal increase in the copy number of a proto-oncogene. An increase in gene copy number is expected to increase the amount of the encoded protein, thereby contributing to malignancy. Gene amplification does not normally happen in mammalian cells, but it is a common occurrence in cancer cells. Robert Gallo and Mark Groudine discovered that c-*myc* was amplified in a human leukemia cell line. Many human cancers are associated with the amplification of particular oncogenes. In such cases, the extent of oncogene amplification may be correlated with the progression of tumors to increasing malignancy. These include the amplification of N-*myc* in neuroblastomas and *erbB-2* in breast carcinomas.

Chromosomal Translocation A third type of genetic alteration that can lead to cancer is a chromosomal translocation. Although structural abnormalities are common in cancer cells, very specific types of chromosomal translocations have been identified in certain types of tumors. In 1960, Peter Nowell and David Hungerford discovered that chronic myelogenous leukemia (CML) is correlated with the presence of a shortened version of chromosome 22, which they called the Philadelphia chromosome after the city where it was discovered. Rather than a deletion, this shortened chromosome is the result of a reciprocal translocation between chromosomes 9 and 22. Later studies revealed that this translocation activates a proto-oncogene, *abl*, in an unusual way (**Figure 24.15**). The reciprocal translocation involves breakpoints within the *abl* and *bcr* genes. Following the reciprocal

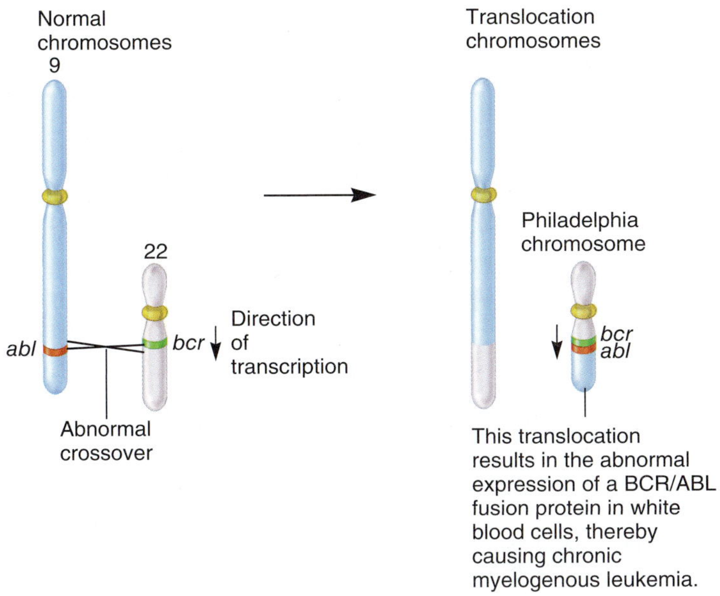

FIGURE 24.15 The reciprocal translocation commonly found in people with chronic myelogenous leukemia.

Genes→Traits In healthy individuals, the *abl* gene is located on chromosome 9, and the *bcr* gene is on chromosome 22. In certain forms of myelogenous leukemia, a reciprocal translocation causes the *abl* gene to fuse with the *bcr* gene. This combined gene, under the control of the *bcr* promoter, encodes an abnormal fusion protein that overexpresses the tyrosine kinase function of the ABL protein in white blood cells and leads to leukemia.

CONCEPT CHECK: Why does this translocation cause leukemia rather than cancer in a different tissue type, such as the lung?

translocation, the coding sequence of the *abl* gene fuses with the promoter and coding sequence of the *bcr* gene. This yields an oncogene that encodes an abnormal fusion protein, which contains the polypeptide sequences encoded from both genes. The *abl* gene encodes a tyrosine kinase enzyme, which uses ATP to attach phosphate groups onto target proteins. This phosphorylation activates certain proteins involved with cell division. Normally, the *abl* gene is highly regulated. However, in the Philadelphia chromosome, the fusion gene is controlled by the *bcr* promoter, which is active in white blood cells. This explains why this fusion causes a type of cancer called a leukemia, which involves a proliferation of white blood cells.

Interestingly, the study of the *abl* gene has led to an effective treatment for CML. Until recently, the only successful treatment was to destroy the patient's bone marrow and then restore blood-cell production by infusing stem cells from the bone marrow of a healthy donor. With knowledge about the function of the ABL protein, researchers have developed the drug imatinib mesylate (Gleevec) that appears to dramatically improve survival rates. This molecule fits into the active site of the ABL protein, preventing ATP from binding there. Without ATP, the ABL protein cannot phosphorylate its target proteins. This prevents the ABL protein from stimulating cell division. In a clinical trial, almost 90% of the CML patients treated with the drug showed no further progression of their disease! In 2001, Gleevec was given FDA approval and is now used to treat CML and certain other types of cancers.

Viral Integration A fourth way that oncogenes can occur is via viral integration. As part of their reproductive cycle, certain viruses integrate their genomes into the chromosomal DNA of their host cell. If the integration occurs next to a proto-oncogene, a viral promoter or enhancer sequence may cause the proto-oncogene to be overexpressed. For example, in certain lymphomas that occur in birds, the genome of the avian leukosis virus has been found to be integrated next to the c-*myc* gene and enhances its level of transcription.

Tumor-Suppressor Genes Play a Role in Preventing the Proliferation of Cancer Cells

Thus far, we have considered how oncogenes promote cancer due to gain-of-function mutations. An oncogene is an abnormally activated gene that leads to uncontrolled cell growth. We now turn our attention to a second category of genes called **tumor-suppressor genes.** As the name suggests, the role of a tumor-suppressor gene is to prevent cancerous growth. Therefore, when a tumor-suppressor gene is inactivated by mutation, cancer is more likely to occur. It is a loss-of-function mutation in a tumor-suppressor gene that promotes cancer.

The first identification of a human tumor-suppressor gene involved studies of retinoblastoma, a tumor that occurs in the retina of the eye. Some people have inherited a predisposition to develop this disease within the first few years of life. By comparison, the noninherited form of retinoblastoma tends to occur later in life but only rarely.

Based on these differences, in 1971, Alfred Knudson proposed a "two-hit" model for retinoblastoma. According to this model, retinoblastoma requires two mutations to occur. People with the hereditary form already have received one mutant gene from one of their parents. They need only one additional mutation in the other copy of this tumor-suppressor gene to develop the disease. Because the retina has more than 1 million cells, it is relatively likely that a mutation may occur in one of these cells at an early age, leading to the disease. However, people with the noninherited form of the disease must have two mutations in the same retinal cell to cause the disease. Because two rare events are much less likely to occur than a single such event, the noninherited form of this disease is expected to occur much later in life and only rarely. Therefore, this hypothesis explains the different populations typically affected by the inherited and noninherited forms of retinoblastoma.

Since Knudson's original hypothesis, molecular studies have confirmed the two-hit hypothesis for retinoblastoma. In this case, the gene in which mutations occur is designated *rb* (for retinoblastoma). This tumor-suppressor gene is found on the long arm of chromosome 13. Most people have two normal copies of the *rb* gene. Persons with hereditary retinoblastoma have inherited one normal and one defective copy. In nontumorous cells throughout the body, they have one functional copy and one defective copy of *rb*. However, in retinal tumor cells, the normal *rb* gene has also suffered the second hit (i.e., a mutation), which renders it defective. Without the tumor-suppressor function, cells are allowed to grow and divide in an unregulated manner, which ultimately leads to cancer.

More recent studies have revealed how the Rb protein suppresses the proliferation of cancer cells (**Figure 24.16**). The Rb protein regulates a transcription factor called E2F, which activates genes required for cell-cycle progression. (The eukaryotic cell cycle is described in Chapter 3; see Figure 3.5.) The binding of the Rb protein to E2F inhibits its activity and prevents the cell from progressing through the cell cycle. As discussed later in this chapter, when a normal cell is supposed to divide, cellular proteins called cyclins bind to cyclin-dependent protein kinases (CDKs). This activates the kinases, which then leads to the phosphorylation of the Rb protein. The phosphorylated form of the Rb protein is released from E2F, thereby allowing E2F to activate genes needed to progress through the cell cycle. What happens when both copies of the *rb* gene are rendered inactive by mutation? The answer is that the E2F protein is always active, which explains why uncontrolled cell division occurs.

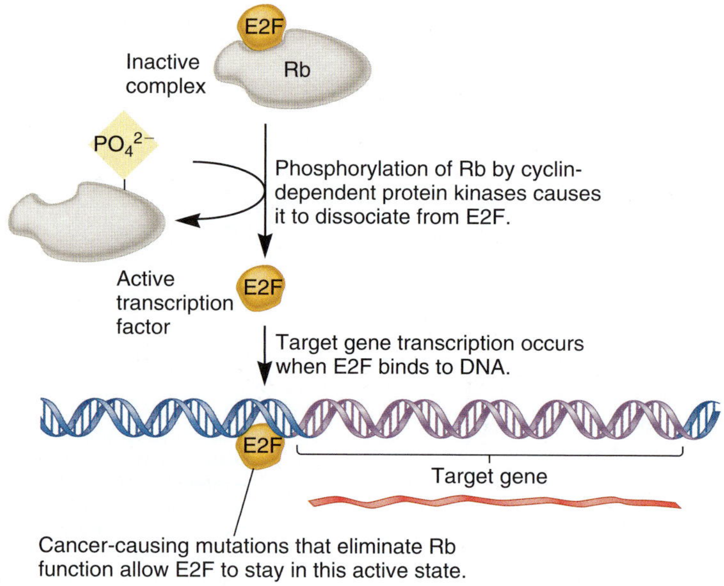

FIGURE 24.16 Interactions between the Rb and E2F proteins. The binding of the Rb protein to the transcription factor E2F inhibits the ability of E2F to function. This prevents cell division. For cell division to occur, cyclins bind to cyclin-dependent protein kinases, which then phosphorylate the Rb protein. The phosphorylated Rb protein is released from E2F. The free form of E2F can activate target genes needed to progress through the cell cycle.

CONCEPT CHECK: If a cell cannot make any Rb protein, how will this affect the function of E2F?

The Vertebrate *p53* Gene Is a Master Tumor-Suppressor Gene That Senses DNA Damage

After the *rb* gene, the second tumor-suppressor gene discovered was the *p53* gene. The *p53* gene is the most commonly altered gene in human cancers. About 50% of all human cancers are associated with defects in *p53*. These include malignant tumors of the lung, breast, esophagus, liver, bladder, and brain as well as sarcomas, lymphomas, and leukemias. For this reason, an enormous amount of research has been aimed at elucidating the function of the p53 protein.

A primary role of the p53 protein is to determine if a cell has incurred DNA damage. The expression of the *p53* gene is caused by the formation of damaged DNA. The event that induces *p53* gene expression appears to be double-stranded breaks in the DNA. The p53 protein functions as a transcription factor. It contains a DNA-binding domain and a transcriptional activation domain. If damage is detected, p53 can promote three types of cellular pathways aimed at preventing the proliferation of cells with damaged DNA.

1. When confronted with DNA damage, the expression of *p53* activates genes involved with DNA repair. This may prevent the accumulation of mutations that activate oncogenes or inactivate tumor-suppressor genes.
2. If a cell is in the process of dividing, it can arrest itself in the cell cycle. By stopping the cell cycle, a cell has more time to repair its DNA and avoid producing two mutant daughter cells. For this to happen, the p53 protein stimulates the expression of another gene termed *p21*. The p21 protein inhibits cyclin/CDK protein complexes that are needed to progress from the G_1 phase of the cell cycle to the S phase.
3. The most drastic event is that the expression of *p53* can initiate a series of events called **apoptosis,** or programmed cell death. In response to DNA-damaging agents, a cell may self-destruct if adequate DNA repair is not possible. Apoptosis is an active process that involves cell shrinkage, chromatin condensation, and DNA degradation. This process is facilitated by proteases known as **caspases,** which are sometimes called the "executioners" of the cell. Caspases digest selected cellular proteins, such as components of the intracellular cytoskeleton. This causes the cell to break down into small vesicles that are eventually phagocytized by cells of the immune system. In this way, an organism can eliminate cells with cancer-causing potential.

Tumor-Suppressor Genes Can No Longer Inhibit Cancer When Their Function Is Lost

During the past three decades, researchers have identified many tumor-suppressor genes that, when defective, contribute to the development and progression of cancer (**Table 24.9**). What are the general functions of these genes? They tend to fall into two broad categories: genes that negatively regulate cell division and genes that maintain genome integrity.

Negative Regulators of Cell Division Some tumor-suppressor genes encode proteins that directly affect the regulation of cell division. The *rb* gene is an example. As mentioned earlier, the Rb protein negatively regulates E2F. If both copies of the *rb* gene are inactivated, the growth of cells is accelerated. Therefore, loss of function of these kinds of negative regulators has a direct effect on the abnormal cell division rates seen in cancer cells. In other words, when the Rb protein is lost, a cell becomes more likely to divide.

Maintenance of Genome Integrity Alternatively, other tumor-suppressor genes play a role in the proper maintenance of the integrity of the genome. The term **genome maintenance** refers to cellular mechanisms that prevent mutations from occurring

TABLE 24.9

Functions of Selected Tumor-Suppressor Genes

Gene	Function
Genes That Negatively Regulate Cell Division	
rb	The Rb protein is a negative regulator of E2F (see Figure 24.16). The inhibition of E2F prevents the transcription of certain genes required for DNA replication and cell division.
p16	A protein kinase that negatively regulates cyclin-dependent kinases. This protein controls the transition from the G_1 phase of the cell cycle to the S phase.
NF1	The NF1 protein stimulates Ras to hydrolyze its GTP to GDP. Loss of *NF1* function causes the Ras protein to be overactive, which promotes cell division.
APC	A negative regulator of a cell-signaling pathway that leads to the activation of genes that promote cell division
Genes That Maintain Genome Integrity	
p53	p53 is a transcription factor that is a checkpoint protein that positively regulates a few specific target genes and negatively regulates others in a general manner. It acts as a sensor of DNA damage. It can prevent the progression through the cell cycle and also can promote apoptosis.
BRCA-1, BRCA-2	BRCA1 and BRCA2 proteins are both involved in the cellular defense against DNA damage. These proteins facilitate DNA repair and can promote apoptosis if repair is not achieved.

and/or prevent mutant cells from surviving or dividing. The proteins encoded by such genes help to ensure that gene mutations or changes in chromosome structure and number do not occur and are not transmitted to daughter cells. The proteins that participate in genome maintenance can be subdivided into two classes: checkpoint proteins and those involved directly with DNA repair.

Figure 24.17 shows a simplified diagram of cell-cycle control. Proteins called cyclins and cyclin-dependent protein kinases (CDKs) are responsible for advancing a cell through the four phases of the cell cycle. For example, an activated G_1-cyclin/CDK complex is necessary to advance from the G_1 phase to the S phase. Human cells produce several types of cyclins and CDKs. The formation of activated cyclin/CDK complexes is regulated by a variety of factors. One way is via **checkpoint proteins** that monitor the state of the cell and stop the progression through the cell cycle if abnormalities, such as DNA breaks and improperly segregated chromosomes, are detected. These proteins are called checkpoint proteins because their role is to check the integrity of the genome and prevent cells from progressing past a certain point in the cell cycle if genetic abnormalities are detected. This provides a mechanism for stopping the accumulation of genetic abnormalities that could produce cancer cells within the body. When checkpoint genes are lost, cell division may not be directly accelerated. However, the loss of checkpoint protein function makes it more likely that undesirable genetic changes occur that could cause cancerous growth.

Several checkpoint proteins regulate the cell cycle of human cells. Figure 24.17 shows three of the major checkpoints where these proteins exert their effects. Both the G_1 and G_2 checkpoints involve the functions of proteins that can sense if the DNA has incurred damage. If so, these checkpoint proteins, such as p53, can prevent the formation of active cyclin/CDK complexes. This stops the progression of the cell cycle. A checkpoint also exists in metaphase. This checkpoint is monitored by proteins that can sense if a chromosome is not correctly attached to the spindle apparatus, making it likely that it will be improperly segregated.

In addition to checkpoint proteins, a second class of proteins involved with genome maintenance consists of DNA repair enzymes, which were discussed in Chapter 18 (refer back to Table 18.7). The genes encoding such enzymes are inactivated in certain forms of cancer. The loss of a DNA repair enzyme makes it more likely for a cell to accumulate mutations that could create an oncogene or eliminate the function of a tumor-suppressor gene. In Chapter 18, we considered how defects in the nucleotide excision repair process are responsible for the disease called xeroderma pigmentosum, which results in a predisposition to developing skin cancer (refer back to Figure 18.19). As discussed later in this chapter, defects in DNA mismatch repair enzymes can contribute to colorectal cancer. In these cases, the loss of a DNA repair enzyme contributes to a higher mutation rate, which makes it more likely for other genes to incur cancer-causing mutations.

Tumor-Suppressor Genes Can Be Silenced in a Variety of Ways

Thus far, we have considered the functions of proteins that are encoded by tumor-suppressor genes. Cancer biologists also want to understand how tumor-suppressor genes are inactivated, because this knowledge may aid in the prevention of cancer. Researchers have identified three common ways that the function of tumor-suppressor genes can be lost.

1. A mutation can occur specifically within a tumor-suppressor gene to inactivate its function. For example, a mutation could inactivate the promoter of a tumor-suppressor gene or introduce an early stop codon in the coding sequence. Either of these would prevent the expression of a functional protein.
2. Tumor-suppressor genes are sometimes inhibited via DNA methylation. As discussed in Chapter 15, DNA methylation usually inhibits the transcription of eukaryotic genes, particularly when it occurs in the vicinity of the promoter. The methylation of CpG islands near the promoters of tumor-suppressor genes has been found in many types of tumors, suggesting that this form of gene inactivation plays an important role in the formation or progression of malignancy. However, further research is needed to determine why tumor-suppressor genes are aberrantly methylated in cancer cells.
3. Many types of cancer are associated with aneuploidy. As discussed in Chapter 8, aneuploidy involves the loss or addition of one or more chromosomes, so the total number of chromosomes is not an even multiple of a set. In some cases, chromosome loss may contribute to the progression of cancer because the lost chromosome carries one or more tumor-suppressor genes.

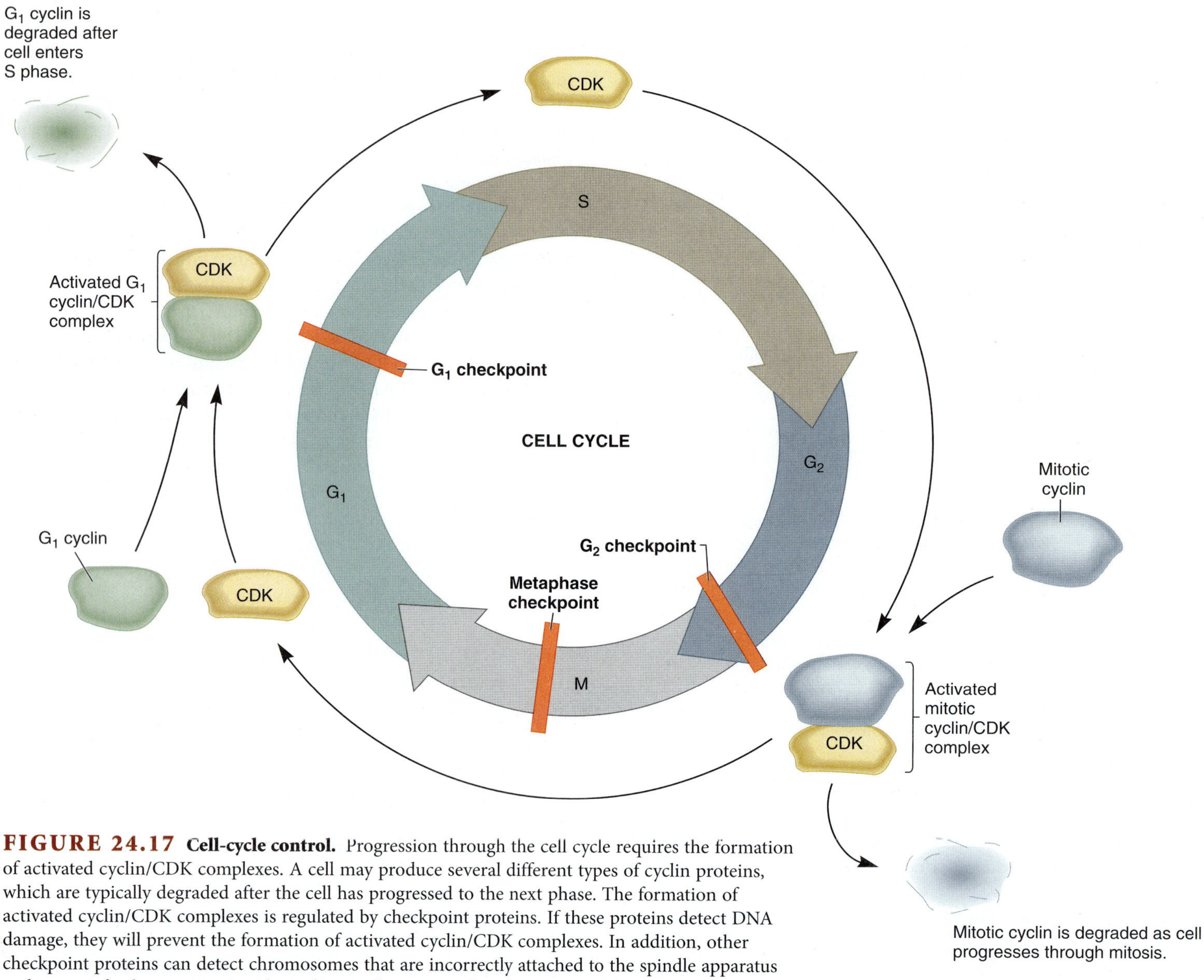

FIGURE 24.17 **Cell-cycle control.** Progression through the cell cycle requires the formation of activated cyclin/CDK complexes. A cell may produce several different types of cyclin proteins, which are typically degraded after the cell has progressed to the next phase. The formation of activated cyclin/CDK complexes is regulated by checkpoint proteins. If these proteins detect DNA damage, they will prevent the formation of activated cyclin/CDK complexes. In addition, other checkpoint proteins can detect chromosomes that are incorrectly attached to the spindle apparatus and prevent further progression through metaphase. Note that this is a simplified diagram of the cell cycle of humans.

CONCEPT CHECK: What is a checkpoint?

Most Forms of Cancer Involve Multiple Genetic Changes Leading to Malignancy

The discovery of oncogenes and tumor-suppressor genes, along with molecular techniques that can detect genetic alterations, has enabled researchers to study the progression of certain forms of cancer at the molecular level. Many cancers begin with a benign genetic alteration that, over time and with additional mutations, progresses eventually to malignancy. Furthermore, a malignancy can continue to accumulate genetic changes that make it even more difficult to treat. For example, some tumors may acquire mutations that cause them to be resistant to chemotherapeutic agents.

In 1990, Eric Fearon and Bert Vogelstein proposed a series of genetic changes that lead to colorectal cancer, the second most common cancer in the United States. As shown in **Figure 24.18**, colorectal cancer is derived from cells in the mucosa of the colon. The loss of function of *APC*, a tumor-suppressor gene on chromosome 5, leads to an increased proliferation of mucosal cells and the development of a benign polyp, a noncancerous growth. Additional genetic changes involving the loss of other tumor-suppressor genes and the activation of an oncogene (namely, *ras*) eventually lead to the development of a carcinoma. In Figure 24.18, the genetic changes that lead to colon cancer are portrayed as occurring in an orderly sequence. Although the growth of a tumor often begins with mutations in *APC*, it is the total number of genetic changes, not their exact order, that is important.

Among different types of tumors, researchers have identified a large number of genes that are mutated in cancer cells. Though not all of these mutant genes have been directly shown to affect the growth rate of cells, such mutations are likely to be found in tumors because they provide some type of growth advantage for the cell population from which the cancer developed. For example, certain mutations may enable cells to metastasize to neighboring locations. These mutations may not affect growth rate, but they provide the growth advantage that cancer cells are not limited to growing in a particular location, but can migrate to new locations.

Researchers have estimated that about 300 different genes may play a role in the development of human cancer. With an approximate genome size of about 22,000 genes, this observation indicates that over 1% of our genes have the potential to promote cancer if their function is altered by a mutation.

In addition to mutations within specific genes, other common genetic changes associated with cancer are abnormalities in chromosome structure and number. **Figure 24.19** compares the chromosome composition of a normal male cell and a tumor cell

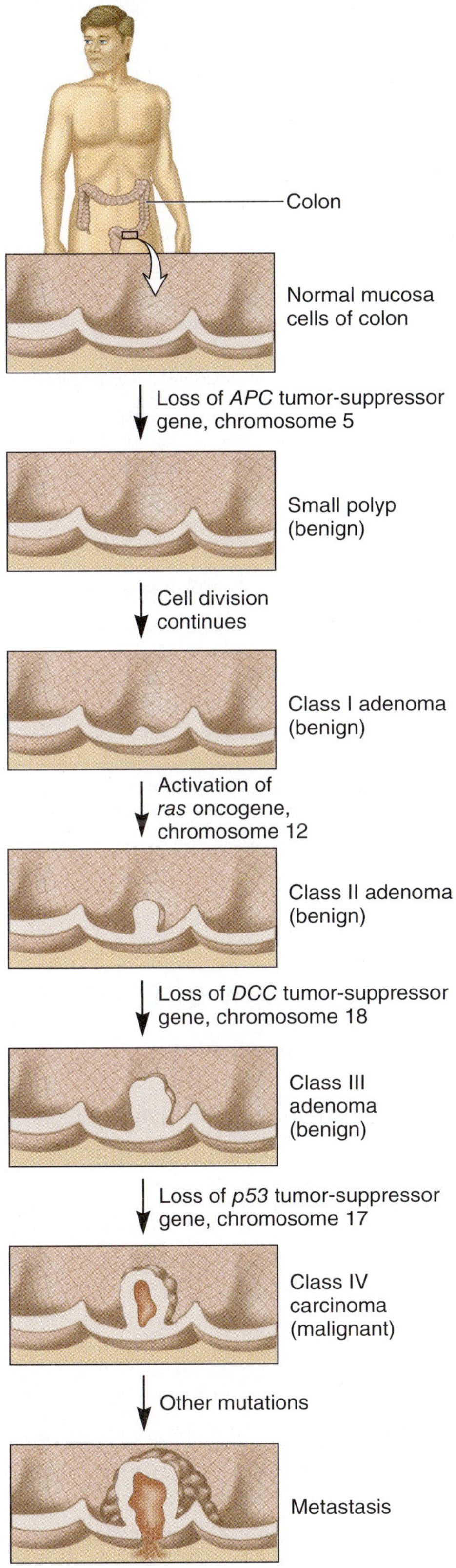

FIGURE 24.18 Multiple genetic changes leading to colorectal cancer.

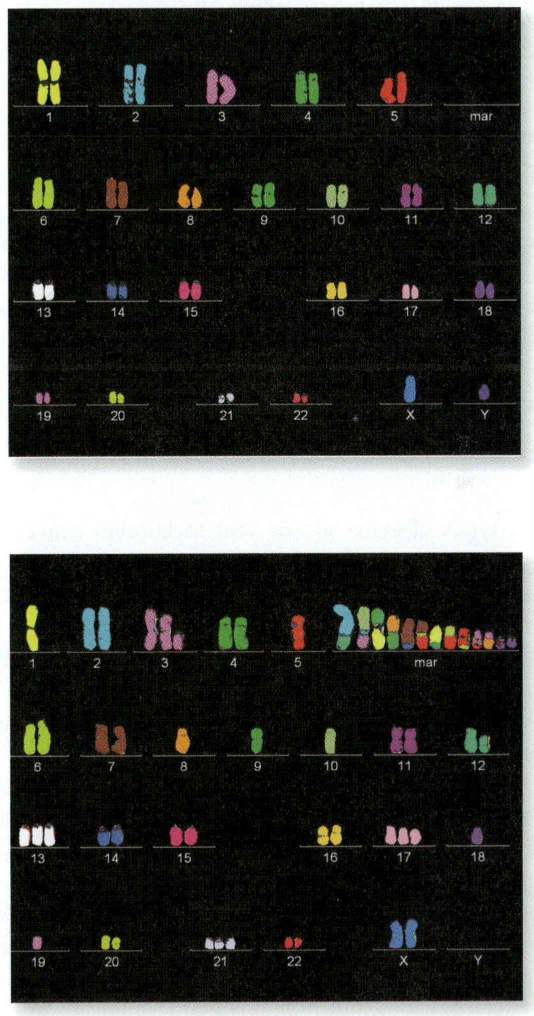

FIGURE 24.19 A comparison between chromosomes found in a normal human cell and a cancer cell from the same person. The bottom set found in a cancer cell is highly abnormal, with extra copies of some chromosomes and lost copies of others. Multiply arranged chromosomes (designated mar in this figure) are made of fused pieces of different chromosomes; they are also common in cancer cells.

taken from the same person. The normal composition for this person is 22 pairs of chromosomes plus two sex chromosomes (X and Y). By comparison, the chromosome composition of the tumor cell is quite bizarre, including the observation that the tumor cell has two X chromosomes, which is characteristic of females. There are many cases in which chromosomes are missing. If tumor-suppressor genes were on these missing chromosomes, their function is lost as well. Figure 24.19 also shows a few cases of extra chromosomes. If these chromosomes contain proto-oncogenes, the expression of those genes may be overactive. Finally, tumor cells often contain chromosomes that have translocations. Such translocations may create fused genes (as in the case of the Philadelphia chromosome discussed earlier in this chapter), or they may place two genes close together so that the regulatory sequences of one gene affect the expression of the other.

Inherited Forms of Cancers May Be Caused by Defects in Tumor-Suppressor Genes

Before we end our discussion of the genetic basis of cancer, let's consider which genes are most likely to be affected in inherited forms of the disease. As mentioned previously, about 5–10% of all cases of cancer involve inherited (germ-line) mutations. These familial forms of cancer occur because people have inherited mutations from one or both parents that give them an increased susceptibility to developing cancer. This does not mean they will definitely get cancer, but they are more likely to develop the disease than are individuals in the general population. When individuals have family members who have developed certain forms of cancer, they may be tested to determine if they also carry a mutant gene. For example, von Hippel-Lindau disease and familial adenomatous polyposis are examples of syndromes for which genetic testing to identify at-risk family members is considered the standard of care.

What types of genes are mutant in familial cancers? Most inherited forms of cancer involve a defect in a tumor-suppressor gene (**Table 24.10**). In these cases, the individual is heterozygous, with one normal and one inactive allele.

At the level of human pedigree, a predisposition for developing cancer is inherited in a dominant fashion because a heterozygote exhibits this predisposition. **Figure 24.20a** shows a pedigree for familial breast cancer. In this case, individuals with the disorder have inherited a loss-of-function mutation in one *BRCA-1* gene. As seen in the pedigree, the development of breast cancer shows a dominant pattern of inheritance with incomplete penetrance. Most affected individuals have an affected parent.

At the cellular level, however, the actual development of cancer is recessive. A cell must be homozygous for a loss-of-function allele to become cancerous. How does this occur? An individual initially is heterozygous, but then a somatic mutation in the normal *BRCA-1* gene eliminates its function. This somatic mutation makes the cell homozyous for two loss-of-function alleles (**Figure 24.20b**). This phenomenon is called **loss of heterozygosity (LOH)**—the loss of function of a normal allele when the other allele was already inactivated.

TABLE 24.10

Inherited Mutant Genes That Confer a Predisposition to Develop Cancer

Gene	Type of Cancer*
Tumor-Suppressor Genes†	
VHL	Causes von Hippel-Lindau disease, which is typically characterized by a clear-cell renal carcinoma
APC	Familial adenomatous polyposis and familial colon cancer
rb	Retinoblastoma
p53	Li-Fraumeni syndrome, which is characterized by a wide spectrum of tumors, including soft-tissue and bone sarcomas, brain tumors, adenocortical tumors, and premenopausal breast cancers
BRCA-1	Familial breast cancer
BRCA-2	Familial breast cancer
NF1	Neurofibromatosis
MSH2	Hereditary malignant melanoma
MLH1	Nonpolyposis colorectal cancer
XP-A to *XP-G*	UV-sensitive forms of cancer such as basal cell carcinoma
Oncogenes	
RET	Multiple endocrine neoplasia type 2

*Many of the genes described in this table are mutated in more than one type of cancer. The cancers listed are those in which it has been firmly established that a predisposition to develop the disease is commonly due to germ-line mutations in the designated gene.
†*MSH2*, *MLH1*, and *XP-A* to *XP-G* encode proteins that are involved in DNA repair.

24.5 COMPREHENSION QUESTIONS

1. An oncogene is produced from a __________ that has acquired a __________.
   a. proto-oncogene, loss-of-function mutation
   b. proto-oncogene, gain-of-function mutation
   c. tumor-suppressor gene, loss-of-function mutation
   d. tumor-suppressor gene, gain-of-function mutation
2. Which of the following is a type of genetic change that could produce an oncogene?
   a. Missense mutation
   b. Gene amplification
   c. Chromosomal translocation
   d. All of the above
3. Tumor-suppressor genes promote cancer when
   a. they are overexpressed.
   b. they are expressed in the wrong cell type.
   c. their function is inactivated.
   d. they are expressed at the wrong stage of development.

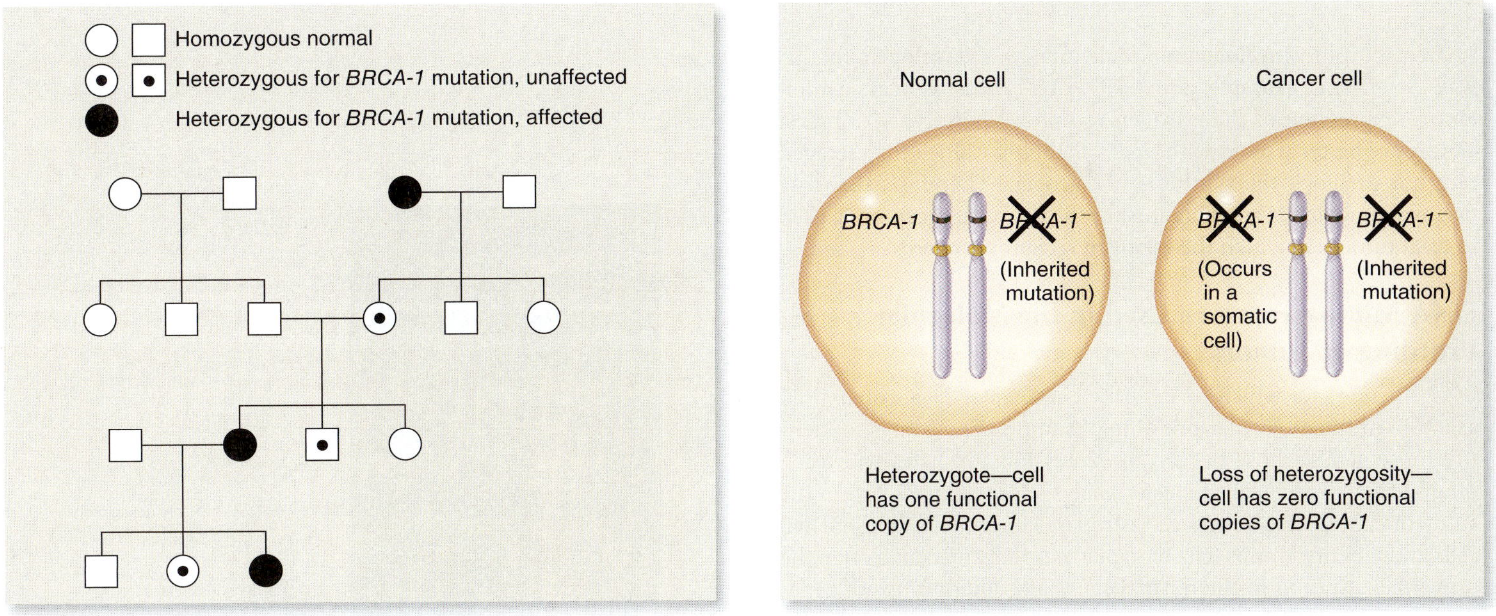

(a) Pedigree for familial breast cancer **(b) Development of cancer at the cellular level**

FIGURE 24.20 Inheritance pattern of familial breast cancer. **(a)** A family pedigree that involves a loss-of-function mutation in the *BRCA-1* gene. The individuals in this pedigree were tested to determine if they carry a mutation in the *BRCA-1* gene. Affected individuals are shown with black symbols. The disorder follows a dominant pattern of inheritance. Note: Males can occasionally develop breast cancer, although it is much more common in females. **(b)** Breast cancer at the cellular level. Normal cells in an affected individual are heterozygous, whereas cancer cells in the same individual have lost their heterozygosity. Therefore, at the cellular level, cancer is recessive because both alleles must be inactivated for it to occur.

CONCEPT CHECK: Explain why familial breast cancer shows a dominant pattern of inheritance in a pedigree even though it is recessive at the cellular level.

4. Normal (nonmutant) tumor-suppressor genes often function
   a. as negative regulators of cell division.
   b. in the maintenance of genome integrity.
   c. in the stimulation of cell division.
   d. both a and b.
5. Most forms of cancer involve
   a. the activation of a single oncogene.
   b. the inactivation of a single tumor-suppressor gene.
   c. the activation of multiple oncogenes.
   d. the activation of multiple oncogenes and the inactivation of multiple tumor-suppressor genes.

24.6 PERSONALIZED MEDICINE

Learning Outcomes:

1. Define personalized medicine and molecular profiling.
2. Describe examples in which personalized medicine affects patient care.

Personalized medicine is the use of information about a patient's genotype and other clinical data in order to select a medication, therapy, or preventative measure that is specifically suited to that patient. As we gain a better understanding of human genes and disease states, researchers expect that personalized medicine will become an increasingly important aspect of health care. In this section, we will begin by examining how personalized medicine can affect treatment options for cancer patients. We will then consider how personalized medicine may play a role in determining the correct dosage for certain types of drugs.

Molecular Profiling Is Increasingly Used to Classify Tumors and Govern Treatment Options

Traditionally, different types of tumors have been classified according to their appearance under a microscope. Although this approach is useful, a major drawback is that two tumors may have a very similar microscopic appearance but yet have very different underlying genetic changes and clinical outcomes. For this reason, researchers and clinicians are turning to methods that enable them to understand the molecular changes that occur in diseases such as cancer. This general approach is called **molecular profiling.**

In cancer biology, molecular profiling involves the identification of the genes that play a role in the development of cancer. Why is this useful? First, molecular profiling can distinguish between tumors that look very similar under the microscope. Second, researchers are optimistic that molecular profiling may lead to improved treatment options. As we gain a better understanding of the genetic changes associated with particular types of cancers, researchers may be able to develop drugs that specifically target the proteins that are encoded by cancer-causing gene mutations. As discussed earlier, the drug imatinib mesylate, which is used to treat chronic myelogenous leukemia, was developed in this way. As another example, about 70% of all breast cancers exhibit an overexpression of the estrogen receptor. These types of breast cancer are better treated with drugs that either block the estrogen receptor or block the synthesis of estrogen. Therefore, the drug tamoxifen, which is an antagonist of the estrogen receptor, is used to treat tumors in which the estrogen receptor is overexpressed.

DNA Microarrays Are Used in the Molecular Profiling of Tumors

DNA microarrays, which are described in Chapter 23, are often used as a tool in the molecular profiling of tumors. The goal is to identify those genes whose pattern of expression correlates with each other—an approach called cluster analysis (see Figure 23.2). In the study of cancer, researchers can compare cancer cells with normal cells and identify groups (clusters) of genes that are turned on in the cancer cells and off in the normal cells, and other groups of genes that are turned off in the cancer cells and on in the normal cells. Likewise, researchers can compare two different types of tumors and identify groups of genes that show different patterns of expression.

As an example, **Figure 24.21a** shows a computer-generated image that illustrates the results of a microarray analysis of 47 samples, most of which came from the tumors of patients with a type of cancer called diffuse large B-cell lymphoma (DLBCL). Each column represents the expression pattern of a set of genes from a particular sample. Genes that are expressed are shown in red; those that are not expressed are shown in green. During the course of these studies, the researchers identified two different patterns of gene expression. The tumor samples on the left side showed a set of genes (next to the orange bar) that tended to be turned on in the tumor and another set of genes (next to the blue bar) that tended to be turned off in the tumor. This pattern of gene expression was similar to the pattern found in a type of B cell called germinal center B cells. In contrast, the tumor samples on the right side showed the opposite pattern. The upper genes tended to be turned off in these patients, and the lower genes were turned on. These samples showed a gene expression pattern found in activated B cells. These results suggest that the two groups of tumors may have originated in B cells at different stages of development—those on the left originated in germinal center B cells, whereas those on the right originated in activated B cells. Furthermore, the patients from whom these tumors were derived also appeared to have very different

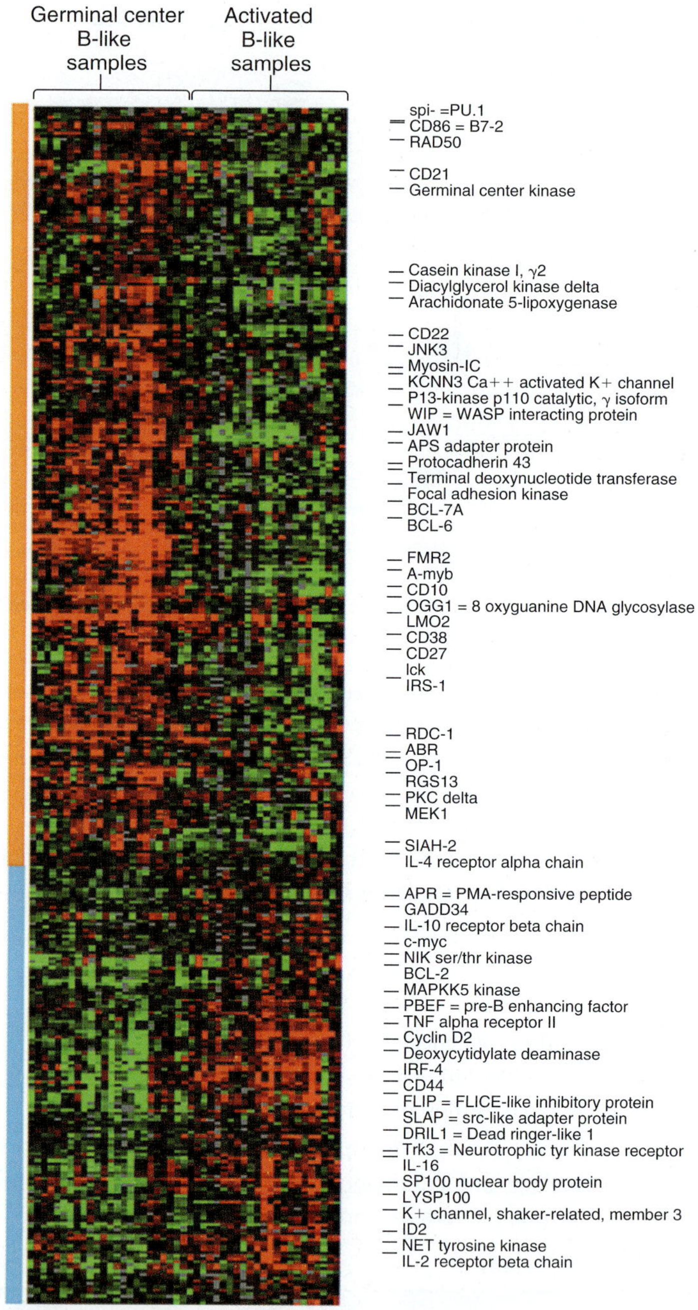

(a) Cluster analysis

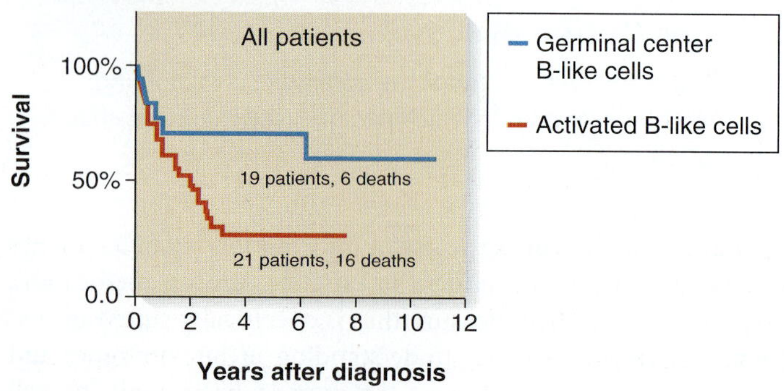

(b) Patient outcomes

FIGURE 24.21 The use of DNA microarrays to classify types of tumors. (a) Forty-seven samples, mostly from patients with diffuse large B-cell lymphoma (DLBCL), were subjected to a DNA microarray analysis. The DNA microarray data were then subjected to a cluster analysis to identify genes that are coordinately expressed. The figure shown here is a graphical illustration of a cluster analysis. Each column represents one sample; each row represents the expression of a particular gene. The names of some of the genes are shown along the right side. (Note: The rows and columns are not easily resolved in this illustration.) Genes highly expressed are shown in red; those not expressed are shown in green. One group of samples had an expression pattern similar to that found in germinal center B cells; the other group had an expression pattern typical of activated B cells. **(b)** Survival of patients with DLBCL.

clinical outcomes (**Figure 24.21b**). The patients whose tumors had a pattern of gene expression similar to activated B cells had a significantly lower overall survival rate than did the other patients.

A Patient's Genotype Is Important in Determining the Proper Dosage of Certain Drugs

Pharmacogenetics is the study or clinical testing of genetic variation that causes differing responses to drugs. The proper dosage for any given drug depends on a variety of factors. Some key factors include the following:

- For drugs that are taken orally, the rate of transport of the drug from the digestive tract into the bloodstream
- The rate of transport of the drug into the appropriate cells where the drug exerts its effect
- The ability of the drug to affect the function of a specific target protein
- The ability of the drug to be taken up and metabolized by the liver
- The rate of excretion of the drug from the body

These five factors are affected by genetics because proteins, which are encoded by genes, are directly involved. For example, transport proteins are often required for the uptake of drugs into the bloodstream, into specific cell types, and into liver cells. Enzymes, which are proteins, are involved in metabolizing drugs into inactive products. Finally, nearly all drugs exert their effects by binding to specific target proteins and altering their function. For example, acetylsalicylic acid, more commonly known as aspirin, binds to a protein called cyclooxygenase, which inhibits its enzymatic function and reduces inflammation. Why is genetics important in the proper dosage of drugs? The answer is that genetic variation in the human population often affects the function of proteins that are involved with drug transport, drug metabolism, and the ability of drugs to affect their target proteins.

To understand how genetic variation can affect the proper dosage of a specific drug, let's consider drug metabolism by the liver. Many drugs are broken down in the liver by a family of related enzymes called cytochrome P450. An example is warfarin (Coumadin), which is used clinically as an anticoagulant, but requires periodic monitoring and is associated with adverse side effects. This drug is metabolized by a cytochrome P450 enzyme designated CYP2C9. Researchers have identified over 80 variants (SNPs) in CYP2C9 in human populations. Some of these variants affect CYP2C9 function. Clinically, the variable activity of CYP2C9 can result in four different levels of warfarin metabolism, which are described as ultrarapid, extensive, intermediate, and poor.

Currently, the dose of warfarin given to patients is either a "one size fits all" approach or it may take into consideration characteristics such as gender, age, and liver function. Adjustments to the dosage are made based on periodic blood tests that measure the level of blood coagulation. Even so, overcoagulation and undercoagulation remain a problem in a significant number of patients. Recently, genetic tests have become available that can help doctors predict the proper warfarin dosage for their patients. For example, persons with an ultrarapid metabolizer genotype would require a higher dosage because the drug would tend to be rapidly broken down in their bodies. By comparison, someone with a poor metabolizer genotype would require a lower dosage. In the future, such genetic tests may be routinely used by doctors to determine the proper drug dosage.

As discussed in Chapter 22, DNA sequencing technologies are progressing to the point where the sequencing of a person's entire genome will be inexpensive enough to be used as a routine diagnostic procedure. Therefore, in the future, many clinicians are predicting that it will become common practice for patients' genome sequences to be determined and analyzed to improve their care. As we gain a better understanding of how genetic variation affects drug action, transport, and metabolism, pharmacogenetics is expected to play an increasingly important role.

24.6 COMPREHENSION QUESTION

1. Personalized medicine may be used
   a. to characterize types of tumors.
   b. to predict the outcome of certain types of cancers.
   c. to determine the proper dosage of drugs.
   d. for all of the above.

KEY TERMS

Page 607. monozygotic (MZ) twins, dizygotic (DZ) twins, concordance
Page 608. age of onset
Page 610. haploinsufficiency, gain-of-function mutations, dominant-negative mutations
Page 612. locus heterogeneity
Page 613. haplotype, founder
Page 614. linkage disequilibrium
Page 615. International HapMap Project, HapMap, genetic testing, genetic screening
Page 617. amniocentesis, chorionic villus sampling, preimplantation genetic diagnosis (PGD), in vitro fertilization (IVF)
Page 618. prion
Page 619. cancer, clonal, benign, malignant, invasive, metastatic, carcinogen
Page 620. transformation, acutely transforming viruses (ACTs), oncogene, focus
Page 623. growth factors, proto-oncogene
Page 625. tumor-suppressor genes
Page 626. apoptosis, caspases, genome maintenance
Page 627. checkpoint proteins
Page 630. loss of heterozygosity (LOH)
Page 631. personalized medicine, molecular profiling
Page 633. pharmacogenetics

CHAPTER SUMMARY

- Approximately 12,000 genetic diseases are known to afflict people.

24.1 Inheritance Patterns of Genetic Diseases

- A genetic basis for a human disease may be suggested by a variety of different observations (see Figure 24.1).
- Thousands of human genetic diseases follow a simple Mendelian pattern of inheritance. These include autosomal recessive, autosomal dominant, X-linked recessive, and X-linked dominant (see Figures 24.2–24.4, Tables 24.1–24.4).
- Recessive diseases are usually caused by loss-of-function alleles, whereas dominant diseases may be caused by haploinsuffiency, gain-of-function mutations, or dominant-negative mutations.
- Many genetic diseases exhibit locus heterogeneity, which means that mutations in more than one gene can cause the disorder.

24.2 Detection of Disease-Causing Alleles via Haplotypes

- The term haplotype refers to the linkage of alleles or molecular markers along a single chromosome (see Figure 24.5).
- Disease-causing mutations may occur in a founder with a specific haplotype (see Figure 24.6).
- Researchers may identify disease-causing alleles by determining their location within a chromosome via haplotype association studies (see Figure 24.7).

24.3 Genetic Testing and Screening

- Genetic testing for abnormalities can be performed in a variety of ways (see Table 24.5).
- Genetic screening is population-wide genetic testing.
- Genetic testing, including amniocentesis, chorionic villus sampling, and preimplantation genetic diagnosis, can be done prior to birth (see Figure 24.8).

24.4 Prions

- Prions are infectious agents composed entirely of protein. They can be acquired by eating infected meat or via a genetic predisposition (see Table 24.6).
- Prions cause disease because the PrP protein in the abnormal (prion) conformation is able to convert normal proteins to the abnormal conformation (see Figure 24.9).

24.5 Genetic Basis of Cancer

- Cancer is clonal in origin. It is usually a multistep process involving several mutations that begins with benign growth and progresses to growth that is invasive and metastatic (see Figure 24.10).
- Certain viruses cause cancer by carrying oncogenes into cells (see Table 24.7).
- Weinberg and colleagues showed that DNA from malignant cells could transform normal mouse cells (see Figures 24.11, 24.12).
- Oncogenes often exert their effects via cell division pathways that control the cell cycle (see Figure 24.13).
- Oncogenes are due to gain-of-function mutations in proto-oncogenes. An example is a mutation in the *ras* gene that prevents the Ras protein from hydrolyzing GTP (see Table 24.8, Figure 24.14).
- Different types of genetic changes can convert proto-oncogenes into oncogenes. These include missense mutations, gene amplification, chromosomal translocation, and viral integration (see Figure 24.15).
- Knudson proposed a two-hit model to explain the occurrence of retinoblastoma (see Figure 24.16).
- The vertebrate *p53* gene senses DNA damage and prevents damaged cells from dividing. It may result in apoptosis.
- Most tumor-suppressor genes encode proteins that are negative regulators of cell division or play a role in genome maintenance (see Table 24.9).
- Checkpoint proteins are a category of proteins that prevent cells that may harbor genetic abnormalities from progressing through the cell cycle (see Figure 24.17).
- Tumor-suppressor genes can be silenced by loss-of-function mutations, DNA methylation, and chromosome loss.
- Most forms of cancer involve multiple genetic changes (see Figure 24.18).
- Alterations in chromosome structure and number are common in cancer cells (see Figure 24.19).
- An inherited predisposition to develop cancer is usually caused by mutations in tumor-suppressor genes (see Table 24.10).
- Familial breast cancer exhibits a dominant pattern of inheritance. At the cellular level, loss of heterozygosity promotes cancer (see Figure 24.20).

24.6 Personalized Medicine

- Personalized medicine is the use of information about a patient's genotype and other clinical data in order to select a medication, therapy, or preventative measure that is specifically suited to that patient.
- Molecular profiling is used to classify tumors, which may affect treatment options. An example is the use of microarrays (see Figure 24.21).
- Pharmacogenetics, which is the study or clinical testing of genetic variation that causes differing responses to drugs, may play an increasing role in the future for selecting the proper dosages of drugs given to patients.

PROBLEM SETS & INSIGHTS

Solved Problems

S1. The pedigree shown below concerns a human disease known as familial hypercholesterolemia.

This disorder is characterized by an elevation of serum cholesterol in the blood. Though relatively rare, this genetic abnormality can be a contributing factor to heart attacks. At the molecular level, this disease is caused by a defective gene that encodes a protein called low-density lipoprotein receptor (LDLR). In the bloodstream, serum cholesterol is bound to a carrier protein known as low-density lipoprotein (LDL). LDL binds to LDLR so that cells can absorb cholesterol. When LDLR is defective, it becomes more difficult for the cells to absorb cholesterol. This explains why the levels of LDL blood cholesterol remain high. Based on the pedigree, what is the most likely pattern of inheritance of this disorder?

Answer: The pedigree is consistent with a dominant pattern of inheritance. An affected individual always has an affected parent. Also, individuals III-8 and III-9, who are both affected, produced unaffected offspring. If this trait was recessive, two affected parents should always produce affected offspring. However, because the trait is dominant, two heterozygous parents can produce homozygous unaffected offspring. The ability of two affected parents to have unaffected offspring is a striking characteristic of dominant inheritance. On average, we would expect that two heterozygous parents should produce 25% unaffected offspring. In the family containing IV-4, IV-5, IV-6, and IV-7, three out of four offspring are actually unaffected. This higher-than-expected proportion of unaffected offspring is not too surprising because the family is a very small group and may deviate substantially from the expected value due to random sampling error.

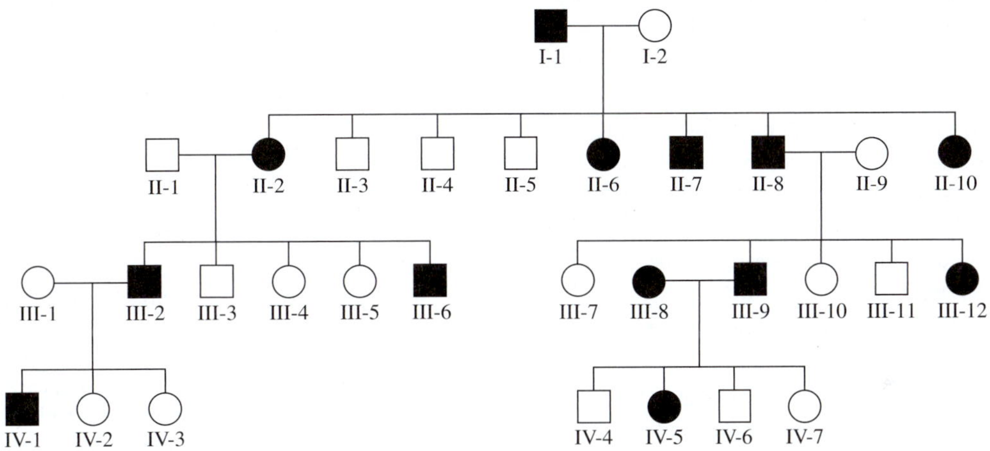

S2. One way to identify a human oncogene is to use human DNA from a malignant cell to transform a mouse cell. A mouse cell that has been transformed by DNA from a malignant human cell may have a human oncogene incorporated into its genome; it also is likely to have *Alu* sequences that are closely linked to this human oncogene. (Note: As discussed in Chapter 19, the human genome contains *Alu* sequences interspersed every 5000–6000 bp. *Alu* sequences are not found in the mouse genome.) Discuss how the *Alu* sequence can provide a way to clone human oncogenes.

Answer: One approach is transposon tagging, described in Chapter 19. When a mouse cell is transformed with a human DNA fragment containing an oncogene, that fragment is likely to contain an *Alu* sequence as well. To clone the human oncogene, the chromosomal DNA can be isolated from the transformed mouse cells, digested with a restriction enzyme, and cloned into vectors to create a library of DNA fragments. The members of the library that carry the human oncogene can be identified using a probe complementary to the *Alu* sequence because this sequence is not found in the mouse genome. Using this strategy, researchers have identified several human oncogenes. In human bladder carcinoma, for example, the human oncogene called *ras* was identified this way.

S3. Oncogenes sometimes result from genetic rearrangements (e.g., translocations) that produce gene fusions. An example is the Philadelphia chromosome, in which a reciprocal translocation between chromosomes 9 and 22 leads to fusion of the first part of the *bcr* gene with the *abl* gene. Suggest two different reasons why a gene fusion could create an oncogene.

Answer: An oncogene is derived from a genetic change that abnormally activates the expression of a gene that plays a role in cell division. A genetic change that creates a gene fusion can abnormally activate the expression of the gene in two ways.

The first way is at the level of transcription. The promoter and part of the coding sequence of one gene may become fused with the coding sequence of another gene. For example, the promoter and part of the coding sequence of the *bcr* gene may fuse with the coding sequence of the *abl* gene. After this has occurred, the *abl* gene is now under the control of the *bcr* promoter, rather than its own normal promoter. Because the *bcr* promoter is turned on in different cells compared with the *abl* promoter, overexpression of the ABL protein occurs in certain cell types compared with its normal level of expression.

A second way that a gene fusion can cause abnormal activation is at the level of protein structure. A fusion protein has parts of two different polypeptides. The first portion of a fusion protein may affect the structure of the second portion of the polypeptide in such a way that the second portion becomes abnormally active, or vice versa.

Conceptual Questions

C1. With regard to pedigree analysis, make a list of the patterns that distinguish among recessive, dominant, and X-linked genetic diseases.

C2. Explain, at the molecular level, why human genetic diseases often follow a simple Mendelian pattern of inheritance, whereas most normal traits, such as the shape of your nose or the size of your head, are governed by multiple gene interactions.

C3. Many genetic disorders exhibit locus heterogeneity. Define and give two examples of locus heterogeneity. How does locus heterogeneity confound a pedigree analysis?

C4. In general, why do changes in chromosome structure or number tend to affect an individual's phenotype? Explain why some changes in chromosome structure, such as reciprocal translocations, do not.

C5. We often speak of diseases such as phenylketonuria (PKU) and achondroplasia as having a "genetic basis." Explain whether the following statements are accurate with regard to the genetic basis of any human disease (not just PKU and achondroplasia).

A. An individual must inherit two copies of a mutant allele to have disease symptoms.

B. A genetic predisposition means that an individual has inherited one or more alleles that make it more likely that he or she will develop disease symptoms than other individuals in a population will.

C. A genetic predisposition to develop a disease may be passed from parents to offspring.

D. The genetic basis for a disease is always more important than the environment.

C6. Figure 24.1 illustrates albinism in two different species. Describe two other genetic disorders found in both humans and animals.

C7. Discuss why a genetic disease might have a particular age of onset. Would an infectious disease have an age of onset? Explain why or why not.

C8. Gaucher disease (type I) is due to a defect in a gene that encodes a protein called acid β-glucosidase. This enzyme plays a role in carbohydrate metabolism within the lysosome. The gene is located on the long arm of chromosome 1. People who inherit two defective copies of this gene exhibit Gaucher disease, the major symptoms of which include an enlarged spleen, bone lesions, and changes in skin pigmentation. Let's suppose a phenotypically unaffected woman, whose father had Gaucher disease, has a child with a phenotypically unaffected man, whose mother had Gaucher disease.

A. What is the probability that this child will have the disease?

B. What is the probability that this child will have two normal copies of this gene?

C. If this couple has five children, what is the probability that one of them will have Gaucher disease and four will be phenotypically unaffected?

C9. Ehler-Danlos syndrome is a relatively rare disorder caused by a mutation in a gene that encodes a protein called collagen (type 3 A1). Collagen is a protein found in the extracellular matrix that plays an important role in the formation of skin, joints, and other connective tissues. People with this syndrome have extraordinarily flexible skin and very loose joints. The pedigree shown here contains several members affected with Ehler-Danlos syndrome, shown with black symbols. Based on this pedigree, does this syndrome appear to be an autosomal recessive, autosomal dominant, X-linked recessive, or X-linked dominant trait? Explain your reasoning.

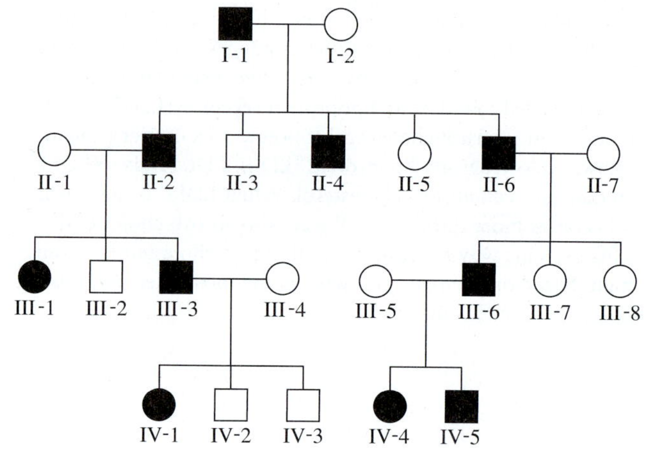

C10. Hurler syndrome is due to a mutation in a gene that encodes a protein called α-L-iduronidase. This protein functions within the lysosome as an enzyme that breaks down mucopolysaccharides (a type of polysaccharide that has many acidic groups attached). When this enzyme is defective, excessive amounts of the mucopolysaccharides dermatan sulfate and heparin sulfate accumulate within the lysosomes, especially in liver cells and connective tissue cells. This leads to symptoms such as an enlarged liver and spleen, bone abnormalities, corneal clouding, heart problems, and severe neurological problems. The pedigree shown here contains three members affected with Hurler syndrome, indicated with black symbols. Based on this pedigree, does this syndrome appear to be an autosomal recessive, autosomal dominant, X-linked recessive, or X-linked dominant trait? Explain your reasoning.

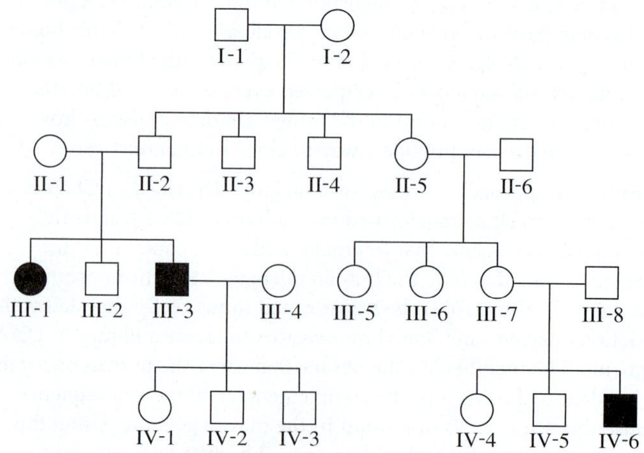

C11. Like Hurler syndrome, Fabry disease involves an abnormal accumulation of substances within lysosomes. However, the lysosomes of individuals with Fabry disease show an abnormal accumulation of lipids. The defective enzyme is α-galactosidase A, which is a lysosomal enzyme that functions in lipid metabolism. This defect causes cell damage, especially to the kidneys, heart, and eyes. The gene that encodes α-galactosidase A is found on the X chromosome. Let's suppose a phenotypically unaffected couple produces two sons with Fabry disease and one phenotypically unaffected daughter. What is the probability that the daughter will have an affected son?

C12. Achondroplasia is a rare form of dwarfism caused by an autosomal dominant mutation that affects the gene that encodes a fibroblast growth factor receptor. Among 1,422,000 live births, the number of babies born with achondroplasia was 31. Among those 31 babies, 18 of them had one parent with achondroplasia. The remaining babies had two unaffected parents. How do you explain these 13 babies, assuming that the mutant allele has 100% penetrance? What are the odds that these 13 individuals will pass this mutant gene to their offspring?

C13. Lesch-Nyhan syndrome is due to a mutation in a gene that encodes a protein called hypoxanthine-guanine phosphoribosyltransferase (HPRT). HPRT is an enzyme that functions in purine metabolism. People afflicted with this syndrome have severe neurodegeneration and loss of motor control. The pedigree shown here contains several members with Lesch-Nyhan syndrome. Affected members are shown with black symbols. Based on this pedigree, does this syndrome appear to be an autosomal recessive, autosomal dominant, X-linked recessive, or X-linked dominant trait? Explain your reasoning.

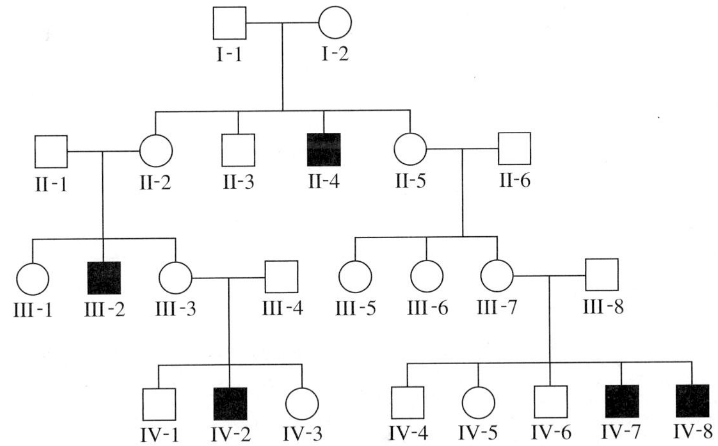

C14. Marfan syndrome is due to a mutation in a gene that encodes a protein called fibrillin-1. It is inherited as a dominant trait. The fibrillin-1 protein is the main constituent of extracellular microfibrils. These microfibrils can exist as individual fibers or associate with a protein called elastin to form elastic fibers. People with the disorder tend to be unusually tall with long limbs, and they may have defects in their heart valves and aorta. Let's suppose a phenotypically unaffected woman has a child with a man who has Marfan syndrome.

A. What is the probability this child will have the disease?

B. If this couple has three children, what is the probability none of them will have Marfan syndrome?

C15. Sandhoff disease is due to a mutation in a gene that encodes a protein called hexosaminidase B. This disease has symptoms that are similar to Tay-Sachs disease. Weakness begins in the first 6 months of life. Individuals exhibit early blindness and progressive mental and motor deterioration. The pedigree shown here contains three members with Sandhoff disease. Affected members are shown with black symbols.

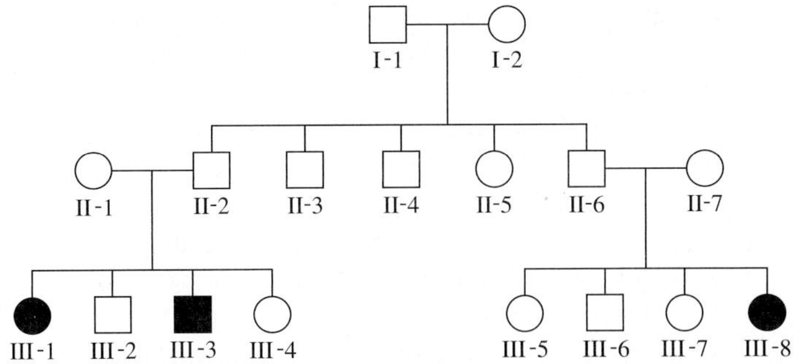

A. Based on this pedigree, does this syndrome appear to be an autosomal recessive, autosomal dominant, X-linked recessive, or X-linked dominant trait? Explain your reasoning.

B. What is the likelihood that II-1, II-2, II-3, II-4, II-5, II-6, and II-7 carry a mutant allele for the hexosaminidase B gene?

C16. Describe the two assumptions that underlie the identification of disease-causing alleles via haplotypes.

C17. What is the purpose of the International HapMap Project? How will it help researchers who study disease-causing alleles?

C18. What is a prion? Explain how a prion relies on normal cellular proteins to cause a disease such as mad cow disease.

C19. Some people have a genetic predisposition for developing prion diseases. Examples are described in Table 24.6. In the case of Gerstmann-Straüssler-Scheinker disease, the age of onset is typically at 30–50 years, and the duration of the disease (which leads to death) is about 5 years. Suggest a possible explanation why someone can live for a relatively long time without symptoms and then succumb to the disease in a relatively short time.

C20. Familial fatal insomnia (described in Table 24.6) is a prion disease inherited as an autosomal dominant trait. Researchers have identified the *PrP* gene, located on the short arm of chromosome 20, and tried to understand the relationship between the *PrP* gene sequence and the molecular basis of familial fatal insomnia.

A. It has been found that a rare mutation at codon 178, changing an aspartic acid to an asparagine, can cause this disease. In addition, codon 129 seems to play a role. The human population is polymorphic at codon 129, which may encode valine or methionine. If codon 178 is the normal aspartic acid (Asp) codon, it does not seem to matter if valine or methionine is found at position 129. In other words, if codon 178 is aspartic acid, Met-129 and Val-129 codons are not associated with disease symptoms. However, if codon 178 specifies asparagine (Asn), then it does matter. Familial fatal insomnia seems to require an asparagine at codon 178 and a methionine at codon 129 in the *PrP* gene sequence. Suggest a possible reason why this is the case.

B. Also, researchers have compared the sequences of the *PrP* gene in many people with familial fatal insomnia. Keep in mind that this is a dominant autosomal trait, so people with this disorder have one mutant copy of the *PrP* gene and one normal copy. People with familial fatal insomnia must have one abnormal copy of the gene that contains Asn-178 and Met-129. The second copy of the gene has Asp-178, and it may have Met-129 or Val-129. Some results suggest that people having a Met-129 codon in this second copy of the *PrP* gene causes the disease to develop more rapidly than in people who have Val-129 in the second copy. Propose an explanation why disease symptoms may occur more rapidly when Met-129 is found in the second copy of the *PrP* gene.

C21. What is the difference between an oncogene and a tumor-suppressor gene? Give two examples.

C22. What is a proto-oncogene? What are the typical functions of proteins encoded by proto-oncogenes? At the level of protein function, what are the general ways that proto-oncogenes can be converted into oncogenes?

C23. What is a retroviral oncogene? Is it necessary for viral infection and proliferation? How have retroviruses acquired oncogenes?

C24. A genetic predisposition to developing cancer is usually inherited as a dominant trait. At the level of cellular function, are the alleles involved actually dominant? Explain why some individuals who have inherited these dominant alleles do not develop cancer during their lifetimes.

C25. Describe the types of genetic changes that commonly convert a proto-oncogene into an oncogene. Give three examples. Explain how the genetic changes are expected to alter the activity of the gene product.

C26. Relatively few inherited forms of cancer involve the inheritance of mutant oncogenes. Instead, most inherited forms of cancer are defects in tumor-suppressor genes. Give two or more reasons why we seldom see inherited forms of cancer involving activated oncogenes.

C27. The *rb* gene encodes a protein that inhibits E2F, a transcription factor that activates several genes involved in cell division. Mutations in *rb* are associated with certain forms of cancer, such as retinoblastoma. Under each of the following conditions, would you expect cancer to occur?

A. One copy of *rb* is defective; both copies of *E2F* are functional.

B. Both copies of *rb* are defective; both copies of *E2F* are functional.

C. Both copies of *rb* are defective; one copy of *E2F* is defective.

D. Both copies of *rb* and *E2F* are defective.

C28. A *p53* knockout mouse in which both copies of *p53* are defective has been produced by researchers. This type of mouse appears normal at birth. However, it is highly sensitive to UV light. Based on your knowledge of *p53*, explain the normal appearance at birth and the high sensitivity to UV light.

C29. With regard to cancer cells, which of the following statements are true?

A. Cancer cells are clonal, which means they are derived from a single mutant cell.

B. To become cancerous, cells usually accumulate multiple genetic changes that eventually result in uncontrolled growth.

C. Most cancers are caused by oncogenic viruses.

D. Cancer cells have lost the ability to properly regulate cell division.

C30. When the DNA of a human cell becomes damaged, the *p53* gene is activated. What is the general function of the p53 protein? Is it an enzyme, transcription factor, cell-cycle protein, or something else? Describe three ways that the synthesis of the p53 protein affects cellular function. Why is it beneficial for these three things to happen when a cell's DNA has been damaged?

Experimental Questions

E1. Which of the following experimental observations would suggest that a disease has a genetic basis?

A. The frequency of the disease is less likely in relatives that live apart compared with relatives that live together.

B. The frequency of the disease is unusually high in a small group of genetically related individuals who live in southern Spain.

C. The disease symptoms usually begin around the age of 40.

D. It is more likely that both monozygotic twins will be affected by the disease than will dizygotic twins.

E2. At the beginning of this chapter, we discussed the types of experimental observations that suggest a disease is inherited. Which of these observations do you find the least convincing? Which do you find the most convincing? Explain your answer.

E3. What is meant by the term genetic testing? What is different between testing at the protein level versus testing at the DNA level? Describe five different techniques used in genetic testing.

E4. A particular disease is found in a group of South American Indians. During the 1920s, many of these people migrated to Central America. In the Central American group, the disease is never found. Discuss whether or not you think the disease has a genetic component. What types of further observations would you make?

E5. Chapter 20 describes a blotting method known as Western blotting that can be used to detect a polypeptide that is translated from a particular mRNA. In this method, a particular polypeptide or protein is detected by an antibody that specifically recognizes a segment of its amino acid sequence. After the antibody binds to the polypeptide within a gel, a secondary antibody (which is labeled) is used to visualize the polypeptide as a dark band. For example, an antibody that recognizes α-galactosidase A could be used to specifically detect the amount of α-galactosidase A protein on a gel. The enzyme α-galactosidase A is defective in individuals with Fabry disease, which shows an X-linked recessive pattern of inheritance. Amy, Nan, and Pete are siblings, and Pete has Fabry disease. Aileen, Jason, and Jerry are brothers and sister, and Jerry has Fabry disease. Amy, Nan, and Pete are not related to Aileen, Jason, and Jerry. Amy, Nan, and Aileen are concerned that they could be carriers of a defective α-galactosidase A gene. A sample of cells was obtained from each of these six individuals and subjected to Western blotting, using an antibody against α-galactosidase A. Samples were also obtained from two unrelated normal individuals (lanes 7 and 8). The results are shown here.

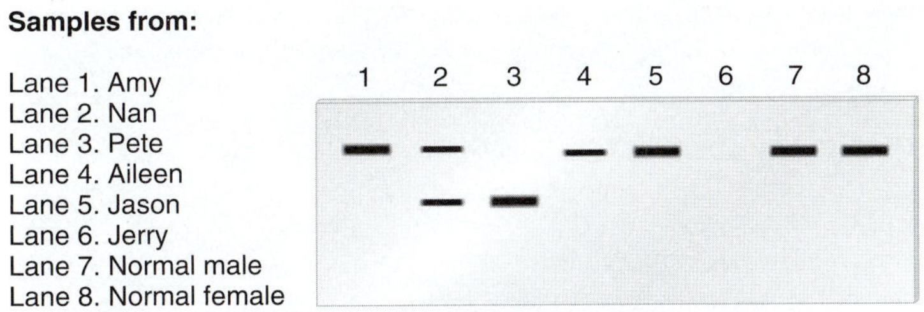

Note: Due to X inactivation in females, the amount of expression of genes on the single X chromosome in males is equal to the amount of expression from genes on both X chromosomes in females.

A. Explain the type of mutation (i.e., missense, nonsense, promoter, etc.) that causes Fabry disease in Pete and Jerry.

B. What would you tell Amy, Nan, and Aileen regarding the likelihood that they are carriers of the mutant allele and the probability of having affected offspring?

E6. An experimental assay for the blood-clotting protein called factor IX is available. A blood sample was obtained from each member of the pedigree shown here. The amount of factor IX protein is shown within the symbol of each member and is expressed as a percent of the amount observed in normal individuals who do not carry a mutant copy of the gene.

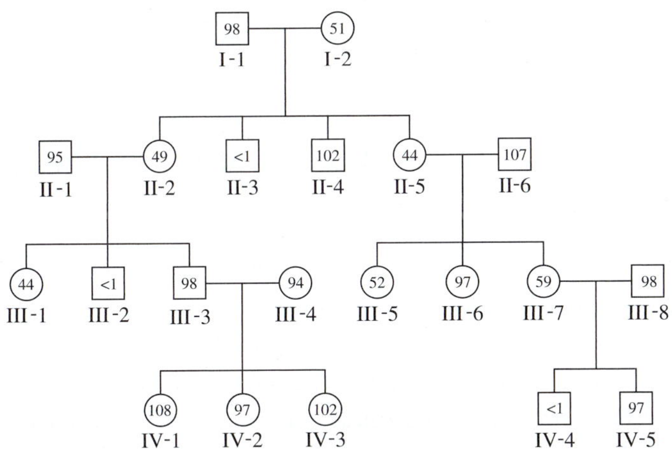

What are the likely genotypes of each member of this pedigree?

E7. Let's suppose a cell line has become malignant because it has accumulated mutations that inactivate two different tumor-suppressor genes. A researcher followed the protocol described in the experiment of Figure 24.12 and isolated DNA from this mutant cell line. The DNA was used to transform normal fibroblast (NIH3T3) cells. What results would you expect? Would you expect a high number of malignant foci or not? Explain your answer.

E8. What is a transformed cell? Describe three different methods to transform cells in a laboratory.

E9. In the experiment of Figure 24.12, what would be the results if the DNA sample had been treated with DNase, RNase, or protease prior to the treatment with calcium and phosphate ions?

E10. Explain how the experimental study of cancer-causing viruses has increased our understanding of cancer.

E11. Discuss ways to distinguish whether a particular form of cancer involves an inherited predisposition or is due strictly to (postzygotic) somatic mutations. In your answer, consider that only one mutation may be inherited, but the cancer might develop only after several somatic mutations.

E12. The codon change (Gly-12 to Val-12) in human *ras*H that converts it to oncogenic *ras*H has been associated with many types of cancers. For this reason, researchers would like to develop drugs to inhibit oncogenic *ras*H. Based on your molecular understanding of the Ras protein, what types of drugs might you develop? In other words, what would be the structure of the drugs, and how would they inhibit Ras protein? How would you test the efficacy of the drugs? What might be some side effects?

E13. Describe how normal mammalian cells grow on solid media in a laboratory and how cancer cells grow. Is a malignant focus derived from a single cancer cell or from many independent cancer cells that happen to be in the same region of a tissue culture plate?

E14. Explain how DNA microarrays are used in the molecular profiling of cancer.

Questions for Student Discussion/Collaboration

1. Make a list of the benefits that may arise from genetic testing as well as possible negative consequences. Discuss the items on your list.

2. Our government has finite funds to devote to cancer research. Discuss which aspects of cancer biology you would spend the most money pursuing.

   A. Identifying and characterizing oncogenes and tumor-suppressor genes

   B. Identifying agents in our environment that cause cancer

   C. Identifying viruses that cause cancer

   D. Devising methods aimed at killing cancer cells in the body

   E. Informing the public of the risks involved in exposure to carcinogens

   In the long run, which of these areas would you expect to be the most effective in decreasing mortality due to human cancer?

Note: All answers appear at the website for this textbook; the answers to the even-numbered questions and all of the Concept Check and Comprehension Questions are in the back of the book.

www.mhhe.com/brookergenetics5e

Visit the website for practice tests, answer keys, and other learning aids for this chapter. Enhance your understanding of genetics with our interactive exercises, quizzes, animations, and much more.

CHAPTER OUTLINE

The fruit fly (Drosophila melanogaster), a model organism in the study of developmental genetics.

25 DEVELOPMENTAL GENETICS

Multicellular organisms, such as animals and plants, begin their lives with a fairly simple organization, namely, a fertilized egg, and then proceed step by step to a much more complex arrangement. As this occurs, cells divide and change their characteristics as they become highly specialized units within a multicellular individual. Each cell in an organism has its own particular role. In animals, for example, muscle cells allow an organism to move, and intestinal cells facilitate the absorption of nutrients. This division of labor among the various cells and organs of the body promotes the survival of the individual.

Developmental genetics is concerned with the roles genes play in orchestrating the changes that occur during development. In this chapter, we will examine how the sequential actions of genes provide a program for the development of an organism from a fertilized egg to an adult. The last couple of decades have seen staggering advances in our understanding of developmental genetics at the molecular level. Scientists have chosen a few experimental organisms, such as the fruit fly (*Drosophila melanogaster*), nematode (*Caenorhabditis elegans*), zebrafish (*Danio rerio*), mouse (*Mus musculus*), and a flowering plant (*Arabidopsis thaliana*), to work toward the identification and characterization of the genes required for running their developmental programs. In certain organisms, notably *Drosophila*, many of the genes that play a critical role in the early stages of development have been identified. Researchers are now studying how the proteins encoded by these genes control the course of development. In this chapter, we will explore the body plans of invertebrates, vertebrates, and plants, and consider several examples in which geneticists understand how the actions of genes govern the developmental process.

25.1 OVERVIEW OF ANIMAL DEVELOPMENT

Learning Outcomes:

1. List the four events that give rise to a body pattern.
2. Define positional information and describe three types.
3. Outline the four overlapping stages of animal development.

Although the details differ widely among various species, the general features and steps in animal development are largely conserved within this kingdom. In this section, we will examine the types of molecules that promote developmental changes and consider how genes play the underlying role in orchestrating a plan of development from a fertilized egg to the adult.

The Generation of a Body Pattern Depends on Four Types of Cellular Events

Multicellular development in animals (and plants) follows a body plan or pattern. The term **pattern** refers to the spatial arrangement of different regions of the body. At the cellular level, the body pattern is due to the arrangement of cells and their specialization.

The progressive growth of a fertilized egg into an adult organism involves four types of cellular events: cell division, cell migration, cell differentiation, and cell death (apoptosis) (**Figure 25.1**). The coordination of these four events leads to the formation of a body with a particular pattern. As we will see, the timing of gene expression and the localization of gene products at precise regions in the fertilized egg and early embryo are the critical phenomena that underlie this coordination.

(a) Cell division

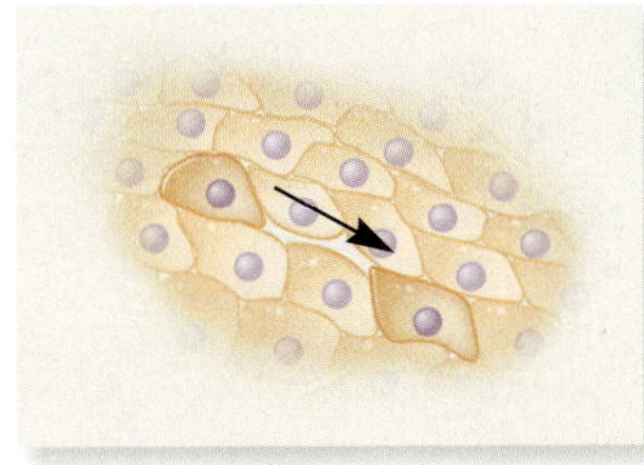

(b) Cell migration

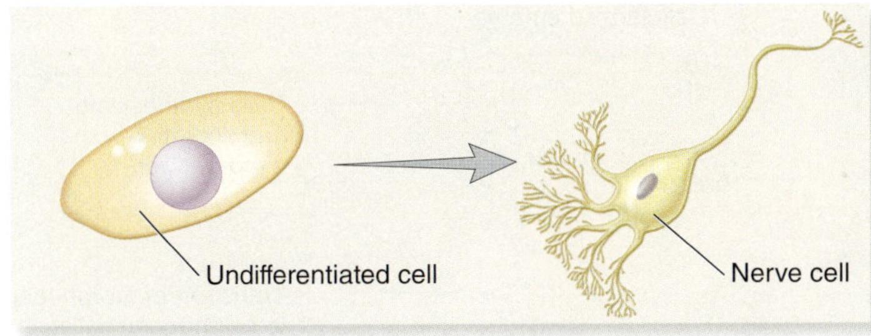

(c) Cell differentiation

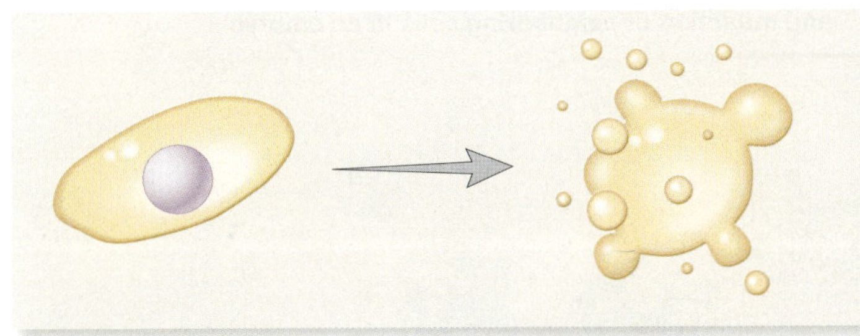

(d) Cell death (apoptosis)

FIGURE 25.1 Four events that may happen to cells during the course of development.

Positional Information Controls How Cells Behave During Development

Before we consider how genes affect development, let's examine a central concept in developmental biology known as **positional information.** For an organism to develop a body pattern with unique morphological and cellular features, each cell of the body must receive signals—positional information—that affect its course of development.

A **morphogen** is a molecule that conveys positional information and promotes developmental changes over time. It is generally a diffusible molecule that acts in a concentration-dependent manner to influence the developmental fate of a cell. Morphogens provide the positional information that stimulates a cell to divide, migrate, differentiate, or die. For example, when a region of a *Drosophila* embryo is exposed to a high concentration of the morphogen known as Bicoid, it develops structures that are characteristic of the anterior region of the body. Many such morphogens function as transcription factors that regulate the expression of many genes. This topic is described in greater detail later. How do morphogens control pattern development? Within an oocyte and during embryonic development, morphogens typically are distributed along a concentration gradient. In other words, the concentration of a morphogen varies from low to high in different regions of the developing organism. A key feature of morphogens is that they act in a concentration-dependent manner. In a particular concentration range, a morphogen or a combination of two or more different morphogens restrict a cell into a specific developmental pathway.

During the earliest stages of development of certain species, morphogenic gradients are preestablished within the oocyte (**Figure 25.2a**). Following fertilization, the zygote subdivides into many smaller cells. Due to the preestablished gradient of morphogens within the oocyte, these smaller cells have higher or lower concentrations of morphogens, depending on their location in the embryo. In this way, the morphogen gradients in the oocyte can provide positional information that is important in establishing the general polarity of an embryo along two main axes: the anteroposterior axis and the dorsoventral axis, which are described later.

A morphogen gradient can also be established in the embryo by secretion and transport to neighboring cells (**Figure 25.2b**). A certain cell or group of cells may synthesize and secrete a morphogen at a specific stage of development. After secretion, the morphogen may be transported to neighboring cells. The concentration of the morphogen is usually highest near the cells that secrete it. The morphogen may then influence the developmental fate of the cells exposed to it. The process by which a cell or group of cells governs the developmental fate of neighboring cells is known as **induction.**

In addition to morphogens, positional information is conveyed by **cell adhesion** (**Figure 25.2c**). Each cell makes its own collection of surface receptors that enable it to adhere to other cells and/or to the extracellular matrix (ECM), which consists primarily of carbohydrates and fibrous proteins. Such receptors are known as **cell adhesion molecules (CAMs).** A cell may gain positional information via the combination of contacts it makes with other cells or with the ECM. The contacts, in turn, may activate intracellular signal transduction pathways that lead to developmental changes.

The phenomenon of cell adhesion, and its role in multicellular development, was first recognized by Henry Wilson in 1907. He took multicellular sponges and passed them through a sieve, which disaggregated them into individual cells. Remarkably, the cells actively migrated until they adhered to one another to form a

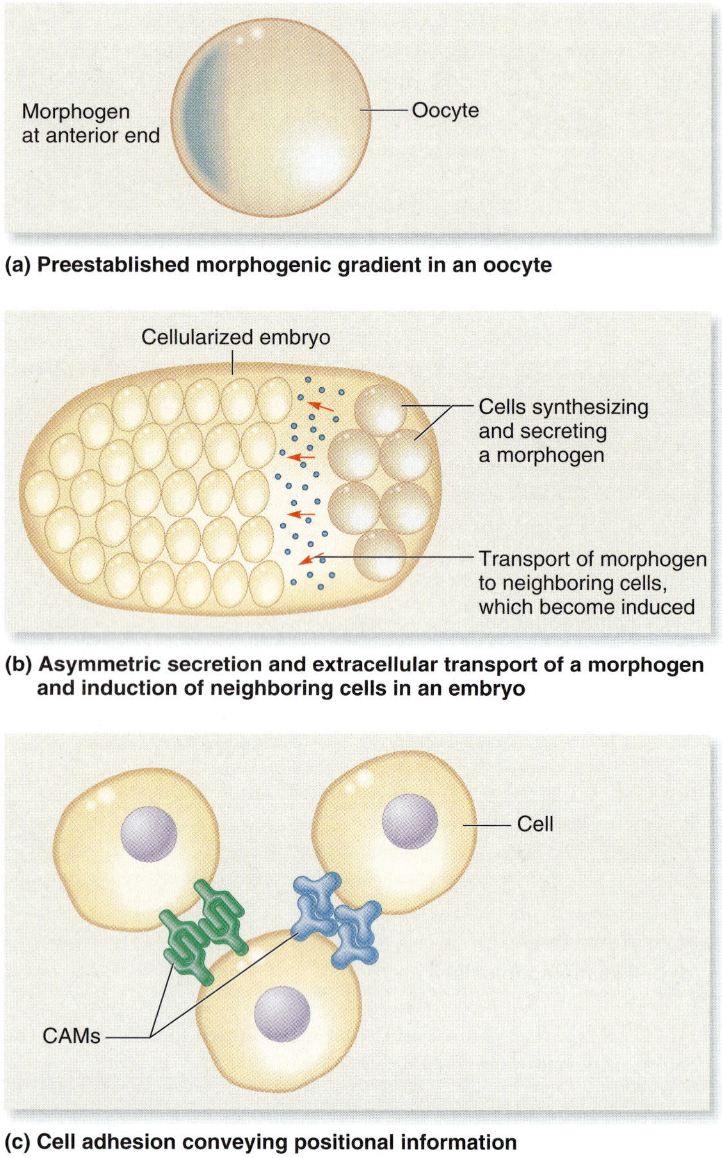

FIGURE 25.2 **Three molecular mechanisms of positional information.**

CONCEPT CHECK: Which of these mechanisms involves diffusible morphogens?

new sponge, complete with the chambers and canals that characterize a sponge's internal structure! When sponge cells from different species were mixed, they sorted themselves properly, adhering only to cells of the same species. Overall, these results indicate that cell adhesion plays an important role in governing the position a cell will adopt during development.

The Study of Mutants with Disrupted Developmental Patterns Has Identified Genes That Control Development

Mutations that alter the course of development in experimental organisms, such as *Drosophila*, have greatly contributed to our understanding of the normal process of development. For example, Figure 25.3 shows a photograph of a fly that carries multiple mutations in a cluster of genes called *bithorax*. This mutant fly has four wings instead of two; the halteres (balancing organs that resemble miniature wings), which are found on the third thoracic segment, are changed into wings, normally found on the second thoracic segment. (The term bithorax refers to the observation that the characteristics of the second thoracic segment are duplicated.)

FIGURE 25.3 **The bithorax mutation in *Drosophila*.**

Genes→Traits A normal fly contains two wings on the second thoracic segment and two halteres on the third thoracic segment. However, this mutant fly contains multiple mutations in a cluster of genes called the *bithorax* complex. In this fly, the third thoracic segment has the same characteristics as the second thoracic segment, thereby producing a fly with four wings instead of the normal number of two.

Edward Lewis, a pioneer in the genetic study of development, became interested in the bithorax phenotype and began investigating it in 1946. Researchers later discovered that the mutant chromosomal region actually contained a complex of three genes involved in specifying developmental pathways in the fly. A gene that plays a central role in specifying the final identity of a body region is called a **homeotic gene.** We will discuss particular examples of homeotic genes later in this chapter.

During the 1960s and 1970s, interest in the relationship between genetics and embryology blossomed as biologists began to appreciate the role of genetics at the molecular and cellular levels. It soon became clear that the genomes of multicellular organisms contain groups of genes that initiate a program of development involving networks of gene regulation. By identifying mutant alleles that disrupt development, geneticists have begun to unravel the pattern of gene expression that underlies the normal pattern of multicellular development.

Animal Development Occurs in Four Overlapping Phases

With the exception of sponges, all animal bodies are organized along axes. Some simple animals, such as jellyfish, exhibit radial symmetry in which the body is circularly symmetrical around an axis. However, most animals, including humans, are **bilaterians,** meaning they have left-right symmetry. Our discussion of animal development in this chapter will focus on the bilaterians. For these species,

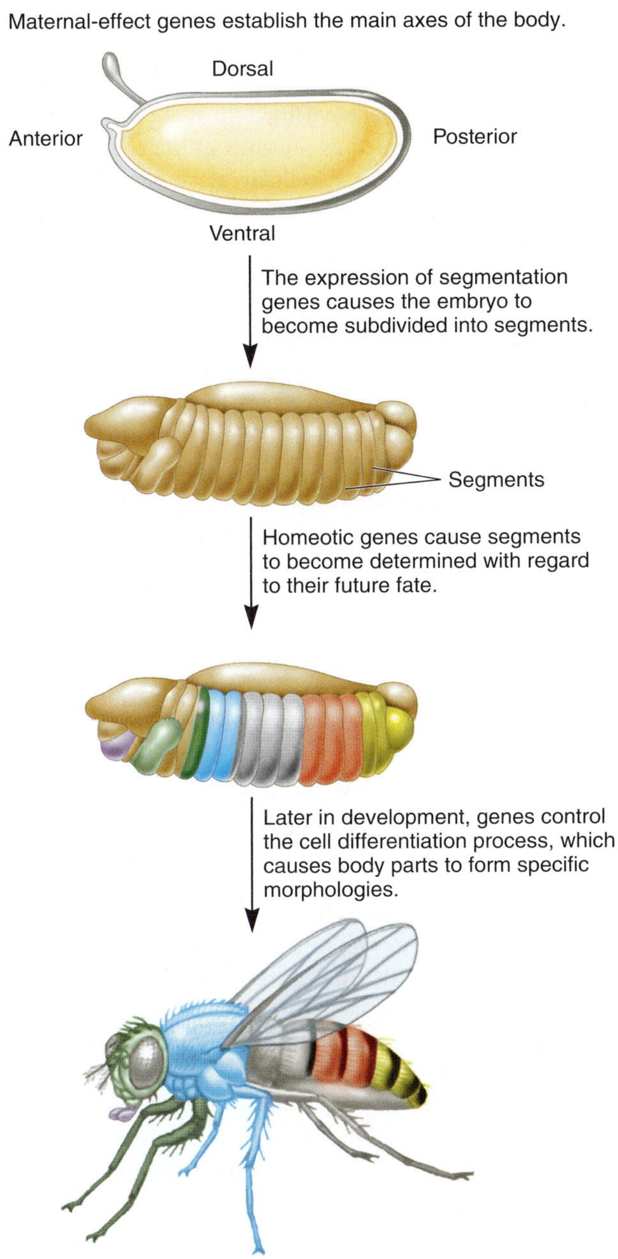

FIGURE 25.4 The four overlapping phases of animal development.

CONCEPT CHECK: How does genetics play a role in development?

the body is typically organized along two major axes: an anteroposterior axis and a dorsoventral axis. We will consider these axes later when we discuss invertebrate development. Left-right symmetry occurs relative to the anteroposterior axis.

By comparing a variety of bilaterians, researchers have discovered that development generally proceeds in four overlapping stages (Figure 25.4).

1. ***Formation of body axes:*** The first phase of pattern development is the establishment of the body axes.
2. ***Segmentation of the body:*** In the second phase, the body is subdivided into segments. In invertebrates, the segments may remain morphologically distinct, whereas in vertebrates, distinct segments are obvious only in the early stages of development.
3. ***Determination of structures within the segments:*** As the segments form, groups of cells become destined to develop into particular structures and cell types. This process, called **determination,** occurs before the structures and cell types have changed their morphologies.
4. ***Cell differentiation:*** Toward the end of development, cells differentiate into particular cell types. This results in tissues and organs with specific morphologies.

The formation of a bilaterian body with a specific pattern is the result of these four overlapping phases. The phases themselves are controlled by sets of genes. One set is expressed very early and controls the formation of body axes. As you will learn later, the genes in this category often exhibit a maternal-effect pattern of inheritance. A second set is expressed next that results in segmentation. In *Drosophila*, these are called segmentation genes. A third set of genes, known as homeotic genes, controls the final fate of groups of cells within segments. Finally, a fourth set of genes controls the cellular differentiation process. In the following two sections, we will explore the phases of animal development in greater detail and discuss examples of genes that play a key role in these four phases.

25.1 COMPREHENSION QUESTIONS

1. Positional information may provide a cue for a cell to
   a. divide.
   b. migrate.
   c. differentiate.
   d. undergo apoptosis.
   e. do all of the above.
2. Molecules that convey positional information include
   a. morphogens.
   b. cell adhesion molecules.
   c. ATP.
   d. both a and b.
3. Which of the following is the correct order for developmental phases in animals?
   A. Segmentation of the body
   B. Determination
   C. Cell differentiation
   D. Formation of body axes
   a. A, B, C, D
   b. A, D, C, B
   c. D, A, B, C
   d. D, A, C, B

25.2 INVERTEBRATE DEVELOPMENT

Learning Outcomes:

1. List the stages of *Drosophila* development.
2. Compare and contrast how maternal-effect genes, gap genes, and homeotic genes affect *Drosophila* development.
3. Explain how an understanding of cell lineages in *Caenorhabditis elegans* aids in the identification of mutations that affect the timing of developmental changes.

We will focus our discussion of invertebrate development by considering two model organisms, *Drosophila melanogaster* and *Caenorhabditis elegans*, that have been pivotal in our understanding of developmental genetics. As we first saw in Chapter 3, *Drosophila* has been a favorite subject of geneticists since 1910, when Thomas Hunt Morgan isolated his first white-eyed mutant. The fruit fly has been used to discover many of the fundamental principles of genetics, including the chromosome theory of inheritance, the random mutation theory, and linkage mapping, to mention only a few.

In developmental genetics, *Drosophila* is also useful for a variety of reasons. First, the techniques for generating and analyzing mutants in this organism are highly advanced, and researchers have identified many mutant strains with altered developmental pathways. Second, at the embryonic and larval stages, *Drosophila* is large enough to conduct experiments in which portions of the body are transplanted to different sites. It is also small enough to examine under the microscope and determine where particular genes are expressed at critical stages of development.

By comparison, *C. elegans* is used by developmental geneticists because of its simplicity. The adult organism is a small transparent worm composed of only about 1000 somatic cells. Starting with the fertilized egg, the pattern of cell division and the developmental fate of each cell are completely known.

In this section, we will begin by examining the stages of *Drosophila* development. We then focus our attention on embryonic development (embryogenesis) because it is during this stage that the overall body plan is determined. We will see how the timing of the expression of particular genes and the localization of gene products within the embryo influence the developmental process. We then briefly consider development in *C. elegans*. In this organism, we will examine how the timing of gene expression plays a key role in determining the developmental fate of particular cells.

Drosophila Progresses Through Several Stages to Become an Adult

Figure 25.5 illustrates the general sequence of events in *Drosophila* development. The oocyte is the cell most critical in determining the pattern of development in the adult organism. It is an elongated cell with preestablished axes (Figure 25.5a). After fertilization takes place, the zygote goes through a series of nuclear divisions that are not accompanied by cytoplasmic division. Initially, the resulting nuclei are scattered throughout the yolk, but eventually they migrate to the periphery of the cytoplasm. This is the syncytial blastoderm stage (Figure 25.5b).

After the nuclei have lined up along the cell membrane, individual cells are formed as portions of the cell membrane surround each nucleus, creating a cellular blastoderm (Figure 25.5c). This structure is composed of a sheet of cells on the outside with yolk in the center. In this arrangement, the cells are distributed asymmetrically. At the posterior end are a group of cells called the pole cells, the primordial germ cells that eventually give rise to gametes in the adult organism. After blastoderm formation is complete, dramatic changes occur during gastrulation (Figure 25.5d). This stage produces three cell layers known as the ectoderm, mesoderm, and endoderm.

As this process occurs, the embryo begins to subdivide into detectable units. Initially, shallow grooves partition the embryo into 14 **parasegments,** which are only transient subdivisions. A short time later, these grooves disappear, and new boundaries are formed that divide the embryo into morphologically discrete **segments.** Figure 25.5e shows the segmented pattern of a *Drosophila* embryo about 10 hours after fertilization. Later in this section, we will explore how the coordination of gene expression underlies the formation of these parasegments and segments.

At the end of **embryogenesis,** which lasts about 18 to 22 hours, a larva hatches from the egg (Figure 25.5f) and begins feeding. *Drosophila* has three larval stages, separated by molts. During molting, the larva sheds its cuticle, a hardened extracellular shell that is secreted by the epidermis. Each larval stage between two molts is also called an instar.

After the third larval stage, *Drosophila* forms a pupa that undergoes metamorphosis. During metamorphosis, groups of cells called imaginal disks that were produced earlier in development differentiate into the structures found in the adult fly, such as the head, wings, legs, and abdomen. The adult fly then emerges from the pupal case. The adult is produced about 10–14 days after fertilization. In *Drosophila*, as in all bilateral animals, the final result of development is an adult body organized along two major axes: the **anteroposterior axis** and the **dorsoventral axis** (Figure 25.5g). The **left-right axis** occurs relative to the anteroposterior axis. An additional axis, used mostly for designating limb parts, is the **proximodistal axis.** Proximal refers to a location that is close to the body, whereas distal is farther away.

The Early Stages of Embryonic Development Determine the Pattern of Structures in the Adult Organism

Although many interesting developmental events occur during the three larval stages and the pupal stage of *Drosophila*, we will focus most of our attention on the events that occur in the oocyte and during embryonic development. Even before hatching, the embryo has developed the basic body plan that is eventually found in the adult organism (see Figure 25.5e). In other words, during the early stages of development, the embryo is divided into segments that correspond to the segments of the larva and adult. Therefore, an understanding of how these segments form is critical to our understanding of pattern formation.

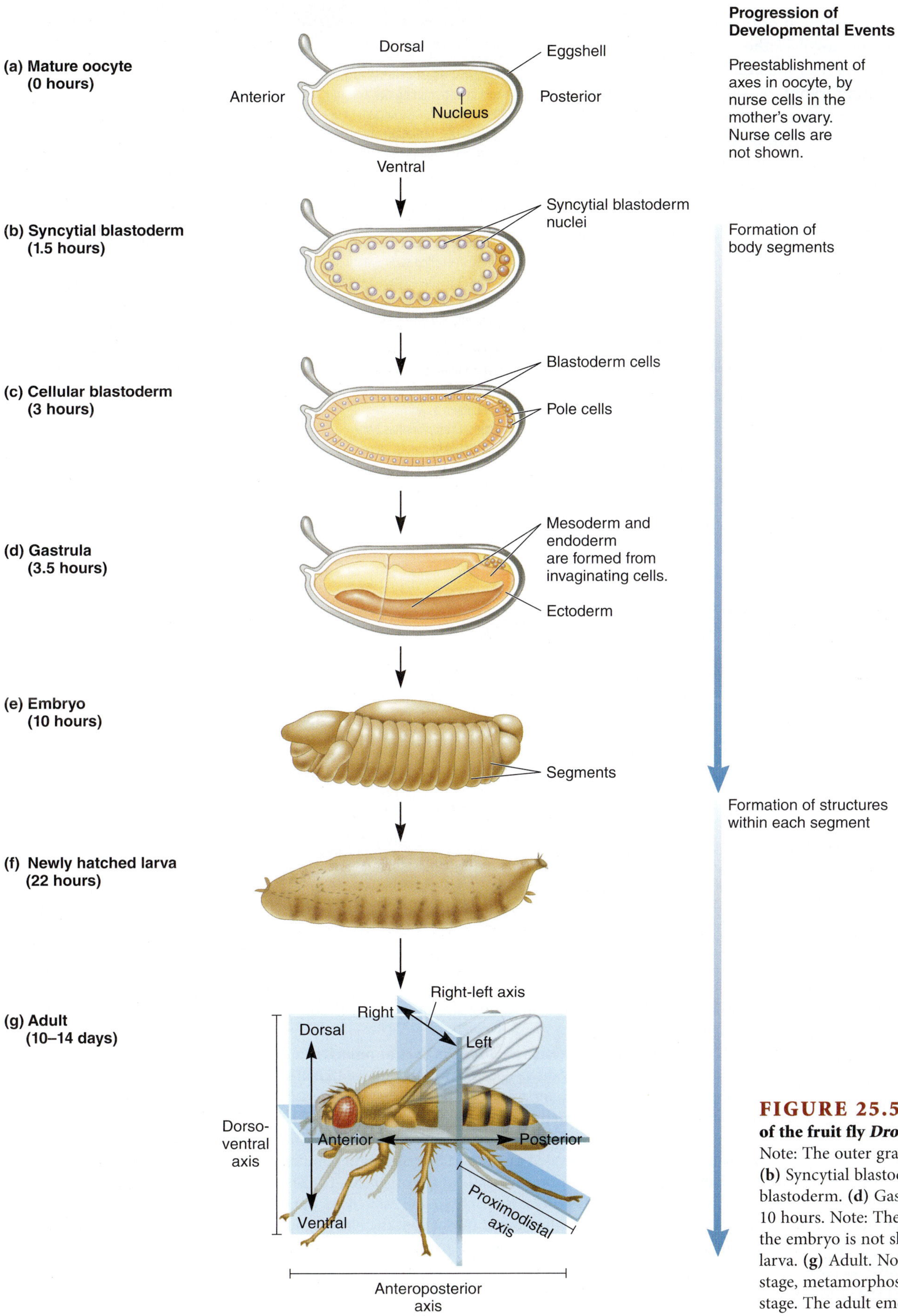

FIGURE 25.5 Developmental stages of the fruit fly *Drosophila.* **(a)** Oocyte. Note: The outer gray layer is an eggshell. **(b)** Syncytial blastoderm. **(c)** Cellular blastoderm. **(d)** Gastrula. **(e)** Embryo at 10 hours. Note: The eggshell surrounding the embryo is not shown. **(f)** Newly hatched larva. **(g)** Adult. Note: Prior to the adult stage, metamorphosis occurs during a pupal stage. The adult emerges from the pupal case.

In *Drosophila*, the establishment of the body axes and division of the body into segments involves the participation of a few dozen genes. **Table 25.1** lists many of the important genes governing pattern formation during embryonic development. These genes were identified by characterizing mutants that had altered development patterns. They are often given interesting names based on the observed phenotype when they are mutant. It is beyond the scope of this textbook to describe how all of these genes exert their effects during embryonic development. Instead, we will consider a few examples that illustrate how the expression of a particular gene and the localization of its gene product have a defined effect on the pattern of development.

TABLE 25.1

Examples of *Drosophila* Genes That Play a Role in Pattern Development

Description	Examples
Maternal-effect genes play a role in determining the axes of development. Also certain genes govern the formation of the extreme terminal (anterior and posterior) regions.	Anterior: *bicoid, exuperantia, swallow, staufen* Posterior: *nanos, cappuccino, oskar pumilio, spire, staufen, tudor, vasa* Terminal: *torso, torsolike, Trunk, NTF-1* Dorsoventral: *Toll, cactus, dorsal, easter, gurken, nudel, pelle, pipe, snake, spatzle*
Some genes play a role in promoting the subdivision of the embryo into segments. These are called segmentation genes. The three types are gap genes, pair-rule genes, and segment-polarity genes.	Gap genes: *empty spiracles, giant, huckebein, hunchback, knirps, Krüppel, tailless, orthodenticle* Pair-rule genes: *even-skipped, hairy, runt, fushi tarazu, paired* Segment-polarity genes: *frizzled, frizzled-2, engrailed, patched, smoothened, hedgehog, wingless, gooseberry*
Some genes play a role in determining the fate of particular segments. These are known as homeotic genes. *Drosophila* has two clusters of homeotic genes known as the *Antennapedia* complex and the *bithorax* complex.	*Antennapedia* complex: *labial, proboscipedia, Deformed, Sex combs reduced, Antennapedia* *Bithorax* complex: *Ultrabithorax, abdominal A, Abdominal B*

The Gene Products of Maternal-Effect Genes Are Deposited Asymmetrically into the Oocyte and Establish the Anteroposterior and Dorsoventral Axes at a Very Early Stage of Development

The first stage in *Drosophila* embryonic pattern development is the establishment of the body axes. This occurs before the embryo becomes segmented. The morphogens necessary to establish these axes are distributed prior to fertilization. During oogenesis, gene products such as mRNA, which are important in early developmental stages, are deposited asymmetrically within the egg. Later, after the egg has been fertilized and development begins, these gene products establish independent developmental programs that govern the formation of the body axes of the embryo. Proper development depends on the products of many different genes.

As shown in **Figure 25.6**, a few gene products act as key morphogens, or receptors for morphogens, that initiate changes in embryonic development. As shown here, these gene products are deposited asymmetrically in the egg. For example, the product of the *bicoid* gene is necessary to initiate development of the anterior structures of the organism. During oogenesis, the mRNA for *bicoid* accumulates in the anterior region of the oocyte. In contrast, the mRNA from the *nanos* gene accumulates in the posterior end. Later in development, the *nanos* mRNA is translated into protein, which functions to influence posterior development. The *nanos* gene is required for the formation of the abdomen.

In addition to morphogens such as Bicoid and Nanos, the development of the structures at the extreme anterior and posterior ends of the embryo are regulated in part by a receptor

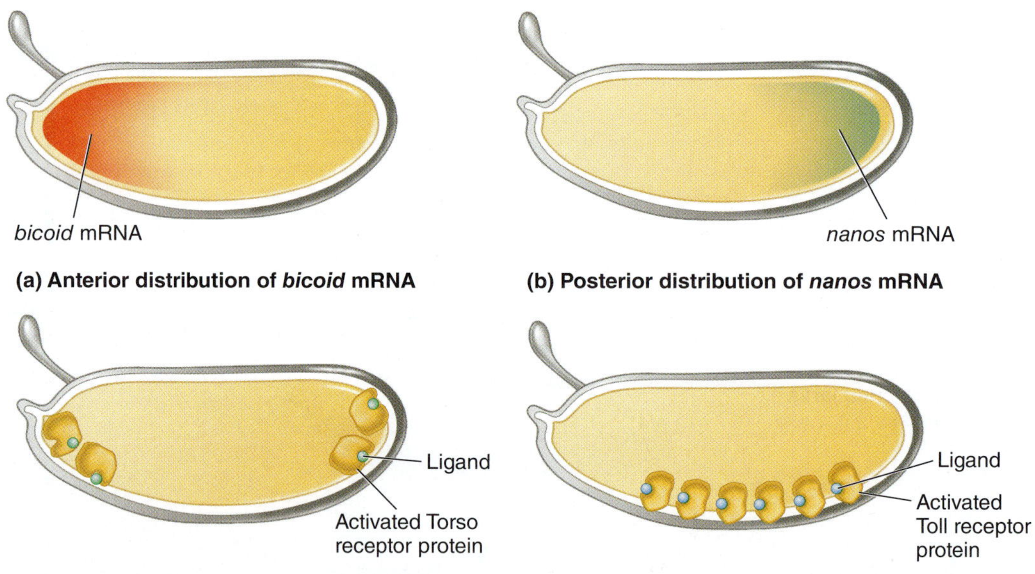

(a) Anterior distribution of *bicoid* mRNA **(b) Posterior distribution of *nanos* mRNA**

(c) Terminal distribution of Torso receptor protein **(d) Ventral distribution of Toll receptor protein**

FIGURE 25.6 The establishment of the axes of polarity in *Drosophila*. This figure shows some of the maternal gene products deposited in the oocyte that are critical in the establishment of the anteroposterior, terminal, and dorsoventral axes. **(a)** *Bicoid* mRNA is distributed in the anterior end of the oocyte and promotes the formation of anterior structures. **(b)** *Nanos* mRNA is localized to the posterior end and promotes the formation of posterior structures. **(c)** The Torso receptor protein is found in the plasma membrane and is activated by ligand binding at either end of the oocyte. It causes the formation of structures that are found only at the ends of the organism. **(d)** The Toll receptor protein is activated by ligand binding at the ventral side of the embryo and establishes the dorsoventral axis. Note: The Torso and Toll receptor proteins are distributed throughout the plasma membranes of the oocyte and embryo, but they are activated by ligand binding only in the regions shown in this figure. Note: The gray region surrounding the oocyte and embryo is the eggshell; the oocyte is inside the eggshell. Torso and Toll proteins are embedded in the plasma membrane of the oocyte and embryonic cells.

CONCEPT CHECK: Describe the orientations of the anteroposterior and dorsoventral axes.

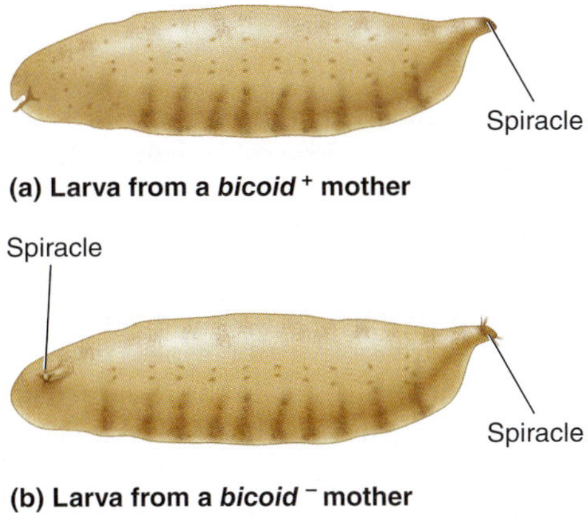

FIGURE 25.7 The *bicoid* mutation in *Drosophila*. **(a)** A larva from a normal *bicoid*$^+$ mother. **(b)** A larva from a homozygous *bicoid*$^-$ mother in which both ends of the larva develop posterior structures. For example, both ends develop a spiracle, which normally is found only at the posterior end. This is a lethal condition.

CONCEPT CHECK: What is the normal function of the Bicoid protein?

protein called Torso. This receptor is activated by the binding of a signaling molecule, called a ligand, which occurs only at the anterior and posterior ends of the egg. This activation is necessary for the formation of the terminal ends of the embryo. The dorsoventral axis is governed by the actions of several proteins, including a receptor protein known as Toll. This receptor is found in plasma membranes of cells throughout the embryo. However, ligand binding is needed to activate Toll, and this occurs only along the ventral midline of the embryo.

Let's now take a closer look at the molecular mechanism of one morphogen, Bicoid. The *bicoid* gene got its name because a larva whose mother is defective in this gene develops with two posterior ends (**Figure 25.7**). This allele exhibits a maternal-effect pattern of inheritance, in which the genotype of the mother determines the phenotypic traits of the offspring (see Chapter 5). A female fly that is phenotypically normal (because its mother was heterozygous for the normal *bicoid* allele), but genotypically homozygous for an inactive *bicoid* allele (because it inherited an inactive allele from both its mother and father), produces 100% abnormal offspring even when mated to a male that is homozygous for the normal *bicoid* allele. In other words, the genotype of the mother determines the phenotype of the offspring. This occurs because the *bicoid* gene product is provided to the oocyte via maternal nurse cells.

In the ovaries of female flies, the nurse cells are localized asymmetrically toward the anterior end of the oocyte. During oogenesis, gene products are transferred from nurse cells into the oocyte via cell-to-cell connections called cytoplasmic bridges. Maternally-encoded gene products enter one side of the oocyte, which eventually becomes the anterior end of the embryo (**Figure 25.8a**). The *bicoid* gene is actively transcribed in the nurse cells, and *bicoid* mRNA is transported into the anterior end of the oocyte. The 3′ end of *bicoid* mRNA contains a signal that is recognized by RNA-binding proteins that are necessary for the transport of this mRNA into the oocyte. After it enters the oocyte, the *bicoid* mRNA is trapped at the anterior end.

How can researchers determine the location of *bicoid* mRNA in the oocyte and resulting zygote? **Figure 25.8b** shows an in situ hybridization experiment in which a *Drosophila* egg was examined via a labeled probe complementary to the *bicoid* mRNA. (The technique of in situ hybridization is described in Chapter 22; see Figure 22.2.) As seen here, the *bicoid* mRNA is highly concentrated near the anterior end of the egg cell. Following fertilization, the *bicoid* mRNA is translated and a gradient of Bicoid protein is established as shown in **Figure 25.8c**.

After fertilization occurs, the Bicoid protein functions as a transcription factor. A remarkable feature of this protein is that its ability to influence gene expression is tuned exquisitely to its concentration. Depending on the distribution of the Bicoid protein, this transcription factor activates genes only in certain regions of

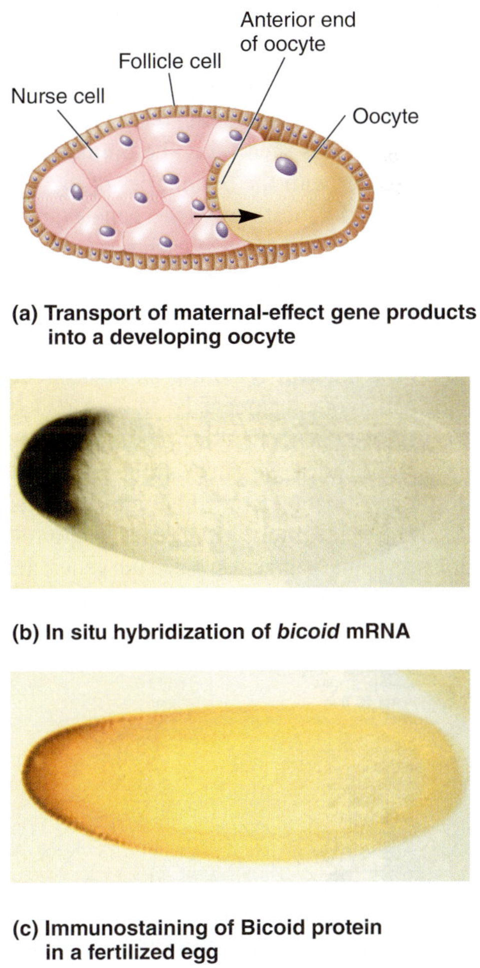

FIGURE 25.8 Asymmetrical localization of gene products during oogenesis in *Drosophila*. **(a)** The nurse cells transport gene products into the anterior (left) end of the developing oocyte. **(b)** An in situ hybridization experiment showing that the *bicoid* mRNA is trapped near the anterior end of the oocyte. **(c)** The *bicoid* mRNA is translated into protein soon after fertilization. The location of the Bicoid protein is revealed by immunostaining using an antibody that specifically recognizes this protein.

CONCEPT CHECK: Where are maternal-effect gene products made first? Where do they go?

the embryo. For example, Bicoid stimulates the transcription of a gene called *hunchback* in the anterior half of the embryo, but its concentration is too low in the posterior half to activate the *hunchback* gene there.

Gap, Pair-Rule, and Segment-Polarity Genes Act Sequentially to Divide the *Drosophila* Embryo into Segments

After the anteroposterior and dorsoventral regions of the embryo have been established by maternal-effect genes, the next developmental process organizes the embryo transiently into parasegments and then permanently into segments. The segmentation pattern of the embryo is shown in **Figure 25.9**. This pattern of positional information is maintained, or "remembered," throughout the rest of development. In other words, each segment of the embryo gives rise to unique morphological features in the adult. For example, T2 becomes a thoracic segment with a pair of legs and a pair of wings, and A8 becomes a segment of the abdomen.

Figure 25.9 shows the overlapping relationship between parasegments and segments. As seen here, the boundaries of the segments are out of register with the boundaries of the parasegments. The parasegments are the locations where gene expression is controlled spatially. The anterior region of each segment coincides with the posterior region of a parasegment; the posterior region of a segment coincides with the anterior region of the next parasegment. The pattern of gene expression that occurs in the posterior region of one parasegment and the anterior region of an adjacent parasegment results in the formation of a segment.

Now that we have a general understanding of the way the *Drosophila* embryo is subdivided, we can examine how particular genes cause it to become segmented into this pattern. Three classes of genes, collectively called **segmentation genes,** play a role in the formation of body segments: gap genes, pair-rule genes, and segment-polarity genes. The expression and activation patterns of these genes in specific regions of the embryo cause it to become segmented.

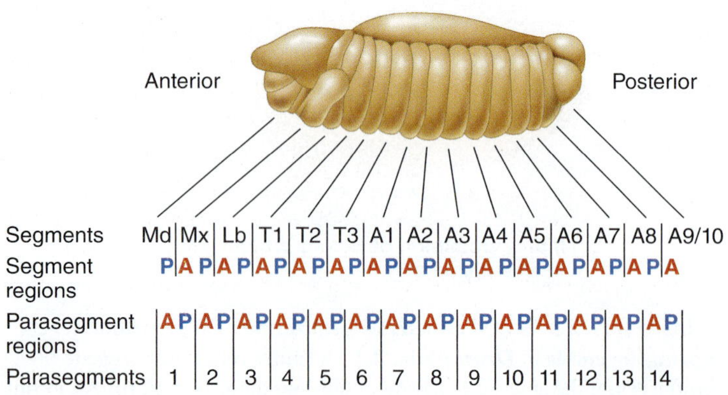

FIGURE 25.9 A comparison of segments and parasegments in the *Drosophila* embryo. Note that the parasegments and segments are out of register. The posterior (P) and anterior (A) regions are shown for each segment. The head segments are Md (mandibular), Mx (maxillary) and Lb (labial). The thoracic and abdominal segments are designated with the letters T and A, respectively.

How were the three classes of segmentation genes discovered? In the 1970s, segmentation genes were identified by Christiane Nüsslein-Volhard and Eric Wieschaus, who undertook a systematic search for mutations affecting embryonic development. Their pioneering effort identified most of the genes required for the embryo to develop a segmented pattern.

Figure 25.10 represents a few of the phenotypic effects that occur when a particular segmentation gene is defective. The gray boxes indicate the regions missing in the resulting larvae. In flies with a mutation in a gap gene known as *Krüppel*, a contiguous section of the larva is missing (Figure 25.10a). In other words, a gap of several segments has occurred. By comparison, a defect in a pair-rule gene causes alternating regions to be deleted (Figure 25.10b). For example, when the *even-skipped* gene is defective, portions of alternating segments in the resulting larva are missing. Finally, segment-polarity mutations cause individual segments to be missing either an anterior or a posterior region. Figure 25.10c shows a mutation in a segmentation gene known as *gooseberry*. When this gene is defective, the anterior portion of each segment is missing from the larva. In the case of segment-polarity mutants, the segments adjacent to the deleted regions exhibit a mirror-image duplication. Overall, the phenotypic effects of mutant segmentation genes provided geneticists with important clues regarding the roles of these genes in the developmental process of segmentation.

Figure 25.11 presents a partial, simplified scheme of the genetic hierarchy that leads to a segmented pattern in the *Drosophila* embryo. Keep in mind that although this figure presents the general sequence of events that occurs during the early stages of embryonic development, many more genes are actually involved in this process (refer back to Table 25.1). As described in this figure, the following steps occur:

1. Maternal-effect gene products, such as *bicoid* mRNA, are deposited asymmetrically into the oocyte. These gene products form a gradient that later influences the formation of axes, such as the anteroposterior axis.
2. In contrast to maternal-effect genes, which are expressed during oogenesis, **zygotic genes** are expressed after fertilization. The first zygotic genes to be activated are the gap genes. Maternal-effect gene products are responsible for activating the gap genes. As shown in step 2 of Figure 25.11, gap genes divide the embryo into a series of broad bands or regions. These bands do not correspond to parasegments or segments within the embryo.
3. Products of the gap genes and maternal-effect genes then activate the pair-rule genes. The photograph shown in step 3 of Figure 25.11 illustrates the alternating pattern of expression of the *even-skipped* gene. The expression of pair-rule genes in stripes defines the boundaries of parasegments. *Even-skipped* is expressed in the odd-numbered parasegments. Solved problem S4 at the end of the chapter examines how the *even-skipped* gene can be expressed in this pattern of alternating stripes. The pair-rule genes divide the broad regions established by gap genes into seven bands or stripes.

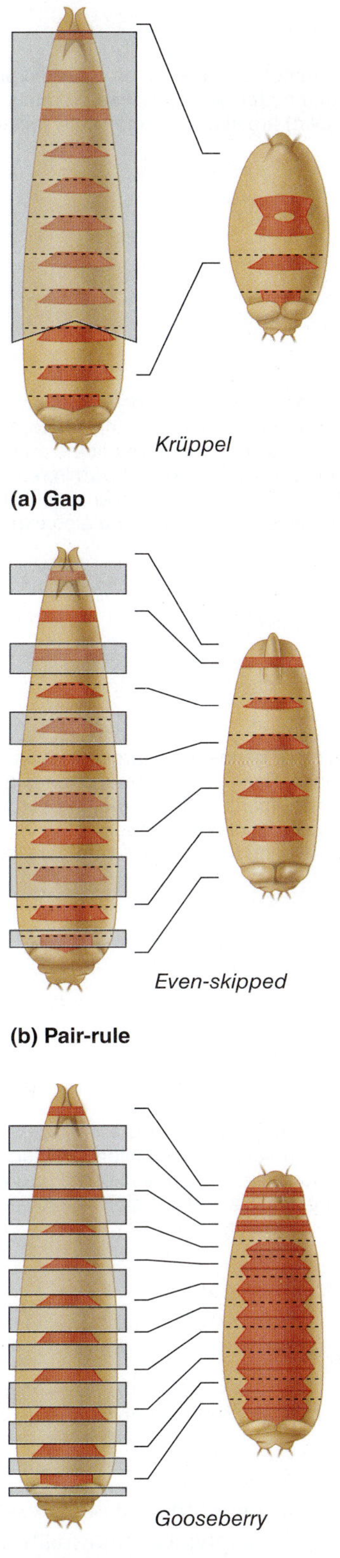

FIGURE 25.10 Phenotypic effects in *Drosophila* larvae with mutations in segmentation genes. The effects shown here are caused by mutations in **(a)** a gap gene, **(b)** a pair-rule gene, and **(c)** a segment-polarity gene. The dotted lines indicate the boundaries between segments.

CONCEPT CHECK: Describe the different effects of a mutation in a gap gene versus a pair-rule gene.

4. Once the pair-rule genes are activated in an alternating banding arrangement, their gene products then regulate the expression of segment-polarity genes. The segment-polarity genes divide the embryo into 14 stripes, one within each parasegment. As shown in step 4 of Figure 25.11, the segment-polarity gene *engrailed* is expressed in the anterior region of each parasegment. Another segment-polarity gene, *wingless*, is expressed in the posterior region. Later in development, the anterior region of one parasegment and the posterior region of another parasegment develop into a segment with particular morphological characteristics.

The Expression of Homeotic Genes Controls the Phenotypic Characteristics of Segments

Thus far, we have considered how the *Drosophila* embryo becomes organized along axes and then into a segmented body pattern. Now let's examine how each segment develops its unique morphological features. Geneticists often use the term **cell fate** to describe the ultimate morphological features that a cell or group of cells adopts. For example, the fate of the cells in segment T2 in the *Drosophila* embryo is to develop into a thoracic segment containing two legs and two wings. In *Drosophila*, the fate of the cells in each segment of the body is determined at a very early stage of embryonic development, long before the morphological features become apparent.

Our understanding of developmental fate has been greatly aided by the identification of mutant genes that alter cell fates. In animals, the first mutant of this type was described by the German entomologist Ernst Gustav Kraatz in 1876. He observed a sawfly (*Cimbex axillaris*) in which part of an antenna was replaced with a leg. During the late nineteenth century, the English zoologist William Bateson collected many of these types of observations and published them in 1894 in a book entitled *Materials for the Study of Variation Treated with Especial Regard to Discontinuity in the Origin of Species*. In this book, Bateson coined the term **homeotic** to describe mutants in which one body part is replaced by another, such as the transformation of the antennae of insects into legs.

As mentioned earlier, Edward Lewis began to study strains of *Drosophila* having homeotic mutations. This work, which began in 1946, was the first systematic study of homeotic genes. Each homeotic gene controls the fate of a particular region of the body. *Drosophila* contains two clusters of homeotic genes called the *Antennapedia* complex and the *bithorax* complex. **Figure 25.12** shows the organization of genes within these complexes. As discussed later in this chapter, these types of genes are found in all animals and referred to as *Hox* genes. The *Antennapedia* complex contains five genes, designated *lab*, *pb*, *Dfd*, *Scr*, and *Antp*. The *bithorax* complex has three genes, *Ubx*, *abd-A*, and *Abd-B*. Both of these complexes are located on chromosome 3 in *Drosophila*, but a large segment of DNA separates them.

As noted in Figure 25.12, the order of these genes along chromosome 3 correlates with their pattern of gene expression along the anteroposterior axis of the body. For example, *lab* is expressed in an anterior segment and governs the formation of mouth structures. The *Antp* gene is expressed strongly in the thoracic region

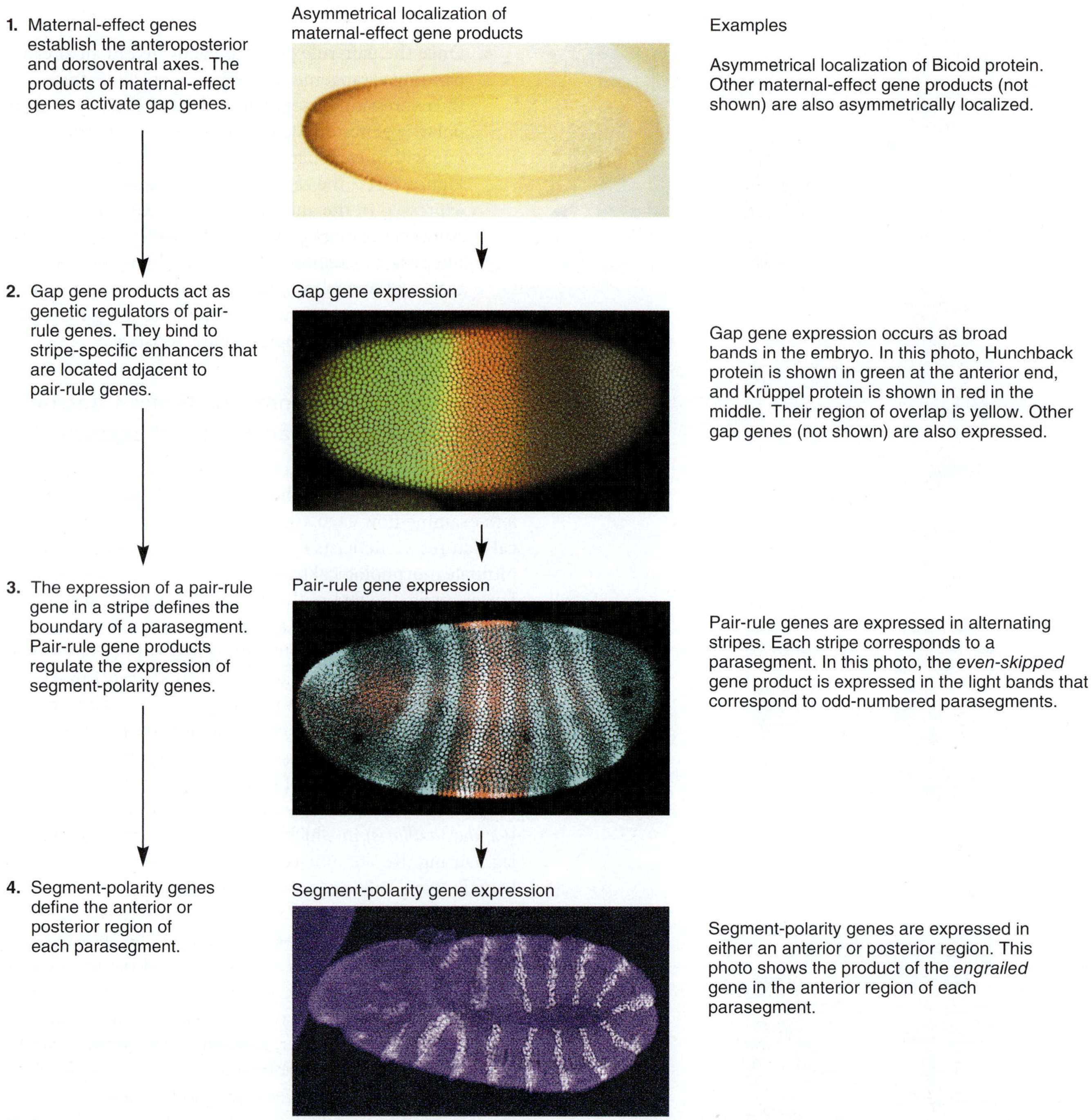

FIGURE 25.11 **Overview of the genetic hierarchy leading to segmentation in *Drosophila*.** Note: When comparing steps 3 and 4, the elongated embryo bends in the middle and folds back on itself.

during embryonic development and controls the formation of thoracic structures. Transcription of the *Abd-B* gene occurs in the posterior region of the embryo. This gene controls the formation of the posterior-most abdominal segments.

The role of homeotic genes in determining the identity of particular segments has been revealed by mutations that alter their function. For example, the *Antp* gene is normally expressed in the thoracic region. The *Antennapedia* mutation causes the *Antp* gene to also be expressed in the region where the antennae are made. These mutant flies have legs in the place of antennae (**Figure 25.13**)! This is an example of a **gain-of-function mutation.** In this case, the *Antp* gene is expressed normally in the thoracic region and also expressed abnormally in the anterior segment that normally gives rise to the antennae.

Investigators have also studied many **loss-of-function mutations** in homeotic genes. In this case, when a particular homeotic gene is defective, the region it normally governs is usually controlled by the homeotic gene that acts in the adjacent anterior region. For example, the *Ubx* gene functions within parasegments 5 and 6. If this gene is missing, this section of the fly becomes converted to the morphological features that would normally be produced from parasegment 4.

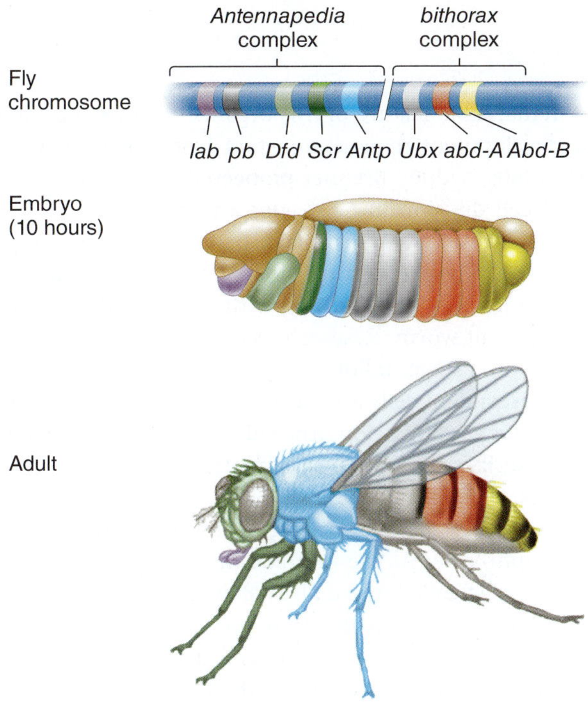

FIGURE 25.12 Expression pattern of homeotic genes in *Drosophila.* These homeotic genes are found in two complexes termed the *Antennapedia* complex and *bithorax* complex. The order of homeotic genes, *labial* (*lab*), *proboscipedia* (*pb*), *Deformed* (*Dfd*), *Sex combs reduced* (*Scr*), *Antennapedia* (*Antp*), *Ultrabithorax* (*Ubx*), *abdominal A* (*abd-A*), and *Abdominal B* (*Abd-B*), correlates with the order of expression in the embryo. The expression pattern of seven of these genes is shown. *Lab* (purple) is expressed in the region that eventually gives rise to mouth structures. *Dfd* (light green) is expressed in the region that will form much of the head. *Scr* (forest green) and *Antp* (light blue) are expressed in embryonic segments that give rise to thoracic segments. *Ubx* (gray), *abd-A* (red), and *Abd-B* (yellow) are expressed in posterior segments that will form the abdomen. The order of gene expression, from anterior to posterior, parallels the order of genes along the chromosome. The expression pattern of the *pb* gene is not well understood.

CONCEPT CHECK: Explain how the physical arrangements of the homeotic genes correlate with their effects on phenotype.

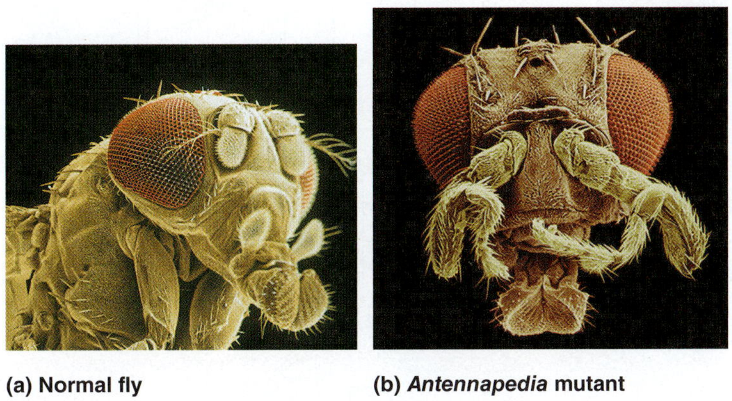

FIGURE 25.13 The *Antennapedia* mutation in *Drosophila.* Genes→Traits **(a)** A normal fly with antennae. **(b)** This mutant fly has a gain-of-function mutation in which the *Antp* gene is expressed in the embryonic segment that normally gives rise to antennae. The expression of *Antp* causes this region to have legs rather than antennae.

Homeotic genes are part of the genetic hierarchy that produces the morphological characteristics of the fly. How are they regulated? Homeotic genes are controlled by gap genes and pair-rule genes, and they are also regulated by interactions among themselves. Epigenetic modifications maintain the patterns of homeotic gene expression after the initial patterns are established by segmentation genes. As described in Chapter 16, trithorax group complexes and polycomb group complexes promote epigenetic modifications that result in gene activation or repression, respectively. In regions of the embryo where the homeotic genes are active, the products of *Trithorax* genes may maintain an open conformation in which transcription can take place. Alternatively, Polycomb group complexes repress the expression of homeotic genes in regions where they should not act. One way this may be accomplished is by changing chromatin structure to convert it to a closed conformation in which transcription is difficult or impossible (refer back to Figure 16.6). Overall, the concerted actions of many gene products cause the homeotic genes to be expressed only in the appropriate region of the embryo, as shown in Figure 25.12.

Because they are part of a genetic hierarchy in which genes activate other genes, perhaps it is not surprising that homeotic genes encode transcription factors. The coding sequence of homeotic genes contains a 180-bp consensus sequence known as a **homeobox** (**Figure 25.14a**). This sequence was first discovered in the *Antp* and *Ubx* genes, and it has since been found in all *Drosophila* homeotic genes and in some other genes affecting pattern development, such as *bicoid.* The protein domain encoded by the homeobox is called a **homeodomain.** The arrangement of α helices within the homeodomain promotes the binding of the protein to the major groove of DNA (**Figure 25.14b**). In this way, homeotic proteins can bind to DNA in a sequence-specific manner. In addition to DNA-binding ability, homeotic proteins also contain a transcriptional activation domain that functions to activate the genes to which the homeodomain can bind.

The transcription factors encoded by homeotic genes activate other genes encoding proteins that produce the morphological characteristics of each segment. Much current research attempts to identify these genes and determine how their expression in particular regions of the embryo leads to morphological changes in the embryo, larva, and adult.

The Developmental Fate of Each Cell in the Nematode *Caenorhabditis elegans* Is Known

We now turn our attention to another invertebrate, *C. elegans*, a nematode that has been the subject of numerous studies in developmental genetics. The embryo develops within the eggshell and hatches when it reaches a size of 550 cells. After hatching, it continues to grow and mature as it passes through four successive larval stages. It takes about 3 days for a fertilized egg to develop into an adult worm about 1 mm in length. With regard to sex, *C. elegans* can be a male (and produce only sperm) or a hermaphrodite (capable of producing sperm and egg cells). An adult male is composed of 1031 somatic cells and produces about 1000 sperm. A hermaphrodite consists of 959 somatic cells and produces about 2000 gametes (both sperm and eggs).

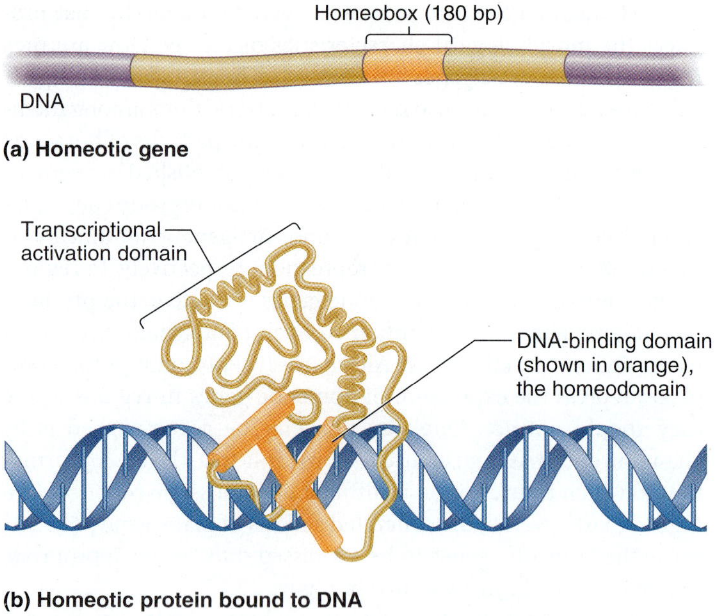

FIGURE 25.14 Molecular features of homeotic proteins. **(a)** A homeotic gene (shown in tan and orange) contains a 180-bp sequence called the homeobox (shown in orange). **(b)** When a homeotic gene is expressed, it produces a protein that functions as a transcription factor. The homeobox encodes a region of the protein called a homeodomain, which binds to the major groove of DNA. These DNA-binding sites are found within genetic regulatory elements (i.e., enhancers). The enhancers are found in the vicinity of promoters that are turned on by homeotic proteins. For this to occur, the homeotic protein also contains a transcriptional activation domain, which activates the transcription of a gene after the homeodomain has bound to the DNA.

A remarkable feature of this organism is that the pattern of cellular development remains constant from worm to worm. In the early 1960s, Sydney Brenner pioneered the effort to study the pattern of cell division in *C. elegans* and establish it as a model organism. Because *C. elegans* is transparent and composed of relatively few cells, researchers can follow cell division step by step under the microscope, beginning with a fertilized egg and ending with an adult worm. Researchers can identify a particular cell at an embryonic stage, follow that cell as it divides, and observe where its descendant cells are located in the adult. An illustration that depicts how cell division proceeds is called a **cell lineage diagram.** It describes the cell division patterns and fates of any cell's descendants.

Figure 25.15 shows a partial cell lineage diagram for a *C. elegans* hermaphrodite. At the first cell division, the egg divides to produce two cells, called AB and P_1. AB then divides into two cells—ABa and ABp; and P_1 divides into two cells—EMS and P_2. The EMS cell then divides into two cells, called MS and E. The cellular descendants of the E cell give rise to the worm's intestine. In other words, the fate of the E cell's descendants is to develop into intestinal cells. This diagram also illustrates the concept of a **cell lineage**—a series of cells that are derived from each other by cell division. For example, the EMS cell, E cell, and intestinal cells are all part of the same cell lineage.

Why is a cell lineage diagram for an organism an important experimental advantage? It allows researchers to investigate how gene expression in any cell, at any stage of development, may affect the outcome of a cell's fate. In the experiment described next, we will see how the timing of gene expression is an important factor in the fate of a cell's descendants.

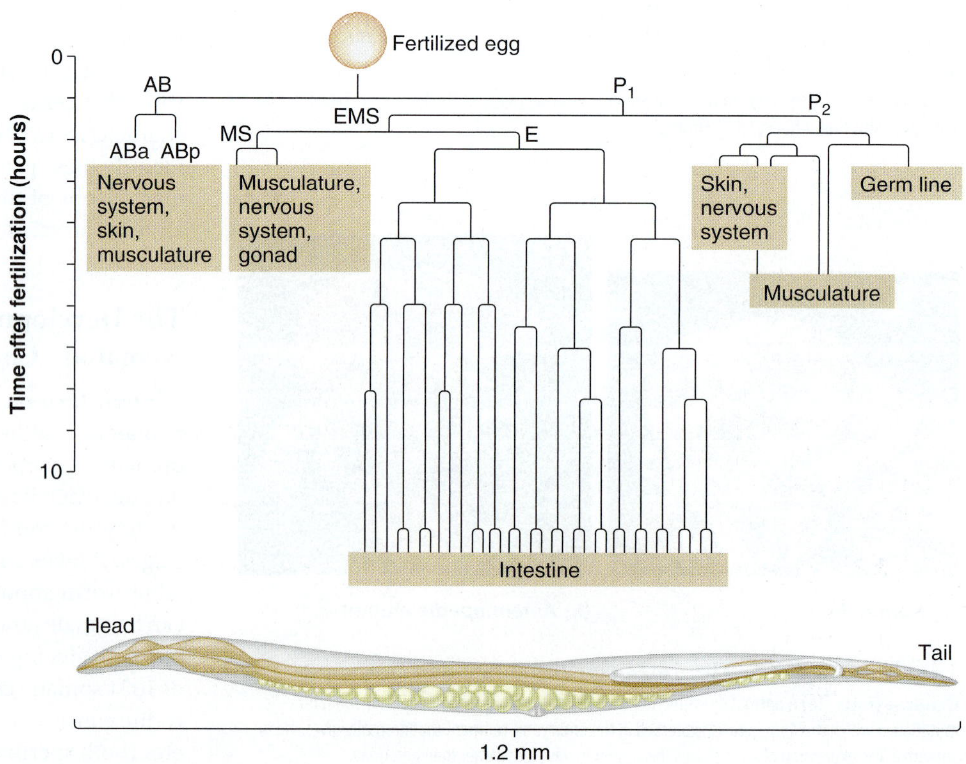

FIGURE 25.15 A cell lineage diagram of the nematode *Caenorhabditis elegans*. This partial cell lineage diagram illustrates how the cells divide to produce different regions of the adult. The fate of the intestinal cell lineage is shown in greater detail than that of other cell lineages. A complete cell lineage diagram is known for this organism, although its level of detail is beyond the scope of this textbook. Note: The lowercase letters "a" and "p" refer to anterior and posterior, respectively, which thereby indicates the relative positions of the daughter cells.

CONCEPT CHECK: What is a cell lineage?

EXPERIMENT 25A

Mutations Can Disrupt the Timing of Developmental Changes in *C. elegans*

Our discussion of *Drosophila* development focused on how the spatial expression and localization of gene products can lead to a particular pattern of embryonic development. An important issue in development is the timing of developmental changes at the cellular level. The cells of a multicellular organism must divide at the proper time and differentiate into the correct cell type. If the timing of these processes is not coordinated, certain tissues will develop too early or too late, disrupting the developmental process.

In *C. elegans*, the timing of developmental events can be examined carefully at the cellular level. As mentioned, the fate of each cell in *C. elegans* has been determined. Using a microscope, a researcher can focus on a particular cell within this transparent worm and watch it divide into two cells, then four cells, and so forth. Therefore, a scientist can judge whether a cell is behaving as it should during the developmental process.

To identify genes that play a role in the timing of cell fates, researchers have searched for mutant alleles that disrupt the normal timing process. In a collaboration in the late 1970s, H. Robert Horvitz and John Sulston set out to identify mutant alleles in *C. elegans* that disrupt cell fates or the timing of cell fates. Using a microscope, they screened thousands of worms for altered morphologies that might indicate an abnormality in development. During this screening process, one of the phenotypic abnormalities they found was a defective egg-laying phenotype. They reasoned that because the egg-laying system depends on a large number of cell types (vulval cells, muscle cells, and nerve cells), an abnormality in any of the cell lineages leading to these cell types might cause an inability to lay eggs.

In *C. elegans*, a defective egg-laying phenotype is easy to identify, because the hermaphrodite is able to fertilize its own eggs but unable to lay them. When this occurs, the eggs actually hatch within the hermaphrodite's body. This leads to the death of the hermaphrodite as it becomes filled with hatching worms. This defective egg-laying phenotype, in which the hermaphrodite becomes filled with its own offspring, is called a "bag of worms." Eventually, the newly hatched larvae eat their way out and can be saved for further study.

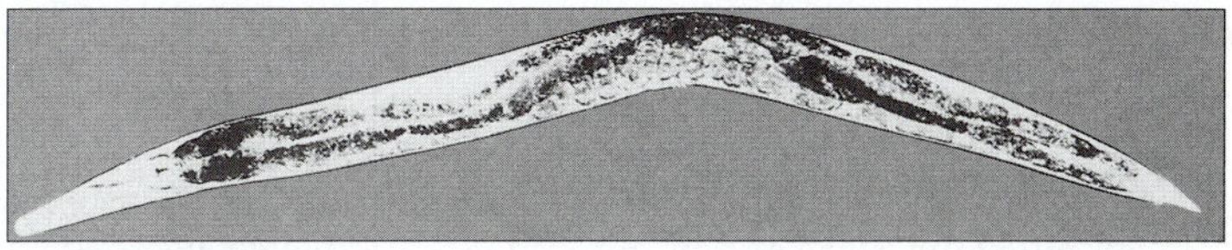

Normal *C. elegans*

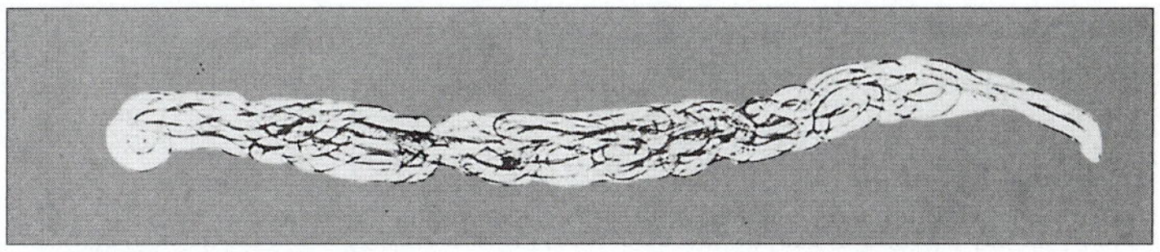

C. elegans with the defective egg-laying phenotype, the "bag of worms"

In their initial study, published in 1980, the defective egg-laying phenotype produced several mutant strains that were defective in particular cell lineages. A few years later, in the experiment described in **Figure 25.16**, Victor Ambros and H. Robert Horvitz took this same approach and were able to identify genes that play a key role in the timing of cell fate. They began with wild-type *C. elegans* and three mutant lines designated *n536*, *n355*, and *n540*. All three of the mutant lines showed an egg-laying defect. Right after hatching, they observed the fates of particular cells via microscopy. This involved spending hours looking at the nuclei of specific cells (which are relatively easy to see in this transparent worm) and timing the patterns of cell division. The patterns in the mutant and wild-type strains were then used to create cell lineage diagrams for particular cells.

THE HYPOTHESIS

Mutations that cause a defective egg-laying phenotype may affect the timing of cell lineages.

TESTING THE HYPOTHESIS

FIGURE 25.16 Identification of mutations that affect the timing of development in *C. elegans*.

Starting material: Prior to this work, many laboratories had screened thousand of *C. elegans* worms and identified many different mutant strains that were defective in egg-laying. (Note: When mutated, many different genes may cause a defective egg-laying phenotype. Only some of them are expected to be genes that alter the timing of cell fate within a particular cell lineage.)

	Experimental level	Conceptual level
1. Obtain a large number of *C. elegans* strains that have a defective egg-laying phenotype. The wild-type strain was also studied as a control.	Normal adult worm: 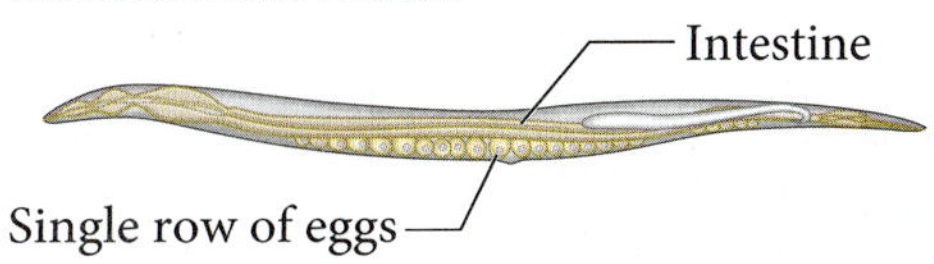	

2. Right after hatching, observe the fate of particular cells via microscopy. In this example, a researcher began watching a cell called the T cell and monitored its division pattern, and the pattern of subsequent daughter cells, during the larval stages. These patterns were examined in both wild-type and defective egg-laying worms.

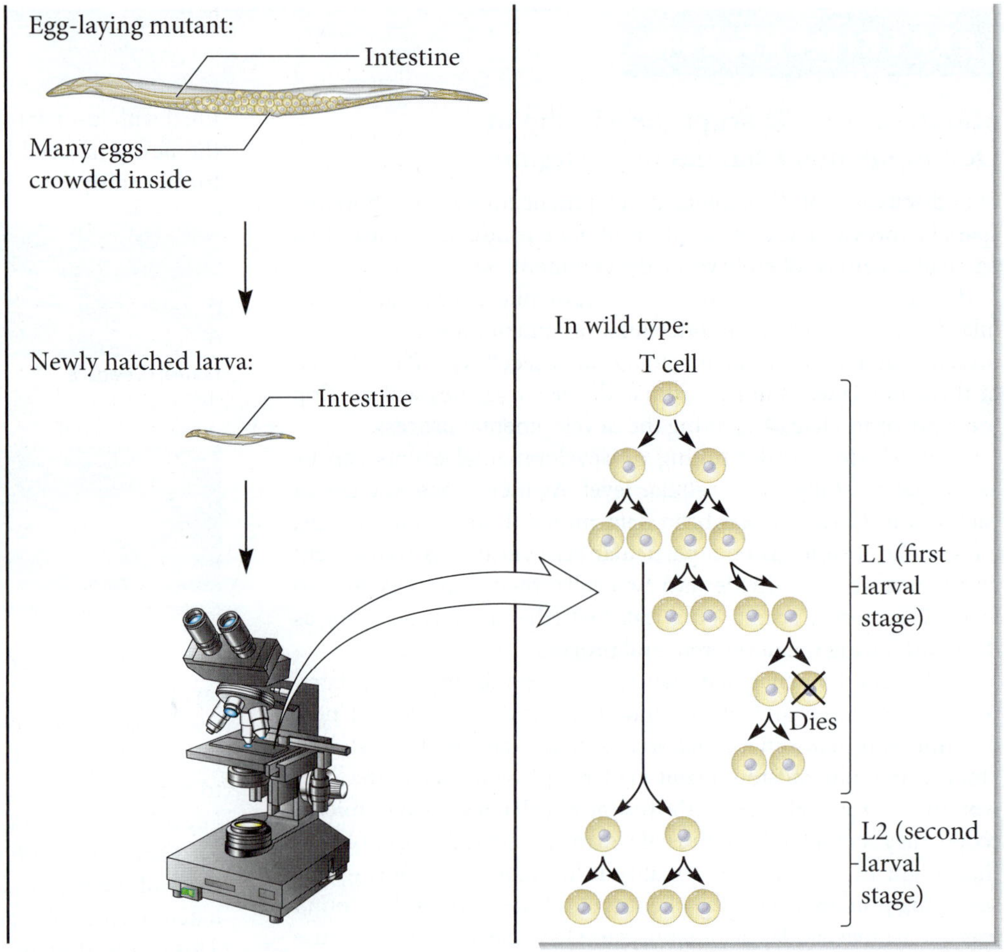

THE DATA

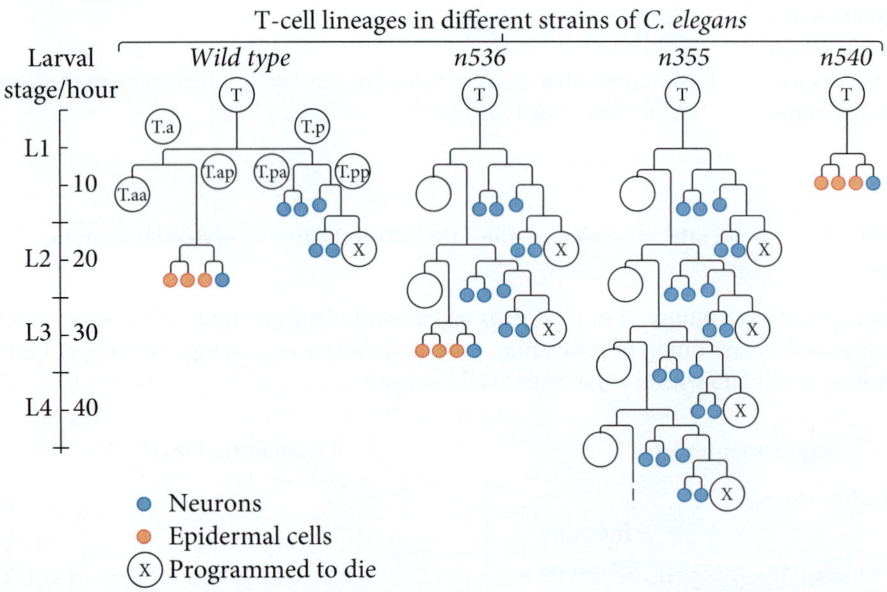

Data from V. Ambros and H.R. Horvitz (1984) Heterochronic mutants of the nematode *Caenorhabditis elegans*. *Science 226*, 409–416.

INTERPRETING THE DATA

Cell lineage diagrams involving cells derived from a cell called a T cell are shown in The Data of Figure 25.16. As seen here, the wild-type strain follows a particular pattern of cell division for the T-cell lineage. Each division event occurs at specific times during the L1 and L2 larval stages. In the normal strain, the T cell divides during the L1 larval stage to produce a T.a and T.p cell. The T.a cell also divides during L1 to produce a T.aa and T.ap cell. The T.p cell divides during L1 to produce T.pa and T.pp. These cells also divide during L1, eventually producing five neurons (labeled in blue) and one cell that is programmed to die (designated with an X). During the L2 larval stage, the T.ap cell resumes division to produce four cells: three epidermal cells (labeled in red) and one neuron.

The other T-cell lineages in The Data are from worms that carry mutations causing an egg-laying defect. Later research revealed that these three mutations are located in a gene called *lin-14*. The allele designated *n536* has caused the reiteration of the normal events of L1 during the L2 larval stage. In L2, the only cell of this lineage that is supposed to divide is T.ap. In worms carrying the *n536* allele, however, this cell behaves as if it were a T cell, rather than a T.ap cell, and produces a group of cells identical to what a T cell produces during the L1 stage. In the L3 stage, the cells in the *n536* strain behave as if they were in L2. In addition to the egg-laying defect, the phenotypic outcome of this irregularity in the timing of cell fates is a worm that has several additional cells derived from the T-cell lineage.

An allele that causes multiple reiterations is the *n355* allele. This strain continues to reiterate the normal events of L1 during the L2, L3, and L4 stages. In contrast, the *n540* allele has an opposite effect on the T-cell lineage. During the L1 larval stage, the T cell behaves as if it were a T.ap cell in the L2 stage. In this case, it skips the divisions and cell fates of the L1 and proceeds directly to cell fates that occur during the L2 stage.

The types of mutations described here are called heterochronic mutations. The term heterochrony refers to a change in the relative timing of developmental events. In **heterochronic mutations,** the timing of the fate of particular cell lineages is not synchronized with the development of the rest of the organism. More recent molecular data have shown that this is due to an irregular pattern of gene expression. In wild-type worms, the lin-14 protein accumulates during the L1 stage and promotes the T-cell division pattern shown for the wild type. During L2, the lin-14 protein diminishes to negligible levels. The *n536* and *n355* alleles are examples of gain-of-function mutations. In strains with these alleles, the lin-14 protein persists during later larval stages. For the *n536* allele, it is made for one additional cell division, whereas the *n355* allele continues to express *lin-14* for several cell divisions. By comparison, the *n540* allele is a loss-of-function mutation. This allele causes lin-14 to be inactive during L1, so it cannot promote the normal L1 pattern of cell division and cell fate.

Overall, the results described in this experiment are consistent with the idea that the precise timing of *lin-14* expression during development is necessary to correctly control the fate of particular cells in *C. elegans*. Mutations that alter the expression of *lin-14* lead to phenotypic abnormality, including the inability to lay eggs. This detrimental phenotypic consequence illustrates the importance of the correct timing for cell division and differentiation during development.

A self-help quiz involving this experiment can be found at www.mhhe.com/brookergenetics5e.

25.2 COMPREHENSION QUESTIONS

1. The expression of maternal-effect genes directly leads to
   a. the establishment of body axes.
   b. segmentation.
   c. determination.
   d. cell differentiation.
2. Which of the following are types of segmentation genes?
   a. Gap genes
   b. Pair-rule genes
   c. Segment-polarity genes
   d. All of the above
3. The expression of homeotic genes leads to
   a. the establishment of body axes.
   b. the formation of segments in the embryo.
   c. the determination of structures within segments.
   d. cell differentiation.
4. Homeotic genes encode proteins that function as
   a. cell-signaling proteins.
   b. transcription factors.
   c. hormones.
   d. all of the above.

25.3 VERTEBRATE DEVELOPMENT

Learning Outcomes:

1. Describe the relationship between homeotic genes in *Drosophila* and mice.
2. Explain how *Hox* gene expression affects vertebrate development.
3. Describe how cell differentiation is controlled in vertebrates.

Biologists have studied the morphological features of development in many vertebrate species. Historically, amphibians and birds have been studied extensively, because their eggs are rather large and easy to manipulate. For example, certain developmental stages of the frog and chicken have been described in great detail. In more recent times, the successes obtained in *Drosophila* have shown the great power of genetic analyses in elucidating the underlying molecular mechanisms that govern biological development. With this knowledge, many researchers are attempting to understand the genetic pathways that govern the development of the more complex body structure found in vertebrate organisms.

Several vertebrate species have been the subject of developmental studies. These include the mouse (*Mus musculus*), the frog (*Xenopus laevis*), and the small aquarium zebrafish (*Danio rerio*). In this section, we will primarily discuss the genes that are important in mammalian development, particularly those that have been characterized in the mouse, one of the best-studied mammals. As we will see, several genes affecting its developmental pathways have been cloned and characterized. We will also examine how these genes affect the course of vertebrate development.

Homeotic Genes Are Found in Vertebrates

Vertebrates typically have long generation times and produce relatively few offspring. Therefore, it is usually not practical to screen large numbers of embryos or offspring in search of mutant phenotypes with developmental defects. As an alternative, a successful way of identifying genes that affect vertebrate development has been the use of molecular techniques to identify vertebrate genes similar to those that control development in simpler organisms such as *Drosophila*.

As discussed in Chapters 23 and 28, species that are evolutionarily related to each other often contain genes with similar DNA sequences. When two or more genes have similar sequences because they are derived from the same ancestral gene, they are called **homologous genes.** Homologous genes found in different species are termed **orthologs.** Experimentally, a DNA strand from one gene hybridizes to a complementary strand of a homolog, because they have similar sequences.

Researchers initially followed a strategy of using cloned *Drosophila* genes as probes for identifying homologous vertebrate genes, which has been quite successful. With this method, researchers have found groups of homeotic genes in vertebrate species that are homologous to those in the fruit fly. These groups of homeotic genes are called ***Hox* complexes.** The term *Hox* is a contraction of homeobox, which is a domain found in all Hox proteins (see Figure 25.14). As shown in **Figure 25.17**, the mouse has four *Hox* complexes, designated *HoxA* (on chromosome 6), *HoxB* (on chromosome 11), *HoxC* (on chromosome 15), and *HoxD* (on chromosome 2). A total of 38 genes are found in the four complexes. Thirteen different types of homeotic genes occur within the four *Hox* complexes, although none of the four complexes contains representatives of all 13 types of genes. The addition of *Hox* genes into the genomes of animals has allowed animals to develop more complex body plans.

Remarkably, several of the homeotic genes in fruit flies and mammals are strikingly similar. Among the first six types of homeotic genes found in mammals, five are homologous to genes found in the *Antennapedia* complex of *Drosophila*. Among the last seven, three are homologous to the genes of the *bithorax* complex. These results indicate that fundamental similarities occur in the ways that animals as different as fruit flies and mammals undergo embryonic development. This suggests that a common body plan underlies animal development.

Like the *Antennapedia* and *bithorax* complexes in *Drosophila*, the arrangement of *Hox* genes along the mouse chromosomes reflects their pattern of expression from the anterior to the posterior end (**Figure 25.18a**). This phenomenon is seen in more detail in **Figure 25.18b**, which shows the expression pattern for a group of *HoxB* genes in a mouse embryo. Overall, these results are consistent with the idea that the *Hox* genes play a role in determining the fates of segments along the anteroposterior axis.

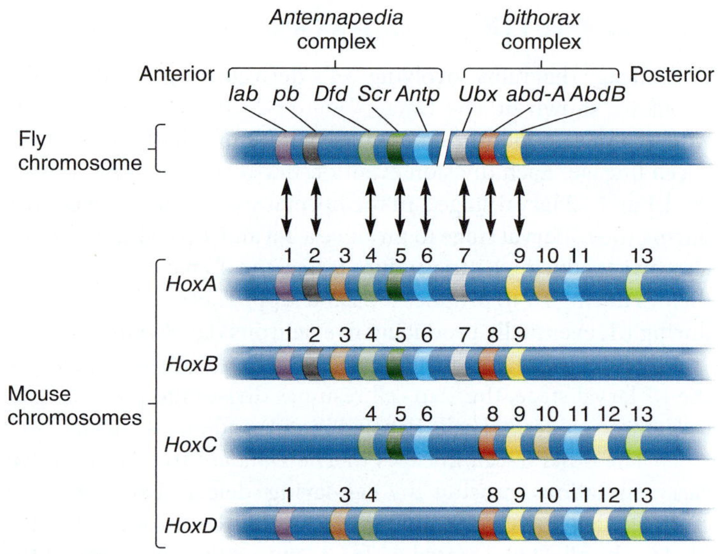

FIGURE 25.17 A comparison of homeotic genes in *Drosophila* and the mouse. The mouse contains four gene clusters, *HoxA–D*, that correspond to certain homeotic genes found in *Drosophila*. Thirteen different types of homeotic genes are found in the mouse, although each *Hox* gene cluster does not contain all 13 genes. In this drawing, orthologous genes are aligned in columns. For example, *lab* is the ortholog to *HoxA-1*, *HoxB-1*, and *HoxD-1*; *Ubx* is orthologous to *HoxA-7* and *HoxB-7*. In *Drosophila*, the homeotic genes are located on chromosome 3. In the mouse, the chromosomal locations are as follows: *HoxA* (chromosome 6), *Hox B* (chromosome 11), *HoxC* (chromosome 15) , and *HoxD* (chromosome 2).

CONCEPT CHECK: What is an ortholog?

Researchers Use Reverse Genetics to Understand *Hox* Gene Function

Currently, researchers are trying to understand the functional roles of the genes within the *Hox* complexes in vertebrate development. In *Drosophila*, great advances in developmental genetics have been made by studying mutant alleles in genes that control development. In mice, however, few natural mutations have been identified that affect development. This has made it difficult to understand the role that genetics plays in the development of the mouse and other vertebrate organisms.

How have researchers overcome this problem? Geneticists are taking an approach known as **reverse genetics.** In this strategy, researchers first identify the wild-type gene using cloning methods. In this case, the *Hox* genes in vertebrates were initially cloned using *Drosophila* genes as probes. The next step is to create a mutant version of a *Hox* gene in vitro. This mutant allele is then reintroduced into a mouse using gene replacement techniques

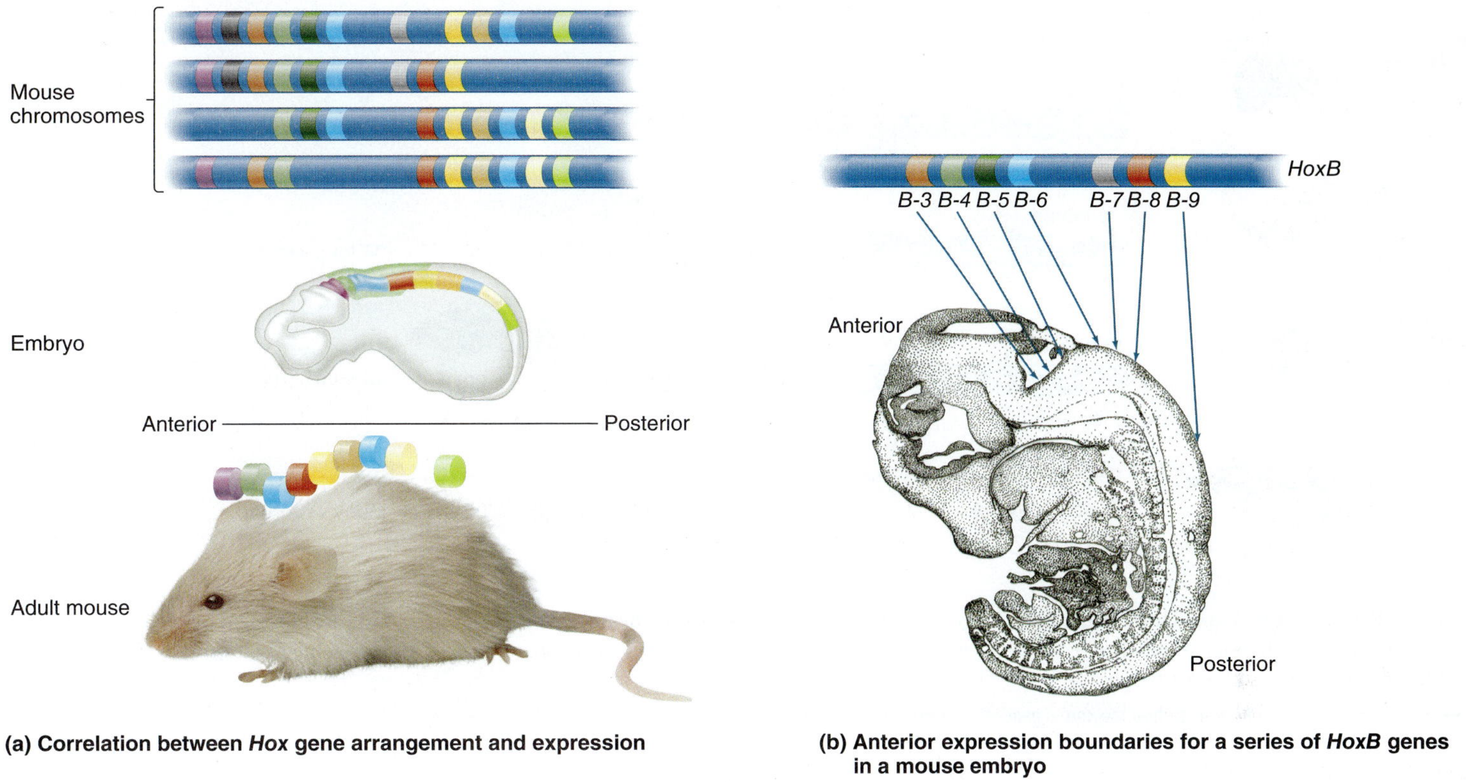

FIGURE 25.18 Expression pattern of *Hox* genes in the mouse. **(a)** A schematic illustration of the *Hox* gene expression in the embryo and the corresponding regions in the adult. **(b)** A more detailed description of *HoxB* expression in a mouse embryo. The arrows indicate the anterior-most boundaries for the expression of *HoxB-3* to *HoxB-9*. The order of *Hox* gene expression, from the anterior end to posterior of the embryo, is in the same order as the genes are found along the chromosome.

described in Chapter 21. Eliminating the function of the wild-type gene produces a **gene knockout.** In this way, researchers can determine how the mutant allele affects the phenotype of the mouse.

The concept of reverse genetics means that the experimental steps occur in an order opposite to that in the conventional approach used in *Drosophila* and other organisms. In the fly, mutant alleles were identified by their phenotype first, and then they were cloned. This is also called **forward genetics.** In contrast, the mouse genes were cloned first, the mutations were made in vitro, and then the mutated genes were introduced into the mouse to observe their phenotypic effects.

In recent years, several laboratories have used a reverse genetic approach to understand how many different genes, including *Hox* genes, affect vertebrate development. In *Drosophila*, loss-of-function alleles for homeotic genes usually show an anterior transformation. This means that the segment where the defective homeotic gene is expressed now exhibits characteristics that resemble the adjacent anterior segment. Similarly, certain gene knockouts (e.g., *HoxA-2*, *B-4*, and *C-8*) also show anterior transformations within particular regions of the mouse. However, knockouts of other *Hox* genes (e.g., *A-11*) have posterior transformations, and knockouts of *A-3* and *A-1* exhibit abnormalities in morphology but no clear homeotic transformations. Interestingly, a *HoxA-5* knockout in mice shows evidence of both anterior and posterior transformations, which are also seen in *Drosophila* when its ortholog, *Scr*, is knocked out.

Overall, the current picture indicates that the *Hox* genes in vertebrates play a key role in patterning the anteroposterior axis. During the evolution of animals, increases in the number of *Hox* genes and changes in their patterns of expression have had an important effect on their morphologies. As an example, let's consider the expression of the *HoxC-6* gene. The *HoxC-6* gene is expressed during embryonic development prior to vertebrae formation. Differences in the relative position of its expression correlate with the number of neck vertebrae produced (**Figure 25.19**). In the mouse, which has a relatively short neck, *HoxC-6* expression occurs posterior to the region of the early embryo that later develops into vertebrae 7. In contrast, *HoxC-6* expression in the chicken and goose occurs much farther back, posterior to vertebrae 14 and 17, respectively. The forelimbs also arise at this boundary in all vertebrates. However, snakes, which have no neck or forelimbs, do not have such a boundary because *HoxC-6* expression begins toward their heads.

Genes That Encode Transcription Factors Also Play a Key Role in Cell Differentiation

Thus far, we have focused our attention on patterns of gene expression that occur during the very early stages of development. These genes control the basic body plan of the organism. As this process occurs, cells become **determined.** As mentioned earlier, this refers to the phenomenon that a cell is destined to become a particular

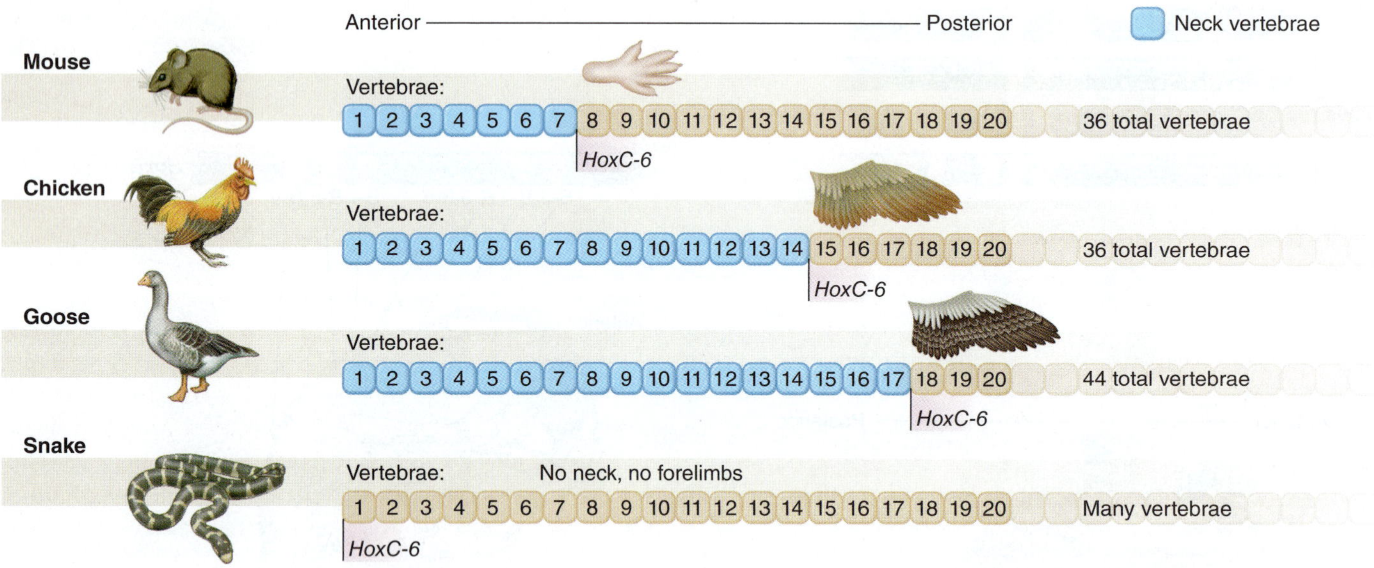

FIGURE 25.19 Expression of the *HoxC-6* gene in different species of vertebrates. This pink region indicates the general area of *HoxC-6* gene expression in the mouse, chicken, goose, and snake. The vertical black line shows the anterior edge of *HoxC-6* gene expression, which defines the posterior boundary of neck vertebrae.

CONCEPT CHECK: Briefly describe how the *HoxC-6* gene affects vertebrate development.

cell type. In other words, its fate has been predetermined to eventually become a particular type of cell, such as a nerve cell. This occurs long before a cell becomes **differentiated.** This later term means that a cell's morphology and function have changed, usually permanently, into a highly specialized cell type. For example, an undifferentiated mesodermal cell may differentiate into a specialized muscle cell, or an undifferentiated ectodermal cell may differentiate into a nerve cell.

At the molecular level, the profound morphological differences between muscle cells and nerve cells arise from gene regulation. Though muscle and nerve cells contain the same set of genes, they regulate the expression of those genes in very different ways. Certain genes that are transcriptionally active in muscle cells are inactive in nerve cells, and vice versa. Therefore, muscle and nerve cells express different proteins, which affect the characteristics of the respective cells in distinct ways. In this manner, differential gene regulation underlies cell differentiation.

Researchers have identified specific genes that cause cells to differentiate into particular cell types. In 1987, Harold Weintraub and colleagues identified a gene, which they called *MyoD.* This gene plays a key role in skeletal muscle-cell differentiation. Experimentally, when the cloned *MyoD* gene was introduced into fibroblast cells in a laboratory, the fibroblasts differentiated into skeletal muscle cells. This result was particularly remarkable because fibroblasts normally differentiate into osteoblasts (bone cells), chondrocytes (cartilage cells), adipocytes (fat cells), and smooth muscle cells, but in vivo, they never differentiate into skeletal muscle cells.

Since this initial discovery, researchers have found that *MyoD* belongs to a small group of genes that initiate muscle development. Besides *MyoD*, these include *Myogenin*, *Myf5*, and *Mrf4*. All four of these genes encode transcription factors that contain a **basic domain** and a **helix-loop-helix domain.** The basic domain is responsible for DNA binding and the activation of skeletal muscle-cell-specific genes. The helix-loop-helix domain is necessary for dimer formation between transcription factor proteins. Because of their common structural features and their role in muscle differentiation, MyoD, Myogenin, Myf5, and Mrf4 constitute a family of proteins called **myogenic bHLH proteins.** (Note: bHLH refers to the basic domain and the helix-loop-helix domain.) They are found in all vertebrates and have been identified in several invertebrates, such as *Drosophila* and *C. elegans.* In all cases, the myogenic bHLH genes are activated during skeletal muscle-cell development.

Two key features enable myogenic bHLH proteins to promote muscle-cell differentiation. First, the basic domain binds to a skeletal muscle-cell-specific enhancer sequence; this sequence is adjacent to genes that are expressed only in muscle cells (**Figure 25.20**). Therefore, when myogenic bHLH proteins are activated, they can bind to these enhancers and activate the expression of many different muscle-cell-specific genes. In this way, myogenic bHLH proteins function as master switches that activate the expression of muscle-specific genes. When the encoded proteins are synthesized, they change the characteristics of an undifferentiated cell into those of a highly specialized skeletal muscle cell.

A second key feature of myogenic bHLH proteins is that their activity is regulated by dimerization, which occurs via the helix-loop-helix domain. As shown in Figure 25.20, heterodimers—dimers formed from two different proteins—may be activating or inhibitory. When a heterodimer forms between a myogenic bHLH protein and an E protein, which also contains a basic domain, the heterodimer binds to the DNA and activates gene expression (Figure 25.20a). However, when a heterodimer forms between a myogenic bHLH protein and a protein called Id (for inhibitor of differentiation), the heterodimer cannot bind to DNA because the

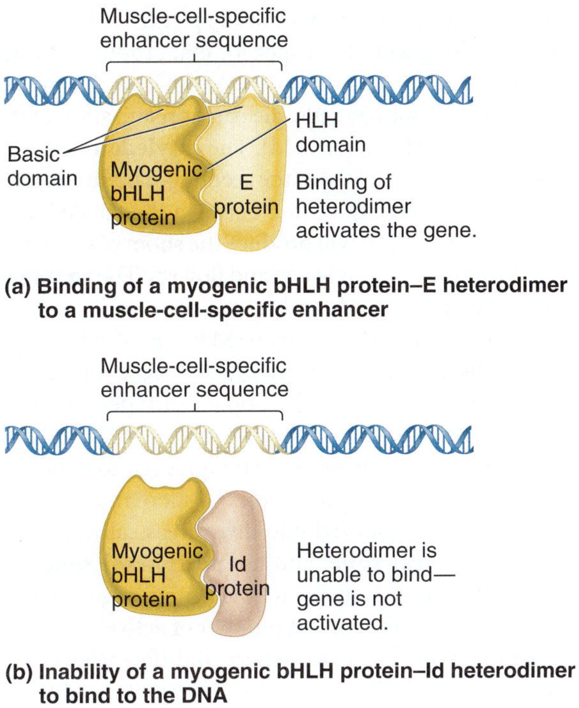

FIGURE 25.20 **Regulation of muscle-cell-specific genes by myogenic bHLH proteins.**

CONCEPT CHECK: At which stage of development does the Id protein function? Why is its function important?

Id protein lacks a basic domain (Figure 25.20b). Two basic domains are needed for the heterodimer to bind to the DNA. The Id protein is produced during early stages of development and prevents myogenic bHLH proteins from promoting muscle differentiation too soon. At later stages of development, the amount of Id protein falls, and myogenic bHLH proteins can then combine with E proteins to induce muscle-cell differentiation.

25.3 COMPREHENSION QUESTIONS

1. *Hox* genes encode transcription factors that
   a. control segmentation.
   b. promote determination.
   c. cause cell differentiation.
   d. do all of the above.
2. A cell that is __________ has a particular morphology and function.
   a. determined
   b. differentiated
   c. undergoing apoptosis
   d. dividing
3. Myogenic bHLH proteins are __________ that promote __________.
   a. cell-signaling proteins, muscle-cell differentiation
   b. cell-signaling proteins, muscle-cell determination
   c. transcription factors, muscle-cell differentiation
   d. transcription factors, muscle-cell determination

25.4 PLANT DEVELOPMENT

Learning Outcomes:

1. Describe how plant growth occurs from meristems.
2. Explain how homeotic genes control flower development.

In developmental plant biology, the model organism for genetic analysis is *Arabidopsis thaliana* (**Figure 25.21**). Unlike most flowering plants, which have long generation times and large genomes, *Arabidopsis* has a generation time of about 2 months and a genome size of 14×10^7 bp, which is similar to *Drosophila* and *C. elegans*. A flowering *Arabidopsis* plant produces a large number of seeds and is small enough to be grown in the laboratory. Like *Drosophila*, *Arabidopsis* can be subjected to mutagens to generate mutations that alter developmental processes. The small genome size of this organism makes it relatively easy to map these mutant alleles and eventually clone the relevant genes (as described in Chapters 20 and 22).

FIGURE 25.21 **The model organism *Arabidopsis*.** The plant is relatively small, making it easy to grow many of them in the laboratory.

The morphological patterns of growth are markedly different between animals and plants. As described previously, animal embryos become organized along anteroposterior, dorsoventral, and left-right axes, and then they subdivide into segments. By comparison, the form of plants has two key features. The first is the root-shoot axis. Most plant growth occurs via cell division near the tips of the shoots and roots.

Second, this growth occurs in a well-defined radial or circular pattern. For example, rings of dividing cells occur in the stems of plants. Stems also produce buds as they grow that give rise to additional branches, leaves, and flowers. Overall, the radial pattern in which a plant shoot gives off the buds that produce branches, leaves, and flowers is an important mechanism that determines much of the general morphology of a plant.

At the cellular level, too, plant development differs markedly from animal development. For example, cell migration does not occur during plant development. In addition, the development of a plant does not rely on morphogens that are deposited asymmetrically in the oocyte. In plants, an entirely new individual can be regenerated from many types of somatic cells. In other words, many plant cells are **totipotent,** meaning that they have the ability to differentiate into every cell type and to produce an entire new individual. By comparison, animal development typically relies on the organization within an oocyte as a starting point for development.

In spite of these apparent differences, the underlying molecular mechanisms of pattern development in plants still share some similarities with those in animals. In this section, we will consider a few examples in which the genes encoding transcription factors play a key role in plant development.

Plant Growth Occurs from Meristems Formed During Embryonic Development

Figure 25.22 illustrates a common sequence of events that takes place in the development of seed plants such as *Arabidopsis.* After fertilization, the first cellular division is asymmetrical and produces a smaller cell, called the apical cell, and a larger cell called a basal cell (Figure 25.22a). The apical cell gives rise to most of the embryo, and it later develops into the shoot of the plant. The basal cell gives rise to the root, along with the suspensor that produces extraembryonic tissue required for seed formation. At the "heart" stage, which is composed of only about 100 cells, the basic organization of the plant has been established (Figure 25.22d). The **cotyledons** are structures that will become the first leaves of the seedling. The **shoot meristem** arises from a group of cells located between the cotyledons. These cells are the precursors that will produce the shoot of the plant, along with lateral structures such as leaves and flowers. The **root meristem** is located on the opposite side and creates the root.

A meristem contains an organized group of actively dividing stem cells. As discussed in Chapter 21, stem cells retain the ability to divide and differentiate into multiple cell types. As they grow, meristems produce offshoots of proliferating cells. On a shoot meristem, for example, these offshoots or buds give rise to structures such as leaves and flowers. The organization of a shoot meristem is shown in **Figure 25.23**. It is organized into three areas called the **organizing center,** the **central zone,** and the **peripheral zone.** The role of the organizing center is to ensure the proper organization of the meristem and preserve the correct number of actively dividing stem cells. The central zone is an area where undifferentiated stem cells are always maintained. The peripheral zone contains dividing cells that eventually differentiate into plant structures. For example, the peripheral zone may form a bud, which will produce a leaf or flower.

In *Arabidopsis*, the organization of a shoot meristem is controlled by two critical genes termed *WUS* and *CLV3*. The *WUS* gene encodes a transcription factor that is expressed in the organizing center (Figure 25.23). The expression of the *WUS* gene induces the adjacent cells in the central zone to become undifferentiated stem cells. These stem cells then turn on the *CLV3* gene, which encodes a secreted protein. The CLV3 protein binds to receptors in the cells of the central and peripheral zones, preventing them from expressing the *WUS* gene. This limits the area of *WUS* gene expression to the underlying organizing center, thereby maintaining a small population of stem cells at the growing tip. A shoot meristem in *Arabidopsis* contains only about 100 cells. The inhibition of *WUS* expression in the peripheral cells also allows them to embark on a path of cell differentiation so they can produce structures such as leaves and flowers.

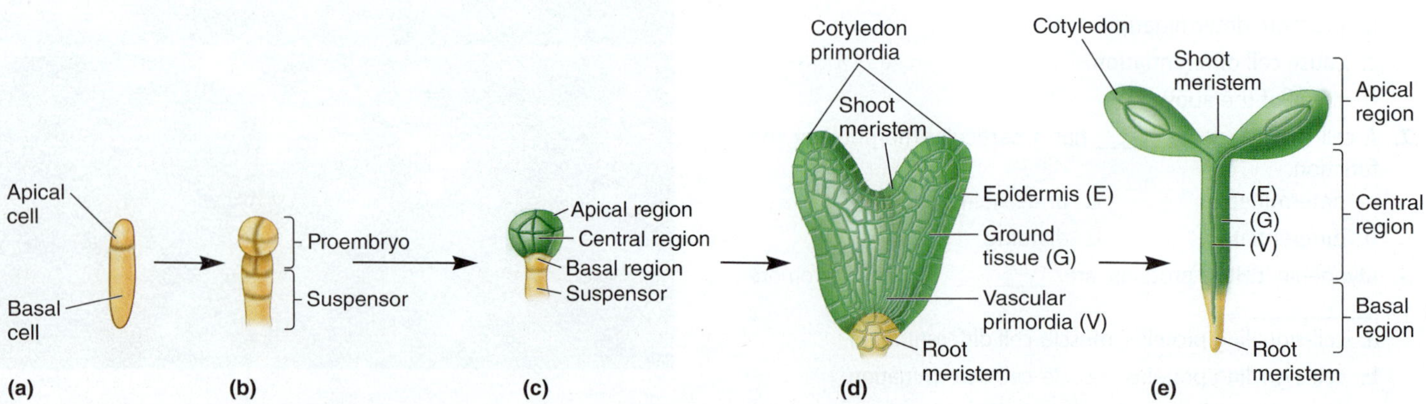

FIGURE 25.22 Developmental steps in the formation of a plant embryo.

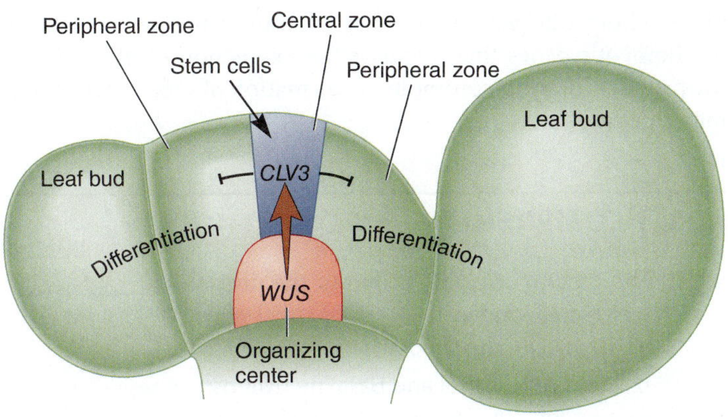

FIGURE 25.23 Organization of a shoot meristem. The organization of a shoot meristem is controlled by the *WUS* and *CLV3* genes, which are abbreviations for *Wuschel* and *clavata*, respectively. The *WUS* gene is expressed in the organizing center and induces the cells in the central zone to become undifferentiated stem cells. The red arrow indicates that the WUS protein induces these central zone stem cells to turn on the *CLV3* gene, which encodes a secreted protein that binds to receptors in the cells of the peripheral zone. The black lines with a vertical slash indicate that the CLV3 protein prevents the cells in the peripheral zone from expressing the *WUS* gene. This limits the area of *WUS* gene expression to the underlying organizing center, thereby maintaining a small population of stem cells at the growing tip. The cells in the peripheral zone are allowed to divide and eventually differentiate into lateral structures such as leaves.

CONCEPT CHECK: Why is it important to maintain the correct number of stem cells in the growing tip?

Plant Homeotic Genes Control Flower Development

Although the term homeotic was coined by William Bateson to describe homeotic mutations in animals, the first known mutations in homeotic genes were observed in plants. In ancient Greece and Rome, for example, double flowers in which stamens were replaced by petals were noted. In current research, geneticists have been studying these types of mutations to better understand developmental pathways in plants. Many homeotic mutations affecting flower development have been identified in *Arabidopsis* and also in the snapdragon (*Antirrhinum majus*).

A normal *Arabidopsis* flower is composed of four concentric whorls of structures (**Figure 25.24a**). The outer whorl contains four sepals, which protect the flower bud before it opens. The second whorl is composed of four petals, and the third whorl contains six stamens, the structures that make the male gametophyte, pollen. Finally, the innermost whorl contains two carpels, which are fused together. The carpel produces the female gametophyte. An example of a homeotic mutant in *Arabidopsis* is shown in **Figure 25.24b**. The sepals have been transformed into carpels, and the petals into stamens.

By analyzing the effects of many different homeotic mutations in *Arabidopsis*, Elliot Meyerowitz and colleagues proposed the **ABC model** for flower development. In this model, three classes of genes, called *A*, *B*, and *C*, govern the formation of sepals, petals, stamens, and carpels. More recently, a fourth category of genes called the *Sepallata* genes (*SEP* genes) have been found to be required for this process. **Figure 25.25** illustrates how these genes

FIGURE 25.24 Examples of homeotic mutations in *Arabidopsis*.

Genes→Traits **(a)** A normal flower. It is composed of four concentric whorls of structures: sepals, petals, stamens, and carpel. **(b)** A homeotic mutant in which the sepals have been transformed into carpels and the petals have been transformed into stamens. **(c)** A triple mutant in which all of the whorls have been changed into leaves.

(a) Normal flower

(b) Single homeotic mutant

(c) Triple mutant

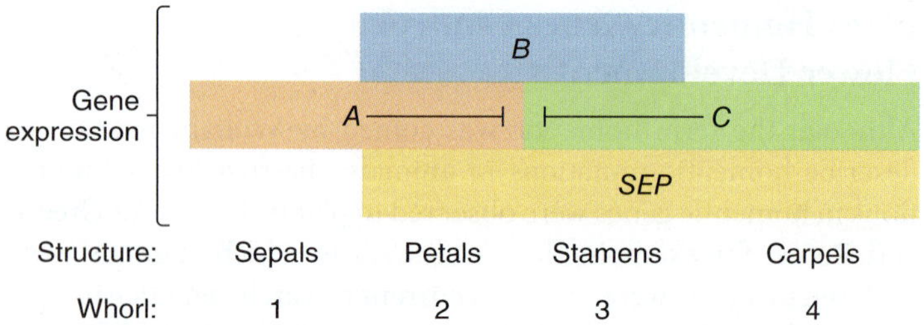

FIGURE 25.25 The ABC model of homeotic gene action in *Arabidopsis.* Note: This is a revised model based on the identification of *SEP* genes. The black lines with a vertical slash indicate that the *A* gene product represses the *C* gene product, and vice versa.

CONCEPT CHECK: What would be the expected result if the *A* gene was inactive?

affect normal flower development in *Arabidopsis*. Gene *A* products are made in tissues that will become the outermost whorl (whorl 1) and promote sepal formation. In tissues that will form whorl 2, gene *A*, gene *B*, and *SEP* gene products are made, which promotes petal formation. The expression of gene *B*, gene *C*, and *SEP* genes causes stamens to be made in whorl 3. Finally, gene *C* and *SEP* genes promote carpel formation in whorl 4.

What is the molecular explanation for how the homeotic mutants in the *A*, *B*, *C*, or *SEP* genes cause their phenotypic effects? In the original ABC model, it was proposed that genes *A* and *C* repress each other's expression, and gene *B* functions independently. In a mutant defective in gene *A* expression, gene *C* would also be expressed in tissues that give rise to whorls 1 and 2. This produces a carpel-stamen-stamen-carpel arrangement. When gene *B* is defective, a flower cannot make petals or stamens. Therefore, a gene *B* defect yields a flower with a sepal-sepal-carpel-carpel arrangement. When gene *C* is defective, gene *A* is expressed in tissues that give rise to all four whorls. This results in a sepal-petal-petal-sepal pattern. If the expression of *SEP* genes is defective, the flower consists entirely of sepals, which is the origin of the gene's name.

What happens if genes *A*, *B*, and *C* are all defective? As shown in **Figure 25.24c**, this produces a "flower" that is composed entirely of leaves! These results indicate that the leaf structure is the default pathway and that the *A*, *B*, *C*, and *SEP* genes cause development to deviate from a leaf structure in order to make something else. In this regard, the sepals, petals, stamens, and carpels can be viewed as modified leaves. Interestingly, the German philosopher and poet Johann Goethe originally proposed this idea—that flower formation comes from modifications of the leaf—over 200 years ago.

Arabidopsis has two types of gene *A* (*apetala1* and *apetala2*), two types of gene *B* (*apetala3* and *pistillata*), one type of gene *C* (*agamous*), and three *SEP* genes (*SEP1*, *SEP2*, and *SEP3*). All of these plant homeotic genes encode transcription factor proteins that contain a DNA-binding domain and a dimerization domain. However, the *Arabidopsis* homeotic genes do not contain a sequence similar to the homeobox found in animal homeotic genes.

Like the *Drosophila* homeotic genes, plant homeotic genes are part of a hierarchy of gene regulation. Genes that are expressed within a flower bud produce proteins that activate the expression of these homeotic genes. Once they are transcriptionally activated, the homeotic genes then regulate the expression of other genes, the products of which promote the formation of sepals, petals, stamens, or carpels.

25.4 COMPREHENSION QUESTIONS

1. The growth of plants is due to the division of __________, which are found in __________.
   a. stem cells, the shoots
   b. stem cells, apical and basal meristems
   c. somatic cells, the shoots
   d. somatic cells, apical and basal meristems
2. Flower development occurs in __________ according to __________.
   a. 3 whorls, maternal-effect genes
   b. 3 whorls, the ABC model
   c. 4 whorls, maternal-effect genes
   d. 4 whorls, the ABC model

25.5 SEX DETERMINATION IN ANIMALS

Learning Outcome:

1. Outline the molecular mechanism of sex determination in *Drosophila* and mammals.

To end our discussion of development, let's consider how genetics plays a role in **sex determination**—the process that governs the development of male and female individuals. In the animal kingdom, the existence of two sexes is nearly universal. As discussed in Chapter 3 (see Figure 3.17) the underlying factors that distinguish female versus male development vary widely. In animals, sex determination is often caused by differences in chromosomal composition. In flies, the ratio of the number of X chromosomes to the number of sets of autosomes determines sex. By comparison, in mammals, the presence of the *SRY* gene on the Y chromosome causes maleness. The adoption of one of two sexual fates is an event that has been studied in great detail in several species. Researchers have discovered that sex determination is a process controlled genetically by a hierarchy of genes that exert their effects in early embryonic development. In this section, we will consider features of these hierarchies in *Drosophila* and mammals.

In *Drosophila*, Sex Determination Involves a Regulatory Cascade That Includes Alternative Splicing

In diploid fruit flies, XX flies develop into females, and X0 flies become males. The ratio of the number of X chromosomes to the number of sets of autosomes is the determining factor. In a diploid fly that carries two sets of chromosomes, this ratio is 1.0 in

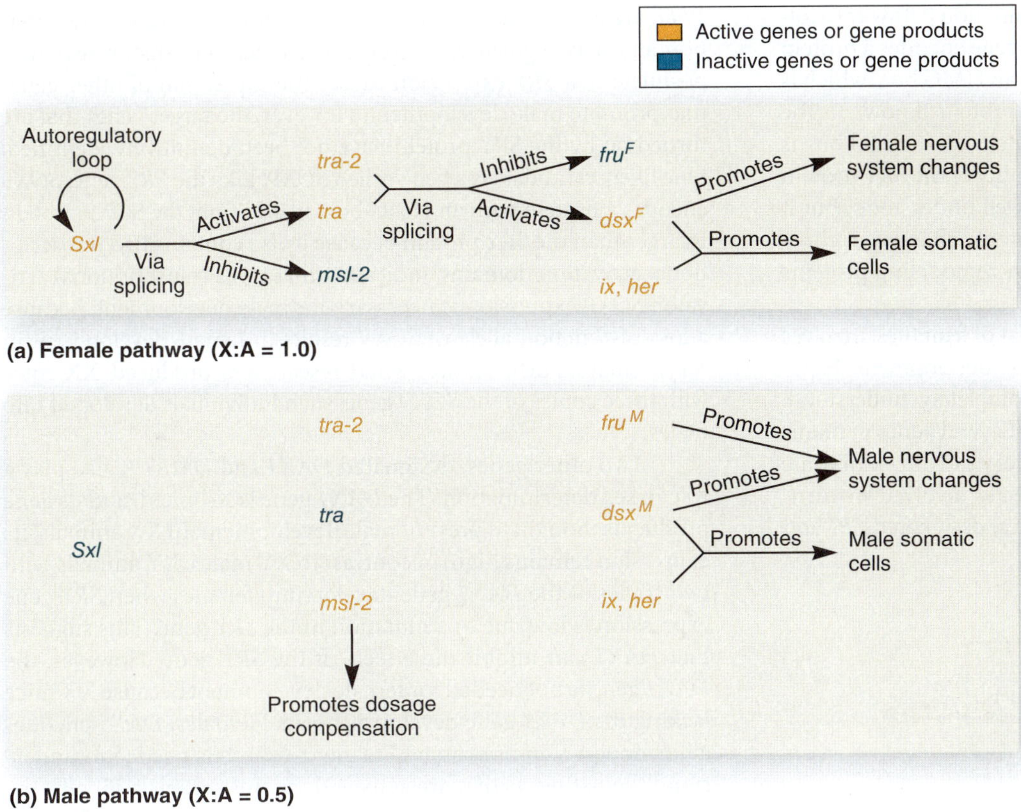

FIGURE 25.26
Sex determination pathway for *Drosophila melanogaster*. Genes or gene products that are functionally expressed are shown in light orange; those that are not expressed are shown in blue. The gene names are acronyms for the phenotypes that result from mutations that cause loss-of-function or aberrant expression. These are as follows: *Sxl* (sex lethal), *msl* (male sex lethal), *tra* (transformer), *dsx* (double sex), *fru* (fruitless), *ix* (intersex), and *her* (hermaphrodite). Note: This is a simplified pathway. More gene products are involved than shown here.

females versus 0.5 in males. Although male fruit flies usually carry a Y chromosome, it is not necessary for male development. The mechanism of sex determination in *Drosophila* begins in early embryonic development and involves a regulatory cascade composed of several genes. Females and males follow one of two alternative pathways. Simplified versions of these pathways are depicted in **Figure 25.26**.

Let's begin with the pathway that produces female flies. In females, the higher ratio of X chromosomes results in the embryonic expression of a gene designated *Sxl* (Figure 25.26a). The *Sxl* gene product is a protein that functions in the splicing of pre-mRNA. In female embryos, the Sxl protein enhances its own expression by splicing its own pre-mRNA, an event termed an **autoregulatory loop.** In addition, it splices the pre-mRNA from two other genes called *msl-2* and *tra*. The Sxl protein promotes the splicing of the *msl-2* pre-mRNA in a way that introduces an early stop codon in the coding sequence, thereby producing a shortened version of the msl-2 protein that is functionally inactive. By comparison, the Sxl protein promotes the splicing of *tra* pre-mRNA to produce an mRNA that is translated into a functional protein. Therefore, *Sxl* activates *tra*.

The *tra* gene product and a constitutively expressed product from a gene called *tra-2* are also splicing factors. In the female, they cause the alternative splicing of the pre-mRNAs that are expressed from the *fru* and *dsx* genes. The tra and tra-2 proteins cause these pre-mRNAs to be spliced into mRNAs designated fru^F and dsx^F, respectively. The female-specific fru^F mRNA is not translated into a sex-specific protein. However, the dsx^F mRNA, together with two other proteins encoded by the *ix* and *her* genes, promotes female sexual development and controls some aspects of female-specific behavior via the central nervous system. The dsx^F protein is known to be a transcription factor that regulates certain genes that promote these changes.

How are males produced? In X0 or XY flies, the *Sxl* gene is transcriptionally activated, but it is spliced in a way that places an early stop codon in the coding sequence. Therefore, a functional Sxl protein is not made (Figure 25.26b). This permits the expression of *msl-2*, which promotes dosage compensation. In fruit flies, dosage compensation is accomplished by turning up the expression of X-linked genes in the male to a level that is twofold higher. Therefore, even though the male has only one X chromosome, the expression of X-linked genes in the male and female is approximately equal. The absence of *Sxl* expression in male embryos also promotes the development of maleness. Without *Sxl*, the *tra* mRNA is not properly spliced, so *tra* is not expressed. Without the tra protein, the *fru* and *dsx* mRNAs are spliced in a different way to produce mRNAs designated fru^M and dsx^M. The dsx^M protein is a transcription factor that regulates several different genes, thereby promoting male development. In addition, the fru^M gene product is necessary for the regulation of genes involved in male-specific behaviors.

In Mammals, the *SRY* Gene on the Y Chromosome Determines Maleness

In most mammals, such as humans, mice, and marsupials, the presence of the *SRY* gene on the Y chromosome determines maleness. In cases of abnormal sex chromosome composition such as XXY, an individual develops into a male. The *SRY* gene, which is located on the Y chromosome, causes the sex determination pathway to follow a male developmental scheme. The *SRY* gene encodes a protein that contains a DNA-binding domain called an HMG box, which is found in a broad category of DNA-binding proteins known as the high-mobility group. In several of these proteins, the HMG box is known to cause DNA bends. Thus far, the ability of the SRY protein to promote male sex determination is not well understood, but it may act as a transcription factor and/or promote changes in chromatin architecture, much like the chromatin-remodeling proteins discussed in Chapters 10 and 15.

Sex determination in mammals, like that of fruit flies, involves a cascade that is initiated in early embryonic development. However, the details of the pathway are not completely understood. **Figure 25.27** illustrates a simplified pathway. Several genes in mammals have been identified that are expressed very early in embryonic development and may be directly or indirectly involved in turning on the *SRY* gene. For example, two genes, designated *SFI* and *WTI*, are expressed in the early embryo prior to sexual differentiation and may regulate *SRY* expression. Once activated, researchers postulate that *SRY* expression causes the expression of other genes that promote male development. However, the target genes that are turned on by the SRY protein have not been definitively identified. One likely candidate is a gene called *SOX9*. Like the *SRY* gene, *SOX9* encodes a protein with an HMG box. In addition, the SOX9 protein is larger than the SRY protein because it also contains two transcriptional activation domains, indicating that it is a transcriptional activator. *SOX9* expression is necessary for male development, because a loss-of-function allele of *SOX9* results in female development in XY animals. Furthermore, when researchers produced XX mice with three copies of the *SOX9* gene, such individuals developed into males.

Two other genes, designated *DAX1* and *DMRT1*, also play a role in sex determination. The *DAX1* gene is X-linked, and its gene product is thought to prevent male development. In XX animals, its expression remains high, in contrast to XY males. XY animals with two copies of the *DAX1* gene develop into females when *SRY* gene expression is low due to a mutation in the *SRY* gene. This suggests that *DAX1* can inhibit the effects of the *SRY* gene. However, the *DAX1* gene is not needed for female development because XX mice lacking the *DAX1* gene develop as normal females. *DAX1* encodes a hormone-receptor protein. Further research is needed to understand the pathways that are activated via this receptor. Finally, the *DMRT1* gene in mammals is evolutionarily related to the *dsx* gene in *Drosophila*. Both genes encode transcription factors involved in the differentiation of the gonads. In mammals, *DMRT1* expression is gonad-specific, and it is expressed at higher levels in the testes. XY mice that are lacking the *DMRT1* gene develop as males, but they are infertile due to severe defects in testis structure and the inability to produce sperm cells.

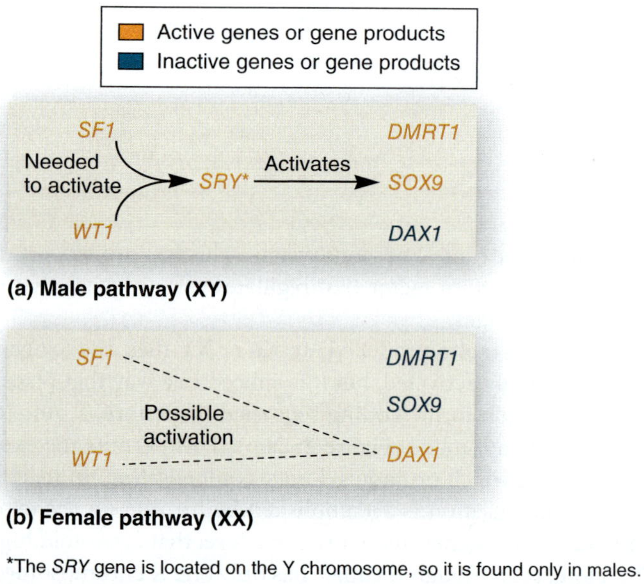

FIGURE 25.27 Sex determination pathway for mammals. Genes or gene products that are expressed at relatively high levels are shown in light orange; those that are not expressed or are inactive are shown in blue. The regulation of the *DAX1* gene is not well understood, but it may involve activation via *SF1* and/or *WT1* gene products. The gene names are acronyms that usually refer to the effects of mutations or their relatedness to other genes or chromosomal regions. These are as follows: *SF1* (steroidogenic factor-1), *WT1* (Wilms tumor gene), *SRY* (sex-determining region of the Y chromosome), *DMRT1* (doublesex- and mab-3-related transcription factor), *SOX9* (SRY-like HMG box), and *DAX1* (dosage-sensitive sex-reversal, adrenal hypoplasia congenital, X chromosome). Note: This is a simplified pathway.

25.5 COMPREHENSION QUESTIONS

1. A key event that initially determines female or male development in *Drosophila* is
   a. the transcription of the *Sxl* gene.
   b. alternative splicing of the *Sxl* RNA.
   c. expression of the *ix* gene.
   d. expression of the *her* gene.
2. An individual who is XY but is missing the *SRY* gene would be expected to develop into
   a. a male.
   b. a female.
   c. a hermaphrodite.
   d. none of the above because sex differentiation would not occur.

KEY TERMS

Page 640. developmental genetics
Page 641. pattern, positional information, morphogen, induction, cell adhesion, cell adhesion molecules (CAMs)
Page 642. homeotic gene, bilaterians
Page 643. determination
Page 644. parasegments, segments, embryogenesis, anteroposterior axis, dorsoventral axis, left-right axis, proximodistal axis
Page 648. segmentation genes, zygotic genes
Page 649. cell fate, homeotic
Page 650. gain-of-function mutation, loss-of-function mutations
Page 651. homeobox, homeodomain
Page 652. cell lineage diagram, cell lineage
Page 655. heterochronic mutations
Page 656. homologous genes, orthologs, *Hox* complexes, reverse genetics
Page 657. gene knockout, forward genetics, determined
Page 658. differentiated, basic domain, helix-loop-helix domain, myogenic bHLH proteins
Page 660. totipotent, cotyledons, shoot meristem, root meristem, organizing center, central zone, peripheral zone
Page 661. ABC model
Page 662. sex determination
Page 663. autoregulatory loop

CHAPTER SUMMARY

- Developmental genetics is concerned with the roles that genes play in orchestrating the changes that occur during development.

25.1 Overview of Animal Development

- Positional information in animals may cause a cell to divide, migrate, differentiate, or undergo apoptosis (see Figure 25.1).
- A morphogen is a molecule that conveys positional information.
- Three molecular mechanisms that convey positional information include (1) a preestablished morphogenic gradient in an oocyte, (2) asymmetrical secretion and induction of neighboring cells in an embryo, and (3) cell adhesion (see Figure 25.2).
- The study of mutants that disrupt development has identified genes that control development (see Figure 25.3).
- Development in animals involves four overlapping phases, which include the formation of axes, segmentation, determination, and cell differentiation (see Figure 25.4).

25.2 Invertebrate Development

- *Drosophila* proceeds through different developmental stages from a fertilized egg to an adult. Various sets of genes are responsible for developmental changes (see Figure 25.5, Table 25.1).
- Maternal-effect genes establish the anteroposterior and dorsoventral axes due to their asymmetrical distribution. An example is *bicoid*, which promotes the formation of anterior structures (see Figures 25.6–25.8).
- The *Drosophila* embryo is divided into segments (see Figure 25.9).
- Three categories of segmentation genes, called gap, pair-rule, and segment-polarity genes, have been identified based on their effects on development when mutant (see Figure 25.10).
- A hierarchy of gene expression, which includes maternal-effect genes, gap genes, pair-rule genes, and segment-polarity genes, gives rise to a segmented embryo (see Figure 25.11).
- Homeotic genes control the developmental fate of particular segments (see Figures 25.12, 25.13).
- Homeotic proteins contain a DNA-binding domain and a transcriptional activation domain (see Figure 25.14).
- In *C. elegans*, a cell lineage diagram describes the cell fate of each cell of the worm's body (see Figure 25.15).
- Heterochronic mutations in *C. elegans* disrupt the timing of developmental changes (see Figure 25.16).

25.3 Vertebrate Development

- Homeotic genes in vertebrates are found in *Hox* complexes (see Figure 25.17).
- *Hox* genes control the fate of regions along the anteroposterior axis (see Figures 25.18, 25.19).
- Transcription factors also control cell differentiation. An example is myoD, which causes cells to differentiate into skeletal muscle cells (see Figure 25.20).

25.4 Plant Development

- *Arabidopsis thaliana* is a model organism for studying plant development (see Figure 25.21).
- Plant growth occurs from shoot and root meristems (see Figure 25.22).
- The expression of *WUS* and *CLV3* controls the correct number of stem cells in the central zone of a shoot meristem (see Figure 25.23).
- The ABC model describes how homeotic genes control flower development in plants (see Figures 25.24, 25.25).

25.5 Sex Determination in Animals

- At the molecular level, sex determination is controlled by pathways that activate specific genes or proteins and inactivate others (see Figures 25.26–25.27).

PROBLEM SETS & INSIGHTS

Solved Problems

S1. Discuss the functional roles of maternal-effect genes, gap genes, pair-rule genes, and segment-polarity genes in *Drosophila*.

Answer: These genes are involved in pattern formation of the *Drosophila* embryo. The asymmetrical distribution of maternal-effect gene products in the oocyte establishes the anteroposterior and dorsoventral axes. These gene products also control the expression of the gap genes, which are expressed as broad bands in certain regions of the embryo. The overlapping expression of maternal-effect genes and gap genes controls the pair-rule genes, which are expressed in alternating stripes. A stripe corresponds to a parasegment. Within each parasegment, the expression of segment-polarity genes defines an anterior and posterior region. With regard to morphology, an anterior region of one parasegment and the posterior region of an adjacent parasegment form a segment of the fly.

S2. With regard to genes affecting development, what are the phenotypic effects of gain-of-function mutations versus loss-of-function mutations?

Answer: Gain-of-function mutations cause a gene to be expressed in the wrong place, at the wrong time, or at an abnormally high level. When the gene is expressed in the wrong place, that region may develop into an inappropriate structure. For example, when *Antp* is abnormally expressed in an anterior segment, this segment develops legs in place of antennae. When gain-of-function mutations cause a gene to be expressed at the wrong time, this can also disrupt the development process. Gain-of-function heterochronic mutations cause cell lineages to be reiterated, thereby altering the course of development. By comparison, loss-of-function mutations result in a defect in the expression of a gene with the result that protein function is reduced or abolished. This usually disrupts the developmental process, because the cells in the region where the gene is normally expressed are not directed to develop along the correct pathway.

S3. Mutations in genes that control the early stages of development are often lethal (e.g., see Figure 25.7b). To circumvent this problem, developmental geneticists may try to isolate temperature-sensitive developmental mutants, or *ts* alleles. If an embryo carries a *ts* allele, it will develop correctly at the permissive temperature (e.g., 25°C) but will fail to develop if incubated at the nonpermissive temperature (e.g., 30°C). In most cases, *ts* alleles are missense mutations that slightly alter the amino acid sequence of a protein, causing a change in its structure that prevents it from working properly at the nonpermissive temperature. *ts* alleles are particularly useful because they can provide insight regarding the stage of development when the protein encoded by the gene is necessary. Researchers can take groups of embryos that carry a *ts* allele and expose them to the permissive and nonpermissive temperature at different stages of development. In the experiment described next, embryos were divided into five groups and exposed to the permissive or nonpermissive temperature at different times after fertilization.

Time After Fertilization (hours):	**Group** 1	2	3	4	5
0–1	25°C	25°C	25°C	25°C	25°C
1–2	25°C	30°C	25°C	25°C	25°C
2–3	25°C	25°C	30°C	25°C	25°C
3–4	25°C	25°C	25°C	30°C	25°C
4–5	25°C	25°C	25°C	25°C	30°C
5–6	25°C	25°C	25°C	25°C	25°C
SURVIVAL:	Yes	Yes	Yes	No	Yes

Explain these results.

Answer: By varying the temperature during different stages of development, researchers can pinpoint the stage when the function of the protein encoded by this *ts* allele is critical. As shown, embryos fail to survive if they are subjected to the nonpermissive temperature between 3 and 4 hours after fertilization, but they do survive if subjected to the nonpermissive temperature at other times of development. These results indicate that this protein plays a crucial role at the 3- to 4-hour stage of development.

S4. An intriguing question in developmental genetics is, how can a particular gene, such as *even-skipped*, be expressed in a multiple banding pattern as seen in Figure 25.11? Another way of asking this question is, how is the positional information within the broad bands of the gap genes able to be deciphered in a way that causes the pair-rule genes to be expressed in this alternating banding pattern? The answer lies in a complex mechanism of genetic regulation. Certain pair-rule genes have several stripe-specific enhancers that are controlled by multiple transcription factors. A stripe-specific enhancer is typically a short segment of DNA, 300–500 bp in length, that contains binding sequences recognized by several different transcription factors. This term is a bit misleading because a stripe-specific enhancer is a regulatory region that contains both enhancer and silencer elements.

In 1992, Michael Levine and his colleagues investigated stripe-specific enhancers located near the promoter of the *even-skipped* gene. A segment of DNA, termed the stripe 2 enhancer, controls the expression of the *even-skipped* gene; this enhancer is responsible for the expression of the *even-skipped* gene in stripe 2, which corresponds to parasegment 3 of the embryo. The stripe 2 enhancer is a segment of DNA that contains binding sites for four transcription factors that are the products of the *Krüppel*, *bicoid*, *hunchback*, and *giant* genes. The Hunchback and Bicoid transcription factors bind to this enhancer and activate the transcription of the *even-skipped* gene. In contrast, the transcription factors encoded by the *Krüppel* and *giant* genes bind to the stripe 2 enhancer and repress transcription. The figure shown next describes the concentrations of these four transcription factor proteins in the region of parasegment 3 (i.e., stripe 2) in the *Drosophila* embryo.

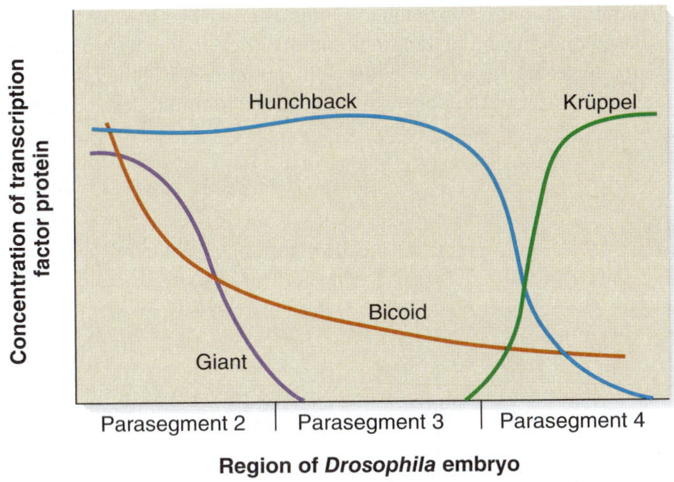

To study stripe-specific enhancers, researchers have constructed artificial genes in which the enhancer is linked to a reporter gene, the expression of which is easy to detect. The next figure shows the results of an experiment in which an artificial gene was made by putting the stripe 2 enhancer next to the β-galactosidase gene. This artificial gene was introduced into *Drosophila*, and then embryos containing this gene were analyzed for β-galactosidase activity. If a region of the embryo is expressing β-galactosidase, the region will stain darkly because β-galactosidase converts a colorless compound into a dark blue compound.

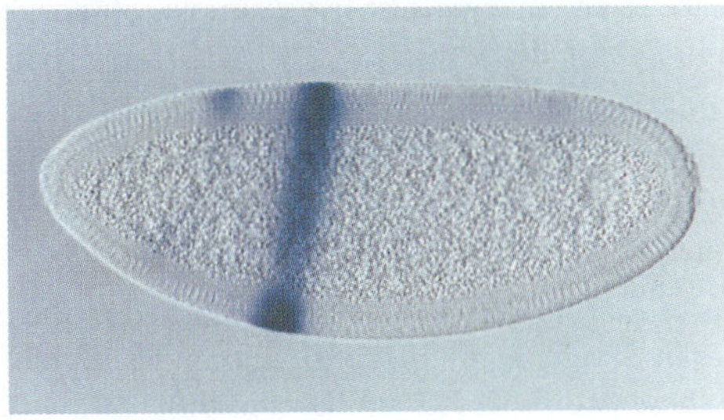

Explain these results.

Answer: As shown in the first figure to this problem, the concentrations of the Hunchback and Bicoid transcription factors are relatively high in the region of the embryo corresponding to stripe 2 (which is parasegment 3). The levels of Krüppel and Giant are very low in this region. Therefore, the high levels of activators and low levels of repressors cause the *even-skipped* gene to be transcribed. In this experiment, β-galactosidase was made only in stripe 2 (i.e., parasegment 3). These results show that the stripe-2-specific enhancer controls gene expression only in parasegment 3. Because we know that the *even-skipped* gene is expressed as several alternating stripes (as seen in Figure 25.11), the *even-skipped* gene must contain other stripe-specific enhancers that allow it to be expressed in these other alternating parasegments.

Conceptual Questions

C1. What four types of cellular processes must occur to enable a fertilized egg to develop into an adult multicellular animal? Briefly discuss the role of each process.

C2. The arrangement of body axes of the fruit fly are shown in Figure 25.5g. Are the following statements true or false with regard to body axes in the mouse?

A. Along the anteroposterior axis, the head is posterior to the tail.

B. Along the dorsoventral axis, the vertebrae of the back are dorsal to the stomach.

C. Along the dorsoventral axis, the feet are dorsal to the hips.

D. Along the proximodistal axis, the feet on the hind legs are distal to the upper parts of the hind legs.

C3. If you observed fruit flies with the following developmental abnormalities, would you guess that a mutation has occurred in a segmentation gene or a homeotic gene? Explain your guess.

A. Three abdominal segments were missing.

B. One abdominal segment had legs.

C. A fly with the correct number of segments had two additional thoracic segments and two fewer abdominal segments.

C4. Which of the following statement(s) is/are true with regard to positional information in *Drosophila*?

A. Morphogens are a type of molecule that conveys positional information.

B. Morphogenetic gradients are established only in the oocyte, prior to fertilization.

C. Cell adhesion molecules also provide a way for a cell to obtain positional information.

C5. Discuss the morphological differences between the parasegments and segments of *Drosophila*. Discuss the evidence, providing specific examples, that suggests the parasegments of the embryo are the subdivisions for the organization of gene expression.

C6. Here are schematic diagrams of mutant larvae.

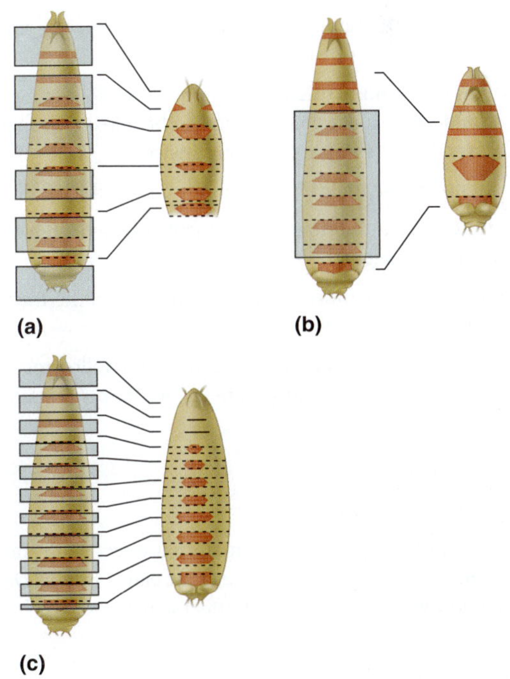

The left side of each pair shows a wild-type larva, with gray boxes showing the sections that are missing in the mutant larva. Which type of gene is defective in each larva: a gap gene, a pair-rule gene, or a segment-polarity gene?

C7. Describe what a morphogen is and how it exerts its effects. What do you expect will happen when a morphogen is expressed in the wrong place in an embryo? List five examples of morphogens that function in *Drosophila*.

C8. What is the meaning of positional information? Discuss three different ways that cells obtain positional information. Which of these three ways do you think is the most important for the formation of a segmented body pattern in *Drosophila*?

C9. Gradients of morphogens can be preestablished in the oocyte. Also, later in development, morphogens can be secreted from cells. How are these two processes similar and different?

C10. Discuss how the anterior portion of the anteroposterior axis is established. What aspects of oogenesis are critical in establishing this axis? What do you think would happen if the *bicoid* mRNA was not trapped at the anterior end but instead diffused freely throughout the oocyte?

C11. Describe the function of the Bicoid protein. Explain how its ability to exert its effects in a concentration-dependent manner is a critical feature of its function.

C12. With regard to development, what are the roles of the maternal-effect genes versus the zygotic genes? Which types of genes are needed earlier in the development process?

C13. Discuss the role of homeotic genes in development. Explain what happens to the phenotype of a fly when a gain-of-function homeotic gene mutation causes the protein to be expressed in an abnormal region of the embryo. What are the consequences of a loss-of-function mutation in such a gene?

C14. Describe the molecular features of the homeobox and homeodomain. Explain how these features are important in the function of homeotic genes.

C15. What would you predict to be the phenotype of a *Drosophila* larva whose mother was homozygous for a loss-of-function allele in the *nanos* gene?

C16. Based on the photographs in Figure 25.13, in which segments would the *Antp* gene normally be expressed?

C17. If a mutation in a homeotic gene produced the following phenotypes, would you expect the mutation to be a loss-of-function or a gain-of-function allele? Explain your answer.

A. An abdominal segment has antennae attached to it.

B. The most anterior abdominal segment resembles the most posterior thoracic segment.

C. The most anterior thoracic segment resembles the most posterior abdominal segment.

C18. Explain how loss-of-function mutations in the following categories of genes would affect the morphologies of *Drosophila* larvae:

A. Gap genes

B. Pair-rule genes

C. Segment-polarity genes

C19. What is the difference between a maternal-effect gene and a zygotic gene? Of the following genes that play a role in *Drosophila* development, which are maternal-effect genes and which are zygotic? Explain your answer.

A. *nanos* C. *bicoid*

B. *Antp* D. *lab*

C20. Cloning of mammals (such as Dolly) is described in Chapter 21. Based on your understanding of animal development, explain why an enucleated egg is needed to clone mammals. In other words, what features of the oocyte are essential for animal development?

C21. A hypothetical cell lineage is shown here.

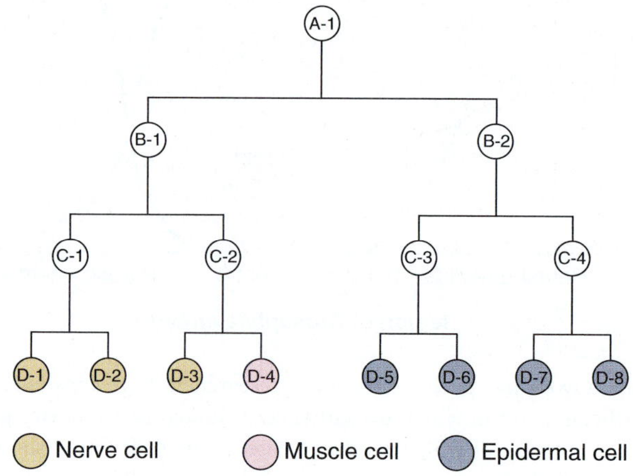

A gene, which we will call gene *X*, is activated in the B-1 cell, so the B-1 cell will progress through the proper developmental stages to produce three nerve cells (D-1, D-2, and D-3) and one muscle cell (D-4). Gene *X* is normally inactivated in A-1, C-1, and C-2 cells, as well as the four D cells. Draw the expected cell lineages if a heterochronic mutation had the following effects:

A. Gene *X* is turned on one cell division too early.

B. Gene *X* is turned on one cell division too late.

C22. What is a heterochronic mutation? How does it affect the phenotypic outcome of an organism? What phenotypic effects would you expect if a heterochronic mutation affected the cell lineage that determines the fates of intestinal cells?

C23. Discuss the similarities and differences between the *bithorax* and *Antennapedia* complexes in *Drosophila* and the *Hox* gene complexes in mice.

C24. What is cell differentiation? Discuss the role of myogenic bHLH proteins in the differentiation of muscle cells. Explain how they work at the molecular level. In your answer, indicate how protein dimerization is a key feature in gene regulation.

C25. The *MyoD* gene in mammals plays a role in skeletal muscle-cell differentiation, whereas the *Hox* genes are homeotic genes that play a role in the differentiation of particular regions of the body. Explain how the functions of these genes are similar and different.

C26. What is a totipotent cell? In the following examples, which cells would be totipotent?

A. Humans B. Corn C. Yeast D. Bacteria

C27. What is a meristem? Explain the role of meristems in plant development.

C28. Discuss the morphological differences between animal and plant development. How are they different at the cellular level? How are they similar at the genetic level?

C29. Predict the phenotypic consequences of each of the following mutations:

A. *apetala1* defective C. *apetala1* and *pistillata* defective

B. *pistillata* defective

C30. Explain how alternative splicing affects sex determination in *Drosophila.*

Experimental Questions

E1. Researchers have used the cloning methods described in Chapter 20 to clone the *bicoid* gene and express large amounts of the Bicoid protein. The Bicoid protein was then injected into the posterior end of a zygote immediately after fertilization. What phenotypic results would you expect? What do you think would happen if the Bicoid protein was injected into a segment of a larva?

E2. Compare and contrast the experimental advantages of *Drosophila* and *C. elegans* in the study of developmental genetics.

E3. What is meant by the term cell fate? What is a cell lineage diagram? Discuss the experimental advantage of having a cell lineage diagram. What is a cell lineage?

E4. Explain why a cell lineage diagram is necessary to determine if a mutation is heterochronic.

E5. Explain the rationale behind the use of the "bag of worms" phenotype as a way to identify heterochronic mutations.

E6. Here are the results of cell lineage analyses of hypodermal cells in wild-type and mutant strains of *C. elegans*.

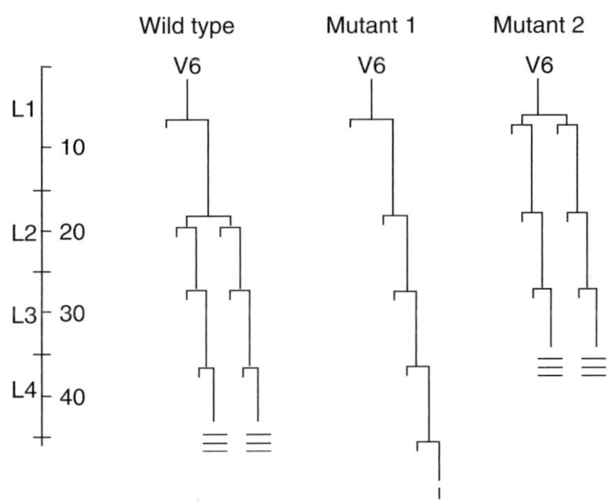

Explain the nature of the mutations in the altered strains.

E7. Take a look at solved problem S3 before answering this question. *Drosophila* embryos carrying a *ts* mutation were exposed to the permissive (25°C) or nonpermissive (30°C) temperature at different stages of development. Explain these results.

Time After Fertilization (hours):	Group 1	2	3	4	5
0–1	25°C	25°C	25°C	25°C	25°C
1–2	25°C	30°C	25°C	25°C	25°C
2–3	25°C	25°C	30°C	25°C	25°C
3–4	25°C	25°C	25°C	30°C	25°C
4–5	25°C	25°C	25°C	25°C	30°C
5–6	25°C	25°C	25°C	25°C	25°C
SURVIVAL:	Yes	No	No	Yes	Yes

E8. All of the homeotic genes in *Drosophila* have been cloned. As discussed in Chapter 20, cloned genes can be manipulated in vitro. They can be subjected to cutting and pasting, site-directed mutagenesis, etc. After *Drosophila* genes have been altered in vitro, they can be inserted into a *Drosophila* transposon vector (i.e., a P element vector), and then the altered gene within a P element can be injected into *Drosophila* embryos. The P element then transposes into the chromosomes, thereby introducing one or more copies of the altered gene into the *Drosophila* genome. This method is termed P element transformation.

With these ideas in mind, how would you make a mutant gene with a gain of function in which the *Antp* gene would be expressed where the *abd-A* gene is normally expressed? What phenotype would you expect for flies that carried this altered gene?

E9. You need to understand solved problem S4 before answering this question. If the artificial gene containing the stripe 2 enhancer and the β-galactosidase gene was found within an embryo that also contained the following loss-of-function mutations, what results would you expect? In other words, would there be a stripe or not? Explain why.

A. *Krüppel*

B. *bicoid*

C. *hunchback*

D. *giant*

E10. Two techniques commonly used to study the expression patterns of genes that play a role in development are Northern blotting and in situ hybridization. As described in Chapter 20, Northern blotting is used to detect RNA that is transcribed from a particular gene. In this method, a specific RNA is detected by using a short segment of cloned DNA as a probe. The DNA probe, which is labeled, is complementary to the RNA that the researcher wishes to detect. After the probe DNA binds to the RNA within a blot of a gel, the RNA is visualized as a labeled band on a nylon membrane. For example, a DNA probe that is complementary to the *bicoid* mRNA could be used to specifically detect the amount and size of the mRNA in a blot.

A second technique, termed fluorescence in situ hybridization (FISH), is used to identify the locations of genes on chromosomes. This technique is also used to locate gene products within oocytes, embryos, and larvae. For this reason, it has been commonly used by developmental geneticists to understand the expression patterns of genes during development. The micrograph in Figure 25.8b is derived from the application of the FISH technique. In this case, the probe was complementary to *bicoid* mRNA.

Now here is the question. Suppose a researcher has three different *Drosophila* strains that have loss-of-function mutations in the *bicoid* gene. We will call them *bicoid-A*, *bicoid-B*, and *bicoid-C*; the wild type is designated *bicoid*$^+$. To study these mutations, phenotypically normal female flies that are homozygous for the *bicoid* mutation were obtained, and their oocytes were analyzed using these two techniques. A wild-type strain was also analyzed as a control. In other words, RNA was isolated from some oocytes and analyzed by Northern blotting, and some oocytes were subjected to in situ hybridization. In both cases, the probe was complementary to the *bicoid* mRNA. The results are shown next.

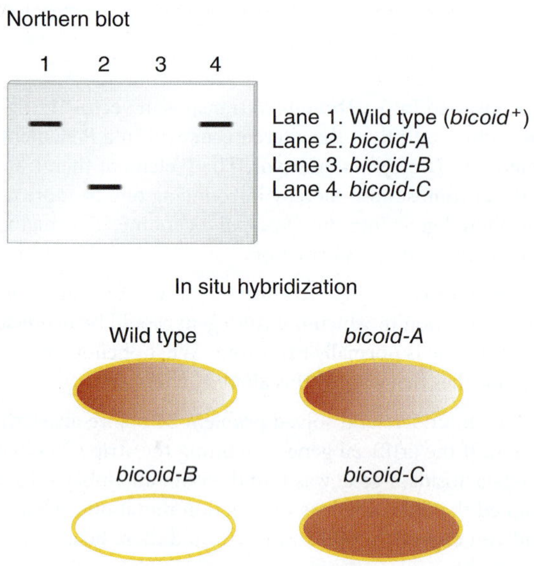

A. How can phenotypically normal female flies be homozygous for a loss-of-function allele in the *bicoid* gene?

B. Explain the type of mutation (e.g., deletion, point mutation, etc.) in each of the three strains. Explain how the mutation may cause a loss of normal function for the *bicoid* gene product.

C. Discuss how the use of both techniques provides more definitive information than the application of just one of the two techniques.

E11. Explain one experimental strategy for determining the functional role of the mouse *HoxD-3* gene.

E12. In the experiment of Figure 25.16, suggest reasons why the *n536*, *n355*, and *n540* strains have an egg-laying defect.

E13. Another way to study the role of proteins (e.g., transcription factors) that function in development is to microinject the mRNA that encodes a protein, or the purified protein itself, into an oocyte or embryo, and then determine how this affects the subsequent development of the embryo, larva, and adult. For example, if Bicoid protein is injected into the posterior region of an oocyte, the resulting embryo will develop into a larva that has anterior structures at both ends. Based on your understanding of the function of these developmental genes, what would be the predicted phenotype if the following proteins or mRNAs were injected into normal oocytes?

A. *Nanos* mRNA injected into the anterior end of an oocyte

B. Antp protein injected into the posterior end of an embryo

C. *Toll* mRNA injected into the dorsal side of an early embryo

E14. Why have geneticists used reverse genetics to study the genes involved in vertebrate development? Explain how this strategy differs from traditional genetic analyses like those done by Mendel.

Questions for Student Discussion/Collaboration

1. Compare and contrast the experimental advantages and disadvantages of *Drosophila*, *C. elegans*, mammals, and *Arabidopsis*.
2. It seems that developmental genetics boils down to a complex network of gene regulation. Try to draw how this network is structured for *Drosophila*. How many genes do you think are necessary to describe a complete developmental network for the fruit fly? How many genes do you think are needed for a network to specify one segment? Do you think it is more difficult to identify genes that are involved in the beginning, middle, or end of this network? Suppose you were trying to identify all of the genes needed for development in a chicken. Knowing what you know about *Drosophila* development, would you first try to identify genes necessary for early development, or would you begin by identifying genes involved in cell differentiation?
3. At the molecular level, how do you think a gain-of-function mutation in a developmental gene might cause it to be expressed in the wrong place or at the wrong time? Explain what type of DNA sequence would be altered.

Note: All answers appear at the website for this textbook; the answers to the even-numbered questions and all of the Concept Check and Comprehension Questions are in the back of the book.

www.mhhe.com/brookergenetics5e

Visit the website for practice tests, answer keys, and other learning aids for this chapter. Enhance your understanding of genetics with our interactive exercises, quizzes, animations, and much more.

CHAPTER OUTLINE

The African cheetah. *This species has a relatively low level of genetic variation because the population was reduced to a small size approximately 10,000–12,000 years ago.*

26 POPULATION GENETICS

Until now, we have primarily focused our attention on genes within individuals and their related family members. In this chapter and Chapters 27 and 28, we turn to the study of genes in a population or species. The field of **population genetics** is concerned with the extent of genetic variation within a group of individuals and how that variation changes over time. The field of population genetics emerged as a branch of genetics in the 1920s and 1930s. Its mathematical foundations were developed by theoreticians who extended the principles of Gregor Mendel and Charles Darwin by deriving formulas to explain the occurrence of genotypes within populations. These foundations can be largely attributed to three scientists: Sir Ronald Fisher, Sewall Wright, and J. B. S. Haldane. As we will see, support for their mathematical theories was provided by several researchers who analyzed the genetic composition of natural and experimental populations. More recently, population geneticists have used techniques to probe genetic variation at the molecular level. In addition, staggering advances in computer technology have aided population geneticists in the analysis of their genetic theories and data. In this chapter, we will explore the genetic variation that occurs in populations and consider the reasons why the genetic composition of populations may change over the course of several generations.

26.1 GENES IN POPULATIONS AND THE HARDY-WEINBERG EQUATION

Learning Outcomes:

1. Define gene pool and population.
2. Describe the extent of polymorphism in natural populations.
3. Use the Hardy-Weinberg equation to calculate allele and genotype frequencies.

Population genetics may seem like a significant departure from other topics in this textbook, but it is a direct extension of our understanding of Mendel's laws of inheritance, molecular genetics, and the ideas of Darwin, which are described in Chapter 28. In the field of population genetics, the focus shifts away from the individual and toward the population of which the individual is a member. Conceptually, all of the alleles of every gene in a population make up the **gene pool.**

One way to view gene pools is to consider them on a generation-per-generation basis. From this viewpoint, individuals of one generation constitute a gene pool. In turn, individuals that reproduce contribute to the gene pool of the next generation. Population geneticists study the genetic variation within gene pools and how such variation changes from one generation to the next. The emphasis is often on allelic variation. In this introductory section, we will examine some of the general features of populations and gene pools.

A Population Is a Group of Interbreeding Individuals That Share a Gene Pool

In genetics, the term population has a very specific meaning. With regard to sexually reproducing species, a **population** is a group of individuals of the same species that occupy the same region and can interbreed with one another. Many species occupy a wide geographic range and are divided into discrete populations. For example, distinct populations of a given species may be located on different continents, or populations on the same continent could be divided by a geographical barrier such as a large mountain range.

A large population usually is composed of smaller groups called **local populations,** or **demes.** The members of a local population are far more likely to breed among themselves than with other members of the general population. Local populations are often separated from each other by moderate geographic barriers. As shown in Figure 26.1, the large ground finch (*Geospiza magnirostris*) is found on a small volcanic island called Daphne Major, which is one of the Galápagos Islands. Daphne Major is located north of the much larger Santa Cruz Island. The population of large ground finches on Daphne Major constitutes a local population of this species. Breeding is much more likely to occur among members of a local population than between members of neighboring populations of the same species. On relatively rare occasions, however, a bird may fly from Daphne Major to Santa Cruz Island, which would make breeding between the two different local populations possible.

Populations typically are dynamic units that change from one generation to the next. A population may change its size, geographic location, and genetic composition. With regard to size, natural populations commonly go through cycles of "feast or famine," during which environmental factors cause the population to swell or shrink. In addition, natural predators or disease may periodically decrease the size of a population to significantly lower levels; the population later may rebound to its original size. Populations or individuals within populations may migrate to a new site and establish a distinct population in this location. The environment at this new geographic location may differ from the original site. What are the consequences of such changes? As population sizes and locations change, their genetic composition generally changes as well. As described later, population geneticists have developed mathematical models that predict how the gene pool will change in response to fluctuations in size, migration, and changes in the environment.

(a) Large ground finch (b) A view of Daphne Major (the small island in the distance) from Santa Cruz Island

FIGURE 26.1 A local population of the large ground finch. **(a)** The large ground finch (*Geospiza magnirostris*) on Daphne Major. **(b)** A view of Daphne Major, one of the Galápagos Islands, from Santa Cruz Island.

CONCEPT CHECK: What does the term *local population* mean?

At the Population Level, Some Genes May Be Monomorphic, But Most Are Polymorphic

In population genetics, the term **polymorphism** (meaning many forms) refers to the observation that many traits display variation within a population. Historically, polymorphism first referred to the variation in traits that are observable with the naked eye. Polymorphisms in color and pattern have long attracted the attention of population geneticists. These include studies involving yellow and red varieties of the elder-flowered orchid, and brown, pink, and yellow land snails, which are discussed later in this chapter. Figure 26.2 illustrates a striking example of polymorphism in the Hawaiian happy-face spider (*Theridion grallator*). The three individuals shown in this figure are from the same species, but they differ in alleles that affect color and pattern.

What is the underlying cause of polymorphism? At the DNA level, polymorphism may be due to two or more alleles that influence the phenotype of the individual that inherits them. In other words, it is due to genetic variation. Geneticists also use the term **polymorphic** to describe a gene that commonly exists as two or more alleles in a population. By comparison, a **monomorphic** gene exists predominantly as a single allele in a population. By convention, when a single allele is found in at least 99% of all cases, the gene is considered monomorphic.

At the level of a particular gene, a polymorphism may involve various types of changes such as a deletion of a significant region of the gene, a duplication of a region, or a change in a single nucleotide. This last phenomenon is called a **single-nucleotide polymorphism (SNP).** SNPs are the smallest type of genetic change that can occur within a given gene and are also the most common. In humans, for example, SNPs represent 90% of all the variation in DNA sequences that occurs among different people. SNPs are found very frequently in genes. In the human population, a gene that is 2000–3000 bp in length, on average, contains 10 different sites that are polymorphic. The high frequency of SNPs indicates that polymorphism is the norm for most human genes. Likewise, relatively large, healthy populations of nearly all species exhibit a high level of genetic variation as evidenced by the occurrence of SNPs within most genes.

Within a population, the alleles of a given gene may arise by different types of genetic changes. Figure 26.3 considers a gene that exists in multiple forms in humans. This example is a short segment of DNA found within the human β-globin gene. The top sequence is an allele designated Hb^A, whereas the middle sequence is called Hb^S. These alleles differ from each other by a single nucleotide, so they are an example of a single-nucleotide polymorphism. As discussed in Chapter 4, the Hb^S allele causes sickle cell disease in a homozygote. The bottom sequence contains a short, 5-bp deletion compared with the other two alleles. This deletion results in a nonfunctional β-globin polypeptide. Therefore, the bottom sequence is an example of a loss-of-function allele.

FIGURE 26.2 Polymorphism in the Hawaiian happy-face spider (*Theridion grallator*).

Genes→Traits These three spiders are members of the same species and carry the same genes. However, several genes that affect pigmentation patterns are polymorphic, meaning that each of these genes exists in multiple alleles within the population. This polymorphism within the Hawaiian happy-face spider population produces members that look quite different from each other.

CONCEPT CHECK: Are polymorphisms common or rare in natural populations?

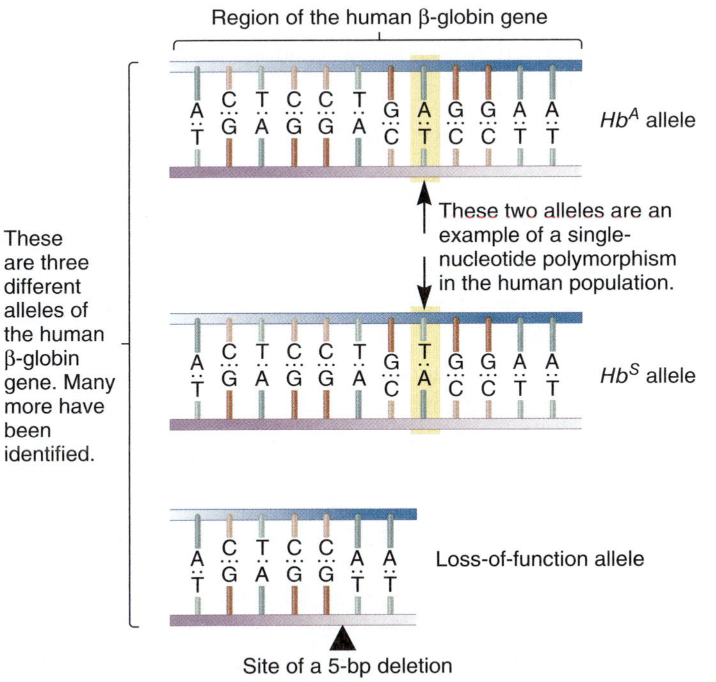

FIGURE 26.3 The relationship between alleles and various types of mutations. The DNA sequence shown here is a small portion of the β-globin gene in humans. Mutations have altered the gene to create the three different alleles in this figure. The top two alleles differ by a single base pair and are referred to as a single-nucleotide polymorphism (SNP). The bottom allele has a 5-bp deletion that begins right after the arrowhead. The deletion results in a nonfunctional polypeptide. It is a loss-of-function allele.

Population Genetics Is Concerned with Allele and Genotype Frequencies

As we have seen, population geneticists want to understand the prevalence of polymorphic genes within populations. Their goal is to identify the causative factors that govern changes in genetic variation. Much of their work evaluates the frequency of alleles in a quantitative way. Two fundamental calculations are central to population genetics: **allele frequencies** and **genotype frequencies.** The allele and genotype frequencies are defined as

$$\text{Allele frequency} = \frac{\text{Number of copies of a specific allele in a population}}{\text{Total number of all types of alleles for that gene in a population}}$$

$$\text{Genotype frequency} = \frac{\text{Number of individuals with a particular genotype in a population}}{\text{Total number of individuals in a population}}$$

Though these two frequencies are related, a clear distinction between them must be kept in mind. As an example, let's consider a hypothetical population of 100 frogs with the following genotypes:

64 dark green frogs with the genotype *GG*
32 medium green frogs with the genotype *Gg*
4 light green frogs with the genotype *gg*

When calculating an allele frequency, homozygous individuals have two copies of an allele, whereas heterozygotes have only one. For example, in tallying the *g* allele, each of the 32 heterozygotes has one copy of the *g* allele, and each light green frog has two copies. The allele frequency for *g* equals

$$g = \frac{32 + 2(4)}{2(64) + 2(32) + 2(4)}$$
$$= \frac{40}{200} = 0.2, \text{ or } 20\%$$

This result tells us that the allele frequency of *g* is 20%. In other words, 20% of the alleles for this gene in the population are the *g* allele.

Let's now calculate the genotype frequency of *gg* (light green) frogs:

$$gg = \frac{4}{64 + 32 + 4}$$
$$= \frac{4}{100} = 0.04, \text{ or } 4\%$$

We see that 4% of the individuals in this population are light green frogs.

Allele and genotype frequencies are always less than or equal to 1 (i.e., ≤100%). If a gene is monomorphic, the allele frequency for the single allele will be equal to or slightly less than a value of 1.0.

For polymorphic genes, if we add up the frequencies for all of the alleles in the population, we should obtain a value of 1.0. In our frog example, the allele frequency of *g* equals 0.2. The frequency of the other allele, *G*, equals 0.8. If we add the two together, we obtain a value of $0.2 + 0.8 = 1.0$.

The Hardy-Weinberg Equation Can Be Used to Calculate Genotype Frequencies Based on Allele Frequencies

Now that we have a general understanding of genes in populations, we can begin to relate these concepts to mathematical expressions as a way to examine whether allele and genotype frequencies are changing over the course of many generations. In 1908, a British mathematician, Godfrey Harold Hardy, and a German physician, Wilhelm Weinberg, independently derived a simple mathematical expression that predicts stability of allele and genotype frequencies from one generation to the next. It is called the **Hardy-Weinberg equilibrium,** because (under a given set of conditions, described later) allele and genotype frequencies do not change over the course of many generations.

Why is the Hardy-Weinberg equilibrium a useful concept? An equilibrium is a null hypothesis, which suggests that evolutionary change is not occurring. In reality, however, populations rarely achieve an equilibrium. Therefore, the main usefulness of the Hardy-Weinberg equilibrium is that it provides a framework on which to understand changes in allele frequencies within a population when such an equilibrium is violated.

To appreciate the Hardy-Weinberg equilibrium, let's return to our hypothetical frog example in which a gene is polymorphic and exists as two different alleles: *G* and *g*. If the allele frequency of *G* is denoted by the variable *p*, and the allele frequency of *g* by *q*, then

$$p + q = 1$$

For example, if $p = 0.8$, then *q* must be 0.2. In other words, if the allele frequency of *G* equals 80%, the remaining 20% of alleles must be *g*, because together they equal 100%.

The Hardy-Weinberg equilibrium relates allele frequencies and genotype frequencies. For a diploid species, each individual inherits two copies of most genes. The Hardy-Weinberg equilibrium assumes that two gametes are chosen at random from the gene pool. Therefore, because gametes are chosen independently, we can use the product rule (discussed in Chapter 2) to predict the likelihood of diploid genotypes:

$$(p + q)(p + q) = 1$$

$$p^2 + 2pq + q^2 = 1 \text{ (Hardy-Weinberg equation)}$$

The Hardy-Weinberg equation applies to a gene in a diploid species that is found in only two alleles. If this equation is related to our hypothetical frog population in which a gene exists in alleles designated *G* and *g*, then

p^2 equals the genotype frequency of *GG*

$2pq$ equals the genotype frequency of *Gg*

q^2 equals the genotype frequency of *gg*

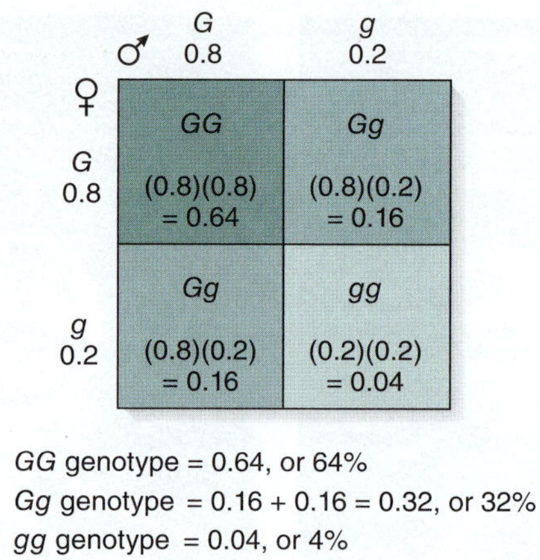

FIGURE 26.4 A comparison between allele frequencies and the union of alleles in a Punnett square. In a population in Hardy-Weinberg equilibrium, the frequency of gametes carrying a particular allele is equal to the allele frequency in the population.

If $p = 0.8$ and $q = 0.2$ and if the population is in Hardy-Weinberg equilibrium, then

$$GG = p^2 = (0.8)^2 = 0.64$$

$$Gg = 2pq = 2(0.8)(0.2) = 0.32$$

$$gg = q^2 = (0.2)^2 = 0.04$$

In other words, if the allele frequency of *G* is 80% and the allele frequency of *g* is 20%, the genotype frequency of *GG* is 64%, *Gg* is 32%, and *gg* is 4%.

To illustrate the relationship between allele frequencies and genotypes, Figure 26.4 compares the Hardy-Weinberg equation with the way that gametes combine randomly with each other to produce offspring. In a population, the frequency of a gamete carrying a particular allele is equal to the allele frequency in that population. In this example, the frequency of a gamete carrying the *G* allele equals 0.8.

We can use the product rule to determine the frequency of genotypes. For example, the frequency of producing a *GG* homozygote is $0.8 \times 0.8 = 0.64$, or 64%. Likewise, the probability of inheriting both *g* alleles is $0.2 \times 0.2 = 0.04$, or 4%. As seen in Figure 26.4, heterozygotes can be produced in two different ways. An offspring could inherit the *G* allele from its father and *g* from its mother, or *G* from its mother and *g* from its father. Therefore, the frequency of heterozygotes is $pq + pq$, which equals $2pq$; in our example, this is $2(0.8)(0.2) = 0.32$, or 32%.

One particularly useful feature of the Hardy-Weinberg equation is that it allows us to estimate the frequency of heterozygotes for recessive genetic diseases, assuming a Hardy-Weinberg equilibrium. As an example, let's consider cystic fibrosis, which involves a gene that encodes a chloride transporter. Persons afflicted with this disorder have an irregularity in salt and water balance. One of the symptoms is thick mucus in the lungs that can contribute to repeated lung infections. In human populations of Northern European descent, the frequency of affected individuals is approximately 1 in 2500. Because this is a recessive disorder, affected individuals are homozygotes. If *q* represents the allele frequency of the disease-causing allele, then

$$q^2 = 1/2500$$

$$q^2 = 0.0004$$

We take the square root to determine q:

$$q = \sqrt{0.0004}$$

$$q = 0.02$$

If p represents the normal allele,

$$p = 1 - q$$

$$p = 1 - 0.02 = 0.98$$

The frequency of heterozygous carriers is

$$2pq = 2(0.98)(0.02) = 0.0392, \text{ or } 3.92\%$$

In the absence of evolutionary changes, the Hardy-Weinberg equation predicts an equilibrium—unchanging allele and genotype frequencies from generation to generation—if certain conditions are met in a population. With regard to the gene of interest, these conditions are as follows:

1. ***No new mutations:*** The gene of interest does not incur any new mutations.
2. ***No genetic drift:*** The population is so large that allele frequencies do not change due to random sampling effects.
3. ***No migration:*** Individuals or their gametes do not travel between different populations. (Note: For some species, gametes can be dispersed via wind or water.)
4. ***No natural selection:*** All of the genotypes have the same reproductive success.
5. ***Random mating:*** With respect to the gene of interest, the members of the population mate with each other without regard to their phenotypes and genotypes.

The Hardy-Weinberg equation provides a quantitative relationship between allele and genotype frequencies in a population. Figure 26.5 describes this relationship for different allele frequencies of *g* and *G*. As expected, when the allele frequency of *g* is very low, the *GG* genotype predominates; when the *g* allele frequency is high, the *gg* homozygote is most prevalent in the population. When the allele frequencies of *g* and *G* are intermediate in value, the heterozygote predominates.

In reality, no population satisfies the Hardy-Weinberg equilibrium completely. Nevertheless, in large natural populations with little migration and negligible natural selection, the Hardy-Weinberg equilibrium may be nearly approximated for certain genes. In addition, the Hardy-Weinberg equilibrium can be extended to situations in which a single gene exists in three or more alleles, as described in solved problem S4 at the end of the chapter.

As discussed in Chapter 2, the chi square test can determine if observed and expected data are in agreement. Therefore, we can use a chi square test to determine whether a population exhibits Hardy-Weinberg equilibrium for a particular gene. To do so, it is necessary to distinguish between the homozygotes and heterozygotes, either phenotypically (due to codominance or incomplete dominance) or at the molecular level by analyzing the gene sequence. This is necessary so that we can determine both the allele and genotype frequencies. As an example, let's consider a human blood type called the MN type. In this case, the blood type is determined by two codominant alleles, *M* and *N*. In an Inuit population

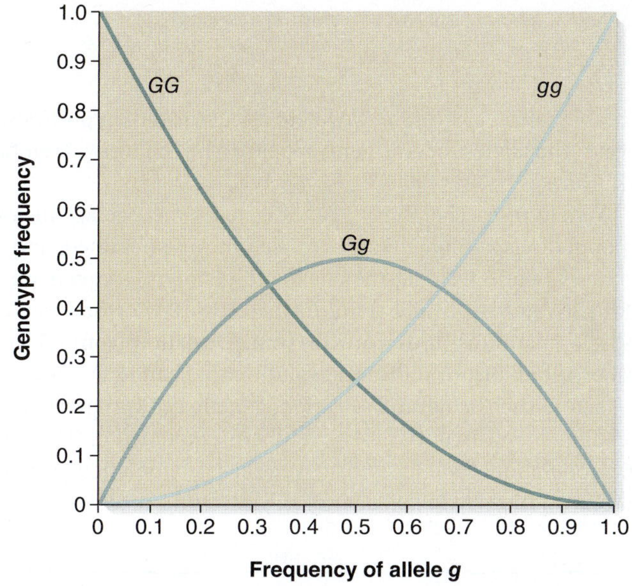

FIGURE 26.5 The relationship between allele frequencies and genotype frequencies according to the Hardy-Weinberg equilibrium. This graph assumes that *g* and *G* are the only two alleles for this gene.

in East Greenland, it was found that among 200 people, 168 were *MM*, 30 were *MN*, and 2 were *NN*. We can use these observed data to calculate the expected number of each genotype based on the Hardy-Weinberg equation.

$$\text{Allele frequency of } M = \frac{2(168) + 30}{400} = 0.915$$

$$\text{Allele frequency of } N = \frac{2(2) + 30}{400} = 0.085$$

Expected frequency of $MM = p^2 = (0.915)^2 = 0.837$

Expected number of MM individuals $= 0.837 \times 200 = 167.4$ (or 167 if rounded to the nearest individual)

Expected frequency of $NN = q^2 = (0.085)^2 = 0.007$

Expected number of NN individuals $= 0.007 \times 200 = 1.4$ (or 1 if rounded to the nearest individual)

Expected frequency of $MN = 2pq = 2(0.915)(0.085) = 0.156$

Expected number of MN individuals $= 0.156 \times 200 = 31.2$ (or 31 if rounded to the nearest individual)

$$\chi^2 = \frac{(O_1 - E_1)^2}{E_1} + \frac{(O_2 - E_2)^2}{E_2} + \frac{(O_3 - E_3)^2}{E_3}$$

$$\chi^2 = \frac{(168 - 167)^2}{167} + \frac{(30 - 31)^2}{31} + \frac{(2 - 1)^2}{1}$$

$$\chi^2 = 1.04$$

Let's refer back to Table 2.1 in Chapter 2 to evaluate the calculated chi square value. For a chi square of 1.04 with 1 degree of freedom (df), the P value is between 0.5 and 0.2, which is well within the acceptable range. Therefore, we fail to reject the null hypothesis. In this case, the alleles for this gene appear to be in Hardy-Weinberg equilibrium.

When researchers have investigated other genes in various populations, a high chi square value may be obtained, and the hypothesis is rejected that the allele and genotype frequencies are in Hardy-Weinberg equilibrium. In these cases, we would say that such a population is in **disequilibrium.** Deviation from a Hardy-Weinberg equilibrium indicates evolutionary change. As discussed next, factors such as natural selection, genetic drift, migration, and inbreeding may disrupt Hardy-Weinberg equilibrium. Therefore, when population geneticists discover that a population is not in equilibrium, they try to determine which factors are at work.

26.1 COMPREHENSION QUESTIONS

1. A gene pool is
   a. all of the genes in a single individual.
   b. all of the genes in the gametes from a single individual.
   c. all of the genes in a population of individuals.
   d. the random mixing of genes during sexual reproduction.
2. In natural populations, most genes are
   a. polymorphic.
   b. monomorphic.
   c. recessive.
   d. both a and c.
3. A gene exists in two alleles designated *D* and *d*. If 48 copies of this gene are the *D* allele and 152 are the *d* allele, what is the allele frequency of *D*?
   a. 0.24
   b. 0.32
   c. 0.38
   d. 0.76
4. The allele frequency of *C* is 0.4 and *c* is 0.6. If the population is in Hardy-Weinberg equilibrium, what is the frequency of heterozygotes?
   a. 0.16
   b. 0.24
   c. 0.26
   d. 0.48

26.2 OVERVIEW OF MICROEVOLUTION

Learning Outcomes:

1. Define microevolution.
2. Explain the role of mutation in microevolution.
3. List the mechanisms that may cause allele and genotype frequencies to significantly change from one generation to the next.

The genetic variation in natural populations typically changes over the course of many generations. The term **microevolution** describes changes in a population's gene pool from generation to generation. Such change is rooted in two related phenomena (**Table 26.1**). First, the introduction of new genetic variation into a population is one essential aspect of microevolution. As discussed later in this chapter, gene variation can originate by a variety of molecular mechanisms. For example, new alleles of preexisting genes can arise by random mutations. Such events provide a continuous source of new variation to populations. However, due to their low rate of occurrence, mutations do not act as a major factor in promoting widespread changes in a population.

Microevolution also involves the action of evolutionary mechanisms that alter the prevalence of a given allele or genotype in a population. These mechanisms are natural selection, genetic drift, migration, and nonrandom mating (see Table 26.1). The collective contributions of these evolutionary mechanisms over the course of many generations have the potential to promote widespread genetic changes in a population. In the following sections, we will examine how these mechanisms can affect the type of genetic variation that occurs when a gene exists in two or more alleles in a population. As you will learn, these mechanisms may cause a particular allele to be favored, or they may create a balance where two or more alleles are maintained in a population.

TABLE 26.1
Factors That Govern Microevolution

Source of New Allelic Variation*	
Mutation	Throughout most of this chapter, we consider allelic variation. Random mutations within preexisting genes introduce new alleles into populations, but at a very low rate. New mutations may be beneficial, neutral, or deleterious. For new alleles to rise to a significant percentage in a population, other evolutionary mechanisms (i.e., natural selection, genetic drift, and/or migration) must operate on them.
Mechanisms That Alter Existing Genetic Variation	
Natural selection	This is the phenomenon in which the environment selects for individuals that possess certain traits that favor reproductive success. For example, natural selection may be related to the survival of members to reproductive age.
Genetic drift	This is a change in genetic variation from generation to generation due to random sampling error. Allele frequencies may change as a matter of chance from one generation to the next. This tends to have a greater effect in a small population.
Migration	Migration can occur between two different populations. The introduction of migrants into a recipient population may change the allele frequencies of that population.
Nonrandom mating	This is the phenomenon in which individuals select mates based on their phenotypes or genetic lineage. This can alter the relative proportion of homozygotes and heterozygotes predicted by the Hardy-Weinberg equation, but does not change allele frequencies.

*Allelic variation is just one source of new genetic variation. Section 26.7 considers a variety of mechanisms through which new genetic variation can occur.

26.2 COMPREHENSION QUESTION

1. Which of the following is a factor that does *not* promote widespread changes in allele or genotype frequencies?
   a. New mutation
   b. Natural selection
   c. Genetic drift
   d. Migration
   e. Nonrandom mating

26.3 NATURAL SELECTION

Learning Outcomes:

1. Explain the process of natural selection.
2. Compare and contrast directional, stabilizing, disruptive, and balancing selection.

In the 1850s, Charles Darwin and Alfred Russel Wallace independently proposed the theory of evolution by **natural selection.** According to this theory, the conditions found in nature result in the selective survival and reproduction of individuals whose characteristics make them better adapted to their environment. These surviving individuals are more likely to reproduce and contribute offspring to the next generation. Natural selection can be related not only to differential survival but also to mating efficiency and fertility.

A modern restatement of the principles of natural selection can relate our knowledge of molecular genetics to the phenotypes of individuals.

1. Within a population, allelic variation arises in various ways, such as through random mutations that cause differences in DNA sequences. A mutation that creates a new allele may alter the amino acid sequence of the encoded protein, which, in turn, may alter the function of the protein.
2. Some alleles may encode proteins that enhance an individual's survival or reproductive capability compared with that of other members of the population. For example, an allele may produce a protein that is more efficient at a higher temperature, conferring on the individual a greater probability of survival in a hot climate.
3. Individuals with beneficial alleles are more likely to survive and contribute to the gene pool of the next generation.
4. Over the course of many generations, allele frequencies of many different genes may change through this process, thereby significantly altering the characteristics of a population or species. The net result of natural selection is a population that is better adapted to its environment and more successful at reproduction. Even so, it should be emphasized that species are not perfectly adapted to their environments, because mutations are random events and because the environment tends to change from generation to generation.

Fisher, Wright, and Haldane developed mathematical formulas to explain the theory of natural selection. As our knowledge of the process of natural selection has increased, it has become apparent that it operates in many different ways. In this section, we will consider a few examples of how natural selection may occur.

Darwinian Fitness Is a Measure of Reproductive Success

To begin our discussion of natural selection, we must examine the concept of **Darwinian fitness**—the relative likelihood that one genotype will contribute to the gene pool of the next generation compared to other genotypes. Natural selection acts on phenotypes that are derived from individuals' genotypes. Although Darwinian fitness often correlates with physical fitness, the two concepts are not identical. Darwinian fitness is a measure of reproductive success. An extremely fertile genotype may have a higher Darwinian fitness than a less fertile genotype that appears more physically fit.

To consider Darwinian fitness, let's use our example of a gene existing in the *A* and *a* alleles. If the three genotypes have the same level of mating success and fertility, we can assign fitness values to each genotype class based on their likelihood of surviving to reproductive age. For example, let's suppose that the relative survival to adulthood of each of the three genotype classes is as follows: For every five *AA* individuals that survive, four *Aa* individuals survive, and one *aa* individual survives. By convention, the genotype with the highest reproductive ability is given a fitness value of 1.0. Relative fitness values are denoted by the variable *w*. The fitness values of the other genotypes are assigned values relative to this 1.0 value:

$$\text{Fitness of } AA\text{: } w_{AA} = 1.0$$

$$\text{Fitness of } Aa\text{: } w_{Aa} = 4/5 = 0.8$$

$$\text{Fitness of } aa\text{: } w_{aa} = 1/5 = 0.2$$

Differences in reproductive success among genotypes may occur for various reasons. In this case, the fittest genotype is more likely to survive to reproductive age. In other situations, the most fit genotype is more likely to mate. For example, a bird with brightly colored feathers may have an easier time attracting a mate than a bird with duller plumage. Finally, a third possibility is that the fittest genotype may be more fertile. It may produce a higher number of gametes or gametes that are more successful at fertilization.

Also keep in mind that the above discussion of relative fitness presents the simplified case in which a single gene affects reproductive success. However, most traits are affected by allelic variation involving multiple genes. In Chapter 27, we will examine quantitative traits, such as weight and height, which are determined by alleles of many different genes. When natural selection acts on quantitative traits, relative fitness values are determined by allelic variation of multiple genes, not just one.

By studying species in their native environments, population geneticists have discovered that natural selection can occur in several ways. The patterns of natural selection depend on the relative fitness values of different genotypes and on the variation of environmental effects. The four patterns of natural selection that we will consider are called directional, stabilizing, disruptive, and balancing selection. In most of the examples described next, natural selection leads to adaptation so that the members of a species are more likely to survive to reproductive age.

Directional Selection Favors the Extreme Phenotype

Directional selection favors individuals at one extreme of a phenotypic distribution that are more likely to survive and reproduce in a particular environment. In some cases, directional selection may act on phenotypes that are largely determined by the alleles of a single gene. For example, the level of resistance to an insecticide in a mosquito population may be determined by alleles of a single gene. In other cases, directional selection may act on phenotypes that are determined by multiple genes. For example, body weight in mammals is influenced by alleles of many different genes. If directional selection favored higher body weight, it would affect the allele frequencies of many different genes.

Different phenomena may initiate the process of directional selection. One way that directional selection may arise is that a new allele may be introduced into a population by mutation, and the new allele may promote a higher fitness in individuals that carry it (**Figure 26.6**). If the homozygote carrying the favored allele has the highest fitness value, directional selection may cause this favored allele to eventually become the predominant allele in the population, perhaps even becoming a monomorphic gene.

Another possibility is that a population may be exposed to a prolonged change in its living environment. Under the new environmental conditions, the relative fitness values may change to favor one or more genotypes, which will promote the elimination of other genotypes. As an example, let's suppose a population of finches on the mainland already has genetic variation in beak size. A small number of birds migrate to an island where the seeds are generally larger than they are on the mainland. In this new environment, birds with larger beaks have a higher fitness because they are better able to crack open the larger seeds and thereby survive to reproductive age. Over the course of many generations, directional selection produces a population of birds carrying alleles that promote larger beak size.

For traits in which fitness is largely determined by alleles of a single gene, we can calculate how directional selection changes allele frequencies in a step-by-step, generation-per-generation way. To appreciate how this occurs, let's take a look at how fitness affects the Hardy-Weinberg equilibrium and allele frequencies. Again, let's suppose a gene exists in two alleles: *A* and *a*. The three fitness values, which are based on relative survival levels, are

$$w_{AA} = 1.0$$

$$w_{Aa} = 0.8$$

$$w_{aa} = 0.2$$

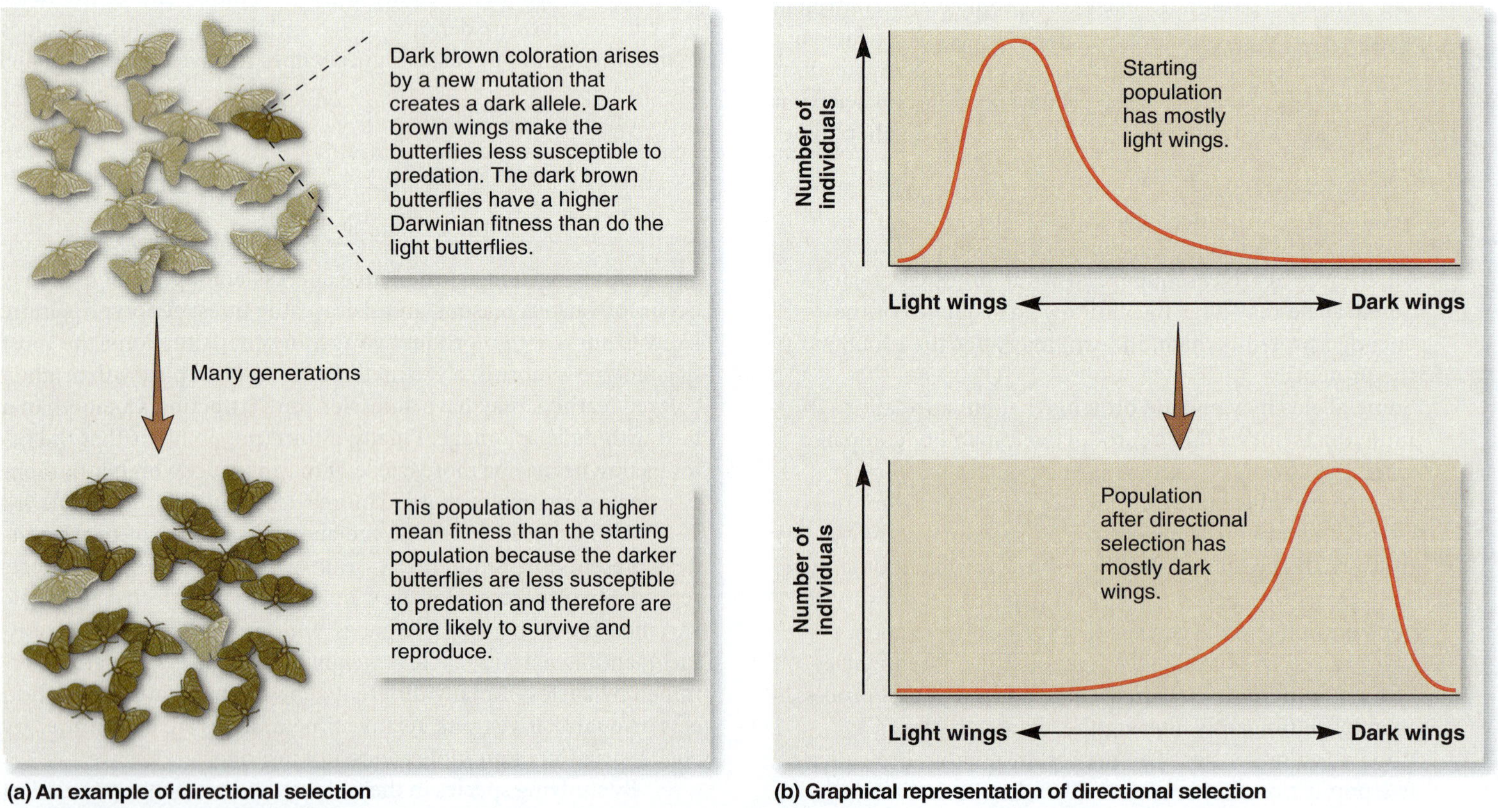

FIGURE 26.6 Directional selection. **(a)** A new mutation arises in a population that confers higher Darwinian fitness. In this example, butterflies with dark wings are more likely to survive and reproduce. Over many generations, directional selection favors the prevalence of darker individuals. **(b)** A graphical representation of directional selection.

CONCEPT CHECK: In this form of natural selection, explain the meaning of the word directional.

In the next generation, we expect that the Hardy-Weinberg equilibrium is modified in the following way due to directional selection:

Frequency of *AA*: p^2w_{AA}

Frequency of *Aa*: $2pqw_{Aa}$

Frequency of *aa*: q^2w_{aa}

In a population that is changing due to natural selection, these three terms may not add up to 1.0, as they do in the Hardy-Weinberg equilibrium. Instead, the three terms sum to a value known as the **mean fitness of the population ($\overline{w}$):**

$$p^2w_{AA} + 2pqw_{Aa} + q^2w_{aa} = \overline{w}$$

Dividing both sides of the equation by the mean fitness of the population,

$$\frac{p^2w_{AA}}{\overline{w}} + \frac{2pqw_{Aa}}{\overline{w}} + \frac{q^2w_{aa}}{\overline{w}} = 1$$

Using this equation, we can calculate the expected genotype and allele frequencies after one generation of directional selection:

Frequency of *AA* genotype: $\dfrac{p^2w_{AA}}{\overline{w}}$

Frequency of *Aa* genotype: $\dfrac{2pqw_{Aa}}{\overline{w}}$

Frequency of *aa* genotype: $\dfrac{q^2w_{aa}}{\overline{w}}$

Allele frequency of *A*: $p_A = \dfrac{p^2w_{AA}}{\overline{w}} + \dfrac{pqw_{Aa}}{\overline{w}}$

Allele frequency of *a*: $q_a = \dfrac{q^2w_{aa}}{\overline{w}} + \dfrac{pqw_{Aa}}{\overline{w}}$

As an example, let's suppose that the starting allele frequencies are $A = 0.5$ and $a = 0.5$, and use fitness values of 1.0, 0.8, and 0.2 for the three genotypes, *AA*, *Aa*, and *aa*, respectively. We begin by calculating the mean fitness of the population:

$$p^2w_{AA} + 2pqw_{Aa} + q^2w_{aa} = \overline{w}$$

$$\overline{w} = (0.5)^2(1) + 2(0.5)(0.5)(0.8) + (0.5)^2(0.2)$$

$$\overline{w} = 0.25 + 0.4 + 0.05 = 0.7$$

After one generation of directional selection,

Frequency of *AA* genotype: $\dfrac{p^2w_{AA}}{\overline{w}} = \dfrac{(0.5)^2(1)}{0.7} = 0.36$

Frequency of *Aa* genotype: $\dfrac{2pqw_{Aa}}{\overline{w}} = \dfrac{2(0.5)(0.5)(0.8)}{0.7} = 0.57$

Frequency of *aa* genotype: $\dfrac{q^2w_{aa}}{\overline{w}} = \dfrac{(0.5)^2(0.2)}{0.7} = 0.07$

Allele frequency of *A*: $p_A = \dfrac{p^2w_{AA}}{\overline{w}} + \dfrac{pqw_{Aa}}{\overline{w}}$

$$= \frac{(0.5)^2(1)}{0.7} + \frac{(0.5)(0.5)(0.8)}{0.7} = 0.64$$

Allele frequency of *a*: $q_a = \dfrac{q^2w_{aa}}{\overline{w}} + \dfrac{pqw_{Aa}}{\overline{w}}$

$$= \frac{(0.5)^2(0.2)}{0.7} + \frac{(0.5)(0.5)(0.8)}{0.7} = 0.36$$

After one generation, the allele frequency of *A* has increased from 0.5 to 0.64, and the frequency of *a* has decreased from 0.5 to 0.36. This has occurred because the *AA* genotype has the highest fitness, whereas the *Aa* and *aa* genotypes have lower fitness values. Another interesting feature of natural selection is that it raises the mean fitness of the population. If we assume the individual fitness values are constant, the mean fitness of this next generation is

$$\overline{w} = p^2w_{AA} + 2pqw_{Aa} + q^2w_{aa}$$

$$= (0.64)^2(1) + 2(0.64)(0.36)(0.8) + (0.36)^2(0.2)$$

$$= 0.80$$

The mean fitness of the population has increased from 0.7 to 0.8.

What are the consequences of natural selection at the population level? This population is better adapted to its environment than the previous one. Another way of viewing this calculation is that the subsequent population has a greater reproductive potential than the previous one. We could perform the same types of calculations to find the allele frequencies and mean fitness value in the next generation. If we assume the individual fitness values remain constant, the frequencies of *A* and *a* in the next generation are 0.85 and 0.15, respectively, and the mean fitness increases to 0.93. As we can see, the general trend is to increase *A*, decrease *a*, and increase the mean fitness of the population.

In the previous example, we considered the effects of natural selection by beginning with allele frequencies at intermediate levels (namely, $A = 0.5$ and $a = 0.5$). **Figure 26.7** illustrates what would happen if a new mutation introduced the *A* allele into a population that was originally monomorphic for the *a* allele. As before, the *AA* homozygote has a fitness of 1.0, the *Aa* heterozygote 0.8, and the recessive *aa* homozygote 0.2. Initially, the *A* allele is at a very low frequency in the population. If it is not lost initially due to genetic drift, its frequency slowly begins to rise and then, at intermediate values, rises much more rapidly.

Eventually, this type of natural selection may lead to fixation of a beneficial allele at a frequency of 1.0, or 100%. However, a new beneficial allele is in a precarious situation when its frequency is very low. As discussed later, genetic drift is likely to eliminate new mutations, even beneficial ones, due to chance fluctuations.

Researchers have identified many examples of directional selection in nature that are related to single genes. As mentioned in Chapter 7, resistance to antibiotics is a growing concern in the treatment of infection. Resistance to antibiotics typically occurs in a directional manner. Similarly, the resistance of insects to pesticides, such as DDT (dichlorodiphenyltrichloroethane), occurs in

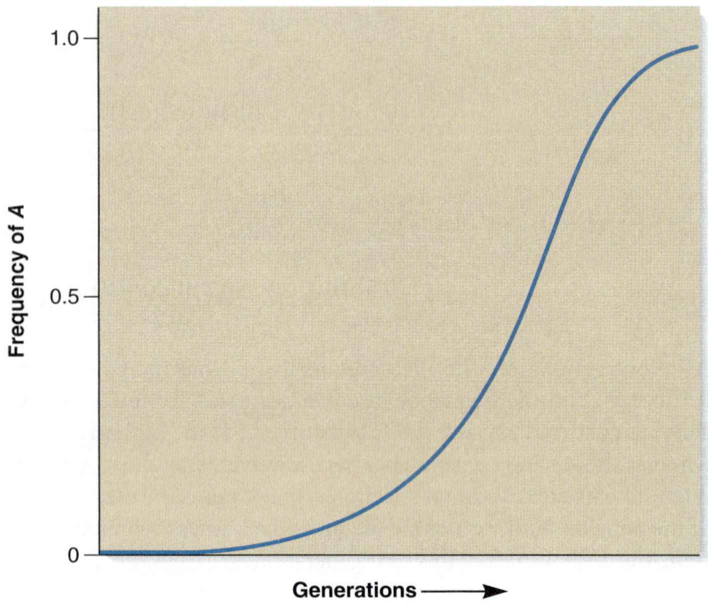

FIGURE 26.7 The fate of a beneficial allele that is introduced into a population as a new mutation. A new allele (*A*) is beneficial in the homozygous condition: w_{AA} = 1.0. The heterozygote, *Aa* (w_{Aa} = 0.8), and homozygote, *aa* (w_{aa} = 0.2), have lower fitness values.

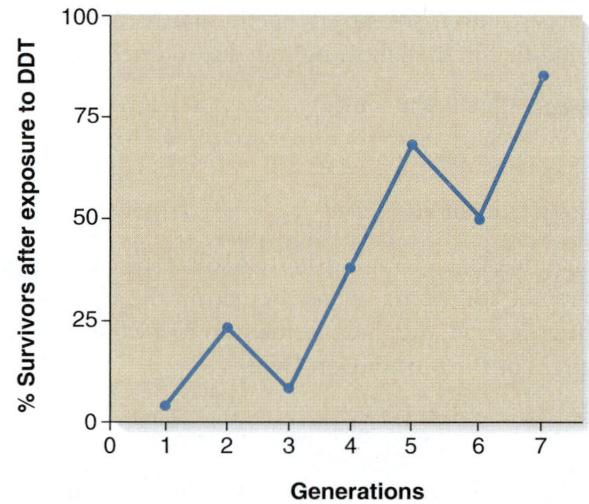

FIGURE 26.8 Directional selection for DDT resistance in a mosquito population. In this experiment, mosquito larvae (*Aedes aegypti*) were exposed to 10 mg/L of DDT. The percentage of survivors was recorded, and then the survivors of each generation were used as parents for the next generation.

CONCEPT CHECK: In this example, is directional selection promoting genetic diversity? Explain.

a directional manner. DDT usage began in the 1940s as a way to decrease the populations of mosquitoes and other insects. Subsequently, certain insect species have become resistant to DDT by a dominant mutation in a single enzyme-encoding gene. The resulting mutant enzyme detoxifies DDT, making it harmless to the insect. Figure 26.8 shows the results of an experiment in which mosquito larvae (*Aedes aegypti*) were exposed to DDT over the course of seven generations. The starting population showed a low level of DDT resistance, as evidenced by the low percentage of survivors after exposure to DDT. By comparison, after seven generations, nearly 100% of the population was DDT-resistant. These results illustrate the power of directional selection in promoting change in a population. Since the 1950s, resistance to nearly every known insecticide has evolved within 10 years of its commercial introduction!

Balanced Polymorphisms May Occur Due to Heterozygote Advantage or Negative Frequency-Dependent Selection

A common misperception is that natural selection always eliminates "weaker alleles" from a population. Researchers have discovered certain patterns of natural selection that actually favor the maintenance of two or more alleles in a population. One example is called **balancing selection.**

For genetic variation involving a single gene, balancing selection may arise when the heterozygote has a higher fitness than either corresponding homozygote, a situation called **heterozygote advantage.** In this case, an equilibrium is reached in which both alleles are maintained in the population. If the fitness values are known for each of the genotypes, the allele frequencies at equilibrium can be calculated. To do so, we must consider the **selection coefficient (*s*),** which measures the degree to which a genotype is selected against.

$$s = 1 - w$$

By convention, the genotype with the highest fitness has an *s* value of zero. Genotypes at a selective disadvantage have *s* values that are greater than 0 but less than or equal to 1.0. An extreme case is a recessive lethal allele. It would have an *s* value of 1.0 in the homozygote, whereas the *s* value in the heterozygote could be 0.

Let's consider the following case of relative fitness, where

$$w_{AA} = 0.7$$

$$w_{Aa} = 1.0$$

$$w_{aa} = 0.4$$

The selection coefficients are

$$s_{AA} = 1 - 0.7 = 0.3$$

$$s_{Aa} = 1 - 1.0 = 0$$

$$s_{aa} = 1 - 0.4 = 0.6$$

The population reaches an equilibrium when

$$s_{AA}p = s_{aa}q$$

If we take this equation, let $q = 1 - p$, and then solve for p:

$$p = \text{Allele frequency of } A = \frac{s_{aa}}{s_{AA} + s_{aa}}$$

$$= \frac{0.6}{0.3 + 0.6} = 0.67$$

If we let $p = 1 - q$ and then solve for q:

$$q = \text{Allele frequency of } a = \frac{s_{AA}}{s_{AA} + s_{aa}}$$

$$= \frac{0.3}{0.3 + 0.6} = 0.33$$

In this example, balancing selection maintains the two alleles in the population at frequencies in which A equals 0.67 and a equals 0.33.

Heterozygote advantage can sometimes explain the high frequency of alleles that are harmful in a homozygous condition. A classic example is the Hb^S allele of the human β-globin gene. A homozygous Hb^SHb^S individual exhibits sickle cell disease, a disorder that leads to the sickling of the red blood cells. The Hb^SHb^S homozygote has a lower fitness than a homozygote with two copies of the more common β-globin allele, Hb^AHb^A. However, the heterozygote, Hb^AHb^S, has a higher level of fitness than either homozygote in areas where malaria is endemic (**Figure 26.9**). Compared with Hb^AHb^A homozygotes, heterozygotes have a 10–15% better chance of survival if infected by the malarial parasite, *Plasmodium falciparum*. Therefore, the Hb^S allele is maintained in populations where malaria is prevalent, even though the allele is detrimental in the homozygous state.

In addition to sickle cell disease, other gene mutations that cause human disease in the homozygous state are thought to be prevalent because of heterozygote advantage. For example, the high prevalence of the allele causing cystic fibrosis may be related to this phenomenon, but the advantage that a heterozygote may possess is not understood.

Negative frequency-dependent selection is a second mechanism of balancing selection. In this pattern of natural selection, the fitness of a genotype decreases when its frequency becomes higher. In other words, rare individuals have a higher fitness than more common individuals. Therefore, rare individuals are more likely to reproduce, whereas common individuals are less likely, thereby producing a balanced polymorphism in which no genotype becomes too rare or too common.

An interesting example of negative frequency-dependent selection involves the elder-flowered orchid, *Dactylorhiza sambucina* (**Figure 26.10**). Throughout its range in central and southern Europe, both yellow- and red-flowered individuals are prevalent. The explanation for this polymorphism is related to its pollinators, which are mainly bumblebees such as *Bombus lapidarius* and *Bombus terrestris*. The pollinators increase their visits to the flower color of *D. sambucina* as it becomes less common in a given area. One reason this may occur is because *D. sambucina* is a rewardless flower; that is, it does not provide its pollinators with any reward for visiting, such as sweet nectar. Pollinators learn that the more common color of *D. sambucina* in a given area does not offer a reward, and they increase their visits to the less-common flower. Thus, the relative fitness of the less-common flower increases.

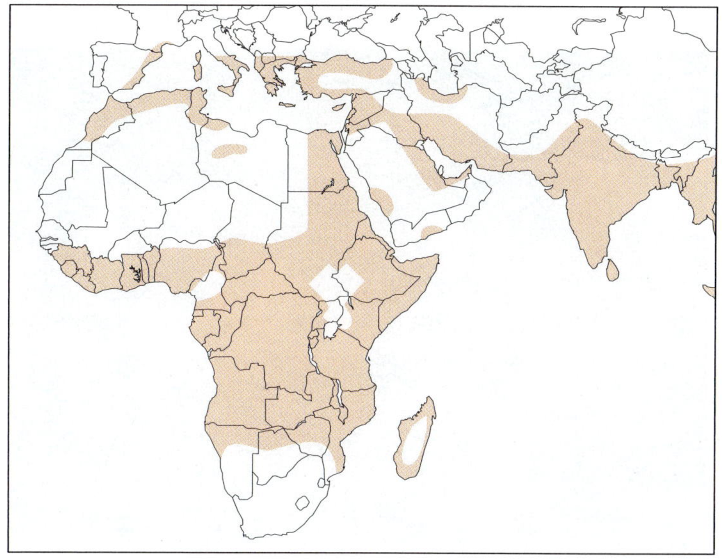

(a) Malaria prevalence

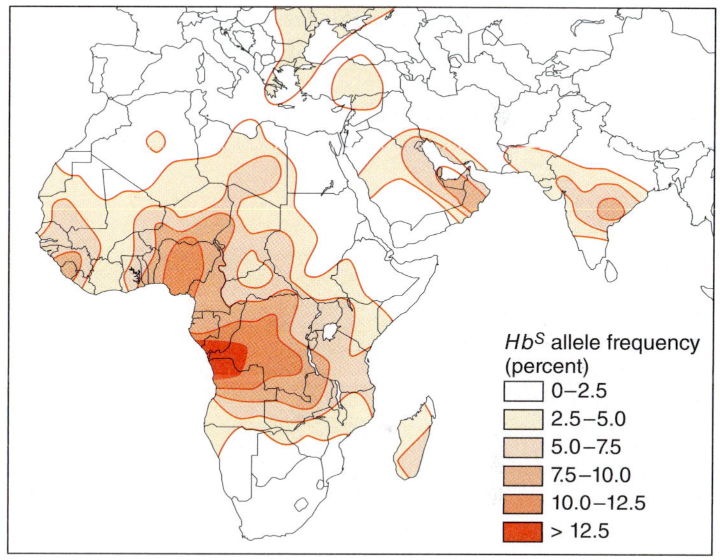

(b) Hb^S allele frequency

FIGURE 26.9 The geographic relationship between malaria and the frequency of the sickle cell allele in human populations. **(a)** The geographic prevalence of malaria in Africa and surrounding areas. **(b)** The frequency of the Hb^S allele in the same areas.

Genes→Traits The sickle cell allele of the β-globin gene is maintained in human populations by balancing selection. In areas where malaria is endemic, the heterozygote carrying one copy of the Hb^S allele confers a greater fitness than either of the corresponding homozygotes (Hb^AHb^A and Hb^SHb^S). Therefore, even though the Hb^SHb^S homozygotes suffer the detrimental consequences of sickle cell disease, this negative aspect is balanced by the beneficial effects of malarial resistance in the heterozygotes.

CONCEPT CHECK: Explain why the Hb^S allele is prevalent in certain regions even though it is detrimental in the homozygous condition.

Disruptive Selection Favors Multiple Phenotypes

As we have just seen, polymorphisms may occur due to balancing selection. Researchers have discovered another pattern of natural selection that favors the maintenance of two or more alleles in a heterogeneous environment. This pattern, called **disruptive selection,** also known as diversifying selection, favors the survival of two or more different phenotypes (**Figure 26.11**). It typically acts on quantitative traits that are determined by multiple genes. In disruptive selection, the fitness values of particular genotypes are higher in one environment and lower in a different one. Disruptive

FIGURE 26.10 The two color variations found in the elder-flowered orchid, *Dactylorhiza sambucina*. The two colors are maintained in the population due to negative frequency-dependent selection.

CONCEPT CHECK: Explain how negative frequency-dependent selection works.

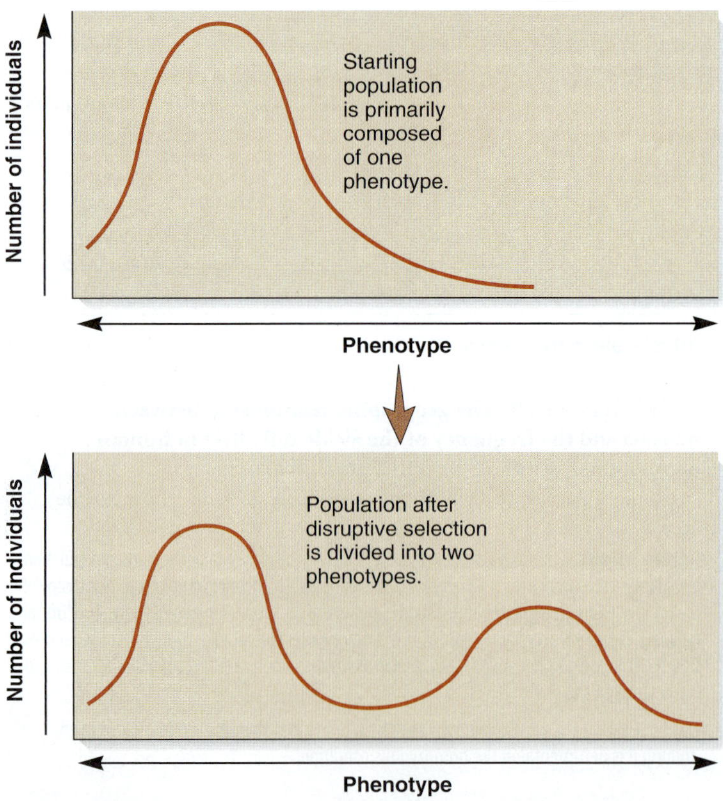

FIGURE 26.11 Disruptive selection. Over time, this form of selection favors two or more phenotypes due to heterogeneous environments.

CONCEPT CHECK: Does this form of natural selection favor polymorphism? Explain why or why not.

(a) Land snails

Habitat	*Brown*	*Pink*	*Yellow*
Beechwoods	0.23	0.61	0.16
Deciduous woods	0.05	0.68	0.27
Hedgerows	0.05	0.31	0.64
Rough herbage	0.004	0.22	0.78

(b) Frequency of snail color

FIGURE 26.12 Polymorphism in the land snail, *Cepaea nemoralis*. **(a)** This species of snail can exist in several different colors and banding patterns. **(b)** Coloration of the snails is correlated with the specific environments where they are located.

Genes→Traits Snail coloration is an example of genetic polymorphism due to heterogeneous environments; the genes governing shell coloration are polymorphic. The predation of snails is correlated with their ability to be camouflaged in their natural environment. Snails with brown shells are most prevalent in beechwoods, where the soil is dark. Pink snails are most abundant in the leaf litter of beechwoods and deciduous woods. Yellow snails are most prevalent in sunnier locations, such as hedgerows and rough herbage.

selection is likely to occur in populations that occupy diverse environments so that some members of the species survive and reproduce in each type of environmental condition.

As an example, **Figure 26.12a** shows a photograph of land snails, *Cepaea nemoralis*, that live in woods and open fields. This snail is polymorphic in color and banding patterns. In 1954, Arthur Cain and Philip Sheppard found that snail color was correlated with the environment. As shown in **Figure 26.12b**, the highest frequency of brown shell color was found in snails in the beechwoods, where there are wide expanses of dark soil. Their frequency was substantially less in other environments. By comparison, pink snails are most common in the leaf litter of both beechwoods and deciduous woods, and the yellow snails are most abundant in the sunny, grassy areas of hedgerows and rough herbage. Researchers have suggested that this disruptive selection can be explained by different levels of predation by thrushes. Depending on the environment, certain snail phenotypes may be more easily seen by their predators than other phenotypes. Migration can occasionally occur among the snail populations, which keeps the polymorphism in balance among these different environments.

Stabilizing Selection Favors Individuals with Intermediate Phenotypes

In **stabilizing selection,** the extreme phenotypes for a trait are selected against, and those individuals with intermediate phenotypes have the highest fitness values. Stabilizing selection is typically directed at quantitative traits, such as body weight and offspring number, which are determined by multiple genes. Stabilizing selection tends to decrease genetic diversity for genes affecting such traits because it eliminates alleles that cause a greater variation in phenotypes. In 1947, David Lack proposed that stabilizing selection may apply to clutch size in birds. Under stabilizing selection, birds that lay too many or too few eggs have lower fitness values than those that lay an intermediate number (**Figure 26.13**). Laying too many eggs may cause many offspring to die due to inadequate parental care and food. In addition, the strain on the parents themselves may decrease their likelihood of survival and therefore their ability to produce more offspring. Having too few offspring, on the other hand, does not contribute many individuals to the next generation. Therefore, the most successful parents are those that produce an intermediate clutch size. In the 1980s, Swedish evolutionary biologist Lars Gustafsson and colleagues examined the phenomenon of stabilizing selection in the collared flycatcher, *Ficedula albicollis*, on the island of Gotland, which is southeast of the mainland of Sweden. They discovered that Lack's hypothesis that clutch size is subject to the action of stabilizing selection appears to be true for this species.

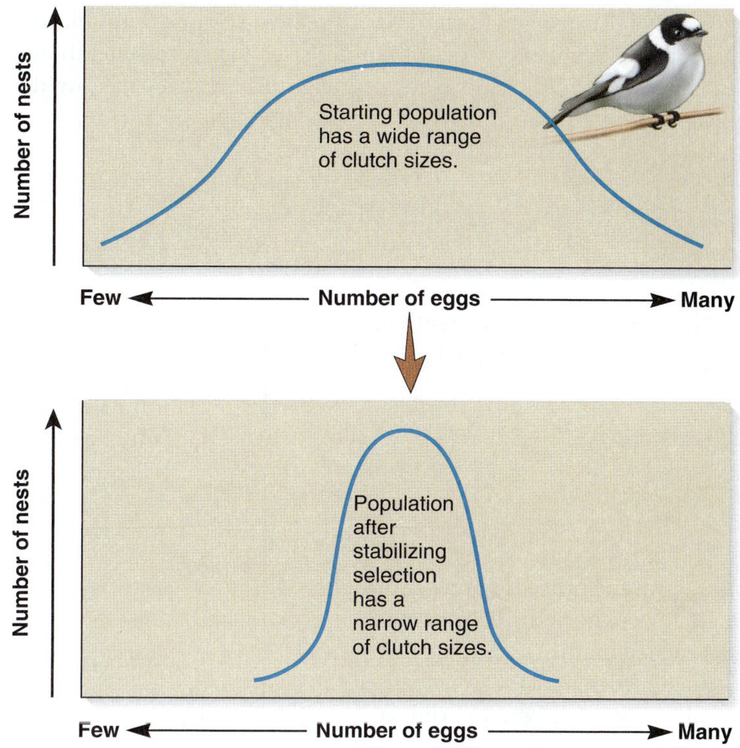

FIGURE 26.13 Stabilizing selection. In this pattern of natural selection, the extremes of a phenotypic distribution are selected against. Those individuals with intermediate traits have the highest fitness. This results in a population with less diversity and more uniform traits.

CONCEPT CHECK: In general, why does stabilizing selection decrease genetic diversity?

EXPERIMENT 26A

The Grants Have Observed Natural Selection in Galápagos Finches

Let's now turn to a study that demonstrates natural selection in action. Since 1973, Peter Grant, Rosemary Grant, and their colleagues have studied the process of natural selection in finches found on the Galápagos Islands. For over 30 years, the Grants have focused much of their research on one of the Galápagos Islands known as Daphne Major. This small island (0.34 km^2) has a moderate degree of isolation (8 km from the nearest island), an undisturbed habitat, and a resident population of finches, including the medium ground finch, *Geospiza fortis* (**Figure 26.14**).

To study natural selection, the Grants observed various traits in the medium ground finch, including beak size, over the course of many years. The medium ground finch has a relatively small crushing beak, suitable for breaking open small, tender seeds. Beak size is an example of a quantitative trait, which is controlled by multiple genes. The Grants quantified beak size among the medium ground finches of Daphne Major by carefully measuring beak depth (a measurement of the beak from top to bottom, at its base) on individual birds and then releasing them. During the course of their studies, they compared the beak sizes of parents and offspring by examining many broods over several years. The depth of the beak was transmitted from parents to offspring, regardless of environmental conditions, indicating that differences in beak sizes are due to genetic differences in the population. In other words, they found that beak depth is a heritable trait.

FIGURE 26.14 The medium ground finch (*Geospiza fortis*), which is found on Daphne Major.

By measuring many birds every year, the Grants were able to assemble a detailed portrait of natural selection from generation to generation. In the study shown in **Figure 26.15**, they measured beak depth in 1976 and 1978.

THE HYPOTHESIS

Beak size will be influenced by natural selection. Environments that produce larger seeds will select for birds with larger beaks.

TESTING THE HYPOTHESIS

FIGURE 26.15 Natural selection in medium ground finches of Daphne Major.

	Experimental level		Conceptual level
1. In 1976, measure beak depth in parents and offspring of the species *G. fortis*.	Capture birds and measure beak depth.		This is a way to measure a trait that may be subject to natural selection.
2. Repeat the procedure on offspring that were born in 1978 and had reached mature size. A drought had occurred in 1977 that caused plants on the island to produce mostly larger seeds and relatively few small seeds.	Capture birds and measure beak depth.		This is a way to measure a trait that may be subject to natural selection.

THE DATA

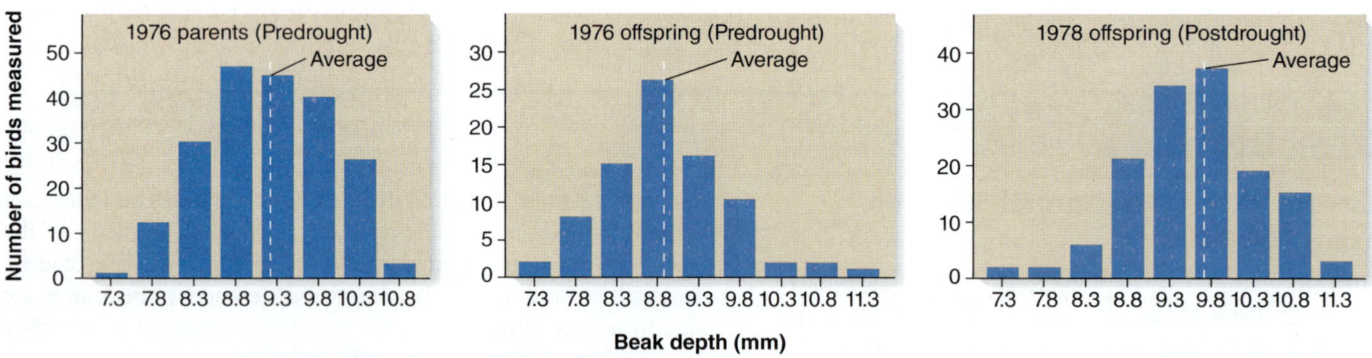

Data from: B. Rosemary Grant and Peter R. Grant (2003) What Darwin's finches can teach us about the evolutionary origin and regulation of biodiversity. *Bioscience 53*, 965–975.

INTERPRETING THE DATA

In the wet year of 1976, the plants of Daphne Major produced the small seeds that these finches were able to eat in abundance. However, a drought occurred in 1977. During this year, the plants on Daphne Major tended to produce few of the smaller seeds, which the finches rapidly consumed. To survive, the finches resorted to eating larger, drier seeds, which were harder to crush. As a result, the birds that survived tended to have larger beaks, because they were better able to break open these large seeds. In the year after the drought, the average beak depth of offspring in the population increased to approximately 9.8 mm because the surviving birds with larger beaks passed this trait on to their offspring. This is likely to be due to directional selection (see Figure 26.6), although genetic drift could also contribute to these data. Overall, these results illustrate the power of natural selection to alter the nature of a trait, in this case, beak depth, in a given population.

A self-help quiz involving this experiment can be found at www.mhhe.com/brookergenetics5e.

26.3 COMPREHENSION QUESTIONS

1. Darwinian fitness is a measure of
   a. survival.
   b. reproductive success.
   c. heterozygosity of the gene pool.
   d. polymorphisms in a population.
2. Within a particular population, darkly colored rats are more likely to survive than more lightly colored individuals. This scenario is likely to result in
   a. directional selection.
   b. stabilizing selection.
   c. disruptive selection.
   d. balancing selection.
3. A population occupies a diverse environment in which the fitness of some genotypes is higher in one environment while the fitness of other individuals is higher in another environment. This scenario is likely to result in
   a. directional selection.
   b. stabilizing selection.
   c. disruptive selection.
   d. balancing selection.
4. A gene exists in two alleles, and the heterozygote has the highest fitness. This scenario is likely to result in
   a. directional selection.
   b. stabilizing selection.
   c. disruptive selection.
   d. balancing selection.

26.4 GENETIC DRIFT

Learning Outcomes:

1. Define genetic drift.
2. Explain and calculate how population size affects genetic drift.
3. Compare and contrast the bottleneck effect and the founder effect.

In the 1930s, geneticist Sewall Wright played a key role in developing the concept of **genetic drift,** which refers to changes in allele frequencies in a population due to random fluctuations. As a matter of chance, the frequencies of alleles found in gametes that unite to form zygotes vary from generation to generation. Over the long run, genetic drift usually results in either the loss of an allele or its fixation at 100% in the population. The process is random with regard to particular alleles. Genetic drift can lead to the loss or fixation of deleterious, neutral, or beneficial alleles. The rate at which this occurs depends on the population size and on the initial allele frequencies. **Figure 26.16** illustrates the potential consequences of genetic drift in one large ($N = 1000$) and five small ($N = 20$) populations. At the beginning of this hypothetical simulation, all of these populations have identical allele frequencies: $A = 0.5$ and $a = 0.5$. In the five small populations, this allele frequency fluctuates substantially from generation to generation. Eventually, one of the alleles is eliminated and the other is fixed at 100%. At this point, the gene has become monomorphic and cannot fluctuate any further. By comparison, the allele frequencies in the large population fluctuate much less, because random sampling error is expected to have a smaller effect. Nevertheless, genetic drift leads to homozygosity even in large populations, but this takes many more generations to occur.

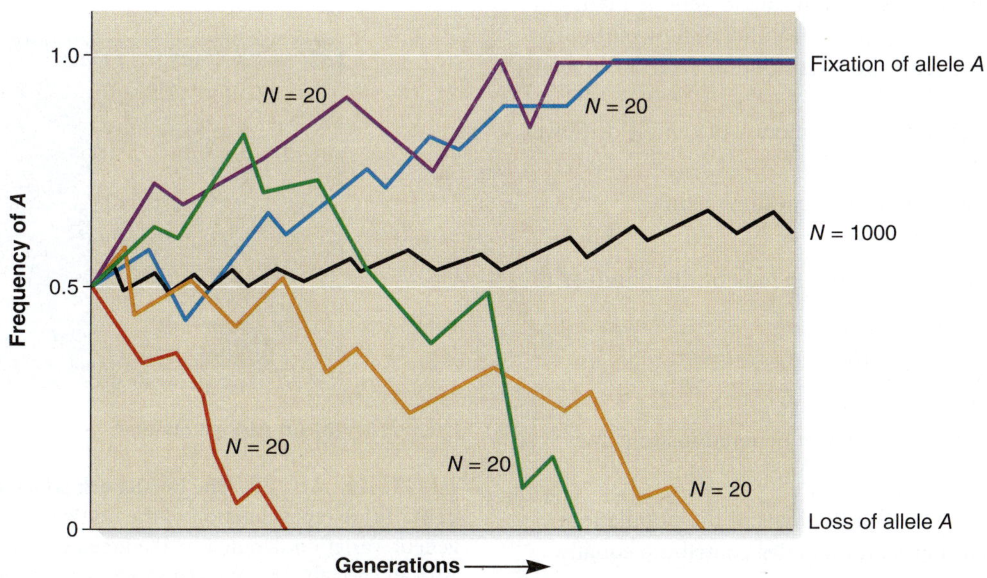

ONLINE ANIMATION

FIGURE 26.16 A hypothetical simulation of genetic drift. In all cases, the starting allele frequencies are $A = 0.5$ and $a = 0.5$. The colored lines illustrate five populations in which $N = 20$; the black line shows a population in which $N = 1000$.

CONCEPT CHECK: How does population size affect genetic drift?

Now let's ask two questions:

1. How many new mutations do we expect in a natural population?
2. How likely is it that any new mutation will be either fixed in, or eliminated from, a population due to genetic drift?

With regard to the first question, the average number of new mutations depends on the mutation rate (μ) and the number of individuals in a population (N). If each individual has two copies of the gene of interest, the expected number of new mutations in this gene is

$$\text{Expected number of new mutations} = 2N\mu$$

From this, we see that a new mutation is more likely to occur in a large population than in a small one. This makes sense, because the larger population has more copies of the gene to be mutated.

With regard to the second question, the probability of fixation of a newly arising allele due to genetic drift is

$$\text{Probability of fixation} = \frac{1}{2N} \text{ (assuming equal numbers of males and females contribute to the next generation)}$$

In other words, the probability of fixation is the same as the initial allele frequency in the population. For example, if $N = 20$, the probability of fixation of a new allele equals $1/(2 \times 20)$, or 2.5%. Conversely, a new allele may be lost from the population.

$$\text{Probability of elimination} = 1 - \text{probability of fixation}$$
$$= 1 - \frac{1}{2N}$$

If $N = 20$, the probability of elimination equals $1 - 1/(2 \times 20)$, or 97.5%. As you may have noticed, the value of N has opposing effects with regard to new mutations and their eventual fixation in a population. When N is very large, new mutations are much more likely to occur. Each new mutation, however, has a greater chance of being eliminated from the population due to genetic drift. On the other hand, when N is small, the probability of new mutations is also small, but if they occur, the likelihood of fixation is relatively large.

Now that we appreciate the phenomenon of genetic drift, we can ask a third question:

3. If fixation of a new allele does occur, how many generations is it likely to take?

The formula for calculating this also depends on the number of individuals in the population:

$$\bar{t} = 4N$$

where

$\bar{t}$ equals the average number of generations to achieve fixation

N equals the number of individuals in the population, assuming that males and females contribute equally to each succeeding generation

As you may have expected, allele fixation takes much longer in large populations. If a population has 1 million breeding members, it takes, on average, 4 million generations, perhaps an insurmountable period of time, to reach fixation. In a small group of 100 individuals, however, fixation takes only 400 generations, on average.

In nature, allele frequencies in small populations are more susceptible to genetic drift. This may commonly occur in the following two ways.

Bottleneck Effect Changes in population size may influence genetic drift via the **bottleneck effect** (**Figure 26.17**). In nature, a population can be reduced dramatically in size by events such as earthquakes, floods, drought, or human destruction of habitat. Such events may randomly eliminate most of the members of the

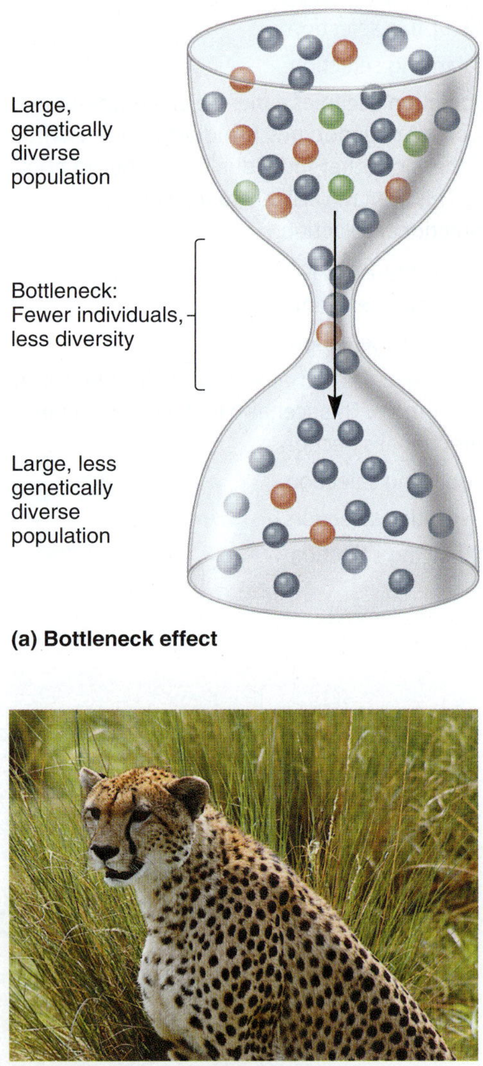

(b) An African cheetah

FIGURE 26.17 The bottleneck effect, an example of genetic drift. **(a)** A representation of the bottleneck effect. Note that the genetic variation denoted by the green balls has been lost. **(b)** The African cheetah. The modern species has low genetic variation due to a genetic bottleneck that is thought to have occurred about 10,000–12,000 years ago.

CONCEPT CHECK: What is happening at the bottleneck? Describe the effect of genetic drift during the bottleneck.

population without regard to its genetic composition. The initial bottleneck may be greatly influenced by genetic drift because the surviving members may have allele frequencies that differ from those of the original population. In addition, allele frequencies are expected to drift substantially during the generations when the population size is small. In extreme cases, alleles may even be eliminated. Eventually, the population with the bottleneck may regain its original size. However, the new population has less genetic variation than the original large population. As an example, the African cheetah population lost a substantial amount of its genetic variation due to a bottleneck effect (Figure 26.17b). DNA analysis by population geneticists has suggested that a severe bottleneck occurred approximately 10,000–12,000 years ago, when the population size was dramatically reduced. The population eventually rebounded, but the bottleneck significantly decreased the genetic variation.

Founder Effect Geography and population size may also influence genetic drift via the **founder effect.** The key difference between the bottleneck effect and the founder effect is that the founder effect involves migration; a small group of individuals separates from a larger population and establishes a colony in a new location. For example, a few individuals may migrate from a large continental population and become the founders of an island population. The founder effect has two important consequences. First, the founding population is expected to have less genetic variation than the original population from which it was derived. Second, as a matter of chance, the allele frequencies in the founding population may differ markedly from those of the original population.

Population geneticists have studied many examples of isolated populations that were started from a few members of another population. In the 1960s, Victor McKusick studied allele frequencies in the Old Order Amish of Lancaster County, Pennsylvania. At that time, this was a group of about 8000 people, descended from just three couples who immigrated to the United States in the 1700s. Among this population of 8000, a genetic disease known as the Ellis-van Creveld syndrome (a recessive form of dwarfism) was found at a frequency of 0.07, or 7%. By comparison, this disorder is extremely rare in other human populations, even the population from which the founding members had originated. The high frequency of dwarfism in the Lancaster County population is a chance occurrence due to the founder effect. The recessive allele can be traced back to one couple who came to the area in 1744.

26.4 COMPREHENSION QUESTIONS

1. Genetic drift is
   a. a change in allele frequencies due to random fluctuations.
   b. likely to result in allele loss or fixation over the long run.
   c. more pronounced in smaller populations.
   d. all of the above.
2. Which of the following types of genetic drift involve the migration of a population from one location to another?
   a. The bottleneck effect c. Both a and b
   b. The founder effect d. None of the above

26.5 MIGRATION

Learning Outcome:

1. Explain and calculate how migration affects allele frequencies between neighboring populations.

We have just seen how migration to a new location by a relatively small group can result in a population with an altered genetic composition via genetic drift. In addition, migration between two different established populations can alter allele frequencies. **Gene flow** refers to the transfer of alleles or genes from one population (a donor population), thereby changing the gene pool of a recipient population. One way this occurs is by the migration of fertile individuals from one population to another population and the successful breeding of such migrants with the members of the recipient population. Gene flow depends not only on migration, but also on the ability of the migrants' alleles to be passed to subsequent generations.

To determine the effects of migration, we need to consider three populations: the original donor population, the original recipient population, and the population that has new members due to migration. We will call this third population a conglomerate because it is composed of members of both the donor and recipient populations. To calculate changes in allele frequencies in the conglomerate population, we must know the original allele frequencies in the donor and recipient populations prior to migration. In addition, we must know the proportion of the conglomerate population that is due to migrants from the donor population. With these data, we begin by calculating the change in allele frequency in the conglomerate population using the following equation:

$$\Delta p_C = m(p_D - p_R)$$

where

Δp_C is the change in allele frequency in the conglomerate population

p_D is the allele frequency in the donor population

p_R is the allele frequency in the original recipient population

m is the proportion of migrants that make up the conglomerate population

$$m = \frac{\text{Number of migrants in the conglomerate population}}{\text{Total number of individuals in the conglomerate population}}$$

As an example, let's suppose the allele frequency of A is 0.7 in the donor population and 0.3 in the recipient population. A group of 20 individuals migrates and joins the recipient population, which originally had 80 members. Thus,

$$m = \frac{20}{20 + 80}$$
$$= 0.2$$
$$\Delta p_C = m(p_D - p_R)$$
$$= 0.2(0.7 - 0.3)$$
$$= 0.08$$

We can now calculate the allele frequency in the conglomerate:

$$p_C = p_R + \Delta p_C$$
$$= 0.3 + 0.08 = 0.38$$

Therefore, in the conglomerate population, the allele frequency of *A* has changed from 0.3 (its value before migration) to 0.38. This increase in allele frequency arises from the higher allele frequency of *A* in the donor population.

In our previous example, we considered the consequences of a unidirectional migration from a donor to a recipient population. In nature, it is common for individuals to migrate in both directions. What are the main consequences of bidirectional migration? Depending on its rate, migration tends to reduce differences in allele frequencies between neighboring populations. Populations that frequently mix their gene pools via migration tend to have similar allele frequencies, whereas isolated populations are expected to have more dissimilar allele frequencies.

Migration can also enhance genetic diversity within a population. As mentioned, new mutations are relatively rare events. Therefore, a particular mutation may arise only in one population. Migration may then introduce this new allele into neighboring populations.

26.5 COMPREHENSION QUESTION

1. Gene flow depends on
   a. migration.
   b. the ability of migrant alleles to be passed to subsequent generations.
   c. genetic drift.
   d. both a and b.

26.6 NONRANDOM MATING

Learning Outcomes:

1. Define assortative mating, inbreeding, and outbreeding.
2. Calculate the inbreeding coefficient for members of pedigree.
3. Explain how inbreeding affects the Hardy-Weinberg equilibrium.

As mentioned earlier, one of the conditions required to establish the Hardy-Weinberg equilibrium is random mating. This means that individuals choose their mates irrespective of their genotypes and phenotypes. In many cases, particularly in human populations, this condition is violated frequently.

When mating is nonrandom in a population, the process is called **assortative mating.** Positive assortative mating occurs when individuals with similar phenotypes choose each other as mates. The opposite situation, where dissimilar phenotypes mate preferentially, is called negative assortative mating.

In addition, individuals may choose a mate that is part of the same genetic lineage. The mating of two genetically related individuals, such as cousins, is called **inbreeding.** It sometimes occurs in human societies and is more likely to take place in nature when population size becomes very limited. In Chapter 27, we will examine how inbreeding is a useful strategy for developing agricultural breeds or strains with desirable characteristics. Conversely, **outbreeding,** which involves mating between unrelated individuals, can create hybrids that are heterozygous for many genes. In the absence of other evolutionary processes, inbreeding and outbreeding do not affect allele frequencies in a population. However, they do alter the relative proportions of homozygotes and heterozygotes that are predicted by the Hardy-Weinberg equation.

Inbreeding involves a smaller gene pool because the parents are related genetically. In the 1940s, Gustave Malécot developed methods to quantify the degree of inbreeding. The **inbreeding coefficient (*F*)** is the probability that two alleles in a particular individual will be identical for a given gene because both copies are due to descent from a common ancestor. An inbreeding coefficient (*F*) can be computed by analyzing the degree of relatedness within a pedigree.

Figure 26.18 illustrates a human pedigree involving a mating between cousins. Individuals III-2 and III-3 are cousins and have produced the daughter labeled IV-1. She is said to be inbred, because her parents are genetically related to each other. Let's determine the inbreeding coefficient for individual IV-1. To begin this problem, we must first identify all of this individual's common ancestors. A common ancestor is anyone who is an ancestor to both of an individual's parents. In Figure 26.18, IV-1 has one common ancestor, I-2, her great-grandfather. I-2 is the grandfather of III-2 and III-3.

Our next step is to determine the inbreeding paths. An inbreeding path for an individual is the shortest path through the pedigree that includes both parents and the common ancestor. In

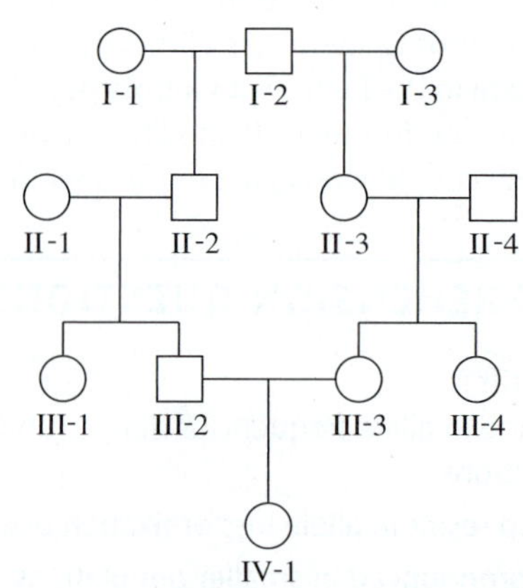

FIGURE 26.18 A human pedigree containing inbreeding. Individual IV-1 is the result of inbreeding because her parents are related.

CONCEPT CHECK: How does inbreeding affect the likelihood of expressing recessive traits? Explain.

a pedigree, there is an inbreeding path for each common ancestor. The length of each inbreeding path is calculated by adding together all of the individuals in the path except the individual of interest. In this case, there is only one path because IV-1 has only one common ancestor. To add the members of the path, we begin with individual IV-1, but we do not count her. We then move to her father (III-2); to her grandfather (II-2); to I-2, her great-grandfather (the common ancestor); back down to her other grandmother (II-3); and finally to her mother (III-3). This path has five members. Finally, to calculate the inbreeding coefficient, we use the following formula:

$$F = \Sigma(1/2)^n(1 + F_A)$$

where

- F is the inbreeding coefficient of the individual of interest
- n is the number of individuals in the inbreeding path, excluding the inbred offspring
- F_A is the inbreeding coefficient of the common ancestor
- Σ indicates that we add together $(1/2)^n(1 + F_A)$ for each inbreeding path

In this case, there is only one common ancestor and, therefore, only one inbreeding path. Also, we do not know anything about the heritage of the common ancestor, so we assume that F_A is zero. Thus, in our example of Figure 26.18,

$$F = \Sigma(1/2)^n(1 + 0)$$

$$= (1/2)^5 = 1/32 = 0.03125, \text{ or } 3.125\%$$

What does this value mean? The inbreeding coefficient, 3.125%, tells us the probability that a gene in the inbred individual (IV-1) is homozygous due to its inheritance from a common ancestor (I-2). In this case, each gene in individual IV-1 has a 3.125% chance of being homozygous because the individual has inherited the same allele twice from her great-grandfather (I-2), once through each parent.

As an example, let's suppose that the common ancestor (I-2) is heterozygous for the gene involved with cystic fibrosis (CF). His genotype would be *Cc*, where *c* is the recessive allele that causes CF. Since *F* is 3.125%, there is a 3.125% probability that the inbred individual (IV-1) is homozygous (*CC* or *cc*) for this gene because she has inherited both copies from her great-grandfather. She has a 1.56% probability of inheriting both normal alleles (*CC*) and a 1.56% probability of inheriting both mutant alleles (*cc*). The inbreeding coefficient is denoted by the letter *F* (for Fixation) because it is the probability that an allele will be fixed in the homozygous condition. The term *fixation* signifies that the homozygous individual can pass only one type of allele to their offspring.

In other pedigrees, an individual may have two or more common ancestors. In this case, the inbreeding coefficient *F* is calculated as the sum of the inbreeding paths. Such an example is described in solved problem S2 at the end of the chapter.

The effects of inbreeding and outbreeding can also be examined at the population level. For example, let's consider the situation in which the frequency of allele $A = p$ and the frequency of allele $a = q$. In a given population, the genotype frequencies are determined in the following way:

$p^2 + fpq$	equals the frequency of *AA*
$2pq(1 - f)$	equals the frequency of *Aa*
$q^2 + fpq$	equals the frequency of *aa*

where *f* is a measure of how much the genotype frequencies deviate from Hardy-Weinberg equilibrium due to nonrandom mating. The value of *f* ranges from −1 to +1. When inbreeding occurs, the value is greater than zero. When outbreeding occurs, the value is less than zero.

As an example, let's suppose that $p = 0.8$, $q = 0.2$, and $f = 0.25$. We can calculate the frequencies of the *AA*, *Aa*, and *aa* genotypes under these conditions as follows:

$$AA = p^2 + fpq = (0.8)^2 + (0.25)(0.8)(0.2) = 0.68$$

$$Aa = 2pq(1 - f) = 2(0.8)(0.2)(1 - 0.25) = 0.24$$

$$aa = q^2 + fpq = (0.2)^2 + (0.25)(0.8)(0.2) = 0.08$$

There will be 68% *AA* homozygotes, 24% heterozygotes, and 8% *aa* homozygotes. If mating had been random (i.e., $f = 0$), the genotype frequencies of *AA* would be p^2, which equals 64%, and *aa* would be q^2, which equals 4%. The frequency of heterozygotes would be $2pq$, which equals 32%. When comparing these numbers, we see that inbreeding raises the proportions of homozygotes and decreases the proportion of heterozygotes. In natural populations, the value of *f* tends to become larger as a population becomes smaller, because each individual has a more limited choice in mate selection.

What are the consequences of inbreeding in a population? From an agricultural viewpoint, it results in a higher proportion of homozygotes, which may exhibit a desirable trait. For example, an animal breeder may use inbreeding to produce animals that are larger because they have become homozygous for alleles promoting larger size. On the negative side, many genetic diseases are inherited in a recessive manner (see Chapter 24). For these disorders, inbreeding increases the likelihood that an individual will be homozygous for the recessive trait and therefore afflicted with the disease. Also, in natural populations, inbreeding lowers the mean fitness of the population if homozygous offspring have a lower fitness value. This can be a serious problem as natural populations become smaller due to human destruction of their habitats. As the population shrinks, inbreeding becomes more likely because individuals have fewer potential mates from which to choose. The inbreeding, in turn, produces homozygotes that are less fit, thereby decreasing the reproductive success of the population. This phenomenon is called **inbreeding depression.** Conservation biologists sometimes try to circumvent this problem by introducing individuals from one population into another. For example, the endangered Florida panther (*Felis concolor coryi*) suffers from inbreeding-related defects, which include poor sperm quality and quantity and morphological abnormalities. To help alleviate these effects, panthers of the same species from Texas have been introduced into the Florida population.

26.6 COMPREHENSION QUESTION

1. Inbreeding refers to mating between individuals that are
   a. homozygous.
   b. heterozygous.
   c. part of the same genetic lineage.
   d. both a and c.

26.7 SOURCES OF NEW GENETIC VARIATION

Learning Outcomes:

1. List different sources of genetic variation.
2. Define mutation rate and calculate how it affects allele frequencies in populations.
3. Outline the processes of exon shuffling and horizontal gene transfer and explain how they produce genetic variation.
4. Describe the occurrence of repetitive sequences and explain how they are used in DNA fingerprinting.

In the previous sections, we primarily focused on genetic variation in which a single gene exists in two or more alleles. These simplified scenarios allow us to appreciate the general principles behind evolutionary mechanisms. However, as researchers have analyzed genetic variation at the molecular, cellular, and population level, they have come to understand that new genetic variation occurs in many ways (**Table 26.2**). Among eukaryotic species, sexual reproduction is an important way that new genetic variation occurs among offspring. In Chapters 2 and 6, we considered how independent assortment and crossing over during sexual reproduction may produce new combinations of alleles among different genes, thereby producing new genetic variation in the resulting offspring. Similarly, in Chapter 28, we will consider how breeding between members of different species may produce hybrid offspring that harbor new combinations of genetic material. Such hybridization events have been important in the evolution of new species, particularly those in the plant kingdom. Though prokaryotic species reproduce asexually, they also possess mechanisms for gene transfer, such as conjugation, transduction, and transformation (see Chapter 7). These mechanisms are important for fostering genetic variation among bacterial and archaeal populations.

Rare mutations in DNA may also give rise to new types of variation (see Table 26.2). As discussed earlier in this chapter (see Figure 26.3) and in Chapter 18, mutations may occur within a particular gene to create new alleles of that gene. Such allelic variation is common in natural populations. Also, as described in Chapter 8, gene duplications may create a gene family; each family member acquires independent mutations and often evolves more specialized functions. An example is the globin gene family (see Figure 8.7). Changes in chromosome structure and number are also important in the evolution of new species (see Chapter 28). In this section, we examine some additional mechanisms through which an organism can acquire new genetic variation. These include exon shuffling, horizontal gene transfer, and changes in repetitive sequences. The diversity of mechanisms that foster genetic variation underscores its profound importance in the evolution of species that are both well adapted to their native environments and successful at reproduction.

TABLE 26.2
Sources of New Genetic Variation That Occur in Populations

Type	Description
Independent assortment	The independent segregation of different homologous chromosomes may give rise to new combinations of alleles in offspring (see Chapter 2).
Crossing over	Recombination (crossing over) between homologous chromosomes can also produce new combinations of alleles that are located on the same chromosome (see Chapter 6).
Interspecies crosses	On occasion, members of different species may breed with each other to produce hybrid offspring. This topic is discussed in Chapter 28.
Prokaryotic gene transfer	Prokaryotic species possess mechanisms of genetic transfer such as conjugation, transduction, and transformation (see Chapter 7).
New alleles	Point mutations can occur within a gene to create single-nucleotide polymorphisms (SNPs). In addition, genes can be altered by small deletions and additions. Gene mutations are also discussed in Chapter 18.
Gene duplications	Events, such as misaligned crossovers, can add additional copies of a gene into a genome and lead to the formation of gene families (see Chapter 8).
Chromosome structure and number	Chromosome structure may be changed by deletions, duplications, inversions, and translocations. Changes in chromosome number result in aneuploid, polyploid, and alloploid offspring. These mechanisms are discussed in Chapters 8 and 28.
Exon shuffling	New genes can be created when exons of preexisting genes are rearranged to make a gene that encodes a protein with a new combination of protein domains.
Horizontal gene transfer	Genes from one species can be introduced into another species and become incorporated into that species' genome.
Changes in repetitive sequences	Short repetitive sequences are common in genomes due to the occurrence of transposable elements and due to tandem arrays. The number and lengths of repetitive sequences tend to show considerable variation in natural populations.

Mutations Provide the Source of Genetic Variation

As discussed in Chapters 8 and 18, mutations involve changes in gene sequences, chromosome structure, and/or chromosome number. Mutations are random events that occur spontaneously at a low rate or are caused by mutagens at a higher rate. In 1926, the Russian geneticist Sergei Chetverikov was the first to suggest that mutational variability provides the raw material for evolution but does not constitute a significant evolutionary change. In other words, mutation can provide new alleles to a population but does not substantially alter allele frequencies. Chetverikov proposed that populations in nature absorb mutations like a sponge and retain them in a heterozygous condition, thereby providing a source of variability for future change.

Population geneticists often consider how new mutations affect the survival and reproductive potential of the individual that inherits them. A new mutation may be deleterious, neutral, or beneficial, depending on its effect. For genes that encode proteins, the effects of new mutations depend on their influence on protein function. Deleterious and neutral mutations are far more likely to occur than beneficial ones. For example, alleles can be altered in many different ways that render an encoded protein defective. As discussed in Chapter 18, deletions and point mutations such as missense mutations, nonsense mutations, and frameshift mutations all may cause a gene to express a protein that is nonfunctional or less functional than the wild-type protein. Also, mutations in noncoding regions can alter gene expression (refer back to Table 18.2). Neutral mutations, which are not acted upon by natural selection, occur in several different ways. For example, a neutral mutation can change the wobble base without affecting the amino acid sequence of the encoded protein, or it can be a missense mutation that has no effect on protein function. Such point mutations happen at specific sites within the coding sequence. Neutral mutations can also occur within introns, the noncoding sequences of genes. By comparison, beneficial mutations are relatively uncommon. To be advantageous, a new mutation could alter the amino acid sequence of a protein to yield a better-functioning product. Such mutations occur less frequently than deleterious or neutral mutations.

The **mutation rate** is defined as the probability that a gene will be altered by a new mutation. The rate is typically expressed as the number of new mutations in a given gene per generation. A common value for the mutation rate is in the range of 1 in 100,000 to 1 in 1,000,000, or 10^{-5} to 10^{-6} per generation. However, mutation rates vary depending on species, cell types, chromosomal location, and gene size. Furthermore, in experimental studies, the mutation rate is usually measured by following the change of a normal (functional) gene to a deleterious (nonfunctional) allele. The mutation rate producing beneficial alleles is expected to be substantially less.

It is clear that new mutations provide genetic variability, but population geneticists also want to know how much the mutation rate affects the allele frequencies in a population. Can random mutations have a large effect on allele frequencies over time? To answer this question, let's take this simple case: A gene exists in an allele, *A*; the allele frequency of *A* is denoted by the variable *p*. A mutation can convert the *A* allele into a different allele called *a*. The allele frequency of *a* is designated by *q*. The conversion of the *A* allele into the *a* allele by mutation occurs at a rate that is designated μ. If we assume that the rate of the reverse mutation (*a* to *A*) is negligible, the change in allele frequency of the *a* allele (Δq) after one generation is

$$\Delta q = \mu p$$

For example, let's consider the following conditions:

$p = 0.8$ (i.e., frequency of *A* is 80%)

$q = 0.2$ (i.e., frequency of *a* is 20%)

$\mu = 10^{-5}$ (a hypothetical mutation rate of converting *A* to *a*)

$$\Delta q = (10^{-5})(0.8) = (0.00001)(0.8) = 0.000008$$

Therefore, in the next generation (designated $n + 1$),

$$q_{n+1} = 0.2 + 0.000008 = 0.200008$$

$$p_{n+1} = 0.8 - 0.000008 = 0.799992$$

As we can see from this calculation, new mutations do not significantly alter the allele frequencies in a single generation.

We can use the following equation to calculate the change in allele frequency after any number of generations:

$$(1 - \mu)^t = \frac{p_t}{p_0}$$

where

μ is the mutation rate of the conversion of *A* to *a*

t is the number of generations

p_0 is the allele frequency of *A* in the starting generation

p_t is the allele frequency of *A* after *t* generations

As an example, let's suppose that the allele frequency of *A* is 0.8, $\mu = 10^{-5}$, and we want to know what the allele frequency will be after 1000 generations ($t = 1000$). Plugging these values into the preceding equation and solving for p_t,

$$(1 - 0.00001)^{1000} = \frac{p_t}{0.8}$$

$$p_t = 0.792$$

Therefore, after 1000 generations, the frequency of *A* has dropped only from 0.8 to 0.792. Again, these results point to how slowly the occurrence of new mutations changes allele frequencies. In natural populations, the rate of new mutation is rarely a significant catalyst in shaping allele frequencies. Instead, other processes, such as natural selection and genetic drift, have far greater effects on allele frequencies.

New Genes Are Produced in Eukaryotes via Exon Shuffling

Sources of new genetic variation are revealed when the parts of genes that encode protein domains are compared within a single species. Many proteins, particularly those found in eukaryotic species, have a modular structure composed of two or more domains with different functions. For example, certain transcription factors have discrete domains involved with hormone binding, dimerization, and DNA binding. As described in Chapter 15, the glucocorticoid receptor has a domain that binds the hormone, a second domain that facilitates protein dimerization, and a third that allows the glucocorticoid receptor to bind to a glucocorticoid response element (GRE) next to particular genes (see Figure 15.7). By comparing the modular structure of eukaryotic proteins with the genes

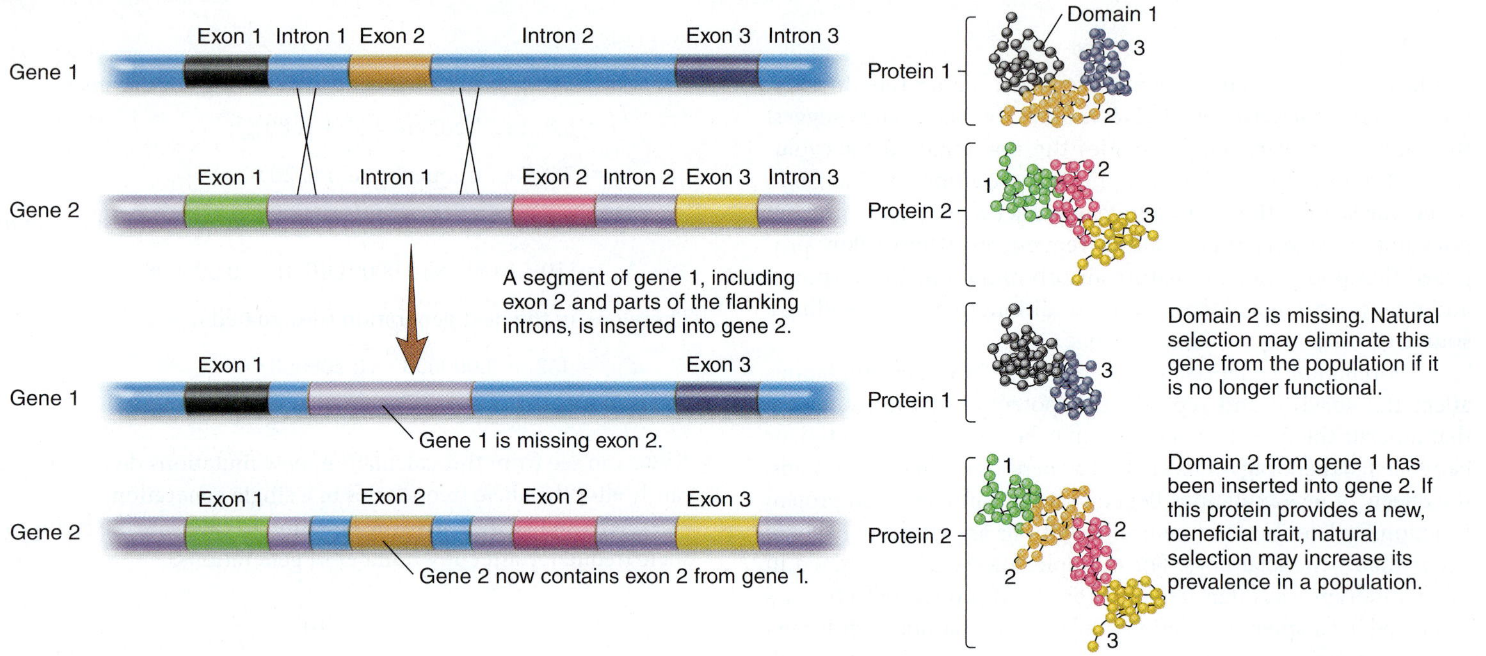

FIGURE 26.19 The process of exon shuffling. In this example, a segment of one gene containing an exon and its flanking introns has been inserted into another gene. A rare, abnormal crossing over event called nonhomologous recombination may cause this to happen. This results in proteins that have new combinations of domains and possibly new functions.

that encode them, geneticists have discovered that each domain tends to be encoded by one coding sequence, or exon, or by a series of two or more adjacent exons.

During the evolution of eukaryotic species, many new genes have been created by a process known as **exon shuffling,** in which an exon and its flanking introns are inserted into a gene, thereby producing a new gene that encodes a protein with an additional domain (**Figure 26.19**). This process may also involve the duplication and rearrangement of exons. Exon shuffling results in novel genes that express proteins with diverse functional modules. Such proteins can then alter traits in the organism and may be acted on by evolutionary processes, such as natural selection and genetic drift.

Exon shuffling may occur by more than one mechanism. Transposable elements, which are described in Chapter 19, may promote the insertion of exons into the coding sequences of genes. Alternatively, an abnormal crossover event could promote the insertion of an exon into another gene (this is the case in Figure 26.19). This is called nonhomologous recombination because the two regions involved in the crossover are not homologous to each other.

New Genes Are Acquired via Horizontal Gene Transfer

Species also accumulate genetic changes by a process called **horizontal gene transfer,** in which an organism incorporates genetic material from another organism without being the offspring of that organism. It often involves the exchange of genetic material between different species. **Figure 26.20** illustrates one possible mechanism for horizontal gene transfer. In this example, a eukaryotic cell has engulfed a bacterium by endocytosis. During the degradation of the bacterium, a bacterial gene escapes to the nucleus of the cell, where it is inserted into one of the chromosomes. In this way, a gene has been transferred from a bacterial species to a eukaryotic species. By analyzing gene sequences among many different species, researchers have discovered that horizontal gene transfer is a common phenomenon. This process can occur from prokaryotes to eukaryotes, from eukaryotes to prokaryotes, between different species of prokaryotes, and between different species of eukaryotes.

Gene transfer among bacterial species is relatively widespread. As discussed in Chapter 7, bacterial species may carry out three natural mechanisms of gene transfer known as conjugation, transduction, and transformation. By analyzing the genomes of bacterial species, scientists have determined that many genes within a given bacterial genome are derived from other species

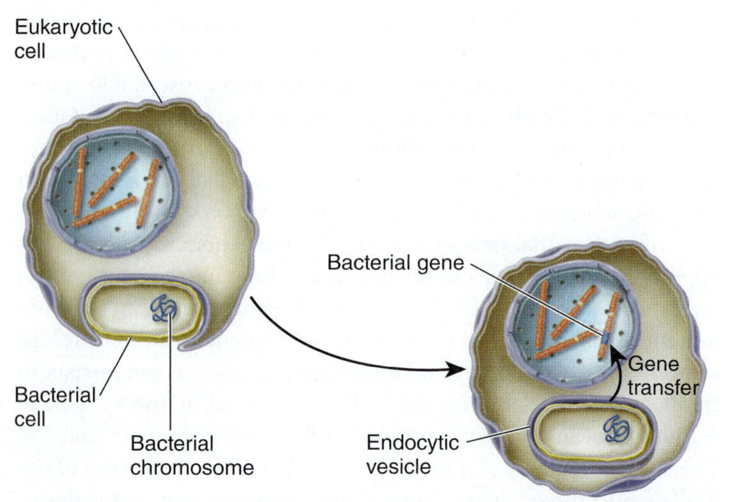

ONLINE ANIMATION

FIGURE 26.20 Horizontal gene transfer from a bacterium to a eukaryote. In this example, a bacterium is engulfed by a eukaryotic cell, and a bacterial gene is transferred to one of the eukaryotic chromosomes.

via horizontal gene transfer. Genome studies have suggested that as much as 20–30% of the variation in the genetic composition of modern prokaryotic species can be attributed to this process. For example, roughly 17% of the genes of *E. coli* and *Salmonella typhimurium* have been acquired from other species via horizontal gene transfer during the past 100 million years. The roles of these acquired genes are quite varied, though they commonly involve functions that are readily acted on by natural selection. These include genes that confer antibiotic resistance, the ability to degrade toxic compounds, and pathogenicity (the ability to cause disease).

Genetic Variation Is Produced via Changes in Repetitive Sequences

Another source of genetic variation comes from changes in **repetitive sequences**—short sequences typically a few base pairs to a few thousand base pairs long that are repeated many times within a species' genome. Repetitive sequences usually come from two types of sources. First, transposable elements (TEs) are genetic sequences that can move from place to place in a species' genome (see Chapter 19). The prevalence and movement of TEs provide a great deal of genetic variation between species and within a single species. In certain eukaryotic species, TEs have become fairly abundant (see Table 19.3).

A second type of repetitive sequence is nonmobile and involves short sequences that are tandemly repeated. In a **microsatellite** (also called short tandem repeats, STRs), the repeat unit is usually 1–6 bp long, and the whole tandem repeat is less than a couple hundred base pairs in length. For example, the most common microsatellite encountered in humans is a sequence $(CA)_N$, where N may range from 5 to more than 50. In other words, this dinucleotide sequence can be tandemly repeated 5–50 or more times. The $(CA)_N$ microsatellite is found, on average, about every 10,000 bases in the human genome. In a **minisatellite,** the repeat unit is typically 6–80 bp in length, and the size of the minisatellite ranges from 1–20 kbp. An example of a minisatellite in humans is telomeric DNA. In a human sperm cell, for example, the repeat unit is 6 bp and the size of a telomere is about 15 kbp. (Note: Tandem repeat sequences are called satellites because they sediment away from the rest of the chromosomal DNA during equilibrium gradient centrifugation.)

Alterations in the sequences of microsatellites often escape the proofreading function of DNA polymerase. Therefore, microsatellites tend to be "hotspots" for mutation in which the number of tandem repeats changes. For example, a microsatellite with a 4-bp repeat unit and a length of 64 bp may undergo a mutation that adds three more repeat units and becomes 76 bp long. How does this occur? One common mechanism to explain this phenomenon is DNA strand slippage that occurs during DNA replication due to the formation of hairpin structures. As discussed in Chapter 18, this mechanism can explain trinucleotide repeat expansion (TNRE) (see Figure 18.11).

Because repetitive sequences tend to vary within a population, they have become a common tool that geneticists use in a variety of ways. For example, as described in Chapter 24, microsatellites are used as molecular markers to map the locations of genes. Likewise, population geneticists analyze microsatellites or minisatellites to study variation at the population level and to determine the relationships among individuals and neighboring populations. The sizes of microsatellites and minisatellites found in closely related individuals tend to be more similar than they are in unrelated individuals. As described next, this phenomenon is the basis for DNA fingerprinting.

DNA Fingerprinting Is Used for Identification and Relationship Testing

The technique of **DNA fingerprinting,** also known as **DNA profiling,** analyzes individuals based on the occurrence of repetitive sequences in their genome. When subjected to traditional DNA fingerprinting, the chromosomal DNA gives rise to a series of bands on a gel (**Figure 26.21**). The sizes and order of bands constitute an

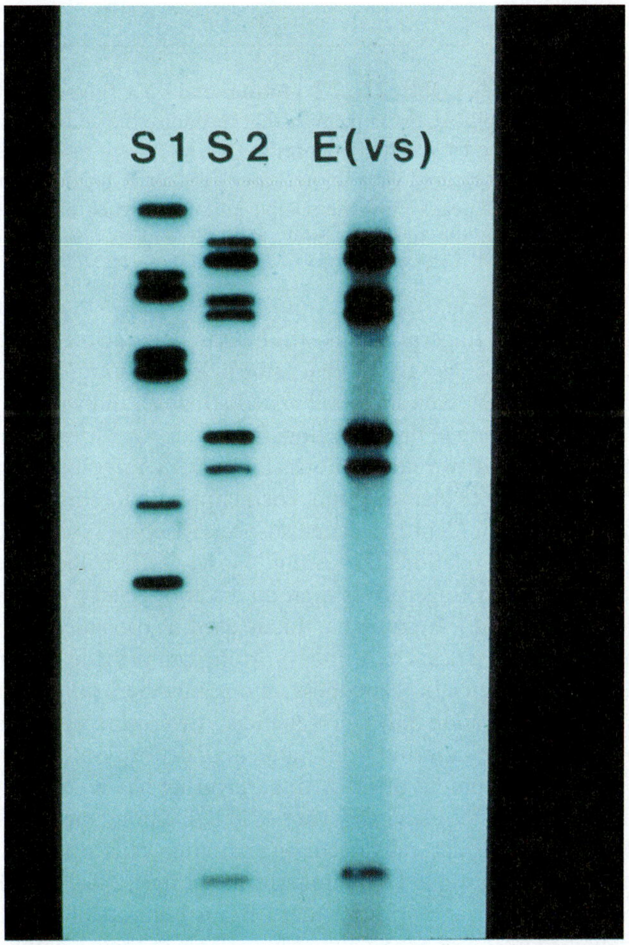

FIGURE 26.21 **A comparison of two DNA fingerprints.** The chromosomal DNA from two different individuals (Suspect 1—S1, and Suspect 2—S2) was subjected to DNA fingerprinting. The DNA evidence at a crime scene, E(vs), was also subjected to DNA fingerprinting. Following the hybridization of a labeled probe, the DNA appears as a series of bands on a gel. The dissimilarity in the pattern of these bands distinguishes different individuals, much as the differences in physical fingerprint patterns can be used for identification. As seen here, S2 matches the DNA found at the crime scene.

CONCEPT CHECK: What are two common applications of DNA fingerprinting?

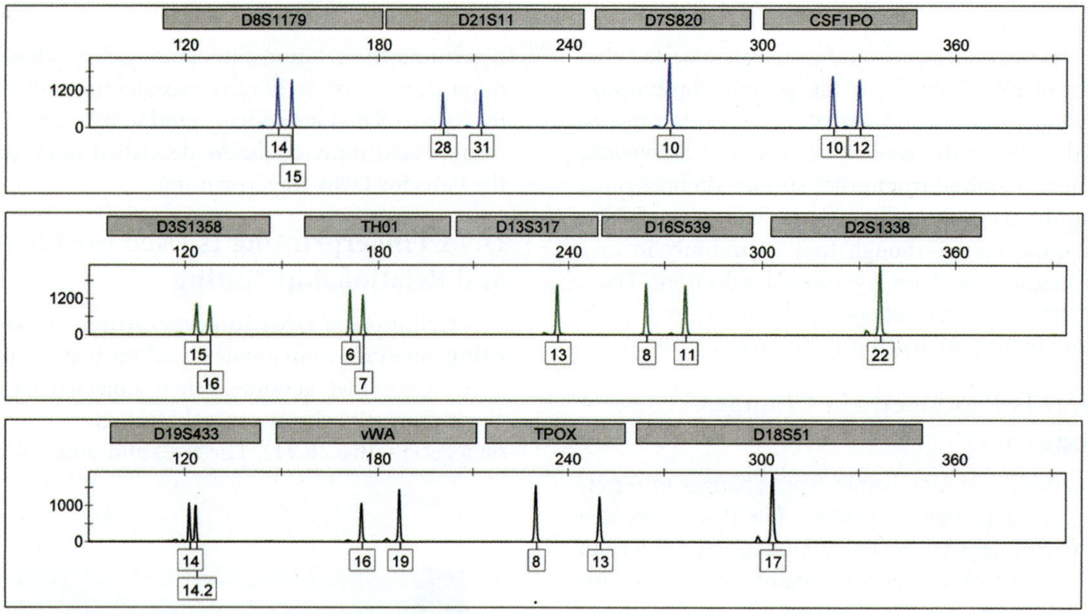

ONLINE ANIMATION

FIGURE 26.22 **Automated DNA fingerprinting.** In automated DNA fingerprinting, microsatellites in a sample of DNA are amplified, using primers that recognize the ends of microsatellites. The microsatellite fragments are fluorescently labeled and then separated by gel electrophoresis. The fluorescent molecules within each microsatellite are excited with a laser, and the amount of fluorescence is measured via a fluorescence detector. A printout from the detector is shown here. The gray boxes indicate the names of specific microsatellites. The peaks show the relative amounts of each microsatellite. The boxes beneath each peak indicate the number of tandem repeats in a given microsatellite. In this example, the individual is heterozygous for certain microsatellites (e.g., D8S1179) and homozygous for others (e.g., D7S820).

individual's DNA fingerprint. Like the human fingerprint, the DNA of each individual has a distinctive pattern. It is the unique pattern of these bands that makes it possible to distinguish individuals.

A comparison of the DNA fingerprints among different individuals has found two applications. First, DNA fingerprinting is used as a method of identification. For example, in forensics, DNA fingerprinting can identify a crime suspect. A second use of DNA fingerprinting is relationship testing. Closely related individuals have more similar fingerprints than do distantly related ones (see solved problem S6). In humans, this is used in paternity testing. In population genetics, DNA fingerprinting can provide evidence regarding the degree of relatedness among members of a population. Such information may help geneticists determine if a population is likely to be suffering from inbreeding depression.

The development of DNA fingerprinting has relied on the identification of DNA sites that vary in length among members of a population. This naturally occurring variation causes each individual to have a unique DNA fingerprint. In the 1980s, Alec Jeffreys and his colleagues found that certain minisatellites within human chromosomes are particularly variable in their lengths. As discussed earlier, minisatellites tend to vary within populations due to changes in the number of tandem repeats at each site.

DNA fingerprinting is now done using the technique of polymerase chain reaction (PCR), which amplifies microsatellites. Like minisatellites, microsatellites are found in multiple sites in the genome of humans and other species and vary in length among different individuals. In this procedure, the microsatellites from a sample of DNA are amplified by PCR using primers that flank the repetitive region. During this process, the amplified microsatellites are fluorescently labeled. They are then separated by gel electrophoresis according to their molecular masses. As in automated DNA sequencing described in Chapter 20, a laser excites the fluorescent molecule within a microsatellite, and a detector records the amount of fluorescence emission for each one. As shown in Figure 26.22, this type of DNA fingerprint yields a series of peaks, each peak having a characteristic molecular mass. In this automated approach, the pattern of peaks rather than bands constitutes an individual's DNA fingerprint.

26.7 COMPREHENSION QUESTIONS

1. The mutation rate is
   a. the likelihood that a new mutation will occur in a given gene.
   b. too low to substantially change allele frequencies in a population.
   c. lower for mutations that create beneficial alleles.
   d. all of the above.
2. The transfer of an antibiotic resistance gene from one bacterial species to a different species is an example of
   a. exon shuffling.
   b. horizontal gene transfer.
   c. genetic drift.
   d. migration.
3. DNA fingerprinting analyzes the DNA from individuals based on the occurrence of ____________ in their genomes.
   a. repetitive sequences
   b. abnormalities in chromosome structure
   c. specific genes
   d. viral insertions

KEY TERMS

Page 671. population genetics, gene pool
Page 672. population, local populations, demes, polymorphism, polymorphic, monomorphic, single-nucleotide polymorphism (SNP)
Page 673. allele frequencies, genotype frequencies
Page 674. Hardy-Weinberg equilibrium, Hardy-Weinberg equation
Page 676. disequilibrium, microevolution
Page 677. natural selection, Darwinian fitness
Page 678. directional selection
Page 679. mean fitness of the population ($\overline{w}$)
Page 680. balancing selection, heterozygote advantage, selection coefficient (s)
Page 681. negative frequency-dependent selection, disruptive selection
Page 683. stabilizing selection
Page 685. genetic drift
Page 686. bottleneck effect
Page 687. founder effect, gene flow
Page 688. assortative mating, inbreeding, outbreeding, inbreeding coefficient (F)
Page 689. inbreeding depression
Page 691. mutation rate
Page 692. exon shuffling, horizontal gene transfer
Page 693. repetitive sequences, microsatellite, minisatellite, DNA fingerprinting, DNA profiling

CHAPTER SUMMARY

- Population genetics is concerned with changes in genetic variation within a population over time.

26.1 Genes in Populations and the Hardy-Weinberg Equation

- All of the alleles of every gene in a population constitute the gene pool.
- For sexually reproducing organisms, a population is a group of individuals of the same species that occupy the same region and can interbreed with one another (see Figure 26.1).
- Genetic polymorphism refers to inherited traits or genes that exhibit variation in a population (see Figure 26.2).
- Single-nucleotide polymorphisms (SNPs) are the most common type of variation in genes (see Figure 26.3).
- Geneticists analyze genetic variation by determining allele and genotype frequencies.
- The Hardy-Weinberg equation can be used to calculate genotype frequencies based on allele frequencies (see Figures 26.4, 26.5).
- Deviation from Hardy-Weinberg equilibrium indicates that evolutionary change is occurring.
- The Hardy-Weinberg equation can be used to estimate the frequency of heterozygous carriers.

26.2 Overview of Microevolution

- Mutations are the source of new genetic variation. However, the occurrence of new mutations does not greatly change allele frequencies because they are relatively rare. Other factors, such as natural selection, genetic drift, migration, and nonrandom mating, may alter allele and/or genotype frequencies (see Table 26.1).

26.3 Natural Selection

- Natural selection is a process that changes allele frequencies from one generation to the next based on fitness, which is the relative reproductive successes of different genotypes.
- Directional selection favors the extreme phenotype (see Figures 26.6–26.8).
- Balancing selection results in a stable polymorphism. Examples include heterozygote advantage and negative-frequency dependent selection (see Figures 26.9, 26.10).
- Disruptive selection favors multiple phenotypes. This may occur due to heterogeneous environments (see Figures 26.11, 26.12).
- Stabilizing selection favors individuals with intermediate phenotypes (see Figure 26.13).
- The Grants observed natural selection in a finch population. The selection involved a change in beak size due to drought conditions that produced larger seeds (see Figures 26.14, 26.15).

26.4 Genetic Drift

- Genetic drift is a change in allele frequency due to chance fluctuations. Over the long run, it often results in allele fixation or loss. The effect of genetic drift is greater in small populations (see Figure 26.16).
- Two examples of genetic drift are the bottleneck effect and the founder effect (see Figure 26.17).

26.5 Migration

- Migration can alter allele frequencies. It tends to reduce differences in allele frequencies between neighboring populations and increase genetic diversity within a population.

26.6 Nonrandom Mating

- Nonrandom mating may alter the genotype frequencies that would be based on a Hardy-Weinberg equilibrium. Inbreeding results in a higher proportion of homozygotes in a population (see Figure 26.18).

26.7 Sources of New Genetic Variation

- A variety of different mechanisms bring about genetic variation (see Table 26.2).
- Mutation rate is the probability that a gene will be altered by a new mutation.
- New genes in eukaryotes are produced by exon shuffling (see Figure 26.19).
- A species may acquire a new gene from another species via horizontal gene transfer (see Figure 26.20).
- A common source of genetic variation in populations involves changes in repetitive sequences, such as microsatellites.
- DNA fingerprinting is a technique that relies on variation in repetitive sequences within a population. It is used as a means of identification and in relationship testing (see Figures 26.21, 26.22).

PROBLEM SETS & INSIGHTS

Solved Problems

S1. The phenotypic frequency of people who cannot taste phenylthiocarbamide (PTC) is approximately 0.3. The inability to taste this bitter substance is due to a recessive allele. If we assume there are only two alleles in the population (namely, tasters, *T*, and nontasters, *t*) and that the population is in Hardy-Weinberg equilibrium, calculate the frequencies of these two alleles.

Answer: Let p = allele frequency of the taster allele and q = the allele frequency of the nontaster allele. The frequency of nontasters is 0.3.

This is the frequency of the genotype *tt*, which in this case is equal to q^2:

$$q^2 = 0.3$$

To determine the frequency q of the nontaster allele, we take the square root of both sides of this equation:

$$q = 0.55$$

With this value, we can calculate the frequency p of the taster allele:

$$p = 1 - q$$
$$= 1 - 0.55 = 0.45$$

S2. In the pedigree shown here, answer the following questions with regard to individual VII-1:

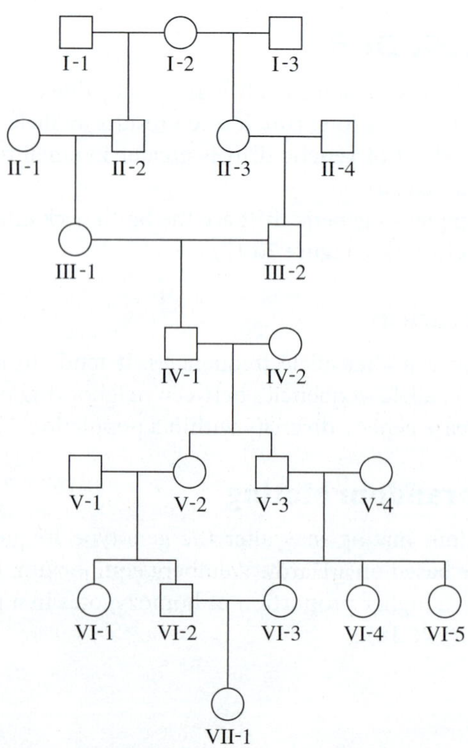

A. Who are the common ancestors of her parents?

B. What is the inbreeding coefficient?

Answer:

A. The common ancestors are IV-1 and IV-2. They are the grandparents of VI-2 and VI-3, who are the parents of VII-1.

B. The inbreeding coefficient is calculated using the formula

$$F = \Sigma(1/2)^n(1 + F_A)$$

In this case, there are two common ancestors, IV-1 and IV-2. Also, IV-1 is inbred, because I-2 is a common ancestor to both of IV-1's parents. The first step is to calculate F_A, the inbreeding coefficient for this common ancestor. The inbreeding path for IV-1 contains five people: III-1, II-2, I-2, II-3, and III-2. Therefore,

$$n = 5$$
$$F_A = (1/2)^5 = 0.03$$

Now we can calculate the inbreeding coefficient for VII-1. Each inbreeding path contains five people: VI-2, V-2, IV-1, V-3, and VI-3; and VI-2, V-2, IV-2, V-3, and VI-3. Thus,

$$F = (1/2)^5 (1 + 0.03) + (1/2)^5 (1 + 0)$$
$$= 0.032 + 0.031 = 0.063$$

S3. The Hardy-Weinberg equation provides a way to predict genotype frequency based on allele frequency. In the case of mammals, males are hemizygous for X-linked genes, whereas females have two copies. Among males, the frequency of any X-linked trait equals the frequency of males with the trait. For example, if an allele frequency for an X-linked disease-causing allele was 5%, then 5% of all males would be affected with the disorder. Female genotype frequencies are computed using the Hardy-Weinberg equation.

As a specific example, let's consider the human X-linked trait known as hemophilia A (see Chapter 24 for a description of this disorder). In human populations, the allele frequency of the hemophilia A allele is approximately 1 in 10,000, or 0.0001. The other allele for this gene is the normal allele. Males can be affected or unaffected, whereas females can be affected, unaffected carriers, or unaffected noncarriers.

A. What are the allele frequencies for the mutant and normal allele in the human population?

B. Among males, what is the frequency of affected individuals?

C. Among females, what is the frequency of affected individuals and heterozygous carriers?

D. Within a population of 100,000 people, what is the expected number of affected males? In this same population, what is the expected number of carrier females?

Answer: Let p represent the normal allele and q represent the allele that causes hemophilia.

A. X^H normal allele, frequency $= 0.9999 = p$

X^h hemophilia allele, frequency $= 0.0001 = q$

B. X^hY genotype frequency of affected males $= q = 0.0001$

C. X^hX^h genotype frequency of affected females $= q^2 = (0.0001)^2 = 0.00000001$

X^HX^h genotype frequency of carrier females $= 2pq = 2(0.9999)(0.0001) = 0.0002$

D. We assume this population is composed of 50% males and 50% females.

Number of affected males $= 50{,}000 \times 0.0001 = 5$

Number of carrier females $= 50{,}000 \times 0.0002 = 10$

S4. The Hardy-Weinberg equation can be modified to include situations of three or more alleles. In its standard (two-allele) form, the Hardy-Weinberg equation reflects the Mendelian notion that each individual inherits two copies of each gene, one from each parent. For a two-allele situation, it can also be written as

$(p + q)^2 = 1$ (Note: The number 2 in this equation reflects the idea that the genotype is due to the inheritance of two alleles, one from each parent.)

This equation can be expanded to include three or more alleles. For example, let's consider a situation in which a gene exists as three alleles: *A1*, *A2*, and *A3*. The allele frequency of *A1* is designated by the letter p, *A2* by the letter q, and *A3* by the letter r. Under these circumstances, the Hardy-Weinberg equation becomes

$$(p + q + r)^2 = 1$$

$$p^2 + q^2 + r^2 + 2pq + 2pr + 2qr = 1$$

where

p^2 is the genotype frequency of *A1A1*

q^2 is the genotype frequency of *A2A2*

r^2 is the genotype frequency of *A3A3*

$2pq$ is the genotype frequency of *A1A2*

$2pr$ is the genotype frequency of *A1A3*

$2qr$ is the genotype frequency of *A2A3*

Now here is the question. As discussed in Chapter 4, the gene that affects human blood type can exist in three alleles. In a Japanese population, the allele frequencies are

$I^A = 0.28$

$I^B = 0.17$

$i = 0.55$

Based on these allele frequencies, calculate the different possible genotype frequencies and blood type frequencies.

Answer: If we let p represent I^A, q represent I^B, and r represent i, then

p^2 is the genotype frequency of I^AI^A, which is type A blood $= (0.28)^2 = 0.08$

q^2 is the genotype frequency of I^BI^B, which is type B blood $= (0.17)^2 = 0.03$

r^2 is the genotype frequency of ii, which is type O blood $= (0.55)^2 = 0.30$

$2pq$ is the genotype frequency of I^AI^B, which is type AB blood $= 2(0.28)(0.17) = 0.09$

$2pr$ is the genotype frequency of I^Ai, which is type A blood $= 2(0.28)(0.55) = 0.31$

$2qr$ is the genotype frequency of I^Bi, which is type B blood $= 2(0.17)(0.55) = 0.19$

Type A $= 0.08 + 0.31 = 0.39$, or 39%

Type B $= 0.03 + 0.19 = 0.22$, or 22%

Type O $= 0.30$, or 30%

Type AB $= 0.09$, or 9%

S5. Let's suppose that pigmentation in a species of insect is controlled by a single gene existing in two alleles, *D* for dark and *d* for light. The heterozygote *Dd* is intermediate in color. In a heterogeneous environment, the allele frequencies are $D = 0.7$ and $d = 0.3$. This polymorphism is maintained because the environment contains some dimly lit forested areas and some sunny fields. During a hurricane, a group of 1000 insects is blown to a completely sunny area. In this environment, the fitness values are $DD = 0.3$, $Dd = 0.7$, and $dd = 1.0$. Calculate the allele frequencies in the next generation.

Answer: The first step is to calculate the mean fitness of the population:

$$p^2w_{DD} + 2pqw_{Dd} + q^2w_{dd} = \overline{w}$$

$$\overline{w} = (0.7)^2(0.3) + 2(0.7)(0.3)(0.7) + (0.3)^2(1.0)$$

$$= 0.15 + 0.29 + 0.09 = 0.53$$

After one generation of selection, we get

$$\text{Allele frequency of } D\text{: } p_D = \frac{p^2w_{DD}}{\overline{w}} + \frac{pqw_{Dd}}{\overline{w}}$$

$$= \frac{(0.7)^2(0.3)}{0.53} + \frac{(0.7)(0.3)(0.7)}{0.53}$$

$$= 0.55$$

$$\text{Allele frequency of } d\text{: } q_d = \frac{q^2w_{dd}}{\overline{w}} + \frac{pqw_{Dd}}{\overline{w}}$$

$$= \frac{(0.3)^2(1.0)}{0.53} + \frac{(0.7)(0.3)(0.7)}{0.53}$$

$$= 0.45$$

After one generation, the allele frequency of *D* has decreased from 0.7 to 0.55, and the frequency of *d* has increased from 0.3 to 0.45.

S6. An important application of DNA fingerprinting is relationship testing. Persons who are related genetically have some bands or peaks in common. The number they share depends on the closeness of their genetic relationship. For example, an offspring is expected to receive half of his or her minisatellites from one parent and the rest from the other. The diagram shown here schematically illustrates a traditional DNA fingerprint of an offspring, mother, and two potential fathers.

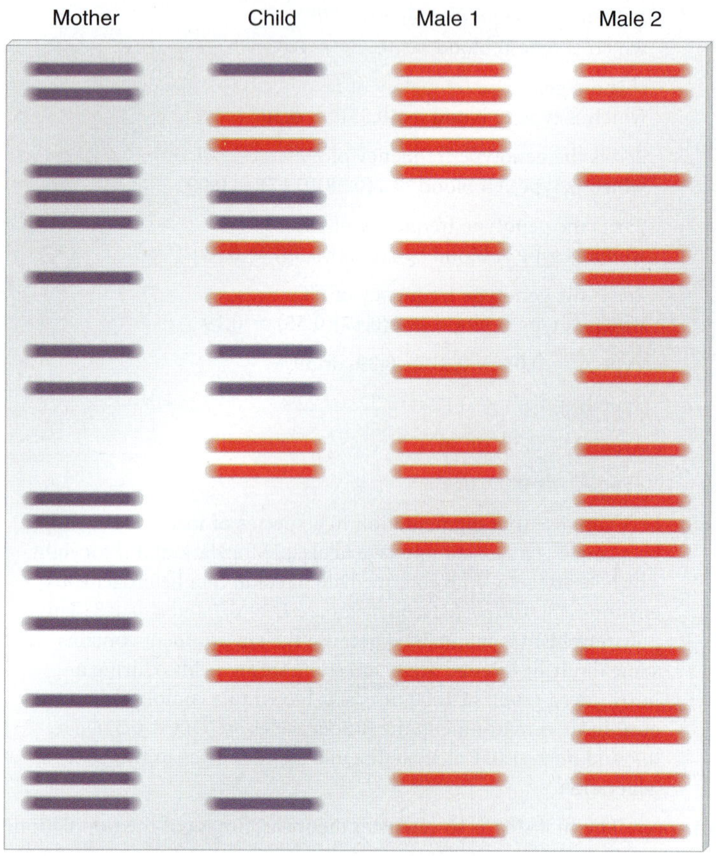

In paternity testing, the offspring's DNA fingerprint is first compared with that of the mother. The bands that the offspring have in common with the mother are depicted in purple. The bands that are not similar between the offspring and the mother must have been inherited from the father. These bands are depicted in red. Which male could be the father?

Answer: Male 2 does not have many of the paternal bands. Therefore, he can be excluded as being the father of this child. However, male 1 has all of the paternal bands. He is very likely to be the father.

Geneticists can calculate the likelihood that the matching bands between the offspring and a prospective father could occur just as a matter of random chance. To do so, they must analyze the frequency of each band in a reference population (e.g., Caucasians living in the United States). For example, let's suppose that DNA fingerprinting analyzed 40 bands. Of these, 20 bands matched with the mother and 20 bands matched with a prospective father. If the probability of each of these bands in a reference population was 1/4, the likelihood of such a match occurring by random chance would be $(1/4)^{20}$, or roughly 1 in 1 trillion. Therefore, a match between two samples is rarely a matter of random chance.

Conceptual Questions

C1. What is the gene pool? How is a gene pool described in a quantitative way?

C2. In genetics, what does the term *population* mean? Pick any species you like and describe how its population might change over the course of many generations.

C3. What is a genetic polymorphism? What is the source of genetic variation?

C4. State for each of the following whether it is an example of an allele, genotype, and/or phenotype frequency:

A. Approximately 1 in 2500 people of Northern European descent is born with cystic fibrosis.

B. The percentage of carriers of the sickle cell allele in West Africa is approximately 13%.

C. The number of new mutations for achondroplasia, a genetic disorder, is approximately 5×10^{-5}.

C5. The term *polymorphism* can refer to both genes and traits. Explain the meaning of a polymorphic gene and a polymorphic trait. If a gene is polymorphic, does the trait that the gene affects also have to be polymorphic? Explain why or why not.

C6. Cystic fibrosis (CF) is a recessive autosomal trait. In certain populations of Northern European descent, the number of people born with this disorder is about 1 in 2500. Assuming Hardy-Weinberg equilibrium for this trait:

A. What are the frequencies for the normal and CF alleles?

B. What are the genotype frequencies of homozygous normal, heterozygous, and homozygous affected individuals?

C. Assuming random mating, what is the probability that two phenotypically unaffected heterozygous carriers will choose each other as mates?

C7. For a gene existing in two alleles, what are the allele frequencies when the heterozygote frequency is at its maximum value, assuming a Hardy-Weinberg equilibrium? What if there are three alleles?

C8. In a population, the frequencies of two alleles are $B = 0.67$ and $b = 0.33$. The genotype frequencies are $BB = 0.50$, $Bb = 0.37$, and $bb = 0.13$. Do these numbers suggest inbreeding? Explain why or why not.

C9. The ability to roll your tongue is inherited as a recessive trait. The frequency of the rolling allele is approximately 0.6, and the dominant (nonrolling) allele is 0.4. What is the frequency of individuals who can roll their tongues?

C10. What evolutionary factors can cause allele frequencies to change and possibly lead to a genetic polymorphism? Discuss the relative importance of each type of process.

C11. What is the difference between a random and an adaptive evolutionary process? Describe two or more examples of each. At the molecular level, explain how mutations can be random or adaptive.

C12. What is Darwinian fitness? What types of characteristics can promote high fitness values? Give several examples.

C13. What is the intuitive meaning of the mean fitness of a population? How does its value change in response to natural selection?

C14. Describe the similarities and differences among directional, balancing, disruptive, and stabilizing selection.

C15. Do the following examples describe directional, disruptive, balancing, or stabilizing selection?

A. Polymorphisms in snail color and banding pattern as described in Figure 26.12

B. Thick fur among mammals exposed to cold climates

C. Birth weight in humans

D. Sturdy stems and leaves among plants exposed to windy climates

C16. In the term genetic drift, what is drifting? Why is this an appropriate term to describe this phenomenon?

C17. Why is genetic drift more significant in small populations? Why does it take longer for genetic drift to cause allele fixation in large populations than in small ones?

C18. A group of four birds flies to a new location and initiates the formation of a new colony. Three of the birds are homozygous *AA*, and one bird is heterozygous *Aa*.

A. What is the probability that the *a* allele will become fixed in the population via genetic drift?

B. If fixation of the *a* allele occurs, how long will it take?

C. How will the growth of the population, from generation to generation, affect the answers to parts A and B? Explain.

C19. Describe what happens to allele frequencies as a result of the bottleneck effect. Discuss the relevance of this effect with regard to species that are approaching extinction.

C20. With regard to genetic drift, are the following statements true or false? If a statement is false, explain why.

A. Over the long run, genetic drift leads to allele fixation or loss.

B. When a new mutation occurs within a population, genetic drift is more likely to cause the loss of the new allele rather than the fixation of the new allele.

C. Genetic drift promotes genetic diversity in large populations.

D. Genetic drift is more significant in small populations.

C21. When two populations frequently intermix due to migration, what are the long-term consequences with regard to allele frequencies and genetic variation?

C22. Two populations of antelope are separated by a mountain range. The antelope are known to occasionally migrate from one population to the other. Migration can occur in either direction. Explain how migration affects the following phenomena:

A. Genetic diversity in the two populations

B. Allele frequencies in the two populations

C. Genetic drift in the two populations

C23. Does inbreeding affect allele frequencies? Why or why not? How does it affect genotype frequencies? With regard to rare recessive diseases, what are the consequences of inbreeding in human populations?

C24. Using the pedigree shown here, answer the following questions for individual VI-1:

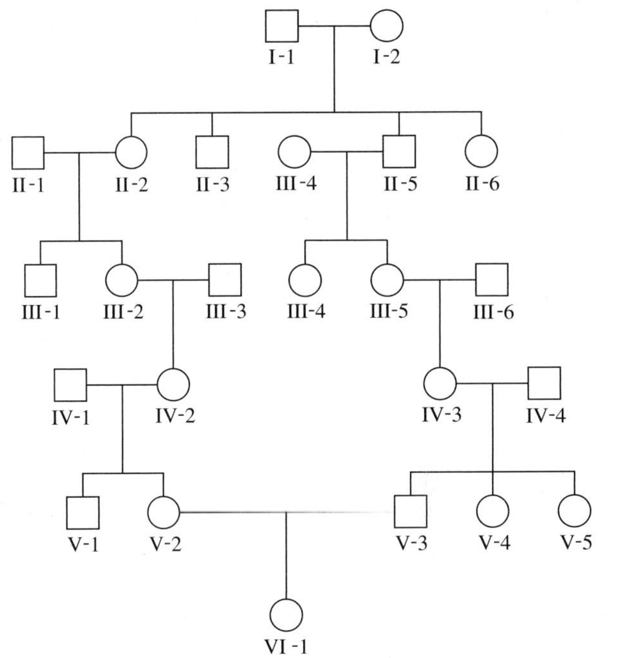

A. Is this individual inbred?

B. If so, who are her common ancestor(s)?

C. Calculate the inbreeding coefficient for VI-1.

D. Are the parents of VI-1 inbred?

C25. A family pedigree is shown here.

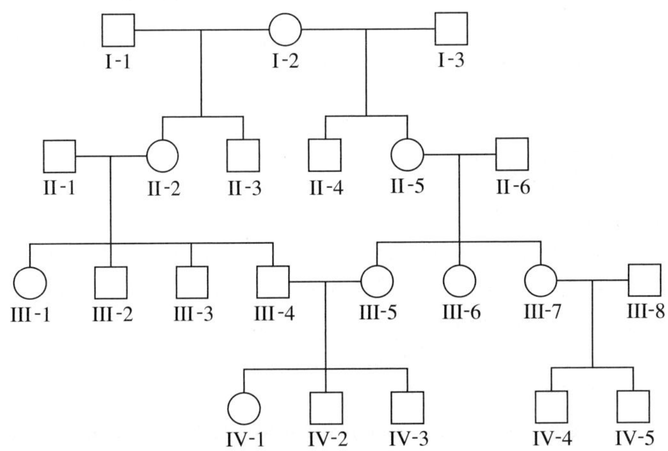

A. What is the inbreeding coefficient for individual IV-3?

B. Based on the data shown in this pedigree, is individual IV-4 inbred?

C26. A family pedigree is shown here.

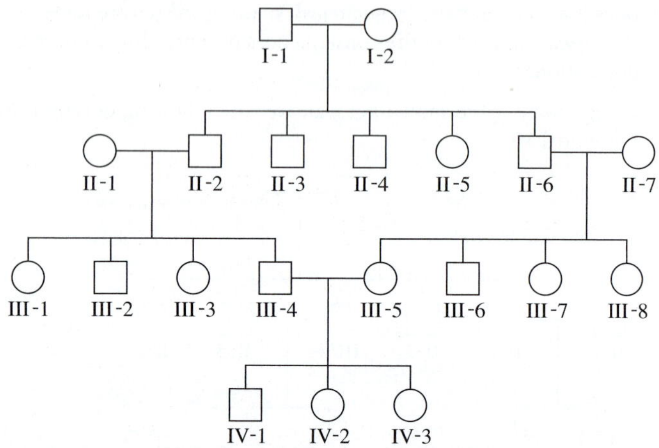

A. What is the inbreeding coefficient for individual IV-2? Who is/are her common ancestors?

B. Based on the data shown in this pedigree, is individual III-4 inbred?

C27. Antibiotics are commonly used to combat bacterial and fungal infections. During the past several decades, however, antibiotic-resistant strains of microorganisms have become alarmingly prevalent. This has undermined the effectiveness of antibiotics in treating many types of infectious disease. Discuss how the following processes that alter allele frequencies may have contributed to the emergence of antibiotic-resistant strains:

A. Random mutation

B. Genetic drift

C. Natural selection

C28. Let's suppose the mutation rate for converting a B allele into a b allele is 10^{-4}. The current allele frequencies are $B = 0.6$ and $b = 0.4$. How long will it take for the allele frequencies to equal each other, assuming that no genetic drift is taking place?

Experimental Questions

E1. You will need to be familiar with the techniques described in Chapter 20 to answer this question. Gene polymorphisms can be detected using a variety of cellular and molecular techniques. Which techniques would you use to detect gene polymorphisms at the following levels?

A. DNA level

B. RNA level

C. Polypeptide level

E2. You will need to understand solved problem S4 to answer this question. The gene for coat color in rabbits can exist in four alleles termed C (full coat color), c^{ch} (chinchilla), c^h (Himalayan), and c (albino). In a population of rabbits in Hardy-Weinberg equilibrium, the allele frequencies are

$C = 0.34$

$c^{ch} = 0.17$

$c^h = 0.44$

$c = 0.05$

Assume that C is dominant to the other three alleles. c^{ch} is dominant to c^h and c, and c^h is dominant to c.

A. What is the frequency of albino rabbits?

B. Among 1000 rabbits, how many would you expect to have a Himalayan coat color?

C. Among 1000 rabbits, how many would be heterozygotes with a chinchilla coat color?

E3. In a large herd of 5468 sheep, 76 animals have yellow fat, compared with the rest of the members of the herd, which have white fat. Yellow fat is inherited as a recessive trait. This herd is assumed to be in Hardy-Weinberg equilibrium.

A. What are the frequencies of the white and yellow fat alleles in this population?

B. Approximately how many sheep with white fat are heterozygous carriers of the yellow allele?

E4. The human MN blood group is determined by two codominant alleles, M and N. The following data were obtained from various human populations:

		Percentages		
Population	**Place**	***MM***	***MN***	***NN***
Inuit	East Greenland	83.5	15.6	0.9
Navajo Indians	New Mexico	84.5	14.4	1.1
Finns	Karajala	45.7	43.1	11.2
Russians	Moscow	39.9	44.0	16.1
Aborigines	Queensland	2.4	30.4	67.2

Data from E.B. Speiss (1990). *Genes in Populations,* 2d ed. Wiley-Liss, New York.

A. Calculate the allele frequencies in these five populations.

B. Which populations appear to be in Hardy-Weinberg equilibrium?

C. Which populations do you think have had significant intermixing due to migration?

E5. You will need to understand solved problem S4 before answering this question. In an island population, the following data were obtained for the numbers of people with each of the four blood types:

Type O	721
Type A	932
Type B	235
Type AB	112

Is this population in Hardy-Weinberg equilibrium? Explain your answer.

E6. Resistance to the poison warfarin is a genetically determined trait in rats. Homozygotes carrying the resistance allele (WW) have a lower fitness because they suffer from vitamin K deficiency, but heterozygotes (Ww) do not. However, the heterozygotes are still resistant to warfarin. In an area where warfarin is applied, the heterozygote has a survival advantage. Due to warfarin resistance, the heterozygote is also more fit than the normal homozygote (*ww*). If the relative fitness values for *Ww*, *WW*, and *ww* individuals are 1.0, 0.37, and 0.19, respectively, in areas where warfarin is applied, calculate the allele frequencies at equilibrium. How would this equilibrium be affected if the rats were no longer exposed to warfarin?

E7. Describe, in as much experimental detail as possible, how you would test the hypothesis that snail color distribution is due to predation.

E8. In the Grants' study of the medium ground finch, do you think the pattern of natural selection was directional, stabilizing, disruptive, or balancing? Explain your answer. If the environment remained dry indefinitely (for many years), what do you think would be the long-term outcome?

E9. A recessive lethal allele has achieved a frequency of 0.22 due to genetic drift in a very small population. Based on natural selection, how would you expect the allele frequencies to change in the next three generations? Note: Your calculation can assume that genetic drift is not altering allele frequencies in either direction.

E10. Among a large population of 2 million gray mosquitoes, one mosquito is heterozygous for a body color gene; this mosquito has one gray allele and one blue allele. There is no selective advantage or disadvantage between gray and blue body color. All of the other mosquitoes carry the gray allele.

A. What is the probability of fixation of the blue allele?

B. If fixation happens to occur, how many generations is it likely to take?

C. Qualitatively, how would the answers to parts A and B be affected if the blue allele conferred a slight survival advantage?

E11. In a donor population, the allele frequencies for the common (Hb^A) and sickle cell (Hb^S) alleles are 0.9 and 0.1, respectively. A group of 550 individuals migrates to a new population containing 10,000 individuals; in the recipient population, the allele frequencies are $Hb^A = 0.99$ and $Hb^S = 0.01$.

A. Calculate the allele frequencies in the conglomerate population.

B. Assuming the donor and recipient populations are each in Hardy-Weinberg equilibrium, calculate the genotype frequencies in the conglomerate population prior to further mating between the donor and recipient populations.

C. What will be the genotype frequencies of the conglomerate population in the next generation, assuming it achieves Hardy-Weinberg equilibrium in one generation?

E12. Here are traditional DNA fingerprints of five people: a child, mother, and three potential fathers:

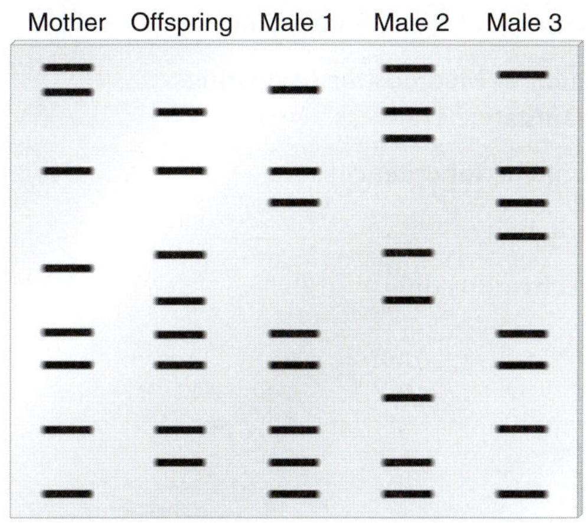

Which males can be ruled out as being the father? Explain your answer. If one of the males could be the father, explain the general strategy for calculating the likelihood that he could match the offspring's DNA fingerprint by chance alone. (See solved problem S6 before answering this question.)

E13. What is DNA fingerprinting? How can it be used in human identification?

E14. When analyzing the automated DNA fingerprints of a father and his biological daughter, a technician examined 50 peaks and found that 30 of them were a perfect match. In other words, 30 out of 50 peaks, or 60%, were a perfect match. Is this percentage too high, or would you expect a value of only 50%? Explain why or why not.

E15. What would you expect to be the minimum percentage of matching peaks in an automated DNA fingerprint for the following pairs of individuals?

A. Mother and son

B. Sister and brother

C. Uncle and niece

D. Grandfather and grandson

Questions for Student Discussion/Collaboration

1. Discuss examples of positive and negative assortative mating in natural populations, human populations, and agriculturally important species.

2. Discuss the role of mutation in the origin of genetic polymorphisms. Suppose that a genetic polymorphism involves two alleles at frequencies of 0.45 and 0.55. Describe three different scenarios to explain these observed allele frequencies. You can propose that the alleles are neutral, beneficial, or deleterious.

3. Most new mutations are detrimental, yet rare beneficial mutations can be adaptive. With regard to the fate of new mutations, discuss whether you think it is more important for natural selection to select against detrimental alleles or to select in favor of beneficial ones. Which do you think is more significant in human populations?

Note: All answers appear at the website for this textbook; the answers to the even-numbered questions and all of the Concept Check and Comprehension Questions are in the back of the book.

www.mhhe.com/brookergenetics5e

Visit the website for practice tests, answer keys, and other learning aids for this chapter. Enhance your understanding of genetics with our interactive exercises, quizzes, animations, and much more.

CHAPTER OUTLINE

Domesticated wheat. *The color of wheat ranges from a dark red to white, which is an example of a complex or quantitative trait.*

27 QUANTITATIVE GENETICS

In this chapter, we will examine **complex traits**—characteristics that are determined by several genes and are significantly influenced by environmental factors. Many complex traits are viewed as **quantitative traits** because they can be described numerically. In humans, quantitative traits include height, the shape of our noses, and the rate at which we metabolize food, to name a few examples. The field of genetics that studies the mode of inheritance of complex or quantitative traits is called **quantitative genetics.** Quantitative genetics is an important branch of genetics for several reasons. In agriculture, most of the key characteristics of interest to plant and animal breeders are quantitative traits. These include traits such as weight, fruit size, resistance to disease, and the ability to withstand harsh environmental conditions. As we will see, genetic techniques have improved our ability to develop strains of agriculturally important species with desirable quantitative traits. In addition, many human diseases are viewed as complex traits that are influenced by several genes.

Quantitative genetics is also important in the study of evolution. Many of the traits that allow a species to adapt to its environment are quantitative. Examples include the swift speed of the cheetah and the sturdiness of tree branches in windy climates. The importance of quantitative traits in the evolution of species will be discussed in Chapter 28. In this chapter, we examine how genes and the environment contribute to the phenotypic expression of quantitative traits. We will begin with an examination of quantitative traits and how to analyze them using statistical techniques. We then look at experimental ways to identify quantitative trait loci (QTLs)—locations on chromosomes containing genes that affect the outcome of quantitative traits. Advances in genetic mapping strategies have enabled researchers to identify these genes. Last, we look at heritability, which is a measure of the amount of phenotypic variation in a population due to genetic variation, and consider various ways of calculating and modifying the genetic variation that affects phenotype.

27.1 OVERVIEW OF QUANTITATIVE TRAITS

Learning Outcomes:

1. List examples of quantitative traits.
2. Explain how quantitative traits may be described with a frequency distribution.

When we compare characteristics among members of the same species, the differences are often quantitative rather than qualitative. Humans, for example, have the same basic anatomical features (two eyes, two ears, and so on), but they differ in quantitative ways. People vary with regard to height, weight, the shape of facial features, skin and hair pigmentation, and many other characteristics. As shown in **Table 27.1**, quantitative traits can be categorized as anatomical, physiological, or behavioral. In addition, many human

TABLE 27.1

Types of Quantitative Traits

Trait	Examples
Anatomical traits	Height, weight, number of bristles in *Drosophila*, ear length in corn, and the degree of pigmentation in flowers and skin
Physiological traits	Metabolic traits, speed of running and flight, ability to withstand harsh temperatures, and milk production in mammals
Behavioral traits	Mating calls, courtship rituals, ability to learn a maze, and the ability to grow or move toward light
Diseases	Predisposition toward heart disease, hypertension, cancer, diabetes, and arthritis

diseases exhibit characteristics and inheritance patterns analogous to those of quantitative traits. Three of the leading causes of death worldwide—heart disease, cancer, and diabetes—are considered quantitative traits.

In many cases, quantitative traits are easily measured and described numerically. Height and weight can be measured in centimeters (or inches) and kilograms (or pounds), respectively. The number of bristles on a fruit fly's body can be counted, and metabolic rate can be assessed as the amount of glucose burned per minute. Behavioral traits can also be quantified. A mating call can be evaluated with regard to its duration, sound level, and pattern. The ability to learn a maze can be described as the time and/or repetitions it takes to learn the skill. Quantitative traits, such as height and weight, are viewed as **continuous traits**—traits that do not fall into discrete categories. Some, such as bristle number in *Drosophila*, are **meristic traits**—traits that can be counted and expressed in whole numbers. Finally, certain diseases, such as diabetes, are viewed as quantitative traits even though the disease itself can also be considered qualitative—either you have it or you don't. Even so, diseases have quantitative characteristics, such as their degree of severity. For example, some people with diabetes require low doses of insulin to prevent adverse symptoms, whereas others require higher doses. For diabetes, heart disease, and cancer, the alleles of several different genes contribute to the likelihood that an individual will develop the disease. A certain threshold must be reached in which the number of disease-causing alleles results in the development of the disease. These are referred to as **threshold traits**—traits that are inherited quantitatively due to the contributions of many genes.

Many Quantitative Traits Exhibit a Continuum of Phenotypic Variation That Follows a Normal Distribution

In Part II of this textbook, we discussed many traits that fall into discrete categories. For example, fruit flies might have white or red eyes, and pea plants might have wrinkled or round seeds. The alleles that govern these traits affect the phenotype in a qualitative way. In analyzing crosses involving these types of traits, each offspring can be put into a particular phenotypic category. Such attributes are called **discontinuous traits.**

In contrast, quantitative traits show a continuum of phenotypic variation within a group of individuals. For such traits, it is often impossible to place organisms into a discrete phenotypic class. For example, **Figure 27.1a** is a classic photograph from 1914 showing the range of heights of 175 students at the Connecticut Agricultural College. Though height is found at minimum and maximum values, the range of heights between these values is fairly continuous.

How do geneticists describe traits that show a continuum of phenotypes? Because most quantitative traits do not naturally fall into a small number of discrete categories, an alternative way to describe them is with a **frequency distribution.** To construct a frequency distribution, the trait is divided arbitrarily into a number of convenient, discrete phenotypic categories. For example, in Figure 27.1, the range of heights is partitioned into 1-inch intervals. Then a graph is made that shows the numbers of individuals found in each of the categories.

Number of students	1	0	0	1	5	7	7	22	25	26	27	17	11	17	4	4	1
Height (inches)	58	59	60	61	62	63	64	65	66	67	68	69	70	71	72	73	74

Albert F. Blakeslee, "CORN AND MEN: The Interacting Influence of Heredity and Environment—Movements for Betterment of Men, or Corn, or Any Other Living Thing, One-sided Unless They Take Both Factors into Account," *Journal of Heredity*, 1914, 5:511–518, by permission of Oxford University Press.

(a)

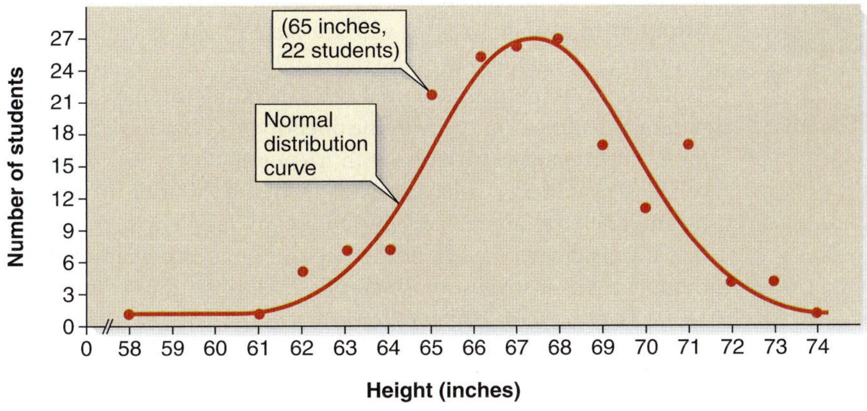

(b)

FIGURE 27.1 Normal distribution of a quantitative trait. (a) The distribution of heights in 175 students at the Connecticut Agricultural College in 1914. (b) A frequency distribution for the heights of students shown in (a).

CONCEPT CHECK: Is height a discontinuous (discrete) trait or does it follow a continuum?

Figure 27.1b shows a frequency distribution for the heights of students pictured in Figure 27.1a. The measurement of height is plotted along the x-axis, and the number of individuals who exhibit that phenotype is plotted on the y-axis. The values along the x-axis are divided into the discrete 1-inch intervals that define the phenotypic categories, even though height is essentially continuous within a group of individuals. For example, in Figure 27.1a, 22 students were between 64.5 and 65.5 inches in height, which is plotted as the point (65 inches, 22 students) on the graph in Figure 27.1b. This type of analysis can be conducted on any group of individuals who vary with regard to a quantitative trait.

The line in the frequency distribution depicts a **normal distribution,** a distribution for a large sample in which the trait of interest varies in a symmetrical way around an average value. The distribution of measurements of many biological characteristics is approximated by a symmetrical bell-shaped curve like that in Figure 27.1b. Normal distributions are common when the phenotype is determined by the cumulative effect of many small independent factors.

27.1 COMPREHENSION QUESTIONS

1. Which of the following is an example of a quantitative trait?
   a. Height
   b. Rate of glucose metabolism
   c. Ability to learn a maze
   d. All of the above
2. Saying that a quantitative trait follows a continuum means that
   a. the numerical value for the trait increases with the age of the individual.
   b. environmental effects are additive.
   c. phenotypes are continuous and do not fall into discrete categories.
   d. the trait continuously changes during the life of an individual.

27.2 STATISTICAL METHODS FOR EVALUATING QUANTITATIVE TRAITS

Learning Outcome:

1. Calculate the mean, variance, standard deviation, and correlation coefficient for quantitative traits and explain their meanings.

In the early 1900s, Francis Galton in England and his student Karl Pearson showed that many traits in humans and domesticated animals are quantitative in nature. To understand the underlying genetic basis of these traits, they founded what became known as the **biometric field** of genetics, which involved the statistical study of biological traits. During this period, Galton and Pearson developed various statistical tools for studying the variation of quantitative traits within groups of individuals. Many of these tools are still in use today. In this section, we will examine how statistical tools are used to analyze the variation of quantitative traits within groups.

Statistical Methods Are Used to Evaluate a Frequency Distribution Quantitatively

Statistical tools can be used to analyze a normal distribution in a number of ways. One measure you are probably familiar with is a parameter called the **mean,** which is the sum of all the values in the group divided by the number of individuals in the group. The mean is computed using the following formula:

$$\overline{X} = \frac{\Sigma X}{N}$$

where

$\overline{X}$	is the mean
ΣX	is the sum of all the values in the group
N	is the number of individuals in the group

A more generalized form of this equation can be used:

$$\overline{X} = \frac{\Sigma f_i X_i}{N}$$

where

$\overline{X}$	is the mean
$\Sigma f_i X_i$	is the sum of all the values in the group; each value in the group is multiplied by its frequency (f_i) in the group
N	is the number of individuals in the group

For example, suppose a bushel of corn had ears with the following lengths (rounded to the nearest centimeter): 15, 14, 13, 14, 15, 16, 16, 17, 15, and 15. Then

$$\overline{X} = \frac{4(15) + 2(14) + 13 + 2(16) + 17}{10}$$

$$\overline{X} = 15 \text{ cm}$$

In genetics, we are often interested in the amount of phenotypic variation in a group. As we will see later in this chapter and in Chapter 28, variation lies at the heart of breeding experiments and evolution. Without variation, selective breeding is impossible, and natural selection cannot favor one phenotype over another. A common way to evaluate variation within a population is with a statistic called the **variance,** which is a measure of the variation around the mean. It helps us appreciate how far a set of numbers is spread out. Variance is determined by adding up the squared deviations from the mean. In relatively small sample sizes, this sum is divided by $N - 1$ to give an estimate of the variance.

$$V_x = \frac{\Sigma f_i (X_i - \overline{X})^2}{N - 1}$$

where

V_X is the variance

$X_i - \overline{X}$ is the difference between each value and the mean

N is the number of observations

For example, if we use the values given previously for the lengths of ears of corn, the variance in this group is calculated as follows:

$$\Sigma f_i (X_i - \overline{X})^2 = 4(15 - 15)^2 + 2(14 - 15)^2 + (13 - 15)^2 + 2(16 - 15)^2 + (17 - 15)^2$$

$$\Sigma f_i (X_i - \overline{X})^2 = 0 + 2 + 4 + 2 + 4$$

$$\Sigma f_i (X_i - \overline{X})^2 = 12 \text{ cm}^2$$

$$V_X = \frac{\Sigma f_i (X_i - \overline{X})^2}{N - 1}$$

$$V_X = \frac{12 \text{ cm}^2}{9}$$

$$V_X = 1.33 \text{ cm}^2$$

Although variance is a measure of the variation around the mean, this statistic may be difficult to understand intuitively because the variance is computed from squared deviations. For example, weight can be measured in grams; the corresponding variance is measured in square grams. Even so, variances are centrally important to the analysis of quantitative traits because they are additive under certain conditions. This means that the variances for different factors that contribute to a quantitative trait, such as genetic and environmental factors, can be added together to predict the total variance for that trait. Later, we will examine how this property is useful in predicting the outcome of genetic crosses.

To gain a more intuitive grasp of variation, we can take the square root of the variance. This statistic is called the **standard deviation (*SD*).** Again, using the same values for length, the standard deviation is

$$SD = \sqrt{V_X} = \sqrt{1.33}$$

$$SD = 1.15 \text{ cm}$$

If the values in a population follow a normal distribution, it is easier to appreciate the amount of variation by considering the standard deviation. Figure 27.2 illustrates the relationship between the standard deviation and the percentages of individuals that deviate from the mean. Approximately 68% of all individuals have values within one standard deviation from the mean, either in the positive or negative direction. About 95% are within two standard deviations, and 99.7% are within three standard deviations. When a quantitative characteristic follows a normal distribution, less than 0.3% of the individuals have values that are more or less than three standard deviations from the mean of the population. In our corn example, three standard deviations equal 3.45 cm. Therefore, we expect that approximately 0.3% of the ears of corn would be less than 11.55 cm or greater than 18.45 cm, assuming that corn ear length follows a normal distribution.

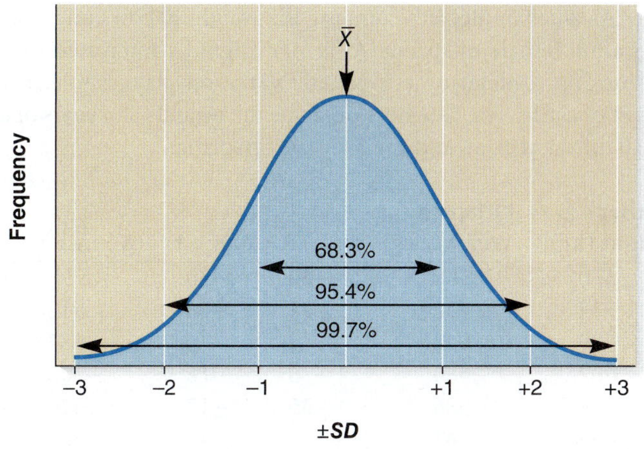

FIGURE 27.2 The relationship between the standard deviation and the proportions of individuals in a normal distribution. For example, approximately 68% of the individuals in a population are between the mean and one standard deviation (1 *SD*) above or below the mean.

CONCEPT CHECK: What percentage of individuals fall above 2 *SDs*?

Some Statistical Methods Compare Two Variables with Each Other

In many biological problems, it is useful to compare two different variables. For example, we may wish to compare the occurrence of two different phenotypic traits. Do obese animals have larger hearts? Are brown eyes more likely to occur in people with dark skin pigmentation? A second type of comparison is between traits and environmental factors. Does insecticide resistance occur more frequently in areas that have been exposed to insecticides? Is heavy body weight more prevalent in colder climates? Finally, a third type of comparison is between traits and genetic relationships. Do tall parents tend to produce tall offspring? Do women with diabetes tend to have brothers with diabetes?

To gain insight into such questions, a statistic known as the correlation coefficient is often applied. To calculate this statistic, we first need to determine the **covariance,** which describes the relationship between two variables within a group. The covariance is similar to the variance, except that we multiply together the deviations of two different variables rather than squaring the deviations from a single factor.

$$CoV_{(X,Y)} = \frac{\Sigma f_i[(X_i - \overline{X})(Y_i - \overline{Y})]}{N - 1}$$

where

$CoV_{(X,Y)}$ is the covariance between X and Y values

X_i represents the values for one variable, and $\overline{X}$ is the mean value in the group

Y_i represents the values for another variable, and $\overline{Y}$ is the mean value in that group

N is the total number of pairs of observations

As an example, let's compare the weight of cows and that of their adult female offspring. A farmer might be interested in this relationship to determine if genetic variation plays a role in the weight of cattle. The data here describe the weights at 5 years of age for 10 different cows and their female offspring.

Mother's Weight (kg)	Offspring's Weight (kg)	$X_i - \overline{X}$	$Y_i - \overline{Y}$	$(X_i - \overline{X})(Y_i - \overline{Y})$
570	568	−26	−30	780
572	560	−24	−38	912
599	642	3	44	132
602	580	6	−18	−108
631	586	35	−12	−420
603	642	7	44	308
599	632	3	34	102
625	580	29	−18	−522
584	605	−12	7	−84
575	585	−21	−13	273
$\overline{X} = 596$	$\overline{Y} = 598$			$\Sigma = 1373$
$SD_X = 21.1$	$SD_Y = 30.5$			

$$CoV_{(X,Y)} = \frac{\Sigma f_i[(X_i - \overline{X})(Y_i - \overline{Y})]}{N - 1}$$

$$CoV_{(X,Y)} = \frac{1373}{10 - 1}$$

$$CoV_{(X,Y)} = 152.6$$

After we calculate the covariance, we can evaluate the strength of the association between the two variables by calculating a **correlation coefficient (*r*).** This value, which ranges between −1 and +1, indicates how two factors vary in relation to each other. The correlation coefficient is calculated as

$$r_{(X,Y)} = \frac{CoV_{(X,Y)}}{SD_X SD_Y}$$

A positive *r* value means that two factors tend to vary in the same way relative to each other; as one factor increases, the other increases with it. A value of zero indicates that the two factors do not vary in a consistent way relative to each other; the values of the two factors are not related. Finally, a negative correlation, in which the correlation coefficient is negative, indicates that the two factors tend to vary in opposite ways to each other; as one factor increases, the other decreases.

Let's use the data of 5-year weights for mother and offspring to calculate a correlation coefficient.

$$r_{(X,Y)} = \frac{152.6}{(21.1)(30.5)}$$

$$r_{(X,Y)} = 0.237$$

The result is a positive correlation between the 5-year weights of mother and offspring. In other words, the positive correlation coefficient suggests that heavy mothers tend to have heavy offspring and lighter mothers have lighter offspring.

How do we evaluate the value of *r*? After a correlation coefficient has been calculated, one must consider whether the *r* value represents a true association between the two variables or if it could simply be due to chance. To accomplish this, we can test the hypothesis that there is no real correlation (i.e., the null hypothesis, $r = 0$). The null hypothesis is that the observed *r* value differs from zero due only to random sampling error. We followed a similar approach in the chi square analysis described in Chapter 2. Like the chi square value, the significance of the correlation coefficient is directly related to sample size and the degrees of freedom (*df*). In testing the significance of correlation coefficients, *df* equals $N - 2$, because two variables are involved. *N* equals the number of paired observations. We will reject the null hypothesis if the correlation coefficient results in a probability that is less than 0.05 (less than 5%) or if the probability is less than 0.01 (less than 1%). These are called the 5% and 1% significance levels, respectively. **Table 27.2** shows the relationship between the *r* values and degrees of freedom at the 5% and 1% significance levels.

TABLE 27.2

Values of *r* at the 5% and 1% Significance Levels

Degrees of Freedom (*df*)*	5%	1%	Degrees of Freedom (*df*)	5%	1%
1	0.997	1.000	24	0.388	0.496
2	0.950	0.990	25	0.381	0.487
3	0.878	0.959	26	0.374	0.478
4	0.811	0.917	27	0.367	0.470
5	0.754	0.874	28	0.361	0.463
6	0.707	0.834	29	0.355	0.456
7	0.666	0.798	30	0.349	0.449
8	0.632	0.765	35	0.325	0.418
9	0.602	0.735	40	0.304	0.393
10	0.576	0.708	45	0.288	0.372
11	0.553	0.684	50	0.273	0.354
12	0.532	0.661	60	0.250	0.325
13	0.514	0.641	70	0.232	0.302
14	0.497	0.623	80	0.217	0.283
15	0.482	0.606	90	0.205	0.267
16	0.468	0.590	100	0.195	0.254
17	0.456	0.575	125	0.174	0.228
18	0.444	0.561	150	0.159	0.208
19	0.433	0.549	200	0.138	0.181
20	0.423	0.537	300	0.113	0.148
21	0.413	0.526	400	0.098	0.128
22	0.404	0.515	500	0.088	0.115
23	0.396	0.505	1000	0.062	0.081

*Note: *df* equals $N - 2$.
From J. T. Spence, B. J. Underwood (1976). *Elementary Statistics*. Prentice-Hall, Englewood Cliffs, New Jersey.

The use of Table 27.2 is valid only if several assumptions are met. First, the values of X and Y in the study must have been obtained by an unbiased sampling of the entire population. In addition, this approach assumes that the values of X and Y follow a normal distribution, like that of Figure 27.1, and that the relationship between X and Y is linear.

To illustrate the use of Table 27.2, let's consider the correlation coefficient we have just calculated for 5-year weights of cows and their female offspring. In this case, we obtained a value of 0.237 for r, and the value of N was 10. Under these conditions, df equals 10 (the number of paired observations) minus 2, which equals 8. To be valid at a 5% significance level, the value of r would have to be 0.632 or higher. Because the value we obtained is much less than this, it is fairly likely that this value could have occurred as a matter of random sampling error. In this case, we cannot reject the null hypothesis, and, therefore, we cannot conclude the positive correlation is due to a true association between the weights of mothers and offspring.

In an actual experiment, however, a researcher would examine many more pairs of cows and offspring, perhaps 500–1000. If a correlation coefficient of 0.237 was observed for $N = 1000$, the value would be significant at the 1% level. In this case, we would reject the null hypothesis that weights are not associated with each other. Instead, we would conclude that a real association occurs between the weights of mothers and their offspring. In fact, these kinds of experiments have been done for cattle weights, and the correlation coefficients between mothers and offspring have often been found to be significant.

If a statistically significant correlation coefficient is obtained, how do we interpret its meaning? An r value that is statistically significant suggests a true association, but it does not necessarily imply a cause-and-effect relationship. When parents and offspring display a significant correlation for a trait, we should not jump to the conclusion that genetics is the underlying cause of the positive association. In many cases, parents and offspring share similar environments, so the positive association might be rooted in environmental factors. In general, correlation coefficients are quite useful in identifying positive or negative associations between two variables. We should use caution, however, because this statistic, by itself, cannot prove that the association is due to cause and effect.

A **regression analysis** may be used when researchers suspect, or when their experimentation has shown, that two variables are related due to cause and effect—that one variable (the independent variable) affects the outcome of another (the dependent variable). Researchers use a regression analysis to predict how much the dependent variable changes in response to the independent variable. This approach is described in solved problem S5 at the end of the chapter.

27.2 COMPREHENSION QUESTIONS

1. The variance is
   a. a measure of the variation around the mean.
   b. computed as a squared deviation.
   c. higher when there is less phenotypic variation.
   d. both a and b are correct.
2. Which of the following statistics is used to compare two variables?
   a. Mean
   b. Correlation coefficient
   c. Variance
   d. Standard deviation

27.3 POLYGENIC INHERITANCE

Learning Outcomes:

1. Define quantitative trait locus.
2. Describe how polygenic inheritance may result in a continuum of phenotypes.
3. Explain how quantitative trait loci are mapped along chromosomes using molecular markers.

Thus far, we have seen that quantitative traits tend to show a continuum of variation and can be analyzed with various statistical tools. At the beginning of the 1900s, a great debate focused on the inheritance of quantitative traits. The biometric school, founded by Francis Galton and Karl Pearson, argued that these types of traits are not controlled by discrete genes that affect phenotypes in a predictable way. To some extent, the biometric school favored the idea of blending inheritance, which proposed that the traits of each parent would mix together to form the traits of the offspring (see Chapter 2).

Alternatively, the followers of Mendel, led by William Bateson in England and William Castle in the United States, held firmly to the idea that traits are governed by genes, which are inherited as discrete units. As we now know, Bateson and Castle were correct in their assertion that genes are inherited as discrete units. However, as we will see in this section, studying quantitative traits is difficult because these traits are controlled by multiple genes and substantially influenced by environmental factors.

Most quantitative traits are polygenic and exhibit a continuum of phenotypic variation. The term **polygenic inheritance** refers to the transmission of a trait governed by two or more different genes. The location on a chromosome that harbors one or more genes that affect the outcome of a quantitative trait is called a **quantitative trait locus (QTL).** As discussed later, QTLs are chromosomal regions identified by genetic mapping. Because such mapping usually locates the QTL to a relatively large chromosomal region, a QTL may contain a single gene or two or more closely linked genes that affect a quantitative trait.

Just a few years ago, it was extremely difficult for geneticists to determine the inheritance patterns for genes underlying quantitative traits, particularly those determined by three or more genes having multiple alleles for each gene. Recently, however, molecular genetic tools (described in Chapters 22 and 24) have greatly enhanced our ability to find regions in the genome where QTLs are likely to reside. This has been a particularly exciting advance in the field of quantitative genetics. In some cases, the identification of QTLs may allow for the improvement of quantitative traits in agriculturally important species.

Polygenic Inheritance and Environmental Factors May Produce a Continuum of Phenotypes

The first experiment demonstrating that continuous variation is related to polygenic inheritance was conducted by the Swedish geneticist Herman Nilsson-Ehle in 1909. He studied the inheritance of red pigment in the hull of bread wheat, *Triticum aestivum* (**Figure 27.3a**). When true-breeding plants with white hulls were crossed to true-breeding plants with red hulls, the F_1 generation had an intermediate color. When the F_1 generation was allowed to self-fertilize, great variation in redness was observed in the F_2 generation, including white, light red, intermediate red, medium red, and dark red. An unsuspecting observer might conclude that this F_2 generation displayed a continuous variation in hull color. However, as shown in **Figure 27.3b**, Nilsson-Ehle carefully categorized the colors of the hulls and discovered that they followed a 1:4:6:4:1 ratio. He concluded that this species has two different genes that control hull color, each gene existing in a red or white allelic form. He hypothesized that these two loci must contribute additively to the color of the hull. In this case, each red allele carried by a given plant contributed to the red color of the hull. Plants carrying more red alleles had a deeper red color.

As we have just seen, Nilsson-Ehle categorized wheat hull colors into several discrete genotypic categories. However, for many quantitative traits, this is difficult or impossible. In general, as the number of genes controlling a trait increases and the influence of the environment increases, the categorization of phenotypes into discrete genotypic classes becomes increasingly difficult, if not impossible. Therefore, a Punnett square cannot be used to analyze most quantitative traits. Instead, statistical methods, which are described later, must be employed.

(a) Red and white hulls of wheat

r1r1r2r2 (White) x *R1R1R2R2* (Dark red)

↓

R1r1R2r2 (Intermediate red)

↓ Self-fertilization

♀ \ ♂	*R1R2*	*R1r2*	*r1R2*	*r1r2*
R1R2	*R1R1R2R2* Dark red	*R1R1R2r2* Medium red	*R1r1R2R2* Medium red	*R1r1R2r2* Intermediate red
R1r2	*R1R1R2r2* Medium red	*R1R1r2r2* Intermediate red	*R1r1R2r2* Intermediate red	*R1r1r2r2* Light red
r1R2	*R1r1R2R2* Medium red	*R1r1R2r2* Intermediate red	*r1r1R2R2* Intermediate red	*r1r1R2r2* Light red
r1r2	*R1r1R2r2* Intermediate red	*R1r1r2r2* Light red	*r1r1R2r2* Light red	*r1r1r2r2* White

(b) Inheritance of hull color in wheat

FIGURE 27.3 The Nilsson-Ehle experiment studying how continuous variation is related to polygenic inheritance in wheat. **(a)** Red (top) and white (bottom) varieties of wheat, *Triticum aestivum*. **(b)** Nilsson-Ehle carefully categorized the colors of the hulls in the F_2 generation and discovered that they followed a 1:4:6:4:1 ratio. This occurs because the contributions of the red alleles are additive.

Genes→Traits In this example, two genes, with two alleles each (red and white), govern hull color. Offspring can display a range of colors, depending on how many copies of the red allele they inherit. If an offspring is homozygous for the red allele of both genes, it will have dark red hulls. By comparison, if it carries three red alleles and one white allele, it will be medium red (which is not quite as deep in color). In this way, this quantitative trait can exhibit a range of phenotypes from dark red to white.

CONCEPT CHECK: What does it mean to say that these alleles are additive?

Figure 27.4 illustrates how genotypes and phenotypes may overlap for quantitative traits. Figure 27.4a considers a situation in which seed weight is controlled by one gene with light (*w*) and heavy (*W*) alleles. A heterozygous plant (*Ww*) is allowed to self-fertilize. Let's consider how environmental factors, such as sunlight, soil richness, and rainfall, may affect phenotypes. First, the environment may have great variation—some plants may be exposed to much more sunlight than others or they may be planted in better soil or receive more rain. Second, such environmental variation may or may not have a great effect on seed weight. For example, in one species of plant, receiving low or high amounts of rain may have a great effect on seed weight, whereas in another species, such variation in rainfall may have a minimal effect. When seed weight is only slightly influenced by the environment, as seen on the left, the light, intermediate, and heavy seeds fall into separate, well-defined categories. When the environment has a greater effect on seed weight, as shown on the right, more phenotypic variation is found in seed weight within each genotypic class. The variation in the frequency distribution on the right is much higher. Even so, most individuals can still be classified into the three main categories.

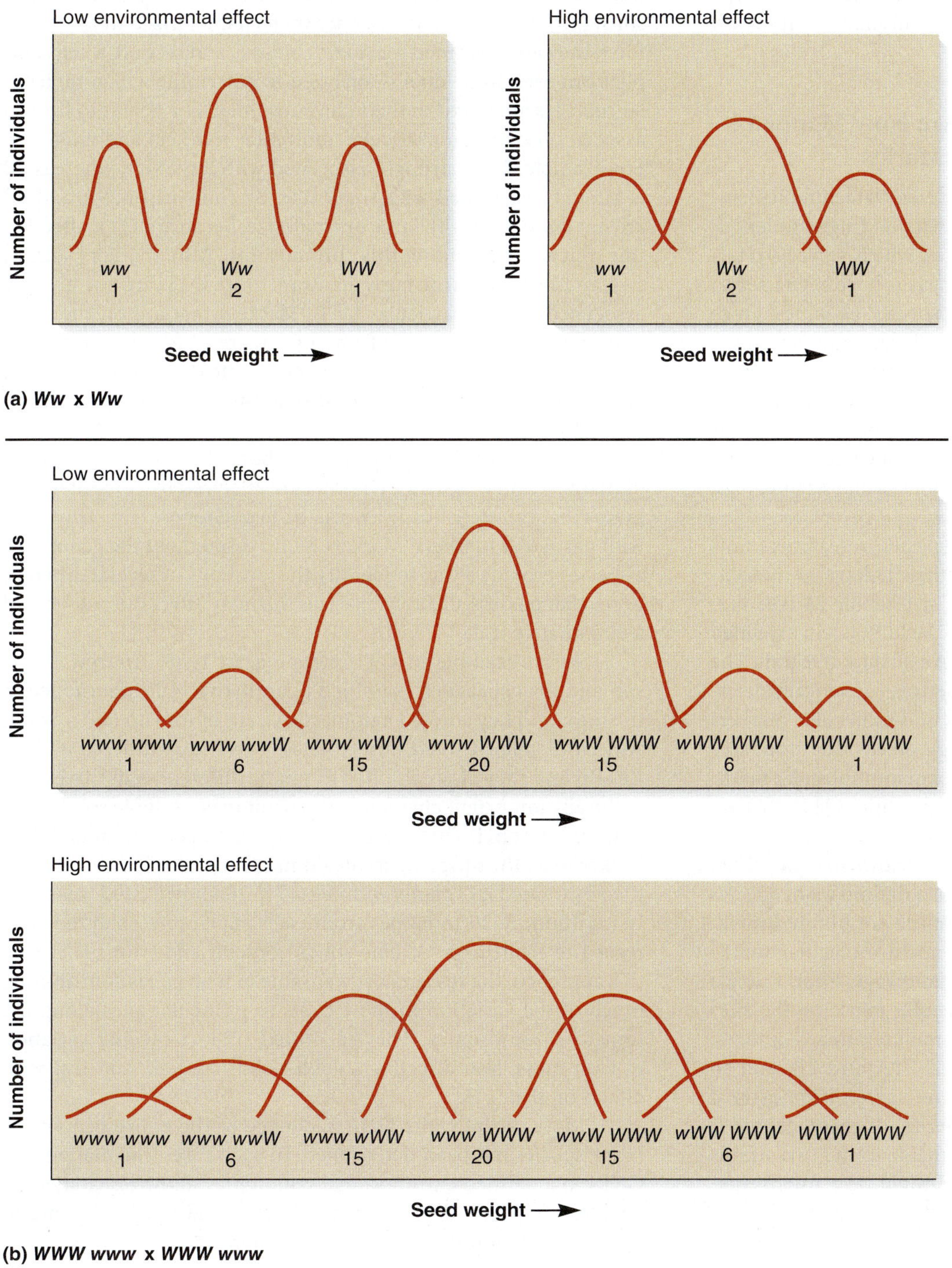

FIGURE 27.4 How genotypes and phenotypes may overlap for quantitative traits. **(a)** Situations in which seed weight is controlled by one gene, existing in light (*w*) and heavy (*W*) alleles. **(b)** Situations in which seed weight is governed by three genes instead of one, each existing in light and heavy alleles. Note: The 1:2:1 and 1:6:15:20:15:6:1 ratios were derived by using a Punnett square and assuming a cross between individuals that are both heterozygous for one or three different genes, respectively.

Genes→Traits The ability of geneticists to correlate genotype and phenotype depends on how many genes are involved and how much the environment causes the phenotype to vary. In **(a)**, a single gene influences seed weight. In the graph on the left side, the environment does not cause much variation in seed weight. This makes it easy to distinguish the three genotypes. There is no overlap in the seed weights of *ww*, *Ww*, and *WW* individuals. In the graph on the right side, the environment causes more variation in seed weight. In this case, a few individuals with *ww* genotypes produce seeds having the same weight as a few individuals with *Ww* genotypes; and a few *Ww* genotypes produce seeds having the same weight as *WW* genotypes. As shown in **(b)**, it becomes even more difficult to distinguish genotype based on phenotype when three genes are involved. The overlaps are minor when the environment does not cause much seed weight variation. However, when the environment causes substantial phenotypic variation, the overlaps between genotypes and phenotypes are very pronounced and greatly confound genetic analysis.

CONCEPT CHECK: Explain how gene number and environmental variation affect the overlaps between phenotypes and different genotypes.

By comparison, Figure 27.4b illustrates a situation in which seed weight is governed by three genes instead of one, each existing in light and heavy alleles. When the variation in environmental factors is low and/or plays a minor role in the outcome of this trait, a cross between two heterozygotes is expected to produce a 1:6:15:20:15:6:1 ratio. As shown in the upper illustration in Figure 27.4b, nearly all individuals fall within a phenotypic category that corresponds to their genotype. When the environment has a greater effect on phenotype, as shown in the lower illustration, the situation becomes more ambiguous. For example, individuals with one *w* allele and five *W* alleles have a phenotype that overlaps with that of individuals having six *W* alleles or two *w* alleles and four *W* alleles. Therefore, it becomes difficult to categorize each phenotype into a unique genotypic class. Instead, the trait displays a continuum ranging from light to heavy seed weight.

Quantitative Trait Loci (QTLs) Are Now Mapped by Linkage to Known Molecular Markers

As discussed in Chapters 22 and 24, newer research techniques have identified molecular markers, such as restriction fragment length polymorphism (RFLPs) and microsatellites, which serve as reference points along chromosomes. These markers have been used to construct detailed genetic maps of many species' genomes. Once a genome map is obtained, it becomes much easier to determine the locations of genes that affect a quantitative trait. In addition to model organisms such as *Drosophila*, *Arabidopsis*, *Caenorhabditis elegans*, and mice, detailed molecular maps have been obtained for many agriculturally important species. These include crops such as corn, rice, and tomatoes, as well as livestock such as cattle, pigs, sheep, and chickens.

To map the genes in a eukaryotic species, researchers now determine their locations by identifying their linkage to molecular markers. This approach is described in Chapter 24 (see Figures 24.5–24.7). In 1989, Eric Lander and David Botstein extended this technique to identify QTLs that govern a quantitative trait. The basis of **QTL mapping** is the association between genetically determined phenotypes for quantitative traits and molecular markers such as RFLPs, microsatellites, and single-nucleotide polymorphisms (SNPs). In this approach, the sites of many molecular markers are already known, and the researcher identifies QTLs that are close to particular markers.

The general strategy for QTL mapping is shown in **Figure 27.5**. This figure depicts two different strains of a diploid plant species with four chromosomes per set. The strains are highly inbred, which means they are homozygous for most molecular markers and genes. They differ in two important ways. First, the two strains differ with regard to many molecular markers; the sites of many markers along each chromosome are already known. These markers are designated 1A and 1B, 2A and 2B, and so forth. The markers 1A and 1B mark the same chromosomal location in this species, namely, the upper tip of chromosome 1. However, the two markers are distinguishable in the two strains at the molecular level. For example, 1A might be a microsatellite that is 148 bp, whereas 1B might be 212 bp. Second, the two strains differ in a quantitative trait of interest. In this example, the strain on the left produces large fruit, whereas the strain on the right produces small fruit. The unknown genes affecting this trait are designated with a black or blue QTL label. A black QTL indicates a site that harbors one or more alleles that promote large fruit. A blue QTL is at the same site but carries alleles that promote small fruit. Prior to conducting their crosses, researchers would not know the chromosomal locations of the QTLs shown in this figure. The purpose of the experiment is to determine their locations.

With these ideas in mind, the protocol shown in Figure 27.5 begins by crossing the two inbred strains to each other and then backcrossing the F_1 offspring to both parental strains. This produces a second generation with a great degree of variation. The offspring from these backcrosses are then characterized in two ways. First, they are examined for their fruit size, and second, a cell sample from each individual is analyzed to determine which molecular markers are found in their chromosomes. The goal is to find an association between particular molecular markers and fruit size. For example, 2A is strongly associated with large size, whereas 2B is strongly associated with small size. By comparison, 9A and 9B are not associated with large or small size, because a QTL affecting this trait is not found on this chromosome. Also, markers such as 14A and 14B, which are fairly far away from a QTL, are not strongly associated with a particular QTL. Markers that are on the same chromosome but far away from a QTL are often separated from a QTL during meiosis in the F_1 heterozygote due to crossing over. Only closely linked markers are strongly associated with a particular QTL.

Overall, QTL mapping involves the analysis of a large number of markers and offspring. The data are analyzed by computer programs that can statistically associate the phenotype (e.g., fruit size) with particular markers. Markers found throughout the genome of a species provide a way to identify the locations of several different genes that possess allelic differences that may affect the outcome of a quantitative trait.

As an example of QTL mapping, in 1988, Andrew Paterson and his colleagues examined quantitative trait inheritance in the tomato. They studied a domestic strain of tomato and a South American green-fruited variety. These two strains differed in their RFLPs, and they also exhibited dramatic differences in three agriculturally important characteristics: fruit mass, soluble solids content, and fruit pH. The researchers crossed the two strains and then backcrossed the offspring to the domestic tomato. The researchers then examined 237 plants with regard to 70 known RFLP markers. In addition, 5–20 tomatoes from each plant were analyzed with regard to fruit mass, soluble solids content, and fruit pH. Using this approach, the researchers were able to map genes contributing much of the variation in these traits to particular sites along the tomato chromosomes. They identified six loci causing variation in fruit mass, four affecting soluble solids content, and five with effects on fruit pH.

More recently, the DNA sequence of the entire genome of many species has been determined. In such cases, the mapping of QTLs to a defined chromosomal region may allow researchers to analyze the DNA sequence in that region and to identify one or more genes that influence the trait of interest.

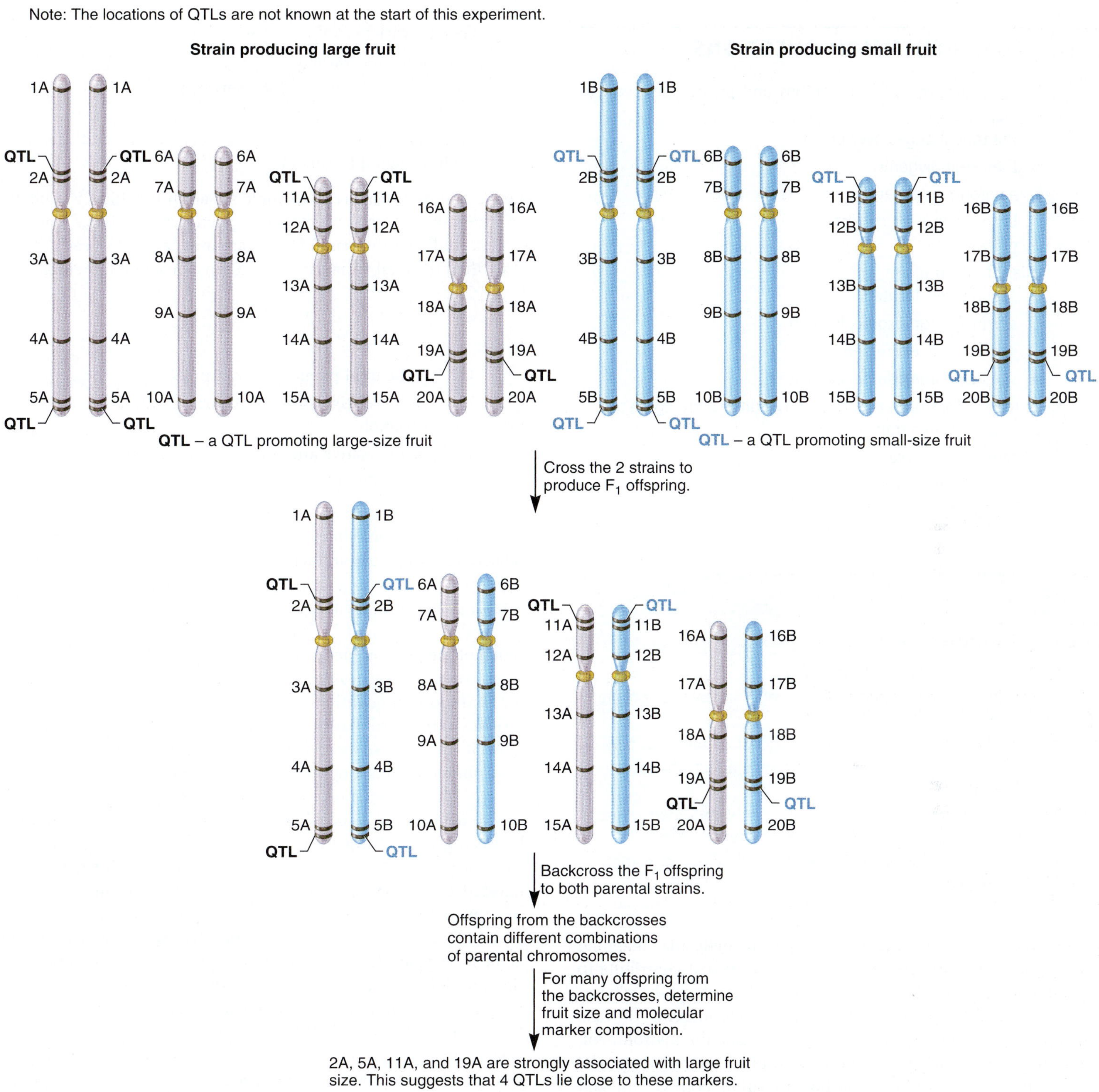

FIGURE 27.5 The general strategy for QTL mapping via molecular markers. Two different inbred strains have four chromosomes per set. The strain on the left produces large fruit, and the strain on the right produces small fruit. The goal of this mapping strategy is to locate the unknown genes affecting this trait, which are designated with a QTL label. A black QTL indicates a site carrying one or more alleles that promote large fruit, and a blue QTL carries alleles that promote small fruit. The two strains differ with regard to many molecular markers designated 1A and 1B, 2A and 2B, and so forth. The two strains are crossed, and then the F_1 offspring are backcrossed to both parental strains. Many offspring from the backcrosses are then examined for their fruit size and to determine which molecular markers are found in their chromosomes. The data are analyzed by computer programs that can statistically associate the phenotype (e.g., fruit size) with particular markers. Markers found throughout the genome of this species provide a way to locate several different genes that may affect the outcome of a single quantitative trait. In this case, the analysis predicts four QTLs promoting heavier fruit weight that are linked to regions of the chromosomes with the following markers: 2A, 5A, 11A, and 19A.

CONCEPT CHECK: What are the two ways that strains A and B differ?

27.3 COMPREHENSION QUESTIONS

1. For quantitative traits, genotypes and phenotypes tend to overlap because
   a. the trait changes over time.
   b. they are polygenic.
   c. environmental variation affects the trait.
   d. both b and c are correct.
2. A QTL is a __________ where one or more genes affecting a quantitative trait are _________.
   a. site in a cell, located
   b. site in a chromosome, located
   c. site in a cell, expressed
   d. site in a chromosome, expressed
3. To map QTLs, strains are crossed that differ with regard to
   a. a quantitative trait.
   b. molecular markers.
   c. a quantitative trait and molecular markers.
   d. a quantitative trait and a discontinuous trait.

27.4 HERITABILITY

Learning Outcomes:

1. Explain the relationship among phenotypic variance, genetic variance, and environmental variance using an equation.
2. Describe how interactions and associations between genotype and environmental factors may affect phenotypic variance.
3. Define heritability, and distinguish between broad-sense heritability and narrow-sense heritability.
4. Calculate narrow-sense heritability using correlation coefficients.

As we have just seen, recent approaches in molecular mapping have enabled researchers to identify the genes that contribute to a quantitative trait. The other key factor that affects the phenotypic outcome of quantitative traits is the environment. All traits of organisms are influenced by both genetics and the environment, and this is particularly pertinent in the study of quantitative traits. Researchers want to understand how variation, both genetic and environmental, affects the phenotypic results. In this section, we examine how geneticists analyze the genetic and environmental components that affect quantitative traits.

Both Genetic Variance and Environmental Variance May Contribute to Phenotypic Variance

Earlier, we examined the amount of phenotypic variation within a group by calculating the variance. Geneticists partition quantitative trait variation into components that are attributable to the following different causes:

Genetic variance (V_G)
Environmental variance (V_E)
Variance due to interactions between genetic and environmental factors ($V_{G \times E}$)
Variance due to associations between genetic and environmental factors ($V_{G \leftarrow\rightarrow E}$)

Let's begin by considering a simple situation in which V_G and V_E are the only factors that determine phenotypic variance, and the genetic and environmental factors are independent of each other. If so, then the total variance for a trait in a group of individuals is

$$V_P = V_G + V_E$$

where

V_P is the total phenotypic variance
V_G is the relative amount of variance due to genetic variation
V_E is the relative amount of variance due to environmental variation

Why is this equation useful? The partitioning of variance into genetic and environmental components allows us to estimate their relative importance in influencing the phenotypic variance within a group. If V_G is very high and V_E is very low, genetics plays a greater role in promoting phenotypic variation within a group. Alternatively, if V_G is low and V_E is high, environmental factors underlie much of the phenotypic variation. As described later in this chapter, a livestock breeder might want to apply selective breeding if V_G for an important quantitative trait is high. In this way, the characteristics of the herd may be improved. Alternatively, if V_G is negligible, it would make more sense to investigate and manipulate the environmental causes of phenotypic variation.

With experimental and domesticated species, one possible way to determine V_G and V_E is by comparing the variation in traits between genetically identical and genetically disparate groups. For example, researchers have followed the practice of **inbreeding** to develop genetically homogeneous strains of mice. Inbreeding in mice involves many generations of brother-sister matings, which eventually produces strains that are **monomorphic** for all or nearly all of their genes. Recall that monomorphic members of a population are all homozygous for the same allele of a given gene. Within such an inbred strain of mice, V_G equals zero. Therefore, all phenotypic variance is due to V_E. When studying quantitative traits such as weight, an experimenter might want to know the genetic and environmental variance for a different, genetically heterogeneous group of mice. To do so, the genetically homogeneous and heterogeneous mice could be raised under the same environmental conditions and their weights measured. The phenotypic variance for weight could then be calculated as described earlier. Let's suppose we obtained the following results:

$V_P = 15\ g^2$ for the group of genetically homogeneous mice
$V_P = 22\ g^2$ for the group of genetically heterogeneous mice

In the case of the homogeneous mice, $V_P = V_E$, because V_G equals 0. Therefore, V_E equals 15 g^2. To estimate V_G for the heterogeneous group of mice, we could assume that V_E is the same for them as it is for the homogeneous mice, because the two groups were raised in identical environments. This assumption allows us to calculate the genetic variance for the heterogeneous mice.

$$V_P = V_G + V_E$$
$$22\ g^2 = V_G + 15\ g^2$$
$$V_G = 7\ g^2$$

This result tells us that some of the phenotypic variance in the genetically heterogeneous group is due to the environment (namely, 15 g^2), and some (7 g^2) is due to genetic variation in alleles that affect weight.

Phenotypic Variance May Also Be Influenced by Interactions and Associations Between Genotype and the Environment

Thus far, we have considered the simple situation in which genetic variance and environmental variance are independent of each other and affect the phenotypic variance in an additive way. As another example, let's suppose that three genotypes, *HH*, *Hh*, and *hh*, affect height, producing tall, intermediate, and short plants, respectively. Greater sunlight makes the plants grow taller regardless of their genotypes. In this case, our assumption that $V_P = V_G + V_E$ would be reasonably valid.

However, let's consider a different environmental factor such as minerals in the soil. As a hypothetical example, let's suppose the *H* allele encodes a fully functional protein involved with mineral uptake from the soil, whereas the *h* allele carries a mutation that causes a decrease in the affinity of this protein for certain minerals. In this case, the *Hh* and *hh* plants are shorter because they do not take up enough minerals to support maximal growth, whereas the *HH* plants are not limited by mineral uptake. According to this hypothetical scenario, adding minerals to the soil enables the defective protein to transport more minerals into the roots, thereby enhancing the growth rate of *hh* plants by a large amount and the *Hh* plants by a smaller amount, because the minerals are more easily taken up by the plants (**Figure 27.6**). The height of *HH* plants is not affected by mineral supplementation. When the environmental effects on phenotype differ according to genotype, this phenomenon is called a **genotype-environment interaction.** Variation due to interactions between genetic and environmental factors is termed $V_{G \times E}$, as noted earlier.

Interactions between genetic and environmental factors are common. As an example, **Table 27.3** shows results from a study conducted in 2000 by Cristina Vieira, Trudy Mackay, and colleagues in which they investigated genotype-environment interaction for QTLs affecting life span in *Drosophila melanogaster*. The data seen in the table compare the life span in days of male and female flies from two different strains of *D. melanogaster* raised at different temperatures. Because males and females differ in their sex chromosomes and gene expression patterns, they can be viewed as having different genotypes. The effects of environmental changes depended greatly on the strain and the sex of the flies. Under standard culture conditions, the females of strain *A* had the longest life span, whereas females of strain *B* had the shortest. In strain *A*, high temperature increased the longevity of males and decreased the longevity of females. In contrast, under hotter conditions, the longevity of males of strain *B* was dramatically reduced, whereas females of this same strain were not significantly affected. Lower growth temperature also had different effects in these two strains. Although low temperature increased the longevity of both strains, the effects were most dramatic in the males of strain *A* and the females of strain *B*. Taken together, these results illustrate the potential complexity of genotype-environmental interaction when measuring a quantitative trait such as life span.

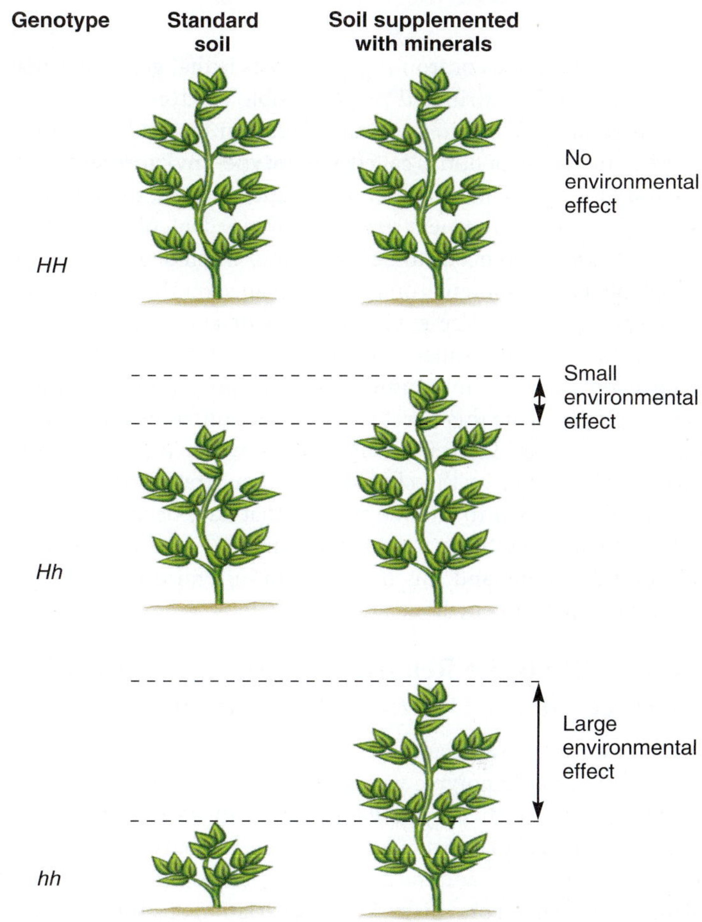

FIGURE 27.6 A schematic example of genotype-environment interaction. When grown in standard soil, the three genotypes *HH*, *Hh*, and *hh* show tall, intermediate, and short heights, respectively. When the soil is supplemented with minerals, a large effect is seen on the *hh* genotype and a smaller effect on the *Hh* genotype. The *HH* genotype is unaffected by the environmental change.

TABLE 27.3
Longevity of Two Strains of *Drosophila melanogaster**

	Strain *A*		Strain *B*	
Temperature	**Male**	**Female**	**Male**	**Female**
Standard	33.6	39.5	37.5	28.9
High	36.3	33.9	23.2	28.6
Low	77.5	48.3	45.8	77.0

*Longevity was measured in the mean number of days of survival. Strains *A* and *B* were inbred strains of *D. melanogaster* called Oregon and 2b, respectively. The standard-, high-, and low-temperature conditions were 25°C, 29°C, and 14°C, respectively.

Another issue confronting geneticists is that genotypes may not be randomly distributed in all possible environments. When certain genotypes are preferentially found in particular environments, this phenomenon is called a **genotype-environment association** ($V_{G\leftarrow\rightarrow E}$). When such an association occurs, the effects of genotype and environment are not independent of each other, and the association needs to be considered when determining the effects of genetic and environmental variance on the total phenotypic variance. Genotype-environment associations are very common in the study of human genetics, in which large families tend to have more similar environments than the population as a whole. One way to evaluate this effect is to compare different genetic relationships, such as identical versus fraternal twins. We will examine this approach later in the chapter. Another strategy that geneticists might follow is to analyze siblings that have been adopted by different parents at birth. Their environmental conditions tend to be more disparate, and this may help to minimize the effects of genotype-environment association.

Heritability Is the Relative Amount of Phenotypic Variance That Is Due to Genetic Variance

Another way to view variance is to focus our attention on the genetic contribution to phenotypic variance. The term **heritability** refers to the amount of phenotypic variance within a specific group of individuals raised in a particular environment that is due to genetic variance. Both genes and the environment are essential to produce all of the traits of an organism. Even so, variation of a trait in a population may be due entirely to environmental variation, entirely to genetic variation, or more commonly to a combination of the two.

If all of the phenotypic variance in a group is due to genetic variance, the heritability would have a value of 1. If all of the phenotypic variation is due to environmental effects, the heritability would equal 0. For most groups of organisms, the heritability for a given trait lies between these two extremes. For example, both genes and diet affect the size an individual will attain. Some individuals inherit alleles that tend to make them larger, and a proper diet also promotes larger size. Other individuals inherit alleles that make them small, and an inadequate diet may contribute to small size. Taken together, both genetics and the environment affect the amount of phenotypic variation in such a population.

If we assume that environment and genetics are independent and the only two factors affecting phenotype, then

$$h_B^2 = V_G/V_P$$

h_B^2 is the heritability in the broad sense

V_G is the variance due to genetics

V_P is the total phenotypic variance, which equals $V_G + V_E$

The heritability defined here, h_B^2, called **broad-sense heritability,** takes into account different types of genetic variation that may affect the phenotype. As we have seen throughout this textbook, genes can affect phenotypes in various ways. As described earlier, the Nilsson-Ehle experiment showed that the alleles determining hull color in wheat affect the phenotype in an additive way. A heterozygote shows a phenotype that is intermediate between the respective homozygotes. Alternatively, alleles affecting other traits may show a dominant/recessive relationship. In this case, the alleles are not strictly additive, because the heterozygote has a phenotype closer to, or perhaps the same as, the homozygote containing two copies of the dominant allele. For example, Mendel discovered that both *PP* and *Pp* pea plants have purple flowers. In addition, another complicating factor is epistasis (described in Chapter 4), in which the alleles for one gene can mask the phenotypic expression of the alleles of another gene. To account for these differences, geneticists usually subdivide V_G into these three different genetic categories:

$$V_G = V_A + V_D + V_I$$

where

V_A is the variance due to the additive effects of alleles

V_D is the variance due to the effects of alleles that follow a dominant/recessive pattern of inheritance

V_I is the variance due to the effects of alleles that interact in an epistatic manner

In analyzing quantitative traits, geneticists may focus on V_A and ignore the contributions of V_D and V_I. They do this for scientific as well as practical reasons. For some quantitative traits, the additive effects of alleles may play a primary role in the phenotypic outcome. In addition, when the alleles behave additively, we can predict the outcomes of crosses based on the quantitative characteristics of the parents. The heritability of a trait due to the additive effects of alleles is called **narrow-sense heritability:**

$$h_N^2 = V_A/V_P$$

For many quantitative traits, the value of V_A may be relatively large compared with V_D and V_I. In such cases, the determination of the narrow-sense heritability provides an estimate of the broad-sense heritability.

How can the narrow-sense heritability be determined? In this chapter, we will consider two common ways. As discussed later, one way to calculate the narrow-sense heritability involves selective breeding practices, which are done with agricultural species. A second common strategy to determine h_N^2 involves the measurement of a quantitative trait among groups of genetically related individuals. For example, agriculturally important traits, such as egg weight in poultry, can be analyzed in this way. To calculate the heritability, a researcher determines the observed egg weights between individuals whose genetic relationships are known, such as a mother and her female offspring. These data can then be used to compute a correlation coefficient between the parent and offspring, using the methods described earlier. The narrow-sense heritability is then calculated as

$$h_N^2 = r_{obs}/r_{exp}$$

where

r_{obs} is the observed phenotypic correlation coefficient between related individuals

r_{exp} is the expected correlation coefficient based on the known genetic relationship

In our example, r_{obs} is the observed phenotypic correlation coefficient between parent and offspring. In particular research studies, the observed phenotypic correlation coefficient for egg weights between mothers and daughters has been found to be about

0.3 (although this varies among strains). The expected correlation coefficient, r_{exp}, is based on the known genetic relationship. A parent and child share 50% of their genetic material, so r_{exp} equals 0.50. So,

$$\begin{aligned} h_N^2 &= r_{obs}/r_{exp} \\ &= 0.3/0.50 \\ &= 0.60 \end{aligned}$$

(Note: For siblings, $r_{exp} = 0.50$; for identical twins, $r_{exp} = 1.0$; and for an aunt-niece relationship, $r_{exp} = 0.25$.)

According to this calculation, about 60% of the phenotypic variance in egg weight is due to additive genetic variance; the other 40% is due to the environment. This calculation assumes that V_D and V_I are negligible.

When calculating heritabilities from correlation coefficients, keep in mind that this computation also assumes that genetics and the environment are independent variables. However, this is not always the case. The environments of parents and offspring are often more similar to each other than they are to those of unrelated individuals. As mentioned earlier, there are several ways to minimize this confounding factor. First, in human studies, one may analyze the heritabilities from correlation coefficients between adopted children and their biological parents. Alternatively, one can examine a variety of relationships (aunt-niece, identical twins versus fraternal twins, and so on) and see if the heritability values are roughly the same in all cases. This approach was applied in the study that is described next.

EXPERIMENT 27A

The Heritability of Dermal Ridge Count in Human Fingerprints Is Very High

Fingerprints are inherited as a quantitative trait. It has long been known that identical twins have fingerprints that are very similar, whereas fraternal twins show considerably less similarity. Galton was the first researcher to study fingerprint patterns, but this trait became more amenable to genetic studies in the 1920s, when Kristine Bonnevie, a Norwegian geneticist, developed a method for counting the number of ridges within a human fingerprint.

As shown in **Figure 27.7**, human fingerprints can be categorized as having an arch, loop, or whorl, or a combination of these patterns. The primary difference among these patterns is the number of triple junctions, each known as a triradius (Figure 27.7b and c). At a triradius, a ridge emanates in three different directions. An arch has zero triradii, a loop has one, and a whorl has two. In Bonnevie's method of counting, a line is drawn from a triradius to the center of the fingerprint. The ridges that touch this line are then counted. (Note: The triradius ridge itself is not counted, and the last ridge is not counted if it forms the center of the fingerprint.) With this method, one can obtain a ridge count for all 10 fingers. Bonnevie conducted a study on a small population and found that ridge count correlation coefficients were relatively high in genetically related individuals.

Sarah Holt, who was also interested in the inheritance of this quantitative trait, carried out a more extensive analysis of ridge counts by examining the fingerprint patterns of a large group of people and their close relatives. In the experiment of **Figure 27.8**, the ridge counts for pairs of related individuals were determined by the method described in Figure 27.7. The correlation coefficients for ridge counts were then calculated among the pairs of related or unrelated individuals. To estimate the narrow-sense heritability, the observed correlation coefficients were then divided by the expected correlation coefficients based on the known genetic relationships.

THE HYPOTHESIS

Dermal ridge count has a genetic component. The goal of this experiment was to determine the contribution of genetics in the variation of dermal ridge counts.

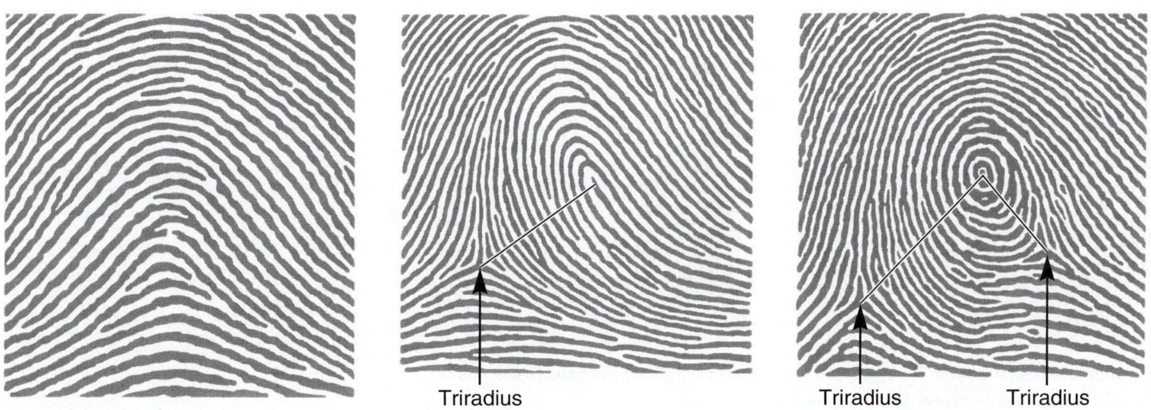

(a) Arch (no triradius) **(b) Loop (1 triradius)** **(c) Whorl (2 triradii)**

FIGURE 27.7
Human fingerprints and the ridge count method of Bonnevie. **(a)** This print has an arch rather than a triradius. The ridge count is zero. **(b)** This print has one triradius. A straight line is drawn from the triradius to the center of the print. The number of ridges dissecting this straight line is 13. **(c)** This print has two triradii. Straight lines are drawn from both triradii to the center. There are 16 ridges touching the left line and 7 touching the right line, giving a total ridge count of 23.

TESTING THE HYPOTHESIS **FIGURE 27.8 Heritability of human fingerprint patterns.**

Starting material: A group of human subjects from Great Britain.

Experimental level | **Conceptual level**

1. Take a person's finger and blot it onto an ink pad.

2. Roll the person's finger onto a recording surface to obtain a print.

3. With a low-power binocular microscope, count the number of ridges, using the method described in Figure 27.7.

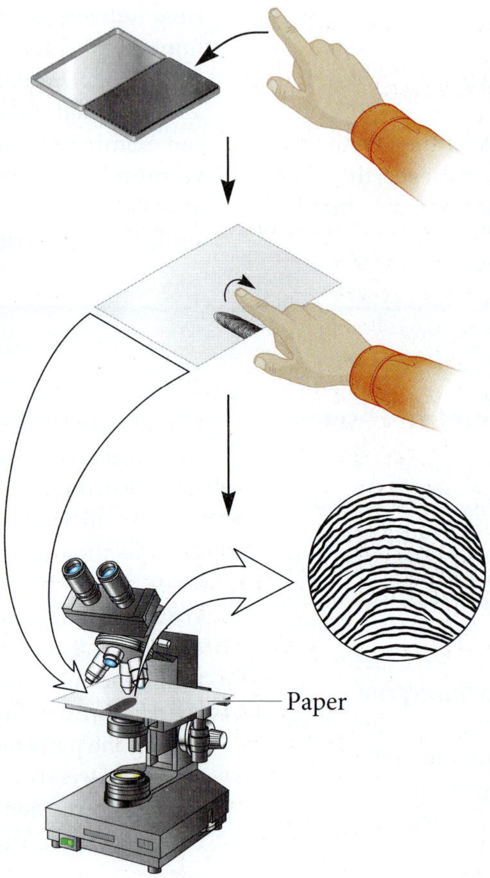

This is a method to measure a quantitative trait.

4. Calculate the correlation coefficients between different pairs of individuals as described earlier in this chapter.

See The Data.

The correlation coefficient provides a way to determine the heritability for the quantitative trait.

THE DATA

Type of Relationship	*Number of Pairs Examined*	*Correlation Coefficient* (r_{obs})	*Heritability* r_{obs}/r_{exp}
Parent-child	810	0.48 ± 0.04*	0.96
Parent-parent	200	0.05 ± 0.07	—†
Sibling-sibling	642	0.50 ± 0.04	1.00
Identical twins	80	0.95 ± 0.01	0.95
Fraternal twins	92	0.49 ± 0.08	0.98
			0.97 average heritability

*± = Standard error of the mean.

†Note: We cannot calculate a heritability value because the value for r_{exp} is not known. Nevertheless, the value for r_{obs} is very low, suggesting that there is a negligible correlation between unrelated individuals.

Adapted from S. B. Holt (1961). Quantitative genetics of fingerprint patterns. *Br Med Bull 17*, 247–250.

INTERPRETING THE DATA

As seen in The Data, the results indicate that genetics plays the major role in explaining the variation in this trait. Genetically unrelated individuals (namely, parent-parent relationships) have a negligible correlation for this trait. By comparison, individuals who are genetically related have a substantially higher correlation coefficient. When the observed correlation coefficient is divided by the expected correlation coefficient based on the known genetic relationships, the average heritability value is 0.97, which is very close to 1.0.

What do these high heritability values mean? They indicate that nearly all of the phenotypic variance in fingerprint pattern is due to genetic variance. Significantly, fraternal and identical twins have substantially different observed correlation coefficients, even though we expect they have been raised in very similar environments. These results support the idea that genetics is playing the major role in promoting variation and the results are not biased heavily by environmental similarities that may be associated with genetically related individuals. From an experimental viewpoint, the results show us how the determination of correlation coefficients between related individuals can provide insight into the relative contributions of genetics and environment to the variation of a quantitative trait.

A self-help quiz involving this experiment can be found at www.mhhe.com/brookergenetics5e.

27.4 COMPREHENSION QUESTIONS

1. In a population of squirrels in North Carolina, the heritability for body weight is high. This means that
   a. body weight is primarily controlled by genes.
   b. the environment has little influence on body weight.
   c. the variance in body weight is mostly due to genetic variation.
   d. both a and b are correct.
2. If two or more different genotypes do not respond to environmental variation in the same way, this is an example of
   a. a genotype-environment association.
   b. a genotype-environment interaction.
   c. the additive effects of alleles.
   d. both a and b.
3. One way to estimate narrow-sense heritability for a given trait is to compare __________ for __________.
   a. variances, related pairs of individuals
   b. correlation coefficients, related pairs of individuals
   c. variances, unrelated pairs of individuals
   d. correlation coefficients, unrelated pairs of individuals

27.5 SELECTIVE BREEDING

Learning Outcomes:

1. Describe the effects of selective breeding.
2. Calculate heritability from the results of selective breeding experiments.
3. Explain how dominance and overdominance may contribute to the beneficial characteristics of hybrids.

The term **selective breeding** refers to programs and procedures designed to modify phenotypes in species of economically important plants and animals. This phenomenon, also called **artificial selection,** is related to natural selection, discussed in Chapter 26. In forming his theory of natural selection, Charles Darwin was influenced by his observations of selective breeding by pigeon fanciers and other breeders. The primary difference between artificial and natural selection is how the parents are chosen. Natural selection is due to natural variation in reproductive success. In artificial selection, the breeder chooses individuals that possess traits that are desirable from a human perspective. In this section, we will examine the effects of selective breeding and consider its relationship to heritability.

Selective Breeding of Species Can Alter Quantitative Traits Dramatically

For centuries, humans have been practicing selective breeding to obtain domestic species with interesting or agriculturally useful characteristics. For example, the common breeds of dogs have been obtained by selective breeding strategies (**Figure 27.9**). As shown here, it is very striking how selective breeding can modify the quantitative traits in a species. When comparing a greyhound with a bulldog, the magnitude of the differences is amazing. They hardly look like members of the same species. A QTL study in 2007 by Nathan Sutter and colleagues indicates that the size of dogs is determined, in part, by alleles in the *Igf1* gene that encodes a growth hormone called insulin-like growth factor 1. A particular allele of this gene was found to be common to all small breeds of dogs and nearly absent from very large breeds, suggesting that this allele is one of several genes that influences body size in small breeds of dogs.

Likewise, most of the food we eat is obtained from species that have been modified profoundly by selective breeding strategies. This includes products such as grains, fruits, vegetables, meat, milk, and juices. **Figure 27.10** illustrates how certain characteristics in the wild mustard plant (*Brassica oleracea*) have been modified by selective breeding to create several varieties of important domesticated crops. This plant is native to Europe and Asia, and plant breeders began to modify its traits approximately 4000 years ago. As seen here, certain quantitative traits in the domestic strains, such as stems and lateral buds, differ considerably from those of the original wild species.

The phenomenon that underlies selective breeding is genetic variation. Within a group of individuals, allelic variation may affect the outcome of quantitative traits. The fundamental strategy of the

FIGURE 27.9 Some common breeds of dogs that have been obtained by selective breeding.

Genes→Traits By selecting parents carrying the alleles that have a desired effect on certain quantitative traits, dog breeders have produced breeds with distinctive sets of traits. For example, the bulldog has alleles that produce short legs and a flat face. By comparison, the corresponding genes in a German shepherd are found in alleles that produce longer legs and a more pointy snout. All of the dogs shown in this figure carry the same kinds of genes (e.g., many genes that affect their sizes, shapes, and fur color). However, the alleles for many of these genes are different among these dogs, thereby producing breeds with strikingly different phenotypes.

CONCEPT CHECK: What are the similarities and differences between natural selection and selective breeding?

selective breeder is to choose parents that will pass on alleles to their offspring that produce desirable phenotypic characteristics. For example, if a breeder wants large cattle, the largest members of the herd are chosen as parents for the next generation. These large cattle will transmit an array of alleles to their offspring that confer large size. The breeder often chooses genetically related individuals (e.g., brothers and sisters) as the parental stock. As mentioned previously, the practice of mating between genetically related individuals is known as inbreeding. Some of the consequences of inbreeding are also described in Chapter 26.

What is the outcome when selective breeding is conducted for a quantitative trait? **Figure 27.11a** shows the results of a program begun at the Illinois Agricultural Experiment Station in 1896,

FIGURE 27.10 Crop plants developed by selective breeding of the wild mustard plant (*Brassica oleracea*).

Genes→Traits The wild mustard plant carries a large amount of genetic (i.e., allelic) variation, which was used by plant breeders to produce modern strains that are agriculturally desirable and economically important. For example, by selecting for alleles that promote the formation of large lateral leaf buds, the strain of Brussels sprouts was created. By selecting for alleles that alter the leaf morphology, kale was developed. Although these six agricultural plants look quite different from each other, they carry many of the same alleles as the wild mustard. However, they differ in alleles affecting the formation of stems, leaves, flower buds, and leaf buds.

CONCEPT CHECK: Discuss the types of traits that have been subjected to selective breeding in this example.

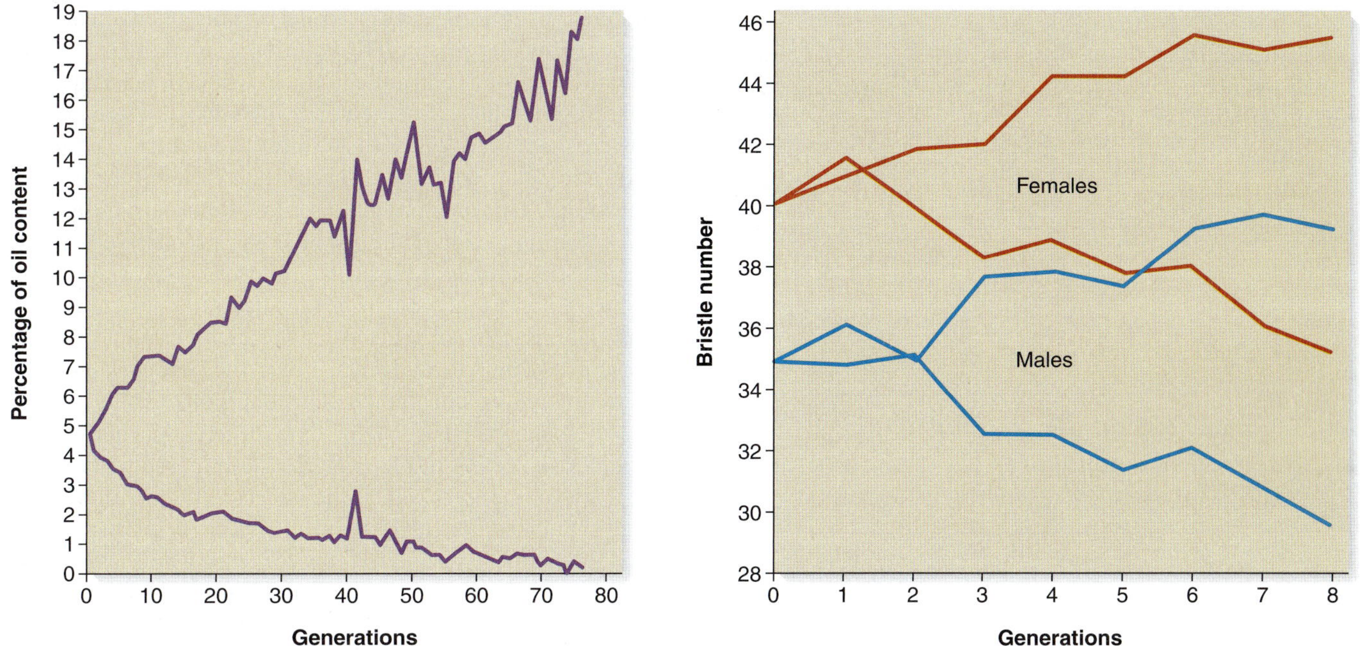

(a) Results of selective breeding for high and low oil content in corn

(b) Results of selective breeding for high and low bristle number in flies

FIGURE 27.11 **Common results of selective breeding for a quantitative trait.**

CONCEPT CHECK: What are two reasons why a selection limit is reached in which artificial selection no longer has an effect?

even before the rediscovery of Mendel's laws. This experiment began with 163 ears of corn with an oil content ranging from 4–6%. In each of 80 succeeding generations, corn plants were divided into two separate groups. In one group, several members with the highest oil content were chosen as parents of the next generation. In the other group, several members with the lowest oil content were chosen. After 80 generations, the oil content in the first group rose to over 18%; in the other group, it dropped to less than 1%. These results show that selective breeding can modify quantitative traits in a very directed manner.

Similar results have been obtained for many other quantitative traits. **Figure 27.11b** shows an experiment by Kenneth Mather conducted in the 1940s, in which flies were selected on the basis of their bristle number. The starting group had an average of 40 bristles for females and 35 bristles for males. After eight generations, the group selected for high bristle number had an average of 46 bristles for females and 39 for males, whereas the group selected for low bristle number had an average of 35 bristles for females and 30 for males.

When comparing the curves in Figure 27.11, keep in mind that quantitative traits are often at an intermediate value in unselected populations. Therefore, artificial selection can increase or decrease the magnitude of the trait. Oil content can go up or down, and bristle number can increase or decrease.

Figure 27.11 also shows the phenomenon known as a **selection limit**—after many generations a plateau is reached where artificial selection is no longer effective. A selection limit may occur for two reasons. Presumably, the starting population possesses a large amount of genetic variation, which contributes to the diversity in phenotypes. By carefully choosing the parents, each succeeding generation has a higher proportion of the desirable alleles. However, after many generations, the population may be nearly monomorphic for all or most of the desirable alleles that affect the trait of interest. At this point, additional selective breeding will have no effect. When this occurs, the heritability for the trait is near zero, because nearly all genetic variation for the trait of interest has been eliminated from the population. Without the introduction of new mutations into the population, further selection is not possible. A second reason for a selection limit is related to fitness. Some alleles that accumulate in a population due to artificial selection may have a negative influence on the population's overall fitness. A selection limit is reached in which the desired effects of artificial selection are balanced by the negative effects on fitness.

Selective Breeding Provides a Way of Estimating Heritability

When conducting artificial selection experiments, the response to selection is a common way to estimate the narrow-sense heritability in a starting population. The narrow-sense heritability measured in this way is also called the **realized heritability.** It is calculated as

$$h_N^2 = \frac{R}{S}$$

where

R is the response in the offspring to selection. It is the difference between the mean of the offspring and the mean of the starting population.

S is the selection differential in the parents. It is the difference between the mean of the parents and the mean of the starting population.

Here,

$$R = \overline{X}_O - \overline{X}$$

$$S = \overline{X}_P - \overline{X}$$

where

$\overline{X}$ is the mean of the starting population

$\overline{X}_O$ is the mean of the offspring

$\overline{X}_P$ is the mean of the parents

So,

$$h_N^2 = \frac{\overline{X}_O - \overline{X}}{\overline{X}_P - \overline{X}}$$

The narrow-sense heritability is the proportion of the variance in phenotype that can be used to predict changes in the population mean when selection is practiced.

As an example, let's suppose we began with a population of fruit flies in which the average bristle number for both sexes was 37.5. The parents chosen from this population had an average bristle number of 40. The offspring of the next generation had an average bristle number of 38.7. With these values, the realized heritability is

$$h_N^2 = \frac{38.7 - 37.5}{40 - 37.5}$$

$$h_N^2 = \frac{1.2}{2.5}$$

$$h_N^2 = 0.48$$

This result tells us that about 48% of the phenotypic variance is due to the additive effects of alleles.

An important aspect of narrow-sense heritabilities is their ability to predict the outcome of selective breeding. In this case, the goal is to predict the mean phenotypes of offspring. If we rearrange our realized heritability equation

$$R = h_N^2 S$$

$$\overline{X}_O - \overline{X} = h_N^2 (\overline{X}_P - \overline{X})$$

This equation is referred to as the breeder's equation, because it is used to calculate the mean phenotypes of offspring based on the mean weights of the parents, the mean weights of the starting population, and the heritability. Solved problem S1 at the end of the chapter illustrates the use of this equation.

Heterosis May Be Explained by Dominance or Overdominance

As we have just seen, selective breeding can alter the phenotypes of domesticated species in a highly directed way. An unfortunate consequence of inbreeding, however, is that it may inadvertently promote homozygosity for deleterious alleles. This phenomenon is called **inbreeding depression.** In addition, genetic drift, described in Chapter 26, may contribute to the loss of beneficial alleles. In agriculture, it is widely observed that when two different inbred strains are crossed to each other, the resulting offspring are often more vigorous (e.g., larger or longer-lived) than either of the inbred parental strains. This phenomenon is called **heterosis,** or **hybrid vigor.**

In modern agricultural breeding practices, many strains of plants and animals are hybrids produced by crossing two different inbred lines. In fact, much of the success of agricultural breeding programs is founded in heterosis. In rice, for example, hybrid strains have a 15–20% yield advantage over the best conventional inbred varieties under similar cultivation conditions.

As shown in Figure 25.12, two different phenomena may contribute to heterosis. In 1908, Charles Davenport developed the dominance hypothesis, in which the effects of dominant alleles explain the favorable outcome in a hybrid. He suggested that highly inbred strains have become homozygous for recessive alleles of one or more genes. In the homozygous state, the recessive alleles are somewhat deleterious (but not lethal). Because the homozygosity occurs by chance, two different inbred strains are likely to be homozygous for recessive alleles in different genes. Therefore, when they are crossed to each other, the resulting hybrids are heterozygous and do not suffer the consequences of homozygosity for deleterious recessive alleles. In other words, the benefit of the dominant alleles explains the observed heterosis. Steven Tanksley, working with colleagues in China, found that heterosis in rice seems to be due to the phenomenon of dominance. This is a common explanation for heterosis.

In 1908, George Shull and Edward East proposed a second hypothesis, known as the overdominance hypothesis (see Figure 25.12). As described in Chapter 4, overdominance occurs when the heterozygote is more vigorous than either corresponding homozygote. According to this idea, heterosis can occur because the resulting hybrids are heterozygous for one or more genes that display overdominance. The heterozygote is more vigorous than

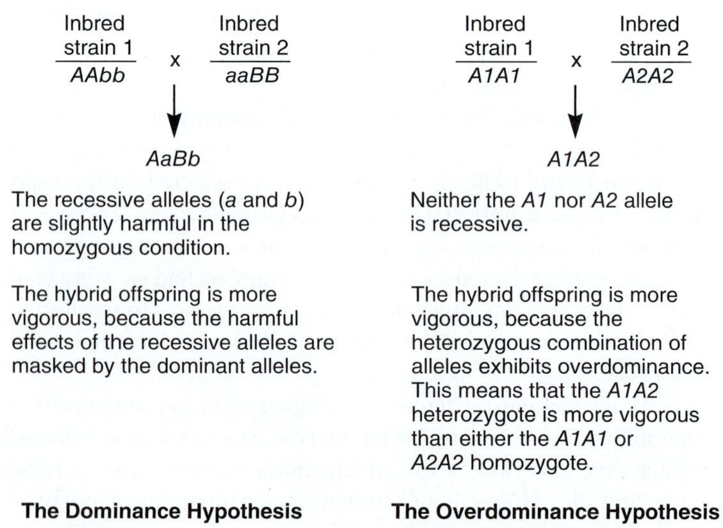

FIGURE 27.12 Mechanisms to explain heterosis. The two common explanations are the dominance hypothesis and the overdominance hypothesis.

either homozygote. In corn, Charles Stuber and his colleagues have found that several QTLs for grain yield support the overdominance hypothesis.

Finally, it should be pointed out that overdominance is very difficult to distinguish from **pseudo-overdominance,** a phenomenon initially suggested by James Crow. To understand pseudo-overdominance, let's consider an example. When a QTL is identified in a mapping experiment, the researcher may not know if the QTL contains one or more than one gene affecting the quantitative trait. Let's consider a scenario in which a QTL actually has two closely linked genes, both affecting the same quantitative trait. In one strain, the alleles of the two genes, *a* and *B*, are closely linked, whereas *A* and *b* are closely linked in another strain. In this example, *A* is dominant to *a* and *B* is dominant to *b*. In a cross between *aaBB* and *AAbb* individuals, the hybrid is heterozygous (*AaBb*) for two different genes. Heterosis is observed because the hybrid has copies of both dominant alleles. If a researcher assumed that only one gene was located at this QTL, she or he may incorrectly conclude that the hybrid is heterozygous for a single gene and overdominance is occurring. This incorrect conclusion is why it is called pseudo-overdominance. Therefore, without very fine mapping, which is rarely done for QTLs, it is hard to distinguish between overdominance and pseudo-overdominance. However, if the genome sequence is available, the identification of candidate genes in the mapped region may make it possible to distinguish between overdominance and pseudo-overdominance.

27.5 COMPREHENSION QUESTIONS

1. For selective breeding to be successful, the starting population must

a. have genetic variation that affects the trait of interest.

b. be very large.

c. be amenable to phenotypic variation caused by environmental effects.

d. have very little phenotypic variation.

2. The mean weight of cows in a population is 520 kg. Animals with a mean weight of 540 kg are used as parents and produce offspring that have a mean weight of 535 kg. What is the narrow-sense heritability (h_N^2) for body weight in this population of cows?

a. 0.25

b. 0.5

c. 0.75

d. 1.0

KEY TERMS

Page 702. complex traits, quantitative traits, quantitative genetics
Page 703. continuous traits, meristic traits, threshold traits, discontinuous traits, frequency distribution
Page 704. normal distribution, biometric field, mean, variance
Page 705. standard deviation (*SD*), covariance
Page 706. correlation coefficient (*r*)
Page 707. regression analysis, polygenic inheritance, quantitative trait locus (QTL)
Page 710. QTL mapping
Page 712. inbreeding, monomorphic
Page 713. genotype-environment interaction
Page 714. genotype-environment association, heritability, broad-sense heritability, narrow-sense heritability
Page 717. selective breeding, artificial selection
Page 719. selection limit, realized heritability
Page 720. inbreeding depression, heterosis, hybrid vigor
Page 721. pseudo-overdominance
Page 723. regression coefficient

CHAPTER SUMMARY

- Quantitative genetics is the field of genetics concerned with complex and quantitative traits.

27.1 Overview of Quantitative Traits

- Quantitative traits can be categorized as anatomical, physiological, or behavioral (see Table 27.1).
- Quantitative traits often exhibit a continuum and may follow a normal distribution (see Figure 27.1).

27.2 Statistical Methods for Evaluating Quantitative Traits

- Statistical methods, including the mean, variance, standard deviation, covariance, and correlation coefficient, are used to analyze quantitative traits (see Figure 27.2, Table 27.2).

27.3 Polygenic Inheritance

- Polygenic inheritance refers to an inheritance pattern in which multiple genes affect a single trait.
- The locations along a chromosome that contain genes affecting a quantitative trait are called quantitative trait loci (QTLs).

- Polygenic inheritance and environmental factors may cause a quantitative trait to fall along a continuum (see Figures 27.3, 27.4).
- QTLs are mapped by their linkage to molecular markers (see Figure 27.5).

27.4 Heritability

- Genetic variance and environmental variance may contribute additively to the total phenotypic variance.
- Genetic variance and environmental variance may exhibit interactions and associations (see Figure 27.6, Table 27.3).
- Heritability is the amount of phenotypic variance within a group of individuals that is due to genetic variance.
- Broad-sense heritability refers to all genetic factors affecting heritability, which includes the additive effects of alleles, effects due to dominant/recessive relationships, and effects due to epistatic interactions.
- Narrow-sense heritability is heritability that is due to the additive effects of alleles.
- Holt determined that dermal ridge count has a very high heritability value in humans (see Figures 27.7, 27.8).

27.5 Selective Breeding

- Selective breeding refers to programs and procedures designed to modify phenotypes in commercially important plants and animals (see Figures 27.9, 27.10).
- When starting with a genetically diverse population, selective breeding can usually modify a trait in a desired direction until a selection limit is reached (see Figure 27.11).
- Heterosis is the phenomenon in which the crossing of different inbred strains produces hybrids that are more vigorous than the inbred strains. This may be due to dominance or overdominance (see Figure 27.12).

PROBLEM SETS & INSIGHTS

Solved Problems

S1. The narrow-sense heritability (h_N^2) for potato weight in a starting population of potatoes is 0.42, and the mean weight is 1.4 pounds. If a breeder crosses plants with average potato weights of 1.9 and 2.1 pounds, respectively, what is the predicted average weight of potatoes in the offspring?

Answer: The mean weight of the parental plants is 2.0 pounds. To solve for the mean weight of the offspring:

$$R = h_N^2 S$$

$$\overline{X}_O - \overline{X} = h_N^2 (\overline{X}_P - \overline{X})$$

$$\overline{X}_O - 1.4 = 0.42\,(2.0 - 1.4)$$

$$\overline{X}_O = 1.65 \text{ pounds}$$

S2. A farmer wants to increase the average body weight in a herd of cattle. She begins with a herd having a mean weight of 595 kg and chooses individuals to breed that have a mean weight of 625 kg. Twenty offspring were obtained, having the following weights in kilograms: 612, 587, 604, 589, 615, 641, 575, 611, 610, 598, 589, 620, 617, 577, 609, 633, 588, 599, 601, and 611. Calculate the realized heritability for body weight in this herd.

Answer:

$$h_N^2 = \frac{R}{S} = \frac{\overline{X}_O - \overline{X}}{\overline{X}_P - \overline{X}}$$

We already know the mean weight of the starting herd (595 kg) and the mean weight of the parents (625 kg). The only value missing is the mean weight of the offspring, X_O.

$$\overline{X}_O = \frac{\text{Sum of the offspring's weights}}{\text{Number of offspring}}$$

$$\overline{X}_O = 604 \text{ kg}$$

$$h_N^2 = \frac{604 - 595}{625 - 595} = 0.3$$

S3. The following data describe the 6-week weights (in grams) of mice and their offspring of the same sex:

Parent (g)	Offspring (g)
24	26
21	24
24	22
27	25
23	21
25	26
22	24
25	24
22	24
27	24

Calculate the correlation coefficient.

Answer: To calculate the correlation coefficient, we first need to calculate the mean and standard deviation for each group:

$$\overline{X}_{\text{parents}} = \frac{24 + 21 + 24 + 27 + 23 + 25 + 22 + 25 + 22 + 27}{10} = 24$$

$$\overline{X}_{\text{offspring}} = \frac{26 + 24 + 22 + 25 + 21 + 26 + 24 + 24 + 24 + 24}{10} = 24$$

$$SD_{\text{parents}} = \sqrt{\frac{0 + 9 + 0 + 9 + 1 + 1 + 4 + 1 + 4 + 9}{9}} = 2.1$$

$$SD_{\text{offspring}} = \sqrt{\frac{4 + 0 + 4 + 1 + 9 + 4 + 0 + 0 + 0 + 0}{9}} = 1.6$$

Next, we need to calculate the covariance.

$$CoV_{\text{(parents, offspring)}} = \frac{\Sigma\left[(X_P - \overline{X}_P)(X_O - \overline{X}_O)\right]}{N - 1}$$

$$= \frac{0 + 0 + 0 + 3 + 3 + 2 + 0 + 0 + 0 + 0}{9}$$

$$= 0.9$$

Finally, we calculate the correlation coefficient:

$$r_{\text{(parents, offspring)}} = \frac{CoV_{(P,O)}}{SD_P SD_O}$$

$$r_{\text{(parents, offspring)}} = \frac{0.9}{(2.1)(1.6)}$$

$$r_{\text{(parent, offspring)}} = 0.27$$

S4. In a wild strain of strawberry plants, the phenotypic variance for fruit weight is 4.1 g^2. In another strain of highly inbred strawberries raised under the same environmental conditions, the phenotypic variance is 2.7 g^2. With regard to the wild strain,

A. Estimate V_G.

B. What is h_B^2?

C. Assuming that all of the genetic variance is additive, what is h_N^2?

Answer:

A. If we assume that the highly inbred strain has no genetic variance:

$$V_G \text{ (for the wild strain)} = 4.1\ g^2 - 2.7\ g^2 = 1.4\ g^2$$

B. $h_B^2 = 1.4\ g^2/4.1\ g^2 = 0.34$

C. It is the same as h_B^2 so it also equals 0.34.

S5. As described in this chapter, the correlation coefficient provides a way to determine the strength of association between two variables. When the variables are related due to cause and effect (i.e., one variable affects the outcome of another), researchers may use a regression analysis to predict how much one variable changes in response to the other. This is easier to understand if we plot the data for two variables. The graph shown here compares mothers' and offspring's body weights in cattle.

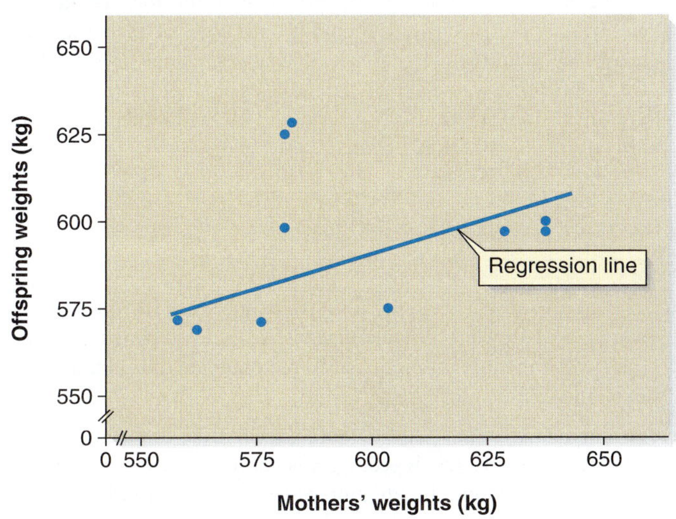

The line running through the data points is called a regression line. It is a line that minimizes the squared vertical distances to all of the points. For many types of data, particularly those involving quantitative traits, the regression line can be represented by the equation

$$Y = bX + a$$

where

b is the regression coefficient

a is a constant

In this example, *X* is the value of a mother's weight, and *Y* is the value of its offspring's weight. The value of *b*, known as the **regression coefficient,** represents the slope of the regression line. The value of *a* is the *y*-intercept (i.e., the value of *Y* when *X* equals zero). The equation shown is very useful because it allows us to predict the value of *Y* at any given value of *X*, and vice versa. To do so, we first need to determine the values of *b* and *a*. This can be accomplished in the following manner:

$$b = \frac{CoV_{(X,Y)}}{V_X}$$

$$a = \overline{Y} - b\overline{X}$$

Once the values of *a* and *b* have been computed, we can use them to predict the values of *Y* or *X* by using the equation

$$Y = bX + a$$

For example, if $b = 0.5$, $a = 2$, and $X = 58$, this equation can be used to compute a value of *Y* that equals 31. It is important to keep in mind that this equation predicts the average value of *Y*. As we see in the preceding figure, the data points tend to be scattered around the regression line. Deviations may occur between the data points and the line. The equation predicts the most likely values. In an actual experiment, however, some deviation will occur between the predicted values and the experimental values due to random sampling error.

Now here is the question. Using the data found in this chapter regarding weight in cattle, what is the predicted weight of an offspring if its mother weighed 660 pounds?

Answer: We first need to calculate *a* and *b*.

$$b = \frac{CoV_{(X,Y)}}{V_X}$$

We need to use the data on page 706 to calculate V_X, which is the variance for the mothers' weights. The variance equals 445.1. The covariance is already calculated on page 706; it equals 152.6.

$$b = \frac{152.6}{445.1}$$

$$b = 0.34$$

$$a = \overline{Y} - b\overline{X}$$

$$a = 598 - (0.34)(596) = 395.4$$

Now we are ready to calculate the predicted weight of the offspring using the equation

$$Y = bX + a$$

In this problem, $X = 660$ pounds.

$$Y = 0.34(660) + 395.4$$

$$Y = 619.8 \text{ pounds}$$

The average weight of the offspring is predicted to be 619.8 pounds.

S6. Are the following statements regarding heritability true or false?

A. Heritability applies to a specific population raised in a particular environment.

B. Heritability in the narrow sense takes into account all types of genetic variance.

C. Heritability is a measure of the amount that genetics contributes to the outcome of a trait.

Answer:

A. True

B. False. Narrow-sense heritability considers only the effects of additive alleles.

C. False. Heritability is a measure of the amount of phenotypic variance that is due to genetic variance; it applies to the variance of a specific population raised in a particular environment.

S7. The correlation coefficients for height were determined for 15 pairs of individuals with the following genetic relationships:

Father-daughter: 0.41

Father-granddaughter: 0.18

Sister-sister: 0.40

Sister-sister (fraternal twins): 0.41

Sister-sister (identical twins): 0.83

What is the average heritability for height in this group of females?

Answer: $h_N^2 = r_{obs}/r_{exp}$

The value for r_{exp} comes from the known genetic relationships:

Father-daughter	$r_{obs} = 0.41$	$r_{exp} = 0.5$	$h_N^2 = 0.82$
Father-granddaughter	$r_{obs} = 0.18$	$r_{exp} = 0.25$	$h_N^2 = 0.72$
Sister-sister	$r_{obs} = 0.40$	$r_{exp} = 0.5$	$h_N^2 = 0.80$
Twin sisters (fraternal)	$r_{obs} = 0.41$	$r_{exp} = 0.5$	$h_N^2 = 0.82$
Twin sisters (identical)	$r_{obs} = 0.83$	$r_{exp} = 1.0$	$h_N^2 = 0.83$

If we take the average of the five h_N^2 values, the average heritability is 0.80.

Conceptual Questions

C1. Give several examples of quantitative traits. How are these quantitative traits described within groups of individuals?

C2. At the molecular level, explain why quantitative traits often exhibit a continuum of phenotypes within a population. How does the environment help produce this continuum?

C3. What is a normal distribution? Discuss this curve with regard to quantitative traits within a population. What is the relationship between the standard deviation and the normal distribution?

C4. Explain the difference between a continuous trait and a discontinuous trait. Give two examples of each. Are quantitative traits likely to be continuous or discontinuous? Explain why.

C5. What is a frequency distribution? Explain how the graph is made for a quantitative trait that is continuous.

C6. The variance for weight in a particular herd of cattle is 484 pounds^2. The mean weight is 562 pounds. How heavy would an animal have to be if it was in the top 2.5% of the herd? The bottom 0.13%?

C7. Two different varieties of potatoes both have the same mean weight of 1.5 pounds. One group has a very low variance, and the other has a much higher variance.

A. Discuss the possible reasons for the differences in variance.

B. If you were a potato farmer, would you rather raise a variety with a low or high variance? Explain your answer from a practical point of view.

C. If you were a potato breeder and you wanted to develop potatoes with a heavier weight, would you choose the variety with a low or high variance? Explain your answer.

C8. If an r value equals 0.5 and $N = 4$, would you conclude that a positive correlation is found between the two variables? Explain your answer. What if $N = 500$?

C9. What does it mean when a correlation coefficient is negative? Can you think of examples?

C10. When a correlation coefficient is statistically significant, what do you conclude about the two variables? What do the results mean with regard to cause and effect?

C11. What is polygenic inheritance? Discuss the issues that make polygenic inheritance difficult to study.

C12. What is a quantitative trait locus (QTL)? Does a QTL contain one gene or multiple genes? What technique is commonly used to identify QTLs?

C13. Let's suppose that weight in a species of mammal is polygenic, and each gene exists as a heavy and light allele. If the allele frequencies in the population were equal for both types of allele (i.e., 50% heavy alleles and 50% light alleles), what percentage of individuals would be homozygous for the light alleles at all of the genes affecting this trait, if the trait was determined by the following number of genes?

A. Two

B. Three

C. Four

C14. The broad-sense heritability for a trait equals 1.0. In your own words, explain what this value means. Would you conclude that the environment is unimportant in the outcome of this trait? Explain your answer.

C15. From an agricultural point of view, discuss the advantages and disadvantages of selective breeding. It is common for plant breeders to take two different, highly inbred strains, which are the product of many generations of selective breeding, and cross them to make hybrids. How does this approach overcome some of the disadvantages of selective breeding?

C16. Many beautiful varieties of roses have been produced, particularly in the last few decades. These newer varieties often have very striking and showy flowers, making them desirable as horticultural specimens. However, breeders and novices alike have noticed that some of these newer varieties are not very fragrant compared with the older, more traditional varieties. From a genetic point of view, suggest an explanation why some of these newer varieties with superb flowers are not as fragrant.

C17. In your own words, explain the meaning of the term heritability. Why is a heritability value valid only for a particular population of individuals raised in a particular environment?

C18. What is the difference between broad-sense heritability and narrow-sense heritability? Why is narrow-sense heritability such a useful concept in the field of agricultural genetics?

C19. The heritability for egg weight in a group of chickens on a farm in Maine is 0.95. Are the following statements regarding heritability true or false? If a statement is false, explain why.

A. The environment in Maine has very little effect on the outcome of this trait.

B. Nearly all of the phenotypic variance for this trait in this group of chickens is due to genetic variance.

C. The trait is polygenic and likely to involve a large number of genes.

D. Based on the observation of the heritability in the Maine chickens, it is reasonable to conclude that the heritability for egg weight in a group of chickens on a farm in Montana is also very high.

C20. In a fairly large population of people living in a commune in the southern United States, everyone cares about good nutrition. All of the members of this population eat very nutritious foods, and their diets are very similar. How do you think the height of individuals in this commune population would compare with that of the general population in the following categories?

A. Mean height

B. Heritability for height

C. Genetic variation for alleles that affect height

C21. When artificial selection is practiced over many generations, it is common for the trait to reach a plateau in which further selection has little effect on the outcome of the trait. This phenomenon is illustrated in Figure 27.11. Explain why.

C22. Discuss whether a natural population of wolves or a domesticated population of German shepherds is more likely to have a higher heritability for the trait of size.

C23. With regard to heterosis, would the following statements be consistent with the dominance hypothesis, the overdominance hypothesis, or both?

A. Strains that have been highly inbred have become monomorphic for one or more recessive alleles that are somewhat detrimental to the organism.

B. Hybrid vigor occurs because highly inbred strains are monomorphic for many genes, whereas hybrids are more likely to be heterozygous for those same genes.

C. If a gene exists in two alleles, hybrids are more vigorous because heterozygosity for the gene is more beneficial than homozygosity of either allele.

Experimental Questions

E1. Here are data for height and weight among 10 male college students.

Height (cm)	Weight (kg)
159	48
162	50
161	52
175	60
174	64
198	81
172	58
180	74
161	50
173	54

A. Calculate the correlation coefficients for this group.

B. Is the correlation coefficient statistically significant? Explain.

E2. The abdomen length (in millimeters) was measured in 15 male *Drosophila*, and the following data were obtained: 1.9, 2.4, 2.1, 2.0, 2.2, 2.4, 1.7, 1.8, 2.0, 2.0, 2.3, 2.1, 1.6, 2.3, and 2.2. Calculate the mean, standard deviation, and variance for this population of male fruit flies.

E3. In one strain of cabbage, you conduct an RFLP analysis of head weight; you determine that seven QTLs affect this trait. In another strain of cabbage, you find that only four QTLs affect this trait. Note that both strains of cabbage are from the same species, although they may have been subjected to different degrees of inbreeding. Explain how one strain can have seven QTLs and another strain four QTLs for exactly the same trait. Is the second strain missing three genes?

E4. From an experimental viewpoint, what does it mean to say that an RFLP is associated with a trait? Let's suppose that two strains of pea plants differ in two RFLPs that are linked to two genes governing pea size. RFLP-1 is found in 2000-bp and 2700-bp bands, and RFLP-2 is found in 3000-bp and 4000-bp bands. The plants producing large peas have RFLP-1 (2000 bp) and RFLP-2 (3000 bp); those producing small peas have RFLP-1 (2700 bp) and RFLP-2 (4000 bp). A cross is made between these two strains, and the F_1 offspring are allowed to self-fertilize. Five phenotypic classes are observed: small peas, small-medium peas, medium peas, medium-large peas, and large peas. We assume that each of the two genes makes an equal contribution to pea size and that the genetic variance is additive. Draw a gel and explain what RFLP banding patterns you would expect to observe for these five phenotypic categories. Note: Certain phenotypic categories may have more than one possible banding pattern.

E5. Let's suppose that two strains of pigs differ in 500 RFLPs. One strain is much larger than the other. The pigs are crossed to each other, and the members of the F_1 generation are also crossed among themselves to produce an F_2 generation. Three distinct RFLPs are associated with F_2 pigs that are larger. How would you interpret these results?

E6. Outline the steps you would follow to determine the number of genes that influence the yield of rice. Describe the results you might get if rice yield is governed by variation in six different genes.

E7. In a wild strain of tomato plants, the phenotypic variance for tomato weight is 3.2 g^2. In another strain of highly inbred tomatoes raised under the same environmental conditions, the phenotypic variance is 2.2 g^2. With regard to the wild strain,

A. Estimate V_G.

B. What is h_B^2?

C. Assuming that all of the genetic variance is additive, what is h_N^2?

E8. The average thorax length in a *Drosophila* population is 1.01 mm. You want to practice selective breeding to make larger *Drosophila*. To do so, you choose 10 parents (5 males and 5 females) of the following sizes: 0.97, 0.99, 1.05, 1.06, 1.03, 1.21, 1.22, 1.17, 1.19, and 1.20. You mate them and then analyze the thorax sizes of 30 offspring (half male and half female): 0.99, 1.15, 1.20, 1.33, 1.07, 1.11, 1.21, 0.94, 1.07, 1.11, 1.20, 1.01, 1.02, 1.05, 1.21, 1.22, 1.03, 0.99, 1.20, 1.10, 0.91, 0.94, 1.13, 1.14, 1.20, 0.89, 1.10, 1.04, 1.01, 1.26. Calculate the realized heritability in this group of flies.

E9. In a strain of mice, the average 6-week body weight is 25 g and the narrow-sense heritability for this trait is 0.21.

A. What would be the average weight of the offspring if parents with a mean weight of 27 g were chosen?

B. What weight of parents would you have to choose to obtain offspring with an average weight of 26.5 g?

E10. You need to understand solved problem S5 before answering this question. The variance in height for fathers (in square inches) was 112, the variance for sons was 122, and the covariance was 144. The mean height for fathers was 68 inches, and the mean height for sons was 69 inches. If a father had a height of 70 inches, what is the most probable height of his son?

E11. A danger in computing heritability values from studies involving genetically related individuals is the possibility that these individuals share more similar environments than do unrelated individuals. In the experiment of Figure 27.8, which data are the most compelling evidence that ridge count is not caused by genetically related individuals sharing common environments? Explain.

E12. A large, genetically heterogeneous group of tomato plants was used as the original breeding stock by two different breeders, named Mary and Hector. Each breeder was given 50 seeds and began an artificial selection strategy, much like the one described in Figure 27.11. The seeds were planted, and the breeders selected the 10 plants with the highest mean tomato weights as the breeding stock for the next generation. This process was repeated over the course of 12 growing seasons, and the following data were obtained:

Mean Weight of Tomatoes (pounds)

Year	Mary's Tomatoes	Hector's Tomatoes
1	0.7	0.8
2	0.9	0.9
3	1.1	1.2
4	1.2	1.3
5	1.3	1.3
6	1.4	1.4
7	1.4	1.5
8	1.5	1.5
9	1.5	1.5
10	1.5	1.5
11	1.5	1.5
12	1.5	1.5

A. Explain these results.

B. Another tomato breeder, named Martin, got some seeds from Mary's and Hector's tomato strains (after 12 generations), grew the plants, and then crossed them to each other. The mean weight of the tomatoes in these hybrids was about 1.7 pounds. For a period of 5 years, Martin subjected these hybrids to the same experimental strategy that Mary and Hector had followed, and he obtained the following results:

Mean Weight of Tomatoes (pounds)

Year	Martin's Tomatoes
1	1.7
2	1.8
3	1.9
4	2.0
5	2.0

Explain Martin's data. Why was Martin able to obtain tomatoes heavier than 1.5 pounds, whereas Mary's and Hector's strains appeared to plateau at this weight?

E13. The correlation coefficients for height were determined for 15 pairs of individuals with the following genetic relationships:

Mother-daughter: 0.36

Mother-granddaughter: 0.17

Sister-sister: 0.39

Sister-sister (fraternal twins): 0.40

Sister-sister (identical twins): 0.77

What is the average heritability for height in this group of females?

E14. An animal breeder had a herd of sheep with a mean weight of 254 pounds at 3 years of age. He chose animals with mean weights of 281 pounds as parents for the next generation. When these offspring reached 3 years of age, their mean weights were 269 pounds.

A. Calculate the narrow-sense heritability for weight in this herd.

B. Using the heritability value that you calculated in part A, what weight of animals would you have to choose to get offspring that weigh 275 pounds (at 3 years of age)?

E15. The trait of blood pressure in humans has a frequency distribution that is similar to a normal distribution. The following graph shows the ranges of blood pressures for a selected population of people. The red line depicts the frequency distribution of the systolic pressures for the entire population. Several individuals with high blood pressure were identified, and the blood pressures of their relatives were determined. This frequency distribution is depicted with a blue line. (Note: The blue line does not include the people who were identified with high blood pressure; it includes only their relatives.)

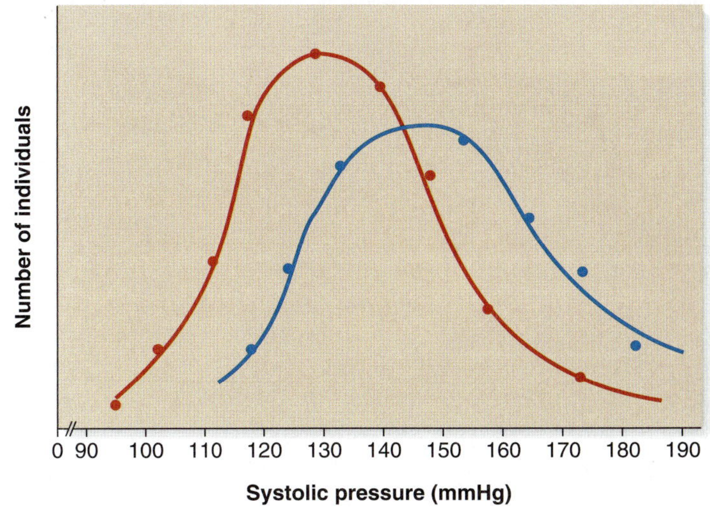

What do these data suggest with regard to a genetic basis for high blood pressure? What statistical approach could you use to determine the heritability for this trait?

Questions for Student Discussion/Collaboration

1. Discuss why heritability is an important phenomenon in agriculture.
2. From a biological viewpoint, speculate as to why many traits seem to fit a normal distribution. Students with a strong background in math and statistics may want to explain how a normal distribution is generated, and what it means. Can you think of biological examples that do not fit a normal distribution?
3. What is heterosis? Discuss whether it is caused by a single gene or several genes. Discuss the two major hypotheses proposed to explain heterosis. Which do you think is more likely to be correct?

Note: All answers appear at the website for this textbook; the answers to the even-numbered questions and all of the Concept Check and Comprehension Questions are in the back of the book.

www.mhhe.com/brookergenetics5e

Visit the website for practice tests, answer keys, and other learning aids for this chapter. Enhance your understanding of genetics with our interactive exercises, quizzes, animations, and much more.

CHAPTER OUTLINE

28.1 Origin of Species

28.2 Phylogenetic Trees

28.3 Molecular Evolution

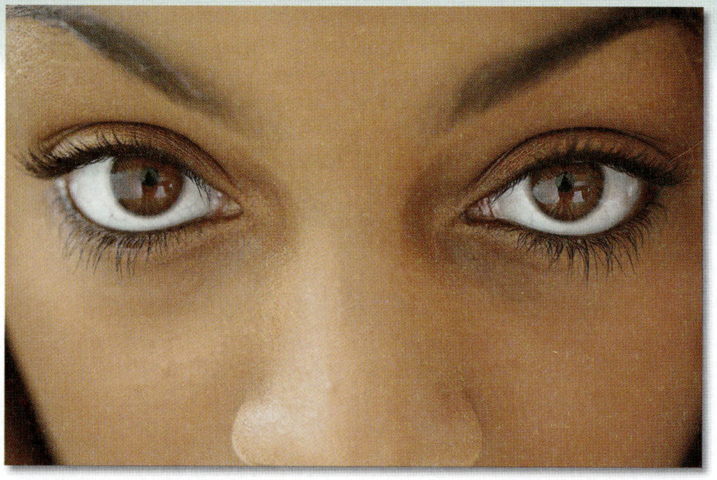

The evolution of eyes. *Developmental biologists have recently discovered that the eyes of many diverse species, including fruit flies, frogs, mice, and people, are under the control of the homologous gene called Pax6, suggesting that eyes may have originated once during the evolution of animals.*

28 EVOLUTIONARY GENETICS

Biological evolution, or simply **evolution,** is a heritable change in one or more characteristics of a population or species from one generation to the next. Evolution is a process. It can be viewed on a small scale as it relates to a single gene or it can be viewed on a larger scale as it relates to the formation of new species. In Chapter 26, we examined several factors that cause allele frequencies to change in populations. This process, also known as **microevolution,** concerns the changing composition of gene pools with regard to particular alleles over measurable periods of time. As we have seen, several evolutionary mechanisms, such as mutation, genetic drift, migration, natural selection, and inbreeding, affect the allele and genotypic frequencies within natural populations. On a microevolutionary scale, evolution can be viewed as a change in allele frequency over time.

A goal of this chapter is to relate phenotypic changes that occur during evolution to the underlying genetic changes that cause them to happen. In the first part of the chapter, we will be concerned with evolution on a large scale, which leads to the origin of new species. The question of how species form has been central to the development of evolutionary theory. The term **macroevolution** refers to large-scale evolutionary changes that create new species and higher taxa. It concerns the establishment of the diversity of organisms over long periods of time through the accumulated evolution and extinction of many species.

In the second part of this chapter, we will link molecular genetics to the evolution of species. Techniques for analyzing chromosomes and DNA sequences have greatly enhanced our understanding of evolutionary processes at the molecular level. The term **molecular evolution** refers to patterns and processes associated with evolutionary change at the molecular level. Such changes may be phenotypically neutral or they may underlie the phenotypic changes associated with evolution. In this chapter, we examine how molecular data can provide information about the phylogenetic relationships among different organisms.

The topic of molecular evolution is a fitting way to end our discussion of genetics because it integrates the ongoing theme of this textbook—the relationship between molecular genetics and traits—in the broadest and most profound ways. Theodosius Dobzhansky, an influential evolutionary scientist, once said, "Nothing in biology makes sense except in the light of evolution." The extraordinarily diverse and seemingly bizarre array of species on our planet can be explained naturally within the context of evolution. An examination of molecular evolution allows us to make sense of the existence of these species at both the population and the molecular levels.

28.1 ORIGIN OF SPECIES

Learning Outcomes:

1. Explain the two factors that lead to adaptive evolution.
2. Outline the characteristics used to distinguish species.
3. Define species concept and give examples.
4. Describe different mechanisms of how new species come into existence.

A key topic in evolutionary genetics is the origin of species. How do new species come into existence? We will begin by considering how the experiences of Darwin allowed him to explain how evolutionary change leads to the formation of new species. We will then explore the ways that modern biologists tackle the problem of distinguishing the enormous array of species on Earth. Finally, we will examine different mechanisms by which new species have come into existence.

Darwin's Experiences Allowed Him to Propose the Theory of Evolution

Charles Darwin, a British naturalist born in 1809, proposed the theory of evolution and provided evidence that existing species have evolved from preexisting ones. Like many great scientists, Darwin had a broad background in science, which enabled him to see connections among different disciplines. His thinking was influenced by the field of geology. According to the geologist Charles Lyell, the processes that alter the Earth are uniform through time. This view, which was known as uniformitarianism, suggested that the Earth is very old and that slow geological processes can lead eventually to substantial changes in the Earth's characteristics.

Darwin's own experimental observations also greatly affected his thinking. His famous voyage on the *HMS Beagle*, which lasted from 1832 to 1836, involved a careful examination of many different species. He observed the similarities among many discrete species, yet noted the differences that enabled them to be adapted to their environmental conditions. He was particularly struck by the distinctive adaptations of island species. For example, the finches found on the Galápagos Islands had unique phenotypic characteristics compared with those of similar finches found on the mainland.

A third important influence on Darwin was a paper published in 1798, "Essay on the Principle of Population," by Thomas Malthus, an English economist. Malthus asserted that the population size of humans has the potential to increase exponentially. However, such potential increases are not realized due to factors that limit population growth such as famine, war, and disease. Therefore, not all offspring are able to survive and reproduce.

With these three ideas in mind, Darwin had largely formulated his theory of evolution by natural selection by the mid-1840s. He then spent several years studying barnacles without having published his ideas. Charles Lyell, who had greatly influenced Darwin's thinking, strongly encouraged Darwin to publish his theory of evolution. In 1856, Darwin began to write a long book to explain his ideas. In 1858, however, Alfred Wallace, a naturalist working in the East Indies, sent Darwin an unpublished manuscript to read prior to its publication. In it, Wallace proposed the same ideas concerning evolution. Darwin therefore quickly excerpted some of his own writings on this subject, and two papers, one by Darwin and one by Wallace, were published in the *Proceedings of the Linnaean Society of London*. These papers were not widely recognized. A short time later, however, Darwin finished his book, *On the Origin of Species by Means of Natural Selection*, which expounded his ideas in greater detail and with experimental support. This book, which received high praise from many scientists and scorn from others, started a great debate concerning natural selection. Although some of his ideas were incomplete because the genetic basis of traits was not understood at that time, Darwin's work represents one of the most important contributions to our understanding of biology.

Darwin called evolution "the theory of descent with modification through variation and natural selection." This form of evolution, which is sometimes called **adaptive evolution,** is based on two fundamental principles: genetic variation and natural selection. A modern interpretation of evolution can view these two principles at the species level (macroevolution) and at the level of genes in populations (microevolution).

1. ***Genetic variation at the species level:*** As we have seen in Chapter 26, genetic variation is a consistent feature of natural populations. Darwin observed that many species exhibit a great amount of phenotypic variation. Although the theory of evolution preceded Mendel's pioneering work in genetics, Darwin (as well as many other people before him) observed that offspring resemble their parents more than they do unrelated individuals. Therefore, he assumed that traits are passed from parent to offspring. However, the genetic basis for the inheritance of traits was not understood at that time.

   At the gene level: Genetic variation can involve allelic differences in genes. These differences are caused by random mutations. Different alleles may affect the functions of the proteins they encode, thereby affecting the phenotype of the organism. Likewise, changes in chromosome structure and number may affect gene expression, thereby influencing the phenotype of the individual.
2. ***Natural selection at the species level:*** Darwin agreed with Malthus that most species produce many more offspring than will survive and reproduce, resulting in an ever-present struggle for existence. Over the course of many generations, those individuals that happen to possess the most favorable traits will dominate the composition of the population. The result of natural selection is to make a species better adapted to its environment and more successful at reproduction.

   At the gene level: Some alleles encode proteins that provide the individual with a selective advantage. Over time, natural selection may change the allele frequencies of genes, thereby leading to the fixation of beneficial alleles and the elimination of detrimental alleles.

Each Species Is Established Using Characteristics and Histories That Distinguish It from Other Species

Before we begin to consider how biologists study the evolution of new species, we need to consider how species are defined and identified. A **species** is a group of organisms that maintains a distinctive set of attributes in nature. How many different species are on Earth? The number is astounding. Currently, about 2 million species have been identified and catalogued. However, this number does not include a vast number of species that have yet to be classified. This is particularly true among bacteria and archaea, which are difficult to categorize into distinct species. Also, new invertebrate and even vertebrate species are still being found in the far reaches of pristine habitats. Common estimates of the total number of species range from 3 to 100 million! A study in 2011 suggested that the Earth has about 8.7 million different species.

When studying natural populations, evolutionary biologists are often confronted with situations in which some differences between two populations are apparent, but it is difficult to decide whether the two populations truly represent separate species. When two or more geographically restricted groups of the same species display one or more traits that are somewhat different but not enough to warrant their placement into different species, biologists sometimes classify such groups as **subspecies.** Similarly, many bacterial species are subdivided into **ecotypes.** Each ecotype is a genetically distinct population adapted to its local environment.

Members of the same species share an evolutionary history that is distinct from other species. Although this may seem like a reasonable way to characterize a given species, evolutionary biologists would agree that the identification of many species is a difficult undertaking. What criteria do we use to distinguish species? How many differences must exist between two populations to classify them as distinct species? Such questions are often difficult to answer.

The characteristics that a biologist uses to identify a species depend, in large part, on the species in question. For example, the traits used to distinguish insect species are quite different from those used to identify different bacterial species. The relatively high level of horizontal gene transfer among bacteria presents special challenges in the grouping of bacterial species. The division of bacteria into separate species is sometimes very difficult and perhaps arbitrary. The most commonly used characteristics to identify species are morphological traits, their inability to interbreed with members of other species, molecular features, ecological factors, and evolutionary relationships. A comparison of these concepts will help you to appreciate the various approaches that biologists use to identify the bewildering array of species on our planet.

Morphological Traits One way to establish that a population constitutes a unique species is based on their physical characteristics. Organisms are classified as the same species if their anatomical traits appear to be very similar. Likewise, microorganisms can be classified according to morphological traits at the cellular level. By comparing many different morphological traits, biologists may decide that certain populations constitute a unique species.

Although an analysis of morphological traits is a common way for biologists to establish that a particular group constitutes a species, this approach has a few drawbacks. First, researchers may have trouble deciding how many traits to consider. In addition, quantitative traits such as size and weight, which vary in a continuous way among members of the same species, are difficult to analyze. Another drawback is that the degree of dissimilarity that distinguishes different species may not show a simple relationship; the members of the same species sometimes look very different, and conversely, members of different species sometimes look remarkably similar to each other. For example, **Figure 28.1a** shows two different frogs of the species *Dendrobates tinctorius*, commonly called the dyeing poison frog. This species exists in many different-colored morphs, which are individuals of the same species that have noticeably dissimilar appearances. In contrast, **Figure 28.1b** shows two different species of meadowlarks, the western meadowlark (*Sturnella neglecta*) and eastern meadowlark (*Sturnella magna*). Both species are nearly identical in shape, coloration, and habitat, and their ranges overlap in the central United States.

Inability to Interbreed Why do biologists describe two species, such as the western and eastern meadowlarks, as being different if they are morphologically similar? One reason is that biologists have discovered that they are unable to breed with each other in nature. In the zone of overlap, very little interspecies mating takes place between western and eastern meadowlarks, largely due to differences in their songs. The song of the western meadowlark is a long series of flutelike gurgling notes that go down the scale. By comparison, the eastern meadowlark's song is a simple series of whistles, typically about four or five notes. These differences in songs enable meadowlarks to recognize potential mates as members of their own species.

Therefore, a second way to identify a species is by their inability to interbreed with other species. In the late 1920s, geneticist Theodosius Dobzhansky proposed that each species is unable to successfully interbreed with other species—a phenomenon called **reproductive isolation.** In 1942, evolutionary biologist Ernst Mayr expanded on the ideas of Dobzhansky to provide a definition of a species. According to Mayr, a key feature of sexually reproducing species is that, in nature, the members of one species have the potential to interbreed with one another to produce viable, fertile offspring but cannot successfully interbreed with members of other species.

Reproductive isolation has been used to distinguish many plant and animal species, especially those that look alike but do not interbreed. How does reproductive isolation occur? **Table 28.1** describes several ways. These are classified as **prezygotic isolating mechanisms,** which prevent the formation of a zygote, and **postzygotic isolating mechanisms,** which prevent the development of a viable and fertile individual after fertilization has taken place. Species that are reproductively isolated in nature may interbreed when kept in captivity. For example, different species of the genus *Drosophila* rarely mate with each other in nature. In the laboratory, however, it is fairly easy to produce interspecies hybrids.

Although reproductive isolation has been commonly used to classify species, it suffers from four main drawbacks. First, in nature, it may be difficult to determine if two populations are

(a) Frogs of the same species

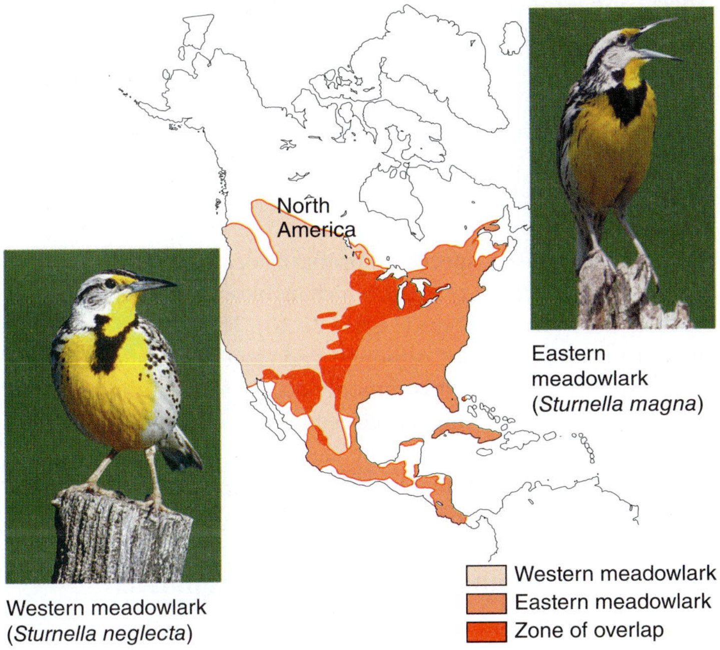

(b) Birds of different species

FIGURE 28.1 Morphological dissimilarities and similarities in species. **(a)** In some cases, two members of the same species can look quite different. These photographs show two members of the same species, the dyeing poison frog (*Dendrobates tinctorius*). **(b)** In comparison, members of different species can look quite similar as illustrated by the western meadowlark (*Sturnella neglecta*) and eastern meadowlark (*Sturnella magna*).

CONCEPT CHECK: Does this figure illustrate the strength or potential problem of using morphological traits to establish species?

TABLE 28.1
Types of Reproductive Isolation Among Different Species

Prezygotic Isolating Mechanisms	
Habitat isolation	Species may occupy different habitats so they never come in contact with each other.
Temporal isolation	Species have different mating or flowering seasons, mate at different times of day, or become sexually active at different times of the year.
Sexual isolation	Sexual attraction between males and females of different animal species is limited due to differences in behavior, physiology, or morphology.
Mechanical isolation	The anatomical structures of genitalia prevent mating between different species.
Gametic isolation	Gametic transfer takes place, but the gametes fail to unite with each other. This can occur because the male and female gametes fail to attract, because they are unable to fuse, or because the male gametes are inviable in the female reproductive tract of another species.
Postzygotic Isolating Mechanisms	
Hybrid inviability	The egg of one species is fertilized by the sperm from another species, but the fertilized egg fails to develop past early embryonic stages.
Hybrid sterility	The interspecies hybrid survives, but it is sterile. For example, the mule, which is sterile, is a cross between a female horse (*Equus caballus*) and a male donkey (*Equus asinus*).
Hybrid breakdown	The F_1 interspecies hybrid is viable and fertile, but succeeding generations (i.e., F_2, etc.) become increasingly inviable. This is usually due to the formation of less fit genotypes by genetic recombination.

reproductively isolated, particularly if they are populations with nonoverlapping geographical ranges. Second, biologists have noted many cases in which two different species can interbreed in nature yet consistently maintain themselves as separate species. For example, different species of yucca plants, such as *Yucca pallida* and *Yucca constricta*, do interbreed in nature yet typically maintain populations with distinct characteristics. For this reason, they are viewed as distinct species. A third drawback of reproductive isolation is that it does not apply to asexual species such as bacteria. Likewise, some species of plants and fungi only reproduce asexually. Finally, a fourth drawback is that it cannot be applied to extinct species. For these reasons, reproductive isolation has been primarily used to distinguish closely related species of modern animals and plants that reproduce sexually.

Molecular Features Molecular features are now commonly used to determine if two populations are different species. Evolutionary biologists often compare DNA sequences within genes, gene order along chromosomes, chromosome structure, and chromosome number as features to identify similarities and differences among different populations. DNA sequence differences are often used to compare populations. For example, researchers may compare the DNA sequence of the 16S rRNA gene between bacterial populations as a way to decide if the two populations represent different species. When the sequences are very similar, such populations are probably judged to be the same species. However, it may be difficult to draw the line when separating groups into different species. Is a 2% difference in their genome sequences sufficient to warrant placement into two different species, or do we need a 5% difference?

Ecological Factors A variety of factors related to an organism's habitat can be used to distinguish one species from another. For example, certain species of warblers can be distinguished by the habitat in which they forage for food. Some species search the ground for food, others forage in bushes or small trees, and some species primarily forage in tall trees. Such habitat differences are used to distinguish different species that look morphologically similar.

Many bacterial species have been categorized as distinct species based on ecological factors. Bacterial cells of the same species are likely to use the same types of resources (e.g., sugars and vitamins) and grow under the same types of conditions (e.g., temperature and pH). However, a drawback of this approach is that different groups of bacteria sometimes display very similar growth characteristics, and even the same species may show great variation in the growth conditions it will tolerate.

Evolutionary Relationships Later in this chapter, we will examine the methods used to produce evolutionary trees that describe the evolutionary relationships among different species. In some cases, such relationships are based on an analysis of the fossil record. Alternatively, another way to establish evolutionary relationships is by the analysis of DNA sequences. Researchers can obtain samples of cells from different individuals and compare the genes within those cells to see how similar or different they are.

A Species Concept Is a Way to Define and Distinguish Different Species

A **species concept** is a way to define what a species is and/or provide an approach for distinguishing one species from another. Several different species concepts have been proposed; we will consider a few examples. In 1942, Ernst Mayr proposed an early species concept called the **biological species concept.** According to this idea, a species is a group of individuals whose members have the potential to interbreed with one another in nature to produce viable, fertile offspring but cannot successfully interbreed with members of other species. The biological species concept emphasizes reproductive isolation as the most important criterion for delimiting species. Since then, over 20 different species concepts have been proposed by evolutionary biologists. Another example is the **evolutionary species concept** proposed by American paleontologist George Gaylord Simpson in 1961. According to this idea, a species should be defined based on the separate evolution of lineages. A third example is the **ecological species concept,** described by American evolutionary biologist Leigh Van Valen in 1976. According to this viewpoint, each species occupies an ecological niche, which is the unique set of habitat resources that a species requires, as well as its influence on the environment and other species.

Most evolutionary biologists agree that different methods are needed to distinguish the vast array of species on Earth. Even so, some evolutionary biologists have questioned whether it is valid to have many different species concepts. In 1998, Kevin de Queiroz suggested that there is only a single general species concept, which concurs with Simpson's evolutionary species concept and includes all previous concepts. According to de Queiroz's **general lineage concept,** each species is a population of an independently evolving lineage. Each species has evolved from a specific series of ancestors and, as a consequence, forms a group of organisms with a particular set of characteristics. Multiple criteria are used to determine if a population is part of an independent evolutionary lineage, and thus a species, which is distinct from others. Typically, researchers use analyses of morphology, reproductive isolation, DNA sequences, and ecology to determine if a population or group of populations is distinct from others. Because of its generality, the general lineage concept has received significant support.

Speciation Usually Occurs via a Branching Process Called Cladogenesis

Speciation is the process by which new species are formed via evolution. One way it can occur is by **anagenesis** (from the Greek *ana*, up, and *genesis*, origin) in which a single species evolves into a different species over the course of many generations (**Figure 28.2a**). However, most evolutionary biologists would argue that anagenesis is not a common mechanism for speciation, though it could occur if an entire species was confined to a single environment for long periods of time.

By comparison, speciation primarily occurs via **cladogenesis** (from the Greek *clados*, branch), which involves the division of a species into two or more species (**Figure 28.2b**). This form of speciation increases species diversity. Although cladogenesis is usually thought of as a splitting process, it commonly occurs as a budding process, which results in the original species plus one or more new species with different characteristics. If we view evolution as a tree, the new species bud from the original species and develop characteristics that prevent them from breeding with the original one.

Cladogenesis Can Be Allopatric, Parapatric, or Sympatric

Depending on the geographic locations of the evolving population(s) and the environment that a species occupies, cladogenesis is categorized as allopatric, parapatric, or sympatric (**Table 28.2**).

Allopatric Speciation **Allopatric speciation** (from the Greek *allos*, other, and Latin *patria*, homeland) is thought to be the most prevalent way for a species to diverge. It happens when members of a species become geographically separated from the other members. This form of speciation can occur by the geographic subdivision of large populations via geological processes. For example, a mountain range may emerge and split a species that occupies the lowland regions, or a creeping glacier may divide a population. **Figure 28.3** shows an interesting example in which geological separation promoted speciation. Two species of antelope squirrels occupy opposite rims of the Grand Canyon. On the south rim is Harris's antelope squirrel (*Ammospermophilus harrisi*), whereas a closely related white-tailed antelope squirrel (*Ammospermophilus leucurus*) is found on the north rim. Presumably, these two species evolved from a common species that existed before the canyon was formed. Over time, the accumulation of genetic changes in the two separated populations led to the formation of two morphologically distinct species. Interestingly, birds that can easily fly across the canyon have not diverged into different species on the opposite rims.

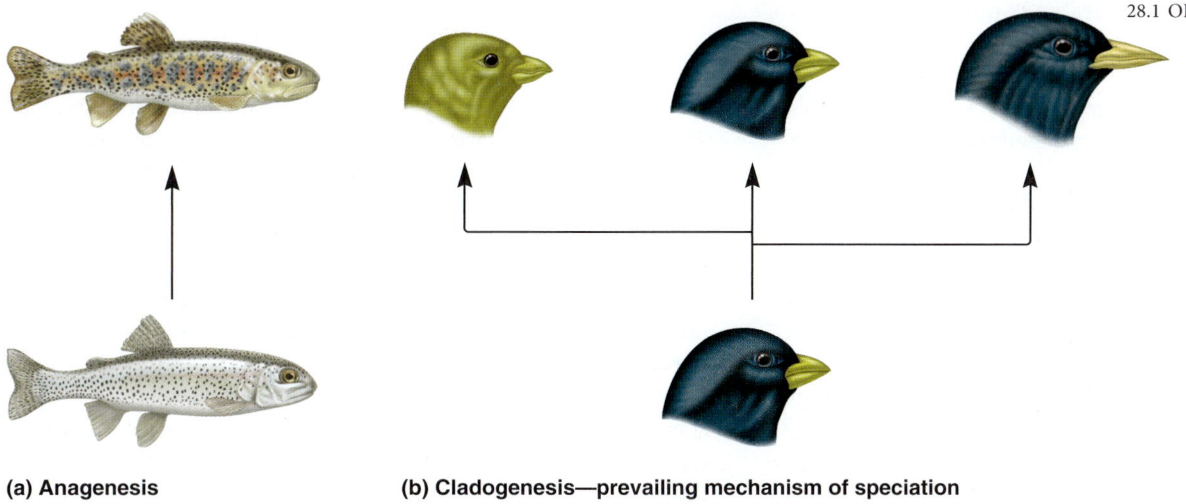

FIGURE 28.2 A comparison between anagenesis and cladogenesis, two patterns of speciation. **(a)** Anagenesis is the change of one species into another. **(b)** Cladogenesis involves a process in which one original species is separated into two or more species.

CONCEPT CHECK: Which is more common: anagenesis or cladogenesis?

TABLE 28.2

Common Genetic Mechanisms That Underlie Allopatric, Parapatric, and Sympatric Speciation

Type of Speciation	Common Genetic Mechanisms Responsible for Speciation
Allopatric—two large populations are separated by geographic barriers	Many small genetic differences may accumulate over a long period, leading to reproductive isolation. Some of these genetic differences may be adaptive, whereas others are neutral.
Allopatric—a small founding population separates from the main population	Genetic drift may lead to the rapid formation of a new species. If the group has moved to an environment that is different from its previous one, natural selection is expected to favor beneficial alleles and eliminate harmful alleles.
Parapatric—two populations occupy overlapping ranges, so a limited amount of interbreeding occurs	A new combination of alleles or chromosomal rearrangement may rapidly limit the amount of gene flow between neighboring populations because hybrid offspring have a very low fitness.
Sympatric—within a population occupying a single habitat in a range, a small group evolves into a reproductively isolated species	An abrupt genetic change leads to reproductive isolation. For example, a mutation may affect gamete recognition. In plants, the formation of a tetraploid often leads to the formation of a new species because the interspecies hybrid (e.g., diploid × tetraploid) is triploid and sterile. Also, members of a population may occupy different local environments that are continuous with each other.

Allopatric speciation can also occur via a second mechanism, known as the **founder effect,** which is thought to be more rapid and frequent than allopatric speciation caused by geological events. The founder effect, which was discussed in Chapter 26, occurs when a small group migrates to a new location that is geographically separated from the main population. For example, a storm may force a small group of birds from the mainland to a distant island. In this case, the migration of individuals between the island and the mainland is a very infrequent event. In a relatively short time, the founding population on the island may evolve into a new species. Two evolutionary mechanisms may contribute to this rapid evolution. First, genetic drift may quickly lead to the random fixation of certain alleles and the elimination of other alleles from the population. Another factor is natural selection. The environment on an island may differ significantly from the mainland environment. For this reason, natural selection on the island may favor different types of alleles.

Parapatric Speciation **Parapatric speciation** (from the Greek *para*, beside) occurs when members of a species are partially separated. In other words, the geographic separation is not complete. For example, members of a given species may invade a new ecological niche at the periphery of an existing population. Alternatively, a mountain range may divide a species into two populations but have breaks in the range where the two groups are connected physically.

FIGURE 28.3 An example of allopatric speciation: two closely related species of antelope squirrels that occupy opposite rims of the Grand Canyon.

Genes→Traits Harris's antelope squirrel (*Ammospermophilus harrisi*) is found on the south rim of the Grand Canyon, whereas the white-tailed antelope squirrel (*Ammospermophilus leucurus*) is found on the north rim. These two species evolved from a common species that existed before the canyon was formed. After the canyon was formed, the two separated populations accumulated genetic changes due to mutation, genetic drift, and natural selection that eventually led to the formation of two distinct species.

In these zones of contact, the members of two populations can interbreed, although this tends to occur infrequently. In addition, parapatric speciation may occur among very sedentary species even though no geographic isolation exists. Certain organisms are so sedentary that as little as 100–1000 m may be sufficient to limit the interbreeding between neighboring groups. Plants, terrestrial snails, rodents, grasshoppers, lizards, and many flightless insects may speciate in a parapatric manner.

During parapatric speciation, **hybrid zones** exist where two populations can interbreed. For speciation to occur, the amount of gene flow within the hybrid zones must become very limited. In other words, selection must occur against the offspring produced in the hybrid zone. This can happen if each of the two parapatric populations accumulates different chromosomal rearrangements, such as inversions and balanced translocations. How do chromosomal rearrangements, such as inversions, prevent interbreeding? As discussed in Chapter 8, if a hybrid individual has one chromosome with a large inversion and one that does not carry the inversion, crossing over during meiosis can lead to the production of grossly abnormal chromosomes. Therefore, such a hybrid individual is substantially less fertile. By comparison, an individual homozygous for two normal chromosomes or for two chromosomes carrying the same inversion is fertile, because crossing over can proceed normally.

Sympatric Speciation **Sympatric speciation** (from the Greek *sym*, together) occurs when a new species arises in the same geographic area as the species from which it was derived. In plants, a common way for sympatric speciation to occur is the formation of polyploids. As discussed in Chapter 8, complete nondisjunction of chromosomes during gamete formation can increase the number of chromosome sets within a single species (autopolyploidy) or between different species (allopolyploidy). Polyploidy is a major form of speciation in plants. In ferns and flowering plants, at least 30% of the species are polyploid. By comparison, polyploidy is much less common in animals, but it can occur. For example, roughly 30 species of reptiles and amphibians have been identified that are polyploids derived from diploid relatives.

The formation of a polyploid can abruptly lead to reproductive isolation. As an example, let's consider the probable events that led to the formation of a natural species of common hemp nettle known as *Galeopsis tetrahit*. This species is thought to be an allotetraploid derived from two diploid species: *Galeopsis pubescens* and *Galeopsis speciosa*. As shown in **Figure 28.4a**, *G. tetrahit* has 32 chromosomes, whereas the two diploid species contain 16 chromosomes each ($2n = 16$). **Figure 28.4b** illustrates the chromosomal composition of offspring derived from crosses between the allotetraploid and the diploid species. The allotetraploid crossed to another allotetraploid produces an allotetraploid. The allotetraploid is fertile, because all of its chromosomes occur in homologous pairs that can segregate evenly during meiosis. However, a cross between an allotetraploid and a diploid produces an offspring that is monoploid for one chromosome set and diploid for the other. These offspring are expected to be sterile, because they produce highly aneuploid gametes that have incomplete sets of chromosomes. This hybrid sterility renders the allotetraploid reproductively isolated from the diploid species.

Another way that sympatric speciation may occur is when members of a population occupy different local environments that are continuous with each other. An example of this type of sympatric speciation was described by Jeffrey Feder, Guy Bush, and colleagues. They studied the North American apple maggot fly (*Rhagoletis pomonella*). This fly originally fed on native hawthorn trees. However, the introduction of apple trees approximately 200 years ago provided a new local environment for this species. The apple-feeding populations of this species develop more rapidly because apples mature more quickly than hawthorn fruit. The result is partial temporal isolation in reproduction (see Table 28.1). Although the two populations—those that feed on apple trees and those that feed on hawthorn trees—are considered subspecies, evolutionary biologists speculate they may eventually become distinct species due to reproductive isolation and the accumulation of independent mutations in the two populations.

28.1 COMPREHENSION QUESTIONS

1. Evolution that results from genotypes with a higher reproductive success is based on
   a. genetic variation.
   b. natural selection.
   c. genetic drift.
   d. both a and b.
2. Characteristics that are used to establish species include
   a. morphological traits.
   b. reproductive isolation.
   c. molecular features.
   d. ecological factors.
   e. evolutionary relationships.
   f. all of the above.
3. A pair of birds flies to a deserted island and establishes a colony. Over time, this population evolves into a new species. This is an example of
   a. allopatric speciation.
   b. parapatric speciation.
   c. sympatric speciation.
   d. all of the above.
4. The formation of polyploids is common in plants and can abruptly produce a new species. This is an example of
   a. allopatric speciation.
   b. parapatric speciation.
   c. sympatric speciation.
   d. all of the above.

28.2 PHYLOGENETIC TREES

Learning Outcomes:

1. Describe the key features of phylogenetic trees.
2. Compare and contrast the approaches of cladistics and phenetics for making phylogenetic trees.
3. Define horizontal gene transfer and explain how it affects the relationships in phylogenetic trees.

Thus far, we have considered the factors that play a role in the formation of new species. In this section, we will examine **phylogeny**—the sequence of events involved in the evolutionary development of a species or group of species. A **phylogenetic tree** is a diagram that describes a phylogeny. Such a tree is a hypothesis of the evolutionary relationships among different species.

(a) Chromosomal composition of 3 *Galeopsis* species

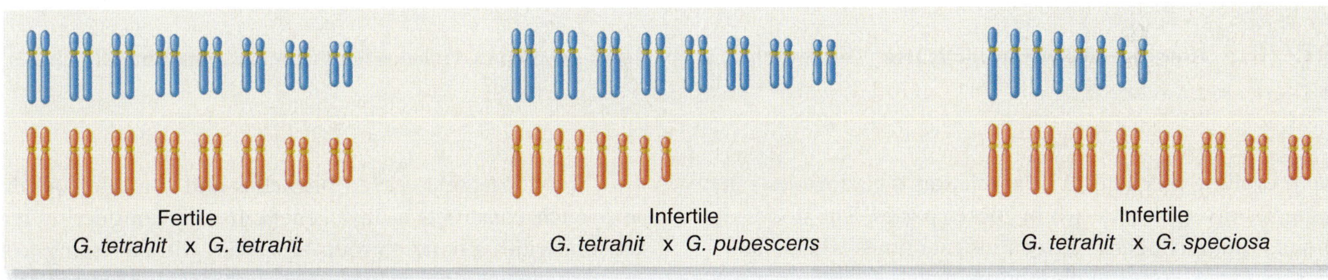

(b) Outcome of intraspecies and interspecies crosses

FIGURE 28.4 A comparison of crosses between three natural species of hemp nettle with different ploidy levels. **(a)** *Galeopsis tetrahit* is an allotetraploid that is thought to be derived from *Galeopsis pubescens* and *Galeopsis speciosa*. **(b)** If *G. tetrahit* is crossed with the other two species, the F_1 hybrid offspring will be monoploid for one chromosome set and diploid for the other. The F_1 offspring are likely to be sterile, because they will produce highly aneuploid gametes.

CONCEPT CHECK: Why is *G. tetrahit* reproductively isolated from the other two species?

Phylogenetic trees are now based on **homology,** which refers to similarities among various species that occur because the species are derived from a common ancestor. Attributes that are the result of homology are said to be **homologous.** For example, the wing of a bat, the arm of a human, and the front leg of a cat are homologous structures. By comparison, a bat wing and insect wing are not homologous; they arose independently of each other. When constructing phylogenetic trees, researchers identify homologous features that are shared by some species but not by others. This allows them to group species based on their shared characteristics. Researchers typically study homology at the level of morphological traits or at the level of genes.

Historically, comparisons of morphological similarities and differences have been used to construct evolutionary trees. In this approach, species that share certain characteristics (i.e., homologous traits) tend to be placed closer together on the tree. In addition, species have been categorized based on physiology, biochemistry, and even behavior. Although these approaches continue to be used, researchers are increasingly using molecular data to infer evolutionary relationships. In 1963, Linus Pauling and Emile Zuckerkandl were the first to suggest the use of molecular data to establish evolutionary relationships. When comparing homologous genes in different species, the DNA sequences from closely related species are more similar to each other than are the sequences from distantly related species. In this section, we will examine the general features of phylogenetic trees, how they can be constructed, and the types of information they reveal.

A Phylogenetic Tree Depicts the Evolutionary Relationships Among Different Species

Let's first take a look at what information is found within a phylogenetic tree and the form in which it is presented. **Figure 28.5** shows a hypothetical phylogenetic tree of the relationships between various butterfly species labeled A through J. The vertical axis represents time, with the oldest species at the bottom.

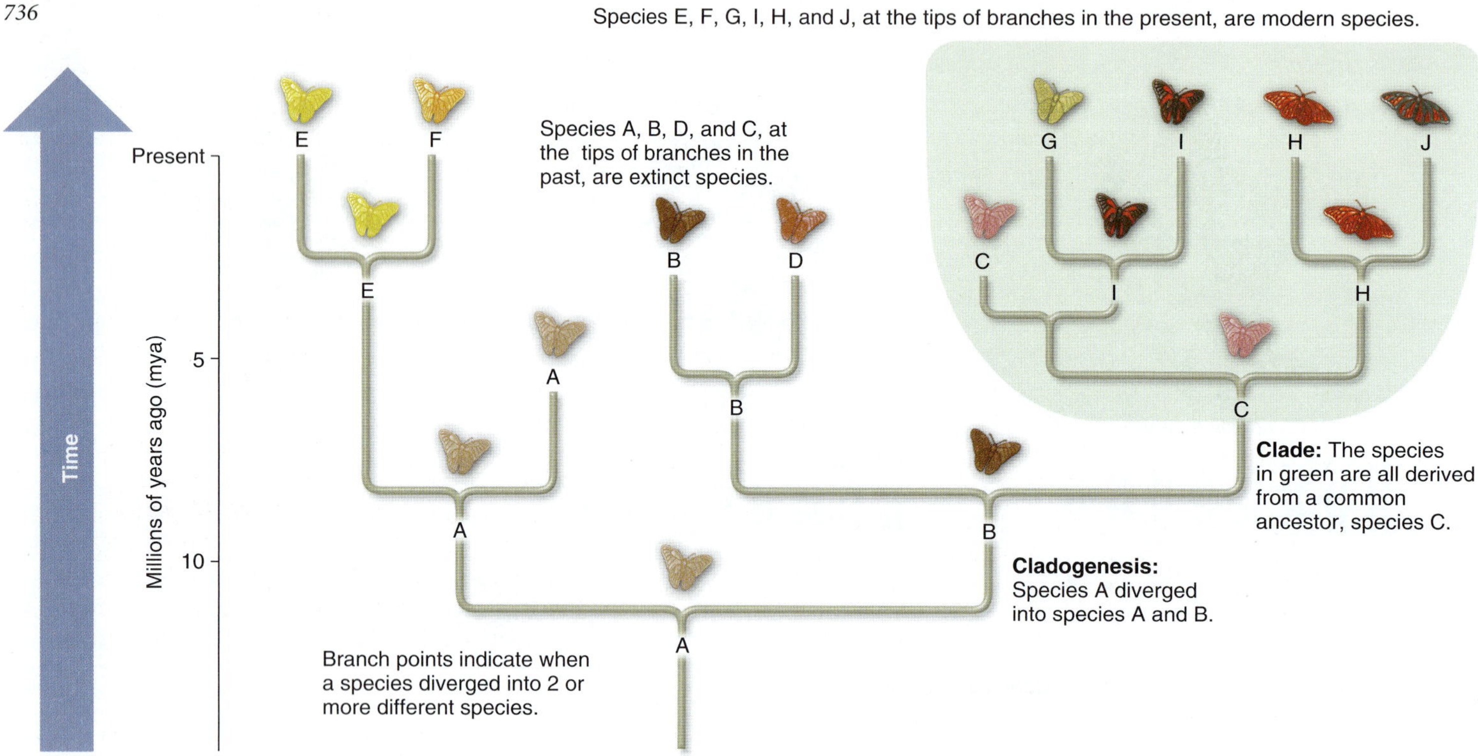

FIGURE 28.5 How to read a phylogenetic tree. This hypothetical tree shows the proposed relationships among various butterfly species.

CONCEPT CHECK: What is a clade?

The prevailing mechanism of speciation is cladogenesis, in which a species diverges into two or more species. The nodes or branch points in a phylogenetic tree illustrate times when cladogenesis has occurred. For example, approximately 12 million years ago (mya), species A diverged into species A and species B by cladogenesis. The tips of branches represent either species that became extinct in the past, such as species A, B, D and C, or modern species, such as E, F, G, I, H, and J, that are at the top of the tree.

By studying the branch points of a phylogenetic tree, researchers can group species according to common ancestry. A **monophyletic group,** also known as a **clade,** is a group of species consisting of all descendants of the group's most common ancestor. For example, the group highlighted in light green is a clade derived from the common ancestor labeled C. The present-day descendants of a common ancestor can also be called a clade. In this case, species G, I, H, and J form a modern clade. Likewise, the entire tree shown in Figure 28.5 forms a clade, with species A as a common ancestor. As we see in this figure, smaller and more recent clades are subsets of larger ones.

The phylogenetic tree in Figure 28.5 includes ancestral species. This type of tree could be obtained by examining the fossil record. However, many phylogenetic trees do not include ancestral species, but instead focus on the relationships among modern species.

A Phylogenetic Tree Can Be Constructed Using Cladistic and Phenetic Approaches

Now that we appreciate the concept of a phylogenetic tree, let's turn our attention to how biologists actually construct them. The most popular methods of building phylogenetic trees can be classified into two broad categories called phenetics and cladistics. A **phenetic approach** constructs a phylogenetic tree by considering the overall similarities among a group of species without trying to understand their evolutionary history. Such trees are called **phenograms.** By comparison, in a **cladistic approach,** a phylogenetic tree is constructed by considering the various possible pathways of evolution and choosing from among the best possible trees. In this approach, a goal is to group species based on knowledge concerning traits that occurred earlier in evolution versus traits that came later. Such trees are called **cladograms.**

Which approach is better? For data involving morphological traits and for the construction of complex evolutionary trees that include many taxonomic levels, the cladistic approach is generally superior. Even so, both approaches are used, and many evolutionary biologists construct their evolutionary trees based on a combination of both cladistics and phenetics. We will briefly consider both types of methods next.

Cladistics In the 1950s, the German entomologist Willi Hennig proposed that evolutionary relationships should be inferred from new features shared by descendants of a common ancestor. A cladistic approach compares features, also called characters, that are either shared or not shared by different species. These can be morphological traits, such as the shapes of birds' beaks, or molecular traits, such as sequences of homologous genes. Such characters may come in different versions called character states. Those that are shared with a distant ancestor are called **ancestral characters** (also called **primitive characters**). Such characters are viewed as being older—ones that arose earlier in evolution. In contrast, a **shared derived character,** or **synapomorphy,** is a trait that is shared by a group of organisms but not by a distant common ancestor. Compared with ancestral characters, shared derived characters are more recent traits on an evolutionary timescale. For example, among mammals, only some species, such as whales and dolphins, have flippers. In this case, flippers were derived from the two front limbs of an ancestral species. The word *derived* refers to the observation that evolution involves the modification of traits in preexisting species. In other words, the features of newer populations of organisms are derived from changes in preexisting populations. The basis of the cladistic approach is to analyze many shared derived characters among groups of species to deduce the pathway that gave rise to the species.

To understand the concept of ancestral versus shared derived characters, let's consider how a cladogram can be constructed based on molecular data such as a sequence of a gene. Our example uses molecular data obtained from seven different hypothetical species called A through G. In these species, the same gene was sequenced from seven different species; a portion of the gene sequence is shown as follows:

A: GATAGTACCC
B: GATAGTTCCC
C: GATAGTTCCG
D: GGTATTACCC
E: GGTATAACCC
F: GGTAGTACCA
G: GGTAGTACCC

In a cladogram, an **ingroup** is a group in which a researcher is interested that is hypothesized to be monophyletic. By comparison, an **outgroup** is a species or group of species that possess characteristics that set it apart as being more distantly related to an ingroup. The root of a cladogram is placed between the outgroup and the ingroup. In the cladogram of **Figure 28.6**, the outgroup is species E. This may have been inferred because the other species share traits that are not found in species E. The other species (A, B, C, D, F, and G) form the ingroup. For these data, a mutation that changes the DNA sequence is analogous to a modification of a characteristic. Species that share such genetic changes possess shared derived characters because the new genetic sequence was derived from a more ancestral sequence.

Now that we understand some of the general principles of cladistics, let's consider the steps a researcher follows to construct a cladogram using this approach.

1. ***Choose the species in whose evolutionary relationships you are interested.*** In a simple cladogram, individual species are compared with each other. In more complex cladograms, species may be grouped into larger taxa (e.g., families) and compared with each other. If such grouping is done, the groups must be clades for the results to be reliable.
2. ***Choose characters for comparing different species.*** As mentioned, a character is a general feature of an organism. Characters may come in different versions called character states. For example, a base at a particular location in a gene can be considered a character, and this character could exist in different character states, such as A, T, G, or C, due to mutations.
3. ***Determine the polarity of character states.*** In other words, determine if a character state is ancestral or derived. Comparisons are made within the ingroup and between the ingroup and the outgroup.
4. ***Group species (or higher taxa) based on shared derived characters.***
5. ***Build a cladogram based on the following principles:***
   - All species (or higher taxa) are placed on tips in the phylogenetic tree, not at branch points. A cladogram does not include ancestral species at branch points.
   - Each cladogram branch point should have a list of one or more shared derived characters that are common to all species above the branch point unless the character is later modified.

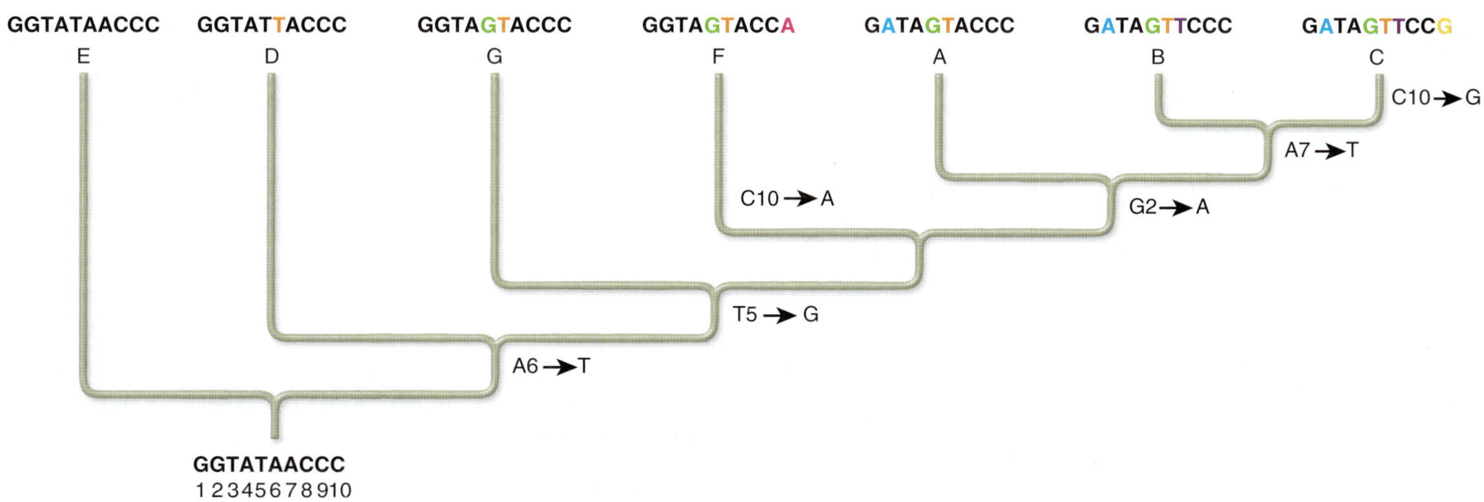

FIGURE 28.6 Shared derived characters involving a molecular trait. This phylogenetic tree illustrates a cladogram of relationships involving homologous gene sequences found in seven species. Mutations that alter an ancestral DNA sequence are shared among certain species, thereby allowing the construction of a cladogram.

CONCEPT CHECK: What is the difference between an ancestral character and a shared derived character?

- All shared derived characters appear together only once in a cladogram unless they independently arose during evolution more than once in the ancestors of different clades.

6. ***Choose the best cladogram among possible options.*** When grouping species (or higher taxa), more than one cladogram is possible. Therefore, analyzing the data and producing the most likely cladogram is a key aspect of this process. As described next, different approaches can be followed to achieve this goal.

The greatest challenge in a cladistic approach is to determine the correct order of events. It may not always be obvious which traits are ancestral and came earlier and which are derived and came later in evolution. Different approaches can be used to deduce the correct order. First, for morphological traits, a common way to deduce the order of events is to analyze fossils and determine the relative dates that certain traits arose. A second strategy assumes that the best hypothesis is the one that requires the fewest number of evolutionary changes. This concept, called the **principle of parsimony,** states that the preferred hypothesis is the one that is the simplest. For example, if two species possess a tail, we would initially assume that a tail arose once during evolution and that both species have descended from a common ancestor with a tail. Such a hypothesis is simpler than assuming that tails arose twice during evolution .

Let's consider a simple example to illustrate how the principle of parsimony can be used. This example involves molecular data obtained from four different hypothetical species called A through D. In these species, a homologous region of DNA was sequenced as shown here.

```
    12345
A   GTACA
B:  GACAG
C:  GTCAA
D:  GACCG
```

Many different trees could be constructed from this information. Three of them are shown in Figure 28.7. These three trees hypothesize that sequence A is ancestral, though other trees are possible. In these examples, tree 1 requires seven mutations, tree 2 requires six, and tree 3 requires only five. Because tree 3 requires the fewest number of mutations, it is considered the most parsimonious. Based on the principle of parsimony, tree 3 is the more likely choice.

Phenetics As mentioned, a phenetic approach constructs a phylogenetic tree based on overall similarities among a group of species. A relatively simple method is the UPGMA method (unweighted pair group method with arithmetic mean). When comparing gene sequences, this method assumes that after two species have diverged from each other, the rate of accumulation of neutral mutations in a homologous region of DNA is approximately the same in the two species. As discussed later in Section 28.3, this means that the molecular clock for neutral mutations is linear. If this assumption is met, closely related sequences (i.e., those that have diverged more recently) have greater sequence similarity compared to distantly related species.

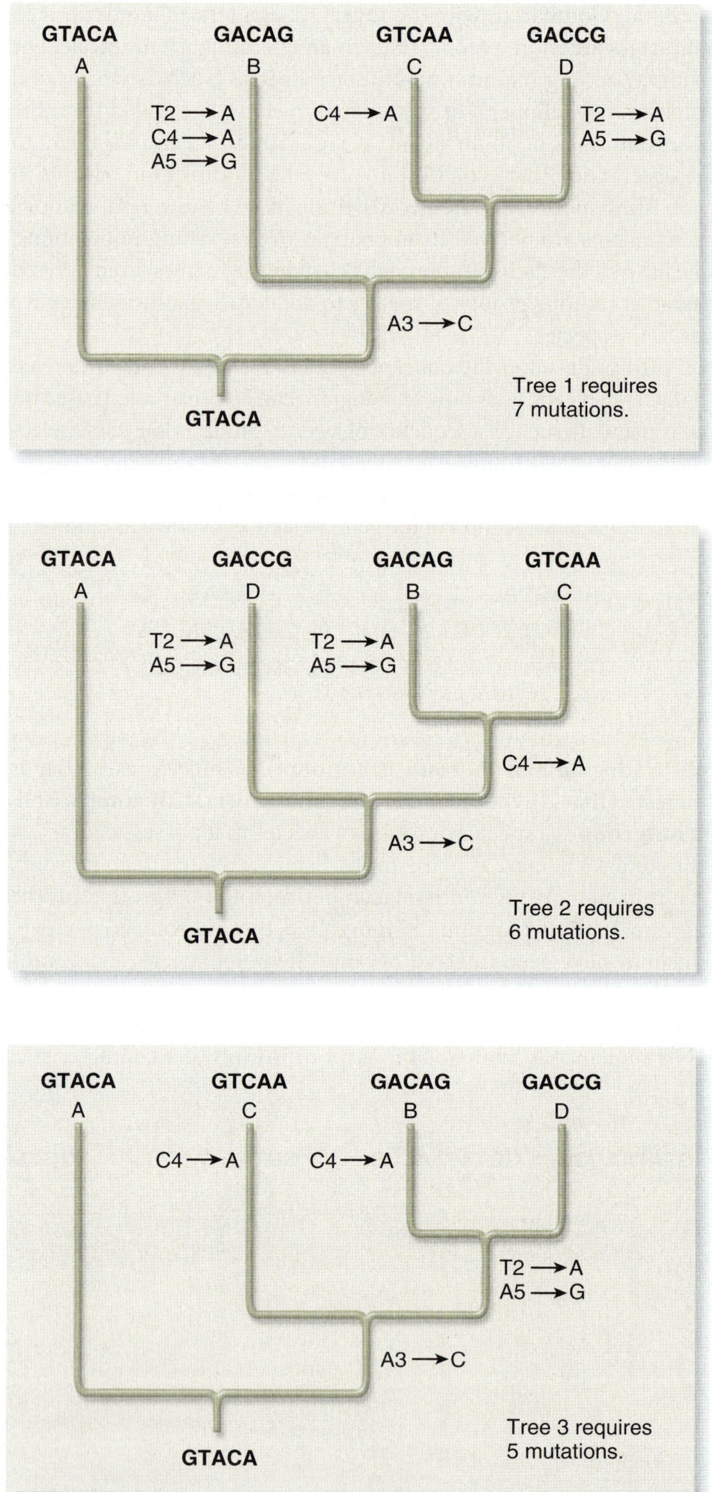

FIGURE 28.7 The cladistic approach: choosing a cladogram from molecular genetic data. This figure shows three different phylogenetic trees for the evolution of a short DNA sequence, but many more are possible. According to the principle of parsimony, the cladogram shown in tree 3 is the more plausible choice because it requires only five mutations. When constructing cladograms based on long genetic sequences, researchers use computers to generate trees with the fewest possible genetic changes.

CONCEPT CHECK: What is the principle of parsimony?

As an example of the UPGMA method, we can consider the nucleotide similarities among five species of primates. In a homologous region containing 10,000 bp, the following numbers of nucleotide substitutions were found:

	Human	Chimpanzee	Gorilla	Orangutan	Rhesus monkey
Human	0	145	151	298	751
Chimpanzee	145	0	157	294	755
Gorilla	151	157	0	304	739
Orangutan	298	294	304	0	710
Rhesus monkey	751	755	739	710	0

To construct a tree, we begin with the pair that is the most closely related. In this case, it is humans and chimpanzees, because that pair has the fewest number of nucleotide substitutions. The number of nucleotide substitutions per 10,000 nucleotides equals 145. We divide 145 by 2 to yield the average number of substitutions, which occurred in each species since the time they diverged. This number equals 72.5. In other words, about 72.5 substitutions occurred in humans and about 72.5 (different) substitutions occurred in chimpanzees since they diverged from a common ancestor and that is why there are 145 nucleotide differences today. This value is expressed as the percentage of nucleotide differences. So,

$$\underset{\text{(humans vs chimpanzee)}}{\text{Percentage of nucleotide differences}} = \frac{72.5 \times 100}{10{,}000} = 0.725$$

Next, we consider the species that is the most closely related to humans and chimpanzees. That is the gorilla: 151 substitutions occurred between gorillas and humans, and 157 occurred between gorillas and chimpanzees. The average number of substitutions between these two pairs equals 151 + 157 divided by 2, which equals 154. We divide 154 by 2 to yield the average number of substitutions that occurred in each species since the time they diverged from a common ancestor. This number equals 77. In other words, about 77 substitutions occurred in humans or chimpanzees and about 77 (different) substitutions occurred in gorillas since they diverged from a common ancestor and that is why there are 154 nucleotide differences today. To express this value as the percentage of nucleotide differences:

$$\underset{\text{(gorillas vs humans/chimpanzees)}}{\text{Percentage of nucleotide substitutions}} = \frac{77 \times 100}{10{,}000} = 0.77$$

Next, we consider the species that is the most closely related to humans, chimpanzees, and gorillas. That is the orangutan: 298 substitutions occurred between orangutans and humans, 294 between orangutans and chimpanzees, and 304 between orangutans and gorillas. The average number of substitutions between these three pairs equals 298 + 294 + 304 divided by 3, which equals 299. We divide 299 by 2 to yield the average number of substitutions that occurred in each species since the time they diverged from a common ancestor. This number approximately equals 149. To express this value as the percentage of nucleotide differences:

$$\underset{\text{(orangutans vs humans/chimpanzees/gorillas)}}{\text{Percentage of nucleotide substitutions}} = \frac{149 \times 100}{10{,}000} = 1.49$$

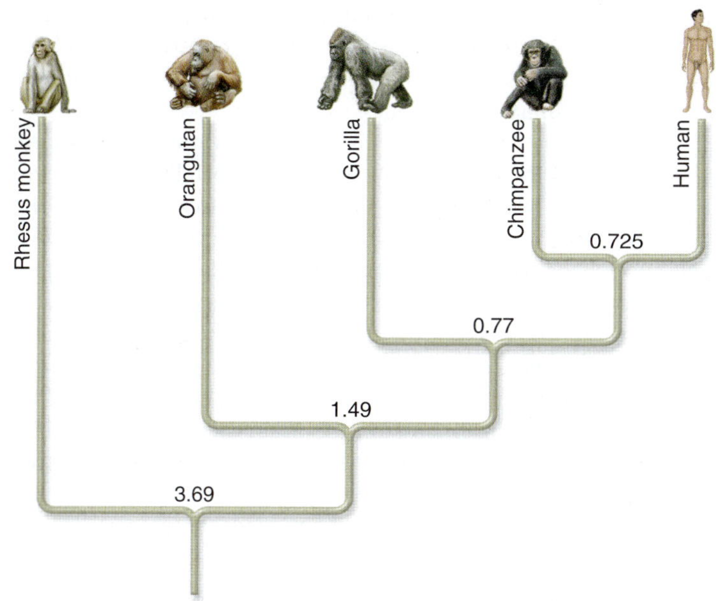

FIGURE 28.8 Reconstruction of a phylogenetic tree using the UPGMA method.

Finally, we consider the most distantly related species, which is the Rhesus monkey: 751 substitutions occurred between Rhesus monkeys and humans, 755 between Rhesus monkeys and chimpanzees, 739 between Rhesus monkeys and gorillas, and 710 between Rhesus monkeys and orangutans. The average number of substitutions between these four pairs equals 751 + 755 + 739 + 710 divided by 4, which equals 739. We divide 739 by 2 to yield the average number of substitutions that occurred in each species since the time they diverged from a common ancestor. This number approximately equals 369. To express this value as the percentage of nucleotide differences:

$$\underset{\text{(Rhesus monkeys vs humans/chimpanzees/gorillas/orangutans)}}{\text{Percentage of nucleotide substitutions}} = \frac{369 \times 100}{10{,}000} = 3.69$$

With these data, we can construct the phylogenetic tree shown in Figure 28.8. This illustration shows the relative evolutionary relationships among these primate species.

Maximum Likelihood and Bayesian Methods Are Used to Discriminate Among Possible Phylogenetic Trees

Evolutionary biologists also apply other approaches, such as maximum likelihood and Bayesian methods, when proposing and evaluating phylogenetic trees. These methods involve the use of an evolutionary model—a set of assumptions about how evolution is likely to happen. For example, mutations affecting the third base in a codon are often neutral because they don't affect the amino acid sequence of the encoded protein and therefore don't affect the fitness of an organism. As discussed later in this chapter, such neutral mutations are more likely to become prevalent in a population than are mutations in the first or second base. Therefore, one possible assumption of an evolutionary model is that neutral mutations are more likely than nonneutral ones.

According to an approach called **maximum likelihood,** researchers ask the question: What is the probability that an evolutionary model and a proposed phylogenetic tree would give rise to the observed data? The rationale is that a phylogenetic tree that gives a higher probability of producing the observed data is preferred to any trees that give a lower probability. By comparison, **Bayesian methods** ask the question: What is the probability that a particular phylogenetic tree is correct, given the observed data and a particular evolutionary model? Though the computational strategies of maximum likelihood and Bayesian methods are different (and beyond the scope of this textbook), the goal of both approaches is to identify one or more trees that are most likely to be correct based on an evolutionary model and the available data.

Phylogenetic Trees Refine Our Understanding of Evolutionary Relationships

For molecular evolutionary studies, the DNA sequences of many genes have been obtained from a wide range of sources. Several different types of gene sequences have been used to construct phylogenetic trees. The gene encoding 16S rRNA (bacterial) or 18S rRNA (eukaryotic), an rRNA found in the small ribosomal subunit, is commonly analyzed. This gene has been sequenced from thousands of different species. Because rRNA is universal in all living organisms, its function was established at an early stage in the evolution of life on this planet, and its sequence has changed fairly slowly. Presumably, most mutations in this gene are deleterious, so few neutral or beneficial alleles can occur. This limitation causes this gene sequence to change very slowly during evolution. Furthermore, 16S and 18S rRNAs are rather large molecules and therefore contain a large amount of sequence information.

In 1977, Carl Woese analyzed 16S and 18S rRNA sequences and proposed three main evolutionary branches called **domains**: the **Bacteria,** the **Archaea,** and the **Eukaryotes** (also called Eukarya). From these types of genetic analyses, it has become apparent that all living organisms are connected through a complex evolutionary tree.

Although the work of Woese was a breakthrough in our appreciation of evolution, more recent molecular genetic data have shed new light on the classification of species. Specifically, biologists once categorized eukaryotic species into four kingdoms: fungi (Fungi), animals (Animalia), plants (Plantae), and protists (Protista). However, recent models propose several major groups, called **supergroups,** as a way to organize eukaryotes into monophyletic groups. Figure 28.9 shows a diagram that hypothesizes seven eukaryotic supergroups. Each supergroup contains distinctive types of protists. In addition, Plantae is found within Land plants and algal relatives, and Fungi and Animalia are within the Opisthokonta supergroup. As seen in this figure, molecular data and newer ways of building trees reveal that protists played a key role in the evolution of many diverse groups of eukaryotic species, producing several large monophyletic supergroups.

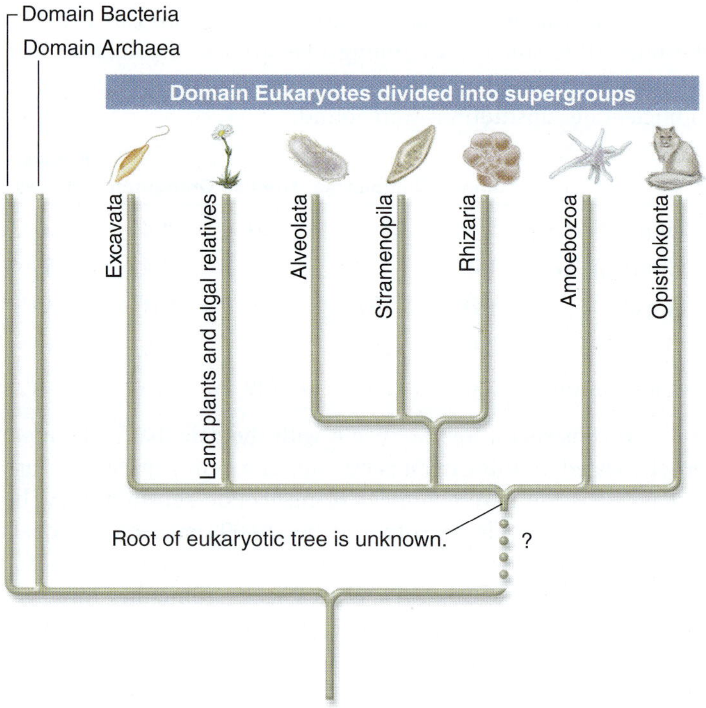

FIGURE 28.9 A modern cladogram for eukaryotes. Each supergroup shown here is hypothesized to be monophyletic. This drawing should be considered a working hypothesis. The arrangement of these supergroups relative to each other is not entirely certain.

CONCEPT CHECK: Discuss where protists are found in this newer grouping of eukaryotic species.

Horizontal Gene Transfer Also Contributes to the Evolution of Species

The types of phylogenetic trees considered thus far are examples of **vertical evolution,** in which species evolve from preexisting species by the accumulation of gene mutations and by changes in chromosome structure and number. Vertical evolution involves genetic changes in a series of ancestors that form a lineage. In addition to vertical evolution, however, species accumulate genetic changes by another mechanism called **horizontal gene transfer.** This refers to a process in which an organism incorporates genetic material from another organism without being the offspring of that organism. It often involves the exchange of genetic material between different species.

An analysis of many genomes suggests that horizontal gene transfer was prevalent during the early stages of evolution, when all organisms were unicellular, but continued even after the divergence of the three major domains of life. With regard to modern organisms, horizontal gene transfer remains prevalent among prokaryotic species. By comparison, this process is less common among eukaryotes, though it does occur. Researchers have speculated that multicellularity and sexual reproduction have presented barriers to horizontal gene transfer in many eukaryotes. For a gene to be transmitted to eukaryotic offspring, it must be transferred into a eukaryotic gamete or a cell that gives rise to gametes.

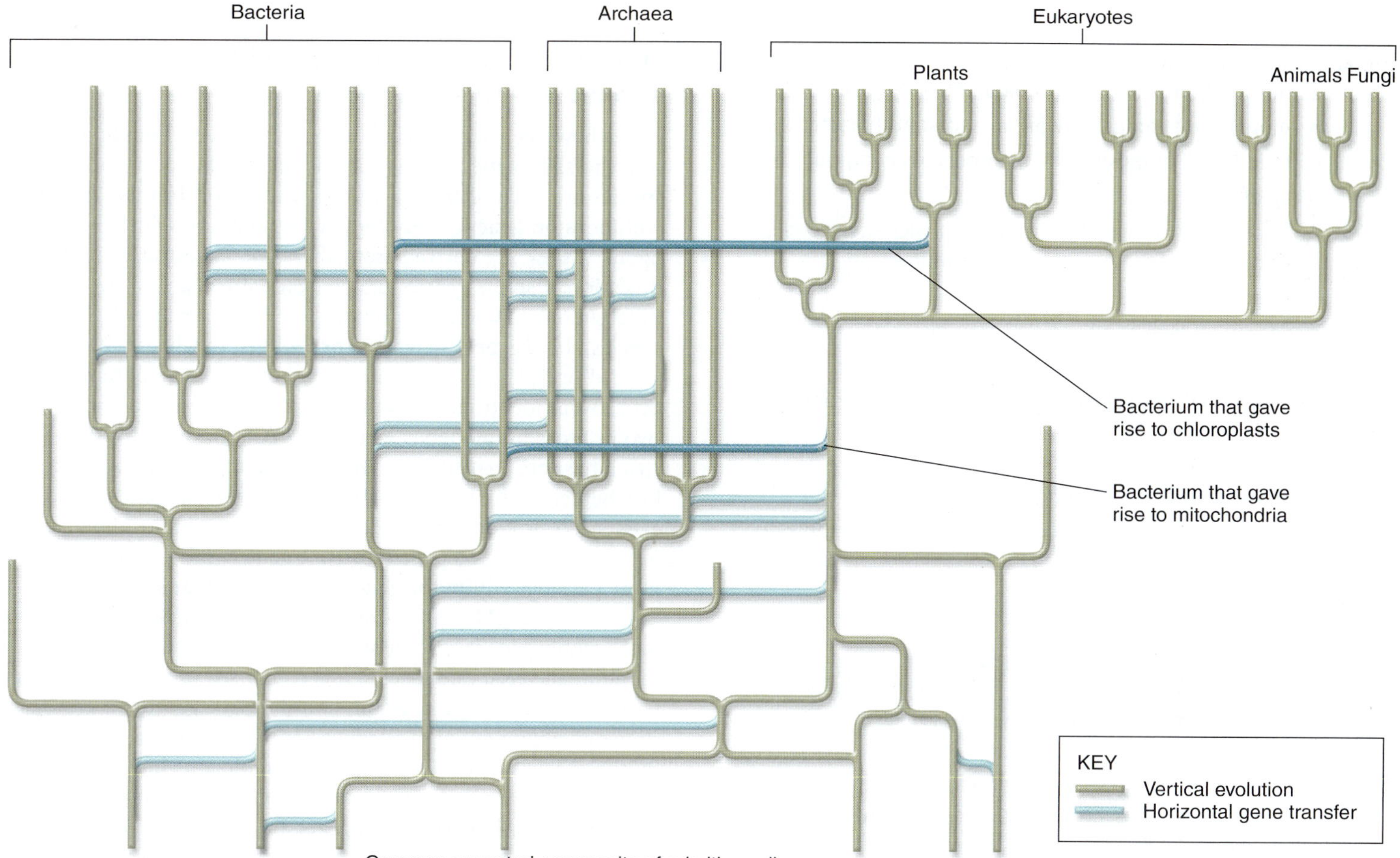

FIGURE 28.10 A revised view regarding the evolution of life, incorporating the concept of horizontal gene transfer. This phylogenetic tree shows a classification of life on Earth that includes the contribution of horizontal gene transfer in the evolution of species on our planet. This phenomenon was prevalent during the early stages of evolution when all organisms were unicellular. Horizontal gene transfer continues to be a prominent factor in the speciation of bacteria and archaea. Note: This tree is meant to be schematic. Figure 28.9 is a more realistic representation of the evolutionary relationships among modern species. In this simplified drawing, the horizontal gene transfer that gave rise to chloroplasts is shown as a single event, though multiple transfers of plastids have occurred among different supergroups during the evolution of life on Earth.

CONCEPT CHECK: What is horizontal gene transfer?

How has horizontal gene transfer affected evolution? In the past few decades, scientists have debated the role of horizontal gene transfer in the earliest stages of evolution, prior to the emergence of the two prokaryotic domains. The traditional viewpoint was that the three domains of life arose from a single type of cell called the universal ancestor. However, genomic research has suggested that horizontal gene transfer may have been particularly common during the early stages of evolution on Earth, when all species were unicellular. Rather than proposing that all life arose from a single type of cell, horizontal gene transfer may have been so prevalent that the universal ancestor may have actually been an ancestral community of cell lineages that evolved as a whole. If that was the case, the tree of life cannot be traced back to a single universal ancestor.

Figure 28.10 illustrates a schematic scenario for the evolution of life on Earth that includes the roles of both vertical evolution and horizontal gene transfer. This has been described as a "web of life" rather than a "tree of life." Instead of a universal ancestor, a web of life began with a community of primitive cells that transferred genetic material in a horizontal fashion. Horizontal gene transfer was also prevalent during the early evolution of bacteria and archaea and when eukaryotes first emerged as unicellular species. Although horizontal gene transfer remains a prominent way to foster evolutionary change in modern bacteria and archaea, the region of the diagram that contains eukaryotic species has a more treelike structure, because horizontal gene transfer has become much less common in these species.

28.2 COMPREHENSION QUESTIONS

1. Phylogenetic trees are based on
   a. natural selection.
   b. genetic drift.
   c. homology.
   d. none of the above.

2. A shared derived character is
   a. derived from an ancestral character.
   b. more recent compared with an ancestral character.
   c. found in all or most of the members of an ingroup.
   d. all of the above.
3. An approach that is used to propose a phylogenetic tree is
   a. cladistics and the principle of parsimony.
   b. phenetics.
   c. maximum likelihood and Bayesian methods.
   d. all of the above.
4. Horizontal gene transfer is a process in which genetic material from an organism is
   a. transferred from cell to cell.
   b. transferred to its offspring.
   c. transferred to another organism that is not the offspring of that organism.
   d. released into the environment.

28.3 MOLECULAR EVOLUTION

Learning Outcomes:

1. Compare and contrast orthologs and paralogs.
2. Describe how researchers analyze orthologs to construct phylogenetic trees.
3. Outline the neutral theory of evolution.
4. Explain the concept of a molecular clock and how it is used to date a phylogenetic tree.
5. Describe how changes in chromosome structure and number are associated with molecular evolution.

Recall from the chapter introduction that molecular evolution refers to patterns and processes associated with evolutionary change at the molecular level. Such changes may be phenotypically neutral or they may underlie the phenotypic changes associated with adaptive evolution. Differences in nucleotide sequences are quantitative and can be analyzed using mathematical principles in conjunction with computer programs. Evolutionary changes at the DNA level can be objectively compared among different species to establish evolutionary relationships. Furthermore, this approach can be used to compare any two existing organisms, no matter how greatly they differ in their morphological traits. For example, we can compare DNA sequences between humans and bacteria, or between plants and fruit flies. Such comparisons would be very difficult at a morphological level. In this section, we will examine how evolution occurs at the molecular level.

Homologous Genes Are Derived from a Common Ancestral Gene

As we have seen, a phylogenetic tree is based on homology, which refers to similarities among various species that occur because the species are derived from a common ancestor. In this section, we will focus on genetic homology. Two genes are said to be homologous if they are derived from the same ancestral gene. During evolution, a single species may become divided into two or more different ones. When two homologous genes are found in different species, these genes are termed **orthologs.** In Chapter 25, we considered orthologs of *Hox* genes. In that case, several homologous genes were identified in the fruit fly and the mouse. In addition, two or more homologous genes can be found within a single species. These are termed paralogous genes, or **paralogs.** As discussed in Chapter 8, this can occur because abnormal gene duplication events can produce multiple copies of a gene and ultimately lead to the formation of a **gene family.** A gene family consists of two or more paralogs within the genome of a particular species.

Figure 28.11a shows examples of both orthologs and paralogs in the globin gene family. Hemoglobin is an oxygen-carrying protein found in all vertebrate species. It is composed of two different subunits, encoded by the α-globin and β-globin genes. Figure 28.11a shows an alignment of the deduced amino acid sequences encoded by these genes. The sequences are homologous between humans and horses because of their evolutionary relationship. We say that the human and horse α-globin genes are orthologs of each other, as are the human and horse β-globin genes. The α-globin and β-globin genes in humans are paralogs of each other.

As shown in **Figure 28.11b**, the sequences of the orthologs are more similar to each other than they are to the paralogs. For example, the sequences of human and horse β globins show 25 differences, whereas the human β globin and human α globins show 84 differences. What do these results mean? They indicate that the gene duplications that created the α-globin and β-globin genes occurred long before the evolutionary divergence that produced different species of mammals. For this reason, a greater amount of time has elapsed for the α- and β-globin genes to accumulate changes compared to the amount of time since the evolutionary divergence of mammalian species. This idea is schematically shown in **Figure 28.12**.

Based on the analysis of genetic sequences, evolutionary biologists have estimated that the gene duplication that produced the α-globin and β-globin gene lineages occurred approximately 400 mya, whereas the speciation events that resulted in different species of mammals occurred less than 200 mya. Therefore, the α-globin and β-globin genes have had much more time to accumulate changes relative to each other.

Hemoglobin

α	1	—	2	3	4	5	6	7	8	9	10	11	12	13	14	15	16	17	18	19	20	21	22	23	24	25	26	27	28	29	30
Human	Val	—	Leu	Ser	Pro	Ala	Asp	Lys	Thr	Asn	Val	Lys	Ala	Ala	Trp	Gly	Lys	Val	Gly	Ala	His	Ala	Gly	Glu	Tyr	Gly	Ala	Glu	Ala	Leu	Glu
Horse	Val	—	Leu	Ser	Ala	Ala	Asp	Lys	Thr	Asn	Val	Lys	Ala	Ala	Trp	Ser	Lys	Val	Gly	Gly	His	Ala	Gly	Glu	Val	Gly	Ala	Glu	Ala	Leu	Glu

β	1	2	3	4	5	6	7	8	9	10	11	12	13	14	15	16	17	18	—	—	19	20	21	22	23	24	25	26	27	28	29
Human	Val	His	Leu	Thr	Pro	Glu	Glu	Lys	Ser	Ala	Val	Thr	Ala	Leu	Trp	Gly	Lys	Val	—	—	Asn	Val	Asp	Glu	Val	Gly	Gly	Glu	Ala	Leu	Gly
Horse	Val	Gln	Leu	Ser	Gly	Glu	Glu	Lys	Ala	Ala	Val	Leu	Ala	Leu	Trp	Asp	Lys	Val	—	—	Asn	Glu	Glu	Glu	Val	Gly	Gly	Glu	Ala	Leu	Gly

α	31	32	33	34	35	36	37	38	39	40	41	42	43	44	45	46	—	47	48	49	50	—	—	—	—	—	51	52	53	54	55
Human	Arg	Met	Phe	Leu	Ser	Phe	Pro	Thr	Thr	Lys	Thr	Tyr	Phe	Pro	His	Phe	—	Asp	Leu	Ser	His	—	—	—	—	—	Gly	Ser	Ala	Gln	Val
Horse	Arg	Met	Phe	Leu	Gly	Phe	Pro	Thr	Thr	Lys	Thr	Tyr	Phe	Pro	His	Phe	—	Asp	Leu	Ser	His	—	—	—	—	—	Gly	Ser	Ala	Gln	Val

β	30	31	32	33	34	35	36	37	38	39	40	41	42	43	44	45	46	47	48	49	50	51	52	53	54	55	56	57	58	59	60
Human	Arg	Leu	Leu	Val	Val	Tyr	Pro	Trp	Thr	Gln	Arg	Phe	Phe	Glu	Ser	Phe	Gly	Asp	Leu	Ser	Thr	Pro	Asp	Ala	Val	Met	Gly	Asn	Pro	Lys	Val
Horse	Arg	Leu	Leu	Val	Val	Tyr	Pro	Trp	Thr	Gln	Arg	Phe	Phe	Asp	Ser	Phe	Gly	Asp	Leu	Ser	Asn	Pro	Gly	Ala	Val	Met	Gly	Asn	Pro	Lys	Val

α	56	57	58	59	60	61	62	63	64	65	66	67	68	69	70	71	72	73	74	75	76	77	78	79	80	81	82	83	84	85	86
Human	Lys	Gly	His	Gly	Lys	Lys	Val	Ala	Asp	Ala	Leu	Thr	Asn	Ala	Val	Ala	His	Val	Asp	Asp	Met	Pro	Asn	Ala	Leu	Ser	Ala	Leu	Ser	Asp	Leu
Horse	Lys	Ala	His	Gly	Lys	Lys	Val	Gly	Asp	Ala	Leu	Thr	Leu	Ala	Val	Gly	His	Leu	Asp	Asp	Leu	Pro	Gly	Ala	Leu	Ser	Asp	Leu	Ser	Asn	Leu

β	61	62	63	64	65	66	67	68	69	70	71	72	73	74	75	76	77	78	79	80	81	82	83	84	85	86	87	88	89	90	91
Human	Lys	Ala	His	Gly	Lys	Lys	Val	Leu	Gly	Ala	Phe	Ser	Asp	Gly	Leu	Ala	His	Leu	Asp	Asn	Leu	Lys	Gly	Thr	Phe	Ala	Thr	Leu	Ser	Glu	Leu
Horse	Lys	Ala	His	Gly	Lys	Lys	Val	Leu	His	Ser	Phe	Gly	Glu	Gly	Val	His	His	Leu	Asp	Asn	Leu	Lys	Gly	Thr	Phe	Ala	Ala	Leu	Ser	Glu	Leu

α	87	88	89	90	91	92	93	94	95	96	97	98	99	100	101	102	103	104	105	106	107	108	109	110	111	112	113	114	115	116	117
Human	His	Ala	His	Lys	Leu	Arg	Val	Asp	Pro	Val	Asn	Phe	Lys	Leu	Leu	Ser	His	Cys	Leu	Leu	Val	Thr	Leu	Ala	Ala	His	Leu	Pro	Ala	Glu	Phe
Horse	His	Ala	His	Lys	Leu	Arg	Val	Asp	Pro	Val	Asn	Phe	Lys	Leu	Leu	Ser	His	Cys	Leu	Leu	Ser	Thr	Leu	Ala	Val	His	Leu	Pro	Asn	Asp	Phe

β	92	93	94	95	96	97	98	99	100	101	102	103	104	105	106	107	108	109	110	111	112	113	114	115	116	117	118	119	120	121	122
Human	His	Cys	Asp	Lys	Leu	His	Val	Asp	Pro	Glu	Asn	Phe	Arg	Leu	Leu	Gly	Asn	Val	Leu	Val	Cys	Val	Leu	Ala	His	His	Phe	Gly	Lys	Glu	Phe
Horse	His	Cys	Asp	Lys	Leu	His	Val	Asp	Pro	Glu	Asn	Phe	Arg	Leu	Leu	Gly	Asn	Val	Leu	Ala	Val	Val	Leu	Ala	Arg	His	Phe	Gly	Lys	Asp	Phe

α	118	119	120	121	122	123	124	125	126	127	128	129	130	131	132	133	134	135	136	137	138	139	140	141
Human	Thr	Pro	Ala	Val	His	Ala	Ser	Leu	Asp	Lys	Phe	Leu	Ala	Ser	Val	Ser	Thr	Val	Leu	Thr	Ser	Lys	Tyr	Arg
Horse	Thr	Pro	Ala	Val	His	Ala	Ser	Leu	Asp	Lys	Phe	Leu	Ser	Ser	Val	Ser	Thr	Val	Leu	Thr	Ser	Lys	Tyr	Arg

β	123	124	125	126	127	128	129	130	131	132	133	134	135	136	137	138	139	140	141	142	143	144	145	146
Human	Thr	Pro	Pro	Val	Gln	Ala	Ala	Tyr	Gln	Lys	Val	Val	Ala	Gly	Val	Ala	Asn	Ala	Leu	Ala	His	Lys	Tyr	His
Horse	Thr	Pro	Glu	Leu	Gln	Ala	Ser	Tyr	Gln	Lys	Val	Val	Ala	Gly	Val	Ala	Asn	Ala	Leu	Ala	His	Lys	Tyr	His

(a) Alignment of human and horse globin polypeptides

Orthologs	Number of Amino Acid Differences
Human α globin vs horse α globin	18 out of 141
Human β globin vs horse β globin	25 out of 146
Paralogs	
Human α globin vs human β globin	84 out of 146
Horse α globin vs horse β globin	81 out of 146

(b) A comparison of amino acid differences between orthologs and paralogs

FIGURE 28.11 A comparison of the α- and β-globin polypeptides from humans and horses. **(a)** An alignment of the deduced amino acid sequences obtained by sequencing the exon portions of the corresponding genes. The gaps indicate where additional amino acids are found in the sequence of myoglobin, another member of this gene family. **(b)** A comparison of amino acid differences between orthologs and paralogs.

CONCEPT CHECK: What is the difference between a paralog and an ortholog?

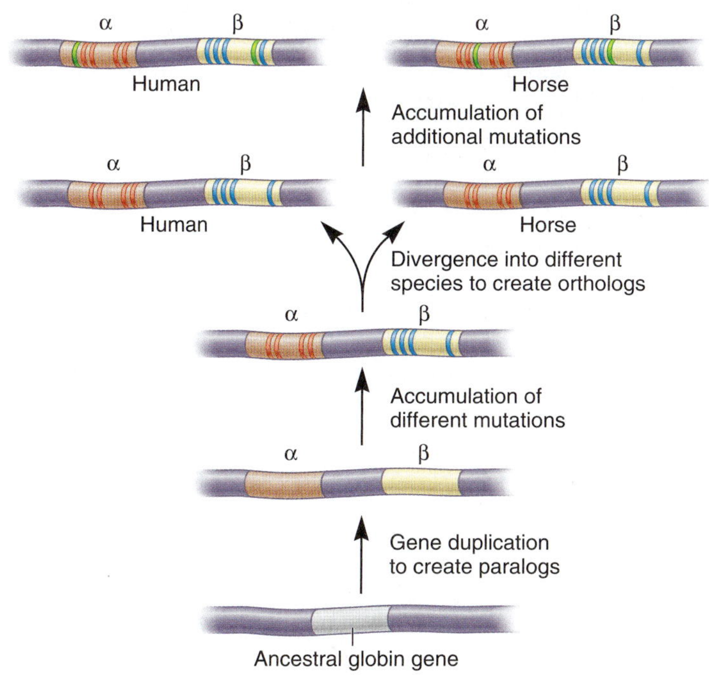

FIGURE 28.12 Evolution of paralogs and orthologs. In this schematic example, the ancestral globin gene duplicated to create the α- and β-globin genes, which are paralogous genes. (Note: Actually, gene duplications occurred several times to produce a large globin gene family, but this is a simplified example that shows only one gene duplication event. Also, chromosomal rearrangements have placed the α- and β-globin genes on different chromosomes.) Over time, these two paralogs accumulated different mutations, designated with red and blue lines. At a later point in evolution, a species divergence occurred to create different mammalian species including humans and horses. Over time, the orthologs also accumulated different mutations, designated with green lines. As noted here, the orthologs have fewer differences than the paralogs because the gene duplication occurred prior to the species divergence.

EXPERIMENT 28A

Scientists Can Compare Orthologs Among Living and Extinct Flightless Birds to Establish Evolutionary Relationships

A common way to construct phylogenetic trees is by comparing orthologs. The majority of such trees have been constructed from molecular data using DNA samples collected from living species. With this approach, we can infer the prehistoric changes that gave rise to present-day DNA sequences. In addition, scientists have discovered that it is occasionally possible to obtain DNA sequence information from species that lived in the past. In 1984, the first successful attempt at determining DNA sequences from an extinct species was accomplished by groups at the University of California at Berkeley and the San Diego Zoo, including Russell Higuchi, Barbara Bowman, Mary Freiberger, Oliver Ryder, and Allan Wilson. They obtained a sample of dried muscle from a museum specimen of the quagga (*Equus quagga*), a zebra-like species that became extinct in 1883. This piece of muscle tissue was obtained from an animal that had died 140 years ago. A sample of its skin and muscle had been preserved in salt in the Museum of Natural History at Mainz, Germany. The researchers extracted DNA from the sample, cloned pieces of it into vectors, and then sequenced hybrid vectors containing the quagga DNA. This pioneering study opened the field of **ancient DNA analysis,** also known as **molecular paleontology.**

Since the mid-1980s, many researchers have become excited about the information that might be derived from sequencing DNA obtained from older specimens. Currently there is debate about how long DNA can remain significantly intact after an organism has died. Over time, the structure of DNA is degraded by hydrolysis and the loss of purines. Nevertheless, under certain conditions (e.g., cold temperature, low oxygen), DNA samples may remain stable for as long as 50,000–100,000 years, and perhaps longer.

In most studies involving prehistoric specimens (in particular, those that are much older than the salt-preserved quagga sample), the ancient DNA is extracted from bone, dried muscle, or preserved skin. These samples are often obtained from museum specimens that have been gathered by archaeologists. However, it is unlikely that enough DNA can be extracted to enable a researcher to directly clone the DNA into a vector. Since 1985, however, the advent of PCR technology, described in Chapter 20, has made it possible to amplify very small amounts of DNA using PCR primers that flank a region within the 12S rRNA gene, a mitochondrial gene. By comparing orthologs of the 12S rRNA among different species, researchers have achieved some success at elucidating the phylogenetic relationships between modern and extinct species.

In the experiment described in **Figure 28.13**, Alan Cooper, Cécile Mourer-Chauviré, Geoffrey Chambers, Arndt von Haeseler, Allan Wilson, and Svante Pääbo investigated the evolutionary relationships among some extinct and modern species of flightless birds. Two groups of flightless birds, the moas and the kiwis, existed in New Zealand during the Pleistocene era. The moas are now extinct, although 11 species were formerly present. In this study, the researchers investigated the phylogenetic relationships among four extinct species of moas that were available as museum samples, three kiwis of New Zealand, and several other (nonextinct) species of flightless birds. These included the emu and the cassowary (found in Australia and New Guinea), the ostrich (found in Africa and formerly Asia), and two rheas (found in South America).

In this work, the researchers wanted to compare orthologs of the 12S rRNA gene. To do so, they collected samples from the various species and used PCR to amplify the 12S rRNA gene from each sample. This provided enough DNA to subject the gene to DNA sequencing. The sequences of the 12S rRNA gene orthologs were aligned using computer programs as described in Chapter 23.

THE GOAL (DISCOVERY-BASED SCIENCE)

Because DNA is a relatively stable molecule, it can be amplified by PCR from a preserved sample of a deceased organism and subjected to DNA sequencing. A comparison of these DNA sequences with modern species may help elucidate the phylogenetic relationships between extinct and modern species.

ACHIEVING THE GOAL

FIGURE 28.13 DNA analysis of orthologs reveals phylogenetic relationships among extinct and modern flightless birds.

Starting material: Tissue samples from four extinct species of moas were obtained from museum specimens. Tissue samples were also obtained from three species of kiwis, one emu, one cassowary, one ostrich, and two species of rhea.

1. For cellular samples, treat with proteinase K (which digests protein) and a detergent that dissolves cell membranes. This releases the DNA from the cells.

2. Individually, mix the DNA samples with a pair of PCR primers that are complementary to the 12S rRNA gene.

3. Subject the samples to PCR. See Chapter 20 (Figure 20.6) for a description of PCR.

4. Subject the amplified DNA fragments to DNA sequencing. See Chapter 20 for a description of DNA sequencing.

5. Align the DNA sequences to each other. Methods of DNA sequence alignment are described in Chapter 23.

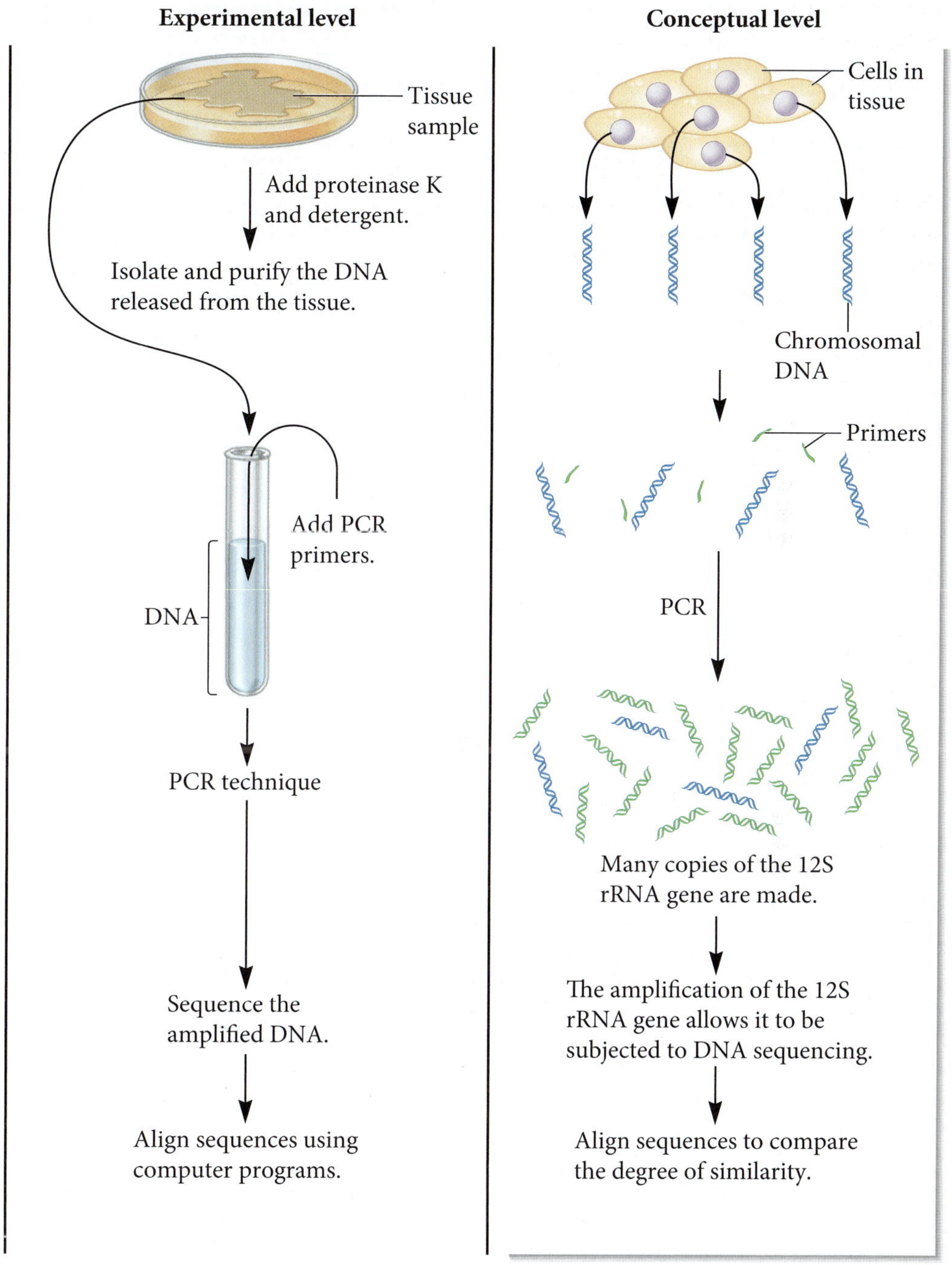

THE DATA

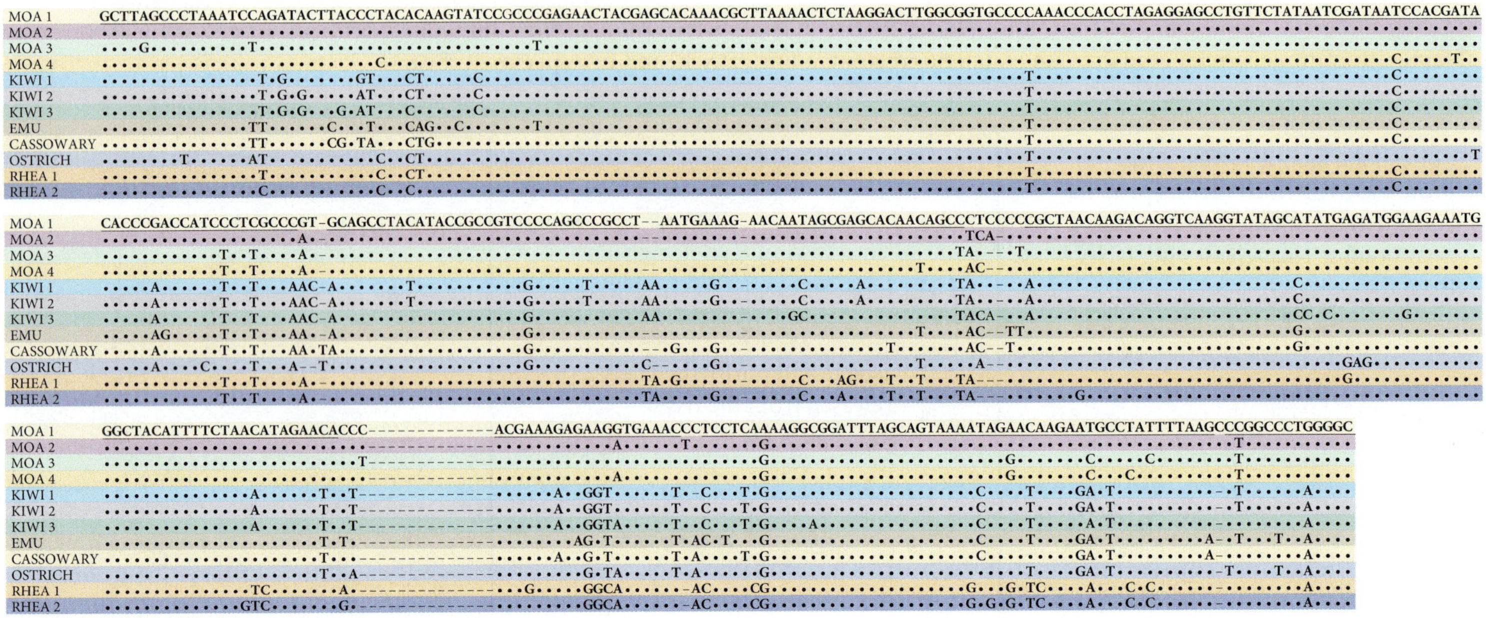

Data from A. Cooper, C. Mourer-Chauviré, G. K. Chambers, et al. (1992) Independent origins of New Zealand moas and kiwis. *Proc Natl Acad Sci USA 89*, 8741–8744.

INTERPRETING THE DATA

The Data of Figure 28.13 illustrate a multiple sequence alignment of the amplified DNA sequences. The first line shows the DNA sequence of one extinct moa species, and underneath it are the sequences of the other species. When the other sequences are identical to the first sequence, a dot is placed in the corresponding position. When the sequences are different, the changed base (A, T, G, or C) is placed there. In a few regions, the genes are different lengths. In these cases, a dash is placed at the corresponding position.

As you can see from the large number of dots, the sequences among these flightless birds are very similar. To establish evolutionary relationships, researchers focus on sites where the gene sequences are not identical. At these sites, base changes have occurred, thereby identifying species with shared derived characters. Some surprising results were obtained. Certain sites in the DNA sequences from the kiwis (a New Zealand species) are the same as the sequence from the ostrich (an African species), but different from those of the moas, which were once found in New Zealand. Likewise, several sites in the DNA sequences of the kiwis are the same as the emu and cassowary (found in Australia and New Guinea), but different from the moas. Contrary to their original expectations, the authors concluded that the kiwis are more closely related to Australian and African flightless birds than they are to the moas. They proposed that New Zealand was colonized twice by ancestors of flightless birds. As shown in **Figure 28.14**, the researchers constructed a new phylogenetic tree to illustrate the relationships among these modern and extinct species.

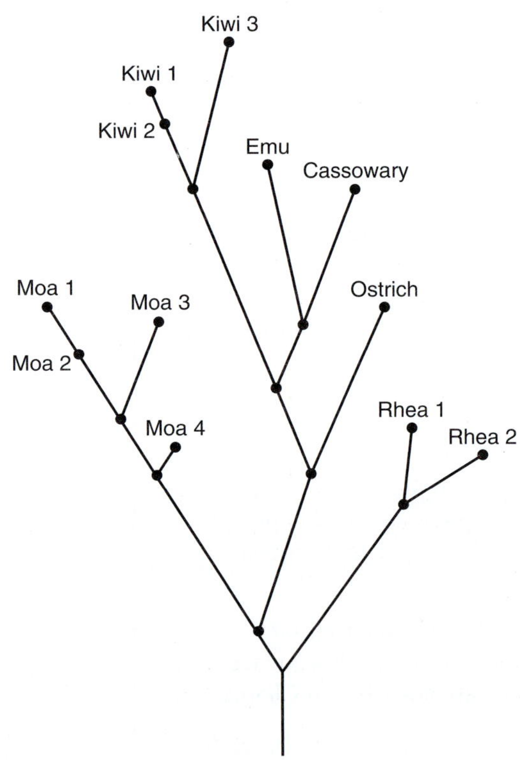

FIGURE 28.14 A revised phylogenetic tree of moas, kiwis, emus, cassowaries, ostriches, and rheas based on a comparison of orthologs.

A self-help quiz involving this experiment can be found at www.mhhe.com/brookergenetics5e.

Genetic Variation at the Molecular Level Is Associated with Neutral Changes in Gene Sequences

As we have seen, the globin genes in mammals and the 12S rRNA genes in flightless birds exhibit variation in their sequences. Is such variation due primarily to mutations favored by natural selection or due to genetic drift?

A **nonneutral mutation** is one that affects the phenotype of the organism and can be acted on by natural selection. Such a mutation may only subtly alter the phenotype of an organism, or it may have a major effect. According to Darwin, natural selection is the agent that leads to evolutionary change in populations. It selects for individuals with the highest Darwinian fitness and often promotes the establishment of beneficial alleles and the elimination of deleterious ones. Therefore, many geneticists have assumed that natural selection is the dominant factor in changing the genetic composition of natural populations, thereby leading to variation.

In opposition to this viewpoint, in 1968, Motoo Kimura proposed the **neutral theory of evolution.** According to this theory, most genetic variation observed in natural populations is due to the accumulation of neutral mutations that do not affect the phenotype of the organism and are not acted on by natural selection. For example, a mutation within a protein-encoding gene that changes a glycine codon from GGG to GGC does not affect the amino acid sequence of the encoded protein. Because neutral mutations do not affect phenotype, they spread throughout a population according to their frequency of appearance and to genetic drift. This theory has been called the "survival of the luckiest" and also **non-Darwinian evolution** to contrast it with Darwin's theory, which focuses on fitness. Kimura agreed with Darwin that natural selection is responsible for adaptive changes in a species during evolution. His main argument is that most modern variation in gene sequences is neutral with respect to natural selection.

In support of the theory, Kimura and his colleague Tomoko Ohta outlined five principles that govern the evolution of genes at the molecular level:

1. For each protein, the rate of evolution, in terms of amino acid substitutions, is approximately constant with regard to neutral substitutions that do not affect protein structure or function.

   Evidence: As an example, the amount of genetic variation between the coding sequence of the human α-globin and β-globin genes is approximately the same as the difference between the α-globin and β-globin genes in the horse (shown earlier in Figure 28.11b). This type of comparison holds true in many different genes compared among many different species.
2. Proteins that are functionally less important for the survival of an organism, or parts of a protein that are less important for its function, tend to evolve faster than more important proteins or regions of a protein. In other words, during evolution, less important proteins or protein domains accumulate amino acid substitutions more rapidly than important ones do.

   Evidence: Certain proteins are critical for survival, and their structure is precisely suited to their function. Examples are the histone proteins necessary for nucleosome formation in eukaryotes. Histone genes tolerate very few mutations and have evolved extremely slowly. By comparison, fibrinopeptides, which bind to fibrinogen to form a blood clot, evolve very rapidly. Presumably, the sequence of amino acids in this polypeptide is not very important for allowing it to aggregate and form a clot.
3. Amino acid substitutions that do not significantly alter the existing structure and function of a protein are found more commonly than disruptive amino acid changes.

   Evidence: When comparing the coding sequences within homologous genes of modern species, nucleotide differences are more likely to be observed in the wobble base than in the first or second base within a codon. Mutations in the wobble base are often silent because they do not change the amino acid sequence of the protein. In addition, conservative substitutions (i.e., a substitution with a similar amino acid, such as a nonpolar amino acid for another nonpolar amino acid) are fairly common. By comparison, nonconservative substitutions—those that significantly alter the structure and function of a protein—are less frequent. Nonsense and frameshift mutations are very rare within the coding sequences of genes. Also, intron sequences evolve more rapidly than exon sequences.
4. Gene duplication often precedes the emergence of a gene having a new function.

   Evidence: When a single copy of a gene exists in a species, it usually plays a functional role similar to that of the homologous gene found in another species. Gene duplications have produced gene families in which each family member can evolve somewhat different functional roles. An example is the globin family described in Chapter 8.
5. Selective elimination of definitely deleterious mutations and the random fixation of selectively neutral or very slightly deleterious alleles occur far more frequently in evolution than selection of advantageous mutants.

   Evidence: As mentioned in principle 3, silent and conservative mutations are much more common than nonconservative substitutions. Presumably these nonconservative mutations usually have a negative effect on the phenotype of the organism, so they are effectively eliminated from the population by natural selection. On rare occasions, however, an amino acid substitution due to a mutation may have a beneficial effect on the phenotype. For example, a nonconservative mutation in the β-globin gene produced the Hb^S allele, which gives an individual resistance to malaria in the heterozygous condition.

In general, the DNA sequencing of hundreds of thousands of different genes from thousands of species has provided compelling support for these five principles of gene evolution at the molecular level. When it was first proposed, the neutral theory sparked a great debate. Some geneticists, called selectionists, strongly opposed the neutral theory of evolution. However, the debate largely cooled after Tomoko Ohta incorporated the concept of nearly neutral mutations into the theory. Nearly neutral mutations have a minimal effect on phenotype—they may be slightly beneficial or slightly detrimental. Ohta suggested that the prevalence of such alleles can depend mostly on natural selection or mostly on genetic drift, depending on the population size.

Why do evolutionary biologists care about neutral or nearly neutral mutations? One reason is that their prevalence is used as a tool to add a time scale to phylogenetic trees. This topic is discussed next.

Molecular Clocks Can Be Used to Date the Divergence of Species

According to the neutral theory of evolution, most of the observed variation is due to neutral mutations. In a sense, the relatively constant rate of neutral or nearly neutral mutations acts as a **molecular clock** on which to measure evolutionary time. According to this idea, neutral mutations become fixed in a population at a rate that is proportional to the rate of mutation per generation. On this basis, the genetic divergence between species that is due to neutral mutations reflects the time elapsed since they diverged from a common ancestor.

Figure 28.15a shows an example of a molecular clock involving a study of superoxide dismutase found in 27 different fruit fly species. (Superoxide dismutase is an enzyme that protects cells against harmful free radicals.) Twenty-three species were in the genus *Drosophila*, two in the genus *Chymomyza*, one in the genus *Scaptodrosophila*, and one in the genus *Ceratitis*. The genus *Ceratitis* is in the family Tephritidae and is more distantly related to the other 26 species, which are in the family Drosophilidae. In this figure, the *y*-axis is a measure of the average number of amino acid differences in superoxide dismutase between pairs of species or between groups of species. The *x*-axis plots the amount of time that has elapsed since a pair or two different groups diverged from a common ancestor. For example, the yellow dot represents the average number of amino acid differences between *Ceratitis* (in family Tephritidae) versus the other species (in family Drosophilidae). Approximately 30 amino acid differences were observed. By comparison, each of the two red dots

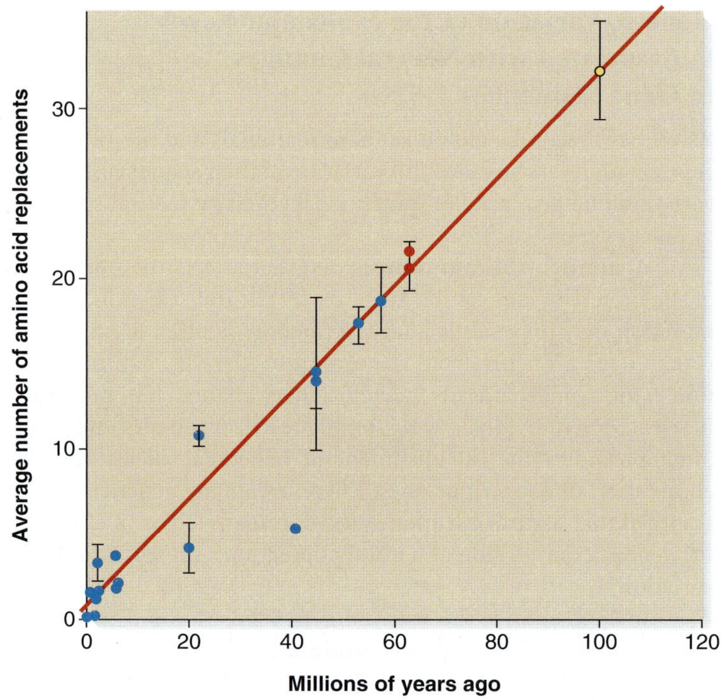

(a) An example of a molecular clock

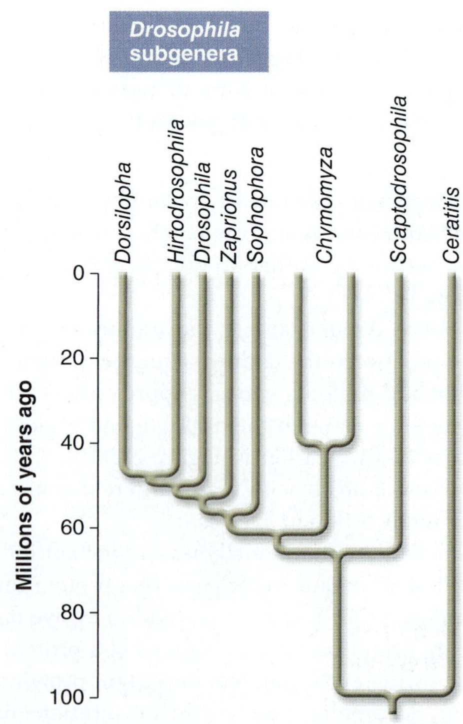

(b) A phylogenetic tree with a timescale

FIGURE 28.15 A molecular clock. According to the concept of a molecular clock, neutral or nearly neutral mutations accumulate over evolutionary time at a fairly constant rate. **(a)** The clock shown here is based on an analysis of superoxide dismutase found in 27 different fruit fly species. Twenty-three species were in the genus *Drosophila*, two in the genus *Chymomyza*, one in the genus *Scaptodrosophila*, and one in the genus *Ceratitis*. When comparing the amino acid sequence of this enzyme between species or groups of species, those that diverged more recently tend to have fewer differences than those whose common ancestor occurred in the very distant past. **(b)** A phylogenetic tree for the species shown in part (a). Parts (a) and (b) are modified from F. J. Ayala (1997) Vagaries of the molecular clock. *Proc Natl Acad Sci USA 94*, 7776–7783.

CONCEPT CHECK: With regard to phylogenetic trees, how is a molecular clock useful?

compares one of the *Chymomyza* species with all of the *Drosophila* species. The blue dots are pairwise comparisons of some *Drosophila* species with each other or with the *Scaptodrosophila* species. As a general trend, species that diverged a longer time ago show a greater number of amino acid differences than species that diverged more recently. The explanation for this phenomenon is that species accumulate independent mutations (e.g., nearly neutral mutations) after they have diverged from each other. A longer period of time since their divergence allows for a greater accumulation of mutations that makes their sequences different.

To further understand the concept of a molecular clock, let's consider how the molecular clock data are related to a phylogenetic tree (**Figure 28.15b**). In this diagram, the *Drosophila* genus is divided into five subgenera. The divergence between the genus *Ceratitis* and the other genera occurred a long time ago, nearly 100 mya. The molecular clock data described in Figure 28.15a showed a relatively large number of amino acid differences (about 30) between *Ceratitis* and the other genera. By comparison, the divergence between *Chymomyza* species and *Drosophila* species occurred more recently—about 65 mya as shown in Figure 28.15b. Likewise the molecular clock data revealed fewer amino acid differences of about 20 between *Chymomyza* and *Drosophila* species (see Figure 28.15a). Finally, the five subgenera of *Drosophila* diverged the most recently and had the fewest number of amino acid differences.

Figure 28.15a suggests a linear relationship between the number of sequence changes and the time of divergence. Such a relationship indicates that the observed rate of neutral or nearly neutral mutations remains constant over millions of years. For example, the linear relationship shown in Figure 28.15a indicates that a pair of species showing 15 amino acid differences diverged from a common ancestor that existed about 50 mya, whereas a pair showing 30 amino acid differences diverged from a common ancestor that existed approximately 100 mya. In other words, twice as many amino acid differences (30 vs 15) correlate with a divergence time that is twice as long (100 mya vs 50 mya). Although actual data sometimes show a relatively linear relationship over a defined period, evolutionary biologists have discovered that molecular clocks are often not linear. This is particularly true over very long periods or when comparing distantly related taxa. When comparing different species or groups of species, several factors can contribute to the nonlinearity of molecular clocks. These include the following:

1. Differences in population sizes that may affect the relative effects of genetic drift and natural selection
2. Differences in mutation rates among different species
3. Differences in the generation times of the species being analyzed
4. Differences in the relative number of sites in a gene or protein that are susceptible to neutral mutations

To produce a timescale, researchers need to calibrate their molecular clocks. How much time does it take to accumulate a certain percentage of nucleotide changes? To perform such a calibration, researchers must have information regarding the date when two species diverged from a common ancestor. Such information could come from the fossil record, for instance. The genetic differences between those species are then divided by the time elapsed since their divergence from a common ancestor to calculate a rate of change. For example, fossil evidence suggests that humans and chimpanzees diverged from a common ancestor approximately 6 mya. The percentage of nucleotide differences in mitochondrial DNA between humans and chimpanzees is 12%. From these data, the molecular clock for changes in mitochondrial DNA sequences of primates is calibrated at roughly 2% nucleotide changes per million years.

Evolution Is Associated with Changes in Chromosome Structure and Number

Thus far, we have focused on mutations that alter the DNA sequences within genes. In addition to gene mutations, other types of changes, such as gene duplications, transpositions, inversions, translocations, and changes in chromosome number, are important features of evolution.

As discussed earlier, changes in chromosome structure and/or number may lead to reproductive isolation and the origin of new species. As an example of variation in chromosome structure among closely related species, **Figure 28.16** compares the banding pattern of the three largest chromosomes in humans and the corresponding chromosomes in chimpanzees, gorillas, and orangutans. The banding patterns are strikingly similar because these species are closely related evolutionarily. However, some interesting differences are observed. Humans have one large chromosome 2, but this chromosome is divided into two separate chromosomes in the other three species. This explains why humans have 23 pairs of chromosomes and the other species have 24. This may have occurred by a fusion of the two smaller chromosomes during the development of the human lineage, which would be an example of a Robertsonian translocation, described in Chapter 8. Another interesting change in chromosome structure is seen in chromosome 3. The banding patterns among humans, chimpanzees, and gorillas are very similar, but the orangutan has a large inversion that flips the arrangement of bands in the centromeric region.

Synteny Groups Contain the Same Group of Linked Genes

With the advent of molecular techniques, researchers can analyze the chromosomes of two or more different species and identify regions that contain the same groups of linked genes, which are called **synteny groups.** Within a particular synteny group, the same types of genes are found in the same order. In 1995, Graham Moore and colleagues analyzed the locations of molecular markers along the chromosomes of several cereal grasses, including rice (*Oryza sativa*), wheat (*Triticum aestivum*), maize (*Zea mays*), foxtail millet (*Setaria italica*), sugarcane (*Saccharum officinarum*), and sorghum (*Sorghum vulgare*). From this analysis, they were able to identify several large synteny groups that are common to most of these species (**Figure 27.17**). As an example, let's compare rice (12 chromosomes per set) with wheat (7 chromosomes per set). In rice, chromosome 6 contains two synteny groups designated R6a and R6b, whereas chromosome 8 contains a single synteny group, R8. In wheat, chromosome 7 consists of R6a and R6b at either end,

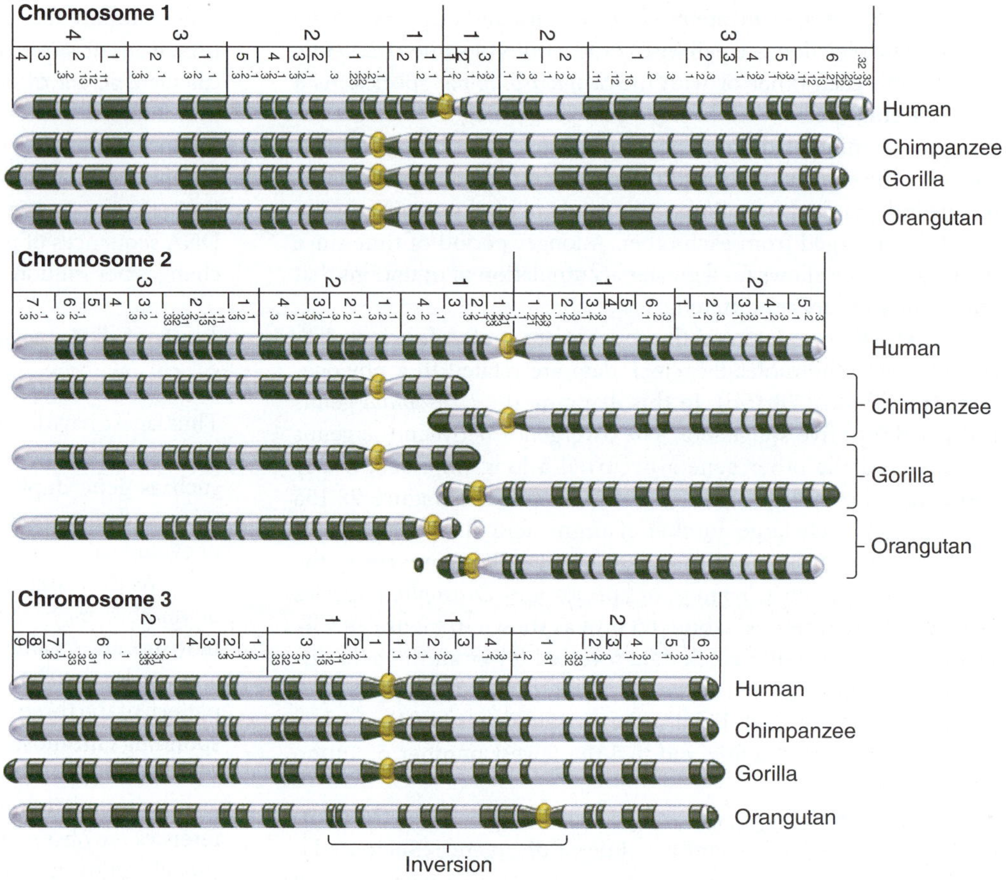

FIGURE 28.16 A comparison of banding patterns among the three largest human chromosomes and the corresponding chromosomes in the chimpanzee, gorilla, and orangutan. This is a schematic drawing of late prophase chromosomes. The conventional numbering system of the banding patterns is shown next to the human chromosomes.

CONCEPT CHECK: Describe two differences among these chromosomes.

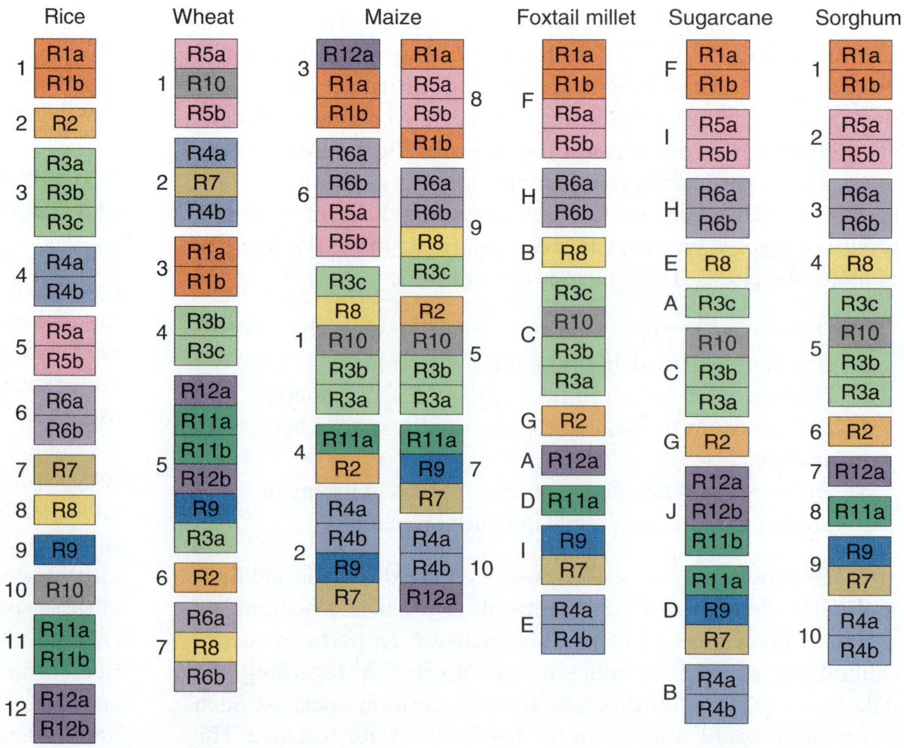

FIGURE 28.17 Synteny groups in cereal grasses. The synteny groups are named according to their relative arrangement in rice, which is shown on the left and has 12 chromosomes per set. In some cases, a synteny group or portion of a group may have incurred an inversion, but these are not shown in this figure. Also, the genome of maize contains two copies of most synteny groups, suggesting that it is derived from an ancestral polyploid species.

and R8 is sandwiched in the middle. One possible explanation for this difference could be a chromosomal rearrangement during the evolution of wheat in which R8 became inserted into the middle of a chromosome containing R6a and R6b. Overall, the evolution of cereal grass species has maintained most of the same types of genes, but many chromosomal rearrangements have occurred. These types of rearrangements promote reproductive isolation.

28.3 COMPREHENSION QUESTIONS

1. Homologous genes found in different species are called
   a. orthologs.
   b. paralogs.
   c. analogs.
   d. none of the above.
2. A molecular clock may not be linear because
   a. mutation rates may differ among different species.
   b. differences in population sizes may affect the relative effects of natural selection and genetic drift.
   c. different species may differ in their generation times.
   d. all of the above may occur.
3. When comparing the chromosomes of closely related species,
   a. the banding patterns are often similar.
   b. a few structural alterations may be seen.
   c. a change in chromosome number may be seen.
   d. all of the above are common.

KEY TERMS

Page 728. biological evolution, evolution, microevolution, macroevolution, molecular evolution
Page 729. adaptive evolution
Page 730. species, subspecies, ecotypes, reproductive isolation, prezygotic isolating mechanisms, postzygotic isolating mechanisms
Page 732. species concept, biological species concept, evolutionary species concept, ecological species concept, general lineage concept, speciation, anagenesis, cladogenesis, allopatric speciation
Page 733. founder effect, parapatric speciation
Page 734. hybrid zones, sympatric speciation, phylogeny, phylogenetic tree
Page 735. homology, homologous
Page 736. monophyletic group, clade, phenetic approach, phenogram, cladistic approach, cladogram, ancestral characters, primitive characters, shared derived characters, synapomorphy
Page 737. ingroup, outgroup
Page 738. principle of parsimony
Page 740. maximum likelihood, Bayesian methods, domains, Bacteria, Archaea, Eukaryotes, supergroups, vertical evolution, horizontal gene transfer
Page 742. orthologs, paralogs, gene family
Page 744. ancient DNA analysis, molecular paleontology
Page 747. nonneutral mutation, neutral theory of evolution, non-Darwinian evolution
Page 748. molecular clock
Page 749. synteny groups

CHAPTER SUMMARY

- Evolution is a heritable change in one or more characteristics of a population or species from one generation to the next.

28.1 Origin of Species

- The most commonly used characteristics to distinguish species are morphological traits, the ability to interbreed, molecular features, ecological factors, and evolutionary relationships. However, each approach has its drawbacks (see Figure 28.1, Table 28.1).
- A species concept is a way to define what a species is and/or provide an approach for distinguishing one species from another. The general lineage concept is receiving wide support.
- Speciation is the process by which new species are formed via evolution. It usually occurs via cladogenesis, in which a species diverges into two or more species (see Figure 28.2).
- Cladogenesis may occur via allopatric, parapatric, or sympatric speciation (see Table 28.2, Figures 28.3, 28.4).

28.2 Phylogenetic Trees

- A phylogenetic tree is a diagram that describes a phylogeny—the sequence of events involved in the evolutionary development of a species or group of species (see Figure 28.5).
- Different methods, including cladistics and phenetics, are used to construct phylogenetic trees. One way to evaluate the

validity of possible trees is by the principle of parsimony (see Figures 28.6–28.8).

- Maximum likelihood and Bayesian methods are also used to discriminate among possible phylogenetic trees.
- Phylogenetic trees depict the relationships among different groups of species such as domains and supergroups (see Figure 28.9).
- In addition to vertical evolution, horizontal gene transfer has played an important role in the evolution of life on Earth (see Figure 28.10).

28.3 Molecular Evolution

- Molecular evolution refers to patterns and processes associated with evolutionary change at the molecular level.
- Homologous genes are derived from the same ancestral gene. They can be orthologs or paralogs (see Figures 28.11, 28.12).
- Phylogenetic trees can be constructed by comparing orthologs from extinct and modern species (see Figures 28.13, 28.14).
- The neutral theory of evolution proposes that most variation in gene sequences is due to mutations that are neutral or nearly neutral with regard to phenotype. Such variation accumulates in populations largely due to genetic drift.
- The relatively constant rate of neutral or nearly neutral mutations acts as a molecular clock on which to measure evolutionary time (see Figure 28.15).
- Evolution also involves changes in chromosome structure and number (see Figures 28.16, 28.17).

PROBLEM SETS & INSIGHTS

Solved Problems

S1. A codon for leucine is UUA. A mutation causing a single-base substitution in a gene can change this codon in the transcribed mRNA into GUA (valine), AUA (isoleucine), CUA (leucine), UGA (stop), UAA (stop), UCA (serine), UUG (leucine), UUC (phenylalanine), or UUU (phenylalanine). According to the neutral theory, which of these mutations would you expect to be the most likely within the genetic variation of a natural population? Explain.

Answer: The neutral theory proposes that neutral mutations accumulate to the greatest extent in a population. Leucine is a nonpolar amino acid. For a UUA codon, single-base changes of CUA and UUG are silent, so they would be the most likely to occur in a natural population. Likewise, conservative substitutions to other nonpolar amino acids such as isoleucine (AUA), valine (GUA), and phenylalanine (UUC and UUU) may not affect protein structure and function, so they may also occur and not be eliminated rapidly by natural selection. The polar amino acid serine (UCA) is a nonconservative substitution; one would predict that it is more likely to disrupt protein function. Therefore, it may be less likely to be found. Finally, the stop codons, UGA and UAA, would be expected to diminish or eliminate protein function. These types of mutations are selected against and, therefore, are not usually found in natural populations.

S2. Explain why homologous genes have sequences that are similar but not identical.

Answer: Homologous genes are derived from the same ancestral gene. Therefore, as a starting point, they had identical sequences. Over time, however, each gene accumulates random mutations that the other homologous genes did not acquire. These random mutations change the gene from its original sequence. Therefore, much of the sequence between homologous genes remains identical, but some of the sequence is altered due to the accumulation of independent random mutations.

S3. Explain why plants are more likely than animals to evolve by sympatric speciation.

Answer: A common way for sympatric speciation to occur is by the formation of polyploids. For example, if one species is diploid ($2n$), nondisjunction could produce an individual that is tetraploid ($4n$). If the tetraploid individual was a plant and if it was monoecious (i.e., produces both pollen and egg cells), the plant could multiply to produce many tetraploid offspring. These offspring would be reproductively isolated from the diploid plants that are found in the same geographic area. This isolation occurs because the offspring of a cross between a diploid and a tetraploid plant would be infertile. For example, if the pollen from a diploid plant fertilized the egg from a tetraploid plant, the offspring would be triploid. The triploid offspring might be viable, but it is likely that triploids would be sterile because they would produce highly aneuploid gametes. The gametes would be aneuploid because they have an odd number of homologous chromosomes. In this case, there would be three copies of each homologous chromosome, and these could not be equally distributed into gametes. Therefore, when a tetraploid is produced, it can immediately become its own unique species that is reproductively isolated from other species.

In contrast, polyploidy rarely occurs in animals. Perhaps the main reason is because animals usually cannot tolerate polyploidy. Tetraploid animals typically die during early stages of development. In addition, most species of animals have male and female sexes. To develop a tetraploid species from a diploid species, nondisjunction would have to produce both a male and a female offspring that could reproduce with each other. The chance of producing one tetraploid offspring is relatively rare. The chances of producing two tetraploid offspring that happen to be male and female and happen to mate with each other to produce many offspring would be extremely rare.

S4. As described in Figure 28.16, evolution is associated with changes in chromosome structure and number. As seen there, chromosome 2 in humans is divided into two distinct chromosomes in chimpanzees, gorillas, and orangutans. In addition, chromosome 3 in the orangutan has a large inversion not found in the other three primates. Discuss the potential role of these types of changes in the evolution of these primate species.

Answer: As discussed in Chapter 8, changes in chromosome structure, such as inversions and balanced translocations, may not have any phenotypic effects. Likewise, the division of a single chromosome into two distinct chromosomes may not have any phenotypic effect as long as the total amount of genetic material remains the same. Overall, the types of changes in chromosome structure and number shown in Figure 28.16 may not have caused any changes in the phenotypes of primates. However, the changes would be expected to promote reproductive isolation. For example, if a gorilla mated with an orangutan, the offspring would be an inversion heterozygote for chromosome 3. As shown in Figure 8.11, crossing over during gamete formation in an inversion heterozygote may produce chromosomes that have too much or too little genetic material. This is particularly likely if the inversion is fairly large (like the one shown in Figure 28.16). The inheritance of too much or too little genetic material is likely to be detrimental or even lethal. For this reason, the hybrid offspring of a gorilla and orangutan would probably be infertile. (Note: In reality, there are several other reasons why interspecies matings between gorillas and orangutans do not produce viable offspring.)

Overall, the primary effect of changes in chromosome structure and number, like the ones shown in Figure 28.16, is to promote reproductive isolation. Once two populations become reproductively isolated, they accumulate different mutations which, over the course of many generations, leads to two different species with distinct characteristics.

Conceptual Questions

C1. Discuss the two principles on which evolution is based.

C2. Evolution, which involves genetic changes in a population of organisms over time, is often described as the unifying theme in biology. Discuss how evolution is unifying at the molecular and cellular levels.

C3. What is a species? What types of observations do researchers analyze when trying to identify species?

C4. What is meant by the term reproductive isolation? Give several examples.

C5. Would the following examples of reproductive isolation be considered a prezygotic or postzygotic mechanism?

A. Horses and donkeys can interbreed to produce mules, but the mules are infertile.

B. Three species of the orchid genus *Dendrobium* produce flowers 8 days, 9 days, and 11 days after a rainstorm. The flowers remain open for 1 day.

C. Two species of fish release sperm and eggs into seawater at the same time, but the sperm of one species do not fertilize the eggs of the other species.

D. Two tree frogs, *Hyla chrysoscelis* (diploid) and *H. versicolor* (tetraploid), can produce viable offspring, but the offspring are sterile.

C6. Distinguish between anagenesis and cladogenesis. Which type of speciation is more prevalent? Why?

C7. Describe three or more genetic mechanisms that may lead to the rapid evolution of a new species. Which of these genetic mechanisms are influenced by natural selection, and which are not?

C8. Explain the type of speciation (allopatric, parapatric, or sympatric) most likely to occur under each of the following conditions:

A. A pregnant female rat is transported by an ocean liner to a new continent.

B. A meadow containing several species of grasses is exposed to a pesticide that promotes nondisjunction.

C. In a very large lake containing several species of fish, the water level gradually falls over the course of several years. Eventually, the large lake becomes subdivided into smaller lakes, some of which are connected by narrow streams.

C9. Alloploids are produced by crosses involving two different species. Explain why alloploids may be reproductively isolated from the two original species from which they were derived. Explain why alloploids are usually sterile, whereas allotetraploids (containing a diploid set from each species) are commonly fertile.

C10. Discuss whether the phenomenon of reproductive isolation applies to bacteria, which reproduce asexually. How would a geneticist divide bacteria into separate species?

C11. Discuss the major differences among allopatric, parapatric, and sympatric speciation.

C12. The following are two DNA sequences from homologous genes:

```
TTGCATAGGCATACCGTATGATATCGAAAACTAGAAAAATAGGGCGATAGCTA
GTATGTTATCGAAAAGTAGCAAAATAGGGCGATAGCTACCCAGACTACCGGAT
```

The two sequences, however, do not begin and end at the same location. Try to line them up according to their homologous regions.

C13. What is meant by the term molecular clock? How is this concept related to the neutral theory of evolution?

C14. Would the rate of deleterious or beneficial mutations be a good molecular clock? Why or why not?

C15. Which would you expect to exhibit a faster rate of evolutionary change, the nucleotide sequence of a gene or the amino acid sequence of the encoded polypeptide of the same gene? Explain your answer.

C16. When comparing the coding region of protein-encoding genes among closely related species, certain regions of the gene are commonly found to have evolved more rapidly (i.e., have tolerated more changes in sequence) than other regions of the gene. Explain why different regions of a protein-encoding gene evolve at different rates.

C17. Plant seeds contain storage proteins that are encoded by plant genes. When the seed germinates, these proteins are rapidly hydrolyzed (i.e., the covalent bonds between amino acids within the polypeptides are broken), which releases amino acids for the developing seedling. Would you expect the genes that encode plant storage proteins to evolve more slowly or more rapidly than genes that encode enzymes? Explain your answer.

C18. Take a look at the α-globin and β-globin sequences in Figure 28.11. Which sequences are more similar, the α globin in humans and the α globin in horses, or the α globin in humans and the β globin in humans? Based on your answer, would you conclude that the gene duplication that gave rise to the α-globin and β-globin genes occurred before or after the divergence of humans and horses? Explain your reasoning.

C19 Compare and contrast the neutral theory of evolution versus the Darwinian (i.e., selectionist) theory of evolution. Explain why the neutral theory of evolution is sometimes called non-Darwinian evolution.

C20. For each of the following examples, discuss whether it would be the result of neutral mutations or mutations that have been acted on by natural selection, or both:

A. When comparing sequences of homologous genes, differences in the coding sequence are most common at the wobble base (i.e., the third base in each codon).

B. For a protein-encoding gene, the regions that encode portions of the polypeptide that are vital for structure and function are less likely to display mutations than other regions of the gene.

C. When comparing the sequences of homologous genes, introns usually have more sequence differences than exons.

C21. As discussed in Chapter 26, genetic variation is prevalent in natural populations. This variation is revealed in the DNA sequencing of genes. According to the neutral theory of evolution, discuss the relative importance of natural selection against detrimental mutations, natural selection in favor of beneficial mutations, and neutral mutations in accounting for the genetic variation we see in natural populations.

C22. If you were comparing the karyotypes of species that are closely related evolutionarily, what types of similarities and differences would you expect to find?

Experimental Questions

E1. Two populations of snakes are separated by a river. The snakes cross the river only on rare occasions. The snakes in the two populations look very similar to each other, except that the members of the population on the eastern bank of the river have a yellow spot on the top of their head, whereas the members of the western population have an orange spot on the top of their head. Discuss two experimental methods that you might follow to determine whether the two populations are members of the same species or members of different ones.

E2. Sympatric speciation by allotetraploidy has been proposed as a common mechanism for speciation. Let's suppose you were interested in the origin of certain grass species in southern California. Experimentally, how would you go about determining if some of the grass species are the result of allotetraploidy?

E3. Two diploid species of closely related frogs, which we will call species A and species B, were analyzed with regard to genes that encode an enzyme called hexokinase. Species A has two distinct copies of this gene: *A1* and *A2*. In other words, this diploid species is *A1A1 A2A2*. The other species has three copies of the hexokinase gene, which we will call *B1*, *B2*, and *B3*. A diploid individual of species B would be *B1B1 B2B2 B3B3*. These hexokinase genes from the two species were subjected to DNA sequencing, and the percentage of sequence identity was compared among these genes. The results are shown here.

Percentage of DNA Sequence Identity

	A1	*A2*	*B1*	*B2*	*B3*
A1	100	62	54	94	53
A2	62	100	91	49	92
B1	54	91	100	67	90
B2	94	49	67	100	64
B3	53	92	90	64	100

If we assume that hexokinase genes were never lost in the evolution of these frog species, how many distinct hexokinase genes do you think there were in the most recent ancestor that preceded the divergence of these two species? Explain your answer. Also explain why species B has three distinct copies of this gene, whereas species A has only two.

E4. A researcher sequenced a portion of a bacterial gene and obtained the following sequence, beginning with the start codon, which is underlined:

<u>ATG</u> CCG GAT TAC CCG GTC CCA AAC AAA ATG ATC GGC CGC CGA ATC TAT CCC

The bacterial strain that contained this gene has been maintained in the laboratory and grown serially for many generations. Recently, another person working in the laboratory isolated DNA from the bacterial strain and sequenced the same region. The following results were obtained:

<u>ATG</u> CCG GAT TAT CCG GTC CCA AAT AAA ATG ATC GGC CGC CGA ATC TAC CCC

Explain why these sequencing differences may have occurred.

E5. F_1 hybrids between two species of cotton, *Gossypium barbadense* and *G. hirsutum*, are very vigorous plants. However, F_1 crosses produce many seeds that do not germinate and a high percentage of very weak F_2 offspring. Suggest two reasons for these observations.

E6. A species of antelope contains 20 chromosomes per set. The species is divided by a mountain range into two separate populations, which we will call the eastern and western population. When comparing the karyotypes of these two populations, it was discovered that the members of the eastern population are homozygous for a large inversion within chromosome 14. How would this inversion affect the interbreeding between the two populations? Could such an inversion play an important role in speciation?

E7. Explain why molecular techniques were needed as a way to provide evidence for the neutral theory of evolution.

E8. Prehistoric specimens often contain minute amounts of ancient DNA. What technique can be used to increase the amount of DNA in an older sample? Explain how this technique is performed and how it increases the amount of a specific region of DNA.

E9. In the experiment of Figure 28.13, explain how we know that the kiwis are more closely related to the emu and cassowary than to the moas. Cite particular regions in the sequences that support your answer.

E10. In Chapter 22, we learned about a technique called fluorescence in situ hybridization (FISH), during which a labeled piece of DNA is hybridized to a set of chromosomes. Let's suppose that we cloned a piece of DNA from *G. pubescens* (see Figure 28.4) and used it as a labeled probe for in situ hybridization. What would you expect to happen if we hybridized it to the *G. speciosa* or *G. tetrahit* chromosomes? Describe your expected results.

E11. A team of researchers has obtained a dinosaur bone (*Tyrannosaurus rex*) and has attempted to extract ancient DNA from it. Using primers to the 12S rRNA mitochondrial gene, they have used PCR and obtained a DNA segment that yields a sequence homologous to crocodile DNA. Other scientists are skeptical that this sequence is really from the dinosaur. Instead, they believe that it may be due to contamination from more recent DNA, such as the remains of a reptile that lived much more recently. What criteria might you use to establish the credibility of the dinosaur sequence?

E12. Discuss how the principle of parsimony can be used in a cladistics approach of constructing a phylogenetic tree.

E13. A homologous DNA region, which was 20,000 bp in length, was sequenced among four different species. The following number of nucleotide differences were obtained:

	Species A	Species B	Species C	Species D
Species A	0	443	719	465
Species B	443	0	744	423
Species C	719	744	0	723
Species D	465	423	723	0

Construct a phylogenetic tree that describes the evolutionary relationships among these four species using the UPGMA method. Your tree should include values that show the percentage of nucleotide substitutions.

E14. As discussed in this chapter and Chapter 26, genes are sometimes transferred between different species via horizontal gene transfer. Discuss how horizontal gene transfer might give misleading results when constructing a phylogenetic tree. How could you overcome this problem?

Questions for Student Discussion/Collaboration

1. The raw material for evolution is random mutation. Discuss whether or not you view evolution as a random process.
2. Compare the forms of speciation that are slow with those that occur more rapidly. Make a list of the slow and fast forms. With regard to mechanisms of genetic change, what features do slow and rapid speciation have in common? What features are different?
3. Do you think that Darwin would object to the neutral theory of evolution?

Note: All answers appear at the website for this textbook; the answers to the even-numbered questions and all of the Concept Check and Comprehension Questions are in the back of the book.

www.mhhe.com/brookergenetics5e

Visit the website for practice tests, answer keys, and other learning aids for this chapter. Enhance your understanding of genetics with our interactive exercises, quizzes, animations, and much more.

Note: Appendix A: Experimental Techniques can be found on the website for this textbook: www.mhhe.com/brookergenetics5e

APPENDIX B SOLUTIONS TO EVEN-NUMBERED PROBLEMS AND ALL COMPREHENSION AND CONCEPT CHECK QUESTIONS

CHAPTER 1

Answers to Comprehension Questions

1.1: d, c, b, b

1.2: d, d, a, c

1.3: d, a, d

Concept Check Questions (in figure legends)

FIGURE 1.1 Understanding our genes may help to diagnose inherited diseases. It may also lead to the development of drugs to combat diseases. Other answers are possible.

FIGURE 1.2 Many ethical issues are associated with human cloning. Is it the wrong thing to do? Does it conflict an individual's religious views? And so on.

FIGURE 1.3 Because females mate only once, sorting the male mosquitos and releasing sterile males into the environment can limit mosquito reproduction.

FIGURE 1.4 DNA is a macromolecule.

FIGURE 1.5 DNA and proteins are found in chromosomes. A small amount of RNA may also be associated with chromosomes when transcription is occurring.

FIGURE 1.6 The information to make a polypeptide is stored in DNA.

FIGURE 1.7 The dark-colored butterfly has a more active pigment-producing enzyme.

FIGURE 1.8 Genetic variation is the reason these frogs look different.

FIGURE 1.9 These are examples of variation in chromosome number.

FIGURE 1.10 A corn gamete would contain 10 chromosomes. (The leaf cells are diploid.)

FIGURE 1.11 The horse populations have become adapted to their environment, which has changed over the course of many years.

FIGURE 1.12 There are several possible examples of other model organisms, including rats and frogs.

End-of-Chapter Questions:

Conceptual Questions

C2. A chromosome is a very long polymer of DNA. A gene is a specific sequence of DNA within that polymer; the sequence of bases creates a gene and distinguishes it from other genes. Genes are located in chromosomes, which are found within living cells.

C4. At the molecular level, a gene (a sequence of DNA) is first transcribed into RNA. The genetic code within the RNA is used to synthesize a protein with a particular amino acid sequence. This second process is called translation.

C6. Genetic variation involves the occurrence of genetic differences within members of the same species or different species. Within any population, variation may occur in the genetic material. Variation may occur in particular genes so that some individuals carry one allele and other individuals carry a different allele. An example would be differences in coat color among mammals. Variation may also occur in chromosome structure and number. In plants, differences in chromosome number can affect disease resistance.

C8. You could pick almost any trait. For example, flower color in petunias would be an interesting choice. Some petunias are red and others are purple. There must be different alleles in a flower color gene that affect this trait in petunias. In addition, the amount of sunlight, fertilizer, and water also affects the intensity of flower color.

C10. A DNA sequence is a sequence of nucleotides. Each nucleotide may have one of four different bases (i.e., A, T, G, or C). When we speak of a DNA sequence, we focus on the sequence of bases.

C12. A. A gene is a segment of DNA. For most genes, the expression of the gene results in the production of a functional protein. The functioning of proteins within living cells affects the traits of an organism.

B. A gene is a segment of DNA that usually encodes the information for the production of a specific protein. Genes are found within chromosomes. Many genes are found within a single chromosome.

C. An allele is an alternative version of a particular gene. For example, suppose a plant has a flower color gene. One allele could produce a white flower, while a different allele could produce an orange flower. The white allele and orange allele are alleles of the flower color gene.

D. A DNA sequence is a sequence of nucleotides. The information within a DNA sequence (which is transcribed into an RNA sequence) specifies the amino acid sequence within a protein.

C14. A. How genes and traits are transmitted from parents to offspring.

B. How the genetic material functions at the molecular and cellular levels.

C. Why genetic variation exists in populations and how it changes over the course of many generations.

Experimental Questions

E2. This would be used primarily by molecular geneticists. The sequence of DNA is a molecular characteristic of DNA. In addition, as we will learn throughout this textbook, the sequence of DNA is interesting to transmission and population geneticists as well.

E4. A. Transmission geneticists. Dog breeders are interested in how genetic crosses affect the traits of dogs.

B. Molecular geneticists. This is a good model organism to study genetics at the molecular level.

C. Both transmission geneticists and molecular geneticists. Fruit flies are easy to cross and study the transmission of genes and traits from parents to offspring. Molecular geneticists have also studied many genes in fruit flies to see how they function at the molecular level.

D. Population geneticists. Most wild animals and plants would be the subject of population geneticists. In the wild, you cannot make controlled crosses. But you can study genetic variation within populations and try to understand its relationship to the environment.

E. Transmission geneticists. Agricultural breeders are interested in how genetic crosses affect the outcome of traits.

CHAPTER 2

Answers to Comprehension Questions

2.1: d, a, b

2.2: c, b

2.3: b, c, a

2.4: b, c

2.5: a, b (use the binomial expansion), c

Concept Check Questions (in figure legends)

FIGURE 2.2 The male gamete is found within pollen grains.

FIGURE 2.3 The white flower is providing the sperm, and the purple flower is providing the eggs.

FIGURE 2.4 A true-breeding strain maintains the same trait over the course of many generations.

FIGURE 2.6 Segregation means that the *T* and *t* alleles separate from each other so that a gamete receives one of them, but not both.

FIGURE 2.7 In this hypothesis, two different genes are linked. The alleles of the same gene are not linked.

FIGURE 2.9 Independent assortment allows for new combinations of alleles among different genes to be found in future generations of offspring.

FIGURE 2.10 Such a parent could make two types of gametes, *Ty* and *ty*, in equal proportions.

FIGURE 2.11 Horizontal lines connect two individuals that have offspring together, and they connect all of the offspring that produced by the same two parents.

End-of-Chapter Questions:

Conceptual Questions

C2. In the case of plants, cross-fertilization occurs when the pollen and eggs come from different plants, whereas in self-fertilization, they come from the same plant.

C4. A homozygote that has two copies of the same allele.

C6. Diploid organisms contain two copies of each type of gene. When they make gametes, only one copy of each gene is found in a gamete. Two alleles cannot stay together within the same gamete.

C8. Genotypes: 1:1 *Tt* and *tt*

Phenotypes: 1:1 Tall and dwarf

C10. *c* is the recessive allele for constricted pods; *Y* is the dominant allele for yellow color. The cross is *ccYy* × *CcYy*. Follow the directions for setting up a Punnett square, as described in Chapter 2. The genotypic ratio is 2 *CcYY* : 4 *CcYy* : 2 *Ccyy* : 2 *ccYY* : 4 *ccYy* : 2 *ccyy*. This 2:4:2:2:4:2 ratio can be reduced to a 1:2:1:1:2:1 ratio.

The phenotypic ratio is 6 smooth pods, yellow seeds : 2 smooth pods, green seeds : 6 constricted pods, yellow seeds : 2 constricted pods, green seeds. This 6:2:6:2 ratio could be reduced to a 3:1:3:1 ratio.

C12. Offspring with a nonparental phenotype are consistent with the idea of independent assortment. If two different traits were always transmitted together as unit, it would not be possible to get nonparental phenotypic combinations. For example, if a true-breeding parent had two dominant traits and was crossed to a true-breeding parent having the two recessive traits, the F_2 generation could not have offspring with one recessive and one dominant phenotype. However, because independent assortment can occur, it is possible for F_2 offspring to have one dominant and one recessive trait.

C14. A. Barring a new mutation during gamete formation, the chance is 100% because they must be heterozygotes in order to produce a child with a recessive disorder.

B. Construct a Punnett square. There is a 50% chance of heterozygous children.

C. Use the product rule. The chance of being phenotypically normal is 0.75 (i.e., 75%), so the answer is 0.75 × 0.75 × 0.75 = 0.422, which is 42.2%.

D. Use the binomial expansion equation, where $n = 3$, $x = 2$, $p = 0.75$, $q = 0.25$. The answer is 0.422, or 42.2%.

C16. First construct a Punnett square. The chances are 75% of producing a solid pup and 25% of producing a spotted pup.

A. Use the binomial expansion equation, where $n = 5$, $x = 4$, $p = 0.75$, $q = 0.25$. The answer is 0.396 = 39.6% of the time.

B. You can use the binomial expansion equation for each litter. For the first litter, $n = 6$, $x = 4$, $p = 0.75$, $q = 0.25$; for the second litter, $n = 5$, $x = 5$, $p = 0.75$, $q = 0.25$. Because the litters are in a specified order, we use the product rule and multiply the probability of the first litter times the probability of the second litter. The answer is 0.070, or 7.0%.

C. To calculate the probability of the first litter, we use the product rule and multiply the probability of the first pup (0.75) times the probability of the remaining four. We use the binomial expansion equation to calculate the probability of the remaining four, where $n = 4$, $x = 3$, $p = 0.75$, $q = 0.25$. The probability of the first litter is 0.316. To calculate the probability of the second litter, we use the product rule and multiply the probability of the first pup (0.25) times the probability of the second pup (0.25) times the probability of the remaining five. To calculate the probability of the remaining five, we use the binomial expansion equation, where $n = 5$, $x = 4$, $p = 0.75$, $q = 0.25$. The probability of the second litter is 0.025. To get the probability of these two litters occurring in this order, we use the product rule and multiply the probability of the first litter (0.316) times the probability of the second litter (0.025). The answer is 0.008, or 0.8%.

D. Because this is a specified order, we use the product rule and multiply the probability of the firstborn (0.75) times the probability of the second born (0.25) times the probability of the remaining four. We use the binomial expansion equation to calculate the probability of the remaining four pups, where $n = 4$, $x = 2$, $p = 0.75$, $q = 0.25$. The answer is 0.040, or 4.0%.

C18. A. Use the product rule:

$$(1/4)(1/4) = 1/16$$

B. Use the binomial expansion equation:

$$n = 4, p = 1/4, q = 3/4, x = 2$$

$$P = 0.21 = 21\%$$

C. Use the product rule:

$$(1/4)(3/4)(3/4) = 0.14, \text{ or } 14\%$$

C20. A. $1/4$

B. 1, or 100%

C. $(3/4)(3/4)(3/4) = 27/64 = 0.42$, or 42%

D. Use the binomial expansion equation, where

$$n = 7, p = 3/4, q = 1/4, x = 3$$

$$P = 0.058, \text{ or } 5.8\%$$

E. The probability that the first plant is tall is $3/4$. To calculate the probability that among the next four, any two will be tall, we use the binomial expansion equation, where $n = 4$, $p = 3/4$, $q = 1/4$, and $x = 2$.

The probability P equals 0.21.

To calculate the overall probability of these two events:

$$(3/4)(0.21) = 0.16, \text{ or } 16\%$$

C22. It violates the law of segregation because two copies of one gene are in the gamete. The two alleles for the *A* gene did not segregate from each other.

C24. Based on this pedigree, it is likely to be dominant inheritance because an affected child always has an affected parent. In fact, it is a dominant disorder.

C26. It is impossible for the F_1 individuals to be true-breeding because they are all heterozygotes.

C28. 2 *TY*, *tY*, 2 *Ty*, *ty*, *TTY*, *TTy*, 2 *TtY*, 2 *Tty*

It may be tricky to think about, but you get 2 *TY* and 2 *Ty* because either of the two *T* alleles could combine with *Y* or *y*. Also, you get 2 *TtY* and 2 *Tty* because either of the two *T* alleles could combine with *t* and then combine with *Y* or *y*.

C30. The genotype of the F_1 plants is *Tt Yy Rr*. According to the laws of segregation and independent assortment, the alleles of each gene will segregate from each other, and the alleles of different genes will randomly assort into gametes. A *Tt Yy Rr* individual could make eight types of gametes: *TYR*, *TyR*, *Tyr*, *TYr*, *tYR*, *tyR*, *tYr*, and *tyr*, in equal proportions (i.e., $^1/_8$ of each type of gamete). To determine genotypes and phenotypes, you could make a large Punnett square that would contain 64 boxes. You would need to line up the eight possible gametes across the top and along the side and then fill in the 64 boxes. Alternatively, you could use one of the two approaches described in solved problem S4. The genotypes and phenotypes would be

1 *TT YY RR*

2 *TT Yy RR*

2 *TT YY Rr*

2 *Tt YY RR*

4 *TT Yy Rr*

4 *Tt Yy RR*

4 *Tt YY Rr*

8 *Tt Yy Rr* = 27 tall, yellow, round

1 *TT yy RR*

2 *Tt yy RR*

2 *TT yy Rr*

4 *Tt yy Rr* = 9 tall, green, round

1 *TT YY rr*

2 *TT Yy rr*

2 *Tt YY rr*

4 *Tt Yy rr* = 9 tall, yellow, wrinkled

1 *tt YY RR*

2 *tt Yy RR*

2 *tt YY Rr*

4 *tt Yy Rr* = 9 dwarf, yellow, round

1 *TT yy rr*

2 *Tt yy rr* = 3 tall, green, wrinkled

1 *tt yy RR*

2 *tt yy Rr* = 3 dwarf, green, round

1 *tt YY rr*

2 *tt Yy rr* = 3 dwarf, yellow, wrinkled

1 *tt yy rr* = 1 dwarf, green, wrinkled

C32. The wooly-haired male is a heterozygote, because he has the trait and his mother did not. (He must have inherited the normal allele from his mother.) Therefore, he has a 50% chance of passing the wooly allele to his offspring; his offspring have a 50% of passing the allele to their offspring; and these grandchildren have a 50% chance of passing the allele to their offspring (the wooly-haired man's great-grandchildren). Because this is an ordered sequence of independent events, we use the product rule: $0.5 \times 0.5 \times 0.5 = 0.125$, or 12.5%. Because no other Scandinavians are on the island, the chance is 87.5% for the offspring being normal (because they could not inherit the wooly hair allele from anyone else). We use the binomial expansion equation to determine the likelihood that one out of eight great-grandchildren will have wooly hair, where $n = 8$, $x = 1$, $p = 0.125$, $q = 0.875$. The answer is 0.393, or 39.3%, of the time.

C34. Use the product rule. If the woman is heterozygous, there is a 50% chance of having an affected offspring: $(0.5)^7 = 0.0078$, or 0.78%, of the time. This is a pretty small probability. If the woman has an eighth child who is unaffected, however, she has to be a heterozygote, because it is a dominant trait. She would have to pass a normal allele to an unaffected offspring. The answer is 100%.

Experimental Questions

E2. The experimental difference depends on where the pollen comes from. In self-fertilization, the pollen and eggs come from the same plant. In cross-fertilization, they come from different plants.

E4. According to Mendel's law of segregation, the genotypic ratio should be 1 homozygote dominant : 2 heterozygotes : 1 homozygote recessive. This data table considers only the plants with a dominant phenotype. The genotypic ratio should be 1 homozygote dominant : 2 heterozygotes. The homozygote dominants would be true-breeding, but the heterozygotes would not be true-breeding. This 1:2 ratio is very close to what Mendel observed.

E6. All three offspring had black fur. The ovaries from the albino female could produce eggs with only the dominant black allele (because they were obtained from a true-breeding black female). The actual phenotype of the albino mother does not matter. Therefore, all offspring would be heterozygotes (*Bb*) and have black fur.

E8. If we construct a Punnett square according to Mendel's laws, we expect a 9:3:3:1 ratio. Because a total of 556 offspring were observed, the expected number of offspring are

$556 \times {}^9/_{16} = 313$ round, yellow

$556 \times {}^3/_{16} = 104$ wrinkled, yellow

$556 \times {}^3/_{16} = 104$ round, green

$556 \times {}^1/_{16} = 35$ wrinkled, green

If we plug the observed and expected values into the chi square equation, we get a value of 0.51. With four categories, our degrees of freedom equal $n - 1$, or 3. If we look up the value of 0.51 in the chi square table (see Table 2.1), we see that it falls between the *P* values of 0.80 and 0.95. This means that the probability is 80% to 95% that any deviation between observed results and expected results was caused by random sampling error. Therefore, we accept the hypothesis. In other words, the results are consistent with the law of independent assortment.

E10. A. If we let c^+ represent normal wings and c represent curved wings, and e^+ represents gray body and e represents ebony body,

Parental Cross: $cce^+e^+ \times c^+c^+ee$

F_1 generation is heterozygous c^+ce^+e

An F_1 offspring crossed to flies with curved wings and ebony bodies is:

$c^+ce^+e \times ccee$

The F_2 offspring would be a 1:1:1:1 ratio of flies:

$c^+ce^+e : c^+cee : cce^+e : ccee$

B. The phenotypic ratio of the F_2 flies would be a 1:1:1:1 ratio of flies:

normal wings, gray body : normal wings, ebony bodies : curved wings, gray bodies : curved wings, ebony bodies

C. From part B, we expect $^1/_4$ of each category. There are a total of 444 offspring. The expected number of each category is $^1/_4 \times$ 444, which equals 111.

$$\chi^2 = \frac{(114-111)^2}{111} + \frac{(105-111)^2}{111} + \frac{(111-111)^2}{111} + \frac{(114-111)^2}{111}$$

$$\chi^2 = 0.49$$

With 3 degrees of freedom, a value of 0.49 or greater is likely to occur between 80% and 95% of the time. Therefore, we accept our hypothesis.

E12. Follow through the same basic chi square strategy as before. We expect a 3:1 ratio, or $^3/_4$ of the dominant phenotype and $^1/_4$ of the recessive phenotype.

The observed and expected values are as follows (rounded to the nearest whole number):

Observed*	Expected	$\frac{(O - E)^2}{E}$
5474	5493	0.066
1850	1831	0.197
6022	6017	0.004
2001	2006	0.012
705	697	0.092
224	232	0.276
882	886	0.018
299	295	0.054
428	435	0.113
152	145	0.338
651	644	0.076
207	215	0.298
787	798	0.152
277	266	0.455
		$\chi^2 = 2.15$

*Due to rounding, the observed and expected values may not add up to precisely the same number.

Because $n = 14$, there are 13 degrees of freedom. If we look up this value in the chi square table, we have to look between 10 and 15 degrees of freedom. In either case, we would expect the value of 2.15 or greater to occur more than 99% of the time. Therefore, we accept the hypothesis.

E14. The dwarf parent with terminal flowers must be homozygous for both genes, because it is expressing these two recessive traits: *ttaa*, where *t* is the recessive dwarf allele, and *a* is the recessive allele for terminal flowers. The phenotype of the other parent is dominant for both traits. However, because this parent was able to produce dwarf offspring with axial flowers, it must have been heterozygous for both genes: *TtAa*.

E16. Our hypothesis is that blue flowers and purple seeds are dominant traits and they are governed by two genes that assort independently. According to this hypothesis, the F_2 generation should yield a ratio of 9 blue flowers, purple seeds : 3 blue flowers, green seeds : 3 white flowers, purple seeds : 1 white flower, green seeds. Because a total of 300 offspring are produced, the expected numbers would be

$^9/_{16} \times 300 = 169$ blue flowers, purple seeds

$^3/_{16} \times 300 = 56$ blue flowers, green seeds

$^3/_{16} \times 300 = 56$ white flowers, purple seeds

$^1/_{16} \times 300 = 19$ white flowers, green seeds

$$\chi^2 = \frac{(103 - 169)^2}{169} + \frac{(49 - 56)^2}{56} + \frac{(44 - 56)^2}{56} + \frac{(104 - 19)^2}{19}$$

$$\chi^2 = 409.5$$

If we look up this value in the chi square table under 3 degrees of freedom, the value is much higher than would be expected 1% of the time by chance alone. Therefore, we reject the hypothesis. The idea that the two genes are assorting independently seems to be incorrect. The F_1 generation supports the idea that blue flowers and purple seeds are dominant traits.

Questions for Student Discussion/Collaboration

2. If we construct a Punnett square, the following probabilities will be obtained:

   tall with axial flowers, $^3/_8$

   dwarf with terminal flowers, $^1/_8$

   The probability of being tall with axial flowers or dwarf with terminal flowers:

   $^3/_8 + {}^1/_8 = {}^4/_8 = {}^1/_2$

   We use the product rule to calculate the ordered events of the first three offspring being tall/axial or dwarf/terminal, and the fourth offspring being tall axial:

   $(^1/_2)(^1/_2)(^1/_2)(^3/_8) = {}^3/_{64} = 0.047 = 4.7\%$

CHAPTER 3

Answers to Comprehension Questions

3.1: c, d, a	3.4: a, c
3.2: d, b, b	3.5: c, d
3.3: d, d	3.6: c. a, a, b

Concept Check Questions (in figure legends)

FIGURE 3.1 Compartmentalization means that cells have membrane-bound compartments.

FIGURE 3.2 The chromosomes would not be spread out very well and would probably be overlapping. It would be difficult to see individual chromosomes.

FIGURE 3.3 Homologs are similar in size, banding pattern, and carry the same types of genes. However, the alleles of a given gene may be different.

FIGURE 3.4 FtsZ assembles into a ring at the future site of the septum and recruits other proteins to this site that produce a cell wall between the two daughter cells.

FIGURE 3.5 The G_1 phase is a phase of the cell cycle when a cell may make the decision to divide. By comparison, the G_0 phase is a phase in which a cell is either not progressing through the cell cycle or has made a decision to never divide again.

FIGURE 3.6 Homologs are genetically similar; one is inherited from the mother and the other from the father. By comparison, chromatids are the product of DNA replication. The chromatids within a pair of sister chromatids are genetically identical.

FIGURE 3.7 One end of a kinetochore microtubule is attached to a kinetochore on a chromosome. The other end is within the centrosome.

FIGURE 3.8 Anaphase.

FIGURE 3.9 Ingression occurs because myosin motor proteins shorten the contractile ring, which is formed from actin proteins.

FIGURE 3.10 The end result of crossing over is that homologous chromosomes have exchanged pieces.

FIGURE 3.11 The cells at the end of meiosis are haploid, whereas the mother cell is diploid.

FIGURE 3.12 In mitosis, each pair of sister chromatids is attached to both poles, whereas in metaphase of meiosis I, each pair of sister chromatids is attached to just one pole.

FIGURE 3.13 Polar bodies are small cells produced during oogenesis that degenerate.

FIGURE 3.14 All of the nuclei in the embryo sac are haploid. The central cell has two haploid nuclei and all of the other cells, including the egg, have just one.

FIGURE 3.15 Homologous chromosomes separate at anaphase of meiosis I.

FIGURE 3.16 If we view the left and right sides as being distinctly different, these chromosomes could line up in eight different ways.

FIGURE 3.17 In the X-Y system, the presence of the Y chromosome causes maleness, whereas in the X-0 system, it is the ratio between the number of X chromosomes and number of sets of autosomes that determines sex. A ratio of 0.5 is male and 1.0 is female.

End-of-Chapter Questions:

Conceptual Questions

C2. The term homolog refers to the members of a chromosome pair. Homologs are usually the same size and carry the same types and order of genes. They may differ in that the genes they carry may be different alleles.

C4. Metaphase is the organization phase, and anaphase is the separation phase.

C6. In metaphase I of meiosis, each pair of chromatids is attached to only one pole via the kinetochore microtubules. In metaphase of mitosis, there are two attachments (i.e., to both poles). If the attachment was lost, a chromosome would not migrate to a pole and may not become enclosed in a nuclear membrane after telophase. If left out in the cytoplasm, it would eventually be degraded.

C8. The reduction occurs because there is a single DNA replication event but two cell divisions. Because of the nature of separation during anaphase I, each cell receives one copy of each type of chromosome.

C10. It means that the maternally derived and paternally derived chromosomes are randomly aligned along the metaphase plate during metaphase I. Refer to Figure 3.17.

C12. There are three pairs of chromosomes. The number of different, random alignments equals 2^n, where n equals the number of chromosomes per set. So the possible number of arrangements equals 2^3, which is 8.

C14. The probability would be much lower because pieces of maternal chromosomes would be mixed with the paternal chromosomes. Therefore, it is unlikely to inherit a chromosome that was completely paternally derived.

C16. During interphase, the chromosomes are greatly extended. In this conformation, they might get tangled up with each other and not sort properly during meiosis and mitosis. The condensation process probably occurs so that the chromosomes easily align along the equatorial plate during metaphase without getting tangled up.

C18. During prophase II, your drawing should show four replicated chromosomes (i.e., four structures that look like X's). Each chromosome is one homolog. During prophase of mitosis, there should be eight replicated chromosomes (i.e., eight X's). During prophase of mitosis, there are pairs of homologs. The main difference is that prophase II has a single copy of each of the four chromosomes, whereas prophase of mitosis has four pairs of homologs. At the end of meiosis I, each daughter cell has received only one copy of a homologous pair, not both. This is due to the alignment of homologs during metaphase I and their separation during anaphase I.

C20. DNA replication does not take place during interphase II. The chromosomes at the end of telophase I have already replicated (i.e., they are found in pairs of sister chromatids). During meiosis II, the sister chromatids separate from each other, yielding individual chromosomes.

C22. A. 20 C. 30

B. 10 D. 20

C24. A. Dark males and light females; reciprocal: all dark offspring

B. All dark offspring; reciprocal: dark females and light males

C. All dark offspring; reciprocal: dark females and light males

D. All dark offspring; reciprocal: dark females and light males

C26. To produce sperm, a spermatogonial cell first goes through mitosis to produce two cells. One of these remains a spermatogonial cell and the other progresses through meiosis. In this way, the testes continue to maintain a population of spermatogonial cells.

C28. A. *A B C, A B c, A b C, A b c, a B C, a b C, a B c, a b c*

B. *A B C, A b C*

C. *A B C, A B c, a B C, a B c*

D. *A b c, a b c*

C30. A. The fly is a male because the ratio of X chromosomes to sets of autosomes is $^1/_2$, or 0.5.

B. The fly is female because the ratio of X chromosomes to sets of autosomes is 1.0.

C. The fly is male because the ratio of X chromosomes to sets of autosomes is 0.5.

D. The fly is female because the ratio of X chromosomes to sets of autosomes is 1.0.

Experimental Questions

E2. Perhaps the most convincing observation was that all of the white-eyed flies of the F_2 generation were males. This suggests a link between sex determination and the inheritance of this trait. Because sex determination in fruit flies is determined by the number of X chromosomes, this suggests a relationship between the inheritance of the X chromosome and the inheritance of this trait.

E4. The basic strategy is to set up a pair of reciprocal crosses. The phenotype of sons is usually the easiest way to discern the two patterns. If it is Y linked, the trait will be passed only from father to son. If it is X linked, the trait will be passed from mother to son.

E6. The 3:1 sex ratio occurs because the female produces 50% gametes that are XX (and must produce female offspring) and 50% that are X (and produce half male and half female offspring). The original female had one X chromosome carrying the red allele and two other X chromosomes carrying the eosin allele. Set up a Punnett square assuming that this female produces the following six types of gametes: $X^{w+}X^{w-e}$, $X^{w+}X^{w-e}$, $X^{w-e}X^{w-e}$, X^{w+}, X^{w-e}, X^{w-e}. The male of this cross is X^wY.

Male gametes

Female gametes ♀ / ♂	X^w	Y
$X^{w^+}X^{w-e}$	$X^{w^+}X^{w-e}X^w$ Red, female	$X^{w^+}X^{w-e}Y$ Red, female
$X^{w^+}X^{w-e}$	$X^{w^+}X^{w-e}X^w$ Red, female	$X^{w^+}X^{w-e}Y$ Red, female
$X^{w-e}X^{w-e}$	$X^{w-e}X^{w-e}X^w$ Eosin female	$X^{w-e}X^{w-e}Y$ Eosin female
X^{w^+}	$X^{w^+}X^w$ Red, female	$X^{w^+}Y$ Red, male
X^{w-e}	$X^{w-e}X^w$ Light-eosin female	$X^{w-e}Y$ Light-eosin male
X^{w-e}	$X^{w-e}X^w$ Light-eosin female	$X^{w-e}Y$ Light-eosin male

E8. If we use the data from the F_1 mating (i.e., F_2 results), there were 3470 red-eyed flies. We would expect a 3:1 ratio between red- and white-eyed flies. Therefore, assuming that all red-eyed offspring survived, there should have been about 1157 (i.e., $^{3470}/_3$) white-eyed flies. However, there were only 782. If we divide 782 by 1157, we get a value of 0.676, or a 67.6% survival rate.

E10. You need to make crosses to understand the pattern of inheritance of traits (determined by genes) from parents to offspring. And you need to microscopically examine cells to understand the pattern of transmission of chromosomes. The correlation between the pattern of transmission of chromosomes during meiosis and Mendel's laws of segregation and independent assortment is what led to the chromosome theory of inheritance.

E12. Originally, individuals who had abnormalities in their composition of sex chromosomes provided important information. In mammals, X0 individuals are females, whereas in flies, X0 individuals are males. In mammals, XXY individuals are males, while in flies, XXY individuals are females. These results indicate that the presence of the Y chromosome causes maleness in mammals, but it does not in flies. A further analysis of flies with abnormalities in the number of sets of autosomes revealed that it is the ratio between the number of X chromosomes and the number of sets of autosomes that determines sex in flies.

Questions for Student Discussion/Collaboration

2. It's not possible to give a direct answer, but the point is for students to be able to draw chromosomes in different configurations and understand the various phases. The chromosomes may or may not be

    1. In homologous pairs
    2. Connected as sister chromatids
    3. Associated in bivalents
    4. Lined up in metaphase
    5. Moving toward the poles

    And so on.

CHAPTER 4

Answers to Comprehension Questions

4.1: d	4.6: b
4.2: d, d, c	4.7: c
4.3: c	4.8: d
4.4: d, a	4.9: b, d
4.5: c, a	

Concept Check Questions (in figure legends)

FIGURE 4.1 Both of these colors are considered wild type because both of them are prevalent in natural populations.

FIGURE 4.2 Yes. The *PP* homozygote probably makes twice as much of the protein that is needed for purple pigment formation.

FIGURE 4.3 Individual III-2 shows incomplete penetrance.

FIGURE 4.4 Genes and the environment determine an organism's traits.

FIGURE 4.5 50% of the functional enzyme is not enough to give a red color.

FIGURE 4.6 It is often easier to observe incomplete dominance at the molecular/cellular level.

FIGURE 4.7 In this case, the heterozygote is resistant to malaria.

FIGURE 4.8 The scenario show in part (a) explains the effects of the sickle cell allele.

FIGURE 4.9 The *i* allele is a loss-of-function allele.

FIGURE 4.10 The key pedigree feature that points to X-linked inheritance is that only males are affected with the disorder. Also, carrier females often have affected brothers.

FIGURE 4.11 The reciprocal cross yields a different result because females carry two copies of an X-linked gene, whereas males have only one.

FIGURE 4.12 Homologous regions of the X and Y chromosome are important for chromosome synapsis (pairing) during meiosis.

FIGURE 4.13 A heterozygous female would not have scurs.

FIGURE 4.14 Certain traits are expressed only in males or females, possibly due to differences in the levels of sex hormones or other factors that differ between the sexes.

FIGURE 4.15 The heterozygote has one normal copy of the gene, which allows for development to proceed in a way that is not too far from normal. Having two mutant copies of the gene probably messes up development to a degree that is incompatible with survival.

FIGURE 4.17 Epistasis means that the alleles of one gene mask the phenotypic effects of the alleles of a different gene. Complementation occurs when two strains exhibiting the same recessive trait produce offspring that show the dominant (wild-type) trait. This usually means the alleles for the recessive trait are in two different genes.

FIGURE 4.20 In some cases, a single gene knockout does not have an effect due to gene redundancy. Other explanations are also possible.

FIGURE 4.21 The two genes are redundant. Having one normal copy of either gene produces a triangular capsule. If both genes are inactive, an ovate capsule is produced.

End-of-Chapter Questions:

Conceptual Questions

C2. Sex-influenced traits are influenced by the sex of the individual even though the gene that governs the trait may be autosomally inherited. Pattern baldness in people is an example. Sex-limited traits are an extreme example of sex influence. The expression of a sex-limited trait is limited to one sex. For example, colorful plumage in certain species of birds is limited to the male sex. Sex-linked traits involve traits whose genes are found on the sex chromosomes. Examples in humans include hemophilia and color blindness.

C4. If the normal allele is dominant, it tells you that one copy of the gene produces a sufficient amount of the protein encoded by the gene. Having twice as much of this protein, as in the normal homozygote, does not alter the phenotype. If the allele is incompletely dominant, this means that one copy of the normal allele does not produce the same trait as the homozygote.

C6. The ratio would be 1 normal : 2 star-eyed individuals.

C8. If individual 1 is *ii*, individual 2 could be I^Ai, I^AI^A, I^Bi, I^BI^B, or I^AI^B.

If individual 1 is I^Ai or I^AI^A, individual 2 could be I^Bi, I^BI^B, or I^AI^B.

If individual 1 is I^Bi or I^BI^B, individual 2 could be I^Ai, I^AI^A, or I^AI^B.

Assuming individual 1 is the parent of individual 2:

If individual 1 is *ii*, individual 2 could be I^Ai or I^Bi.

If individual 1 is I^Ai, individual 2 could be I^Bi or I^AI^B.

If individual 1 is I^AI^A, individual 2 could be I^AI^B.

If individual 1 is I^Bi, individual 2 could be I^Ai or I^AI^B.

If individual 1 is I^BI^B, individual 2 could be I^AI^B.

C10. The father could not be I^AI^B, I^BI^B, or I^AI^A. He is contributing the O allele to his offspring. Genotypically, he could be I^Ai, I^Bi, or *ii* and have type A, B, or O blood, respectively.

C12. Perhaps it should be called codominant at the "hair level" because one or the other allele is dominant with regard to a single hair. However, this is not the same as codominance in blood types, in which every cell can express both alleles.

C14. A. X-linked recessive (unaffected mothers transmit the trait to sons).

B. Autosomal recessive (affected daughters and sons are produced from unaffected parents).

C16. First set up the following Punnett square:

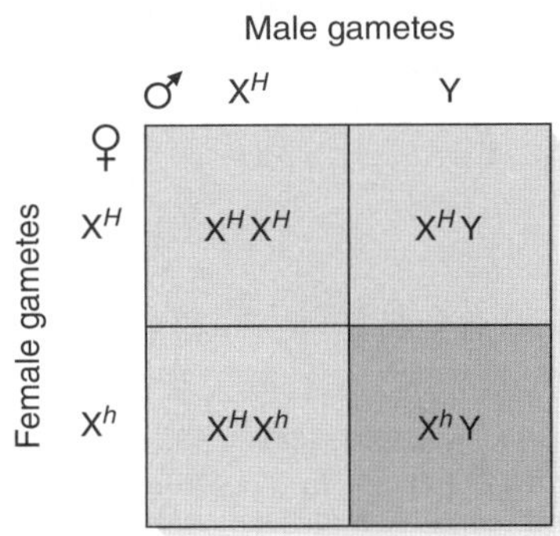

There is a $^1/_4$ probability of each type of offspring.

A. $^1/_4$

B. $(^3/_4)(^3/_4)(^3/_4)(^3/_4) = {^{81}/_{256}}$

C. $^3/_4$

D. The probability of an affected offspring is $^1/_4$, and the probability of an unaffected offspring is $^3/_4$. For this problem, you use the binomial expansion equation where $x = 2$, $n = 5$, $p = {^1/_4}$, and $q = {^3/_4}$. The answer is 0.26, or 26%, of the time.

C18. We know that the parents must be heterozygotes for both genes.

The genotypic ratio of their offspring is 1 *ScSc* : 2 *Scsc* : 1 *scsc*

The phenotypic ratio depends on sex: 1 *ScSc* male with scurs : 1 *ScSc* female with scurs : 2 *Scsc* males with scurs : 2 *Scsc* hornless females : 1 *scsc* hornless male : 1 *scsc* hornless female

A. 50%

B. $^1/8$, or 12.5%

C. $(^3/8)(^3/8)(^3/8) = {^{27}/512} = 0.05$, or 5%

C20. It probably occurred in the summer. In the Siamese cat, dark fur occurs in cooler regions of the body. If the fur grows during the summer, these regions are likely to be somewhat warmer, and therefore the fur will be lighter.

C22. First, you would cross heterozygous birds to each other. This would yield an F_1 generation consisting of a ratio of 1 *HH* : 2 *Hh* : 1 *hh*. The male offspring that are *hh* would have cock-feathering. All the female offspring would have hen-feathering. You would then take cock-feathered males and cross them to F_1 females. If all of the males within a brood were cock-feathered, it is likely that the mother was *hh*. If so, all of the offspring would be *hh*. The offspring from such a brood could be crossed to each other. If they are truly *hh*, the males of the F_3 generation should all be cock-feathered.

C24. A. Could be.

B. No, because an affected female has an unaffected son.

C. Could be.

D. No, because an affected male has an unaffected daughter.

E. No, because it affects both sexes.

C26. Molecular: The β-globin gene for Hb^A homozygotes encodes a β-globin polypeptide with a normal amino acid sequence compared with Hb^S homozygotes, whose β-globin genes encode a polypeptide that has an abnormal structure. The abnormal structure affects the ability of hemoglobin to carry oxygen.

Cellular: Under conditions of low oxygen, Hb^SHb^S cells form a sickle shape compared with the normal biconcave disk shape of Hb^AHb^A cells.

Organism: In Hb^SHb^S individuals, the sickle shape decreases the life span of the red blood cell, which causes anemia. Also, the clogging of red blood cells in the capillaries causes tissue damage and painful crises. This does not occur in Hb^AHb^A individuals.

C28.

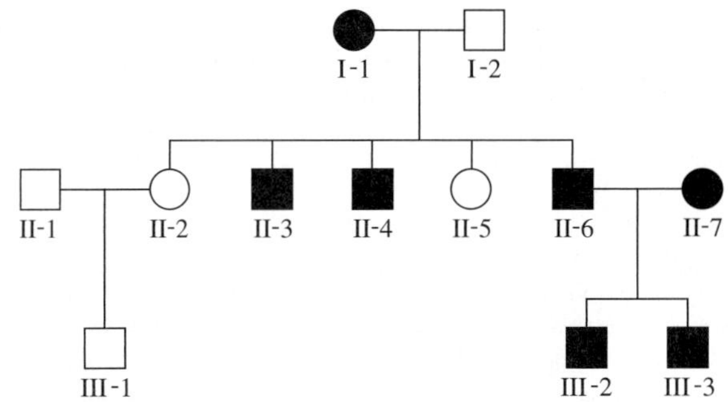

Experimental Questions

E2. Two redundant genes are involved in feathering. The unfeathered Buff Rocks are homozygous recessive for the two genes. The Black Langhans are homozygous dominant for both genes. In the F_2 generation (which is a double heterozygote crossed to another double heterozygote), 1 out of 16 offspring will be doubly homozygous for both recessive genes. All the others will have at least one dominant allele for one of the two (redundant) genes.

E4. The reason why all the puppies have black hair is because albino alleles are found in two different genes. If we let the letters *A* and *B* represent the two different pigmentation genes, then one of the dogs is *AAbb*, and the other is *aaBB*. Their offspring are *AaBb* and therefore are not albinos because they have one dominant copy of each gene.

E6. In general, you cannot distinguish between autosomal and pseudoautosomal inheritance from a pedigree analysis. Mothers and fathers have an equal probability of passing the alleles to sons and daughters. However, if an offspring had a chromosomal abnormality, you might be able to tell. For example, in a family tree involving the *Mic2* allele, an offspring that was X0 would have less of the gene product, and an offspring that was XXX or XYY or XXY would have extra amounts of the gene products. This may lead you to suspect that the gene is located on the sex chromosomes.

E8. One parent must be *RRPp*. The other parent could be *RRPp* or *RrPp*. All the offspring would inherit (at least) one dominant *R* allele. With regard to the other gene, $^3/_4$ would inherit at least one copy of the dominant *P* allele. These offspring would have a walnut comb. The other $^1/_4$ would be homozygous *pp* and have a rose comb (because they would also have a dominant *R* allele).

E10. Let's use the letters *A* and *B* for these two genes. Gene *A* exists in two alleles, which we will call *A* and *a*. Gene *B* exists in two alleles, *B* and *b*. The uppercase alleles are dominant to the lowercase alleles. The true-breeding long-shaped squash is *aabb*, and the true-breeding disk-shaped squash is *AABB*. The F_1 offspring are *AaBb*. You can construct a Punnett square, with 16 boxes, to determine the outcome of self-fertilization of the F_1 plants.

To get the disk-shaped phenotype, an offspring must inherit at least one dominant allele from both genes.

1 *AABB* + 2 *AaBB* + 2 *AABb* + 4 *AaBb* = 9 disk-shaped offspring

To get the round phenotype, an offspring must inherit at least one dominant allele for one of the two genes but must be homozygous recessive for only one of the two genes.

1 *aaBB* + 1 *AAbb* + 2 *aaBb* + 2 *Aabb* = 6 round-shaped offspring

To get the long phenotype, an offspring must inherit all recessive alleles:

1 *aabb* = 1 long-shaped offspring

E12. The results obtained when crossing two F_1 offspring appear to yield a 9:3:3:1 ratio, which would be expected if eye color is affected by two different genes that exist in dominant and recessive alleles. Neither gene is X-linked. Let pr^+ represent the red allele of the first gene and *pr* the purple allele. Let sep^+ represent the red allele of the second gene and *sep* the sepia allele.

The first cross is $pr\,pr\ sep^+sep^+ \times pr^+pr^+sep\ sep$

All the F_1 offspring would be $pr^+pr\ sep^+sep$. They have red eyes because they have a dominant red allele for each gene. When the F_1 offspring are crossed to each other, the following results would be obtained:

♀ \ ♂	♂ pr^+sep^+	pr^+sep	$pr\ sep^+$	$pr\ sep$
pr^+sep^+	pr^+pr^+ sep^+sep^+ Red	pr^+pr^+ sep^+sep Red	pr^+pr sep^+sep^+ Red	pr^+pr sep^+sep Red
pr^+sep	pr^+pr^+ sep^+sep Red	pr^+pr^+ $sep\ sep$ Sepia	pr^+pr sep^+sep Red	pr^+pr $sep\ sep$ Sepia
$pr\ sep^+$	pr^+pr sep^+sep^+ Red	pr^+pr sep^+sep Red	$pr\ pr$ sep^+sep^+ Purple	$pr\ pr$ sep^+sep Purple
$pr\ sep$	pr^+pr sep^+sep Red	pr^+pr $sep\ sep$ Sepia	$pr\ pr$ sep^+sep Purple	$pr\ pr$ $sep\ sep$ Pur/Sepia

In this case, one gene exists as the red (dominant) or purple (recessive) allele, and the second gene exists as the red (dominant) or sepia (recessive) allele. If an offspring is homozygous for the purple allele, it will have purple eyes. Similarly, if an offspring is homozygous for the sepia allele, it will have sepia eyes. An offspring homozygous for both recessive alleles has purplish sepia eyes. To have red eyes, it must have at least one copy of the dominant red allele for both genes. Based on an expected 9 red : 3 purple : 3 sepia : 1 purplish sepia, the observed and expected numbers of offspring are as follows:

Observed	Expected
146 purple eyes	148 purple eyes ($791 \times ^3/_{16}$)
151 sepia eyes	148 sepia eyes ($791 \times ^3/_{16}$)
50 purplish sepia eyes	49 purplish sepia eyes ($791 \times ^1/_{16}$)
444 red eyes	445 red eyes ($791 \times ^9/_{16}$)
791 total offspring	

If we plug the observed and expected values into our chi square formula, we obtain a chi square value of about 0.11. With 3 degrees of freedom, this is well within our expected range of values, so we cannot reject our hypothesis that purple and sepia alleles are in two different genes and that these recessive alleles are epistatic to each other.

E14. To see if the allele is X linked, the pink-eyed male could be crossed to a red-eyed female. All the offspring would have red eyes, assuming that the pink allele is recessive. When crossed to red-eyed males, the F_1 females will produce $^1/_2$ red-eyed daughters, $^1/_4$ red-eyed sons, and $^1/_4$ pink-eyed sons if the pink allele is X-linked.

If the pink allele is X linked, then one could determine if it is in the same X-linked gene as the white and eosin alleles by crossing pink-eyed males to white-eyed females. (Note: We already know that white and eosin are alleles of the same gene.) If the pink and white alleles are in the same gene, the F_1 female offspring should have pink eyes (assuming that the pink allele is dominant over white). However, if the pink and white alleles are in different genes, the F_1 females will have red eyes (assuming that pink is recessive to red). This is because the F_1 females will be heterozygous for two genes, $X^{w+p}X^{wp+}$, in which the X^{w+} and X^{p+} alleles are the dominant wild-type alleles that produce red eyes, and the X^w and X^p alleles are recessive alleles for these two different genes, which produce white eyes and pink eyes, respectively.

E16. The total number of offspring shown in the data was 209. Rounded to the nearest whole number, the expected numbers would be as follows:

red-eyed females = $^8/_{16} \times 209 = 105$

red-eyed males = $^4/_{16} \times 209 = 52$

light eosin-eyed males = $^3/_{16} \times 209 = 39$

cream-eyed males = $^1/_{16} \times 209 = 13$

If we plug the observed and expected values into a chi square equation (see Chapter 2), the chi square value equals 1.2. If we look this value up in the chi square table (Table 2.1) with 3 degrees of freedom, the value lies between a probability of 0.8 and 0.5. Therefore, we cannot reject the hypothesis that cream is autosomal and modifier of eosin, and eosin and red are X-linked genes.

Questions for Student Discussion/Collaboration

2. Perhaps the easiest way to solve this problem is to take one trait at a time. With regard to combs, all the F_1 generation would be $RrPp$ or walnut comb. With regard to shanks, they would all be feathered, because they would inherit one dominant copy of a feathered allele. With regard to hen- or cock-feathering: 1 male cock-feathered : 1 male hen-feathered : 2 females hen-feathered. Overall then, we would have a 1:1:2 ratio of

   walnut comb/feathered shanks/cock-feathered males

   walnut comb/feathered shanks/hen-feathered males

   walnut comb/feathered shanks/hen-feathered females

CHAPTER 5

Answers to Comprehension Questions

5.1: d, d 5.3: d, c, a, b

5.2: b, a, d 5.4: b, a, c, b

Concept Check Questions (in figure legends)

FIGURE 5.1 The offspring are all dextral because all of the F_1 mothers are *Dd*, and the genotype of the mother determines the phenotype of the offspring.

FIGURE 5.2 The egg cell will receive both *D* and *d* gene products.

FIGURE 5.3 The Barr body is more brightly staining because it is very compact.

FIGURE 5.4 X-chromosome inactivation initially occurs during embryonic development.

FIGURE 5.5 They migrate differently because their amino acid sequences are slightly different.

FIGURE 5.7 The *Tsix* gene is involved in choosing the X chromosome that is not inactivated.

FIGURE 5.8 Only the maintenance phase occurs in an adult female.

FIGURE 5.9 All of the offspring would be normal because they would inherit an active copy from their father.

FIGURE 5.10 Erasure allows eggs to transmit unmethylated copies of the gene to the offspring.

FIGURE 5.11 Maintenance methylation is automatic methylation that occurs when a methylated gene replicates and is transferred to daughter cells. It occurs in somatic cells. De novo methylation is the methylation of a gene that is not already methylated. It occurs in germ-line cells.

FIGURE 5.12 The offspring on the left did not receive a copy of either gene from their mother. Their father silences the *AS* gene, so they have AS because they do not have an active copy of the *AS* gene. By comparison, they receive an active *PWS* gene from their father, which prevents them from having PWS.

FIGURE 5.13 A nucleoid is not surrounded by a membrane as is the cell nucleus.

FIGURE 5.14 Mitochondria need rRNA and tRNA genes to translate proteins within the mitochondrial matrix.

FIGURE 5.15 A reciprocal cross is a cross in which the sexes and phenotypes of the parents are reversed compared to a first cross.

FIGURE 5.16 No. Once a patch of tissue is white, it has lost all of the normal chloroplasts, so it could not produce a patch of green tissue.

FIGURE 5.17 Chloroplast and mitochondrial genomes have lost most of their genes during evolution. Many of these have been transferred to the cell nucleus.

End-of-Chapter Questions:

Conceptual Questions

C2. A maternal effect gene is one in which the genotype of the mother determines the phenotype of the offspring. At the cellular level, this happens because maternal effect genes are expressed in diploid nurse cells and then the gene products are transported into the egg. These gene products play key roles in the early steps of embryonic development.

C4. The genotype of the mother must be $bic^- \ bic^-$. That is why it produces abnormal offspring. Because the mother is alive and able to produce offspring, its mother (the maternal grandmother) must have been $bic^+ \ bic^-$ and passed the bic^- allele to its daughter (the mother in this problem). The maternal grandfather also must have passed the bic^- allele to its daughter. The maternal grandfather could be either $bic^+ \ bic^-$ or $bic^- \ bic^-$.

C6. The mother must be heterozygous. She is phenotypically abnormal because her mother must have been homozygous for the abnormal recessive allele. However, because she produces all normal offspring, she must have inherited the normal dominant allele from her father. She produces all normal offspring because this is a maternal effect gene, and the gene product of the normal dominant allele is transferred to the egg.

C8. Maternal effect genes exert their effects because the gene products are transferred from nurse cells to eggs. The gene products, mRNA and proteins, do not last a very long time before they are eventually degraded. Therefore, they can exert their effects only during early stages of embryonic development.

C10. Dosage compensation refers to the phenomenon that the level of expression of genes on the sex chromosomes is the same in males and females, even though they have different numbers of sex chromosomes. In many species it seems necessary so that the balance of gene expression between the autosomes and sex chromosomes is similar between the two sexes.

C12. In mammals, one of the X chromosomes is inactivated in females; in *Drosophila*, the level of transcription on the X chromosome in males is doubled; in *C. elegans*, the level of transcription of the X chromosome in hermaphrodites is decreased by 50% of that of males.

C14. X-chromosome inactivation begins with the counting of Xics. If there are two X chromosomes, in the process of initiation, one is targeted for inactivation. During embryogenesis, this inactivation begins at the Xic locus and spreads to both ends of the X chromosome until it becomes a highly condensed Barr body. The *Tsix* gene may play a role in the choice of the X chromosome that remains active. The *Xist* gene, which is located in the Xic region, remains transcriptionally active on the inactivated X chromosome. It is thought to play an important role in X-chromosome inactivation by coating the inactive X chromosome. After X-chromosome inactivation is established, it is maintained in the same X chromosome in somatic cells during subsequent cell divisions. In germ cells, however, the X chromosomes are not inactivated, so an egg can transmit either copy of an active (noncondensed) X chromosome.

C16. A. One C. Two

B. Zero D. Zero

C18. The offspring inherited X^B from its mother and X^O and Y from its father. It is an XXY animal, which is male (but somewhat feminized).

C20. Erasure and reestablishment of the imprint occurs during gametogenesis. It is necessary to erase the imprint because each sex will transmit either inactive or active alleles of a gene. In somatic cells, the two alleles for a gene are imprinted according to the sex of the parent from which the allele was inherited.

C22. A person born with paternal uniparental disomy 15 would have Angelman syndrome, because this individual would not have an active copy of the *AS* gene; the paternally inherited copies of the *AS* gene are silenced. This individual would have normal offspring, because she does not have a deletion in either copy of chromosome 15.

C24. In some species, such as marsupials, X-chromosome inactivation depends on the sex. This is similar to imprinting. Also, once X-chromosome inactivation occurs during embryonic development, it is remembered throughout the rest of the life of the organism, which is also similar to imprinting. X-chromosome inactivation in mammals is different from genomic imprinting, in that it is not sex dependent. The X chromosome that is inactivated could be inherited from the mother or the father. There was no marking process on the X chromosome that occurred during gametogenesis. In contrast, genomic imprinting always involves a marking process during gametogenesis.

C26. The term reciprocal cross refers to two parallel crosses that involve the same genotypes of the two parents, but their sexes are opposite in the two crosses. For example, the reciprocal cross of female *BB* × male *bb* is the cross female *bb* × male *BB*. Autosomal inheritance gives the same result because the autosomes are transmitted from parent to offspring in the same way for both sexes. However, for extranuclear inheritance, the mitochondria and plastids are not transmitted via the gametes in the same way for both sexes. For maternal inheritance, the reciprocal crosses would show that the gene is always inherited from the mother.

C28. Paternal leakage means that, in a small percentage of cases, an organelle is inherited from the paternal parent. If paternal leakage was 3%, then 3% of the time the offspring would inherit the organelles from the father. If the father was transmitting a dominant allele in the organellar genome (and the mother did not), then 3% of the offspring would exhibit the trait. Thus, among a total of 200 offspring, 6 would be expected to inherit paternal mitochondria.

C30. The mitochondrial and chloroplast genomes are composed of a circular chromosome found in one or more copies in a region of the organelle known as the nucleoid. The number of genes per chromosome varies from species to species. Chloroplast genomes tend to be larger than mitochondria genomes. See Table 5.3 for examples of the variation among mitochondrial and chloroplast genomes.

C30. Mitochondria and chloroplasts evolved from an endosymbiotic relationship in which bacteria took up residence within a primordial eukaryotic cell. Throughout evolution, there has been a movement of genes out of the organellar genomes and into the nuclear genome. The genomes of modern mitochondria and chloroplasts only contain a fraction of genes necessary for organellar structure and function. Nuclear genes encode most of the proteins that function within chloroplasts and mitochondria. Long ago, these genes were originally in the mitochondrial and chloroplasts genomes but have been subsequently transferred to the nuclear genome.

C32. Superficially, the tendency to develop this form of leukemia seems to be inherited from mother to offspring, much like the inheritance of mitochondria. To show that it is not, one could separate newborn mice from their mothers and place them with mothers that do not carry AMLV. These offspring would not be expected to develop leukemia, even though their mother would.

Experimental Questions

E2. The first type of observation was based on cytological studies. The presence of the Barr body in female cells was consistent with the idea that one of the X chromosomes was highly condensed. The second type of observation was based on genetic mutations. A variegated phenotype that is found only in females is consistent with the idea that certain patches express one allele and other patches express the other allele. This variegated phenotype would occur only if the inactivation of one X chromosome happened at an early stage of embryonic development and was inherited permanently thereafter.

E4. The pattern of inheritance is consistent with imprinting. In every cross, the allele that is inherited from the father is expressed in the offspring, but the allele inherited from the mother is not.

E6. We assume that the snails in the large colony on the second island are true-breeding, *DD*. Let the male snail from the deserted island mate with a female snail from the large colony. Then let the F_1 snails mate with each other to produce an F_2 generation. Then let the F_2 generation mate with each other to produce an F_3 generation. Here are the expected results:

Female *DD* × Male *DD*

All F_1 snails coil to the right.

All F_2 snails coil to the right.

All F_3 snails coil to the right.

Female *DD* × Male *Dd*

All F_1 snails coil to the right.

All F_2 snails coil to the right because all of the F_1 females are *DD* or *Dd*.

$^{15}/_{16}$ of F_3 snails coil to the right; $^{1}/_{16}$ of F_3 snails coil to the left (because $^{1}/_{16}$ of the F_2 females are *dd*).

Female *DD* × Male *dd*

All F_1 snails coil to the right.

All F_2 snails coil to the right because all of the F_1 females are *Dd*.

$^{3}/_{4}$ of F_3 snails coil to the right; $^{1}/_{4}$ of F_3 snails coil to the left (because $^{1}/_{4}$ of the F_2 females are *dd*).

E8. Let's first consider the genotypes of male A and male B. Male A must have two normal copies of the *Igf2* gene. We know this because male A's mother was *Igf2 Igf2*; the father of male A must have been a heterozygote *Igf2 Igf2*$^-$ because half of the litter that contained male A also contained dwarf offspring. But because male A was not dwarf, it must have inherited the normal allele from its father. Therefore, male A must be *Igf2 Igf2*. We cannot be completely sure of the genotype of male B. It must have inherited the normal *Igf2* allele from its father because male B is phenotypically normal. We do not know the genotype of male B's mother, but she could be either *Igf2*$^-$ *Igf2*$^-$ or *Igf2 Igf2*$^-$. In either case, the mother of male B could pass the *Igf2*$^-$ allele to an offspring, but we do not know for sure if she did. So, male B could be either *Igf2 Igf2*$^-$ or *Igf2 Igf2*.

For the *Igf2* gene, we know that the maternal allele is inactivated. Therefore, the genotypes and phenotypes of females A and B are irrelevant. The phenotype of the offspring is determined only by the allele that is inherited from the father. Because we know that male A has to be *Igf2 Igf2*, we know that it can produce only normal offspring. Because both females A and B both produced dwarf offspring, male A cannot be the father. In contrast, male B could be either *Igf2 Igf2* or *Igf2 Igf2*$^-$. Because both females gave birth to dwarf babies (and because male A and male B were the only two male mice in the cage), we conclude that male B must be *Igf2 Igf2*$^-$ and is the father of both litters.

E10. In fruit flies, the expression of a male's X-linked genes is turned up twofold. In mice, one of the two X chromosomes is inactivated; that is why females and males produce the same total amount of mRNA for most X-linked genes. In *C. elegans*, the expression of hermaphrodite X-linked genes is turned down twofold. Overall, the total amount of expression of X-linked genes is the same in males and females (or hermaphrodites) of these three species. In fruit flies and *C. elegans*, heterozygous females and hermaphrodites express 50% of each allele compared with a homozygous male, so that heterozygous females and hermaphrodites produce the same total amount of mRNA from X-linked genes compared with males. Note: In heterozygous females of fruit flies, mice, and *C. elegans*, there is 50% of each gene product (compared to hemizygous males and homozygous females).

A.

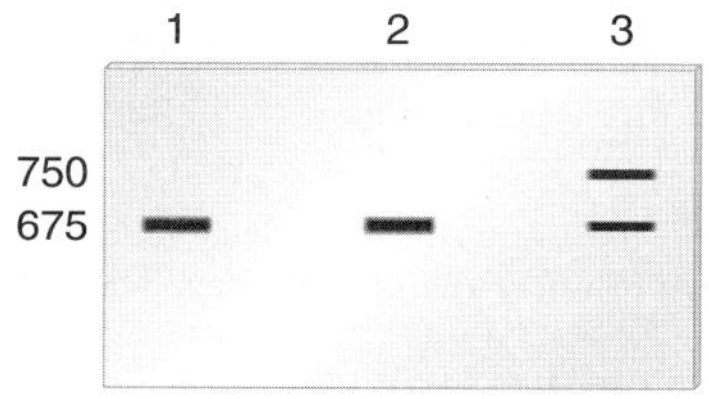

B.

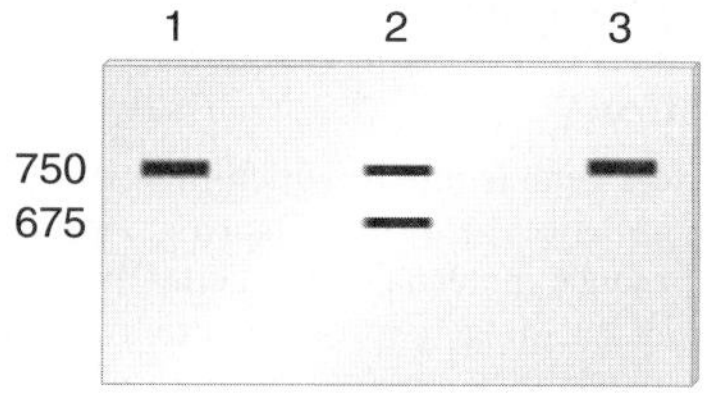

C.

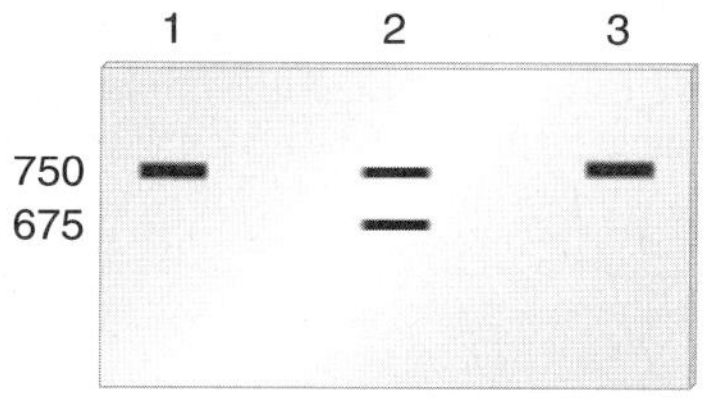

Questions for Student Discussion/Collaboration

2. Most of the genes originally within mitochondria and chloroplasts have been transferred to the nucleus. Therefore, mitochondria and chloroplasts have lost most of the genes that would be needed for them to survive as independent organisms.

CHAPTER 6

Answers to Comprehension Questions

6.1: b, d

6.2: a, a, b

6.3: d, d, b

6.4: c, b

6.5: b

Concept Check Questions (in figure legends)

FIGURE 6.1 The offspring found in excess are those with purple flowers, long pollen, and red flowers, round pollen.

FIGURE 6.2 No, such a crossover would not change the arrangements of these alleles.

FIGURE 6.3 A single crossover can produce offspring with: gray body, red eyes, miniature wings; gray body, white eyes, miniature wings; yellow body, red eyes, long wings; and yellow body, white eyes, long wings.

FIGURE 6.4 When genes are relatively close together, a crossover is relatively unlikely to occur between them. Therefore, the nonrecombinant offspring are more common.

FIGURE 6.5 The reason is because the *w* and *m* genes are farther apart than the *y* and *w* genes.

FIGURE 6.6 Crossing over can change the combination of kernel phenotypes and also it can change the morphologies of the chromosomes compared to the parental chromosomes.

FIGURE 6.8 Genetic mapping is useful to: (1) understand the complexity and genetic organization of a species; (2) understand the underlying basis of inherited traits; (3) clone genes; (4) understand evolution; (5) diagnose and treat diseases; (6) predict the likelihood of a couple having offspring with genetic diseases; (7) agricultural breeders of livestock and crops.

FIGURE 6.9 Crossing over occurred during oogenesis in the female parent of the recombinant offspring.

FIGURE 6.10 Multiple crossovers prevent the maximum percentage of recombinant offspring from exceeding 50%.

FIGURE 6.15 Mitotic recombination occurs in somatic cells.

End-of-Chapter Questions:

Conceptual Questions

C2. An independent assortment hypothesis is used because it enables us to calculate the expected values based on Mendel's ratios. Using the observed and expected values, we can calculate whether or not the deviations between the observed and expected values are too large to occur as a matter of chance. If the deviations are very large, we reject the hypothesis of independent assortment.

C4.

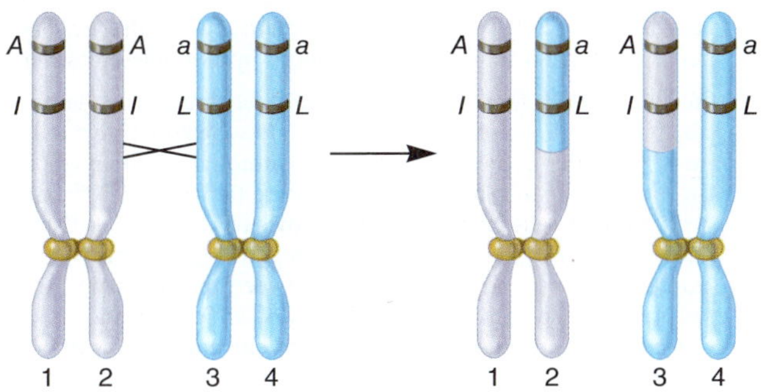

If the chromosomes (on the right side) labeled 2 and 4 move into one daughter cell, that will lead to a patch that is albino and has long fur. The other cell will receive chromosomes 1 and 3, which will produce a patch that has dark, short fur.

C6. A single crossover produces *A B C*, *A b c*, *a B C*, and *a b c*.

A. Between 2 and 3, between genes *B* and *C*

B. Between 1 and 4, between genes *A* and *B*

C. Between 1 and 4, between genes *B* and *C*

D. Between 2 and 3, between genes *A* and *B*

C8. The likelihood of scoring a basket would be greater if the basket was larger. Similarly, the chances of a crossover initiating in a region between two genes is proportional to the size of the region between the two genes. There are a finite number (usually a few) of crossovers that occur between homologous chromosomes during meiosis, and the likelihood that a crossover will occur in a region between two genes depends on how big that region is.

C10. The pedigree suggests a linkage between the dominant allele causing nail-patella syndrome and the I^B allele of the ABO blood type gene. In every case, the individual who inherits the I^B allele also inherits this disorder.

C12. *Ass-1* 43 *Sdh-1* 5 *Hdc* 9 *Hao-1* 6 *Odc-2* 8 *Ada-1*

C14. The inability to detect double crossovers causes the map distance to be underestimated. In other words, more crossovers occur in the region than we realize. When we have a double crossover, we do not get a recombinant offspring (in a dihybrid cross). Therefore, the second crossover cancels out the effects of the first crossover.

C16. The key feature is that all the products of a single meiosis are contained within a single sac. The spores in this sac can be dissected, and then their genetic traits can be analyzed individually.

C18. In an unordered ascus, the products of meiosis are free to move around. In an ordered octad (or tetrad), they are lined up according to their relationship to each other during meiosis and mitosis.

Experimental Questions (includes most mapping questions)

E2. They could have used a strain with two abnormal chromosomes. In this case, the recombinant chromosomes would either look normal or have abnormalities at both ends.

E4. A gene on the Y chromosome in mammals would be transmitted only from father to son. It would be difficult to genetically map Y-linked genes because a normal male has only one copy of the Y chromosome, so you do not get any crossing over between two Y chromosomes. Occasionally, abnormal males (XYY) are born with two Y chromosomes. If such males were heterozygous for alleles of Y-linked genes, one could examine the normal male offspring of XYY fathers and determine if crossing over has occurred.

E6. The answer is explained in solved problem S5. We cannot get more than 50% recombinant offspring because the pattern of multiple crossovers can yield an average maximum value of only 50%. When a testcross does yield a value of 50% recombinant offspring, it can mean two different things. Either the two genes are on different chromosomes or the two genes are on the same chromosome but at least 50 mu apart.

E8. If two genes are at least 50 mu apart, you would need to map genes between them to show that the two genes were actually in the same linkage group. For example, if gene *A* was 55 mu from gene *B*, there might be a third gene (e.g., gene *C*) that was 20 mu from *A* and 35 mu from *B*. These results would indicate that *A* and *B* are 55 mu apart, assuming dihybrid testcrosses between genes *A* and *B* yielded 50% recombinant offspring.

E10. A. Because they are 12 mu apart, we expect 12% (or 120) recombinant offspring. This would be approximately 60 *Aabb* and 60 *aaBb* plus 440 *AaBb* and 440 *aabb*.

B. We would expect 60 *AaBb*, 60 *aabb*, 440 *Aabb*, and 440 *aaBb*.

E12. Due to the large distance between the two genes, they will assort independently even though they are actually on the same chromosome. According to independent assortment, we expect 50% parental and 50% recombinant offspring. Therefore, this cross will produce 150 offspring in each of the four phenotypic categories.

E14. A. If we hypothesize two genes independently assorting, then the predicted ratio is 1:1:1:1. There are a total of 390 offspring. The expected number of offspring in each category is about 98. Plugging the figures into our chi square formula,

$$\chi^2 = \frac{(117-98)^2}{98} + \frac{(115-98)^2}{98} + \frac{(78-98)^2}{98} + \frac{(80-98)^2}{98}$$

$$\chi^2 = 3.68 + 2.95 + 4.08 + 3.31$$

$$\chi^2 = 14.02$$

Looking up this value in the chi square table under 1 degree of freedom, we reject our hypothesis, because the chi square value is above 7.815.

B. Map distance:

$$\text{Map distance} = \frac{78 + 80}{117 + 115 + 78 + 80}$$

$$= 40.5 \text{ mu}$$

Because the value is relatively close to 50 mu, it is probably a significant underestimate of the true distance between these two genes.

E16. The percentage of recombinants for the green, yellow and wide, narrow is 7%, or 0.07; there will be 3.5% of the green, narrow and 3.5% of the yellow, wide. The remaining 93% parentals will be 46.5% green, wide and 46.5% yellow, narrow. The third gene assorts independently. There will be 50% long and 50% short with respect to each of the other two genes. To calculate the number of offspring out of a total of 800, we multiply 800 by the percentages in each category.

(0.465 green, wide)(0.5 long)(800) = 186 green, wide, long

(0.465 yellow, narrow)(0.5 long)(800) = 186 yellow, narrow, long

(0.465 green, wide)(0.5 short)(800) = 186 green, wide, short

(0.465 yellow, narrow)(0.5 short)(800) = 186 yellow, narrow, short

(0.035 green, narrow)(0.5 long)(800) = 14 green, narrow, long

(0.035 yellow, wide)(0.5 long)(800) = 14 yellow, wide, long

(0.035 green, narrow)(0.5 short)(800) = 14 green, narrow, short

(0.035 yellow, wide)(0.5 short)(800) = 14 yellow, wide, short

E18. Let's use the following symbols: *G* for green pods, *g* for yellow pods, *S* for green seedlings, *s* for bluish green seedlings, *C* for normal plants, *c* for creepers. The parental cross is *GG SS CC* crossed to *gg ss cc*.

The F_1 plants would all be *Gg Ss Cc*. If the genes are linked, the alleles *G*, *S*, and *C* would be linked on one chromosome, and the alleles *g*, *s*, and *c* would be linked on the homologous chromosome.

The testcross is F_1 plants, which are *Gg Ss Cc* crossed to *gg ss cc*.

To measure the distances between the genes, we can separate the data into gene pairs.

Pod color, seedling color

2210 green pods, green seedlings—nonrecombinant

296 green pods, bluish green seedlings—recombinant

2198 yellow pods, bluish green seedlings—nonrecombinant

293 yellow pods, green seedlings—recombinant

$$\text{Map distance} = \frac{296 + 293}{2210 + 296 + 2198 + 293} \times 100 = 11.8 \text{ mu}$$

Pod color, plant stature

2340 green pods, normal—nonrecombinant

166 green pods, creeper—recombinant

2323 yellow pods, creeper—nonrecombinant

168 yellow pods, normal—recombinant

$$\text{Map distance} = \frac{166 + 168}{2340 + 166 + 2323 + 168} \times 100 = 6.7 \text{ mu}$$

Seedling color, plant stature

2070 green seedlings, normal—nonrecombinant

433 green seedlings, creeper—recombinant

2056 bluish green seedlings, creeper—nonrecombinant

438 bluish green seedlings, normal—recombinant

$$\text{Map distance} = \frac{433 + 438}{2070 + 433 + 2056 + 438} \times 100 = 17.4 \text{ mu}$$

The order of the genes is seedling color, pod color, and plant stature (or you could say the opposite order). Pod color is in the middle. If we use the two shortest distances to construct our map:

S ___ 11.8 ___ *G* ___ 6.7 ___ *C*

E20. To answer this question, we can consider genes in pairs. Let's consider the two gene pairs that are closest together. The distance between the wing length and eye color genes is 12.5 mu. From this cross, we expect 87.5% to have long wings and red eyes or short wings and purple eyes, and 12.5% to have long wings and purple eyes or short wings and red eyes. Therefore, we expect 43.75% to have long wings and red eyes, 43.75% to have short wings and purple eyes, 6.25% to have long wings and purple eyes, and 6.25% to have short wings and red eyes. If we have 1000 flies, we expect 438 to have long wings and red eyes, 438 to have short wings and purple eyes, 62 to have long wings and purple eyes, and 62 to have short wings and red eyes (rounding to the nearest whole number).

The distance between the eye color and body color genes is 6 mu. From this cross, we expect 94% to have a parental combination (red eyes and gray body or purple eyes and black body) and 6% to have a nonparental combination (red eyes and black body or purple eyes and gray body). Therefore, of our 438 flies with long wings and red eyes, we expect 94% of them (or about 412) to have long wings, red eyes, and gray body and 6% of them (or about 26) to have long wings, red eyes, and black bodies. Of our 438 flies with short wings and purple eyes, we expect about 412 to have short wings, purple eyes, and black bodies and 26 to have short wings, purple eyes, and gray bodies.

Of the 62 flies with long wings and purple eyes, we expect 94% of them (or about 58) to have long wings, purple eyes, and black bodies and 6% of them (or about 4) to have long wings, purple eyes, and gray bodies. Of the 62 flies with short wings and red eyes, we expect 94% (or about 58) to have short wings, red eyes, and gray bodies and 6% (or about 4) to have short wings, red eyes, and black bodies.

In summary,

Long wings, red eyes, gray body	412
Long wings, purple eyes, gray body	4
Long wings, red eyes, black body	26
Long wings, purple eyes, black body	58
Short wings, red eyes, gray body	58
Short wings, purple eyes, gray body	26
Short wings, red eyes, black body	4
Short wings, purple eyes, black body	412

The flies with long wings, purple eyes, and gray bodies, or short wings, red eyes, and black bodies, are produced by a double-crossover event.

E22. Yes. Begin with females that have one X chromosome that is X^{Nl} and the other X chromosome that is X^{nL}. These females have to be mated to $X^{NL}Y$ males because a living male cannot carry the *n* or *l* allele. In the absence of crossing over, a mating between $X^{Nl}X^{nL}$ females to $X^{NL}Y$ males should not produce any surviving male offspring. However, during oogenesis in these heterozygous female mice, there could be a crossover in the region between the two genes, which would produce an X^{NL} chromosome and an X^{nl} chromosome. Male offspring inheriting these recombinant chromosomes will be either $X^{NL}Y$ or $X^{nl}Y$ (whereas nonrecombinant males will be $X^{nL}Y$ or $X^{Nl}Y$). Only the male mice that inherit $X^{NL}Y$ will live. The living males represent only half of the recombinant offspring. (The other half are $X^{nl}Y$, which are born dead.)

To compute map distance:

$$\text{Map distance} = \frac{2(\text{Number of male living offspring})}{\text{Number of males born dead} + \text{Number of males born alive}} \times 100$$

Questions for Student Discussion/Collaboration

2. The X and Y chromosomes are not completely distinct linkage groups. One might describe them as overlapping linkage groups having some genes in common, but most genes are not common to both.

CHAPTER 7

Answers to Comprehension Questions

7.1: c	7.5: b, c
7.2: a, d	7.6: d
7.3: c, b	7.7: a, b
7.4: a, b	7.8: b

Concept Check Questions (in figure legends)

FIGURE 7.1 To grow, the colonies must have functional copies of all five genes. This could occur by the transfer of the met^+ and bio^+ genes to the met^- bio^- thr^+ leu^+ thi^+ strain or the transfer of the thr^+, leu^+, and thi^+ genes to the met^+ bio^+ thr^- leu^- thi^- strain.

FIGURE 7.2 Because bacteria are too small to pass through the filter, the apparatus can determine if direct cell-to-cell contact is necessary for gene transfer to occur.

FIGURE 7.3 It would be found in an F^+ cell.

FIGURE 7.4 Relaxase is a part of the relaxosome, which is needed for the cutting of the F factor and its transfer to the recipient cell. The coupling factor guides the DNA strand to the exporter, which transports it to the recipient cell.

FIGURE 7.5 An F′ factor carries a portion of the bacterial chromosome, whereas an F factor does not.

FIGURE 7.6 Because conjugation occurred for a longer period of time, *pro*$^+$ was transferred in the conjugation experiment shown in the bottom right.

FIGURE 7.8 This type of map is based on the time of gene transfer in conjugation experiments, which is determined in minutes.

FIGURE 7.9 The *lacZ* gene is closer to the origin of transfer; its transfer began at 16 minutes.

FIGURE 7.10 The normal process is for bacteriophage DNA to be incorporated into a phage coat. In transduction, a segment of bacterial chromosomal DNA is incorporated into a phage coat.

FIGURE 7.11 If the two genes were very far apart, the *arg*$^+$ gene would never be cotransduced with the *met*$^+$ gene.

FIGURE 7.12 If the recipient cell did not have a *lys*$^-$ gene, the *lys*$^+$ gene could be incorporated by nonhomologous recombination.

FIGURE 7.13 The genetic material that determines the structure of the virus is found inside the virus itself.

FIGURE 7.14 As more viruses are made, they are released from host cells and eventually cause neighboring cells to lyse. This make the plaque grow larger.

FIGURE 7.16 Noncomplementation indicates that two mutations are in different genes, whereas complementation indicates that the two mutations are in the same gene.

FIGURE 7.18 The mutant phages that infected *E. coli* K12(λ) would not be able to propagate and form plaques unless intragenic recombination produced a wild-type phage. This intragenic recombination would be a very rare event, which is why more plaques were used to infect this strain.

End-of-Chapter Questions:

Conceptual Questions

C2. It is not a form of sexual reproduction, in which two distinct parents produce gametes that unite to form a new individual. However, conjugation is similar to sexual reproduction in the sense that the genetic material from two cells are somewhat mixed. In conjugation, there is not the mixing of two genomes, one from each gamete. Instead, genetic material from one cell is transferred to another. This transfer can alter the combination of genetic traits in the recipient cell.

C4. An F^+ strain contains a separate, circular piece of DNA that has its own origin of transfer. An *Hfr* strain has its origin of transfer integrated into the bacterial chromosome. An F^+ strain can transfer only the DNA contained on the F factor. If given enough time, an *Hfr* strain can actually transfer the entire bacterial chromosome to the recipient cell.

C6. Sex pili promote the binding of donor and recipient cells.

C8. Though exceptions are common, interspecies genetic transfer via conjugation is not as likely because the cell surfaces do not interact correctly. Interspecies genetic transfer via transduction is also not very likely because each species of bacteria is sensitive to particular bacteriophages. The correct answer is transformation. A consequence of interspecies genetic transfer is that new genes can be introduced into a bacterial species from another species. For example, interspecies genetic transfer could provide the recipient bacterium with a new trait, such as resistance to an antibiotic. Evolutionary biologists call this horizontal gene transfer, while the passage of genes from parents to offspring is termed vertical gene transfer.

C10. Cotransduction is the transduction of two or more genes. The distance between the genes determines the frequency of cotransduction. When two genes are close together, the cotransduction frequency would be higher compared to two genes that are relatively farther apart.

C12. If a site that frequently incurred a breakpoint was between two genes, the cotransduction frequency of these two genes would be much less than expected. This is because the site where the breakage occurred would separate the two genes from each other.

C14. The transfer of conjugative plasmids such as F factor DNA.

C16. A. If it occurred in a single step, transformation is the most likely mechanism because conjugation does not usually occur between different species, particularly distantly related species, and different species are not usually infected by the same bacteriophages.

B. It could occur in a single step, but it may be more likely to have involved multiple steps.

C. The use of antibiotics selects for the survival of bacteria that have resistance genes. If a population of bacteria is exposed to an antibiotic, those carrying resistance genes will survive, and their relative numbers will increase in subsequent generations.

C18. The term *allele* means alternative forms of the same gene. Therefore, mutations in the same gene among different phages are alleles of each other; the mutations may be at different positions within the same gene. When we map the distance between mutations in the same gene, we are mapping the distance between the mutations that create different alleles of the same gene. An intragenic map describes the locations of mutations within the same gene.

Experimental Questions

E2. Mix the two strains together and then put some of them on plates containing streptomycin and some of them on plates without streptomycin. If mated colonies are present on both types of plates, then the *thr*$^+$, *leu*$^+$, and *thi*$^+$ genes were transferred to the *met*$^+$ *bio*$^+$ *thr*$^-$ *leu*$^-$ *thi*$^-$ strain. If colonies are found only on the plates that lack streptomycin, then the *met*$^+$ and *bio*$^+$ genes are being transferred to the *met*$^-$ *bio*$^-$ *thr*$^+$ *leu*$^+$ *thi*$^+$ strain. This answer assumes a one-way transfer of genes from a donor to a recipient strain.

E4. An interrupted mating experiment is a procedure in which two bacterial strains are allowed to mate, and then the mating is interrupted at various time points. The interruption occurs by agitation of the solution in which the bacteria are found. This type of study is used to map the locations of genes. It is necessary to interrupt mating so that you can vary the time and obtain information about the order of transfer; which gene transferred first, second, and so on.

E6. Mate unknown strains *A* and *B* to the F^- strain in your lab that is resistant to streptomycin and cannot use lactose. This is done in two separate tubes (i.e., strain *A* plus your F^- strain in one tube, and strain *B* plus your F^- strain in the other tube). Plate the mated cells on growth media containing lactose plus streptomycin. If you get growth of colonies, the unknown strain had to be strain *A*, the F^+ strain that had lactose utilization genes on its F factor.

E8. A. If we extrapolate these lines back to the *x*-axis, the *hisE* intersects at about 3 minutes, and the *pheA* intersects at about 24 minutes. These are the values for the times of entry. Therefore, the distance between these two genes is 21 minutes (i.e., 24 minus 3).

B.

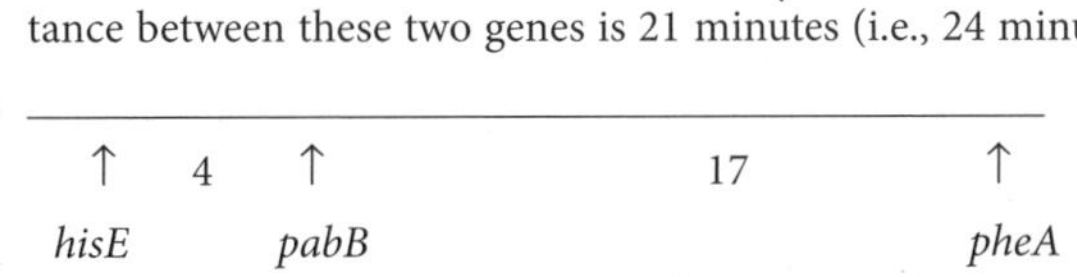

E10. One possibility is that you could treat the P1 lysate with DNase I, an enzyme that digests DNA. (Note: If DNA were digested with DNase I, the function of any genes within the DNA would be destroyed.) If the DNA were within a P1 phage, it would be protected from DNase I digestion. This would allow you to distinguish between transformation (which would be inhibited by DNase I) versus transduction (which would not be inhibited by DNase I). Another possibility is that you could try to fractionate the P1 lysate. Naked DNA would be smaller than a P1 phage carrying DNA. You could try to filter the lysate to remove naked DNA, or you could subject the lysate to centrifugation and remove the lighter fractions that contain naked DNA.

E12. Cotransduction frequency $= (1 - d/L)^3$

For the normal strain,

Cotransduction frequency $= (1 - 0.7/2)^3 = 0.275$, or 27.5%

For the new strain,

Cotransduction frequency $= (1 - 0.7/5)^3 = 0.64$, or 64%

The experimental advantage is that you could map genes that are farther than 2 minutes apart. You could map genes that are up to 2 minutes apart.

E14. Cotransduction frequency $= (1 - {}^{d}/_{L})^3$

$$0.53 = (1 - {}^{d}/_{2}\text{ minutes})^3$$
$$(1 - {}^{d}/_{2}\text{ minutes}) = \sqrt[3]{0.53}$$
$$(1 - {}^{d}/_{2}\text{ minutes}) = 0.81$$
$$d = 0.38\text{ minutes}$$

E16. A. We first need to calculate the cotransformation frequency, which equals $^{2}/_{70}$, or 0.029.

$$\text{Cotransformation frequency} = (1 - {}^{d}/_{L})^3$$
$$0.029 = (1 - {}^{d}/_{2}\text{ minutes})^3$$
$$d = 1.4\text{ minutes}$$

B. Cotransformation frequency $= (1 - {}^{d}/_{L})^3$

$$= (1 - {}^{1.4}/_{4})^3$$
$$= 0.27$$

As you may have expected, the cotransformation frequency is much higher when the transformation involves larger pieces of DNA.

E18. Benzer could use this observation as a way to evaluate if intragenic recombination had occurred. If two *rII* mutations recombined to make a wild-type gene, the phage would produce plaques in this *E. coli* K12(λ) strain.

E20. *rIIA*: L47 and L92

rIIB: L33, L40, L51, L62, L65, and L91

Questions for Student Discussion/Collaboration

2. Flower color in sweet peas is another example. When two different genes affect flower color, the white allele of one gene may be epistatic to the purple allele of another gene. A heterozygote for both genes would have purple flowers. In other words, the two purple alleles are complementing the two white alleles. Other similar examples are discussed in Chapter 4 in the section on Gene Interactions.

CHAPTER 8

Answers to Comprehension Questions

8.1: c, d
8.2: c
8.3: a, d
8.4: b, d, d
8.5: b
8.6: a, d
8.7: c, b
8.8: c, b

Concept Check Questions (in figure legends)

FIGURE 8.1 The staining of chromosomes results in banding patterns that make it easier to distinguish chromosomes that are similar in size and have similar centromere locations.

FIGURE 8.2 Deletions and duplications alter the total amount of genetic material.

FIGURE 8.3 If a chromosomal fragment does not contain a centromere, it will not segregate properly. If it remains outside the nucleus, it will be degraded.

FIGURE 8.5 Nonallelic homologous recombination occurs because of the pairing of homologous sites that duplicated within the chromosomes. This pairing causes the chromosomes to be misaligned.

FIGURE 8.11 These chromosomes form an inversion loop because the homologous regions are pairing with each other. For the inverted and noninverted regions to pair, a loop must form.

FIGURE 8.12 The mechanism shown in part (b) may occur if transposable elements are found in different chromosomes. These elements may promote the pairing between nonhomologous chromosomes and a subsequent crossover could occur.

FIGURE 8.13 Two out of six gametes (the two on the left) will produce a viable offspring with a normal phenotype. Therefore, the probability is $^{2}/_{6}$, or $^{1}/_{3}$.

FIGURE 8.14 These chromosomes form a translocation cross because homologous regions are pairing with each other.

FIGURE 8.15 Aneuploid, monosomic, monosomy 3.

FIGURE 8.16 The genes on chromosome 2 would be found in single copies, whereas the genes on the other chromosomes would be found in two copies. The expression of genes on chromosome 2 would be less (perhaps 50%) relative to a normal individual. This creates an imbalance between genes on chromosome 2 and those on the other chromosomes.

FIGURE 8.19 About 512.

FIGURE 8.20 Polyploid plants are often more vigorous and disease resistant. They may have larger flowers and produce more fruit.

FIGURE 8.21 During meiosis in a triploid individual, the homologs cannot pair properly. This results in highly aneuploid gametes, which are usually nonviable. Also, if aneuploid gametes participate in fertilization, the offspring are usually nonviable.

FIGURE 8.22 In this case, nondisjunction means that pairs of chromosomes are not separating from each other properly.

FIGURE 8.24 In autopolyploidy, multiple sets of chromosomes come from the same species. In allopolyploidy, multiple sets of chromosomes come from at least two different species.

End-of-Chapter Questions:

Conceptual Questions

C2. Small deletions and duplications are less likely to affect phenotype simply because they usually involve fewer genes. If a small deletion did have a phenotypic effect, you would conclude that a gene or genes in this region is required to have a normal phenotype.

C4. A gene family is a group of genes that are derived from the process of gene duplication. They have similar sequences, but the sequences have some differences due to the accumulation of mutations over many generations. The members of a gene family usually encode proteins with similar but specialized functions. The specialization may occur in different cells or at different stages of development.

C6. It has a pericentric inversion.

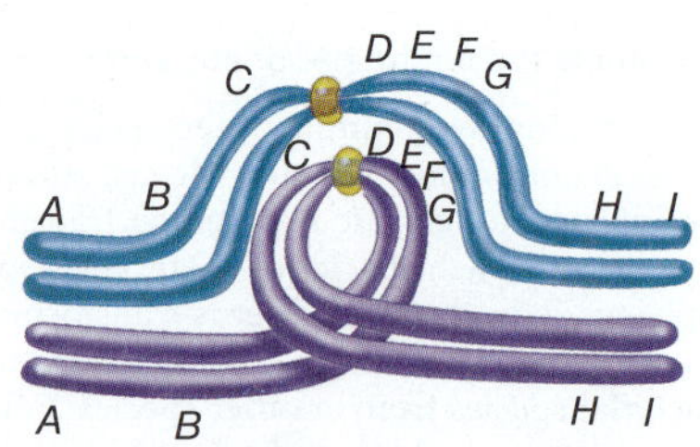

C8. There are four products from meiosis. One would be a normal chromosome, and one would contain the inversion shown in the drawing to conceptual question C8. The other two chromosomes would be dicentric or acentric with the following order of genes:

centromere		centromere	
A ↓	B C D E F G H I J D C B	↓ A	Dicentric
	M L K J I H G F E K L M		Acentric

C10. In the absence of crossing over, alternate segregation would yield half of the cells with two normal chromosomes and half with a balanced translocation. For adjacent-1 segregation, all cells will be unbalanced. Two cells would be

$\underline{A\ B\ C\ D\ E} + \underline{A\ I\ J\ K\ L\ M}$

And the other two cells would be

$\underline{H\ B\ C\ D\ E} + \underline{H\ I\ J\ K\ L\ M}$

C12. One of the parents may carry a balanced translocation between chromosomes 5 and 7. The phenotypically abnormal offspring has inherited an unbalanced translocation due to the segregation of translocated chromosomes during meiosis.

C14. A deletion and an unbalanced translocation are more likely to have phenotypic effects because they create genetic imbalances. For a deletion, there are too few copies of several genes, and for an unbalanced translocation, there are too many. Pericentric inversions and reciprocal translocations are not likely to affect phenotype unless they cause a position effect, but they may affect fertility and cause problems in future offspring.

C16. It is due to a crossover within the inverted region. You should draw the inversion loop (as is done in Figure 8.11a). The crossover occurred between *P* and *U*.

C18. This person has a total of 46 chromosomes. However, this person would be considered aneuploid rather than euploid, because one of the sets is missing a sex chromosome and one set has an extra copy of chromosome 21.

C20. It may be related to genetic balance. In aneuploidy, there is an imbalance in gene expression between the chromosomes found in their normal copy number versus those that are either too many or too few. In polyploidy, the balance in gene expression is still maintained.

C22. The male offspring is the result of nondisjunction during oogenesis. The female produced an egg without any sex chromosomes. The male parent transmitted a single X chromosome carrying the red allele. This produces an X0 male offspring with red eyes.

C24. Trisomies 13, 18, and 21 survive because the chromosomes are small and probably contain fewer genes compared to the larger chromosomes. Individuals with abnormal numbers of X chromosomes can survive because the extra copies are converted to transcriptionally inactive Barr bodies. The other trisomies are lethal because they cause a great amount of imbalance between the level of gene expression on the normal diploid chromosomes relative to the chromosomes that are trisomic.

C26. Endopolyploidy means that a particular somatic tissue is polyploid even though the rest of the organism is not. The biological significance is not entirely understood, although it has been speculated that an increase in chromosome number in certain cells may enhance their ability to produce specific gene products that are needed in great abundance.

C28. In certain types of cells, such as salivary cells, the homologous chromosomes pair with each other and then replicate approximately nine times to produce a polytene chromosome. The centromeres from each chromosome aggregate with each other at the chromocenter. This structure has six arms that arise from the single arm of two telocentric chromosomes (the X and 4) and two arms each from the metacentric chromosomes 2 and 3.

C30. The turtles are two distinct species that appear phenotypically identical. The turtles with 48 chromosomes are polyploid relatives (i.e., tetraploids) of the species with 24 chromosomes. In animals, it is somewhat hard to imagine how this could occur because most animals cannot self-fertilize, so there had to be two animals (i.e., one male and one female) that became tetraploids. It is easy to imagine how one animal could become a tetraploid; complete nondisjunction could occur during the first cell division of a fertilized egg, thereby creating a tetraploid cell that continued to develop into a tetraploid animal. This would have to happen independently (i.e., in two individuals of opposite sex) to create a tetraploid species. If you mated a tetraploid turtle with a diploid turtle, the offspring would be triploid and probably phenotypically normal. However, the triploid offspring would be sterile because they would make highly aneuploid gametes.

C32. Polyploid, triploid, and euploid should not be used.

C34. The boy carries a translocation involving chromosome 21: probably a translocation in which nearly all of chromosome 21 is translocated to chromosome 14. He would have one normal copy of chromosome 14, one normal copy of chromosome 21, and the translocated chromosome that contains both chromosome 14 and chromosome 21. This boy is phenotypically normal because the total amount of genetic material is normal, although the total number of chromosomes is 45 (because chromosome 14 and chromosome 21 are fused into a single chromosome). His sister has familial Down syndrome because she has inherited the translocated chromosome, but she also must have one copy of chromosome 14 and two copies of chromosome 21. She has the equivalent of three copies of chromosome 21 (i.e., two normal copies and one copy fused with chromosome 14). This is why she has familial Down syndrome. One of the parents of these two children is probably normal with regard to karyotype (i.e., the parent has 46 normal chromosomes). The other parent would have a karyotype that would be like the phenotypically normal boy.

C36. Nondisjunction is a mechanism whereby the chromosomes do not segregate equally into the two daughter cells. This can occur during meiosis to produce cells with altered numbers of chromosomes, or it can occur during mitosis to produce a genetic mosaic individual. A third way to alter chromosome number is by interspecies crosses that produce an alloploid.

C38. A mutation occurred during early embryonic development to create the blue patch of tissue. One possibility is a mitotic nondisjunction in which the two chromosomes carrying the *b* allele went to one cell and the two chromosomes carrying the *B* allele went to the other daughter cell. A second possibility is that the chromosome carrying the *B* allele was lost. A third possibility is that the *B* allele was deleted. This would cause the recessive *b* allele to exhibit pseudodominance.

C40. Homeologous chromosomes are chromosomes from two species that are evolutionarily related to each other. For example, chromosome 1 in the sable antelope and the roan antelope are homeologous; they carry many of the same genes.

C42. In general, Turner syndrome could be due to nondisjunction during either oogenesis or spermatogenesis. However, the Turner individual with color blindness is due to nondisjunction during spermatogenesis. The sperm lacked a sex chromosome, due to nondisjunction, and the egg carried an X chromosome with the recessive color blindness allele. This X chromosome had to be inherited from the mother, because the father was not color blind. The mother must be heterozygous for the recessive color blindness allele, and the father is hemizygous for the normal allele. Therefore, the mother must have transmitted a single X chromosome carrying the color blindness allele to her offspring, indicating that nondisjunction did not occur during oogenesis.

Experimental Questions

E2. Due to the mix up, the ratio of green/red fluorescence would be 0.5 in regions where the cancer cells had a normal amount of DNA. If a duplication occurred on both chromosomes of cancer cells, the ratio would be 1.0. If a deletion occurred on a single chromosome, the ratio would be 0.25.

E4. Colchicine interferes with the spindle apparatus and thereby causes nondisjunction. At high concentrations, it can cause complete nondisjunction and produce polyploid cells.

E6. First, you would cross the two strains together. It is difficult to predict the phenotype of the offspring. Nevertheless, you would keep crossing offspring to each other and backcrossing them to the parental strains until you obtained a great-tasting tomato strain that was resistant to heat and the viral pathogen. You could then make this strain tetraploid by treatment with colchicine. If you crossed the tetraploid strain with your great-tasting diploid strain that was resistant to heat and the viral pathogen, you may get a triploid that had these characteristics. This triploid would probably be seedless.

E8. A polytene chromosome is formed when a chromosome replicates many times, and the chromatids lie side by side, as shown in Figure 8.19. The homologous chromosomes also lie side by side. Therefore, if there is a deletion, there will be a loop. The loop is the segment that is not deleted from one of the two homologs.

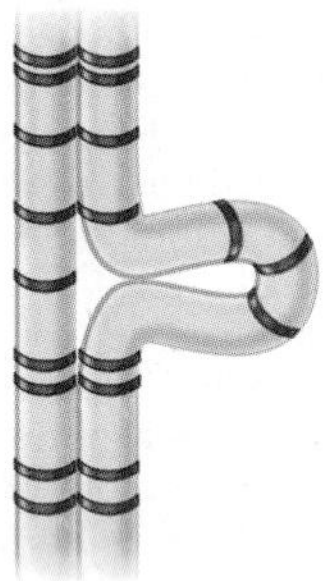

Questions for Student Discussion/Collaboration

2. There are many possibilities. The students could look in agriculture and botany books to find many examples. In the insect world, there are interesting examples of euploidy affecting gender determination. Among amphibians and reptiles, there are also several examples of closely related species that have euploid variation.
4. 1. Polyploids are often more robust and disease resistant.
   2. Allopolyploids may have useful combinations of traits.
   3. Hybrids are often more vigorous.
   4. Strains with an odd number of chromosome sets (e.g., triploids) are usually seedless.

CHAPTER 9

Answers to Comprehension Questions

9.1: d, b
9.2: c
9.3: c, d
9.4: a
9.5: d, c
9.6: d, a, a
9.7: d

Concept Check Questions (in figure legends)

FIGURE 9.1 In this experiment, the type R bacteria had taken up genetic material from the heat-killed type S bacteria, which converted the type R bacteria into type S. This enabled them to proliferate within the mouse and kill it.

FIGURE 9.2 RNase and protease were added to the DNA extract to rule out the possibility that small amounts of contaminating RNA or protein was responsible for converting the type R bacteria into type S.

FIGURE 9.4 It must be injected into the bacterial cytoplasm.

FIGURE 9.7 Ribose and uracil are not found in DNA.

FIGURE 9.10 Deoxyribose and phosphate form the backbone of a DNA strand.

FIGURE 9.11 Modeling is useful because it shows how molecules can fit together in a complicated three-dimensional structure.

FIGURE 9.15 Hydrogen bonding between base pairs and base stacking holds the DNA strands together.

FIGURE 9.16 The major and minor grooves are the indentations where the bases make contact with water. The major groove is wider than the minor groove.

FIGURE 9.17 B DNA is a right-handed helix and the backbone is helical, whereas Z DNA is a left-handed helix and the backbone appears to zigzag slightly. Z DNA has the bases tilted relative to the central axis, whereas they are perpendicular in B DNA. There are also minor differences in the number of bases pairs per turn.

FIGURE 9.19 Covalent bonds hold nucleotides together in an RNA strand.

FIGURE 9.20 A bonds with U and G bonds with C.

End-of-Chapter Questions:

Conceptual Questions

C2. The transformation process is described in Chapter 7.
   1. A fragment of DNA binds to the cell surface.
   2. It penetrates the cell wall/cell membrane.
   3. It enters the cytoplasm.
   4. It recombines with the chromosome.
   5. The genes within the DNA are expressed (i.e., transcription and translation).
   6. The gene products create a capsule. That is, they are enzymes that synthesize a capsule using cellular molecules as building blocks.

C4. The building blocks of a nucleotide are a sugar (ribose or deoxyribose), a nitrogenous base, and a phosphate group. In a nucleotide, the phosphate is already linked to the 5′ position on the sugar. When two nucleotides are hooked together, a phosphate on one nucleotide forms a covalent bond with the 3′ hydroxyl group on another nucleotide.

C6. The structure is a phosphate group connecting two sugars at the 3′ and 5′ positions, as shown in Figure 9.10.

C8. 3′—CCGTAATGTGATCCGGA—5′

C10. A drawing of a DNA helix with 10 bp per turn would look like Figure 9.15 in the textbook. To make 15 bp per turn, you would have to add 5 more base pairs, but the helix should still make only one complete turn.

C12. The nucleotide bases occupy the major and minor grooves. Phosphate and sugar are found in the backbone. If a DNA-binding protein does not recognize a nucleotide sequence, it probably is not binding in the grooves but instead is binding to the DNA backbone (i.e., sugar-phosphate sequence). DNA-binding proteins that recognize a base sequence must bind into a major or minor groove of the DNA, which is where the bases would be accessible to a DNA-binding protein. Most DNA-binding proteins that recognize a base sequence fit into the major groove. By comparison, other DNA-binding proteins, such as histones, which do not recognize a base sequence, bind to the DNA backbone.

C14. The structure is shown in Figure 9.7. You begin numbering at the carbon that is to the right of the ring oxygen and continue to number the carbon atoms in a clockwise direction. Antiparallel means that the backbones are running in the opposite direction. In one strand, the sugar carbons are oriented in a 3′ to 5′ direction, while in the other strand they are oriented in a 5′ to 3′ direction.

C16. Double-stranded RNA and DNA both form a helical structure due to base pairing. They different in that the number of base pairs per turn is slightly different and RNA follows an AU/GC base pairing rule whereas DNA follows AT/GC base pairing.

C18. Its nucleotide base sequence.

C20. G = 32%, C = 32%, A = 18%, T = 18%

C22. One possibility is a sequential mechanism. First, the double helix could unwind and replicate itself as described in Chapter 11. This would produce two double helices. Next, the third strand (bound in the major groove) could replicate itself via a semiconservative mechanism. This new strand could be copied to make a copy that is identical to the strand that lies in the major groove. At this point, you would have two double helices and two strands that could lie in the major groove. These could assemble to make two triple helices.

C24. Lysines and arginines, and also polar amino acids.

C26. This DNA molecule contains 280 bp. There are 10 bp per turn, so there are 28 complete turns.

C28. A hydroxyl group is at the 3′ end, and a phosphate group is at the 5′ end.

C30. Not necessarily. The AT/GC rule is required only of double-stranded DNA molecules.

C32. The first thing we need to do is to determine how many base pairs are in this DNA molecule. The linear length of 1 bp is 0.34 nm, which equals 0.34×10^{-9} m. One centimeter equals 10^{-2} meters.

$$\frac{10^{-2}}{0.34 \times 10^{-9}} = 2.9 \times 10^{7}\,\text{bp}$$

There are approximately 2.9×10^7 bp in this DNA molecule, which equals 5.8×10^7 nucleotides. If 15% are adenine, then 15% must also be thymine. This leaves 70% for cytosine and guanine. Because cytosine and guanine bind to each other, there must be 35% cytosine and 35% guanine. If we multiply 5.8×10^7 times 0.35, we get

$(5.8 \times 10^7)(0.35) = 2.0 \times 10^7$ cytosines, or about 20 million cytosines

C34. The methyl group is not attached to one of the atoms that hydrogen bonds with guanine, so methylation would not directly affect hydrogen bonding. It could indirectly affect hydrogen bonding if it perturbed the structure of DNA. Methylation may affect gene expression because it could alter the ability of proteins to recognize DNA sequences. For example, a protein might bind into the major groove by interacting with a sequence of bases that includes one or more cytosines. If the cytosines are methylated, this may prevent a protein from binding into the major groove properly. Alternatively, methylation could enhance protein binding. In Chapter 5, we considered DNA-binding proteins that were influenced by the methylation of DMRs (differentially methylated regions) that occur during genomic imprinting.

Experimental Questions

E2. A. There are different possible reasons why most of the cells were not transformed.

1. Most of the cells did not take up any of the type S DNA.
2. The type S DNA was usually degraded after it entered the type R bacteria.
3. The type S DNA was usually not expressed in the type R bacteria.

B. The antibody/centrifugation steps were used to remove the bacteria that had not been transformed. It enabled the researchers to determine the phenotype of the bacteria that had been transformed. If this step was omitted, there would have been so many colonies on the plate it would have been difficult to identify any transformed bacterial colonies, because they would have represented a very small proportion of the total number of bacterial colonies.

C. They were trying to demonstrate that it was really the DNA in their DNA extract that was the genetic material. It was possible that the extract was not entirely pure and could contain contaminating RNA or protein. However, treatment with RNase and protease did not prevent transformation, indicating that RNA and protein were not the genetic material. In contrast, treatment with DNase blocked transformation, confirming that DNA is the genetic material.

E4. A. There are several possible explanations why about 35% of the DNA is in the supernatant. One possibility is that not all of the DNA was injected into the bacterial cells. Alternatively, some of the cells may have been broken during the shearing procedure, thereby releasing the DNA.

B. If the radioactivity in the pellet had been counted instead of the supernatant, the following figure would be produced:

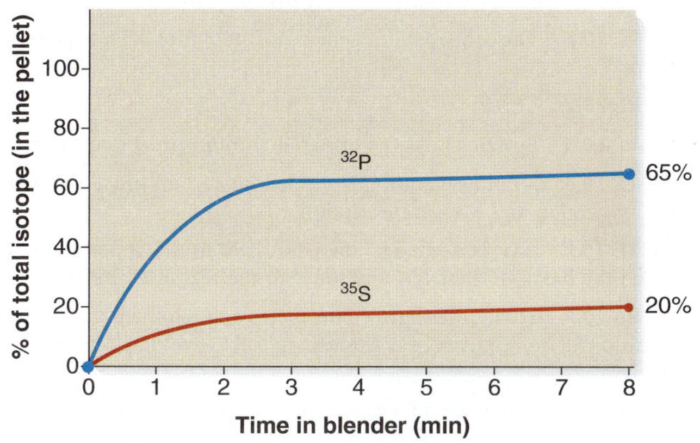

C. ^{32}P and ^{35}S were chosen as radioisotopes to label the phages because phosphorous is found in nucleic acids, while sulfur is found only in proteins.

D. There are multiple reasons why less than 100% of the phage protein is removed from the bacterial cells during the shearing process. For example, perhaps the shearing just is not strong enough to remove all of the phages, or perhaps the tail fibers remain embedded in the bacterium and only the head region is sheared off.

E6. This is really a matter of opinion. The Avery, MacLeod, and McCarty experiment seems to indicate directly that DNA is the genetic material, because DNase prevented transformation and RNase and protease did not. However, one could argue that the DNA is required for the rough bacteria to take up some other contaminant in the DNA preparation. It would seem that the other contaminant would not be RNA or protein. The Hershey and Chase experiments indicate that DNA is being injected into bacteria, although quantitatively the results are not entirely convincing. Some ^{35}S-labeled protein was not sheared off, so the results do not definitely rule out the possibility that protein could be the genetic material. But the results do indicate that DNA is the more likely candidate.

E8. A. The purpose of chromatography was to separate the different types of bases.

B. It was necessary to separate the bases and determine the total amount of each type of base. In a DNA strand, all the bases are found within a single molecule, so it is difficult to measure the total amount of each type of base. When the bases are removed from the strand, each type can be purified, and then the total amount of each type of base can be measured by spectroscopy.

C. Chargaff's results would probably not be very convincing if done on a single species. The strength of his data was that all species appeared to conform to the AT/GC rule, suggesting that this is a consistent feature of DNA structure. In a single species, the observation is that A = T and G = C could occur as a matter of chance.

Questions for Student Discussion/Collaboration

2. There are many possibilities. You could use a DNA-specific chemical and show that it causes heritable mutations. Perhaps you could inject an oocyte with a piece of DNA and produce a mouse with a new trait.

CHAPTER 10

Answers to Comprehension Questions

10.1: d	10.4: a
10.2: d, b, d	10.5: b, b, d, c
10.3: e	10.6: d, c

Concept Check Questions (in figure legends)

FIGURE 10.1 The sequences of genes constitute most of a bacterial genome.

FIGURE 10.2 Three.

FIGURE 10.3 Proteins hold the bacterial chromosomal loops in place.

FIGURE 10.6 Strand separation is needed for certain processes such as DNA replication and RNA transcription.

FIGURE 10.7 ATP is needed so the DNA held in the upper jaws can pass through the break in the DNA and move to the region of the lower jaws.

FIGURE 10.8 Eukaryotes have centromeres and telomeres, which bacterial chromosomes do not. Also, eukaryotes typically have much more repetitive sequences.

FIGURE 10.9 One reason for variation in genome size is that the number of genes among different eukaryotes varies. A second reason is that the amount of repetitive sequences varies.

FIGURE 10.11 It is 11 nm at its widest point.

FIGURE 10.14 The solenoid model depicts the nucleosomes in a repeating, spiral arrangement, whereas the zigzag model depicts a more irregular and dynamic arrangement of nucleosomes.

FIGURE 10.15 The nuclear matrix helps to organize and compact the chromosomes within the cell nucleus and also aids in their condensation during cell division.

FIGURE 10.16 A chromosome territory is a discrete region in the cell nucleus that is occupied by a single chromosome.

FIGURE 10.17 Active genes are found in more loosely packed euchromatic regions.

FIGURE 10.18 In the conversion of a 300-nm fiber to a 700-nm fiber, the radial loop domains become more tightly packed.

FIGURE 10.22 At the beginning of prophase, cohesin lines the region between sister chromatids. It is first removed from the arms of the sister chromatids but remains in the centromeric region. At anaphase, the cohesin at the centromeres is degraded, which allows the sister chromatids to separate.

End-of-Chapter Questions:

Conceptual Questions

C2. A bacterium with two nucleoids is similar to a diploid eukaryotic cell because it would have two copies of each gene. The bacterium is different, however, with regard to alleles. A eukaryotic cell can have two different alleles for the same gene. For example, a cell from a pea plant could be heterozygous, *Tt*, for the gene that affects height. By comparison, a bacterium with two nucleoids has two identical chromosomes. Therefore, a bacterium with two nucleoids is homozygous for its chromosomal genes. Note: As discussed in Chapter 7, a bacterium can contain another piece of DNA, called an F′ factor, that can carry a few genes. The alleles on an F′ factor can be different from the alleles on the bacterial chromosome.

C4. A. One loop is 40,000 bp. One base pair is 0.34 nm, which equals 0.34×10^{-3} μm. If we multiply the two together:

$$(40{,}000)(0.34 \times 10^{-3}) = 13.6\ \mu m$$

B. Circumference = πD

$$13.6\ \mu m = \pi D$$

$$D = 4.3\ \mu m$$

C. No, it is too big to fit inside of *E. coli*. Supercoiling is needed to make the loops more compact.

C6. These drugs would diminish the amount of negative supercoiling in DNA. Negative supercoiling is needed to compact the chromosomal DNA, and it also aids in strand separation. Bacteria might not be able to survive and/or transmit their chromosomes to daughter cells if their DNA was not compacted properly. Also, because negative supercoiling aids in strand separation, these drugs would make it more difficult for the DNA strands to separate. Therefore, the bacteria would have a difficult time transcribing their genes and replicating their DNA, because both processes require strand separation. As discussed in Chapter 11, DNA replication is needed to make new copies of the genetic material to transmit from mother to daughter cells. If DNA replication was inhibited, the bacteria could not grow and divide into new daughter cells. As discussed in Chapters 12–14, gene transcription is necessary for bacterial cells to make proteins. If gene transcription was inhibited, the bacteria could not make many proteins that are necessary for survival.

C8.

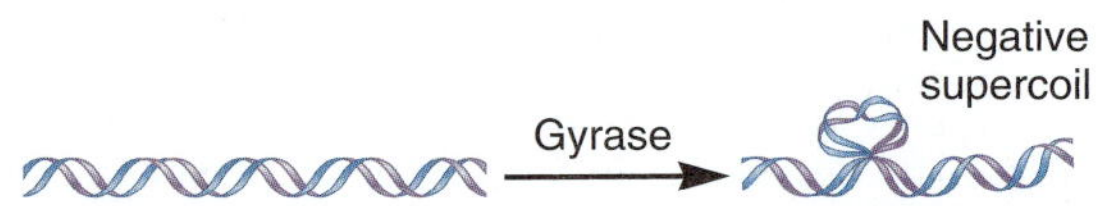

C10. The centromere is the attachment site for the kinetochore, which attaches to the spindle. If a chromosome is not attached to the spindle, it is free to "float around" within the cell, and it may not be near a pole when the nuclear membrane re-forms during telophase. If a chromosome is left outside of the nucleus, it is degraded during interphase. That is why the chromosome without a centromere may not be found in daughter cells.

C12. Highly repetitive DNA, as its name suggests, is a DNA sequence that is repeated many times, from tens of thousands to millions of times throughout the genome. It can be interspersed in the genome or found clustered in a tandem array, in which a short nucleotide sequence is repeated many times in a row. In DNA renaturation studies, highly repetitive DNA renatures at a much faster rate because there are many copies of the complementary sequences.

C14. During interphase (i.e., G_1, S, and G_2), the euchromatin is found primarily as a 30-nm fiber in a radial loop configuration. Most interphase chromosomes also have some heterochromatic regions where the radial loops are more highly compacted. During M phase, each chromosome becomes entirely heterochromatic, which is needed for the proper sorting of the chromosomes during nuclear division.

C16.

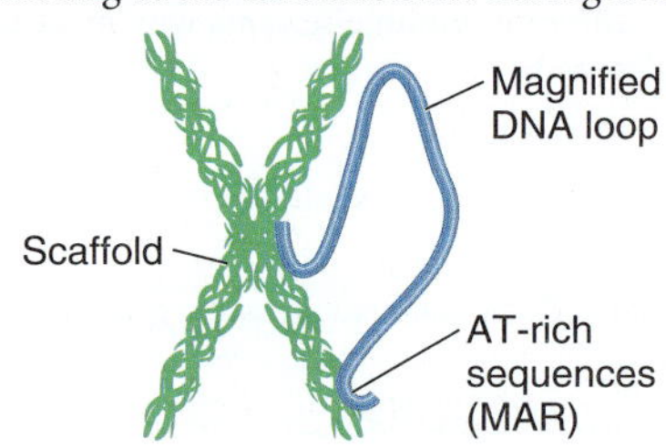

C18. During interphase, the chromosomes are found within the cell nucleus. They are less tightly packed and are transcriptionally active. Segments of chromosomes are anchored to the nuclear matrix. During M phase, the chromosomes become highly condensed, and the nuclear membrane is fragmented into vesicles. The chromosomal DNA remains anchored to a scaffold, formed from the nuclear matrix. The chromosomes eventually become attached to the spindle apparatus via microtubules that attach to the kinetochore, which is attached to the centromere.

C20. There are 146 bp around the core histones. If the linker region is 54 bp, we expect 200 bp of DNA (i.e., 146 + 54) for each nucleosome and linker region. If we divide 46,000 bp by 200 bp, we get 230. Because there are two molecules of H2A for each nucleosome, there would be 460 molecules of H2A in a 46,000-bp sample of DNA.

C22. The role of the core histones is to form the nucleosomes. In a nucleosome, the DNA is wrapped 1.65 times around the core histones. Histone H1 binds to the linker region. It may play a role in compacting the DNA into a 30-nm fiber.

C24. The answer is B and E. A Barr body is composed of a type of highly compacted chromatin called heterochromatin. Euchromatin is not so compacted. A Barr body is not composed of euchromatin. A Barr body is one chromosome, the X chromosome. The term genome refers to all the types of chromosomes that make up the genetic composition of an individual.

C26. SMC stands for structural maintenance of chromosomes. SMC proteins use energy from ATP to catalyze changes in chromosome structure. Together with topoisomerases, SMC proteins have been shown to promote major changes in DNA structure. Two examples of SMC proteins are condensin and cohesin, which play different roles in metaphase chromosome structure. The function of condensin is to promote the proper compaction of metaphase chromosomes, while the function of cohesin is to promote the binding (i.e., cohesion) between sister chromatids.

Experimental Questions

E2. Supercoiled DNA would look curled up into a relatively compact structure. You could add different purified topoisomerases and see how they affect the structure via microscopy. For example, DNA gyrase relaxes positive supercoils, and topoisomerase I relaxes negative supercoils. If you added topoisomerase I to a DNA preparation and it became less compacted, then the DNA was negatively supercoiled.

E4. DNase I cuts the DNA in the linker region. A single nucleosome plus one half of a linker region on each side of it contains about 200 bp of DNA. At lower DNase I concentration, sometimes the linker region was uncut. Therefore, a segment of DNA could be derived from 2 or more nucleosomes and would be in multiples of 200 bp.

E6. You would get DNA fragments of about 446 to 496 bp (i.e., 146 bp plus 300 to 350 bp).

E8. Histones are positively charged, and DNA is negatively charged. They bind to each other by these ionic interactions. Salt is composed of positively charged ions and negatively charged ions. For example, when dissolved in water, NaCl becomes individual ions of Na^+ and Cl^-. When chromatin is exposed to a salt such as NaCl, the positively charged Na^+ could bind to the DNA, and the negatively charged Cl^- could bind to the histones. This would prevent the histones and DNA from binding to each other.

Questions for Student Discussion/Collaboration

2. This is a matter of opinion. It seems strange to have so much DNA that seems to have no obvious function. It's a waste of energy. Perhaps it has a function that we don't know about yet. On the other hand, evolution does allow bad things to accumulate within genomes, such as genes that cause diseases, etc. Perhaps this is just another example of the negative consequences of evolution.

CHAPTER 11

Answers to Comprehension Questions

11.1: c, d, b
11.2: b, d
11.3: a, d
11.4: a, d
11.5: b, d, c, b

Concept Check Questions (in figure legends)

FIGURE 11.1 The two features that allow DNA to be replicated are its double-stranded structure and the base pairing between A and T and between G and C.

FIGURE 11.5 The DnaA boxes are recognized by DnaA protein, which binds to them and causes the DNA to unwind at the AT-rich region.

FIGURE 11.6 Two replication forks are formed at the origin.

FIGURE 11.7 Primase is needed for DNA replication because DNA polymerase cannot initiate DNA replication on a bare template strand.

FIGURE 11.8 The template strand is read in the 3′ to 5′ direction.

FIGURE 11.10 The leading strand is made as one, long continuous strand in the same direction that the replication fork is moving. The lagging strand is made as Okazaki fragments in the direction away from the replication fork.

FIGURE 11.12 Having the replication machinery in a complex allows the coordination of different enzymes. For example, helicase and primase can work together to make multiple primers for the lagging strand.

FIGURE 11.14 Yes, so the strands can move relative to each and untangle the catenane.

FIGURE 11.16 The oxygen in the newly made ester bond comes from the sugar.

FIGURE 11.19 Because eukaryotic chromosomes are so large, they need multiple origins so the DNA can be replicated in a reasonable length of time.

FIGURE 11.23 Six times (36 divided by 6).

End-of-Chapter Questions:

Conceptual Questions

C2. Bidirectional replication refers to DNA replication in both directions starting from one origin.

C4. A.

```
TTGGHTGUTGG
HHUUTHUGHUU
```

B.

```
                  TTGGHTGUTGG
                  HHUUTHUGHUU
                       ↓
      TTGGHTGUTGG  CCAAACACCAA
      AACCCACAACC  HHUUTHUGHUU
                       ↓
TTGGHTGUTGG  TTGGGTGTTGG  CCAAACACCAA  CCAAACACCAA
AACCCACAACC  AACCCACAACC  GGTTTGTGGTT  HHUUTHUGHUU
```

C6. Let's assume there are 4,600,000 bp of DNA and that DNA replication is bidirectional at a rate of 750 nucleotides per second.

If there were just a single replication fork,

4,600,000/750 = 6133 seconds, or 102.2 minutes.

Because replication is bidirectional, $^{102.2}/_{2} = 51.1$ minutes.

Actually, this is an average value based on a variety of growth conditions. Under optimal growth conditions, replication can occur substantially faster.

With regard to errors, if we assume an error rate of one mistake per 100,000,000 nucleotides,

$4{,}600{,}000 \times 1000$ bacteria = 4,600,000,000 nucleotides of replicated DNA

4,600,000,000/100,000,000 = 46 mistakes

When you think about it, this is pretty amazing. In this population, DNA polymerase would cause only 46 single mistakes in a total of 1000 bacteria, each containing 4.6 million bp of DNA.

C8. DNA polymerase would slide from right to left. The new strand would be 3′—CTAGGGCTAGGCGTATGTAAATGGTCTAGTGGTGG—5′

C10. A. When looking at Figure 11.5, the first, second, and fourth DnaA boxes are running in the same direction, and the third and fifth are running in the opposite direction. Once you realize that, you can see the sequences are very similar to each other.

B. According to the direction of the first DnaA box, the consensus sequence is

```
TGTGGATAA
ACACCTATT
```

C. This sequence is 9 nucleotides long. Because there are four kinds of nucleotides (i.e., A, T, G, and C), the chance of this sequence occurring by random chance is 4^{-9}, which equals once every

262,144 nucleotides. Because the *E. coli* chromosome is more than 10 times longer than this, it is fairly likely that this consensus sequence occurs elsewhere. The reason why there are not multiple origins, however, is because the origin has five copies of the consensus sequence very close together. The chance of having five copies of this consensus sequence occurring close together (as a matter of random chance) is very small.

C12. 1. According to the AT/GC rule, a pyrimidine always hydrogen bonds with a purine. A transition still involves a pyrimidine hydrogen bonding to a purine, but a transversion causes a purine to hydrogen bond with a purine or a pyrimidine to hydrogen bond with a pyrimidine. The structure of the double helix makes it much more difficult for this latter type of hydrogen bonding to occur.

2. The induced-fit phenomenon of the active site of DNA polymerase makes it unlikely for DNA polymerase to catalyze covalent bond formation if the wrong nucleotide is bound to the template strand. A transition mutation creates a somewhat bad interaction between the bases in opposite strands, but it is not as bad as the fit caused by a transversion mutation. In a transversion, a purine is opposite another purine, or a pyrimidine is opposite a pyrimidine. This is a very bad fit.

3. The proofreading function of DNA polymerase is able to detect and remove an incorrect nucleotide that has been incorporated into the growing strand. A transversion is going to cause a larger distortion in the structure of the double helix and make it more likely to be detected by the proofreading function of DNA polymerase.

C14. Primase and DNA polymerase are able to knock the single-strand binding proteins off the template DNA.

C16. A. The right Okazaki fragment was made first. It is farthest away from the replication fork. The fork (not seen in the strand diagram in the original question) would be to the left of the three Okazaki fragments, and moving from right to left.

B. The RNA primer in the right Okazaki fragment would be removed first. DNA polymerase would begin by elongating the DNA strand of the middle Okazaki fragment and removing the right RNA primer with its 5′ to 3′ exonuclease activity. DNA polymerase I would use the 3′ end of the DNA of the middle Okazaki fragment as a primer to synthesize DNA in the region where the right RNA primer is removed. If the middle fragment was not present, DNA polymerase could not fill in this DNA (because it needs a primer).

C. You need DNA ligase only at the right arrow. DNA polymerase I begins at the end of the left Okazaki fragment and synthesizes DNA to fill in the region as it removes the middle RNA primer. At the left arrow, DNA polymerase I is simply extending the length of the left Okazaki fragment. No ligase is needed here. When DNA polymerase I has extended the left Okazaki fragment through the entire region where the RNA primer has been removed, it hits the DNA of the middle Okazaki fragment. This occurs at the right arrow. At this point, the DNA of the middle Okazaki fragment has a 5′ end that is a monophosphate. DNA ligase is needed to connect this monophosphate with the 3′ end of the region where the middle RNA primer has been removed.

D. As mentioned in the answer to part C, the 5′ end of the DNA in the middle Okazaki fragment is a monophosphate. It is a monophosphate because it was previously connected to the RNA primer by a phosphoester bond. At the location of the right arrow, there was only one phosphate connecting this deoxyribonucleotide to the last ribonucleotide in the RNA primer. For DNA polymerase to function, the energy to connect two nucleotides comes from the hydrolysis of the incoming triphosphate. In this location shown at the right arrow, however, the nucleotide is already present at the 5′ end of the DNA, and it is a monophosphate. DNA ligase needs energy to connect this nucleotide with the left Okazaki fragment. It obtains energy from the hydrolysis of ATP or NAD^+.

C18. The picture would depict a ring of helicase proteins traveling along a DNA strand and separating the two helices, as shown in Figure 11.6.

C20. The leading strand is primed once, at the origin, and then DNA polymerase III synthesizes DNA continuously in the direction of the replication fork. In the lagging strand, many short pieces of DNA (Okazaki fragments) are made. This requires many RNA primers. The primers are removed by DNA polymerase I, which then fills in the gaps with DNA. DNA ligase then covalently connects the Okazaki fragments. Having the enzymes within a complex such as a primosome or replisome provides coordination among the different steps in the replication process and thereby allows it to proceed faster and more efficiently.

C22. A processive enzyme is one that remains clamped to one of its substrates. In the case of DNA polymerase, it remains clamped to the template strand as it makes a new daughter strand. This is important to ensure a fast rate of DNA synthesis.

C24. The opposite strand is made in the conventional way by DNA polymerase using the strand made via telomerase as a template.

C26. The ends labeled B and C could not be replicated by DNA polymerase. DNA polymerase makes a strand in the 5′ to 3′ direction using a template strand that is running in the 3′ to 5′ direction. Also, DNA polymerase requires a primer. At the ends labeled B and C, there is no place (upstream) for a primer to be made.

C28. As shown in Figure 11.23, the first step involves a binding of telomerase to the telomere. The 3′ overhang binds to the complementary RNA in telomerase. For this reason, a 3′ overhang is necessary for telomerase to replicate the telomere.

Experimental Questions

E2. A. You would probably still see a band of DNA, but you would only see a heavy band.

B. You would probably not see a band because the DNA would not be released from the bacteria. The bacteria would sediment to the bottom of the tube.

C. You would not see a band. UV light is needed to see the DNA, which absorbs light in the UV region.

E4. Rapid-stop mutants immediately stop DNA replication when the cells are shifted to a nonpermissive temperature. Slow-stop mutants finish the round of DNA replication that is occurring when the temperature is shifted, but they cannot initiate a new round of DNA replication. Rapid-stop mutants affect proteins that are needed to make DNA at a replication fork, whereas slow-stop mutants affect proteins that are needed to initiate DNA replication at an origin of replication.

E6. A. Heat is used to separate the DNA strands, so you do not need helicase.

B. Each primer must be a sequence that is complementary to one of the DNA strands. There are two types of primers, and each type binds to one of the two complementary strands.

C. A thermophilic DNA polymerase is used because DNA polymerases isolated from nonthermophilic species would be permanently inactivated during the heating phase of the PCR cycle. Remember that DNA polymerase is a protein, and most proteins are denatured by heating. However, proteins from thermophilic organisms have evolved to withstand heat, which is how thermophilic organisms survive at high temperatures.

D. With each cycle, the amount of DNA is doubled. Because there are initially 10 copies of the DNA, there will be 10×2^{27} copies after 27 cycles. $10 \times 2^{27} = 1.34 \times 10^9 = 1.34$ billion copies of DNA. As you can see, PCR can amplify the amount of DNA by a staggering amount!

Questions for Student Discussion/Collaboration

2. DNA synthesis at a replication fork is very similar in bacteria and eukaryotes, which involves helicase, topoisomerase, single-strand binding protein, primase, DNA polymerase, and ligase. Eukaryotes have more types of DNA polymerases, and different ones are involved with synthesizing the leading and lagging strands. In bacteria, the RNA primers are removed by DNA polymerase I, whereas in eukaryotes, they are removed by flap endonuclease. Bacteria have a single origin of replication that is recognized DnaA protein. Eukaryotes have multiple origins of replications that are recognized by ORC. After ORC binds, other proteins including MCM helicase complete a process called DNA replication licensing. Another difference is that the ends of eukaryotic chromosomes are replicated by telomerase.

CHAPTER 12

Answers to Comprehension Questions

12.1: a, c, a 12.3: b, c, d, b 12.5: d

12.2: d, b, a, b 12.4: d, c, b, d

Concept Check Questions (in figure legends)

FIGURE 12.2 The mutation would not affect the length of the RNA, because it would not terminate transcription. However, the encoded polypeptide would be shorter.

FIGURE 12.5 When comparing a group of related sequences, the consensus sequence is the most common base found at each location within that sequence.

FIGURE 12.6 Parts of σ factor must fit into the major groove so it can recognize a base sequence of a promoter.

FIGURE 12.7 The –10 is AT rich, which has fewer hydrogen bonds compared to a region with a lot of GC base pairs.

FIGURE 12.10 The mutation would prevent ρ-dependent termination of transcription.

FIGURE 12.11 NusA helps RNA polymerase to pause, which facilitates transcriptional termination.

FIGURE 12.13 The TATA box provides a precise starting point for the transcription of eukaryotic protein-encoding genes.

FIGURE 12.14 The phosphorylation of CTD allows RNA polymerase to proceed to the elongation phase of transcription.

FIGURE 12.17 An endonuclease can cleave within a strand, whereas an exonuclease digests a strand, one nucleotide at a time, starting at one end.

FIGURE 12.20 Splicing via a spliceosome is very common in eukaryotes.

FIGURE 12.22 snRNPs are involved in recognizing the intron boundaries, cutting out the intron, and connecting the two adjacent exons together.

FIGURE 12.23 The 7-methylguanosine cap is important for the proper splicing of pre-mRNA, the exit of mRNA from the nucleus, and the binding of mRNA to a ribosome.

FIGURE 12.25 A consequence of RNA editing is that the amino acid sequence of an encoded polypeptide may be changed, which can affect its function.

End-of-Chapter Questions:

Conceptual Questions

C2. The release of σ factor marks the transition to the elongation stage of transcription.

C4. GGCATTGTCA

C6. The most highly conserved positions are the first, second, and sixth. In general, when promoter sequences are conserved, they are more likely to be important for binding. That explains why changes are not found at these positions; if a mutation altered a conserved position, the promoter would probably not work very well. By comparison, changes are occasionally tolerated at the fourth position and frequently at the third and fifth positions. The positions that tolerate changes are less important for binding by σ factor.

C8. This will not affect transcription. However, it will affect translation by preventing the initiation of polypeptide synthesis.

C10. Sigma factor can slide along the major groove of the DNA. In this way, it is able to recognize base sequences that are exposed in the groove. When it encounters a promoter sequence, hydrogen bonding between the bases and the σ factor protein can promote a tight and specific interaction.

C12. DNA-G/RNA-C

DNA-C/RNA-G

DNA-A/RNA-U

DNA-T/RNA-A

The template strand is

3′—CCGTACGTAATGCCGTAGTGTGATCCCTAG—5′

and the coding strand is

5′—GGCATGCATTACGGCATCACACTAGGGATC—3′

The promoter would be to the left (in the 3′ direction) of the template strand.

C14. Transcriptional termination occurs when the hydrogen bonding is broken between the DNA and the part of the newly made RNA transcript that is located in the open complex.

C16. DNA helicase and ρ protein bind to a nucleic acid strand and travel in the 5′ to 3′ direction. When they encounter a double-stranded region, they break the hydrogen bonds between complementary strands. ρ protein is different from DNA helicase in that it moves along an RNA strand, while DNA helicase moves along a DNA strand. The purpose of DNA helicase function is to promote DNA replication; the purpose of ρ protein function is to promote transcriptional termination.

C18. A. Mutations that alter the uracil-rich region by introducing guanines and cytosines, and mutations that prevent the formation of the stem-loop structure.

B. Mutations that alter the termination sequence, and mutations that alter the ρ recognition site.

C. Eventually, somewhere downstream from the gene, another transcriptional termination sequence would be found, and transcription would terminate there. This second termination sequence might be found randomly, or it might be at the end of an adjacent gene.

C20. Eukaryotic promoters are somewhat variable with regard to the pattern of sequence elements that may be found. In the case of protein-encoding genes that are transcribed by RNA polymerase II, it is common to have a TATA box, which is about 25 bp upstream from a transcriptional start site. The TATA box is important in the identification of the transcriptional start site and the assembly of RNA polymerase and various transcription factors. The transcriptional start site defines where transcription actually begins.

C22. The two models are described in Figure 12.15. The allosteric model is more like ρ-independent termination whereas the torpedo model is more like ρ-dependent termination. In the torpedo model, a protein knocks RNA polymerase off the DNA, much like the effects of ρ protein.

C24. Hydrogen bonding is usually the predominant type of interaction when proteins and DNA follow an assembly and disassembly process. In addition, ionic bonding and hydrophobic interactions could occur. Covalent interactions would not occur. High temperature and low salt concentrations tend to break hydrogen bonds. Therefore, high temperature and low salt would inhibit assembly and stimulate disassembly.

C26. In bacteria, the 5′ end of the tRNA is cleaved by RNase P. The 3′ end is cleaved by a different endonuclease, and then a few nucleotides are digested away by an exonuclease that removes nucleotides until it reaches a CCA sequence.

C28. A ribozyme has a catalytic part that is composed of RNA. Examples are RNase P and self-splicing group I and II introns. It is thought that the spliceosome may contain catalytic RNAs as well.

C30. Self-splicing means that an RNA molecule can splice itself without the aid of a protein. Group I and II introns can be self-splicing, although proteins can also enhance the rate of splicing.

C32. In alternative splicing, variation occurs in the pattern of splicing, so the resulting mRNAs contain alternative combinations of exons. The biological significance is that two or more different proteins can be produced from a single gene. This is a more efficient use of the genetic material. In multicellular organisms, alternative splicing is often used in a cell-specific manner.

C34. As shown at the left side of Figure 12.20, the guanosine, which binds to the guanosine-binding site, does not have a phosphate group attached to it. This guanosine is the nucleoside that winds up at the 5′ end of the intron. Therefore, the intron does not have a phosphate group at its 5′ end.

C36. U5

Experimental Questions

E2. An R loop is a loop of DNA that occurs when RNA is hybridized to double-stranded DNA. While the RNA is hydrogen bonding to one of the DNA strands, the other strand does not have a partner to hydrogen bond with, so it bubbles out as a loop. RNA is complementary to the template strand, so that is the strand it binds to.

E4. The 1100-nucleotide band would be observed from a normal individual (lane 1). A deletion that removed the −50 to −100 region would greatly diminish transcription, so the homozygote would produce hardly any of the transcript (just a faint amount, as shown in lane 2), and the heterozygote would produce roughly half as much of the 1100-nucleotide transcript (lane 3) compared with a normal individual. A nonsense codon would not have an effect on transcription; it affects only translation. So the individual with this mutation would produce a normal amount of the 1100-nucleotide transcript (lane 4). A mutation that removed the splice acceptor site would prevent splicing. Therefore, this individual would produce a 1550-nucleotide transcript (actually, 1547 to be precise, 1550 minus 3). The Northern blot is shown here:

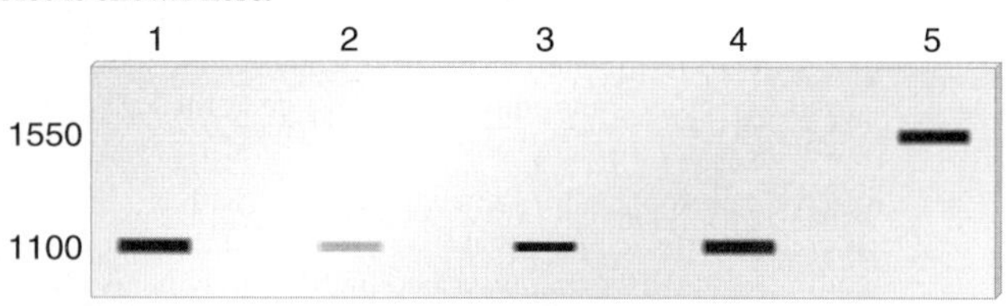

E6. A. It would not be retarded because ρ protein would not bind to the mRNA encoded by a gene that is terminated in a ρ-independent manner. The mRNA from such genes does not contain the sequence near the 3′ end that acts as a recognition site for the binding of ρ protein.

B. It would be retarded because ρ protein would bind to the mRNA.

C. It would be retarded because U1 would bind to the pre-mRNA.

D. It would not be retarded because U1 would not bind to mRNA that has already had its introns removed. U1 binds only to pre-mRNA.

E8. A. mRNA molecules would bind to this column because they have a polyA tail. The string of adenine nucleotides in the polyA tail is complementary to a stretch of thymines in the poly-dT column, so the two would hydrogen bond to each other. To purify mRNAs, one begins with a sample of cells; the cells need to be broken open by some technique such as homogenization or sonication. This would release the RNAs and other cellular macromolecules. The large cellular structures (organelles, membranes, etc.) could be removed from the cell extract by a centrifugation step. The large cellular structures would be found in the pellet, whereas soluble molecules such as RNA and proteins would stay in the supernatant. At this point, you would want the supernatant to contain a high salt concentration and neutral pH. The supernatant would then be poured over the poly-dT column. The mRNAs would bind to the poly-dT column and other molecules (i.e., other types of RNAs and proteins) would flow through the column. Because the mRNAs would bind to the poly-dT column via hydrogen bonds, to break the bonds, you could add a solution that contains a low salt concentration and/or a high pH. This would release the mRNAs, which would then be collected in a low salt/high pH solution as it dripped from the column.

B. The basic strategy is to attach a short stretch of DNA nucleotides to the column matrix that is complementary to the type of RNA that you want to purify. For example, if an rRNA contained a sequence 5′–AUUCCUCCA–3′, a researcher could chemically synthesize an oligonucleotide with the sequence 3′–TAAGGAGGT–5′ and attach it to the column matrix. To purify rRNA, one would use this 3′–TAAGGAGGT–5′ column and follow the general strategy described in part A.

Questions for Student Discussion/Collaboration

2. RNA transcripts come in two basic types: those that function as RNA (e.g., tRNA, rRNA, etc.) versus those that are translated (i.e., mRNA). As described in this chapter, they play a myriad of functional roles. RNAs that form complexes with proteins carry out some interesting roles. In some cases, the role is to bind other types of RNA molecules. For example, rRNA in bacteria plays a role in binding mRNA. In other cases, the RNA plays a catalytic role. An example is RNaseP. The structure and function of RNA molecules may be enhanced by forming a complex with proteins.

CHAPTER 13

Answers to Comprehension Questions

13.1: a, c	13.4: b, c, b
13.2: a, d, b, c	13.5: d, d
13.3: c, b	13.6: d, a, c, c

Concept Check Questions (in figure legends)

FIGURE 13.1 Alkaptonuria occurs when homogentisic acid oxidase is defective.

FIGURE 13.2 The strain 2 mutants are unable to convert O-acetylhomoserine into cystathionine.

FIGURE 13.3 The role of DNA is to store the information that specifies the amino acid sequence of a polypeptide. It is a storage role.

FIGURE 13.5 Tryptophan and phenylalanine are the least soluble in water.

FIGURE 13.6 Hydrogen bonding promotes the formation of secondary structures in proteins.

FIGURE 13.8 Only one radiolabeled amino acid was found in each sample. If radioactivity was trapped on the filter, this meant that the codon triplet specified that particular amino acid.

FIGURE 13.9 A tRNA has an antiocodon that recognizes a codon in the mRNA. It also has a 3′ acceptor site where the correct amino acid is attached.

FIGURE 13.11 A charged tRNA has an amino acid attached to it.

FIGURE 13.12 The wobble rules allow for a smaller population of tRNAs to recognize all of the possible mRNA codons.

FIGURE 13.14 A site in mRNA promotes the binding of the mRNA to the ribosome. The codons are needed during elongation to specify the polypeptide sequence. The stop codon is needed to terminate transcription.

FIGURE 13.16 The Shine-Dalgarno sequence in mRNA is complementary to a region in the 16S rRNA within the small ribosomal subunit. These complementary sequences hydrogen bond with each other.

FIGURE 13.17 Peptidyl transferase catalyzes the peptide bond formation between amino acids in the growing polypeptide chain.

FIGURE 13.18 The structures of release factors, which are proteins, resemble the structures of tRNAs.

End-of-Chapter Questions:

Conceptual Questions

C2. When we say the genetic code is degenerate, it means that more than one codon can specify the same amino acid. For example, GGG, GGC, GGA, and GGU all specify glycine. In general, the genetic code is nearly universal, because it is used in the same way by viruses, prokaryotes, fungi, plants, and animals. As discussed in Table 13.3, there are a few exceptions, which occur primarily in protists and yeast and mammalian mitochondria.

C4. A. This mutant tRNA would recognize glycine codons in the mRNA but would put in tryptophan amino acids where glycine amino acids are supposed to be in the polypeptide chain.

B. This mutation tells us that the aminoacyl-tRNA synthetase is primarily recognizing other regions of the tRNA molecule besides the anticodon region. In other words, tryptophanyl-tRNA synthetase (the aminoacyl-tRNA synthetase that attaches tryptophan) primarily recognizes other regions of the $tRNA^{Trp}$ sequence (that is, other than the anticodon region), such as the T- and D-loops. If aminoacyl-tRNA synthetases recognized only the anticodon region, we would expect glycyl-tRNA synthetase to recognize this mutant tRNA and attach glycine. That is not what happens.

C6. A. The answer is three. There are six leucine codons: UUA, UUG, CUU, CUC, CUA, and CUG. The anticodon AAU would recognize UUA and UUG. You would need two other tRNAs to efficiently recognize the other four leucine codons. These could be GAG and GAU or GAA and GAU.

B. The answer is one. There is only one codon, AUG, so you need only one tRNA with the anticodon UAC.

C. The answer is three. There are six serine codons: AGU, AGC, UCU, UCC, UCA, and UCG. You would need only one tRNA to recognize AGU and AGC. This tRNA could have the anticodon UCG or UCA. You would need two tRNAs to efficiently recognize the other four tRNAs. These could be AGG and AGU or AGA and AGU.

C8. 3′–CUU–5′ or 3′–CUC–5′

C10. It can recognize 5′–GGU–3′, 5′–GGC–3′, and 5′–GGA–3′. All of these specify glycine.

C12. All tRNA molecules have some basic features in common. They all have a cloverleaf structure with three stem-loop structures. The second stem-loop contains the anticodon sequence that recognizes the codon sequence in mRNA. At the 3′ end, there is an acceptor stem, with the sequence CCA, that serves as an attachment site for an amino acid. Most tRNAs also have base modifications that occur within their nucleotide sequences.

C14. The role of aminoacyl-tRNA synthetase is to specifically recognize tRNA molecules and attach the correct amino acid to them. This ability is sometimes described as the second genetic code because the specificity of the attachment is a critical step in deciphering the genetic code. For example, if a tRNA has a 3′–GGG–5′ anticodon, it will recognize a 5′–CCC–3′ codon, which should specify proline. It is essential that the aminoacyl-tRNA synthetase known as prolyl-tRNA-synthetase recognizes this tRNA and attaches proline to the 3′ end. The other aminoacyl-tRNA synthetases should not recognize this tRNA.

C16. Bases that have been chemically modified can occur at various locations throughout the tRNA molecule. The significance of all of these modifications is not entirely known. However, within the anticodon region, base modification may alter base pairing to allow the anticodon to recognize two or more different bases within the codon.

C18. No, it is not. Due to the wobble rules, the 5′ base in the anticodon of a tRNA can recognize two or more bases in the third (3′) position of the mRNA. Therefore, any given cell type synthesizes far fewer than 61 types of tRNAs.

C20. The assembly process is very complex at the molecular level. In eukaryotes, 33 proteins and one rRNA assemble to form a 40S subunit, and 49 proteins and three rRNAs assemble to form a 60S subunit. This assembly occurs within the nucleolus.

C22. A. On the surface of the 30S subunit and at the interface between the two subunits.

B. Within the 50S subunit.

C. From the 50S subunit.

D. To the 30S subunit.

C24. Most bacterial mRNAs contain a Shine-Dalgarno sequence, which is necessary for the binding of the mRNA to the small ribosomal subunit. This sequence, UUAGGAGGU, is complementary to a sequence in the 16S rRNA. Due to this complementarity, these sequences will hydrogen bond to each other during the initiation stage of translation.

C26. The ribosome binds at the 5′ end of the mRNA and then scans in the 3′ direction in search of an AUG start codon. If it finds one that reasonably obeys Kozak's rules, it will begin translation at that site. Aside from an AUG start codon, two other important features are a guanosine at the +4 position and a purine at the −3 position.

C28. The A (aminoacyl) site is the location where a tRNA carrying a single amino acid initially binds. The only exception is the initiator tRNA, which binds to the P (peptidyl) site. The growing polypeptide chain is removed from the tRNA in the P site and transferred to the amino acid attached to the tRNA in the A site. The ribosome translocates in the 3′ direction, with the result that the two tRNAs in the P and A sites are moved to the E (exit) and P sites, and the uncharged tRNA in the E site is released.

C30. The initiation phase involves the binding of the Shine-Dalgarno sequence to the rRNA in the 30S subunit. The elongation phase involves the binding of anticodons in tRNA to codons in mRNA.

C32. A. The E site and P site. (Note: A tRNA without an amino acid attached is only briefly found in the P site, just before translocation occurs.)

B. The P site and A site. (Note: A tRNA with a polypeptide chain attached is only briefly found in the A site, just before translocation occurs.)

C. Usually the A site, except the initiator tRNA, which can be found in the P site.

C34. The tRNAs bind to the mRNA because their anticodon and codon sequences are complementary. When the ribosome translocates in the 5′ to 3′ direction, the tRNAs remain bound to their complementary codons, and the two tRNAs shift from the A site and P site to the P site and E site. If the ribosome moved in the 3′ direction, it would have to dislodge the tRNAs and drag them to a new position where they would not (necessarily) be complementary to the mRNA.

C36. 52

Experimental Questions

E2. The initiation phase of translation is very different in bacteria and in eukaryotes, so they would not be translated very efficiently. A bacterial mRNA would not be translated very efficiently in a eukaryotic translation system, because it lacks a cap structure attached to its 5′ end. A eukaryotic mRNA would not have a Shine-Dalgarno sequence near its 5′ end, so it would not be translated very efficiently in a bacterial translation system.

E4. The term *polysome* refers to a structure in which a single mRNA is being translated into polypeptides by many ribosomes.

E6.

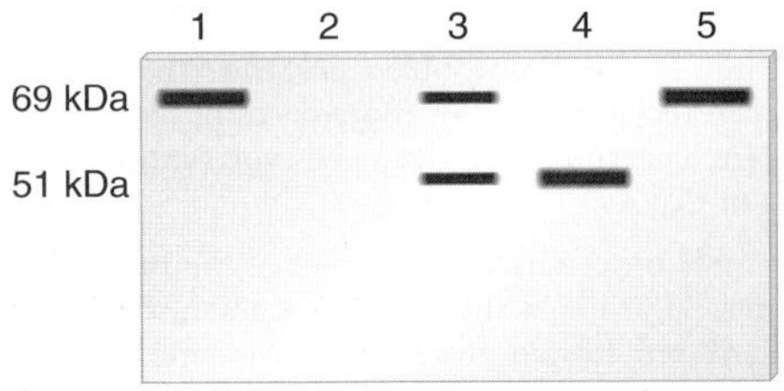

E8. A. If codon usage were significantly different between kangaroo and yeast cells, this would inhibit the translation process. For example, if the preferred leucine codon in kangaroos was CUU, translation would probably be slow in a yeast translation system. We would expect the cell-free translation system from yeast cells to primarily contain leucine tRNAs with an anticodon sequence that is AAC, because this $tRNA^{Leu}$ would match the preferred yeast leucine codon, which is UUG. In a yeast translation system, there probably would not be a large amount of tRNA with an anticodon of GAA, which would match the preferred leucine codon, CUU, of kangaroos. For this reason, kangaroo mRNA would not be translated very well in a yeast translation system, but it probably would be translated to some degree.

B. The advantage of codon bias is that a cell can rely on a smaller population of tRNA molecules to efficiently translate its proteins. A disadvantage is that mutations, which do not change the amino acid sequence but do change a codon (e.g., UUG to UUA), may inhibit the production of a polypeptide if a preferred codon is changed to a nonpreferred codon.

Questions for Student Discussion/Collaboration

2. This could be a very long list. There are similarities along several lines:

   1. There is a lot of molecular recognition going on, either between two nucleic acid molecules or between proteins and nucleic acid molecules. Students may see these as similarities or differences, depending on their point of view.
   2. There is biosynthesis going on in both processes. Small building blocks are being connected together. This requires an input of energy.
   3. There are genetic signals that determine the beginning and ending of these processes.

   There are also many differences:

   1. Transcription produces an RNA molecule with a similar structure to the DNA, whereas translation produces a polypeptide with a structure that is very different from RNA.
   2. Depending on your point of view, it seems that translation is more biochemically complex, requiring more proteins and RNA molecules to accomplish the task.

CHAPTER 14

Answers to Comprehension Questions

14.1: c, a

14.2: b, d, b, d

14.3: a, b, a

14.4: d, d

14.5: a

Concept Check Questions (in figure legends)

FIGURE 14.1 Gene regulation is more efficient. A cell does not waste energy making RNAs and proteins it does not need.

FIGURE 14.2 Activators and repressors are regulatory proteins. Inducers, corepressors, and inhibitors are small effector molecules.

FIGURE 14.3 The *lacZ*, *lacY*, and *lacA* genes are under the control of the *lac* promoter.

FIGURE 14.5 The lac repressor is bound to the *lac* operon when it is not exposed to lactose—when allolactose is not bound to the repressor.

FIGURE 14.8 The repressor protein allows the cell to avoid turning on the operon in the absence of lactose. The activator protein allows the cell to choose between glucose and lactose.

FIGURE 14.9 The data in which O_2 and O_3 are deleted indicate that O_1, by itself, is not very strong in repressing the *lac* operon.

FIGURE 14.11 Tryptophan acts as a corepressor that causes the trp repressor to bind to the *trp* operon and repress transcription.

FIGURE 14.12 Hydrogen bonding between complementary sequences causes the stem-loops to form.

FIGURE 14.13 The presence of tryptophan causes the *trpL* gene to be translated to its stop codon. This blocks both regions 1 and 2, which allows a 3–4 stem-loop to form. The 3–4 stem-loop causes transcriptional termination.

FIGURE 14.14 The *micF* antisense RNA binds to the *ompF* mRNA and inhibits its translation.

FIGURE 14.15 Feedback inhibition prevents a bacterium from overproducing the product of a metabolic pathway.

FIGURE 14.16 The RNA conformation with an antiterminator stem-loop favors transcription.

FIGURE 14.17 The RNA conformation with the Shine-Dalgarno antisequestor favors translation.

End-of-Chapter Questions:

Conceptual Questions

C2. In bacteria, gene regulation greatly enhances the efficiency of cell growth. It takes a lot of energy to transcribe and translate genes. Therefore, a cell is much more efficient and better at competing in its environment if it expresses genes only when the gene product is needed. For example, a bacterium will express only the genes that are necessary for lactose metabolism when a bacterium is exposed to lactose. When the environment is missing lactose, these genes are turned off. Similarly, when tryptophan levels are high within the cytoplasm, the genes required for tryptophan biosynthesis are repressed.

C4. A. Regulatory protein
B. Effector molecule
C. DNA segment
D. Effector molecule
E. Regulatory protein
F. DNA segment
G. Effector molecule

C6. A mutation that has a *cis*-effect is within a genetic regulatory sequence, such as an operator site, that affects the binding of a genetic regulatory protein. A *cis*-effect mutation affects only the adjacent genes that the genetic regulatory sequence controls. A mutation having a *trans*-effect is usually in a gene that encodes a genetic regulatory protein. A *trans*-effect mutation can be complemented in a merozygote experiment by the introduction of a normal gene that encodes the regulatory protein.

C8. A. No transcription would take place. The *lac* operon could not be expressed.

B. No regulation would take place. The operon would be continuously turned on.

C. The rest of the operon would function normally, but none of the transacetylase would be made.

C10. Diauxic growth refers to the phenomenon in which a cell first uses up one type of sugar (such as glucose) before it begins to metabolize a second sugar (e.g., lactose). In this case, it is caused by gene regulation. When a bacterial cell is exposed to both sugars, the uptake of glucose causes the cAMP levels in the cell to fall. When this occurs, the catabolite activator protein (CAP) is removed from the *lac* operon, so it is not able to be activated by CAP.

C12. A mutation that prevented the lac repressor from binding to the operator would make the *lac* operon constitutive only in the absence of glucose. However, this mutation would not be entirely constitutive because transcription would be inhibited in the presence of glucose. The disadvantage of constitutive expression of the *lac* operon is that the bacterial cell would waste a lot of energy transcribing the genes and translating the mRNA when lactose was not present.

C14. A. Attenuation will not occur because loop 2–3 will form.

B. Attenuation will occur because 2–3 cannot form, so 3–4 will form.

C. Attenuation will not occur because 3–4 cannot form.

D. Attenuation will not occur because 3–4 cannot form.

C16. The addition of Gs and Cs into the U-rich sequence would prevent attenuation. The U-rich sequence promotes the dissociation of the mRNA from the DNA, when the terminator stem-loop forms. This causes RNA polymerase to dissociate from the DNA and thereby causes transcriptional termination. The UGGUUGUC sequence would probably not dissociate because of the Gs and Cs. Remember that GC base pairs have three hydrogen bonds and are more stable than AU base pairs, which have only two hydrogen bonds.

C18. It takes a lot of cellular energy to translate mRNA into a protein. A cell wastes less energy if it prevents the initiation of translation rather than a later stage such as elongation or termination.

C20. One mechanism is that histidine could act as corepressor that shuts down the transcription of the histidine synthetase gene. A second mechanism would be that histidine could act as an inhibitor via feedback inhibition. A third possibility is that histidine inhibits the ability of the mRNA encoding histidine synthetase to be translated. Perhaps it induces a gene that encodes an antisense RNA. If the amount of histidine synthetase protein was identical in the presence and absence of extracellular histidine, a feedback inhibition mechanism is favored, because this affects only the activity of the histidine synthetase enzyme, not the amount of the enzyme. The other two mechanisms would diminish the amount of this protein.

C22. The two proteins are similar in that both bind to a segment of DNA and repress transcription. They are different in three ways. (1) They recognize different effector molecules (i.e., the lac repressor recognizes allolactose, and the trp repressor recognizes tryptophan). (2) Allolactose causes the lac repressor to release from the operator, while tryptophan causes the trp repressor to bind to its operator. (3) The sequences of the operator sites that these two proteins recognize are different from each other. Otherwise, the lac repressor could bind to the *trp* operator, and the trp repressor could bind to the *lac* operator.

Experimental Questions

E2. In samples loaded in lanes 1 and 4, we expect the repressor to bind to the operator because no lactose is present. In the sample loaded into lane 4, the CAP protein could still bind cAMP because there is no glucose. However, there really is no difference between lanes 1 and 4, so it does not look like the CAP can activate transcription when the lac repressor is bound. If we compare samples loaded into lanes 2 and 3, the lac repressor would not be bound in either case, and the CAP would not be bound in the sample loaded into lane 3. There is less transcription in lane 3 compared to lane 2, but because there is some transcription seen in lane 3, we can conclude that the removal of the CAP (because cAMP levels are low) is not entirely effective at preventing transcription. Overall, the results indicate that the binding of the lac repressor is more effective at preventing transcription of the *lac* operon compared to the removal of the CAP.

E4. A. Yes, if you do not sonicate, then β-galactosidase will not be released from the cell, and not much yellow color will be observed. (Note: You may observe a little yellow color because some β-ONPG may be taken into the cell.)

B. No, you should still get yellow color in the first two tubes even if you forgot to add lactose because the unmated strain does not have a functional lac repressor.

C. Yes, if you forgot to add β-ONPG, you could not get yellow color because the cleavage of β-ONPG by β-galactosidase is what produces the yellow color.

E6. You could mate a strain that has an F′ factor carrying a normal *lac* operon and a normal *lacI* gene to this mutant strain. Because the mutation is in the operator site, you would still continue to get expression of β-galactosidase, even in the absence of lactose.

Questions for Student Discussion/Collaboration

2. A DNA loop may inhibit transcription by preventing RNA polymerase from recognizing the promoter. Or, it may inhibit transcription by preventing the formation of the open complex. Alternatively, a bend may enhance transcription by exposing the base sequence that the σ factor of RNA polymerase recognizes. The bend may expose the major groove in such a way that this base sequence is more accessible to binding by σ factor and RNA polymerase.

CHAPTER 15

Answers to Comprehension Questions

15.1: b, b, b, d 15.3: d, b 15.5: b

15.2: d, a, b, d 15.4: c

Concept Check Questions (in figure legends)

FIGURE 15.1 Transcriptional regulation is the most energy efficient, because a cell avoids wasting energy making RNA or protein.

FIGURE 15.3 An α helix can bind into the major groove of DNA and recognize a specific sequence of bases.

FIGURE 15.4 If TFIID cannot bind to the TATA box, RNA polymerase will not be recruited to the core promoter, and therefore transcription will not begin.

FIGURE 15.5 When an activator interacts with mediator, it causes mediator to phosphorylate CTD, which causes RNA polymerase to proceed to the elongation phase of transcription.

FIGURE 15.7 The glucocorticoid receptor binds only next to genes that have a GRE by their core promoters.

FIGURE 15.9 Nucleosome eviction may allow certain proteins access to binding to particular sites in the DNA.

FIGURE 15.10 Histone modifications may directly affect the interaction between histones and the DNA, or they may affect the binding of other proteins to the chromatin.

FIGURE 15.12 An NFR is needed at the core promoter so that activators can recognize enhancers and then the preinitiation complex can form.

FIGURE 15.13 Histone eviction or displacement is needed for elongation because RNA polymerase cannot transcribe though nucleosomes. It needs to unwind the DNA for transcription to take place.

FIGURE 15.15 In part (a), DNA methylation is preventing an activator protein from binding to the DNA. This prevents transcriptional activation.

FIGURE 15.16 De novo methylation occurs on unmethylated DNA, whereas maintenance methylation occurs on hemimethylated DNA.

FIGURE 15.17 Insulators prevent one gene from regulating a neighboring gene. This allows each gene to control its own regulation.

End-of-Chapter Questions:

Conceptual Questions

C2. Regulatory elements are relatively short genetic sequences that are recognized by regulatory transcription factors. After the regulatory transcription factor has bound to the regulatory element, it will affect the rate of transcription, either activating it or repressing it, depending on the action of the regulatory protein. Regulatory elements are typically located in the upstream region near the promoter, but they can be located almost anywhere (i.e., upstream and downstream) and even quite far from the promoter.

C4. Transcriptional activation occurs when a regulatory transcription factor binds to a response element and activates transcription. Such proteins, called activators, may interact with TFIID and/or mediator to promote the assembly of RNA polymerase and general transcription factors at the promoter region. They also could alter the structure of chromatin so that RNA polymerase and transcription factors are able to gain access to the promoter. Transcriptional inhibition occurs when a regulatory transcription factor inhibits transcription. Such repressors also may interact with TFIID and/or mediator to inhibit RNA polymerase.

C6. A. DNA binding.

B. DNA binding.

C. Protein dimerization.

C8. For the glucocorticoid receptor to bind to a GRE, a steroid hormone must first enter the cell. The hormone then binds to the glucocorticoid receptor, which releases HSP90. The release of HSP90 exposes a nuclear localization signal (NLS) within the receptor, which enables it to dimerize and then enter the nucleus. Once inside the nucleus, the dimer binds to a GRE, which activates transcription of the adjacent genes.

C10. Phosphorylation of the CREB protein causes it to act as a transcriptional activator. The unphosphorylated CREB protein can still bind to CREs, but it does not stimulate transcription via CBP.

C12. A. Eventually, the glucocorticoid hormone will be degraded by the cell. The glucocorticoid receptor binds the hormone with a certain affinity. The binding is a reversible process. Once the concentration of the hormone falls below the affinity of the hormone for the receptor, the receptor will no longer have the glucocorticoid hormone bound to it. When the hormone is released, the glucocorticoid receptor will change its conformation, and it will no longer bind to the DNA.

B. An enzyme known as a phosphatase will eventually cleave the phosphate groups from the CREB protein. When the phosphates are removed, the CREB protein will stop activating transcription.

C14. The enhancer found in A would work, but the ones found in B and C would not. The sequence that is recognized by the transcriptional activator is 5′–GTAG–3′ in one strand and 3′–CATC–5′ in the opposite strand. This is the same arrangement found in A. In B and C, however, the arrangement is 5′–GATG–3′ and 3′–CATC–5′. In the arrangement found in B and C, the two middle bases (i.e., A and T) are not in the correct order.

C16.

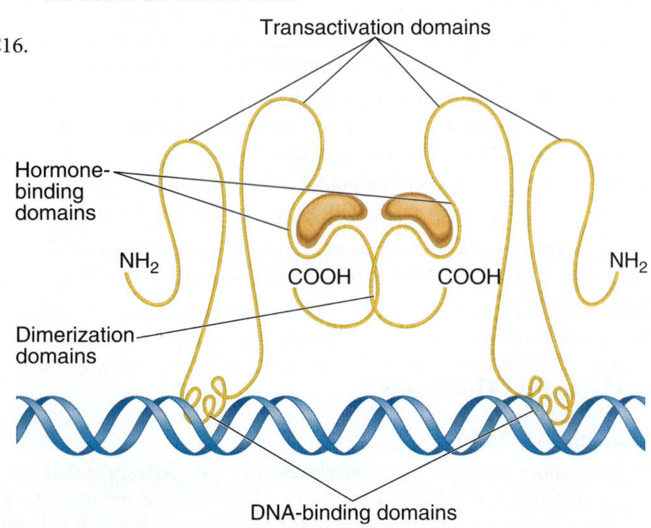

This is one hypothetical drawing of the glucocorticoid receptor. As discussed in your textbook, the glucocorticoid receptor forms a homodimer. The dimer shown here has rotational symmetry. If you flip the left side around, it has the same shape as the right side. The hormone-binding, DNA-binding, dimerization, and transactivation domains are labeled. The glucocorticoid hormone is shown in orange.

C18. ATP-dependent chromatin remodeling complexes may change the positions of nucleosomes, evict histones, and/or replace histones with histone variants.

C20. The attraction between DNA and histones occurs because the histones are positively charged and the DNA is negatively charged. The covalent attachment of acetyl groups decreases the amount of positive charge on the histone proteins and thereby may decrease the binding of the DNA. In addition, histone acetylation may attract proteins to the region that loosen chromatin compaction.

C22. A nucleosome-free region (NFR) is a location in the genome where nucleosomes are missing. They are typically found at the beginning and ends of genes. An NFR at the beginning of a gene is thought to be important so that genes can be activated. The NFR at the end of a gene may be important for its proper termination.

C24. An insulator is a segment of DNA that functions as a boundary between two adjacent genes. An insulator may act as a barrier to changes in chromatin structure or block the effects of a neighboring enhancer.

C26. Perhaps the methyltransferase is responsible for methylating and inhibiting a gene that causes a cell to become a muscle cell. The methyltransferase is inactivated by the mutation.

Experimental Questions

E2. These results indicate that the fibroblasts undergo maintenance methylation because they can replicate and methylate DNA if it has already been methylated. However, the cells do not undergo de novo methylation, because if the donor DNA was unmethylated, the DNA in the daughter cells remains unmethylated.

E4. Based on these results, there are enhancers that are located in regions A, D, and E. When these enhancers are deleted, the level of transcription is decreased. There also appears to be a silencer in region B, because a deletion of this region increases the rate of transcription. There do not seem to be any response elements in region C, or at least not any that function in muscle cells. Region F contains the core promoter, so the deletion of this region inhibits transcription.

E6. The results indicate that protein X binds to the DNA fragment and retards its mobility (lanes 3 and 4). However, the hormone is not required for DNA binding. Because we already know that the hormone is needed for transcriptional activation, it must play some other role. Perhaps the hormone activates a signaling pathway that leads to the phosphorylation of the transcription factor, and phosphorylation is necessary for translational activation. This situation would be similar to the CREB protein, which is activated by phosphorylation. The CREB protein can bind to the DNA whether or not it is phosphorylated.

Questions for Student Discussion/Collaboration

2. Probably the most efficient method would be to systematically make deletions of progressively smaller sizes. For example, you could begin by deleting 20,000 bp on either side of the gene and see if that affects transcription. If you found that only the deletion on the 5′ end of the gene had an effect, you could then start making deletions from the 5′ end, perhaps in 10,000-bp or 5000-bp increments until you localized response elements. You would then make smaller deletions in the putative region until it was down to a hundred or a few dozen nucleotides. At this point, you might conduct site-directed mutagenesis, as described in Chapter 20, as a way to specifically identify the regulatory element sequence.

CHAPTER 16

Answers to Comprehension Questions

16.1: f, c, b, d   16.3: a, d

16.2: c, b, d   16.4: b, d, d, a

Concept Check Questions (in figure legends)

FIGURE 16.2 Following DNA replication, each DNA strand is hemimethylated. As discussed in Chapter 15, hemimethylated DNA becomes fully methylated via maintenance methylation, which is a routine event in mammalian and plant cells. If another copy of the same gene in the cell was not methylated to begin with, it would need to undergo de novo methylation, which is not a routine event. De novo methylation is highly regulated and typically only occurs during gamete formation or embryonic development.

FIGURE 16.3 The *Igf2* gene follows a *cis*-epigenetic pattern.

FIGURE 16.5 The symmetry break results in a shift of all of the pluripotency factors and CTCFs to one of the X chromosomes, which becomes the active X chromosome. The other one becomes inactivated. This symmetry break occurs only during embryonic development.

FIGURE 16.6 A knot-like structure would prevent the binding of proteins, such as TFIID and RNA polymerase, to the gene, thereby inhibiting transcription.

FIGURE 16.9 Exon 3 would be spliced out. Exon 4 would still be within the mRNA.

FIGURE 16.12 A RISC binds to a specific mRNA because the small RNA within RISC is complementary to that mRNA. The bonding is hydrogen bonding between complementary bases.

FIGURE 16.14 The mutation would cause the overproduction of ferritin, because ferritin synthesis would occur even if iron levels were low.

End-of-Chapter Questions:

Conceptual Questions

C2. 1. DNA methylation—the attachment of methyl groups to cytosines in DNA. This often silences transcription.

2. Covalent histone modification—the covalent attachment of groups to the amino terminal tails of histones. This may silence or activate genes.

3. Chromatin remodeling—changes in the positions of nucleosomes. This may lead to a closed or open conformation for transcription.

4. Histone variants—replacement of standard histones for histone variants. This may silence or activate genes.

5. Feedback loop—the activation of a gene encoding a transcription factor. After the transcription factor is made, it continues to activate its own expression as well as the expression of other genes.

C4. In a *cis*-epigenetic mechanism, the pattern of gene modification is maintained when two or more copies of a gene are found in the same cell. One copy may be silenced and the other is active. In a *trans*-epigenetic mechanism, soluble proteins, such as transcription factors, are responsible for maintaining gene activation. In this case, all copies of the gene will be active. The imprinting of *Igf2* is a *cis*-epigenetic mechanism.

C6. During oogenesis, the ICR and DMR are not methylated. This allows CTCF to bind to these sites and to each other to form a loop. This loop prevents a nearby enhancer from activating the *Igf2* gene. During sperm formation, methylation of these sites prevents CTCF binding, thereby preventing loop formation. The nearby enhancer activates the *Igf2* gene. Therefore, the copy of the *Igf2* gene inherited from the father via sperm is active.

C8. See Figure 16.5. The symmetry break is the step that chooses which X chromosome will be active and which one will be inactivated.

C10. Epigenetic modifications lead to changes in gene expression that persist from early development to adulthood. This allows some genes to be expressed in certain cell types but not in others. The trithorax and polycomb group complexes are involved with causing such epigenetic changes.

C12. See Figure 16.6.

C14. Dietary factors can affect the occurrence of epigenetic changes, particularly those that happen during early stages of development. For example, dietary factors that affect the level of DNA methylation appear to be responsible or partly responsible for altering coat color in mice and female bee development. These dietary factors may affect the level of *S*-adenosyl methionine and/or the function of DNA methyltransferase and thereby influence if particular genes are methylated or not. If they are methylated, this typically inactivates genes.

C16. Alternative splicing is the splicing of a pre-mRNA in two or more different ways. An advantage is that an organism can carry fewer genes in its genome and yet express proteins that are specialized to function in particular cell types. A disadvantage is that this process uses energy to remove the introns. Also, energy is used to synthesize DNA that contains introns. In general, having introns and removing them during splicing uses a fair amount of energy.

C18. As shown in Figure 16.8, the unique feature of the smooth muscle mRNA for α-tropomyosin is that it contains exon 2. Splicing factors that are found only in smooth muscle cells may recognize the splice junction at the 3′ end of intron 1 and the 5′ end of intron 2 and promote splicing at these sites. This would cause exon 2 to be included in the mRNA. Furthermore, because smooth muscle mRNA does not contain exon 3, a splicing suppressor may bind to the 3′ end of intron 2. This would promote exon skipping, so that exon 3 would not be contained in the mRNA.

C20. Translational control is faster because it acts closer to the step of protein synthesis. If a cell quickly needs ferritin, translational control will allow its synthesis to occur more quickly. By comparison, transcriptional control would require the synthesis and processing of ferritin mRNA before more ferritin protein could be made.

C22. RNA interference refers to the phenomenon in which the presence of a double-stranded RNA molecule leads to the silencing of a complementary mRNA. For this to occur, the double-stranded RNA is processed by dicer into small RNA fragments (miRNA or siRNA). The miRNA or siRNA associates with a complex called the RNA-induced silencing complex (RISC). The RISC complex then binds to a complementary mRNA. This either leads to mRNA degradation or inhibits translation.

C24. If mRNA stability is low, this means that it is degraded more rapidly. Therefore, low stability results in a low mRNA concentration. The length of the polyA tail is one factor that affects stability. A longer tail makes mRNA more stable. Certain mRNAs have sequences that affect their half-lives. For example, AU-rich elements (AREs) are found in many short-lived mRNAs. The AREs are recognized by cellular proteins that cause the mRNAs to be rapidly degraded.

Experimental Questions

E2. The coat color relies on the expression of the *Agouti* gene. In this case, the *Agouti* gene is under the control of a promoter that is found in a transposable element, causing it to be overexpressed. When the promoter is not methylated and the gene is overexpressed, the coat color is yellow because the *Agouti* gene product promotes the synthesis of yellow pigment. When the promoter is methylated, this inhibits the *Agouti* gene and coat color is darker because less yellow pigment is made.

E4. The agent in cigarette smoke may have caused the methylation of CpG islands near the promoter of the *p53* gene, thereby inhibiting transcription. Another possibility is that it may affect covalent histone modifications or chromatin remodeling in a way that causes the gene to be in a closed conformation.

E6. They were injecting pre-siRNA. In this case, it was a perfect match to the *mex-3* RNA and caused its degradation.

E8. The gel shown in (a) is correct for ferritin mRNA regulation. The presence and absence of iron does not affect the amount of mRNA; it affects the ability of the mRNA to be translated. The gel shown in (b) is correct for transferrin receptor mRNA. In the absence of iron, the IRP binds to the IRE in the mRNA and stabilizes it, so the mRNA level is high. In the presence of iron, the iron binds to IRP and IRP is released from the IRE, and the transferrin receptor mRNA is degraded.

Questions for Student Discussion/Collaboration

2. Both epigenetic variation and variation in DNA sequences are similar in that they can affect gene expression, thereby affecting phenotype. One key difference is reversibility. Barring rare mutations, changes in DNA sequences are not usually reversible. In contrast, epigenetic modifications are reversible though they may be permanent during the life of a particular individual.

CHAPTER 17

Answers to Comprehension Questions

17.1: c, d
17.2: c, d
17.3: b, b
17.4: b, c, c

Concept Check Questions (in figure legends)

FIGURE 17.1 One way that viruses vary is in their genomes. Viral genomes vary with regard to size, RNA versus DNA, and single stranded versus double stranded. The structures of viruses also vary with regard to the complexity of their capsids and whether or not they contain an envelope.

FIGURE 17.3 A virus may remain latent following its integration into the genome of the host cell.

FIGURE 17.4 The lytic cycle produces new phage particles.

FIGURE 17.6 Emerging viruses are viruses that have arisen recently and are more likely to cause infection compared with previous strains.

FIGURE 17.8 Neither cycle could be followed. The N protein is needed to make a longer transcript from P_L for the lysogenic cycle and also to make a longer transcript from P_R for the lytic cycle.

FIGURE 17.9 This promoter is used to express the λ repressor during the maintenance of the lysogenic cycle.

FIGURE 17.10 An adequate supply of nutrients or UV light favor a switch to the lytic cycle.

FIGURE 17.12 The two enzymatic functions are the synthesis of DNA using RNA or DNA as a template and the function of digesting RNA via the RNase H component.

FIGURE 17.14 The fully spliced HIV RNA is needed.

End-of-Chapter Questions:

Conceptual Questions

C2. All viruses have a nucleic acid genome and a capsid composed of protein. Some eukaryotic viruses are surrounded by an envelope that is composed of a membrane with embedded proteins.

C4. A viral envelope is composed of a membrane with embedded proteins. It is made when the virus buds from the host cell, taking with it a portion of the host cell's plasma membrane.

C6. The attachment step usually involves the binding of the virus to a specific protein on the surface of the host cell. Only certain cell types will make that specific protein.

C8. Reverse transcriptase is used to copy the viral RNA into DNA so it can be integrated into a chromosome of the host cell.

C10. A temperate phage can follow either the lytic or lysogenic cycle, whereas a virulent phage can follow only the lytic cycle.

C12. In the lytic cycle, the virus directs the bacterial cell to make more virus particles until eventually the cell lyses and releases them. In the lysogenic cycle, the viral genome is incorporated into the host cell's genome as a prophage. It remains there in a dormant state until some stimulus causes it to excise itself from the bacterial chromosome and enter the lytic cycle.

C14. The O_R region contains three operator sites and two promoters. P_{RM} and P_R transcribe in opposite directions. The λ repressor will first bind to O_{R1} and then O_{R2}. The binding of the λ repressor to O_{R1} and O_{R2} inhibits transcription from P_R, thereby switching off the lytic cycle. Early in the lysogenic cycle, the λ repressor protein concentration may become so high that it will occupy O_{R3}. Later, when the λ repressor concentration begins to drop, it will first be removed from O_{R3}. This allows transcription from P_{RM} and maintains the lysogenic cycle. By comparison, the cro protein has its highest affinity for O_{R3}, and so it binds there first. This blocks transcription from P_{RM} and thereby switches off the lysogenic cycle. The cro protein has a similar affinity for O_{R2} and O_{R1}, and so it may occupy either of these sites next. It will bind to both O_{R2} and O_{R1}. This turns down the expression from P_R, which is not needed in the later stages of the lytic cycle.

C16. A genetic switch is a DNA sequence that governs a choice between two genetic pathways. The segment of DNA usually contains several operator sites that are recognized by two or more transcriptional regulatory proteins. A genetic switch is similar to a simple operator in that the binding of proteins to the DNA is the underlying event that controls transcription. The main difference between a genetic switch and a simple operator is the degree of complexity. A simple operator usually controls a single gene or a single operon, whereas a genetic switch often has multiple operator sites and controls the choice between two (or more) promoters.

C18. The host-cell tRNA provides a primer to initiate the process of reverse transcription.

C20. See Figure 17.13.

C22. After the 3′ ends of the HIV DNA are covalently attached to the 5′ ends of the host DNA, short gaps occur because of the removal of two nucleotides from each end of the HIV DNA and because integrase makes staggered cuts in the host DNA. These two gaps are filled in by gap repair synthesis.

C24. Gag polyprotein plays the central role in the assembly of the components that make up an HIV particle. It interacts with the plasma membrane and promotes budding, which is how HIV acquires its envelope. Gag polyprotein is also responsible for binding to components that are found within the HIV particle, such as HIV RNA, as well as certain HIV proteins and host proteins.

C26. HIV protease cleaves Gag polyprotein and Gag-pol polyprotein into their individual proteins, which allows the proteins to function as separate units. For example, the cleavage of Gag polyproteins allows the capsid proteins to assemble with each other and form a capsid.

Experimental Questions

E2. Electron microscopy must be used to visualize a virus.

E4. Following infection, the features found on the infected leaves correlated with the RNA that was initially found in the reconstituted virus. Also, the biochemical composition of the capsid protein in newly made viruses was consistent with the RNA that was found in the reconstituted virus.

E6. Drugs that bind to and inhibit HIV protease would prevent the cleavage of Gag and Gag-pol polyproteins. This cleavage is needed for the maturation of HIV.

E8. A cell that has a λ prophage is making a significant amount of the λ repressor. If another phage infects the cell, the λ repressor inhibits transcription from P_R and P_L, thereby inhibiting the early steps that are required for either the lytic or lysogenic cycles.

E10. If the F^- strain is lysogenic for phage λ, the λ repressor is already being made in that cell. If the F^- strain receives genetic material from an *Hfr* strain, you would not expect it to have an effect on the lysogenic cycle, which is already established in the F^- cell. However, if the *Hfr* strain is lysogenic for λ and the F^- strain is not, the *Hfr* strain could transfer the integrated λ DNA (i.e., the prophage) to the F^- strain. The cytoplasm of the F^- strain would not contain any λ repressor. Therefore, this λ DNA could choose between the lytic and lysogenic cycle. If it follows the lytic cycle, the F^- recipient bacterium will lyse.

Questions for Student Discussion/Collaboration

2. Step 1. Without the λ repressor around, P_R will be turned on, which will allow the synthesis of cII and cro.

   Step 2. cII will activate P_L, which will lead to the synthesis of integrase and exisionase so that the λ genome will be excised from the *E. coli* genome.

   Step 3. Without the λ repressor around, cro will be able to activate P_R; this will eventually allow the synthesis of the operons that encode proteins required for λ replication and coat protein synthesis.

CHAPTER 18

Answers to Comprehension Questions

18.1: b, c, a, d
18.2: b
18.3: c, d, c
18.4: b, b, b, a
18.5: c, d, a, d, b

Concept Check Questions (in figure legends)

FIGURE 18.2 Position effect means that the expression of a gene has been altered due to a change in its position along a chromosome.

FIGURE 18.3 In part (b), the sequence of the eye color gene is not changed. The change in expression is due to a position effect.

FIGURE 18.5 A somatic mutation is not passed from parent to offspring.

FIGURE 18.7 The probability is 75%.

FIGURE 18.8 The deamination of methylcytosine is more difficult to repair because thymine is a normal base found in DNA. This makes it difficult for DNA repair enzymes to distinguish between the correct and altered strand.

FIGURE 18.10 A reactive oxygen species is a chemical that contains oxygen and may react with cellular molecules in a way that may cause harm.

FIGURE 18.13 5-Bromouracil causes a transition.

FIGURE 18.14 Thymine dimer formation is often the result of exposure to UV light. It most commonly occurs in skin cells.

FIGURE 18.15 Enzymes within the rat liver extract may convert non-mutagenic molecules into a mutagenic form. Adding the extract allows researchers to identify molecules that may be mutagenic in people.

FIGURE 18.16 Photolyase is particularly valuable to plants, which are exposed to sun and susceptible to thymine dimer formation.

FIGURE 18.18 Cuts are made on both sides of the DNA damage so the region can be removed by UvrD.

FIGURE 18.20 MutH distinguishes between the parental strand and the newly made daughter strand, which ensures that the daughter strand is repaired.

FIGURE 18.22 An advantage of this DNA repair system is that it can occur at any stage of the cell cycle. A disadvantage is that it may be imprecise and result in a short deletion in the DNA.

End-of-Chapter Questions:

Conceptual Questions

C2. It is a gene mutation, a point mutation, a base substitution, a transition mutation, a deleterious mutation, a mutant allele, a nonsense mutation, a conditional mutation, and a temperature-sensitive lethal mutation.

C4. A. It would probably inhibit protein function, particularly if it was not near the end of the coding sequence.

B. It may or may not affect protein function, depending on the nature of the amino acid substitution and whether the substitution is in a critical region of the protein.

C. It would increase the amount of functional protein.

D. It may affect protein function if the alteration in splicing changes an exon in the mRNA that results in a protein with a perturbed structure.

C6. A. Not appropriate, because the second mutation is at a different codon.

B. Appropriate.

C. Not appropriate, because the second mutation is in the same gene as the first mutation.

D. Appropriate.

C8. A. Silent, because the same amino acid (glycine) is encoded by GGA and GGT.

B. Missense, because a different amino acid is encoded by CGA compared with GGA.

C. Missense, because a different amino acid is encoded by GTT compared with GAT.

D. Frameshift, because an extra base is inserted into the sequence.

C10. One possibility is that a translocation may move a gene next to a heterochromatic region of another chromosome, thereby diminishing its expression, or it could be moved next to a euchromatic region and increase its expression. Another possibility is that the translocation breakpoint may move the gene next to a new promoter or regulatory sequences that may now influence the gene's expression.

C12. A. No; the position (i.e., chromosomal location) of a gene has not been altered.

B. Yes; the expression of a gene has been altered because it has been moved to a new chromosomal location.

C. Yes; the expression of a gene has been altered because it has been moved to a new chromosomal location.

C14. If a mutation within the germ line is passed to an offspring, all of the cells of the offspring's body will carry the mutation. A somatic mutation affects only the somatic cell in which it originated and all of the daughter cells that the somatic cell produces. If a somatic mutation occurs early during embryonic development, it may affect a fairly large region of the organism. Because germ-line mutations affect the entire organism, they are potentially more harmful (or beneficial), but this is not always the case. Somatic mutations can cause quite harmful effects, such as cancer.

C16. A thymine dimer can interfere with DNA replication because DNA polymerase cannot slide past the dimer and add bases to the newly growing strand. Alkylating mutagens such as nitrous acid will cause DNA replication to make mistakes in the base pairing. For example, an alkylated cytosine will base-pair with adenine during DNA replication, thereby creating a mutation in the newly made strand. A third example is 5-bromouracil, which is a thymine analog. It may base-pair with guanine instead of adenine during DNA replication.

C18. During TNRE, a trinucleotide repeat sequence gets longer. If someone was mildly affected with a TNRE disorder, he or she might be concerned that an expansion of the repeat might occur during gamete formation, yielding offspring more severely affected with the disorder, a phenomenon called anticipation. This phenomenon may depend on the sex of the parent with the TNRE.

C20. According to the random mutation theory, spontaneous mutations can occur in any gene and do not involve exposure of the organism to a particular environment that selects for specific types of mutation. However, the structure of chromatin may cause certain regions of the DNA to be more susceptible to random mutations. For example, DNA in an open conformation may be more accessible to mutagens and more likely to incur mutations. Similarly, hot spots—certain regions of a gene that are more likely to mutate than other regions—can occur within a single gene. Also, another reason that some genes mutate at a higher rate is that some genes are larger than others, which provides a greater chance for mutation.

C22. Excision repair systems could fix this damage. Also, homologous recombination repair could fix the damage.

C24. Anticipation means that the TNRE expands even further in future generations. Anticipation may depend on the sex of the parent with the TNRE.

C26. The mutation frequency is the total number of mutant alleles divided by the total number of alleles in the population. If there are 1,422,000 babies, there are 2,844,000 copies of this gene (because each baby has two copies). The mutation frequency is $^{31}/_{2,844,000}$, which equals 1.09×10^{-5}. The mutation rate is the number of new mutations per generation. There are 13 babies who did not have a parent with achondroplasia; thus, 13 is the number of new mutations. If we calculate the mutation rate as the number of new mutations in a given gene per generation, then we should divide 13 by 2,844,000. In this case, the mutation rate would be 4.6×10^{-6}.

C28. The effects of mutations are cumulative. If one mutation occurs in a cell, this mutation will be passed to the daughter cells. If a mutation occurs in the daughter cell, now there will be two mutations. These two mutations will be passed to the next generation of daughter cells, and so forth. The accumulation of many mutations eventually kills the cells. That is why mutagens are more effective at killing dividing cells compared to nondividing cells. It is because the number of mutations accumulates to a lethal level.

There are two main side effects to this treatment. First, some normal (noncancerous) cells of the body, particularly skin cells and intestinal cells, are actively dividing. These cells are also killed by chemotherapy and radiation therapy. Secondly, it is possible that the therapy may produce mutations that will cause noncancerous cells to become cancerous. For these reasons, a maximal dose of chemotherapy or radiation therapy is recommended.

C30. A. Yes.

B. No; the albino trait affects the entire individual.

C. No; the early apple-producing trait affects the entire tree.

D. Yes.

C32. Mismatch repair is aimed at eliminating mismatches that may have occurred during DNA replication. In this case, the wrong base is in the newly made strand. The binding of MutH, which occurs on a hemimethylated sequence, provides a sensing mechanism to distinguish between the unmethylated and methylated strands. In other words, MutH binds to the hemimethylated DNA in a way that allows the mismatch repair system to distinguish which strand is methylated and which is not.

C34. Because sister chromatids are genetically identical, an advantage of homologous recombination is that it can be an error-free mechanism to repair a DSB. A disadvantage, however, is that it occurs only during the S and G_2 phase of the cell cycle in eukaryotes or following DNA replication in bacteria. An advantage of NHEJ is that it doesn't involve the participation of a sister chromatid, so it can occur at any stage of the cell cycle. However, a disadvantage is that NHEJ can result in small deletions in the region that has been repaired. Overall, NHEJ is a quick but error-prone repair mechanism, while HR is a more accurate method of repair that is limited to certain stages of the cell cycle.

C36. The underlying genetic defect that causes xeroderma pigmentosum is a defect in one of the genes that encode a polypeptide involved with nucleotide excision repair. These individuals are defective in repairing DNA abnormalities such as thymine dimers and abnormal bases. Therefore, they are very sensitive to environmental agents such as UV light. Because they are defective at repair, UV light is more likely to cause mutations in these people compared to unaffected individuals. For this reason, people with XP develop pigmentation abnormalities and premalignant lesions and have a high predisposition to skin cancer.

C38. Both types of repair systems recognize an abnormality in the DNA and excise the abnormal strand. The normal strand is then used as a template to synthesize a complementary strand of DNA. The systems differ in the types of abnormalities they detect. The mismatch repair system detects base pair mismatches, while the excision repair system recognizes thymine dimers, chemically modified bases, missing bases, and certain types of cross-links. The mismatch repair system operates immediately after DNA replication, allowing it to distinguish between the daughter strand (which contains the wrong base) and the parental strand. The excision repair system can operate at any time in the cell cycle.

Experimental Questions

E2. To show that antibiotic resistance is due to random mutation, one could follow the same basic strategy shown in Figure 18.6, except the secondary plates would contain the antibiotic instead of T1 phage. If the antibiotic resistance arose as a result of random mutation on the master plate, one would expect the antibiotic-resistant colonies to appear at the same locations on two different secondary plates.

E4. You would expose the bacteria to the physical agent. You could also expose the bacteria to the rat liver extract, but it is probably not necessary for two reasons. First, a physical mutagen is not something that a person would eat. Therefore, the actions of digestion via the liver are probably irrelevant if you are concerned that the agent might be a mutagen. Second, the rat liver extract would not be expected to alter the properties of a physical mutagen.

E6. The results suggest that the strain is defective in excision repair. If we compare the normal and mutant strains that have been incubated for 2 hours at 37°C, much of the radioactivity in the normal strain has been transferred to the soluble fraction because it has been excised. In the mutant strain, however, less of the radioactivity has been transferred to the soluble fraction, suggesting that it is not as efficient at removing thymine dimers.

Questions for Student Discussion/Collaboration

2. The worst time to be exposed to mutagens would be at very early stages of embryonic development. An early embryo is most sensitive to mutation because it will affect a large region of the body. Adults must also worry about mutagens for several reasons. Mutations in somatic cells can cause cancer, a topic discussed in Chapter 22. Also, adults should be careful to avoid mutagens that may affect the ovaries or testes because these mutations could be passed to offspring.

CHAPTER 19

Answers to Comprehension Questions

19.1: a, d, a, d

19.2: a, b

19.3: c, d, b

Concept Check Questions (in figure legends)

FIGURE 19.1 An advantage of genetic recombination is that it may foster genetic diversity, which may produce organisms that have reproductive advantages over other organisms.

FIGURE 19.2 A heteroduplex region may be produced after branch migration because the two strands are not perfectly complementary.

FIGURE 19.3 During recombination, a D loop is formed when one DNA strand from one homolog invades the homologous region of the other homolog.

FIGURE 19.5 During gene conversion, the region of the gene that contained the genetic variation that created the *b* allele was digested away. The same region from the homolog, which carried the *B* allele, was then used as a template to make another copy of the *B* allele.

FIGURE 19.10 Retrotransposition always causes the TE to increase in number. Transposition by DNA transpososons, by itself, does not increase the TE number. However, as discussed later, such transposition can increase the numbers of a TE if it occurs around the time of DNA replication.

FIGURE 19.14 Reverse transcriptase uses RNA as a template to make a strand of DNA.

End-of-Chapter Questions:

Conceptual Questions

C2. Branch migration will not create a heteroduplex during SCE because the sister chromatids are genetically identical. There should not be any mismatches between the complementary strands. Gene conversion cannot take place because the sister chromatids carry alleles that are already identical to each other.

C4. The two molecular mechanisms that can explain the phenomenon of gene conversion are mismatch DNA repair and gap repair synthesis. Both mechanisms could occur in the double-strand break model.

C6. A recombinant chromosome is one that has been derived from a crossover and contains a combination of alleles that is different from the parental chromosomes. A recombinant chromosome is a hybrid of the parental chromosomes.

C8. Gene conversion occurs when a pair of different alleles is converted to a pair of identical alleles. For example, a pair of *Bb* alleles could be converted to *BB* or *bb*.

C10. Gene conversion is likely to take place near the breakpoint. According to the double-strand break model, a gap may be created by the digestion of one DNA strand in the double helix. Gap repair synthesis may result in gene conversion. A second way that gene conversion can occur is by mismatch repair. A heteroduplex may be created after DNA strand migration and may be repaired in such a way as to cause gene conversion.

C12. No, it does not necessarily involve DNA mismatch repair; it could also involve gap repair synthesis. The double-strand break model involves a migration of DNA strands and the digestion of a gap. Therefore, in this gap region, only one chromatid is providing the DNA strands. As seen in Figure 17.7, the top chromosome is using the top DNA strand from the bottom chromosome in the gap region. The bottom chromosome is using the bottom DNA strand from the bottom chromosome. After DNA synthesis, both chromosomes may have the same allele.

C14. First, gene rearrangement of V, D, and J domains occurs within the light- and heavy-chain genes. Second, within a given B cell, different combinations of light and heavy chains are possible. And third, imprecise fusion may occur between the V, D, and J domains.

C16. One segment, which includes some variable (V) domains and perhaps one or more joining (J) domain, of DNA is removed from the κ light-chain gene. One segment, which may include one or more J domains and the region between the J domain and C domain, is removed during pre-mRNA splicing.

C18. The ends of a short region would be flanked by direct repeats. This is a universal characteristic of all transposable elements. In addition, many elements contain IRs or LTRs that are involved in the transposition process. One might also look for the presence of a transposase or reverse transcriptase gene, although this is not an absolute requirement, because nonautonomous transposable elements typically lack transposase or reverse transcriptase.

C20. Direct repeats occur because transposase or integrase produces staggered cuts in the two strands of chromosomal DNA. The transposable element is then inserted into this site, which temporarily leaves two gaps. The gaps are filled in by DNA polymerase. Because this gap filling occurs at sites that are complementary, the two gaps end up with the exact same sequence, thereby creating direct repeats.

C22. Transposable elements are mutagens, because they alter (disrupt) the sequences of chromosomes and genes within chromosomes. They do this by inserting themselves into genes.

C24. A. LTR and non-LTR retrotransposons.

B. Insertion elements and simple transposons.

C. All of them have direct repeats.

D. Insertion elements and simple transposons.

C26. A. As shown in solved problem S2a, a crossover between direct repeats (DRs) will excise the region between the repeats. In this case, it will excise a large chromosomal region, which does not contain a centromere, and will be lost. The remaining portion of the chromosome is shown here. It has one direct repeat that is formed partly from DR-1 and partly from DR-4. It is designated DR-1/4.

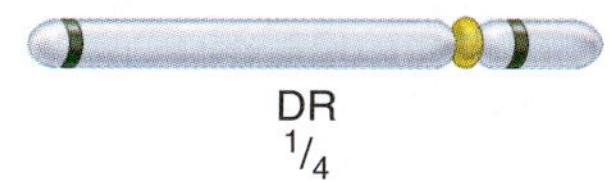

B. A crossover between IR-1 and IR-4 will cause an inversion between the intervening region. This is a similar effect to the crossover within a single transposable element, as described in solved problem S2b. The difference is that the crossover causes a large chromosomal region to be inverted.

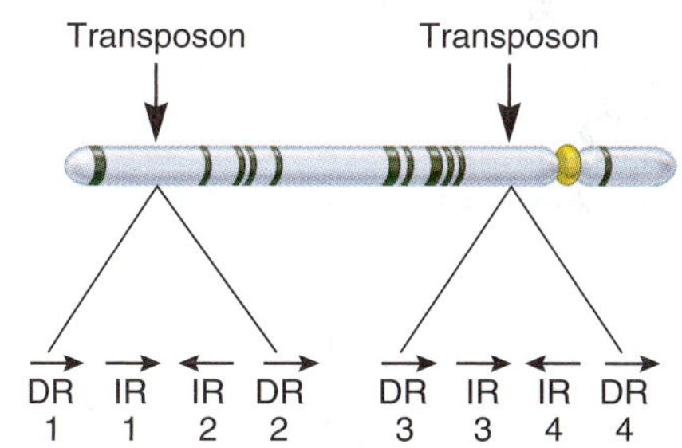

C28. At a frequency of about 1 in 10,000, recombination occurs between the inverted repeats within this transposon. As shown in solved problem S2b, recombination between inverted repeats causes the sequence within the transposon to be reversed. If this occurred in a strain that expressed the *H2/rH1* operon, the promoter would be flipped in the opposite direction, so the *H2* gene and the *rH1* gene would not be expressed. The H2 flagellar protein would not be made, and the H1 repressor protein would not be made. The H1 flagellar protein would be made because the expression of the *H1* gene would not be repressed. A strain expressing the *H1* gene could also "switch" at a frequency of about 1 in 10,000 by the same mechanism. If a crossover occurred between the inverted repeats within the transposable element, the promoter would be flipped again, and the *H2* and *rH1* genes would be expressed again.

Experimental Questions

E2. It is due to the excision of the transposable element and chromosomal rejoining. This conclusion was based on the reversion of the phenotype (namely, colorless to red), which suggests a restoration of gene function and lack of mutability of the phenotype.

E4. Transposon tagging is an experimental method that is aimed at cloning genes. In this approach, a transposon is introduced into a strain, and the researcher identifies individuals in which the function of a gene of interest has been inactivated. In many cases, the loss of function has occurred because the transposon has been inserted into the gene. If so, DNA can be isolated, digested into fragments, and cloned into viral vectors, creating a DNA library. The library is then probed using a radiolabeled fragment of DNA that is complementary to the transposon in order to identify clones in which the transposon has been inserted.

E6. First, you would inject the plasmid carrying the P element and *rosy*$^+$ gene into embryos that were homozygous for the recessive *rosy* allele (i.e., *rosy rosy*). Allow these embryos to mature into adult flies, and then mate them to *rosy rosy* flies. Analyze the color of the F_1 offspring. If the eyes are red, at least one copy of the P element has hopped into the chromosome of these flies. Possibly, the P element has hopped into a gene that is involved with wing development. However, at this point, such a fly would have one normal copy of the gene and one copy that has been inactivated by the P element. If the inactivation of the gene were inherited as a dominant phenotype, these flies would have deformed wings. Alternatively (and more likely), however, the inactivation of one copy of most genes would be inherited as a recessive trait. If this were the case, the wings of the red-eyed flies would still appear normal.

To identify these recessive genes, take each F_1 offspring with red eyes (and normal wings) and mate them to *rosy rosy* flies. About 1/4 of the F_2 offspring should have red eyes (assuming that one P element has hopped into the genome of the red-eyed flies). Take the F_2 offspring with red eyes (from a single cross) and mate them to each other. In other words, mate F_2 brothers and sisters to each other. From these crosses, 1/4 of the F_3 flies should have rosy eyes, 1/2 should have one copy of the *rosy*$^+$ P element and red eyes, and 1/4 should be homozygous for the P element and have red eyes. If the P element has inserted into a gene that is required for wing development, 1/4 of the offspring should have deformed wings and red eyes. In this way, you could identify genes that are required for wing development.

Note: This experiment is a lot of work! You would have to identify thousands of flies with red eyes and set up thousands of crosses to identify a few genes that play a role in wing development. Nevertheless, this approach has been successful in the identification of genes that play a role in many aspects of *Drosophila* development.

Questions for Student Discussion/Collaboration

2. Beneficial consequences: You wouldn't get (as many) translocations, inversions, and the accumulation of selfish DNA.

   Harmful consequences: The level of genetic diversity would be decreased, because linked combinations of alleles would not be able to recombine. You wouldn't be able to produce antibody diversity in the same way. Gene duplication could not occur, so the evolution of new genes would be greatly inhibited.

CHAPTER 20

Answers to Comprehension Questions

20.1: d, b, c, a, a	20.4: c
20.2: c, b, d	20.5: d, b, b
20.3: b	20.6: b, c

Concept Check Questions (in figure legends)

FIGURE 20.1 Eight

FIGURE 20.2 In this experiment, the selectable marker gene selects for the growth of bacteria that have taken up a plasmid.

FIGURE 20.3 Reverse transcriptase refers to the idea that this enzyme catalyzes the opposite of transcription. It uses an RNA template to make DNA, whereas during transcription a DNA template is used to make RNA.

FIGURE 20.4 The advantage of a DNA library is that it lacks introns. This is useful if a researcher wants to focus their attention on the coding sequence of a gene or if they want to express the gene in cells that do not splice out introns.

FIGURE 20.6 The products that predominate are the DNA fragments that are flanked by primers. This occurs because these products can be used as templates to make products only like themselves.

FIGURE 20.8 The probe must be cleaved to separate the quencher from the reporter.

FIGURE 20.13 Some examples are: to study the core promoter of a gene; to study regulatory elements; to identify splice sites; and to study the importance of particular amino acids with regard to protein function.

FIGURE 20.15 Conditions of low stringency reveal gene families.

FIGURE 20.17 The secondary antibody is labeled, which provides a way to detect the protein of interest.

FIGURE 20.19 The binding of a protein to DNA prevents DNase I from cleaving the region where the protein is bound.

End-of-Chapter Questions:

Conceptual Questions

C2. A restriction enzyme recognizes and binds to a specific DNA sequence and then cleaves a (covalent) ester bond in each of two DNA strands.

C4. The term cDNA refers to DNA that is made using RNA as the starting material. Compared to genomic DNA, it lacks introns.

Experimental Questions

E2. Remember that AT base pairs form two hydrogen bonds, whereas GC base pairs form three hydrogen bonds. The order (from stickiest to least sticky) would be

*Bam*HI = *Pst*I = *Sau*3AI > *Eco*RI

E4. In conventional gene cloning, many copies are made because the vector replicates to a high copy number within the cell, and the cells divide to produce many more cells. In PCR, the replication of the DNA to produce many copies is facilitated by primers, deoxyribonucleoside triphosphates (dNTPs), and *Taq* polymerase.

E6. A recombinant vector is a vector that has a piece of foreign DNA inserted into it. The foreign DNA came from somewhere else, such as the chromosomal DNA of some organism. To construct a recombinant vector, the vector and source of foreign DNA are digested with the same restriction enzyme. The complementary ends of the fragments are allowed to hydrogen bond to each other (i.e., sticky ends are allowed to bind), and then DNA ligase is added to create covalent bonds. In some cases, a piece of the foreign DNA will become ligated to the vector, thereby creating a recombinant vector. In other cases, the two ends of the vector ligate back together, restoring the vector to its original structure.

As described in Figure 20.2, the insertion of foreign DNA can be detected using X-Gal. As seen here, the insertion of the foreign DNA causes the inactivation of the *lacZ* gene. The *lacZ* gene encodes the enzyme β-galactosidase, which converts the colorless compound X-Gal to a blue compound. If the *lacZ* gene is inactivated by the insertion of foreign DNA, the enzyme will not be produced, and the bacterial colonies will be white. If the vector has simply recircularized and the *lacZ* gene remains intact, the enzyme will be produced, and the colonies will be blue.

E8. It would be necessary to use cDNA so that the gene would not carry any introns. Bacterial cells do not contain spliceosomes (which are described in Chapter 12). To express a eukaryotic protein in bacteria, a researcher would clone cDNA into bacteria, because the cDNA does not contain introns.

E10. Initially, the mRNA would be mixed with reverse transcriptase and nucleotides to create a complementary strand of DNA. Reverse transcriptase also needs a primer. This could be a primer that is known to be complementary to the β-globin mRNA. Alternatively, mature mRNAs have a polyA tail, so one could add a primer that consists of many Ts, called a poly-dT primer. After the complementary DNA strand has been made, the sample would then be mixed with primers, *Taq* polymerase, and nucleotides and subjected to the standard PCR protocol. Note: The PCR reaction would have two kinds of primers. One primer would be complementary to the 5′ end of the mRNA and would be unique to the β-globin sequence. The other primer would be complementary to the 3′ end. This second primer could be a poly-dT primer, or it could be a unique primer that would bind slightly upstream from the polyA-tail region.

E12. The exponential phase of PCR is chosen because it is during this phase that the amount of PCR product is proportional to the amount of the original DNA in the sample.

E14. AGGTCGGTTGCCATCGCAATAATTTCTGCCTGAACCCAATA

E16. There are lots of different strategies one could follow. For example, you could mutate every other base and see what happens. It would be best to make very nonconservative mutations such as a purine for a pyrimidine or a pyrimidine for a purine. If the mutation prevents protein binding in a gel retardation assay, then the mutation is probably within the response element. If the mutation has no effect on protein binding, it probably is outside the response element.

E18. One interpretation would be that the gene is part of a gene family. In this case, the family would contain four homologous members. At high stringency, the probe binds only to the gene that is its closest match, but at low stringency it recognizes the three other homologous genes.

E20. The purpose of gel electrophoresis is to separate the many DNA fragments, RNA molecules, or proteins that were obtained from the sample you want to probe. This separation is based on molecular mass and allows you to identify the molecular mass of the DNA fragment, RNA molecule, or protein that is being recognized by the probe.

E20. It appears that this mRNA is alternatively spliced to create a higher molecular mass and a lower molecular mass product. Nerve cells produce a very large amount of the larger mRNA, whereas spleen cells produce a moderate amount of the smaller mRNA. Both types are produced in small amounts by the muscle cells. It appears that kidney cells do not transcribe this gene.

E22. The Northern blot is shown here. The female mouse expresses the same total amount of this mRNA compared with the male. In the heterozygous female, there would be 50% of the 900 bp band and 50% of the 825 bp band.

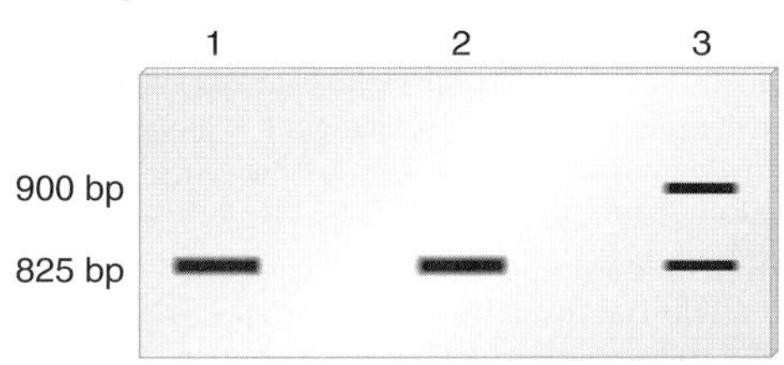

E24. Restriction enzymes recognize many sequences throughout the chromosomal DNA. If two fragments from different samples have the same molecular mass in a Southern blot, it is likely (though not certain) that the two fragments are found at the same chromosomal site in the genome. In this Southern blot, most of the transposable elements are found at the same sites within the genomes of these different yeast strains. However, a couple of bands are different among the three strains. These results indicate that the *Ty* element may occasionally transpose to a new location or that chromosomal changes (point mutations, chromosomal rearrangements, deletions, and so on) may have slightly changed the genomes among these three strains of yeast in a way that changes the distances between the restriction sites.

E26. Lane 1 shows that β-globin is made in normal red blood cells. In red blood cells from a thalassemia patient (lane 2), however, very little is made. Perhaps this person is homozygous for a down promoter mutation, which diminishes the transcription of the gene. As shown in lanes 3 and 4, β-globin is not made in muscle cells.

E28. Western blotting.

E30. You would first make a DNA library using chromosomal DNA from pig cells. You would then radiolabel a portion of a strand from the human β-globin gene and use it as a probe in a colony hybridization experiment; each colony would contain a different cloned piece of the pig genome. You would identify "hot" colonies that hybridize to the human β-globin probe. You would then go back to the master plate and pick these colonies and grow them in a test tube. You would then isolate the plasmid DNA from the colonies and subject the DNA to DNA sequencing. By comparing the DNA sequences of the human β-globin gene and the putative clones, you could determine if the putative clones were homologous to the human clone and likely to be the pig homolog of the β-globin gene.

E32. The rationale behind a gel retardation assay is that a segment of DNA with a protein bound to it will migrate more slowly through a gel than will the same DNA without any bound protein. A shift in a DNA band to a higher molecular mass provides a way to identify DNA-binding proteins.

E34. TFIID can bind to this DNA fragment by itself, as seen in lane 2. However, TFIIB and RNA polymerase II cannot bind to the DNA by themselves (lanes 3 and 4). As seen in lane 5, TFIIB can bind, if TFIID is also present, because the mobility shift is higher than TFIID alone (compare lanes 2 and 5). In contrast, RNA polymerase II cannot bind to the DNA when only TFIID is present. The mobility shift in lane 6 is the same as that found in lane 2, indicating that only TFIID is bound. Finally, in lane 7, when all three components are present, the mobility shift is higher than when both TFIIB and TFIID are present (compare lanes 5 and 7). These results mean that all three proteins are bound to the DNA. Taken together, the results indicate that TFIID can bind by itself, TFIIB needs TFIID to bind, and RNA polymerase II needs both proteins to bind to the DNA.

E36. The region of the gel from about 350 bp to 175 bp does not contain any bands. This is the region being covered up; the "footprint" is about 175 bp long.

Questions for Student Discussion/Collaboration

2. A. Does a particular amino acid within a protein sequence play a critical role in the protein's structure or function?

   B. Does a DNA sequence function as a promoter?

   C. Does a DNA sequence function as a regulatory site?

   D. Does a DNA sequence function as a splicing junction?

   E. Is a sequence important for correct translation?

   F. Is a sequence important for RNA stability?

   And many others . . .

CHAPTER 21

Answers to Comprehension Questions

21.1: d, b, c
21.2: a, b, b
21.3: c, d
21.4: b, d
21.5: d, c

Concept Check Questions (in figure legends)

FIGURE 21.1 Cyanogen bromide separates β-galactosidase from the A or B chains.

FIGURE 21.3 In gene addition, a cloned gene is added to an organism's genome. In gene replacement, a cloned gene is swapped with a gene that is already present in the organism's genome.

FIGURE 21.5 The *Neo*^{R} gene selects for cells that have taken up the cloned gene.

FIGURE 21.6 A gene knockout replaces a normal gene with a gene that is inactivated. A gene knockin inserts a gene into a noncritical site in the genome.

FIGURE 21.7 The β-lactoglobulin promoter is used because it is expressed in mammary cells.

FIGURE 21.8 The nucleus of the oocyte is removed so the resulting organism contains (nuclear) genetic material only from the somatic cell.

FIGURE 21.9 No. Carbon Copy has not received genetic material from a different species.

FIGURE 21.10 When stem cells divide, they produce one cell that remains a stem cell and another cell that differentiates. This pattern maintains a population of stem cells.

FIGURE 21.12 Multipotent.

FIGURE 21.16 Only the T DNA within the T-DNA vector is transferred to a plant.

FIGURE 21.17 The liposome method is relatively safe in that it doesn't evoke an immune response. A disadvantage is that it may be inefficient at getting the cloned gene into many cells. The retrovirus method is typically very good at getting the cloned gene into cells. A disadvantage is that it may cause a potentially harmful immune response.

End-of-Chapter Questions:

Conceptual Questions

C2. *Agrobacterium radiobacter* synthesizes an antibiotic that kills *Agrobacterium tumefaciens*. The genes, which are necessary for antibiotic biosynthesis and resistance, are plasmid encoded and can be transferred during interspecies conjugation. If *A. tumefaciens* received this plasmid during conjugation, it would be resistant to killing. Therefore, the conjugation-deficient strain prevents the occurrence of *A. tumefaciens*–resistant strains.

C4. A biological control agent is an organism that prevents the harmful effects of some other agent in the environment. Examples include *Bacillus thuringiensis*, a bacterium that synthesizes compounds that act as toxins to kill insects, and the use of *A. radiobacter* to prevent crown gall disease caused by *A. tumefaciens*.

C6. A mouse model is a strain of mice that carries a mutation in a mouse gene that is analogous to a mutation in a human gene that causes disease. These mice can be used to study the disease and to test potential therapeutic agents.

C8. The T DNA gets transferred to the plant cell; it then is incorporated into the plant cell's genome.

C10. A. With regard to maternal effect genes, the phenotype would depend on the animal that donated the egg. It is the cytoplasm of the egg that accumulates the gene products of maternal effect genes.

B. The extranuclear traits depend on the mitochondrial genome. Mitochondria are found in the egg and in the somatic cell. So, theoretically, both cells could contribute extranuclear traits. In reality, however, researchers have found that the mitochondria in Dolly were from the animal that donated the egg. It is not clear why she had no mitochondria from the mammary cell.

C. The cloned animal would be genetically identical to the animal that donated the nucleus with regard to traits that are determined by nuclear genes, which are expressed during the lifetime of the organism. The cloned animal would/could differ from the animal that donated the nucleus with regard to traits that are determined by maternal effect genes and mitochondrial genes. Such an animal is not a true clone, but it is likely that it would greatly resemble the animal that donated the nucleus, because the vast majority of genes are found in the cell nucleus.

C12. Some people are concerned with the release of genetically engineered microorganisms into the environment. The fear is that such organisms may continue to proliferate and it may not be possible to "stop them." A second concern involves the use of genetically engineered organisms in the food we eat. Some people are worried that genetically engineered organisms may pose an unknown health risk. A third issue is ethics. Some people feel that it is morally wrong to tamper with the genetics of organisms. This opinion may also apply to genetic techniques such as cloning, stem cell research, and gene therapy.

Experimental Questions

E2. One possibility is to clone the toxin-producing genes from *B. thuringiensis* and introduce them into *Pseudomonas syringae*. This bacterial strain would have the advantage of not needing repeated applications. However, it would be a recombinant strain and might be viewed in a negative light by people who are hesitant to use recombinant organisms in the field. By comparison, *B. thuringiensis* is a naturally occurring species.

E4. Basically, one can follow the strategy described in Figure 21.5. If homologous recombination occurs, only the *Neo*^{R} gene is incorporated into the genome. The cells will be neomycin resistant and also resistant to gancyclovir. If gene addition occurs, the cells will usually be sensitive to gancyclovir. By growing the cells in the presence of neomycin and gancyclovir, one can select for homologous recombinants. The chimeras are identified by the observation that they have a mixture of light and dark fur.

E6. 1. *A. tumefaciens* (T DNA-mediated gene transfer): Genes are cloned into the T DNA of the Ti plasmid. The T DNA from this plasmid is transferred to the plant cell when it is infected by the bacterium.

2. Biolistic gene transfer: DNA is coated on microprojectiles that are "shot" into the plant cell.

3. Microinjection: A microscopic needle containing a solution of DNA is used to inject DNA into plant cells.

4. Electroporation: An electric current is used to introduce DNA into plant cells.

5. Also, DNA can be introduced into protoplasts by treatment with polyethylene glycol and calcium phosphate. This method is similar to bacterial transformation procedures.

E8. In Mendel's work, and the work of many classical geneticists, an altered (mutant) phenotype is the initial way to identify a gene. For example, Mendel recognized a gene that affects plant height by the identification of tall and dwarf plants. The transmission of this gene could be followed in genetic crosses, and eventually, the gene could be cloned using molecular techniques. Reverse genetics uses the opposite sequence of steps. The gene is cloned first, and a phenotype for the gene is discovered later, by making a transgenic animal with a gene knockout.

E10. A chimera is an organism that is composed of cells from two different individuals (usually of the same species). Chimeras are made by mixing together embryonic cells from two individuals and allowing the cells to organize themselves and develop into a single individual.

E12. You would use Southern blotting to determine the number of gene copies. As described in Chapter 20, you will observe multiple bands in a Southern blot if there are multiple copies of a gene. You need to know this information to predict the outcome of crosses. For example, if there are four integrated copies at different sites in the genome, an offspring could inherit anywhere from zero to four copies of the gene. You would want to understand this so you could predict the phenotypes of the offspring.

You would use Northern blotting or Western blotting to monitor the level of gene expression. A Northern blot will indicate if the gene is transcribed into mRNA, and a Western blot will indicate if the mRNA is translated into protein.

E14. Reproductive cloning means the cloning of entire multicellular organisms. In plants, this is easy. Most species of plants can be cloned by asexual cuttings. In animals, cloning occurs naturally, as in identical twins. Identical twins are genetic replicas of each other because they begin from the same fertilized egg. (Note: There could be some somatic mutations that occur in identical twins that would make them slightly different.) Recently, as in the case of Dolly, reproductive cloning has become possible by fusing somatic cells with enucleated eggs. The advantage, from an agricultural point of view, is that reproductive cloning could allow one to choose the best animal in a herd and make many clones from it. Breeding would no longer be necessary. Also, breeding may be less reliable because the offspring inherit traits from both the mother and father.

E16. Ex vivo therapy involves the removal of living cells from the body and their modification after they have been removed. The modified cells are then reintroduced back into a person's body. This approach works well for cells such as blood cells that are easily removed and replaced. By comparison, this approach would not work very well for many cell types. For example, lung cells cannot be removed and put back again. In this case, in vivo approaches must be sought.

Questions for Student Discussion/Collaboration

2. From a genetic viewpoint, the recombinant and nonrecombinant strains are very similar. The main difference is their history. The recombinant strain has been subjected to molecular techniques to eliminate a particular gene. The nonrecombinant strain has had the same gene eliminated by a spontaneous mutation or via mutagens. The nonrecombinant strain has the advantage of a better public perception. People are less worried about releasing nonrecombinant strains into the environment.

CHAPTER 22

Answers to Comprehension Questions

22.1: a   22.4: b, d, b

22.2: d   22.5: a, b, b

22.3: c, d, a   22.6: c

Concept Check Questions (in figure legends)

FIGURE 22.1 A genetic map is a chart that describes the order and relative distances between genes and other sites along a chromosome.

FIGURE 22.2 The probe binds only to a specific site that has a complementary sequence.

FIGURE 22.5 Microsatellites are polymorphic when the number of repeat sequences varies.

FIGURE 22.7 A contig is a series of clones that have overlapping fragments of DNA that span a region of chromosome or an entire chromosome.

FIGURE 22.10 Their advantage is they can be used to clone large fragments of DNA.

FIGURE 22.12 Chromosome walking is used to identify a gene that has already been mapped to a site along a chromosome.

FIGURE 22.15 Yes. The sequence is determined as the DNA is synthesized.

End-of-Chapter Questions:

Conceptual Questions

C2. A. Yes.

B. No; this is only one chromosome in the genome.

C. Yes.

D. Yes.

Experimental Questions

E2. They are complementary to each other.

E4. Because normal cells contain two copies of chromosome 14, one would expect that a probe would bind to complementary DNA sequences on both of these chromosomes. If a probe recognized only one of two chromosomes, this means that one of the copies of chromosome 14 has been lost, or it has suffered a deletion in the region where the probe binds. With regard to cancer, the loss of this genetic material may be related to the uncontrollable cell growth.

E6. After the cells and chromosomes have been fixed to the slide, it is possible to add two or more different probes that recognize different sequences (i.e., different sites) within the genome. Each probe has a different fluorescence emission wavelength. Usually, a researcher will use computer imagery that recognizes the wavelength of each probe and then assigns that probe a bright color. The color seen by the researcher is not the actual color emitted by the probe; it is a secondary color assigned by the computer. In a sense, the probes, with the aid of a computer, are "painting" the regions of the chromosomes that are recognized by a probe. An example of chromosome painting is shown in Figure 22.3. In this example, human chromosome 5 is painted with six different colors.

E8. A contig is a collection of clones that contain overlapping segments of DNA that span a particular region of a chromosome. To determine if two clones are overlapping, one could conduct a Southern blotting experiment. In this approach, one of the clones is used as a probe. If it is overlapping with the second clone, it will bind to it in a Southern blot. Therefore, the second clone is run on a gel, and the first clone is used as a probe. If the band corresponding to the second clone is labeled, this means that the two clones are overlapping.

E10. BAC cloning vectors have the replication properties of a bacterial chromosome and the cloning properties of a plasmid. To replicate like a chromosome, the BAC vector contains an origin of replication from an F factor. Therefore, in a bacterial cell, a BAC can behave as a chromosome. Like a plasmid, BACs also contain selectable markers and convenient cloning sites for the insertion of large segments of DNA. The primary advantage is the ability to clone very large pieces of DNA.

E12. *Deduced Outcome*

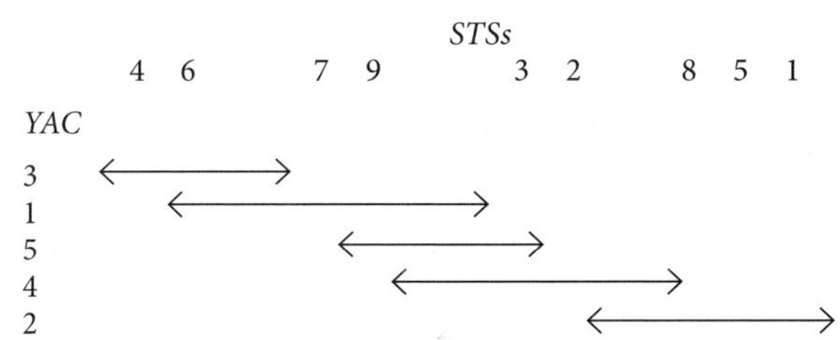

E14. A. One homolog contains the STS-1 that is 289 bp and STS-2 that is 422 bp, whereas the other homolog contains STS-1 that is 211 bp and STS-2 that is 115 bp. This is based on the observation that 28 of the sperm have either the 289-bp and 422-bp bands or the 211-bp and 115-bp bands.

B. There are two recombinant sperm; see lanes 12 and 18. Because there are two recombinant sperm out of a total of 30,

$$\text{Map distance} = \frac{2}{30} \times 100$$
$$= 6.7 \text{ mu}$$

C. In theory, this method could be used. However, there is not enough DNA in one sperm to carry out an RFLP analysis unless the DNA is amplified by PCR.

E16. The proper order is A, C, D, B.

1. Clone large fragments of DNA to make a BAC library.
2. Subclone BAC fragments to make a cosmid library.
3. Subclone cosmid fragments for DNA sequencing.
4. Determine the DNA sequence of subclones from a cosmid library.

E18. A sequence-tagged site (STS) is a segment of DNA, usually quite short (e.g., 100–400 bp in length), that serves as a unique site in the genome. STSs are identified using primers in a PCR reaction. STSs serve as molecular markers in genetic mapping studies. Sometimes the region within an STS may contain a microsatellite. A microsatellite is a short DNA segment that is variable in length, usually due to a short repeating sequence. When a microsatellite is within an STS, the length of the STS will vary among different individuals or even the same individual may be heterozygous for the STS. This makes the STS polymorphic. Polymorphic STSs can be used in linkage analysis, because their transmission can be followed in family pedigrees and through crosses of experimental organisms.

E20. The first piece of information you would start with is the location of a gene or marker that is known from previous mapping studies to be close to the gene of interest. You would begin with a clone containing this gene (or marker) and follow the procedure of chromosome walking to eventually reach the gene of interest. A contig would make this much easier because you would not have to conduct a series of subcloning experiments to reach your gene. Instead, you could simply analyze the members of the contig.

E22. The overall advantage of next-generation sequencing is that a large amount of DNA sequence can be obtained in a short period of time, and at a lower cost. One common innovation is that the need to subclone fragments of DNA into vectors is no longer necessary. Also, some of the methods involve the parallel analysis of an enormous number of samples simultaneously.

E24. First, researchers obtain a sample from the environment. This could be a soil sample, a water sample from the ocean, or a fecal sample from a person. After the sample is filtered, the cells within the sample are lysed and the DNA is extracted and purified. During this procedure, the purified DNA is sheared and contains fragments of different sizes. The DNA fragments are then randomly inserted into a cloning vector and transformed into a host cell. The result is a DNA library of thousands or tens of thousands of cells, each carrying a DNA fragment from the metagenome. All cells together constitute the metagenomic library. The members of the library are then subjected to shotgun DNA sequencing to identify the genes they may contain. Note: The cloning step may be skipped if certain types of DNA sequencing methods, such as pyrosequencing, are used.

Questions for Student Discussion/Collaboration

2. This is a matter of opinion. Many people would say that the ability to identify many human genetic diseases is the most important outcome. In addition, we will better understand how humans are constructed at the molecular level. As we gain a greater understanding of our genetic makeup, some people are worried that this may lead to greater discrimination. Insurance or healthcare companies could refuse people who are known to carry genetic abnormalities. Similarly, employers could make their employment decisions based on the genetic makeup of potential employees rather than their past accomplishments. At the family level, genetic information may affect how people choose mates, and whether or not they decide to have children.

CHAPTER 23

Answers to Comprehension Questions

23.1: c, b, c, a

23.2: d, d, a, d

23.3: a, d, c

Concept Check Questions (in figure legends)

FIGURE 23.1 A key point is that mRNA is made only when a gene is expressed. In this experiment, mRNA is first isolated and then used to make cDNA, which is fluorescently labeled. The fluorescent spots indicate which genes have been transcribed into mRNA.

FIGURE 23.3 The antibody recognizes the protein of interest that is covalently crosslinked to DNA. The antibody is also attached to heavy beads, which makes it easy to separate it from the rest of the cellular components by centrifugation. This provides an easy way to purify the protein of interest along with its attached DNA.

FIGURE 23.4 Each of these mechanisms alters the structure and possibly the function of proteins, thereby increasing protein diversity.

FIGURE 23.6 The purpose of tandem mass spectrometry is to determine the amino acid sequence of a peptide fragment.

FIGURE 23.8 These homologous genes have similar sequences because they were derived from a common ancestral gene. They are not identical because they have accumulated random mutations after they diverged from each other.

FIGURE 23.11 Yes, you would expect them to be similar because their RNA sequences would also be similar.

End-of-Chapter Questions:

Conceptual Questions

C2. There are two main reasons why the proteome is larger than the genome. The first reason involves the processing of pre-mRNA, a phenomenon that occurs primarily in eukaryotic species. RNA splicing and editing can alter the codon sequence of mRNA and thereby produce alternative forms of proteins that have different amino acid sequences. The second reason for protein diversity is posttranslational modifications. There are many ways that a given protein's structure can be covalently modified by cellular enzymes. These include proteolytic processing, disulfide bond formation, glycosylation, attachment of lipids, phosphorylation, methylation, and acetylation, to name a few.

C4. Centromeric sequences, origins of replication, telomeric sequences, repetitive sequences, and enhancers. Other examples are possible.

C6. There are a few interesting trends. Sequences 1 and 2 are similar to each other, as are sequences 3 and 4. There are a few places where amino acid residues are conserved among all five sequences. These amino acids may be particularly important with regard to function.

C8. A. Correct.

B. Correct.

C. This is not correct. These are short genetic sequences that happen to be similar to each other. The *lac* operon and *trp* operon are not derived from the same primordial operon.

D. This is not correct. The two genes are homologous to each other. It is correct to say that their sequences are 60% identical.

Experimental Questions

E2. The cDNA labeled with a green dye is derived from mRNA obtained from cells at an early time point, when glucose levels were high. The other samples of cDNA were derived from cells collected at later time points, when glucose levels were falling and when the diauxic shift was occurring. These were labeled with a red dye. The green fluorescence provides a baseline for gene expression when glucose is high. At later time points, if the red : green ratio is high (i.e., >1), this means that a gene is induced as glucose levels fall, because there is more red cDNA compared to green cDNA. If the ratio is low (i.e., <1), this means that a gene is being repressed.

E4. In the first dimension (i.e., in the tube gel), proteins migrate to the point in the gel where their net charge is zero. In the second dimension (i.e., the slab gel), proteins are coated with SDS and separated according to their molecular mass.

E6. In tandem mass spectroscopy, the first spectrometer determines the mass of a peptide fragment from a protein of interest. The second spectrometer determines the masses of progressively smaller fragments derived from that peptide. Because the masses of each amino acid are known, the molecular masses of these smaller fragments reveal the amino acid sequence of the peptide. With peptide sequence information, it is possible to use the genetic code and produce DNA sequences that could encode such a peptide. More than one sequence is possible due to the degeneracy of the genetic code. These sequences are used as query sequences to search a genomic database. This program will (hopefully) locate a match. The genomic sequence can then be analyzed to determine the entire coding sequence for the protein of interest.

E8. One strategy is a search by signal approach, which relies on known sequences such as promoters, start and stop codons, and splice sites to help predict whether or not a DNA sequence contains a protein-encoding gene. It attempts to identify a region that contains a promoter sequence, then a start codon, a coding sequence, and a stop codon. A second strategy is a search-by-content approach. The goal is to identify sequences whose nucleotide content differs significantly from a random distribution, which is usually due to codon bias. A search-by-content approach attempts to locate coding regions by identifying regions where the nucleotide content displays a bias. A third method to locate protein-encoding genes is to search for long open reading frames within a DNA sequence. An open reading frame is a sequence that does not contain any stop codons.

E10. By searching a database, one can identify genetic sequences that are homologous to a newly determined sequence. In most cases, homologous sequences carry out identical or very similar functions. Therefore, if one identifies a homologous member of a database whose function is already understood, this provides an important clue regarding the function of the newly determined sequence.

E12. The basis for secondary structure prediction is that certain amino acids tend to be found more frequently in α helices or β sheets. This information is derived from the statistical frequency of amino acids within secondary structures that have already been crystallized. Predictive methods are perhaps 60–70% accurate, which is not very good.

E14. The BACKTRANSLATE program works by using the genetic code. Each amino acid has one or more codons (i.e., 3-base sequences) that are specified by the genetic code. This program would produce a sequence file that was a nucleotide base sequence. The BACKTRANSLATE program would produce a degenerate base sequence because the genetic code is degenerate. For example, lysine can be specified by AAA or AAG. The program would probably store a single file that had degeneracy at particular positions. For example, if the amino acid sequence was lysine–methionine–glycine–glutamine, the program would produce the following sequence:

```
5′—AA(A/G)ATGGG(T/C/A/G)CA(A/G)
```

The bases found in parentheses are the possible bases due to the degeneracy of the genetic code.

E16. A. To identify a specific transposable element, a program would use sequence recognition. The sequence of P elements is already known. The program would be supplied with this information and scan a sequence file looking for a match.

B. To identify a stop codon, a program would use sequence recognition. There are three stop codons that are specific 3-base sequences. The program would be supplied with these three sequences and scan a sequence file to identify a perfect match.

C. To identify an inversion of any kind, a program would use pattern recognition. In this case, the program would be looking for a pattern in which the same sequence was running in opposite directions in a comparison of the two sequence files.

D. A search by signal approach uses both sequence recognition and pattern recognition as a means to identify genes. It looks for an organization of sequence elements that would form a functional gene. A search-by-content approach identifies genes based on patterns, not on specific sequence elements. This approach looks for a pattern in which the nucleotide content is different from a random distribution. The third approach to identify a gene is to scan a genetic sequence for long open reading frames. This approach is a combination of sequence recognition and pattern recognition. The program is looking for specific sequence elements (i.e., stop codons), but it is also looking for a pattern in which the stop codons are far apart.

E18. A. The amino acids that are most conserved (i.e., the same in all of the family members) are most likely to be important for structure and/or function. This is because a mutation that changed the amino acid might disrupt structure and function, and these kinds of mutations would be selected against. Completely conserved amino acids are found at the following positions: 101, 102, 105, 107, 108, 116, 117, 124, 130, 134, 139, 143, and 147.

B. The amino acids that are least conserved are probably not very important because changes in the amino acid does not seem to inhibit function. (If it did inhibit function, natural selection would eliminate such a mutation.) At one location, position 118, there are five different amino acids.

E20. This sequence is within the *lacY* gene of the *lac* operon of *E. coli*. It is found on page 593, nucleotides 801–850.

E22. A. This sequence has two regions that are about 20 amino acids long and very hydrophobic. Therefore, it is probable that this polypeptide has two transmembrane segments.

B.

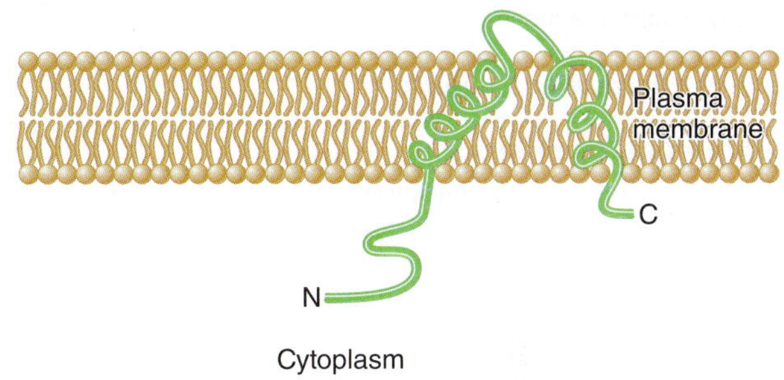

E24. A. True.

B. False. The programs are only about 60–70% accurate.

C. True.

Questions for Student Discussion/Collaboration

2. This is a very difficult question. In 20 years, we may have enough predictive information so that the structure of macromolecules can be predicted from their genetic sequences. If so, it would be better to be a mathematical theoretician with some genetics background. If not, it is probably better to be a biophysicist with some genetics background.

CHAPTER 24

Answers to Comprehension Questions

24.1: b, a, b, b	24.4: c
24.2: c, c	24.5: b, d, c, d, d
24.3: a, c	24.6: d

Concept Check Questions (in figure legends)

FIGURE 24.2 The key feature is that affected offspring have both parents who are unaffected by the disease. The parents are heterozygous carriers.

FIGURE 24.3 The key feature is that affected offspring have an affected parent.

FIGURE 24.4 The key feature is that all of the affected individuals are males. Furthermore, these males all have mothers who were descendants of Queen Victoria.

FIGURE 24.5 Haplotype refers to the linkage of alleles or molecular markers along a chromosome.

FIGURE 24.7 In this example, the original mutation that caused the Huntington allele occurred in a germ-line cell or gamete of a founder such that the Huntington allele was linked to the G8-C marker.

FIGURE 24.9 The PrP^{Sc} protein may come from eating infected meat or some people have a genetic predisposition that occasionally causes the PrP^{C} protein to convert to the PrP^{Sc} protein.

FIGURE 24.14 Such a mutation would keep the Ras protein in an active state and thereby promote cancerous growth. The EGF pathway would be turned on.

FIGURE 24.15 The *bcr* gene is expressed in white blood cells. This translocation causes the *abl* gene to be under the control of the *bcr* promoter. The abnormal expression of *abl* in white blood cells causes leukemia.

FIGURE 24.16 E2F will be active all of the time, and this will lead to cancer.

FIGURE 24.17 A checkpoint is a point in the cell cycle in which proteins detect if the cell is in the proper condition to divide. If an abnormality such as DNA damage is detected, the checkpoint proteins will halt the cell cycle.

FIGURE 24.20 An individual with a predisposition for familial breast cancer inherits only one copy of the mutant allele. Therefore, it shows a dominant pattern of inheritance. However, to actually get breast cancer, the other allele must become mutant in somatic cells.

End-of-Chapter Questions:

Conceptual Questions

C2. When a disease-causing allele affects a trait, it is causing a deviation from normality, but the gene involved is not usually the only gene that governs the trait. For example, an allele causing hemophilia prevents the normal blood-clotting pathway from operating correctly. It follows a simple Mendelian pattern because a single gene affects the phenotype. Even so, it is known that normal blood clotting is due to the actions of many genes.

C4. Changes in chromosome number and unbalanced changes in chromosome structure tend to affect phenotype because they create an imbalance of gene expression. For example, in Down syndrome, there are three copies of chromosome 21 and, therefore, three copies of all the genes on chromosome 21. This leads to a relative overexpression of genes that are located on chromosome 21 compared with the other chromosomes. Balanced translocations and inversions often are without phenotypic consequences because the total amount of genetic material is not altered, and the level of gene expression is not significantly changed.

C6. There are lots of possible answers; here are a few. Dwarfism occurs in people and dogs. Breeds like the dachshund and basset hound are types of dwarfism in dogs. There are diabetic people and mice. There are forms of inherited obesity in people and mice. Hip dysplasia is found in people and dogs.

C8. A. Because a person must inherit two defective copies of this gene and it is known to be on chromosome 1, the mode of transmission is autosomal recessive. Both members of this couple must be heterozygous, because they have one affected parent (who had to transmit the mutant allele to them) and their phenotypes are unaffected (so they must have received the normal allele from their other parent). Because both parents are heterozygotes, there is a $^1/_4$ chance of producing an affected child (a homozygote) with Gaucher disease. If we let *G* represent the nonmutant allele and *g* the mutant allele:

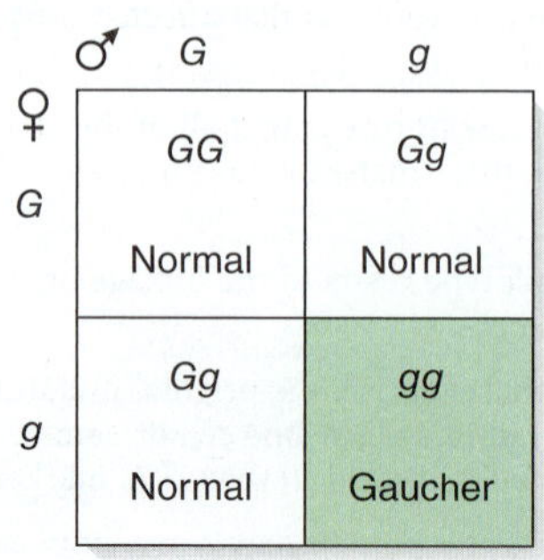

B. From this Punnett square, we can also see that there is a $^1/_4$ chance of producing a homozygote with both normal copies of the gene.

C. We need to apply the binomial expansion equation to solve this problem (see Chapter 2 for a description of this equation). In this problem, $n = 5$, $x = 1$, $p = 0.25$, $q = 0.75$. The answer is 0.396, or 39.6%.

C10. The mode of transmission is autosomal recessive. All of the affected individuals do not have affected parents. Also, the disorder is found in both males and females. If it were X-linked recessive, individual III-1 would have to have an affected father, which she does not.

C12. The 13 babies have acquired a new mutation. In other words, during spermatogenesis or oogenesis, or after the egg was fertilized, a new mutation occurred in the fibroblast growth factor gene. These 13 individuals have the same chances of passing the mutant allele to their offspring as the 18 individuals who inherited the mutant allele from a parent. The chance is 50%.

C14. Because this is a dominant trait, the mother must have two normal copies of the gene, and the father (who is affected) is most likely to be a heterozygote. (Note: The father could be a homozygote, but this is extremely unlikely because the dominant allele is very rare.) If we let *M* represent the mutant Marfan allele and *m* the normal allele, the following Punnett square can be constructed:

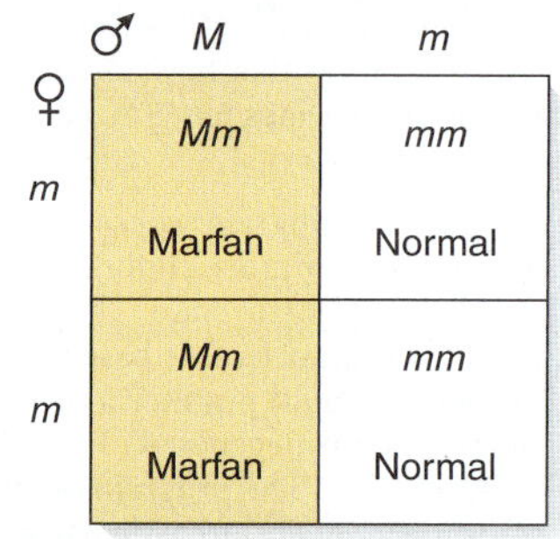

A. There is a 50% chance that this couple will have an affected child.

B. We use the product rule. The odds of having an unaffected child are 50%. So if we multiply 0.5 × 0.5 × 0.5, this equals 0.125, or a 12.5% chance of having three unaffected offspring.

C16. 1. The disease-causing allele had its origin in a single individual known as a founder, who lived many generations ago. Since that time, the allele has spread throughout portions of the human population.

2. When the disease-causing allele originated in the founder, it occurred in a region with a particular haplotype. The haplotype is not likely to have changed over the course of several generations if the disease-causing allele and markers in this region are very close together.

C18. A prion is a protein that behaves like an infectious agent. The infectious form of the prion protein has an abnormal conformation. This abnormal conformation is termed PrP^{Sc}, and the normal conformation of the protein is termed PrP^{C}. An individual can be "infected" with the abnormal conformation of the protein by eating products from another animal that had the disease, or the prion protein may convert spontaneously to the abnormal conformation. A prion protein in the PrP^{Sc} conformation can bind to a prion protein in the PrP^{C} conformation and convert it to the PrP^{Sc} form. An accumulation of prions in the PrP^{Sc} form is what causes the disease symptoms.

C20. A. Keep in mind that a conformational change is a stepwise process. It begins in one region of a protein and involves a series of small changes in protein structure. The entire conformational change from PrP^{C} to PrP^{Sc} probably involves many small changes in protein structure that occur in a stepwise manner. Perhaps the conformational change (from PrP^{C} to PrP^{Sc}) begins in the vicinity of position 178 and then proceeds throughout the rest of the protein. If there is a methionine at position 129, the complete conformational change (from PrP^{C} to PrP^{Sc}) can take place. However, perhaps the valine at position 129 somehow blocks one of the steps needed to complete the conformational change.

B. Once the PrP^{Sc} conformational change is completed, a PrP^{Sc} protein can bind to another prion protein in the PrP^{C} conformation. Perhaps it begins to convert it (to the PrP^{Sc} conformation) by initiating a small change in protein structure in the vicinity of position 178. The conversion would then proceed in a stepwise manner until the PrP^{Sc} conformation has been achieved. If a valine is at position 129, this could somehow inhibit one of the steps that are needed to complete the conformational change. If an individual had Val-129 in the polypeptide encoded by the second *PrP* gene, half of their prion proteins would be less sensitive to conversion by PrP^{Sc} compared to the proteins of individuals who had Met-129. This would explain why individuals with Val-129 in half of the prion proteins would have disease symptoms that would progress more slowly.

C22. A proto-oncogene is a normal cellular gene that typically plays a role in cell division. It can be altered by mutation to become an oncogene and thereby cause cancer. At the level of protein function, a proto-oncogene can become an oncogene by synthesizing too much of a protein or synthesizing the same amount of a protein that is abnormally active.

C24. The predisposition to develop cancer is inherited in a dominant fashion because the heterozygote has the higher predisposition. The mutant allele is actually recessive at the cellular level. But because we have so many cells in our bodies, it becomes relatively likely that a defective mutation will occur in the single normal gene and lead to a cancerous cell. Some heterozygous individuals may not develop the disease as a matter of chance. They may be lucky and not get a defective mutation in the normal gene, or perhaps their immune system is better at destroying cancerous cells once they arise.

C26. If an oncogene was inherited, it may cause uncontrollable cell growth at an early stage of development, thereby causing an embryo to develop improperly. This could lead to an early spontaneous abortion, which would explain why we do not observe individuals with inherited oncogenes. Another possibility is that inherited oncogenes may adversely affect gamete survival, which would make it difficult for them to be passed from parent to offspring. A third possibility would be that oncogenes could affect the fertilized zygote in a way that would prevent the correct implantation in the uterus.

C28. The role of *p53* is to sense DNA damage and prevent damaged cells from proliferating. Perhaps, prior to birth, the fetus is in a protected environment, so DNA damage may be minimal. In other words, the fetus may not really need *p53*. After birth, agents such as UV light may cause DNA damage. At this point, *p53* is important. A *p53* knockout is more sensitive to UV light because it cannot repair its DNA properly in response to this DNA-damaging agent, and it cannot kill cells that have become irreversibly damaged.

C30. The p53 protein is a regulatory transcription factor; it binds to DNA and influences the transcription rate of nearby genes. This transcription factor (1) activates genes that promote DNA repair; (2) activates a protein that inhibits cyclin/CDK protein complexes that are required for cell division; and (3) activates genes that promote apoptosis. If a cell is exposed to DNA damage, it has a greater potential to become malignant. Therefore, an organism wants to avoid the proliferation of such a cell. When exposed to an agent that causes DNA damage, a cell will try to repair the damage. However, if the damage is too extensive, the p53 protein will stop the cell from dividing and program it to die. This helps to prevent the proliferation of cancer cells in the body.

Experimental Questions

E2. Perhaps the least convincing is the higher incidence of the disease in particular populations. Because populations living in specific geographic locations are exposed to their own unique environment, it is difficult to distinguish genetic versus environmental causes for a particular disease. The most convincing evidence might be the higher incidence of a disease in related individuals and/or the ability to correlate a disease with the presence of a mutant gene. Overall, however, the reliability that a disease has a genetic component should be based on as many observations as possible.

E4. You would probably conclude that it is less likely to have a genetic component. If it were rooted primarily in genetics, it would be likely to be found in the Central American population. Of course, there is a chance that very few or none of the people who migrated to Central America were carriers of the mutant gene, which is somewhat unlikely for a large migrating population. By comparison, one might suspect that an environmental agent present in South America but not present in Central America may underlie the disease. Researchers could try to search for this environmental agent (e.g., a pathogenic organism).

E6. Males I-1, II-1, II-4, II-6, III-3, III-8, and IV-5 have a normal copy of the gene. Males II-3, III-2, and IV-4 are hemizygous for an inactive mutant allele. Females III-4, III-6, IV-1, IV-2, and IV-3 have two normal copies of the gene, whereas females I-2, II-2, II-5, III-1, III-5, and III-7 are heterozygous carriers of a mutant allele.

E8. A transformed cell is one that has become malignant. In a laboratory, this can be done in three ways. First, the cells could be treated with a mutagen that would convert a proto-oncogene into an oncogene. Second, cells could be exposed to the DNA from a malignant cell line. Under the appropriate conditions, this DNA can be taken up by the cells and integrated into their genome so that they become malignant. A third way to transform cells is by exposure to an oncogenic virus.

E10. By comparing oncogenic viruses with strains that have lost their oncogenicity, researchers have been able to identify particular genes that cause cancer. This has led to the identification of many oncogenes. From this work, researchers have also learned that normal cells contain proto-oncogenes that usually play a role in cell division. This suggests that oncogenes exert their effects by upsetting the cell division process. In particular, it appears that oncogenes are abnormally active and keep the cell division cycle in a permanent "on" position.

E12. One possible category of drugs would be GDP analogs (i.e., compounds that resemble the structure of GDP). Perhaps one could find a GDP analog that binds to the Ras protein and locks it in the inactive conformation.

One way to test the efficacy of such a drug would be to incubate the drug with a type of cancer cell that is known to have an overactive Ras protein and then plate the cells on solid media. If the drug locked the Ras protein in the inactive conformation, it should inhibit the formation of malignant growth or malignant foci.

There are possible side effects of such drugs. First, they might block the growth of normal cells, because Ras protein plays a role in normal cell proliferation. Second (if you have taken a cell biology course), there are many GTP/GDP-binding proteins in cells, and the drugs could somehow inhibit cell growth and function by interacting with these proteins.

E14. Molecular profiling is used to identify gene expression patterns in tumors. Sometimes tumors that look morphologically similar have very different expression patterns. Molecular profiling makes it possible to distinguish such tumors. This may be useful in deciding on treatment options and in the prognosis of tumor progression.

Questions for Student Discussion/Collaboration

2. There isn't a clearly correct answer to this question, but it should stimulate a large amount of discussion.

CHAPTER 25

Answers to Comprehension Questions

25.1: e, d, c

25.2: a, d, c, b

25.3: b, b, c

25.4: b, d

25.5: b, b

Concept Check Questions (in figure legends)

FIGURE 25.2 The mechanisms in parts (a) and (b) involve diffusible morphogens.

FIGURE 25.4 Genes encode proteins that control the changes that cause development to happen. It is largely a hierarchy in which certain sets of genes control other sets of genes.

FIGURE 25.6 The anteroposterior axis is from head to tail. The dorsoventral axis is from back (spine in vertebrates) to front.

FIGURE 25.7 The bicoid protein promotes the formation anterior structures in the fruit fly.

FIGURE 25.8 Maternal effect gene products are first made in nurse cells and are then transported into the oocyte.

FIGURE 25.10 A mutation in a gap gene causes several adjacent segments to be missing, whereas a mutation in a pair-rule gene causes regions in alternating segments to be missing.

FIGURE 25.12 The arrangement of homeotic genes correlates with the anteroposterior axis.

FIGURE 25.15 A cell lineage is a series of cells that are derived from each other via cell division.

FIGURE 25.17 An ortholog is a gene that is homologous to another gene in a different species.

FIGURE 25.19 The expression of the *HoxC-6* gene appears to determine the boundary between the neck and thoracic region in vertebrates.

FIGURE 25.20 The ID protein functions during early stages of embryonic development. It prevents myogenic bHLH proteins from promoting muscle differentiation too soon.

FIGURE 25.23 The correct number of stem cells is needed to promote proper growth and development.

FIGURE 25.25 The whorls would be carpel-stamen-stamen-carpel.

End-of-Chapter Questions:

Conceptual Questions

C2. A. False; the head is anterior to the tail.

B. True.

C. False; the feet are ventral to the hips.

D. True.

C4. A. True.

B. False; because gradients are also established after fertilization during embryonic development.

C. True.

C6. A. This is a mutation in a pair-rule gene (*runt*).

B. This is a mutation in a gap gene (*knirps*).

C. This is a mutation in a segment-polarity gene (*patched*).

C8. Positional information refers to the phenomenon whereby the spatial locations of morphogens and cell adhesion molecules (CAMs) provide a cell with information regarding its position relative to other cells. In *Drosophila*, the formation of a segmented body pattern relies initially on the spatial location of maternal gene products. These gene products lead to the sequential activation of the segmentation genes. Positional information can come from morphogens that are found within the oocyte, from morphogens secreted from cells during development, and from cell-to-cell contact. Although all three are important, morphogens in the oocyte have the greatest influence on the overall body structure.

C10. The anterior portion of the anteroposterior axis is established by the action of Bicoid. During oogenesis, the mRNA for Bicoid enters the anterior end of the oocyte and is sequestered there to establish an anterior (high) to posterior (low) gradient. Later, when the mRNA is translated, the Bicoid protein in the anterior region establishes a genetic hierarchy that leads to the formation of anterior structures. If Bicoid was not trapped in the anterior end, it is likely that anterior structures would not form.

C12. Maternal effect gene products influence the formation of the main body axes, including the anteroposterior, dorsoventral, and terminal regions. They are expressed during oogenesis and needed very early in development. Zygotic genes, particularly the three classes of the segmentation genes, are necessary after the axes have been established. The segmentation genes are expressed after fertilization.

C14. The coding sequence of homeotic genes contains a 180-bp consensus sequence known as a homeobox. The protein domain encoded by the homeobox is called a homeodomain. The homeodomain contains three conserved sequences that are folded into α-helical conformations. The arrangement of these α helices promotes the binding of the protein to the major groove of the DNA. Helix III is called the recognition helix because it recognizes a particular nucleotide sequence within the major groove. In this way, homeotic proteins are able to bind to DNA in a sequence-specific manner and thereby activate particular genes.

C16. It would normally be expressed in the three thoracic segments that have legs (T1, T2, and T3).

C18. A. When a mutation inactivates a gap gene, a contiguous section of the larva is missing.

B. When a mutation inactivates a pair-rule gene, some regions that are derived from alternating parasegments are missing.

C. When a mutation inactivates a segment-polarity gene, portions are missing at either the anterior or posterior end of the segments.

C20. Proper development in mammals is likely to require the products of maternal effect genes that play a key role in initiating embryonic development. The adult body plan is merely an expansion of the embryonic body plan, which is established in the oocyte. Because the starting point for the development of an embryo is the oocyte, this explains why an enucleated oocyte is needed to clone mammals.

C22. A heterochronic mutation is one that alters the timing when a gene involved in development is normally expressed. The gene may be expressed too early or too late, which causes certain cell lineages to be out of sync with the rest of the animal. If a heterochronic mutation affected the intestine, the animal may end up with too many intestinal cells if it is a gain-of-function mutation or too few if it is a loss-of-function mutation. In either case, the effects might be detrimental because the growth of the intestine must be coordinated with the growth of the rest of the animal.

C24. Cell differentiation is the specialization of a cell into a particular cell type. In the case of skeletal muscle cells, the bHLH proteins play a key role in the initiation of cell differentiation. When bHLH proteins are activated, they are able to bind to enhancers and activate the expression of many different muscle-specific genes. In this way, myogenic bHLH proteins turn on the expression of many muscle-specific proteins. When these proteins are synthesized, they change the characteristics of the cell into those of a muscle cell. Myogenic bHLH proteins are regulated by dimerization. When a heterodimer forms between a myogenic bHLH protein and an E protein, it activates gene expression. However, when a heterodimer forms between myogenic bHLH proteins and a protein called Id, the heterodimer is unable to bind to DNA. The Id protein is produced during early stages of development and prevents myogenic bHLH proteins from promoting muscle differentiation too soon. At later stages of development, the amount of Id protein falls, and myogenic bHLH proteins can combine with E proteins to induce muscle differentiation.

C26. A totipotent cell is a cell that has the potential to create a complete organism.

A. In humans, a fertilized egg is totipotent, and the cells during the first few embryonic divisions are totipotent. However, after several divisions, embryonic cells lose their totipotency and, instead, are determined to become particular tissues within the body.

B. In plants, many living cells are totipotent.

C. Because yeast are unicellular, one cell is a complete individual. Therefore, yeast cells are totipotent; they can produce new individuals by cell division.

D. Because bacteria are unicellular, one cell is a complete individual. Therefore, bacteria are totipotent; they can produce new individuals by cell division.

C28. Animals begin their development from an egg and then form anteroposterior and dorsoventral axes. The formation of an adult organism is an expansion of the embryonic body plan. Plants grow primarily from two meristems: shoot and root meristems. At the cellular level, plant development is different in that it does not involve cell migration, and most plant cells are totipotent. Animals require the organization within an oocyte to begin development. At the genetic level, however, animal and plant development are similar in that a genetic hierarchy of transcription factors governs pattern formation and cell specialization.

C30. The *tra* and *tra-2* gene products are splicing factors. In the female, they cause the alternative splicing of the pre-mRNAs that are expressed from the *fru* and *dsx* genes. The tra and tra-2 proteins cause these pre-mRNAs to be spliced into mRNAs designated fru^F and dsx^F. The absence of *Sxl* expression in the male prevents *tra* from being expressed. Without the tra protein, the *fru* and *dsx* mRNAs are spliced in a different way to produce mRNAs designated fru^M and dsx^M.

Experimental Questions

E2. *Drosophila* has an advantage in that researchers have identified many more mutant alleles that alter development in specific ways. The hierarchy of gene regulation is particularly well understood in the fruit fly. *C. elegans* has the advantage of simplicity and a complete knowledge of cell fate. This enables researchers to explore how the timing of gene expression is critical to the developmental process.

E4. To determine that a mutation is affecting the timing of developmental decisions, a researcher needs to know the normal time or stage of development when cells are supposed to divide and what types of cells will be produced. A lineage diagram provides this. With this information, one can then determine if particular mutations alter the timing when cell division occurs.

E6. Mutant 1 is a gain-of-function allele; it keeps reiterating the L1 pattern of division. Mutant 2 is a loss-of-function allele; it skips the L1 pattern and immediately follows an L2 pattern.

E8. As discussed in Chapter 15, most eukaryotic genes have a core promoter that is adjacent to the coding sequence; regulatory elements that control the transcription rate at the promoter are typically upstream from the core promoter. Therefore, to get the *Antp* gene product expressed where the *abd-A* gene product is normally expressed, you would link the upstream genetic regulatory region of the *abd-A* gene to the coding sequence of the *Antp* gene. This construct would be inserted into the middle of a P element (see next). The construct shown here would then be introduced into an embryo by P element transformation.

P element	*abd-A* regulatory region	/	*Antp* coding sequence	P element

The *Antp* gene product is normally expressed in the thoracic region and produces segments with legs, as illustrated in Figure 25.12. Therefore, because the *abd-A* gene product is normally expressed in the anterior abdominal segments, one might predict that the genetic construct shown above would produce a fly with legs attached to the segments that are supposed to be the anterior abdominal segments. In other words, the anterior abdominal segments might resemble thoracic segments with legs.

E10. A. The female flies must have had mothers that were heterozygous for a (dominant) normal allele and the mutant allele. Their fathers were either homozygous for the mutant allele or heterozygous. The female flies inherited a mutant allele from both their father and mother. Nevertheless, because their mother was heterozygous for the normal (dominant) allele and mutant allele, and because this is a maternal effect gene, their phenotype is based on the genotype of their mother. The normal allele is dominant, so they have a normal phenotype.

B. *Bicoid-A* appears to have a deletion that removes part of the sequence of the gene, resulting in a shorter mRNA. *Bicoid-B* could also have a deletion that removes all of the sequence of the *bicoid* gene, or it could have a promoter mutation that prevents the expression of the *bicoid* gene. *Bicoid-C* seems to have a point mutation that does not affect the amount of the bicoid mRNA.

C. With regard to function, all three mutations are known to be loss-of-function mutations. *Bicoid-A* probably eliminates function by truncating the Bicoid protein. The Bicoid protein is a transcription factor. The *bicoid-A* mutation probably shortens this protein, thereby inhibiting its function. The *bicoid-B* mutation prevents expression of the bicoid mRNA. Therefore, none of the Bicoid protein would be made, and this would explain the loss of function. The *bicoid-C* mutation seems to prevent the proper localization of the *bicoid* mRNA in the oocyte. There must be proteins within the oocyte that recognize specific sequences in the *bicoid* mRNA and trap it in the anterior end of the oocyte. This mutation must change these sequences and prevent these proteins from recognizing the bicoid mRNA.

E12. An egg-laying defect is somehow related to an abnormal anatomy. The *n540* strain has fewer neurons compared to a normal worm. Perhaps the *n540* strain is unable to lay eggs because it is missing neurons that are needed for egg laying. The *n536* and *n355* strains have an abnormal abundance of neurons. Perhaps this overabundance also interferes with the proper neural signals needed for egg laying.

E14. Geneticists interested in mammalian development have used reverse genetics because it has been difficult for them to identify mutations in developmental genes based on phenotypic effects in the embryo. This is because it is difficult to screen a large number of mammalian embryos in search of abnormal ones that carry mutant genes. It is easy to have thousands of flies in a laboratory, but it is not easy to have thousands of mice. Instead, it is easier to clone the normal gene based on its homology to invertebrate genes and then make mutations in vitro. These mutations can be introduced into a mouse to create a gene knockout. This strategy is opposite to that of Mendel, who characterized genes by first identifying phenotypic variants (e.g., tall versus dwarf, green seeds versus yellow seeds, etc.).

Questions for Student Discussion/Collaboration

2. In this problem, the students should try to make a flow diagram that begins with maternal-effect genes, then gap genes, pair-rule genes, and segment polarity genes. These genes then lead to homeotic genes and finally genes that encode proteins that affect cell structure and function. It's almost impossible to make an accurate flow diagram because there are so many gene interactions, but it is instructive to think about developmental genetics in this way. It is probably easier to identify mutant phenotypes that affect later stages of development because they are less likely to be lethal. However, modern methods can screen for conditional mutants as described in solved problem S3. To identify all of the genes necessary for chicken development, you may begin with early genes, but this assumes you have some way to identify them. If they had been identified, you would then try to identify the genes that they stimulate or repress. This could be done using molecular methods described in Chapters 14, 15, and 20.

CHAPTER 26

Answers to Comprehension Questions

26.1: c, a, a, d

26.2: a

26.3: b, a, c, d

26.4: d, b

26.5: d

26.6: c

26.7: d, b, a

Concept Check Questions (in figure legends)

FIGURE 26.1 A local population is a group of individuals that are more likely to interbreed with each other compared with more distant populations.

FIGURE 26.2 Polymorphisms are very common in nearly all natural populations.

FIGURE 26.6 The word *directional* means that selection is favoring a particular phenotype. Selection is moving the population in the direction in which that phenotype will predominate.

FIGURE 26.8 No, because directional selection is eliminating individuals that are sensitive to DDT.

FIGURE 26.9 The Hb^S allele is an advantage in the heterozygous condition because it confers resistance to malaria. The heterozygote advantage outweighs the homozygote disadvantage.

FIGURE 26.10 In negative frequency-dependent selection, the rarer phenotype has a higher fitness, which improves its reproductive success.

FIGURE 26.11 Yes, this selection fosters polymorphism. The fitness values of phenotypes depend on the environment. Some phenotypes are the fittest in one environment whereas other phenotypes are the fittest in another environment.

FIGURE 26.13 Stabilizing selection decreases genetic diversity because it eliminates individuals that carry alleles that promote more extreme phenotypes.

FIGURE 26.16 Genetic drift tends to have a greater effect in small populations. It can lead to the rapid loss or fixation of an allele.

FIGURE 26.17 At the bottleneck, genetic diversity may be lower because there are fewer individuals. Also, during the time when the bottleneck occurs, genetic drift may promote the loss of certain alleles and the fixation of other alleles, thereby diminishing genetic diversity.

FIGURE 26.18 Inbreeding increases the likelihood of homozygosity, and therefore tends to increase the likelihood that an individual will exhibit a recessive trait. This occurs because an individual can inherit both copies of the same allele from a common ancestor.

FIGURE 26.21 DNA fingerprinting is used for identification and for relationship testing.

End-of-Chapter Questions:

Conceptual Questions

C2. A population is a group of interbreeding individuals. Let's consider a squirrel population in a forested area. Over the course of many generations, several things could happen to this population. A forest fire, for example, could dramatically decrease the number of individuals and thereby cause a bottleneck. This would decrease the genetic diversity of the population. A new predator may enter the region and natural selection may select for the survival of squirrels that are best able to evade the predator. Another possibility is that a group of squirrels within the population may migrate to a new region and found a new squirrel population.

C4. A. Phenotype frequency and genotype frequency.

B. Genotype frequency.

C. Allele frequency.

C6. A. The genotype frequency for the *CF* homozygote is $^1/_{2500}$, or 0.004. This would equal q^2. The allele frequency is the square root of this value, which equals 0.02. The frequency of the corresponding normal allele equals $1 - 0.02 = 0.98$.

B. The frequency for the *CF* homozygote is 0.004; for the unaffected homozygote, $(098)^2 = 0.96$; and for the heterozygote, 2(0.98)(0.02), which equals 0.039.

C. If a person is known to be a heterozygous carrier, the chances that this particular person will happen to choose another as a mate is equal to the frequency of heterozygous carriers in the population, which equals 0.039, or 3.9%. The chances that two randomly chosen individuals will choose each other as mates equals $0.039 \times 0.039 = 0.0015$, or 0.15%.

C8. If we apply the Hardy-Weinberg equation:

$BB = (0.67)^2 = 0.45$, or 45%

$Bb = 2(0.67)(0.33) = 0.44$, or 44%

$bb = (0.33)^2 = 0.11$, or 11%

C10. Migration, genetic drift, and natural selection are the main factors that alter allele frequencies within a population. Natural selection acts to eliminate harmful alleles and promote beneficial alleles. Genetic drift involves random changes in allele frequencies that may eventually lead to elimination or fixation of alleles. It is thought to be important in the establishment of neutral alleles in a population. Migration is important because it introduces new alleles into neighboring populations. According to the neutral theory, genetic drift is largely responsible for the variation seen in natural populations.

C12. Darwinian fitness is the relative likelihood that a genotype will survive and contribute to the gene pool of the next generation as compared to other genotypes. The genotype with the highest reproductive ability is given a value of 1.0. Characteristics that promote survival, ability to attract a mate, or an enhanced fertility would be expected to promote Darwinian fitness. Examples are the thick fur of a polar bear, which helps it to survive in a cold climate; the bright plumage of male birds, which helps them to attract a mate; and the high number of gametes released by certain species of fish, which enhances their fertility.

C14. In all cases, these forms of natural selection favor one or more phenotypes because such phenotypes have a reproductive advantage. However, the patterns differ with regard to whether a single phenotype or multiple phenotypes are favored, and whether the phenotype that is favored is in the middle of the phenotypic range or at one or both extremes. Directional selection favors one phenotype at a phenotypic extreme. Over time, natural selection is expected to favor the fixation of alleles that cause these phenotypic characteristics. Disruptive selection favors two or more phenotypic categories. It will lead to a population with a balanced polymorphism for the trait. Examples of balancing selection are heterozygote advantage and negative-frequency dependent selection. These promote a stable polymorphism in a population. Stabilizing selection favors individuals with intermediate phenotypes. It tends to decrease genetic diversity because alleles that favor extreme phenotypes are eliminated.

C16. In genetic drift, allele frequencies are drifting. Genetic drift is an appropriate term because the word drift implies a random process. Nevertheless, drift can be directional. A boat may drift from one side of a lake to another. It would not drift in a straight path, but the drifting process will alter its location. Similarly, allele frequencies can drift up and down and eventually lead to the elimination or fixation of particular alleles within a population.

C18. A. Probability of fixation $= {}^1/_{2N} = {}^1/_{2(4)} = {}^1/_8$, or 0.125

B. $t = 4N = 4(4) = 16$

C. The preceding calculations assume a constant population size. If the population grows after it has been founded by these four individuals, the probability of fixation will be lower and the time it takes to reach fixation will be longer.

C20. A. True.

B. True.

C. False; it causes allele loss or fixation, which results in less diversity.

D. True.

C22. A. Migration will increase the genetic diversity in both populations. A random mutation could occur in one population to create a new allele. This new allele could be introduced into the other population via migration.

B. The allele frequencies between the two populations will tend to be similar to each other, due to the intermixing of their alleles.

C. Genetic drift depends on population size. When two populations intermix, this has the effect of increasing the overall population size. In a sense, the two smaller populations behave somewhat like one big population. Therefore, the effects of genetic drift are lessened when the individuals in two populations can migrate. The net effect is that allele loss and allele fixation are less likely to occur due to genetic drift.

C24. A. Yes.

B. The common ancestors are I-1 and I-2.

C. $F = \Sigma(^1/_2)^n(1 + F_A)$

$F = (^1/_2)^9 + (^1/_2)^9$

$F = {}^1/_{512} + {}^1/_{512} = {}^2/_{512} = 0.0039$

D. Not that we can tell.

C26. A. The inbreeding coefficient is calculated using the formula:

$F = \Sigma(^1/_2)^n(1 + F_A)$

In this case there are two common ancestors, I-1 and I-2. Because we have no prior history on I-1 or I-2, we assume they are not inbred, which makes $F_A = 0$. The two inbreeding loops for IV-2 contain five people, III-4, II-2, I-1, II-6, and III-5, and III-4, II-2, I-2, II-6, and III-5. Therefore, $n = 5$ for both loops.

$F = (^1/_2)^5\ (1 + 0) + (^1/_2)5\ (1 + 0)$

$F = 0.031 + 0.031 = 0.062$

B. Based on the data shown in this pedigree, individual III-4 is not inbred.

C28. We can use the following equation to calculate the change in allele frequency after any number of generations:

$$(1 - u)^t = \frac{p_t}{p_o}$$

$(1 - 10^{-4})^t = 0.5 / 0.6 = 0.833$

$(0.9999)t = 0.833$

$t = 1827$ generations

Experimental Questions

E2. Solved problem S4 shows how the Hardy-Weinberg equation can be modified to include situations of three or more alleles. In this case:

$(p + q + r + s)^2 = 1$

$p^2 + q^2 + r^2 + s^2 + 2pq + 2qr + 2qs + 2rp + 2rs + 2sp = 1$

Let $p = C$, $q = c^{ch}$, $r = c^h$, and $s = c$.

A. The frequency of albino rabbits is s^2:

$s^2 = (0.05) = 0.0025 = 0.25\%$

B. Himalayan is dominant to albino but recessive to full and chinchilla. Therefore, Himalayan rabbits would be represented by r^2 and by $2rs$:

$r^2 + 2rs = (0.44)^2 + 2(0.44)(0.05) = 0.24 = 24\%$

Among 1000 rabbits, about 240 would have a Himalayan coat color.

C. Chinchilla is dominant to Himalayan and albino but recessive to full coat color. Therefore, heterozygotes with chinchilla coat color would be represented by $2qr$ and by $2qs$:

$2qr + 2qs = 2(0.17)(0.44) + 2(0.17)(0.05) = 0.17$, or 17%

Among 1000 rabbits, about 170 would be heterozygotes with chinchilla fur.

E4. A.

Inuit	$M = 0.913$	$N = 0.087$
Navajo	$M = 0.917$	$N = 0.083$
Finns	$M = 0.673$	$N = 0.327$
Russians	$M = 0.619$	$N = 0.381$
Aborigines	$M = 0.176$	$N = 0.824$

B. To determine if these populations are in equilibrium, we can use the Hardy-Weinberg equation and calculate the expected number of individuals with each genotype. For example:

Inuit $MM = (0.913)^2 = 0.833 = 83.3\%$

$MN = 2(0.913)(0.087) = 0.159 = 15.9\%$

$NN = (0.087)^2 = 0.0076 = 0.76\%$

In general, the values agree pretty well with an equilibrium. The same is true for the other four populations.

C. Based on similar allele frequencies, the Inuit and Navajo Indians seem to have interbred, as well as the Finns and Russians.

E6. The selection coefficients are:

$s_{ww} = 1 - 0.19 = 0.81$

$s_{WW} = 1 - 0.37 = 0.63$

If the rats are not exposed to warfarin, the equilibrium will no longer exist, and natural selection will tend to eliminate the warfarin-resistance allele because the homozygotes are vitamin K deficient.

E8. This is an example of directional selection. Over the long run, directional selection may lead to the loss of certain alleles and the fixation of others. In this case, alleles promoting smaller beak size might be lost from the population, while alleles promoting larger beak size could become fixed.

E10. A. Probability of fixation $= {}^1/_2N$ (Assuming equal numbers of males and females contributing to the next generation)

Probability of fixation $= {}^1/_2(2{,}000{,}000)$

$= 1$ in 4,000,000 chance

B. $\bar{t} = 4N$

Where

$\bar{t}$ = the average number of generations to achieve fixation

N = the number of individuals in population, assuming that males and females contribute equally to each succeeding generation

$\bar{t} = 4(2\text{ million}) = 8$ million generations

C. If the blue allele had a selective advantage, the value calculated in part A would be slightly larger; there would be a higher chance of allele fixation. The value calculated in part B would be smaller; it would take a shorter period of time to reach fixation.

E12. Male 2 is the potential father, because he contains the bands found in the offspring but not found in the mother. To calculate the probability, one would have to know the probability of having each of the types of bands that match. In this case, for example, male 2 and the offspring have four bands in common. As a simple calculation, we could eliminate the four bands the offspring shares with the mother. If the probability of having each paternal band is $^1/_4$, the chances that this person is not the father are $(^1/_4)^4$.

E14. This percentage is not too high. Based on their genetic relationship, we expect that a father and daughter must share at least 50% of the same bands in a DNA fingerprint. However, the value can be higher than that because the mother and father may have some bands in common, even though they are not genetically related. For example, at one site in the genome, the father may be heterozygous for a 4100-bp and 5200-bp minisatellite, and the mother may also be heterozygous in this same region and have 4100-bp and 4700-bp minisatellites. The father could pass the 5200-bp band to his daughter, and the mother could pass the 4100-bp band. Thus, the daughter would inherit the 4100-bp and 5200-bp bands. This would be a perfect match to both of the father's bands, even though the father transmitted only the 5200-bp band to his daughter. The 4100-bp band matches because the father and mother happened to have a minisatellite in common. Therefore, the 50% estimate of matching bands in a DNA fingerprint based on genetic relationships is a minimum estimate. The value can be higher than that.

Questions for Student Discussion/Collaboration

2. Mutation is responsible for creating new alleles, but the rate of new mutations is so low that it cannot explain allele frequencies in this range. Let's call the two alleles *B* and *b* and assume that *B* was the original allele and *b* is a more recent allele that arose as a result of mutation. Three scenarios to explain the allele frequencies:
    1. The *b* allele is neutral and reached its present frequency by genetic drift. It hasn't reached elimination or fixation yet.
    2. The *b* allele is beneficial, and its frequency is increasing due to natural selection. However, there hasn't been enough time to reach fixation.
    3. The *Bb* heterozygote is at a selective advantage, leading to a balanced polymorphism.

CHAPTER 27

Answers to Comprehension Questions

27.1: d, c

27.2: d, b

27.3: d, b, c

27.4: c, b, b

27.5: a, c

Concept Check Questions (in figure legends)

FIGURE 27.1 In most populations (like this one), height follows a continuum.

FIGURE 27.2 About 95.4% are within 2 standard deviations, which means that 4.6% are outside of this range. Half of them, or 2.3%, fall above 2 standard deviations.

FIGURE 27.3 When alleles are additive, these means they contribute in an incremental way to the outcome of a trait. Having three heavy alleles will make an individual heavier than having two heavy alleles.

FIGURE 27.4 Increases in gene number and more environmental variation tend to cause greater overlaps between different genotypes and the same phenotype.

FIGURE 27.5 These two strains differ with regard to a quantitative trait and they differ in their molecular markers.

FIGURE 27.9 Both natural selection and selective breeding affect allele frequencies due to differences in reproductive success. In the case of natural selection, reproductive success is determined by environmental conditions. For selective breeding, reproductive success is determined by the people who choose the parents for breeding.

FIGURE 27.10 The strains differ with regard to the relative sizes of different parts of the plants. These include the stems, leaves, leaf buds, and flower buds.

FIGURE 27.11 A selection limit may be reached because a population has become monomorphic for all of the desirable alleles. A second reason is because the desired effects of artificial selection are balanced by the negative effects on fitness.

End-of-Chapter Questions:

Conceptual Questions

C2. At the molecular level, quantitative traits often exhibit a continuum of phenotypic variation because they are usually influenced by multiple genes that exist as multiple alleles. A large amount of environmental variation will also increase the overlap between genotypes and phenotypes for polygenic traits.

C4. A discontinuous trait is one that falls into discrete categories. Examples include brown eyes versus blue eyes in humans and purple versus white flowers in pea plants. A continuous trait is one that does not fall into discrete categories. Examples include height in humans and fruit weight in tomatoes. Most quantitative traits are continuous; the trait falls within a range of values. The reason why quantitative traits are continuous is because they are usually polygenic and greatly influenced by the environment. As shown in Figure 25.4b, this tends to create ambiguities between genotypes and a continuum of phenotypes.

C6. To be in the top 2.5% is about 2 standard deviation units. If we take the square root of the variance, the standard deviation would be 22 lb. To be in the top 2.5%, an animal would have to weigh at least 44 lb heavier than the mean, which equals 606 lb. To be in the bottom 0.13%, an animal would have to be 3 standard deviations lighter, which would be at least 66 lb lighter than the mean, or 496 lb.

C8. There is a positive correlation, but it could have occurred as a matter of chance alone. According to Table 27.2, this value could have occurred through random sampling error. You would need to conduct more experimentation to determine if there is a significant correlation, such as examining a greater number of pairs of individuals. If $N = 500$, the correlation would be statistically significant, and you would conclude that the correlation did not occur as a matter of random chance. However, you could not conclude cause and effect.

C10. When a correlation coefficient is statistically significant, it means that the association is likely to have occurred for reasons other than random sampling error. It may indicate cause and effect but not necessarily. For example, large parents may have large offspring due to genetics (cause and effect). However, the correlation may be related to the sharing of similar environments rather than cause and effect.

C12. Quantitative trait loci are sites within chromosomes that contain genes that affect a quantitative trait. It is possible for a QTL to contain one gene, or it may contain two or more closely linked genes. QTL mapping, which involves linkage to known molecular markers, is commonly used to determine the locations of QTLs.

C14. If the broad sense heritability equals 1.0, it means that all of the variation in the population is due to genetic variation rather than environmental variation. It does not mean that the environment is unimportant in the outcome of the trait. Under another set of environmental conditions, the trait may have turned out quite differently.

C16. When a species is subjected to selective breeding, the breeder is focusing his/her attention on improving one particular trait. In this case, the rose breeder is focused on the size and quality of the flowers. Because the breeder usually selects a small number of individuals (e.g., the ones with best flowers) as the breeding stock for the next generation, this may lead to a decrease in the allelic diversity at other genes. For example, several genes affect flower fragrance. In an unselected population, these genes may exist as "fragrant alleles" and "nonfragrant alleles." After many generations of breeding for large flowers, the fragrant alleles may be lost from the population, just as a matter of random chance. This is a common problem of selective breeding. As you select for an improvement in one trait, you may inadvertently diminish the quality of an unselected trait.

Others have suggested that the lack of fragrance may be related to flower structure and function. Perhaps the amount of energy that a flower uses to make beautiful petals somehow diminishes its capacity to make fragrance.

C18. Broad-sense heritability takes into account all genetic factors that affect the phenotypic variation in a trait. Narrow-sense heritability considers only alleles that behave in an additive fashion. In many cases, the alleles affecting quantitative traits appear to behave additively. More importantly, if a breeder assumes that the heritability of a trait is due to the additive effects of alleles, it is possible to predict the outcome of selective breeding. This is also termed the realized heritability.

C20. A. Because of their good nutrition, you may speculate that they would grow to be taller.

B. If the environment is rather homogeneous, then heritability values tend to be higher because the environment contributes less to the amount of variation in the trait. Therefore, in the commune, the heritability might be higher, because they uniformly practice good nutrition. On the other hand, because the commune is a smaller size than the general population, the amount of genetic variation might be less, so this would make the heritability lower. However, because the problem states that the commune population is large, we would probably assume that the amount of genetic variation is similar to that in the general population. Overall, the best guess would be that the heritability in the commune population is higher because of the uniform nutrition standards.

C. As stated in part B, the amount of variation would probably be similar, because the commune population is large. As a general answer, larger populations tend to have more genetic variation. Therefore, the general population probably has a bit more variation.

C22. A natural population of animals is more likely to have a higher genetic diversity compared to a domesticated population. This is because domesticated populations have been subjected to many generations of selective breeding, which decreases the genetic diversity. Therefore, V_G is likely to be higher for the natural population. The other issue is the environment. It is difficult to say which group would have a more homogeneous environment. In general, natural populations tend to have a more heterogeneous environment, but not always. If the environment is more heterogeneous, this tends to cause more phenotypic variation, which makes V_E higher.

$$\text{Heritability} = V_G/V_T = V_G/(V_G + V_E)$$

When V_G is high, heritability increases. When V_E is high, heritability decreases. In the natural wolf population, we would expect that V_G would be high. In addition, we would guess that V_E might be high as well (but that is less certain). Nevertheless, if this were the case, the heritability of the wolf population might be similar to the domestic population. This is because the high V_G in the wolf population is balanced by its high V_E. On the other hand, if V_E is not that high in the wolf population, or if it is fairly high in the domestic population, then the wolf population would have a higher heritability for this trait.

Experimental Questions

E2. To calculate the mean, we add the values together and divide by the total number.

$$\text{Mean} = \frac{1.9 + 2(2.4) + 2(2.1) + 3(2.0) + 2(2.2) + 1.7 + 1.8 + 2(2.3) + 1.6}{15}$$

$$\text{Mean} = 2.1$$

The variance is the sum of the squared deviations from the mean divided by $N - 1$. The mean value of 2.1 must be subtracted from each value, and then the square is taken. These 15 values are added together and then divided by 14 (which is $N - 1$).

$$\text{Variance} = \frac{0.85}{14} = 0.061$$

The standard deviation is the square root of the variance.

Standard deviation = 0.25

E4. When we say an RFLP is associated with a trait, we mean that a gene that influences a trait is closely linked to an RFLP. At the chromosomal level, the gene of interest is so closely linked to the RFLP that a crossover almost never occurs between them.

Note: Each plant inherits four RFLPs, but it may be homozygous for one or two of them.

Small: 2700 and 4000 (homozygous for both)

Small–medium: 2700 (homozygous), 3000, and 4000; or 2000, 2700, and 4000 (homozygous)

Medium: 2000 and 4000 (homozygous for both); or 2700 and 3000 (homozygous for both); or 2000, 2700, 3000, and 4000

Medium–large: 2000 (homozygous), 3000, and 4000; or 2000, 2700, and 3000 (homozygous)

Large: 2000 and 3000 (homozygous for both)

E6. Let's assume there is an extensive molecular marker map for the rice genome. We would begin with two strains of rice, one with a high yield and one with a low yield, that greatly differ with regard to the molecular markers they carry. We would make a cross between these two strains to get F_1 hybrids. We would then backcross the F_1 hybrids to either of the parental strains and then examine hundreds of offspring with regard to their rice yields and molecular markers. In this case, our expected results would be that six different markers in the high-producing strain would be correlated with offspring that produce higher yields. We might get fewer than six bands if some of these genes are closely linked and associate with the same marker. We also might get fewer than six if the two parental strains have the same marker that is associated with one or more of the genes that affect yield.

E8. $h_N^2 = \frac{R}{S}$

$R = \overline{X}_O - \overline{X}$

$S = \overline{X}_P - \overline{X}$

In this problem:

$\overline{X}$ equals 1.01 (as given in the problem)

$\overline{X}_O$ equals 1.09 (by calculating the mean for the offspring)

$\overline{X}_P$ equals 1.11 (by calculating the mean for the parents)

$R = 1.09 - 1.01 = 0.08$

$S = 1.11 - 1.01 = 0.10$

$h_N^2 = 0.08/0.10 = 0.8$ (which is a pretty high heritability value)

E10. We first need to calculate a and b. In this calculation, X represents the height of fathers, and Y represents the height of sons.

$$b = \frac{144}{112} = 1.29$$

$$a = 69 - (1.29)(68) = -18.7$$

For a father who is 70 in. tall,

$$Y = (1.29)(70) + (-18.7) = 71.6$$

The most likely height of the son would be 71.6 in.

E12. A. After six or seven generations, the selective breeding seems to have reached a plateau. This suggests that the tomato plants have become monomorphic for the alleles that affect tomato weight.

B. There does seem to be heterosis, because the first generation has a weight of 1.7 lb, which is heavier than either Mary's or Hector's tomatoes. This partially explains why Martin has obtained tomatoes heavier than 1.5 lb. However, heterosis is not the whole story; it does not explain why Martin obtained tomatoes that weigh 2 lb. Even though Mary's and Hector's tomatoes were selected for heavier weight, they may not have all of the "heavy alleles" for each gene that controls weight. For example, let's suppose there are 20 genes that affect weight, with each gene existing in a light and heavy allele. During the early stages of selective breeding, when Mary and Hector picked their 10 plants as seed producers for the next generation, as a matter of random chance, some of these plants may have been homozygous for the light alleles at a few of the 20 genes that control weight. Therefore, just as a matter of chance, they probably "lost" a few of the heavy alleles that affect weight. So, after 12 generations of breeding, they have predominantly heavy alleles but also have light alleles for some of the genes. If we represent heavy alleles with a capital letter and light alleles with a lowercase letter, Mary's and Hector's strains could be the following:

Mary's strain: *AA BB cc DD EE FF gg hh II JJ KK LL mm NN OO PP QQ RR ss TT*

Hector's strain: *AA bb CC DD EE ff GG HH II jj kk LL MM NN oo PP QQ RR SS TT*

As we see here, Mary's strain is homozygous for the heavy allele at 15 of the genes but carries the light allele at the other 5. Similarly, Hector's strain is homozygous for the heavy allele at 15 genes and carries the light allele at the other 5. It is important to note, however, that the light alleles in Mary's and Hector's strains are not in the same genes. Therefore, when Martin crosses them together, he will initially get the following:

Martin's F_1 offspring: *AA Bb Cc DD EE Ff Gg Hh II Jj Kk LL Mm NN Oo PP QQ RR Ss TT*

If the alleles are additive and contribute equally to the trait, we would expect about the same weight (1.5 lb), because this hybrid has a total of 10 light alleles. However, if heterosis is occurring, genes (which were homozygous recessive in Mary's and Hector's strains) will become heterozygous in the F_1 offspring, and this may make the plants healthier and contribute to a higher weight. If Martin's F_1 strain is subjected to selective breeding, the 10 genes that are heterozygous in the F_1 offspring may eventually become homozygous for the heavy allele. This would explain why Martin's tomatoes achieved a weight of 2.0 lb after five generations of selective breeding.

E14. A. $h_N^2 = \frac{\overline{X}_O - \overline{X}}{\overline{X}_P - \overline{X}}$

$h_N^2 = \frac{269 - 254}{281 - 254} = 0.56$

B. $0.56 = \frac{275 - 254}{\overline{X}_P - 254}$

$\overline{X}_P = 291.5$ lb

Questions for Student Discussion/Collaboration

2. Most traits depend on the influence of many genes. Also, genetic variation is a common phenomenon in most populations. Therefore, most individuals have a variety of alleles that contribute to a given trait. For quantitative traits, some alleles may make the trait bigger, and other alleles may make the trait turn out smaller. If a population contains many different genes and alleles that govern a quantitative trait, most individuals will have an intermediate phenotype because they will have inherited some large and some small alleles. Fewer individuals will inherit a predominance of large alleles or a predominance of small alleles. An example of a quantitative trait that does not fit a normal distribution is snail pigmentation. The dark snails and light snails are favored rather than the intermediate colors because they are less susceptible to predation.

CHAPTER 28

Answers to Comprehension Questions

28.1: d, f, a, c

28.2: c, d, d, c

28.3: a, d, d

Concept Check Questions (in figure legends)

FIGURE 28.1 It illustrates the potential problem of using morphological traits to establish species.

FIGURE 28.2 Cladogenesis is much more common.

FIGURE 28.4 The offspring that are produced from a cross between *Galeopsis tetrahit* and either of the other two species will have two sets of chromosomes from one species plus one set from the other. The chromosomes that exist as a single set cannot segregate evenly during meiosis thereby causing sterility. This sterility is why *G. tetrahit* is reproductively isolated from the other two species.

FIGURE 28.5 A clade is a group of species consisting of all of the descendants of the group's most common ancestor.

FIGURE 28.6 A shared derived character is more recent (on an evolutionary time scale) compared to a primitive character. A shared derived character is shared by a group of organisms but not by a distant common ancestor.

FIGURE 28.7 The idea that the preferred hypothesis is the one that is the simplest.

FIGURE 28.9 The protists are distributed within seven supergroups.

FIGURE 28.10 Horizontal gene transfer is the transfer of genetic material to another individual that is not an offspring of that individual. It can occur between different species.

FIGURE 28.11 Orthologs are homologous genes found in different species, whereas paralogs are homologous genes found in the same species.

FIGURE 28.15 A molecular clock allows researchers to put a time scale on their phylogenetic trees.

FIGURE 28.16 In humans, chromosome 2 is a single chromosome but it is divided into two chromosomes in the other species. With regard to chromosome 3, the orangutan has a large inversion that the other species do not have.

End-of-Chapter Questions:

Conceptual Questions

C2. Evolution is unifying because all living organisms on this planet evolved from an interrelated group of common ancestors. At the molecular level, all organisms have a great deal in common. With the exception of some viruses, they all use DNA as their genetic material. This DNA is found within chromosomes, and the sequence of the DNA is organized into units called genes. Most genes are protein-encoding genes that encode the amino acid sequence of polypeptides. Polypeptides fold to form functional units called proteins. At the cellular level, all living organisms also share many similarities. For example, living cells share many of the same basic features including a plasma membrane, ribosomes, enzymatic pathways, and so on. In addition, as discussed in Chapter 5, the mitochondria and chloroplasts of eukaryotic cells are evolutionarily derived from bacterial cells.

C4. Reproductive isolation occurs when two species are unable to mate and produce viable offspring. As discussed in Table 28.1, several prezygotic and postzygotic mechanisms can prevent interspecies matings.

C6. Anagenesis is the evolution of one species into another, whereas cladogenesis is the divergence of one species into two or more species. Of the two, cladogenesis is more prevalent. There may be many reasons why. It is common for an abrupt genetic change such as alloploidy to produce a new species from a preexisting one. Also, migrations of a few members of species into a new region may lead to the formation of a new species in the region (i.e., allopatric speciation).

C8. A. Allopatric.

B. Sympatric.

C. At first, it may involve parapatric speciation with a low level of intermixing. Eventually, when smaller lakes are formed, allopatric speciation will occur.

C10. Reproductive isolation does not really apply to bacteria. Two different bacteria of the same species do not produce gametes that have to fuse to produce an offspring, although bacteria can exchange genetic material (as described in Chapter 7). For this reason, it becomes more difficult to distinguish different species of bacteria. A geneticist would probably divide bacteria into different species based on their cellular traits and the sequences of their DNA. Historically, bacteria were first categorized as different species based on morphological and physiological differences. Later, when genetic tools such as DNA sequencing became available, the previously identified species could be categorized based on genetic sequences. One issue that makes categorization rather difficult is that a species of bacteria can exist as closely related strains that may have a small number of genetic differences.

C12. Line up the sequences where the two Gs are underlined.

TTGCATAGGCATACCGTATGATATCGAAAACTAGAAAAATAGGGCGATAGCTA

GTATGTTATCGAAAAGTAGCAAAATAGGGCGATAGCTACCCAGACTACCGGAT

C14. The rate of deleterious and beneficial mutations would probably not be a good molecular clock. Their rate of formation might be relatively constant, but their rate of elimination or fixation would probably be quite variable. These alleles are acted upon by natural selection. As environmental conditions change, the degree to which natural selection would favor beneficial alleles and eliminate deleterious alleles would also change. For example, natural selection favors the sickle cell allele in regions where malaria is prevalent but not in other regions. Therefore, the prevalence of this allele does not depend solely on its rate of formation and random genetic drift.

C16. Some regions of a polypeptide are particularly important for the structure or function of a protein. For example, a region of a polypeptide may form the active site of an enzyme. The amino acids that are found within the active site are likely to be precisely located for the binding of the enzyme's substrate and/or for catalysis. Changes in the amino acid sequence of the active site usually have a detrimental effect on the enzyme's functions. Therefore, these types of polypeptide sequences (like those found in active sites) are not likely to change. If they did change, natural selection would tend to prevent the change from being transmitted to future generations. In contrast, other regions of a polypeptide are less important. These other regions would be more tolerant of changes in amino acid sequence and therefore would evolve more rapidly. When comparing related protein sequences, regions that are important for function can often be identified based on less sequence variation.

C18. The α-globin sequences in humans and horses are more similar to each other, compared to the α-globin in humans and the β-globin in humans. This suggests that the gene duplication that produced the α-globin and β-globin genes occurred first. After this gene duplication occurred, each gene accumulated several different mutations that caused the sequences of the two genes to diverge. At a much later time, during the evolution of mammals, a split occurred that produced different branches in the evolutionary tree of mammals. One branch eventually led to the formation of horses and a different branch led to the formation of humans. During the formation of these mammalian branches (which has been more recent), some additional mutations occurred in the α- and β-globin genes. This explains why the α-globin gene in humans and horses is not exactly the same. However, it is more similar than the α- and β-globin genes within humans because the divergence of humans and horses occurred much more recently than the gene duplication that produced the α- and β-globin genes. In other words, there has been much less time for the α-globin gene in humans to diverge from the α-globin gene in horses.

C20. A. This is an example of neutral mutation. Mutations in the wobble base are neutral when they do not affect the amino acid sequence.

B. This is an example of natural selection. Random mutations that occur in vital regions of a polypeptide sequence are likely to inhibit function. Therefore, these types of mutations are eliminated by natural selection. That is why they are relatively rare.

C. This is a combination of neutral mutation and natural selection. The prevalence of mutations in introns is due to the accumulation of neutral mutations. Most mutations within introns do not have any effect on the expression of the exons, which contain the polypeptide sequence. In contrast, mutations within the exons are more likely to be affected by natural selection. As mentioned in the answer to part B, mutations in vital regions are likely to inhibit function. Natural selection tends to eliminate these mutations. Therefore, mutations within exons are less likely than mutations within introns.

C22. Generally, one would expect a similar number of chromosomes with very similar banding patterns. However, there may be a few notable differences. An occasional translocation could change the size or chromosomal number between two different species. Also, an occasional inversion may alter the banding pattern between two species.

Experimental Questions

E2. Perhaps the easiest way to determine allotetraploidy is by the chromosomal examination of closely related species. A researcher could karyotype the chromosomes from many different species and look for homologous chromosomes that have similar banding patterns. This may enable them to identify allotetraploids that contain a diploid set of chromosomes from two different species.

E4. The mutations that have occurred in this sequence are neutral mutations. In all cases, the wobble base has changed, and this change would not affect the amino acid sequence of the encoded polypeptide. Therefore, a reasonable explanation is that the gene has accumulated random neutral mutations over the course of many generations. This observation would be consistent with the neutral theory of evolution. A second explanation would be that one of these two researchers made a few experimental mistakes when determining the sequence of this region.

E6. Inversions do not affect the total amount of genetic material. Usually, inversions do not affect the phenotype of the organism. Therefore, if members of the two populations were to interbreed, the offspring would probably be viable because they would have inherited a normal amount of genetic material from each parent. However, such offspring would be inversion heterozygotes. As described in Chapter 8 (see Figure 8.11), crossing over during meiosis may create chromosomes that have too much or too little genetic material. If these unbalanced chromosomes are passed to the next generation of offspring, the offspring may not survive. For this reason, inversion heterozygotes (that are phenotypically normal) may not be very fertile because many of their offspring will die. Because inversion heterozygotes are less fertile, this would tend to keep the eastern and western populations reproductively isolated. Over time, this would aid in the independent evolution of the two populations and would ultimately promote the evolution of the two populations into separate species.

E8. The technique of PCR is used to amplify the amount of DNA in a sample. To accomplish this, one must use oligonucleotide primers that are complementary to the region that is to be amplified. For example, as described in the experiment of Figure 28.13, PCR primers that were complementary to and flank the 12S rRNA gene can be used to amplify the 12S rRNA gene. The technique of PCR is described in Chapter 20.

E10. We would expect the probe to hybridize to the natural *G. tetrahit* and also the artificial *G. tetrahit*, because both of these strains contain two sets of chromosomes from *G. pubescens*. We would expect two bright spots in the in situ experiment. Depending on how closely related *G. pubescens* and *G. speciosa* are, the probe may also hybridize to two sites in the *G. speciosa* genome, but this is difficult to predict a priori. If so, the *G. tetrahit* species would show four spots.

E12. The principle of parsimony chooses a phylogenetic tree that requires the fewest number of evolutionary changes. When using molecular data, researchers can use computer programs that compare DNA sequences from homologous genes of different species and construct a tree that requires the fewest numbers of mutations. Such a tree is the most likely pathway for the evolution of such species.

E14. If you constructed a phylogenetic tree using a gene sequence that was the result of horizontal gene transfer, the results might be misleading. For example, if a bacterial gene was transferred to a protist, and you used that gene to construct your tree, the results would suggest that the bacterium and protist are very closely related, which they are not. Evolutionary biologists can overcome this problem by choosing several different genes in the construction of their phylogenetic trees, assuming that most of them are not transferred horizontally.

Questions for Student Discussion/Collaboration

2. The founder effect and allotetraploidy are examples of rapid forms of evolution. In addition, some single gene mutations may have a great impact on phenotype and lead to the rapid evolution of new species by cladogenesis. Geological processes may promote the slower accumulation of alleles and alter a species' characteristics more gradually. In this case, it is the accumulation of many phenotypically minor genetic changes that ultimately leads to reproductive isolation. Slow and fast mechanisms of evolution have the common theme that they result in reproductive isolation. This is a prerequisite for the evolution of new species. Fast mechanisms tend to involve small populations and a few number of genetic changes. Slower mechanisms may involve larger populations and involve the accumulation of a large number of genetic changes that each contributes in a small way.

GLOSSARY

GLOSSARY

A

A an abbreviation for adenine.
ABC model a model for flower development.
acentric describes a chromosome without a centromere.
acentric fragment a fragment of a chromosome that lacks a centromere.
acquired antibiotic resistance the acquisition of antibiotic resistance because a bacterium has taken up a gene or plasmid from another bacterial strain.
acridine dye a type of chemical mutagen that causes frameshift mutations.
acrocentric a chromosome with the centromere significantly off center, but not at the very end.
activator a transcriptional regulatory protein that increases the rate of transcription.
acutely transforming virus (ACT) a virus that readily transforms normal cells into malignant cells, when grown in a laboratory.
adaptive evolution see *natural selection.*
adaptor hypothesis a hypothesis that proposes a tRNA has two functions: recognizing a three-base codon sequence in mRNA and carrying an amino acid that is specific for that codon.
adenine a purine base found in DNA and RNA. It base-pairs with thymine in DNA.
AFLP see *amplified restriction fragment length polymorphism.*
age of onset for alleles that cause genetic diseases, the time of life at which disease symptoms appear.
alkaptonuria a human genetic disorder involving the accumulation of homogentisic acid due to a defect in homogentisic acid oxidase.
alkyltransferase an enzyme that can remove methyl or ethyl groups from guanine bases.
allele an alternative form of a specific gene.
allele frequency the number of copies of a particular allele in a population divided by the total number of all alleles for that gene in the population.
allelic variation genetic variation in a population that involves the occurrence of two or more different alleles for a particular gene.
allodiploid an organism that contains one set of chromosomes from two different species.
allopatric speciation an evolutionary phenomenon in which speciation occurs when members of a species become geographically separated from the other members.
alloploid an organism that contains chromosomes from two or more different species.
alloploidy the phenomenon in which a cell or organism contains sets of chromosomes from two or more different species.
allopolyploid an organism that contains two (or more) sets of chromosomes from two (or more) species.
allosteric enzyme an enzyme that contains two binding sites: a catalytic site and a regulatory site.
allosteric regulation the phenomenon in which an effector molecule binds to a noncatalytic site on a protein and causes a conformational change that regulates its function.
allosteric site the site on a protein where a small effector molecule binds to regulate the function of the protein.
allotetraploid an organism that contains two sets of chromosomes from two different species.
allozymes two or more enzymes (encoded by the same type of gene) with alterations in their amino acid sequences, which may affect their gel mobilities.
α helix a type of secondary structure found in proteins.
alternative exon an exon that is not always found in mRNA. It is only found in certain types of alternatively spliced mRNAs.
alternative splicing refers to the phenomenon in which a pre-mRNA can be spliced in more than one way.
Ames test a test using strains of a bacterium, *Salmonella typhimurium,* to determine if a substance is a mutagen.
amino acid a building block of polypeptides and proteins. It contains an amino group, a carboxyl group, and a side chain.
aminoacyl site (A site) a site on the ribosome where a charged tRNA initially binds.
aminoacyl-tRNA a tRNA molecule that has an amino acid covalently attached to its 3′ end.
aminoacyl-tRNA synthetase an enzyme that catalyzes the attachment of a specific amino acid to the correct tRNA.
2-aminopurine a base analog that acts as a chemical mutagen.
amino terminal end see *amino-terminus.*
amino-terminus the location of the first amino acid in a polypeptide chain. The amino acid at the amino-terminus still retains a free amino group that is not covalently attached to the second amino acid.
amniocentesis a method of obtaining cellular material from a fetus for the purpose of genetic testing.
amplified restriction fragment length polymorphism (AFLP) an RFLP that is amplified via PCR.
anabolic enzyme an enzyme involved in connecting organic molecules to create larger molecules.
anagenesis the evolutionary phenomenon in which a single species is transformed into a different species over the course of many generations.
anaphase the fourth stage of M phase. As anaphase proceeds, half of the chromosomes move to one pole, and the other half move to the other pole.
ancestral character see *primitive character.*
ancient DNA analysis analysis of DNA that is extracted from the remains of extinct species.
aneuploid not euploid. Refers to a variation in chromosome number such that the total number of chromosomes is not an exact multiple of a set or *n* number.
aneuploidy a cell or organism that is aneuploid.
annealing the process in which two complementary segments of DNA bind to each other.
annotated in files involving genetic sequences, annotation is a description of the known function and features of the sequence, as well as other pertinent information.
anteroposterior axis in animals, the axis that runs from the head (anterior) to the tail or base of the spine or abdomen (posterior).
anther the structure in flowering plants that gives rise to pollen grains.
anther culture the generation of monoploid plants by cold-shock treatment of anthers.
antibiotic any substance produced by a microorganism that inhibits the growth of other microorganics, such as pathogenic bacteria.
antibodies proteins produced by the B cells of the immune system that recognize foreign substances (namely, viruses, bacteria, and so forth) and target them for destruction.
antibody microarray a small silica, glass, or plastic slide that is dotted with many different antibodies, which recognize particular amino acid sequences within proteins.
anticipation the phenomenon in which the severity of an inherited disease tends to get worse in future generations.
anticodon a three-nucleotide sequence in tRNA that is complementary to a codon in mRNA.
antigens foreign substances that are recognized by antibodies.
antiparallel an arrangement in a double helix in which one strand is running in the 5′ to 3′ direction, while the other strand runs 3′ to 5′.
antisense RNA an RNA strand that is complementary to a strand of mRNA.
antisense strand also called the template strand. It is the strand of DNA that is used as a template for RNA synthesis.
antitermination the function of certain proteins, such as N protein in bacteria, that prevents transcriptional termination.
AP endonuclease a DNA repair enzyme that recognizes a DNA region that is missing a base and makes a cut in the DNA backbone near that site.
apical–basal-patterning gene one of several plant genes that play a role in embryonic development.
apical region in plants, the region that produces the leaves and flowers.
apoptosis programmed cell death.
apurinic site a site in DNA that is missing a purine base.
Archaea also called archaebacteria; one of the three domains of life. Archaea are prokaryotic species. They tend to live in extreme environments and are less common than bacteria.
ARS elements DNA sequences found in yeast that function as origins of replication.
artificial chromosomes cloning vectors that can accommodate large DNA inserts and behave like chromosomes when inside of living cells.

artificial selection see *selective breeding.*
artificial transformation transformation of bacteria that occurs via experimental treatments.
ascus (pl. **asci**) a sac that contains haploid spores of fungi (i.e., yeast or molds).
asexual reproduction a form of reproduction that does not involve the union of gametes; at the cellular level, a preexisting cell divides to produce two new cells.
A site see *aminoacyl site.*
assortative mating breeding in which individuals preferentially mate with each other based on their phenotypes.
aster microtubule the microtubule that emanates outward from the centrosome toward the plasma membrane.
AT/GC rule in DNA, the phenomenon in which an adenine base in one strand always hydrogen bonds with a thymine base in the opposite strand, and a guanine base always hydrogen bonds with a cytosine.
ATP-dependent chromatin remodeling see *chromatin remodeling.*
attachment site a site in a host cell chromosome where a virus integrates during site-specific recombination.
attenuation a mechanism of genetic regulation, seen in the *trp* operon, in which a short RNA is made but its synthesis is terminated before RNA polymerase can transcribe the rest of the operon.
attenuator sequence a sequence found in certain operons (e.g., *trp* operon) in bacteria that stops transcription soon after it has begun.
AU-rich element (ARE) a sequence found in many short-lived mRNAs that contains the consensus sequence AUUUA.
automated DNA sequencing the use of fluorescently labeled dideoxyribonucleotides and a fluorescence detector to sequence DNA.
autonomous elements see *autonomous transposable element.*
autonomous transposable element a transposable element that contains all of the information necessary for transposition or retroposition to take place.
autopolyploid a polyploid produced within a single species due to nondisjunction.
autoradiography a technique that involves the use of X-ray film to detect the location of radioisotopes as they are found in macromolecules or cells. It is used to detect a particular band on a gel or to map the location of a gene within an intact chromosome.
autoregulatory loop a form of gene regulation in which a protein, such as a splicing factor or a transcription factor, regulates its own expression.
autosomes chromosomes that are not sex chromosomes.
auxotroph a strain that cannot synthesize a particular nutrient and needs that nutrient supplemented in its growth medium or diet.

B

BAC see *bacterial artificial chromosome.*
backbone the portion of a DNA or RNA strand that is composed of the repeated covalent linkage of the phosphates and sugar molecules.
backcross in genetics, this usually refers to a cross of F_1 hybrids to individuals that have genotypes of the parental generation.
Bacteria one of the three domains of living organisms. Also called eubacteria. Bacteria are prokaryotic species.
bacterial artificial chromosome (BAC) a cloning vector that propagates in bacteria and is used to clone large fragments of DNA.
bacteriophages (or **phages**) viruses that infect bacteria.
balanced polymorphism when natural selection favors the maintenance of two or more alleles in a population.
balanced translocation a translocation, such as a reciprocal translocation, in which the total amount of genetic material is normal or nearly normal.
balancing selection a pattern of natural selection that favors the maintenance of two or more alleles. Examples include heterozygote advantage and negative frequency-dependent selection.
band shift assay see *gel retardation assay.*
Barr body a structure in the interphase nuclei of somatic cells of female mammals that is a highly condensed X chromosome.
basal region in plants, the region that produces the roots.
basal transcription in eukaryotes, a low level of transcription via the core promoter. The binding of transcription factors to enhancer elements may increase transcription above the basal level.
basal transcription apparatus the minimum number of proteins needed to transcribe a gene.
base a nitrogen-containing molecule that is a portion of a nucleotide in DNA or RNA. Examples of bases are adenine, thymine, guanine, cytosine, and uracil.
base excision repair a type of DNA repair in which a modified base is removed from a DNA strand. Following base removal, a short region of the DNA strand is removed, which is then resynthesized using the complementary strand as a template.
base pair (bp) the structure in which two nucleotides in opposite strands of DNA hydrogen bond with each other. For example, an AT base pair is a structure in which an adenine-containing nucleotide in one DNA strand hydrogen bonds with a thymine-containing nucleotide in the complementary strand.
base pair mismatch when two bases opposite each other in a double helix do not conform to the AT/GC rule. For example, if A were opposite C, that would be a base mismatch.
base substitution a point mutation in which one base is substituted for another.
basic domain a protein domain containing several basic amino acids, which is often involved in binding to DNA.
Bayesian methods with regard to phylogenetic trees, this method asks the question: What is the probability that a particular phylogenetic tree is correct given the observed data and a particular evolutionary model?
B DNA the predominant form of DNA in living cells. It is a right-handed DNA helix with 10 bp per turn.
behavioral trait a trait that involves behavior. An example is the ability to learn a maze.
beneficial mutation a mutation that has a beneficial effect on phenotype.
benign a noncancerous tumor that is not invasive and cannot metastasize.
β sheet a type of secondary structure found in proteins.
bHLH a structure found in transcription factor proteins with a basic domain involved in DNA binding and a helix-loop-helix domain involved in dimerization.
bidirectionally the phenomenon in which two replication forks move in opposite directions, outward from the origin.
bidirectional replication the phenomenon in which two DNA replication forks emanate in both directions from an origin of replication.
bilateral gynandromorph an animal in which one side is phenotypically male and the other side is female.
bilateralism an animal with right-left symmetry.
binary fission the physical process whereby a bacterial cell divides into two daughter cells. During this event, the two daughter cells become divided by the formation of a septum.
binomial expansion equation an equation used to solve genetic problems involving two types of unordered events.
biodegradation the breakdown of a larger molecule into a smaller molecule via cellular enzymes.
bioinformatics the study of biological information. Recently, this term has been associated with the analysis of genetic sequences, using computers and computer programs.
biolistic gene transfer the use of microprojectiles to introduce DNA into plant cells.
biological control the use of microorganisms or products from microorganisms to alleviate plant diseases or damage from undesirable environmental conditions (e.g., frost damage).
biological evolution the accumulation of genetic changes in a species or population over the course of many generations.
biological species concept definition of a species as a group of individuals whose members have the potential to interbreed with one another in nature to produce viable, fertile offspring, but that cannot interbreed successfully with members of other species.
biometric field a field of genetics that involves the statistical study of biological traits.
bioremediation the use of microorganisms to decrease pollutants in the environment.
biotechnology technologies that involve the use of living organisms, or products from living organisms, as a way to benefit humans.
biotransformation the conversion of one molecule into another via cellular enzymes. This term is often used to describe the conversion of a toxic molecule into a nontoxic one.
bivalent a structure in which two pairs of homologous sister chromatids have synapsed (i.e., aligned) with each other.
BLAST (basic local alignment search tool) a computer program that can start with a particular genetic sequence and then locate homologous sequences within a large database.
blending hypothesis of inheritance an early, incorrect hypothesis of heredity. According to this view, the seeds that dictate hereditary traits are able to blend together from generation to generation. The blended traits would then be passed to the next generation.
bottleneck effect a type of genetic drift that occurs when most members of a population are eliminated without any regard to their genetic composition.
box in genetics, a term used to describe a sequence with a specialized function.
branch migration the lateral movement of a Holliday junction.
breakpoint the region where two chromosome pieces break and rejoin with other chromosome pieces.

broad-sense heritability heritability that takes into account all genetic factors.
5-bromodeoxyuridine (BrdU) a base analog that can be incorporated into chromosomes during DNA replication. The presence of this analog can affect the ability of the chromosomes to absorb certain dyes. This is the basis for the staining of harlequin chromosomes.
5-bromouracil a base analog that acts as a chemical mutagen.

C

C an abbreviation for cytosine.
cAMP see *cyclic-AMP.*
cAMP response element (CRE) a short DNA sequence found next to certain eukaryotic genes that is recognized by the cAMP response element-binding (CREB) protein.
cAMP response element-binding protein see *CREB protein.*
cancer a disease characterized by uncontrolled cell division.
cancer cell a cell that has lost its normal growth control. Cancer cells are invasive (i.e., they can invade normal tissues) and metastatic (i.e., they can migrate to other parts of the body).
CAP an abbreviation for the catabolite activator protein, a genetic regulatory protein found in bacteria.
capping the covalent attachment of a 7-methylguanosine nucleotide to the 5′ end of mRNA in eukaryotes.
capsid the protein coat of a virus, which encloses its genome.
CAP site the sequence of DNA that is recognized by CAP.
carbohydrate organic molecules with the general formula $C(H_2O)$. An example of a simple carbohydrate is the sugar glucose. Large carbohydrates are composed of multiple sugar units.
carboxyl-terminus the location of the last amino acid in a polypeptide chain. The amino acid at the carboxyl-terminus still retains a free carboxyl group that is not covalently attached to another amino acid.
carcinogen an agent that can cause cancer.
caspases proteolytic enzymes that play a role in apoptosis.
catabolic enzyme an enzyme that is involved in the breakdown of organic molecules into smaller units.
catabolite activator protein see *CAP.*
catabolite repression the phenomenon in which a catabolite (e.g., glucose) represses the expression of certain genes (e.g., the *lac* operon).
catenane interlocked circular molecules.
cDNA see *complementary DNA.*
cDNA library a DNA library made from a collection of cDNAs.
cell adhesion when the surfaces of cells bind to each other or to the extracellular matrix.
cell adhesion molecule (CAM) a molecule (e.g., surface protein or carbohydrate) that plays a role in cell adhesion.
cell culture refers to the growth of cells in a laboratory.
cell cycle in eukaryotic cells, a series of stages through which a cell progresses in order to divide. The phases are G for growth, S for synthesis (of the genetic material), and M for mitosis. There are two G phases, G_1 and G_2.
cell fate the final morphological features that a cell or group of cells will adopt.
cell-free translation system an experimental mixture that can synthesize polypeptides.
cell fusion the process in which individual cells are mixed together and made to fuse with each other.
cell lineage a series of cells that are descended from a cell or group of cells by cell division.
cell lineage diagram an illustration of the cell division patterns and fates of any cell's descendants.
cell plate the structure that forms between two daughter plant cells that leads to the separation of the cells by the formation of an intervening cell wall.
cellular level refers to an observation or experimentation at the level of individual cells.
cellular trait a trait that is observed at the cellular level. An example is the shape of a cell.
centiMorgans (cM) (same as a map unit) a unit of map distance obtained from genetic crosses. Named in honor of Thomas Hunt Morgan.
central dogma of genetics the idea that the usual flow of genetic information is from DNA to RNA to polypeptide (protein). In addition, DNA replication serves to copy the information so that it can be transmitted from cell to cell and from parent to offspring.
central region in plants, it is the region that creates the stem. It is the radial pattern of cells in the central region that causes the radial growth observed in plants.
central zone in plants, an area in the meristem where undifferentiated stem cells are always maintained.
centrifugation a method to separate cell organelles and macromolecules in which samples are placed in tubes and spun very rapidly. The rate at which particles move toward the bottom of the tube depends on their densities, sizes, shapes, and the viscosity of the medium.
centrifuge a machine that contains a motor, which causes a rotor holding centrifuge tubes to spin very rapidly.
centriole a pair of structures within the centrosome of animal cells.
centromere a segment of eukaryotic chromosomal DNA that provides an attachment site for the kinetochore.
centrosome a cellular structure from which microtubules emanate.
chain termination the stoppage of growth of a DNA strand, RNA strand, or polypeptide sequence.
chaperone a protein that aids in the folding of polypeptides.
character in genetics, a general characteristic such as eye color.
Chargaff's rule the observation that in DNA the amount of A equals T and the amount of G equals C.
charged tRNA a tRNA that has an amino acid attached to its 3′ end by an ester bond.
checkpoint protein a protein that monitors the conditions of DNA and chromosomes and may prevent a cell from progressing through the cell cycle if an abnormality is detected.
chiasma (pl. **chiasmata**) the site where crossing over occurs between two chromosomes. It resembles the Greek letter chi, χ.
chimera an organism composed of cells that are embryonically derived from two different individuals.
ChIP-chip assay a form of chromatin immunoprecipitation that utilizes a microarray to determine where in the genome a particular protein binds.
chi square (χ^2) test a commonly used statistical method for determining the goodness of fit. This method can be used to analyze population data in which the members of the population fall into different categories.
chloroplast DNA (cpDNA) the genetic material found within a chloroplast.
chorionic villus sampling a method for obtaining cellular material from a fetus for the purpose of genetic testing.
chromatid following chromosomal replication in eukaryotes, the two copies that remain attached to each other in the form of sister chromatids.
chromatin the association between DNA and proteins that is found within chromosomes.
chromatin immunoprecipitation (ChIP) a method for determining whether proteins bind to a particular region of DNA. This method analyzes DNA-protein interactions as they occur in the chromatin of living cells.
chromatin immunoprecipitation sequencing (ChIP-Seq) a technique that is used to determine where in a genome a particular protein binds to the DNA.
chromatin remodeling a change in chromatin structure that alters the degree of compaction and/or the spacing and histone composition of nucleosomes.
chromatography a method of separating different macromolecules and small molecules based on their chemical and physical properties. A sample is dissolved in a liquid solvent and exposed to some type of matrix, such as a gel, a column containing beads, or a thin strip of paper.
chromocenter the central point where polytene chromosomes aggregate.
chromomere a dark band within a polytene chromosome.
chromosome the structures within living cells that contain the genetic material. Genes are physically located within the structure of chromosomes. Biochemically, chromosomes contain a very long segment of DNA, which is the genetic material, and proteins, which are bound to the DNA and provide it with an organized structure.
chromosome map a chart that depicts the linear arrangement of genes along a chromosome.
chromosome painting the use of probes to identify particular regions of chromosomes. The probes are usually assigned a computer-generated color.
chromosome territory in the cell nucleus, each chromosome occupies a nonoverlapping region called a chromosome territory.
chromosome theory of inheritance a theory of Sutton and Boveri that the inheritance patterns of traits can be explained by the transmission patterns of chromosomes during gametogenesis and fertilization.
chromosome walking a common method used in positional cloning in which a mapped gene or molecular marker provides a starting point to molecularly "walk" toward a gene of interest via overlapping clones.
***cis*-acting element** a sequence of DNA, such as a regulatory element, that exerts a *cis*-effect.
***cis*-effect** an effect on gene expression due to genetic sequences that are within the same chromosome and often are immediately adjacent to the gene of interest.
***cis*-epigenetic mechanism** an epigenetic mechanism that may affect only one copy of a gene in a cell that has two copies of the gene. It may be caused by DNA methylation or changes in chromatin structure.

cistron the smallest genetic unit that produces a positive result in a complementation experiment. A cistron is equivalent to a gene.
clade a group of species consisting of all descendents of the group's most common ancestor.
cladistic approach a way to construct a phylogenetic tree, also called a cladogram, by considering the various possible pathways of evolution and then choosing the most plausible tree.
cladogenesis during evolution, a form of speciation that involves the division of a species into two or more species.
cladogram a phylogenetic tree that has been constructed using a cladistic approach.
cleavage furrow a constriction that causes the division of two animal cells during cytokinesis.
clonal an adjective to describe a clone. For example, a clonal population of cells is a group of cells that are derived from the same cell.
clone the general meaning of this term is to make many copies of something. In genetics, this term has several meanings: (1) a single cell that has divided to produce a colony of genetically identical cells; (2) an individual that has been produced from a somatic cell of another individual, such as the sheep Dolly; (3) many copies of a DNA fragment that are propagated within a vector or produced by PCR.
closed complex the complex between transcription factors, RNA polymerase, and a promoter before the DNA has denatured to form an open complex.
closed conformation a tightly packed conformation of chromatin that cannot be transcribed.
cluster analysis the analysis of microarray data to determine if certain groups (i.e., clusters) of genes are expressed under the same conditions.
cM an abbreviation for centiMorgans; also see *map unit*.
coactivators proteins that increase the rate of transcription but do not directly bind to the DNA itself.
coding strand the strand in DNA that is not used as a template for mRNA synthesis.
codominance a pattern of inheritance in which two alleles are both expressed in the heterozygous condition. For example, a person with the genotype I^AI^B has the blood type AB and expresses both surface antigens A and B.
codon a sequence of three nucleotides in mRNA that functions in translation. A start codon, which usually specifies methionine, initiates translation, and a stop codon terminates translation. The other codons specify the amino acids within a polypeptide sequence according to the genetic code.
codon bias in a given species, the phenomenon in which certain codons are used more frequently than others.
coefficient of inbreeding (F) see *inbreeding coefficient*.
cohesin a protein complex that facilitates the alignment of sister chromatids.
colinearity the correspondence between the sequence of codons in the DNA coding strand and the amino acid sequence of a polypeptide.
colony hybridization a technique in which a probe is used to identify bacterial colonies that contain a hybrid vector with a gene of interest.
combinatorial control the phenomenon common in eukaryotes in which the combination of many factors determines the expression of any given gene.
common ancestor someone who is an ancestor to both of an individual's parents.
comparative genomic hybridization (CGH) a hybridization technique to determine if cells (e.g., cancer cells) have changes in chromosome structure, such as deletions or duplications.
comparative genomics using information from genome projects to understand the genetic variation between different populations and evolutionary relationships among different species.
competence factors proteins that are needed for bacterial cells to become naturally transformed by extracellular DNA.
competence-stimulating peptide (CSP) a peptide secreted by certain species of bacteria that allows them to become competent for transformation.
competent cells cells that can be transformed by extracellular DNA.
complementary describes sequences in two DNA strands that match each other according to the AT/GC rule. For example, if one strand has the sequence of ATGGCGGATTT, then the complementary strand must be TACCGCCTAAA.
complementary DNA (cDNA) DNA that is made from an RNA template by the action of reverse transcriptase.
complementation a phenomenon in which the presence of two different mutant alleles in the same organism produces a wild-type phenotype. It usually happens because the two mutations are in different genes, so the organism carries one copy of each mutant allele and one copy of each wild-type allele.
complementation test an experimental procedure in which the goal is to determine if two different mutations that affect the same trait are in the same gene or in two different genes.
complete nondisjunction during meiosis or mitosis, when all of the chromosomes fail to disjoin and remain in one of the two daughter cells.
complete transposable element see *autonomous transposable element.*
complex traits characteristics that are determined by several genes and are significantly influenced by environmental factors.
composite transposon a transposon that contains additional genes, such as antibiotic resistance genes, that are not necessary for transposition per se.
computer data file a file (a collection of information) stored by a computer.
computer program a series of operations that can analyze data in a defined way.
concordance in genetics, the degree to which pairs of individuals (e.g., identical twins or fraternal twins) exhibit the same trait.
condensation a change in chromatin structure to become more compact.
condense refers to chromosomes forming a more compact structure.
condensin a protein complex that plays a role in the condensation of interphase chromosomes to become metaphase chromosomes.
conditional alleles alleles in which the phenotypic expression depends on the environmental conditions. An example is temperature-sensitive alleles, which affect the phenotype only at a particular temperature.
conditional lethal allele an allele that is lethal, but only under certain environmental conditions.
conditional mutant a mutant whose phenotype depends on the environmental conditions, such as a temperature-sensitive mutant.
conglomerate a population composed of members of an original population plus new members that have migrated from another population.
conjugation a form of genetic transfer between bacteria that involves direct physical interaction between two bacterial cells. One bacterium acts as donor and transfers genetic material to a recipient cell.
conjugation bridge a connection between two bacterial cells that provides a passageway for DNA during conjugation.
conjugative plasmid a plasmid that can be transferred to a recipient cell during conjugation.
consensus sequence the most commonly occurring bases within a sequence element.
conservative model an incorrect model in which both strands of parental DNA remain together following DNA replication.
conservative transposition see *simple transposition.*
constitutive exon an exon that is always found in mRNA following splicing.
constitutive gene a gene that is not regulated and has essentially constant levels of expression over time.
constitutive heterochromatin regions of chromosomes that are always heterochromatic and are permanently transcriptionally inactive.
contig a series of clones that contain overlapping pieces of chromosomal DNA.
continuous traits quantitative traits that fall along a continuum rather than occurring in discrete categories.
control element see *regulatory sequence* or *element.*
coordinate regulation a group of genes that are regulated in the same way.
copy number variation (CNV) a type of structural variation in which a segment of DNA that is 1000 bp or more in length commonly exhibits copy number differences among members of the same species.
core enzyme the subunits of an enzyme that are needed for catalytic activity, as in the core enzyme of RNA polymerase.
corepressor a small effector molecule that binds to a repressor protein, thereby causing the repressor protein to bind to DNA and inhibit transcription.
core promoter a DNA sequence that is absolutely necessary for transcription to take place. It provides the binding site for general transcription factors and RNA polymerase.
correlation coefficient (*r*) a statistic with a value that ranges between -1 and 1. It describes how two factors vary relative to each other.
cosmid a vector that is a hybrid between a plasmid vector and phage λ. Cosmid DNA can replicate in a cell like a plasmid or be packaged into a protein coat like a phage. Cosmid vectors can accept fragments of DNA that are typically tens of thousands of base pairs in length.
cotransduction the phenomenon in which a bacteriophage transfers a piece of DNA carrying two closely linked genes.
cotransformation the phenomenon in which bacterial transformation transfers a piece of DNA carrying two closely linked genes.
cotranslational events that occur during translation.
cotyledon a food-storing structure that will become the first leaves of a seedling.
covalent histone modification the covalent attachment of functional groups onto histone proteins. These modifications may affect chromatin structure and constitute a histone code.
covariance a statistic that describes the degree of variation between two variables within a group.
cpDNA an abbreviation for chloroplast DNA.

CpG island a group of CG sequences that may be clustered near a promoter region of a gene. The methylation of the cytosine bases usually inhibits transcription.
CREB protein (cAMP response element-binding protein) a regulatory transcription factor that becomes activated in response to specific cell-signaling molecules that cause the synthesis of cAMP.
cross a mating between two distinct individuals. An analysis of their offspring may be conducted to understand how traits are passed from parent to offspring.
cross-fertilization same meaning as *cross.* It requires that the male and female gametes come from separate individuals.
crossing over a physical exchange of chromosome pieces that most commonly occurs during prophase of meiosis I.
C-terminus see *carboxyl-terminus.*
cycle threshold method (C_t method) in quantitative PCR, a method of determining the starting amount of DNA based on a threshold level at which the accumulation of fluorescence is significantly greater than the background level.
cyclic-AMP (cAMP) in bacteria, a small effector molecule that binds to CAP (catabolite activator protein). In eukaryotes, cAMP functions as a second messenger in a variety of intracellular signaling pathways.
cyclin a type of protein that plays a role in the regulation of the eukaryotic cell cycle.
cyclin-dependent protein kinases (CDKs) enzymes that are regulated by cyclins and can phosphorylate other cellular proteins by covalently attaching a phosphate group.
cytogeneticist a scientist who studies chromosomes under the microscope.
cytogenetic mapping the mapping of genes or genetic sequences using microscopy.
cytogenetics the field of genetics that involves the microscopic examination of chromosomes.
cytokinesis the division of a single cell into two cells. The two nuclei produced in M phase are segregated into separate daughter cells during cytokinesis.
cytological mapping see *cytogenetic mapping.*
cytoplasmic inheritance (also known as *extranuclear inheritance*) the inheritance of genetic material that is not found within the cell nucleus.
cytosine a pyrimidine base found in DNA and RNA. It base-pairs with guanine in DNA.

D

Dam see *DNA adenine methyltransferase.*
Darwinian fitness the relative likelihood that a genotype will survive and contribute to the gene pool of the next generation as compared with other genotypes.
database a computer storage facility that stores many data files, such as those containing genetic sequences.
daughter strand in DNA replication, the newly made strand of DNA.
deamination the removal of an amino group from a molecule. For example, the removal of an amino group from cytosine produces uracil.
decoding function the ability of the 16S rRNA to detect when an incorrect tRNA is bound at the A site and prevent elongation until the mispaired tRNA is released from the A site.
decondensed chromosomes that form a less compact structure.
deficiency condition in which a segment of chromosomal material is missing.
degeneracy in genetics, this term means that more that one codon specifies the same amino acid. For example, the codons GGU, GGC, GGA, and GGG all specify the amino acid glycine.
degrees of freedom in a statistical analysis, the number of categories that are independent of each other.
deleterious mutation a mutation that is detrimental with regard to its effect on phenotype.
deletion condition in which a segment of DNA is missing.
deletion mapping the use of strains carrying deletions within a defined region to map a mutation of unknown location.
deme see *local population.*
de novo methylation the methylation of DNA that has not been previously methylated. This is usually a highly regulated event.
deoxyribonucleic acid (DNA) the genetic material. It is a double-stranded structure, with each strand composed of repeating units of deoxyribonucleotides.
deoxyribose the sugar found in DNA.
depurination the removal of a purine base from DNA.
determination the process in which a cell or region of an organism has adopted a particular fate that will occur later in development.
determined cell a cell that is destined to differentiate into a specific cell type.
development in multicellular species, the series of genetically programmed stages in which a fertilized egg becomes an embryo and eventually develops into an adult.
developmental genetics the area of genetics concerned with the roles of genes in orchestrating the changes that occur during development.
diakinesis the fifth stage of prophase of meiosis I.
diauxic growth the sequential use of two sugars by a bacterium.
dicentric describes a chromosome with two centromeres.
dicentric bridge the region between the two centromeres in a dicentric chromosome.
dicer an endonuclease that makes a cut in double-stranded RNA.
dideoxyribonucleotide (ddNTP) a nucleotide used in DNA sequencing that is missing the 3′—OH group. If a dideoxyribonucleotide is incorporated into a DNA strand, it stops any further growth of the strand.
dideoxy sequencing a method of DNA sequencing that uses dideoxyribonucleotides to terminate the growth of DNA strands.
differential centrifugation a form of centrifugation involving a series of centrifugation steps in which the supernatant or pellet is used in each subsequent centrifugation step.
differentially methylated region (DMR) in the case of imprinting, a site that is methylated during spermatogenesis or oogenesis, but not both.
differentiated cell a cell that has become a specialized type of cell within a multicellular organism.
dihybrid testcross a cross in which an experimenter crosses an individual that is heterozygous for two genes to an individual that is homozygous recessive for the same two genes.
dimeric DNA polymerase a complex of two DNA polymerase proteins that move as a unit during DNA replication.
dioecious in plants, a species in which male and female gametophytes are produced on a single (sporophyte) individual.
diploid an organism or cell that contains two copies of each type of chromosome.
diplotene the fourth stage of prophase of meiosis I.
directionality in DNA and RNA, refers to the 5′ to 3′ arrangement of nucleotides in a strand; in proteins, refers to the linear arrangement of amino acids from the N-terminal to C-terminal ends.
directional selection natural selection that favors an extreme phenotype. This usually leads to the fixation of the favored allele.
direct repeat (DR) short DNA sequences that flank transposable elements in which the DNA sequence is repeated in the same direction.
discontinuous trait a trait in which each offspring can be put into a particular phenotypic category.
discovery-based science experimentation that does not require a preconceived hypothesis. In some cases, the goal is to collect data to be able to formulate a hypothesis.
disequilibrium in population genetics, refers to a population that is not in Hardy-Weinberg equilibrium.
dispersive model an incorrect model for DNA replication in which segments of parental DNA and newly made DNA are interspersed in both strands following the replication process.
disruptive selection natural selection that favors both extremes of a phenotypic category. This results in a balanced polymorphism.
dizygotic twins also known as fraternal twins; twins formed from separate pairs of sperm and egg cells.
DNA see *deoxyribonucleic acid.*
DnaA box sequence serves as a recognition site for the binding of the DnaA protein, which is involved in the initiation of bacterial DNA replication.
DNA adenine methyltransferase (Dam) an enzyme in bacteria that attaches methyl groups to the adenine base in DNA that is found within the sequence GATC.
DnaA protein a protein that binds to the dnaA box sequence at the origin of replication in bacteria and initiates DNA replication.
DNA fingerprinting a technology for identifying a particular individual based on the properties of his or her DNA.
DNA gap repair synthesis the synthesis of DNA in a region where a DNA strand has been previously removed, usually by a DNA repair enzyme or by an enzyme involved in homologous recombination.
DNA gyrase also known as topoisomerase II; an enzyme that introduces negative supercoils into DNA using energy from ATP. Gyrase can also relax positive supercoils when they occur.
DNA helicase an enzyme that separates the two strands of DNA.
DNA library a collection of many hybrid vectors, each carrying a particular fragment of DNA from a larger source. For example, each hybrid vector in a DNA library might carry a small segment of chromosomal DNA from a particular species.
DNA ligase an enzyme that catalyzes a covalent bond between two DNA fragments.
DNA methylation the phenomenon in which an enzyme covalently attaches a methyl group ($—CH_3$) to a base (usually adenine or cytosine) in DNA.
DNA methyltransferase the enzyme that attaches methyl groups to adenine or cytosine bases.

DNA microarray a small silica, glass, or plastic slide that is dotted with many different sequences of DNA, corresponding to short sequences within known genes.

DNA-*N*-glycosylase an enzyme that can recognize an abnormal base and cleave the bond between it and the sugar in the DNA backbone.

DNA polymerase an enzyme that catalyzes the covalent attachment of nucleotides together to form a strand of DNA.

DNA primase an enzyme that synthesizes a short RNA primer for DNA replication.

DNA probe in a hybridization experiment, a single-stranded DNA fragment with a base sequence that is complementary to a gene of interest. The DNA probe is labeled to detect a gene of interest.

DNA profiling see *DNA fingerprinting.*

DNA replication the process in which original DNA strands are used as templates for the synthesis of new DNA strands.

DNA replication licensing in eukaryotes, occurs when MCM helicase is bound at an origin, enabling the formation of two replication forks.

DNA translocase an ATPase that moves along the DNA. It is a component of chromatin-remodeling complexes.

DNase an enzyme that cuts the sugar-phosphate backbone in DNA.

DNase I an endonuclease that cleaves DNA.

DNase I footprinting a method for studying protein-DNA interactions in which the binding of a protein to DNA protects the DNA from digestion by DNase I.

DNase I sensitivity the phenomenon in which DNA is in an open conformation and able to be digested by DNase I.

DNA sequencing a method for determining the base sequence in a segment of DNA.

DNA supercoiling the formation of additional coils in DNA due to twisting forces.

DNA uptake signal sequences DNA sequences found in certain species of bacteria that are needed for a DNA fragment to be taken up during transformation.

domain a segment of a protein that has a specific function.

dominant describes an allele that determines the phenotype in the heterozygous condition. For example, if a plant is *Tt* and has a tall phenotype, the *T* (tall) allele is dominant over the *t* (dwarf) allele.

dominant-negative mutation a mutation that produces an altered gene product that acts antagonistically to the normal gene product. Shows a dominant pattern of inheritance.

dorsoventral axis in animals, the axis from the upper side of the back (e.g., the spine in humans) to the opposite side (e.g., the stomach in humans).

dosage compensation the phenomenon that in species with sex chromosomes, one of the sex chromosomes is altered so that males and females have similar levels of gene expression, even though they do not contain the same complement of sex chromosomes.

double helix the arrangement in which two strands of DNA (and sometimes RNA) interact with each other to form a double-stranded helical structure.

double-strand break model a model for homologous recombination in which the event that initiates recombination is a double-strand break in one of the double helices.

down promoter mutation a mutation in a promoter that inhibits the rate of transcription.

down regulation genetic regulation that leads to a decrease in gene expression.

duplication a segment of DNA that is repeated more than once within a genome.

dyad see *sister chromatids.*

dynamic mutation a type of mutation in which a repeated sequence, such as three nucleotides, can readily increase in number from one generation to the next.

E

ecological species concept a species concept in which each species occupies an ecological niche, which is the unique set of habitat resources that a species requires, as well as the species' influence on the environment and other species.

ecotypes genetically distinct populations that are adapted to their local environments.

editosome a complex that catalyzes RNA editing.

egg cell also known as an ovum; it is a female gamete that is usually very large and nonmotile.

electrophoresis the migration of ions or charged molecules in response to an electric field.

electroporation the use of electric current that creates transient pores in the plasma membrane of a cell to allow entry of DNA.

elongation (1) in transcription, the synthesis of RNA using DNA as a template; (2) in translation, the synthesis of a polypeptide using the information within mRNA.

embryogenesis an early stage of animal and plant development leading to the production of an embryo with organized tissue layers and a body plan organization.

embryonic germ cell (EG cell) a type of pluripotent stem cell found in the gonads of the fetus.

embryonic stem cell (ES cell) a type of pluripotent stem cell found in the early blastocyst.

embryo sac in flowering plants, the female gametophyte that contains an egg cell.

emerging virus a virus that has arisen recently and that is more likely to cause infection.

empirical approach a strategy in which experiments are designed to determine quantitative relationships as a way to derive laws that govern biological, chemical, or physical phenomena.

endonuclease an enzyme that can cut in the middle of a DNA strand.

endopolyploidy in a diploid individual, the phenomenon in which certain cells of the body may be polyploid.

endosperm in flowering plants, the material in the seed, which is $3n$, that nourishes the developing embryo.

endosymbiosis a symbiotic relationship in which the symbiont actually lives inside (*endo*) the larger of the two species.

endosymbiosis theory the theory that the ancient origin of plastids and mitochondria was the result of certain species of bacteria taking up residence within a primordial eukaryotic cell.

enhancer a DNA sequence that functions as a regulatory element. The binding of a regulatory transcription factor to the enhancer increases the level of transcription.

environment the surroundings an organism experiences.

enzyme a protein that functions to accelerate chemical reactions within the cell.

enzyme adaptation the phenomenon in which a particular enzyme appears within a living cell after the cell has been exposed to the substrate for that enzyme.

epigenetic inheritance an epigenetic change that is passed from parent to offspring.

epigenetics the study of mechanisms that lead to changes in gene expression that can be passed from cell to cell and are reversible, but do not involve a change in the DNA sequence.

episome a segment of bacterial DNA that can exist as an F factor and also integrate into the chromosome.

epistasis an inheritance pattern where one gene can mask the phenotypic effects of a different gene.

epitope the structure on the surface of an antigen that is recognized by an antibody.

equilibrium density centrifugation a form of centrifugation in which the particles sediment through the gradient, reaching a position where the density of the particle matches the density of the solution.

error-prone replication a form of DNA replication carried out by lesion-replicating DNA polymerases that results in a high rate of mutation.

E site see *exit site.*

essential gene a gene that is essential for survival.

ethyl methanesulfonate (EMS) a type of chemical mutagen that alkylates bases (i.e., attaches methyl or ethyl groups).

eubacteria see *Bacteria.*

euchromatin DNA that is not highly compacted and may be transcriptionally active.

Eukarya see *Eukaryotes.*

Eukaryotes one of the three domains of life. A defining feature of these organisms is that their cells contain nuclei bounded by cell membranes. Some simple eukaryotic species are single-celled protists and yeast; more complex multicellular species include fungi, plants, and animals.

euploid describes an organism in which the chromosome number is an exact multiple of a chromosome set.

E-value with regard to the BLAST program, the number of times that the match or a better one would be expected to occur purely by random chance in a search of an entire database.

evolution see *biological evolution.*

evolutionary species concept a species concept in which each species is defined based on the separate evolution of lineages.

excisionase an enzyme that excises a prophage from a host cell's chromosome.

exit site (E site) a site on the ribosome from which an uncharged tRNA exits.

exon a segment of RNA that is contained within the RNA after splicing has occurred. In mRNA, the coding sequence of a polypeptide is contained within the exons.

exon shuffling the phenomenon in which exons are transferred between different genes during evolution. Some researchers believe that transposable elements have played a role in this phenomenon.

exon skipping when an exon is spliced out of a pre-mRNA.

exonuclease an enzyme that digests an RNA or DNA strand from the end.

expression vector a cloning vector that contains a promoter so that the gene of interest is transcribed into RNA when the vector is introduced into a host cell.

expressivity the degree to which a trait is expressed. For example, flowers with deep red color have a high expressivity of the red allele.

extranuclear inheritance (also known as *cytoplasmic inheritance*) the inheritance of genetic material that is not found within the nucleus.

ex vivo approach in the case of gene therapy, refers to genetic manipulations that occur outside the body.

F

F see *inbreeding coefficient.*
facultative heterochromatin heterochromatin that is derived from the conversion of euchromatin to heterochromatin.
fate map a diagram that depicts how cell division proceeds in an organism.
feedback inhibition the phenomenon in which the final product of a metabolic pathway inhibits an enzyme that acts early in the pathway.
feedback loop a mechanism in which a gene that encodes a transcription is regulated by that same transcription factor.
fertilization the union of gametes (e.g., sperm and egg) to begin the life of a new organism.
F factor a fertility factor found in certain strains of bacteria in addition to their circular chromosome. Strains of bacteria that contain an F factor are designated F^+; strains without F factors are F^-.
F′ factors an F factor that also carries genes derived from the bacterial chromosome.
F_1 generation the offspring produced from a cross of the parental generation.
F_2 generation the offspring produced from a cross of the F_1 generation.
fidelity a term used to describe the accuracy of a process. If there are few mistakes, a process has a high fidelity.
fine-structure mapping also known as intragenic mapping; the aim of fine-structure mapping is to ascertain the distances between two (or more) different mutations within the same gene.
fitness see *Darwinian fitness.*
flap endonuclease an enzyme that removes small pieces of RNA flaps that are generated by the action of DNA polymerase δ. It removes RNA primers during DNA replication.
fluorescence in situ hybridization (FISH) a form of in situ hybridization in which the probe is fluorescently labeled.
focus with regard to cancer cells, it means a clump of raised cells that grow without regard to contact inhibition.
footprinting see *DNase I footprinting.*
fork see *replication fork.*
forked-line method a method to solve independent assortment problems in which lines are drawn to connect particular genotypes.
forward genetics the traditional method of genetics in which mutants are first identified by their effect on phenotype, and then the mutant gene is cloned.
forward mutation a mutation that changes the wild-type genotype into some new variation.
founder with regard to genetic diseases, an individual who lived many generations ago and was the person in which a genetic disease originated.
founder effect changes in allele frequencies that occur when a small group of individuals separates from a larger population and establishes a colony in a new location.
fraction (1) following centrifugation, a portion of the liquid contained within a centrifuge tube; (2) following column chromatography, a portion of the liquid that has been eluted from a column.
frameshift mutation a mutation that involves the addition or deletion of nucleotides not in a multiple of three and thereby shifts the reading frame of the codon sequence downstream from the mutation.
frequency distribution a graph that describes the numbers of individuals that are found in each of several phenotypic categories.
functional genomics the study of gene function at the genome level. It involves the study of many genes simultaneously.
functional protein microarray a type of protein microarray that monitors a particular kind of protein function, such as the ability to bind a specific drug.

G

G an abbreviation for guanine.
gain-of-function mutation a mutation that causes a gene to be expressed in an additional place where it is not normally expressed or during a stage of development when it is not normally expressed.
gamete a reproductive cell (usually haploid) that can unite with another reproductive cell to create a zygote. Sperm and egg cells are types of gametes.
gametogenesis the production of gametes (e.g., sperm or egg cells).
gametophyte the haploid generation of plants.
gap gene one category of segmentation genes.
G bands the chromosomal banding pattern that is observed when the chromosomes have been treated with the chemical dye Giemsa.
gel electrophoresis a method that combines chromatography and electrophoresis to separate molecules and macromolecules. Samples are loaded into wells at one end of the gel, and an electric field is applied across the gel that causes charged molecules to migrate from one side of the gel to the other.
gel mobility shift assay see *gel retardation assay.*
gel retardation assay a technique for studying protein-DNA interactions in which the binding of protein to a DNA fragment retards its mobility during gel electrophoresis.
gene a unit of heredity that may influence the outcome of an organism's traits. At the molecular level, a gene contains the information to make a functional product, either RNA or protein.
gene addition the addition of a cloned gene into a site in a chromosome of a living cell.
gene amplification an increase in the copy number of a gene.
gene chip see *DNA microarray.*
gene cloning the production of many copies of a gene using molecular methods, such as PCR or the introduction of a gene into a vector that replicates in a host cell.
gene conversion the phenomenon in which one allele is converted to another due to genetic recombination and DNA repair.
gene dosage effect the phenomenon when the number of copies of a gene affects the phenotypic expression of a trait.
gene duplication an increase in the copy number of a gene. Can lead to the evolution of gene families.
gene expression the process in which the information within a gene is accessed, first to synthesize RNA and usually proteins, and eventually to affect the phenotype of the organism.
gene family two or more different genes within a single species that are homologous to each other because they were derived from the same ancestral gene.
gene flow changes in allele frequencies due to migration.
gene interaction when two or more different genes influence the outcome of a single trait.
gene knockin a type of gene addition in which a gene of interest has been added to a noncritical site in the mouse genome.
gene knockout when both copies of a normal gene have been replaced by an inactive mutant gene.
gene modifier effect when the allele of one gene modifies the phenotypic effect of the allele of a different gene.
gene mutation a relatively small mutation that affects only a single gene.
gene pool the totality of all genes within a particular population.
gene rearrangement a rearrangement in segments of a gene, as occurs in antibody precursor genes.
gene redundancy the phenomenon in which an inactive gene is compensated for by another gene with a similar function.
gene regulation the phenomenon in which the level of gene expression can vary under different conditions.
gene replacement the swapping of a cloned gene made experimentally with a normal chromosomal gene found in a living cell.
gene therapy the introduction of cloned genes into living cells in an attempt to cure or alleviate disease.
generalized transduction a form of transduction in which any piece of the bacterial chromosomal DNA can be incorporated into a phage.
general lineage concept a species concept which states that each species is a population of an independently evolving lineage.
general transcription factor (GTF) one of several proteins that are necessary for basal transcription at the core promoter.
genetically modified organism (GMO) an organism that has received genetic material via recombinant DNA technology.
genetic approach in research, the study of mutant genes that have abnormal function. By studying mutant genes, researchers may better understand normal genes and normal biological processes.
genetic code the correspondence between a codon (i.e., a sequence of three bases in an mRNA molecule) and the functional role that the codon plays during translation. Each codon specifies a particular amino acid or the end of translation.
genetic cross a mating between two individuals and the analysis of their offspring in an attempt to understand how traits are passed from parent to offspring.
genetic drift random changes in allele frequencies due to sampling error.
genetic linkage see *linkage.*
genetic linkage map see *genetic map.*
genetic map a chart that describes the relative locations of genes or other DNA segments along a chromosome.
genetic mapping any method used to determine the linear order of genes as they are linked to each other along the same chromosome. This term is also used to describe the use of genetic crosses to determine the linear order of genes. See also *linkage mapping.*
genetic marker any genetic sequence that is used to mark a specific location on a chromosome.
genetic mosaic see *mosaicism.*
genetic polymorphism when two or more alleles occur in population; each allele is found at a frequency of 1% or higher.
genetic recombination (1) the process in which chromosomes are broken and then rejoined to form a novel genetic combination; (2) the process in which alleles are assorted and passed to offspring in combinations that are different from the parents.

genetics the study of heredity; the branch of biology that deals with heredity and variation.
genetic screening the use of testing methods at the population level to determine if individuals are heterozygous carriers for or have a genetic disease.
genetic testing the analysis of individuals with regard to their genes or gene products. In many cases, the goal is to determine if an individual carries a mutant gene.
genetic transfer the physical transfer of genetic material from one bacterial cell to another.
genetic variation genetic differences among members of the same species or among different species.
genome all of the chromosomes and DNA sequences that an organism or species can possess.
genome database a database that focuses on the genetic sequences and characteristics of a single species.
genome maintenance refers to cellular mechanisms that prevent mutations from occurring and/or prevent mutant cells from surviving or dividing.
genome-sequencing projects research endeavors that have the ultimate goal of determining the sequence of DNA bases of the entire genome of a given species.
genomic clone a clone made from the digestion and cloning of chromosomal DNA.
genomic imprinting a pattern of inheritance that involves a change in a single gene or chromosome during gamete formation. Depending on whether the modification occurs during spermatogenesis or oogenesis, imprinting governs whether an offspring will express a gene that has been inherited from its mother or father.
genomic library a DNA library made from chromosomal DNA fragments.
genomics the molecular analysis of the entire genome of a species.
genotype the genetic composition of an individual, especially in terms of the alleles for particular genes.
genotype-environment association when certain genotypes are preferentially found in particular environments.
genotype-environment interaction when the environmental effects on phenotype differ according to genotype.
genotype frequency the number of individuals with a particular genotype in a population divided by the total number of individuals in the population.
germ cells the gametes (i.e., sperm and egg cells).
germ line a lineage of cells that gives rise to gametes.
germ-line mutation a mutation in a cell of the germ line.
GloFish genetically modified aquarium fish that glow due to the introduction of genes that encode fluorescent proteins.
glucocorticoid receptor a type of steroid receptor that functions as a regulatory transcription factor.
goodness of fit the degree to which the observed data and expected data are similar to each other. If the observed and predicted data are very similar, the goodness of fit is high.
gradualism an evolutionary hypothesis suggesting that each new species evolves continuously over long spans of time. The principal idea is that large phenotypic differences that cause the divergence of species are due to the accumulation of many small genetic changes.
grande normal (large-sized) yeast colonies.
grooves in DNA, the indentations where the atoms of the bases are in contact with the surrounding water. In B DNA, there is a smaller minor groove and a larger major groove.
group I intron a type of intron found in self-splicing RNA that uses free guanosine in its splicing mechanism.
group II intron a type of intron found in self-splicing RNA that uses an adenine nucleotide within the intron itself in its splicing mechanism.
growth factors protein factors that influence cell division.
guanine a purine base found in DNA and RNA. It base-pairs with cytosine in DNA.
gyrase see *DNA gyrase.*

H

haplodiploid a species, such as certain bees, in which one sex is haploid (e.g., male) and the other sex is diploid (e.g., female).
haploid describes the phenomenon that gametes contain half the genetic material found in somatic cells. For a species that is diploid, a haploid gamete contains a single set of chromosomes.
haploinsufficiency the phenomenon in which a person has only a single functional copy of a gene, and that single functional copy does not produce a normal phenotype. Shows a dominant pattern of inheritance.
haplotype the linkage of particular alleles or molecular markers along a single chromosome.
HapMap a catalog of common human genetic variations.
Hardy-Weinberg equation $p^2 + 2pq + q^2 = 1$.
Hardy-Weinberg equilibrium the phenomenon by which, under certain conditions, allele frequencies are maintained in a stable condition and genotypes can be predicted according to the Hardy-Weinberg equation.
helicase see *DNA helicase.*
helix-loop-helix domain a domain found in transcription factors that enables them to dimerize.
helix-turn-helix motif a structure found in transcription factor proteins that promotes binding to the major groove of DNA.
hemizygous describes the single copy of an X-linked gene in the male. A male mammal is said to be hemizygous for X-linked genes.
heritability the amount of phenotypic variation within a particular group of individuals that is due to genetic factors.
heterochromatin highly compacted DNA. It is usually transcriptionally inactive.
heterochronic mutation a mutation that alters the timing of expression of a gene and thereby alters the outcome of cell fates.
heterodimer when two polypeptides encoded by different genes bind to each other to form a dimer.
heteroduplex a double-stranded region of DNA that contains one or more base mismatches.
heterogametic sex in species with two types of sex chromosomes, the heterogametic sex produces two types of gametes. For example, in mammals, the male is the heterogametic sex, because a sperm can contain either an X or a Y chromosome.
heterogamous describes a species that produces two morphologically different types of gametes (i.e., sperm and eggs).
heterogeneity see *locus heterogeneity.*
heterogeneous nuclear RNA (hnRNA) same as pre-mRNA.
heteroplasmy when a cell contains variation in a particular type of organelle. For example, a plant cell could contain some chloroplasts that make chlorophyll and other chloroplasts that do not.
heterosis the phenomenon in which hybrids display traits superior to either corresponding parental strain. Heterosis is usually different from overdominance, because the hybrid may be heterozygous for many genes, not just a single gene, and because the superior phenotype may be due to the masking of deleterious recessive alleles.
heterozygote an individual who is heterozygous.
heterozygote advantage see *overdominance.* A pattern of inheritance in which a heterozygote has a higher Darwinian fitness compared with either of the corresponding homozygotes.
heterozygous describes a diploid individual that has different versions (i.e., two different alleles) of the same gene.
Hfr **strain** (for **high frequency of recombination**) a bacterial strain in which an F factor has become integrated into the bacterial chromosome. During conjugation, an *Hfr* strain can transfer segments of the bacterial chromosome.
highly repetitive sequences sequences that are found tens of thousands or even millions of times throughout the genome.
high stringency refers to highly selective hybridization conditions that promote the binding of DNA or RNA fragments that are perfect or almost perfect matches.
high-throughput sequencing the ability to sequence large amounts of DNA in a short period of time. It usually involves the sequencing of many samples at the same time.
highly repetitive sequences base sequences that are found tens of thousands or even millions of times throughout a genome.
histone acetyltransferase an enzyme that attaches acetyl groups to the amino terminal tails of histone proteins.
histone code hypothesis the hypothesis that the pattern of histone modification acts much like a language or code in specifying alterations in chromatin structure.
histone deacetylase an enzyme that removes acetyl groups from the amino terminal tails of histone proteins.
histone variants histone proteins whose amino acid sequences are slightly different from the standard histones. They often play a specialized role in chromatin structure and function.
histones a group of proteins involved in forming the nucleosome structure of eukaryotic chromatin.
hnRNA an abbreviation for heterogeneous nuclear RNA.
holandric gene a gene on the Y chromosome.
Holliday junction a site where an unresolved crossover has occurred between two homologous chromosomes.
Holliday model a model to explain the molecular mechanism of homologous recombination.
holoenzyme an enzyme containing all of its subunits, as in the holoenzyme of RNA polymerase that has σ factor along with the core enzyme.
homeobox a 180-bp consensus sequence found in homeotic genes.
homeodomain the protein domain encoded by the homeobox. The homeodomain promotes the binding of the protein to the DNA.
homeologous describes the homologous chromosomes from closely related species.

homeotic an adjective that was originally used to describe mutants in which one body part is replaced by another.
homeotic gene a gene that functions in governing the developmental fate of a particular region of the body.
homoallelic two or more alleles in different organisms that are due to mutations at exactly the same base pair within a gene.
homodimer when two polypeptides encoded by the same gene bind to each other to form a dimer.
homogametic sex in species with two types of sex chromosomes, the homogametic sex produces only one type of gamete. For example, in mammals, the female is the homogametic sex, because an egg can only contain an X chromosome.
homolog one of the chromosomes in a pair of homologous chromosomes.
homologous in the case of genes, this term describes two genes that are derived from the same ancestral gene. Homologous genes have similar DNA sequences. In the case of chromosomes, the two homologs of a chromosome pair are said to be homologous to each other.
homologous recombination the exchange of DNA segments between homologous chromosomes.
homologous recombination repair (HRR) also called homology-directed repair, occurs when the DNA strands from a sister chromatid are used to repair a lesion in the other sister chromatid.
homology structures that are similar to each other because they evolved from a common ancestor.
homology-based modeling the modeling of a three-dimensional structure of a molecule based on its homology to another molecule whose structure is already known.
homozygous describes a diploid individual that has two identical alleles of a particular gene.
horizontal gene transfer the transfer of genes from one individual to another individual that is not its offspring.
host cell a cell that is infected with a virus or bacterium.
host range the spectrum of host species that a virus or other pathogen can infect.
hot spots sites within a gene that are more likely to be mutated than other locations.
housekeeping gene a gene that encodes a protein required in most cells of a multicellular organism.
***Hox* complexes** a group of several *Hox* genes located in a particular chromosomal region.
***Hox* genes** mammalian genes that play a role in development. They are homologous to homeotic genes found in *Drosophila*.
Human Genome Project a worldwide collaborative project that provided a detailed map of the human genome and obtained a complete DNA sequence of the human genome.
human immunodeficiency virus (HIV) an enveloped animal virus containing two copies of single-stranded RNA that is responsible for acquired immunodeficiency syndrome (AIDS).
hybrid (1) an offspring obtained from a hybridization experiment; (2) a cell produced from a cell fusion experiment in which the two separate nuclei have fused to make a single nucleus.
hybrid dysgenesis a syndrome involving defective *Drosophila* offspring, due to the phenomenon that P elements can transpose freely.
hybridization (1) the mating of two organisms of the same species with different characteristics; (2) the phenomenon in which two single-stranded molecules renature together to form a hybrid molecule.
hybrid vector a cloning vector that contains an insert of foreign DNA.
hybrid vigor see *heterosis.*
hybrid zones during parapatric speciation, places where two populations can interbreed.
hypothesis testing using statistical tests to determine if the data from experimentation are consistent with a hypothesis.

I

ICR see *imprinting control region.*
illegitimate recombination see *nonhomologous recombination.*
immunoglobulin (IgG) see *antibodies.*
immunoprecipitation the use of antibodies to cause other molecules, such as proteins, to precipitate, which allows them to be collected by centrifugation.
imprinting see *genomic imprinting.*
imprinting control region (ICR) a DNA region that is differentially methylated and plays a role in genomic imprinting.
inborn error of metabolism a genetic disease that involves a defect in a metabolic enzyme.
inbreeding the practice of mating between genetically related individuals.
inbreeding coefficient (F) the probability that two alleles in a particular individual will be identical for a given gene because both copies are due to descent from a common ancestor.
inbreeding depression the phenomenon in which inbreeding produces homozygotes that are less fit, thereby decreasing the reproductive success of a population.
incomplete dominance a pattern of inheritance in which a heterozygote that carries two different alleles exhibits a phenotype that is intermediate to the corresponding homozygous individuals. For example, an *Rr* heterozygote may be pink, whereas the *RR* and *rr* homozygotes are red and white, respectively.
incomplete penetrance a pattern of inheritance in which a dominant allele does not always control the phenotype of the individual.
incomplete transposable element see *nonautonomous element.*
induced a gene that has been transcriptionally activated by an inducer.
induced mutation a mutation caused by environmental agents.
inducer a small effector molecule that binds to a genetic regulatory protein and thereby increases the rate of transcription.
inducible gene a gene that is regulated by an inducer, which is a small effector molecule that causes transcription to increase.
induction (1) the effects of an inducer on increasing the transcription of a gene; (2) the process by which a cell or group of cells governs the developmental fate of neighboring cells.
ingroup in cladistics, a group of species in which a researcher is interested.
inhibitor a small effector molecule that binds to an activator protein, causing the protein to be released from the DNA, thereby inhibiting transcription.
initiation (1) in transcription, the stage that involves the initial binding of RNA polymerase to the promoter in order to begin RNA synthesis; (2) in translation, the formation of a complex between mRNA, the initiator tRNA, and the ribosomal subunits.
initiator tRNA during translation, the tRNA that binds to the start codon.
insertion elements (IS elements) the simplest transposable elements. They are commonly found in bacteria.
insertional mutagenesis introducing a transposon into a cell for the purpose of inactivating a gene's function.
in situ hybridization a technique used to cytologically map the locations of genes or other DNA sequences within large eukaryotic chromosomes. In this method, a complementary probe is used to detect the location of a gene within a set of chromosomes.
insulator a segment of DNA that insulates a gene from the effects of nearby regulatory elements, such as enhancers.
integrase an enzyme that functions in the integration of viral DNA or retroelements into a chromosome.
interference see *positive interference.*
intergenic mapping the determination of the distance between two different genes.
intergenic region in a chromosome, a region of DNA that lies between two different genes.
intergenic suppressor a suppressor mutation that is in a different gene from the gene that contains the first mutation.
internal nuclear matrix a network of irregular protein fibers and other proteins that is connected to the nuclear lamina and fills the interior of the nucleus.
International HapMap Project worldwide effort to identify common human genetic variation.
interphase the series of phases G_1, S, and G_2, during which a eukaryotic cell spends most of its life.
interrupted mating a method used in conjugation experiments in which the length of time that the bacteria spend conjugating is stopped by a blender treatment or other type of harsh agitation.
interstitial deficiency see *interstitial deletion.*
interstitial deletion when an internal segment is lost from a linear chromosome.
intervening sequence also known as an *intron.* A segment of RNA that is removed during RNA splicing.
intragenic mapping see *fine-structure mapping.*
intragenic suppressor a suppressor mutation that is within the same gene as the first mutation that it suppresses.
intrinsic termination transcriptional termination that does not require the function of the rho protein.
intron intervening sequences that are found between exons. Introns are spliced out of the RNA prior to translation.
invasive refers to a tumor that can invade surrounding tissue.
inversion a change in the orientation of genetic material along a chromosome such that a segment is flipped in the reverse direction.
inversion heterozygote a diploid individual that carries one normal chromosome and a homologous chromosome with an inversion.
inversion loop the loop structure that is formed when the homologous chromosomes of an inversion heterozygote attempt to align themselves (i.e., synapse) during meiosis.
inverted repeats DNA sequences found in transposable elements that are identical (or very similar) but run in opposite directions.

in vitro fertilization (IVF) in the case of humans, the fertilization of an egg outside of a female's body.
iron regulatory protein (IRP) a translational regulatory protein that recognizes iron response elements that are found in specific mRNAs. It may inhibit translation or stabilize the mRNA.
iron response element (IRE) an RNA sequence that is recognized by the iron regulatory protein.
isoacceptor tRNAs two different tRNAs that can recognize the same codon.
isoelectric focusing a form of gel electrophoresis in which a protein migrates to the point in the gel where its net charge is zero.
isogamous describes a species that makes morphologically similar gametes.
isolating mechanism a mechanism that favors reproductive isolation. These can be prezygotic or postzygotic.

K

karyotype a photographic representation of all the chromosomes within a cell. It reveals how many chromosomes are found within an actively dividing somatic cell.
kinetochore a group of cellular proteins that attach to the centromere during meiosis and mitosis.
kinetochore microtuble a microtubule connected to a kinetochore.
knockin see *gene knockin.*
knockout see *gene knockout.*
Kozak's rules a set of rules that describes the most favorable types of bases that flank a eukaryotic start codon.

L

lac repressor a protein that binds to the operator site of the *lac* operon and inhibits transcription.
lagging strand a strand during DNA replication that is synthesized as short Okazaki fragments in the direction away from the replication fork.
lariat an excised intron structure composed of a circle and a tail.
latent refers to a virus that is within a cell but is not making new viruses. A virus may remain latent for a long time.
law of independent assortment see *Mendel's law of independent assortment.*
law of segregation see *Mendel's law of segregation.*
leading strand a strand during DNA replication that is synthesized continuously toward the replication fork.
left-right axis in bilaterians, the axis that determines left-right symmetry.
leptotene the first stage of prophase of meiosis I.
lesion-replicating polymerase a type of DNA polymerase that can replicate over a DNA region that contains an abnormal structure (i.e., a lesion).
lethal allele an allele that may cause the death of an organism.
lethal mutation a mutation that produces a lethal allele that may cause the death of an organism.
library see *DNA library.*
ligase see *DNA ligase.*
lineage diagram in developmental biology, a depiction of a cell and all of its descendants that are produced by cell division.
LINEs in mammals, long interspersed elements that are usually 1 to 5 kbp in length and found in 20,000 to 1,000,000 copies per genome.
linkage the occurrence of two or more genes along the same chromosome.
linkage disequilibrium when alleles and molecular markers are associated with each other at a frequency that is significantly higher than expected by random chance.
linkage group a group of genes that are linked together because they are found on the same chromosome.
linkage mapping the mapping of genes or other genetic sequences along a chromosome by analyzing the outcome of crosses.
lipid a general name given to an organic molecule that is insoluble in water. Cell membranes contain a large amount of lipid.
liposome a vesicle that is surrounded by a phospholipid bilayer.
local population a segment of a population that is slightly isolated. Members of a local population are more likely to breed with each other than with members that are outside of the local population.
locus (pl. **loci**) the physical location of a gene within a chromosome.
locus control region (LCR) a segment of DNA that is involved in the regulation of chromatin opening and closing.
locus heterogeneity the phenomenon in which a particular type of disease or trait may be caused by mutations in two or more different genes.
LOH see *loss of heterozygosity.*
long terminal repeats (LTRs) sequences containing many short segments that are tandemly repeated. They are found in retroviruses and viral-like retroelements.
loop domain a segment of chromosomal DNA that is anchored by proteins, so it forms a loop.
loss-of-function allele an allele of a gene that encodes an RNA or protein that is nonfunctional or compromised in function.
loss-of-function mutation a change in a genetic sequence that creates a loss-of-function allele.
loss of heterozygosity (LOH) when a heterozygous somatic cell incurs a genetic change that inactivates the single functional allele.
low stringency refers to hybridization conditions in which DNA or RNA fragments that have some mismatches are still able to recognize and bind to each other.
LTR retrotransposons a type of retrotransposon that is derived from a virus and has long terminal repeats.
LTRs see *long terminal repeats.*
Lyon hypothesis a hypothesis to explain the pattern of X-chromosome inactivation seen in mammals. Initially, both X chromosomes are active. However, at an early stage of embryonic development, one of the two X chromosomes is randomly inactivated in each somatic cell.
lysis cell breakage.
lysogenic refers to a phage that can follow the lysogenic cycle.
lysogenic cycle a type of growth cycle for a phage in which the phage integrates its genetic material into the chromosome of the bacterium. This integrated phage DNA can exist in a dormant state for a long time, during which no new bacteriophages are made.
lytic cycle a type of growth cycle for a phage in which the phage directs the synthesis of many copies of the phage genetic material and coat proteins. These components then assemble to make new phages. When synthesis and assembly is completed, the bacterial host cell is lysed, and the newly made phages are released into the environment.

M

macroevolution evolutionary changes at or above the species level involving relatively large changes in form and function that are sufficient to produce new species and higher taxa.
macromolecule a large organic molecule composed of smaller building blocks. Examples include DNA, RNA, proteins, and large carbohydrates.
MADS box a domain found in several transcription factors that play a role in plant development.
maintenance methylation the methylation of hemimethylated DNA following DNA replication.
major groove a wide indentation in the DNA double helix in which the bases have access to water.
malignant describes a tumor composed of cancerous cells.
mammalian cloning experimentally, this refers to the use of somatic cells and enucleated eggs to create a clone of a mammal.
map distance the relative distance between sites (e.g., genes) along a single chromosome.
mapping the experimental process of determining the relative locations of genes or other segments of DNA along individual chromosomes.
map unit (mu) a unit of map distance obtained from genetic crosses. One map unit is equivalent to 1% recombinant offspring in a testcross.
mass spectrometry a technique to accurately measure the mass of a molecule, such as a peptide fragment.
maternal effect an inheritance pattern for certain nuclear genes in which the genotype of the mother directly determines the phenotypic traits of her offspring.
maternal inheritance inheritance of DNA that occurs through the cytoplasm of the egg.
matrix-attachment region (MAR) a site in the chromosomal DNA that is anchored to the nuclear matrix or scaffold.
maturase a protein that enhances the rate of splicing of Group I and II introns.
maximum likelihood a strategy for choosing the best phylogenetic tree based on assumptions about the most likely ways that DNA incurs mutations.
MCM helicase a group of eukaryotic proteins needed to complete a process called DNA replication licensing, which is necessary for the formation of two replication forks at an origin of replication.
mean the sum of all the values in a group divided by the number of individuals in the group.
mean fitness of the population ($\overline{w}$) the average fitness of a population that is calculated by considering the frequencies and fitness values for all genotypes.
mediator a protein complex that interacts with RNA polymerase II and various regulatory transcription factors. Depending on its interactions with regulatory transcription factors, mediator may stimulate or inhibit RNA polymerase II.
meiosis a form of nuclear division in which the sorting process results in the production of haploid cells from a diploid cell.
meiotic nondisjunction the event in which chromosomes do not segregate equally during meiosis.

Mendelian inheritance the common pattern of inheritance observed by Mendel, which involves the transmission of eukaryotic genes that are located on the chromosomes found within the cell nucleus.

Mendel's law of independent assortment two different genes randomly assort their alleles during gamete formation (if they are not linked).

Mendel's law of segregation the two copies of a gene segregate from each other during transmission from parent to offspring.

meristem in plants, an organized group of actively dividing cells.

meristic traits traits that can be counted and expressed in whole numbers.

merozygote a partial diploid strain of bacteria containing F′ factor genes.

messenger RNA (mRNA) a type of RNA that contains the information for the synthesis of a polypeptide.

metacentric describes a chromosome with the centromere in the middle.

metagenome a collection of genes from an environmental sample.

metagenomics the study of a complex mixture of genetic material obtained from an environmental sample.

metaphase the third stage of M phase. The chromosomes align along the center of the spindle apparatus, and the formation of the spindle apparatus is complete.

metaphase plate the plane at which chromosomes align during metaphase.

metastatic describes cancer cells that migrate to other parts of the body.

methylation see *DNA methylation.*

methyl-CpG-binding protein a protein that binds to a CpG island when it is methylated.

methyl-directed mismatch repair a DNA repair system that detects a mismatch and specifically removes the segment from the newly made daughter strand.

methyltransferase see *DNA methyltransferase.*

microarray see *DNA microarray, protein microarray,* or *antibody microarray.*

microevolution changes in the gene pool with regard to particular alleles that occur over the course of many generations.

microinjection the use of microscopic-sized needles or pipettes to inject a substance into cells, such as DNA.

microRNAs (miRNAs) RNAs that are transcribed from eukaryotic genes that are endogenous—genes that are normally found in the genome. They silence the expression of mRNAs via RNA interference.

microsatellite composed of many repeated copies of short sequences. Microsatellites are interspersed throughout a genome and are quite variable in length among different individuals. They can be amplified by PCR.

microscopy the use of a microscope to view cells or subcellular structures.

microtubule-organizing center (MTOC) a site in a cell where microtubules begin to grow.

minichromosomes structures formed from many copies of circular DNA molecules.

minimal medium a type of growth medium for microorganisms that contains a mixture of reagents that are required for growth; nothing additional has been added.

minisatellite a repetitive sequence that was formerly used in DNA fingerprinting. Its use has been largely superseded by smaller repetitive sequences called microsatellites.

minor groove a narrow indentation in the DNA double helix in which the bases have access to water.

minute a unit of measure in bacterial conjugation experiments. This unit refers to the relative time it takes for genes to first enter a recipient strain during conjugation.

mismatch repair system see *methyl-directed mismatch repair.*

missense mutation a base substitution that leads to a change in the amino acid sequence of the encoded polypeptide.

mitochondrial DNA (mtDNA) the DNA found within mitochondria.

mitosis a type of nuclear division into two nuclei, such that each daughter cell receives the same complement of chromosomes.

mitotic nondisjunction an event in which chromosomes do not segregate equally during mitosis.

mitotic recombination crossing over that occurs during mitosis.

mitotic spindle apparatus (also known as the *mitotic spindle*) the structure that organizes and separates the chromosomes during M phase of the eukaryotic cell cycle.

model organism an organism studied by many scientists so that researchers can more easily compare their results and begin to unravel the properties of a given species.

moderately repetitive sequences sequences that are found a few hundred to several thousand times in the genome.

molecular clock the phenomenon by which the rate of neutral mutations can be used as a tool to measure evolutionary time.

molecular evolution the molecular changes in the genetic material that underlie the phenotypic changes associated with evolution.

molecular genetics an examination of DNA structure and function at the molecular level.

molecular level the perspective of viewing life at the level of individual molecules.

molecular marker a segment of DNA that is found at a specific site in the genome and has properties that enable it to be uniquely recognized using molecular tools, such as gel electrophoresis.

molecular paleontology the analysis of DNA sequences from extinct species.

molecular pharming a recombinant technology that involves the production of medically important proteins in the mammary glands of livestock.

molecular profiling methods that enable researchers to understand the molecular changes that occur in diseases, such as cancer.

molecular trait a trait that is observed at the molecular level. An example is the amount of a given protein in a cell.

monad the single chromatid within a dyad.

monoallelic expression in the case of imprinting, refers to the phenomenon that only one of the two alleles of a given gene is transcriptionally expressed.

monohybrid an individual produced from a single-factor cross in which the parents had different variants for a single character.

monomorphic a term used to describe a gene that is found as only one allele in a population.

monophyletic group see *clade.*

monoploid an organism with a single set of chromosomes within its somatic cells.

monosomic a diploid cell that is missing a chromosome (i.e., $2n - 1$).

monozygotic twins twins that are genetically identical because they were formed from the same sperm and egg.

morph a form or phenotype in a population. For example, red eyes and white eyes are different eye color morphs.

morphogen a molecule that conveys positional information and promotes developmental changes.

morphological trait a trait that affects the morphology (physical form) of an organism. An example is eye color.

mosaicism when the cells of part of an organism differ genetically from the rest of the organism.

motif the name given to a domain or amino acid sequence that functions in a similar manner in many different proteins.

M phase a general name given to nuclear division that can apply to mitosis or meiosis. It is divided into prophase, prometaphase, metaphase, anaphase, and telophase.

mRNA see *messenger RNA.*

mtDNA an abbreviation for mitochondrial DNA.

MTOC see *microtubule-organizing center.*

multicellularity the property by which an organism consists of more than one cell.

multinomial expansion equation an equation to solve genetic problems involving three or more types of unordered events.

multiple alleles when the same gene exists in two or more alleles within a population.

multiple-sequence alignment an alignment of two or more genetic sequences based on their homology to each other.

multiplication method a method for solving independent assortment problems in which the probabilities of the outcome for each gene are multiplied together.

multipotent a type of stem cell that can differentiate into several different types of cells.

mutable site a site in a chromosome that tends to break at a fairly high rate due to the presence of a transposable element.

mutagen an agent that causes alterations in the structure of DNA.

mutant alleles alleles that have been created by altering a wild-type allele by mutation.

mutation a permanent change in the genetic material that can be passed from cell to cell or from parent to offspring.

mutation frequency the number of mutant genes divided by the total number of genes within the population.

mutation rate the likelihood that a gene will be altered by a new mutation.

myogenic bHLH protein a type of transcription factor involved in muscle cell differentiation.

N

n an abbreviation that designates the number of chromosomes in a set. In humans, $n = 23$, and a diploid cell has $2n = 46$ chromosomes.

narrow-sense heritability heritability that takes into account only those genetic factors that are additive.

natural selection refers to the process whereby differential fitness acts on the gene pool. When a mutation creates a new allele that is beneficial, the allele may become prevalent within future generations because the individuals possessing this allele are more likely to reproduce and pass the beneficial allele to their offspring.

natural transformation a natural process of transformation that occurs in certain strains of bacteria.

negative control transcriptional regulation by repressor proteins.

negative frequency-dependent selection a pattern of natural selection in which the fitness of a genotype decreases when its frequency becomes higher.

neutral mutation a mutation that has no detectable effect on protein function or no detectable effect on the survival of the organism.
neutral theory of evolution the theory that most genetic variation observed in natural populations is due to the accumulation of neutral mutations.
next-generation sequencing technologies newer DNA sequencing technologies that are more rapid and inexpensive.
nick translation the phenomenon in which DNA polymerase uses its 5′ to 3′ exonuclease activity to remove a region of DNA and, at the same time, replaces it with new DNA.
nitrogen mustard an alkylating agent that can cause mutations in DNA.
nitrous acid a type of chemical mutagen that deaminates bases, thereby changing amino groups to keto groups.
nonallelic homologous recombination recombination that occurs at sites within chromosomes due to the occurrence of repetitive sequences.
nonautonomous element a transposable element that lacks a gene, such as transposase or reverse transcriptase, that is necessary for transposition.
noncoding RNAs RNAs that do not encode proteins.
noncoding strand the strand of DNA within a protein-encoding gene that is complementary to the mRNA. The noncoding strand is used as a template to make mRNA.
noncomplementation the phenomenon in which two mutant alleles in the same organism do not produce a wild-type phenotype.
non-Darwinian evolution see *neutral theory of evolution.*
nondisjunction event in which chromosomes do not segregate properly during mitosis or meiosis.
nonessential genes genes that are not absolutely required for survival, although they are likely to be beneficial to the organism.
nonhomologous end-joining (NHEJ) the joining of DNA fragment ends that are different.
nonhomologous end-joining (NHEJ) protein a protein that joins the ends of DNA fragments that are not homologous. This occurs during site-specific recombination of immunoglobulin genes.
nonhomologous recombination the exchange of DNA between nonhomologous segments of chromosomes or plasmids.
non-LTR retrotransposon a type of retrotransposon that does not have long terminal repeats.
nonneutral mutation a mutation that affects the phenotype of the organism and can be acted on by natural selection.
nonparental see *recombinant.*
nonparental ditype (NPD) an ascus that contains cells that all have a nonparental combination of alleles.
nonrecombinant in a testcross, refers to a phenotype or arrangement of alleles on a chromosome that is found in the parental generation.
nonrecombinant offspring for genes that are linked, nonrecombinant offspring are not the result of a crossover.
nonsense codon a stop codon.
nonsense mutation a mutation that involves a change from a sense codon to a stop codon.
nontemplate strand a strand of DNA that is not used as a template during transcription.
normal distribution a distribution for a large sample in which the trait of interest varies in a symmetrical way around an average value.
norm of reaction the effects of environmental variation on an individual's traits.
Northern blotting a technique used to detect a specific RNA within a mixture of many RNA molecules.
N-terminus see *amino-terminus.*
nuclear genes genes that are located on chromosomes found in the cell nucleus of eukaryotic cells.
nuclear lamina a collection of fibers that line the inner nuclear membrane.
nuclear matrix (or **nuclear scaffold**) a group of proteins that anchor the loops found in eukaryotic chromosomes.
nucleic acid RNA or DNA. A macromolecule that is composed of repeating nucleotide units.
nucleoid a darkly staining region that contains the genetic material of mitochondria, chloroplasts, or bacteria.
nucleolus a region within the nucleus of eukaryotic cells where the assembly of ribosomal subunits occurs.
nucleoprotein a complex of DNA (or RNA) and protein.
nucleoside structure in which a base is attached to a sugar, but no phosphate is attached to the sugar.
nucleosome the repeating structural unit within eukaryotic chromatin. It is composed of double-stranded DNA wrapped around an octamer of histone proteins.
nucleosome-free region (NFR) a region within a eukaryotic chromosome where nucleosomes are not found.
nucleotide the repeating structural unit of nucleic acids, composed of a sugar, phosphate, and base.
nucleotide excision repair (NER) a DNA repair system in which several nucleotides in the damaged strand are removed from the DNA and the undamaged strand is used as a template to resynthesize a normal strand.
nucleus a membrane-bound organelle in eukaryotic cells where the linear sets of chromosomes are found.
null hypothesis a hypothesis that assumes there is no real difference between the observed and expected values.

O

octad a group of eight fungal spores contained within an ascus.
Okazaki fragments short segments of DNA that are synthesized in the lagging strand during DNA replication.
oligonucleotide a short strand of DNA, typically a few or a few dozen nucleotides in length.
oncogene a mutant gene that promotes cancer.
one-gene/one-enzyme hypothesis the idea, which later needed to be expanded, that one gene encodes one enzyme.
oogenesis the production of egg cells.
open complex the region of separation of two DNA strands produced by RNA polymerase during transcription.
open conformation a loosely packed chromatin structure that is capable of transcription.
open reading frame (ORF) a genetic sequence that does not contain stop codons.
operator (or **operator site**) a sequence of nucleotides in bacterial DNA that provides a binding site for a genetic regulatory protein.
operon an arrangement in DNA in which two or more genes are found within a regulatory unit that is under the transcriptional control of a single promoter.
ordered octad an ascus composed of eight cells whose order depends on crossing over during meiosis.
ordered tetrad an ascus composed of four cells whose order depends on crossing over during meiosis.
ORF see *open reading frame.*
organelle a large specialized structure within a cell, which is often surrounded by a single or double membrane.
organism level when the level of observation or experimentation involves a whole organism.
organizing center in plants, a region of the meristem that ensures the proper organization of the meristem and preserves the correct number of actively dividing stem cells.
orientation-independent refers to certain types of genetic regulatory elements that can function in the forward or reverse direction. Certain enhancers are orientation independent.
origin of replication a nucleotide sequence that functions as an initiation site for the assembly of several proteins required for DNA replication.
origin of transfer the location on an F factor or within the chromosome of an *Hfr* strain that is the initiation site for the transfer of DNA from one bacterium to another during conjugation.
origin recognition complex (ORC) a complex of six proteins found in eukaryotes that is necessary to initiate DNA replication.
orthologs homologous genes in different species that were derived from the same ancestral gene.
outbreeding mating between genetically unrelated individuals.
outgroup in cladistics, a species or group of species that is less related compared to members of the ingroup.
ovary (1) in plants, the structure in which the ovules develop; (2) in animals, the structure that produces egg cells and female hormones.
overdominance an inheritance pattern in which a heterozygote is more vigorous than either of the corresponding homozygotes.
ovule the structure in higher plants where the female gametophyte (i.e., embryo sac) is produced.
ovum a female gamete, also known as an egg cell.
oxidative DNA damage changes in DNA structure that are caused by reactive oxygen species (ROS).
oxidative stress an imbalance between the production of reactive oxygen species (ROS) and an organism's ability to break them down.

P

p an abbreviation for the short arm of a chromosome.
P1 artificial chromosomes (PACs) cloning vector developed from P1 bacteriophage chromosomes that can contain 300,000 bp or more of DNA.
PACs See *P1 artificial chromosomes.*
pachytene the third stage of prophase of meiosis I.
pair-rule gene one category of segmentation genes.
palindromic when a sequence is the same in the forward and reverse directions.
pangenesis an incorrect hypothesis of heredity. It suggested that hereditary traits could be modified depending on the lifestyle of the individual. For example, it was believed that a person who practiced a particular skill would produce offspring that would be better at that skill.

paracentric inversion an inversion in which the centromere is found outside of the inverted region.
paralogs homologous genes within a single species that constitute a gene family.
parapatric speciation a form of speciation that occurs when members of a species are only partially separated or when a species is very sedentary.
parasegments transient subdivisions that occur in the *Drosophila* embryo prior to the formation of segments.
parental in a testcross, refers to a phenotype or arrangement of alleles on a chromosome that is the same as one or both members of the parental generation.
parental ditype (PD) an ascus that contains four spores with the parental combinations of alleles.
parental generation in a genetic cross, the first generation in the experiment. In Mendel's studies, the parental generation was true-breeding with regard to particular traits.
parental strand in DNA replication, the DNA strand that is used as a template.
parthenogenesis the formation of an individual from an unfertilized egg.
particulate theory of inheritance a theory proposed by Mendel. It states that traits are inherited as discrete units that remain unchanged as they are passed from parent to offspring.
paternal leakage the phenomenon in which maternal inheritance is generally observed, but the male parent may, on rare occasions, provide mitochondria or chloroplasts to the zygote.
pattern the spatial arrangement of different regions of the body. At the cellular level, the body pattern is due to the arrangement of cells and their specialization.
pattern recognition in bioinformatics, this term refers to a program that recognizes a pattern of symbols.
PCR see *polymerase chain reaction.*
pedigree analysis a genetic analysis using information contained within family trees. In this approach, the aim is to determine the type of inheritance pattern that a gene follows.
pedigrees charts representing family relationships.
pellet a collection of particles found at the bottom of a centrifuge tube.
peptide bond a covalent bond formed between the carboxyl group in one amino acid in a polypeptide and the amino group in the next amino acid.
peptidyl site (P site) a site on the ribosome that carries a tRNA along with a polypeptide chain.
peptidyl transfer the step during the elongation stage of translation in which the polypeptide is removed from the tRNA in the P site and transferred to the amino acid at the A site.
peptidyl transferase a complex that functions during translation to catalyze the formation of a peptide bond between the amino acid in the A site of the ribosome and the growing polypeptide chain.
pericentric inversion an inversion in which the centromere is located within the inverted region of the chromosome.
peripheral zone in plants, an area in the meristem that contains dividing cells that eventually differentiate into plant structures.
personalized medicine the application of genetic or molecular data in the treatment of disease.
petites mutant strains of yeast that form small colonies due to defects in mitochondrial function.
PGD see *preimplantation genetic diagnosis.*
P generation the parental generation in a genetic cross.
phage see *bacteriophage.*
phage λ a bacteriophage that infects *E. coli.*
pharmacogenetics the study or clinical testing of genetic variation that causes differing responses to drugs.
pharming see *molecular pharming.*
phenetic approach the construction of a phylogenetic tree based on the overall similarities of a group of species without understanding their evolutionary history.
phenogram a phylogenetic tree based on the phenetic approach.
phenotype the observable traits of an organism.
phenylketonuria (PKU) a human genetic disorder arising from a defect in phenylalanine hydroxylase.
phosphodiester linkage in a DNA or RNA strand, a linkage in which a phosphate group connects two sugar molecules together.
photolyase an enzyme found in yeast and plants that can repair thymine dimers by splitting the dimers, which returns the DNA to its original condition.
photoreactivation a type of DNA repair of thymine dimers via photolyase that requires light.
phylogenetic tree a diagram that describes the evolutionary relationships among different species.
phylogeny the sequence of events involved in the evolutionary development of a species or group of species.
physical mapping the mapping of genes or other genetic sequences using DNA cloning methods.
physiological trait a trait that affects a cellular or body function. An example is the rate of glucose metabolism.
PKU see *phenylketonuria.*
plaque a clear zone within a bacterial lawn on a petri plate. It is due to repeated cycles of viral infection and bacterial lysis.
plasmid a general name used to describe circular pieces of DNA that exist independently of the chromosomal DNA. Some plasmids are used as vectors in cloning experiments.
pleiotropy the multiple effects of a single gene on the phenotype of an organism.
pluripotency factors transcription factors that stimulate the expression of the *Tsix* gene.
pluripotent a type of stem cell that can differentiate into all or nearly all the types of cells of the adult organism.
point mutation a change in a single base pair within DNA.
polar microtubule a microtubule that projects toward the region where the chromosomes will be found during mitosis.
polar mutation a mutation in an operon that affects the expression of another gene that is found downstream in the operon.
polarity in genetics, the phenomenon in which a nonsense mutation in one gene affects the translation of a downstream gene in an operon.
pollen also, pollen grain; the male gametophyte of flowering plants.
polyA-binding protein a protein that binds to the 3′ polyA tail of mRNAs and protects the mRNA from degradation.
polyadenylation the process of attaching a string of adenine nucleotides to the 3′ end of eukaryotic mRNAs.
polyA tail the string of adenine nucleotides at the 3′ end of eukaryotic mRNAs.
polycistronic mRNA an mRNA transcribed from an operon that encodes two or more proteins.
polycomb group (PcG) protein complexes that are key regulators of epigenetic changes that are programmed during development. They cause gene repression.
polycomb response element (PRE) a DNA element that is initially recognized by a PRE-binding protein, which then recruits PRC2 to the site.
polygenic inheritance the transmission of traits that are governed by two or more different genes.
polymerase chain reaction (PCR) the method for amplifying a DNA region involving the sequential use of oligonucleotide primers and *Taq* polymerase.
polymerase switch during DNA replication, when one type of DNA polymerase (e.g., α) is switched for another type (e.g., β).
polymorphic a term used to describe a trait or gene that is found in two or more forms in a population.
polymorphism (1) the prevalence of two or more phenotypic forms in a population; (2) the phenomenon in which a gene exists in two or more alleles within a population.
polypeptide a sequence of amino acids that is the product of mRNA translation. One or more polypeptides fold and associate with each other to form a functional protein.
polyploid an organism or cell with three or more sets of chromosomes.
polyribosome an mRNA transcript that has many bound ribosomes in the act of translation.
polysome see *polyribosome.*
polytene chromosomes chromosomes that are found in certain cells, such as *Drosophila* salivary cells, in which the chromosomes have replicated many times and the copies lie side by side.
population a group of individuals of the same species that are capable of interbreeding with one another.
population genetics the field of genetics that is primarily concerned with the prevalence of genetic variation within populations.
population level when the level of observation or experimentation involves a population of organisms.
positional cloning a cloning strategy in which a gene is cloned based on its mapped position along a chromosome.
positional information chemical substances and other environmental cues that enable a cell to deduce its position relative to other cells.
position effect a change in phenotype that occurs when the position of a gene is changed from one chromosomal site to a different location.
positive control genetic regulation by activator proteins.
positive interference the phenomenon in which a crossover that occurs in one region of a chromosome decreases the probability that another crossover will occur nearby.
posttranslational describes events that occur after translation is completed.
posttranslational covalent modification the covalent attachment of a molecule to a protein after it has been synthesized via ribosomes.
postzygotic isolating mechanism a mechanism of reproductive isolation that prevents an offspring from being viable or fertile.
preimplantation genetic diagnosis (PGD) a form of genetic testing in which an embryo obtained via in vitro fertilization is tested for genetic abnormalities.

preinitiation complex the stage of transcription in which the assembly of RNA polymerase and general transcription factors occurs at the core promoter, but the DNA has not yet started to unwind.

preintegration complex in the case of HIV, refers to a complex of proteins and double-stranded HIV DNA that will be integrated into the host chromosomal DNA.

pre-mRNA in eukaryotes, the transcription of protein-encoding genes produces a long transcript known as pre-mRNA, which is located within the nucleus. This pre-mRNA is usually altered by splicing and other modifications before it exits the nucleus.

prereplication complex (preRC) in eukaryotes, an assembly of at least 14 different proteins, including a group of 6 proteins called the origin recognition complex (ORC), that acts as the initiator of eukaryotic DNA replication.

prezygotic isolating mechanism a mechanism for reproductive isolation that prevents the formation of a zygote.

Pribnow box the TATAAT sequence that is often found at the −10 region of bacterial promoters.

primary structure with regard to proteins, the linear sequence of amino acids.

primase an enzyme that synthesizes a short RNA primer during DNA replication.

primer a short segment of DNA or RNA that initiates DNA replication.

primer extension in PCR, the step during which DNA is made by extending the length of a DNA primer.

primitive character a trait that is shared with a distant ancestor.

primosome a multiprotein complex composed of DNA helicase, primase, and several accessory proteins.

principle of parsimony the principle that the preferred hypothesis is the one that is the simplest.

prion an infectious particle that causes several types of neurodegenerative diseases affecting humans and livestock. It is composed entirely of protein.

probability the chance that an event will occur in the future.

probe a molecule, such as an oligonucleotide, that is used to identify another molecule or a site within a chromosome.

processing body (P-body) the cellular structure where the RISC-mRNA complex is stored and may be later reused.

processive enzyme an enzyme, such as RNA and DNA polymerase, which glides along the DNA and does not dissociate from the template strand as it catalyzes the covalent attachment of nucleotides.

product rule the probability that two or more independent events will occur is equal to the products of their individual probabilities.

proflavin a type of chemical mutagen that causes frameshift mutations.

prokaryotes another name for bacteria and archaea. The term refers to the fact that their chromosomes are not contained within a separate nucleus of the cell.

prometaphase the second phase of M phase. During this phase, the nuclear membrane vesiculates, and the mitotic spindle is completely formed.

promoter a sequence within a gene that initiates (i.e., promotes) transcription.

promoter bashing the approach of making deletions in the vicinity of a promoter as a way to identify the core promoter and regulatory elements.

proofreading function the ability of DNA polymerase to remove mismatched bases from a newly made strand.

prophage phage DNA that has been integrated into the bacterial chromosome.

prophase the first stage of M phase. The chromosomes have already replicated and begin to condense. The mitotic spindle starts to form.

protease an enzyme that digests the polypeptide backbone found in proteins.

protein a functional unit composed of one or more polypeptides.

protein-encoding genes genes that produce mRNA and encode polypeptides; also called structural genes.

protein microarray a small silica, glass, or plastic slide that is dotted with many different proteins.

proteome the collection of all proteins that a given cell or species can make.

proteomics the study of protein function at the genome level. It involves the study of many proteins simultaneously.

proto-oncogene a normal cellular gene that does not cause cancer but which may incur a gain-of-function mutation or become incorporated into a viral genome and thereby lead to cancer.

protoplast a plant cell without a cell wall.

prototroph a strain that does not need a particular nutrient supplemented in its growth medium or diet.

provirus viral DNA that has been incorporated into the chromosome of a host cell.

proximodistal axis in animals, an axis involving limbs in which the part of the limb attached to the trunk is located proximal, whereas the end of the limb is located distal.

pseudoautosomal inheritance the inheritance pattern of genes that are found on both the X and Y chromosomes. Even though such genes are located physically on the sex chromosomes, their pattern of inheritance is identical to that of autosomal genes.

pseudodominance a pattern of inheritance that occurs when a single copy of a recessive allele is phenotypically expressed because the second copy of the gene has been deleted from the homologous chromosome.

pseudo-overdominance occurs when two closely linked genes exhibit heterosis. The heterosis is really due to the masking of recessive alleles.

P site see *peptidyl site.*

pulse-chase experiment an experimental strategy in which cells are given a radiolabeled compound for a brief period of time—a pulse—followed by an excess amount of unlabeled compound—a chase.

Punnett square a diagrammatic method in which the gametes that two parents can produce are aligned next to a square grid as a way to predict the types of offspring the parents will produce and in what proportions.

purine a type of nitrogenous base that has a double-ring structure. Examples are adenine and guanine.

***P* value** in a chi square table, the probability that the deviations between observed and expected values are due to random chance.

pyrimidine a type of nitrogenous base that has a single-ring structure. Examples are cytosine, thymine, and uracil.

pyrosequencing a type of next-generation DNA sequencing.

pyrrolysine an amino acid that may be incorporated into polypeptides during translation.

Q

q an abbreviation for the long arm of a chromosome.

QTL see *quantitative trait loci.*

QTL mapping the determination of the location of QTLs using mapping methods, such as genetic crosses coupled with the analysis of molecular markers.

quantitative genetics the area of genetics concerned with traits that can be described in a quantitative way.

quantitative trait a trait, usually polygenic in nature, that can be described with numbers.

quantitative trait loci (QTLs) the locations on chromosomes where the genes that influence quantitative traits reside.

quaternary structure in proteins, refers to the binding of two or more polypeptides to each other to form a protein.

R

radial loop domains organization of chromatin during interphase into loops, often 25,000 to 200,000 bp in size, which are anchored to the nuclear matrix at matrix-attachment regions.

radioimmunoassay a method for measuring the amount of an antigen in a biological sample.

RAG1/RAG2 enzymes that recognize recombination signal sequences and make double-stranded cuts. In the case of V/J recombination in immunoglobulin genes, a cut is made at the end of one V region and the beginning of one J region.

random genetic drift see *genetic drift.*

random mutation theory according to this theory, mutations are a random process—they can occur in any gene and do not involve exposure of an organism to a particular condition that selects for specific types of mutations.

random sampling error the deviation between the observed and expected outcomes due to chance.

reactive oxygen species (ROS) products of oxygen metabolism in all aerobic organisms that can damage cellular molecules, including DNA, proteins, and lipids.

reading frame a series of codons beginning with the start codon as a frame of reference.

real-time PCR a method of PCR in which the synthesis of DNA is monitored in real time. It can quantitate the starting amount of DNA in a sample.

realized heritability a form of narrow-sense heritability that is observed when selective breeding is practiced.

recessive a trait or gene that is masked by the presence of a dominant trait or gene.

recessive epistasis a form of epistasis in which an individual must be homozygous for either recessive allele to mask a particular phenotype.

reciprocal crosses a pair of crosses in which the traits of the two parents differ with regard to sex. For example, one cross could be a red-eyed female fly and a white-eyed male fly, and the reciprocal cross would be a red-eyed male fly and a white-eyed female fly.

reciprocal translocation when two different chromosomes exchange pieces.

recombinant (1) refers to combinations of alleles or traits that are not found in the parental generation; (2) describes DNA molecules that are produced by molecular techniques in which segments of DNA are joined to each other in ways that differ from their original arrangement in their native chromosomal sites. The cloning of DNA into vectors is an example.
recombinant cells for linked genes, refers to cells that are the product of a crossover.
recombinant DNA molecules molecules that are produced in a test tube by covalently linking DNA fragments from two different sources.
recombinant DNA technology the use of in vitro molecular techniques to isolate and manipulate different pieces of DNA.
recombinant offspring for linked genes, refers to offspring that are the product of a crossover.
recombinant vector a vector that contains an inserted fragment of DNA, such as a gene from a chromosome.
recombination see *genetic recombination.*
recombinational repair DNA repair via homologous recombination.
recombination signal sequence a specific DNA sequence that is involved in site-specific recombination. Such sequences are found in immunoglobulin genes.
redundancy see *gene redundancy.*
regression analysis the approach of using correlated data to predict the outcome of one variable when given the value of a second variable.
regression coefficient in a regression analysis, a coefficient that describes the slope of the regression line.
regulatory sequence, or element a sequence of DNA (or possibly RNA) that binds a regulatory protein and thereby influences gene expression. Bacterial operator sites and eukaryotic enhancers and silencers are examples.
regulatory transcription factor a protein or protein complex that binds to a regulatory element and influences the rate of transcription via RNA polymerase.
relative fitness the fitness value of one genotype relative to another genotype.
relaxosome a protein complex that recognizes the origin of transfer in F factors and other conjugative plasmids, cuts one DNA strand, and aids in the transfer of the T DNA.
release factor a protein that recognizes a stop codon and promotes translational termination and the release of the completed polypeptide.
repetitive sequences DNA sequences that are present in many copies in the genome.
replica plating a technique in which a replica of bacterial colonies is transferred to a new petri plate.
replication see *DNA replication.*
replication fork the region in which two DNA strands have separated and new strands are being synthesized.
replisome a complex that contains a primosome and dimeric DNA polymerase.
repressible gene a gene that is regulated by a corepressor or inhibitor, which are small effector molecules that cause transcription to decrease.
repressor a regulatory protein that binds to DNA and inhibits transcription.
reproductive cloning the cloning of a eukaryotic organism, which may involve the use of genetic material from somatic cells.
reproductive isolation the inability of a species to successfully interbreed with other species.
resolution the last stage of homologous recombination, in which the entangled DNA strands become resolved into two separate structures.
resolvase an endonuclease that makes the final cuts in DNA during the resolution phase of recombination.
response element see *regulatory sequence,* or *element.*
restriction endonuclease (or **restriction enzyme**) an endonuclease that cleaves DNA. The restriction enzymes used in cloning experiments bind to specific base sequences and then cleave the DNA backbone at two defined locations, one in each strand.
restriction fragment length polymorphism (RFLP) genetic variation within a population in the lengths of DNA fragments that are produced when chromosomes are digested with particular restriction enzymes.
restriction mapping a technique for determining the locations of restriction endonuclease sites within a segment of DNA.
restriction point a point in the G_1 phase of the cell cycle that causes a cell to progress to cell division.
retroelement a type of transposable element that moves via an RNA intermediate.
retroposon see *retroelement.*
retrospective testing a procedure in which a researcher formulates a hypothesis and then collects observations as a way to confirm or refute the hypothesis. This approach relies on the observation and testing of already existing materials, which themselves are the result of past events.
retrotransposition a form of transposition in which the element is transcribed into RNA. The RNA is then used as a template via reverse transcriptase to synthesize a DNA molecule that is integrated into a new region of the genome via integrase.
retrotransposon see *retroelement.*
reverse genetics an experimental strategy in which researchers first identify the wild-type gene using cloning methods. The next step is to make a mutant version of the wild-type gene, introduce it into an organism, and see how the mutant gene affects the phenotype of the organism.
reverse mutation see *reversion.*
reverse transcriptase an enzyme that uses an RNA template to make a complementary strand of DNA.
reverse transcriptase PCR a modification of PCR in which the first round of replication involves the use of RNA and reverse transcriptase to make a complementary strand of DNA.
reversion a mutation that returns a mutant allele back to the wild-type allele.
R factor a type of plasmid found commonly in bacteria that confers resistance to a toxic substance, such as an antibiotic.
RFLP see *restriction fragment length polymorphism.*
RFLP map a genetic map composed of many RFLP markers.
RFLP mapping the mapping of a gene or other genetic sequence relative to the known locations of RFLPs within a genome.
R group the side chain of an amino acid.
rho (ρ) a protein that is involved in transcriptional termination for certain bacterial genes.
rho-dependent termination transcriptional termination that requires the function of the rho protein.
rho-independent termination transcription termination that does not require the rho protein. It is also known as intrinsic termination.
ribonucleic acid (RNA) a nucleic acid that is composed of ribonucleotides. In living cells, RNA is synthesized via the transcription of DNA.
ribose the sugar found in RNA.
ribosome a large macromolecular structure that acts as the catalytic site for polypeptide synthesis. The ribosome allows the mRNA and tRNAs to be positioned correctly as the polypeptide is made.
ribosome-binding site a sequence in bacterial mRNA that is needed to bind to the ribosome and initiate translation.
riboswitch a mechanism for regulating transcription, translation, RNA stability, and splicing in which an RNA molecule can switch between two secondary conformations based on whether or not a small molecule, such as TPP, binds to the RNA.
ribozyme an RNA molecule with enzymatic activity.
right-left axis in bilateral animals, this axis determines the two sides of the body relative to the anteroposterior axis.
R loop experimentally, a DNA loop that is formed because RNA is displacing it from its complementary DNA strand.
RNA see *ribonucleic acid.*
RNA editing the process in which a change occurs in the nucleotide sequence of an RNA molecule that involves additions or deletions of particular bases, or a conversion of one type of base to a different type.
RNA-induced silencing complex (RISC) the complex that mediates RNA interference.
RNA interference the phenomenon that double-stranded RNA targets complementary RNAs within the cell for silencing or degradation.
RNA polymerase an enzyme that synthesizes a strand of RNA using a DNA strand as a template.
RNA primer a short strand of RNA, made by DNA primase, that is used to elongate a strand of DNA during DNA replication.
RNase an enzyme that cuts the sugar-phosphate backbone in RNA.
RNaseP a bacterial enzyme that is an endonuclease and cuts precursor tRNA molecules. RNaseP is a ribozyme, which means that its catalytic ability is due to the action of RNA.
RNA splicing the process in which pieces of RNA are removed and the remaining pieces are covalently attached to each other.
Robertsonian translocation the structure produced when two telocentric chromosomes fuse at their short arms.
root meristem an actively dividing group of cells that gives rise to root structures.
RT-PCR see *reverse transcriptase PCR* or *real-time PCR.*

S

satellite DNA in a density centrifugation experiment, a peak of DNA that is separated from the majority of the chromosomal DNA. It is usually composed of highly repetitive sequences.
SBS see *sequencing by synthesis.*
scaffold a collection of proteins that holds the DNA in place and gives chromosomes their characteristic shapes.

scaffold-attachment region (SAR) a site in the chromosomal DNA that is anchored to the nuclear matrix or scaffold.
SCE see *sister chromatid exchange.*
science a way of knowing about our natural world. The science of genetics allows us to understand how the expression of genes produces the traits of an organism.
scientific method a basis for conducting science. It is a process that scientists typically follow so they may reach verifiable conclusions about the world in which they live.
scintillation counting a technique that permits a researcher to count the number of radioactive emissions from a sample containing a population of radioisotopes.
search by content in bioinformatics, an approach to predict the location of a gene because the nucleotide content of a particular region differs significantly (due to codon bias) from a random distribution.
search by signal in bioinformatics, an approach that relies on known sequences such as promoters, start and stop codons, and splice sites to help predict whether or not a DNA sequence contains a gene.
secondary structure a regular repeating pattern of molecular structure, such as the DNA double helix or the α helix and β sheet found in proteins.
sedimentation coefficient a measure of centrifugation that is normally expressed in Svedberg units (S). A sedimentation coefficient has the units of seconds: $1\ S = 1 \times 10^{-13}$ seconds.
segmental duplication a small segment of a chromosome that has a tandem duplication.
segmentation gene in animals, a gene whose encoded product is involved in the development of body segments.
segment polarity gene one category of segmentation genes.
segments anatomical subdivisions that occur during the development of species, such as *Drosophila.*
segregate when two things are placed in separate locations. For example, homologous chromosomes segregate into different gametes.
selectable marker a gene that provides a selectable phenotype in a cloning experiment. Many selectable markers are genes that confer antibiotic resistance.
selection coefficient one minus the fitness value.
selectionists scientists who oppose the neutral theory of evolution.
selection limit the phenomenon in which several generations of artificial selection results in a plateau where artificial selection is no longer effective.
selective breeding programs and procedures designed to modify the phenotypes in economically important species of plants and animals.
selenocysteine an amino acid that may be incorporated into polypeptides during translation.
self-fertilization fertilization that involves the union of male and female gametes derived from the same parent.
selfish DNA hypothesis the idea that transposable elements exist because they possess characteristics that allow them to multiply within the host cell DNA and inhabit the host without offering any selective advantage.
self-splicing refers to RNA molecules that can remove their own introns without the aid of other proteins or RNA.
semiconservative model the correct model for DNA replication in which the newly made double-stranded DNA contains one parental strand and one daughter strand.
semiconservative replication the net result of DNA replication in which the DNA contains one original strand and one newly made strand.
semilethal alleles lethal alleles that kill some individuals but not all.
semisterility when an individual has a lowered fertility.
senescent a cell that is no longer capable of dividing.
sense codon a codon that encodes a specific amino acid.
sense strand the strand of DNA within a protein-encoding gene that has the same sequence as mRNA except that T is found in the DNA instead of U.
sequence complexity the number of times a particular base sequence appears throughout the genome of a given species.
sequence element in genetics, a sequence with a specialized function.
sequence recognition in bioinformatics, this term refers to a program that recognizes particular sequence elements.
sequence-tagged site (STS) a short segment of DNA, usually between 100 and 400 bp long, the base sequence of which is found to be unique within the entire genome. Sequence-tagged sites are identified by PCR.
sequencing see *DNA sequencing.*
sequencing by synthesis (SBS) a next-generation form of DNA sequencing in which the synthesis of DNA is directly monitored to deduce the base sequence.
sequencing ladder a series of bands on a gel that can be followed in order (e.g., from the bottom of the gel to the top of the gel) to determine the base sequence of DNA.
sex chromosomes a pair of chromosomes (e.g., X and Y in mammals) that determines sex in a species.
sex determination the factors that determine whether an organism develops into a male or a female.
sex-influenced inheritance an inheritance pattern in which an allele is dominant in one sex but recessive in the opposite sex.
sex-limited inheritance an inheritance pattern in which a trait is found in only one of the two sexes. An example would be beard development in men.
sex-limited traits traits that occur in only one of the two sexes.
sex linkage the phenomenon that certain genes are found on one of the two types of sex chromosomes but not both.
sex-linked gene a gene that is located on one of the sex chromosomes.
sex pilus (pl. **pili**) a structure on the surface of bacterial cells that acts as an attachment site to promote the binding of bacteria to each other.
sexual dimorphism species in which the males and females are morphologically distinct.
sexual reproduction the process whereby parents make gametes (e.g., sperm and egg) that fuse with each other in the process of fertilization to begin the life of a new organism.
sexual selection natural selection that acts to promote characteristics that give individuals a greater chance of reproducing.
shared derived character a trait shared by a group of organisms but not by a distant common ancestor.
Shine-Dalgarno sequence a sequence in bacterial mRNAs that functions as a ribosomal binding site.
shoot meristem an actively dividing group of cells that gives rise to shoot structures.
short-interfering RNAs (siRNAs) RNAs that originate from sources that are exogenous, which means they are not normally made by cells. They silence mRNAs via RNA interference.
short tandem repeat sequences (STRs) short DNA sequences that are repeated many times in a row. Often found in centromeric and telomeric regions.
shotgun sequencing a genome sequencing strategy in which DNA fragments to be sequenced are randomly generated from larger DNA fragments.
shuttle vector a cloning vector that can propagate in two or more different species, such as *E. coli* and yeast.
side chain in an amino acid, the chemical structure that is attached to the carbon atom (i.e., the α carbon) that is located between the amino group and carboxyl group.
sigma (σ) factor a transcription factor that recognizes bacterial promoter sequences and facilitates the binding of RNA polymerase to the promoter.
silencer a DNA sequence that functions as a regulatory element. The binding of a regulatory transcription factor to the silencer decreases the level of transcription.
silent mutation a mutation that does not alter the amino acid sequence of the encoded polypeptide even though the nucleotide sequence has changed.
similarity with regard to DNA, refers to a comparison of DNA sequences that have regions where the bases match up. Similarity may be due to homology.
simple Mendelian inheritance an inheritance pattern involving a simple, dominant/recessive relationship that produces observed ratios in the offspring that readily obey Mendel's laws.
simple sequence repeat (SSR) a short base sequence found within a chromosome that is repeated many times in a row.
simple translocation when one piece of a chromosome becomes attached to a different chromosome.
simple transposition a cut-and-paste mechanism for transposition in which a transposable element is removed from one site and then inserted into another.
simple transposon a small genetic element that carries one or two genes that are not required for transposition.
SINEs in mammals, short interspersed elements that are less than 500 bp in length.
single-factor cross a cross in which an experimenter is following the outcome of only a single trait.
single-nucleotide polymorphism (SNP) a genetic polymorphism within a population in which two alleles of the gene differ by a single nucleotide.
single-strand binding protein a protein that binds to both of the single strands of DNA during DNA replication and prevents them from re-forming a double helix.
siRNA see *short-interfering RNAs.*
sister chromatid exchange (SCE) the phenomenon in which crossing over occurs between sister chromatids, thereby exchanging identical genetic material.

sister chromatids pairs of replicated chromosomes that are attached to each other at the centromere. Sister chromatids are genetically identical.
site-directed mutagenesis a technique that enables scientists to change the sequence of cloned DNA segments.
site-specific recombination when two different DNA segments break and rejoin with each other at a specific site. This occurs during the integration of certain viruses into the host chromosome and during the rearrangement of immunoglobulin genes.
SMC proteins proteins that use energy from ATP to catalyze changes in chromosome structure.
snRNP a complex containing small nuclear RNA and a set of proteins, which are components of the spliceosome.
somatic cell any cell of the body except for germ-line cells that give rise to gametes.
somatic mutation a mutation in a somatic cell.
Southern blotting a technique used to detect the presence of a particular genetic sequence within a mixture of many chromosomal DNA fragments.
speciation the process by which new species are formed via evolution.
species a group of organisms that maintains a distinctive set of attributes in nature.
species concepts different approaches for distinguishing species.
spectrophotometer a device used by researchers to determine how much radiation at various wavelengths a sample can absorb.
spermatids immature sperm cells produced from spermatogenesis.
spermatogenesis the production of sperm cells.
sperm cell a male gamete. Sperm are small and usually travel relatively far distances to reach the female gamete.
spindle see *mitotic spindle apparatus.*
spindle pole during cell division in eukaryotes, one of two sites in the cell where microtubules originate.
spliceosome a multisubunit complex that functions in the splicing of eukaryotic pre-mRNA.
splicing see *RNA splicing.*
splicing factor a protein that regulates the process of RNA splicing.
spontaneous mutation a change in DNA structure that results from random abnormalities in biological processes.
spores haploid cells that are produced by certain species, such as fungi (i.e., yeast and molds).
sporophyte the diploid generation of plants.
SR protein a type of splicing factor.
SSR see *simple sequence repeat.*
stabilizing selection natural selection that favors individuals with an intermediate phenotype.
stamen the structure found in flowering plants that produces the male gametophyte (i.e., pollen).
standard deviation a statistic that is computed as the square root of the variance.
start codon a 3-base sequence in mRNA that initiates translation. It is usually 5′-AUG-3′ and encodes methionine.
stem cell a cell that has the capacity to divide and to differentiate into one or more specific cell types.
steroid receptor a category of transcription factors that respond to steroid hormones. An example is the glucocorticoid receptor.
stigma the structure in flowering plants on which the pollen land and the pollen tube starts to grow so that sperm cells can reach the egg cells.
stop codon a 3-base sequence in mRNA that signals the end of translation of a polypeptide. The three stop codons are 5′–UAA–3′, 5′–UAG–3′, and 5′–UGA–3′.
strain within a given species, a group that displays one or more genetic differences compared to another group.
strand in DNA or RNA, nucleotides covalently linked together to form a long, linear polymer.
stripe-specific enhancer in *Drosophila*, a regulatory region that controls the expression of a gene so that it occurs only in a particular parasegment during early embryonic development.
STRs see *short tandem repeat sequences.*
structural gene see *protein-encoding genes.*
STS see *sequence-tagged site.*
subcloning the procedure of making smaller DNA clones from a larger one.
submetacentric describes a chromosome in which the centromere is slightly off center.
subspecies a population within a species that has some distinct characteristics that differ from other members of the species.
subunit a component of a larger complex. In a protein, each subunit is a single polypeptide.
supercoiling see *DNA supercoiling.*
supergroups a relatively recent way for evolutionary biologists to subdivide the eukaryotic domain.
supernatant following centrifugation, the fluid that is found above the pellet.
suppressor (or **suppressor mutation**) a mutation at a second site that suppresses the phenotypic effects of another mutation.
SWI/SNF family a group of related proteins that catalyze chromatin remodeling.
symmetry break an event during X-chromosome inactivation in which the pluripotency factors and CTCFs that were previously bound to both X chromosomes shift entirely to one of the X chromosomes.
sympatric speciation a form of speciation that occurs when members of a species diverge while occupying the same habitat within the same range.
synapomorphy see *shared derived character.*
synapsis the event in which homologous chromosomes recognize each other and then align themselves along their entire lengths.
synaptonemal complex a complex of proteins that promote the interconnection between homologous chromosomes during meiosis.
synonymous codons two different codons that specify the same amino acid.
synteny group a group of genes that are found in the same order on the chromosomes of different species.
systematists biologists who study the evolutionary relationships among different species.

T

T an abbreviation for thymine.
tandem array (or **tandem repeat**) a short nucleotide sequence that is repeated many times in a row.
tandem mass spectrometry the sequential use of two mass spectrometers. It can be used to determine the sequence of amino acids in a polypeptide.
***Taq* polymerase** a thermostable form of DNA polymerase used in PCR experiments.
target-site primed reverse transcription (TPRT) a mechanism of DNA synthesis that occurs during the movement of certain types of retrotransposons.
TATA box a sequence found within eukaryotic core promoters that determines the starting site for transcription. The TATA box is recognized by a TATA-binding protein, which is a component of TFIID.
tautomer the forms of certain small molecules, such as bases, which can spontaneously interconvert between chemically similar forms.
tautomeric shift a change in chemical structure, such as an alternation between the keto and enol forms of the bases that are found in DNA.
T DNA a segment of DNA found within a Ti plasmid that is transferred from a bacterium to infected plant cells. The T DNA from the Ti plasmid becomes integrated into the chromosomal DNA of the plant cell by recombination.
T-DNA vectors vectors that carry T DNA that are used to introduce cloned genes into a plant cell.
TE see *transposable element.*
telocentric describes a chromosome with its centromere at one end.
telomerase the enzyme that recognizes telomeric sequences at the ends of eukaryotic chromosomes and synthesizes additional numbers of telomeric repeat sequences.
telomerase reverse transcriptase (TERT) the enzyme within telomerase that uses RNA as a template to make DNA.
telomeres specialized DNA sequences found at the ends of linear eukaryotic chromosomes.
telophase the fifth stage of M phase. The chromosomes have reached their respective poles and decondense.
temperate phage a bacteriophage that usually exists in the lysogenic cycle.
temperature-sensitive allele an allele in which the resulting phenotype depends on the environmental temperature.
temperature-sensitive (ts) lethal allele an allele that is lethal at a certain environmental temperature.
temperature-sensitive (ts) mutant a mutant that has a normal phenotype at a permissive temperature, but a different phenotype, such as failure to grow, at the nonpermissive temperature.
template DNA a strand of DNA that is used to synthesize a complementary strand of DNA or RNA.
template strand see *template DNA.*
terminal deficiency see *terminal deletion.*
terminal deletion when a segment is lost from the end of a linear chromosome.
termination (1) in transcription, the release of the newly made RNA transcript and RNA polymerase from the DNA; (2) in translation, the release of the polypeptide and the last tRNA and the disassembly of the ribosomal subunits and mRNA.
termination codon see *stop codon.*
termination sequences (ter sequences) in *E. coli*, a pair of sequences in the chromosome that bind a protein known as the termination utilization substance (Tus), which stops the movement of the replication forks.
terminator a sequence within a gene that signals the end of transcription.
TERT see *telomerase reverse transcriptase.*
tertiary structure the three-dimensional structure of a macromolecule, such as the tertiary structure of a polypeptide.
testcross an experimental cross between a recessive individual and an individual whose genotype the experimenter wishes to determine.
tetrad (1) the association among four sister chromatids during meiosis; (2) a group of four fungal spores contained within an ascus.

tetraploid having four sets of chromosomes (i.e., 4*n*).
tetratype (T) an ascus that has two parental cells and two nonparental cells.
TFIID a type of general transcription factor in eukaryotes that is needed for RNA polymerase II function. It binds to the TATA box and recruits RNA polymerase II to the core promoter.
thermocycler a device that automates the timing of temperature changes in each cycle of a PCR experiment.
30-nm fiber the association of nucleosomes to form a more compact structure that is 30 nm in diameter.
three-factor cross a cross in which the experimenter follows the outcome of three different traits.
threshold traits traits that are inherited quantitatively due to the contributions of many genes.
thymine a pyrimidine base found in DNA. It base-pairs with adenine in DNA.
thymine dimer a DNA lesion involving a covalent linkage between two adjacent thymine bases in a DNA strand.
Ti plasmid a tumor-inducing plasmid found in *Agrobacterium tumefaciens.* It is responsible for promoting tumor formation after a plant has been infected.
tissue-specific gene a gene that is highly regulated and is expressed in a particular cell type.
TNRE see *trinucleotide repeat expansion.*
topoisomerase an enzyme that alters the degree of supercoiling in DNA.
topoisomerase II during DNA replication, this protein travels in front of DNA helicase and alleviates positive supercoiling.
topoisomers DNA conformations that differ with regard to supercoiling.
totipotent a cell that possesses the genetic potential to produce an entire individual. A somatic plant cell or a fertilized egg is totipotent.
TPRT see *target-site primed reverse transcription.*
trait any characteristic that an organism displays. Morphological traits affect the appearance of an organism. Physiological traits affect the ability of an organism to function. A third category of traits are those that affect an organism's behavior (behavioral traits).
***trans*-acting factor** a regulatory protein that binds to a regulatory element in the DNA and exerts a *trans* effect.
transactivation domain a domain in a transcription factor that activates RNA polymerase.
transcription the process of synthesizing RNA from a DNA template.
transcriptional start site the site in a gene where transcription begins.
transcription factors a broad category of proteins that influence the ability of RNA polymerase to transcribe DNA into RNA.
transduction a form of genetic transfer between bacterial cells in which a bacteriophage transfers bacterial DNA from one bacterium to another.
***trans*-effect** an effect on gene expression that occurs even though two DNA segments are not physically adjacent to each other. *Trans*-effects are mediated through diffusible genetic regulatory proteins.
***trans*-epigenetic mechanism** an epigenetic mechanism that affects all of the genes of a given type within a cell. It is usually caused by diffusible transcription factors that constitute a feedback loop.
transfection (1) when a viral vector is introduced into a bacterial cell; (2) the introduction of any type of recombinant DNA into a eukaryotic cell.
transfer RNA (tRNA) a type of RNA used in translation that carries an amino acid. The anticodon in tRNA is complementary to a codon in the mRNA.
transformation (1) when a plasmid vector or segment of chromosomal DNA is introduced into a bacterial cell; (2) when a normal cell is converted into a malignant cell.
transgene a gene from one species that is introduced into another species.
transgenerational epigenetic inheritance epigenetic inheritance that is transmitted from parent to offspring.
transgenic organism an organism that has DNA from another species incorporated into its genome via recombinant DNA techniques.
transition a point mutation involving a change of a pyrimidine to another pyrimidine (e.g., C to T) or a purine to another purine (e.g., A to G).
translation the synthesis of a polypeptide using the codon information within mRNA.
translational regulatory protein a protein that regulates translation.
translational repressor a protein that binds to mRNA and inhibits its ability to be translated.
translesion synthesis (TLS) the synthesis of DNA over a template strand that harbors some type of DNA damage. This occurs via lesion-replicating polymerases.
translocation (1) when one segment of a chromosome breaks off and becomes attached to a different chromosome; (2) when a ribosome moves from one codon in an mRNA to the next codon.
translocation cross the structure that is formed when the chromosomes of a reciprocal translocation attempt to synapse during meiosis. This structure contains two normal (nontranslocated) chromosomes and two translocated chromosomes. A total of eight chromatids are found within the cross.
transposable element (TE) a small genetic element that can move to multiple locations within the chromosomal DNA.
transposase the enzyme that catalyzes the movement of some types of transposable elements.
transposition the phenomenon in which a transposable element moves to a new location.
transposon a type of transposable element that moves via transposase.
transposon tagging a technique for cloning genes in which a transposon inserts into a gene and inactivates it. The transposon-tagged gene is then cloned using a complementary transposon as a probe to identify the gene.
transversion a point mutation in which a purine is interchanged with a pyrimidine, or vice versa.
trithorax group (TrxG) protein complexes that are key regulators of epigenetic changes that are programmed during development. They cause gene activation.
trimethylation the attachment of three methyl groups to a single amino acid, such as lysine. Occurs via polycomb and trithorax group complexes.
trinucleotide repeat expansion (TNRE) a type of mutation that involves an increase in the number of tandemly repeated trinucleotide sequences.
triplex DNA a double-stranded DNA that has a third strand wound around it to form a triple-stranded structure.
triploid an organism or cell that contains three sets of chromosomes.
trisomic a diploid cell with one extra chromosome (i.e., $2n + 1$).
tRNA see *transfer RNA.*
trp repressor a protein that binds to the operator site of the *trp* operon and inhibits transcription.
true-breeding line a strain of a particular species that continues to produce the same trait after several generations of self-fertilization (in plants) or inbreeding.
true-breeding strain see *true-breeding line.*
tumor-suppressor gene a gene that functions to inhibit cancerous growth.
two-dimensional gel (2D-gel) electrophoresis a technique to separate proteins that involves isoelectric focusing in the first dimension and SDS-gel electrophoresis in the second dimension.
two-factor cross a cross in which an experimenter follows the outcome of two different traits.

U

U an abbreviation for uracil.
unbalanced translocation a translocation in which a cell has too much genetic material compared with a normal cell.
unipotent a type of stem cell that can differentiate into only a single type of cell.
universal in genetics, the phenomenon that nearly all organisms use the same genetic code with just a few exceptions.
unordered octad an ascus composed of eight unordered cells.
unordered tetrad an ascus composed of four unordered cells.
3′-untranslated region (3′-UTR) the region of an mRNA that is downstream from the stop codon.
up promoter mutation a mutation in a promoter that increases the rate of transcription.
up regulation genetic regulation that leads to an increase in gene expression.
uracil a pyrimidine base found in RNA.
UTR an abbreviation for the untranslated region of mRNA.
U-tube a U-shaped tube that has a filter at the bottom of the U. The pore size of the filter allows the passage of small molecules (e.g., DNA molecules) from one side of the tube to the other, but restricts the passage of bacterial cells.

V

variance the sum of the squared deviations from the mean divided by the degrees of freedom.
variants individuals of the same species that exhibit different traits. An example is tall and dwarf pea plants.
V(D)J recombination site-specific recombination that occurs within immunoglobulin genes.
vector a small segment of DNA that is used as a carrier of another segment of DNA. Vectors are used in DNA cloning experiments.
vertical evolution the phenomenon that species evolve from preexisting species by the accumulation of gene mutations and by changes in chromosome structure and number. Vertical evolution involves genetic changes in a series of ancestors that form a lineage.
vertical gene transfer the transfer of genetic material from parents to offspring or from mother cell to daughter cell.

viral envelope a lipid bilayer that is derived from the plasma membrane of the host cell and embedded with viral-encoded spike glycoproteins. The envelope encloses the capsid.

viral genome the genetic material of a virus.

viral reproductive cycle the series of steps that results in the production of new viruses.

viral vector a vector used in gene cloning that is derived from a naturally occurring virus.

virulent phage a phage that follows the lytic cycle.

virus a small infectious particle that contains nucleic acid as its genetic material, surrounded by a capsid of proteins. Some viruses also have an envelope consisting of a membrane embedded with spike proteins.

VNTRs also called minisatellites; segments of DNA that are located in several places in a genome and have a variable number of tandem repeats. The pattern of VNTRs was originally used in DNA fingerprinting.

W

Western blotting a technique used to detect a specific protein among a mixture of proteins.

whole-genome shotgun sequencing a genome sequencing strategy that bypasses the mapping step. The whole genome is subjected to shotgun sequencing.

wild type the relatively prevalent genotype in a natural population.

wild-type allele an allele that is fairly prevalent in a natural population, generally greater than 1% of the population. For polymorphic genes, there may be more than one wild-type allele.

wobble base the first base (from the 5′ end) in an anticodon. This term suggests that the first base in the anticodon can wobble a bit to recognize the third base in the mRNA.

wobble rules rules that govern the binding specificity between the third base in a codon and the first base in an anticodon.

X

X-chromosome inactivation (XCI) a process in which mammals equalize the expression of X-linked genes by randomly turning off one X chromosome in the somatic cells of females.

xenotransplantation the transplantation of cells, tissues, or organs from one animal species to another.

X inactivation see *X-chromosome inactivation.*

X-inactivation center (Xic) a site on the X chromosome that appears to play a critical role in X-chromosome inactivation.

X-linked genes (alleles) genes (or alleles of genes) that are physically located within the X chromosome.

X-linked inheritance an inheritance pattern in certain species that involves genes that are located only on the X chromosome.

X-linked recessive an allele or trait in which the gene is found on the X chromosome and the allele is recessive relative to a corresponding dominant allele.

Y

YAC see *yeast artificial chromosome.*

yeast artificial chromosome (YAC) a cloning vector propagated in yeast that can reliably contain very large insert fragments of DNA.

Y-linked genes (alleles) genes (or alleles of genes) that are located only on the Y chromosome.

Z

Z DNA a left-handed DNA double helix that is found occasionally in the DNA of living cells.

zygote a cell formed from the union of a sperm and an egg.

zygotene the second stage of prophase of meiosis I.

zygotic gene a gene that is expressed after fertilization.

CREDITS

Photographs

Chapter 1

Opener: © Corbis; 1.2: Photo courtesy of The Roslin Institute, The University of Edinburgh; 1.3a: © Advanced Cell Technology, Inc., Worcester, Massachusetts; 1.3b: Photo taken by Flaminia Catteruccia, Jason Benton and Andrea Crisanti, and assembled by www.luciariccidesign.com; 1.4(chromosome): © Biophoto Associates/Science Source; 1.5: © CNRI/SPL/Science Source; 1.8(left): © Mark Smith/Science Source; 1.8(right): © Dante Fenolio/Science Source; 1.9a: © Joseph Sohm/Visions of America/Corbis; 1.9b: © Paul Edmondson/Photodisc Red/Getty Images RF; 1.12a: © George Musil/Visuals Unlimited; 1.12b: © SciMAT/Science Source; 1.12c: © Steve Hopkin/ardea.com; 1.12d: © Brad Mogen/Visuals Unlimited; 1.12e: © J-M. Labat/Science Source; 1.12f: © Wally Eberhart/Visuals Unlimited.

Chapter 2

Opener: © Chris Martin Bahr/SPL/Science Source; 2.1: © SPL/Science Source; 2.2b: © Nigel Cattlin/Science Source.

Chapter 3

Opener: © Photomicrograph by Dr. Conly L. Rieder, Wadsworth Center, Albany, New York 12201-0509; 3.2b: © David Parker/Science Source; 3.2c, 3.6a(top left & bottom inset): © Leonard Lessin/Science Source; 3.6a(right): © Biophoto Associates/Science Source; 3.8a–f: © Photomicrographs by Dr. Conly L. Rieder, Wadsworth Center, Albany, New York 12201-0509; 3.9a: © Dr. David M. Phillips/Visuals Unlimited; 3.9b: © Ed Reschke.

Chapter 4

Opener: © Robert Calentine/Visuals Unlimited; 4.1: © blickwinkel/Alamy; 4.3b: © Bob Shanley/The Palm Beach Post; 4.4a(left): © Peter Weimann/Animals Animals; 4.4a(right): © Tom Walker/Visuals Unlimited; 4.4b: © Sally Haugen/Virginia Schuett, www.pkunews.org; 4.4c: © Phototake, Inc./Alamy; 4.7a–b: © Stan Flegler/Visuals Unlimited; 4.11a: © Fondazione San Raffaele del Monte Tabor/AP Photo; 4.13: Courtesy of Sheila M. Schmutz, Ph.D.; 4.14a–b: © Robert Maier/Animals Animals; 4.15a: © Jane Burton/naturepl.com.

Chapter 5

Opener: © John Mendenhall Institute for Cellular and Molecular Biology, University of Texas at Austin; 5.3a(left, right): © Courtesy of I. Solovei, University of Munich (LMU); 5.3b: © Tim Davis/Science Source; 5.6: © G.W. Willis, MD/Visuals Unlimited; p. 109(bottom left): Ronald G. Davidson, Harold M. Nitowsky, and Barton Childs (1963), "Demonstration of Two Populations of Cells in the Human Female Heterozygous for Glucose-6-Phosphate Dehydrogenase Variants," *PNAS*, *50*(3):481–485, Fig. 2. Courtesy Harold M. Nitowsky. Reproduced with author permission; 5.9: © Courtesy of Dr. Argiris Efstratiadis; 5.13a: Reproduced with permission from *The Journal of Cell Biology*, 72:687–694 © 1977 The Rockefeller University Press; 5.13b: S.P. Gibbs, R. Mak, R. Ng, & T. Slankis (1974), "The chloroplast nucleoid in *Ochromonas danica*. II. Evidence for an increase in plastid DNA during greening," *J Cell Sci.*, *16*(3):579–591. Fig. 1. © The Company of Biologists Limited. dev.biologists.org.

Chapter 6

Opener: © Educational Images Ltd./CustomMedical.

Chapter 7

Opener: © Eye of Science/Science Source; 7.4b: © Dr. L. Caro/SPL/Science Source; 7.15a–b: © Carolina Biological Supply/Phototake.

Chapter 8

Opener: © BSIP/Phototake; 8.1a(left): © Scott Camazine/Science Source; 8.1a(middle): © Michael Abbey/Science Source; 8.1a(right): © Carlos R. Carvalho/Universidade Federal de Viçosa; 8.1c: © CNRI/Phototake; 8.4a: © Biophoto Associates/Science Source; 8.4b: © Jeff Noneley; 8.13b: © Paul Benke/University of Miami School of Medicine; 8.13c: © Denys Kuvaiev/Alamy RF; 8.18a–b: © A.B. Sheldon; 8.19b: © Dr. David M. Phillips/Visuals Unlimited; 8.20: © James Steinberg/Science Source; 8.25: T.J. Robinson & E.H. Harley (1995), "Absence of geographic chromosomal variation in the roan and sable antelope and the cytogenetics of a naturally occurring hybrid," *Cytogenet. Cell Genet.*, *71*(4):363–369. Permission granted by S. Karger AG, Basel.

Chapter 9

Opener: © Ken Eward/Science Source; 9.3: © Omikron/Science Source; 9.11b: © Tom Hollyman/Science Source; 9.12a: © Science Source; 9.14: © Barrington Brown/Science Source; 9.16b: © Laguna Design/SPL/Science Source; 9.21b: © Alfred Pasieka/SPL/Science Source.

Chapter 10

Opener: © Dr. Gopal Murti/Visuals Unlimited; 10.2: C. Robinow & E. Kellenberger (1994), "The bacterial nucleoid revisited," *Microbiol. Mol. Biol. Rev.*, *58*(2):211–232. Reproduced with permission from American Society for Microbiology; 10.9b: © Ann & Rob Simpson; 10.9c: © William Leonard; 10.13a–b: © Fritz Thoma, ETH Zurich; 10.14a: © Jerome Rattner/University of Calgary; 10.15b–c: Nickerson et al., "The nuclear matrix revealed by eluting chromatin from a cross-linked nucleus," *PNAS*, *94*(9):4446–4450. Fig. 3. © 1997 National Academy of Sciences, USA; 10.16a–b: Courtesy of Felix Habermann and Irina Solovei, University of Munich (LMU, Biocenter); 10.18(1): © Dr. Gopal Murti/Visuals Unlimited; 10.18(2): © Ada L. Olins and Donald E. Olins/Biological Photo Service; 10.18(3): Jerome Rattner/University of Calgary; 10.18(4): © Dr. James Paulson, Ph.D.; 10.18(5–6), 10.19a: © Peter Engelhardt/Department of Virology, Haartman Institute; 10.19b: © Dr. Donald Fawcett/Visuals Unlimited.

Chapter 11

Opener: © Clive Freeman/The Royal Institution/Science Source; p. 265: M. Meselson & F. Stahl (1958), "The Replication of DNA in *Escherichia coli*," *PNAS*, *44*(7):671–682, Fig. 4A. Courtesy of M. Meselson; 11.4b: From: Cold Spring Harbor, *Symposia of Quantitative Biology* (1963), 28:43 © Cold Spring Harbor Laboratory Press; 11.8b: From: Cold Spring Harbor, *DNA Replication and Human Disease* (2006), DePamphilis, Ch. 5, Fig. 3. © Cold Spring Harbor Laboratory Press; 11.18: Courtesy of Dr. Joel A. Huberman.

Chapter 12

Opener: From: Patrick Cramer, David A. Bushnell, Roger D. Kornberg (2001), "Structural Basis of Transcription: RNA Polymerase II at 2.8 Ångstrom Resolution," *Science*, *292*(5523):1863–1876. Reprinted with permission of AAAS; 12.6: © Laguna Design/SPL/Science Source; 12.12a(left): Courtesy of Dr. Seth Darst; 12.12a(right): From: Patrick Cramer, David A. Bushnell, Roger D. Kornberg (2001), "Structural Basis of Transcription: RNA Polymerase II at 2.8 Ångstrom Resolution," *Science*, *292*(5523):1863–1876, Fig. 1. Image courtesy of David A. Bushnell; p. 304: From: Tilghman et al. (1978), "Intervening sequence of DNA identified in the structural portion of a mouse β-glovin gene," *PNAS*, *75*(2): 725–729, Fig. 2.

Chapter 13

Opener: © Tom Pantages; 13.13a: © Dr. Elena Kiseleva/SPL/Science Source; 13.13b: © Tom Pantages.

Chapter 14

Opener & 14.10b: © SPL/Science Source.

Chapter 15

Opener: Courtesy Song Tan, Penn State University, www.bmb.psu.edu/faculty/tan/lab.

Chapter 16

Opener: © Andia/Alamy; 16.7b: From: R.A. Waterland & R.L. Jirtle (2003), "Transposable elements: Targets for early nutritional effects on epigenetic gene regulation," *Mol. Cell. Biol.*, *23*(15):5293–5300, Fig. 2A. Reproduced with permission from American Society for Microbiology.

Chapter 17

Opener: © Nigel Cattlin/Science Source; 17.1a: © Robley C. Williams/Biological Photo Service; 17.1b: © Dr. Linda M. Stannard, University of Cape Town/SPL/Science Source; 17.1c: © Chris Bjornberg/Science Source; 17.1d: © Omikron/Science Source; 17.5: C.S. Goldsmith and A. Balish/CDC; 17.6: Cynthia Goldsmith/CDC; 17.15b: © Eye of Science/Science Source.

Chapter 18
Opener: © Robert Brooker; 18.1a(left, right): © Phototake, Inc./Alamy; 18.3a–b: From: N. Aulner, C. Monod, G. Mandicourt et al. (2002), "The AT-Hook protein D1 is essential for *Drosophila melanogaster* development and is implicated in position-effect variegation," *Mol. Cell. Biol.*, *22*(4):1218–1232, Fig. 7. Reproduced with permission from American Society for Microbiology; 18.5: © Otero/gtphoto; 18.19: © Anne Chadwick Williams/Sacramento Bee/ZUMAPRESS.com/Newscom.

Chapter 19
Opener: © Matt Meadows/Getty Images; 19.2b: © Dr. John D. Cunningham/ Visuals Unlimited.

Chapter 20
Opener: © Argus Fotoarchive.

Chapter 21
Opener: © Najlah Feanny/Corbis SABA; 21.2: © R.L. Brinster and R.E. Hammer, School of Veterinary Medicine, University of Pennsylvania; 21.4: Glofish © www.glofish.com; 21.5(bottom): © Alan Handyside/Wellcome Images; 21.9: © Corbis; 21.13: Courtesy Monsanto; 21.14a: © Bill Barksdale/AgStock Images; 21.15b: © Jack Bostrack/Visuals Unlimited.

Chapter 22
Opener: © Dr. Peter Lansdorp/Visuals Unlimited; 22.3: From: T. Ried, A. Baldini, T.C. Rand, & D.C. Ward (2002), "Simultaneous visualization of seven different DNA probes by in situ hybridization using combinatorial fluorescence and digital imaging microscopy," *PNAS*, *89*(4):1388–1392. Courtesy Thomas Ried; 22.16(left): © Joseph Sohm/Visions of America/Getty Images RF; 22.16(right): © Joseph Gareri/Getty Images RF.

Chapter 23
Opener: © Alfred Pasieka/SPL/Science Source; p. 584(bottom): Courtesy Joseph DeRisi; 23.5b: © Medical School/University of Newcastle upon Tyne/Simon Fraser/Science Source.

Chapter 24
Opener: © UK Stock Images Ltd./Alamy RF; 24.1(top): © Friedrich Stark/Alamy; 24.1(bottom): © Mitch Reardon/Science Source; 24.4(top left): © Hulton-Deutsch Collection/Corbis; 24.8(karyotyping): © CNRI/SPL/Science Source; 24.11: © Dr. Geoffrey M. Cooper; 24.19(top, bottom): Courtesy Dr. Ruhong Li, Molecular and Cell Biology, University of California at Berkeley. Permission granted by The Duesberg Laboratory. Appeared in Jean Marx (2002), "Debate Surges Over the Origins of Genomic Defects in Cancer," *Science*, *297*(5581, 26):544–546; 24.21: © Dr. Pat Brown.

Chapter 25
Opener: © N.A. Callow/NHPA/Photoshot; 25.3: Courtesy of E.B. Lewis, California Institute of Technology; 25.8b–c, 25.11(1): Christiane Nusslein-Volhard, *Development*, Supplement 1, 1991. © The Company of Biologists Limited. dev.biologists.org; 25.11(2–4): © Jim Langeland, Steve Paddock, and Sean Carroll/University of Wisconsin–Madison; 25.13a: © Juergen Berger/SPL/ Science Source; 25.13b: © F.R. Turner, Indiana University/Visuals Unlimited; p. 653(normal and defective *C. elegans*): From: H.R. Horvitz & J. Sulston (1980), "Isolation and genetic characterization of cell lineage mutants of the nematode *Caenorhabditis elegans*," *Genetics*, *96*(2):435–454, Fig. 1A&B. Courtesy Dr. Horvitz; 25.18: © Photodisc/Object Series/Vol. 50/Getty Images RF; 25.21: © Jeremy Burgess/SPL/Science Source; 25.24a–c: Elliott Meyerowitz and John Bowman (1991), *Development*, 112:1–20. Courtesy of Elliott Meyerowitz; p. 667(top right): Courtesy Stephen Small/New York University.

Chapter 26
Opener: © Photodisc/Getty Images RF; 26.1a: © Miguel Castro/Science Source; 26.1b: © Deborah Freund; 26.2(all): © Geoff Oxford; 26.10: © Paul Harcourt Davies/SPL/Science Source; 26.12: © R. Koenig/agefotostock; 26.14: © Gerald & Buff Corsi/Visuals Unlimited; 26.17b: © Corbis RF; 26.21: © Leonard Lessin.

Chapter 27
Opener: © Photodisc/Vol. 74/Getty Images RF; 27.1a: From: Albert F. Blakeslee (1914), "CORN AND MEN: The Interacting Influence of Heredity and Environment—Movements for Betterment of Men, or Corn, or Any Other Living Thing, One-sided Unless They Take Both Factors into Account," *Journal of Heredity*, *5*(11):511–518, by permission of Oxford University Press; 27.3a(top): © Nigel Cattlin/Visuals Unlimited; 27.3a(bottom): © irin-k/agefotostock RF; 27.9(greyhound): © Henry Ausloos/Animals Animals; 27.9(German shepherd): © Roger Tidman/Corbis; 27.9(bulldog): © Philip Gould/Corbis; 27.9(cocker spaniel): © Juniors Bildarchiv/Alamy; 27.10(wild mustard): © Inga Spence/Visuals Unlimited; 27.10(kohlrabi): © Nigel Cattlin/Science Source; 27.10(kale): © Valerie Giles/Science Source; 27.10(broccoli, Brussels sprouts, cabbage, cauliflower): © Michael P. Gadomski/Science Source.

Chapter 28
Opener: © LWA/Stephen Welstead/Blend Images/Corbis RF; 28.1a(left): © Mark Smith/Science Source; 28.1a(right): © Pascal Goetgheluck/ardea.com; 28.1b(left): © Rod Planck/Science Source; 28.1b(right): © Ron Austing/Science Source; 28.3(left): © Paul & Joyce Berquist/Animals Animals; 28.3(right): © Gerald & Buff Corsi/Visuals Unlimited.

Appendix A
A.1a: © Michael Gabridge/Visuals Unlimited; A.1b: © Fred Hossler/Visuals Unlimited; A.8: © Richard Wehr/CustomMedical.

Line Art

Chapter 9
9.17: Illustration, Irving Geis. Rights owned by Howard Hughes Medical Institute. Not to be reproduced without permission.

INDEX

Note: Page numbers followed by *f* and *t* indicate figures and tables, respectively.

A

B

G

H

I

N

Q

R

S

T

U

V

W

X

Y

Z